SOLUTIONS MANUAL

MECHANICS of MATERIALS

FOURTH EDITION

R.C. HIBBELER

Prentice Hall, Upper Saddle River, NJ 07458

Acquisitions Editor: Eric Svendsen
Supplement Editor: Kristen Blanco
Special Projects Manager: Barbara A. Murray
Production Editor: Barbara A. Till
Supplement Cover Manager: Paul Gourhan
Supplement Cover Designer: PM Workshop Inc.
Manufacturing Buyer: Pat Brown

Upper Saddle River, NJ 07458

Printed in the United States of America

10 9 8 7 6 5 4 3 2 1

ISBN 0-13-016615-4

Prentice-Hall International (UK) Limited, London
Prentice-Hall of Australia Pty. Limited, Sydney
Prentice-Hall Canada, Inc., Toronto
Prentice-Hall Hispanoamericana, S.A., Mexico
Prentice-Hall of India Private Limited, New Delhi
Pearson Education Asia Pte. Ltd., Singapore
Prentice-Hall of Japan, Inc., Tokyo
Editora Prentice-Hall do Brazil, Ltda., Rio de Janeiro

CONTENTS

To the Instructor

This manual contains the solutions to all the problems in *Mechanics of Materials, 4th edition.* In any group of problems, the easiest problems are presented first, followed by more the difficult problems. When assigning problems, note that the answers to all but every forth problem, which is indicated by an asterisk (*) before the problem number, are listed in the back of the text. Also, those problems indicated by a square (■) will require additional numerical work. The solutions may be readily determined using a numerical analysis method on a pocket calculator or desktop computer.

You may wish to use one of the lists of homework problems given on the following pages. Here you will find three lists for which the answers are in the back of the book, a fourth list for problems without answers, and a fifth sheet which can be used to develop your own personal syllabus. The prepared lists generally represent assignments with an easy, moderate, and sometimes a more challenging problem.

If you have any questions regarding the solutions in this manual, I would greatly appreciate hearing from you.

R. C. Hibbeler

hibbeler@bell south.net

Section	Topic	Assignment List 1
1.1-1.2	Equilibrium of a Deformable Body	1.3,1.7,1.18
1.3-1.5	Average Normal and Shear Stress	1.34,1.43,1.46,1.63
1.6-1.7	Design of Simple Connections	1.78,1.83,1.90,1.97
2.1-2.2	Strain	2.1,2.10,2.21
3.1-3.5	The Stress Strain Diagram	3.2,3.9,3.14
3.6-3.8	Poisson's ratio, Shear Stress-Strain Diagram	3.26,3.30,3.34
4.1-4.2	Deformation of an Axially Loaded Member	4.9,4.15,4.19,4.35
4.3-4.5	Statically Indeterminate Member	4.37,4.45,4.49,4.59
4.6	Thermal Stresses	4.75,4.81,4.90
4.7	Stress Concentrations	4.95,4.101,4.102
4.8-4.9	Inelastic Deformation and Residual Stresses	4.107,4.109,4.111
5.1-5.3	Torsion Stress and Power	5.2,5.9,5.23,5.31
5.4	Angle of Twist	5.45,5.53,5.59,5.65
5.5	Statically Indeterminate Members	5.75,5.83,5.85
5.6	Noncircular Shafts	5.91,5.97,5.98
5.7	Thin-Walled Tubes	5.101,5.105
5.8	Stress Concentrations	5.113,5.114
5.9-5.10	Inelastic Torsion and Residual Stresses	5.119,5.123,5.127
6.1-6.2	Shear and Moment Diagrams	6.2,6.7,6.13
		6.22,6.25,6.30
6.3-6.4	Bending Stress	6.42,6.45,6.51,6.63,6.85
6.5	Unsymmetrical Bending	6.98,6.102,6.107
6.6-6.7	Composite Beams	6.115,6.118,6.122
6.8	Curved Beams	6.130,6.134,6.139
6.9	Stress Concentrations	6.143,6.146,6.149
6.10-6.11	Inelastic Bending	6.154,6.162,6.165,6.173
7.1-7.3	Shear Stress	7.2,7.7,7.13,7.23
7.4	Shear Flow in Built-up Members	7.34,7.39,7.45
7.5-6	Shear Center	7.57,7.62,7.70
8.1	Thin-Walled Pressure Vessels	8.3,8.7,8.10
8.2	Stress Due to Combined Loadings	8.19,8.29,8.31
		8.43,8.47,8.51
9.1-9.3	Stress Transformation	9.6,9.9,9.17
		9.23,9.31,9.37 *
9.4-9.6	Mohr's Circle	9.53,9.61,9.67
		9.73,9.77,9.82 *
9.7	Absolute Maximum Shear Stress	9.86,9.87
		9.93 *
10.1-10.2	Strain Transformation	10.2,10.5,10.6
10.3	Mohr's Circle	10.13,10.14,10.17
10.4-10.5	Abs. Maximum Shear Strain, Strain Rosettes	10.22,10.25,10.29
10.6	Material Property Relations	10.38,10.47,10.54
10.7	Theories of Failure	10.75,10.87,10.91
11.1-11.2	Prismatic Beam Design	11.5,11.13,11.22
11.3	Fully Stressed Beams	11.30,11.34
11.4	Shaft Design	11.38,11.41
12.1-12.2	Slope and Displacement by Integration	12.2,12.15,12.18
12.3	Discontinuity Functions	12.23,12.29,12.31
12.4	Moment -Area Theorems	12.49,12.53,12.63
12.5	Method of Superposition	12.77,12.82,12.86
12.6-12.7	Indet. Beams-Method of Integration	12.90,12.94,12.95
12.8	Indet. Beams-Mom. Area Theorems	12.99,12.102
12.10	Indet. Beams-Method of Superposition	12.109,12.113,12.117
13.1-13.3	Buckling of an Ideal Column	13.5,13.17,13.33
13.4-13.5	The Secant Formula, Inelastic Buckling	13.49,13.53,13.69
13.6	Design of Columns for Concentric Loading	13.77,13.85,13.97
13.7	Design of Columns for Eccentric Loading	13.102,13.109,13.111
14.1-14.2	Elastic Strain Energy	14.3,14.11,14.15
14.3	Conservation of Energy	14.22,14.25,14.27
14.4	Impact	14.46,14.49,14.51
14.5-14.6	Principle of Virtual Forces-Trusses	14.61,14.67,14.73
14.7	Principle of Virtual Forces-Beams	14.79,14.85,14.102
14.8	Castigliano's Theorem-Trusses	14.115,14.123,14.129
14.9	Castigliano's Theorem-Beams	14.135,14.138,14.141

* These problems require coverage of Chapters 4-8.

Section	Topic	Assignment List 2
1.1-1.2	Equilibrium of a Deformable Body	1.5,1.10,1.23
1.3-1.5	Average Normal and Shear Stress	1.39,1.45,1.47,1.59
1.6-1.7	Design of Simple Connections	1.81,1.87,1.93,1.98
2.1-2.2	Strain	2.5,2.13,2.17
3.1-3.5	The Stress Strain Diagram	3.5,3.7,3.13
3.6-3.8	Poisson's ratio, Shear Stress-Strain Diagram	3.25,3.33,3.37
4.1-4.2	Deformation of an Axially Loaded Member	4.6,4.13,4.17,4.25
4.3-4.5	Statically Indeterminate Member	4.41,4.46,4.54,4.61
4.6	Thermal Stresses	4.77,4.83,4.86
4.7	Stress Concentrations	4.97,4.98,4.103
4.8-4.9	Inelastic Deformation and Residual Stresses	4.105,4.106,4.107
5.1-5.3	Torsion Stress and Power	5.3,5.13,5.19,5.27
5.4	Angle of Twist	5.42,5.50,5.55,5.61
5.5	Statically Indeterminate Members	5.78,5.81,5.87
5.6	Noncircular Shafts	5.91,5.94,5.97
5.7	Thin-Walled Tubes	5.103,5.109
5.8	Stress Concentrations	5.111,5.114
5.9-5.10	Inelastic Torsion and Residual Stresses	5.122,5.129,5.131
6.1-6.2	Shear and Moment Diagrams	6.3,6.9,6.14
		6.21,6.26,6.29
6.3-6.4	Bending Stress	6.41,6.46,6.50,6.61,6.73
6.5	Unsymmetrical Bending	6.97,6.101,6.109
6.6-6.7	Composite Beams	6.117,6.119,6.121
6.8	Curved Beams	6.129,6.135,6.138
6.9	Stress Concentrations	6.141,6.145,6.150
6.10-6.11	Inelastic Bending	6.153,6.157,6.161,6.167
7.1-7.3	Shear Stress	7.3,7.5, 7.11,7.22
7.4	Shear Flow in Built-up Members	7.35,7.37,7.41
7.5-6	Shear Center	7.55,7.63,7.67
8.1	Thin-Walled Pressure Vessels	8.1,8.5,8.9
8.2	Stress Due to Combined Loadings	8.21,8.31,8.38
		8.42,8.46,8.54
9.1-9.3	Stress Transformation	9.2,9.11,9.18
		9.21,9.39,9.43 *
9.4-9.6	Mohr's Circle	9.51,9.59,9.65
		9.71,9.75,9.81 *
9.7	Absolute Maximum Shear Stress	9.85,9.87
		9.91 *
10.1-10.2	Strain Transformation	10.1,10.3,10.10
10.3	Mohr's Circle	10.11,10.17,10.19
10.4-10.5	Abs. Maximum Shear Strain, Strain Rosettes	10.21,10.25,10.27
10.6	Material Property Relations	10.34,10.41,10.57
10.7	Theories of Failure	10.73,10.77,10.85
11.1-11.2	Prismatic Beam Design	11.6,11.19,11.27
11.3	Fully Stressed Beams	11.31,11.33
11.4	Shaft Design	11.39,11.43
12.1-12.2	Slope and Displacement by Integration	12.3,12.9,12.15
12.3	Discontinuity Functions	12.25,12.37,12.39
12.4	Moment -Area Theorems	12.47,12.57,12.71
12.5	Method of Superposition	12.75,12.81,12.85
12.6-12.7	Indet. Beams-Method of Integration	12.89,12.95,12.97
12.8	Indet. Beams-Mom. Area Theorems	12.98,12.105
12.10	Indet. Beams-Method of Superposition	12.106,12.115,12.118
13.1-13.3	Buckling of an Ideal Column	13.7,13.15,13.27
13.4-13.5	The Secant Formula, Inelastic Buckling	13.55,13.63,13.70
13.6	Design of Columns for Concentric Loading	13.75,13.87,13.93
13.7	Design of Columns for Eccentric Loading	13.103,13.106,13.118
14.1-14.2	Elastic Strain Energy	14.1,14.9,14.17
14.3	Conservation of Energy	14.21,14.27,14.29
14.4	Impact	14.45,14.53,14.57
14.5-14.6	Principle of Virtual Forces-Trusses	14.59,14.69,14.71
14.7	Principle of Virtual Forces-Beams	14.77,14.83,14.101
14.8	Castigliano's Theorem-Trusses	14.113,14.125,14.127
14.9	Castigliano's Theorem-Beams	14.133,14.137,14.145

* These problems require coverage of Chapters 4-8.

Section	Topic	Assignment List 3
1.1-1.2	Equilibrium of a Deformable Body	1.9,1.17,1.22
1.3-1.5	Average Normal and Shear Stress	1.37,1.41,1.50,1.61
1.6-1.7	Design of Simple Connections	1.71,1.85,1.91,1.94
2.1-2.2	Strain	2.3,2.9,2.18
3.1-3.5	The Stress Strain Diagram	3.3,3.11,3.15
3.6-3.8	Poisson's ratio, Shear Stress-Strain Diagram	3.27,3.31,3.35
4.1-4.2	Deformation of an Axially Loaded Member	4.7,4.14,4.21,4.26
4.3-4.5	Statically Indeterminate Member	4.37,4.43,4.51,4.55
4.6	Thermal Stresses	4.79,4.82,4.89
4.7	Stress Concentrations	4.99,4.102,4.103
4.8-4.9	Inelastic Deformation and Residual Stresses	4.106,4.107,4.110
5.1-5.3	Torsion Stress and Power	5.1,5.7,5.21,5.29
5.4	Angle of Twist	5.43,5.51,5.545.63
5.5	Statically Indeterminate Members	5.77,5.82,5.86
5.6	Noncircular Shafts	5.91,5.95,5.105
5.7	Thin-Walled Tubes	5.106,5.110
5.8	Stress Concentrations	5.111,5.113
5.9-5.10	Inelastic Torsion and Residual Stresses	5.117,5.121,5.123
6.1-6.2	Shear and Moment Diagrams	6.1,6.6,6.11
		6.19,6.23,6.37
6.3-6.4	Bending Stress	6.39,6.43,6.53,6.66,6.86
6.5	Unsymmetrical Bending	6.99,6.103,6.106
6.6-6.7	Composite Beams	6.114,6.121,6.126
6.8	Curved Beams	6.131,6.133,6.136
6.9	Stress Concentrations	6.142,6.146,6.149
6.10-6.11	Inelastic Bending	6.151,6.155,6.163,6.171
7.1-7.3	Shear Stress	7.1,7.9,7.14,7.25
7.4	Shear Flow in Built-up Members	7.33,7.38,7.43
7.5-6	Shear Center	7.59,7.61,7.73
8.1	Thin-Walled Pressure Vessels	8.2,8.6,8.11
8.2	Stress Due to Combined Loadings	8.15,8.18,8.27
		8.47,8.53,8.58
9.1-9.3	Stress Transformation	9.3,9.10,9.15
		9.25,9.34,9.43 *
9.4-9.6	Mohr's Circle	9.57,9.62,9.69
		9.74,9.77,9.83 *
9.7	Absolute Maximum Shear Stress	9.85,9.87
		9.93 *
10.1-10.2	Strain Transformation	10.1,10.3,10.7
10.3	Mohr's Circle	10.11,10.17,10.18
10.4-10.5	Abs. Maximum Shear Strain, Strain Rosettes	10.21,10.29,10.31
10.6	Material Property Relations	10.35,10.41,10.61
10.7	Theories of Failure	10.79,10.86,10.93
11.1-11.2	Prismatic Beam Design	11.1,11.15,11.25
11.3	Fully Stressed Beams	11.29,11.35
11.4	Shaft Design	11.37,11.47
12.1-12.2	Slope and Displacement by Integration	12.1,12.9,12.17
12.3	Discontinuity Functions	12.29,12.33,12.38
12.4	Moment -Area Theorems	12.43,12.59,12.67
12.5	Method of Superposition	12.78,12.83,12.87
12.6-12.7	Indet. Beams-Method of Integration	12.91,12.93,12.97
12.8	Indet. Beams-Mom. Area Theorems	12.101,12.103
12.10	Indet. Beams-Method of Superposition	12.107,12.111,12.114
13.1-13.3	Buckling of an Ideal Column	13.9,13.11,13.35
13.4-13.5	The Secant Formula, Inelastic Buckling	13.47,13.57,13.67
13.6	Design of Columns for Concentric Loading	13.78,13.83,13.91
13.7	Design of Columns for Eccentric Loading	13.101,13.107,13.115
14.1-14.2	Elastic Strain Energy	14.2,14.5,14.11
14.3	Conservation of Energy	14.23,14.27,14.29
14.4	Impact	14.43,14.45,14.54
14.5-14.6	Principle of Virtual Forces-Trusses	14.63,14.65,14.75
14.7	Principle of Virtual Forces-Beams	14.81,14.87,14.103
14.8	Castigliano's Theorem-Trusses	14.117,14.121,14.126
14.9	Castigliano's Theorem-Beams	14.134,14.139,14.143

* These problems require coverage of Chapters 4-8.

Section	Topic	Assignment List Without Book Ans.
1.1-1.2	Equilibrium of a Deformable Body	1.4,1.8,1.24
1.3-1.5	Average Normal and Shear Stress	1.36,1.44,1.52,1.56
1.6-1.7	Design of Simple Connections	1.80,1.92,1.100,1.104
2.1-2.2	Strain	2.8,2.16,2.24
3.1-3.5	The Stress Strain Diagram	3.8,3.12,3.16
3.6-3.8	Poisson's ratio, Shear Stress-Strain Diagram	3.24,3.32,3.36
4.1-4.2	Deformation of an Axially Loaded Member	4.4,4.12,4.24,4.32
4.3-4.5	Statically Indeterminate Member	4.40,4.52,4.56,4.60
4.6	Thermal Stresses	4.76,4.84,4.92
4.7	Stress Concentrations	4.96,4.100
4.8-4.9	Inelastic Deformation and Residual Stresses	4.104,4.108
5.1-5.3	Torsion Stress and Power	5.4,5.16,5.24,5.36
5.4	Angle of Twist	5.48,5.56,5.60
5.5	Statically Indeterminate Members	5.80,5.84,5.88
5.6	Noncircular Shafts	5.92,5.96
5.7	Thin-Walled Tubes	5.104,5.108
5.8	Stress Concentrations	5.112,5.116
5.9-5.10	Inelastic Torsion and Residual Stresses	5.120,5.128,5.132
6.1-6.2	Shear and Moment Diagrams	6.4,6.8,6.16
		6.20,6.28,6.36
6.3-6.4	Bending Stress	6.44,6.48,6.56,6.60,6.80
6.5	Unsymmetrical Bending	6.96,6.100,6.108
6.6-6.7	Composite Beams	6.116,6.120,6.124
6.8	Curved Beams	6.128,6.132,6.136
6.9	Stress Concentrations	6.140,6.144,6.148
6.10-6.11	Inelastic Bending	6.152,6.160,6.168,6.172
7.1-7.3	Shear Stress	7.4,7.7,16.20,7.24
7.4	Shear Flow in Built-up Members	7.36,7.40,7.48
7.5-6	Shear Center	7.56,7.64,7.68
8.1	Thin-Walled Pressure Vessels	8.4,8.8,8.12
8.2	Stress Due to Combined Loadings	8.16,8.28,8.32
		8.40,8.48,8.56
9.1-9.3	Stress Transformation	9.4,9.12,9.20
		9.24,9.32,9.40 *
9.4-9.6	Mohr's Circle	9.52,9.60,9.68
		9.72,9.76,9.80 *
9.7	Absolute Maximum Shear Stress	9.88,9.96
		9.92 *
10.1-10.2	Strain Transformation	10.4,10.8
10.3	Mohr's Circle	10.1210.16
10.4-10.5	Abs. Maximum Shear Strain, Strain Rosettes	10.20,10.24,10.28
10.6	Material Property Relations	10.40,10.48,10.56
10.7	Theories of Failure	10.72,10.80,10.88
11.1-11.2	Prismatic Beam Design	11.8,11.16,11.28
11.3	Fully Stressed Beams	11.32,11.36
11.4	Shaft Design	11.40,11.44
12.1-12.2	Slope and Displacement by Integration	12.4,12.12,12.16
12.3	Discontinuity Functions	12.24,12.32,12.36
12.4	Moment -Area Theorems	12.44,12.52,12.60
12.5	Method of Superposition	12.72,12.76,12.84
12.6-12.7	Indet. Beams-Method of Integration	12.88,12.92,12.96
12.8	Indet. Beams-Mom. Area Theorems	12.100,12.104
12.10	Indet. Beams-Method of Superposition	12.108,12.112,12.116
13.1-13.3	Buckling of an Ideal Column	13.8,13.16,13.28
13.4-13.5	The Secant Formula, Inelastic Buckling	13.52,13.60,13.68
13.6	Design of Columns for Concentric Loading	13.72,13.80,13.92
13.7	Design of Columns for Eccentric Loading	13.104,13.112,13.116
14.1-14.2	Elastic Strain Energy	14.4,14.12,14.16
14.3	Conservation of Energy	14.20,14.24,14.28
14.4	Impact	14.40,14.48,14.52
14.5-14.6	Principle of Virtual Forces-Trusses	14.60,14.64,14.72
14.7	Principle of Virtual Forces-Beams	14.80,14.88,14.96
14.8	Castigliano's Theorem-Trusses	14.112,14.116,14.128
14.9	Castigliano's Theorem-Beams	14.132,14.136,14.144

*These problems require coverage of Chapters 4-8.

Section	Topic	Assignment
1.1-1.2	Equilibrium of a Deformable Body	
1.3-1.5	Average Normal and Shear Stress	
1.6-1.7	Design of Simple Connections	
2.1-2.2	Strain	
3.1-3.5	The Stress Strain Diagram	
3.6-3.8	Poisson's ratio, Shear Stress-Strain Diagram	
4.1-4.2	Deformation of an Axially Loaded Member	
4.3-4.5	Statically Indeterminate Member	
4.6	Thermal Stresses	
4.7	Stress Concentrations	
4.8-4.9	Inelastic Deformation and Residual Stresses	
5.1-5.3	Torsion Stress and Power	
5.4	Angle of Twist	
5.5	Statically Indeterminate Members	
5.6	Noncircular Shafts	
5.7	Thin-Walled Tubes	
5.8	Stress Concentrations	
5.9-5.10	Inelastic Torsion and Residual Stresses	
6.1-6.2	Shear and Moment Diagrams	
6.3-6.4	Bending Stress	
6.5	Unsymmetrical Bending	
6.6-6.7	Composite Beams	
6.8	Curved Beams	
6.9	Stress Concentrations	
6.10-6.11	Inelastic Bending	
7.1-7.3	Shear Stress	
7.4	Shear Flow in Built-up Members	
7.5-6	Shear Center	
8.1	Thin-Walled Pressure Vessels	
8.2	Stress Due to Combined Loadings	
9.1-9.2	Stress Transformation	
9.3	Princ. Stress and Max. In-Plane Shear Stress	
9.4-9.6	Mohr's Circle	
9.7	Absolute Maximum Shear Stress	
10.1-10.2	Strain Transformation	
10.3	Mohr's Circle	
10.4-10.5	Abs. Maximum Shear Strain, Strain Rosettes	
10.6	Material Property Relations	
10.7	Theories of Failure	
11.1-11.2	Prismatic Beam Design	
11.3	Fully Stressed Beams	
11.4	Shaft Design	
12.1-12.2	Slope and Displacement by Integration	
12.3	Discontinuity Functions	
12.4	Moment -Area Theorems	
12.5	Method of Superposition	
12.6-12.7	Indet. Beams-Method of Integration	
12.8	Indet. Beams-Mom. Area Theorems	
12.10	Indet. Beams-Method of Superposition	
13.1-13.3	Buckling of an Ideal Column	
13.4-13.5	The Secant Formula, Inelastic Buckling	
13.6	Design of Columns for Concentric Loading	
13.7	Design of Columns for Eccentric Loading	
14.1-14.2	Elastic Strain Energy	
14.3	Conservation of Energy	
14.4	Impact	
14.5-14.6	Principle of Virtual Forces-Trusses	
14.7	Principle of Virtual Forces-Beams	
14.8	Castigliano's Theorem-Trusses	
14.9	Castigliano's Theorem-Beams	

1–1. Determine the resultant internal torque acting on the cross sections through points C and D of the shaft. The support bearings at A and B allow free turning of the shaft.

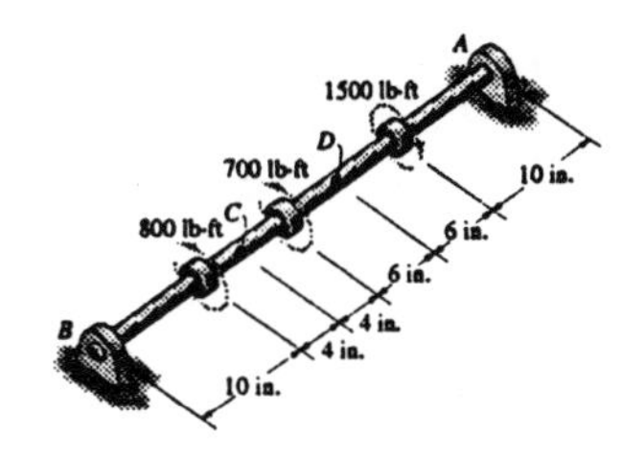

Equations of Equilibrium :

$\circlearrowleft + \quad 800 - T_C = 0 \qquad T_C = 800 \text{ lb}\cdot\text{ft}$ **Ans**

$\circlearrowleft + \quad T_D - 1500 = 0 \qquad T_D = 1500 \text{ lb}\cdot\text{ft}$ **Ans**

1–2. Determine the resultant internal torque acting on the cross sections through points C and D of the shaft. The shaft is fixed at B.

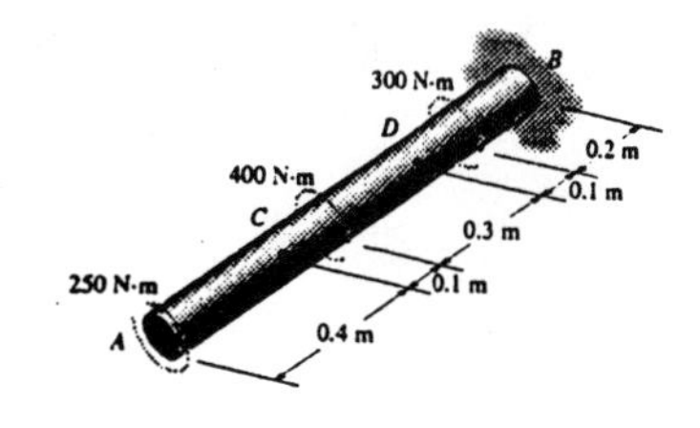

Equations of Equilibrium :

$\circlearrowleft + \quad 250 - T_C = 0 \qquad T_C = 250 \text{ N}\cdot\text{m}$ **Ans**

$\circlearrowleft + \quad 250 - 400 + T_D = 0 \qquad T_D = 150 \text{ N}\cdot\text{m}$ **Ans**

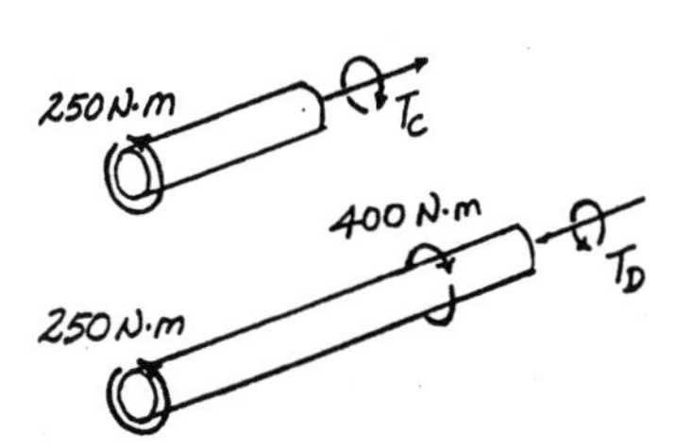

1–3. Determine the resultant internal loadings on the cross sections through points C and D of the beam.

Support Reactions :

$\circlearrowleft + \Sigma M_A = 0;\quad B_y(7) - 6(1.5) - 2(3) - 8(5) - 3(7) = 0$

$B_y = 10.86 \text{ kN}$

$+\uparrow \Sigma F_y = 0;\quad A_y - 4 - 6 - 2 - 8 - 3 + 10.86 = 0$

$A_y = 12.14 \text{ kN}$

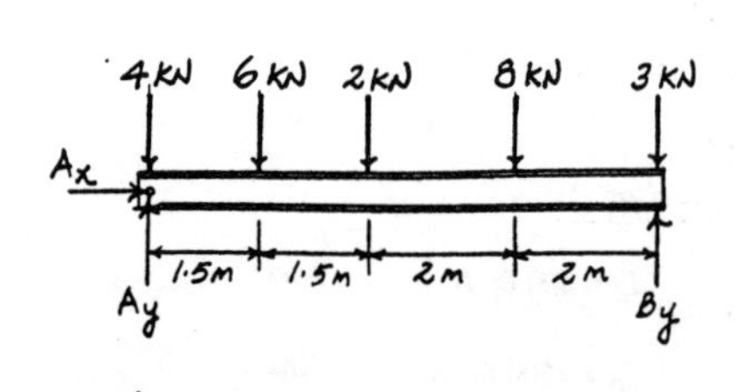

$\overset{+}{\rightarrow} \Sigma F_x = 0;\quad A_x = 0$

Equations of Equilibrium : For point C

$\overset{+}{\rightarrow} \Sigma F_x = 0;\quad N_C = 0$ **Ans**

$+\uparrow \Sigma F_y = 0;\quad 12.14 - 4 - V_C = 0$

$V_C = 8.14 \text{ kN}$ **Ans**

$\circlearrowleft + \Sigma M_C = 0;\quad M_C + 4\,(0.75) - 12.14\,(0.75) = 0$

$M_C = 6.11 \text{ kN}\cdot\text{m}$ **Ans**

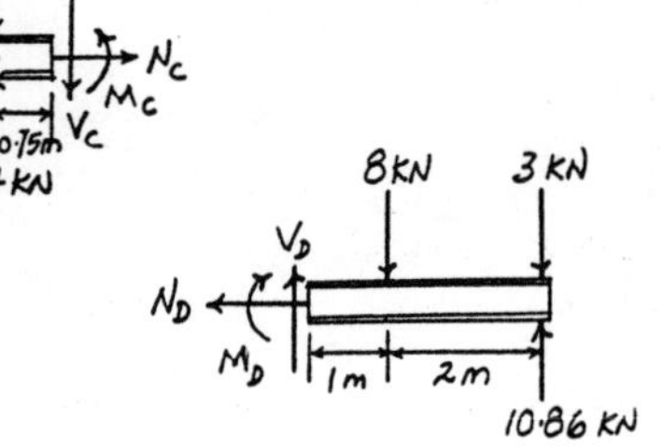

Equations of equilibrium : For point D

$\overset{+}{\rightarrow} \Sigma F_x = 0;\quad N_D = 0$ **Ans**

$+\uparrow \Sigma F_y = 0;\quad V_D + 10.86 - 8 - 3 = 0$

$V_D = 0.143 \text{ kN}$ **Ans**

$\circlearrowleft + \Sigma M_D = 0;\quad 10.86(3) - 8(1) - 3(3) - M_D = 0$

$M_D = 15.6 \text{ kN}\cdot\text{m}$ **Ans**

***1–4.** A force of 80 N is supported by the bracket as shown. Determine the resultant internal loadings acting on the section through point A.

Equations of Equilibrium :

$+\nearrow \Sigma F_{x'} = 0;\quad N_A - 80\cos 15^\circ = 0$

$N_A = 77.3 \text{ N}$ **Ans**

$\nwarrow + \Sigma F_{y'} = 0;\quad V_A - 8\,0\sin 15^\circ = 0$

$V_A = 20.7 \text{ N}$ **Ans**

$\circlearrowleft + \Sigma M_A = 0;\quad M_A + 80\cos 45^\circ(0.3\cos 30^\circ)$

$- 80\sin 45^\circ(0.1 + 0.3\sin 30^\circ) = 0$

$M_A = -0.555 \text{ N}\cdot\text{m}$ **Ans**

or

$\circlearrowleft + \Sigma M_A = 0;\quad M_A + 80\sin 15^\circ(0.3 + 0.1\sin 30^\circ)$

$- 80\cos 15^\circ(0.1\cos 30^\circ) = 0$

$M_A = -0.555 \text{ N}\cdot\text{m}$ **Ans**

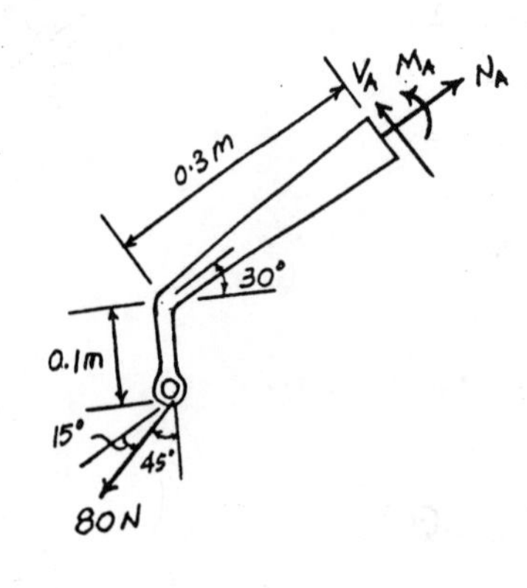

Negative sign indicates that M_A acts in the opposite direction to that shown on FBD.

1–5. Determine the resultant internal loadings on the cross section through point D on member AB.

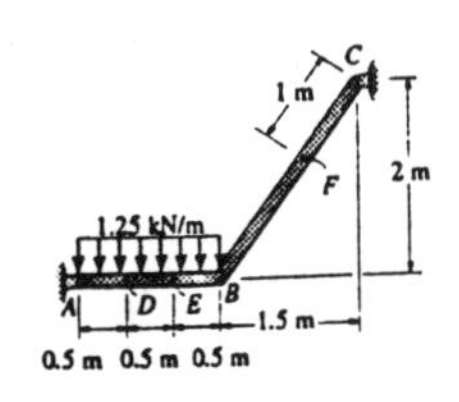

Support Reactions : Member BC is the two force member.

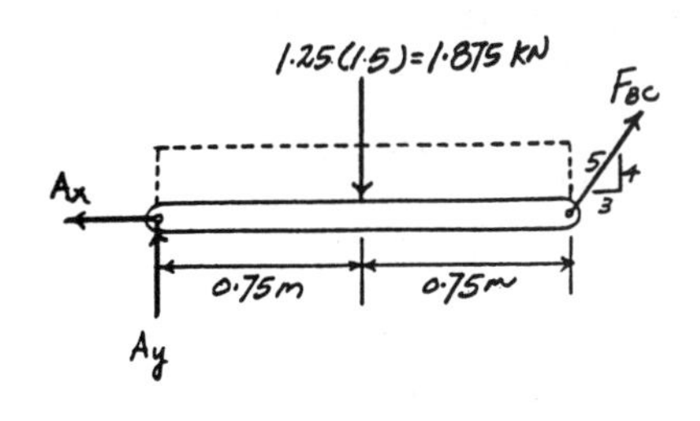

$\curvearrowleft + \Sigma M_A = 0; \quad \frac{4}{5}F_{BC}(1.5) - 1.875(0.75) = 0$

$F_{BC} = 1.1719 \text{ kN}$

$+\uparrow \Sigma F_y = 0; \quad A_y + \frac{4}{5}(1.1719) - 1.875 = 0$

$A_y = 0.9375 \text{ kN}$

$\xrightarrow{+} \Sigma F_x = 0; \quad \frac{3}{5}(1.1719) - A_x = 0$

$A_x = 0.7031 \text{ kN}$

Equations of Equilibrium : For point D

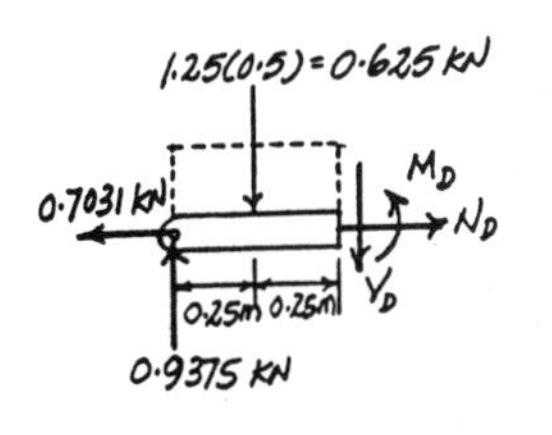

$\xrightarrow{+} \Sigma F_x = 0; \quad N_D \ 0.7031 = 0$

$N_D = 0.703 \text{ kN}$ **Ans**

$+\uparrow \Sigma F_y = 0; \quad 0.9375 - 0.625 - V_D = 0$

$V_D = 0.3125 \text{ kN}$ **Ans**

$\curvearrowleft + \Sigma M_D = 0; \quad M_D + 0.625(0.25) - 0.9375(0.5) = 0$

$M_D = 0.3125 \text{ kN} \cdot \text{m}$ **Ans**

1–6. Determine the resultant internal loadings at cross sections through points E and F on the assembly.

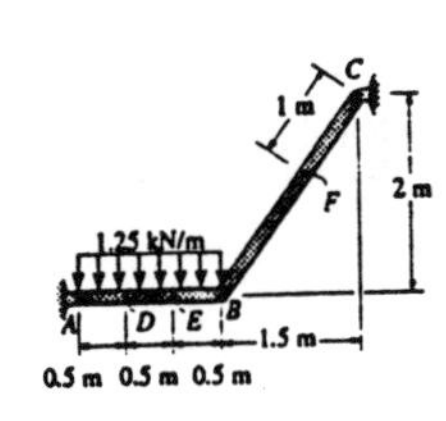

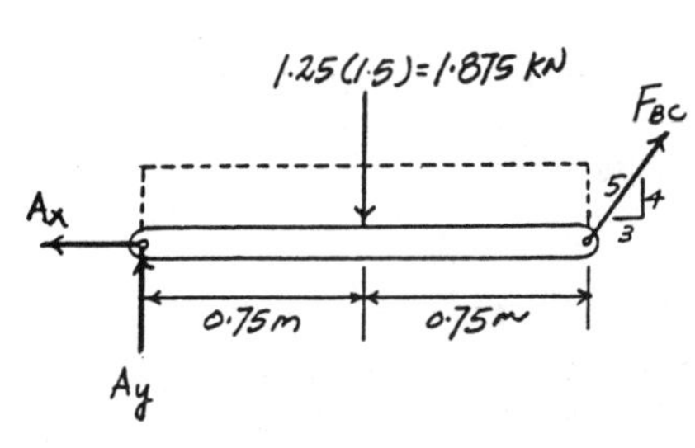

Support Reactions : Member BC is the two - force member.

$\curvearrowleft + \Sigma M_A = 0; \quad \frac{4}{5}F_{BC}(1.5) - 1.875(0.75) = 0$

$F_{BC} = 1.1719 \text{ kN}$

$+\uparrow \Sigma F_y = 0; \quad A_y + \frac{4}{5}(1.1719) - 1.875 = 0$

$A_y = 0.9375 \text{ kN}$

$\xrightarrow{+} \Sigma F_x = 0; \quad \frac{3}{5}(1.1719) - A_x = 0$

$A_x = 0.7031 \text{ kN}$

Equations of Equilibrium : For point F

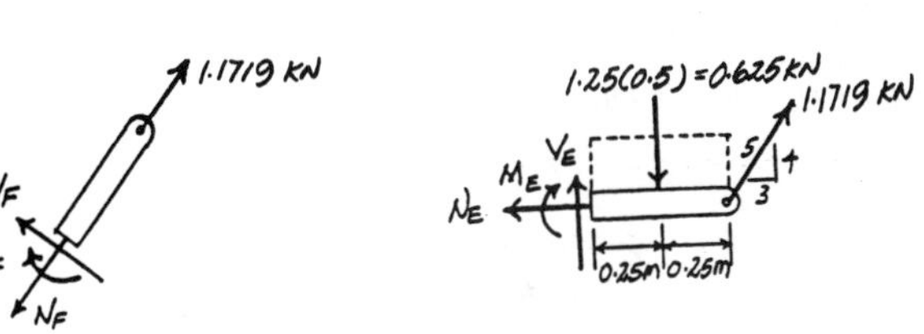

$+\nearrow \Sigma F_{x'} = 0; \quad N_F - 1.1719 = 0$

$N_F = 1.17 \text{ kN}$ **Ans**

$+\nwarrow \Sigma F_{y'} = 0; \quad V_F = 0$ **Ans**

$\curvearrowleft + \Sigma M_F = 0; \quad M_F = 0$ **Ans**

Equations of Equilibrium : For point E

$\xleftarrow{+} \Sigma F_x = 0; \quad N_E - \frac{3}{5}(1.1719) = 0$

$N_E = 0.703 \text{ kN}$ **Ans**

$+\uparrow \Sigma F_y = 0; \quad V_E - 0.625 + \frac{4}{5}(1.1719) = 0$

$V_E = -0.3125 \text{ kN}$ **Ans**

$\curvearrowleft + \Sigma M_E = 0; \quad -M_E - 0.625(0.25) + \frac{4}{5}(1.1719)(0.5) = 0$

$M_E = 0.3125 \text{ kN} \cdot \text{m}$ **Ans**

Negative sign indicates that V_E acts in the opposite direction to that shown on FBD.

1–7. The beam supports the distributed load shown. Determine the resultant internal loadings on the cross section through point C. Assume the reactions at the supports A and B are vertical.

Support Reactions :

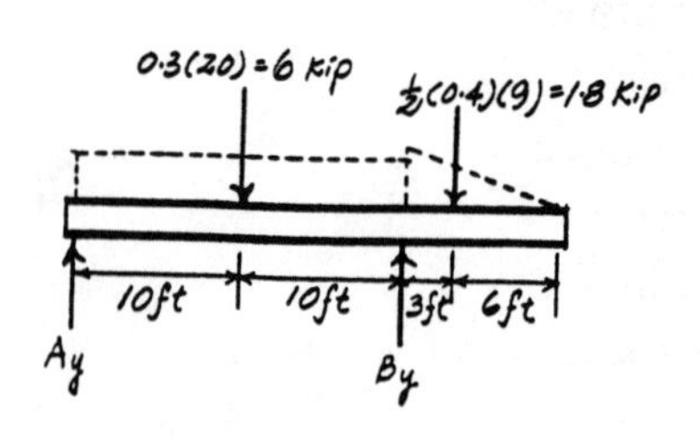

$\curvearrowleft + \Sigma M_A = 0;\quad B_y(20) - 6(10) - 1.8(23) = 0$

$B_y = 5.07 \text{ kip}$

$+\uparrow \Sigma F_y = 0;\quad A_y + 5.07 - 6 - 1.8 = 0$

$A_y = 2.73 \text{ kip}$

Equations of Equilibrium : For point C

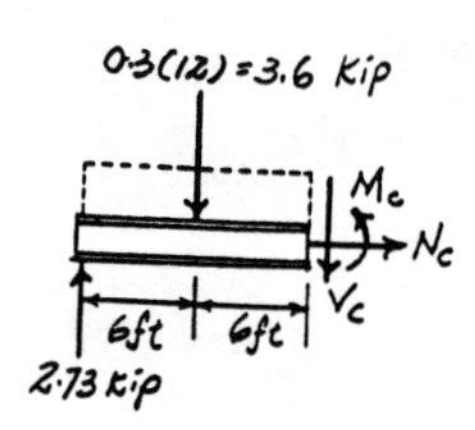

$\xrightarrow{+} \Sigma F_x = 0;\quad N_C = 0$ **Ans**

$+\uparrow \Sigma F_y = 0;\quad 2.73 - 3.60 - V_C = 0$

$V_C = -0.870 \text{ kip}$ **Ans**

$\curvearrowleft + \Sigma M_C = 0;\quad M_C + 3.60(6) - 2.73\,(12) = 0$

$M_C = 11.2 \text{ kip}\cdot\text{ft}$ **Ans**

Negative sign indicates that $\mathbf{V_C}$ acts in the opposite direction to that shown on FBD.

***1–8.** The beam supports the distributed load shown. Determine the resultant internal loadings on the cross sections through points D and E. Assume the reactions at the supports A and B are vertical.

Support Reactions :

$\curvearrowleft + \Sigma M_A = 0;\quad B_y(20) - 6(10) - 1.8(23) = 0$

$B_y = 5.07 \text{ kip}$

$+\uparrow \Sigma F_y = 0;\quad A_y + 5.07 - 6 - 1.8 = 0$

$A_y = 2.73 \text{ kip}$

Equations of Equilibrium : For point D

$\xrightarrow{+} \Sigma F_x = 0;\quad N_D = 0$ **Ans**

$+\uparrow \Sigma F_y = 0;\quad 2.73 - 1.8 - V_D = 0$

$V_D = 0.930 \text{ kip}$ **Ans**

$\curvearrowleft + \Sigma M_D = 0;\quad M_D + 1.8(3) - 2.73(6) = 0$

$M_D = 11.0 \text{ kip}\cdot\text{ft}$ **Ans**

Equations of Equilibrium : For point E

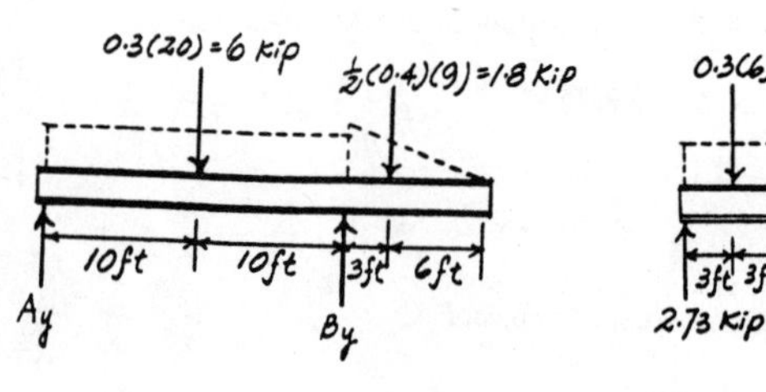

$\xleftarrow{+} \Sigma F_x = 0;\quad N_E = 0$ **Ans**

$+\uparrow \Sigma F_y = 0;\quad V_E - 0.45 = 0$

$V_E = 0.450 \text{ kip}$ **Ans**

$\curvearrowleft + \Sigma M_E = 0;\quad -M_E - 0.45(1.5) = 0$

$M_E = -0.675 \text{ kip}\cdot\text{ft}$ **Ans**

Negative sign indicates that $\mathbf{M_E}$ acts in the opposite direction to that shown on FBD.

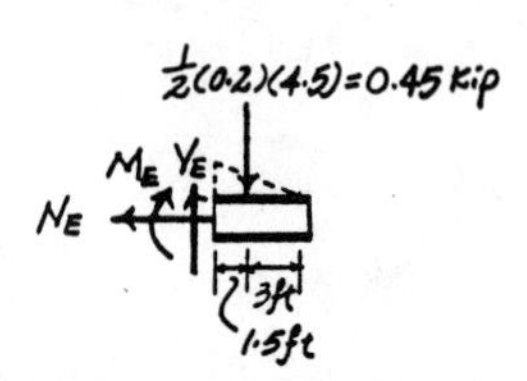

1–9. The boom DF of the jib crane and the column DE have a uniform weight of 50 lb/ft. If the hoist and load weigh 300 lb, determine the resultant internal loadings in the crane on cross sections through points A, B, and C.

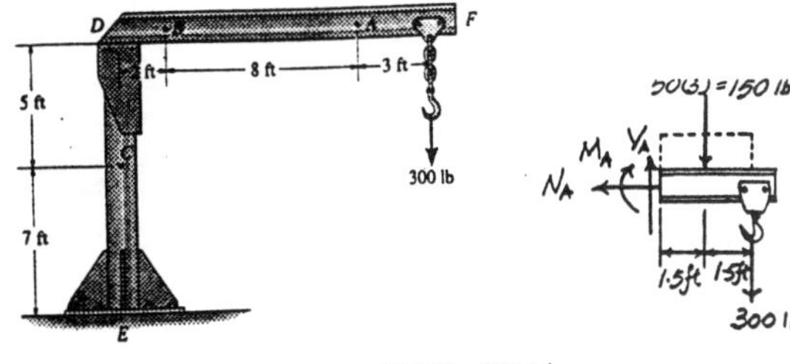

Equations of Equilibrium : For point A

$\overset{+}{\leftarrow} \Sigma F_x = 0;\qquad N_A = 0$ **Ans**

$+\uparrow \Sigma F_y = 0;\qquad V_A - 150 - 300 = 0$

$V_A = 450\text{ lb}$ **Ans**

$\circlearrowleft + \Sigma M_A = 0;\qquad -M_A - 150(1.5) - 300(3) = 0$

$M_A = -1125\text{ lb}\cdot\text{ft} = -1.125\text{ kip}\cdot\text{ft}$ **Ans**

Negative sign indicates that $\mathbf{M}_A$ acts in the opposite direction to that shown on FBD.

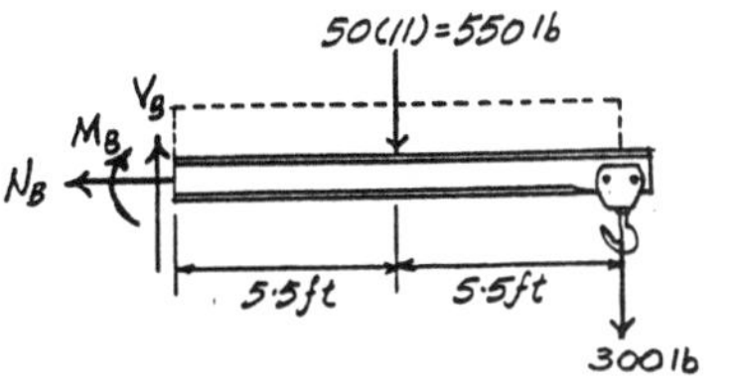

Equations of Equilibrium : For point B

$\overset{+}{\leftarrow} \Sigma F_x = 0;\qquad N_B = 0$ **Ans**

$+\uparrow \Sigma F_y = 0;\qquad V_B - 550 - 300 = 0$

$V_B = 850\text{ lb}$ **Ans**

$\circlearrowleft + \Sigma M_B = 0;\qquad -M_B - 550(5.5) - 300(11) = 0$

$M_B = -6325\text{ lb}\cdot\text{ft} = -6.325\text{ kip}\cdot\text{ft}$ **Ans**

Negative sign indicates that $\mathbf{M}_B$ acts in the opposite direction to that shown on FBD.

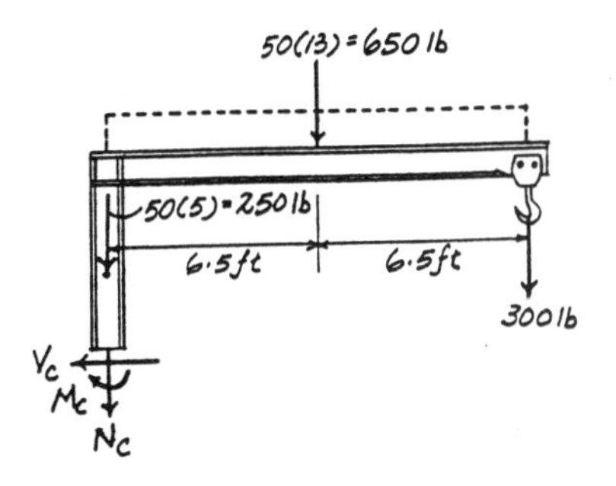

Equations of Equilibrium : For point C

$\overset{+}{\leftarrow} \Sigma F_x = 0;\qquad V_C = 0$ **Ans**

$+\uparrow \Sigma F_y = 0;\qquad -N_C - 250 - 650 - 300 = 0$

$N_C = -1200\text{ lb} = -1.20\text{ kip}$ **Ans**

$\circlearrowleft + \Sigma M_C = 0;\qquad -M_C - 650(6.5) - 300(13) = 0$

$M_C = -8125\text{ lb}\cdot\text{ft} = -8.125\text{ kip}\cdot\text{ft}$ **Ans**

Negative signs indicate that $\mathbf{N}_C$ and $\mathbf{M}_C$ act in the opposite direction to that shown on FBD.

1–10. The force $F = 80$ lb acts on the gear tooth. Determine the resultant internal loadings on the root of the tooth, i.e., at the centroid point A of section a–a.

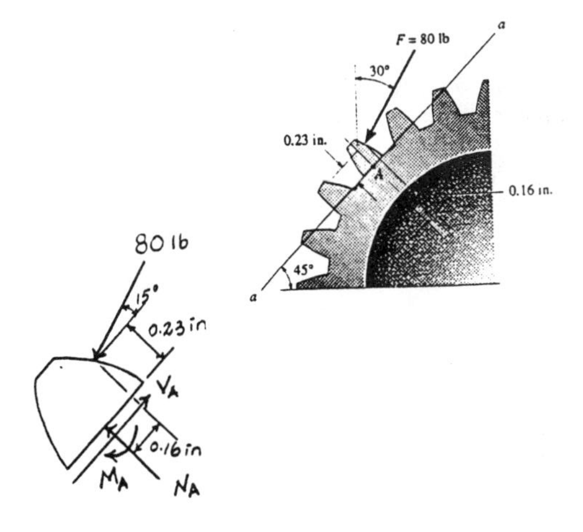

Equations of Equilibrium : For section $a-a$

$+\nearrow \Sigma F_{x'} = 0;\qquad V_A - 80\cos 15° = 0$

$V_A = 77.3\text{ lb}$ **Ans**

$\nwarrow + \Sigma F_{y'} = 0;\qquad N_A - 80\sin 15° = 0$

$N_A = 20.7\text{ lb}$ **Ans**

$\circlearrowleft + \Sigma M_A = 0;\qquad -M_A - 80\sin 15°(0.16) + 80\cos 15°(0.23) = 0$

$M_A = 14.5\text{ lb}\cdot\text{in.}$ **Ans**

1–11. The telescoping boom is constructed from three segments. When extended to its full position as shown, AB weighs 80 lb/ft, BC weighs 55 lb/ft, and CD weighs 25 lb/ft. If the platform and worker have a total weight of 275 lb and center of gravity at G, determine the resultant internal loadings acting on the cross section through point E.

Equations of Equilibrium : For point E

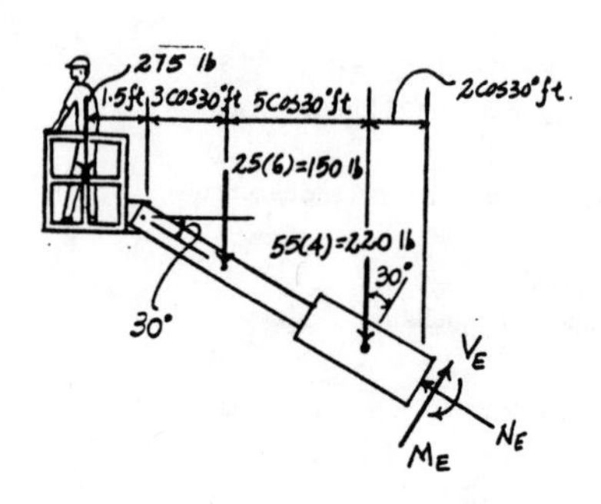

$\nwarrow + \Sigma F_{x'} = 0;\quad N_E - 275\sin 30° - 150\sin 30° - 220\sin 30° = 0$

$N_E = 322.5 \text{ lb}$ **Ans**

$\nearrow + \Sigma F_{y'} = 0;\quad V_E - 275\cos 30° - 150\cos 30° - 220\cos 30° = 0$

$V_E = 559 \text{ lb}$ **Ans**

$\circlearrowleft + \Sigma M_E = 0;\quad -M_E + 275(1.5 + 10\cos 30°) + 150(7\cos 30°) + 220(2\cos 30°) = 0$

$M_E = 4084 \text{ lb}\cdot\text{ft} = 4.08 \text{ kip}\cdot\text{ft}$ **Ans**

***1–12.** Solve Prob. 1–11 for the internal loadings on the cross section through point F.

Equations of Equilibrium : For point F

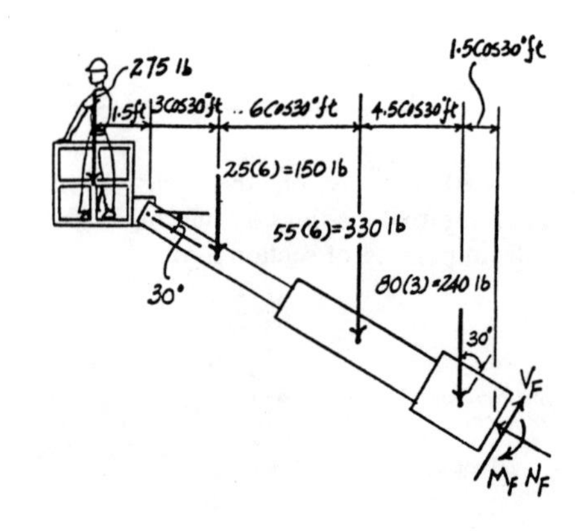

$\nwarrow + \Sigma F_{x'} = 0;\quad N_F - 275\sin 30° - 150\sin 30° - 330\sin 30° - 240\sin 30° = 0$

$N_F = 497.5 \text{ lb}$ **Ans**

$\nearrow + \Sigma F_{y'} = 0;\quad V_F - 275\cos 30° - 150\cos 30° - 330\cos 30° - 240\cos 30° = 0$

$V_F = 862 \text{ lb}$ **Ans**

$\circlearrowleft + \Sigma M_F = 0;\quad -M_F + 275(1.5 + 15\cos 30°) + 150(12\cos 30°) + 330(6\cos 30°) + 240(1.5\cos 30°) = 0$

$M_F = 7570 \text{ lb.ft} = 7.57 \text{ kip}\cdot\text{ft}$ **Ans**

1–13. The "stick" on a long front excavator is 5.4 m long and supports a bucket of soil that has a mass of 200 kg and center of mass at G. Determine the resultant internal loadings acting on the cross section through C when the stick is held in the position shown. Neglect the mass of the stick.

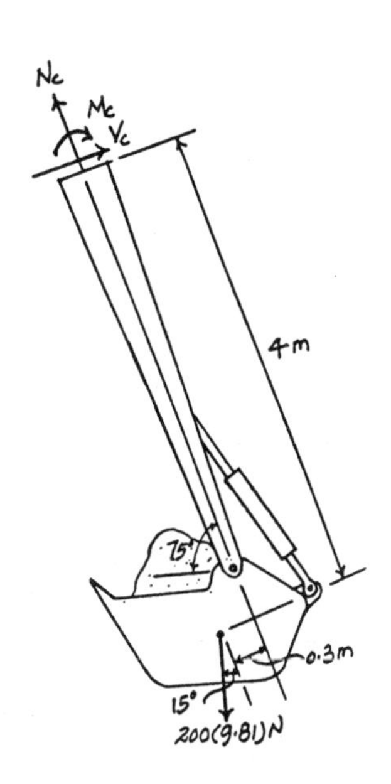

Equations of Equilibrium :

$+\nearrow \Sigma F_{x'} = 0;\quad V_C - 200(9.81)\sin 15° = 0$

$V_C = 508 \text{ N}$ **Ans**

$+\nwarrow \Sigma F_{y'} = 0;\quad N_C - 200(9.81)\cos 15° = 0$

$N_C = 1895 \text{ N} = 1.90 \text{ kN}$ **Ans**

$\curvearrowleft + \Sigma M_C = 0;\quad -M_C + 200(9.81)\cos 15°(0.3) - 200(9.81)\sin 15°(4) = 0$

$M_C = -1463 \text{ N}\cdot\text{m} = -1.46 \text{ kN}\cdot\text{m}$ **Ans**

Negative sign indicates that $\mathbf{M}_C$ acts in the opposite direction to that shown on FBD.

1–14. Determine the resultant internal normal and shear forces in the member at (*a*) section *a–a* and (*b*) section *b–b*, each of which passes through point *A*. Take $\theta = 60°$. The 650-N load is applied along the centroidal axis of the member.

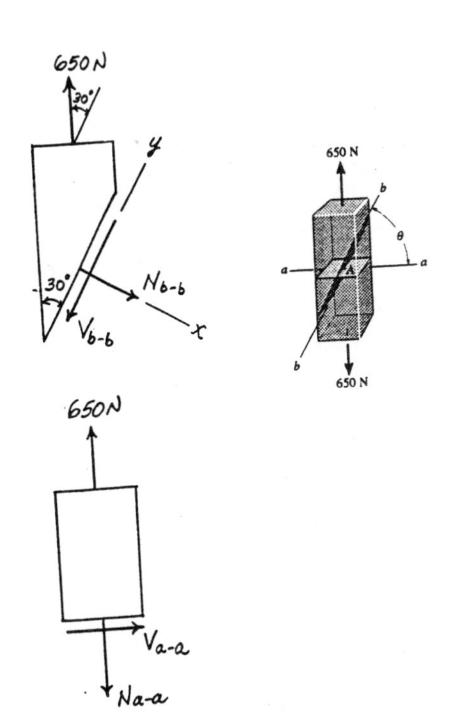

Equations of Equilibrium : For section *a-a*

$+\uparrow \Sigma F_y = 0;\quad 650 - N_{a-a} = 0$

$N_{a-a} = 650 \text{ N}$ **Ans**

$\xrightarrow{+} \Sigma F_x = 0;\quad V_{a-a} = 0$ **Ans**

Equations of Equilibrium : For section *b-b*

$+\nearrow \Sigma F_y = 0;\quad 650\cos 30° - V_{b-b} = 0$

$V_{b-b} = 563 \text{ N}$ **Ans**

$\searrow + \Sigma F_x = 0;\quad N_{b-b} - 650 \sin 30° = 0$

$N_{b-b} = 325 \text{ N}$ **Ans**

1–15. Determine the resultant internal normal and shear forces in the member at section b–b, each as a function of θ. Plot these results for $0° \le \theta \le 90°$. The 650-N load is applied along the centroidal axis of the member.

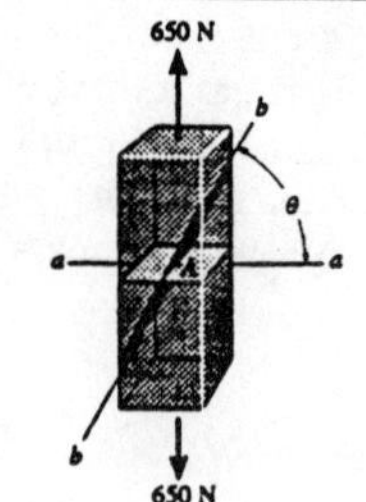

Equations of Equilibrium : For section $b-b$

$$\nearrow\!+ \Sigma F_x = 0; \quad N_{b-b} - 650 \cos\theta = 0$$

$$N_{b-b} = 650 \cos\theta \qquad \textbf{Ans}$$

$$+\!\nearrow \Sigma F_y = 0; \quad -V_{b-b} + 650 \sin\theta = 0$$

$$V_{b-b} = 650 \sin\theta \qquad \textbf{Ans}$$

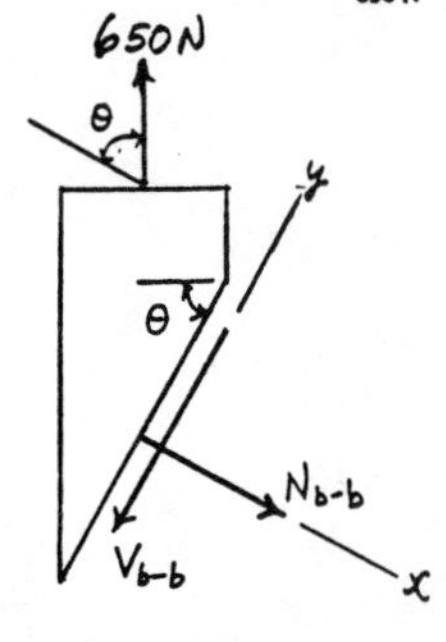

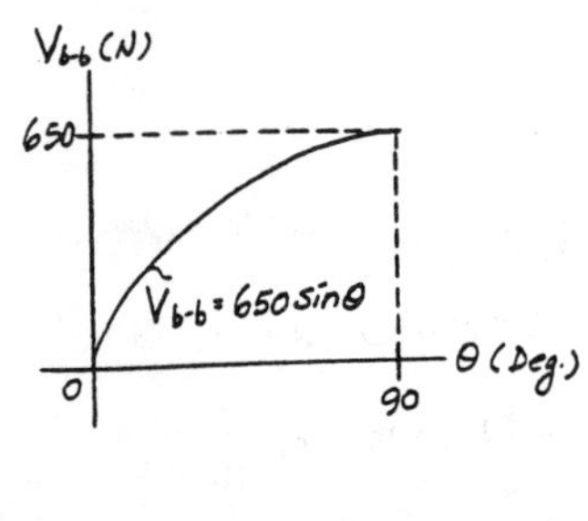

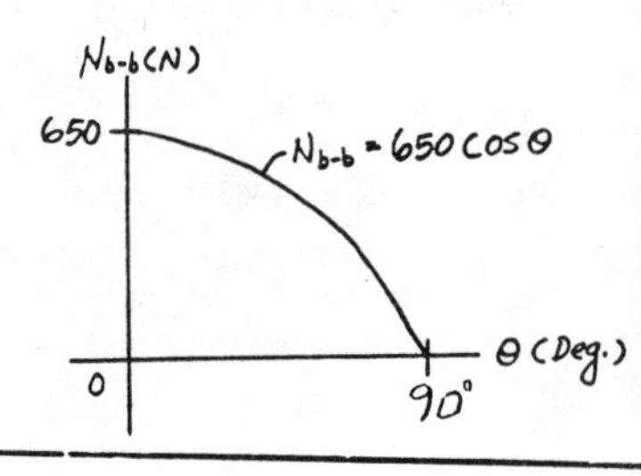

***1–16.** The sky hook is used to support the cable of a scaffold over the side of a building. If it consists of a smooth rod that contacts the parapet of a wall at points A, B, and C, determine the normal force, shear force, and moment on the cross section at points D and E.

Support Reactions :

$$+\uparrow \Sigma F_y = 0; \quad N_B - 18 = 0 \qquad N_B = 18.0 \text{ kN}$$

$$\circlearrowleft\!+ \Sigma M_C = 0; \quad 18(0.7) - 18.0(0.2) - N_A(0.1) = 0$$

$$N_A = 90.0 \text{ kN}$$

$$\xrightarrow{+} \Sigma F_x = 0; \quad N_C - 90.0 = 0 \qquad N_C = 90.0 \text{ kN}$$

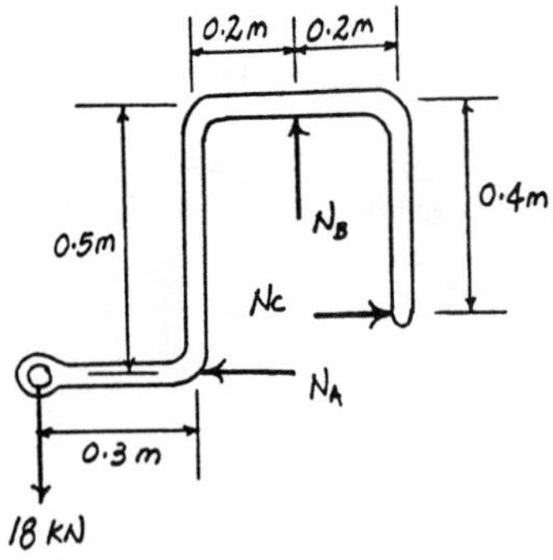

Equations of Equilibrium : For point D

$$\xrightarrow{+} \Sigma F_x = 0; \quad V_D - 90.0 = 0$$

$$V_D = 90.0 \text{ kN} \qquad \textbf{Ans}$$

$$+\uparrow \Sigma F_y = 0; \quad N_D - 18 = 0$$

$$N_D = 18.0 \text{ kN} \qquad \textbf{Ans}$$

$$\circlearrowleft\!+ \Sigma M_D = 0; \quad M_D + 18(0.3) - 90.0(0.3) = 0$$

$$M_D = 21.6 \text{ kN} \cdot \text{m} \qquad \textbf{Ans}$$

Equations of Equilibrium : For point E

$$\xrightarrow{+} \Sigma F_x = 0; \quad 90.0 - V_E = 0$$

$$V_E = 90.0 \text{ kN} \qquad \textbf{Ans}$$

$$+\uparrow \Sigma F_y = 0; \quad N_E = 0 \qquad \textbf{Ans}$$

$$\circlearrowleft\!+ \Sigma M_E = 0; \quad 90.0(0.2) - M_E = 0$$

$$M_E = 18.0 \text{ kN} \cdot \text{m} \qquad \textbf{Ans}$$

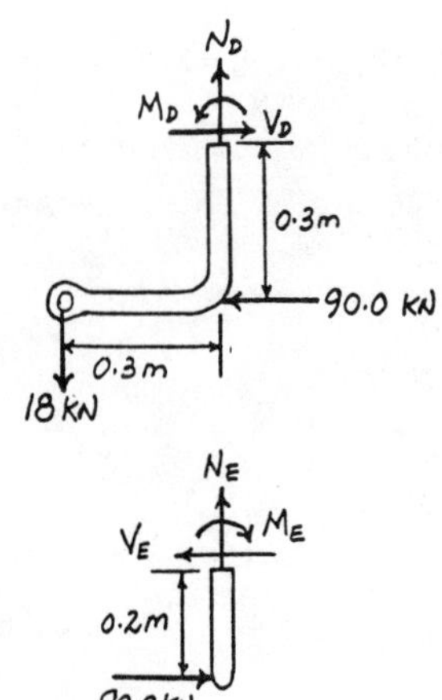

1–17. Determine the normal force, shear force, and moment at a section through point C. Take $P = 8$ kN.

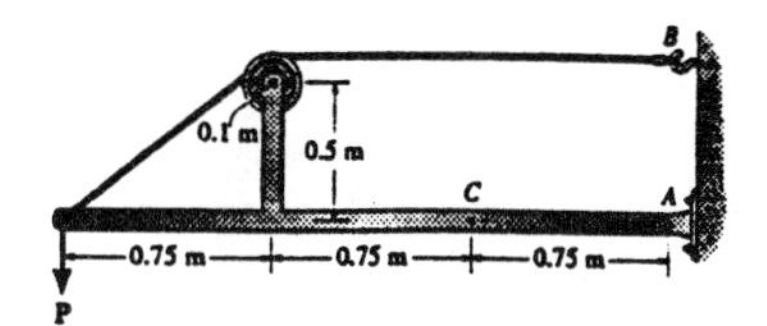

Support Reactions :

$\circlearrowleft + \Sigma M_A = 0;\quad 8(2.25) - T(0.6) = 0 \qquad T = 30.0 \text{ kN}$

$\xrightarrow{+} \Sigma F_x = 0;\quad 30.0 - A_x = 0 \qquad A_x = 30.0 \text{ kN}$

$+\uparrow \Sigma F_y = 0;\quad A_y - 8 = 0 \qquad A_y = 8.00 \text{ kN}$

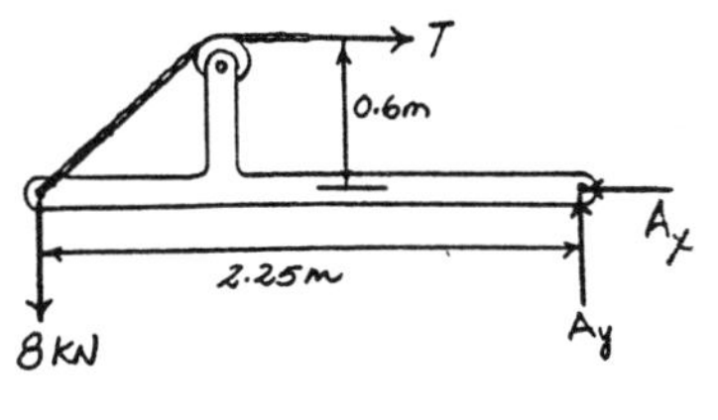

Equations of Equilibrium : For point C

$\xrightarrow{+} \Sigma F_x = 0;\quad -N_C - 30.0 = 0$

$N_C = -30.0 \text{ kN}$ **Ans**

$+\uparrow \Sigma F_y = 0;\quad V_C + 8.00 = 0$

$V_C = -8.00 \text{ kN}$ **Ans**

$\circlearrowleft + \Sigma M_C = 0;\quad 8.00(0.75) - M_C = 0$

$M_C = 6.00 \text{ kN} \cdot \text{m}$ **Ans**

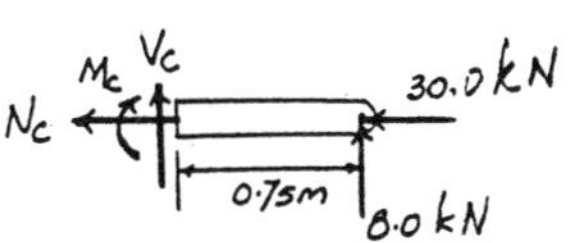

Negative signs indicate that N_C and V_C act in the opposite direction to that shown on FBD.

1–18. The cable will fail when subjected to a tension of 2 kN. Determine the largest vertical load P the frame will support and calculate the internal normal force, shear force, and moment at the cross section through point C for this loading.

Support Reactions :

$\circlearrowleft + \Sigma M_A = 0;\quad P(2.25) - 2(0.6) = 0$

$P = 0.5333 \text{ kN} = 0.533 \text{ kN}$ **Ans**

$\xrightarrow{+} \Sigma F_x = 0;\quad 2 - A_x = 0 \qquad A_x = 2.00 \text{ kN}$

$+\uparrow \Sigma F_y = 0;\quad A_y - 0.5333 = 0 \qquad A_y = 0.5333 \text{ kN}$

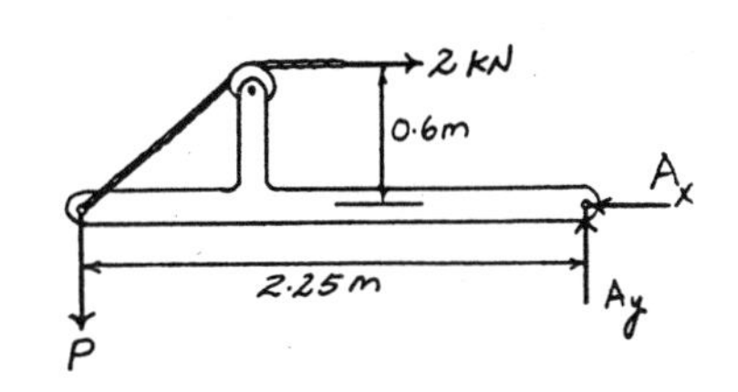

Equations of Equilibrium : For point C

$\xrightarrow{+} \Sigma F_x = 0;\quad -N_C - 2.00 = 0$

$N_C = -2.00 \text{ kN}$ **Ans**

$+\uparrow \Sigma F_y = 0;\quad V_C + 0.5333 = 0$

$V_C = -0.533 \text{ kN}$ **Ans**

$\circlearrowleft + \Sigma M_C = 0;\quad 0.5333(0.75) - M_C = 0$

$M_C = 0.400 \text{ kN} \cdot \text{m}$ **Ans**

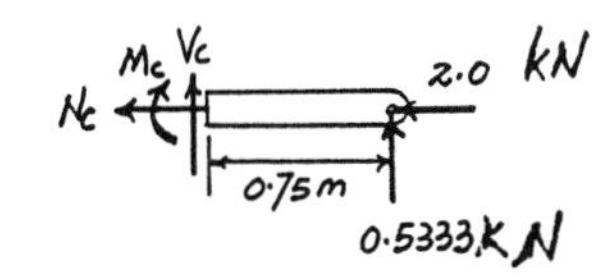

Negative signs indicate that N_C and V_C act in the opposite direction to that shown on FBD.

1–19. Determine the resultant internal loadings on the cross sections through points D and E on the frame.

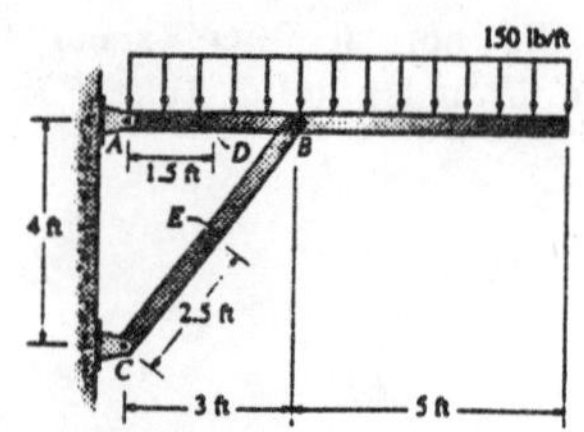

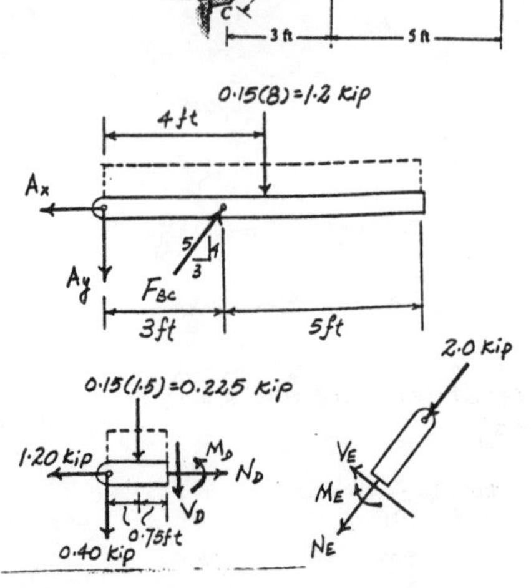

Support Reactions : Member BC is a two force member.

$\circlearrowleft + \Sigma M_A = 0;\quad \frac{4}{5}F_{BC}(3) - 1.20(4) = 0$

$F_{BC} = 2.00$ kip

$\xrightarrow{+} \Sigma F_x = 0;\quad \frac{3}{5}(2.00) - A_x = 0 \quad A_x = 1.20$ kip

$+\uparrow \Sigma F_y = 0;\quad \frac{4}{5}(2.00) - 1.20 - A_y = 0$

$A_y = 0.400$ kip

Equations of Equilibrium : For point D

$\xrightarrow{+} \Sigma F_x = 0;\quad N_D - 1.20 = 0 \quad N_D = 1.20$ kip **Ans**

$+\uparrow \Sigma F_y = 0;\quad -V_D - 0.225 - 0.400 = 0$

$V_D = -0.625$ kip **Ans**

$\circlearrowleft + \Sigma M_D = 0;\quad M_D + 0.225(0.75) + 0.400(1.5) = 0$

$M_D = -0.769$ kip · ft **Ans**

Negative signs indicate that V_D and M_D act in the opposite direction to that shown on FBD.

Equations of Equilibrium : For point E

$\nearrow + \Sigma F_{x'} = 0;\quad -N_E - 2.00 = 0$

$N_E = -2.00$ kip **Ans**

$\nwarrow + \Sigma F_{y'} = 0;\quad V_E = 0$ **Ans**

$\circlearrowleft + \Sigma M_E = 0;\quad M_E = 0$ **Ans**

Negative sign indicates that N_E acts in the opposite direction to that shown on FBD.

***1–20.** The drum lifter suspends the 500-lb drum. The linkage is pin connected to the plate at A and B. The gripping action on the drum chime is such that only horizontal and vertical forces are exerted on the drum at G and H. Determine the resultant internal loadings on the cross section through point I.

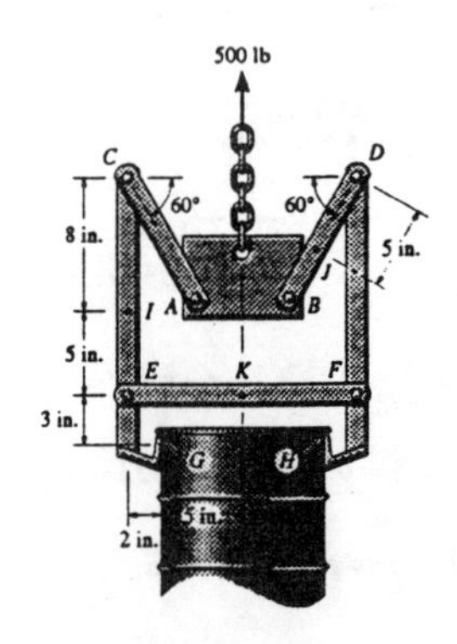

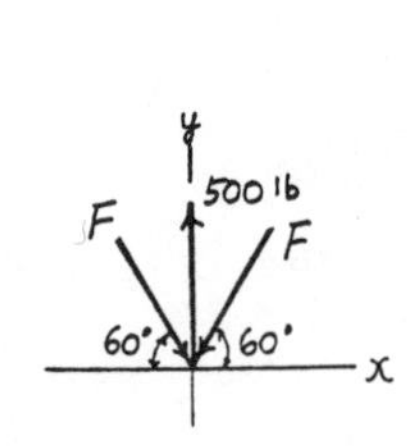

Equations of Equilibrium : Members AC and BD are two-force members.

$+\uparrow \Sigma F_y = 0;\quad 500 - 2F\sin 60° = 0$

$F = 288.7$ lb

Equations of Equilibrium : For point I

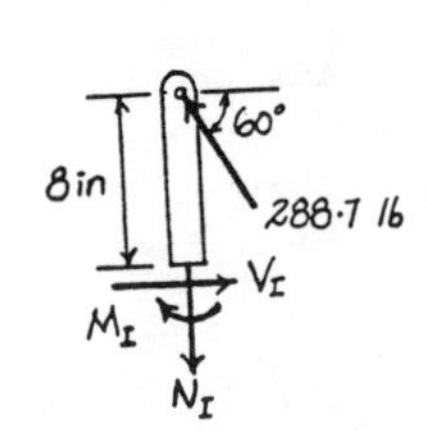

$\xrightarrow{+} \Sigma F_x = 0;\quad V_I - 288.7\cos 60° = 0$

$V_I = 144$ lb **Ans**

$+\uparrow \Sigma F_y = 0;\quad 288.7\sin 60° - N_I = 0$

$N_I = 250$ lb **Ans**

$\circlearrowleft + \Sigma M_I = 0;\quad 288.7\cos 60°(8) - M_I = 0$

$M_I = 1154.7$ lb · in. $= 1.15$ kip · in. **Ans**

1–21. Determine the resultant internal loadings on the cross sections through points K and J on the drum lifter in Prob. 1–20.

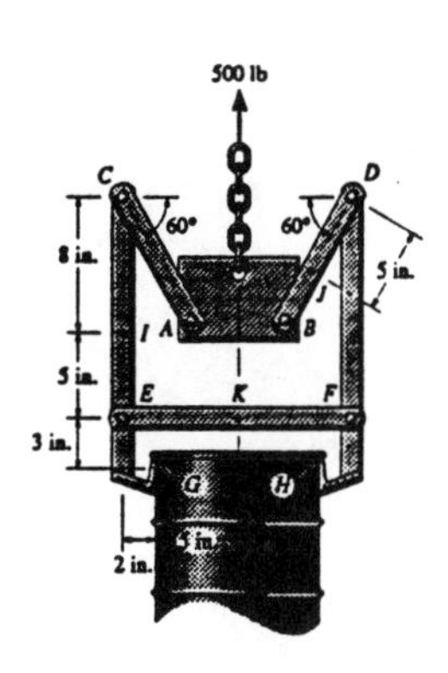

Equations of Equilibrium : Members AC and BD are two force members.

$+\uparrow \Sigma F_y = 0; \qquad 500 - 2F \sin 60° = 0$

$F = 288.7 \text{ lb}$

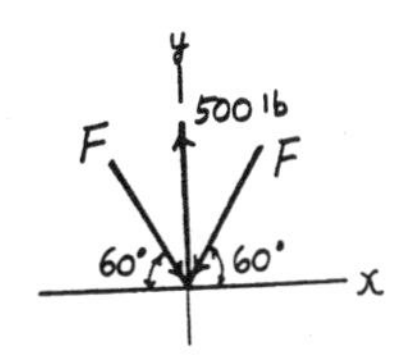

Equations of Equilibrium : For point J

$\Sigma F_{y'} = 0; \qquad V_J = 0 \qquad$ **Ans**

$\Sigma F_{x'} = 0; \qquad 288.7 + N_J = 0 \qquad N_J = -289 \text{ lb} \qquad$ **Ans**

$\circlearrowleft + \Sigma M_J = 0; \qquad M_J = 0 \qquad$ **Ans**

Negative sign indicates that N_J acts in the opposite direction to that shown on FBD.

Support Reactions : For member DFH

$\circlearrowleft + \Sigma M_H = 0; \qquad F_{EF}(3) - 288.7\cos 60°(16) + 288.7 \sin 60°(2) = 0$

$F_{EF} = 603.1 \text{ lb}$

Equations of Equilibrium : For point K

$\overset{+}{\leftarrow} \Sigma F_x = 0; \qquad N_K - 603. = 0 \qquad N_K = 603 \text{ lb} \qquad$ **Ans**

$+\uparrow \Sigma F_y = 0; \qquad V_K = 0 \qquad$ **Ans**

$\circlearrowleft + \Sigma M_K = 0; \qquad M_K = 0 \qquad$ **Ans**

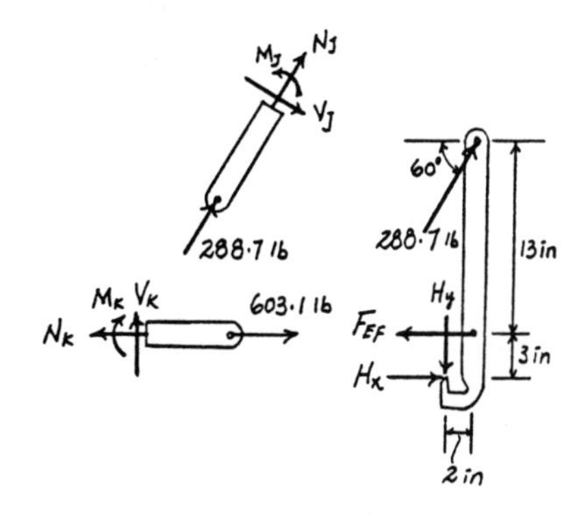

1–22. Determine the resultant internal loadings in the beam at cross sections through points D and E. Point E is just to the right of the 3-kip load.

Support Reactions : For member AB

$\curvearrowleft + \Sigma M_B = 0; \quad 9.00(4) - A_y(12) = 0 \quad A_y = 3.00 \text{ kip}$

$\xrightarrow{+} \Sigma F_x = 0; \quad B_x = 0$

$+\uparrow \Sigma F_y = 0; \quad B_y + 3.00 - 9.00 = 0 \quad B_y = 6.00 \text{ kip}$

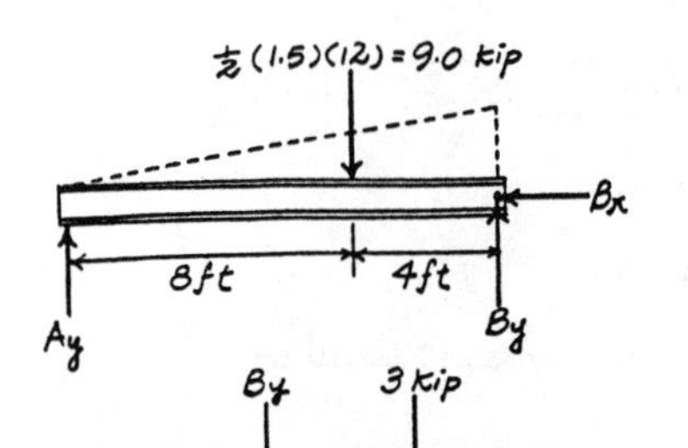

Equations of Equilibrium : For point D

$\xrightarrow{+} \Sigma F_x = 0; \quad N_D = 0$ **Ans**

$+\uparrow \Sigma F_y = 0; \quad 3.00 - 2.25 - V_D = 0$

$V_D = 0.750 \text{ kip}$ **Ans**

$\curvearrowleft + \Sigma M_D = 0; \quad M_D + 2.25(2) - 3.00(6) = 0$

$M_D = 13.5 \text{ kip} \cdot \text{ft}$ **Ans**

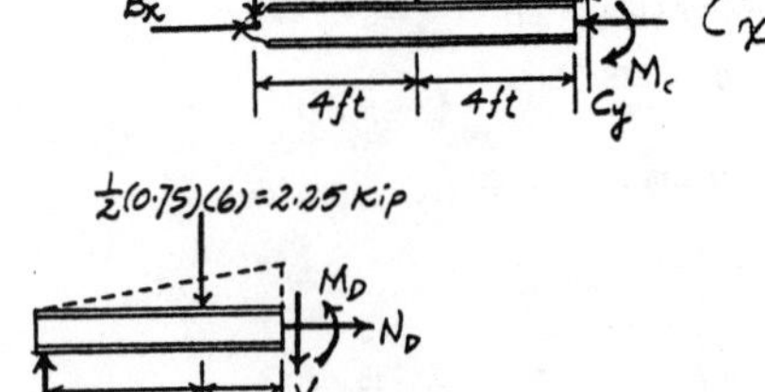

Equations of Equilibrium : For point E

$\xrightarrow{+} \Sigma F_x = 0; \quad N_E = 0$ **Ans**

$+\uparrow \Sigma F_y = 0; \quad -6.00 - 3 - V_E = 0$

$V_E = -9.00 \text{ kip}$ **Ans**

$\curvearrowleft + \Sigma M_E = 0; \quad M_E + 6.00(4) = 0$

$M_E = -24.0 \text{ kip} \cdot \text{ft}$ **Ans**

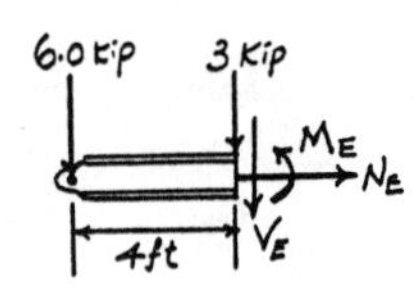

Negative signs indicate that M_E and V_E act in the opposite direction to that shown on FBD.

1–23. The wishbone construction of the power pole supports the three lines, each exerting a force of 800 lb on the bracing struts. If the struts are pin connected at A, B, and C, determine the resultant internal loadings at cross sections through points D, E, and F.

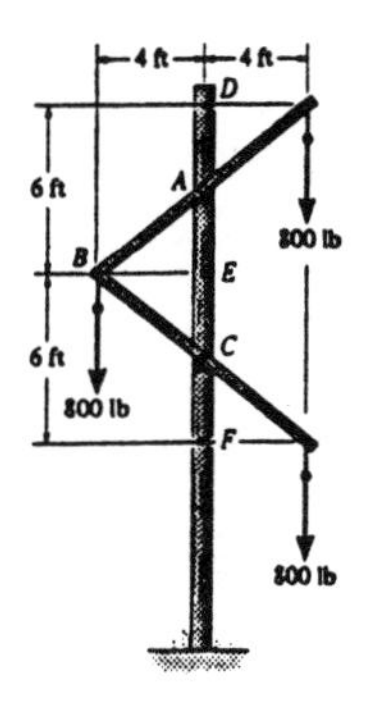

Support Reaction : FBD(a) and (b).

$$\circlearrowleft+\Sigma M_A = 0; \quad B_y(4) + B_x(3) - 800(4) = 0 \qquad [1]$$

$$\circlearrowleft+\Sigma M_C = 0; \quad B_x(3) + 800(4) - B_y(4) - 800(4) = 0 \qquad [2]$$

Solving Eq. [1] and [2] yields

$$B_y = 400.0 \text{ lb} \qquad B_x = 533.33 \text{ lb}$$

From FBD (a)

$$\overset{+}{\rightarrow}\Sigma F_x = 0; \quad 533.33 - A_x = 0 \quad A_x = 533.33 \text{ lb}$$
$$+\uparrow\Sigma F_y = 0; \quad A_y - 800 - 400.0 = 0 \quad A_y = 1200 \text{ lb}$$

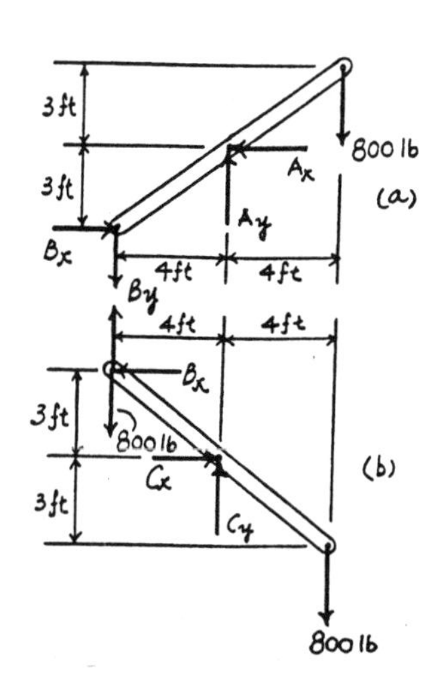

From FBD (b)

$$\overset{+}{\rightarrow}\Sigma F_x = 0; \quad C_x - 533.33 = 0 \quad C_x = 533.33 \text{ lb}$$
$$+\uparrow\Sigma F_y = 0; \quad C_y + 400.0 - 800 - 800 = 0 \quad C_y = 1200 \text{ lb}$$

Equations of Equilibrium : For point D [FBD(c)].

$$\overset{+}{\rightarrow}\Sigma F_x = 0; \quad V_D = 0 \qquad \textbf{Ans}$$

$$+\uparrow\Sigma F_y = 0; \quad N_D = 0 \qquad \textbf{Ans}$$

$$\circlearrowleft+\Sigma M_D = 0; \quad M_D = 0 \qquad \textbf{Ans}$$

For point E [FBD(d)].

$$\overset{+}{\rightarrow}\Sigma F_x = 0; \quad 533.33 - V_E = 0 \quad V_E = 533 \text{ lb} \qquad \textbf{Ans}$$

$$+\uparrow\Sigma F_y = 0; \quad N_E - 1200 = 0 \quad N_E = 1200 \text{ lb} \qquad \textbf{Ans}$$

$$\circlearrowleft+\Sigma M_E = 0; \quad M_E - 533.33(3) = 0 \quad M_E = 1600 \text{ lb}\cdot\text{ft} \qquad \textbf{Ans}$$

For point F [FBD(e)].

$$\overset{+}{\rightarrow}\Sigma F_x = 0; \quad V_F + 533.33 - 533.33 = 0 \quad V_F = 0 \qquad \textbf{Ans}$$

$$+\uparrow\Sigma F_y = 0; \quad N_F - 1200 - 1200 = 0 \quad N_F = 2400 \text{ lb} \qquad \textbf{Ans}$$

$$\circlearrowleft+\Sigma M_F = 0; \quad M_F - 533.33(6) = 0 \quad M_F = 3200 \text{ lb}\cdot\text{ft} \qquad \textbf{Ans}$$

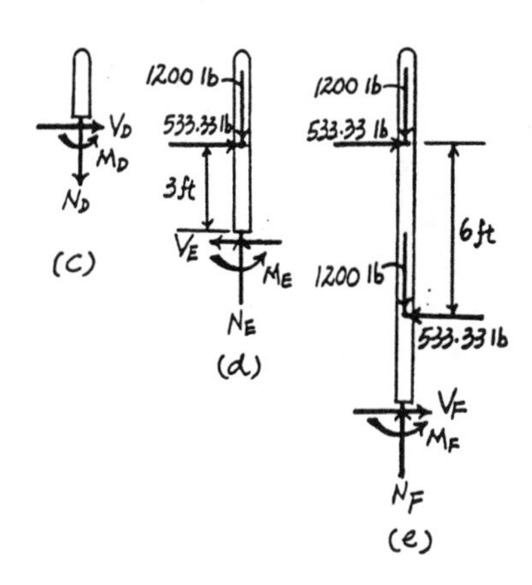

***1–24.** Determine the resultant internal loadings acting on section a–a through the centroid, point C on the beam.

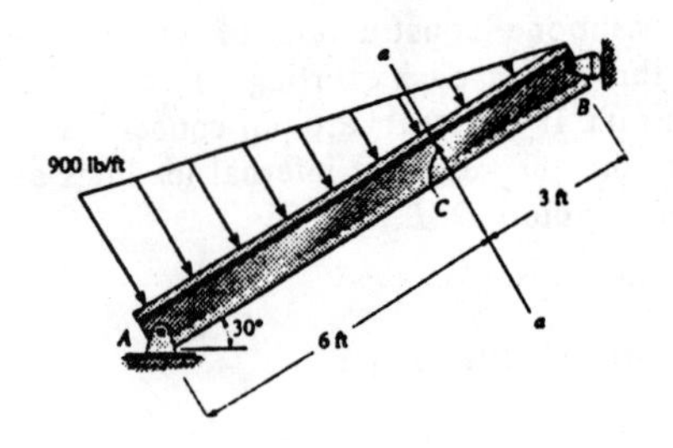

Support Reaction :

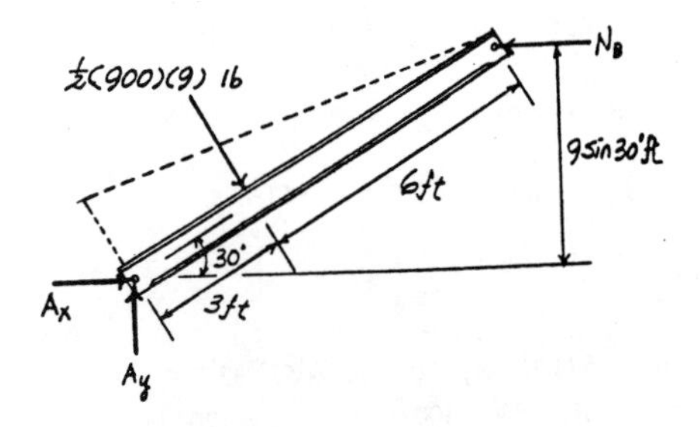

$$\curvearrowleft + \Sigma M_A = 0; \quad N_B(9\sin 30°) - \frac{1}{2}(900)(9)(3) = 0$$

$$N_B = 2700 \text{ lb}$$

Equations of Equilibrium : For section $a-a$

$$\nearrow + \Sigma F_{x'} = 0; \quad -N_{a-a} - 2700\cos 30° = 0$$

$$N_{a-a} = -2338 \text{ lb} = -2.34 \text{ kip} \quad \textbf{Ans}$$

$$\nwarrow + \Sigma F_{y'} = 0; \quad V_{a-a} - \frac{1}{2}(300)(3) + 2700\sin 30° = 0$$

$$V_{a-a} = -900 \text{ lb} = -0.900 \text{ kip} \quad \textbf{Ans}$$

$$\curvearrowleft + \Sigma M_C = 0; \quad -M_{a-a} - \frac{1}{2}(300)(3)(1)$$

$$+2700\sin 30°(3) = 0$$

$$M_{a-a} = 3600 \text{ lb}\cdot\text{ft} = 3.60 \text{ kip}\cdot\text{ft} \quad \textbf{Ans}$$

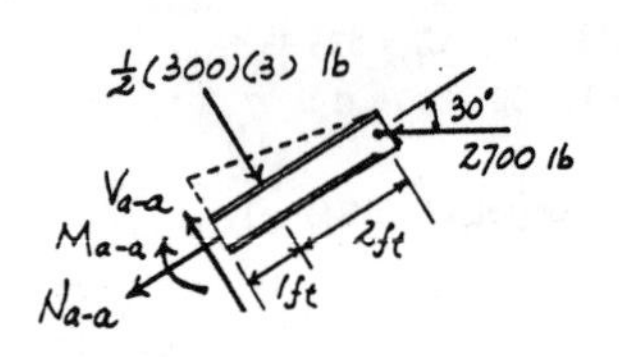

Negative signs indicate that N_{a-a} and V_{a-a} act in the opposite direction to that shown on FBD.

1–25. Determine the resultant internal loadings acting on section b–b through the centroid, point C on the beam.

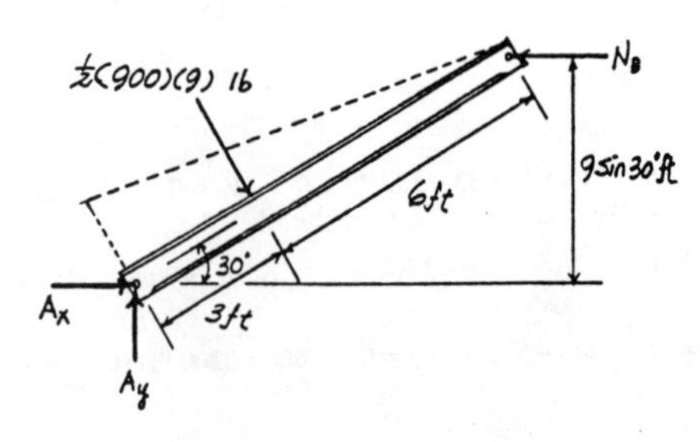

Support Reaction :

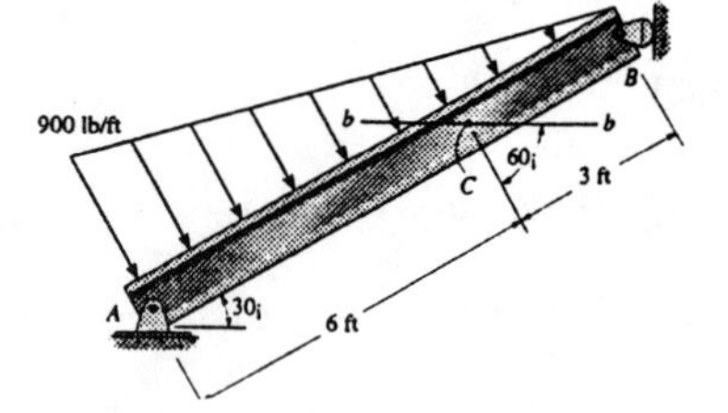

$$\curvearrowleft + \Sigma M_A = 0; \quad N_B(9\sin 30°) - \frac{1}{2}(900)(9)(3) = 0$$

$$N_B = 2700 \text{ lb}$$

Equations of Equilibrium : For section $b-b$

$$\xrightarrow{+} \Sigma F_x = 0; \quad V_{b-b} + \frac{1}{2}(300)(3)\sin 30° - 2700 = 0$$

$$V_{b-b} = 2475 \text{ lb} = 2.475 \text{ kip} \quad \textbf{Ans}$$

$$+\uparrow \Sigma F_y = 0; \quad N_{b-b} - \frac{1}{2}(300)(3)\cos 30° = 0$$

$$N_{b-b} = 389.7 \text{ lb} = 0.390 \text{ kip} \quad \textbf{Ans}$$

$$\curvearrowleft + \Sigma M_C = 0; \quad 2700(3\sin 30°)$$

$$-\frac{1}{2}(300)(3)(1) - M_{b-b} = 0$$

$$M_{b-b} = 3600 \text{ lb}\cdot\text{ft} = 3.60 \text{ kip}\cdot\text{ft} \quad \textbf{Ans}$$

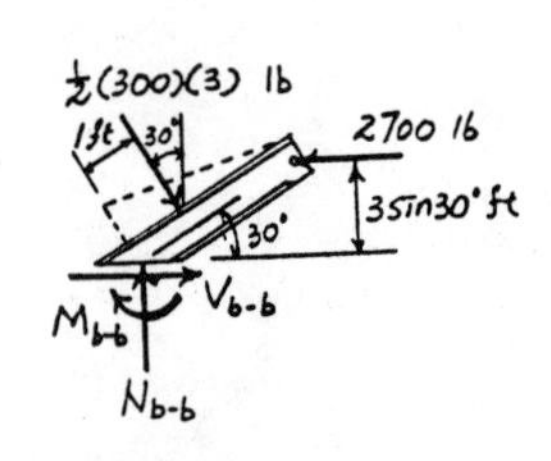

1–26. The support has a constant width w and a specific weight (weight/volume) γ. Determine the resultant internal loads acting at the centroid C of a cross section located a distance s from the apex.

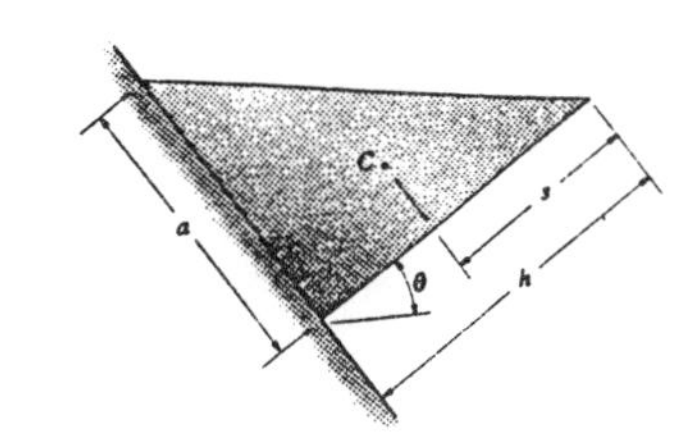

Equations of Equilibrium : The weight of the sectioned frustrum is W.

$$W = \gamma V = \gamma\left[\frac{1}{2}\left(\frac{s}{h}a\right)ws\right] = \frac{aws^2}{2h}\gamma$$

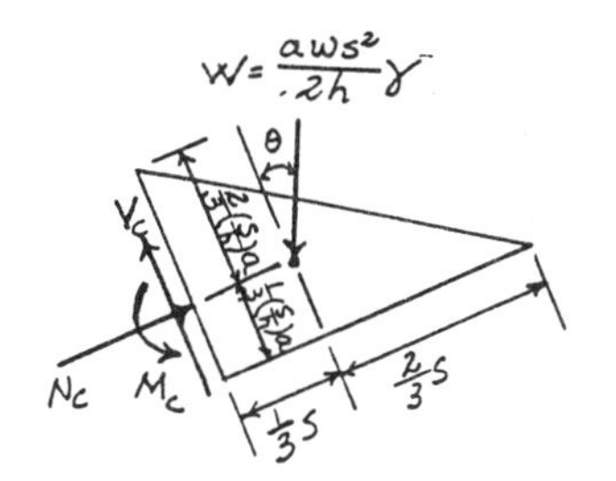

$+\Sigma F_{x'} = 0;\quad N_C - \frac{aws^2}{2h}\gamma(\sin\theta) = 0$

$$N_C = \frac{aws^2}{2h}\gamma\sin\theta \qquad \textbf{Ans}$$

$+\Sigma F_{y'} = 0;\quad V_C - \frac{aws^2}{2h}\gamma(\cos\theta) = 0$

$$V_C = \frac{aws^2}{2h}\gamma\cos\theta \qquad \textbf{Ans}$$

$+\Sigma M_C = 0;\quad M_C - \frac{aws^2}{2h}\gamma(\cos\theta)\left(\frac{1}{3}s\right) - \frac{aws^2}{2h}\gamma(\sin\theta)\left[\frac{1}{6}\left(\frac{s}{h}\right)a\right] = 0$

$$M_C = \frac{aws^3}{12h^2}\gamma(2h\cos\theta + a\sin\theta) \qquad \textbf{Ans}$$

1–27. The shaft is supported at its ends by two bearings A and B and is subjected to the forces applied to the pulleys fixed to the shaft. Determine the resultant internal loadings acting on the cross section through point C. The 400-N forces act in the $-z$ direction and the 200-N and 80-N forces act in the $+y$ direction. The journal bearings at A and B exert only y and z components of force on the shaft.

Support Reactions :

$\Sigma M_z = 0;\quad 160(0.4) + 400(0.7) - A_y(1.4) = 0$
$$A_y = 245.71\text{ N}$$

$\Sigma F_y = 0;\quad -245.71 - B_y + 400 + 160 = 0$
$$B_y = 314.29\text{ N}$$

$\Sigma M_y = 0;\quad 800(1.1) - A_z(1.4) = 0 \qquad A_z = 628.57\text{ N}$

$\Sigma F_z = 0;\quad B_z + 628.57 - 800 = 0 \qquad B_z = 171.43\text{ N}$

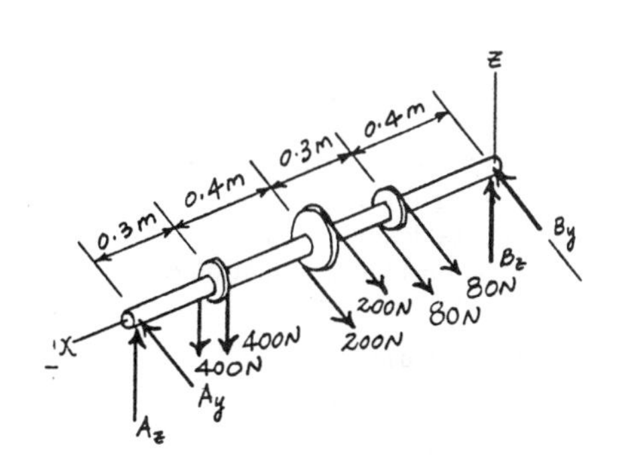

Equations of Equilibrium : For point C

$\Sigma F_x = 0;\quad (N_C)_x = 0 \qquad \textbf{Ans}$

$\Sigma F_y = 0;\quad -245.71 + (V_C)_y = 0$
$$(V_C)_y = 246\text{ N} \qquad \textbf{Ans}$$

$\Sigma F_z = 0;\quad 628.57 - 800 + (V_C)_z = 0$
$$(V_C)_z = 171\text{ N} \qquad \textbf{Ans}$$

$\Sigma M_x = 0;\quad (T_C)_x = 0 \qquad \textbf{Ans}$

$\Sigma M_y = 0;\quad (M_C)_y - 628.57(0.5) + 800(0.2) = 0$
$$(M_C)_y = 154\text{ N}\cdot\text{m} \qquad \textbf{Ans}$$

$\Sigma M_z = 0;\quad (M_C)_z - 245.71(0.5) = 0$
$$(M_C)_z = 123\text{ N}\cdot\text{m} \qquad \textbf{Ans}$$

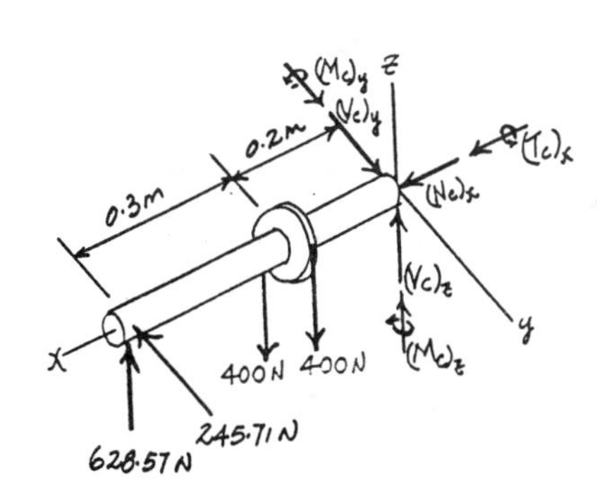

***1–28.** The shaft is supported at its ends by two bearings A and B and is subjected to the forces applied to the pulleys fixed to the shaft. Determine the resultant internal loadings acting on the cross section through point D. The 400-N forces act in the $-z$ direction and the 200-N and 80-N forces act in the $+y$ direction. The journal bearings at A and B exert only y and z components of force on the shaft.

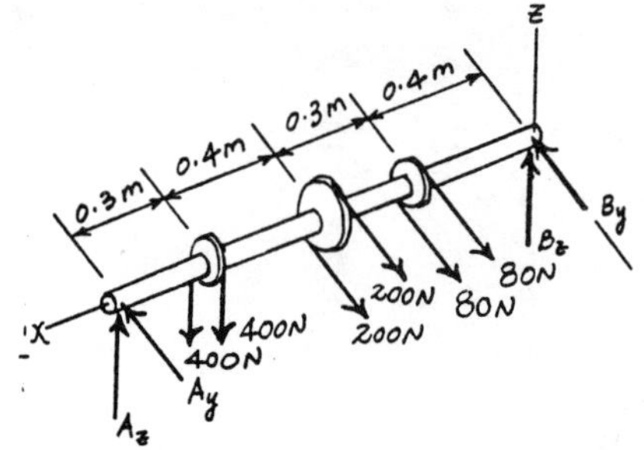

Support Reactions :

$\Sigma M_z = 0;\quad 160(0.4) + 400(0.7) - A_y(1.4) = 0$
$A_y = 245.71\text{ N}$

$\Sigma F_y = 0;\quad -245.71 - B_y + 400 + 160 = 0$
$B_y = 314.29\text{ N}$

$\Sigma M_y = 0;\quad 800(1.1) - A_z(1.4) = 0 \qquad A_z = 628.57\text{ N}$

$\Sigma F_z = 0;\quad B_z + 628.57 - 800 = 0 \qquad B_z = 171.43\text{ N}$

Equations of Equilibrium : For point D

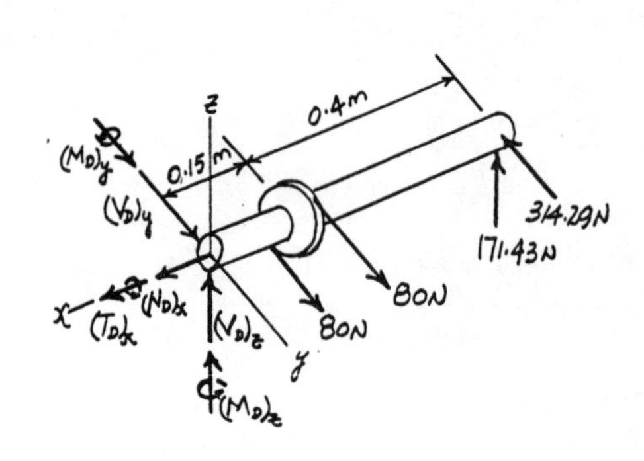

$\Sigma F_x = 0;\quad (N_D)_x = 0$ Ans

$\Sigma F_y = 0;\quad (V_D)_y - 314.29 + 160 = 0$
$(V_D)_y = 154\text{ N}$ Ans

$\Sigma F_z = 0;\quad 171.43 + (V_D)_z = 0$
$(V_D)_z = -171\text{ N}$ Ans

$\Sigma M_x = 0;\quad (T_D)_x = 0$ Ans

$\Sigma M_y = 0;\quad 171.43\,(0.55) + (M_D)_y = 0$
$(M_D)_y = -94.3\text{ N}\cdot\text{m}$ Ans

$\Sigma M_z = 0;\quad 314.29(0.55) - 160(0.15) + (M_D)_z = 0$
$(M_D)_z = -149\text{ N}\cdot\text{m}$ Ans

1–29. The sign has a weight of 1500 lb and center of gravity at G. If it is subjected to the uniform wind load of 60 lb/ft^2, determine the resultant internal loadings acting on the cross section of the post at A. The post has a weight of 100 lb/ft.

Equations of Equilibrium : For point A

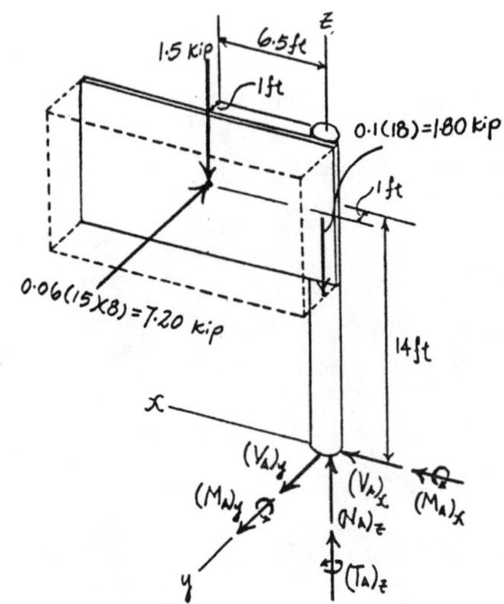

$\Sigma F_x = 0;\quad (V_A)_x = 0$ Ans

$\Sigma F_y = 0;\quad (V_A)_y - 7.20 = 0$
$(V_A)_y = 7.20\text{ kip}$ Ans

$\Sigma F_z = 0;\quad (N_A)_z - 1.5 - 1.8 = 0$
$(N_A)_z = 3.30\text{ kip}$ Ans

$\Sigma M_x = 0;\quad (M_A)_x + 7.20(14) - 1.5(1) = 0$
$(M_A)_x = -99.3\text{ kip}\cdot\text{ft}$ Ans

$\Sigma M_y = 0;\quad (M_A)_y + 1.5(6.5) = 0$
$(M_A)_y = -9.75\text{ kip}\cdot\text{ft}$ Ans

$\Sigma M_z = 0;\quad (T_A)_z - 7.20(6.5) = 0$
$(T_A)_z = 46.8\text{ kip}\cdot\text{ft}$ Ans

1–30. The pipe has a mass of 12 kg/m. If it is fixed to the wall at A, determine the resultant internal loadings acting on the cross section through B.

Equations of Equilibrium : For point B

$\Sigma F_x = 0;\quad (V_B)_x = 0$ **Ans**

$\Sigma F_y = 0;\quad (N_B)_y + \frac{4}{5}(750) = 0$

$(N_B)_y = -600\text{ N}$ **Ans**

$\Sigma F_z = 0;\quad (V_B)_z - 235.44 - 235.44 - \frac{3}{5}(750) = 0$

$(V_B)_z = 921\text{ N}$ **Ans**

$\Sigma M_x = 0;\quad (M_B)_x - 235.44(1) - 235.44(2) - \frac{3}{5}(750)(2) = 0$

$(M_B)_x = 1606\text{ N}\cdot\text{m}$ **Ans**

$\Sigma M_y = 0;\quad (T_B)_y = 0$ **Ans**

$\Sigma M_z = 0;\quad (M_B)_z + 800 = 0$

$(M_B)_z = -800\text{ N}\cdot\text{m}$ **Ans**

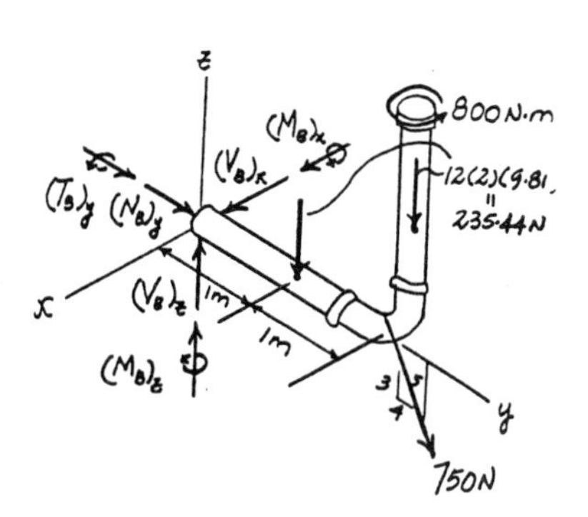

1–31. The curved rod has a radius r and is fixed to the wall at B. Determine the resultant internal loadings acting on the cross section through A which is located at an angle θ from the horizontal.

Equations of Equilibrium : For point A

$+\Sigma F_x = 0;\quad P\cos\theta - N_A = 0$

$N_A = P\cos\theta$ **Ans**

$+\Sigma F_y = 0;\quad V_A - P\sin\theta = 0$

$V_A = P\sin\theta$ **Ans**

$+\Sigma M_A = 0;\quad M_A - P[r(1-\cos\theta)] = 0$

$M_A = Pr(1-\cos\theta)$ **Ans**

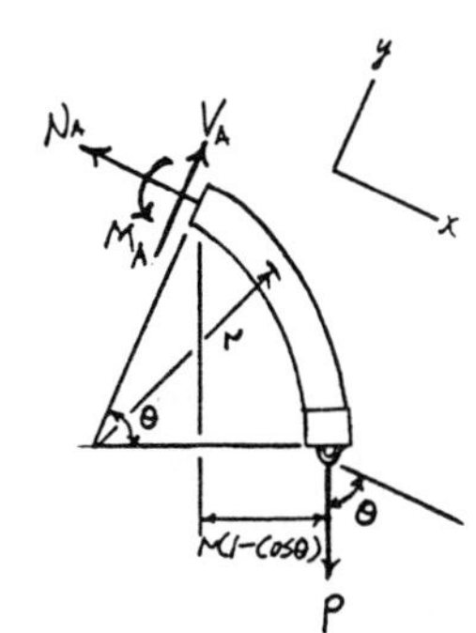

***1–32.** The distributed loading $w = w_0 \sin\theta$, measured per unit length, acts on the curved rod. Determine the normal force, shear force, and moment in the rod at $\theta = 45°$. Neglect the weight of the rod.

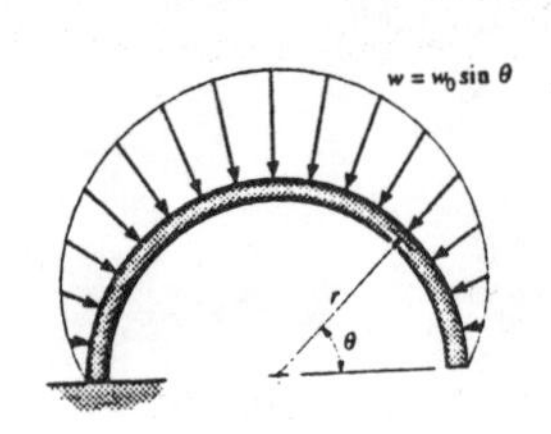

Resultant Force :

$$dA = \omega ds \quad \text{where } ds = rd\theta$$
$$dA = (\omega_0 \sin\theta)(rd\theta) = \omega_0 r\sin\theta d\theta$$

$$F_R = \int dA = \omega_0 r\int_0^{\frac{\pi}{4}} \sin\theta d\theta$$
$$= \omega_0 r(-\cos\theta)\Big|_0^{\frac{\pi}{4}}$$
$$= 0.2929\omega_0 r$$

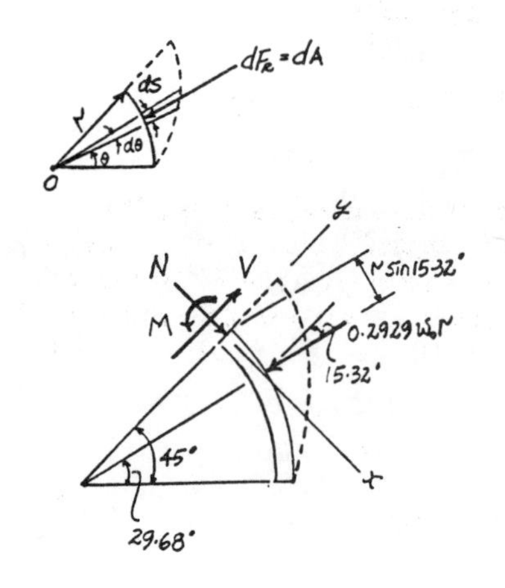

Location of Resultant Force :

$$F_R\bar{s} = \int \tilde{s}dA \qquad \text{where} \quad \bar{s} = r\bar{\theta}$$

$$(0.2929\omega_0 r)\, r\bar{\theta} = \int_0^{\frac{\pi}{4}} r\theta(\omega_0 \sin\theta r d\theta)$$

$$0.2929\omega_0 r^2\bar{\theta} = \omega_0 r^2\int_0^{\frac{\pi}{4}} \theta\sin\theta d\theta$$

$$0.2929\bar{\theta} = \int_0^{\frac{\pi}{4}} \theta\sin\theta d\theta$$

$$0.2929\bar{\theta} = (\sin\theta - \theta\cos\theta)\Big|_0^{\frac{\pi}{4}}$$

$$\bar{\theta} = 0.5181 \text{ rad} = 29.68°$$

Equations of Equilibrium :

$+\Sigma F_x = 0;\qquad -N + 0.2929\omega_0 r\sin 15.32° = 0$

$N = 0.0774\omega_0 r$ **Ans**

$+\Sigma F_y = 0;\qquad V - 0.2929\omega_0 r\cos 15.32° = 0$

$V = 0.282\omega_0 r$ **Ans**

$+\Sigma M = 0;\qquad M - 0.2929\omega_0 r(r\sin 15.32°) = 0$

$M = 0.0774\omega_0 r^2$ **Ans**

1–33. The column is subjected to an axial force of 80 kN, which is applied through the centroid C of the cross-sectional area. Determine the distance $\bar{y}$ to the centroid and the average normal stress acting at section a–a. Show this distribution of stress acting over the area's cross section.

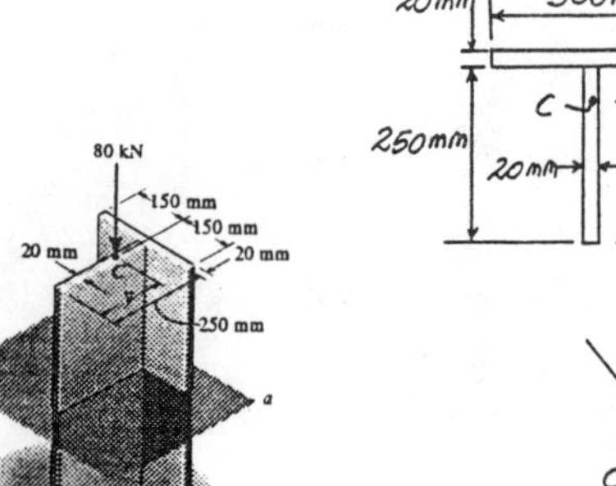

20mm 300mm 250mm 20mm C $\bar{y}$

Section Properties :

$$\bar{y} = \frac{125(250)(20)\ 260(300)(20)}{250(20) + 300(20)}$$
$$= 198.63 \text{ mm} = 199 \text{ mm}$$ **Ans**

$A = 0.25(0.02) + 0.3(0.02)\ 0.011 \text{ m}^2$

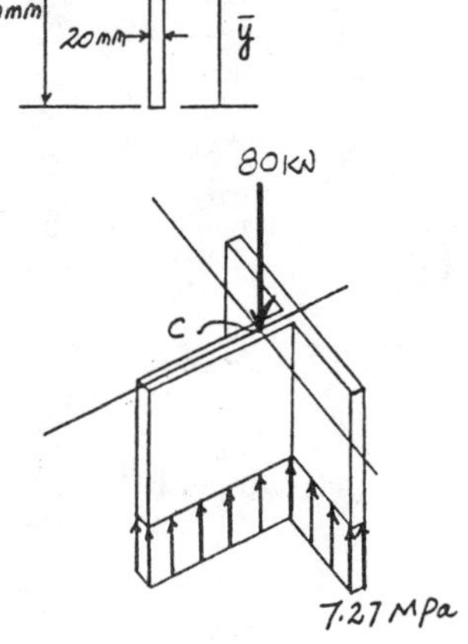

Average Normal Stress :

$$\sigma = \frac{P}{A} = \frac{80(10^3)}{0.011}\ 7.27 \text{ MPa}$$ **Ans**

1–34. The cinder block has the dimensions shown. If the material fails when the average normal stress reaches 120 psi, determine the largest centrally applied vertical load **P** it can support.

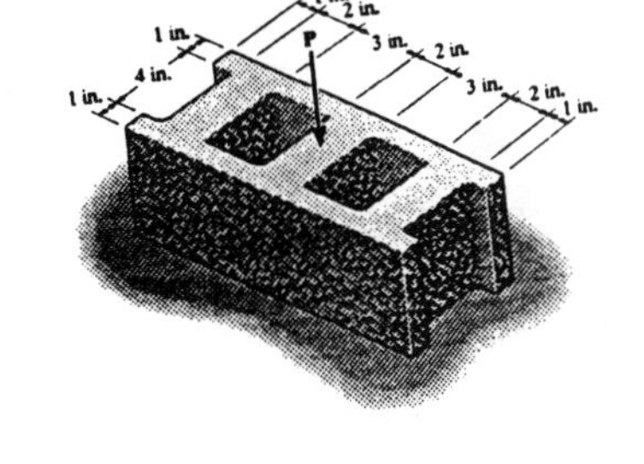

Cross section Area :

$$A = 6(14) - 2[4(1) + 3(4)] = 52 \text{ in}^2$$

Average Normal Stress :

$$\sigma_{allow} = \frac{P_{allow}}{A}; \qquad 120 = \frac{P_{allow}}{52}$$

$$P_{allow} = 6240 \text{ lb} = 6.24 \text{ kip} \qquad \textbf{Ans}$$

1–35. The cinder block has the dimensions shown. If it is subjected to a centrally applied force of $P = 800$ lb, determine the average normal stress in the material. Show the result acting on a differential volume element of the material.

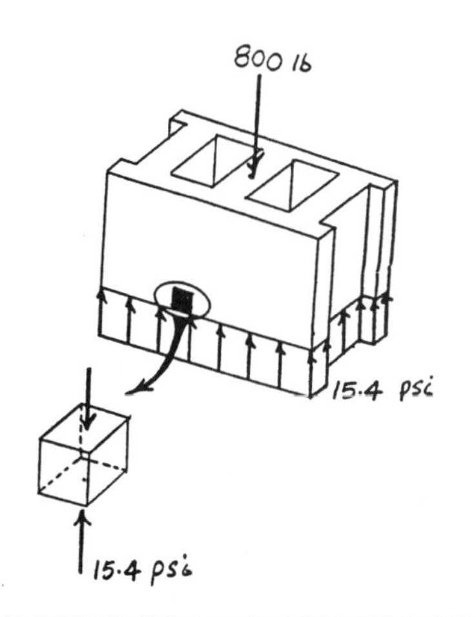

Cross section Area :

$$A = 14(6) - 4(8) = 52 \text{ in}^2$$

Average Normal Stress :

$$\sigma = \frac{P}{A} = \frac{800}{52} = 15.4 \text{ psi} \qquad \textbf{Ans}$$

***1–36.** The thrust bearing is subjected to the loads shown. Determine the average normal stress developed on cross sections through points *B*, *C*, and *D*. Sketch the results on a differential volume element located at each section.

Average Normal Stress :

$$\sigma_B = \frac{500}{\frac{\pi}{4}(\frac{65}{1000})^2} = 151 \text{ kPa} \qquad \textbf{Ans}$$

$$\sigma_C = \frac{500}{\frac{\pi}{4}(\frac{140}{1000})^2} = 32.5 \text{ kPa} \qquad \textbf{Ans}$$

$$\sigma_D = \frac{200}{\frac{\pi}{4}(\frac{100}{1000})^2} = 25.5 \text{ kPa} \qquad \textbf{Ans}$$

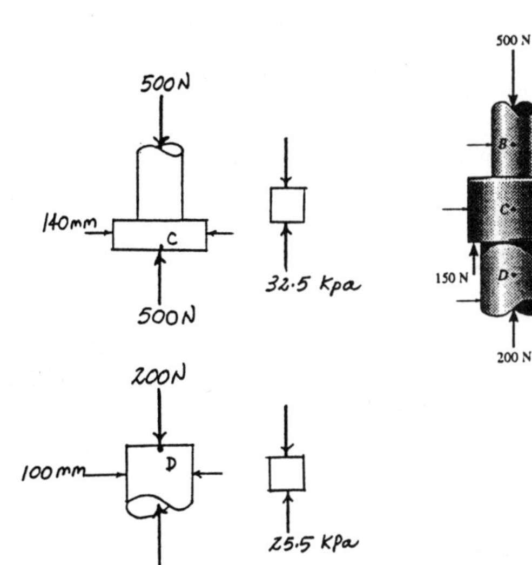

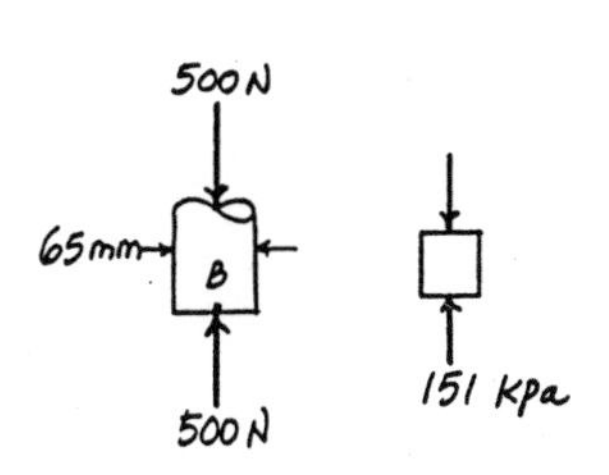

1–37. The 50-lb lamp is supported by two steel rods connected by a ring at A. Determine which rod is subjected to the greater average normal stress and compute its value. Take $\theta = 60°$. The diameter of each rod is given in the figure.

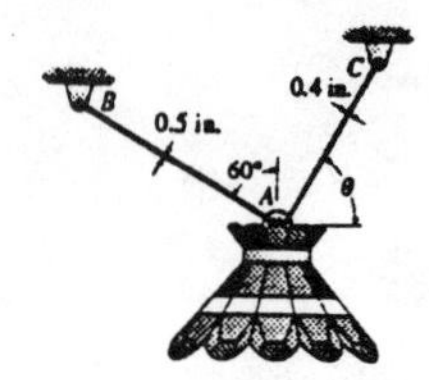

Method of Joints :

$\nwarrow + \Sigma F_x = 0;\quad F_{AB} - 50 \sin 30° = 0 \quad F_{AB} = 25.0 \text{ lb}$

$\nearrow + \Sigma F_y = 0;\quad F_{AC} - 50 \cos 30° = 0 \quad F_{AC} = 43.30 \text{ lb}$

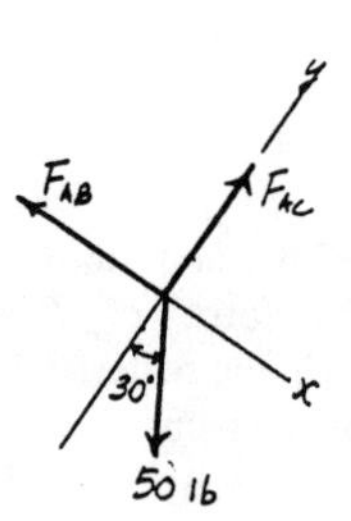

Average Normal Stress : The largest normal stress is in rod AC, since it has the smallest diameter and is subjected to the greatest force.

$$\sigma_{max} = \sigma_{AC} = \frac{F_{AC}}{A_{AC}} = \frac{43.30}{\frac{\pi}{4}(0.4)^2} = 345 \text{ psi} \qquad \textbf{Ans}$$

1–38. Solve Prob. 1–37 for $\theta = 45°$.

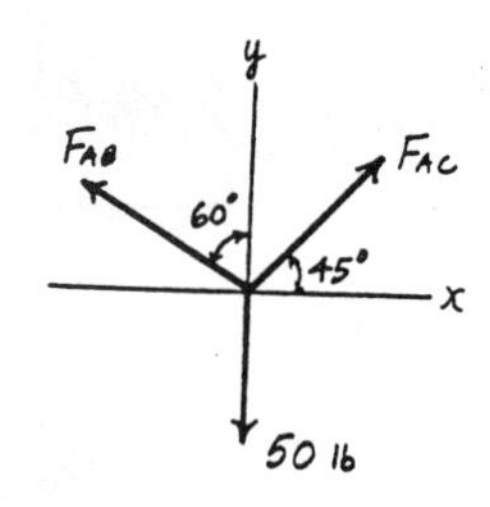

Method of Joints :

$\overset{+}{\leftarrow} \Sigma F_x = 0;\quad F_{AB} \sin 60° - F_{AC} \cos 45° = 0 \qquad [1]$

$+\uparrow \Sigma F_y = 0;\quad F_{AB} \cos 60° + F_{AC} \sin 45° - 50 = 0 \qquad [2]$

Solving Eqs. [1] and [2] yields :

$F_{AC} = 44.83 \text{ lb} \qquad F_{AB} = 36.60 \text{ lb}$

Average Normal Stress : The largest normal stress is in rod AC, since it has the smallest diameter and is subjected to the greatest force.

$$\sigma_{max} = \sigma_{AC} = \frac{F_{AC}}{A_{AC}} = \frac{44.83}{\frac{\pi}{4}(0.4^2)} = 357 \text{ psi} \qquad \textbf{Ans}$$

1–39. The 50-lb lamp is supported by two steel rods connected by a ring at A. Determine the angle of orientation θ of AC such that the average normal stress in rod AC is twice the average normal stress in rod AB. What is the magnitude of this stress in each rod? The diameter of each rod is given in the figure.

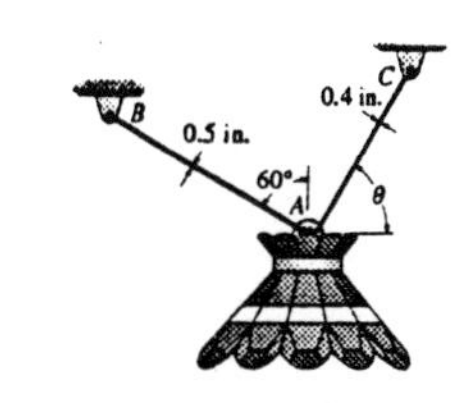

Method of Joints :

$\overset{+}{\leftarrow} \Sigma F_x = 0;\quad F_{AB}\sin 60° - F_{AC}\cos\theta = 0$ [1]

$+\uparrow \Sigma F_y = 0;\quad F_{AB}\cos 60° + F_{AC}\sin\theta - 50 = 0$ [2]

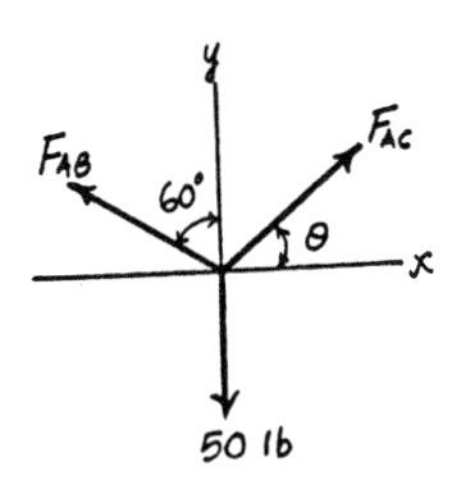

Average Normal Stress : Require, $\sigma_{AC} = 2\sigma_{AB}$

$$\frac{F_{AC}}{\frac{\pi}{4}(0.4^2)} = 2\left[\frac{F_{AB}}{\frac{\pi}{4}(0.5^2)}\right];\quad F_{AC} = 1.28F_{AB} \quad [3]$$

Solving Eqs. [1] to [3] yields :

$\theta = 47.4°$ **Ans**

$F_{AB} = 34.66$ lb $\quad F_{AC} = 44.37$ lb

$$\sigma_{AC} = \frac{F_{AC}}{A_{AC}} = \frac{44.37}{\frac{\pi}{4}(0.4^2)} = 353 \text{ psi} \quad \textbf{Ans}$$

$$\sigma_{AB} = \frac{F_{AB}}{A_{AB}} = \frac{34.66}{\frac{\pi}{4}(0.5^2)} = 177 \text{ psi} \quad \textbf{Ans}$$

***1–40.** The shaft is subjected to the axial force of 30 kN. If the shaft passes through the 53 - mm diameter hole in the fixed support A. determine the bearing stress acting on the collar C. Also, what is the average shear stress acting along the inside surface of the collar where it is fixed connected to the 52 - mm diameter shaft?

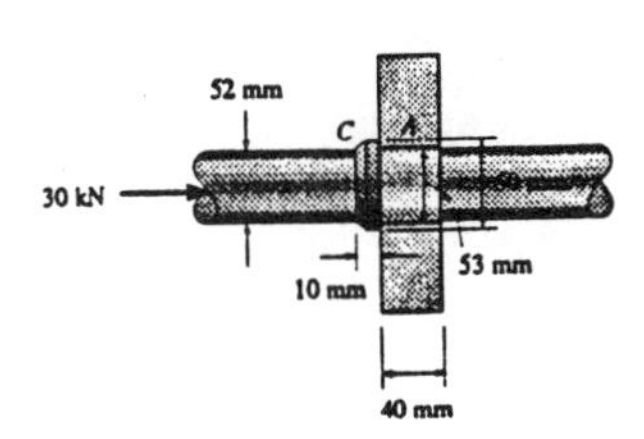

Bearing Stress :

$$\sigma_b = \frac{P}{A} = \frac{30(10^3)}{\frac{\pi}{4}(0.06^2 - 0.053^2)} = 48.3 \text{ MPa} \quad \textbf{Ans}$$

Average Shear Stress :

$$\tau_{avg} = \frac{V}{A} = \frac{30(10^3)}{\pi(0.052)(0.01)} = 18.4 \text{ MPa} \quad \textbf{Ans}$$

1–41. The control link is held in equilibrium. Determine the average shear stress in the pins at A, B, and C. Pin A is subjected to double shear and pins B and C are subjected to single shear. Each pin has a diameter of 10 mm.

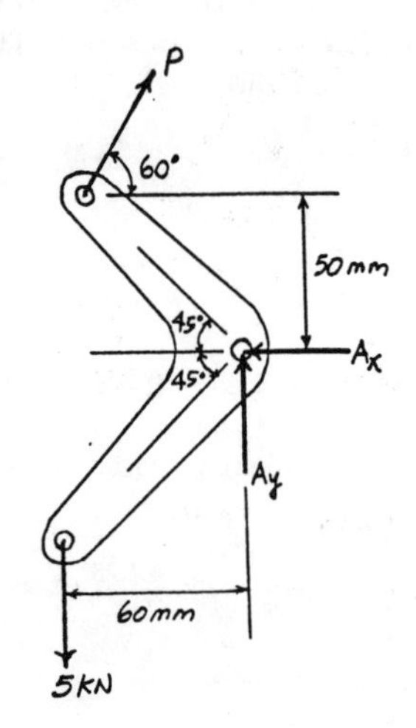

Support Reactions :

$$\circlearrowleft + \Sigma M_A = 0; \quad 5(60) - P\cos 60°(50) - P\sin 60°\left(\frac{50}{\tan 45°}\right) = 0$$

$$P = 4.392 \text{ kN}$$

$$+\uparrow \Sigma F_y = 0; \quad A_y + 4.392\sin 60° - 5 = 0 \qquad A_y = 1.196 \text{ kN}$$

$$\overset{+}{\leftarrow} \Sigma F_x = 0; \quad A_x - 4.392\cos 60° = 0 \qquad A_x = 2.196 \text{ kN}$$

Average Shear Stress : Pin A is subjected to double shear and acted upon by $F_A = \sqrt{2.196^2 + 1.196^2} = 2.501$ kN.

$$(\tau_A)_{avg} = \frac{V_A}{A_A} = \frac{2.501(10^3)/2}{\frac{\pi}{4}(0.01^2)} = 15.9 \text{ MPa} \qquad \textbf{Ans}$$

$$(\tau_B)_{avg} = \frac{V_B}{A_B} = \frac{5(10^3)}{\frac{\pi}{4}(0.01^2)} = 63.7 \text{ MPa} \qquad \textbf{Ans}$$

$$(\tau_C)_{avg} = \frac{V_C}{A_C} = \frac{4.392(10^3)}{\frac{\pi}{4}(0.01^2)} = 55.9 \text{ MPa} \qquad \textbf{Ans}$$

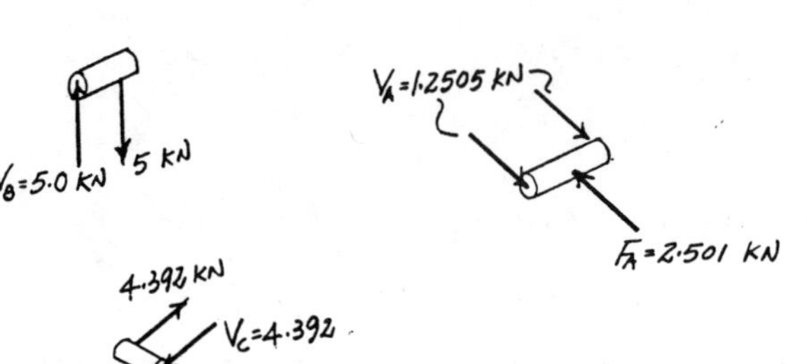

1–42. The J hanger is used to support the pipe such that the force on the vertical bolt is 775 N. Determine the average normal stress developed in the bolt BC if the bolt has a diameter of 8 mm. Assume A is a pin.

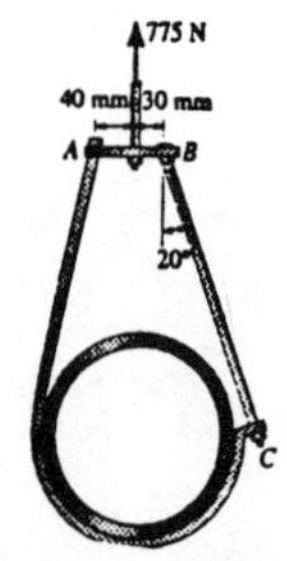

Support Reaction :

$$\circlearrowleft + \Sigma M_A = 0; \quad 775(40) - F_{BC}\cos 20°(70) = 0$$

$$F_{BC} = 471.28 \text{ N}$$

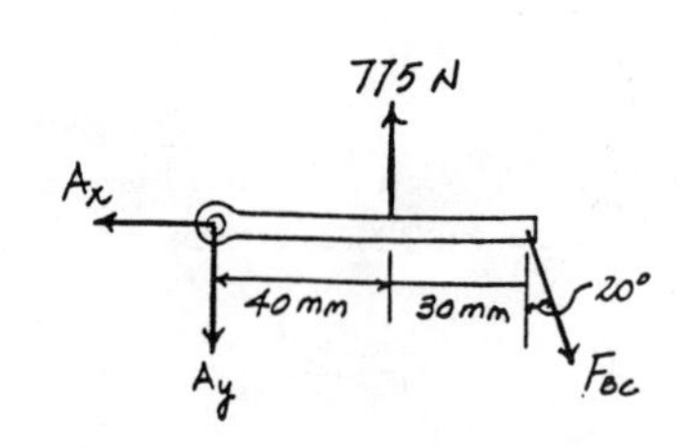

Average Normal Stress :

$$\sigma_{BC} = \frac{F_{BC}}{A_{BC}} = \frac{471.28}{\frac{\pi}{4}(0.008^2)} = 9.38 \text{ MPa} \qquad \textbf{Ans}$$

1–43. The pins on the frame at B and C each have a diameter of 0.25 in. If these pins are subjected to *double shear*, determine the average shear stress in each pin.

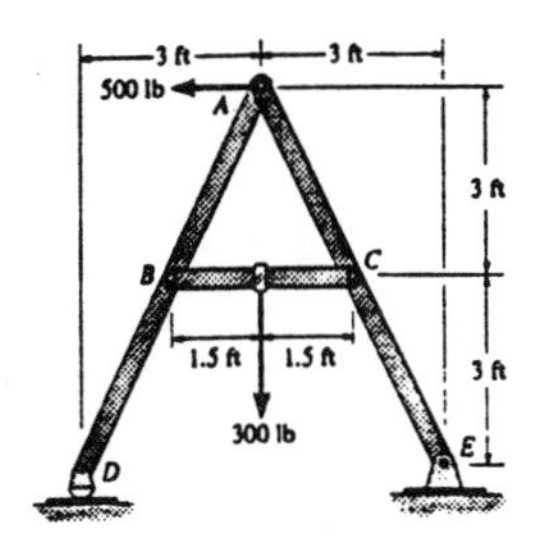

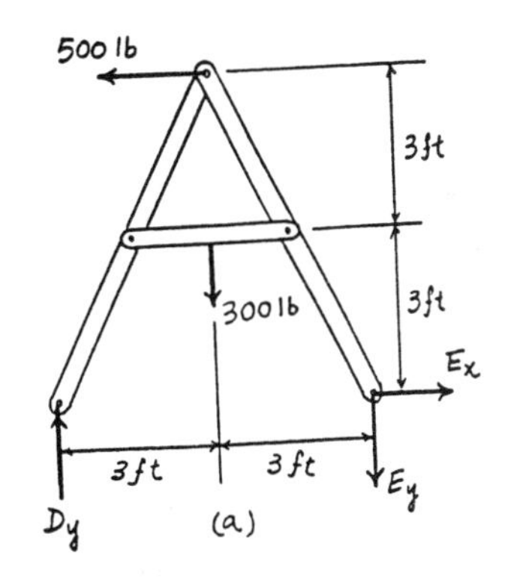

Support Reactions : FBD (a)

$$\circlearrowleft + \Sigma M_E = 0; \quad 500(6) + 300(3) - D_y(6) = 0$$
$$D_y = 650 \text{ lb}$$

$$\overset{+}{\leftarrow} \Sigma F_x = 0; \quad 500 - E_x = 0 \qquad E_x = 500 \text{ lb}$$

$$+\uparrow \Sigma F_y = 0; \quad 650 - 300 - E_y = 0 \qquad E_y = 350 \text{ lb}$$

From FBD (c),

$$\circlearrowleft + \Sigma M_B = 0; \quad C_y(3) - 300(1.5) = 0 \qquad C_y = 150 \text{ lb}$$

$$+\uparrow \Sigma F_y = 0; \quad B_y + 150 - 300 = 0 \qquad B_y = 150 \text{ lb}$$

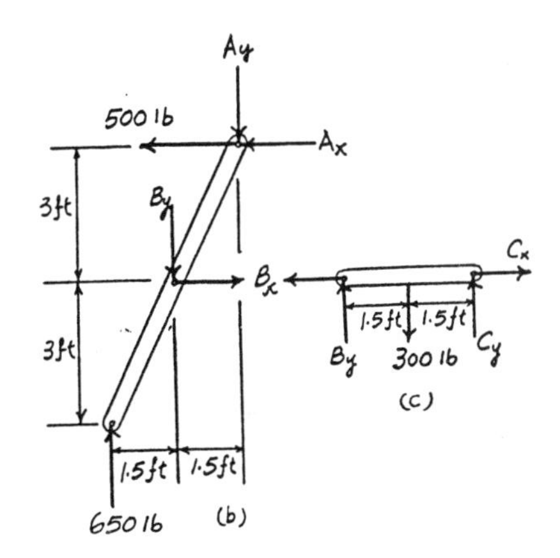

From FBD (b)

$$\circlearrowleft + \Sigma M_A = 0; \quad 150(1.5) + B_x(3) - 650(3) = 0$$
$$B_x = 575 \text{ lb}$$

From FBD (c),

$$\overset{+}{\rightarrow} \Sigma F_x = 0; \quad C_x - 575 = 0 \qquad C_x = 575 \text{ lb}$$

Hence, $F_B = F_C = \sqrt{575^2 + 150^2} = 594.24$ lb

Average shear stress : Pins B and C are subjected to double shear as shown on FBD (d)

$$(\tau_B)_{avg} = (\tau_C)_{avg} = \frac{V}{A} = \frac{297.12}{\frac{\pi}{4}(0.25^2)}$$
$$= 6053 \text{ psi} = 6.05 \text{ ksi} \qquad \textbf{Ans}$$

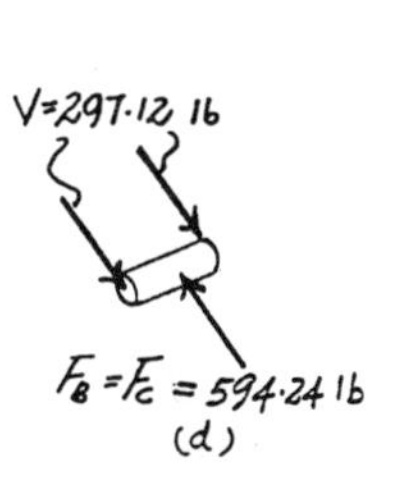

***1–44.** Solve Prob. 1–43 assuming that pins B and C are subjected to *single shear*.

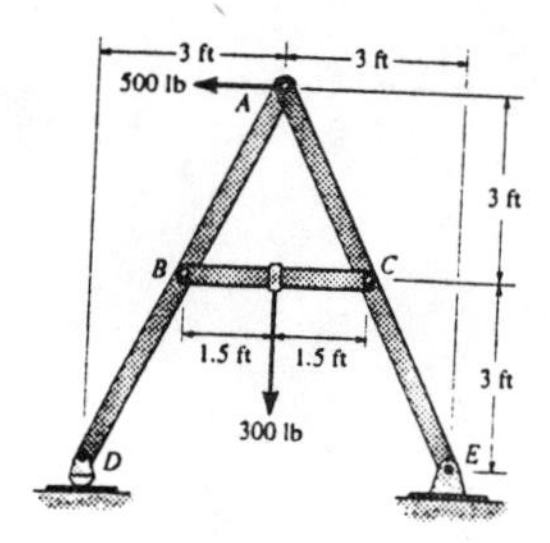

Support Reactions : FBD (a)

$$\curvearrowleft + \Sigma M_E = 0; \quad 500(6) + 300(3) - D_y(6) = 0$$
$$D_y = 650 \text{ lb}$$

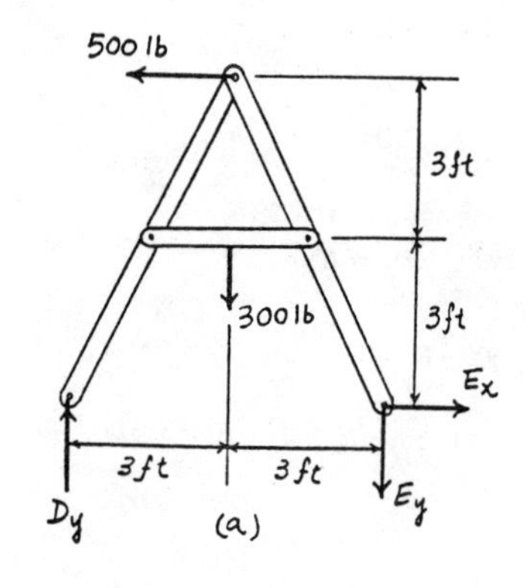

$$\overset{+}{\leftarrow} \Sigma F_x = 0; \quad 500 - E_x = 0 \qquad E_x = 500 \text{ lb}$$

$$+\uparrow \Sigma F_y = 0; \quad 650 - 300 - E_y = 0 \qquad E_y = 350 \text{ lb}$$

From FBD (c),

$$\curvearrowleft + \Sigma M_B = 0; \quad C_y(3) - 300(1.5) = 0 \qquad C_y = 150 \text{ lb}$$

$$+\uparrow \Sigma F_y = 0; \quad B_y + 150 - 300 = 0 \qquad B_y = 150 \text{ lb}$$

From FBD (b)

$$\curvearrowleft + \Sigma M_A = 0; \quad 150(1.5) + B_x(3) - 650(3) = 0$$
$$B_x = 575 \text{ lb}$$

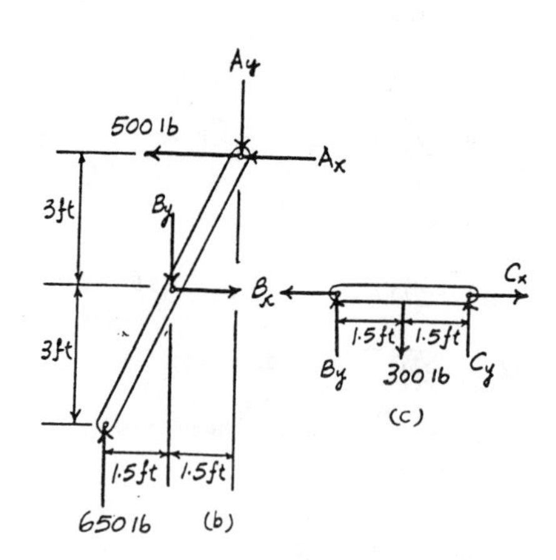

From FBD (c),

$$\overset{+}{\rightarrow} \Sigma F_x = 0; \quad C_x - 575 = 0 \qquad C_x = 575 \text{ lb}$$

Hence, $F_B = F_C = \sqrt{575^2 + 150^2} = 594.24$ lb

Average shear stress : Pins B and C are subjected to single shear as shown on FBD (d)

$$(\tau_B)_{avg} = (\tau_C)_{avg} = \frac{V}{A} = \frac{594.24}{\frac{\pi}{4}(0.25^2)}$$
$$= 12106 \text{ psi} = 12.1 \text{ ksi} \qquad \textbf{Ans}$$

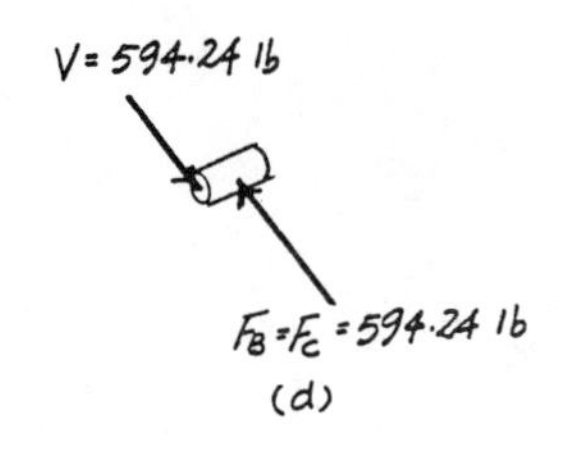

1–45. The pins on the frame at D and E each have a diameter of 0.25 in. If these pins are subjected to *double shear*, determine the average shear stress in each pin.

Support Reactions : FBD (a)

$\circlearrowleft + \Sigma M_E = 0; \quad 500(6) + 300(3) - D_y(6) = 0$

$D_y = 650 \text{ lb}$

$\overset{+}{\leftarrow} \Sigma F_x = 0; \quad 500 - E_x = 0 \qquad E_x = 500 \text{ lb}$

$+\uparrow \Sigma F_y = 0; \quad 650 - 300 - E_y = 0 \qquad E_y = 350 \text{ lb}$

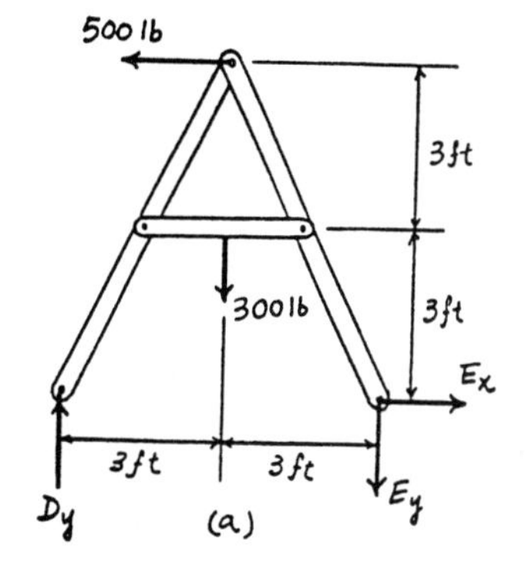

Average shear stress : Pins D and E are subjected to double shear as shown on FBD (b) and (c).

For Pin D, $F_D = D_y = 650$ lb then $V_D = \frac{F_D}{2} = 325$ lb

$$(\tau_D)_{avg} = \frac{V_D}{A_D} = \frac{325}{\frac{\pi}{4}(0.25)^2}$$

$= 6621 \text{ psi} = 6.62 \text{ ksi}$ **Ans**

For Pin E, $F_E = \sqrt{500^2 + 350^2} = 610.32$ lb then $V_E = \frac{F_E}{2} = 305.16$ lb

$$(\tau_E)_{avg} = \frac{V_E}{A_E} = \frac{305.16}{\frac{\pi}{4}(0.25^2)}$$

$= 6217 \text{ psi} = 6.22 \text{ ksi}$ **Ans**

$V_D = 325$ lb

(b)

$F_D = 650$ lb

$F_E = 610.32$ lb

(c)

$V_E = 305.16$ lb

1–46. Solve Prob. 1–45 assuming that pins D and E are subjected to *single shear*.

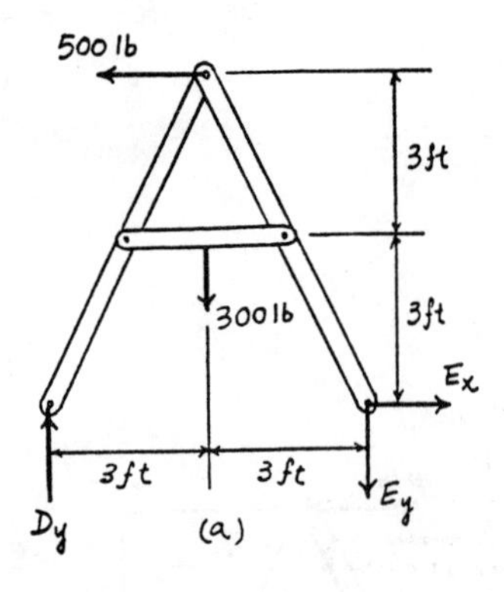

Support Reactions : FBD (a)

$$\curvearrowleft + \Sigma M_E = 0; \quad 500(6) + 300(3) - D_y(6) = 0$$
$$D_y = 650 \text{ lb}$$

$$\overset{+}{\leftarrow} \Sigma F_x = 0; \quad 500 - E_x = 0 \qquad E_x = 500 \text{ lb}$$

$$+\uparrow \Sigma F_y = 0; \quad 650 - 300 - E_y = 0 \qquad E_y = 350 \text{ lb}$$

Average shear stress : Pins D and E are subjected to single shear as shown on FBD (b) and (c).

For Pin D, $\quad V_D = F_D = D_y = 650$ lb

$$(\tau_D)_{avg} = \frac{V_D}{A_D} = \frac{650}{\frac{\pi}{4}(0.25^2)}$$
$$= 13242 \text{ psi} = 13.2 \text{ ksi} \qquad \textbf{Ans}$$

For Pin E, $\quad V_E = F_E = \sqrt{500^2 + 350^2} = 610.32$ lb

$$(\tau_E)_{avg} = \frac{V_E}{A_E} = \frac{610.32}{\frac{\pi}{4}(0.25^2)}$$
$$= 12433 \text{ psi} = 12.4 \text{ ksi} \qquad \textbf{Ans}$$

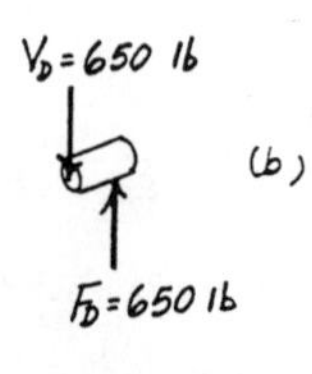

F_E = 610.32 lb

(c)

V_E = 610.32 lb

1–47. The beam is supported by a pin at A and a short link BC. If $w = 2$ kN/m, determine the average shear stress developed in the pins at A, B, and C. All pins are subjected to double shear as shown, and each has a diameter of 18 mm.

Support Reactions : FBD (a)

$$\circlearrowleft + \Sigma M_A = 0; \quad 10(2.5) - F_{BC}\sin 30°(5) = 0$$
$$F_{BC} = 10.0 \text{ kN}$$

$$\overset{+}{\rightarrow} \Sigma F_x = 0; \quad A_x - 10.0\cos 30° = 0 \quad A_x = 8.660 \text{ kN}$$

$$+\uparrow \Sigma F_y = 0; \quad A_y + 10.0\sin 30° - 10.0 = 0$$
$$A_y = 5.00 \text{ kN}$$

Hence, $F_A = \sqrt{5.00^2 + 8.660^2} = 10.0$ kN

Average shear stress : Pins A, B, and C are subjected to double shear as shown on FBD (b) and (c). Hence, $V_B = V_C = V = \frac{F_{BC}}{2} = 5.00$ kN and $V_A = \frac{F_A}{2} = 5.00$ kN.

$$(\tau_A)_{avg} = \frac{V_A}{A} = \frac{5.00(10^3)}{\frac{\pi}{4}(0.018^2)} = 19.6 \text{ MPa} \qquad \textbf{Ans}$$

$$(\tau_C)_{avg} = (\tau_B)_{avg} = \frac{V}{A} = \frac{5.00(10^3)}{\frac{\pi}{4}(0.018^2)}$$
$$= 19.6 \text{ MPa} \qquad \textbf{Ans}$$

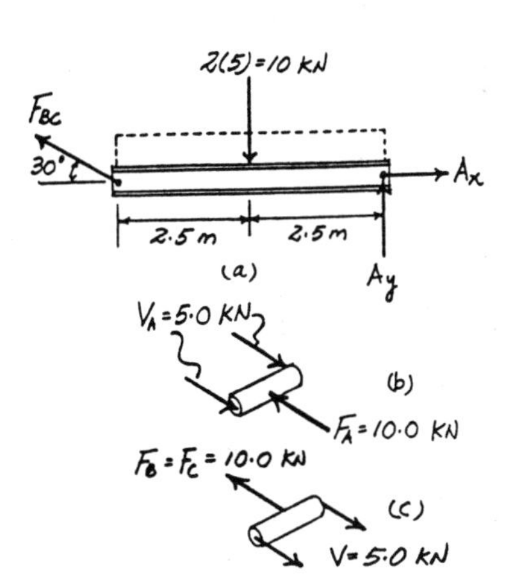

***1–48.** The beam is supported by a pin at A and a short link BC. Determine the maximum distributed load w the beam will support if the average shear stress in each pin is not to exceed 80 MPa. All pins are subjected to double shear as shown, and each has a diameter of 18 mm.

Support Reactions : FBD (a)

$$\circlearrowleft + \Sigma M_A = 0; \quad 5w(2.5) - F_{BC}\sin 30°(5) = 0$$
$$F_{BC} = 5.00w$$

$$\overset{+}{\rightarrow} \Sigma F_x = 0; \quad -5.00w\cos 30° + A_x = 0$$
$$A_x = 4.330w$$

$$+\uparrow \Sigma F_y = 0; \quad A_y - 5w + 5.00w \sin 30° = 0$$
$$A_y = 2.50w$$

Hence

$F_B = 5.00w$; $F_A = \sqrt{(4.330w)^2 + (2.50w)^2} = 5.00w$

Average shear stress : All pins are subjected to double shear as shown on FBD(b) and experience the same shear force of $\frac{5.00w}{2} = 2.50w$.

$$\tau_{allow} = \frac{V}{A}; \quad 80(10^6) = \frac{2.50w}{\frac{\pi}{4}(0.018^2)}$$

$$w = 8143 \text{ N/m } 8.14 \text{ kN/m} \qquad \textbf{Ans}$$

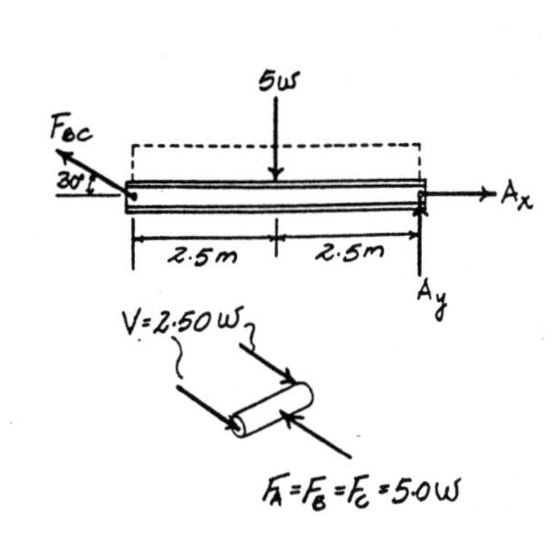

1–49. The open square butt joint is used to transmit a force of 50 kip from one plate to the other. Determine the average normal and average shear stress components that this loading creates on the face of the weld, section AB.

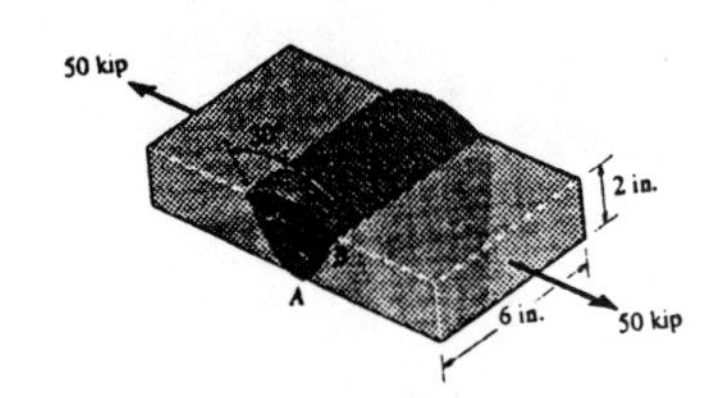

Equations of Equilibrium :

$\nwarrow\!+ \Sigma F_y = 0;\quad N - 50\cos 30° = 0 \qquad N = 43.30 \text{ kip}$

$\nearrow\!+ \Sigma F_x = 0;\quad -V + 50\sin 30° = 0 \qquad V = 25.0 \text{ kip}$

Average Normal and Shear Stress :

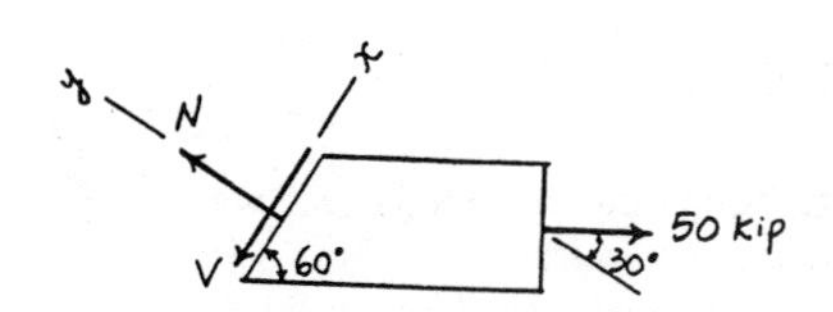

$$A' = \left(\frac{2}{\sin 60°}\right)(6) = 13.86 \text{ in}^2$$

$$\sigma = \frac{N}{A'} = \frac{43.30}{13.86} = 3.125 \text{ ksi} \qquad \textbf{Ans}$$

$$\tau_{avg} = \frac{V}{A'} = \frac{25.0}{13.86} = 1.80 \text{ ksi} \qquad \textbf{Ans}$$

1–50. The yoke is subjected to the force and couple moment. Determine the average shear stress in the bolt acting on the cross sections through A and B. The bolt has a diameter of 0.25 in. *Hint:* The couple moment is resisted by a set of couple forces developed in the shank of the bolt.

At A force on bolt shank is zero, then

$$\tau_A = 0 \qquad \textbf{Ans}$$

Equations of Equilibrium : Force on bolt shank at B.

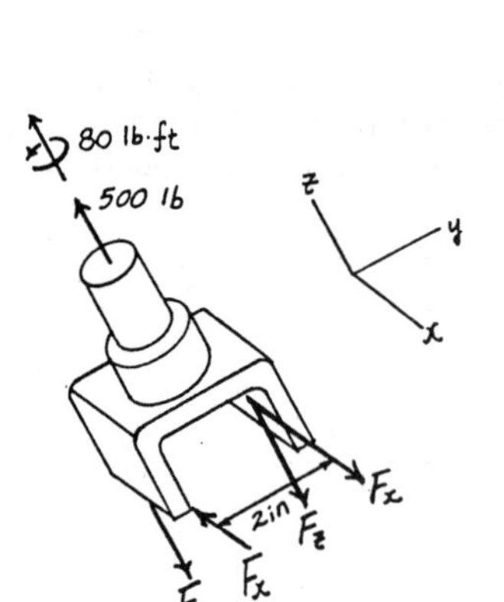

$\Sigma F_z = 0;\quad 500 - 2F_z = 0 \qquad F_z = 250 \text{ lb}$

$\Sigma M_z = 0;\quad 80 \text{ lb}\cdot\text{ft}\left(\frac{12 \text{ in}}{\text{ft}}\right) - F_x(2 \text{ in.}) = 0$

$F_x = 480 \text{ lb}$

Average Shear Stress : The bolt shank subjected to a shear force of $V_B = F_B = \sqrt{250^2 + 480^2} = 541.2$ lb.

$$(\tau_B)_{avg} = \frac{541.2}{\frac{\pi}{4}(0.25)^2} = 11.0 \text{ ksi} \qquad \textbf{Ans}$$

1–51. The row of staples AB contained in the stapler is glued together so that the maximum shear stress the glue can withstand is $\tau_{max} = 12$ psi. Determine the minimum force **F** that must be placed on the plunger in order to shear off a staple from its row and allow it to exit undeformed through the groove at C. The outer dimensions of the staple are shown in the figure. It has a thickness of 0.05 in. Assume all the other parts are rigid and neglect friction.

Average Shear Stress :

$$A = 0.5(0.3) - 0.4(0.25) = 0.05 \text{ in}^2$$

$$\tau_{max} = \frac{V}{A};\quad 12 = \frac{V}{0.05}$$

$$F_{min} = V = 0.60 \text{ lb} \qquad \textbf{Ans}$$

***1–52.** The plug is used to close the end of the cylindrical tube that is subjected to an internal pressure of $p = 650$ Pa. Determine the average shear stress which the glue exerts on the sides of the tube needed to hold the cap in place.

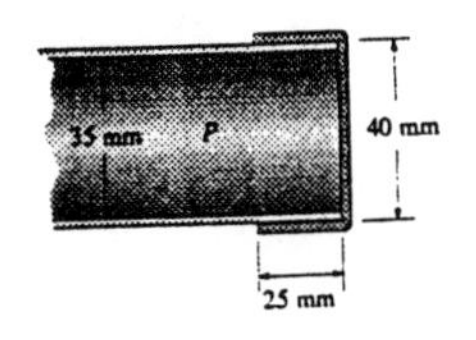

Average Shear Stress :

$$V = 650\left[\frac{\pi}{4}\left(0.035^2\right)\right] = 0.6254 \text{ N}$$

$$\tau_{avg} = \frac{V}{A} = \frac{0.6254}{\pi(0.04)(0.025)} = 199 \text{ Pa} \qquad \textbf{Ans}$$

1–53. The anchor bolt was pulled out of the concrete wall and the failure surface formed part of a frustum and cylinder. This indicates a shear failure occurred along the cylinder BC and tension failure along the frustum AB. If the shear and normal stresses along these surfaces have the magnitudes shown, determine the force **P** that must have been applied to the bolt.

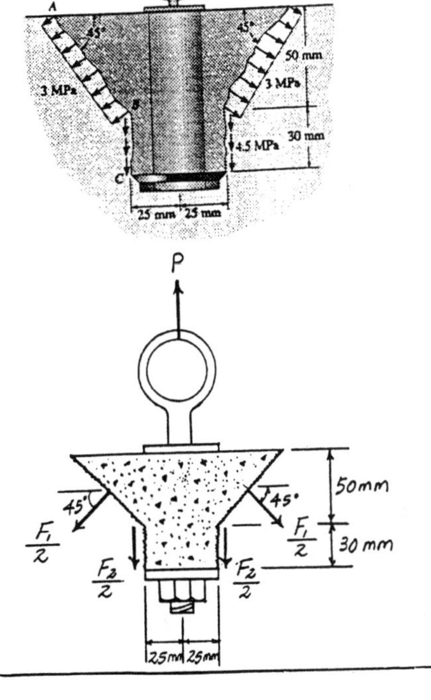

Average Normal Stress :

For the frustum, $\quad A = 2\pi\bar{x}L = 2\pi(0.025 + 0.025)\left(\sqrt{0.05^2 + 0.05^2}\right)$

$$= 0.02221 \text{ m}^2$$

$$\sigma = \frac{P}{A}; \quad 3\left(10^6\right) = \frac{F_1}{0.02221}$$

$$F_1 = 66.64 \text{ kN}$$

Average Shear Stress :

For the cylinder, $\quad A = \pi(0.05)(0.03) = 0.004712 \text{ m}^2$

$$\tau_{avg} = \frac{V}{A}; \quad 4.5\left(10^6\right) = \frac{F_2}{0.004712}$$

$$F_2 = 21.21 \text{ kN}$$

Equation of Equilibrium :

$$+\uparrow \Sigma F_y = 0; \quad P - 21.21 - 66.64\sin 45° = 0$$

$$P = 68.3 \text{ kN} \qquad \textbf{Ans}$$

1–54. Member A of the timber step joint for a truss is subjected to a compressive force of 5 kN. Determine the average normal stress acting in the hanger rod C which has a diameter of 10 mm and in member B which has a thickness of 30 mm.

Equations of Equilibrium :

$$\overset{+}{\rightarrow} \Sigma F_x = 0; \quad 5\cos 60° - F_B = 0 \quad F_B = 2.50 \text{ kN}$$

$$+\uparrow \Sigma F_y = 0; \quad F_c - 5\sin 60° = 0 \quad F_c = 4.330 \text{ kN}$$

Average Normal Stress :

$$\sigma_B = \frac{F_B}{A_B} = \frac{2.50(10^3)}{(0.04)(0.03)} = 2.08 \text{ MPa} \qquad \textbf{Ans}$$

$$\sigma_C = \frac{F_C}{A_C} = \frac{4.330(10^3)}{\frac{\pi}{4}(0.01^2)} = 55.1 \text{ MPa} \qquad \textbf{Ans}$$

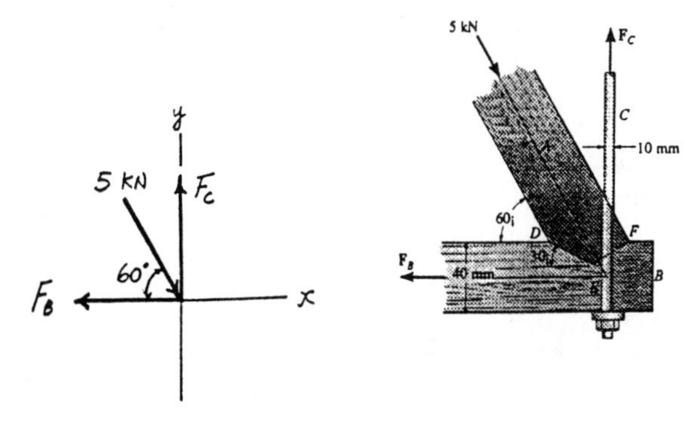

1–55. The crimping tool is used to crimp the end of the wire E. If a force of 20 lb is applied to the handles, determine the average shear stress in the pin at A. The pin is subjected to double shear and has a diameter of 0.2 in. Only a vertical force is exerted on the wire.

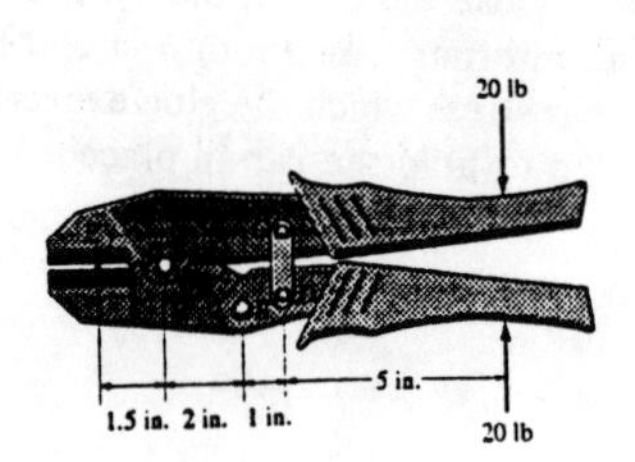

Support Reactions :

From FBD (a)

$+\Sigma M_D = 0;\quad 20(5) - B_y(1) = 0 \qquad B_y = 100 \text{ lb}$

$\xrightarrow{+} \Sigma F_x = 0; \qquad B_x = 0$

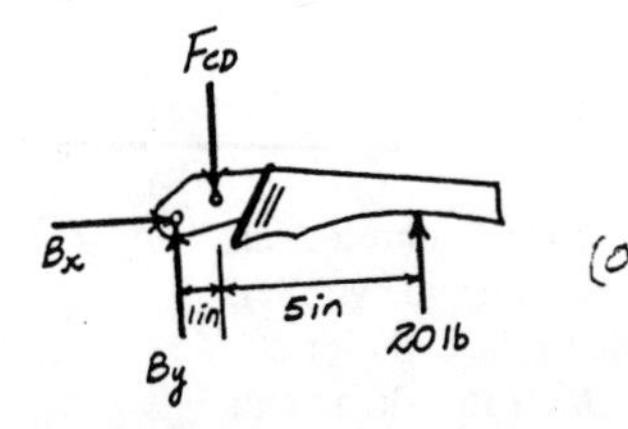

From FBD (b)

$\xrightarrow{+} \Sigma F_x = 0; \qquad A_x = 0$

$+\Sigma M_E = 0;\quad A_y(1.5) - 100(3.5) = 0$

$A_y = 233.33 \text{ lb}$

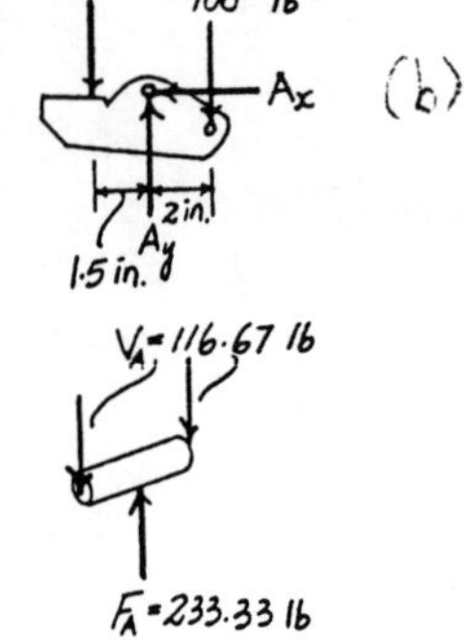

Average Shear Stress : Pin A is subjected to double shear. Hence,

$$V_A = \frac{F_A}{2} = \frac{A_y}{2} = 116.67 \text{ lb}$$

$$(\tau_A)_{avg} = \frac{V_A}{A_A} = \frac{116.67}{\frac{\pi}{4}(0.2^2)}$$

$$= 3714 \text{ psi} = 3.71 \text{ ksi} \qquad \textbf{Ans}$$

***1–56.** Solve Prob. 1–55 for pin B. The pin is subjected to double shear and has a diameter of 0.2 in.

Support Reactions :

From FBD (a)

$+\Sigma M_D = 0;\quad 20(5) - B_y(1) = 0 \qquad B_y = 100 \text{ lb}$

$\xrightarrow{+} \Sigma F_x = 0; \qquad B_x = 0$

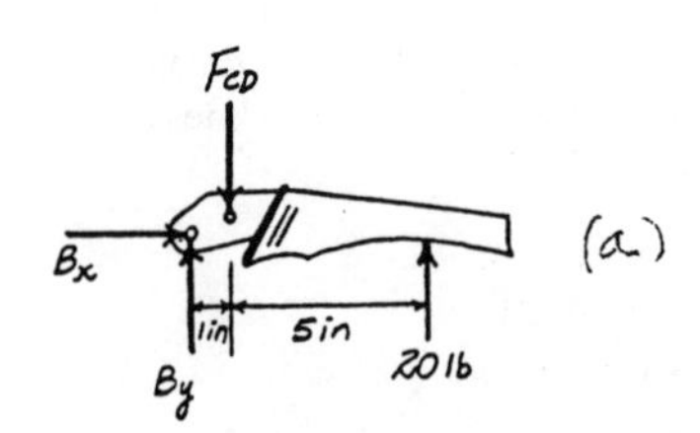

Average Shear Stress : Pin B is subjected to double shear. Hence,

$$V_B = \frac{F_B}{2} = \frac{B_y}{2} = 50.0 \text{ lb}$$

$$(\tau_B)_{avg} = \frac{V_B}{A_B} = \frac{50.0}{\frac{\pi}{4}(0.2^2)}$$

$$= 1592 \text{ psi} = 1.59 \text{ ksi} \qquad \textbf{Ans}$$

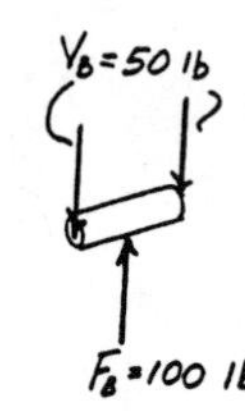

1–57. The frame is subjected to the load of 200 lb. Determine the average shear stress in the bolt at A as a function of the bar angle θ. Plot this function, $0 \le \theta \le 90°$, and indicate the values of θ for which this stress is a minimum. The bolt has a diameter of 0.25 in. and is subjected to single shear.

Support Reactions :

$$\curvearrowleft + \Sigma M_C = 0; \quad F_{AB}\cos\theta\,(0.5) + F_{AB}\sin\theta\,(2) - 200(3.5) = 0$$

$$F_{AB}(0.5\cos\theta + 2\sin\theta) = 700$$

$$F_{AB} = \frac{700}{0.5\cos\theta + 2\sin\theta}$$

Average Shear Stress : Pin A is subjected to single shear. Hence, $V_A = F_{AB}$

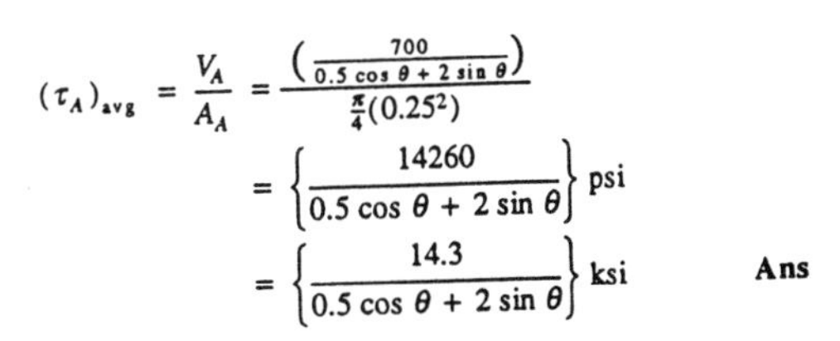

$$(\tau_A)_{avg} = \frac{V_A}{A_A} = \frac{\left(\frac{700}{0.5\cos\theta + 2\sin\theta}\right)}{\frac{\pi}{4}(0.25^2)}$$

$$= \left\{\frac{14260}{0.5\cos\theta + 2\sin\theta}\right\} \text{psi}$$

$$= \left\{\frac{14.3}{0.5\cos\theta + 2\sin\theta}\right\} \text{ksi} \qquad \textbf{Ans}$$

$$\frac{d\tau}{d\theta} = 0$$

$$\frac{(0.5\cos\theta + 2\sin\theta)(0) - (-0.5\sin\theta + 2\cos\theta)(14260)}{(0.5\cos\theta + 2\sin\theta)^2} = 0$$

$$0.5\sin\theta - 2\cos\theta = 0$$

$$\tan\theta = 4; \qquad \theta_{min} = 76.0° \qquad \textbf{Ans}$$

1–58. The bar has a cross-sectional area A and is subjected to the axial load P. Determine the average normal and average shear stresses acting over the shaded section, which is oriented at θ from the horizontal. Plot the variation of these stresses as a function of θ $(0 \le \theta \le 90°)$.

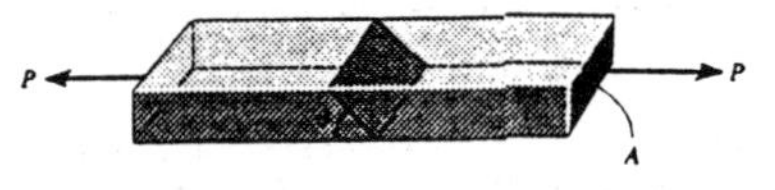

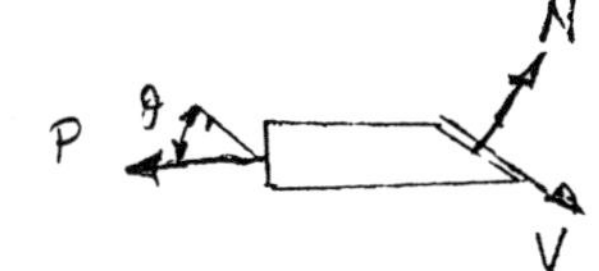

Equations of Equilibrium :

$$\searrow + \Sigma F_x = 0; \quad V - P\cos\theta = 0 \qquad V = P\cos\theta$$

$$\nearrow + \Sigma F_y = 0; \quad N - P\sin\theta = 0 \qquad N = P\sin\theta$$

Average Normal Stress and Shear Stress :

Area at θ plane, $A' = \dfrac{A}{\sin\theta}$.

$$\sigma = \frac{N}{A'} = \frac{P\sin\theta}{\frac{A}{\sin\theta}} = \frac{P}{A}\sin^2\theta \qquad \textbf{Ans}$$

$$\tau_{avg} = \frac{V}{A'} = \frac{P\cos\theta}{\frac{A}{\sin\theta}}$$

$$= \frac{P}{A}\sin\theta\cos\theta = \frac{P}{2A}\sin 2\theta \qquad \textbf{Ans}$$

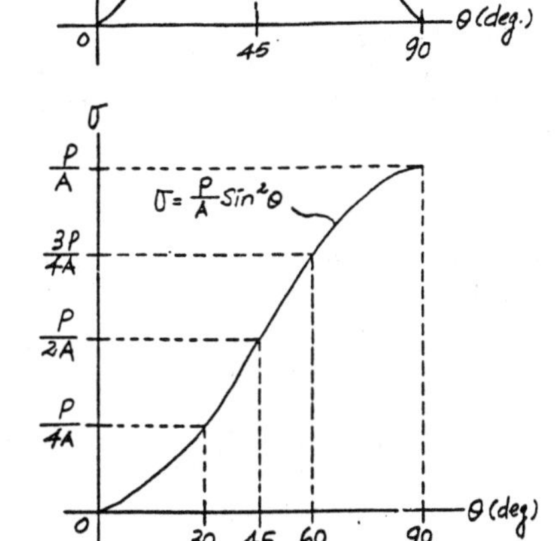

1–59. The specimen has a cross-sectional area A and is subjected to the axial load P. Determine the maximum average shear stress in the material and associated angle θ of the plane over which it acts.

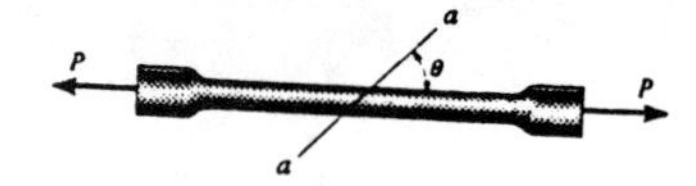

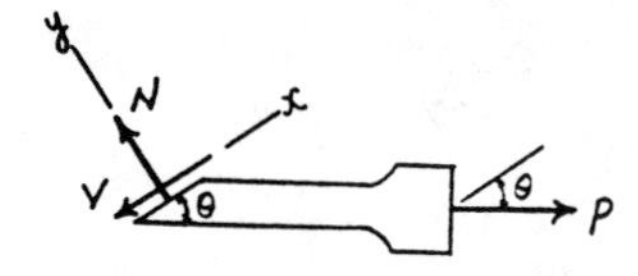

Equations of Equilibrium :

$+\nearrow \Sigma F_x = 0; \quad P\cos\theta - V = 0 \quad V = P\cos\theta$

$+\nwarrow \Sigma F_y = 0; \quad N - P\sin\theta = 0 \quad N = P\sin\theta$

Average Normal Stress and Shear Stress :

Area at θ plane, $A' = \dfrac{A}{\sin\theta}$.

$$\sigma = \frac{N}{A} = \frac{P\sin\theta}{\frac{A}{\sin\theta}} = \frac{P}{A}\sin^2\theta$$

$$\tau_{avg} = \frac{V}{A'} = \frac{P\cos\theta}{\frac{A}{\sin\theta}} = \frac{P}{A}\sin\theta\cos\theta = \frac{P}{2A}\sin 2\theta$$

$$\frac{d\tau_{avg}}{d\theta} = \frac{P}{A}\cos 2\theta = 0$$

$2\theta = 90°; \quad \theta = 45°$ **Ans**

Therefore, $(\tau_{avg})_{max} = \dfrac{P}{2A}\sin 2(45°) = \dfrac{P}{2A}$ **Ans**

Also, when $\theta = 45°$, $\sigma = \dfrac{P}{A}\sin^2 45° = \dfrac{P}{2A}$

***1–60.** The bars of the truss each have a cross-sectional area of 1.25 in². Determine the average normal stress in each member due to the loading $P = 6$ kip. State whether the stress is tensile or compressive.

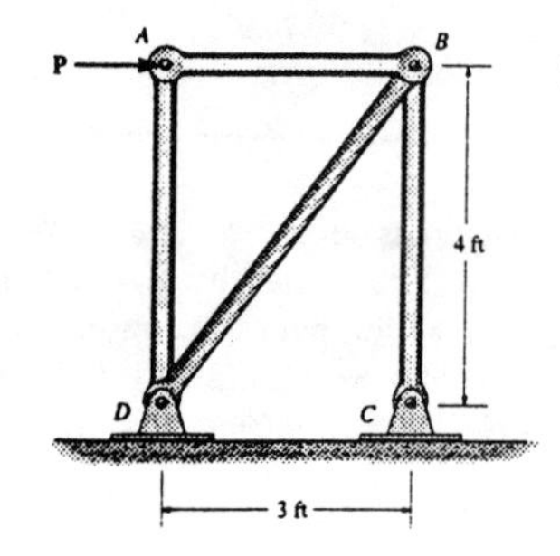

Method of Joints :

Joint A

$\overset{+}{\rightarrow}\Sigma F_x = 0; \quad 6 - F_{AB} = 0 \qquad F_{AB} = 6.00$ kip (C)

$+\uparrow \Sigma F_y = 0; \quad F_{AD} = 0$

Joint B

$\overset{+}{\rightarrow}\Sigma F_x = 0; \quad 6.00 - \frac{3}{5}F_{BD} = 0 \qquad F_{BD} = 10.0$ kip (T)

$+\uparrow \Sigma F_y = 0; \quad F_{BC} - \frac{4}{5}(10.0) = 0 \qquad F_{BC} = 8.00$ kip (C)

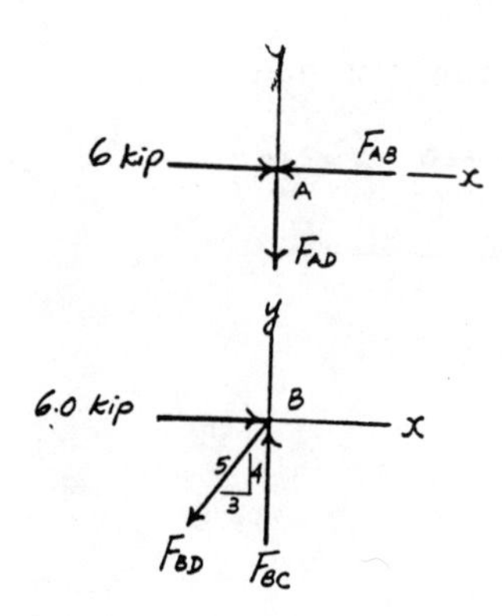

Average normal stress :

$\sigma_{AB} = \dfrac{F_{AB}}{A} = \dfrac{6.00}{1.25} = 4.80$ ksi (C) **Ans**

$\sigma_{AD} = \dfrac{F_{AD}}{A} = 0$ **Ans**

$\sigma_{BD} = \dfrac{F_{BD}}{A} = \dfrac{10.0}{1.25} = 8$ ksi (T) **Ans**

$\sigma_{BC} = \dfrac{F_{BC}}{A} = \dfrac{8.00}{1.25} = 6.40$ ksi (C) **Ans**

1–61. The bars of the truss each have a cross-sectional area of 1.25 in². If the maximum average normal stress in any bar is not allowed to exceed 20 ksi, determine the maximum load **P** that can be applied to the truss.

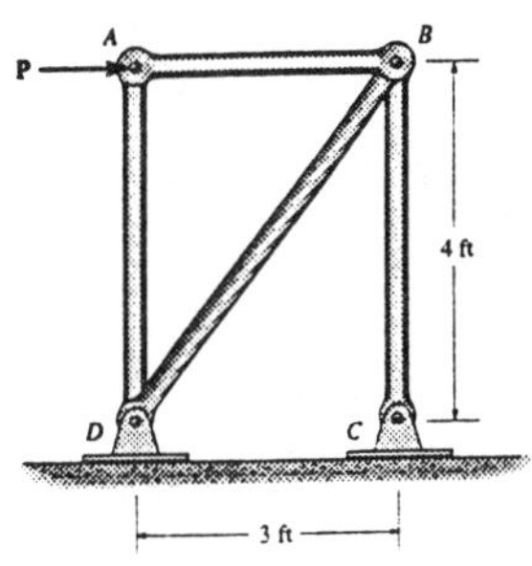

Method of Joint :

Joint A

$\xrightarrow{+} \Sigma F_x = 0; \quad -F_{AB} + P = 0 \qquad F_{AB} = 1.00P$

$+\uparrow \Sigma F_y = 0; \quad F_{AD} = 0$

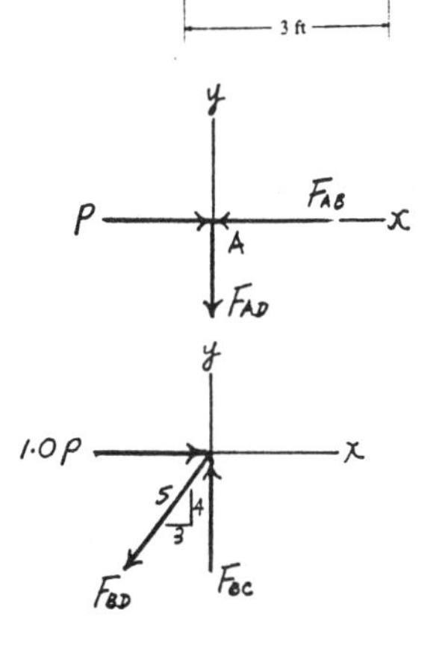

Joint B

$\xrightarrow{+} \Sigma F_x = 0; \quad -\frac{3}{5}F_{BD} + 1.00P = 0 \qquad F_{BD} = 1.667P$

$+\uparrow \Sigma F_y = 0; \quad F_{BC} - \frac{4}{5}(1.667P) = 0 \qquad F_{BC} = 1.333P$

Average normal stress : Member *BD* is subjected to the greatest force. Therefore, it is the critical member.

$$\sigma_{allow} = \frac{F_{BD}}{A_{BD}}; \qquad 20 = \frac{1.667P}{1.25}$$

$$P = 15.0 \text{ kip} \qquad \textbf{Ans}$$

1–62. The bar is subjected to a uniform distributed axial loading of 10 kN/m and a concentrated force of 1.5 kN at its midpoint as shown. Determine the maximum average normal stress in the bar and its location *x*.

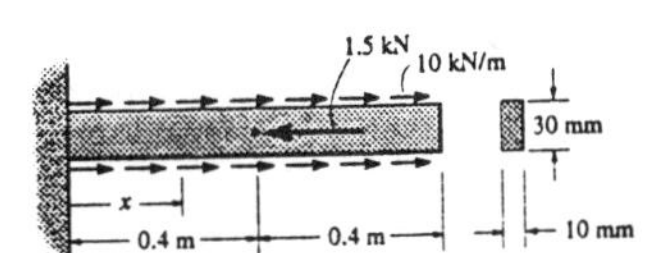

Equations of Equilibrium :

For $0 \le x < 0.4$ m

$\xrightarrow{+} \Sigma F_x = 0; \quad 10(0.8 - x) - 1.5 - N = 0$

$N = (6.50 - 10.0x)$ kN

For $0.4\text{m} < x \le 0.8\text{m}$

$\xrightarrow{+} \Sigma F_x = 0; \quad 10(0.8 - x) - N = 0$

$N = (8.00 - 10.0x)$ kN

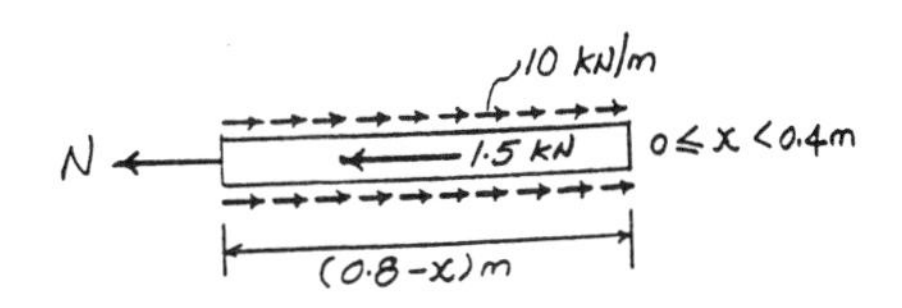

Average Normal Stress : By inspection, maximum average normal stress occurs at region $0 \le x < 0.4$m where $\boldsymbol{x = 0}$. Therefore,

$N = 6.50 - 10.0(0) = 6.50$ kN

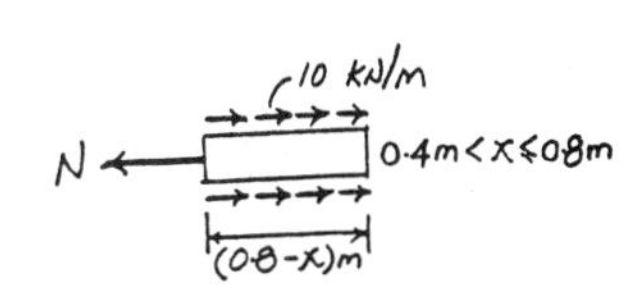

$$\sigma_{max} = \frac{N}{A} = \frac{6.50(10^3)}{(0.01)(0.03)} = 21.7 \text{ MPa} \qquad \textbf{Ans}$$

1–63. The prismatic bar has a cross-sectional area A. If it is subjected to a distributed axial loading that increases linearly from $w = 0$ at $x = 0$ to $w = w_0$ at $x = a$, and then decreases linearly to $w = 0$ at $x = 2a$, determine the average normal stress in the bar as a function of x for $0 \leq x < a$.

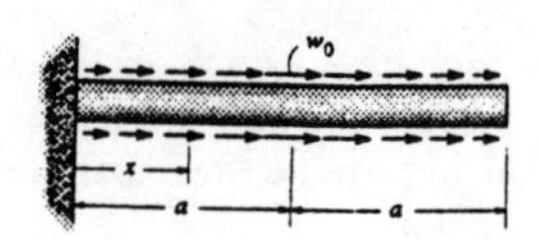

Equation of Equilibrium :

$$\overset{+}{\rightarrow} \Sigma F_x = 0; \qquad -N + \frac{1}{2}\left(\frac{w_0}{a}x + w_0\right)(a - x) + \frac{1}{2}w_0 a = 0$$

$$N = \frac{w_0}{2a}\left(2a^2 - x^2\right)$$

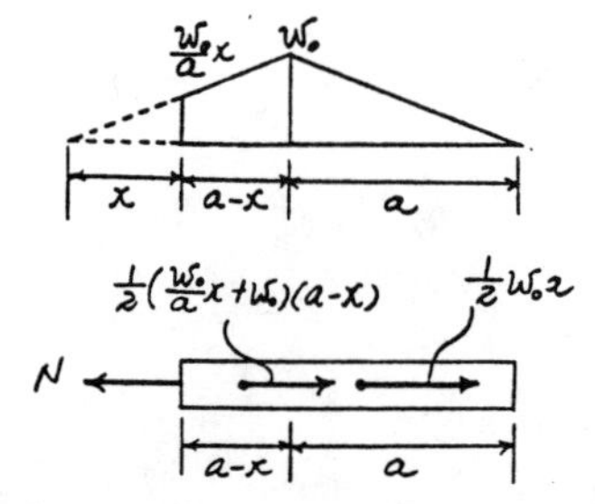

Average Normal Stress :

$$\sigma = \frac{N}{A} = \frac{\frac{w_0}{2a}(2a^2 - x^2)}{A} = \frac{w_0}{2aA}\left(2a^2 - x^2\right) \qquad \textbf{Ans}$$

***1–64.** The prismatic bar has a cross-sectional area A. If it is subjected to a distributed axial loading that increases linearly from $w = 0$ at $x = 0$ to $w = w_0$ at $x = a$, and then decreases linearly to $w = 0$ at $x = 2a$, determine the average normal stress in the bar as a function of x for $a < x \leq 2a$.

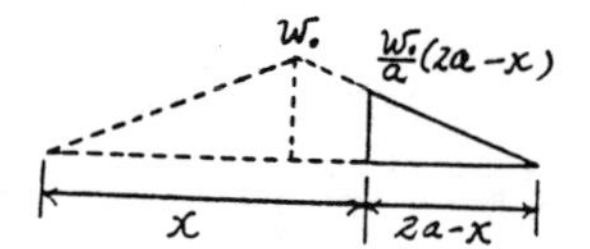

Equation of Equilibrium :

$$\overset{+}{\rightarrow} \Sigma F_x = 0; \qquad -N + \frac{1}{2}\left[\frac{w_0}{a}(2a - x)\right](2a - x) = 0$$

$$N = \frac{w_0}{2a}(2a - x)^2$$

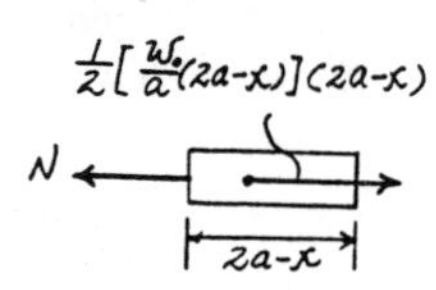

Average Normal Stress :

$$\sigma = \frac{N}{A} = \frac{\frac{w_0}{2a}(2a - x)^2}{A} = \frac{w_0}{2aA}(2a - x)^2 \qquad \textbf{Ans}$$

1–65. The bar has a cross-sectional area of $400\ (10^{-6})\ \text{m}^2$. If it is subjected to a uniform axial distributed loading along its length and to two concentrated loads as shown, determine the average normal stress in the bar as a function of x for $0 < x < 0.5$ m.

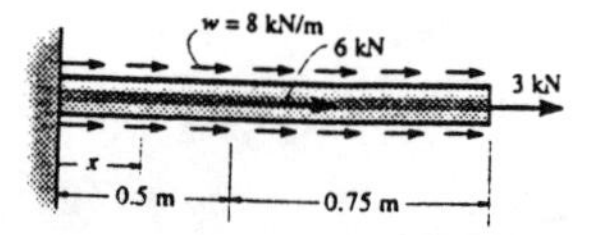

Equation of Equilibrium :

$$\overset{+}{\rightarrow} \Sigma F_x = 0; \qquad -N + 3 + 6 + 8(1.25 - x) = 0$$

$$N = (19.0 - 8.00x)\ \text{kN}$$

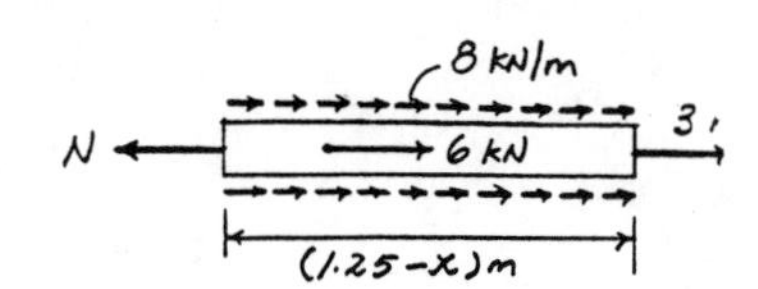

Average Normal Stress :

$$\sigma = \frac{N}{A} = \frac{(19.0 - 8.00x)(10^3)}{400(10^{-6})}$$

$$= (47.5 - 20.0x)\ \text{MPa} \qquad \textbf{Ans}$$

1–66. The bar has a cross-sectional area of $400\,(10^{-6})\ \text{m}^2$. If it is subjected to a uniform axial distributed loading along its length and to two concentrated loads as shown, determine the average normal stress in the bar as a function of x for $0.5\ \text{m} < x < 1.25\ \text{m}$.

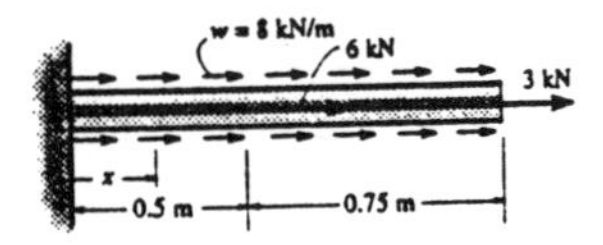

Equation of Equilibrium :

$$\xrightarrow{+} \Sigma F_x = 0; \qquad -N + 3 + 8(1.25 - x) = 0$$

$$N = (13.0 - 8.00x)\ \text{kN}$$

Average Normal Stress :

$$\sigma = \frac{N}{A} = \frac{(13.0 - 8.00x)(10^3)}{400(10^{-6})}$$

$$= (32.5 - 20.0x)\ \text{MPa} \qquad \textbf{Ans}$$

1–67. The boom has a uniform weight of 600 lb and is hoisted into position using the cable BC. If the cable has a diameter of 0.5 in., plot the average normal stress in the cable as a function of the boom position θ for $0° \leq \theta \leq 90°$.

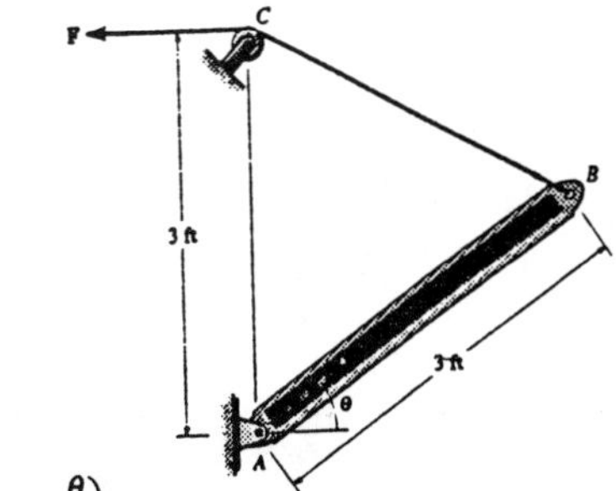

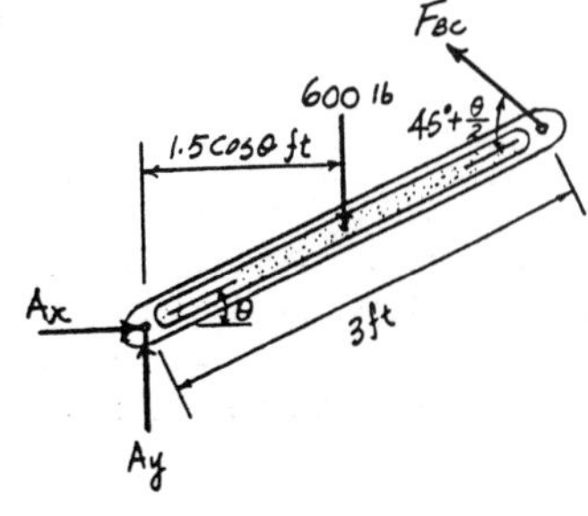

Support Reactions :

$$\circlearrowleft + \Sigma M_A = 0; \qquad F_{BC}\sin\left(45° + \frac{\theta}{2}\right)(3) - 600(1.5\cos\theta) = 0$$

$$F_{BC} = \frac{300\cos\theta}{\sin\left(45° + \frac{\theta}{2}\right)}$$

Average Normal Stress :

$$\sigma_{BC} = \frac{F_{BC}}{A_{BC}} = \frac{\frac{300\cos\theta}{\sin\left(45° + \frac{\theta}{2}\right)}}{\frac{\pi}{4}(0.5^2)}$$

$$= \left\{\frac{1.528\cos\theta}{\sin\left(45° + \frac{\theta}{2}\right)}\right\}\ \text{ksi} \qquad \textbf{Ans}$$

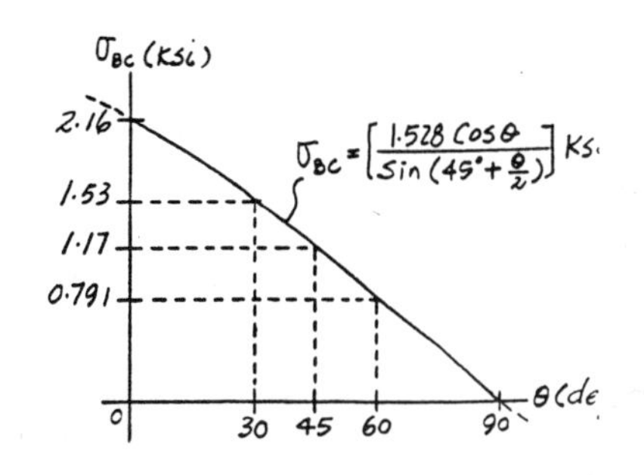

***1–68.** The two-member frame is subjected to the distributed loading shown. Determine the average normal stress and average shear stress acting at sections a–a and b–b. Member CB has a square cross section of 30 mm on each side. Take $w = 6$ kN/m.

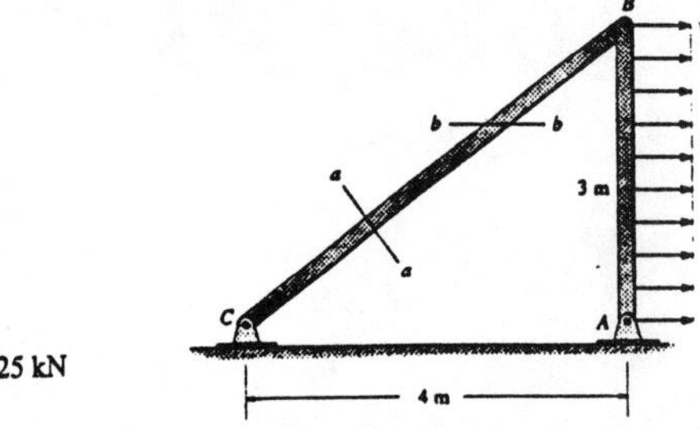

Support Reactions : FBD (a)

$$\circlearrowleft + \Sigma M_A\ 0; \quad \frac{4}{5}F_{BC}(3) - 6(3)(1.5) = 0 \qquad F_{BC} = 11.25 \text{ kN}$$

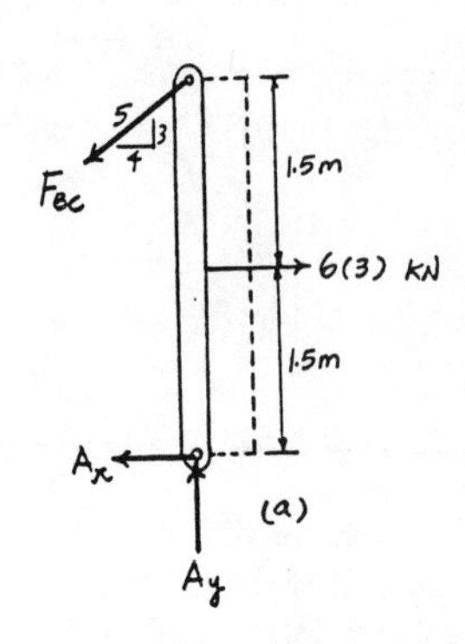

Equations of Equilibrium : For section a-a, FBD (b)

$$+\swarrow \Sigma F_x = 0; \qquad 11.25 - N_{a\text{-}a} = 0 \qquad N_{a\text{-}a} = 11.25 \text{ kN}$$

$$\nwarrow + \Sigma F_y = 0; \qquad V_{a\text{-}a} = 0$$

Average Normal Stress And Shear Sress : The cross-sectional area of section a-a, $A = 0.03(0.03) = 0.90(10^{-3})$ m^2

$$\sigma_{a\text{-}a} = \frac{N_{a\text{-}a}}{A} = \frac{11.25(10^3)}{0.9(10^{-3})} = 12.5 \text{ MPa} \qquad \textbf{Ans}$$

$$(\tau_{a\text{-}a})_{avg} = \frac{V_{a\text{-}a}}{A} = 0 \qquad \textbf{Ans}$$

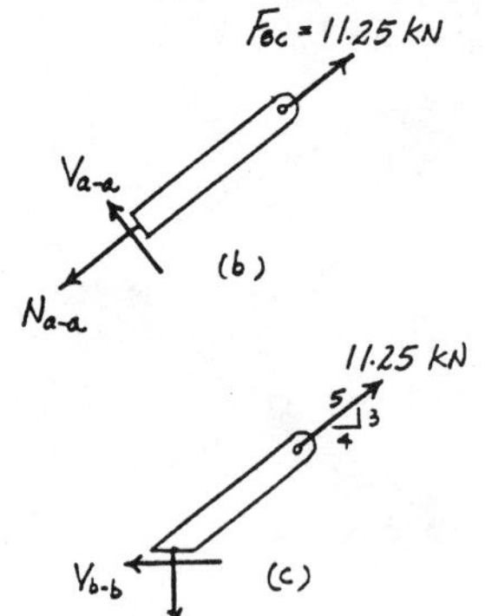

Equations of Equilibrium : For section b-b, FBD (c)

$$\xrightarrow{+} \Sigma F_x = 0; \qquad \frac{4}{5}(11.25) - V_{b\text{-}b} = 0 \qquad V_{b\text{-}b} = 9.00 \text{ kN}$$

$$+\uparrow \Sigma F_y = 0; \qquad \frac{3}{5}(11.25) - N_{b\text{-}b} = 0 \qquad N_{b\text{-}b} = 6.75 \text{ kN}$$

Average Normal Stress And Shear Sress : The cross-sectional area of section b-b, $A' = \frac{5A}{3} = 1.50(10^{-3})$ m^2

$$\sigma_{b\text{-}b} = \frac{N_{b\text{-}b}}{A'} = \frac{6.75(10^3)}{1.50(10^{-3})} = 4.50 \text{ MPa} \qquad \textbf{Ans}$$

$$(\tau_{b\text{-}b})_{avg} = \frac{V_{b\text{-}b}}{A'} = \frac{9.00(10^3)}{1.50(10^{-3})} = 6.00 \text{ MPa} \qquad \textbf{Ans}$$

1–69. The two-member frame is subjected to the distributed loading shown. Determine the largest intensity w of the uniform loading that can be applied to the frame without causing either the average normal stress or the average shear stress at section b–b to exceed $\sigma = 15$ MPa and $\tau = 16$ MPa, respectively. Member CB has a square cross-section of 30 mm on each side.

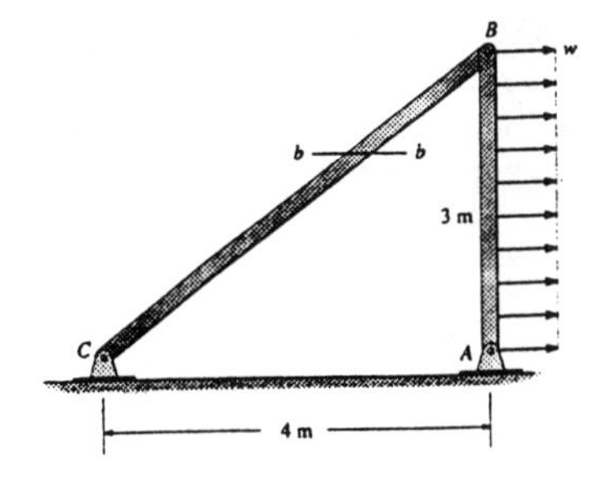

Support Reactions : FBD (a)

$$\circlearrowleft + \Sigma M_A = 0; \quad \frac{4}{5}F_{BC}(3) - 3w(1.5) = 0 \quad F_{BC} = 1.875w$$

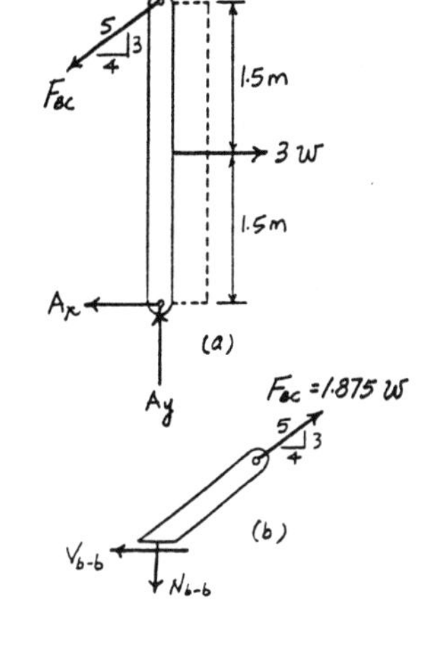

Equations of Equilibrium : For section b - b, FBD (b)

$$\xrightarrow{+} \Sigma F_x = 0; \quad \frac{4}{5}(1.875w) - V_{b-b} = 0 \quad V_{b-b} = 1.50w$$

$$+\uparrow \Sigma F_y = 0; \quad \frac{3}{5}(1.875w) - N_{b-b} = 0 \quad N_{b-b} = 1.125w$$

Average Normal Stress And Shear Sress : The cross - sectional area of section b - b, $A' = \frac{5A}{3}$; where $A = (0.03)(0.03) = 0.90(10^{-3})\ \text{m}^2$.

Then $A' = \frac{5}{3}(0.90)(10^{-3}) = 1.50(10^{-3})\ \text{m}^2$.

Assume failure due to normal stress.

$$(\sigma_{b-b})_{\text{Allow}} = \frac{N_{b-b}}{A'}; \quad 15(10^6) = \frac{1.125w}{1.50(10^{-3})}$$

$$w = 20000\ \text{N/m} = 20.0\ \text{kN/m}$$

Assume failure due to shear stress.

$$(\tau_{b-b})_{\text{Allow}} = \frac{V_{b-b}}{A'}; \quad 16(10^6) = \frac{1.50w}{1.50(10^{-3})}$$

$w = 16000\ \text{N/m} = 16.0\ \text{kN/m}$ ***(Controls !)*** **Ans**

1–70. The uniform beam is supported by two rods AB and CD that have cross-sectional areas of 10 mm² and 15 mm², respectively. If $d = 1$ m, determine the average normal stress in each rod.

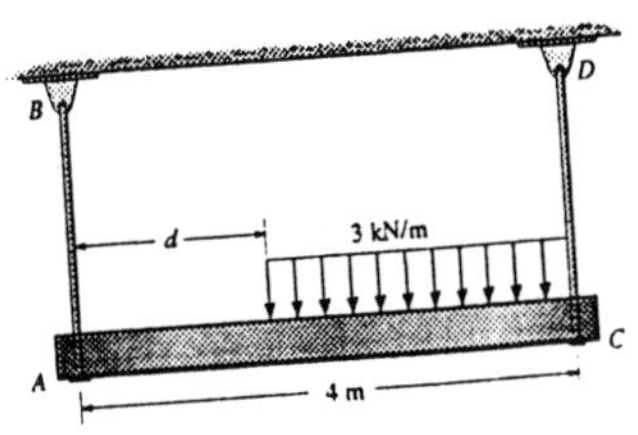

Support Reactions :

$$\circlearrowleft + \Sigma M_A = 0; \quad F_{CD}(4) - 3(3)(2.5) = 0$$
$$F_{CD} = 5.625\ \text{kN}$$

$$\circlearrowleft + \Sigma M_C = 0; \quad 3(3)(1.5) - F_{AB}(4) = 0$$
$$F_{AB} = 3.375\ \text{kN}$$

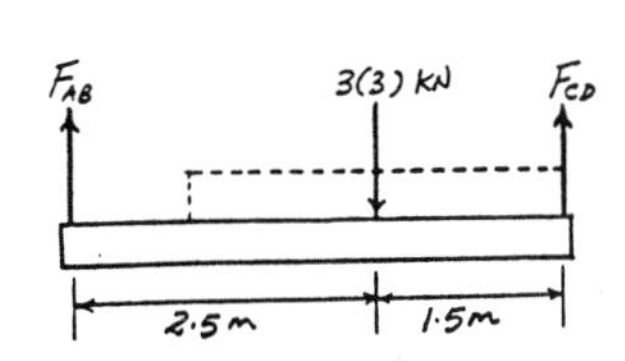

Average Normal Stress :

$$\sigma_{AB} = \frac{F_{AB}}{A_{AB}} = \frac{3.375(10^3)}{10(10^{-6})} = 337.5\ \text{MPa} \qquad \textbf{Ans}$$

$$\sigma_{CD} = \frac{F_{CD}}{A_{CD}} = \frac{5.625(10^3)}{15(10^{-6})} = 375\ \text{MPa} \qquad \textbf{Ans}$$

1–71. The uniform beam is supported by two rods AB and CD that have cross-sectional areas of 10 mm² and 15 mm², respectively. Determine the position d of the distributed load so that the average normal stress in each rod is the same.

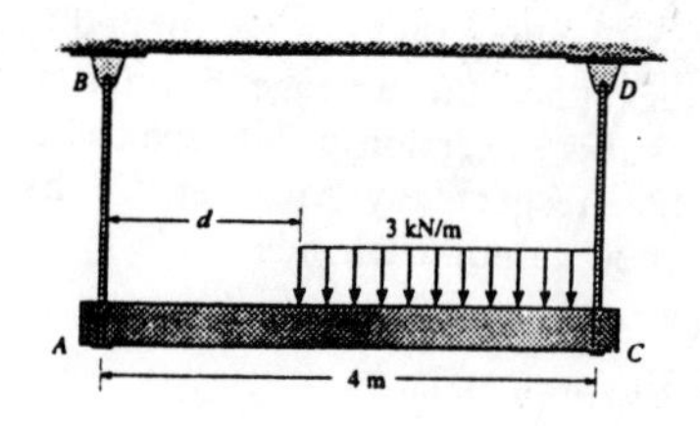

Support Reactions :

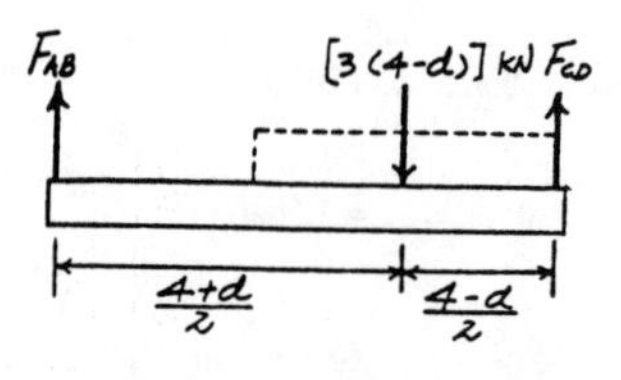

$$\circlearrowleft +\Sigma M_A = 0; \qquad F_{CD}(4) - 3(4-d)\left(\frac{4+d}{2}\right) = 0$$

$$4F_{CD} - 24 + \frac{3d^2}{2} = 0 \qquad [1]$$

$$\circlearrowleft +\Sigma M_C = 0; \qquad 3(4-d)\left(\frac{4-d}{2}\right) - F_{AB}(4) = 0$$

$$24 - 12d + \frac{3d^2}{2} - 4F_{AB} = 0 \qquad [2]$$

Average Normal Stress :

$$\sigma_{AB} = \sigma_{CD}\,; \qquad \frac{F_{AB}}{10} = \frac{F_{CD}}{15}$$

$$F_{CD} = 1.5\,F_{AB} \qquad [3]$$

Solving Eqs. [1] to [3] yields :

$d = 0.800$ m or $d = 4.00$ m **Ans**

$F_{CD} = 5.76$ kN or $F_{CD} = 0$

$F_{AB} = 3.84$ kN or $F_{AB} = 0$

***1–72.** The pedestal in the shape of a frustum of a cone is made of concrete having a specific weight of γ. Determine the bearing stress acting at its base. *Hint:* The volume of a cone of radius r and height h is $V = \frac{1}{3}\pi r^2 h$.

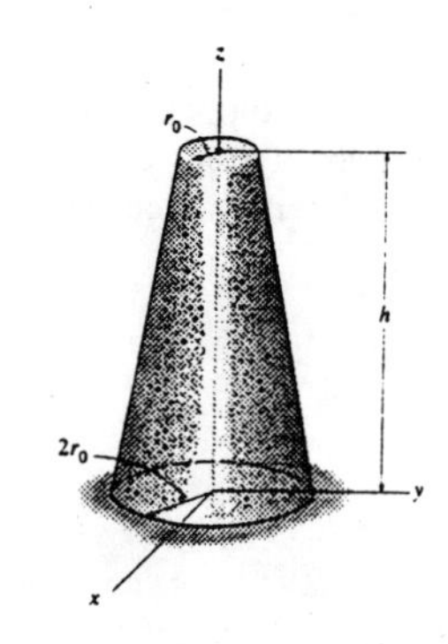

Geometrical Properties :

$$\frac{h'}{r_0} = \frac{h'+h}{2\,r_0}; \qquad h' = h$$

$$V = \frac{1}{3}\pi(2r_0)^2(2h) - \frac{1}{3}\pi r_0^2 h = \frac{7}{3}\pi r_0^2 h$$

$$W = \gamma V = \frac{7}{3}\pi r_0^2 h\gamma$$

$$A = \pi(2r_0)^2 = 4\pi r_0^2$$

Bearing Stress :

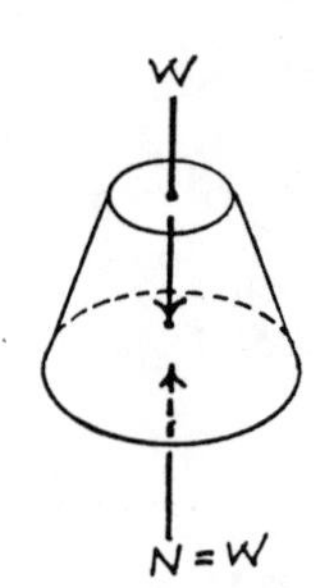

$$\sigma = \frac{N}{A} = \frac{\frac{7}{3}\pi r_0^2 h\gamma}{4\pi r_0^2} = \frac{7\gamma h}{12} \qquad \textbf{Ans}$$

1–73. The pedestal in the shape of a frustum of a cone is made of concrete having a specific weight of γ. Determine the bearing stress acting in the pedestal at its midheight, $z = h/2$. *Hint:* The volume of a cone of radius r and height h is $V = \frac{1}{3}\pi r^2 h$.

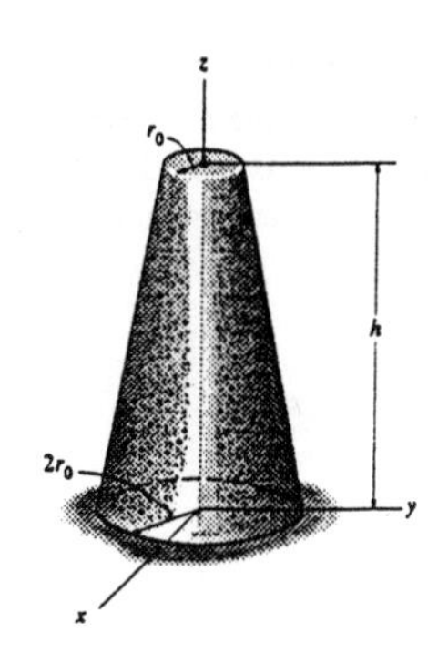

Geometrical Properties :

$$\frac{h'}{r_0} = \frac{h' + h}{2r_0}; \qquad h' = h$$

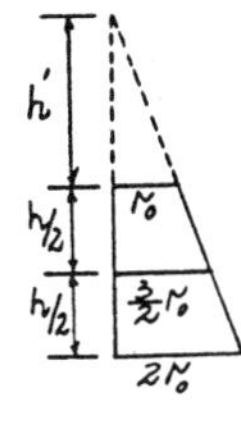

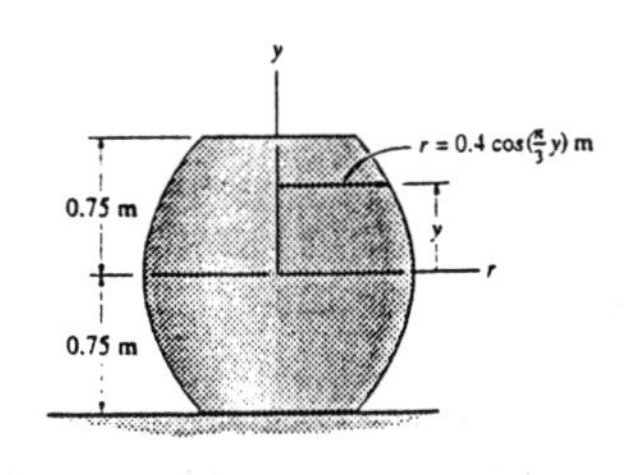

$$V = \frac{1}{3}\pi\left(\frac{3}{2}r_0\right)^2\left(\frac{3}{2}h\right) - \frac{1}{3}\pi r_0{}^2 h = \frac{19}{24}\pi r_0^2 h$$

$$W = \gamma V = \frac{19}{24}\pi r_0^2 h\gamma$$

$$A = \pi\left(\frac{3}{2}r_0\right)^2 = \frac{9}{4}\pi r_0^2$$

Bearing Stress :

$$\sigma = \frac{N}{A} = \frac{\frac{19}{24}\pi r_0^2 h\gamma}{\frac{9}{4}\pi r_0{}^2} = \frac{19\gamma h}{54} \qquad \textbf{Ans}$$

1–74. The shape has a radius that is defined by $r = 0.4\cos(\pi y/3)$ m. Determine the average normal stress at the support if the material has a density of $\rho = 3\ \text{Mg/m}^3$.

Geometrical Properties :

$$A = \pi\left(0.2828^2\right) = 0.2513\ \text{m}^2$$

$$dV = \pi r^2 dy = \pi\left[0.4\cos\frac{\pi}{3}y\right]^2 dy$$

$$V = \int dV = 0.16\pi\int_{-0.75\text{m}}^{0.75\text{m}}\cos^2\left(\frac{\pi}{3}y\right)dy$$

However, $\cos^2\left(\frac{\pi}{3}y\right) = \frac{\cos\frac{2\pi}{3}y + 1}{2}$

Therefore,

$$V = 0.08\pi\int_{-0.75\text{m}}^{0.75\text{m}}\left(\cos\frac{2\pi}{3}y + 1\right)dy$$
$$= 0.08\pi\left[\frac{3}{2\pi}\sin\frac{2\pi}{3}y + y\right]\Bigg|_{-0.75\text{m}}^{0.75\text{m}}$$
$$= 0.6170\ \text{m}^3$$

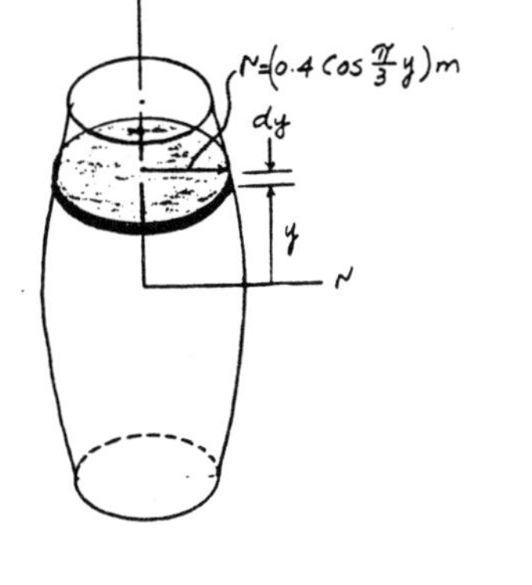

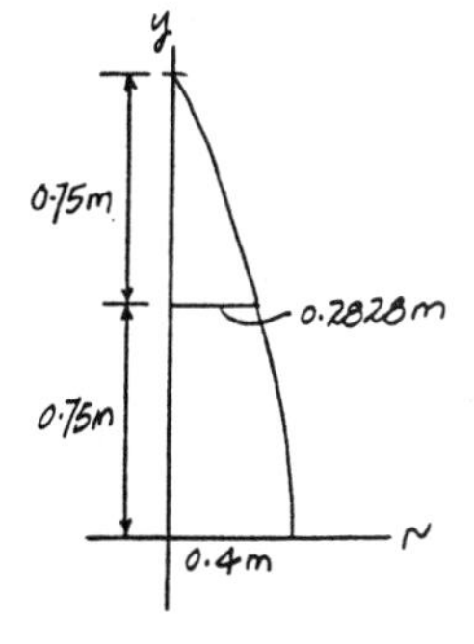

$$W = \rho g V = (3000)(9.81)(0.6170) = 18.158\ \text{kN}$$

Average Normal Stress :

$$\sigma = \frac{W}{A} = \frac{18.158}{0.2513} = 72.2\ \text{kPa} \qquad \textbf{Ans}$$

1–75. The uniform bar, having a cross-sectional area of A and mass per unit length of m, is pinned at its center. If it is rotating in the horizontal plane at a constant angular rate of ω, determine the average normal stress in the bar as a function of x.

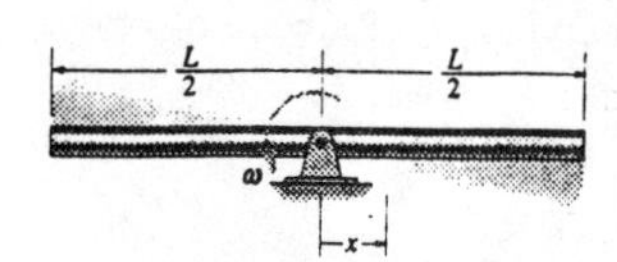

Equation of Motion :

$$\overset{+}{\leftarrow} \Sigma F_x = ma_N; \quad N = m\left[\frac{1}{2}(L-2x)\right]\omega^2\left[\frac{1}{4}(L+2x)\right]$$

$$= \frac{m\omega^2}{8}\left(L^2 - 4x^2\right)$$

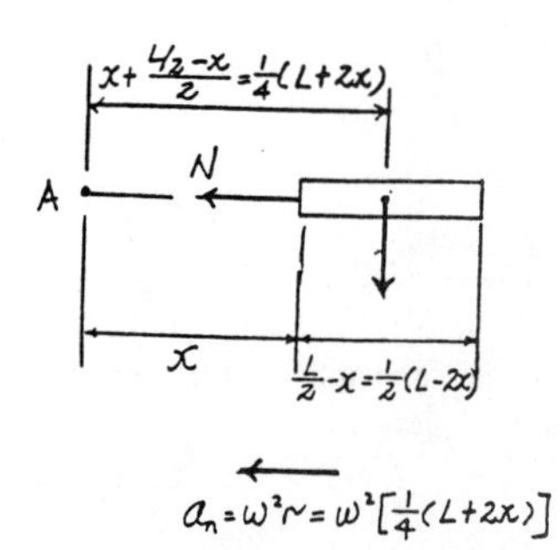

Average Normal Stress :

$$\sigma = \frac{N}{A} = \frac{m\omega^2}{8A}\left(L^2 - 4x^2\right) \qquad \textbf{Ans}$$

***1–76.** The block is subjected to the centrally applied axial load of 60 kip. Determine the magnitudes of the normal and shear stresses acting on an inclined plane which contain the edges AB and BC. *Hint:* Use vector analysis to show that the outward normal to the plane can be defined by the unit vector $n = 0.4193\mathbf{i} + 0.5447\mathbf{j} + 0.7263\mathbf{k}$. The stress components act along and perpendicular to this vector.

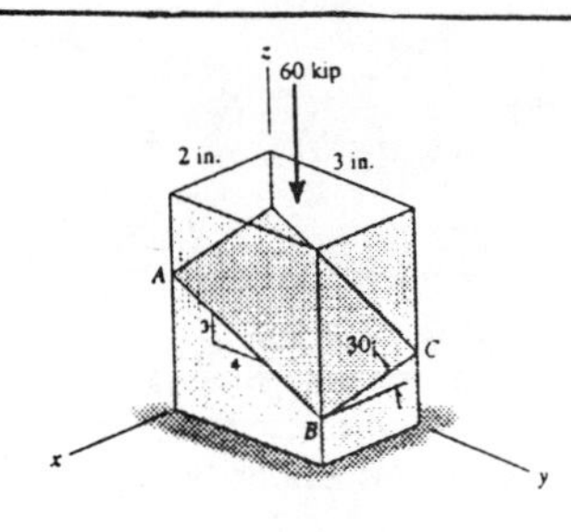

Geometrical Properties :

$$\mathbf{r}_{BC} = \{-2\mathbf{i} + 2\tan 30°\mathbf{k}\} \text{ in.}$$

$$\mathbf{u}_{BC} = \frac{\mathbf{r}_{BC}}{r_{BC}} = \frac{-2\mathbf{i} + 2\tan 30°\mathbf{k}}{\sqrt{(-2)^2 + (2\tan 30°)^2}} = -0.8660\mathbf{i} + 0.50\mathbf{k}$$

$$\mathbf{r}_{BA} = \{-3\mathbf{j} + 2.25\mathbf{k}\} \text{ in.}$$

$$\mathbf{u}_{BA} = \frac{\mathbf{r}_{BA}}{r_{BA}} = \frac{-3\mathbf{j} + 2.25\mathbf{k}}{\sqrt{(-3)^2 + 2.25^2}} = -0.80\mathbf{j} + 0.60\mathbf{k}$$

$$\mathbf{r}_{BC} \times \mathbf{r}_{BA} = (-2\mathbf{i} + 2\tan 30°\mathbf{k}) \times (-3\mathbf{j} + 2.25\mathbf{k})$$
$$= \{6\tan 30°\mathbf{i} + 4.50\mathbf{j} + 6.00\mathbf{k}\} \text{ in.}$$

$$\left|\mathbf{r}_{BC} \times \mathbf{r}_{BA}\right| = \sqrt{(6\tan 30°)^2 + 4.50^2 + 6.00^2} = 8.2614 \text{ in.}$$

$$\mathbf{n} = \frac{\mathbf{r}_{BC} \times \mathbf{r}_{BA}}{\left|\mathbf{r}_{BC} \times \mathbf{r}_{BA}\right|} = \frac{6\tan 30°\mathbf{i} + 4.50\mathbf{j} + 6.00\mathbf{k}}{8.2614}$$
$$= 0.4193\mathbf{i} + 0.5447\mathbf{j} + 0.7263\mathbf{k}$$

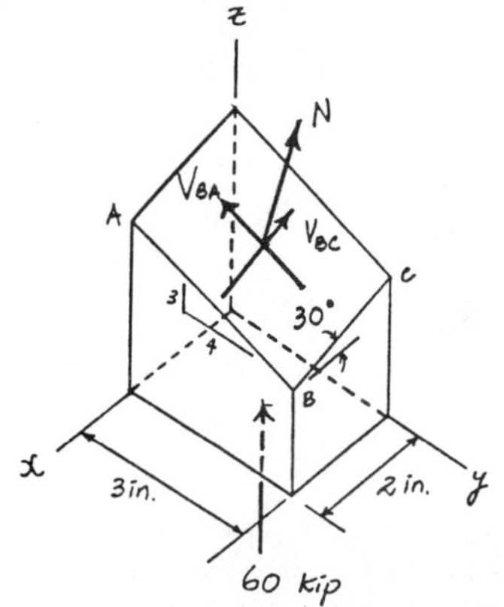

Equations of Equilibrium :

$$\Sigma F_n = 0; \quad N + (60\text{k}) \cdot (0.4193\mathbf{i} + 0.5447\mathbf{j} + 0.7263\mathbf{k}) = 0$$
$$N = -43.58 \text{ kip}$$

$$\Sigma(F_t)_{BC} = 0; \quad V_{BC} + (60\text{k}) \cdot (-0.8660\mathbf{i} + 0.50\mathbf{k}) = 0$$
$$V_{BC} = -30.0 \text{ kip}$$

$$\Sigma(F_t)_{BA} = 0; \quad V_{BA} + (60\text{k}) \cdot (-0.80\mathbf{j} + 0.60\mathbf{k}) = 0$$
$$V_{BA} = -36.0 \text{ kip}$$

Average Normal Stress And Shear Stress: The cross-sectional area for the inclined surface is $A' = \left(\frac{2}{\cos 30°}\right)\left(\frac{3}{4/5}\right) = 8.6603 \text{ in}^2$.

$$\sigma = \frac{N}{A} = \frac{43.58}{8.6603} = 5.03 \text{ ksi} \qquad \textbf{Ans}$$

$$V = \sqrt{V_{BC}^2 + V_{BA}^2} = \sqrt{30.0^2 + 36.0^2} = 46.86 \text{ kip}$$

$$\tau_{avg} = \frac{V}{A} = \frac{46.86}{8.6603} = 5.41 \text{ ksi} \qquad \textbf{Ans}$$

1–77. The joint is fastened together using two bolts. Determine the required diameter of the bolts if the failure shear stress for the bolts is $\tau_{fail} = 350$ MPa. Use a factor of safety for shear of F.S. = 2.5.

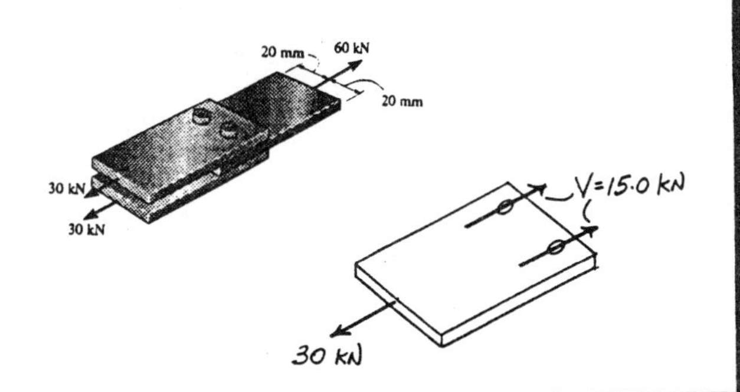

Allowable Shear Stress :

$$\tau_{allow} = \frac{\tau_{fail}}{F.S} = \frac{V}{A}; \qquad \frac{350(10^6)}{2.5} = \frac{15.0(10^3)}{\frac{\pi}{4}d^2}$$

$d = 0.01168$ m = 11.7 mm **Ans**

1–78. The rods *AB* and *CD* are made of steel having a failure tensile stress of $\sigma_{fail} = 510$ MPa. Using a factor of safety of F.S. = 1.75 for tension, determine their smallest diameter so that they can support the load shown. The beam is assumed to be pin connected at *A* and *C*.

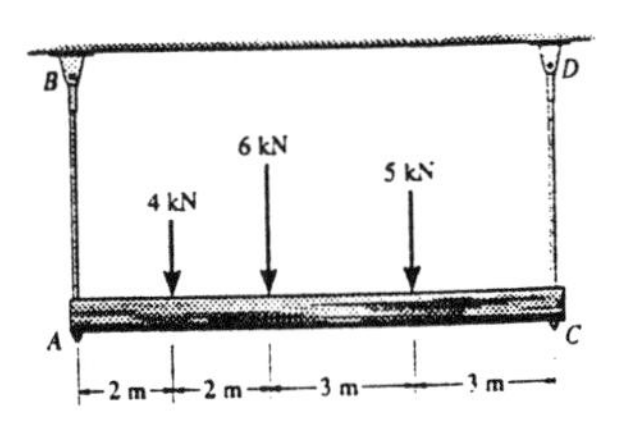

Support Reactions :

$$+\Sigma M_A = 0; \qquad F_{CD}(10) - 5(7) - 6(4) - 4(2) = 0$$

$$F_{CD} = 6.70 \text{ kN}$$

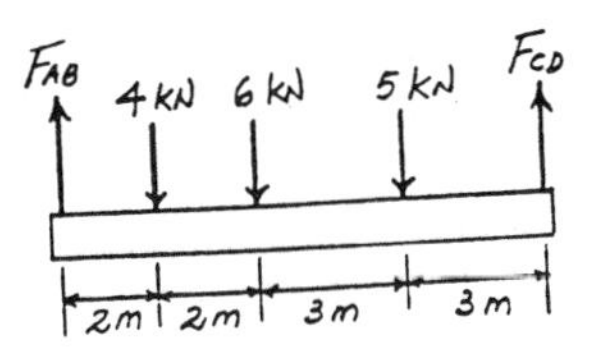

$$+\Sigma M_C = 0; \qquad 4(8) + 6(6) + 5(3) - F_{AB}(10) = 0$$

$$F_{AB} = 8.30 \text{ kN}$$

Allowable Normal Stress : Design of rod sizes

For rod AB

$$\sigma_{allow} = \frac{\sigma_{fail}}{F.S} = \frac{F_{AB}}{A_{AB}}; \qquad \frac{510(10^6)}{1.75} = \frac{8.30(10^3)}{\frac{\pi}{4}d_{AB}^2}$$

$d_{AB} = 0.006022$ m = 6.02 mm **Ans**

For rod CD

$$\sigma_{allow} = \frac{\sigma_{fail}}{F.S} = \frac{F_{CD}}{A_{CD}}; \qquad \frac{510(10^6)}{1.75} = \frac{6.70(10^3)}{\frac{\pi}{4}d_{CD}^2}$$

$d_{CD} = 0.005410$ m = 5.41 mm **Ans**

1–79. Member *A* of the timber step joint for a truss is subjected to a compressive force of 5 kN. Determine the required diameter *d* of the steel rod at *C* and the height *h* of member *B* if the allowable normal stress for the steel is $(\sigma_{allow})_{st} = 157$ MPa and the allowable normal stress for the wood is $(\sigma_{allow})_w = 2$ MPa. Member *B* is 50 mm thick.

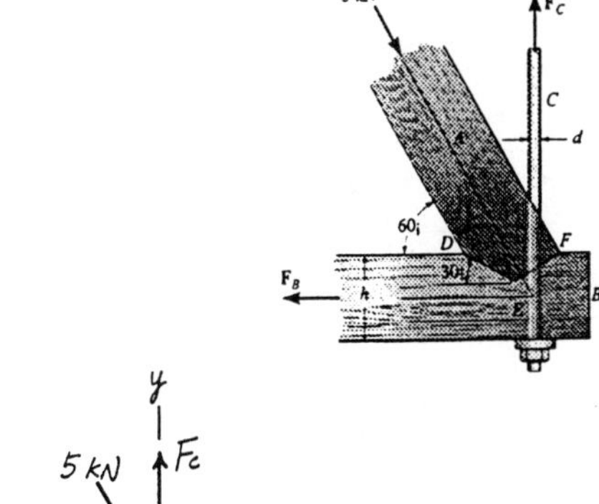

Equations of Equilibrium :

$$\xrightarrow{+} \Sigma F_x = 0; \qquad 5\cos 60° - F_B = 0 \qquad F_B = 2.50 \text{ kN}$$

$$+\uparrow \Sigma F_y = 0; \qquad F_C - 5\sin 60° = 0 \qquad F_C = 4.330 \text{ kN}$$

Allowable Normal Stress : Design of member sizes

$$(\sigma_{allow})_w = \frac{F_B}{A_B}; \qquad 2(10^6) = \frac{2.50(10^3)}{(0.05)h}$$

$h = 0.025$ m = 25.0 mm **Ans**

$$(\sigma_{allow})_{st} = \frac{F_C}{A_C}; \qquad 157(10^6) = \frac{4.330(10^3)}{\frac{\pi}{4}d^2}$$

$d = 0.005926$ m = 5.93 mm **Ans**

***1–80.** The eye bolt is used to support the load of 5 kip. Determine its diameter d to the nearest $\frac{1}{8}$ in. and the required thickness h to the nearest $\frac{1}{8}$ in. of the support so that the washer will not penetrate or shear through it. The allowable normal stress for the bolt is $\sigma_{\text{allow}} = 21$ ksi and the allowable shear stress for the supporting material is $\tau_{\text{allow}} = 5$ ksi.

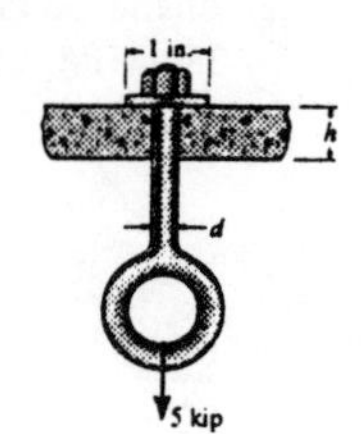

Allowable Normal Stress : Design of bolt size

$$\sigma_{\text{allow}} = \frac{P}{A_b}; \quad 21.0(10^3) = \frac{5(10^3)}{\frac{\pi}{4}d^2}$$

$$d = 0.5506 \text{ in.}$$

$$\text{Use } d = \frac{5}{8} \text{ in.} \qquad \textbf{Ans}$$

Allowable Shear Stress : Design of support thickness

$$\tau_{\text{allow}} = \frac{V}{A}; \quad 5(10^3) = \frac{5(10^3)}{\pi(1)(h)}$$

$$h = 0.3183 \text{ in.}$$

$$\text{Use } h = \frac{3}{8} \text{ in.} \qquad \textbf{Ans}$$

1–81. Determine the required cross-sectional area of member BC and the diameter of the pins at A and B if the allowable normal stress for BC is $\sigma_{\text{allow}} = 10$ ksi and the allowable shear stress for the pins is $\tau_{\text{allow}} = 3$ ksi.

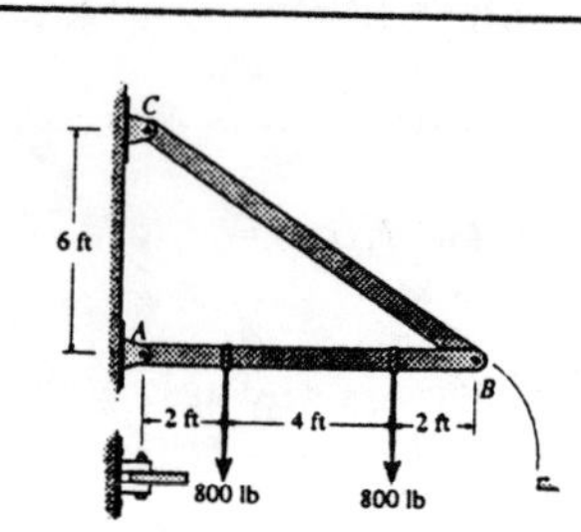

Support Reactions :

$$\circlearrowleft +\Sigma M_A = 0; \quad \frac{3}{5}F_{BC}(8) - 800(6) - 800(2) = 0$$

$$F_{BC} = 1333.33 \text{ lb}$$

$$\xrightarrow{+} \Sigma F_x = 0; \quad A_x - \frac{4}{5}(1333.33) = 0 \qquad A_x = 1066.67 \text{ lb}$$

$$+\uparrow \Sigma F_y = 0; \quad A_y + \frac{3}{5}(1333.33) - 800 - 800 = 0$$

$$A_y = 800 \text{ lb}$$

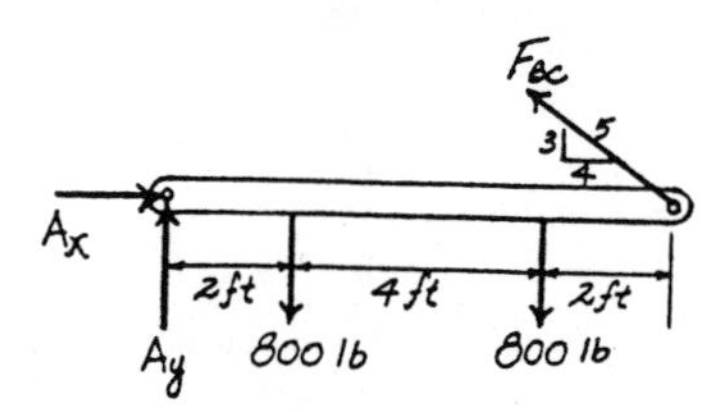

Allowable Normal Stress : Design of member BC

$$\sigma_{\text{allow}} = \frac{F_{BC}}{A_{BC}}; \quad 10(10^3) = \frac{1333.33}{A_{BC}}$$

$$A_{BC} = 0.133 \text{ in}^2 \qquad \textbf{Ans}$$

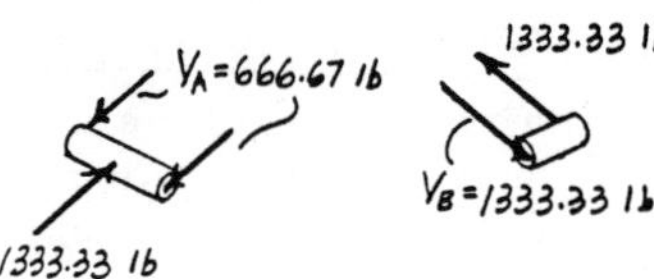

Allowable Shear Stress : Design of pin sizes.

For pin A

Pin A is subjected to double shear and

$F_A = \sqrt{1066.67^2 + 800^2} = 1333.33$ lb.

Therefore, $V_A = \frac{F_A}{2} = 666.67$ lb

$$\tau_{\text{allow}} = \frac{V_A}{A_A}; \quad 3(10^3) = \frac{666.67}{\frac{\pi}{4}d_A^2}$$

$$d_A = 0.532 \text{ in.} \qquad \textbf{Ans}$$

For pin B

Pin B is subjected to single shear. Therefore,

$V_B = F_{BC} = 1333.33$ lb

$$\tau_{\text{allow}} = \frac{V_B}{A_B}; \quad 3(10^3) = \frac{1333.33}{\frac{\pi}{4}d_B^2}$$

$$d_B = 0.752 \text{ in.} \qquad \textbf{Ans}$$

1–82. Determine the required cross-sectional area of member BC and the diameter of the pins at A and B if the allowable normal stress for BC is $\sigma_{allow} = 8$ ksi and the allowable shear stress for the pins is $\tau_{allow} = 5$ ksi.

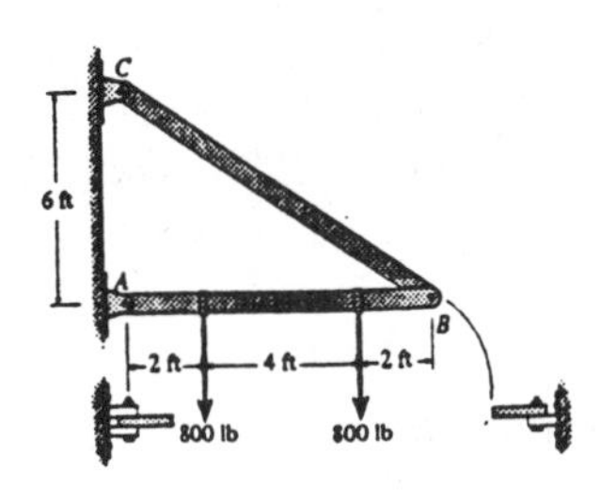

Support Reactions :

$$\curvearrowleft + \Sigma M_A = 0; \quad \frac{3}{5}F_{BC}(8) - 800(6) - 800(2) = 0$$

$$F_{BC} = 1333.33 \text{ lb}$$

$$\xrightarrow{+} \Sigma F_x = 0; \quad A_x - \frac{4}{5}(1333.33) = 0 \quad A_x = 1066.67 \text{ lb}$$

$$+\uparrow \Sigma F_y = 0; \quad A_y + \frac{3}{5}(1333.33) - 800 - 800 = 0$$

$$A_y = 800 \text{ lb}$$

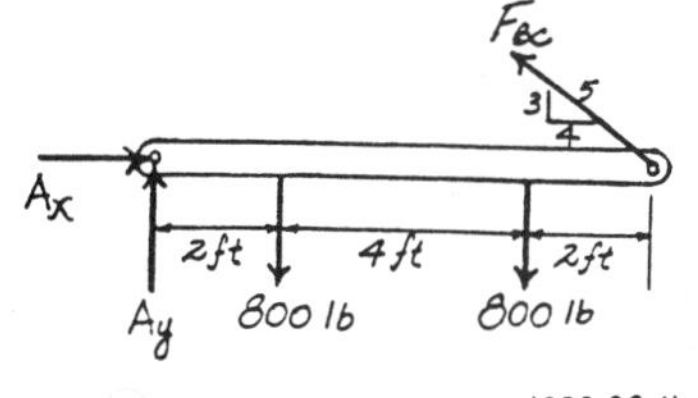

Allowable Normal Stress : Design of member BC

$$\sigma_{allow} = \frac{F_{BC}}{A_{BC}}; \quad 8(10^3) = \frac{1333.33}{A_{BC}}$$

$$A_{BC} = 0.167 \text{ in}^2 \qquad \textbf{Ans}$$

Allowable Shear Stress : Design of pin sizes

For pin A

Pin A is subjected to double shear and

$F_A = \sqrt{1066.67^2 + 800^2} = 1333.33$ lb.

Therefore, $V_A = \frac{F_A}{2} = 666.67$ lb

$$\tau_{allow} = \frac{V_A}{A_A}; \quad 5(10^3) = \frac{666.67}{\frac{\pi}{4}d_A^2}$$

$$d_A = 0.412 \text{ in.} \qquad \textbf{Ans}$$

For pin B

Pin B is subjected to single shear. Therefore,

$V_B = F_{BC} = 1333.33$ lb

$$\tau_{allow} = \frac{V_B}{A_B}; \quad 5(10^3) = \frac{1333.33}{\frac{\pi}{4}d_B^2}$$

$$d_B = 0.583 \text{ in.} \qquad \textbf{Ans}$$

1–83. The hanger is supported using the rectangular pin. Determine the required thickness t of the hanger, and dimensions a and b if the suspended load is $P = 60$ kN. The allowable tensile stress is $(\sigma_t)_{allow} = 150$ MPa, the allowable bearing stress is $(\sigma_b)_{allow} = 290$ MPa, and the allowable shear stress is $(\tau_s)_{allow} = 125$ MPa.

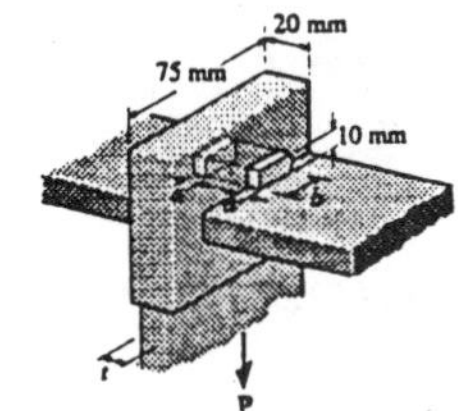

Allowable Normal Stress : For the hanger

$$(\sigma_t)_{allow} = \frac{P}{A}; \quad 150(10^6) = \frac{60(10^3)}{(0.075)t}$$

$$t = 0.005333 \text{ m} = 5.33 \text{ mm} \qquad \textbf{Ans}$$

Allowable Shear Stress : For the pin

$$\tau_{allow} = \frac{V}{A}; \quad 125(10^6) = \frac{30(10^3)}{(0.01)b}$$

$$b = 0.0240 \text{ m} = 24.0 \text{ mm} \qquad \textbf{Ans}$$

Allowable Bearing Stress : For the bearing area

$$(\sigma_b)_{allow} = \frac{P}{A}; \quad 290(10^6) = \frac{30(10^3)}{(0.0240)a}$$

$$a = 0.00431 \text{ m} = 4.31 \text{ mm} \qquad \textbf{Ans}$$

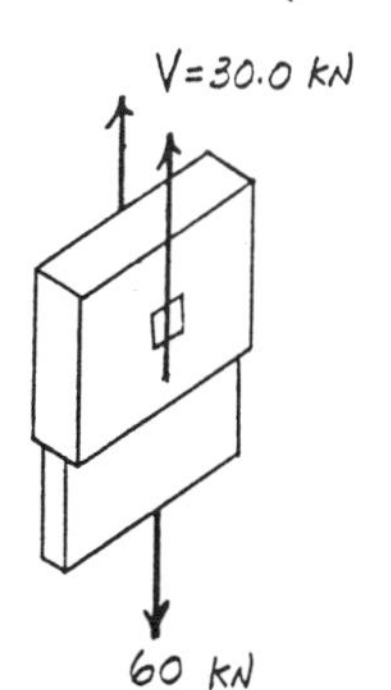

***1–84.** The hanger is supported using the rectangular pin. Determine the magnitude of the allowable suspended load **P** if the allowable bearing stress is $(\sigma_b)_{allow} = 220$ MPa, the allowable tensile stress is $(\sigma_t)_{allow} = 150$ MPa, and the allowable shear stress is $(\tau_s)_{allow} = 130$ MPa. Take $t = 6$ mm, $a = 5$ mm, and $b = 25$ mm.

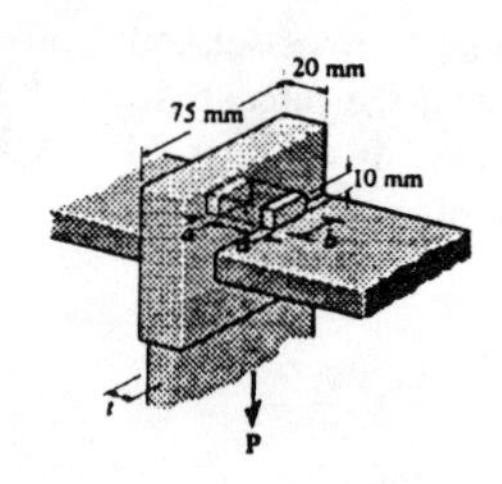

Allowable Normal Stress : For the hanger

$$(\sigma_t)_{allow} = \frac{P}{A}; \quad 150(10^6) = \frac{P}{(0.075)(0.006)}$$
$$P = 67.5 \text{ kN}$$

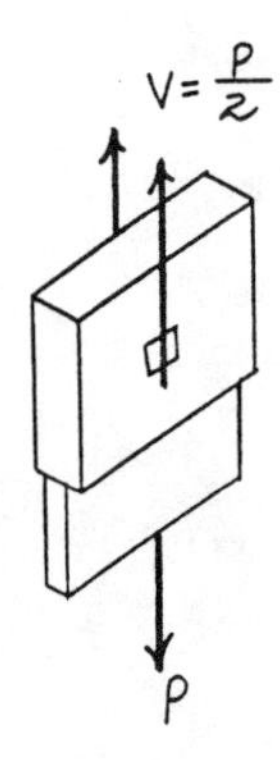

Allowable Shear Stress : The pin is subjected to double shear. Therefore, $V = \frac{P}{2}$

$$\tau_{allow} = \frac{V}{A}; \quad 130(10^6) = \frac{P/2}{(0.01)(0.025)}$$
$$P = 65.0 \text{ kN}$$

Allowable Bearing Stress : For the bearing area

$$(\sigma_b)_{allow} = \frac{P}{A}; \quad 220(10^6) = \frac{P/2}{(0.005)(0.025)}$$
$$P = 55.0 \text{ kN } (\textit{Controls !}) \quad \textbf{Ans}$$

1–85. The frame is subjected to the load of 1.5 kip. Determine the required diameter of the pins at A and B if the allowable shear stress for the material is $\tau_{allow} = 6$ ksi. Pin A is subjected to double shear, whereas pin B is subjected to single shear.

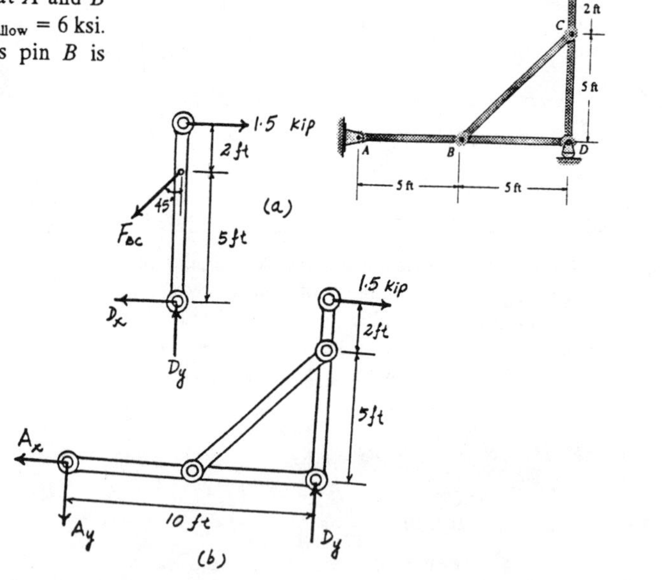

Support Reactions : From FBD (a),

$$\circlearrowleft + \Sigma M_D = 0; \quad F_{BC}(\sin 45°)(5) - 1.5(7) = 0$$
$$F_{BC} = 2.970 \text{ kip}$$

From FBD (b),

$$\circlearrowleft + \Sigma M_A = 0; \quad D_y(10) - 1.5(7) = 0 \quad D_y = 1.05 \text{ kip}$$

$$\overset{+}{\leftarrow} \Sigma F_x = 0; \quad A_x - 1.5 = 0 \quad A_x = 1.50 \text{ kip}$$

$$+\uparrow \Sigma F_y = 0; \quad 1.05 - A_y = 0 \quad A_y = 1.05 \text{ kip}$$

Allowable Shear Stress : Design of pin sizes

For pin A

Pin A is subjected to double shear and $F_A = \sqrt{1.50^2 + 1.05^2} = 1.831$ kip.

Therefore, $V_A = \frac{F_A}{2} = 0.9155$ kip

$$\tau_{allow} = \frac{V_A}{A_A}; \quad 6 = \frac{0.9155}{\frac{\pi}{4} d_A^2}$$
$$d_A = 0.441 \text{ in.} \quad \textbf{Ans}$$

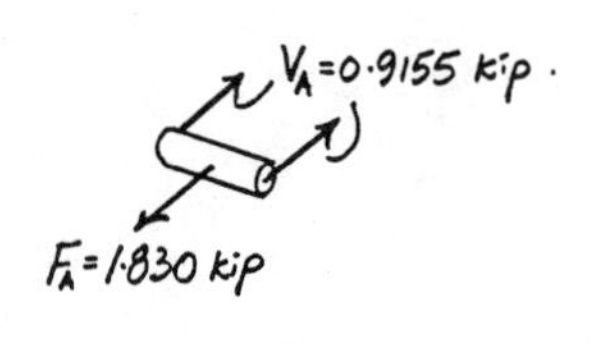

For pin B

Pin B is subjected to single shear. Therefore, $V_B = F_B = F_{BC} = 2.970$ kip

$$\tau_{allow} = \frac{V_B}{A_B}; \quad 6 = \frac{2.970}{\frac{\pi}{4} d_B^2}$$
$$d_B = 0.794 \text{ in.} \quad \textbf{Ans}$$

1–86. The frame is subjected to the distributed loading of 2 kN/m. Determine the required diameter of the pins at A and B if the allowable shear stress for the material is τ_{allow} = 100 MPa. Both pins are subjected to double shear.

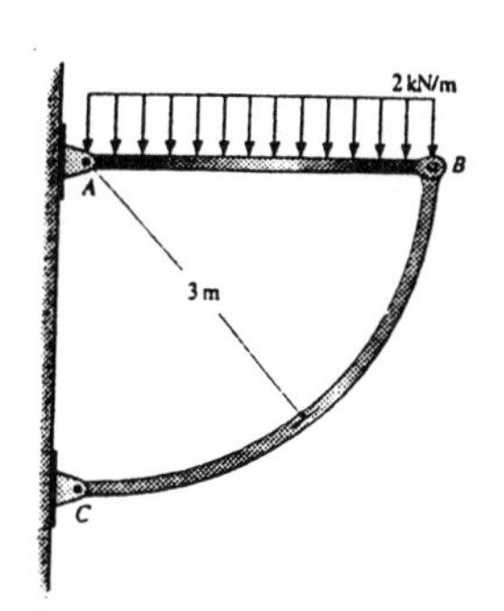

Support Reactions : Member BC is a two force member.

$\circlearrowleft + \Sigma M_A = 0;\quad F_{BC}\sin 45°(3) - 6(1.5) = 0$

$F_{BC} = 4.243$ kN

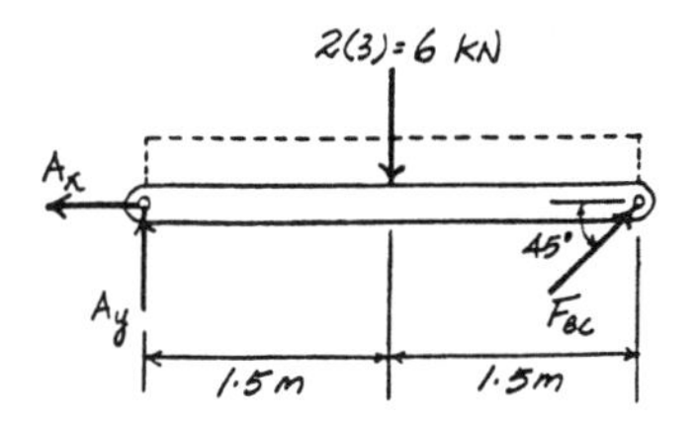

$+\uparrow \Sigma F_y = 0;\quad A_y + 4.243\sin 45° - 6 = 0$

$A_y = 3.00$ kN

$\overset{+}{\leftarrow} \Sigma F_x = 0;\quad A_x - 4.243\cos 45° = 0$

$A_x = 3.00$ kN

Allowable Shear Stress : Pin A and pin B are subjected to double shear.

$F_A = \sqrt{3.00^2 + 3.00^2} = 4.243$ kN and

$F_B = F_{BC} = 4.243$ kN.

Therefore,

$V_A = V_B = \dfrac{4.243}{2} = 2.1215$ kN

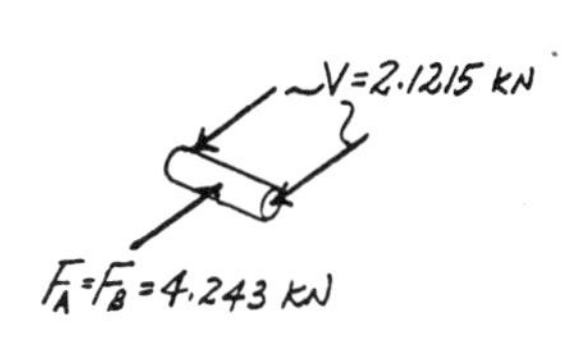

$\tau_{allow} = \dfrac{V}{A};\quad 100(10^6) = \dfrac{2.1215(10^3)}{\frac{\pi}{4}d^2}$

$d = 0.005197$ m $= 5.20$ mm

$d_A = d_B = d = 5.20$ mm **Ans**

1–87. The connection is made using a bolt and nut and two washers. If the allowable bearing stress for the boards is $(\sigma_b)_{allow}$ = 2 ksi, and the allowable tensile stress for the bolt shank S is $(\sigma_t)_{allow}$ = 18 ksi, determine the maximum allowable tension in the bolt shank. The bolt shank has a diameter of 0.31 in., and the washers have an outer diameter of 0.75 in. and inner diameter (hole) of 0.50 in.

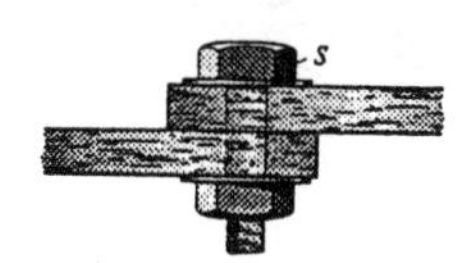

Allowable Normal Stress : Assume tension failure

$\sigma_{allow} = \dfrac{P}{A};\quad 18 = \dfrac{P}{\frac{\pi}{4}(0.31^2)}$

$P = 1.36$ kip

Allowable Bearing Stress : Assume bearing failure

$(\sigma_b)_{allow} = \dfrac{P}{A};\quad 2 = \dfrac{P}{\frac{\pi}{4}(0.75^2 - 0.50^2)}$

$P = 0.491$ kip ***(controls!)*** **Ans**

***1–88.** The two plates are joined together using a rectangular pin. Determine the smallest width w of the plates and the dimension b of the pin if the applied load is $P = 40$ kN. The allowable tensile stress, shear stress, and bearing stress for the material is $(\sigma_t)_{allow} = 175$ MPa, $\tau_{allow} = 125$ MPa, and, $(\sigma_b)_{allow} = 380$ MPa.

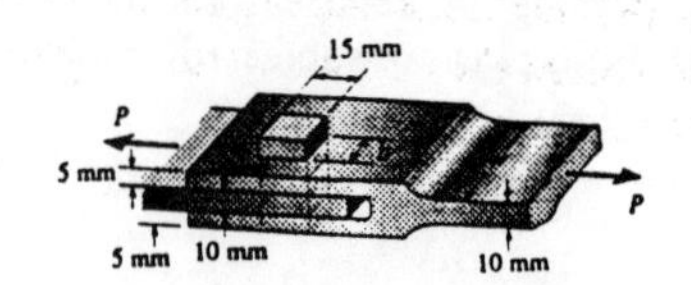

Allowable Normal Stress : For the plate

$$(\sigma_t)_{allow} = \frac{P}{A}; \quad 175(10^6) = \frac{40(10^3)}{(0.01)w}$$

$w = 0.02286$ m $= 22.9$ mm **Ans**

Allowable Shear Stress : For the rectangular pin

$$\tau_{allow} = \frac{V}{A}; \quad 125(10^6) = \frac{20(10^3)}{(0.015)b}$$

$b = 0.01067$ m $= 10.7$ mm ***(controls !)*** **Ans**

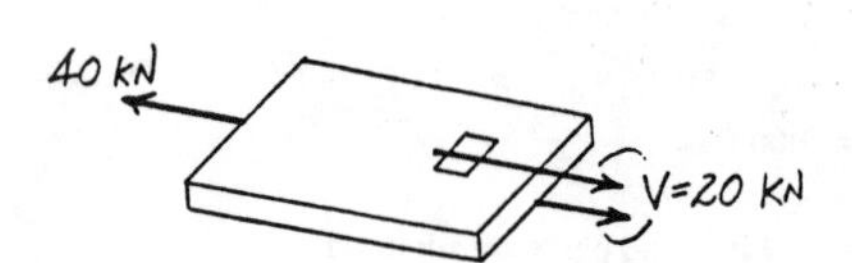

Allowable Bearing Stress : For the bearing area

$$(\sigma_b)_{allow} = \frac{P}{A}; \quad 380(10^6) = \frac{20(10^3)}{(0.005)b}$$

$b = 0.01053$ m $= 10.5$ mm

1–89. The two plates are joined together using a square pin. Determine the maximum applied load P if the allowable shear stress for the material is $\tau_{allow} = 125$ MPa, the allowable bearing stress is $(\sigma_b)_{allow} = 400$ MPa, and the allowable tensile stress is $(\sigma_t)_{allow} = 175$ MPa. Take $w = 25$ mm, $b = 10$ mm..

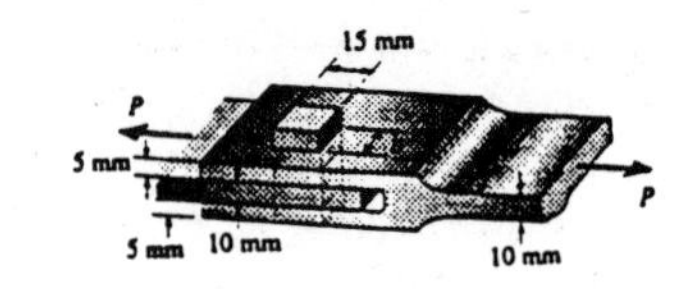

Solution

Allowable Normal Stress : For the plate.

$$(\sigma_t)_{allow} = \frac{P}{A}; \quad 175(10^6) = \frac{P}{0.025(0.01)}$$

$P = 43.75$ kN

Allowable Shear Stress : For the rectangular pin.

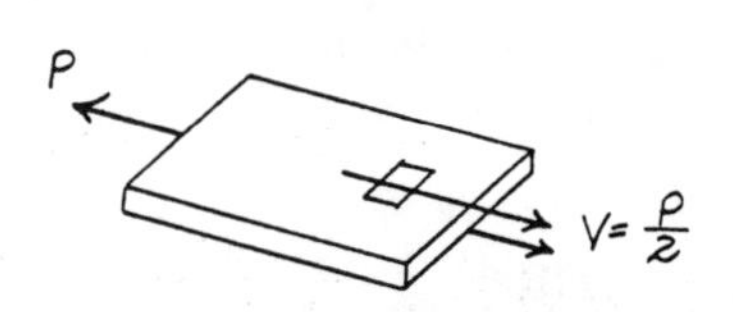

$$\tau_{allow} = \frac{V}{A}; \quad 125(10^6) = \frac{P/2}{0.015(0.01)}$$

$P = 37.5$ kN ***(controls !)*** **Ans**

Allowable Bearing Stress : For the bearing area.

$$(\sigma_b)_{allow} = \frac{P}{A}; \quad 400(10^6) = \frac{P/2}{(0.01)(0.005)}$$

$P = 40.0$ kN

1–90. The aluminum bracket A is used to support the centrally applied load of 8 kip. If it has a constant thickness of 0.5 in., determine the smallest height h in order to prevent a shear failure. The failure shear stress is $\tau_{fail} = 23$ ksi. Use a factor of safety for shear of F.S. = 2.5.

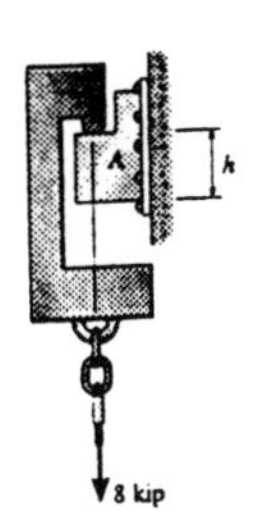

Equation of Equilibrium :

$+\uparrow \Sigma F_y = 0; \quad V - 8 = 0 \quad V = 8.00 \text{ kip}$

Allowable Shear Stress : Design of the support size

$$\tau_{allow} = \frac{\tau_{fail}}{\text{F.S}} = \frac{V}{A}; \quad \frac{23(10^3)}{2.5} = \frac{8.00(10^3)}{h(0.5)}$$

$h = 1.74$ in. **Ans**

1–91. Determine the smallest dimensions of the circular shaft and circular end cap if the load it is required to support is $P = 150$ kN. The allowable tensile stress, bearing stress, and shear stress is $(\sigma_t)_{allow} = 175$ MPa, $(\sigma_b)_{allow} = 275$ MPa, and $\tau_{allow} = 115$ MPa.

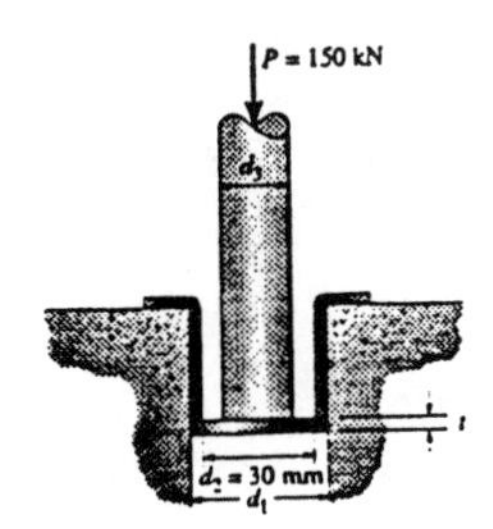

Allowable Normal Stress : Design of end cap outer diameter

$$(\sigma_t)_{allow} = \frac{P}{A}; \quad 175\left(10^6\right) = \frac{150(10^3)}{\frac{\pi}{4}\left(d_1^2 - 0.03^2\right)}$$

$d_1 = 0.04462 \text{ m} = 44.6 \text{ mm}$ **Ans**

Allowable Bearing Stress : Design of circular shaft diameter

$$(\sigma_b)_{allow} = \frac{P}{A}; \quad 275\left(10^6\right) = \frac{150(10^3)}{\frac{\pi}{4}d_3^2}$$

$d_3 = 0.02635 \text{ m} = 26.4 \text{ mm}$ **Ans**

Allowable Shear Stress : Design of end cap thickness

$$\tau_{allow} = \frac{V}{A}; \quad 115\left(10^6\right) = \frac{150(10^3)}{\pi(0.02635)\,t}$$

$t = 0.01575 \text{ m} = 15.8 \text{ mm}$ **Ans**

***1–92.** The assembly consists of three disks A, B, and C that are used to support the load of 140 kN. Determine the smallest diameter d_1 of the top disk, the diameter d_2 within the support space, and the diameter d_3 of the hole in the bottom disk. The allowable bearing stress for the material is $(\sigma_{allow})_b = 350$ MPa and allowable shear stress is $\tau_{allow} = 125$ MPa.

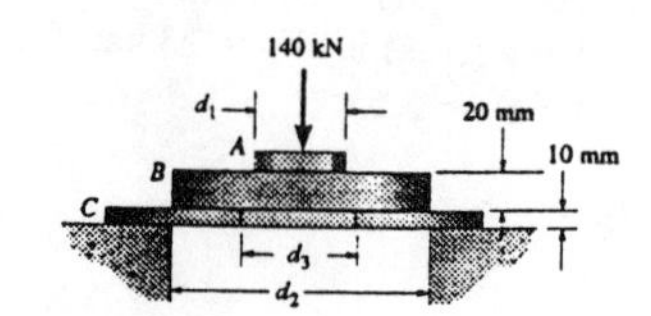

Solution

Allowable Shear Stress : Assume shear failure for disk C.

$$\tau_{allow} = \frac{V}{A}; \quad 125(10^6) = \frac{140(10^3)}{\pi d_2\,(0.01)}$$

$$d_2 = 0.03565 \text{ m} = 35.7 \text{ mm} \qquad \textbf{Ans}$$

Allowable Bearing Stress : Assume bearing failure for disk C.

$$(\sigma_b)_{allow} = \frac{P}{A}; \quad 350(10^6) = \frac{140(10^3)}{\frac{\pi}{4}(0.03565^2 - d_3^2)}$$

$$d_3 = 0.02760 \text{ m} = 27.6 \text{ mm} \qquad \textbf{Ans}$$

Allowable Bearing Stress : Assume bearing failure for disk B.

$$(\sigma_b)_{allow} = \frac{P}{A}; \quad 350(10^6) = \frac{140(10^3)}{\frac{\pi}{4}d_1^2}$$

$$d_1 = 0.02257 \text{ m} = 22.6 \text{ mm}$$

Since $d_3 = 27.6 \text{ mm} > d_1 = 22.6 \text{ mm}$, disk B might fail due to shear.

$$\tau = \frac{V}{A} = \frac{140(10^3)}{\pi(0.02257)(0.02)} = 98.7 \text{ MPa} < \tau_{allow} = 125 \text{ MPa } (O.K!)$$

Therefore, $\qquad d_1 = 22.6 \text{ mm} \qquad$ **Ans**

1–93. Strips A and B are to be glued together using the two strips C and D. Determine the required thickness t of C and D so that all strips will fail simultaneously. The width of strips A and B is 1.5 times that of strips C and D.

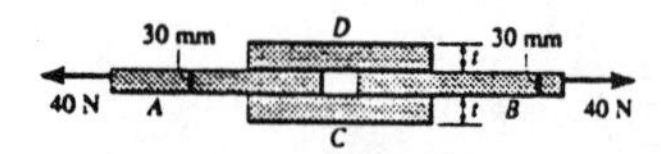

Average Normal Stress : Requires,

$$\sigma_A = \sigma_B = \sigma_C; \quad \frac{40}{(0.03)(1.5w)} = \frac{20}{wt}$$

$$t = 0.0225 \text{ m} = 22.5 \text{ mm} \qquad \textbf{Ans}$$

$P_A = 40N \qquad P_C = P_D = 20N$

1–94. The wood specimen is subjected to the pull of 10 kN in a tension testing machine. If the allowable normal stress for the wood is $(\sigma_t)_{allow} = 12$ MPa and the allowable shear stress is $\tau_{allow} = 1.2$ MPa, determine the required dimensions b and t so that the specimen reaches these stresses simultaneously. The specimen has a width of 25 mm.

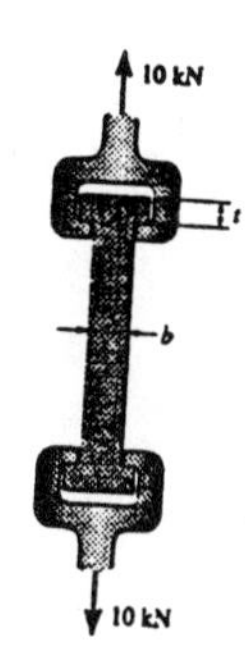

Allowable Shear Stress : Shear limitation

$$\tau_{allow} = \frac{V}{A}; \quad 1.2(10^6) = \frac{5.00(10^3)}{(0.025)\,t}$$

$t = 0.1667$ m $= 167$ mm **Ans**

Allowable Normal Stress : Tension limitation

$$\sigma_{allow} = \frac{P}{A}; \quad 12.0(10^6) = \frac{10(10^3)}{(0.025)\,b}$$

$b = 0.03333$ m $= 33.3$ mm **Ans**

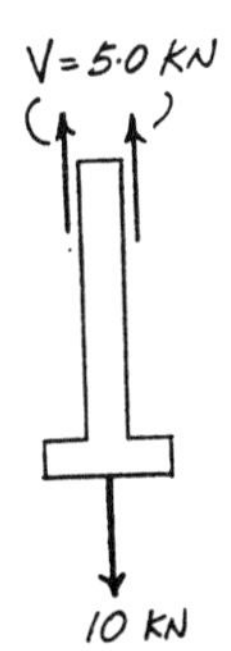

1–95. The cotter is used to hold the two rods together. Determine the smallest thickness t of the cotter and the smallest diameter d of the rods. All parts are made of steel for which the failure tensile stress is $\sigma_{fail} = 500$ MPa and the failure shear stress is $\tau_{fail} = 375$ MPa. Use a factor of safety of $(F.S.)_t = 2.50$ in tension and $(F.S.)_s = 1.75$ in shear.

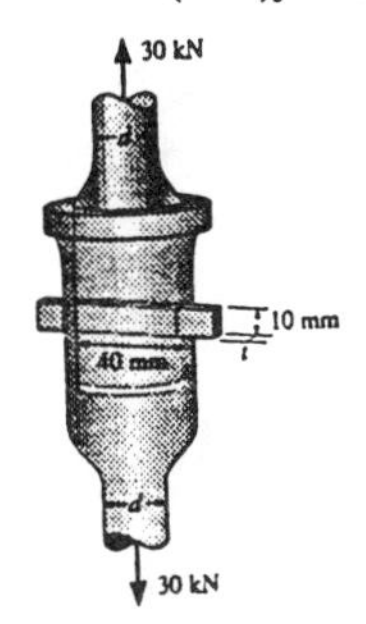

Allowable Normal Stress : Design of rod size

$$\sigma_{allow} = \frac{\sigma_{fail}}{F.S} = \frac{P}{A}; \quad \frac{500(10^6)}{2.5} = \frac{30(10^3)}{\frac{\pi}{4}d^2}$$

$d = 0.01382$ m $= 13.8$ mm **Ans**

Allowable Shear Stress : Design of cotter size.

$$\tau_{allow} = \frac{\tau_{fail}}{F.S} = \frac{V}{A}; \quad \frac{375(10^6)}{1.75} = \frac{15.0(10^3)}{(0.01)\,t}$$

$t = 0.0070$ m $= 7.00$ mm **Ans**

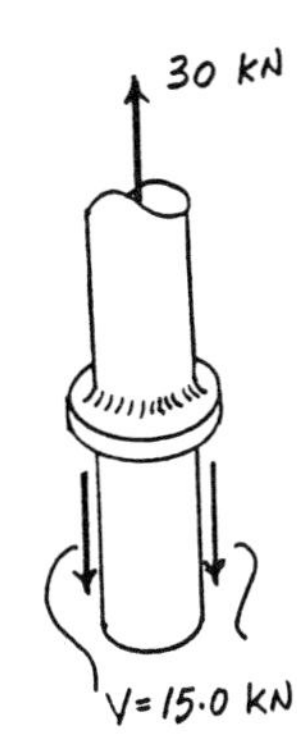

***1–96.** If the allowable bearing stress for the material under the supports at A and B is $(\sigma_b)_{allow} = 400$ psi, determine the size of *square* bearing plates A' and B' required to support the load. Dimension the plates to the nearest $\frac{1}{2}$ in. The reactions at the supports are vertical. Take $P = 1500$ lb.

Support Reactions :

$$\circlearrowleft + \Sigma M_A = 0; \quad F_B(15) - 9.00(7.5) - 1.50(22.5) = 0$$
$$F_B = 6.75 \text{ kip}$$

$$+\uparrow \Sigma F_y = 0; \quad F_A + 6.75 - 9.00 - 1.50 = 0$$
$$F_A = 3.75 \text{ kip}$$

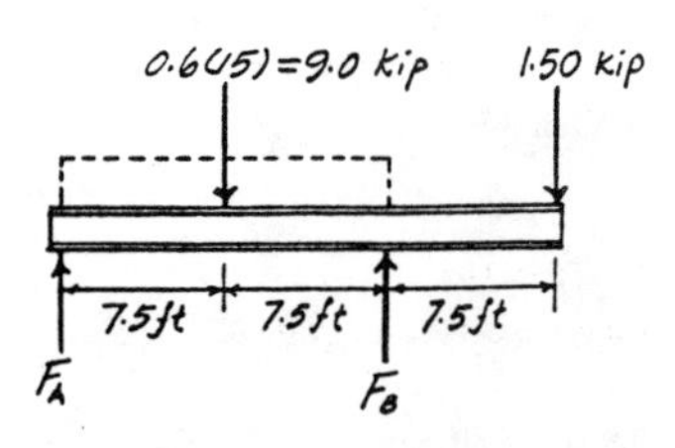

Allowable Bearing Stress : Design of bearing plates

For plate A'

$$(\sigma_b)_{allow} = \frac{F_A}{A_{A'}}; \quad 400 = \frac{3.75(10^3)}{L_{A'}^2} \quad L_{A'} = 3.06 \text{ in.}$$

Use $3\frac{1}{2}$ in. × $3\frac{1}{2}$ in. plate **Ans**

For plate B'

$$(\sigma_b)_{allow} = \frac{F_B}{A_{B'}}; \quad 400 = \frac{6.75(10^3)}{L_{B'}^2} \quad L_{B'} = 4.11 \text{ in.}$$

Use $4\frac{1}{2}$ in. × $4\frac{1}{2}$ in. plate **Ans**

1–97. If the allowable bearing stress for the material under the support at A and B is $(\sigma_b)_{allow} = 400$ psi, determine the maximum load **P** that can be applied to the beam. The bearing plates A' and B' have square cross sections of 2 in. × 2 in. and 4 in. × 4 in., respectively.

Support Reactions :

$$\circlearrowleft + \Sigma M_A = 0; \quad F_B(15) - 9.00(7.5) - P(22.5) = 0$$
$$15F_B - 22.5P = 67.5 \qquad [1]$$

$$\circlearrowleft + \Sigma M_B = 0; \quad 9.00(7.5) - P(7.5) - F_A(15) = 0$$
$$15F_A + 7.5P = 67.5 \qquad [2]$$

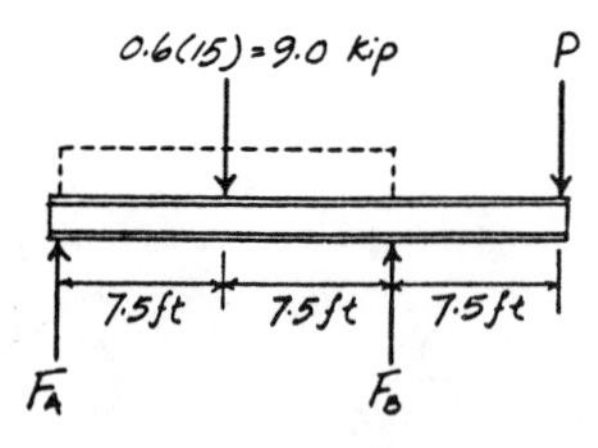

Allowable Bearing Stress : Assume failure of material occurs under plate A'

$$(\sigma_b)_{allow} = \frac{F_A}{A_{A'}}; \quad 400 = \frac{F_A}{2(2)}$$
$$F_A = 1600 \text{ lb} = 1.60 \text{ kip}$$

From Eq. [2]

$$P = 5.80 \text{ kip}$$

Assume failure of material occurs under B'

$$(\sigma_b)_{allow} = \frac{F_B}{A_{B'}}; \quad 400 = \frac{F_B}{4(4)}$$
$$F_B = 6400 \text{ lb} = 6.40 \text{ kip}$$

From Eq. [1]

$$P = 1.27 \text{ kip}$$

Choose the ***smallest*** value $P = 1.27$ kip **Ans**

1–98. The lapbelt assembly is to be subjected to a force of 800 N. Determine (a) the required thickness t of the belt if the allowable tensile stress for the material is $(\sigma_t)_{allow} = 10$ MPa, (b) the required lap length d_l if the glue can sustain an allowable shear stress of $(\tau_{allow})_g = 0.75$ MPa, and (c) the required diameter d_r of the pin if the allowable shear stress for the pin is $(\tau_{allow})_p = 30$ MPa.

Allowable Normal Stress : Design of belt thickness.

$$(\sigma_t)_{allow} = \frac{P}{A}; \quad 10(10^6) = \frac{800}{(0.045)t}$$

$$t = 0.001778 \text{ m} = 1.78 \text{ mm} \qquad \textbf{Ans}$$

Allowable Shear Stress : Design of lap length.

$$(\tau_{allow})_g = \frac{V_A}{A}; \quad 0.750(10^6) = \frac{400}{(0.045)\,d_l}$$

$$d_l = 0.01185 \text{ m} = 11.9 \text{ mm} \qquad \textbf{Ans}$$

Allowable Shear Stress : Design of pin size.

$$(\tau_{allow})_p = \frac{V_B}{A}; \quad 30(10^6) = \frac{400}{\frac{\pi}{4}d_r^2}$$

$$d_r = 0.004120 \text{ m} = 4.12 \text{ mm} \qquad \textbf{Ans}$$

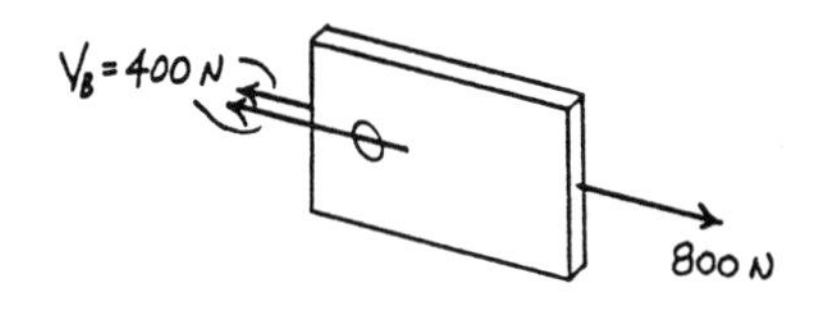

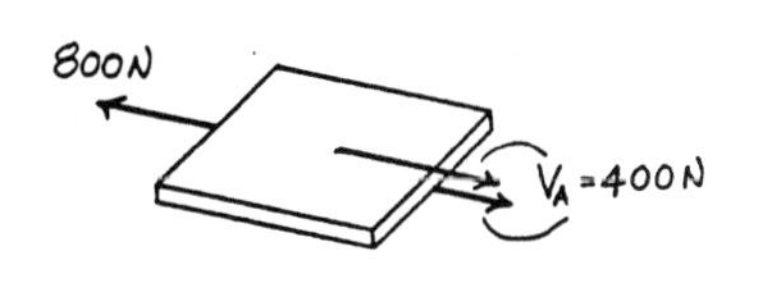

1–99. The pin is subjected to double shear since it is used to connect the three links together. Due to wear, the load is distributed over the top and bottom of the pin as shown on the free-body diagram. Determine the diameter d of the pin if the allowable shear stress is $\tau_{allow} = 10$ ksi and the load $P = 8$ kip. Also, determine the load intensities w_1 and w_2.

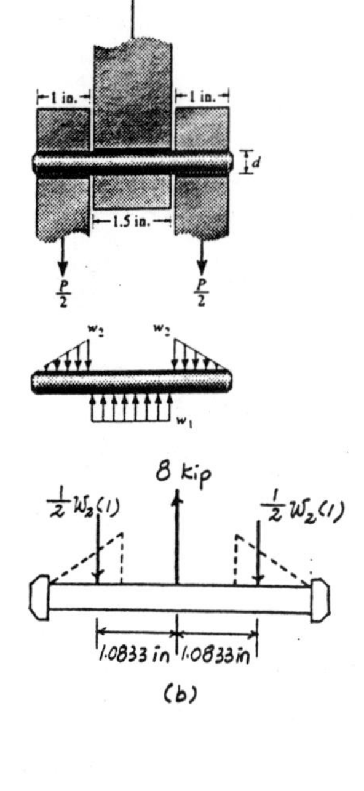

Equations of Equilibrium :
FBD (a)

$$+\uparrow \Sigma F_y = 0; \quad 8 - w_1(1.5) = 0$$

$$w_1 = 5.33 \text{ kip / in.} \qquad \textbf{Ans}$$

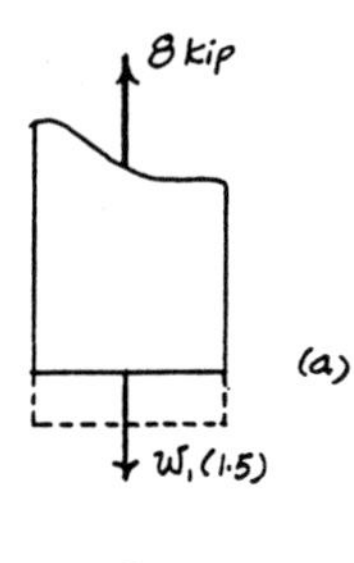

FBD (b)

$$+\uparrow \Sigma F_y = 0; \quad -2\left(\frac{1}{2}w_2\right)(1) + 8 = 0$$

$$w_2 = 8.00 \text{ kip / in.} \qquad \textbf{Ans}$$

Allowable Shear Stress : Design of pin size

$$\tau_{allow} = \frac{V}{A}; \quad 10 = \frac{4.00}{\frac{\pi}{4}d^2}$$

$$d = 0.714 \text{ in.} \qquad \textbf{Ans}$$

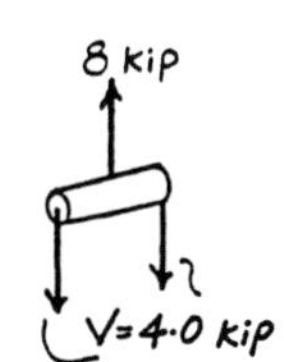

***1–100.** The clevis C and pin are made of steel having an allowable normal stress of $\sigma_{allow} = 21$ ksi and an allowable shear stress of $\tau_{allow} = 12$ ksi. Determine to the nearest $\frac{1}{4}$ in. the diameters of the rod, d_r, and the pin, d_p, needed to support the load. Notice that the screw at the end of the rod has "upset" threads, so that the rod will *not* fail in this region.

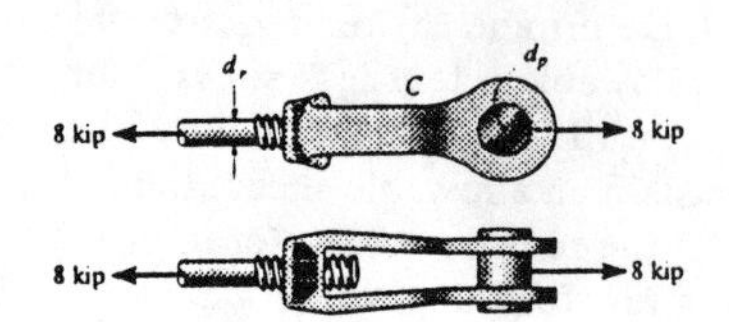

Allowable Normal Stress : Design of rod size

$$\sigma_{allow} = \frac{P}{A}; \quad 21(10^3) = \frac{8(10^3)}{\frac{\pi}{4}d_r^2} \quad d_r = 0.696 \text{ in.}$$

$$\text{Use } d_r = \frac{3}{4} \text{ in.} \qquad \textbf{Ans}$$

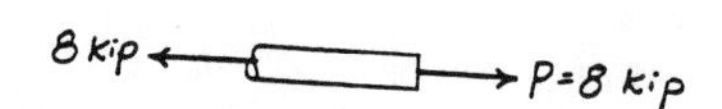

Allowable Shear Stress : Design of pin size

$$\tau_{allow} = \frac{V}{A}; \quad 12(10^3) = \frac{4(10^3)}{\frac{\pi}{4}d_p^2} \quad d_p = 0.651 \text{ in.}$$

$$\text{Use } d_p = \frac{3}{4} \text{ in.} \qquad \textbf{Ans}$$

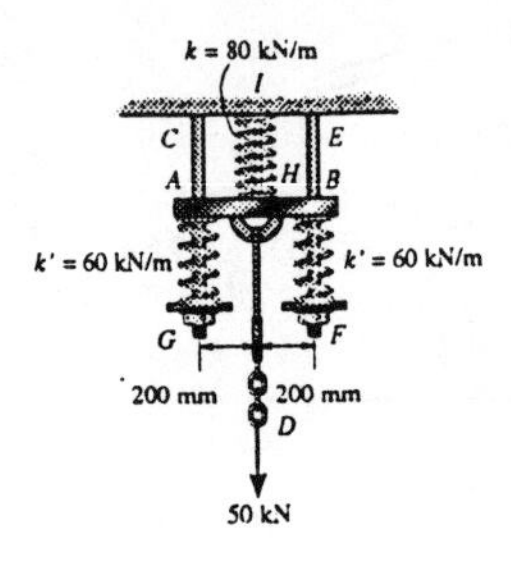

1–101. The spring mechanism is used as a shock absorber for a load applied to the drawbar AB. Determine the force in each spring when the 50-kN force is applied. Each spring is originally unstretched and the drawbar slides along the smooth guide posts CG and EF. The ends of all springs are attached to their respective members. Also, what is the required diameter of the shank of bolts CG and EF if the allowable stress for the bolts is $\sigma_{allow} = 150$ MPa.

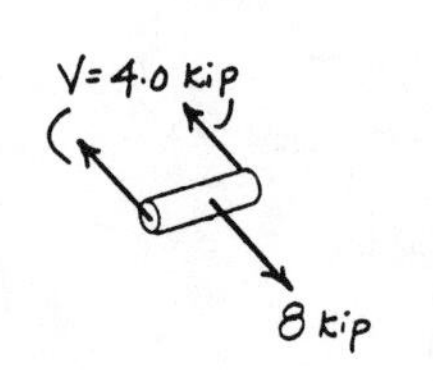

Solution

Equations of Equilibrium :

$$\circlearrowleft + \Sigma M_H = 0; \quad -F_{BF}(200) + F_{AG}(200) = 0$$

$$F_{BF} = F_{AG} = F$$

$$+\uparrow \Sigma F_y = 0; \quad 2F + F_{HI} - 50 = 0 \qquad [1]$$

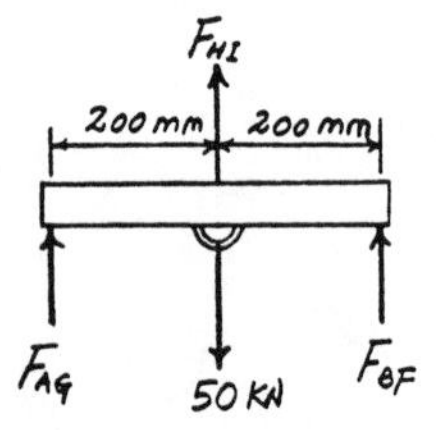

Required,

$$\Delta_H = \Delta_B; \quad \frac{F_{HI}}{80} = \frac{F}{60}$$

$$F = 0.75 F_{HI} \qquad [2]$$

Solving Eqs. [1] and [2] yields,

$$F_{HI} = 20.0 \text{ kN} \qquad \textbf{Ans}$$

$$F_{BF} = F_{AG} = F = 15.0 \text{ kN} \qquad \textbf{Ans}$$

Allowable Normal Stress : Design of bolt shank size.

$$\sigma_{allow} = \frac{P}{A}; \quad 150(10^6) = \frac{15.0(10^3)}{\frac{\pi}{4}d^2}$$

$$d = 0.01128 \text{ m} = 11.3 \text{ mm}$$

$$d_{EF} = d_{CG} = 11.3 \text{ mm} \qquad \textbf{Ans}$$

1–102. The compound wooden beam is connected by a bolt at B. Assuming that the connections at A, B, C, and D exert only vertical forces on the beam, determine the required diameter of the bolt and the required outer diameter of the washers, each to the nearest $\frac{1}{8}$ in., if the allowable tensile stress for the bolt is $(\sigma_t)_{allow} = 21$ ksi and the allowable bearing stress for the wood is $(\sigma_b)_{allow} = 4$ ksi. Assume that the hole in the washers is the same size as the bolt diameter.

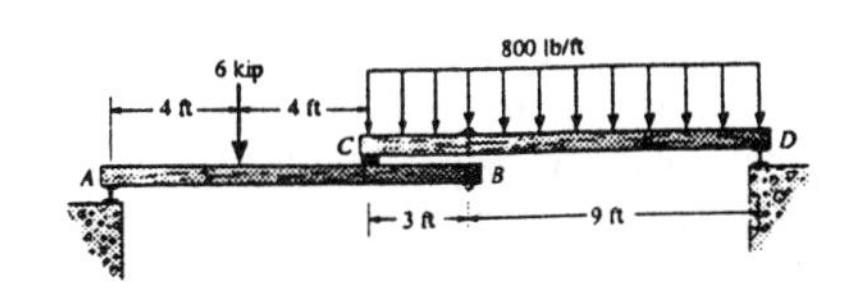

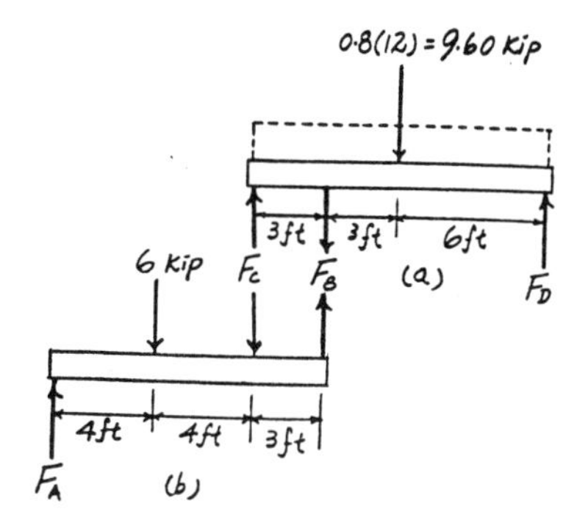

Equations of Equilibrium :

From FBD (a)

$$\circlearrowleft +\Sigma M_D = 0; \quad 9.60(6) + F_B(9) - F_C(12) = 0$$

$$9F_B - 12F_C + 57.6 = 0 \qquad [1]$$

From FBD (b)

$$\circlearrowleft +\Sigma M_A = 0; \quad F_B(11) - F_C(8) - 6(4) = 0$$

$$11F_B - 8F_C - 24 = 0 \qquad [2]$$

Solving Eqs. [1] and [2] yields :

$$F_B = 12.48 \text{ kip} \qquad F_C = 14.16 \text{ kip}$$

Allowable Normal Stress : Design of bolt size

$$(\sigma_t)_{allow} = \frac{F_B}{A_b}; \quad 21.0(10^3) = \frac{12.48(10^3)}{\frac{\pi}{4}d_b^{\,2}}$$

$$d_b = 0.8699 \text{ in.}$$

$$\text{Use } d_b = \frac{7}{8} \text{ in.} \qquad \textbf{Ans}$$

Allowable Bearing Stress : Design of washer size

$$(\sigma_b)_{allow} = \frac{F_B}{A_w}; \quad 4(10^3) = \frac{12.48(10^3)}{\frac{\pi}{4}[d_w^2 - (\frac{7}{8})^2]}$$

$$d_w = 2.177 \text{ in.}$$

$$\text{Use } d_w = 2\frac{1}{4} \text{ in.} \qquad \textbf{Ans}$$

1–103. The assembly is used to support the distributed loading of $w = 500$ lb/ft. Determine the factor of safety with respect to yielding for the steel rod BC and the pins at B and C if the yield stress for the steel in tension is $\sigma_y = 36$ ksi and in shear $\tau_y = 18$ ksi. The rod has a diameter of 0.4 in., and the pins each have a diameter of 0.30 in.

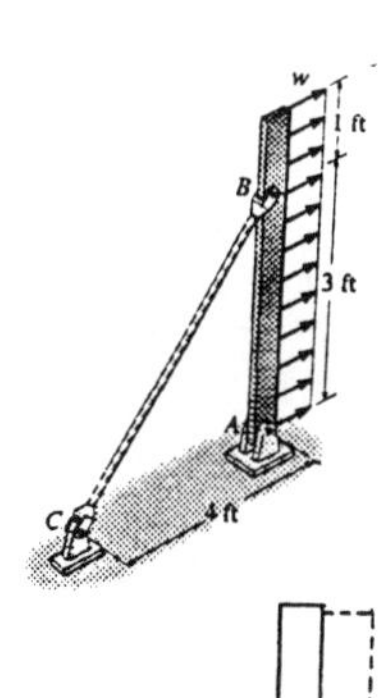

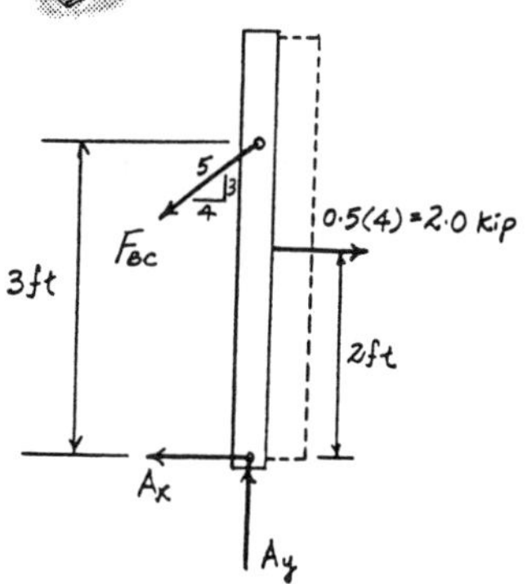

Support Reactions :

$$\circlearrowleft +\Sigma M_A = 0; \quad \frac{4}{5}F_{BC}(3) - 2.00(2) = 0 \qquad F_{BC} = 1.667 \text{ kip}$$

Factor of Safety :

For Rod BC

$$\sigma_{BC} = \frac{F_{BC}}{A_{BC}} = \frac{1.667}{\frac{\pi}{4}(0.4)^2} = 13.263 \text{ ksi}$$

$$\text{F.S} = \frac{\sigma_y}{\sigma_{BC}} = \frac{36}{13.263} = 2.71 \qquad \textbf{Ans}$$

For pins B and C

Pins B and C are subjected to double shear and $F_B = F_C = F_{BC} = 1.667$ kip. Therefore,

$$V_B = V_C = V = \frac{1.667}{2} = 0.8333 \text{ kip}$$

$$\tau_{avg} = \frac{V}{A} = \frac{0.8333}{\frac{\pi}{4}(0.3^2)} = 11.789 \text{ ksi}$$

$$\text{F.S} = \frac{\tau_y}{\tau_{avg}} = \frac{18}{11.789} = 1.53 \qquad \textbf{Ans}$$

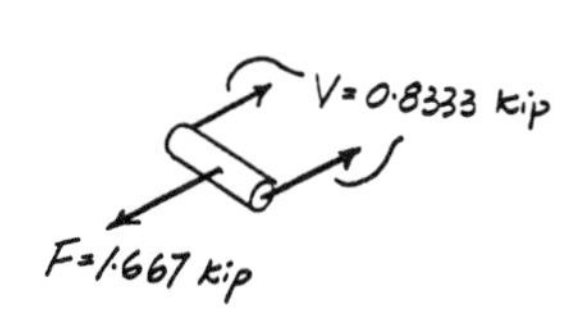

***1–104.** If the allowable shear stress for each of the 0.3-in.-diameter steel pins at A, B, and C is $\tau_{allow} = 12.5$ ksi, and the allowable normal stress for the 0.40-in.-diameter rod BC is $\sigma_{allow} = 22$ ksi, determine the largest intensity w of the uniform distributed load that can be supported by the beam.

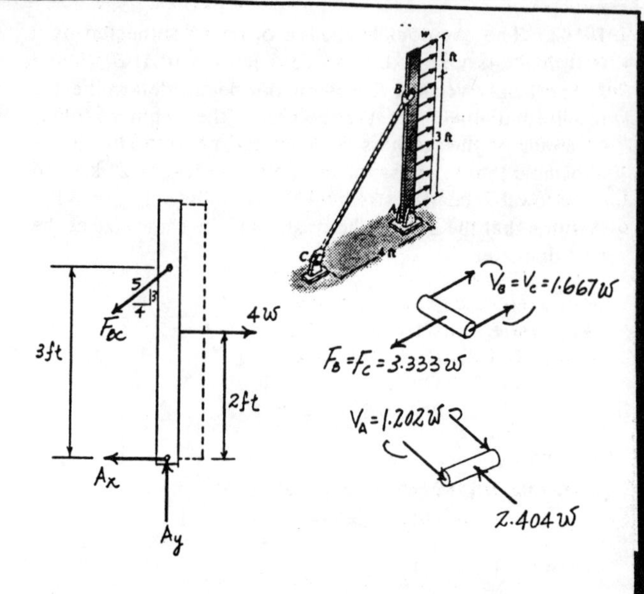

Support Reactions :

$$\circlearrowleft + \Sigma M_A = 0; \quad \frac{4}{5}F_{BC}(3) - 4w(2) = 0 \qquad F_{BC} = 3.333\,w$$

$$\xrightarrow{+} \Sigma F_x = 0; \quad 4w - \frac{4}{5}(3.333w) - A_x = 0 \qquad A_x = 1.333\,w$$

$$+\uparrow \Sigma F_y = 0; \quad A_y - \frac{3}{5}(3.333w) = 0 \qquad A_y = 2.00\,w$$

Allowable Normal Stress : Assume failure of rod BC

$$\sigma_{allow} = \frac{F_{BC}}{A_{BC}}; \quad 22.0 = \frac{3.333w}{\frac{\pi}{4}(0.4^2)}$$

$$w = 0.829 \text{ kip/ft}$$

Allowable Shear Stress : Assume failure of pins B and C. Pins B and C are subjected to double shear and $F_B = F_C = F_{BC} = 3.333w$. Therefore,

$$V_B = V_C = V = \frac{3.333w}{2} = 1.667w$$

$$\tau_{allow} = \frac{V}{A}; \quad 12.5 = \frac{1.667w}{\frac{\pi}{4}(0.3^2)}$$

$$w = 0.530 \text{ kip/ft}$$

Assume failure of pin A. Pin A is subjected to double shear and

$$F_A = \sqrt{(1.333w)^2 + (2.00w)^2} = 2.404w$$

Therefore,

$$V_A = \frac{F_A}{2} = 1.202w$$

$$\tau_{allow} = \frac{V_A}{A}; \quad 12.5 = \frac{1.202w}{\frac{\pi}{4}(0.3^2)}$$

$$w = 0.735 \text{ kip/ft}$$

Choose the ***smallest*** value of $w = 0.530$ kip/ft **Ans**

1–105. The bar is supported by the pin. If the allowable tensile stress for the bar is $(\sigma_t)_{allow} = 21$ ksi, and the allowable shear stress for the pin is $\tau_{allow} = 12$ ksi, determine the diameter of the pin for which the load P will be a maximum. What is this maximum load? Assume the hole in the bar has the same diameter d as the pin. Take $t = \frac{1}{4}$ in. and $w = 2$ in.

Allowable Normal Stress : The effective cross - sectional area A' for the bar must be considered here by taking into account the reduction in cross sectional area introduced by the hole. Here $A' = (2-d)(\frac{1}{4})$.

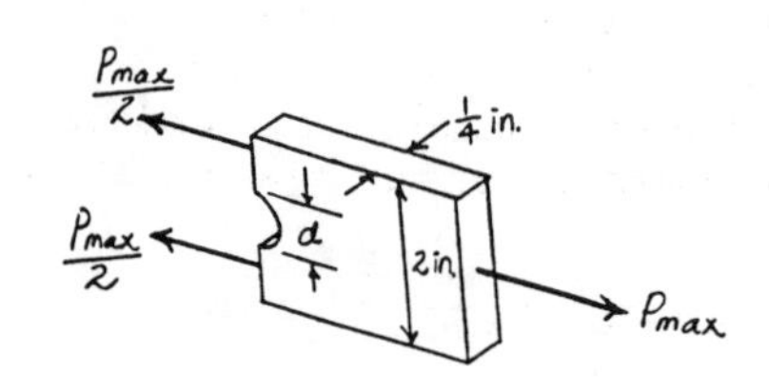

$$(\sigma_t)_{allow} = \frac{P}{A'}; \quad 21(10^3) = \frac{P_{max}}{(2-d)(\frac{1}{4})} \qquad [1]$$

Allowable Shear Stress : The pin is subjected to double shear and therefore, $V = \frac{P_{max}}{2}$.

$$\tau_{allow} = \frac{V}{A}; \quad 12(10^3) = \frac{P_{max}/2}{\frac{\pi}{4}d^2} \qquad [2]$$

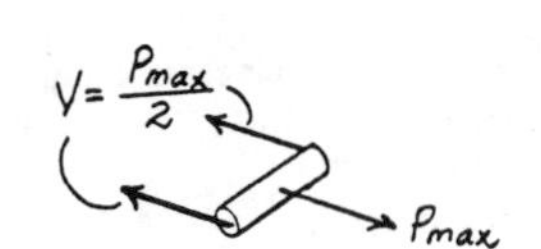

Solving Eq.[1] and[2] yields :

$d = 0.620$ in. **Ans**

$P_{max} = 7.25$ kip **Ans**

1–106. The bar is connected to the support using a pin having a diameter of $d = 1$ in. If the allowable tensile stress for the bar is $(\sigma_t)_{allow} = 20$ ksi, and the allowable bearing stress between the pin and the bar is $(\sigma_b)_{allow} = 30$ ksi, determine the dimensions w and t such that the gross area of the cross section is $wt = 2$ in^2 and the load P is a maximum. What is this maximum load? Assume the hole in the bar has the same diameter as the pin.

Allowable Normal Stress : The effective cross - sectional area A' for the bar must be considered here by taking into account the reduction in cross - sectional area introduced by the hole. Here $A' = (w-1)t = wt - t = (2-t)\text{ in}^2$ where $wt = 2\text{ in}^2$.

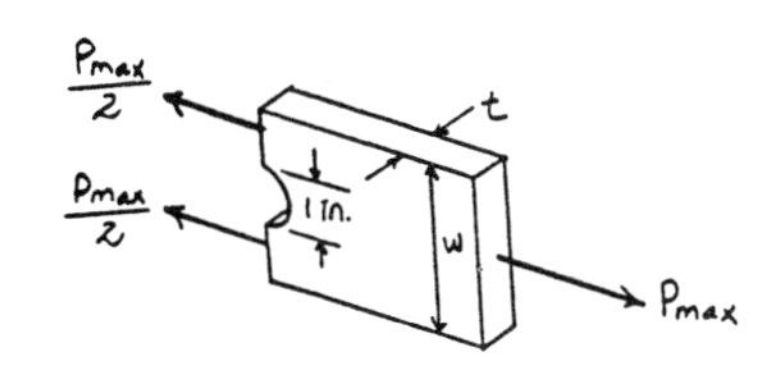

$$(\sigma_t)_{allow} = \frac{P}{A'};\quad 20(10^3) = \frac{P_{max}}{2-t} \qquad [1]$$

Allowable Bearing Stress : The projected area $A_p = (1)t = t\text{ in}^2$.

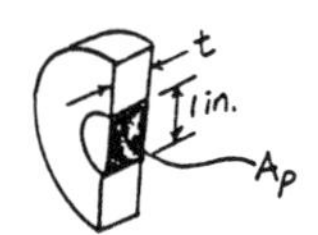

$$(\sigma_b)_{allow} = \frac{P}{A_p};\quad 30(10^3) = \frac{P_{max}}{t} \qquad [2]$$

Solving Eq. [1] and [2] yields :

$t = 0.800$ in. **Ans**

$P_{max} = 24.0$ kip **Ans**

And $w = 2.50$ in. **Ans**

1–107. A force of 8 kN is applied at the *center* of the wooden post. If the post is placed at the corner of its base plate B, can the bearing stress that the base plate exerts on the slab S be assumed uniformly distributed? Why or why not? What is the average compressive stress in the wooden post?

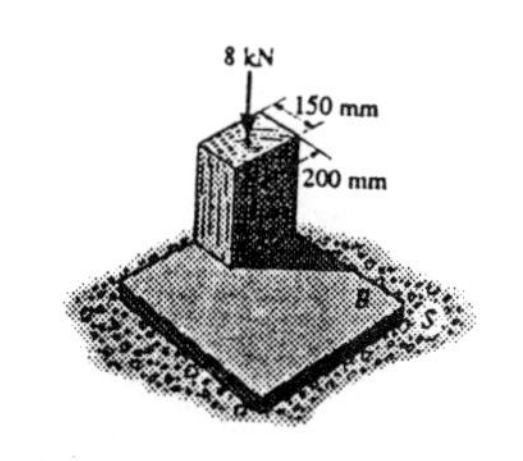

No. The force $P = 8$ kN must act at the centroid of the plate to produce uniform bearing sress.

Average Normal Stress :

$$\sigma = \frac{P}{A} = \frac{8(10^3)}{0.15(0.20)} = 267\text{ kPa} \qquad \textbf{Ans}$$

***1–108.** The column is made of concrete having a density of 2.30 Mg/m^3. At its top B it is subjected to an axial compressive force of 15 kN. Determine the compressive stress in the column as a function of the distance z measured from its base.

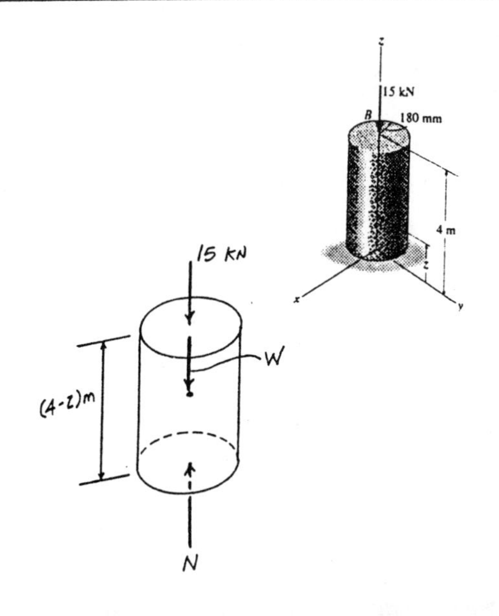

Equation of Equilibrium :

$$W = \rho g V = \left[2.30(9.81)\left(\frac{\pi}{4}\right)(0.36^2)(4-z)\right]\text{ kN}$$

$$+\uparrow \Sigma F_y = 0;\quad N - 15 - \left[2.30(9.81)\left(\frac{\pi}{4}\right)(0.36^2)(4-z)\right] = 0$$

$$N = [\,24.1865 - 2.2966\,z]\text{ kN}$$

Average Normal Stress :

$$\sigma = \frac{N}{A} = \frac{[24.1865 - 2.2966\,z]\text{ kN}}{\frac{\pi}{4}(0.36^2)}$$

$$= (238 - 22.6\,z)\text{ kPa} \qquad \textbf{Ans}$$

1–109. The column is made of concrete having a density of 2.30 Mg/m³. At its top B it is subjected to an axial compressive force of 15 kN. Determine the compressive stress in the column at z = 1 m, 2 m, and 3 m.

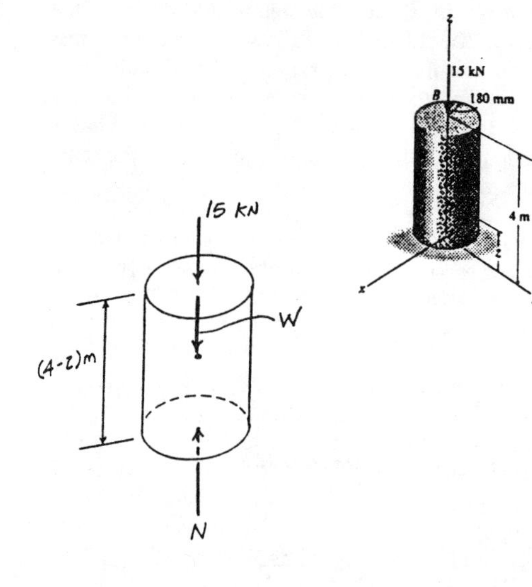

Equation of Equilibrium :

$$W = \rho g V = \left[2.30(9.81)\left(\frac{\pi}{4}\right)\left(0.36^2\right)(4-z)\right] \text{kN}$$

$$+\uparrow \Sigma F_y = 0; \quad N - 15 - \left[2.30(9.81)\left(\frac{\pi}{4}\right)\left(0.36^2\right)(4-z)\right] = 0$$

$$N = [\,24.1865 - 2.2966\,z]\ \text{kN}$$

Average Normal Stress :

$$\sigma = \frac{N}{A} = \frac{[24.1865 - 2.2966\,z]\ \text{kN}}{\frac{\pi}{4}(0.36^2)}$$

$$= (237.62 - 22.563\,z)\ \text{kPa}$$

Evaluate the result at

$z = 1$ m; $\quad \sigma = 237.62 - 22.563(1) = 215$ kPa **Ans**

$z = 2$ m; $\quad \sigma = 237.62 - 22.563(2) = 192$ kPa **Ans**

$z = 3$ m; $\quad \sigma = 237.62 - 22.563(3) = 170$ kPa **Ans**

1–110. The bearing pad consists of a 150 mm by 150 mm block of aluminum that supports a compressive load of 6 kN. Determine the average normal and shear stress acting on the plane through section a–a. Show the results on a differential volume element located on the plane.

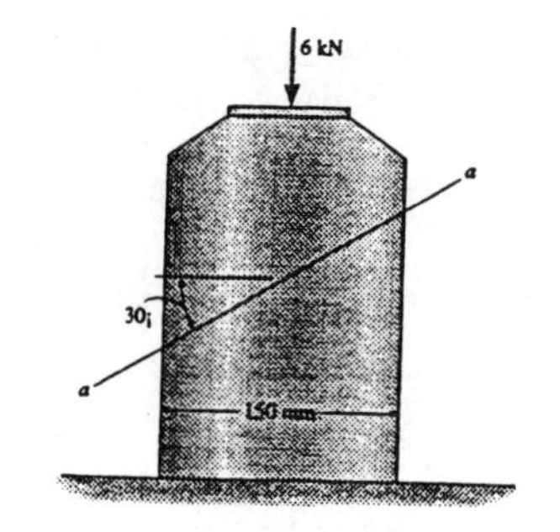

Equations of Equilibrium :

$\nearrow + \Sigma F_x = 0; \quad V_{a-a} - 6\cos 60° = 0 \quad V_{a-a} = 3.00$ kN

$\nwarrow + \Sigma F_y = 0; \quad N_{a-a} - 6\sin 60° = 0 \quad N_{a-a} = 5.196$ kN

Average Normal Stress And Shear Stress : The cross sectional Area at section a-a is $A = \left(\frac{0.15}{\sin 60°}\right)(0.15) = 0.02598\ \text{m}^2$.

$$\sigma_{a-a} = \frac{N_{a-a}}{A} = \frac{5.196(10^3)}{0.02598} = 200\ \text{kPa} \qquad \textbf{Ans}$$

$$\tau_{a-a} = \frac{V_{a-a}}{A} = \frac{3.00(10^3)}{0.02598} = 115\ \text{kPa} \qquad \textbf{Ans}$$

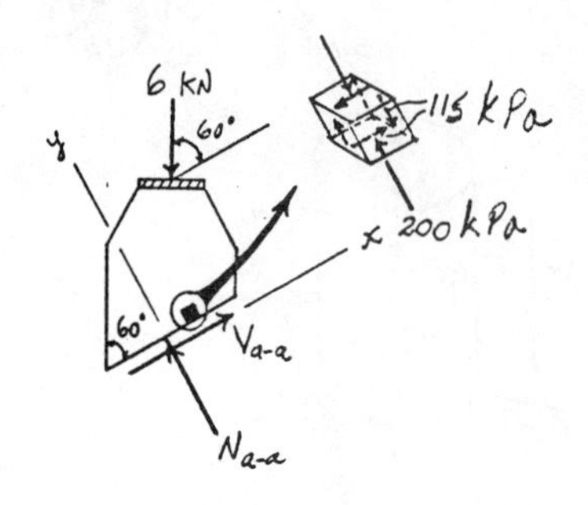

1–111. The circular punch B exerts a force of 2 kN on the top of the plate A. Determine the average shear stress in the plate due to this loading.

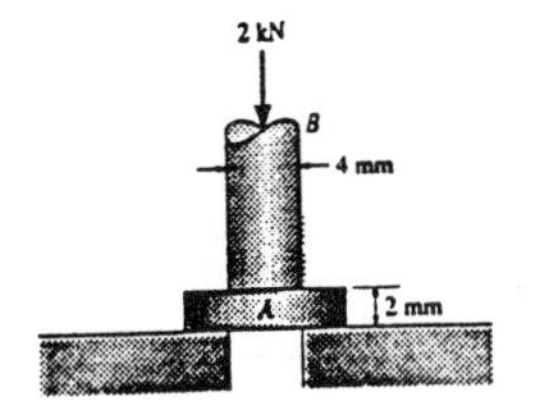

Average Shear Stress : The shear area $A = \pi(0.004)(0.002) = 8.00(10^{-6})\pi\ \text{m}^2$

$$\tau_{avg} = \frac{V}{A} = \frac{2(10^3)}{8.00(10^{-6})\pi} = 79.6\ \text{MPa} \qquad \textbf{Ans}$$

***1–112.** The column has a cross-sectional area of $16(10^3)\ \text{mm}^2$. It is subjected to an axial force of 80 kN. If the base plate to which the column is attached has a length of 175 mm, determine its width d so that the average bearing stress under the plate at the ground is two-thirds of the average compressive stress in the column. Sketch the stress distributions acting over the column's cross-sectional area and at the bottom of the base plate.

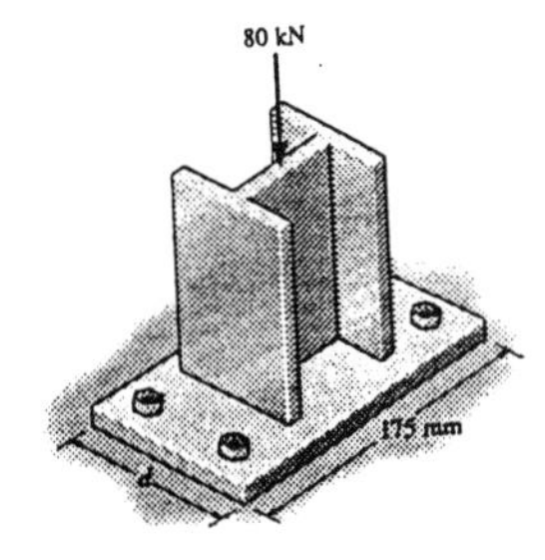

Average Normal Stress : For column

$$\sigma = \frac{P}{A} = \frac{80(10^3)}{0.016} = 5.00\ \text{MPa}$$

Average Bearing Stress : For base plate

$$\sigma_b = \frac{P}{A}; \qquad \frac{2}{3}(5.00)(10^6) = \frac{80(10^3)}{0.175(d)}$$

$$d = 0.1371\ \text{m} = 137\ \text{mm} \qquad \textbf{Ans}$$

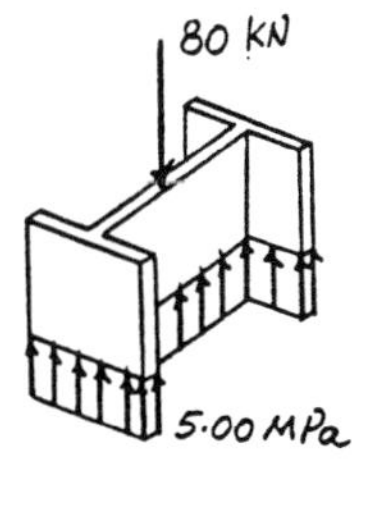

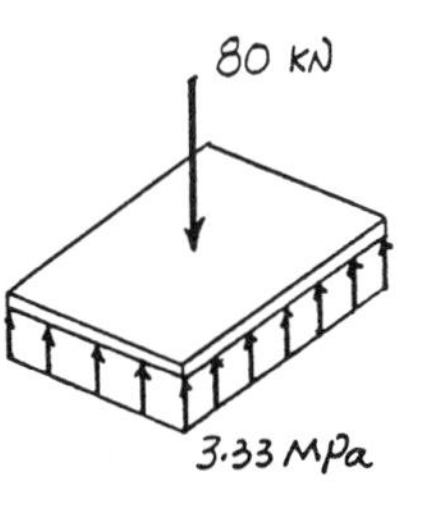

1–113. The cable has a specific weight γ (weight/volume) and cross-sectional area A. If the sag s is small, so that its length is approximately L and its weight can be distributed uniformly along the horizontal axis, determine the average normal stress in the cable at its lowest point C.

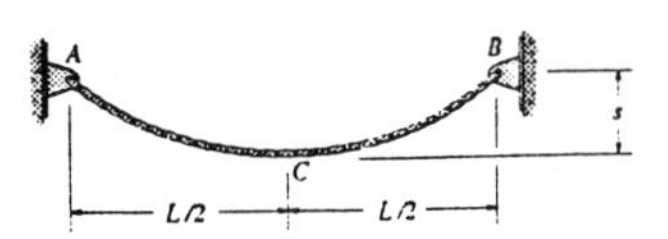

Equation of Equilibrium :

$$\curvearrowleft +\Sigma M_A = 0; \qquad T\,s - \frac{\gamma A L}{2}\left(\frac{L}{4}\right) = 0$$

$$T = \frac{\gamma A L^2}{8\,s}$$

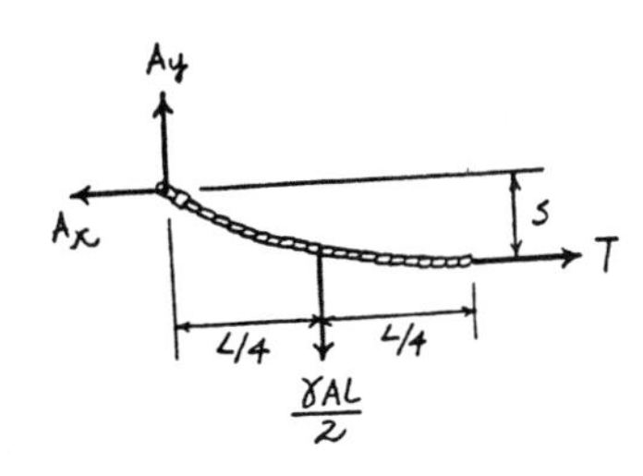

Average Normal Stress :

$$\sigma = \frac{T}{A} = \frac{\frac{\gamma A L^2}{8\,s}}{A} = \frac{\gamma L^2}{8\,s} \qquad \textbf{Ans}$$

1–114. Member B is subjected to a compressive force of 650 lb. If A and B are both made of wood and are $\frac{3}{8}$ in. thick, determine to the nearest $\frac{1}{4}$ in. the smallest dimension d of the support so that the average shear stress along section C does not exceed $\tau_{allow} = 300$ psi.

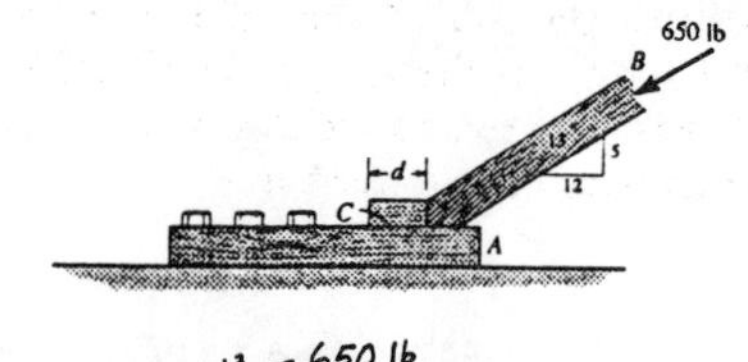

Equation of Equilibrium :

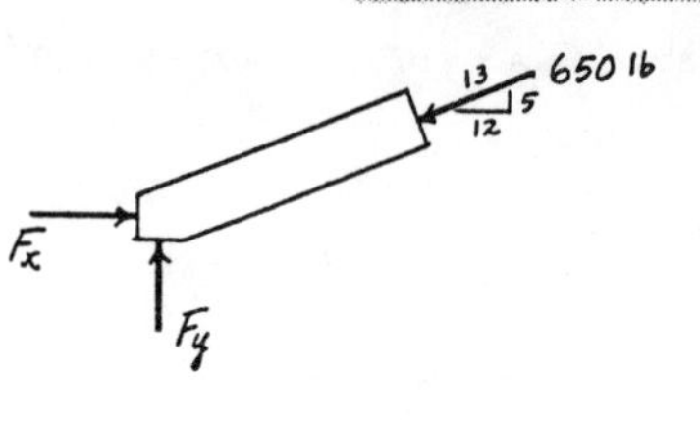

$$\overset{+}{\rightarrow}\Sigma F_x = 0; \quad F_x - 650\left(\frac{12}{13}\right) = 0 \quad F_x = 600 \text{ lb}$$

Allowable Shear Stress :

$$\tau_{allow} = \frac{V_C}{A_C}; \quad 300 = \frac{600}{\left(\frac{3}{8}\right)d}$$

$$d = 5.333 \text{ in.}$$

$$\text{Use } d = 5\frac{1}{2} \text{ in.} \qquad \textbf{Ans}$$

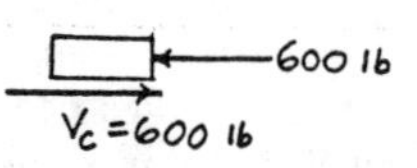

1–115. Member B is subjected to a compressive force of 650 lb. If A and B are both made of wood and are $\frac{3}{8}$ in. thick, and $d = 6$ in., determine the average shear stress along section C.

Equation of Equilibrium :

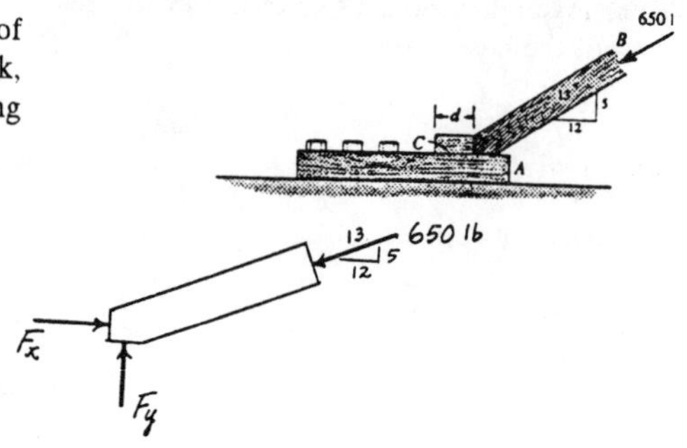

$$\overset{+}{\rightarrow}\Sigma F_x = 0; \quad F_x - 650\left(\frac{12}{13}\right) = 0 \quad F_x = 600 \text{ lb}$$

Average Shear Stress :

$$\tau_{avg} = \frac{V}{A} = \frac{600}{\left(\frac{3}{8}\right)(6)} = 267 \text{ psi} \qquad \textbf{Ans}$$

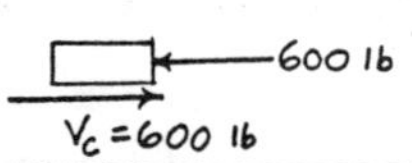

***1–116.** The 3-Mg concrete pipe is suspended by the three wires. If BD and CD have a diameter of 10 mm and AD has a diameter of 7 mm, determine the average normal stress in each wire.

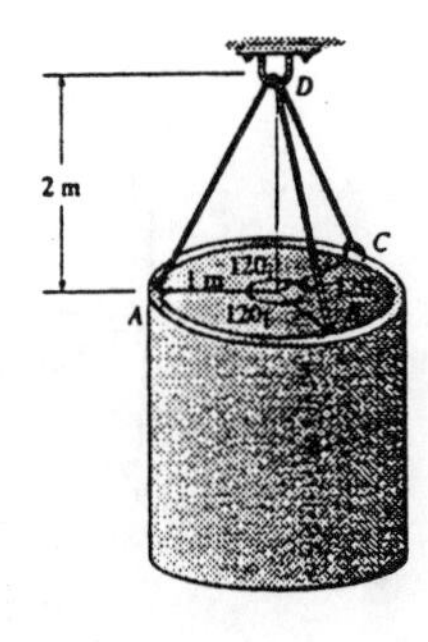

Equations of Equilibrium :

$$\Sigma M_x = 0; \quad F_{BD}(1\sin 60°) - F_{CD}(1\sin 60°) = 0$$
$$F_{BD} = F_{CD} = F$$

$$\Sigma M_y = 0; \quad 2F(1\cos 60°) - F_{AD}(1) = 0$$
$$F_{AD} = F$$

$$\Sigma F_z = 0; \quad 3\left[F\left(\frac{2}{\sqrt{5}}\right)\right] - 29.43 = 0$$
$$F = 10.97 \text{ kN}$$

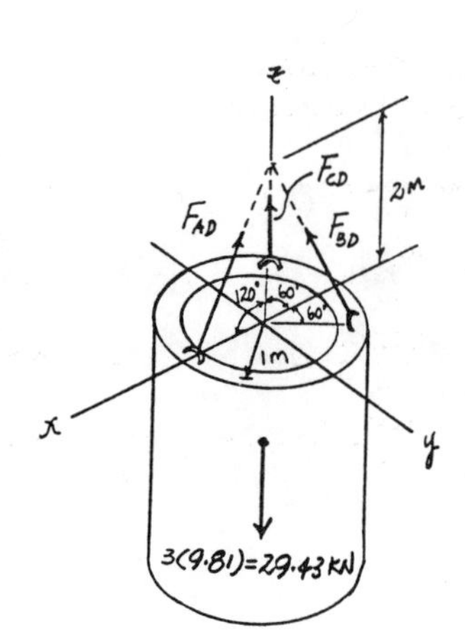

Average Normal Stress :

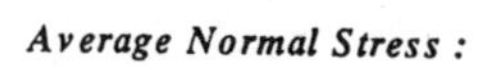

$$\sigma_{BD} = \sigma_{CD} = \frac{F}{A_{BD}} = \frac{10.97(10^3)}{\frac{\pi}{4}(0.01^2)} = 140 \text{ MPa} \qquad \textbf{Ans}$$

$$\sigma_{AD} = \frac{F}{A_{AD}} = \frac{10.97(10^3)}{\frac{\pi}{4}(0.007^2)} = 285 \text{ MPa} \qquad \textbf{Ans}$$

2–1. The center portion of the rubber balloon has a diameter of $d = 4$ in. If the air pressure within it causes the balloon's diameter to become $d = 5$ in., determine the average normal strain in the rubber.

Average Normal Strain :

$$\varepsilon_{avg} = \frac{\pi d - \pi d_0}{\pi d_0} = \frac{d - d_0}{d_0} = \frac{5-4}{4} = 0.250 \text{ in./in.} \quad \textbf{Ans}$$

2–2. A rubber band has an unstretched length of 10 in. If it is stretched around a pole having an outer diameter of 6 in., determine the average normal strain in the band.

Average Normal Strain :

$$L_0 = 10 \text{ in.} \qquad L = 6\pi \text{ in.}$$

$$\varepsilon = \frac{L - L_0}{L_0} = \frac{6\pi - 10}{10} = 0.885 \text{ in./in.} \quad \textbf{Ans}$$

2–3. Bar ABC is originally in a horizontal position. If loads cause the end A to be displaced downwards $\Delta_A = 0.002$ in. and the bar rotates $\theta = 0.2°$, determine the average normal strain in the rods AD, BE, and CF.

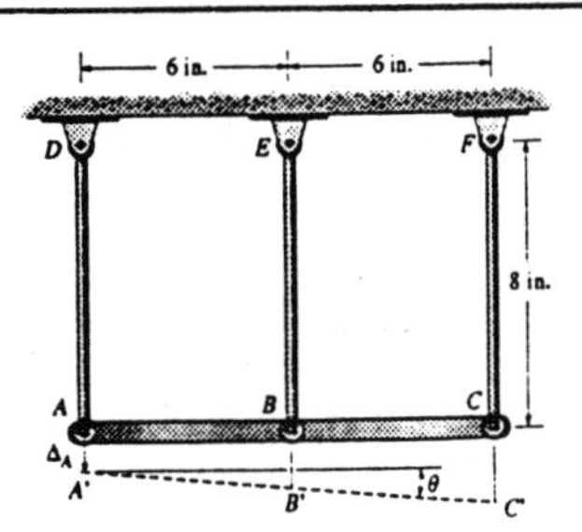

Geometry :

$$\Delta_B = 0.002 + 6\tan 0.2° = 0.02294 \text{ in.}$$
$$\Delta_C = 0.002 + 12\tan 0.2° = 0.04389 \text{ in.}$$

Average Normal Strain :

$$(\varepsilon_{AD})_{avg} = \frac{\Delta_A}{L_{AD}} = \frac{0.002}{8} = 0.250(10^{-3}) \text{ in./in.} \quad \textbf{Ans}$$

$$(\varepsilon_{BE})_{avg} = \frac{\Delta_B}{L_{BE}} = \frac{0.02294}{8} = 2.87(10^{-3}) \text{ in./in.} \quad \textbf{Ans}$$

$$(\varepsilon_{CF})_{avg} = \frac{\Delta_C}{L_{CF}} = \frac{0.04389}{8} = 5.49(10^{-3}) \text{ in./in.} \quad \textbf{Ans}$$

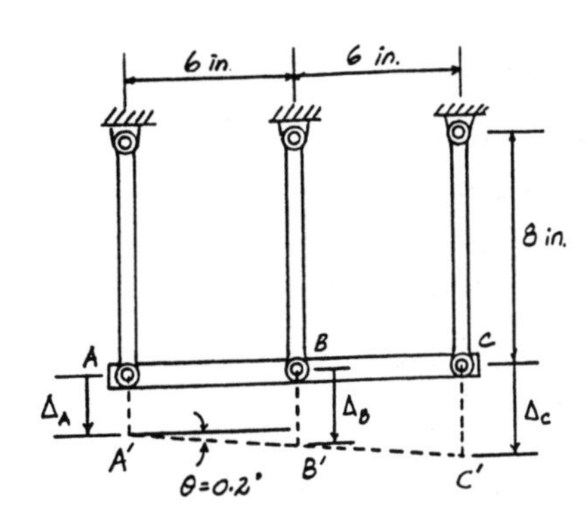

***2–4.** The rigid beam is supported by a pin at A and wires BD and CE. If the load **P** on the beam is displaced 10 mm downward, determine the normal strain developed in wires CE and BD.

Geometry :

$$\frac{\Delta_{BD}}{3} = \frac{\Delta_{CE}}{5} = \frac{10}{7}$$

$\Delta_{BD} = 4.2857$ mm $\qquad \Delta_{CE} = 7.1429$ mm

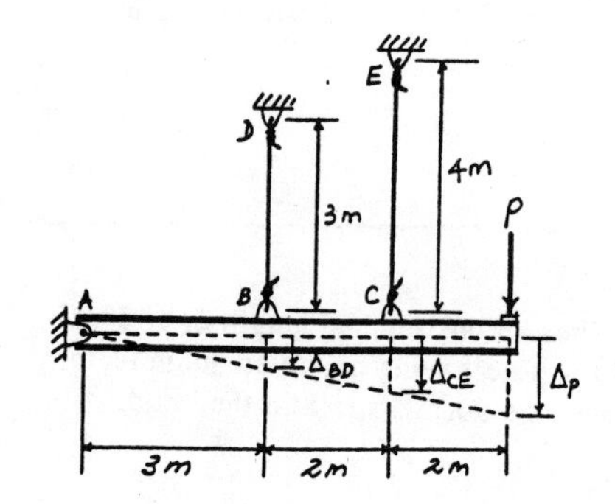

Average Normal Strain :

$$(\varepsilon_{CE})_{avg} = \frac{\Delta_{CE}}{L_{CE}} = \frac{7.1429}{4000} = 1.79(10^{-3})\ \text{mm/mm} \qquad \textbf{Ans}$$

$$(\varepsilon_{BD})_{avg} = \frac{\Delta_{BD}}{L_{BD}} = \frac{4.2857}{3000} = 1.43(10^{-3})\ \text{mm/mm} \qquad \textbf{Ans}$$

2–5. The rigid beam is supported by a pin at A and wires BD and CE. If the maximum allowable normal strain in each wire is $\epsilon_{max} = 0.002$, determine the maximum vertical displacement of the load **P**.

Geometry :

$$\frac{\Delta_{BD}}{3} = \frac{\Delta_{CE}}{5} = \frac{\Delta_P}{7}$$

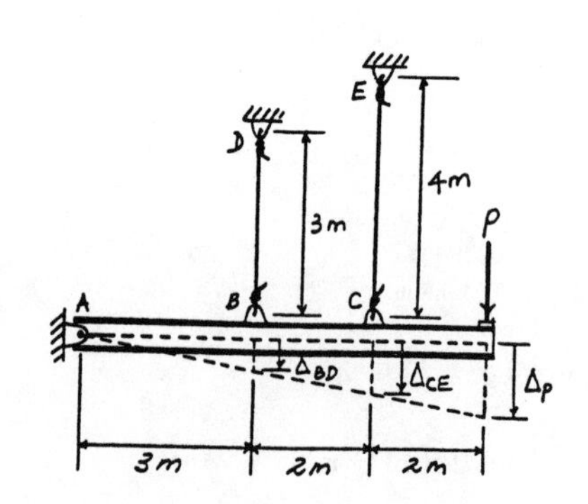

Average Normal Strain :

$$\Delta_{BD} = \varepsilon_{max} L_{BD} = 0.002(3000) = 6.00\ \text{mm}$$

$$\Delta_P = \frac{7}{3}\Delta_{BD} = \frac{7}{3}(6.00) = 14.0\ \text{mm}$$

$$\Delta_{CE} = \varepsilon_{max} L_{CE} = 0.002(4000) = 8.00\ \text{mm}$$

$$\Delta_P = \frac{7}{5}\Delta_{CE} = \frac{7}{5}(8.00) = 11.2\ \text{mm} \quad \textit{(controls!)} \qquad \textbf{Ans}$$

2–6. Determine the approximate average normal strain in the wire AB as a function of the rotation θ of the rigid bar CA by assuming θ is small. What is this value if $\theta = 2°$?

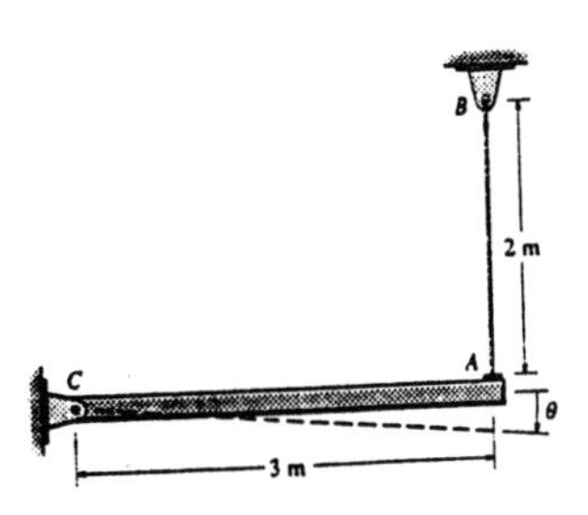

$$\alpha = \left(\frac{180-\theta}{2}\right) + 90° = 180° - \frac{\theta}{2}$$

$$AA' = 2\left(3\sin\frac{\theta}{2}\right) = 6\sin\frac{\theta}{2}$$

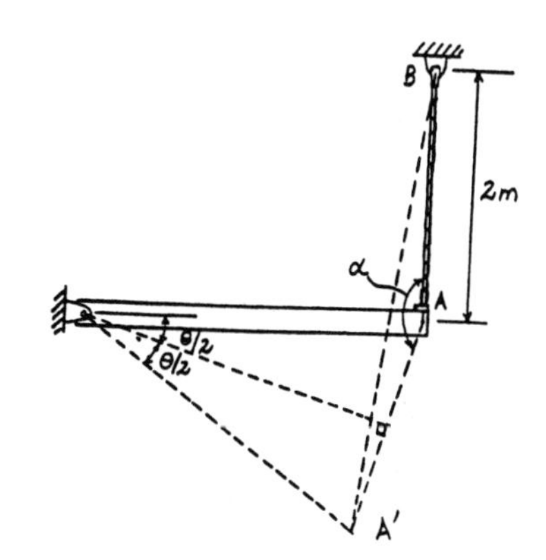

$$BA' = \sqrt{2^2 + \left(6\sin\frac{\theta}{2}\right)^2 - 2(2)\left(6\sin\frac{\theta}{2}\right)\cos\left(180° - \frac{\theta}{2}\right)}$$

$$= \sqrt{4 + 36\sin^2\frac{\theta}{2} + 24\sin\frac{\theta}{2}\cos\frac{\theta}{2}}$$

$$= \sqrt{4 + 36\left(\frac{1-\cos\theta}{2}\right) + 12\sin\theta}$$

$$= \sqrt{4 + 18 - 18\cos\theta + 12\sin\theta}$$

$$= \sqrt{22 - 18\cos\theta + 12\sin\theta}$$

Average Normal Strain :

$$\varepsilon_{avg} = \frac{BA' - BA}{BA} = \frac{\sqrt{22 - 18\cos\theta + 12\sin\theta} - 2}{2}$$

$$= \sqrt{5.5 - 4.5\cos\theta + 3\sin\theta} - 1$$

If θ is small then, $\cos\theta \approx 1$ and $\sin\theta \approx \theta$

$$(\varepsilon_{avg})_{Approx} \approx \sqrt{5.5 - 4.5 + 3\theta} - 1$$

$$= \sqrt{1 + 3\theta} - 1 = 1 + \frac{1}{2}(3\theta) - 1$$

$$= 1.50\,\theta \qquad \textbf{Ans}$$

$$(\varepsilon_{avg})_{approx} = 1.5\,(2°)\left(\frac{\pi}{180}\right) = 0.05235988 \text{ m/m}$$

$$= 52.4\left(10^{-3}\right) \text{ m/m} \qquad \textbf{Ans}$$

2–7. The two wires are connected together at A. If the force **P** causes point A to be displaced vertically 3 mm, determine the normal strain developed in each wire.

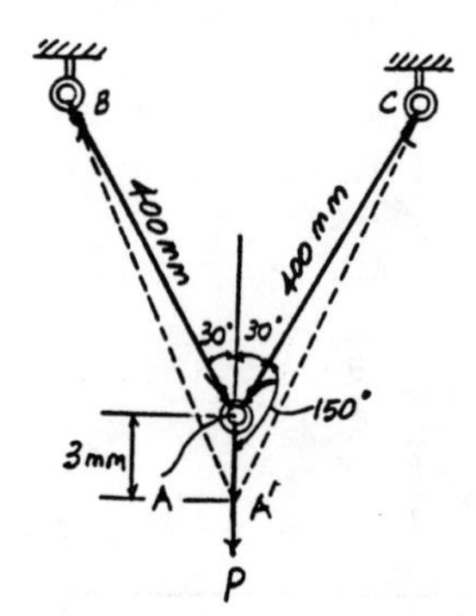

Geometry :

$$L_{A'C} = L_{A'B} = \sqrt{3^2 + 400^2 - 2(3)(400)\cos 150^\circ}$$
$$= 402.601 \text{ mm}$$

Average Normal Strain :

$$(\varepsilon_{AB})_{avg} = (\varepsilon_{AC})_{avg} = \frac{L_{A'C} - L_{AC}}{L_{AC}}$$
$$= \frac{402.601 - 400}{400}$$
$$= 6.50(10^{-3}) \text{ mm/mm} \qquad \textbf{Ans}$$

***2–8.** Two bars are used to support a load. When unloaded, AB is 5 in. long, AC is 8 in. long, and the ring at A has coordinates (0, 0). If a load **P** acts on the ring at A, the normal strain in AB becomes $\epsilon_{AB} = 0.02$ in./in., and the normal strain in AC becomes $\epsilon_{AC} = 0.035$ in./in. Determine the coordinate position of the ring due to the load.

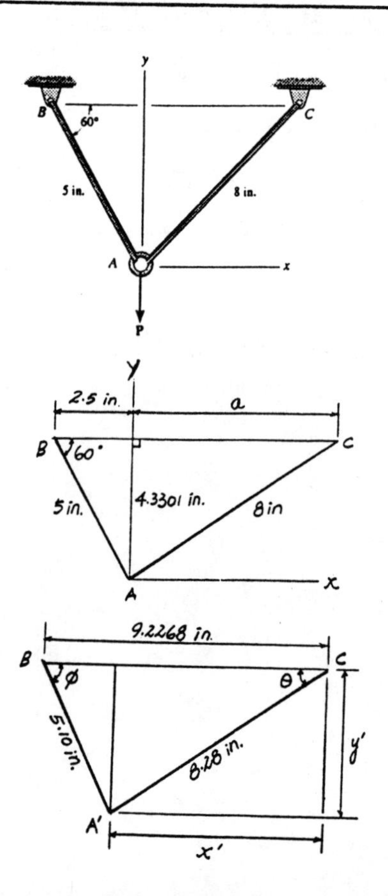

Average Normal Strain :

$$L'_{AB} = L_{AB} + \varepsilon_{AB} L_{AB} = 5 + (0.02)(5) = 5.10 \text{ in.}$$
$$L'_{AC} = L_{AC} + \varepsilon_{AC} L_{AC} = 8 + (0.035)(8) = 8.28 \text{ in.}$$

Geometry :

$$a = \sqrt{8^2 - 4.3301^2} = 6.7268 \text{ in.}$$

$$5.10^2 = 9.2268^2 + 8.28^2 - 2(9.2268)(8.28)\cos\theta$$
$$\theta = 33.317^\circ$$

$$x' = 8.28 \cos 33.317^\circ = 6.9191 \text{ in.}$$
$$y' = 8.28 \sin 33.317^\circ = 4.5480 \text{ in.}$$

$$x = -(x' - a)$$
$$= -(6.9191 - 6.7268) = -0.192 \text{ in.} \qquad \textbf{Ans}$$

$$y = -(y' - 4.3301)$$
$$= -(4.5480 - 4.3301) = -0.218 \text{ in.} \qquad \textbf{Ans}$$

2–9. Two bars are used to support a load **P**. When unloaded, AB is 5 in. long, AC is 8 in. long, and the ring at A has coordinates (0, 0). If a load is applied to the ring at A, so that it moves it to the coordinate position (0.25 in., −0.73 in.), determine the normal strain in each bar.

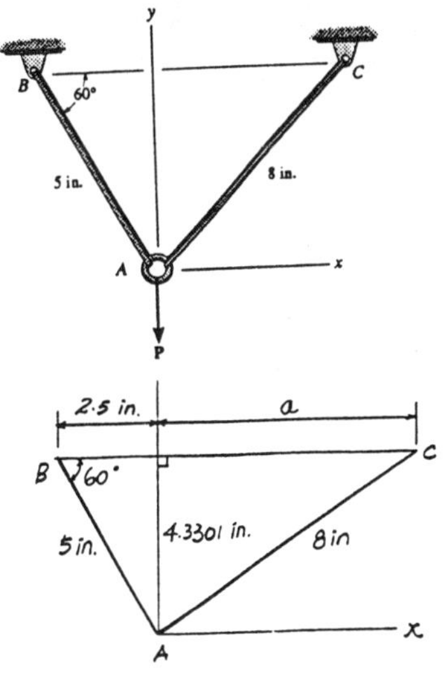

Geometry :

$$a = \sqrt{18^2 - 4.3301^2} = 6.7268 \text{ in.}$$

$$L_{A'B} = \sqrt{(2.5 + 0.25)^2 + (4.3301 + 0.73)^2} = 5.7591 \text{ in.}$$

$$L_{A'C} = \sqrt{(6.7268 - 0.25)^2 + (4.3301 + 0.73)^2} = 8.2191 \text{ in.}$$

Average Normal Strain :

$$\varepsilon_{AB} = \frac{L_{A'B} - L_{AB}}{L_{AB}} = \frac{5.7591 - 5}{5} = 0.152 \text{ in./in.} \qquad \textbf{Ans}$$

$$\varepsilon_{AC} = \frac{L_{A'C} - L_{AC}}{L_{AC}} = \frac{8.2191 - 8}{8} = 0.0274 \text{ in./in.} \qquad \textbf{Ans}$$

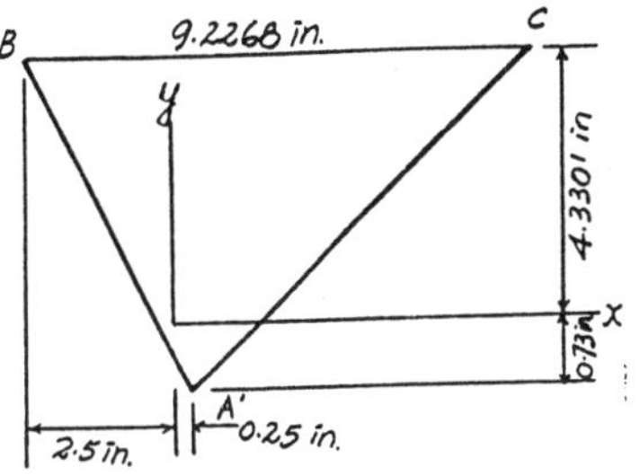

2–10. The guy wire AB of a building frame is originally unstretched. Due to an earthquake, the two columns of the frame tilt $\theta = 2°$. Determine the approximate normal strain in the wire when the frame is in this position. Assume the columns are rigid and rotate about their lower supports.

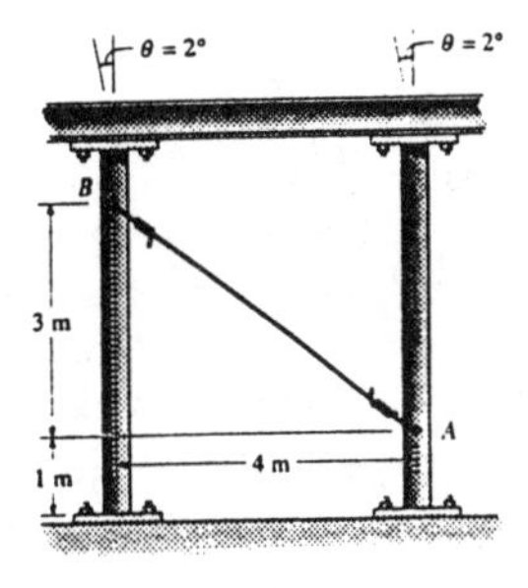

Geometry : The vertical displacement is negligible.

$$x_A = (1)\left(\frac{2°}{180°}\right)\pi = 0.03491 \text{ m}$$

$$x_B = (4)\left(\frac{2°}{180°}\right)\pi = 0.13963 \text{ m}$$

$$x = 4 + x_B - x_A = 4.10472 \text{ m}$$

$$A'B' = \sqrt{3^2 + 4.10472^2} = 5.08416 \text{ m}$$

$$AB = \sqrt{3^2 + 4^2} = 5.00 \text{ m}$$

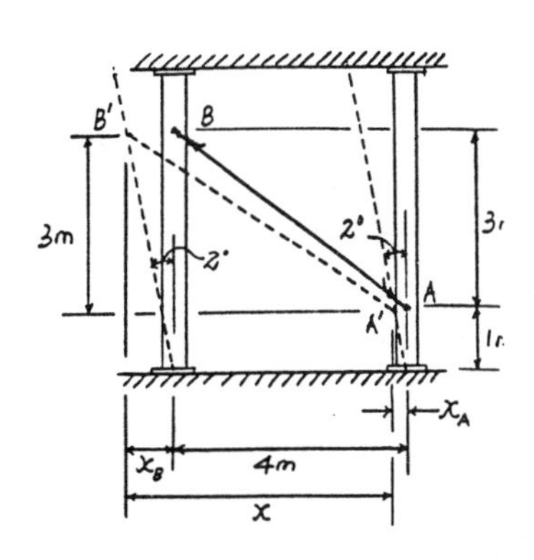

Average Normal Stress :

$$\varepsilon_{AB} = \frac{A'B' - AB}{AB} = \frac{5.08416 - 5}{5} = 16.8\left(10^{-3}\right) \text{ m/m} \qquad \textbf{Ans}$$

2–11. Due to its weight, the rod is subjected to a normal strain that varies along its length such that $\epsilon = kz$, where k is a constant. Determine the displacement ΔL of its end B when it is suspended as shown.

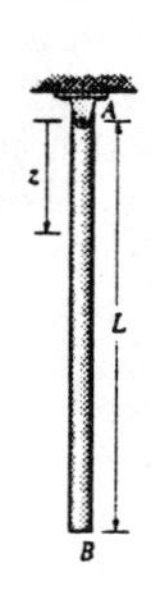

Normal Strain :

$$\varepsilon_z = \frac{\delta z}{dz} = k\,z$$

$$\int_0^{\Delta L} \delta z = k \int_0^{L} z\,dz$$

$$\Delta L = \frac{k\,L^2}{2} \qquad \textbf{Ans}$$

***2–12.** The rectangular membrane has an unstretched length L_1 and width L_2. If the sides are increased by small amounts ΔL_1 and ΔL_2, determine the normal strain along the diagonal AB.

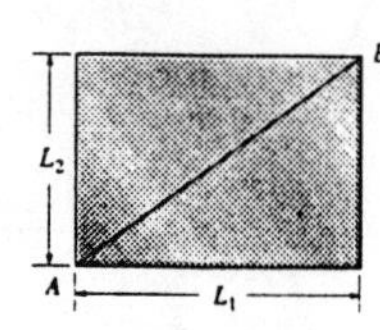

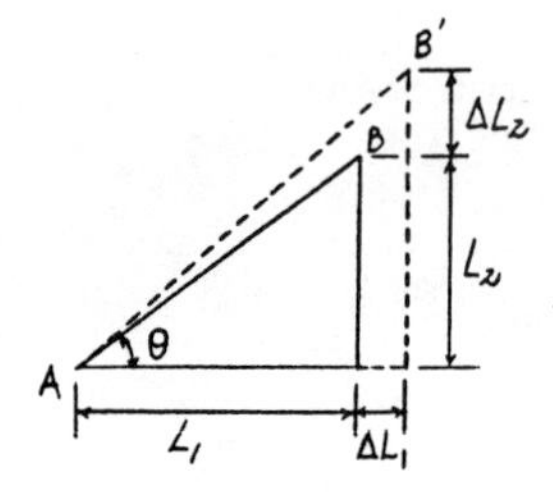

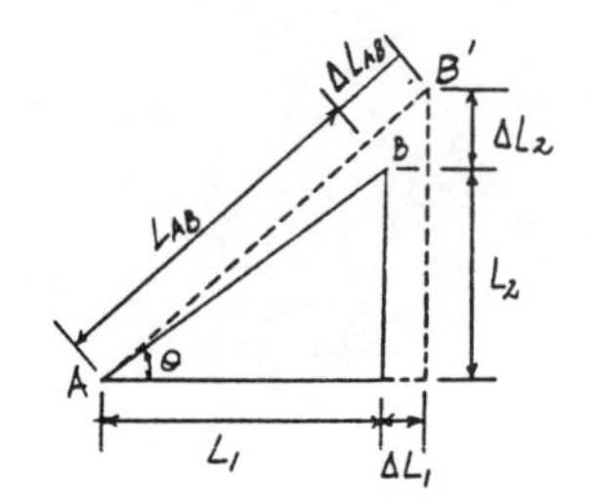

Geometry :

$$L_{AB} = \sqrt{L_1^2 + L_2^2}$$

$$L_{AB'} = \sqrt{(L_1 + \Delta L_1)^2 + (L_2 + \Delta L_2)^2}$$

$$= \sqrt{L_1^2 + L_2^2 + \Delta L_1^2 + \Delta L_2^2 + 2L_1 \Delta L_1 + 2L_2 \Delta L_2}$$

Average Normal Strain :

$$\varepsilon_{AB} = \frac{L_{AB'} - L_{AB}}{L_{AB}}$$

$$\varepsilon_{AB} = \sqrt{\frac{L_1^2 + L_2^2 + \Delta L_1^2 + \Delta L_2^2 + 2L_1 \Delta L_1 + 2L_2 \Delta L_2}{L_1^2 + L_2^2}} - 1$$

$$= \sqrt{1 + \frac{\Delta L_1^2 + \Delta L_2^2 + 2L_1 \Delta L_1 + 2L_2 \Delta L_2}{L_{AB}^2}} - 1$$

Neglecting the higher order terms

$$\varepsilon_{AB} = \left(1 + \frac{2L_1 \Delta L_1 + 2L_2 \Delta L_2}{L_{AB}^2}\right)^{\frac{1}{2}} - 1$$

Using the binomial theorem,

$$\varepsilon_{AB} = 1 + \frac{1}{2}\left(\frac{2L_1 \Delta L_1}{L_{AB}^2} + \frac{2L_2 \Delta L_2}{L_{AB}^2}\right) + \ldots - 1$$

$$\varepsilon_{AB} \approx \frac{L_1 \Delta L_1}{L_{AB}^2} + \frac{L_2 \Delta L_2}{L_{AB}^2}$$

However, $L_{AB} = \dfrac{L_1}{\cos\theta} = \dfrac{L_2}{\sin\theta}$

$$\varepsilon_{AB} = \frac{\Delta L_1}{L_1}\cos^2\theta + \frac{\Delta L_2}{L_2}\sin^2\theta \qquad \textbf{Ans}$$

Also

Geometry :

$$(L_1 + \Delta L_1) = (1 + \varepsilon_x)L_1 = (1 + \varepsilon_x)L_{AB}\cos\theta$$
$$(L_2 + \Delta L_2) = (1 + \varepsilon_y)L_2 = (1 + \varepsilon_y)L_{AB}\sin\theta$$

Thus,

$$(L_{AB} + \Delta L_{AB})^2 = (1 + \varepsilon_x)^2 L_{AB}^2 \cos^2\theta + (1 + \varepsilon_y)^2 L_{AB}^2 \sin^2\theta$$
$$(L_{AB} + \Delta L_{AB})^2 = L_{AB}^2(\cos^2\theta + 2\varepsilon_x\cos^2\theta + \varepsilon_x^2\cos^2\theta$$
$$+ \sin^2\theta + 2\varepsilon_y \sin^2\theta + \varepsilon_y^2 \sin^2\theta)$$

For small strain $\varepsilon_x^2 = \varepsilon_y^2 \approx 0$

$$L_{AB} + \Delta L_{AB} = L_{AB}\left(1 + 2\varepsilon_x\cos^2\theta + 2\varepsilon_y\sin^2\theta\right)^{\frac{1}{2}}$$

Using the binomial theorem,

$$L_{AB} + \Delta L_{AB} = L_{AB}\left(1 + \varepsilon_x\cos^2\theta + \varepsilon_y\sin^2\theta + \ldots\right)$$

Average Normal Strain :

$$\varepsilon_{AB} = \frac{(L_{AB} + \Delta L_{AB}) - L_{AB}}{L_{AB}} = \varepsilon_x\cos^2\theta + \varepsilon_y\sin^2\theta$$

However, $\varepsilon_x = \dfrac{\Delta L_1}{L_1}$ and $\varepsilon_y = \dfrac{\Delta L_2}{L_2}$

$$\varepsilon_{AB} = \frac{\Delta L_1}{L_1}\cos^2\theta + \frac{\Delta L_2}{L_2}\sin^2\theta \qquad \textbf{Ans}$$

2–13. The rectangular plate is subjected to the deformation shown by the dashed line. Determine the average shear strain γ_{xy} of the plate.

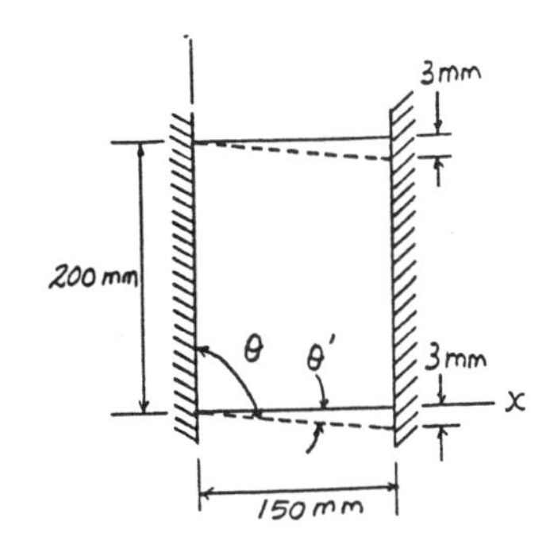

Geometry :

$$\theta' = \tan^{-1}\frac{3}{150} = 0.0200 \text{ rad}$$

$$\theta = \left(\frac{\pi}{2} + 0.0200\right) \text{ rad}$$

Shear Strain :

$$\gamma_{xy} = \frac{\pi}{2} - \theta = \frac{\pi}{2} - \left(\frac{\pi}{2} + 0.0200\right)$$
$$= -0.0200 \text{ rad} \qquad \text{Ans}$$

2–14. The rectangular plate is subjected to the deformation shown by the dashed line. Determine the shear strain $\gamma_{x'y'}$ in the plate. The x' axis is directed from A through point B.

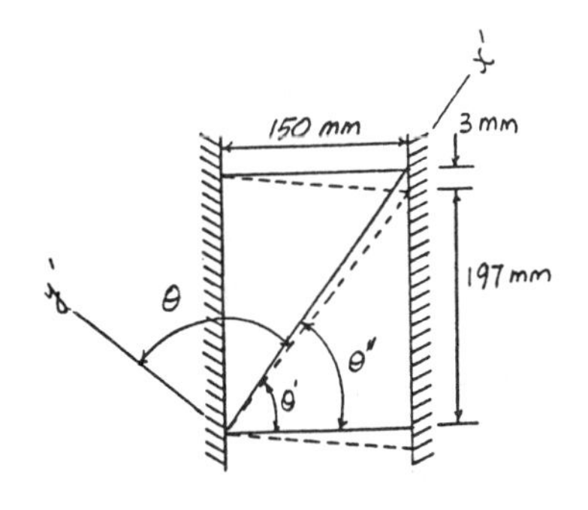

Geometry :

$$\theta' = \tan^{-1}\frac{197}{150} = 52.7136^\circ \qquad \theta'' = \tan^{-1}\frac{200}{150} = 53.1301^\circ$$

$$\theta = 90^\circ + (53.1301^\circ - 52.7136^\circ) = 90.4165^\circ = 1.57807 \text{ rad}$$

Shear Strain :

$$\gamma_{x'y'} = \frac{\pi}{2} - \theta = \frac{\pi}{2} - 1.57807$$
$$= -7.27(10^{-3}) \text{ rad} \qquad \text{Ans}$$

2–15. The rectangular plate is subjected to the deformation shown by the dashed line. Determine the normal strains ϵ_x, ϵ_y, $\epsilon_{x'}$ $\epsilon_{y'}$,

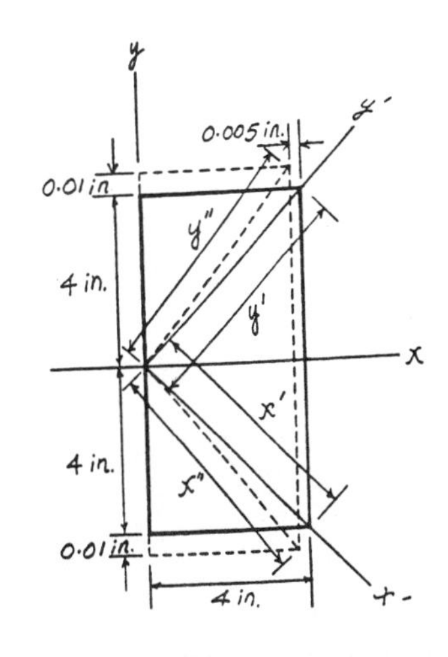

Geometry :

$$x'' = y'' = \sqrt{(4+0.01)^2 + (4-0.005)^2} = 5.6604 \text{ in.}$$

$$x' = y' = \sqrt{4^2 + 4^2} = 5.6569 \text{ in.}$$

Average Normal Strian :

$$\varepsilon_y = \frac{0.01}{4} = 2.50\left(10^{-3}\right) \text{ in./in.} \qquad \text{Ans}$$

$$\varepsilon_x = \frac{-0.005}{4} = -1.25\left(10^{-3}\right) \text{ in./in.} \qquad \text{Ans}$$

$$\varepsilon_{x'} = \varepsilon_{y'} = \frac{x'' - x'}{x'} = \frac{5.6604 - 5.6569}{5.6569}$$
$$= 0.627\left(10^{-3}\right) \text{ in./in.} \qquad \text{Ans}$$

***2–16.** The rectangular plate is subjected to the deformation shown by the dashed line. Determine the shear strains γ_{xy} and $\gamma_{x'y'}$ developed at point A.

Shear Strain :

Since the right angle of the element in the $x-y$ plane does not distort, then

$$(\gamma_A)_{xy} = 0 \qquad \textbf{Ans}$$

$$\tan\theta = \frac{4.01}{3.995}; \qquad \theta = 0.787272 \text{ rad}$$

$$(\gamma_A)_{x'y'} = \frac{\pi}{2} - 2\theta$$

$$= \frac{\pi}{2} - 2(0.787272)$$

$$= -3.75(10^{-3}) \text{ rad} \qquad \textbf{Ans}$$

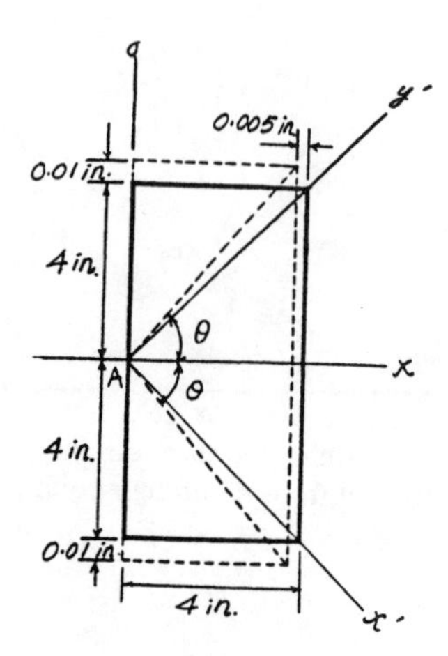

2–17. The piece of plastic is originally rectangular. Determine the shear strain γ_{xy} at corners A and B if the plastic distorts as shown by the dashed lines.

Geometry : For small angles,

$$\alpha = \psi = \frac{2}{302} = 0.00662252 \text{ rad}$$

$$\beta = \theta = \frac{2}{403} = 0.00496278 \text{ rad}$$

Shear Strain :

$$(\gamma_B)_{xy} = \alpha + \beta$$

$$= 0.0116 \text{ rad} = 11.6(10^{-3}) \text{ rad} \qquad \textbf{Ans}$$

$$(\gamma_A)_{xy} = -(\theta + \psi)$$

$$= -0.0116 \text{ rad} = -11.6(10^{-3}) \text{ rad} \qquad \textbf{Ans}$$

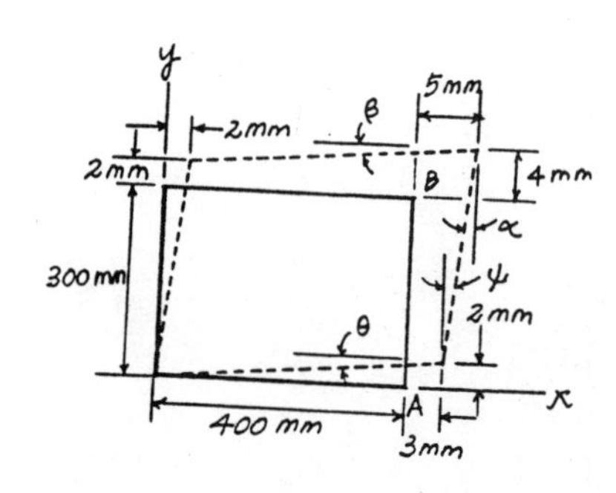

2–18. The piece of plastic is originally rectangular. Determine the shear strain γ_{xy} at corners D and C if the plastic distorts as shown by the dashed lines.

Geometry : For small angles,

$$\alpha = \psi = \frac{2}{403} = 0.00496278 \text{ rad}$$

$$\beta = \theta = \frac{2}{302} = 0.00662252 \text{ rad}$$

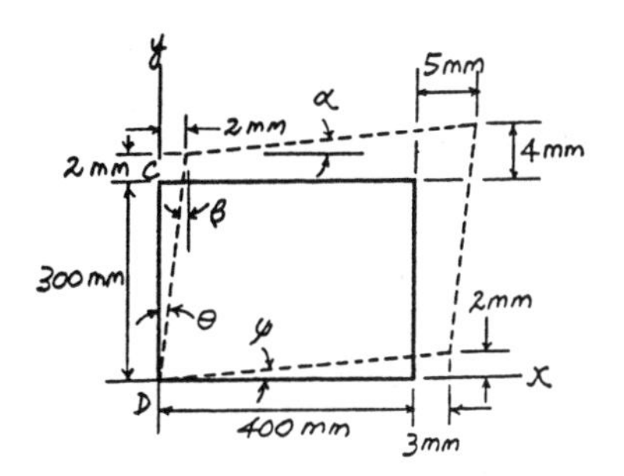

Shear Strain :

$$(\gamma_C)_{xy} = -(\alpha + \beta)$$
$$= -0.0116 \text{ rad} = -11.6(10^{-3}) \text{ rad} \qquad \textbf{Ans}$$

$$(\gamma_D)_{xy} = \theta + \psi$$
$$= 0.0116 \text{ rad} = 11.6(10^{-3}) \text{ rad} \qquad \textbf{Ans}$$

2–19. The piece of plastic is originally rectangular. Determine the average normal strain that occurs along the diagonals AC and DB.

Geometry :

$$AC = DB = \sqrt{400^2 + 300^2} = 500 \text{ mm}$$
$$DB' = \sqrt{405^2 + 304^2} = 506.4 \text{ mm}$$
$$A'C' = \sqrt{401^2 + 300^2} = 500.8 \text{ mm}$$

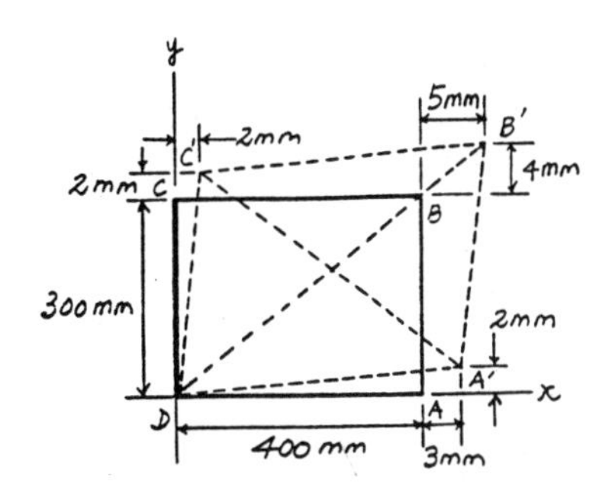

Average Normal Strain :

$$\varepsilon_{AC} = \frac{A'C' - AC}{AC} = \frac{500.8 - 500}{500}$$
$$= 0.00160 \text{ mm/mm} = 1.60(10^{-3}) \text{ mm/mm} \qquad \textbf{Ans}$$

$$\varepsilon_{DB} = \frac{DB' - DB}{DB} = \frac{506.4 - 500}{500}$$
$$= 0.0128 \text{ mm/mm} = 12.8(10^{-3}) \text{ mm/mm} \qquad \textbf{Ans}$$

***2–20.** A square piece of material is deformed into the dashed parallelogram. Determine the average normal strain that occurs along the diagonals AC and BD.

Geometry :

$$AC = BD = \sqrt{15^2 + 15^2} = 21.2132 \text{ mm}$$

$$AC' = \sqrt{15.18^2 + 15.24^2 - 2(15.18)(15.24)\cos 90.3^\circ}$$
$$= 21.5665 \text{ mm}$$

$$B'D' = \sqrt{15.18^2 + 15.24^2 - 2(15.18)(15.24)\cos 89.7^\circ}$$
$$= 21.4538 \text{ mm}$$

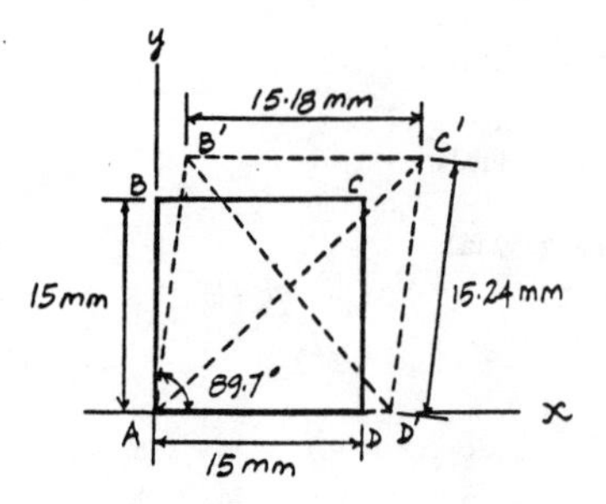

Average Normal Strain :

$$\varepsilon_{AC} = \frac{AC' - AC}{AC} = \frac{21.5665 - 21.2132}{21.2132}$$
$$= 0.01665 \text{ mm/mm} = 16.7(10^{-3}) \text{ mm/mm} \qquad \textbf{Ans}$$

$$\varepsilon_{BD} = \frac{B'D' - BD}{BD} = \frac{21.4538 - 21.2132}{21.2132}$$
$$= 0.01134 \text{ mm/mm} = 11.3(10^{-3}) \text{ mm/mm} \qquad \textbf{Ans}$$

2–21. A square piece of material is deformed into the dashed position. Determine the shear strain γ_{xy} at corners A and D.

Shear Strain :

$$(\gamma_A)_{xy} = \frac{\pi}{2} - \left(\frac{89.7^\circ}{180^\circ}\right)\pi$$
$$= 5.24(10^{-3}) \text{ rad} \qquad \textbf{Ans}$$

$$(\gamma_D)_{xy} = \frac{\pi}{2} - \left(\frac{90.3^\circ}{180^\circ}\right)\pi$$
$$= -5.24(10^{-3}) \text{ rad} \qquad \textbf{Ans}$$

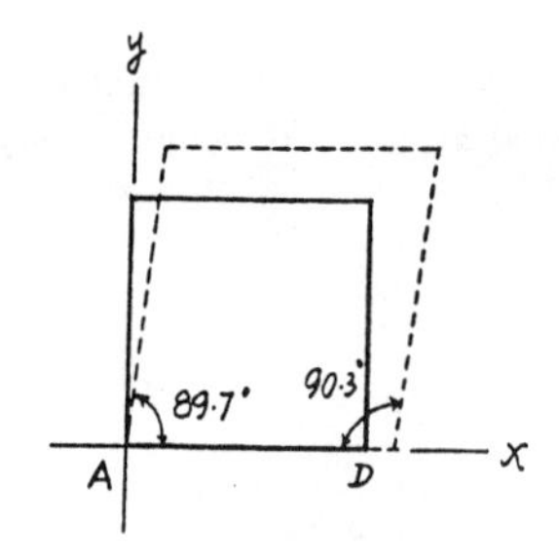

2–22. A square piece of material is deformed into the dashed position. Determine the shear strain γ_{xy} at corners B and C.

Shear Strain :

$$(\gamma_B)_{xy} = \frac{\pi}{2} - \left(\frac{90.3^\circ}{180^\circ}\right)\pi$$
$$= -5.24(10^{-3}) \text{ rad} \qquad \textbf{Ans}$$

$$(\gamma_C)_{xy} = \frac{\pi}{2} - \left(\frac{89.7^\circ}{180^\circ}\right)\pi$$
$$= 5.24(10^{-3}) \text{ rad} \qquad \textbf{Ans}$$

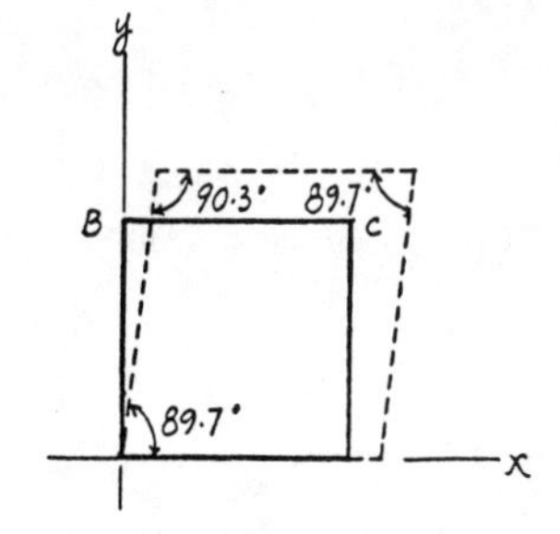

2–23. The square deforms into the position shown by the dashed lines. Determine the average normal strain along each diagonal, AB and CD. Side $D'B'$ remains horizontal.

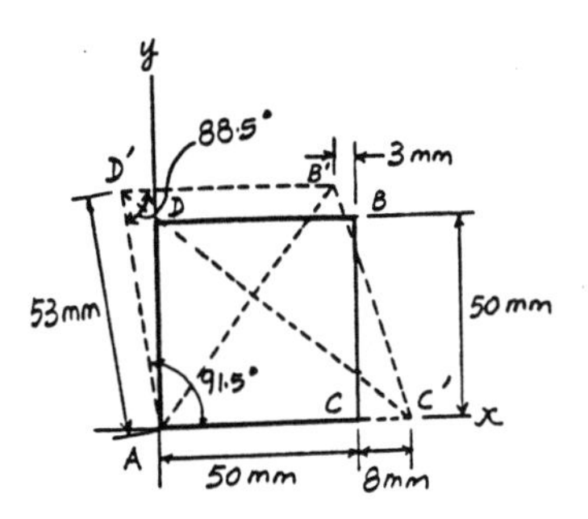

Geometry :

$$AB = CD = \sqrt{50^2 + 50^2} = 70.7107 \text{ mm}$$

$$C'D' = \sqrt{53^2 + 58^2 - 2(53)(58)\cos 91.5^\circ} = 79.5860 \text{ mm}$$

$$B'D' = 50 + 53\sin 1.5^\circ - 3 = 48.3874 \text{ mm}$$

$$AB' = \sqrt{53^2 + 48.3874^2 - 2(53)(48.3874)\cos 88.5^\circ} = 70.8243 \text{ mm}$$

Average Normal Strain :

$$\varepsilon_{AB} = \frac{AB' - AB}{AB} = \frac{70.8243 - 70.7107}{70.7107} = 1.61\left(10^{-3}\right) \text{ mm/mm} \qquad \textbf{Ans}$$

$$\varepsilon_{CD} = \frac{C'D' - CD}{CD} = \frac{79.5860 - 70.7107}{70.7107} = 126\left(10^{-3}\right) \text{ mm/mm} \qquad \textbf{Ans}$$

***2–24.** The square deforms into the position shown by the dashed lines. Determine the shear strain at each of its corners, A, B, C, and D. Side $D'B'$ remains horizontal.

Geometry :

$$B'C' = \sqrt{(8+3)^2 + (53\sin 88.5^\circ)^2} = 54.1117 \text{ mm}$$

$$C'D' = \sqrt{53^2 + 58^2 - 2(53)(58)\cos 91.5^\circ} = 79.5860 \text{ mm}$$

$$B'D' = 50 + 53\sin 1.5^\circ - 3 = 48.3874 \text{ mm}$$

$$\cos\theta = \frac{(B'D')^2 + (B'C')^2 - (C'D')^2}{2(B'D')(B'C')} = \frac{48.3874^2 + 54.1117^2 - 79.5860^2}{2(48.3874)(54.1117)} = -0.20328$$

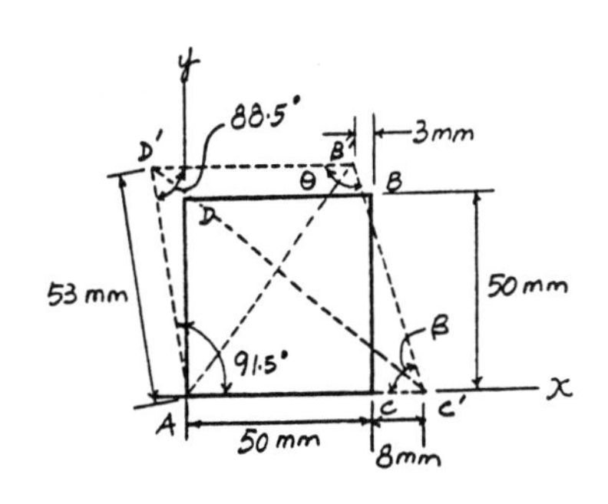

$$\theta = 101.73^\circ$$
$$\beta = 180^\circ - \theta = 78.27^\circ$$

Shear Strain :

$$(\gamma_A)_{xy} = \frac{\pi}{2} - \pi\left(\frac{91.5^\circ}{180^\circ}\right) = -0.0262 \text{ rad} \qquad \textbf{Ans}$$

$$(\gamma_B)_{xy} = \frac{\pi}{2} - \theta = \frac{\pi}{2} - \pi\left(\frac{101.73^\circ}{180^\circ}\right) = -0.205 \text{ rad} \qquad \textbf{Ans}$$

$$(\gamma_C)_{xy} = \frac{\pi}{2} - \beta = \frac{\pi}{2} - \pi\left(\frac{78.27^\circ}{180^\circ}\right) = 0.205 \text{ rad} \qquad \textbf{Ans}$$

$$(\gamma_D)_{xy} = \frac{\pi}{2} - \pi\left(\frac{88.5^\circ}{180^\circ}\right) = 0.0262 \text{ rad} \qquad \textbf{Ans}$$

2–25. The block is deformed into the position shown by the dashed lines. Determine the average normal strain along line AB

Geometry :

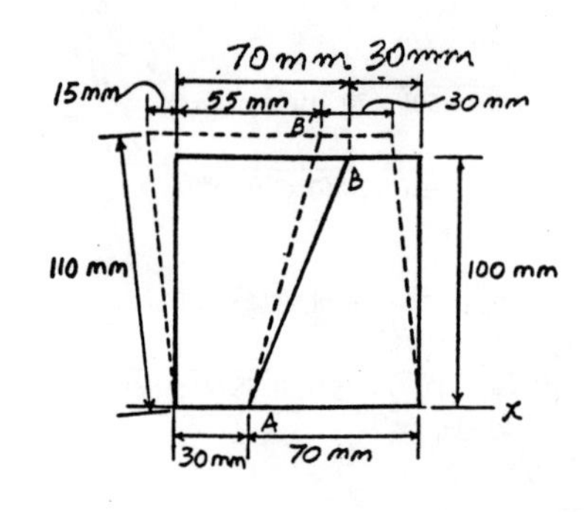

$$AB = \sqrt{100^2 + (70-30)^2} = 107.7033 \text{ mm}$$

$$AB' = \sqrt{(70-30-15)^2 + (110^2 - 15^2)} = 111.8034 \text{ mm}$$

Average Normal Strain :

$$\varepsilon_{AB} = \frac{AB' - AB}{AB}$$
$$= \frac{111.8034 - 107.7033}{107.7033}$$
$$= 0.0381 \text{ mm/mm} = 38.1\left(10^{-3}\right) \text{ mm} \qquad \textbf{Ans}$$

2–26. The block is deformed into the position shown by the dashed lines. Determine the shear strain at corners C and D

Geometry :

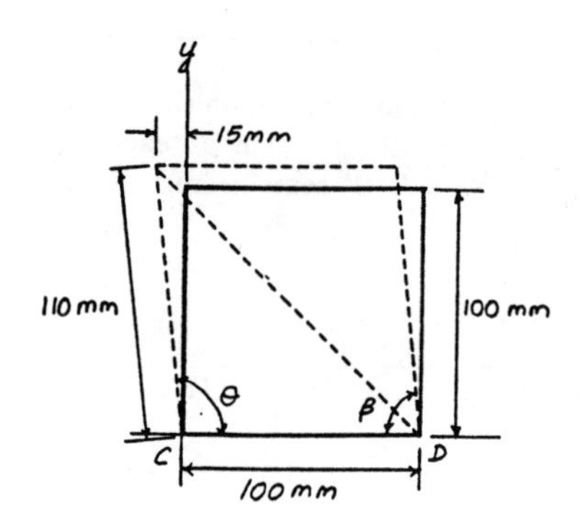

$$\theta = 90° + \sin^{-1}\left(\frac{15}{110}\right) = 97.84° = 1.70759 \text{ rad}$$
$$\beta = \pi - 1.70759 = 1.43401 \text{ rad}$$

Shear Strain :

$$(\gamma_C)_{xy} = \frac{\pi}{2} - \theta = \frac{\pi}{2} - 1.70759 = -0.137 \text{ rad} \qquad \textbf{Ans}$$
$$(\gamma_D)_{xy} = \frac{\pi}{2} - \beta = \frac{\pi}{2} - 1.43401 = 0.137 \text{ rad} \qquad \textbf{Ans}$$

2–27. The bar is originally 300 mm long when it is flat. If it is subjected to a shear strain defined by $\gamma_{xy} = 0.02x$, where x is in millimeters, determine the displacement Δy at the end of its bottom edge. It is distorted into the shape shown, where no elongation of the bar occurs in the x direction.

Shear Strain :

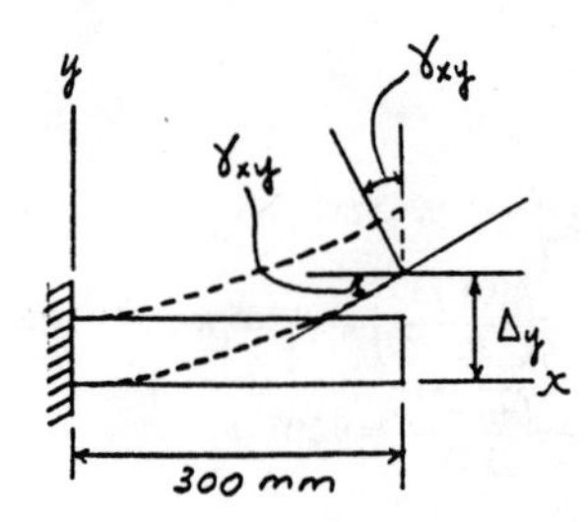

$$\frac{dy}{dx} = \tan \gamma_{xy} ; \qquad \frac{dy}{dx} = \tan(0.02\,x)$$

$$\int_0^{\Delta y} dy = \int_0^{300 \text{ mm}} \tan(0.02\,x)\,dx$$

$$\Delta y = -50[\ln \cos(0.02x)]\big|_0^{300\text{mm}}$$
$$= 2.03 \text{ mm} \qquad \textbf{Ans}$$

*2–28. The rubber band AB has an unstretched length of 1 ft. If it is fixed at B and attached to the surface at point A', determine the average normal strain in the band. The surfac is defined by the function $y = (x^2)$ ft, where x is in feet.

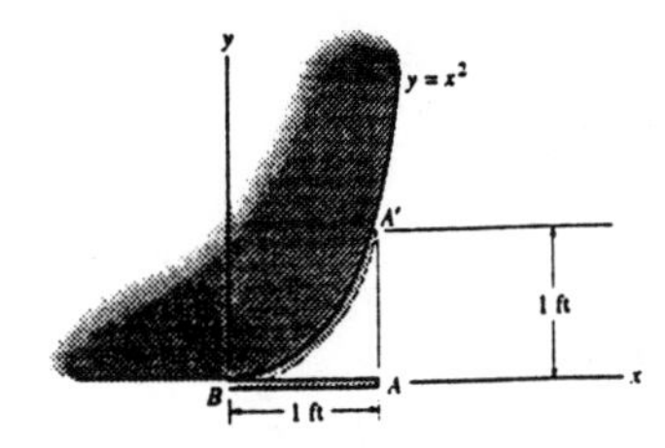

Geometry :

$$L = \int_0^{1\text{ ft}} \sqrt{1 + \left(\frac{dy}{dx}\right)^2}\, dx$$

However $\quad y = x^2 \quad$ then $\quad \dfrac{dy}{dx} = 2x$

$$L = \int_0^{1\text{ ft}} \sqrt{1 + 4x^2}\, dx$$

$$= \frac{1}{4}\left[2x\sqrt{1+4x^2} + \ln\left(2x + \sqrt{1+4x^2}\right)\right]\Bigg|_0^{1\text{ ft}}$$

$$= 1.47894 \text{ ft}$$

Average Normal Strain :

$$\varepsilon_{avg} = \frac{L - L_0}{L_0} = \frac{1.47894 - 1}{1} = 0.479 \text{ ft/ft} \qquad \textbf{Ans}$$

2–29. A thin wire is wrapped along a surface having the form $y = 0.5x^2$, where x and y are in inches. Originally the end B is at $x = 10$ in. If the wire undergoes a normal strain along its length of $\epsilon = 0.005x$, determine the change in length of the wire. Hint: For the curve, $y = f(x)$, $ds = \sqrt{1 + (dy/dx)^2}\, dx$.

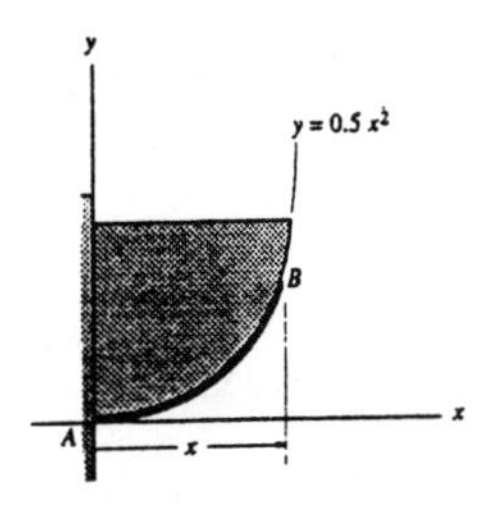

Normal Strain :

$\delta = \varepsilon\, ds \quad$ where $\quad \varepsilon = 0.005x$ and $ds = \sqrt{1 + \left(\dfrac{dy}{dx}\right)^2}\, dx$

However, $\quad y = 0.5x^2 \quad$ Then $\quad \dfrac{dy}{dx} = x$

$$\Delta = \int \varepsilon\, ds = \int_0^{10\text{ in.}} (0.005x)\sqrt{1+x^2}\,dx$$

$$= 0.005\int_0^{10\text{ in.}} x\sqrt{1+x^2}\,dx$$

$$= \frac{0.005}{3}\left[\left(1+x^2\right)^{\frac{3}{2}}\right]\Bigg|_0^{10\text{ in.}}$$

$$= 1.69 \text{ in.} \qquad \textbf{Ans}$$

2–30. The fiber AB has a length L and orientation θ. If its ends A and B undergo very small displacements u_A and v_B, respectively, determine the normal strain in the fiber when it is in position $A'B'$.

Geometry :

$$L_{A'B'} = \sqrt{(L\cos\theta - u_A)^2 + (L\sin\theta + v_B)^2}$$
$$= \sqrt{L^2 + u_A^2 + v_B^2 + 2L(v_B\sin\theta - u_A\cos\theta)}$$

Average Normal strain :

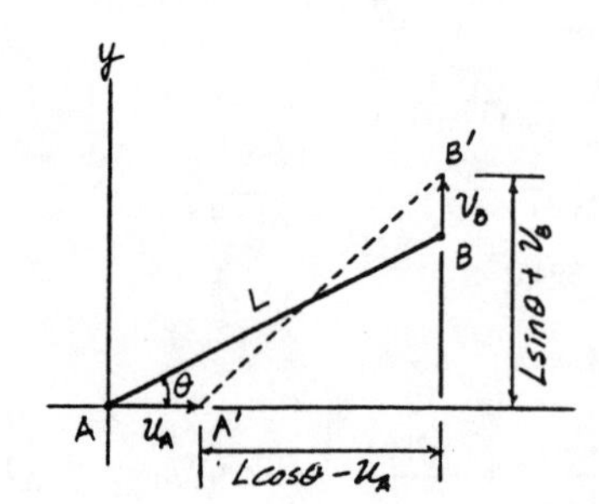

$$\varepsilon_{AB} = \frac{L_{A'B'} - L}{L}$$
$$= \sqrt{1 + \frac{u_A^2 + v_B^2}{L^2} + \frac{2(v_B\sin\theta - u_A\cos\theta)}{L}} - 1$$

Neglecting higher terms u_A^2 and v_B^2

$$\varepsilon_{AB} = \left[1 + \frac{2(v_B\sin\theta - u_A\cos\theta)}{L}\right]^{\frac{1}{2}} - 1$$

Using the binomial theorem :

$$\varepsilon_{AB} = 1 + \frac{1}{2}\left(\frac{2v_B\sin\theta}{L} - \frac{2u_A\cos\theta}{L}\right) + \dots - 1$$
$$= \frac{v_B\sin\theta}{L} - \frac{u_A\cos\theta}{L} \qquad \textbf{Ans}$$

2–31. If the normal strain is defined in reference to the final length, that is,

$$\epsilon_n' = \lim_{P\to P'}\left(\frac{\Delta s' - \Delta s}{\Delta s'}\right)$$

instead of in reference to the original length, Eq. 2–2, show that the difference in these strains is represented as a second-order term, namely, $\epsilon_n - \epsilon_n' = \epsilon_n\epsilon_n'$.

$$\varepsilon_n = \frac{\Delta S' - \Delta S}{\Delta S}$$

$$\varepsilon_n - \varepsilon_n' = \frac{\Delta S' - \Delta S}{\Delta S} - \frac{\Delta S' - \Delta S}{\Delta S'}$$
$$= \frac{\Delta S'^2 - \Delta S\Delta S' - \Delta S'\Delta S + \Delta S^2}{\Delta S\Delta S'}$$
$$= \frac{\Delta S'^2 + \Delta S^2 - 2\Delta S'\Delta S}{\Delta S\Delta S'}$$
$$= \frac{(\Delta S' - \Delta S)^2}{\Delta S\Delta S'} = \left(\frac{\Delta S' - \Delta S}{\Delta S}\right)\left(\frac{\Delta S' - \Delta S}{\Delta S'}\right)$$
$$= \varepsilon_n\varepsilon_n' \quad (Q.E.D)$$

3–1. A concrete cylinder having a diameter of 6.00 in. and gauge length of 12 in. is tested in compression. The results of the test are reported in the table as load versus contraction. Draw the stress–strain diagram using scales of 1 in. = 0.5 ksi and 1 in. = $0.2(10^{-3})$ in./in. From the diagram, determine approximately the modulus of elasticity.

Load (kip)	Contraction (in.)
0	0
5.0	0.0006
9.5	0.0012
16.5	0.0020
20.5	0.0026
25.5	0.0034
30.0	0.0040
34.5	0.0045
38.5	0.0050
46.5	0.0062
50.0	0.0070
53.0	0.0075

Stress and Strain :

$$\sigma = \frac{P}{A} \text{ (ksi)} \qquad \varepsilon = \frac{\delta L}{L} \text{ (in./in.)}$$

σ	ε
0	0
0.177	0.00005
0.336	0.00010
0.584	0.000167
0.725	0.000217
0.902	0.000283
1.061	0.000333
1.220	0.000375
1.362	0.000417
1.645	0.000517
1.768	0.000583
1.874	0.000625

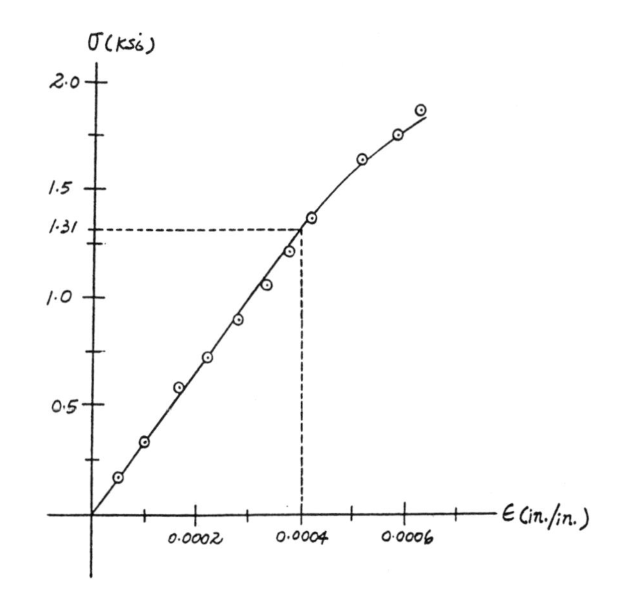

Modulus of Elasticity : From the stress – strain diagram

$$E_{\text{approx}} = \frac{1.31-0}{0.0004-0} = 3.275\left(10^3\right) \text{ ksi} \qquad \textbf{Ans}$$

3–2. Data taken from a stress–strain test for a ceramic is given in the table. The curve is linear between the origin and the first point. Plot the diagram, and determine the modulus of elasticity and the modulus of resilience.

σ (ksi)	ε (in./in.)
0	0
33.2	0.0006
45.5	0.0010
49.4	0.0014
51.5	0.0018
53.4	0.0022

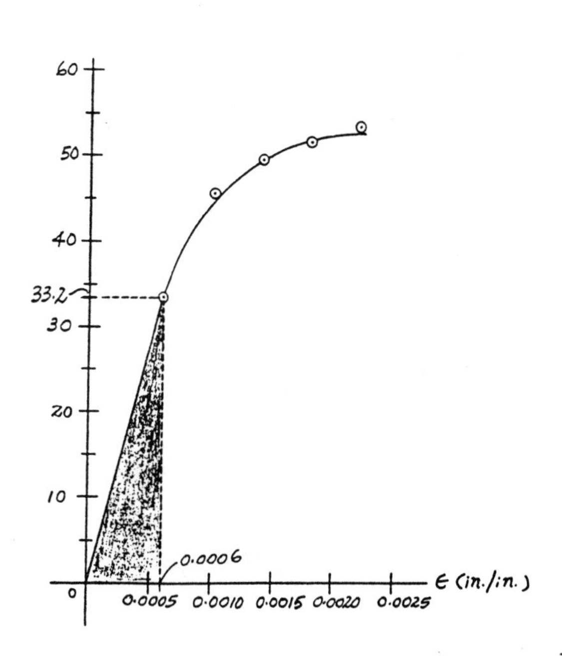

Modulus of Elasticity : From the stress – strain diagram

$$E = \frac{33.2-0}{0.0006-0} = 55.3\left(10^3\right) \text{ ksi} \qquad \textbf{Ans}$$

Modulus of Resilience : The modulus of resilience is equal to the area under the *linear portion* of the stress – strain diagram (shown shaded).

$$u_r = \frac{1}{2}(33.2)\left(10^3\right)\left(\frac{\text{lb}}{\text{in}^2}\right)\left(0.0006\,\frac{\text{in.}}{\text{in.}}\right) = 9.96\,\frac{\text{in}\cdot\text{lb}}{\text{in}^3} \qquad \textbf{Ans}$$

3–3. Data taken from a stress–strain test for a ceramic is given in the table. The curve is linear between the origin and the first point. Plot the diagram, and determine approximately the modulus of toughness. The rupture stress is $\sigma_r = 53.4$ ksi.

σ (ksi)	ϵ (in./in.)
0	0
33.2	0.0006
45.5	0.0010
49.4	0.0014
51.5	0.0018
53.4	0.0022

Modulus of Toughness : The modulus of toughness is equal to the area under the stress – strain diagram (shown shaded).

$$(u_t)_{\text{approx}} = \frac{1}{2}(33.2)(10^3)\left(\frac{\text{lb}}{\text{in}^2}\right)(0.0004+0.0010)\left(\frac{\text{in.}}{\text{in.}}\right)$$

$$+45.5(10^3)\left(\frac{\text{lb}}{\text{in}^2}\right)(0.0012)\left(\frac{\text{in.}}{\text{in.}}\right)$$

$$+\frac{1}{2}(7.90)(10^3)\left(\frac{\text{lb}}{\text{in}^2}\right)(0.0012)\left(\frac{\text{in.}}{\text{in.}}\right)$$

$$+\frac{1}{2}(12.3)(10^3)\left(\frac{\text{lb}}{\text{in}^2}\right)(0.0004)\left(\frac{\text{in.}}{\text{in.}}\right)$$

$$= 85.0\ \frac{\text{in}\cdot\text{lb}}{\text{in}^3} \qquad \textbf{Ans}$$

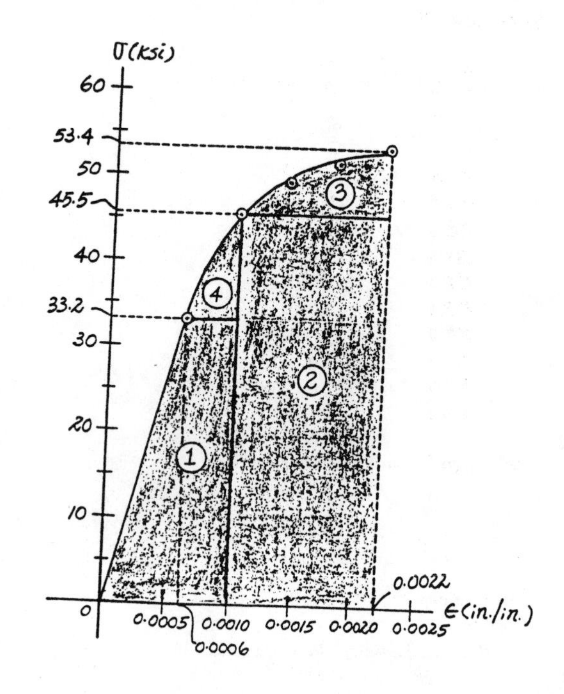

***3–4.** Data taken from a stress–strain test is given in the table. The curve is linear between the origin and the first point. Plot the diagram, and determine the modulus of elasticity and the modulus of resilience.

σ (ksi)	ϵ (in./in.)
0	0
32.0	0.0016
33.5	0.0018
40.0	0.0030
41.2	0.0050

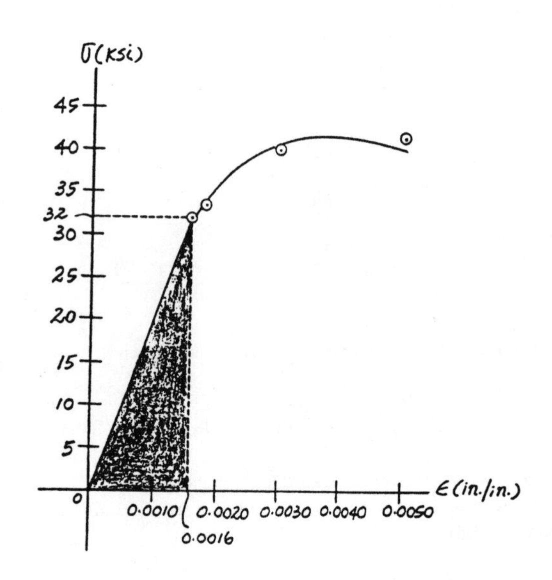

Modulus of Elasticity : From the stress – strain diagram

$$E = \frac{32.0-0}{0.0016-0} = 20.0(10^3)\ \text{ksi} \qquad \textbf{Ans}$$

Modulus of Resilience : The modulus of resilience is equal to the area under the *linear portion* of the stress – strain diagram (shown shaded).

$$u_r = \frac{1}{2}(32.0)(10^3)\left(\frac{\text{lb}}{\text{in}^2}\right)(0.0016)\left(\frac{\text{in.}}{\text{in.}}\right) = 25.6\ \frac{\text{in}\cdot\text{lb}}{\text{in}^3} \qquad \textbf{Ans}$$

3–5. A tension test was performed on a specimen having an original diameter of 12.5 mm and a gauge length of 50 mm. The data is listed in the table. Plot the stress–strain diagram, and determine approximately the modulus of elasticity, the ultimate stress, and the fracture stress. Use a scale of 20 mm = 50 MPa and 20 mm = 0.05 mm/mm. Redraw the linear-elastic region, using the same stress scale but a strain scale of 20 mm = 0.001 mm/mm.

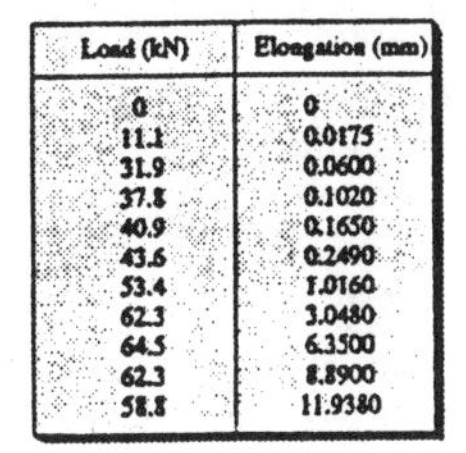

Load (kN)	Elongation (mm)
0	0
11.1	0.0175
31.9	0.0600
37.8	0.1020
40.9	0.1650
43.6	0.2490
53.4	1.0160
62.3	3.0480
64.5	6.3500
62.3	8.8900
58.8	11.9380

Stress and Strain :

$$\sigma = \frac{P}{A}\ \text{(MPa)} \qquad \varepsilon = \frac{\delta L}{L}\ \text{(mm/mm)}$$

σ (MPa)	ε (mm/mm)
0	0
90.45	0.00035
259.9	0.00120
308.0	0.00204
333.3	0.00330
355.3	0.00498
435.1	0.02032
507.7	0.06096
525.6	0.12700
507.7	0.17780
479.1	0.23876

Modulus of Elasticity : From the stress – strain diagram

$$(E)_{\text{approx}} = \frac{228.75(10^3) - 0}{0.001 - 0} = 229\ \text{GPa} \qquad \textbf{Ans}$$

Ultimate and Fracture Stress : From the stress – strain diagram

$$(\sigma_u)_{\text{approx}} = 528\ \text{MPa} \qquad \textbf{Ans}$$

$$(\sigma_f)_{\text{approx}} = 479\ \text{MPa} \qquad \textbf{Ans}$$

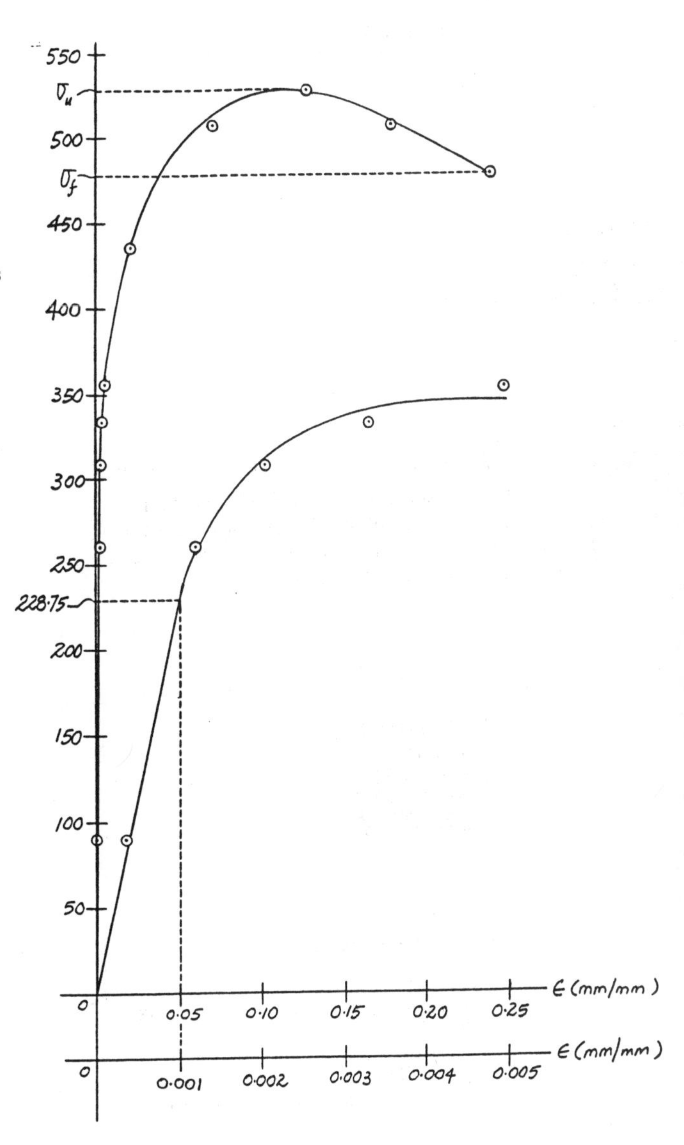

3–6. A tension test was performed on a steel specimen having an original diameter of 12.5 mm and gauge length of 50 mm. Using the data listed in the table, plot the stress–strain diagram, and determine approximately the modulus of toughness. Use a scale of 20 mm = 50 MPa and 20 mm = 0.05 mm/mm.

Load (kN)	Elongation (mm)
0	0
11.1	0.0175
31.9	0.0600
37.8	0.1020
40.9	0.1650
43.6	0.2490
53.4	1.0160
62.3	3.0480
64.5	6.3500
62.3	8.8900
58.8	11.9380

Stress and Strain :

$\sigma = \frac{P}{A}$ (MPa)	$\varepsilon = \frac{\delta L}{L}$ (mm/mm)
0	0
90.45	0.00035
259.9	0.00120
308.0	0.00204
333.3	0.00330
355.3	0.00498
435.1	0.02032
507.7	0.06096
525.6	0.12700
507.7	0.17780
479.1	0.23876

Modulus of Toughness : The modulus of toughness is equal to the total area under the stress – strain diagram and can be approximated by counting the number of squares. The total number of squares is 187.

$$(u_t)_{\text{approx}} = 187(25)\left(10^6\right)\left(\frac{\text{N}}{\text{m}^2}\right)\left(0.025\ \frac{\text{m}}{\text{m}}\right) = 117\ \text{MJ/m}^3 \quad \textbf{Ans}$$

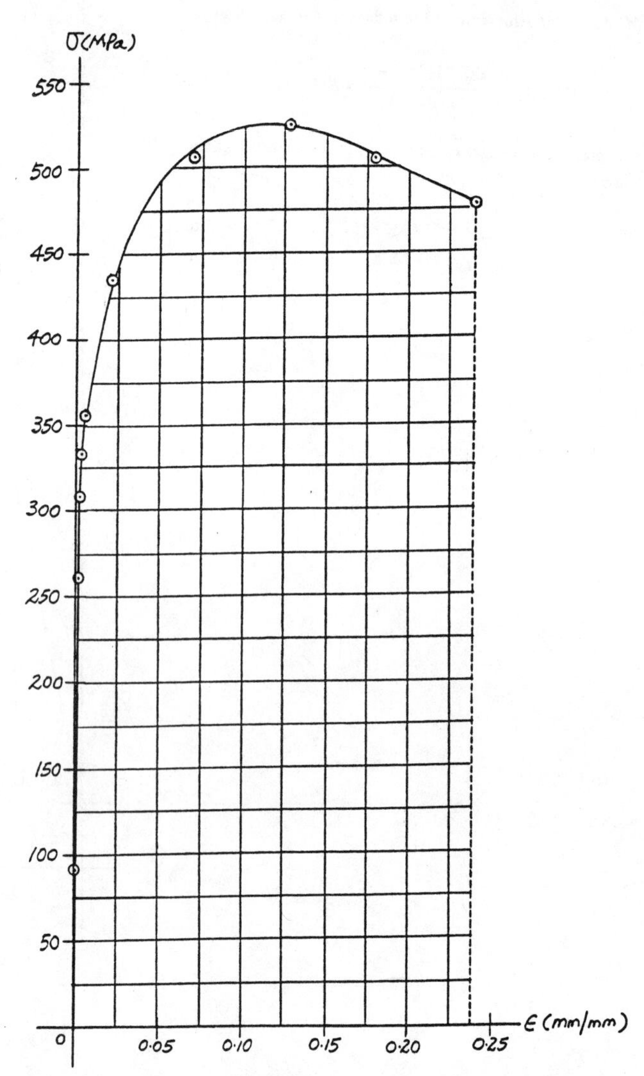

3–7. The stress–strain diagram for a steel alloy having an original diameter of 0.5 in. and a gauge length of 2 in. is given in the figure. Determine approximately the modulus of elasticity for the material, the load on the specimen that causes yielding, and the ultimate load the specimen will support.

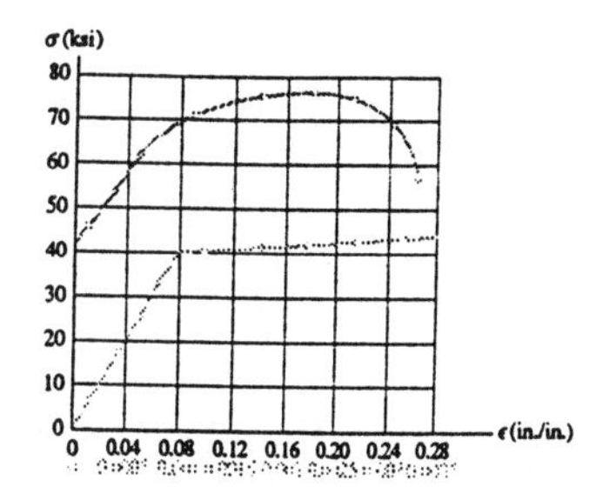

Modulus of Elasticity : From the stress – strain diagram, $\sigma = 40$ ksi when $\varepsilon = 0.001$ in./in.

$$E_{\text{approx}} = \frac{40-0}{0.001-0} = 40.0(10^3)\ \text{ksi} \qquad \textbf{Ans}$$

Yield Load : From the stress – strain diagram, $\sigma_Y = 40.0$ ksi.

$$P_Y = \sigma_Y A = 40.0\left[\left(\frac{\pi}{4}\right)(0.5^2)\right] = 7.85\ \text{kip} \qquad \textbf{Ans}$$

Ultimate Load : From the stress – strain diagram, $\sigma_u = 76.25$ ksi.

$$P_u = \sigma_u A = 76.25\left[\left(\frac{\pi}{4}\right)(0.5^2)\right] = 15.0\ \text{kip} \qquad \textbf{Ans}$$

***3–8.** The stress–strain diagram for a steel alloy having an original diameter of 0.5 in. and a gauge length of 2 in. is given in the figure. If the specimen is loaded until it is stressed to 70 ksi, determine the approximate amount of elastic recovery and the permanent increase in the gauge length after it is unloaded.

Modulus of Elasticity : From the stress – strain diagram, $\sigma = 40$ ksi when $\varepsilon = 0.001$ in./in.

$$E = \frac{40-0}{0.001-0} = 40.0(10^3)\ \text{ksi}$$

Elastic Recovery :

$$\text{Elastic recovery} = \frac{\sigma}{E} = \frac{70}{40.0(10^3)} = 0.00175\ \text{in./in.}$$

Thus,

The amount of Elastic Recovery = 0.00175(2) = 0.00350 in. **Ans**

Permanent Set :

$$\text{Permanent set} = 0.08 - 0.00175 = 0.07825\ \text{in./in.}$$

Thus,

Permanent elongation = 0.07825(2) = 0.1565 in. **Ans**

3–9. The stress–strain diagram for a steel alloy having an original diameter of 0.5 in. and a gauge length of 2 in. is given in the figure. Determine approximately the modulus of resilience and the modulus of toughness for the material.

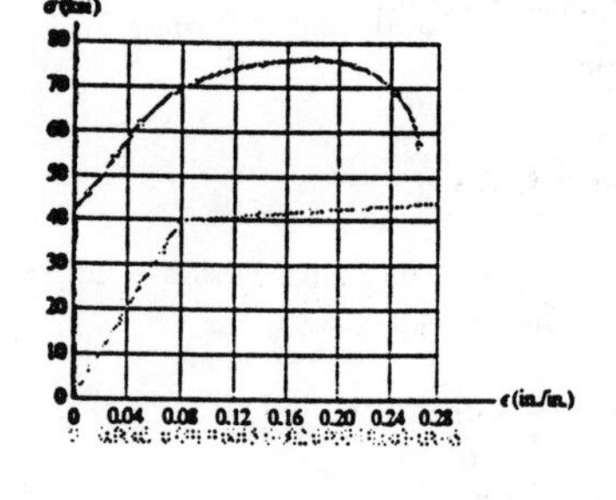

Modulus of Resilience : The modulus of resilience is equal to the area under the *linear portion* of the stress – strain diagram.

$$(u_r)_{\text{approx}} = \frac{1}{2}(40.0)(10^3)\left(\frac{\text{lb}}{\text{in}^2}\right)\left(0.001\ \frac{\text{in.}}{\text{in.}}\right) = 20.0\ \frac{\text{in}\cdot\text{lb}}{\text{in}^3}$$

Modulus of Toughness : The modulus of toughness is equal to the total area under the stress – strain diagram and can be approximated by counting the number of squares. The total number of squares is 45.

$$(u_t)_{\text{approx}} = 45\left(10\ \frac{\text{kip}}{\text{in}^2}\right)\left(0.04\ \frac{\text{in.}}{\text{in.}}\right) = 18.0\ \frac{\text{in}\cdot\text{kip}}{\text{in}^3} \qquad \textbf{Ans}$$

3–10. The stress–strain diagram for a bar of steel alloy is shown in the figure. Determine approximately the modulus of elasticity, the proportional limit, the ultimate stress, and the modulus of resilience. If the bar is loaded until it is stressed to 360 MPa, determine the elastic strain recovery and the permanent set or strain in the bar when it is unloaded.

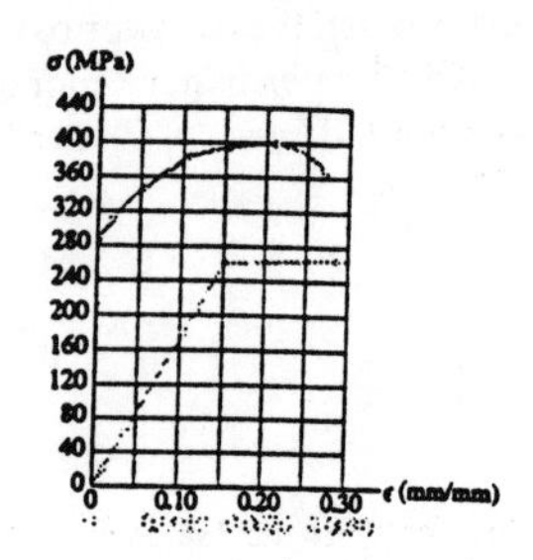

Modulus of Elasticity : From the stress – strain diagram, $\sigma = 260$ MPa when $\varepsilon = 0.0015$ mm/mm.

$$E_{\text{approx}} = \frac{260(10^6) - 0}{0.0015 - 0} = 173.33\ \text{GPa} = 173\ \text{GPa} \qquad \textbf{Ans}$$

Proportional Limit and Ultimate Stress : From the stress – strain diagram

$$\sigma_{pl} = 260\ \text{MPa} \qquad \textbf{Ans}$$
$$\sigma_u = 400\ \text{MPa} \qquad \textbf{Ans}$$

Modulus of Resilience : The modulus of resilience is equal to the area under the *linear portion* of the stress – strain diagram.

$$u_r = \frac{1}{2}(260)(10^6)\left(\frac{\text{N}}{\text{m}^2}\right)\left(0.0015\ \frac{\text{m}}{\text{m}}\right) = 195\ \text{kJ/m}^3 \qquad \textbf{Ans}$$

Elastic Recovery :

$$\text{Elastic recovery} = \frac{\sigma}{E} = \frac{360(10^6)}{173.33(10^9)}$$
$$= 0.002077\ \text{mm/mm}$$
$$= 0.00208\ \text{mm/mm} \qquad \textbf{Ans}$$

Permanent Set : From the stress – strain diagram, the strain is 0.075 mm/mm when the bar is stressed 360 MPa.

$$\text{Permanent set} = 0.075 - 0.002077 = 0.0729\ \text{mm/mm} \qquad \textbf{Ans}$$

3–11. An A-36 steel bar has a length of 50 in. and cross-sectional area of 0.7 in^2. Determine the length of the bar if it is subjected to an axial tension of 5000 lb. The material has linear-elastic behavior.

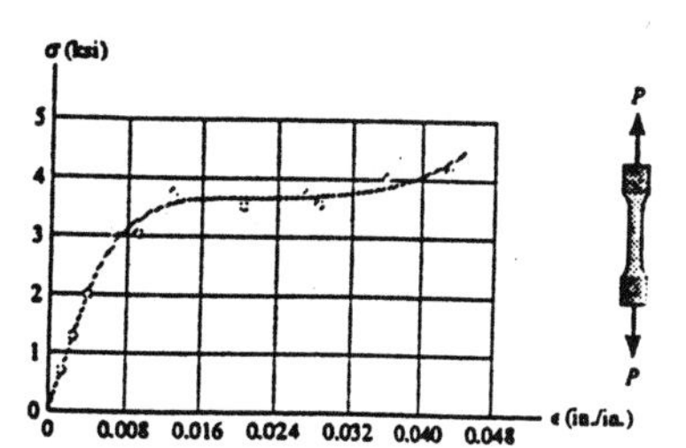

Normal Stress :

$$\sigma = \frac{P}{A} = \frac{5}{0.7} = 7.143 \text{ ksi} < \sigma_Y = 36.0 \text{ ksi}$$

Hence Hooke's law is still valid.

Normal Strain :

$$\varepsilon = \frac{\sigma}{E} = \frac{7.143}{29.0(10^3)} = 0.2463(10^{-3}) \text{ in./in.}$$

Thus,

$$\delta L = \varepsilon L_0 = 0.2463(10^{-3})(50) = 0.01231 \text{ in.}$$

$$L = L_0 + \delta L = 50.0123 \text{ in.} \qquad \textbf{Ans}$$

***3–12.** The stress–strain diagram for polyethylene, which is used to sheath coaxial cables, is determined from testing a specimen that has a gauge length of 10 in. If a load P on the specimen develops a strain of $\epsilon = 0.024$ in./in., determine the approximate length of the specimen, measured between the gauge points, when the load is removed. Assume the specimen recovers elastically.

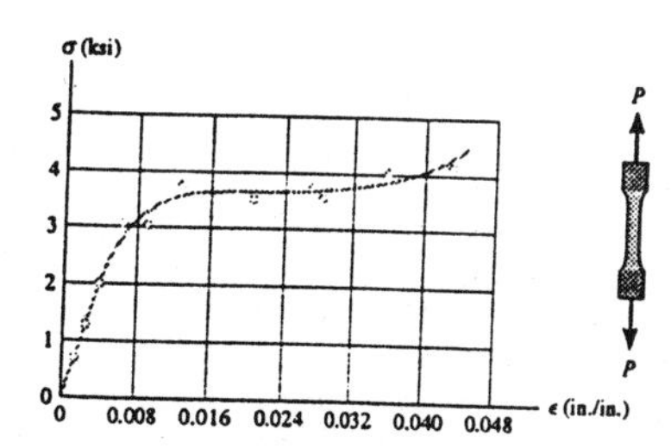

Modulus of Elasticity : From the stress – strain diagram, $\sigma = 2$ ksi when $\varepsilon = 0.004$ in./in..

$$E = \frac{2-0}{0.004-0} = 0.500(10^3) \text{ ksi}$$

Elastic Recovery : From the stress – strain diagram, $\sigma = 3.70$ ksi when $\varepsilon = 0.024$ in./in..

$$\text{Elastic recovery} = \frac{\sigma}{E} = \frac{3.70}{0.500(10^3)} = 0.00740 \text{ in./in.}$$

Permanent Set :

$$\text{Permanent set} = 0.024 - 0.00740 = 0.0166 \text{ in./in.}$$

Thus,

$$\text{Permanent elongation} = 0.0166(10) = 0.166 \text{ in.}$$

$$\begin{aligned} L &= L_0 + \text{permanent elongation} \\ &= 10 + 0.166 \\ &= 10.17 \text{ in.} \end{aligned} \qquad \textbf{Ans}$$

3–13. The change in weight of an airplane is determined from reading the strain gauge A mounted in the plane's aluminum wheel strut. *Before* the plane is loaded, the strain-gauge reading in a strut is $\epsilon_1 = 0.00100$ in./in., whereas after loading $\epsilon_2 = 0.00243$ in./in. Determine the change in the force on the strut if the cross-sectional area of the strut is 3.5 in². $E_{al} = 10(10^3)$ ksi.

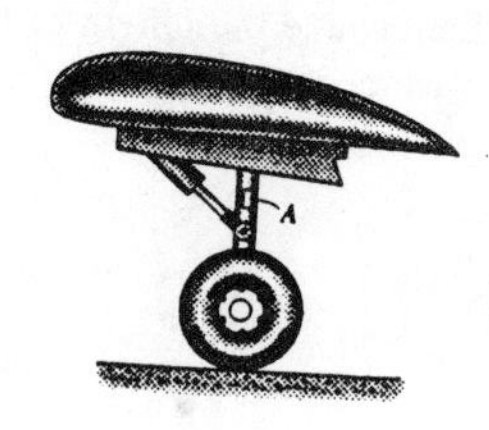

Stress - Strain Relationship : Applying Hooke's law $\sigma = E\varepsilon$.

$$\sigma_1 = 10(10^3)(0.00100) = 10.0 \text{ ksi}$$
$$\sigma_2 = 10(10^3)(0.00243) = 24.3 \text{ ksi}$$

Normal Force : Applying equation $\sigma = \frac{P}{A}$.

$$P_1 = 10.0(3.5) = 35.0 \text{ kip}$$
$$P_2 = 24.3(3.5) = 85.05 \text{ kip}$$

$$\Delta P = P_2 - P_1 = 85.05 - 35.0 = 50.0 \text{ kip} \qquad \textbf{Ans}$$

3–14. By adding plasticizers to polyvinyl chloride it is possible to reduce its stiffness. The stress–strain diagrams for three types of this material showing this effect are given below. Specify the type that should be used in the manufacture of a rod having a length of 5 in. and a diameter of 2 in., that is required to support at least an axial load of 20 kip and also be able to stretch at most $\frac{1}{4}$ in.

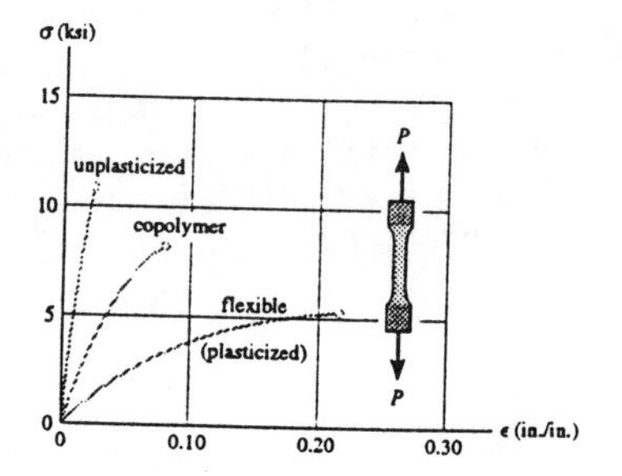

Normal Stress :

$$\sigma = \frac{P}{A} = \frac{20}{\frac{\pi}{4}(2^2)} = 6.366 \text{ ksi}$$

Normal Strain :

$$\varepsilon = \frac{0.25}{5} = 0.0500 \text{ in./in.}$$

From the stress – strain diagram, the ***copolymer*** will satisfy both stress and strain requirements. **Ans**

3–15. A bar having a length of 5 in. and cross-sectional area of 0.7 in² is subjected to an axial force of 8000 lb. If the bar stretches 0.002 in., determine the modulus of elasticity of the material. The material has linear-elastic behavior.

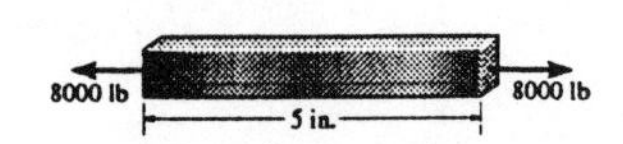

Normal Stress and Strain :

$$\sigma = \frac{P}{A} = \frac{8.00}{0.7} = 11.43 \text{ ksi}$$
$$\varepsilon = \frac{\delta L}{L} = \frac{0.002}{5} = 0.000400 \text{ in./in.}$$

Modulus of Elasticity :

$$E = \frac{\sigma}{\varepsilon} = \frac{11.43}{0.000400} = 28.6(10^3) \text{ ksi} \qquad \textbf{Ans}$$

***3–16.** A structural member in a nuclear reactor is made of a zirconium alloy. If an axial load of 4 kip is to be supported by the member, determine its required cross-sectional area. Use a factor of safety of 3 relative to yielding. What is the load on the member if it is 3 ft long and its elongation is 0.02 in.? $E_{zr} = 14(10^3)$ ksi, $\sigma_Y = 57.5$ ksi. The material has elastic behavior.

Allowable Normal Stress :

$$\text{F.S.} = \frac{\sigma_y}{\sigma_{\text{allow}}}$$

$$3 = \frac{57.5}{\sigma_{\text{allow}}}$$

$$\sigma_{\text{allow}} = 19.17 \text{ ksi}$$

$$\sigma_{\text{allow}} = \frac{P}{A}$$

$$19.17 = \frac{4}{A}$$

$$A = 0.2087 \text{ in}^2 = 0.209 \text{ in}^2 \qquad \textbf{Ans}$$

Stress - Strain Relationship : Applying Hooke's law with

$$\varepsilon = \frac{\delta}{L} = \frac{0.02}{3\,(12)} = 0.000555 \text{ in./in.}$$

$$\sigma = E\varepsilon = 14\left(10^3\right)(0.000555) = 7.778 \text{ ksi}$$

Normal Force : Applying equation $\sigma = \frac{P}{A}$.

$$P = \sigma A = 7.778\,(0.2087) = 1.62 \text{ kip} \qquad \textbf{Ans}$$

3–17. A specimen is originally 1 ft long, has a diameter of 0.5 in., and is subjected to a force of 500 lb. When the force is increased from 500 lb to 1800 lb, the specimen elongates 0.009 in. Determine the modulus of elasticity for the material if it remains linear elastic.

Normal Stress and Strain : Applying $\sigma = \frac{P}{A}$ and $\varepsilon = \frac{\delta L}{L}$.

$$\sigma_1 = \frac{0.500}{\frac{\pi}{4}(0.5^2)} = 2.546 \text{ ksi}$$

$$\sigma_2 = \frac{1.80}{\frac{\pi}{4}(0.5^2)} = 9.167 \text{ ksi}$$

$$\Delta\varepsilon = \frac{0.009}{12} = 0.000750 \text{ in./in.}$$

Modulus of Elasticity :

$$E = \frac{\Delta\sigma}{\Delta\varepsilon} = \frac{9.167 - 2.546}{0.000750} = 8.83\left(10^3\right) \text{ ksi} \qquad \textbf{Ans}$$

3–18. The steel wires AB and AC support the 200-kg mass. If the allowable axial stress for the wires is $\sigma_{allow} = 130$ MPa, determine the required diameter of each wire. Also, what is the new length of wire AB after the load is applied? Take the unstretched length of AB to be 750 mm. $E_{st} = 200$ GPa.

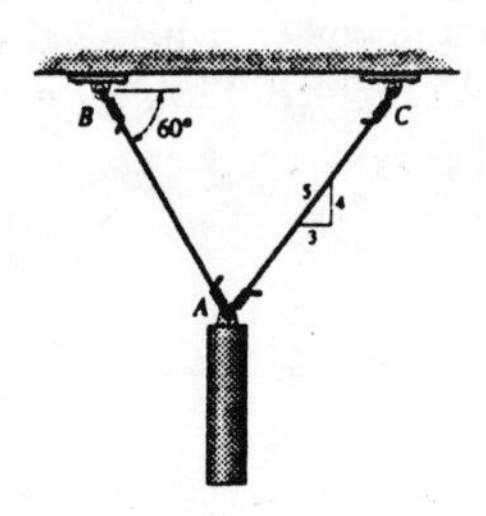

Axial Force : The axial forces exerted by wires AB and AC are shown on FBD.

Allowable Normal Stress :

For wire AB

$$\sigma_{allow} = 130(10^6) = \frac{1280.10}{\frac{\pi}{4}(d_{AB}^2)}$$

$$d_{AB} = 0.003541 \text{ m} = 3.54 \text{ mm} \qquad \textbf{Ans}$$

For wire AC

$$\sigma_{allow} = 130(10^6) = \frac{1066.75}{\frac{\pi}{4}(d_{AC}^2)}$$

$$d_{AC} = 0.003232 \text{ m} = 3.23 \text{ mm} \qquad \textbf{Ans}$$

Stress – Strain Relationship : Applying Hooke's law

$$\varepsilon_{AB} = \frac{\sigma}{E} = \frac{130(10^6)}{200(10^9)} = 0.000650 \text{ mm/mm}$$

Thus,

$$L_{AB} = (L_{AB})_0 + \varepsilon_{AB}(L_{AB})_0$$
$$= 750 + 750(0.000650) = 750.49 \text{ mm} \qquad \textbf{Ans}$$

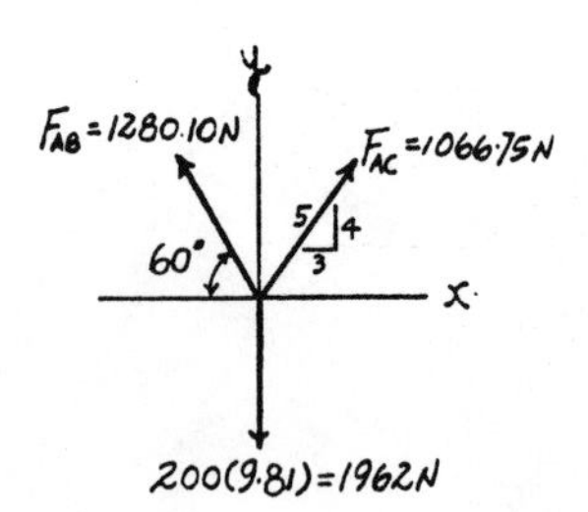

3–19. The bar DBA is rigid and is originally held in the horizontal position. When the weight W is supported from D,it causes the end D to displace downward 0.025 in. Determine the strain in wires CD and BE. Also, if the wires are made of A-36 steel and have a cross-sectional area of 0.002 in^2, determine the weight W

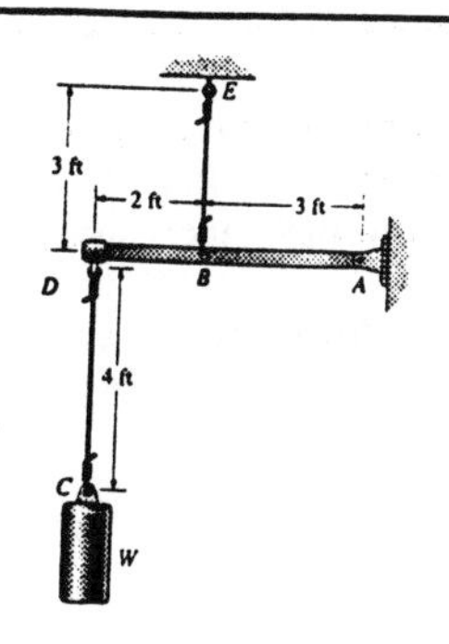

Normal Stress and Strain : For wire BE

$$\varepsilon_{BE} = \frac{\delta_B}{L_{BE}} = \frac{0.015}{3(12)} = 0.0004167 \text{ in./in.} = 0.000417 \text{ in./in.} \quad \textbf{Ans}$$

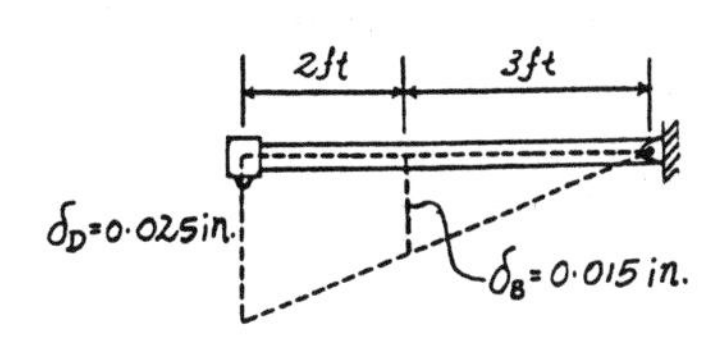

Applying Hooke's law

$$\sigma_{BE} = E\varepsilon_{BE} = 29(10^3)(0.0004167) = 12.083 \text{ ksi}$$

Thus,

$$F_{BE} = \sigma_{BE}A_{BE} = 12.083(0.002) = 0.02417 \text{ kip} = 24.17 \text{ lb}$$

Equations of Equilibrium : From FBD (a)

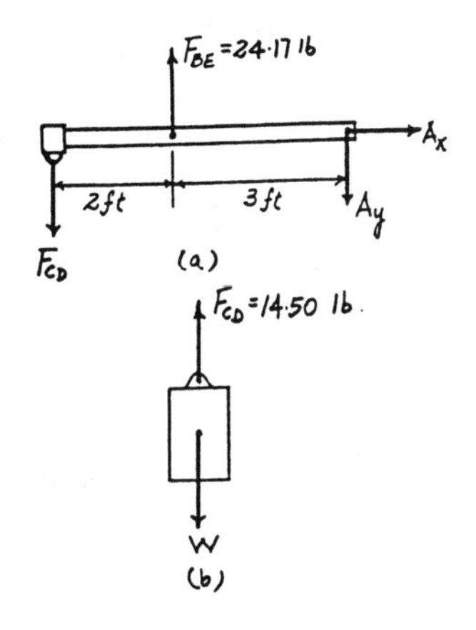

$$\circlearrowleft +\Sigma M_A = 0; \quad F_{CD}(5) - 24.17(3) = 0 \quad F_{CD} = 14.50 \text{ lb}$$

From FBD (b)

$$+\uparrow \Sigma F_y = 0; \quad W - 14.50 = 0 \quad W = 14.5 \text{ lb} \quad \textbf{Ans}$$

Normal Stress and Strain : For wire CD

$$\sigma_{CD} = \frac{F_{CD}}{A_{CD}} = \frac{14.5}{0.002} = 7.25 \text{ ksi}$$

Applying Hooke's law

$$\varepsilon_{CD} = \frac{\sigma_{CD}}{E} = \frac{7.25}{29(10^3)} = 0.000250 \text{ in./in.} \quad \textbf{Ans}$$

***3–20.** The stress–strain diagram for many metal alloys can be described analytically using the Ramberg-Osgood three parameter equation $\epsilon = \sigma/E + k\sigma^n$, where E, k, and n are determined from measurements taken from the diagram. Using the stress–strain diagram shown in the figure, take $E = 30(10^3)$ ksi and determine the other two parameters k and n, and thereby obtain an analytical expression for the curve.

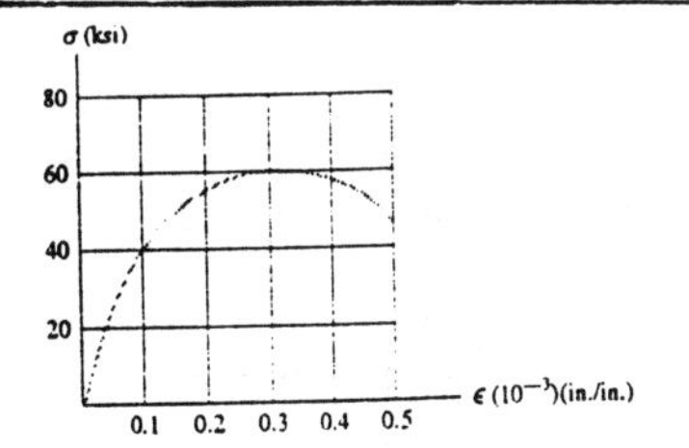

Stress – Strain Relationship : From the stress – strain diagram, when $\sigma = 40$ ksi, $\varepsilon = 0.1(10^{-3})$ in./in. and $\sigma = 60$ ksi, $\varepsilon = 0.3(10^{-3})$ in./in. Substituting the above data into the Ramberg – Osgood equation, we have

$$0.1(10^{-3}) = \frac{40}{30(10^3)} + k(40^n)$$

$$-0.0012333 = k(40^n) \quad [1]$$

$$0.3(10^{-3}) = \frac{60}{30(10^3)} + k(60^n)$$

$$-0.001700 = k(60^n) \quad [2]$$

Solving Eq. [1] and [2] yields :

$$n = 0.791 \qquad k = -66.5(10^{-6}) \quad \textbf{Ans}$$

3–21. Direct tension indicators are sometimes used instead of torque wrenches to insure that a bolt has a prescribed tension when used for connections. If a nut on the bolt is tightened so that the six heads of the indicator that were originally 3 mm high, are crushed 0.3 mm, leaving a contact area on each head of 1.5 mm², determine the tension in the bolt shank. The material has the stress–strain diagram shown.

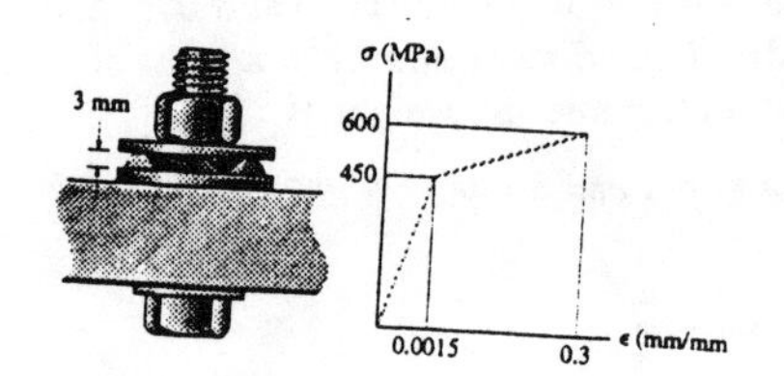

Stress – Strain Relatioship : From the stress – strain diagram with

$$\varepsilon = \frac{0.3}{3} = 0.1 \text{ mm/mm} > 0.0015 \text{ mm/mm}$$

$$\frac{\sigma - 450}{0.1 - 0.0015} = \frac{600 - 450}{0.3 - 0.0015}$$

$$\sigma = 499.497 \text{ MPa}$$

Axial Force : For each head

$$P = \sigma A = 499.4971\left(10^{6}\right)(1.5)\left(10^{-6}\right) = 749.24 \text{ N}$$

Thus, the tension in the bolt is

$$T = 6P = 6(749.24) = 4495 \text{ N} = 4.50 \text{ kN} \qquad \textbf{Ans}$$

3–22. The $\sigma - \epsilon$ diagram for elastic fibers that make up human skin and muscle is shown. Determine the modulus of elasticity of the fibers, and estimate their modulus of toughness and modulus of resilience.

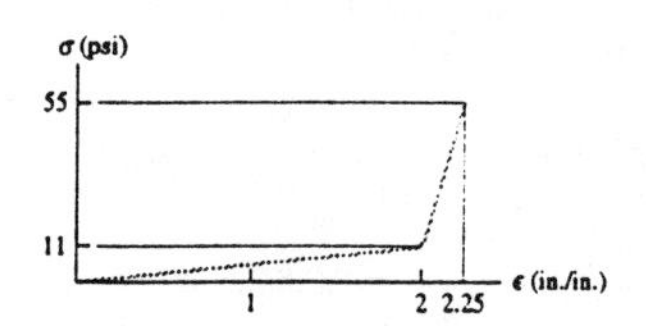

Modulus of Elasticity : From the stress – strain diagram, when $\varepsilon = 2$ in./in., $\sigma = 11$ psi

$$E = \frac{11 - 0}{2 - 0} = 5.50 \text{ psi} \qquad \textbf{Ans}$$

Modulus of Resilience : The modulus of resilience is equal to the area under the *linear portion* of the stress – strain diagram.

$$u_r = \frac{1}{2}(11)\left(\frac{\text{lb}}{\text{in}^2}\right)(2)\left(\frac{\text{in.}}{\text{in.}}\right) = 11.0\,\frac{\text{in.}\cdot\text{lb}}{\text{in}^3} \qquad \textbf{Ans}$$

Modulus of Toughness : The modulus of toughness is equal to the total area under the stress – strain diagram.

$$u_t = \frac{1}{2}(11)\left(\frac{\text{lb}}{\text{in}^2}\right)(2)\left(\frac{\text{in.}}{\text{in.}}\right) + \frac{1}{2}(11 + 55)\left(\frac{\text{lb}}{\text{in}^2}\right)(0.25)\left(\frac{\text{in.}}{\text{in.}}\right)$$

$$= 19.25\,\frac{\text{in.}\cdot\text{lb}}{\text{in}^3} \qquad \textbf{Ans}$$

3–23. The A-36 steel wire AB has a cross-sectional area of $10\ \text{mm}^2$ and is unstretched when $\theta = 60°$. Determine the applied load P needed to cause $\theta = 59.95°$.

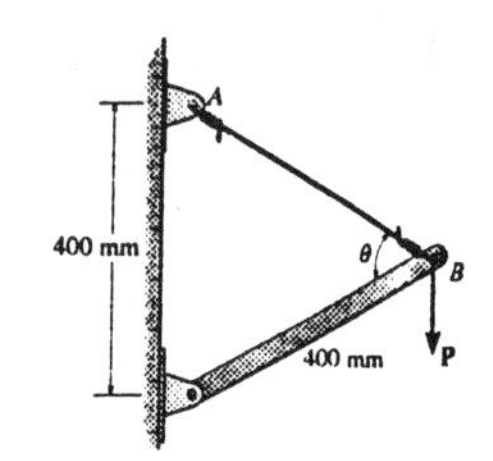

Geometry : From triangle $AB'C$

$$\frac{L_{AB'}}{\sin 60.10°} = \frac{400}{\sin 59.95°} \qquad L_{AB'} = 400.6044\ \text{mm}$$

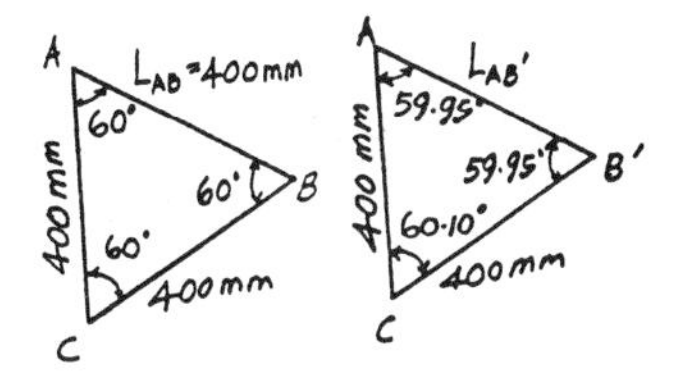

Normal Strain :

$$\varepsilon_{AB} = \frac{L_{AB'} - L_{AB}}{L_{AB}} = \frac{400.6044 - 400}{400}$$
$$= 0.001511\ \text{mm/mm}$$

Applying Hooke's law

$$\sigma_{AB} = E\,\varepsilon_{AB} = 200(10^9)\,(0.001511) = 302.22\ \text{MPa}$$

Thus,

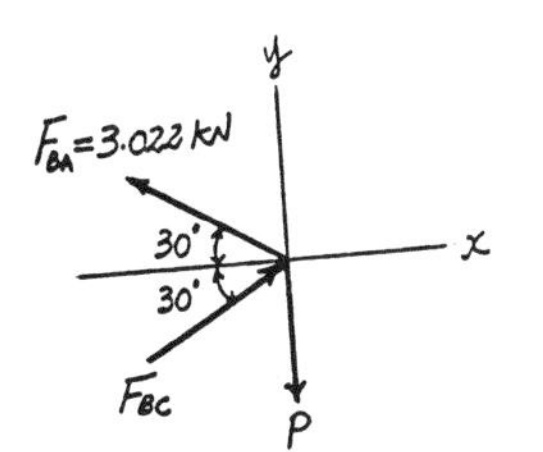

$$F_{AB} = \sigma_{AB} A_{AB} = 302.22(10^6)\,(10)\,(10^{-6})$$
$$= 3022.24\ \text{N} = 3.022\ \text{kN}$$

Equations of Equilibrium :

$$\overset{+}{\rightarrow} \Sigma F_x = 0; \qquad F_{BC} \cos 30° - 3.022 \cos 30° = 0$$
$$F_{BC} = 3.022\ \text{kN}$$

$$+\uparrow \Sigma F_y = 0; \qquad 3.022 \sin 30° + 3.022 \sin 30° - P = 0$$

$$P = 3.02\ \text{kN} \qquad \textbf{Ans}$$

***3–24.** The brake pads for a bicycle tire are made of rubber. If a frictional force of 50 N is applied to each side of the tires, determine the average shear strain in the rubber. Each pad has cross-sectional dimensions of 20 mm and 50 mm. $G_r = 0.20$ MPa.

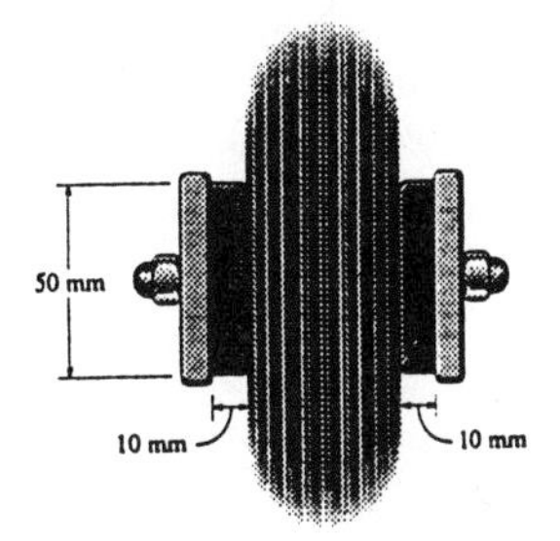

Average Shear Stress : The shear force is $V = 50$ N.

$$\tau = \frac{V}{A} = \frac{50}{0.02(0.05)} = 50.0\ \text{kPa}$$

Shear Stress – Strain Relationship : Applying Hooke's law for shear

$$\tau = G\,\gamma$$
$$50.0(10^3) = 0.2(10^6)\,\gamma$$
$$\gamma = 0.250\ \text{rad} \qquad \textbf{Ans}$$

3–25. The plastic rod is made of Kevlar 49 and has a diameter of 10 mm. If an axial load of 80 kN is applied to it, determine the change in its length and the change in diameter.

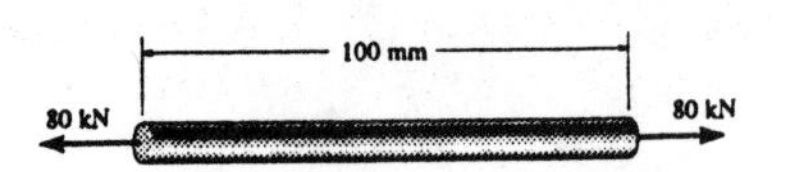

Normal Stress :

$$\sigma = \frac{P}{A} = \frac{80(10^3)}{\frac{\pi}{4}(0.01^2)} = 1.0186 \text{ GPa}$$

Normal Strain : Applying Hooke's law

$$\varepsilon = \frac{\sigma}{E} = \frac{1.0186(10^9)}{131(10^9)} = 7.7755\left(10^{-3}\right) \text{ mm/mm}$$

Thus,

$$\delta L = \varepsilon L = 7.7755\left(10^{-3}\right)(100) = 0.778 \text{ mm} \qquad \textbf{Ans}$$

Poisson's Ratio : The lateral and longitudinal strain can be related using Poisson's ratio.

$$\varepsilon_{\text{lat}} = -\nu\varepsilon_{\text{long}} = -0.34(7.7755)\left(10^{-3}\right)$$
$$= -2.64367\left(10^{-3}\right) \text{ mm/mm}$$

Thus,

$$\delta d = \varepsilon_{\text{lat}} d = -2.6436\left(10^{-3}\right)(10) = -0.0264 \text{ mm} \qquad \textbf{Ans}$$

3–26. A short cylindrical block of bronze C86100, having an original diameter of 1.5 in. and a length of 3 in., is placed in a compression machine and squeezed until its length becomes 2.98 in. Determine the new diameter of the block.

Normal Strain :

$$\varepsilon = \frac{L - L_0}{L_0} = \frac{2.98 - 3}{3} = -0.006667 \text{ in./in.}$$

Poisson's Ratio : The lateral and longitudinal can be related using poisson's ratio.

$$\varepsilon_{\text{lat}} = -\nu\varepsilon_{\text{long}} = -0.34(-0.006667)$$
$$= 0.002267 \text{ in./in.}$$

Thus,

$$\delta d = \varepsilon_{\text{lat}} d = 0.002267(1.5) = 0.00340 \text{ in}$$

$$d = d_0 + \delta d = 1.5 + 0.00340 = 1.5034 \text{ in} \qquad \textbf{Ans}$$

3–27. A short cylindrical block of 6061-T6 aluminum, having an original diameter of 20 mm and a length of 75 mm, is placed in a compression machine and squeezed until the axial load applied is 5 kN. Determine (a) the decrease in its length and (b) its new diameter.

a) ***Normal Stress :***

$$\sigma = \frac{P}{A} = \frac{-5(10^3)}{\frac{\pi}{4}(0.02^2)} = -15.915 \text{ MPa}$$

Normal Strain : Applying Hooke's law

$$\varepsilon = \frac{\sigma}{E} = \frac{-15.915(10^9)}{68.9(10^9)} = -0.2310(10^{-3}) \text{ mm/mm}$$

Thus,

$$\delta L = \varepsilon L = -0.2310(10^{-3})(75) = -0.0173 \text{ mm} \qquad \textbf{Ans}$$

b) ***Poisson's Ratio :*** The lateral and longitudinal strain can be related using Poisson's ratio.

$$\varepsilon_{lat} = -\nu\varepsilon_{long} = -0.35(-0.2310)(10^{-3})$$
$$= 80.848(10^{-6}) \text{ mm/mm}$$

Thus,

$$\delta d = \varepsilon_{lat} d = 80.848(10^{-6})(20) = 0.001617 \text{ mm}$$

$$d = d_0 + \delta d = 20 + 0.001617 = 20.00162 \text{ mm} \qquad \textbf{Ans}$$

***3–28.** The block is made of titanium Ti-6A1-4V and is subjected to a compression of 0.06 in. along the y axis, and its shape is given a tilt of $\theta = 89.7°$. Determine ϵ_x, ϵ_y, and γ_{xy}.

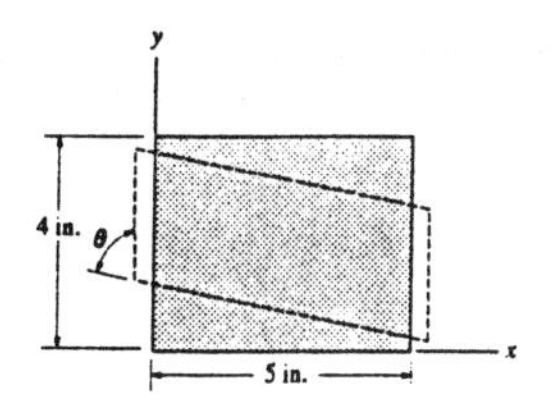

Normal Strain :

$$\varepsilon_y = \frac{\delta L_y}{L_y} = \frac{-0.06}{4} = -0.0150 \text{ in./in.} \qquad \textbf{Ans}$$

Poisson's Ratio : The lateral and longitudinal strain can be related using Poisson's ratio.

$$\varepsilon_x = -\nu\varepsilon_y = -0.36(-0.0150)$$
$$= 0.00540 \text{ in./in.} \qquad \textbf{Ans}$$

Shear Strain :

$$\beta = 180° - 89.7° = 90.3° = 1.576032 \text{ rad}$$

$$\gamma_{xy} = \frac{\pi}{2} - \beta = \frac{\pi}{2} - 1.576032 = -0.00524 \text{ rad} \qquad \textbf{Ans}$$

3–29. The elastic portion of the stress–strain diagram for a steel alloy is shown in the figure. The specimen from which it was obtained had an original diameter of 13 mm and a gauge length of 50 mm. When the applied load on the specimen is 50 kN, the diameter is 12.99265 mm. Determine Poisson's ratio for the material.

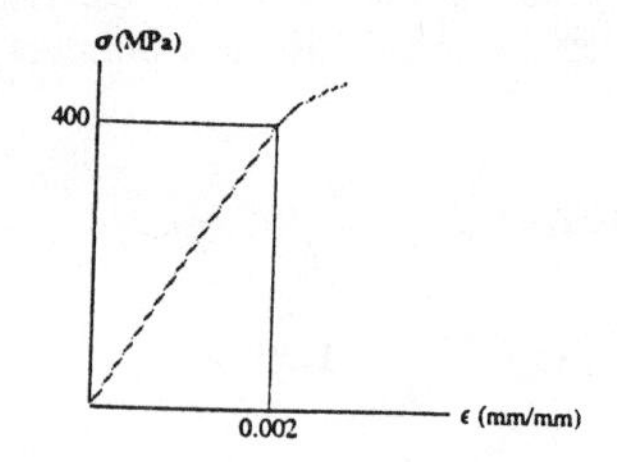

Normal Stress :

$$\sigma = \frac{P}{A} = \frac{50(10^3)}{\frac{\pi}{4}(0.013^2)} = 376.70 \text{ Mpa}$$

Normal Strain : From the swtress – strain diagram, the modulus of elasticity $E = \dfrac{400(10^6)}{0.002} = 200$ GPa. Applying Hooke's law

$$\varepsilon_{long} = \frac{\sigma}{E} = \frac{376.70(10^6)}{200(10^9)} = 1.8835\left(10^{-3}\right) \text{ mm/mm}$$

$$\varepsilon_{lat} = \frac{d - d_0}{d_0} = \frac{12.99265 - 13}{13} = -0.56538\left(10^{-3}\right) \text{ mm/mm}$$

Poisson's Ratio : The lateral and longitudinal strain can be related using Poisson's ratio.

$$v = -\frac{\varepsilon_{lat}}{\varepsilon_{long}} = -\frac{-0.56538(10^{-3})}{1.8835(10^{-3})} = 0.300 \qquad \textbf{Ans}$$

3–30. The elastic portion of the stress–strain diagram for a steel alloy is shown in the figure. The specimen from which it was obtained had an original diameter of 13 mm and a gauge length of 50 mm. If a load of $P = 20$ kN is applied to the specimen, determine its diameter and gauge length. Take $v = 0.4$.

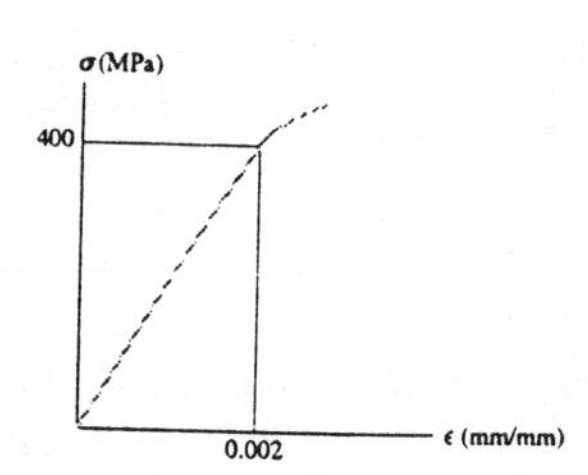

Normal Stress :

$$\sigma = \frac{P}{A} = \frac{20(10^3)}{\frac{\pi}{4}(0.013^2)} = 150.68 \text{Mpa}$$

Normal Strain : From the Stress – Strain diagram, the modulus of elasticity $E = \dfrac{400(10^6)}{0.002} = 200$ GPa. Applying Hooke's Law

$$\varepsilon_{long} = \frac{\sigma}{E} = \frac{150.68(10^6)}{200(10^9)} = 0.7534\left(10^{-3}\right) \text{ mm/mm}$$

Thus,

$$\delta L = \varepsilon_{long} L_0 = 0.7534\left(10^{-3}\right)(50) = 0.03767 \text{ mm}$$

$$L = L_0 + \delta L = 50 + 0.03767 = 50.0377 \text{ mm} \qquad \textbf{Ans}$$

Poisson's Ratio : The lateral and longitudinal can be related using poisson's ratio.

$$\varepsilon_{lat} = -v\varepsilon_{long} = -0.4(0.7534)\left(10^{-3}\right)$$
$$= -0.3014\left(10^{-3}\right) \text{ mm/mm}$$

$$\delta d = \varepsilon_{lat} d = -0.3014\left(10^{-3}\right)(13) = -0.003918 \text{ mm}$$

$$d = d_0 + \delta d = 13 + (-0.003918) = 12.99608 \text{ mm} \qquad \textbf{Ans}$$

3–31. The shear stress–strain diagram for a steel alloy is shown in the figure. If a bolt having a diameter of 0.25 in. is made of this material and used in the lap joint, determine the modulus of elasticity E and the force P required to cause the material to yield. Take $\nu = 0.3$.

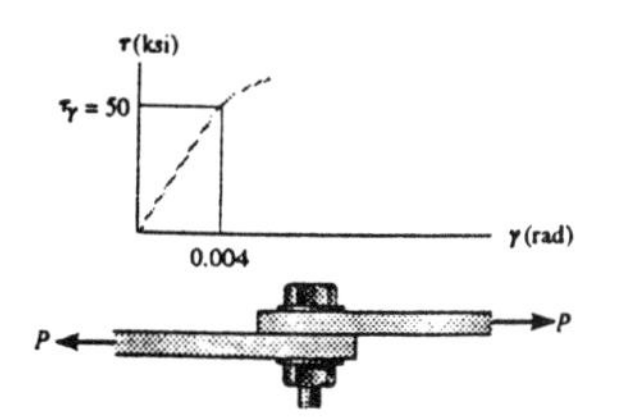

Modulus of Rigidity : From the stress – strain diagram,

$$G = \frac{50}{0.004} = 12.5(10^3)\text{ ksi}$$

Modulus of Elasticity :

$$G = \frac{E}{2(1+\nu)}$$

$$12.5(10^3) = \frac{E}{2(1+0.3)}$$

$$E = 32.5(10^3)\text{ ksi} \qquad \textbf{Ans}$$

Yielding Shear : The bolt is subjected to a yielding shear of $V_Y = P$. From the stress – strain diagram, $\tau_Y = 50$ ksi

$$\tau_Y = \frac{V_Y}{A}$$

$$50 = \frac{P}{\frac{\pi}{4}(0.25^2)}$$

$$P = 2.45\text{ kip} \qquad \textbf{Ans}$$

***3–32.** A shear spring is made by bonding the rubber annulus to a rigid fixed ring and a plug. When an axial load **P** is placed on the plug, show that the slope at point A in the rubber is $dy/dr = -\tan\gamma = -\tan(P/(2\pi hGr))$. For small angles we can write $dy/dr = -P/(2\pi hGr)$. Integrate this expression and evaluate the constant of integration using the condition that $y = 0$ at $r = r_o$. From the result compute the deflection $y = \delta$ of the plug.

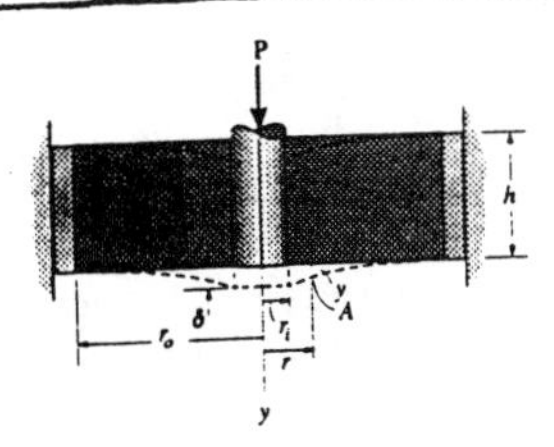

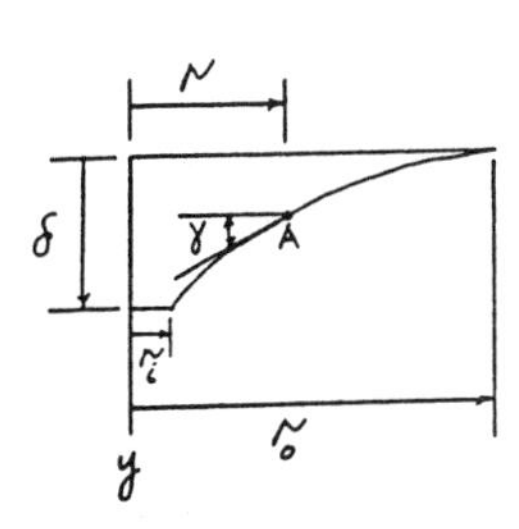

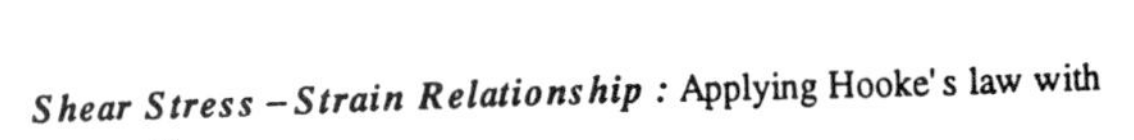

Shear Stress – Strain Relationship : Applying Hooke's law with $\tau_A = \dfrac{P}{2\pi r h}$.

$$\gamma = \frac{\tau_A}{G} = \frac{P}{2\pi h G r}$$

$$\frac{dy}{dr} = -\tan\gamma = -\tan\left(\frac{P}{2\pi h G r}\right) \qquad (Q.E.D)$$

If γ is small, then $\tan\gamma \approx \gamma$. Therefore,

$$\frac{dy}{dr} = -\frac{P}{2\pi h G r}$$

$$y = -\frac{P}{2\pi h G}\int\frac{dr}{r}$$

$$y = -\frac{P}{2\pi h G}\ln r + C$$

At $r = r_o$, $\quad y = 0$

$$0 = -\frac{P}{2\pi h G}\ln r_o + C$$

$$C = \frac{P}{2\pi h G}\ln r_o$$

Then, $\quad y = \dfrac{P}{2\pi h G}\ln\dfrac{r_o}{r}$

At $r = r_i$, $\quad y = \delta$

$$\delta = \frac{P}{2\pi h G}\ln\frac{r_o}{r_i} \qquad \textbf{Ans}$$

3–33. A shear spring is made from two blocks of rubber, each having a height h, width b, and thickness a. The blocks are bonded to three plates as shown. If the plates are rigid and the shear modulus of the rubber is G, determine the displacement of plate A if a vertical load **P** is applied to this plate. Assume that the displacement is small so that $\delta = a \tan \gamma \approx a\gamma$.

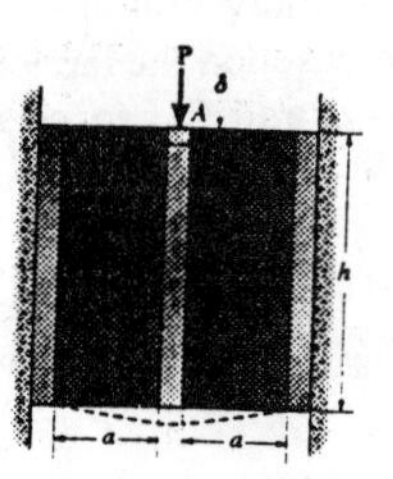

Average Shear Stress : The rubber block is subjected to a shear force of $V = \frac{P}{2}$.

$$\tau = \frac{V}{A} = \frac{\frac{P}{2}}{b\,h} = \frac{P}{2\,b\,h}$$

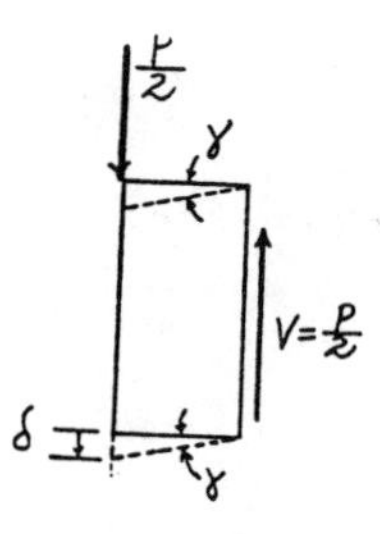

Shear Strain : Applying Hooke's law for shear

$$\gamma = \frac{\tau}{G} = \frac{\frac{P}{2\,b\,h}}{G} = \frac{P}{2\,b\,h\,G}$$

Thus,

$$\delta = a\,\gamma = \frac{P\,a}{2\,b\,h\,G} \qquad \textbf{Ans}$$

3–34. The head H is connected to the cylinder of a compressor using six steel bolts. If the clamping force in each bolt is 800 lb, determine the normal strain in the bolts. Each bolt has a diameter of $\frac{3}{16}$ in. If $\sigma_Y = 40$ ksi and $E_{st} = 29(10^3)$ ksi, what is the strain in each bolt when the nut is unscrewed so that the clamping force is released?

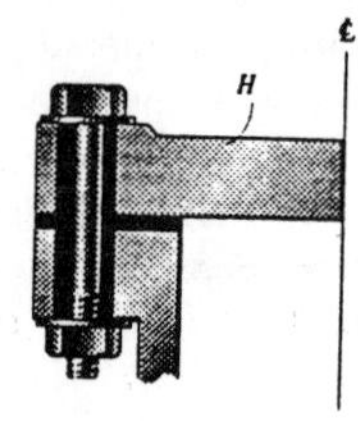

Normal Stress :

$$\sigma = \frac{P}{A} = \frac{800}{\frac{\pi}{4}\left(\frac{3}{16}\right)^2} = 28.97\text{ ksi} < \sigma_Y = 40\text{ ksi}$$

Normal Strain : Since $\sigma < \sigma_Y$, Hooke's law is still valid.

$$\varepsilon = \frac{\sigma}{E} = \frac{28.97}{29(10^3)} = 0.000999\text{ in./in.} \qquad \textbf{Ans}$$

If the nut is unscrewed, the load is zero. Therefore, the strain $\varepsilon = 0$ **Ans**

3–35. The stone has a mass of 800 kg and center of gravity at G. It rests on a pad at A and a roller at B. The pad is fixed to the ground and has a compressed height of 30 mm, a width of 140 mm, and a length of 150 mm. If the coefficient of static friction between the pad and the stone is $\mu_s = 0.8$, determine the approximate horizontal displacement of the stone, caused by the shear strains in the pad, before the stone begins to slip. Assume the normal force at A acts 1.5 m from G as shown. The pad is made from a material having $E = 4$ MPa and $\nu = 0.35$.

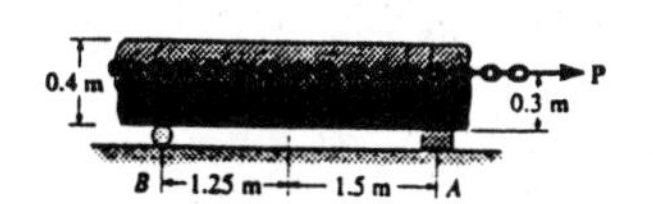

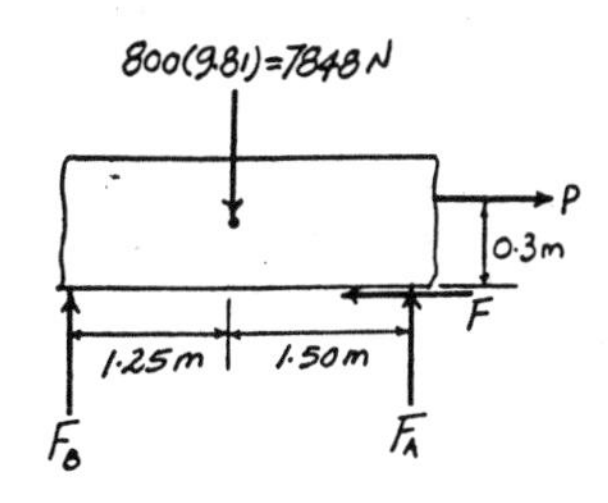

Equations of Equilibrium :

$$\circlearrowleft + \Sigma M_B = 0; \quad F_A(2.75) - 7848(1.25) - P(0.3) = 0 \qquad [1]$$

$$\xrightarrow{+} \Sigma F_x = 0; \quad P - F = 0 \qquad [2]$$

Note : The normal force at A does not act exactly at A. It has to shift due to friction.

Friction Equation :

$$F = \mu_s F_A = 0.8 F_A \qquad [3]$$

Solving Eqs. [1], [2] and [3] yields :

$$F_A = 3908.37 \text{ N} \qquad F = P = 3126.69 \text{ N}$$

Average Shear Stress : The pad is subjected to a shear force of $V = F = 3126.69$ N.

$$\tau = \frac{V}{A} = \frac{3126.69}{(0.14)(0.15)} = 148.89 \text{ kPa}$$

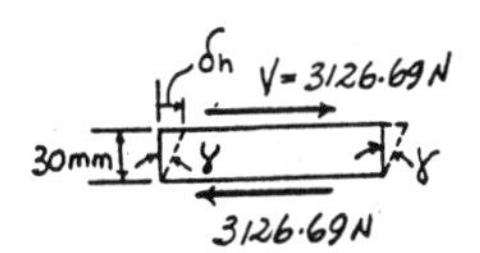

Modulus of Rigidity :

$$G = \frac{E}{2(1+\nu)} = \frac{4}{2(1+0.35)} = 1.481 \text{ MPa}$$

Shear Strain : Applying Hooke's law for shear

$$\gamma = \frac{\tau}{G} = \frac{148.89(10^3)}{1.481(10^6)} = 0.1005 \text{ rad}$$

Thus,

$$\delta_h = h\gamma = 30(0.1005) = 3.02 \text{ mm} \qquad \textbf{Ans}$$

***3–36.** The rigid pipe is supported by a pin at C and an A-36 steel guy wire AB. If the wire has a diameter of 0.2 in., determine how much it stretches when a load of $P = 300$ lb acts on the pipe. The material remains elastic.

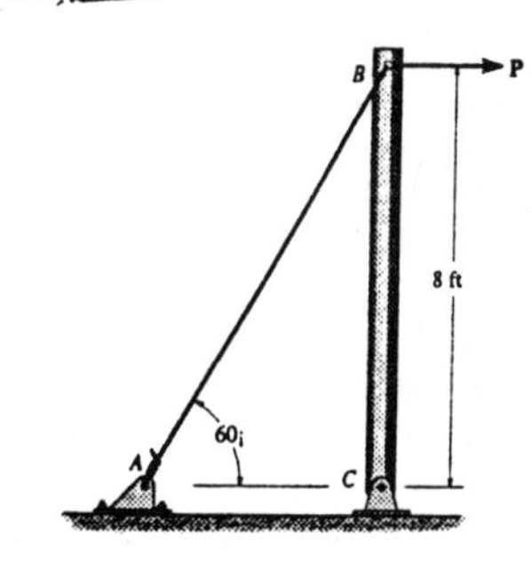

Equation of Equilibrium :

$$\circlearrowleft + \Sigma M_C = 0; \quad F_{AB} \sin 30°(8) - 300(8) = 0$$
$$F_{AB} = 600 \text{ lb}$$

Normal Stress and Strain :

$$\sigma_{AB} = \frac{F_{AB}}{A} = \frac{600}{\frac{\pi}{4}(0.2^2)} = 19.10 \text{ ksi}$$

Applying Hooke's law

$$\varepsilon_{AB} = \frac{\sigma_{AB}}{E} = \frac{19.10}{29.0(10^3)} = 0.6586(10^{-3}) \text{ in./in.}$$

Thus,

$$\delta L_{AB} = \varepsilon_{AB} L_{AB}$$
$$= 0.6586(10^{-3})\left[\frac{8(12)}{\cos 30°}\right]$$
$$= 0.0730 \text{ in.} \qquad \textbf{Ans}$$

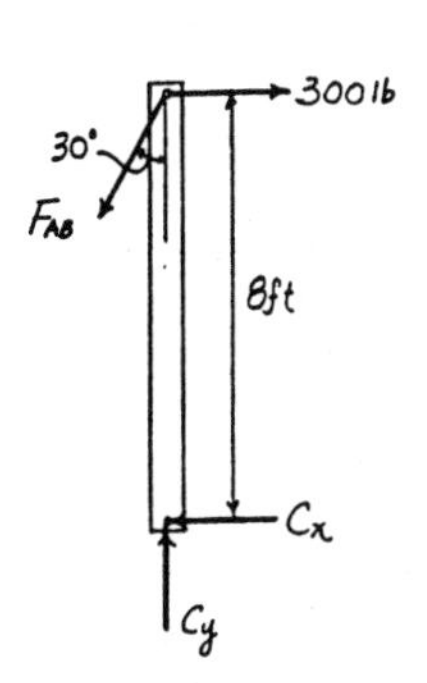

3–37. The rigid pipe is supported by a pin at C and an A-36 guy wire AB. If the wire has a diameter of 0.2 in., determine the load **P** if the end B is displaced 0.10 in. to the right. $E_s = 29(10^3)$ ksi.

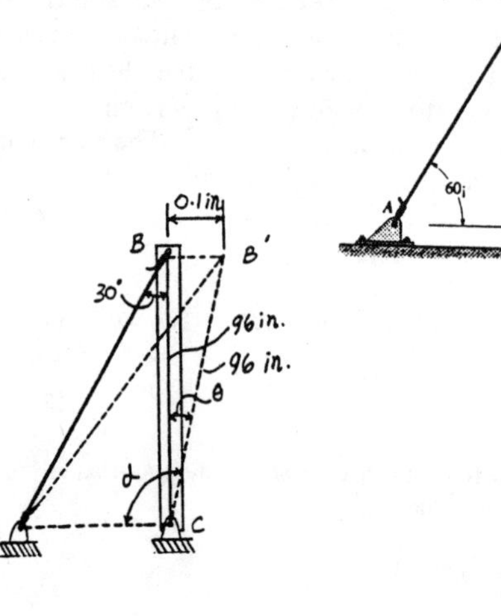

Geometry :

$$\sin\theta = \frac{0.1}{96} \qquad \theta = 0.05968°$$

$$\alpha = 90° + 0.05968° = 90.05968°$$

$$AC = 96\tan 30° = 55.4256 \text{ in}$$

$$AB = \frac{96}{\cos 30°} = 110.8513 \text{ in}$$

$$AB' = \sqrt{96^2 + 55.4256^2 - 2(96)(55.4256)\cos 90.05968°}$$
$$= 110.9012 \text{ in.}$$

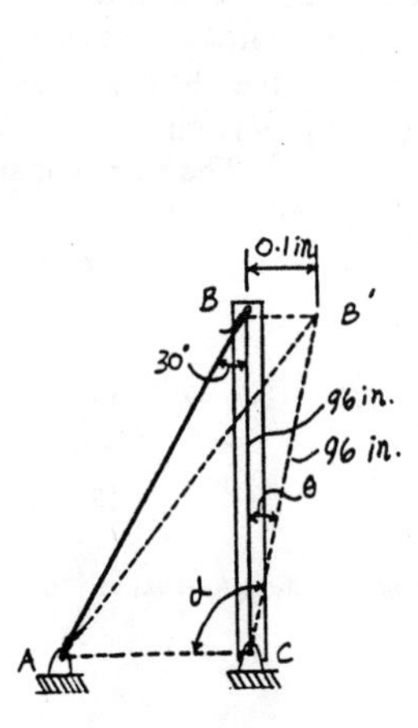

Normal Stress and Strain :

$$\varepsilon_{AB} = \frac{AB' - AB}{AB}$$
$$= \frac{110.9012 - 110.8513}{110.8513}$$
$$= 0.4510(10^{-3}) \text{ in./in.}$$

Applying Hooke's law

$$\sigma_{AB} = E\varepsilon_{AB} = 29.0(10^3)\,0.4510(10^{-3}) = 13.08 \text{ ksi}$$

Thus,

$$F_{AB} = \sigma_{AB}A = 13.08(10^3)\left[\frac{\pi}{4}(0.2^2)\right] = 410.85 \text{ lb}$$

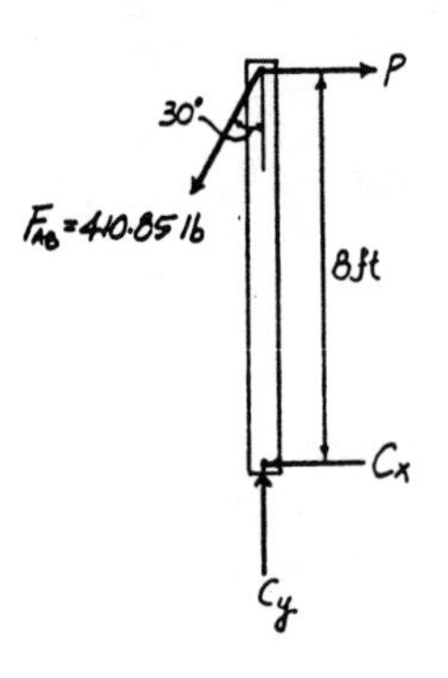

Equation of Equilibrium :

$$\circlearrowleft + \Sigma M_C = 0; \qquad 410.85\sin 30°(8) - P(8) = 0$$

$$P = 205 \text{ lb} \qquad \textbf{Ans}$$

3–38. The 8-mm-diameter bolt is made of an aluminum alloy. It fits through a magnesium sleeve that has an inner diameter of 12 mm and an outer diameter of 20 mm. If the original lengths of the bolt and sleeve are 80 mm and 50 mm, respectively, determine the strains in the sleeve and the bolt if the nut on the bolt is tightened so that the tension in the bolt is 8 kN. Assume the material at A is rigid. $E_{al} = 70$ GPa, $E_{mg} = 45$ GPa.

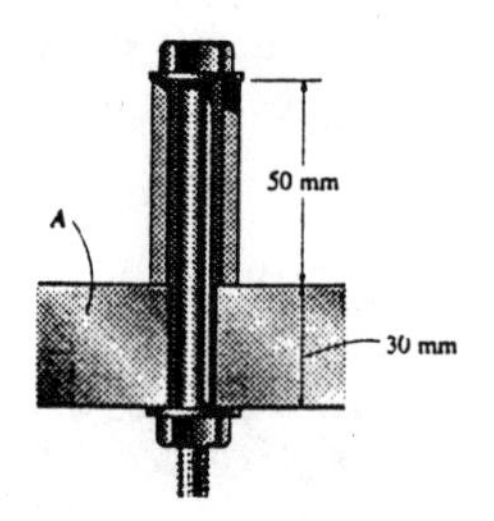

Normal Stress :

$$\sigma_b = \frac{P}{A_b} = \frac{8(10^3)}{\frac{\pi}{4}(0.008^2)} = 159.15 \text{ MPa}$$

$$\sigma_s = \frac{P}{A_s} = \frac{8(10^3)}{\frac{\pi}{4}(0.02^2 - 0.012^2)} = 39.79 \text{ MPa}$$

Normal Strain : Applying Hooke's Law

$$\varepsilon_b = \frac{\sigma_b}{E_{al}} = \frac{159.15(10^6)}{70(10^9)} = 0.00227 \text{ mm/mm} \qquad \textbf{Ans}$$

$$\varepsilon_s = \frac{\sigma_s}{E_{mg}} = \frac{39.79(10^6)}{45(10^9)} = 0.000884 \text{ mm/mm} \qquad \textbf{Ans}$$

3–39. A tension test was performed on a steel specimen having an original diameter of 0.503 in. and gauge length of 2.00 in. The data is listed in the table below. Plot the stress–strain diagram and determine approximately the modulus of elasticity, the yield stress, the ultimate stress, and the rupture stress. Use a scale of 1 in. = 20 ksi and 1 in. = 0.05 in./in. Redraw the elastic region, using the same stress scale but a strain scale of 1 in. = 0.001 in./in.

Load (kip)	Elongation (in.)
0	0
1.50	0.0005
4.60	0.0015
8.00	0.0025
11.00	0.0035
11.80	0.0050
11.80	0.0080
12.00	0.0200
16.60	0.0400
20.00	0.1000
21.50	0.2800
19.50	0.4000
18.50	0.4600

Stress and Strain :

$\sigma = \dfrac{P}{A}$ (ksi)	$\varepsilon = \dfrac{\delta L}{L}$ (in./in.)
0	0
7.55	0.00025
23.15	0.00075
40.26	0.00125
55.36	0.00175
59.38	0.00250
59.38	0.00400
60.39	0.01000
83.54	0.02000
100.65	0.05000
108.20	0.14000
98.13	0.20000
93.10	0.23000

Modulus of Elasticity : From the stress – strain diagram

$$E_{\text{approx}} = \frac{48.0-0}{0.0015-0} = 32.0\left(10^3\right)\ \text{ksi} \qquad \textbf{Ans}$$

Yield, Ultimate and Fracture Stress : From the stress – strain diagram

$$(\sigma_Y)_{\text{approx}} = 55.4\ \text{ksi} \qquad \textbf{Ans}$$

$$\sigma_u = 110\ \text{ksi} \qquad \textbf{Ans}$$

$$\sigma_f = 93.1\text{ksi} \qquad \textbf{Ans}$$

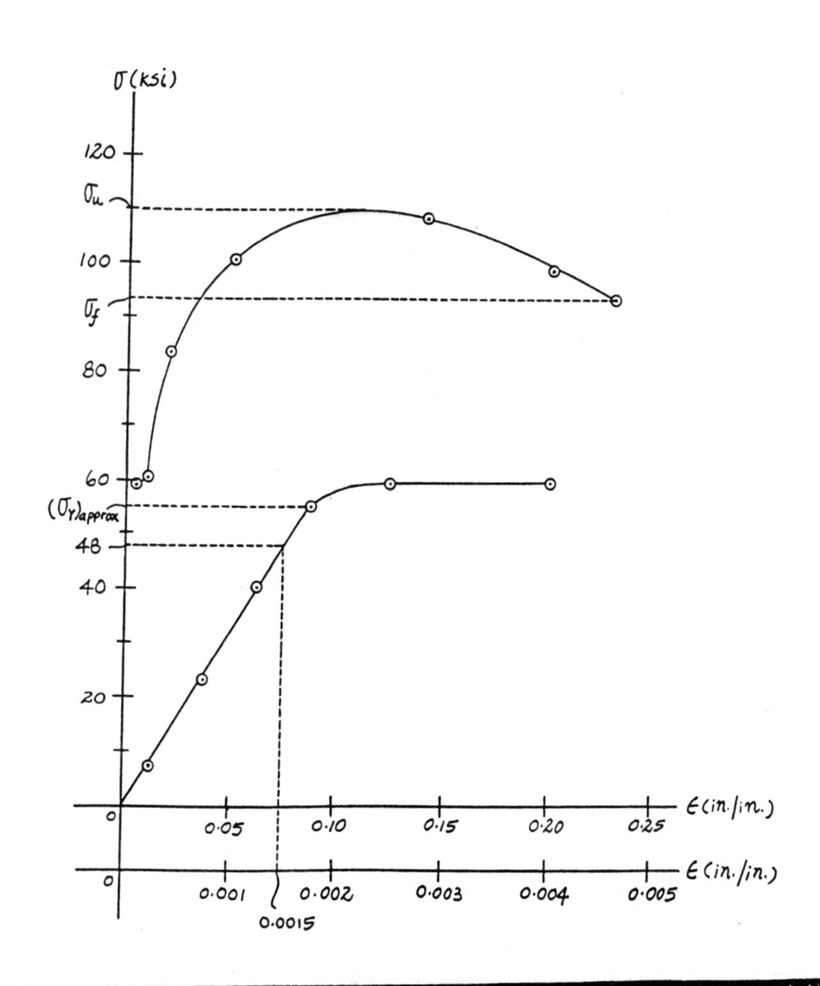

***3–40.** The stress–strain diagram for a polyester resin is given in the figure. If the rigid beam is supported by a link *AB* and post *CD*, both made of this material, and subjected to a load of $P = 80$ kN, determine the angle of tilt of the beam when the load is applied. The diameter of the link is 40 mm and the diameter of the post is 80 mm.

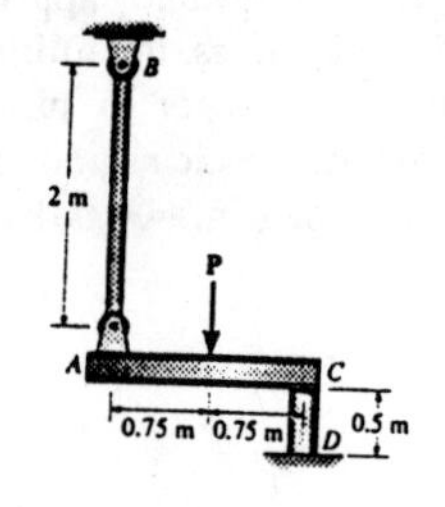

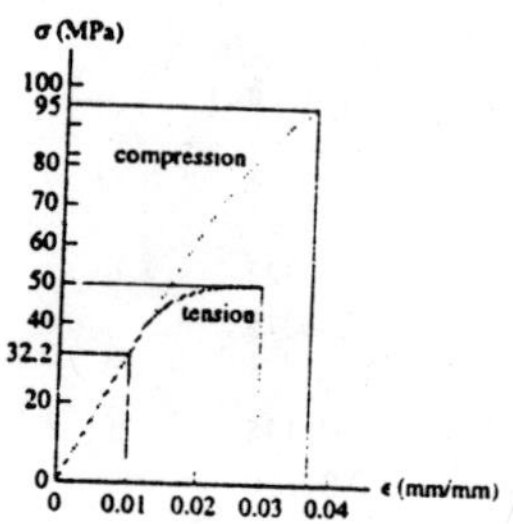

Support Reactions : As shown on FBD.

Normal Stress :

$$\sigma_{AB} = \frac{F_{AB}}{A_{AB}} = \frac{40(10^3)}{\frac{\pi}{4}(0.04^2)} = 31.83 \text{ MPa} < 32.2 \text{ MPa}$$

$$\sigma_{CD} = \frac{F_{CD}}{A_{CD}} = \frac{40(10^3)}{\frac{\pi}{4}(0.08^2)} = 7.958 \text{ MPa} < 32.2 \text{ MPa}$$

Normal Strain : From the stress - strain diagram,

$E = \dfrac{32.2(10^6)}{0.01} = 3.22$ GPa. Applying Hooke's law

$$\varepsilon_{AB} = \frac{\sigma_{AB}}{E} = \frac{31.83(10^6)}{3.22(10^9)} = 0.009885 \text{ mm/mm}$$

$$\varepsilon_{CD} = \frac{\sigma_{CD}}{E} = \frac{7.958(10^6)}{3.22(10^9)} = 0.002471 \text{ mm/mm}$$

Thus,

$$\delta L_{AB} = \varepsilon_{AB} L_{AB} = 0.009885(2000) = 19.771 \text{ mm}$$

$$\delta L_{CD} = \varepsilon_{CD} L_{CD} = 0.002471(500) = 1.236 \text{ mm}$$

Angle of tilt :

$$\tan\alpha = \frac{18.536}{1500} \qquad \alpha = 0.708° \qquad \textbf{Ans}$$

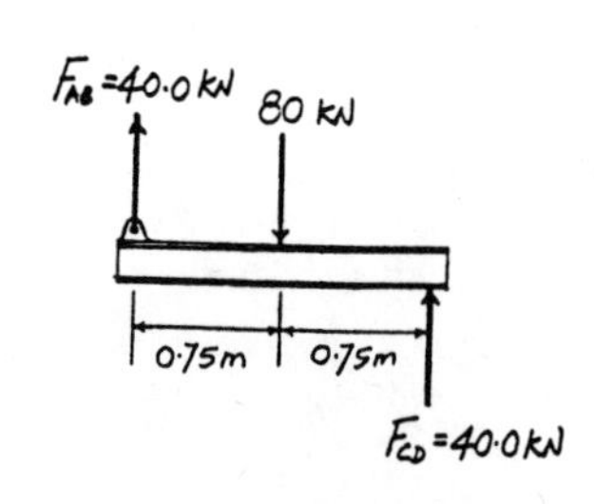

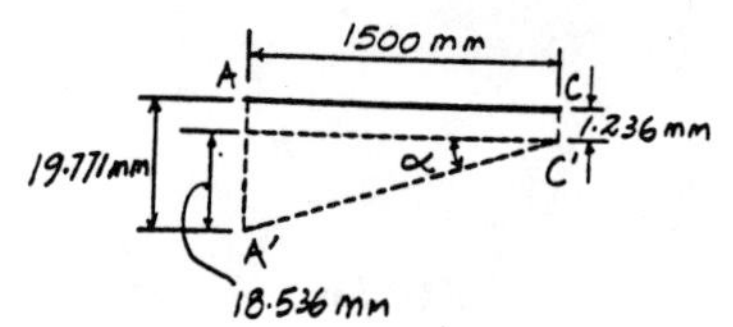

3–41. The stress–strain diagram for a polyester resin is given in the figure. If the rigid beam is supported by a link AB and post CD made of this material, determine the largest load P that can be applied to the beam before either AB or CD fractures. The diameter of the link is 12 mm and the diameter of the post is 40 mm.

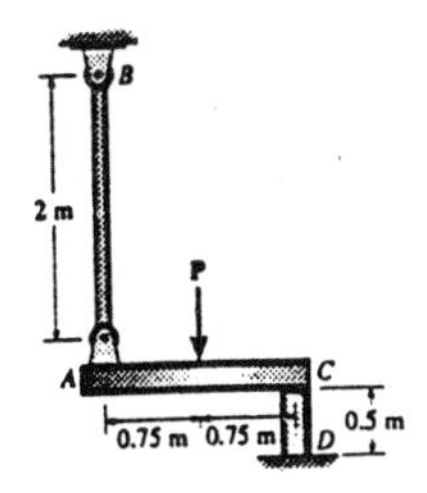

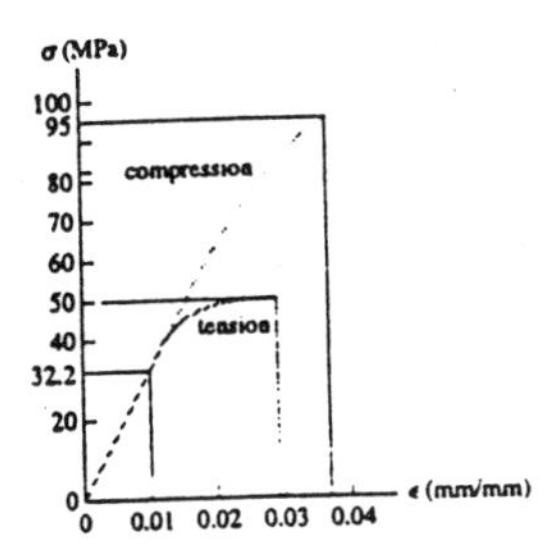

Fracture Stress :

Fracture of strut AB

$$(\sigma_f)_{AB} = \frac{F_{AB}}{A_{AB}}$$

$$50(10^6) = \frac{\frac{P}{2}}{\frac{\pi}{4}(0.012^2)}$$

$$P = 11.3 \text{ kN } (\textbf{\textit{Controls!}})$$ **Ans**

Fracture of post CD

$$(\sigma_f)_{CD} = \frac{F_{CD}}{A_{CD}}$$

$$95(10^6) = \frac{\frac{P}{2}}{\frac{\pi}{4}(0.04^2)}$$

$$P = 238.8 \text{ kN}$$

4–1. The ship is pushed through the water using an A-36 steel propeller shaft that is 8 m long, measured from the propeller to the thrust bearing D at the engine. If it has an outer diameter of 400 mm and a wall thickness of 50 mm, determine the amount of axial contraction of the shaft when the propeller exerts a force on the shaft of 5 kN. The bearings at B and C are journal bearings.

Internal Force : As shown on FBD.

Displacement :

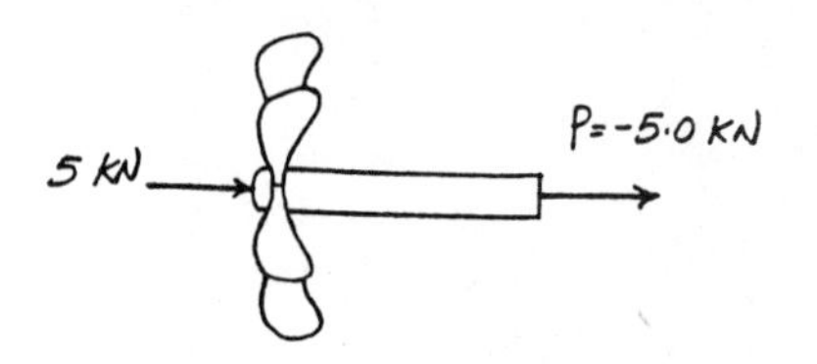

$$\delta_A = \frac{PL}{AE} = \frac{-5.00\,(10^3)(8)}{\frac{\pi}{4}(0.4^2 - 0.3^2)\,200(10^9)}$$

$$= -3.638(10^{-6})\text{ m}$$

$$= -3.64\left(10^{-3}\right)\text{ mm} \qquad \textbf{Ans}$$

Negative sign indicates that end A moves towards end D.

4–2. The A-36 steel column is used to support the symmetric loads from the two floors of a building. Determine the vertical displacement of its top A if $P_1 = 40$ kip, $P_2 = 62$ kip, and the column has a cross-sectional area of 23.4 in^2.

Internal Forces : As shown on FBD (a) and (b)

Displacement :

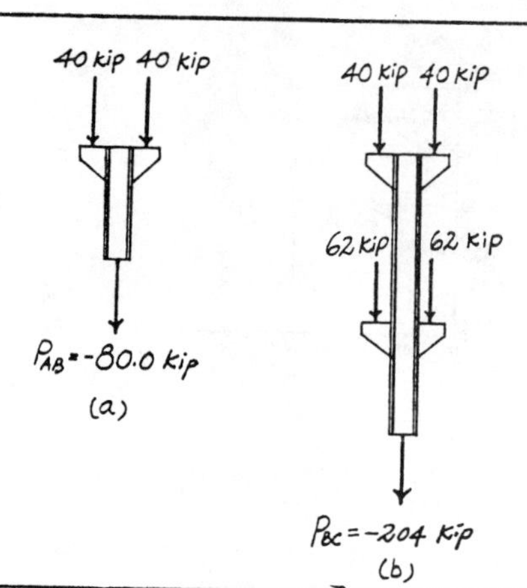

$$\delta_A = \sum \frac{PL}{AE} = \frac{-80.0\,(12)(12)}{23.4\,(29.0)(10^3)} + \frac{(-204)\,(12)(12)}{23.4\,(29.0)(10^3)}$$

$$= -\,0.0603\text{ in.} \qquad \textbf{Ans}$$

Negative sign indicates that end A moves toward end C.

4–3. The A-36 steel column is used to support the symmetric loads from the two floors of a building. Determine the loads P_1 and P_2 if A moves downward 0.12 in. and B moves downward 0.09 in. when the loads are applied. The column has a cross-sectional area of 23.4 in^2.

Internal Forces : As shown on FBD.

Displacement :

For point A

$$\delta_A = \sum \frac{PL}{AE}; \qquad -0.12\text{ in.} = \frac{-\,2P_1\,(144)}{(23.4)\,29.0(10^3)} + \frac{-2(P_1 + P_2)\,(12)(12)}{(23.4)\,29.0(10^3)}$$

$$282.75 = 2P_1 + P_2 \qquad [1]$$

For point B

$$\delta_B = \sum \frac{PL}{AE}; \qquad -0.09\text{ in.} = \frac{-2\,(P_1 + P_2)\,(12)(12)}{(23.4)\,29.0(10^3)}$$

$$212.0625 = P_1 + P_2 \qquad [2]$$

Solving Eqs. [1] and [2] yields :

$P_1 = 70.7$ kip **Ans**

$P_2 = 141$ kip **Ans**

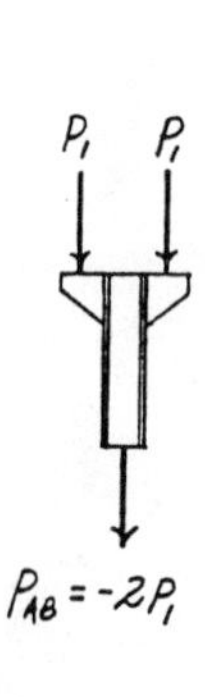

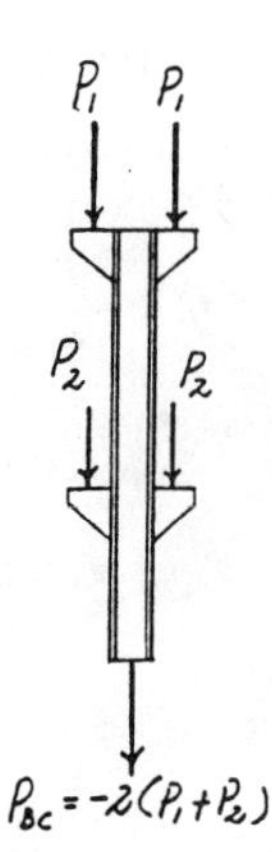

***4–4.** The bronze C86100 shaft is subjected to the axial loads shown. Determine the displacement of end A with respect to end D if the diameters of each segment are $d_{AB} = 0.75$ in., $d_{BC} = 2$ in., and $d_{CD} = 0.5$ in.

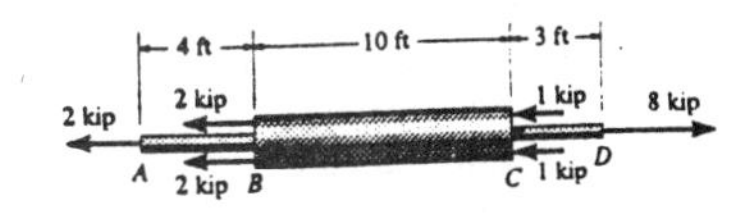

Internal Forces : As shown on FBD.
Displacement :

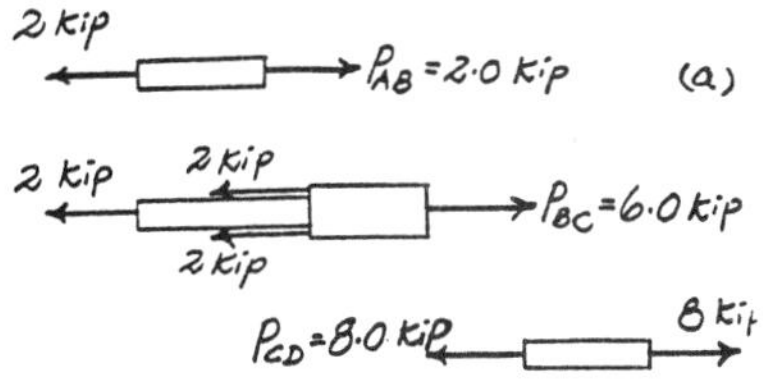

$$\delta_{A/D} = \sum \frac{PL}{AE} = \frac{2.00(4)(12)}{\frac{\pi}{4}(0.75^2)(15.0)(10^3)} + \frac{6.00(10)(12)}{\frac{\pi}{4}(2^2)\,(15.0)(10^3)} + \frac{8.00(3)(12)}{\frac{\pi}{4}(0.5^2)(15.0)(10^3)}$$

$$= 0.128 \text{ in.} \qquad \textbf{Ans}$$

Positive sign indicates that end A moves away from end D.

4–5. Determine the displacement of end A with respect to end C of the shaft in Prob. 4–4.

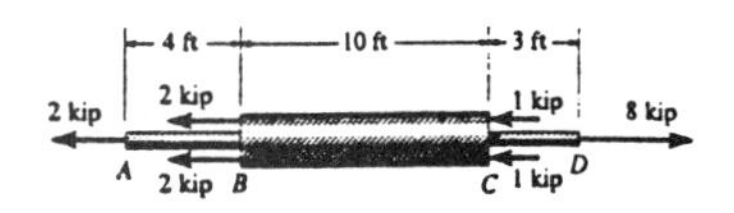

Internal Forces : As shown on FBD.
Displacement :

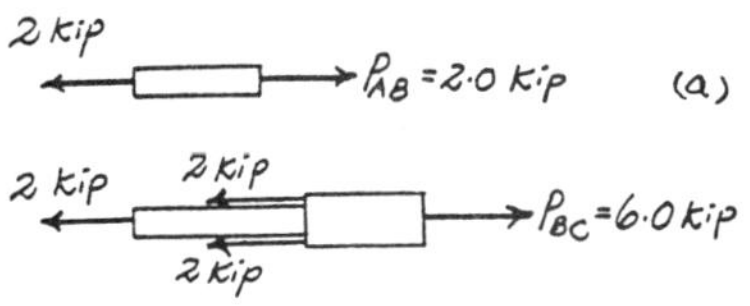

$$\delta_{A/C} = \sum \frac{PL}{AE} = \frac{2.00(4)(12)}{\frac{\pi}{4}(0.75^2)(15.0)(10^3)} + \frac{6.00(10)(12)}{\frac{\pi}{4}(2^2)\,(15.0)(10^3)}$$

$$= 0.0298 \text{ in.} \qquad \textbf{Ans}$$

Positive sign indicates that end A moves away from C.

4–6 The assembly consists of an A-36 steel rod CB and a 6061-T6 aluminum rod BA, each having a diameter of 1 in. If the rod is subjected to the axial loading $P_1 = 12$ kip at A and $P_2 = 18$ kip at the coupling B, determine the displacement of the coupling B and the end A. The unstretched length of each segment is shown in the figure. Neglect the size of the connections at B and C, and assume that they are rigid.

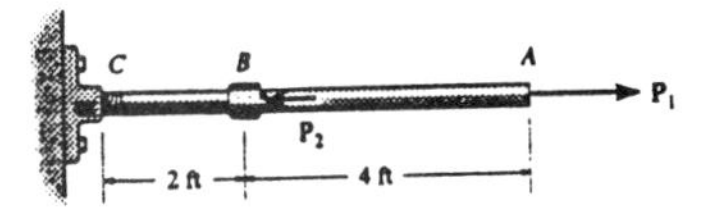

Internal Forces : As shown on FBD.
Displacement :

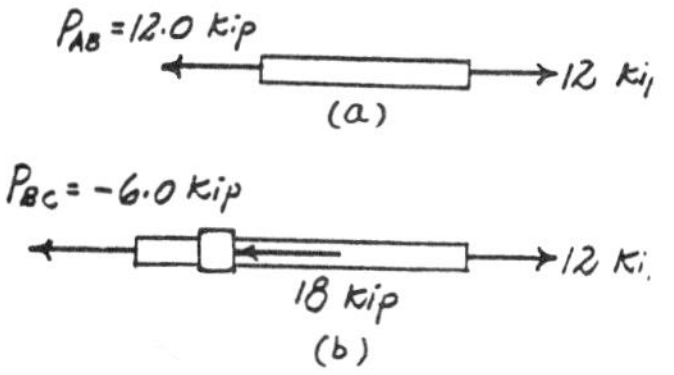

$$\delta_B = \frac{P_{BC}L_{BC}}{A_{BC}E_{st}} = \frac{-6.00(2)(12)}{\frac{\pi}{4}(1^2)(29.0)(10^3)}$$

$$= -0.00632 \text{ in.} \qquad \textbf{Ans}$$

Negative sign indicates coupling B moves towards end C.

$$\delta_A = \sum \frac{PL}{AE} = \frac{12.0(4)(12)}{\frac{\pi}{4}(1^2)(10.0)(10^3)} + \frac{-6.00(2)(12)}{\frac{\pi}{4}(1)^2(29.0)(10^3)}$$

$$= 0.0670 \text{ in.} \qquad \textbf{Ans}$$

Positive sign indicates end A moves away from end C.

4–7. The assembly consists of an A-36 steel rod CB and a 6061-T6 aluminum rod BA, each having a diameter of 1 in. Determine the applied loads P_1 and P_2 if A is displaced 0.08 in. to the right and B is displaced 0.02 in. to the left when the loads are applied. The unstretched length of each segment is shown in the figure. Neglect the size of the connections at B and C, and assume that they are rigid.

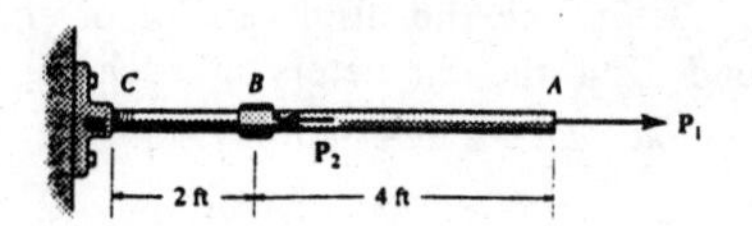

Internal Forces : As shown on FBD.
Displacement :

For point A

$$\delta_A = \sum \frac{PL}{AE}; \quad 0.08 = \frac{P_1(4)(12)}{\frac{\pi}{4}(1^2)(10.0)(10^3)} + \frac{(P_1 - P_2)(2)(12)}{\frac{\pi}{4}(1^2)(29.0)(10^3)}$$

$$2.618 = 0.2344\,P_1 - 0.03448 P_2 \qquad [1]$$

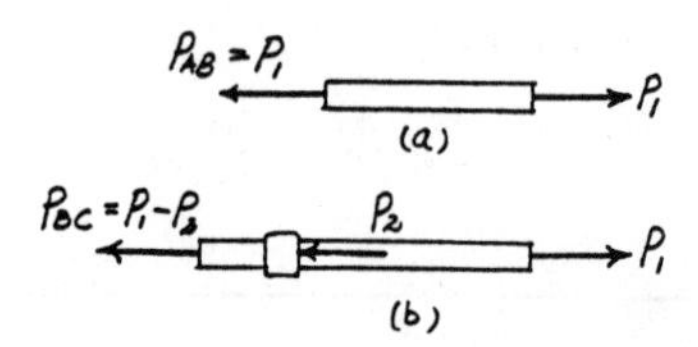

For point B

$$\delta_B = \sum \frac{PL}{AE}; \quad -0.02 = \frac{(P_1 - P_2)(2)(12)}{\frac{\pi}{4}(1^2)(29.0)(10^3)}$$

$$18.980 = P_2 - P_1 \qquad [2]$$

Solving Eqs. [1] and [2] yields :

$P_1 = 16.4$ kip **Ans**
$P_2 = 35.3$ kip **Ans**

***4–8.** The joint is made from three A-36 steel plates that are bonded together at their seams. Determine the displacement of end A with respect to end D when the joint is subjected to the axial loads shown. Each plate has a thickness of 6 mm.

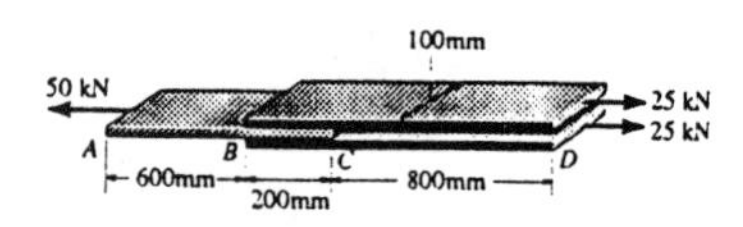

Internal Forces : As shown on FBD.
Displacement :

50 kN $P_{AB} = 50.0$ kN

50 kN $P_{BC} = 50.0$ kN

25 kN $P_{CD} = 25.0$ kN

$$\delta_{A/D} = \sum \frac{PL}{AE} = \frac{50.0(10^3)(0.6)}{(0.1)(0.006)(200)(10^9)} + \frac{50.0(10^3)(0.2)}{(0.1)(0.018)(200)(10^9)} + \frac{25.0(10^3)(0.8)}{(0.1)(0.006)(200)(10^9)}$$

$= 0.0004444$ m $= 0.444$ mm **Ans**

Positive sign indicates that end A moves away from end D.

4–9. The 15-mm-diameter A-36 steel shaft AC is supported by a rigid collar, which is fixed to the shaft at B. If it is subjected to an axial load of 80 kN at its end, determine the uniform pressure distribution p on the collar required for equilibrium. Also, what is the elongation on segment BC and segment BA?

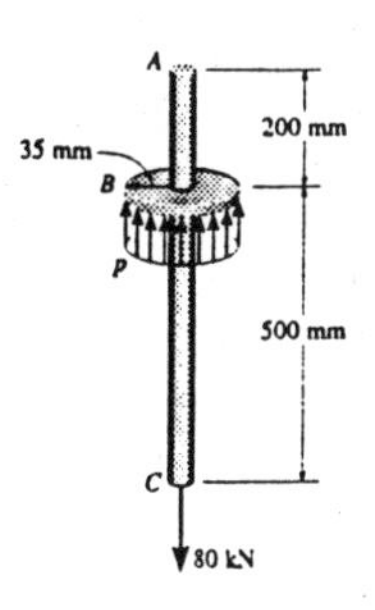

Equations of Equilibrium : FBD (a)

$$+\uparrow \Sigma F_y = 0; \quad P\left[\frac{\pi}{4}\left(0.07^2 - 0.015^2\right)\right] - 80(10^3) = 0$$

$$P = 21.79(10^6)\ \text{Pa} = 21.8\ \text{MPa} \qquad \textbf{Ans}$$

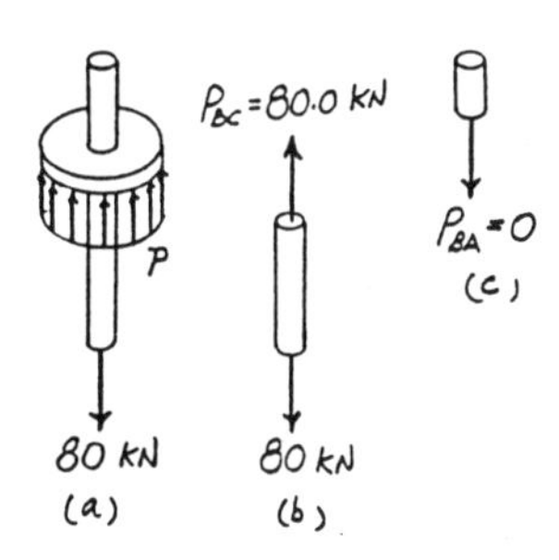

Internal Forces : As shown on FBD (b) and (c).

Displacement :

$$\delta_{BC} = \frac{P_{BC} L_{BC}}{A_{BC} E} = \frac{80.0(10^3)(500)}{\frac{\pi}{4}(0.015^2)(200)(10^9)}$$

$$= 1.13\ \text{mm} \qquad \textbf{Ans}$$

$$\delta_{BA} = \frac{P_{BA} L_{BA}}{A_{BA} E} = 0 \qquad \textbf{Ans}$$

4–10. The truss is made from three A-36 steel members, each having a cross-sectional area of 400 mm². Determine the vertical displacement of the roller at C when the truss supports the load of $P = 10$ kN.

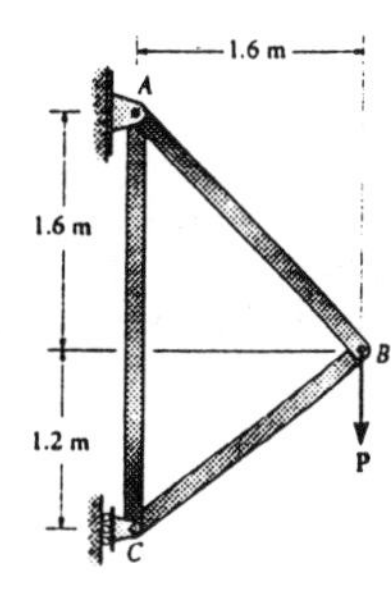

Method of Joints :

Joint B

$$\overset{+}{\rightarrow} \Sigma F_x = 0; \quad \frac{4}{5}F_{BC} - F_{BA}\cos 45^\circ = 0 \qquad [1]$$

$$+\uparrow \Sigma F_y = 0; \quad \frac{3}{5}F_{BC} + F_{BA}\sin 45^\circ - 10 = 0 \qquad [2]$$

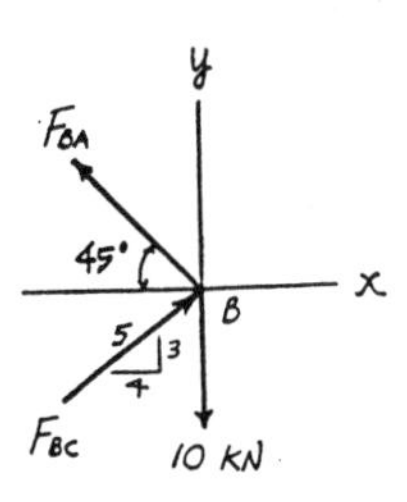

Solving Eqs. [1] and [2] yields :

$$F_{BA} = 8.081\ \text{kN} \qquad F_{BC} = 7.143\ \text{kN}$$

Joint C

$$+\uparrow \Sigma F_y = 0; \quad F_{CA} - \frac{3}{5}(7.143) = 0$$

$$F_{CA} = 4.2857\ \text{kN}$$

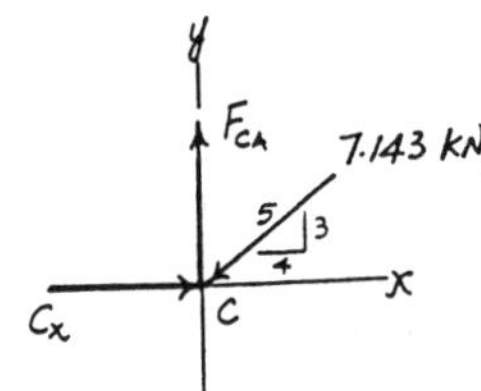

Displacement : By observation only member AC resists the vertical displacement of roller C.

$$\delta_{C_y} = \frac{F_{CA} L_{CA}}{A_{CA} E}$$

$$= \frac{4.2857(10^3)(2.8)(10^3)}{400(10^{-6})(200)(10^9)} = 0.150\ \text{mm} \qquad \textbf{Ans}$$

4–11. The truss is made from three A-36 steel members, each having a cross-sectional area of 400 mm^2. Determine the load P required to displace the roller downward 0.2 mm.

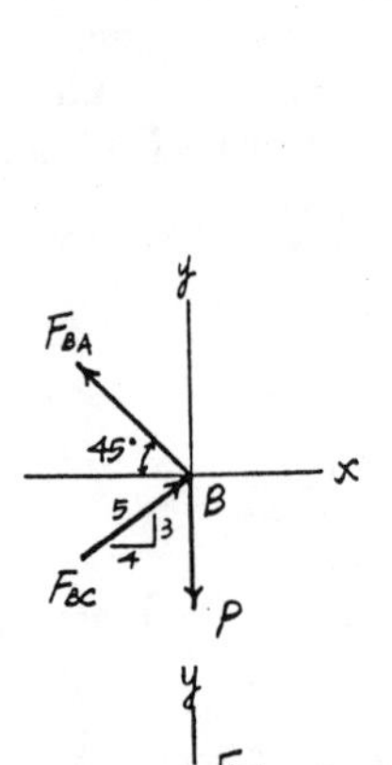

Method of Joints :

Joint B

$$\xrightarrow{+} \Sigma F_x = 0; \quad \frac{4}{5}F_{BC} - F_{BA}\cos 45^\circ = 0 \qquad [1]$$

$$+\uparrow \Sigma F_y = 0; \quad \frac{3}{5}F_{BC} + F_{BA}\sin 45^\circ - P = 0 \qquad [2]$$

Solving Eqs. [1] and [2] yields :

$$F_{BA} = 0.8081P \qquad F_{BC} = 0.7143P$$

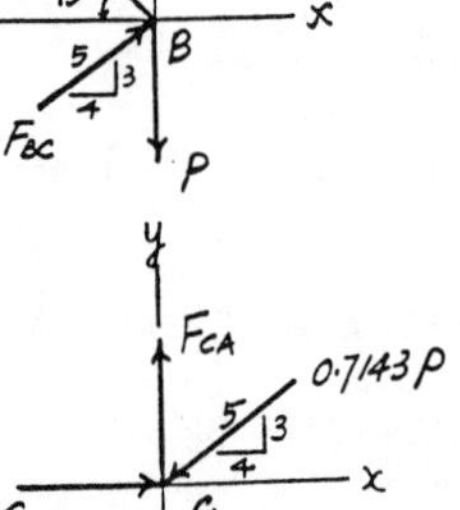

Joint C

$$+\uparrow \Sigma F_y = 0; \quad F_{CA} - \frac{3}{5}(0.7143P) = 0$$

$$F_{CA} = 0.4286P$$

Displacement : By observation only member AC resists the vertical displacement of roller C.

$$\delta_{C_v} = \frac{PL}{AE}; \qquad 0.0002 = \frac{(0.4286P)(2.8)}{400(10^{-6})(200)(10^9)}$$

$$P = 13.3 \text{ kN} \qquad \textbf{Ans}$$

***4–12.** The linkage is made from three pin-connected 304 stainless steel members, each having a cross-sectional area of 0.75 in^2. If a horizontal force of P = 6 kip is applied to the end B of member AB, determine the horizontal displacement of point B.

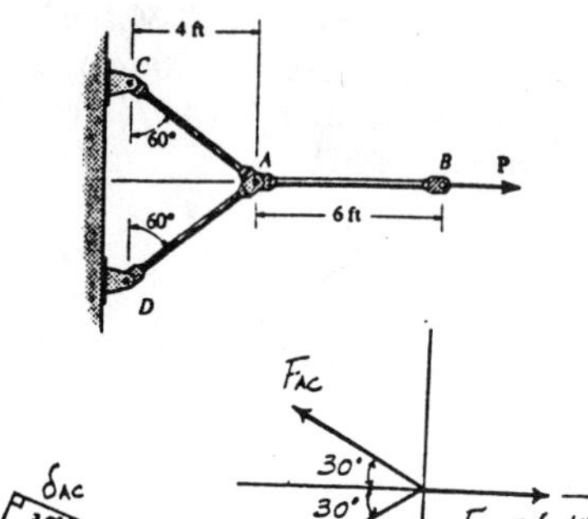

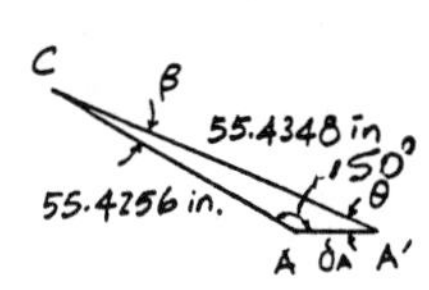

Method of Joints :

$$+\uparrow \Sigma F_y = 0; \quad F_{AC}\sin 30^\circ - F_{AD}\sin 30^\circ = 0$$

$$F_{AC} = F_{AD} = F$$

$$\xrightarrow{+} \Sigma F_x = 0; \quad 2F\cos 30^\circ - 6 = 0$$

$$F = 3.4641 \text{ kip}$$

Displacement :

$$\delta_{B/A} = \frac{F_{AB}L_{AB}}{A_{AB}E} = \frac{6(6)(12)}{(0.75)(28.0)(10^3)} = 0.02057 \text{ in.}$$

$$\delta_{AC} = \delta_{AD} = \frac{F_{AC}L_{AC}}{A_{AC}E} = \frac{3.4641\left(\frac{4}{\sin 60^\circ}\right)(12)}{0.75(28.0)(10^3)} = 0.009143 \text{ in.}$$

Geometry : From triangle $AA'C$

$$\frac{\sin\theta}{55.4256} = \frac{\sin 150^\circ}{55.4348}; \qquad \theta = 29.994544^\circ$$

$$\beta = 180^\circ - 150^\circ - 29.994544^\circ = 0.005455^\circ$$

$$\frac{\delta_A}{\sin 0.005455^\circ} = \frac{55.4348}{\sin 150^\circ}; \qquad \delta_A = 0.01056 \text{ in.}$$

or

$$\delta_A = \frac{\delta_{AC}}{\cos 30^\circ} = \frac{0.009143}{\cos 30^\circ} = 0.01056 \text{ in.}$$

$$\delta_B = \delta_A + \delta_{B/A} = 0.01056 + 0.02057$$
$$= 0.0311 \text{ in.} \qquad \textbf{Ans}$$

4–13. The linkage is made from three pin-connected 304 stainless steel members, each having a cross-sectional area of 0.75 in^2. Determine the magnitude of the force **P** needed to displace point B 0.08 in. to the right.

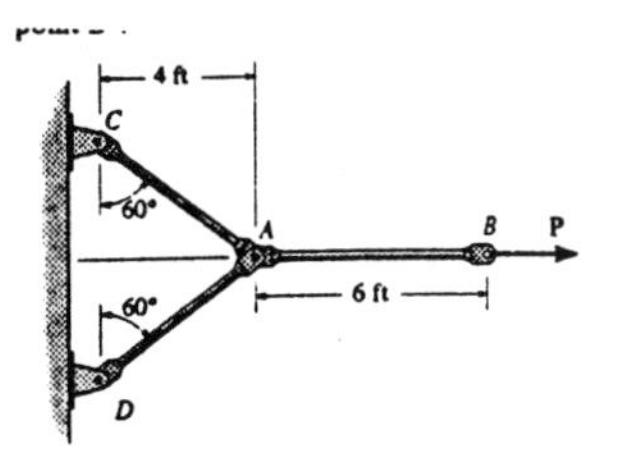

Method of Joints :

$$+\uparrow \Sigma F_y = 0; \quad F_{AC}\sin 30^\circ - F_{AD}\sin 30^\circ = 0$$

$$F_{AC} = F_{AD} = F$$

$$\xrightarrow{+} \Sigma F_x = 0; \quad P - 2F\cos 30^\circ = 0 \quad F = 0.5774P$$

Displacement :

For AB

$$\delta_{B/A} = \frac{F_{AB}L_{AB}}{A_{AB}E} = \frac{P(6)(12)}{0.75(28.0)(10^3)} = 0.003429P$$

For AC

$$\delta_{AC} = \frac{F_{AC}L_{AC}}{A_{AC}E} = \frac{0.5774P\left(\frac{4}{\sin 60^\circ}\right)(12)}{0.75(28.0)(10^3)} = 0.001524P$$

Geometry :

$$\delta_A = \frac{\delta_{AC}}{\cos 30^\circ} = \frac{0.001524P}{\cos 30^\circ} = 0.001760P$$

Require,

$$\delta_{B/A} + \delta_A = 0.08\,; \quad 0.003429P + 0.001760P = 0.08$$

$$P = 15.4 \text{ kip} \qquad \textbf{Ans}$$

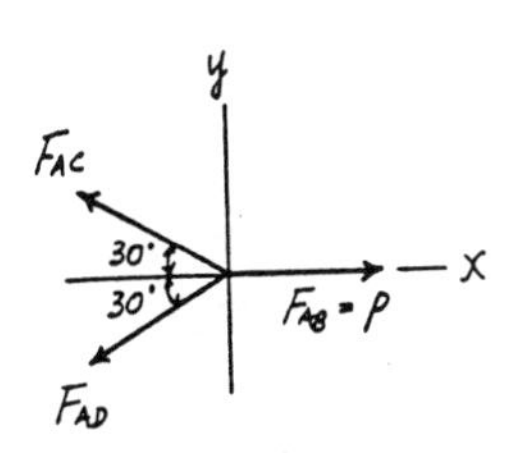

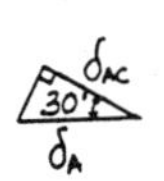

4–14. The load is supported by the four 304 stainless steel wires that are connected to the rigid members AB and DC. Determine the vertical displacement of the 500-lb load if the members were horizontal when the load was originally applied. Each wire has a cross-sectional area of 0.025 in^2.

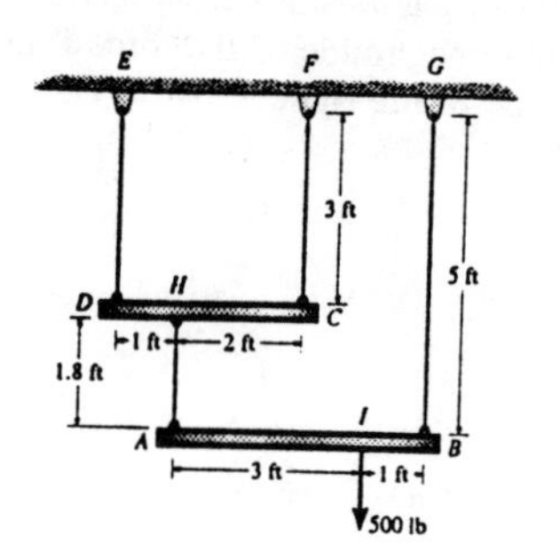

Internal Forces in the wires :

FBD (b)

$+\Sigma M_A = 0;\quad F_{BG}(4) - 500(3) = 0 \quad F_{BG} = 375.0 \text{ lb}$

$+\uparrow \Sigma F_y = 0;\quad F_{AH} + 375.0 - 500 = 0 \quad F_{AH} = 125.0 \text{ lb}$

FBD (a)

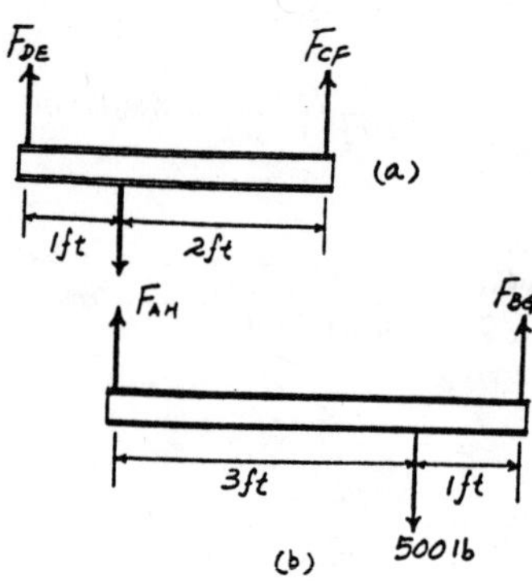

$+\Sigma M_D = 0;\quad F_{CF}(3) - 125.0(1) = 0 \quad F_{CF} = 41.67 \text{ lb}$

$+\uparrow \Sigma F_y = 0;\quad F_{DE} + 41.67 - 125.0 = 0 \quad F_{DE} = 83.33 \text{ lb}$

Displacement :

$$\delta_D = \frac{F_{DE}L_{DE}}{A_{DE}E} = \frac{83.33(3)(12)}{0.025(28.0)(10^6)} = 0.0042857 \text{ in.}$$

$$\delta_C = \frac{F_{CF}L_{CF}}{A_{CF}E} = \frac{41.67(3)(12)}{0.025(28.0)(10^6)} = 0.0021429 \text{ in.}$$

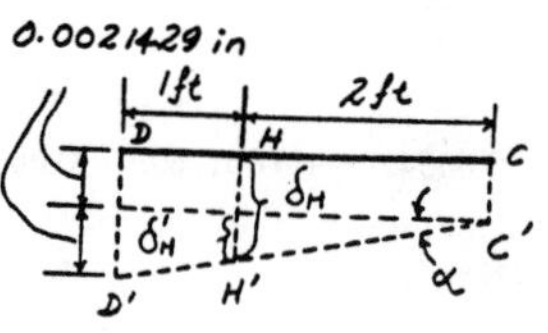

$$\frac{\delta'_H}{2} = \frac{0.0021429}{3}; \qquad \delta'_H = 0.0014286 \text{ in.}$$

$$\delta_H = 0.0014286 + 0.0021429 = 0.0035714 \text{ in.}$$

$$\delta_{A/H} = \frac{F_{AH}L_{AH}}{A_{AH}E} = \frac{125.0(1.8)(12)}{0.025(28.0)(10^6)} = 0.0038571 \text{ in.}$$

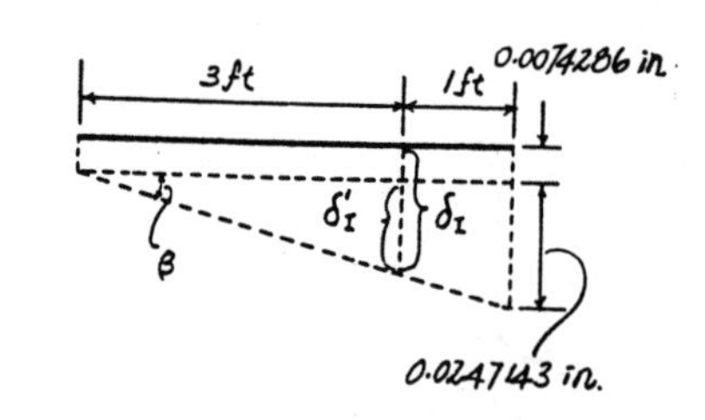

$$\delta_A = \delta_H + \delta_{A/H} = 0.0035714 + 0.0038571 = 0.0074286 \text{ in.}$$

$$\delta_B = \frac{F_{BG}L_{BG}}{A_{BG}E} = \frac{375.0(5)(12)}{0.025(28.0)(10^6)} = 0.0321428 \text{ in.}$$

$$\frac{\delta'_I}{3} = \frac{0.0247143}{4}; \qquad \delta'_I = 0.0185357 \text{ in.}$$

$$\delta_I = 0.0074286 + 0.0185357 = 0.0260 \text{ in.} \qquad \textbf{Ans}$$

4–15. The load is supported by the four 304 stainless steel wires that are connected to the rigid members *AB* and *DC*. Determine the angle of tilt of each member after the 500-lb load is applied. The members were originally horizontal, and each wire has a cross-sectional area of 0.025 in^2.

Internal Forces in the wires :

FBD (b)

$\circlearrowleft + \Sigma M_A = 0; \quad F_{BG}(4) - 500(3) = 0 \quad F_{BG} = 375.0 \text{ lb}$

$+\uparrow \Sigma F_y = 0; \quad F_{AH} + 375.0 - 500 = 0 \quad F_{AH} = 125.0 \text{ lb}$

FBD (a)

$\circlearrowleft + \Sigma M_D = 0; \quad F_{CF}(3) - 125.0(1) = 0 \quad F_{CF} = 41.67 \text{ lb}$

$+\uparrow \Sigma F_y = 0; \quad F_{DE} + 41.67 - 125.0 = 0 \quad F_{DE} = 83.33 \text{ lb}$

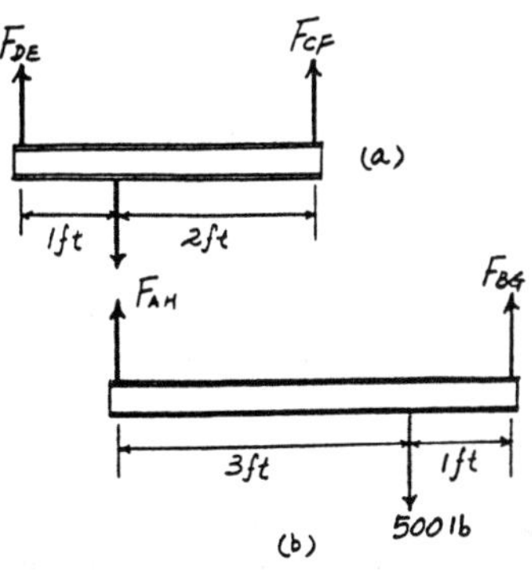

Displacement :

$$\delta_D = \frac{F_{DE}L_{DE}}{A_{DE}E} = \frac{83.33(3)(12)}{0.025(28.0)(10^6)} = 0.0042857 \text{ in.}$$

$$\delta_C = \frac{F_{CF}L_{CF}}{A_{CF}E} = \frac{41.67(3)(12)}{0.025(28.0)(10^6)} = 0.0021429 \text{ in.}$$

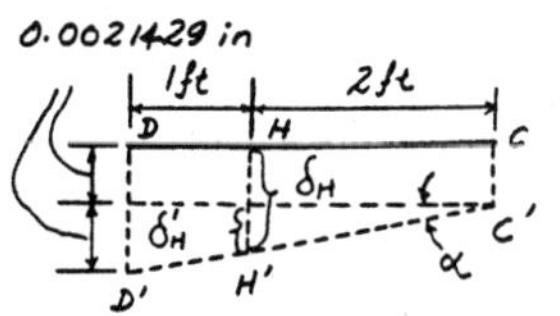

$$\frac{\delta_H'}{2} = \frac{0.0021429}{3}; \qquad \delta_H' = 0.0014286 \text{ in.}$$

$$\delta_H = \delta_H' + \delta_C = 0.0014286 + 0.0021429 = 0.0035714 \text{ in.}$$

$$\tan\alpha = \frac{0.0021429}{36}; \qquad \alpha = 0.00341° \qquad \textbf{Ans}$$

$$\delta_{A/H} = \frac{F_{AH}L_{AH}}{A_{AH}E} = \frac{125.0(1.8)(12)}{0.025(28.0)(10^6)} = 0.0038571 \text{ in.}$$

$$\delta_A = \delta_H + \delta_{A/H} = 0.0035714 + 0.0038571 = 0.0074286 \text{ in.}$$

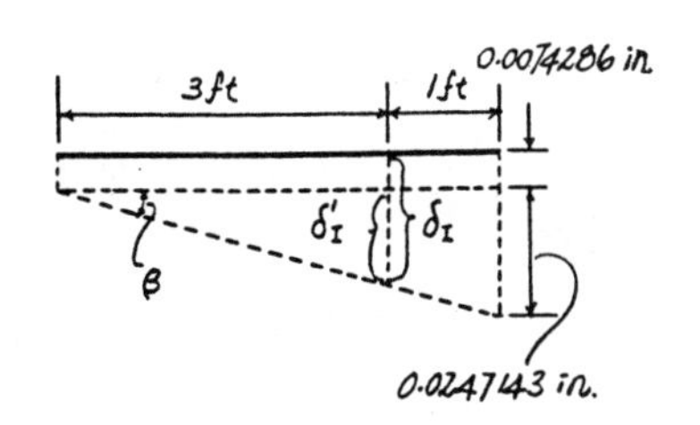

$$\delta_B = \frac{F_{BG}L_{BG}}{A_{BG}E} = \frac{375.0(5)(12)}{0.025(28.0)(10^6)} = 0.0321428 \text{ in.}$$

$$\tan\beta = \frac{0.0247143}{48}; \qquad \beta = 0.0295° \qquad \textbf{Ans}$$

***4–16.** The assembly consists of three titanium (Ti-6A1-4V) rods and a rigid bar AC. The cross-sectional area of each rod is given in the figure. If a force of 6 kip is applied to the ring F, determine the horizontal displacement of point F.

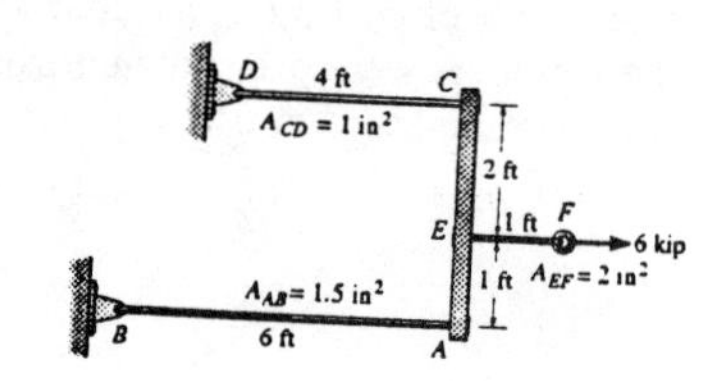

Internal Force in the Rods :

$$\circlearrowleft + \Sigma M_A = 0; \quad F_{CD}(3) - 6(1) = 0 \quad F_{CD} = 2.00 \text{ kip}$$

$$\xrightarrow{+} \Sigma F_x = 0; \quad 6 - 2.00 - F_{AB} = 0 \quad F_{AB} = 4.00 \text{ kip}$$

Displacement :

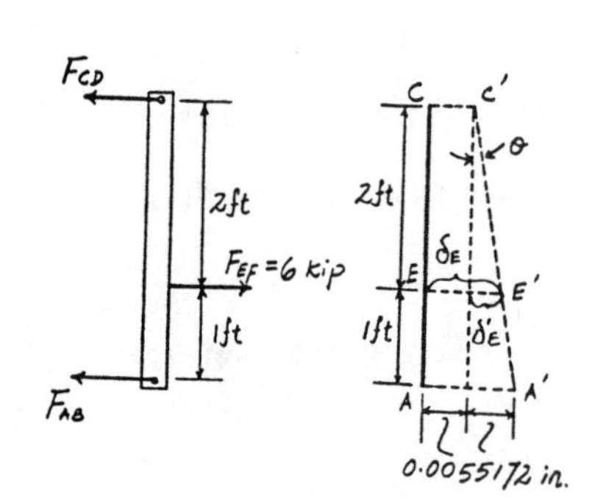

$$\delta_C = \frac{F_{CD}L_{CD}}{A_{CD}E} = \frac{2.00(4)(12)}{(1)(17.4)(10^3)} = 0.0055172 \text{ in.}$$

$$\delta_A = \frac{F_{AB}L_{AB}}{A_{AB}E} = \frac{4.00(6)(12)}{(1.5)(17.4)(10^3)} = 0.0110344 \text{ in.}$$

$$\delta_{F/E} = \frac{F_{EF}L_{EF}}{A_{EF}E} = \frac{6.00(1)(12)}{(2)(17.4)(10^3)} = 0.0020690 \text{ in.}$$

$$\frac{\delta'_E}{2} = \frac{0.0055172}{3}; \quad \delta'_E = 0.0036782 \text{ in.}$$

$$\delta_E = \delta_C + \delta'_E = 0.0055172 + 0.0036782 = 0.0091954 \text{ in.}$$

$$\delta_F = \delta_E + \delta_{F/E}$$
$$= 0.0091954 + 0.0020690 = 0.0113 \text{ in.} \quad \textbf{Ans}$$

4–17. The assembly consists of three titanium (Ti-6A1-4V) rods and a rigid bar AC. The cross-sectional area of each rod is given in the figure. If a force of 6 kip is applied to the ring F, determine the angle of tilt in radians of bar AC

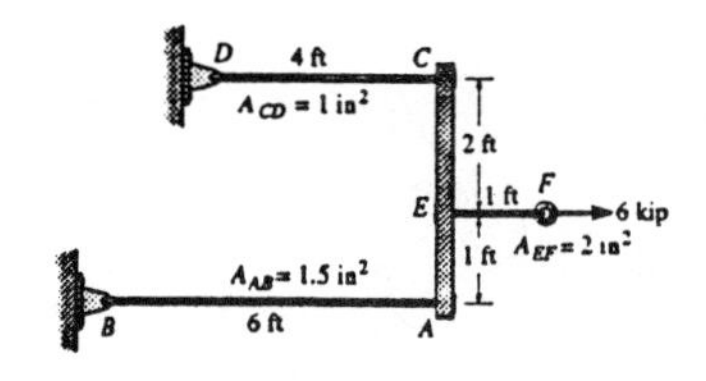

Internal Force in the Rods :

$$\circlearrowleft + \Sigma M_A = 0; \quad F_{CD}(3) - 6(1) = 0 \quad F_{CD} = 2.00 \text{ kip}$$

$$\xrightarrow{+} \Sigma F_x = 0; \quad 6 - 2.00 - F_{AB} = 0 \quad F_{AB} = 4.00 \text{ kip}$$

Displacement :

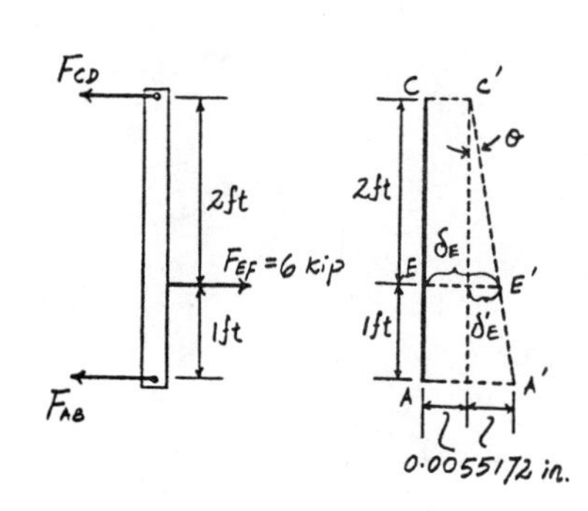

$$\delta_C = \frac{F_{CD}L_{CD}}{A_{CD}E} = \frac{2.00(4)(12)}{(1)(17.4)(10^3)} = 0.0055172 \text{ in.}$$

$$\delta_A = \frac{F_{AB}L_{AB}}{A_{AB}E} = \frac{4.00(6)(12)}{(1.5)(17.4)(10^3)} = 0.0110344 \text{ in.}$$

$$\theta = \tan^{-1}\frac{\delta_A - \delta_C}{3(12)} = \tan^{-1}\frac{0.0110344 - 0.0055172}{3(12)}$$
$$= 0.00878° \quad \textbf{Ans}$$

4–18. A spring-supported pipe hanger consists of two springs which are originally unstretched and have a stiffness of $k = 60$ kN/m, three 304 stainless steel rods, AB and CD which have a diameter of 5 mm and EF which has a diameter of 12 mm, and a rigid beam GH. If the pipe and the fluid it carries have a total weight of 4 kN, determine the displacement of the pipe when it is attached to the support.

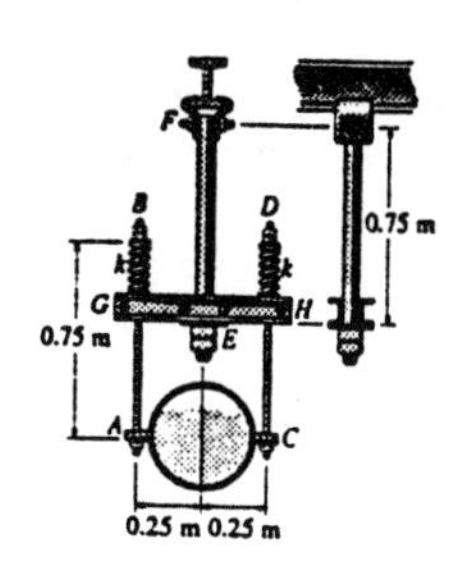

Internal Force in the Rods :

FBD (a)

$$\curvearrowleft + \Sigma M_A = 0; \quad F_{CD}(0.5) - 4(0.25) = 0 \quad F_{CD} = 2.00 \text{ kN}$$

$$+\uparrow \Sigma F_y = 0; \quad F_{AB} + 2.00 - 4 = 0 \quad F_{AB} = 2.00 \text{ kN}$$

FBD (b)

$$+\uparrow \Sigma F_y = 0; \quad F_{EF} - 2.00 - 2.00 = 0 \quad F_{EF} = 4.00 \text{ kN}$$

Displacement :

$$\delta_D = \delta_E = \frac{F_{EF}L_{EF}}{A_{EF}E} = \frac{4.00(10^3)(750)}{\frac{\pi}{4}(0.012)^2(193)(10^9)} = 0.1374 \text{ mm}$$

$$\delta_{A/B} = \delta_{C/D} = \frac{P_{CD}L_{CD}}{A_{CD}E} = \frac{2(10^3)(750)}{\frac{\pi}{4}(0.005)^2(193)(10^9)} = 0.3958 \text{ mm}$$

$$\delta_C = \delta_D + \delta_{C/D} = 0.1374 + 0.3958 = 0.5332 \text{ mm}$$

Displacement of the spring

$$\delta_{sp} = \frac{F_{sp}}{k} = \frac{2.00}{60} = 0.0333333 \text{ m} = 33.3333 \text{ mm}$$

$$\delta_{tot} = \delta_C + \delta_{sp}$$
$$= 0.5332 + 33.3333 = 33.87 \text{ mm} \qquad \textbf{Ans}$$

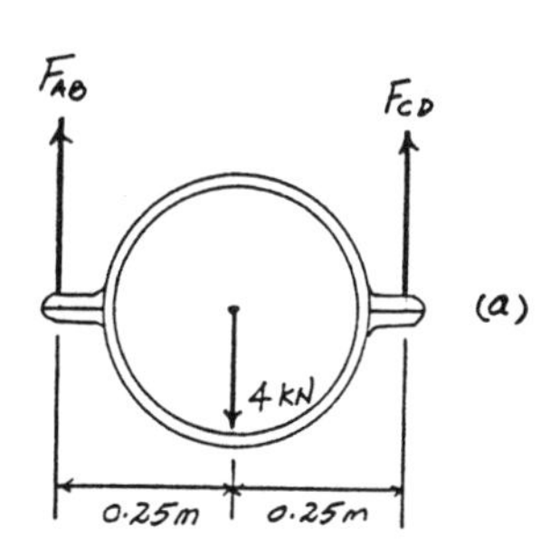

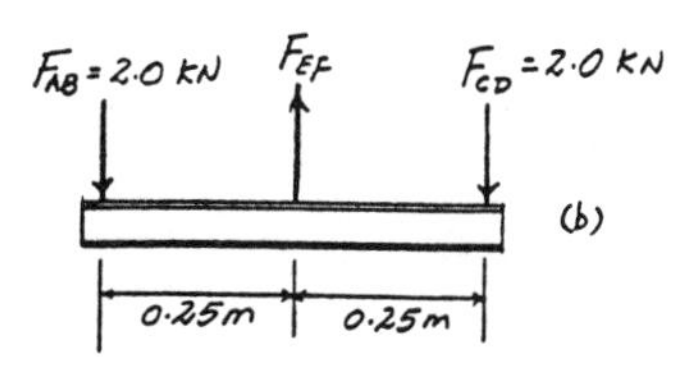

4–19. A spring-supported pipe hanger consists of two springs which are originally unstretched and have a stiffness of $k = 60$ kN/m, three 304 stainless steel rods, AB and CD which have a diameter of 5 mm and EF which has a diameter of 12 mm, and a rigid beam GH. If the pipe is displaced 82 mm when it is filled with fluid, determine the weight of the fluid.

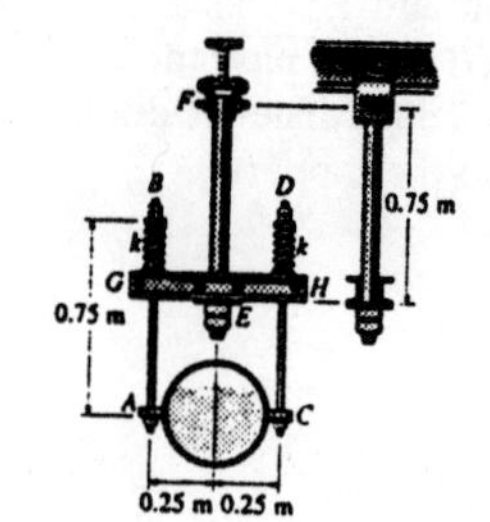

Internal Force in the Rods :

FBD (a)

$$\circlearrowleft + \Sigma M_A = 0; \quad F_{CD}(0.5) - W(0.25) = 0 \quad F_{CD} = \frac{W}{2}$$

$$+\uparrow \Sigma F_y = 0; \quad F_{AB} + \frac{W}{2} - W = 0 \quad F_{AB} = \frac{W}{2}$$

FBD (b)

$$+\uparrow \Sigma F_y = 0; \quad F_{EF} - \frac{W}{2} - \frac{W}{2} = 0 \quad F_{EF} = W$$

Displacement :

$$\delta_D = \delta_E = \frac{F_{EF}L_{EF}}{A_{EF}E} = \frac{W(750)}{\frac{\pi}{4}(0.012)^2(193)(10^9)} = 34.35988(10^{-6})W$$

$$\delta_{A/B} = \delta_{C/D} = \frac{F_{CD}L_{CD}}{A_{CD}E} = \frac{\frac{W}{2}(750)}{\frac{\pi}{4}(0.005)^2(193)(10^9)} = 98.95644(10^{-6})W$$

$$\delta_C = \delta_D + \delta_{C/D}$$
$$= 34.35988(10^{-6})W + 98.95644(10^{-6})W$$
$$= 0.133316(10^{-3})W$$

Displacement of the spring

$$\delta_{sp} = \frac{F_{sp}}{k} = \frac{\frac{W}{2}}{60(10^3)}(1000) = 0.008333\,W$$

$$\delta_{tot} = \delta_C + \delta_{sp}$$
$$82 = 0.133316(10^{-3})W + 0.008333W$$

$$W = 9685 \text{ N} = 9.69 \text{ kN}$$ **Ans**

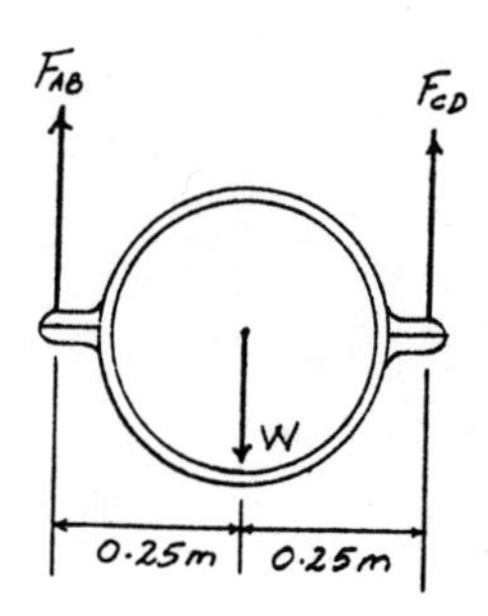

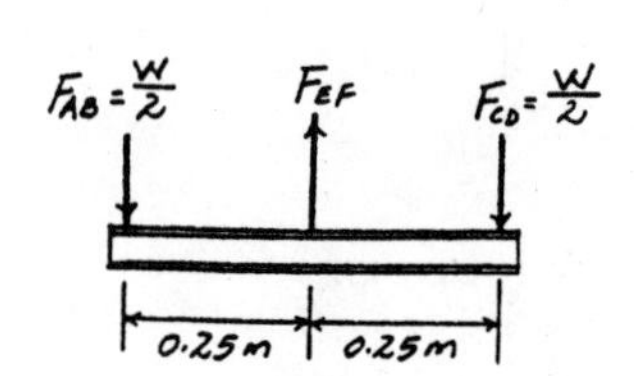

***4–20.** The rigid beam is supported at its ends by two A-36 steel tie rods. If the allowable stress for the steel is $\sigma_{allow} = 16.2$ ksi, the load $w = 3$ kip/ft, and $x = 4$ ft, determine the diameter of each rod so that the beam remains in the horizontal position when it is loaded.

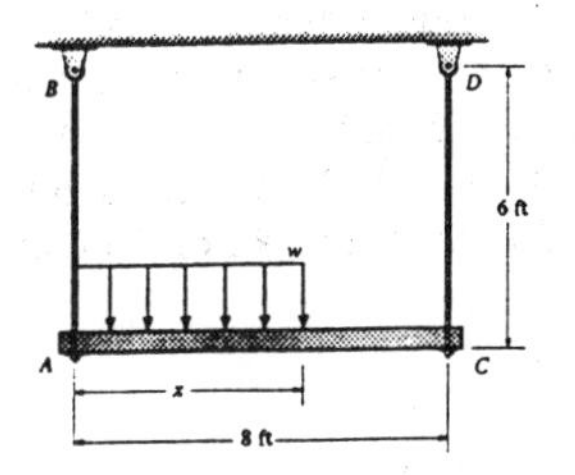

Internal Force in the Rods :

$$+\Sigma M_A = 0; \quad F_{CD}(8) - 12.0(2) = 0 \quad F_{CD} = 3.00 \text{ kip}$$

$$+\uparrow \Sigma F_y = 0; \quad F_{AB} + 3.00 - 12.0 = 0 \quad F_{AB} = 9.00 \text{ kip}$$

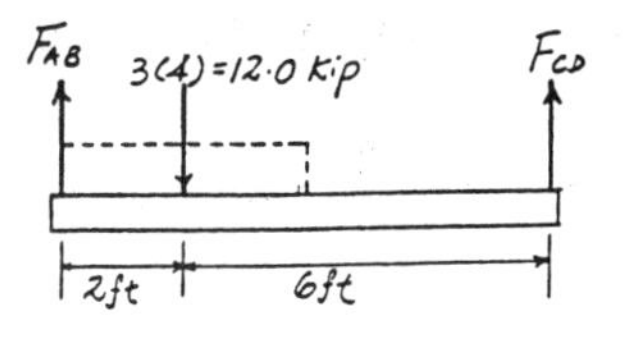

Displacement : To maintain the rigid beam in the horizontal position, the elongation of both rods AB and CD must be the same.

$$\delta_{AB} = \delta_{CD}$$

$$\frac{9.00(6)(12)}{\frac{\pi}{4}d_{AB}^2 E} = \frac{3.00(6)(12)}{\frac{\pi}{4}d_{CD}^2 E};$$

$$9d_{CD}^2 = 3d_{AB}^2 ; \quad d_{AB} = \sqrt{3}\, d_{CD} \qquad [1]$$

Allowable Normal Stress : Assume failure of rod AB

$$\sigma_{allow} = \frac{F_{AB}}{A_{AB}}; \quad 16.2 = \frac{9.00}{\frac{\pi}{4}d_{AB}^2}$$

$$d_{AB} = 0.841 \text{ in.} \qquad \textbf{Ans}$$

From Eq. [1] $\quad d_{CD} = 0.486$ in. **Ans**

Assume failure of rod CD

$$\sigma_{allow} = \frac{F_{CD}}{A_{AB}}; \quad 16.2 = \frac{3.00}{\frac{\pi}{4}d_{CD}^2}$$

$$d_{CD} = 0.486 \text{ in.} \qquad \textbf{Ans}$$

From Eq. [1] $\quad d_{AB} = 0.841$ in. **Ans**

4–21. The rigid beam is supported at its ends by two A-36 steel tie rods. The rods have diameters $d_{AB} = 0.5$ in. and $d_{CD} = 0.3$ in. If the allowable stress for the steel is $\sigma_{allow} = 16.2$ ksi, determine the intensity of the distributed load w and its length x on the beam so that the beam remains in the horizontal position when it is loaded.

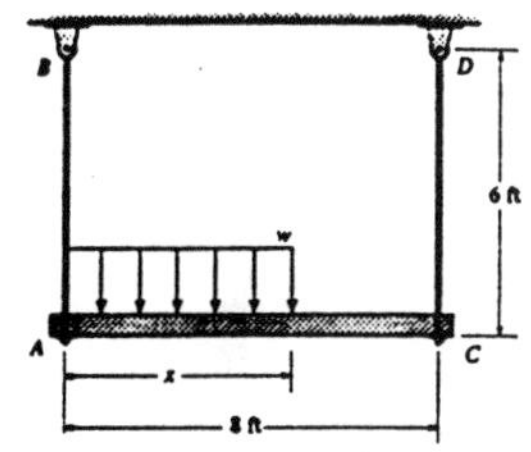

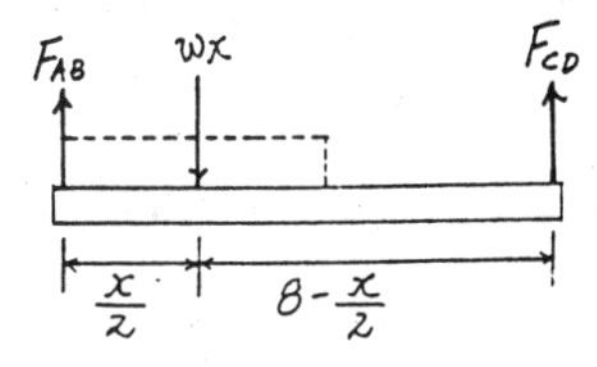

Internal Force in the Rods :

$$\curvearrowleft + \Sigma M_A = 0; \quad F_{CD}(8) - wx\left(\frac{x}{2}\right) = 0$$

$$8F_{CD} - \frac{wx^2}{2} = 0 \qquad [1]$$

$$\curvearrowleft + \Sigma M_C = 0; \quad -F_{AB}(8) + wx\left(8 - \frac{x}{2}\right) = 0$$

$$8wx - \frac{wx^2}{2} - 8F_{AB} = 0 \qquad [2]$$

Displacement : To maintain the rigid beam in the horizontal position, both elongations of rods AB and CD must be the same.

$$\delta_{AB} = \delta_{CD}$$

$$\frac{F_{AB}(6)(12)}{\frac{\pi}{4}(0.5^2)E} = \frac{F_{CD}(6)(12)}{\frac{\pi}{4}(0.3^2)E}$$

$$F_{CD} = 0.360\, F_{AB} \qquad [3]$$

Allowable Normal Stress : Assume failure of rod AB

$$\sigma_{allow} = \frac{F_{AB}}{A_{AB}}; \quad 16.2 = \frac{F_{AB}}{\frac{\pi}{4}(0.5^2)} \quad F_{AB} = 3.1809 \text{ kip}$$

Using $F_{AB} = 3.1809$ kip and solving Eqs. [1] to [3] yields :

$$F_{CD} = 1.1451 \text{ kip}$$

$$x = 4.24 \text{ ft} \qquad \textbf{Ans}$$

$$w = 1.02 \text{ kip / ft} \qquad \textbf{Ans}$$

Assume failure of rod CD

$$\sigma_{allow} = \frac{F_{CD}}{A_{CD}}; \quad 16.2 = \frac{F_{CD}}{\frac{\pi}{4}(0.3^2)} \quad F_{CD} = 1.1451 \text{ kip}$$

Therefore, rods AB and CD fail simultaneously.

4–22. The rigid frame is supported with A-36 steel wires AB and DE. If each wire has a diameter of 12 mm, determine the horizontal displacement of point E when the 50-kN load is applied to the frame.

Internal Force in the Wires : From FBD(a)

$$\circlearrowleft + \Sigma M_C = 0; \quad F_{AB}(1) - 50(1) = 0 \quad F_{AB} = 50.0 \text{ kN}$$

From FBD(b)

$$+\uparrow \Sigma F_y = 0; \quad \frac{3}{\sqrt{10}}(F_{EC}) - 50 = 0 \quad F_{EC} = 52.705 \text{ kN}$$

$$\xrightarrow{+} \Sigma F_x = 0; \quad \frac{1}{\sqrt{10}}(52.705) - F_{ED} = 0 \quad F_{ED} = 16.67 \text{ kN}$$

Displacement :

$$\delta_{AB} = \frac{F_{AB}L_{AB}}{A_{AB}E} = \frac{50.0(10^3)(2000)}{\frac{\pi}{4}(0.012^2)\,200(10^9)} = 4.4210 \text{ mm}$$

$$(\delta_B)_h = \frac{\delta_{AB}}{\tan 26.565^\circ} = 8.8419 \text{ mm}$$

$$\delta_{DE} = \frac{F_{DE}L_{DE}}{A_{DE}E} = \frac{16.67(10^3)(1000)}{\frac{\pi}{4}(0.012^2)\,200(10^9)} = 0.7368 \text{ mm}$$

$$(\delta_E)_h = (\delta_D)_h + \delta_{DE} = \frac{3}{2}(\delta_B)_h + \delta_{DE} = \frac{3}{2}(8.8419) + 0.7368 = 14.0 \text{ mm} \quad \textbf{Ans}$$

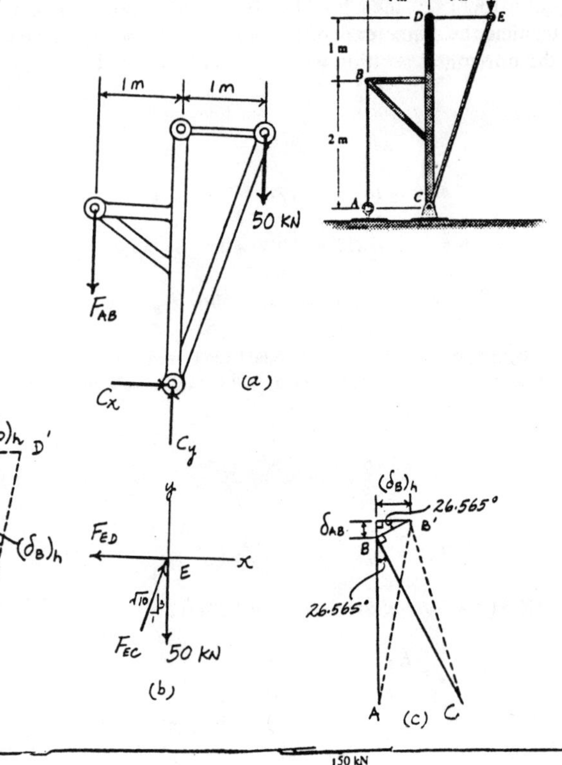

4–23. The rigid frame is constructed with A-36 steel wires AB and DE. If each wire has a diameter of 12 mm, determine the vertical displacement of point E when the 50-kN load is applied to the frame.

Internal Force in the Wires :
FBD (a)

$$\circlearrowleft + \Sigma M_C = 0; \quad F_{AB}(1) - 50(1) = 0 \quad F_{AB} = 50.0 \text{ kN}$$

FBD (b)

$$+\uparrow \Sigma F_y = 0; \quad \frac{3}{\sqrt{10}}(F_{EC}) - 50 = 0 \quad F_{EC} = 52.705 \text{ kN}$$

$$\xrightarrow{+} \Sigma F_x = 0; \quad \frac{1}{\sqrt{10}}(52.705) - F_{ED} = 0 \quad F_{ED} = 16.67 \text{ kN}$$

Displacement :

$$\delta_{AB} = \frac{F_{AB}L_{AB}}{A_{AB}E} = \frac{50.0(10^3)(2000)}{\frac{\pi}{4}(0.012^2)\,200(10^9)} = 4.4210 \text{ mm}$$

$$(\delta_B)_h = \frac{\delta_{AB}}{\tan 26.565^\circ} = 8.8419 \text{ mm}$$

$$\delta_{DE} = \frac{F_{DE}L_{DE}}{A_{DE}E} = \frac{16.67(10^3)(1000)}{\frac{\pi}{4}(0.012^2)\,200(10^9)} = 0.7368 \text{ mm}$$

$$(\delta_E)_h = (\delta_D)_h + \delta_{DE} = \frac{3}{2}(\delta_B)_h + \delta_{DE} = \frac{3}{2}(8.8419) + 0.7368 = 13.9997 \text{ mm}$$

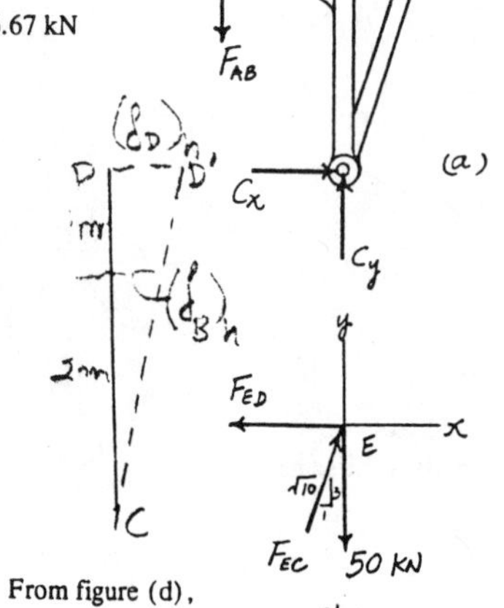

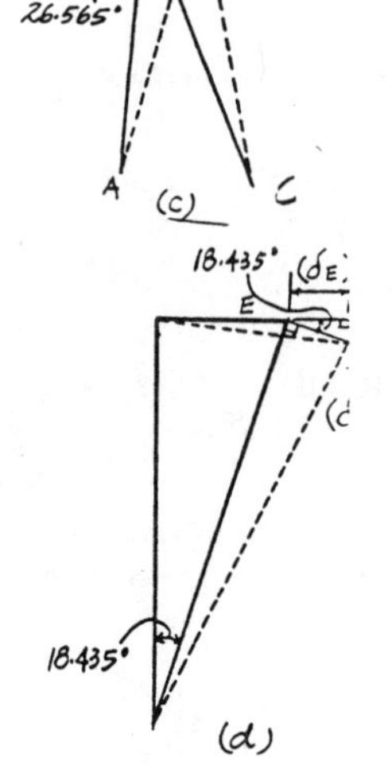

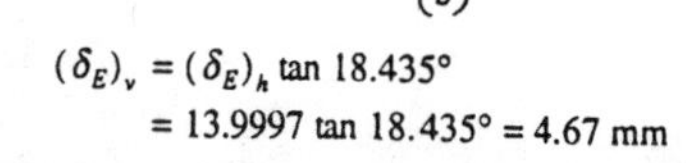

From figure (d),

$$(\delta_E)_v = (\delta_E)_h \tan 18.435^\circ = 13.9997 \tan 18.435^\circ = 4.67 \text{ mm} \quad \textbf{Ans}$$

***4–24.** The post is made of Douglas fir and has a diameter of 60 mm. If it is subjected to the load of 20 kN and the soil provides a frictional resistance that is uniformly distributed along its sides of $w = 4$ kN/m, determine the force **F** at its bottom needed for equilibrium. Also, what is the displacement of the top of the post A with respect to its bottom B? Neglect the weight of the post.

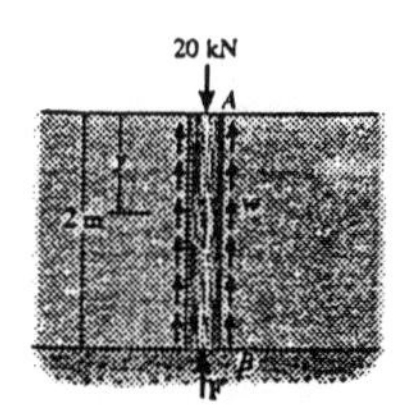

Equation of Equilibrium : For entire post [FBD (a)]

$$+\uparrow \Sigma F_y = 0; \quad F + 8.00 - 20 = 0 \quad F = 12.0 \text{ kN} \quad \textbf{Ans}$$

Internal Force : FBD (b)

$$+\uparrow \Sigma F_y = 0; \quad -F(y) + 4y - 20 = 0$$

$$F(y) = \{4y - 20\} \text{ kN}$$

Displacement :

$$\delta_{A/B} = \int_0^L \frac{F(y)\,dy}{A(y)E} = \frac{1}{AE}\int_0^{2\text{ m}} (4y - 20)\,dy$$

$$= \frac{1}{AE}\left(2y^2 - 20y\right)\Big|_0^{2\text{ m}}$$

$$= -\frac{32.0 \text{ kN}\cdot\text{m}}{AE}$$

$$= -\frac{32.0(10^3)}{\frac{\pi}{4}(0.06^2)\,13.1(10^9)}$$

$$= -0.8639\left(10^{-3}\right) \text{ m}$$

$$= -0.864 \text{ mm} \qquad \textbf{Ans}$$

Negative sign indicates that end A moves toward end B.

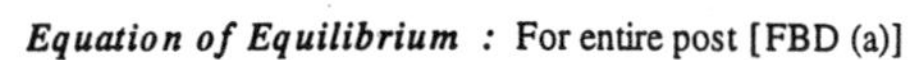

4–25. The post is made of Douglas fir and has a diameter of 60 mm. If it is subjected to the load of 20 kN and the soil provides a frictional resistance that is distributed along its length and varies linearly from $w = 0$ at $y = 0$ to $w = 3$ kN/m at $w = 2$ m, determine the force **F** at its bottom needed for equilibrium. Also, what is the displacement of the top of the post A with respect to its bottom B? Neglect the weight of the post.

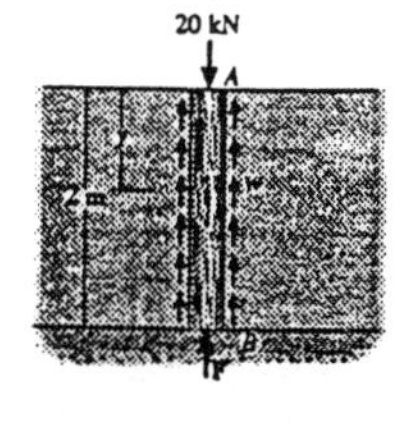

Equation of Equilibrium : For entire post [FBD (a)]

$$+\uparrow \Sigma F_y = 0; \quad F + 3.00 - 20 = 0 \quad F = 17.0 \text{ kN} \quad \textbf{Ans}$$

Internal Force : FBD (b)

$$+\uparrow \Sigma F_y = 0; \quad -F(y) + \frac{1}{2}\left(\frac{3y}{2}\right)y - 20 = 0$$

$$F(y) = \left\{\frac{3}{4}y^2 - 20\right\} \text{ kN}$$

Displacement :

$$\delta_{A/B} = \int_0^L \frac{F(y)\,dy}{A(y)E} = \frac{1}{AE}\int_0^{2\text{ m}} \left(\frac{3}{4}y^2 - 20\right)dy$$

$$= \frac{1}{AE}\left(\frac{y^3}{4} - 20y\right)\Big|_0^{2\text{ m}}$$

$$= -\frac{32.0 \text{ kN}\cdot\text{m}}{AE}$$

$$= -\frac{38.0(10^3)}{\frac{\pi}{4}(0.06^2)\,13.1(10^9)}$$

$$= -1.026\left(10^{-3}\right) \text{ m}$$

$$= -1.03 \text{ mm} \qquad \textbf{Ans}$$

Negative sign indicates that end A moves toward end B.

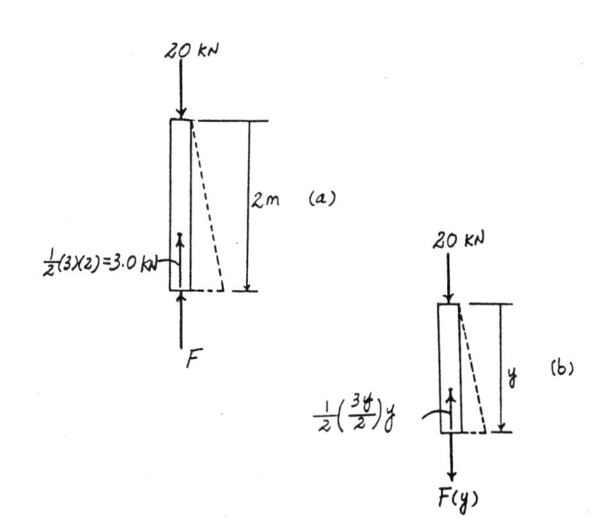

4–26. The segments of pipe and couplings used for drilling an oil well 15 000 ft deep are made of A-36 steel weighing 20 lb/ft. They have an outer diameter of 5.50 in. and an inner diameter of 4.75 in. In order to prevent buckling or sidesway of the pipe due to its own weight, it is partially supported at its top by the drawworks of the rig. If this force is $P = 299$ kip, determine the force **F** of the ground on the drill pipe and the elongation of the pipe for this condition.

Equation of Equilibrium : For the entire pipe [FBD (a)]

$$+\uparrow \Sigma F_y = 0; \quad 299 + F - 0.02(15000) = 0$$

$$F = 1.00 \text{ kip} \qquad \textbf{Ans}$$

Internal Force : FBD (b).

$$+\uparrow \Sigma F_y = 0; \quad P(y) - 0.02y + 1.00 = 0$$

$$P(y) = \{0.02y - 1.00\} \text{ kip}$$

Displacement :

$$\delta = \int_0^L \frac{P(y)\,dy}{A(y)\,E} = \frac{1}{AE}\int_0^{15000\text{ ft}} (0.02y - 1.00)\,dy$$

$$= \frac{2.235(10^6)\text{ kip}\cdot\text{ft}}{AE}$$

$$= \frac{2.235(10^6)}{\frac{\pi}{4}(5.50^2 - 4.75^2)(29.0)(10^3)}$$

$$= 12.8 \text{ ft} \qquad \textbf{Ans}$$

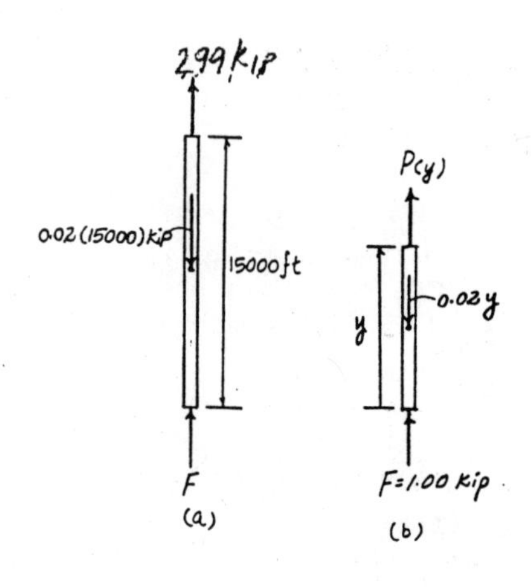

4–27. The segments of pipe and couplings used for drilling an oil well 15 000 ft deep are made of A-36 steel weighing 20 lb/ft. They have an outer diameter of 5.50 in. and an inner diameter of 4.75 in. Determine the force **P** needed to remove the pipe, excluding friction along its sides and requiring $F = 0$. Also, what is the elongation of the pipe as it just begins to be lifted out?

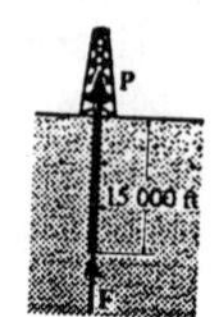

Equation of Equilibrium : For the entire pipe [FBD (a)]

$$+\uparrow \Sigma F_y = 0; \quad P - 0.02(15000) = 0$$

$$P = 300 \text{ kip} \qquad \textbf{Ans}$$

Internal Force : FBD(b)

$$+\uparrow \Sigma F_y = 0; \quad P(y) - 0.02y = 0 \qquad P(y) = \{0.02y\} \text{ kip}$$

Displacement :

$$\delta = \int_0^L \frac{P(y)\,dy}{A(y)\,E} = \frac{1}{AE}\int_0^{15000\text{ ft}} (0.02y)\,dy$$

$$= \frac{2.25(10^6)\text{ kip}\cdot\text{ft}}{AE}$$

$$= \frac{2.25(10^6)}{\frac{\pi}{4}(5.50^2 - 4.75^2)(29.0)(10^3)}$$

$$= 12.9 \text{ ft} \qquad \textbf{Ans}$$

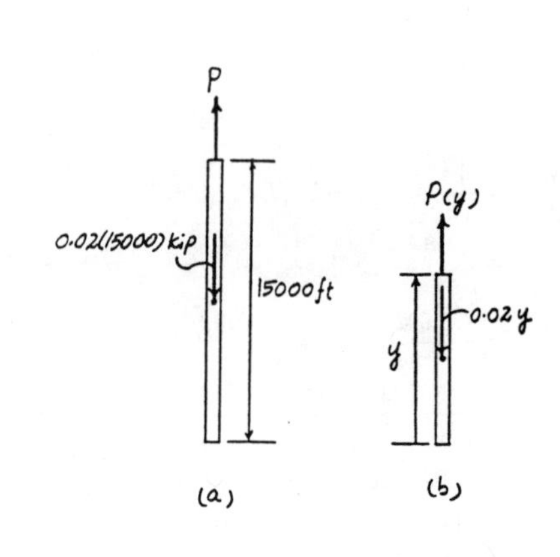

***4–28.** The rod has a slight taper and length L. It is suspended from the ceiling and supports a load $\mathbf{P}$ at its end. Show that the displacement of its end due to this load is $\delta = PL/(\pi E r_2 r_1)$. Neglect the weight of the material. The modulus of elasticity is E.

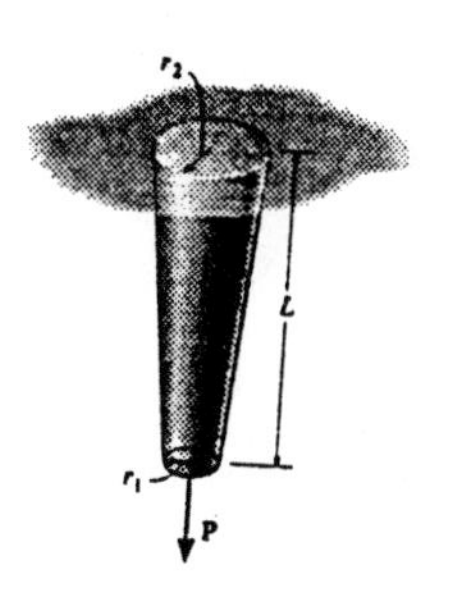

Geometry :

$$r(y) = r_1 + \frac{r_2 - r_1}{L}y = \frac{r_1 L + (r_2 - r_1)y}{L}$$

$$A(y) = \frac{\pi}{L^2}\left[r_1 L + (r_2 - r_1)y\right]^2$$

Displacement :

$$\delta = \int_0^L \frac{P(y)\,dy}{A(y)E} = \frac{PL^2}{\pi E}\int_0^L \frac{dy}{\left[r_1 L + (r_2 - r_1)y\right]^2}$$

$$= -\frac{PL^2}{\pi E}\left[\frac{1}{(r_2 - r_1)\left[r_1 L + (r_2 - r_1)y\right]}\right]\Bigg|_0^L$$

$$= -\frac{PL^2}{\pi E(r_2 - r_1)}\left[\frac{1}{r_1 L + (r_2 - r_1)L} - \frac{1}{r_1 L}\right]$$

$$= -\frac{PL^2}{\pi E(r_2 - r_1)}\left[\frac{1}{r_2 L} - \frac{1}{r_1 L}\right]$$

$$= -\frac{PL^2}{\pi E(r_2 - r_1)}\left[\frac{r_1 - r_2}{r_2 r_1 L}\right]$$

$$= \frac{PL^2}{\pi E(r_2 - r_1)}\left[\frac{r_2 - r_1}{r_2 r_1 L}\right]$$

$$= \frac{PL}{\pi E r_2 r_1} \qquad \textit{(Q.E.D)}$$

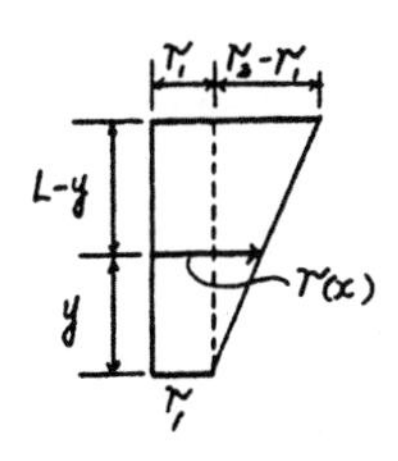

4–29. Solve Prob. 4–28 by including the weight of the material, considering its specific weight to be γ (weight/volume).

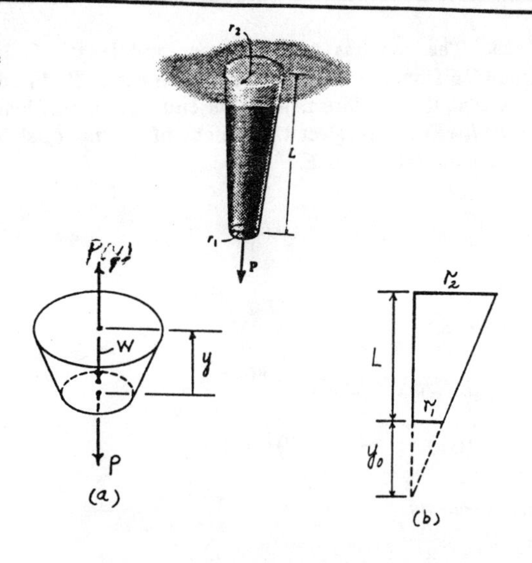

Internal Forces **:** FBD(a).

$$+\uparrow \Sigma F_y = 0; \qquad P(y) - P - W = 0 \qquad P(y) = P + W$$

Geometry **:** From diagram (b)

$$\frac{L + y_0}{r_2} = \frac{y_0}{r_1}; \qquad y_0 = \frac{L\, r_1}{r_2 - r_1}$$

From diagram (c)

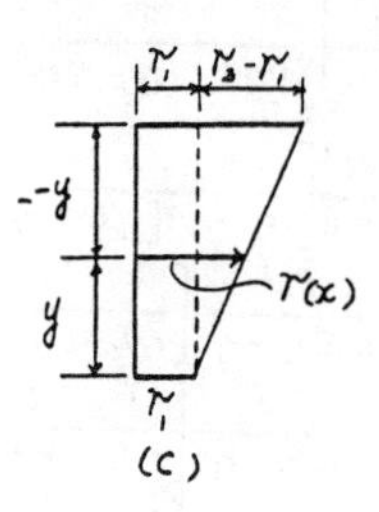

$$r(y) = r_1 + \frac{r_2 - r_1}{L} y = \frac{r_1 L + (r_2 - r_1) y}{L}$$

$$A(y) = \frac{\pi}{L^2}\left[r_1 L + (r_2 - r_1) y\right]^2$$

$$W = \frac{\gamma \pi}{3L^2}\left[r_1 L + (r_2 - r_1) y\right]^2 \left[y + \frac{L\, r_1}{r_2 - r_1}\right] - \frac{\gamma \pi}{3}\left(r_1^2\right)\left(\frac{L\, r_1}{r_2 - r_1}\right)$$

$$= \frac{\gamma \pi}{3L^2 (r_2 - r_1)}\left\{\left[r_1 L + (r_2 - r_1) y\right]^3 - r_1^3 L^3\right\}$$

Displacement :

$$\delta = \int_0^L \frac{P(y)\, dy}{A(y) E} = \int_0^L \frac{P dy}{A(y) E} + \int_0^L \frac{W dy}{A(y) E}$$

From prob. 4 - 28, $\qquad \delta_P = \int_0^L \frac{P dy}{A(y) E} = \frac{PL}{\pi E\, r_2 r_1}$

$$\delta_W = \int_0^L \frac{W dy}{A(y) E}$$

$$= \frac{\gamma}{3E(r_2 - r_1)} \int_0^L \frac{\left[r_1 L + (r_2 - r_1) y\right]^3 - r_1^3 L^3}{\left[r_1 L + (r_2 - r_1) y\right]^2} dy$$

$$= \frac{\gamma}{3E(r_2 - r_1)} \int_0^L \left[r_1 L + (r_2 - r_1) y\right] dy - \frac{\gamma r_1^3 L^3}{3E(r_2 - r_1)} \int_0^L \frac{dy}{\left[r_1 L + (r_2 - r_1) y\right]^2}$$

$$= \frac{\gamma}{3E(r_2 - r_1)}\left[r_1 L y + \frac{(r_2 - r_1) y^2}{2}\right]\Bigg|_0^L + \frac{\gamma r_1^3 L^3}{3E(r_2 - r_1)^2}\left[\frac{1}{r_1 L + (r_2 - r_1) y}\right]\Bigg|_0^L$$

$$= \frac{\gamma}{3E(r_2 - r_1)}\left[r_1 L^2 + \frac{(r_2 - r_1) L^2}{2}\right] + \frac{\gamma r_1^3 L^3}{3E(r_2 - r_1)^2}\left[\frac{1}{r_2 L} - \frac{1}{r_1 L}\right]$$

$$= \frac{\gamma}{6E(r_2 - r_1)}\left[2 r_1 L^2 + r_2 L^2 - r_1 L^2\right] + \frac{\gamma r_1^3 L^3}{3E(r_2 - r_1)^2}\left[\frac{-(r_2 - r_1)}{r_2 r_1 L}\right]$$

$$= \frac{\gamma L^2 (r_2 + r_1)}{6E(r_2 - r_1)} - \frac{\gamma L^2 r_1^2}{3E\, r_2 (r_2 - r_1)}$$

Therefore,

$$\delta = \delta_W + \delta_P = \frac{PL}{\pi E\, r_2 r_1} + \frac{\gamma L^2 (r_2 + r_1)}{6E(r_2 - r_1)} - \frac{\gamma L^2 r_1^2}{3E\, r_2 (r_2 - r_1)} \qquad \textbf{Ans}$$

4-30. Determine the relative displacement of one end of the tapered plate with respect to the other end when it is subjected to an axial load P. Neglect the weight of the plate.

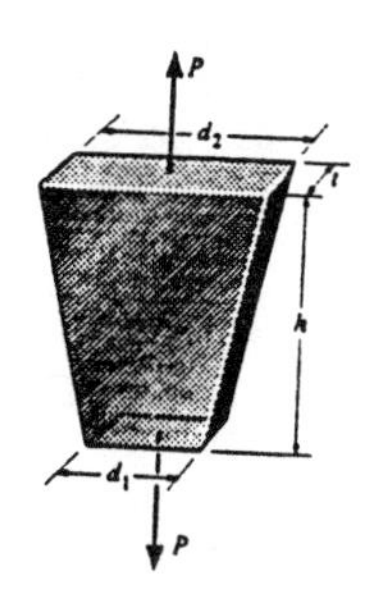

Geometry :

$$d(y) = d_1 + \frac{d_2 - d_1}{h}y = \frac{d_1 h + (d_2 - d_1)y}{h}$$

$$A(y) = \left[\frac{d_1 h + (d_2 - d_1)y}{h}\right]t$$

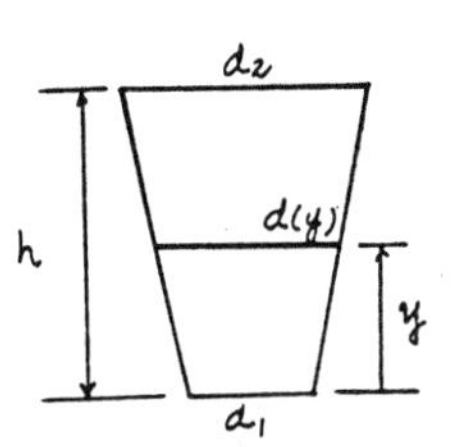

Displacement :

$$\delta = \int_0^L \frac{P(y)\,dy}{A(y)E}$$

$$= \frac{P}{E}\int_0^h \frac{dy}{\left[\frac{d_1 h + (d_2 - d_1)y}{h}\right]t}$$

$$= \frac{Ph}{Et}\int_0^h \frac{dy}{d_1 h + (d_2 - d_1)y}$$

$$= \frac{Ph}{Etd_1 h}\int_0^h \frac{dy}{1 + \frac{d_2 - d_1}{d_1 h}y}$$

$$= \frac{Ph}{Etd_1 h}\left(\frac{d_1 h}{d_2 - d_1}\right)\left[\ln\left(1 + \frac{d_2 - d_1}{d_1 h}y\right)\right]\Bigg|_0^h$$

$$= \frac{Ph}{Et(d_2 - d_1)}\left[\ln\left(1 + \frac{d_2 - d_1}{d_1}\right)\right]$$

$$= \frac{Ph}{Et(d_2 - d_1)}\left[\ln\left(\frac{d_1 + d_2 - d_1}{d_1}\right)\right]$$

$$= \frac{Ph}{Et(d_2 - d_1)}\ln\frac{d_2}{d_1} \qquad \textbf{Ans}$$

■4-31. The bar is made of 2014-T6 aluminum alloy and has a variable diameter that can be expressed as $d = (x^2 + 0.02)$ m. Determine the relative displacement of end A with respect to end B.

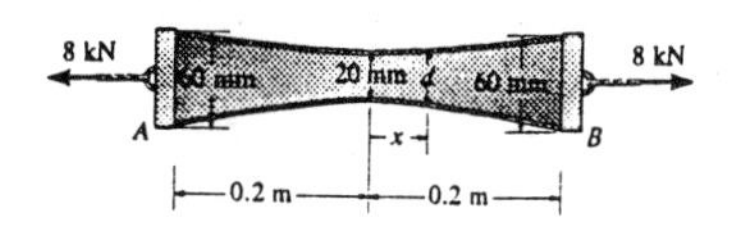

Internal Forces : As shown on FBD.

Displacement :

$$\delta_{A/B} = \int_0^L \frac{P(x)\,dx}{A(x)E} \qquad \text{Where} \quad A(x) = \frac{\pi}{4}\left(x^2 + 0.02\right)^2$$

$$= 2\int_0^{0.2\text{ m}} \frac{8.00(10^3)}{\frac{\pi}{4}(x^2 + 0.02)^2(73.1)(10^9)}dx$$

$$= 0.27868\left(10^{-6}\right)\int_0^{0.2\text{m}} \frac{dx}{(x^2 + 0.02)^2} = 0.0703 \text{ mm} \quad \textbf{Ans}$$

***4–32.** The bar has a variable radius of $r = (0.1/(x+1))$ m and a variable modulus of elasticity of $E = 125(1+x)$ MPa. Determine the displacement of end A when it is subjected to the axial load of 4 kN.

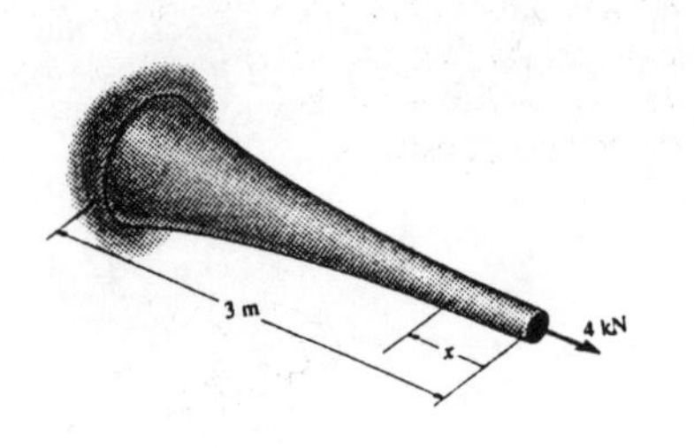

Internal Forces : As shown on FBD.

Displacement :

$$\delta = \int_0^L \frac{P\,dx}{A(x)E(x)} \qquad \text{Where} \quad A = \pi\left(\frac{0.1}{x+1}\right)^2$$

$$= 4(10^3)\int_0^{3\,\text{m}} \frac{dx}{\pi\left(\frac{0.1}{x+1}\right)^2 125(10^6)(1+x)}$$

$$= \frac{0.032}{\pi}\int_0^{3\,\text{m}} (x+1)\,dx = 0.007639 \text{ m}$$

$$= 7.64 \text{ mm}$$

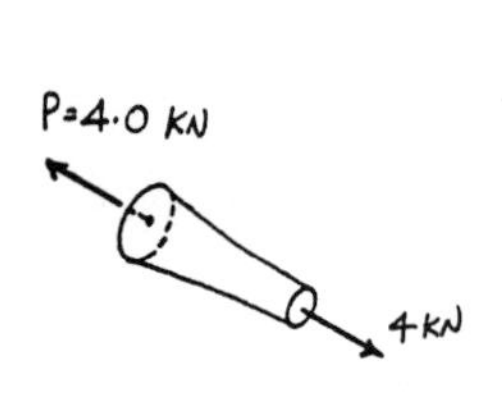

4–33. The support is made by cutting off the two opposite sides of a sphere that has a radius r_0. If the original height of the support is $r_0/2$, determine how far it shortens when it supports a load **P**. The modulus of elasticity is E.

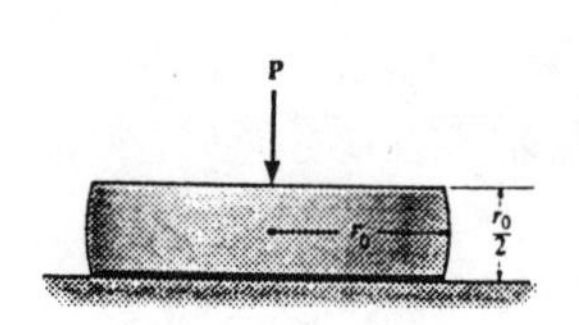

Geometry :

$$A = \pi r^2 = \pi(r_0\cos\theta)^2 = \pi r_0^2\cos^2\theta$$

$$y = r_0\sin\theta\,; \qquad dy = r_0\cos\theta\,d\theta$$

Displacement :

$$\delta = \int_0^L \frac{P(y)\,dy}{A(y)\,E}$$

$$= 2\left[\frac{P}{E}\int_0^{\theta} \frac{r_0\cos\theta\,d\theta}{\pi r_0^2\cos^2\theta}\right] = 2\left[\frac{P}{\pi r_0 E}\int_0^{\theta} \frac{d\theta}{\cos\theta}\right]$$

$$= \frac{2P}{\pi r_0 E}[\ln(\sec\theta + \tan\theta)]\Big|_0^{\theta}$$

$$= \frac{2P}{\pi r_0 E}[\ln(\sec\theta + \tan\theta)]$$

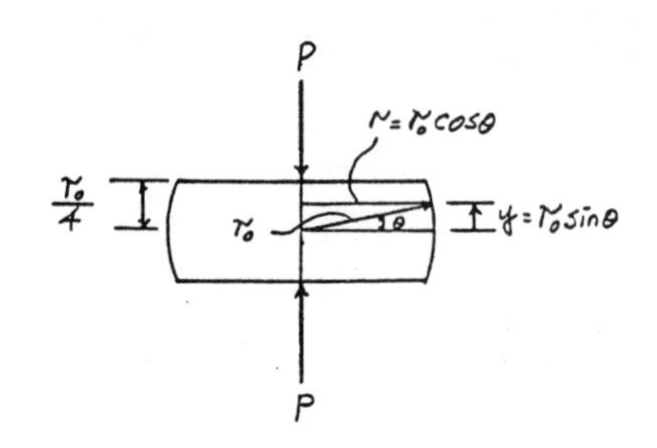

When $y = \frac{r_0}{4}$; $\quad \theta = 14.48°$

$$\delta = \frac{2P}{\pi r_0 E}[\ln(\sec 14.48° + \tan 14.48°)]$$

$$= \frac{0.511P}{\pi r_0 E} \qquad \textbf{Ans}$$

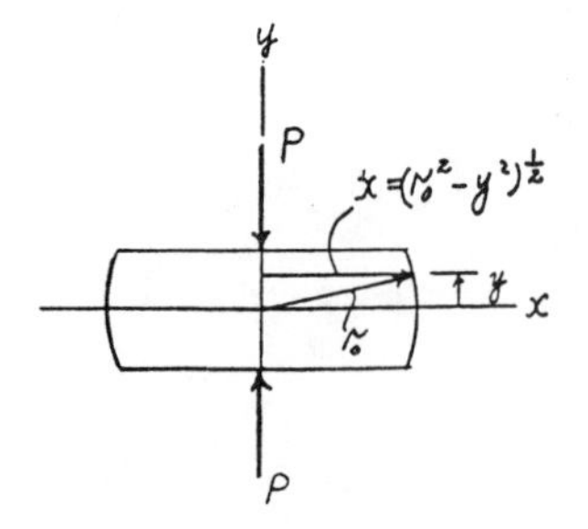

Also,

Geometry :

$$A(y) = \pi x^2 = \pi(r_0^2 - y^2)$$

Displacement :

$$\delta = \int_0^L \frac{P(y)\,dy}{A(y)\,E}$$

$$= \frac{2P}{\pi E}\int_0^{\frac{r_0}{4}} \frac{dy}{r_0^2 - y^2} = \frac{2P}{\pi E}\left[\frac{1}{2r_0}\ln\frac{r_0+y}{r_0-y}\right]\Bigg|_0^{\frac{r_0}{4}}$$

$$= \frac{P}{\pi r_0 E}[\ln 1.667 - \ln 1]$$

$$= \frac{0.511\,P}{\pi r_0 E} \qquad \textbf{Ans}$$

4–34. The casting is made of a material that has a specific weight γ and modulus of elasticity E. If it is formed into a pyramid having the dimensions shown, determine how far its end is displaced due to gravity when it is suspended in the vertical position.

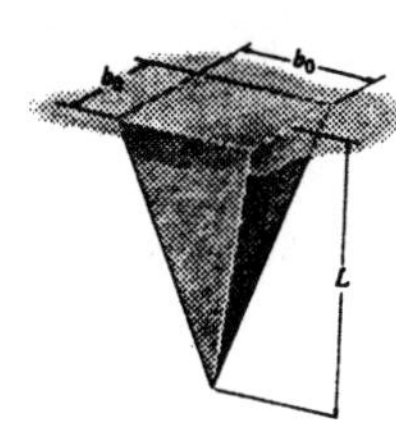

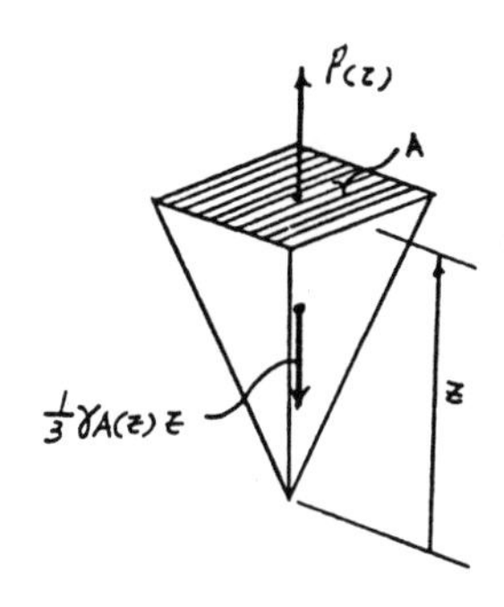

Internal Forces :

$$+\uparrow \Sigma F_z = 0; \quad P(z) - \frac{1}{3}\gamma A z = 0 \qquad P(z) = \frac{1}{3}\gamma A z$$

Displacement :

$$\delta = \int_0^L \frac{P(z)\,dz}{A(z)\,E}$$

$$= \int_0^L \frac{\frac{1}{3}\gamma A z}{AE}\,dz$$

$$= \frac{\gamma}{3E}\int_0^L z\,dz$$

$$= \frac{\gamma L^2}{6E} \qquad \textbf{Ans}$$

4–35. The ball is truncated at its ends and is used to support the bearing load **P**. If the modulus of elasticity for the material is E, determine the decrease in its height when the load is applied.

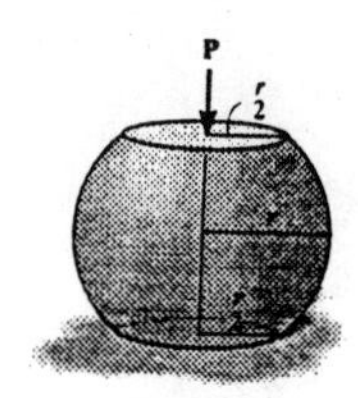

Displacement :

Geometry :

$$A(y) = \pi x^2 = \pi (r^2 - y^2)$$

Displacement : When $x = \frac{r}{2}$, $\quad y = \pm\frac{\sqrt{3}}{2}r$

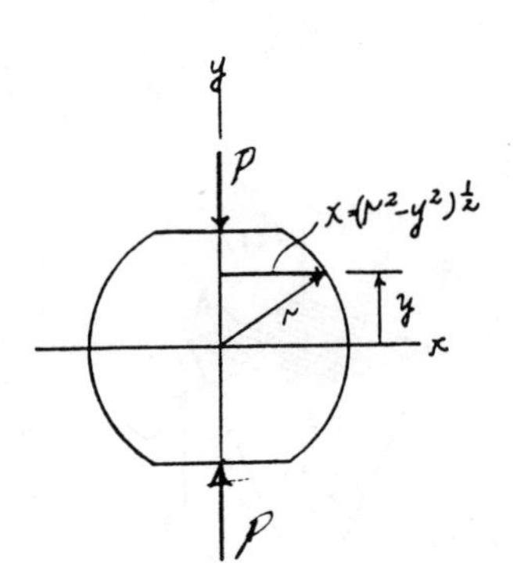

$$\delta = \int_0^L \frac{P(y)\,dy}{A(y)\,E}$$

$$= \frac{P}{\pi E}\int_{-\frac{\sqrt{3}}{2}r}^{\frac{\sqrt{3}}{2}r} \frac{dy}{r^2 - y^2}$$

$$= \frac{P}{\pi E}\left[\frac{1}{2r}\ln\frac{r+y}{r-y}\right]\Bigg|_{-\frac{\sqrt{3}}{2}r}^{\frac{\sqrt{3}}{2}r}$$

$$= \frac{P}{2\pi rE}[\ln 13.9282 - \ln 0.07180]$$

$$= \frac{2.63\,P}{\pi rE} \qquad \textbf{Ans}$$

4–37. The column is constructed from high-strength concrete and six A-36 steel reinforcing rods. If it is subjected to an axial force of 30 kip, determine the average normal stress in the concrete and in each rod. Each rod has a diameter of 0.75 in.

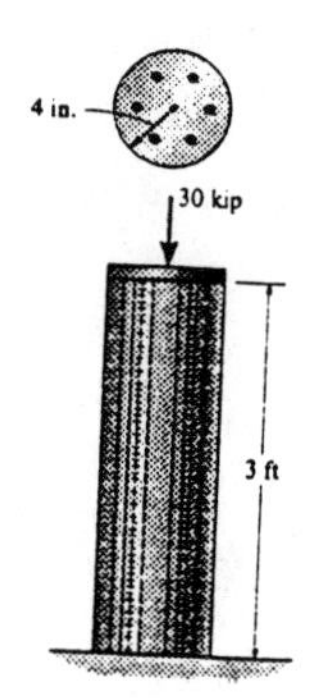

Equations of Equilibrium :

$$+\uparrow \Sigma F_y = 0; \qquad 6P_{st} + P_{con} - 30 = 0 \qquad [1]$$

Compatibility :

$$\delta_{st} = \delta_{con}$$

$$\frac{P_{st}(3)(12)}{\frac{\pi}{4}(0.75^2)(29.0)(10^3)} = \frac{P_{con}(3)(12)}{[\frac{\pi}{4}(8^2) - 6(\frac{\pi}{4})(0.75)^2](4.20)(10^3)}$$

$$P_{st} = 0.064065\,P_{con} \qquad [2]$$

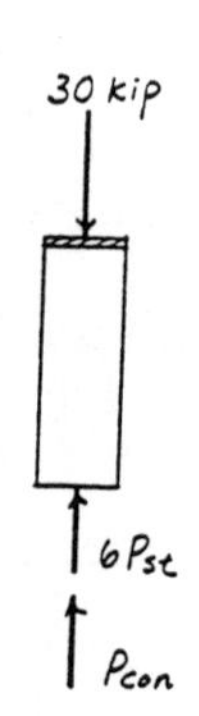

Solving Eqs. [1] and [2] yields :

$$P_{st} = 1.388 \text{ kip} \qquad P_{con} = 21.670 \text{ kip}$$

Average Normal Stress :

$$\sigma_{st} = \frac{P_{st}}{A_{st}} = \frac{1.388}{\frac{\pi}{4}(0.75^2)} = 3.14 \text{ ksi} \qquad \textbf{Ans}$$

$$\sigma_{con} = \frac{P_{con}}{A_{con}} = \frac{21.670}{\frac{\pi}{4}(8^2) - 6\left(\frac{\pi}{4}\right)(0.75^2)} = 0.455 \text{ ksi} \qquad \textbf{Ans}$$

4–38. The column is constructed from high-strength concrete and six A-36 steel reinforcing rods. If it is subjected to an axial force of 30 kip, determine the required diameter of each rod so that one-fourth of the load is carried by the concrete and three-fourths by the steel.

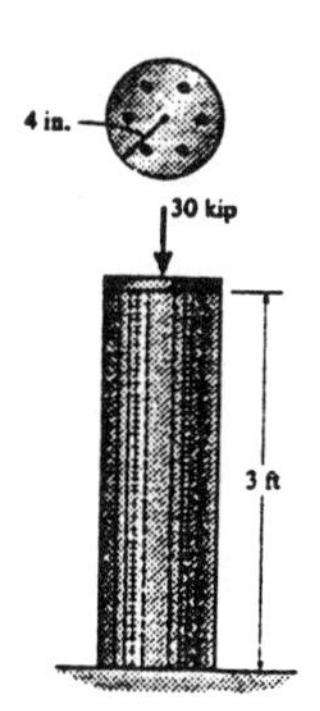

Equilibrium : The force of 30 kip is required to distribute in such a manner that 3/4 of the force is carried by steel and 1/4 of the force is carried by concrete. Hence

$$P_{st} = \frac{3}{4}(30) = 22.5 \text{ kip} \qquad P_{con} = \frac{1}{4}(30) = 7.50 \text{ kip}$$

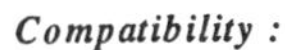

Compatibility :

$$\delta_{st} = \delta_{con}$$

$$\frac{P_{st}L}{A_{st}E_{st}} = \frac{P_{con}L}{A_{con}E_{con}}$$

$$A_{st} = \frac{22.5A_{con}E_{con}}{7.50\ E_{st}}$$

$$6\left(\frac{\pi}{4}\right)d^2 = \frac{3\left[\frac{\pi}{4}(8^2) - 6\left(\frac{\pi}{4}\right)d^2\right](4.20)(10^3)}{29.0(10^3)}$$

$$d = 1.80 \text{ in.} \qquad \textbf{Ans}$$

4–39. The A-36 steel pipe has a 6061-T6 aluminum core. It is subjected to a tensile force of 200 kN. Determine the average normal stress in the aluminum and the steel due to this loading. The pipe has an outer diameter of 80 mm and an inner diameter of 70 mm.

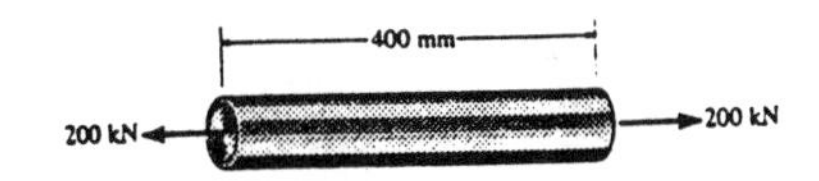

Equations of Equilibrium :

$$\overset{+}{\leftarrow} \Sigma F_x = 0; \qquad P_{al} + P_{st} - 200 = 0 \qquad [1]$$

Compatibility :

$$\delta_{al} = \delta_{st}$$

$$\frac{P_{al}(400)}{\frac{\pi}{4}(0.07^2)(68.9)(10^9)} = \frac{P_{st}(400)}{\frac{\pi}{4}(0.08^2 - 0.07^2)(200)(10^9)}$$

$$P_{al} = 1.125367\ P_{st} \qquad [2]$$

Solving Eqs. [1] and [2] yields :

$$P_{st} = 94.10 \text{ kN} \qquad P_{al} = 105.90 \text{ kN}$$

Average Normal Stress :

$$\sigma_{al} = \frac{P_{al}}{A_{al}} = \frac{105.90(10^3)}{\frac{\pi}{4}(0.07^2)} = 27.5 \text{ MPa} \qquad \textbf{Ans}$$

$$\sigma_{st} = \frac{P_{st}}{A_{st}} = \frac{94.10(10^3)}{\frac{\pi}{4}(0.08^2 - 0.07^2)} = 79.9 \text{ MPa} \qquad \textbf{Ans}$$

***4–40.** The concrete column is reinforced using four steel reinforcing rods, each having a diameter of 18 mm. Determine the average normal stress in the concrete and the steel if the column is subjected to an axial load of 800 kN. E_{st} = 200 GPa, E_c = 25 GPa.

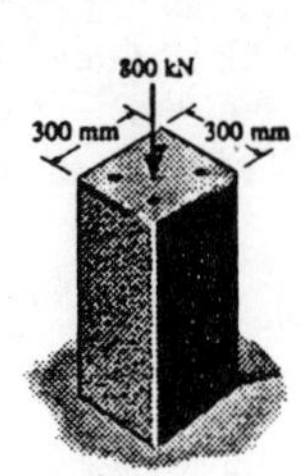

Equilibrium :

$$+\uparrow \Sigma F_y = 0; \qquad P_{st} + P_{con} - 800 = 0 \qquad [1]$$

Compatibility :

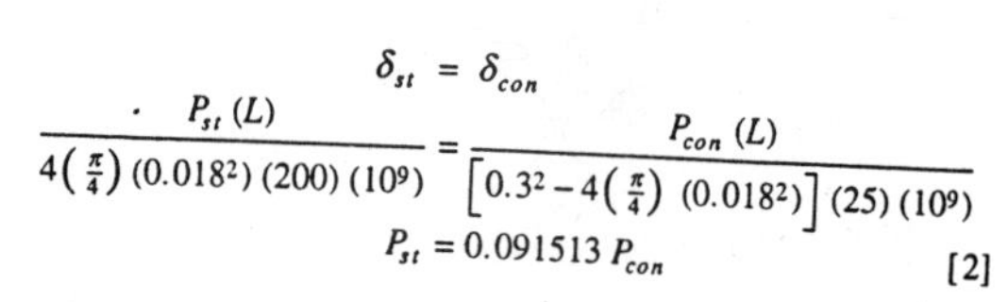

$$\delta_{st} = \delta_{con}$$

$$\frac{P_{st}\,(L)}{4\left(\frac{\pi}{4}\right)(0.018^2)\,(200)\,(10^9)} = \frac{P_{con}\,(L)}{\left[0.3^2 - 4\left(\frac{\pi}{4}\right)(0.018^2)\right](25)\,(10^9)}$$

$$P_{st} = 0.091513\,P_{con} \qquad [2]$$

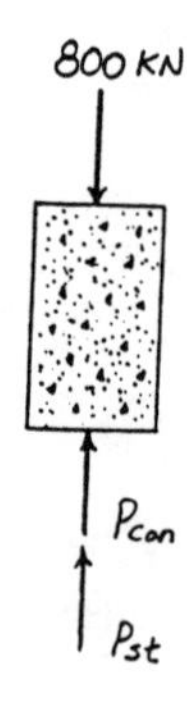

Solving Eqs. [1] and [2] yields :

$$P_{st} = 67.072 \text{ kN} \qquad P_{con} = 732.928 \text{ kN}$$

Average Normal Sress :

$$\sigma_{st} = \frac{67.072(10^3)}{4\left(\frac{\pi}{4}\right)(0.018^2)} = 65.9 \text{ MPa} \qquad \textbf{Ans}$$

$$\sigma_{con} = \frac{732.928(10^3)}{\left[0.3^2 - 4\left(\frac{\pi}{4}\right)(0.018^2)\right]} = 8.24 \text{ MPa} \qquad \textbf{Ans}$$

4–41. The column is constructed from high-strength concrete and four A-36 steel reinforcing rods. If it is subjected to an axial force of 800 kN, determine the required diameter of each rod so that one-fourth of the load is carried by the steel and three-fourths by the concrete.

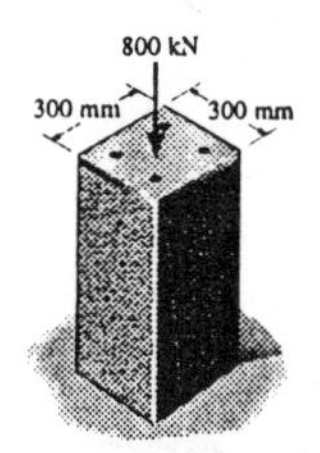

Equilibrium : Require $P_{st} = \frac{1}{4}(800) = 200$ kN and $P_{con} = \frac{3}{4}(800) = 600$ kN.

Compatibility :

$$\delta_{con} = \delta_{st}$$

$$\frac{P_{con}L}{(0..3^2 - A_{st})(29.0)(10^9)} = \frac{P_{st}L}{A_{st}(200)(10^9)}$$

$$A_{st} = \frac{0.09P_{st}}{6.8966P_{con} + P_{st}}$$

$$4\left[\left(\frac{\pi}{4}\right)d^2\right] = \frac{0.09(200)}{6.8966(600) + 200}$$

$$d = 0.03634 \text{ m} = 36.3 \text{ mm} \qquad \textbf{Ans}$$

4–42. The 304 stainless steel post A has a diameter of $d = 2$ in. and is surrounded by a red brass C83400 tube B. Both rest on the rigid surface. If a force of 5 kip is applied to the rigid cap, determine the average normal stress developed in the post and the tube.

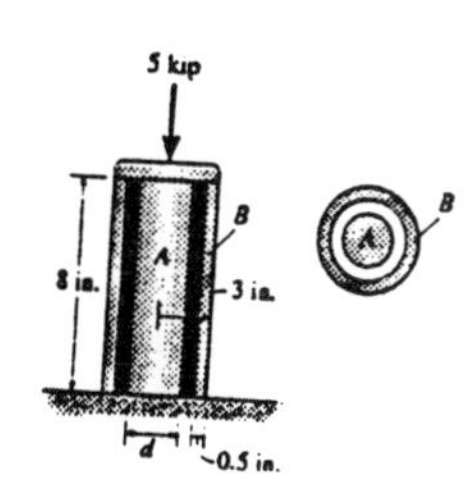

Equations of Equilibrium :

$$+\uparrow \Sigma F_y = 0; \qquad P_{st} + P_{br} - 5 = 0 \qquad [1]$$

Compatibility :

$$\delta_{st} = \delta_{br}$$

$$\frac{P_{st}(8)}{\frac{\pi}{4}(2^2)(28.0)(10^3)} = \frac{P_{br}(8)}{\frac{\pi}{4}(6^2 - 5^2)(14.6)(10^3)}$$

$$P_{st} = 0.69738\, P_{br} \qquad [2]$$

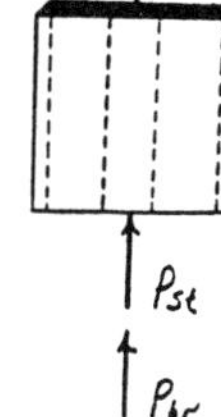

Solving Eqs. [1] and [2] yields :

$$P_{br} = 2.9457 \text{ kip} \qquad P_{st} = 2.0543 \text{ kip}$$

Average Normal Stress :

$$\sigma_{br} = \frac{P_{br}}{A_{br}} = \frac{2.9457}{\frac{\pi}{4}(6^2 - 5^2)} = 0.341 \text{ ksi} \qquad \textbf{Ans}$$

$$\sigma_{st} = \frac{P_{st}}{A_{st}} = \frac{2.0543}{\frac{\pi}{4}(2^2)} = 0.654 \text{ ksi} \qquad \textbf{Ans}$$

4–43. The 304 stainless steel post A is surrounded by a red brass C83400 tube B. Both rest on the rigid surface. If a force of 5 kip is applied to the rigid cap, determine the required diameter d of the steel post so that the load is shared equally between the post and tube.

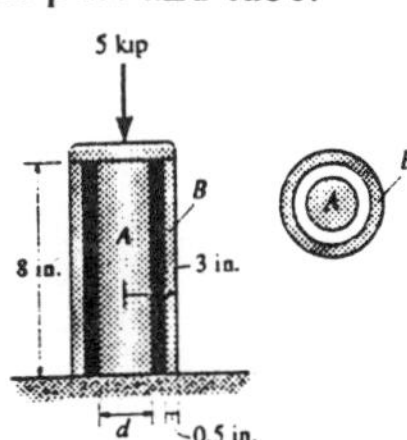

Equilibrium : The force of 60 kip is shared equally by the brass and steel. Hence

$$P_{st} = P_{br} = P = 2.50 \text{ kip}$$

Compatibility :

$$\delta_{st} = \delta_{br}$$

$$\frac{PL}{A_{st}E_{st}} = \frac{PL}{A_{br}E_{br}}$$

$$A_{st} = \frac{A_{br}E_{br}}{E_{st}}$$

$$\left(\frac{\pi}{4}\right)d^2 = \frac{\frac{\pi}{4}(6^2 - 5^2)(14.6)(10^3)}{28.0(10^3)}$$

$$d = 2.39 \text{ in.} \qquad \textbf{Ans}$$

***4–44.** The two pipes are made of the same material and are connected as shown. If the cross-sectional area of BC is A and that of CD is $2A$, determine the reactions at B and D when a force $\mathbf{P}$ is applied at the junction C.

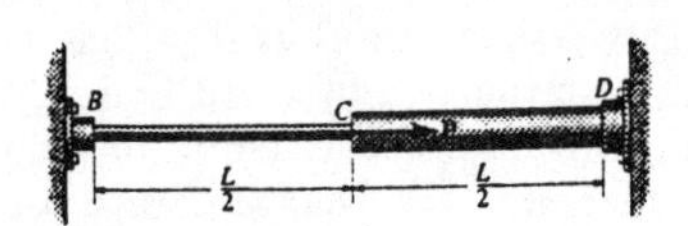

Equations of Equilibrium :

$$\overset{+}{\leftarrow} \Sigma F_x = 0; \quad F_B + F_D - P = 0 \qquad [1]$$

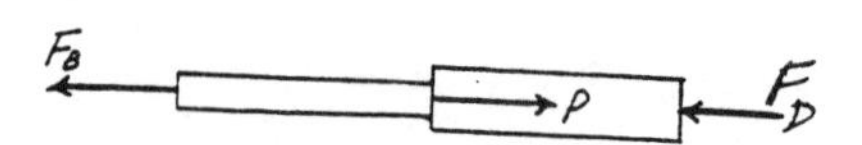

Compatibility :

$$(\overset{+}{\rightarrow}) \quad 0 = \delta_P - \delta_B$$

$$0 = \frac{P(\frac{L}{2})}{2AE} - \left[\frac{F_B(\frac{L}{2})}{AE} + \frac{F_B(\frac{L}{2})}{2AE}\right]$$

$$0 = \frac{PL}{4AE} - \frac{3F_B L}{4AE}$$

$$F_B = \frac{P}{3} \qquad \text{Ans}$$

From Eq. [1]

$$F_D = \frac{2}{3}P \qquad \text{Ans}$$

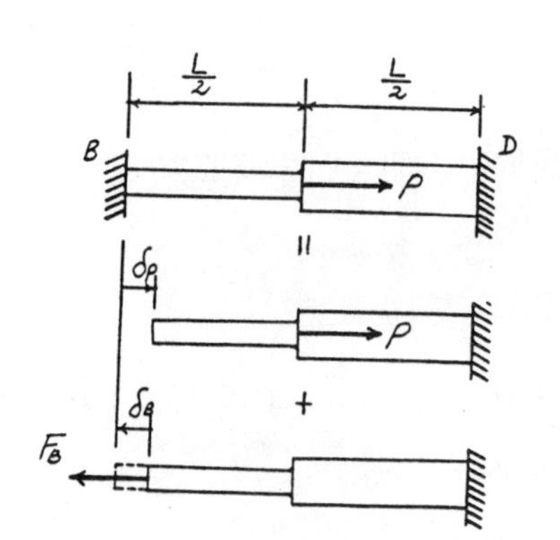

4–45. The uniform bar is subjected to the load $\mathbf{P}$ at collar B. Determine the reactions at the pins A and C. Neglect the size of the collar.

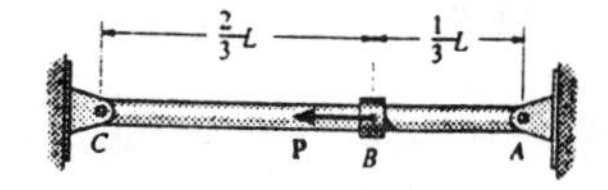

Equations of Equilibrium :

$$\overset{+}{\rightarrow} \Sigma F_x = 0; \quad F_A + F_C - P = 0 \qquad [1]$$

Compatibility :

$$(\overset{+}{\leftarrow}) \quad 0 = \delta_P - \delta_C$$

$$0 = \frac{P(\frac{L}{3})}{AE} - \frac{F_C(L)}{AE}$$

$$F_C = \frac{P}{3} \qquad \text{Ans}$$

From Eq. [1]

$$F_A = \frac{2}{3}P \qquad \text{Ans}$$

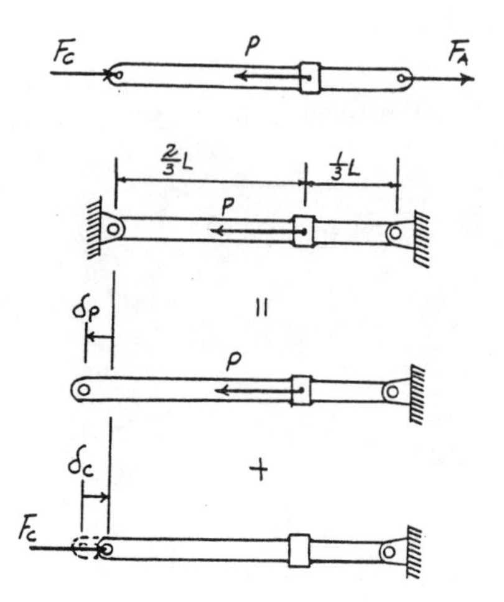

4–46. The bolt AB has a diameter of 20 mm and passes through a sleeve that has an inner diameter of 40 mm and an outer diameter of 50 mm. The bolt and sleeve are made of A-36 steel and are secured to the rigid brackets as shown. If the bolt length is 220 mm and the sleeve length is 200 mm, determine the tension in the bolt when a force of 50 kN is applied to the brackets.

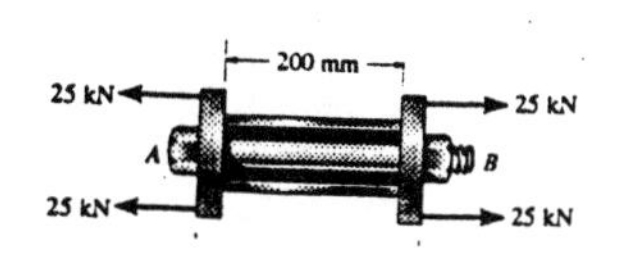

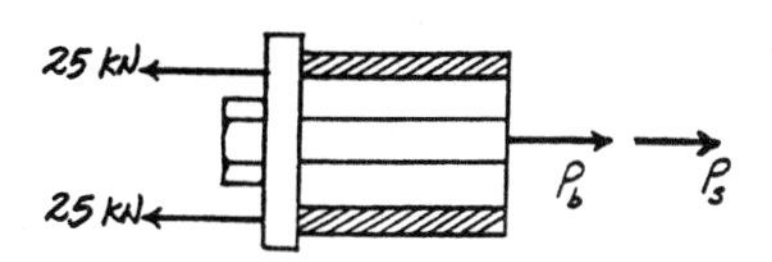

Equation of Equilibrium :

$$\xrightarrow{+} \Sigma F_x = 0; \quad P_b + P_s - 25 - 25 = 0$$

$$P_b + P_s - 50 = 0 \qquad [1]$$

Compatibility :

$$\delta_b = \delta_s$$

$$\frac{P_b(220)}{\frac{\pi}{4}(0.02^2)200(10^9)} = \frac{P_s(200)}{\frac{\pi}{4}(0.05^2 - 0.04^2)(200)(10^9)}$$

$$P_b = 0.40404\,P_s \qquad [2]$$

Solving Eqs. [1] and [2] yields :

$P_s = 35.61$ kN

$P_b = 14.4$ kN **Ans**

4–47. The load of 1500 lb is to be supported by the two vertical A-36 steel wires. If originally wire AB is 50 in. long and wire AC is 50.1 in. long, determine the force developed in each wire after the load is suspended. Each wire has a cross-sectional area of 0.02 in^2.

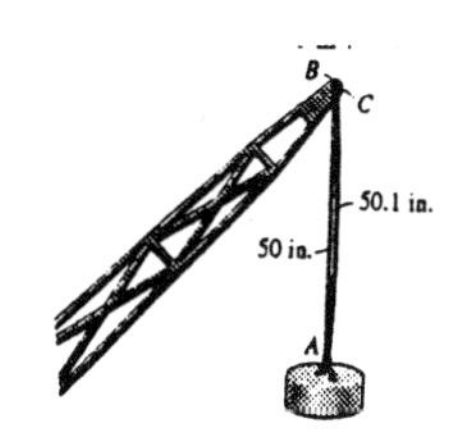

Equations of Equilibrium :

$$+\uparrow \Sigma F_y = 0; \quad T_{AC} + T_{AB} - 1.50 = 0 \qquad [1]$$

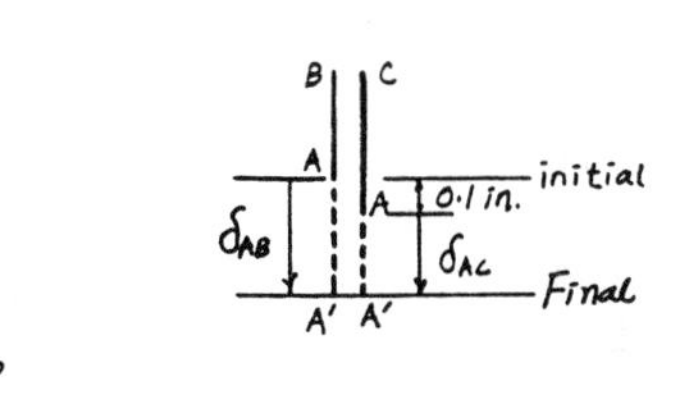

Compatibility :

$(+\downarrow)$

$$\delta_{AC} + 0.1 = \delta_{AB}$$

$$\frac{T_{AC}(50.1)}{0.02(29.0)(10^3)} + 0.1 = \frac{T_{AB}(50)}{0.02(29.0)(10^3)}$$

$$50.1T_{AC} - 50T_{AB} + 58.0 = 0 \qquad [2]$$

Solving Eqs. [1] and [2] yields :

$T_{AC} = 0.170$ kip **Ans**

$T_{AB} = 1.33$ kip **Ans**

***4–48.** The load of 15 kip is to be supported by the two vertical A-36 steel wires. If originally wire AB is 50 in. long and wire AC is 50.1 in. long, determine the cross-sectional area of AB if the load is to be shared equally between both wires. Wire AC has a cross-sectional area of 0.02 in^2.

Equilibrium : The force of 15.0 kip is shared equally by the two wires. Hence

$$T_{AB} = T_{AC} = \frac{15.0}{2} = 7.50 \text{ kip}$$

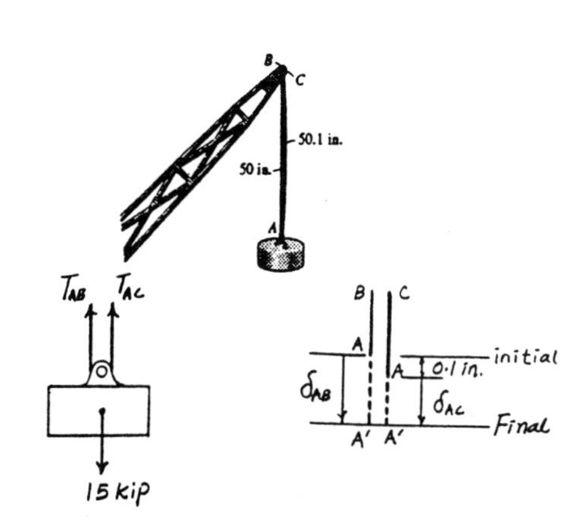

Compatibility :

$(+\downarrow)$

$$\delta_{AC} + 0.1 = \delta_{AB}$$

$$\frac{7.50(50.1)}{(0.02)(29.0)(10^3)} + 0.1 = \frac{7.50(50)}{A_{AB}(29.0)(10^3)}$$

$$A_{AB} = 0.0173 \text{ in}^2 \qquad \textbf{Ans}$$

4–49. The rigid link is supported by a pin at A, a steel wire BC having an unstretched length of 200 mm and cross-sectional area of 22.5 mm^2, and a short aluminum block having an unloaded length of 50 mm and cross-sectional area of 40 mm^2. If the link is subjected to the vertical load shown, determine the average normal stress in the wire and the block. E_{st} = 200 GPa, E_{al} = 70 GPa.

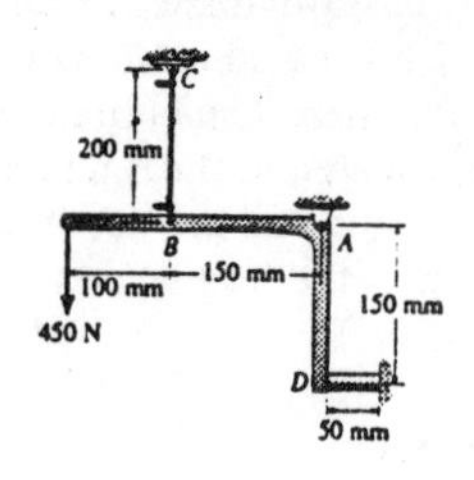

Equations of Equilibrium :

$$\circlearrowleft + \Sigma M_A = 0; \quad 450(250) - F_{BC}(150) - F_D(150) = 0$$

$$750 - F_{BC} - F_D = 0 \qquad [1]$$

Compatibility :

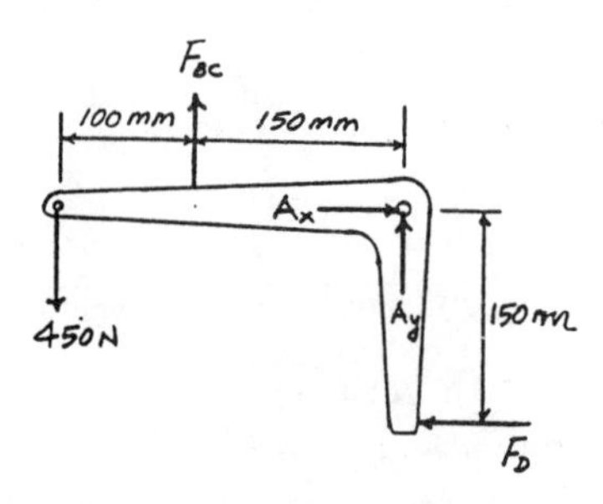

$$\delta_{BC} = \delta_D$$

$$\frac{F_{BC}(200)}{22.5(10^{-6})200(10^9)} = \frac{F_D(50)}{40(10^{-6})70(10^9)}$$

$$F_{BC} = 0.40177\, F_D \qquad [2]$$

Solving Eqs. [1] and [2] yields :

$$F_D = 535.03 \text{ N} \qquad F_{BC} = 214.97 \text{ N}$$

Average Normal Stress :

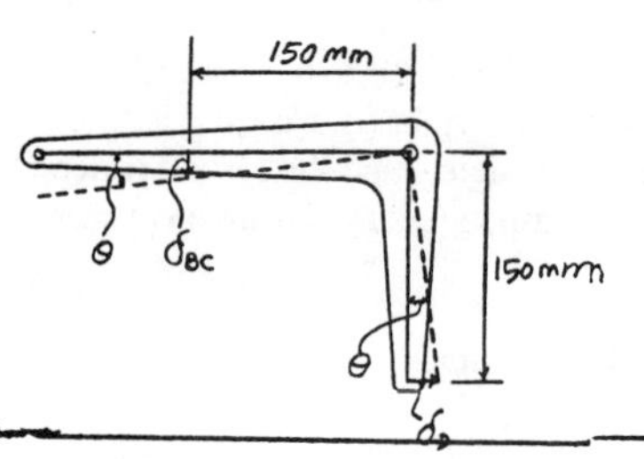

$$\sigma_D = \frac{F_D}{A_D} = \frac{535.03}{40(10^{-6})} = 13.4 \text{ MPa} \qquad \textbf{Ans}$$

$$\sigma_{BC} = \frac{F_{BC}}{A_{BC}} = \frac{214.97}{22.5(10^{-6})} = 9.55 \text{ MPa} \qquad \textbf{Ans}$$

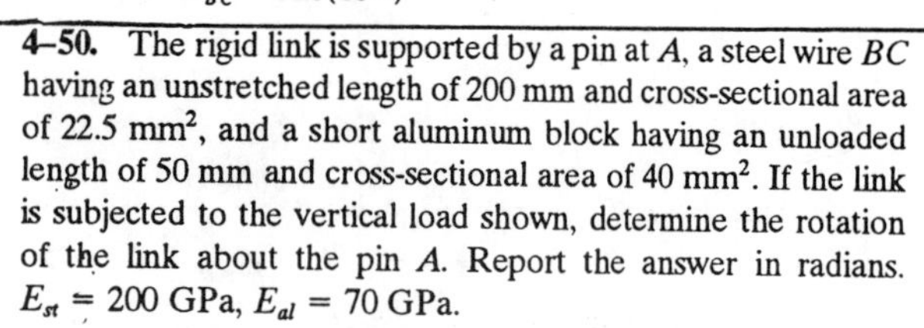

4–50. The rigid link is supported by a pin at A, a steel wire BC having an unstretched length of 200 mm and cross-sectional area of 22.5 mm^2, and a short aluminum block having an unloaded length of 50 mm and cross-sectional area of 40 mm^2. If the link is subjected to the vertical load shown, determine the rotation of the link about the pin A. Report the answer in radians. E_{st} = 200 GPa, E_{al} = 70 GPa.

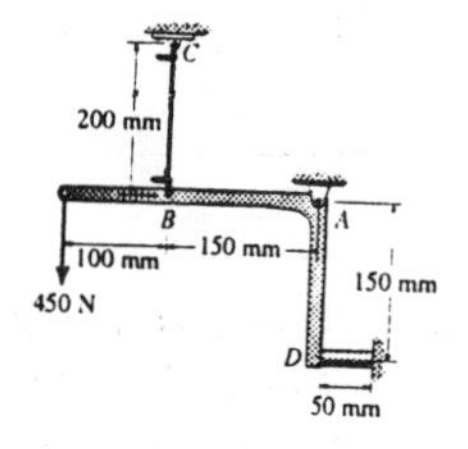

Equations of Equilibrium :

$$\circlearrowleft + \Sigma M_A = 0; \quad 450(250) - F_{BC}(150) - F_D(150) = 0$$

$$750 - F_{BC} - F_D = 0 \qquad [1]$$

Compatibility :

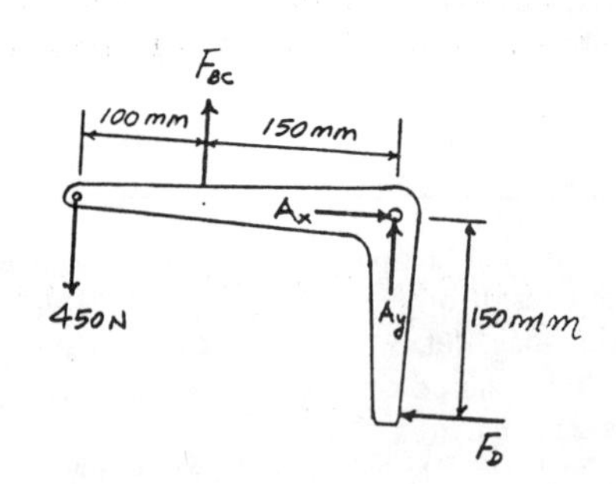

$$\delta_{BC} = \delta_D$$

$$\frac{F_{BC}(200)}{22.5(10^{-6})200(10^9)} = \frac{F_D(50)}{40(10^{-6})70(10^9)}$$

$$F_{BC} = 0.40177\, F_D \qquad [2]$$

Solving Eqs. [1] and [2] yields :

$$F_D = 535.03 \text{ N} \qquad F_{BC} = 214.97 \text{ N}$$

Displacement :

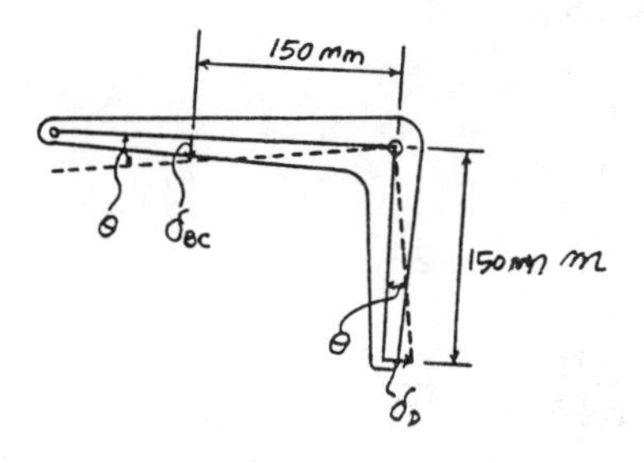

$$\delta_D = \frac{F_D L_D}{A_D E_{al}} = \frac{535.03(50)}{40(10^{-6})(70)(10^9)} = 0.009554 \text{ mm}$$

$$\tan\theta = \frac{\delta_D}{150} = \frac{0.009554}{150}$$

$$\theta = 63.7(10^{-6}) \text{ rad} \qquad \textbf{Ans}$$

4–51. The three A-36 steel wires each have a diameter of 2 mm and unloaded lengths of $L_{CA} = 1.60$ m and $L_{AB} = L_{AD} = 2.00$ m. Determine the force in each wire after the 150-kg mass is suspended from the ring at A.

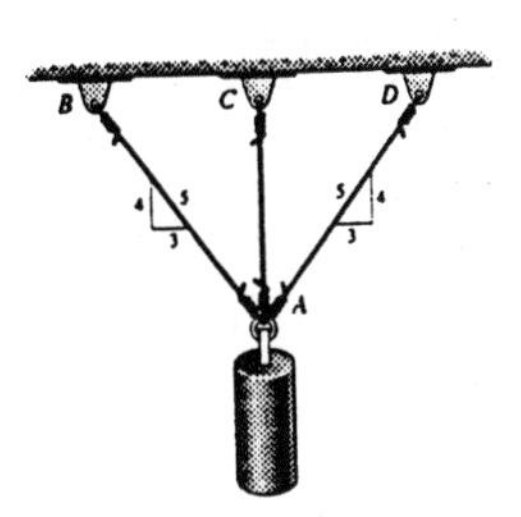

Equations of Equilibrium :

$$\xrightarrow{+} \Sigma F_x = 0; \quad \frac{3}{5}F_{AD} - \frac{3}{5}F_{AB} = 0 \quad F_{AD} = F_{AB} = F$$

$$+\uparrow \Sigma F_y = 0; \quad 2\left(\frac{4}{5}F\right) + F_{AC} - 150(9.81) = 0$$

$$1.6F + F_{AC} - 1471.5 = 0 \qquad [1]$$

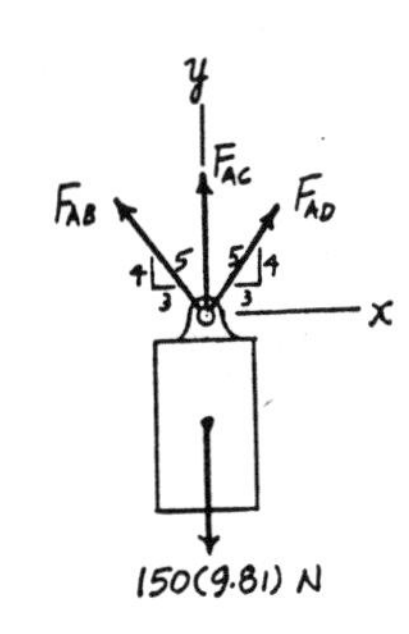

Compatibility :

$$\delta_{AD} = \delta_{AC} \cos\theta$$

Since the displacement is very small, $\cos\theta = \frac{4}{5}$

$$\delta_{AD} = \frac{4}{5}\delta_{AC}$$

$$\frac{F(2)}{AE} = \frac{4}{5}\left[\frac{F_{AC}(1.6)}{AE}\right]$$

$$F = 0.640\,F_{AC} \qquad [2]$$

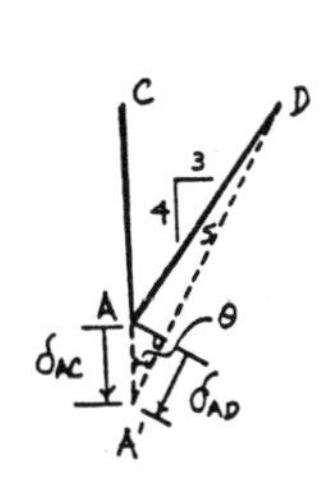

Solving Eqs. [1] and [2] yields :

$F_{AC} = 727$ N **Ans**

$F_{AB} = F_{AD} = F = 465$ N **Ans**

***4–52.** The A-36 steel wires AB and AD each have a diameter of 2 mm and the unloaded lengths of each wire are $L_{CA} = 1.60$ m and $L_{AB} = L_{AD} = 2.00$ m. Determine the required diameter of wire AC so that each wire is subjected to the same force caused by the 150-kg mass suspended from the ring at A.

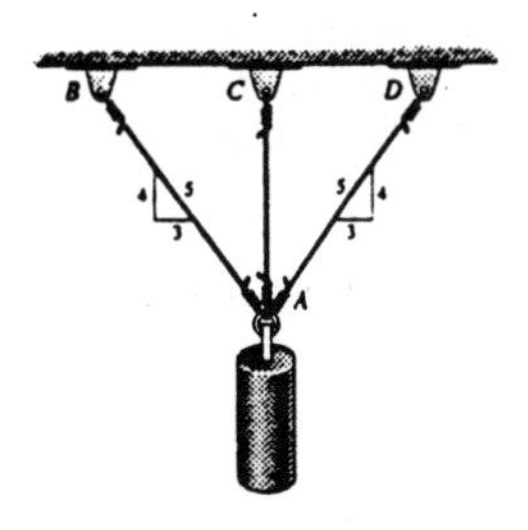

Equations of Equilibrium : Each wire is required to carry the same amount of load. Hence

$$F_{AB} = F_{AC} = F_{AD} = F$$

Compatibility :

$$\delta_{AD} = \delta_{AC} \cos\theta$$

Since the displacement is very small, $\cos\theta = \frac{4}{5}$

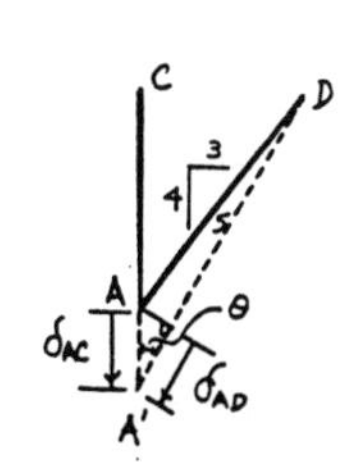

$$\delta_{AD} = \frac{4}{5}\delta_{AC}$$

$$\frac{F(2)}{\frac{\pi}{4}(0.002^2)E} = \frac{F(1.6)}{\frac{\pi}{4}d_{AC}^2 E}$$

$d_{AC} = 0.001789 \text{ m} = 1.79 \text{ mm}$ **Ans**

4–53. The three bars are pinned together and subjected to the force **P**. If each bar has the same length L, cross-sectional area A, and modulus of elasticity E, determine the force in each bar

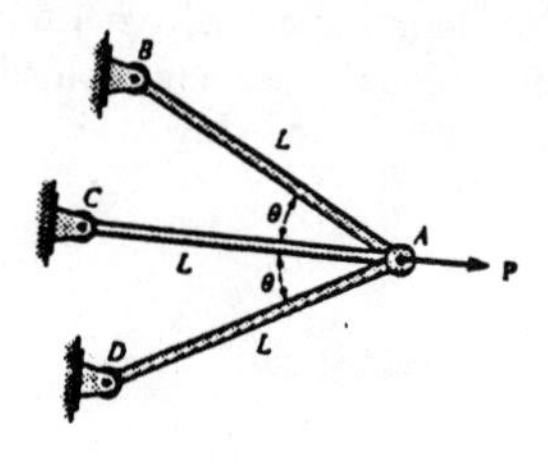

Equations of Equilibrium :

$+\uparrow \Sigma F_y = 0; \quad F_{AB}\sin\theta - F_{AD}\sin\theta = 0$

$F_{AB} = F_{AD} = F$

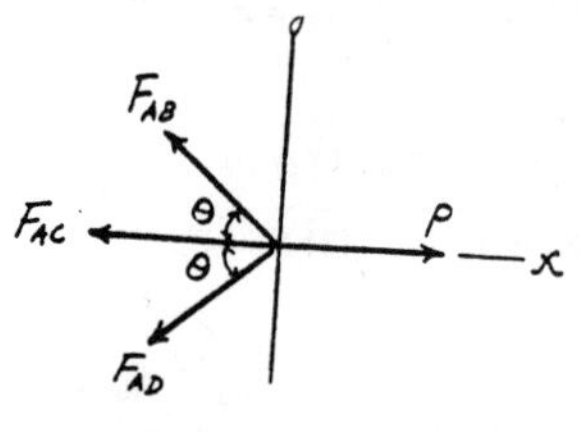

$\xrightarrow{+} \Sigma F_x = 0; \quad P - F_{AC} - 2F\cos\theta = 0 \qquad [1]$

Compatibility :

$\delta_{AB} = \delta_{AC}\cos\theta$

$\frac{FL}{AE} = \frac{F_{AC}L}{AE}\cos\theta$

$F = F_{AC}\cos\theta \qquad [2]$

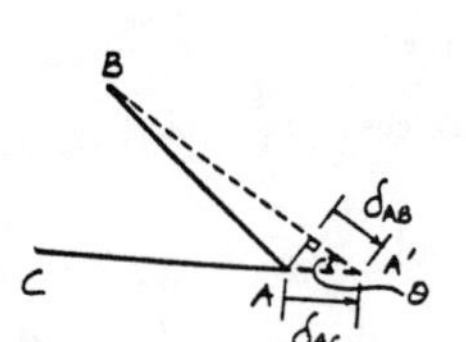

Solving Eqs. [1] and [2] yields :

$F_{AC} = \frac{P}{1+2\cos^2\theta}$ **Ans**

$F_{AB} = F_{AD} = F = \frac{P\cos\theta}{1+2\cos^2\theta}$ **Ans**

4–54. The assembly consists of an A-36 steel bolt and a C83400 red brass tube. If the nut is drawn up snug against the tube so that $L = 75$ mm, then turned an additional amount so that it advances 0.02 mm on the bolt, determine the force in the bolt and the tube. The bolt has a diameter of 7 mm and the tube has a cross-sectional area of 100 mm²

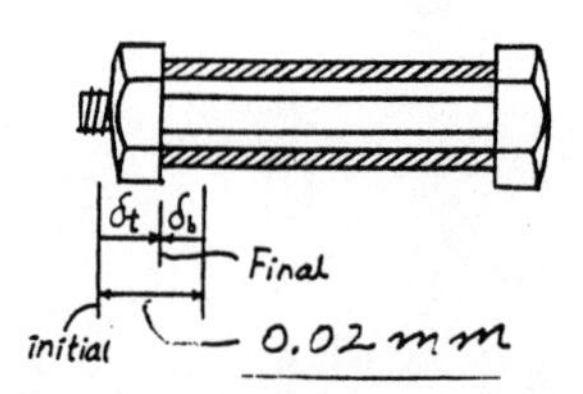

Equilibrium : Since no external load is applied, the force acting on the tube and the bolt is the same.

Compatibility :

$0.02 = \delta_t + \delta_b$

$0.02 = \frac{P(75)}{100(10^{-6})(101)(10^9)} + \frac{P(75)}{\frac{\pi}{4}(0.007^2)(200)(10^9)}$

$P = 1164.82\text{ N} = 1.16\text{ kN}$ **Ans**

4–55. The assembly consists of an A-36 steel bolt and a C83400 red brass tube. The nut is drawn up snug against the tube so that $L = 75$ mm. Determine the maximum additional amount of advance of the nut on the bolt so that none of the material will yield. The bolt has a diameter of 7 mm and the tube has a cross-sectional area of 100 mm^2.

Allowable Normal Stress :

$$(\sigma_Y)_{st} = 250\,(10^6) = \frac{P_{st}}{\frac{\pi}{4}(0.007)^2}$$

$$P_{st} = 9.621 \text{ kN}$$

$$(\sigma_Y)_{br} = 70.0(10^6) = \frac{P_{br}}{100(10^{-6})}$$

$$P_{br} = 7.00 \text{ kN}$$

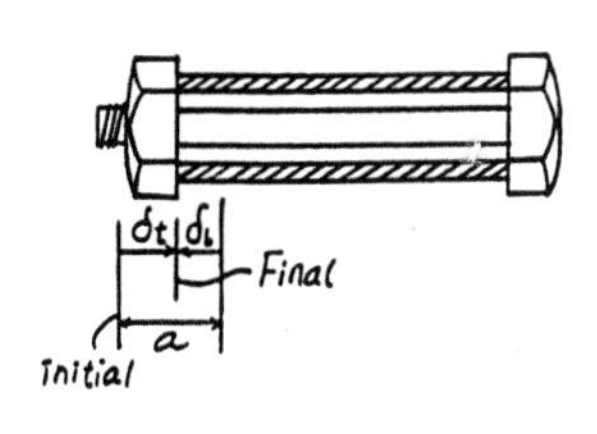

Since $P_{st} > P_{br}$, by comparison the brass will yield first.

Compatibility :

$$a = \delta_t + \delta_b$$

$$= \frac{7.00(10^3)(75)}{100(10^{-6})(101)(10^9)} + \frac{7.00(10^3)(75)}{\frac{\pi}{4}(0.007)^2(200)(10^9)}$$

$$= 0.120 \text{ mm} \qquad \textbf{Ans}$$

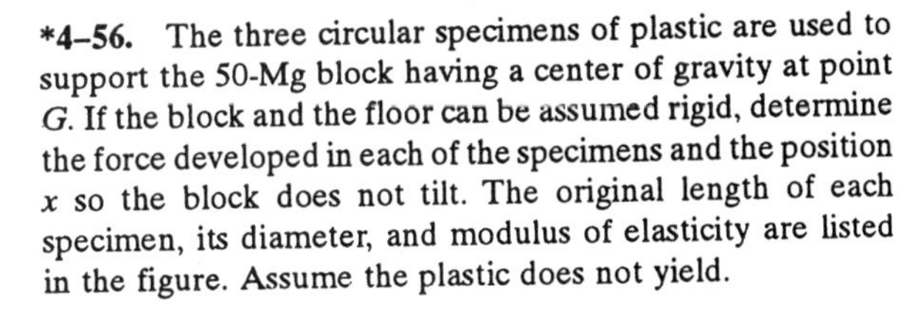

***4–56.** The three circular specimens of plastic are used to support the 50-Mg block having a center of gravity at point G. If the block and the floor can be assumed rigid, determine the force developed in each of the specimens and the position x so the block does not tilt. The original length of each specimen, its diameter, and modulus of elasticity are listed in the figure. Assume the plastic does not yield.

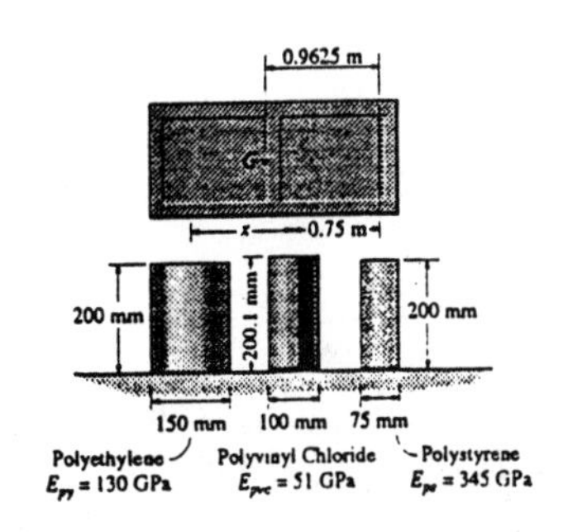

Equations of Equilibrium :

$$\curvearrowleft + \Sigma M_B = 0; \quad F_{ps}(0.75) - F_{py}(x) + 50(10^3)(9.81)(0.2125) = 0$$

$$0.75F_{ps} - F_{py}x + 104231.25 = 0 \qquad [1]$$

$$+\uparrow \Sigma F_y = 0; \quad F_{py} + F_{pvc} + F_{ps} - 50(10^3)(9.81) = 0 \qquad [2]$$

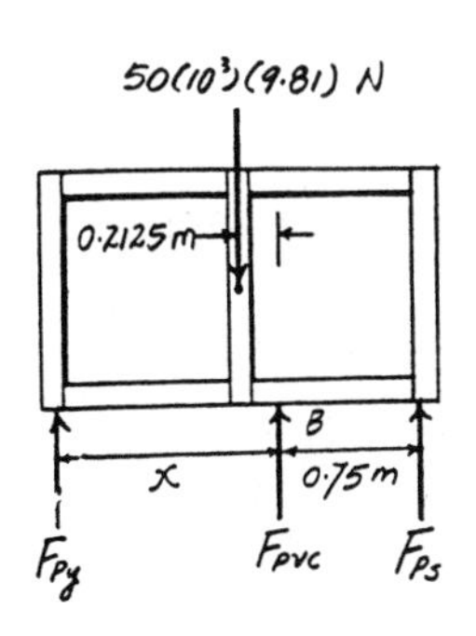

Compatibility : For the block to remain in the horizontal position then,

$$\delta_{py} = \delta_{ps}$$

$$\frac{F_{py}(200)}{\frac{\pi}{4}(0.15)^2(130)(10^9)} = \frac{F_{ps}(200)}{\frac{\pi}{4}(0.075^2)(345)(10^9)}$$

$$F_{py} = 1.5072\,F_{ps} \qquad [3]$$

$$\delta_{py} = \delta_{pvc} - 0.1 \text{ mm}$$

$$\frac{F_{py}(200)}{\frac{\pi}{4}(0.15)^2(130)(10^9)} = \frac{F_{pvc}(200.1)}{\frac{\pi}{4}(0.1^2)(51)(10^9)} - 0.1$$

$$0.08706F_{py} = 0.49956F_{pvc} - 0.1(10^6) \qquad [4]$$

Solving Eqs. [1] to [4] yields :

$$F_{ps} = 104878.3 \text{ N} = 105 \text{ kN} \qquad \textbf{Ans}$$

$$F_{py} = 158077.5 \text{ N} = 158 \text{ kN} \qquad \textbf{Ans}$$

$$F_{pvc} = 227544.2 \text{ N} = 228 \text{ kN} \qquad \textbf{Ans}$$

$$x = 1.16 \text{ m} \qquad \textbf{Ans}$$

4–57. The three suspender bars are made of the s material and have equal cross-sectional areas A. Detern the force in each bar if the rigid beam ACE is subjecte the uniform distributed loading w.

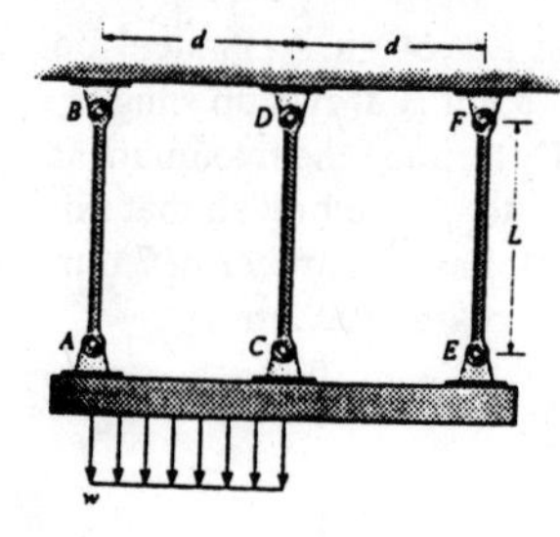

Equations of Equilibrium :

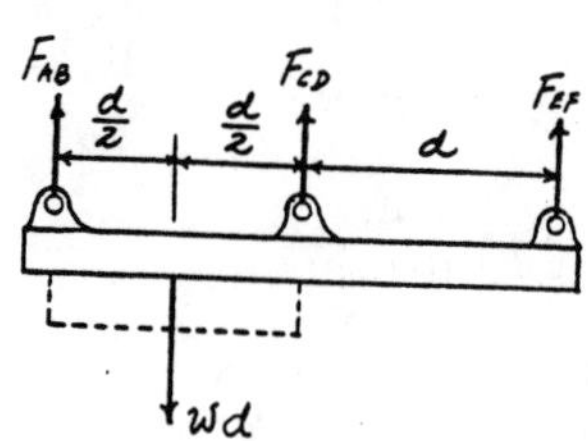

$$\circlearrowleft + \Sigma M_A = 0; \quad F_{EF}(2\,d) + F_{CD}(d) - w\,d\left(\frac{d}{2}\right) = 0$$

$$2\,F_{EF} + F_{CD} = \frac{w\,d}{2} \qquad [1]$$

$$+\uparrow \Sigma F_y = 0; \quad F_{AB} + F_{CD} + F_{EF} - w\,d = 0 \qquad [2]$$

Compatibility :

$$\frac{\delta_C - \delta_E}{d} = \frac{\delta_A - \delta_E}{2\,d}$$

$$2\delta_C - \delta_E = \delta_A$$

$$\frac{2\,F_{CD}(L)}{AE} - \frac{F_{EF}(L)}{AE} = \frac{F_{AB}(L)}{AE}$$

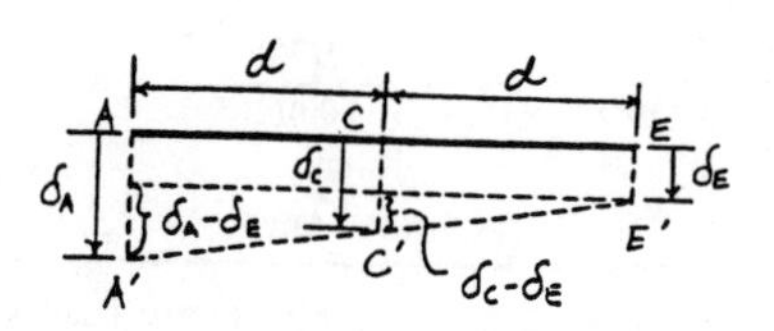

$$2\,F_{CD} - F_{EF} = F_{AB} \qquad [3]$$

Solving Eqs. [1] to [3] yields :

$$F_{EF} = \frac{1}{12}\,w\,d \qquad \textbf{Ans}$$

$$F_{AB} = \frac{7}{12}\,w\,d \qquad \textbf{Ans}$$

$$F_{CD} = \frac{1}{3}\,w\,d \qquad \textbf{Ans}$$

4–58. The bar is pinned at A and supported by two aluminum rods, each having a diameter of 1 in. and a modulus of elasticity $E_{al} = 10(10^3)$ ksi. If the bar is assumed to be rigid and initially vertical, determine the displacement of the end B when the force of 2 kip is applied.

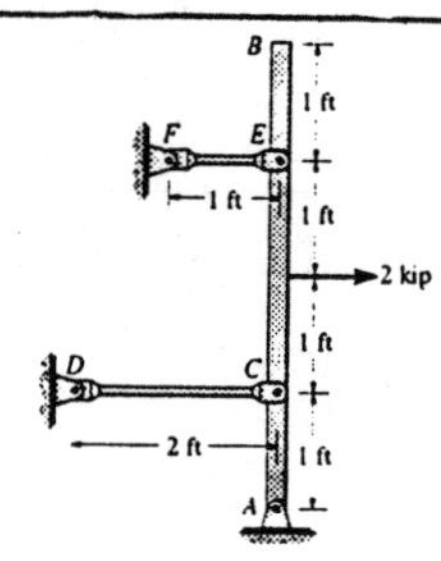

Equations of Equilibrium :

$$\circlearrowleft + \Sigma M_A = 0; \quad F_{CD}(1) + F_{EF}(3) - 2(2) = 0 \qquad [1]$$

Compatibility :

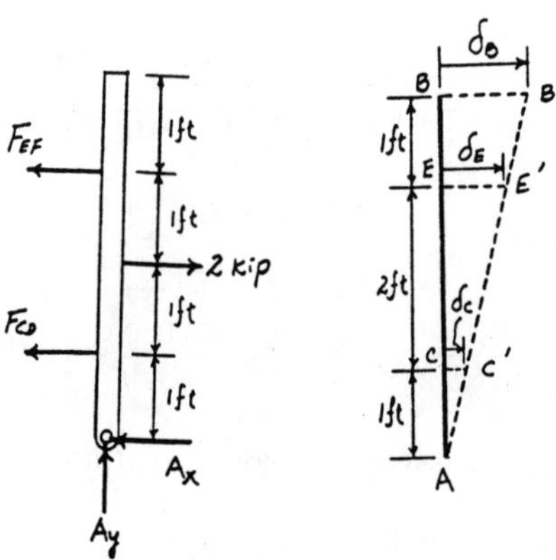

$$\delta_C = \frac{\delta_E}{3}$$

$$\frac{F_{CD}(2)(12)}{AE} = \frac{1}{3}\left[\frac{F_{EF}(1)(12)}{AE}\right]$$

$$F_{EF} = 6F_{CD} \qquad [2]$$

Solving Eqs. [1] and [2] yields :

$$F_{CD} = 0.21053 \text{ kip} \qquad F_{EF} = 1.2632 \text{ kip}$$

Displacement : Point B

$$\frac{\delta_B}{4} = \frac{\delta_E}{3}$$

$$\delta_B = \frac{4}{3}\delta_E = \frac{4}{3}\left[\frac{1.2632(1)(12)}{\frac{\pi}{4}(1^2)(10)(10^3)}\right] = 0.00257 \text{ in.} \qquad \textbf{Ans}$$

4–59. The bar is pinned at A and supported by two aluminum rods, each having a diameter of 1 in. and a modulus of elasticity $E_{al} = 10(10^3)$ ksi. If the bar is assumed to be rigid and initially vertical, determine the force in each rod when the 2-kip load is applied.

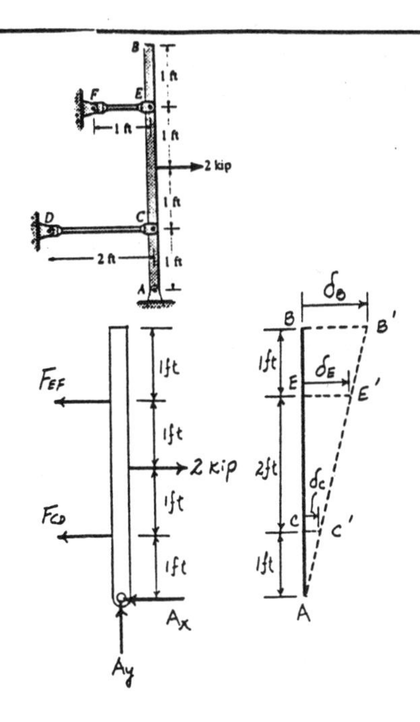

Equations of Equilibrium :

$$\circlearrowleft + \Sigma M_A = 0; \quad F_{CD}(1) + F_{EF}(3) - 2(2) = 0 \qquad [1]$$

Compatibility :

$$\delta_C = \frac{\delta_E}{3}$$

$$\frac{F_{CD}(2)(12)}{AE} = \frac{1}{3}\left[\frac{F_{EF}(1)(12)}{AE}\right]$$

$$F_{EF} = 6F_{CD} \qquad [2]$$

Solving Eqs. [1] and [2] yields :

$F_{CD} = 0.211$ kip **Ans**

$F_{EF} = 1.26$ kip **Ans**

***4–60.** The support consists of a rigid cap attached to a solid cylindrical piece of 2014-T6 aluminum having an unloaded length of 100.1 mm. The outer tube is made of magnesium alloy Am 1004-T61 and acts as a safety back-up if the load is increased. Determine the load that each metal supports if (a) P = 20 kN and (b) P = 30 kN.

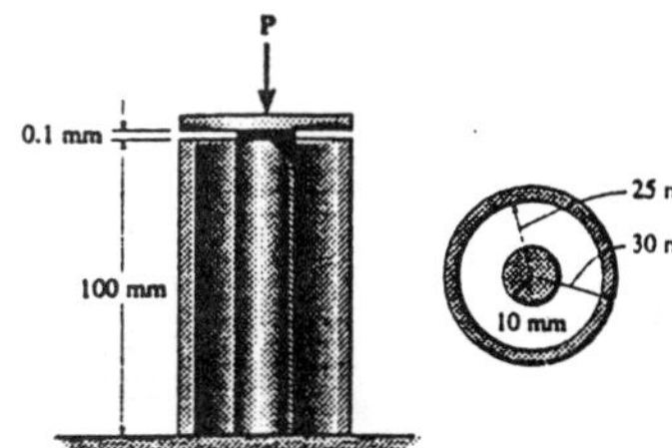

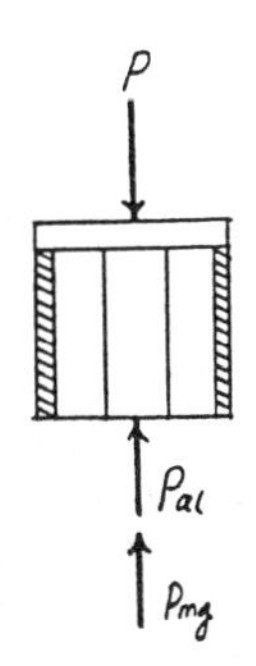

Equations of Equilibrium .

$$+\uparrow \Sigma F_y = 0; \quad P_{al} + P_{mg} - P = 0 \qquad [1]$$

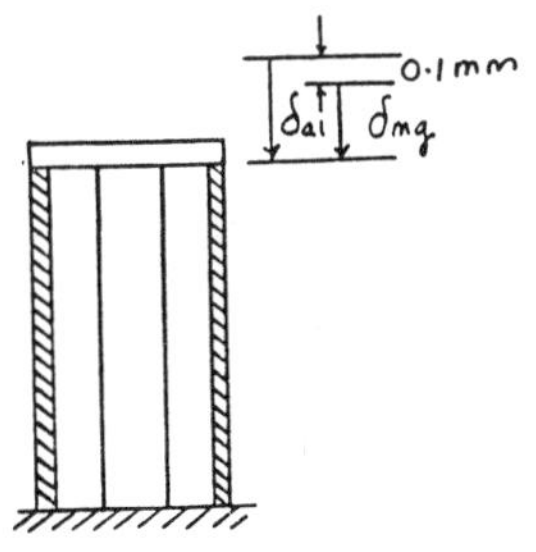

a) When the cap is loaded with P = 20 kN

$$\delta_{al} = \frac{20(10^3)(100.1)}{\frac{\pi}{4}(0.02^2)(73.1)(10^9)} = 0.0872 \text{ mm} < 0.1 \text{ mm}$$

Therefore, the cap does not touch the magnesium tube. Hence

$P_{mg} = 0$ **Ans**

From Eq. [1],

$P_{al} = P = 20.0$ kN **Ans**

b) When the cap is loaded with P = 30 kN

$$\delta_{al} = \frac{30(10^3)(100.1)}{\frac{\pi}{4}(0.02^2)(73.1)(10^9)} = 0.131 \text{ mm} > 0.1 \text{ mm}$$

Therefore, the magnesium tube will be deformed.

Compatibility :

$$\delta_{al} = \delta_{mg} + 0.1$$

$$\frac{P_{al}(100.1)}{\frac{\pi}{4}(0.02^2)(73.1)(10^9)} = \frac{P_{mg}(100)}{\frac{\pi}{4}(0.06^2 - 0.05^2)(44.7)(10^9)} + 0.1$$

$$4.3588\, P_{al} = 2.5895\, P_{mg} + 0.1(10^6) \qquad [2]$$

Solving Eqs. [1] and [2] with P = 30 kN

$P_{al} = 25.6$ kN **Ans**

$P_{mg} = 4.43$ kN **Ans**

4–61. The support consists of a rigid cap attached to a solid cylindrical piece of 2014-T6 aluminum having an unloaded length of 100.1 mm. The outer tube is made of magnesium alloy Am 1004-T61 and acts as a safety back-up if the load is increased. Determine the force **P** that should be applied to the cap so that the aluminum and magnesium carry an equal portion of the load.

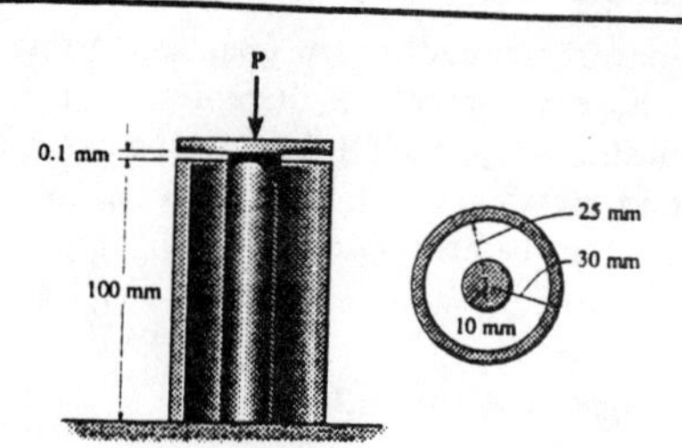

Equations of Equilibrium : The aluminum and magnesium are required to carry the same amount of load. Hence,

$$P_{al} = P_{mg} = \frac{P}{2}$$

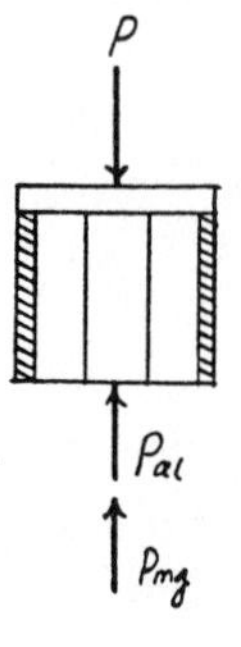

Compatibility :

$$\delta_{al} = \delta_{mg} + 0.1$$

$$\frac{\frac{P}{2}(100.1)}{\frac{\pi}{4}(0.02^2)(73.1)(10^9)} = \frac{\frac{P}{2}(100)}{\frac{\pi}{4}(0.06^2 - 0.05^2)(44.7)(10^9)} + 0.1$$

$$P = 113036.8 \text{ N} = 113 \text{ kN}$$ **Ans**

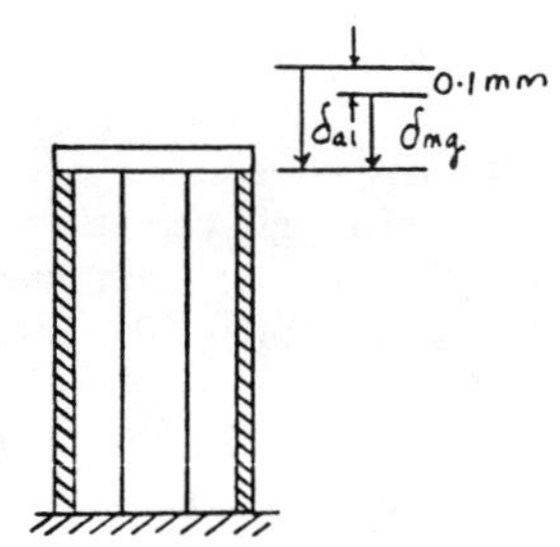

4–62. The assembly consists of two posts made from material 1 having a modulus of elasticity of E_1 and each a cross-sectional area A_1, and a material 2 having a modulus of elasticity E_2 and cross-sectional area A_2. If a central load **P** is applied to the rigid cap, determine the force in each post. The support is also rigid.

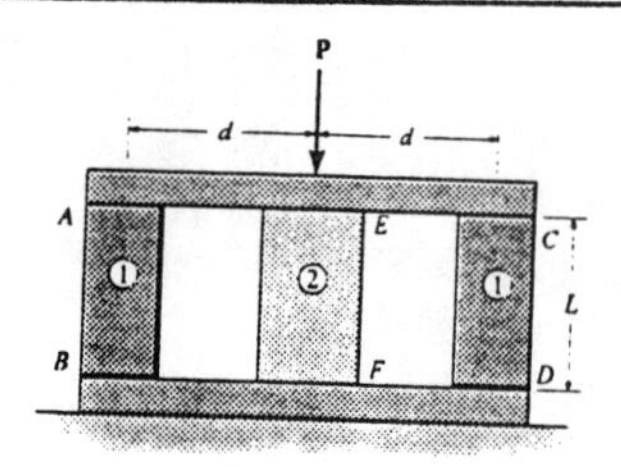

Equilibrium :

$$+\uparrow \Sigma F_y = 0; \quad 2F_1 + F_2 - P = 0 \qquad [1]$$

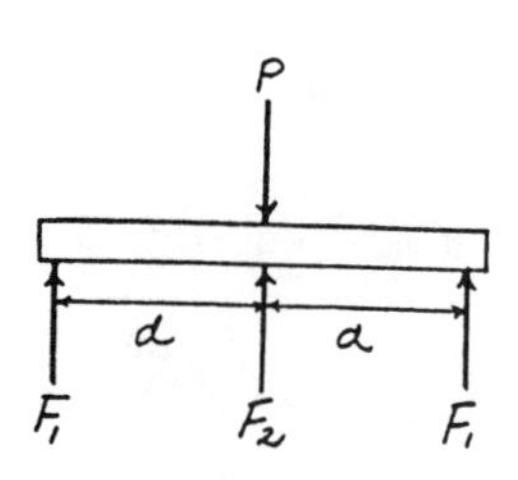

Compatibility :

$$\delta = \delta_1 = \delta_2$$

$$\frac{F_1 L}{A_1 E_1} = \frac{F_2 L}{A_2 E_2} \qquad F_1 = \left(\frac{A_1 E_1}{A_2 E_2}\right) F_2 \qquad [2]$$

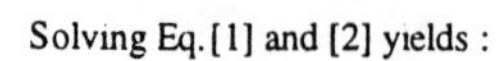

Solving Eq. [1] and [2] yields :

$$F_1 = \left(\frac{A_1 E_1}{2A_1 E_1 + A_2 E_2}\right) P$$ **Ans**

$$F_2 = \left(\frac{A_2 E_2}{2A_1 E_1 + A_2 E_2}\right) P$$ **Ans**

4–63. The assembly consists of two posts AB and CD made from material 1 having a modulus of elasticity of E_1 and each a cross-sectional area A_1, and a central post EF made from material 2 having a modulus of elasticity E_2 and a cross-sectional area A_2. If posts AB and CD are to be replaced by those having a material 2, determine the required cross-sectional area of these new posts so that both assemblies deform the same amount when loaded. The support is also rigid.

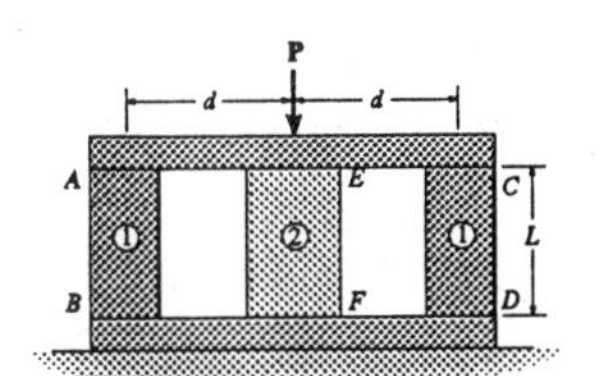

Equilibrium :

$$+\uparrow \Sigma F_y = 0; \quad 2F_1 + F_2 - P = 0 \qquad [1]$$

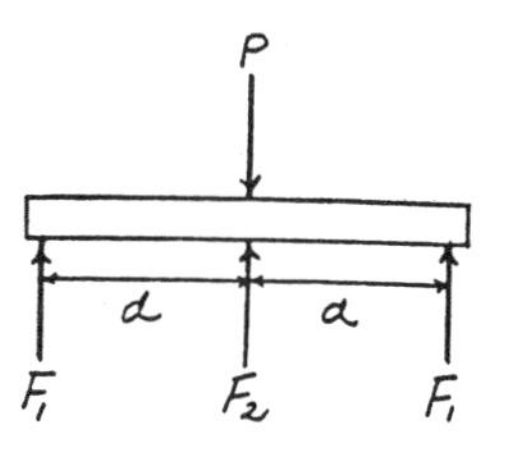

Compatibility :

$$\delta_{in} = \delta_1 = \delta_2$$

$$\frac{F_1 L}{A_1 E_1} = \frac{F_2 L}{A_2 E_2} \qquad F_1 = \left(\frac{A_1 E_1}{A_2 E_2}\right) F_2 \qquad [2]$$

Solving Eq. [1] and [2] yields :

$$F_1 = \left(\frac{A_1 E_1}{2A_1 E_1 + A_2 E_2}\right) P \qquad F_2 = \left(\frac{A_2 E_2}{2A_1 E_1 + A_2 E_2}\right) P$$

$$\delta_{in} = \frac{F_2 L}{A_2 E_2} = \frac{\left(\frac{A_2 E_2}{2A_1 E_1 + A_2 E_2}\right) PL}{A_2 E_2} = \frac{PL}{2A_1 E_1 + A_2 E_2}$$

Compatibility : When material 1 has been replaced by material 2 for two side posts, then

$$\delta_{final} = \delta_1 = \delta_2$$

$$\frac{F_1 L}{A_1' E_2} = \frac{F_2 L}{A_2 E_2} \qquad F_1 = \left(\frac{A_1'}{A_2}\right) F_2 \qquad [3]$$

Solving for F_2 from Eq. [1] and [3]

$$F_2 = \left(\frac{A_2}{2A_1' + A_2}\right) P$$

$$\delta_{final} = \frac{F_2 L}{A_2 E_2} = \frac{\left(\frac{A_2}{2A_1' + A_2}\right) PL}{A_2 E_2} = \frac{PL}{E_2 (2A_1' + A_2)}$$

Requires, $\delta_{in} = \delta_{final}$

$$\frac{PL}{2A_1 E_1 + A_2 E_2} = \frac{PL}{E_2 (2A_1' + A_2)}$$

$$A_1' = \left(\frac{E_1}{E_2}\right) A_1 \qquad \textbf{Ans}$$

***4–64.** The assembly consists of two posts AB and CD made from material 1 having a modulus of elasticity of E_1 and each a cross-sectional area A_1, and a central post EF made from material 2 having a modulus of elasticity E_2 and a cross-sectional area A_2. If post EF is to be replaced by one having a material 1, determine the required cross-sectional area of this new post so that both assemblies deform the same amount when loaded. The support is also rigid.

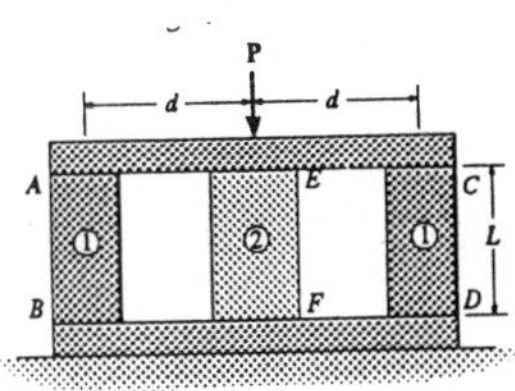

$$+\uparrow \Sigma F_y = 0; \quad 2F_1 + F_2 - P = 0 \qquad [1]$$

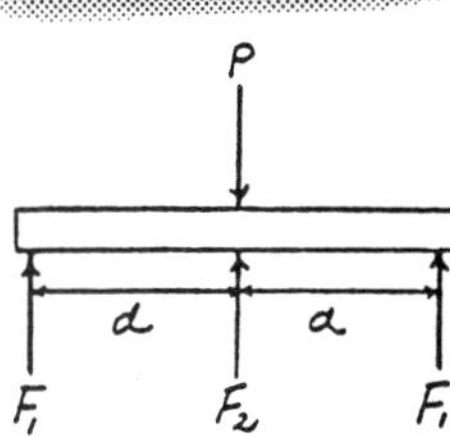

Compatibility :

$$\delta_{in} = \delta_1 = \delta_2$$

$$\frac{F_1 L}{A_1 E_1} = \frac{F_2 L}{A_2 E_2} \qquad F_1 = \left(\frac{A_1 E_1}{A_2 E_2}\right) F_2 \qquad [2]$$

Solving Eq. [1] and [2] yields :

$$F_1 = \left(\frac{A_1 E_1}{2A_1 E_1 + A_2 E_2}\right) P \qquad F_2 = \left(\frac{A_2 E_2}{2A_1 E_1 + A_2 E_2}\right) P$$

$$\delta_{in} = \frac{F_2 L}{A_2 E_2} = \frac{\left(\frac{A_2 E_2}{2A_1 E_1 + A_2 E_2}\right) P}{A_2 E_2} = \frac{PL}{2A_1 E_1 + A_2 E_2}$$

Compatibility : When material 2 has been replaced by material 1 for central posts, then

$$\delta_{final} = \delta_1 = \delta_2$$

$$\frac{F_1 L}{A_1 E_1} = \frac{F_2 L}{A_2' E_1} \qquad F_2 = \left(\frac{A_2'}{A_1}\right) F_1 \qquad [3]$$

Solving for F_1 from Eq. [1] and [3]

$$F_1 = \left(\frac{A_1}{2A_1 + A_2'}\right) P$$

$$\delta_{final} = \frac{F_1 L}{A_1 E_1} = \frac{\left(\frac{A_1}{2A_1 + A_2'}\right) PL}{A_1 E_1} = \frac{PL}{E_1 (2A_1 + A_2')}$$

Requires, $\delta_{in} = \delta_{final}$

$$\frac{PL}{2A_1 E_1 + A_2 E_2} = \frac{PL}{E_1 (2A_1 + A_2')}$$

$$A_2' = \left(\frac{E_2}{E_1}\right) A_2 \qquad \textbf{Ans}$$

4–65. The bracket is held to the wall using three A-36 steel bolts at B, C, and D. Each bolt has a diameter of 0.5 in. and an unstretched length of 2 in. If a force of 800 lb is placed on the bracket as shown, determine the force developed in each bolt. For the calculation, assume that the bolts carry no shear; rather, the vertical force of 800 lb is supported by the toe at A. Also, assume that the wall and bracket are rigid. A greatly exaggerated deformation of the bolts is shown.

$\circlearrowleft + \Sigma M_A = 0; \quad F_D(3.5) + F_C(1.5) + F_B(0.5) - 800(2) = 0 \quad [1]$

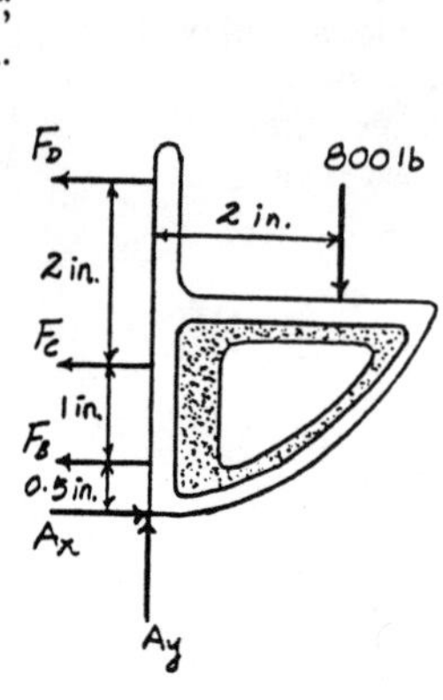

Compatibility :

$$\frac{\delta_D}{3.5} = \frac{\delta_C}{1.5}$$

$$\delta_D = 2.3333\ \delta_C$$

$$\frac{F_D L}{AE} = 2.333\frac{F_C L}{AE}; \quad F_D = 2.333\ F_C \quad [2]$$

$$\frac{\delta_B}{0.5} = \frac{\delta_C}{1.5}$$

$$\delta_B = 0.3333\ \delta_C$$

$$\frac{F_B L}{AE} = 0.3333\frac{F_C L}{AE}; \quad F_B = 0.3333\ F_C \quad [3]$$

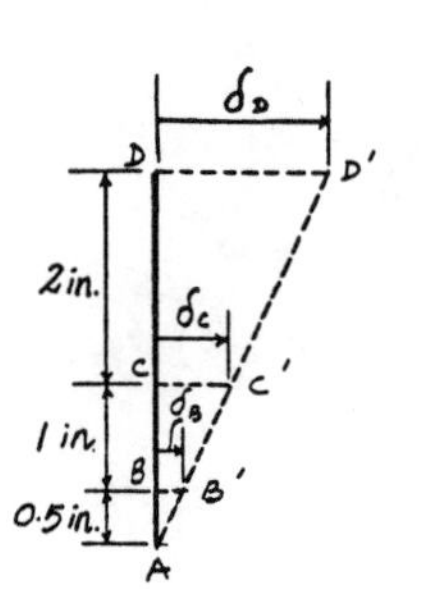

Solving Eqs. [1] to [3] yields :

$F_C = 163$ lb **Ans**

$F_D = 380$ lb **Ans**

$F_B = 54.2$ lb **Ans**

4–66. The bracket is held to the wall using three A-36 steel bolts at B, C, and D. Each bolt has a diameter of 0.5 in. and an unstretched length of 2 in. If a force of 800 lb is placed on the bracket as shown, determine how far s the top bracket at bolt D moves away from the wall. For the calculation, assume that the bolts carry no shear; rather, the vertical force of 800 lb is supported by the toe at A. Also, assume that the wall and bracket are rigid. A greatly exaggerated deformation of the bolts is shown.

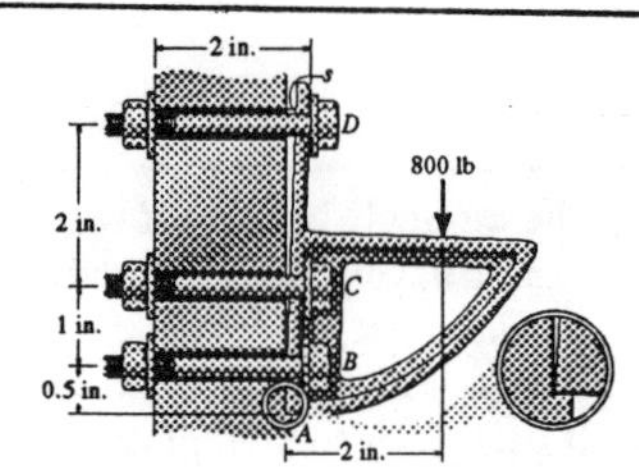

Equations of Equilibrium :

$\circlearrowleft + \Sigma M_A = 0; \quad F_D(3.5) + F_C(1.5) + F_B(0.5) - 800(2) = 0 \quad [1]$

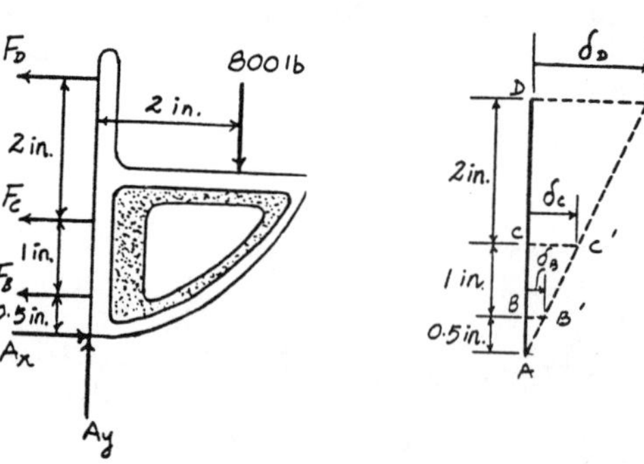

Compatibility :

$$\frac{\delta_D}{3.5} = \frac{\delta_C}{1.5}$$

$$\delta_D = 2.3333\ \delta_C$$

$$\frac{F_D L}{AE} = 2.333\frac{F_C L}{AE}; \quad F_D = 2.333\ F_C \quad [2]$$

$$\frac{\delta_B}{0.5} = \frac{\delta_C}{1.5}$$

$$\delta_B = 0.3333\ \delta_C$$

$$\frac{F_B L}{AE} = 0.3333\frac{F_C L}{AE}; \quad F_B = 0.3333\ F_C \quad [3]$$

Solving Eqs. [1] to [3] yields :

$F_C = 162.71$ lb $\quad F_D = 379.66$ lb $\quad F_B = 54.24$ lb

Displacement :

$$s = \delta_D = \frac{F_D L}{AE}$$

$$= \frac{379.66(2)}{\frac{\pi}{4}(0.5^2)(29.0)(10^6)} = 0.133(10^{-3}) \text{ in.} \quad \textbf{Ans}$$

4–67. The post is made of 6061-T6 aluminum and has a diameter of 50 mm. It is fixed supported at A and B and at its center C there is a coiled spring attached to the rigid collar which is fixed to the post. If the spring is originally uncompressed, determine the reactions at A and B when the force $P = 40$ kN is applied to the collar.

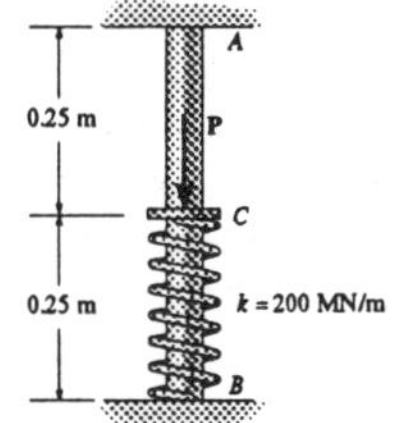

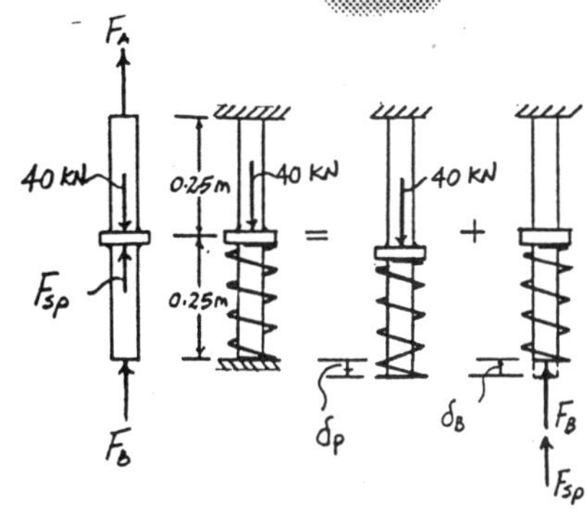

Equations of Equilibrium :

$$+\uparrow \Sigma F_y = 0; \quad F_A + F_B + F_{sp} - 40(10^3) = 0 \qquad [1]$$

Compatibility :

$$0 = \delta_P - \delta_B$$

$$0 = \frac{40(10^3)(0.25)}{\frac{\pi}{4}(0.05^2)68.9(10^9)} - \left[\frac{(F_B + F_{sp})(0.25)}{\frac{\pi}{4}(0.05^2)68.9(10^9)} + \frac{F_B + F_{sp}}{\frac{\frac{\pi}{4}(0.05^2)68.9(10^9)}{0.25} + 200(10^6)}\right]$$

$$F_B + F_{sp} = 23119.45 \qquad [2]$$

Also,

$$\delta_{sp} = \delta_{BC}$$

$$\frac{F_{sp}}{200(10^6)} = \frac{F_B + F_{sp}}{\frac{\frac{\pi}{4}(0.05^2)68.9(10^9)}{0.25} + 200(10^6)}$$

$$F_B = 2.7057 F_{sp} \qquad [3]$$

Solving Eq.[2] and [3] yields

$$F_{sp} = 6238.9 \text{ N}$$

$$F_B = 16880.6 \text{ N} = 16.9 \text{ kN} \qquad \textbf{Ans}$$

Substitute the results into Eq.[1]

$$F_A = 16880.6 \text{ N} = 16.9 \text{ kN} \qquad \textbf{Ans}$$

***4–68.** The post is made of 6061-T6 aluminum and has a diameter of 50 mm. It is fixed supported at A and B and at its center C there is a coiled spring attached to the rigid collar which is fixed to the post. If the spring is originally uncompressed, determine the compression in the spring when the load of $P = 50$ kN is applied to the collar.

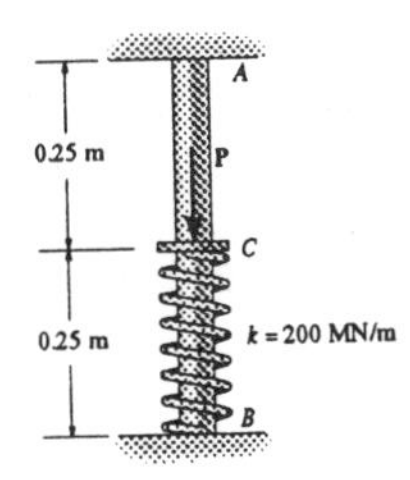

Compatibility :

$$0 = \delta_P - \delta_B$$

$$0 = \frac{50(10^3)(0.25)}{\frac{\pi}{4}(0.05^2)68.9(10^9)} - \left[\frac{(F_B + F_{sp})(0.25)}{\frac{\pi}{4}(0.05^2)68.9(10^9)} + \frac{F_B + F_{sp}}{\frac{\frac{\pi}{4}(0.05^2)68.9(10^9)}{0.25} + 200(10^6)}\right]$$

$$F_B + F_{sp} = 28899.31 \qquad [1]$$

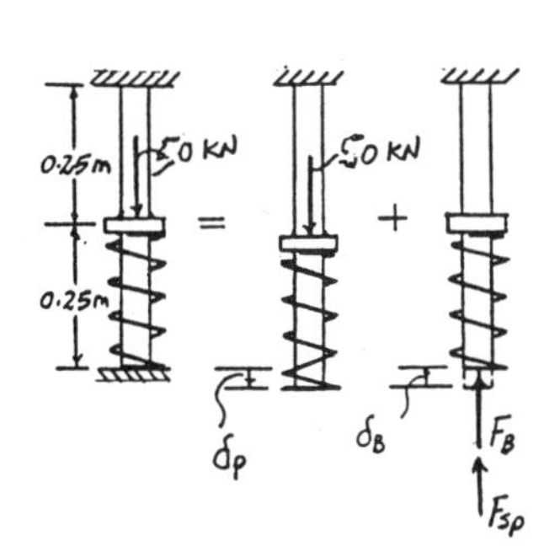

Also,

$$\delta_{sp} = \delta_{BC}$$

$$\frac{F_{sp}}{200(10^6)} = \frac{F_B + F_{sp}}{\frac{\frac{\pi}{4}(0.05^2)68.9(10^9)}{0.25} + 200(10^6)}$$

$$F_B = 2.7057 F_{sp} \qquad [2]$$

Solving Eqs.[1] and [2] yield

$$F_{sp} = 7798.6 \text{ N} \qquad F_B = 21100.7 \text{ N}$$

Thus,

$$\delta_{sp} = \frac{F_{sp}}{k} = \frac{7798.6}{200(10^6)}$$

$$= 0.0390(10^{-3}) \text{ m} = 0.0390 \text{ mm} \qquad \textbf{Ans}$$

4–69. The rigid bar is supported by the two short white spruce wooden posts and a spring. If each of the posts has an unloaded length of 1 m and a cross-sectional area of 600 mm², and the spring has a stiffness of $k = 2$ MN/m and an unstretched length of 1.02 m, determine the force in each post after the load is applied to the bar.

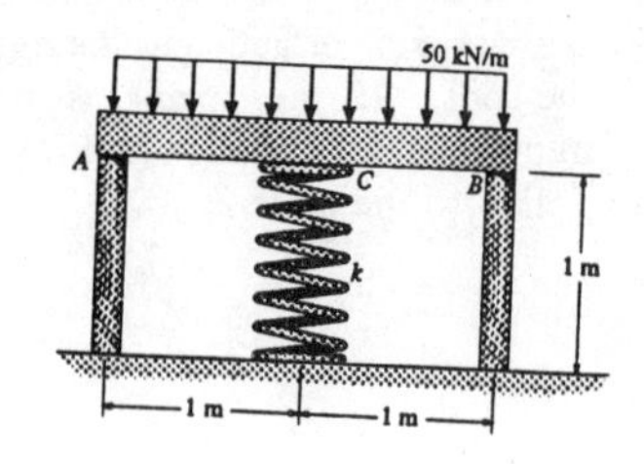

Equations of Equilibrium :

$$\circlearrowleft + \Sigma M_C = 0; \quad F_B(1) - F_A(1) = 0 \quad F_A = F_B = F$$

$$+\uparrow \Sigma F_y = 0; \quad 2F + F_{sp} - 100(10^3) = 0 \qquad [1]$$

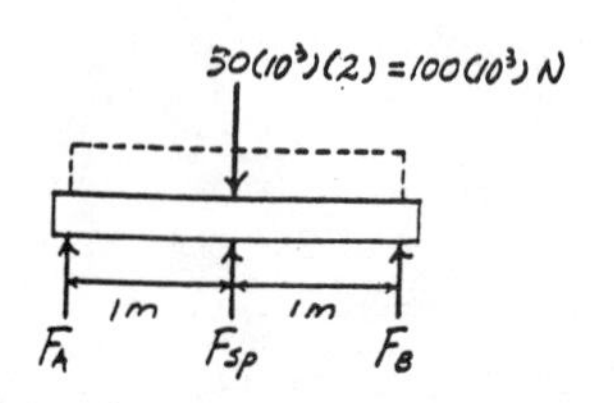

Compatibility :

$$(+\downarrow) \qquad \delta_A + 0.02 = \delta_{sp}$$

$$\frac{F(1)}{600(10^{-6})9.65(10^9)} + 0.02 = \frac{F_{sp}}{2.0(10^6)}$$

$$0.1727F + 20(10^3) = 0.5\,F_{sp} \qquad [2]$$

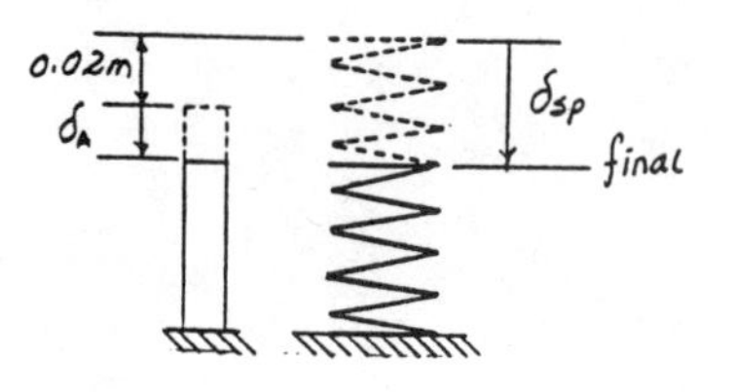

Solving Eqs. [1] and [2] yields :

$F_A = F_B = F = 25581.7 \text{ N} = 25.6 \text{ kN}$ **Ans**

$F_{sp} = 48\,836.5 \text{ N}$

4–70. The rigid bar is supported by the two short white spruce wooden posts and a spring. If each of the posts has an unloaded length of 1 m and a cross-sectional area of 600 mm², and the spring has a stiffness of $k = 2$ MN/m and an unstretched length of 1.02 m, determine the vertical displacement of A and B after the load is applied to the bar.

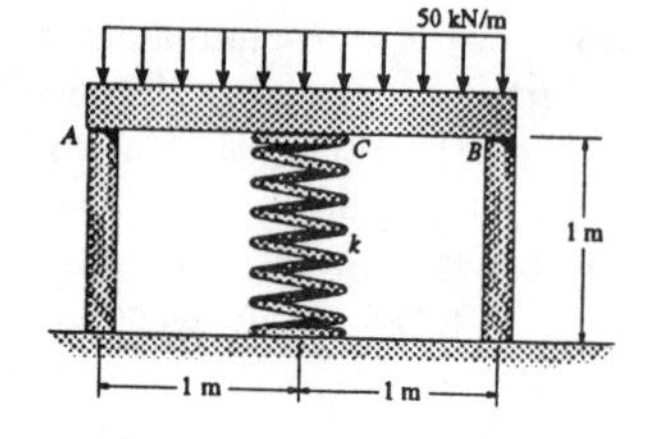

Equations of Equilibrium :

$$\circlearrowleft + \Sigma M_C = 0; \quad F_B(1) - F_A(1) = 0 \quad F_A = F_B = F$$

$$+\uparrow \Sigma F_y = 0; \quad 2F + F_{sp} - 100(10^3) = 0 \qquad [1]$$

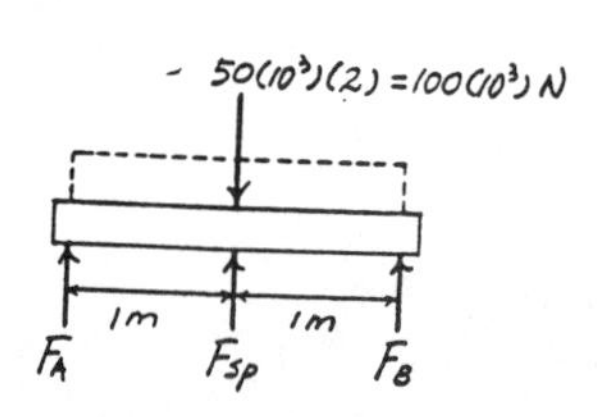

Compatibility :

$$(+\downarrow) \qquad \delta_A + 0.02 = \delta_{sp}$$

$$\frac{F(1)}{600(10^{-6})9.65(10^9)} + 0.02 = \frac{F_{sp}}{2.0(10^6)}$$

$$0.1727F + 20(10^3) = 0.5\,F_{sp} \qquad [2]$$

Solving Eqs. [1] and [2] yields :

$F = 25\,581.7 \text{ N} \qquad F_{sp} = 48\,836.5 \text{ N}$

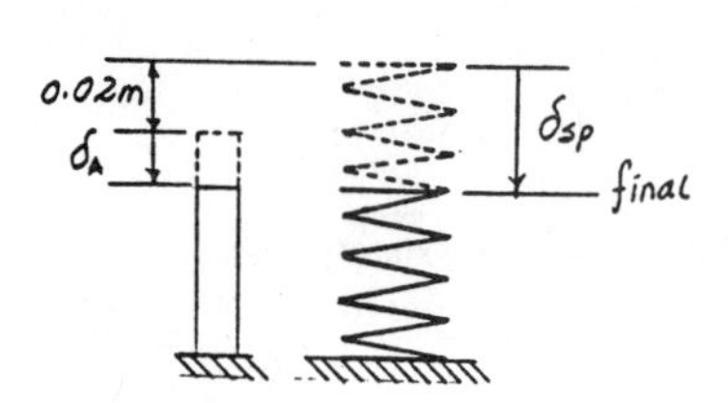

Displacement :

$$\delta_A = \delta_B = \frac{FL}{AE}$$

$$= \frac{25\,581.7(1000)}{600(10^{-6})(9.65)(10^9)} = 4.42 \text{ mm} \qquad \textbf{Ans}$$

4–71. The rigid member is held in the position shown by three A-36 steel tie rods. Each rod has an unstretched length of 0.75 m and a cross-sectional area of 125 mm². Determine the forces in the rods if a turnbuckle on rod EF undergoes one full turn. The lead of the screw is 1.5 mm. Neglect the size of the turnbuckle and assume that it is rigid. *Note:* The lead would cause the rod, when *unloaded*, to shorten 1.5 mm when the turnbuckle is rotated one revolution.

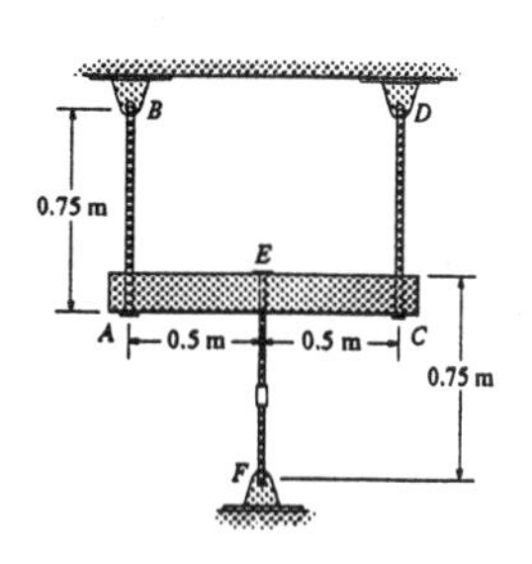

Equations of Equilibrium :

$$\curvearrowleft + \Sigma M_E = 0; \quad -F_{AB}(0.5) + F_{CD}(0.5) = 0$$

$$F_{AB} = F_{CD} = F \qquad [1]$$

$$+\uparrow \Sigma F_y = 0; \quad -F_{EF} + 2F = 0$$

$$F_{EF} = 2F \qquad [2]$$

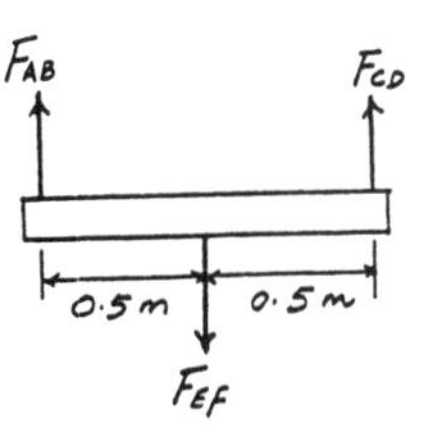

Compatibility : Rod EF shortens 1.5 mm causing AB (and DC) to elongate. Thus,

$$0.0015 = \delta_{AB} + \delta_{EF}$$

$$0.0015 = \frac{F(0.75)}{(125)(10^{-6})(200)(10^9)} + \frac{2F(0.75)}{(125)(10^{-6})(200)(10^9)}$$

$$F = 16666.67 \text{ N}$$

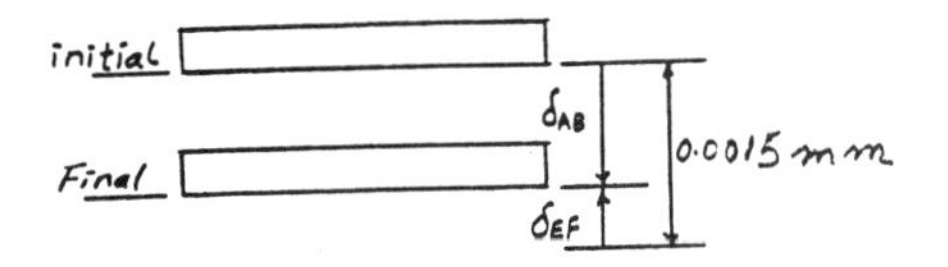

From Eq. [1] $\quad F_{AB} = F_{CD} = F = 16.7 \text{ kN}$ **Ans**

From Eq. [2] $\quad F_{EF} = 2F = 33.3 \text{ kN}$ **Ans**

***4–72.** The bolt is made of A-36 steel and the single threaded screw on the bolt has a lead of 0.100 mm. Determine the maximum average normal stress in the bolt if the nut is given one full turn after it first fits snug against the rigid plate. Also, assume the washer, the bolt head, and nut are rigid. *Note:* The lead represents the distance the screw advances along its axis for one complete turn of the screw.

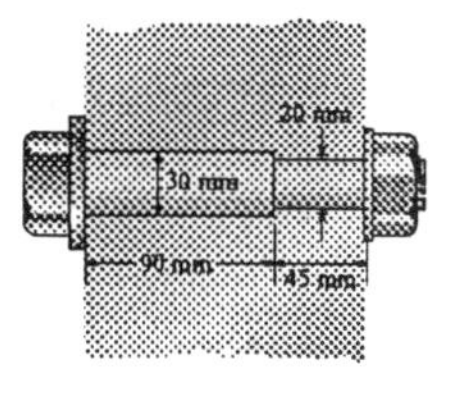

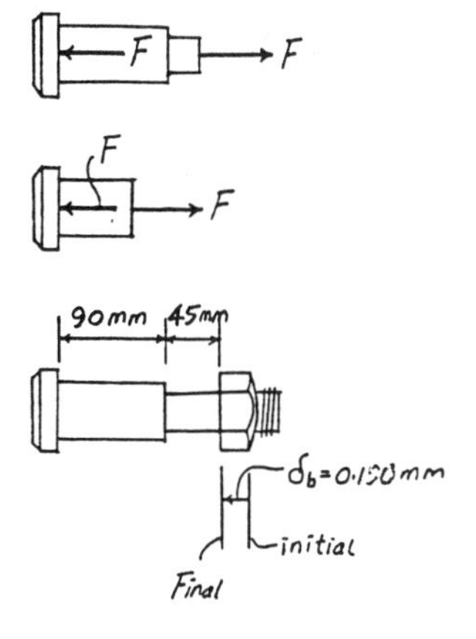

Compatibility :

$$0.100 \text{ mm} = \delta_b$$

$$0.100 = \frac{F(90)}{\frac{\pi}{4}(0.03^2)200(10^9)} + \frac{F(45)}{\frac{\pi}{4}(0.02^2)200(10^9)}$$

$$F = 73.92(10^3) \text{ N}$$

Maximum Average Normal Stress :

$$\sigma_{\max} = \frac{F}{A} = \frac{73.92(10^3)}{\frac{\pi}{4}(0.02^2)} = 235 \text{ MPa} \qquad \textbf{Ans}$$

4–73. The wheel is subjected to a force of 18 kN from the axial. Determine the force in each of the three spokes. Assume the rim is rigid and the spokes are made of the same material, and each has the same cross-sectional area.

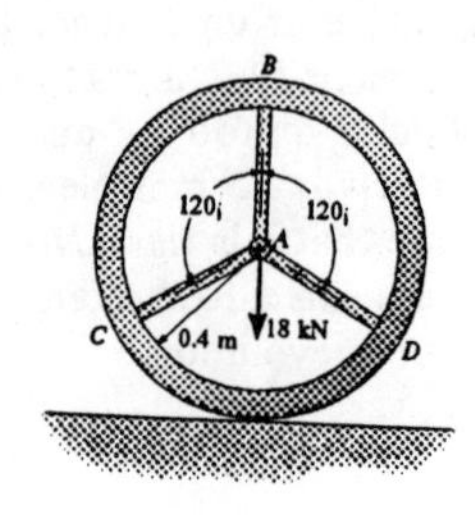

Equations of Equilibrium :

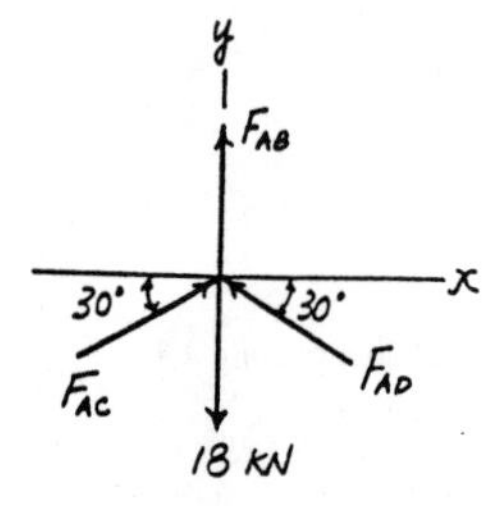

$$\xrightarrow{+} \Sigma F_x = 0; \quad F_{AC}\cos 30° - F_{AD}\cos 30° = 0$$
$$F_{AC} = F_{AD} = F$$

$$+\uparrow \Sigma F_y = 0; \quad F_{AB} + 2F\sin 30° - 18 = 0$$
$$F_{AB} + F = 18 \qquad [1]$$

Compatibility :

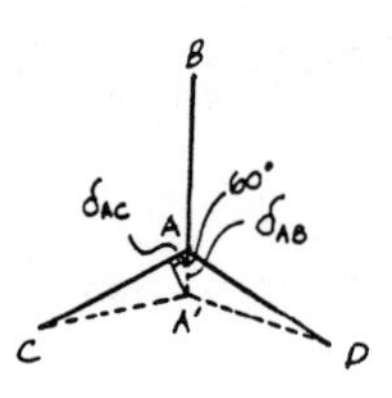

$$\delta_{AC} = \delta_{AB}\cos 60°$$
$$\frac{F(0.4)}{AE} = \frac{F_{AB}(0.4)}{AE}\cos 60°$$
$$F = 0.5F_{AB} \qquad [2]$$

Solving Eq.[1] and [2] yields :

$F_{AB} = 12.0 \text{ kN (T)}$ **Ans**

$F_{AC} = F_{AD} = F = 6.00 \text{ kN (C)}$ **Ans**

4–74. The electrical switch closes when the linkage rods CD and AB heat up, causing the rigid arm BDE both to translate and rotate until contact is made at F. Originally, BDE is vertical and the temperature is 20°C. If AB is made of bronze C86100 and CD is made of aluminum 6061-T6, determine the gap s required so that the switch will close when the temperature becomes 110°C.

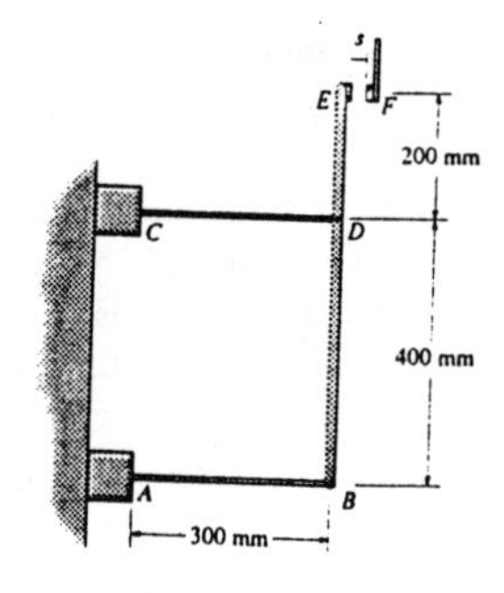

Thermal Expansion :

$$\delta_{AB} = \alpha_{cu}\Delta TL = 17.0(10^{-6})(110-20)(300) = 0.4590 \text{ mm}$$
$$\delta_{CD} = \alpha_{al}\Delta TL = 24.0(10^{-6})(110-20)(300) = 0.6480 \text{ mm}$$

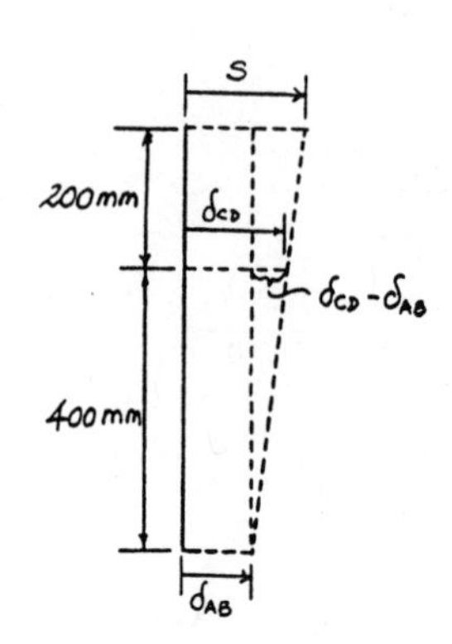

Geometry :

$$s = \delta_{AB} + (\delta_{CD} - \delta_{AB})\left(\frac{600}{400}\right)$$
$$= 0.4590 + (0.6480 - 0.4590)\left(\frac{600}{400}\right)$$
$$= 0.7425 \text{ mm}$$ **Ans**

4–75. A 6-ft-long steam pipe is made of A-36 steel and is connected directly to two turbines A and B as shown. The pipe has an outer diameter of 4 in. and a wall thickness of 0.25 in. The connection was made at $T_1 = 70°F$. If the turbines' points of attachment are assumed rigid, determine the force the pipe exerts on the turbines when the steam and thus the pipe reach a temperature of $T_2 = 275°F$.

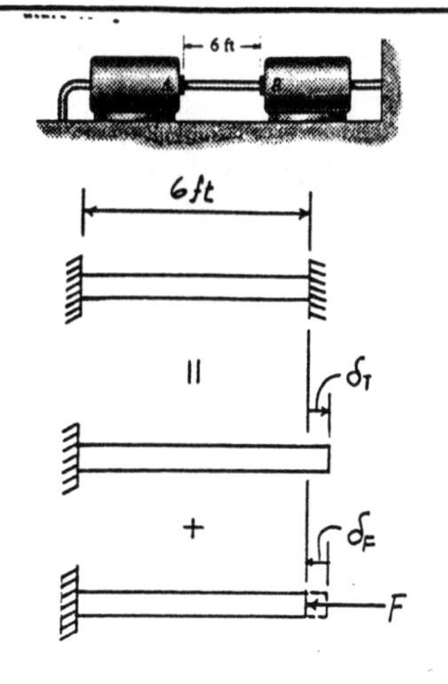

Compatibility :

$$(\overset{+}{\rightarrow}) \quad 0 = \delta_T - \delta_F$$

$$0 = 6.60(10^{-6})(275-70)(6)(12) - \frac{F(6)(12)}{\frac{\pi}{4}(4^2 - 3.5^2)(29.0)(10^3)}$$

$F = 116$ kip **Ans**

***4–76.** A 6-ft-long steam pipe is made of A-36 steel and is connected directly to two turbines A and B as shown. The pipe has an outer diameter of 4 in. and a wall thickness of 0.25 in. The connection was made at $T_1 = 70°F$. If the turbines' points of attachment are assumed to have a stiffness of $k = 80(10^3)$ kip/in., determine the force the pipe exerts on the turbines when the steam and thus the pipe reach a temperature of $T_2 = 275°F$.

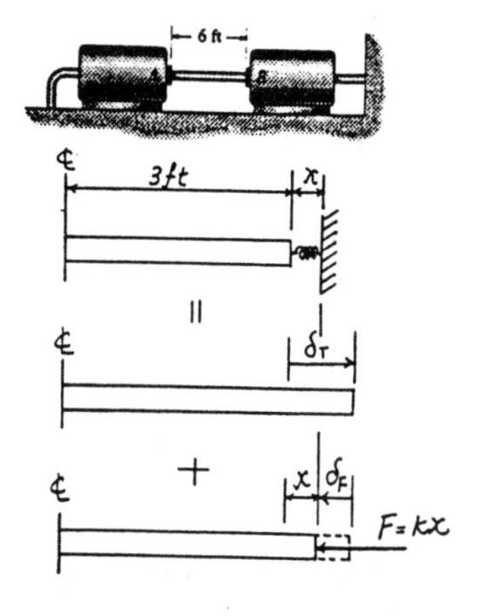

Compatibility :

$$x = \delta_T - \delta_F$$

$$x = 6.60(10^{-6})(275-70)(3)(12) - \frac{80(10^3)(x)(3)(12)}{\frac{\pi}{4}(4^2 - 3.5^2)(29.0)(10^3)}$$

$x = 0.001403$ in.

$F = kx = 80(10^3)(0.001403) = 112$ kip **Ans**

4–77. The 40-ft-long A-36 steel rails on a train track are laid with a small gap between them to allow for thermal expansion. Determine the required gap δ so that the rails just touch one another when the temperature is increased from $T_1 = -20°F$ to $T_2 = 90°F$. Using this gap, what would be the axial force in the rails if the temperature were to rise to $T_3 = 110°F$? The cross-sectional area of each rail is 5.10 in^2.

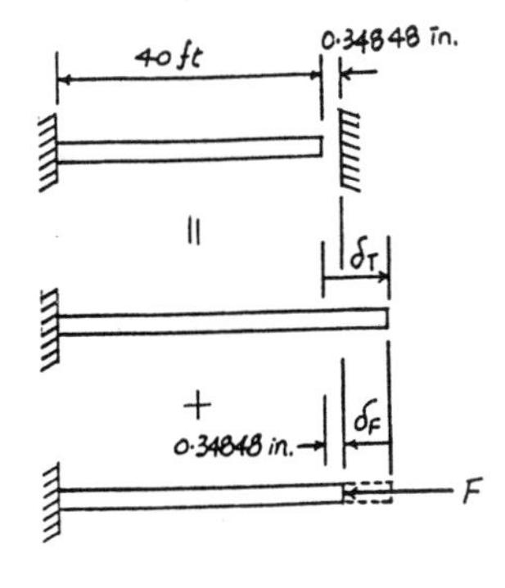

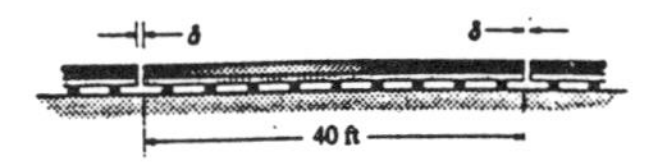

Thermal Expansion : Note that since adjacent rails expand, each rail will be required to expand $\frac{\delta}{2}$ on each end, or δ for the entire rail.

$$\delta = \alpha \Delta T L = 6.60(10^{-6})[90 - (-20)](40)(12)$$
$$= 0.34848 \text{ in.} = 0.348 \text{ in.}$$ **Ans**

Compatibility :

$$(\overset{+}{\rightarrow}) \quad 0.34848 = \delta_T - \delta_F$$

$$0.34848 = 6.60(10^{-6})[110 - (-20)](40)(12) - \frac{F(40)(12)}{5.10(29.0)(10^3)}$$

$F = 19.5$ kip **Ans**

4–78. The 0.4-in.-diameter A-36 steel bolt is used to hold the (rigid) assembly together. Determine the clamping force that must be provided by the bolt when $T_1 = 90°F$ so that the clamping force it exerts when $T_2 = 175°F$ is 500 lb.

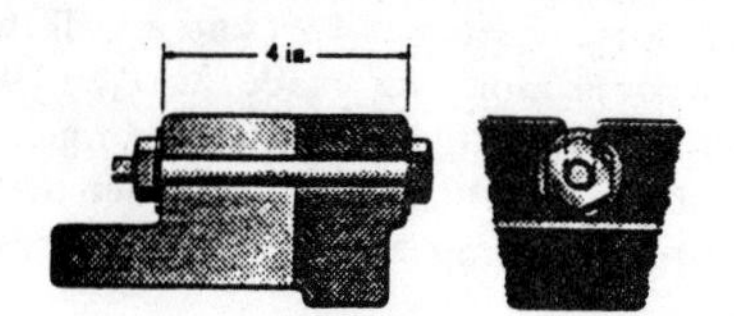

Compatibility :

$$\delta_F = \delta + \delta_T$$

$$\frac{F(4)}{\frac{\pi}{4}(0.4^2)(29.0)(10^3)} = \frac{0.5(4)}{\frac{\pi}{4}(0.4^2)(29.0)(10^3)} + 6.60(10^{-6})(175-90)(4)$$

$$F = 2.54 \text{ kip} \qquad \textbf{Ans}$$

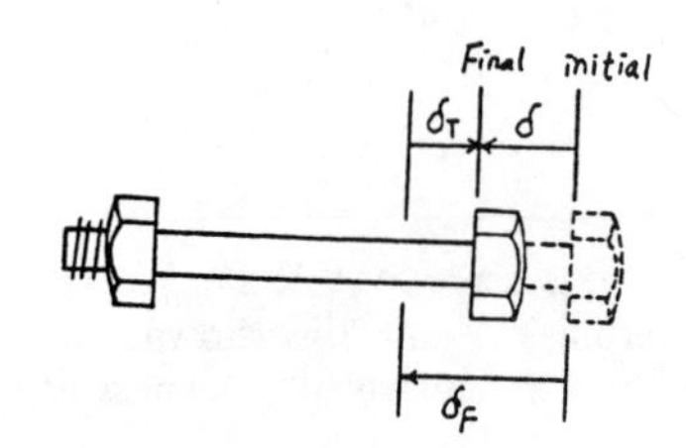

Also,

Thermal Expansion :

$$\delta_T = \alpha \Delta T L = 6.60(10^{-6})(175-90)(4) = 0.002244 \text{ in.}$$

The force when the bolt expands δ_T is

$$\delta_T = 0.002244 = \frac{F_T(4)}{\frac{\pi}{4}(0.4^2)(29.0)(10^3)}$$

$$F_T = 2.044 \text{ kip}$$

The initial clamping force is therefore

$$F = F_T + 0.5 = 2.54 \text{ kip} \qquad \textbf{Ans}$$

4–79. The 0.4-in.-diameter A-36 steel bolt is used to hold the (rigid) assembly together. If the nut is snug (no axial force in the bolt) when $T = 90°F$, determine the clamping force it exerts on the assembly when $T = 20°F$.

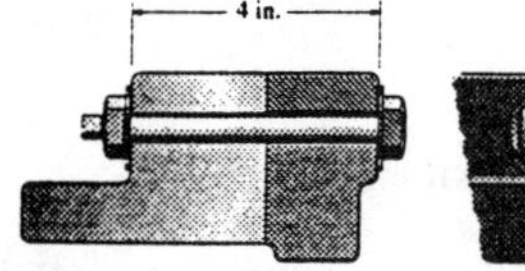

Compatibility :

$$(\overset{+}{\rightarrow}) \qquad 0 = \delta_F - \delta_T$$

$$0 = \frac{F(4)}{\frac{\pi}{4}(0.4^2)(29.0)(10^6)} - 6.60(10^{-6})(70)(4)$$

$$F = 1.68 \text{ kip} \qquad \textbf{Ans}$$

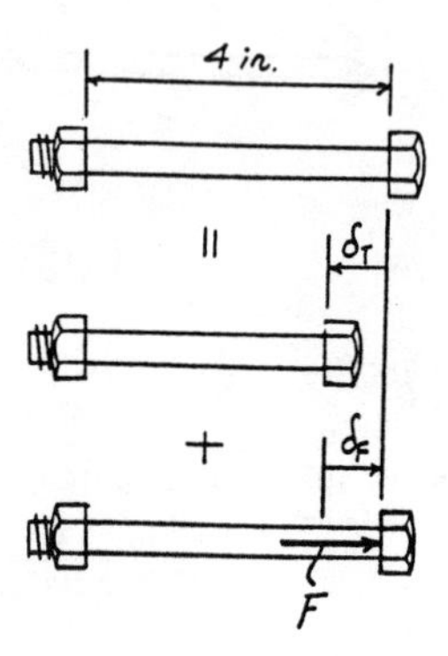

***4–80.** The shaft is held within the two rigid bearing supports. If it becomes overheated such that its temperature rises $\Delta T = 18°C$, determine the average normal stress developed within the shaft. Assume the supports maintain their original temperature. The shaft is made from L2 tool steel.

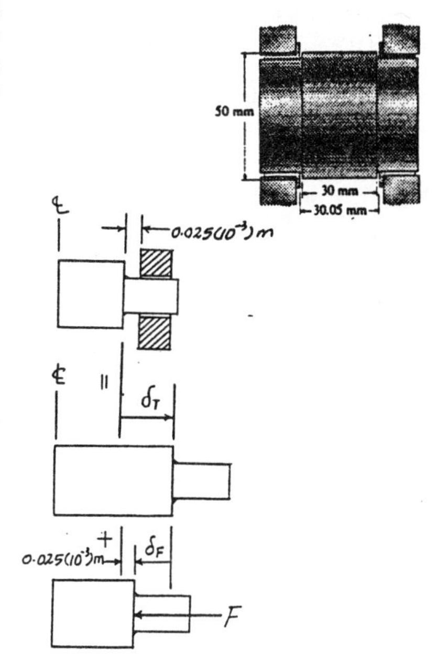

Compatibility :

$$(\overset{+}{\rightarrow}) \qquad 0.025\left(10^{-3}\right) = \delta_T - \delta_F$$

$$0.025\left(10^{-3}\right) = 12.0\left(10^{-6}\right)(180)(0.015) - \frac{F(0.015)}{\frac{\pi}{4}(0.05^2)\,200(10^9)}$$

$$F = 193\,731.5 \text{ N}$$

Average Normal Stress :

$$\sigma = \frac{F}{A} = \frac{193\,731.5}{\frac{\pi}{4}(0.05^2)} = 98.7 \text{ MPa} \qquad \textbf{Ans}$$

4–81. The A-36 steel pipe having a cross-sectional area of 0.5 in^2 is connected to fixed supports and carries a liquid that causes the pipe to be subjected to a temperature drop of $\Delta T = -0.2x^{3/2}$ °F, where x is in inches. Determine the axial force developed in the pipe.

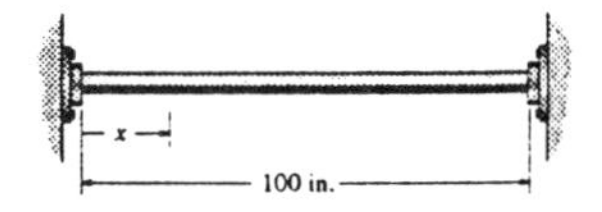

Compatibility :

$$(\overset{+}{\rightarrow}) \qquad 0 = \delta_F + \delta_T \qquad \text{where} \qquad \delta_T = \alpha\int \Delta T\,dx$$

$$0 = \frac{F(100)}{0.5(29.0)(10^3)} + 6.60\left(10^{-6}\right)\int_0^{100\text{in.}}\left(-0.2x^{\frac{3}{2}}\right)dx$$

$$F = 7.66 \text{ kip} \qquad \textbf{Ans}$$

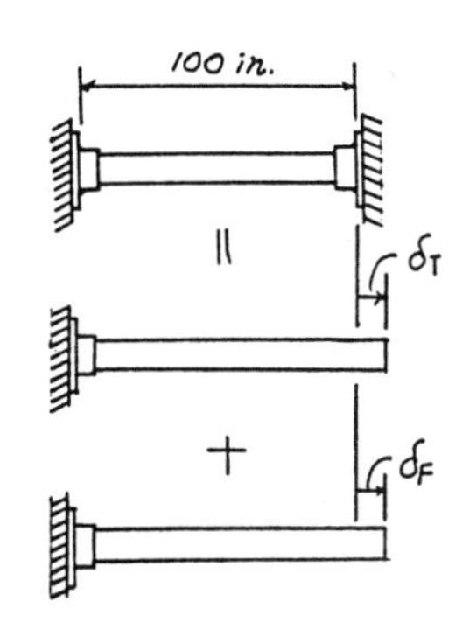

4–82. The A-36 steel pipe having a cross-sectional area of 0.5 in² is connected to fixed supports and carries a liquid that causes the pipe to be subjected to a temperature drop of $\Delta T = -0.2x^{1/2}$ °F, where x is in inches. Determine the maximum and minimum normal strain.

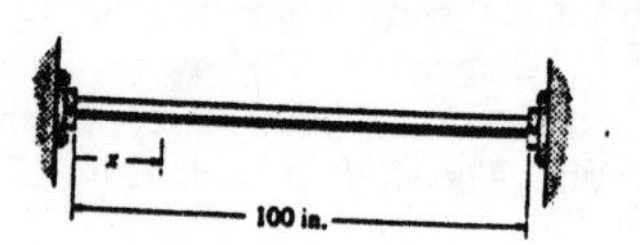

Compatibility :

$$(\overset{+}{\rightarrow}) \qquad 0 = \delta_F + \delta_T \qquad \text{where} \qquad \delta_T = \alpha \int \Delta T dx$$

$$0 = \frac{F(100)}{0.5(29.0)(10^3)} + 6.60(10^{-6}) \int_0^{100\text{in.}} \left(-0.2x^{\frac{1}{2}}\right) dx$$

$$F = 0.1276 \text{ kip}$$

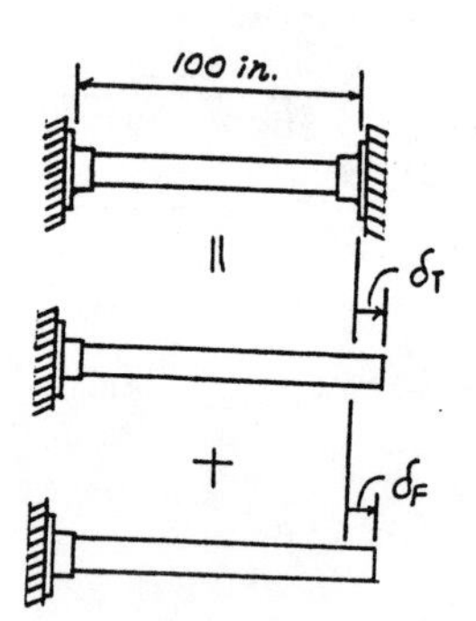

Normal Strain : The normal strain developed in the pipe is contributed by normal stress and thermal strain. Therefore,

$$\varepsilon = \frac{\sigma}{E} + \alpha \Delta T \qquad \text{where} \qquad \sigma = \frac{F}{A} = \frac{0.1276}{0.5} = 0.2552 \text{ ksi}$$

$$= \frac{0.2552}{29.0(10^3)} + 6.60(10^{-6})\left(-0.2x^{\frac{1}{2}}\right)$$

$$= 8.80(10^{-6}) - 1.32(10^{-6})x^{\frac{1}{2}}$$

By observation, maximum and minimum normal strain occurs at $x = 0$ and $x = 100$ in. respectively. Then

$$\varepsilon_{max} = 8.80(10^{-6}) - 0$$

$$= 8.80(10^{-6}) \text{ in./in.} \qquad \textbf{Ans}$$

$$\varepsilon_{min} = 8.80(10^{-6}) - 1.32(10^{-6})\left(100^{\frac{1}{2}}\right)$$

$$= -4.40(10^{-6}) \text{ in./in.} \qquad \textbf{Ans}$$

4–83. The A-36 steel pipe having a cross-sectional area of 0.5 in² is connected to fixed supports and carries a liquid that causes the pipe to be subjected to a temperature variation of $\Delta T = (20 - x)$°F, where x is in inches. Determine the average normal stress developed in the pipe.

Compatibility :

$$(\overset{+}{\rightarrow}) \qquad 0 = \delta_F + \delta_T \qquad \text{where} \qquad \delta_T = \alpha \int \Delta T dx$$

$$0 = \frac{F(100)}{0.5(29.0)(10^3)} + 6.60(10^{-6}) \int_0^{100\text{in.}} (20 - x)\, dx$$

$$F = 2.871 \text{ kip}$$

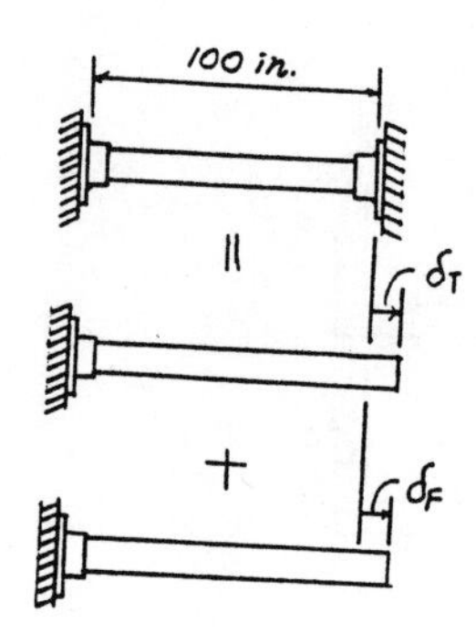

Average Normal Stress :

$$\sigma = \frac{F}{A} = \frac{2.871}{0.5} = 5.74 \text{ ksi} \qquad \textbf{Ans}$$

***4–84.** Two bars, each made of a different material, are connected and placed between two walls when the temperature is $T_1 = 10°C$. Determine the force exerted on the (rigid) supports when the temperature becomes $T_2 = 20°C$. The material properties and cross-sectional area of each bar are given in the figure.

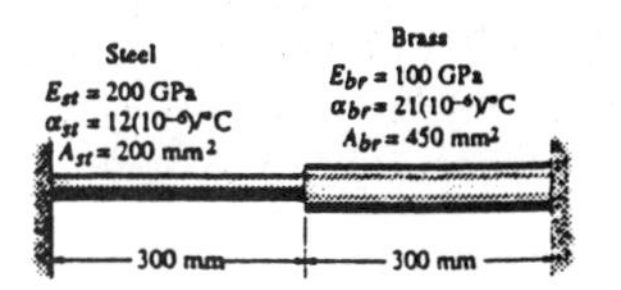

Compatibility :

$$(\overset{+}{\leftarrow}) \quad 0 = \delta_T - \delta_F$$

$$0 = 12(10^{-6})(20-10)(0.3) + 21(10^{-6})(20-10)(0.3) - \frac{F(0.3)}{200(10^{-6})(200)(10^9)} - \frac{F(0.3)}{450(10^{-6})(100)(10^9)}$$

$F = 6988.2$ N $= 6.99$ kN **Ans**

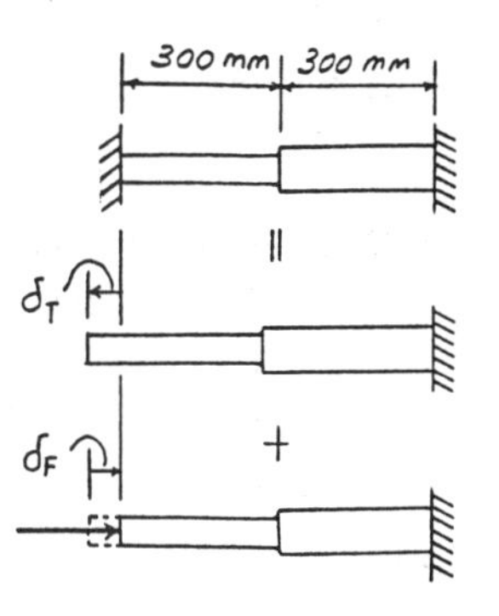

4–85. The two circular rod segments, one of aluminum and the other of copper, are fixed to the rigid walls such that there is a gap of 0.008 in. between them when $T_1 = 60°F$. What larger temperature T_2 is required in order to just close the gap? Each rod has a diameter of 1.25 in. $\alpha_{al} = 13(10^{-6})/°F$, $E_{al} = 10(10^3)$ ksi, $\alpha_{cu} = 9.4(10^{-6})/°F$, $E_{cu} = 18(10^3)$ ksi. Determine the average normal stress in each rod if $T_2 = 200°F$.

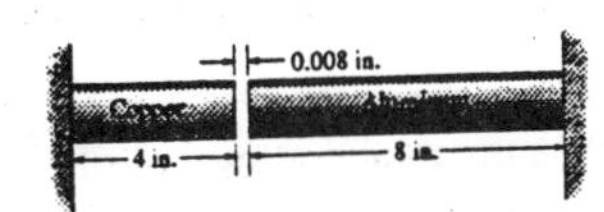

Thermal Expansion : To close the gap

$$\delta_T = \alpha_{al}\Delta T\, L_{al} + \alpha_{cu}\Delta T\, L_{cu}$$

$$0.008 = 13(10^{-6})(T_2 - 60)(8) + 9.4(10^{-6})(T_2 - 60)(4)$$

$T_2 = 116$ °F **Ans**

Compatibility :

$$0.008 = (\delta_T)_{cu} - (\delta_F)_{cu} + (\delta_T)_{al} - (\delta_F)_{al}$$

$$0.008 = 9.4(10^{-6})(200-60)(4) - \frac{F(4)}{\frac{\pi}{4}(1.25^2)(18)(10^3)} + 13(10^{-6})(200-60)(8) - \frac{F(8)}{\frac{\pi}{4}(1.25^2)(10)(10^3)}$$

$F = 14.195$ kip

Average Normal Stress :

$$\sigma_{al} = \sigma_{cu} = \frac{F}{A} = \frac{14.195}{\frac{\pi}{4}(1.25^2)} = 11.6 \text{ ksi}$$ **Ans**

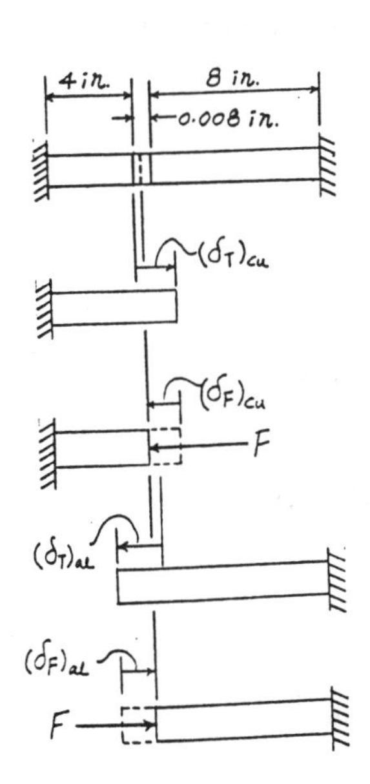

4–86. The two circular rod segments, one of aluminum and the other of copper, are fixed to the rigid walls such that there is a gap of 0.008 in. between them when $T_1 = 60°F$. Each rod has a diameter of 1.25 in. $\alpha_{al} = 13(10^{-6})/°F$, $E_{al} = 10(10^3)$ ksi, $\alpha_{cu} = 9.4(10^{-6})/°F$, $E_{cu} = 18(10^3)$ ksi. Determine the average normal stress in each rod if $T_2 = 300°F$, and also calculate the new length of the aluminum segment.

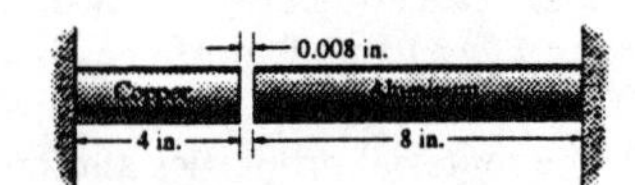

Compatibility :

$$0.008 = (\delta_T)_{cu} - (\delta_F)_{cu} + (\delta_T)_{al} - (\delta_F)_{al}$$

$$0.008 = 9.4(10^{-6})(300-60)(4) - \frac{F(4)}{\frac{\pi}{4}(1.25^2)(18)(10^3)}$$

$$+13(10^{-6})(300-60)(8) - \frac{F(8)}{\frac{\pi}{4}(1.25^2)(10)(10^3)}$$

$$F = 31.194 \text{ kip}$$

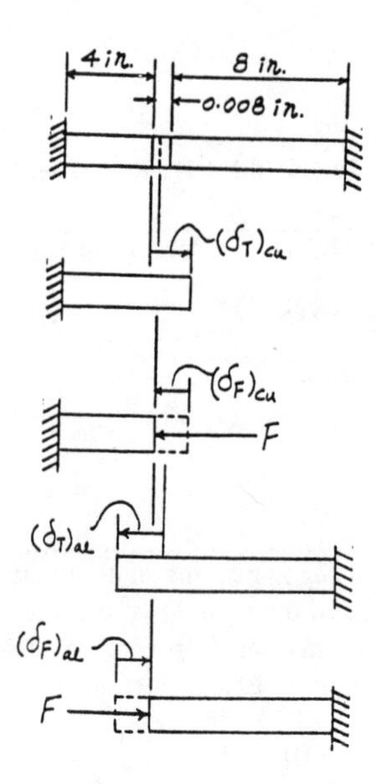

Average Normal Stress :

$$\sigma_{al} = \sigma_{cu} = \frac{F}{A} = \frac{31.194}{\frac{\pi}{4}(1.25^2)} = 25.4 \text{ ksi} \qquad \textbf{Ans}$$

Displacement :

$$\delta_{al} = (\delta_T)_{al} - (\delta_F)_{al}$$

$$= 13(10^{-6})[300-60](8) - \frac{31.194(8)}{\frac{\pi}{4}(1.25^2)(10)(10^3)}$$

$$= 0.0046247 \text{ in.}$$

$$L'_{al} = L_{al} + \delta_{al} = 8 + 0.0046247 = 8.00462 \text{ in.} \qquad \textbf{Ans}$$

4–87. The pipe is made of A-36 steel and is connected to the collars at A and B. When the temperature is 60°F, there is no axial load in the pipe. If hot gas traveling through the pipe causes its temperature to rise by $\Delta T = (40 + 15x)°F$, where x is in feet, determine the average normal stress in the pipe. The inner diameter is 2 in., the wall thickness is 0.15 in.

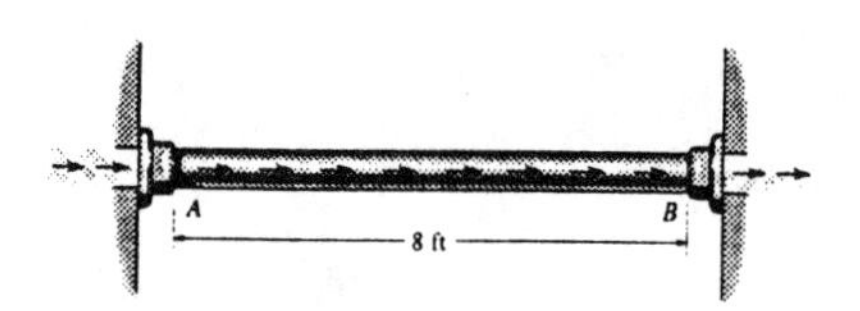

Compatibility :

$$0 = \delta_T - \delta_F \qquad \text{Where} \qquad \delta_T = \int_0^L \alpha \, \Delta T \, dx$$

$$0 = 6.60(10^{-6})\int_0^{8\text{ft}} (40 + 15x)\,dx - \frac{F(8)}{A(29.0)(10^3)}$$

$$0 = 6.60(10^{-6})\left[40(8) + \frac{15(8)^2}{2}\right] - \frac{F(8)}{A(29.0)(10^3)}$$

$$F = 19.14\,A$$

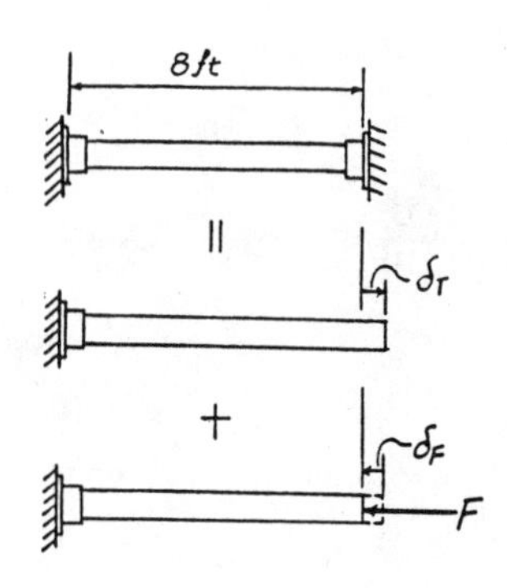

Average Normal Stress :

$$\sigma = \frac{19.14\,A}{A} = 19.1 \text{ ksi} \qquad \textbf{Ans}$$

***4–88.** The bronze 86100 pipe has an inner radius of 0.5 in. and a wall thickness of 0.2 in. If the gas flowing through it changes the temperature of the pipe uniformly from $T_A = 200°F$ at A to $T_B = 60°F$ at B, determine the axial force it exerts on the walls. The pipe was fitted between the walls when $T = 60°F$.

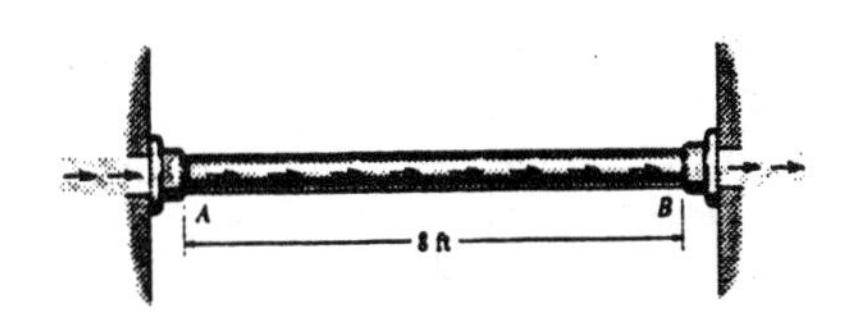

Temperature Gradient :

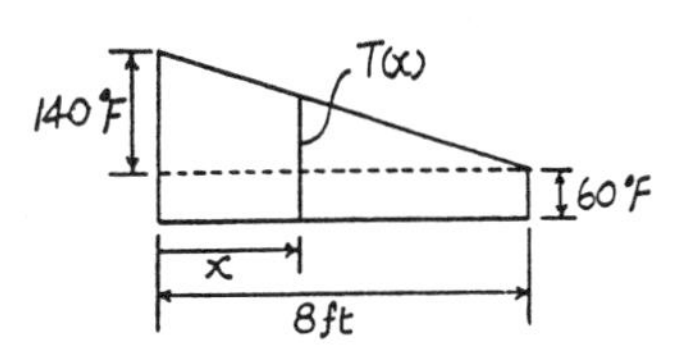

$$T(x) = 60 + \left(\frac{8-x}{8}\right)140 = 200 - 17.5x$$

Compatibility :

$$0 = \delta_T - \delta_F \quad \text{Where} \quad \delta_T = \int \alpha \Delta T dx$$

$$0 = 9.60(10^{-6})\int_0^{8\,ft}[(200 - 17.5x) - 60]\,dx - \frac{F(8)}{\frac{\pi}{4}(1.4^2 - 1^2)\,15.0(10^3)}$$

$$0 = 9.60(10^{-6})\int_0^{8\,ft}(140 - 17.5x)\,dx - \frac{F(8)}{\frac{\pi}{4}(1.4^2 - 1^2)\,15.0(10^3)}$$

$F = 7.60$ kip **Ans**

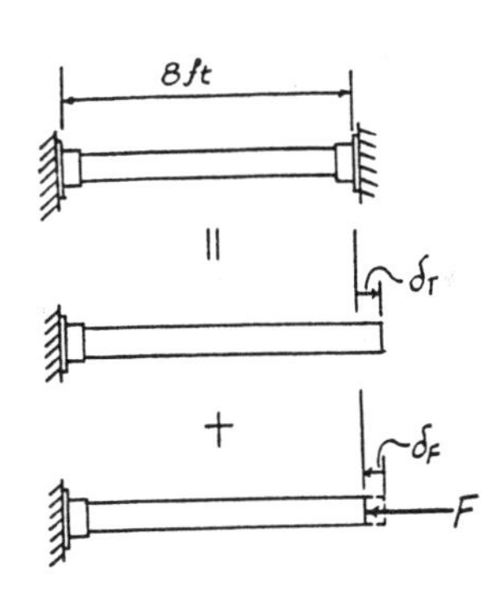

4–89. The rigid block has a weight of 80 kip and is to be supported by posts A and B, which are made of A-36 steel, and the post C, which is made of C83400 red brass. If all the posts have the same original length before they are loaded, determine the average normal stress developed in each post when post C is heated so that its temperature is increased by 20°F. Each post has a cross-sectional area of 8 in^2.

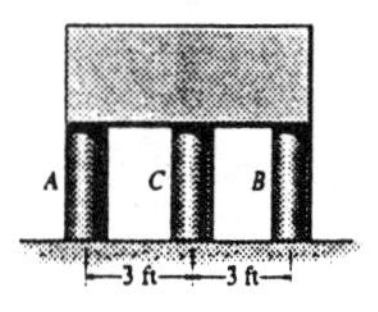

Equations of Equilibrium :

$$\curvearrowleft + \Sigma M_C = 0; \quad F_B(3) - F_A(3) = 0 \quad F_A = F_B = F$$

$$+\uparrow \Sigma F_y = 0; \quad 2F + F_C - 80 = 0 \qquad [1]$$

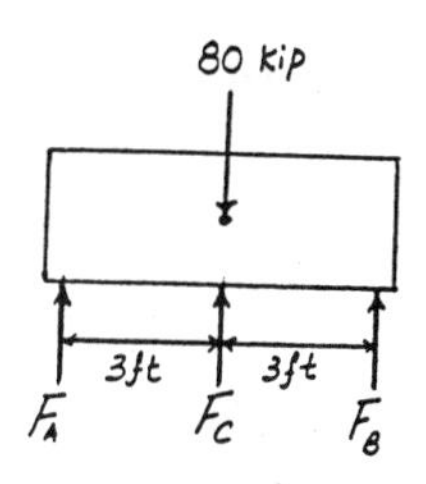

Compatibility :

$$(+\downarrow) \qquad (\delta_C)_F - (\delta_C)_T = \delta_F$$

$$\frac{F_C L}{8(14.6)(10^3)} - 9.80(10^{-6})(20)L = \frac{FL}{8(29.0)(10^3)}$$

$$8.5616\,F_C - 4.3103\,F = 196 \qquad [2]$$

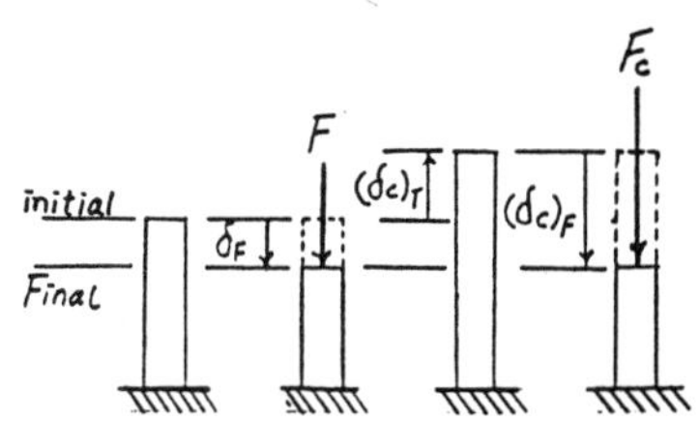

Solving Eqs. [1] and [2] yields :

$F = 22.81$ kip $\qquad F_C = 34.38$ kip

average Normal Sress :

$$\sigma_A = \sigma_B = \frac{F}{A} = \frac{22.81}{8} = 2.85 \text{ ksi} \qquad \textbf{Ans}$$

$$\sigma_C = \frac{F_C}{A} = \frac{34.38}{8} = 4.30 \text{ ksi} \qquad \textbf{Ans}$$

4–90. The aluminum 2014-T6 pipe CD is placed within the clamp and the screws on the clamp are tightened snug. If the assembly experiences a temperature increase of $\Delta T = 50°C$, determine the average normal stress developed within the pipe and screw. Assume the heads on the clamp are rigid and the screws are made of A-36 steel. The screws have a diameter of 14 mm, and the pipe has an outer diameter of 35 mm and a wall thickness of 2 mm.

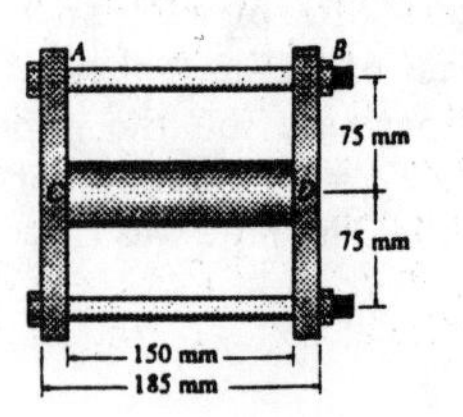

Equations of Equilibrium :

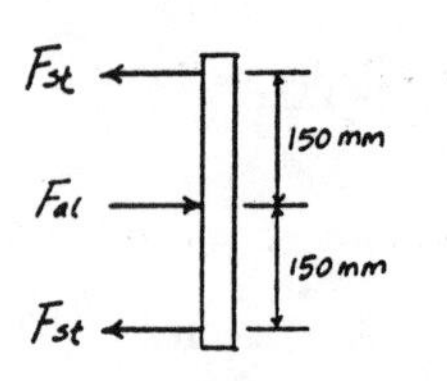

$$\overset{+}{\rightarrow} \Sigma F_x = 0; \quad F_{al} - 2F_{st} = 0 \qquad [1]$$

Compatibility :

$$(\overset{+}{\rightarrow}) \quad (\delta_{st})_T + (\delta_{st})_F = (\delta_{al})_T - (\delta_{al})_F$$

$$12.0(10^{-6})(50)(0.185) + \frac{F_{st}(0.185)}{\frac{\pi}{4}(0.014^2)\,200(10^9)}$$

$$= 23.0(10^{-6})(50)(0.150) - \frac{F_{al}(0.150)}{\frac{\pi}{4}(0.035^2 - 0.031^2)\,73.1(10^9)}$$

$$0.006009F_{st} + 0.009896F_{al} = 61.5 \qquad [2]$$

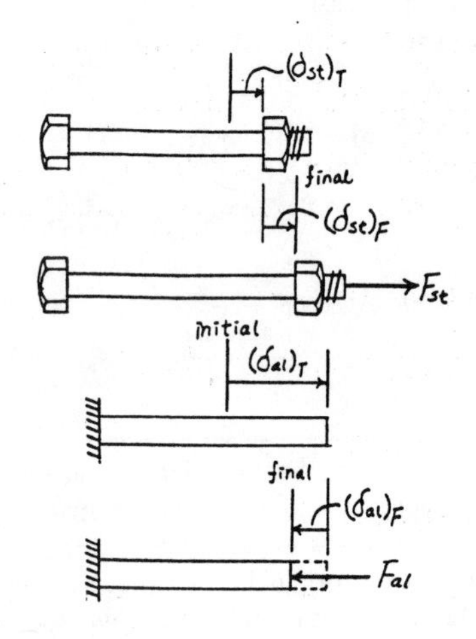

Solving Eq. [1] and [2] yields :

$$F_{st} = 2383.6 \text{ N} \qquad F_{al} = 4767.1 \text{ N}$$

Average Normal Stress :

$$\sigma_{al} = \frac{F_{al}}{A} = \frac{4767.1}{\frac{\pi}{4}(0.035^2 - 0.031^2)} = 23.0 \text{ MPa} \qquad \textbf{Ans}$$

$$\sigma_{st} = \frac{F_{st}}{A} = \frac{2383.6}{\frac{\pi}{4}(0.014^2)} = 15.5 \text{ MPa} \qquad \textbf{Ans}$$

4–91. The steel bolt has a diameter of 7 mm and fits through an aluminum sleeve as shown. The sleeve has an inner diameter of 8 mm and an outer diameter of 10 mm. The nut at A is adjusted so that it just presses up against the sleeve. If the assembly is originally at a temperature of $T_1 = 20°C$ and then is heated to a temperature of $T_2 = 100°C$, determine the average normal stress in the bolt and the sleeve. $E_{st} = 200$ GPa, $E_{al} = 70$ GPa, $\alpha_{st} = 14(10^{-6})/°C$, $\alpha_{al} = 23(10^{-6})/°C$.

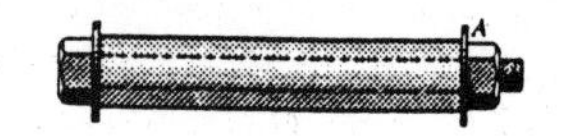

Compatibility :

$$(\delta_s)_T - (\delta_b)_T = (\delta_s)_F + (\delta_b)_F$$

$$23(10^{-6})(100-20)L - 14(10^{-6})(100-20)L$$

$$= \frac{FL}{\frac{\pi}{4}(0.01^2 - 0.008^2)70(10^9)} + \frac{FL}{\frac{\pi}{4}(0.007^2)200(10^9)}$$

$$F = 1133.54 \text{ N}$$

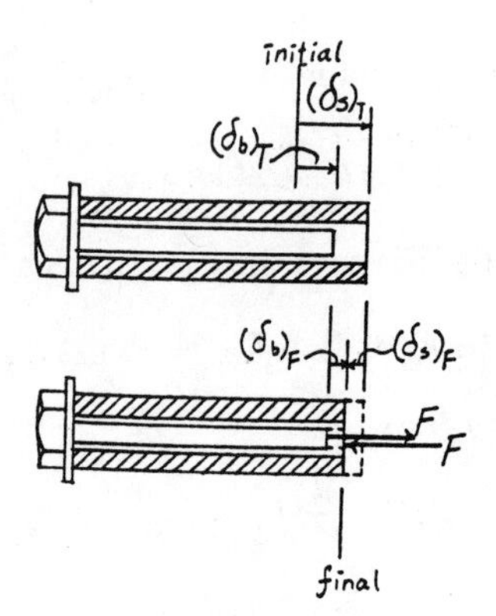

Average Normal Stress :

$$\sigma_s = \frac{F}{A_s} = \frac{1133.54}{\frac{\pi}{4}(0.01^2 - 0.008^2)} = 40.1 \text{ MPa} \qquad \textbf{Ans}$$

$$\sigma_b = \frac{F}{A_b} = \frac{1133.54}{\frac{\pi}{4}(0.007^2)} = 29.5 \text{ MPa} \qquad \textbf{Ans}$$

***4–92.** The wires AB and AC are made of steel and wire AD is made of copper. Before the 150-lb force is applied, AB and AC are each 60 in. long and AD is 40 in. long. If the temperature is increased by 80°F, determine the force in each wire needed to support the load. Take $E_{st} = 29(10^3)$ ksi, $E_{cu} = 17(10^3)$ ksi, $\alpha_{st} = 8(10^{-6})/°\text{F}$, $\alpha_{cu} = 9.60(10^{-6})/°\text{F}$. Each wire has a cross-sectional area of 0.0123 in².

Equations of Equilibrium :

$$\xrightarrow{+} \Sigma F_x = 0; \quad F_{AC}\cos 45° - F_{AB}\cos 45° = 0$$

$$F_{AC} = F_{AB} = F$$

$$+\uparrow \Sigma F_y = 0; \quad 2F\sin 45° + F_{AD} - 150 = 0 \qquad [1]$$

Compatibility :

$$(\delta_{AC})_T = 8.0(10^{-6})(80)(60) = 0.03840 \text{ in.}$$

$$(\delta_{AC})_{T_v} = \frac{(\delta_{AC})_T}{\cos 45°} = \frac{0.03840}{\cos 45°} = 0.05431 \text{ in.}$$

$$(\delta_{AD})_T = 9.60(10^{-6})(80)(40) = 0.03072 \text{ in.}$$

$$\delta_0 = (\delta_{AC})_{T_v} - (\delta_{AD})_T = 0.05431 - 0.03072 = 0.02359 \text{ in.}$$

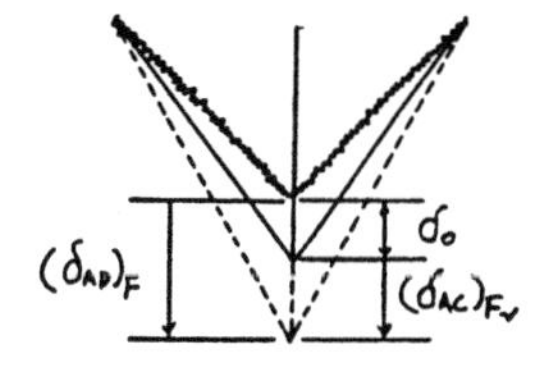

$$(\delta_{AD})_F = (\delta_{AC})_{F_v} + \delta_0$$

$$\frac{F_{AD}(40)}{0.0123(17.0)(10^6)} = \frac{F(60)}{0.0123(29.0)(10^6)\cos 45°} + 0.02359$$

$$0.1913F_{AD} - 0.2379F = 23.5858 \qquad [2]$$

Solving Eq. [1] and [2] yields :

$F_{AC} = F_{AB} = F = 10.0$ lb **Ans**

$F_{AD} = 136$ lb **Ans**

4–93. The rod is made of A-36 steel and has a diameter of 0.25 in. If the springs are compressed 0.5 in. when the temperature of the rod is $T = 40°\text{F}$, determine the force in the rod when its temperature is $T = 160°\text{F}$.

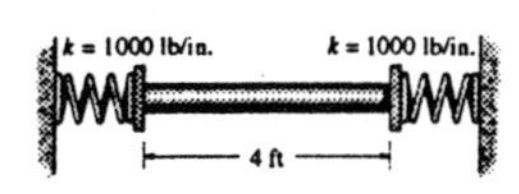

Compatibility :

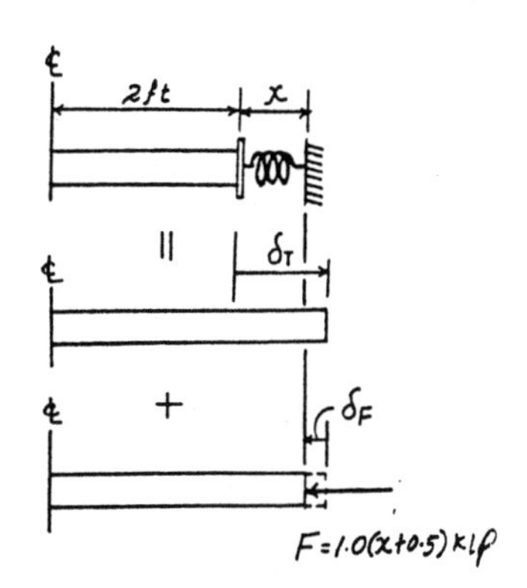

$$(\xrightarrow{+}) \quad x = \delta_T - \delta_F$$

$$x = 6.60(10^{-6})(160-40)(2)(12) - \frac{1.00(x + 0.5)(2)(12)}{\frac{\pi}{4}(0.25^2)(29.0)(10^3)}$$

$$x = 0.01040 \text{ in.}$$

$F = 1.00(0.01040 + 0.5) = 0.510$ kip **Ans**

4–94. The bar has a cross-sectional area A, length L, modulus of elasticity E, and coefficient of thermal expansion α. The temperature of the bar changes uniformly along its length from an original temperature of T_A at A to T_B at B so that at any point x along the bar $T = T_A + x(T_B - T_A)/L$. Determine the force the bar exerts on the rigid walls. Initially no axial force is in the bar.

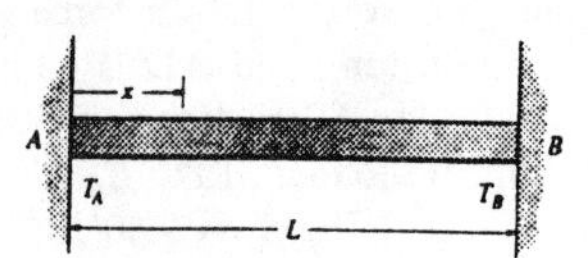

Compatibility :

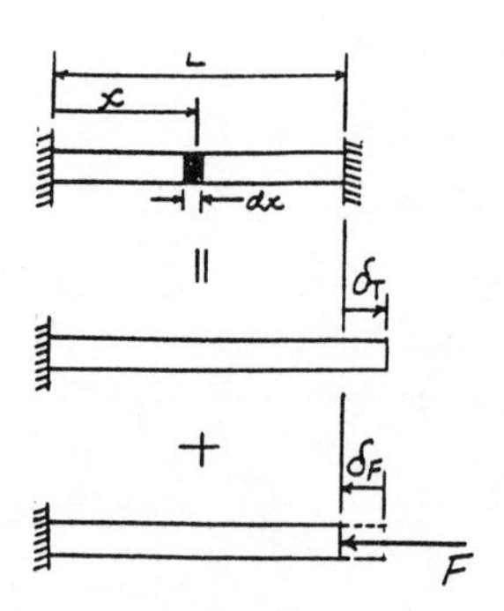

$(\xrightarrow{+})$ $\quad 0 = \delta_T - \delta_F \quad$ Where $\quad \delta_T = \int_0^L \alpha \Delta T dx$

$$0 = \alpha \int_0^L \left(T_A + \frac{T_B - T_A}{L} x - T_A \right) dx - \frac{FL}{AE}$$

$$0 = \alpha \int_0^L \left(\frac{T_B - T_A}{L} x \right) dx - \frac{FL}{AE}$$

$$0 = \alpha \left(\frac{T_B - T_B}{2L} \right) x^2 \Big|_0^L - \frac{FL}{AE}$$

$$F = \frac{\alpha AE}{2} (T_B - T_A) \qquad \textbf{Ans}$$

4–95. The steel bar has the dimensions shown. Determine the maximum axial force P that can be applied so as not to exceed an allowable tensile stress of $\sigma_{allow} = 150$ MPa.

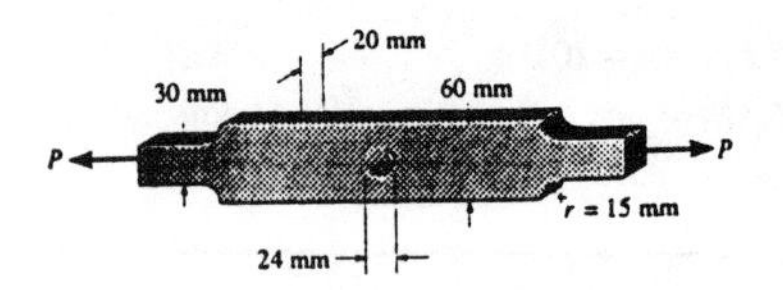

Assume failure occurs at the fillet :

$$\frac{w}{h} = \frac{60}{30} = 2 \quad \text{and} \quad \frac{r}{h} = \frac{15}{30} = 0.5$$

From the text, $\quad K = 1.4$

$$\sigma_{max} = \sigma_{allow} = K\sigma_{avg}$$

$$150(10^6) = 1.4\left[\frac{P}{0.03(0.02)}\right]$$

$$P = 64.3 \text{ kN}$$

Assume failure occurs at the hole :

$$\frac{r}{w} = \frac{12}{60} = 0.2$$

From the text. $\quad K = 2.45$

$$\sigma_{max} = \sigma_{allow} = K\sigma_{avg}$$

$$150(10^6) = 2.45\left[\frac{P}{(0.06 - 0.024)(0.02)}\right]$$

$$P = 44.1 \text{ kN} \quad \textit{(controls!)} \qquad \textbf{Ans}$$

***4–96.** Determine the maximum axial force P that can be applied to the bar. The bar is made of steel and has an allowable stress of $\sigma_{allow} = 21$ ksi.

Assume failure occurs at the fillet :

$$\frac{r}{h} = \frac{0.25}{1.25} = 0.2 \quad \text{and} \quad \frac{w}{h} = \frac{1.875}{1.25} = 1.5$$

From the text, $K = 1.75$

$$\sigma_{max} = \sigma_{allow} = K\sigma_{avg}$$

$$21 = 1.75\left[\frac{P}{1.25\,(0.125)}\right]$$

$$P = 1.875 \text{ kip}$$

Assume failure occurs at the hole :

$$\frac{r}{w} = \frac{0.375}{1.875} = 0.20$$

From the text, $K = 2.45$

$$\sigma_{max} = \sigma_{allow} = K\sigma_{avg}$$

$$21 = 2.45\left[\frac{P}{(1.875 - 0.75)(0.125)}\right]$$

$P = 1.21$ kip ***(controls!)*** **Ans**

4–97. Determine the maximum normal stress developed in the bar when it is subjected to a tension of $P = 2$ kip.

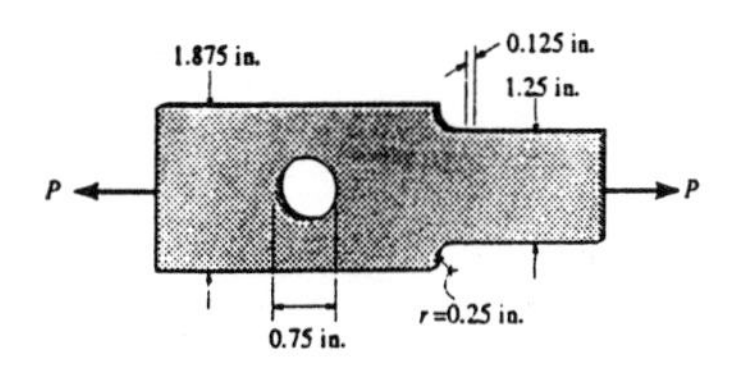

Maximum Normal Stress at fillet :

$$\frac{r}{h} = \frac{0.25}{1.25} = 0.2 \quad \text{and} \quad \frac{w}{h} = \frac{1.875}{1.25} = 1.5$$

From the text, $K = 1.75$

$$\sigma_{max} = K\left(\frac{P}{A}\right) = 1.75\left[\frac{2}{1.25(0.125)}\right] = 22.4 \text{ ksi}$$

Maximum Normal Stress at hole :

$$\frac{r}{w} = \frac{0.375}{1.875} = 0.20$$

From the text, $K = 2.45$

$$\sigma_{max} = 2.45\left[\frac{2}{(1.875 - 0.75)(0.125)}\right]$$

$= 34.8$ ksi ***(Controls!)*** **Ans**

4–98. Determine the maximum normal stress developed in the bar when it is subjected to a tension of $P = 8$ kN.

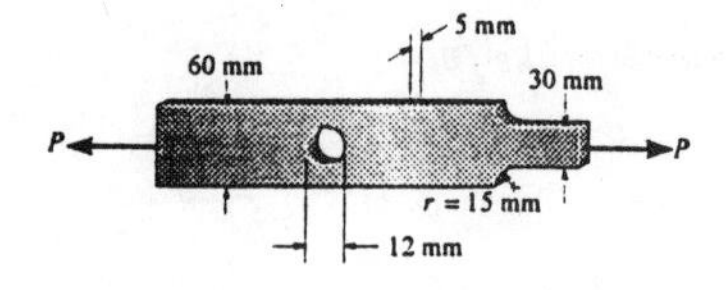

Maximum Normal Stress at fillet :

$$\frac{r}{h} = \frac{15}{30} = 0.5 \quad \text{and} \quad \frac{w}{h} = \frac{60}{30} = 2$$

From the text, $K = 1.4$

$$\sigma_{max} = K\sigma_{avg} = K\frac{P}{h\,t}$$

$$= 1.4\left[\frac{8(10^3)}{(0.03)(0.005)}\right] = 74.7 \text{ MPa}$$

Maximum Normal Stress at the hole :

$$\frac{r}{w} = \frac{6}{60} = 0.1$$

From the text, $K = 2.65$

$$\sigma_{max} = K\sigma_{avg} = K\frac{P}{(w - 2r)\,t}$$

$$= 2.65\left[\frac{8(10^3)}{(0.06 - 0.012)(0.005)}\right]$$

$= 88.3$ MPa ***(controls!)*** **Ans**

4–99. If the allowable normal stress for the bar is $\sigma_{allow} = 120$ MPa, determine the maximum axial force P that can be applied to the bar.

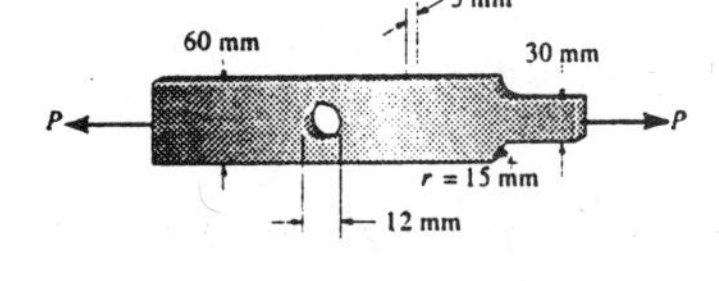

Assume failure occurs at fillet :

$$\frac{r}{h} = \frac{15}{30} = 0.5 \quad \text{and} \quad \frac{w}{h} = \frac{60}{30} = 2$$

From the text, $K = 1.4$

$$\sigma_{max} = \sigma_{allow} = K\sigma_{avg}$$

$$120(10^6) = 1.4\left[\frac{P}{0.03(0.005)}\right]$$

$P = 12.9$ kN

Assume failure occurs at hole :

$$\frac{r}{w} = \frac{6}{60} = 0.1$$

From the text, $K = 2.65$

$$\sigma_{max} = \sigma_{allow} = K\sigma_{avg}$$

$$120(10^6) = 2.65\left[\frac{P}{(0.06 - 0.012)(0.005)}\right]$$

$P = 10.9$ kN ***(controls!)*** **Ans**

***4-100.** The A-36 steel plate has a thickness of 12 mm. If there are shoulder fillets at B and C, and $\sigma_{allow} = 150$ MPa, determine the maximum axial load P that it can support. Compute its elongation neglecting the effect of the fillets.

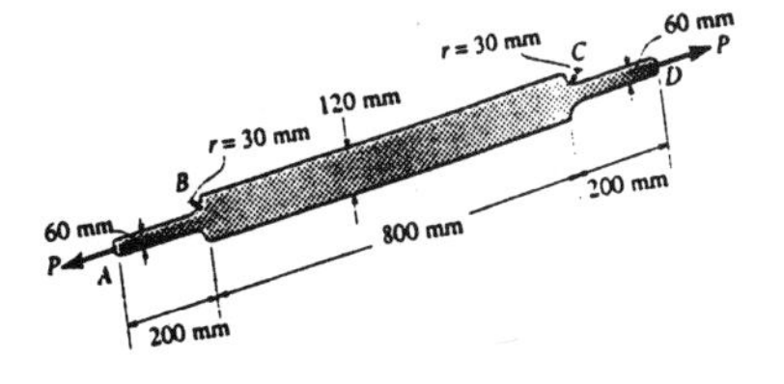

Maximum Normal Stress at fillet :

$$\frac{r}{h} = \frac{30}{60} = 0.5 \quad \text{and} \quad \frac{w}{h} = \frac{120}{60} = 2$$

From the text, $K = 1.4$

$$\sigma_{max} = \sigma_{allow} = K\sigma_{avg}$$

$$150(10^6) = 1.4\left[\frac{P}{0.06(0.012)}\right]$$

$$P = 77142.86 \text{ N} = 77.1 \text{ kN} \qquad \textbf{Ans}$$

Displacement :

$$\delta = \Sigma\frac{PL}{AE}$$

$$= \frac{77142.86(400)}{(0.06)(0.012)(200)(10^9)} + \frac{77142.86(800)}{(0.12)(0.012)(200)(10^9)}$$

$$= 0.429 \text{ mm} \qquad \textbf{Ans}$$

4-101. The resulting stress distribution along the section AB of the bar is shown in the figure. From this distribution, determine the approximate resultant axial force P applied to the bar. Also, what is the stress-concentration factor for this geometry?

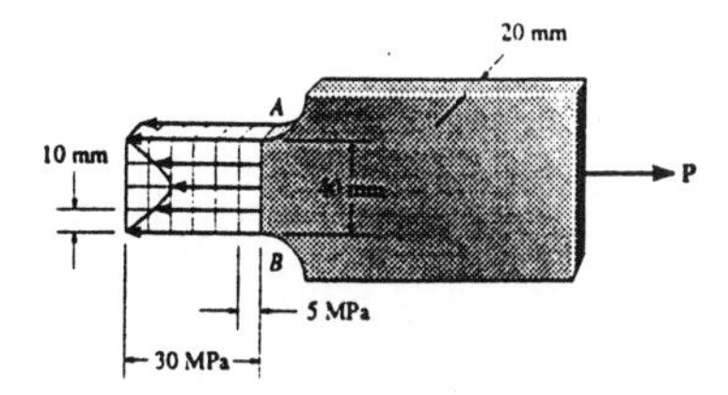

Resultant Force : Number of squares is approximately 20.

$$P = 20(5)(10^6)(0.01)(0.02) = 20.0 \text{ kN} \qquad \textbf{Ans}$$

Stress Concentration Factor :

$$\sigma_{avg} = \frac{P}{h\,t} = \frac{20.0(10^3)}{0.04(0.02)} = 25.0 \text{ MPa}$$

$$K = \frac{\sigma_{max}}{\sigma_{avg}} = \frac{30}{25.0} = 1.20 \qquad \textbf{Ans}$$

4-102. The resulting stress distribution along the section AB of the bar is shown in the figure. From this distribution, determine the approximate resultant axial force P applied to the bar. Also, what is the stress-concentration factor for this geometry?

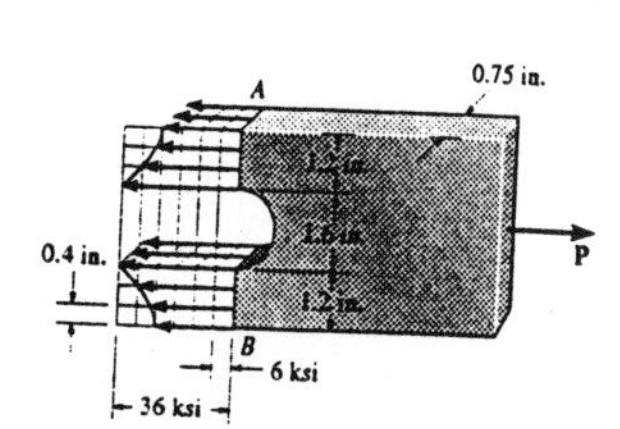

Resultant Force : Number of squares is approximately 28.

$$P = 28(6)(0.4)(0.75) = 50.4 \text{ kip} \qquad \textbf{Ans}$$

Stress Concentration Factor :

$$\sigma_{avg} = \frac{P}{(w - 2r)\,t} = \frac{50.4}{(4 - 1.6)(0.75)} = 28.0 \text{ ksi}$$

$$K = \frac{\sigma_{max}}{\sigma_{avg}} = \frac{36}{28.0} = 1.29 \qquad \textbf{Ans}$$

4–103. The 300-kip weight is slowly set on the top of a post made of 2014-T6 aluminum with an A-36 steel core. If both materials can be considered elastic perfectly plastic, determine the stress in each material.

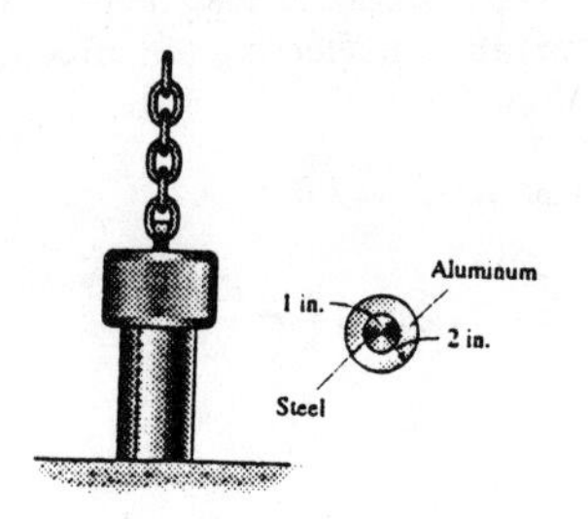

Equations of Equilibrium :

$$+\uparrow \Sigma F_y = 0; \qquad P_{st} + P_{al} - 300 = 0 \qquad [1]$$

Elastic Analysis : Assume both materials still behave elastically under the load.

$$\delta_{st} = \delta_{al}$$

$$\frac{P_{st} L}{\frac{\pi}{4}(2)^2(29)(10^3)} = \frac{P_{al} L}{\frac{\pi}{4}(4^2 - 2^2)(10.6)(10^3)}$$

$$P_{st} = 0.9119\, P_{al}$$

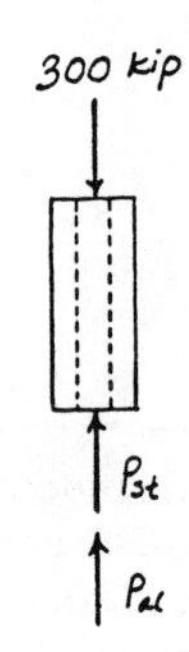

Solving Eqs. [1] and [2] yields :

$$P_{al} = 156.91 \text{ kip} \qquad P_{st} = 143.09 \text{ kip}$$

Average Normal Stress :

$$\sigma_{al} = \frac{P_{al}}{A_{al}} = \frac{156.91}{\frac{\pi}{4}(4^2 - 2^2)}$$
$$= 16.65 \text{ ksi} < (\sigma_Y)_{al} = 60.0 \text{ ksi} \quad \textit{(OK!)}$$

$$\sigma_{st} = \frac{P_{st}}{A_{st}} = \frac{143.09}{\frac{\pi}{4}(2^2)}$$
$$= 45.55 \text{ ksi} > (\sigma_Y)_{st} = 36.0 \text{ ksi}$$

Therefore, the steel core yields and so the elastic analysis is invalid. The stress in the steel is

$$\sigma_{st} = (\sigma_Y)_{st} = 36.0 \text{ ksi} \qquad \textbf{Ans}$$

$$P_{st} = (\sigma_Y)_{st} A_{st} = 36.0\left(\frac{\pi}{4}\right)\left(2^2\right) = 113.10 \text{ kip}$$

From Eq. [1] $\quad P_{al} = 186.90$ kip

$$\sigma_{al} = \frac{P_{al}}{A_{al}} = \frac{186.90}{\frac{\pi}{4}(4^2 - 2^2)} = 19.83 \text{ ksi} < (\sigma_Y)_{al} = 60.0 \text{ ksi}$$

Then $\qquad \sigma_{al} = 19.3$ ksi $\qquad$ **Ans**

***4–104.** The weight is suspended from steel and aluminum wires, each having the same initial length of 3 m and cross-sectional area of 4 mm^2. If the materials can be assumed to be elastic perfectly plastic, with $(\sigma_Y)_{st} = 120$ MPa and $(\sigma_Y)_{al} = 70$ MPa, determine the force in each wire if the weight is (*a*) 600 N and (*b*) 720 N. $E_{al} = 70$ GPa, $E_{st} = 200$ GPa.

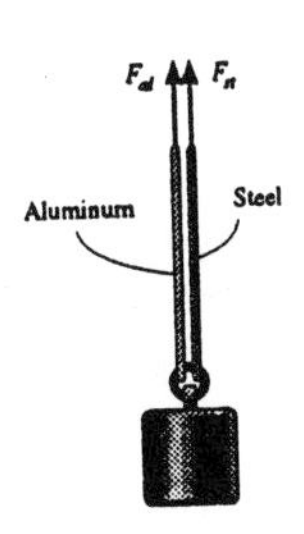

Equations of Equilibrium :

$$+\uparrow \Sigma F_y = 0; \quad F_{al} + F_{st} - w = 0 \qquad [1]$$

Elastic Analysis : Assume both wires behave elastically.

$$\delta_{al} = \delta_{st}$$

$$\frac{F_{al}L}{A(70)(10^9)} = \frac{F_{st}L}{A(200)(10^9)}$$

$$F_{al} = 0.350\, F_{st} \qquad [2]$$

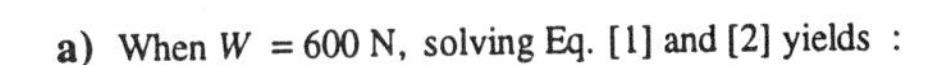

a) When $W = 600$ N, solving Eq. [1] and [2] yields :

$F_{st} = 444.44\text{ N} = 444\text{ N}$ **Ans**

$F_{al} = 155.55\text{ N} = 156\text{ N}$ **Ans**

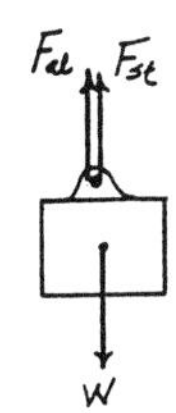

Average Normal Stress :

$$\sigma_{al} = \frac{F_{al}}{A_{al}} = \frac{155.55}{4.00(10^{-6})} = 38.88\text{ MPa} < (\sigma_Y)_{al} = 70.0\text{ MPa} \quad \textit{(OK!)}$$

$$\sigma_{st} = \frac{F_{st}}{A_{st}} = \frac{444.44}{4.00(10^{-6})} = 111.11\text{ MPa} < (\sigma_Y)_{st} = 120\text{ MPa} \quad \textit{(OK!)}$$

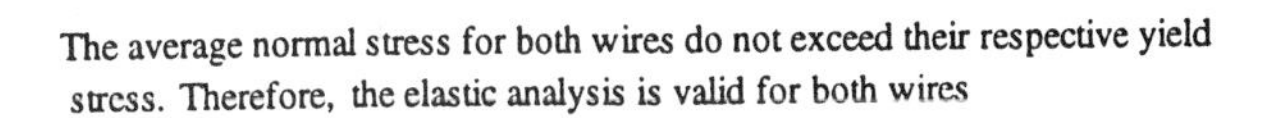

The average normal stress for both wires do not exceed their respective yield stress. Therefore, the elastic analysis is valid for both wires

b) When $W = 720$ N, solving Eq. [1] and [2] yields :

$F_{st} = 533.33$ N $\qquad F_{al} = 186.67$ N

Average Nomal Stress :

$$\sigma_{al} = \frac{F_{al}}{A_{al}} = \frac{186.67}{4.00(10^{-6})} = 46.67\text{ MPa} < (\sigma_Y)_{al} = 70.0\text{ MPa} \quad \textit{(OK!)}$$

$$\sigma_{st} = \frac{F_{st}}{A_{st}} = \frac{533.33}{4.00(10^{-6})} = 133.33\text{ MPa} > (\sigma_Y)_{st} = 120\text{ MPa}$$

Therefore, the steel wire yields. Hence,

$F_{st} = (\sigma_Y)_{st}A_{st} = 120(10^6)(4.00)(10^{-6}) = 480$ N **Ans**

From Eq. [1], $F_{al} = 240$ N **Ans**

$$\sigma_{al} = \frac{240}{4.00(10^{-6})} = 60.00\text{ MPa} < (\sigma_Y)_{al} \quad \textit{(OK!)}$$

4–105. The bar has a cross-sectional area of 0.5 in² and is made of a material that has a stress–strain diagram that can be approximated by the two line segments shown. Determine the elongation of the bar due to the applied loading.

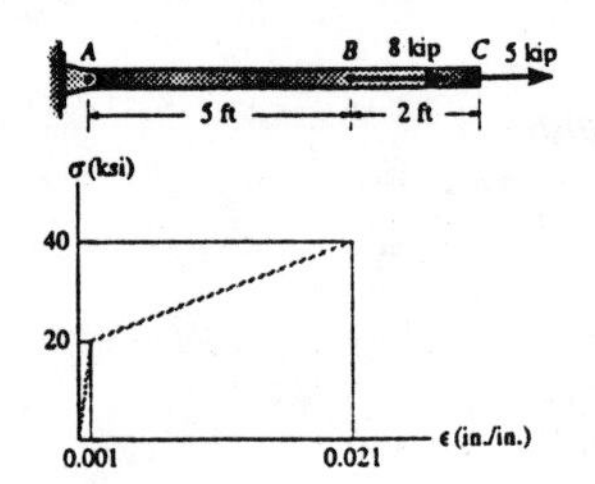

Average Normal Stress and Strain : For segment BC

$$\sigma_{BC} = \frac{P_{BC}}{A_{BC}} = \frac{5}{0.5} = 10.0 \text{ ksi}$$

$$\frac{10.0}{\varepsilon_{BC}} = \frac{20}{0.001}; \quad \varepsilon_{BC} = \frac{0.001}{20}(10.0) = 0.00050 \text{ in./in.}$$

Average Normal Stress and Strain : For segment AB

$$\sigma_{AB} = \frac{P_{AB}}{A_{AB}} = \frac{13}{0.5} = 26.0 \text{ ksi}$$

$$\frac{26.0 - 20}{\varepsilon_{AB} - 0.001} = \frac{40 - 20}{0.021 - 0.001}$$

$$\varepsilon_{AB} = 0.0070 \text{ in./in.}$$

Elongation :

$$\delta_{BC} = \varepsilon_{BC} L_{BC} = 0.00050(2)(12) = 0.0120 \text{ in.}$$
$$\delta_{AB} = \varepsilon_{AB} L_{AB} = 0.0070(5)(12) = 0.420 \text{ in.}$$

$$\delta_{Tot} = \delta_{BC} + \delta_{AB} = 0.0120 + 0.420 = 0.432 \text{ in.}$$ **Ans**

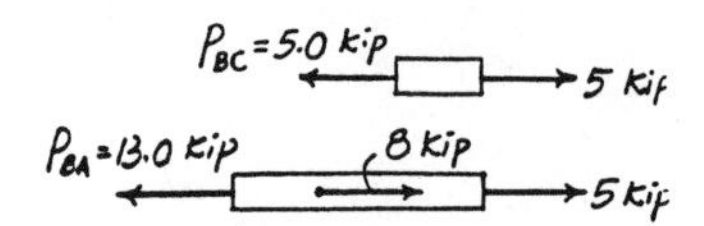

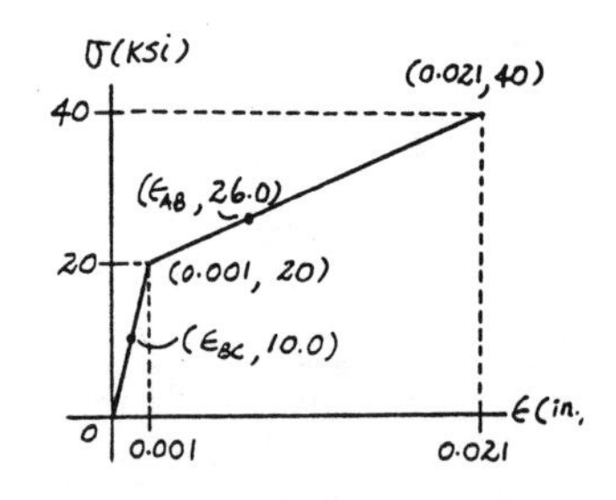

4–106. The wire BC has a diameter of 0.125 in. and the material has the stress–strain characteristics shown in the figure. Determine the vertical displacement of the handle at D if the pull at the grip is slowly increased and reaches a magnitude of (*a*) $P = 450$ lb, (*b*) $P = 600$ lb.

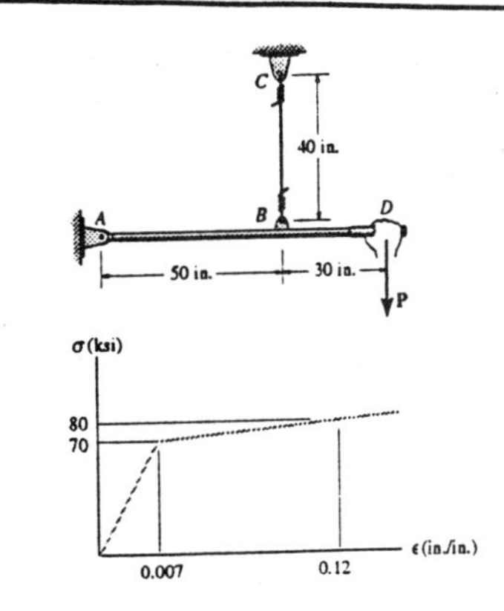

Equations of Equilibrium :

$$\curvearrowleft + \Sigma M_A = 0; \quad F_{BC}(50) - P(80) = 0 \qquad [1]$$

a) From Eq. [1] when $P = 450$ lb, $\quad F_{BC} = 720$ lb

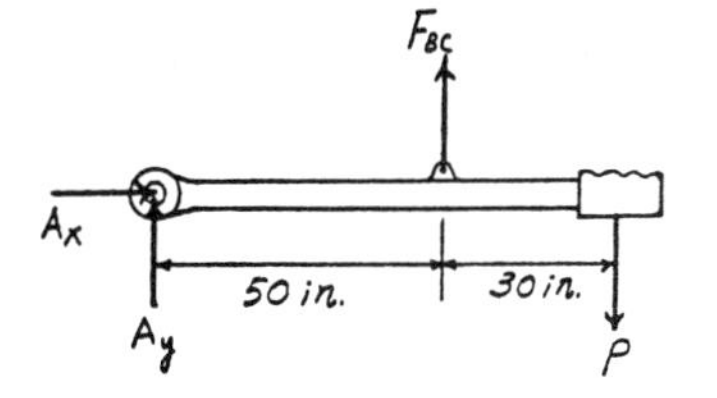

Average Normal Stress and Strain :

$$\sigma_{BC} = \frac{F_{BC}}{A_{BC}} = \frac{720}{\frac{\pi}{4}(0.125^2)} = 58.67 \text{ ksi}$$

From the Stress – Strain diagram

$$\frac{58.67}{\varepsilon_{BC}} = \frac{70}{0.007}; \quad \varepsilon_{BC} = 0.0055867 \text{ in./in.}$$

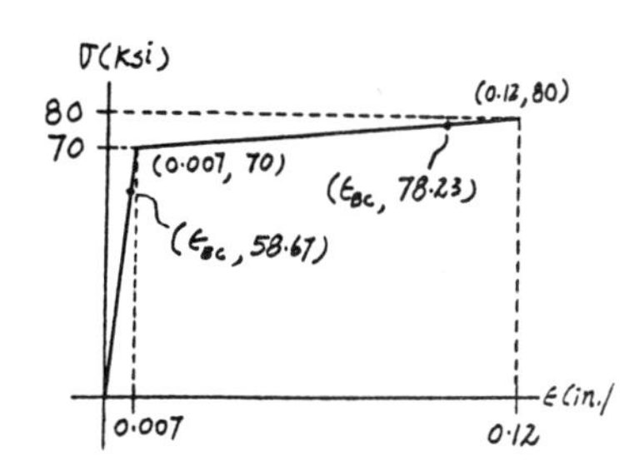

Displacement :

$$\delta_{BC} = \varepsilon_{BC} L_{BC} = 0.005867(40) = 0.2347 \text{ in.}$$

$$\frac{\delta_D}{80} = \frac{\delta_{BC}}{50}; \quad \delta_D = \frac{8}{5}(0.2347) = 0.375 \text{ in.} \qquad \textbf{Ans}$$

b) From Eq. [1] when $P = 600$ lb, $\quad F_{BC} = 960$ lb

Average Normal Stress and Strain :

$$\sigma_{BC} = \frac{F_{BC}}{A_{BC}} = \frac{960}{\frac{\pi}{4}(0.125)^2} = 78.23 \text{ ksi}$$

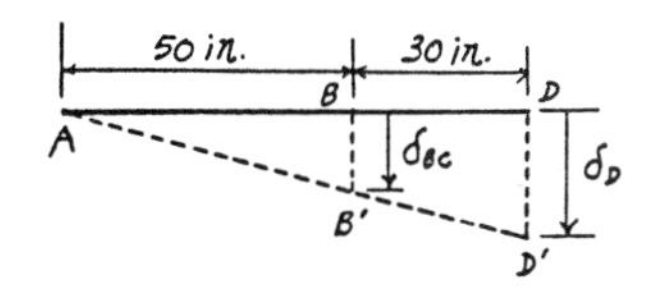

From Stess – Strain diagram

$$\frac{78.23 - 70}{\varepsilon_{BC} - 0.007} = \frac{80 - 70}{0.12 - 0.007} \qquad \varepsilon_{BC} = 0.09997 \text{ in./in.}$$

Displacement :

$$\delta_{BC} = \varepsilon_{BC} L_{BC} = 0.09997(40) = 3.9990 \text{ in.}$$

$$\frac{\delta_D}{80} = \frac{\delta_{BC}}{50}; \quad \delta_D = \frac{8}{5}(3.9990) = 6.40 \text{ in.} \qquad \textbf{Ans}$$

4–107. The rigid beam is supported by the three rods A, B, and C of equal length. Rods A and C have a diameter of 25 mm and are made of aluminum, for which $E_{al} = 70$ GPa and $(\sigma_Y)_{al} = 20$ MPa. Rod B has a diameter of 10 mm and is made of brass, for which $E_{br} = 100$ GPa and $(\sigma_Y)_{br} = 590$ MPa. Determine the smallest magnitude of **P** so that (*a*) only rods A and C yield and (*b*) all the rods yield.

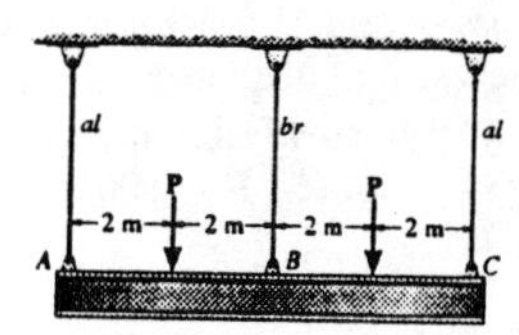

Equations of Equilibrium :

$\curvearrowleft + \Sigma M_B = 0; \quad F_A = F_C = F_{al}$

$+\uparrow \Sigma F_y = 0: \quad F_{br} + 2F_{al} - 2P = 0 \qquad [1]$

a) ***Plastic analysis*** : Rod A and C are required to yield

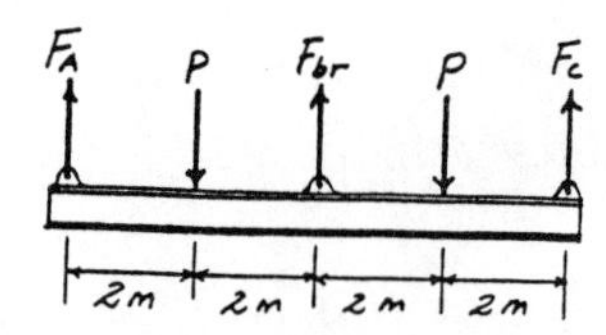

$$F_{al} = (\sigma_Y)_{al} A$$
$$= 20(10^6)\left(\frac{\pi}{4}\right)(0.025^2) = 9.817 \text{ kN}$$

$$(\varepsilon_{al})_Y = \frac{(\sigma_{al})_Y}{E_{al}} = \frac{20(10^6)}{70(10^9)} = 0.0002857 \text{ mm/mm}$$

Compatibility : The beam will remain horizontal after the displacement since the loading and the system are symmetrical.

$$\delta_{br} = \delta_{al}$$

$$\frac{F_{br}(L)}{\frac{\pi}{4}(0.01^2)(100)(10^9)} = 0.0002857\,L$$

$$F_{br} = 2.244 \text{ kN}$$

$$\sigma_{br} = \frac{2.244(10^3)}{\frac{\pi}{4}(0.01^2)}$$
$$= 28.57 \text{ MPa} < (\sigma_Y)_{br} = 590 \text{ MPa} \quad (OK!)$$

From Eq. [1] $\qquad P = 10.9$ kN $\qquad$ **Ans**

b) ***Plastic Analysis*** : All the rods are required to yield

$$F_{br} = (\sigma_Y)_{br} A$$
$$= (590)(10^6)\left(\frac{\pi}{4}\right)(0.01^2) = 46.338 \text{ kN}$$

$$F_{al} = (\sigma_Y)_{al} A$$
$$= 20(10^6)\left(\frac{\pi}{4}\right)(0.025^2) = 9.817 \text{ kN}$$

From Eq. [1] $\qquad P = 33.0$ kN $\qquad$ **Ans**

***4–108.** The rigid beam is supported by the three rods A, B, and C of equal length. Rods A and C have a diameter of 15 mm and are made of aluminum, for which $E_{al} = 70$ GPa and $(\sigma_y)_{al} = 20$ MPa. Rod B is made of brass, for which $E_{br} = 100$ GPa and $(\sigma_Y)_{br} = 590$ MPa. If $P = 130$ kN, determine the largest diameter of rod B so that all the rods yield at the same time.

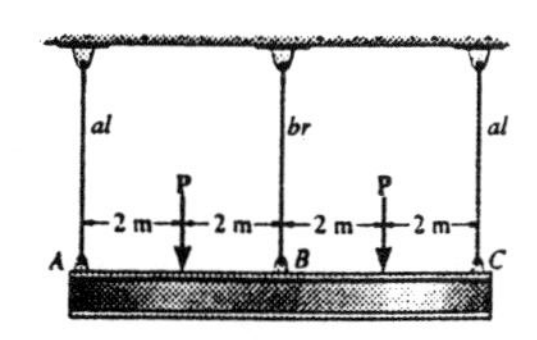

Equations of Equilibrium :

$$\curvearrowleft + \Sigma M_B = 0; \quad F_A = F_C = F_{al}$$

$$+\uparrow \Sigma F_y = 0; \quad F_{br} + 2F_{al} - 260 = 0 \qquad [1]$$

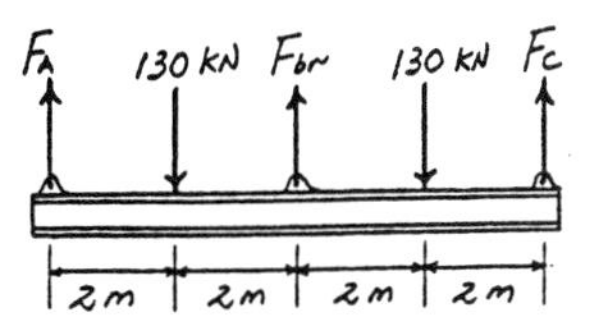

Plastic Analysis : All the rods are required to yield

$$F_{al} = (\sigma_Y)_{al}A$$
$$= 20(10^6)\left(\frac{\pi}{4}\right)(0.015^2) = 3.534 \text{ kN}$$

$$F_{br} = (\sigma_Y)_{br}A$$
$$= (590)(10^6)\left(\frac{\pi}{4}\right)(d_{br}^2) = (463384.91 d_{br}^2) \text{ kN}$$

Substitute the results into Eq. [1]

$$463384.91 d_{br}^2 + 2(3.534) - 260 = 0$$
$$d_{br} = 0.02336 \text{ m} = 23.4 \text{ mm} \qquad \textbf{Ans}$$

4–109. The rigid beam is supported by three A-36 steel wires, each having a length of 4 ft. The cross-sectional area of AB and EF is 0.015 in^2, and the cross-sectional area of CD is 0.006 in^2. Determine the largest distributed load w that can be supported by the beam before any of the wires begin to yield. If the steel is assumed to be elastic perfectly plastic, determine how far the beam is displaced downward just before all the wires begin to yield.

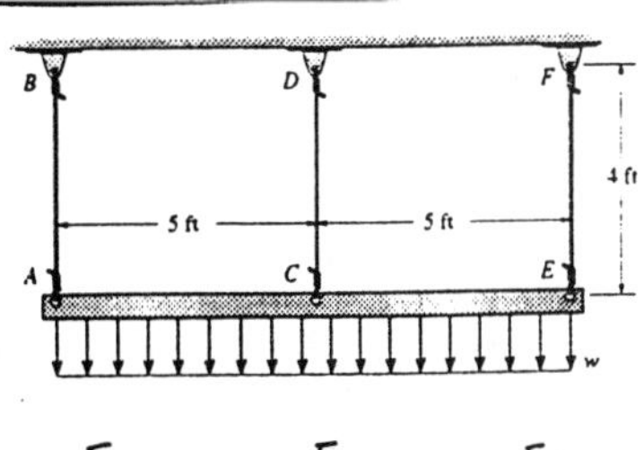

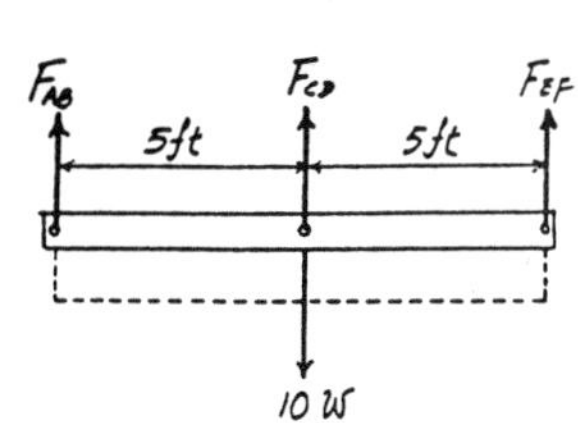

Equations of Equilibrium :

$$\curvearrowleft + \Sigma M_C = 0; \quad F_{EF}(5) - F_{AB}(5) = 0 \quad F_{EF} = F_{AB} = F$$

$$+\uparrow \Sigma F_y = 0; \quad 2F + F_{CD} - 10w = 0 \qquad [1]$$

Compatibility : The beam will remain horizontal after the displacement since the loading and the system are symmetrical.

$$\delta_{AB} = \delta_{CD}$$
$$\frac{F(L)}{0.015\,E} = \frac{F_{CD}(L)}{0.006\,E}$$
$$F = 2.50\,F_{CD} \qquad [2]$$

Plastic Analysis : Assume wire AB yields

$$F = (\sigma_Y)_{st}A = 36.0(0.015) = 0.540 \text{ kip}$$

Using $F = 0.540$ kip and solving Eqs. [1] and [2] yields :

$$F_{CD} = 0.216 \text{ kip} \qquad w = 0.1296 \text{ kip/ft}$$

Plastic Analysis : Assume wire CD yields

$$F_{CD} = (\sigma_Y)_{st}A = 36.0(0.006) = 0.216 \text{ kip}$$

Using $F_{CD} = 0.216$ kip and solving Eqs. [1] and [2] yields :

$$F = 0.540 \text{ kip} \qquad w = 0.1296 \text{ kip/ft}$$

The three wires AB, CD and EF yield simultaneously. Hence,

$$w = 0.130 \text{ kip/ft} \qquad \textbf{Ans}$$

Displacement :

$$\delta = \frac{F_{CD}L}{A_{CD}E} = \frac{0.216(4)(12)}{0.006(29)(10^3)} = 0.0596 \text{ in.} \qquad \textbf{Ans}$$

4–110. The rigid bar is supported by a pin at A and two steel wires, each having a diameter of 4 mm. If the yield stress for the wires is $\sigma_Y = 530$ MPa, and $E_{st} = 200$ GPa, determine the intensity of the distributed load w that can be placed on the beam and will just cause wire EB to yield. What is the displacement of point G for this case? For the calculation, assume that the steel is elastic perfectly plastic.

Equations of Equilibrium :

$$\circlearrowleft + \Sigma M_A = 0; \quad F_{BE}(0.4) + F_{CD}(0.65) - 0.8w\,(0.4) = 0$$

$$0.4\,F_{BE} + 0.65 F_{CD} = 0.32w \qquad [1]$$

Plastic Analysis : Wire CD will yield first followed by wire BE. When both wires yield

$$F_{BE} = F_{CD} = (\sigma_Y)A$$
$$= 530(10^6)\left(\frac{\pi}{4}\right)(0.004^2) = 6.660 \text{ kN}$$

Substituting the results into Eq.[1] yields :

$$w = 21.9 \text{ kN/m} \qquad \textbf{Ans}$$

Displacement : When wire BE achieves yield stress, the corresponding yield strain is

$$\varepsilon_Y = \frac{\sigma_Y}{E} = \frac{530(10^6)}{200(10^9)} = 0.002650 \text{ mm/mm}$$
$$\delta_{BE} = \varepsilon_Y L_{BE} = 0.002650(800) = 2.120 \text{ mm}$$

From the geometry

$$\frac{\delta_G}{0.8} = \frac{\delta_{BE}}{0.4}$$
$$\delta_G = 2\delta_{BE} = 2(2.120) = 4.24 \text{ mm} \qquad \textbf{Ans}$$

4–111. The rigid bar is supported by a pin at A and two steel wires, each having a diameter of 4 mm. If the yield stress for the wires is $\sigma_Y = 530$ MPa, and $E_{st} = 200$ GPa, determine a) the intensity of the distributed load w that can be placed on the beam that will cause only one of the wires to start to yield and b) the smallest intensity of the distributed load that will cause both wires to yield. For the calculation, assume that the steel is elastic perfectly plastic.

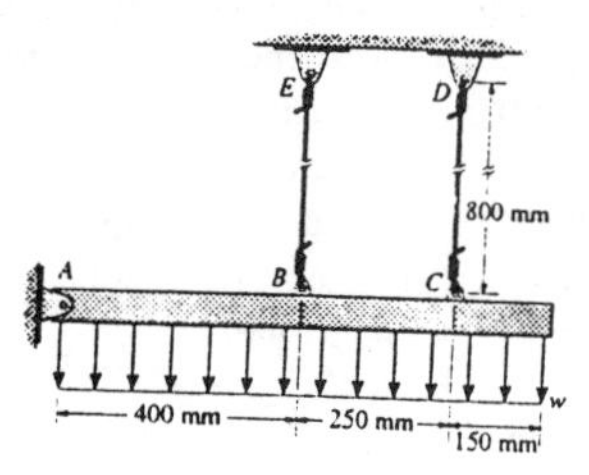

Equations of Equilibrium :

$$\circlearrowleft + \Sigma M_A = 0; \quad F_{BE}(0.4) + F_{CD}(0.65) - 0.8w\,(0.4) = 0$$

$$0.4\,F_{BE} + 0.65\,F_{CD} = 0.32w \qquad [1]$$

a) By observation, wire CD will yield first.

Then $F_{CD} = \sigma_Y A = 530(10^6)\left(\frac{\pi}{4}\right)(0.004^2) = 6.660$ kN.

From the geometry

$$\frac{\delta_{BE}}{0.4} = \frac{\delta_{CD}}{0.65}; \quad \delta_{CD} = 1.625\delta_{BE}$$
$$\frac{F_{CD}L}{AE} = 1.625\frac{F_{BE}L}{AE}$$
$$F_{CD} = 1.625\,F_{BE} \qquad [2]$$

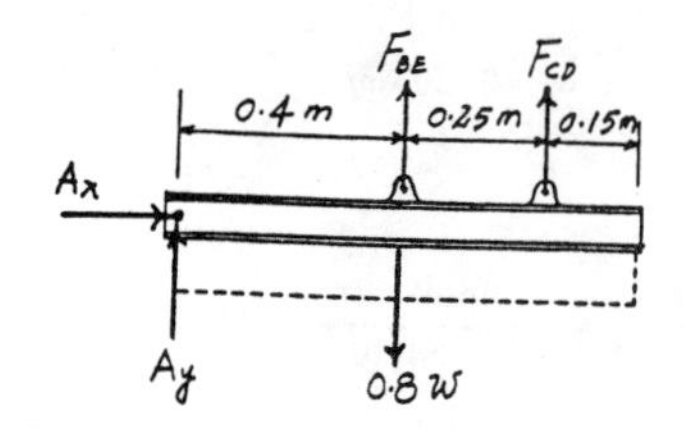

Using $F_{CD} = 6.660$ kN and solving Eqs. [1] and [2] yields :

$$F_{BE} = 4.099 \text{ kN}$$
$$w = 18.7 \text{ kN/m} \qquad \textbf{Ans}$$

b) When both wires yield

$$F_{BE} = F_{CD} = (\sigma_Y)A$$
$$= 530(10^6)\left(\frac{\pi}{4}\right)(0.004^2) = 6.660 \text{ kN}$$

Substituting the results into Eq.[1] yields :

$$w = 21.9 \text{ kN/m} \qquad \textbf{Ans}$$

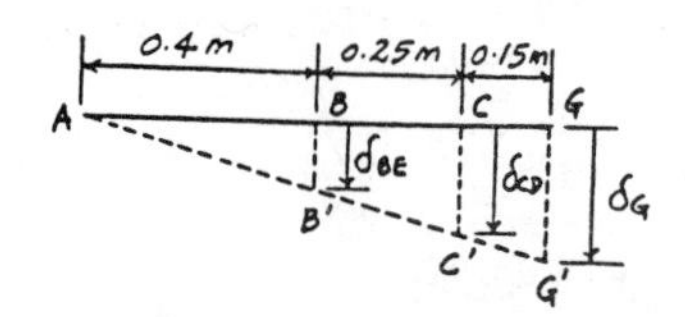

***4–112.** The rigid link is supported by a pin at A and two A - 36 steel wires, each having an unstretched length of 12 in. and cross-sectional area of 0.0125 in^2. Determine the force developed in the wires when the link supports the vertical load of 350 lb.

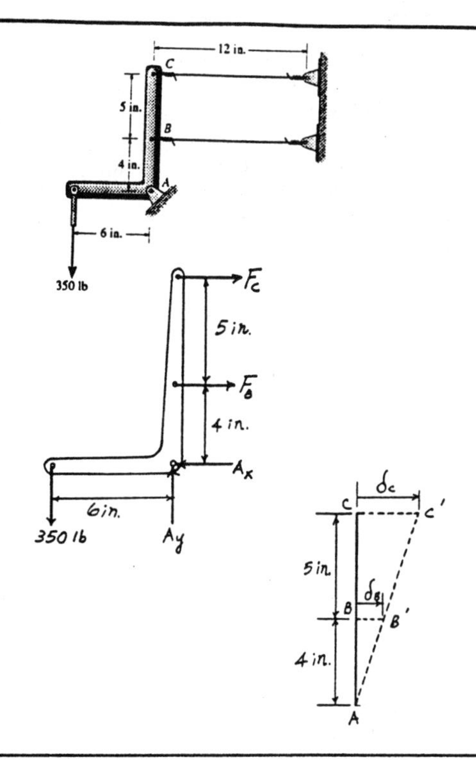

Equations of Equilibrium :

$$\curvearrowleft +\Sigma M_A = 0; \qquad -F_C(9) - F_B(4) + 350(6) = 0 \qquad [1]$$

Compatibility :

$$\frac{\delta_B}{4} = \frac{\delta_C}{9}$$

$$\frac{F_B(L)}{4AE} = \frac{F_C(L)}{9AE}$$

$$9F_B - 4F_C = 0 \qquad [2]$$

Solving Eqs. [1] and [2] yields :

$$F_B = 86.6 \text{ lb} \qquad \textbf{Ans}$$
$$F_C = 195 \text{ lb} \qquad \textbf{Ans}$$

4–113. The 50-mm-diameter cylinder is made of Am 1004-T61 magnesium and is placed in the clamp when the temperature is $T_1 = 15°C$. The two 304-stainless-steel carriage bolts of the clamp each have a diameter of 10 mm, and they hold the cylinder snug with negligible force against the rigid jaws. Determine the temperature at which the average normal stress in either the magnesium or steel does not exceed $\tau_{allow} = 2$ MPa.

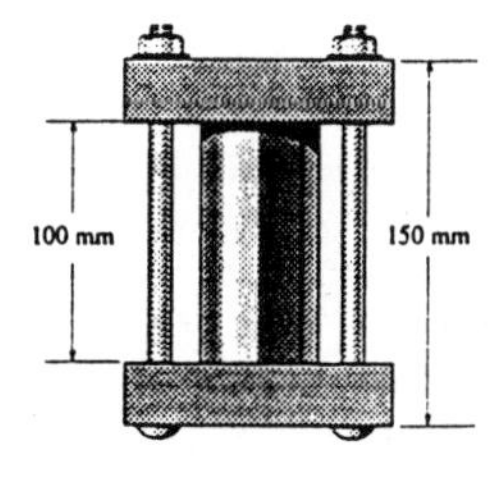

Equations of Equilibrium :

$$+\uparrow \Sigma F_y = 0; \qquad -2F_{st} + F_{mg} = 0 \qquad [1]$$

Compatibility :

$$(\delta_{mg})_T - (\delta_{mg})_F = (\delta_{st})_T + (\delta_{st})_F$$

$$26(10^{-6})(0.1)(\Delta T) - \frac{F_{mg}(0.1)}{\frac{\pi}{4}(0.05^2)\,44.7(10^9)}$$

$$= 17(10^{-6})(0.150)(\Delta T) + \frac{F_{st}(0.150)}{\frac{\pi}{4}(0.01^2)\,193(10^9)}$$

$$0.009896F_{st} + 0.001139F_{mg} = 0.05\Delta T \qquad [2]$$

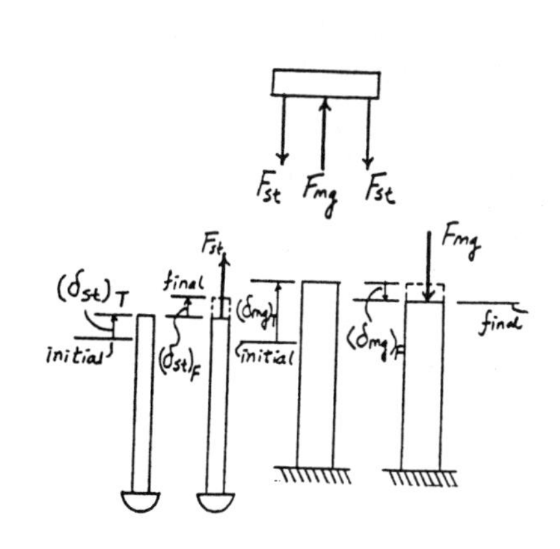

Allowable normal stress : Since the steel has a smaller cross – sectional area, it will achieve the allowable normal stress first.

$$F_{st} = \sigma_{\text{allow}}A = 12(10^6)\left(\frac{\pi}{4}\right)\left(0.01^2\right) = 942.48 \text{ N}$$

Using this result $F_{st} = 942.48$ N and solving Eq.[1] and [2] yields :

$$\Delta T = 229.47 \text{ °C} \qquad F_{mg} = 1884.96 \text{ N}$$

Thus,

$$T_2 = 229.47 \text{ °C} + 15 \text{ °C} = 244 \text{ °C} \qquad \textbf{Ans}$$

4–114. The assembly consists of a 2014-T6-aluminum cylinder having an outer diameter of 200 mm and inner diameter of 150 mm, together with a concentric solid inner cylinder of Am 1004-T61 magnesium, having a diameter of 125 mm. If the clamping force in the bolts AB and CD is 4 kN when the temperature is $T_1 = 16°C$, determine the clamping force in the bolts when the temperature becomes $T_2 = 48°C$. Assume the bolts and the restraining bars are rigid

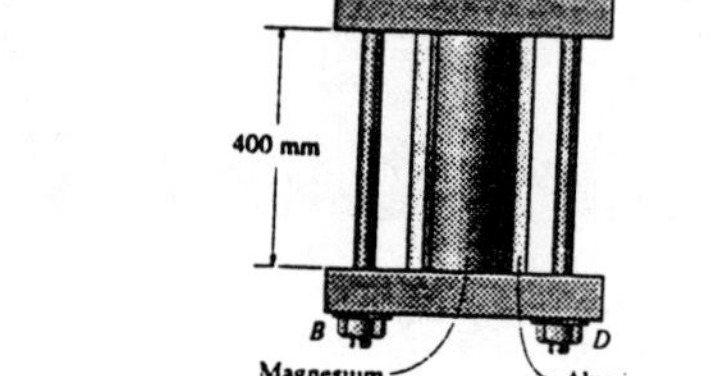

Equations of Equilibrium :

$$+\uparrow \Sigma F_y = 0; \quad 2F - \left(F_{al} + F_{mg}\right) = 0 \qquad [1]$$

Compatibility : For aluminum cylinder

$$\delta_F - \delta_T = 0$$

$$\frac{F_{al}(0.4)}{\frac{\pi}{4}(0.2^2 - 0.15^2)\,73.1(10^9)} - 23.0(10^{-6})(48 - 16)(0.4) = 0$$

$$F_{al} = 739.47 \text{ kN}$$

For magnesium core

$$\delta_F - \delta_T = 0$$

$$\frac{F_{mg}(0.4)}{\frac{\pi}{4}(0.125^2)(44.7)(10^9)} - 26.0\left(10^{-6}\right)(48 - 16)(0.4) = 0$$

$$F_{mg} = 456.39 \text{ kN}$$

Substituting the results into Eq.[1] yields :

$$F = 597.93 \text{ kN}$$

Thus, the final clamping force is

$$F_1 = F_0 + F = 4 + 597.93 = 602 \text{ kN} \qquad \textbf{Ans}$$

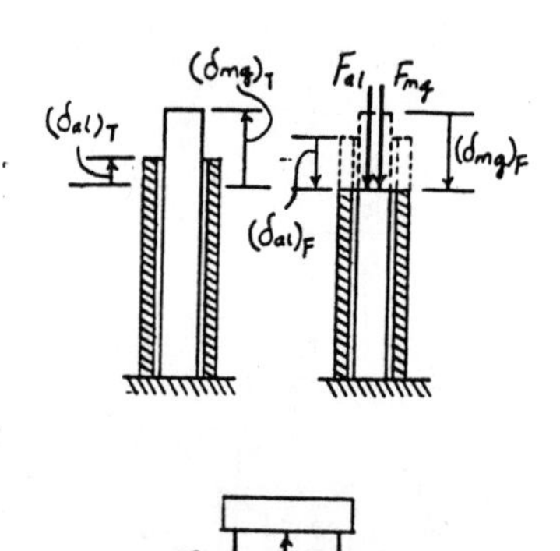

4–115. The assembly consists of two bars AB and CD of the same material having a modulus of elasticity E_1 and coefficient of thermal expansion α_1, and a bar EF having a modulus of elasticity E_2 and coefficient of thermal expansion α_2. All the bars have the same length L and cross-sectional area A. If the rigid beam is originally horizontal at temperature T_1, determine the angle it makes with the horizontal when the temperature is increased to T_2.

Equations of Equilibrium :

$$\curvearrowleft + \Sigma M_C = 0; \qquad F_{AB} = F_{EF} = F$$

$$+\uparrow \Sigma F_y = 0; \qquad F_{CD} - 2F = 0 \qquad [1]$$

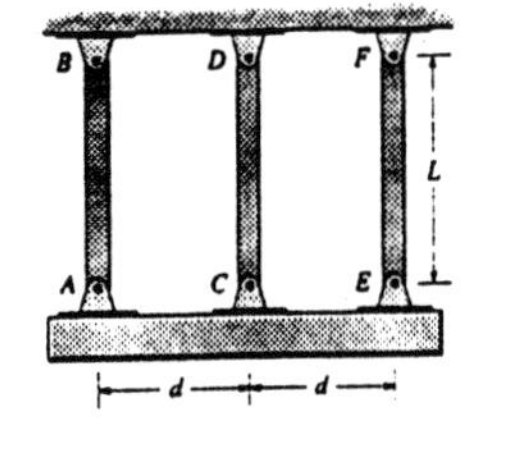

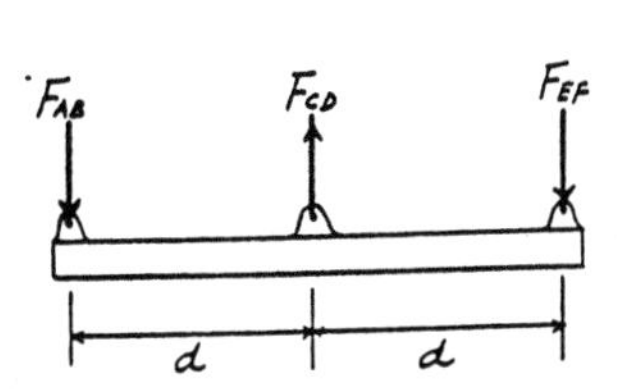

Compatibility :

$$\delta_{AB} = (\delta_{AB})_T - (\delta_{AB})_F \qquad \delta_{CD} = (\delta_{CD})_T + (\delta_{CD})_F$$

$$\delta_{EF} = (\delta_{EF})_T - (\delta_{EF})_F$$

From the geometry

$$\frac{\delta_{CD} - \delta_{AB}}{d} = \frac{\delta_{EF} - \delta_{AB}}{2d}$$

$$2\delta_{CD} = \delta_{EF} + \delta_{AB}$$

$$2\left[(\delta_{CD})_T + (\delta_{CD})_F\right] = (\delta_{EF})_T - (\delta_{EF})_F + (\delta_{AB})_T - (\delta_{AB})_F$$

$$2\left[\alpha_1 (T_2 - T_1) L + \frac{F_{CD}(L)}{AE_1}\right] = \alpha_2 (T_2 - T_1) L - \frac{F(L)}{AE_2} + \alpha_1 (T_2 - T_1) L - \frac{F(L)}{AE_1} \qquad [2]$$

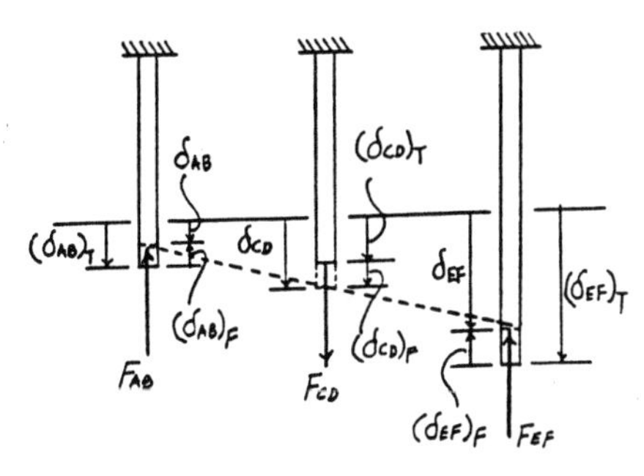

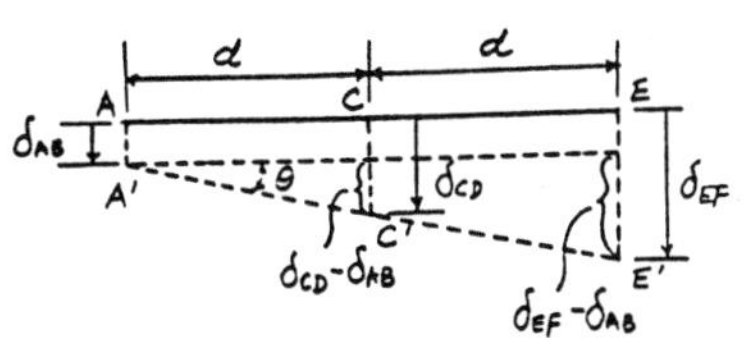

Substitute Eq. [1] into [2].

$$2\alpha_1 (T_2 - T_1) L + \frac{4FL}{AE_1} = \alpha_2 (T_2 - T_1) L - \frac{FL}{AE_2} + \alpha_1 (T_2 - T_1) L - \frac{FL}{AE_1}$$

$$\frac{5F}{AE_1} + \frac{F}{AE_2} = \alpha_2 (T_2 - T_1) - \alpha_1 (T_2 - T_1)$$

$$F\left(\frac{5E_2 + E_1}{AE_1 E_2}\right) = (T_2 - T_1)(\alpha_2 - \alpha_1)\ ; \qquad F = \frac{AE_1 E_2 (T_2 - T_1)(\alpha_2 - \alpha_1)}{5E_2 + E_1}$$

$$(\delta_{EF})_T = \alpha_2 (T_2 - T_1) L$$

$$(\delta_{EF})_F = \frac{AE_1 E_2 (T_2 - T_1)(\alpha_2 - \alpha_1)(L)}{AE_2 (5E_2 + E_1)} = \frac{E_1 (T_2 - T_1)(\alpha_2 - \alpha_1)(L)}{5E_2 + E_1}$$

$$\delta_{EF} = (\delta_{EF})_T - (\delta_{EF})_F = \frac{\alpha_2 L (T_2 - T_1)(5E_2 + E_1) - E_1 L (T_2 - T_1)(\alpha_2 - \alpha_1)}{5E_2 + E_1}$$

$$(\delta_{AB})_T = \alpha_1 (T_2 - T_1) L$$

$$(\delta_{AB})_F = \frac{AE_1 E_2 (T_2 - T_1)(\alpha_2 - \alpha_1)(L)}{AE_1 (5E_2 + E_1)} = \frac{E_2 (T_2 - T_1)(\alpha_2 - \alpha_1)(L)}{5E_2 + E_1}$$

$$\delta_{AB} = (\delta_{AB})_T - (\delta_{AB})_F = \frac{\alpha_1 L (5E_2 + E_1)(T_2 - T_1) - E_2 L (T_2 - T_1)(\alpha_2 - \alpha_1)}{5E_2 + E_1}$$

$$\delta_{EF} - \delta_{AB} = \frac{L(T_2 - T_1)}{5E_2 + E_1}[\alpha_2 (5E_2 + E_1) - E_1 (\alpha_2 - \alpha_1) - \alpha_1 (5E_2 + E_1) + E_2 (\alpha_2 - \alpha_1)]$$

$$= \frac{L(T_2 - T_1)}{5E_2 + E_1}[(5E_2 + E_1)(\alpha_2 - \alpha_1) + (\alpha_2 - \alpha_1)(E_2 - E_1)]$$

$$= \frac{L(T_2 - T_1)(\alpha_2 - \alpha_1)}{5E_2 + E_1}(5E_2 + E_1 + E_2 - E_1)$$

$$= \frac{L(T_2 - T_1)(\alpha_2 - \alpha_1)(6E_2)}{5E_2 + E_1}$$

$$\theta = \frac{\delta_{EF} - \delta_{AB}}{2d} = \frac{3E_2 L (T_2 - T_1)(\alpha_2 - \alpha_1)}{d(5E_2 + E_1)} \qquad \textbf{Ans}$$

***4–116.** The composite shaft, consisting of aluminum, copper, and steel sections, is subjected to the loading shown. Determine the displacement of end A with respect to end D and the average normal stress in each section. The cross-sectional area and modulus of elasticity for each section are shown. Neglect the size of the collars at B and C.

Internal Forces : As shown on FBD.

Average Normal Stress :

$$\sigma_{AB} = \frac{P_{AB}}{A_{AB}} = \frac{2.00}{0.09} = 22.2 \text{ ksi (T)} \qquad \textbf{Ans}$$

$$\sigma_{BC} = \frac{P_{BC}}{A_{BC}} = \frac{-5.00}{0.12} = -41.67 \text{ ksi} = 41.7 \text{ ksi (C)} \qquad \textbf{Ans}$$

$$\sigma_{CD} = \frac{P_{CD}}{A_{CD}} = \frac{-1.50}{0.06} = -25.00 \text{ ksi} = 25.0 \text{ ksi (C)} \qquad \textbf{Ans}$$

Displacement :

$$\delta_{A/D} = \sum \frac{PL}{AE} = \frac{2.00(18)}{(0.09)(10)(10^3)} + \frac{(-5.00)(12)}{(0.12)(18)(10^3)} + \frac{(-1.50)(16)}{(0.06)(29)(10^3)}$$

$$= -0.00157 \text{ in.} \qquad \textbf{Ans}$$

Negative sign indicates that end A moves towards end D.

4–117. Determine the displacement of B with respect to C of the composite shaft in Prob. 4–116.

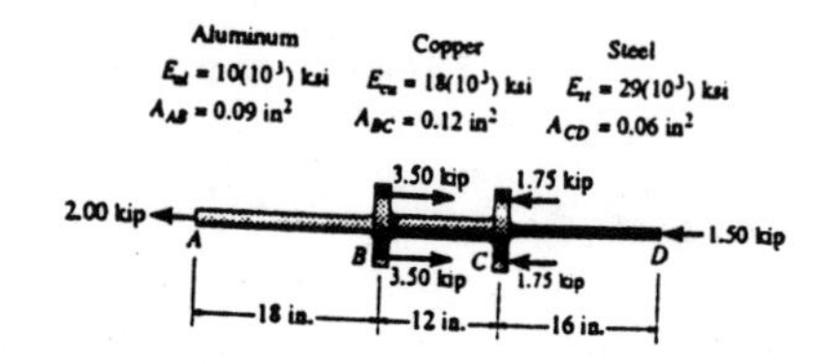

Internal Force : As shown on FBD.

Displacement :

$$\delta_{B/C} = \frac{P_{BC}L_{BC}}{A_{BC}E_{BC}}$$

$$= \frac{(-5.00)(12)}{(0.12)(18)(10^3)} = -0.0278 \text{ in.} \qquad \textbf{Ans}$$

Negative sign indicates that end B moves towards end C.

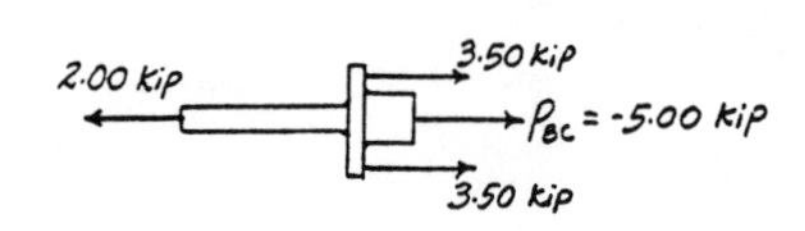

4–118. The A-36 steel column, having a cross-sectional area of 18 in^2, is encased in high-strength concrete as shown. If an axial force of 60 kip is applied to the column, determine the average compressive stress in the concrete and in the steel. How far does the column shorten? It has an original length of 8 ft.

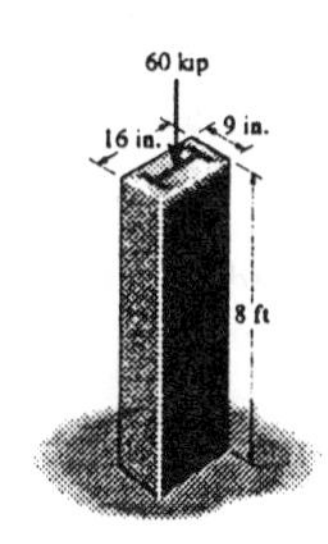

Equations of Equilibrium :

$$+\uparrow \Sigma F_y = 0; \quad P_{st} + P_{con} - 60 = 0 \qquad [1]$$

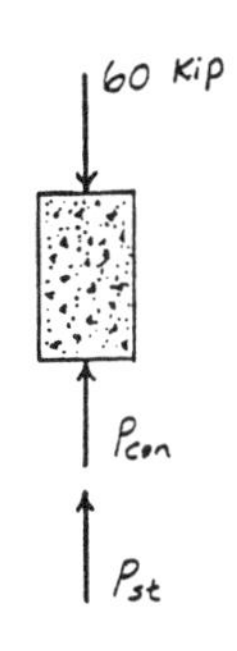

Compatibility :

$$\delta_{st} = \delta_{con}$$

$$\frac{P_{st}(8)(12)}{18(29.0)(10^3)} = \frac{P_{con}(8)(12)}{[(9)(16) - 18](4.20)(10^3)}$$

$$P_{st} = 0.98639\, P_{con} \qquad [2]$$

Solving Eqs. [1] and [2] yields :

$$P_{st} = 29.795 \text{ kip} \qquad P_{con} = 30.205 \text{ kip}$$

Average Normal Stress :

$$\sigma_{st} = \frac{P_{st}}{A_{st}} = \frac{29.795}{18} = 1.66 \text{ ksi} \qquad \textbf{Ans}$$

$$\sigma_{con} = \frac{P_{con}}{A_{con}} = \frac{30.205}{[9(16) - 18]} = 0.240 \text{ ksi} \qquad \textbf{Ans}$$

Displacement : Either the concrete or steel can be used for the calculation.

$$\delta = \frac{P_{st}L}{A_{st}E} = \frac{29.795(8)(12)}{18(29.0)(10^3)} = 0.00548 \text{ in.} \qquad \textbf{Ans}$$

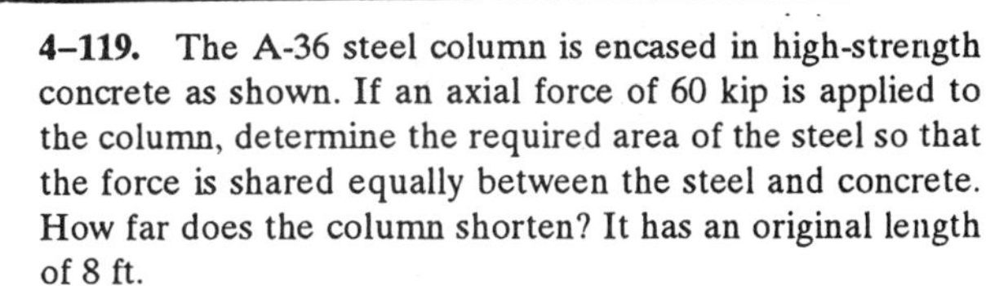

4–119. The A-36 steel column is encased in high-strength concrete as shown. If an axial force of 60 kip is applied to the column, determine the required area of the steel so that the force is shared equally between the steel and concrete. How far does the column shorten? It has an original length of 8 ft.

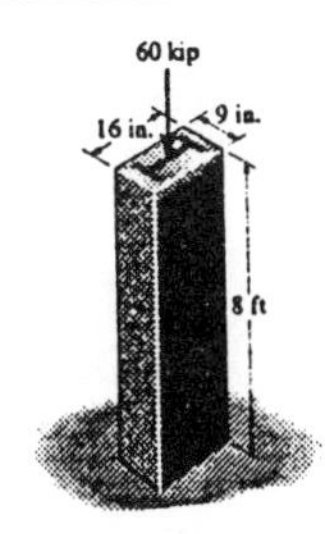

Equilibrium : The force of 60 kip is shared equally by the concrete and steel. Hence

$$P_{st} = P_{con} = P = 30.0 \text{ kip}$$

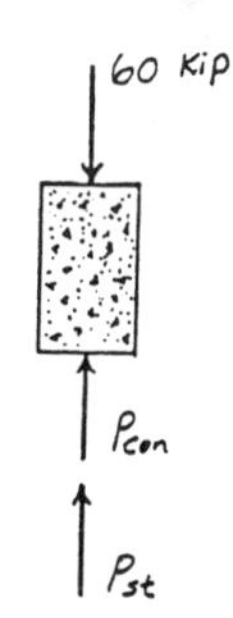

Compatibility :

$$\delta_{con} = \delta_{st}$$

$$\frac{PL}{A_{con}E_{con}} = \frac{PL}{A_{st}E_{st}}$$

$$A_{st} = \frac{A_{con}E_{con}}{E_{st}} = \frac{[9(16) - A_{st}]\, 4.20(10^3)}{29.0(10^3)}$$

$$A_{st} = 18.22 \text{ in}^2 = 18.2 \text{ in}^2 \qquad \textbf{Ans}$$

Displacement : Either the concrete or steel can be used for the calculation.

$$\delta = \frac{P_{st}L}{A_{st}E_{st}} = \frac{30.0(8)(12)}{18.22(29.0)(10^3)} = 0.00545 \text{ in.} \qquad \textbf{Ans}$$

*4–120. The brass plug is force-fitted into the rigid casting. The uniform normal bearing pressure on the plug is estimated to be 15 MPa. If the coefficient of static friction between the plug and casting is $\mu_s = 0.3$, determine the axial force P needed to pull the plug out. Also, calculate the displacement of end B relative to end A just before the plug starts to slip out. $E_{br} = 98$ GPa.

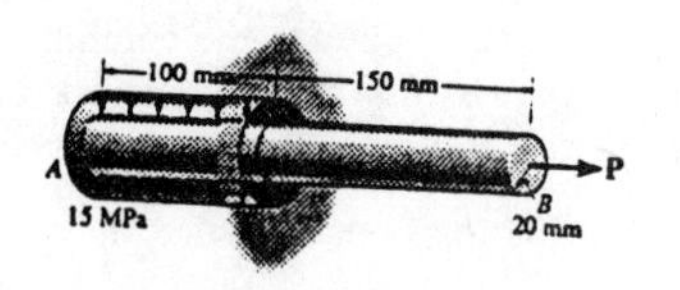

Equations of Equilibrium :

$$\xrightarrow{+} \Sigma F_x = 0; \quad P - 4.50(10^6)(2)(\pi)(0.02)(0.1) = 0$$

$$P = 56.549 \text{ kN} = 56.5 \text{ kN} \quad \textbf{Ans}$$

Displacement :

$$\delta_{B/A} = \sum \frac{PL}{AE}$$

$$= \frac{56.549(10^3)(0.15)}{\pi(0.02^2)(98)(10^9)} + \int_0^{0.1\text{ m}} \frac{0.56549(10^6)\,x\,dx}{\pi(0.02^2)(98)(10^9)}$$

$$= 0.00009184 \text{ m} = 0.0918 \text{ mm} \quad \textbf{Ans}$$

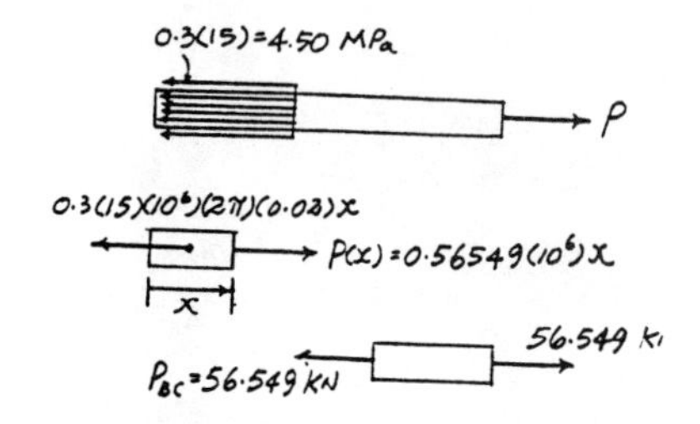

4–121. The assembly consists of a 30-mm-diameter aluminum bar ABC with fixed collar at B and a 10-mm-diameter steel rod CD. Determine the displacement of point D when the assembly is loaded as shown. Neglect the size of the collar at B and the connection at C. $E_{st} = 200$ GPa, $E_{al} = 70$ GPa.

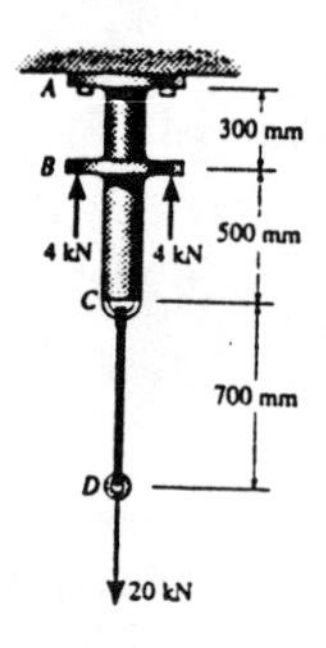

Internal Force : As shown on FBD.
Displacement :

$$\delta_D = \frac{20.0(10^3)(700)}{\frac{\pi}{4}(0.01)^2(200)(10^9)} + \frac{20.0(10^3)(500)}{\frac{\pi}{4}(0.03^2)(70)(10^9)} + \frac{12.0(10^3)(300)}{\frac{\pi}{4}(0.03^2)(70)(10^9)}$$

$$= 1.17 \text{ mm} \quad \textbf{Ans}$$

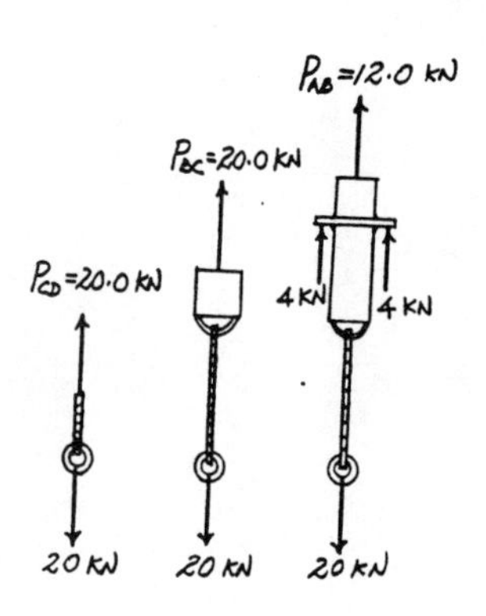

5–1. A shaft is made of a steel alloy having an allowable shear stress of $\tau_{allow} = 12$ ksi. If the diameter of the shaft is 1.5 in., determine the maximum torque **T** that can be transmitted. What would be the maximum torque **T′** if a 1-in.-diameter hole is bored through the shaft? Sketch the shear-stress distribution along a radial line in each case.

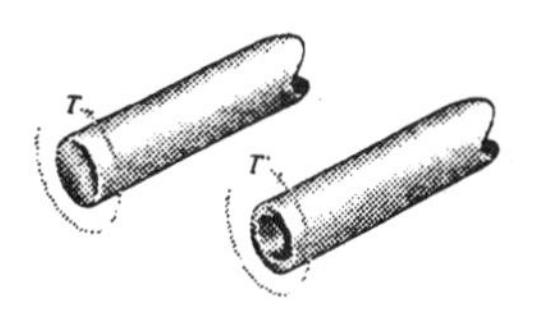

a) ***Allowable Shear Stress :*** Applying the torsion formula

$$\tau_{max} = \tau_{allow} = \frac{Tc}{J}$$

$$12 = \frac{T\,(0.75)}{\frac{\pi}{2}(0.75^4)}$$

$$T = 7.95 \text{ kip} \cdot \text{in.} \qquad \textbf{Ans}$$

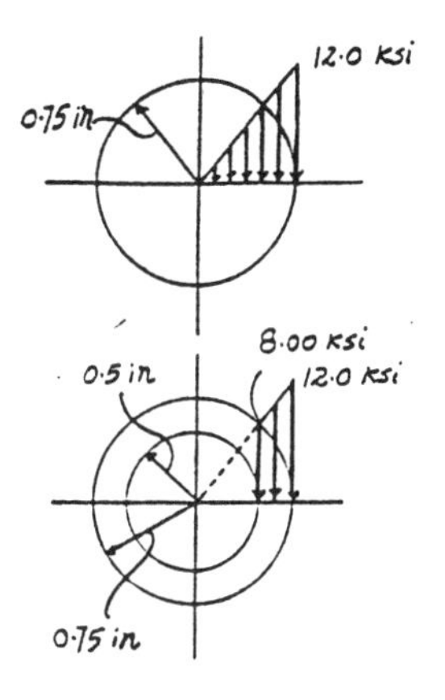

b) ***Allowable Shear Stress :*** Applying the torsion formula

$$\tau_{max} = \tau_{allow} = \frac{T'c}{J}$$

$$12 = \frac{T'\,(0.75)}{\frac{\pi}{2}(0.75^4 - 0.5^4)}$$

$$T' = 6.381 \text{ kip} \cdot \text{in.} = 6.38 \text{ kip} \cdot \text{in.} \qquad \textbf{Ans}$$

$$\tau_{\rho=0.5\text{in}} = \frac{T'\rho}{J} = \frac{6.381(0.5)}{\frac{\pi}{2}(0.75^4 - 0.5^4)} = 8.00 \text{ ksi}$$

5–2. The tube is subjected to a torque of 750 N · m. Determine the amount of this torque that is resisted by the grey shaded section. Solve the problem two ways: (*a*) by using the torsion formula, (*b*) by finding the resultant of the shear-stress distribution.

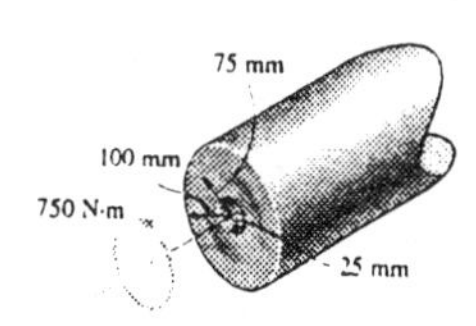

a) ***Applying Torsion Formula :***

$$\tau_{max} = \frac{Tc}{J} = \frac{750(0.1)}{\frac{\pi}{2}(0.1^4 - 0.025^4)} = 0.4793 \text{ MPa}$$

$$\tau_{max} = 0.4793(10^6) = \frac{T'(0.1)}{\frac{\pi}{2}(0.1^4 - 0.075^4)}$$

$$T' = 515 \text{ N} \cdot \text{m} \qquad \textbf{Ans}$$

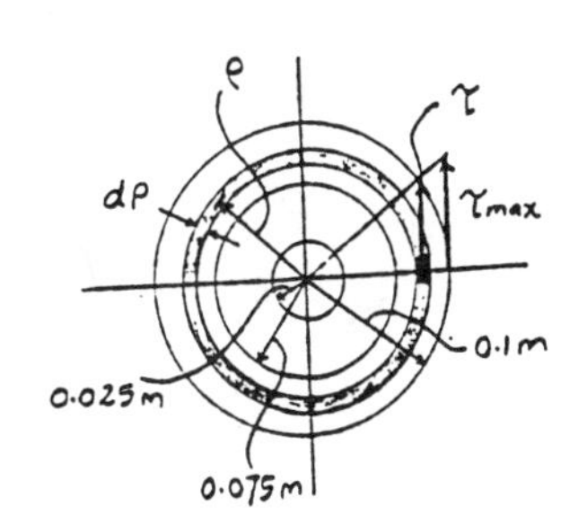

b) ***Integration Method :***

$$\tau = \left(\frac{\rho}{c}\right)\tau_{max} \quad \text{and} \quad dA = 2\pi\rho\, d\rho$$

$$dT' = \rho\tau\, dA = \rho\tau(2\pi\rho\, d\rho) = 2\pi\tau\rho^2\, d\rho$$

$$T' = \int 2\pi\tau\rho^2\, d\rho = 2\pi\int_{0.075\text{m}}^{0.1\text{m}} \tau_{max}\left(\frac{\rho}{c}\right)\rho^2\, d\rho$$

$$= \frac{2\pi\tau_{max}}{c}\int_{0.075\text{m}}^{0.1\text{m}} \rho^3\, d\rho$$

$$= \frac{2\pi(0.4793)(10^6)}{0.1}\left[\frac{\rho^4}{4}\right]\Big|_{0.075\text{m}}^{0.1\text{m}}$$

$$= 515 \text{ N} \cdot \text{m} \qquad \textbf{Ans}$$

5–3. The copper shaft has an outer diameter of 1.5 in. and an inner diameter of 1 in. If it is subjected to the applied torques as shown, determine the maximum shear stress developed in the shaft. The shaft is supported by two smooth journal bearings at A and B.

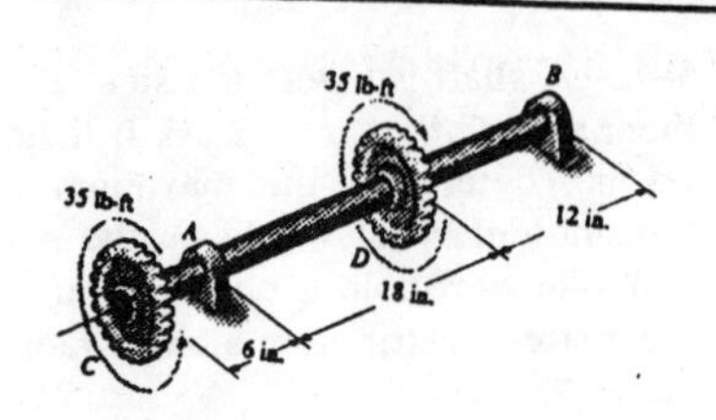

Internal Torque : As shown on FBD.
Maximum Shear Stress : Applying the torsion formula

35 lb·ft   T = 35.0 lb·ft

$$\tau_{max} = \frac{Tc}{J} = \frac{35.0(12)(0.75)}{\frac{\pi}{2}(0.75^4 - 0.5^4)}$$

$$= 790 \text{ psi} \qquad \textbf{Ans}$$

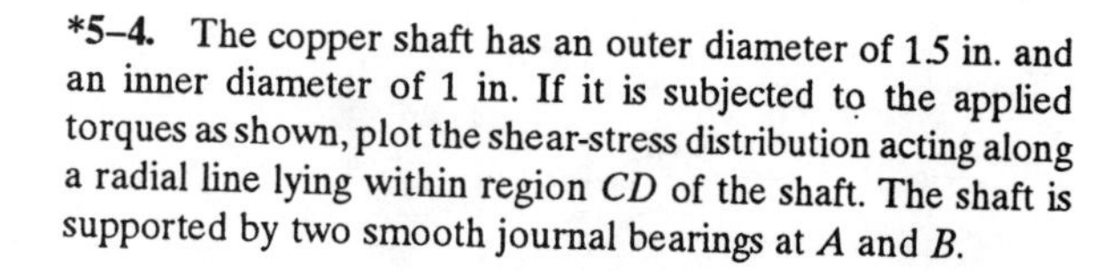

***5–4.** The copper shaft has an outer diameter of 1.5 in. and an inner diameter of 1 in. If it is subjected to the applied torques as shown, plot the shear-stress distribution acting along a radial line lying within region CD of the shaft. The shaft is supported by two smooth journal bearings at A and B.

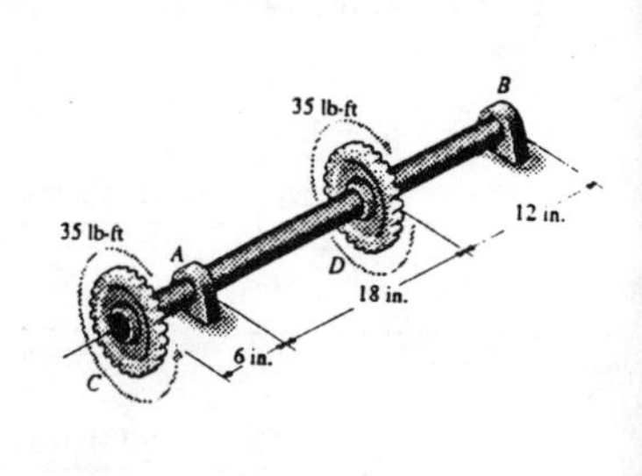

Internal Torque : As shown on FBD.
Maximum Shear Stress : Applying torsion formula.

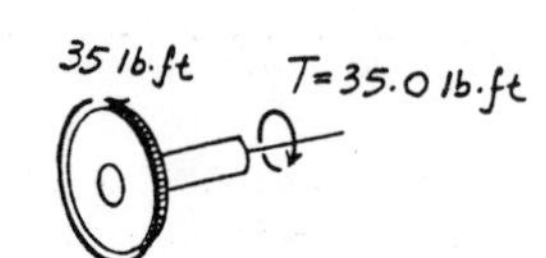

$$\tau_{max} = \frac{Tc}{J} = \frac{35.0(12)(0.75)}{\frac{\pi}{2}(0.75^4 - 0.5^4)}$$

$$= 790 \text{ psi}$$

At $\rho = 0.5$ in.

$$\tau_{\rho=0.5in} = \frac{T\rho}{J} = \frac{35.0(12)(0.5)}{\frac{\pi}{2}(0.75^4 - 0.5^4)} = 527 \text{ psi}$$

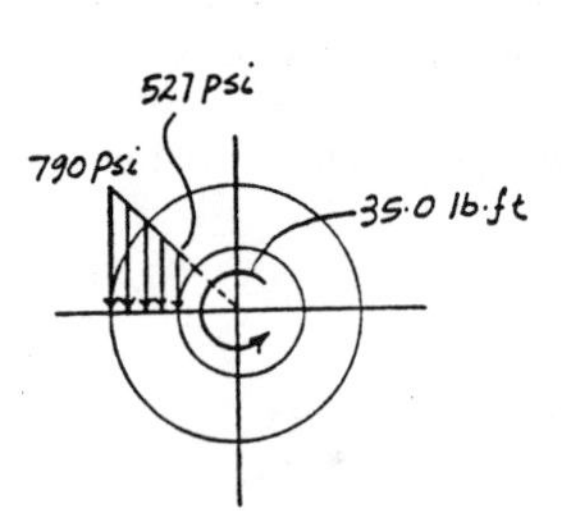

5–5. The solid 30-mm-diameter shaft is used to transmit the torques applied to the gears. Determine the shear stress developed in the shaft at points C and D. Indicate the shear stress on volume elements located at these points.

Internal Torque : As shown on FBD.
Maximum Shear Stress : Applying torsion formula

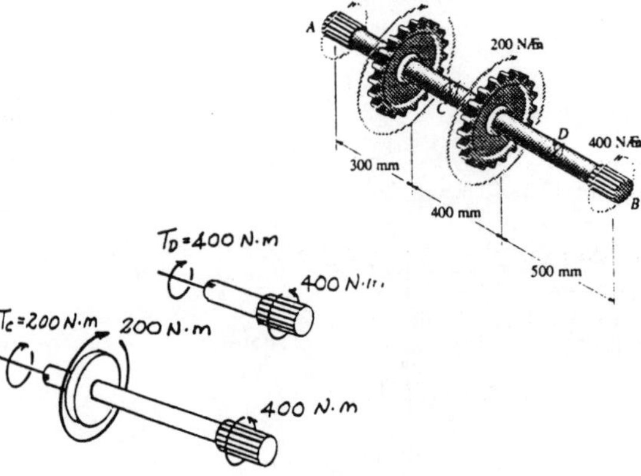

$$\tau_C = \frac{T_C\, c}{J} = \frac{200(0.015)}{\frac{\pi}{2}(0.015^4)}$$

$$= 37.7 \text{ MPa} \qquad \textbf{Ans}$$

$$\tau_D = \frac{T_D\, c}{J} = \frac{400(0.015)}{\frac{\pi}{2}(0.015^4)}$$

$$= 75.5 \text{ MPa} \qquad \textbf{Ans}$$

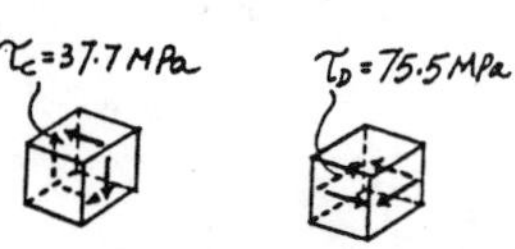

5–6. The solid 30-mm-diameter shaft is used to transmit the torques applied to the gears. Determine the absolute maximum shear stress on the shaft.

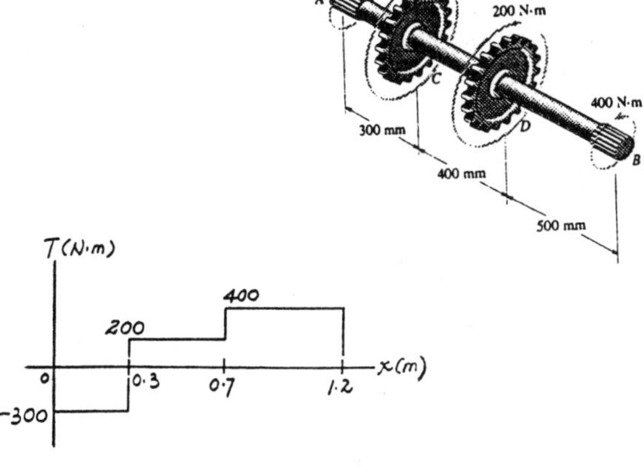

Internal Torque : As shown on torque diagram.

Maximum Shear Stress : From the torque diagram $T_{max} = 400$ N · m. Then, applying torsion Formula.

$$\tau_{\substack{max \\ ABS}} = \frac{T_{max}\,c}{J}$$

$$= \frac{400(0.015)}{\frac{\pi}{2}(0.015^4)} = 75.5 \text{ MPa} \qquad \textbf{Ans}$$

5–7. The solid aluminum shaft has a diameter of 50 mm and an allowable shear stress of $\tau_{allow} = 6$ MPa. Determine the largest torque T_1 that can be applied to the shaft if it is also subjected to the other torsional loadings. It is required that $\mathbf{T}_1$ act in the direction shown. Also, determine the maximum shear stress within regions CD and DE.

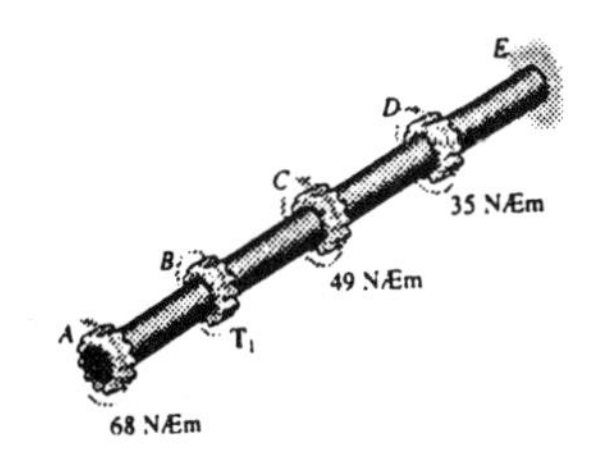

Internal Torque : As shown on FBD.

Maximum Shear Stress : Applying the torsion Formula and assume failure at region BC.

$$\tau_{max} = \tau_{allow} = \frac{T_{BC}\,c}{J}$$

$$6(10^6) = \frac{(T_1 - 68)(0.025)}{\frac{\pi}{2}(0.025^4)}$$

$$T_1 = 215.26 \text{ N} \cdot \text{m} = 215 \text{ N} \cdot \text{m} \qquad \textbf{Ans}$$

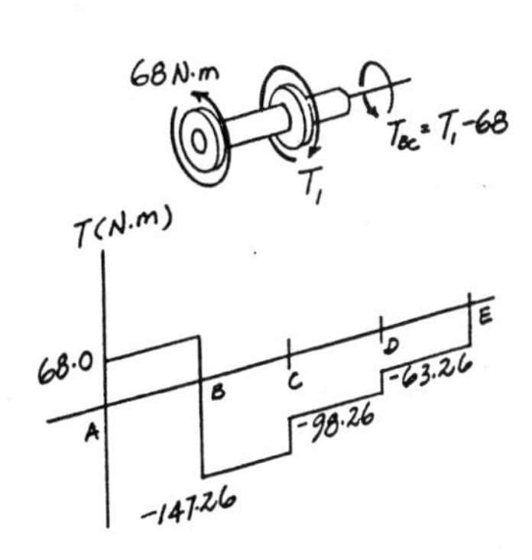

Maximum torque occurs within region BC as indicated on the torque diagram.

Maximum Shear Stresses at Other Region : From the torque diagram, the internal torque at region CD and DE are $T_{CD} = 98.26$ N · m and $T_{DE} = 63.26$ N · m respectively.

$$(\tau_{max})_{CD} = \frac{T_{CD}\,c}{J} = \frac{98.26(0.025)}{\frac{\pi}{2}(0.025^4)} = 4.00 \text{ MPa} \qquad \textbf{Ans}$$

$$(\tau_{max})_{DE} = \frac{T_{DE}\,c}{J} = \frac{63.26(0.025)}{\frac{\pi}{2}(0.025^4)} = 2.58 \text{ MPa} \qquad \textbf{Ans}$$

***5–8.** The solid aluminum shaft has a diameter of 50 mm. Determine the absolute maximum shear stress in the shaft and sketch the shear-stress distribution along a radial line of the shaft where the shear stress is maximum. Set $T_1 = 20\ \text{N}\cdot\text{m}$.

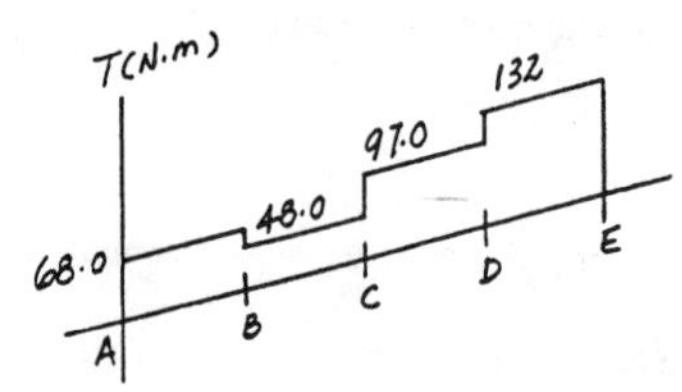

Internal Torque : From the torque diagram, the critical section is segment DE and $T_{DE} = 132\ \text{N}\cdot\text{m}$.

Maximum Shear Stress : Applying the torsion formula

$$\tau_{\substack{abs\\max}} = \frac{T_{DE}\,c}{J}$$

$$= \frac{132(0.025)}{\frac{\pi}{2}(0.025^4)} = 5.38\ \text{MPa} \qquad \textbf{Ans}$$

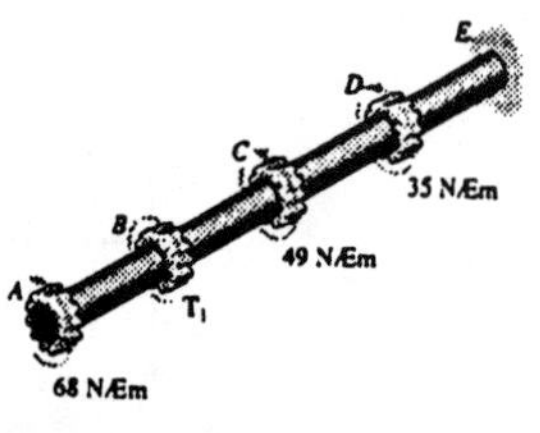

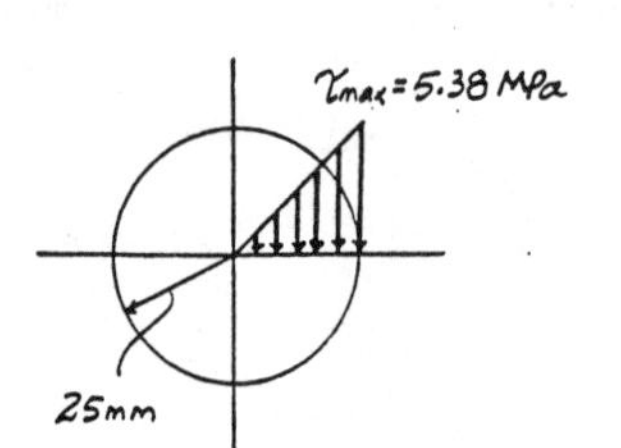

5–9. The motor delivers a torque of 50 N · m to the shaft AB. This torque is transmitted to shaft CD using the gears at E and F. Determine the equilibrium torque $\mathbf{T}'$ on shaft CD and the maximum shear stress in each shaft. The bearings B, C, and D allow free rotation of the shafts.

Equilibrium :

$$\circlearrowleft + \Sigma M_E = 0; \quad 50 - F(0.05) = 0 \quad F = 1000\ \text{N}$$

$$\circlearrowleft + \Sigma M_F = 0; \quad T' - 1000(0.125) = 0$$

$$T' = 125\ \text{N}\cdot\text{m} \qquad \textbf{Ans}$$

Internal Torque : As shown on FBD.

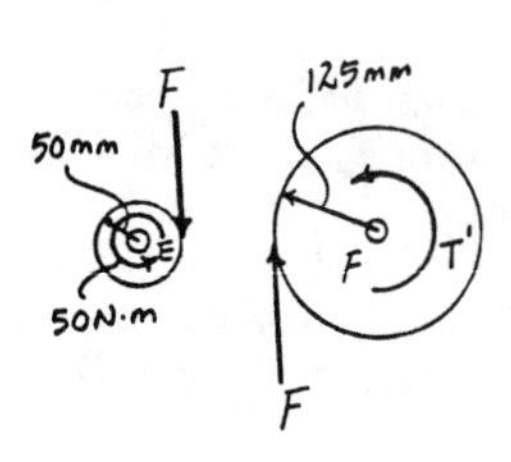

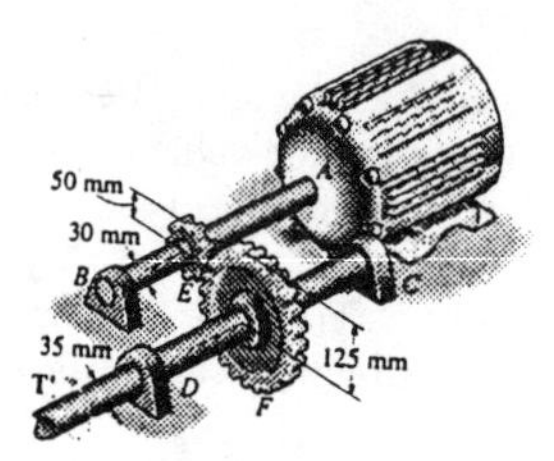

Maximum Shear Stress : Applying torsion Formula.

$$(\tau_{AB})_{max} = \frac{T_{AB}\,c}{J} = \frac{50.0(0.015)}{\frac{\pi}{2}(0.015^4)} = 9.43\ \text{MPa} \qquad \textbf{Ans}$$

$$(\tau_{CD})_{max} = \frac{T_{CD}\,c}{J} = \frac{125(0.0175)}{\frac{\pi}{2}(0.0175^4)} = 14.8\ \text{MPa} \qquad \textbf{Ans}$$

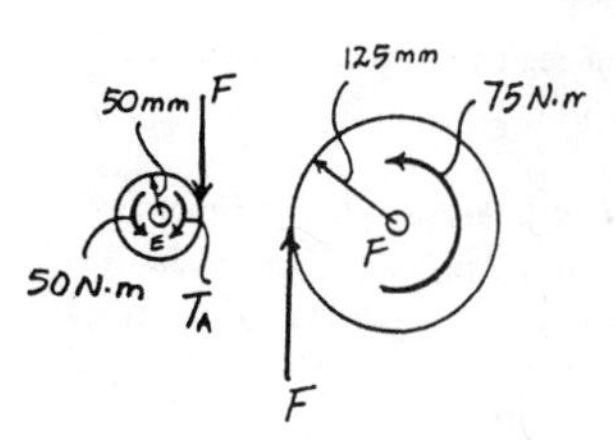

5–10. If the applied torque on shaft CD is $T' = 75\ \text{N}\cdot\text{m}$, determine the absolute maximum shear stress in each shaft. The bearings B, C, and D allow free rotation of the shafts, and the motor holds the shafts fixed from rotating.

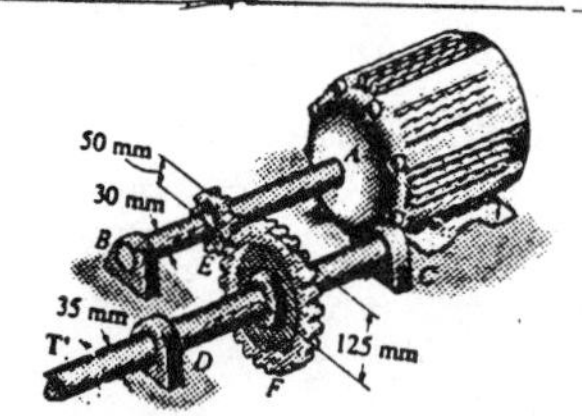

Equilibrium :

$$\circlearrowleft + \Sigma M_F = 0; \quad 75 - F(0.125) = 0; \quad F = 600\ \text{N}$$

$$\circlearrowleft + \Sigma M_E = 0; \quad 50 - 600(0.05) - T_A = 0$$

$$T_A = 20.0\ \text{N}\cdot\text{m}$$

Internal Torque : As shown on FBD.

Maximum Shear Stress : Applying the torsion formula

$$(\tau_{EA})_{max} = \frac{T_{EA}\,c}{J} = \frac{30.0(0.015)}{\frac{\pi}{2}(0.015^4)} = 5.66\ \text{MPa} \qquad \textbf{Ans}$$

$$(\tau_{CD})_{max} = \frac{T_{CD}\,c}{J} = \frac{75.0(0.0175)}{\frac{\pi}{2}(0.0175^4)} = 8.91\ \text{MPa} \qquad \textbf{Ans}$$

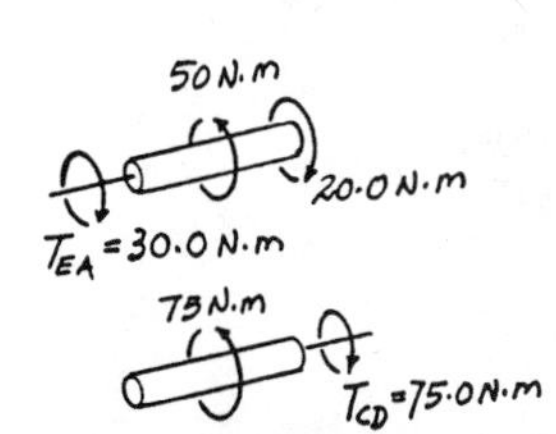

5–11. The rod has a diameter of 0.5 in. and a weight of 3 lb/ft. Determine the maximum shear stress due to torsion at a section through A.

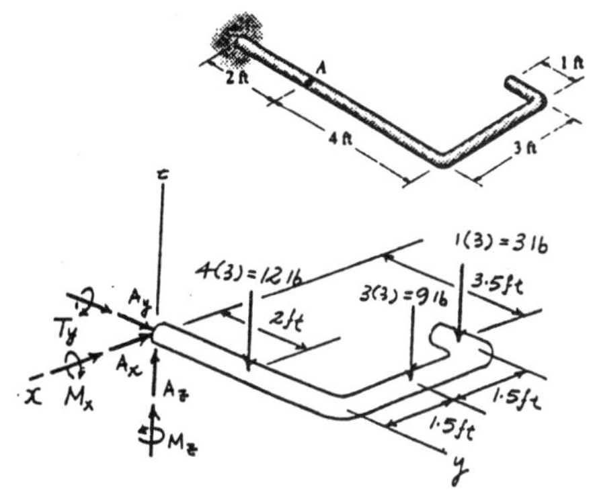

Internal Torque :

$$\Sigma M_y = 0; \quad T_y - 9(1.5) - 3(3) = 0 \quad T_y = 22.5 \text{ lb} \cdot \text{ft}$$

Maximum Shear Stress : Applying the torsion formula

$$(\tau_A)_{\max} = \frac{Tc}{J} = \frac{22.5(12)(0.25)}{\frac{\pi}{2}(0.25^4)}$$

$$= 11.0 \text{ ksi} \qquad \textbf{Ans}$$

5–13. The copper pipe has an outer diameter of 2.50 in. and an inner diameter of 2.30 in. If it is tightly secured to the wall at C and a uniformly distributed torque is applied to it as shown, determine the shear stress developed at points A and B. These points lie on the pipe's outer surface. Sketch the shear stress on volume elements located at A and B.

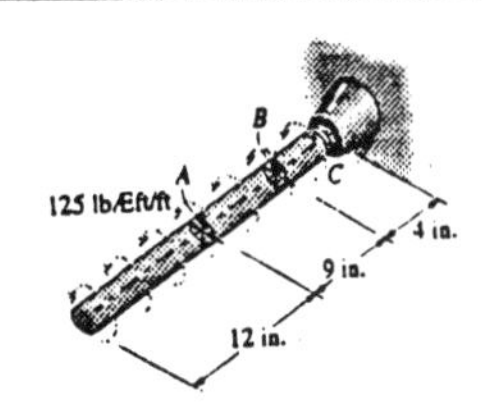

Internal Torque : As shown on FBD.

Maximum Shear Stress : Applying the torsion formula

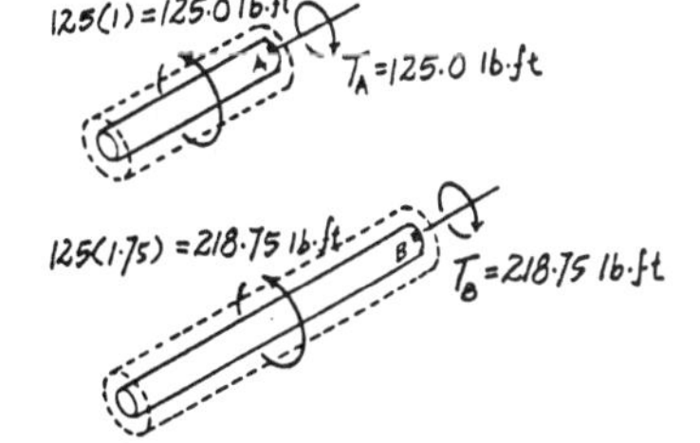

$$\tau_A = \frac{T_A c}{J}$$

$$= \frac{125.0(12)(1.25)}{\frac{\pi}{2}(1.25^4 - 1.15^4)} = 1.72 \text{ ksi} \qquad \textbf{Ans}$$

$$\tau_B = \frac{T_B c}{J}$$

$$= \frac{218.75(12)(1.25)}{\frac{\pi}{2}(1.25^4 - 1.15^4)} = 3.02 \text{ ksi} \qquad \textbf{Ans}$$

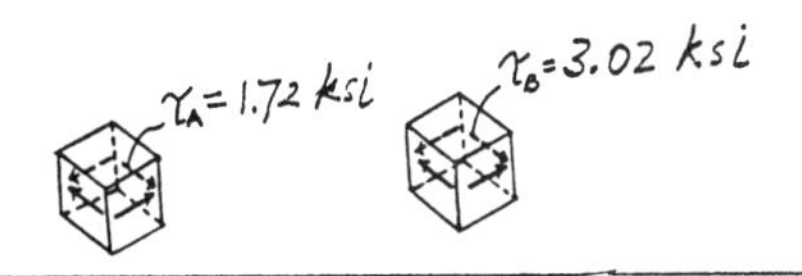

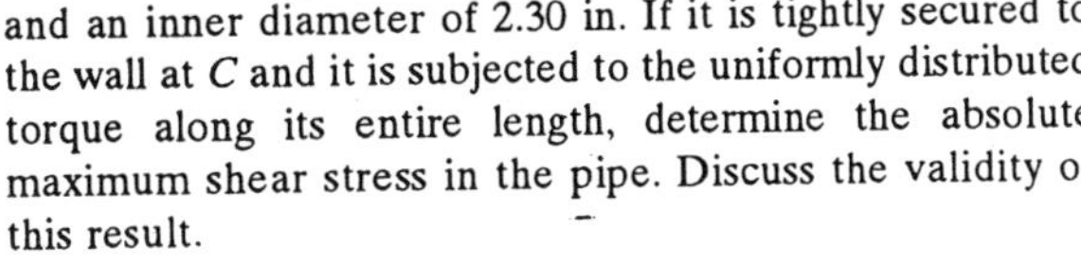

5–14. The copper pipe has an outer diameter of 2.50 in. and an inner diameter of 2.30 in. If it is tightly secured to the wall at C and it is subjected to the uniformly distributed torque along its entire length, determine the absolute maximum shear stress in the pipe. Discuss the validity of this result.

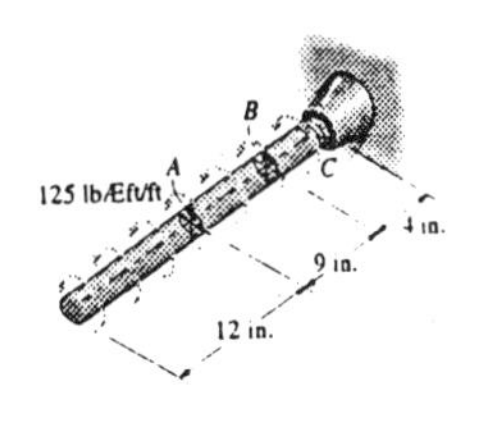

Internal Torque : The maximum torque occurs at the support C.

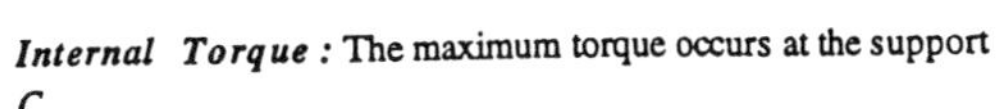

$$T_{\max} = (125 \text{ lb} \cdot \text{ft/ft})\left(\frac{25 \text{ in.}}{12 \text{ in./ft}}\right) = 260.42 \text{ lb} \cdot \text{ft}$$

Maximum Shear Stress : Applying the torsion formula

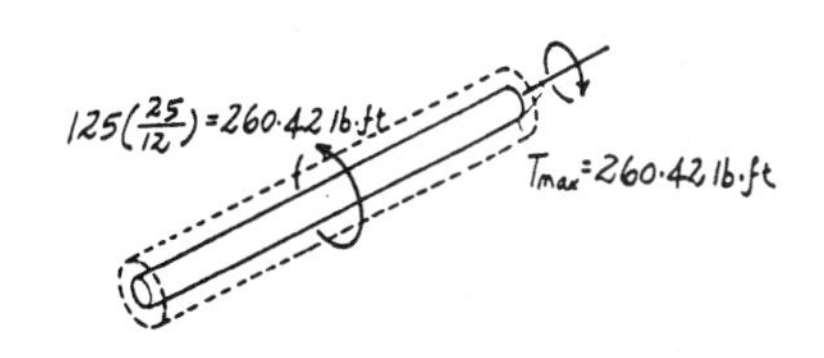

$$\tau_{\substack{\text{abs}\\ \max}} = \frac{T_{\max} c}{J}$$

$$= \frac{260.42(12)(1.25)}{\frac{\pi}{2}(1.25^4 - 1.15^4)} = 3.59 \text{ ksi} \qquad \textbf{Ans}$$

According to Saint-Venant's principle, application of the torsion formula should be at points sufficiently removed from the supports or points of concentrated loading.

5–15. The 60-mm-diameter solid shaft is subjected to the distributed and concentrated torsional loadings shown. Determine the shear stress at points A and B and sketch the shear stress on volume elements located at these points.

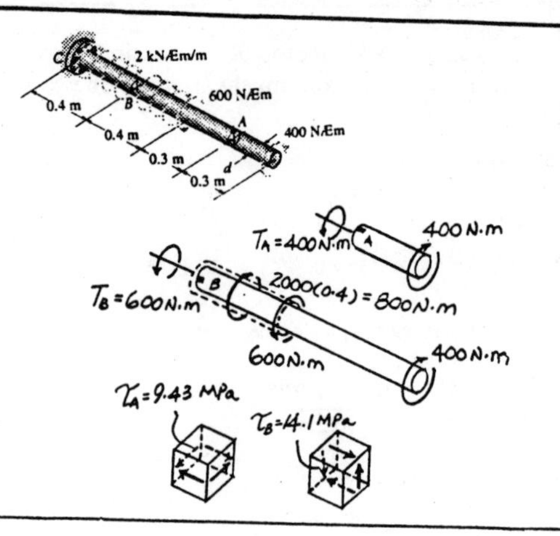

Internal Torque : As shown on FBD.

Maximum Shear Stress : Applying the torsion formula

$$\tau_A = \frac{T_A c}{J} = \frac{400(0.03)}{\frac{\pi}{2}(0.03^4)} = 9.43 \text{ MPa} \qquad \textbf{Ans}$$

$$\tau_B = \frac{T_B c}{J} = \frac{600(0.03)}{\frac{\pi}{2}(0.03^4)} = 14.1 \text{ MPa} \qquad \textbf{Ans}$$

***5–16.** The 60-mm diameter solid shaft is subjected to the distributed and concentrated torsional loadings shown. Determine the absolute maximum and minimum shear stresses in the shaft and specify their locations, measured from the fixed end.

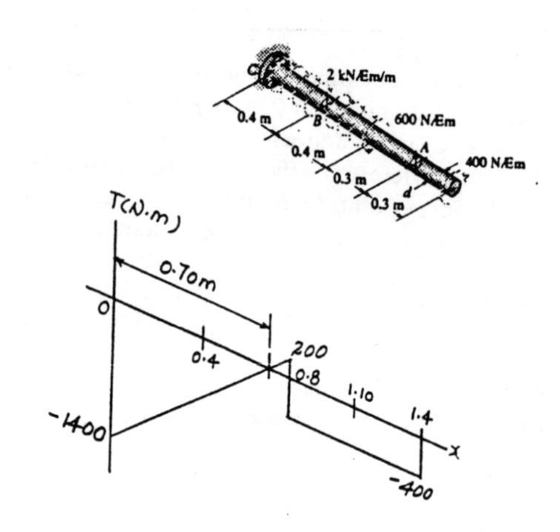

Internal Torque : From the torque diagram, the maximum torque $T_{max} = 1400$ N · m occurs at the fixed support and the minimum torque $T_{min} = 0$ occurs at $x = 0.700$ m.

Shear Stress : Applying the torsion formula

$$\tau_{\substack{abs\\min}} = \frac{T_{min} c}{J} = 0 \quad \text{occurs at} \quad x = 0.700 \text{ m} \qquad \textbf{Ans}$$

$$\tau_{\substack{abs\\max}} = \frac{T_{max} c}{J} = \frac{1400(0.03)}{\frac{\pi}{2}(0.03^4)} = 33.0 \text{ MPa} \qquad \textbf{Ans}$$

occurs at $x = 0$ **Ans**

According to Saint - Venant's principle, application of the torsion formula should be at points sufficiently removed from the supports or points of concentrated loading. Therefore, $\tau_{\substack{abs\\max}}$ obtained ***is not valid.***

5–17. The solid shaft is subjected to the distributed and concentrated torsional loadings shown. Determine the required diameter d of the shaft if the allowable shear stress for the material is $\tau_{allow} = 175$ MPa.

Internal Torque : From the torque diagram, the maximum torque $T_{max} = 1400$ N · m occurs at the fixed support.

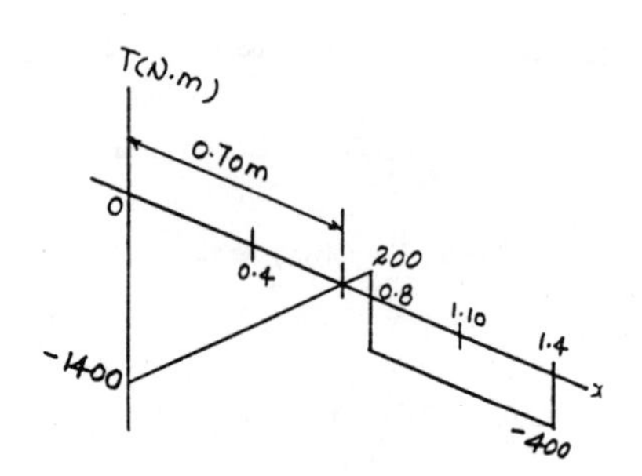

Allowable Shear Stress : Applying the torsion formula

$$\tau_{\substack{abs\\max}} = \tau_{allow} = \frac{T_{max} c}{J}$$

$$175(10^6) = \frac{1400\left(\frac{d}{2}\right)}{\frac{\pi}{2}\left(\frac{d}{2}\right)^4}$$

$$d = 0.03441 \text{ m} = 34.4 \text{ mm} \qquad \textbf{Ans}$$

According to Saint Venant's principle, application of the torsion formula should be at points sufficiently removed from the supports or points of concentrated loading. Therefore, the above analysis ***is not valid.***

5–18. The wooden post, which is half buried in the ground, is subjected to a torsional moment of 50 N · m that causes the post to rotate at constant angular velocity. This moment is resisted by a *linear distribution* of torque developed by soil friction, which varies from zero at the ground to t_0 N · m/m at its base. Determine the equilibrium value for t_0, and then calculate the shear stress at points *A* and *B*, which lie on the outer surface of the post.

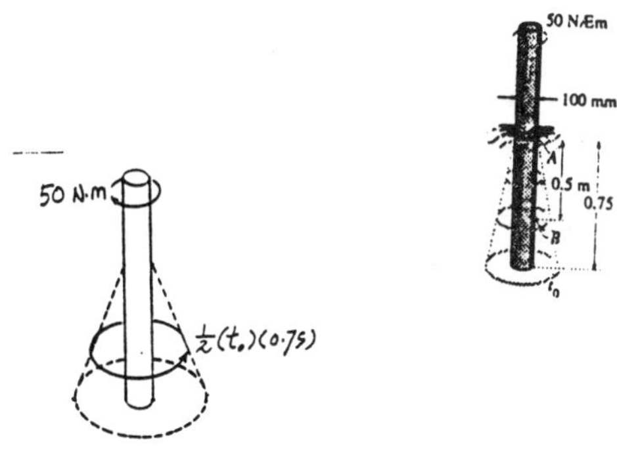

Equilibrium :

$$\Sigma M_z = 0; \quad \frac{1}{2}t_0(0.75) - 50 = 0$$

$$t_0 = 133.33 \text{ N}\cdot\text{m/m} = 133 \text{ N}\cdot\text{m/m} \qquad \textbf{Ans}$$

Internal Torque : As shown on FBD.

Maximum Shear Stress : Applying the torsion formula

$$\tau_A = \frac{T_A c}{J} = \frac{50.0(0.05)}{\frac{\pi}{2}(0.05^4)} = 0.255 \text{ MPa} \qquad \textbf{Ans}$$

$$\tau_B = \frac{T_B c}{J} = \frac{27.78(0.05)}{\frac{\pi}{2}(0.05^4)} = 0.141 \text{ MPa} \qquad \textbf{Ans}$$

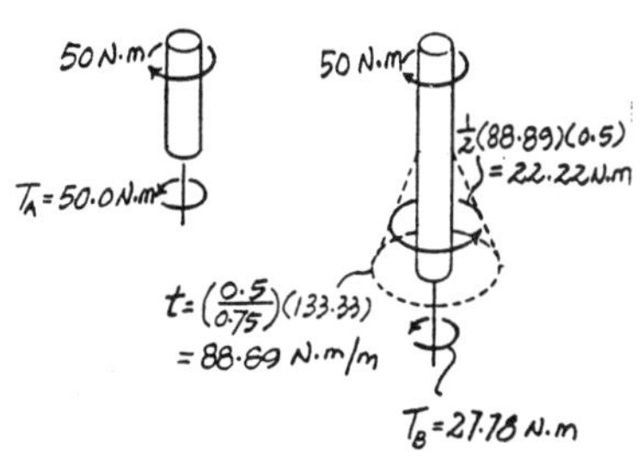

5–19. Determine the absolute maximum shear stress in the post in Prob. 5–18.

Equilibrium :

$$\Sigma M_z = 0; \quad \frac{1}{2}t_0(0.75) - 50 = 0 \qquad t_0 = 133.33 \text{ N}\cdot\text{m/m}$$

Internal Torque : From the torque diagram, $T_{max} = 50.0$ N · m.

Maximum Shear Stress : Applying torsion Formula.

$$\tau_{\substack{max\\ABS}} = \frac{T_{max}c}{J} = \frac{50.0(0.05)}{\frac{\pi}{2}(0.05^4)} = 0.255 \text{ MPa} \qquad \textbf{Ans}$$

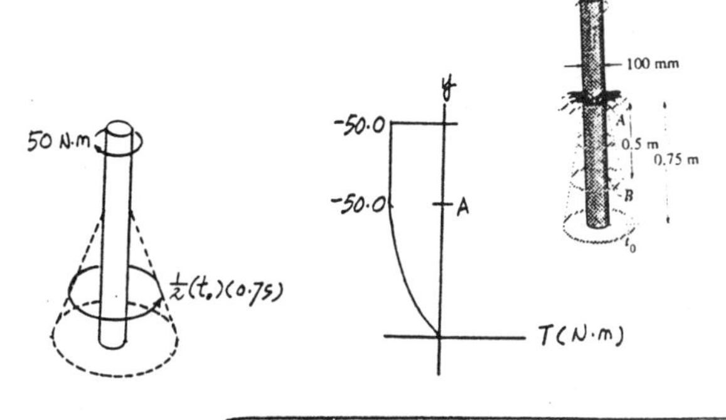

***5–20.** The solid shaft is subjected to the distributed and concentrated torsional loadings shown. Determine the shear stress at points *A* and *B* and sketch the shear stress on volume elements located at these points. The distributed torque from *D* to *C* varies from zero to 900 N · m/m. Take *d* =40mm.

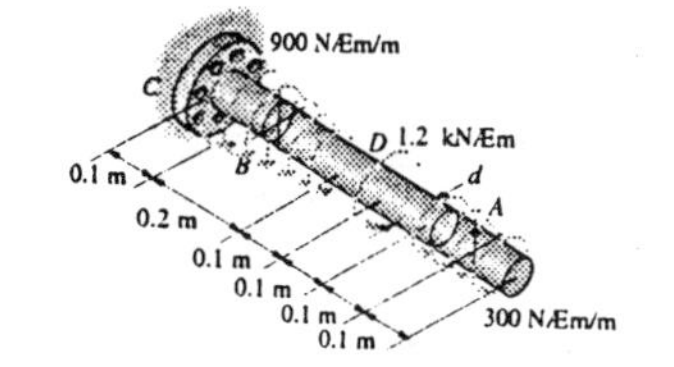

Internal Torque : As shown on FBD.

Maximum Shear Stress : Applying the torsion formula

$$\tau_A = \frac{T_A c}{J} = \frac{30.0(0.02)}{\frac{\pi}{2}(0.02^4)} = 2.39 \text{ MPa} \qquad \textbf{Ans}$$

$$\tau_B = \frac{T_B c}{J} = \frac{1080(0.02)}{\frac{\pi}{2}(0.02^4)} = 85.9 \text{ MPa} \qquad \textbf{Ans}$$

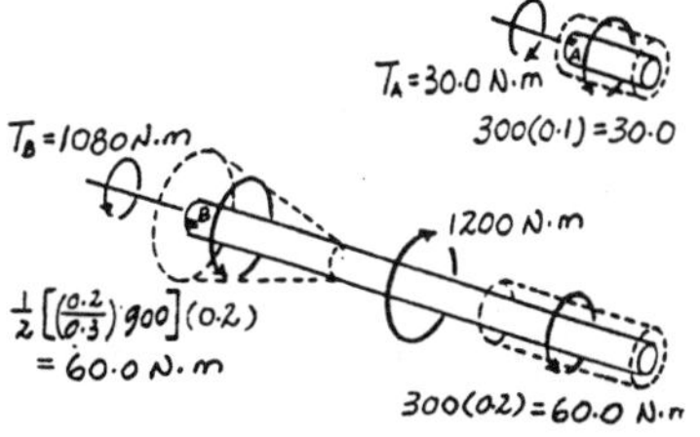

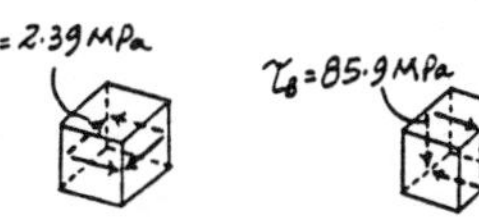

5–21. The solid shaft is subjected to the distributed and concentrated torsional loadings shown. Determine the absolute maximum shear stress in the shaft and specify its location, measured from the fixed end C. Take d = 40 mm.

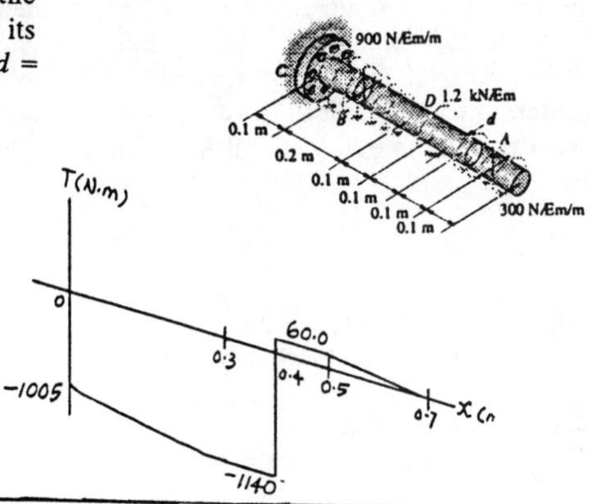

Internal Torque : From the torque diagram, the maximum torque T_{max} = 1140 N · m occurs in the region 0.3 m > x > 0.4 m.

Shear Stress : Applying the torsion formula

$$\tau_{\substack{abs\\max}} = \frac{T_{max}\,c}{J} = \frac{1140(0.02)}{\frac{\pi}{2}(0.02^4)} = 90.7 \text{ MPa} \qquad \textbf{Ans}$$

in the region 0.3 m > x > 0.4 m **Ans**

5–22. The solid shaft is subjected to the distributed and concentrated torsional loadings shown. Determine the required diameter d of the shaft if the allowable shear stress for the material is τ_{allow} = 150 MPa.

Internal Torque : From the torque diagram, the maximum torque T_{max} = 1140 N · m occurs in the region 0.3 m > x > 0.4 m.

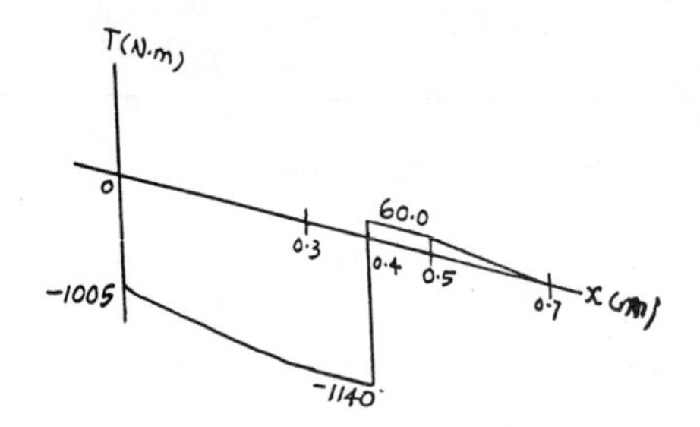

Allowable Shear Stress : Applying the torsion formula

$$\tau_{\substack{abs\\max}} = \tau_{allow} = \frac{T_{max}\,c}{J}$$

$$150(10^6) = \frac{1140\left(\frac{d}{2}\right)}{\frac{\pi}{2}\left(\frac{d}{2}\right)^4}$$

$$d = 0.03383 \text{ m} = 33.8 \text{ mm} \qquad \textbf{Ans}$$

5–23. Determine to the nearest $\frac{1}{8}$ in. the diameter of a shaft that is required to transmit 150 hp at 4000 rev/min. The material has an allowable shear stress of τ_{allow} = 8 ksi.

Internal Torque :

$$\omega = 4000\,\frac{\text{rev}}{\text{min}}\left(\frac{2\pi \text{ rad}}{\text{rev}}\right)\frac{1 \text{ min}}{60 \text{ s}} = 133.33\pi \text{ rad/s}$$

$$P = 150 \text{ hp}\left(\frac{550 \text{ ft}\cdot\text{lb/s}}{1 \text{ hp}}\right) = 82500 \text{ ft}\cdot\text{lb/s}$$

$$T = \frac{P}{\omega} = \frac{82500}{133.33\pi} = 196.95 \text{ lb}\cdot\text{ft}$$

Allowable Shear Stress : Applying the torsion formula

$$\tau_{max} = \tau_{allow} = \frac{T\,c}{J}$$

$$8(10^3) = \frac{196.95(12)\left(\frac{d}{2}\right)}{\frac{\pi}{2}\left(\frac{d}{2}\right)^4}$$

$$d = 1.146 \text{ in.}$$

Use $d = 1\frac{1}{4}$ in. diameter of shaft **Ans**

***5–24.** The drilling pipe on an oil rig is made from steel pipe having an outer diameter of 4.5 in. and a thickness of 0.25 in. If the pipe is turning at 650 rev/min while being powered by a 15-hp motor, determine the maximum shear stress in the pipe.

Internal Torque:

$$\omega = 650\ \frac{\text{rev}}{\text{min}}\left(\frac{2\pi\ \text{rad}}{\text{rev}}\right)\frac{1\ \text{min}}{60\ \text{s}} = 21.67\pi\ \text{rad/s}$$

$$P = 15\ \text{hp}\left(\frac{550\ \text{ft}\cdot\text{lb/s}}{1\ \text{hp}}\right) = 8250\ \text{ft}\cdot\text{lb/s}$$

$$T = \frac{P}{\omega} = \frac{8250}{21.67\pi} = 121.20\ \text{lb}\cdot\text{ft}$$

Maximum Shear Stress: Applying the torsion formula

$$\tau_{\max} = \frac{Tc}{J}$$

$$= \frac{121.20(12)(2.25)}{\frac{\pi}{2}(2.25^4 - 2^4)} = 216\ \text{psi} \qquad \textbf{Ans}$$

5–25. The gear motor can develop 1/10 hp when it turns at 300 rev/min. If the shaft has a diameter of $\frac{1}{2}$ in., determine the maximum shear stress that will be developed in the shaft.

Internal Torque:

$$\omega = 300\ \frac{\text{rev}}{\text{min}}\left(\frac{2\pi\ \text{rad}}{\text{rev}}\right)\frac{1\ \text{min}}{60\ \text{s}} = 10.0\pi\ \text{rad/s}$$

$$P = \frac{1}{10}\ \text{hp}\left(\frac{550\ \text{ft}\cdot\text{lb/s}}{1\ \text{hp}}\right) = 55.0\ \text{ft}\cdot\text{lb/s}$$

$$T = \frac{P}{\omega} = \frac{55.0}{10.0\pi} = 1.751\ \text{lb}\cdot\text{ft}$$

Maximum Shear Stress: Applying the torsion formula

$$\tau_{\max} = \frac{Tc}{J}$$

$$= \frac{1.751(12)(0.25)}{\frac{\pi}{2}(0.25^4)} = 856\ \text{psi} \qquad \textbf{Ans}$$

5–26. The gear motor can develop 1/10 hp when it turns at 80 rev/min. If the allowable shear stress for the shaft is $\tau_{\text{allow}} = 4$ ksi, determine the smallest diameter of the shaft to the nearest $\frac{1}{8}$ in. that can be used.

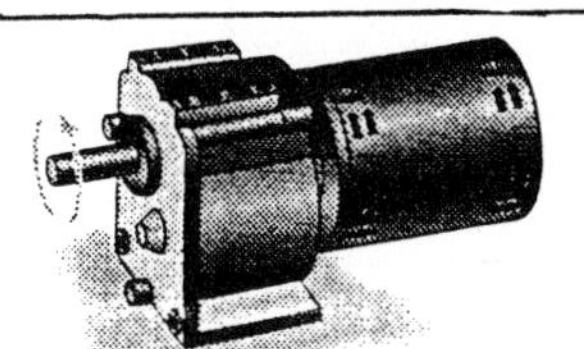

Internal Torque:

$$\omega = 80\ \frac{\text{rev}}{\text{min}}\left(\frac{2\pi\ \text{rad}}{\text{rev}}\right)\frac{1\ \text{min}}{60\ \text{s}} = 2.667\pi\ \text{rad/s}$$

$$P = \frac{1}{10}\ \text{hp}\left(\frac{550\ \text{ft}\cdot\text{lb/s}}{1\ \text{hp}}\right) = 55.0\ \text{ft}\cdot\text{lb/s}$$

$$T = \frac{P}{\omega} = \frac{55.0}{2.667\pi} = 6.565\ \text{lb}\cdot\text{ft}$$

Allowable Shear Stress: Applying the torsion formula

$$\tau_{\max} = \tau_{\text{allow}} = \frac{Tc}{J}$$

$$4(10^3) = \frac{6.565(12)\left(\frac{d}{2}\right)}{\frac{\pi}{2}\left(\frac{d}{2}\right)^4}$$

$$d = 0.4646\ \text{in.}$$

Use $d = \frac{1}{2}$ in. diameter of shaft. **Ans**

5–27. The 3-hp reducer motor can turn at 330 rev/min. If the shaft has a diameter of $\frac{3}{4}$ in., determine the maximum shear stress that will be developed in the shaft.

Internal Torque :

$$\omega = 330\,\frac{\text{rev}}{\text{min}}\left(\frac{2\pi\ \text{rad}}{\text{rev}}\right)\frac{1\ \text{min}}{60\ \text{s}} = 11.0\pi\ \text{rad/s}$$

$$P = 3\ \text{hp}\left(\frac{550\ \text{ft}\cdot\text{lb/s}}{1\ \text{hp}}\right) = 1650\ \text{ft}\cdot\text{lb/s}$$

$$T = \frac{P}{\omega} = \frac{1650}{11.0\pi} = 47.75\ \text{lb}\cdot\text{ft}$$

Maximum Shear Stress : Applying the torsion formula

$$\tau_{max} = \frac{Tc}{J}$$

$$= \frac{47.75(12)\left(\frac{3}{8}\right)}{\frac{\pi}{2}\left(\frac{3}{8}\right)^4} = 6.92\ \text{ksi} \qquad \textbf{Ans}$$

***5–28.** The 3-hp reducer motor can turn at 330 rev/min. If the allowable shear stress for the shaft is $\tau_{allow} = 8$ ksi, determine the smallest diameter of the shaft to the nearest $\frac{1}{8}$ in. that can be used.

Internal Torque :

$$\omega = 330\,\frac{\text{rev}}{\text{min}}\left(\frac{2\pi\ \text{rad}}{\text{rev}}\right)\frac{1\ \text{min}}{60\ \text{s}} = 11.0\pi\ \text{rad/s}$$

$$P = 3\ \text{hp}\left(\frac{550\ \text{ft}\cdot\text{lb/s}}{1\ \text{hp}}\right) = 1650\ \text{ft}\cdot\text{lb/s}$$

$$T = \frac{P}{\omega} = \frac{1650}{11.0\pi} = 47.75\ \text{lb}\cdot\text{ft}$$

Allowable Shear Stress : Applying torsion formula

$$\tau_{max} = \tau_{allow} = \frac{Tc}{J}$$

$$8(10^3) = \frac{47.75(12)\left(\frac{d}{2}\right)}{\frac{\pi}{2}\left(\frac{d}{2}\right)^4}$$

$$d = 0.7145\ \text{in.}$$

Use $d = \frac{3}{4}$ in. diameter of shaft **Ans**

5–29. A motor delivers 500 hp to the steel shaft which is tubular and has an outer diameter of 2 in. If the shaft is rotating at 200 rad/s, determine its largest inner diameter to the nearest $\frac{1}{8}$ in. if the allowable shear stress for the material is $\tau_{allow} = 25$ ksi.

Internal Torque :

$$P = 500\ \text{hp}\left(\frac{550\ \text{ft}\cdot\text{lb/s}}{1\ \text{hp}}\right) = 275\ 000\ \text{ft}\cdot\text{lb/s}$$

$$T = \frac{P}{\omega} = \frac{275\ 000}{200} = 1375\ \text{lb}\cdot\text{ft}$$

Allowable Shear Stress : Applying torsion formula.

$$\tau_{max} = \tau_{allow} = \frac{Tc}{J}$$

$$25(10^3) = \frac{1375(12)(1)}{\frac{\pi}{2}\left[1^4 - \left(\frac{d_i}{2}\right)^4\right]}$$

$$d_i = 1.745\ \text{in.}$$

Use inner diameter $d_i = 1\frac{5}{8}$ in. **Ans**

5–30. The pump operates using the motor that has a power of 85 W. If the impeller at B is turning at 150 rev/min, determine the maximum shear stress developed in the 20-mm-diameter transmission shaft at A.

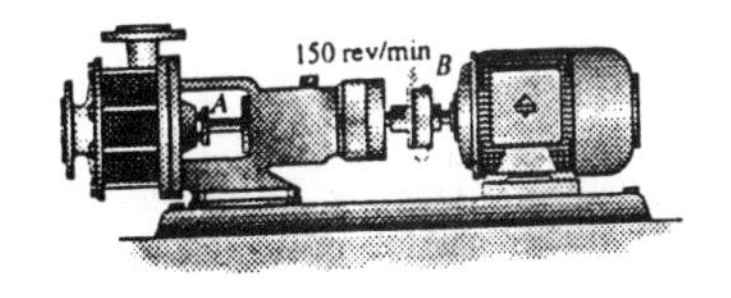

Internal Torque :

$$\omega = 150\ \frac{\text{rev}}{\text{min}}\left(\frac{2\pi\ \text{rad}}{\text{rev}}\right)\frac{1\ \text{min}}{60\ \text{s}} = 5.00\pi\ \text{rad/s}$$

$$P = 85\ \text{W} = 85\ \text{N}\cdot\text{m/s}$$

$$T = \frac{P}{\omega} = \frac{85}{5.00\pi} = 5.411\ \text{N}\cdot\text{m}$$

Maximum Shear Stress : Applying torsion formula

$$\tau_{max} = \frac{Tc}{J}$$

$$= \frac{5.411(0.01)}{\frac{\pi}{2}(0.01^4)} = 3.44\ \text{MPa} \qquad \textbf{Ans}$$

5–31. A steel tube having an outer diameter of d_1 = 2.5 in. is used to transmit 35 hp when turning at 2700 rev/min. Determine the inner diameter d_2 of the tube to the nearest $\frac{1}{8}$ in. if the allowable shear stress is τ_{allow} = 10 ksi.

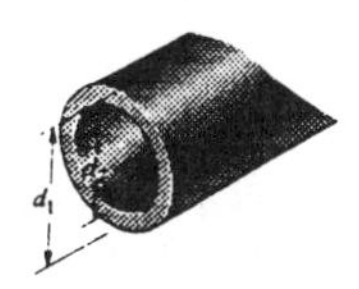

Internal Torque :

$$\omega = 2700\ \frac{\text{rev}}{\text{min}}\left(\frac{2\pi\ \text{rad}}{1\ \text{rev}}\right)\frac{1\ \text{min}}{60\ \text{s}} = 90.0\pi\ \text{rad/s}$$

$$P = 35\ \text{hp}\left(\frac{550\ \text{ft}\cdot\text{lb/s}}{1\ \text{hp}}\right) = 19250\ \text{ft}\cdot\text{lb}$$

$$T = \frac{P}{\omega} = \frac{19250}{90.0\pi} = 68.08\ \text{lb}\cdot\text{ft}$$

Allowable Shear Stress : Applying torsion formula

$$\tau_{max} = \tau_{allow} = \frac{Tc}{J}$$

$$10(10^3) = \frac{68.08(12)(1.25)}{\frac{\pi}{2}\left[1.25^4 - \left(\frac{d_2}{2}\right)^4\right]}$$

$$d_2 = 2.483\ \text{in.}$$

$$\text{Use} \quad d_2 = 2\frac{3}{8}\ \text{in.} \qquad \textbf{Ans}$$

Note that the internal diameter must be decreased.

***5–32.** A steel tube having an outer diameter of d_1 = 2.5 in. and an inner diameter of d_2 = 2 in. is used to transmit 45 hp. Determine its maximum rate of rotation if the allowable shear stress is τ_{allow} = 12 ksi.

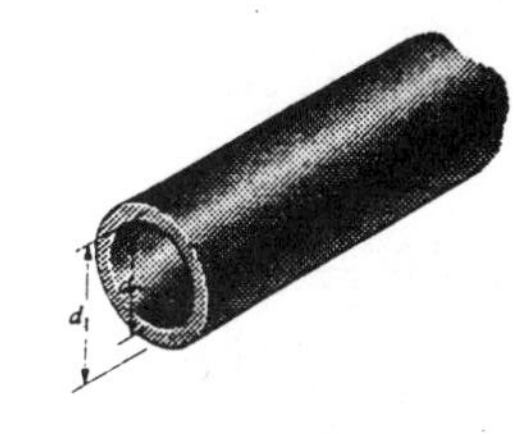

Allowable Shear Stress : Applying torsion formula

$$\tau_{max} = \tau_{allow} = \frac{Tc}{J}$$

$$12(10^3) = \frac{T(1.25)}{\frac{\pi}{2}(1.25^4 - 1^4)}$$

$$T = 21\,735.9\ \text{lb}\cdot\text{in.} = 1811.3\ \text{lb}\cdot\text{ft}$$

Internal Torque :

$$P = T\omega$$

$$45\ \text{hp}\left(\frac{550\ \text{ft}\cdot\text{lb/s}}{1\ \text{hp}}\right) = (1811.3\ \text{lb}\cdot\text{ft})\,\omega$$

$$\omega = 13.66\ \text{rad/s}$$

$$\omega = 13.66\ \text{rad/s}\left(\frac{1\ \text{rev}}{2\pi\ \text{rad}}\right)\left(\frac{60\ \text{s}}{1\ \text{min}}\right) = 130\ \text{rpm} \qquad \textbf{Ans}$$

5–33. A ship has a propeller drive shaft that is turning at 1500 rev/min while developing 1800 hp. If it is 8 ft long and has a diameter of 4 in., determine the maximum shear stress in the shaft caused by torsion.

Internal Torque :

$$\omega = 1500\ \frac{\text{rev}}{\text{min}}\left(\frac{2\pi\ \text{rad}}{1\ \text{rev}}\right)\frac{1\ \text{min}}{60\ \text{s}} = 50.0\ \pi\ \text{rad/s}$$

$$P = 1800\ \text{hp}\left(\frac{550\ \text{ft}\cdot\text{lb/s}}{1\ \text{hp}}\right) = 990\ 000\ \text{ft}\cdot\text{lb/s}$$

$$T = \frac{P}{\omega} = \frac{990\ 000}{50.0\pi} = 6302.54\ \text{lb}\cdot\text{ft}$$

Maximum Shear Stress : Applying torsion formula

$$\tau_{\max} = \frac{T\,c}{J} = \frac{6302.54(12)(2)}{\frac{\pi}{2}(2^4)}$$
$$= 6018\ \text{psi} = 6.02\ \text{ksi} \qquad \textbf{Ans}$$

5–34. The motor A develops a power of 300 W and turns its connected pulley at 90 rev/min. Determine the required diameters of the steel shafts on the pulleys at A and B if the allowable shear stress is $\tau_{\text{allow}} = 85$ MPa.

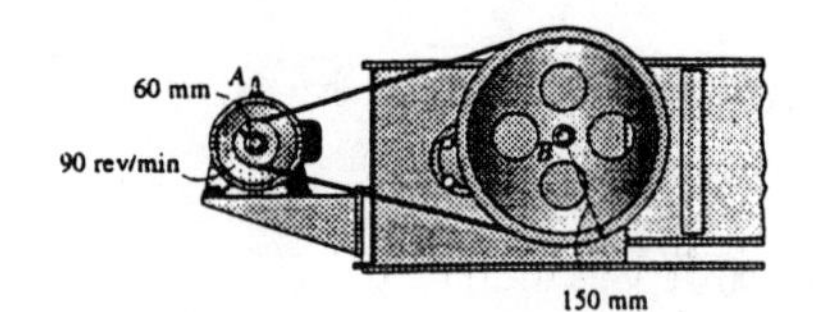

Internal Torque : For shafts A and B

$$\omega_A = 90\ \frac{\text{rev}}{\text{min}}\left(\frac{2\pi\ \text{rad}}{\text{rev}}\right)\frac{1\ \text{min}}{60\ \text{s}} = 3.00\pi\ \text{rad/s}$$
$$P = 300\ \text{W} = 300\ \text{N}\cdot\text{m/s}$$
$$T_A = \frac{P}{\omega_A} = \frac{300}{3.00\pi} = 31.83\ \text{N}\cdot\text{m}$$

$$\omega_B = \omega_A\left(\frac{r_A}{r_B}\right) = 3.00\pi\left(\frac{0.06}{0.15}\right) = 1.20\pi\ \text{rad/s}$$
$$P = 300\ \text{W} = 300\ \text{N}\cdot\text{m/s}$$
$$T_B = \frac{P}{\omega_B} = \frac{300}{1.20\pi} = 79.58\ \text{N}\cdot\text{m}$$

Allowable Shear Stress : For shaft A

$$\tau_{\max} = \tau_{\text{allow}} = \frac{T_A\,c}{J}$$
$$85(10^6) = \frac{31.83\left(\frac{d_A}{2}\right)}{\frac{\pi}{2}\left(\frac{d_A}{2}\right)^4}$$

$$d_A = 0.01240\ \text{m} = 12.4\ \text{mm} \qquad \textbf{Ans}$$

For shaft B

$$\tau_{\max} = \tau_{\text{allow}} = \frac{T_B\,c}{J}$$
$$85(10^6) = \frac{79.58\left(\frac{d_B}{2}\right)}{\frac{\pi}{2}\left(\frac{d_B}{2}\right)^4}$$

$$d_B = 0.01683\ \text{m} = 16.8\ \text{mm} \qquad \textbf{Ans}$$

5–35. The 4-hp motor turns the shaft AC at a rate of 1300 rev/min. Determine the maximum shear stress developed in the shaft within region AB, where it has a diameter of 0.5 in., and within region BC, where it has a diameter of 0.65 in.

Internal Torque :

$$\omega = 1300 \frac{\text{rev}}{\text{min}}\left(\frac{2\pi \text{ rad}}{\text{rev}}\right)\frac{1 \text{ min}}{60 \text{ s}} = 43.33\pi \text{ rad/s}$$

$$P = 4 \text{ hp}\left(\frac{550 \text{ ft}\cdot\text{lb/s}}{1 \text{ hp}}\right) = 2200 \text{ ft}\cdot\text{lb/s}$$

$$T = \frac{P}{\omega} = \frac{2200}{43.33\pi} = 16.16 \text{ lb}\cdot\text{ft}$$

Maximum Shear Stress : Applying torsion formula

$$(\tau_{AB})_{\max} = \frac{16.16(12)(0.25)}{\frac{\pi}{2}(0.25^4)} = 7.90 \text{ ksi} \qquad \textbf{Ans}$$

$$(\tau_{BC})_{\max} = \frac{16.16(12)(0.325)}{\frac{\pi}{2}(0.325^4)} = 3.60 \text{ ksi} \qquad \textbf{Ans}$$

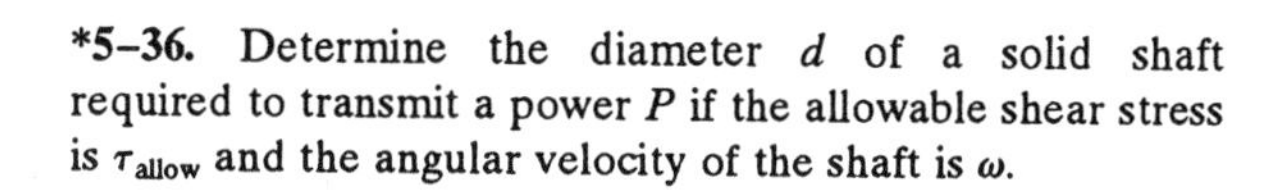

***5–36.** Determine the diameter d of a solid shaft required to transmit a power P if the allowable shear stress is τ_{allow} and the angular velocity of the shaft is ω.

Internal Torque :

$$T = \frac{P}{\omega}$$

Allowable Shear Stress : Applying torsion formula

$$\tau_{\max} = \tau_{\text{allow}} = \frac{T c}{J}$$

$$\tau_{\text{allow}} = \frac{\frac{P}{\omega}\left(\frac{d}{2}\right)}{\frac{\pi}{2}\left(\frac{d}{2}\right)^4}$$

$$\tau_{\text{allow}} = \frac{16P}{\pi\,\omega\,d^3}$$

$$d = \left(\frac{16P}{\pi\,\omega\,\tau_{\text{allow}}}\right)^{\frac{1}{3}} \qquad \textbf{Ans}$$

5–37. The assembly consists of a solid 15-mm-diameter rod connected to the inside of a tube using a rigid disk at B. Determine the absolute maximum shear stress in the rod and in the tube. The tube has an outer diameter of 30 mm and a wall thickness of 3 mm

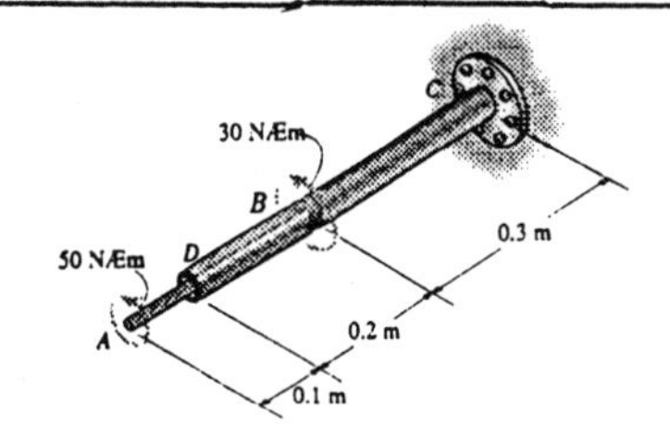

Internal Torque : As shown on FBD.
Maximum Shear Stress : Applying torsion formula

$$(\tau_r)_{\max} = \frac{(T_r)_{\max}\,c}{J}$$

$$= \frac{50.0(0.0075)}{\frac{\pi}{2}(0.0075^4)} = 75.5 \text{ MPa} \qquad \textbf{Ans}$$

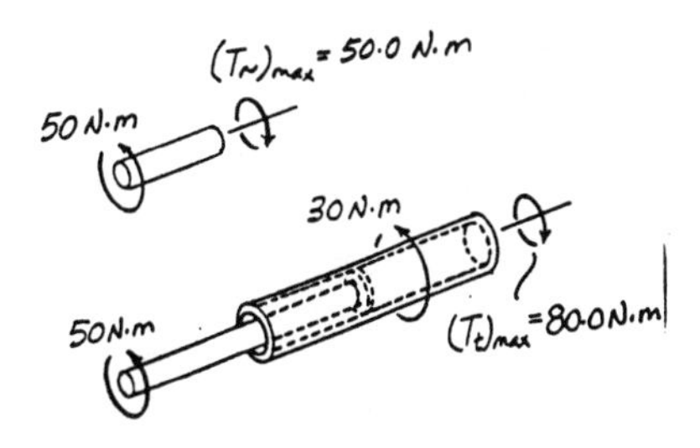

$$(\tau_t)_{\max} = \frac{(T_t)_{\max}\,c}{J}$$

$$= \frac{80.0(0.015)}{\frac{\pi}{2}(0.015^4 - 0.012^4)} = 25.6 \text{ MPa} \qquad \textbf{Ans}$$

5–38. The drive shaft of an automobile is made of a steel tube having an allowable shear stress of $\tau_{allow} = 8$ ksi. If the outer diameter is 2.5 in. and the engine delivers 200 hp to the shaft when it is turning at 1140 rev/min, determine the minimum required thickness of the shaft wall.

Internal Torque :

$$\omega = 1140 \frac{\text{rev}}{\text{min}}\left(\frac{2\pi \text{ rad}}{\text{rev}}\right)\frac{1 \text{ min}}{60 \text{ s}} = 38.0\pi \text{ rad/s}$$

$$P = 200 \text{ hp}\left(\frac{550 \text{ ft}\cdot\text{lb/s}}{1 \text{ hp}}\right) = 110\,000 \text{ ft}\cdot\text{lb/s}$$

$$T = \frac{P}{\omega} = \frac{110\,000}{38.0\pi} = 921.42 \text{ lb}\cdot\text{ft}$$

Allowable Shear Stress : Applying torsion formula

$$\tau_{max} = \tau_{allow} = \frac{Tc}{J}$$

$$8(10^3) = \frac{921.42(12)(1.25)}{\frac{\pi}{2}(1.25^4 - r_i^4)}$$

$$r_i = 1.0762 \text{ in.}$$

$$t = r_o - r_i = 1.25 - 1.0762 = 0.174 \text{ in.} \qquad \textbf{Ans}$$

5–39. The drive shaft of an automobile is to be designed as a thin-walled tube. The engine delivers 150 hp when the shaft is turning at 1500 rev/min. Determine the minimum thickness of the tube's wall if the outer diameter is 2.5 in. The material has an allowable shear stress of $\tau_{allow} = 7$ ksi.

Internal Torque :

$$\omega = 1500 \frac{\text{rev}}{\text{min}}\left(\frac{2\pi \text{ rad}}{\text{rev}}\right)\frac{1 \text{ min}}{60 \text{ s}} = 50.0\,\pi \text{ rad/s}$$

$$P = 150 \text{ hp}\left(\frac{550 \text{ ft}\cdot\text{lb/s}}{1 \text{ hp}}\right) = 82\,500 \text{ ft}\cdot\text{lb/s}$$

$$T = \frac{P}{\omega} = \frac{82\,500}{50.0\pi} = 525.21 \text{ lb}\cdot\text{ft}$$

Allowable Shear Stress : Applying torsion formula

$$\tau_{max} = \tau_{allow} = \frac{Tc}{J}$$

$$7(10^3) = \frac{525.21(12)(1.25)}{\frac{\pi}{2}(1.25^4 - r_i^4)}$$

$$r_i = 1.1460 \text{ in.}$$

$$t = r_o - r_i = 1.25 - 1.1460 = 0.104 \text{ in.} \qquad \textbf{Ans}$$

***5–40.** A motor delivers 500 hp to the steel shaft AB, which is tubular and has an outer diameter of 2 in. and an inner diameter of 1.84 in. Determine the *smallest* angular velocity at which it can rotate if the allowable shear stress for the material is $\tau_{allow} = 25$ ksi.

Internal Torque :

$$P = 500\text{ hp}\left(\frac{550\text{ ft}\cdot\text{lb/s}}{1\text{ hp}}\right) = 275\ 000\text{ ft}\cdot\text{lb/s}$$

$$T = \frac{P}{\omega} = \frac{275\ 000}{\omega}$$

Allowable Shear Stress : Applying torsion formula

$$\tau_{max} = \tau_{allow} = \frac{Tc}{J}$$

$$25(10^3) = \frac{\left(\frac{275\ 000}{\omega}\right)(12)(1)}{\frac{\pi}{2}(1^4 - 0.92^4)}$$

$$\omega = 296\text{ rad/s} \qquad \textbf{Ans}$$

5–41. A cylindrical spring consists of a rubber annulus bonded to a rigid ring and shaft. If the ring is held fixed and a torque **T** is applied to the shaft, determine the maximum shear stress in the rubber.

Equilibrium :

$$\circlearrowleft + \Sigma M_O = 0; \qquad T - F(r) = 0 \qquad F = \frac{T}{r}$$

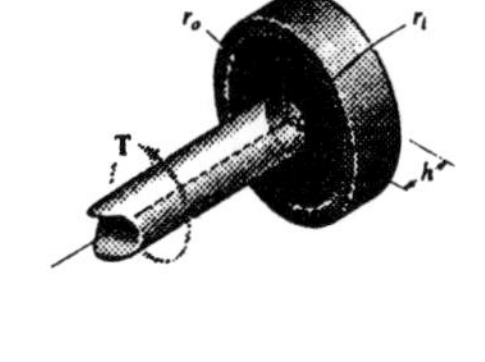

Maximum Shear Stress :

$$\tau_{avg} = \frac{F}{A} = \frac{\frac{T}{r}}{2\pi r h} = \frac{T}{2\pi r^2 h}$$

Shear stress is maximum when r is the smallest, i.e. $r = r_i$. Hence,

$$(\tau_{avg})_{max} = \frac{T}{2\pi r_i^2 h} \qquad \textbf{Ans}$$

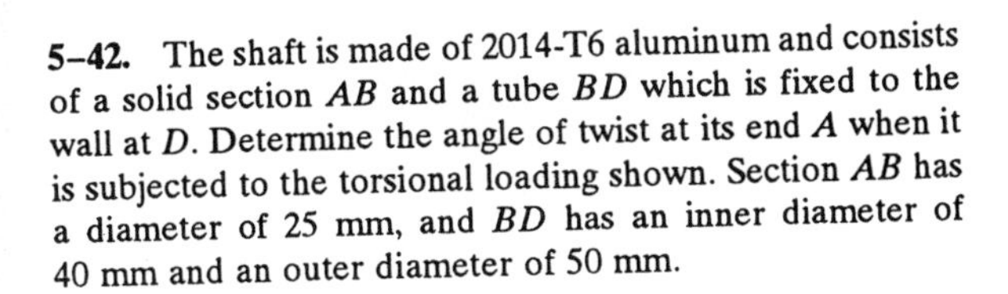

5–42. The shaft is made of 2014-T6 aluminum and consists of a solid section AB and a tube BD which is fixed to the wall at D. Determine the angle of twist at its end A when it is subjected to the torsional loading shown. Section AB has a diameter of 25 mm, and BD has an inner diameter of 40 mm and an outer diameter of 50 mm.

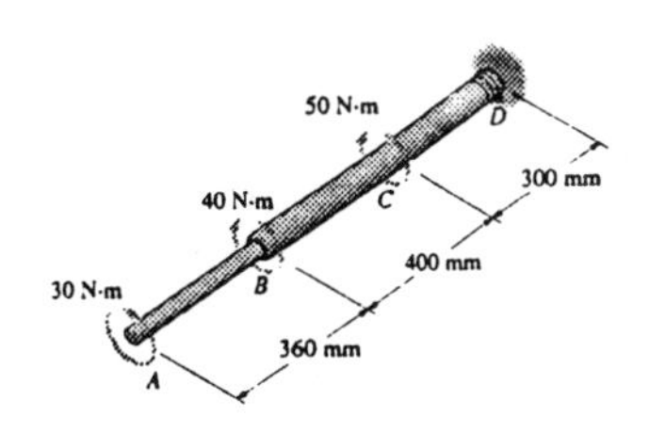

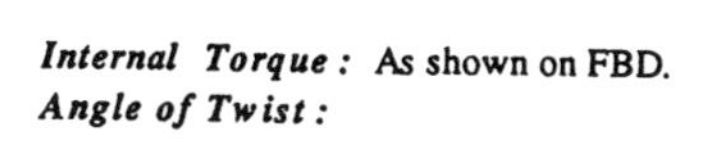

Internal Torque : As shown on FBD.
Angle of Twist :

$$\phi_A = \sum\frac{TL}{JG} = \frac{30.0(0.36)}{\frac{\pi}{2}(0.0125^4)\,27.0(10^9)} + \frac{-10.0(0.40)}{\frac{\pi}{2}(0.025^4 - 0.02^4)\,27.0(10^9)} + \frac{-60.0(0.30)}{\frac{\pi}{2}(0.025^4 - 0.02^4)\,27.0(10^9)}$$

$$= 0.008181\text{ rad} = 0.469° \qquad \textbf{Ans}$$

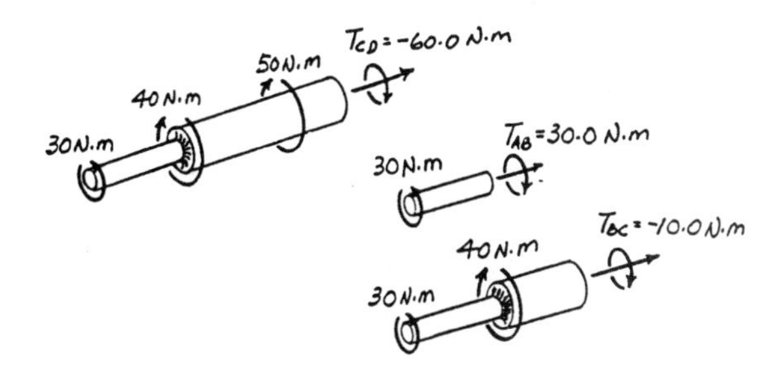

5–43. The shaft is made of 2014-T6 aluminum and consists of a solid section AB and a tube BD which is fixed to the wall at D. Determine the angle of twist at its end A relative to C when it is subjected to the torsional loading shown. Section AB has a diameter of 25 mm, and BD has an inner diameter of 40 mm and an outer diameter of 50 mm.

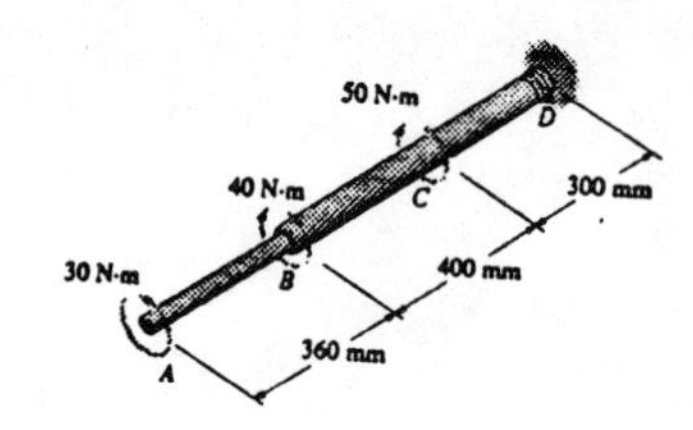

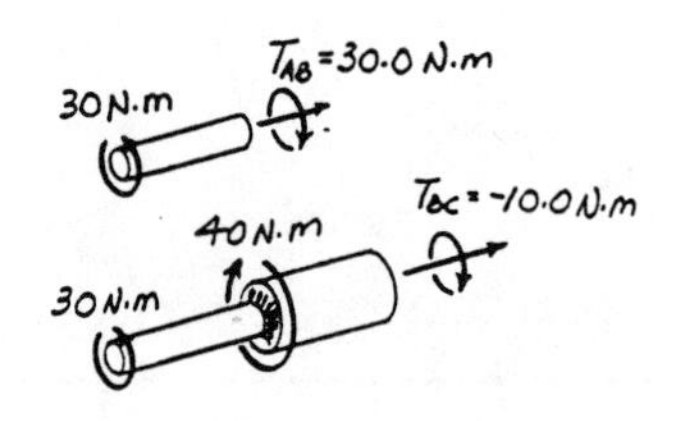

Internal Torque: As shown on FBD.

Angle of Twist:

$$\phi_{A/C} = \sum \frac{TL}{JG} = \frac{30.0(0.36)}{\frac{\pi}{2}(0.0125^4)\,27.0(10^9)} + \frac{-10.0(0.40)}{\frac{\pi}{2}(0.025^4 - 0.02^4)\,27.0(10^9)}$$

$$= 0.01002 \text{ rad} = 0.574° \qquad \textbf{Ans}$$

***5–44.** The hydrofoil boat has an A-36 steel propeller shaft that is 100 ft long. It is connected to an in-line diesel engine that delivers a maximum power of 2500 hp and causes the shaft to rotate at 1700 rpm. If the outer diameter of the shaft is 8 in. and the wall thickness is $\frac{3}{8}$ in., determine the maximum shear stress developed in the shaft. Also, what is the "wind up" or angle of twist in the shaft at full power?

Internal Torque :

$$\omega = 1700\,\frac{\text{rev}}{\text{min}}\left(\frac{2\pi\ \text{rad}}{\text{rev}}\right)\frac{1\ \text{min}}{60\ \text{s}} = 56.67\pi\ \text{rad/s}$$

$$P = 2500\ \text{hp}\left(\frac{550\ \text{ft}\cdot\text{lb/s}}{1\ \text{hp}}\right) = 1\,375\,000\ \text{ft}\cdot\text{lb/s}$$

$$T = \frac{P}{\omega} = \frac{1\,375\,000}{56.67\pi} = 7723.7\ \text{lb}\cdot\text{ft}$$

Maximum Shear Stress : Applying torsion Formula.

$$\tau_{\max} = \frac{Tc}{J}$$

$$= \frac{7723.7(12)(4)}{\frac{\pi}{2}(4^4 - 3.625^4)} = 2.83\ \text{ksi} \qquad \textbf{Ans}$$

Angle of Twist :

$$\phi = \frac{TL}{JG} = \frac{7723.7(12)(100)(12)}{\frac{\pi}{2}(4^4 - 3.625^4)11.0(10^6)}$$

$$= 0.07725\ \text{rad} = 4.43^\circ \qquad \textbf{Ans}$$

5–45. The tubular drive shaft for the propeller of a hovercraft is 6 m long. If the motor delivers 4 MW of power to the shaft when the propellers rotate at 25 rad/s, determine the required inner diameter of the shaft if the outer diameter is 250 mm. What is the angle of twist of the shaft when it is operating? Take $\tau_{\text{allow}} = 90$ MPa and $G = 75$ GPa.

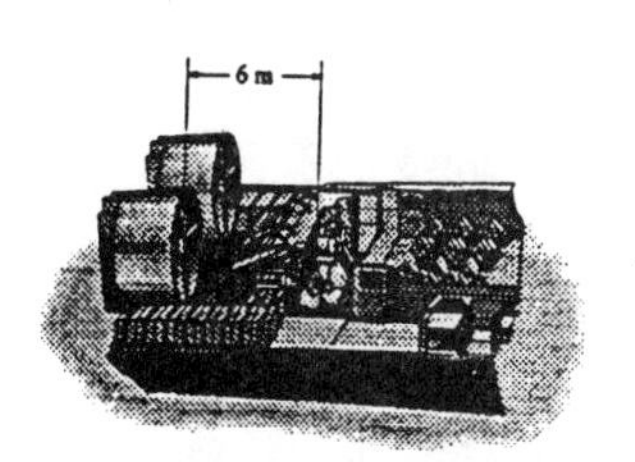

Internal Torque :

$$P = 4(10^6)\ \text{W} = 4(10^6)\ \text{N}\cdot\text{m/s}$$

$$T = \frac{P}{\omega} = \frac{4(10^6)}{25} = 160(10^3)\ \text{N}\cdot\text{m}$$

Maximum Shear Stress : Applying torsion Formula.

$$\tau_{\max} = \tau_{\text{allow}} = \frac{Tc}{J}$$

$$90(10^6) = \frac{160(10^3)(0.125)}{\frac{\pi}{2}\left[0.125^4 - \left(\frac{d_i}{2}\right)^4\right]}$$

$$d_i = 0.2013\ \text{m} = 201\ \text{mm} \qquad \textbf{Ans}$$

Angle of Twist :

$$\phi = \frac{TL}{JG} = \frac{160(10^3)(6)}{\frac{\pi}{2}(0.125^4 - 0.10065^4)75(10^9)}$$

$$= 0.0576\ \text{rad} = 3.30^\circ \qquad \textbf{Ans}$$

5–46. The A-36 steel axle is made from tubes AB and CD and a solid section BC. It is supported on smooth bearings that allow it to rotate freely. If the ends are subjected to 85 N · m torques, determine the angle of twist of end A relative to end D. The tubes have an outer diameter of 30 mm and an inner diameter of 20 mm. The solid section has a diameter of 40 mm.

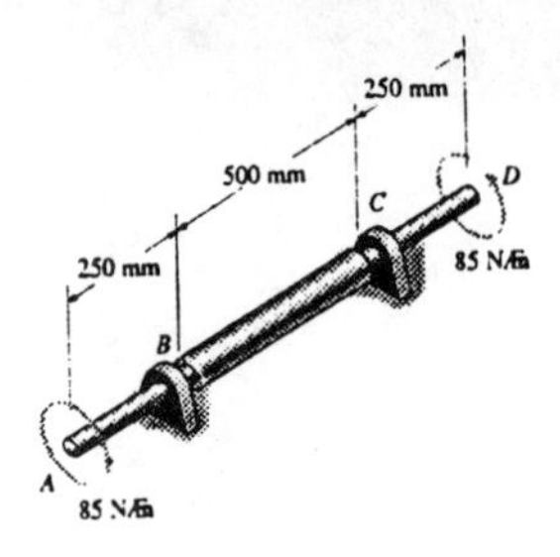

Internal Torque : As shown on FBD.

Angle of Twist :

$$\phi_{A/D} = \sum \frac{TL}{JG} = \left[\frac{-85.0(0.25)}{\frac{\pi}{2}(0.015^4 - 0.01^4)75.0(10^9)}\right](2) + \frac{-85.0(0.5)}{\frac{\pi}{2}(0.02^4)75.0(10^9)}$$

$$= -0.01113 \text{ rad} = |\ 0.638°\ | \qquad \textbf{Ans}$$

5–47. The A-36 steel axle is made from tubes AB and CD and a solid section BC. It is supported on smooth bearings that allow it to rotate freely. If ends A and D are subjected to 85 N · m torques, determine the angle of twist of end B of the solid section relative to end C. The tubes have an outer diameter of 30 mm and an inner diameter of 20 mm. The solid section has a diameter of 40 mm.

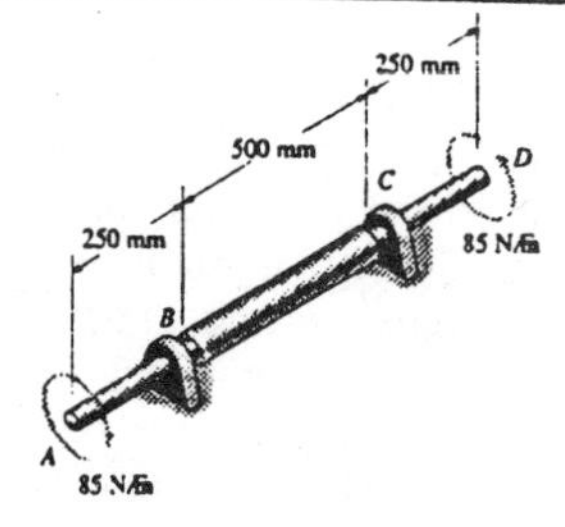

Internal Torque : As shown on FBD.

Angle of Twist :

$$\phi_{B/C} = \frac{T_{BC}L_{BC}}{JG} = \frac{-85.0(0.5)}{\frac{\pi}{2}(0.02^4)75.0(10^9)}$$

$$= -002255 \text{ rad} = |\ 0.129°\ | \qquad \textbf{Ans}$$

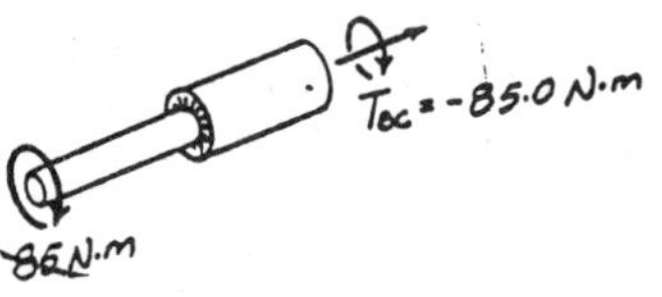

***5–48.** The gears attached to the 304 stainless steel shaft are subjected to the torques shown. Determine the angle of twist of end A with respect to end B. The shaft has a diameter of 1.5 in.

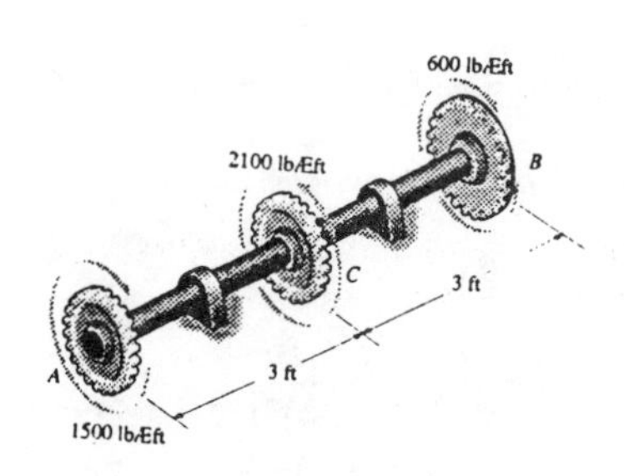

Internal Torque : As shown on FBD.

Angle of Twist :

$$\phi_{A/B} = \sum \frac{TL}{JG} = \frac{-1500(12)(3)(12)}{\frac{\pi}{2}(0.75^4)(11.0)(10^6)} + \frac{600(12)(3)(12)}{\frac{\pi}{2}(0.75^4)(11.0)(10^6)}$$

$$= -0.07112 \text{ rad} = |4.07°| \qquad \textbf{Ans}$$

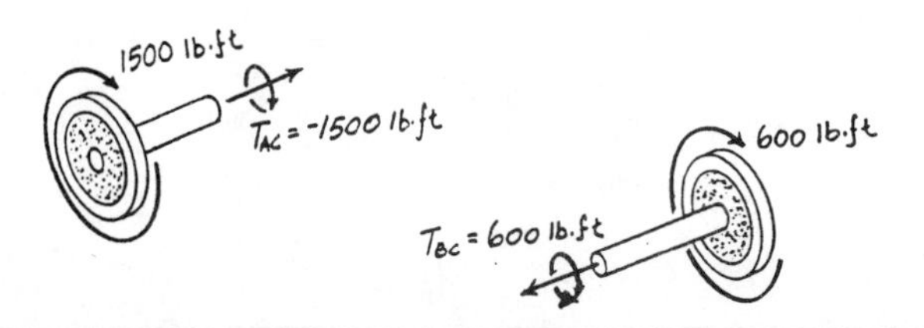

5–49. The gears attached to the 304 stainless steel shaft are subjected to the torques shown. Determine the angle of twist of gear C with respect to gear B. The shaft has a diameter of 1.5 in.

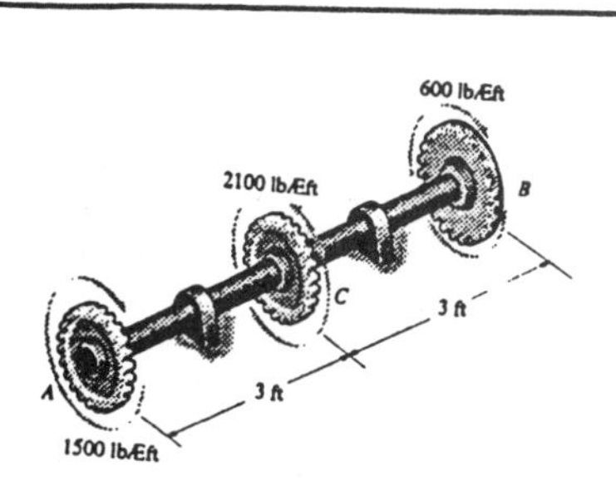

Internal Torque : As shown on FBD.

Angle of Twist :

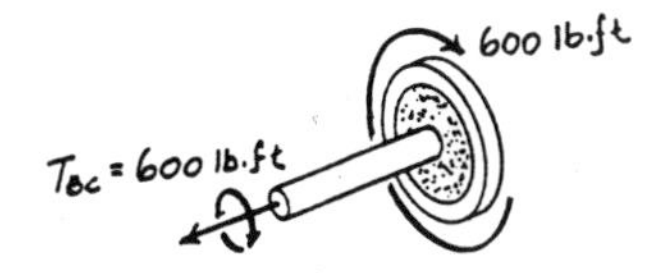

$$\phi_{C/B} = \frac{T_{BC}L_{BC}}{JG} = \frac{600(12)(3)(12)}{\frac{\pi}{2}(0.75^4)(11.0)(10^6)}$$

$$= 0.04741 \text{ rad} = 2.72° \qquad \textbf{Ans}$$

5–50. The 20-mm-diameter A-36 steel shaft is subjected to the torques shown. Determine the angle of twist of the end B.

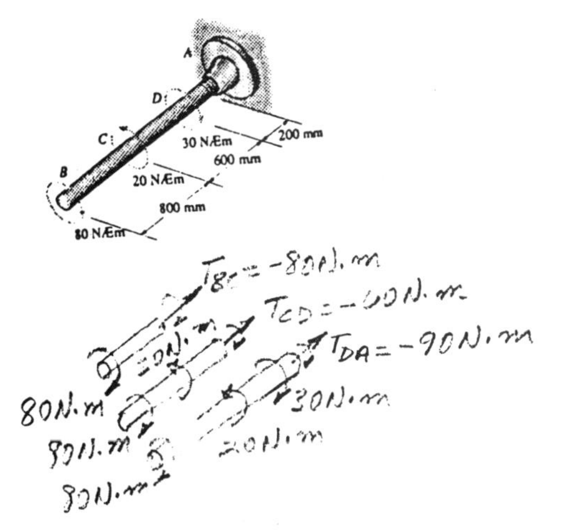

Internal Torque : As shown on FBD.

Angle of Twist :

$$\phi_B = \sum \frac{TL}{JG}$$

$$= \frac{1}{\frac{\pi}{2}(0.01^4)(75.0)(10^9)}[-80.0(0.8) + (-60.0)(0.6) + (-90.0)(0.2)]$$

$$= -0.1002 \text{ rad} = |\ 5.74°\ | \qquad \textbf{Ans}$$

5–51. The motor delivers 40 hp to the 304 stainless steel shaft while it rotates at 20 Hz. The shaft is supported on smooth bearings at A and B, which allow free rotation of the shaft. The gears C and D fixed to the shaft remove 25 hp and 15 hp, respectively. Determine the diameter of the shaft to the nearest $\frac{1}{8}$ in. if the allowable shear stress is τ_{allow} = 8 ksi and the allowable angle of twist of C with respect to D is 0.20°.

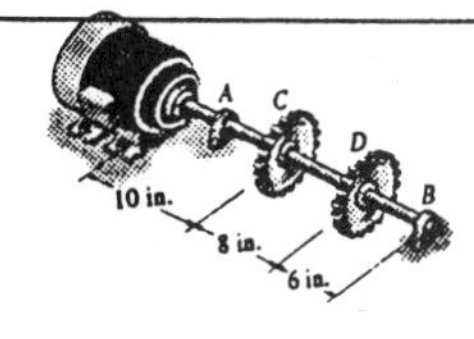

External Applied Torque : Applying $T = \frac{P}{2\pi f}$, we have

$$T_M = \frac{40(550)}{2\pi(20)} = 175.07 \text{ lb}\cdot\text{ft} \qquad T_C = \frac{25(550)}{2\pi(20)} = 109.42 \text{ lb}\cdot\text{ft}$$

$$T_D = \frac{15(550)}{2\pi(20)} = 65.65 \text{ lb}\cdot\text{ft}$$

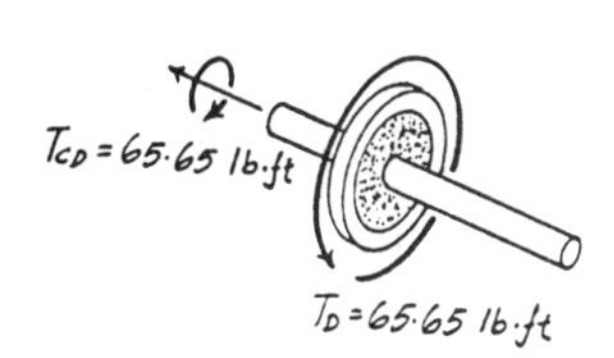

T_{CD} = 65.65 lb·ft

T_D = 65.65 lb·ft

Internal Torque : As shown on FBD.

Allowable Shear Stress : Assume failure due to shear stress. By observation, section AC is the critical region.

$$\tau_{max} = \tau_{allow} = \frac{Tc}{J}$$

$$8(10^3) = \frac{175.07(12)\left(\frac{d}{2}\right)}{\frac{\pi}{2}\left(\frac{d}{2}\right)^4}$$

$$d = 1.102 \text{ in.}$$

Angle of Twist : Assume failure due to angle of twist limitation.

$$\phi_{C/D} = \frac{T_{CD}L_{CD}}{JG}$$

$$\frac{0.2(\pi)}{180} = \frac{65.65(12)(8)}{\frac{\pi}{2}\left(\frac{d}{2}\right)^4(11.0)(10^6)}$$

$d = 1.137$ in. (***controls !***)

$$\text{Use } d = 1\frac{1}{4} \text{ in.} \qquad \textbf{Ans}$$

***5–52.** The motor delivers 40 hp to the 304 stainless steel solid shaft while it rotates at 20 Hz. The shaft has a diameter of 1.5 in. and is supported on smooth bearings at A and B, which allow free rotation of the shaft. The gears C and D fixed to the shaft remove 25 hp and 15 hp, respectively. Determine the absolute maximum stress in the shaft and the angle of twist of gear C with respect to gear D.

External Applied Torque : Applying $T = \dfrac{P}{2\pi f}$, we have

$$T_M = \frac{40(550)}{2\pi(20)} = 175.07\ \text{lb}\cdot\text{ft} \qquad T_C = \frac{25(550)}{2\pi(20)} = 109.42\ \text{lb}\cdot\text{ft}$$

$$T_D = \frac{15(550)}{2\pi(20)} = 65.65\ \text{lb}\cdot\text{ft}$$

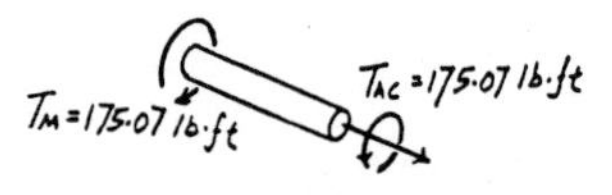

Internal Torque : As shown on FBD.

Allowable Shear Stress : The maximum torque occurs within region AC of the shaft where $T_{max} = T_{AC} = 175.07$ lb · ft.

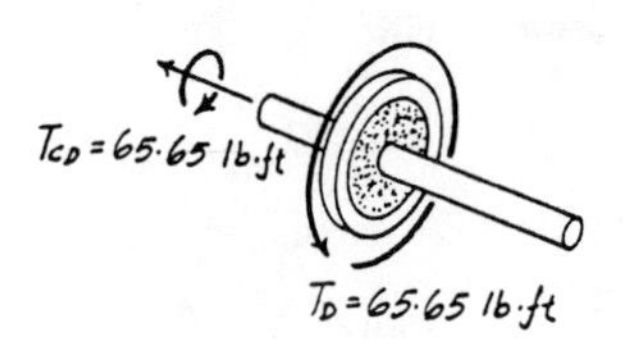

$$\tau_{abs\ max} = \frac{T_{max}c}{J} = \frac{175.07(12)(0.75)}{\frac{\pi}{2}(0.75^4)} = 3.17\ \text{ksi} \qquad \textbf{Ans}$$

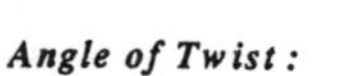

Angle of Twist :

$$\phi_{C/D} = \frac{T_{CD}L_{CD}}{JG}$$

$$= \frac{65.65(12)(8)}{\frac{\pi}{2}(0.75^4)(11.0)(10^6)}$$

$$= 0.001153\ \text{rad} = 0.0661^\circ \qquad \textbf{Ans}$$

5–53. The A-36 steel shaft rotates at $\omega = 1000$ rev/min and transmits the power shown. Determine the absolute maximum shear stress in the shaft and the angle of twist of C with respect to F if the inner and outer diameters are $d_i = 40$ mm and $d_o = 50$ mm. ,respectively.

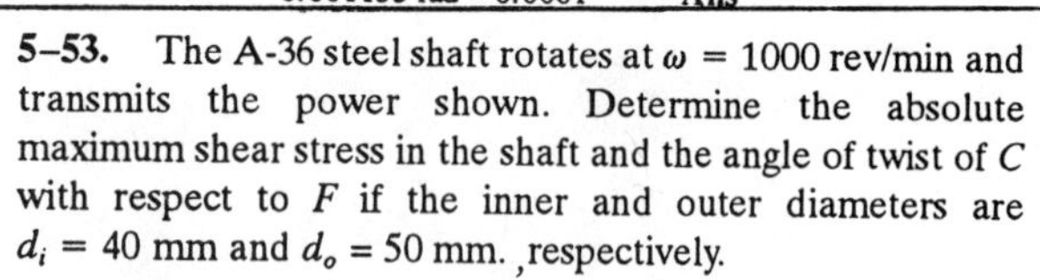

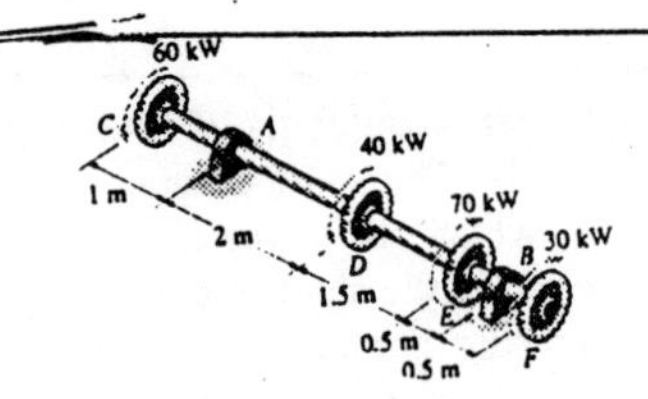

External Applied Torque : Applying $T = \dfrac{P}{\omega}$ where

$$\omega = \frac{1000(2\pi)}{60} = 33.33\pi\ \text{rad/s, we have}$$

$$T_C = \frac{60(10^3)}{33.33\pi} = 572.96\ \text{N}\cdot\text{m} \qquad T_F = \frac{30(10^3)}{33.33\pi} = 286.48\ \text{N}\cdot\text{m}$$

$$T_E = \frac{70(10^3)}{33.33\pi} = 668.45\ \text{N}\cdot\text{m}$$

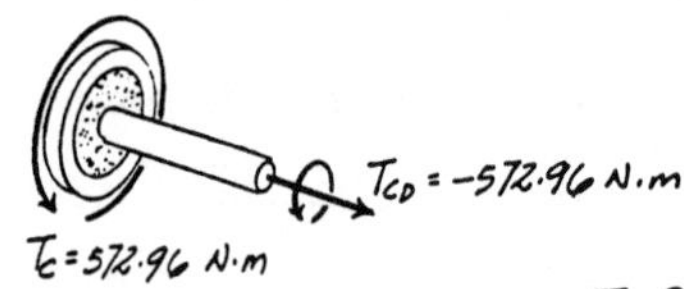

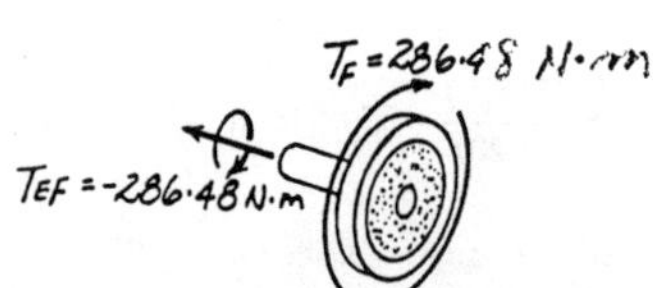

Internal Torque : As shown on FBD.

Maximum Shear Stress : The maximum torque occurs within region DE of the shaft where $T_{max} = T_{DE} = 954.93$ N · m

$$\tau_{abs\ max} = \frac{T_{max}c}{J} = \frac{954.93(0.025)}{\frac{\pi}{2}(0.025^4 - 0.02^4)} = 65.9\ \text{MPa} \qquad \textbf{Ans}$$

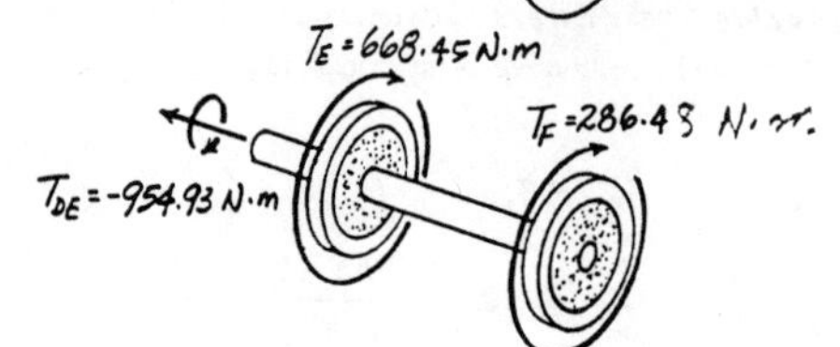

Angle of Twist :

$$\phi_{C/F} = \sum \frac{TL}{JG}$$

$$= \frac{1}{\frac{\pi}{2}(0.025^4 - 0.02^4)(75.0)(10^9)}[(-572.96)(3)$$

$$+ (-954.93)(1.5) + (-286.48)(1)]$$

$$= -0.1265\ \text{rad} = |\ 7.25^\circ\ | \qquad \textbf{Ans}$$

5–54. The A-36 steel shaft rotates at $\omega = 1000$ rev/min and transmits the power shown. Determine the inner and outer diameters of the shaft if $d_i/d_o = 0.8$, the allowable shear stress is $\tau_{allow} = 145$ MPa, and the allowable angle of twist is $\phi_{allow} = 1.5°/\text{m}$.

External Applied Torque : Applying $T = \dfrac{P}{\omega}$ where

$\omega = \dfrac{1000(2\pi)}{60} = 33.33\pi$ rad/s, we have

$$T_C = \frac{60(10^3)}{33.33\pi} = 572.96 \text{ N}\cdot\text{m} \qquad T_F = \frac{30(10^3)}{33.33\pi} = 286.48 \text{ N}\cdot\text{m}$$

$$T_E = \frac{70(10^3)}{33.33\pi} = 668.45 \text{ N}\cdot\text{m}$$

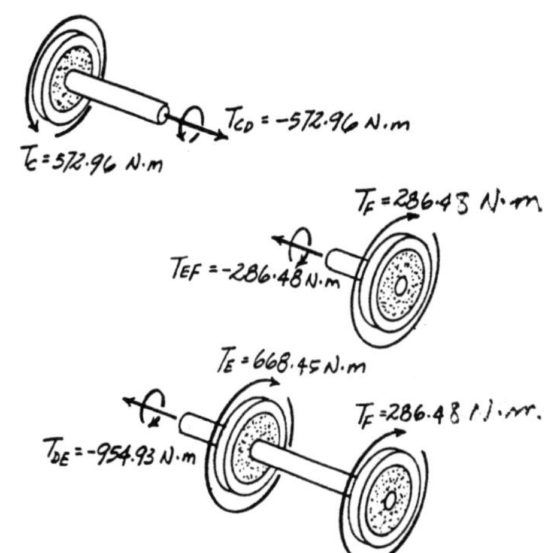

Internal Torque : As shown on FBD.

Allowable Shear Stress : Assume failure due to shear stress. By observation, section DE is the critical region.

$$\tau_{max} = \tau_{allow} = \frac{T_{DE}\, c}{J}$$

$$145(10^6) = \frac{954.93\left(\frac{d_o}{2}\right)}{\frac{\pi}{2}\left[\left(\frac{d_o}{2}\right)^4 - \left(\frac{0.8d_o}{2}\right)^4\right]}$$

$$d_o = 0.03844 \text{ m}$$

Angle of Twist : Assume failure due to angle of twist limitation. By observation $\phi_{C/F}$ is critical.

$$\phi_{C/F} = \sum \frac{TL}{JG}$$

$$-\frac{1.5(5.5)\pi}{180} = \frac{1}{\frac{\pi}{2}\left[\left(\frac{d_o}{2}\right)^4 - \left(\frac{0.8d_o}{2}\right)^4\right](75.0)(10^9)}[(-572.96)(3)$$

$$+(-954.93)(1.5) + (-286.48)(1)]$$

$$d_o = 0.04841 \text{ m} = 48.4 \text{ mm } (\textbf{\textit{controls !}}) \qquad \textbf{Ans}$$

$$d_i = 0.8d_o = 0.8(0.04841) = 0.03873 \text{ m} = 38.7 \text{ mm} \qquad \textbf{Ans}$$

5–55. The A-36 steel shaft rotates at $\omega = 125$ rad/s and transmits the power shown. Determine the absolute maximum shear stress in the shaft and the angle of twist of C with respect to F. The inner and outer diameters of the shaft are $d_i = 30$ mm and $d_o = 40$ mm. The journal bearings at B and G are smooth.

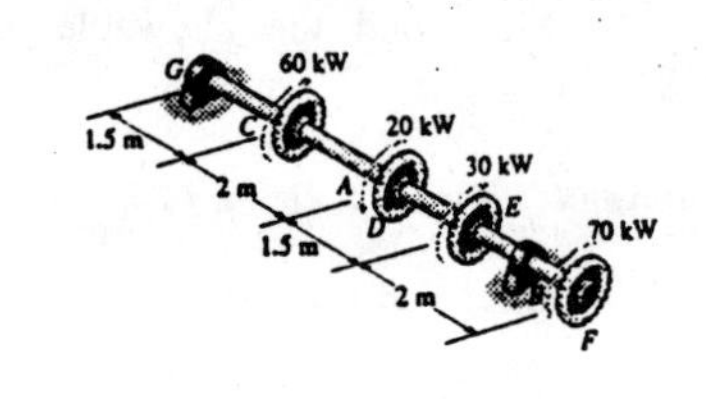

External Applied Torque : Applying $T = \frac{P}{\omega}$ where, we have

$$T_C = \frac{60(10^3)}{125} = 480 \text{ N}\cdot\text{m} \qquad T_D = \frac{20(10^3)}{125} = 160 \text{ N}\cdot\text{m}$$

$$T_F = \frac{70(10^3)}{125} = 560 \text{ N}\cdot\text{m}$$

Internal Torque : As shown on FBD.

Maximum Shear Stress : The maximum torque occurs within region $EF$of the shaft where $T_{max} = T_{EF} = 560$ N · m

$$\tau_{\substack{abs\\max}} = \frac{T_{max}\,c}{J} = \frac{560(0.02)}{\frac{\pi}{2}(0.02^4 - 0.015^4)} = 65.2 \text{ MPa} \qquad \textbf{Ans}$$

Angle of Twist :

$$\phi_{C/F} = \sum \frac{TL}{JG}$$

$$= \frac{1}{\frac{\pi}{2}(0.02^4 - 0.015^4)(75.0)(10^9)}[(480)(2) + (320)(1.5) + (560)(2)]$$

$$= 0.1987 \text{ rad} = 11.4° \qquad \textbf{Ans}$$

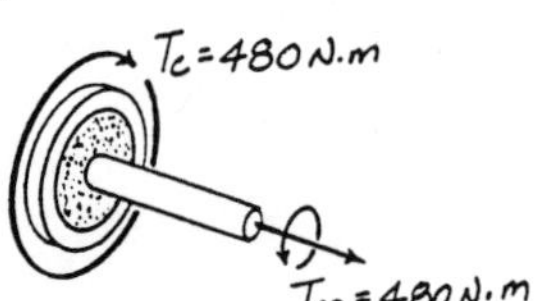

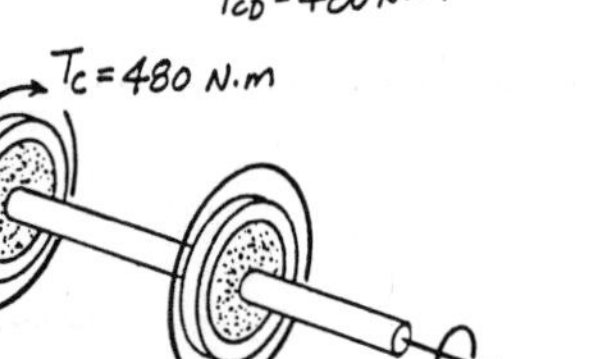

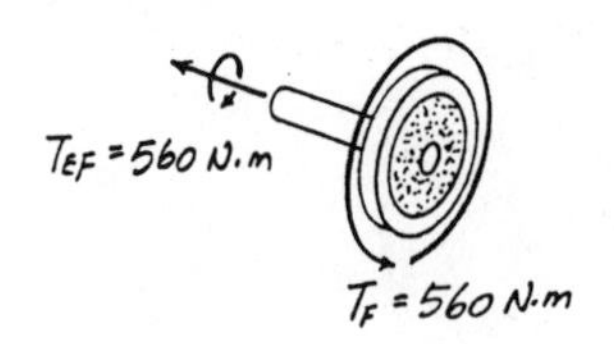

***5–56.** The A-36 steel shaft rotates at $\omega = 125$ rad/s and transmits the power shown. Determine the inner and outer diameters of the shaft if $d_i/d_o = 0.6$, the allowable shear stress is $\tau_{allow} = 150$ MPa, and the allowable relative angle of twist is $\phi_{allow} = 2°/\text{m}$. The journal bearings at B and G are smooth.

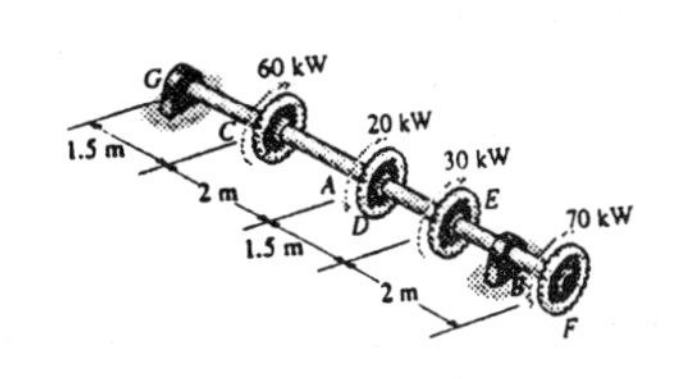

External Applied Torque : Applying $T = \frac{P}{\omega}$, we have

$$T_C = \frac{60(10^3)}{125} = 480 \text{ N}\cdot\text{m} \qquad T_D = \frac{20(10^3)}{125} = 160 \text{ N}\cdot\text{m}$$

$$T_F = \frac{70(10^3)}{125} = 560 \text{ N}\cdot\text{m}$$

Internal Torque : As shown on FBD.

Allowable Shear Stress : Assume failure due to shear stress. By observation, section EF is the critical region.

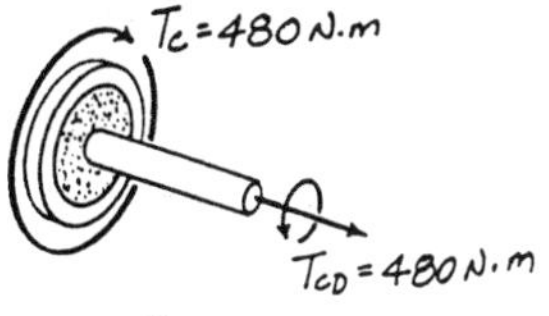

$$\tau_{max} = \tau_{allow} = \frac{T_{EF}\, c}{J}$$

$$150(10^6) = \frac{560\left(\frac{d_o}{2}\right)}{\frac{\pi}{2}\left[\left(\frac{d_o}{2}\right)^4 - \left(\frac{0.6d_o}{2}\right)^4\right]}$$

$$d_o = 0.02795 \text{ m}$$

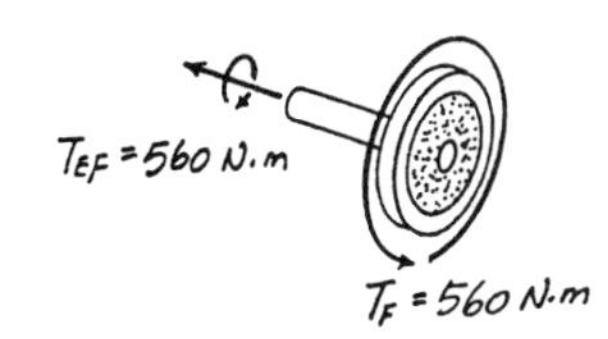

Angle of Twist : Assume failure due to angle of twist limitation. By observation $\phi_{C/F}$ is critical.

$$\phi_{C/F} = \sum \frac{TL}{JG}$$

$$\frac{2(5.5)\pi}{180°} = \frac{1}{\frac{\pi}{2}\left[\left(\frac{d_o}{2}\right)^4 - \left(\frac{0.6d_o}{2}\right)^4\right](75.0)(10^9)}[(480)(2) + (320)(1.5) + (560)(2)]$$

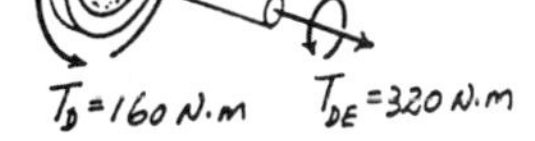

$$d_o = 0.03798 \text{ m} = 38.0 \text{ mm } (\textit{controls !}) \qquad \textbf{Ans}$$

$$d_i = 0.6d_o = 0.6(0.03798) = 0.02279 \text{ m} = 22.8 \text{ mm} \qquad \textbf{Ans}$$

5–57. The-30-mm-diameter shafts are made of L2 tool steel and are supported on journal bearings that allow the shaft to rotate freely. If the motor at A develops a torque of $T = 45$ N · m on the shaft AB, while the turbine at E is fixed from turning, determine the amount of rotation of gears B and C.

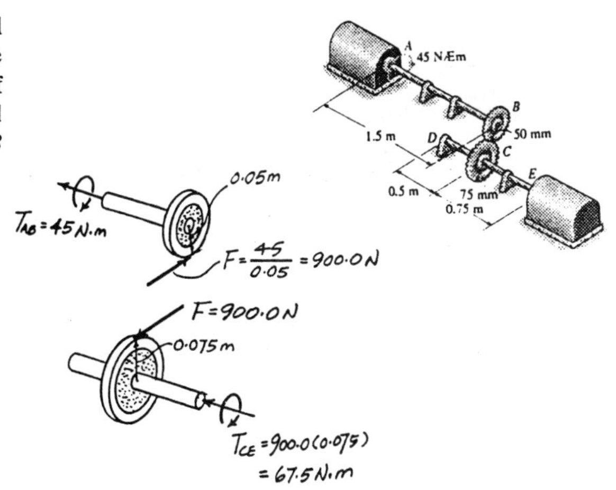

Internal Torque : As shown on FBD.

Angle of Twist :

$$\phi_B = \frac{T_{AB}L_{AB}}{JG} = \frac{45.0(1.5)}{\frac{\pi}{2}(0.015^4)75.0(10^9)}$$

$$= 0.01132 \text{ rad} = 0.648° \qquad \textbf{Ans}$$

$$\phi_C = \frac{T_{CE}L_{CE}}{JG} = \frac{67.5(0.75)}{\frac{\pi}{2}(0.015^4)75.0(10^9)}$$

$$= 0.008488 \text{ rad} = 0.486° \qquad \textbf{Ans}$$

5–58. The two shafts are made of A-36 steel. Each has a diameter of 1 in., and they are supported by bearings at A, B, and C, which allow free rotation. If the support at D is fixed, determine the angle of twist of end B when the torques are applied to the assembly as shown.

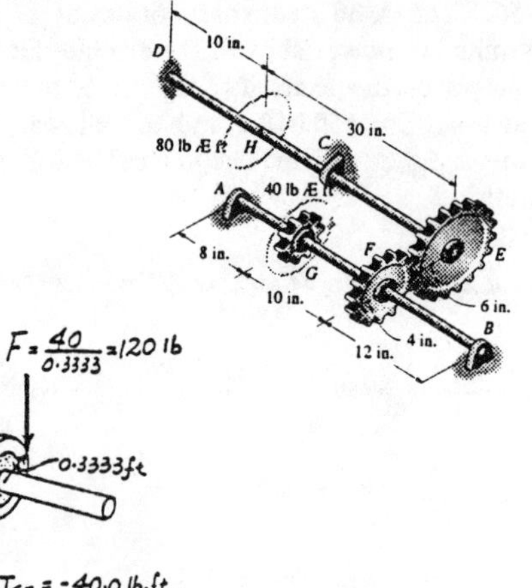

Internal Torque : As shown on FBD.

Angle of Twist :

$$\phi_E = \sum \frac{TL}{JG}$$

$$= \frac{1}{\frac{\pi}{2}(0.5^4)(11.0)(10^6)}[-60.0(12)(30) + 20.0(12)(10)]$$

$$= -0.01778 \text{ rad} = 0.01778 \text{ rad}$$

$$\phi_F = \frac{6}{4}\phi_E = \frac{6}{4}(0.01778) = 0.02667 \text{ rad}$$

Since there is no torque applied between F and B then

$$\phi_B = \phi_F = 0.02667 \text{ rad} = 1.53° \qquad \textbf{Ans}$$

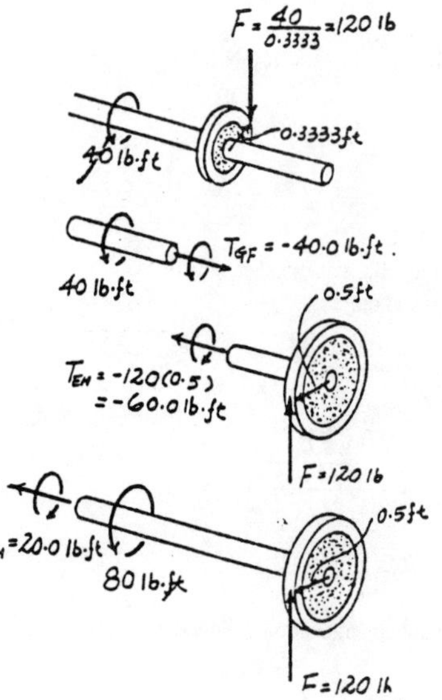

5–59. The two shafts are made of A-36 steel. Each has a diameter of 1 in., and they are supported by bearings at A, B, and C, which allow free rotation. If the support at D is fixed, determine the angle of twist of end A when the torques are applied to the assembly as shown.

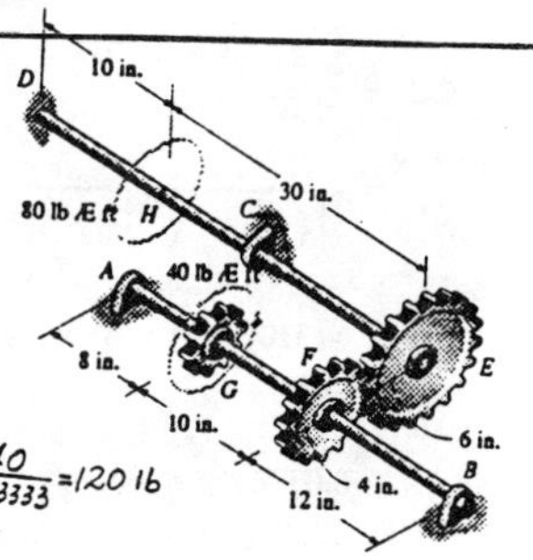

Internal Torque : As shown on FBD.

Angle of Twist :

$$\phi_E = \sum \frac{TL}{JG}$$

$$= \frac{1}{\frac{\pi}{2}(0.5^4)(11.0)(10^6)}[-60.0(12)(30) + 20.0(12)(10)]$$

$$= -0.01778 \text{ rad} = 0.01778 \text{ rad}$$

$$\phi_F = \frac{6}{4}\phi_E = \frac{6}{4}(0.01778) = 0.02667 \text{ rad}$$

$$\phi_{A/F} = \frac{T_{GF}L_{GF}}{JG}$$

$$= \frac{-40(12)(10)}{\frac{\pi}{2}(0.5^4)(11.0)(10^6)}$$

$$= -0.004445 \text{ rad} = 0.004445 \text{ rad}$$

$$\phi_A = \phi_F + \phi_{A/F}$$

$$= 0.02667 + 0.004445$$

$$= 0.03111 \text{ rad} = 1.78° \qquad \textbf{Ans}$$

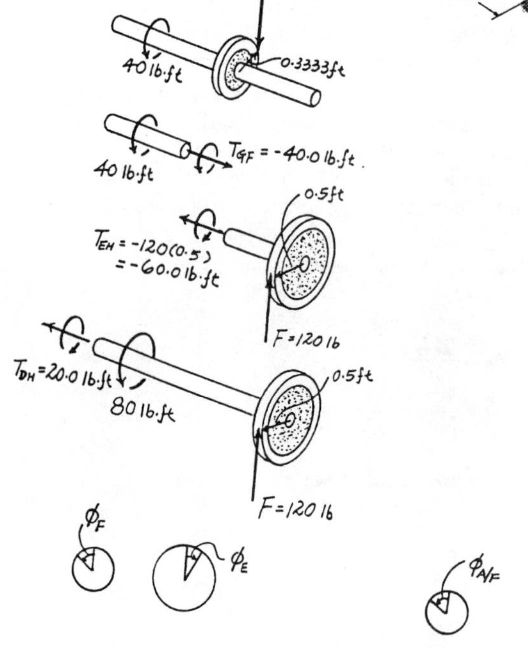

***5–60.** The two A-36 steel shafts AC and FD are coupled together using the meshed-gear arrangement shown. If A is a fixed support, determine the angle of twist at end F due to the applied torsional loading. Each shaft has a diameter of 30 mm, and the journal bearings at F, B, and G allow free rotation.

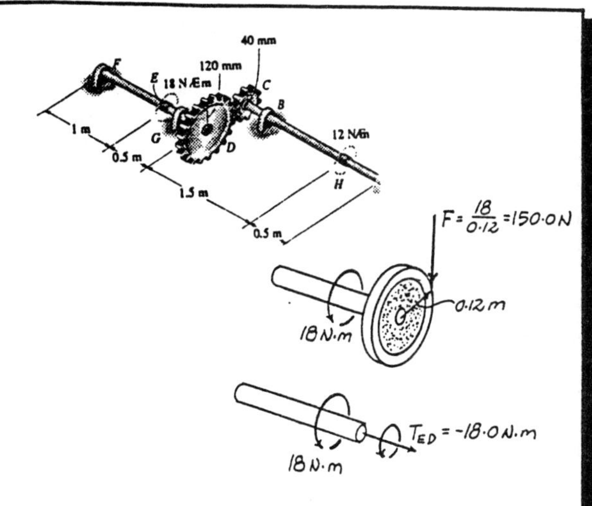

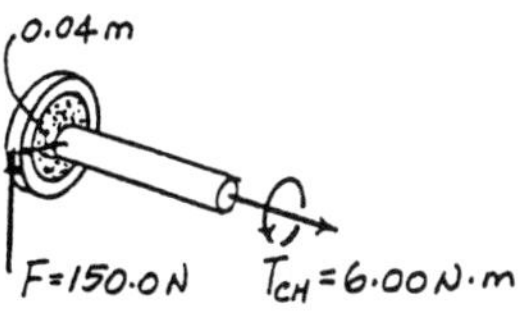

Internal Torque : As shown on FBD.
Angle of Twist :

$$\phi_C = \sum \frac{TL}{JG} = \frac{1}{JG}[6(1.5) + (-6)(0.5)] = \frac{6.00\ \text{N}\cdot\text{m}^2}{JG}$$

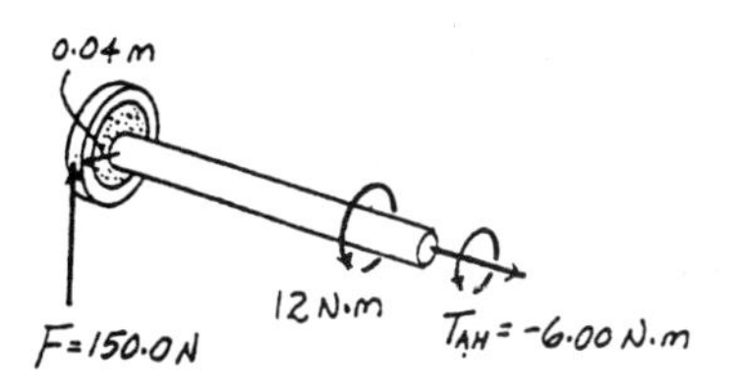

$$\phi_D = \frac{0.04}{0.12}\phi_C = \frac{2.00}{JG}$$

$$\phi_{E/D} = \frac{T_{ED}L_{ED}}{JG} = \frac{-18.0(0.5)}{JG} = \frac{-9.00\ \text{N}\cdot\text{m}^2}{JG} = \frac{9.00\ \text{N}\cdot\text{m}^2}{JG}$$

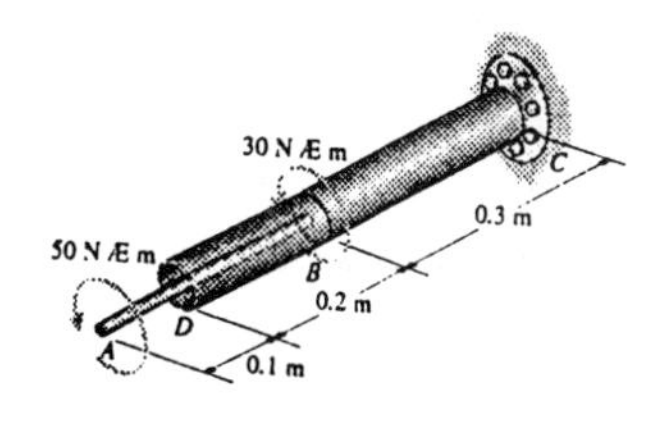

$$\phi_E = \phi_D + \phi_{E/D}$$
$$= \frac{2.00}{JG} + \frac{9.00}{JG} = \frac{11.0\ \text{N}\cdot\text{m}^2}{JG}$$

Since there is no torque between E and F, then

$$\phi_F = \phi_E = \frac{11.00}{\frac{\pi}{2}(0.015)^4(75.0)(10^9)}$$
$$= 0.001844\ \text{rad} = 0.106° \qquad \textbf{Ans}$$

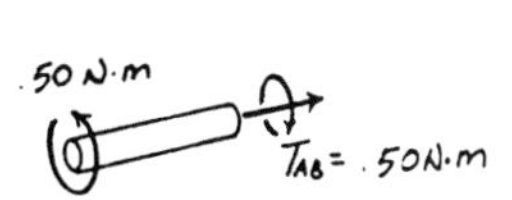

5–61. The assembly is made of A-36 steel and consists of a solid rod 15 mm in diameter connected to the inside of a tube using a rigid disk at B. Determine the angle of twist at A. The tube has an outer diameter of 30 mm and wall thickness of 3 mm.

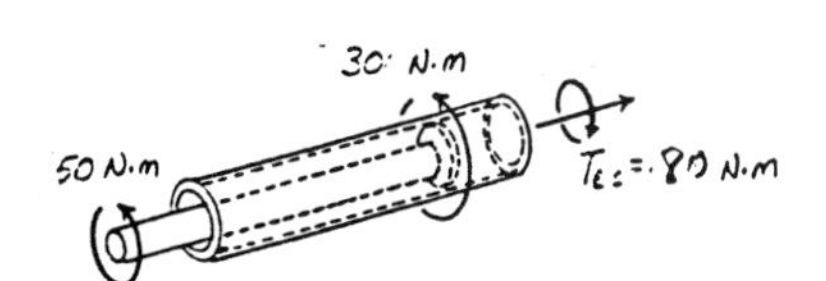

Internal Torque : As shown on FBD.
Angle of Twist :

$$\phi_A = \sum \frac{TL}{JG}$$
$$= \frac{50(0.3)}{\frac{\pi}{2}(0.0075^4)\,75.0(10^9)} + \frac{80(0.3)}{\frac{\pi}{2}(0.015^4 - 0.012^4)\,75.0(10^9)}$$

$$= 0.04706\ \text{rad} = 2.70° \qquad \textbf{Ans}$$

5–62. The assembly is made of A-36 steel and consists of a solid rod 15 mm in diameter connected to the inside tube using a rigid disk at B. Determine the angle of twist at D. The tube has an outer diameter of 30 mm and wall thickness of 3 mm.

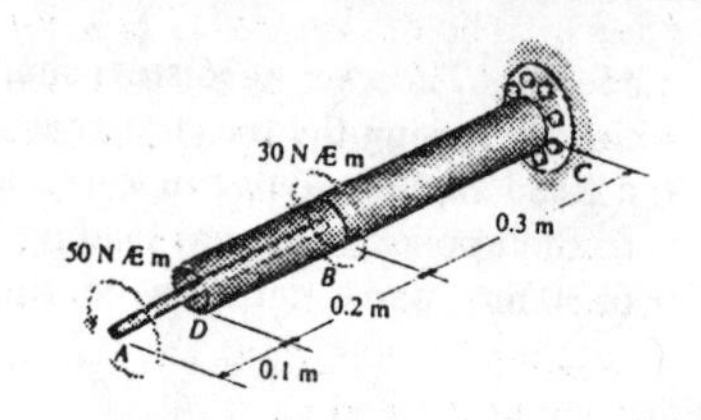

Internal Torque : As shown on FBD.

Angle of Twist : Since there is no torque between B and D, then

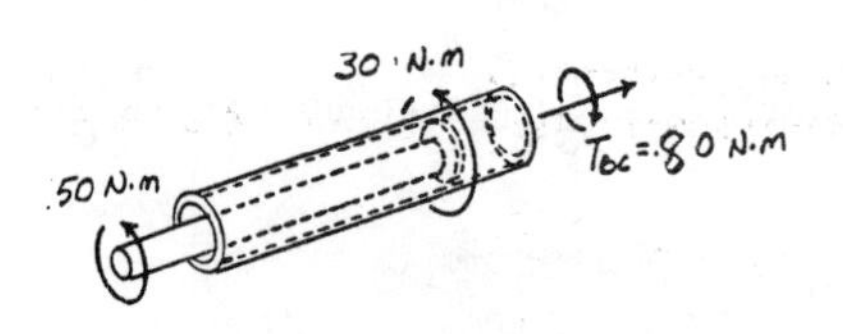

$$\phi_D = \phi_B = \frac{T_{BC}L_{BC}}{JG}$$

$$= \frac{80(0.3)}{\frac{\pi}{2}(0.015^4 - 0.012^4)\,75.0(10^9)}$$

$$= 006816 \text{ rad} = 0.391° \qquad \textbf{Ans}$$

5–63. The device serves as a compact torsional spring. It is made of A-36 steel and consists of a solid inner shaft CB which is surrounded by and attached to a tube AB using a rigid ring at B. The ring at A can also be assumed rigid and is fixed from rotating. If a torque of T = 2k · in. is applied to the shaft, determine the angle of twist at the end C and the maximum shear stress in the tube and shaft.

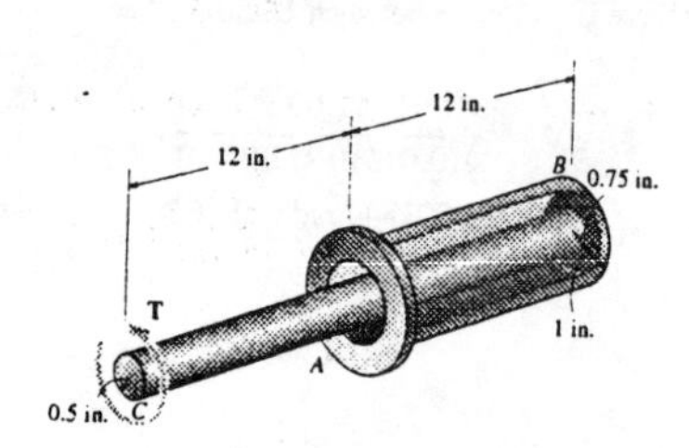

Internal Torque : As shown on FBD.

Maximum Shear Stress :

$$(\tau_{BC})_{max} = \frac{T_{BC}c}{J} = \frac{2.00(0.5)}{\frac{\pi}{2}(0.5^4)} = 10.2 \text{ ksi} \qquad \textbf{Ans}$$

$$(\tau_{BA})_{max} = \frac{T_{BA}c}{J} = \frac{2.00(1)}{\frac{\pi}{2}(1^4 - 0.75^4)} = 1.86 \text{ ksi} \qquad \textbf{Ans}$$

2 Kip·in

T_{BC} = 2.0 Kip·in

2 Kip·in

T_{BA} = -2.0 Kip·in.

Angle of Twist :

$$\phi_B = \frac{T_{BA}L_{BA}}{JG}$$

$$= \frac{(2.00)(12)}{\frac{\pi}{2}(1^4 - 0.75^4)\,11.0(10^3)} = 0.002032 \text{ rad}$$

$$\phi_{C/B} = \frac{T_{BC}L_{BC}}{JG}$$

$$= \frac{2.00(24)}{\frac{\pi}{2}(0.5^4)\,11.0(10^3)} = 0.044448 \text{ rad}$$

$$\phi_C = \phi_B + \phi_{C/B}$$

$$= 0.002032 + 0.044448$$

$$= 0.04648 \text{ rad} = 2.66° \qquad \textbf{Ans}$$

*5–64. The device serves as a compact torsion spring. It is made of A-36 steel and consists of a solid inner shaft CB which is surrounded by and attached to a tube AB using a rigid ring at B. The ring at A can also be assumed rigid and is fixed from rotating. If the allowable shear stress for the material is $\tau_{allow} = 12$ ksi and the angle of twist at C is limited to $\phi_{allow} = 3°$, determine the maximum torque T that can be applied at the end C.

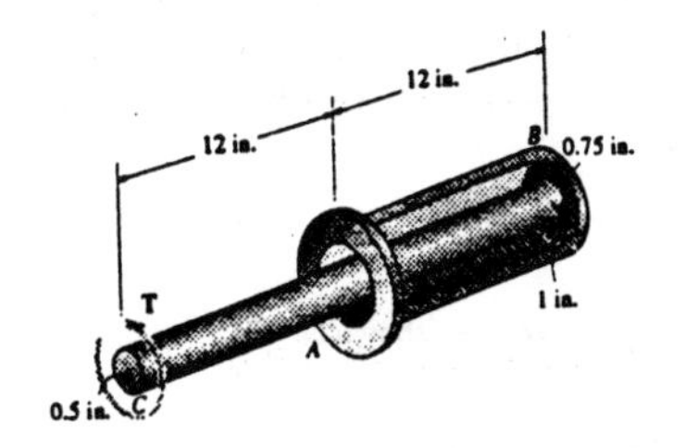

Internal Torque : As shown on FBD.

Allowable Shear Stress : Assume failure due to shear stress.

$$\tau_{max} = \tau_{allow} = \frac{T_{BC}\, c}{J}$$

$$12.0 = \frac{T\,(0.5)}{\frac{\pi}{2}(0.5^4)}$$

$$T = 2.356 \text{ kip}\cdot\text{in}$$

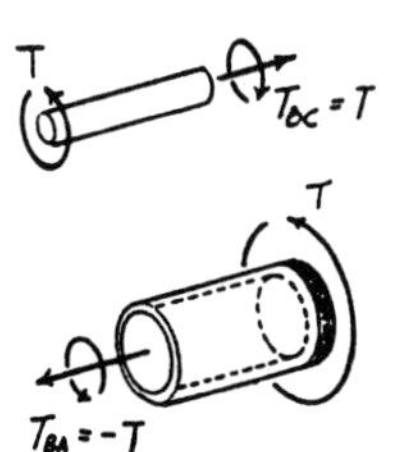

$$\tau_{max} = \tau_{allow} = \frac{T_{BA}\, c}{J}$$

$$12.0 = \frac{T\,(1)}{\frac{\pi}{2}(1^4 - 0.75^4)}$$

$$T = 12.89 \text{ kip}\cdot\text{in}$$

Angle of Twist : Assume failure due to angle of twist limitation.

$$\phi_B = \frac{T_{BA} L_{BA}}{JG} = \frac{T(12)}{\frac{\pi}{2}(1^4 - 0.75^4)\, 11.0(10^3)}$$

$$= 0.001016T$$

$$\phi_{C/B} = \frac{T_{BC} L_{BC}}{JG} = \frac{T(24)}{\frac{\pi}{2}(0.5^4)\, 11.0(10^3)}$$

$$= 0.022224T$$

$$(\phi_C)_{allow} = \phi_B + \phi_{C/B}$$

$$\frac{3(\pi)}{180} = 0.001016T + 0.022224T$$

$$T = 2.25 \text{ kip}\cdot\text{in} \quad (\textit{controls!})$$ **Ans**

5–65. The contour of the surface of the shaft is defined by the equation $y = e^{ax}$, where a is a constant. If the shaft is subjected to a torque T at its ends, determine the angle of twist of end A with respect to end B. The shear modulus is G.

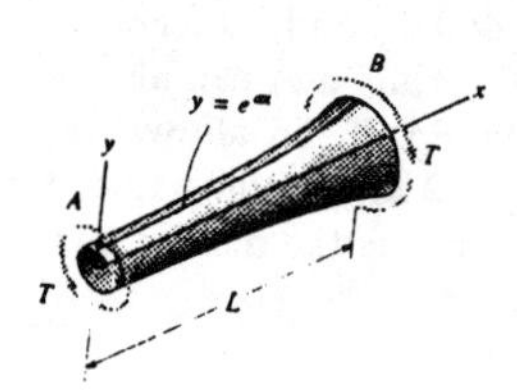

Internal Torque : As shown on FBD.

Angle of Twist : The polar moment of inertia is $J(x) = \frac{\pi}{2}(e^{ax})^4 = \frac{\pi}{2}e^{4ax}$.

$$\phi = \int_0^L \frac{T(x)\,dx}{J(x)G}$$
$$= \frac{2T}{\pi G}\int_0^L \frac{dx}{e^{4ax}}$$
$$= \frac{2T}{\pi G}\left[-\frac{1}{4a}e^{-4ax}\right]_0^L$$
$$= \frac{T}{2a\pi G}\left(1 - e^{-4aL}\right) \qquad \textbf{Ans}$$

5–66. The tapered shaft is made of 2014-T6 aluminum alloy and has a radius which can be described by the function $r = 0.02(1 + x^{3/2})$ m, where x is in meters. Determine the angle of twist of its end A if it is subjected to a torque of 450 N · m.

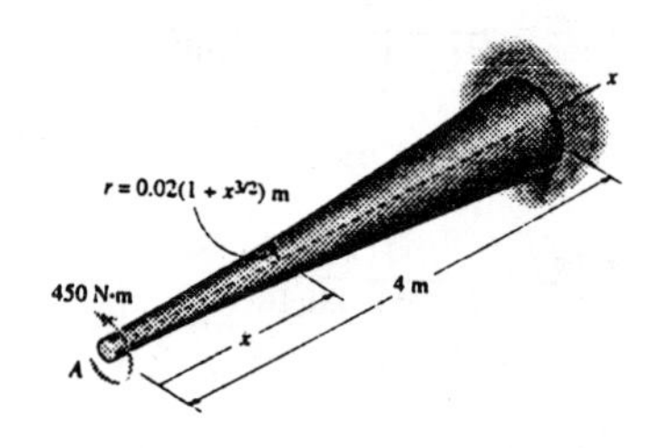

Internal Torque : As shown on FBD.

Angle of Twist : The polar moment of inertia is $J(x) = \frac{\pi}{2}\left[0.02\left(1+x^{3/2}\right)\right]^4 = 80\pi\left(10^{-9}\right)\left(1+x^{3/2}\right)^4$.

$$\phi_A = \int_0^L \frac{T(x)\,dx}{J(x)\,G}$$
$$= \int_0^{4\,\text{m}} \frac{450\,dx}{80\pi(10^{-9})(1+x^{3/2})^4(27.0)(10^9)}$$
$$= 0.066315\int_0^{4\,\text{m}} \frac{dx}{(1+x^{3/2})^4}$$

Evaluating the integral using Simpson's rule, we have

$$\phi_A = 0.066315(0.4179)\ \text{rad}$$
$$= 0.0277\ \text{rad} = 1.59° \qquad \textbf{Ans}$$

5–67. The shaft has a radius c and is subjected to a torque per unit length of t_0, which is distributed uniformly over the shaft's entire length L. If it is fixed at its far end A, determine the angle of twist ϕ of end B. The shear modulus is G.

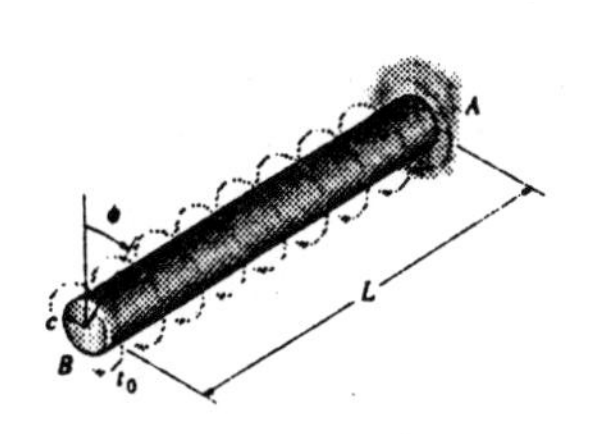

Internal Torque : As shown on FBD.

Angle of Twist :

$$\phi = \int \frac{T(x)\,dx}{JG} = \frac{-t_0}{JG}\int_0^L x\,dx$$

$$= \frac{-t_0}{JG}\left[\frac{x^2}{2}\right]_0^L$$

$$= \frac{-t_0 L^2}{2JG}$$

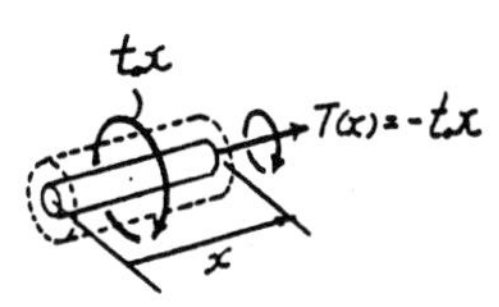

However, $J = \frac{\pi}{2}c^4$

$$\phi = \frac{t_0 L^2}{\pi c^4 G} \qquad \textbf{Ans}$$

***5–68.** The glass tube is confined within a rubber stopper, so that when the tube is twisted at constant angular velocity the stopper creates a *constant distribution* of frictional torque along the contacting length AB of the tube. If the tube has an inner diameter of 2 mm and an outer diameter of 4 mm, determine the shear stress developed at a point located at its inner and outer walls at a section through level C. Show the shear-stress distribution acting along a radial line segment at this section. Also, determine the angle of twist at A with respect to B. $G_g = 10$ GPa.

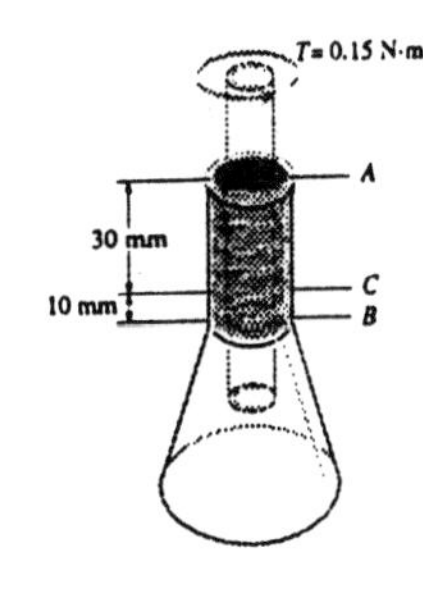

Equilibrium : FBD (a).

$\Sigma M_z = 0;\quad 0.15 - t(0.04) = 0 \qquad t = 3.75\ \text{N.m / m}$

Internal Torque : As shown on FBD (b) and (c).

Shear Stress : Applying torsion formula, we have

$$\tau_o = \frac{0.0375(0.002)}{\frac{\pi}{2}(0.002^4 - 0.001^4)} = 3.18\ \text{MPa} \qquad \textbf{Ans}$$

$$\tau_i = \frac{0.0375(0.001)}{\frac{\pi}{2}(0.002^4 - 0.001^4)} = 1.59\ \text{MPa} \qquad \textbf{Ans}$$

Angle of Twist :

$$\phi_{A/B} = \int_0^L \frac{T(x)\,dx}{JG}$$

$$= \frac{1}{\frac{\pi}{2}(0.002^4 - 0.001^4)\,10(10^9)}\int_0^{0.04\text{m}} 3.75x\,dx$$

$$= 0.01273\ \text{rad} = 0.730^\circ \qquad \textbf{Ans}$$

5–69. The A-36 steel shaft has a diameter of 50 mm and is subjected to the distributed and concentrated loadings shown. Determine the absolute maximum shear stress in the shaft and plot a graph of the angle of twist of the shaft in radians versus x.

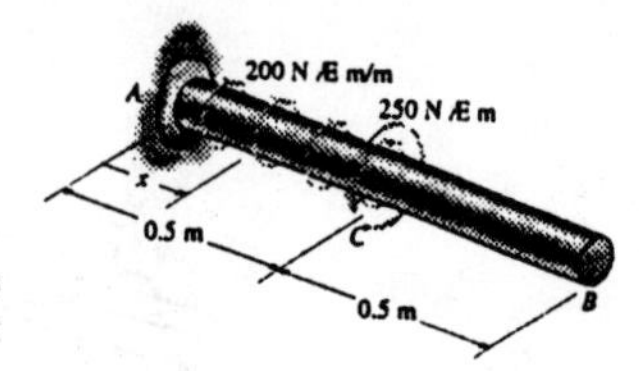

Internal Torque : As shown on FBD.

Maximum Shear Stress : The maximum torque occurs at $x = 0.5$ m where $T_{max} = 150 + 200(0.5) = 250$ N · m.

$$\tau_{\substack{max\\ABS}} = \frac{T_{max}\,c}{J} = \frac{250(0.025)}{\frac{\pi}{2}(0.025^4)} = 10.2 \text{ MPa} \qquad \textbf{Ans}$$

Angle of Twist :

For $0 \le x < 0.5$ m

$$\phi(x) = \int_0^L \frac{T(x)\,dx}{JG}$$

$$= \int_0^x \frac{(150 + 200x)\,dx}{JG}$$

$$= \frac{150x + 100x^2}{JG}$$

$$= \frac{150x + 100x^2}{\frac{\pi}{2}(0.025^4)\,75.0(10^9)}$$

$$= \left[3.26x + 2.17x^2\right](10^{-3}) \text{ rad}$$

At $x = 0.5$ m, $\quad \phi = \phi_C = 0.00217$ rad

For $0.5 \text{ m} < x < 1$ m $\quad$ Since $T(x) = 0$, then

$$\phi(x) = \phi_C = 0.00217 \text{ rad}$$

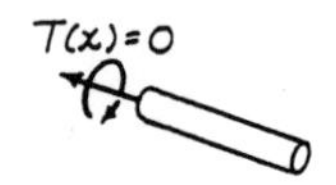

For 0.5m < x ≤ 1m

T(x) = 150 + 200x

200(0.5-x)

0.5-x

250 N·m

For 0 ≤ x < 0.5m

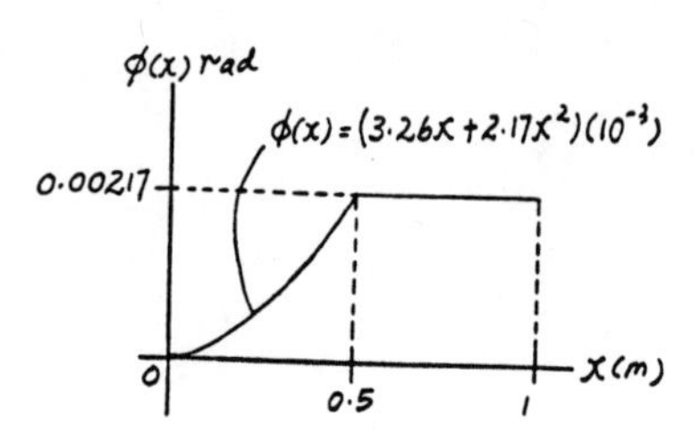

5–70. The 60-mm-diameter solid shaft is made of A-36 steel and is subjected to the distributed and concentrated torsional loadings shown. Determine the angle of twist at the free end A of the shaft due to these loadings.

Internal Torque : As shown on FBD.

Angle of Twist :

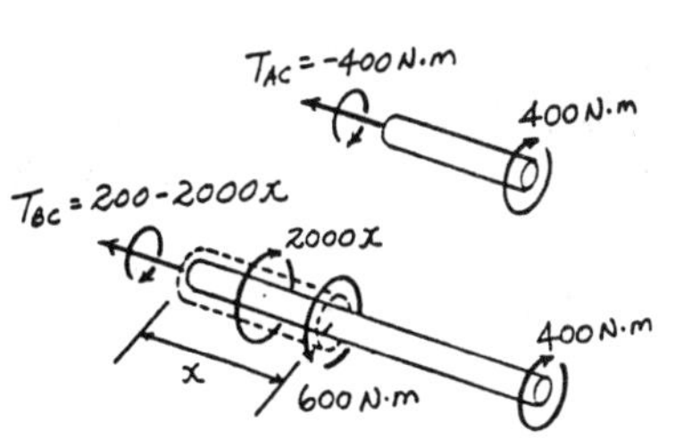

$$\phi_A = \sum \frac{TL}{JG}$$

$$= \frac{-400(0.6)}{\frac{\pi}{2}(0.03^4)75.0(10^9)} + \int_0^{0.8\text{m}} \frac{(200-2000x)\,dx}{\frac{\pi}{2}(0.03^4)75.0(10^9)}$$

$$= -0.007545 \text{ rad} = |\ 0.432°\ | \qquad \textbf{Ans}$$

5–71. The tapered shaft has a length L and a radius r at end A and $2r$ at end B. If it is fixed at end B and is subjected to a torque T, determine the angle of twist of end A. The shear modulus is G.

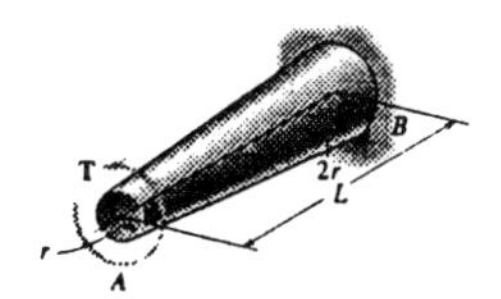

Geometry :

$$r(x) = r + \frac{r}{L}x = \frac{rL + rx}{L}$$

$$J(x) = \frac{\pi}{2}\left(\frac{rL + rx}{L}\right)^4 = \frac{\pi r^4}{2L^4}(L + x)^4$$

Angle of Twist :

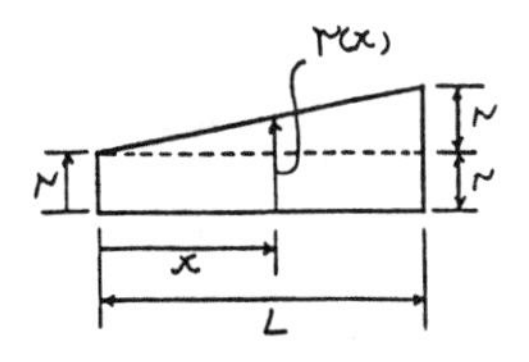

$$\phi = \int_0^L \frac{T\,dx}{J(x)G}$$

$$= \frac{2TL^4}{\pi r^4 G}\int_0^L \frac{dx}{(L + x)^4}$$

$$= \frac{2TL^4}{\pi r^4 G}\left[-\frac{1}{3(L + x)^3}\right]\Bigg|_0^L$$

$$= \frac{7TL}{12\pi r^4 G} \qquad \textbf{Ans}$$

***5–72.** The solid steel shaft consists of two tapered ends, AB and CD, and a central portion BC having a constant diameter. It is supported by two smooth bearings, which allow it to rotate freely. If the gears fixed to its ends are subjected to counterbalancing torques of 1700 lb · in., determine the angle of twist of one end of the shaft relative to the other end. *Hint:* Use the results of Prob. 5–71. $G_{st} = 12(10^3)$ ksi.

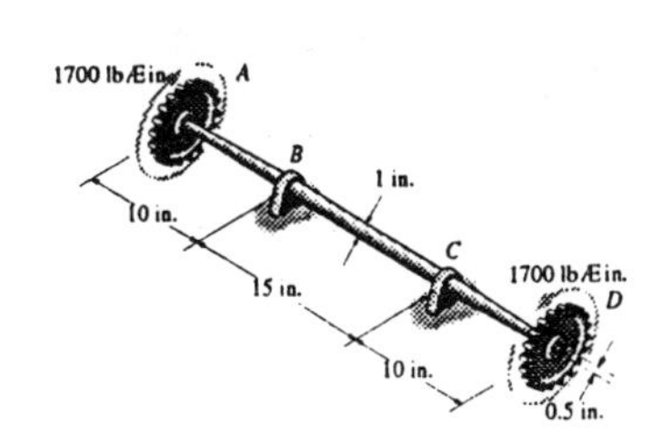

Internal Torque : As shown on FBD.

Angle of Twist : Use the formula developed in Prob. 5 – 71.

$$\phi_{A/D} = \frac{TL}{JG} + 2\left(\frac{7TL}{12\pi r^4 G}\right) \quad \text{where } r = 0.25 \text{ in.}$$

$$= \frac{1700(15)}{\frac{\pi}{2}(0.5^4)(12)(10^6)} + 2\left[\frac{(7)(1700)(10)}{12\pi(0.25^4)(12)(10^6)}\right]$$

$$= 0.1563 \text{ rad} = 8.96° \qquad \textbf{Ans}$$

5–73. Show that when the bar is subjected to the torque T, the length L shortens by an amount $L(1 - \sqrt{1 - (2T/\pi c^3 G)^2})$. The shaft is made of material having a shear modulus G.

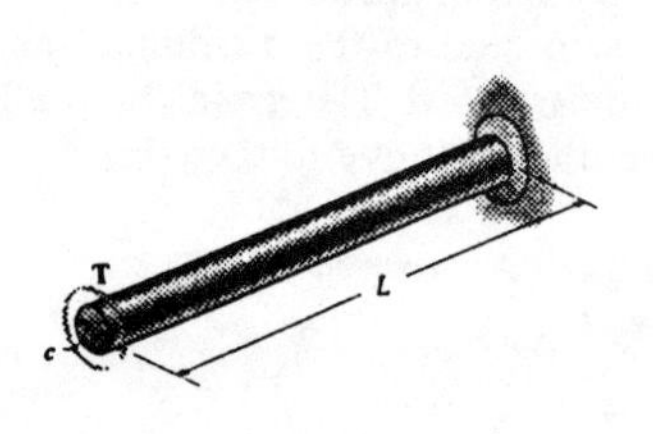

Angle of Twist :

$$\phi = \frac{TL}{JG} = \frac{TL}{\left(\frac{\pi}{2}c^4\right)G} = \frac{2TL}{\pi c^4 G} \qquad [1]$$

Geometry :

$$\delta L = L - L' \qquad \text{where } L' = \sqrt{L^2 - (c\phi)^2}$$

$$= L - \sqrt{L^2 - (c\phi)^2} \qquad [2]$$

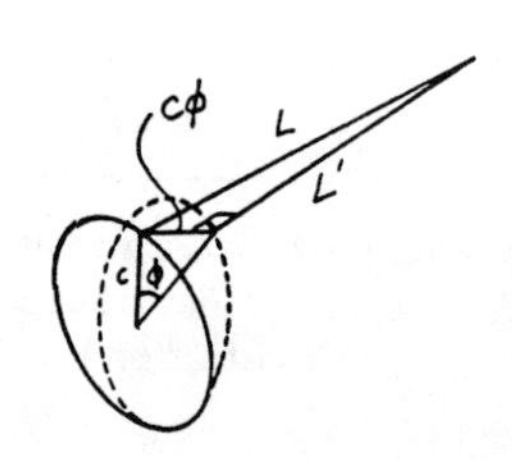

Substitute Eq. [1] into [2] yields :

$$\delta L = L - \sqrt{L^2 - \left[c\left(\frac{2TL}{\pi c^4 G}\right)\right]^2}$$

$$= L - \sqrt{L^2 - \left(\frac{2TL}{\pi c^3 G}\right)^2}$$

$$= L - \sqrt{L^2\left[1 - \left(\frac{2T}{\pi c^3 G}\right)^2\right]}$$

$$= L - L\sqrt{1 - \left(\frac{2T}{\pi c^3 G}\right)^2}$$

$$= L\left[1 - \sqrt{1 - \left(\frac{2T}{\pi c^3 G}\right)^2}\right] \qquad (Q.E.D.)$$

5–74. A cylindrical spring consists of a rubber annulus bonded to a rigid ring and shaft. If the ring is held fixed and a torque T is applied to the rigid shaft, determine the angle of twist of the shaft. The shear modulus of the rubber is G. *Hint:* As shown in the figure, the deformation of the element at radius r can be determined from $r\,d\theta = dr\,\gamma$. Use this expression along with $\tau = T/(2\pi r^2 h)$, from Prob. 5–41, to obtain the result.

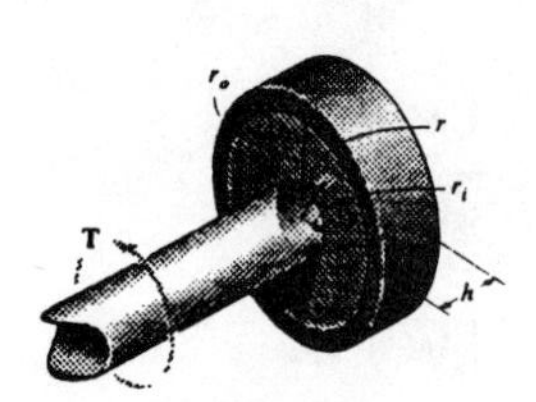

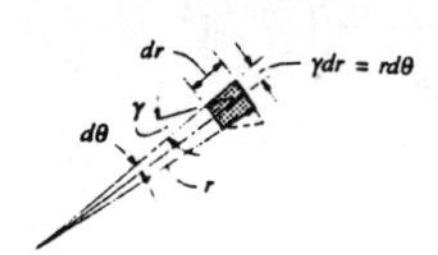

Geometry :

$$r\,d\theta = \gamma\,dr \qquad d\theta = \frac{\gamma dr}{r} \qquad [1]$$

Shear Stress and Shear Strain Relationship : From Prob. 5 – 41, we have $\tau = \dfrac{T}{2\pi r^2 h}$ and $\gamma = \dfrac{\tau}{G}$. Then

$$\gamma = \frac{T}{2\pi r^2 hG} \qquad [2]$$

Angle of Twist : Substituting Eq. [2] into [1] yields

$$d\theta = \frac{T}{2\pi hG}\frac{dr}{r^3}$$

$$\theta = \frac{T}{2\pi hG}\int_{r_i}^{r_o}\frac{dr}{r^3}$$

$$= \frac{T}{2\pi hG}\left[-\frac{1}{2r^2}\right]\Bigg|_{r_i}^{r_o}$$

$$= \frac{T}{2\pi hG}\left[-\frac{1}{2r_o^2} + \frac{1}{2r_i^2}\right]$$

$$= \frac{T}{4\pi hG}\left[\frac{1}{r_i^2} - \frac{1}{r_o^2}\right] \qquad \textbf{Ans}$$

5–75. The A-36 steel shaft has a diameter of 50 mm and is fixed at its ends A and B. If it is subjected to the couple, determine the maximum shear stress in regions AC and CB of the shaft.

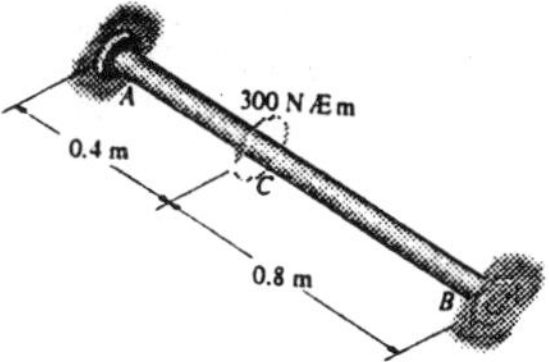

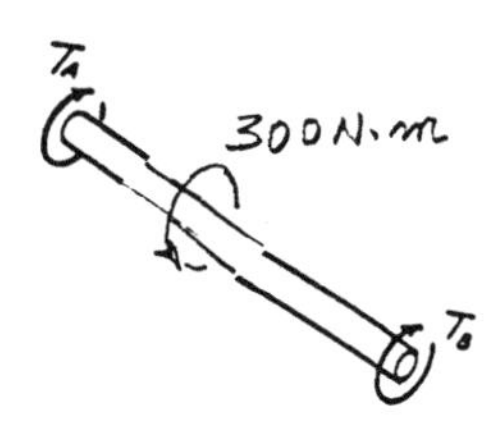

Equilibrium :

$$T_A + T_B - 300 = 0 \qquad [1]$$

Compatibility :

$$\phi_{C/A} = \phi_{C/B}$$

$$\frac{T_A(0.4)}{JG} = \frac{T_B(0.8)}{JG}$$

$$T_A = 2.00T_B \qquad [2]$$

Solving Eqs. [1] and [2] yields :

$$T_A = 200 \text{ N}\cdot\text{m} \qquad T_B = 100 \text{ N}\cdot\text{m}$$

Maximum Shear stress :

$$(\tau_{AC})_{\max} = \frac{T_A c}{J} = \frac{200(0.025)}{\frac{\pi}{2}(0.025^4)} = 8.15 \text{ MPa} \qquad \textbf{Ans}$$

$$(\tau_{CB})_{\max} = \frac{T_B c}{J} = \frac{100(0.025)}{\frac{\pi}{2}(0.025^4)} = 4.07 \text{ MPa} \qquad \textbf{Ans}$$

***5–76.** The shaft is made of L2 tool steel, has a diameter of 40 mm, and is fixed at its ends A and B. If it is subjected to the couple, determine the maximum shear stress in regions AC and CB.

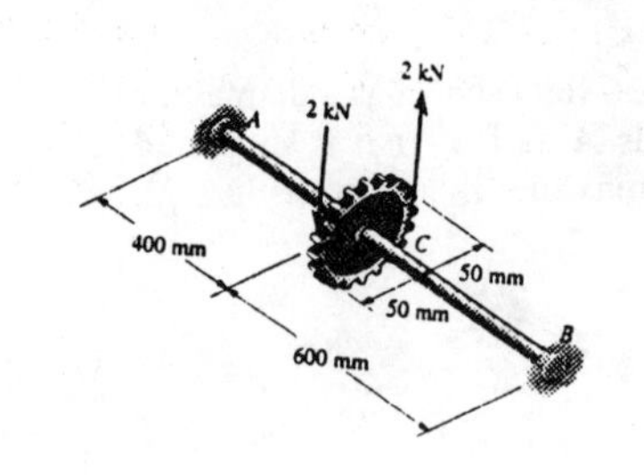

Equilibrium :

$$T_A + T_B - 2(0.1) = 0 \qquad [1]$$

Compatibility :

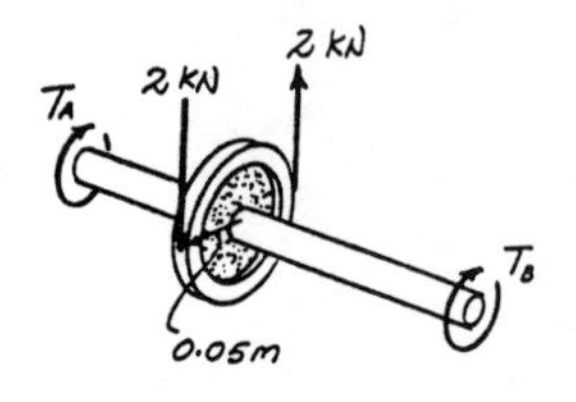

$$\phi_{C/A} = \phi_{C/B}$$

$$\frac{T_A(0.4)}{JG} = \frac{T_B(0.6)}{JG}$$

$$T_A = 1.50 T_B \qquad [2]$$

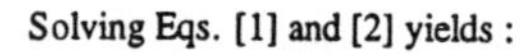

Solving Eqs. [1] and [2] yields :

$$T_B = 0.080 \text{ kN}\cdot\text{m} \qquad T_A = 0.120 \text{ kN}\cdot\text{m}$$

Maximum Shear stress :

$$(\tau_{AC})_{max} = \frac{T_A c}{J} = \frac{0.12(10^3)(0.02)}{\frac{\pi}{2}(0.02^4)} = 9.55 \text{ MPa} \qquad \textbf{Ans}$$

$$(\tau_{CB})_{max} = \frac{T_B c}{J} = \frac{0.08(10^3)(0.02)}{\frac{\pi}{2}(0.02^4)} = 6.37 \text{ MPa} \qquad \textbf{Ans}$$

5–77. The bronze C86100 pipe has an outer diameter of 1.5 in. and a thickness of 0.125 in. The coupling on it at C is being tightened using a wrench. If the torque developed at A is 125 lb · in., determine the magnitude F of the couple forces. The pipe is fixed supported at end B.

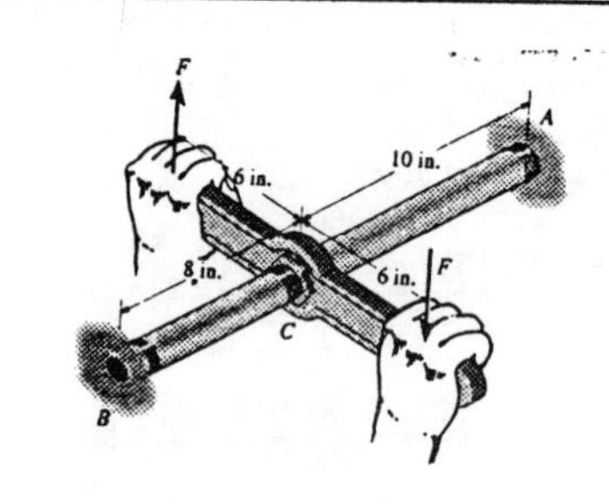

Equilibrium :

$$F(12) - T_B - 125 = 0 \qquad [1]$$

Compatibility :

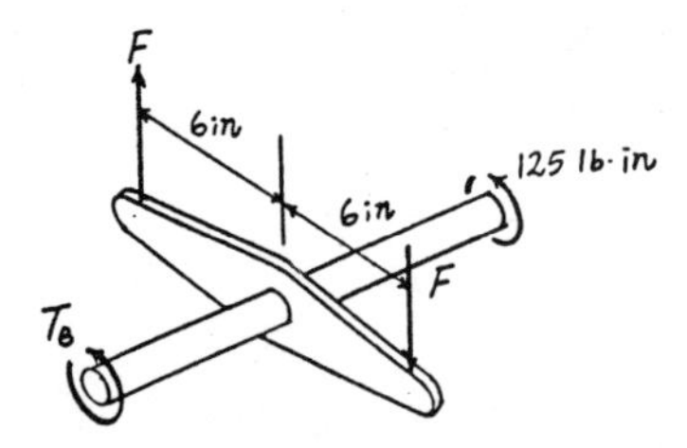

$$\phi_{C/B} = \phi_{C/A}$$

$$\frac{T_B(8)}{JG} = \frac{125(10)}{JG}$$

$$T_B = 156.25 \text{ lb}\cdot\text{in.}$$

From Eq. [1],

$$F(12) - 156.25 - 125 = 0$$

$$F = 23.4 \text{ lb} \qquad \textbf{Ans}$$

5–78. The bronze C86100 pipe has an outer diameter of 1.5 in. and a thickness of 0.125 in. The coupling on it at C is being tightened using a wrench. If the applied force is $F = 20$ lb, determine the maximum shear stress in the pipe.

Equilibrium :

$$T_A + T_B - 20(12) = 0 \qquad [1]$$

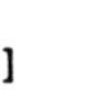

Compatibility :

$$\phi_{C/B} = \phi_{C/A}$$

$$\frac{T_B(8)}{JG} = \frac{T_A(10)}{JG}$$

$$T_B = 1.25T_A \qquad [2]$$

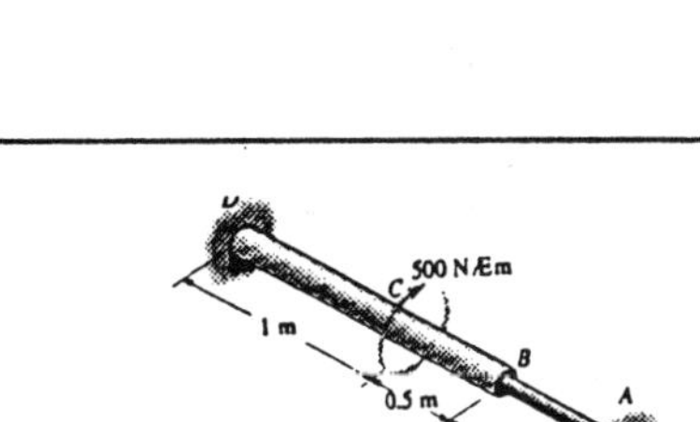

Solving Eqs. [1] and [2] yields :

$$T_A = 106.67 \text{ lb}\cdot\text{in.} \qquad T_B = 133.33 \text{ lb}\cdot\text{in.}$$

Maximum shear stress :

$$\tau_{max} = \frac{T_B\, c}{J} = \frac{133.33(0.75)}{\frac{\pi}{2}(0.75^4 - 0.625^4)} = 389 \text{ psi} \qquad \textbf{Ans}$$

5–79. A rod is made from two segments: AB is A-36 steel and has a diameter of 30 mm and BD is C83400 red brass and has a diameter of 50 mm. It is fixed at its ends and subjected to a torque of $T = 500$ N · m. Determine the torsional reactions at the walls A and D.

Equilibrium :

$$T_A + T_D - 500 = 0 \qquad [1]$$

Compatibility :

$$\phi_{C/D} = \phi_{C/A}$$

$$\frac{T_D(1)}{\frac{\pi}{2}(0.025^4)(37.0)(10^9)} = \frac{T_A(0.75)}{\frac{\pi}{2}(0.015^4)(75.0)(10^9)} + \frac{T_A(0.5)}{\frac{\pi}{2}(0.025^4)(37.0)(10^9)}$$

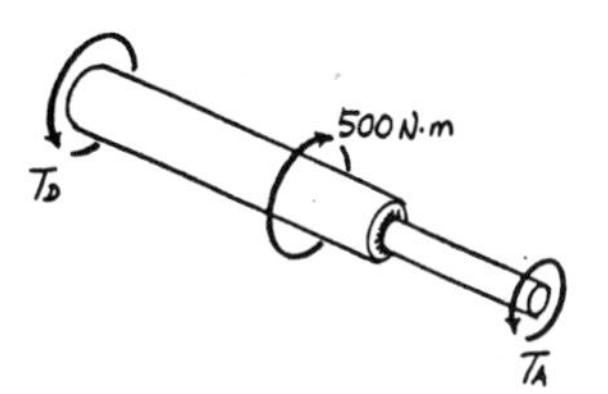

$$T_A = 0.29807T_D \qquad [2]$$

Solving Eqs. [1] and [2] yields :

$$T_D = 385 \text{ N}\cdot\text{m} \qquad \textbf{Ans}$$
$$T_A = 115 \text{ N}\cdot\text{m} \qquad \textbf{Ans}$$

***5–80.** Determine the absolute maximum shear stress in the shaft of Prob. 5–79.

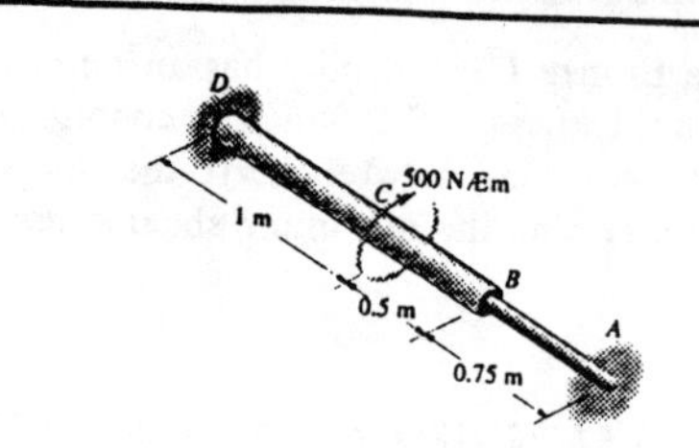

Equilibrium :

$$T_A + T_D - 500 = 0 \qquad [1]$$

Compatibility :

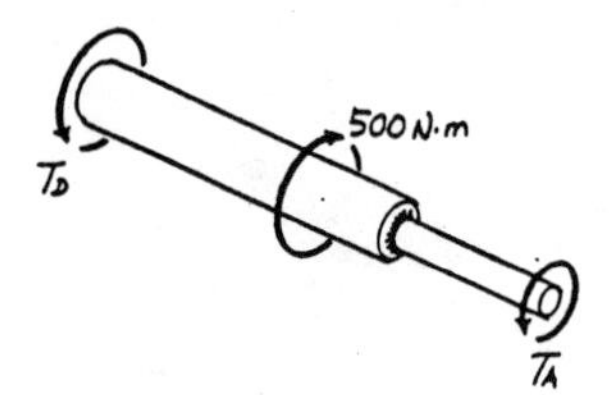

$$\phi_{C/D} = \phi_{C/A}$$

$$\frac{T_D(1)}{\frac{\pi}{2}(0.025^4)(37.0)(10^9)} = \frac{T_A(0.75)}{\frac{\pi}{2}(0.015^4)(75.0)(10^9)} + \frac{T_A(0.5)}{\frac{\pi}{2}(0.025^4)(37.0)(10^9)}$$

$$T_A = 0.29807T_D \qquad [2]$$

Solving Eqs. [1] and [2] yields :

$$T_D = 385.19 \text{ N}\cdot\text{m} \qquad T_A = 114.81 \text{ N}\cdot\text{m}$$

Maximum Shear Stress :

$$(\tau_{CD})_{\max} = \frac{T_D c}{J} = \frac{385.19(0.025)}{\frac{\pi}{2}(0.025^4)} = 15.7 \text{ MPa}$$

$$\tau_{\substack{\text{abs}\\\text{max}}} = (\tau_{AB})_{\max} = \frac{T_A c}{J} = \frac{114.81(0.015)}{\frac{\pi}{2}(0.015^4)} = 21.7 \text{ MPa} \qquad \textbf{Ans}$$

5–81. The A-36 steel shaft is made from two segments: AC has a diameter of 1 in. and CB has a diameter of 2 in. If it is fixed at its ends A and B and subjected to a torque of $T = 500$ lb · ft, determine the absolute maximum shear stress in the shaft.

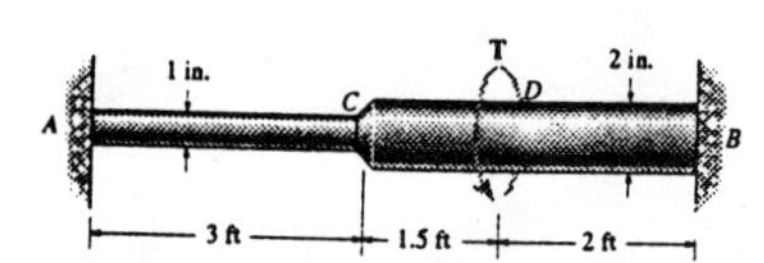

Equilibrium :

$$T_A + T_B - 500 = 0 \qquad [1]$$

Compatibility :

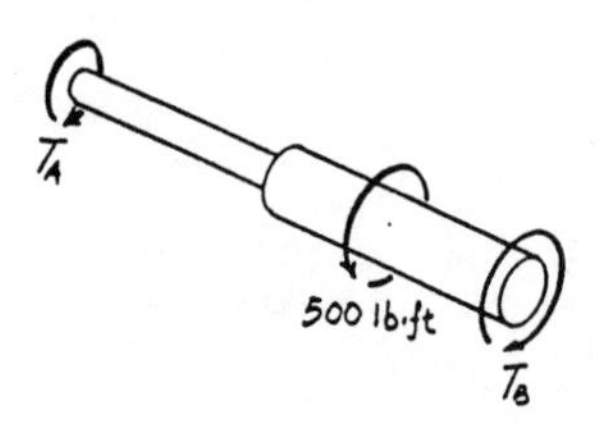

$$\phi_{D/A} = \phi_{D/B}$$

$$\frac{T_A(3)(12)}{\frac{\pi}{2}(0.5^4)G} + \frac{T_A(1.5)(12)}{\frac{\pi}{2}(1^4)G} = \frac{T_B(2)(12)}{\frac{\pi}{2}(1^4)G}$$

$$T_B = 24.75T_A \qquad [2]$$

Solving Eqs. [1] and [2] yields :

$$T_A = 19.417 \text{ lb}\cdot\text{ft} \qquad T_B = 480.583 \text{ lb}\cdot\text{ft}$$

Maximum Shear Stress :

$$(\tau_{AC})_{\max} = \frac{T_A c}{J} = \frac{19.417(12)(0.5)}{\frac{\pi}{2}(0.5^4)} = 1.187 \text{ ksi}$$

$$\tau_{\substack{\text{abs}\\\text{max}}} = (\tau_{BD})_{\max} = \frac{T_B c}{J} = \frac{480.583(12)(1)}{\frac{\pi}{2}(1^4)} = 3.67 \text{ ksi} \qquad \textbf{Ans}$$

5–82. Determine the torsional reactions at the ends A and B of the shaft in Prob. 5–81.

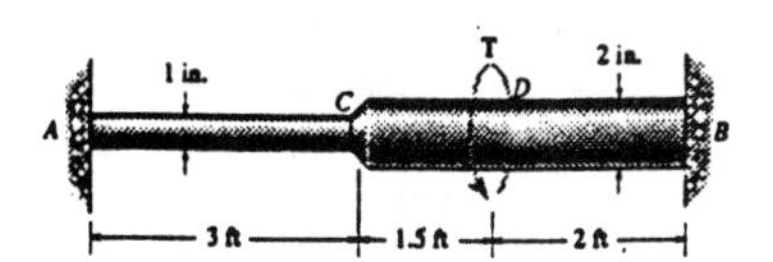

Equilibrium :

$$T_A + T_B - 500 = 0 \qquad [1]$$

Compatibility :

$$\phi_{D/A} = \phi_{D/B}$$

$$\frac{T_A(3)(12)}{\frac{\pi}{2}(0.5^4)G} + \frac{T_A(1.5)(12)}{\frac{\pi}{2}(1^4)G} = \frac{T_B(2)(12)}{\frac{\pi}{2}(1^4)G}$$

$$T_B = 24.75 T_A \qquad [2]$$

Solving Eqs. [1] and [2] yields :

$$T_A = 19.4 \text{ lb} \cdot \text{ft} \qquad \textbf{Ans}$$
$$T_B = 481 \text{ lb} \cdot \text{ft} \qquad \textbf{Ans}$$

5–83. The A-36 steel shaft is made from two segments: AC has a diameter of 1 in. and CB has a diameter of 2 in. If it is fixed at its ends A and B, determine the magnitude of the applied torque T if the reaction at A is 50 lb · ft.

Equilibrium :

$$50 + T_B - T = 0 \qquad [1]$$

Compatibility :

$$\phi_{D/B} = \phi_{D/A}$$

$$\frac{T_B(2)(12)}{\frac{\pi}{2}(1^4)G} = \frac{50(12)(3)(12)}{\frac{\pi}{2}(0.5^4)G} + \frac{50(12)(1.5)(12)}{\frac{\pi}{2}(1^4)G}$$

$$T_B = 14850 \text{ lb} \cdot \text{in.} = 1237.5 \text{ lb} \cdot \text{ft}$$

From Eq. [1]

$$T = 1287.5 \text{ lb} \cdot \text{ft} = 1.29 \text{ kip} \cdot \text{ft} \qquad \textbf{Ans}$$

***5–84.** The tube is made of 2014-T6 aluminum and has an inner diameter of 35 mm and an outer diameter of 60 mm. A torque of $T = 900\ \text{N}\cdot\text{m}$ is applied to its end B. If it is filled with an A-36 steel plug AC, determine the maximum shear stress in the aluminum and the steel and the rotation of the end B.

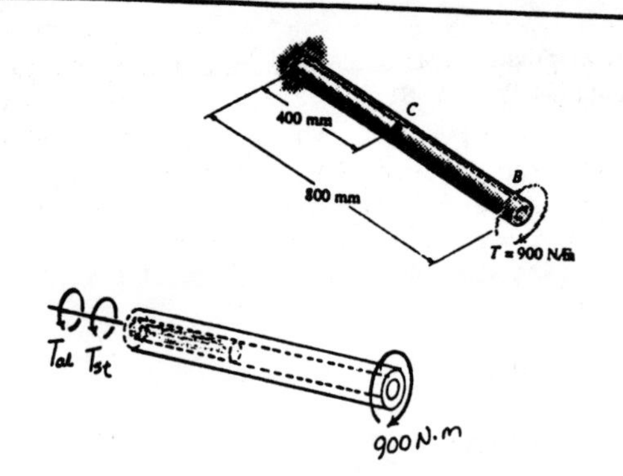

Equilibrium :

$$T_{al} + T_{st} - 900 = 0 \qquad [1]$$

Compatibility : Both the aluminum tube and the steel core undergo the same angle of twist ϕ_C

$$\phi_C = \frac{T_{al}(L)}{\frac{\pi}{2}(0.03^4 - 0.0175^4)27.0(10^9)} = \frac{T_{st}(L)}{\frac{\pi}{2}(0.0175^4)(75.0)(10^9)}$$

$$T_{al} = 2.7491 T_{st} \qquad [2]$$

Solving Eqs. [1] and [2] yields :

$$T_{st} = 240.06\ \text{N}\cdot\text{m} \qquad T_{al} = 659.94\ \text{N}\cdot\text{m}$$

Maximum Shear Stress :

$$(\tau_{al})_{\max} = \frac{Tc}{J} = \frac{900(0.03)}{\frac{\pi}{2}(0.03^4 - 0.0175^4)} = 24.0\ \text{MPa} \qquad \textbf{Ans}$$

$$(\tau_{st})_{\max} = \frac{T_{st}c}{J} = \frac{240.06(0.0175)}{\frac{\pi}{2}(0.0175^4)} = 28.5\ \text{MPa} \qquad \textbf{Ans}$$

Angle of Twist :

$$\phi_B = \sum \frac{TL}{JG}$$

$$= \frac{240.06(0.4)}{\frac{\pi}{2}(0.0175^4)(75.0)(10^9)} + \frac{900(0.4)}{\frac{\pi}{2}(0.03^4 - 0.0175^4)(27.0)(10^9)}$$

$$= 0.02054\ \text{rad} = 1.18^\circ \qquad \textbf{Ans}$$

5–85. The two shafts are made of A-36 steel. Each has a diameter of 25 mm and they are connected using the gears fixed to their ends. Their other ends are attached to fixed supports at A and B. They are also supported by journal bearings at C and D, which allow free rotation of the shafts along their axes. If a torque of 500 N · m is applied to the gear at E as shown, determine the reactions at A and B.

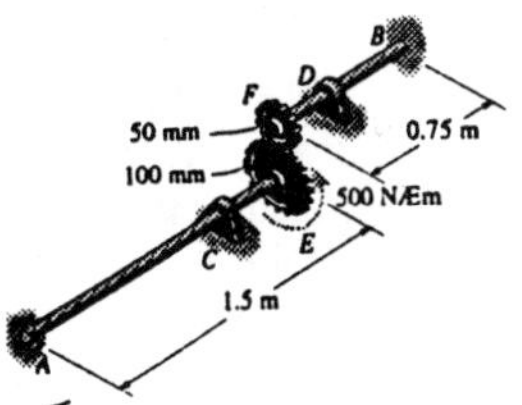

Equilibrium :

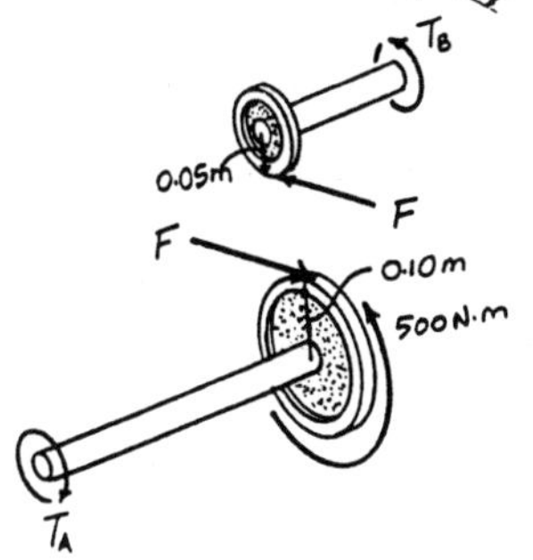

$$T_A + F(0.1) - 500 = 0 \qquad [1]$$

$$T_B - F(0.05) = 0 \qquad [2]$$

From Eqs. [1] and [2]

$$T_A + 2T_B - 500 = 0 \qquad [3]$$

Compatibility :

$$0.1\phi_E = 0.05\phi_F$$

$$\phi_E = 0.5\phi_F$$

$$\frac{T_A(1.5)}{JG} = 0.5\left[\frac{T_B(0.75)}{JG}\right]$$

$$T_A = 0.250T_B \qquad [4]$$

Solving Eqs. [3] and [4] yields :

$T_B = 222 \text{ N}\cdot\text{m}$ **Ans**

$T_A = 55.6 \text{ N}\cdot\text{m}$ **Ans**

5–86. Determine the rotation of the gear at E in Prob. 5–85.

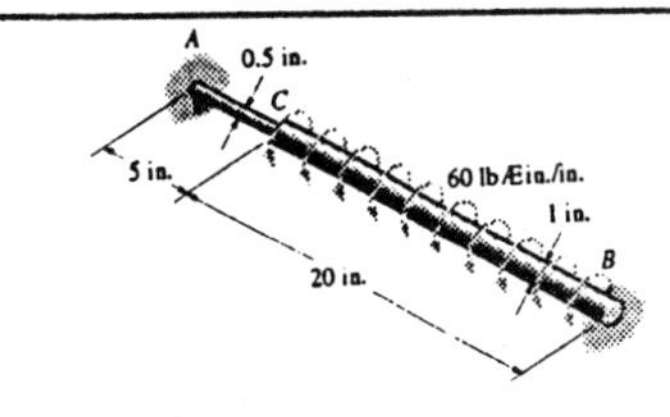

Equilibrium :

$$T_A + F(0.1) - 500 = 0 \qquad [1]$$

$$T_B - F(0.05) = 0 \qquad [2]$$

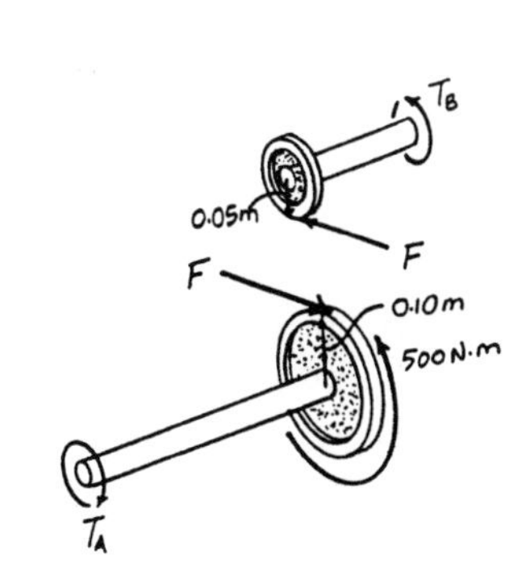

From Eqs. [1] and [2]

$$T_A + 2T_B - 500 = 0 \qquad [3]$$

Compatibility :

$$0.1\phi_E = 0.05\phi_F$$

$$\phi_E = 0.5\phi_F$$

$$\frac{T_A(1.5)}{JG} = 0.5\left[\frac{T_B(0.75)}{JG}\right]$$

$$T_A = 0.250T_B \qquad [4]$$

Solving Eqs. [3] and [4] yields :

$T_B = 222.22 \text{ N}\cdot\text{m}$ $\quad T_A = 55.56 \text{ N}\cdot\text{m}$

Angle of Twist :

$$\phi_E = \frac{T_A L}{JG} = \frac{55.56(1.5)}{\frac{\pi}{2}(0.0125^4)(75.0)(10^9)}$$

$$= 0.02897 \text{ rad} = 1.66°$$ **Ans**

5–87. The A-36 steel shaft is made from two segments: AC has a diameter of 0.5 in. and CB has a diameter of 1 in. If the shaft is fixed at its ends A and B and subjected to a uniform distributed torque of 60 lb · in./in. along segment CB, determine the absolute maximum shear stress in the shaft.

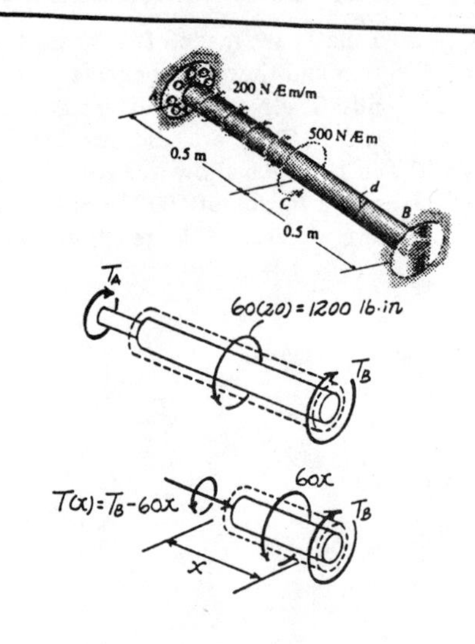

Equilibrium :

$$T_A + T_B - 1200 = 0 \qquad [1]$$

Compatibility :

$$\phi_{C/B} = \int \frac{T(x)\,dx}{JG} = \int_0^{20\text{in.}} \frac{(T_B - 60x)\,dx}{\frac{\pi}{2}(0.5^4)(11.0)(10^6)}$$

$$= 18.52(10^{-6})T_B - 0.011112$$

$$\phi_{C/A} = \phi_{C/B}$$

$$\frac{T_A(5)}{\frac{\pi}{2}(0.25^4)(11.0)(10^6)} = 18.52(10^{-6})T_B - 0.011112$$

$$18.52(10^{-6})T_B - 74.08(10^{-6})T_A = 0.011112$$

$$18.52T_B - 74.08T_A = 11112 \qquad [2]$$

Solving Eqs. [1] and [2] yields :

$$T_A = 120.0 \text{ lb}\cdot\text{in.} \qquad T_B = 1080.0 \text{ lb}\cdot\text{in.}$$

Maximum Shear Stress :

$$\tau_{\substack{abs\\max}} = (\tau_{BC})_{max} = \frac{T_B\,c}{J} = \frac{1080.0(0.5)}{\frac{\pi}{2}(0.5^4)} = 5.50 \text{ ksi} \qquad \textbf{Ans}$$

$$(\tau_{AC})_{max} = \frac{T_A\,c}{J} = \frac{120.0(0.25)}{\frac{\pi}{2}(0.25^4)} = 4.89 \text{ ksi}$$

***5–88.** The shaft is subjected to the distributed and concentrated loadings shown. Determine its required diameter d if the allowable shear stress is $\tau_{allow} = 135$ MPa.

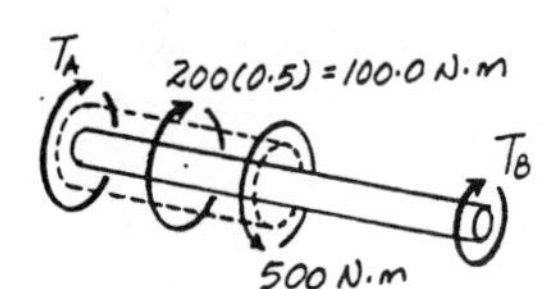

Equilibrium :

$$T_A + T_B + 100 - 500 = 0; \qquad T_A + T_B = 400 \qquad [1]$$

Compatibility :

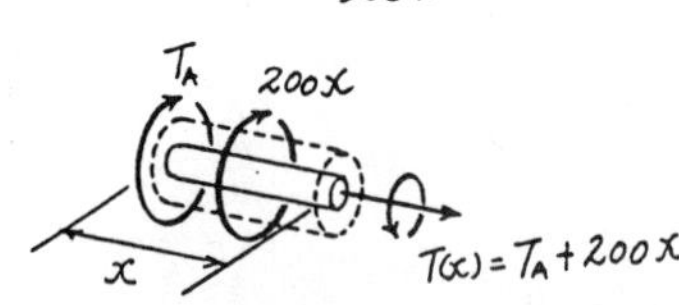

$$\phi_{C/A} = \int \frac{T(x)\,dx}{JG} = \int_0^{0.5\text{m}} \frac{(T_A + 200x)\,dx}{JG}$$

$$= \frac{0.5T_A + 25.0}{JG}$$

$$\phi_{C/A} = \phi_{C/B}$$

$$\frac{0.5T_A + 25.0}{JG} = \frac{T_B(0.5)}{JG}$$

$$T_B - T_A = 50.0 \qquad [2]$$

Solving Eqs. [1] and [2] yields :

$$T_A = 175.0 \text{ N}\cdot\text{m} \qquad T_B = 225.0 \text{ N}\cdot\text{m}$$

Allowable Shear Stress : The maximum torque occurs at $x = 0.5$ m in segment AC where

$$T_{max} = 175 + 200(0.5) = 275.0 \text{ N}\cdot\text{m}$$

$$\tau_{max} = \tau_{allow} = \frac{T_{max}\,c}{J}$$

$$135(10^6) = \frac{275.0\left(\frac{d}{2}\right)}{\frac{\pi}{2}\left(\frac{d}{2}\right)^4}$$

$$d = 0.02181 \text{ m} = 21.8 \text{ mm} \qquad \textbf{Ans}$$

5–89. The tapered shaft is confined by the fixed supports at A and B. If a torque $\mathbf{T}$ is applied at its midpoint, determine the reactions at the supports.

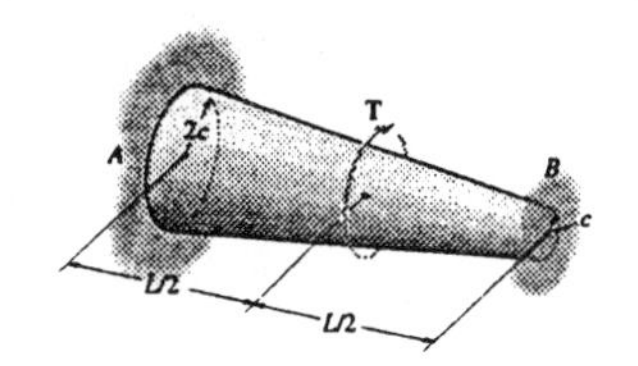

Equilibrium :

$$T_A + T_B - T = 0 \qquad [1]$$

Section Properties :

$$r(x) = c + \frac{c}{L}x = \frac{c}{L}(L+x)$$

$$J(x) = \frac{\pi}{2}\left[\frac{c}{L}(L+x)\right]^4 = \frac{\pi c^4}{2L^4}(L+x)^4$$

Angle of Twist :

$$\phi_T = \int \frac{T dx}{J(x)G} = \int_{\frac{L}{2}}^{L} \frac{T dx}{\frac{\pi c^4}{2L^4}(L+x)^4 G}$$

$$= \frac{2TL^4}{\pi c^4 G}\int_{\frac{L}{2}}^{L} \frac{dx}{(L+x)^4}$$

$$= -\frac{2TL^4}{3\pi c^4 G}\left[\frac{1}{(L+x)^3}\right]\Bigg|_{\frac{L}{2}}^{L}$$

$$= \frac{37TL}{324\pi c^4 G}$$

$$\phi_B = \int \frac{T dx}{J(x)G} = \int_0^L \frac{T_B dx}{\frac{\pi c^4}{2L^4}(L+x)^4 G}$$

$$= \frac{2T_B L^4}{\pi c^4 G}\int_0^L \frac{dx}{(L+x)^4}$$

$$= -\frac{2T_B L^4}{3\pi c^4 G}\left[\frac{1}{(L+x)^3}\right]\Bigg|_0^L$$

$$= \frac{7T_B L}{12\pi c^4 G}$$

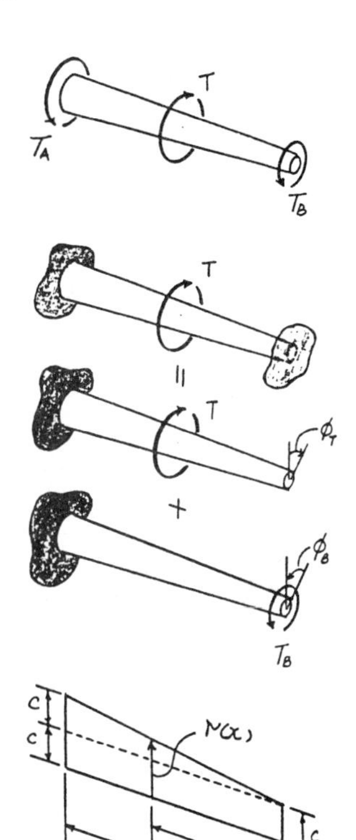

Compatibility :

$$0 = \phi_T - \phi_B$$

$$0 = \frac{37TL}{324\pi c^4 G} - \frac{7T_B L}{12\pi c^4 G}$$

$$T_B = \frac{37}{189}T \qquad \textbf{Ans}$$

Substituting the result into Eq. [1] yields :

$$T_A = \frac{152}{189}T \qquad \textbf{Ans}$$

5–90. Compare the values of the maximum elastic shear stress and the angle of twist developed in 304 stainless steel shafts having circular and square cross sections. Each shaft has the same cross-sectional area of 9 in², length of 36 in., and is subjected to a torque of 4000 lb · in.

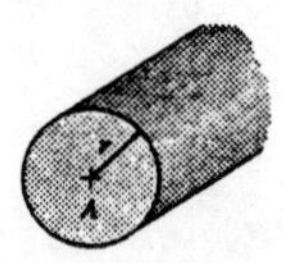

Maximum Shear Stress :
For circular shaft

$$A = \pi c^2 = 9 ; \qquad c = \left(\frac{9}{\pi}\right)^{\frac{1}{2}}$$

$$(\tau_c)_{\max} = \frac{Tc}{J} = \frac{Tc}{\frac{\pi}{2}c^4} = \frac{2T}{\pi c^3} = \frac{2(4000)}{\pi\left(\frac{9}{\pi}\right)^{\frac{3}{2}}} = 525 \text{ psi} \qquad \textbf{Ans}$$

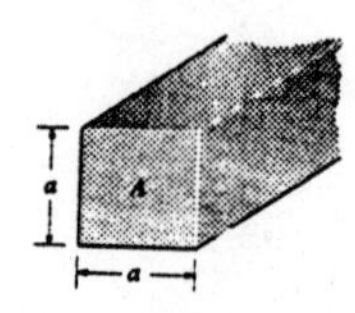

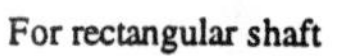
For rectangular shaft

$$A = a^2 = 9 ; \qquad a = 3 \text{ in.}$$

$$(\tau_r)_{\max} = \frac{4.81T}{a^3} = \frac{4.81(4000)}{3^3} = 713 \text{ psi} \qquad \textbf{Ans}$$

Angle of Twist :
For circular shaft

$$\phi_c = \frac{TL}{JG} = \frac{4000(36)}{\frac{\pi}{2}\left(\frac{9}{\pi}\right)^2 11.0(10^6)}$$
$$= 0.001015 \text{ rad} = 0.0582^\circ \qquad \textbf{Ans}$$

For rectangular shaft

$$\phi_r = \frac{7.10\,TL}{a^4 G} = \frac{7.10(4000)(36)}{3^4(11.0)(10^6)}$$
$$= 0.001147 \text{ rad} = 0.0657^\circ \qquad \textbf{Ans}$$

The rectangular shaft has a greater maximum shear stress and angle of twist.

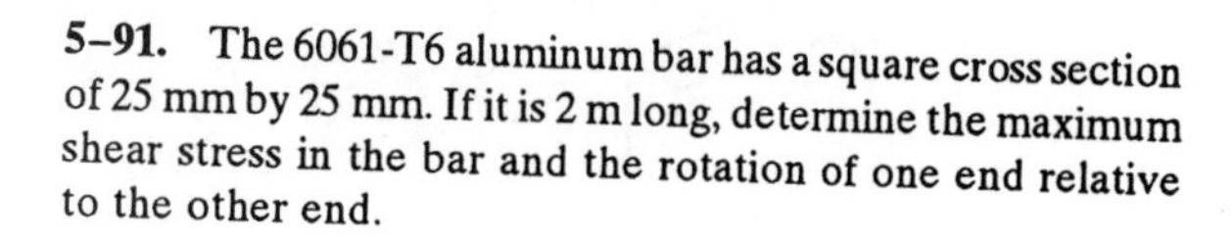
5–91. The 6061-T6 aluminum bar has a square cross section of 25 mm by 25 mm. If it is 2 m long, determine the maximum shear stress in the bar and the rotation of one end relative to the other end.

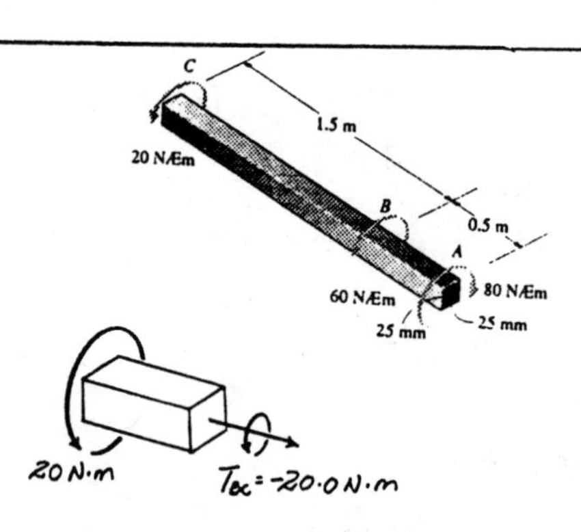

Maximum Shear Stress :

$$\tau_{\max} = \frac{4.81T_{\max}}{a^3} = \frac{4.81(80.0)}{(0.025^3)} = 24.6 \text{ MPa} \qquad \textbf{Ans}$$

Angle of Twist :

$$\phi_{A/C} = \sum \frac{7.10TL}{a^4 G} = \frac{7.10(-20.0)(1.5)}{(0.025^4)(26.0)(10^9)} + \frac{7.10(-80.0)(0.5)}{(0.025^4)(26.0)(10^9)}$$

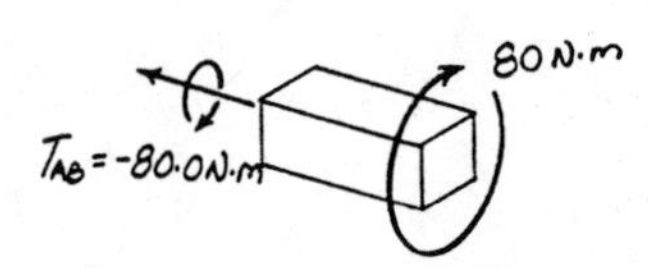

$$= -0.04894 \text{ rad} = |\,2.80^\circ\,| \qquad \textbf{Ans}$$

***5–92.** The shaft is made of red brass C83400 and has an elliptical cross section. If it is subjected to the torsional loading shown, determine the maximum shear stress within regions AC and BC, and the angle of twist ϕ of end B relative to end A.

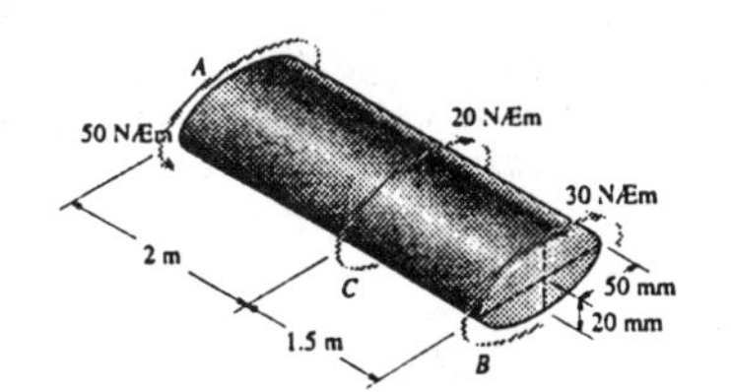

Maximum Shear Stress :

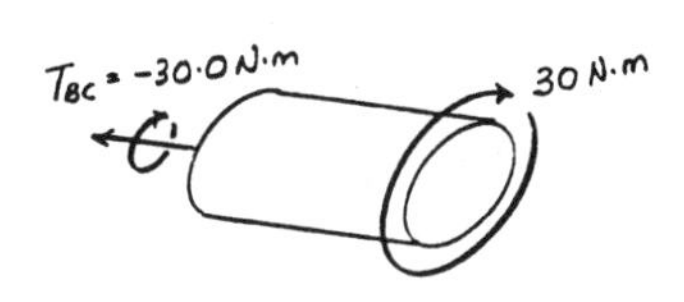

$$(\tau_{BC})_{max} = \frac{2T_{BC}}{\pi a b^2} = \frac{2(30.0)}{\pi(0.05)(0.02^2)}$$
$$= 0.955 \text{ MPa} \qquad \textbf{Ans}$$

$$(\tau_{AC})_{max} = \frac{2T_{AC}}{\pi a b^2} = \frac{2(50.0)}{\pi(0.05)(0.02^2)}$$
$$= 1.59 \text{ MPa} \qquad \textbf{Ans}$$

Angle of Twist :

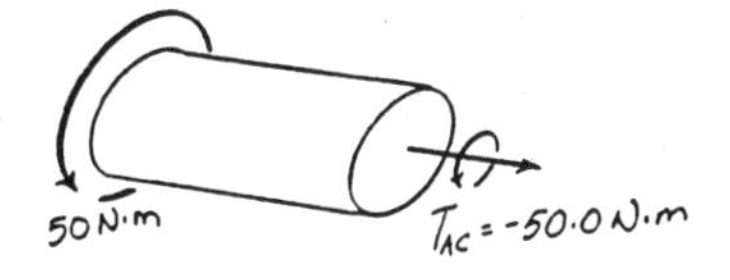

$$\phi_{B/A} = \sum \frac{(a^2+b^2)TL}{\pi a^3 b^3 G}$$
$$= \frac{(0.05^2+0.02^2)}{\pi(0.05^3)(0.02^3)(37.0)(10^9)}[(-30.0)(1.5)+(-50.0)(2)]$$

$$= -0.003618 \text{ rad} = 0.207° \qquad \textbf{Ans}$$

5–93. Solve Prob. 5–92 for the maximum shear stress within regions AC and BC, and the angle of twist ϕ of end B relative to C.

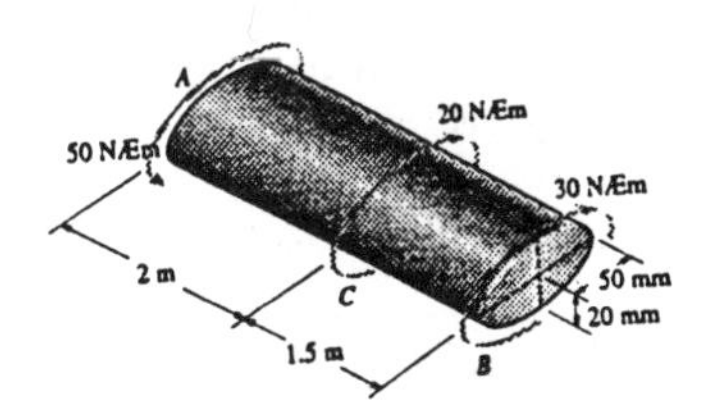

Maximum Shear Stress :

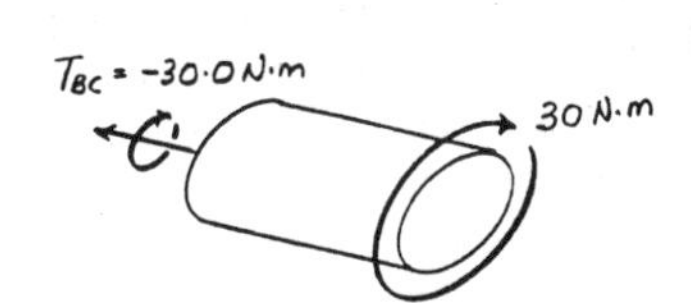

$$(\tau_{BC})_{max} = \frac{2T_{BC}}{\pi a b^2} = \frac{2(30.0)}{\pi(0.05)(0.02^2)}$$
$$= 0.955 \text{ MPa} \qquad \textbf{Ans}$$

$$(\tau_{AC})_{max} = \frac{2T_{AC}}{\pi a b^2} = \frac{2(50.0)}{\pi(0.05)(0.02^2)}$$
$$= 1.59 \text{ MPa} \qquad \textbf{Ans}$$

Angle of Twist :

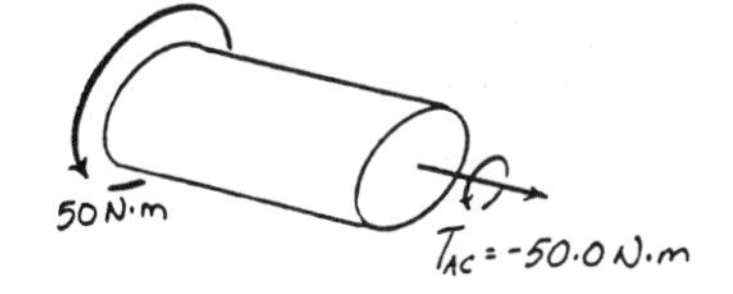

$$\phi_{B/C} = \frac{(a^2+b^2)T_{BC}L}{\pi a^3 b^3 G}$$
$$= \frac{(0.05^2+0.02^2)(-30.0)(1.5)}{\pi(0.05^3)(0.02^3)(37.0)(10^9)}$$

$$= -0.001123 \text{ rad} = |\, 0.0643° |\qquad \textbf{Ans}$$

5-94. The shaft is made of a ceramic material having an allowable shear stress of $\tau_{allow} = 7$ ksi and shear modulus of $G_c = 2.8(10^3)$ ksi. If it has a triangular cross section as shown, determine the dimension a of its sides so that it can resist a torque of 15 lb · in. Also, determine the angle of twist of its end B.

Allowable Shear Stress :

$$\tau_{max} = \tau_{allow} = \frac{20T}{a^3}$$

$$7(10^3) = \frac{20(15)}{a^3}$$

$$a = 0.350 \text{ in.} \qquad \textbf{Ans}$$

Angle of Twist :

$$\phi = \frac{46TL}{a^4 G}$$

$$= \frac{46(15)(40)}{(0.350^4)(2.8)(10^6)}$$

$$= 0.6572 \text{ rad} = 37.7° \qquad \textbf{Ans}$$

5-95. The uniform C86100 bronze rod has an elliptical cross section. If it is subjected to the five torques shown, determine the dimension a of its cross section so that the shear stress in the rod does not exceed $\tau_{allow} = 85$ MPa. Also, what is the rotation of end A with respect to end B?

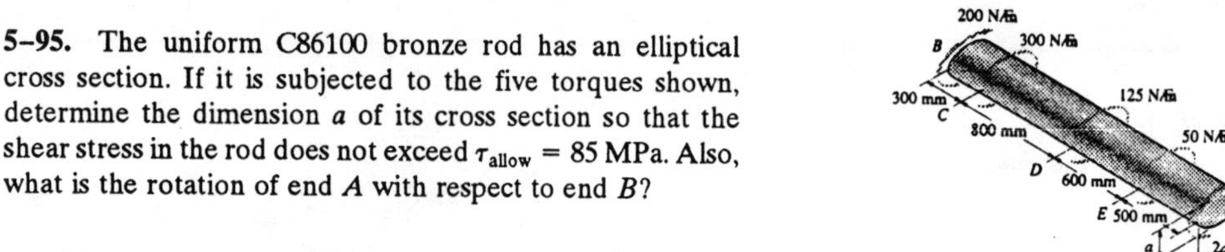

Allowable Shear Stress : The torque diagram shows that section AE is the critical section.

$$\tau_{max} = \tau_{allow} = \frac{2T}{\pi ab^2}$$

$$85(10^6) = \frac{2(275)}{\pi(a)\left(\frac{a}{2}\right)^2}$$

$$a = 0.02020 \text{ m} = 20.2 \text{ mm} \qquad \textbf{Ans}$$

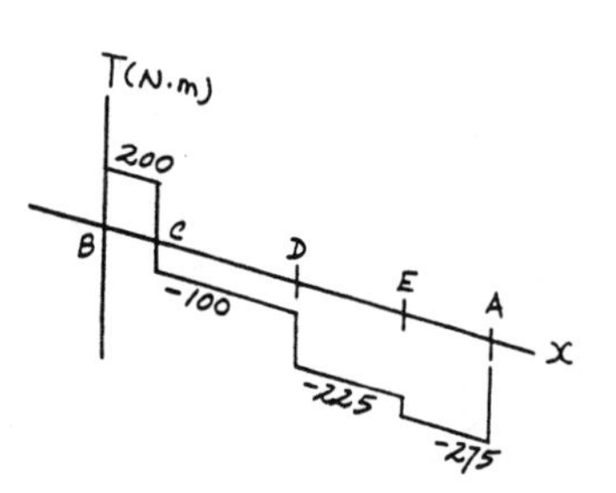

Angle of Twist :

$$\phi_{A/B} = \sum \frac{(a^2 + b^2)TL}{\pi a^3 b^3 G}$$

$$= \frac{(0.02020^2 + 0.01010^2)}{\pi(0.02020^3)(0.01010^3)(38.0)(10^9)}[(-275)(0.5) + (-225)(0.6) + (-100)(0.8) + 200(0.3)]$$

$$= -0.1472 \text{ rad} = |\,8.44°| \qquad \textbf{Ans}$$

***5–96.** The uniform C86100 bronze rod has an elliptical cross section. If it is subjected to the five torques shown, determine the maximum shear stress in the rod and the angle of twist of end A with respect to end B. Take a = 20 mm.

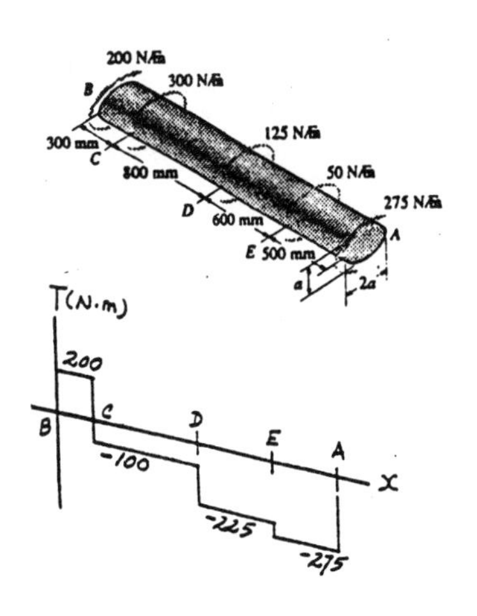

Maximum Shear Stress : The torque diagram shows that the maximum torque occurs at section AE.

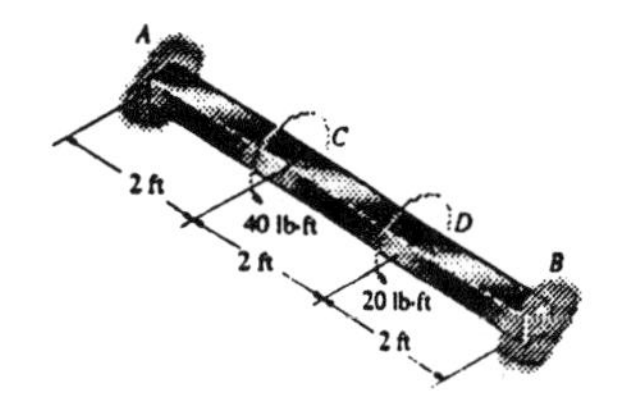

$$\tau_{max} = \frac{2T}{\pi a b^2} = \frac{2(275)}{\pi(0.02)(0.01)^2} = 87.5 \text{ MPa} \qquad \textbf{Ans}$$

Angle of Twist :

$$\phi_{A/B} = \sum \frac{(a^2+b^2)TL}{\pi a^3 b^3 G}$$

$$= \frac{(0.02^2+0.01^2)}{\pi(0.02^3)(0.01^3)(38.0)(10^9)}[(-275)(0.5) + (-225)(0.6) + (-100)(0.8) + 200(0.3)]$$

$$= -0.1531 \text{ rad} = |\,8.77°| \qquad \textbf{Ans}$$

5–97. The 2014-T6 aluminum strut is fixed between the two walls at A and B. If it has a 2 in. by 2 in. square cross section, and it is subjected to the torsional loading shown, determine the reactions at the fixed supports. Also, what is the angle of twist at C?

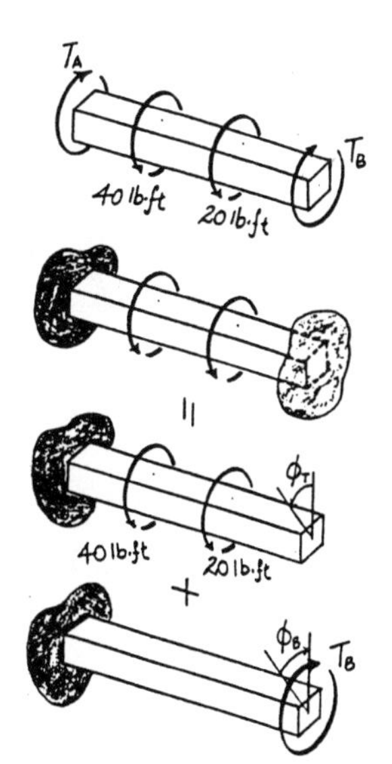

Equilibrium :

$$T_A + T_B - 60 = 0 \qquad [1]$$

Compatibility :

$$0 = \phi_T - \phi_B$$

$$0 = \frac{7.10(60)(12)(2)(12)}{a^4 G} + \frac{7.10(20)(12)(2)(12)}{a^4 G} - \frac{7.10(T_B)(6)(12)}{a^4 G} = 0$$

$$T_B = 320 \text{ lb}\cdot\text{in.} = 26.7 \text{ lb}\cdot\text{ft} \qquad \textbf{Ans}$$

Substituting the results into Eq.[1] yields :

$$T_A = 33.3 \text{ lb}\cdot\text{ft} \qquad \textbf{Ans}$$

Angle of Twist :

$$\phi_C = \frac{7.10 T_A L}{a^4 G} = \frac{7.10(33.33)(12)(2)(12)}{(2^4)(3.90)(10^6)} = 0.001092 \text{ rad} = 0.0626° \qquad \textbf{Ans}$$

5–98. The brass wire has a triangular cross section, 2 mm on a side. If the yield stress for brass is $\tau_Y = 205$ MPa, determine the maximum torque T to which it can be subjected so that the wire will not yield. If this torque is applied to a segment 4 m long, determine the greatest angle of twist of one end of the wire relative to the other end that will not cause permanent damage to the wire. $G_{br} = 37$ GPa.

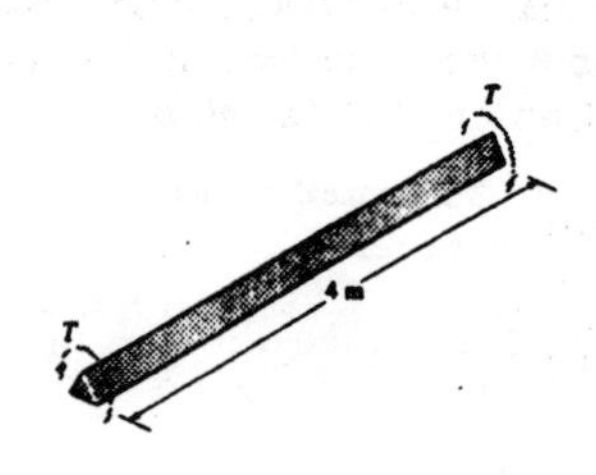

Allowable Shear Stress :

$$\tau_{\max} = \tau_Y = \frac{20T}{a^3}$$

$$205(10^6) = \frac{20T}{0.002^3}$$

$$T = 0.0820 \text{ N}\cdot\text{m} \qquad \textbf{Ans}$$

Angle of Twist :

$$\phi = \frac{46TL}{a^4G} = \frac{46(0.0820)(4)}{(0.002^4)(37)(10^9)}$$
$$= 25.5 \text{ rad} \qquad \textbf{Ans}$$

5–99. The 2014-T6 aluminum tube has a thickness of 5 mm and the mean cross-sectional dimensions shown. Determine the absolute maximum average shear stress in the tube. If the tube has a length of 4 m, determine the angle of twist at A. Set $T = 200$ N · m. Neglect the stress concentrations at the corners.

Section Properties :

$$A_m = \frac{1}{2}(0.1)(0.15) = 0.0075 \text{ m}^2$$

$$\int ds = 2\left(\sqrt{0.05^2 + 0.15^2}\right) + 0.1 = 0.41623 \text{ m}$$

Average Shear Stress :

$$(\tau_{avg})_{\text{abs max}} = \frac{T}{2tA_m} = \frac{300}{2(0.005)(0.0075)}$$
$$= 4.00 \text{ MPa} \qquad \textbf{Ans}$$

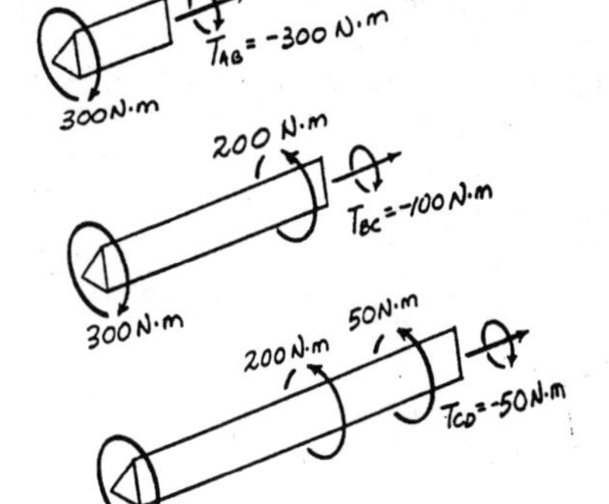

Angle of Twist :

$$\phi = \frac{TL}{4A_m^2G}\int\frac{ds}{t}$$

$$= \frac{0.41623}{4(0.0075^2)(27.0)(10^9)(0.005)}[(-300)(2) + (-100)(1) + (-50)(1)]$$

$$= -0.01028 \text{ rad} = |\,0.589°| \qquad \textbf{Ans}$$

***5–100.** The 2014-T6 aluminum tube has a thickness of 5 mm and the mean cross-sectional dimensions shown. Determine the largest magnitude of **T** so that the average shear stress in the tube does not exceed $\tau_{\text{allow}} = 150$ MPa. Neglect the stress concentrations at the corners.

$$(\tau_{avg})_{\max} = \tau_{\text{allow}} = \frac{T_{CD}}{2tA_m}$$

$$150(10^6) = \frac{T - 250}{2(0.005)(0.0075)}$$

$$T = 11500 \text{ N}\cdot\text{m} = 11.5 \text{ kN}\cdot\text{m} \qquad \textbf{Ans}$$

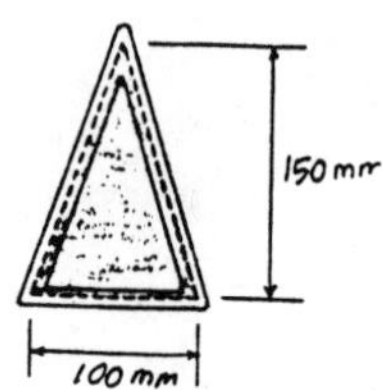

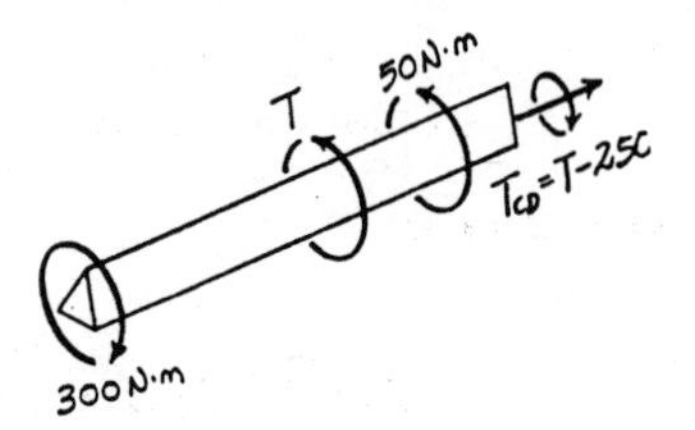

5–101. Consider a thin-walled tube of mean radius r and thickness t. Show that the maximum shear stress in the tube due to an applied torque **T** approaches the average shear stress computed from Eq. 5–18 as r/t' ∞

Section Properties :

$$r_o = r + \frac{t}{2} = \frac{2r + t}{2} \qquad r_i = r - \frac{t}{2} = \frac{2r - t}{2}$$

$$J = \frac{\pi}{2}\left[\left(\frac{2r + t}{2}\right)^4 - \left(\frac{2r - t}{2}\right)^4\right]$$
$$= \frac{\pi}{32}\left[(2r + t)^4 - (2r - t)^4\right]$$
$$= \frac{\pi}{32}\left(64 r^3 t + 16 r t^3\right)$$

Average Shear Stress :

$$\tau_{max} = \frac{Tc}{J} \qquad \text{where} \qquad c = r_o = \frac{2r + t}{2}$$

$$= \frac{T\left(\frac{2r + t}{2}\right)}{\frac{\pi}{32}\left[64 r^3 t + 16 r t^3\right]}$$
$$= \frac{T\left(\frac{2r + t}{2}\right)}{2\pi r t\left[r^2 + \frac{1}{4}t^2\right]}$$
$$= \frac{T\left(\frac{2r}{2r^2} + \frac{t}{2r^2}\right)}{2\pi r t\left(\frac{r^2}{r^2} + \frac{1}{4}\frac{t^2}{r^2}\right)}$$

As $\frac{r}{t} \to \infty$, then $\frac{t}{r} \to 0$

$$\tau_{max} = \frac{T\left(\frac{1}{r} + 0\right)}{2\pi r t(1 + 0)}$$
$$= \frac{T}{2\pi r^2 t}$$
$$= \frac{T}{2 t A_m} \qquad \textit{(Q.E.D.)}$$

5–102. The 304 stainless steel tube has a thickness of 10 mm. If the allowable shear stress is $\tau_{allow} = 80$ MPa, determine the maximum torque T that it can transmit. Also, what is the angle of twist of one end of the tube with respect to the other if the tube is 4 m long? Neglect the stress concentrations at the corners. The mean dimensions are shown.

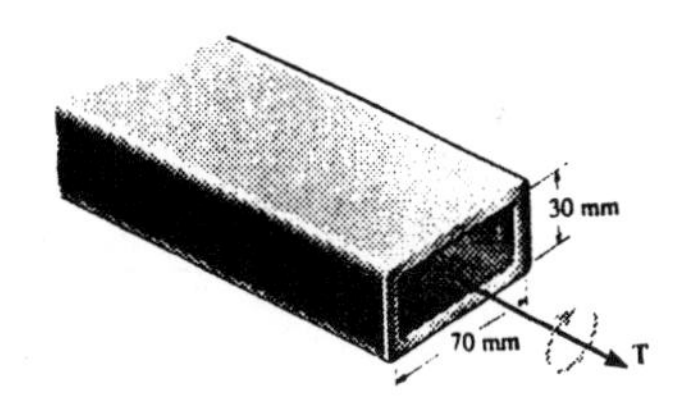

Section Properties :

$$A_m = 0.07(0.03) = 0.00210 \text{ m}^2$$

$$\int ds = 2(0.07) + 2(0.03) = 0.200 \text{ m}$$

Allowable Average Shear Stress :

10 mm

30 mm

70 mm

$$\tau_{avg} = \tau_{allow} = \frac{T}{2 t A_m}$$

$$80(10^6) = \frac{T}{2(0.01)(0.00210)}$$

$$T = 3360 \text{ N}\cdot\text{m} = 3.36 \text{ kN}\cdot\text{m} \qquad \textbf{Ans}$$

Angle of Twist :

$$\phi = \frac{TL}{4 A_m^2 G}\int \frac{ds}{t} = \frac{3360(4)(0.200)}{4(0.00210^2)(75.0)(10^9)(0.01)}$$

$$= 0.2032 \text{ rad} = 11.6° \qquad \textbf{Ans}$$

5–103. The 304 stainless steel tube has a thickness of 10 mm. If the applied torque is $T = 50\ \text{N}\cdot\text{m}$, determine the average shear stress in the tube. Neglect the stress concentrations at the corners. The mean dimensions are shown.

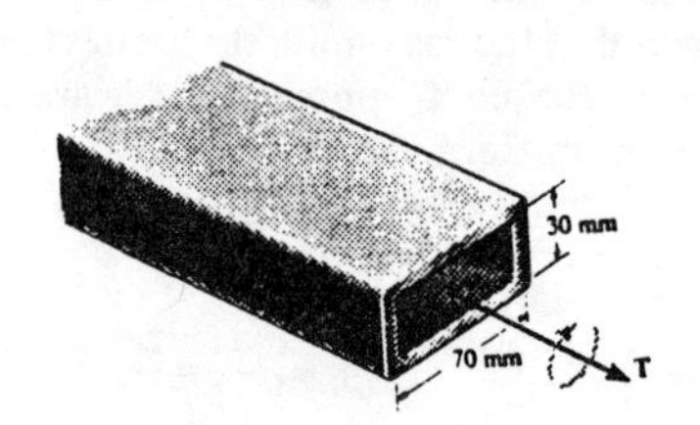

Section Properties :

$$A_m = 0.07(0.03) = 0.00210\ \text{m}^2$$

Average Shear Stress :

$$\tau_{avg} = \frac{T}{2tA_m} = \frac{50}{2(0.01)(0.00210)} = 1.19\ \text{MPa} \qquad \text{Ans}$$

***5–104.** A tube having the dimensions shown is subjected to a torque of $T = 50\ \text{N}\cdot\text{m}$. Neglecting the stress concentrations at its corners, determine the average shear stress in the tube at points A and B. Show the shear stress on volume elements located at these points.

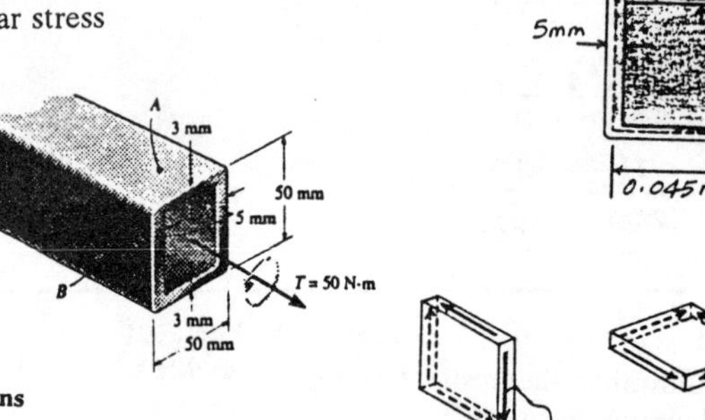

Section Properties :

$$A_m = 0.047(0.045) = 0.002115\ \text{m}^2$$

Average Shear Stress :

$$(\tau_A)_{avg} = \frac{T}{2tA_m} = \frac{50}{2(0.003)(0.002115)} = 3.94\ \text{MPa} \qquad \text{Ans}$$

$$(\tau_B)_{avg} = \frac{T}{2tA_m} = \frac{50}{2(0.005)(0.002115)} = 2.36\ \text{MPa} \qquad \text{Ans}$$

5–105. The steel tube has an elliptical cross section of the mean dimensions shown and a constant thickness of $t = 0.2$ in. If the allowable shear stress is $\tau_{allow} = 8$ ksi, determine the necessary dimension b needed to resist the torque shown. The mean area A_m for the ellipse is $\pi b(0.5b)$.

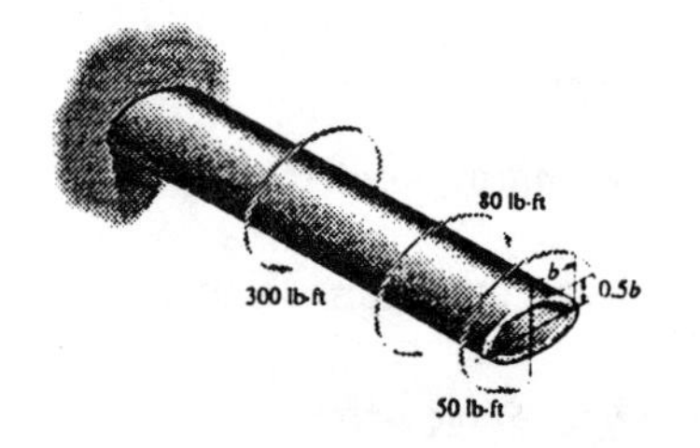

Internal Torque : As shown on FBD the maximum torque is $T_{max} = 270\ \text{lb}\cdot\text{ft}$.

Allowable Average Shear Stress :

$$\tau_{avg} = \tau_{allow} = \frac{T_{max}}{2tA_m}$$

$$8(10^3) = \frac{270(12)}{2(0.2)(\pi b)(0.5b)}$$

$$b = 0.803\ \text{in.} \qquad \text{Ans}$$

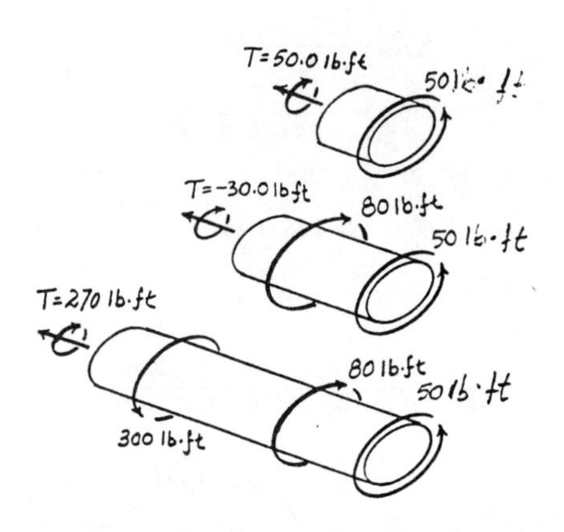

5–106. A 6061-T6 aluminum tube is subjected to a torque of 75 kN · m. Determine the smallest dimension a if the thickness of the material is 7 mm and the allowable shear stress is $\tau_{allow} = 125$ MPa. If the tube has a length of 2 m, determine the angle of twist of one end relative to the other end.

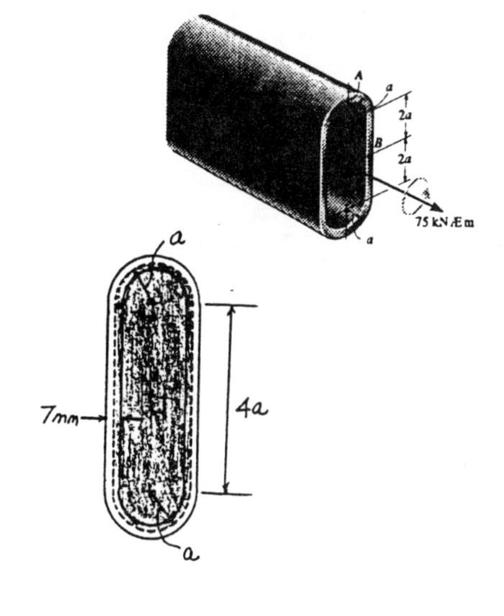

Section Properties :

$$A_m = 4a(2a) + \pi a^2 = (8 + \pi)a^2$$

$$\int ds = (4a)(2) + 2\pi a = (8 + 2\pi)a$$

Allowable Average Shear Stress :

$$\tau_{avg} = \tau_{allow} = \frac{T}{2tA_m}$$

$$125(10^6) = \frac{75(10^3)}{2(0.007)(8+\pi)a^2}$$

$$a = 0.06202 \text{ m} = 62.0 \text{ mm} \qquad \textbf{Ans}$$

Angle of Twist :

$$\phi = \frac{TL}{4A_m^2 G}\int \frac{ds}{t}$$

$$= \frac{75(10^3)(2)(8+2\pi)(0.06202)}{4[(8+\pi)(0.06202^2)]^2(26.0)(10^9)(0.007)}$$

$$= 0.09937 \text{ rad} = 5.69° \qquad \textbf{Ans}$$

5–107. For a given average shear stress, determine the factor by which the torque-carrying capacity is increased if the half-circular sections are reversed from the dashed-line positions to the section shown. The tube is 0.1 in. thick.

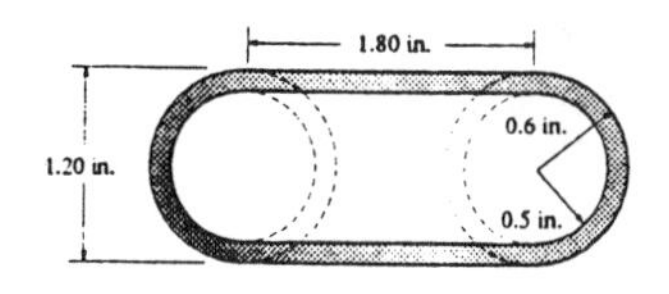

Section Properties :

$$A'_m = (1.1)(1.8) - \left[\frac{\pi(0.55^2)}{2}\right](2) = 1.02967 \text{ in}^2$$

$$A_m = (1.1)(1.8) + \left[\frac{\pi(0.55^2)}{2}\right](2) = 2.93033 \text{ in}^2$$

Average Shear Stress :

$$\tau_{avg} = \frac{T}{2tA_m}; \qquad T = 2tA_m\tau_{avg}$$

Hence,

$$T' = 2tA'_m\tau_{avg}$$

$$\text{The factor of increase} = \frac{T}{T'} = \frac{A_m}{A'_m} = \frac{2.93033}{1.02967}$$

$$= 2.85 \qquad \textbf{Ans}$$

***5–108.** Due to a fabrication error the inner circle of the tube is eccentric with respect to the outer circle. By what percentage is the torsional strength reduced when the eccentricity e is one-fourth of the difference in the radii?

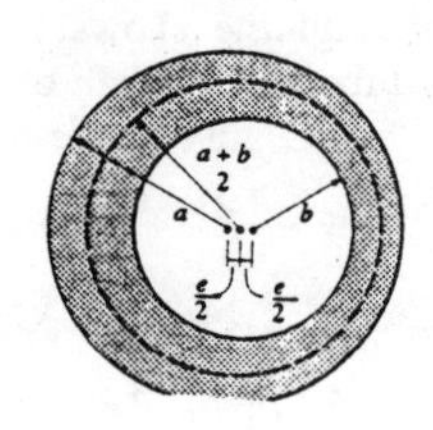

Average Shear Stress :
For the aligned tube

$$\tau_{avg} = \frac{T}{2\,t A_m} = \frac{T}{2(a-b)(\pi)\left(\frac{a+b}{2}\right)^2}$$

$$T = \tau_{avg}(2)(a-b)(\pi)\left(\frac{a+b}{2}\right)^2$$

For the eccentric tube

$$\tau_{avg} = \frac{T'}{2\,t A_m}$$

$$t = a - \frac{e}{2} - \left(\frac{e}{2} + b\right) = a - e - b$$

$$= a - \frac{1}{4}(a-b) - b = \frac{3}{4}(a-b)$$

$$T' = \tau_{avg}(2)\left[\frac{3}{4}(a-b)\right](\pi)\left(\frac{a+b}{2}\right)^2$$

$$\text{Factor} = \frac{T'}{T} = \frac{\tau_{avg}(2)\left[\frac{3}{4}(a-b)\right](\pi)\left(\frac{a+b}{2}\right)^2}{\tau_{avg}(2)(a-b)(\pi)\left(\frac{a+b}{2}\right)^2} = \frac{3}{4}$$

Percent reduction in strength $= \left(1 - \frac{3}{4}\right) \times 100\,\% = 25\,\%$ **Ans**

5–109. A torque of 2 kip · in. is applied to the tube. The mean dimensions are shown and the wall thickness is 0.1 in. Determine the average shear stress in the tube. Neglect stress concentrations at the corners.

Section Properties :

$$A_m = \frac{\pi\,(2^2)}{2} = 2\pi \text{ in}^2$$

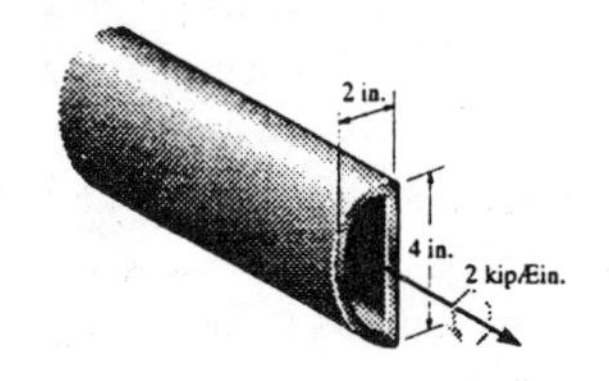

Average Shear Stress :

$$\tau_{avg} = \frac{T}{2\,t A_m}$$

$$= \frac{2(10^3)}{2(0.1)(2\pi)} = 1591 \text{ psi} = 1.59 \text{ ksi} \quad \textbf{Ans}$$

5–110. The plastic hexagonal tube is subjected to a torque of 150 N · m. Determine the mean dimension a of its sides if the allowable shear stress is $\tau_{allow} = 60$ MPa. Each side has a thickness of $t = 3$ mm.

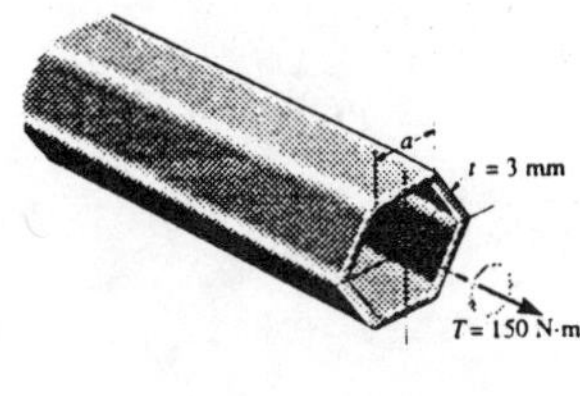

Section Properties :

$$A_m = 6\left[\frac{1}{2}(a)(a\sin 60°)\right] = 2.5981a^2$$

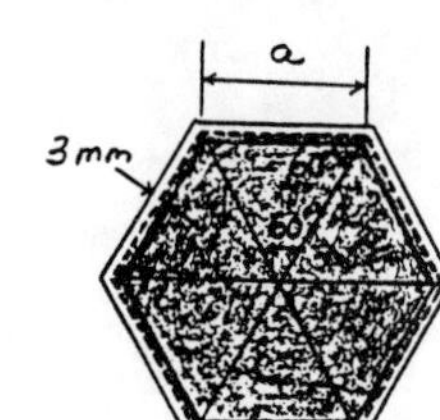

Average Shear Stress :

$$\tau_{avg} = \tau_{allow} = \frac{T}{2\,t A_m}$$

$$60(10^6) = \frac{150}{(2)(0.003)(2.5981a^2)}$$

$$a = 0.01266 \text{ m} = 12.7 \text{ mm} \quad \textbf{Ans}$$

5–111. The steel used for the shaft has an allowable shear stress of τ_{allow} = 8 MPa. If the members are connected with a fillet weld of radius r = 4 mm, determine the maximum torque T that can be applied.

Allowable Shear Stress :

$$\frac{D}{d} = \frac{50}{20} = 2.5 \quad \text{and} \quad \frac{r}{d} = \frac{4}{20} = 0.20$$

From the text, $K = 1.25$

$$\tau_{max} = \tau_{allow} = K\frac{Tc}{J}$$

$$8(10)^6 = 1.25\left[\frac{\frac{T}{2}(0.01)}{\frac{\pi}{2}(0.01^4)}\right]$$

$$T = 20.1 \text{ N}\cdot\text{m} \qquad \textbf{Ans}$$

***5–112.** The members are connected with a fillet weld of radius r = 4 mm. Determine the maximum shear stress in the shaft if T = 10 N · m.

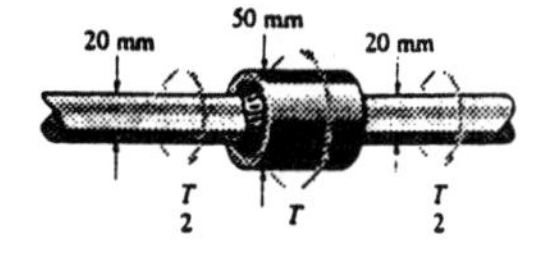

Maximum Shear Stress :

$$\frac{D}{d} = \frac{50}{20} = 2.5 \quad \text{and} \quad \frac{r}{d} = \frac{4}{20} = 0.20$$

From the text, $K = 1.25$

$$\tau_{max} = K\frac{Tc}{J} = 1.25\left[\frac{\frac{10}{2}(0.01)}{\frac{\pi}{2}(0.01^4)}\right] = 3.98 \text{ MPa} \qquad \textbf{Ans}$$

5–113. The shaft is used to transmit 9 hp while turning at 600 rpm. Determine the maximum shear stress in the shaft. The segments are connected using a fillet weld having a radius of 0.15 in.

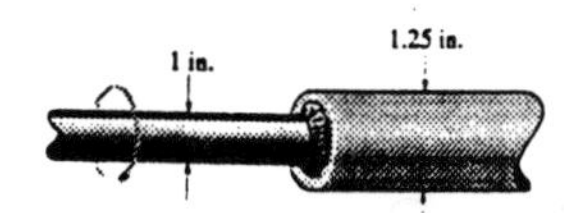

Internal Torque :

$$\omega = 600\,\frac{\text{rev}}{\text{min}}\left(\frac{2\pi \text{ rad}}{1 \text{ rev}}\right)\frac{1 \text{ min}}{60 \text{ s}} = 20.0\,\pi \text{ rad/s}$$

$$P = 9 \text{ hp}\left(\frac{550 \text{ ft}\cdot\text{lb/s}}{1 \text{ hp}}\right) = 4950 \text{ ft}\cdot\text{lb/s}$$

$$T = \frac{P}{\omega} = \frac{4950}{20.0\pi} = 78.78 \text{ lb}\cdot\text{ft}$$

Maximum Shear Stress :

$$\frac{D}{d} = \frac{1.25}{1} = 1.25 \quad \text{and} \quad \frac{r}{d} = \frac{0.15}{1} = 0.15$$

From the text, $K = 1.25$

$$\tau_{max} = K\frac{Tc}{J} = 1.25\left[\frac{78.78(12)(0.5)}{\frac{\pi}{2}(0.5^4)}\right]$$

$$= 6.02 \text{ ksi} \qquad \textbf{Ans}$$

5–114. The shaft is fixed to the wall at A and is subjected to the torques shown. Determine the maximum shear stress in the shaft. A fillet weld having a radius of 4.5 mm is used to connect the shafts at B.

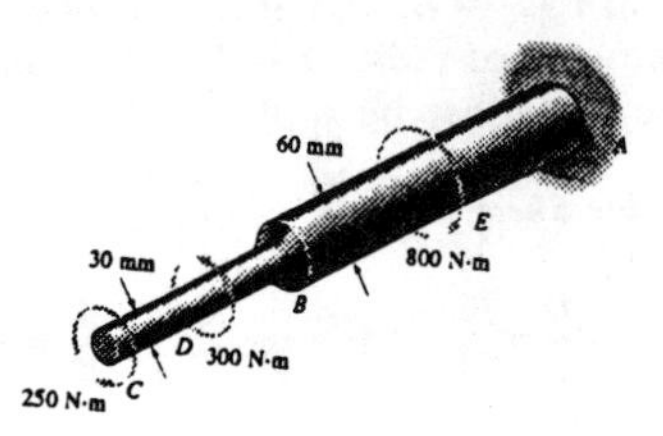

Maximum Shear Stress :
For segment CD

$$(\tau_{CD})_{max} = \frac{T_{CD}\,c}{J} = \frac{250(0.015)}{\frac{\pi}{2}(0.015^4)}$$
$$= 47.2 \text{ MPa} \qquad \textit{(Max)} \qquad \textbf{Ans}$$

$T_{CD} = 250$ N·m, 250 N·m

For segment EA

$$(\tau_{EA})_{max} = \frac{T_{EA}\,c}{J} = \frac{750(0.03)}{\frac{\pi}{2}(0.03^4)} = 17.68 \text{ MPa}$$

300 N·m, $T_{DB} = 50.0$ N·m, 250 N·m

300 N·m, $T_{EA} = 750$ N·m, 800 N·m, 250 N·m

For the fillet

$$\frac{D}{d} = \frac{60}{30} = 2 \qquad \text{and} \qquad \frac{r}{d} = \frac{4.5}{30} = 0.15$$

From the text, $\quad K = 1.3$

$$(\tau_{max})_f = K\,\frac{T_{DB}\,c}{J}$$
$$= 1.3\left[\frac{50(0.015)}{\frac{\pi}{2}(0.015^4)}\right] = 12.26 \text{ MPa}$$

5–115. The steel used for the shaft has an allowable shear stress of $\tau_{allow} = 8$ MPa. If the members are connected together with a fillet weld of radius $r = 2.25$ mm, determine the maximum torque T that can be applied.

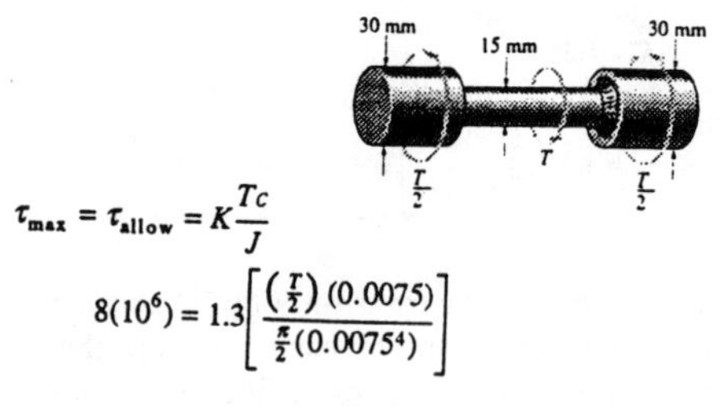

Allowable Shear Stress :

$$\frac{D}{d} = \frac{30}{15} = 2 \qquad \text{and} \qquad \frac{r}{d} = \frac{2.25}{15} = 0.15$$

From the text, $\quad K = 1.30$

$$\tau_{max} = \tau_{allow} = K\frac{Tc}{J}$$

$$8(10^6) = 1.3\left[\frac{\left(\frac{T}{2}\right)(0.0075)}{\frac{\pi}{2}(0.0075^4)}\right]$$

$$T = 8.16 \text{ N}\cdot\text{m} \qquad \textbf{Ans}$$

***5–116.** The assembly is subjected to a torque of 1800 lb · in. If the allowable shear stress for the material is $\tau_{allow} = 12$ ksi, determine the radius of the smallest size fillet that can be used to transmit the torque.

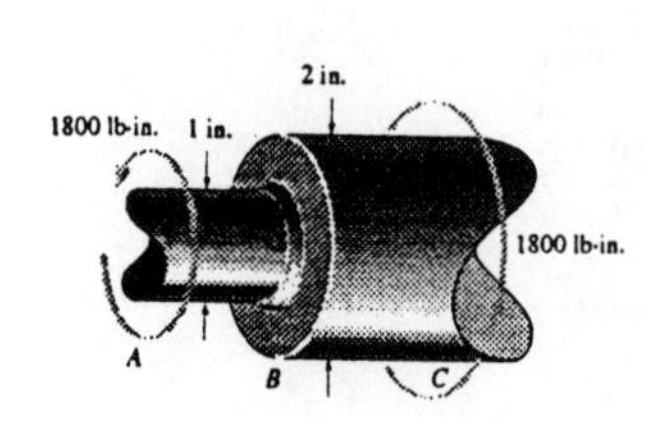

Allowable Shear Stress :

$$\tau_{max} = \tau_{allow} = K\,\frac{Tc}{J}$$

$$12(10^3) = K\left[\frac{1800(0.5)}{\frac{\pi}{2}(0.5^4)}\right]$$

$$K = 1.31$$

For $\frac{D}{d} = \frac{2}{1} = 2$, from the text; $\quad \frac{r}{d} = 0.15$

$$\frac{r}{1} = 0.15; \qquad r = 0.15 \text{ in.} \qquad \textbf{Ans}$$

Check : $\frac{D-d}{2} = \frac{2-1}{2} = 0.5 \text{ in.} > 0.15 \text{ in.} \qquad \textit{(OK!)}$

5–117. A solid shaft is subjected to the torque **T**, which causes the material to yield. If the material is elastic perfectly plastic, show that the torque can be expressed in terms of the angle of twist ϕ of the shaft as $T = \frac{4}{3}T_Y(1 - \phi_Y^3/4\phi^3)$, where T_Y and ϕ_Y are the torque and angle of twist when the material begins to yield.

Elastic - Plastic Torque :

$$\phi = \frac{\gamma L}{\rho} = \frac{\gamma_Y}{\rho_Y}L$$

$$\rho_Y = \frac{\gamma_Y L}{\phi} \qquad [1]$$

When $\rho_Y = c$, $\phi = \phi_Y$, then from Eq. [1]

$$c = \frac{\gamma_Y L}{\phi_Y} \qquad [2]$$

Dividing Eq. [1] by Eq. [2] yields

$$\frac{\rho_Y}{c} = \frac{\phi_Y}{\phi} \qquad [3]$$

Useing Eq. 5 – 26 from the text

$$T = \frac{\pi\,\tau_Y}{6}\left(4\,c^3 - \rho_Y^3\right) = \frac{2\pi\,\tau_Y c^3}{3}\left(1 - \frac{\rho_Y^3}{4\,c^3}\right)$$

Using Eq. 5 – 24, $\quad T_Y = \frac{\pi}{2}\tau_Y c^3$, from the text and Eq. [3]

$$T = \frac{4}{3}T_Y\left(1 - \frac{\phi_Y^3}{4\,\phi^3}\right) \qquad \textit{(Q.E.D.)}$$

5–118. A bar having a circular cross section of 3 in. diameter is subjected to a torque of 100 kip · in. If the material is elastic perfectly plastic, with $\tau_Y = 16$ ksi, determine the radius of the elastic core.

Elastic - Plastic Torque : Applying Eq. 5 – 26 from the text

$$T = \frac{\pi\,\tau_Y}{6}\left(4\,c^3 - \rho_Y^3\right)$$

$$100 = \frac{\pi\,(16)}{6}\left[4\left(1.5^3\right) - \rho_Y^3\right]$$

$\rho_Y = 1.16$ in. **Ans**

5–119. A solid shaft having a diameter of 2 in. is made of elastic-perfectly plastic material having a yield stress of $\tau_Y = 16$ ksi and shear modulus of $G = 12(10^3)$ ksi. Determine the torque required to develop an elastic core in the shaft having a diameter of 1 in. Also, what is the plastic torque?

Elastic - Plastic Torque : Applying Eq. 5 – 26 from the text

$$T = \frac{\pi\,\tau_Y}{6}\left(4\,c^3 - \rho_Y{}^3\right)$$

$$= \frac{\pi\,(16)}{6}\left[4\left(1^3\right) - 0.5^3\right]$$

$= 32.46$ kip · in. $= 2.71$ kip · ft **Ans**

Plastic Torque : Using Eq. 5 – 27 from the text

$$T_P = \frac{2\pi}{3}\tau_Y c^3$$

$$= \frac{2\pi}{3}(16)(1^3)$$

$= 33.51$ kip · in. $= 2.79$ kip · ft **Ans**

***5–120.** A tubular shaft has an inner diameter of 20 mm, an outer diameter of 40 mm, and a length of 1 m. It is made of an elastic perfectly plastic material having a yield stress of $\tau_Y = 100$ MPa. Determine the maximum torque it can transmit. What is the angle of twist of one end with respect to the other end if the shear strain on the inner surface of the tube is about to yield? $G = 80$ GPa.

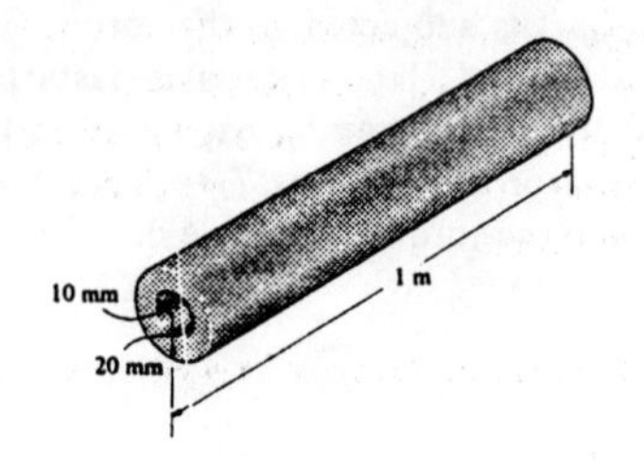

Plastic Torque :

$$T_P = 2\pi\int_{c_i}^{c_o} \tau_Y\, \rho^2\, d\rho$$

$$= 2\pi\tau_Y\left[\frac{\rho^3}{3}\right]\Bigg|_{c_i}^{c_o}$$

$$= \frac{2\pi\tau_Y}{3}\left(c_o^3 - c_i^3\right)$$

$$= \frac{2\pi(100)(10^6)}{3}\left(0.02^3 - 0.01^3\right)$$

$$= 1466\ \text{N}\cdot\text{m} = 1.47\ \text{kN}\cdot\text{m} \qquad \textbf{Ans}$$

Angle of Twist :

$$\gamma_Y = \frac{\tau_Y}{G} = \frac{100(10^6)}{80(10^9)} = 0.00125\ \text{rad}$$

$$\phi = \frac{\gamma_Y}{\rho_Y}L = \left(\frac{0.00125}{0.01}\right)(1) = 0.125\ \text{rad} = 7.16^\circ \qquad \textbf{Ans}$$

5–121. Determine the torque needed to twist a short 3-mm-diameter steel wire through several revolutions if it is made of steel assumed to be elastic perfectly plastic and having a yield stress of $\tau_Y = 80$ MPa. Assume that the material becomes fully plastic.

Plastic Torque : When the material becomes fully plastic, then

$$T_P = 2\pi\int_0^c \tau_Y\, \rho^2 d\rho$$

$$= \frac{2\pi}{3}\tau_Y c^3$$

$$= \frac{2\pi}{3}(80)\left(10^6\right)\left(0.0015^3\right)$$

$$= 0.565\ \text{N}\cdot\text{m} \qquad \textbf{Ans}$$

5–122. A solid shaft has a diameter of 40 mm and length of 1 m. It is made of an elastic perfectly plastic material having a yield stress of $\tau_Y = 100$ MPa. Determine the maximum elastic torque T_Y and the corresponding angle of twist. What is the angle of twist if the torque is increased to $T = 1.2T_Y$? $G = 80$ GPa.

Maximum Elastic Torque :

$$\tau_Y = \frac{T_Y\, c}{J}$$

$$T_Y = \frac{\tau_Y J}{c}$$

$$= \frac{100(10^6)\left(\frac{\pi}{2}\right)(0.02^4)}{0.02}$$

$$= 1256.64\ \text{N}\cdot\text{m} = 1.26\ \text{kN}\cdot\text{m} \qquad \textbf{Ans}$$

Angle of twist :

$$\gamma_Y = \frac{\tau_Y}{G} = \frac{100(10^6)}{80(10^9)} = 0.00125\ \text{rad}$$

$$\phi = \frac{\gamma_Y}{\rho_Y}L = \left(\frac{0.00125}{0.02}\right)(1) = 0.0625\ \text{rad} = 3.58^\circ \qquad \textbf{Ans}$$

Or

$$\phi = \frac{T_Y L}{JG} = \frac{1256.64(1)}{\frac{\pi}{2}(0.02^4)80(10^9)} = 0.0625\ \text{rad}$$

Elastic - Plastic Torque : Applying Eq. 5–26 of the text,

$$T = \frac{\pi\,\tau_Y}{6}\left(4c^3 - \rho_Y^3\right)$$

$$1.2(1256.64) = \frac{\pi(100)(10^6)}{6}\left[4\left(0.02^3\right) - \rho_Y^3\right]$$

$$\rho_Y = 0.01474\ \text{m}$$

$$\phi = \frac{\gamma_Y}{\rho_Y}L = \left(\frac{0.00125}{0.01474}\right)(1) = 0.08483\ \text{rad} = 4.86^\circ \qquad \textbf{Ans}$$

5–123. The stepped shaft is subjected to a torque **T** that produces yielding on the surface of the larger diameter segment. Determine the radius of the elastic core produced in the smaller diameter segment. Neglect the stress concentration at the fillet.

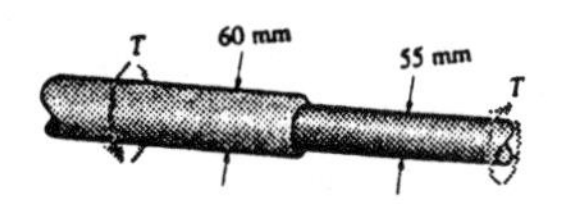

Maximum Elastic Torque : For the larger diameter segment

$$\tau_Y = \frac{T_Y\, c}{J}$$

$$T_Y = \frac{\tau_Y J}{c}$$

$$= \frac{\tau_Y \left(\frac{\pi}{2}\right)(0.03^4)}{0.03}$$

$$= 13.5\left(10^{-6}\right)\pi\tau_Y$$

Elastic - Plastic Torque : For the smaller diameter segment

$T_P = \frac{2\pi}{3}\tau_Y c^3 = \frac{2\pi}{3}\tau_Y\left(0.0275^3\right) = 13.86\left(10^{-6}\right)\pi\tau_Y > 13.5\left(10^{-6}\right)\pi\tau_Y$.

Applying Eq. 5 – 26 of the text, we have

$$T = \frac{\pi\,\tau_Y}{6}\left(4\,c^3 - \rho_Y^3\right)$$

$$13.5\left(10^{-6}\right)\pi\tau_Y = \frac{\pi\,\tau_Y}{6}\left[4\left(0.0275^3\right) - \rho_Y^3\right]$$

$$\rho_Y = 0.01298 \text{ m} = 13.0 \text{ mm} \qquad \textbf{Ans}$$

***5–124.** The tube has a length of 2 m and is made of an elastic plastic material as shown. Determine the torque needed to just cause the material to become fully plastic. What is the permanent angle of twist of the tube when this torque is removed?

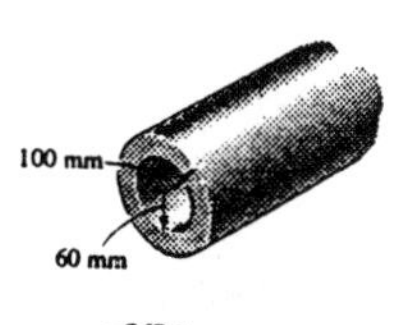

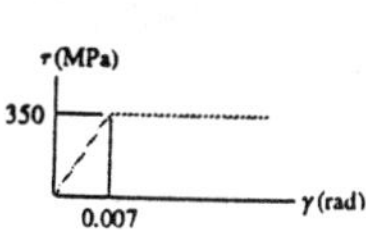

Plastic Torque :

$$T_P = 2\pi\int_{c_i}^{c_o} \tau_Y\, \rho^2\, d\rho$$

$$= \frac{2\,\pi\,\tau_Y}{3}\left(c_o^3 - c_i^3\right)$$

$$= \frac{2\pi\,(350)(10^6)}{3}\left(0.05^3 - 0.03^3\right)$$

$$= 71837.75 \text{ N}\cdot\text{m} = 71.8 \text{ kN}\cdot\text{m} \qquad \textbf{Ans}$$

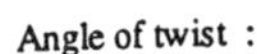

Angle of twist :

$$\phi_P = \frac{\gamma_Y}{\rho_Y}L \qquad \text{Where } \rho_Y = c_i = 0.03 \text{ m}$$

$$= \left(\frac{0.007}{0.03}\right)(2) = 0.4667 \text{ rad}$$

When a reverse $T_P = 71837.75$ N · m is applied,

$$G = \frac{350(10^6)}{0.007} = 50 \text{ GPa}$$

$$\phi'_P = \frac{T_P L}{JG} = \frac{71837.75(2)}{\frac{\pi}{2}(0.05^4 - 0.03^4)\,50(10^9)} = 0.3363 \text{ rad}$$

Permanent angle of twist :

$$\phi_r = \phi_P - \phi'_P = 0.4667 - 0.3363$$

$$= 0.1304 \text{ rad} = 7.47^\circ \qquad \textbf{Ans}$$

5–125. The A-36 steel shaft has a diameter of 2 in. and is subjected to a torque of 3500 lb·ft. Determine the angle of twist for the shaft if $\tau_Y = 80$ ksi and the shaft is 1.5 ft long.

Elastic - Plastic Torque : The maximum elastic torque is $T_Y = \frac{\pi}{2}\tau_Y c^3 = \frac{\pi}{2}(22)\left(1^3\right) = 34.56$ kip · in = 2879.8 lb · ft and the plastic torque is $T_P = \frac{2\pi}{3}\tau_Y c^3 = \frac{2\pi}{3}(22)\left(1^3\right) = 46.08$ kip · in = 3839.7 lb · ft. Since $T_Y < T < T_P$, applying Eq. 5-26 from the text, we have

$$T = \frac{\pi\,\tau_Y}{6}\left(4c^3 - \rho_Y^3\right)$$

$$3500(12) = \frac{\pi(22)(10^3)}{6}\left[4\left(1^3\right) - \rho_Y^3\right]$$

$$\rho_Y = 0.7073 \text{ in.}$$

Angle of Twist :

$$\gamma_Y = \frac{\tau_Y}{G} = \frac{22(10^3)}{11.0(10^6)} = 0.002 \text{ rad}$$

$$\phi = \frac{\gamma_Y}{\rho_Y}L = \left(\frac{0.002}{0.7073}\right)(1.5)(12)$$
$$= 0.05089 \text{ rad} = 2.92° \qquad \text{Ans}$$

5–126. The 2-m-long tube is made of an elastic perfectly plastic material as shown. Determine the applied torque T which subjects the material at the tube's outer edge to a shear strain of $\gamma_{max} = 0.006$ rad. What would be the permanent angle of twist of the tube when this torque is removed? Sketch the residual stress distribution in the tube.

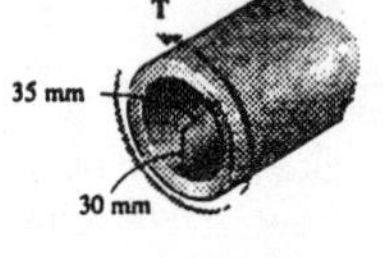

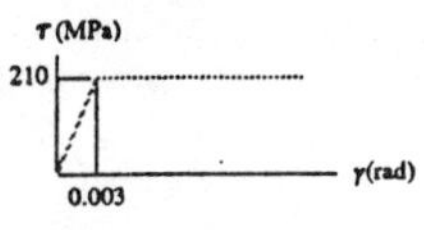

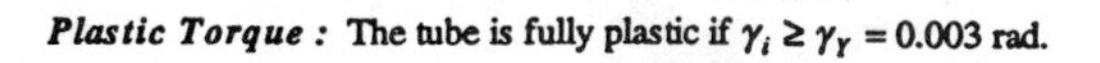

Plastic Torque : The tube is fully plastic if $\gamma_i \geq \gamma_Y = 0.003$ rad.

$$\frac{\gamma}{0.03} = \frac{0.006}{0.035}; \qquad \gamma = 0.005143 \text{ rad}$$

Therefore the tube is fully plastic.

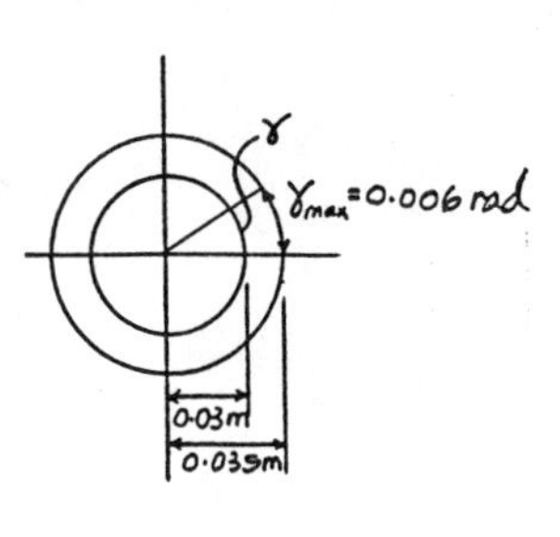

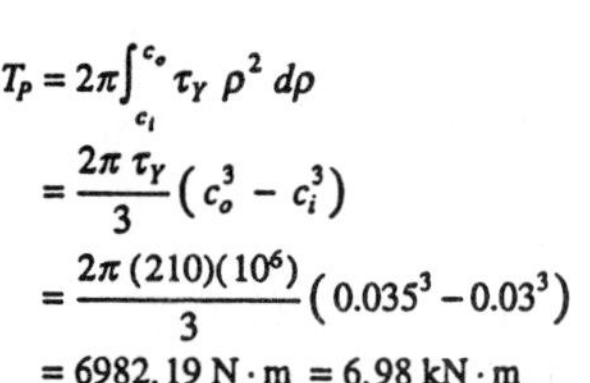

$$T_P = 2\pi\int_{c_i}^{c_o} \tau_Y\,\rho^2\,d\rho$$
$$= \frac{2\pi\,\tau_Y}{3}\left(c_o^3 - c_i^3\right)$$
$$= \frac{2\pi(210)(10^6)}{3}\left(0.035^3 - 0.03^3\right)$$
$$= 6982.19 \text{ N}\cdot\text{m} = 6.98 \text{ kN}\cdot\text{m} \qquad \text{Ans}$$

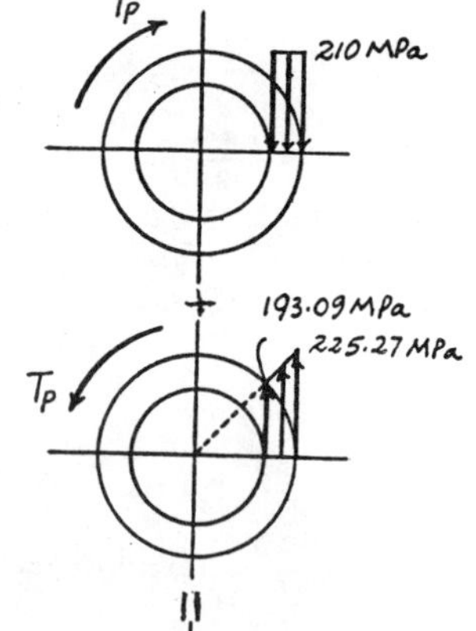

Angle of Twist :

$$\phi_P = \frac{\gamma_{max}}{c_o}L = \left(\frac{0.006}{0.035}\right)(2) = 0.34286 \text{ rad}$$

When a reverse torque of $T_P = 6982.19$ N · m is applied,

$$G = \frac{\tau_Y}{\gamma_Y} = \frac{210(10^6)}{0.003} = 70 \text{ GPa}$$

$$\phi_P' = \frac{T_P L}{JG} = \frac{6982.19(2)}{\frac{\pi}{2}(0.035^4 - 0.03^4)(70)(10^9)} = 0.18389 \text{ rad}$$

Permanent angle of twist,

$$\phi_r = \phi_P - \phi_P'$$
$$= 0.34286 - 0.18389 = 0.1590 \text{ rad} = 9.11° \qquad \text{Ans}$$

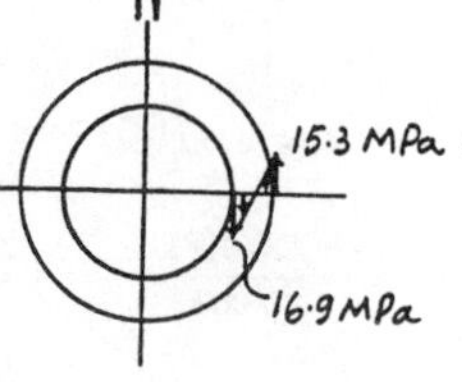

Residual Shear Stress :

$$\tau_{P_o}' = \frac{T_P\,c}{J} = \frac{6982.19(0.035)}{\frac{\pi}{2}(0.035^4 - 0.03^4)} = 225.27 \text{ MPa}$$

$$\tau_{P_i}' = \frac{T_P\,\rho}{J} = \frac{6982.19(0.03)}{\frac{\pi}{2}(0.035^4 - 0.03^4)} = 193.09 \text{ MPa}$$

$$(\tau_r)_o = -\tau_Y + \tau_{P_o}' = -210 + 225.27 = 15.3 \text{ MPa}$$
$$(\tau_r)_i = -\tau_Y + \tau_{P_i}' = -210 + 193.09 = -16.9 \text{ Mpa}$$

5–127. The shaft consists of two sections that are rigidly connected. If the material is elastic perfectly plastic as shown, determine the largest torque T that can be applied to the shaft. Also, draw the shear-stress distribution over a radial line for each section. Neglect the effect of stress concentration.

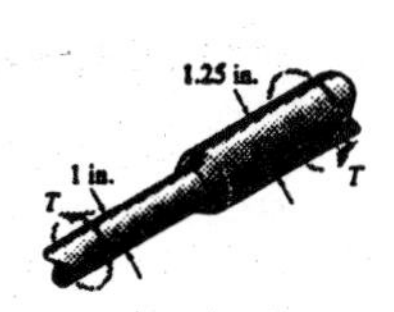

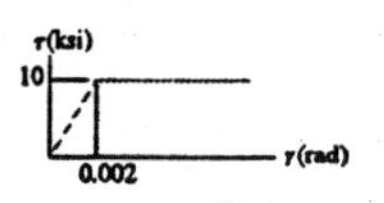

Plastic Torque : For the smaller - diameter segment

$$T_P = 2\pi\int_0^c \tau_Y \rho^2 d\rho$$
$$= \frac{2\pi}{3}\tau_Y c^3$$
$$= \frac{2\pi}{3}(10)(0.5^3)$$
$$= 2.618 \text{ kip}\cdot\text{in.} = 218 \text{ lb}\cdot\text{ft} \qquad \textbf{Ans}$$

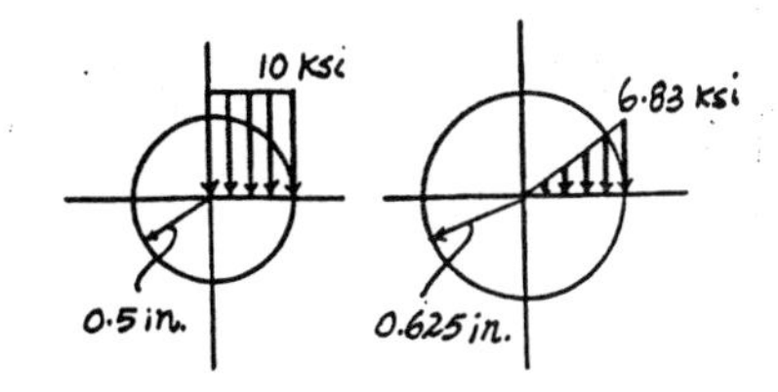

Maximum Shear Stress : For the bigger - diameter segment

$$\tau_{max} = \frac{Tc}{J} = \frac{2.618(0.625)}{\frac{\pi}{2}(0.625^4)}$$
$$= 6.83 \text{ ksi} < \tau_Y = 10 \text{ ksi} \quad \textit{(OK!)}$$

***5–128.** The solid shaft is made of an elastic-perfectly plastic material as shown. Determine the torque T needed to form an elastic core in the shaft having a radius of $\rho_Y = 20$ mm. If the shaft is 3 m long, through what angle does one end of the shaft twist with respect to the other end? When the torque is removed, determine the residual stress distribution in the shaft and the permanent angle of twist.

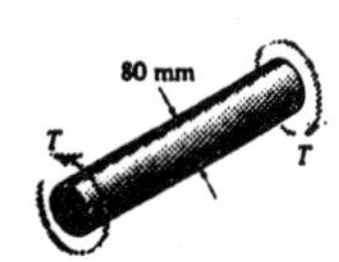

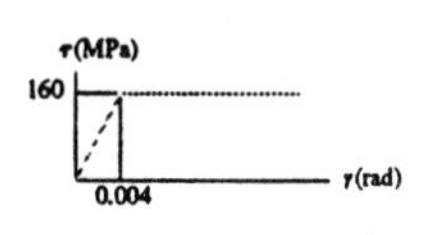

Elastic - Plastic Torque : Applying Eq. 5 – 26 from the text

$$T = \frac{\pi \tau_Y}{6}\left(4c^3 - \rho_Y^3\right)$$
$$= \frac{\pi(160)(10^6)}{6}\left[4\left(0.04^3\right) - 0.02^3\right]$$
$$= 20776.40 \text{ N}\cdot\text{m} = 20.8 \text{ kN}\cdot\text{m} \qquad \textbf{Ans}$$

Angle of Twist :

$$\phi = \frac{\gamma_Y}{\rho_Y}L = \left(\frac{0.004}{0.02}\right)(3) = 0.600 \text{ rad} = 34.4° \qquad \textbf{Ans}$$

When the reverse $T = 20776.4$ N · m is applied,

$$G = \frac{160(10^6)}{0.004} = 40 \text{ GPa}$$

$$\phi' = \frac{TL}{JG} = \frac{20776.4(3)}{\frac{\pi}{2}(0.04^4)(40)(10^9)} = 0.3875 \text{ rad}$$

The permanent angle of twist is,

$$\phi_r = \phi - \phi'$$
$$= 0.600 - 0.3875 = 0.2125 \text{ rad} = 12.2° \qquad \textbf{Ans}$$

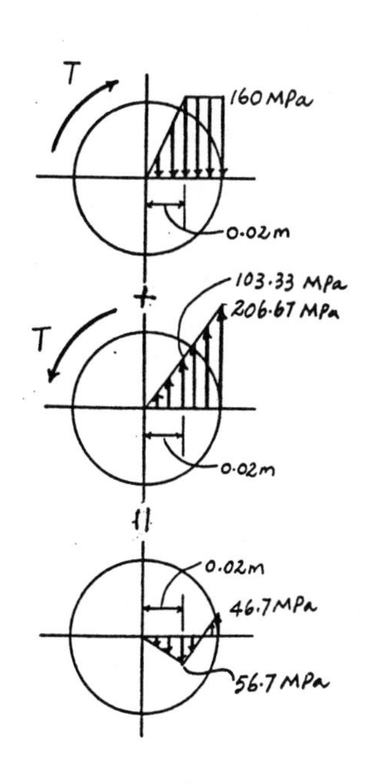

Residual Shear Stress :

$$(\tau')_{\rho=c} = \frac{Tc}{J} = \frac{20776.4(0.04)}{\frac{\pi}{2}(0.04^4)} = 206.67 \text{ MPa}$$
$$(\tau')_{\rho=0.02\text{m}} = \frac{Tc}{J} = \frac{20776.4(0.02)}{\frac{\pi}{2}(0.04^4)} = 103.33 \text{ MPa}$$

$$(\tau_r)_{\rho=c} = -160 + 206.67 = 46.7 \text{ MPa}$$
$$(\tau_r)_{\rho=0.02\text{m}} = -160 + 103.33 = -56.7 \text{ MPa}$$

5–129. The shaft is made of an elastic-perfectly plastic material as shown. Plot the shear-stress distribution acting along a radial line if it is subjected to a torque of $T = 2\text{ kN}\cdot\text{m}$. What is the residual stress distribution in the shaft when the torque is removed?

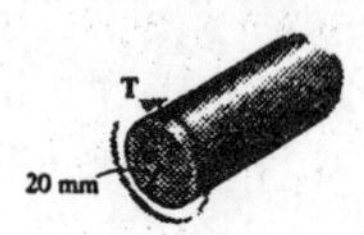

Elastic - Plastic Torque : The maximum elastic torque is

$T_Y = \frac{\pi}{2}\tau_Y c^3 = \frac{\pi}{2}(150)(10^6)(0.02^3) = 1.885\text{ kN}\cdot\text{m}$ and

the plastic torque is $T_P = \frac{2\pi}{3}\tau_Y c^3 = \frac{2\pi}{3}(150)(10^6)(0.02^3)$
$= 2.513\text{ kN}\cdot\text{m}$. Since $T_Y < T < T_P$, applying Eq. 5-26 from the text, we have

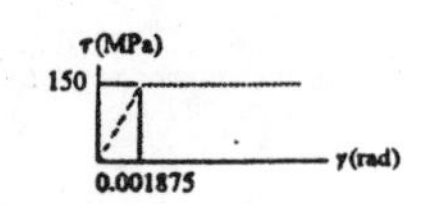

$$T = \frac{\pi\,\tau_Y}{6}\left(4c^3 - \rho_Y^3\right)$$

$$2(10^3) = \frac{\pi(150)(10^6)}{6}\left[4\left(0.02^3\right) - \rho_Y^3\right]$$

$$\rho_Y = 0.01870\text{ m} = 18.7\text{ mm}$$

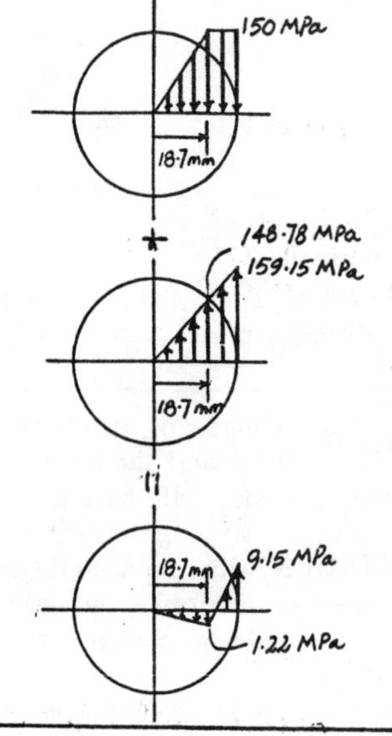

Residual Shear Stress : When the reverse torque $T = 2.0\text{ kN}\cdot\text{m}$ is applied,

$$(\tau')_{\rho=c} = \frac{Tc}{J} = \frac{2000(0.02)}{\frac{\pi}{2}(0.02^4)} = 159.15\text{ MPa}$$

$$(\tau')_{\rho=0.0187\text{m}} = \frac{Tc}{J} = \frac{2000(0.01870)}{\frac{\pi}{2}(0.02^4)} = 148.78\text{ MPa}$$

$$(\tau_r)_{\rho=c} = -150 + 159.15 = 9.15\text{ MPa}$$

$$(\tau_r)_{\rho=0.0187m} = -150 + 148.78 = -1.22\text{ MPa}$$

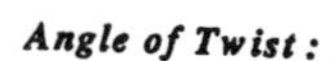

5–130. The shaft is made of an elastic-perfectly plastic material as shown. Determine the torque that the shaft can transmit if the allowable angle of twist is 0.375 rad. Also, determine the permanent angle of twist once the torque is removed. The shaft is 2 m long.

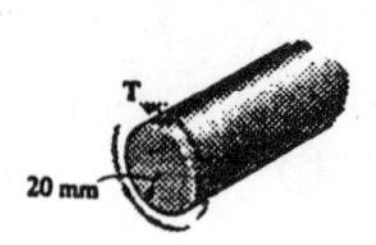

Angle of Twist :

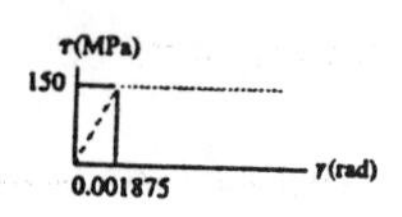

$$\gamma_{max} = \frac{\phi c}{L} = \frac{0.375(0.02)}{2} = 0.00375\text{ rad}$$

$$\frac{\rho_Y}{0.001875} = \frac{0.02}{0.00375}; \quad \rho_Y = 0.01\text{ m}$$

Elastic - Plastic Torque : Applying Eq. 5–26 from the text

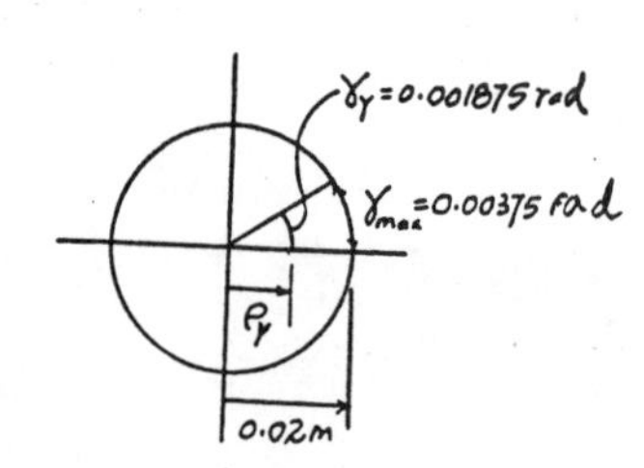

$$T = \frac{\pi\,\tau_Y}{6}\left(4c^3 - \rho_Y^3\right)$$

$$= \frac{\pi(150)(10^6)}{6}\left[4\left(0.02^3\right) - 0.01^3\right]$$

$$= 2434.73\text{ N}\cdot\text{m} = 2.43\text{ kN}\cdot\text{m} \qquad \textbf{Ans}$$

Permanent Angle of Twist : When the reverse torque $T = 2434.73\text{ N}\cdot\text{m}$ is applied,

$$G = \frac{150(10^6)}{0.001875} = 80\text{ GPa}$$

$$\phi' = \frac{TL}{JG} = \frac{2434.73(2)}{\frac{\pi}{2}(0.02^4)(80)(10^9)} = 0.2422\text{ rad}$$

$$\phi_r = \phi - \phi' = 0.375 - 0.2422$$

$$= 0.1328\text{ rad} = 7.61° \qquad \textbf{Ans}$$

5–131. The shear stress–strain diagram for a solid 50-mm-diameter shaft can be approximated as shown in the figure. Determine the torque T required to cause a maximum shear stress in the shaft of 125 MPa. If the shaft is 1.5 m long, what is the corresponding angle of twist?

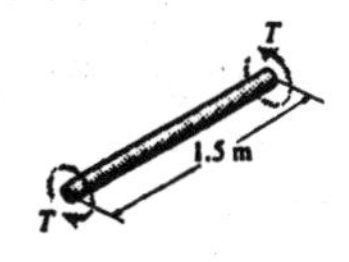

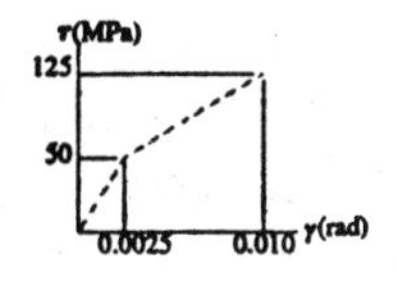

Strain Diagram :

$$\frac{\rho_\gamma}{0.0025} = \frac{0.025}{0.01}; \quad \rho_\gamma = 0.00625 \text{ m}$$

Stress Diagram :

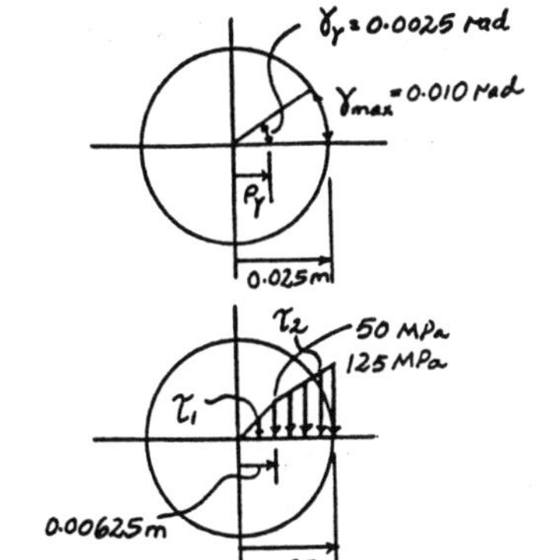

$$\tau_1 = \frac{50(10^6)}{0.00625}\rho = 8(10^9)\,\rho$$

$$\frac{\tau_2 - 50(10^6)}{\rho - 0.00625} = \frac{125(10^6) - 50(10^6)}{0.025 - 0.00625}$$

$$\tau_2 = 4(10^9)\,\rho + 25(10^6)$$

The Ultimate Torque :

$$T = 2\pi\int_0^c \tau\rho^2 d\rho$$

$$= 2\pi\int_0^{0.00625\text{m}} 8(10^9)\,\rho^3\,d\rho$$

$$+2\pi\int_{0.00625\text{m}}^{0.025\text{m}} \left[4(10^9)\,\rho + 25(10^6)\right]\rho^2 d\rho$$

$$= 2\pi\left[2(10^9)\,\rho^4\right]\Big|_0^{0.00625\text{m}}$$

$$+ 2\pi\left[1(10^9)\,\rho^4 + \frac{25(10^6)\rho^3}{3}\right]\Bigg|_{0.00625\text{m}}^{0.025\text{m}}$$

$$= 3269.30 \text{ N}\cdot\text{m} = 3.27 \text{ kN}\cdot\text{m} \qquad \textbf{Ans}$$

Angle of Twist :

$$\phi = \frac{\gamma_{\max}}{c}L = \left(\frac{0.01}{0.025}\right)(1.5) = 0.60 \text{ rad} = 34.4° \qquad \textbf{Ans}$$

***5–132.** A torque is applied to the shaft having a radius of 100 mm. If the material obeys a shear stress–strain relation of $\tau = 20\gamma^{1/3}$ MPa, determine the torque that must be applied to the shaft so that the maximum shear strain becomes 0.005 rad.

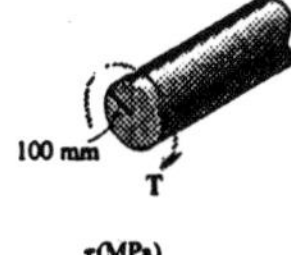

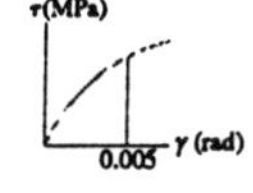

τ – ρ Function :

$$\gamma = \frac{\rho}{c}\gamma_{\max} = \frac{\rho}{0.1}(0.005) = 0.05\rho$$

$$\tau = 20(10^6)\,(0.05\rho)^{\frac{1}{3}} = 7.3681(10^6)\,\rho^{\frac{1}{3}}$$

The Ultimate Torque :

$$T = 2\pi\int_0^c \tau\rho^2 d\rho$$

$$= 2\pi\,(7.3681)(10^6)\int_0^{0.1\text{m}} \rho^{\frac{7}{3}}\,d\rho$$

$$= 46.2949(10^6)\left[0.3\rho^{\frac{10}{3}}\right]\Big|_0^{0.1\text{m}}$$

$$= 6446.46 \text{ N}\cdot\text{m} = 6.45 \text{ kN}\cdot\text{m} \qquad \textbf{Ans}$$

5–133. The 304 stainless steel shaft is 3 m long and has an outer diameter of 60 mm. When it is rotating at 60 rad/s, it transmits 30 kW of power from the engine E to the generator G. Determine the smallest thickness of the shaft if the allowable shear stress is $\tau_{allow} = 150$ MPa and the shaft is restricted not to twist more than 0.08 rad.

Internal Torque :

$$P = 30(10^3)\ \text{W}\left(\frac{1\ \text{N}\cdot\text{m/s}}{\text{W}}\right) = 30(10^3)\ \text{N}\cdot\text{m/s}$$

$$T = \frac{P}{\omega} = \frac{30(10^3)}{60} = 500\ \text{N}\cdot\text{m}$$

Allowable Shear Stress : Assume failure due to shear stress.

$$\tau_{max} = \tau_{allow} = \frac{Tc}{J}$$

$$150(10^6) = \frac{500(0.03)}{\frac{\pi}{2}(0.03^4 - r_i^4)}$$

$$r_i = 0.0293923\ \text{m} = 29.3923\ \text{mm}$$

Angle of Twist : Assume failure due to angle of twist limitation.

$$\phi = \frac{TL}{JG}$$

$$0.08 = \frac{500(3)}{\frac{\pi}{2}\left(0.03^4 - r_i^4\right)(75.0)(10^9)}$$

$$r_i = 0.0284033\ \text{m} = 28.4033\ \text{mm}$$

Choose the smallest value of $r_i = 28.4033$ mm

$$t = r_o - r_i = 30 - 28.4033 = 1.60\ \text{mm} \qquad \textbf{Ans}$$

5–134. The 304 stainless solid steel shaft is 3 m long and has a diameter of 50 mm. It is required to transmit 40 kW of power from the engine E to the generator G. Determine the smallest angular velocity the shaft can have if it is restricted not to twist more than 1.5°.

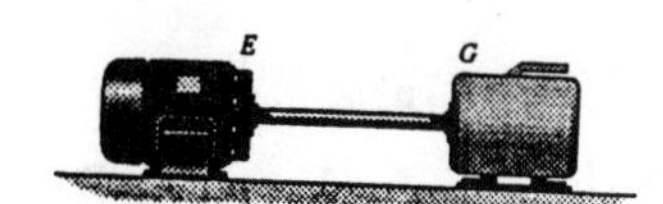

Angle of Twist :

$$\phi = \frac{TL}{JG}$$

$$\frac{1.5\pi}{180} = \frac{T(3)}{\frac{\pi}{2}(0.025^4)(75.0)(10^9)}$$

$$T = 401.60\ \text{N}\cdot\text{m}$$

Angular Velocity :

$$\omega = \frac{P}{T} = \frac{40(10^3)}{401.60} = 99.6\ \text{rad/s} \qquad \textbf{Ans}$$

5–135. The engine of the helicopter is delivering 800 hp to the rotor shaft AB when the blade is rotating at 1500 rev/min. Determine to the nearest $\frac{1}{8}$ in. the diameter of the shaft AB if the allowable shear stress is $\tau_{allow} =$ 8 ksi and the vibrations limit the angle of twist of the shaft to 0.05 rad. The shaft is 2 ft long and made of L2 tool steel.

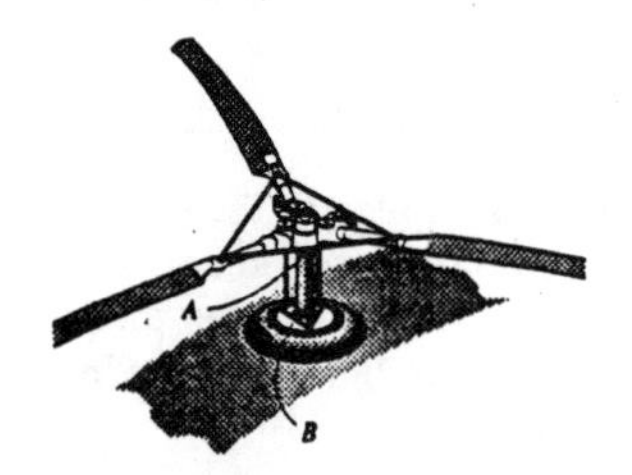

Internal Torque :

$$\omega = 1500\frac{\text{rev}}{\text{min}}\left(\frac{2\pi\ \text{rad}}{1\ \text{rev}}\right)\frac{1\ \text{min}}{60\ \text{s}} = 50.0\,\pi\ \text{rad/s}$$

$$P = 800\ \text{hp}\left(\frac{550\ \text{ft}\cdot\text{lb/s}}{1\ \text{hp}}\right) = 440\,000\ \text{ft}\cdot\text{lb/s}$$

$$T = \frac{P}{\omega} = \frac{440\,000}{50.0\,\pi} = 2801.13\ \text{lb.ft}$$

Allowable Shear Stress : Assume failure due to shear stress

$$\tau_{max} = \frac{Tc}{J}$$

$$8(10^3) = \frac{2801.13(12)\left(\frac{d}{2}\right)}{\frac{\pi}{2}\left(\frac{d}{2}\right)^4}$$

$$d = 2.776\ \text{in.}$$

Angle of Twist : Assume failure due to angle of twist limitation

$$\phi = \frac{TL}{JG}$$

$$0.05 = \frac{2801.13(12)(2)(12)}{\frac{\pi}{2}\left(\frac{d}{2}\right)^4(11.0)(10^6)}$$

$$d = 1.966\ \text{in.}$$

Shear stress failure controls the design. Hence,

Use $d = 2\frac{7}{8}$ in. diameter shaft. **Ans**

***5–136.** The engine of the helicopter is delivering 800 hp to the rotor shaft AB when the blade is rotating at 1500 rev/min. Determine to the nearest $\frac{1}{8}$ in. the diameter of the shaft AB if the allowable shear stress is $\tau_{allow} = 10.5$ ksi and the vibrations limit the angle of twist of the shaft to 0.03 rad. The shaft is 2 ft long and made of L2 tool steel.

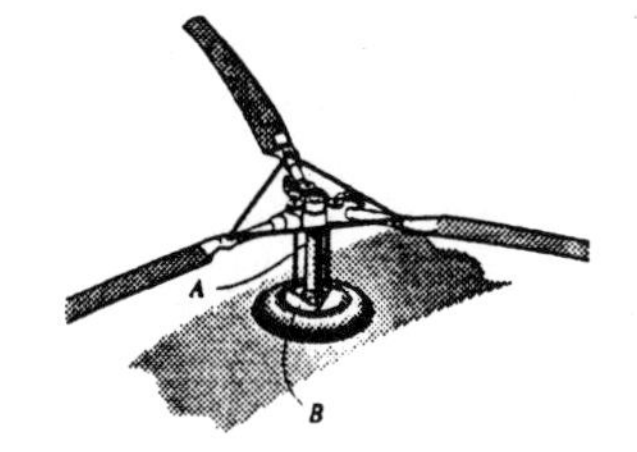

Internal Torque :

$$\omega = 1500\frac{\text{rev}}{\text{min}}\left(\frac{2\pi\ \text{rad}}{1\ \text{rev}}\right)\frac{1\ \text{min}}{60\ \text{s}} = 50.0\ \pi\ \text{rad/s}$$

$$P = 800\ \text{hp}\left(\frac{550\ \text{ft}\cdot\text{lb/s}}{1\ \text{hp}}\right) = 440\ 000\ \text{ft}\cdot\text{lb/s}$$

$$T = \frac{P}{\omega} = \frac{440\ 000}{50.0\ \pi} = 2801.13\ \text{lb.ft}$$

Allowable Shear Stress : Assume failure due to shear stress

$$\tau_{max} = \frac{Tc}{J}$$

$$10.5(10^3) = \frac{2801.13(12)\left(\frac{d}{2}\right)}{\frac{\pi}{2}\left(\frac{d}{2}\right)^4}$$

$$d = 2.536\ \text{in.}$$

Angle of Twist : Assume failure due to angle of twist limitation

$$\phi = \frac{TL}{JG}$$

$$0.03 = \frac{2801.13(12)(2)(12)}{\frac{\pi}{2}\left(\frac{d}{2}\right)^4(11.0)(10^6)}$$

$$d = 2.234\ \text{in.}$$

Shear stress failure controls the design. Hence,

Use $d = 2\frac{5}{8}$ in. diameter shaft. **Ans**

5–137. The device shown is used to mix soils in order to provide in-situ stabilization. If the mixer is connected to an A-36 steel tubular shaft that has an inner diameter of 3 in. and an outer diameter of 4.5 in., determine the angle of twist of the shaft of A relative to C if each mixing blade is subjected to the torques shown.

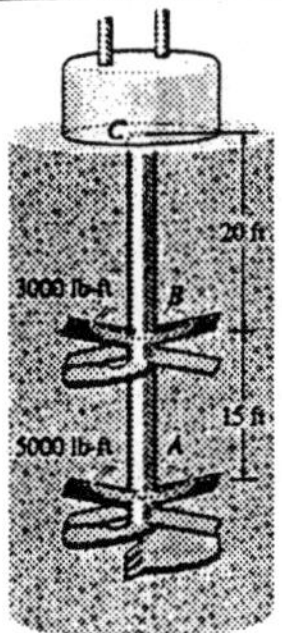

Angle of Twist :

$$\phi_{A/C} = \sum\frac{TL}{JG}$$

$$= \frac{5000(12)(15)(12)}{\frac{\pi}{2}(2.25^4 - 1.5^4)(11.0)(10^6)} + \frac{8000(12)(20)(12)}{\frac{\pi}{2}(2.25^4 - 1.5^4)(11.0)(10^6)}$$

$$= 0.09523\ \text{rad} = 5.46° \quad \textbf{Ans}$$

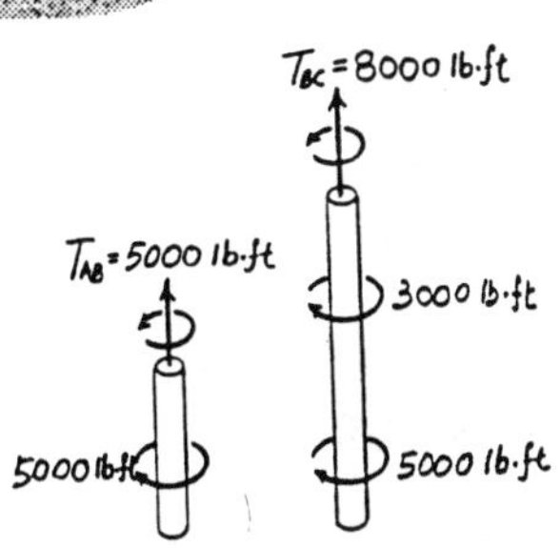

5–138. The rotating flywheel-and-shaft is brought to a sudden stop at D when the bearing freezes. This causes the flywheel to oscillate clockwise–counterclockwise so that a point A on the outer edge of the flywheel is displaced through a 10-mm arc. Determine the maximum shear stress developed in the tubular 304 stainless steel shaft due to this oscillation. The shaft has an inner diameter of 25 mm and an outer diameter of 35 mm. The journal bearings at B and C allow the shaft to rotate freelv.

Angle of Twist :

$$\phi = \frac{TL}{JG} \qquad \text{Where} \quad \phi = \frac{10}{80} = 0.125 \text{ rad}$$

$$0.125 = \frac{T(2)}{\frac{\pi}{2}(0.0175^4 - 0.0125^4)(75.0)(10^9)}$$

$$T = 510.82 \text{ N}\cdot\text{m}$$

Maximum Shear Stress :

$$\tau_{max} = \frac{Tc}{J} = \frac{510.82(0.0175)}{\frac{\pi}{2}(0.0175^4 - 0.0125^4)} = 82.0 \text{ MPa} \qquad \textbf{Ans}$$

5–139. The motor of a fan delivers 150 W of power to the blade when it is turning at 18 rev/s. Determine the smallest diameter of shaft A that can be used to connect the fan blade to the motor if the allowable shear stress is $\tau_{allow} = 80$ MPa.

Internal Torque :

$$\omega = 18\,\frac{\text{rev}}{\text{s}}\left(\frac{2\pi \text{ rad}}{1 \text{ rev}}\right) = 36.0\pi \text{ rad/s}$$

$$P = 150 \text{ W}\left(\frac{1 \text{ N}\cdot\text{m/s}}{1 \text{ W}}\right) = 150 \text{ N}\cdot\text{m/s}$$

$$T = \frac{P}{\omega} = \frac{150}{36.0\pi} = 1.3263 \text{ N}\cdot\text{m}$$

Allowable Shear Stress :

$$\tau_{max} = \tau_{allow} = \frac{Tc}{J}$$

$$80(10^6) = \frac{1.3262\left(\frac{d}{2}\right)}{\frac{\pi}{2}\left(\frac{d}{2}\right)^4}$$

$$d = 0.004387 \text{ m} = 4.39 \text{ mm} \qquad \textbf{Ans}$$

***5–140.** The motor A develops a torque at gear B of 500 lb · ft, which is applied along the axis of the 2-in.-diameter A-36 steel shaft CD. This torque is to be transmitted to the pinion gears at E and F. If these gears are temporarily fixed, determine the maximum shear stress in segments CB and BD of the shaft. Also, what is the angle of twist of each of these segments? The bearings at C and D only exert force reactions on the shaft.

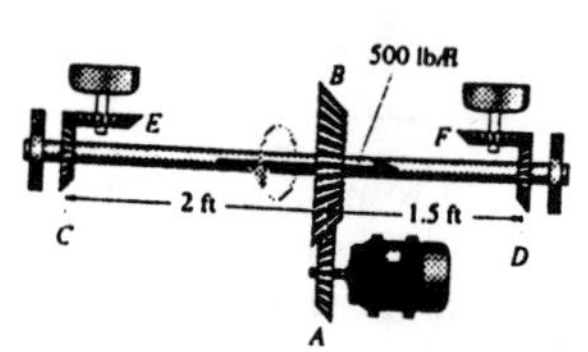

Equilibrium :

$$T_C + T_D - 500 = 0 \qquad [1]$$

Compatibility :

$$\phi_{B/C} = \phi_{B/D}$$

$$\frac{T_C(2)}{JG} = \frac{T_D(1.5)}{JG}$$

$$T_C = 0.75T_D \qquad [2]$$

Solving Eqs. [1] and [2] yields :

$$T_D = 285.71 \text{ lb}\cdot\text{ft} \qquad T_C = 214.29 \text{ lb}\cdot\text{ft}$$

Maximum Shear Stress :

$$(\tau_{CB})_{max} = \frac{T_C c}{J} = \frac{214.29(12)(1)}{\frac{\pi}{2}(1^4)} = 1.64 \text{ ksi} \qquad \textbf{Ans}$$

$$(\tau_{BD})_{max} = \frac{T_D c}{J} = \frac{285.71(12)(1)}{\frac{\pi}{2}(1^4)} = 2.18 \text{ ksi} \qquad \textbf{Ans}$$

Angle of Twist :

$$\phi_{CB} = \phi_{BD} = \frac{T_D L_{BD}}{JG}$$

$$= \frac{285.71(12)(1.5)(12)}{\frac{\pi}{2}(1^4)(11.0)(10^6)}$$

$$= 0.003572 \text{ rad} = 0.205° \qquad \textbf{Ans}$$

5–141. The coupling consists of two disks fixed to separate shafts, each 25 mm in diameter. The shafts are supported on journal bearings that allow free rotation. In order to limit the torque **T** that can be transmitted, a "shear pin" P is used to connect the disks together. If this pin can sustain an *average* shear force of 550 N before it fails, determine the maximum constant torque **T** that can be transmitted from one shaft to the other. Also, what is the maximum shear stress in each shaft when the "shear pin" is about to fail?

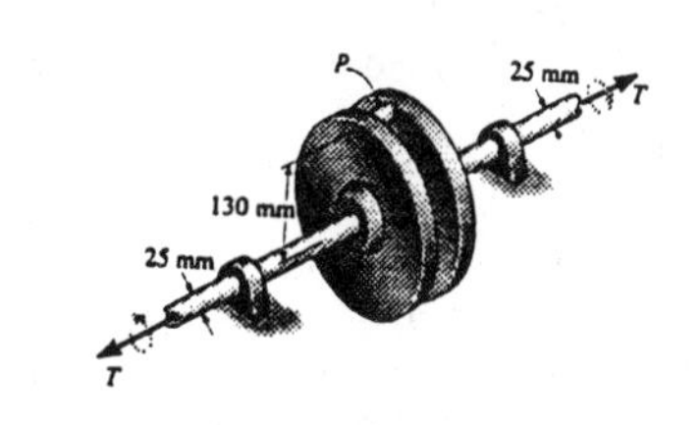

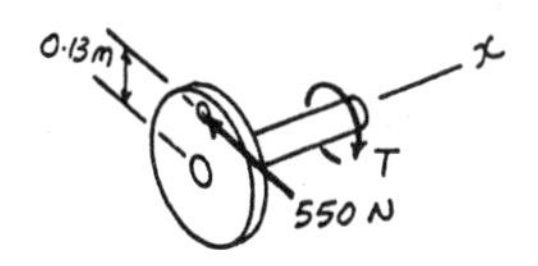

Equilibrium :

$$\Sigma M_x = 0; \quad T - 550(0.13) = 0$$
$$T = 71.5 \text{ N}\cdot\text{m} \qquad \textbf{Ans}$$

Maximum Shear Stress :

$$\tau_{max} = \frac{Tc}{J} = \frac{71.5(0.0125)}{\frac{\pi}{2}(0.0125^4)} = 23.3 \text{ MPa} \qquad \textbf{Ans}$$

5–142. The A-36 steel circular tube is subjected to a torque of 10 kN · m. Determine the shear stress at the mean radius $\rho = 60$ mm and compute the angle of twist of the tube if it is 4 m long and fixed at its far end. Solve the problem using Eqs. 5–7 and 5–15 and by using Eqs. 5–18 and 5–20.

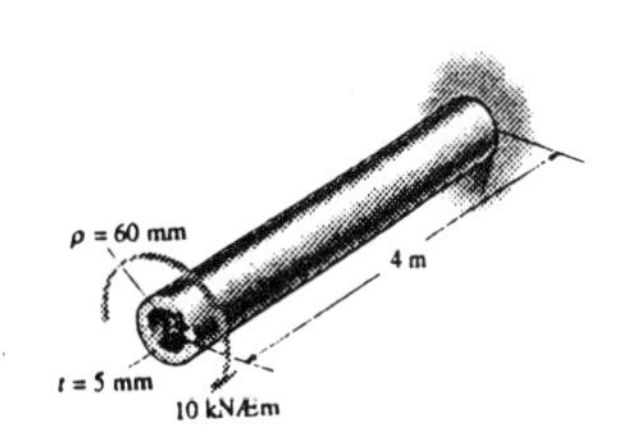

Shear Stress :

Applying Eq. 5 – 7,

$$r_o = 0.06 + \frac{0.005}{2} = 0.0625 \text{ m} \qquad r_i = 0.06 - \frac{0.005}{2} = 0.0575 \text{ m}$$

$$\tau_{\rho=0.06\text{m}} = \frac{T\rho}{J} = \frac{10(10^3)(0.06)}{\frac{\pi}{2}(0.0625^4 - 0.0575^4)} = 88.3 \text{ MPa} \qquad \textbf{Ans}$$

Applying Eq. 5 – 15,

$$\tau_{avg} = \frac{T}{2tA_m} = \frac{10(10^3)}{2(0.005)(\pi)(0.06^2)} = 88.4 \text{ MPa} \qquad \textbf{Ans}$$

Angle of Twist :

Applying Eq. 5 – 18,

$$\phi = \frac{TL}{JG}$$
$$= \frac{10(10^3)(4)}{\frac{\pi}{2}(0.0625^4 - 0.0575^4)(75.0)(10^9)}$$
$$= 0.0785 \text{ rad} \qquad \textbf{Ans}$$

Applying Eq. 5 – 20,

$$\phi = \frac{TL}{4A_m^2 G}\int \frac{ds}{t}$$
$$= \frac{TL}{4A_m^2 G\, t}\int ds \qquad \text{Where} \qquad \int ds = 2\pi\rho$$
$$= \frac{2\pi TL\rho}{4A_m^2 G\, t}$$
$$= \frac{2\pi(10)(10^3)(4)(0.06)}{4[(\pi)(0.06^2)]^2(75.0)(10^9)(0.005)}$$
$$= 0.0786 \text{ rad} \qquad \textbf{Ans}$$

6–1. Draw the shear and moment diagrams for the beam, and determine the shear and moment throughout the beam as a function of x.

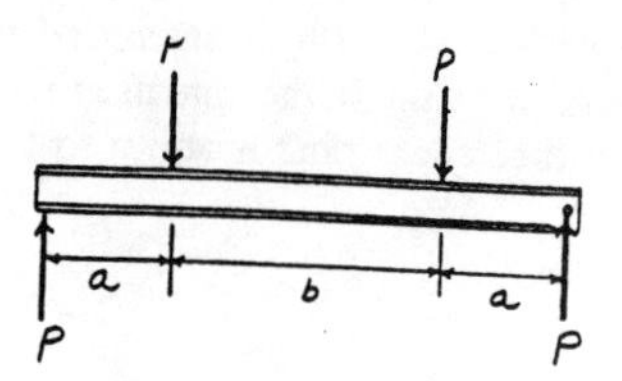

Support Reactions : As shown on FBD.
Shear and Moment Function :

For $0 \le x < a$

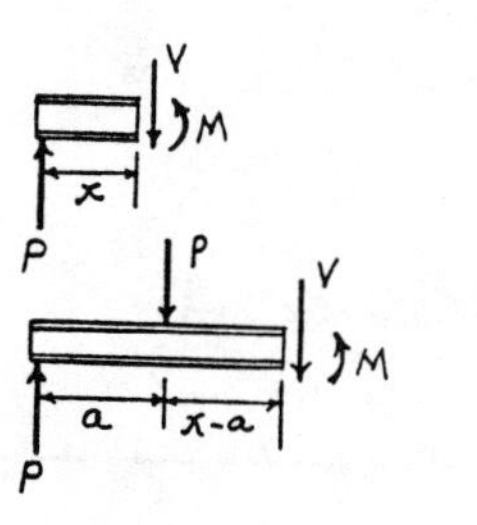

$+\uparrow \Sigma F_y = 0; \quad P - V = 0 \quad V = P$ **Ans**

$\curvearrowleft + \Sigma M_{NA} = 0; \quad M - Px = 0 \quad M = Px$ **Ans**

For $a < x < a+b$

$+\uparrow \Sigma F_y = 0; \quad P - P - V = 0 \quad V = 0$ **Ans**

$\curvearrowleft + \Sigma M_{NA} = 0; \quad M - Pa = 0 \quad M = Pa$ **Ans**

For $a+b < x \le 2a+b$

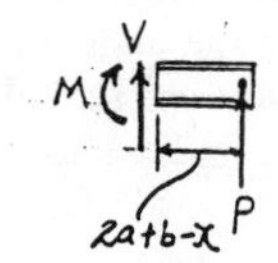

$+\uparrow \Sigma F_y = 0; \quad V + P = 0 \quad V = -P$ **Ans**

$\curvearrowleft + \Sigma M_{NA} = 0; \quad P(2a+b-x) - M = 0$

$M = P(2a+b-x)$ **Ans**

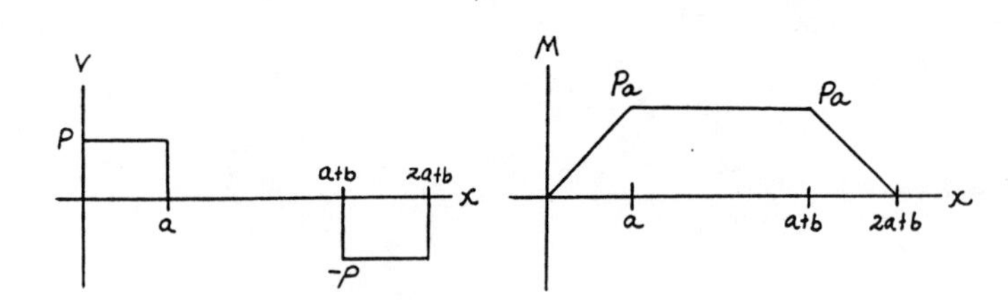

6–2. The shaft is subjected to the loadings caused by belts passing over the two pulleys. Draw the shear and moment diagrams. The bearings at A and B exert only vertical reactions on the shaft.

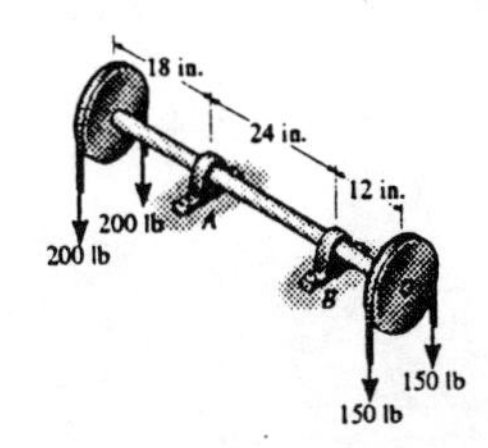

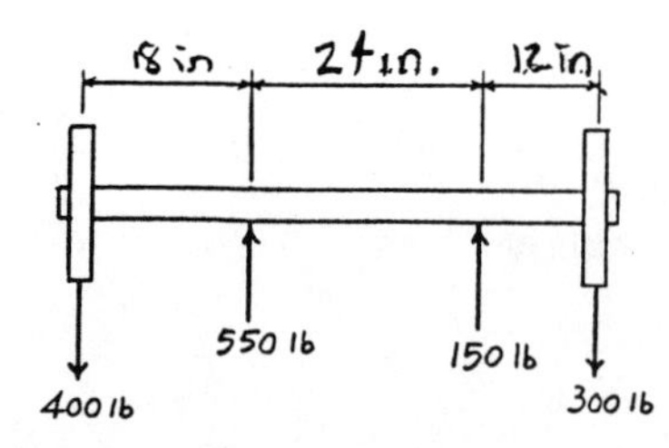

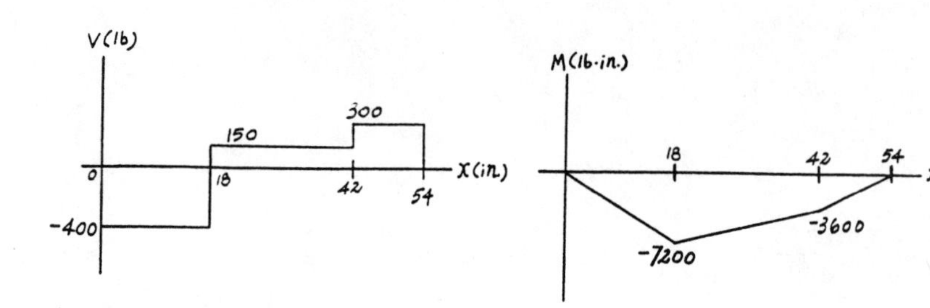

6–3. The three traffic lights each have a mass of 10 kg, and the cantilevered pipe AB has a mass of 1.5 kg/m. Draw the shear and moment diagrams for the pipe. Neglect the mass of the sign.

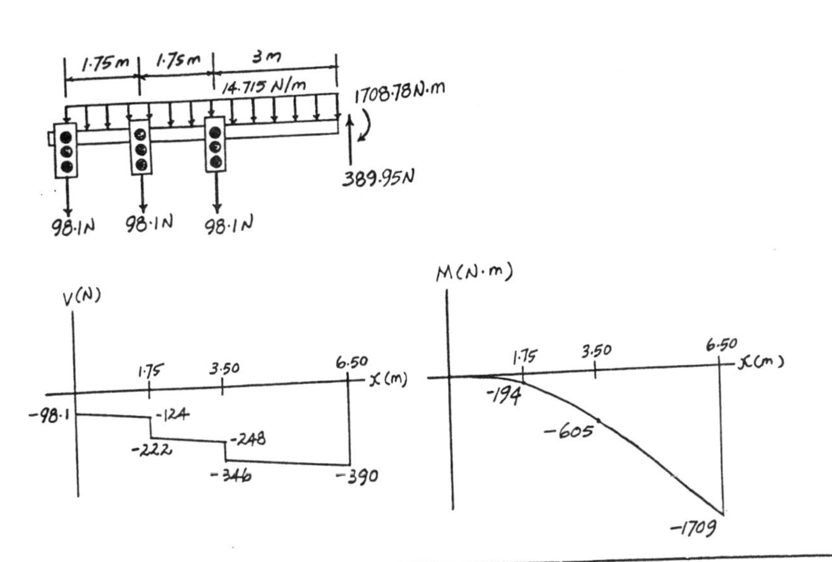

***6–4.** A reinforced concrete pier is used to support the stringers for a bridge deck. Draw the shear and moment diagrams for the pier when it is subjected to the stringer loads shown. Assume the columns at A and B exert only vertical reactions on the pier.

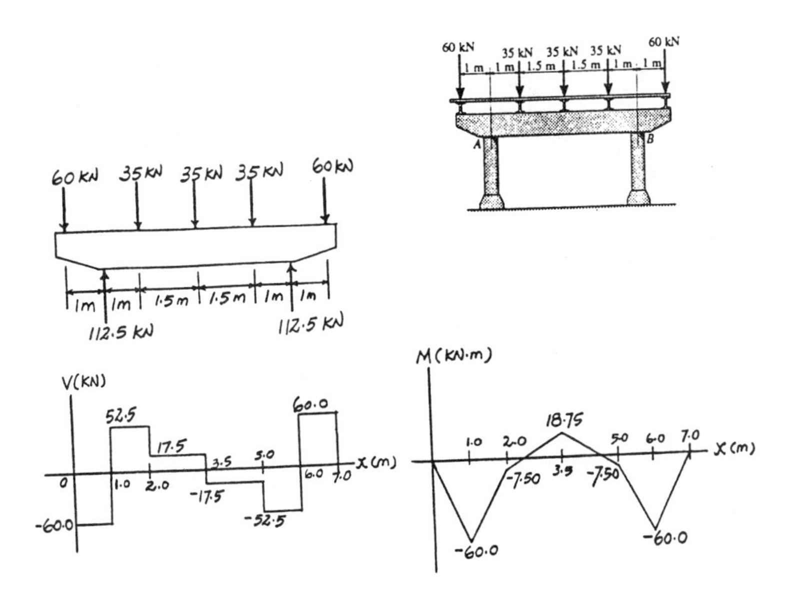

6–5. The steam pipe P rests on the roller guide CD to allow for its thermal expansion. Draw the shear and moment diagrams for the beam AB if the pipe weighs 800 lb. The bearing supports at C and D exert only vertical forces on the beam

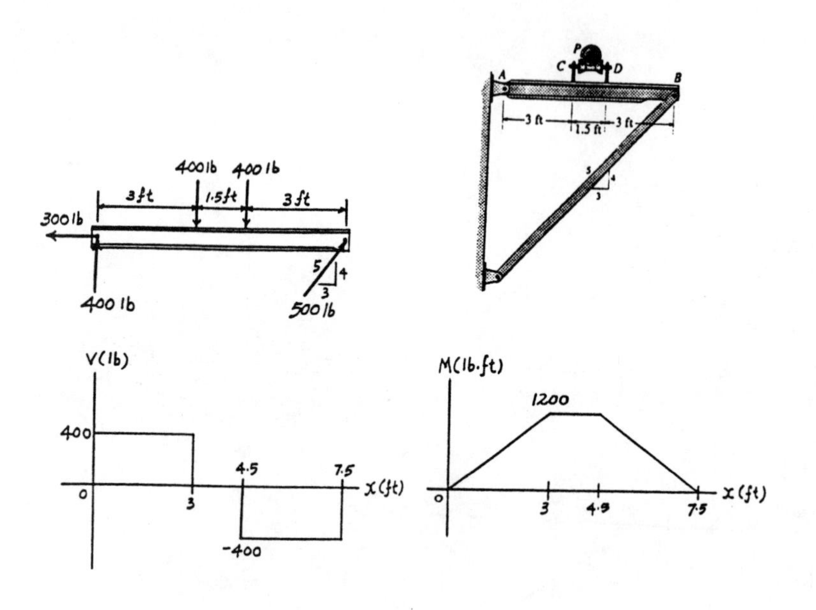

6–6. Draw the shear and moment diagrams for the shaft. The bearings at A and B exert only vertical reactions on the shaft.

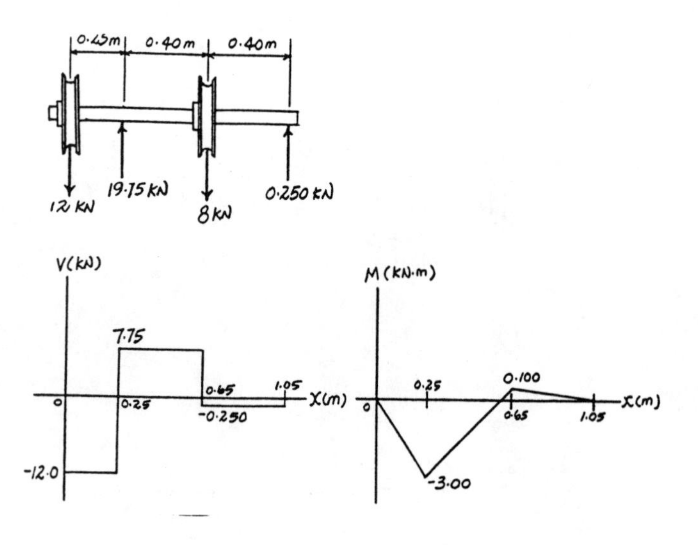

6–7. Draw the shear and moment diagrams for the shaft and determine the shear and moment throughout the shaft as a function of x. The bearings at A and B exert only vertical reactions on the shaft.

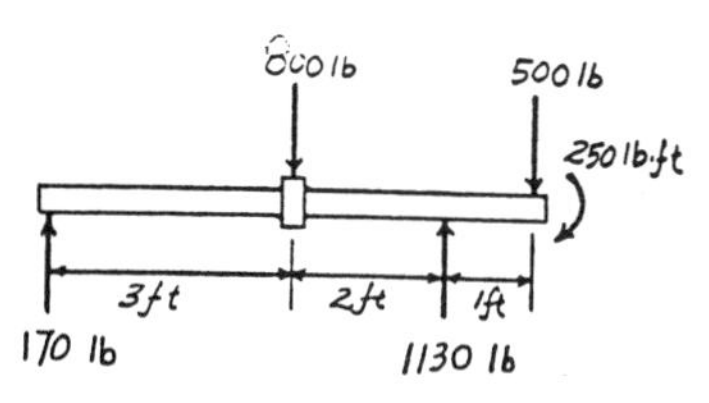

For $0 < x < 3$ ft

$+\uparrow \Sigma F_y = 0; \quad 170 - V = 0 \quad V = 170 \text{ lb}$ **Ans**

$\curvearrowleft + \Sigma M_{NA} = 0; \quad M - 170x = 0$

$M = \{170x\} \text{ lb}\cdot\text{ft}$ **Ans**

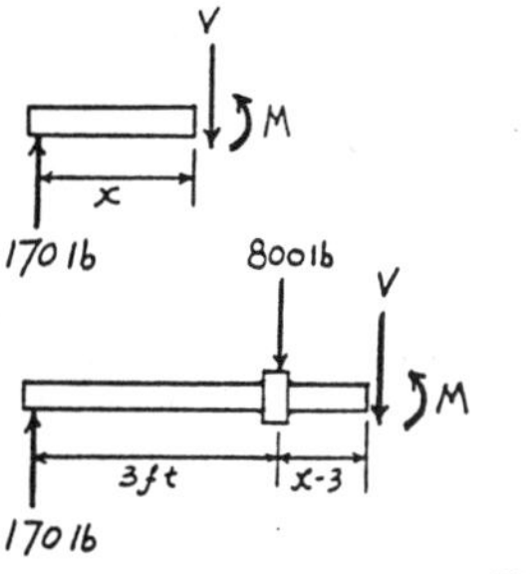

For $3 \text{ ft} < x < 5$ ft

$+\uparrow \Sigma F_y = 0; \quad 170 - 800 - V = 0$

$V = -630 \text{ lb}$ **Ans**

$\curvearrowleft + \Sigma M_{NA} = 0; \quad M + 800(x-3) - 170x = 0$

$M = \{-630x + 2400\} \text{ lb}\cdot\text{ft}$ **Ans**

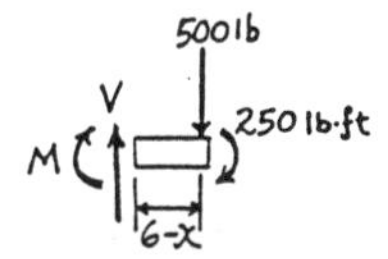

For $5 \text{ ft} < x \leq 6$ ft

$+\uparrow \Sigma F_y = 0; \quad V - 500 = 0 \quad V = 500 \text{ lb}$ **Ans**

$\curvearrowleft + \Sigma M_{NA} = 0; \quad -M - 500(6-x) - 250 = 0$

$M = \{500x - 3250\} \text{ lb}\cdot\text{ft}$ **Ans**

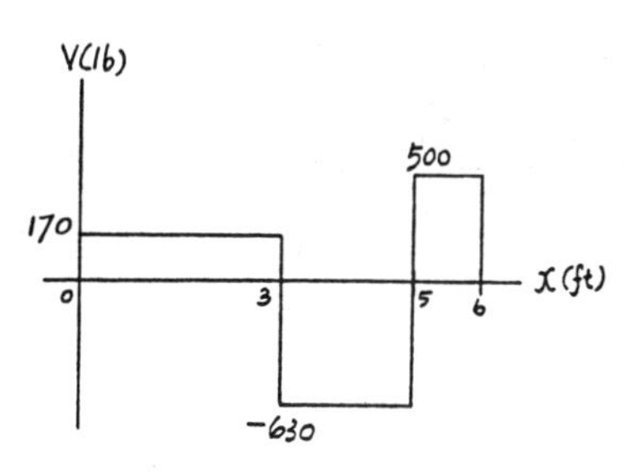

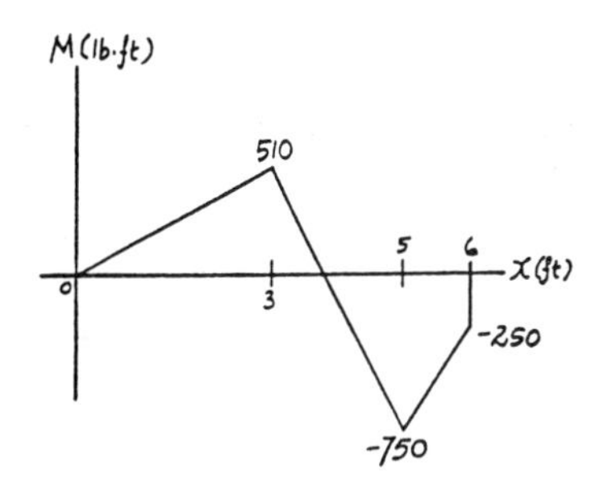

***6–8.** Draw the shear and moment diagrams for the rod segment ABC. The end A is subjected to a force of 5 kN. *Hint:* The reactions at pin D must be replaced by equivalent loadings at C on the axis of the rod. The journal bearing at B exerts only a vertical force on the rod.

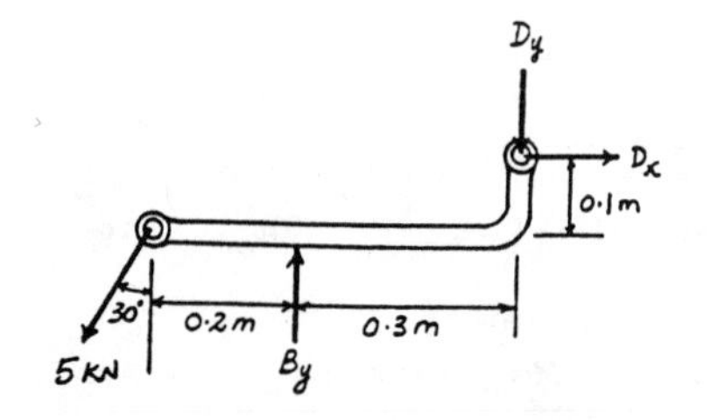

Support Reactions :

$\circlearrowleft + \Sigma M_D = 0;\quad 5\cos 30°(0.5) - 5\sin 30°(0.1) - B_y(0.3) = 0$

$B_y = 6.384\text{ kN}$

$+\uparrow \Sigma F_y = 0;\quad 6.384 - 5\cos 30° - D_y = 0 \quad D_y = 2.053\text{ kN}$

$\xrightarrow{+} \Sigma F_x = 0;\quad D_x - 5\sin 30° = 0 \quad D_x = 2.50\text{ kN}$

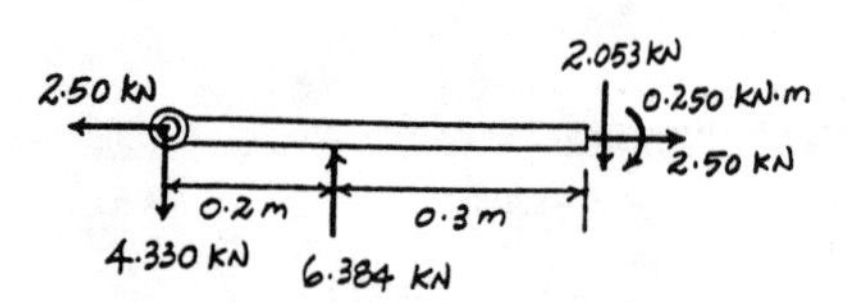

Shear and Moment Diagram :

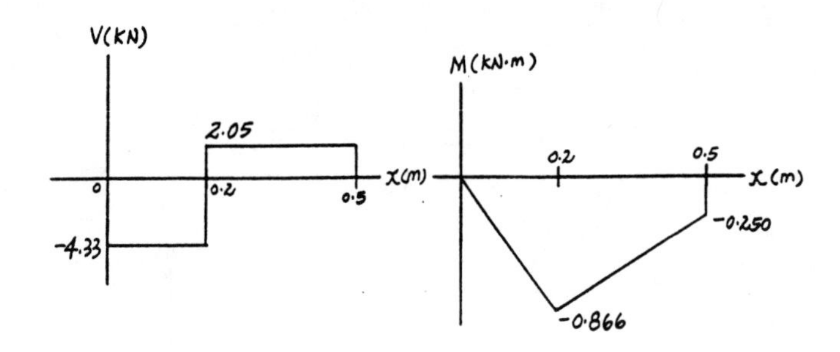

6–9. Draw the shear and moment diagrams for the beam. It is supported by a smooth plate at B which slides within the groove and so it cannot support a vertical force, although it can support a moment and axial load.

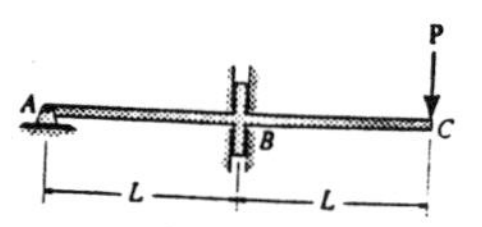

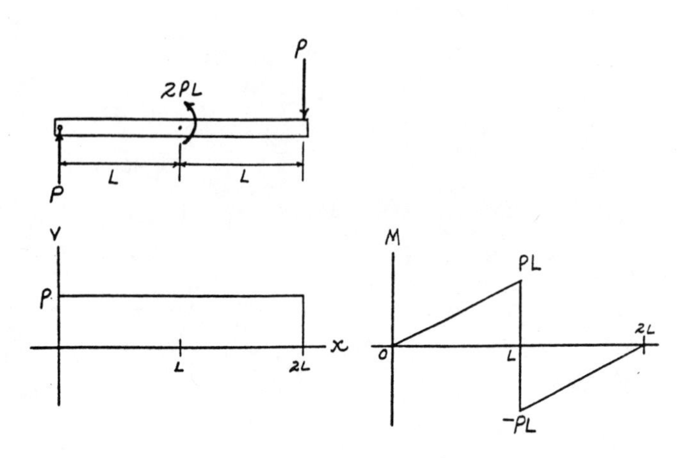

6–10. The walking beam of an oil-field pumping unit is subjected to the draw force of 800 lb. Determine the force P in the pitman arm and the reactions at the pin C. Then draw the shear and moment diagrams for portion AB of the beam. *Hint:* The reactions at C must be replaced by equivalent loadings at point C' on the axis of the beam.

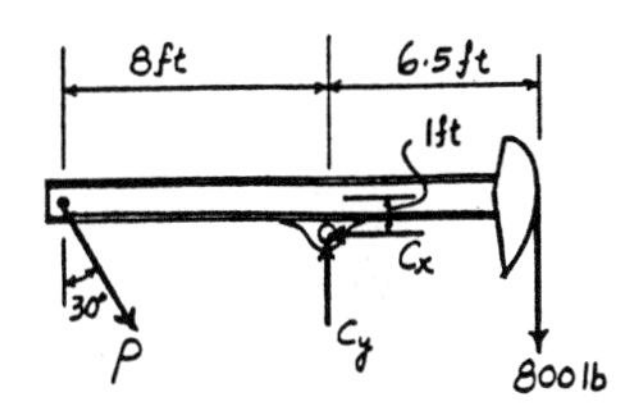

$+\Sigma M_C = 0;\quad P\cos 30°(8) - P\sin 30°(1) - 800(6.5) = 0$

$P = 808.94 \text{ lb} = 809 \text{ lb}$ **Ans**

$+\uparrow \Sigma F_y = 0;\quad C_y - 808.93\cos 30° - 800 = 0$

$C_y = 1500.56 \text{ lb} = 1501 \text{ lb}$ **Ans**

$\xrightarrow{+} \Sigma F_x = 0;\quad 808.93\sin 30° - C_x = 0$

$C_x = 404.47 \text{ lb} = 404 \text{ lb}$ **Ans**

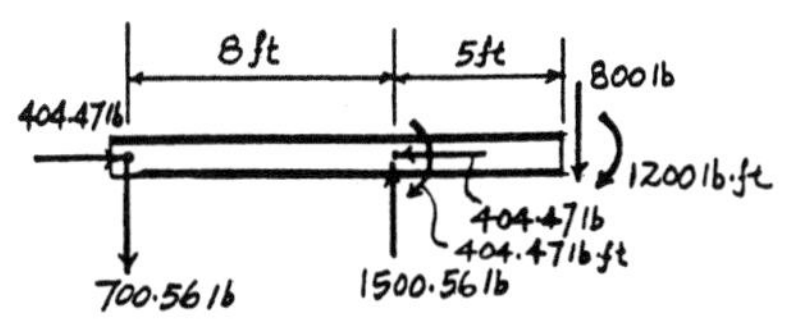

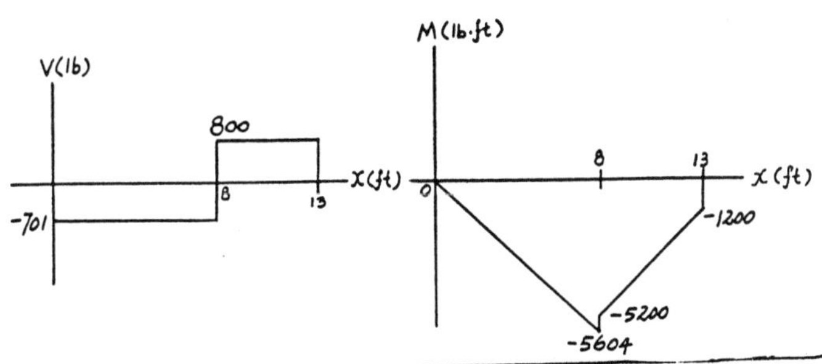

6–11. The handle clamp is intended to develop a maximum clamping force of 1.25 kN at the jaws C. Determine the axial force in each of the two screws A and B, and then draw the shear and moment diagrams for the top jaw beam CD.

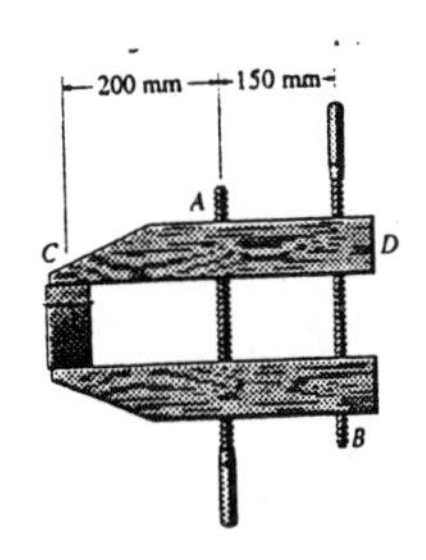

$+\Sigma M_B = 0;\quad F_A(0.15) - 1.25(0.35) = 0$

$F_A = 2.9167 \text{ kN} = 2.92 \text{ kN}$ **Ans**

$+\uparrow \Sigma F_y = 0;\quad 1.25 - 2.9167 + F_B = 0$

$F_B = 1.6667 \text{ kN} = 1.67 \text{ kN}$ **Ans**

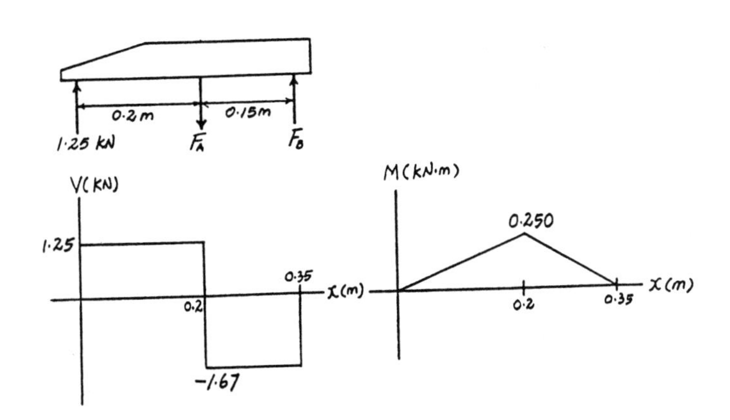

***6–12.** The overhanging beam has been fabricated with a projected arm BD on it. Draw the shear and moment diagrams for the beam ABC if it supports a load of 800 lb. *Hint:* The loading in the supporting strut DE must be replaced by equivalent loads at point B on the axis of the beam.

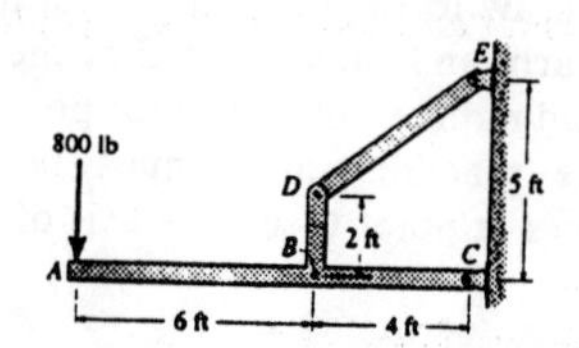

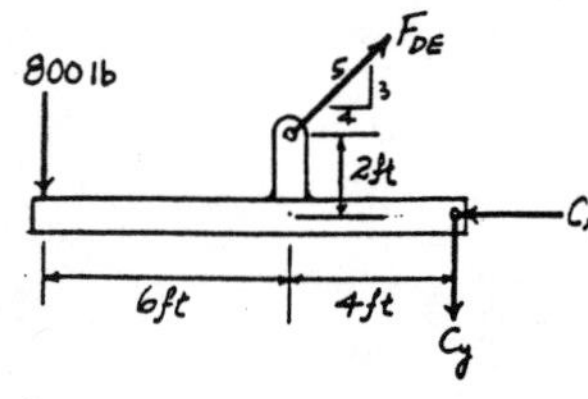

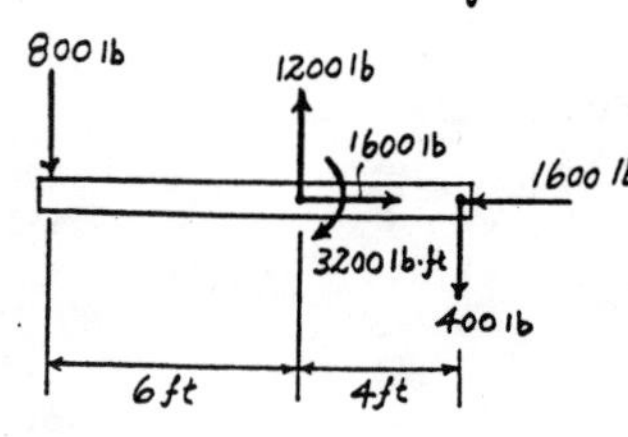

Support Reactions :

$$\circlearrowleft + \Sigma M_C = 0; \quad 800(10) - \frac{3}{5}F_{DE}(4) - \frac{4}{5}F_{DE}(2) = 0$$

$$F_{DE} = 2000 \text{ lb}$$

$$+\uparrow \Sigma F_y = 0; \quad -800 + \frac{3}{5}(2000) - C_y = 0 \qquad C_y = 400 \text{ lb}$$

$$\overset{+}{\rightarrow} \Sigma F_x = 0; \quad -C_x + \frac{4}{5}(2000) = 0 \qquad C_x = 1600 \text{ lb}$$

Shear and Moment Diagram :

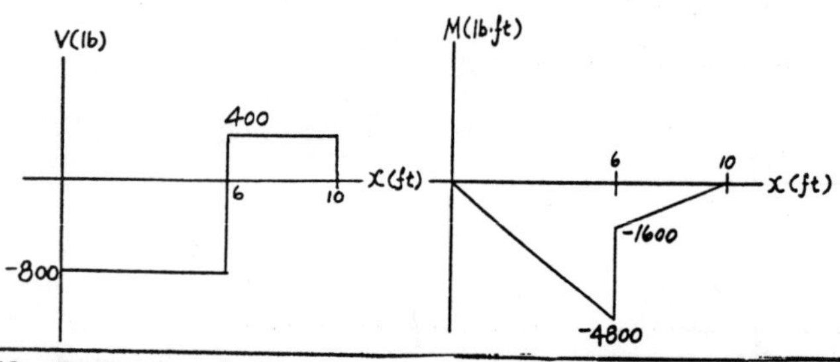

6–13. Draw the shear and moment diagrams for the compound beam. It is supported by a smooth plate at A which slides within the groove and so it cannot support a vertical force, although it can support a moment and axial load.

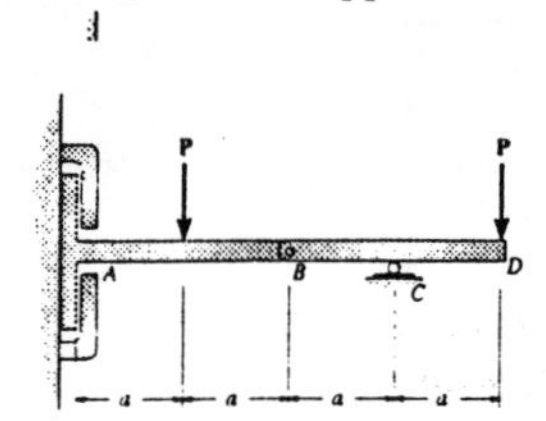

Support Reactions :
From the FBD of segment BD

$$\circlearrowleft + \Sigma M_C = 0; \quad B_y(a) - P(a) = 0 \qquad B_y = P$$

$$+\uparrow \Sigma F_y = 0; \quad C_y - P - P = 0 \qquad C_y = 2P$$

$$\overset{+}{\rightarrow} \Sigma F_x = 0; \qquad B_x = 0$$

From the FBD of segment AB

$$+\Sigma M_A = 0; \quad P(2a) - P(a) - M_A = 0 \qquad M_A = Pa$$

$$+\uparrow \Sigma F_y = 0; \quad P - P = 0 \text{ (equilibrium is statisfied!)}$$

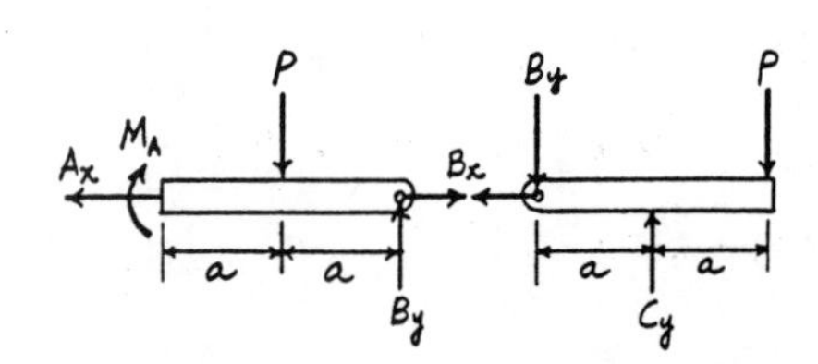

Shear and Moment Diagram :

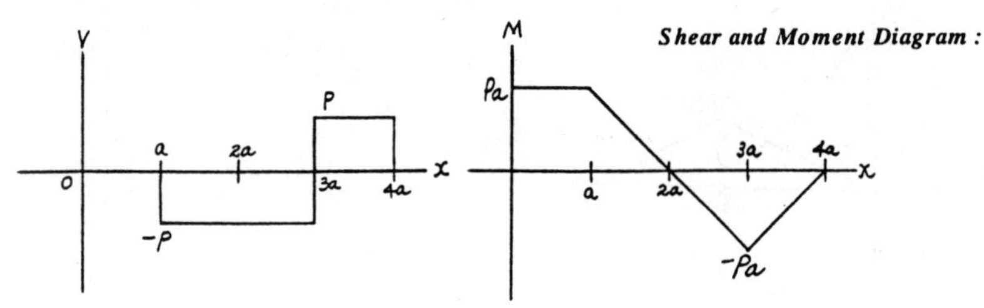

6–14. Draw the shear and moment diagrams for the beam.

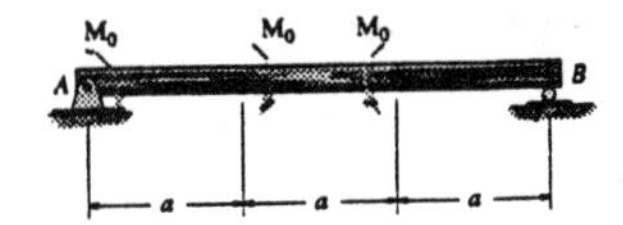

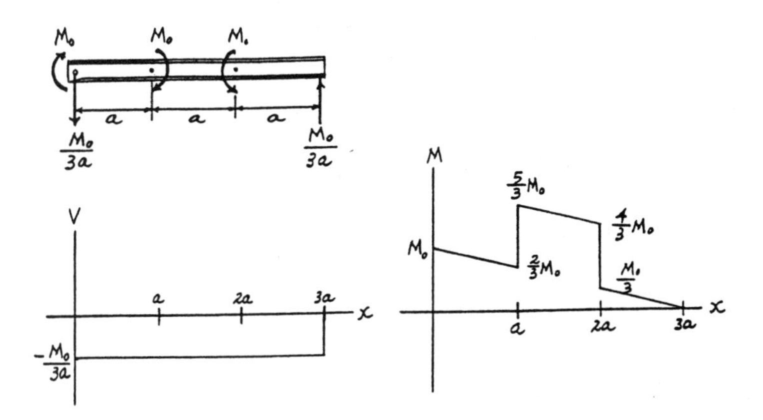

6–15. The beam is subjected to the uniformly distributed moment m (moment / length). Draw the shear and moment diagrams for the beam

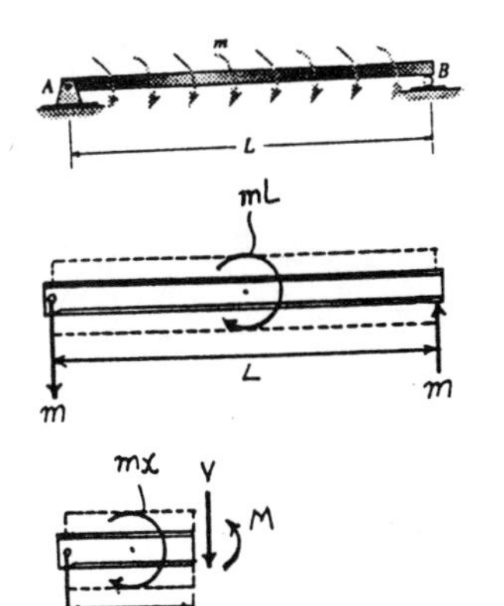

Support Reactions : As shown on FBD.

Shear and Moment Function :

$$+\uparrow \Sigma F_y = 0; \qquad -m - V = 0 \qquad V = -m$$

$$+\Sigma M_{NA} = 0; \qquad M + m(x) - mx = 0 \qquad M = 0$$

Shear and Moment Diagram :

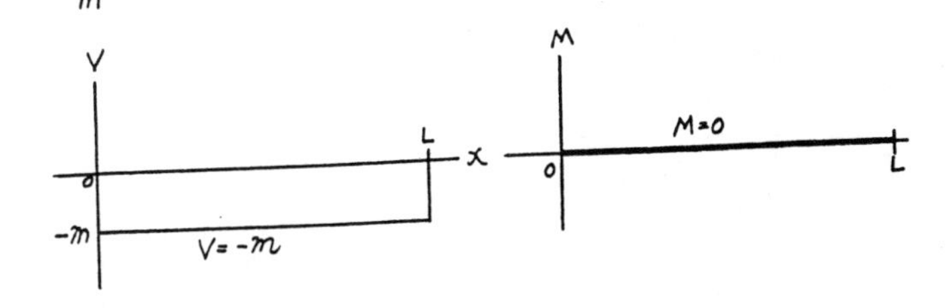

*6–16. The beam is subjected to the uniformly distributed moment m (moment / length). Draw the shear and moment diagrams for the beam

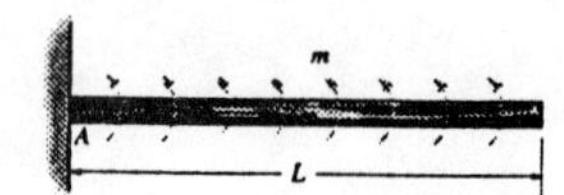

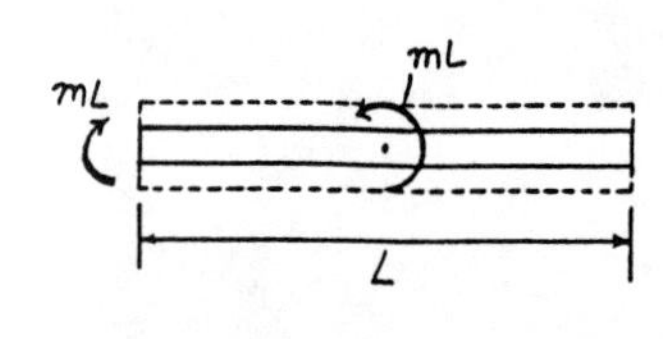

Support Reactions : As shown on FBD.

Shear and Moment Function :

$+\uparrow \Sigma F_y = 0; \qquad V = 0$

$+\Sigma M_{NA} = 0; \qquad M + mx - mL = 0 \qquad M = m(L - x)$

Shear and Moment Diagram :

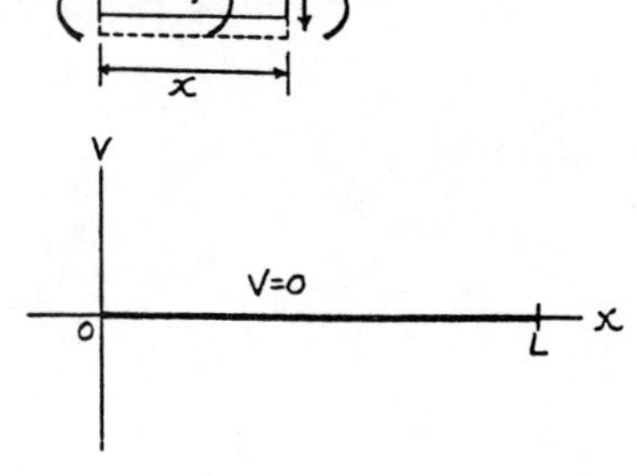

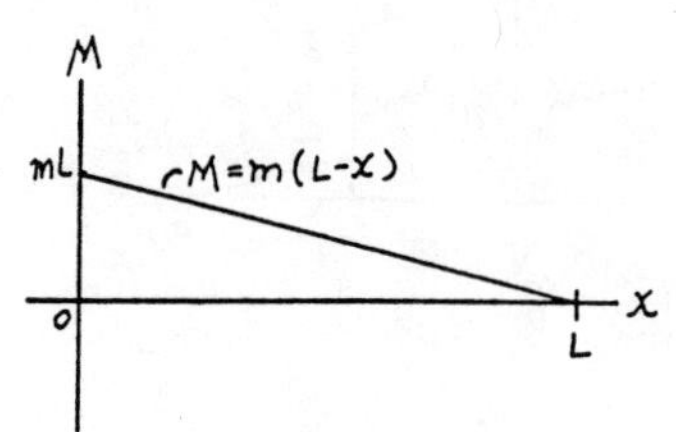

6–17. Draw the shear and moment diagrams for the beam.The bearings at A and B only exert vertical reaction on the beam.

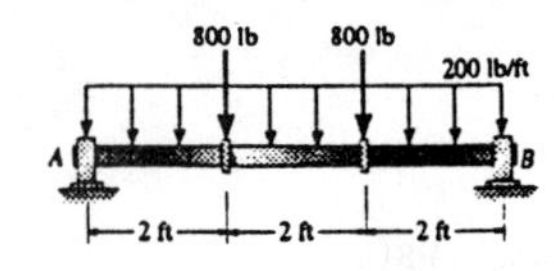

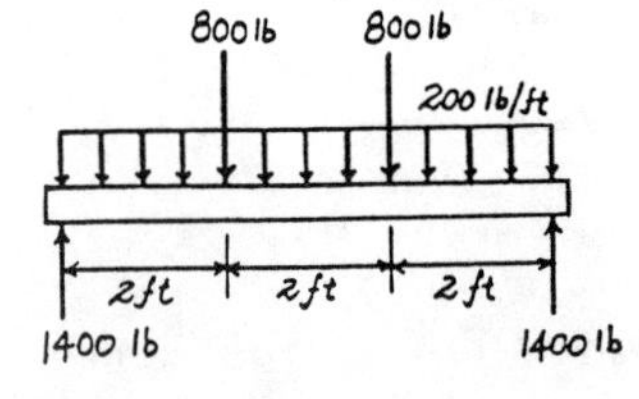

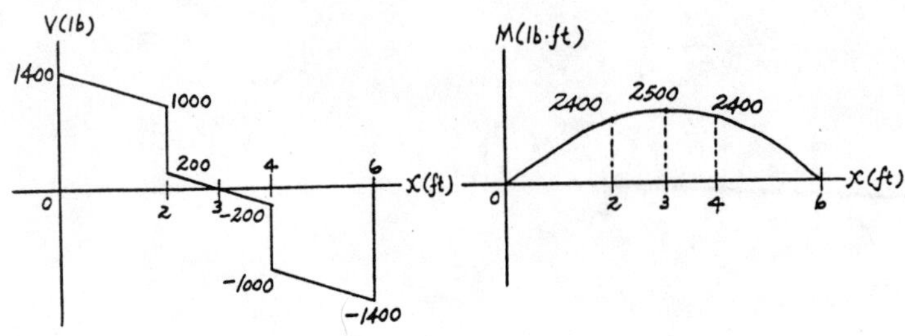

6–18. Draw the shear and moment diagrams for the beam. It is supported by a smooth plate at A which slides within the groove and so it cannot support a vertical force, although it can support a moment and axial load

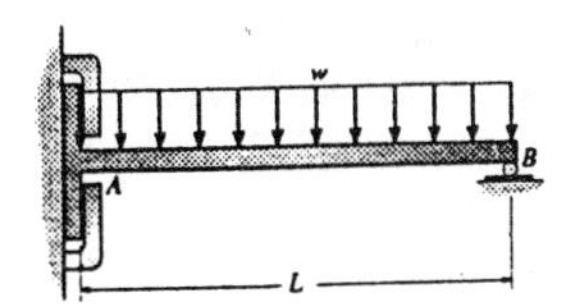

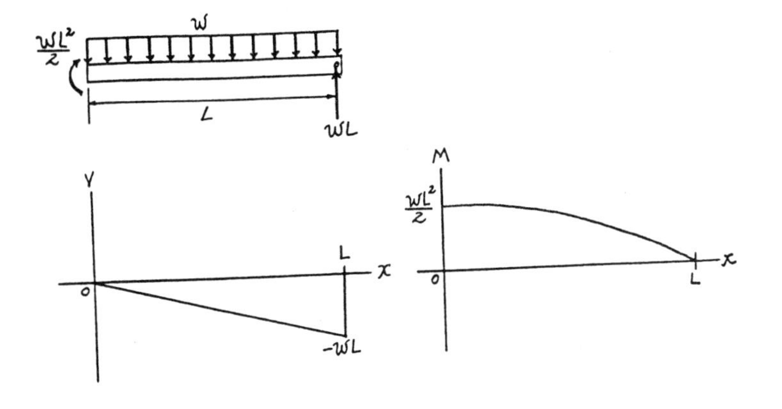

6–19. Draw the shear and moment diagrams for the compound beam.

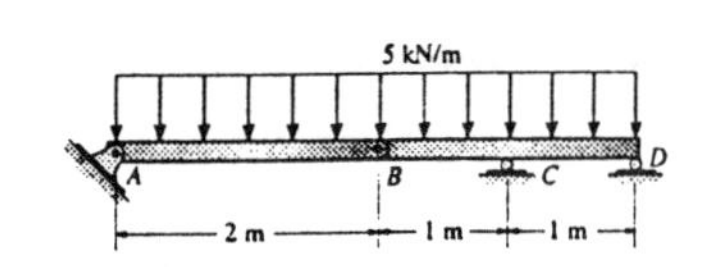

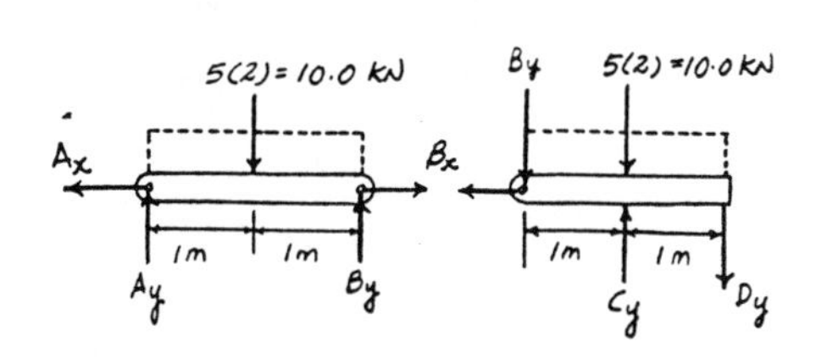

Support Reactions :

From the FBD of segment AB

$$\curvearrowleft + \Sigma M_A = 0; \quad B_y(2) - 10.0(1) = 0 \quad B_y = 5.00 \text{ kN}$$

$$+\uparrow \Sigma F_y = 0; \quad A_y - 10.0 + 5.00 = 0 \quad A_y = 5.00 \text{ kN}$$

From the FBD of segment BD

$$\curvearrowleft + \Sigma M_C = 0; \quad 5.00(1) + 10.0(0) - D_y(1) = 0$$
$$D_y = 5.00 \text{ kN}$$

$$+\uparrow \Sigma F_y = 0; \quad C_y - 5.00 - 5.00 - 10.0 = 0$$
$$C_y = 20.0 \text{ kN}$$

$$\xrightarrow{+} \Sigma F_x = 0; \quad B_x = 0$$

From the FBD of segment AB

$$\xrightarrow{+} \Sigma F_x = 0; \quad A_x = 0$$

Shear and Moment Diagram :

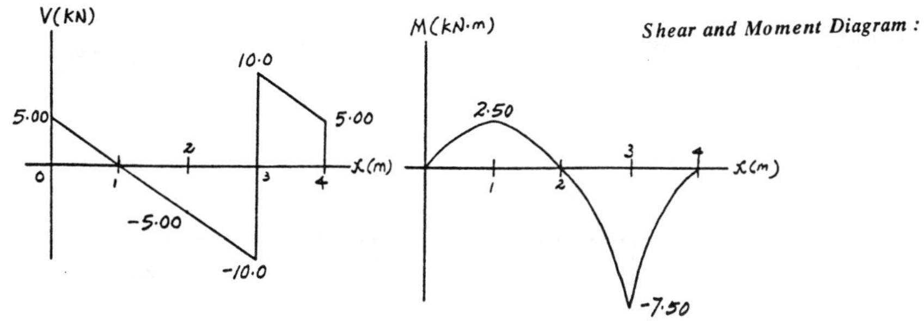

***6–20.** Draw the shear and moment diagrams for the beam, and determine the shear and moment throughout the beam as functions of x.

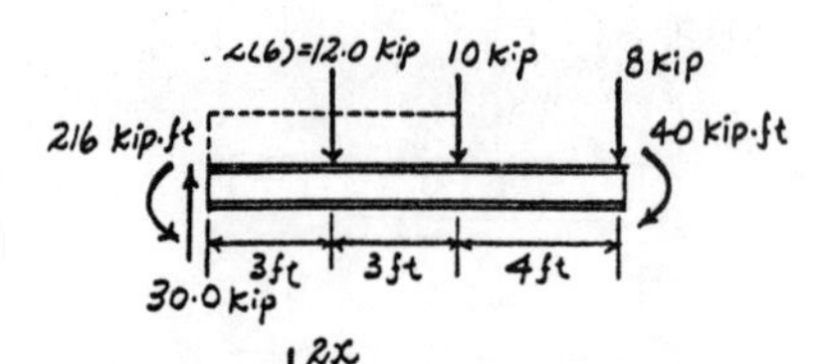

Support Reactions : As shown on FBD.
Shear and Moment Function :

For $0 \le x < 6$ ft

$+\uparrow \Sigma F_y = 0; \quad 30.0 - 2x - V = 0$

$V = \{30.0 - 2x\}$ kip **Ans**

$\circlearrowleft + \Sigma M_{NA} = 0; \quad M + 216 + 2x\left(\frac{x}{2}\right) - 30.0x = 0$

$M = \{-x^2 + 30.0x - 216\}$ kip · ft **Ans**

For 6 ft $< x \le 10$ ft

$+\uparrow \Sigma F_y = 0; \quad V - 8 = 0 \quad V = 8.00$ kip **Ans**

$\circlearrowleft + \Sigma M_{NA} = 0; \quad -M - 8(10 - x) - 40 = 0$

$M = \{8.00x - 120\}$ kip · ft **Ans**

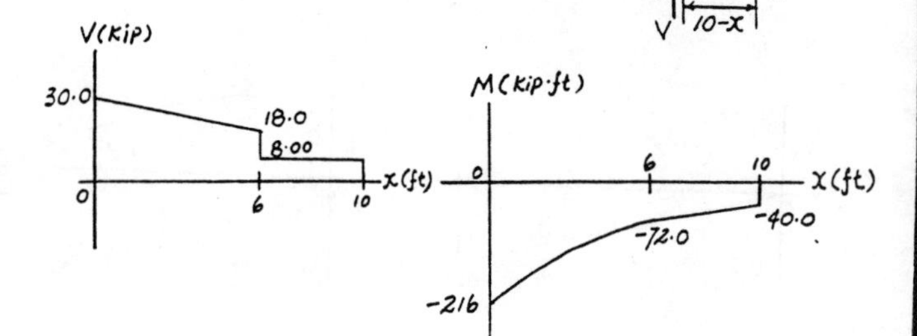

6–21. Draw the shear and moment diagrams for the beam, and determine the shear and moment throughout the beam as functions of x.

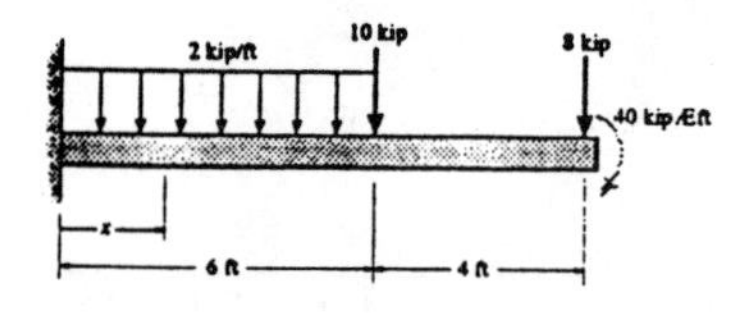

Support Reactions : As shown on FBD.
Shear and Moment Function :

For $0 \le x < 8$ ft

$+\uparrow \Sigma F_y = 0; \quad 7.60 - 0.800x - V = 0$

$V = \{7.60 - 0.800x\}$ kip **Ans**

$\circlearrowleft + \Sigma M_{NA} = 0; \quad M + 44.8 + 0.800x\left(\frac{x}{2}\right) - 7.60x = 0$

$M = \{-0.400x^2 + 7.60x - 44.8\}$ kip · ft **Ans**

For 8 ft $< x \le 16$ ft

$+\uparrow \Sigma F_y = 0; \quad V - 1.20 = 0 \quad V = 1.20$ kip **Ans**

$\circlearrowleft + \Sigma M_{NA} = 0; \quad -M - 1.20(16 - x) = 0$

$M = \{1.20x - 19.2\}$ kip · ft **Ans**

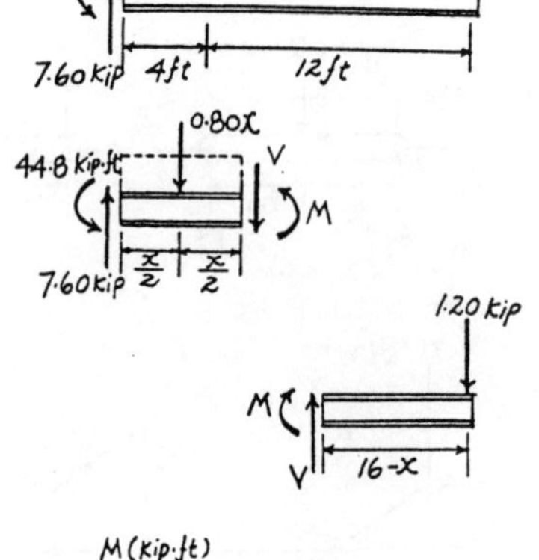

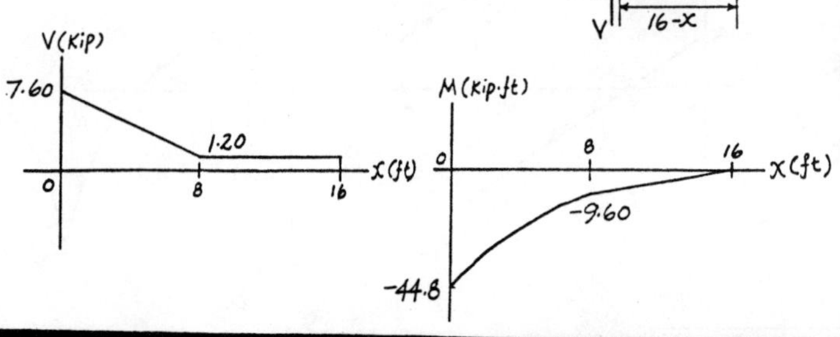

6–22. Draw the shear and moment diagrams for the beam.

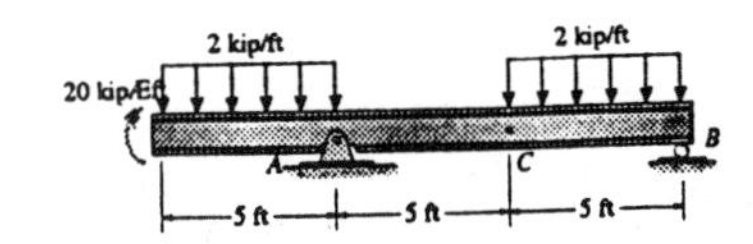

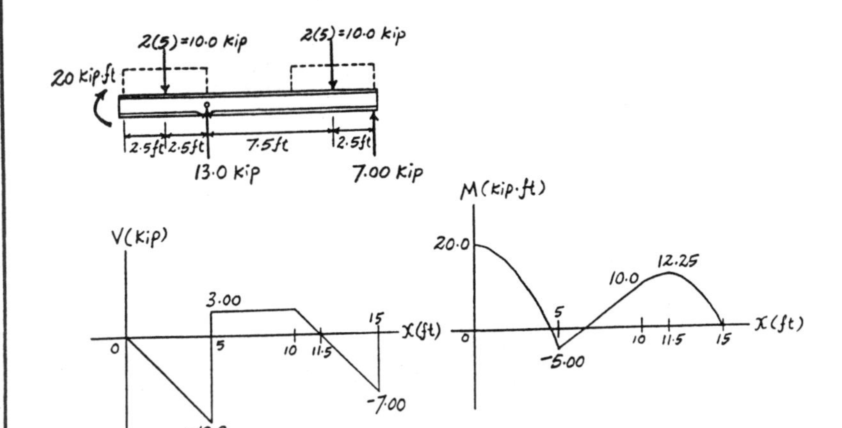

6–23. Draw the shear and moment diagrams for the beam.

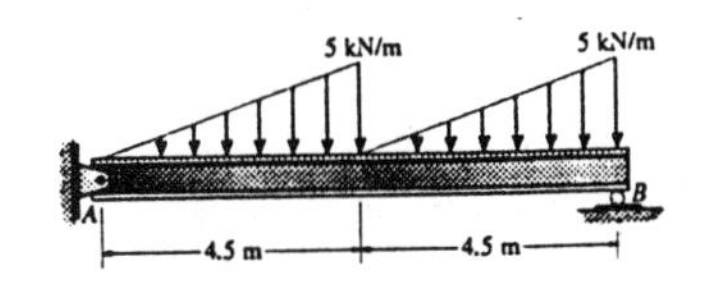

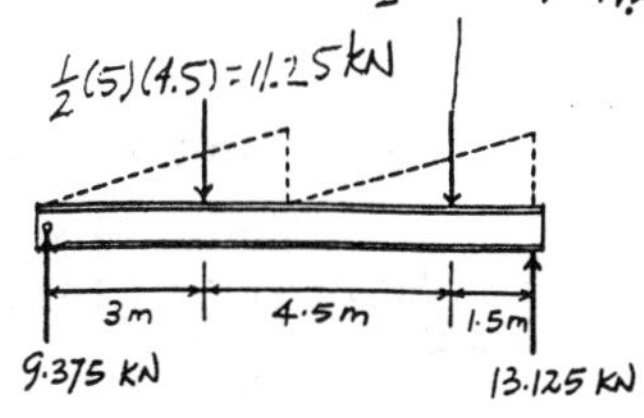

From FBD (a)

$$+\uparrow \Sigma F_y = 0; \quad 9.375 - 0.5556x^2 = 0 \quad x = 4.108 \text{ m}$$

$$\circlearrowleft + \Sigma M_{NA} = 0; \quad M + (0.5556)\left(4.108^2\right)\left(\frac{4.108}{3}\right) - 9.375(4.108) = 0$$

$$M = 25.67 \text{ kN} \cdot \text{m}$$

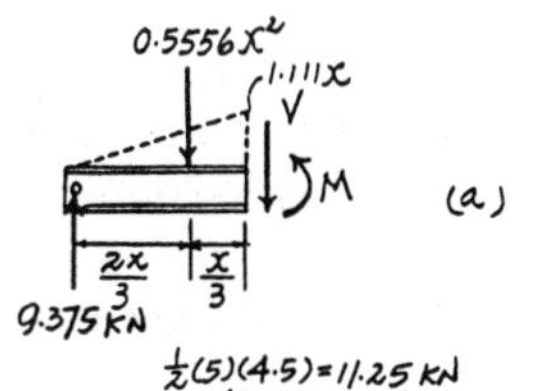

From FBD (b)

$$\circlearrowleft + \Sigma M_{NA} = 0: \quad M + 11.25(1.5) - 9.375(4.5) = 0$$

$$M = 25.31 \text{ kN} \cdot \text{m}$$

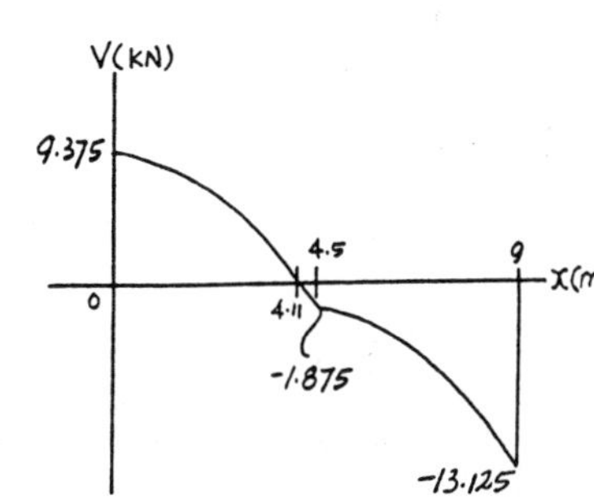

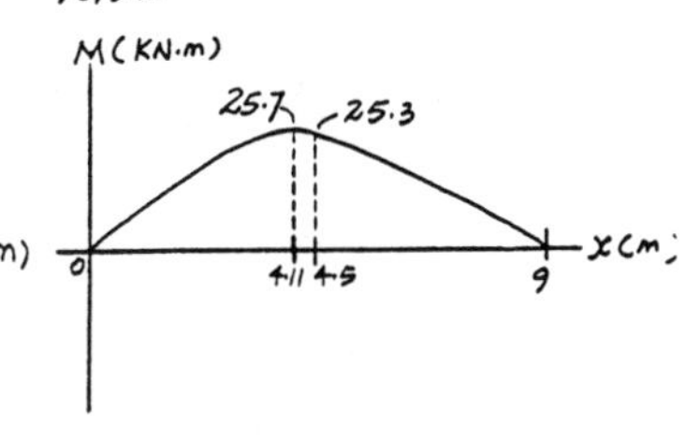

***6–24.** Determine the equilibrium torque T acting on the propeller, and then draw the shear and moment diagrams for the propeller.

200 lb/in.   T   200 lb/in.   30 in.   30 in.

Equilibrium : FBD (a)

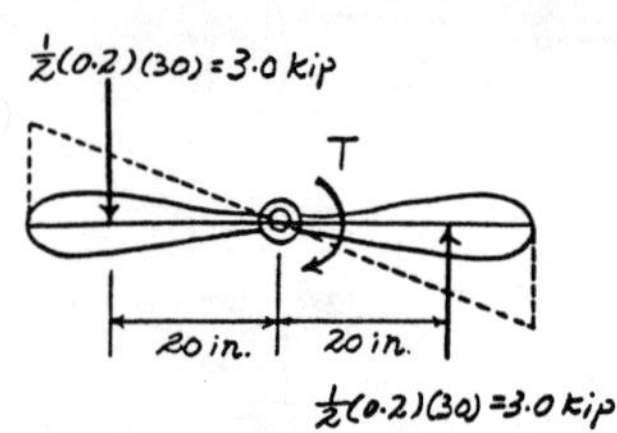

$\circlearrowleft + \Sigma M = 0;\quad 3(40) - T = 0 \qquad T = 120 \text{ kip}\cdot\text{in.} \qquad$ **Ans**

Shear and Moment Diagram : The shear and moment at $x = 30^-$ in. is determined as follows,

$+\uparrow \Sigma F_y = 0;\quad -3 - V = 0 \qquad V = -3.00 \text{ kip}$

$\circlearrowleft + \Sigma M_{NA} = 0;\quad M + 3.0(20) = 0 \qquad M = -60.0 \text{ kip}\cdot\text{in.}$

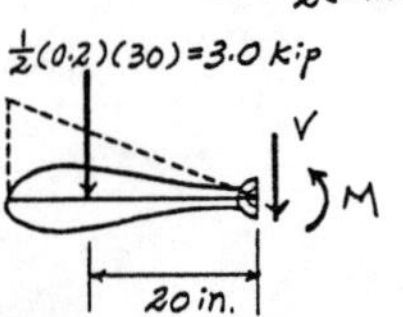

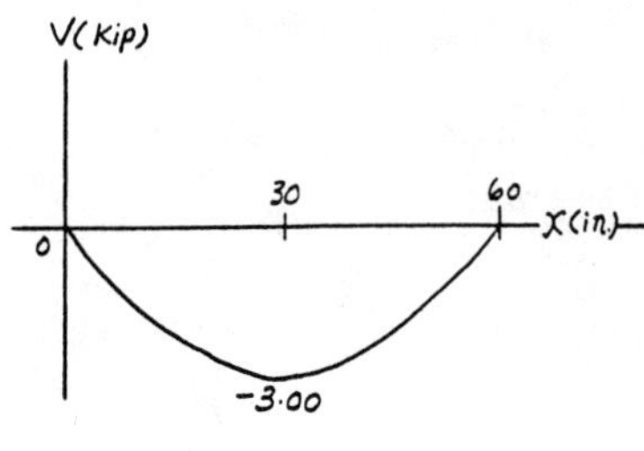
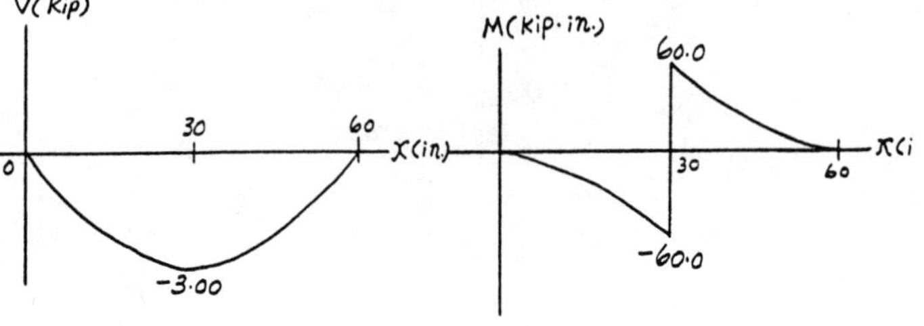

6–25. Draw the shear and moment diagrams for the compound beam.

150 lb/ft   150 lb/ft   A   B   C   6 ft   3 ft

Support Reactions :
From the FBD of segment AB

$\circlearrowleft + \Sigma M_B = 0;\quad 450(4) - A_y(6) = 0 \qquad A_y = 300.0 \text{ lb}$

$+\uparrow \Sigma F_y = 0;\quad B_y - 450 + 300.0 = 0 \qquad B_y = 150.0 \text{ lb}$

$\xrightarrow{+} \Sigma F_x = 0;\quad B_x = 0$

From the FBD of segment BC

$\circlearrowleft + \Sigma M_C = 0;\quad 225(1) + 150.0(3) - M_C = 0$
$M_C = 675.0 \text{ lb}\cdot\text{ft}$

$+\uparrow \Sigma F_y = 0;\quad C_y - 150.0 - 225 = 0 \qquad C_y = 375.0 \text{ lb}$

$\xrightarrow{+} \Sigma F_x = 0;\quad C_x = 0$

Shear and Moment Diagram : The maximum positive moment occurs when $V = 0$.

$+\uparrow \Sigma F_y = 0;\quad 150.0 - 12.5x^2 = 0 \qquad x = 3.464 \text{ ft}$

$\circlearrowleft + \Sigma M_{NA} = 0;\quad 150(3.464)$

$$- 12.5\left(3.464^2\right)\left(\frac{3.464}{3}\right) - M_{max} = 0$$

$$M_{max} = 346.4 \text{ lb}\cdot\text{ft}$$

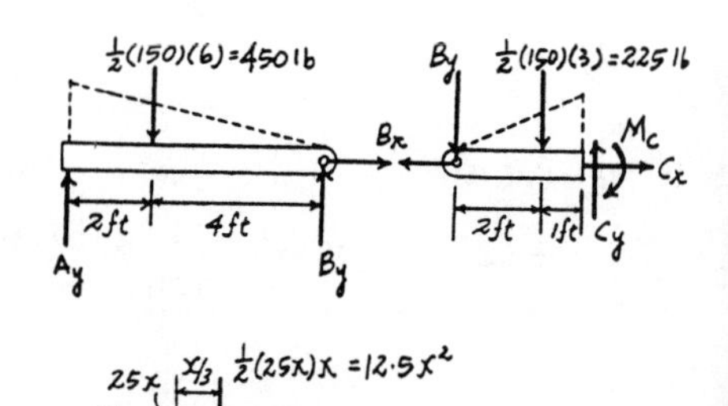

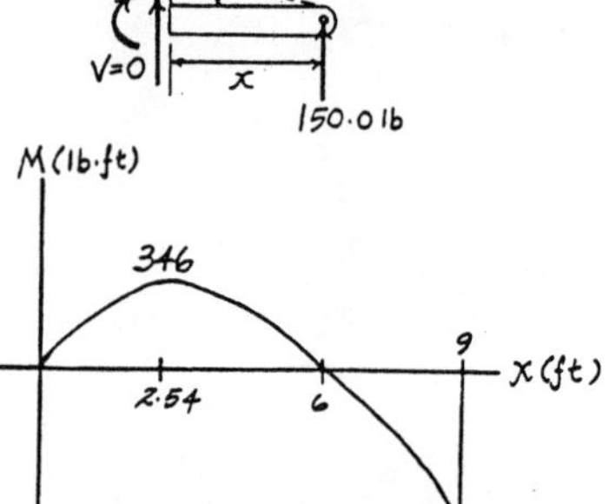

V(lb)   300   0   2.54   6   9   X(ft)   -150   -375

M(lb·ft)   346   2.54   6   9   X(ft)   -675

6–26. The dead-weight loading along the centerline of the airplane wing is shown. If the wing is fixed to the fuselage at A, determine the reactions at A, and then draw the shear and moment diagram for the wing.

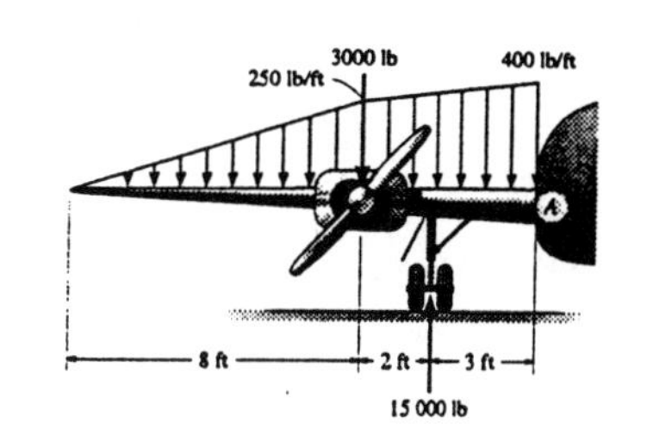

Support Reactions :

$$+\uparrow \Sigma F_y = 0; \quad -1.00 - 3 + 15 - 1.25 - 0.375 - A_y = 0$$

$$A_y = 9.375 \text{ kip} \qquad \textbf{Ans}$$

$$\circlearrowleft + \Sigma M_A = 0; \quad 1.00(7.667) + 3(5) - 15(3) + 1.25(2.5) + 0.375(1.667) + M_A = 0$$

$$M_A = 18.583 \text{ kip} \cdot \text{ft} = 18.6 \text{ kip} \cdot \text{ft} \qquad \textbf{Ans}$$

$$\xrightarrow{+} \Sigma F_x = 0; \quad A_x = 0 \qquad \textbf{Ans}$$

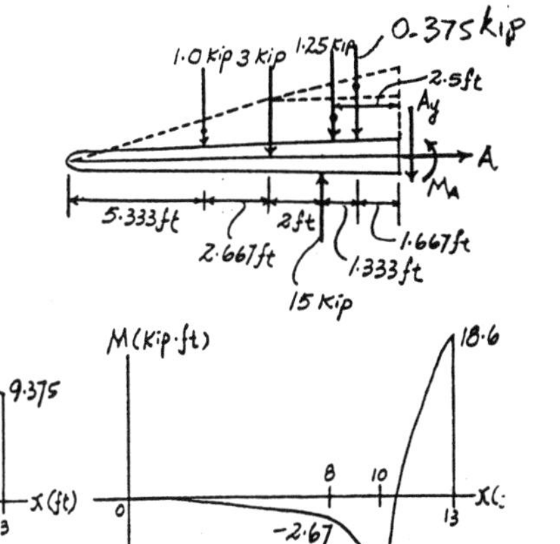

Shear and Moment Diagram :

6–27. Draw the shear and moment diagrams for the beam.

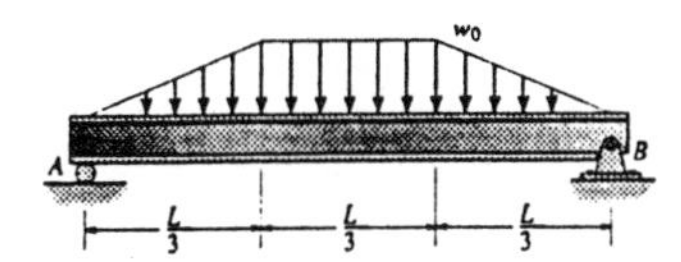

Support Reactions : As shown on FBD.
Shear and Moment Diagram : Shear and moment at $x = L/3$ can be determined using the method of sections.

$$+\uparrow \Sigma F_y = 0; \quad \frac{w_0 L}{3} - \frac{w_0 L}{6} - V = 0 \qquad V = \frac{w_0 L}{6}$$

$$\circlearrowleft + \Sigma M_{NA} = 0; \quad M + \frac{w_0 L}{6}\left(\frac{L}{9}\right) - \frac{w_0 L}{3}\left(\frac{L}{3}\right) = 0$$

$$M = \frac{5 w_0 L^2}{54}$$

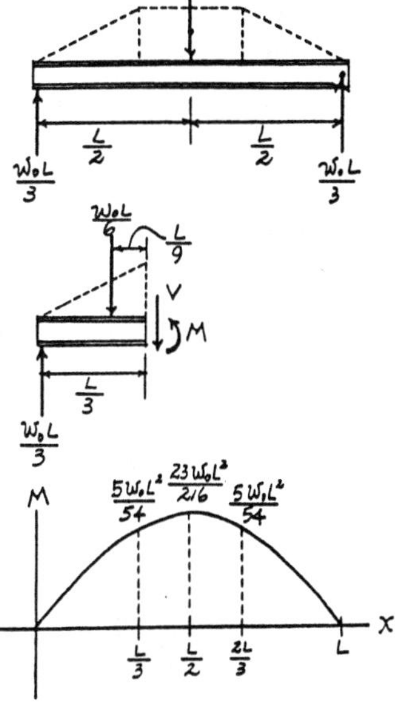

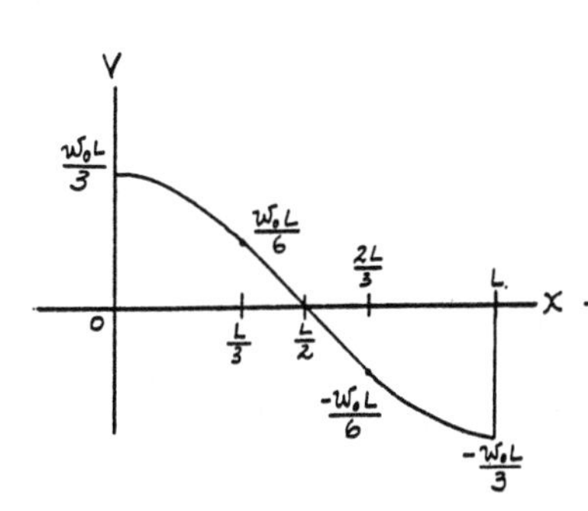

***6–28.** Draw the shear and moment diagrams for the beam, and determine the shear and moment in the beam as functions of x.

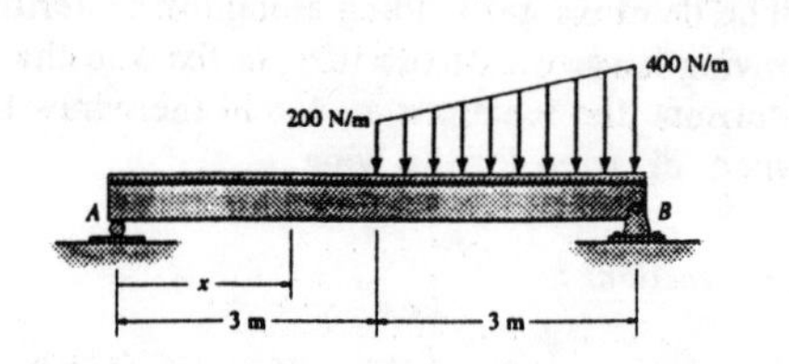

Support Reactions : As shown on FBD.

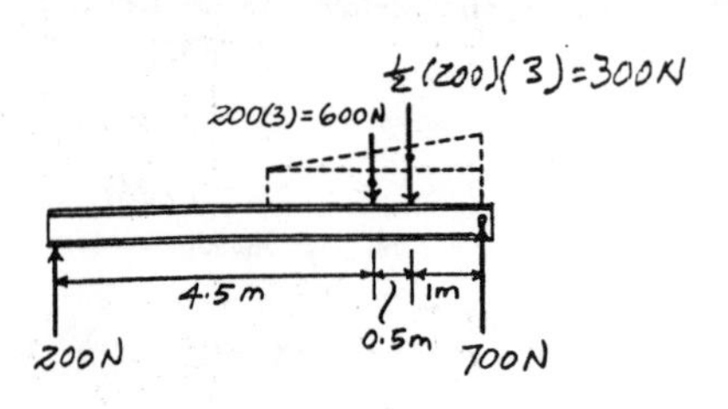

Shear and Moment Functions :

For $0 \le x < 3$ m

$+\uparrow \Sigma F_y = 0; \quad 200 - V = 0 \quad V = 200\text{ N}$ **Ans**

$\circlearrowleft + \Sigma M_{NA} = 0; \quad M - 200x = 0$

$M = \{200x\}\text{ N}\cdot\text{m}$ **Ans**

For $3\text{ m} < x \le 6\text{ m}$

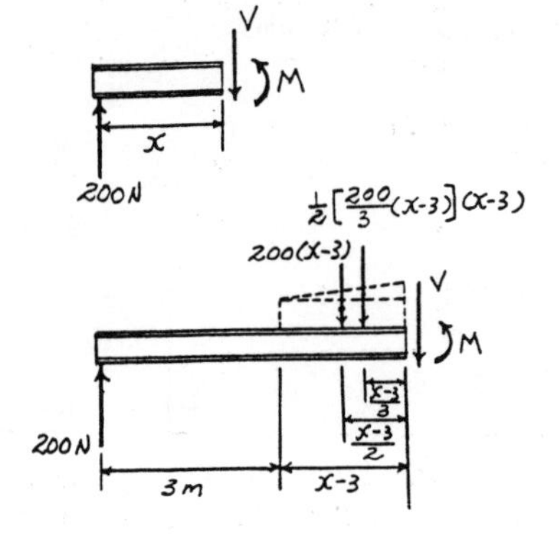

$$+\uparrow \Sigma F_y = 0; \quad 200 - 200(x-3) - \frac{1}{2}\left[\frac{200}{3}(x-3)\right](x-3) - V = 0$$

$$V = \left\{-\frac{100}{3}x^2 + 500\right\}\text{ N} \quad \textbf{Ans}$$

Set $V = 0$, $\quad x = 3.873$ m

$$\circlearrowleft + \Sigma M_{NA} = 0; \quad M + \frac{1}{2}\left[\frac{200}{3}(x-3)\right](x-3)\left(\frac{x-3}{3}\right) + 200(x-3)\left(\frac{x-3}{2}\right) - 200x = 0$$

$$M = \left\{-\frac{100}{9}x^3 + 500x - 600\right\}\text{ N}\cdot\text{m} \quad \textbf{Ans}$$

Substitute $x = 3.87$ m, $\quad M = 691\text{ N}\cdot\text{m}$

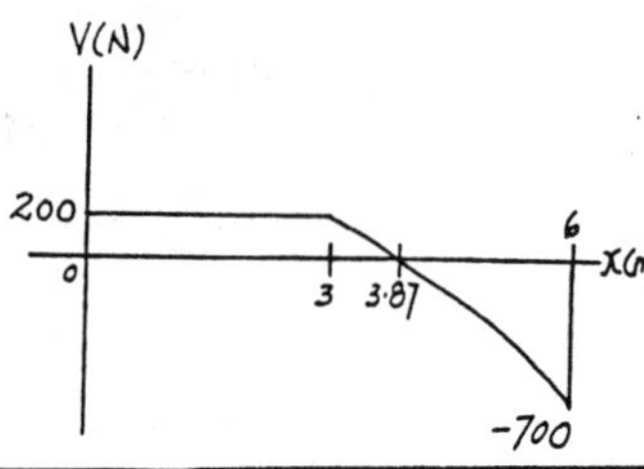

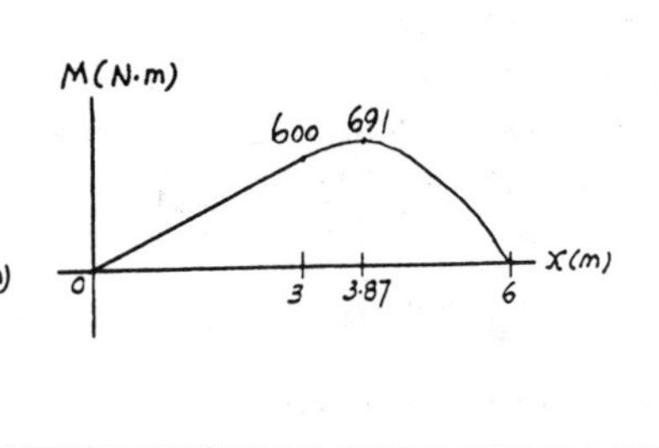

6–29. The T-beam is subjected to the loading shown. Draw the shear and moment diagrams.

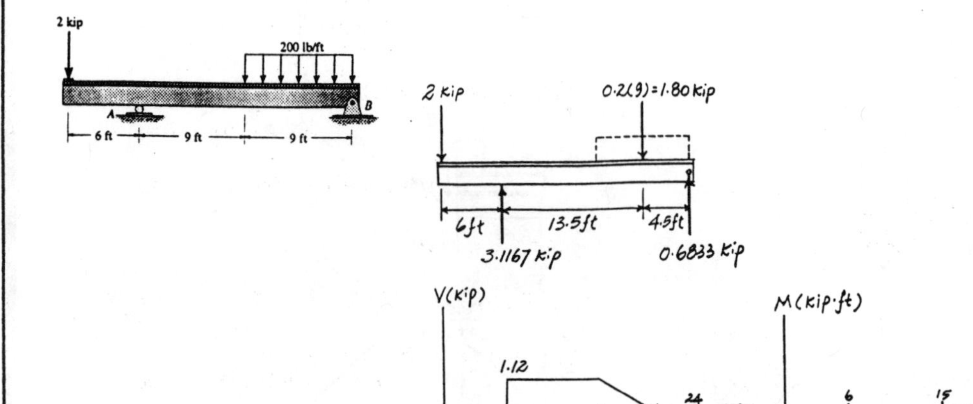

6–30. The boom ABC has a weight of 30 lb/ft and is used to lift the load of 2000 lb. Draw the shear and moment diagrams of the boom when it is in the horizontal position shown.

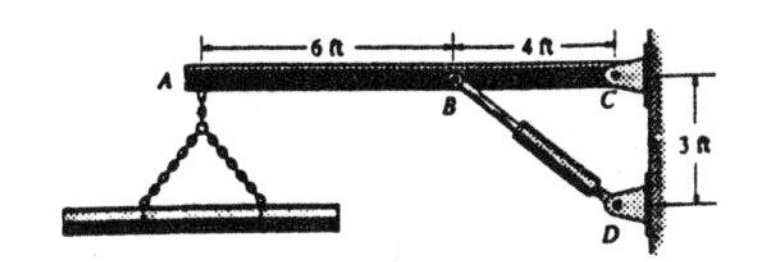

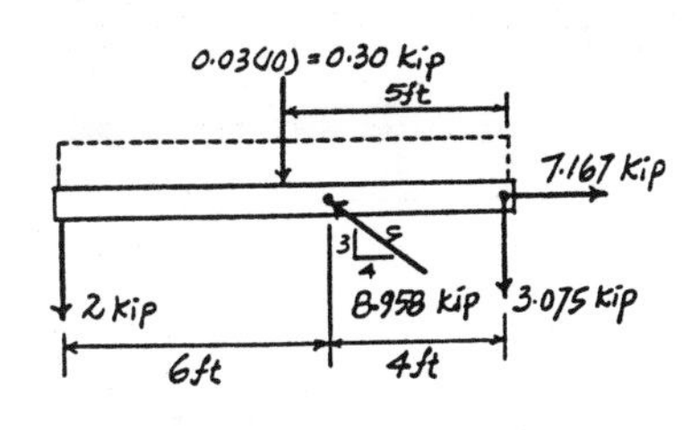

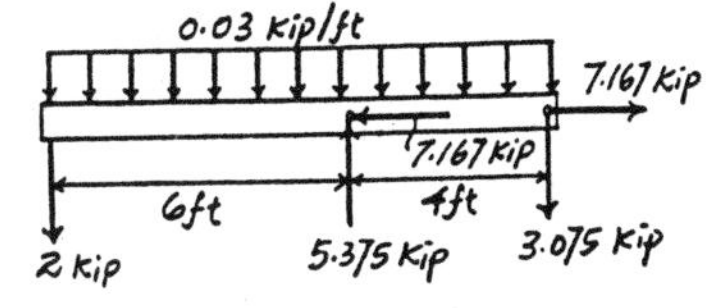

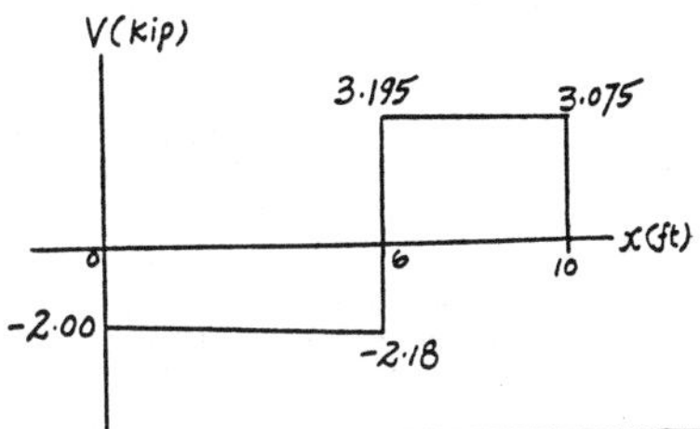

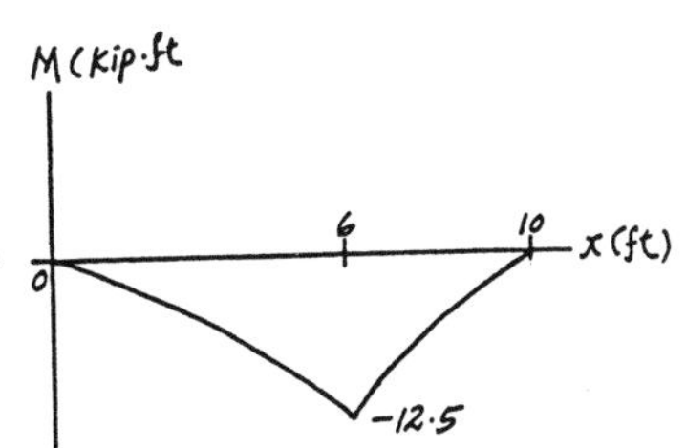

6–31. Draw the shear and moment diagrams for the wood beam, and determine the shear and moment throughout the beam as functions of x.

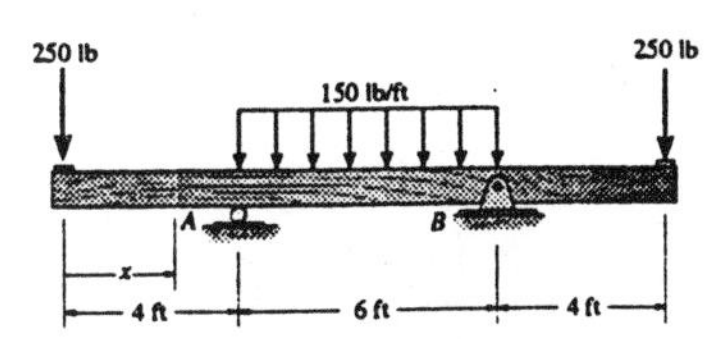

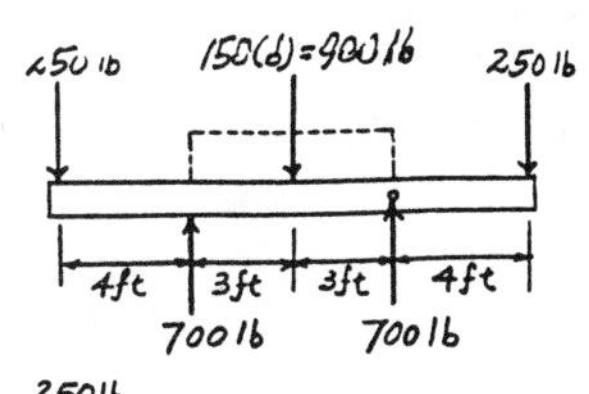

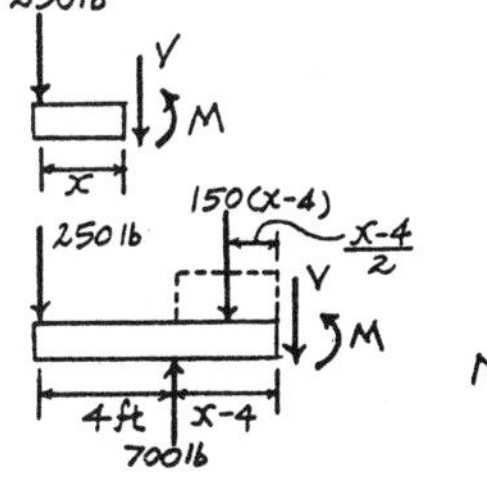

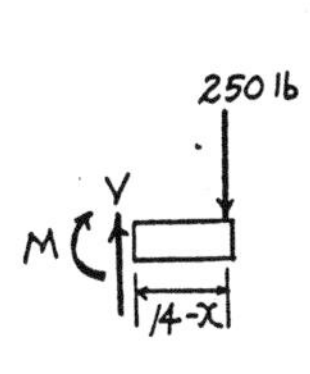

Support Reactions : As shown on FBD.

Shear and Moment Functions :

For $0 \le x < 4$ ft

$+\uparrow \Sigma F_y = 0; \quad -250 - V = 0 \quad V = -250 \text{ lb}$ **Ans**

$\curvearrowleft + \Sigma M_{NA} = 0; \quad M + 250x = 0$

$M = \{-250x\} \text{ lb} \cdot \text{ft}$ **Ans**

For $4 \text{ ft} < x < 10$ ft

$+\uparrow \Sigma F_y = 0; \quad -250 + 700 - 150(x-4) - V = 0$

$V = \{1050 - 150x\} \text{ lb}$ **Ans**

$\curvearrowleft + \Sigma M_{NA} = 0; \quad M + 150(x-4)\left(\frac{x-4}{2}\right) + 250x - 700(x-4) = 0$

$M = \{-75x^2 + 1050x - 4000\} \text{ lb} \cdot \text{ft}$ **Ans**

For $10 \text{ ft} < x \le 14$ ft

$+\uparrow \Sigma F_y = 0; \quad V - 250 = 0 \quad V = 250 \text{ lb}$ **Ans**

$\curvearrowleft + \Sigma M_{NA} = 0; \quad -M - 250(14 - x) = 0$

$M = \{250x - 3500\} \text{ lb} \cdot \text{ft}$ **Ans**

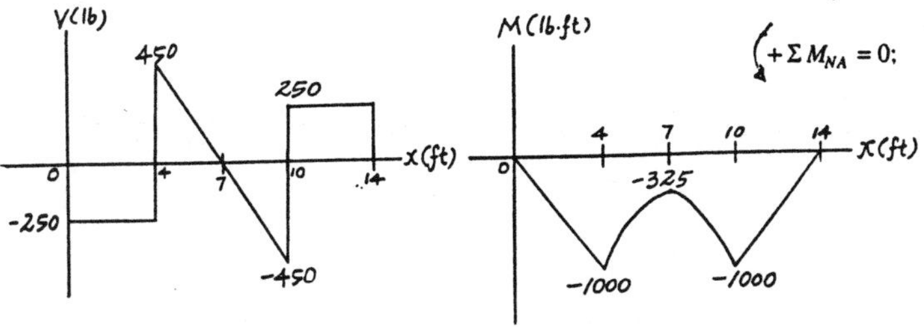

***6–32.** Draw the shear and moment diagrams for the beam, and determine the shear and moment in the beam as functions of x.

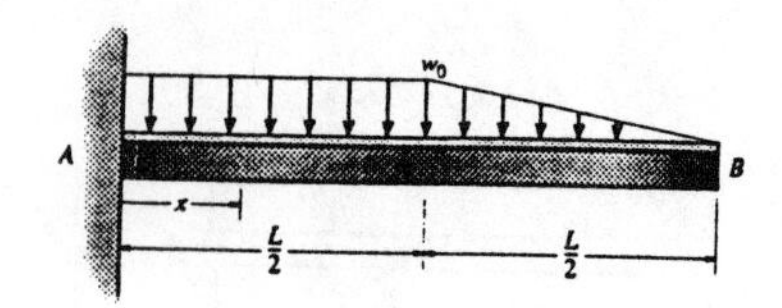

Support Reactions : As shown on FBD.
Shear and Moment Functions :

For $0 \leq x < L/2$

$$+\uparrow \Sigma F_y = 0; \quad \frac{3w_0 L}{4} - w_0 x - V = 0$$

$$V = \frac{w_0}{4}(3L - 4x) \qquad \text{Ans}$$

$$\circlearrowleft + \Sigma M_{NA} = 0; \quad \frac{7w_0 L^2}{24} - \frac{3w_0 L}{4}x + w_0 x\left(\frac{x}{2}\right) + M = 0$$

$$M = \frac{w_0}{24}\left(-12x^2 + 18Lx - 7L^2\right) \qquad \text{Ans}$$

For $L/2 < x \leq L$

$$+\uparrow \Sigma F_y = 0; \quad V - \frac{1}{2}\left[\frac{2w_0}{L}(L-x)\right](L-x) = 0$$

$$V = \frac{w_0}{L}(L-x)^2 \qquad \text{Ans}$$

$$\circlearrowleft + \Sigma M_{NA} = 0; \quad -M - \frac{1}{2}\left[\frac{2w_0}{L}(L-x)\right](L-x)\left(\frac{L-x}{3}\right) = 0$$

$$M = -\frac{w_0}{3L}(L-x)^3 \qquad \text{Ans}$$

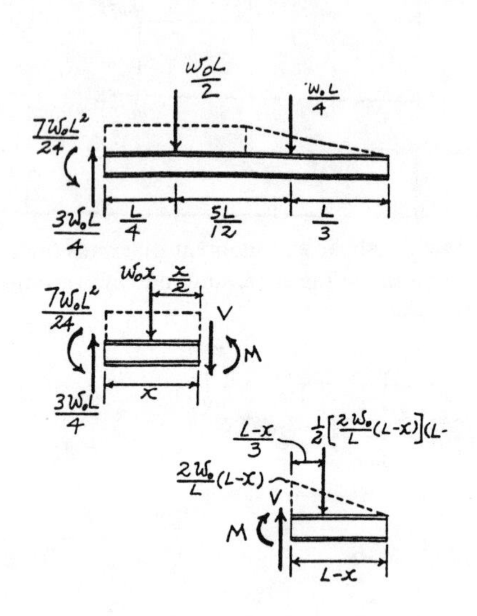

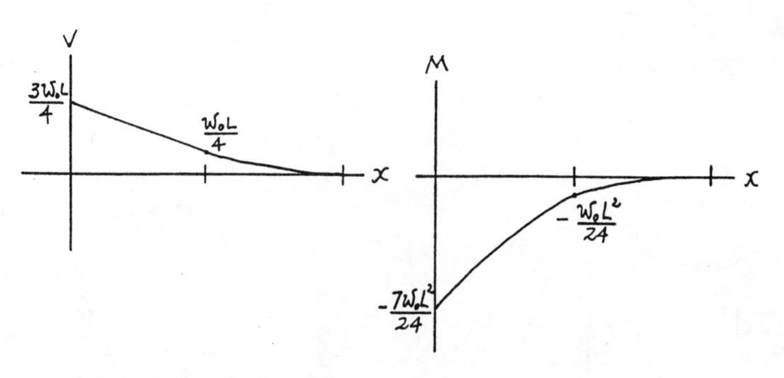

6–33. Draw the shear and moment diagrams for the beam. There is a short vertical link at B and a pin at C.

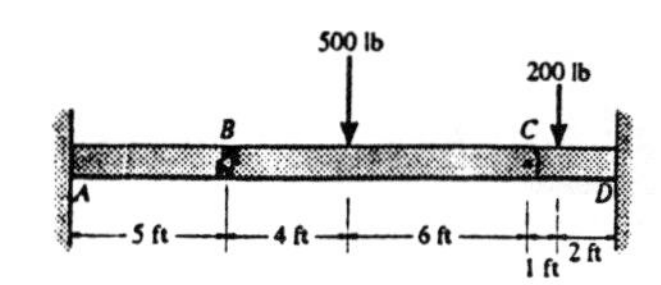

Support Reactions :

From the FBD of segment BC

$\circlearrowleft + \Sigma M_C = 0; \quad -B_y(10) - 500(6) = 0 \quad B_y = 300.0 \text{ lb}$

$+\uparrow \Sigma F_y = 0; \quad C_y - 500 + 300.0 = 0 \quad C_y = 200.0 \text{ lb}$

$\xrightarrow{+} \Sigma F_x = 0; \quad C_x = 0$

From the FBD of segment AB

$\circlearrowleft + \Sigma M_A = 0; \quad M_A - 300.0(5) = 0 \quad M_A = 1500 \text{ lb} \cdot \text{ft}$

$+\uparrow \Sigma F_y = 0; \quad A_y - 300.0 = 0 \quad A_y = 300.0 \text{ lb}$

$\xrightarrow{+} \Sigma F_x = 0; \quad A_x = 0$

From the FBD of segment CD

$\circlearrowleft + \Sigma M_D = 0; \quad -M_D + 200(2) + 200.0(3) = 0$
$M_D = 1000 \text{ lb} \cdot \text{ft}$

$+\uparrow \Sigma F_y = 0; \quad D_y - 200.0 - 200 = 0 \quad D_y = 400.0 \text{ lb}$

$\xrightarrow{+} \Sigma F_x = 0; \quad D_x = 0$

Shear and Moment Diagram :

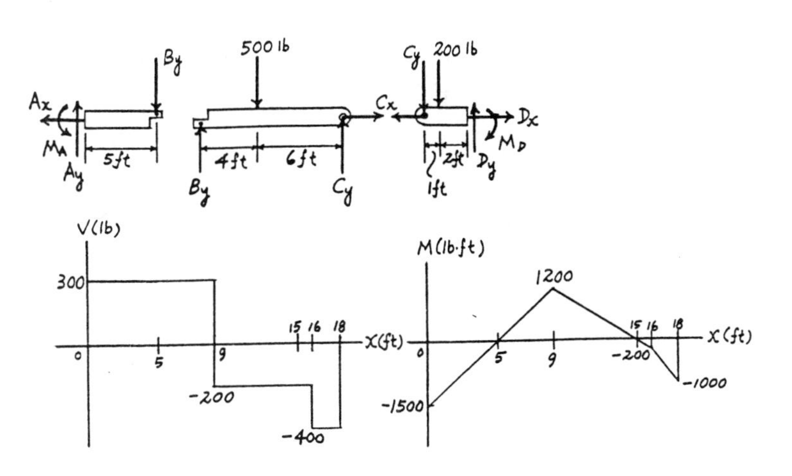

6–34. Draw the shear and moment diagrams for the beam.

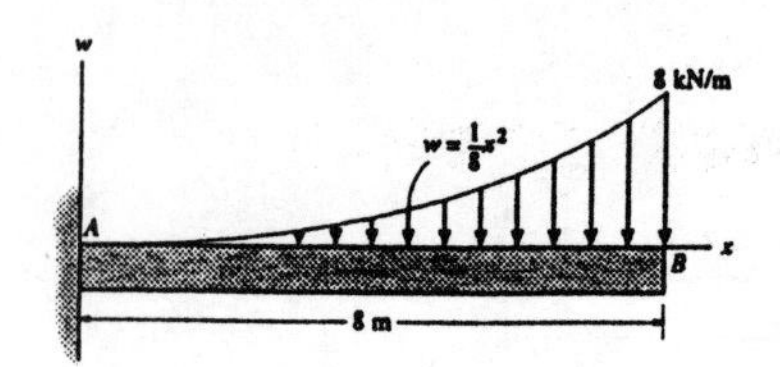

Resultant Force and its Location :

$$F_R = \int_0^L w\,dx = \frac{1}{8}\int_0^{8\,\text{m}} x^2\,dx = 21.33\text{ kN}$$

$$\bar{x} = \frac{\int_0^L wx\,dx}{\int_0^L w\,dx} = \frac{\frac{1}{8}\int_0^{8\,\text{m}} x^3\,dx}{21.33} = 6.00\text{ m}$$

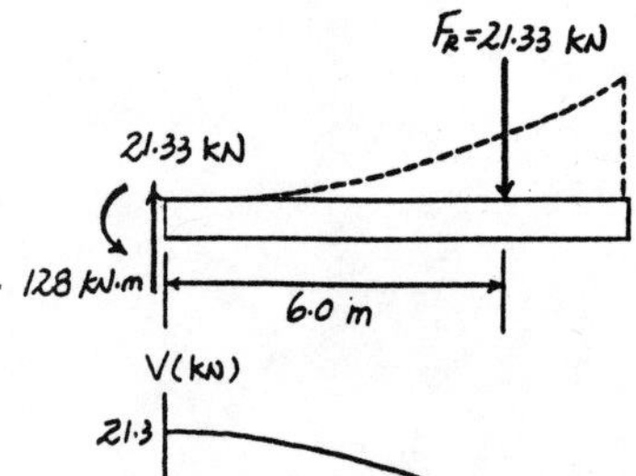

Support Reactions : As shown on FBD.
Shear and Moment Diagram :

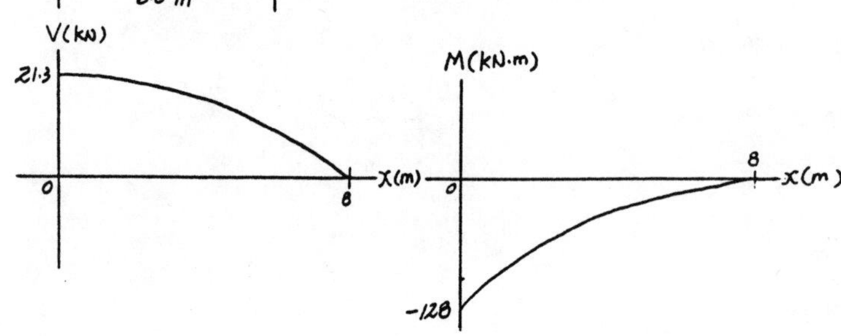

6–35. Draw the shear and moment diagrams for the beam.

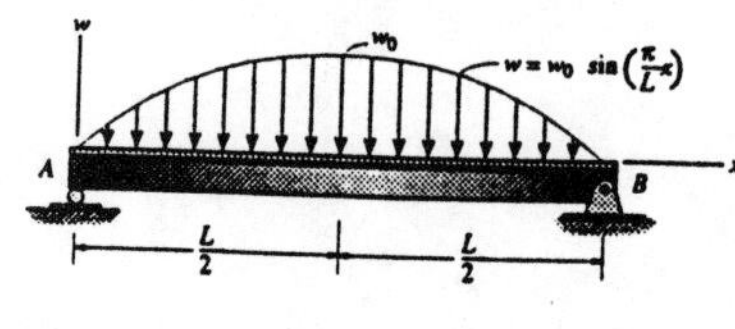

Resultant Force :

$$F_R = \int_0^L w\,dx$$

$$= w_0 \int_0^L \sin\left(\frac{\pi}{L}x\right)dx$$

$$= w_0\left[-\frac{L}{\pi}\cos\left(\frac{\pi}{L}x\right)\right]\Bigg|_0^L$$

$$= \frac{2w_0 L}{\pi}$$

Support Reactions : As shown on FBD.
Shear and Moment Diagram : The location of the resultant force $\bar{x}$ is obtained first.

$$\bar{x} = \frac{\int_0^{L/2} wx\,dx}{\int_0^{L/2} w\,dx} = \frac{w_0\int_0^{\frac{L}{2}} x\sin\left(\frac{\pi}{L}x\right)dx}{w_0\int_0^{\frac{L}{2}}\sin\left(\frac{\pi}{L}x\right)dx} = \frac{\frac{w_0 L^2}{\pi^2}}{\frac{w_0 L}{\pi}} = \frac{L}{\pi}$$

$$\circlearrowleft + \Sigma M_{NA} = 0; \quad M - \frac{w_0 L}{\pi}\left(\frac{L}{\pi}\right) = 0 \quad M = \frac{w_0 L^2}{\pi^2}$$

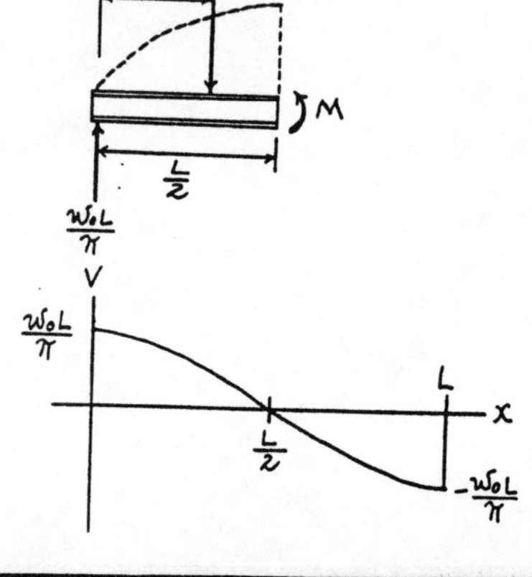

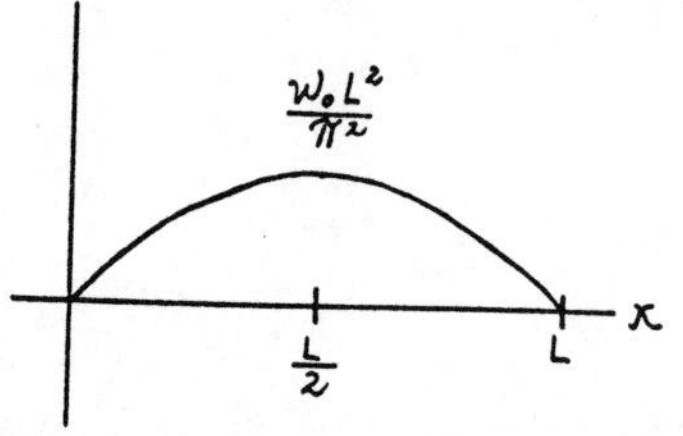

***6–36.** Determine the placement distance a of the roller support so that the largest absolute value of the moment is a minimum. Draw the shear and moment diagrams for this condition.

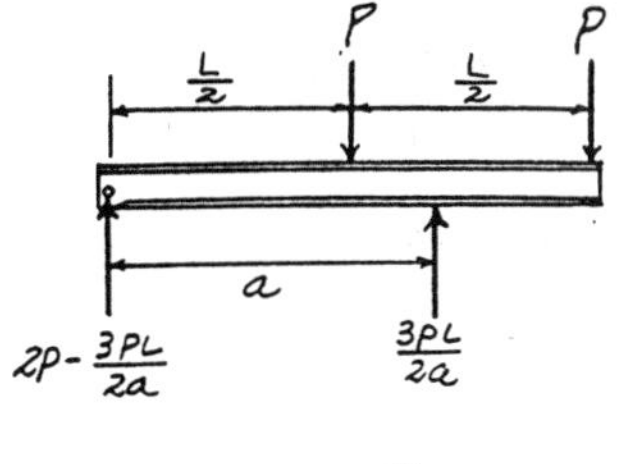

Support Reactions : As shown on FBD.

Absolute Minimum Moment : In order to get the absolute minimum moment, the maximum positive and maximum negative moment must be equal that is $M_{\max(+)} = M_{\min(-)}$.

For the positive moment

$$\curvearrowleft + \Sigma M_{NA} = 0; \quad M_{\max(+)} - \left(2P - \frac{3PL}{2a}\right)\left(\frac{L}{2}\right) = 0$$

$$M_{\max(+)} = PL - \frac{3PL^2}{4a}$$

For the negative moment

$$\curvearrowleft + \Sigma M_{NA} = 0; \quad M_{\max(-)} - P(L-a) = 0$$

$$M_{\max(-)} = P(L-a)$$

$$M_{\max(+)} = M_{\max(-)}$$

$$PL - \frac{3PL^2}{4a} = P(L-a)$$

$$4aL - 3L^2 = 4aL - 4a^2$$

$$a = \frac{\sqrt{3}}{2}L = 0.866L \qquad \textbf{Ans}$$

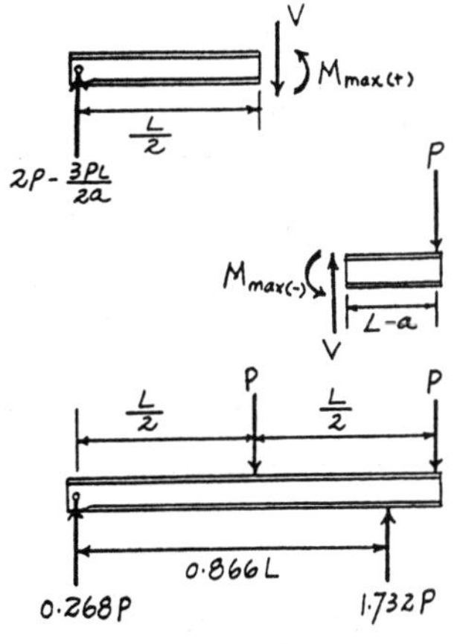

Shear and Moment Diagram :

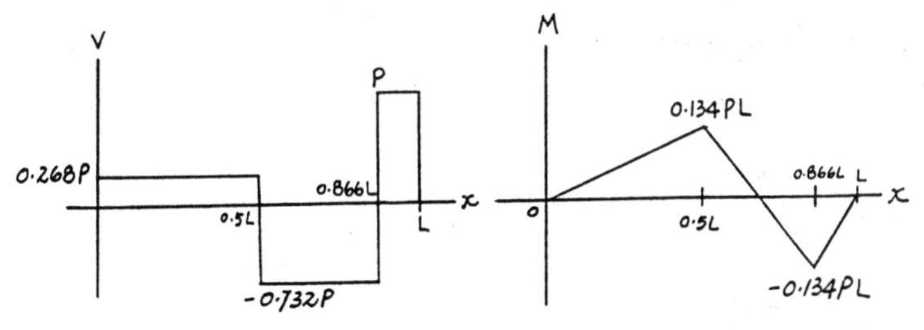

6–37. The truck is to be used to transport the concrete column. If the column has a uniform weight of w (force/length), determine the equal placement a of the supports from the ends so that the absolute maximum bending moment in the column is as small as possible. Also, draw the shear and moment diagrams for the column.

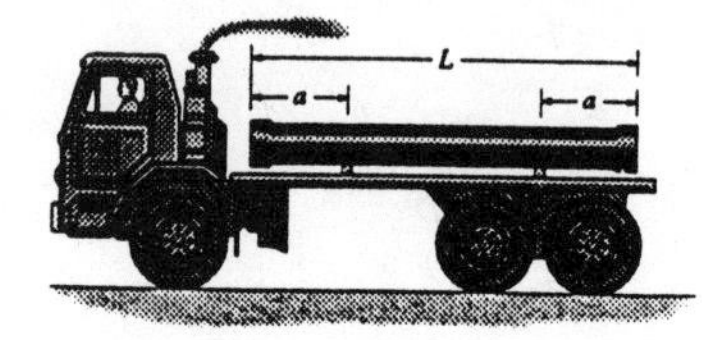

Support Reactions : As shown on FBD.

Absolute Minimum Moment : In order to get the absolute minimum moment, the maximum positive and maximum negative moment must be equal that is $M_{max(+)} = M_{min(-)}$.

For the positive moment

$$\circlearrowleft + \Sigma M_{NA} = 0; \quad M_{max(+)} + \frac{wL}{2}\left(\frac{L}{4}\right) - \frac{wL}{2}\left(\frac{L}{2} - a\right) = 0$$

$$M_{max(+)} = \frac{wL^2}{8} - \frac{waL}{2}$$

For the negative moment

$$\circlearrowleft + \Sigma M_{NA} = 0; \quad wa\left(\frac{a}{2}\right) - M_{max(-)} = 0$$

$$M_{max(-)} = \frac{wa^2}{2}$$

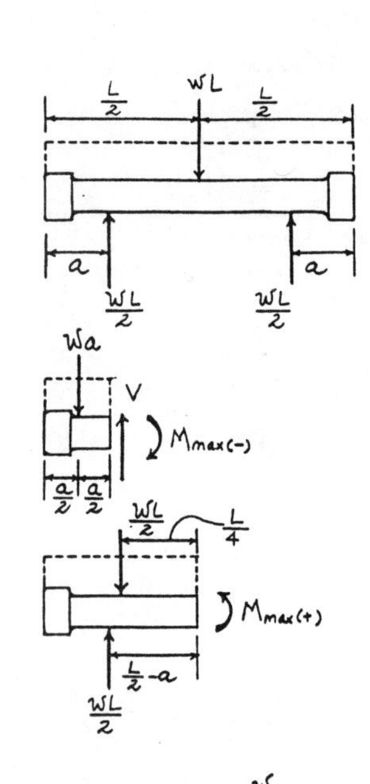

$$M_{max(+)} = M_{max(-)}$$

$$\frac{wL^2}{8} - \frac{wL}{2}a = \frac{wa^2}{2}$$

$$4a^2 + 4La - L^2 = 0$$

$$a = \frac{-4L \pm \sqrt{16L^2 - 4(4)(-L^2)}}{2(4)}$$

$$a = 0.207L \qquad \textbf{Ans}$$

Shear and Moment Diagram :

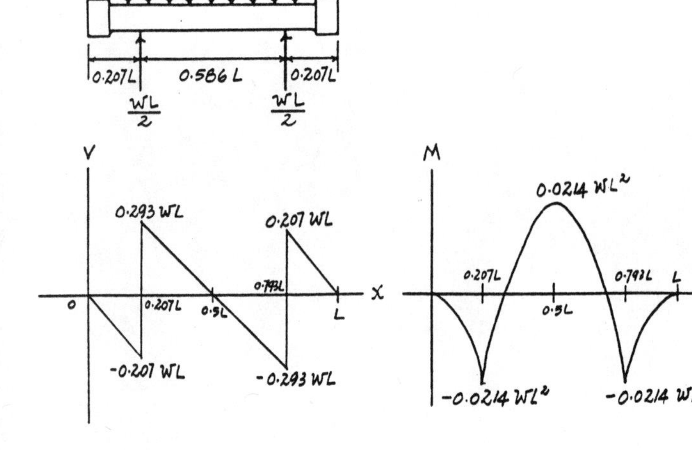

6–38. The beam is subjected to a moment M. Determine the percentage of this moment that is resisted by the stresses acting on both the top and bottom boards, A and B, of the beam.

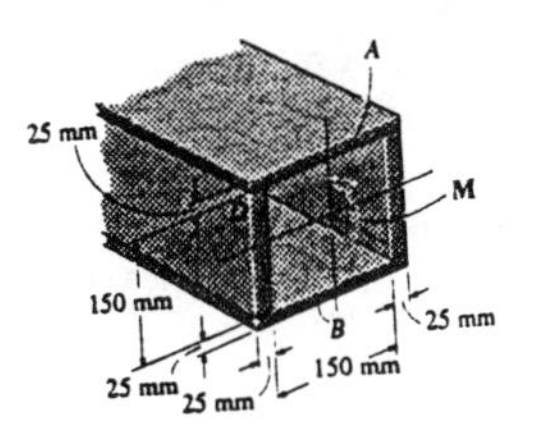

Section Property :

$$I = \frac{1}{12}(0.2)(0.2^3) - \frac{1}{12}(0.15)(0.15^3) = 91.14583(10^{-6})\ \text{m}^4$$

Bending Stress : Applying the flexure formula

$$\sigma = \frac{My}{I}$$

$$\sigma_E = \frac{M(0.1)}{91.14583(10^{-6})} = 1097.143\,M$$

$$\sigma_D = \frac{M(0.075)}{91.14583(10^{-6})} = 822.857\,M$$

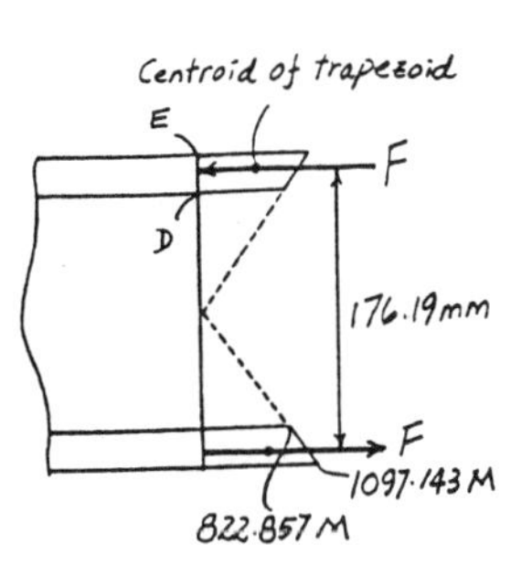

Resultant Force and Moment : For board A or B

$$F = 822.857M(0.025)(0.2) + \frac{1}{2}(1097.143M - 822.857M)(0.025)(0.2) = 4.800\,M$$

$$M' = F(0.17619) = 4.80M(0.17619) = 0.8457\,M$$

$$\%\left(\frac{M'}{M}\right) = 0.8457(100\%) = 84.6\ \%$$ **Ans**

6–39. Determine the moment M that should be applied to the beam in order to create a compressive stress at point D of $\sigma_D = 30$ MPa. Also sketch the stress distribution acting over the cross section and compute the maximum stress developed in the beam.

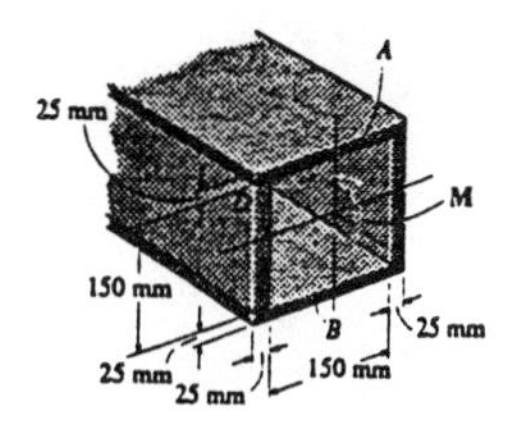

Section Property :

$$I = \frac{1}{12}(0.2)(0.2^3) - \frac{1}{12}(0.15)(0.15^3) = 91.14583(10^{-6})\ \text{m}^4$$

Bending Stress : Applying the flexure formula

$$\sigma = \frac{My}{I}$$

$$30(10^6) = \frac{M(0.075)}{91.14583(10^{-6})}$$

$$M = 36458\ \text{N}\cdot\text{m} = 36.5\ \text{kN}\cdot\text{m}$$ **Ans**

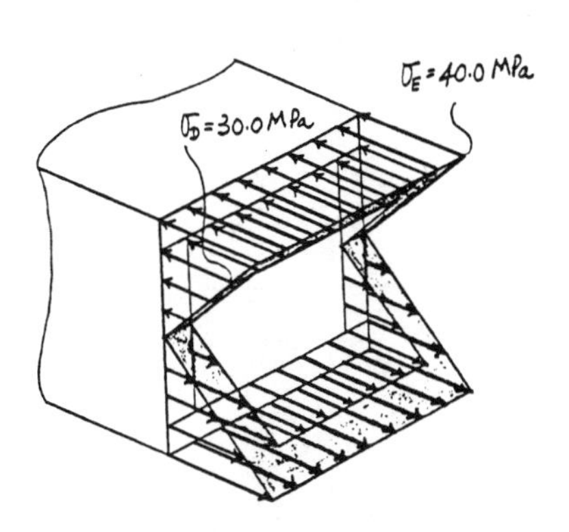

$$\sigma_{max} = \frac{Mc}{I} = \frac{36458(0.1)}{91.14583(10^{-6})} = 40.0\ \text{MPa}$$ **Ans**

***6–40.** The slab of marble, which can be assumed a linear elastic brittle material, has a specific weight of 150 lb/ft³ and a thickness of 0.75 in. Calculate the maximum bending stress in the slab if it is supported (a) on its side and (b) on its edges. If the fracture stress is $\sigma_f = 200$ psi, explain the consequences of supporting the slab in each position.

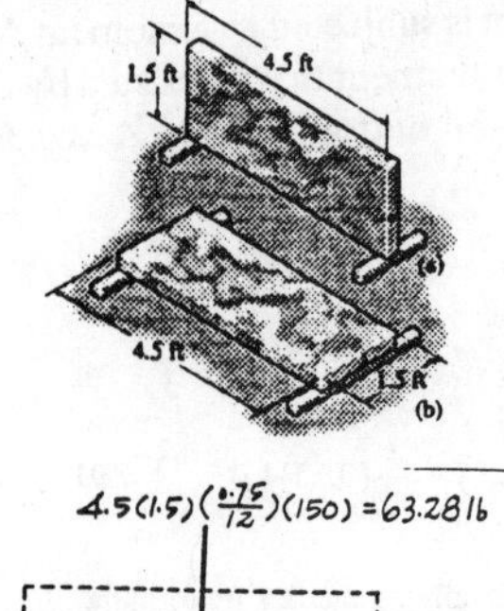

Support Reactions : As shown on FBD(a).
Maximum Moment : In both cases, the maximum moment occurs at midspan as shown on FBD(b).

Maximum Bending Stress : Applying the flexure formula $\sigma_{max} = \frac{Mc}{I}$

a)

$$\sigma_{max} = \frac{35.60(12)(9)}{\frac{1}{12}(0.75)(18^3)} = 10.5 \text{ psi} \qquad \textbf{Ans}$$

b)

$$\sigma_{max} = \frac{35.60(12)(0.375)}{\frac{1}{12}(18)(0.75^3)} = 253 \text{ psi} \qquad \textbf{Ans}$$

The marble slab will break if it is supported as in case (b).

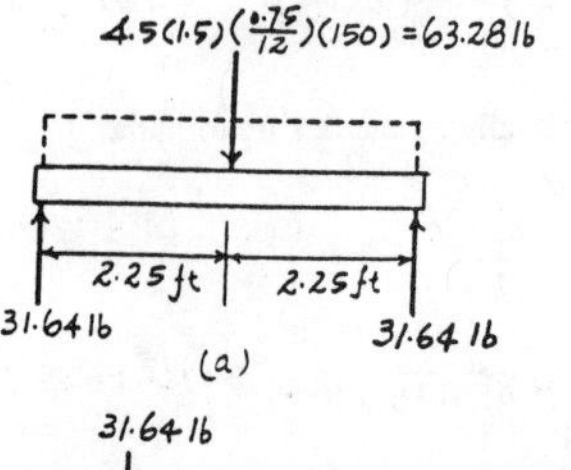
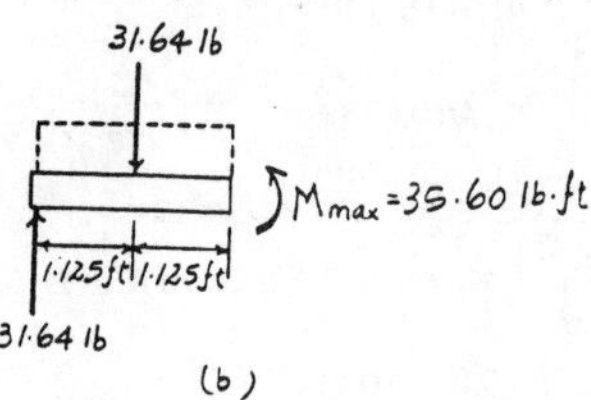

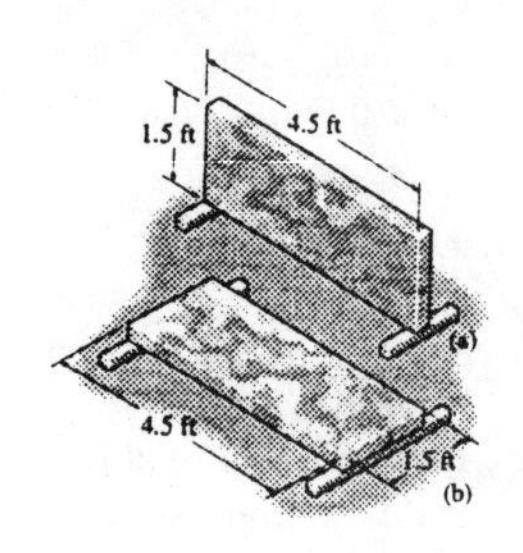

6–41. The slab of marble, which can be assumed a linear elastic brittle material, has a specific weight of 150 lb/ft³. If it is supported on its edges as shown in (b), determine the minimum thickness t it should have without causing it to break. The fracture stress is $\sigma_f = 200$ psi.

Support Reactions : As shown on FBD(a).
Maximum Moment : The maximum moment occurs at midspan as shown on FBD(b).

Maximum Bending Stress : Applying the flexure formula

$$\sigma_{max} = \frac{Mc}{I}$$

$$200 = \frac{569.53\,t(12)\left(\frac{t}{2}\right)(12)}{\frac{1}{12}(18)t^3(12^3)}$$

$$t = 0.07910 \text{ ft} = 0.949 \text{ in.} \qquad \textbf{Ans}$$

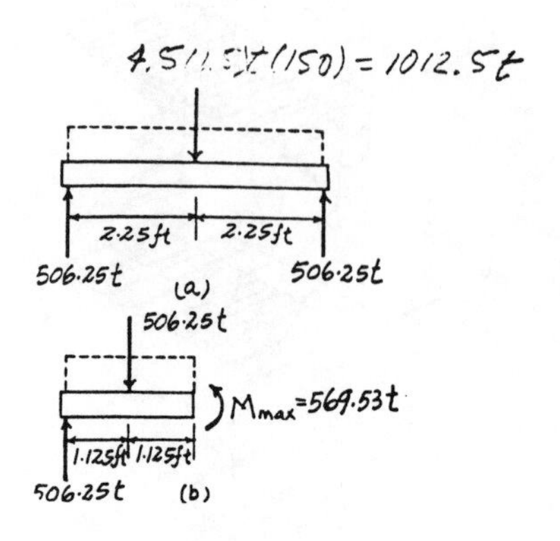

6–42. Two considerations have been proposed for the design of a beam. Determine which one will support a moment of $M = 150\ \text{kN}\cdot\text{m}$ with the least amount of bending stress. What is that stress? By what percentage is it more effective?

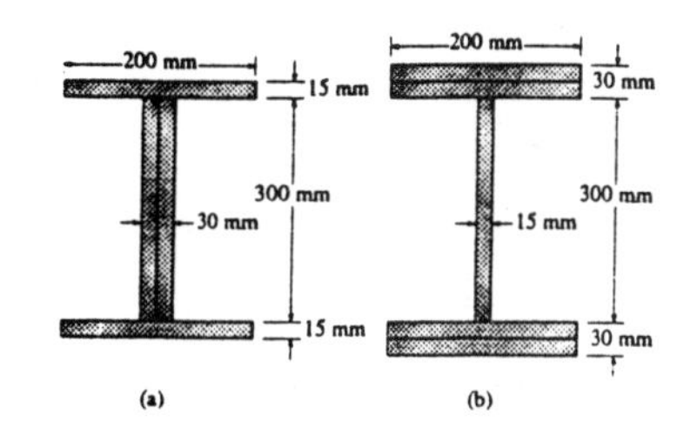

Section Property :

For section (a)

$$I = \frac{1}{12}(0.2)\left(0.33^3\right) - \frac{1}{12}(0.17)(0.3) = 0.21645(10^{-3})\ \text{m}^4$$

For section (b)

$$I = \frac{1}{12}(0.2)\left(0.36^3\right) - \frac{1}{12}(0.185)\left(0.3^3\right) = 0.36135(10^{-3})\ \text{m}^4$$

Maximum Bending Stress : Applying the flexure formula $\sigma_{max} = \frac{Mc}{I}$

For section (a)

$$\sigma_{max} = \frac{150(10^3)(0.165)}{0.21645(10^{-3})} = 114.3\ \text{MPa}$$

For section (b)

$$\sigma_{max} = \frac{150(10^3)(0.18)}{0.36135(10^{-3})} = 74.72\ \text{MPa} = 74.7\ \text{MPa} \qquad \textbf{Ans}$$

By comparison, section (b) will have the least amount of bending stress.

$$\%\text{ of effectiveness} = \frac{114.3 - 74.72}{74.72} \times 100\% = 53.0\ \% \qquad \textbf{Ans}$$

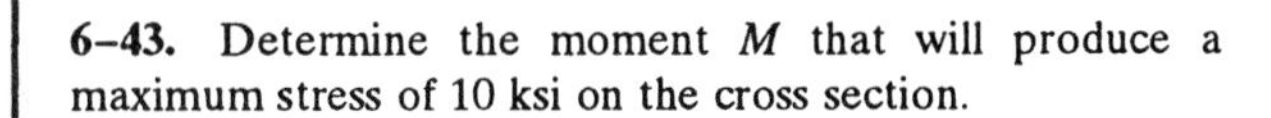

6–43. Determine the moment M that will produce a maximum stress of 10 ksi on the cross section.

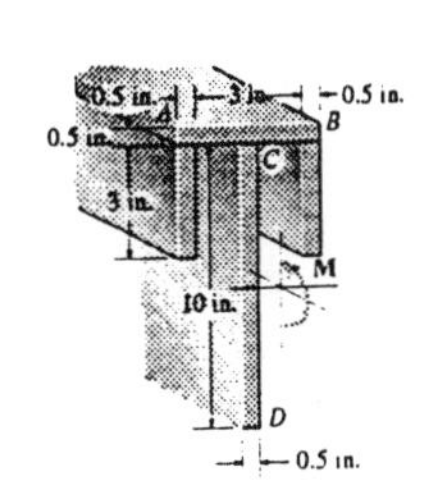

Section Properties :

$$\bar{y} = \frac{\Sigma \tilde{y} A}{\Sigma A} = \frac{0.25(4)(0.5) + 2[2(3)(0.5)] + 5.5(10)(0.5)}{4(0.5) + 2[(3)(0.5)] + 10(0.5)} = 3.40\ \text{in.}$$

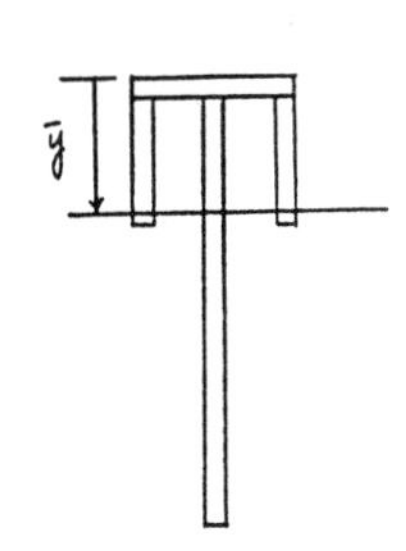

$$I_{NA} = \frac{1}{12}(4)\left(0.5^3\right) + 4(0.5)(3.40 - 0.25)^2 + 2\left[\frac{1}{12}(0.5)(3^3) + 0.5(3)(3.40 - 2)^2\right] + \frac{1}{12}(0.5)\left(10^3\right) + 0.5(10)(5.5 - 3.40)^2$$

$$= 91.73\ \text{in}^4$$

Maximum Bending Stress : Applying the flexure formula

$$\sigma_{max} = \frac{Mc}{I}$$

$$10 = \frac{M(10.5 - 3.4)}{91.73}$$

$$M = 129.2\ \text{kip}\cdot\text{in} = 10.8\ \text{kip}\cdot\text{ft} \qquad \textbf{Ans}$$

***6–44.** Determine the maximum tensile and compressive bending stress in the beam if it is subjected to a moment of $M = 4$ kip · ft.

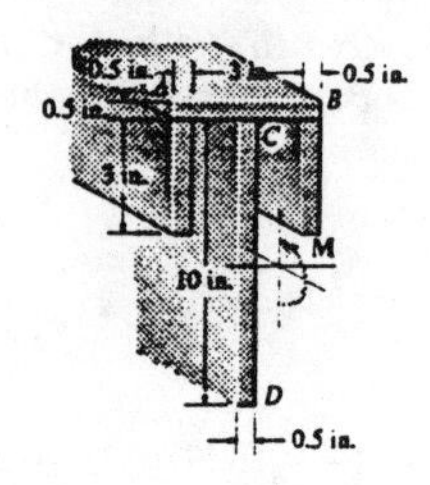

Section Properties :

$$\bar{y} = \frac{\Sigma \tilde{y} A}{\Sigma A}$$

$$= \frac{0.25(4)(0.5) + 2[2(3)(0.5)] + 5.5(10)(0.5)}{4(0.5) + 2[(3)(0.5)] + 10(0.5)} = 3.40 \text{ in.}$$

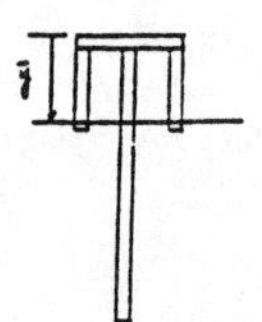

$$I_{NA} = \frac{1}{12}(4)\left(0.5^3\right) + 4(0.5)(3.40 - 0.25)^2 + 2\left[\frac{1}{12}(0.5)(3^3) + 0.5(3)(3.40 - 2)^2\right] + \frac{1}{12}(0.5)\left(10^3\right) + 0.5(10)(5.5 - 3.40)^2$$

$$= 91.73 \text{ in}^4$$

Maximum Bending Stress : Applying the flexure formula $\sigma_{max} = \frac{Mc}{I}$

$$(\sigma_t)_{max} = \frac{4(10^3)(12)(10.5 - 3.40)}{91.73} = 3715.12 \text{ psi} = 3.72 \text{ ksi} \qquad \textbf{Ans}$$

$$(\sigma_c)_{max} = \frac{4(10^3)(12)(3.40)}{91.73} = 1779.07 \text{ psi} = 1.78 \text{ ksi} \qquad \textbf{Ans}$$

6–45. Determine the resultant force the bending stresses produce on the horizontal top flange plate AB of the beam if $M = 4$ kip · ft.

Section Properties :

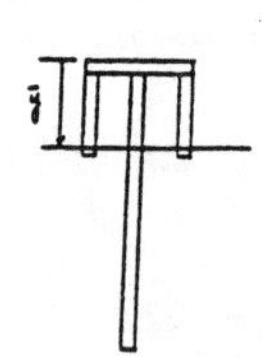

$$\bar{y} = \frac{\Sigma \tilde{y} A}{\Sigma A}$$

$$= \frac{0.25(4)(0.5) + 2[2(3)(0.5)] + 5.5(10)(0.5)}{4(0.5) + 2[(3)(0.5)] + 10(0.5)} = 3.40 \text{ in.}$$

$$I_{NA} = \frac{1}{12}(4)\left(0.5^3\right) + 4(0.5)(3.40 - 0.25)^2 + 2\left[\frac{1}{12}(0.5)(3^3) + 0.5(3)(3.40 - 2)^2\right] + \frac{1}{12}(0.5)\left(10^3\right) + 0.5(10)(5.5 - 3.40)^2$$

$$= 91.73 \text{ in}^4$$

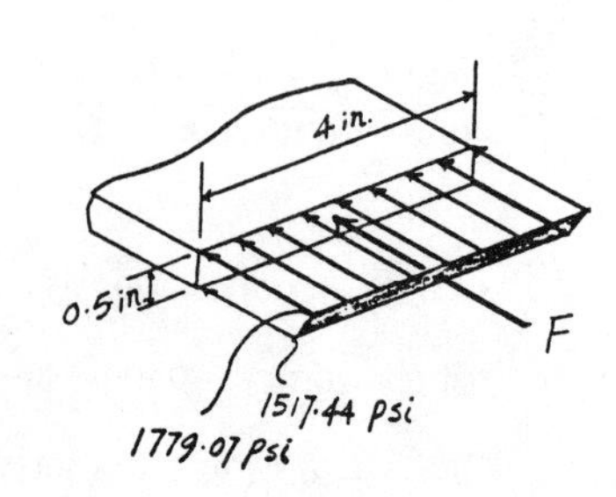

Bending Stress : Applying the flexure formula $\sigma = \frac{My}{I}$ and $\sigma_{max} = \frac{Mc}{I}$

$$\sigma_{y=2.9\text{in.}} = \frac{4(10^3)(12)(2.90)}{91.73} = 1517.44 \text{ psi}$$

$$(\sigma_c)_{max} = \frac{4(10^3)(12)(3.40)}{91.73} = 1779.07 \text{ psi}$$

Resultant Force : For the flange plate AB

$$F = \frac{1}{2}(1779.07 + 1517.44)(4)(0.5)$$

$$= 3296.5 \text{ lb} = 3.30 \text{ kip} \qquad \textbf{Ans}$$

6–46. Determine the resultant force the bending stresses produce on the web *CD* of the beam if $M = 4$ kip · ft.

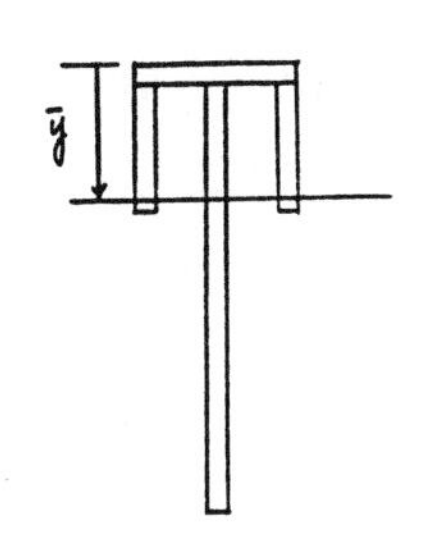

Section Properties :

$$\bar{y} = \frac{\Sigma \tilde{y} A}{\Sigma A}$$

$$= \frac{0.25(4)(0.5) + 2[2(3)(0.5)] + 5.5(10)(0.5)}{4(0.5) + 2[(3)(0.5)] + 10(0.5)} = 3.40 \text{ in.}$$

$$I_{NA} = \frac{1}{12}(4)\left(0.5^3\right) + 4(0.5)(3.40 - 0.25)^2 + 2\left[\frac{1}{12}(0.5)(3^3) + 0.5(3)(3.40-2)^2\right] + \frac{1}{12}(0.5)\left(10^3\right) + 0.5(10)(5.5-3.40)^2$$

$$= 91.73 \text{ in}^4$$

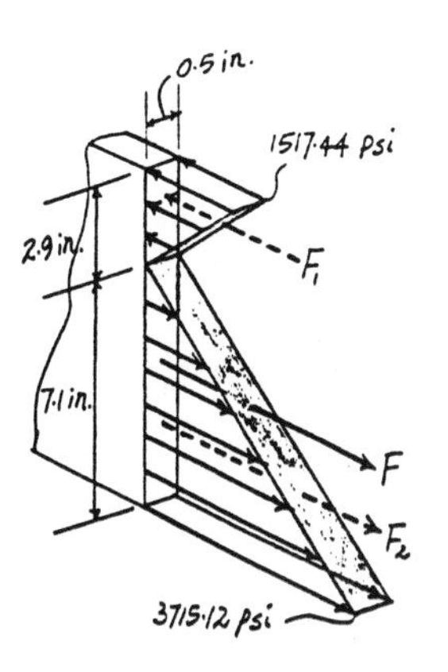

Bending Stress : Applying the flexure formula $\sigma = \frac{My}{I}$ and $\sigma_{max} = \frac{Mc}{I}$

$$\sigma_{y=2.9\text{in.}} = \frac{4(10^3)(12)(2.90)}{91.73} = 1517.44 \text{ psi}$$

$$(\sigma_t)_{max} = \frac{4(10^3)(12)(7.10)}{91.73} = 3715.12 \text{ psi}$$

Resultant Force : For the web plate *CD*.

$$F = F_2 - F_1 = \frac{1}{2}(3715.12)(7.10)(0.5) - \frac{1}{2}(1517.44)(2.90)(0.5)$$

$$= 5494.19 \text{ lb} = 5.49 \text{ kip} \qquad \textbf{Ans}$$

6–47. The aluminum strut has a cross-sectional area in the form of a cross. If it is subjected to the moment $M = 8$ kN · m, determine the bending stress acting at points *A* and *B*, and show the results acting on volume elements located at these points.

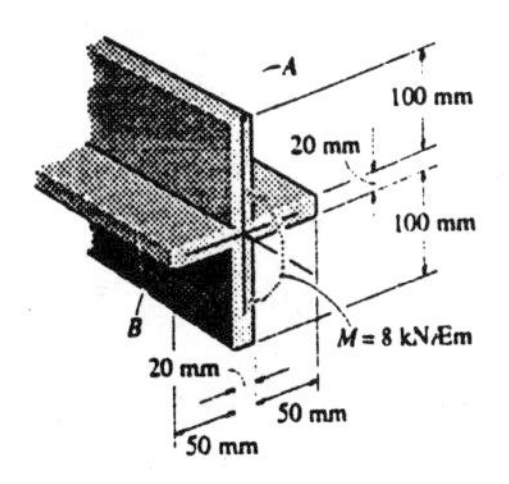

Section Property :

$$I = \frac{1}{12}(0.02)\left(0.22^3\right) + \frac{1}{12}(0.1)\left(0.02^3\right) = 17.8133\left(10^{-6}\right) \text{ m}^4$$

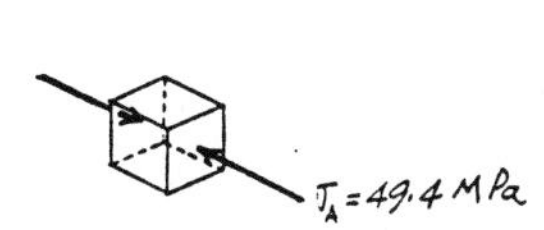

Bending Stress : Applying the flexure formula $\sigma = \frac{My}{I}$

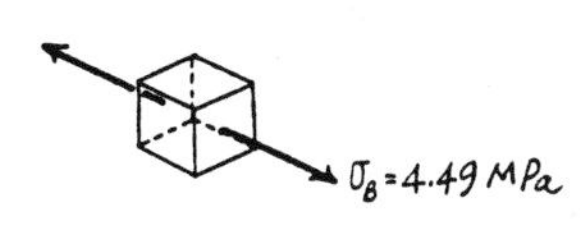

$$\sigma_A = \frac{8(10^3)(0.11)}{17.8133(10^{-6})} = 49.4 \text{ MPa (C)} \qquad \textbf{Ans}$$

$$\sigma_B = \frac{8(10^3)(0.01)}{17.8133(10^{-6})} = 4.49 \text{ MPa (T)} \qquad \textbf{Ans}$$

***6–48.** The aluminum strut has a cross-sectional area in the form of a cross. If it is subjected to the moment $M = 8 \text{ kN} \cdot \text{m}$, determine the maximum bending stress in the beam, and sketch a three-dimensional view of the stress distribution acting over the entire cross-sectional area.

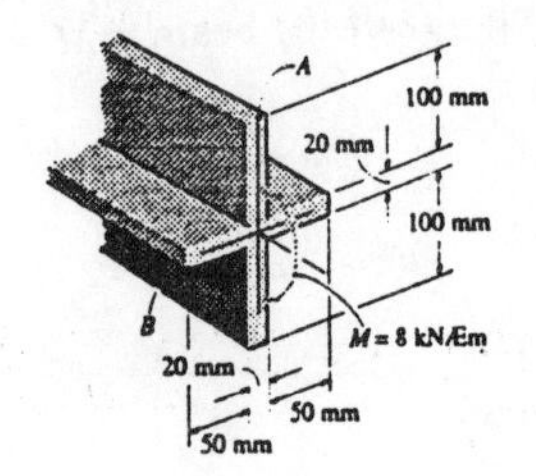

Section Property :

$$I = \frac{1}{12}(0.02)\left(0.22^3\right) + \frac{1}{12}(0.1)\left(0.02^3\right) = 17.8133\left(10^{-6}\right)\ \text{m}^4$$

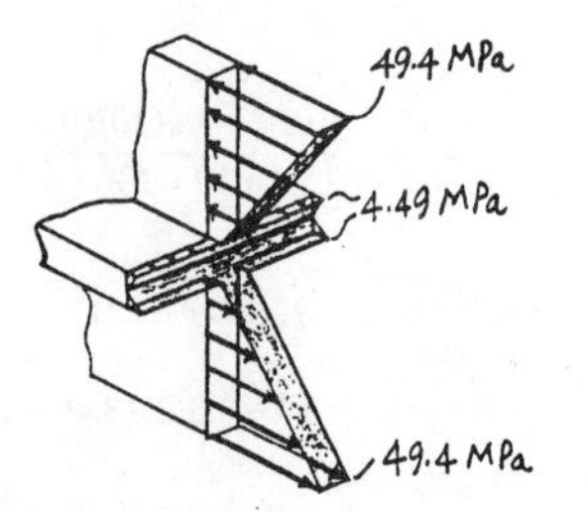

Bending Stress : Applying the flexure formula $\sigma_{\max} = \frac{Mc}{I}$ and $\sigma = \frac{My}{I}$.

$$\sigma_{\max} = \frac{8(10^3)(0.11)}{17.8133(10^{-6})} = 49.4 \text{ MPa} \qquad \textbf{Ans}$$

$$\sigma_{y=0.01\,\text{m}} = \frac{8(10^3)(0.01)}{17.8133(10^{-6})} = 4.49 \text{ MPa}$$

6–49. The beam is made from three boards nailed together as shown. If the moment acting on the cross section is $M = 1 \text{ kip} \cdot \text{ft}$, determine the maximum bending stress in the beam. Sketch a three-dimensional view of the stress distribution acting over the cross section.

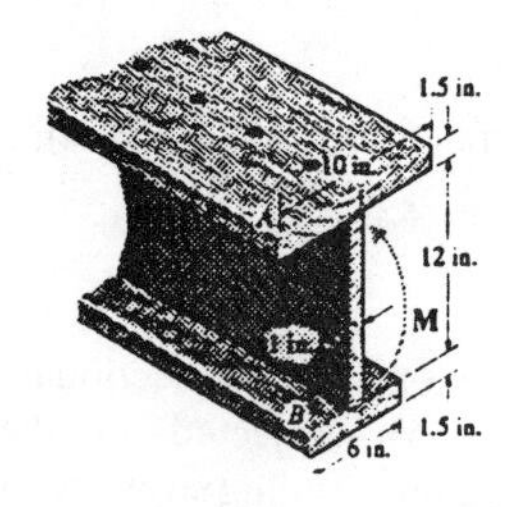

Section Properties :

$$\bar{y} = \frac{\Sigma \tilde{y} A}{\Sigma A} = \frac{0.75(10)(1.5) + 7.5(1)(12) + 14.25(6)(1.5)}{10(1.5) + 1(12) + 6(1.5)} = 6.375 \text{ in.}$$

$$\begin{aligned} I &= \frac{1}{12}(10)\left(1.5^3\right) + 10(1.5)(6.375 - 0.75)^2 \\ &+ \frac{1}{12}(1)\left(12^3\right) + 1(12)(7.5 - 6.375)^2 \\ &+ \frac{1}{12}(6)\left(1.5^3\right) + 6(1.5)(14.25 - 6.375)^2 \\ &= 1196.4375 \text{ in}^4 \end{aligned}$$

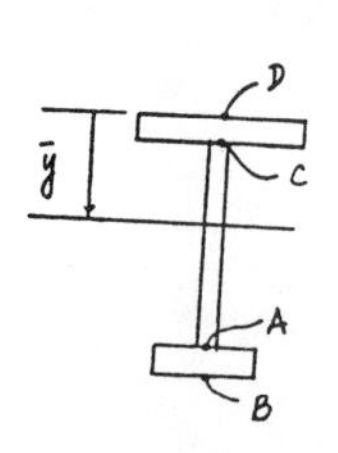

Bending Stress : Maximum bending stresses occurs at point B. Applying the flexure formula

$$\sigma_{\max} = \sigma_B = \frac{Mc}{I} = \frac{1000(12)(15 - 6.375)}{1196.4375} = 86.5 \text{ psi} \qquad \textbf{Ans}$$

$$\sigma_A = \frac{M y_A}{I} = \frac{1000(12)(13.5 - 6.375)}{1196.4375} = 71.5 \text{ psi}$$

$$\sigma_C = \frac{M y_C}{I} = \frac{1000(12)(6.375 - 1.5)}{1196.4375} = 48.9 \text{ psi}$$

$$\sigma_D = \frac{M y_D}{I} = \frac{1000(12)(6.375)}{1196.4375} = 63.9 \text{ psi}$$

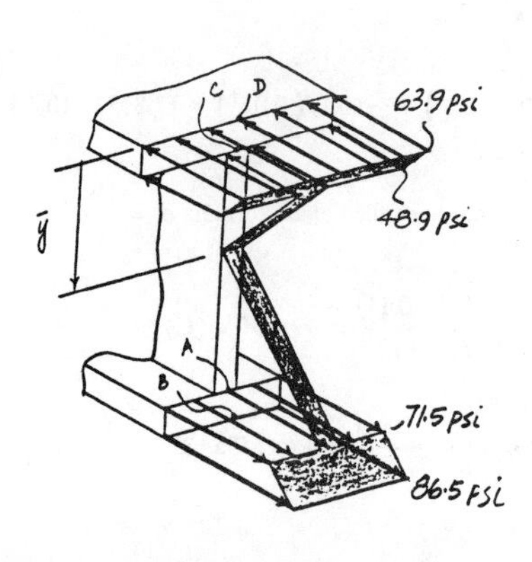

6–50. Determine the resultant force the bending stresses produce on the top board A of the beam if $M = 1$ kip · ft.

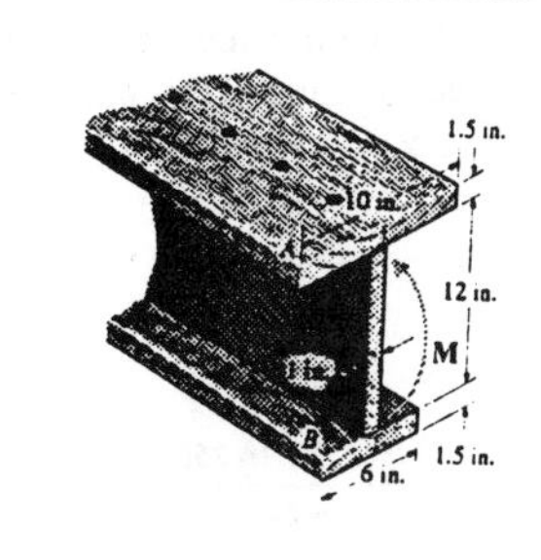

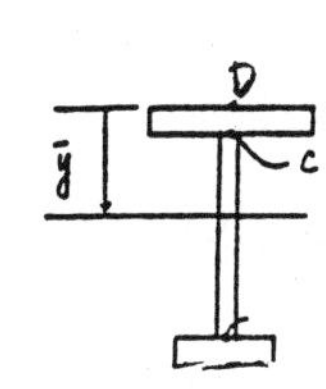

Section Properties :

$$\bar{y} = \frac{\Sigma \tilde{y} A}{\Sigma A} = \frac{0.75(10)(1.5) + 7.5(1)(12) + 14.25(6)(1.5)}{10(1.5) + 1(12) + 6(1.5)}$$

$$= 6.375 \text{ in.}$$

$$I = \frac{1}{12}(10)\left(1.5^3\right) + 10(1.5)(6.375 - 0.75)^2$$
$$+ \frac{1}{12}(1)\left(12^3\right) + 1(12)(7.5 - 6.375)^2$$
$$+ \frac{1}{12}(6)\left(1.5^3\right) + 6(1.5)(14.25 - 6.375)^2$$
$$= 1196.4375 \text{ in}^4$$

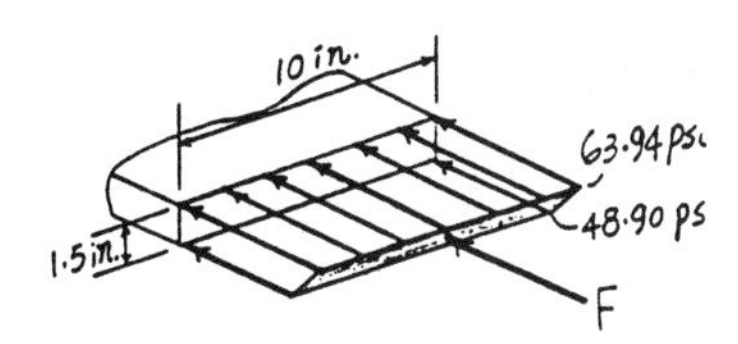

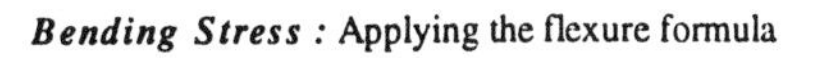

Bending Stress : Applying the flexure formula

$$\sigma_C = \frac{M y_C}{I} = \frac{1000(12)(6.375 - 1.5)}{1196.4375} = 48.90 \text{ psi}$$

$$\sigma_D = \frac{M y_D}{I} = \frac{1000(12)(6.375)}{1196.4375} = 63.94 \text{ psi}$$

The Resultant Force : For top board A

$$F = \frac{1}{2}(63.94 + 48.90)(10)(1.5) = 846 \text{ lb} \qquad \textbf{Ans}$$

6–51. Determine the resultant force the bending stresses produce on the bottom board B of the beam if $M = 1$ kip · ft.

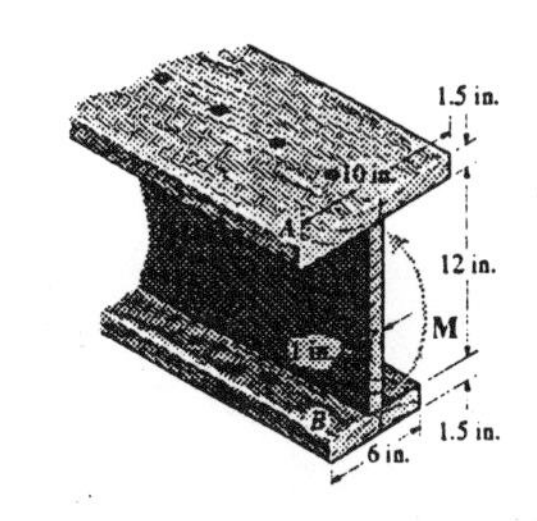

Section Properties :

$$\bar{y} = \frac{\Sigma \tilde{y} A}{\Sigma A} = \frac{0.75(10)(1.5) + 7.5(1)(12) + 14.25(6)(1.5)}{10(1.5) + 1(12) + 6(1.5)}$$

$$= 6.375 \text{ in.}$$

$$I = \frac{1}{12}(10)\left(1.5^3\right) + 10(1.5)(6.375 - 0.75)^2$$
$$+ \frac{1}{12}(1)\left(12^3\right) + 1(12)(7.5 - 6.375)^2$$
$$+ \frac{1}{12}(6)\left(1.5^3\right) + 6(1.5)(14.25 - 6.375)^2$$
$$= 1196.4375 \text{ in}^4$$

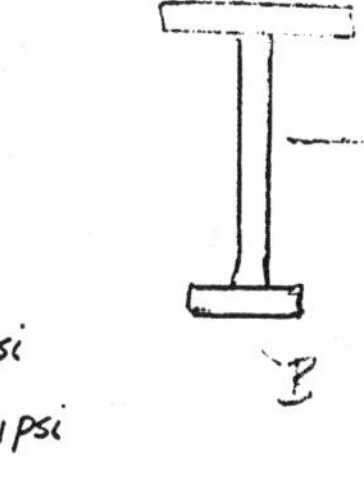

6 in.

1.5 in.

71.46 psi

86.51 psi

F

Bending Stress : Applying the flexure formula

$$\sigma_A = \frac{M y_A}{I} = \frac{1000(12)(13.5 - 6.375)}{1196.4375} = 71.46 \text{ psi}$$

$$\sigma_B = \frac{M y_B}{I} = \frac{1000(12)(15 - 6.375)}{1196.4375} = 86.51 \text{ psi}$$

The Resultant Force : For bottom board B

$$F = \frac{1}{2}(86.51 + 71.46)(6)(1.5) = 711 \text{ lb} \qquad \textbf{Ans}$$

***6–52.** The beam is made from three boards nailed together as shown. Determine the maximum moment M that can be applied to the beam so that the bending stress does not exceed $\sigma_{allow} = 650$ psi.

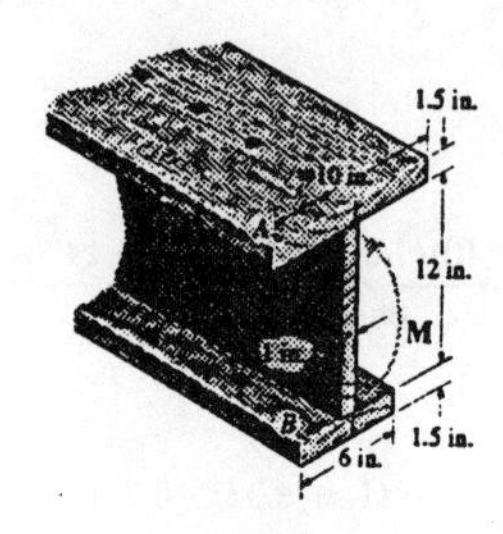

Section Properties :

$$\bar{y} = \frac{\Sigma \tilde{y} A}{\Sigma A} = \frac{0.75(10)(1.5) + 7.5(1)(12) + 14.25(6)(1.5)}{10(1.5) + 1(12) + 6(1.5)}$$
$$= 6.375 \text{ in.}$$

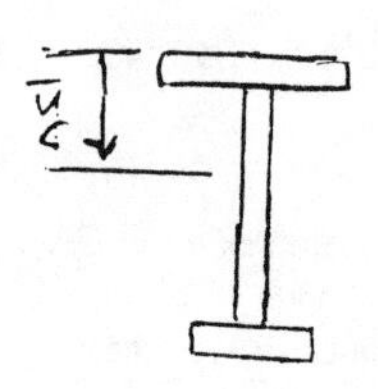

$$I = \frac{1}{12}(10)\left(1.5^3\right) + 10(1.5)(6.375 - 0.75)^2$$
$$+ \frac{1}{12}(1)\left(12^3\right) + 1(12)(7.5 - 6.375)^2$$
$$+ \frac{1}{12}(6)\left(1.5^3\right) + 6(1.5)(14.25 - 6.375)^2$$
$$= 1196.4375 \text{ in}^4$$

Maximum Bending Stress : Applying the flexure formula

$$\sigma_{max} = \sigma_{allow} = \frac{Mc}{I}$$
$$650 = \frac{M(15 - 6.375)}{1196.4375}$$
$$M = 90166.3 \text{ lb} \cdot \text{in} = 7.51 \text{ kip} \cdot \text{ft} \qquad \textbf{Ans}$$

6–53. Determine the largest bending stress developed in the member if it is subjected to an internal bending moment of $M = 40$ kN · m.

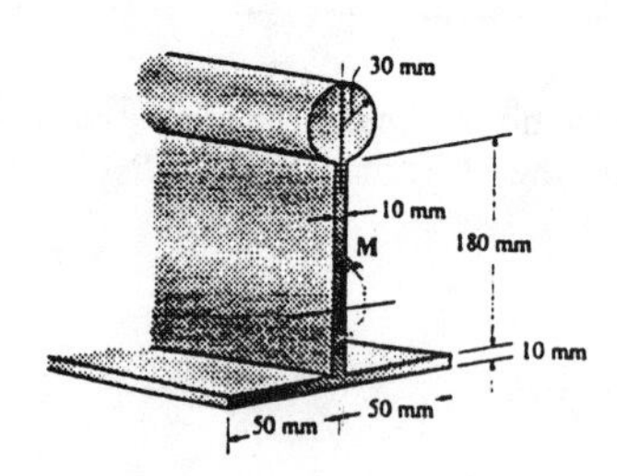

Section Properties :

$$\bar{y} = \frac{\Sigma \tilde{y} A}{\Sigma A}$$
$$= \frac{0.005(0.1)(0.01) + 0.1(0.18)(0.01) + 0.22(\pi)(0.03^2)}{(0.1)(0.01) + (0.18)(0.01) + (\pi)(0.03^2)}$$
$$= 0.143411 \text{ m}$$

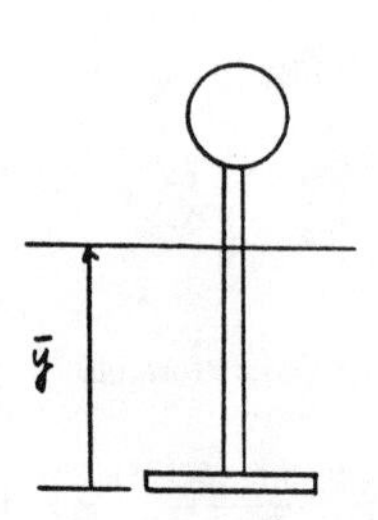

$$I = \frac{1}{12}(0.1)\left(0.01^3\right) + (0.1)(0.01)(0.143411 - 0.005)^2$$
$$+ \frac{1}{12}(0.01)\left(0.18^3\right) + (0.01)(0.18)(0.143411 - 0.1)^2$$
$$+ \frac{1}{4}\pi\left(0.03^4\right) + \pi\left(0.03^2\right)(0.22 - 0.143411)^2$$
$$= 44.64\left(10^{-6}\right) \text{ m}^4$$

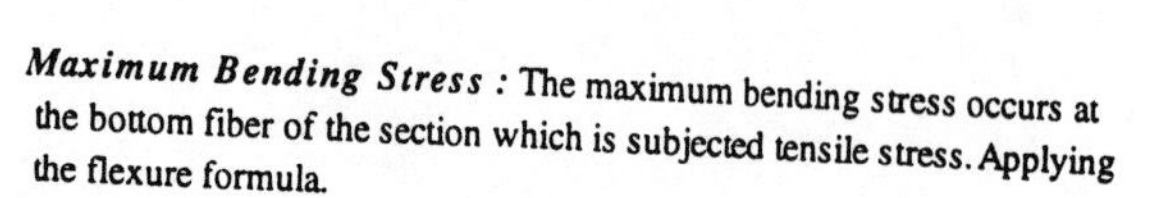

Maximum Bending Stress : The maximum bending stress occurs at the bottom fiber of the section which is subjected tensile stress. Applying the flexure formula.

$$\sigma_{max} = \frac{Mc}{I} = \frac{40(10^3)(0.143411)}{44.64(10^{-6})} = 129 \text{ MPa} \qquad \textbf{Ans}$$

6–54. The member has a cross section with the dimensions shown. Determine the largest internal moment M that can be applied without exceeding allowable tensile and compressive stresses of $(\sigma_t)_{\text{allow}} = 150$ MPa and $(\sigma_t)_{\text{allow}} = 100$ MPa, respectively.

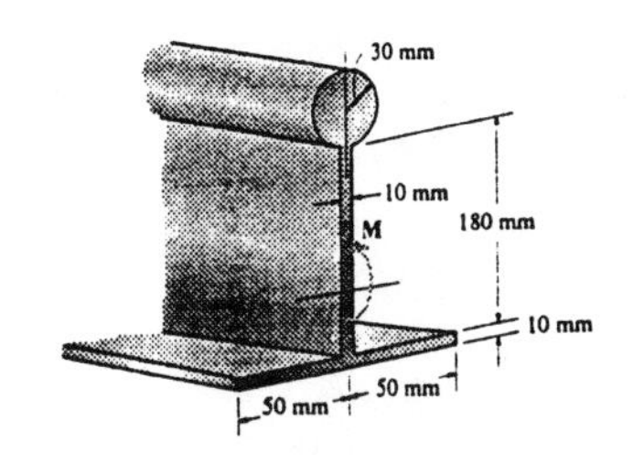

Section Properties :

$$\bar{y} = \frac{\Sigma \bar{y} A}{\Sigma A}$$
$$= \frac{0.005(0.1)(0.01) + 0.1(0.18)(0.01) + 0.22(\pi)(0.03^2)}{(0.1)(0.01) + (0.18)(0.01) + (\pi)(0.03^2)}$$
$$= 0.143411 \text{ m}$$

$$I = \frac{1}{12}(0.1)(0.01^3) + (0.1)(0.01)(0.143411 - 0.005)^2$$
$$+ \frac{1}{12}(0.01)(0.18^3) + (0.01)(0.18)(0.143411 - 0.1)^2$$
$$+ \frac{1}{4}\pi(0.03^4) + \pi(0.03^2)(0.22 - 0.143411)^2$$
$$= 44.64(10^{-6}) \text{ m}^4$$

Maximum Bending Stress : Applying the flexure formula.

Assume failure due to tensile stress

$$\sigma_{\max} = (\sigma_t)_{\text{allow}} = \frac{Mc}{I}$$
$$150(10^6) = \frac{M(0.143411)}{44.64(10^{-6})}$$
$$M = 46\,690 \text{ N}\cdot\text{m}$$

Assume failure due to compreesive stress

$$\sigma_{\max} = (\sigma_c)_{\text{allow}} = \frac{Mc}{I}$$
$$100(10^6) = \frac{M(0.25 - 0.143411)}{44.64(10^{-6})}$$
$$M = 41\,880 \text{ N}\cdot\text{m} \quad \textit{(controls)}$$
$$= 41.9 \text{ kN}\cdot\text{m}$$

Ans

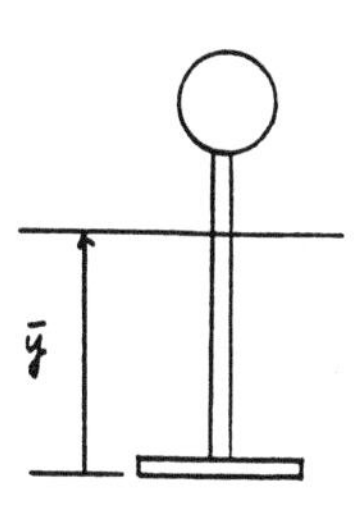

6–55. The member has a cross section with the dimensions shown. Determine the largest bending stress developed in the member if it is subjected to an internal bending moment of $M = 85$ kN · m.

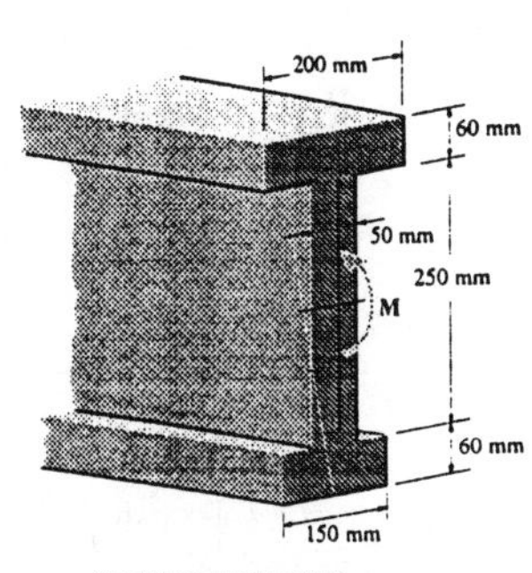

Section Properties :

$$\bar{y} = \frac{\Sigma \bar{y} A}{\Sigma A} = \frac{0.03(0.06)(0.2) + 0.185(0.250)(0.05) + 0.34(0.06)(0.150)}{0.06(0.2) + 0.250(0.05) + 0.06(0.150)}$$
$$= 0.17112 \text{ m}$$

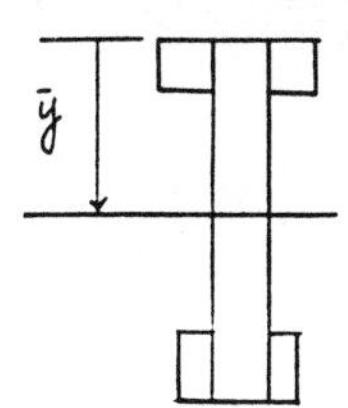

$$I = \frac{1}{12}(0.2)(0.06^3) + (0.2)(0.06)(0.17112 - 0.03)^2$$
$$+ \frac{1}{12}(0.05)(0.25)^3 + (0.05)(0.25)(0.185 - 0.17112)^2$$
$$+ \frac{1}{12}(0.150)(0.06^3) + (0.150)(0.06)(0.34 - 0.17112)^2$$
$$= 569.475(10^{-6}) \text{ m}^4$$

Maximum Bending Stress : The maximum bending stress occurs at the bottom fiber of the section. Applying the flexure formula

$$\sigma_{\max} = \frac{Mc}{I}$$
$$= \frac{85(10^3)(0.37 - 0.17112)}{569.475(10^{-6}) \text{ m}^4}$$
$$= 29.7 \text{ MPa}$$

Ans

***6–56.** The member has a cross section with the dimensions shown. Determine the largest internal moment M that can be applied without exceeding an allowable bending stress of $\sigma_{allow} = 190$ MPa.

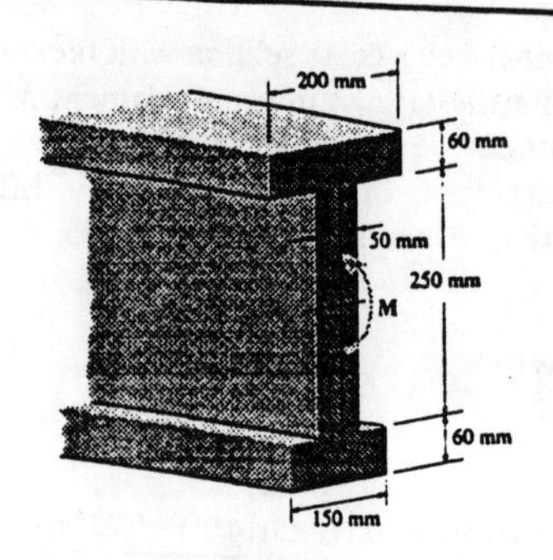

Section Properties :

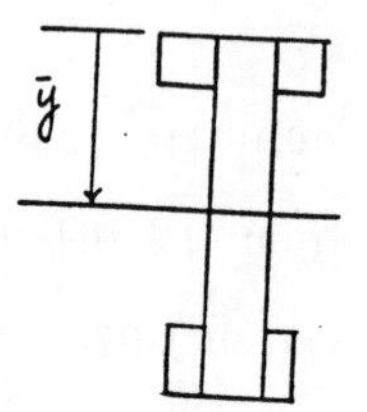

$$\bar{y} = \frac{\Sigma \tilde{y} A}{\Sigma A} = \frac{0.03(0.06)(0.2) + 0.185(0.250)(0.05) + 0.34(0.06)(0.150)}{0.06(0.2) + 0.250(0.05) + 0.06(0.150)}$$
$$= 0.17112 \text{ m}$$

$$I = \frac{1}{12}(0.2)\left(0.06^3\right) + (0.2)(0.06)(0.17112 - 0.03)^2$$
$$+ \frac{1}{12}(0.05)(0.25)^3 + (0.05)(0.25)(0.185 - 0.17112)^2$$
$$+ \frac{1}{12}(0.150)\left(0.06^3\right) + (0.150)(0.06(0.34 - 0.17112)^2$$
$$= 569.475\left(10^{-6}\right) \text{ m}^4$$

Maximum Bending Stress : The maximum bending stress occurs at the bottom of the section. Applying the flexure formula

$$\sigma_{max} = \sigma_{allow} = \frac{Mc}{I}$$
$$190\left(10^6\right) = \frac{M(0.37 - 0.17112)}{569.475(10^{-6}) \text{ m}^4}$$
$$M = 544\ 046 \text{ N} \cdot \text{m} = 544 \text{ kN} \cdot \text{m} \qquad \textbf{Ans}$$

6–57. The bolster or main supporting girder of a truck body is subjected to the uniform distributed load. Determine the bending stress at points A and B.

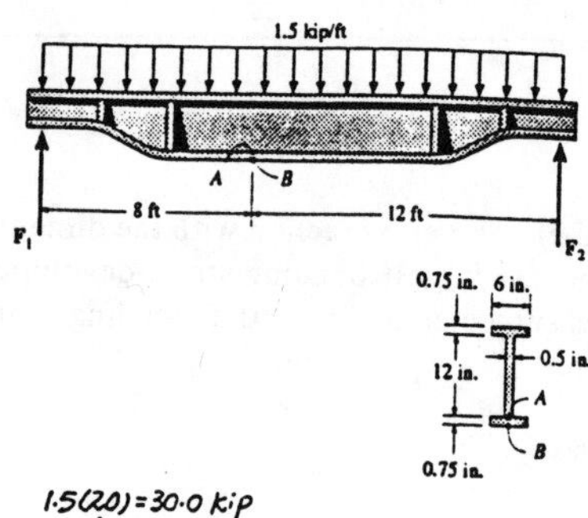

Support Reactions : As shown on FBD.
Internal Moment : Using the method of sections.

$$+\Sigma M_{NA} = 0: \quad M + 12.0(4) - 15.0(8) = 0$$
$$M = 72.0 \text{ kip} \cdot \text{ft}$$

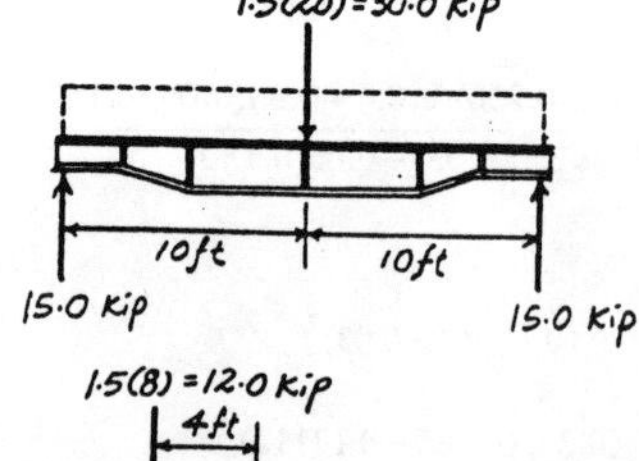

Section Property :

$$I = \frac{1}{12}(6)\left(13.5^3\right) - \frac{1}{12}(5.5)\left(12^3\right) = 438.1875 \text{ in}^4$$

Bending Stress : Applying the flexure formula $\sigma = \frac{My}{I}$

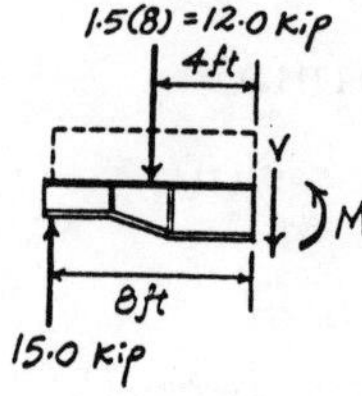

$$\sigma_B = \frac{72.0(12)(6.75)}{438.1875} = 13.3 \text{ ksi} \qquad \textbf{Ans}$$

$$\sigma_A = \frac{72.0(12)(6)}{438.1875} = 11.8 \text{ ksi} \qquad \textbf{Ans}$$

6–58. If the shaft has a diameter of 1.5 in., determine the absolute maximum bending stress in the shaft.

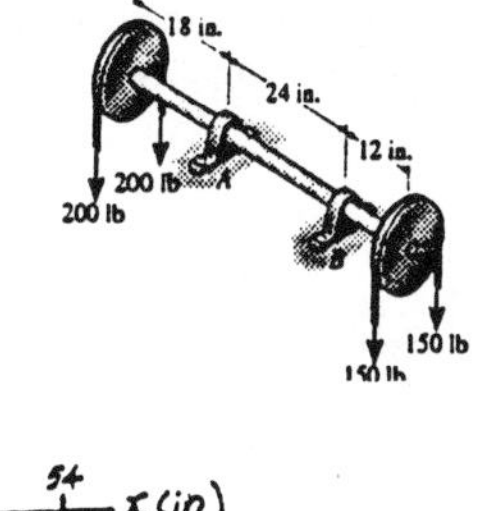

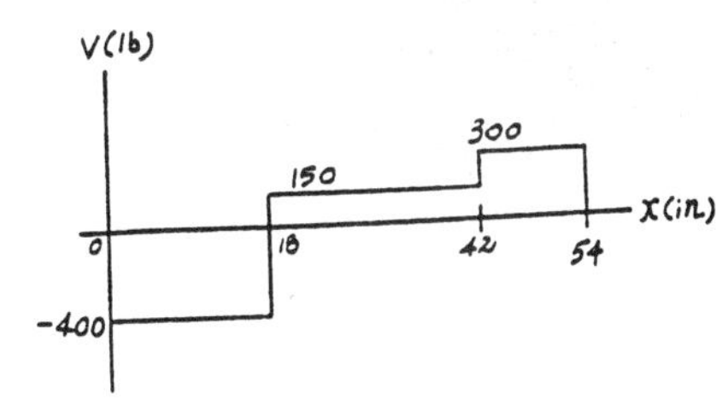

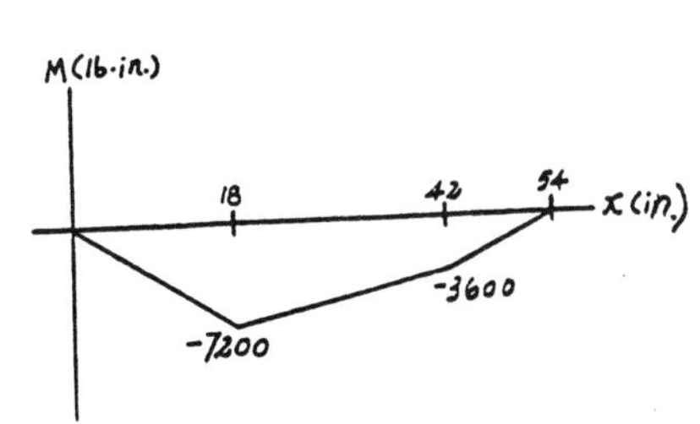

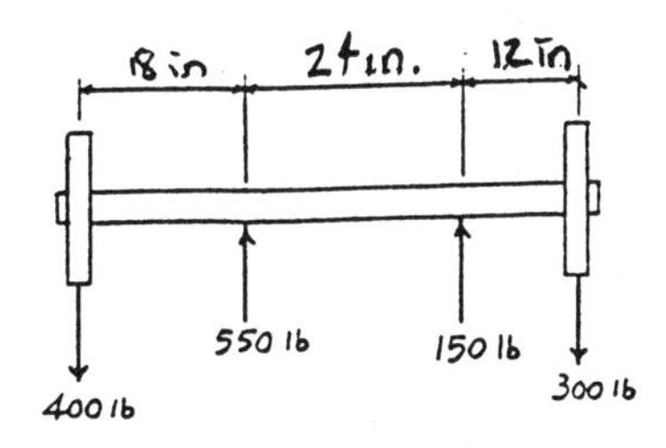

Absolute Maximum Bending Stress : The maximum moment is $M_{max} = 7200$ lb · in as indicated on moment diagram. Applying the flexure formula

$$\sigma_{max} = \frac{M_{max}\,c}{I} = \frac{7200(0.75)}{\frac{\pi}{4}(0.75^4)} = 21.7 \text{ ksi} \qquad \textbf{Ans}$$

6–59. If the shaft has a diameter of 50 mm, determine the absolute maximum bending stress in the shaft.

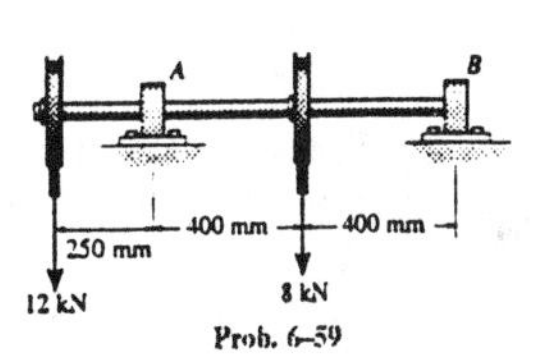

Prob. 6–59

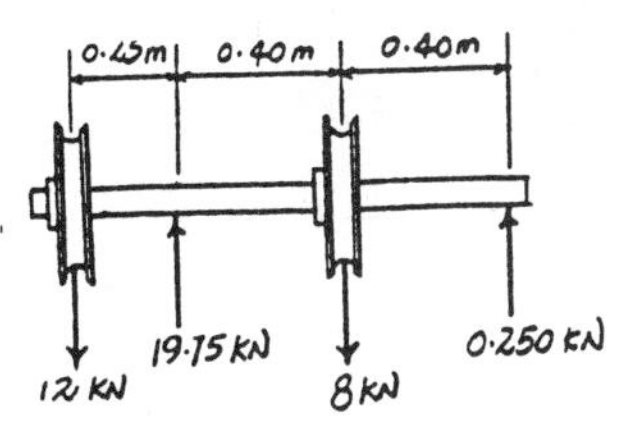

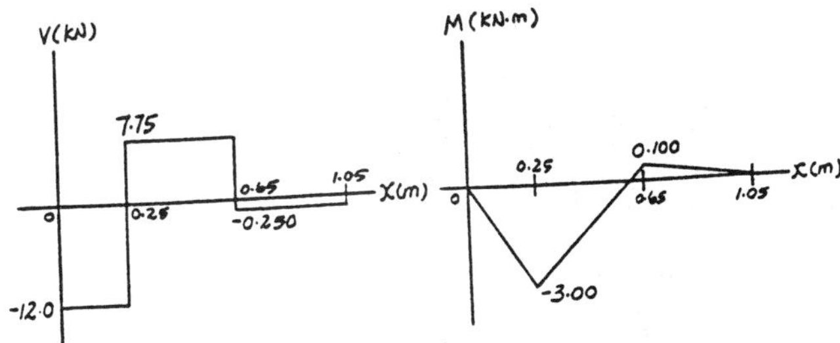

Absolute Maximum Bending Stress : The maximum moment is $M_{max} = 3.00$ kN · m as indicated on moment diagram. Applying the flexure formula

$$\sigma_{max} = \frac{M_{max}\,c}{I} = \frac{3.00(10^3)(0.025)}{\frac{\pi}{4}(0.025^4)} = 244 \text{ MPa} \qquad \textbf{Ans}$$

***6–60.** If the beam in Prob. 6–20 has a rectangular cross section with a width of 8 in. and a height of 16 in., determine the absolute maximum bending stress in the beam.

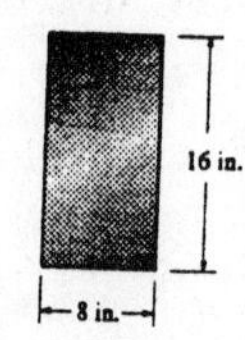

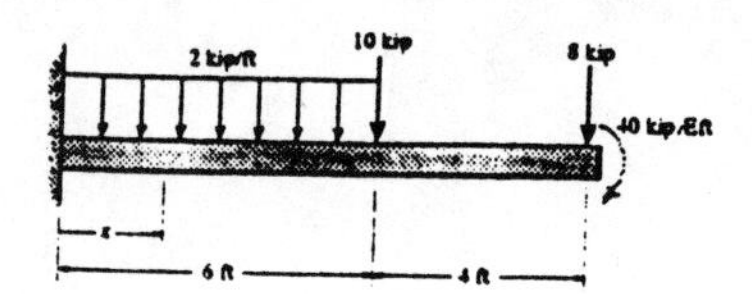

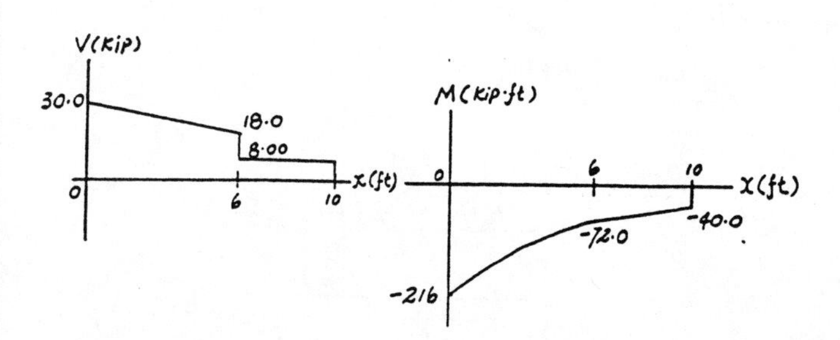

Absolute Maximum Bending Stress : The maximum moment is $M_{max} = 216$ kip · ft as indicated on moment diagram. Applying the flexure formula

$$\sigma_{max} = \frac{M_{max}\,c}{I} = \frac{216(12)(8)}{\frac{1}{12}(8)(16^3)} = 7.59 \text{ ksi} \qquad \textbf{Ans}$$

6–61. If the beam has a square cross section of 9 in. on each side, determine the absolute maximum bending stress in the beam.

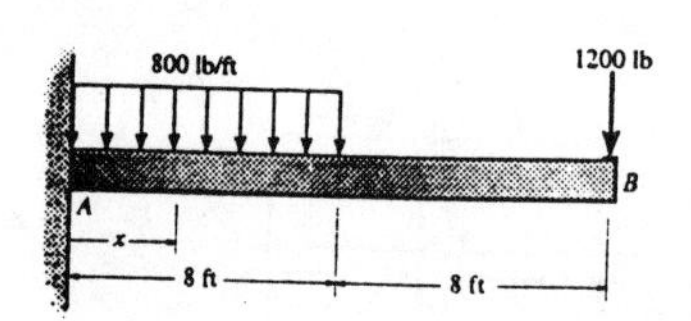

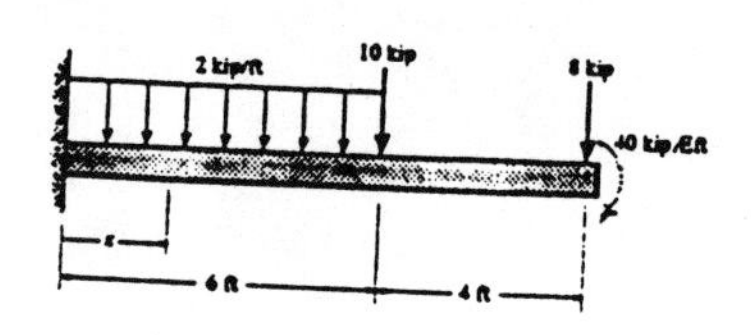

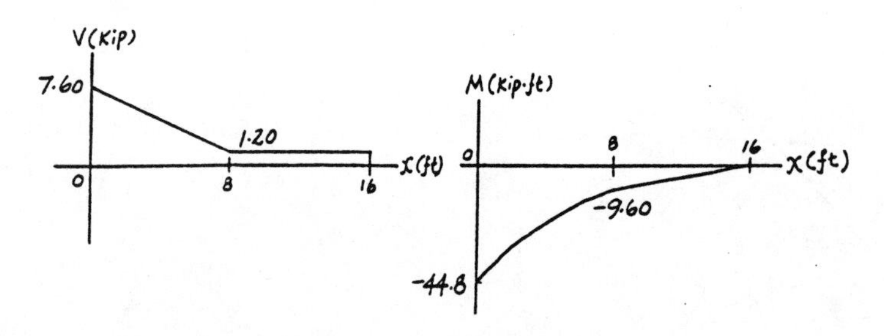

Absolute Maximum Bending Stress : The maximum moment is $M_{max} = 44.8$ kip · ft as indicated on moment diagram. Applying the flexure formula

$$\sigma_{max} = \frac{M_{max}\,c}{I} = \frac{44.8(12)(4.5)}{\frac{1}{12}(9)(9^3)} = 4.42 \text{ ksi} \qquad \textbf{Ans}$$

6–62. If the beam in Prob. 6–22 has a cross section as shown, determine the absolute maximum bending stress in the beam.

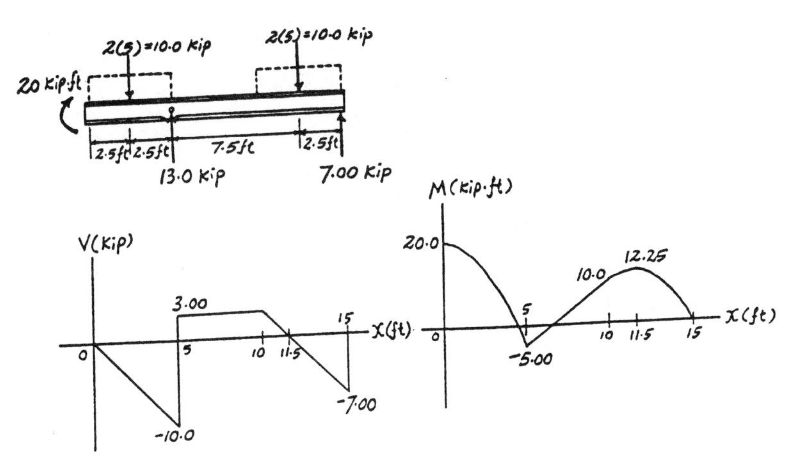

$$I = \frac{1}{12}(4)\left(8^3\right) - \frac{1}{12}(3.75)\left(7^3\right) = 63.479\ \text{in}^4$$

Absolute Maximum Bending Stress : The maximum moment is $M_{max} = 20.0$ kip · ft as indicated on moment diagram. Applying the flexure formula

$$\sigma_{max} = \frac{M_{max}\,c}{I} = \frac{20.0(12)(4)}{63.479} = 15.1\ \text{ksi} \qquad \textbf{Ans}$$

6–63. Determine the absolute maximum bending stress in the beam in Prob. 6–23. The cross section of the beam is as shown.

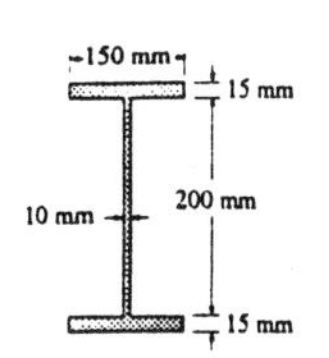

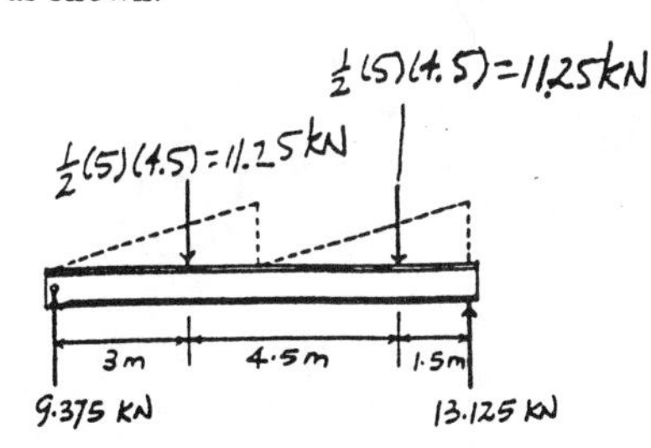

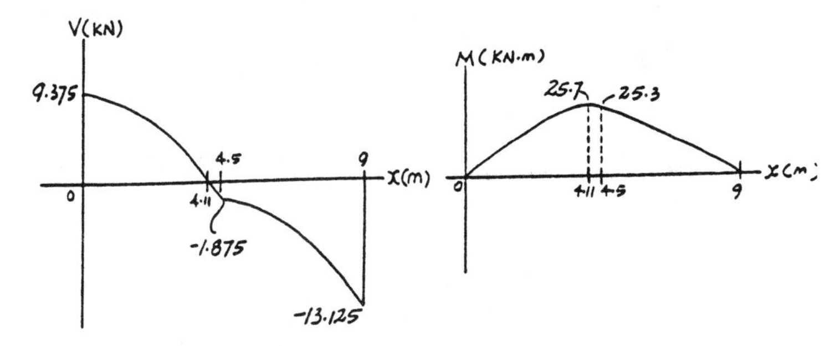

$$I = \frac{1}{12}(0.15)\left(0.23^3\right) - \frac{1}{12}(0.14)\left(0.2^3\right) = 58.754\left(10^{-6}\right)\ \text{m}^4$$

Absolute Maximum Bending Stress : The maximum moment is $M_{max} = 25.7$ kN · m as indicated on the moment diagram. Applying the flexure formula

$$\sigma_{max} = \frac{M_{max}\,c}{I} = \frac{25.7(10^3)(0.115)}{58.754(10^{-6})} = 50.3\ \text{MPa} \qquad \textbf{Ans}$$

***6–64.** If the beam in Prob. 6–27 has a rectangular cross section with a width b and a height h, determine the absolute maximum bending stress in the beam.

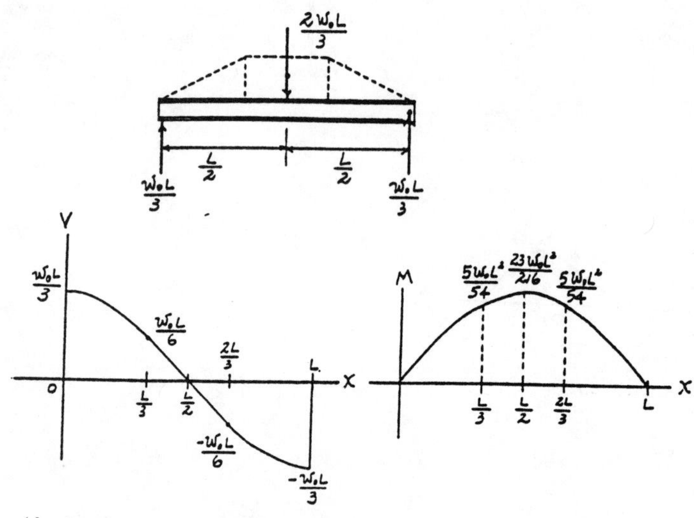

Absolute Maximum Bending Stress : The maximum moment is $M_{max} = \dfrac{23w_0L^2}{216}$ as indicated on the moment diagram. Applying the flexure formula

$$\sigma_{max} = \frac{M_{max}\,c}{I} = \frac{\frac{23w_0L^2}{216}\left(\frac{h}{2}\right)}{\frac{1}{12}bh^3} = \frac{23w_0L^2}{36bh^2} \qquad \textbf{Ans}$$

6–65. If the compound beam in Prob. 6–19 has a square cross section, determine its dimension a if the allowable bending stress is $\sigma_{allow} = 150$ MPa.

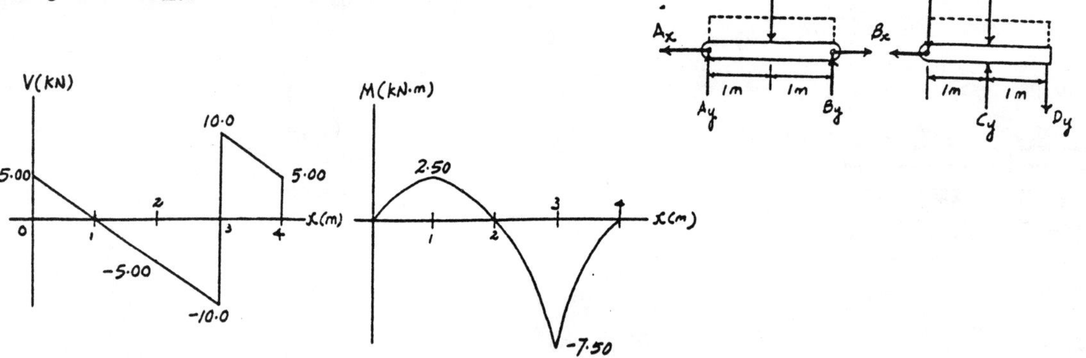

Allowable Bending Stress : The maximum moment is $M_{max} = 7.50$ kN · m as indicated on moment diagram. Applying the flexure formula

$$\sigma_{max} = \sigma_{allow} = \frac{M_{max}\,c}{I}$$

$$150\left(10^6\right) = \frac{7.50\left(10^3\right)\left(\frac{a}{2}\right)}{\frac{1}{12}a^4}$$

$$a = 0.06694 \text{ m} = 66.9 \text{ mm} \qquad \textbf{Ans}$$

6–66. If the compound beam in Prob. 6–25 has a square cross section, determine its required height and width, a, if the allowable bending stress for the material is $\sigma_{allow} = 1.50$ ksi.

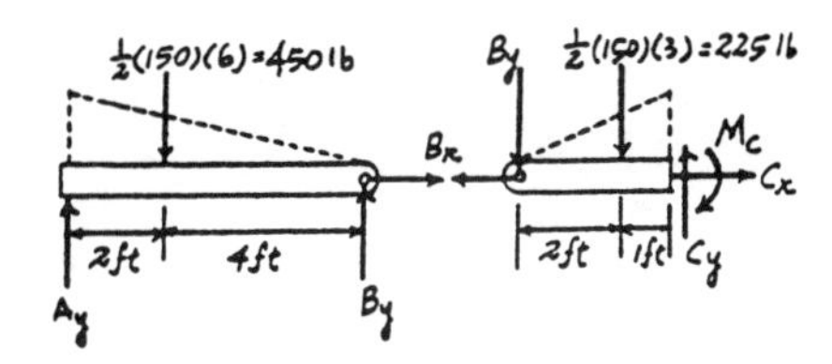

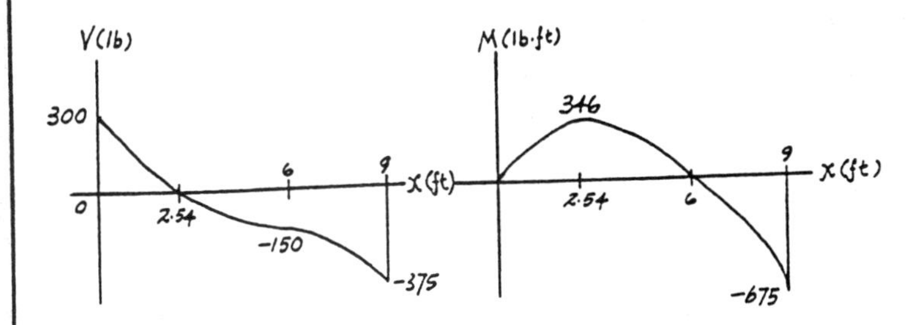

Allowable Bending Stress : The maximum moment is $M_{max} = 675$ lb · ft as indicated on the moment diagram. Applying the flexure formula

$$\sigma_{max} = \sigma_{allow} = \frac{M_{max}\, c}{I}$$

$$1.5(10^3) = \frac{675(12)\left(\frac{a}{2}\right)}{\frac{1}{12}a^4}$$

$$a = 3.19 \text{ in.}$$ **Ans**

6–67. Determine the absolute maximum bending stress in the beam in Prob. 6–28. The cross section of the beam is as shown.

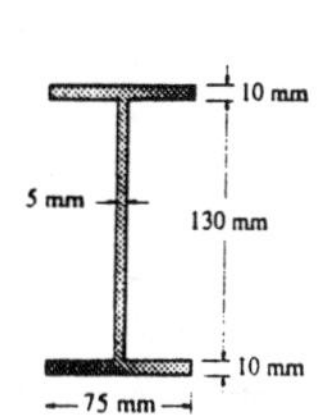

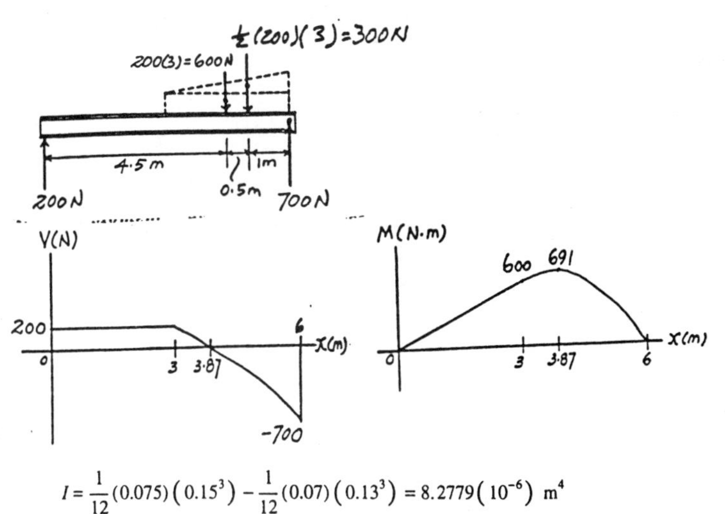

$$I = \frac{1}{12}(0.075)\left(0.15^3\right) - \frac{1}{12}(0.07)\left(0.13^3\right) = 8.2779\left(10^{-6}\right) \text{ m}^4$$

Absolute Maximum Bending Stress : The maximum moment is $M_{max} = 691$ N · m as indicated on the moment diagram. Applying the flexure formula

$$\sigma_{max} = \frac{M_{max}\, c}{I} = \frac{691(0.075)}{8.2779(10^{-6})} = 6.26 \text{ MPa}$$ **Ans**

***6–68.** Determine the absolute maximum bending stress in the T-beam in Prob. 6–29. The cross section of the beam is as shown.

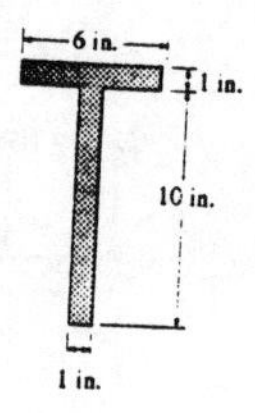

Support Reactions : As shown on FBD.

Internal Moment : As shown on the moment diagram.

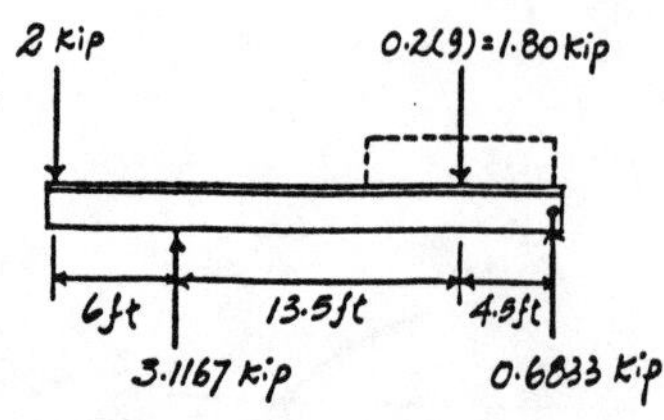

Section Properties :

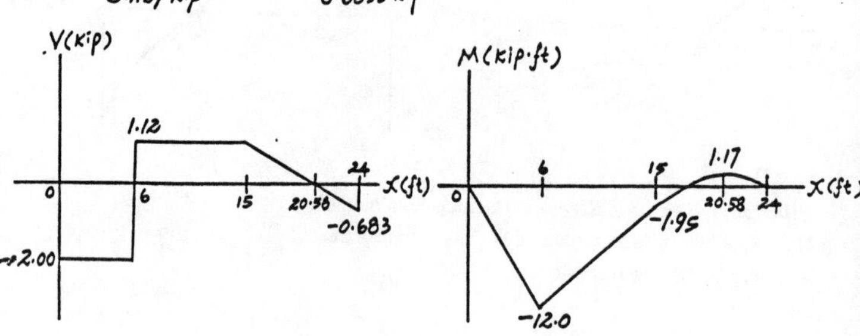

$$\bar{y} = \frac{\Sigma \tilde{y} A}{\Sigma A} = \frac{0.5(6)(1) + 6(10)(1)}{6(1) + 10(1)} = 3.9375 \text{ in.}$$

$$I = \frac{1}{12}(6)(1^3) + 6(1)(3.9375 - 0.5)^2 + \frac{1}{12}(1)(10^3) + 1(10)(6 - 3.9375)^2$$

$$= 197.27 \text{ in}^4$$

Absolute Maximum Bending Stress : The maximum moment is $M_{max} = 12.0$ kip · ft as indicated on the moment diagram. Applying the flexure formula

$$\sigma_{max} = \frac{M_{max}\, c}{I} = \frac{12.0(12)(11 - 3.9375)}{197.27} = 5.16 \text{ ksi} \qquad \textbf{Ans}$$

6–69. If the beam in Prob. 6–17 has a rectangular cross section, determine its required height h and width b if $h = 2b$. The allowable bending stress for the material is $\sigma_{allow} = 24$ ksi.

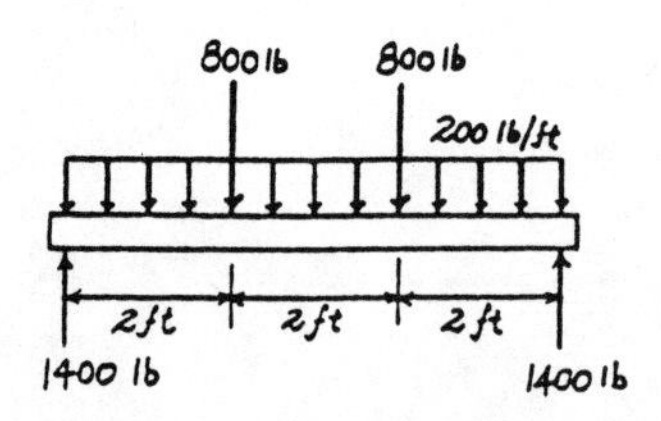

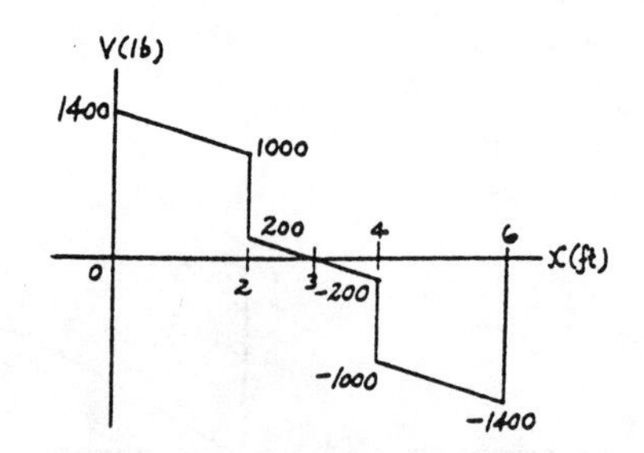

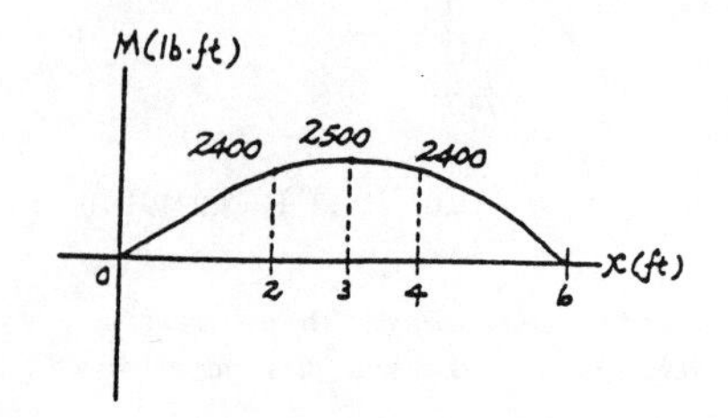

Allowable Bending Stress : The maximum moment is $M_{max} = 2500$ lb · ft as indicated on the moment diagram. Applying the flexure formula

$$\sigma_{max} = \sigma_{allow} = \frac{M_{max}\, c}{I}$$

$$24(10^3) = \frac{2500(12)(b)}{\frac{1}{12}(b)(2b)^3}$$

$$b = 1.23 \text{ in.} \qquad \textbf{Ans}$$

Thus. $\qquad h = 2b = 2.47$ in. $\qquad$ **Ans**

6–70. If the beam in Prob. 6–17 is to be a pipe having an outer diameter of 3 in., determine its inner diameter if the allowable bending stress for the material is $\sigma_{allow} = 24$ ksi.

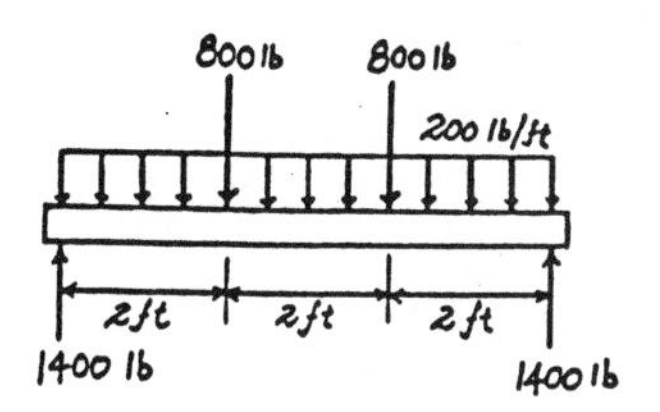

Support Reactions : As shown on FBD.

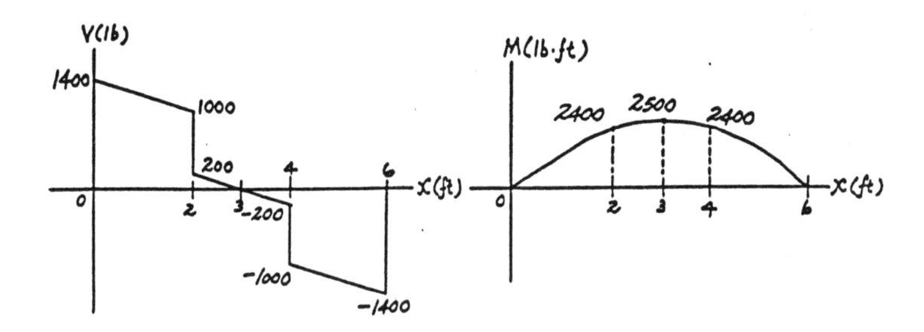

Allowable Bending Stress : The maximum moment is $M_{max} = 2500$ lb · ft as indicated on the moment diagram. Applying the flexure formula

$$\sigma_{max} = \sigma_{allow} = \frac{M_{max}c}{I}$$

$$24(10^3) = \frac{2500(12)(1.5)}{\frac{\pi}{4}\left[1.5^4 - \left(\frac{d_i}{2}\right)^4\right]}$$

$$d_i = 2.56 \text{ in} \qquad \textbf{Ans}$$

6–71. The steel beam has the cross-sectional area shown. Determine the largest intensity of distributed load w_0 that it can support so that the maximum bending stress in the beam does not exceed $\sigma_{max} = 22$ ksi.

Support Reactions : As shown on FBD.
Internal Moment : The maximum moment occurs at mid span. The maximum moment is determined using the method of sections.

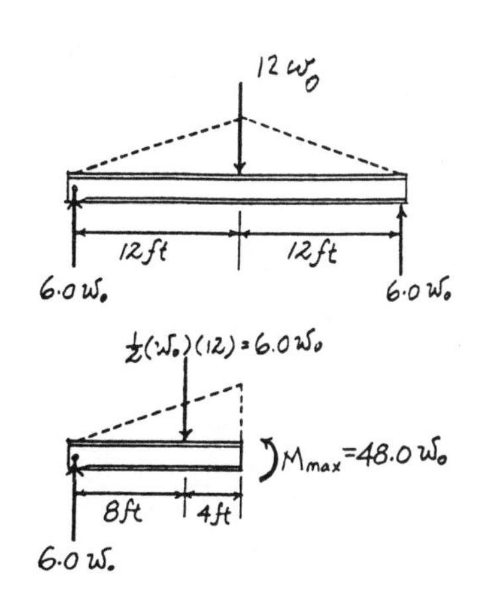

Section Property :

$$I = \frac{1}{12}(8)(10.6^3) - \frac{1}{12}(7.7)(10^3) = 152.344 \text{ in}^4$$

Absolute Maximum Bending Stress : The maximum moment is $M_{max} = 48.0w_0$ as indicated on the FBD. Applying the flexure formula

$$\sigma_{max} = \frac{M_{max}c}{I}$$

$$22 = \frac{48.0w_0(12)(5.30)}{152.344}$$

$$w_0 = 1.10 \text{ kip/ft} \qquad \textbf{Ans}$$

*6–72. The steel beam has the cross-sectional area shown. If w_o = 0.5 kip/ft, determine the maximum bending stress in the beam.

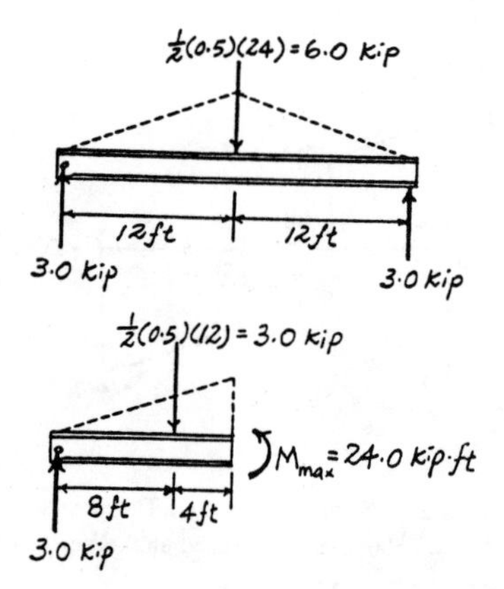

Support Reactions : As shown on FBD.
Internal Moment : The maximum moment occurs at mid span. The maximum moment is determined using the method of sections.

Section Property :

$$I = \frac{1}{12}(8)\left(10.6^3\right) - \frac{1}{12}(7.7)\left(10^3\right) = 152.344\ \text{in}^4$$

Absolute Maximum Bending Stress : The maximum moment is M_{max} = 24.0 kip · ft as indicated on the FBD. Applying the flexure formula

$$\sigma_{max} = \frac{M_{max}\,c}{I} = \frac{24.0(12)(5.30)}{152.344} = 10.0\ \text{ksi}$$

Ans

6–73. The beam has a rectangular cross section as shown. Determine the largest load P that can be supported on its overhanging ends so that the bending stress in the beam does not exceed σ_{max} = 10 MPa.

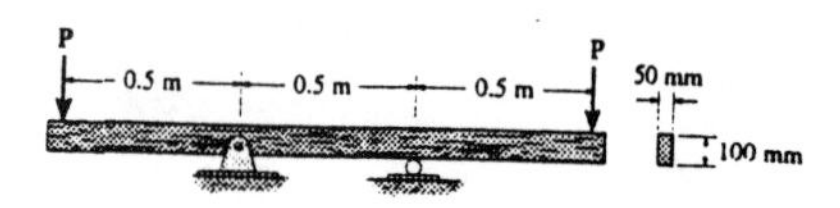

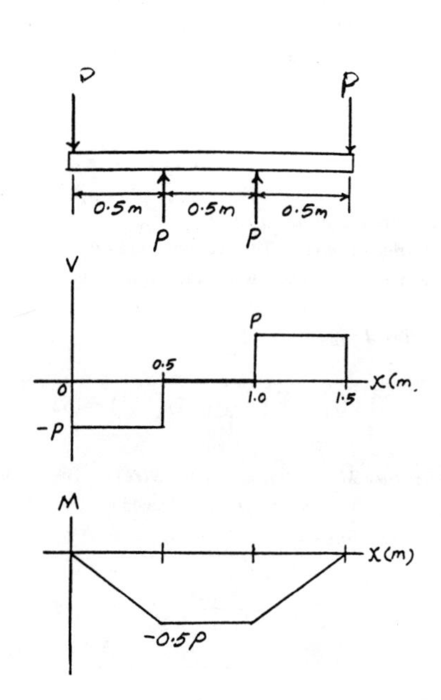

Absolute Maximum Bending Stress : The maximum moment is M_{max} = 0.5P as indicated on the moment diagram. Applying the flexure formula

$$\sigma_{max} = \frac{M_{max}\,c}{I}$$

$$10\left(10^6\right) = \frac{0.5P(0.05)}{\frac{1}{12}(0.05)(0.1^3)}$$

$$P = 1666.7\ \text{N} = 1.67\ \text{kN}$$

Ans

6–74. The beam has the rectangular cross section shown. If $P = 15$ kN, determine the maximum bending stress in the beam. Sketch the stress distribution acting over the cross section.

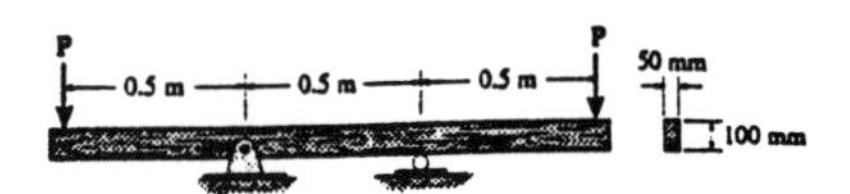

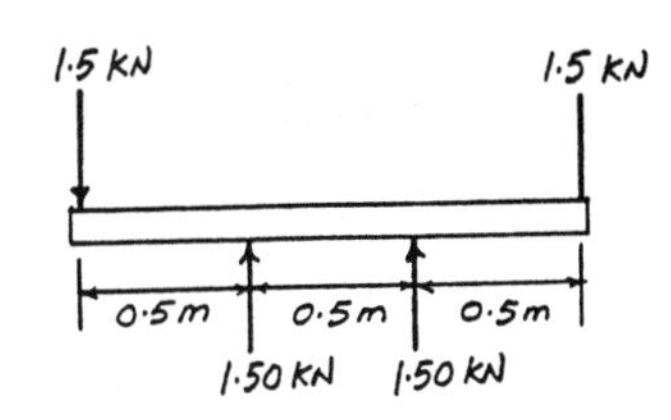

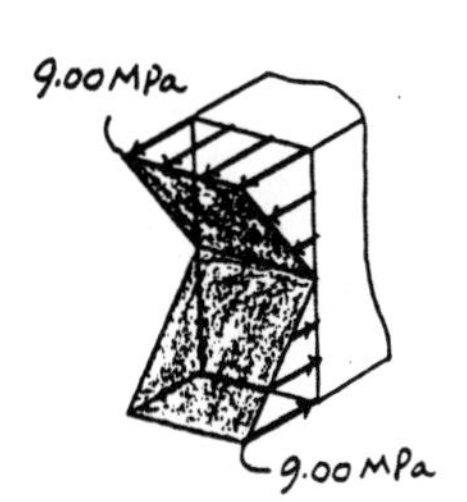

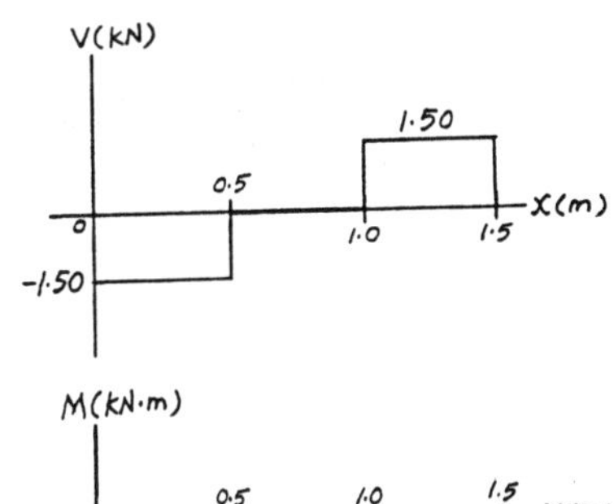

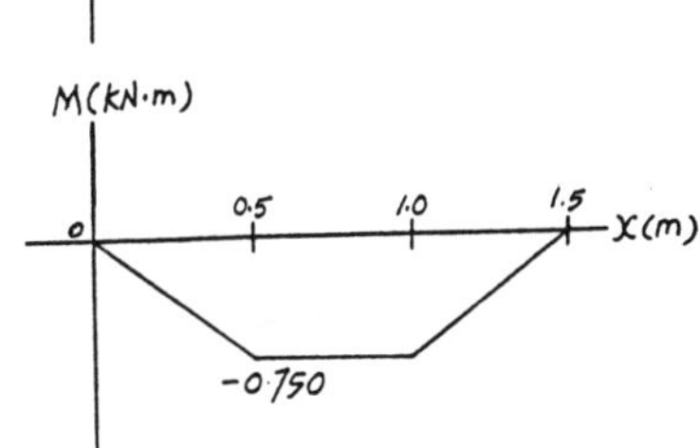

Absolute Maximum Bending Stress : The maximum moment is $M_{max} = 0.750$ kN · m as indicated on moment diagram. Applying the flexure formula

$$\sigma_{max} = \frac{M_{max}\,c}{I} = \frac{0.750(10^3)(0.05)}{\frac{1}{12}(0.05)(0.1^3)} = 9.00 \text{ MPa} \qquad \textbf{Ans}$$

6–75. The load limiter is a device that is used to monitor the tension in cables and thereby provide a warning of overload in elevator, crane, and towing cables. As increased tension P is applied to the cable, it tends to increase the force F, and this causes bending stresses in member AB. Normal strain measurements are then measured electronically at point C in AB, and this is correlated to the load P on the cable. If the normal strain is measured to be $\epsilon = 0.001$ mm/mm, determine the force F developed on the cross beam at C caused by the tension in the cable. The modulus of elasticity for member AB is $E = 135$ GPa. Assume the supports at A and B exert only vertical reactions on the cord

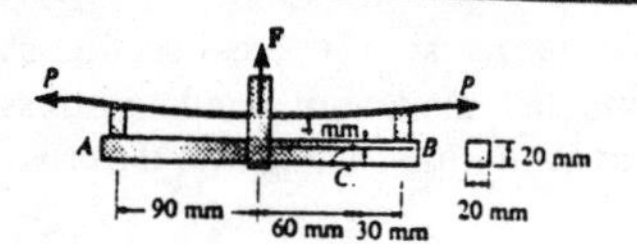

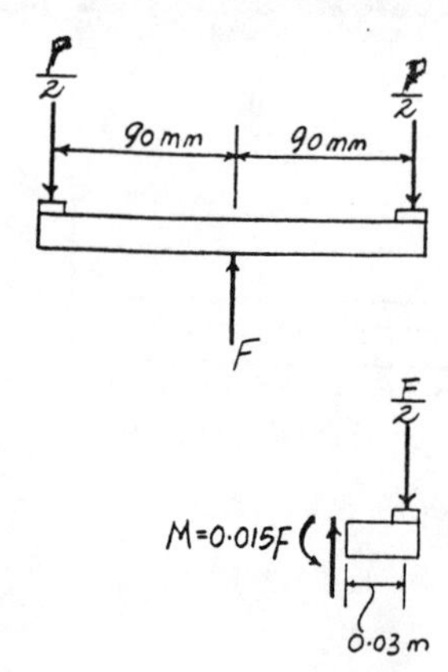

Support Reactions : As shown on FBD.

Internal Moment : The internal moment at a section through point C can be determined using the method of sections.

Absolute Maximum Bending Stress : The internal moment is $M = 0.015F$ as indicated on FBD and the bending stress developed at point C is given by $\sigma = E\varepsilon = 135(10^9)(0.001) = 135$ MPa. Applying the flexure formula

$$\sigma = \frac{My}{I}$$

$$135(10^6) = \frac{0.015F(0.006)}{\frac{1}{12}(0.02)(0.02^3)}$$

$$F = 20\,000 \text{ N} = 20.0 \text{ kN}$$

Ans

***6–76.** The beam is subjected to the loading shown. If its cross-sectional dimension $a = 180$ mm, determine the absolute maximum bending stress in the beam.

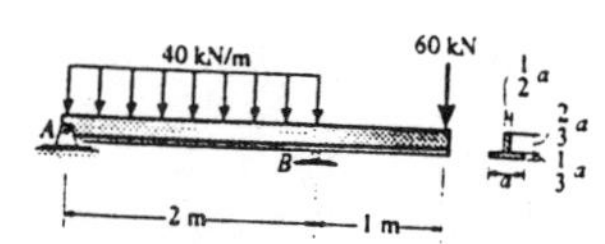

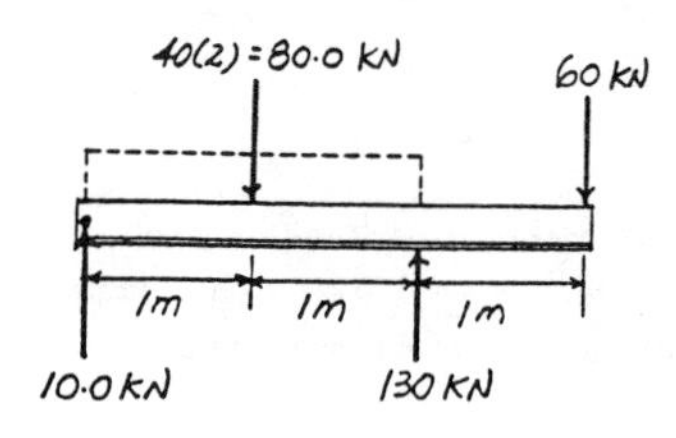

Section Properties :

$$\bar{y} = \frac{\Sigma \tilde{y}A}{\Sigma A} = \frac{0.03(0.18)(0.06) + 0.12(0.12)(0.09)}{(0.18)(0.06) + (0.12)(0.09)} = 0.075 \text{ m}$$

$$I = \frac{1}{12}(0.18)(0.06^3) + 0.18(0.06)(0.075 - 0.03)^2 + \frac{1}{12}(0.09)(0.12^3) + 0.09(0.12)(0.12 - 0.075)^2$$

$$= 59.94(10^{-6}) \text{ m}^4$$

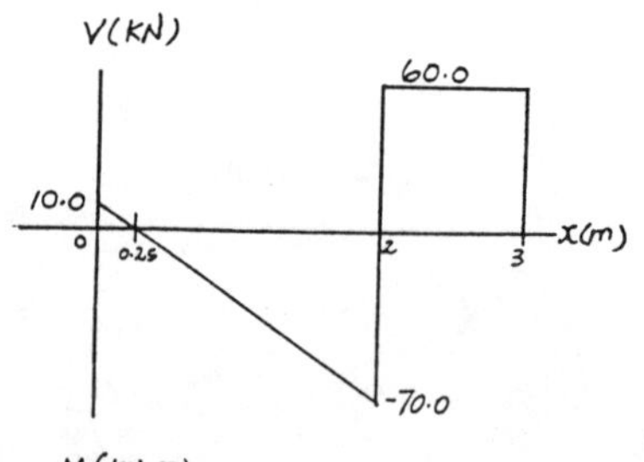

Allowable Bending Stress : The maximum moment is $M_{max} = 60.0$ kN · m as indicated on the moment diagram. Applying the flexure formula

$$\sigma_{max} = \frac{M_{max}c}{I}$$

$$= \frac{60.0(10^3)(0.18 - 0.075)}{59.94(10^{-6})}$$

$$= 105 \text{ MPa}$$

Ans

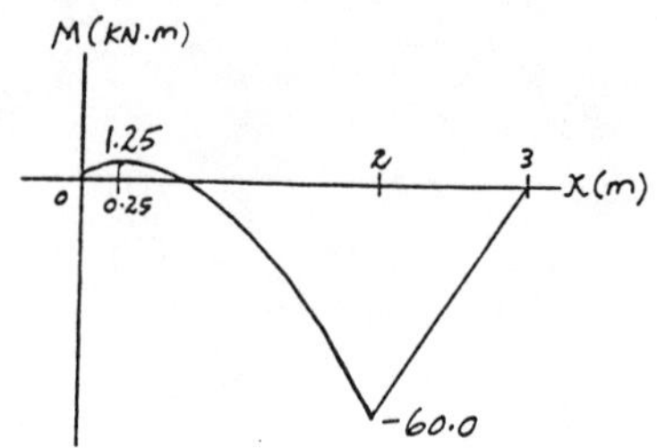

6–77. The beam is subjected to the loading shown. Determine its required cross-sectional dimension a, if the allowable bending stress for the material is $\sigma_{allow} = 150$ MPa.

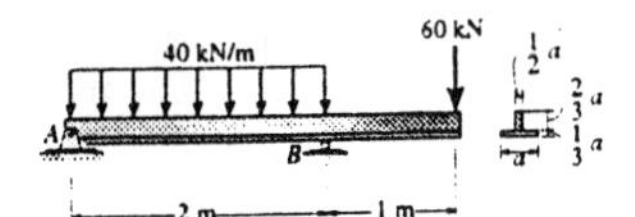

Support Reactions : As shown on FBD.

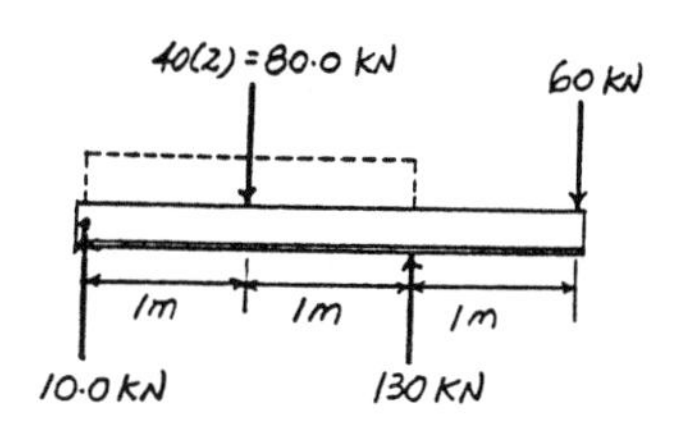

Section Properties :

$$\bar{y} = \frac{\Sigma \tilde{y} A}{\Sigma A} = \frac{\frac{1}{6}a\left(\frac{1}{3}a\right)a + \frac{2}{3}a\left(\frac{2}{3}a\right)\left(\frac{1}{2}a\right)}{\left(\frac{1}{3}a\right)a + \left(\frac{2}{3}a\right)\left(\frac{1}{2}a\right)} = \frac{5}{12}a$$

$$I = \frac{1}{12}(a)\left(\frac{1}{3}a\right)^3 + a\left(\frac{1}{3}a\right)\left(\frac{5}{12}a - \frac{1}{6}a\right)^2 + \frac{1}{12}\left(\frac{1}{2}a\right)\left(\frac{2}{3}a\right)^3 + \frac{1}{2}a\left(\frac{2}{3}a\right)\left(\frac{2}{3}a - \frac{5}{12}a\right)^2$$

$$= \frac{37}{648}a^4$$

Internal Moment : As shown on the moment diagram.

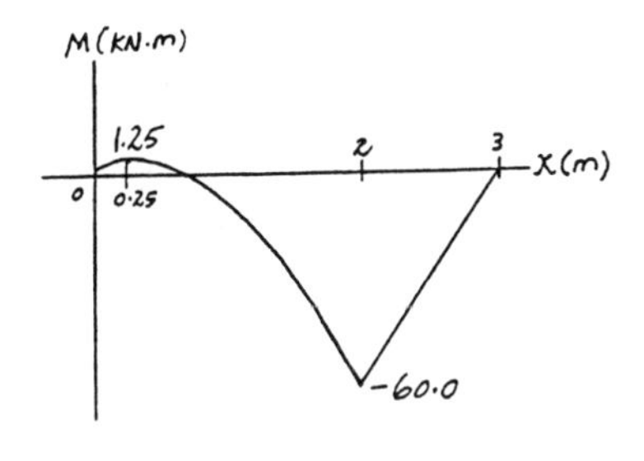

Allowable Bending Stress : The maximum moment is $M_{max} = 60.0$ kN · m as indicated on the moment diagram. Applying the flexure formula

$$\sigma_{max} = \sigma_{allow} = \frac{M_{max}\,c}{I}$$

$$150\left(10^6\right) = \frac{60.0(10^3)\left(a - \frac{5}{12}a\right)}{\frac{37}{648}a^4}$$

$$a = 0.1599 \text{ m} = 160 \text{ mm} \qquad \textbf{Ans}$$

6–78. The simply supported beam is subjected to the load of $P = 1.5$ kN. Determine the maximum bending stress in the beam.

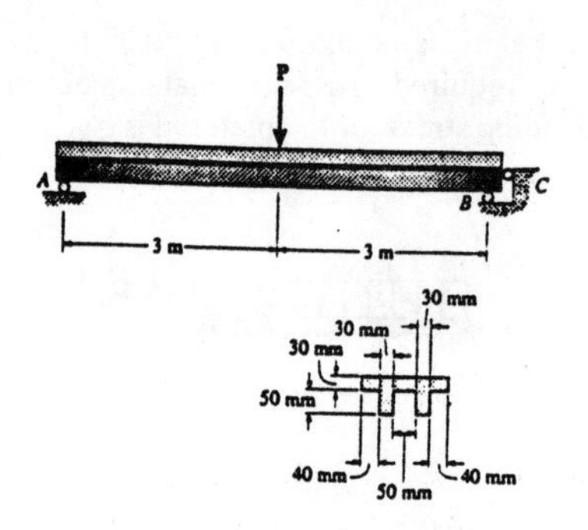

Support Reactions : As shown on FBD.
Internal Moment : The maximum moment occurs at mid span. The maximum moment is determined using the method of sections.

Section Property :

$$\bar{y} = \frac{\Sigma \tilde{y}A}{\Sigma A} = \frac{0.015(0.19)(0.03) + 2[0.055(0.05)(0.03)]}{0.19(0.03) + 2(0.05)(0.03)}$$
$$= 0.028793 \text{ m}$$

$$I = \frac{1}{12}(0.19)(0.03^3) + 0.19(0.03)(0.028793 - 0.015)^2$$
$$+ \frac{1}{12}(0.06)(0.05^3) + 0.06(0.05)(0.055 - 0.028793)^2$$
$$= 4.19733(10^{-6}) \text{ m}^4$$

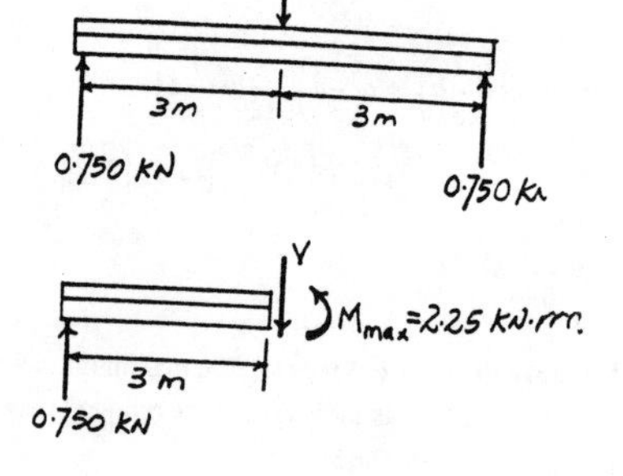

Absolute Maximum Bending Stress : The maximum moment is $M_{max} = 2.25$ kN · m as indicated on the FBD. Applying the flexure formula

$$\sigma_{max} = \frac{M_{max}\,c}{I}$$
$$= \frac{2.25(10^3)(0.08 - 0.028793)}{4.19733(10^{-6})}$$
$$= 27.4 \text{ MPa} \qquad \textbf{Ans}$$

6–79. Determine the magnitude of the maximum load **P** that can be applied to the beam that is made of a material having an allowable bending stress of $\sigma_{allow} = 12$ MPa.

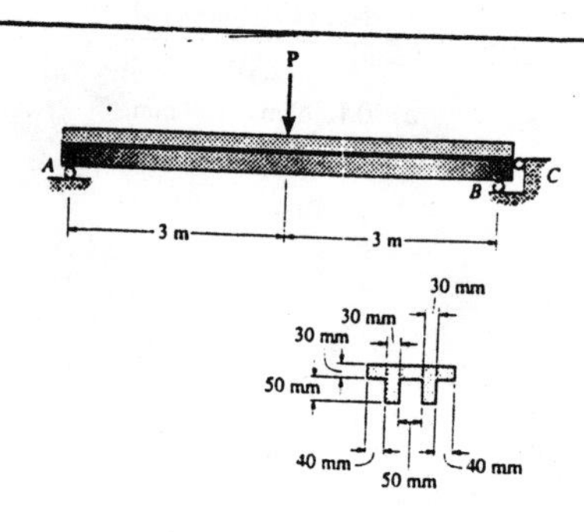

Support Reactions : As shown on FBD.
Internal Moment : The maximum moment occurs at mid span. The maximum moment is determined using the method of sections.

Section Property :

$$\bar{y} = \frac{\Sigma \tilde{y}A}{\Sigma A} = \frac{0.015(0.19)(0.03) + 2[0.055(0.05)(0.03)]}{0.19(0.03) + 2(0.05)(0.03)}$$
$$= 0.028793 \text{ m}$$

$$I = \frac{1}{12}(0.19)(0.03^3) + 0.19(0.03)(0.028793 - 0.015)^2$$
$$+ \frac{1}{12}(0.06)(0.05^3) + 0.06(0.05)(0.055 - 0.028793)^2$$
$$= 4.19733(10^{-6}) \text{ m}^4$$

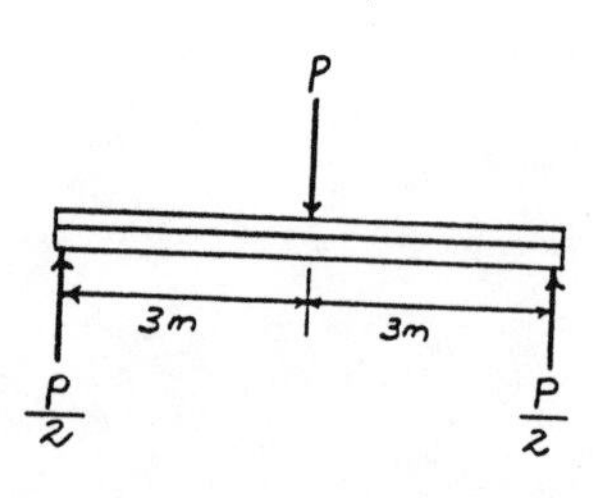

Absolute Maximum Bending Stress : The maximum moment is $M_{max} = \frac{3P}{2}$ as indicated on the FBD. Applying the flexure formula

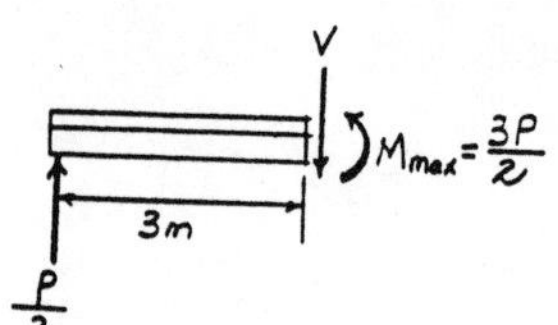

$$\sigma_{max} = \sigma_{allow} = \frac{M_{max}\,c}{I}$$
$$12(10^6) = \frac{\left(\frac{3P}{2}\right)(0.08 - 0.028793)}{4.19733(10^{-6})}$$
$$= 656 \text{ N} \qquad \textbf{Ans}$$

***6–80.** The two solid steel rods are bolted together along their length and support the loading shown. Assume the support at A is a pin and B is a roller. Determine the required diameter d of each of the rods if the allowable bending stress is $\sigma_{allow} = 130$ MPa.

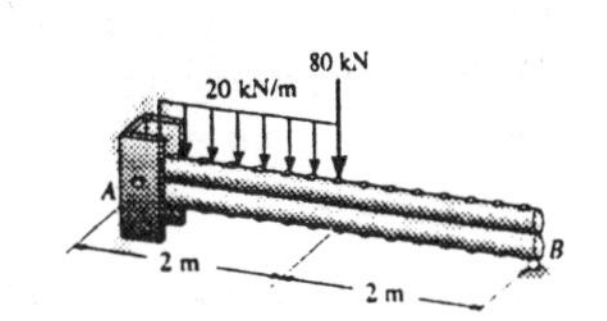

Section Property :

$$I = 2\left[\frac{\pi}{4}\left(\frac{d}{2}\right)^4 + \frac{\pi}{4}d^2\left(\frac{d}{2}\right)^2\right] = \frac{5\pi}{32}d^4$$

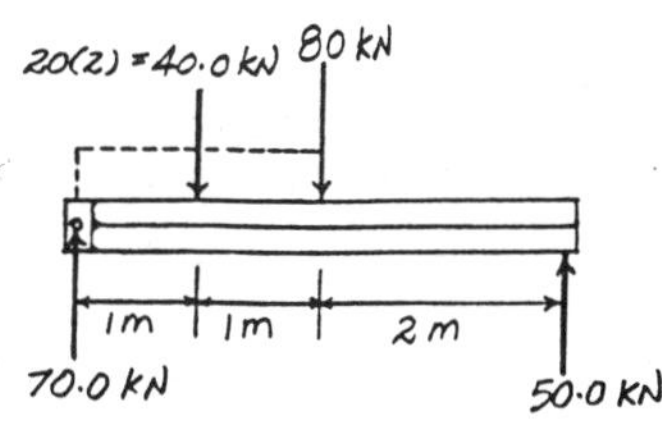

Allowable Bending Stress : The maximum moment is $M_{max} = 100$ kN · m as indicated on moment diagram. Applying the flexure formula

$$\sigma_{max} = \sigma_{allow} = \frac{M_{max}\,c}{I}$$

$$130(10^6) = \frac{100(10^3)(d)}{\frac{5\pi}{32}d^4}$$

$$d = 0.1162 \text{ m} = 116 \text{ mm} \qquad \textbf{Ans}$$

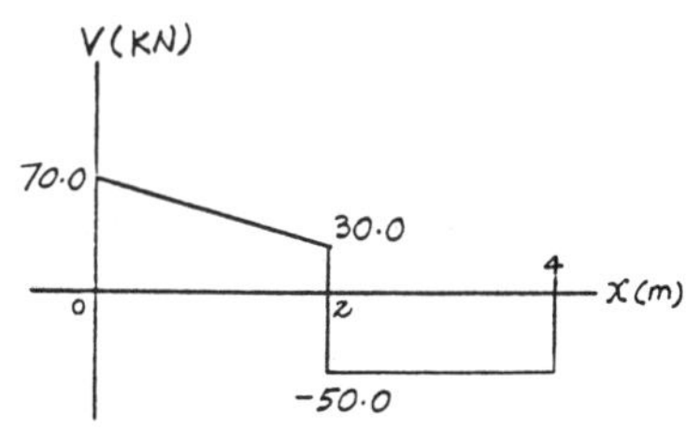

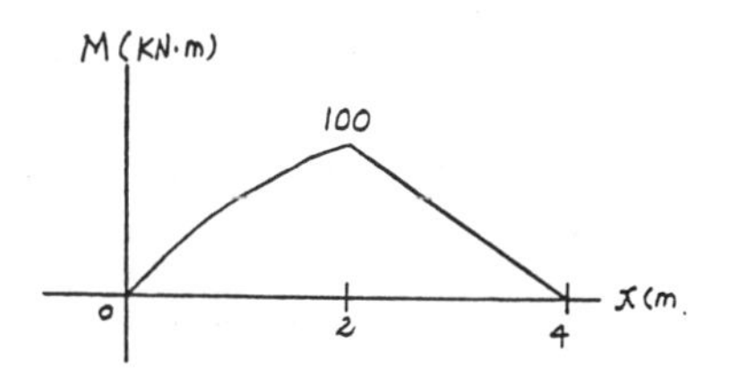

6–81. Solve Prob. 6–80 if the rods are rotated 90° so that both rods rest on the supports at A (pin) and B (roller).

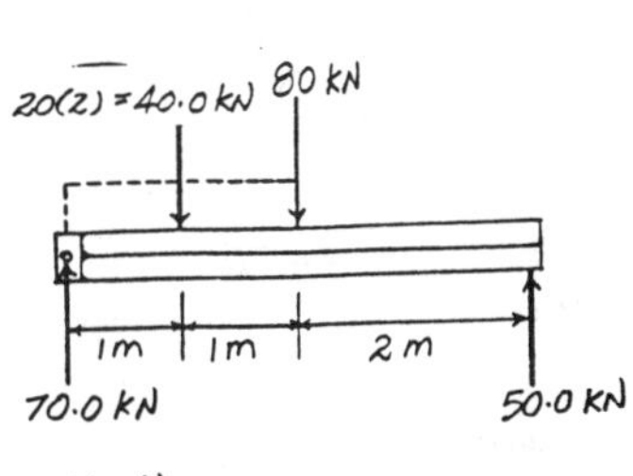

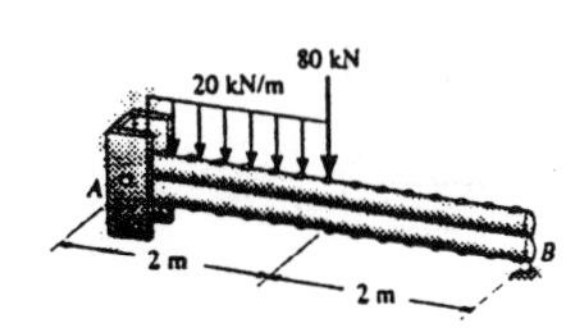

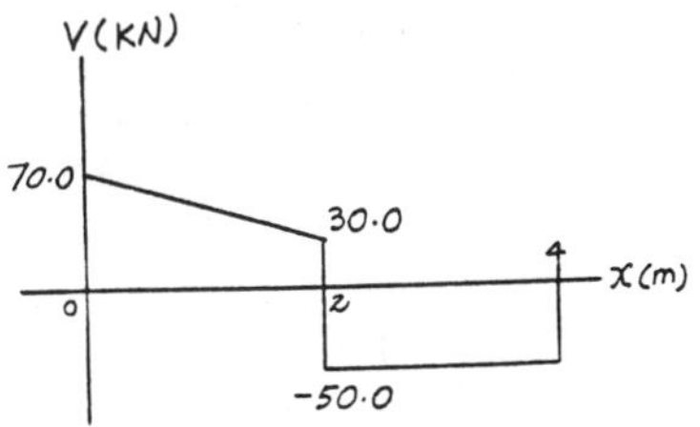

Section Property :

$$I = 2\left[\frac{\pi}{4}\left(\frac{d}{2}\right)^4\right] = \frac{\pi}{32}d^4$$

Allowable Bending Stress : The maximum moment is $M_{max} = 100$ kN · m as indicated on the moment diagram. Applying the flexure formula

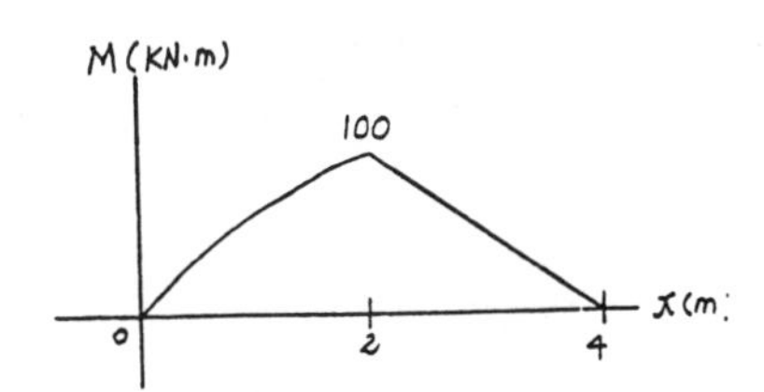

$$\sigma_{max} = \sigma_{allow} = \frac{M_{max}\,c}{I}$$

$$130(10^6) = \frac{100(10^3)(d)}{\frac{\pi}{32}d^4}$$

$$d = 0.1986 \text{ m} = 199 \text{ mm} \qquad \textbf{Ans}$$

6–82. The beam supports the load of 5 kip. Determine the absolute maximum bending stress in the beam if the sides of its triangular cross section are $a = 6$ in.

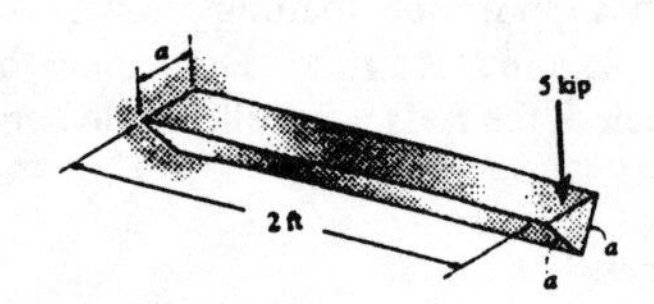

Support Reactions : As shown on FBD.
Section Property :

$$I = \frac{1}{36}(6)(6\sin 60°)^3 = 23.383 \text{ in}^4$$

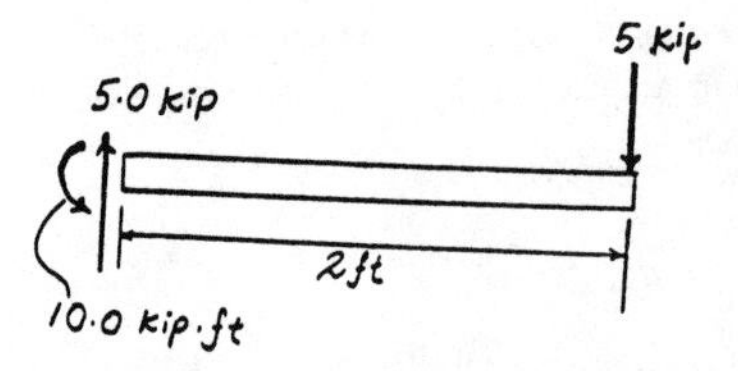

Absolute Maximum Bending Stress : The maximum moment occurs at the fixed support where $M_{max} = 10.0$ kip · ft as indicated on the FBD. Applying the flexure formula

$$\sigma_{max} = \frac{M_{max}\, c}{I}$$

$$= \frac{10.0(12)\left[\frac{2}{3}(6\sin 60°)\right]}{23.383}$$

$$= 17.8 \text{ ksi} \qquad \textbf{Ans}$$

6–83. The beam supports the load of 5 kip.. Determine the required size a of the sides of its triangular cross section if the allowable bending stress is $\sigma_{allow} = 18$ ksi.

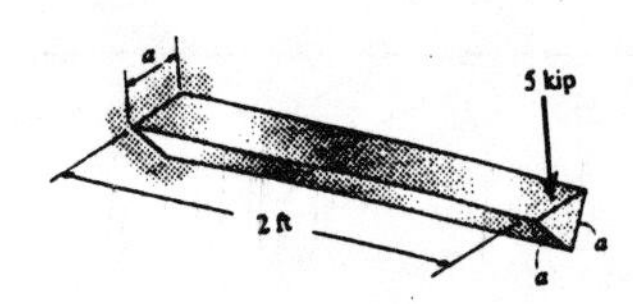

Support Reactions : As shown on FBD.
Section Property :

$$I = \frac{1}{36}(a)(a\sin 60°)^3 = \frac{\sqrt{3}}{96}a^4$$

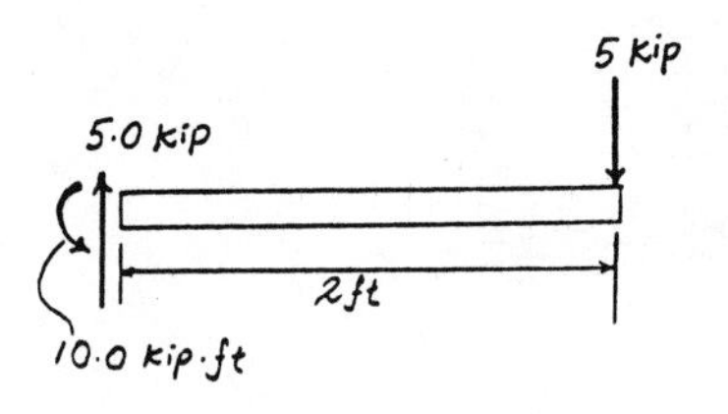

Allowable Bending Stress : The maximum moment occurs at the fixed support where $M_{max} = 10.0$ kip · ft as indicated on the FBD. Applying the flexure formula

$$\sigma_{max} = \sigma_{allow} = \frac{M_{max}\, c}{I}$$

$$18 = \frac{10.0(12)\left(\frac{2}{3}a\sin 60°\right)}{\frac{\sqrt{3}}{96}a^4}$$

$$a = 5.98 \text{ in.} \qquad \textbf{Ans}$$

***6–84.** Determine the absolute maximum bending stress in the tubular shaft if d_i = 160 mm and d_o = 200 mm.

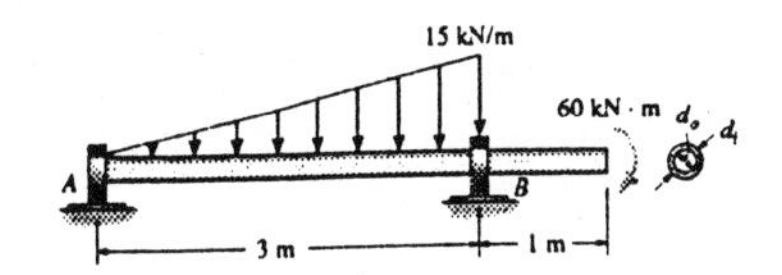

Section Property :

$$I = \frac{\pi}{4}\left(0.1^4 - 0.08^4\right) = 46.370\left(10^{-6}\right)\ \text{m}^4$$

Absolute Maximum Bending Stress : The maximum moment is M_{max} = 60.0 kN · m as indicated on the moment diagram. Applying the flexure formula

$$\sigma_{max} = \frac{M_{max}\,c}{I}$$

$$= \frac{60.0(10^3)(0.1)}{46.370(10^{-6})}$$

$$= 129 \text{ MPa} \qquad \textbf{Ans}$$

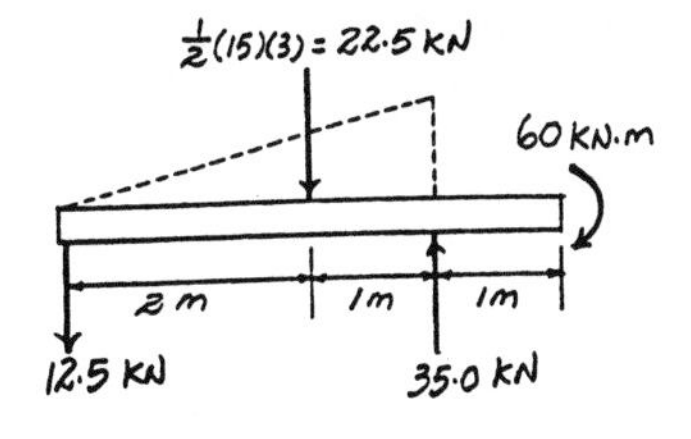

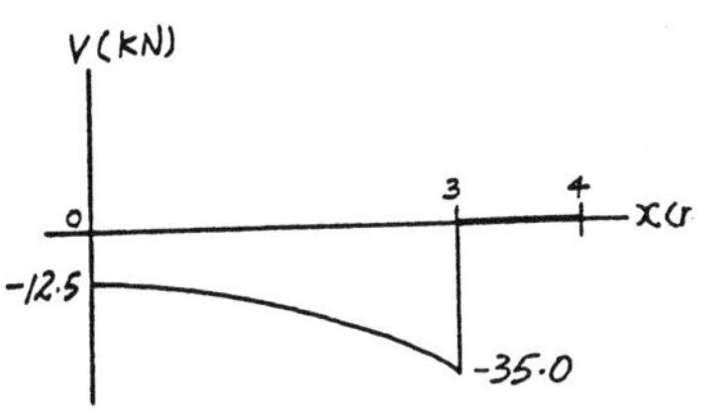

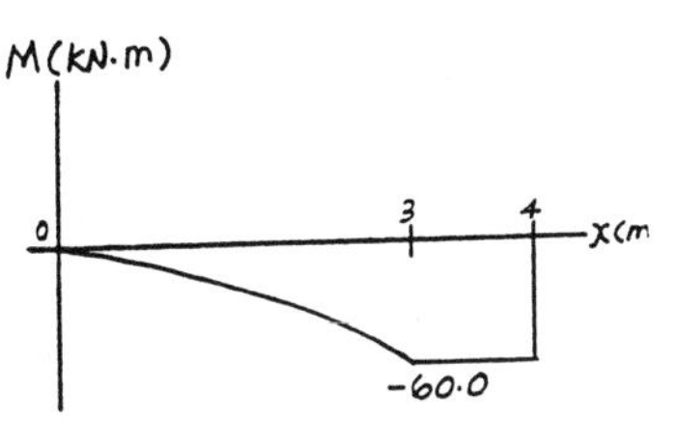

6–85. The tubular shaft is to have a cross section such that its inner diameter and outer diameter are related by $d_i = 0.8d_o$. Determine these required dimensions if the allowable bending stress is σ_{allow} = 155 MPa.

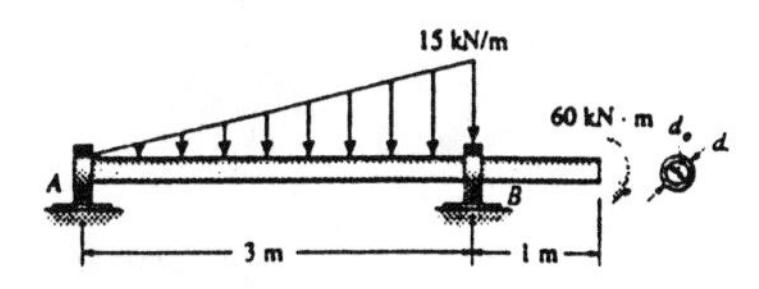

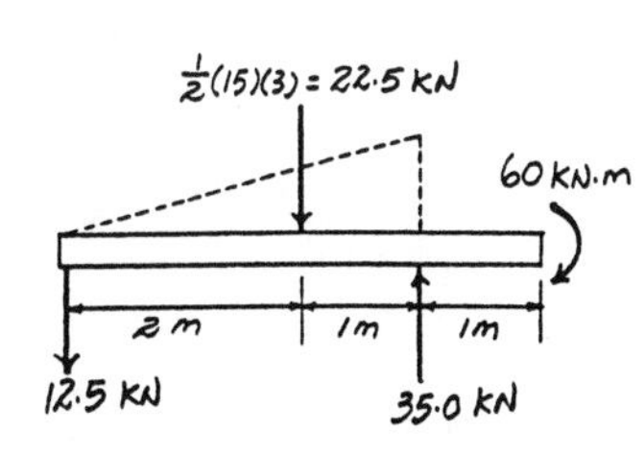

Section Property :

$$I = \frac{\pi}{4}\left[\left(\frac{d_o}{2}\right)^4 - \left(\frac{d_i}{2}\right)^4\right] = \frac{\pi}{4}\left[\frac{d_o^{\,4}}{16} - \left(\frac{0.8d_o}{2}\right)^4\right] = 0.009225\pi d_o^4$$

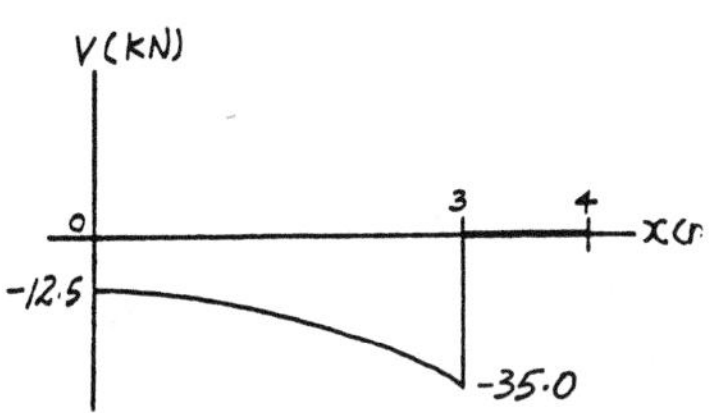

Allowable Bending Stress : The maximum moment is M_{max} = 60.0 kN · m as indicated on the moment diagram. Applying the flexure formula

$$\sigma_{max} = \sigma_{allow} = \frac{M_{max}\,c}{I}$$

$$155\left(10^6\right) = \frac{60.0(10^3)\left(\frac{d_o}{2}\right)}{0.009225\pi d_o^4}$$

$$d_o = 0.1883 \text{ m} = 188 \text{ mm} \qquad \textbf{Ans}$$

Thus, $\quad d_i = 0.8d_o = 151 \text{ mm} \qquad$ **Ans**

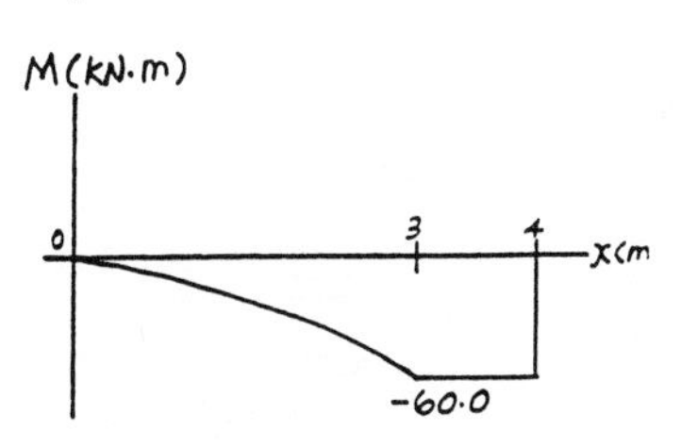

6–86. The man has a mass of 78 kg and stands motionless at the end of the diving board. If the board has the cross section shown, determine the maximum bending stress developed in the board. Assume A is a pin and B is a roller.

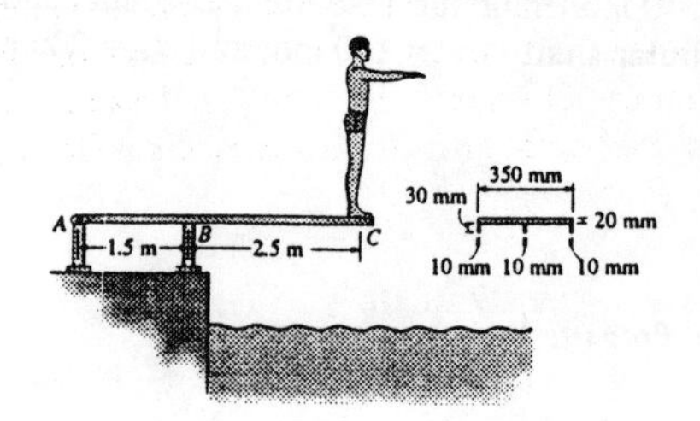

Support Reactions : As shown on FBD.

Internal Moment : The maximum moment occurs at support B. The maximum moment is determined using the method of sections.

Section Property :

$$\bar{y} = \frac{\Sigma \tilde{y} A}{\Sigma A}$$

$$= \frac{0.01(0.35)(0.02) + 0.035(0.03)(0.03)}{0.35(0.02) + 0.03(0.03)} = 0.012848 \text{ m}$$

$$I = \frac{1}{12}(0.35)\left(0.02^3\right) + 0.35(0.02)(0.012848 - 0.01)^2$$

$$+ \frac{1}{12}(0.03)\left(0.03^3\right) + 0.03(0.03)(0.035 - 0.012848)^2$$

$$= 0.79925\left(10^{-6}\right) \text{ m}^4$$

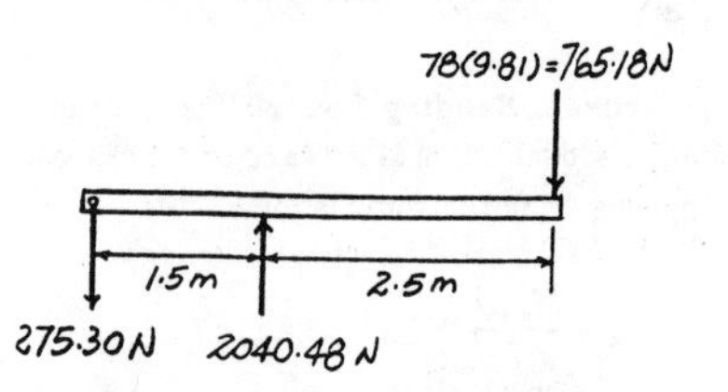

$M_{max} = 1912.95$ N·m

1.5m

1275.30 N

V

Absolute Maximum Bending Stress : The maximum moment is $M_{max} = 1912.95$ N · m as indicated on the FBD. Applying the flexure formula

$$\sigma_{max} = \frac{M_{max}\, c}{I}$$

$$= \frac{1912.95(0.05 - 0.012848)}{0.79925(10^{-6})}$$

$$= 88.9 \text{ MPa} \qquad \textbf{Ans}$$

6–87. The man has a mass of 78 kg and stands motionless at the end of the diving board. If the board has the cross section shown, determine the maximum normal strain developed in the board. The modulus of elasticity for the material is $E = 125$ GPa. Assume A is a pin and B is a roller.

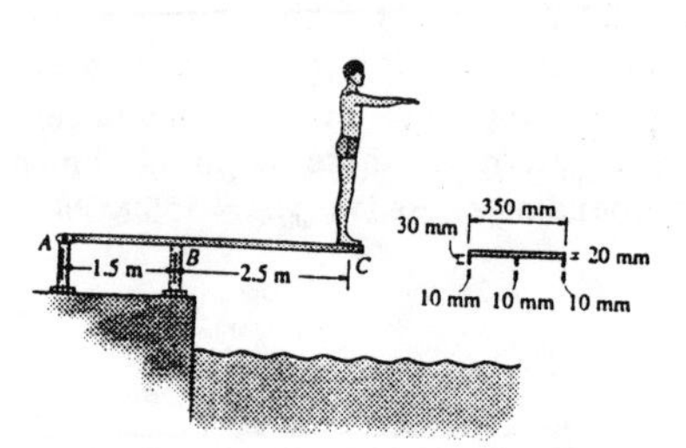

Internal Moment : The maximum moment occurs at support B. The maximum moment is determined using the method of sections.

Section Property :

$$\bar{y} = \frac{\Sigma \tilde{y} A}{\Sigma A}$$

$$= \frac{0.01(0.35)(0.02) + 0.035(0.03)(0.03)}{0.35(0.02) + 0.03(0.03)} = 0.012848 \text{ m}$$

$$I = \frac{1}{12}(0.35)\left(0.02^3\right) + 0.35(0.02)(0.012848 - 0.01)^2$$

$$+ \frac{1}{12}(0.03)\left(0.03^3\right) + 0.03(0.03)(0.035 - 0.012848)$$

$$= 0.79925\left(10^{-6}\right) \text{ m}^4$$

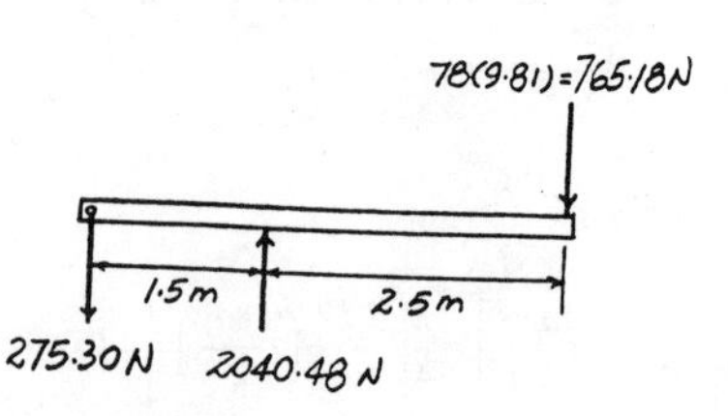

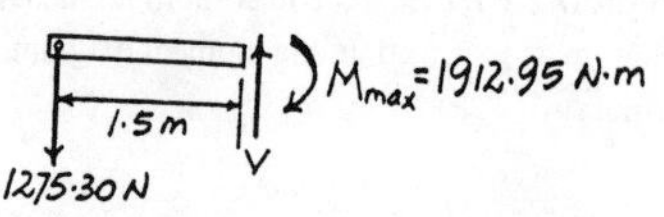

Absolute Maximum Bending Stress : The maximum moment is $M_{max} = 1912.95$ N · m as indicated on the FBD. Applying the flexure formula

$$\sigma_{max} = \frac{M_{max}\, c}{I}$$

$$= \frac{1912.95(0.05 - 0.012848)}{0.79925(10^{-6})}$$

$$= 88.92 \text{ MPa}$$

Absolute Maximum Normal Strain : Applying Hooke's law, we have

$$\varepsilon_{max} = \frac{\sigma_{max}}{E} = \frac{88.92(10^6)}{125(10^9)} = 0.711\left(10^{-3}\right) \text{ mm/mm} \quad \textbf{Ans}$$

***6–88.** The rod is supported by smooth journal bearings at A and B that only exert vertical reactions on the shaft. If $d = 90$ mm, determine the absolute maximum bending stress in the beam, and sketch the stress distribution acting over the cross section.

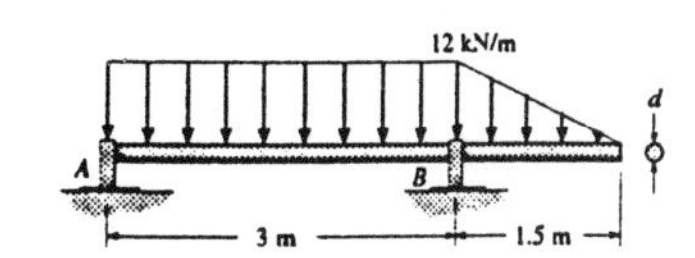

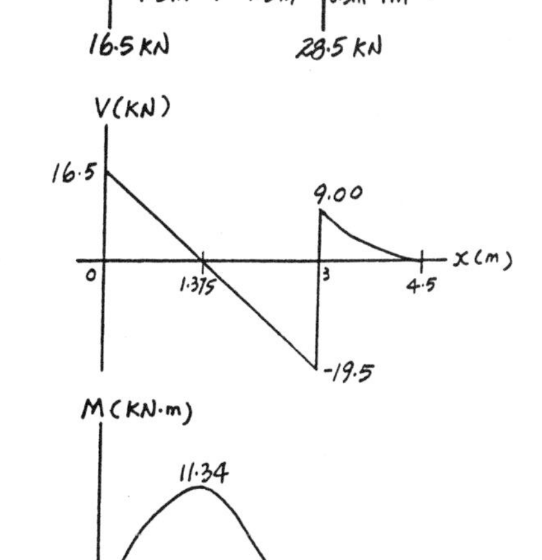

Absolute Maximum Bending Stress : The maximum moment is $M_{max} = 11.34$ kN · m as indicated on the moment diagram. Applying the flexure formula

$$\sigma_{max} = \frac{M_{max}\,c}{I}$$

$$= \frac{11.34(10^3)(0.045)}{\frac{\pi}{4}(0.045^4)}$$

$$= 158 \text{ MPa}$$ **Ans**

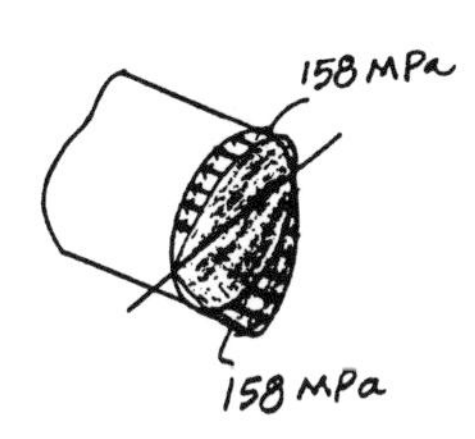

6–89. The rod is supported by smooth journal bearings at A and B that only exert vertical reactions on the shaft. Determine its smallest diameter d if the allowable bending stress is $\sigma_{allow} = 180$ MPa.

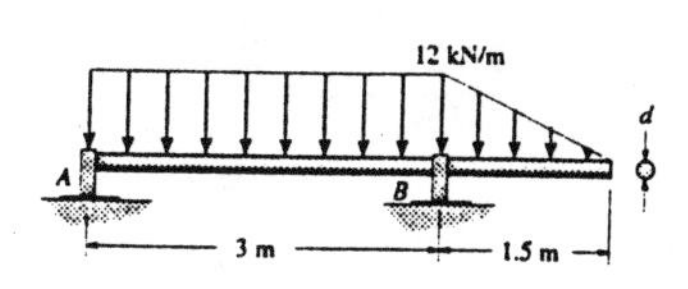

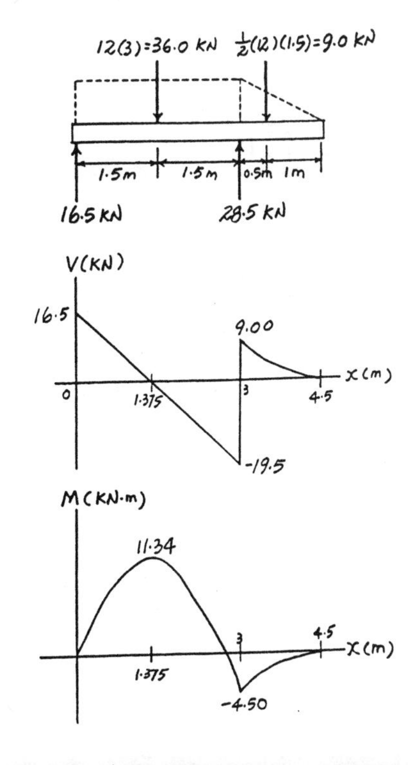

Allowable Bending Stress : The maximum moment is $M_{max} = 11.34$ kN · m as indicated on the moment diagram. Applying the flexure formula

$$\sigma_{max} = \sigma_{allow} = \frac{M_{max}\,c}{I}$$

$$180(10^6) = \frac{11.34(10^3)\left(\frac{d}{2}\right)}{\frac{\pi}{4}\left(\frac{d}{2}\right)^4}$$

$$d = 0.08626 \text{ m} = 86.3 \text{ mm}$$ **Ans**

6–90. The wood beam has a rectangular cross section for which $b = 60$ mm. Determine the absolute maximum bending stress in the beam and sketch the stress distribution acting over the cross section.

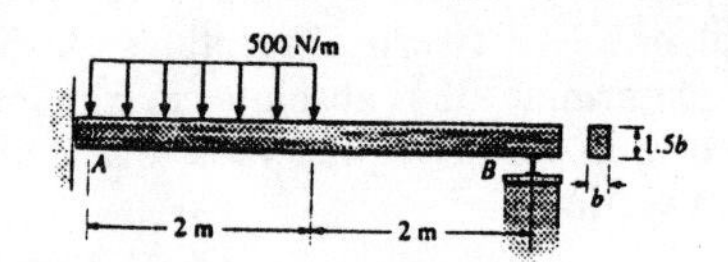

Absolute Maximum Bending Stress : The maximum moment is $M_{max} = 562.5$ N · m as indicated on the moment diagram. Applying the flexure formula

$$\sigma_{max} = \frac{M_{max}\,c}{I}$$

$$= \frac{562.5(0.045)}{\frac{1}{12}(0.06)(0.09^3)}$$

$= 6.94$ MPa **Ans**

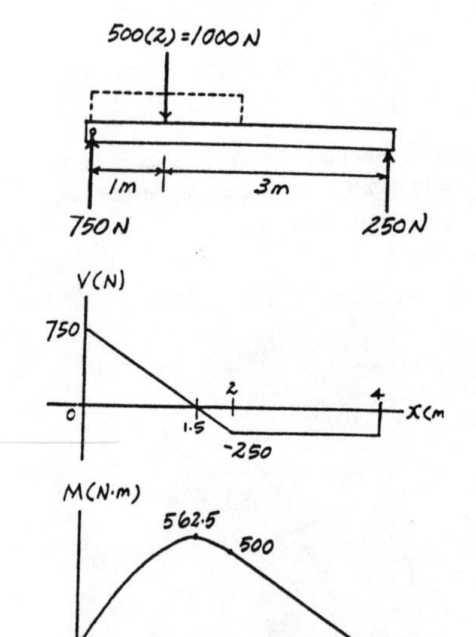

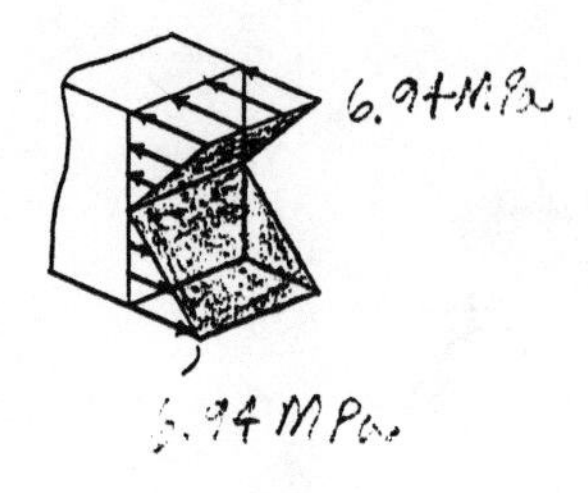

6–91. The wood beam has a rectangular cross section in the proportion shown. Determine its required dimension b if the allowable bending stress is $\sigma_{allow} = 10$ MPa.

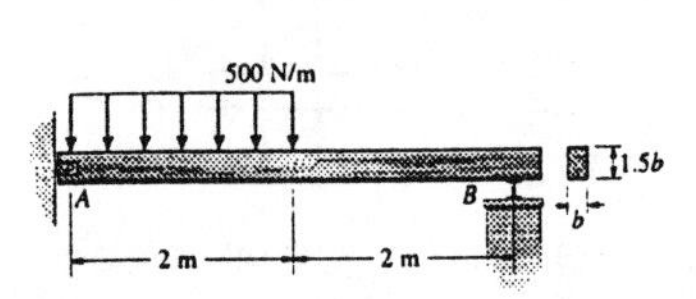

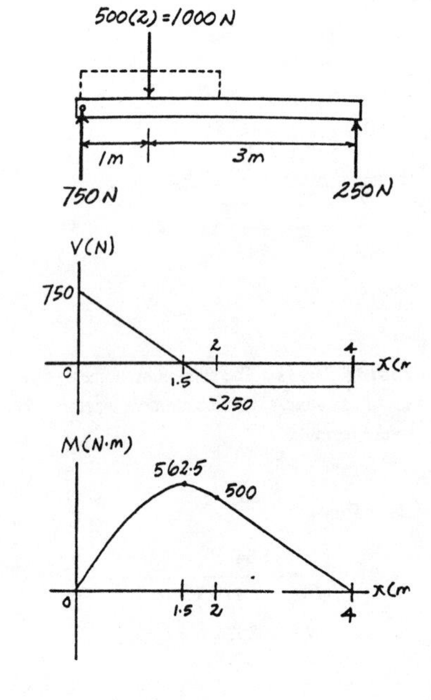

Allowable Bending Stress : The maximum moment is $M_{max} = 562.5$ N · m as indicated on the moment diagram. Applying the flexure formula

$$\sigma_{max} = \sigma_{allow} = \frac{M_{max}\,c}{I}$$

$$10(10^6) = \frac{562.5(0.75b)}{\frac{1}{12}(b)(1.5b)^3}$$

$b = 0.05313$ m $= 53.1$ mm **Ans**

■*6–92. The beam is made of fiberglass for which the allowable bending stress is $\sigma_{allow} = 13$ MPa. Determine the required dimension b needed to support the applied load of $w = 2.5$ kN/m.

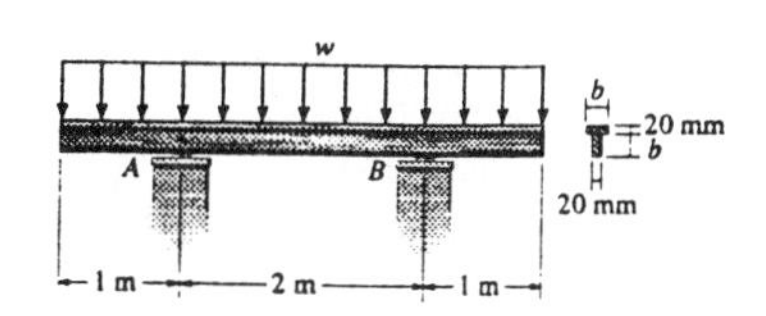

Section Properties :

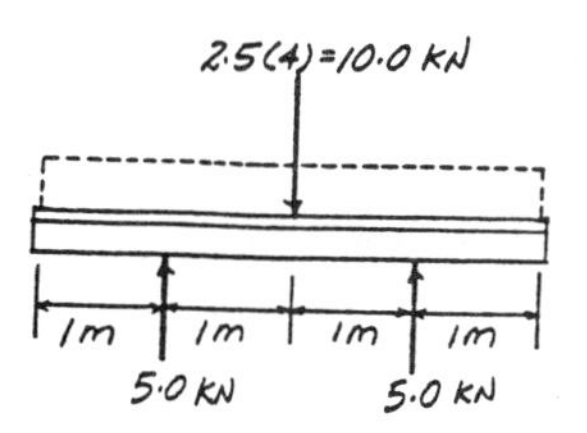

$$\bar{y} = \frac{\Sigma \tilde{y}A}{\Sigma A}$$

$$= \frac{(0.01)(b)(0.02) + \left(\frac{b}{2} + 0.02\right)(0.02)(b)}{b(0.02) + 0.02(b)}$$

$$= 0.015 + 0.25b$$

$$I = \frac{1}{12}(b)\left(0.02^3\right) + b(0.02)\left[(0.015 + 0.25b) - 0.01\right]^2$$

$$+ \frac{1}{12}(0.02)\left(b^3\right) + 0.02(b)\left[(0.5b + 0.02) - (0.015 + 0.25b)\right]^2$$

$$= 0.0041667b^3 + 0.0001b^2 + 1.66667\left(10^{-6}\right)b$$

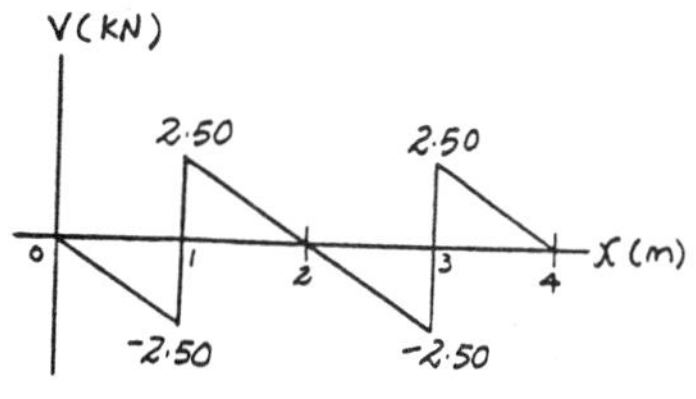

Allowable Bending Stress : The maximum moment is $M_{max} = 1.25$ kN · m as indicated on moment diagram. Applying the the flexure formula

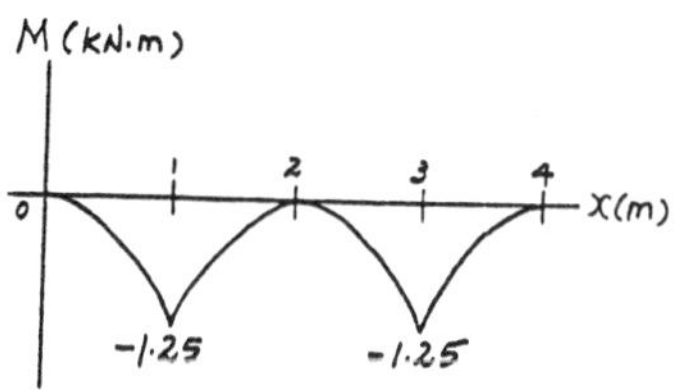

$$\sigma_{max} = \sigma_{allow} = \frac{M_{max}\, c}{I}$$

$$13\left(10^6\right) = \frac{1.25(10^3)\left[(b + 0.02) - (0.015 + 0.25b)\right]}{0.0041667b^3 + 0.0001b^2 + 1.66667(10^{-6})b}$$

$$54166.658b^3 + 1300b^2 - 915.833b - 6.25 = 0$$

Solving the cubic equation by trial and error

$$b = 0.12215 \text{ m} = 122 \text{ mm} \qquad \textbf{Ans}$$

6–93. Determine the maximum allowable uniform load w that can be supported by the fiberglass beam if $b = 125$ mm and the allowable bending stress for the material is $\sigma_{allow} = 13$ MPa.

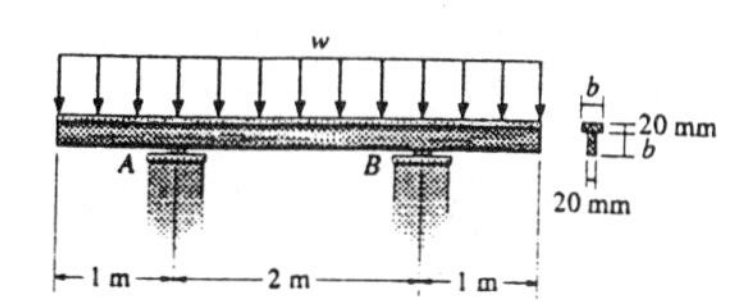

Section Properties :

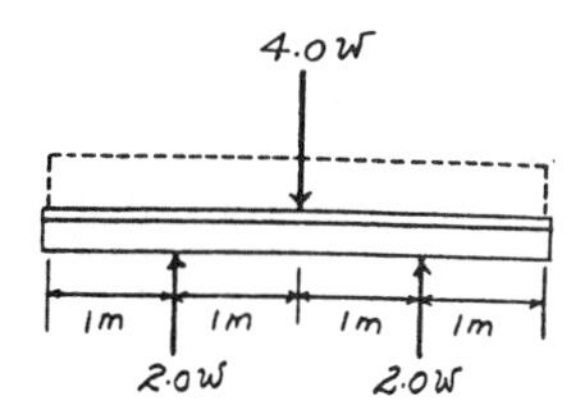

$$\bar{y} = \frac{\Sigma \tilde{y}A}{\Sigma A}$$

$$= \frac{(0.01)(0.125)(0.02) + (0.0825)(0.02)(0.125)}{0.125(0.02) + 0.02(0.125)}$$

$$= 0.04625 \text{ m}$$

$$I = \frac{1}{12}(0.125)\left(0.02^3\right) + 0.125(0.02)(0.04625 - 0.01)^2$$

$$+ \frac{1}{12}(0.02)\left(0.125^3\right) + 0.02(0.125)(0.0825 - 0.04625)^2$$

$$= 9.90885\left(10^{-6}\right) \text{ m}^4$$

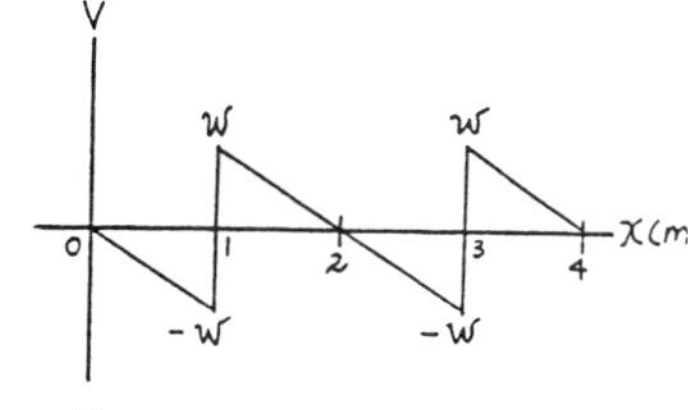

Allowable Bending Stress : The maximum moment is $M_{max} = \frac{w}{2}$ as indicated on moment diagram. Applying the flexure formula

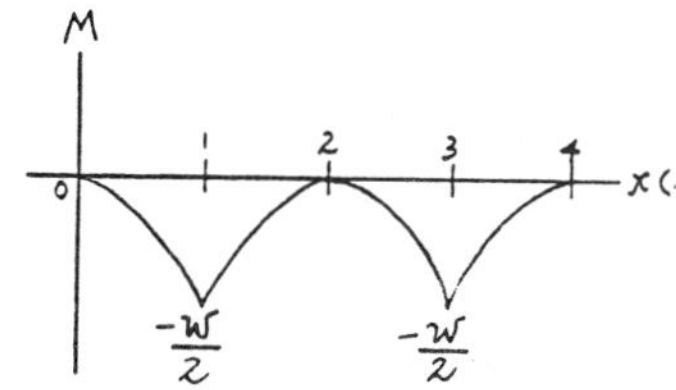

$$\sigma_{max} = \sigma_{allow} = \frac{M_{max}\, c}{I}$$

$$13\left(10^6\right) = \frac{\frac{w}{2}(0.145 - 0.04625)}{9.90885(10^{-6})}$$

$$w = 2609 \text{ N/m} = 2.61 \text{ kN/m} \qquad \textbf{Ans}$$

6–94. The wood beam is subjected to the uniform load of $w = 200$ lb/ft. If the allowable bending stress for the material is $\sigma_{\text{allow}} = 1.40$ ksi, determine the required dimension b of its cross section. Assume the support at A is a pin and B is a roller.

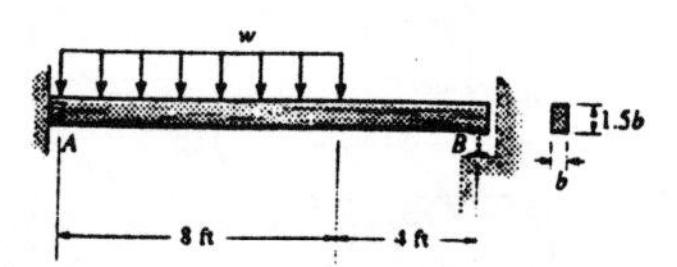

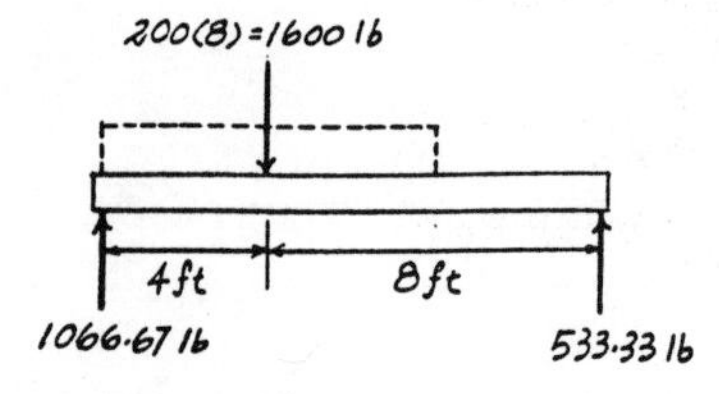

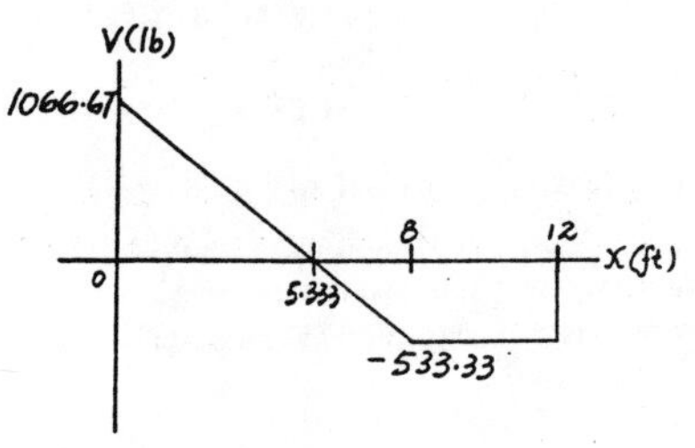

Allowable Bending Stress : The maximum moment is $M_{\max} = 5688.89$ lb · ft as indicated on moment diagram. Applying the flexure formula

$$\sigma_{\max} = \sigma_{\text{allow}} = \frac{M_{\max}c}{I}$$

$$1.40(10^3) = \frac{2844.44(12)(0.75b)}{\frac{1}{12}(b)(1.5b)^3}$$

$$b = 4.02 \text{ in.} \qquad \textbf{Ans}$$

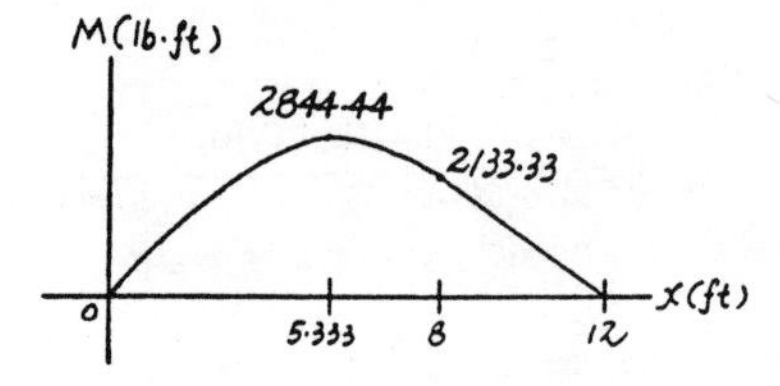

6–95. Determine the maximum allowable uniform load w that can be supported by the wood beam if $b = 4$ in. and the allowable bending stress for the material is $\sigma_{\text{allow}} = 1.40$ ksi. Assume the support at A is a pin and B is a roller.

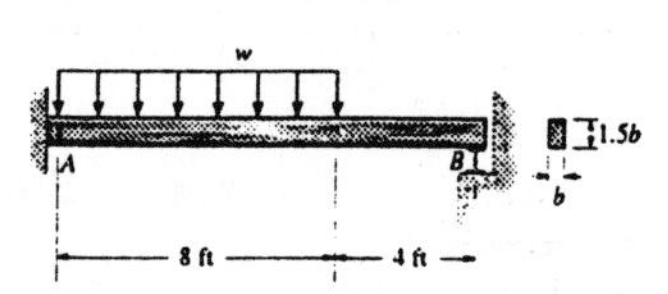

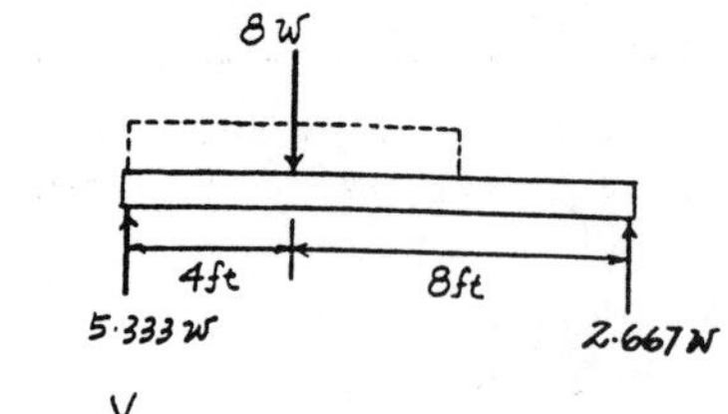

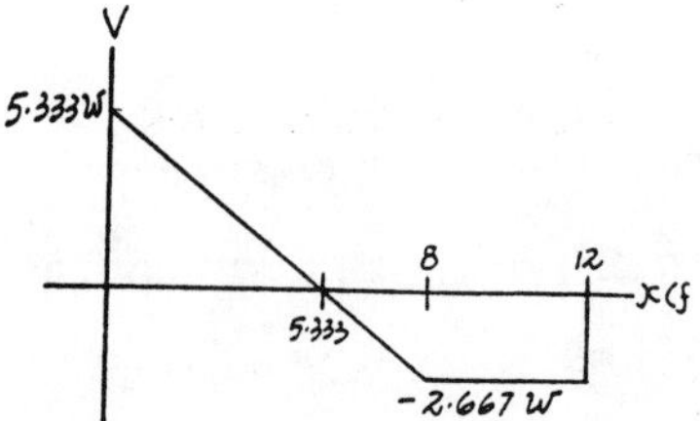

Allowable Bending Stress : The maximum moment is $M_{\max} = 14.222w$ as indicated on moment diagram. Applying the flexure formula

$$\sigma_{\max} = \sigma_{\text{allow}} = \frac{M_{\max}c}{I}$$

$$1.40(10^3) = \frac{14.222w(12)(3)}{\frac{1}{12}(4)(6^3)}$$

$$w = 197 \text{ lb/ft} \qquad \textbf{Ans}$$

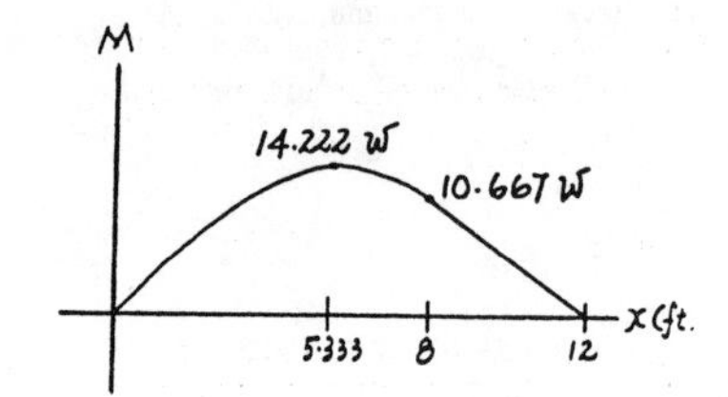

***6–96.** The box beam is subjected to a bending moment of $M = 15$ kip · ft directed as shown. Determine the maximum bending stress in the beam and the orientation of the neutral axis.

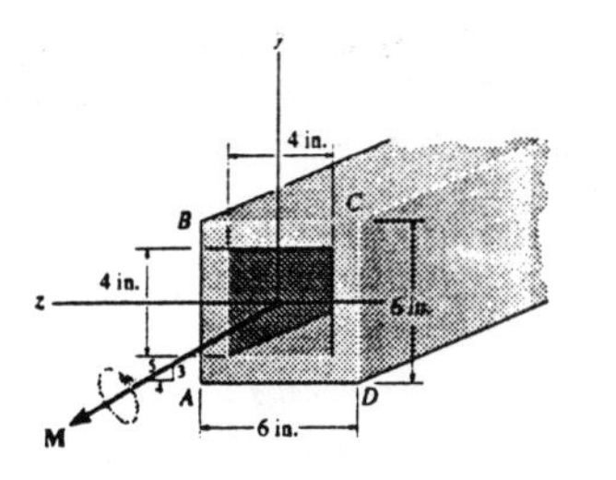

Internal Moment Components :

$$M_y = -\frac{3}{5}(15) = -9.00 \text{ kip} \cdot \text{ft}$$

$$M_z = \frac{4}{5}(15) = 12.0 \text{ kip} \cdot \text{ft}$$

Section Property :

$$I_y = I_z = \frac{1}{12}(6)\left(6^3\right) - \frac{1}{12}(4)\left(4^3\right) = 86.67 \text{ in}^4$$

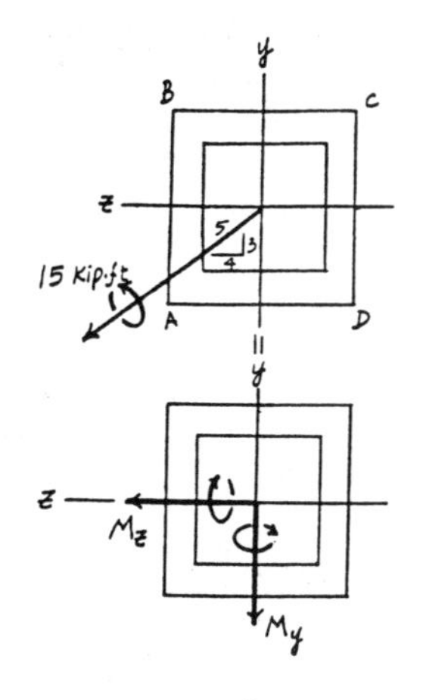

Maximum Bending Stress : By inspection, maximum bending stress occurs at B and D. Applying the flexure formula for biaxial bending

$$\sigma = -\frac{M_z y}{I_z} + \frac{M_y z}{I_y}$$

$$\sigma_D = -\frac{12.0(12)(-3)}{86.67} + \frac{-9.00(12)(-3)}{86.67}$$
$$= 8.72 \text{ ksi (T)} \quad \textbf{(max)} \qquad \textbf{Ans}$$

$$\sigma_B = -\frac{12.0(12)(3)}{86.67} + \frac{-9.00(12)(3)}{86.67}$$
$$= -8.723 \text{ ksi} = 8.72 \text{ ksi(C)} \quad \textbf{(max)} \qquad \textbf{Ans}$$

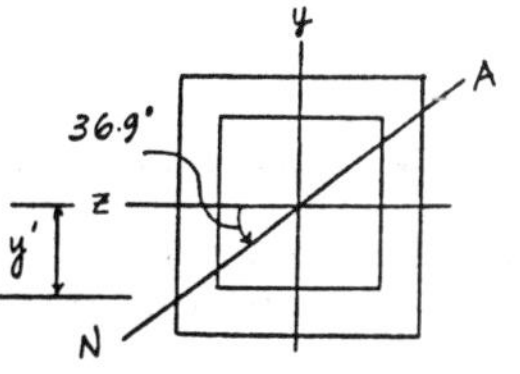

Orientation of Neutral Axis :

$$\tan\alpha = \frac{I_z}{I_y}\tan\theta$$

$$\tan\alpha = (1)(-0.75) \qquad \alpha = -36.9° \qquad \textbf{Ans}$$

$$y' = 3\tan\alpha = 2.25 \text{ in.}$$

6–97. Determine the maximum magnitude of the bending moment **M** so that the bending stress in the member does not exceed 15 ksi.

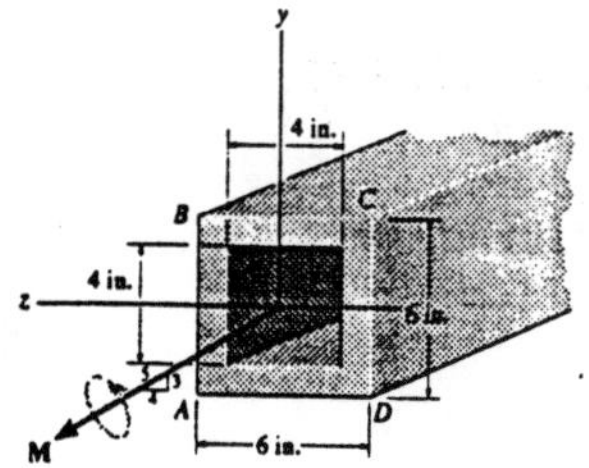

Internal Moment Components :

$$M_y = -\frac{3}{5}(M) = -0.600M \qquad M_z = \frac{4}{5}(M) = 0.800M$$

Section Property :

$$I_y = I_z = \frac{1}{12}(6)\left(6^3\right) - \frac{1}{12}(4)\left(4^3\right) = 86.67 \text{ in}^4$$

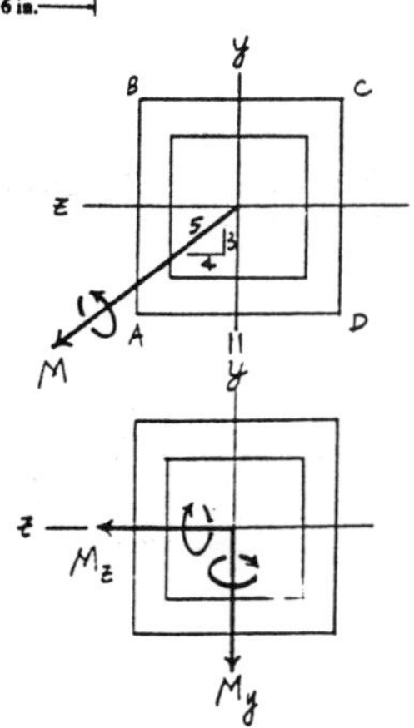

Allowable Bending Stress : By Inspection, maximum bending stress occurs at points B and D. Applying the flexure formula for biaxial bending at either points B or D

$$\sigma_D = \sigma_{\text{allow}} = -\frac{M_z y}{I_z} + \frac{M_y z}{I_y}$$

$$15 = -\frac{0.800M(12)(-3)}{86.67} + \frac{-0.600M(12)(-3)}{86.67}$$

$$M = 25.8 \text{ kip} \cdot \text{ft} \qquad \textbf{Ans}$$

6–98. Determine the maximum magnitude of the bending moment **M** so that the bending stress in the member does not exceed 24 ksi. The location $\bar{y}$ of the centroid C must be determined.

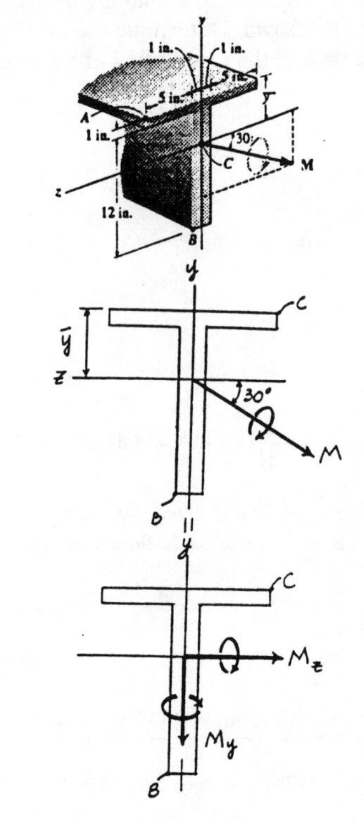

Internal Moment Components :

$$M_y = -M\sin 30° = -0.500M$$
$$M_z = -M\cos 30° = -0.8660M$$

$$\bar{y} = \frac{\Sigma \tilde{y}A}{\Sigma A} = \frac{0.5(12)(1) + 7(2)(12)}{12(1) + 2(12)}$$
$$= 4.833 \text{ in.} = 4.83 \text{ in.} \qquad \textbf{Ans}$$

$$I_z = \frac{1}{12}(12)(1^3) + 12(1)(4.833 - 0.5)^2 + \frac{1}{12}(2)(12^3) + 2(12)(7 - 4.833)^2$$
$$= 627.0 \text{ in}^4$$

$$I_y = \frac{1}{12}(1)(12^3) + \frac{1}{12}(12)(2^3) = 152.0 \text{ in}^4$$

Allowable Bending Stress : By Inspection, maximum Bending Stress can occurs either at points B and C. Applying the flexure formula for biaxial bending at points B and C

$$\sigma_B = \sigma_{\text{allow}} = -\frac{M_z y}{I_z} + \frac{M_y z}{I_y}$$
$$-24 = -\frac{-0.8660M(12)(-8.167)}{627.0} + \frac{-0.500M(12)(1)}{152.0}$$
$$M = 137.3 \text{ kip}\cdot\text{ft}$$

$$\sigma_C = \sigma_{\text{allow}} = -\frac{M_z y}{I_z} + \frac{M_y z}{I_y}$$
$$24 = -\frac{-0.8660M(12)(4.833)}{627.0} + \frac{-0.500M(12)(-6)}{152.0}$$
$$M = 75.7 \text{ kip}\cdot\text{ft} \ (\textit{Controls !}) \qquad \textbf{Ans}$$

6–99. The moment acting on the cross section of the T-beam has a magnitude of $M = 15$ kip · ft and is directed as shown. Determine the bending stress at points A and B. The location $\bar{y}$ of the centroid C must be determined.

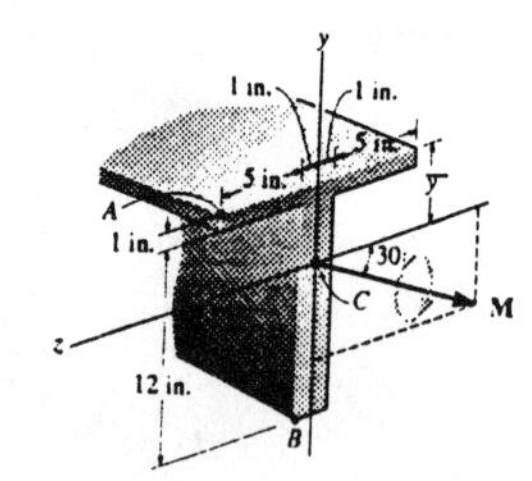

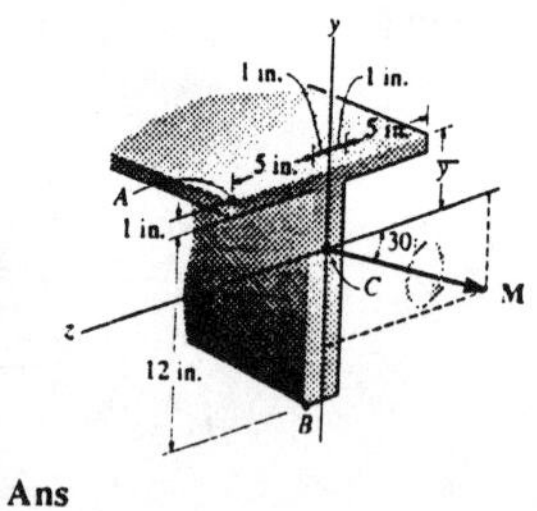

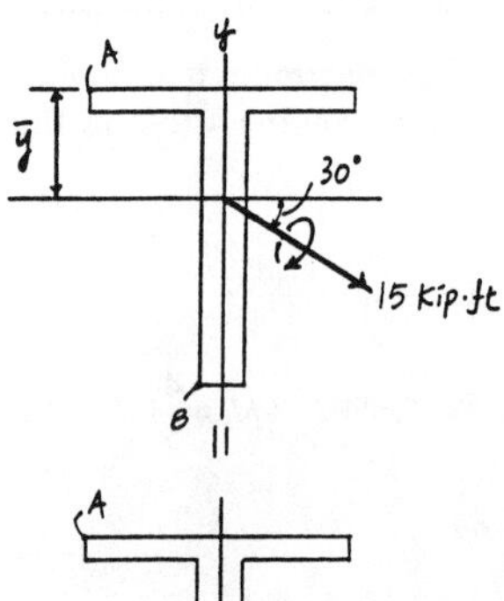

Internal Moment Components :

$$M_y = -15\sin 30° = -7.50 \text{ kip}\cdot\text{ft}$$
$$M_z = -15\cos 30° = -12.990 \text{ kip}\cdot\text{ft}$$

$$\bar{y} = \frac{\Sigma \tilde{y}A}{\Sigma A} = \frac{0.5(12)(1) + 7(2)(12)}{12(1) + 2(12)}$$
$$= 4.833 \text{ in.} = 4.83 \text{ in.} \qquad \textbf{Ans}$$

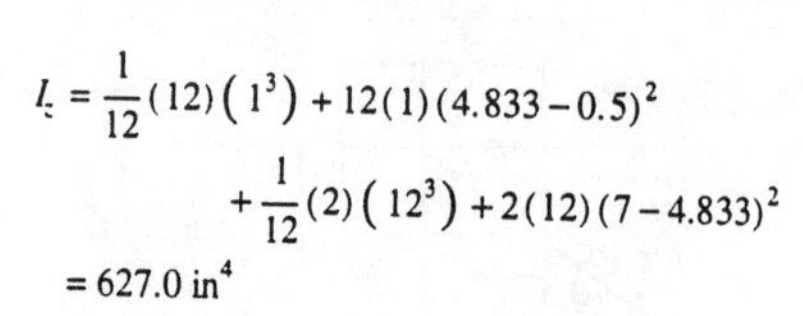

$$I_z = \frac{1}{12}(12)(1^3) + 12(1)(4.833 - 0.5)^2 + \frac{1}{12}(2)(12^3) + 2(12)(7 - 4.833)^2$$
$$= 627.0 \text{ in}^4$$

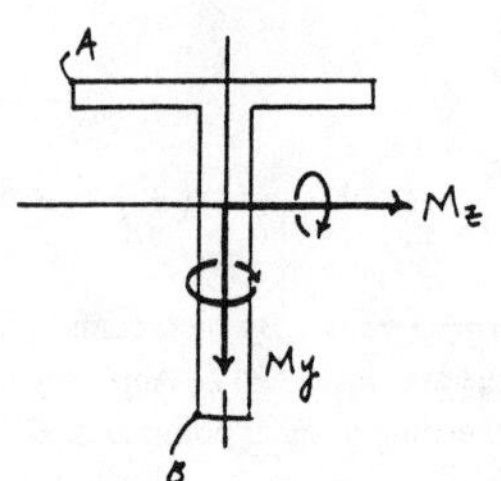

$$I_y = \frac{1}{12}(1)(12^3) + \frac{1}{12}(12)(2^3) = 152.0 \text{ in}^4$$

Bending Stress : Applying the flexure formula for biaxial bending at points A and B

$$\sigma = -\frac{M_z y}{I_z} + \frac{M_y z}{I_y}$$

$$\sigma_A = -\frac{-12.990(12)(4.833)}{627.0} + \frac{-7.50(12)(6)}{152.0}$$
$$= -2.35 \text{ ksi} = 2.35 \text{ksi (C)} \qquad \textbf{Ans}$$

$$\sigma_B = -\frac{-12.990(12)(-8.167)}{627.0} + \frac{-7.50(12)(1)}{152.0}$$
$$= -2.62 \text{ ksi} = 2.62 \text{ksi (C)} \qquad \textbf{Ans}$$

***6–100.** If the resultant internal moment acting on the cross section of the aluminum strut has a magnitude of $M = 520\ \text{N}\cdot\text{m}$ and is directed as shown, determine the bending stress at points A and B. The location $\bar{y}$ of the centroid C of the strut's cross-sectional area must be determined. Also, specify the orientation of the neutral axis.

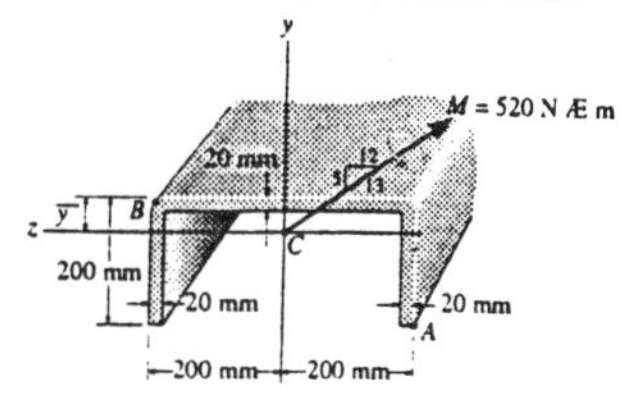

Internal Moment Components :

$$M_z = -\frac{12}{13}(520) = -480\ \text{N}\cdot\text{m} \qquad M_y = \frac{5}{13}(520) = 200\ \text{N}\cdot\text{m}$$

Section Properties :

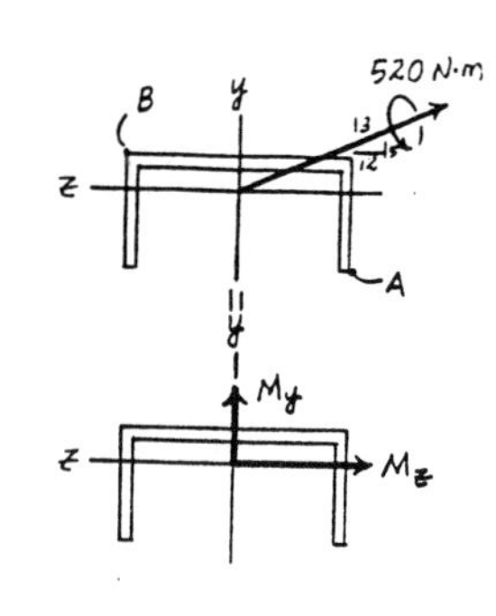

$$\bar{y} = \frac{\Sigma \tilde{y} A}{\Sigma A} = \frac{0.01(0.4)(0.02) + 2[(0.110)(0.18)(0.02)]}{0.4(0.02) + 2(0.18)(0.02)}$$
$$= 0.057368\ \text{m} = 57.4\ \text{mm} \qquad \textbf{Ans}$$

$$I_z = \frac{1}{12}(0.4)\left(0.02^3\right) + (0.4)(0.02)(0.057368 - 0.01)^2$$
$$+ \frac{1}{12}(0.04)\left(0.18^3\right) + 0.04(0.18)(0.110 - 0.057368)^2$$
$$= 57.6014\left(10^{-6}\right)\ \text{m}^4$$

$$I_y = \frac{1}{12}(0.2)\left(0.4^3\right) - \frac{1}{12}(0.18)\left(0.36^3\right) = 0.366827\left(10^{-3}\right)\ \text{m}^4$$

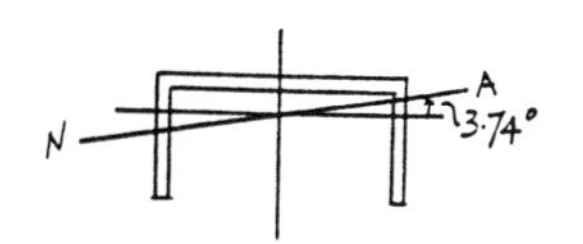

Maximum Bending Stress : Applying the flexure formula for biaxial bending at points A and B

$$\sigma = -\frac{M_z y}{I_z} + \frac{M_y z}{I_y}$$

$$\sigma_A = -\frac{-480(-0.142632)}{57.6014(10^{-6})} + \frac{200(-0.2)}{0.366827(10^{-3})}$$
$$= -1.298\ \text{MPa} = 1.30\ \text{MPa (C)} \qquad \textbf{Ans}$$

$$\sigma_B = -\frac{-480(0.057368)}{57.6014(10^{-6})} + \frac{200(0.2)}{0.366827(10^{-3})}$$
$$= 0.587\ \text{MPa (T)} \qquad \textbf{Ans}$$

Orientation of Neutral Axis :

$$\tan\alpha = \frac{I_z}{I_y}\tan\theta$$

$$\tan\alpha = \frac{57.6014(10^{-6})}{0.366827(10^{-3})}\tan(-22.62^\circ)$$
$$\alpha = -3.74^\circ \qquad \textbf{Ans}$$

6–101. The resultant internal moment acting on the cross section of the aluminum strut has a magnitude of $M = 520\ \text{N}\cdot\text{m}$ and is directed as shown. Determine the maximum bending stress in the strut. The location $\bar{y}$ of the centroid C of the strut's cross-sectional area must be determined. Also, specify the orientation of the neutral axis.

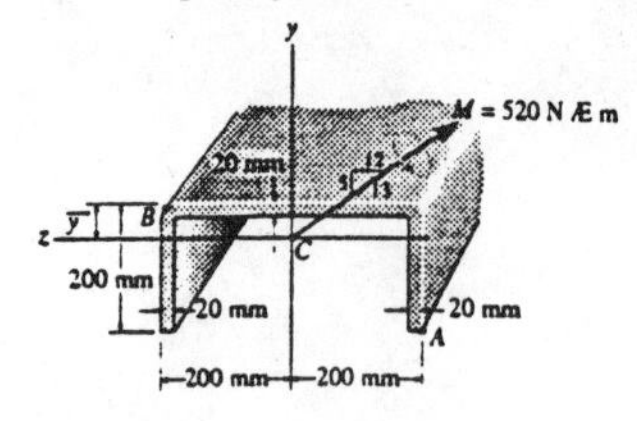

Internal Moment Components :

$$M_z = -\frac{12}{13}(520) = -480\ \text{N}\cdot\text{m} \qquad M_y = \frac{5}{13}(520) = 200\ \text{N}\cdot\text{m}$$

Section Properties :

$$\bar{y} = \frac{\Sigma \tilde{y} A}{\Sigma A} = \frac{0.01(0.4)(0.02) + 2[(0.110)(0.18)(0.02)]}{0.4(0.02) + 2(0.18)(0.02)}$$
$$= 0.057368\ \text{m} = 57.4\ \text{mm} \qquad \textbf{Ans}$$

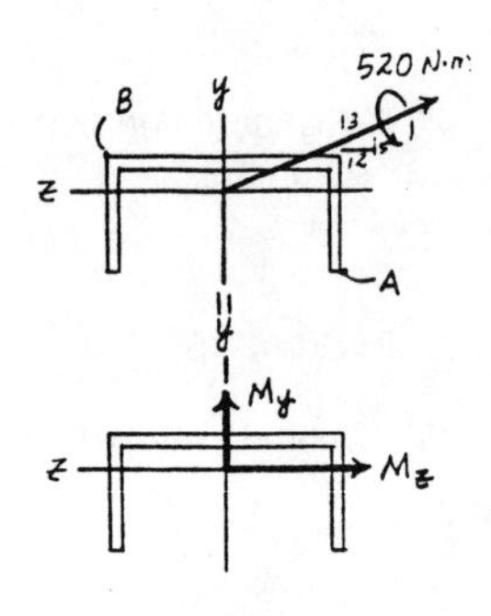

$$I_z = \frac{1}{12}(0.4)\left(0.02^3\right) + (0.4)(0.02)(0.057368 - 0.01)^2$$
$$+ \frac{1}{12}(0.04)\left(0.18^3\right) + 0.04(0.18)(0.110 - 0.057368)^2$$
$$= 57.6014\left(10^{-6}\right)\ \text{m}^4$$

$$I_y = \frac{1}{12}(0.2)\left(0.4^3\right) - \frac{1}{12}(0.18)\left(0.36^3\right) = 0.366827\left(10^{-3}\right)\ \text{m}^4$$

Maximum Bending Stress : By inspection, the maximum bending stress can occur at either point A or B. Applying the flexure formula for biaxial bending at points A and B

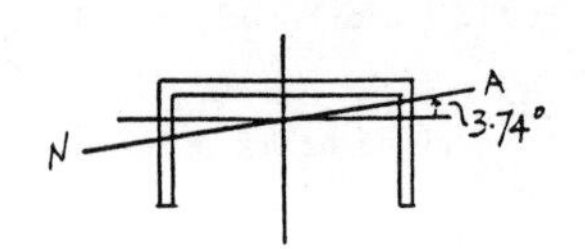

$$\sigma = -\frac{M_z y}{I_z} + \frac{M_y z}{I_y}$$

$$\sigma_A = -\frac{-480(-0.142632)}{57.6014(10^{-6})} + \frac{200(-0.2)}{0.366827(10^{-3})}$$
$$= -1.298\ \text{MPa} = 1.30\ \text{MPa (C)}\ \textbf{(Max)} \qquad \textbf{Ans}$$

$$\sigma_B = -\frac{-480(0.057368)}{57.6014(10^{-6})} + \frac{200(0.2)}{0.366827(10^{-3})}$$
$$= 0.587\ \text{MPa (T)}$$

Orientation of Neutral Axis :

$$\tan\alpha = \frac{I_z}{I_y}\tan\theta$$

$$\tan\alpha = \frac{57.6014(10^{-6})}{0.366827(10^{-3})}\tan(-22.62°)$$
$$\alpha = -3.74° \qquad \textbf{Ans}$$

6–102. The cantilevered wide-flange steel beam is subjected to the concentrated force **P** at its end. Determine the largest magnitude of this force so that the bending stress developed at section A does not exceed $\sigma_{\text{allow}} = 180$ MPa.

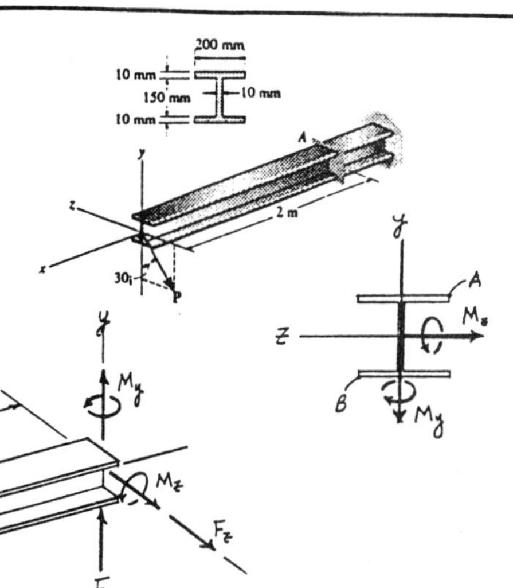

Internal Moment Components : Using method of section

$\Sigma M_z = 0; \quad M_z + P\cos 30°(2) = 0 \quad M_z = -1.732P$

$\Sigma M_y = 0; \quad M_y + P\sin 30°(2) = 0; \quad M_y = -1.00P$

Section Properties :

$$I_z = \frac{1}{12}(0.2)\left(0.17^3\right) - \frac{1}{12}(0.19)\left(0.15^3\right) = 28.44583(10^{-6})\ \text{m}^4$$

$$I_y = 2\left[\frac{1}{12}(0.01)\left(0.2^3\right)\right] + \frac{1}{12}(0.15)\left(0.01^3\right) = 13.34583(10^{-6})\ \text{m}^4$$

Allowable Bending Stress : By inspection, maximum bending stress occurs at points A and B. Applying the flexure formula for biaxial bending at point A,

$$\sigma_A = \sigma_{\text{allow}} = -\frac{M_z y}{I_z} + \frac{M_y z}{I_y}$$

$$180\left(10^6\right) = -\frac{(-1.732P)(0.085)}{28.44583(10^{-6})} + \frac{-1.00P(-0.1)}{13.34583(10^{-6})}$$

$P = 14208$ N $= 14.2$ kN **Ans**

6–103. The cantilevered wide-flange steel beam is subjected to the concentrated force of $P = 600$ N at its end. Determine the maximum bending stress developed in the beam at section A.

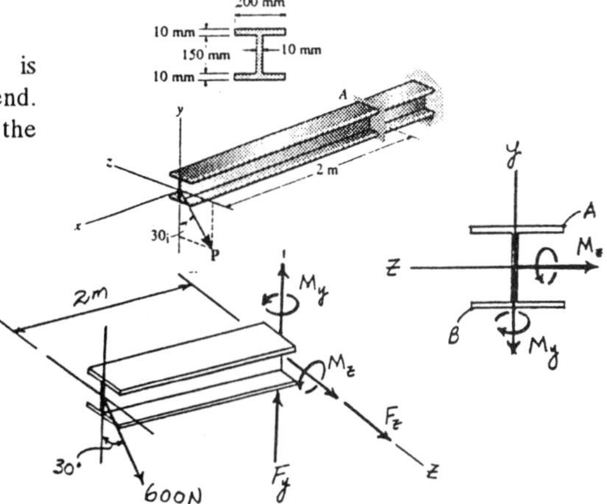

Internal Moment Components : Using method of sections

$\Sigma M_z = 0; \quad M_z + 600\cos 30°(2) = 0 \quad M_z = -1039.23\ \text{N}\cdot\text{m}$

$\Sigma M_y = 0; \quad M_y + 600\sin 30°(2) = 0; \quad M_y = -600.0\ \text{N}\cdot\text{m}$

Section Properties :

$$I_z = \frac{1}{12}(0.2)\left(0.17^3\right) - \frac{1}{12}(0.19)\left(0.15^3\right) = 28.44583(10^{-6})\ \text{m}^4$$

$$I_y = 2\left[\frac{1}{12}(0.01)\left(0.2^3\right)\right] + \frac{1}{12}(0.15)\left(0.01^3\right) = 13.34583(10^{-6})\ \text{m}^4$$

Maximum Bending Stress : By inspection, maximum bending stress occurs at A and B. Applying the flexure formula for biaxial bending at points A and B

$$\sigma = -\frac{M_z y}{I_z} + \frac{M_y z}{I_y}$$

$$\sigma_A = -\frac{-1039.32(0.085)}{28.44583(10^{-6})} + \frac{-600.0(-0.1)}{13.34583(10^{-6})}$$

$= 7.60$ MPa **(T)** **(Max)** **Ans**

$$\sigma_B = -\frac{(-1039.32)(-0.085)}{28.44583(10^{-6})} + \frac{-600.0(0.1)}{13.34583(10^{-6})}$$

$= -7.60$ MPa $= 7.60$ MPa **(C)** **(Max)** **Ans**

***6–104.** The rafter on a roof is subjected to the vertical loadings shown. If it can be considered simply supported at A and B, determine its required dimension b and the position of the neutral axis at the critical section. The allowable bending stress is $\sigma_{allow} = 15$ MPa.

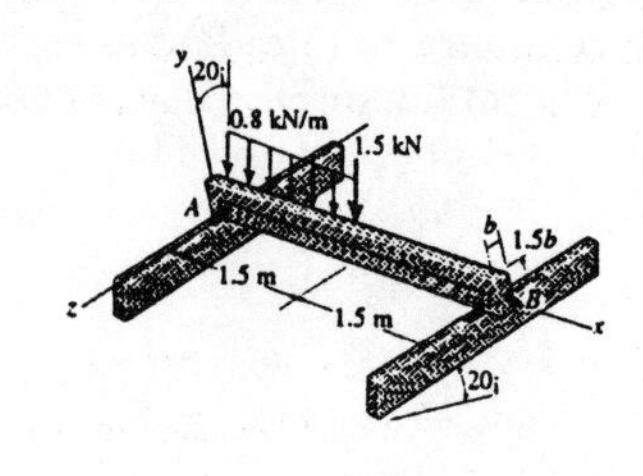

Internal Moment Components : The rafter is subjected to two moment components M_y and M_z. The maximum moment components are shown on their respective moment diagram.

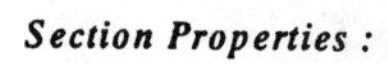

Section Properties :

$$I_z = \frac{1}{12}(b)(1.5b)^3 = 0.28125b^4$$

$$I_y = \frac{1}{12}(1.5b)(b^3) = 0.125b^4$$

Allowable Bending Stress : By inspection, maximum bending stress occurs at points A and B. Applying the flexure formula for biaxial bending at either point A or B, we have

$$\sigma_{allow} = -\frac{M_z y}{I_z} + \frac{M_y z}{I_y}$$

$$15(10^6) = -\frac{1.480(10^3)(-0.75b)}{0.28125b^4} + \frac{0.5387(10^3)(0.5b)}{0.125b^4}$$

$$b = 0.07409 \text{ m} = 74.1 \text{ mm} \qquad \textbf{Ans}$$

Orientation of Neutral Axis :

$$\tan\alpha = \frac{I_z}{I_y}\tan\theta, \qquad \text{Where } \theta = \tan^{-1}\frac{M_y}{M_z} = \tan^{-1}\frac{0.5387}{1.480} = 20.0^\circ$$

$$\tan\alpha = \frac{0.28125b^4}{0.125b^4}\tan 20.0^\circ$$

$$\alpha = 39.3^\circ \qquad \textbf{Ans}$$

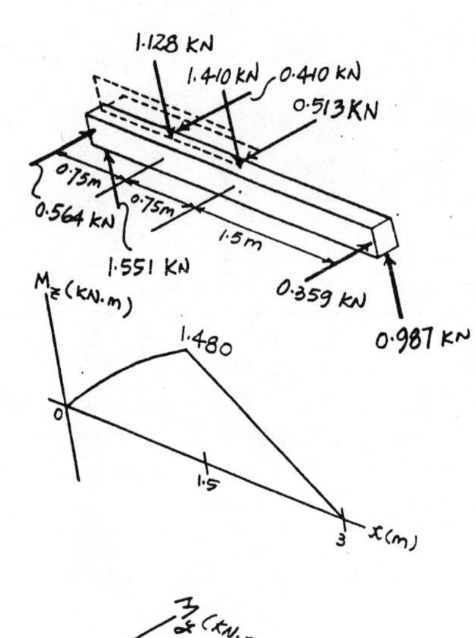

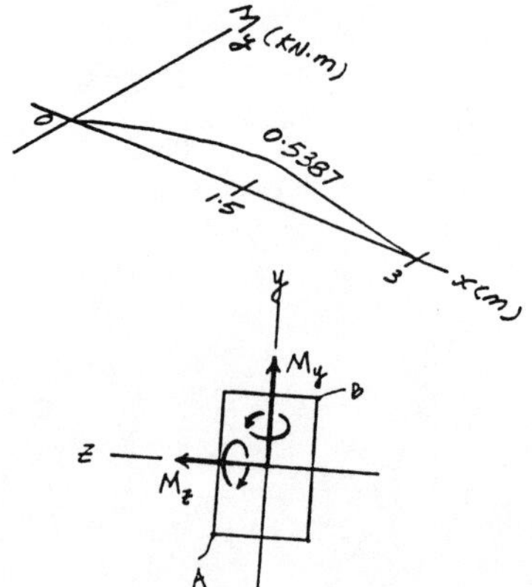

6–105. The rafter on a roof is subjected to the vertical loadings shown. If it can be considered simply suppported at A and B, determine the absolute maximum bending stress developed in the rafter and the orientation of the neutral axis if $b = 75$ mm.

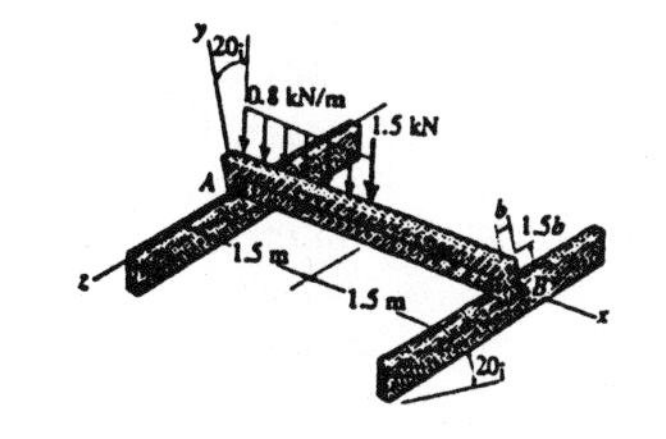

Internal Moment Components : The rafter is subjected to two moment components M_y and M_z. The maximum moment components are shown on their respective moment diagram.

Section Properties :

$$I_z = \frac{1}{12}(0.075)\left(0.1125^3\right) = 8.898926\left(10^{-6}\right)\ \text{m}^4$$

$$I_y = \frac{1}{12}(0.1125)\left(0.075^3\right) = 3.955078\left(10^{-6}\right)\ \text{m}^4$$

Absolute MaximumBending Stress : By inspection, the maximum bending stress occurs at points A and B. Applying the flexure formula for biaxial bending at either point A or B, we have

$$\sigma_{\max} = -\frac{M_z y}{I_z} + \frac{M_y z}{I_y}$$

$$= -\frac{1.480(10^3)(-0.05625)}{8.898926(10^{-6})} + \frac{0.5387(10^3)(0.0375)}{3.955078(10^{-6})}$$

$= 14.5$ MPa **Ans**

Orientation of Neutral Axis :

$$\tan\alpha = \frac{I_z}{I_y}\tan\theta \quad \text{Where } \theta = \tan^{-1}\frac{M_y}{M_z} = \tan^{-1}\frac{0.5387}{1.480} = 20.0°$$

$$\tan\alpha = \frac{8.898926(10^{-6})}{3.955078(10^{-6})}\tan 20.0°$$

$\alpha = 39.3°$ **Ans**

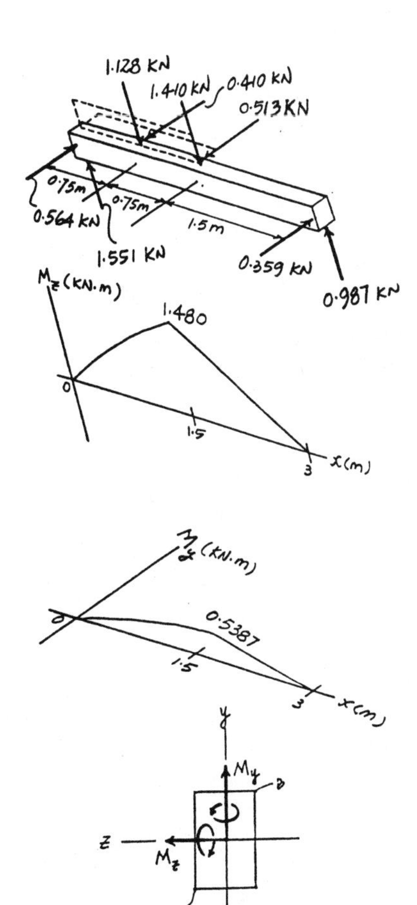

6–106. The 30-mm-diameter shaft is subjected to the vertical and horizontal loadings of two pulleys as shown. It is supported on two journal bearings at A and B which offer no resistance to axial loading. Furthermore, the coupling to the motor at C can be assumed not to offer any support to the shaft. Determine the maximum bending stress developed in the shaft.

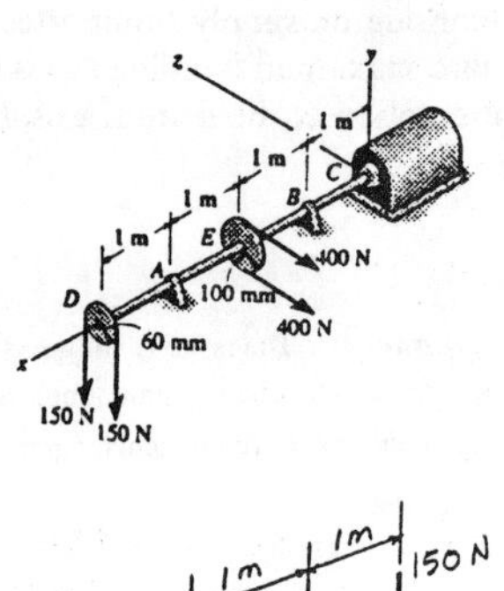

Support Reactions : As shown on FBD.

Internal Moment Components : The shaft is subjected to two bending moment components M_y and M_z. The moment diagram for each component is drawn.

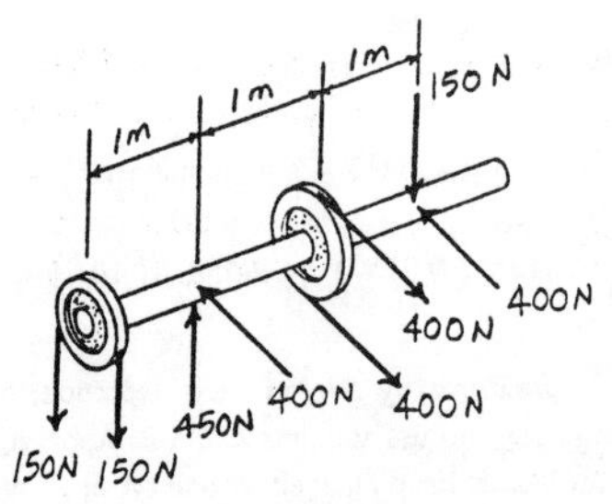

Maximum Bending Stress : Since all the axes through the circle's center for circular shaft are principal axis, then the resultant moment $M = \sqrt{M_y^2 + M_z^2}$ can be used to determine the maximum bending stress. The maximum resultant moment occures at E $M_{max} = \sqrt{400^2 + 150^2} = 427.2$ N · m.

Applying the flexure formula

$$\sigma_{max} = \frac{M_{max}\,c}{I}$$

$$= \frac{427.2(0.015)}{\frac{\pi}{4}(0.015^4)}$$

$$= 161 \text{ MPa} \qquad \textbf{Ans}$$

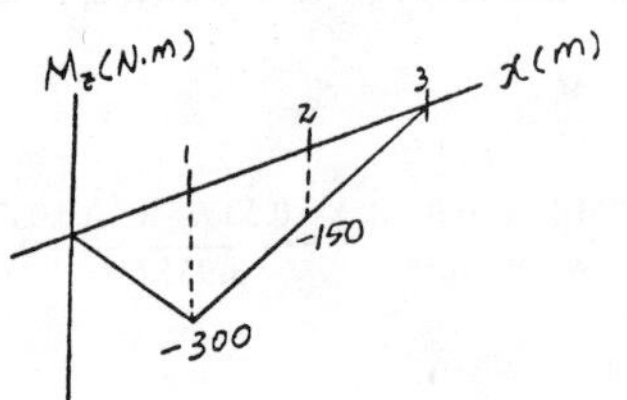

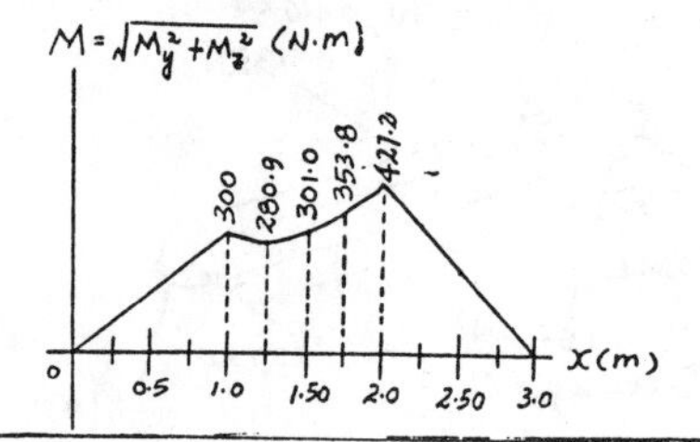

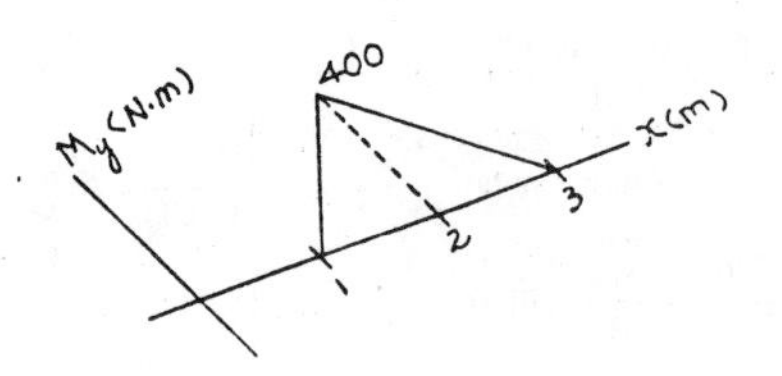

6–107. The shaft is subjected to the vertical and horizontal loadings of two pulleys D and E as shown. It is supported on two journal bearings at A and B which offer no resistance to axial loading. Furthermore, the coupling to the motor at C can be assumed not to offer any support to the shaft. Determine the required diameter d of the shaft if the allowable bending stress for the material is $\sigma_{allow} = 180$ MPa.

Support Reactions : As shown on FBD.

Internal Moment Components : The shaft is subjected to two bending moment components M_y and M_z. The moment diagram for each component is drawn.

Allowable Bending Stress : Since all the axes through the circle's center for a circular shaft are principal axes, then the resultant moment $M = \sqrt{M_y^2 + M_z^2}$ can be used for the design. The maximum resultant moment is $M_{max} = \sqrt{400^2 + 150^2} = 427.2$ N · m.

Applying the flexure formula

$$\sigma_{max} = \sigma_{allow} = \frac{M_{max}\,c}{I}$$

$$180(10^6) = \frac{427.2\left(\frac{d}{2}\right)}{\frac{\pi}{4}\left(\frac{d}{2}\right)^4}$$

$$d = 0.02891 \text{ m} = 28.9 \text{ mm} \qquad \textbf{Ans}$$

***6–108.** The 65-mm-diameter steel shaft is subjected to the two loads that act in the directions shown. If the journal bearings at A and B do not exert an axial force on the shaft, determine the absolute maximum bending stress developed in the shaft.

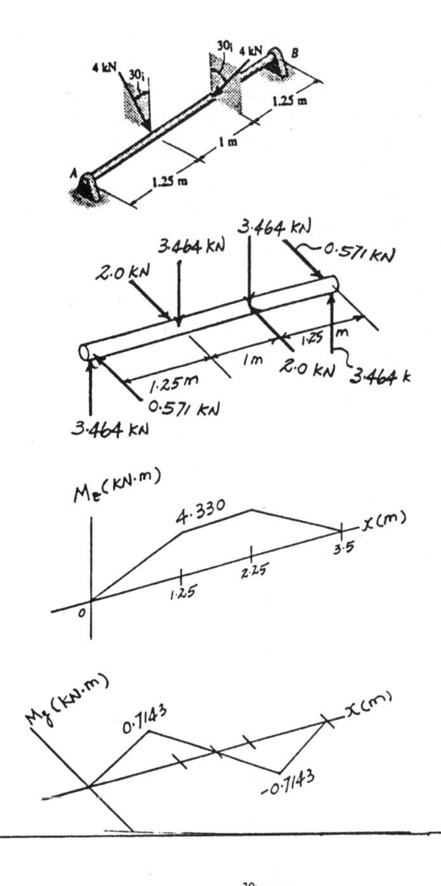

Support Reactions : As shown on FBD.

Internal Moment Components : The shaft is subjected to two bending moment components M_y and M_z. The moment diagram for each component is drawn.

Maximum Bending Stress : Since all the axes through the circle's center for a circular shaft are principal axes, then the resultant moment $M = \sqrt{M_y^2 + M_z^2}$ can be used to determine the maximum bending stress. The maximum resultant moment is $M_{max} = \sqrt{4.330^2 + 0.7143^2} = 4.389$ kN · m. Applying the flexure formula.

$$\sigma_{max} = \frac{M_{max}\, c}{I}$$

$$= \frac{4.389(10^3)(0.0325)}{\frac{\pi}{4}(0.0325^4)}$$

$$= 163 \text{ MPa} \qquad \textbf{Ans}$$

6–109. The steel shaft is subjected to the two loads that act in the directions shown. If the journal bearings at A and B do not exert an axial force on the shaft, determine the required diameter of the shaft if the allowable bending stress is $\sigma_{allow} = 180$ MPa.

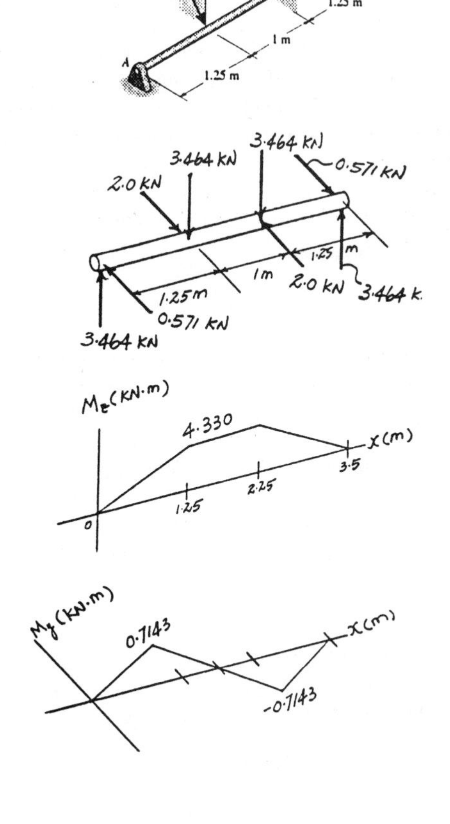

Support Reactions : As shown on FBD.

Internal Moment Components : The shaft is subjected to two bending moment components M_y and M_z. The moment diagram for each component is drawn.

Allowable Bending Stress : Since all the axes through the circle's center for a circular shaft are principal axes, then the resultant moment $M = \sqrt{M_y^2 + M_z^2}$ can be used for the design. The maximum resultant moment is $M_{max} = \sqrt{4.330^2 + 0.7143^2}$ = 4.389 kN · m. Applying the flexure formula,

$$\sigma_{max} = \sigma_{allow} = \frac{M_{max}\, c}{I}$$

$$180(10^6) = \frac{4.389(10^3)\left(\frac{d}{2}\right)}{\frac{\pi}{4}\left(\frac{d}{2}\right)^4}$$

$$d = 0.06286 \text{ m} = 62.9 \text{ mm} \qquad \textbf{Ans}$$

6–110 Consider the general case of a prismatic beam subjected to bending moment components $\mathbf{M}_x$ and $\mathbf{M}_y$, as shown, where the x, y, z axes pass through the centroid of the cross section. If the material is linear elastic, the normal stress in the beaam is a linear function of position such that Using the equilibrium conditions , , , determine the constants a, b, c, and show that the normal stress can be determined from the equation , where the moments and products of inertia are defined in Appendix A.

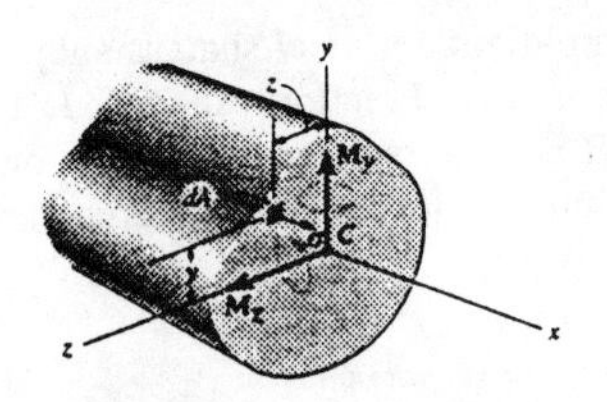

Equilibrium Condition : $\sigma_x = a + by + cz$

$$0 = \int_A \sigma_x dA$$

$$0 = \int_A (a + by + cz)\, dA$$

$$0 = a\int_A dA + b\int_A y\, dA + c\int_A z\, dA \qquad [1]$$

$$M_y = \int_A z\,\sigma_x\, dA$$

$$= \int_A z(a + by + cz)\, dA$$

$$= a\int_A z\, dA + b\int_A yz\, dA + c\int_A z^2\, dA \qquad [2]$$

$$M_z = \int_A -y\,\sigma_x dA$$

$$= \int_A -y(a + by + cz)\, dA$$

$$= -a\int_A y dA - b\int_A y^2\, dA - c\int_A yz\, dA \qquad [3]$$

Section Properties : The integrals are defined in Appendix A. Note that $\int_A y\, dA = \int_A z\, dA = 0$. Thus,

From Eq. [1] $\qquad Aa = 0$

From Eq. [2] $\qquad M_y = bI_{yz} + cI_y$

From Eq. [3] $\qquad M_z = -bI_z - cI_{yz}$

Solving for a, b, c :

$$a = 0 \text{ (Since } A \neq 0)$$

$$b = -\left(\frac{M_z I_y + M_y I_{yz}}{I_y I_z - I_{yz}^2}\right) \qquad c = \frac{M_y I_z + M_z I_{yz}}{I_y I_z - I_{yz}^2}$$

Thus, $$\sigma_x = -\left(\frac{M_z I_y + M_y I_{yz}}{I_y I_z - I_{yz}^2}\right)y + \left(\frac{M_y I_z + M_z I_{yz}}{I_y I_z - I_{yz}^2}\right)z \qquad (Q.E.D.)$$

6–111 Using the techniques outlined in Appendix A, Example A–5 or Example A–6, the Z section has principal moments of inertia $I_{y'} = 29(10^{-6})\ \text{m}^4$ and $I_{z'} = 117(10^{-6})\ \text{m}^4$, computed about the principal axes of inertia y and z, respectively. If the section is subjected to a moment of $M = 2$ kN · m directed horizontally as shown, determine the bending stress produced at point A. Solve the problem using Eq. 6–17.

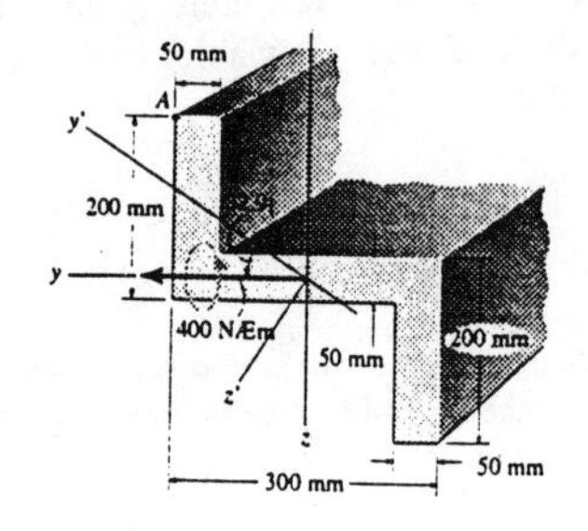

Internal Moment Components :

$$M_{y'} = 250 \cos 32.9° = 209.9\ \text{N} \cdot \text{m}$$
$$M_{z'} = 250 \sin 32.9° = 135.8\ \text{N} \cdot \text{m}$$

Section Property :

$$y' = 0.15 \cos 32.9° + 0.175 \sin 32.9° = 0.2210\ \text{m}$$
$$z' = 0.15 \sin 32.9° - 0.175 \cos 32.9° = -0.06546\ \text{m}$$

Bending Stress : Applying the flexure formula for biaxial bending

$$\sigma = -\frac{M_{z'} y'}{I_{z'}} + \frac{M_{y'} z'}{I_{y'}}$$

$$\sigma_A = -\frac{135.8(0.2210)}{0.471(10^{-3})} + \frac{209.9(-0.06546)}{0.060(10^{-3})}$$

$$= -293\ \text{kPa} = 293\ \text{kPa (C)} \qquad \textbf{Ans}$$

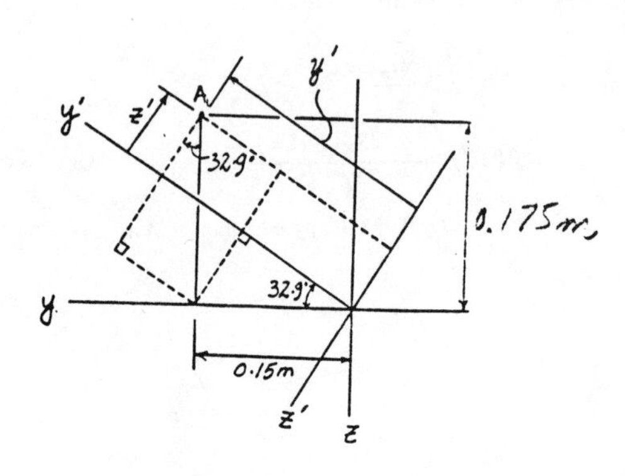

***6–112** Solve Prob. 6–111 using the equation developed in Prob. 6–110.

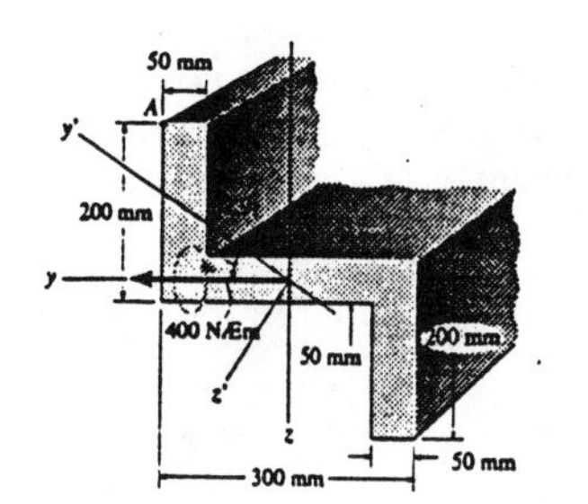

Internal Moment Components :

$$M_y = 250 \text{ N}\cdot\text{m} \qquad M_z = 0$$

Section Properties :

$$I_y = \frac{1}{12}(0.3)\left(0.05^3\right) + 2\left[\frac{1}{12}(0.05)\left(0.15^3\right) + 0.05(0.15)\left(0.1^2\right)\right]$$
$$= 0.18125\left(10^{-3}\right) \text{ m}^4$$

$$I_z = \frac{1}{12}(0.05)\left(0.3^3\right) + 2\left[\frac{1}{12}(0.15)\left(0.05^3\right) + 0.15(0.05)\left(0.125^2\right)\right]$$
$$= 0.350\left(10^{-3}\right) \text{ m}^4$$

$$I_{yz} = 0.15(0.05)(0.125)(-0.1) + 0.15(0.05)(-0.125)(0.1)$$
$$= -0.1875\left(10^{-3}\right) \text{ m}^4$$

Bending Stress : Using formula developed in Prob. 6 - 110

$$\sigma = \frac{-(M_z I_y + M_y I_{yz})y + (M_y I_z + M_z I_{yz})z}{I_y I_z - I_{yz}^2}$$

$$\sigma_A = \frac{-[0 + 250(-0.1875)(10^{-3})](0.15) + [250(0.350)(10^{-3}) + 0](-0.175)}{0.18125(10^{-3})(0.350)(10^{-3}) - [0.1875(10^{-3})]^2}$$

$$= -293 \text{ kPa} = 293 \text{ kPa (C)} \qquad \textbf{Ans}$$

6–113 Using the techniques outlined in Appendix A, Example A–5 or Example A–6, the Z section has principal moments of inertia of $I_{y'} = 29(10^{-6})\text{ m}^4$ and $I_{z'} = 117(10^{-6})\text{ m}^4$, computed about the principal axes of inertia y and z, respectively. If the section is subjected to a moment of $M = 2$ kN · m directed horizontally as shown, determine the bending stress produced at point B. Solve the problem using Eq. 6–17.

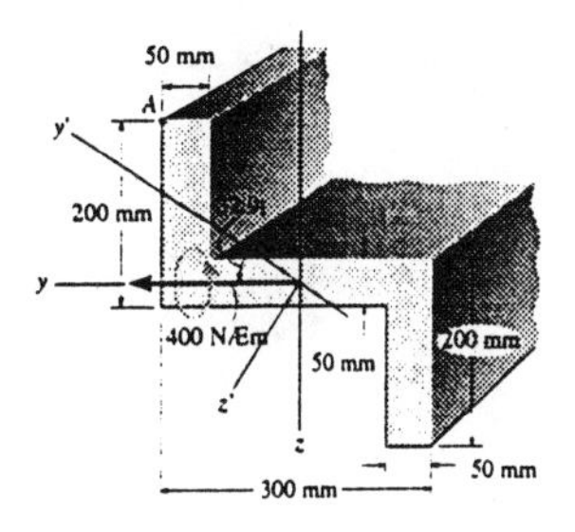

Internal Moment Components :

$$M_{y'} = 250 \cos 32.9° = 209.9 \text{ N}\cdot\text{m}$$
$$M_{z'} = 250 \sin 32.9° = 135.8 \text{ N}\cdot\text{m}$$

Section Property :

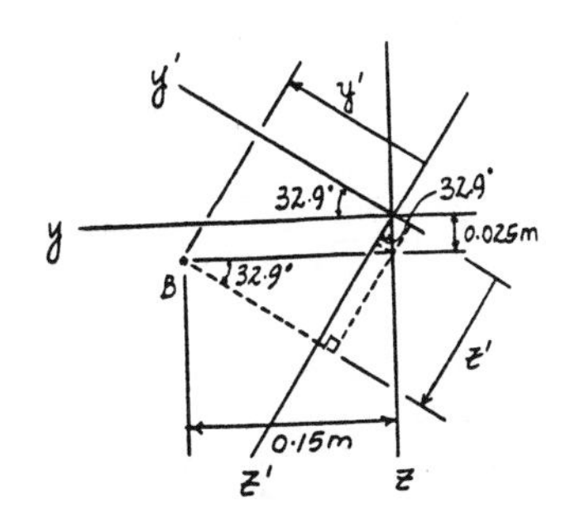

$$y' = 0.15 \cos 32.9° - 0.025 \sin 32.9° = 0.112364 \text{ m}$$
$$z' = 0.15 \sin 32.9° + 0.025 \cos 32.9° = 0.102467 \text{ m}$$

Bending Stress : Applying the flexure formula for biaxial bending

$$\sigma = -\frac{M_{z'} y'}{I_{z'}} + \frac{M_{y'} z'}{I_{y'}}$$

$$\sigma_A = -\frac{135.8(0.112364)}{0.471(10^{-3})} + \frac{209.9(0.102467)}{0.060(10^{-3})}$$

$$= 326 \text{ kPa (T)} \qquad \textbf{Ans}$$

6–114. The composite beam is made of 6061-T6 aluminum (A) and C83400 red brass (B). Determine the dimension h of the brass strip so that the neutral axis of the beam is located at the seam of the two metals. What maximum moment will this beam support if the allowable bending stress for the aluminum is $(\sigma_{allow})_{al} = 128$ MPa and for the brass $(\sigma_{allow})_{br} = 35$ MPa?

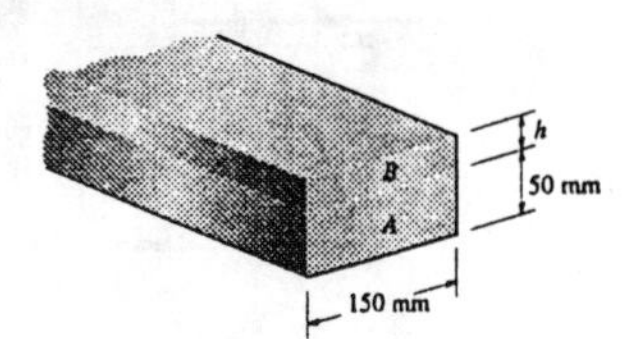

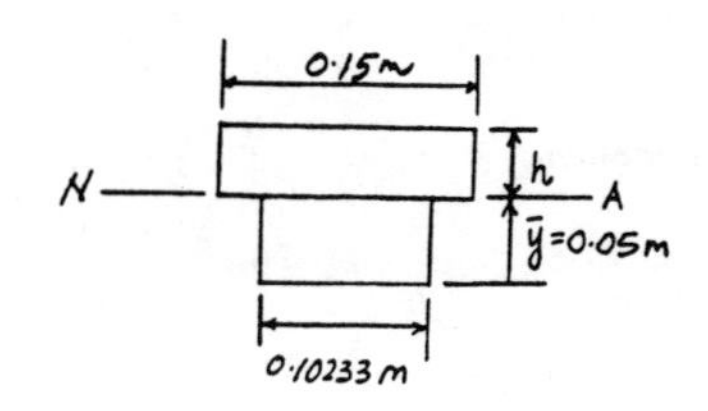

Section Properties :

$$n = \frac{E_{al}}{E_{br}} = \frac{68.9(10^9)}{101(10^9)} = 0.68218$$

$$b_{br} = nb_{al} = 0.68218(0.15) = 0.10233 \text{ m}$$

$$\bar{y} = \frac{\Sigma \tilde{y} A}{\Sigma A}$$

$$0.05 = \frac{0.025(0.10233)(0.05) + (0.05 + 0.5h)(0.15)h}{0.10233(0.05) + (0.15)h}$$

$$h = 0.04130 \text{ m} = 41.3 \text{ mm} \qquad \text{Ans}$$

$$I_{NA} = \frac{1}{12}(0.10233)(0.05^3) + 0.10233(0.05)(0.05 - 0.025)^2 + \frac{1}{12}(0.15)(0.04130^3) + 0.15(0.04130)(0.070649 - 0.05)^2 = 7.7851(10^{-6}) \text{ m}^4$$

Allowable Bending Stress : Applying the flexure formula

Assume failure of red brass

$$(\sigma_{allow})_{br} = \frac{Mc}{I_{NA}}$$

$$35(10^6) = \frac{M(0.04130)}{7.7851(10^{-6})}$$

$$M = 6598 \text{ N}\cdot\text{m} = 6.60 \text{ kN}\cdot\text{m} \; (\textit{controls!}) \qquad \text{Ans}$$

Assume failure of aluminium

$$(\sigma_{allow})_{al} = n\frac{Mc}{I_{NA}}$$

$$128(10^6) = 0.68218\left[\frac{M(0.05)}{7.7851(10^{-6})}\right]$$

$$M = 29215 \text{ N}\cdot\text{m} = 29.2 \text{ kN}\cdot\text{m}$$

6–115. The composite beam is made of 6061-T6 aluminum (A) and C83400 red brass (B). If the height $h = 40$ mm, determine the maximum moment that can be applied to the beam if the allowable bending stress for the aluminum is $(\sigma_{allow})_{al} = 128$ MPa and for the brass $(\sigma_{allow})_{br} = 35$ MPa.

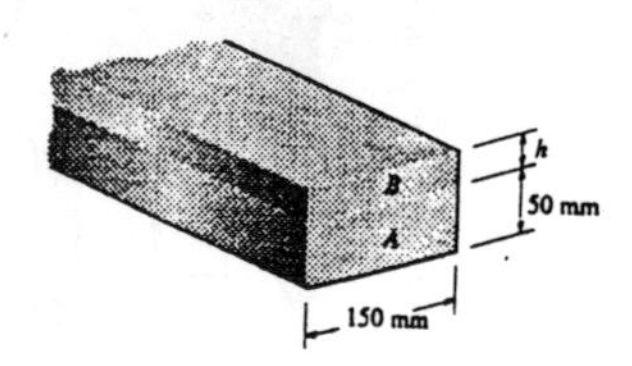

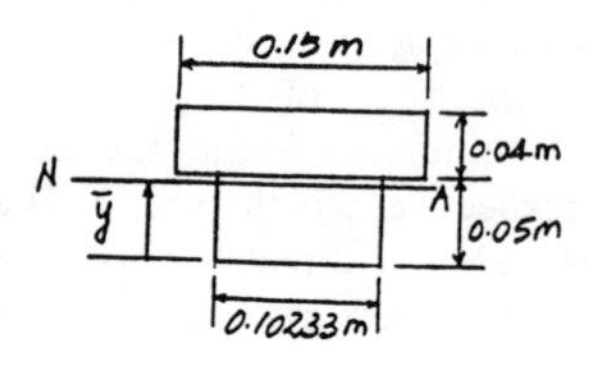

Section Properties : For transformed section.

$$n = \frac{E_{al}}{E_{br}} = \frac{68.9(10^9)}{101.0(10^9)} = 0.68218$$

$$b_{br} = nb_{al} = 0.68218(0.15) = 0.10233 \text{ m}$$

$$\bar{y} = \frac{\Sigma \tilde{y} A}{\Sigma A} = \frac{0.025(0.10233)(0.05) + (0.07)(0.15)(0.04)}{0.10233(0.05) + 0.15(0.04)} = 0.049289 \text{ m}$$

$$I_{NA} = \frac{1}{12}(0.10233)(0.05^3) + 0.10233(0.05)(0.049289 - 0.025)^2 + \frac{1}{12}(0.15)(0.04^3) + 0.15(0.04)(0.07 - 0.049289)^2 = 7.45799(10^{-6}) \text{ m}^4$$

Allowable Bending Stress : Applying the flexure formula

Assume failure of red brass

$$(\sigma_{allow})_{br} = \frac{Mc}{I_{NA}}$$

$$35(10^6) = \frac{M(0.09 - 0.049289)}{7.45799(10^{-6})}$$

$$M = 6412 \text{ N}\cdot\text{m} = 6.41 \text{ kN}\cdot\text{m} \; (\textit{controls!}) \qquad \text{Ans}$$

Assume failure of aluminium

$$(\sigma_{allow})_{al} = n\frac{Mc}{I_{NA}}$$

$$128(10^6) = 0.68218\left[\frac{M(0.049289)}{7.45799(10^{-6})}\right]$$

$$M = 28391 \text{ N}\cdot\text{m} = 28.4 \text{ kN}\cdot\text{m}$$

***6–116.** The member has a brass core bonded to a steel casing. If a load of $P = 12$ kN is applied at its end, determine the maximum bending stress in the member. $E_{br} = 100$ GPa, $E_{st} = 200$ GPa.

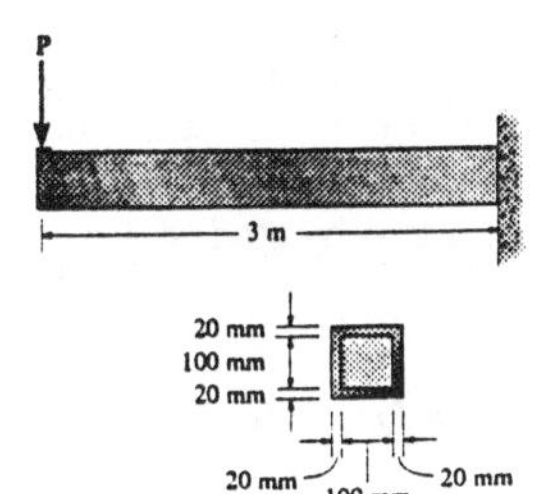

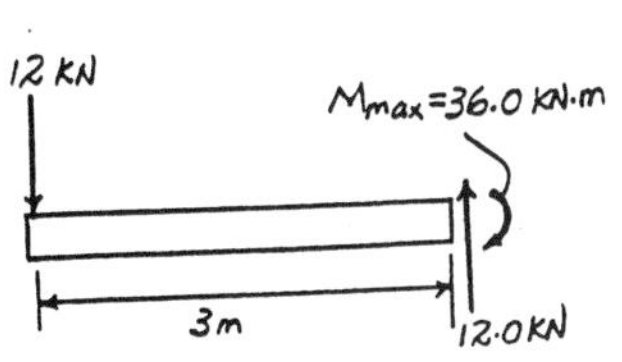

Support Reactions : As shown on FBD.
Section Properties : For transformed section.

$$b_{st} = n\,b_{br} = \frac{100}{200}(0.1) = 0.05 \text{ m}$$

$$I_{NA} = \frac{1}{12}(0.14)\left(0.14^3\right) - \frac{1}{12}(0.05)\left(0.1^3\right)$$
$$= 27.8467(10^{-6}) \text{ m}^4$$

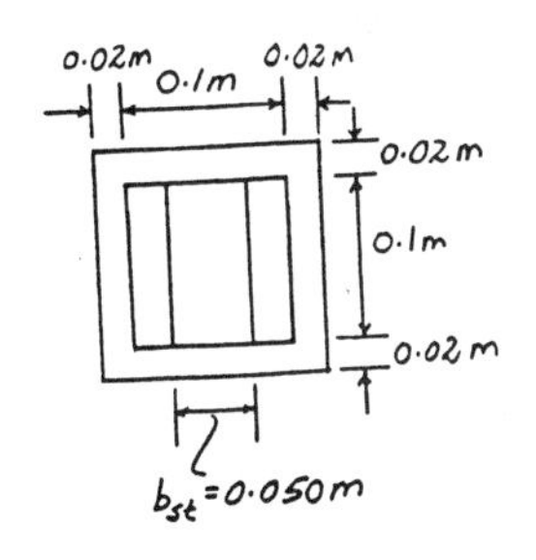

Maximum Bending Stress : The maximum moment occurs at fixed support that is $M_{max} = 36.0$ kN · m and the maximum bending stress is in the material located on the top and bottom surfaces of the steel casing. Applying the flexure formula, we have

$$\sigma_{max} = \sigma_{st} = \frac{Mc}{I} = \frac{36.0(10^3)(0.07)}{27.8467(10^{-6})} = 90.5 \text{ MPa} \qquad \textbf{Ans}$$

6–117. The member has a brass core bonded to a steel casing. If the allowable bending stress for the brass is $(\sigma_{allow})_{br} = 40$ MPa and for the steel $(\sigma_{allow})_{st} = 180$ MPa, determine the maximum load P the member will support. $E_{br} = 100$ GPa, $E_{st} = 200$ GPa.

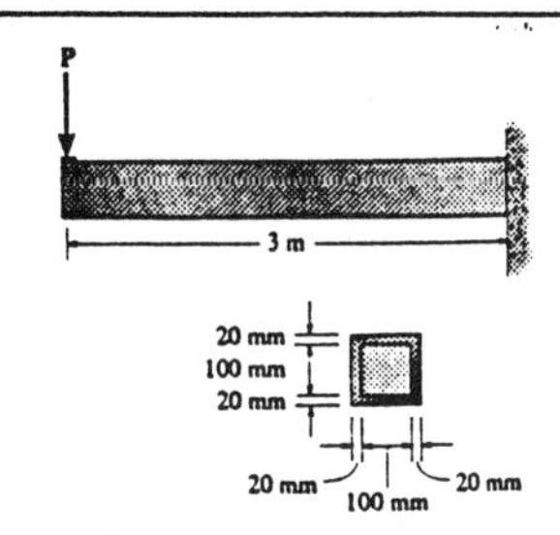

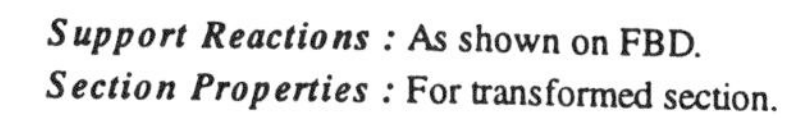

Support Reactions : As shown on FBD.
Section Properties : For transformed section.

$$b_{st} = n\,b_{br} = \frac{100}{200}(0.1) = 0.05 \text{ m}$$

$$I_{NA} = \frac{1}{12}(0.14)\left(0.14^3\right) - \frac{1}{12}(0.05)\left(0.1^3\right)$$
$$= 27.8467(10^{-6}) \text{ m}^4$$

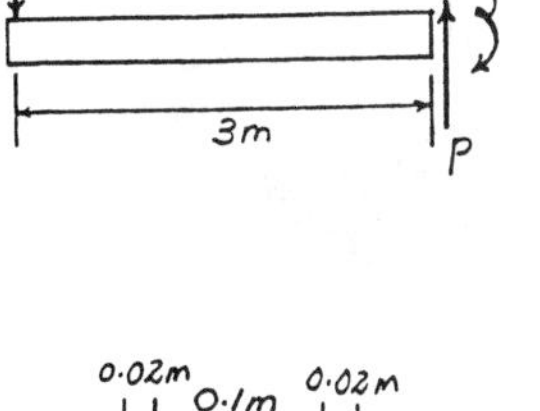

Allowable Bending Stress : The maximum moment occurs at fixed support, that is $M_{max} = 3.00P$. Applying the flexure formula

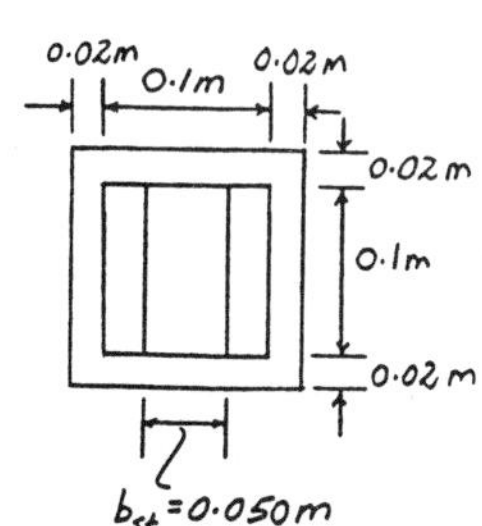

Assume failure of steel

$$(\sigma_{allow})_{st} = \frac{Mc}{I}$$

$$180\left(10^6\right) = \frac{3.00P(0.07)}{27.8467(10^{-6})}$$

$$P = 23869 \text{ N} = 23.87 \text{ kN}$$

Assume failure of brass

$$(\sigma_{allow})_{br} = n\,\frac{My}{I}$$

$$40(10^6) = \frac{100}{200}\left[\frac{3.00P(0.05)}{27.8467(10^{-6})}\right]$$

$$P = 14852 \text{ N} = 14.9 \text{ kN} \quad \textit{(Controls!)} \quad \textbf{Ans}$$

6–118. The top plate is made of 2014-T6 aluminum and is used to reinforce a Kevlar 49 plastic beam. Determine the maximum stress in the aluminum and in the Kevlar if the beam is subjected to a moment of $M = 900$ lb · ft.

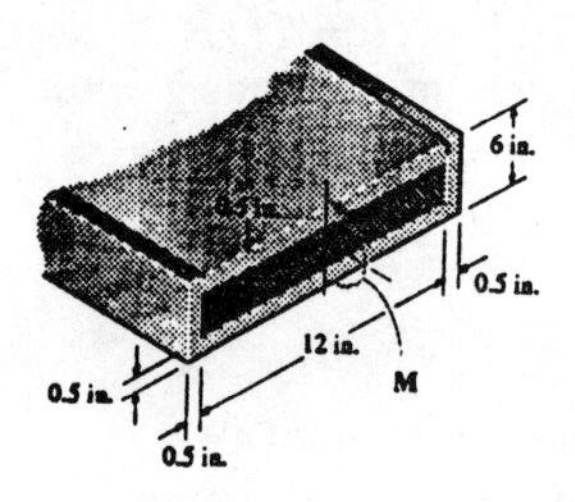

Section Properties :

$$n = \frac{E_{al}}{E_k} = \frac{10.6(10^3)}{19.0(10^3)} = 0.55789$$
$$b_k = n\,b_{al} = 0.55789(12) = 6.6947 \text{ in.}$$

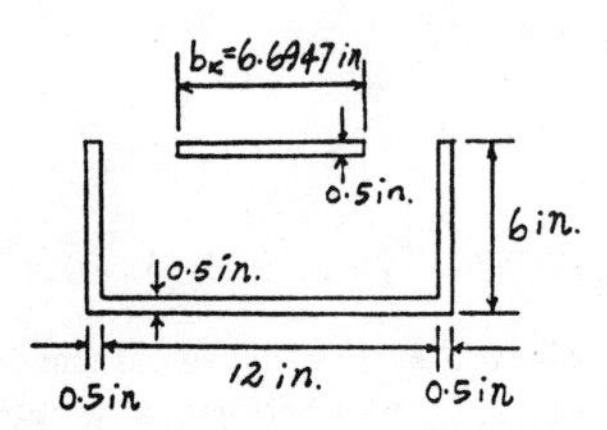

$$\bar{y} = \frac{\Sigma \bar{y} A}{\Sigma A} = \frac{0.25(13)(0.5) + 2[(3.25)(5.5)(0.5)] + 5.75(6.6947)(0.5)}{13(0.5) + 2(5.5)(0.5) + 6.6947(0.5)}$$
$$= 2.5247 \text{ in.}$$

$$I_{NA} = \frac{1}{12}(13)\left(0.5^3\right) + 13(0.5)(2.5247 - 0.25)^2$$
$$+ \frac{1}{12}(1)\left(5.5^3\right) + 1(5.5)(3.25 - 2.5247)^2$$
$$+ \frac{1}{12}(6.6947)\left(0.5^3\right) + 6.6947(0.5)(5.75 - 2.5247)^2$$
$$= 85.4170 \text{ in}^4$$

Maximum Bending Stress : Applying the flexure formula

$$(\sigma_{max})_{al} = n\frac{Mc}{I} = 0.55789\left[\frac{900(12)(6 - 2.5247)}{85.4170}\right] = 245 \text{ psi} \quad \textbf{Ans}$$

$$(\sigma_{max})_k = \frac{Mc}{I} = \frac{900(12)(6 - 2.5247)}{85.4168} = 439 \text{ psi} \quad \textbf{Ans}$$

6–119. The top plate made of 2014-T6 aluminum is used to reinforce a Kevlar 49 plastic beam. If the allowable bending stress for the aluminum is $(\sigma_{allow})_{al} = 40$ ksi and for the Kevlar $(\sigma_{allow})_k = 8$ ksi, determine the maximum moment M that can be applied to the beam.

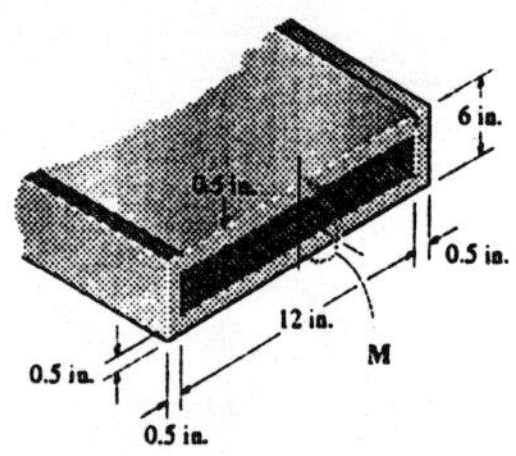

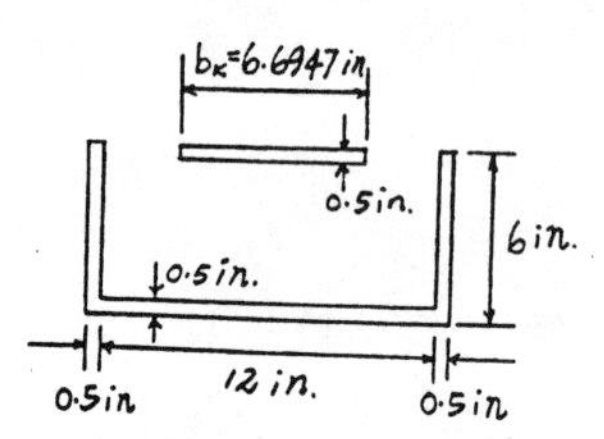

Section Properties :

$$n = \frac{E_{al}}{E_k} = \frac{10.6(10^3)}{19.0(10^3)} = 0.55789$$
$$b_k = n\,b_{al} = 0.55789(12) = 6.6947 \text{ in.}$$

$$\bar{y} = \frac{\Sigma \bar{y} A}{\Sigma A} = \frac{0.25(13)(0.5) + 2[(3.25)(5.5)(0.5)] + 5.75(6.}{13(0.5) + 2(5.5)(0.5) + 6.6947(0.5}$$
$$= 2.5247 \text{ in.}$$

$$I_{NA} = \frac{1}{12}(13)\left(0.5^3\right) + 13(0.5)(2.5247 - 0.25)^2$$
$$+ \frac{1}{12}(1)\left(5.5^3\right) + 1(5.5)(3.25 - 2.5247)^2$$
$$+ \frac{1}{12}(6.6947)\left(0.5^3\right) + 6.6947(0.5)(5.75 - 2.$$
$$= 85.4170 \text{ in}^4$$

Maximum Bending Stress : Applying the flexure formula

Assume failure of aluminium

$$(\sigma_{allow})_{al} = n\frac{Mc}{I}$$
$$40 = 0.55789\left[\frac{M(6 - 2.5247)}{85.4170}\right]$$
$$M = 1762 \text{ kip} \cdot \text{in} = 146.9 \text{ kip} \cdot \text{ft}$$

Assume failure of Kevlar 49

$$(\sigma_{allow})_k = \frac{Mc}{I}$$
$$8 = \frac{M(6 - 2.5247)}{85.4170}$$
$$M = 196.62 \text{ kip} \cdot \text{in}$$
$$= 16.4 \text{ kip} \cdot \text{ft} \quad \textit{(Controls!)} \quad \textbf{Ans}$$

***6–120.** The composite beam consists of a Douglas Fir wood core and three A-36 steel plates. If the allowable bending stress for the wood is $(\sigma_{allow})_w$ = 20 MPa and for the steel $(\sigma_{allow})_{st}$ = 130 MPa, determine the maximum moment that can be applied to the beam.

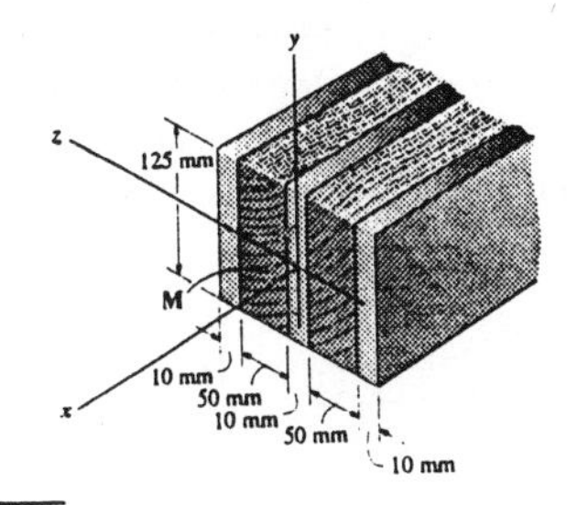

Section Properties : For transformed section

$$n = \frac{E_w}{E_{st}} = \frac{13.1(10^9)}{200(10^9)} = 0.0655$$

$$b_{st} = nb_w = 0.0655(0.1) = 0.00655 \text{ m}$$

$$I_{NA} = \frac{1}{12}(0.03)\left(0.125^3\right) + \frac{1}{12}(0.00655)\left(0.125^3\right)$$

$$= 5.9489\left(10^{-6}\right)\ \text{m}^4$$

N A 0.125m
0.03m b_{st}=0.00655m

Maximum Bending Stress : Applying the flexure formula

Assume failure of steel

$$(\sigma_{allow})_{st} = \frac{Mc}{I}$$

$$130(10^6) = \frac{M(0.0625)}{5.9489(10^{-6})}$$

$$M = 12374 \text{ N}\cdot\text{m}$$

$$= 12.4 \text{ kN}\cdot\text{m} \quad (\textit{Controls!}) \qquad \textbf{Ans}$$

Assume failure of wood

$$(\sigma_{allow})_w = n\frac{Mc}{I}$$

$$20(10^6) = 0.0655\left[\frac{M(0.0625)}{5.9489(10^{-6})}\right]$$

$$M = 29063 \text{ N}\cdot\text{m} = 29.1 \text{ kN . m}$$

6–121. Solve Prob. 6–120 if the moment is applied about the y axis instead of the z axis as shown.

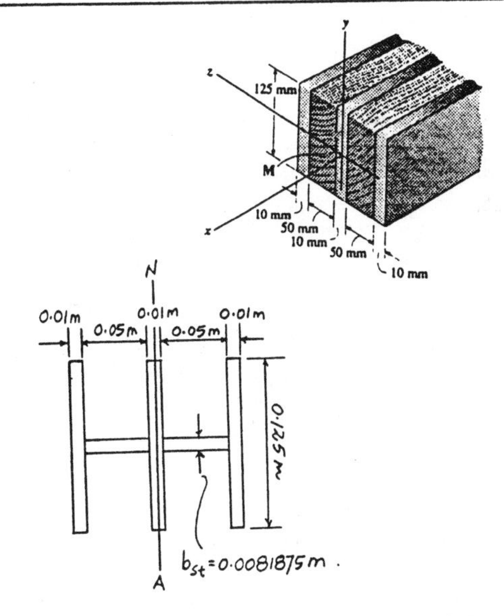

Section Properties : For the transformed section

$$n = \frac{E_w}{E_{st}} = \frac{13.1(10^9)}{200(10^9)} = 0.0655$$

$$b_{st} = nb_w = 0.0655(0.125) = 0.0081875 \text{ m}$$

$$I_{NA} = \frac{1}{12}(0.125)\left(0.13^3\right) - \frac{1}{12}(0.125 - 0.0081875)\left(0.11^3\right)$$

$$+ \frac{1}{12}(0.125 - 0.0081875)\left(0.01^3\right)$$

$$= 9.9387\left(10^{-6}\right)\ \text{m}^4$$

Allowable Bending Stress : Applying the flexure formula

Assume failure of steel

$$(\sigma_{allow})_{st} = \frac{Mc}{I}$$

$$130\left(10^6\right) = \frac{M(0.065)}{9.9387(10^{-6})}$$

$$M = 19877 \text{ N}\cdot\text{m}$$

$$= 19.9 \text{ kN}\cdot\text{m} \quad (\textit{Controls!}) \qquad \textbf{Ans}$$

Assume failure of wood

$$(\sigma_{allow})_w = n\frac{Mz}{I}$$

$$20\left(10^6\right) = 0.0655\left[\frac{M(0.055)}{9.9387(10^{-6})}\right]$$

$$M = 55177 \text{ N}\cdot\text{m} = 55.2 \text{ kN}\cdot\text{m}$$

6–122. A White Spruce beam is reinforced with A-36 steel straps at its top and bottom as shown. Determine the maximum stress developed in the wood and steel if the beam is subjected to a bending moment of $M = 8$ kip · ft. Sketch the stress distribution acting over the cross section.

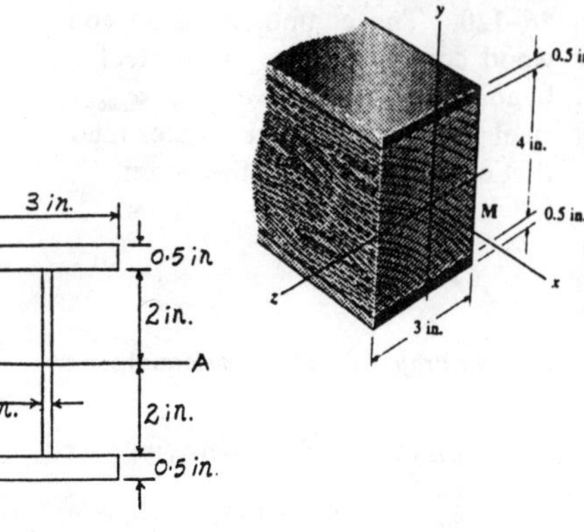

Section Properties : For the transformed section.

$$n = \frac{E_w}{E_{st}} = \frac{1.40(10^3)}{29.0(10^3)} = 0.048276$$

$$b_{st} = nb_w = 0.048276(3) = 0.14483 \text{ in.}$$

$$I_{NA} = \frac{1}{12}(3)\left(5^3\right) - \frac{1}{12}(3 - 0.14483)\left(4^3\right) = 16.0224 \text{ in}^4$$

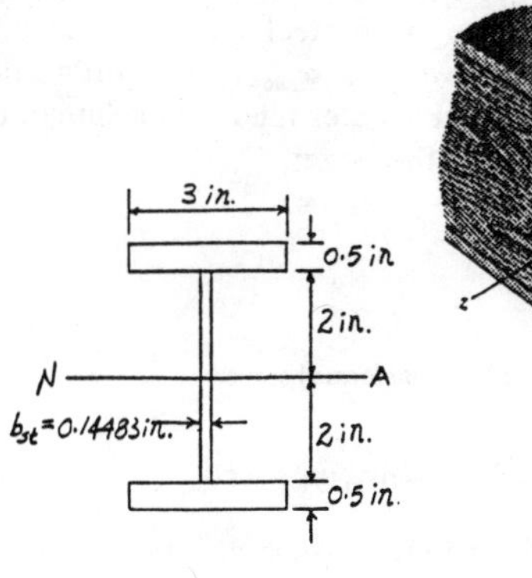

Maximum Bending Stress : Applying the flexure formula

$$(\sigma_{max})_{st} = \frac{Mc}{I}$$

$$= \frac{8(12)(2.5)}{16.0224} = 15.0 \text{ ksi} \qquad \textbf{Ans}$$

$$(\sigma_{max})_w = n\frac{My}{I}$$

$$= 0.048276\left[\frac{8(12)(2)}{16.0224}\right] = 0.578 \text{ ksi} \qquad \textbf{Ans}$$

$$(\sigma_B)_{st} = \frac{My}{I} = \frac{8(12)(2)}{16.0224} = 12.0 \text{ ksi}$$

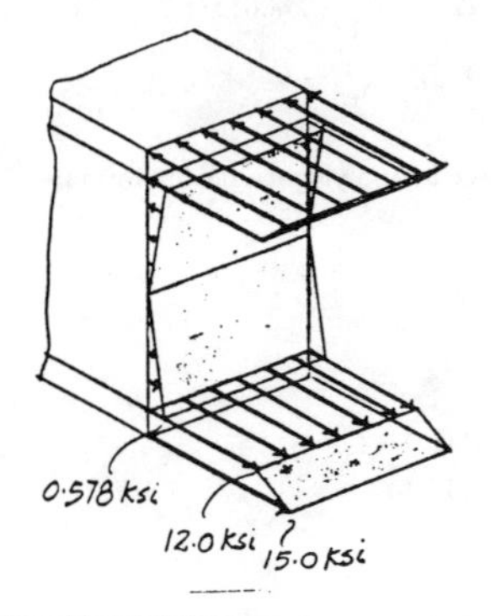

6–123. A White Spruce beam is reinforced with A-36 steel straps at its top and bottom as shown. Determine the bending moment M it can support if $(\sigma_{allow})_{st} = 22$ ksi and $(\sigma_{allow})_w = 2.0$ ksi.

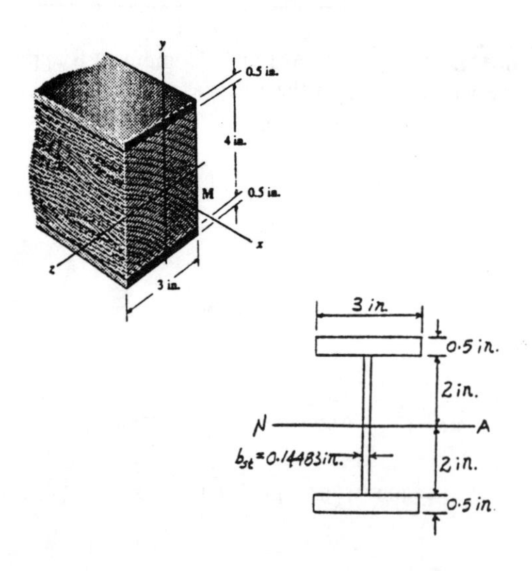

Section Properties : For the transformed section.

$$n = \frac{E_w}{E_{st}} = \frac{1.40(10^3)}{29.0(10^3)} = 0.048276$$

$$b_{st} = nb_w = 0.048276(3) = 0.19310 \text{ in.}$$

$$I_{NA} = \frac{1}{12}(3)\left(5^3\right) - \frac{1}{12}(3 - 0.14483)\left(4^3\right) = 16.0224 \text{ in}^4$$

Allowable Bending Stress : Applying the flexure formula

Assume failure of steel

$$(\sigma_{allow})_{st} = \frac{Mc}{I}$$

$$22 = \frac{M(2.5)}{16.0224}$$

$$M = 141.0 \text{ kip} \cdot \text{in}$$

$$= 11.7 \text{ kip} \cdot \text{ft } (\textit{Controls!}) \qquad \textbf{Ans}$$

Assume failure of wood

$$(\sigma_{allow})_w = n\frac{My}{I}$$

$$2.0 = 0.048276\left[\frac{M(2)}{16.0224}\right]$$

$$M = 331.9 \text{ kip} \cdot \text{in} = 27.7 \text{ kip} \cdot \text{ft}$$

***6–124.** The Douglas Fir beam is reinforced with A-36 steel straps at its center and sides. Determine the maximum stress developed in the wood and steel if the beam is subjected to a bending moment of $M_z = 7.50$ kip · ft. Sketch the stress distribution acting over the cross section.

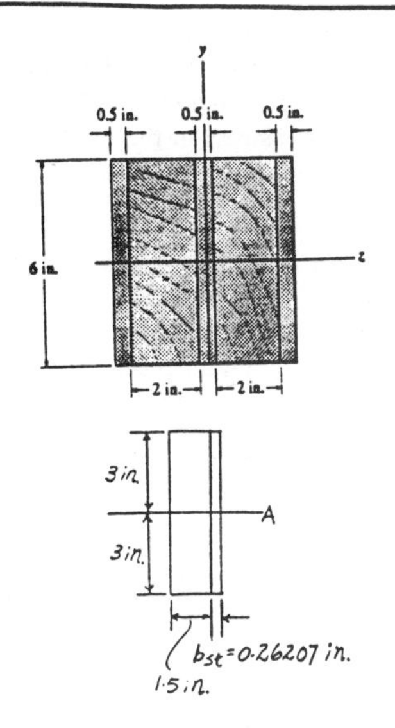

Section Properties : For the transformed section.

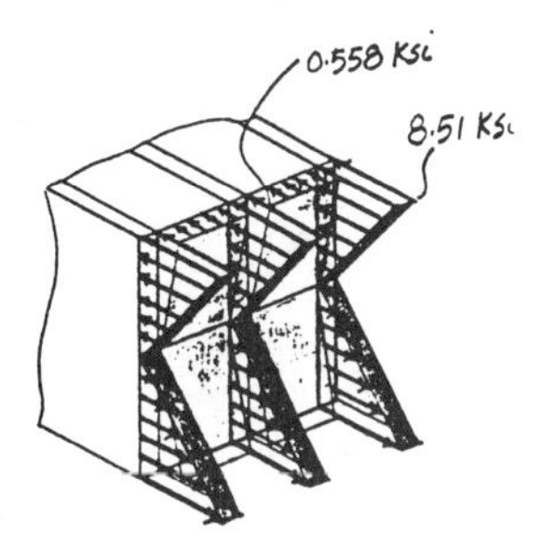

$$n = \frac{E_w}{E_{st}} = \frac{1.90(10^3)}{29.0(10^3)} = 0.065517$$

$$b_{st} = nb_w = 0.065517(4) = 0.26207 \text{ in.}$$

$$I_{NA} = \frac{1}{12}(1.5 + 0.26207)\left(6^3\right) = 31.7172 \text{ in}^4$$

Maximum Bending Stress : Applying the flexure formula

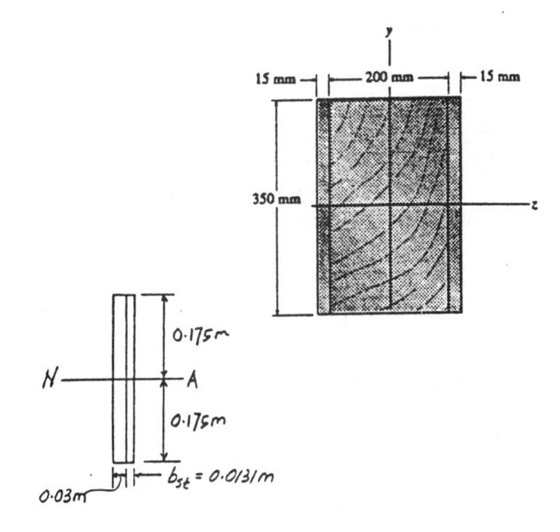

$$(\sigma_{\max})_{st} = \frac{Mc}{I} = \frac{7.5(12)(3)}{31.7172} = 8.51 \text{ ksi} \qquad \textbf{Ans}$$

$$(\sigma_{\max})_w = n\frac{Mc}{I} = 0.065517\left[\frac{7.5(12)(3)}{31.7172}\right] = 0.558 \text{ ksi} \qquad \textbf{Ans}$$

6–125. The Douglas Fir beam is reinforced with A-36 steel straps at its sides. Determine the maximum stress developed in the wood and steel if the beam is subjected to a bending moment of $M_z = 4$ kN · m. Sketch the stress distribution acting over the cross section.

Section Properties : For the transformed section.

$$n = \frac{E_w}{E_{st}} = \frac{13.1(10^9)}{200(10^9)} = 0.0655$$

$$b_{st} = nb_w = 0.0655(0.2) = 0.0131 \text{ m}$$

$$I_{NA} = \frac{1}{12}(0.03 + 0.0131)\left(0.35^3\right) = 153.99\left(10^{-6}\right) \text{ m}^4$$

Maximum Bending Stress : Applying the flexure formula

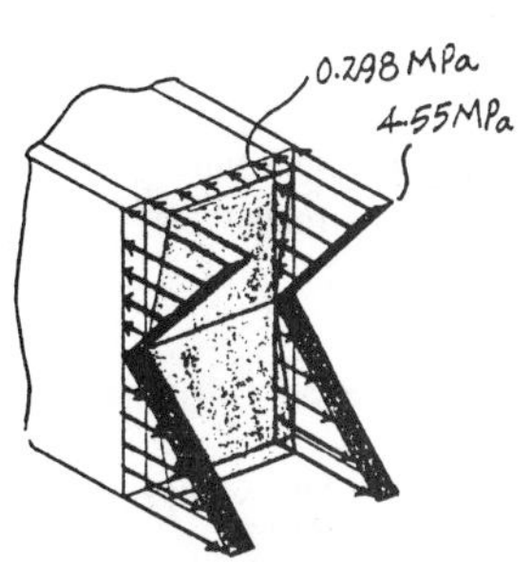

$$(\sigma_{\max})_{st} = \frac{Mc}{I} = \frac{4(10^3)(0.175)}{153.99(10^{-6})} = 4.55 \text{ MPa} \qquad \textbf{Ans}$$

$$(\sigma_{\max})_w = n\frac{Mc}{I} = 0.0655\left[\frac{4(10^3)(0.175)}{153.99(10^{-6})}\right] = 0.298 \text{ MPa} \qquad \textbf{Ans}$$

6–126. The composite beam is made of A-36 steel (A) bonded to C83400 red brass (B) and has the cross section shown. If it is subjected to a moment of $M = 6.5$ kN · m, determine the maximum stress in the brass and steel. Also, what is the stress in each material at the seam where they are bonded together?

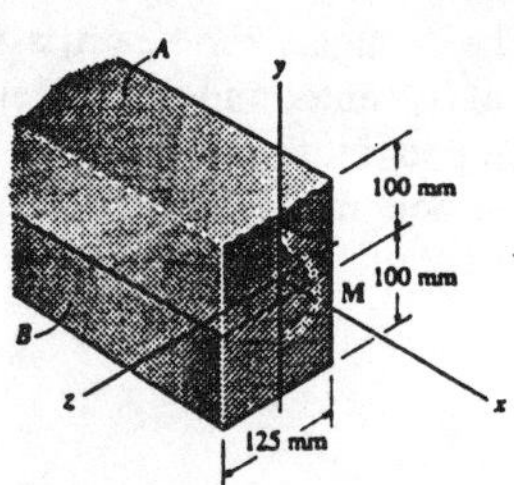

Section Properties : For the transformed section.

$$n = \frac{E_{br}}{E_{st}} = \frac{101(10^9)}{200(10^9)} = 0.505$$

$$b_{st} = n b_{br} = 0.505(0.125) = 0.063125 \text{ m}$$

$$\bar{y} = \frac{\Sigma \tilde{y} A}{\Sigma A} = \frac{0.05(0.125)(0.1) + 0.15(0.1)(0.063125)}{0.125(0.1) + 0.1(0.063125)} = 0.08355 \text{ m}$$

$$I_{NA} = \frac{1}{12}(0.125)(0.1^3) + 0.125(0.1)(0.08355 - 0.05)^2 + \frac{1}{12}(0.063125)(0.1^3) + 0.063125(0.1)(0.15 - 0.08355)^2 = 57.62060(10^{-6}) \text{ m}^4$$

Maximum Bending Stress : Applying the flexure formula

$$(\sigma_{max})_{st} = \frac{My}{I} = \frac{6.5(10^3)(0.08355)}{57.62060(10^{-6})} = 9.42 \text{ MPa} \qquad \textbf{Ans}$$

$$(\sigma_{max})_{br} = n\frac{Mc}{I} = 0.505\left[\frac{6.5(10^3)(0.2 - 0.08355)}{57.62060(10^{-6})}\right] = 6.63 \text{ MPa} \qquad \textbf{Ans}$$

Bending Stress : At seam

$$\sigma_{st} = \frac{My}{I} = \frac{6.5(10^3)(0.1 - 0.08355)}{57.62060(10^{-6})} = 1.855 \text{ MPa} = 1.86 \text{ MPa} \qquad \textbf{Ans}$$

$$\sigma_{br} = n\frac{My}{I} = 0.505(1.855) = 0.937 \text{ MPa} \qquad \textbf{Ans}$$

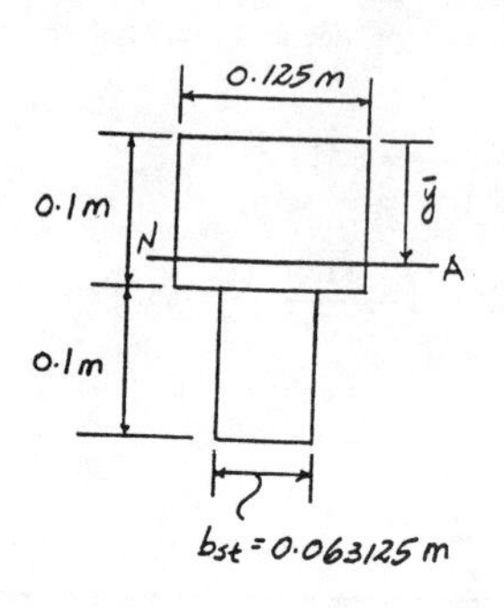

6–127. The composite beam is made of A-36 steel (A) bonded to C83400 red brass (B) and has the cross section shown. If the allowable bending stress for the steel is $(\sigma_{allow})_{st} = 180$ MPa and for the brass $(\sigma_{allow})_{br} = 60$ MPa, determine the maximum moment M that can be applied to the beam..

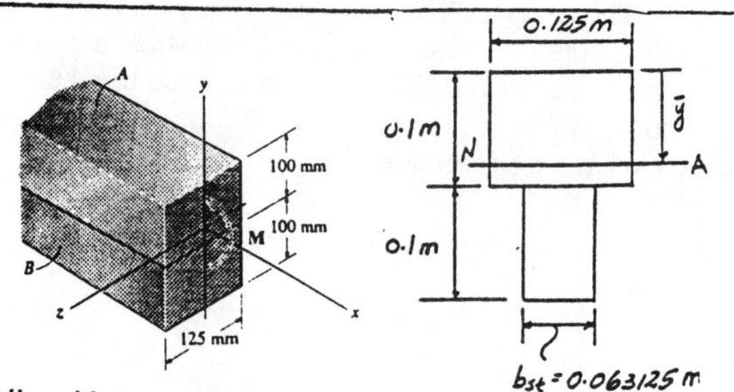

Section Properties : For the transformed section.

$$n = \frac{E_{br}}{E_{st}} = \frac{101(10^9)}{200(10^9)} = 0.505$$

$$b_{st} = n b_{br} = 0.505(0.125) = 0.063125 \text{ m}$$

$$\bar{y} = \frac{\Sigma \tilde{y} A}{\Sigma A} = \frac{0.05(0.125)(0.1) + 0.15(0.1)(0.063125)}{0.125(0.1) + 0.1(0.063125)} = 0.08355 \text{ m}$$

$$I_{NA} = \frac{1}{12}(0.125)(0.1^3) + 0.125(0.1)(0.08355 - 0.05)^2 + \frac{1}{12}(0.063125)(0.1^3) + 0.063125(0.1)(0.15 - 0.08355)^2 = 57.62060(10^{-6}) \text{ m}^4$$

Allowable Bending Stress : Applying the flexure formula

Assume failure of steel

$$(\sigma_{max})_{st} = (\sigma_{allow})_{st} = \frac{My}{I}$$

$$180(10^6) = \frac{M(0.08355)}{57.62060(10^{-6})}$$

$$M = 124130 \text{ N} \cdot \text{m} = 124 \text{ kN} \cdot \text{m}$$

Assume failure of brass

$$(\sigma_{max})_{br} = (\sigma_{allow})_{br} = n\frac{Mc}{I}$$

$$60(10^6) = 0.505\left[\frac{M(0.2 - 0.08355)}{57.62060(10^{-6})}\right]$$

$$M = 58792 \text{ N} \cdot \text{m} = 58.8 \text{ kN} \cdot \text{m} \ \textit{(Controls!)} \qquad \textbf{Ans}$$

***6–128.** Determine the greatest magnitude of the applied forces P if the allowable bending stress is $(\sigma_{allow})_c = 50$ MPa in compression and $(\sigma_{allow})_t = 120$ MPa in tension.

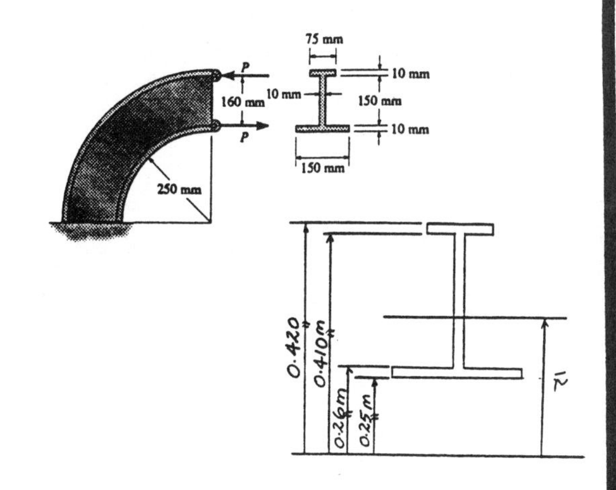

Internal Moment : $M = 0.160P$ is positive since it tends to increase the beam's radius of curvature.

Section Properties :

$$\bar{r} = \frac{\Sigma \bar{y} A}{\Sigma A}$$

$$= \frac{0.255(0.15)(0.01) + 0.335(0.15)(0.01) + 0.415(0.075)(0.01)}{0.15(0.01) + 0.15(0.01) + 0.075(0.01)}$$

$$= 0.3190 \text{ m}$$

$$A = 0.15(0.01) + 0.15(0.01) + 0.075(0.01) = 0.00375 \text{ m}^2$$

$$\Sigma \int_A \frac{dA}{r} = 0.15 \ln \frac{0.26}{0.25} + 0.01 \ln \frac{0.41}{0.26} + 0.075 \ln \frac{0.42}{0.41}$$

$$= 0.012245 \text{ m}$$

$$R = \frac{A}{\Sigma \int_A \frac{dA}{r}} = \frac{0.00375}{0.012245} = 0.306243 \text{ m}$$

$$\bar{r} - R = 0.319 - 0.306243 = 0.012757 \text{ m}$$

Allowable Normal Stress : Applying the curved - beam formula

Assume tension failure

$$(\sigma_{allow})_t = \frac{M(R - r)}{Ar(\bar{r} - R)}$$

$$120(10^6) = \frac{0.16P(0.306243 - 0.25)}{0.00375(0.25)(0.012757)}$$

$$P = 159482 \text{ N} = 159.5 \text{ kN}$$

Assume compression failure

$$(\sigma_{allow})_c = \frac{M(R - r)}{Ar(\bar{r} - R)}$$

$$-50(10^6) = \frac{0.16P(0.306243 - 0.42)}{0.00375(0.42)(0.012757)}$$

$P = 55195 \text{ N} = 55.2 \text{ kN}$ (***Controls!***) **Ans**

6–129. If $P = 6$ kN, determine the maximum tensile and compressive bending stresses in the beam.

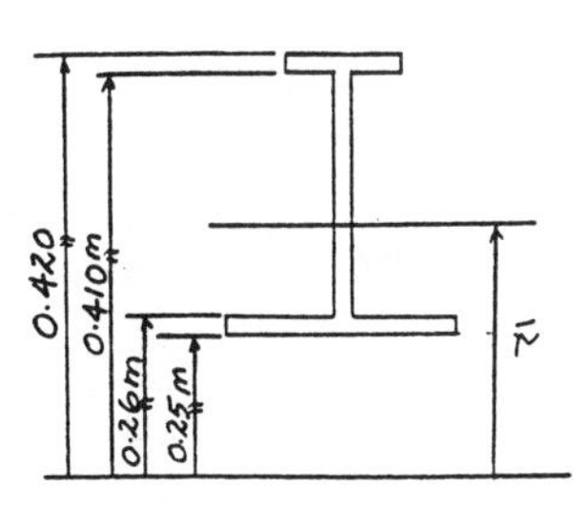

Internal Moment : $M = 0.160(6) = 0.960 \text{ kN} \cdot \text{m}$ is positive since it tends to increase the beam's radius of curvature.

Section Properties :

$$\bar{r} = \frac{\Sigma \bar{y} A}{\Sigma A}$$

$$= \frac{0.255(0.15)(0.01) + 0.335(0.15)(0.01) + 0.415(0.075)(0}{0.15(0.01) + 0.15(0.01) + 0.075(0.01)}$$

$$= 0.3190 \text{ m}$$

$$A = 0.15(0.01) + 0.15(0.01) + 0.075(0.01) = 0.00375 \text{ m}^2$$

$$\Sigma \int_A \frac{dA}{r} = 0.15 \ln \frac{0.26}{0.25} + 0.01 \ln \frac{0.41}{0.26} + 0.075 \ln \frac{0.42}{0.41}$$

$$= 0.012245 \text{ m}$$

$$R = \frac{A}{\Sigma \int_A \frac{dA}{r}} = \frac{0.00375}{0.012245} = 0.306243 \text{ m}$$

$$\bar{r} - R = 0.319 - 0.306243 = 0.012757 \text{ m}$$

Normal Stress : Applying the curved - beam formula

$$(\sigma_{max})_t = \frac{M(R - r)}{Ar(\bar{r} - R)}$$

$$= \frac{0.960(10^3)(0.306243 - 0.25)}{0.00375(0.25)(0.012757)}$$

$= 4.51$ MPa **Ans**

$$(\sigma_{max})_c = \frac{M(R - r)}{Ar(\bar{r} - R)}$$

$$= \frac{0.960(10^3)(0.306243 - 0.42)}{0.00375(0.42)(0.012757)}$$

$= -5.44$ MPa **Ans**

6–130. The curved beam is subjected to a bending moment of $M = 900\ \text{N}\cdot\text{m}$ as shown. Determine the stress at points A and B, and show the stress on a volume element located at each of these points.

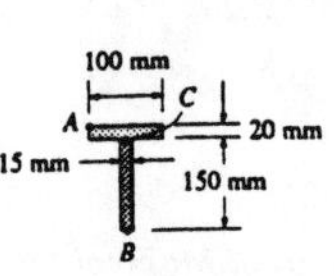

Internal Moment : $M = -900\ \text{N}\cdot\text{m}$ is negative since it tends to decrease the beam's radius of curvature.

Section Properties :

$$\Sigma A = 0.15(0.015) + 0.1(0.02) = 0.00425\ \text{m}^2$$

$$\Sigma \bar{r}A = 0.475(0.15)(0.015) + 0.56(0.1)(0.02) = 2.18875(10^{-3})\ \text{m}^3$$

$$\bar{r} = \frac{\Sigma \bar{r}A}{\Sigma A} = \frac{2.18875\,(10^{-3})}{0.00425} = 0.5150\ \text{m}$$

$$\Sigma \int_A \frac{dA}{r} = 0.015\ \ln\frac{0.55}{0.4} + 0.1\ \ln\frac{0.57}{0.55} = 8.348614(10^{-3})\ \text{m}$$

$$R = \frac{A}{\Sigma\int_A \frac{dA}{r}} = \frac{0.00425}{8.348614(10^{-3})} = 0.509067\ \text{m}$$

$$\bar{r} - R = 0.515 - 0.509067 = 5.933479(10^{-3})\ \text{m}$$

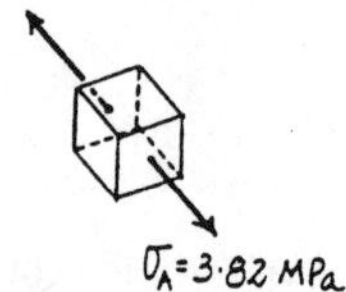

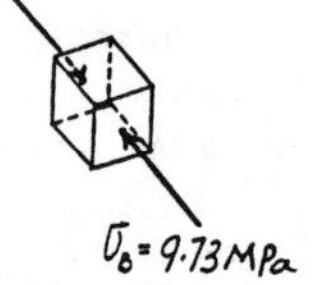

Normal Stress : Applying the curved-beam formula

$$\sigma_A = \frac{M(R - r_A)}{Ar_A(\bar{r} - R)} = \frac{-900(0.509067 - 0.57)}{0.00425(0.57)(5.933479)(10^{-3})}$$

$= 3.82$ MPa (T) **Ans**

$$\sigma_B = \frac{M(R - r_B)}{Ar_B(\bar{r} - R)} = \frac{-900(0.509067 - 0.4)}{0.00425(0.4)(5.933479)(10^{-3})}$$

$= -9.73$ MPa $= 9.73$ MPa (C) **Ans**

6–131. The curved beam is subjected to a bending moment of $M = 900\ \text{N}\cdot\text{m}$. Determine the stress at point C.

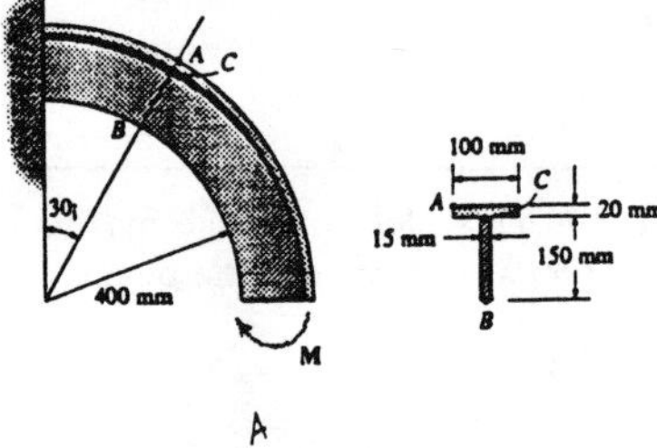

Internal Moment : $M = -900\ \text{N}\cdot\text{m}$ is negative since it tends to decrease the beam's radius of curvature.

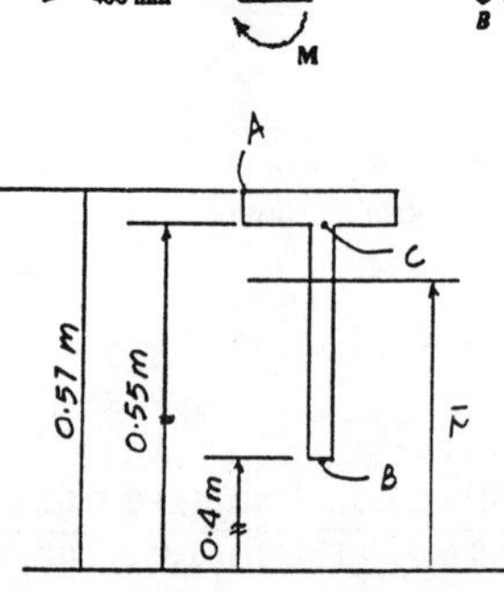

Section Properties :

$$\Sigma A = 0.15(0.015) + 0.1(0.02) = 0.00425\ \text{m}^2$$

$$\Sigma \bar{r}A = 0.475(0.15)(0.015) + 0.56(0.1)(0.02) = 2.18875(10^{-3})\ \text{m}^3$$

$$\bar{r} = \frac{\Sigma \bar{r}A}{\Sigma A} = \frac{2.18875\,(10^{-3})}{0.00425} = 0.5150\ \text{m}$$

$$\Sigma \int_A \frac{dA}{r} = 0.015\ \ln\frac{0.55}{0.4} + 0.1\ \ln\frac{0.57}{0.55} = 8.348614(10^{-3})\ \text{m}$$

$$R = \frac{A}{\Sigma\int_A \frac{dA}{r}} = \frac{0.00425}{8.348614(10^{-3})} = 0.509067\ \text{m}$$

$$\bar{r} - R = 0.515 - 0.509067 = 5.933479(10^{-3})\ \text{m}$$

Normal Stress : Applying the curved-beam formula

$$\sigma_C = \frac{M(R - r_C)}{Ar_C(\bar{r} - R)} = \frac{-900(0.509067 - 0.55)}{0.00425(0.55)(5.933479)(10^{-3})}$$

$= 2.66$ MPa (T) **Ans**

***6–132.** The circular spring clamp produces a compressive force of 3 N on the plates. Determine the maximum bending stress produced in the spring at *A*. The spring has a rectangular cross section as shown.

Internal Moment : As shown on FBD, $M = 0.660\ \text{N}\cdot\text{m}$ is positive since it tends to increase the beam's radius of curvature.

Section Properties :

$$\bar{r} = \frac{0.200+0.210}{2} = 0.205\ \text{m}$$

$$\int_A \frac{dA}{r} = b\ln\frac{r_2}{r_1} = 0.02\ln\frac{0.21}{0.20} = 0.97580328(10^{-3})\ \text{m}$$

$$A = (0.01)(0.02) = 0.200(10^{-3})\ \text{m}^2$$

$$R = \frac{A}{\int_A \frac{dA}{r}} = \frac{0.200(10^{-3})}{0.97580328(10^{-3})} = 0.204959343\ \text{m}$$

$$\bar{r} - R = 0.205 - 0.204959343 = 0.040657(10^{-3})\ \text{m}$$

Maximum Normal Stress : Applying the curved-beam formula

$$\sigma_C = \frac{M(R - r_2)}{Ar_2(\bar{r} - R)} = \frac{0.660(0.204959343 - 0.21)}{0.200(10^{-3})(0.21)(0.040657)(10^{-3})}$$

$$= -1.95\text{MPa} = 1.95\ \text{MPa (C)}$$

$$\sigma_t = \frac{M(R - r_1)}{Ar_1(\bar{r} - R)} = \frac{0.660(0.204959343 - 0.2)}{0.200(10^{-3})(0.2)(0.040657)(10^{-3})}$$

$$= 2.01\ \text{MPa (T)} \quad \textbf{(Max)}$$

Ans

6–133. Determine the maximum compressive force the spring clamp can exert on the plates if the allowable bending stress for the clamp is $\sigma_{allow} = 4$ MPa.

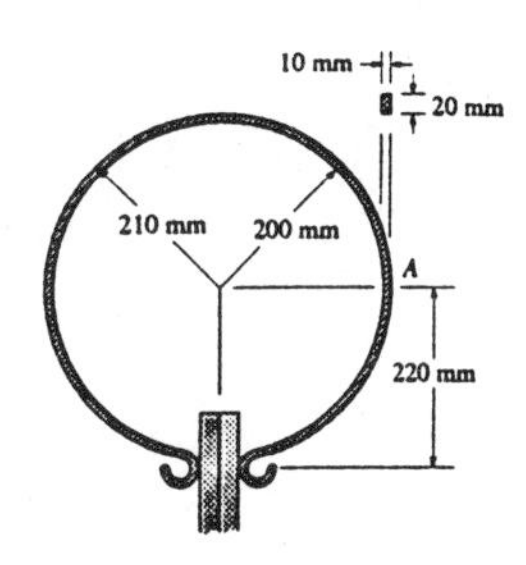

Section Properties :

$$\bar{r} = \frac{0.200+0.210}{2} = 0.205\ \text{m}$$

$$\int_A \frac{dA}{r} = b\ln\frac{r_2}{r_1} = 0.02\ln\frac{0.21}{0.20} = 0.97580328(10^{-3})\ \text{m}$$

$$A = (0.01)(0.02) = 0.200(10^{-3})\ \text{m}^2$$

$$R = \frac{A}{\int_A \frac{dA}{r}} = \frac{0.200(10^{-3})}{0.97580328(10^{-3})} = 0.204959\ \text{m}$$

$$\bar{r} - R = 0.205 - 0.204959343 = 0.040657(10^{-3})\ \text{m}$$

Internal Moment : The internal moment must be computed about the neutral axis as shown on FBD. $M_{max} = 0.424959P$ is positive since it tends to increase the beam's radius of curvature.

Allowable Normal Stress : Applying the curved-beam formula

Assume compression failure

$$\sigma_c = \sigma_{allow} = \frac{M(R - r_2)}{Ar_2(\bar{r} - R)}$$

$$-4(10^6) = \frac{0.424959P(0.204959 - 0.21)}{0.200(10^{-3})(0.21)(0.040657)(10^{-3})}$$

$$P = 3.189\ \text{N}$$

Assume tension failure

$$\sigma_t = \sigma_{allow} = \frac{M(R - r_1)}{Ar_1(\bar{r} - R)}$$

$$4(10^6) = \frac{0.424959P(0.204959 - 0.2)}{0.200(10^{-3})(0.2)(0.040657)(10^{-3})}$$

$$P = 3.09\ \text{N}\ (\textit{Controls!})$$

Ans

6–134. The steel rod has a circular cross section. If it is gripped at its ends and a couple moment of $M = 12$ lb · in. is developed at each grip, determine the stress acting at points A and B and at the centroid C.

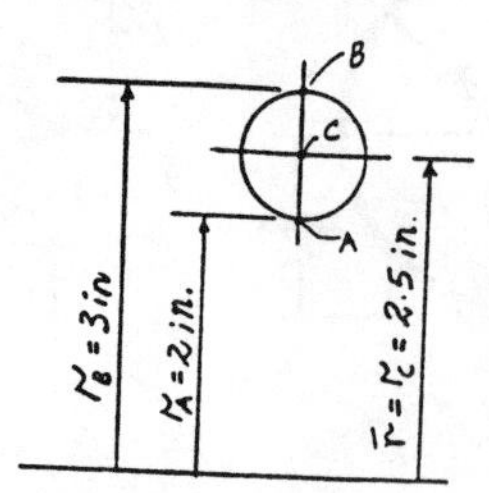

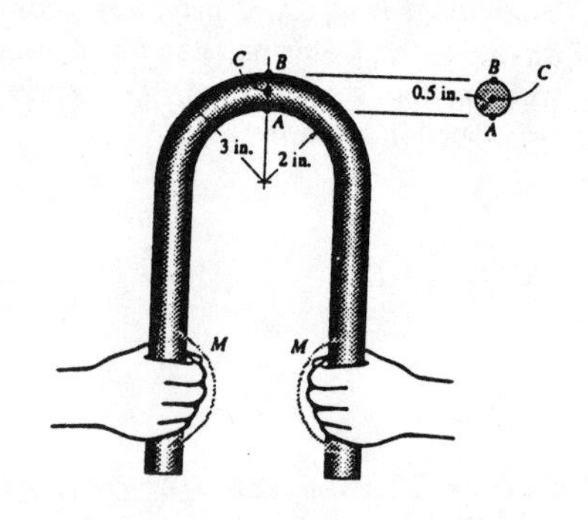

Internal Moment : $M = 12$ lb · in is positive since it tends to increase the beam's radius of curvature.

Section Properties :

$$\int_A \frac{dA}{r} = 2\pi\left(\bar{r} - \sqrt{\bar{r}^2 - c^2}\right)$$

$$= 2\pi\left(2.5 - \sqrt{2.5^2 - 0.5^2}\right) = 0.317365 \text{ in.}$$

$$A = \pi c^2 = \pi\left(0.5^2\right) = 0.25\pi$$

$$R = \frac{A}{\int_A \frac{dA}{r}} = \frac{0.25\pi}{0.317365} = 2.474745 \text{ in.}$$

$$\bar{r} - R = 2.5 - 2.474745 = 0.025255 \text{ in.}$$

Normal Stress : Applying the curved - beam formula

$$\sigma_A = \frac{M(R - r_A)}{A r_A(\bar{r} - R)}$$

$$= \frac{12(2.474745 - 2)}{0.25\pi(2)(0.025255)} = 144 \text{ psi (T)} \qquad \textbf{Ans}$$

$$\sigma_B = \frac{M(R - r_B)}{A r_B(\bar{r} - R)}$$

$$= \frac{12(2.474745 - 3)}{0.25\pi(3)(0.025255)} = -106 \text{ psi} = 106 \text{ psi (C)} \qquad \textbf{Ans}$$

$$\sigma_C = \frac{M(R - r_C)}{A r_C(\bar{r} - R)}$$

$$= \frac{12(2.474745 - 2.5)}{0.25\pi(2.5)(0.025255)} = -6.11 \text{ psi} = 6.11 \text{ psi (C)} \qquad \textbf{Ans}$$

6–135. The ceiling-suspended C-arm is used to support the X-ray camera used in medical diagnoses. If the camera has a mass of 150 kg, with center of mass at G, determine the maximum bending stress at section A.

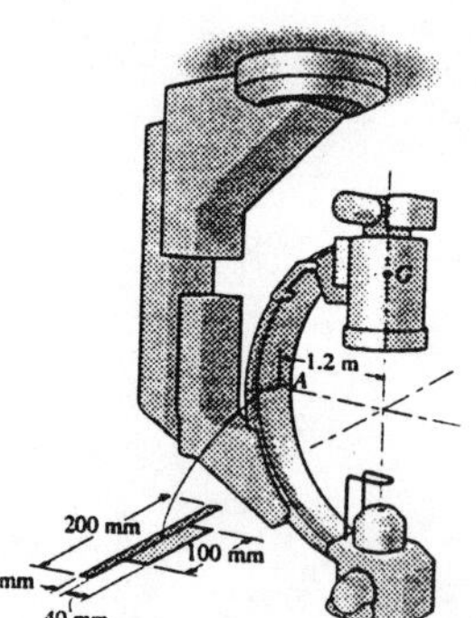

150(9.81) = 1471.5 N

M = 1816.93 N·m

R = 1.234749 m

1471.5 N

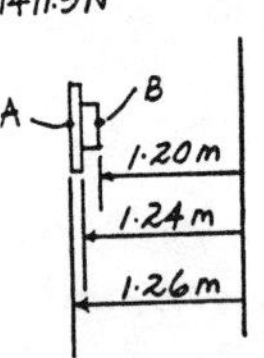

Section Properties :

$$\bar{r} = \frac{\Sigma \bar{r} A}{\Sigma A} = \frac{1.22(0.1)(0.04) + 1.25(0.2)(0.02)}{0.1(0.04) + 0.2(0.02)} = 1.235 \text{ m}$$

$$\Sigma \int_A \frac{dA}{r} = 0.1 \ln \frac{1.24}{1.20} + 0.2 \ln \frac{1.26}{1.24} = 6.479051\left(10^{-3}\right) \text{ m}$$

$$A = 0.1(0.04) + 0.2(0.02) = 0.008 \text{ m}^2$$

$$R = \frac{A}{\int_A \frac{dA}{r}} = \frac{0.008}{6.479051(10^{-3})} = 1.234749 \text{ m}$$

$$\bar{r} - R = 1.235 - 1.234749 = 0.251183\left(10^{-3}\right) \text{ m}$$

Internal Moment : The internal moment must be computed about the neutral axis as shown on FBD. $M = -1816.93$ N · m is negative since it tends to decrease the beam's radius of curvature.

Maximum Normal Stress : Applying the curved - beam formula

$$\sigma_A = \frac{M(R - r_A)}{A r_A(\bar{r} - R)}$$

$$= \frac{-1816.93(1.234749 - 1.26)}{0.008(1.26)(0.251183)(10^{-3})}$$

$$= 18.1 \text{MPa (T)}$$

$$\sigma_B = \frac{M(R - r_B)}{A r_B(\bar{r} - R)}$$

$$= \frac{-1816.93(1.234749 - 1.20)}{0.008(1.20)(0.251183)(10^{-3})}$$

$$= -26.2 \text{ MPa} = 26.2 \text{ MPa (C)} \qquad \text{(Max)} \qquad \textbf{Ans}$$

***6–136.** The curved bar used on a machine has a rectangular cross section. If the bar is subjected to a couple as shown, determine the maximum tensile and compressive stresses acting at section *a–a*. Sketch the stress distribution on the section in three dimensions.

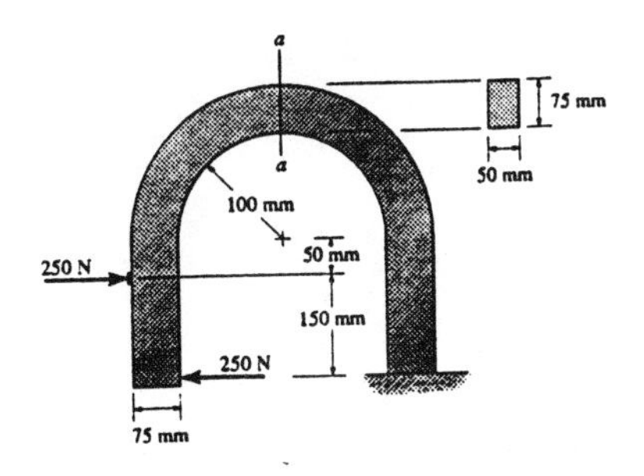

Internal Moment : $M = 37.5$ N · m is positive since it tends to increase the beam's radius of curvature.

Section Properties :

$$\bar{r} = \frac{0.1 + 0.175}{2} = 0.1375 \text{ m}$$

$$A = 0.075(0.05) = 0.00375 \text{ m}^2$$

$$\int_A \frac{dA}{r} = 0.05 \ln \frac{0.175}{0.1} = 0.027981 \text{ m}$$

$$R = \frac{A}{\int_A \frac{dA}{r}} = \frac{0.00375}{0.027981} = 0.134021 \text{ m}$$

$$\bar{r} - R = 0.1375 - 0.134021 = 3.479478(10^{-3}) \text{ m}$$

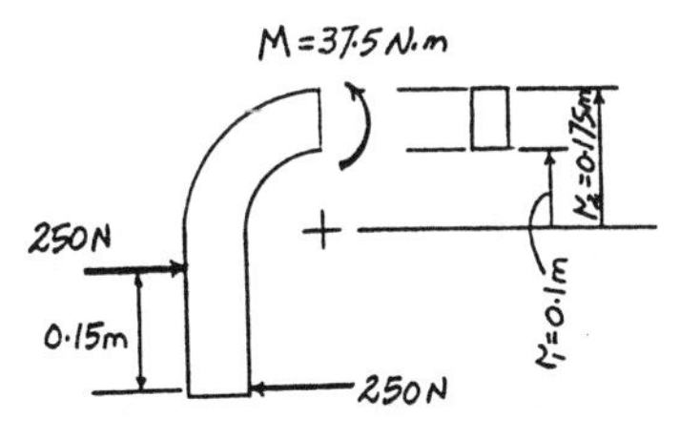

Normal Stress : Applying the curved-beam formula

$$(\sigma_{max})_t = \frac{M(R - r_1)}{Ar_1(\bar{r} - R)}$$
$$= \frac{37.5(0.134021 - 0.1)}{0.00375(0.1)(3.479478)(10^{-3})}$$
$$= 0.978 \text{ MPa (T)} \qquad \textbf{Ans}$$

$$(\sigma_{max})_c = \frac{M(R - r_2)}{Ar_2(\bar{r} - R)}$$
$$= \frac{37.5(0.134021 - 0.175)}{0.00375(0.175)(3.479478)(10^{-3})}$$
$$= -0.673 \text{ MPa} = 0.673 \text{ MPa (C)} \qquad \textbf{Ans}$$

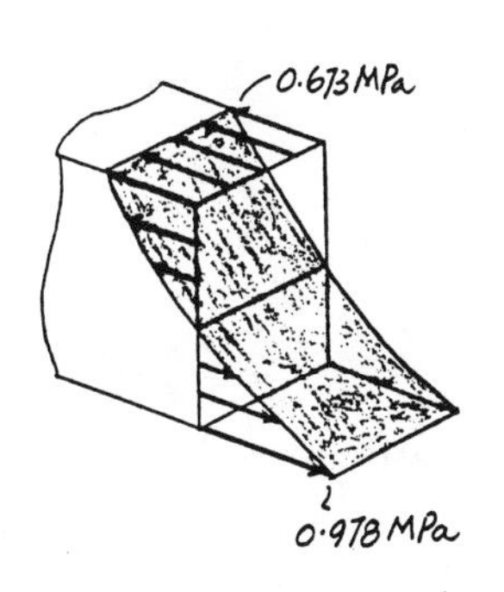

6–137. For the curved beam in Fig. 6–44*a*, show that when the radius of curvature approaches infinity, the curved-beam formula, Eq. 6–24, reduces to the flexure formula, Eq. 6–13.

Normal Stress : Curved - beam formula

$$\sigma = \frac{M(R-r)}{Ar(\bar{r}-R)} \quad \text{where } A' = \int_A \frac{dA}{r} \quad \text{and } R = \frac{A}{\int_A \frac{dA}{r}} = \frac{A}{A'}$$

$$\sigma = \frac{M(A - rA')}{Ar(\bar{r}A' - A)} \qquad [1]$$

$$r = \bar{r} + y \qquad [2]$$

$$\bar{r}A' = \bar{r}\int_A \frac{dA}{r} = \int_A \left(\frac{\bar{r}}{\bar{r}+y} - 1 + 1\right) dA$$

$$= \int_A \left(\frac{\bar{r} - \bar{r} - y}{\bar{r}+y} + 1\right) dA$$

$$= A - \int_A \frac{y}{\bar{r}+y}\, dA \qquad [3]$$

Denominator of Eq. [1] becomes,

$$Ar(\bar{r}A' - A) = Ar\left(A - \int_A \frac{y}{\bar{r}+y}\, dA - A\right) = -Ar\int_A \frac{y}{\bar{r}+y}\, dA$$

Using Eq. [2],

$$Ar(\bar{r}A' - A) = -A\int_A \left(\frac{\bar{r}y}{\bar{r}+y} + y - y\right) dA - Ay\int_A \frac{y}{\bar{r}+y}\, dA$$

$$= A\int_A \frac{y^2}{\bar{r}+y}\, dA - A\int_A y\, dA - Ay\int_A \frac{y}{\bar{r}+y}\, dA$$

$$= \frac{A}{\bar{r}}\int_A \left(\frac{y^2}{1+\frac{y}{\bar{r}}}\right) dA - A\int_A y\, dA - \frac{Ay}{\bar{r}}\int_A \left(\frac{y}{1+\frac{y}{\bar{r}}}\right) dA$$

But, $\int_A y\, dA = 0,$ as $\frac{y}{\bar{r}} \to 0$

Then, $Ar(\bar{r}A' - A) \to \frac{A}{\bar{r}}I$

Eq. [1] becomes $\sigma = \frac{M\bar{r}}{AI}(A - rA')$

Using Eq. [2], $\sigma = \frac{M\bar{r}}{I}(A - \bar{r}A - yA')$

Using Eq. [3],

$$\sigma = \frac{M\bar{r}}{AI}\left[A - \left(A - \int_A \frac{y}{\bar{r}+y}\, dA\right) - y\int_A \frac{dA}{\bar{r}+y}\right]$$

$$= \frac{M\bar{r}}{AI}\left[\int_A \frac{y}{\bar{r}+y}\, dA - y\int_A \frac{dA}{\bar{r}+y}\right]$$

$$= \frac{M\bar{r}}{AI}\left[\int_A \left(\frac{\frac{y}{\bar{r}}}{1+\frac{y}{\bar{r}}}\right) dA - \frac{y}{\bar{r}}\int_A \left(\frac{dA}{1+\frac{y}{\bar{r}}}\right)\right]$$

As $\frac{y}{\bar{r}} \to 0$

$$\int_A \left(\frac{\frac{y}{\bar{r}}}{1+\frac{y}{\bar{r}}}\right) dA = 0 \quad \text{and} \quad \frac{y}{\bar{r}}\int_A \left(\frac{dA}{1+\frac{y}{\bar{r}}}\right) = \frac{y}{\bar{r}}\int_A dA = \frac{yA}{\bar{r}}$$

Therefore, $\sigma = \frac{M\bar{r}}{AI}\left(-\frac{yA}{\bar{r}}\right) = -\frac{My}{I}$ **(Q. E. D.)**

6–138. The member has an elliptical cross section. If it is subjected to a moment of $M = 100\text{ N}\cdot\text{m}$, determine the stress at points A and B. Is the stress at point A', which is located on the member near the wall, the same as that at A? Explain.

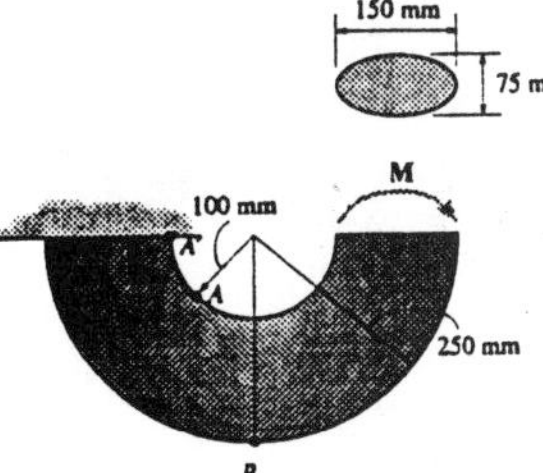

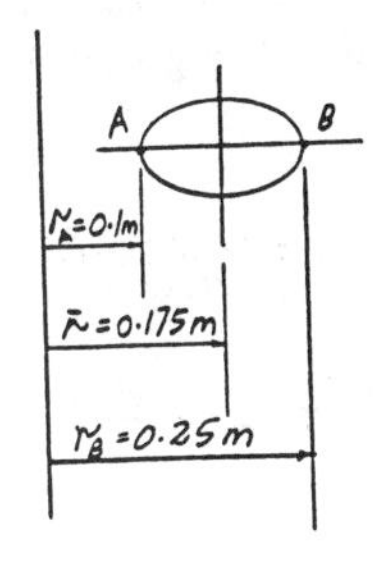

B

Internal Moment : $M = 100\text{ N}\cdot\text{m}$ is positive since it t increase the beam's radius of curvature.

Section Properties :

$a = 0.075\text{ m}$ and $b = 0.0375\text{ m}$

$$A = \pi(0.075)(0.0375) = 0.0028125\pi\text{ m}^2$$

$$\int_A \frac{dA}{r} = \frac{2\pi b}{a}\left(\bar{r} - \sqrt{\bar{r}^2 - a^2}\right)$$

$$= \frac{2\pi(0.0375)}{0.075}\left(0.175 - \sqrt{0.175^2 - 0.075^2}\right)$$

$$= 0.053049\text{ m}$$

$$R = \frac{A}{\int_A \frac{dA}{r}} = \frac{0.0028125\pi}{0.053049} = 0.166557\text{ m}$$

$$\bar{r} - R = 0.175 - 0.166557 = 8.4430(10^{-3})\text{ m}$$

Maximum Normal Stress : Applying the curved - beam formula

$$\sigma_A = \frac{M(R - r_A)}{A r_A(\bar{r} - R)}$$

$$= \frac{100(0.166557 - 0.1)}{0.0028125\pi(0.1)(8.4430)(10^{-3})}$$

$$= 0.892\text{ MPa (T)} \qquad \textbf{Ans}$$

$$\sigma_B = \frac{M(R - r_B)}{A r_B(\bar{r} - R)}$$

$$= \frac{100(0.166557 - 0.25)}{0.0028125\pi(0.25)(8.4430)(10^{-3})}$$

$$= -0.447\text{ MPa} = 0.447\text{ MPa (C)} \qquad \textbf{Ans}$$

No! This is a consequence of Saint - Venant's principle. **Ans**

6–139. The member has an elliptical cross section. If the allowable bending stress is $\sigma_{allow} = 125\text{ MPa}$, determine the maximum moment M that can be applied to the member.

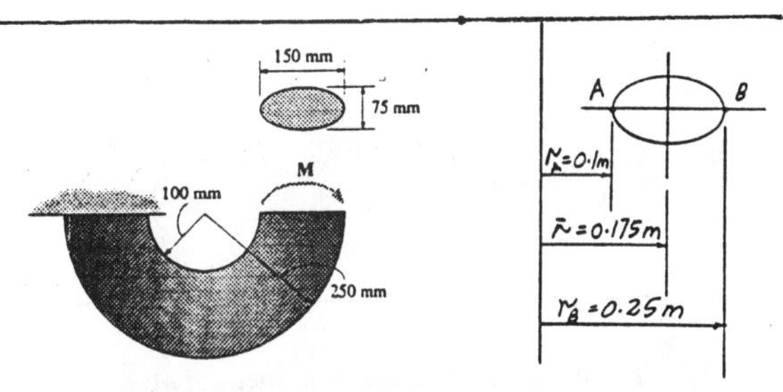

Internal Moment : $M = 100\text{ N}\cdot\text{m}$ is positive since it tends to increase the beam's radius of curvature.

Section Properties :

$a = 0.075\text{ m}$ and $b = 0.0375\text{ m}$

$$A = \pi(0.075)(0.0375) = 0.0028125\pi\text{ m}^2$$

$$\int_A \frac{dA}{r} = \frac{2\pi b}{a}\left(\bar{r} - \sqrt{\bar{r}^2 - a^2}\right)$$

$$= \frac{2\pi(0.0375)}{0.075}\left(0.175 - \sqrt{0.175^2 - 0.075^2}\right)$$

$$= 0.053049\text{ m}$$

$$R = \frac{A}{\int_A \frac{dA}{r}} = \frac{0.0028125\pi}{0.053049} = 0.166557\text{ m}$$

$$\bar{r} - R = 0.175 - 0.166557 = 8.4430(10^{-3})\text{ m}$$

Allowable Normal Stress : Applying the curved - beam formula

Assume compression failure

$$\sigma_B = \sigma_{allow} = \frac{M(R - r_B)}{A r_B(\bar{r} - R)}$$

$$-125(10^6) = \frac{M(0.166557 - 0.25)}{0.0028125\pi(0.25)(8.4430)(10^{-3})}$$

$$M = 27938\text{ N}\cdot\text{m} = 27.94\text{ kN}\cdot\text{m}$$

Assume tension failure

$$\sigma_A = \sigma_{allow} = \frac{M(R - r_A)}{A r_A(\bar{r} - R)}$$

$$125(10^6) = \frac{M(0.166557 - 0.1)}{0.0028125\pi(0.1)(8.4430)(10^{-3})}$$

$$M = 14011\text{ N}\cdot\text{m}$$

$$= 14.0\text{ kN}\cdot\text{m}\ (\textit{controls!}) \qquad \textbf{Ans}$$

***6–140.** The bar is subjected to a moment of $M = 40\text{ N}\cdot\text{m}$. Determine the smallest radius r of the fillets so that an allowable bending stress of $\sigma_{allow} = 124\text{ MPa}$ is not exceeded.

Allowable Bending Stress :

$$\sigma_{allow} = K\frac{Mc}{I}$$

$$124\left(10^{6}\right) = K\left[\frac{40(0.01)}{\frac{1}{12}(0.007)(0.02^{3})}\right]$$

$$K = 1.45$$

Stress Concentration Factor : From the graph in the text with $\frac{w}{h} = \frac{80}{20} = 4$ and $K = 1.45$, then $\frac{r}{h} = 0.25$.

$$\frac{r}{20} = 0.25$$

$$r = 5.00\text{ mm} \qquad \textbf{Ans}$$

6–141. The bar is subjected to a moment of $M = 17.5\text{ N}\cdot\text{m}$. If $r = 5$ mm, determine the maximum bending stress in the material.

Stress Concentration Factor : From the graph in the text with $\frac{w}{h} = \frac{80}{20} = 4$ and $\frac{r}{h} = \frac{5}{20} = 0.25$, then $K = 1.45$.

Maximum Bending Stress :

$$\sigma_{max} = K\frac{Mc}{I}$$

$$= 1.45\left[\frac{17.5(0.01)}{\frac{1}{12}(0.007)(0.02^{3})}\right]$$

$$= 54.4\text{ MPa} \qquad \textbf{Ans}$$

6–142. The bar is subjected to a moment of $M = 20\text{ N}\cdot\text{m}$. Determine the maximum bending stress in the bar and sketch, approximately, how the stress varies over the critical section.

Stress Concentration Factor : From the graph in the text with $\frac{w}{h} = \frac{30}{10} = 3$ and $\frac{r}{h} = \frac{1.5}{10} = 0.15$, then $K = 1.6$.

Maximum Bending Stress :

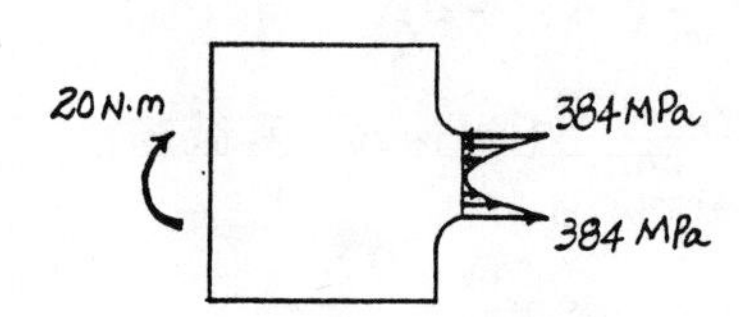

$$\sigma_{max} = K\frac{Mc}{I}$$

$$= 1.6\left[\frac{20(0.005)}{\frac{1}{12}(0.005)(0.01^{3})}\right]$$

$$= 384\text{ MPa} \qquad \textbf{Ans}$$

6–143. The allowable bending stress for the bar is $\sigma_{allow} = 175$ MPa. Determine the maximum moment M that can be applied to the bar.

Stress Concentration Factor : From the graph in the text with $\frac{w}{h} = \frac{30}{10} = 3$ and $\frac{r}{h} = \frac{1.5}{10} = 0.15$, then $K = 1.6$.

Maximum Bending Stress :

$$\sigma_{max} = \sigma_{allow} = K\frac{Mc}{I}$$

$$175(10^6) = 1.6\left[\frac{M(0.005)}{\frac{1}{12}(0.005)(0.01^3)}\right]$$

$$M = 9.11 \text{ N}\cdot\text{m} \qquad \textbf{Ans}$$

***6–144.** Determine the length L of the center portion of the bar so that the maximum bending stress at sections A, B, and C is the same. The bar has a thickness of 10 mm.

Support Reactions : As shown on FBD.

Internal Moments : Using the method of sections

$$\circlearrowleft + \Sigma M_A = 0; \qquad M_A - 175L(0.3) = 0$$

$$M_A = 52.5\,L$$

$$\circlearrowleft + \Sigma M_B = 0; \qquad M_B = 175L\left(0.3 + \frac{L}{4}\right)$$

$$M_B = 52.5\,L + 43.75L^2$$

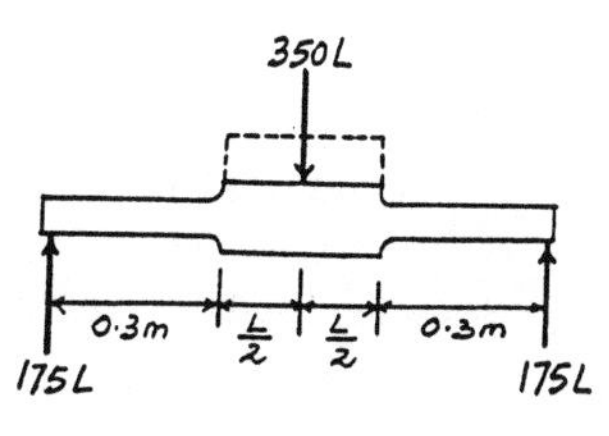

Stress Concentration Factor : From the graph in the texxt with $\frac{w}{h} = \frac{60}{40} = 1.5$ and $\frac{r}{h} = \frac{8}{40} = 0.2$, then $K = 1.45$.

Maximum Bending Stress :

At either section A or C

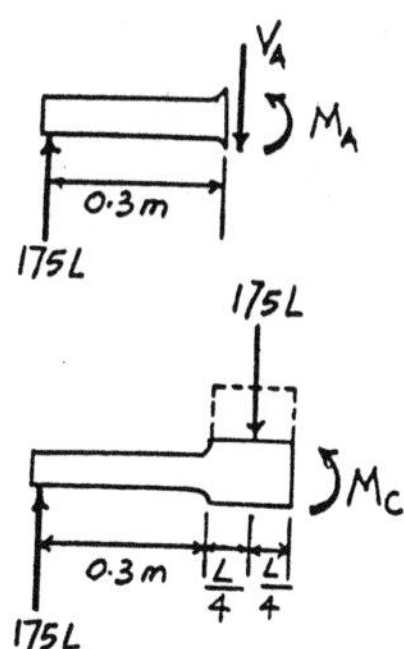

$$(\sigma_{max})_A = K\frac{M_A c}{I}$$

$$= 1.45\left[\frac{52.5L(0.02)}{\frac{1}{12}(0.01)(0.04^3)}\right]$$

$$= 28\,546\,875\,L$$

At section C

$$(\sigma_{max})_C = \frac{M_C c}{I}$$

$$= \frac{(52.5L + 43.75L^2)(0.03)}{\frac{1}{12}(0.01)(0.06^3)}$$

$$= 8\,750\,000L + 7\,291\,666.67L^2$$

Require,

$$(\sigma_{max})_A = (\sigma_{max})_C$$

$$28\,546\,875\,L = 8\,750\,000L + 7\,291\,666.67L^2$$

$$L = 2.715 \text{ m} \qquad \textbf{Ans}$$

6–145. If the radius of each notch on the plate is $r = 10$ mm, determine the largest moment M that can be applied. The allowable bending stress for the material is $\sigma_{allow} = 175$ MPa.

Stress Concentration Factor : From the graph in the text with $\frac{b}{r} = \frac{20}{10} = 2$ and $\frac{r}{h} = \frac{10}{125} = 0.08$, then $K = 2.1$.

Allowable Bending Stress :

$$\sigma_{max} = \sigma_{allow} = K\frac{Mc}{I}$$

$$180(10^6) = 2.1\left[\frac{M(0.0625)}{\frac{1}{12}(0.02)(0.125^3)}\right]$$

$$M = 4464 \text{ N}\cdot\text{m} = 4.46 \text{ kN}\cdot\text{m} \qquad \textbf{Ans}$$

6–146. The symmetric notched plate is subjected to bending. If the radius of each notch is $r = 10$ mm and the applied moment is $M = 2$ kN · m, determine the maximum bending stress in the plate.

Stress Concentration Factor : From the graph in the text with $\frac{b}{r} = \frac{20}{10} = 2$ and $\frac{r}{h} = \frac{10}{125} = 0.08$, then $K = 2.1$.

Allowable Bending Stress :

$$\sigma_{max} = K\frac{Mc}{I}$$

$$= 2.1\left[\frac{2(10^3)(0.0625)}{\frac{1}{12}(0.02)(0.125^3)}\right]$$

$$= 80.6 \text{ MPa} \qquad \textbf{Ans}$$

6–147. The bar has a thickness of 0.5 in. and is made of a material having an allowable bending stress of $\sigma_{allow} = 20$ ksi. Determine the maximum moment M that can be applied.

Stress Concentration Factor : From the graph in the text with $\frac{w}{h} = \frac{6}{2} = 3$ and $\frac{r}{h} = \frac{0.3}{2} = 0.15$, then $K = 1.6$.

Allowable Bending Stress :

$$\sigma_{max} = \sigma_{allow} = K\frac{Mc}{I}$$

$$20 = 1.6\left[\frac{M(1)}{\frac{1}{12}(0.5)(2^3)}\right]$$

$$M = 4.167 \text{ kip}\cdot\text{in} = 347 \text{ lb}\cdot\text{ft} \qquad \textbf{Ans}$$

***6–148.** The bar has a thickness of 0.5 in. and is subjected to a moment of 600 lb · ft. Determine the maximum bending stress in the bar.

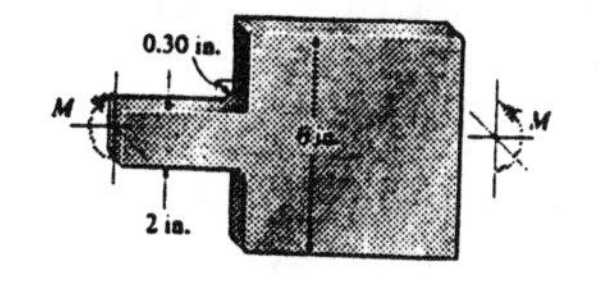

Stress Concentration Factor : From the graph in the text with $\frac{w}{h} = \frac{6}{2} = 3$ and $\frac{r}{h} = \frac{0.3}{2} = 0.15$, then $K = 1.6$.

Maximum Bending Stress :

$$\sigma_{\max} = K\frac{Mc}{I}$$

$$= 1.6\left[\frac{600(12)(1)}{\frac{1}{12}(0.5)(2^3)}\right]$$

$$= 34.6 \text{ ksi} \qquad \textbf{Ans}$$

6–149. The stepped bar has a thickness of 15 mm. Determine the maximum moment that can be applied to its ends if it is made of a material having an allowable bending stress of $\sigma_{\text{allow}} = 200$ MPa.

Stress Concentration Factor :

For the smaller section with $\frac{w}{h} = \frac{30}{10} = 3$ and $\frac{r}{h} = \frac{6}{10} = 0.6$, we have $K = 1.2$ obtained from the graph in the text.

For the larger section with $\frac{w}{h} = \frac{45}{30} = 1.5$ and $\frac{r}{h} = \frac{3}{30} = 0.3$, we have $K = 1.75$ obtained from the graph in the text.

Allowable Bending Stress :

For the smaller section

$$\sigma_{\max} = \sigma_{\text{allow}} = K\frac{Mc}{I}$$

$$200(10^6) = 1.2\left[\frac{M(0.005)}{\frac{1}{12}(0.015)(0.01^3)}\right]$$

$$M = 41.7 \text{ N}\cdot\text{m} \ \textit{(Controls!)} \qquad \textbf{Ans}$$

For the larger section

$$\sigma_{\max} = \sigma_{\text{allow}} = K\frac{Mc}{I}\ ;$$

$$200(10^6) = 1.75\left[\frac{M(0.015)}{\frac{1}{12}(0.015)(0.03^3)}\right]$$

$$M = 257 \text{ N}\cdot\text{m}$$

6–150. The stepped bar has a thickness of 15 mm. If $M = 50\ \text{N}\cdot\text{m}$, determine the maximum bending stress in the bar. Specify where this stress occurs, and sketch the stress distribution over the cross section.

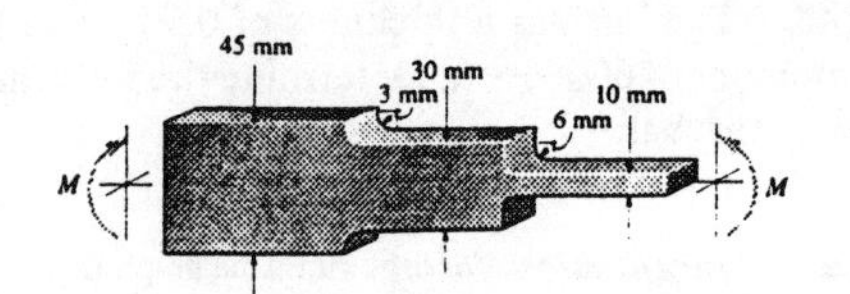

Stress Concentration Factor :

For the smaller section with $\frac{w}{h} = \frac{30}{10} = 3$ and $\frac{r}{h} = \frac{6}{10} = 0.6$, we have $K = 1.2$ obtained from the graph in the text.

For the larger section with $\frac{w}{h} = \frac{45}{30} = 1.5$ and $\frac{r}{h} = \frac{3}{30} = 0.3$, we have $K = 1.75$ obtained from the graph in the text.

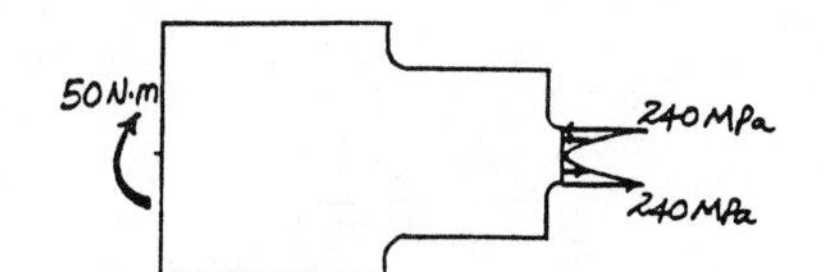

Maximum Bending Stress :

For the smaller section

$$\sigma_{\max} = K\frac{Mc}{I}$$
$$= 1.2\left[\frac{50(0.005)}{\frac{1}{12}(0.015)(0.01^3)}\right]$$
$$= 240\ \text{MPa}\quad \textbf{(Max)}\qquad \textbf{Ans}$$

The maximum bending stress occurs at the **smaller section.** **Ans**

For the larger section

$$\sigma_{\max} = K\frac{Mc}{I}$$
$$= 1.75\left[\frac{50(0.015)}{\frac{1}{12}(0.015)(0.03^3)}\right]$$
$$= 38.89\ \text{MPa}$$

6–151. The channel strut is made of an elastic-perfectly plastic material for which $\sigma_Y = 250$ MPa. Determine the maximum elastic moment and the plastic moment that can be applied to the cross section.

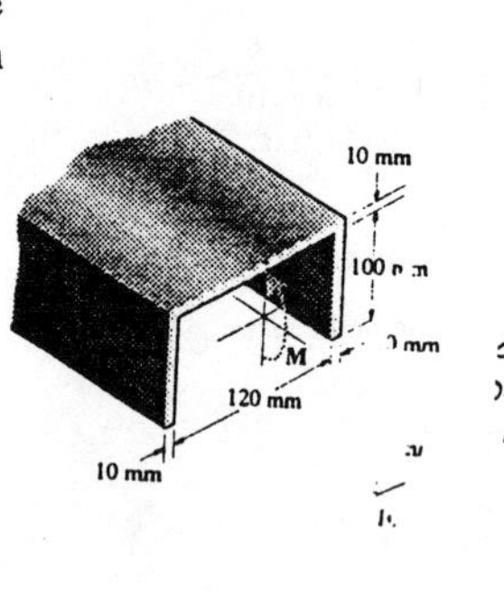

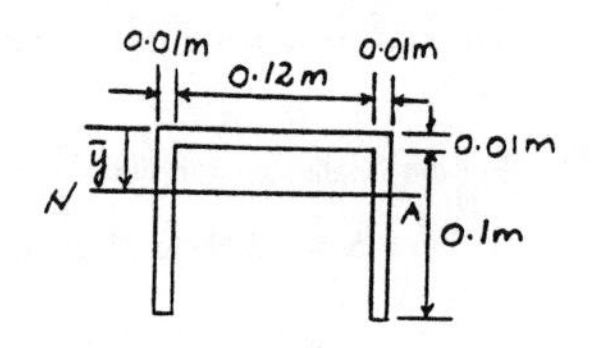

Maximum Elastic Moment : The centroid and the moment of inertia about neutral axis must be determined first.

$$\bar{y} = \frac{\Sigma\tilde{y}A}{\Sigma A} = \frac{0.005(0.14)(0.01) + 2[(0.06)(0.1)(0.01)]}{0.14(0.01) + 2(0.1)(0.01)}$$
$$= 0.03735\ \text{m}$$

$$I_{NA} = \frac{1}{12}(0.14)\left(0.01^3\right) + 0.14(0.01)(0.03735 - 0.005)^2$$
$$+ \frac{1}{12}(0.02)\left(0.1^3\right) + 0.02(0.1)(0.06 - 0.03735)^2$$
$$= 4.16951\left(10^{-6}\right)\ \text{m}^4$$

Applying the flexure formula with $\sigma = \sigma_Y$, we have

$$\sigma_Y = \frac{M_Y c}{I}$$
$$M_Y = \frac{\sigma_Y I}{c} = \frac{250(10^6)(4.16951)(10^{-6})}{(0.11 - 0.03735)}$$
$$= 14349\ \text{N}\cdot\text{m} = 14.3\ \text{kN}\cdot\text{m}\qquad \textbf{Ans}$$

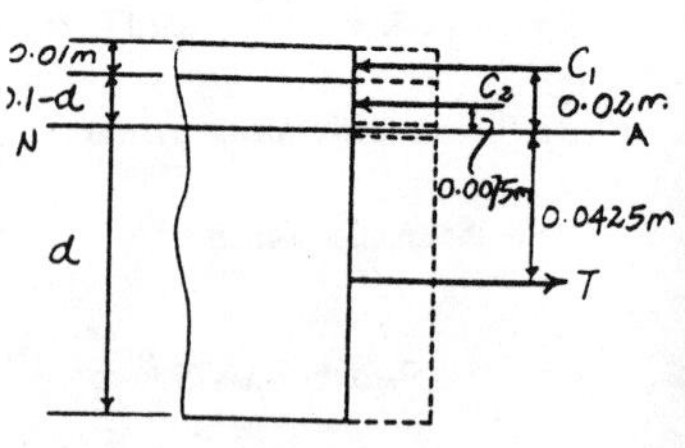

Plastic Moment :

$$\int \sigma\, dA = 0;\quad C_1 + C_2 - T = 0$$
$$\sigma_Y(0.14)(0.01) + \sigma_Y(0.1 - d)(0.02) - \sigma_Y(d)(0.02) = 0$$

$$d = 0.0850\ \text{m} < 0.1\ \text{m}$$

$$M_P = 250\left(10^6\right)(0.01)(0.14)(0.02)$$
$$+ 250\left(10^6\right)(0.015)(0.02)(0.0075)$$
$$+ 250\left(10^6\right)(0.085)(0.02)(0.0425)$$
$$= 25625\ \text{N}\cdot\text{m} = 25.6\ \text{kN}\cdot\text{m}\qquad \textbf{Ans}$$

***6–152.** A rectangular A-36 steel bar has a width of 1 in. and height of 3 in. Determine the moment applied about the horizontal axis that will cause half the bar to yield.

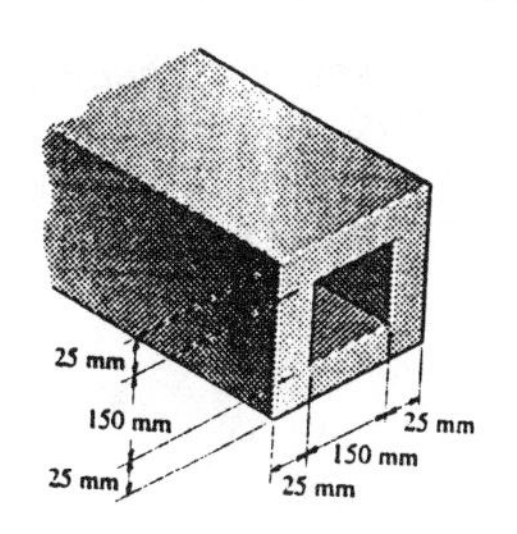

Elastic - Plastic Moment :

$$M = 36(0.75)(1)(2.25) + 36\left(\frac{1}{2}\right)(0.75)(1)(1)$$
$$= 74.25 \text{ kip} \cdot \text{in}$$
$$= 6.19 \text{ kip} \cdot \text{ft} \qquad \textbf{Ans}$$

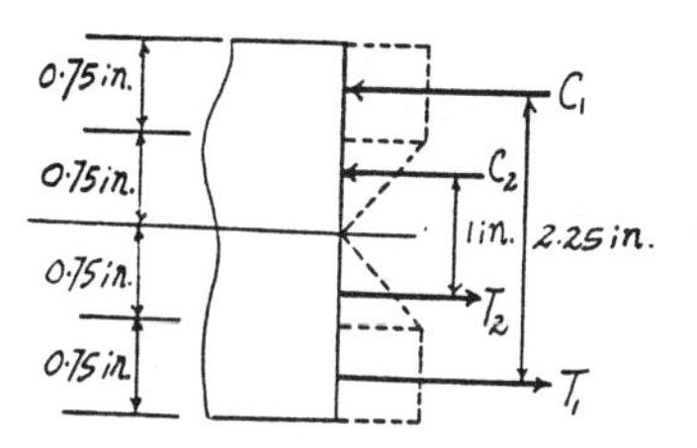

6–153. Determine the shape factor of the beam's cross section.

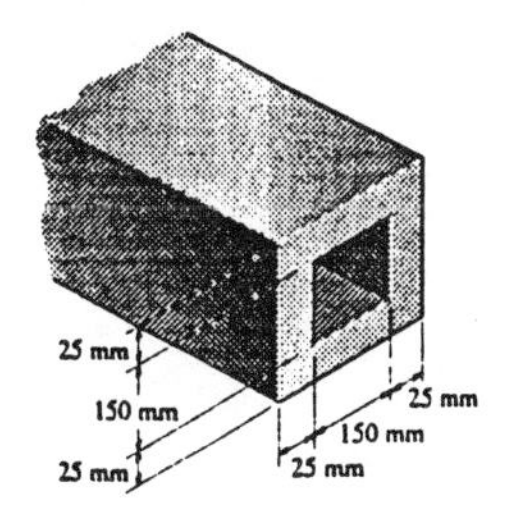

Maximum Elastic Moment : The moment of inertia about neutral axis must be determined first.

$$I_{NA} = \frac{1}{12}(0.2)\left(0.2^3\right) - \frac{1}{12}(0.15)\left(0.15^3\right)$$
$$= 91.14583\left(10^{-6}\right) \text{ m}^4$$

Applyingthe flexure formula with $\sigma = \sigma_Y$, we have

$$\sigma_Y = \frac{M_Y c}{I}$$
$$M_Y = \frac{\sigma_Y I}{c}$$
$$= \frac{\sigma_Y (91.14583)(10^{-6})}{0.1}$$
$$= 0.9114583\left(10^{-3}\right)\sigma_Y$$

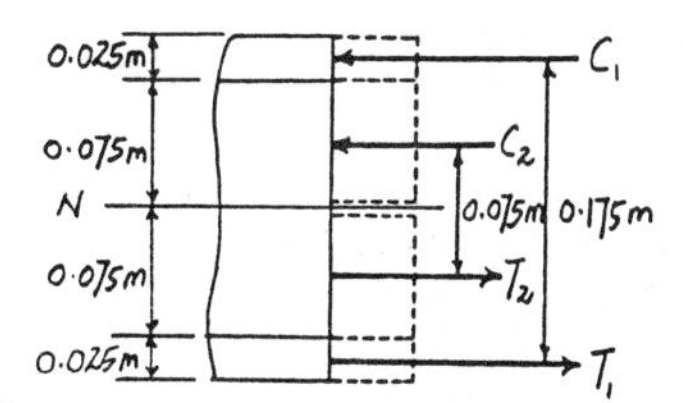

Plastic Moment :

$$M_P = \sigma_Y(0.2)(0.025)(0.175) + \sigma_Y(0.075)(0.05)(0.075)$$
$$= 1.15625\left(10^{-3}\right)\sigma_Y$$

Shape Factor :

$$K = \frac{M_P}{M_Y} = \frac{1.15625(10^{-3})\sigma_Y}{0.9114583(10^{-3})\sigma_Y} = 1.27 \qquad \textbf{Ans}$$

6–154. The box beam is made of an elastic-perfectly plastic material for which $\sigma_Y = 250$ MPa. Determine the residual stress in the top and bottom of the beam after the plastic moment $\mathbf{M}_p$ is applied and then released.

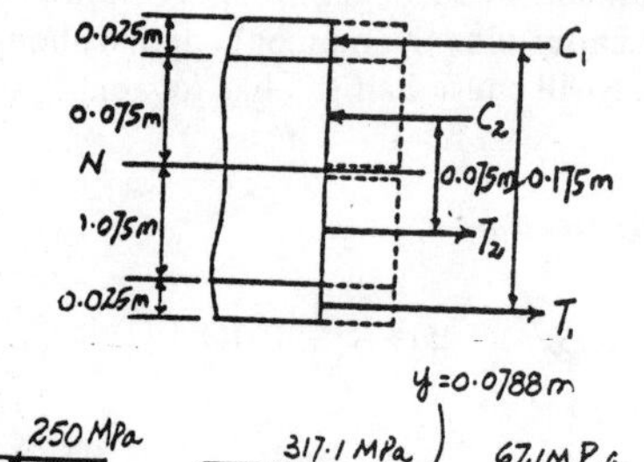

Plastic Moment :

$$M_P = 250(10^6)(0.2)(0.025)(0.175) + 250(10^6)(0.075)(0.05)(0.075) = 289062.5 \text{ N}\cdot\text{m}$$

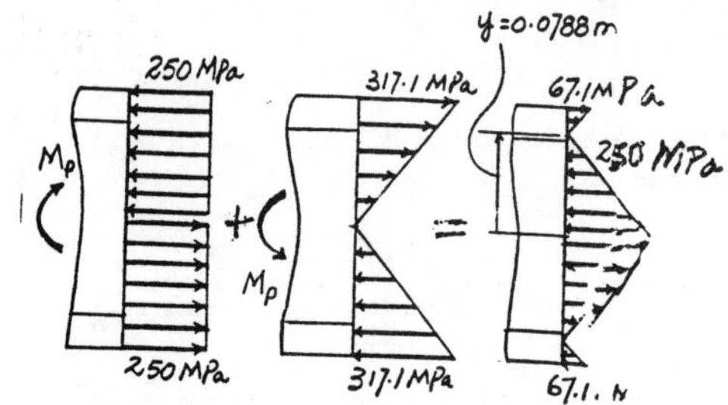

Modulus of Rupture : The modulus of rupture σ_r can be determined using the flexure formula with the application of rever. plastic moment $M_P = 289062.5$ N · m.

$$I = \frac{1}{12}(0.2)(0.2^3) - \frac{1}{12}(0.15)(0.15^3) = 91.14583(10^{-6}) \text{ m}^4$$

$$\sigma_r = \frac{M_P\,c}{I} = \frac{289062.5\,(0.1)}{91.14583(10^{-6})} = 317.14 \text{ MPa}$$

Residual Bending Stress : As shown on the diagram.

$$\sigma'_{top} = \sigma'_{bot} = \sigma_r - \sigma_Y = 317.14 - 250 = 67.1 \text{ MPa} \qquad \textbf{Ans}$$

6–155. Determine the plastic section modulus and the shape factor of the beam's cross section.

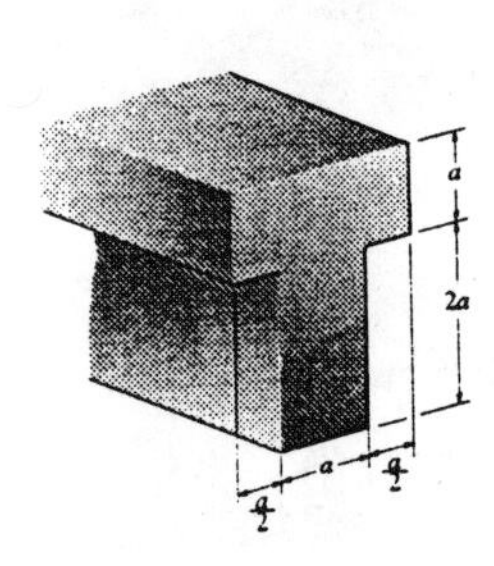

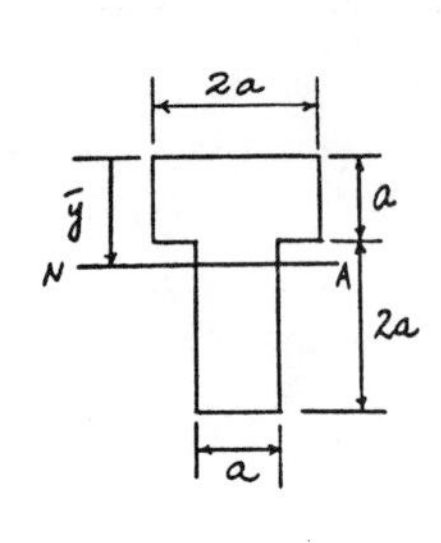

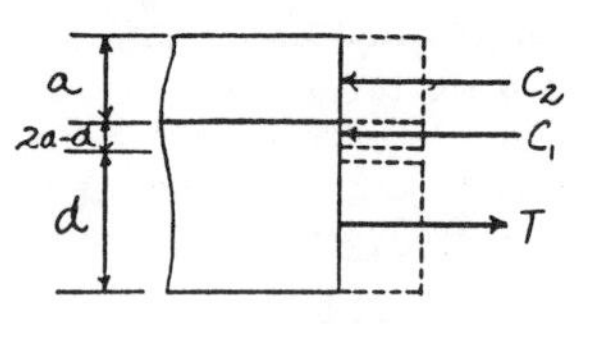

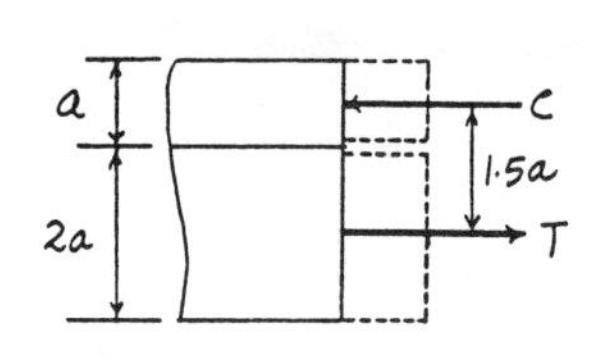

Maximum Elastic Moment : The centroid and the moment of inertia about neutral axis must be determined first.

$$\bar{y} = \frac{\Sigma\tilde{y}A}{\Sigma A} = \frac{0.5a(a)(2a) + 2a(2a)(a)}{a(2a) + 2a(a)} = 1.25a$$

$$I_{NA} = \frac{1}{12}(2a)(a^3) + 2a(a)(1.25a - 0.5a)^2 + \frac{1}{12}(a)(2a)^3 + a(2a)(2a - 1.25a)^2 = 3.0833a^4$$

Applying the flexure formula with $\sigma = \sigma_Y$, we have

$$\sigma_Y = \frac{M_Y c}{I}$$

$$M_Y = \frac{\sigma_Y I}{c} = \frac{\sigma_Y(3.0833a^4)}{(3a - 1.25a)} = 1.7619a^3\sigma_Y$$

Plastic Moment :

$$\int_A \sigma dA = 0; \qquad T - C_1 - C_2 = 0$$

$$\sigma_Y(d)(a) - \sigma_Y(2a - d)(a) - \sigma_Y(a)(2a) = 0$$

$$d = 2a$$

$$M_P = \sigma_Y(2a)(a)(1.5a) = 3.00a^3\sigma_Y$$

Shape Factor :

$$k = \frac{M_P}{M_Y} = \frac{3.00a^3\sigma_Y}{1.7619a^3\sigma_Y} = 1.70 \qquad \textbf{Ans}$$

Plastic Section Modulus :

$$Z = \frac{M_P}{\sigma_Y} = \frac{3.00a^3\sigma_Y}{\sigma_Y} = 3.00a^3 \qquad \textbf{Ans}$$

***6–156.** The beam is made of elastic-perfectly plastic material. Determine the maximum elastic moment and the plastic moment that can be applied to the cross section. Take $a = 50$ mm and $\sigma_Y = 230$ MPa.

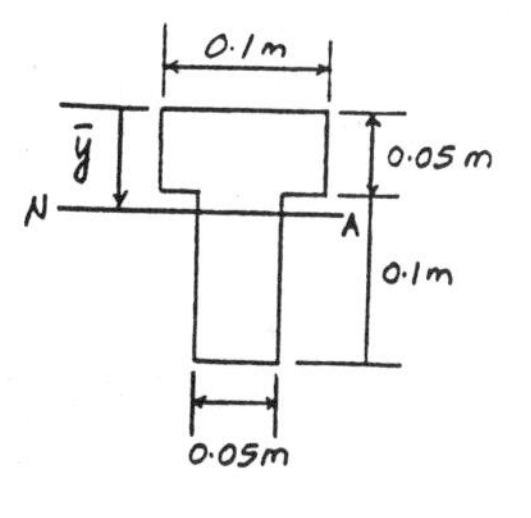

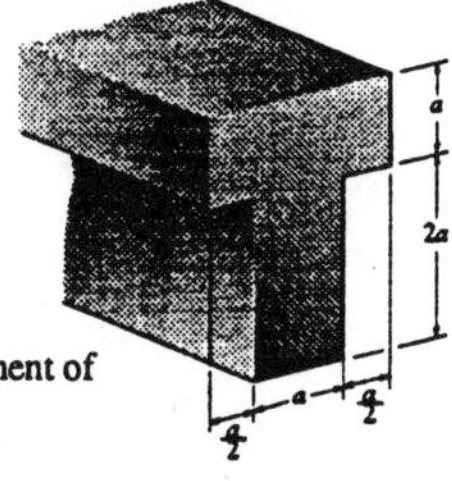

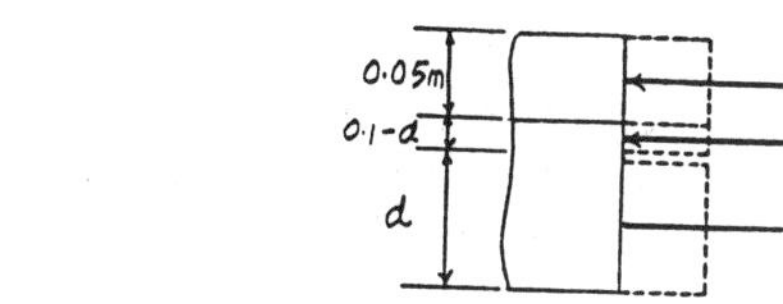

Maximum Elastic Moment : The centroid and the moment of inertia about neutral axis must be determined first.

$$\bar{y} = \frac{\Sigma \tilde{y}A}{\Sigma A} = \frac{0.025(0.05)(0.1) + 0.1(0.1)(0.05)}{0.05(0.1) + 0.1(0.05)} = 0.0625 \text{ m}$$

$$I_{NA} = \frac{1}{12}(0.1)(0.05^3) + 0.1(0.05)(0.0625 - 0.025)^2 + \frac{1}{12}(0.05)(0.1^3) + 0.05(0.1)(0.1 - 0.0625)^2$$
$$= 19.2709(10^{-6}) \text{ m}^4$$

Applying the flexure formula with $\sigma = \sigma_Y$, we have

$$\sigma_Y = \frac{M_Y c}{I}$$

$$M_Y = \frac{\sigma_Y I}{c} = \frac{230(10^6)(19.2709)(10^{-6})}{(0.15 - 0.0625)}$$
$$= 50654.8 \text{ N}\cdot\text{m} = 50.7 \text{ kN}\cdot\text{m} \quad \textbf{Ans}$$

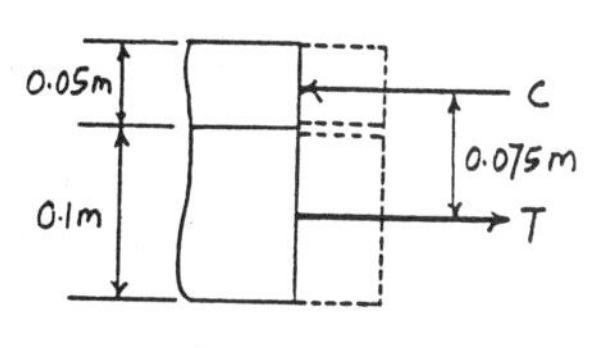

Plastic Moment :

$$\int_A \sigma dA = 0; \quad T - C_1 - C_2 = 0$$

$$\sigma_Y(d)(0.05) - \sigma_Y(0.1 - d)(0.05) - \sigma_Y(0.05)(0.1) = 0$$

$$d = 0.100 \text{ m}$$

$$M_P = 230(10^6)(0.100)(0.05)(0.075)$$
$$= 86250 \text{ N}\cdot\text{m} = 86.25 \text{ kN}\cdot\text{m} \quad \textbf{Ans}$$

6–157. The T-beam is made of an elastic-perfectly plastic material. Determine the maximum elastic moment and the plastic moment that can be applied to the cross section. $\sigma_Y = 36$ ksi.

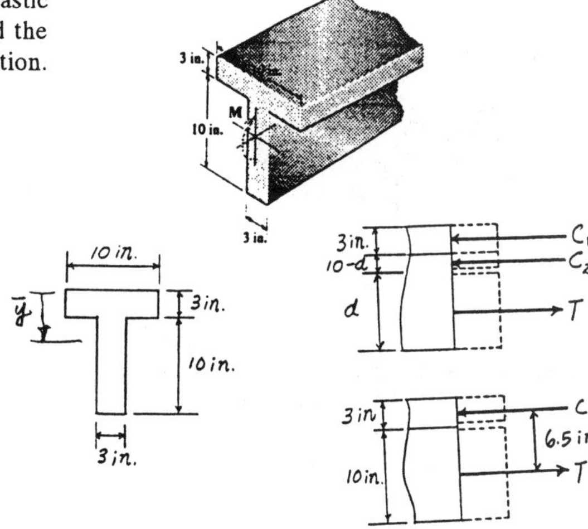

Maximum Elastic Moment : The centroid and the moment of inertia about neutral axis must be determined first.

$$\bar{y} = \frac{\Sigma \tilde{y}A}{\Sigma A} = \frac{1.5(10)(3) + 8(10)(3)}{10(3) + 10(3)} = 4.75 \text{ in.}$$

$$I_{NA} = \frac{1}{12}(10)(3^3) + 10(3)(4.75 - 1.5)^2 + \frac{1}{12}(3)(10^3) + 3(10)(8 - 4.75)^2 = 906.25 \text{ in}^4$$

Applying the flexure formula with $\sigma = \sigma_Y$, we have

$$\sigma_Y = \frac{M_Y c}{I}$$

$$M_Y = \frac{\sigma_Y I}{c} = \frac{36(906.25)}{(13 - 4.75)}$$
$$= 3954.5 \text{ kip}\cdot\text{in} = 330 \text{ kip}\cdot\text{ft} \quad \textbf{Ans}$$

Plastic Moment :

$$\int_A \sigma\, dA = 0; \quad C_1 + C_2 - T = 0$$

$$\sigma_Y(10)(3) + \sigma_Y(3)(10 - d) - \sigma_Y(3)(d) = 0$$

$$d = 10.0 \text{ in.}$$

$$M_P = 36(10.0)(3)(6.5)$$
$$= 7020 \text{ kip}\cdot\text{in} = 585 \text{ kip}\cdot\text{ft} \quad \textbf{Ans}$$

6–158. Determine the plastic section modulus and the shape factor of its cross section.

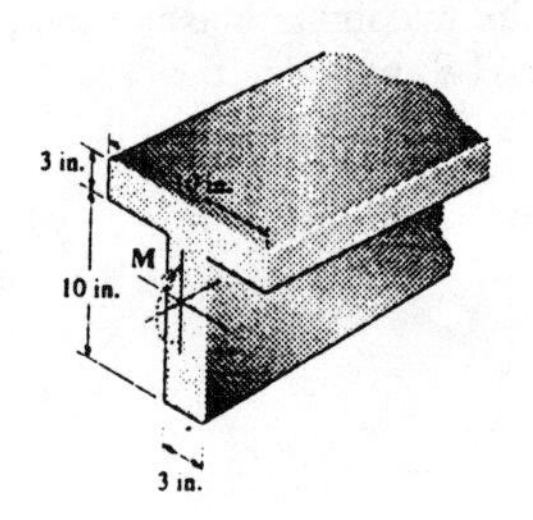

Maximum Elastic Moment : The centroid and the moment of inertia about neutral axis must be determined first.

$$\bar{y} = \frac{\Sigma \tilde{y}A}{\Sigma A} = \frac{1.5(10)(3) + 8(10)(3)}{10(3) + 10(3)} = 4.75 \text{ in.}$$

$$I_{NA} = \frac{1}{12}(10)(3^3) + 10(3)(4.75 - 1.5)^2 + \frac{1}{12}(3)(10^3) + 3(10)(8 - 4.75)^2 = 906.25 \text{ in}^4$$

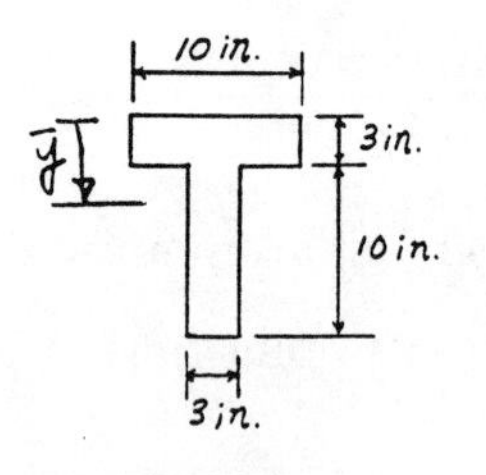

Applying Flexure Formula with $\sigma = \sigma_Y$, we have

$$\sigma_Y = \frac{M_Y c}{I}$$

$$M_Y = \frac{\sigma_Y I}{c} = \frac{\sigma_Y (906.25)}{(13 - 4.75)} = 109.85\sigma_Y$$

Plastic Moment :

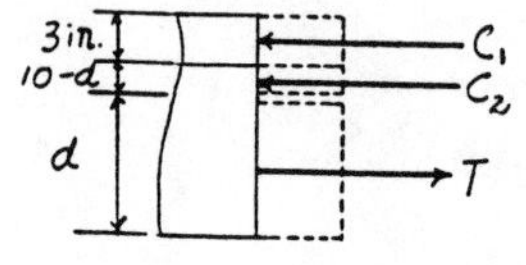

$$\int_A \sigma \, dA = 0; \quad C_1 + C_2 - T = 0$$

$$\sigma_Y(10)(3) + \sigma_Y(3)(10 - d) - \sigma_Y(3)(d) = 0$$

$$d = 10.0 \text{ in.}$$

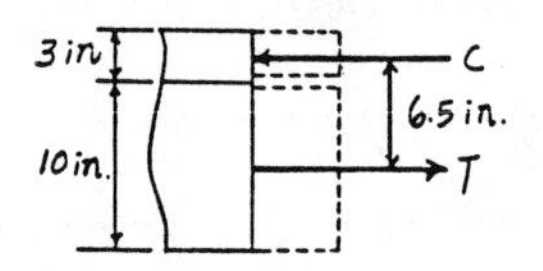

$$M_P = \sigma_Y(10.0)(3)(6.5) = 195\sigma_Y$$

Plastic Section Modulus :

$$Z = \frac{M_P}{\sigma_Y} = \frac{195\sigma_Y}{\sigma_Y} = 195 \text{ in}^3 \qquad \textbf{Ans}$$

Shape Factor :

$$k = \frac{M_P}{M_Y} = \frac{195\sigma_Y}{109.85\sigma_Y} = 1.78 \qquad \textbf{Ans}$$

6–159. The beam is made of an elastic-perfectly plastic material. Determine the plastic moment M_p that can be supported by a beam having the cross section shown. $\sigma_Y = 30$ ksi.

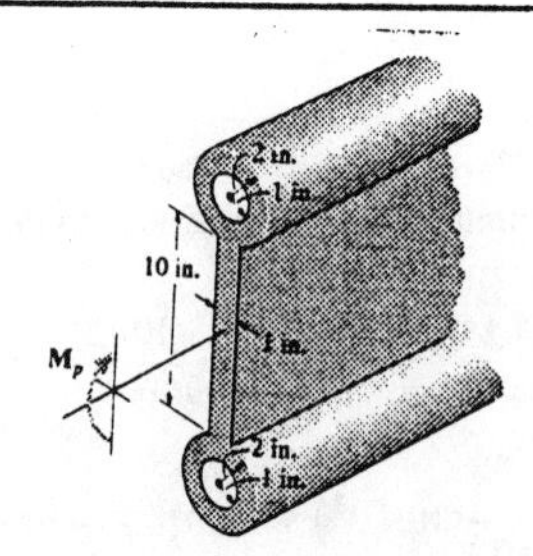

Plastic Moment :

$$T_1 = C_1 = 30[\pi(2^2) - \pi(1^2)] = 90\pi \text{ kip}$$

$$T_2 = C_2 = 30(1)(5) = 150 \text{ kip}$$

$$M_P = 90\pi(14) + 150(5) = 4708.41 \text{ kip} \cdot \text{in} = 392 \text{ kip} \cdot \text{ft} \qquad \textbf{Ans}$$

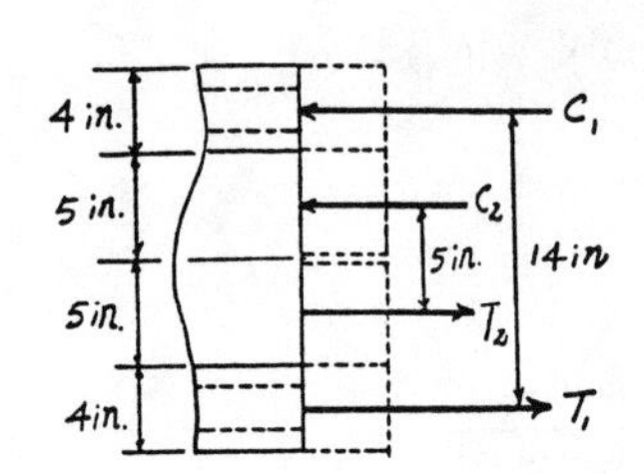

***6–160.** Determine the plastic section modulus and the shape factor of the cross section.

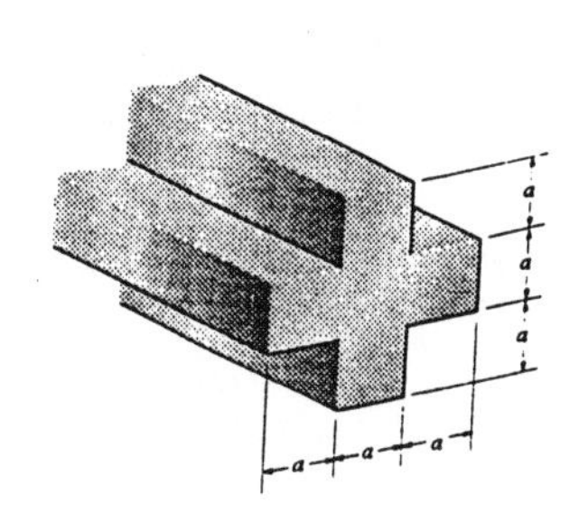

Maximum Elastic Moment : The moment of inertia about neutral axis must be determined first.

$$I_{NA} = \frac{1}{12}(a)(3a)^3 + \frac{1}{12}(2a)\left(a^3\right) = 2.41667a^4$$

Applying the flexure formula with $\sigma = \sigma_Y$, we have

$$\sigma_Y = \frac{M_Y c}{I}$$

$$M_Y = \frac{\sigma_Y I}{c} = \frac{\sigma_Y (2.41667a^4)}{1.5a} = 1.6111a^3\sigma_Y$$

Plastic Moment :

$$M_P = \sigma_Y (a)(a)(2a) + \sigma_Y (0.5a)(3a)(0.5a)$$
$$= 2.75a^3\sigma_Y$$

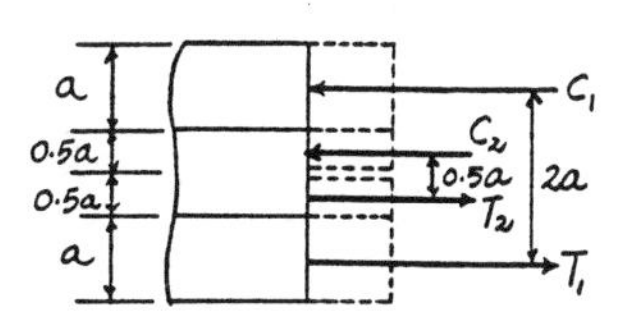

Plastic Section Modulus :

$$Z = \frac{M_P}{\sigma_Y} = \frac{2.75a^3\sigma_Y}{\sigma_Y} = 2.75a^3 \qquad \textbf{Ans}$$

Shape Factor :

$$k = \frac{M_P}{M_Y} = \frac{2.75a^3\sigma_Y}{1.6111a^3\sigma_Y} = 1.71 \qquad \textbf{Ans}$$

6–161. The beam is made of elastic-perfectly plastic material. Determine the maximum elastic moment and the plastic moment that can be applied to the cross section. Take $a = 2$ in. and $\sigma_Y = 36$ ksi.

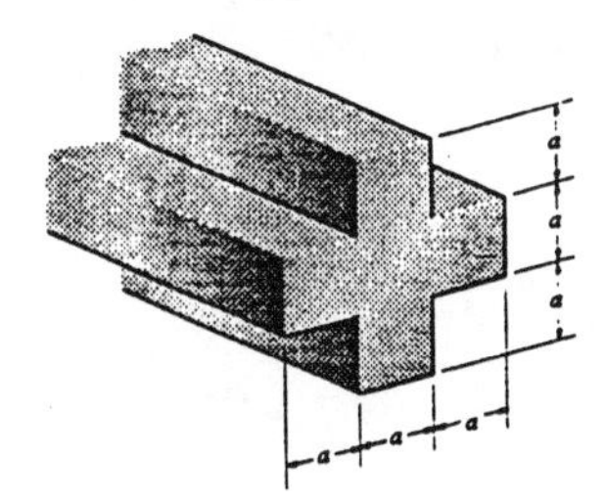

Maximum Elastic Moment : The moment of inertia about neutral axis must be determined first.

$$I_{NA} = \frac{1}{12}(2)\left(6^3\right) + \frac{1}{12}(4)\left(2^3\right) = 38.667 \text{ in}^4$$

Applying the flexure formula with $\sigma = \sigma_Y$, we have

$$\sigma_Y = \frac{M_Y c}{I}$$

$$M_Y = \frac{\sigma_Y I}{c} = \frac{36(38.667)}{3}$$
$$= 464 \text{ kip} \cdot \text{in} = 38.7 \text{ kip} \cdot \text{ft} \qquad \textbf{Ans}$$

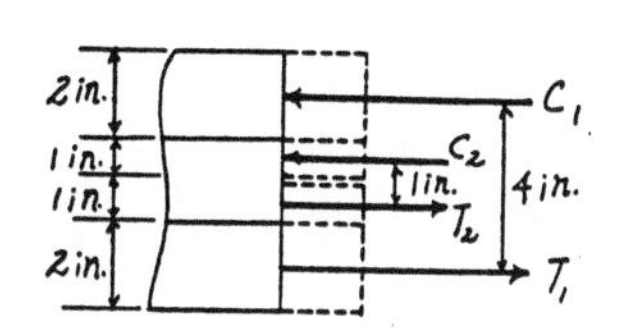

Plastic Moment :

$$M_P = 36(2)(2)(4) + 36(1)(6)(1)$$
$$= 792 \text{ kip} \cdot \text{in} = 66.0 \text{ kip} \cdot \text{ft} \qquad \textbf{Ans}$$

6–162. Determine the plastic section modulus and the shape factor for the member having the box cross section.

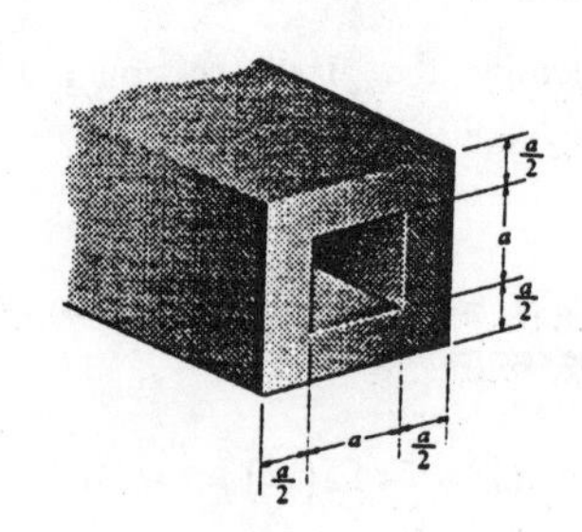

Maximum Elastic Moment : The moment of inertia about the neutral axis must be determined first.

$$I_{NA} = \frac{1}{12}(2a)(2a)^3 - \frac{1}{12}(a)\left(a^3\right) = 1.25a^4$$

Applying the flexure formula with $\sigma = \sigma_Y$, we have

$$\sigma_Y = \frac{M_Y c}{I}$$

$$M_Y = \frac{\sigma_Y I}{c} = \frac{\sigma_Y (1.25a^4)}{a} = 1.25a^3 \sigma_Y$$

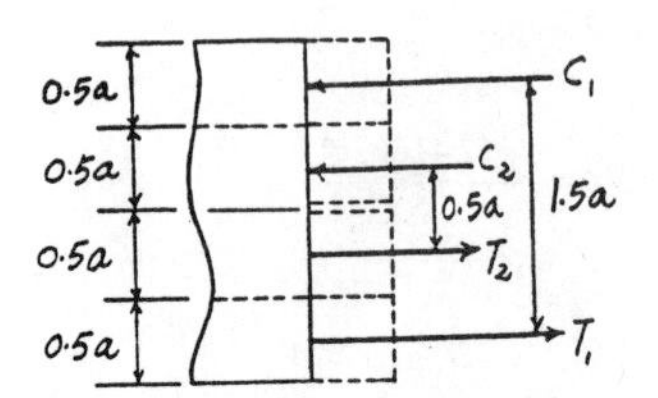

Plastic Moment :

$$M_P = \sigma_Y (0.5a)(2a)(1.5a) + \sigma_Y (0.5a)(a)(0.5a)$$

$$= 1.75a^3 \sigma_Y$$

Plastic Section Modulus :

$$Z = \frac{M_P}{\sigma_Y} = \frac{1.75a^3 \sigma_Y}{\sigma_Y} = 1.75a^3 \qquad \textbf{Ans}$$

Shape Factor :

$$k = \frac{M_P}{M_Y} = \frac{1.75a^3 \sigma_Y}{1.25a^3 \sigma_Y} = 1.40 \qquad \textbf{Ans}$$

6–163. The box beam is made of elastic-perfectly plastic material. Determine the maximum elastic moment and the plastic moment that can be applied to the cross section. Take $a = 100$ mm and $\sigma_Y = 250$ MPa.

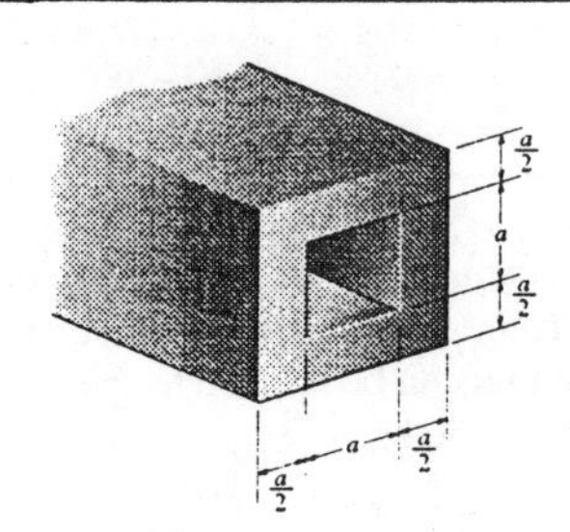

Maximun Elastic Moment : The moment of inertia about neutral axis must be determined first.

$$I_{NA} = \frac{1}{12}(0.2)\left(0.2^3\right) - \frac{1}{12}(0.1)\left(0.1^3\right) = 0.125\left(10^{-3}\right)\ \text{m}^4$$

Applying the flexure formula with $\sigma = \sigma_Y$, we have

$$\sigma_Y = \frac{M_Y c}{I}$$

$$M_Y = \frac{\sigma_Y I}{c}$$

$$= \frac{250(10^6)0.125(10^{-3})}{0.1}$$

$$= 312500\ \text{N}\cdot\text{m} = 312.5\ \text{kN}\cdot\text{m} \qquad \textbf{Ans}$$

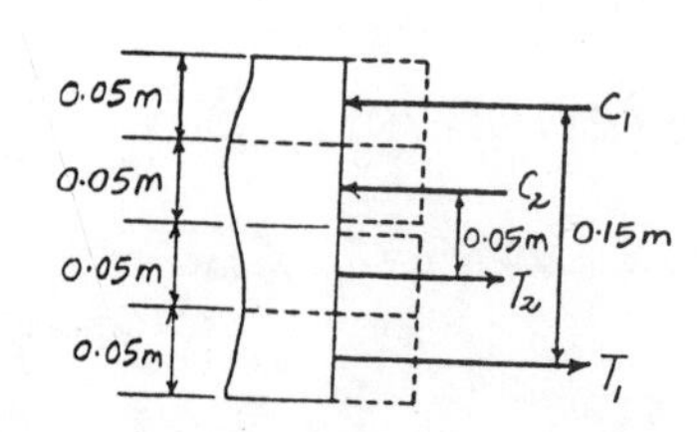

Plastic Moment :

$$M_P = 250\left(10^6\right)(0.05)(0.2)(0.15) + 250\left(10^6\right)(0.05)(0.1)(0.05)$$

$$= 437500\ \text{N}\cdot\text{m} = 437.5\ \text{kN}\cdot\text{m} \qquad \textbf{Ans}$$

***6–164.** Determine the plastic section modulus and the shape factor for the member having the tubular cross section.

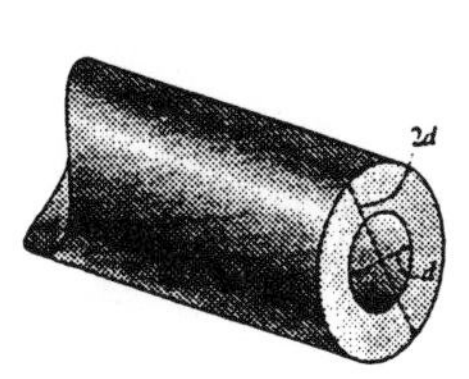

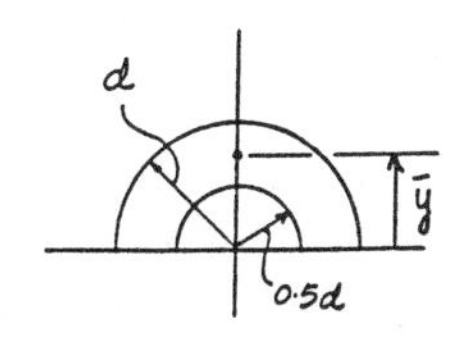

Maximum Elastic Moment : The moment of inertia about neutral axis must be determined first.

$$I_{NA} = \frac{\pi}{4}d^4 - \frac{\pi}{4}\left(\frac{d}{2}\right)^4 = \frac{15\pi}{64}d^4$$

Applying the flexure formula with $\sigma = \sigma_Y$, we have

$$\sigma_Y = \frac{M_Y c}{I}$$

$$M_Y = \frac{\sigma_Y I}{c} = \frac{\sigma_Y\left(\frac{15\pi}{64}d^4\right)}{d} = \frac{15\pi}{64}d^3\sigma_Y$$

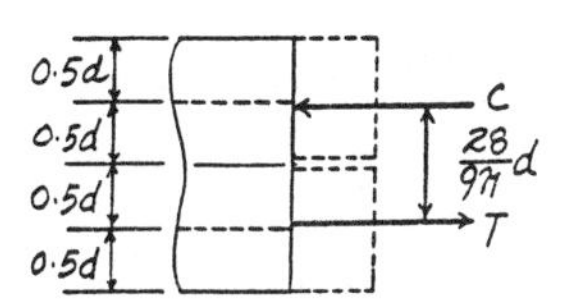

Plastic Moment :

$$\bar{y} = \frac{\Sigma\tilde{y}A}{\Sigma A} = \frac{\frac{4d}{3\pi}\left(\frac{\pi d^2}{2}\right) - \frac{4\left(\frac{d}{2}\right)}{3\pi}\left(\frac{\frac{\pi}{4}d^2}{2}\right)}{\frac{\pi d^2}{2} - \frac{\frac{\pi}{4}d^2}{2}} = \frac{14}{9\pi}d$$

$$M_P = \sigma_Y\left(\frac{\pi d^2}{2} - \frac{\frac{\pi}{4}d^2}{2}\right)\frac{28}{9\pi}d = \frac{7}{6}d^3\sigma_Y$$

Plastic Section Modulus :

$$Z = \frac{M_P}{\sigma_Y} = \frac{\frac{7}{6}d^3\sigma_Y}{\sigma_Y} = \frac{7}{6}d^3 \qquad \textbf{Ans}$$

Shape Factor :

$$k = \frac{M_P}{M_Y} = \frac{\frac{7}{6}d^3\sigma_Y}{\frac{15\pi}{64}d^3\sigma_Y} = 1.58 \qquad \textbf{Ans}$$

6–165. Determine the plastic section modulus and the shape factor for the member.

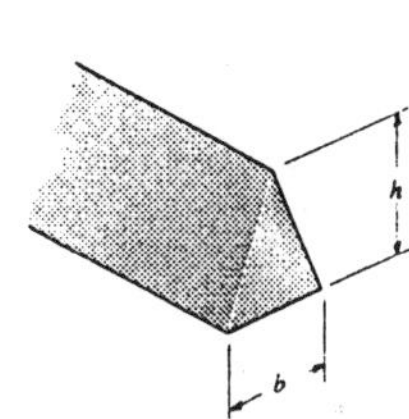

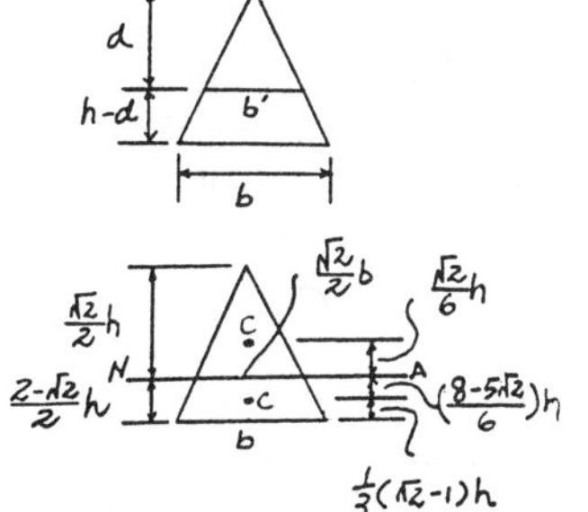

Maximum Elastic Moment : Applying the flexure formula with $\sigma = \sigma_Y$, we have

$$\sigma_Y = \frac{M_Y c}{I}$$

$$M_Y = \frac{\sigma_Y I}{c} = \frac{\sigma_Y\left(\frac{1}{36}bh^3\right)}{\frac{2}{3}h} = \frac{1}{24}bh^2\sigma_Y$$

Plastic Moment : From the geometry $b' = \frac{d}{h}b$

$$\int_A \sigma dA = 0; \qquad T - C = 0$$

$$\sigma_Y\left[\frac{1}{2}\left(\frac{d}{h}b + b\right)(h-d)\right] - \sigma_Y\left[\frac{1}{2}\left(\frac{d}{h}b\right)d\right] = 0$$

$$d = \frac{\sqrt{2}}{2}h$$

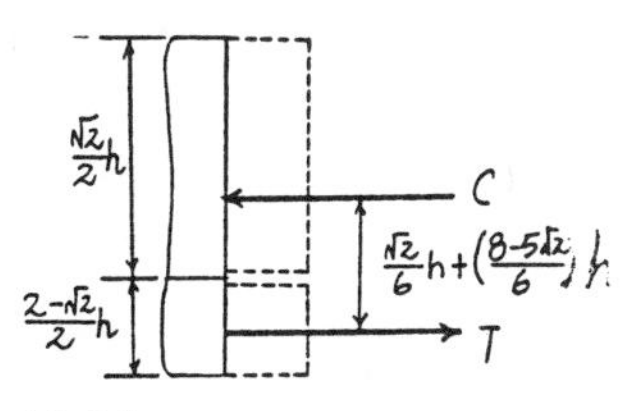

$$M_P = \sigma_Y\left[\frac{1}{2}\left(\frac{\sqrt{2}}{2}b\right)\left(\frac{\sqrt{2}}{2}h\right)\right]\left[\frac{\sqrt{2}}{6}h + \left(\frac{8-5\sqrt{2}}{6}\right)h\right]$$

$$= \frac{2-\sqrt{2}}{6}bh^2\sigma_Y$$

Plastic Section Modulus :

$$Z = \frac{M_P}{\sigma_Y} = \frac{\frac{2-\sqrt{2}}{6}bh^2\sigma_Y}{\sigma_Y} = \frac{2-\sqrt{2}}{6}bh^2 = 0.0976bh^2 \qquad \textbf{Ans}$$

Shape Factor :

$$k = \frac{M_P}{M_Y} = \frac{\frac{2-\sqrt{2}}{6}bh^2\sigma_Y}{\frac{1}{24}bh^2\sigma_Y} = 4\left(2-\sqrt{2}\right) = 2.34 \qquad \textbf{Ans}$$

6–166. The member is made of elastic-perfectly plastic material for which $\sigma_Y = 230$ MPa. Determine the maximum elastic moment and the plastic moment that can be applied to the cross section. Take $b = 50$ mm and $h = 80$ mm.

Maximum Elastic Moment : Applying the flexure formula with $\sigma = \sigma_Y$, we have

$$\sigma_Y = \frac{M_Y c}{I}$$

$$M_Y = \frac{\sigma_Y I}{c} = \frac{230(10^6)\left[\frac{1}{36}(0.05)(0.08^3)\right]}{\frac{2}{3}(0.08)}$$

$$= 3067 \text{ N}\cdot\text{m} = 3.07 \text{ kN}\cdot\text{m} \quad \textbf{Ans}$$

Plastic Moment : From the geometry $b' = \left(\frac{0.05}{0.08}\right)d = 0.625d$

$$\int_A \sigma dA = 0; \qquad T - C = 0$$

$$\sigma_Y\left[\frac{1}{2}(0.625d + 0.05)(0.08 - d)\right] - \sigma_Y\left[\frac{1}{2}(0.625d)\,d\right] = 0$$

$$d = 0.056569 \text{ m}$$

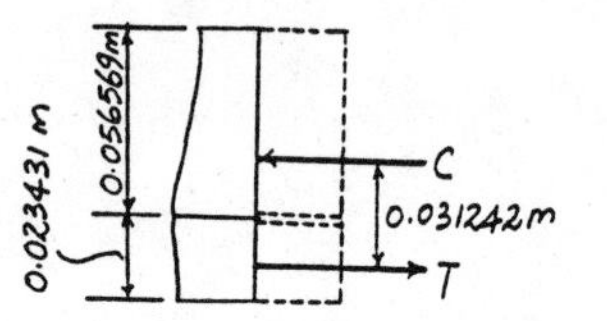

$$M_P = 230\left(10^6\right)\left[\frac{1}{2}(0.035355)(0.056569)\right](0.031242)$$

$$= 7186 \text{ N}\cdot\text{m} = 7.19 \text{ kN}\cdot\text{m} \quad \textbf{Ans}$$

6–167. The beam is made of phenolic, a structural plastic, that has the stress–strain curve shown. If a portion of the curve can be represented by the equation $\sigma = (5(10^6)\epsilon)^{1/2}$ MPa, determine the magnitude w of the distributed load that can be applied to the beam without causing the maximum strain in its fibers at the critical section to exceed $\epsilon_{max} = 0.005$ mm/mm.

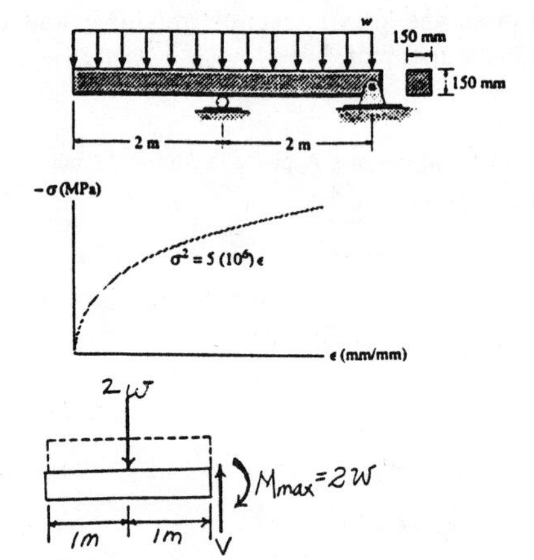

Resultant Internal Forces : The resultant internal forces T and C can be evaluated from the volume of the stress block which is a paraboloid. When $\varepsilon = 0.005$ mm/mm, then

$$\sigma = \sqrt{5(10^6)(0.005)} = 158.11 \text{ MPa}$$

$$T = C = \frac{2}{3}\left[158.11\left(10^6\right)(0.075)\right](0.150) = 1.1859 \text{ MN}$$

$$d = 2\left[\frac{3}{5}(0.075)\right] = 0.090 \text{ m}$$

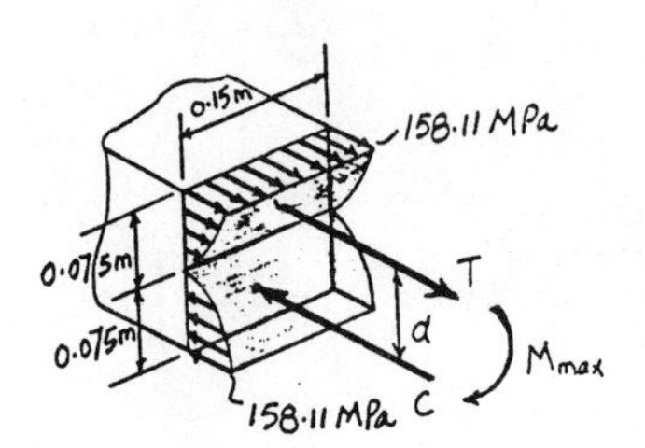

Maximum Internal Moment : The maximum internal moment $M = 2w$ occurs at the overhang support as shown on FBD.

$$M_{max} = Td$$

$$2w = 1.1859\left(10^6\right)(0.090)$$

$$w = 53363 \text{ N/m} = 53.4 \text{ kN/m} \quad \textbf{Ans}$$

***6–168.** The beam is made of an elastic-perfectly plastic material for which $\sigma_Y = 30$ ksi. Determine the intensity of the distributed load w that causes the critical sectoin of the beam to be subjected to be (*a*) the largest elastic moment and (*b*) the largest plastic moment.

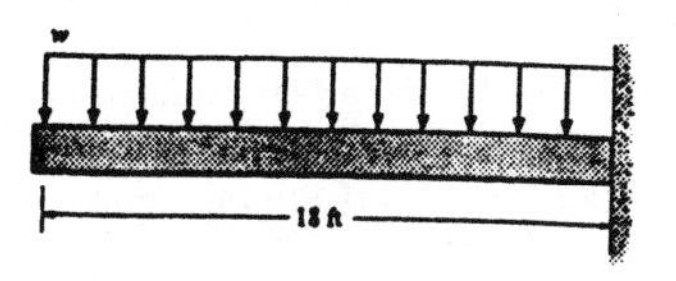

Maximum Internal Moment : The maximum internal moment $M = 162w$ occurs at the fixed support as shown on FBD.

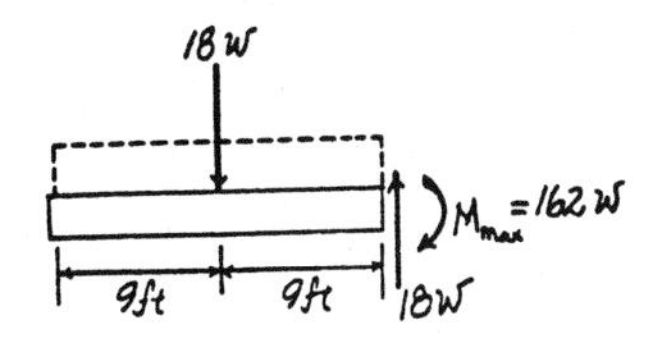

a) ***Maximun Elastic Moment :*** Applying Flexure Formula with $\sigma = \sigma_Y$, we have $\sigma_Y = \dfrac{M_Y c}{I}$

$$M_Y = M_{\max} = \frac{\sigma_Y I}{c}$$

$$162w(12) = \frac{30\left[\frac{1}{12}(6)(6^3)\right]}{3}$$

$$w = 0.556 \text{ kip/ft} \qquad \textbf{Ans}$$

b) ***Plastic Moment :***

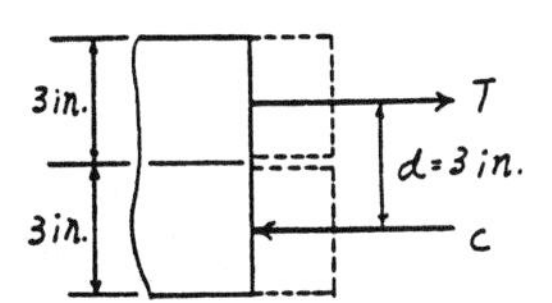

$$M_P = M_{\max} = Td$$

$$162w(12) = 30(3)(6)(3)$$

$$w = 0.833 \text{ kip/ft} \qquad \textbf{Ans}$$

6–169. The beam is made of a polyester that has the stress–strain curve shown. If the curve can be represented by the equation $\sigma = [20\tan^{-1}(15\epsilon)]$ ksi, where $\tan^{-1}(15\epsilon)$ is in radians, determine the magnitude of the force **P** that can be applied to the beam without causing the maximum strain in its fibers at the critical section to exceed $\epsilon_{\max} = 0.003$ in./in.

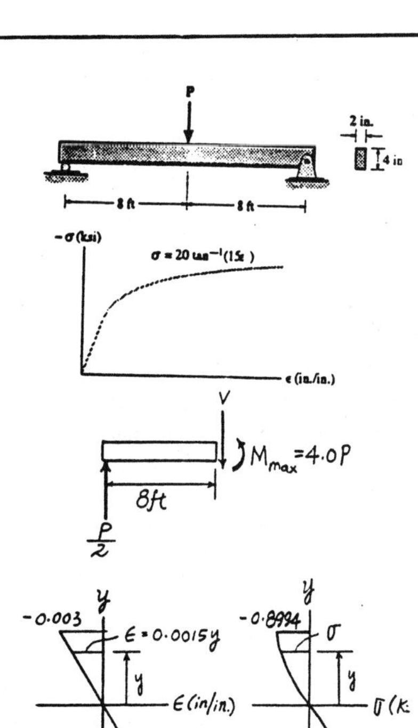

Maximum Internal Moment : The maximum internal moment $M = 4.00P$ occurs at the mid span as shown on FBD.

Stress - Strain Relationship : Using the stress – strain relationship, the bending stress can be expressed in terms of y using $\varepsilon = 0.0015y$.

$$\sigma = 20\tan^{-1}(15\varepsilon)$$
$$= 20\tan^{-1}[15(0.0015y)]$$
$$= 20\tan^{-1}(0.0225y)$$

When $\varepsilon_{\max} = 0.003$ in./in., $y = 2$ in. and $\sigma_{\max} = 0.8994$ ksi

Resultant Internal Moment : The resultant internal moment M can be evaluated from the integal $\int_A y\sigma dA$.

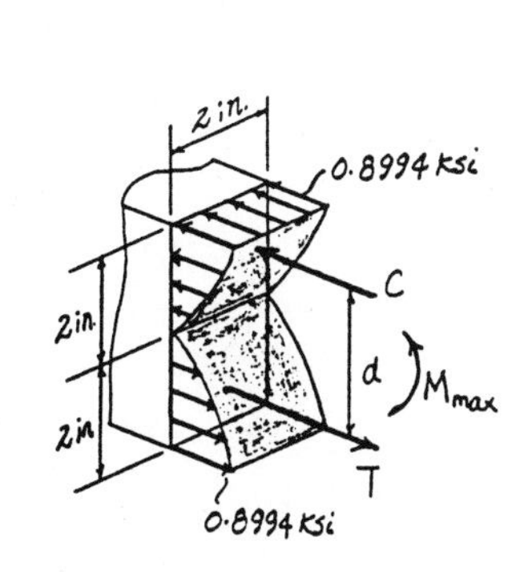

$$M = 2\int_A y\sigma dA$$
$$= 2\int_0^{2\text{in}} y\left[20\tan^{-1}(0.0225y)\right](2dy)$$
$$= 80\int_0^{2\text{in}} y\tan^{-1}(0.0225y)\,dy$$
$$= 80\left[\frac{1+(0.0225)^2y^2}{2(0.0225)^2}\tan^{-1}(0.0225y) - \frac{y}{2(0.0225)}\right]\Bigg|_0^{2\text{in.}}$$
$$= 4.798 \text{ kip}\cdot\text{in}$$

Equating $\qquad M = 4.00P(12) = 4.798$

$$P = 0.100 \text{ kip} = 100 \text{ lb} \qquad \textbf{Ans}$$

6–170. The stress–strain diagram for a titanium alloy can be approximated by the two straight lines. If a strut made of this material is subjected to bending, determine the moment resisted by the strut if the maximum stress reaches a value of (*a*) σ_A and (*b*) σ_B.

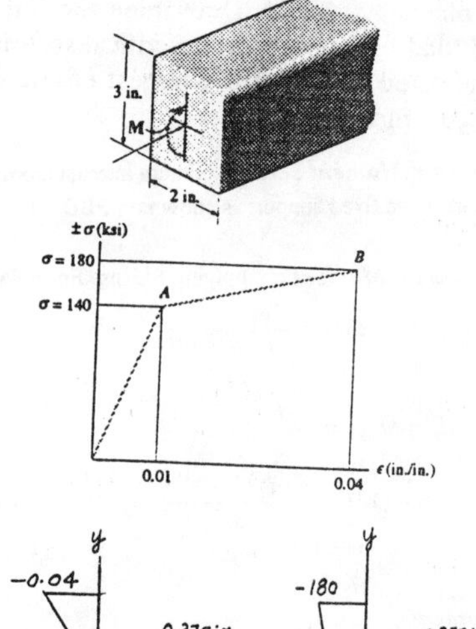

a) ***Maximum Elastic Moment :*** Since the stress is linearly related to strain up to point A. The flexure formula can be applied.

$$\sigma_A = \frac{Mc}{I}$$

$$M = \frac{\sigma_A I}{c}$$

$$= \frac{140\left[\frac{1}{12}(2)(3^3)\right]}{1.5}$$

$$= 420 \text{ kip}\cdot\text{in} = 35.0 \text{ kip}\cdot\text{ft} \qquad \textbf{Ans}$$

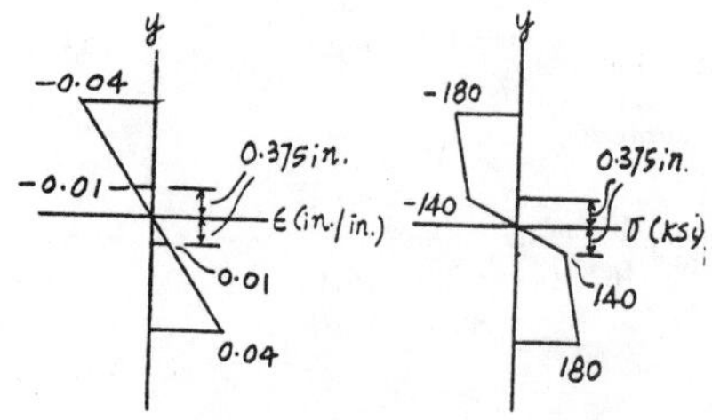

b) ***The Ultimate Moment :***

$$C_1 = T_1 = \frac{1}{2}(140 + 180)(1.125)(2) = 360 \text{ kip}$$

$$C_2 = T_2 = \frac{1}{2}(140)(0.375)(2) = 52.5 \text{ kip}$$

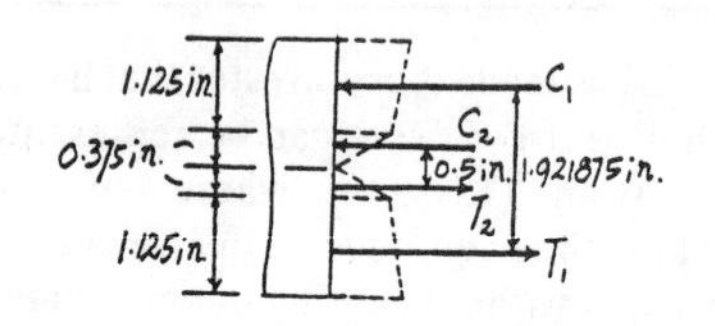

$$M = 360(1.921875) + 52.5(0.5)$$

$$= 718.125 \text{ kip}\cdot\text{in} = 59.8 \text{ kip}\cdot\text{ft} \qquad \textbf{Ans}$$

Note : The centroid of a trapezodial area was used in calculation of moment.

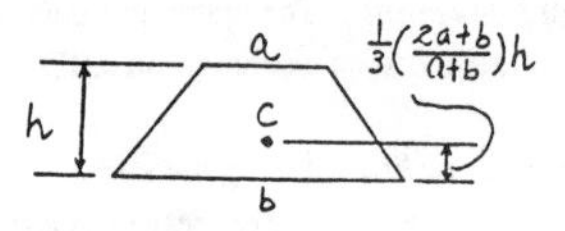

6–171. The beam is made of a material that can be assumed perfectly plastic in tension and elastic-perfectly plastic in compression. Determine the maximum bending moment M that can be supported by the beam so that the compressive material at the outer edge starts to yield.

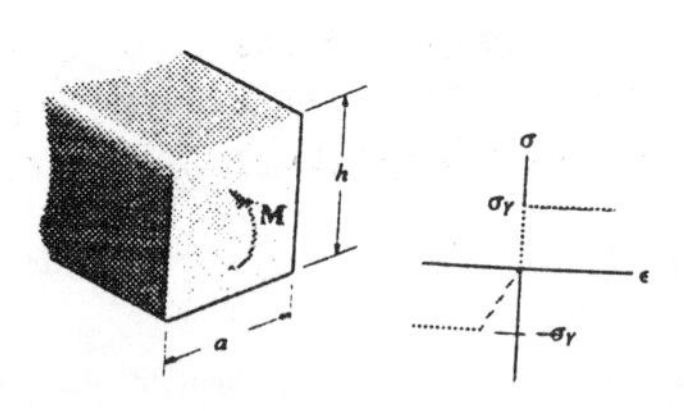

$$\int_A \sigma dA = 0; \qquad C - T = 0$$

$$\frac{1}{2}\sigma_Y(d)(a) - \sigma_Y(h - d)a = 0$$

$$d = \frac{2}{3}h$$

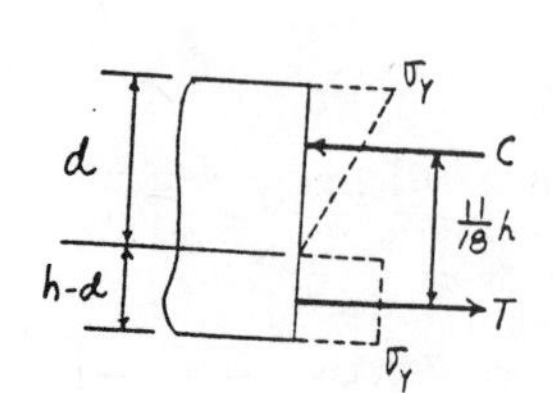

$$M = \frac{1}{2}\sigma_Y\left(\frac{2}{3}h\right)(a)\left(\frac{11}{18}h\right) = \frac{11ah^2}{54}\sigma_Y \qquad \textbf{Ans}$$

***6–172.** The plexiglass bar has a stress–strain curve that can be approximated by the straight-line segments shown. Determine the largest moment M that can be applied to the bar before it fails.

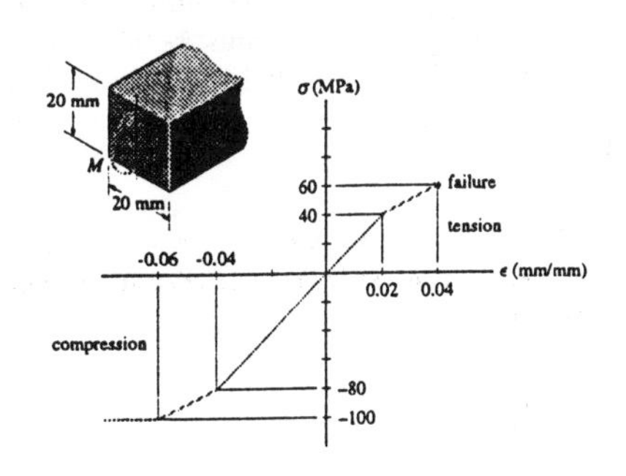

Ultimate Moment :

$$\int_A \sigma\, dA = 0; \qquad C - T_2 - T_1 = 0$$

$$\sigma\left[\frac{1}{2}(0.02-d)(0.02)\right] - 40(10^6)\left[\frac{1}{2}\left(\frac{d}{2}\right)(0.02)\right] - \frac{1}{2}(60+40)(10^6)\left[(0.02)\frac{d}{2}\right] = 0$$

$$\sigma - 50\sigma d - 3500(10^6)d = 0$$

Assume. $\sigma = 74.833$ MPa; $d = 0.010334$ m

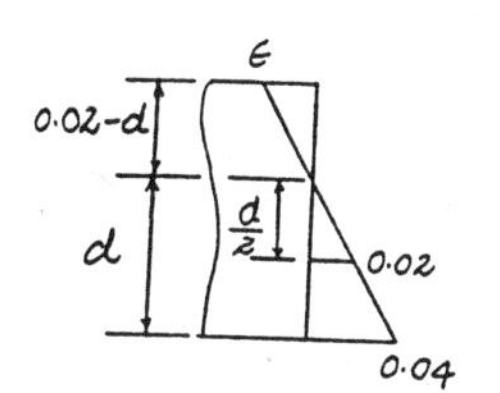

From the strain diagram,

$$\frac{\varepsilon}{0.02-0.010334} = \frac{0.04}{0.010334} \qquad \varepsilon = 0.037417 \text{ mm/mm}$$

From the stress – strain diagram,

$$\frac{\sigma}{0.037417} = \frac{80}{0.04} \qquad \sigma = 74.833 \text{ MPa (**OK!** Close to assumed value)}$$

Therefore,

$$C = 74.833(10^6)\left[\frac{1}{2}(0.02-0.010334)(0.02)\right] = 7233.59 \text{ N}$$

$$T_1 = \frac{1}{2}(60+40)(10^6)\left[(0.02)\left(\frac{0.010334}{2}\right)\right] = 5166.85 \text{ N}$$

$$T_2 = 40(10^6)\left[\frac{1}{2}(0.02)\left(\frac{0.010334}{2}\right)\right] = 2066.74 \text{ N}$$

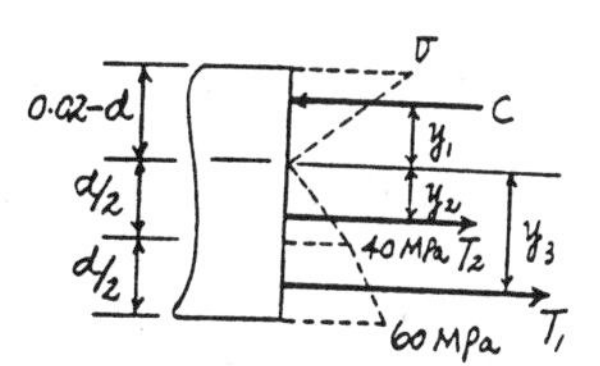

$$y_1 = \frac{2}{3}(0.02-0.010334) = 0.0064442 \text{ m}$$

$$y_2 = \frac{2}{3}\left(\frac{0.010334}{2}\right) = 0.0034445 \text{ m}$$

$$y_3 = \frac{0.010334}{2} + \left[1 - \frac{1}{3}\left(\frac{2(40)+60}{40+60}\right)\right]\left(\frac{0.010334}{2}\right) = 0.0079225 \text{ m}$$

$$M = 7233.59(0.0064442) + 2066.74(0.0034445) + 5166.85(0.0079225)$$
$$= 94.7 \text{ N}\cdot\text{m} \qquad \textbf{Ans}$$

6–1173. Determine the residual stress at the top and bottom of an A-36 steel bar having a circular cross section that has been unloaded from a fully plastic moment. The radius of the bar is 4 in.

Plastic Moment :

$$T = C = 36\left(\frac{1}{2}\right)(\pi)(4^2) = 288\pi \text{ kip}$$

$$M_P = 288\pi\ (3.3953) = 3072 \text{ kip}\cdot\text{in.}$$

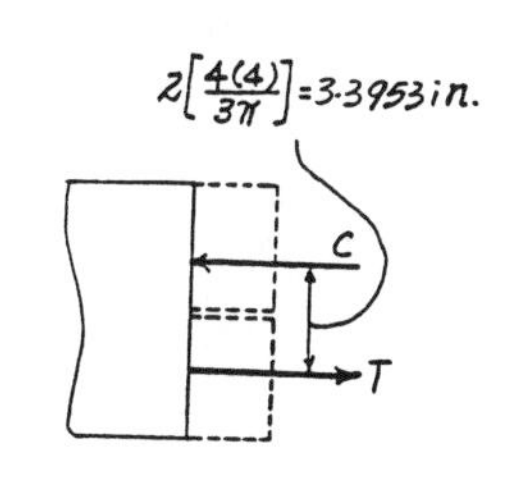

Modulus of Rupture : The modulus of rupture σ_r can be determined using the flexure formula with the application of reverse plastic moment $M_P = 3072$ kip · in.

$$\sigma_r = \frac{M_P c}{I} = \frac{3072(4)}{\frac{\pi}{4}(4^4)} = 61.12 \text{ ksi}$$

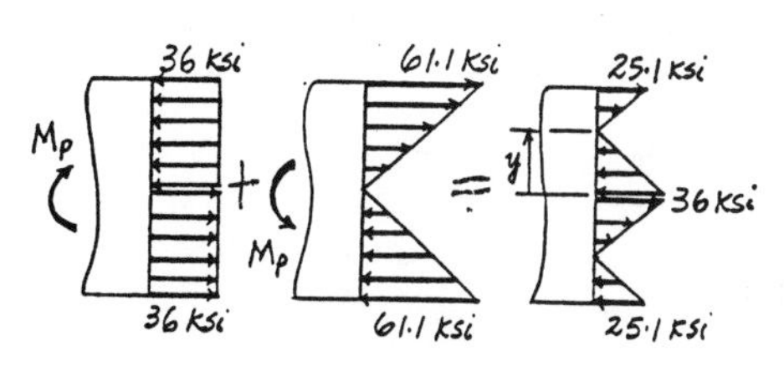

Residual Bending Stress : As shown on the diagram.

$$\sigma'_{top} = \sigma'_{bot} = \sigma_r - \sigma_Y$$
$$= 61.12 - 36.0 = 25.1 \text{ ksi} \qquad \textbf{Ans}$$

$$\frac{y}{36.0} = \frac{4}{61.12}; \qquad y = 2.356 \text{ in.}$$

6–174. Draw the shear and moment diagrams for the beam. *Hint:* The 20-kip load must be replaced by equivalent loadings at point C on the axis of the beam.

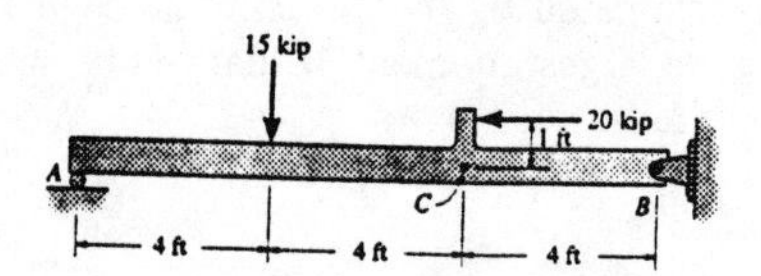

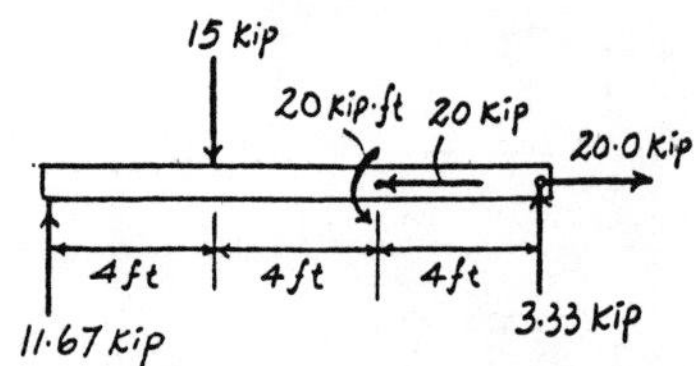

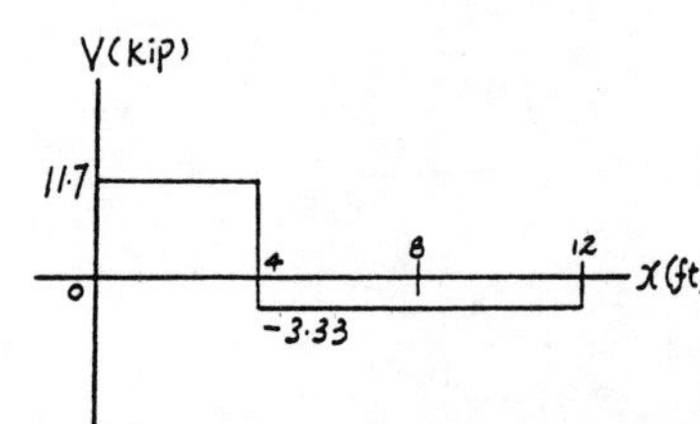

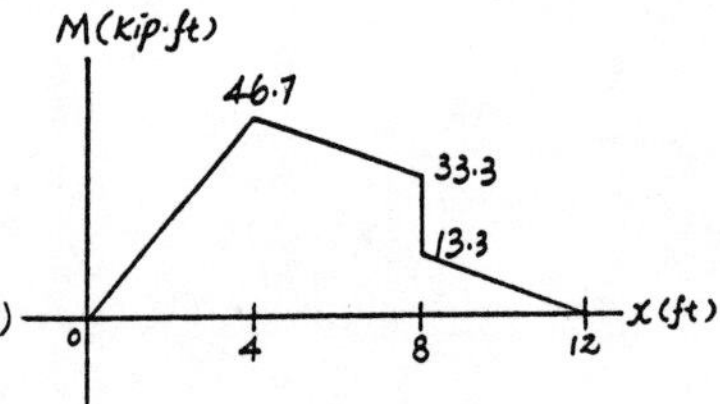

6–175.. The member has a square cross section. If it is made of an elastic-perfectly plastic material, determine the shape factor and the plastic section modulus Z.

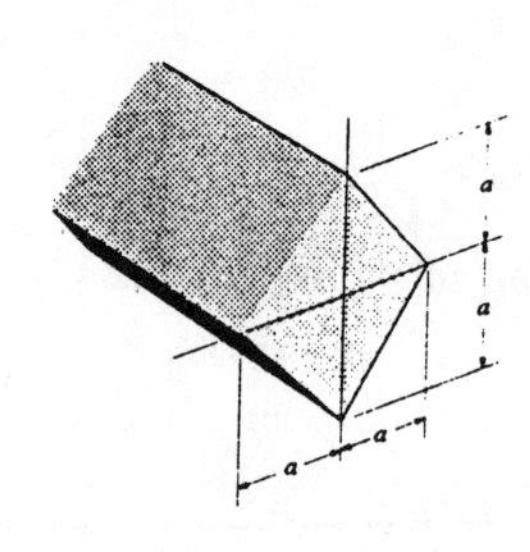

Maximum Elastic Moment : The moment of inertia about neutral axis must be determined first.

$$I_{NA} = 2\left[\frac{1}{36}(2a)\left(a^3\right) + \frac{1}{2}(2a)(a)\left(\frac{a}{3}\right)^2\right] = \frac{a^4}{3}$$

Applying the flexure formula with $\sigma = \sigma_Y$, we have

$$\sigma_Y = \frac{M_Y c}{I}$$

$$M_Y = \frac{\sigma_Y I}{c} = \frac{\sigma_Y\left(\frac{a^4}{3}\right)}{a} = \frac{a^3}{3}\sigma_Y$$

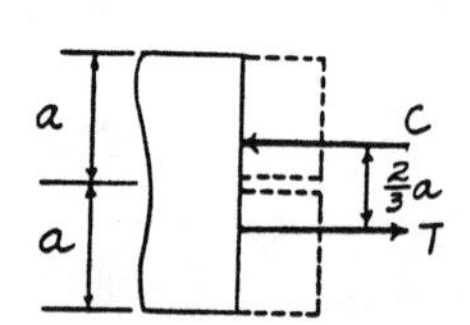

Plastic Moment :

$$C = T = \sigma_Y\left[\frac{1}{2}(2a)(a)\right] = a^2\sigma_Y$$

$$M_P = a^2\sigma_Y\left(\frac{2}{3}a\right) = \frac{2}{3}a^3\sigma_Y$$

Plastic Section Modulus :

$$Z = \frac{M_P}{\sigma_Y} = \frac{\frac{2}{3}a^3\sigma_Y}{\sigma_Y} = \frac{2}{3}a^3 \qquad \textbf{Ans}$$

Shape Factor :

$$K = \frac{M_P}{M_Y} = \frac{\frac{2}{3}a^3\sigma_Y}{\frac{a^3}{3}\sigma_Y} = 2.00 \qquad \textbf{Ans}$$

***6–176.** The beam is made from three boards nailed together as shown. If the moment acting on the cross section is $M = 650\ \text{N}\cdot\text{m}$, determine the resultant force the bending stress produces on the top board.

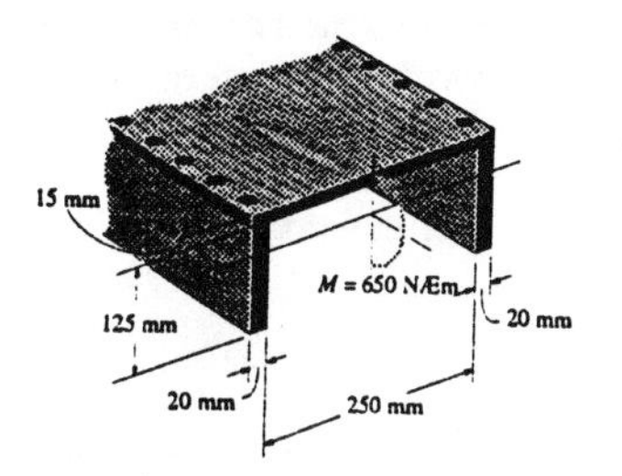

Section Properties :

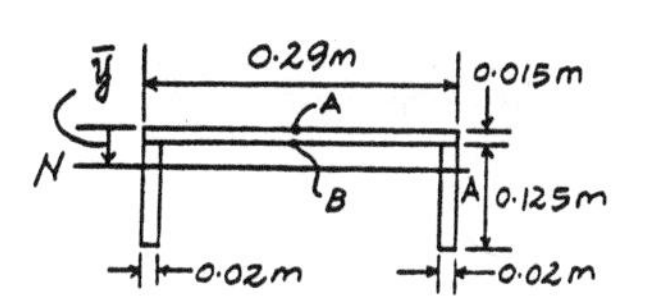

$$\bar{y} = \frac{0.0075(0.29)(0.015) + 2[0.0775(0.125)(0.02)]}{0.29(0.015) + 2(0.125)(0.02)} = 0.044933\ \text{m}$$

$$I_{NA} = \frac{1}{12}(0.29)\left(0.015^3\right) + 0.29(0.015)(0.044933 - 0.0075)^2 + \frac{1}{12}(0.04)\left(0.125^3\right) + 0.04(0.125)(0.0775 - 0.044933)^2 = 17.99037\left(10^{-6}\right)\ \text{m}^4$$

Bending Stress : Applying the flexure formula $\sigma = \frac{My}{I}$

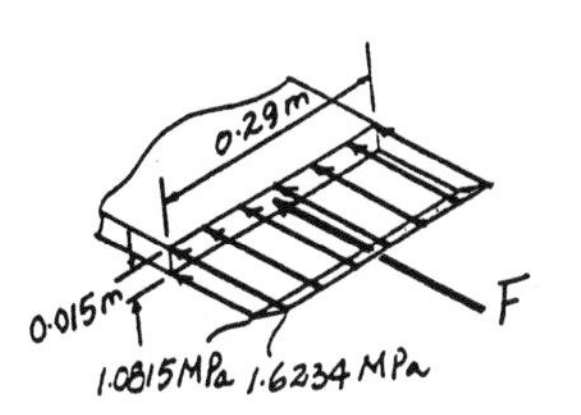

$$\sigma_B = \frac{650(0.044933 - 0.015)}{17.99037(10^{-6})} = 1.0815\ \text{MPa}$$

$$\sigma_A = \frac{650(0.044933)}{17.99037(10^{-6})} = 1.6234\ \text{MPa}$$

Resultant Force :

$$F_R = \frac{1}{2}(1.0815 + 1.6234)\left(10^6\right)(0.015)(0.29) = 5883\ \text{N} = 5.88\ \text{kN} \qquad \textbf{Ans}$$

6–177. The beam is made from three boards nailed together as shown. Determine the maximum tensile and compressive stresses in the beam.

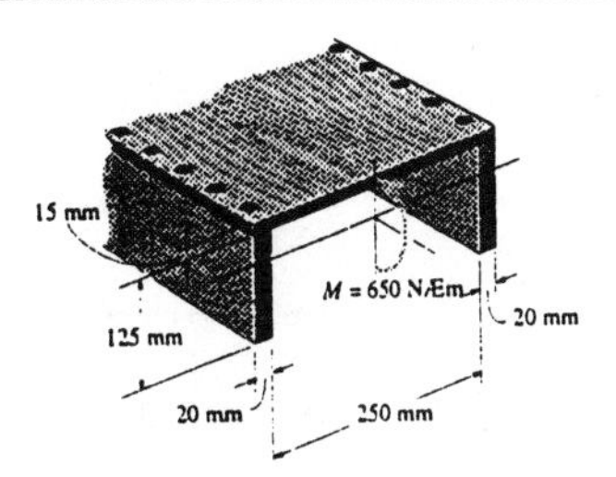

Section Properties :

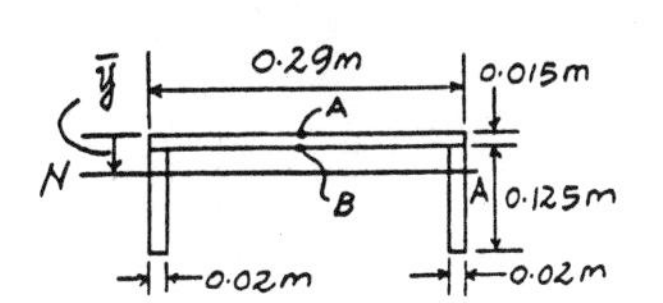

$$\bar{y} = \frac{0.0075(0.29)(0.015) + 2[0.0775(0.125)(0.02)]}{0.29(0.015) + 2(0.125)(0.02)} = 0.044933\ \text{m}$$

$$I_{NA} = \frac{1}{12}(0.29)\left(0.015^3\right) + 0.29(0.015)(0.044933 - 0.0075)^2 + \frac{1}{12}(0.04)\left(0.125^3\right) + 0.04(0.125)(0.0775 - 0.044933)^2 = 17.99037\left(10^{-6}\right)\ \text{m}^4$$

Maximum Bending Stress : Applying the flexure formula $\sigma = \frac{My}{I}$

$$(\sigma_{\max})_t = \frac{650(0.14 - 0.044933)}{17.99037(10^{-6})} = 3.43\ \text{MPa (T)} \qquad \textbf{Ans}$$

$$(\sigma_{\max})_c = \frac{650(0.044933)}{17.99037(10^{-6})} = 1.62\ \text{MPa (C)} \qquad \textbf{Ans}$$

6–178. The beam is constructed from four pieces of wood, glued together as shown. If the internal bending moment is $M = 80$ kip · ft, determine the maximum bending stress in the beam. Sketch a three-dimensional view of the stress distribution acting over the cross section.

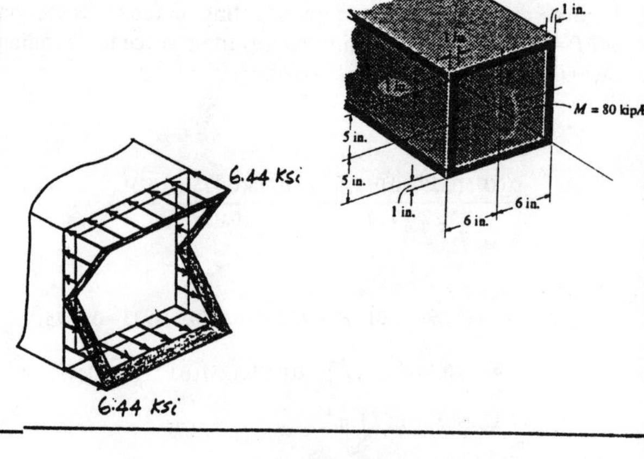

Section Property :

$$I = \frac{1}{12}(12)(12^3) - \frac{1}{12}(10)(10^3) = 894.67 \text{ in}^4$$

Maximum Bending Stress : Applying the flexure formula

$$\sigma_{max} = \frac{Mc}{I} = \frac{80(12)(6)}{894.67} = 6.44 \text{ ksi} \qquad \textbf{Ans}$$

6–179. The beam is constructed from four pieces of wood, glued together as shown. If the internal bending moment is $M = 80$ kip · ft, determine the resultant force the bending moment exerts on the top and bottom boards of the beam.

Section Property :

$$I = \frac{1}{12}(12)(12^3) - \frac{1}{12}(10)(10^3) = 894.67 \text{ in}^4$$

Maximum Bending Stress : Applying the flexure formula

$$\sigma_{max} = \frac{Mc}{I} = \frac{80(12)(6)}{894.67} = 6.438 \text{ ksi}$$

$$\sigma = \frac{My}{I} = \frac{80(12)(5)}{894.67} = 5.365 \text{ ksi}$$

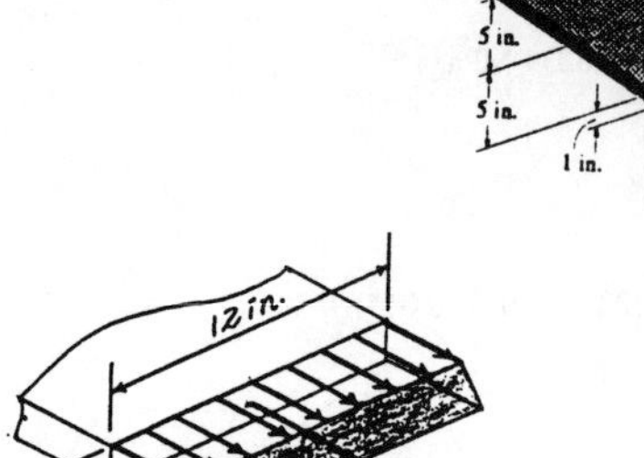

Resultant Force :

$$F = \left(\frac{6.438 + 5.365}{2}\right)(12)(1) = 70.8 \text{ kip} \qquad \textbf{Ans}$$

***6–180.** Draw the shear and moment diagrams for the beam.

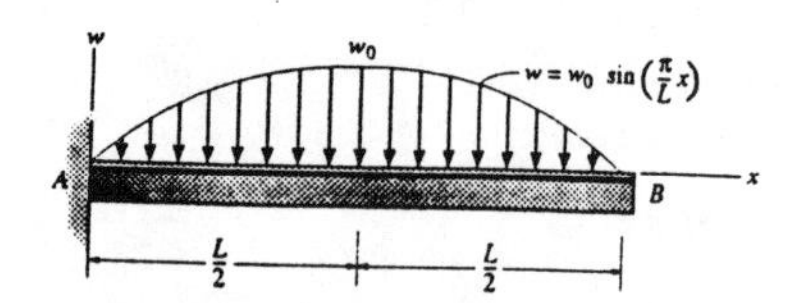

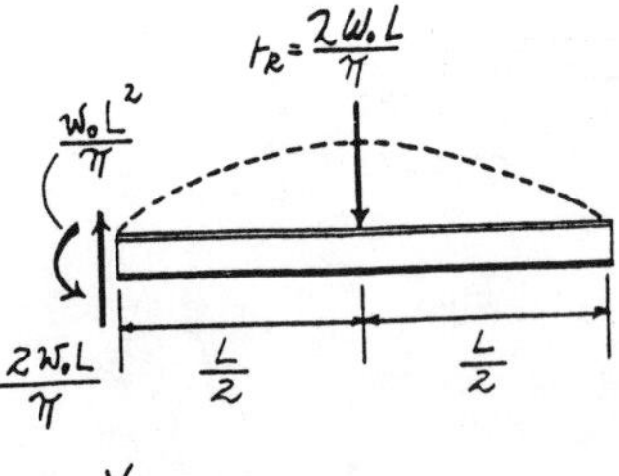

Resultant Force :

$$F_R = \int_0^L w\,dx$$

$$= w_0 \int_0^L \sin\left(\frac{\pi}{L}x\right)dx$$

$$= w_0\left[-\frac{L}{\pi}\cos\left(\frac{\pi}{L}x\right)\right]_0^L$$

$$= \frac{2w_0L}{\pi}$$

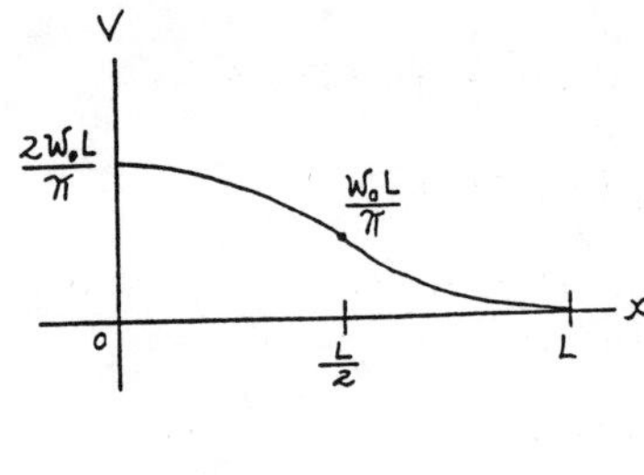

M
0
$\frac{L}{2}$
L
x
$-\frac{w_0L^2}{\pi}$

6–181. The strut has a square cross section a by a and is subjected to the bending moment **M** applied at an angle θ as shown. Determine the maximum bending stress in terms of a, M, and θ. What angle θ will give the largest bending stress in the strut? Specify the orientation of the neutral axis for this case.

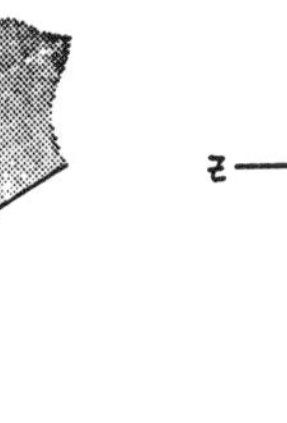

Internal Moment Components :

$$M_z = -M\cos\theta \qquad M_y = -M\sin\theta$$

Section Property :

$$I_y = I_z = \frac{1}{12}a^4$$

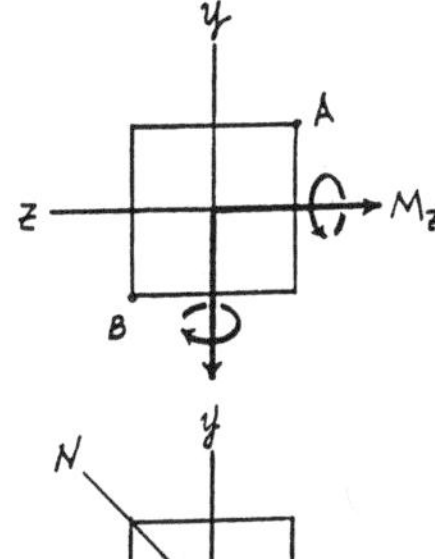

Maximum Bending Stress : By Inspection, Maximum bending stress occurs at A and B. Applying the flexure formula for biaxial bending at point A

$$\sigma = -\frac{M_z y}{I_z} + \frac{M_y z}{I_y}$$

$$= -\frac{-M\cos\theta\,(\frac{a}{2})}{\frac{1}{12}a^4} + \frac{-M\sin\theta\,(-\frac{a}{2})}{\frac{1}{12}a^4}$$

$$= \frac{6M}{a^3}(\cos\theta + \sin\theta) \qquad \textbf{Ans}$$

$$\frac{d\sigma}{d\theta} = \frac{6M}{a^3}(-\sin\theta + \cos\theta) = 0$$

$$\cos\theta - \sin\theta = 0$$

$$\theta = 45° \qquad \textbf{Ans}$$

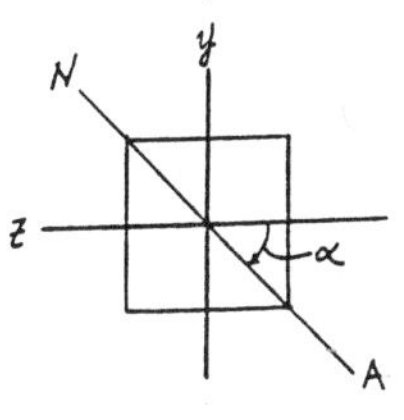

Orientation of Neutral Axis :

$$\tan\alpha = \frac{I_z}{I_y}\tan\theta$$

$$\tan\alpha = (1)\tan(-45°)$$

$$\alpha = -45°$$

6–182. For the section, $I_z = 114(10^{-6})\ \text{m}^4$, $I_y = 31.7(10^{-6})\ \text{m}^4$, $I_{yz} = 15.1(10^{-6})\ \text{m}^4$. Using the techniques outlined in Appendix A, the member's cross-sectional area has principal moments of inertia of $I_{y'} = 29(10^{-6})\ \text{m}^4$ and $I_{z'} = 117(10^{-6})\ \text{m}^4$, computed about the principal axes of inertia y' and z', respectively. If the section is subjected to a moment of $M = 2\ \text{kN}\cdot\text{m}$ directed as shown, determine the stress produced at point A, (a) using Eq. 6–11 and (b) using the equation developed in Prob. 6–110.

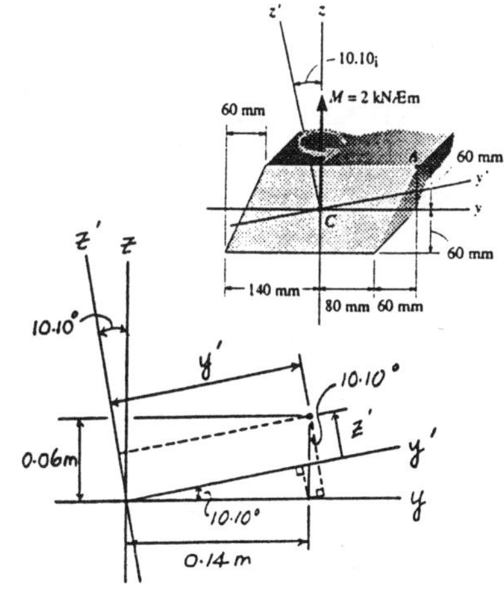

a)
Internal Moment Components :

$$M_{z'} = 2000\cos 10.10° = 1969.0\ \text{N}\cdot\text{m}$$
$$M_{y'} = 2000\sin 10.10° = 350.73\ \text{N}\cdot\text{m}$$

Section Property :

$$y' = 0.14\cos 10.10° + 0.06\sin 10.10° = 0.14835\ \text{m}$$
$$z' = 0.06\cos 10.10° - 0.14\sin 10.10° = 0.034519\ \text{m}$$

Bending Stress : Applying the flexure formula for biaxial bending

$$\sigma = -\frac{M_{z'}y'}{I_{z'}} + \frac{M_{y'}z'}{I_{y'}}$$

$$\sigma_A = -\frac{1969.0(0.14835)}{117(10^{-6})} + \frac{350.73(0.034519)}{29.0(10^{-6})}$$

$$= -2.08\ \text{MPa} = 2.08\ \text{MPa (C)} \qquad \textbf{Ans}$$

b)
Internal Moment Components :

$$M_z = 200\ \text{N}\cdot\text{m} \qquad M_y = 0$$

Bending Stress : Using formula developed in Prob. 6 - 110

$$\sigma = \frac{-(M_z I_y + M_y I_{yz})y + (M_y I_z + M_z I_{yz})z}{I_y I_z - I_{yz}^2}$$

$$= \frac{-[2000(31.7)(10^{-6}) + 0](0.14) + [0 + 2000(15.1)(10^{-6})](0.06)}{31.7(10^{-6})(114)(10^{-6}) - [15.1(10^{-6})]^2}$$

$$= -2.08\ \text{MPa} = 2.08\ \text{MPa (C)} \qquad \textbf{Ans}$$

6–183. A shaft is made of a polymer having a parabolic cross section. If it resists an internal moment of $M = 125\ \text{N}\cdot\text{m}$, determine the maximum bending stress developed in the material (*a*) using the flexure formula and (*b*) using integration. Sketch a three-dimensional view of the stress distribution acting over the cross-sectional area. ***Hint:*** The moment of inertia is determined using Eq. *A*-3 of Appendix *A*.

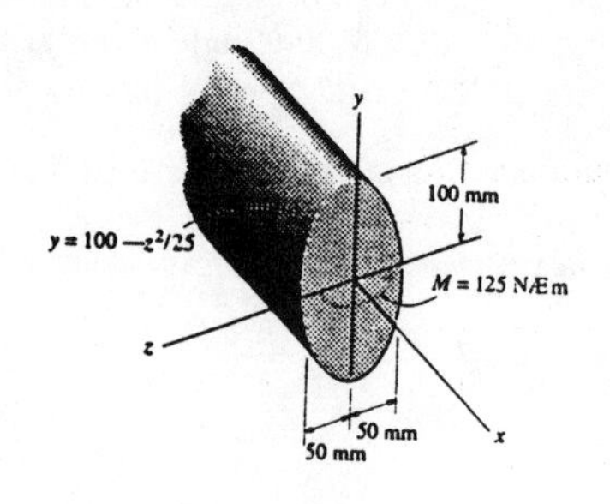

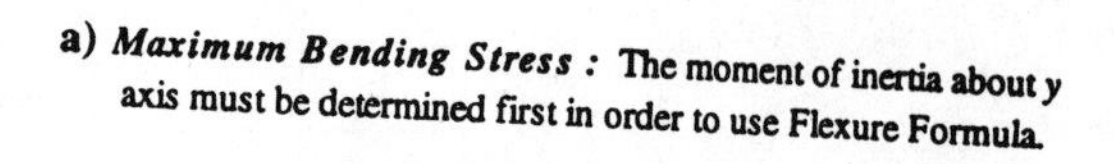

a) ***Maximum Bending Stress :*** The moment of inertia about y axis must be determined first in order to use Flexure Formula.

$$I = \int_A y^2\, dA$$

$$= 2\int_0^{100\text{mm}} y^2\,(2z)\,dy$$

$$= 20\int_0^{100\text{mm}} y^2\sqrt{100-y}\,dy$$

$$= 20\left[-\frac{3}{2}y^2(100-y)^{\frac{3}{2}} - \frac{8}{15}y(100-y)^{\frac{5}{2}} - \frac{16}{105}(100-y)^{\frac{7}{2}}\right]\Bigg|_0^{100\text{mm}}$$

$$= 30.4762\left(10^6\right)\ \text{mm}^4 = 30.4762\left(10^{-6}\right)\ \text{m}^4$$

Thus,

$$\sigma_{\max} = \frac{Mc}{I} = \frac{125(0.1)}{30.4762(10^{-6})} = 0.410\ \text{MPa} \qquad \textbf{Ans}$$

b) ***Maximum Bending Stress :*** Using integration

$$dM = 2[y(\sigma\, dA)] = 2\left\{y\left[\left(\frac{\sigma_{\max}}{100}\right)y\right](2z\,dy)\right\}$$

$$M = \frac{\sigma_{\max}}{5}\int_0^{100\text{mm}} y^2\sqrt{100-y}\,dy$$

$$125\left(10^3\right) = \frac{\sigma_{\max}}{5}\left[-\frac{3}{2}y^2(100-y)^{\frac{3}{2}} - \frac{8}{15}y(100-y)^{\frac{5}{2}} - \frac{16}{105}(100-y)^{\frac{7}{2}}\right]\Bigg|_0^{100\text{mm}}$$

$$125\left(10^3\right) = \frac{\sigma_{\max}}{5}(1.5238)\left(10^6\right)$$

$$\sigma_{\max} = 0.410\ \text{N/mm}^2 = 0.410\ \text{MPa} \qquad \textbf{Ans}$$

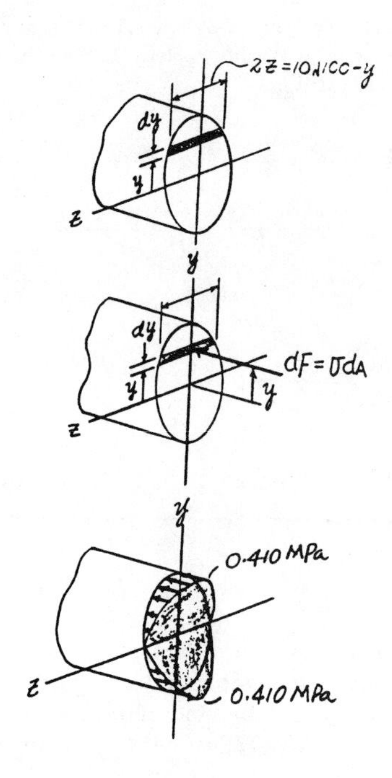

***6–184.** A shaft is made of a polymer having a parabolic cross section. If it resists an internal moment of $M = 125\ \text{N}\cdot\text{m}$, determine the maximum bending stress developed in the material (*a*) using the flexure formula and (*b*) using integration. Sketch a three-dimensional view of the stress distribution acting over the cross-sectional area. *Hint:* The moment of inertia is determined using Eq. *A*-3 of Appendix *A*.

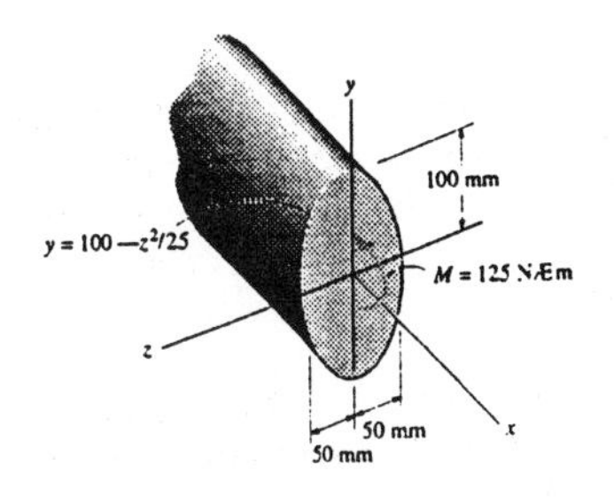

a) *Maximum Bending Stress :* The moment of inertia about the y axis must be determined first in order to use the flexure formula.

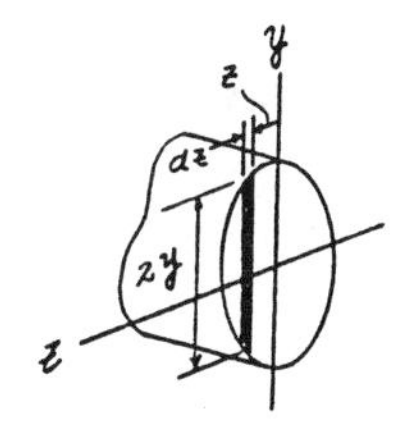

$$I = \int_A z^2\,dA$$

$$= 2\int_0^{50\text{mm}} z^2\,(2y\,dz)$$

$$= 4\int_0^{50\text{mm}} z^2\left(100 - \frac{z^2}{25}\right)dz$$

$$= 4\int_0^{50\text{mm}} \left(100z^2 - \frac{z^4}{25}\right)dz$$

$$= 4\left[100\left(\frac{z^3}{3}\right) - \frac{z^5}{125}\right]\Bigg|_0^{50\text{mm}}$$

$$= 6.667\left(10^6\right)\ \text{mm}^4 = 6.667\left(10^{-6}\right)\ \text{m}^4$$

Thus,

$$\sigma_{\max} = \frac{Mc}{I} = \frac{125(0.05)}{6.667(10^{-6})} = 0.9375\ \text{MPa} \qquad \textbf{Ans}$$

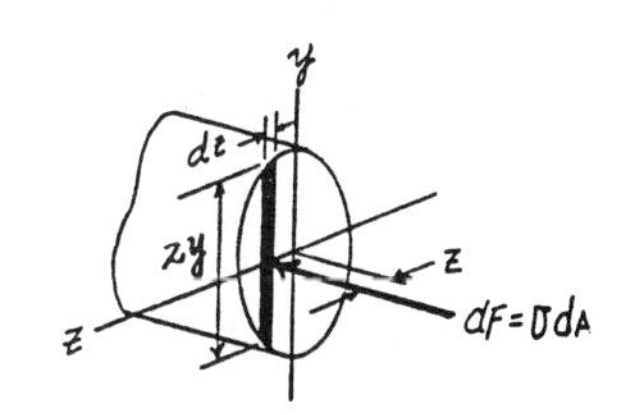

b) *Maximum Bending Stress :* Using integration,

$$dM = 2[z(\sigma\,dA)] = 2\left\{z\left[\left(\frac{\sigma_{\max}}{50}\right)z\right](2y\,dz)\right\}$$

$$M = \frac{2\sigma_{\max}}{25}\int_0^{50\text{mm}} z^2\left(100 - \frac{z^2}{25}\right)dz$$

$$125\left(10^3\right) = \frac{2\sigma_{\max}}{25}\left[100\left(\frac{z^3}{3}\right) - \frac{z^5}{125}\right]\Bigg|_0^{50\text{mm}}$$

$$\sigma_{\max} = 0.9375\ \text{N/mm}^2 = 0.9375\ \text{MPa} \qquad \textbf{Ans}$$

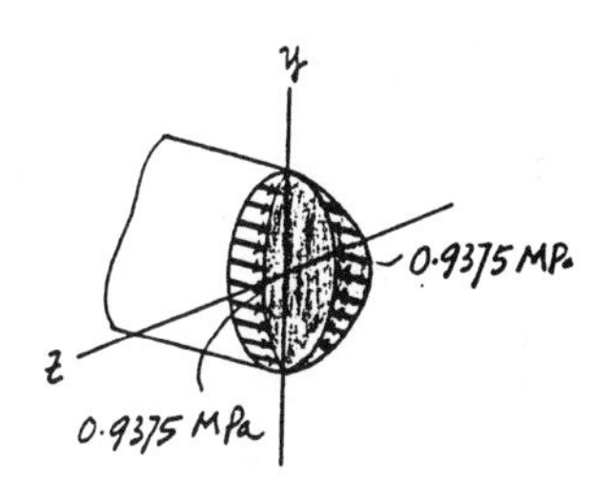

7–1. If the wide-flange beam is subjected to a shear of $V = 25$ kip, determine the maximum shear stress in the beam.

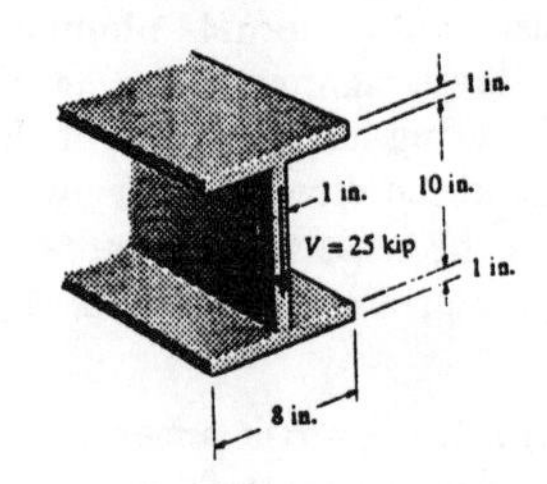

Section Properties :

$$I = \frac{1}{12}(8)\left(12^3\right) - \frac{1}{12}(7)\left(10^3\right) = 568.67 \text{ in}^4$$

$$Q_{max} = \Sigma \bar{y}'A' = 5.5(1)(8) + 2.5(1)(5) = 56.5 \text{ in}^3$$

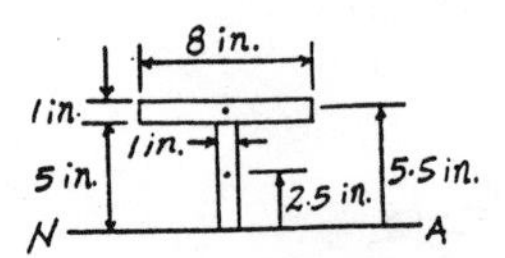

Maximum Shear Stress : Maximum shear stress occurs at the point where the neutral axis passes through the section. Applying the shear formula

$$\tau_{max} = \frac{VQ_{max}}{It} = \frac{25(56.5)}{568.67(1)} = 2.48 \text{ ksi} \qquad \textbf{Ans}$$

7–2. If the wide-flange beam is subjected to a shear of $V = 25$ kip, determine the shear force resisted by the web of the beam.

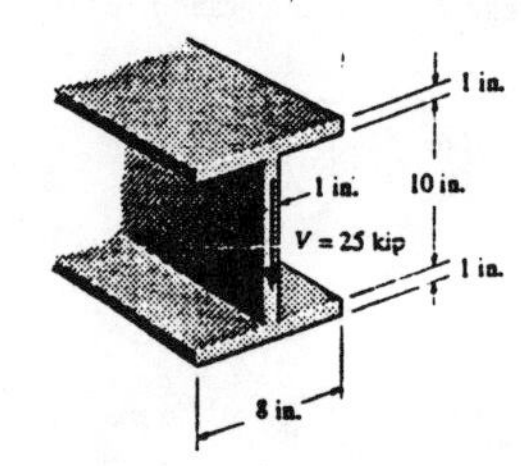

Section Properties :

$$I = \frac{1}{12}(8)\left(12^3\right) - \frac{1}{12}(7)\left(10^3\right) = 568.67 \text{ in}^4$$

$$Q = \Sigma \bar{y}'A' = 5.5(8)(1) + (0.5y + 2.5)(5 - y)(1)$$
$$= 56.5 - 0.5y^2$$

8 in.
1 in
5-y
5.5 in.
0.5y+2.5
1 in.
y
N
A

Shear Stress : Applying the shear formula

$$\tau = \frac{VQ}{It} = \frac{25(56.5 - 0.5y^2)}{568.67(1)} = 2.484 - 0.02198y^2$$

Resultant Shear Force : For the web

$$V_W = \int_A \tau dA$$
$$= \int_{-5\text{in}}^{5\text{in}} \left(2.484 - 0.02198y^2\right)(1)\,dy$$
$$= 23.0 \text{ kip} \qquad \textbf{Ans}$$

7–3. If the T-beam is subjected to a vertical shear of $V = 12$ kip, determine the maximum shear stress in the beam. Also, compute the shear-stress jump at the flange–web junction AB. Sketch the variation of the shear-stress intensity over the entire cross section.

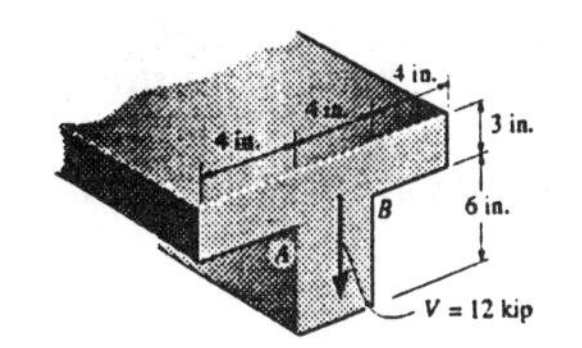

Section Properties :

$$\bar{y} = \frac{\Sigma \tilde{y}A}{\Sigma A} = \frac{1.5(12)(3) + 6(4)(6)}{12(3) + 4(6)} = 3.30 \text{ in.}$$

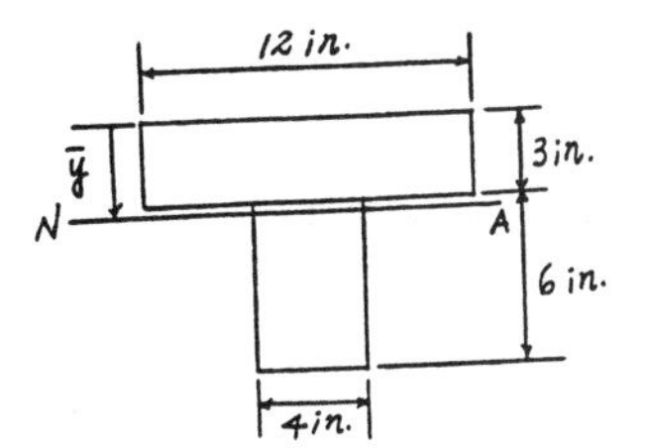

$$I_{NA} = \frac{1}{12}(12)\left(3^3\right) + 12(3)(3.30 - 1.5)^2 + \frac{1}{12}(4)\left(6^3\right) + 4(6)(6 - 3.30)^2 = 390.60 \text{ in}^4$$

$$Q_{\max} = \bar{y}_1' A' = 2.85(5.7)(4) = 64.98 \text{ in}^3$$

$$Q_{AB} = \bar{y}_2' A' = 1.8(3)(12) = 64.8 \text{ in}^3$$

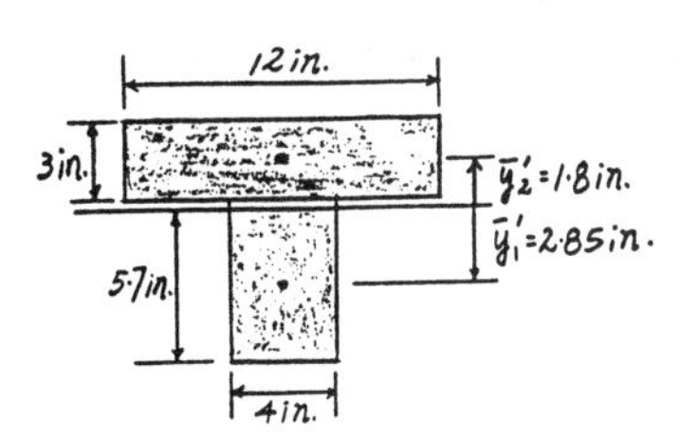

Shear Stress : Applying the shear formula $\tau = \frac{VQ}{It}$

$$\tau_{\max} = \frac{VQ_{\max}}{It} = \frac{12(64.98)}{390.60(4)} = 0.499 \text{ ksi} \qquad \textbf{Ans}$$

$$(\tau_{AB})_F = \frac{VQ_{AB}}{I t_F} = \frac{12(64.8)}{390.60(12)} = 0.166 \text{ ksi} \qquad \textbf{Ans}$$

$$(\tau_{AB})_W = \frac{VQ_{AB}}{I t_W} = \frac{12(64.8)}{390.60(4)} = 0.498 \text{ ksi} \qquad \textbf{Ans}$$

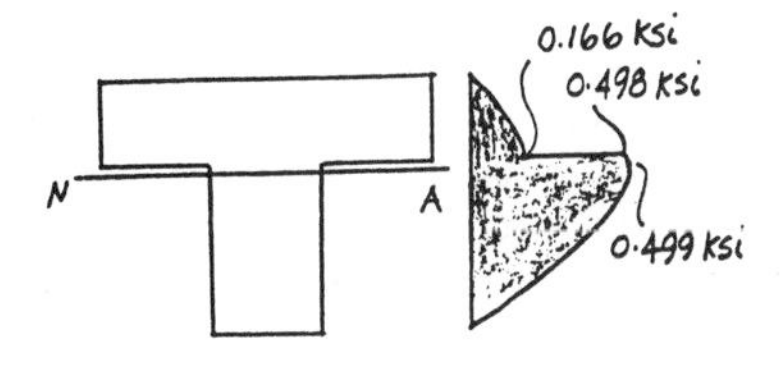

***7–4.** If the T-beam is subjected to a vertical shear of $V = 12$ kip, determine the vertical shear force resisted by the flange.

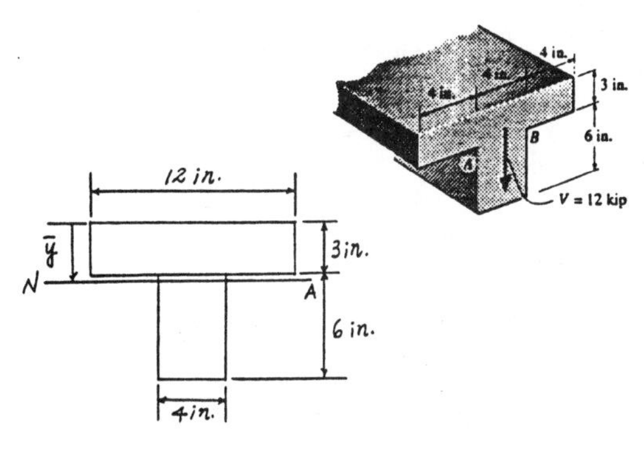

Section Properties :

$$\bar{y} = \frac{\Sigma \tilde{y}A}{\Sigma A} = \frac{1.5(12)(3) + 6(4)(6)}{12(3) + 4(6)} = 3.30 \text{ in.}$$

$$I_{NA} = \frac{1}{12}(12)\left(3^3\right) + 12(3)(3.30 - 1.5)^2 + \frac{1}{12}(4)\left(6^3\right) + 6(4)(6 - 3.30)^2 = 390.60 \text{ in}^4$$

$$Q = \bar{y}'A' = (1.65 + 0.5y)(3.3 - y)(12) = 65.34 - 6y^2$$

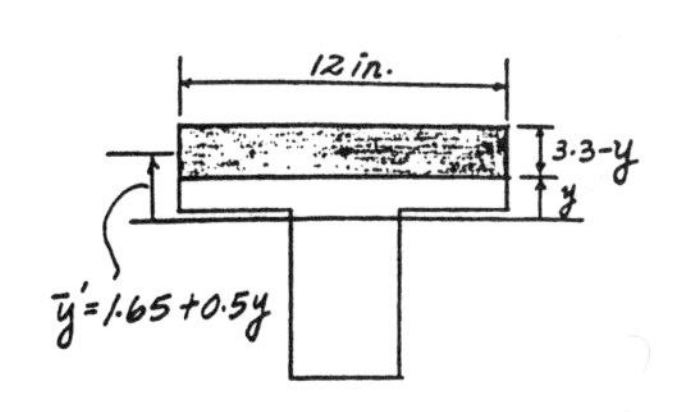

Shear Stress : Applying the shear formula

$$\tau = \frac{VQ}{It} = \frac{12(65.34 - 6y^2)}{390.60(12)} = 0.16728 - 0.01536y^2$$

Resultant Shear Force : For the flange

$$V_f = \int_A \tau dA = \int_{0.3\text{in}}^{3.3\text{in}} \left(0.16728 - 0.01536y^2\right)(12dy) = 3.82 \text{ kip} \qquad \textbf{Ans}$$

7–5. The beam has a rectangular cross section and is made of wood having an allowable shear stress of $\tau_{allow} = 1.6$ ksi. If it is subjected to a shear of $V = 4$ kip, determine the smallest dimension a of its bottom and $1.5a$ of its sides.

Section Properties :

$$I = \frac{1}{12}(a)(1.5a)^3 = 0.28125\,a^4$$

$$Q_{max} = \bar{y}'A' = (0.375a)(0.75a)(a) = 0.28125a^3$$

Allowable Shear Stress : Applying the shear formula

$$\tau_{max} = \tau_{allow} = \frac{VQ_{max}}{It}$$

$$1.6 = \frac{4(0.28125a^3)}{0.28125a^4(a)}$$

$$a = 1.58 \text{ in.} \qquad \textbf{Ans}$$

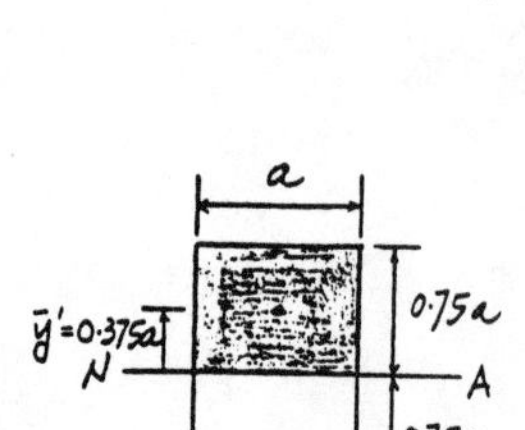

7–6. The beam has a rectangular cross section and is made of wood. If it is subjected to a shear of $V = 4$ kip, and $a = 10$ in., determine the maximum shear stress and plot the shear-stress variation over the cross section. Sketch the result in three dimensions.

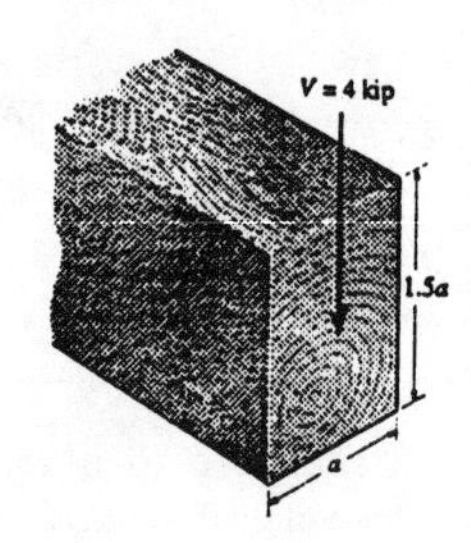

Section Properties :

$$I = \frac{1}{12}(10)\left(15^3\right) = 2812.5 \text{ in}^4$$

$$Q_{max} = \bar{y}'A' = 3.75(7.5)(10) = 281.25 \text{ in}^3$$

Maximum Shear Stress : Maximum shear stress occurs at the point where the neutral axis passes through the section. Applying the shear formula

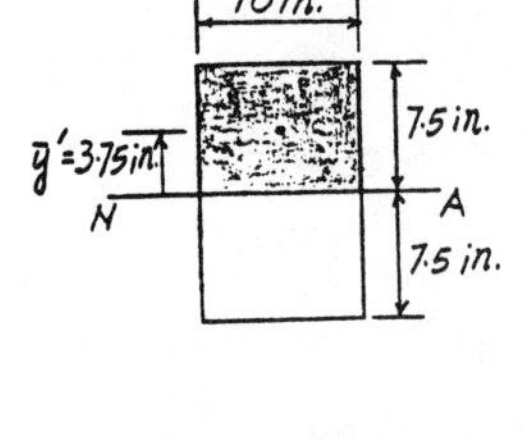

$$\tau_{max} = \frac{VQ_{max}}{It}$$

$$= \frac{4(281.25)}{2812.5(10)} = 0.040 \text{ ksi} = 40.0 \text{ psi} \qquad \textbf{Ans}$$

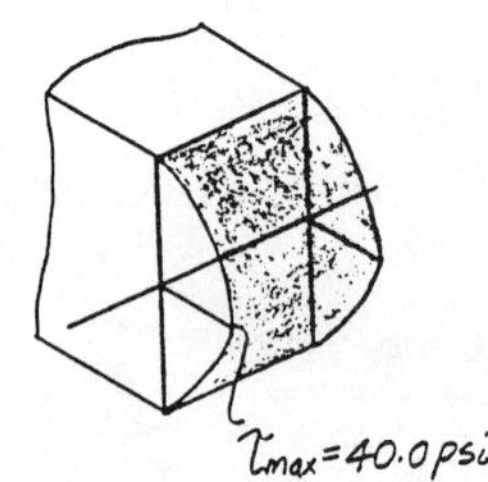

7–7. Determine the maximum shear stress in the member if it is subjected to a shear force of $V = 500$ kN.

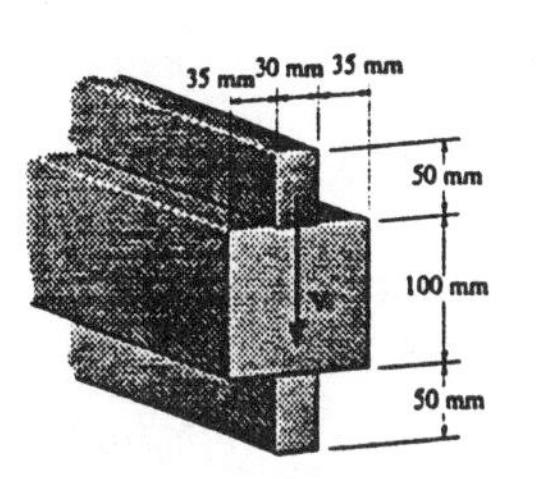

Section Properties :

$$I_{NA} = \frac{1}{12}(0.03)\left(0.2^3\right) + \frac{1}{12}(0.07)\left(0.1^3\right)$$
$$= 25.833\left(10^{-6}\right)\ \text{m}^4$$

$$Q_A = \bar{y}_1{}'A' = 0.075(0.03)(0.05) = 0.1125\left(10^{-3}\right)\ \text{m}^3$$

$$Q_B = \Sigma\bar{y}'A' = 0.075(0.03)(0.05) + 0.025(0.1)(0.05)$$
$$= 0.2375\left(10^{-3}\right)\ \text{m}^3$$

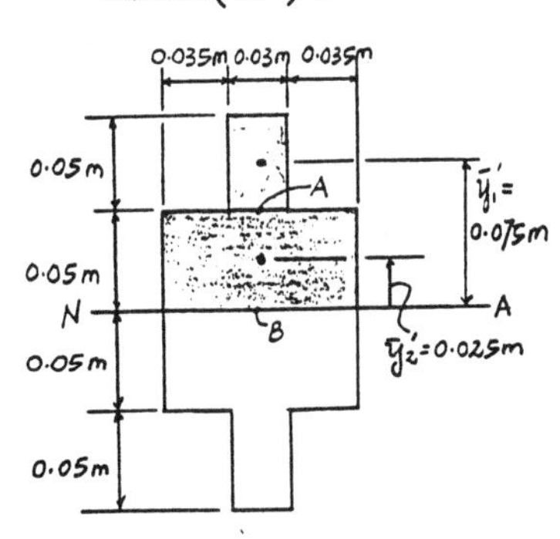

Allowable Shear Stress : Maximum shear stress can occurs at either point *A* or *B*. Applying the shear formula at points A and *B*,

$$\tau_A = \frac{VQ_A}{It_A}$$
$$= \frac{500(10^3)(0.1125)(10^{-3})}{25.833(10^{-6})(0.03)}$$
$$= 72.6\ \text{MPa}\ \textbf{(Max)} \qquad \textbf{Ans}$$

$$\tau_B = \frac{VQ_B}{It_B}$$
$$= \frac{500(10^3)(0.2375)(10^{-3})}{25.833(10^{-6})(0.1)}$$
$$= 45.97\ \text{MPa}$$

***7–8.** Determine the maximum shear force V that the member can support if the allowable shear stress for the material is $\tau_{allow} = 75$ MPa. Also, plot the shear-stress variation over the cross section

Section Properties :

$$I_{NA} = \frac{1}{12}(0.03)\left(0.2^3\right) + \frac{1}{12}(0.07)\left(0.1^3\right)$$
$$= 25.833\left(10^{-6}\right)\ \text{m}^4$$

$$Q_A = \bar{y}_1{}'A' = 0.075(0.03)(0.05) = 0.1125\left(10^{-3}\right)\ \text{m}^3$$

$$Q_B = \Sigma\bar{y}'A' = 0.075(0.03)(0.05) + 0.025(0.1)(0.05)$$
$$= 0.2375\left(10^{-3}\right)\ \text{m}^3$$

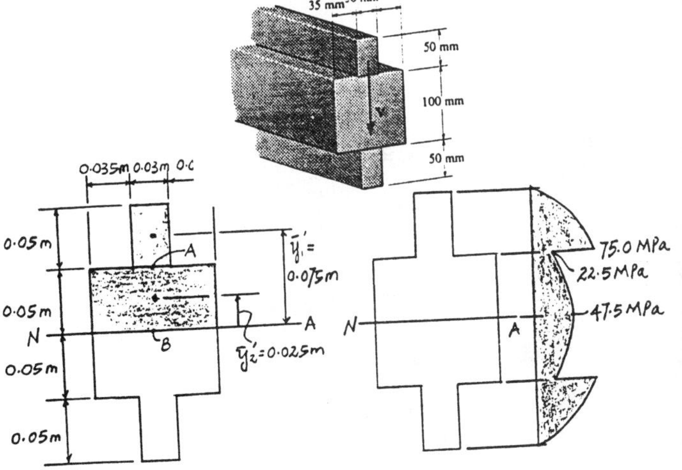

Allowable Shear Stress : Maximum shear stress can occur at either point *A* or *B*. Applying the shear formula at point A

$$\tau_A = \tau_{allow} = \frac{VQ_A}{It_A}$$
$$75\left(10^6\right) = \frac{V(0.1125)(10^{-3})}{25.833(10^{-6})(0.03)}$$

$$V = 516\,667\ \text{N} = 517\ \text{kN}\ (\textit{\textbf{controls!}}) \qquad \textbf{Ans}$$

Applying the shear formula at point *B*

$$\tau_B = \tau_{allow} = \frac{VQ_B}{It_B}$$
$$75\left(10^6\right) = \frac{V(0.2375)(10^{-3})}{25.833(10^{-6})(0.1)}$$

$$V = 815\,789\ \text{N} = 815.8\ \text{kN}$$

Shear Stress : With the applied shear force of 516.67 kN, apply shear formula at point *A* with $t_A = 0.1$ m and at point *B* with $t_B = 0.1$ m.

$$\tau_A = \frac{VQ_A}{It_A} = \frac{516.67(10^3)(0.1125)(10^{-3})}{25.833(10^{-6})(0.1)}$$
$$= 22.5\ \text{MPa}$$

$$\tau_B = \frac{VQ_B}{It_B} = \frac{516.67(10^3)(0.2375)(10^{-3})}{25.833(10^{-6})(0.1)}$$
$$= 47.5\ \text{MPa}$$

7–9. If the beam is subjected to a shear of $V = 15$ kN, determine the web's shear stress at A and B. Indicate the shear-stress components on a volume element located at these points.

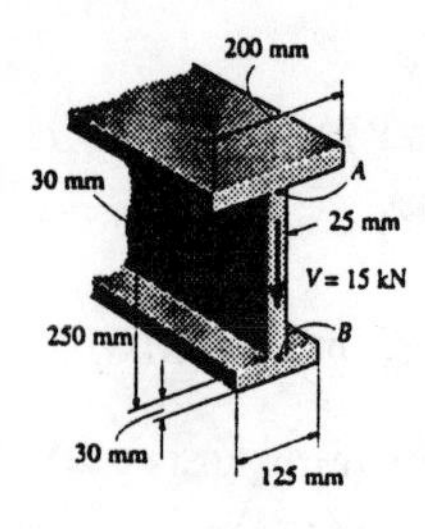

Section Properties :

$$\bar{y} = \frac{\Sigma \tilde{y}A}{\Sigma A}$$

$$= \frac{0.015(0.125)(0.03) + 0.155(0.025)(0.25) + 0.295(0.2)(0.03)}{0.125(0.03) + 0.025(0.25) + 0.2(0.03)}$$

$$= 0.1747 \text{ m}$$

$$I_{NA} = \frac{1}{12}(0.125)(0.03^3) + 0.125(0.03)(0.1747 - 0.015)^2$$

$$+ \frac{1}{12}(0.025)(0.25^3) + 0.025(0.25)(0.1747 - 0.155)^2$$

$$+ \frac{1}{12}(0.2)(0.03^3) + 0.2(0.03)(0.295 - 0.1747)^2$$

$$= 0.218182(10^{-3}) \text{ m}^4$$

$$Q_A = \bar{y}_1'A' = 0.1203(0.2)(0.03) = 0.721875(10^{-3}) \text{ m}^3$$

$$Q_B = \bar{y}_2'A' = 0.1597(0.125)(0.03) = 0.598828(10^{-3}) \text{ m}^3$$

Shear Stress : Applying the shear formula $\tau = \dfrac{VQ}{It}$ at points A and B,

$$\tau_A = \frac{15(10^3)(0.721875)(10^{-3})}{0.218182(10^{-3})(0.025)} = 1.99 \text{ MPa} \qquad \textbf{Ans}$$

$$\tau_B = \frac{15(10^3)(0.598828)(10^{-3})}{0.218182(10^{-3})(0.025)} = 1.65 \text{ MPa} \qquad \textbf{Ans}$$

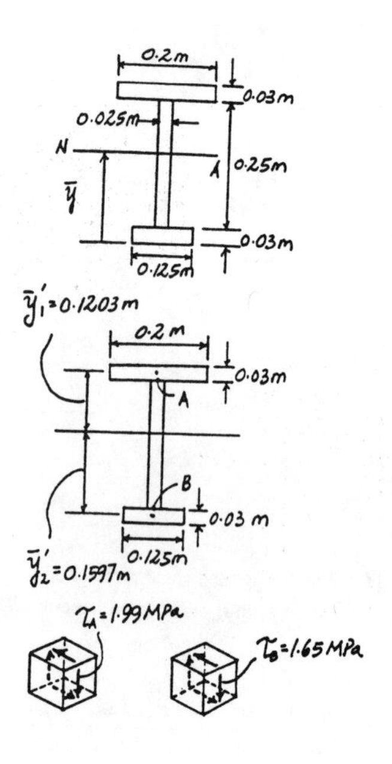

7–10. Determine the maximum shear force V that the box member can support if the allowable shear stress for the material is $\tau_{allow} = 60$ MPa. Also, plot the vertical shear stress variation over the cross sectional area.

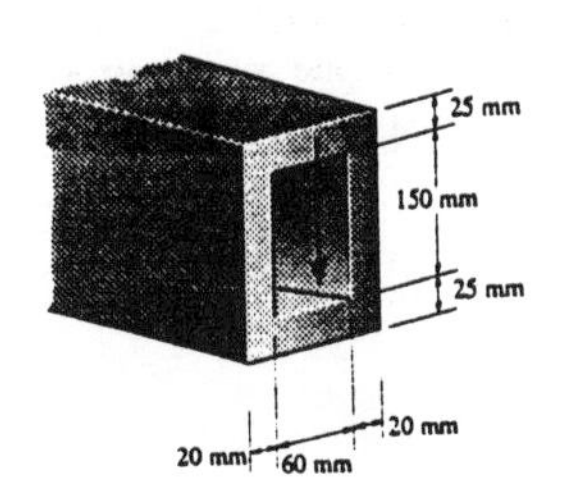

Section Properties :

$$I_{NA} = \frac{1}{12}(0.1)\left(0.2^3\right) - \frac{1}{12}(0.06)\left(0.15^3\right)$$
$$= 49.79167\left(10^{-6}\right)\ \text{m}^4$$

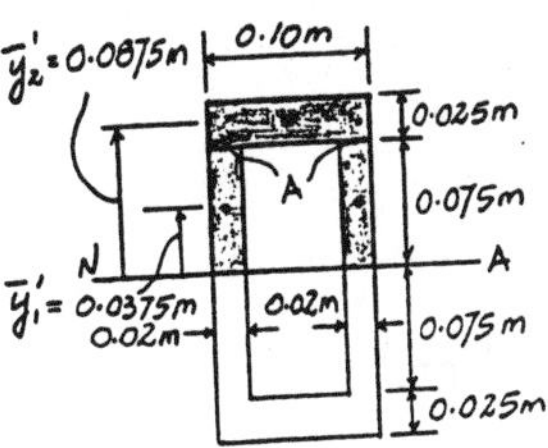

$$Q_A = \bar{y}_2'A' = 0.0875(0.1)(0.025) = 0.21875\left(10^{-3}\right)\ \text{m}^3$$

$$Q_{max} = \Sigma\bar{y}'A'$$
$$= 0.0875(0.1)(0.025) + 0.0375(0.075)(0.04)$$
$$= 0.33125\left(10^{-3}\right)\ \text{m}^3$$

Allowable Shear Stress : Maximum shear stress occurs at the point where the neutral axis passes through the section. Applying the shear formula

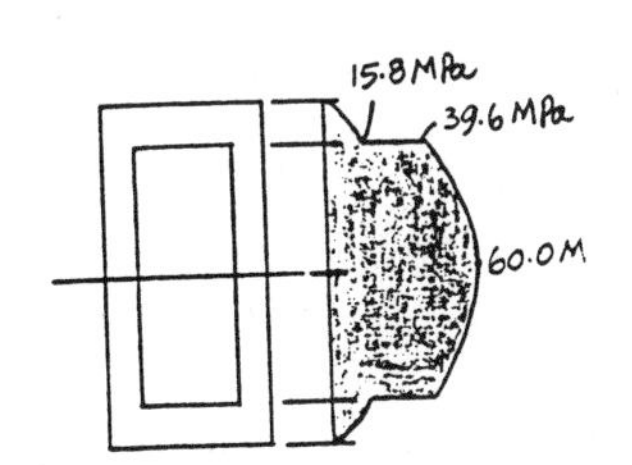

$$\tau_{max} = \tau_{allow} = \frac{VQ_{max}}{It}$$

$$60\left(10^6\right) = \frac{V(0.33125)(10^{-3})}{49.79167(10^{-6})(0.04)}$$

$$V = 360\ 755\ \text{N} = 361\ \text{kN} \qquad \textbf{Ans}$$

Shear Stress : With the applied shear force of 360.755kN, apply the shear formula at point A with the web thickness of $t_w = 0.04$ m and flange thickness of $t_f = 0.1$ m

$$(\tau_A)_w = \frac{VQ_A}{It_w} = \frac{360.755(10^3)(0.21875)(10^{-3})}{49.79167(10^{-6})(0.04)}$$
$$= 39.6\ \text{MPa}$$

$$(\tau_A)_f = \frac{VQ_A}{It_f} = \frac{360.755(10^3)(0.21875)(10^{-3})}{49.79167(10^{-6})(0.1)}$$
$$= 15.8\ \text{MPa}$$

7–11. Determine the maximum shear stress in the box member if it is subjected to a shear force of $V = 350$ kN.

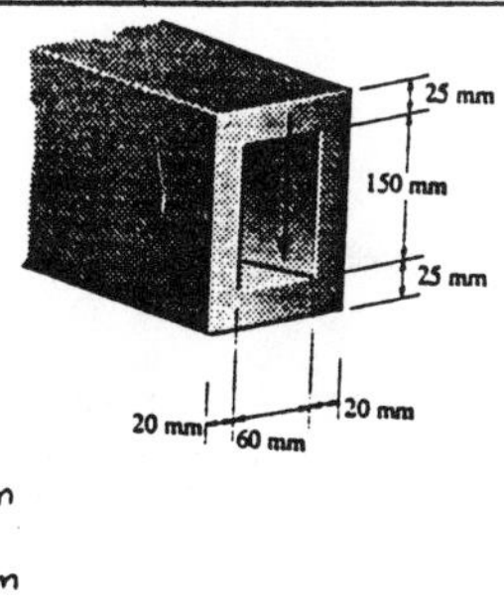

Section Properties :

$$I_{NA} = \frac{1}{12}(0.1)\left(0.2^3\right) - \frac{1}{12}(0.06)\left(0.15^3\right)$$
$$= 49.79167\left(10^{-6}\right)\ \text{m}^4$$

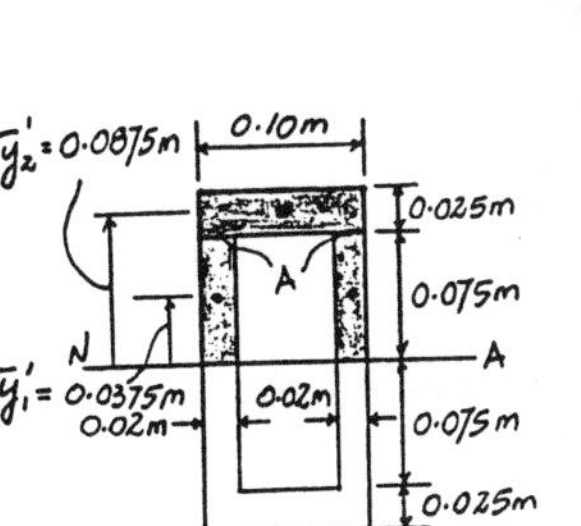

$$Q_{max} = \Sigma\bar{y}'A'$$
$$= 0.0875(0.1)(0.025) + 0.0375(0.075)(0.04)$$
$$= 0.33125\left(10^{-3}\right)\ \text{m}^3$$

Maximum Shear Stress : Maximum shear stress occurs at the point where the neutral axis passes through the section. Applying the shear formula

$$\tau_{max} = \frac{VQ_{max}}{It}$$
$$= \frac{350(10^3)(0.33125)(10^{-3})}{49.79167(10^{-6})(0.04)}$$
$$= 58.2\ \text{MPa} \qquad \textbf{Ans}$$

***7–12.** Determine the maximum shear stress in the strut if it is subjected to a shear force of $V = 20$ kN.

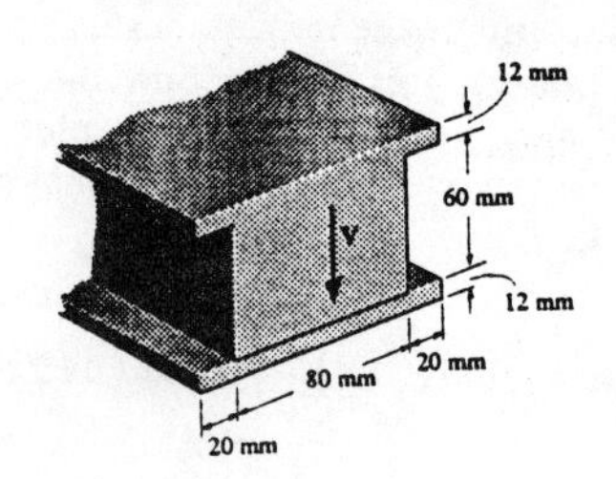

Section Properties :

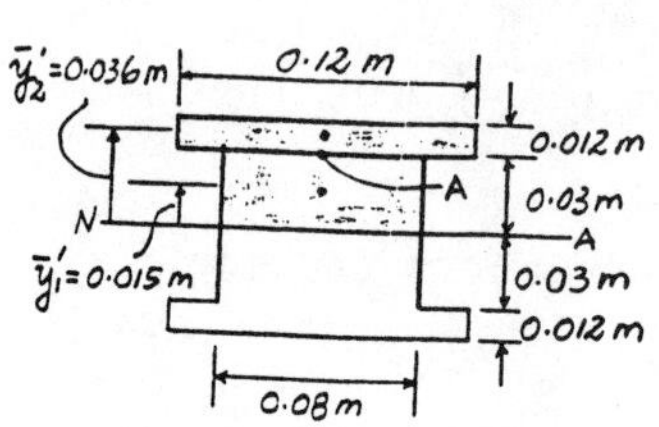

$$I_{NA} = \frac{1}{12}(0.12)\left(0.084^3\right) - \frac{1}{12}(0.04)\left(0.06^3\right)$$
$$= 5.20704\left(10^{-6}\right) \text{ m}^4$$

$$Q_{max} = \Sigma \bar{y}'A'$$
$$= 0.015(0.08)(0.03) + 0.036(0.012)(0.12)$$
$$= 87.84\left(10^{-6}\right) \text{ m}^3$$

Maximum Shear Stress : Maximum shear stress occurs at the point where the neutral axis passes through the section. Applying the shear formula

$$\tau_{max} = \frac{VQ_{max}}{It}$$
$$= \frac{20(10^3)(87.84)(10^{-6})}{5.20704(10^{-6})(0.08)}$$
$$= 4.22 \text{ MPa} \qquad \textbf{Ans}$$

7–13. Determine the maximum shear force V that the strut can support if the allowable shear stress for the material is $\tau_{allow} = 40$ MPa.

Section Properties :

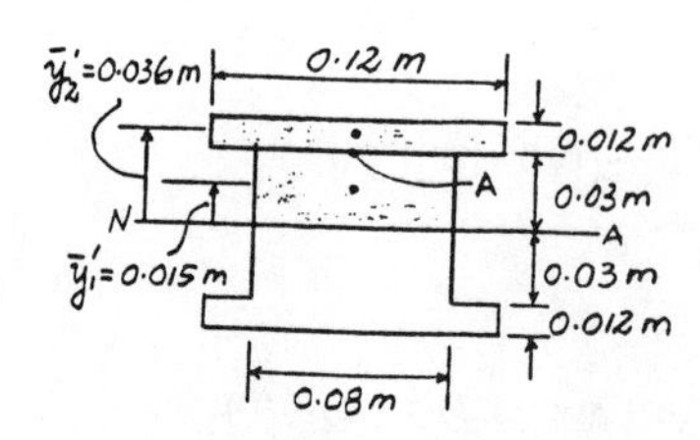

$$I_{NA} = \frac{1}{12}(0.12)\left(0.084^3\right) - \frac{1}{12}(0.04)\left(0.06^3\right)$$
$$= 5.20704\left(10^{-6}\right) \text{ m}^4$$

$$Q_{max} = \Sigma \bar{y}'A'$$
$$= 0.015(0.08)(0.03) + 0.036(0.012)(0.12)$$
$$= 87.84\left(10^{-6}\right) \text{ m}^3$$

Allowable Shear Stress : Maximum shear stress occurs at the point where the neutral axis passes through the section. Applying the shear formula

$$\tau_{max} = \tau_{allow} = \frac{VQ_{max}}{It}$$
$$40\left(10^6\right) = \frac{V(87.84)(10^{-6})}{5.20704(10^{-6})(0.08)}$$
$$= 189\,692 \text{ N} = 190 \text{ kN} \qquad \textbf{Ans}$$

7–14. Plot the intensity of the shear stress distributed over the cross section of the strut if it is subjected to a shear force of $V = 15$ kN.

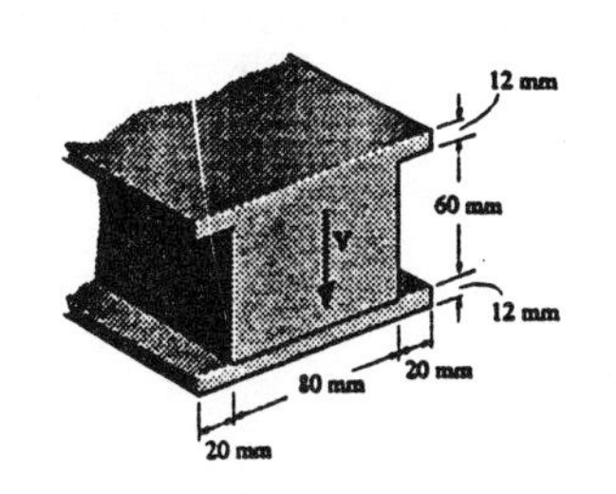

Section Properties :

$$I_{NA} = \frac{1}{12}(0.12)\left(0.084^3\right) - \frac{1}{12}(0.04)\left(0.06^3\right)$$
$$= 5.20704\left(10^{-6}\right)\ \text{m}^4$$

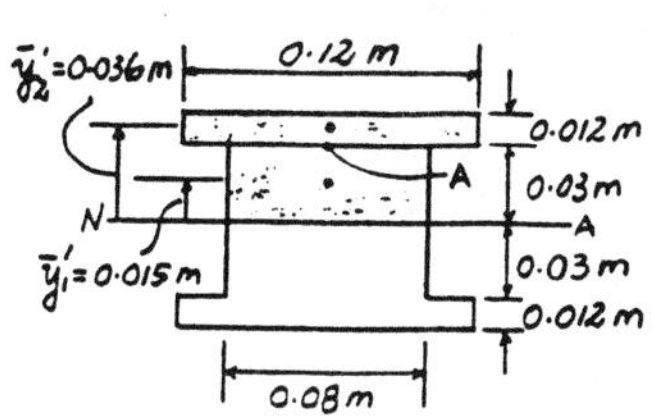

$$Q_A = \bar{y}_2'A' = 0.036(0.012)(0.12) = 51.84\left(10^{-6}\right)\ \text{m}^3$$
$$Q_{max} = \Sigma\bar{y}'A'$$
$$= 0.015(0.08)(0.03) + 0.036(0.012)(0.12)$$
$$= 87.84\left(10^{-6}\right)\ \text{m}^3$$

Shear Stress : Maximum shear stress occurs at the point where the neutral axis passes through the section. Shear stress at point A on the seam can be determined using the shear formula with the web thickness of $t_w = 0.08$ m and flange thickness of $t_f = 0.12$ m. Applying the shear formula at these points

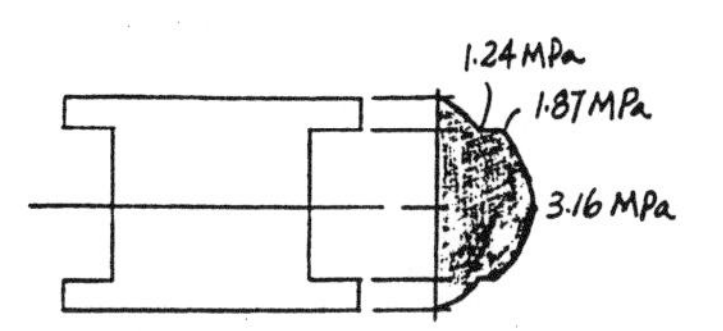

$$\tau_{max} = \frac{VQ_{max}}{It} = \frac{15(10^3)(87.84)(10^{-6})}{5.20704(10^{-6})(0.08)} = 3.16\ \text{MPa}$$

$$(\tau_A)_w = \frac{VQ_A}{It_w} = \frac{15(10^3)(51.84)(10^{-6})}{5.20704(10^{-6})(0.08)} = 1.87\ \text{MPa}$$

$$(\tau_A)_f = \frac{VQ_A}{It_f} = \frac{15(10^3)(51.84)(10^{-6})}{5.20704(10^{-6})(0.12)} = 1.24\ \text{MPa}$$

7–15. Determine the maximum shear stress in the shaft, which has a circular cross section of radius r, and is subjected to the shear force **V**. State the answer in terms of the area A of the cross section.

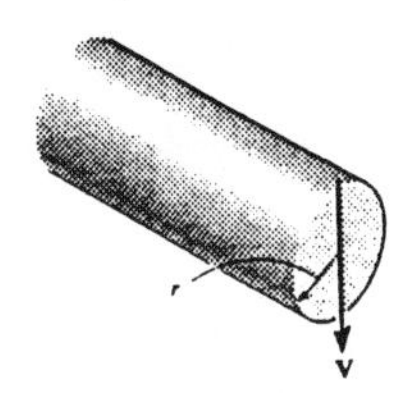

Section Properties :

$$I = \frac{\pi}{4}r^4$$

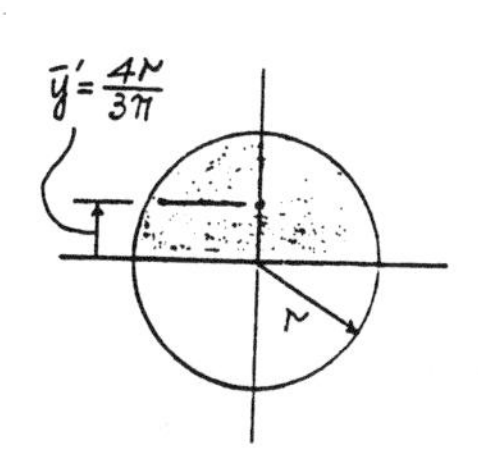

$$Q_{max} = \bar{y}'A' = \frac{4r}{3\pi}\left(\frac{\pi r^2}{2}\right) = \frac{2r^3}{3}$$

Maximum Shear Stress : Maximum shear stress occurs at the point where the neutral axis passes through the section. Applying the shear formula

$$\tau_{max} = \frac{VQ_{max}}{It} = \frac{V\left(\frac{2r^3}{3}\right)}{\left(\frac{\pi}{4}r^4\right)(2r)} = \frac{4V}{3\pi r^2}$$

However $A = \pi r^2$, hence

$$\tau_{max} = \frac{4V}{3A} \qquad \textbf{Ans}$$

***7–16.** If the pipe is subjected to a shear of $V = 15$ kip, determine the maximum shear stress in the pipe.

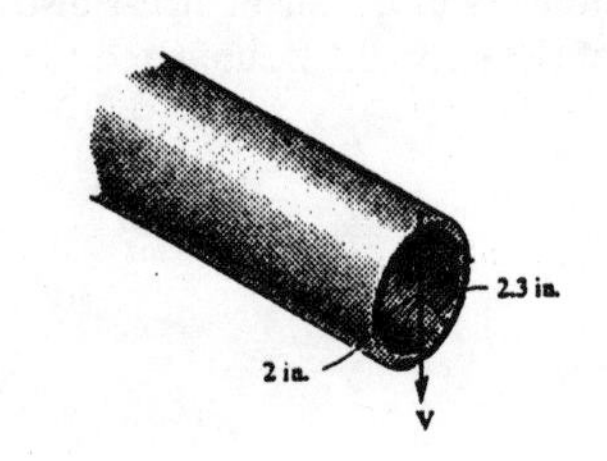

Section Properties :

$$I = \frac{\pi}{4}\left(2.3^4 - 2^4\right) = 9.4123 \text{ in}^4$$

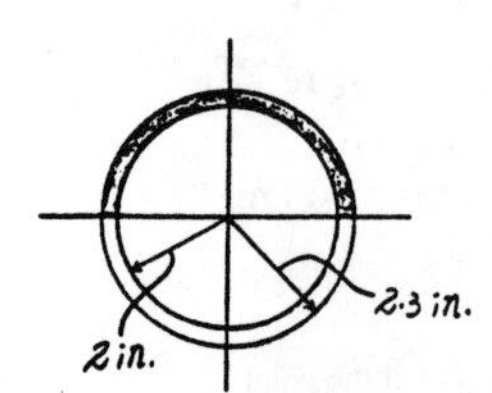

$$Q_{max} = \Sigma \bar{y}' A'$$

$$= \frac{4(2.3)}{3\pi}\left[\frac{\pi(2.3^2)}{2}\right] - \frac{4(2)}{3\pi}\left[\frac{\pi(2^2)}{2}\right]$$

$$= 2.778 \text{ in}^3$$

Maximum Shear Stress : Maximum shear stress occurs at the point where the neutral axis passes through the section. Applying the shear formula

$$\tau_{max} = \frac{VQ_{max}}{It} = \frac{15(2.778)}{9.4123(0.6)} = 7.38 \text{ ksi} \qquad \textbf{Ans}$$

7–17. Determine the largest end forces P that the member can support if the allowable shear stress is $\tau_{allow} = 10$ ksi. The supports at A and B only exert vertical reactions on the beam.

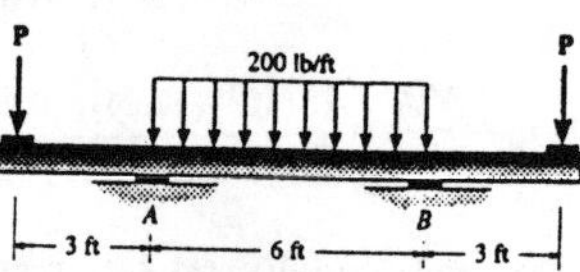

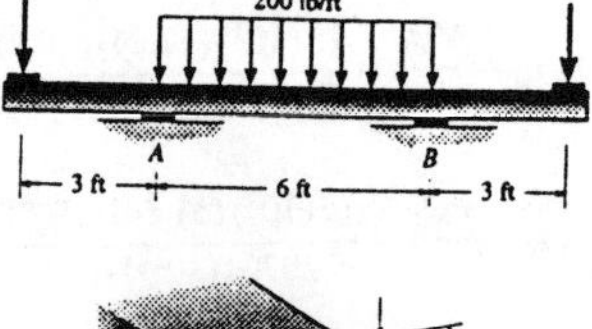

Support Reactions : As shown on FBD.

Internal Shear Force : As shown on Shear diagram, $V_{max} = P$.

Section Properties :

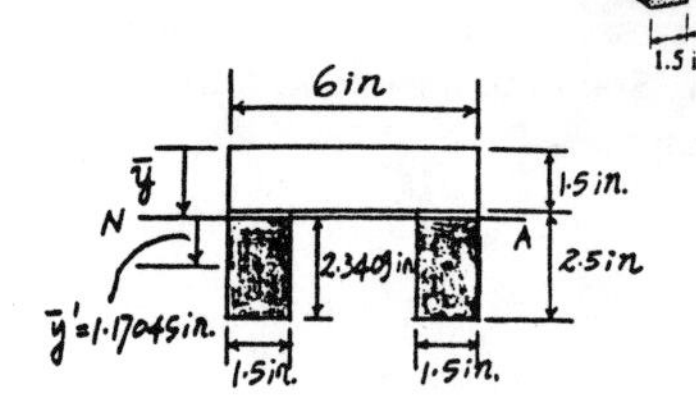

$$\bar{y} = \frac{\Sigma \bar{y}A}{\Sigma A} = \frac{2.75(2.5)(3) + 0.75(6)(1.5)}{2.5(3) + 6(1.5)} = 1.6591 \text{ in.}$$

$$I_{NA} = \frac{1}{12}(3)\left(2.5^3\right) + 3(2.5)(2.75 - 1.6591)^2 + \frac{1}{12}(6)\left(1.5^3\right) + 6(1.5)(1.6591 - 0.75)^2$$

$$= 21.9574 \text{ in}^4$$

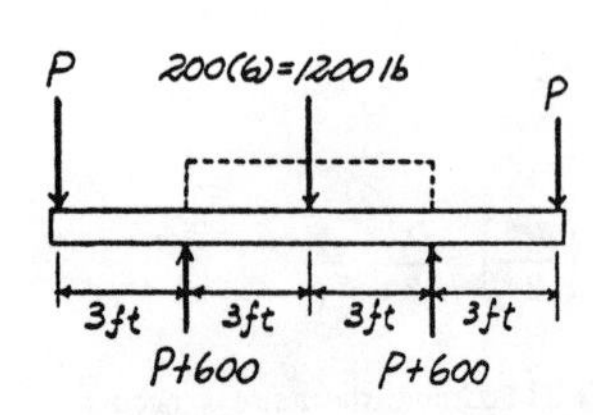

$$Q_{max} = \bar{y}'A' = 1.17045(2.3409)(3) = 8.2198 \text{ in}^3$$

Allowable Shear Stress : Maximum shear stress occurs at the point where the neutral axis passes through the section. Applying the shear formula

$$\tau_{max} = \tau_{allow} = \frac{VQ_{max}}{It}$$

$$10 = \frac{P(8.2198)}{21.9574(3)}$$

$$P = 80.1 \text{ kip} \qquad \textbf{Ans}$$

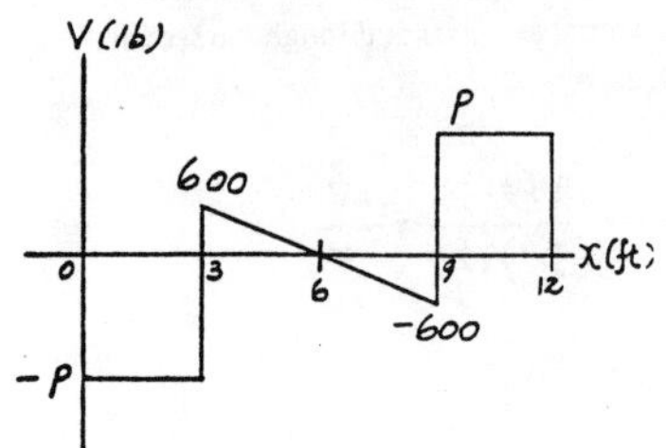

7–18. If the force $P = 800$ lb, determine the maximum shear stress in the beam at the critical section. The supports at A and B only exert vertical reactions on the beam.

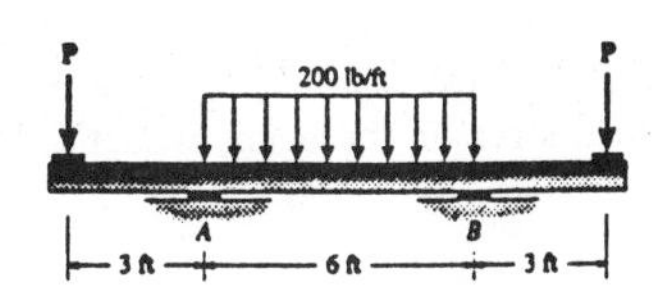

Support Reactions : As shown on FBD.

Internal Shear Force : As shown on Shear diagram, $V_{max} = 800$ lb.

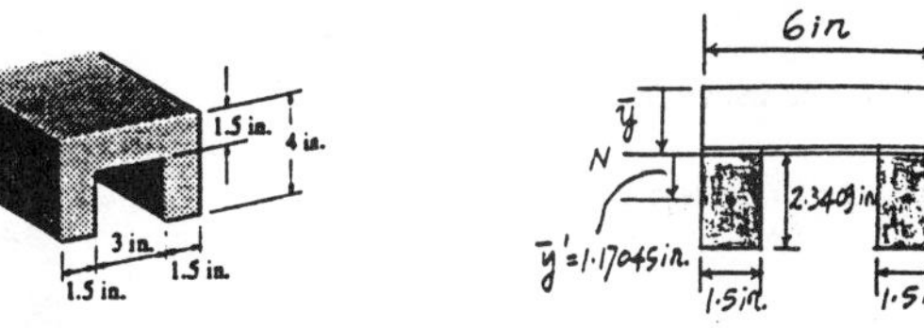

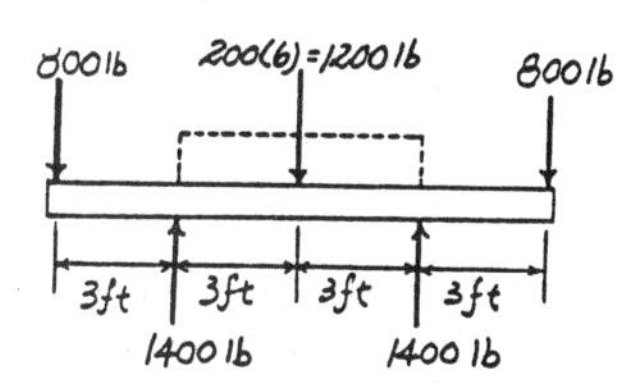

Section Properties :

$$\bar{y} = \frac{\Sigma \tilde{y}A}{\Sigma A} = \frac{2.75(2.5)(3) + 0.75(6)(1.5)}{2.5(3) + 6(1.5)} = 1.6591 \text{ in.}$$

$$I_{NA} = \frac{1}{12}(3)\left(2.5^3\right) + 3(2.5)(2.75 - 1.6591)^2 + \frac{1}{12}(6)\left(1.5^3\right) + 6(1.5)(1.6591 - 0.75)^2 = 21.9574 \text{ in}^4$$

$$Q_{max} = \bar{y}'A' = 1.17045(2.3409)(3) = 8.2198 \text{ in}^3$$

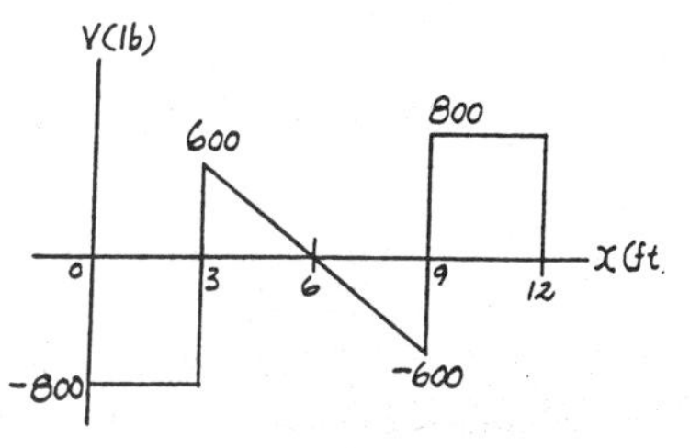

Maximum Shear Stress : Maximum shear stress occurs at the point where the neutral axis passes through the section. Applying the shear formula

$$\tau_{max} = \frac{VQ_{max}}{It} = \frac{800(8.2198)}{21.9574(3)} = 99.8 \text{ psi} \qquad \textbf{Ans}$$

7–19. The T-beam is subjected to the loading shown. Determine the maximum transverse shear stress in the beam at the critical section.

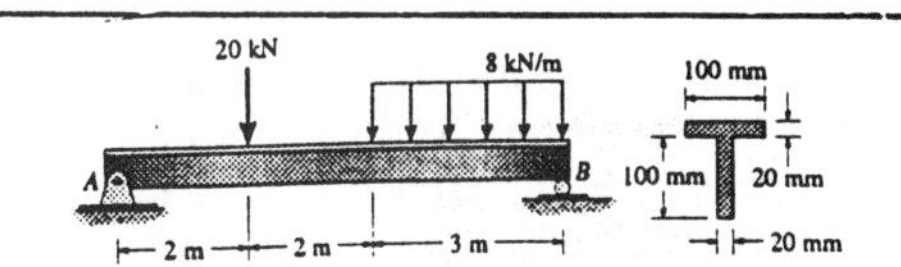

Support Reactions : As shown on FBD.

Internal Shear Force : As shown on Shear diagram, $V_{max} = 24.57$ kN.

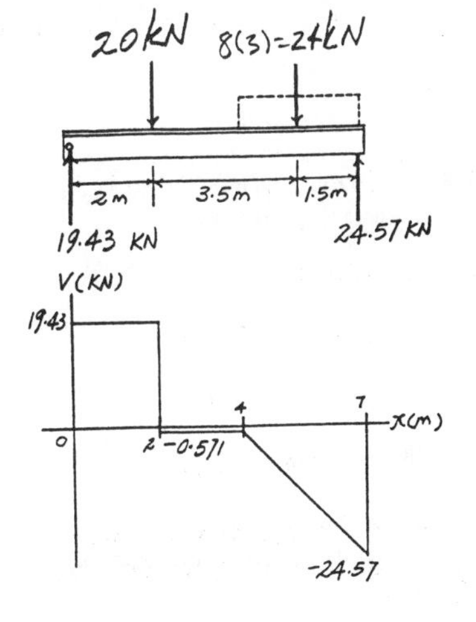

Section Properties :

$$\bar{y} = \frac{\Sigma \tilde{y}A}{\Sigma A} = \frac{0.01(0.1)(0.02) + 0.07(0.1)(0.02)}{0.1(0.02) + 0.1(0.02)} = 0.0400 \text{ m}$$

$$I_{NA} = \frac{1}{12}(0.1)\left(0.02^3\right) + 0.1(0.02)(0.0400 - 0.01)^2 + \frac{1}{12}(0.02)\left(0.1^3\right) + (0.02)(0.1)(0.07 - 0.0400)^2 = 5.3333\left(10^{-6}\right) \text{ m}^4$$

$$Q_{max} = \bar{y}'A' = 0.04(0.02)(0.08) = 64.0\left(10^{-6}\right) \text{ m}^3$$

Maximum Shear Stress : Maximum shear stress occurs at the point where the neutral axis passes through the section. Applying the shear formula

$$\tau_{max} = \frac{VQ_{max}}{It} = \frac{24.57(10^3)64.0(10^{-6})}{5.3333(10^{-6})(0.02)} = 14.7 \text{ MPa} \qquad \textbf{Ans}$$

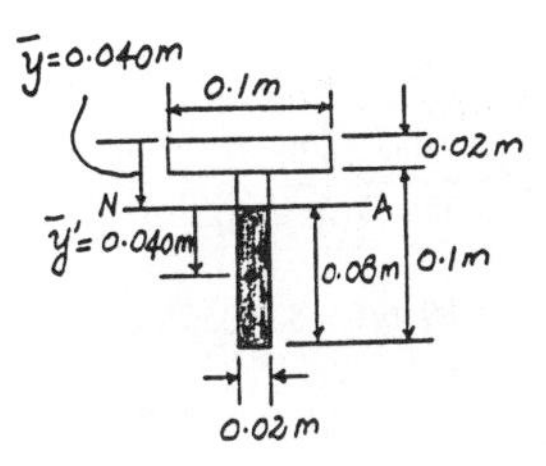

***7–20.** The supports at A and B exert vertical reactions on the wood beam. If the distributed load $w = 4$ kip/ft, determine the maximum shear stress in the beam at section a–a.

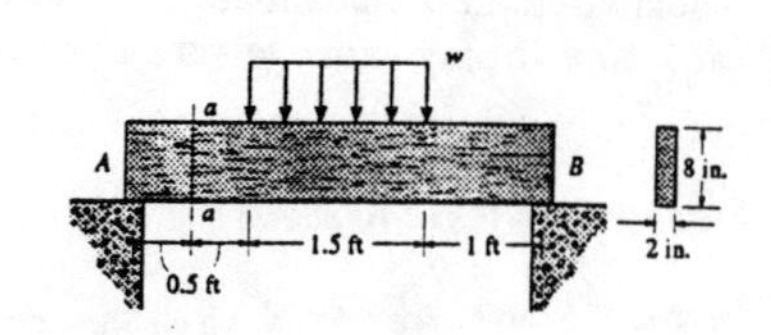

Support Reactions : As shown on FBD.

Internal Shear Force : The shear force at section $a-a$, $V_{a-a} = 3.00$ kip.

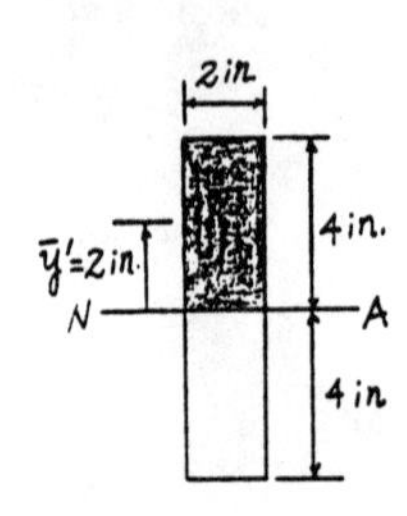

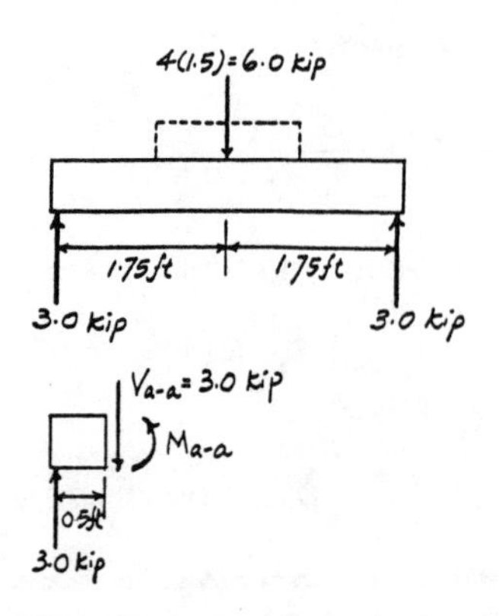

Section Properties :

$$I_{NA} = \frac{1}{12}(2)\left(8^3\right) = 85.333 \text{ in}^4$$

$$Q_{max} = \bar{y}'A' = 2(4)(2) = 16.0 \text{ in}^3$$

Maximum Shear Stress : Maximum shear stress occurs at the point where the neutral axis passes through the section. Applying the shear formula

$$\tau_{max} = \frac{VQ_{max}}{It}$$

$$= \frac{3.00(10^3)\,16.0}{85.333(2)} = 281 \text{ psi} \qquad \textbf{Ans}$$

7–21. The supports at A and B exert vertical reactions on the wood beam. If the allowable shear stress is $\tau_{allow} = 400$ psi, determine the intensity w of the largest distributed load that can be applied to the beam.

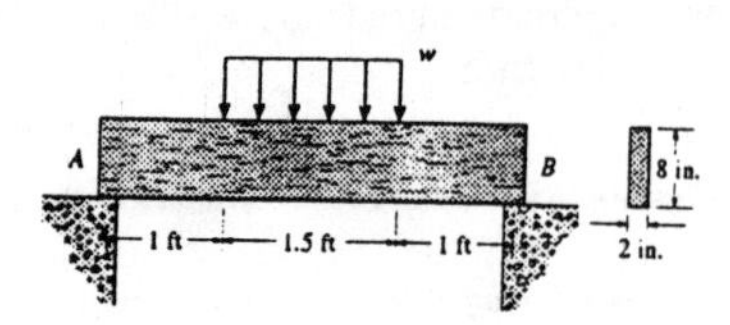

Support Reactions : As shown on FBD.

Internal Shear Force : The maximum shear force occurs at the region $0 \le x < 1$ft where $V_{max} = 0.750w$.

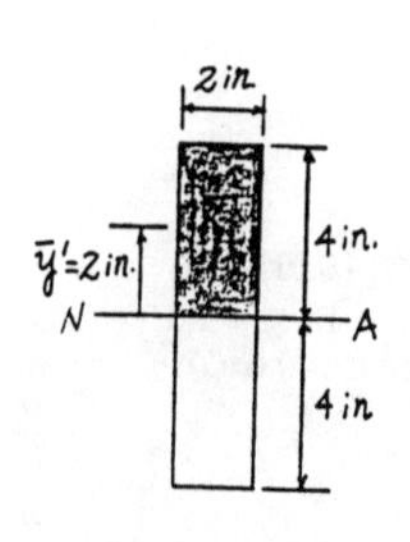

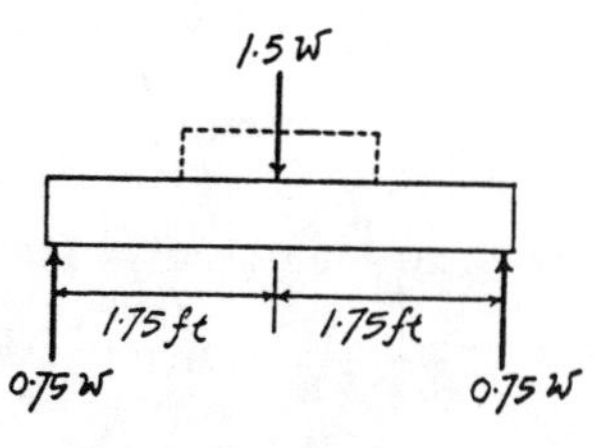

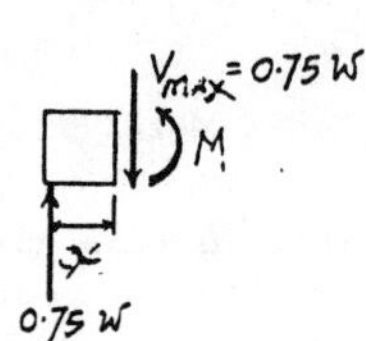

Section Properties :

$$I_{NA} = \frac{1}{12}(2)\left(8^3\right) = 85.333 \text{ in}^4$$

$$Q_{max} = \bar{y}'A' = 2(4)(2) = 16.0 \text{ in}^3$$

Allowable Shear Stress : Maximum shear stress occurs at the point where the neutral axis pass through the section. Applying shear formula

$$\tau_{max} = \tau_{allow} = \frac{VQ_{max}}{It}$$

$$400 = \frac{0.750w(16.0)}{85.333(2)}$$

$$w = 5689 \text{ lb/ft} = 5.69 \text{ kip/ft} \qquad \textbf{Ans}$$

7–22. Railroad ties must be designed to resist large shear loadings. If the tie is subjected to the 30-kip rail loadings and the gravel bed exerts a distributed reaction as shown, determine the intensity w for equilibrium, and find the maximum shear stress in the tie.

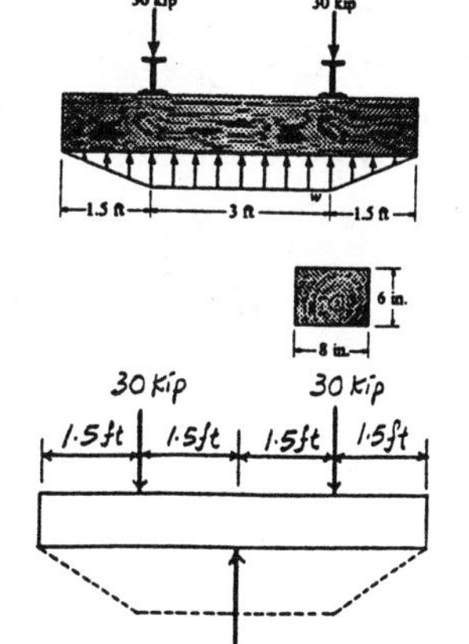

Equations of Equilibrium :

$$+\uparrow \Sigma F_y = 0; \quad 4.50w - 30 - 30 = 0$$

$$w = 13.33 \text{ kip/ft} = 13.3 \text{ kip/ft} \quad \textbf{Ans}$$

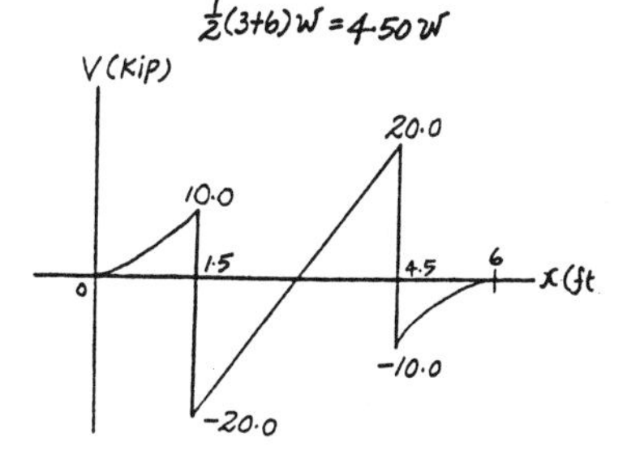

Internal Shear Force : As shown. $\quad V_{max} = 20.0$ kip.

Section Properties :

$$I_{NA} = \frac{1}{12}(8)(6^3) = 144 \text{ in}^4$$

$$Q_{max} = \bar{y}'A' = 1.5(8)(3) = 36.0 \text{ in}^3$$

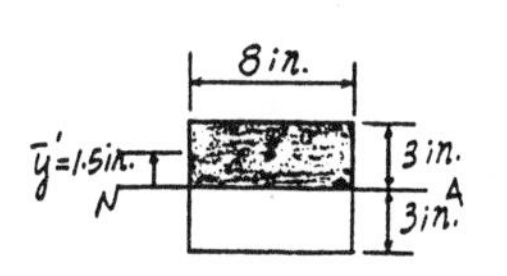

Maximum Shear Stress : Maximum shear stress occurs at the point where the neutral axis passes through the section. Applying the shear formula

$$\tau_{max} = \frac{VQ_{max}}{It}$$

$$= \frac{20.0(10^3)36.0}{144(8)} = 625 \text{ psi} \quad \textbf{Ans}$$

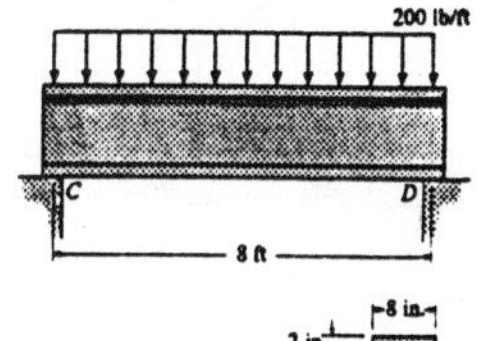

7–23. The beam is made from three plastic pieces glued together at the seams A and B. If it is subjected to the loading shown, determine the shear stress developed in the glued joints at the critical section. The supports at C and D exert only vertical reactions on the beam.

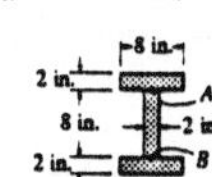

Support Reactions : As shown on FBD.

Internal Shear Force : As shown on shear diagram, $V_{max} = 800$ lb.

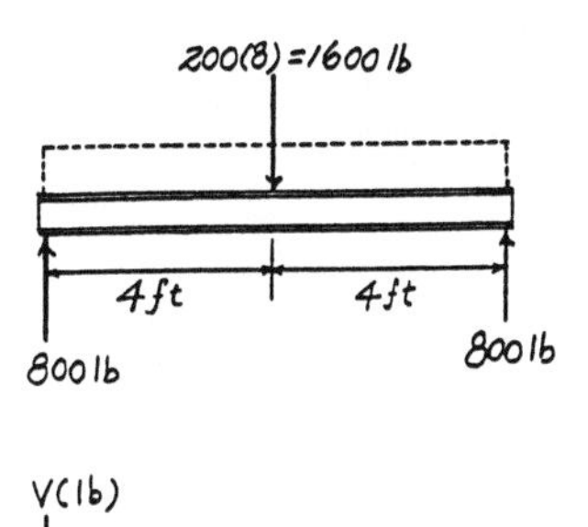

Section Properties :

$$I_{NA} = \frac{1}{12}(8)(12^3) - \frac{1}{12}(6)(8^3) = 896 \text{ in}^4$$

$$Q_A = \bar{y}'A' = 5(8)(2) = 80.0 \text{ in}^3$$

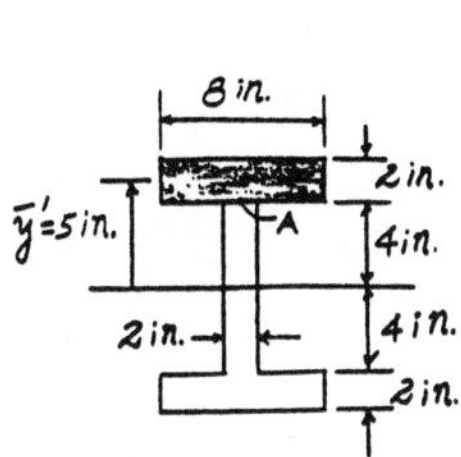

Shear Stress : Applying the shear formula

$$\tau_A = \frac{VQ_A}{It} = \frac{800(80.0)}{896(2)} = 35.7 \text{ psi} \quad \textbf{Ans}$$

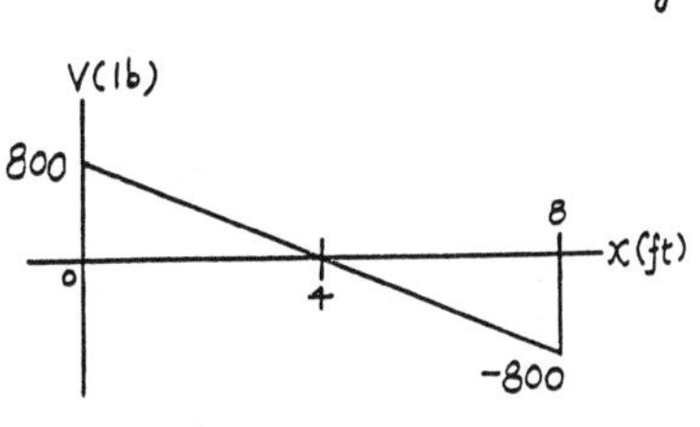

***7–24.** The beam is made from three plastic pieces glued together at the seams A and B. If it is subjected to the loading shown, determine the vertical shear force resisted by the top flange of the beam at the critical section. The supports at C and D exert only vertical reactions on the beam.

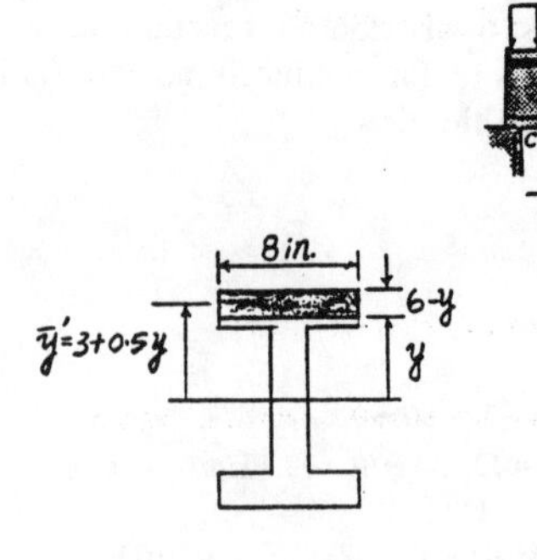

Support Reactions : As shown on FBD.

Internal Shear Force : As shown on shear diagram, $V_{max} = 800$ lb.

Section Properties :

$$I_{NA} = \frac{1}{12}(8)(12^3) - \frac{1}{12}(6)(8^3) = 896 \text{ in}^4$$

$$Q = \bar{y}'A' = (3 + 0.5y)(6 - y)(8) = 144 - 4y^2$$

Shear Stress : Applying the shear formula

$$\tau = \frac{VQ}{It} = \frac{800(144 - 4y^2)}{896(8)} = 16.071 - 0.4464y^2$$

Resultant Shear Force : For the flange

$$V_f = \int_A \tau dA$$

$$= \int_{4\text{in}}^{6\text{in}} (16.071 - 0.4464y^2)(8)\,dy$$

$$= 76.2 \text{ lb}$$

Ans

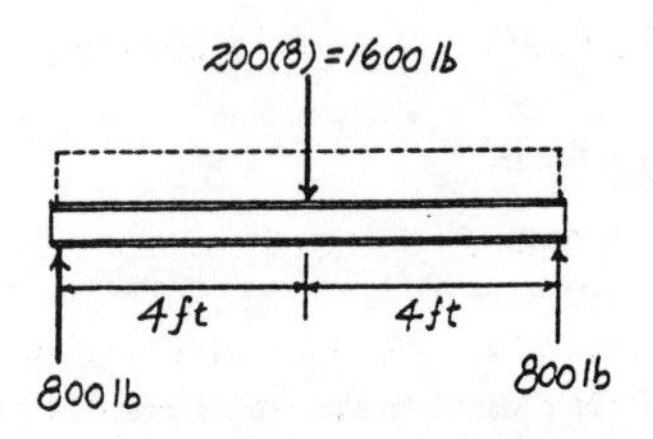

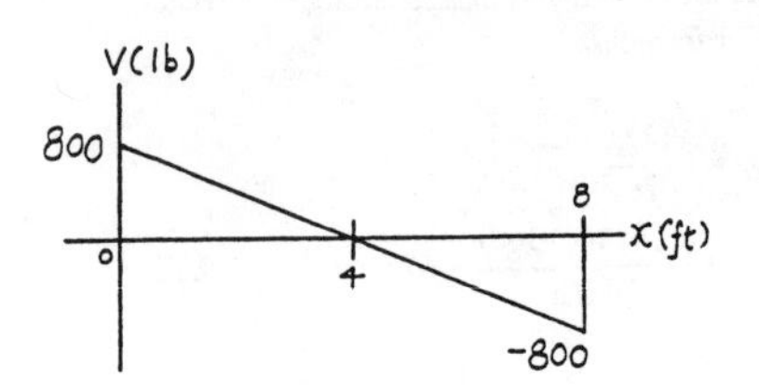

7–25. Determine the shear stress at points B and C located on the web of the fiberglass beam.

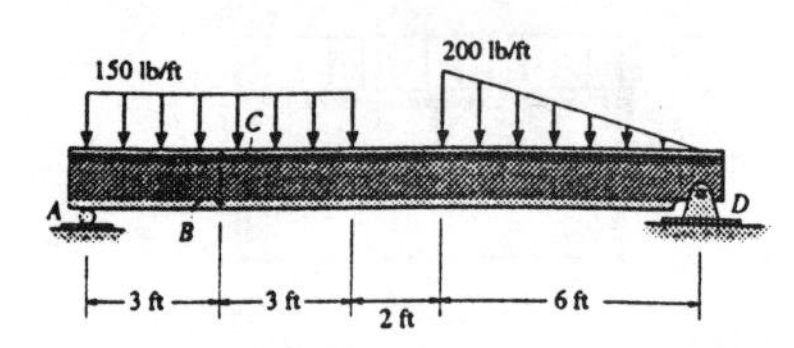

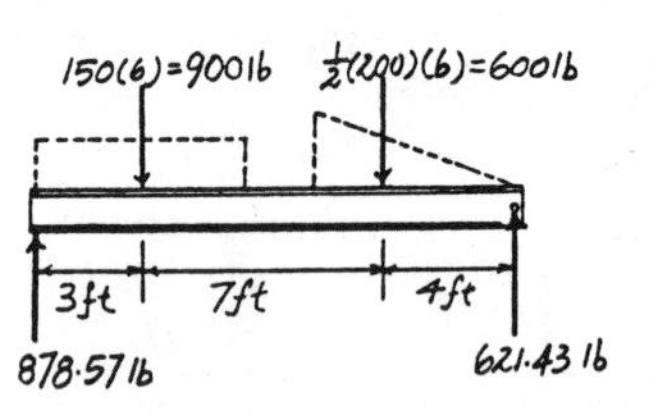

Support Reactions : As shown on FBD.

Internal Shear Force : The shear force at section $a-a$, $V_{a-a} = 428.57$ lb.

Section Properties :

$$I_{NA} = \frac{1}{12}(4)(7.5^3) - \frac{1}{12}(3.5)(6^3) = 77.625 \text{ in}^4$$

$$Q_C = \bar{y}'A' = 3.375(4)(0.75) = 10.125 \text{ in}^3$$

Shear Stress : Applying the shear formula

$$\tau_B = \tau_C = \frac{VQ_C}{It}$$

$$= \frac{428.57(10.125)}{77.625(0.5)} = 112 \text{ psi}$$

Ans

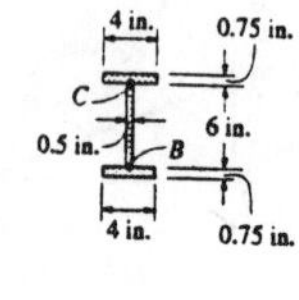

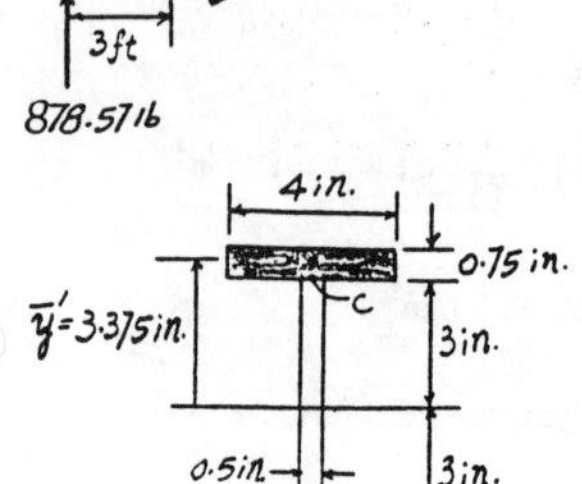

7–26. Determine the maximum shear stress acting in the fiberglass beam at the critical section.

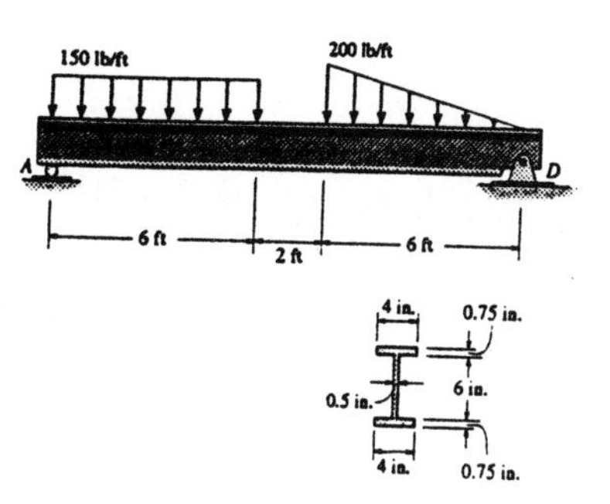

Support Reactions : As shown on FBD.

Internal Shear Force : As shown on shear diagram, $V_{max} = 878.57$ lb.

Section Properties :

$$I_{NA} = \frac{1}{12}(4)\left(7.5^3\right) - \frac{1}{12}(3.5)\left(6^3\right) = 77.625 \text{ in}^4$$

$$\begin{aligned} Q_{max} &= \Sigma \bar{y}' A' \\ &= 3.375(4)(0.75) + 1.5(3)(0.5) = 12.375 \text{ in}^3 \end{aligned}$$

Maximum Shear Stress : Maximum shear stress occurs at the point where the neutral axis passes through the section. Applying the shear formula

$$\begin{aligned} \tau_{max} &= \frac{VQ_{max}}{It} \\ &= \frac{878.57(12.375)}{77.625(0.5)} = 280 \text{ psi} \end{aligned}$$ **Ans**

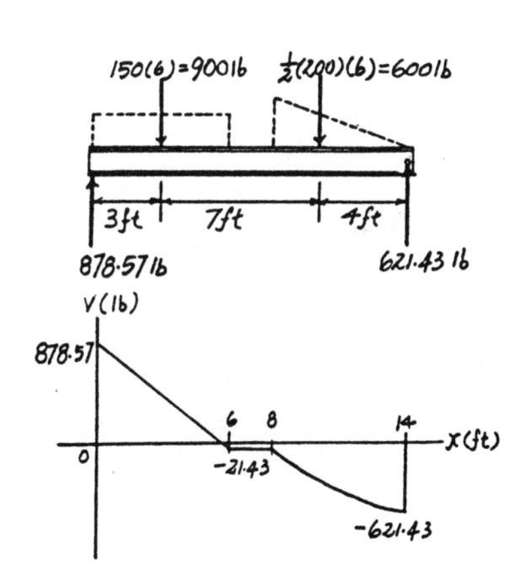

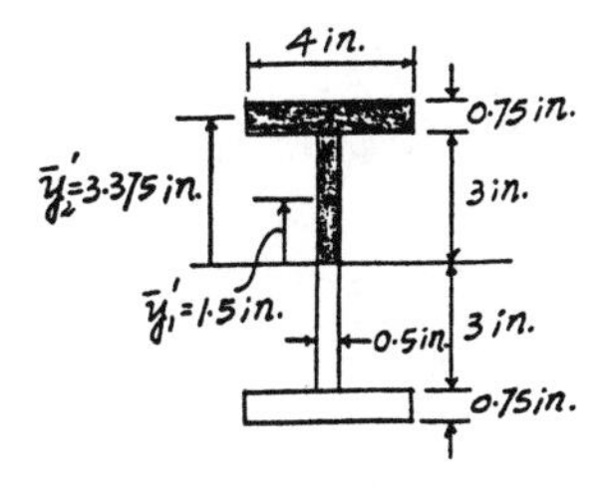

7–27. Determine the variation of the shear stress over the cross section of a hollow rivet. What is the maximum shear stress in the rivet? Also, show that if $r_i \to r_o$, then $\tau_{max} = 2(V/A)$.

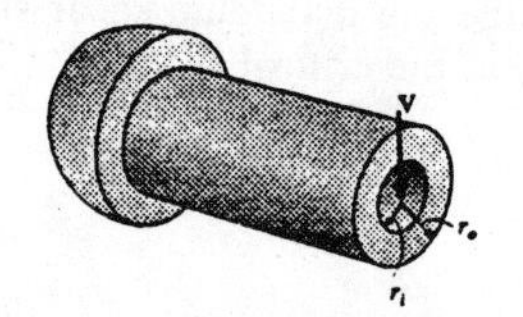

Geometry : Using the equation for a circle, $x = (r^2 - y^2)^{\frac{1}{2}}$.

$$\frac{t_1}{2} = x_2 - x_1$$

$$t_1 = 2(x_2 - x_1) = 2\left[(r_o^2 - y^2)^{\frac{1}{2}} - (r_i^2 - y^2)^{\frac{1}{2}}\right]$$

$$t_2 = 2x_2 = 2(r_o^2 - y^2)^{\frac{1}{2}}$$

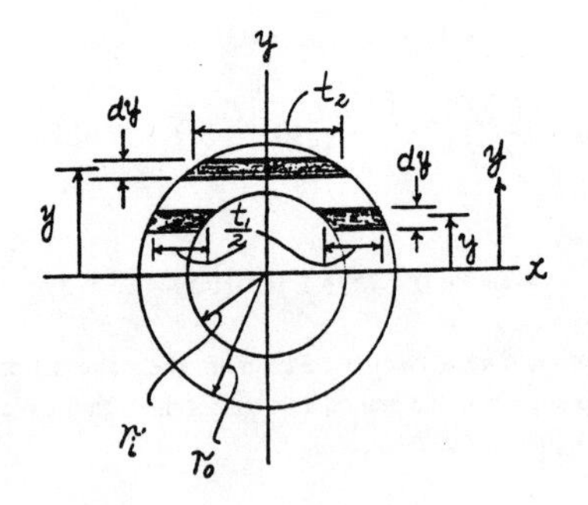

Section Properties :

$$I = \frac{\pi}{4}(r_o^4 - r_i^4)$$

For $0 \le y < r_i$

$$Q = \int_A y\,dA$$

$$= \int_y^{r_i} y t_1\,dy + \int_{r_i}^{r_o} y t_2\,dy$$

$$= \int_y^{r_i} 2y\left[(r_o^2 - y^2)^{\frac{1}{2}} - (r_i^2 - y^2)^{\frac{1}{2}}\right]dy + \int_{r_i}^{r_o} 2y(r_o^2 - y^2)^{\frac{1}{2}}dy$$

$$= \frac{2}{3}\left\{\left[-(r_o^2 - y^2)^{\frac{3}{2}} + (r_i^2 - y^2)^{\frac{3}{2}}\right]\Big|_y^{r_i} + \left[-(r_o^2 - y^2)^{\frac{3}{2}}\right]\Big|_{r_i}^{r_o}\right\}$$

$$= \frac{2}{3}\left[(r_o^2 - y^2)^{\frac{3}{2}} - (r_i^2 - y^2)^{\frac{3}{2}}\right]$$

For $r_i < y \le r_o$

$$Q = \int_A y\,dA = \int_y^{r_o} y t_2\,dy = \int_y^{r_o} 2y(r_o^2 - y^2)^{\frac{1}{2}}dy = \frac{2}{3}\left[(r_o^2 - y^2)^{\frac{3}{2}}\right]$$

Shear Stress : Applying the shear formula $\tau = \frac{VQ}{It}$

For $0 \le y < r_i$

$$\tau = \frac{\frac{2}{3}V\left[(r_o^2 - y^2)^{\frac{3}{2}} - (r_i^2 - y^2)^{\frac{3}{2}}\right]}{\frac{\pi}{4}(r_o^4 - r_i^4)(2)\left[(r_o^2 - y^2)^{\frac{1}{2}} - (r_i^2 - y^2)^{\frac{1}{2}}\right]}$$

$$= \frac{4V}{3\pi}\left[\frac{(r_o^2 - y^2)^{\frac{3}{2}} - (r_i^2 - y^2)^{\frac{3}{2}}}{(r_o^4 - r_i^4)\left[(r_o^2 - y^2)^{\frac{1}{2}} - (r_i^2 - y^2)^{\frac{1}{2}}\right]}\right] \quad \textbf{Ans}$$

For $r_i < y \le r_o$

$$\tau = \frac{\frac{2}{3}V\left[(r_o^2 - y^2)^{\frac{3}{2}}\right]}{\frac{\pi}{4}(r_o^4 - r_i^4)(2)(r_o^2 - y^2)^{\frac{1}{2}}}$$

$$= \frac{4V}{3\pi}\left(\frac{r_o^2 - y^2}{r_o^4 - r_i^4}\right) \quad \textbf{Ans}$$

The maximum shear stress occurs at $y = 0$. Hence

$$\tau_{max} = \frac{4V}{3\pi}\left[\frac{r_o^3 - r_i^3}{(r_o^4 - r_i^4)(r_o - r_i)}\right]$$

However, $r_o^3 - r_i^3 = (r_o^2 + r_o r_i + r_i^2)(r_0 - r_i)$ then

$$\tau_{max} = \frac{4V}{3\pi}\left[\frac{r_o^2 + r_o r_i + r_i^2}{(r_o^4 - r_i^4)}\right] \quad \textbf{Ans}$$

Substitute $A = \pi(r_o^2 - r_i^2)$ into τ_{max}. This yields

$$\tau_{max} = \frac{4V}{3\pi}\left[\frac{r_o^2 + r_o r_i + r_i^2}{(r_o^2 + r_i^2)(r_o^2 - r_i^2)}\right] = \frac{4V}{3A}\left[\frac{r_o^2 + r_o r_i + r_i^2}{(r_o^2 + r_i^2)}\right]$$

As $r_i \to r_o$

$$\tau_{max} = \frac{4V}{3A}\left(\frac{3r_o^2}{2r_o^2}\right) = 2\left(\frac{V}{A}\right) \quad (\textit{Q.E.D.})$$

***7–28.** The beam has a square cross section and is subjected to the shear force **V**. Sketch the shear-stress distribution over the cross section and specify the maximum shear stress. Also, from the neutral axis, locate where a crack along the member will first start to appear due to shear.

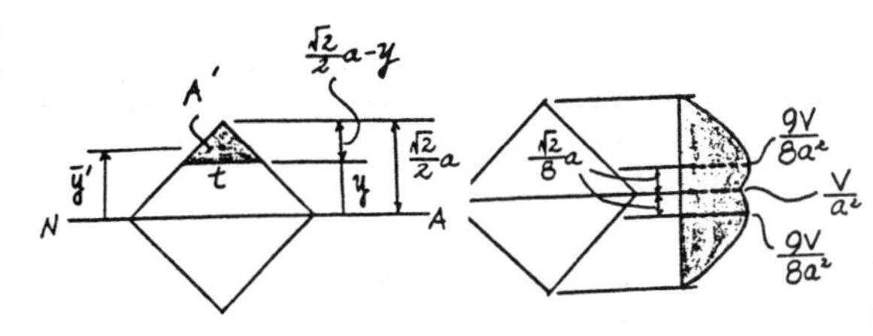

Section Properties :

$$A' = \frac{1}{2}\left(\frac{\sqrt{2}}{2}a - y\right)t = \left(\frac{\sqrt{2}}{4}a - \frac{y}{2}\right)t$$

$$\bar{y}' = y + \frac{1}{3}\left(\frac{\sqrt{2}}{2}a - y\right) = \frac{2}{3}y + \frac{\sqrt{2}}{6}a$$

$$Q = \bar{y}'A' = \left(\frac{2}{3}y + \frac{\sqrt{2}}{6}a\right)\left(\frac{\sqrt{2}}{4}a - \frac{y}{2}\right)t$$

$$= \left(-\frac{y^2}{3} + \frac{\sqrt{2}}{12}ay + \frac{a^2}{12}\right)t$$

$$= \frac{t}{12}\left(-4y^2 + \sqrt{2}ay + a^2\right)$$

$$I = 2\left[\frac{1}{12}\left(\sqrt{2}a\right)\left(\frac{\sqrt{2}}{2}a\right)^3\right] = \frac{a^4}{12}$$

Shear Stress : Applying the shear formula

$$\tau = \frac{VQ}{It} = \frac{V\left[\frac{t}{12}\left(-4y^2 + \sqrt{2}ay + a^2\right)\right]}{\left(\frac{a^4}{12}\right)(t)}$$

$$= \frac{V}{a^4}\left(-4y^2 + \sqrt{2}ay + a^2\right)$$

Maximum Shear Stress : The crack will appear first where the maximum shear stress occurs. For $\tau = \tau_{max}$, $\frac{d\tau}{dy} = 0$.

$$\frac{d\tau}{dy} = \frac{V}{a^4}\left(-8y + \sqrt{2}a\right) = 0 \qquad y = \frac{\sqrt{2}}{8}a \qquad \textbf{Ans}$$

Hence, the maximum shear stress is

$$\tau_{max} = \frac{V}{a^4}\left[-4\left(\frac{\sqrt{2}}{8}a\right)^2 + \sqrt{2}a\left(\frac{\sqrt{2}}{8}a\right) + a^2\right] = \frac{9V}{8a^2} \qquad \textbf{Ans}$$

When $y = 0$; $\qquad \tau = \frac{V}{a^2}$

7–29. The beam has a rectangular cross section and is subjected to a load P that is just large enough to develop a fully plastic moment $M_p = PL$ at the fixed support. If the material is elastic plastic, then at a distance $x < L$ the moment $M = Px$ creates a region of plastic yielding with an associated elastic core having a height $2y'$. This situation has been described by Eq. 6–30 and the moment **M** is distributed over the cross section as shown in Fig. 6–54*e*. Prove that the maximum shear stress developed in the beam is given by $\tau_{max} = \frac{3}{2}(P/A')$, where $A' = 2y'b$, the cross-sectional area of the elastic core.

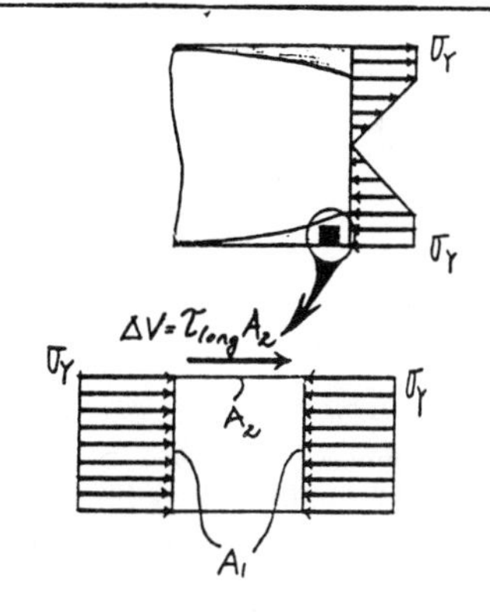

Force Equilibrium : The shaded area indicates the plastic zone. Isolate an element in the plastic zone and write the equation of equilibrium.

$$\xrightarrow{+} \Sigma F_x = 0; \qquad \tau_{long}A_2 + \sigma_Y A_1 - \sigma_Y A_1 = 0$$

$$\tau_{long} = 0$$

This proves that the longitudinal shear stress, τ_{long}, is equal to zero. Hence the corresponding transverse stress, τ_{trans}, is also equal to zero in the plastic zone. Therefore, the shear force $V = P$ is carried by the material only in the elastic zone.

Section Properties :

$$I_{NA} = \frac{1}{12}(b)(2y')^3 = \frac{2}{3}by'^3$$

$$Q_{max} = \bar{y}'A' = \frac{y'}{2}(y')(b) = \frac{y'^2 b}{2}$$

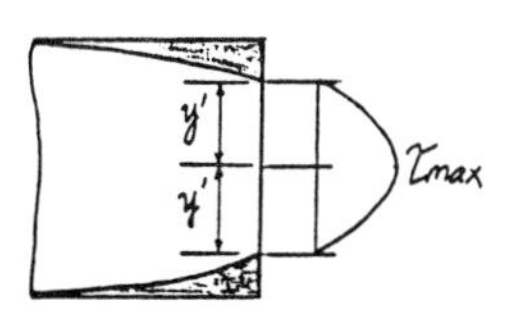

Maximum Shear Stress : Applying the shear formula

$$\tau_{max} = \frac{VQ_{max}}{It} = \frac{V\left(\frac{y'^2 b}{2}\right)}{\left(\frac{2}{3}by'^3\right)(b)} = \frac{3P}{4by'}$$

However, $\quad A' = 2by' \quad$ hence

$$\tau_{max} = \frac{3P}{2A'} \qquad (Q.E.D.)$$

7–30. The beam in Fig. 6–54f is subjected to a fully plastic moment $\mathbf{M}_p$. Prove that the longitudinal and transverse shear stresses in the beam are zero. *Hint:* Consider an element of the beam as shown in Fig. 7–4d.

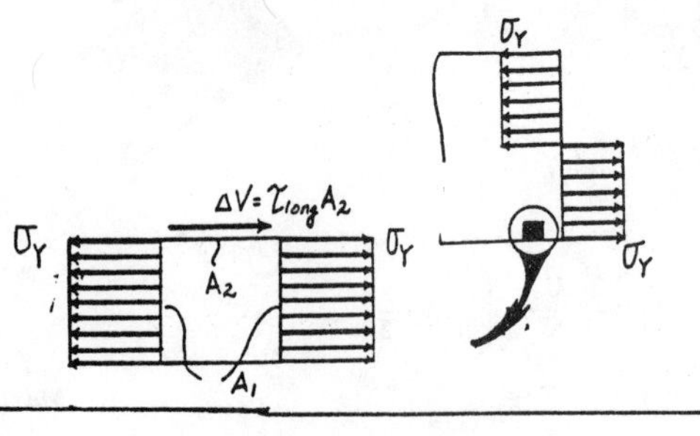

Force Equilibrium : If a fully plastic moment acts on the cross section, then an element of the material taken from the top or bottom of the cross section is subjected to the loading shown. For equilibrium

$$\xrightarrow{+} \Sigma F_x = 0; \qquad \sigma_Y A_1 + \tau_{long} A_2 - \sigma_Y A_1 = 0$$
$$\tau_{long} = 0$$

Thus no shear stress is developed on the longitudinal or transverse plane of the element. (*Q. E. D.*)

***7–32.** The beam is constructed from two boards fastened together at the top and bottom with two rows of nails spaced every 6 in. If each nail can support a 500-lb shear force, determine the maximum shear force V that can be applied to the beam.

Section Properties :

$$I = \frac{1}{12}(6)\left(4^3\right) = 32.0 \text{ in}^4$$
$$Q = \bar{y}'A' = 1(6)(2) = 12.0 \text{ in}^4$$

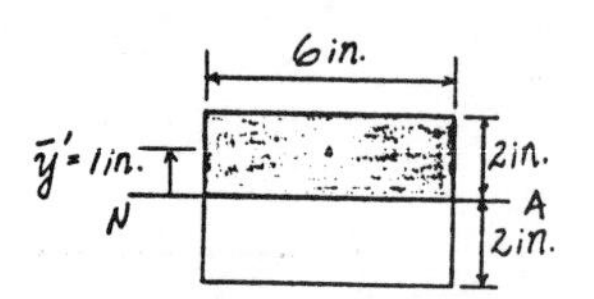

Shear Flow : There are two rows of nails. Hence, the allowable shear flow $q = \dfrac{2(500)}{6} = 166.67$ lb/in.

$$q = \frac{VQ}{I}$$
$$166.67 = \frac{V(12.0)}{32.0}$$

$$V = 444 \text{ lb} \qquad \textbf{Ans}$$

7–33. The beam is constructed from two boards fastened together at the top and bottom with two rows of nails spaced every 6 in. If an internal shear force of $V = 600$ lb is applied to the boards, determine the shear force resisted by each nail.

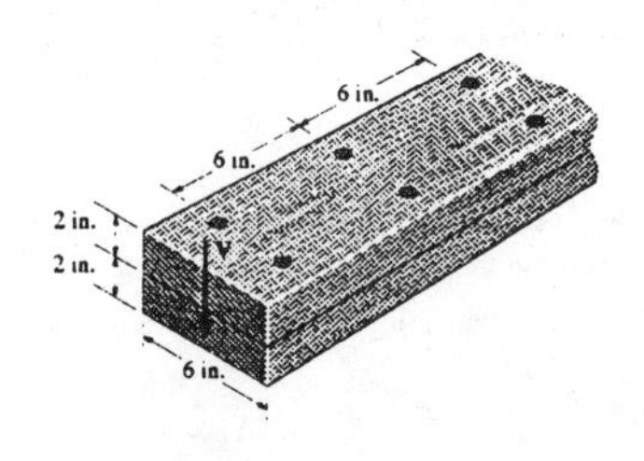

Section Properties :

$$I = \frac{1}{12}(6)\left(4^3\right) = 32.0 \text{ in}^4$$
$$Q = \bar{y}'A' = 1(6)(2) = 12.0 \text{ in}^4$$

Shear Flow :

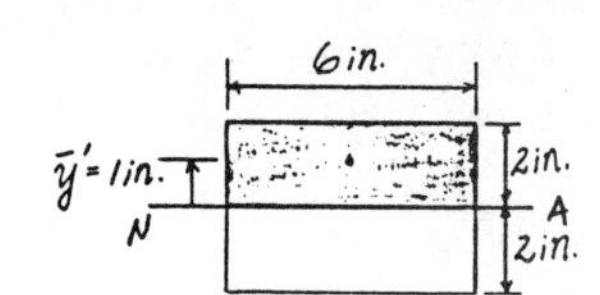

$$q = \frac{VQ}{I} = \frac{600(12.0)}{32.0} = 225 \text{ lb/in.}$$

There are two rows of nails. Hence, the shear force resisted by each nail is

$$F = \left(\frac{q}{2}\right)s = \left(\frac{225 \text{ lb/in.}}{2}\right)(6 \text{ in.}) = 675 \text{ lb} \qquad \textbf{Ans}$$

7–34. A beam is constructed from five boards bolted together as shown. Determine the shear force developed along each shear plane of each bolt if the bolts are spaced $s = 250$ mm apart and the applied shear is $V = 35$ kN.

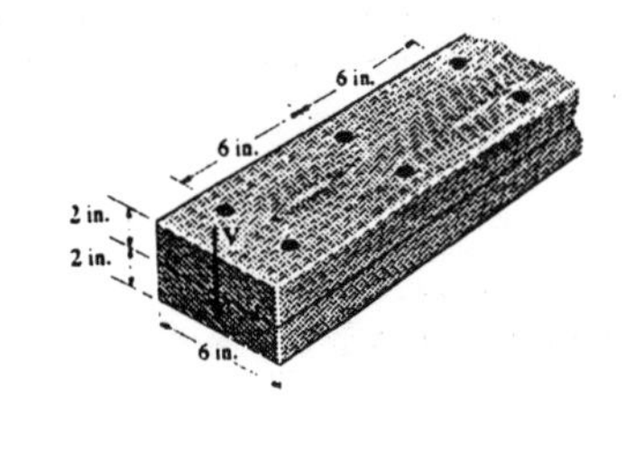

Section Properties :

$$\bar{y} = \frac{\Sigma \tilde{y}A}{\Sigma A} = \frac{0.175(0.075)(0.35) + 0.325(0.25)(0.05)}{0.075(0.35) + 0.25(0.05)}$$
$$= 0.2234 \text{ m}$$

$$I_{NA} = \frac{1}{12}(0.075)\left(0.35^3\right) + 0.075(0.35)(0.2234 - 0.175)^2$$
$$+ \frac{1}{12}(0.05)\left(0.25^3\right) + 0.05(0.25)(0.325 - 0.2234)^2$$
$$= 0.5236\left(10^{-3}\right) \text{ m}^4$$

$$Q = \bar{y}'A' = 0.1016(0.05)(0.25) = 1.2702\left(10^{-3}\right) \text{ m}^3$$

Shear Flow :

$$q = \frac{VQ}{I} = \frac{35(10^3)(1.2702)(10^{-3})}{0.5236(10^{-3})} = 84.90 \text{ kN/m}$$

There are four shear planes on the bolt. Hence, the shear force resisted by each shear plane of the bolt is

$$F = \left(\frac{q}{4}\right)s = \left(\frac{84.90 \text{ kN/m}}{4}\right)(0.25 \text{ m}) = 5.31 \text{ kN} \qquad \textbf{Ans}$$

7–35. A beam is constructed from five boards bolted together as shown. Determine the maximum spacing s of the bolts if they can each resist a shear of 20 kN and the applied shear is $V = 45$ kN.

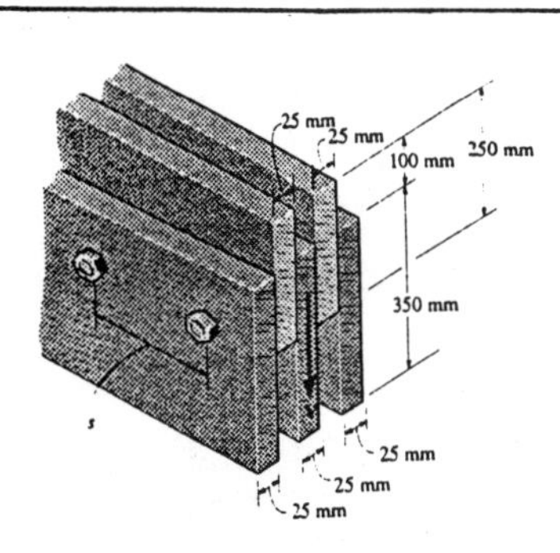

Section Properties :

$$\bar{y} = \frac{\Sigma \tilde{y}A}{\Sigma A} = \frac{0.175(0.075)(0.35) + 0.325(0.25)(0.05)}{0.075(0.35) + 0.25(0.05)}$$
$$= 0.2234 \text{ m}$$

$$I_{NA} = \frac{1}{12}(0.075)\left(0.35^3\right) + 0.075(0.35)(0.2234 - 0.175)^2$$
$$+ \frac{1}{12}(0.05)\left(0.25^3\right) + 0.05(0.25)(0.325 - 0.2234)^2$$
$$= 0.5236\left(10^{-3}\right) \text{ m}^4$$

$$Q = \bar{y}'A' = 0.1016(0.05)(0.25) = 1.2702\left(10^{-3}\right) \text{ m}^3$$

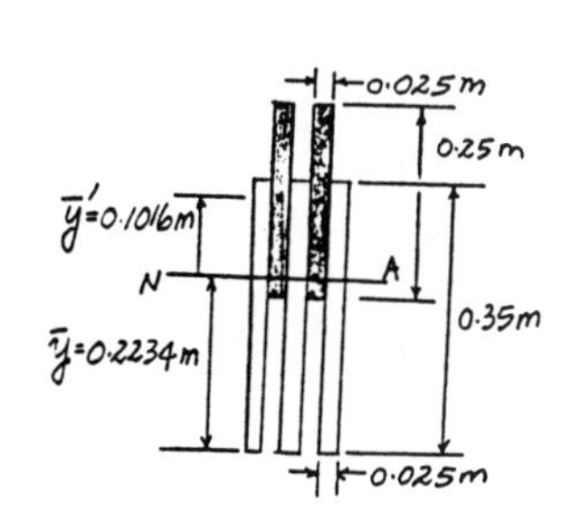

Shear Flow : Since there are four shear planes on the bolt, the allowable shear flow is $q = \dfrac{4(20)(10^3)}{s} = \dfrac{80(10^3)}{s}$

$$q = \frac{VQ}{I}$$

$$\frac{80(10^3)}{s} = \frac{45(10^3)(1.2702)(10^{-3})}{0.5236(10^{-3})}$$

$$s = 0.7329 \text{ m} = 733 \text{ mm} \qquad \textbf{Ans}$$

***7–36.** A beam is constructed from three boards bolted together as shown. Determine the shear force developed along the shear plane between the boards for each bolt if the bolts are spaced $s = 250$ mm apart and the applied shear is $V = 35$ kN.

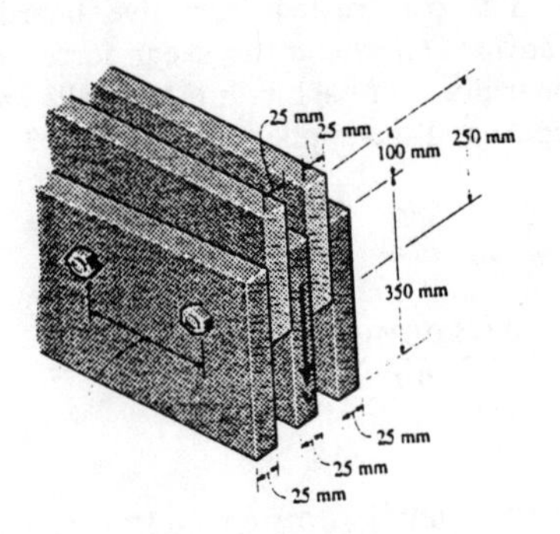

Section Properties :

$$\bar{y} = \frac{\Sigma\tilde{y}A}{\Sigma A} = \frac{0.15(0.05)(0.3) + 0.3(0.025)(0.2)}{0.05(0.3) + 0.025(0.2)}$$
$$= 0.1875 \text{ m}$$

$$I_{NA} = \frac{1}{12}(0.05)\left(0.3^3\right) + 0.05(0.3)(0.1875 - 0.15)^2$$
$$+ \frac{1}{12}(0.025)\left(0.2^3\right) + 0.025(0.2)(0.3 - 0.1875)^2$$
$$= 0.21354(10^{-3}) \text{ m}^4$$

$$Q = \bar{y}'A' = 0.1125(0.025)(0.2) = 0.5625\left(10^{-3}\right) \text{ m}^3$$

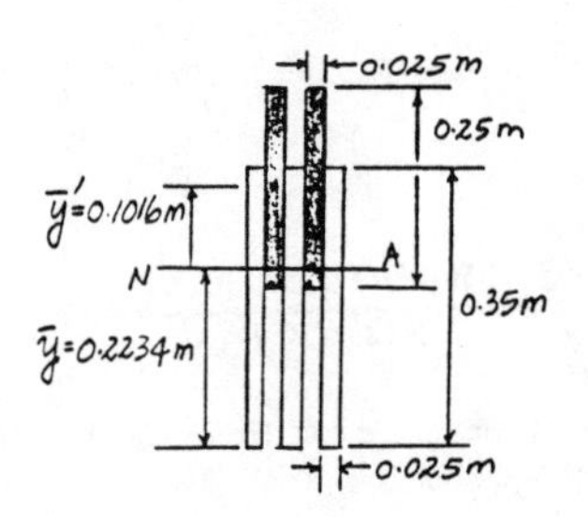

Shear Flow :

$$q = \frac{VQ}{I} = \frac{35(10^3)(0.5625)(10^{-3})}{0.21354(10^{-3})} = 92.20 \text{ kN/m}$$

There are two shear planes on the bolt. Hence, the shear force resiste by each shear plane of the bolt is

$$F = \left(\frac{q}{2}\right)s = \left(\frac{92.20 \text{ kN/m}}{2}\right)(0.25 \text{ m}) = 11.5 \text{ kN} \qquad \textbf{Ans}$$

7–37. The beam is fabricated from two equivalent structural tees and two plates. Each plate has a height of 6 in. and a thickness of 0.5 in. If a shear of $V = 50$ kip is applied to the cross section, determine the maximum spacing of the bolts. Each bolt can resist a shear force of 15 kip.

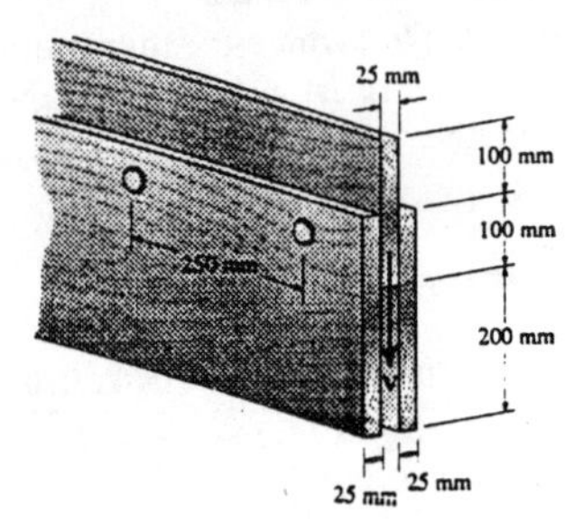

Section Properties :

$$I_{NA} = \frac{1}{12}(3)\left(9^3\right) - \frac{1}{12}(2.5)\left(8^3\right)$$
$$- \frac{1}{12}(0.5)\left(2^3\right) + \frac{1}{12}(1)\left(6^3\right)$$
$$= 93.25 \text{ in}^4$$

$$Q = \Sigma\bar{y}'A' = 2.5(3)(0.5) + 4.25(3)(0.5) = 10.125 \text{ in}^3$$

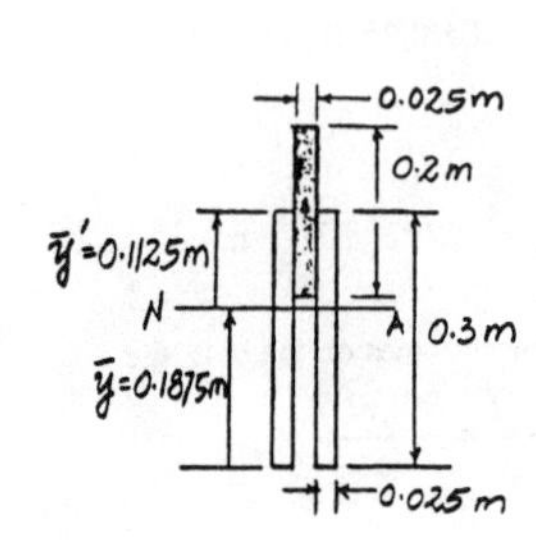

Shear Flow : Since there are two shear planes on the bolt, the allowable shear flow is $q = \frac{2(15)}{s} = \frac{30}{s}$.

$$q = \frac{VQ}{I}$$

$$\frac{30}{s} = \frac{50(10.125)}{93.25}$$

$$s = 5.53 \text{ in.} \qquad \textbf{Ans}$$

7–38. The beam is fabricated from two equivalent structural tees and two plates. Each plate has a height of 6 in. and a thickness of 0.5 in. If the bolts are spaced at $s = 8$ in., determine the maximum shear force **V** that can be applied to the cross section. Each bolt can resist a shear force of 15 kip.

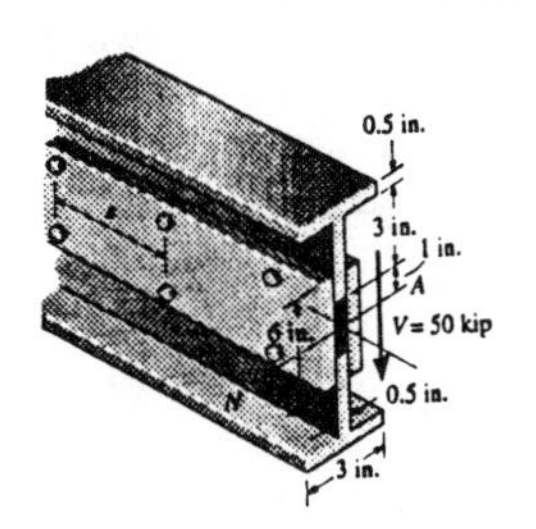

Section Properties :

$$I_{NA} = \frac{1}{12}(3)(9^3) - \frac{1}{12}(2.5)(8^3) - \frac{1}{12}(0.5)(2^3) + \frac{1}{12}(1)(6^3)$$

$$= 93.25 \text{ in}^4$$

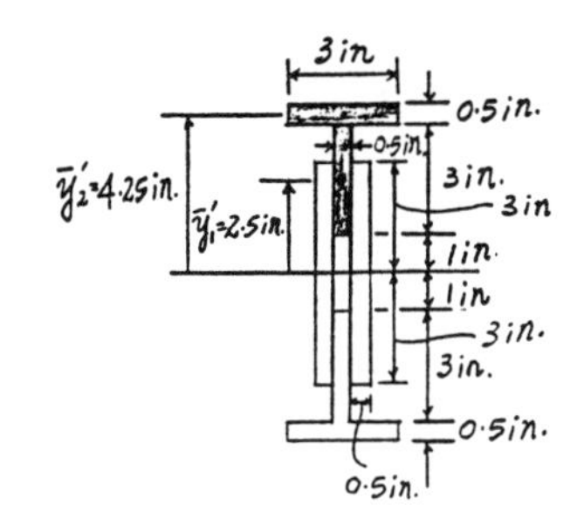

$$Q = \Sigma \bar{y}'A' = 2.5(3)(0.5) + 4.25(3)(0.5) = 10.125 \text{ in}^3$$

Shear Flow : Since there are two shear planes on the bolt, the allowable shear flow is $q = \dfrac{2(15)}{8} = 3.75$ kip/in.

$$q = \frac{VQ}{I}$$

$$3.75 = \frac{V(10.125)}{93.25}$$

$$V = 34.5 \text{ kip} \qquad \textbf{Ans}$$

7–39. The beam is made from four boards nailed together as shown. If the nails can each support a shear force of 100 lb., determine their required spacings s' and s if the beam is subjected to a shear of $V = 700$ lb.

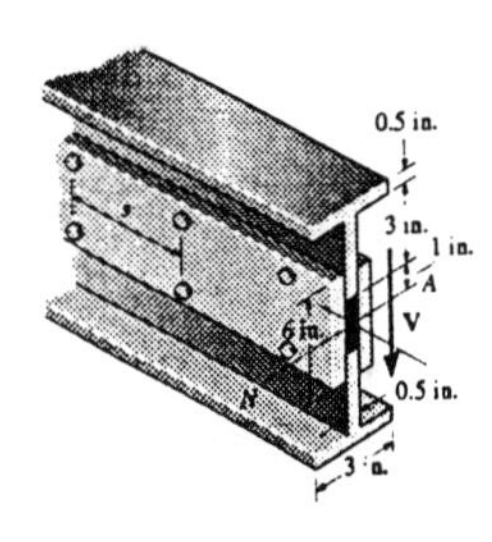

Section Properties :

$$\bar{y} = \frac{\Sigma \bar{y}A}{\Sigma A} = \frac{0.5(10)(1) + 1.5(2)(3) + 6(1.5)(10)}{10(1) + 2(3) + 1.5(10)}$$

$$= 3.3548 \text{ in}$$

$$I_{NA} = \frac{1}{12}(10)(1^3) + 10(1)(3.3548 - 0.5)^2 + \frac{1}{12}(2)(3^3) + 2(3)(3.3548 - 1.5)^2 + \frac{1}{12}(1.5)(10^3) + 1.5(10)(6 - 3.3548)^2$$

$$= 337.43 \text{ in}^4$$

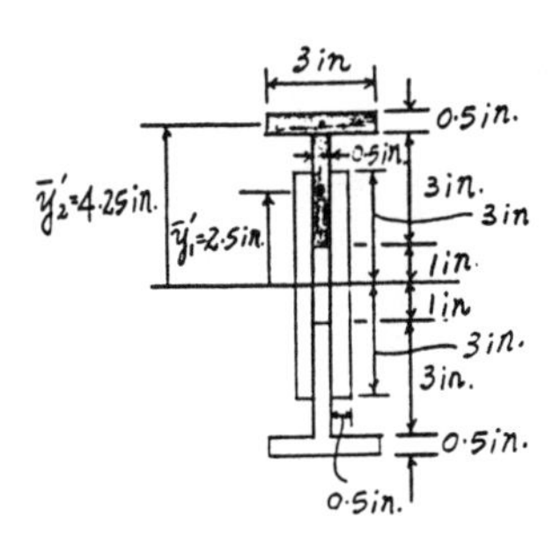

$$Q_A = \bar{y}_1'A' = 1.8548(3)(1) = 5.5645 \text{ in}^3$$

$$Q_B = \bar{y}_2'A' = 2.6452(10)(1.5) = 39.6774 \text{ in}^3$$

Shear Flow : The allowable shear flow at points A and B is $q_A = \dfrac{100}{s}$ and $q_B = \dfrac{100}{s'}$, respectively.

$$q_A = \frac{VQ_A}{I}$$

$$\frac{100}{s} = \frac{700(5.5645)}{337.43}$$

$$s = 8.66 \text{ in.} \qquad \textbf{Ans}$$

$$q_B = \frac{VQ_B}{I}$$

$$\frac{100}{s'} = \frac{700(39.6774)}{337.43}$$

$$s' = 1.21 \text{ in.} \qquad \textbf{Ans}$$

***7–40.** The beam is fabricated from two equivalent channels and two plates. Each plate has a height of 6 in. and a thickness of 0.5 in. If a shear of $V = 50$ kip is applied to the cross section, determine the maximum spacing of the bolts. Each bolt can resist a shear force of 15 kip.

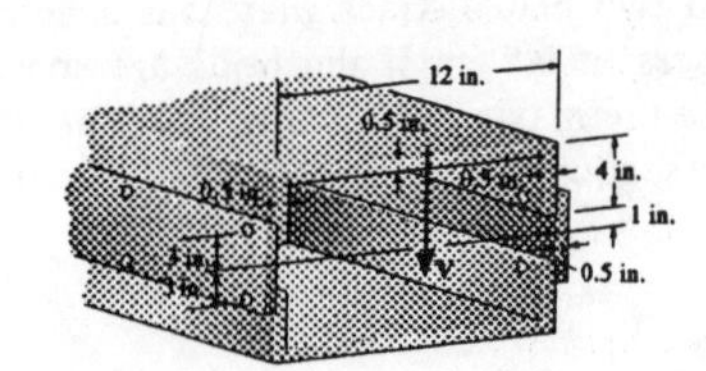

Section Properties :

$$I_{NA} = \frac{1}{12}(12)\left(10^3\right) - \frac{1}{12}(11)\left(9^3\right) - \frac{1}{12}(1)\left(2^3\right) + \frac{1}{12}(1)\left(6^3\right)$$

$$= 349.0833 \text{ in}^4$$

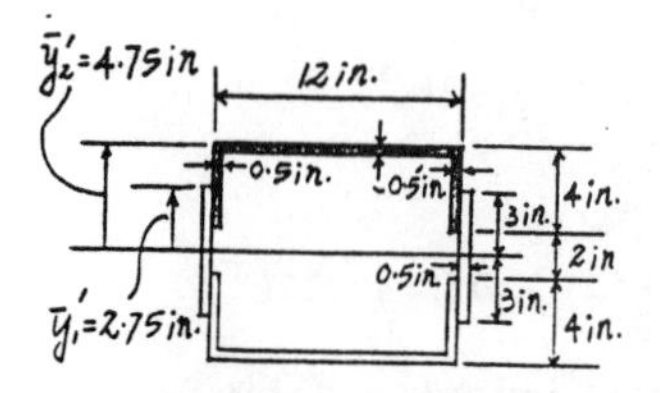

$$Q = \Sigma \bar{y}'A' = 2.75(1)(3.5) + 4.75(12)(0.5) = 38.125 \text{ in}^3$$

Shear Flow : The allowable shear flow is $q = \dfrac{15(2)}{s} = \dfrac{30}{s}$.

$$q = \frac{VQ}{I}$$

$$\frac{30}{s} = \frac{50(38.125)}{349.0833}$$

$$s = 5.49 \text{ in.} \qquad \textbf{Ans}$$

7–41. The strut is constructed from three pieces of plastic that are glued together as shown. If the allowable shear stress for the plastic is $\tau_{allow} = 800$ psi and each glue joint can withstand 250 lb/in., determine the largest allowable distributed loading w that can be applied to the strut.

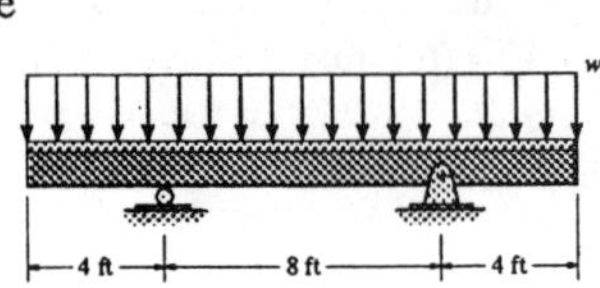

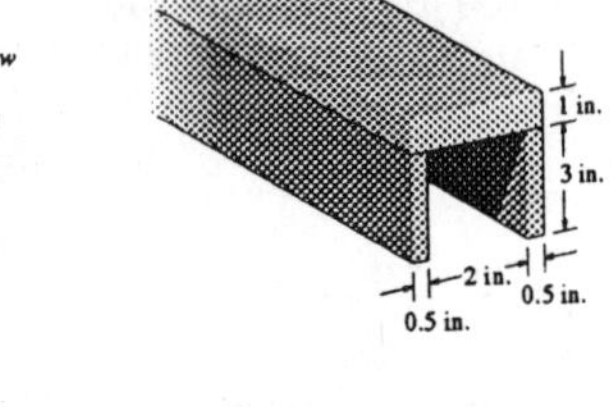

Support Reactions : As shown on FBD.

Internal Shear Force : As shown on shear diagram, $V_{max} = 4.00w$.

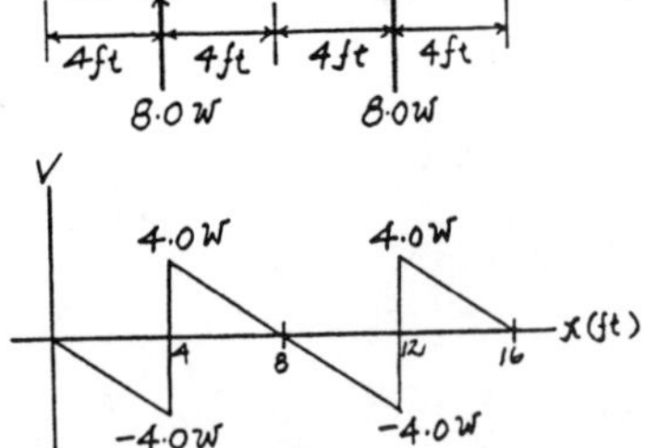

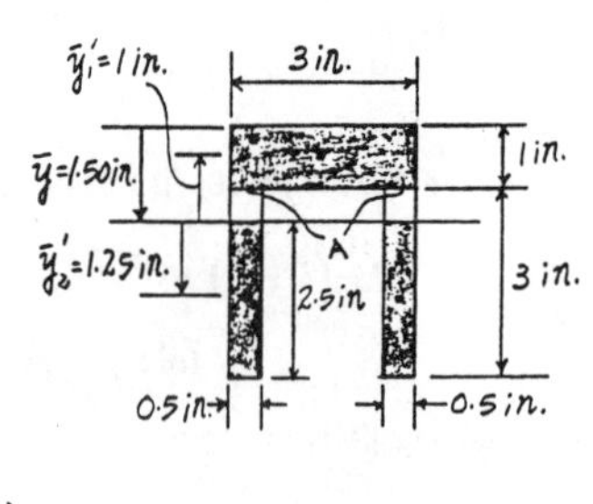

Section Properties :

$$\bar{y} = \frac{\Sigma \bar{y}A}{\Sigma A} = \frac{0.5(3)(1) + 2.5(1)(3)}{3(1) + 1(3)} = 1.50 \text{ in.}$$

$$I_{NA} = \frac{1}{12}(3)\left(1^3\right) + 3(1)(1.5 - 0.5)^2 + \frac{1}{12}(1)\left(3^3\right) + 1(3)(2.5 - 1.5)^2$$

$$= 8.50 \text{ in}^4$$

$$Q_{max} = \bar{y}_2'A' = 1.25(1)(2.5) = 3.125 \text{ in}^3$$

$$Q_A = \bar{y}_1'A' = 1(3)(1) = 3.00 \text{ in}^3$$

Allowable Shear Stress : Assume the beam fails due to shear stress.

$$\tau_{max} = \tau_{allow} = \frac{VQ_{max}}{It}$$

$$800 = \frac{4.00w(3.125)}{8.50(1)}$$

$$w = 544 \text{ lb/ft}$$

Shear Flow : Assume the beam fails at the glue joint and the allowable shear flow is $q = 2(250) = 500$ lb/in.

$$q = \frac{VQ_A}{I}$$

$$500 = \frac{4.00w(3.00)}{8.50}$$

$$w = 354 \text{ lb/ft} \quad \textit{(Controls!)} \qquad \textbf{Ans}$$

7–42. The strut is constructed from three pieces of plastic that are glued together as shown. If the distributed load $w = 200$ lb/ft, determine the shear stress that must be resisted by each glue joint.

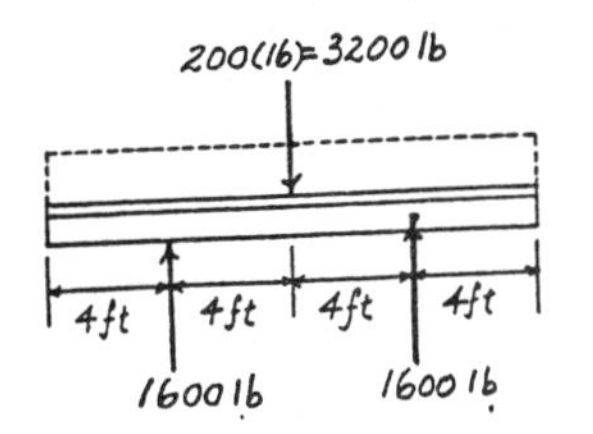

Support Reactions : As shown on FBD.

Internal Shear Force : As shown on shear diagram, $V_{max} = 800$ lb.

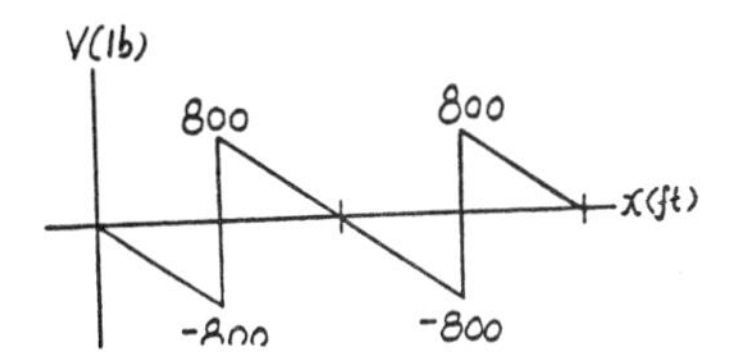

Section Properties :

$$\bar{y} = \frac{\Sigma \bar{y} A}{\Sigma A} = \frac{0.5(3)(1) + 2.5(1)(3)}{3(1) + 1(3)} = 1.50 \text{ in.}$$

$$I_{NA} = \frac{1}{12}(3)(1^3) + 3(1)(1.5 - 0.5)^2 + \frac{1}{12}(1)(3^3) + 1(3)(2.5 - 1.5)^2 = 8.50 \text{ in}^4$$

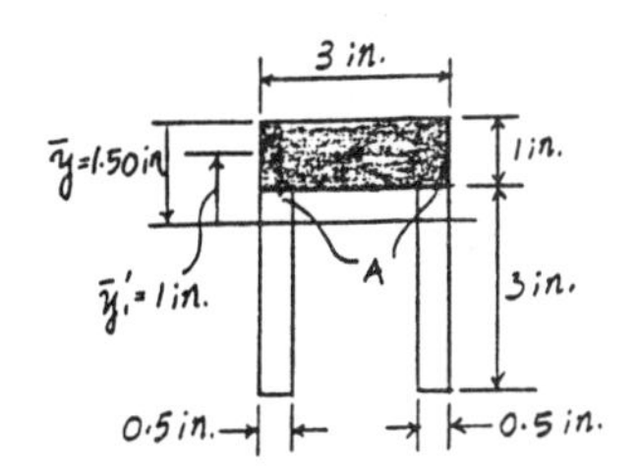

$$Q_A = \bar{y}_1' A' = 1(3)(1) = 3.00 \text{ in}^3$$

Shear Flow : Since there are two glue joints, hence

$$q = \frac{1}{2}\left(\frac{VQ}{I}\right) = \frac{1}{2}\left[\frac{800(3.00)}{8.50}\right] = 141 \text{ lb/in.} \qquad \textbf{Ans}$$

7–43. The beam is subjected to the loading shown, where $P = 7$ kN. Determine the average shear developed in the nails within region AB of the beam. The nails are located on each side of the beam and are spaced 100 mm apart. Each nail has a diameter of 5 mm.

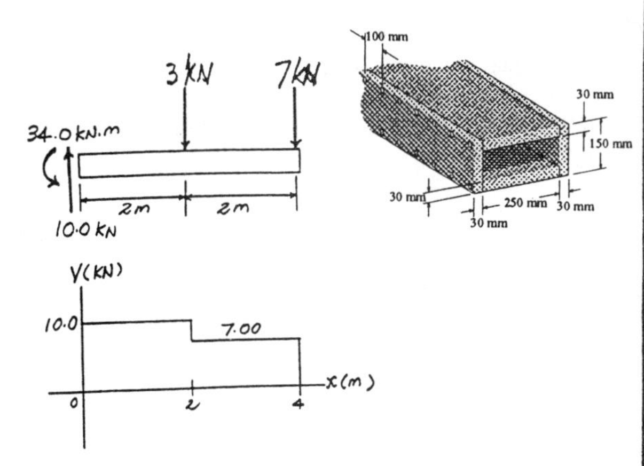

Support Reactions : As shown on FBD.

Internal Shear Force : As shown on shear diagram. $V_{AB} = 10.0$ kN.

Section Properties :

$$I_{NA} = \frac{1}{12}(0.31)(0.15^3) - \frac{1}{12}(0.25)(0.09^3) = 72.0(10^{-6}) \text{ m}^4$$

$$Q = \bar{y}'A' = 0.06(0.25)(0.03) = 0.450(10^{-3}) \text{ m}^3$$

Shear Flow :

$$q = \frac{VQ}{I} = \frac{10.0(10^3)0.450(10^{-3})}{72.0(10^{-6})} = 62.5 \text{ kN/m}$$

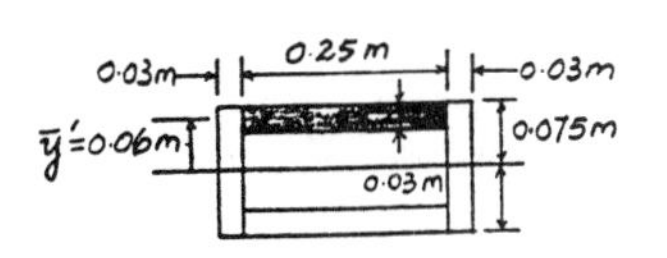

There are two rows of nails. Hence, the shear force resisted by each nail is

$$F = \left(\frac{q}{2}\right)s = \left(\frac{62.5 \text{ kN/m}}{2}\right)(0.1 \text{ m}) = 3.125 \text{ kN}$$

Average Shear Stress : For the nail

$$\tau_{nail} = \frac{F}{A} = \frac{3.125(10^3)}{\frac{\pi}{4}(0.005^2)} = 159 \text{ MPa} \qquad \textbf{Ans}$$

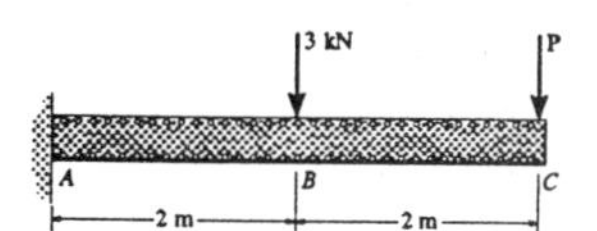

***7–44.** The beam is constructed from four boards which are nailed together. If the nails are on both sides of the beam and each can resist a shear of 3 kN, determine the maximum load P that can be applied to the end of the beam.

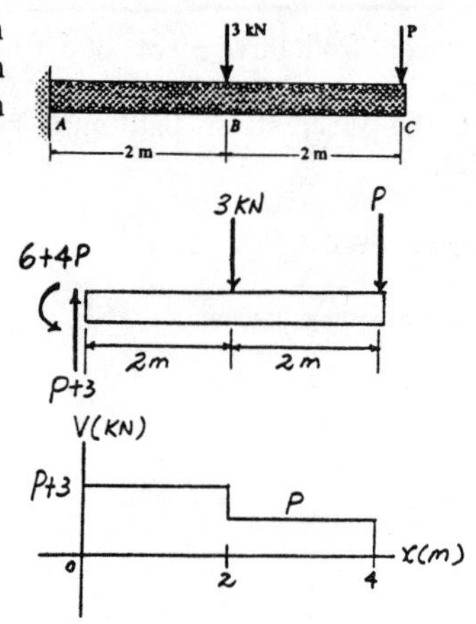

Support Reactions : As shown on FBD.

Internal Shear Force : As shown on shear diagram, $V_{AB} = (P+3)$ kN.

Section Properties :

$$I_{NA} = \frac{1}{12}(0.31)(0.15^3) - \frac{1}{12}(0.25)(0.09^3) = 72.0(10^{-6})\ \text{m}^4$$

$$Q = \bar{y}'A' = 0.06(0.25)(0.03) = 0.450(10^{-3})\ \text{m}^3$$

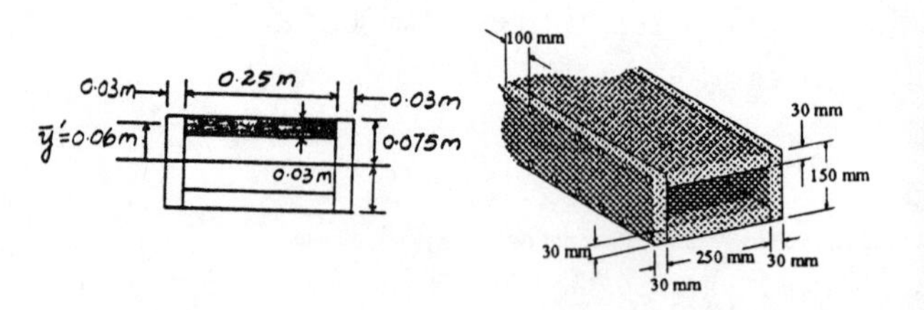

Shear Flow : There are two rows of nails. Hence, the allowable shear flow is $q = \frac{3(2)}{0.1} = 60.0$ kN/m.

$$q = \frac{VQ}{I}$$

$$60.0(10^3) = \frac{(P+3)(10^3)0.450(10^{-3})}{72.0(10^{-6})}$$

$p = 6.60$ kN **Ans**

7–45. The box beam is built up from four members and is subjected to a shear force of $V = 25$ kip. If the glue used at each seam can support 600 lb/in., determine the width b of the top and bottom plates.

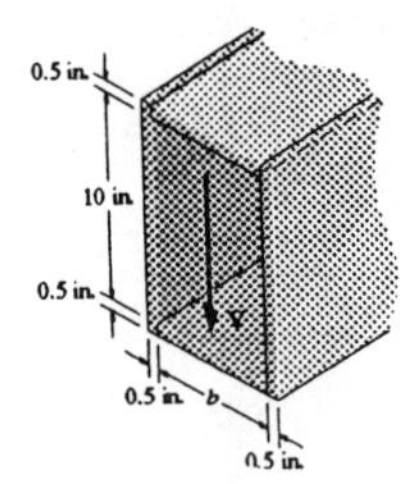

Section Properties :

$$I_{NA} = \frac{1}{12}(b+1)(11^3) - \frac{1}{12}(b)(10^3) = 27.5833b + 110.9167$$

$$Q = \bar{y}'A' = 5.25(b)(0.5) = 2.625b$$

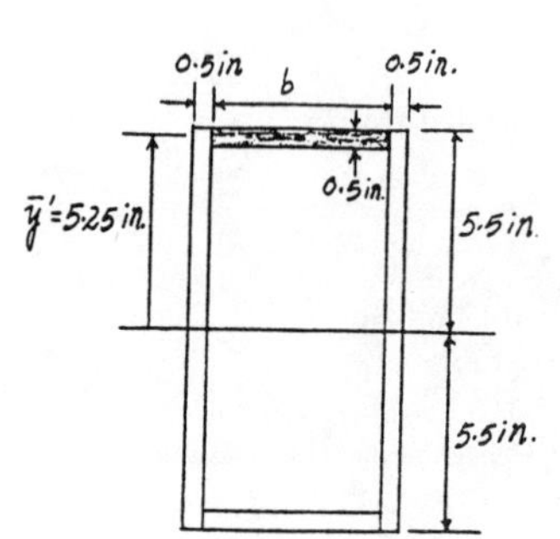

Shear Flow : There are two seams on the section. Hence, the allowable shear flow is $q = 2(600) = 1200$ lb/in.

$$q = \frac{VQ}{I}$$

$$1200 = \frac{25(10^3)(2.625b)}{27.5833b + 110.9167}$$

$b = 4.09$ in. **Ans**

7–46. The box beam is subjected to a shear force of $V = 25$ kip. If the glue used at each seam can support 600 lb/in., determine the width b of the top and bottom plates so that the shear stress at the seams is equal to one half the maximum shear stress in the beam.

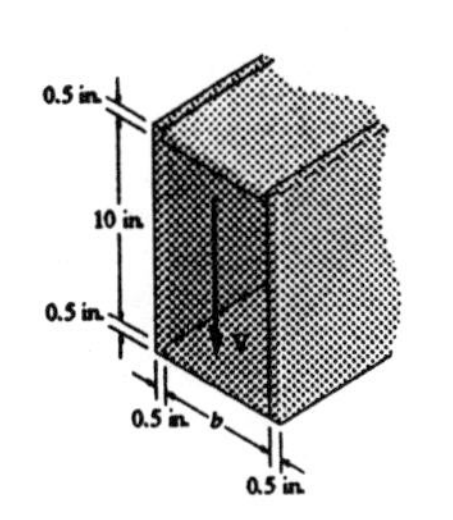

Section Properties :

$$I_{NA} = \frac{1}{12}(b+1)(11^3) - \frac{1}{12}(b)(10^3)$$
$$= 27.5833b + 110.9167$$

$$Q = \bar{y}_2' A' = 5.25(b)(0.5) = 2.625b$$
$$Q_{max} = \Sigma \bar{y}' A' = 5.25(b)(0.5) + 2.75(5.5)(1)$$
$$= 2.625b + 15.125$$

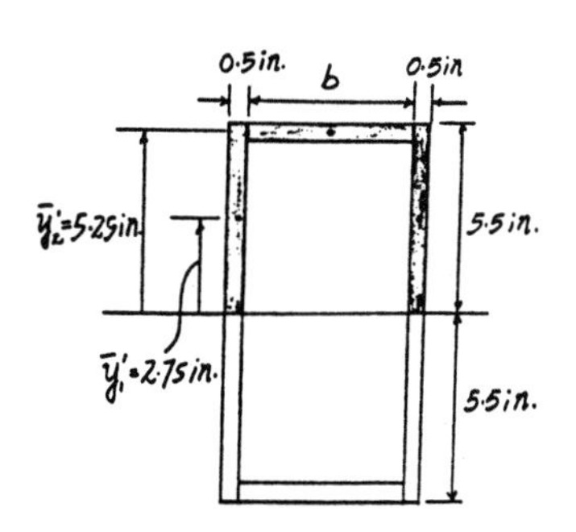

Maximum Shear Stress : Applying the shear formula

$$\tau_{max} = \frac{VQ_{max}}{It} = \frac{25(2.625b + 15.125)}{(27.5833b + 110.9167)(1)}$$

Shear Flow : The shear stress at the seam can be determined using

$$\tau_{seam} = \frac{q}{t} = \frac{VQ}{It} = \frac{25(2.625b)}{(27.5833b + 110.9167)(1)}$$

Requires,

$$\tau_{seam} = \frac{1}{2}\tau_{max}$$
$$\frac{25(2.625b)}{(27.5833b + 110.9167)(1)} = \frac{1}{2}\left[\frac{25(2.625b + 15.125)}{(27.5833b + 110.9167)(1)}\right]$$

$$b = 5.76 \text{ in.}$$ **Ans**

7–47. The double-web girder is constructed from two plywood sheets that are secured to wood members at its top and bottom. If each fastener can support 600 lb in single shear, determine the required spacing s of the fasteners needed to support the loading $P = 3000$ lb. Assume A is pinned and B is a roller.

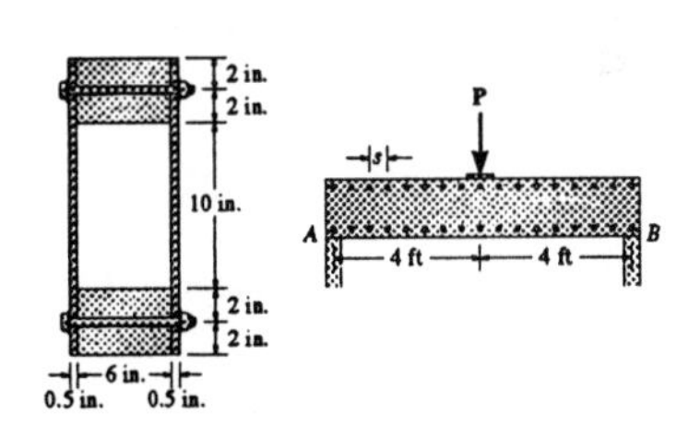

Support Reactions : As shown on FBD.

Internal Shear Force : As shown on shear diagram, $V_{max} = 1500$ lb.

Section Properties :

$$I_{NA} = \frac{1}{12}(7)(18^3) - \frac{1}{12}(6)(10^3) = 2902 \text{ in}^4$$

$$Q = \bar{y}' A' = 7(4)(6) = 168 \text{ in}^3$$

Shear Flow : Since there are two shear planes on the bolt, the allowable shear flow is $q = \frac{2(600)}{s} = \frac{1200}{s}$.

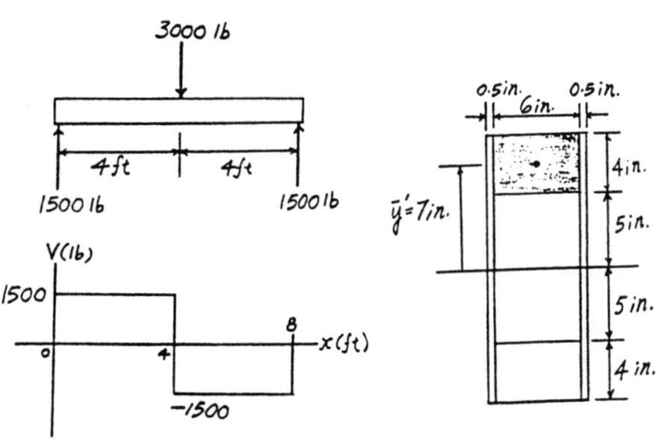

$$q = \frac{VQ}{I}$$
$$\frac{1200}{s} = \frac{1500(168)}{2902}$$

$$s = 13.8 \text{ in.}$$ **Ans**

***7–48.** The double-web girder is constructed from two plywood sheets that are secured to wood members at its top and bottom. The allowable bending stress for the wood is $\sigma_{allow} = 2$ ksi and the allowable shear stress is $\tau_{allow} = 800$ psi. If the fasteners are spaced $s = 6$ in. and each fastener can support 600 lb in single shear, determine the maximum load P that can be applied to the beam.

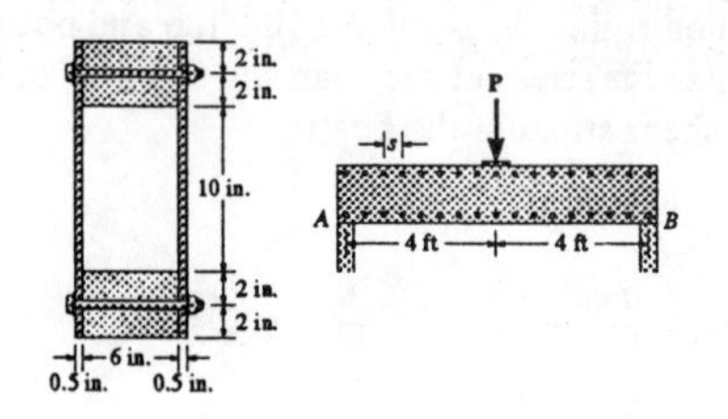

Support Reactions : As shown on FBD.

Internal Shear Force and Moment : As shown on shear and moment diagram, $V_{max} = 0.500P$ and $M_{max} = 2.00P$.

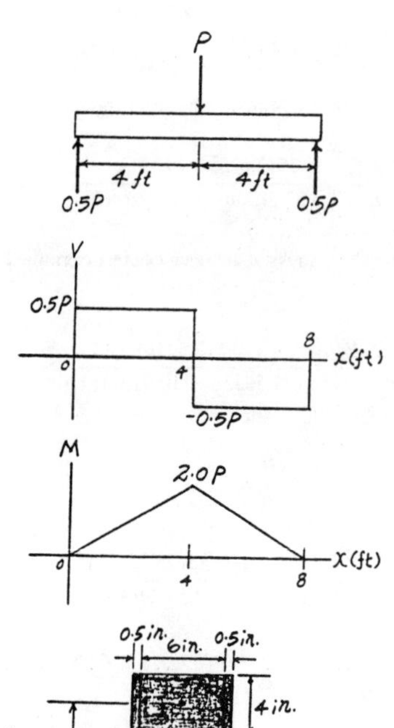

Section Properties :

$$I_{NA} = \frac{1}{12}(7)(18^3) - \frac{1}{12}(6)(10^3) = 2902 \text{ in}^4$$

$$Q = \bar{y}_2' A' = 7(4)(6) = 168 \text{ in}^3$$

$$Q_{max} = \Sigma \bar{y}' A' = 7(4)(6) + 4.5(9)(1) = 208.5 \text{ in}^3$$

Shear Flow : Assume bolt failure. Since there are two shear planes on the bolt, the allowable shear flow is $q = \dfrac{2(600)}{6} = 200$ lb/in.

$$q = \frac{VQ}{I}$$

$$200 = \frac{0.500P(168)}{2902}$$

$$P = 6910 \text{ lb} = 6.91 \text{ kip } (\textit{Controls!}) \qquad \textbf{Ans}$$

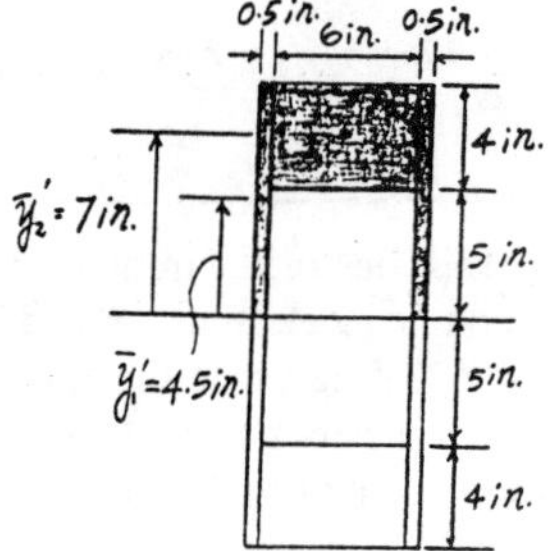

Shear Stress : Assume failure due to shear stress.

$$\tau_{max} = \tau_{allow} = \frac{VQ_{max}}{It}$$

$$800 = \frac{0.500P(208.5)}{2902(1)}$$

$$P = 22270 \text{ lb} = 22.27 \text{ kip}$$

Bending Stress : Assume failure due to bending stress.

$$\sigma_{max} = \sigma_{allow} = \frac{Mc}{I}$$

$$2(10^3) = \frac{2.00P(12)(9)}{2902}$$

$$P = 26870 \text{ lb} = 26.87 \text{ kip}$$

7–51. The member is subjected to a shear force of $V = 10$ kip. Sketch the shear-flow distribution along the vertical plate AB. Indicate numerical values of all peaks.

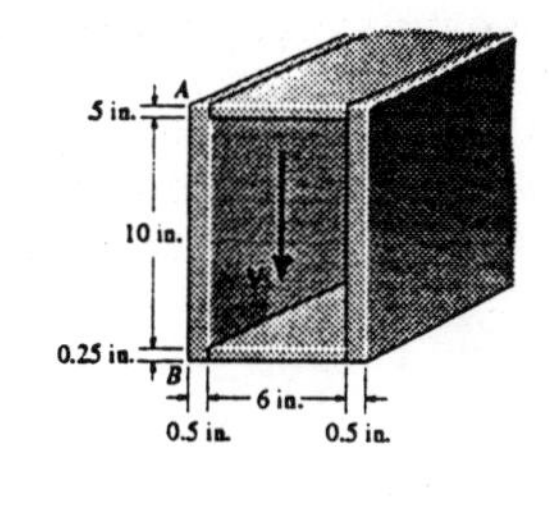

Section Properties :

$$I_{NA} = \frac{1}{12}(7)\left(10.5^3\right) - \frac{1}{12}(6)\left(10^3\right) = 175.28125 \text{ in}^4$$

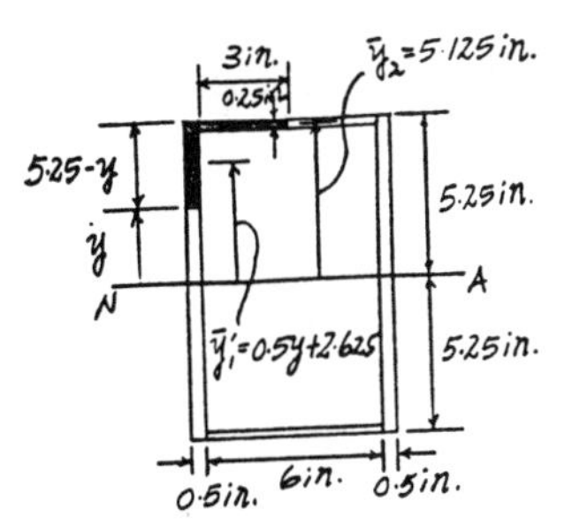

$$Q = \Sigma \bar{y}'A'$$
$$= 5.125(3)(0.25) + (0.5y + 2.625)(5.25 - y)(0.5)$$
$$= 10.734375 - 0.25y^2$$

Shear Flow :

$$q = \frac{VQ}{I} = \frac{10(10^3)(10.734375 - 0.25y^2)}{175.28125}$$
$$= \{612.41 - 14.26y^2\} \text{ lb/in.}$$

At $\quad y = 0, \qquad q = q_{max} = 612$ lb/in. **Ans**

At $\quad y = 5$ in., $\qquad q = 256$ lb/in. **Ans**

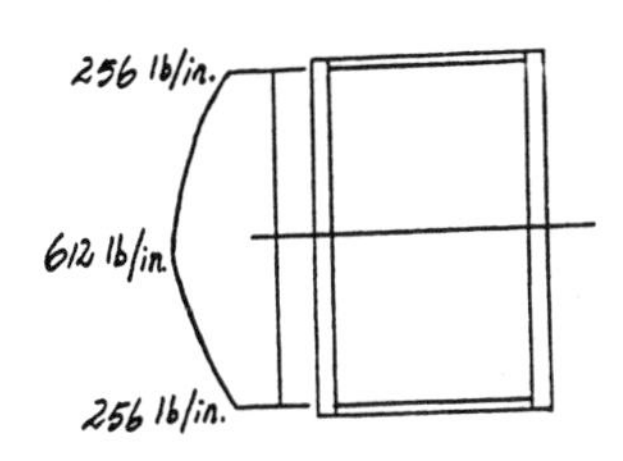

***7–52.** The beam supports a vertical shear of $V = 7$ kip. Determine the resultant force developed in segment AB of the beam.

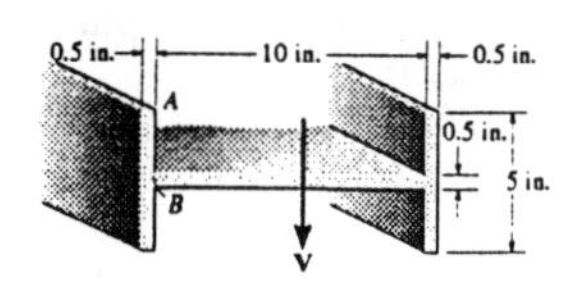

Section Properties :

$$I_{NA} = \frac{1}{12}(1)\left(5^3\right) + \frac{1}{12}(10)\left(0.5^3\right) = 10.52083 \text{ in}^4$$

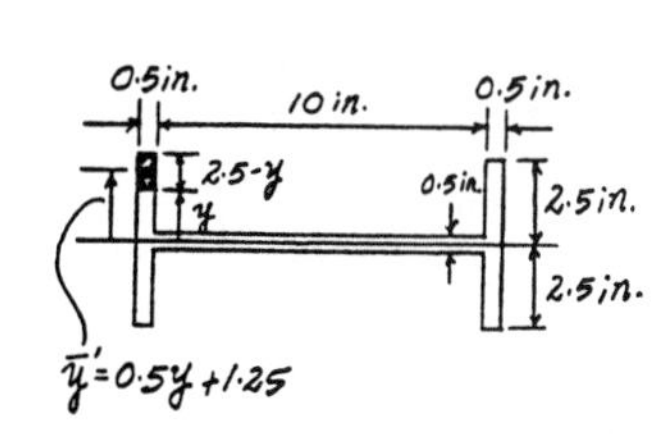

$$Q = \bar{y}'A' = (0.5y + 1.25)(2.5 - y)(0.5)$$
$$= 1.5625 - 0.25y^2$$

Shear Flow :

$$q = \frac{VQ}{I} = \frac{7(10^3)(1.5625 - 0.25y^2)}{10.52083}$$
$$= \{1039.60 - 166.34y^2\} \text{ lb/in.}$$

Resultant Shear Force : For web AB

$$V_{AB} = \int_{0.25\text{in.}}^{2.5\text{in.}} q\,dy$$
$$= \int_{0.25\text{in.}}^{2.5\text{in.}} \left(1039.60 - 166.34y^2\right) dy$$
$$= 1474 \text{ lb} = 1.47 \text{ kip}$$ **Ans**

7–53. A shear force of $V = 18$ kN is applied to the symmetric box girder. Determine the shear flow at A and B.

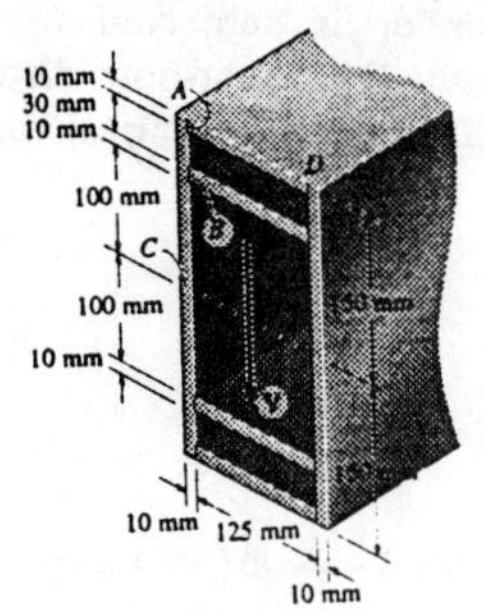

Section Properties :

$$I_{NA} = \frac{1}{12}(0.145)\left(0.3^3\right) - \frac{1}{12}(0.125)\left(0.28^3\right)$$
$$+2\left[\frac{1}{12}(0.125)\left(0.01^3\right) + 0.125(0.01)\left(0.105^2\right)\right]$$
$$= 125.17\left(10^{-6}\right)\ \text{m}^4$$

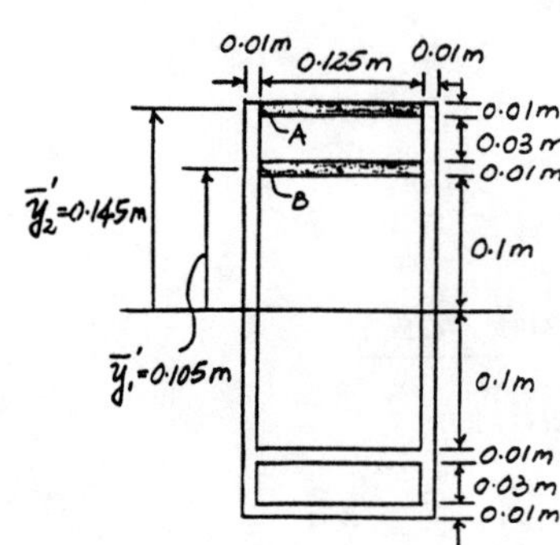

$$Q_A = \bar{y}_2' A' = 0.145(0.125)(0.01) = 0.18125\left(10^{-3}\right)\ \text{m}^3$$

$$Q_B = \bar{y}_1' A' = 0.105(0.125)(0.01) = 0.13125\left(10^{-3}\right)\ \text{m}^3$$

Shear Flow :

$$q_A = \frac{1}{2}\left[\frac{VQ_A}{I}\right]$$
$$= \frac{1}{2}\left[\frac{18(10^3)(0.18125)(10^{-3})}{125.17(10^{-6})}\right]$$
$$= 13033\ \text{N/m} = 13.0\ \text{kN/m}$$

Ans

$$q_B = \frac{1}{2}\left[\frac{VQ_B}{I}\right]$$
$$= \frac{1}{2}\left[\frac{18(10^3)(0.13125)(10^{-3})}{125.17(10^{-6})}\right]$$
$$= 9437\ \text{N/m} = 9.44\ \text{kN/m}$$

Ans

7–54. A shear force of $V = 18$ kN is applied to the box girder. Determine the shear flow at C.

Section Properties :

$$I_{NA} = \frac{1}{12}(0.145)\left(0.3^3\right) - \frac{1}{12}(0.125)\left(0.28^3\right)$$
$$+2\left[\frac{1}{12}(0.125)\left(0.01^3\right) + 0.125(0.01)\left(0.105^2\right)\right]$$
$$= 125.17\left(10^{-6}\right)\ \text{m}^4$$

$$Q_C = \Sigma \bar{y}' A'$$
$$= 0.145(0.125)(0.01) + 0.105(0.125)(0.01) + 0.075(0.15)(0.02)$$
$$= 0.5375\left(10^{-3}\right)\ \text{m}^3$$

Shear Flow :

$$q_C = \frac{1}{2}\left[\frac{VQ_C}{I}\right]$$
$$= \frac{1}{2}\left[\frac{18(10^3)(0.5375)(10^{-3})}{125.17(10^{-6})}\right]$$
$$= 38648\ \text{N/m} = 38.6\ \text{kN/m}$$

Ans

7–55. The beam is subjected to a shear of $V = 25$ kN. Determine the shear flow at points A and B. There is a very small gap at D and E.

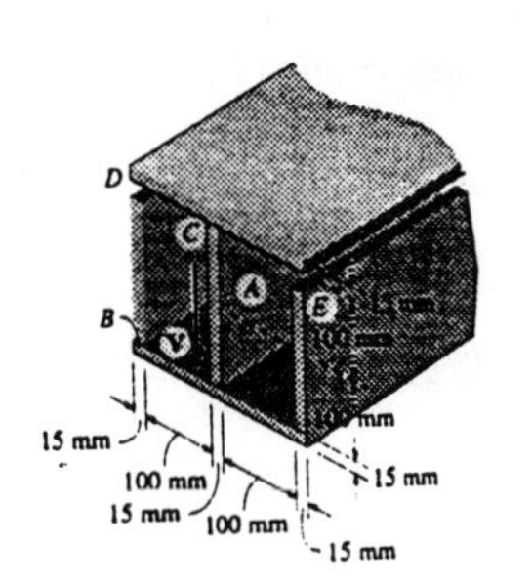

Section Properties :

$$I_{NA} = \frac{1}{12}(0.245)\left(0.23^3\right) - \frac{1}{12}(0.2)\left(0.2^3\right)$$
$$= 115.08\left(10^{-6}\right)\ \text{m}^4$$

$$Q_A = \Sigma \bar{y}'A'$$
$$= 0.05(0.1)(0.015) + 0.1075(0.245)(0.015)$$
$$= 0.4700625\left(10^{-3}\right)\ \text{m}^3$$

$$Q_B = \bar{y}_3'A' = 0.1075(0.245)(0.015)$$
$$= 0.3950625\left(10^{-3}\right)\ \text{m}^3$$

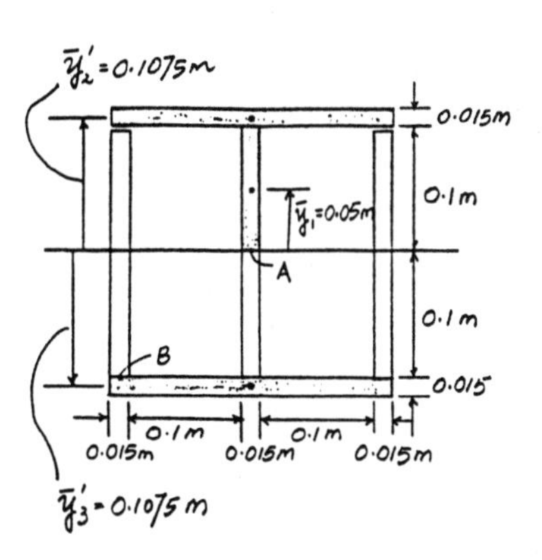

Shear Flow :

$$q_A = \frac{VQ_A}{I}$$
$$= \frac{25(10^3)0.4700625(10^{-3})}{115.08(10^{-6})}$$
$$= 102120\ \text{N/m} = 102\ \text{kN/m} \qquad \textbf{Ans}$$

$$q_B = \frac{1}{3}\left[\frac{VQ_B}{I}\right]$$
$$= \frac{1}{3}\left[\frac{25(10^3)0.3950625(10^{-3})}{115.08(10^{-6})}\right]$$
$$= 28609\ \text{N/m} = 28.6\ \text{kN/m} \qquad \textbf{Ans}$$

***7–56.** A shear force of $V = 18$ kN is applied to the box girder. Determine the position d of the stiffner plates BE and FG so that the shear flow at A is twice as great as the shear flow at B. Use the centerline dimensions for the calculation. All plates are 10 mm thick.

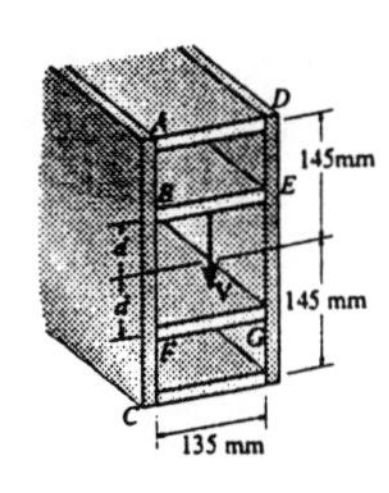

Section Properties :

$$Q_A = \bar{y}_1'A' = 0.145(0.125)(0.01)$$

$$Q_B = \bar{y}_2'A' = d(0.125)(0.01)$$

Shear Flow : Requires

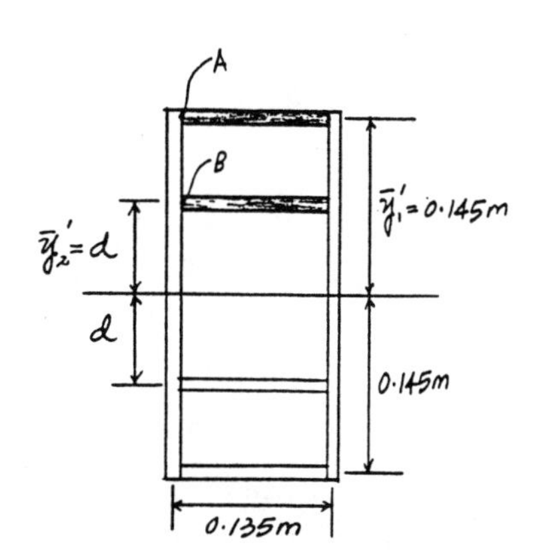

$$q_A = 2q_B$$
$$\frac{1}{2}\left(\frac{VQ_A}{I}\right) = 2\left[\frac{1}{2}\left(\frac{VQ_B}{I}\right)\right]$$
$$Q_A = 2Q_B$$
$$0.145(0.125)(0.01) = 2[d(0.125)(0.01)]$$

$$d = 0.0725\ \text{m} = 72.5\ \text{mm} \qquad \textbf{Ans}$$

7–57. The pipe is subjected to a shear force of $V = 8$ kip. Determine the shear flow in the pipe at points A and B.

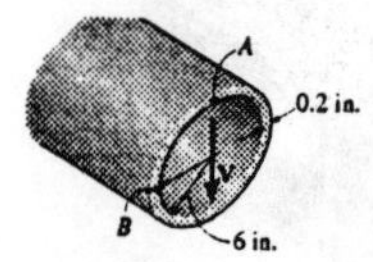

Section Properties :

$$I = \frac{\pi}{4}\left(6.2^4 - 6^4\right) = 45.4084\pi \text{ in}^4$$

Since $A' \to 0$ then $Q_A = 0$

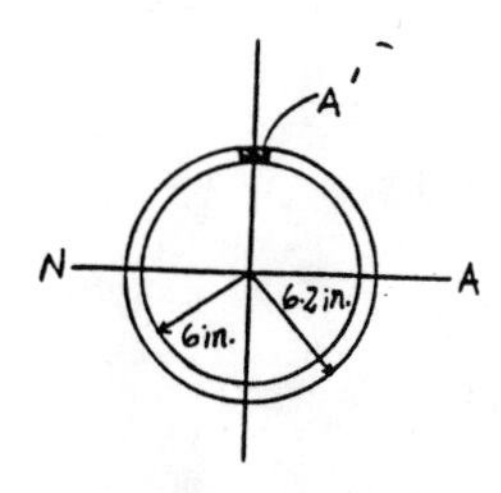

$$Q_B = \Sigma \bar{y}'A' = \frac{4(6.2)}{3\pi}\left[\frac{\pi(6.2^2)}{2}\right] - \frac{4(6)}{3\pi}\left[\frac{\pi(6^2)}{2}\right]$$
$$= 14.8853 \text{ in}^4$$

Shear Flow :

$$q_A = \frac{VQ_A}{I} = 0 \qquad \textbf{Ans}$$

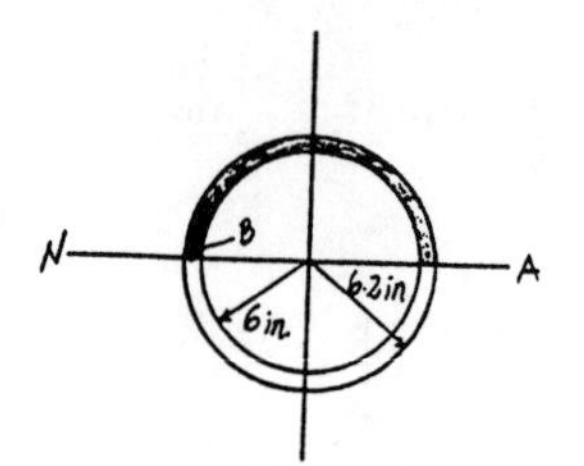

$$q_B = \frac{1}{2}\left(\frac{VQ_B}{I}\right)$$
$$= \frac{1}{2}\left[\frac{8(10^3)(14.8853)}{45.4084\pi}\right] = 417 \text{ lb/in.} \qquad \textbf{Ans}$$

7–58. The square tube is subjected to a shear force of $V = 50$ kN. Determine the maximum shear flow in the tube. The tube is 10 mm thick.

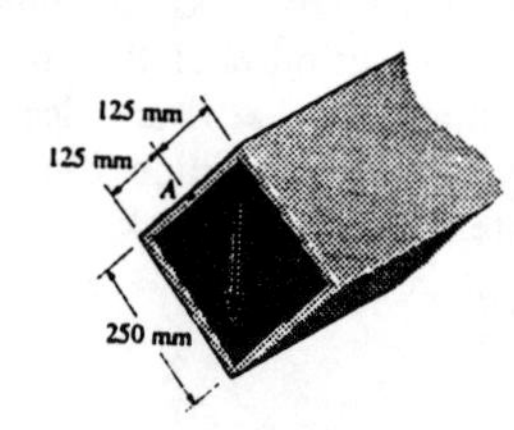

Section Properties :

$$b = \frac{0.01}{\sin 45°} = 0.014142 \text{ m} \qquad h = 2(0.25\sin 45°) = 0.35355 \text{ m}$$

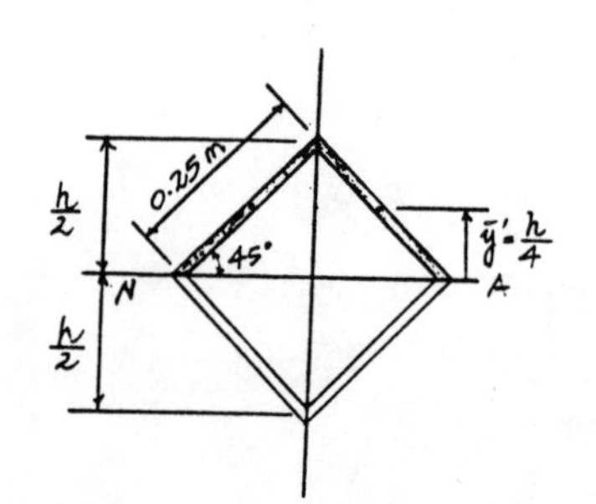

$$I_{NA} = 2\left[\frac{1}{12}(0.014142)\left(0.35355^3\right)\right] = 0.104167\left(10^{-3}\right) \text{ m}^4$$

$$Q_{max} = \bar{y}'A' = 2\left[\left(\frac{0.35355}{4}\right)(0.25)(0.01)\right]$$
$$= 0.44194\left(10^{-3}\right) \text{ m}^3$$

Shear Flow :

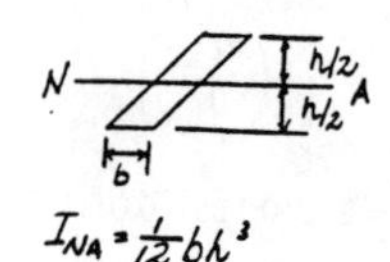

$$q_{max} = \frac{1}{2}\left(\frac{VQ_{max}}{I}\right)$$
$$= \frac{1}{2}\left[\frac{50(10^3)0.44194(10^{-3})}{0.104167(10^{-3})}\right]$$
$$= 106\ 066 \text{ N/m} = 106 \text{ kN/m} \qquad \textbf{Ans}$$

7–59. The square tube is subjected to a shear force of $V = 50$ kN. Determine the shear flow in the tube at point A. The tube is 10 mm thick.

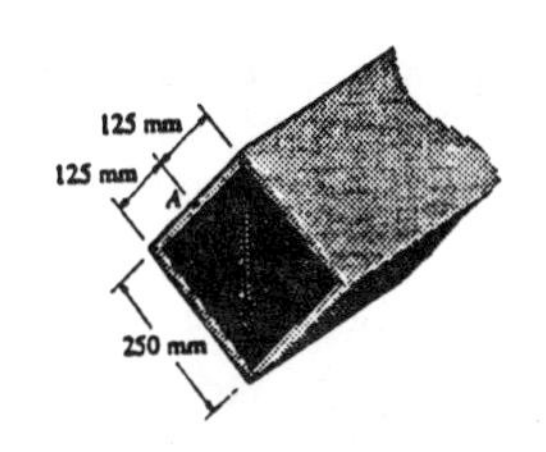

Section Properties :

$$b = \frac{0.01}{\sin 45°} = 0.014142 \text{ m} \qquad h = 2(0.25\sin 45°) = 0.35355 \text{ m}$$

$$I_{NA} = 2\left[\frac{1}{12}(0.014142)\left(0.35355^3\right)\right] = 0.104167\left(10^{-3}\right) \text{ m}^4$$

$$Q_A = \bar{y}'A' = 2\left[\frac{3}{8}(0.35355)(0.125)(0.01)\right]$$
$$= 0.33146\left(10^{-3}\right) \text{ m}^3$$

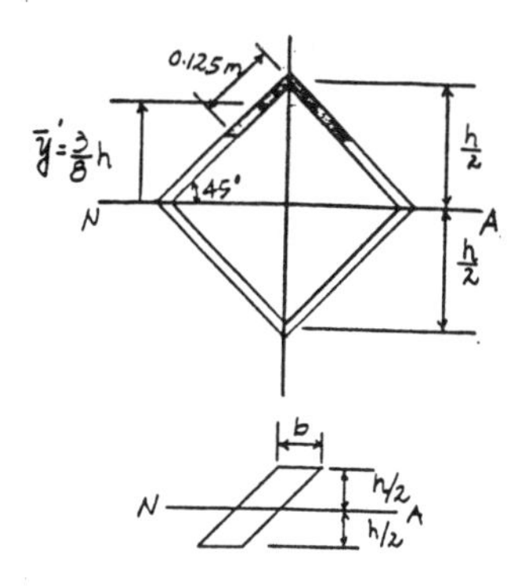

Shear Flow :

$$q_A = \frac{1}{2}\left(\frac{VQ_A}{I}\right)$$
$$= \frac{1}{2}\left[\frac{50(10^3)0.33146(10^{-3})}{0.104167(10^{-3})}\right]$$
$$= 79\,550 \text{ N/m} = 79.5 \text{ kN/m} \qquad \textbf{Ans}$$

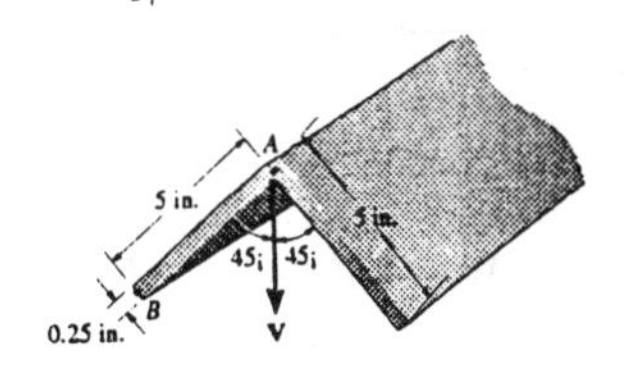

***7–60.** The angle is subjected to a shear of $V = 2$ kip. Sketch the distribution of shear flow along the leg AB. Indicate numerical values at all peaks.

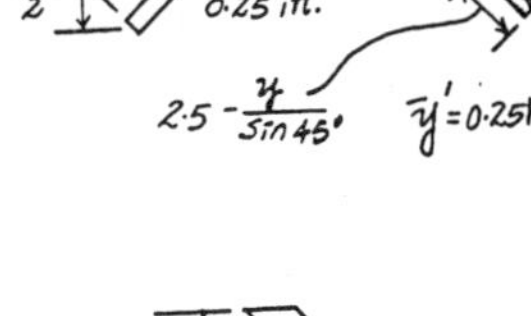

Section Properties :

$$b = \frac{0.25}{\sin 45°} = 0.035355 \text{ in.}$$
$$h = 5\cos 45° = 3.53553 \text{ in.}$$

$$I_{NA} = 2\left[\frac{1}{12}(0.035355)\left(3.53553^3\right)\right] = 2.604167 \text{ in}^4$$

$$Q = \bar{y}'A' = [0.25(3.53553) + 0.5y]\left(2.5 - \frac{y}{\sin 45°}\right)(0.25)$$
$$= 0.55243 - 0.17678y^2$$

Shear Flow :

$$q = \frac{VQ}{I}$$
$$= \frac{2(10^3)(0.55243 - 0.17678y^2)}{2.604167}$$
$$= \{424 - 136y^2\} \text{ lb/in.} \qquad \textbf{Ans}$$

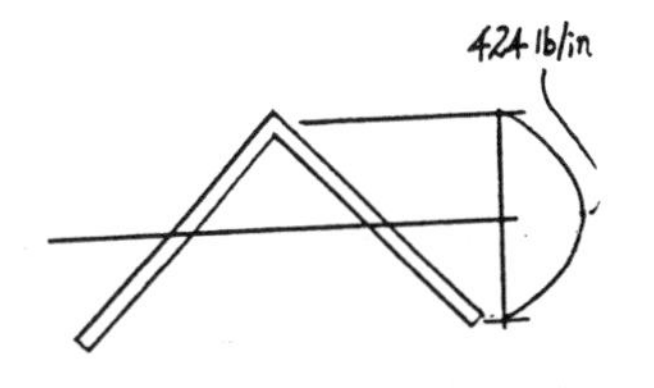

At $y = 0$, $\quad q = q_{max} = 424$ lb/in. $\qquad$ **Ans**

7–61. Determine the location e of the shear center, point O, for the thin-walled member having the cross section shown. The member has a thickness of 0.25 in.

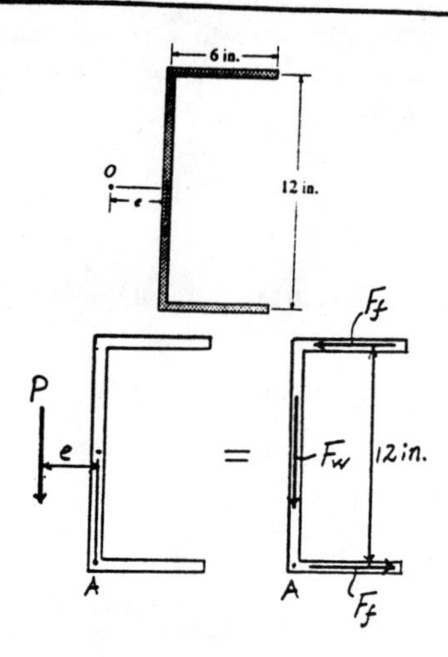

Section Properties :

$$I_{NA} = \frac{1}{12}(t)(12^3) + 2\left[6(t)(6^2)\right] = 576t$$

$$Q = \bar{y}'A' = 6(t)x = 6tx$$

Shear Flow Resultant : For flange

$$q = \frac{VQ}{I} = \frac{P(6tx)}{576t} = \frac{Px}{96}$$

$$F_f = \int_0^{6\text{ in.}} q\,dx = \int_0^{6\text{ in.}} \frac{Px}{96}\,dx = \frac{3}{16}P$$

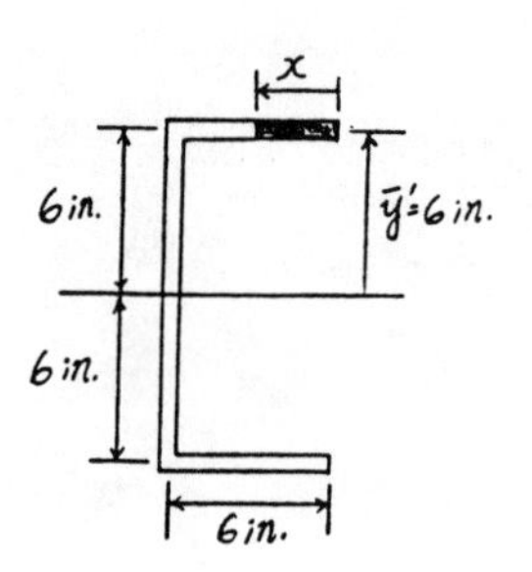

Shear Center : Summing moments about point A.

$$Pe = F_f(12)$$

$$Pe = \frac{3}{16}P(12)$$

$$e = 2\frac{1}{4}\text{ in.} \qquad \textbf{Ans}$$

7–62. Determine the location e of the shear center, point O, for the thin-walled member having the cross section shown. The member segments have the same thickness t.

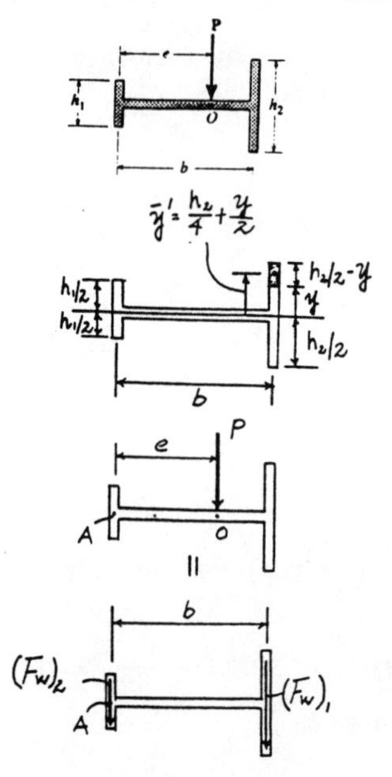

Section Properties :

$$I_{NA} = \frac{1}{12}(t)(h_1^3) + \frac{1}{12}(t)(h_2^3) = \frac{t}{12}(h_1^3 + h_2^3)$$

$$Q = \bar{y}'A' = \left(\frac{h_2}{4} + \frac{y}{2}\right)\left(\frac{h_2}{2} - y\right)t = \left(\frac{h_2^2}{8} - \frac{y^2}{2}\right)t$$

Shear Flow Resultant : For flange

$$q = \frac{VQ}{I} = \frac{P\left(\frac{h_2^2}{8} - \frac{y^2}{2}\right)t}{\frac{t}{12}(h_1^3 + h_2^3)} = \frac{3P}{2(h_1^3 + h_2^3)}(h_2^2 - 4y^2)$$

$$(F_w)_1 = \int_{-\frac{h_2}{2}}^{\frac{h_2}{2}} q\,dy$$

$$= \frac{3P}{2(h_1^3 + h_2^3)}\int_{-\frac{h_2}{2}}^{\frac{h_2}{2}} (h_2^2 - 4y^2)\,dy$$

$$= \frac{3P}{2(h_1^3 + h_2^3)}\left(h_2^2 y - \frac{4y^3}{3}\right)\Bigg|_{-\frac{h_2}{2}}^{\frac{h_2}{2}}$$

$$= \left(\frac{h_2^3}{h_1^3 + h_2^3}\right)P$$

Shear Center : Summing moments about point A.

$$Pe = (F_w)_1\,b$$

$$Pe = \left(\frac{h_2^3}{h_1^3 + h_2^3}\right)Pb$$

$$e = \left(\frac{h_2^3}{h_1^3 + h_2^3}\right)b \qquad \textbf{Ans}$$

7–63. Determine the location e of the shear center, point O, for the thin-walled member having the cross section shown. The member segments have the same thickness t.

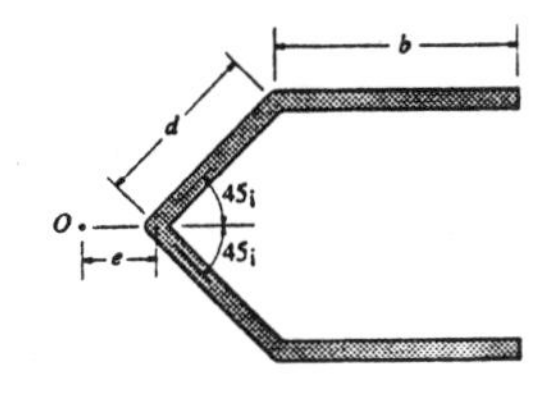

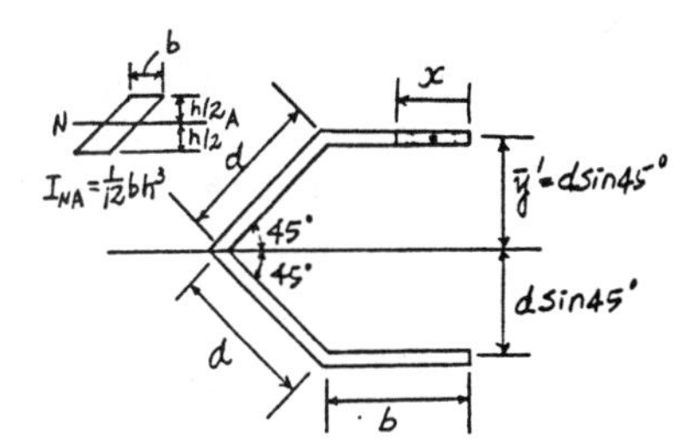

Section Properties :

$$I = \frac{1}{12}\left(\frac{t}{\sin 45°}\right)(2d\sin 45°)^3 + 2\left[bt(d\sin 45°)^2\right]$$
$$= \frac{td^2}{3}(d+3b)$$

$$Q = \bar{y}'A' = d\sin 45°(xt) = (td\sin 45°)x$$

Shear Flow Resultant :

$$q_f = \frac{VQ}{I} = \frac{P(td\sin 45°)x}{\frac{td^2}{3}(d+3b)} = \frac{3P\sin 45°}{d(d+3b)}x$$

$$F_f = \int_0^b q_f dx = \frac{3P\sin 45°}{d(d+3b)}\int_0^b x dx = \frac{3b^2\sin 45°}{2d(d+3b)}P$$

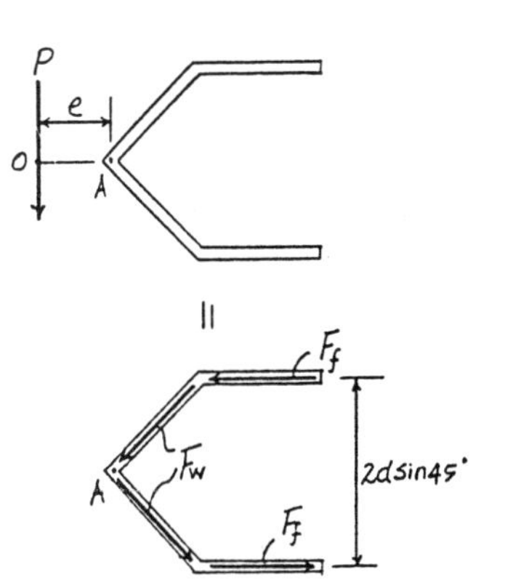

Shear Center : Summing moments about point ,

$$Pe = F_f(2d\sin 45°)$$
$$Pe = \left[\frac{3b^2\sin 45°}{2d(d+3b)}P\right](2d\sin 45°)$$

$$e = \frac{3b^2}{2(d+3b)} \qquad \textbf{Ans}$$

***7–64.** Determine the location e of the shear center, point O, for the thin-walled member having the cross section shown. The member segments have the same thickness t.

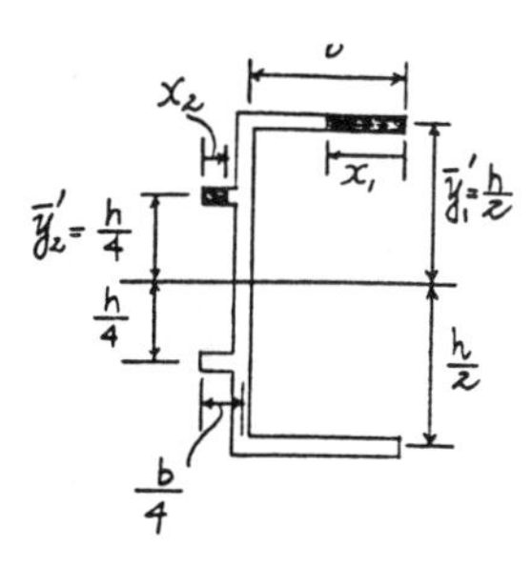

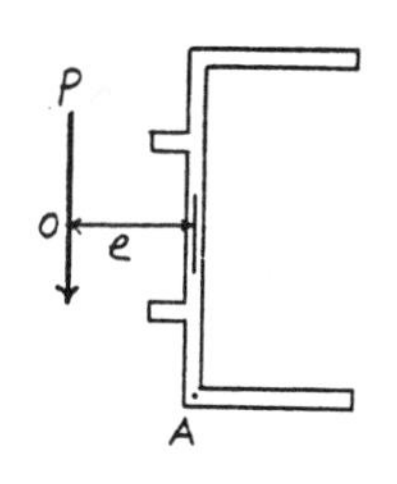

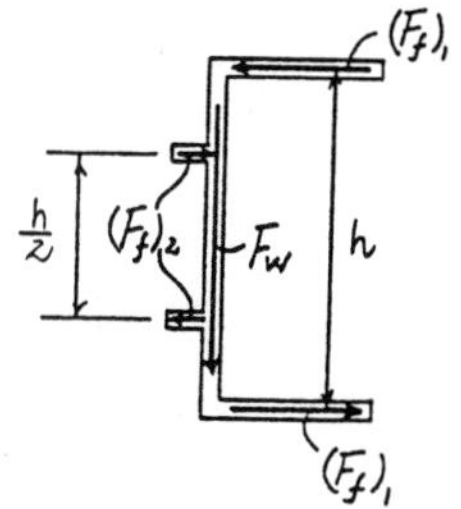

Section Properties :

$$I = \frac{1}{12}(t)(h^3) + 2\left[bt\left(\frac{h}{2}\right)^2\right] + 2\left[\frac{b}{4}t\left(\frac{h}{4}\right)^2\right]$$
$$= \frac{th^2}{96}(8h+51b)$$

$$Q_1 = \bar{y}_1'A' = \frac{h}{2}(x_1 t) = \frac{th}{2}x_1$$
$$Q_2 = \bar{y}_2'A' = \frac{h}{4}(x_2 t) = \frac{th}{4}x_2$$

Shear Flow Resultant :

$$(q_f)_1 = \frac{VQ_1}{I} = \frac{P\left(\frac{th}{2}x_1\right)}{\frac{th^2}{96}(8h+51b)} = \frac{48P}{h(8h+51b)}x_1$$

$$(q_f)_2 = \frac{VQ_2}{I} = \frac{P\left(\frac{th}{4}x_2\right)}{\frac{th^2}{96}(8h+51b)} = \frac{24P}{h(8h+51b)}x_2$$

$$(F_f)_1 = \int_0^b (q_f)_1 dx_1 = \frac{48P}{h(8h+51b)}\int_0^b x_1 dx_1$$
$$= \frac{24b^2}{h(8h+51b)}P$$

$$(F_f)_2 = \int_0^{\frac{b}{4}} (q_f)_2 dx_2 = \frac{24P}{h(8h+51b)}\int_0^{\frac{b}{4}} x_2 dx_2$$
$$= \frac{3b^2}{4h(8h+51b)}P$$

Shear Center : Summing moments about point A,

$$Pe = (F_f)_1 h - (F_f)_2\left(\frac{h}{2}\right)$$
$$Pe = \left[\frac{24b^2}{h(8h+51b)}P\right]h - \left[\frac{3b^2}{4h(8h+51b)}P\right]\left(\frac{h}{2}\right)$$
$$e = \frac{189b^2}{8(8h+51b)} \qquad \textbf{Ans}$$

7–65. Determine the location e of the shear center, point O, for the thin-walled member having the cross section shown. The member segments have the same thickness t.

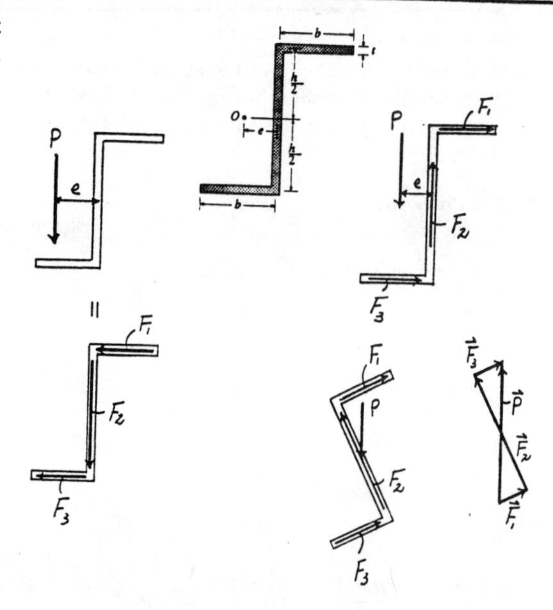

Shear Flow Resultant : The shear force flows through as indicated by F_1, F_2, and F_3 on FBD (b). Hence, The horizontal force equilibrium is not satisfied ($\Sigma F_x \neq 0$). In order to satisfy this equilibrium requirement, F_1 and F_2 must be equal to zero.

Shear Center : Summing moments about point A,

$$Pe = F_2(0) \qquad e = 0 \qquad \textbf{Ans}$$

Also,

The shear flows through the section as indicated by F_1, F_2, F_3.

However, $\underset{\rightarrow}{+}\Sigma F_x \neq 0$

To satisfy this equation, the section must tip so that the resultant of $\vec{F}_1 + \vec{F}_2 + \vec{F}_3 = \vec{P}$

Also, due to the geometry, for calculating F_1 and F_3, we require $F_1 = F_3$.

Hence, $e = 0$ **Ans**

We would evaluate the same thing if the load P was applied along a horizontal axis.

7–66. Determine the location e of the shear center, point O, for the thin-walled member having the cross-sectional area shown, where $b_2 > b_1$. The member segments have the same thickness t.

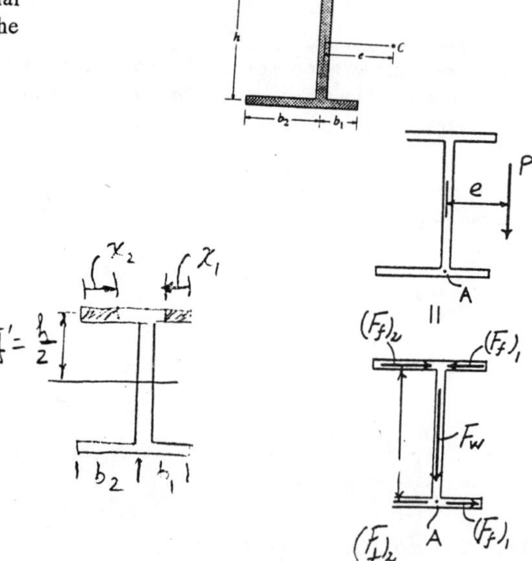

Section Properties :

$$I = \frac{1}{12}t h^3 + 2\left[(b_1 + b_2)t\left(\frac{h}{2}\right)^2\right] = \frac{t h^2}{12}\left[h + 6(b_1 + b_2)\right]$$

$$Q_1 = \bar{y}'A' = \frac{h}{2}(x_1)t = \frac{h t}{2}x_1$$

$$Q_2 = \bar{y}'A' = \frac{h}{2}(x_2)t = \frac{h t}{2}x_2$$

Shear Flow Resultant :

$$q_1 = \frac{VQ_1}{I} = \frac{P\left(\frac{ht}{2}x_1\right)}{\frac{th^2}{12}\left[h + 6(b_1 + b_2)\right]} = \frac{6P}{h\left[h + 6(b_1 + b_2)\right]}x_1$$

$$q_2 = \frac{VQ_2}{I} = \frac{P\left(\frac{ht}{2}x_2\right)}{\frac{th^2}{12}\left[h + 6(b_1 + b_2)\right]} = \frac{6P}{h\left[h + 6(b_1 + b_2)\right]}x_2$$

$$(F_f)_1 = \int_0^{b_1} q_1\,dx_1 = \frac{6P}{h\left[h + 6(b_1 + b_2)\right]}\int_0^{b_1} x_1\,dx_1 = \frac{3Pb_1^2}{h\left[h + 6(b_1 + b_2)\right]}$$

$$(F_f)_2 = \int_0^{b_2} q_2\,dx_2 = \frac{6P}{h\left[h + 6(b_1 + b_2)\right]}\int_0^{b_2} x_2\,dx_2 = \frac{3Pb_2^2}{h\left[h + 6(b_1 + b_2)\right]}$$

Shear Center : Summing moment about point A.

$$Pe = (F_f)_2\,h - (F_f)_1\,h$$

$$Pe = \frac{3Pb_2^2}{h\left[h + 6(b_1 + b_2)\right]}(h) - \frac{3Pb_1^2}{h\left[h + 6(b_1 + b_2)\right]}(h)$$

$$e = \frac{3(b_2^2 - b_1^2)}{h + 6(b_1 + b_2)} \qquad \textbf{Ans}$$

Note that if $b_2 = b_1$, $e = 0$ (I shape).

7–67. Determine the location e of the shear center, point O, for the thin-walled member having the cross section shown. The member segments have the same thickness t.

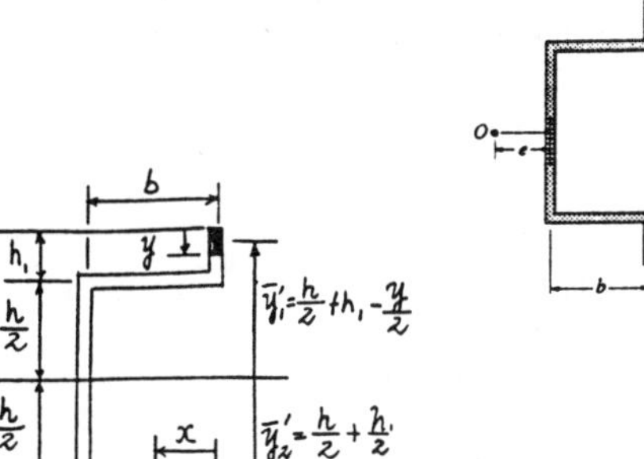

Section Properties :

$$I = \frac{1}{12}t(h + 2h_1)^3 + 2\left[bt\left(\frac{h}{2}\right)^2\right]$$
$$= \frac{1}{12}t(h + 2h_1)^3 + \frac{bth^2}{2}$$
$$= \frac{t}{12}\left[(h+2h_1)^3 + 6bh^2\right]$$

$$Q_1 = \bar{y}'_1 A' = \left(\frac{h}{2} + h_1 - \frac{y}{2}\right)(y\,t) = t\left(\frac{h}{2}y + h_1 y - \frac{y^2}{2}\right)$$
$$Q_2 = \Sigma\bar{y}'A' = \left(\frac{h}{2} + \frac{h_1}{2}\right)h_1 t + \frac{h}{2}(x)t$$
$$= \frac{t}{2}\left[h_1(h + h_1) + hx\right]$$

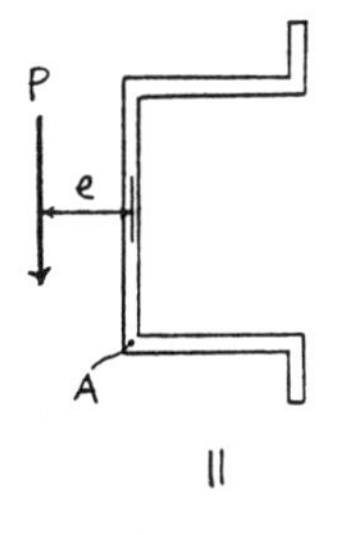

Shear Flow :

$$q_1 = \frac{VQ_1}{I} = \frac{Pt\left(\frac{h}{2}y + h_1 y - \frac{y^2}{2}\right)}{I}$$
$$q_2 = \frac{VQ_2}{I} = \frac{Pt\left[h_1(h + h_1) + hx\right]}{2I}$$

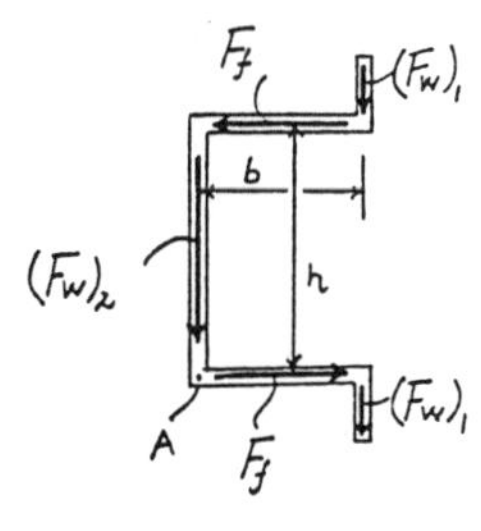

Shear Flow Resultant :

$$(F_w)_1 = \int_0^{h_1} q_1\,dy$$
$$= \frac{Pt}{I}\int_0^{h_1}\left(\frac{h}{2}y + h_1 y - \frac{y^2}{2}\right)dy$$
$$= \frac{Pt}{12I}\left(3hh_1^2 + 4h_1^3\right)$$

$$F_f = \int_0^b q_2\,dx$$
$$= \frac{Pt}{2I}\int_0^b\left[h_1(h + h_1) + hx\right]dx$$
$$= \frac{Pt}{4I}\left[2bh_1(h + h_1) + b^2h\right]$$

Shear Center : Summing moments about point A,

$$Pe = F_f h - 2(F_w)_1 b$$
$$Pe = \frac{Pt}{4I}\left[2bh_1(h + h_1) + b^2h\right]h - 2\left[\frac{Pt}{12I}\left(3hh_1^2 + 4h_1^3\right)\right]b$$
$$e = \frac{tb}{4I}\left[2h_1h(h + h_1) + bh^2\right] - \frac{tb}{6I}\left(3h_1^2h + 4h_1^3\right)$$
$$= \frac{tb}{12I}\left[6h_1h(h + h_1) + 3bh^2\right] - \frac{tb}{12I}\left(6h_1^2h + 8h_1^3\right)$$
$$= \frac{tb}{12I}\left(6h_1h^2 + 6h_1^2h + 3bh^2 - 6h_1^2h - 8h_1^3\right)$$
$$= \frac{tb}{12I}\left(6h_1h^2 + 3bh^2 - 8h_1^3\right)$$
$$= \frac{tb(6h_1h^2 + 3bh^2 - 8h_1^3)}{12\left\{\frac{t}{12}\left[(h+2h_1)^3 + 6bh^2\right]\right\}}$$
$$= \frac{b(6h_1h^2 + 3bh^2 - 8h_1^3)}{(h + 2h_1)^3 + 6bh^2} \qquad \textbf{Ans}$$

***7–68.** Determine the location e of the shear center, point O, for the thin-walled member having the cross section shown. The member segments have the same thickness t.

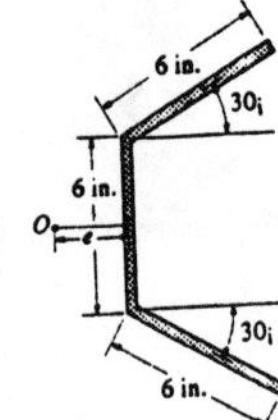

Section Properties :

$$I = \frac{1}{12}t\left(6^3\right) + 2\left[\frac{1}{12}\left(\frac{t}{\sin 30°}\right)(6 \sin 30°)^3 + (6t)(3 + 3\sin 30°)^2\right]$$

$$= 270t$$

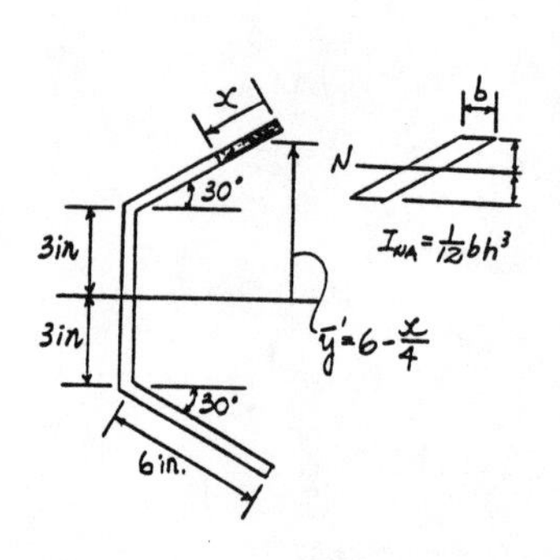

$$\bar{y}' = 3 + 6\sin 30° - \frac{x}{2}\sin 30° = 6 - \frac{x}{4}$$

$$Q = \bar{y}'A' = \left(6 - \frac{x}{4}\right)(x)(t) = t\left(6x - \frac{x^2}{4}\right)$$

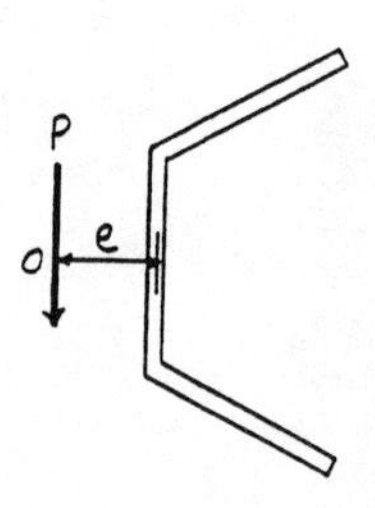

Shear Flow Resultant :

$$q = \frac{VQ}{I} = \frac{P\,t\left(6x - \frac{x^2}{4}\right)}{270\,t} = \frac{P\left(6x - \frac{x^2}{4}\right)}{270}$$

$$F_1 = \int_0^{6\text{in}} q\,dx = \frac{P}{270}\int_0^{6\text{in}}\left(6x - \frac{x^2}{4}\right)dx = \frac{1}{3}P$$

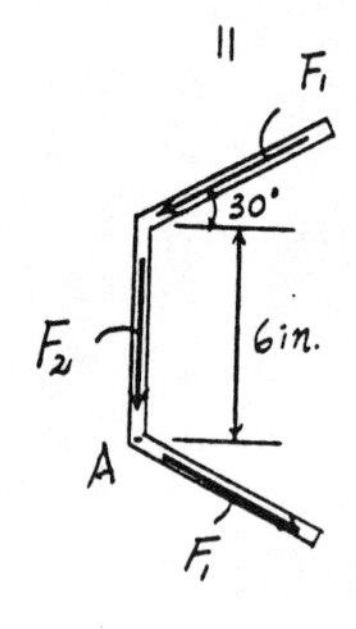

Shear Center : Summing moments about point A,

$$Pe = F_1 \cos 30°(6)$$

$$Pe = \frac{1}{3}P\cos 30°(6)$$

$$e = 1.73 \text{ in.} \qquad \textbf{Ans}$$

7–69. Determine the location e of the shear center, point O, for the thin-walled member having the cross section shown. The member has a thickness t.

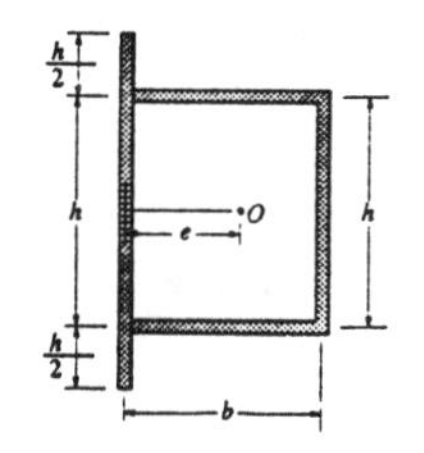

Section Properties :

$$I = \frac{1}{12}(t)(2h)^3 + \frac{1}{12}th^3 + 2\left[bt\left(\frac{h}{2}\right)^2\right] = \frac{th^2}{4}(3h+2b)$$

$$Q_1 = \bar{y}_1' A' = \left(\frac{h}{2} + \frac{y_1}{2}\right)(h - y_1)t = \frac{1}{2}\left(h^2 - y_1^2\right)t$$

$$Q_2 = \Sigma \bar{y}' A'$$

$$= \left(\frac{h}{2} + \frac{y_2}{2}\right)(h - y_2)t + \frac{h}{2}(bt) + \left(\frac{h}{4} + \frac{y_2}{2}\right)\left(\frac{h}{2} - y_2\right)t$$

$$= \frac{t}{8}\left(5h^2 + 4bh - 8y_2^2\right)$$

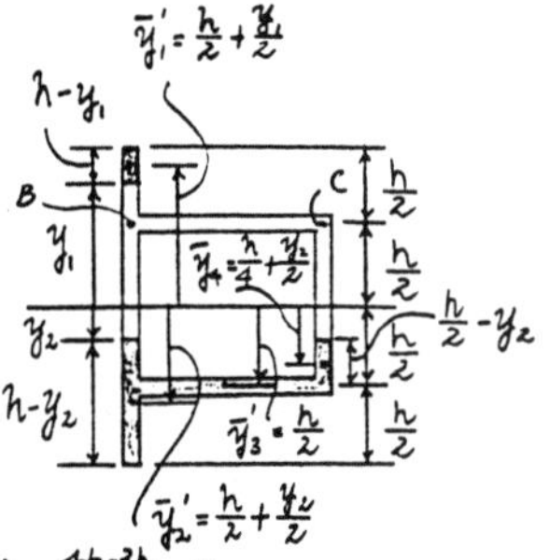

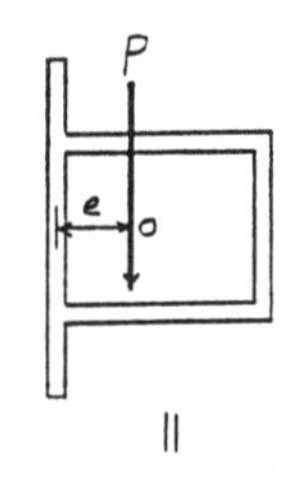

Shear Flow :

$$q_1 = \frac{VQ_1}{I} = \frac{P\left[\frac{1}{2}(h^2 - y_1^2)t\right]}{\frac{th^2}{4}(3h+2b)} = \frac{2P(h^2 - y_1^2)}{h^2(3h+2b)}$$

$$q_2 = \frac{1}{2}\left(\frac{VQ_2}{I}\right)$$

$$= \frac{1}{2}\left[\frac{\frac{Pt}{8}(5h^2 + 4bh - 8y_2^2)}{\frac{th^2}{4}(3h+2b)}\right] = \frac{P(5h^2 + 4bh - 8y_2^2)}{4h^2(3h+2b)}$$

At $y_1 = \frac{h}{2}$, $\quad q_1 = (q_w)_{B-} = \frac{3}{2(3h+2b)}P$

At $y_2 = \frac{h}{2}$, $\quad q_2 = (q_w)_{B+} = \frac{3h+4b}{4h(3h+2b)}P$

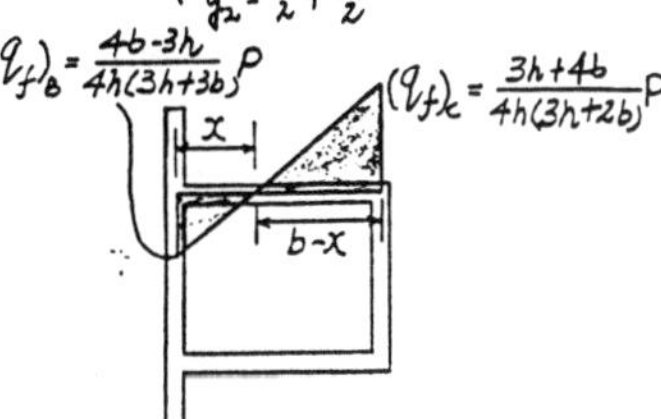

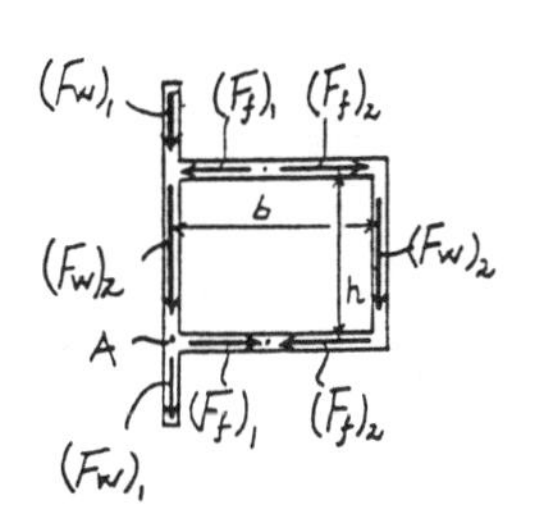

At junction B, q "flows " into lower web from upper web and portion of the top flange. Hence,

$$(q_w)_{B+} = (q_f)_B + (q_w)_{B-}$$

$$\frac{3h+4b}{4h(3h+2b)}P = (q_f)_B + \frac{3}{2(3h+2b)}P$$

$$(q_f)_B = \frac{4b-3h}{4h(3h+2b)}P$$

$$(q_f)_C = (q_w)_{B+} = \frac{3h+4b}{4h(3h+2b)}P$$

Shear Flow Resultant :

For the flange

$$\frac{x}{\frac{4b-3h}{4h(3h+2b)}P} = \frac{b-x}{\frac{3h+4b}{4h(3h+2b)}P}$$

$$x = \frac{4b-3h}{8}$$

$$(F_f)_1 = \frac{1}{2}\left[\frac{4b-3h}{4h(3h+2b)}P\right]\left(\frac{4b-3h}{8}\right) = \frac{(4b-3h)^2}{64h(3h+2b)}P$$

$$(F_f)_2 = \frac{1}{2}\left[\frac{4b+3h}{4h(3h+2b)}P\right]\left(\frac{4b+3h}{8}\right) = \frac{(4b+3h)^2}{64h(3h+2b)}P$$

$$(F_f)_2 - (F_f)_1 = \frac{(4b+3h)^2}{64h(3h+2b)}P - \frac{(4b-3h)^2}{64h(3h+2b)}P$$

$$= \frac{3b}{4(3h+2b)}P$$

For the web

$$(F_w)_2 = \int_{-\frac{h}{2}}^{\frac{h}{2}} q_2\, dy_2$$

$$= \int_{-\frac{h}{2}}^{\frac{h}{2}} \frac{P(5h^2 + 4bh - 8y_2^2)\, dy_2}{4h^2(3h+2b)}$$

$$= \frac{13h+12b}{12(3h+2b)}P$$

Shear Center : Summing moments about point A,

$$Pe = \left[(F_f)_2 - (F_f)_1\right]h + (F_w)_2\, b$$

$$Pe = \left[\frac{3b}{4(3h+2b)}P\right]h + \left[\frac{13h+12b}{12(3h+2b)}P\right]b$$

$$e = \frac{b(22h+12b)}{12(3h+2b)} \qquad \textbf{Ans}$$

7–70. Determine the location e of the shear center, point O, for the thin-walled member having a slit along its side. Each element has a constant thickness t.

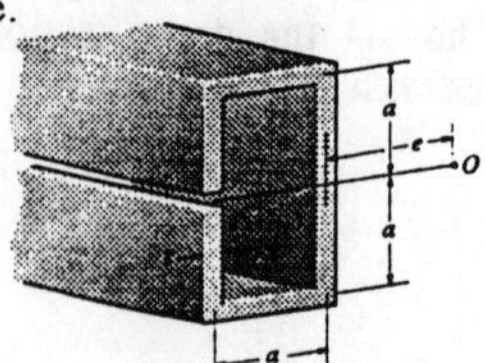

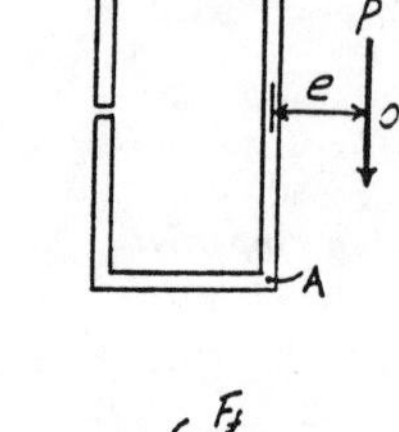

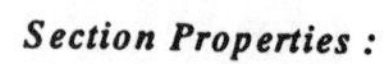

Section Properties :

$$I = \frac{1}{12}(2t)(2a)^3 + 2\left[at\left(a^2\right)\right] = \frac{10}{3}a^3 t$$

$$Q_1 = \bar{y}_1' A' = \frac{y}{2}(yt) = \frac{t}{2}y^2$$

$$Q_2 = \Sigma \bar{y}' A' = \frac{a}{2}(at) + a(xt) = \frac{at}{2}(a+2x)$$

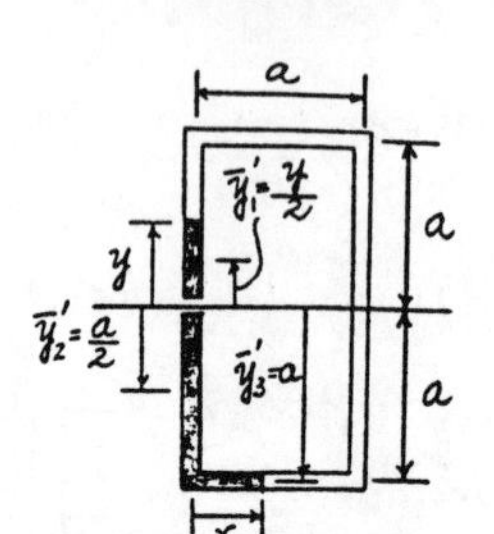

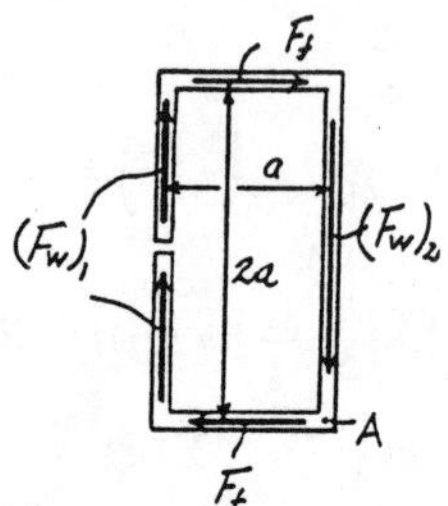

Shear Flow Resultant :

$$q_1 = \frac{VQ_1}{I} = \frac{P\left(\frac{t}{2}y^2\right)}{\frac{10}{3}a^3 t} = \frac{3P}{20a^3}y^2$$

$$q_2 = \frac{VQ_2}{I} = \frac{P\left[\frac{at}{2}(a+2x)\right]}{\frac{10}{3}a^3 t} = \frac{3P}{20a^2}(a+2x)$$

$$(F_w)_1 = \int_0^a q_1\, dy = \frac{3P}{20a^3}\int_0^a y^2\, dy = \frac{P}{20}$$

$$F_f = \int_0^a q_2\, dx = \frac{3P}{20a^2}\int_0^a (a+2x)\, dx = \frac{3}{10}P$$

Shear Center : Summing moments about point A,

$$Pe = 2(F_w)_1\,(a) + F_f(2a)$$

$$Pe = 2\left(\frac{P}{20}\right)a + \left(\frac{3}{10}P\right)2a$$

$$e = \frac{7}{10}a \qquad \textbf{Ans}$$

7–71. Determine the location e of the shear center, point O, for the thin-walled member having the cross section shown.

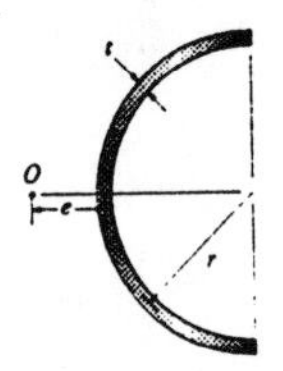

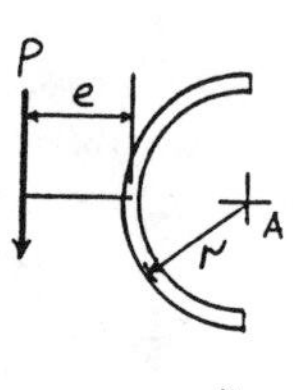

Section Properties :

$$y = r\cos\theta \qquad dA = t\,ds$$

$$dI = y^2 dA = r^2\cos^2\theta(t\,ds) \qquad \text{However;} \qquad ds = r\,d\theta$$

Then.

$$I = r^3 t\int_0^{\pi}\cos^2\theta\, d\theta$$

$$= r^3 t\int_0^{\pi}\left(\frac{\cos 2\theta + 1}{2}\right)d\theta = \frac{\pi r^3 t}{2}$$

$$dQ = \bar{y}' A' = r\cos\theta(t\,r\,d\theta) = r^2 t\cos\theta\, d\theta$$

$$Q = r^2 t\int_0^{\theta}\cos\theta\, d\theta = r^2 t\sin\theta$$

Shear Flow Resultant :

$$q = \frac{VQ}{I} = \frac{p(r^2 t\sin\theta)}{\frac{1}{2}\pi r^3 t} = \frac{2P}{\pi r}\sin 6$$

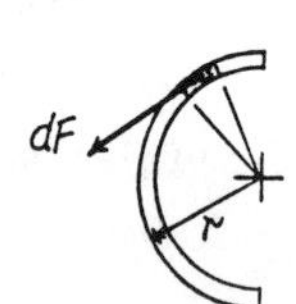

$$F = \int q\,ds = \int qr\,d\theta$$

$$= \int_0^{\pi}\left(\frac{2P}{\pi r}\sin\theta\right)r\,d\theta$$

$$= \frac{2P}{\pi}\int_0^{\pi}\sin\theta\, d\theta$$

$$= \frac{4P}{\pi}$$

Shear Center : Summing moments about point A,

$$P(e + r) = Fr$$

$$P(e + r) = \frac{4Pr}{\pi}$$

$$e = \frac{4r}{\pi} - r = \left(\frac{4-\pi}{\pi}\right)r \qquad \textbf{Ans}$$

***7–72.** Determine the location e of the shear center, point O, for the thin-walled member having the cross-sectional area shown. Two segments of the beam are straight and the other is semicircular. The member segments have the same thickness t.

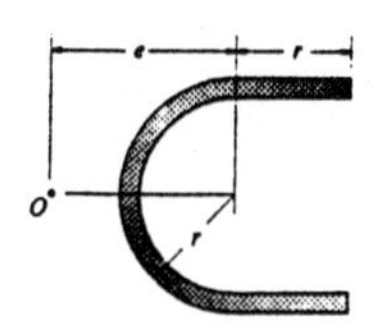

Section Properties :
For the arc

$$y = r\cos\theta \qquad dA = t\,ds$$

$$dI = y^2 dA = r^2\cos^2\theta(t\,ds) \qquad \text{However,} \qquad ds = r\,d\theta$$

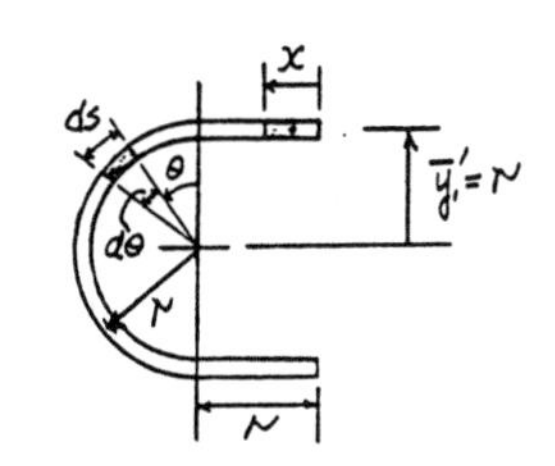

Then,

$$I_{arc} = r^3 t\int_0^{\pi}\cos^2\theta\,d\theta = r^3 t\int_0^{\pi}\left(\frac{\cos 2\theta + 1}{2}\right)d\theta = \frac{1}{2}\pi r^3 t$$

The moment of inertia for the section is

$$I = I_{arc} + 2\left[(rt)\left(r^2\right)\right] = \frac{1}{2}\pi r^3 t + 2r^3 t = \frac{r^3 t}{2}(\pi + 4)$$

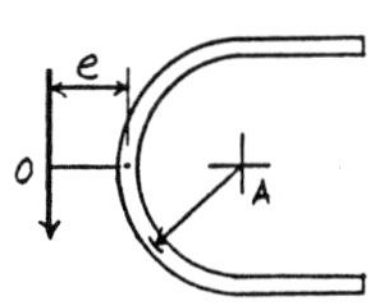

$$Q_1 = \bar{y}_1' A' = r(xt) = rtx$$

$$Q_2 = \Sigma\bar{y}'A' = r(rt) + \int dQ$$

where $\quad dQ = \bar{y}'A' = r\cos\theta(t\,r\,d\theta) = r^2 t\cos\theta\,d\theta$

$$Q_2 = r^2 t + r^2 t\int_0^{\theta}\cos\theta\,d\theta = r^2 t(1 + \sin\theta)$$

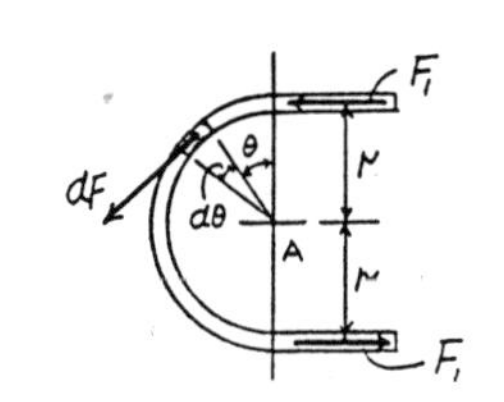

Shear Flow Resultant :

$$q_1 = \frac{VQ_1}{I} = \frac{Prt}{\frac{r^3 t}{2}(\pi + 4)}x = \frac{2P}{r^2(\pi + 4)}x$$

$$q_2 = \frac{VQ_2}{I} = \frac{Pr^2 t(1 + \sin\theta)}{\frac{r^3 t}{2}(\pi + 4)} = \frac{2P(1 + \sin\theta)}{r(\pi + 4)}$$

$$F_1 = \int_0^r q_1\,dx = \frac{2P}{r^2(\pi + 4)}\int_0^r x\,dx = \frac{P}{\pi + 4}$$

$$F = \int q_2\,ds = \int_0^{\pi} q_2\,r\,d\theta = \frac{2P}{\pi + 4}\int_0^{\pi}(1 + \sin\theta)\,d\theta = \frac{2P(\pi + 2)}{\pi + 4}$$

Shear Center : Summing moments about point A,

$$P(e + r) = 2F_1(r) + Fr$$

$$P(e + r) = 2\left(\frac{P}{\pi + 4}\right)r + \left[\frac{2P(\pi + 2)}{\pi + 4}\right]r$$

$$e = \left(\frac{\pi + 2}{\pi + 4}\right)r \qquad \textbf{Ans}$$

7–73. Determine the location e of the shear center, point O, for the tube having a slit along its length.

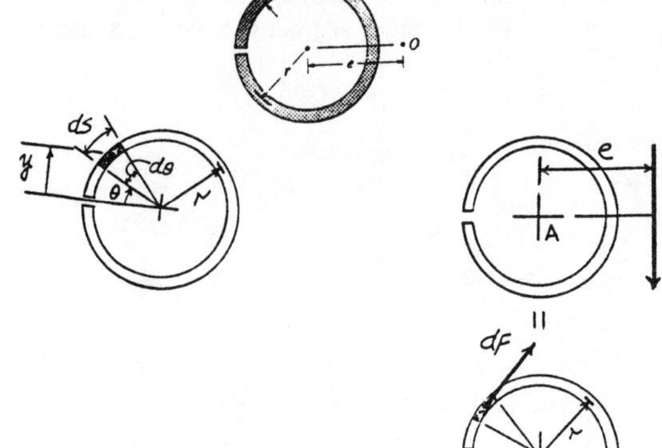

Section Properties :

$$dA = t\,ds = t\,r\,d\theta \qquad y = r\sin\theta$$
$$dI = y^2 dA = r^2\sin^2\theta(t\,r\,d\theta) = r^3 t\sin^2\theta\,d\theta$$

$$I = r^3 t\int_0^{2\pi}\sin^2\theta\,d\theta$$
$$= r^3 t\int_0^{2\pi}\left(\frac{1-\cos 2\theta}{2}\right)d\theta = \pi r^3 t$$

$$dQ = \bar{y}'A' = y\,dA = r\sin\theta(t\,r\,d\theta) = r^2 t\sin\theta\,d\theta$$
$$Q = r^2 t\int_0^{\theta}\sin\theta\,d\theta = r^2 t(1-\cos\theta)$$

Shear Flow Resultant :

$$q = \frac{VQ}{I} = \frac{P\,r^2 t(1-\cos\theta)}{\pi r^3 t} = \frac{P}{\pi r}(1-\cos\theta)$$

$$F = \int_0^{2\pi} q\,ds = \int_0^{2\pi}\frac{P}{\pi r}(1-\cos\theta)\,r\,d\theta$$
$$= \frac{P}{\pi}\int_0^{2\pi}(1-\cos\theta)\,d\theta$$
$$= 2P$$

Shear Center : Summing moments about point A.

$$Pe = Fr$$
$$Pe = 2Pr$$

$$e = 2r \qquad \textbf{Ans}$$

7–74. Determine the location e of the shear center, point O, for the thin-walled tube having the cross section shown. The member segments have the same thickness t.

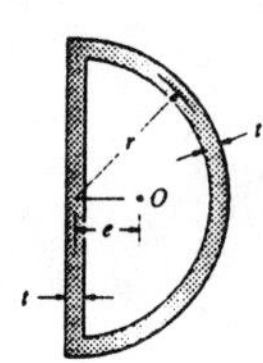

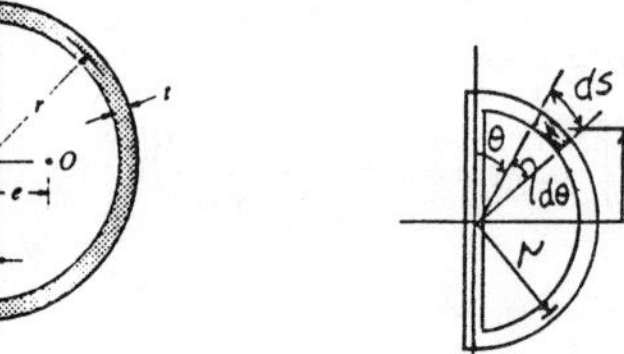

Section Properties :
For the arc

$$y = r\cos\theta \qquad dA = t\,ds$$
$$dI = y^2 dA = r^2\cos^2\theta(t\,ds) \qquad \text{However;} \qquad ds = r\,d\theta$$

Then,

$$I_{arc} = r^3 t\int_0^{\pi}\cos^2\theta\,d\theta = r^3 t\int_0^{\pi}\left(\frac{\cos 2\theta + 1}{2}\right)d\theta = \frac{1}{2}\pi r^3 t$$

The moment of inertia for the section is

$$I = I_{arc} + \frac{1}{12}(t)\left(2r^3\right) = \frac{1}{2}\pi r^3 t + \frac{2}{3}r^3 t = \frac{r^3 t}{6}(3\pi + 4)$$

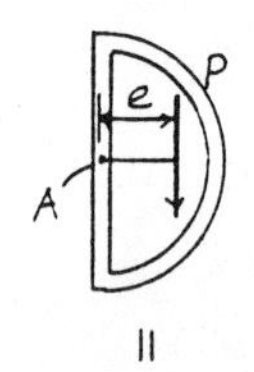

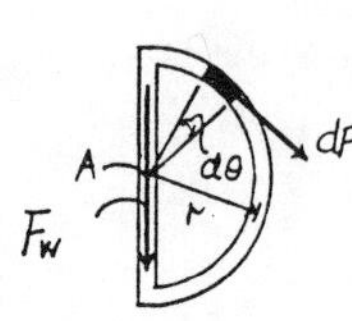

$$dQ = \bar{y}'A' = r\cos\theta(t\,r\,d\theta) = r^2 t\cos\theta\,d\theta$$
$$Q = r^2 t\int_0^{\theta}\cos\theta\,d\theta = r^2 t\sin\theta$$

Shear Flow Resultant :

$$q = \frac{VQ}{I} = \frac{Pr^2 t\sin\theta}{\frac{r^3 t}{6}(3\pi+4)} = \frac{6P}{r(3\pi+4)}\sin\theta$$

$$F = \int q\,ds = \int_0^{\pi} q\,r\,d\theta = \frac{6P}{3\pi+4}\int_0^{\pi}\sin\theta\,d\theta = \frac{12P}{3\pi+4}$$

Shear Center : Summing moments about point A.

$$Pe = Fr$$
$$Pe = \left(\frac{12P}{3\pi+4}\right)r$$
$$e = \left(\frac{12}{3\pi+4}\right)r \qquad \textbf{Ans}$$

7–75. The beam is fabricated from four boards nailed together as shown. Determine the shear force each nail along the sides C and the top D must resist if the nails are uniformly spaced at $s = 3$ in. The beam is subjected to a shear of $V = 4.5$ kip.

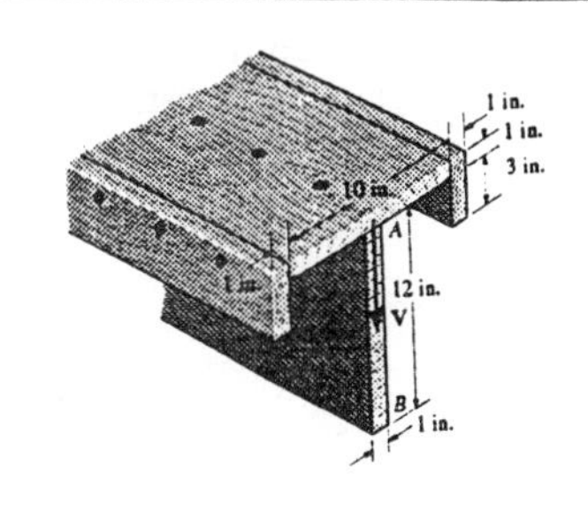

Section Properties :

$$\bar{y} = \frac{\Sigma \tilde{y} A}{\Sigma A} = \frac{0.5(10)(1) + 2(4)(2) + 7(12)(1)}{10(1) + 4(2) + 12(1)} = 3.50 \text{ in.}$$

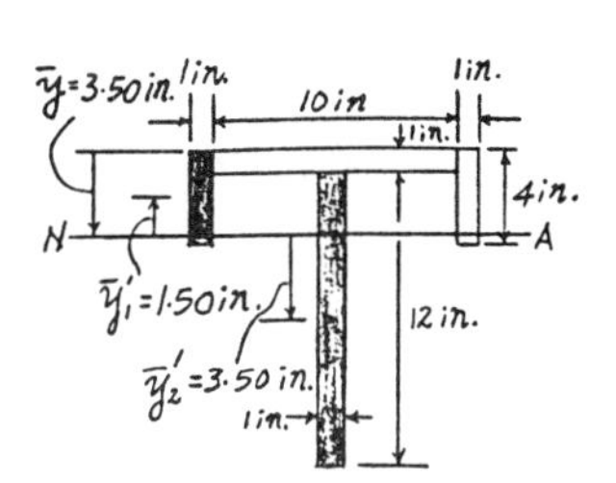

$$I_{NA} = \frac{1}{12}(10)(1^3) + (10)(1)(3.50-0.5)^2 + \frac{1}{12}(2)(4^3) + 2(4)(3.50-2)^2 + \frac{1}{12}(1)(12^3) + 1(12)(7-3.50)^2$$

$$= 410.5 \text{ in}^4$$

$$Q_C = \bar{y}'_1 A' = 1.5(4)(1) = 6.00 \text{ in}^3$$

$$Q_D = \bar{y}'_2 A' = 3.50(12)(1) = 42.0 \text{ in}^3$$

Shear Flow :

$$q_C = \frac{VQ_C}{I} = \frac{4.5(10^3)(6.00)}{410.5} = 65.773 \text{ lb/in.}$$

$$q_D = \frac{VQ_D}{I} = \frac{4.5(10^3)(42.0)}{410.5} = 460.41 \text{ lb/in.}$$

Hence, the shear force resisted by each nail is

$$F_C = q_C s = (65.773 \text{ lb/in.})(3 \text{ in.}) = 197 \text{ lb} \qquad \textbf{Ans}$$

$$F_D = q_D s = (460.41 \text{ lb/in.})(3 \text{ in.}) = 1.38 \text{ kip} \qquad \textbf{Ans}$$

***7–76.** The beam is fabricated from four boards nailed together as shown. Determine the maximum shear stress developed in the beam's web AB. The beam is subjected to a shear of $V = 4.5$ kip.

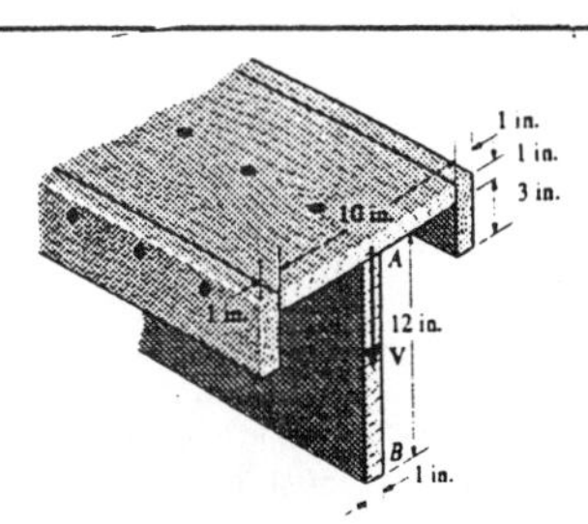

Section Properties :

$$\bar{y} = \frac{\Sigma \tilde{y} A}{\Sigma A} = \frac{0.5(10)(1) + 2(4)(2) + 7(12)(1)}{10(1) + 4(2) + 12(1)} = 3.50 \text{ in.}$$

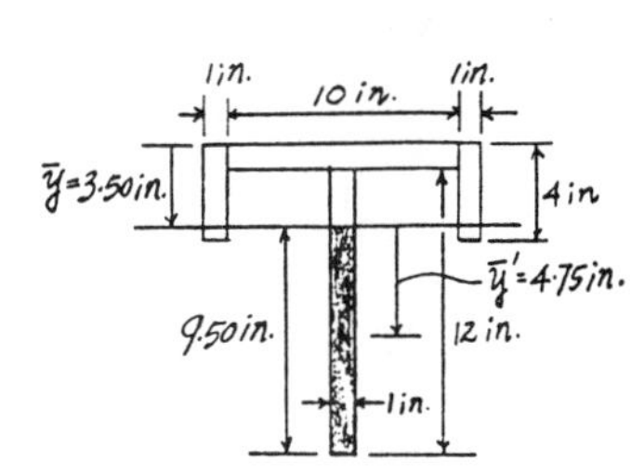

$$I_{NA} = \frac{1}{12}(10)(1^3) + (10)(1)(3.50-0.5)^2 + \frac{1}{12}(2)(4^3) + 2(4)(3.50-2)^2 + \frac{1}{12}(1)(12^3) + 1(12)(7-3.50)^2$$

$$= 410.5 \text{ in}^4$$

$$Q_{max} = \bar{y}'A' = 4.75(9.50)(1) = 45.125 \text{ in}^3$$

Maximum Shear Stress : Maximum shear stress occurs at the point where the neutral axis passes through the section. Applying the shear formula

$$\tau_{max} = \frac{VQ_{max}}{It} = \frac{4.5(10^3)(45.125)}{410.5(1)} = 495 \text{ psi} \qquad \textbf{Ans}$$

7–77. The beam is constructed from four boards glued together at their seams. If the glue can withstand 75 lb/in., what is the maximum vertical shear V that the beam can support? .

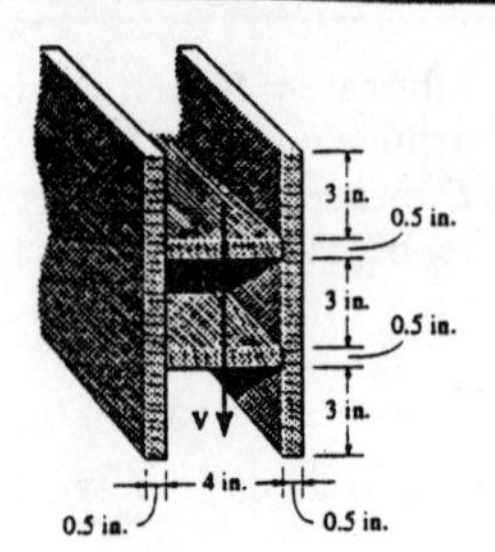

Section Properties :

$$I_{NA} = \frac{1}{12}(1)(10^3) + 2\left[\frac{1}{12}(4)(0.5^3) + 4(0.5)(1.75^2)\right]$$
$$= 95.667 \text{ in}^4$$

$$Q = \bar{y}'A' = 1.75(4)(0.5) = 3.50 \text{ in}^3$$

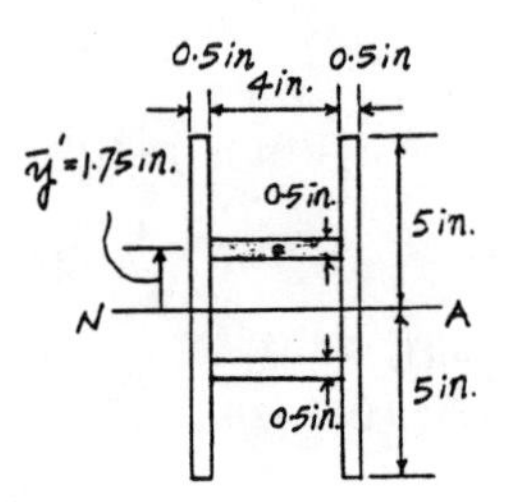

Shear Flow : There are two glue joints in this case, hence the allowable shear flow is 2(75) = 150 lb/in.

$$q = \frac{VQ}{I}$$

$$150 = \frac{V(3.50)}{95.667}$$

$$V = 4100 \text{ lb} = 4.10 \text{ kip} \qquad \textbf{Ans}$$

7–78. Solve Prob. 7–77 if the beam is rotated 90° from the position shown.

Section Properties :

$$I_{NA} = \frac{1}{12}(10)(5^3) - \frac{1}{12}(9)(4^3) = 56.167 \text{ in}^4$$

$$Q = \bar{y}'A' = 2.25(10)(0.5) = 11.25 \text{ in}^3$$

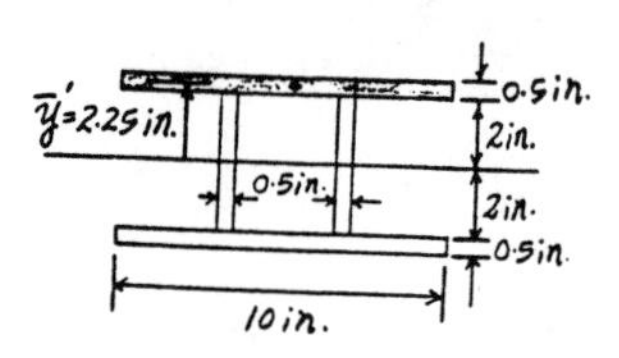

Shear Flow : There are two glue joints in this case, hence the allowable shear flow is 2(75) = 150 lb/in.

$$q = \frac{VQ}{I}$$

$$150 = \frac{V(11.25)}{56.167}$$

$$V = 749 \text{ lb} \qquad \textbf{Ans}$$

7–79. The member is subjected to a shear force of $V = 2$ kN. Determine the shear flow at points A, B, and C. The thickness of each thin-walled segment is 15 mm.

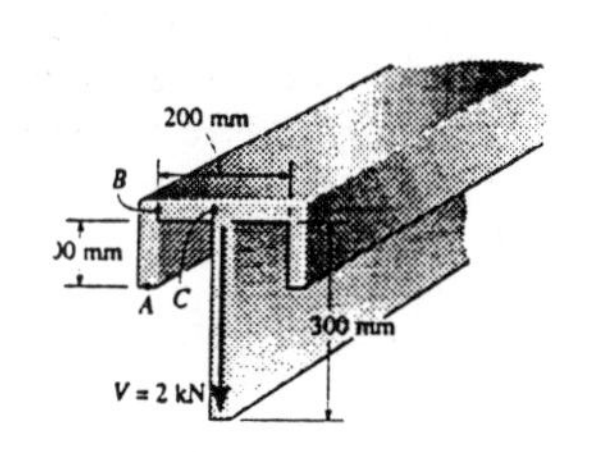

Section Properties :

$$\bar{y} = \frac{\Sigma \tilde{y} A}{\Sigma A}$$

$$= \frac{0.0075(0.2)(0.015) + 0.0575(0.115)(0.03) + 0.165(0.3)(0.015)}{0.2(0.015) + 0.115(0.03) + 0.3(0.015)}$$

$$= 0.08798 \text{ m}$$

$$I_{NA} = \frac{1}{12}(0.2)\left(0.015^3\right) + 0.2(0.015)(0.08798 - 0.0075)^2$$

$$+ \frac{1}{12}(0.03)\left(0.115^3\right) + 0.03(0.115)(0.08798 - 0.0575)^2$$

$$+ \frac{1}{12}(0.015)\left(0.3^3\right) + 0.015(0.3)(0.165 - 0.08798)^2$$

$$= 86.93913\left(10^{-6}\right) \text{ m}^4$$

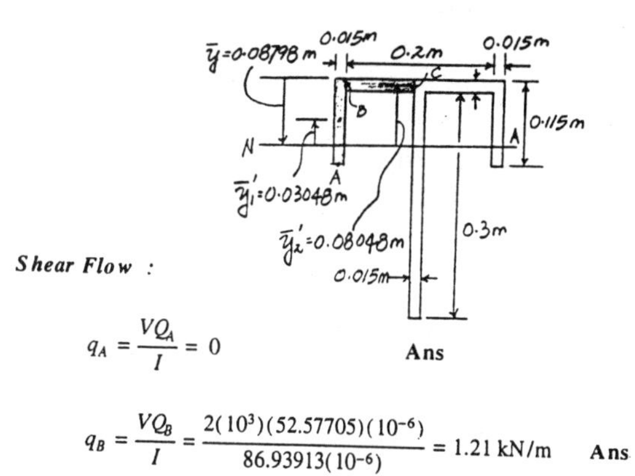

$Q_A = 0$

$Q_B = \bar{y}'_1 A' = 0.03048(0.115)(0.015) = 52.57705\left(10^{-6}\right) \text{ m}^3$

$Q_C = \Sigma \bar{y}' A'$

$= 0.03048(0.115)(0.015) + 0.08048(0.0925)(0.015)$

$= 0.16424\left(10^{-3}\right) \text{ m}^3$

Shear Flow :

$$q_A = \frac{VQ_A}{I} = 0 \qquad \textbf{Ans}$$

$$q_B = \frac{VQ_B}{I} = \frac{2(10^3)(52.57705)(10^{-6})}{86.93913(10^{-6})} = 1.21 \text{ kN/m} \qquad \textbf{Ans}$$

$$q_C = \frac{VQ_C}{I} = \frac{2(10^3)(0.16424)(10^{-3})}{86.93913(10^{-6})} = 3.78 \text{ kN/m} \qquad \textbf{Ans}$$

***7–80.** The member is subjected to a shear force of $V = 2$ kN. Determine the maximum shear flow in the member. All segments of the cross section are 15 mm thick.

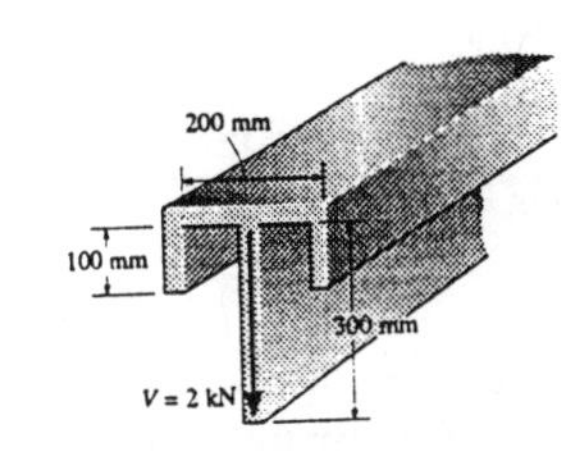

Section Properties :

$$\bar{y} = \frac{\Sigma \tilde{y} A}{\Sigma A}$$

$$= \frac{0.0075(0.2)(0.015) + 0.0575(0.115)(0.03) + 0.165(0.3)(0.015)}{0.2(0.015) + 0.115(0.03) + 0.3(0.015)}$$

$$= 0.08798 \text{ m}$$

$$I_{NA} = \frac{1}{12}(0.2)\left(0.015^3\right) + 0.2(0.015)(0.08798 - 0.0075)^2$$

$$+ \frac{1}{12}(0.03)\left(0.115^3\right) + 0.03(0.115)(0.08798 - 0.0575)^2$$

$$+ \frac{1}{12}(0.015)\left(0.3^3\right) + 0.015(0.3)(0.165 - 0.08798)^2$$

$$= 86.93913\left(10^{-6}\right) \text{ m}^4$$

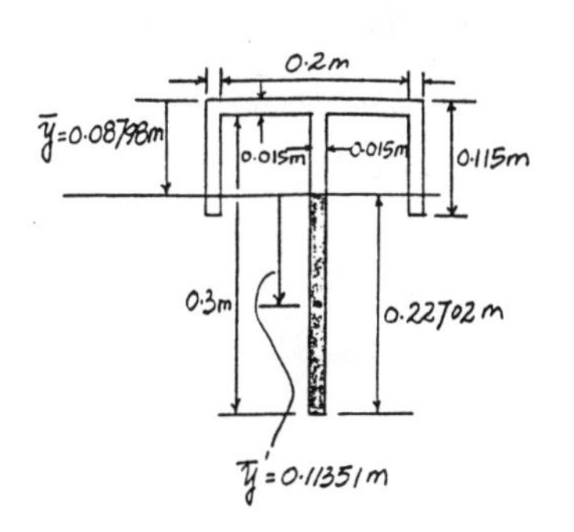

$$Q_{max} = \bar{y}' A' = 0.11351(0.22702)(0.015) = 0.38654\left(10^{-6}\right) \text{ m}^3$$

Maximum Shear Flow : Maximum shear flow occurs at the point where the neutral axis passes through the section.

$$q_{max} = \frac{VQ_{max}}{I} = \frac{2(10^3)0.38654(10^{-6})}{86.93913(10^{-6})} = 8.89 \text{ kN/m} \qquad \textbf{Ans}$$

7–81. Sketch the intensity of the shear-stress distribution acting over the beam's cross-sectional area, and determine the resultant shear force acting on the segment AB. The shear acting at the section is $V = 35$ kip. Show that $I_{NA} = 872.49$ in^4.

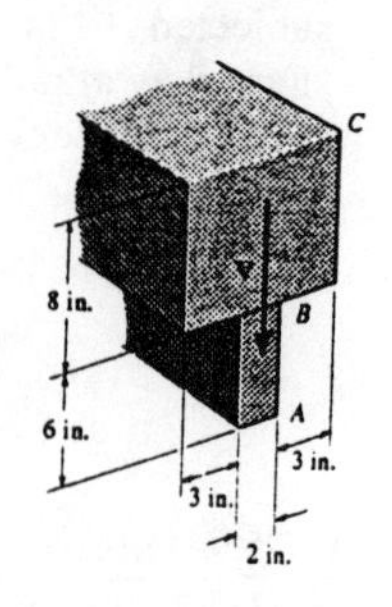

Section Properties :

$$\bar{y} = \frac{\Sigma \tilde{y}A}{\Sigma A} = \frac{4(8)(8) + 11(6)(2)}{8(8) + 6(2)} = 5.1053 \text{ in.}$$

$$I_{NA} = \frac{1}{12}(8)\left(8^3\right) + 8(8)(5.1053 - 4)^2 + \frac{1}{12}(2)\left(6^3\right) + 2(6)(11 - 5.1053)^2$$
$$= 872.49 \text{ in}^4 \qquad (Q.E.D)$$

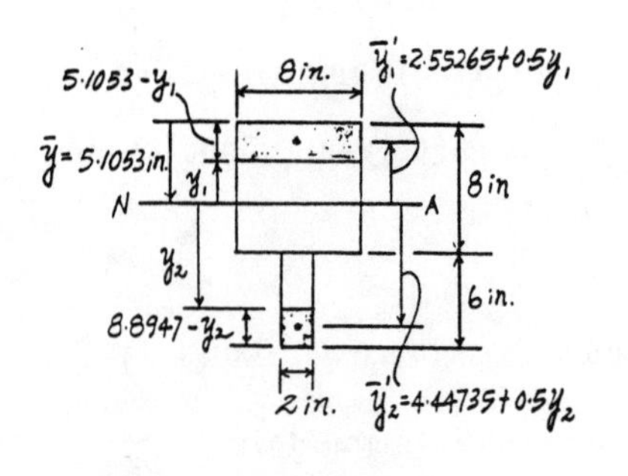

$$Q_1 = \bar{y}_1' A' = (2.55265 + 0.5y_1)(5.1053 - y_1)(8)$$
$$= 104.25 - 4y_1^2$$

$$Q_2 = \bar{y}_2' A' = (4.44735 + 0.5y_2)(8.8947 - y_2)(2)$$
$$= 79.12 - y_2^2$$

Shear Stress : Applying the shear formula $\tau = \frac{VQ}{It}$,

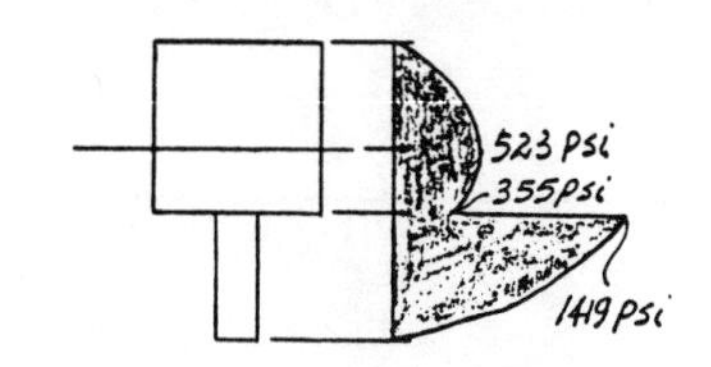

$$\tau_{CB} = \frac{VQ_1}{It} = \frac{35(10^3)(104.25 - 4y_1^2)}{872.49(8)}$$
$$= \{522.77 - 20.06y_1^2\} \text{ psi}$$

At $y_1 = 0$, $\tau_{CB} = 523$ psi

At $y_1 = -2.8947$ in. $\tau_{CB} = 355$ psi

$$\tau_{AB} = \frac{VQ_2}{It} = \frac{35(10^3)(79.12 - y_2^2)}{872.49(2)}$$
$$= \{1586.88 - 20.06y_2^2\} \text{ psi}$$

At $y_2 = 2.8947$ in. $\tau_{AB} = 1419$ psi

Resultant Shear Force : For segment AB.

$$V_{AB} = \int \tau_{AB}\, dA$$
$$= \int_{2.8947\text{in}}^{8.8947\text{in}} \left(1586.88 - 20.06y_2^2\right)(2dy)$$
$$= \int_{2.8947\text{in}}^{8.8947\text{in}} \left(3173.76 - 40.12y_2^2\right) dy$$
$$= 9957 \text{ lb} = 9.96 \text{ kip} \qquad \textbf{Ans}$$

7–82. A steel plate having a thickness of 0.25 in. is formed into the thin-walled section shown. If it is subjected to a shear force of $V = 250$ lb, determine the shear stress at points A and C. Indicate the results on volume elements located at these points.

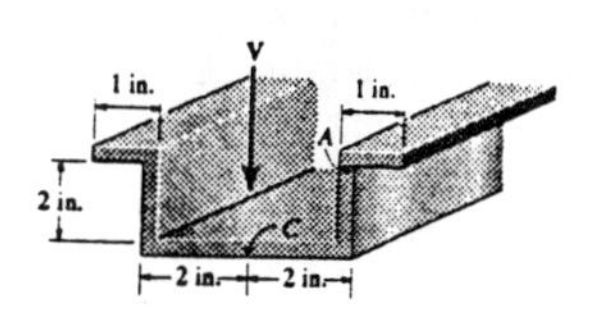

Section Properties :

$$\bar{y} = \frac{\Sigma \tilde{y} A}{\Sigma A}$$
$$= \frac{0.125(4)(0.25) + 1.25(2)(0.5) + 2.375(2)(0.25)}{4(0.25) + 2(0.5) + 2(0.25)}$$
$$= 1.025 \text{ in.}$$

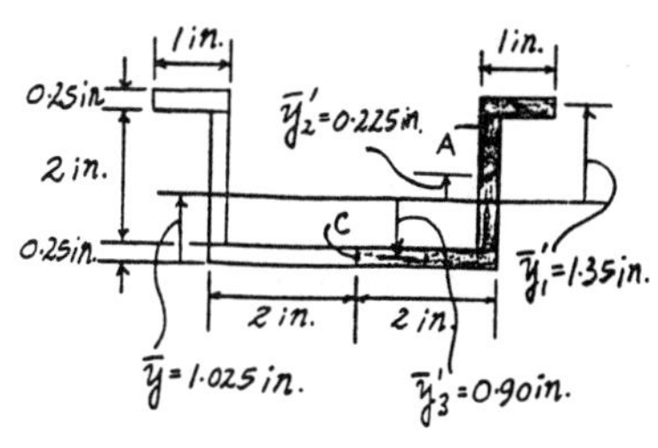

$$I_{NA} = \frac{1}{12}(4)\left(0.25^3\right) + 4(0.25)(1.025 - 0.125)^2$$
$$+ \frac{1}{12}(0.5)\left(2^3\right) + 0.5(2)(1.25 - 1.025)^2$$
$$+ \frac{1}{12}(2)\left(0.25^3\right) + 2(0.25)(2.375 - 1.025)^2$$
$$= 2.1130 \text{ in}^4$$

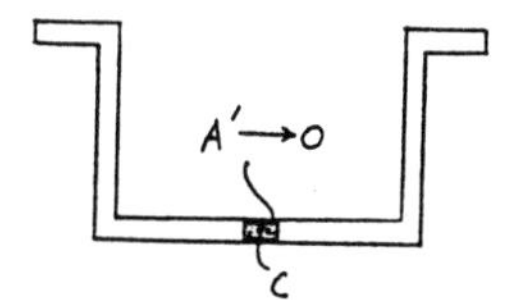

$$Q_A = \bar{y}'_1 A' = 1.35(1)(0.25) = 0.3375 \text{ in}^3$$

$$Q_C = \Sigma \bar{y}' A'$$
$$= 1.35(1)(0.25) + 0.225(2)(0.25) - 0.9(2)(0.25) = 0$$

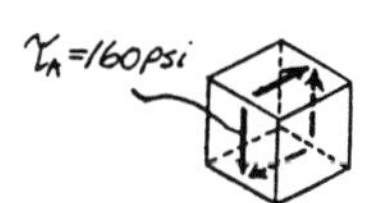

Or since $A' \rightarrow 0$, $Q_C = 0$

Shear Stresses : Applying the shear formula

$$\tau_A = \frac{VQ_A}{It} = \frac{250(0.3375)}{2.1130(0.25)} = 160 \text{ psi} \qquad \textbf{Ans}$$

$$\tau_C = \frac{VQ_C}{It} = 0 \qquad \textbf{Ans}$$

7–83. A steel plate having a thickness of 0.25 in. is formed into the thin-walled section shown. If it is subjected to a shear force of $V = 250$ lb., determine the shear stress at point B.

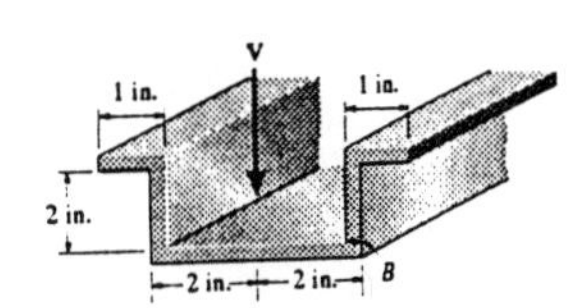

Section Properties :

$$\bar{y} = \frac{\Sigma \tilde{y} A}{\Sigma A}$$
$$= \frac{0.125(4)(0.25) + 1.25(2)(0.5) + 2.375(2)(0.25)}{4(0.25) + 2(0.5) + 2(0.25)}$$
$$= 1.025 \text{ in.}$$

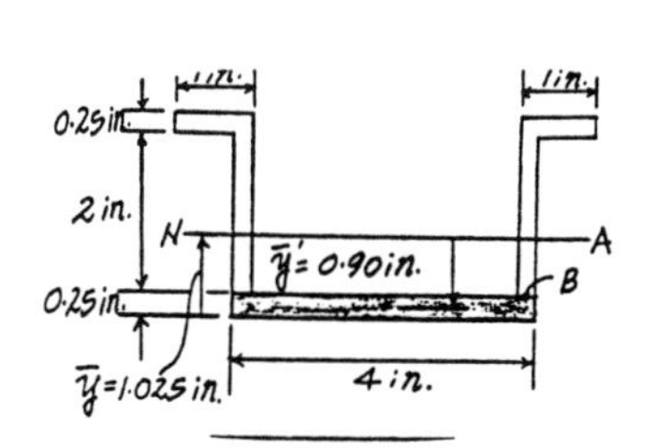

$$I_{NA} = \frac{1}{12}(4)\left(0.25^3\right) + 4(0.25)(1.025 - 0.125)^2$$
$$+ \frac{1}{12}(0.5)\left(2^3\right) + 0.5(2)(1.25 - 1.025)^2$$
$$+ \frac{1}{12}(2)\left(0.25^3\right) + 2(0.25)(2.375 - 1.025)^2$$
$$= 2.1130 \text{ in}^4$$

$$Q_B = \bar{y}' A' = 0.900(4)(0.25) = 0.900 \text{ in}^3$$

Shear Stress : Applying the shear formula

$$\tau_B = \frac{VQ_B}{It} = \frac{250(0.900)}{2.1130(0.5)} = 213 \text{ psi} \qquad \textbf{Ans}$$

***7–84.** A steel plate having a thickness of 0.25 in. is formed into the thin-walled section shown. If it is subjected to a shear force of $V = 250$ lb., determine the maximum shear stress in the plate.

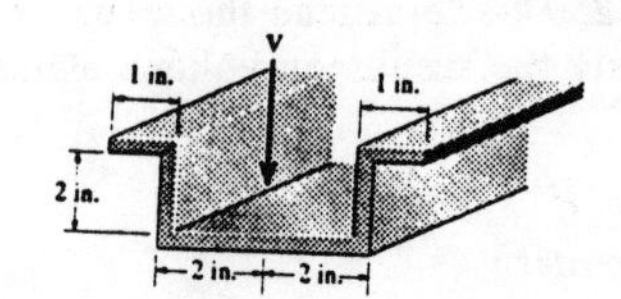

Section Properties :

$$\bar{y} = \frac{\Sigma \tilde{y}A}{\Sigma A}$$

$$= \frac{0.125(4)(0.25) + 1.25(2)(0.5) + 2.375(2)(0.25)}{4(0.25) + 2(0.5) + 2(0.25)}$$

$$= 1.025 \text{ in.}$$

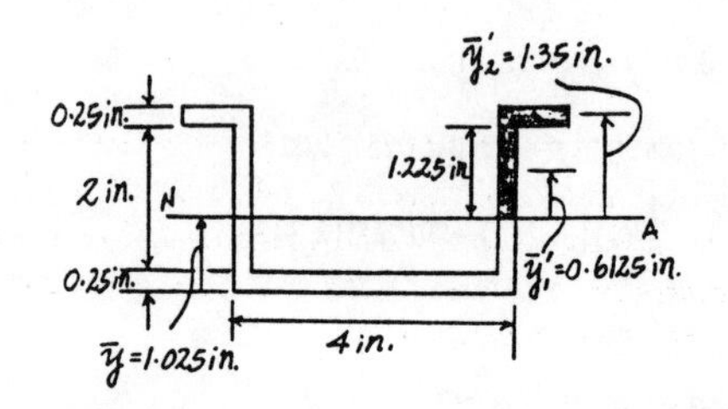

$$I_{NA} = \frac{1}{12}(4)\left(0.25^3\right) + 4(0.25)(1.025 - 0.125)^2$$

$$+ \frac{1}{12}(0.5)\left(2^3\right) + 0.5(2)(1.25 - 1.025)^2$$

$$+ \frac{1}{12}(2)\left(0.25^3\right) + 2(0.25)(2.375 - 1.025)^2$$

$$= 2.1130 \text{ in}^4$$

$$Q_{max} = \Sigma \bar{y}'A'$$

$$= 0.6125(1.225)(0.25) + 1.35(1)(0.25) = 0.52508 \text{ in}^3$$

Maximum Shear Stress : Maximum shear stress occurs at the point where the neutral axis passes through the section. Applying the shear formula

$$\tau_{max} = \frac{VQ_{max}}{It} = \frac{250(0.52508)}{2.1130(0.25)} = 248 \text{ psi} \qquad \textbf{Ans}$$

8–1. A spherical gas tank has an inner radius of r = 1.5 m. If it is subjected to an internal pressure of p = 300 kPa, determine its required thickness if the maximum normal stress is not to exceed 12 MPa.

Spherical Vessel : Applying Eq. 8 – 3

$$\sigma_2 = \sigma_{allow} = \frac{pr}{2t}$$

$$12(10^6) = \frac{300(10^3)(1.5)}{2t}$$

$$t = 0.01875 \text{ m} = 18.75 \text{ mm} \qquad \textbf{Ans}$$

Since $\frac{r}{t} = \frac{1500}{18.75} = 80 > 10$, the *thin wall* analysis is valid.

8–2. A pressurized spherical tank is to be made of 0.5-in-thick steel. If it is subjected to an internal pressure of p = 200 psi, determine its outer radius if the maximum normal stress is not to exceed 15 ksi.

Spherical Vessel : Applying Eq. 8 – 3

$$\sigma_2 = \sigma_{allow} = \frac{pr}{2t}$$

$$15(10^3) = \frac{200r}{2(0.5)}$$

$$r = 75.0 \text{ in.}$$

Since $\frac{r}{t} = \frac{75}{0.5} = 150 > 10$, the *thin wall* analysis is still valid.

$$r_o = r + t = 75.0 + 0.5 = 75.5 \text{ in.} \qquad \textbf{Ans}$$

8–3. The tank of the air compressor is subjected to an internal pressure of 90 psi. If the internal diameter of the tank is 22 in., and the wall thickness is 0.25 in., determine the stress components acting at point A. Draw a volume element of the material at this point, and show the results on the element.

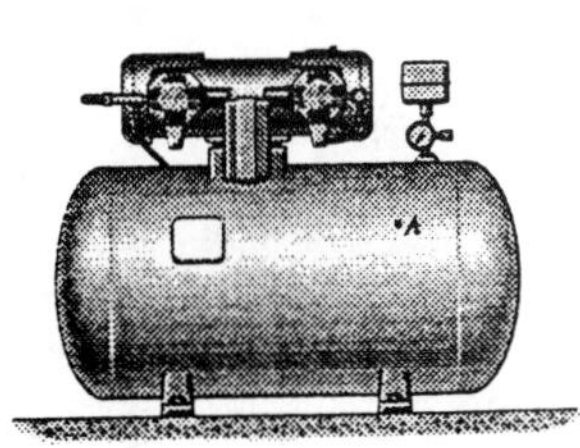

Hoop Stress for Cylindrical Vessels : Since $\frac{r}{t} = \frac{11}{0.25} = 44 > 10$, then *thin wall* analysis can be used. Applying Eq. 8 – 1

$$\sigma_1 = \frac{pr}{t} = \frac{90(11)}{0.25} = 3960 \text{ psi} = 3.96 \text{ ksi} \qquad \textbf{Ans}$$

Longitudinal Stress for Cylindrical Vessels : Applying Eq. 8 – 2

$$\sigma_2 = \frac{pr}{2t} = \frac{90(11)}{2(0.25)} = 1980 \text{ psi} = 1.98 \text{ ksi} \qquad \textbf{Ans}$$

$\sigma_1 = 3.96$ ksi

$\sigma_2 = 1.98$ ksi

***8–4.** The open-ended pipe has a wall thickness of 2 mm and an internal diameter of 40 mm. Calculate the pressure that ice exerted on the interior wall of the pipe to cause it to burst in the manner shown. The maximum stress that the material can support at freezing temperatures is $\sigma_{max} = 360$ MPa. Show the stress acting on a small element of material just before the pipe fails.

Hoop Stress for Cylindrical Pipe : Since $\frac{r}{t} = \frac{20}{2} = 10$, then *thin wall* analysis can be used. Applying Eq. 8 – 1

$$\sigma_1 = \sigma_{allow} = \frac{pr}{t}$$

$$360(10^6) = \frac{p(0.02)}{0.002}$$

$$p = 36.0 \text{ MPa} \qquad \textbf{Ans}$$

$\sigma_1 = 360$ MPa

$\sigma_2 = 0$

Longitudinal Stress for Cylindrical Pipe : Since the pipe is open at both ends, then

$$\sigma_2 = 0$$

8–5. Two hemispheres having an inner radius of 2 ft and wall thickness of 0.25 in. are fitted together, and the inside gauge pressure is reduced to −10 psi. If the coefficient of static friction is $\mu_s = 0.5$ between the hemispheres, determine (a) the torque T needed to initiate the rotation of the top hemisphere relative to the bottom one, (b) the vertical force needed to pull the top hemisphere off the bottom one, and (c) the horizontal force needed to slide the top hemisphere off the bottom one.

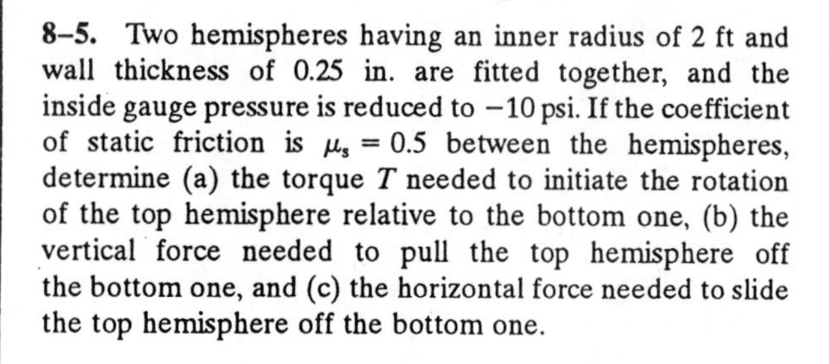

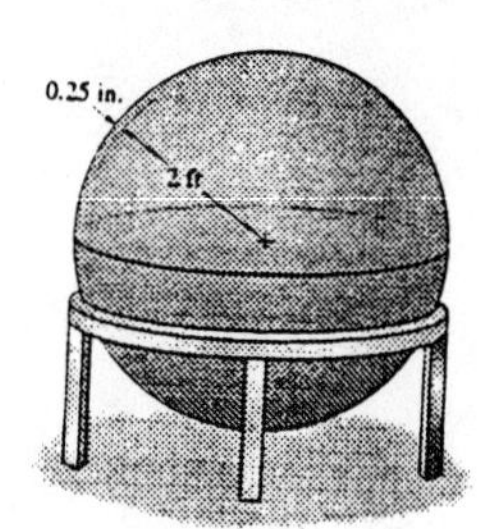

Normal Pressure : Vertical force equilibrium for FBD(a).

$$+\uparrow \Sigma F_y = 0; \quad 10\left[\pi(24^2)\right] - N = 0 \quad N = 5760\pi \text{ lb}$$

The Friction Force : Applying friction formula

$$F_f = \mu_s N = 0.5(5760\pi) = 2880\pi \text{ lb}$$

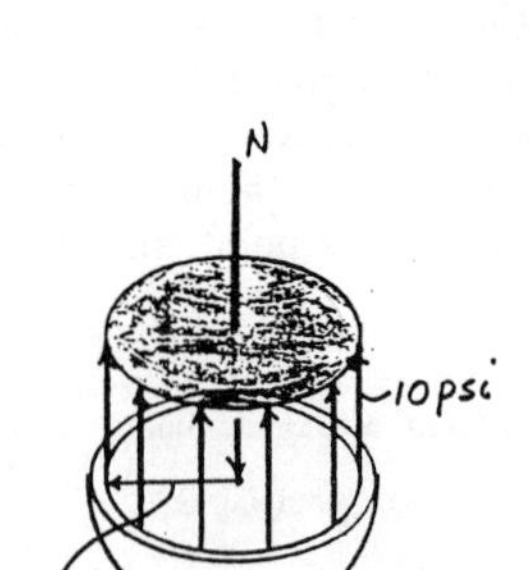

a) ***The Required Torque :*** In order to initiate rotation of the two hemispheres relative to each other, the torque must overcome the moment produced by the friction force about the center of the sphere.

$$T = F_f r = 2880\pi(2 + 0.125/12) = 18\,190 \text{ lb}\cdot\text{ft} = 18.2 \text{ kip}\cdot\text{ft} \qquad \textbf{Ans}$$

b) ***The Required Vertical Force :*** In order to just pull the two hemispheres apart, the vertical force P must overcome the normal force.

$$P = N = 5760\pi = 18096 \text{ lb} = 18.1 \text{ kip} \qquad \textbf{Ans}$$

c) ***The Required Horizontal Force :*** In order to just cause the two hemispheres to slide relative to each other, the horizontal force F must overcome the friction force.

$$F = F_f = 2880\pi = 9048 \text{ lb} = 9.05 \text{ kip} \qquad \textbf{Ans}$$

8–6. The 304 stainless steel band initially fits snugly around the smooth rigid cylinder. If the band is then subjected to a nonlinear temperature drop of $\Delta T = 20\sin^2\theta$ °F, where θ is in radians, determine the circumferential stress in the band.

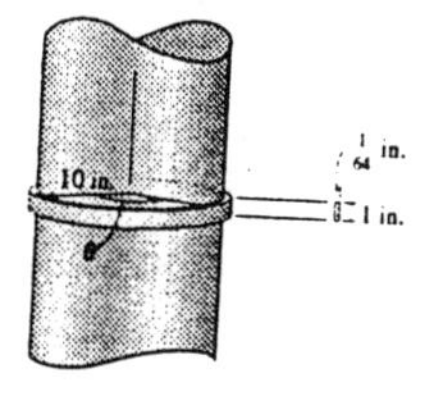

Compatibility : Since the band is fitted to a rigid cylinder (it does not deform under load), then

$$\delta_F - \delta_T = 0$$

$$\frac{P(2\pi r)}{AE} - \int_0^{2\pi} \alpha \Delta T r d\theta = 0$$

$$\frac{2\pi r}{E}\left(\frac{P}{A}\right) = 20\alpha r \int_0^{2\pi} \sin^2\theta d\theta \qquad \text{however, } \frac{P}{A} = \sigma_c$$

$$\frac{2\pi}{E}\sigma_c = 10\alpha \int_0^{2\pi} (1 - \cos 2\theta)\, d\theta$$

$$\sigma_c = 10\alpha E$$

$$= 10(9.60)\left(10^{-6}\right)28.0\left(10^3\right) = 2.69 \text{ ksi} \qquad \textbf{Ans}$$

8–7. The open-ended polyvinyl chloride pipe has an inner diameter of 4 in. and thickness of 0.2 in. If it carries flowing water at 60 psi pressure, determine the state of stress in the walls of the pipe. Neglect the weight of the water.

Hoop Stress for Cylindrical Pipe : Since $\frac{r}{t} = \frac{2}{0.2} = 10$, then *thin wall* analysis can be used. Applying Eq. 8 – 1

$$\sigma_1 = \frac{pr}{t} = \frac{60(2)}{0.2} = 600 \text{ psi} \qquad \textbf{Ans}$$

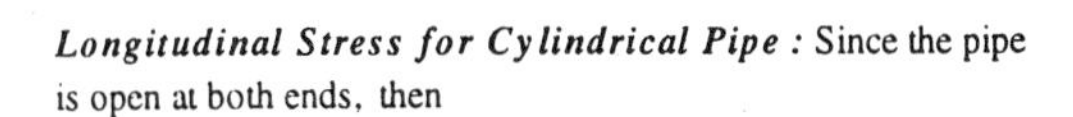

Longitudinal Stress for Cylindrical Pipe : Since the pipe is open at both ends, then

$$\sigma_2 = 0 \qquad \textbf{Ans}$$

***8–8.** The polyvinyl chloride pipe has an inner diameter of 4 in. and thickness of 0.2 in. If the flow of water within the pipe is stopped due to the closing of a valve and the water pressure is 60 psi, determine the state of stress in the walls of the pipe. Neglect the weight of the water.

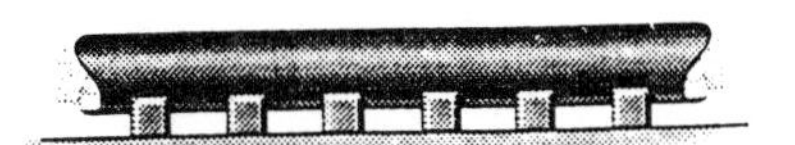

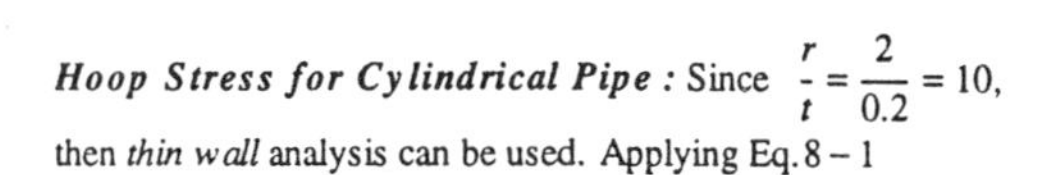

Hoop Stress for Cylindrical Pipe : Since $\frac{r}{t} = \frac{2}{0.2} = 10$, then *thin wall* analysis can be used. Applying Eq. 8 – 1

$$\sigma_1 = \frac{pr}{t} = \frac{60(2)}{0.2} = 600 \text{ psi} \qquad \textbf{Ans}$$

Longitudinal Stress for Cylindrical Pipe : Since the valve for the pipe is closed, then the water pressure exerted on the valve will in turn induce longitudinal stress in the pipe. Applying Eq. 8 – 2

$$\sigma_2 = \frac{pr}{2t} = \frac{60(2)}{2(0.2)} = 300 \text{ psi} \qquad \textbf{Ans}$$

8–9. A wood pipe having an inner diameter of 3 ft is bound together using steel hoops having a cross-sectional area of 0.2 in^2. If the allowable stress for the hoops is $\sigma_{allow} = 12$ ksi, determine their maximum spacing s along the section of pipe so that the pipe can resist an internal gauge pressure of 4 psi. Assume each hoop supports the pressure loading acting along the length s of the pipe.

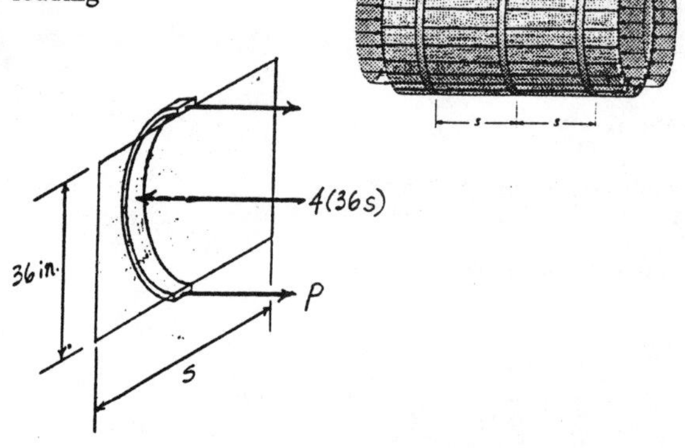

Equilibrium for the Steel Hoop : From the FBD

$$\xrightarrow{+} \Sigma F_x = 0; \quad 2P - 4(36s) = 0 \quad P = 72.0s$$

Hoop Stress for the Steel Hoop :

$$\sigma_1 = \sigma_{allow} = \frac{P}{A}$$

$$12(10^3) = \frac{72.0s}{0.2}$$

$$s = 33.3 \text{ in.} \qquad \textbf{Ans}$$

8–10. The barrel is filled to the top with water. Determine the distance s that the top hoop should be placed from the bottom hoop so that the tensile force in each hoop is the same. Also, what is the force in each hoop? The barrel has an inner diameter of 4 ft. Neglect its wall thickness. Assume that only the hoops resist the water pressure. *Note:* Water develops pressure in the barrel according to Pascal's law, $p = (62.4z)$ lb/ft^2, where z is the depth from the surface of the water in feet.

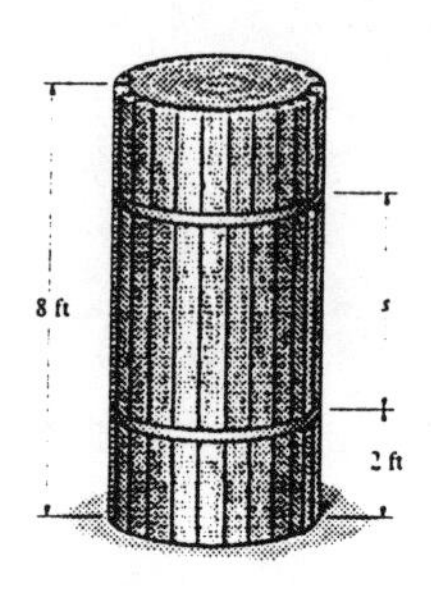

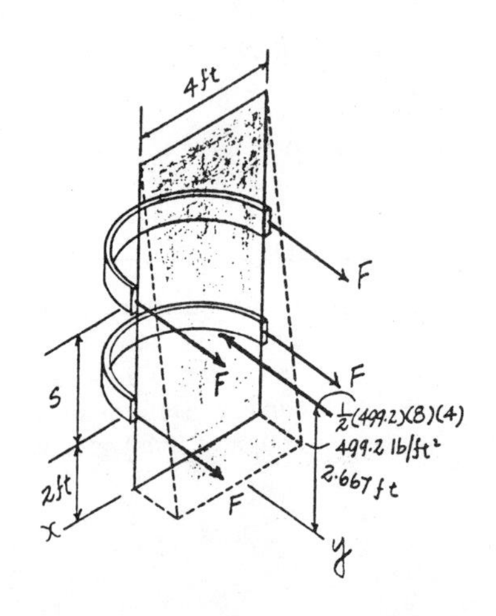

Equilibrium for the Steel Hoop : From the FBD

$$\Sigma F_y = 0; \quad 4F - \frac{1}{2}(499.2)(8)(4) = 0$$

$$F = 1996.8 \text{ lb} = 2.00 \text{ kip} \qquad \textbf{Ans}$$

$$\Sigma M_x = 0; \quad \left[\frac{1}{2}(499.2)(8)(4)\right]2.667$$

$$-2(1996.8)(2) - 2(1996.8)(s+2) = 0$$

$$s = 1.33 \text{ ft} \qquad \textbf{Ans}$$

8–11. The ring, having the dimensions shown, is placed over a flexible membrane which is pumped up with a pressure p. Determine the change in the internal radius of the ring after this pressure is applied. The modulus of elasticity for the ring is E.

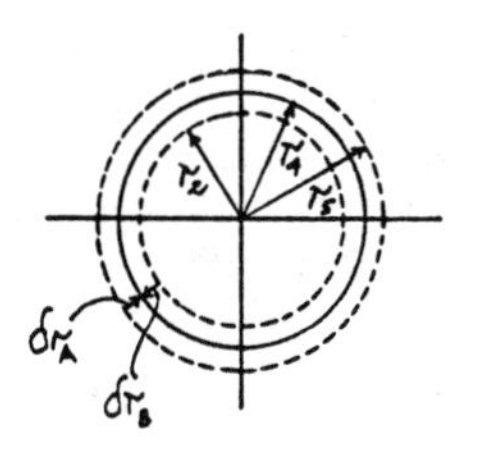

Equilibrium for the Ring : From the FBD

$$\xrightarrow{+} \Sigma F_x = 0; \quad 2P - 2pr_i w = 0 \quad P = pr_i w$$

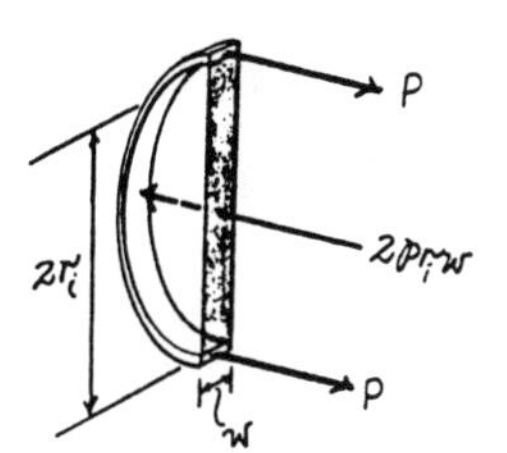

Hoop Stress and Strain for the Ring :

$$\sigma_1 = \frac{P}{A} = \frac{pr_i w}{(r_o - r_i)w} = \frac{pr_i}{r_o - r_i}$$

Using Hooke's Law

$$\varepsilon_1 = \frac{\sigma_1}{E} = \frac{pr_i}{E(r_o - r_i)} \qquad [1]$$

However, $\varepsilon_1 = \frac{2\pi (r_i)_1 - 2\pi r_i}{2\pi r} = \frac{(r_i)_1 - r_i}{r_i} = \frac{\delta r_i}{r_i}.$

Then, from Eq.[1]

$$\frac{\delta r_i}{r_i} = \frac{pr_i}{E(r_o - r_i)}$$

$$\delta r_i = \frac{pr_i^2}{E(r_o - r_i)} \qquad \textbf{Ans}$$

***8–12.** The inner ring A has an inner radius r_1 and outer radius r_2. Before heating, the outer ring B has an inner radius r_3 and an outer radius r_4. and $r_2 > r_3$. If the outer ring is heated and then fitted over the inner ring, determine the pressure between the two rings when ring B reaches the temperature of the inner ring. The material has a modulus of elasticity of E and a coefficient of thermal expansion of α.

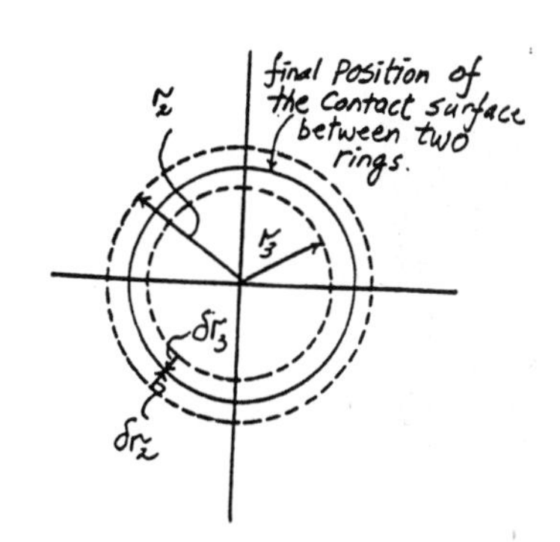

Equilibrium for the Ring : From the FBD

$$\xrightarrow{+} \Sigma F_x = 0; \quad 2P - 2pr_i w = 0 \quad P = pr_i w$$

Hoop Stress and Strain for the Ring :

$$\sigma_1 = \frac{P}{A} = \frac{pr_i w}{(r_o - r_i)w} = \frac{pr_i}{r_o - r_i}$$

Using Hooke's Law

$$\varepsilon_1 = \frac{\sigma_1}{E} = \frac{pr_i}{E(r_o - r_i)} \qquad [1]$$

However, $\varepsilon_1 = \frac{2\pi (r_i)_1 - 2\pi r_i}{2\pi r} = \frac{(r_i)_1 - r_i}{r_i} = \frac{\delta r_i}{r_i}.$

Then, from Eq.[1]

$$\frac{\delta r_i}{r_i} = \frac{pr_i}{E(r_o - r_i)}$$

$$\delta r_i = \frac{pr_i^2}{E(r_o - r_i)}$$

Compatibility : The pressure between the rings requires

$$\delta r_2 + \delta r_3 = r_2 - r_3 \qquad [1]$$

From the result obtained above

$$\delta r_2 = \frac{pr_2^2}{E(r_2 - r_1)} \qquad \delta r_3 = \frac{pr_3^2}{E(r_4 - r_3)}$$

Substitute into Eq.[1]

$$\frac{pr_2^2}{E(r_2 - r_1)} + \frac{pr_3^2}{E(r_4 - r_3)} = r_2 - r_3$$

$$p = \frac{E(r_2 - r_3)}{\frac{r_2^2}{r_2 - r_1} + \frac{r_3^2}{r_4 - r_3}} \qquad \textbf{Ans}$$

8–13. In order to increase the strength of the pressure vessel, filament winding of the same material is wrapped around the circumference of the vessel as shown. If the pretension in the filament is T and the vessel is subjected to an internal pressure p, determine the hoop stresses in the filament and in the wall of the vessel. Use the free-body diagram shown, and assume the filament winding has a thickness t' and width w for a corresponding length of the vessel.

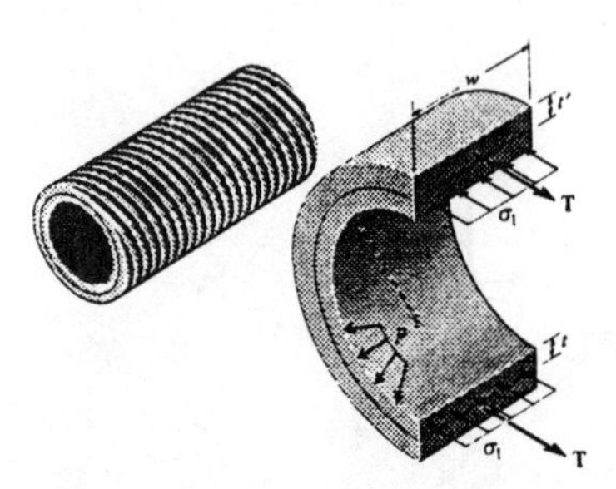

Normal Stress in the Wall and Filament Before the Internal Pressure is Applied : The entire length w of wall is subjected to pretension filament force T. Hence, from equilibrium, the normal stress in the wall at this state is

$$2T-(\sigma_1')_w(2wt)=0 \qquad (\sigma_1')_w=\frac{T}{wt}$$

and for the filament the normal stress is

$$(\sigma_1')_{fil}=\frac{T}{wt'}$$

Normal Stress in the Wall and Filament After the Internal Pressure is Applied : The stress in the filament becomes

$$\sigma_{fil}=\sigma_1+(\sigma_1')_{fil}=\frac{pr}{(t+t')}+\frac{T}{wt'} \qquad \textbf{Ans}$$

And for the wall ,

$$\sigma_w=\sigma_1-(\sigma_1')_w=\frac{pr}{(t+t')}-\frac{T}{wt} \qquad \textbf{Ans}$$

8–14. A closed-ended pressure vessel is fabricated by cross winding glass filaments over a mandrel, so that the wall thickness t of the vessel is composed entirely of filament and an epoxy binder as shown. Consider a segment of the vessel of width w and wrapped at an angle θ. If the vessel is subjected to an internal pressure p, show that the force in the segment is $F_\theta = \sigma_0 wt$, where σ_0 is the stress in the filaments. Also, show that the stresses in the hoop and longitudinal directions are $\sigma_h = \sigma_0 \sin^2\theta$ and $\sigma_l = \sigma_0 \cos^2\theta$, respectively. At what angle θ (optimum winding angle) would the filaments have to be wound so that the hoop and longitudinal stresses are equivalent?

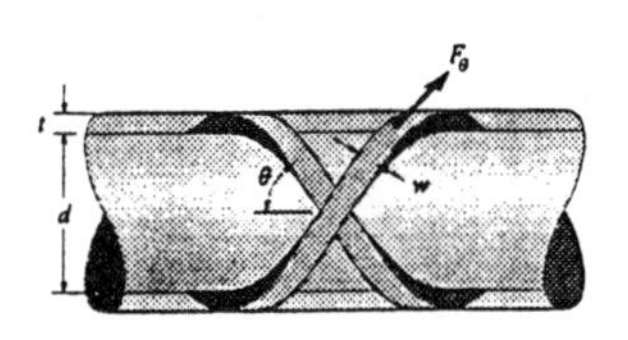

The Hoop and Longitudinal Stresses : Applying Eq. 8 – 1 and Eq. 8 – 2

$$\sigma_1 = \frac{pr}{t} = \frac{p\left(\frac{d}{2}\right)}{t} = \frac{pd}{2t}$$

$$\sigma_2 = \frac{pr}{2t} = \frac{p\left(\frac{d}{2}\right)}{2t} = \frac{pd}{4t}$$

The Hoop and Longitudinal Force for Filament :

$$F_h = \sigma_1 A = \frac{pd}{2t}\left(\frac{w}{\sin\theta}t\right) = \frac{pdw}{2\sin\theta}$$

$$F_l = \sigma_2 A = \frac{pd}{4t}\left(\frac{w}{\cos\theta}t\right) = \frac{pdw}{4\cos\theta}$$

Hence,

$$F_\theta = \sqrt{F_h^2 + F_l^2}$$

$$= \sqrt{\left(\frac{pdw}{2\sin\theta}\right)^2 + \left(\frac{pdw}{4\cos\theta}\right)^2}$$

$$= \frac{pdw}{4}\sqrt{\frac{4}{\sin^2\theta} + \frac{1}{\cos^2\theta}}$$

$$= \frac{pdw}{4}\sqrt{\frac{4\cos^2\theta + \sin^2\theta}{\sin^2\theta\cos^2\theta}}$$

$$= \frac{pdw}{2\sqrt{2}\sin 2\theta}\sqrt{3\cos 2\theta + 5}$$

$$\sigma_\theta = \frac{F_\theta}{A} = \frac{\frac{pdw}{2\sqrt{2}\sin 2\theta}\sqrt{3\cos 2\theta + 5}}{wt}$$

$$= \frac{pd}{2\sqrt{2}t}\left(\frac{\sqrt{3\cos 2\theta + 5}}{\sin 2\theta}\right) \qquad (Q.E.D.)$$

$\frac{d\sigma_\theta}{d\theta} = 0$ when σ_θ is minimum.

$$\frac{d\sigma_\theta}{d\theta} = \frac{pd}{2\sqrt{2}t}\left[-\frac{2\cos 2\theta}{\sin^2 2\theta}\left(\sqrt{3\cos 2\theta + 5}\right) - \frac{3}{\sqrt{3\cos 2\theta + 5}}\right] = 0$$

$$\frac{2\cos 2\theta}{\sin^2 2\theta}\left(\sqrt{3\cos 2\theta + 5}\right) + \frac{3}{\sqrt{3\cos 2\theta + 5}} = 0$$

$$\left(\sqrt{3\cos 2\theta + 5}\right)\left(\frac{2\cos\theta}{\sin^2 2\theta} + \frac{3}{3\cos 2\theta + 5}\right) = 0$$

$$\left(\sqrt{3\cos 2\theta + 5}\right)\left[\frac{3\cos^2 2\theta + 10\cos 2\theta + 3}{\sin^2 2\theta(3\cos 2\theta + 5)}\right] = 0$$

However, $\sqrt{3\cos 2\theta + 5} \neq 0$. Therefore,

$$\frac{3\cos^2 2\theta + 10\cos 2\theta + 3}{\sin^2 2\theta(3\cos 2\theta + 5)} = 0$$

$$3\cos^2 2\theta + 10\cos 2\theta + 3 = 0$$

$$\cos 2\theta = \frac{-10 \pm \sqrt{10^2 - 4(3)(3)}}{2(3)}$$

$$\cos 2\theta = -0.3333$$

$$\theta = 54.7° \qquad \textbf{Ans}$$

Force in θ Direction : Consider a portion of the cylinder. For a filament wire the cross - sectional area is $A = wt$, then

$$F_\theta = \sigma_0 w t \qquad (Q.E.D.)$$

Hoop Stress : The force in hoop direction is $F_h = F_\theta \sin\theta = \sigma_0 wt\sin\theta$ and the area is $A = \frac{wt}{\sin\theta}$. Then due to the internal pressure p,

$$\sigma_h = \frac{F_h}{A} = \frac{\sigma_0 wt\sin\theta}{wt/\sin\theta}$$

$$= \sigma_0 \sin^2\theta \qquad (Q.E.D.)$$

Longitudinal Stress : The force in the longitudinal direction is $F_l = F_\theta\cos\theta = \sigma_0 wt\cos\theta$ and the area is $A = \frac{wt}{\cos\theta}$. Then due to the internal pressure p,

$$\sigma_l = \frac{F_h}{A} = \frac{\sigma_0 wt\cos\theta}{wt/\cos\theta}$$

$$= \sigma_0 \cos^2\theta \qquad (Q.E.D.)$$

Optimum Wrap Angle : This require $\frac{\sigma_h}{\sigma_l} = \frac{pd/2t}{pd/4t} = 2$. Then

$$\frac{\sigma_h}{\sigma_l} = \frac{\sigma_0 \sin^2\theta}{\sigma_0\cos^2\theta} = 2$$

$$\tan^2\theta = 2$$

$$\theta = 54.7° \qquad \textbf{Ans}$$

8–15. The steel bracket is used to connect the ends of two cables. If the allowable normal stress for the steel is $\sigma_{allow} = 24$ ksi, determine the largest tensile force P that can be applied to the cables. The bracket has a thickness of 0.5 in. and a width of 0.75 in.

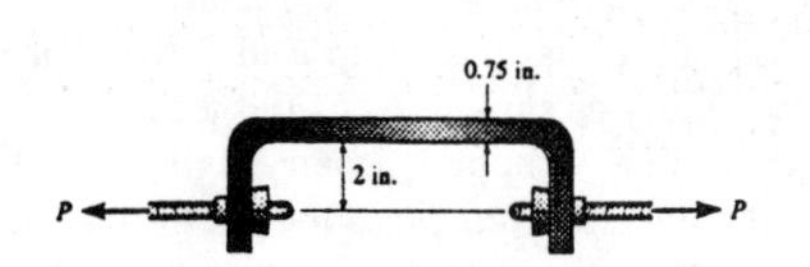

Internal Force and Moment : As shown on FBD.

Section Properties :

$$A = 0.5(0.75) = 0.375 \text{ in}^2$$
$$I = \frac{1}{12}(0.5)\left(0.75^3\right) = 0.01758 \text{ in}^4$$

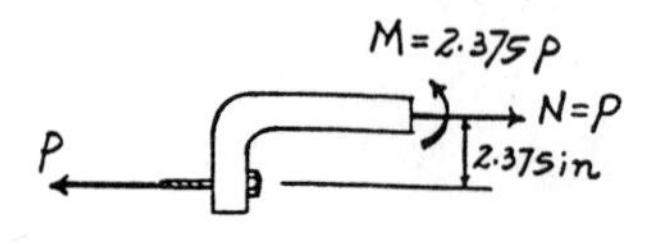

Allowable Normal Stress : The maximum normal stress occurs at the bottom of the steel bracket.

$$\sigma_{max} = \sigma_{allow} = \frac{N}{A} + \frac{Mc}{I}$$
$$24\left(10^3\right) = \frac{P}{0.375} + \frac{2.375P(0.375)}{0.01758}$$

$$P = 450 \text{ lb} \qquad \textbf{Ans}$$

***8–16.** The steel bracket is used to connect the ends of two cables. If the applied force $P = 500$ lb, determine the maximum normal stress in the bracket. The bracket has a thickness of 0.5 in. and a width of 0.75 in.

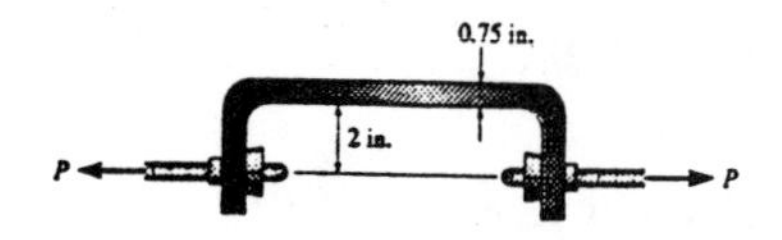

Internal Force and Moment : As shown on FBD.

Section Properties :

$$A = 0.5(0.75) = 0.375 \text{ in}^2$$
$$I = \frac{1}{12}(0.5)\left(0.75^3\right) = 0.01758 \text{ in}^4$$

Maximum Normal Stress : The maximum normal stress occurs at the bottom of the steel bracket.

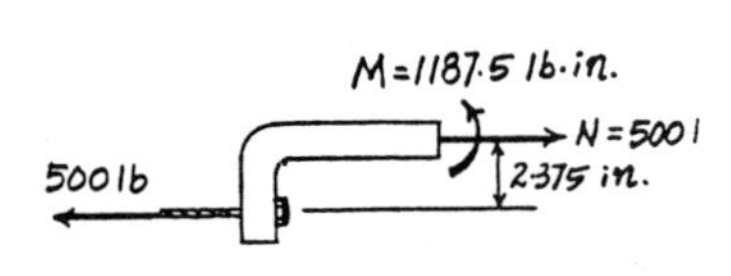

$$\sigma_{max} = \frac{N}{A} + \frac{Mc}{I}$$
$$= \frac{500}{0.375} + \frac{1187.5(0.375)}{0.01758}$$
$$= 26.7 \text{ ksi} \qquad \textbf{Ans}$$

8–17. The joint is subjected to a force of 250 lb as shown. Sketch the normal-stress distribution acting over section a–a if the member has a rectangular cross section of width 0.5 in. and thickness 0.75 in.

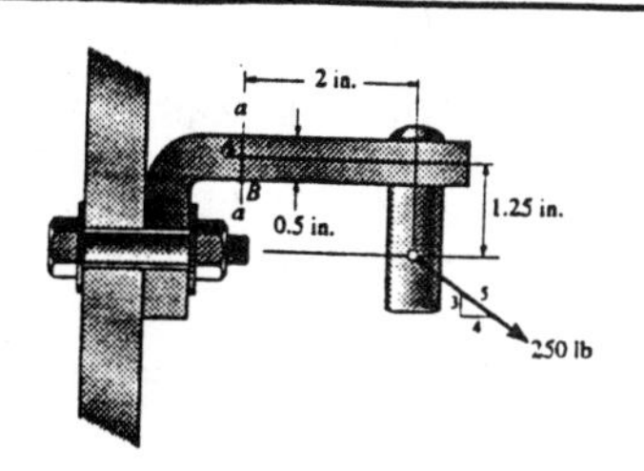

Internal Forces and Moment : As shown on FBD.

$$\xrightarrow{+} \Sigma F_x = 0; \quad \frac{4}{5}(250) - N = 0 \qquad N = 200 \text{ lb}$$

$$+\uparrow \Sigma F_y = 0; \quad V - \frac{3}{5}(250) = 0 \qquad V = 150 \text{ lb}$$

$$\circlearrowleft + \Sigma M_A = 0; \quad M + \frac{4}{5}(250)(1.25) - \frac{3}{5}(250)(2) = 0$$

$$M = 50.0 \text{ lb} \cdot \text{in.}$$

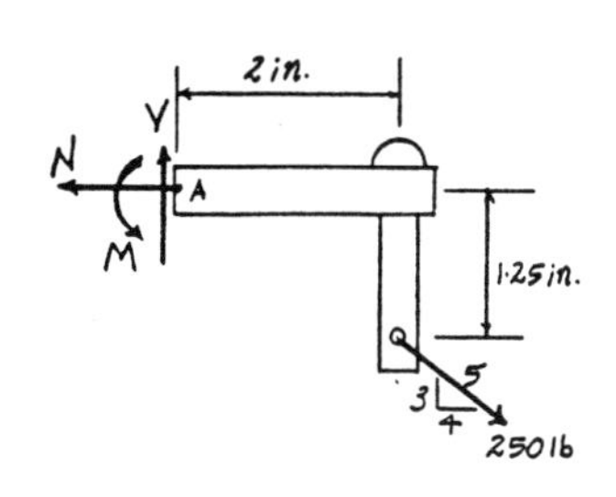

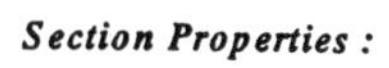

Section Properties :

$$A = 0.5(0.75) = 0.375 \text{ in}^2$$

$$I = \frac{1}{12}(0.75)\left(0.5^3\right) = 0.0078125 \text{ in}^4$$

Normal Stress :

$$\sigma = \frac{N}{A} \pm \frac{Mc}{I}$$

$$= \frac{200}{0.375} \pm \frac{50.0(0.25)}{0.0078125}$$

$$\sigma_C = 533.33 + 1600 = 2133.33 \text{ psi} = 2.13 \text{ ksi (T)}$$

$$\sigma_B = 533.33 - 1600 = -1066.67 \text{ psi} = 1.07 \text{ ksi (C)}$$

$$\frac{0.5 - y}{1066.67} = \frac{y}{2133.33} \qquad y = 0.333 \text{ in.}$$

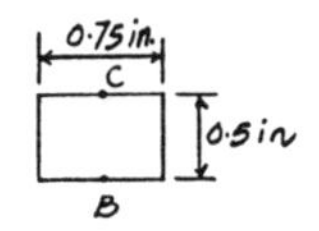

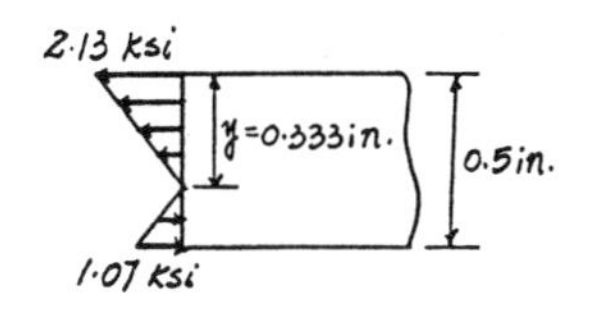

8–18. The joint is subjected to a force of 250 lb as shown. Determine the state of stress at points A and B, and sketch the results on differential elements located at these points. The member has a rectangular cross-sectional area of width 0.5 in. and thickness 0.75 in.

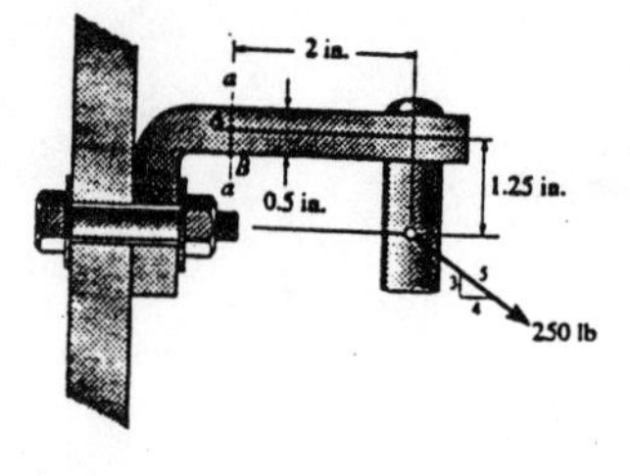

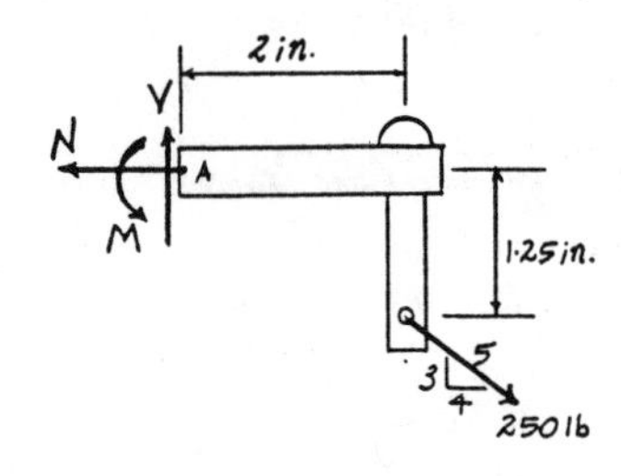

Internal Forces and Moment : As shown on FBD.

$$\xrightarrow{+} \Sigma F_x = 0; \quad \frac{4}{5}(250) - N = 0 \quad N = 200 \text{ lb}$$

$$+\uparrow \Sigma F_y = 0; \quad V - \frac{3}{5}(250) = 0 \quad V = 150 \text{ lb}$$

$$\circlearrowleft + \Sigma M_A = 0; \quad M + \frac{4}{5}(250)(1.25) - \frac{3}{5}(250)(2) = 0$$

$$M = 50.0 \text{ lb} \cdot \text{in.}$$

Section Properties :

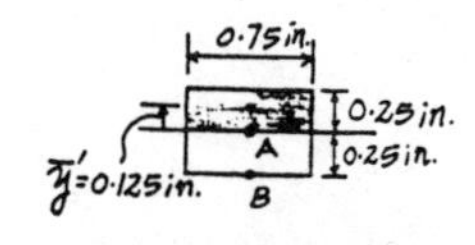

$$A = 0.5(0.75) = 0.375 \text{ in}^2$$

$$I = \frac{1}{12}(0.75)\left(0.5^3\right) = 0.0078125 \text{ in}^4$$

$$Q_A = \bar{y}'A' = 0.125(0.25)(0.75) = 0.0234375 \text{ in}^3$$

$$Q_B = 0$$

Normal Stress :

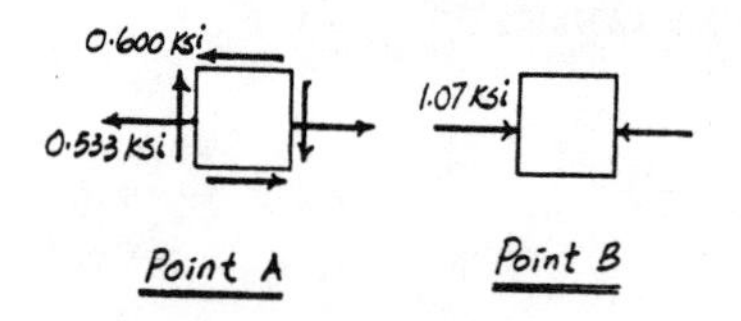

$$\sigma_A = \frac{N}{A} + \frac{My}{I}$$

$$= \frac{200}{0.375} + 0$$

$$= 533.33 \text{ psi} = 0.533 \text{ ksi (T)} \qquad \textbf{Ans}$$

$$\sigma_B = \frac{N}{A} - \frac{Mc}{I}$$

$$= \frac{200}{0.375} - \frac{50.0(0.25)}{0.0078125}$$

$$= -1066.67 \text{ psi} = 1.07 \text{ ksi (C)} \qquad \textbf{Ans}$$

Shear Stress : Applying the shear formula

$$\tau_A = \frac{VQ_A}{It} = \frac{150(0.0234375)}{0.0078125(0.75)} = 600 \text{ psi} = 0.600 \text{ ksi} \qquad \textbf{Ans}$$

$$\tau_B = \frac{VQ_B}{It} = 0 \qquad \textbf{Ans}$$

8–19. The offset link supports the loading of $P = 40$ kN. Determine its required width w if the allowable normal stress is $\sigma_{\text{allow}} = 75$ MPa. The link has a thickness of 40 mm.

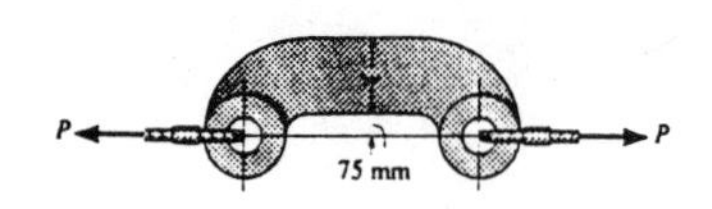

Internal Force and Moment : As shown on FBD.

Section Properties :

$$A = 0.04w$$

$$I = \frac{1}{12}(0.04)\left(w^3\right) = 0.003333w^3$$

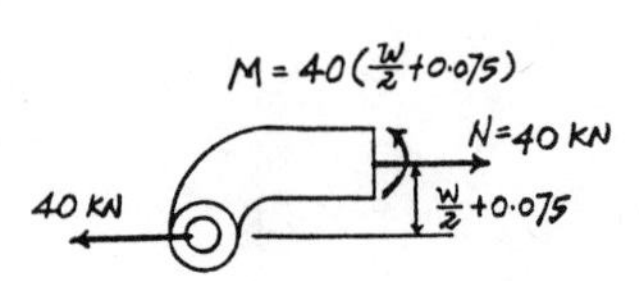

Allowable Normal Stress : The maximum normal stress occurs at the bottom of the link.

$$\sigma_{\max} = \sigma_{\text{allow}} = \frac{N}{A} + \frac{Mc}{I}$$

$$75\left(10^6\right) = \frac{40(10^3)}{0.04w} + \frac{40(10^3)\left(\frac{w}{2} + 0.075\right)\left(\frac{w}{2}\right)}{0.003333w^3}$$

$$75w^2 - 4w - 0.45 = 0$$

$$w = 0.1086 \text{ m} = 109 \text{ mm} \qquad \textbf{Ans}$$

***8–20.** The offset link has a width of w = 200 mm and a thickness of t = 40 mm. If the allowable normal stress is σ_{allow} = 75 MPa, determine the maximum load P that can be applied to the cables.

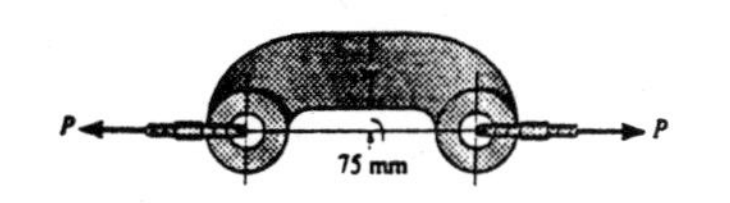

Internal Force and Moment : As shown on FBD.

Section Properties :

$$A = 0.2(0.04) = 0.00800 \text{ m}^2$$

$$I = \frac{1}{12}(0.04)\left(0.2^3\right) = 26.667\left(10^{-6}\right) \text{ m}^4$$

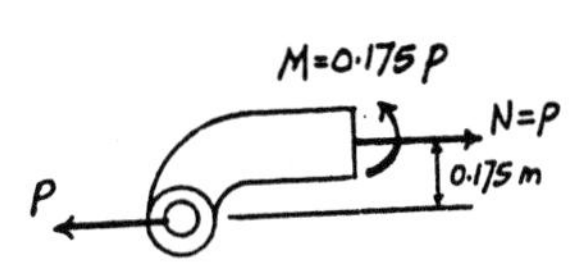

Allowable Normal Stress : The maximum normal stress occurs at the bottom of the link section.

$$\sigma_{max} = \sigma_{allow} = \frac{N}{A} + \frac{Mc}{I}$$

$$75\left(10^6\right) = \frac{P}{0.00800} + \frac{0.175P(0.1)}{26.667(10^{-6})}$$

$$P = 96000 \text{ N} = 96.0 \text{ kN} \qquad \textbf{Ans}$$

8–21. Determine the maximum and minimum normal stress in the bracket at section a when the load is applied at $x = 0$.

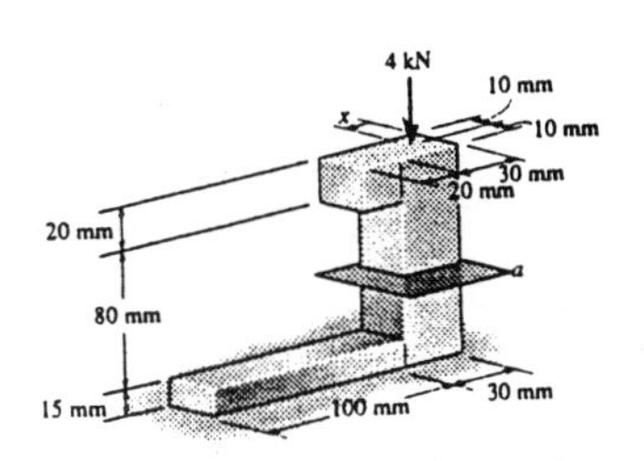

Internal Force and Moment : As shown on FBD.

Section Properties :

$$A = 0.03(0.02) = 0.600\left(10^{-3}\right) \text{ m}^2$$

$$I = \frac{1}{12}(0.02)\left(0.03^3\right) = 45.0\left(10^{-9}\right) \text{ m}^4$$

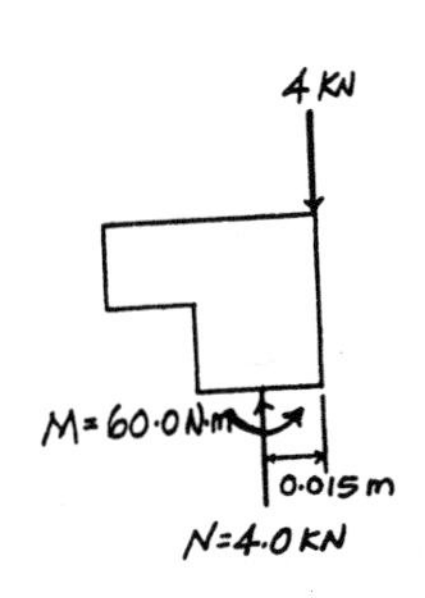

Normal Stress :

$$\sigma = \frac{N}{A} \pm \frac{Mc}{I}$$

$$= \frac{-4.00(10^3)}{0.600(10^{-3})} \pm \frac{60.0(0.015)}{45.0(10^{-9})}$$

$$\sigma_{max} = -26.67 \text{ MPa} = 26.7 \text{ MPa (C)} \qquad \textbf{Ans}$$

$$\sigma_{min} = 13.3 \text{ MPa (T)} \qquad \textbf{Ans}$$

8–22. Determine the maximum and minimum normal stress in the bracket at section a when the load is applied at $x = 50$ mm.

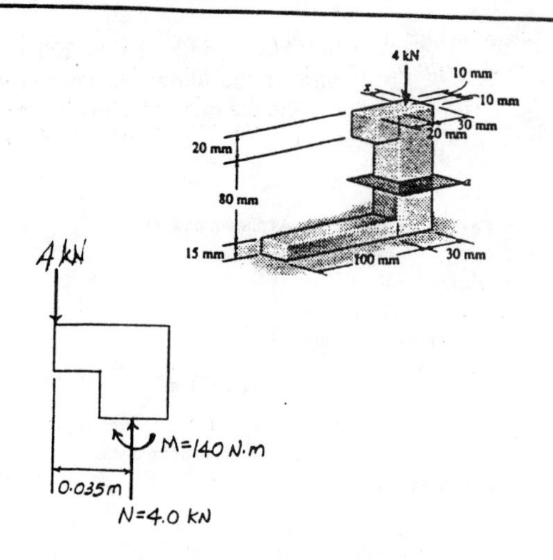

Internal Force and Moment : As shown on FBD.

Section Properties :

$$A = 0.03(0.02) = 0.600\left(10^{-3}\right)\ \text{m}^2$$

$$I = \frac{1}{12}(0.02)\left(0.03^3\right) = 45.0\left(10^{-9}\right)\ \text{m}^4$$

Normal Stress :

$$\sigma = \frac{N}{A} \pm \frac{Mc}{I}$$

$$= \frac{-4.00(10^3)}{0.600(10^{-3})} \pm \frac{140(0.015)}{45.0(10^{-9})}$$

$\sigma_{\max} = -53.33\ \text{MPa} = 53.3\ \text{MPa (C)}$ **Ans**

$\sigma_{\min} = 40.0\ \text{MPa (T)}$ **Ans**

8–23. The bent link is subjected to the cable load of $P = 500$ N. Determine its required diameter d if the allowable normal stress for the material is $\sigma_{\text{allow}} =$ 175 MPa. Consider the critical section to be at A.

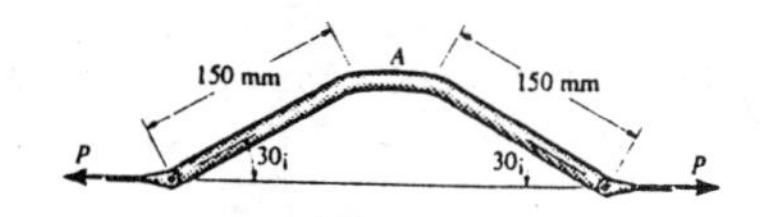

Internal Force and Moment :

$$+\nearrow \Sigma F_{x'} = 0; \quad N - 500\cos 30° = 0 \quad N = 433.01\ \text{N}$$

$$(+\Sigma M_A = 0; \quad M - 500\sin 30°(0.15) = 0 \quad M = 37.5\ \text{N}\cdot\text{m}$$

Section Properties :

$$A = \frac{\pi}{4}d^2 \qquad I = \frac{\pi}{4}r^4 = \frac{\pi}{4}\left(\frac{d}{2}\right)^4 = \frac{\pi}{64}d^4$$

Allowable Normal Stress :

$$\sigma_{\max} = \sigma_{\text{allow}} = \frac{N}{A} + \frac{Mc}{I}$$

$$175\left(10^6\right) = \frac{433.01}{\frac{\pi}{4}d^2} + \frac{37.5\left(\frac{d}{2}\right)}{\frac{\pi}{64}d^4}$$

$$175\left(10^6\right)d^3 - 551.33d - 381.97 = 0$$

$$d = 0.013053\ \text{m} = 13.1\ \text{mm}$$ **Ans**

***8–24.** The bent link has a diameter of $d = 15$ mm and is made of a material having an allowable normal stress of $\sigma_{allow} = 175$ MPa. Determine the maximum load P it will safely support. Consider the critical section to be at A.

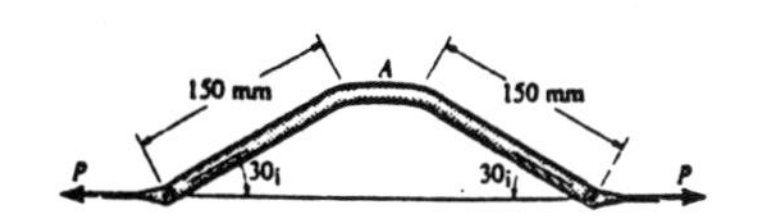

Internal Force and Moment :

$+ \Sigma F_{x'} = 0; \quad N - P\cos 30° = 0 \quad N = 0.8660P$

$\circlearrowleft + \Sigma M_A = 0; \quad M - P\sin 30°(0.15) = 0 \quad M = 0.0750P$

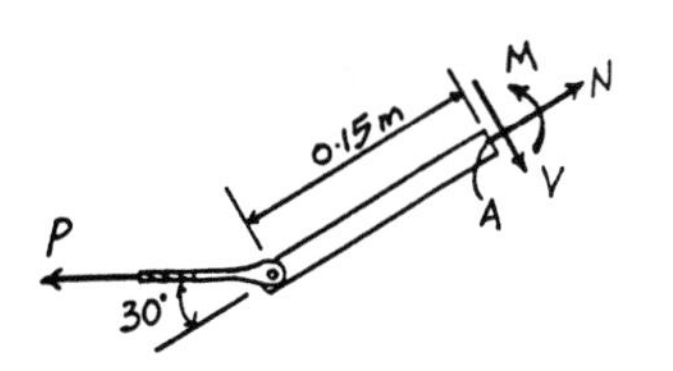

Section Properties :

$$A = \frac{\pi}{4}\left(0.015^2\right) = 0.17671\left(10^{-3}\right)\ \text{m}^2$$

$$I = \frac{\pi}{4}\left(\frac{0.015}{2}\right)^4 = 2.48505\left(10^{-9}\right)\ \text{m}^4$$

Allowable Normal Stress :

$$\sigma_{max} = \sigma_{allow} = \frac{N}{A} + \frac{Mc}{I}$$

$$175\left(10^6\right) = \frac{0.8660P}{0.17671(10^{-3})} + \frac{0.0750P(0.0075)}{2.48505(10^{-9})}$$

$P = 757$ N **Ans**

8–25. The vertical force **P** acts on the bottom of the plate having a negligible weight. Determine the maximum distance d to the edge of the plate at which it can be applied so that it produces no compressive stresses on the plate at section a–a. The plate has a thickness of 10 mm and **P** acts along the centerline of this thickness.

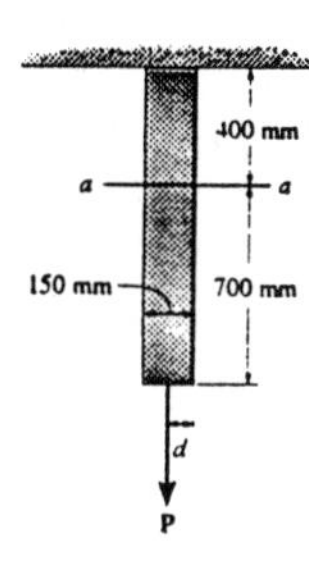

Internal Force and Moment : As shown on FBD.

Section Properties :

$$A = 0.01(0.15) = 0.0015\ \text{m}^4$$

$$I = \frac{1}{12}(0.01)\left(0.15^3\right) = 2.8125\left(10^{-6}\right)\ \text{m}^4$$

Normal Stress : Require $\sigma_A = 0$

$$\sigma_A = 0 = \frac{N}{A} - \frac{Mc}{I}$$

$$0 = \frac{P}{0.0015} - \frac{P(d - 0.075)(0.075)}{2.8125(10^{-6})}$$

$d = 0.100$ m $= 100$ mm **Ans**

N=P
M=P(d-0.075)
A
d-0.075
P

8–26. The vertical force $P = 600$ N acts on the bottom of the plate having a negligible weight. The plate has a thickness of 10 mm and **P** acts along the centerline of this thickness such that $d = 100$ mm. Plot the distribution of normal stress acting along section $a-a$.

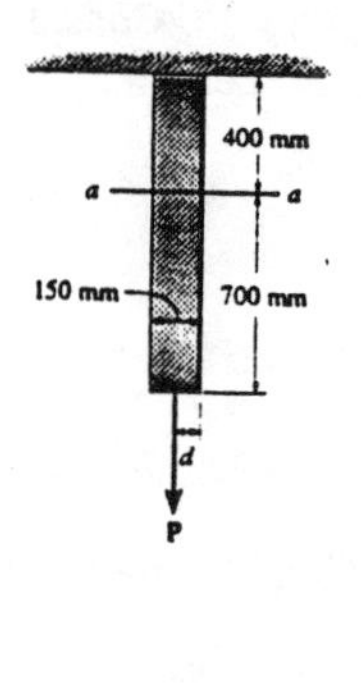

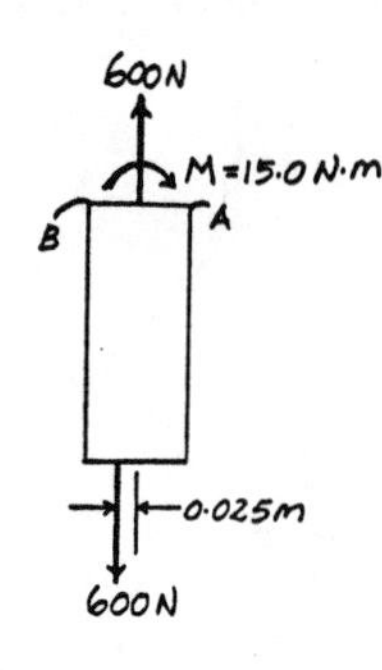

Internal Force and Moment : As shown on FBD.

Section Properties :

$$A = 0.01(0.15) = 0.0015 \text{ m}^4$$

$$I = \frac{1}{12}(0.01)\left(0.15^3\right) = 2.8125\left(10^{-6}\right) \text{ m}^4$$

Normal Stress : Require $\sigma_A = 0$

$$\sigma = \frac{N}{A} \pm \frac{Mc}{I}$$

$$= \frac{600}{0.0015} \pm \frac{15.0(0.075)}{2.8125(10^{-6})}$$

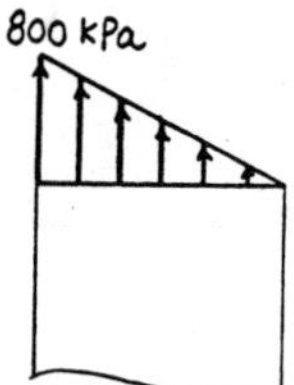

$$\sigma_A = 400\left(10^3\right) - 400\left(10^3\right) = 0$$

$$\sigma_B = 400\left(10^3\right) + 400\left(10^3\right) = 800 \text{ kPa}$$

8–27. The stepped support is subjected to the bearing load of 50 kN. Determine the maximum and minimum compressive stress in the material.

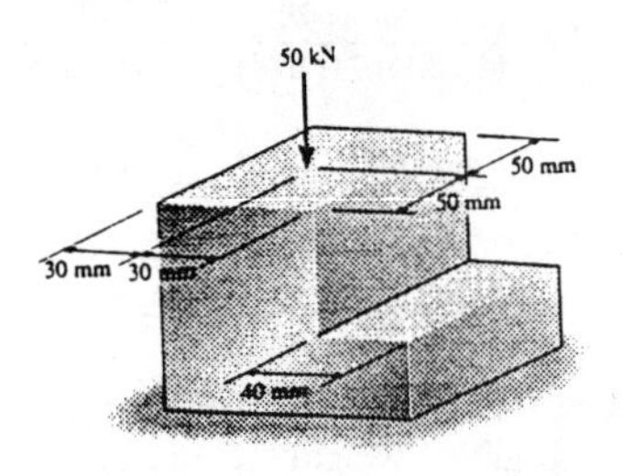

Internal Force and Moment : As shown on FBD.

Section Properties : For the top portion of the stepped support.

$$A = 0.06(0.1) = 0.00600 \text{ m}^2$$

For the bottom portion of the stepped support.

$$A = 0.1(0.1) = 0.0100 \text{ m}^2$$

$$I = \frac{1}{12}(0.1)\left(0.1^3\right) = 8.333\left(10^{-6}\right) \text{ m}^4$$

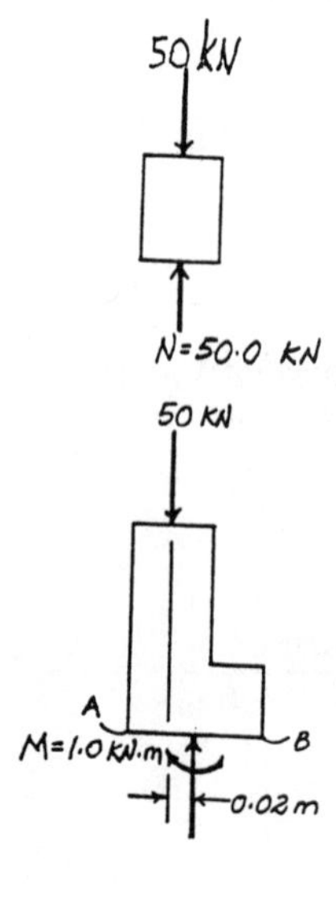

Normal Stress : For the top portion of the stepped support.

$$\sigma = \frac{N}{A} = \frac{-50.0(10^3)}{0.00600} = -8.33 \text{ MPa} = 8.33 \text{ MPa (C)}$$

For the bottom portion of the stepped support.

$$\sigma = \frac{N}{A} \pm \frac{Mc}{I}$$

$$= \frac{-50.0(10^3)}{0.0100} \pm \frac{1.00(10^3)(0.05)}{8.333(10^{-6})}$$

$$\sigma_A = -5.00\left(10^6\right) - 6.00\left(10^6\right)$$
$$= -11.0 \text{ MPa} = 11.0 \text{ MPa (C)}$$

$$\sigma_B = -5.00\left(10^6\right) + 6.00\left(10^6\right)$$
$$= 1.00 \text{ MPa (T)}$$

Therefore, the maximum and minimum compressive stress is

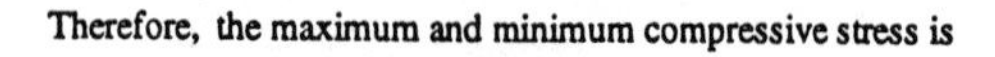

$(\sigma_C)_{max} = 11.0$ MPa **Ans**

$(\sigma_C)_{min} = 8.33$MPa **Ans**

***8–28.** The stepped support is subjected to a pair of axial loads as shown. Determine the maximum and minimum compressive stress in the material.

Internal Force and Moment : As shown on FBD.

Section Properties : For the top portion of the stepped support.

$$A = 0.08(0.03) = 0.00240\ \text{m}^2$$

For the bottom portion of the stepped support.

$$A = 0.08(0.09) = 0.00720\ \text{m}^2$$

$$I = \frac{1}{12}(0.08)\left(0.09^3\right) = 4.860\left(10^{-6}\right)\ \text{m}^4$$

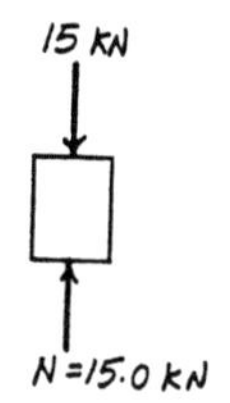

Normal Stress : For the top portion of the stepped support.

$$\sigma = \frac{N}{A} = \frac{-15.0(10^3)}{0.00240} = -6.25\ \text{MPa} = 6.25\ \text{MPa (C)}$$

For the bottom portion of the stepped support.

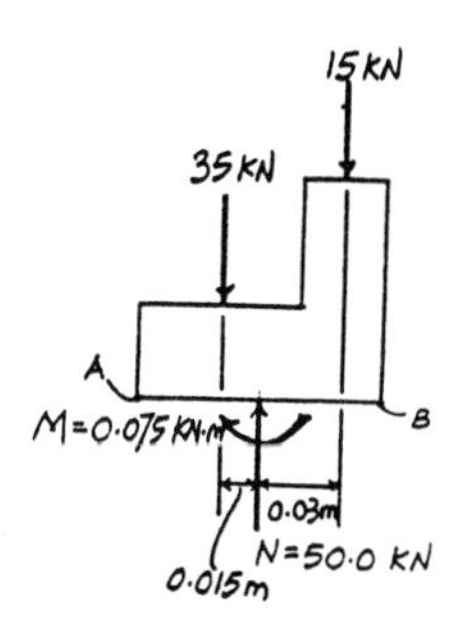

$$\sigma = \frac{N}{A} \pm \frac{Mc}{I}$$

$$= \frac{-50.0(10^3)}{0.00720} \pm \frac{0.0750(10^3)(0.045)}{4.860(10^{-6})}$$

$$\sigma_A = -6.9444\left(10^6\right) - 0.6944\left(10^6\right)$$
$$= -7.639\ \text{MPa} = 7.64\ \text{MPa (C)}$$

$$\sigma_B = -6.9444\left(10^6\right) + 0.6944\left(10^6\right)$$
$$= -6.25\ \text{MPa} = 6.25\ \text{MPa (C)}$$

Therefore, the maximum and minimum compressive stress is

$(\sigma_C)_{max} = 7.64$ MPa **Ans**

$(\sigma_C)_{min} = 6.25$ MPa **Ans**

8–29. Since concrete can support little or no tension, this problem can be avoided by using wires or rods to *prestress* the concrete once it is formed. Consider the simply supported beam shown, which has a rectangular cross section of 18 in. by 12 in. If concrete has a specific weight of 150 lb/ft³, determine the required tension in rod AB, which runs through the beam so that no tensile stress is developed in the concrete at its center section a–a. Neglect the size of the rod and any deflection of the beam.

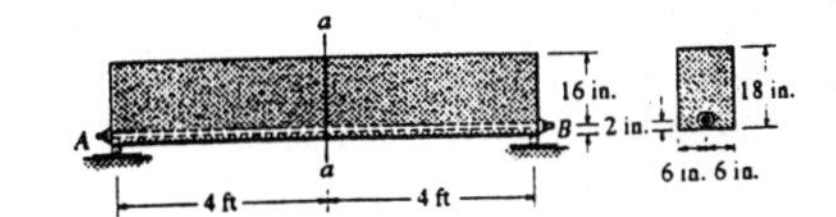

Support Reactions : As shown on FBD.

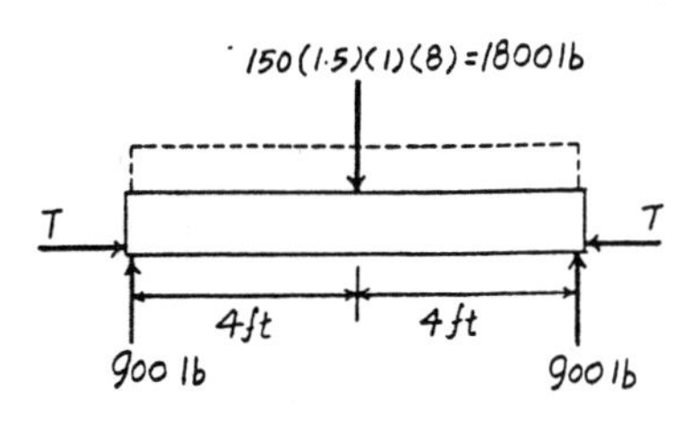

Internal Force and Moment :

$$\xrightarrow{+} \Sigma F_x = 0; \quad T - N = 0 \quad N = T$$

$$\curvearrowleft + \Sigma M_O = 0; \quad M + T(7) - 900(24) = 0$$
$$M = 21600 - 7T$$

Section Properties :

$$A = 18(12) = 216\ \text{in}^2$$

$$I = \frac{1}{12}(12)\left(18^3\right) = 5832\ \text{in}^4$$

Normal Stress : Requires $\sigma_A = 0$

$$\sigma_A = 0 = \frac{N}{A} + \frac{Mc}{I}$$

$$0 = \frac{-T}{216} + \frac{(21600 - 7T)(9)}{5832}$$

$T = 2160$ lb $= 2.16$ kip **Ans**

8–30. Solve Prob. 8–29 if the rod has a diameter of 0.5 in. Use the transformed area method discussed in Sec. 6–6. $E_{st} = 29(10^3)$ ksi, $E_c = 3.60(10^3)$ ksi.

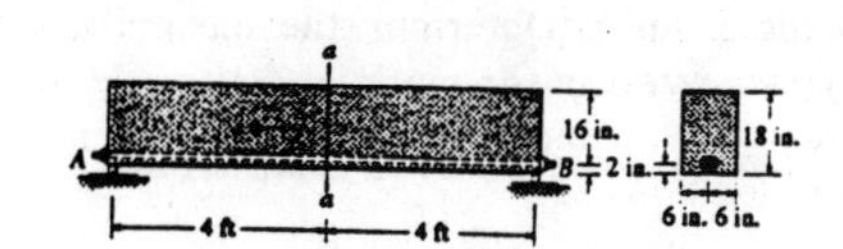

Support Reactions : As shown on FBD.

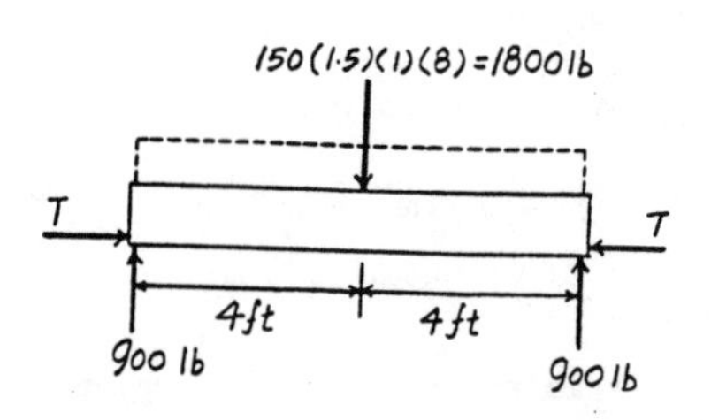

Section Properties :

$$n = \frac{E_{st}}{E_{con}} = \frac{29(10^3)}{3.6(10^3)} = 8.0556$$

$$A_{con} = (n-1)A_{st} = (8.0556-1)\left(\frac{\pi}{4}\right)(0.5^2) = 1.3854 \text{ in}^2$$

$$A = 18(12) + 1.3854 = 217.3854 \text{ in}^2$$

$$\bar{y} = \frac{\Sigma \tilde{y}A}{\Sigma A} = \frac{9(18)(12) + 16(1.3854)}{217.3854} = 9.04461 \text{ in.}$$

$$I = \frac{1}{12}(12)(18^3) + 12(18)(9.04461-9)^2 + 1.3854(16-9.04461)^2$$

$$= 5899.45 \text{ in}^4$$

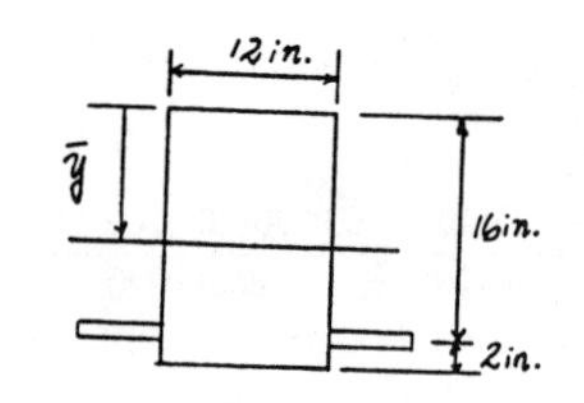

Internal Force and Moment :

$$\xrightarrow{+} \Sigma F_x = 0; \quad T - N = 0 \quad N = T$$

$$+\Sigma M_O = 0; \quad M + T(6.9554) - 900(24) = 0$$
$$M = 21600 - 6.9554T$$

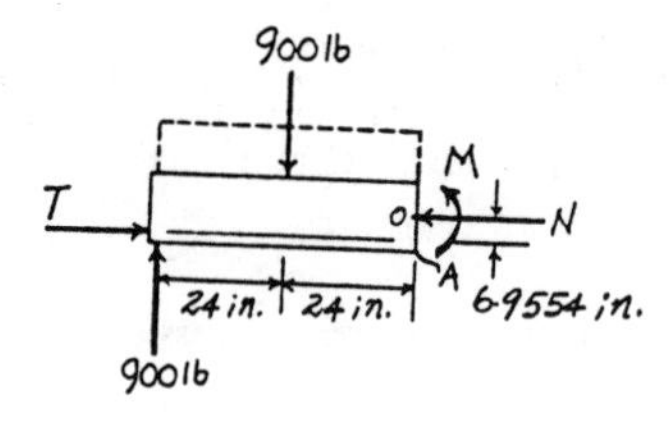

Normal Stress : Requires $\sigma_A = 0$

$$\sigma_A = 0 = \frac{N}{A} + \frac{Mc}{I}$$

$$0 = \frac{-T}{217.3854} + \frac{(21600 - 6.9554T)(8.9554)}{5899.45}$$

$$T = 2163.08 \text{ lb} = 2.16 \text{ kip} \qquad \textbf{Ans}$$

8–31. The chimney is subjected to the uniform wind loading of $w = 150$ lb/ft and has a weight of 2200 lb/ft. If the mortar between the bricks cannot support a tensile stress, determine if the chimney is safe. Take $d = 6$ ft. The thickness of the brick wall is 1 ft.

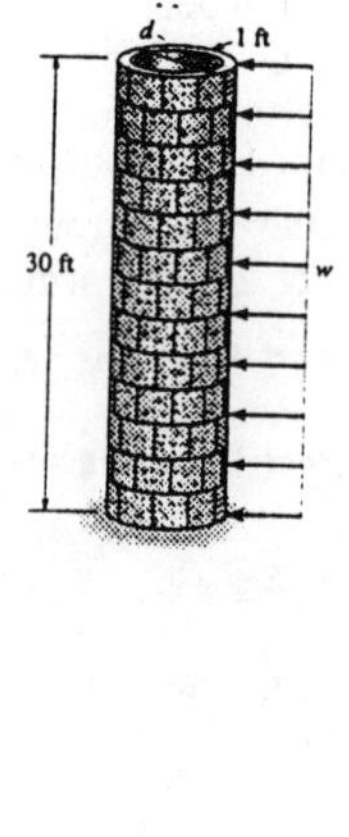

Internal Force and Moment : As shown on FBD.

Section Properties :

$$A = \frac{\pi}{4}(6^2 - 4^2) = 15.7080 \text{ ft}^2$$

$$I = \frac{\pi}{4}(3^4 - 2^4) = 51.0509 \text{ ft}^4$$

Normal Stress : In order for the chimney to be safe, σ_B must remain compressive.

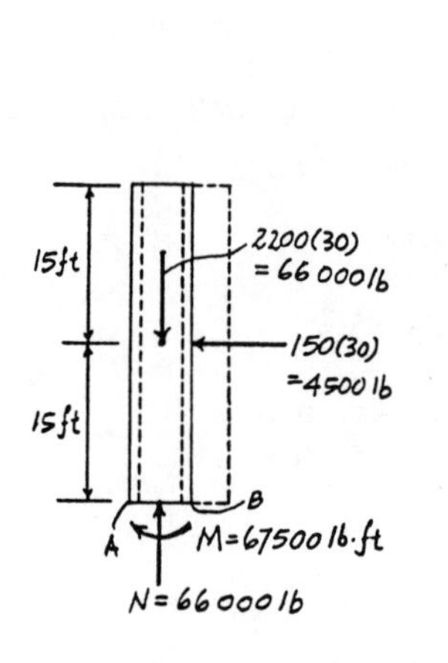

$$\sigma_B = \frac{N}{A} + \frac{Mc}{I}$$
$$= \frac{-66\,000}{15.7080} + \frac{67\,500(3)}{51.0509}$$
$$= -4201.7 + 3966.6$$
$$= -235.1 \text{ psf} = 235.1 \text{ psf (C)}$$

Since σ_B is in compression, the chimney is ***safe***. **Ans**

***8–32.** The chimney is subjected to the uniform wind pressure of $p = 25$ lb/ft^2. It is to be constructed with 1-ft-thick brick walls. If the bricks and mortar have a specific weight of 145 lb/ft^3, determine the smallest outer diameter d of the chimney so that no tensile stress is developed in the material. The wind loading can be approximated by $w = pa$.

Internal Force and Moment : As shown on FBD.

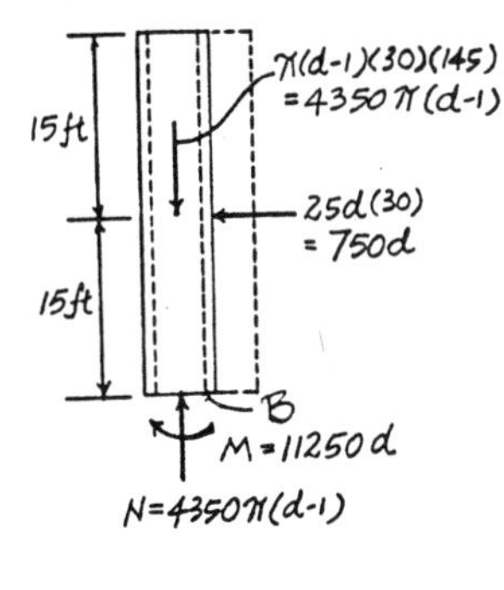

Section Properties :

$$A = \frac{\pi}{4}\left[d^2 - (d-2)^2\right] = \pi(d-1)$$

$$I = \frac{\pi}{64}\left[d^4 - (d-2)^4\right] = \frac{\pi}{8}\left(d^3 - 3d^2 + 4d - 2\right)$$

Normal Stress : Require $\sigma_B = 0$.

$$\sigma_B = 0 = \frac{N}{A} + \frac{Mc}{I}$$

$$0 = \frac{-4350\pi(d-1)}{\pi(d-1)} + \frac{11\,250d\left(\frac{d}{2}\right)}{\frac{\pi}{8}(d^3 - 3d^2 + 4d - 2)}$$

$$4350\pi d^3 - 85997.78d^2 + 17400\pi d - 8700\pi = 0$$

By trial and error,

$$d = 5.65 \text{ ft} \qquad \textbf{Ans}$$

8–33. The block is subjected to the two axial loads shown. Determine the normal stress developed at points A and B. Neglect the weight of the block.

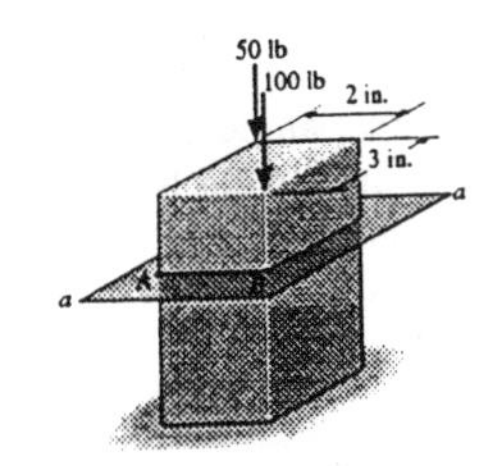

Internal Force and Moment :

$$\Sigma F_x = 0; \quad 50 + 100 + N = 0 \quad N = -150 \text{ lb}$$

$$\Sigma M_z = 0; \quad M_z + 50(1.5) - 100(1.5) = 0$$
$$M_z = 75.0 \text{ lb}\cdot\text{in.}$$

$$\Sigma M_y = 0; \quad M_y + 50(1) - 100(1) = 0$$
$$M_y = 50.0 \text{ lb}\cdot\text{in.}$$

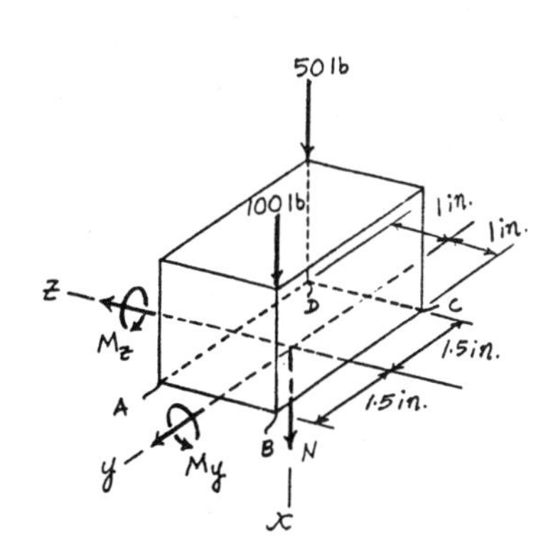

Section Properties :

$$A = 3(2) = 6.00 \text{ in}^2$$

$$I_z = \frac{1}{12}(2)\left(3^3\right) = 4.50 \text{ in}^4$$

$$I_y = \frac{1}{12}(3)\left(2^3\right) = 2.00 \text{ in}^4$$

Normal Stresses :

$$\sigma = \frac{N}{A} - \frac{M_z y}{I_z} + \frac{M_y z}{I_y}$$

$$\sigma_A = \frac{-150}{6.00} - \frac{75.0(1.5)}{4.50} + \frac{50.0(1)}{2.00}$$
$$= -25.0 \text{ psi} = 25.0 \text{ psi (C)} \qquad \textbf{Ans}$$

$$\sigma_B = \frac{-150}{6.00} - \frac{75.0(1.5)}{4.50} + \frac{50.0(-1)}{2.00}$$
$$= -75.0 \text{ psi} = 75.0 \text{ psi (C)} \qquad \textbf{Ans}$$

8–34. The block is subjected to the two axial loads shown. Sketch the normal stress distribution acting over the cross section at section a–a. Neglect the weight of the block.

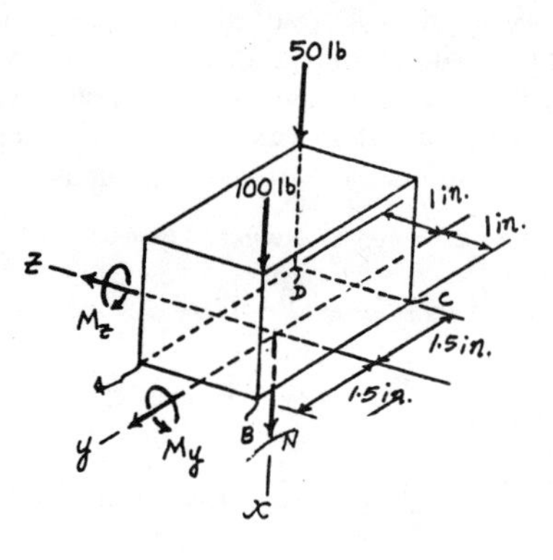

Internal Force and Moment :

$\Sigma F_x = 0;\quad 50 + 100 + N = 0 \quad N = -150$ lb

$\Sigma M_z = 0;\quad M_z + 50(1.5) - 100(1.5) = 0$

$M_z = 75.0$ lb · in.

$\Sigma M_y = 0;\quad M_y + 50(1) - 100(1) = 0$

$M_y = 50.0$ lb · in.

Section Properties :

$$A = 3(2) = 6.00 \text{ in}^2$$

$$I_z = \frac{1}{12}(2)\left(3^3\right) = 4.50 \text{ in}^4$$

$$I_y = \frac{1}{12}(3)\left(2^3\right) = 2.00 \text{ in}^4$$

Normal Stresses :

$$\sigma = \frac{N}{A} - \frac{M_z y}{I_z} + \frac{M_y z}{I_y}$$

$$\sigma_A = \frac{-150}{6.00} - \frac{75.0(1.5)}{4.50} + \frac{50.0(1)}{2.00}$$
$$= -25.0 \text{ psi} = 25.0 \text{ psi (C)}$$

$$\sigma_B = \frac{-150}{6.00} - \frac{75.0(1.5)}{4.50} + \frac{50.0(-1)}{2.00}$$
$$= -75.0 \text{ psi} = 75.0 \text{ psi (C)}$$

$$\sigma_C = \frac{-150}{6.00} - \frac{75.0(-1.5)}{4.50} + \frac{50.0(-1)}{2.00}$$
$$= -25.0 \text{ psi} = 25.0 \text{ psi (C)}$$

$$\sigma_D = \frac{-150}{6.00} - \frac{75.0(-1.5)}{4.50} + \frac{50.0(1)}{2.00}$$
$$= 25.0 \text{ psi (T)}$$

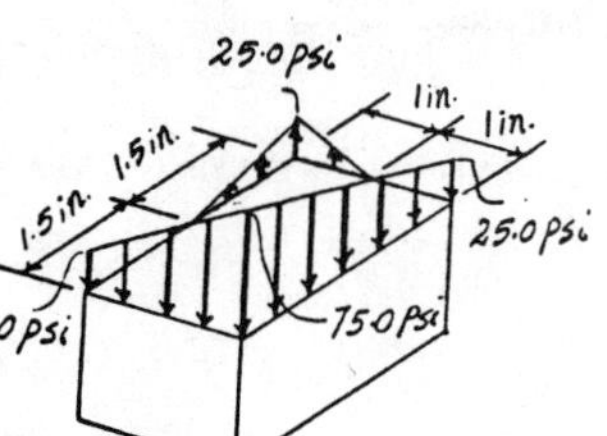

8–35. A bar having a square cross section of 30 mm by 30 mm is 2 m long and is held upward. If it has a mass of 5 kg/m, determine the largest angle θ, measured from the vertical, at which it can be supported before it is subjected to a tensile stress near the grip.

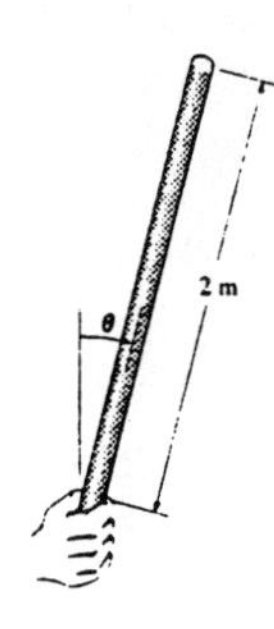

Internal Force and Moment :

$\nearrow + \Sigma F_{x'} = 0;\quad N - 5(2)(9.81)\cos\theta = 0$

$N = 98.1\cos\theta$

$\circlearrowleft + \Sigma M_O = 0;\quad M - 5(2)(9.81)\sin\theta(1) = 0°$

$M = 98.1\sin\theta$

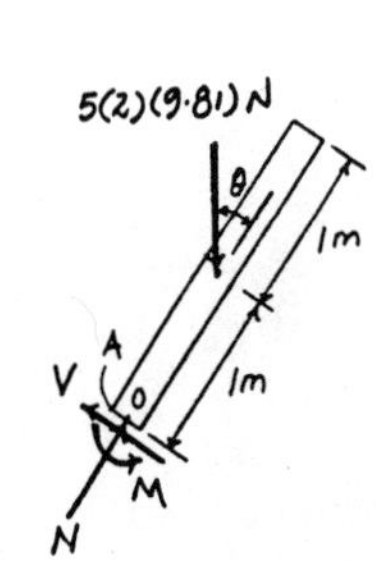

Section Properties :

$$A = 0.03(0.03) = 0.900\left(10^{-3}\right) \text{ m}^2$$

$$I = \frac{1}{12}(0.03)\left(0.03^3\right) = 67.5\left(10^{-9}\right) \text{ m}^4$$

Normal Stress : Require $\sigma_A = 0$.

$$\sigma_A = 0 = \frac{N}{A} + \frac{Mc}{I}$$

$$0 = \frac{-98.1\cos\theta}{0.900(10^{-3})} + \frac{98.1\sin\theta(0.015)}{67.5(10^{-9})}$$

$\tan\theta = 0.0050 \qquad \theta = 0.286°$ **Ans**

***8–36.** Solve Prob. 8–35 if the bar has a circular cross section of 30-mm diameter.

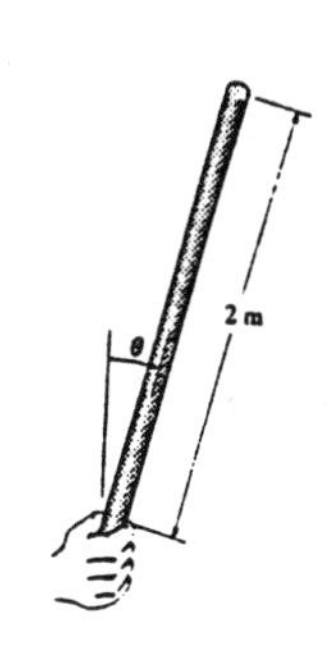

Internal Force and Moment :

$$+\Sigma F_{x'} = 0; \quad N - 5(2)(9.81)\cos\theta = 0$$
$$N = 98.1\cos\theta$$

$$+\Sigma M_O = 0; \quad M - 5(2)(9.81)\sin\theta(1) = 0°$$
$$M = 98.1\sin\theta$$

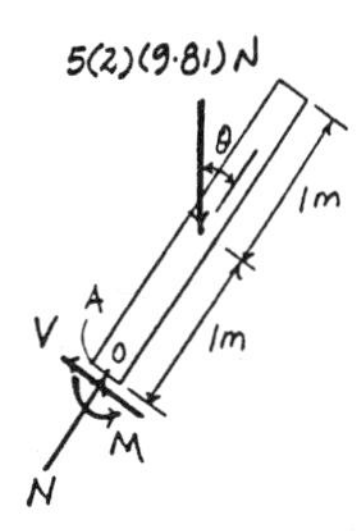

Section Properties :

$$A = \frac{\pi}{4}\left(0.03^2\right) = 0.225\pi\left(10^{-3}\right)\ \text{m}^2$$
$$I = \frac{\pi}{4}\left(0.015^4\right) = 12.65625\pi\left(10^{-9}\right)\ \text{m}^4$$

Normal Stress : Require $\sigma_A = 0$.

$$\sigma_A = 0 = \frac{N}{A} + \frac{Mc}{I}$$
$$0 = \frac{-98.1\cos\theta}{0.225\pi(10^{-3})} + \frac{98.1\sin\theta(0.015)}{12.65625\pi(10^{-9})}$$

$\tan\theta = 0.00375$ $\qquad\theta = 0.215°$ $\qquad$ **Ans**

8–37. The cylindrical pole is subjected to the horizontal load of $P = 400$ N using a sling of negligible thickness. Determine the state of stress at points A and B. Show the results on a differential volume element located at each of these points.

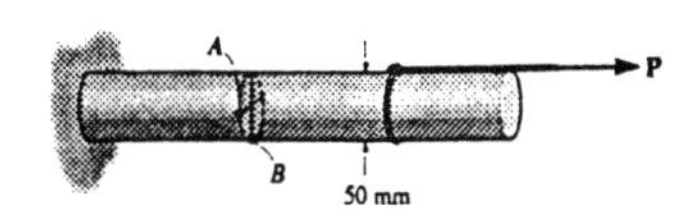

Internal Forces and Moment : As shown on FBD.

Section Properties :

$$A = \frac{\pi}{4}\left(0.05^2\right) = 0.625\pi\left(10^{-3}\right)\ \text{m}^2$$
$$I = \frac{\pi}{4}\left(0.025^4\right) = 97.65625\pi\left(10^{-9}\right)\ \text{m}^4$$

Normal Stress :

$$\sigma_A = \frac{N}{A} + \frac{My}{I}$$
$$= \frac{400}{0.625\pi(10^{-3})} + 0$$
$$= 204\ \text{kPa (T)}$$ **Ans**

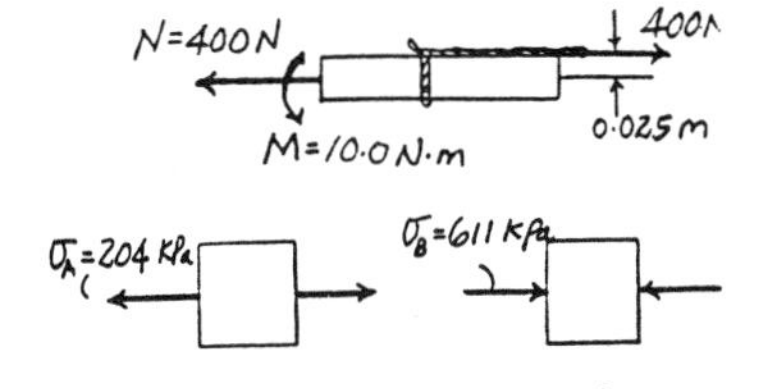

$$\sigma_B = \frac{N}{A} - \frac{Mc}{I}$$
$$= \frac{400}{0.625\pi(10^{-3})} - \frac{10.0(0.025)}{97.65625\pi(10^{-9})}$$
$$= -611\text{kPa} = 611\ \text{kPa (C)}$$ **Ans**

8–38. Determine the maximum load P that can be applied to the sling having a negligible thickness so that the normal stress in the post does not exceed σ_{allow} = 30 MPa.

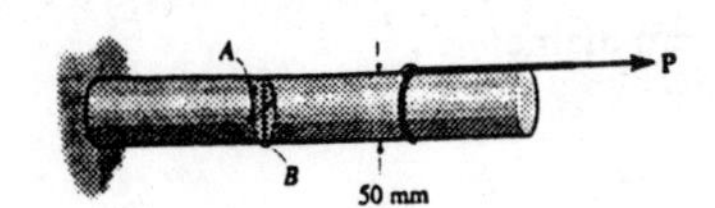

Internal Forces and Moment : As shown on FBD.

Section Properties :

$$A = \frac{\pi}{4}\left(0.05^2\right) = 0.625\pi\left(10^{-3}\right)\ \text{m}^2$$

$$I = \frac{\pi}{4}\left(0.025^4\right) = 97.65625\pi\left(10^{-9}\right)\ \text{m}^4$$

C
P
1=0.025P
0.025m

Normal Stress : Maximum normal stress occurs at point C.

$$\sigma_{max} = \sigma_{allow} = \frac{N}{A} + \frac{Mc}{I}$$

$$30\left(10^6\right) = \frac{P}{0.625\pi(10^{-3})} + \frac{0.025P(0.025)}{97.65625\pi(10^{-9})}$$

$$P = 11781\ \text{N} = 11.8\ \text{kN} \qquad \textbf{Ans}$$

8–39. The cylinder of negligible weight rests on a smooth floor. Determine the eccentric distance e_y at which the load can be placed so that the normal stress at point A is zero.

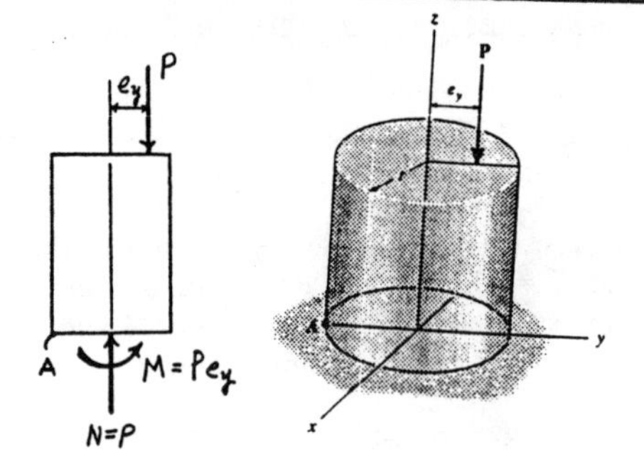

Internal Force and Moment : As shown on FBD.

Section Properties :

$$A = \pi r^2 \qquad I = \frac{\pi}{4} r^4$$

Normal Stress : Require $\sigma_A = 0$

$$\sigma_A = 0 = \frac{N}{A} + \frac{Mc}{I}$$

$$0 = \frac{-P}{\pi r^2} + \frac{Pe_y\,(r)}{\frac{\pi}{4}r^4}$$

$$e_y = \frac{r}{4} \qquad \textbf{Ans}$$

***8–40.** Determine the state of stress at point A when the beam is subjected to the cable force of 4 kN. Indicate the result as a differential volume element.

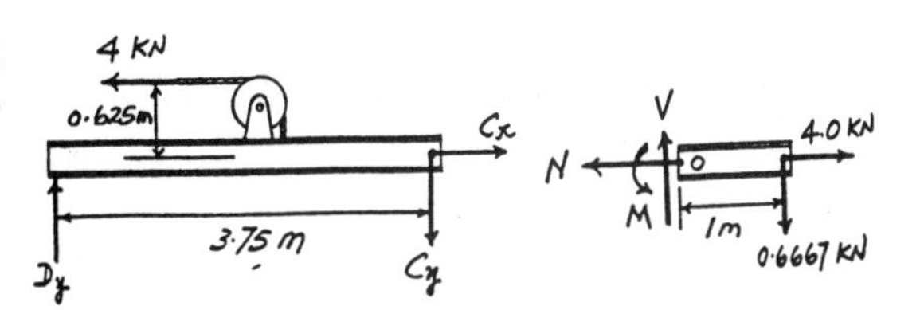

Support Reactions :

$$\curvearrowleft + \Sigma M_D = 0; \quad 4(0.625) - C_y(3.75) = 0$$
$$C_y = 0.6667 \text{ kN}$$
$$\xrightarrow{+} \Sigma F_x = 0; \quad C_x - 4 = 0 \quad C_x = 4.00 \text{ kN}$$

Internal Forces and Moment :

$$\xrightarrow{+} \Sigma F_x = 0; \quad 4.00 - N = 0 \quad N = 4.00 \text{ kN}$$
$$+\uparrow \Sigma F_y = 0; \quad V - 0.6667 = 0 \quad V = 0.6667 \text{ kN}$$
$$\curvearrowleft + \Sigma M_O = 0; \quad M - 0.6667(1) = 0 \quad M = 0.6667 \text{ kN}\cdot\text{m}$$

Section Properties :

$$A = 0.24(0.15) - 0.2(0.135) = 9.00(10^{-3}) \text{ m}^2$$
$$I = \frac{1}{12}(0.15)(0.24^3) - \frac{1}{12}(0.135)(0.2^3) = 82.8(10^{-6}) \text{ m}^4$$
$$Q_A = \Sigma \bar{y}' A' = 0.11(0.15)(0.02) + 0.05(0.1)(0.015)$$
$$= 0.405(10^{-3}) \text{ m}^3$$

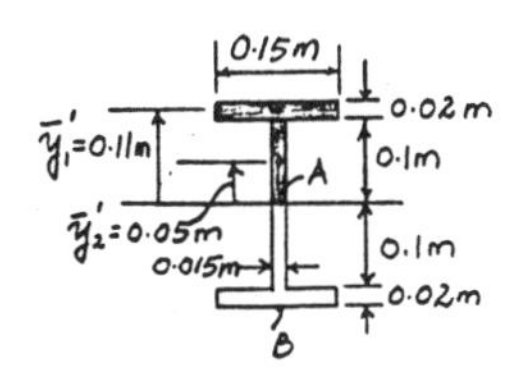

Normal Stress :

$$\sigma = \frac{N}{A} \pm \frac{My}{I}$$
$$\sigma_A = \frac{4.00(10^3)}{9.00(10^{-3})} + \frac{0.6667(10^3)(0)}{82.8(10^{-6})}$$
$$= 0.444 \text{ MPa (T)}$$

Ans

Shear Stress : Applying shear formula.

$$\tau_A = \frac{VQ_A}{It}$$
$$= \frac{0.6667(10^3)[0.405(10^{-3})]}{82.8(10^{-6})(0.015)} = 0.217 \text{ MPa}$$

Ans

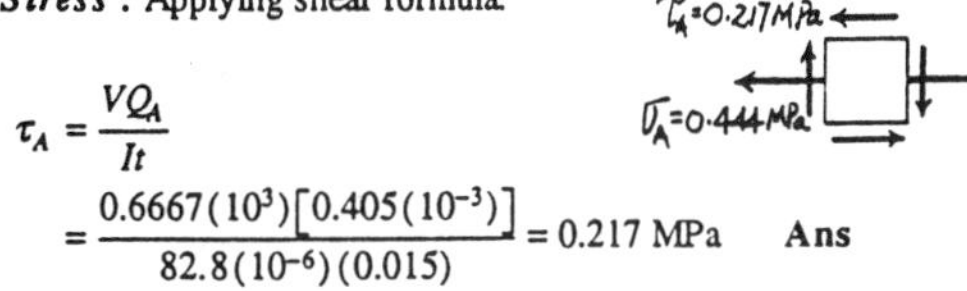

8–41. Determine the state of stress at point B when the beam is subjected to the cable force of 4 kN. Indicate the result as a differential volume element.

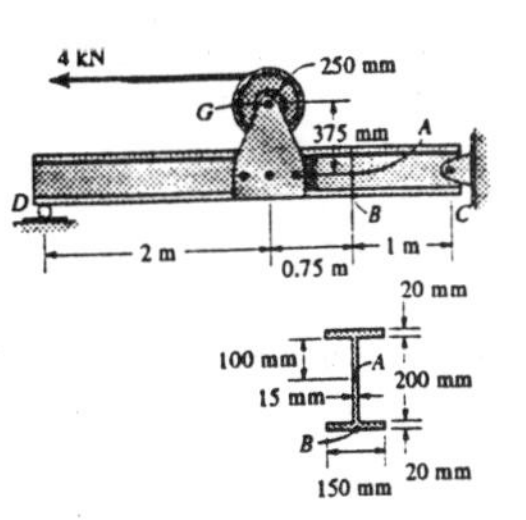

Support Reactions :

$$\curvearrowleft + \Sigma M_D = 0; \quad 4(0.625) - C_y(3.75) = 0$$
$$C_y = 0.6667 \text{ kN}$$
$$\xrightarrow{+} \Sigma F_x = 0; \quad C_x - 4 = 0 \quad C_x = 4.00 \text{ kN}$$

Internal Forces and Moment :

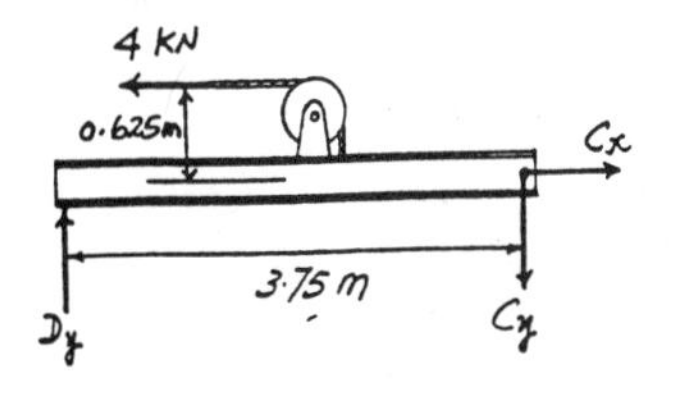

$$\xrightarrow{+} \Sigma F_x = 0; \quad 4.00 - N = 0 \quad N = 4.00 \text{ kN}$$
$$+\uparrow \Sigma F_y = 0; \quad V - 0.6667 = 0 \quad V = 0.6667 \text{ kN}$$
$$\curvearrowleft + \Sigma M_O = 0; \quad M - 0.6667(1) = 0 \quad M = 0.6667 \text{ kN}\cdot\text{m}$$

Section Properties :

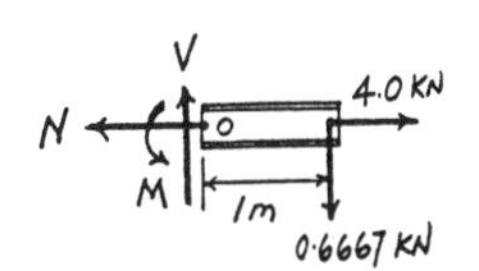

$$A = 0.24(0.15) - 0.2(0.135) = 9.00(10^{-3}) \text{ m}^2$$
$$I = \frac{1}{12}(0.15)(0.24^3) - \frac{1}{12}(0.135)(0.2^3) = 82.8(10^{-6}) \text{ m}$$
$$Q_B = 0$$

Normal Stress :

$$\sigma = \frac{N}{A} \pm \frac{My}{I}$$
$$\sigma_B = \frac{4.00(10^3)}{9.00(10^{-3})} - \frac{0.6667(10^3)(0.12)}{82.8(10^{-6})}$$
$$= -0.522 \text{ MPa} = 0.522 \text{ MPa (C)}$$

Ans

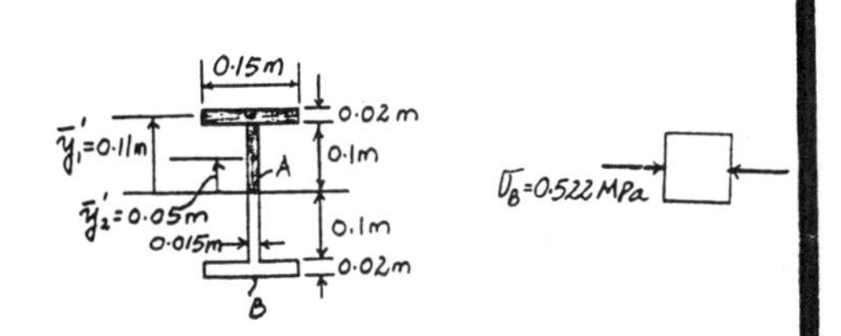

Shear Stress : Since $Q_B = 0$, then

$$\tau_B = 0$$

Ans

8–42. The uniform sign has a weight of 1500 lb and is supported by the pipe AB, which has an inner radius of 2.75 in. and an outer radius of 3.00 in. If the face of the sign is subjected to a uniform wind pressure of p = 150 lb/ft², determine the state of stress at points C and D. Show the results on a differential volume element located at each of these points. Neglect the thickness of the sign, and assume that it is supported along the outside edge of the pipe.

Internal Forces and Moments : As shown on FBD.

$\Sigma F_x = 0;$	$1.50 + N_x = 0$	$N_x = -1.50$ kip
$\Sigma F_y = 0;$	$V_y - 10.8 = 0$	$V_y = 10.8$ kip
$\Sigma F_z = 0;$	$V_z = 0$	
$\Sigma M_x = 0;$	$T_x - 10.8(6) = 0$	$T_x = 64.8$ kip · ft
$\Sigma M_y = 0;$	$M_y - 1.50(6) = 0$	$M_y = 9.00$ kip · ft
$\Sigma M_z = 0;$	$10.8(6) + M_z = 0$	$M_z = -64.8$ kip · ft

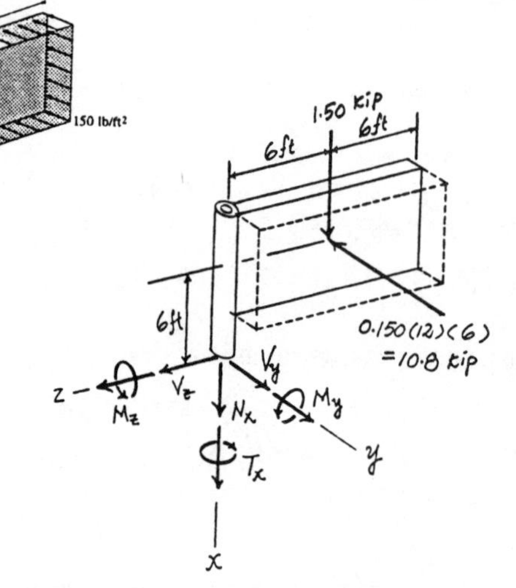

Section Properties :

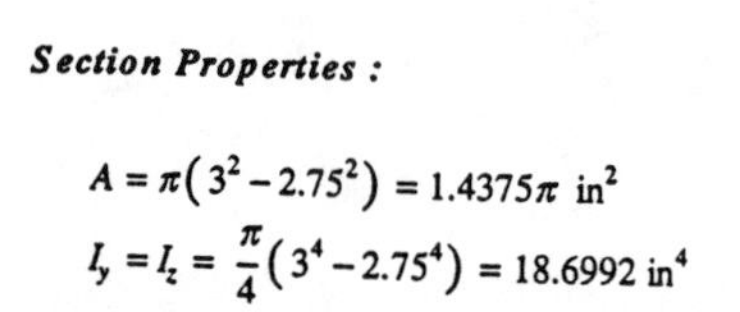

$$A = \pi(3^2 - 2.75^2) = 1.4375\pi \text{ in}^2$$

$$I_y = I_z = \frac{\pi}{4}(3^4 - 2.75^4) = 18.6992 \text{ in}^4$$

$$(Q_C)_z = (Q_D)_y = 0$$

$$(Q_C)_y = (Q_D)_z = \frac{4(3)}{3\pi}\left[\frac{1}{2}(\pi)(3^2)\right] - \frac{4(2.75)}{3\pi}\left[\frac{1}{2}(\pi)(2.75^2)\right]$$

$$= 4.13542 \text{ in}^3$$

$$J = \frac{\pi}{2}(3^4 - 2.75^4) = 37.3984 \text{ in}^4$$

Normal Stress :

$$\sigma = \frac{N}{A} - \frac{M_z y}{I_z} + \frac{M_y z}{I_y}$$

$$\sigma_C = \frac{-1.50}{1.4375\pi} - \frac{(-64.8)(12)(0)}{18.6992} + \frac{9.00(12)(2.75)}{18.6992}$$
$$= 15.6 \text{ ksi (T)} \qquad \textbf{Ans}$$

$$\sigma_D = \frac{-1.50}{1.4375\pi} - \frac{(-64.8)(12)(3)}{18.6992} + \frac{9.00(12)(0)}{18.6992}$$
$$= 124 \text{ ksi (T)} \qquad \textbf{Ans}$$

Shear Stress : The tranverse shear stress in the z and y directions and the torsional shear stress can be obtained using the shear formula and the torsion formula, $\tau_V = \frac{VQ}{It}$ and $\tau_{\text{twist}} = \frac{T\rho}{J}$, respectively.

$$(\tau_{xz})_D = \tau_{\text{twist}} = \frac{64.8(12)(3)}{37.3984} = 62.4 \text{ ksi} \qquad \textbf{Ans}$$

$$(\tau_{xy})_D = \tau_{V_y} = 0 \qquad \textbf{Ans}$$

$$(\tau_{xy})_C = \tau_{V_y} - \tau_{\text{twist}}$$
$$= \frac{10.8(4.13542)}{18.6992(2)(0.25)} - \frac{64.8(12)(2.75)}{37.3984}$$
$$= -52.4 \text{ ksi} \qquad \textbf{Ans}$$

$$(\tau_{xz})_C = \tau_{V_z} = 0 \qquad \textbf{Ans}$$

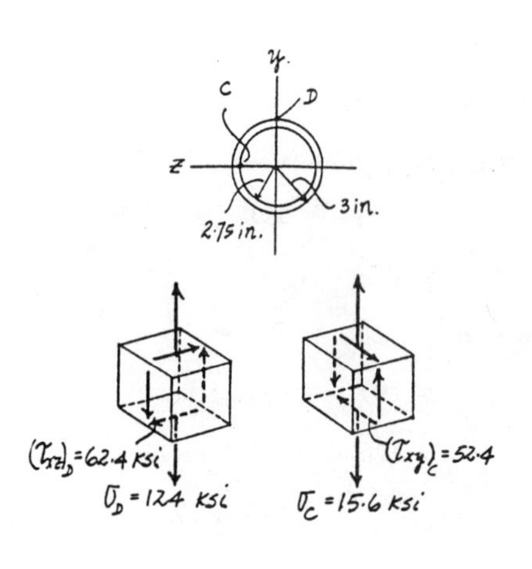

8–43. Solve Prob. 8–42 for points E and F.

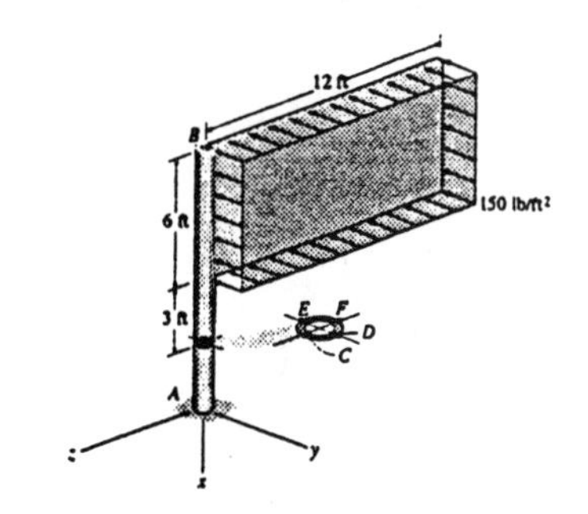

Internal Forces and Moments : As shown on FBD.

$$\Sigma F_x = 0; \quad 1.50 + N_x = 0 \quad N_x = -1.50 \text{ kip}$$
$$\Sigma F_y = 0; \quad V_y - 10.8 = 0 \quad V_y = 10.8 \text{ kip}$$
$$\Sigma F_z = 0; \quad V_z = 0$$
$$\Sigma M_x = 0; \quad T_x - 10.8(6) = 0 \quad T_x = 64.8 \text{ kip}\cdot\text{ft}$$
$$\Sigma M_y = 0; \quad M_y - 1.50(6) = 0 \quad M_y = 9.00 \text{ kip}\cdot\text{ft}$$
$$\Sigma M_z = 0; \quad 10.8(6) + M_z = 0 \quad M_z = -64.8 \text{ kip}\cdot\text{ft}$$

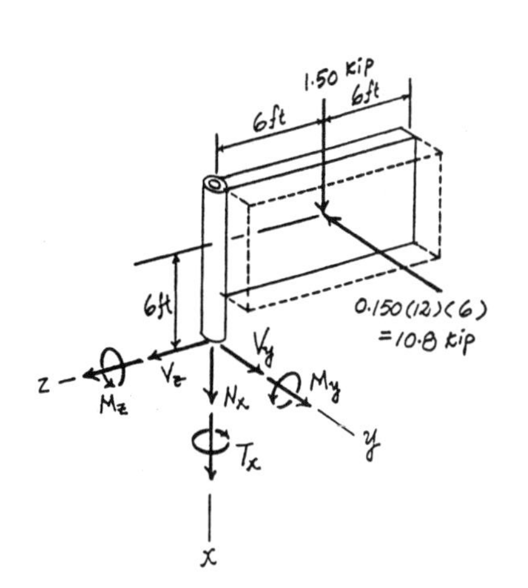

Section Properties :

$$A = \pi\left(3^2 - 2.75^2\right) = 1.4375\pi \text{ in}^2$$

$$I_y = I_z = \frac{\pi}{4}\left(3^4 - 2.75^4\right) = 18.6992 \text{ in}^4$$

$$(Q_C)_z = (Q_D)_y = 0$$

$$(Q_C)_y = (Q_D)_z = \frac{4(3)}{3\pi}\left[\frac{1}{2}(\pi)\left(3^2\right)\right] - \frac{4(2.75)}{3\pi}\left[\frac{1}{2}(\pi)\left(2.75^2\right)\right]$$
$$= 4.13542 \text{ in}^3$$

$$J = \frac{\pi}{2}\left(3^4 - 2.75^4\right) = 37.3984 \text{ in}^4$$

Normal Stress :

$$\sigma = \frac{N}{A} - \frac{M_z y}{I_z} + \frac{M_y z}{I_y}$$

$$\sigma_F = \frac{-1.50}{1.4375\pi} - \frac{(-64.8)(12)(0)}{18.6992} + \frac{9.00(12)(-3)}{18.6992}$$
$$= -17.7 \text{ ksi} = 17.7 \text{ ksi (C)} \qquad \textbf{Ans}$$

$$\sigma_E = \frac{-1.50}{1.4375\pi} - \frac{(-64.8)(12)(-3)}{18.6992} + \frac{9.00(12)(0)}{18.6992}$$
$$= -125 \text{ ksi} = 125 \text{ ksi (C)} \qquad \textbf{Ans}$$

Shear Stress : The tranverse shear stress in the z and y directions and the torsional shear stress can be obtained using the shear formula and the torsion formula, $\tau_V = \frac{VQ}{It}$ and $\tau_{\text{twist}} = \frac{T\rho}{J}$, respectively.

$$(\tau_{xz})_E = -\tau_{\text{twist}} = -\frac{64.8(12)(3)}{37.3984} = -62.4 \text{ ksi} \qquad \textbf{Ans}$$

$$(\tau_{xy})_E = \tau_{V_y} = 0 \qquad \textbf{Ans}$$

$$(\tau_{xy})_F = \tau_{V_y} + \tau_{\text{twist}}$$
$$= \frac{10.8(4.13542)}{18.6992(2)(0.25)} + \frac{64.8(12)(3)}{37.3984}$$
$$= 67.2 \text{ ksi} \qquad \textbf{Ans}$$

$$(\tau_{xz})_F = \tau_{V_z} = 0 \qquad \textbf{Ans}$$

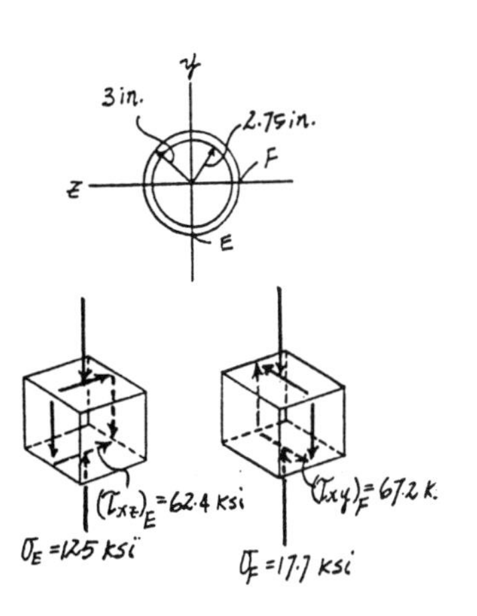

***8–44.** The solid rod is subjected to the loading shown. Determine the state of stress developed in the material at point A, and show the results on a differential volume element at this point.

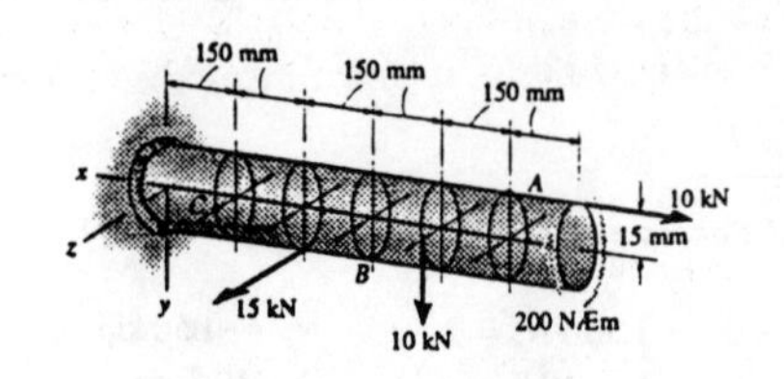

Internal Forces and Moments :

$$\Sigma F_x = 0; \quad N_x - 10 = 0 \quad N_x = 10.0 \text{ kN}$$
$$\Sigma F_y = 0; \quad V_y = 0$$
$$\Sigma F_z = 0; \quad V_z = 0$$
$$\Sigma M_x = 0; \quad T_x + 0.200 = 0 \quad T_x = -0.200 \text{ kN}\cdot\text{m}$$
$$\Sigma M_y = 0; \quad M_y = 0$$
$$\Sigma M_z = 0; \quad M_z - 10(0.03) = 0 \quad M_z = 0.300 \text{ kN}\cdot\text{m}$$

Section Properties :

$$A = \pi\left(0.03^2\right) = 0.900\left(10^{-3}\right)\pi \text{ m}^2$$
$$I_z = I_y = \frac{\pi}{4}\left(0.03^4\right) = 0.2025\left(10^{-6}\right)\pi \text{ m}^4$$
$$J = \frac{\pi}{2}\left(0.03^4\right) = 0.405\left(10^{-6}\right)\pi \text{ m}^4$$
$$(Q_A)_y = 0$$
$$(Q_A)_z = \frac{4(0.03)}{3\pi}\left[\frac{1}{2}\pi\left(0.03^2\right)\right] = 18.0\left(10^{-6}\right) \text{ m}^3$$

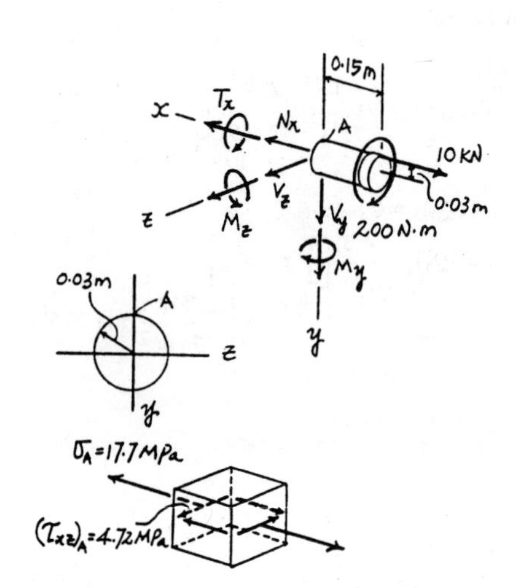

Normal Stress :

$$\sigma = \frac{N}{A} - \frac{M_z y}{I_z} + \frac{M_y z}{I_y}$$

$$\sigma_A = \frac{10.0(10^3)}{0.900(10^{-3})\pi} - \frac{0.300(10^3)(-0.03)}{0.2025(10^{-6})\pi} + 0$$
$$= 17.7 \text{ MPa (T)} \qquad \textbf{Ans}$$

Shear Stress : The tranverse shear stress in the z and y directions and the torsional shear stress can be obtained using the shear formula and the torsion formula, $\tau_V = \dfrac{VQ}{It}$ and $\tau_{twist} = \dfrac{T\rho}{J}$, respectively.

$$(\tau_{xz})_A = \tau_{twist} + \tau_{V_z}$$
$$= \frac{0.200(10^3)(0.03)}{0.405(10^{-6})\pi} + 0 = 4.72 \text{ MPa} \qquad \textbf{Ans}$$

$$(\tau_{xy})_A = \tau_{V_y} = 0 \qquad \textbf{Ans}$$

8–45. The solid rod is subjected to the loading shown. Determine the state of stress at point B, and show the results on a differential volume element located at this point.

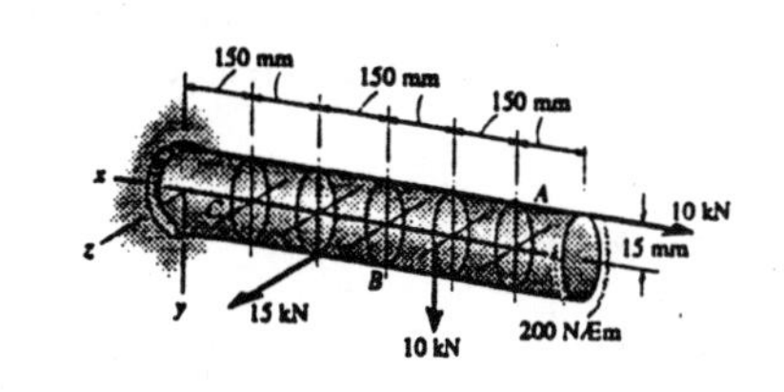

Internal Forces and Moments :

$\Sigma F_x = 0;\quad N_x - 10 = 0 \quad N_x = 10.0 \text{ kN}$

$\Sigma F_y = 0;\quad V_y + 10 = 0 \quad V_y = -10.0 \text{ kN}$

$\Sigma F_z = 0;\quad V_z = 0$

$\Sigma M_x = 0;\quad T_x + 0.200 - 10(0.03) = 0$

$T_x = 0.100 \text{ kN} \cdot \text{m}$

$\Sigma M_y = 0;\quad M_y = 0$

$\Sigma M_z = 0;\quad M_z - 10(0.03) - 10(0.15) = 0$

$M_z = 1.80 \text{ kN} \cdot \text{m}$

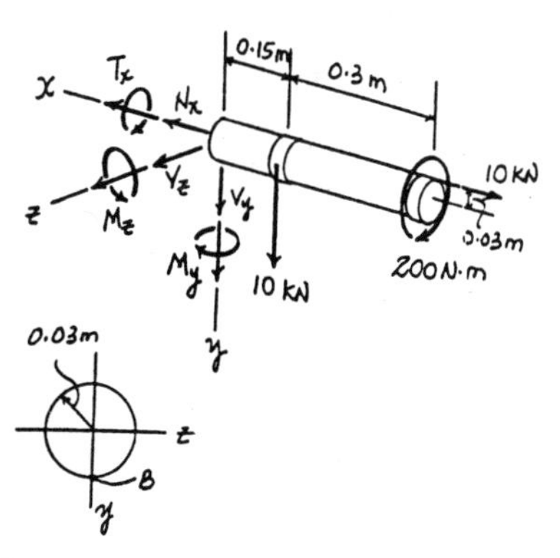

Section Properties :

$$A = \pi\left(0.03^2\right) = 0.900\left(10^{-3}\right)\pi \text{ m}^2$$

$$I_z = I_y = \frac{\pi}{4}\left(0.03^4\right) = 0.2025\left(10^{-6}\right)\pi \text{ m}^4$$

$$J = \frac{\pi}{2}\left(0.03^4\right) = 0.405\left(10^{-6}\right)\pi \text{ m}^4$$

$$(Q_B)_y = 0$$

$$(Q_B)_z = \frac{4(0.03)}{3\pi}\left[\frac{1}{2}\pi\left(0.03^2\right)\right] = 18.0\left(10^{-6}\right) \text{ m}^3$$

Normal Stress :

$$\sigma = \frac{N}{A} - \frac{M_z y}{I_z} + \frac{M_y z}{I_y}$$

$$\sigma_B = \frac{10.0(10^3)}{0.900(10^{-3})\pi} - \frac{1.80(10^3)(0.03)}{0.2025(10^{-6})\pi} + 0$$

$$= -81.3 \text{ MPa} = 81.3 \text{ MPa (C)} \qquad \textbf{Ans}$$

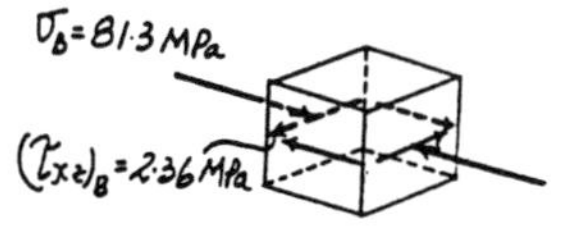

Shear Stress : The tranverse shear stress in the z and y directios and the torsional shear stress can be obtained using the shear formula and the torsion formula, $\tau_V = \frac{VQ}{It}$ and $\tau_{twist} = \frac{T\rho}{J}$, respectively.

$$(\tau_{xz})_B = \tau_{twist} + \tau_{V_z}$$

$$= \frac{0.100(10^3)(0.03)}{0.405(10^{-6})\pi} + 0 = 2.36 \text{ MPa} \qquad \textbf{Ans}$$

$$(\tau_{xy})_B = \tau_{V_y} = 0 \qquad \textbf{Ans}$$

8–46. The solid rod is subjected to the loading shown. Determine the state of stress at point C, and show the results on a differential volume element located at this point.

Internal Forces and Moment :

$\Sigma F_x = 0; \quad N_x - 10 = 0 \quad N_x = 10.0 \text{ kN}$

$\Sigma F_y = 0; \quad V_y + 10 = 0 \quad V_y = -10.0 \text{ kN}$

$\Sigma F_z = 0; \quad V_z + 15 = 0 \quad V_z = -15.0 \text{ kN}$

$\Sigma M_x = 0; \quad T_x + 0.200 - 10(0.03) + 15(0.03) = 0$

$T_x = -0.350 \text{ kN}\cdot\text{m}$

$\Sigma M_y = 0; \quad M_y + 15(0.15) = 0 \quad M_y = -2.25 \text{ kN}\cdot\text{m}$

$\Sigma M_z = 0; \quad M_z - 10(0.03) - 10(0.45) = 0$

$M_z = 4.80 \text{ kN}\cdot\text{m}$

Section Properties :

$$A = \pi(0.03^2) = 0.900(10^{-3})\,\pi \text{ m}^2$$

$$I_z = I_y = \frac{\pi}{4}(0.03^4) = 0.2025(10^{-6})\,\pi \text{ m}^4$$

$$J = \frac{\pi}{2}(0.03^4) = 0.405(10^{-6})\,\pi \text{ m}^4$$

$$(Q_C)_z = 0$$

$$(Q_C)_y = \frac{4(0.03)}{3\pi}\left[\frac{1}{2}\pi(0.03^2)\right] = 18.0(10^{-6}) \text{ m}^3$$

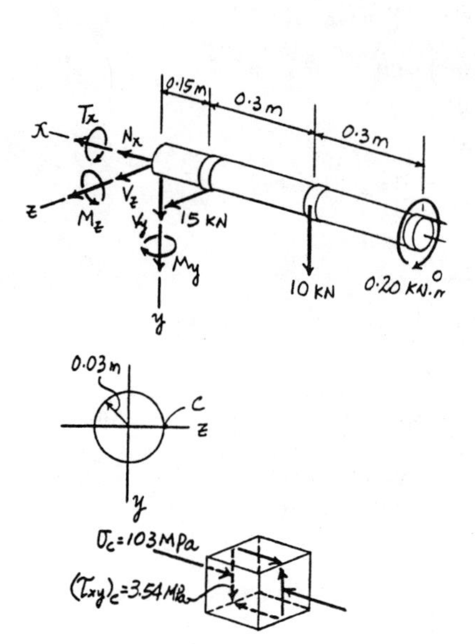

Normal Stress :

$$\sigma = \frac{N}{A} - \frac{M_z y}{I_z} + \frac{M_y z}{I_y}$$

$$\sigma_C = \frac{10.0(10^3)}{0.900(10^{-3})\,\pi} - \frac{4.80(10^3)(0)}{0.2025(10^{-6})\,\pi} + \frac{-2.25(10^3)(0.03)}{0.2025(10^{-6})\,\pi}$$

$= -103 \text{ MPa} = 103 \text{ MPa (C)}$ **Ans**

Shear Stress : The tranverse shear stress in the z and y directions and the torsional shear stress can be obtained using the shear formula and the torsion formula, $\tau_V = \frac{VQ}{It}$ and $\tau_{twist} = \frac{T\rho}{J}$, respectively.

$$(\tau_{xy})_C = \tau_{twist} - \tau_{V_y}$$

$$= \frac{0.350(10^3)(0.03)}{0.405(10^{-6})\,\pi} - \frac{10.0(10^3)\,18.0(10^{-6})}{0.2025(10^{-6})\,\pi(0.06)}$$

$= 3.54 \text{ MPa}$ **Ans**

$(\tau_{xz})_C = \tau_{V_z} = 0$ **Ans**

8–47. The bar has a diameter of 40 mm. If it is subjected to the two force components at its end as shown, determine the state of stress at point A and show the results on a differential volume element located at this point.

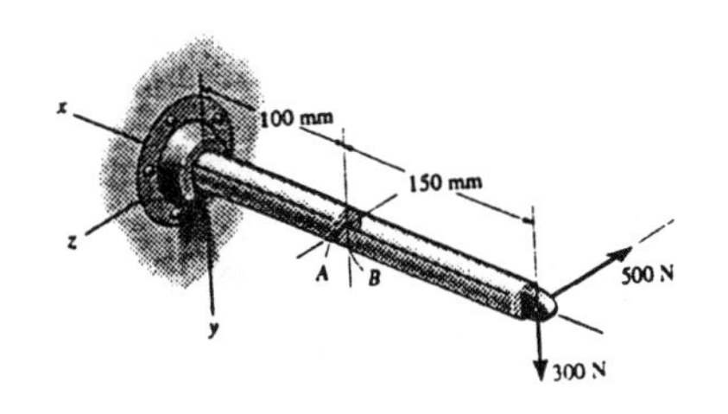

Internal Forces and Moment :

$\Sigma F_x = 0;\quad N_x = 0$

$\Sigma F_y = 0;\quad V_y + 300 = 0\quad V_y = -300\text{ N}$

$\Sigma F_z = 0;\quad V_z - 500 = 0\quad V_z = 500\text{ N}$

$\Sigma M_x = 0;\quad T_x = 0$

$\Sigma M_y = 0;\quad M_y - 500(0.15) = 0\quad M_y = 75.0\text{ N}\cdot\text{m}$

$\Sigma M_z = 0;\quad M_z - 300(0.15) = 0\quad M_z = 45.0\text{ N}\cdot\text{m}$

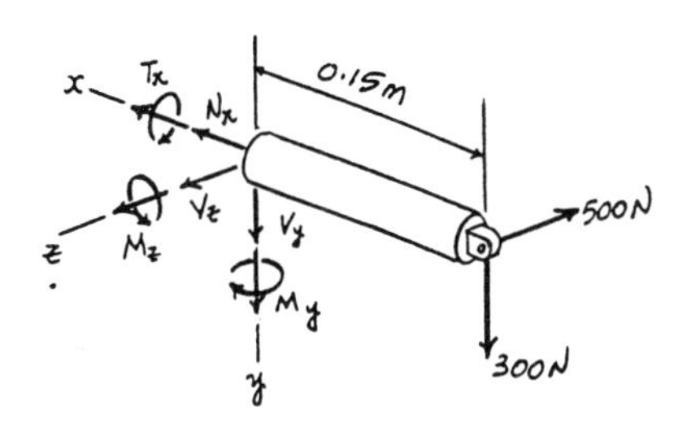

Section Properties :

$$A = \pi\left(0.02^2\right) = 0.400\left(10^{-3}\right)\pi\text{ m}^2$$

$$I_z = I_y = \frac{\pi}{4}\left(0.02^4\right) = 40.0\left(10^{-9}\right)\pi\text{ m}^4$$

$$J = \frac{\pi}{2}\left(0.02^4\right) = 80.0\left(10^{-9}\right)\pi\text{ m}^4$$

$$(Q_A)_z = 0$$

$$(Q_A)_y = \frac{4(0.02)}{3\pi}\left[\frac{1}{2}\pi\left(0.02^2\right)\right] = 5.333\left(10^{-6}\right)\text{ m}^3$$

Normal Stress :

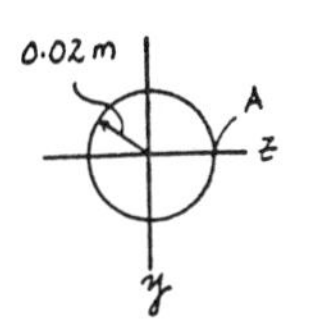

$$\sigma = \frac{N}{A} - \frac{M_z y}{I_z} + \frac{M_y z}{I_y}$$

$$\sigma_A = 0 - \frac{45.0(0)}{40.0(10^{-9})\pi} + \frac{75.0(0.02)}{40.0(10^{-9})\pi}$$

$$= 11.9\text{ MPa (T)}\qquad\textbf{Ans}$$

Shear Stress : The tranverse shear stress in the z and y directions can be obtained using the shear formula, $\tau_V = \frac{VQ}{It}$.

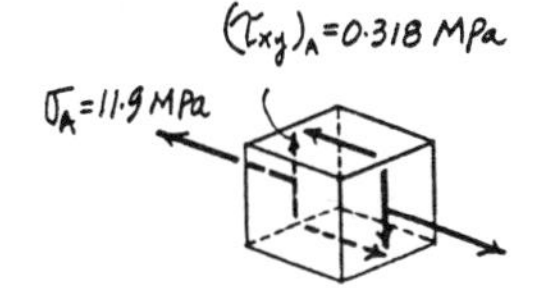

$$(\tau_{xy})_A = -\tau_{V_y} = -\frac{300\left[5.333(10^{-6})\right]}{40.0(10^{-9})\pi\ (0.04)}$$

$$= -0.318\text{ MPa}\qquad\textbf{Ans}$$

$$(\tau_{xz})_A = \tau_{V_z} = 0\qquad\textbf{Ans}$$

***8–48.** Solve Prob. 8–47 for point B.

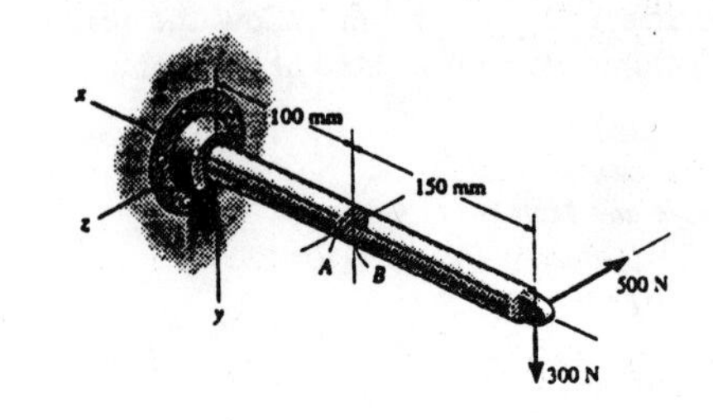

Internal Forces and Moment :

$\Sigma F_x = 0;\quad N_x = 0$

$\Sigma F_y = 0;\quad V_y + 300 = 0 \qquad V_y = -300\text{ N}$

$\Sigma F_z = 0;\quad V_z - 500 = 0 \qquad V_z = 500\text{ N}$

$\Sigma M_x = 0;\quad T_x = 0$

$\Sigma M_y = 0;\quad M_y - 500(0.15) = 0 \qquad M_y = 75.0\text{ N}\cdot\text{m}$

$\Sigma M_z = 0;\quad M_z - 300(0.15) = 0 \qquad M_z = 45.0\text{ N}\cdot\text{m}$

Section Properties :

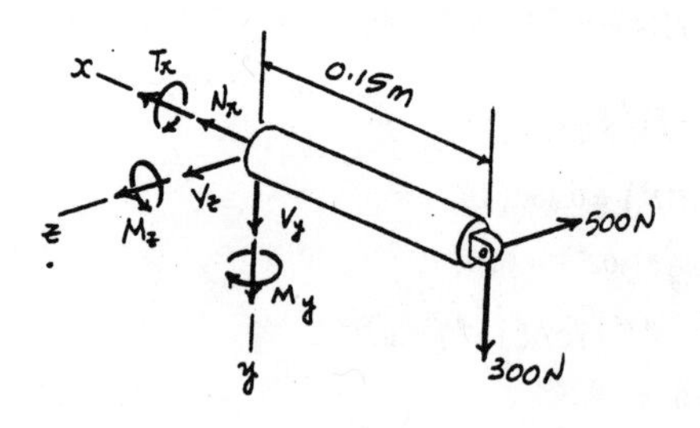

$$A = \pi(0.02^2) = 0.400(10^{-3})\,\pi\text{ m}^2$$

$$I_x = I_y = \frac{\pi}{4}(0.02^4) = 40.0(10^{-9})\,\pi\text{ m}^4$$

$$J = \frac{\pi}{2}(0.02^4) = 80.0(10^{-9})\,\pi\text{ m}^4$$

$$(Q_B)_y = 0$$

$$(Q_B)_z = \frac{4(0.02)}{3\pi}\left[\frac{1}{2}\pi(0.02^2)\right] = 5.333(10^{-6})\text{ m}^3$$

Normal Stress :

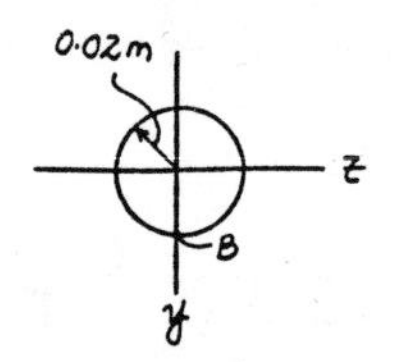

$$\sigma = \frac{N}{A} - \frac{M_z y}{I_z} + \frac{M_y z}{I_y}$$

$$\sigma_B = 0 - \frac{45.0(0.02)}{40.0(10^{-9})\,\pi} + \frac{75.0(0)}{40.0(10^{-9})\,\pi}$$

$$= -7.16\text{ MPa} = 7.16\text{ MPa (C)} \qquad \textbf{Ans}$$

Shear Stress : The tranverse shear stress in the z and y directions can be obtained using the shear formula, $\tau_V = \dfrac{VQ}{It}$.

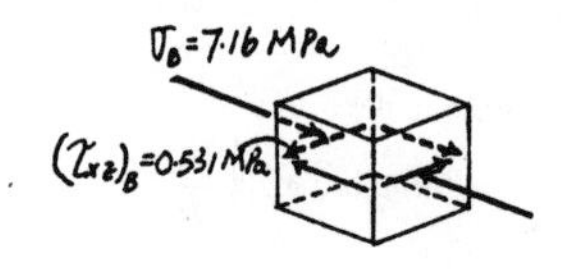

$$(\tau_{xz})_B = \tau_{V_z} = \frac{500[5.333(10^{-6})]}{40.0(10^{-9})\,\pi\ (0.04)}$$

$$= 0.531\text{ MPa} \qquad \textbf{Ans}$$

$$(\tau_{xy})_B = \tau_{V_y} = 0 \qquad \textbf{Ans}$$

8–49. The caster wheel supports a reactive load of 180 N. Determine the state of stress at points A and B on one of the two supporting leaves. Show the results on a differential volume element located at each point.

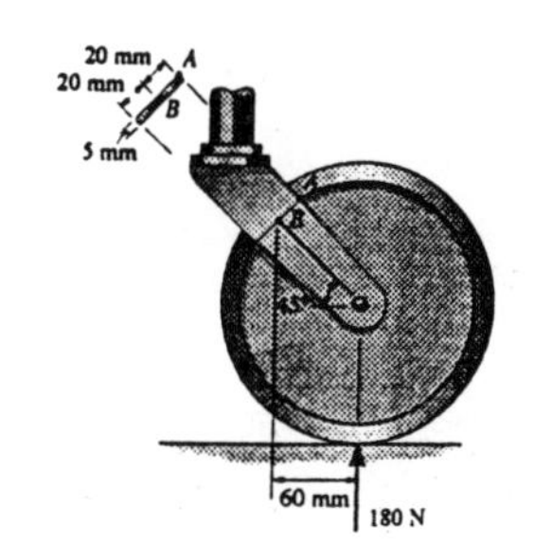

Internal Forces and Moment :

$\circlearrowleft + \Sigma M_B = 0;\quad 180(0.06) - M = 0 \qquad M = 10.8\ \text{N}\cdot\text{m}$

$\nwarrow + \Sigma F_{y'} = 0;\quad N + 180\sin 45° = 0 \qquad N = -127.28\ \text{N}$

$\nearrow + \Sigma F_{x'} = 0;\quad 180\cos 45° - V = 0 \qquad V = 127.28\ \text{N}$

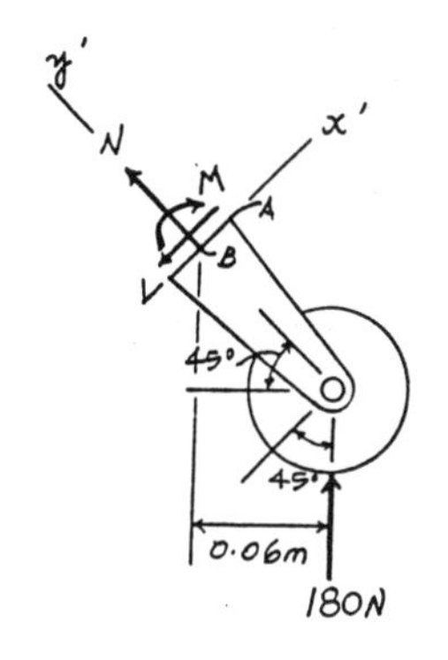

Section Properties :

$$A = 0.01(0.04) = 0.400\left(10^{-3}\right)\ \text{m}^2$$

$$I = \frac{1}{12}(0.01)\left(0.04^3\right) = 53.333\left(10^{-9}\right)\ \text{m}^4$$

$$Q_A = 0$$

$$Q_B = \bar{y}'A' = 0.01(0.01)(0.02) = 2.00\left(10^{-6}\right)\ \text{m}^3$$

Normal Stress :

$$\sigma = \frac{N}{A} \pm \frac{My}{I}$$

$$\sigma_A = \frac{-127.28}{0.400(10^{-3})} - \frac{10.8(0.02)}{53.333(10^{-9})}$$

$$= -4.37\ \text{MPa} = 4.37\ \text{MPa (C)} \qquad \textbf{Ans}$$

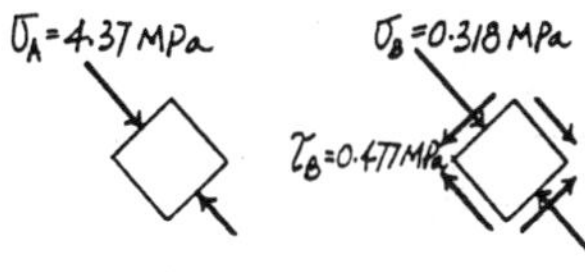

$$\sigma_B = \frac{-127.28}{0.400(10^{-3})} + \frac{10.8(0)}{53.333(10^{-9})}$$

$$= -0.318\ \text{MPa} = 0.318\ \text{MPa (C)} \qquad \textbf{Ans}$$

Shear Stress : Applying the shear formula,

$$\tau = \frac{VQ}{It}$$

$$\tau_A = \frac{127.28(0)}{53.333(10^{-9})(0.01)} = 0 \qquad \textbf{Ans}$$

$$\tau_B = \frac{127.28\left[2.00(10^{-6})\right]}{53.333(10^{-9})(0.01)} = 0.477\ \text{MPa} \qquad \textbf{Ans}$$

8–50. The crane boom is subjected to the load of 500 lb. Determine the state of stress at points A and B. Show the results on a differential volume element located at each of these points.

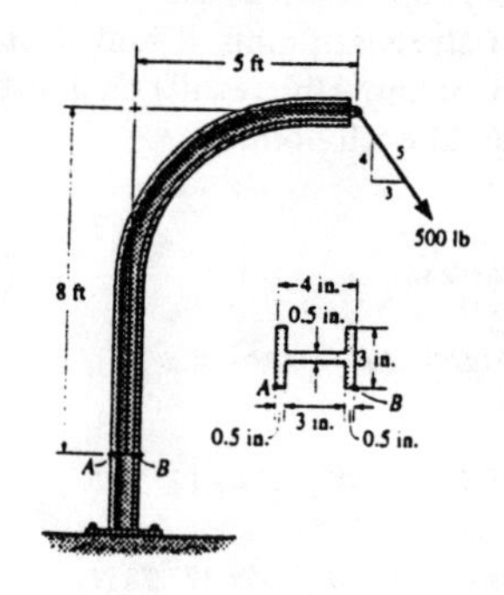

Internal Forces and Moment :

$$\circlearrowleft + \Sigma M_O = 0; \quad M - \frac{3}{5}(500)(8) - \frac{4}{5}(500)(5) = 0$$

$$M = 4400 \text{ lb} \cdot \text{ft}$$

$$\xrightarrow{+} \Sigma F_x = 0; \quad V - \frac{3}{5}(500) = 0 \qquad V = 300 \text{ lb}$$

$$+\uparrow \Sigma F_y = 0; \quad N + \frac{4}{5}(500) = 0 \qquad N = -400 \text{ lb}$$

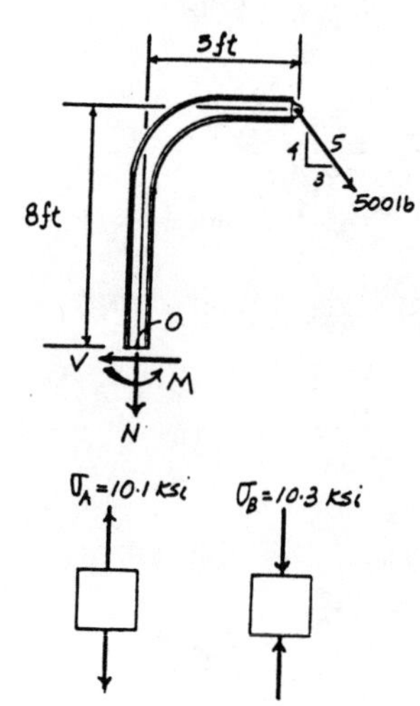

Section Properties :

$$A = 4(3) - 3(2.5) = 4.50 \text{ in}^2$$

$$I = \frac{1}{12}(3)(4^3) - \frac{1}{12}(2.5)(3^3) = 10.375 \text{ in}^4$$

$$Q_A = Q_B = 0$$

Normal Stress :

$$\sigma = \frac{N}{A} \pm \frac{My}{I}$$

$$\sigma_A = \frac{-400}{4.50} + \frac{4400(12)(2)}{10.375}$$

$$= 10089 \text{ psi} = 10.1 \text{ ksi (T)} \qquad \textbf{Ans}$$

$$\sigma_B = \frac{-400}{4.50} - \frac{4400(12)(2)}{10.375}$$

$$= -10267 \text{ psi} = 10.3 \text{ ksi (C)} \qquad \textbf{Ans}$$

Shear Stress : Since $Q_A = Q_B = 0$, then

$$\tau_A = \tau_B = 0 \qquad \textbf{Ans}$$

8–51. The beam supports the loading shown. Determine the state of stress at points E and F at section a–a, and represent the results on a differential volume element located at each of these points.

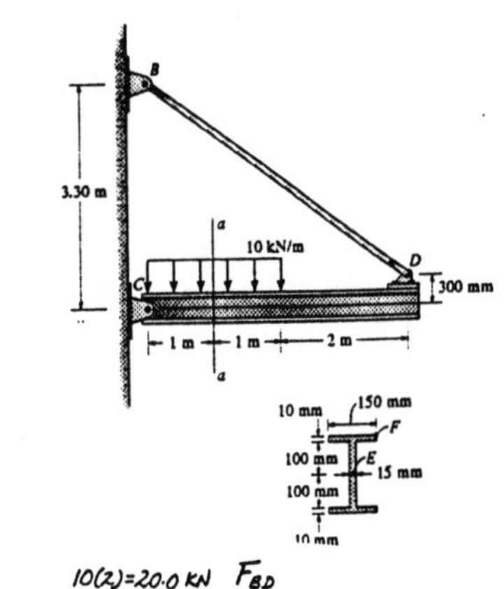

Support Reactions :

$$\curvearrowleft + \Sigma M_C = 0; \quad \frac{3}{5}F_{BD}(4) + \frac{4}{5}F_{BD}(0.3) - 20.0(1) = 0$$

$$F_{BD} = 7.5758 \text{ kN}$$

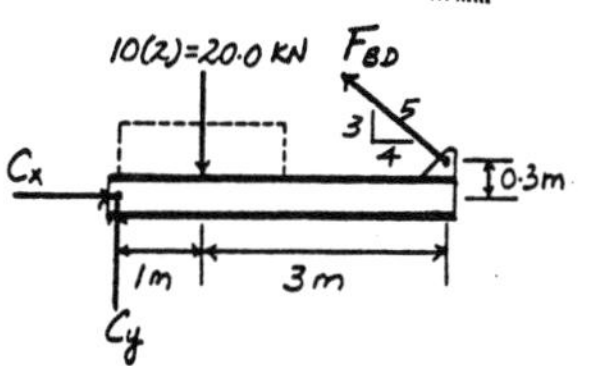

Internal Forces and Moment :

$$\xrightarrow{+} \Sigma F_x = 0; \quad -\frac{4}{5}(7.5758) - N = 0 \quad N = 6.0606 \text{ kN}$$

$$+\uparrow \Sigma F_y = 0; \quad V + \frac{3}{5}(7.5758) - 10 = 0 \quad V = 5.4545 \text{ kN}$$

$$\curvearrowleft + \Sigma M_B = 0; \quad \frac{4}{5}(7.5758)(0.3) + \frac{3}{5}(7.5758)(3) - 10.0(0.5) - M = 0$$

$$M = 10.4545 \text{ kN} \cdot \text{m}$$

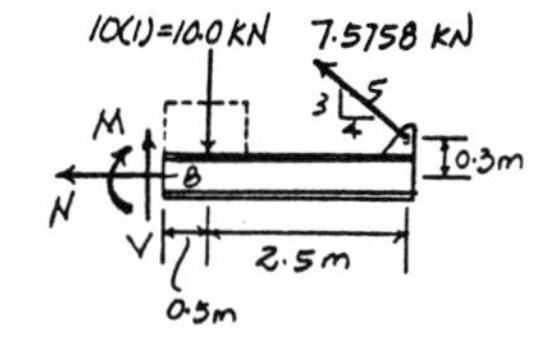

Section Properties :

$$A = 0.22(0.15) - 0.2(0.135) = 6.00\left(10^{-3}\right) \text{ m}^2$$

$$I = \frac{1}{12}(0.15)\left(0.22^3\right) - \frac{1}{12}(0.135)\left(0.2^3\right) = 43.1\left(10^{-6}\right) \text{ m}^4$$

$$Q_E = \Sigma \bar{y}'A' = 0.105(0.15)(0.01) + 0.05(0.1)(0.015)$$

$$= 0.2325\left(10^{-3}\right) \text{ m}^3$$

$$Q_F = 0$$

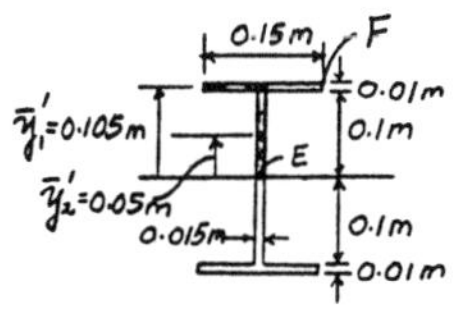

Normal Stress :

$$\sigma = \frac{N}{A} \pm \frac{My}{I}$$

$$\sigma_E = \frac{-6.0606(10^3)}{6.00(10^{-3})} + \frac{10.4545(10^3)(0)}{43.1(10^{-6})}$$

$$= -1.01 \text{ MPa} = 1.01 \text{ MPa (C)} \qquad \textbf{Ans}$$

$$\sigma_F = \frac{-6.0606(10^3)}{6.00(10^{-3})} - \frac{10.4545(10^3)(0.11)}{43.1(10^{-6})}$$

$$= -27.7 \text{ MPa} = 27.7 \text{ MPa (C)} \qquad \textbf{Ans}$$

Shear Stress : Applying the shear formula,

$$\tau = \frac{VQ}{It}$$

$$\tau_E = \frac{5.4545(10^3)\left[0.2325(10^{-3})\right]}{43.1(10^{-6})(0.015)} = 1.96 \text{MPa} \qquad \textbf{Ans}$$

$$\tau_F = 0 \qquad \textbf{Ans}$$

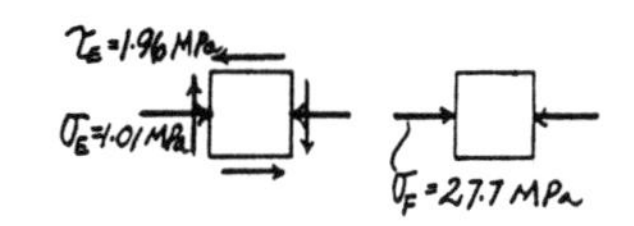

*8–52. The symmetrically loaded spreader bar is used to lift the 2000-lb tank. Determine the state of stress at points A and B, and indicate the results on a differential volume elements.

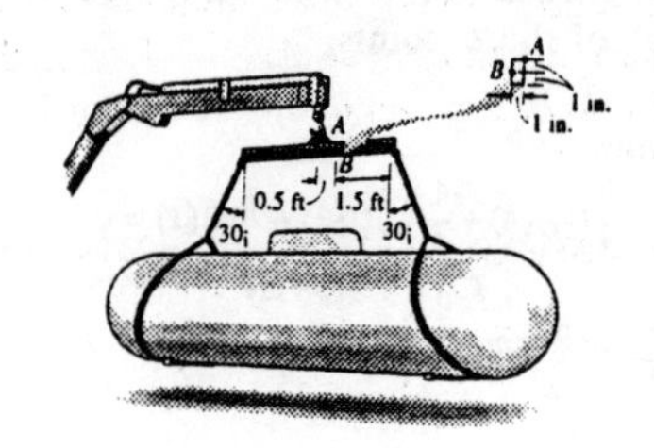

Support Reactions :

$$+\uparrow \Sigma F_y = 0; \quad 2000 - 2F\cos 30° = 0 \quad F = 1154.70 \text{ lb}$$

Internal Forces and Moment :

$$\overset{+}{\rightarrow} \Sigma F_x = 0; \quad 1154.70\sin 30° - N = 0 \quad N = 577.35 \text{ lb}$$

$$+\uparrow \Sigma F_y = 0; \quad V - 1154.70\cos 30° = 0 \quad V = 1000 \text{ lb}$$

$$\circlearrowleft + \Sigma M_B = 0; \quad M - 1154.70\cos 30°(1.5) = 0$$

$$M = 1500 \text{ lb} \cdot \text{ft}$$

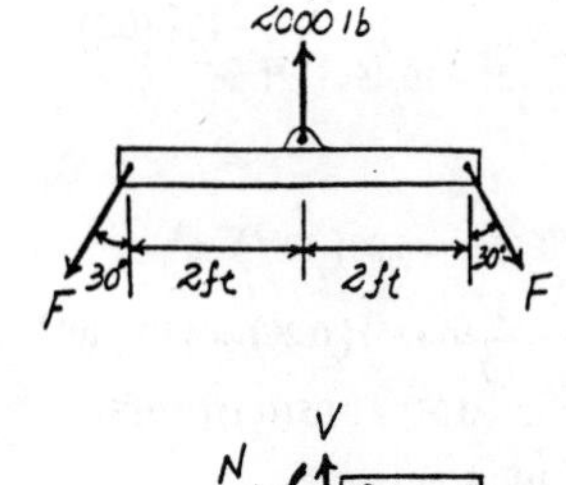

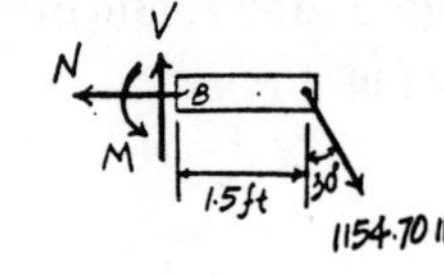

Section Properties :

$$A = 1(2) = 2.00 \text{ in}^2$$

$$I = \frac{1}{12}(1)(2^3) = 0.6666 \text{ in}^4$$

$$Q_B = \bar{y}'A' = 0.5(1)(1) = 0.500 \text{ in}^3$$

$$Q_A = 0$$

Normal Stress :

$$\sigma = \frac{N}{A} \pm \frac{My}{I}$$

$$\sigma_A = \frac{577.35}{2.00} + \frac{1500(12)(1)}{0.6666}$$

$$= 27\,300 \text{ psi} = 27.3 \text{ ksi (T)} \qquad \textbf{Ans}$$

$$\sigma_B = \frac{577.35}{2.00} + \frac{1500(12)(0)}{0.6666}$$

$$= 289 \text{ psi} = 0.289 \text{ ksi (T)} \qquad \textbf{Ans}$$

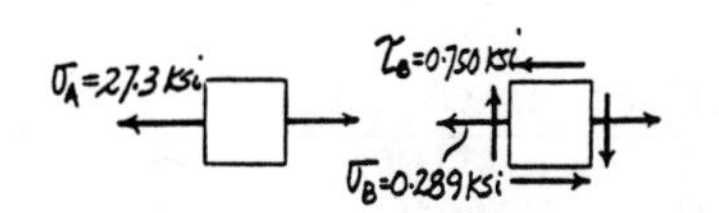

Shear Stress : Applying the shear formula,

$$\tau = \frac{VQ}{It}$$

$$\tau_A = 0 \qquad \textbf{Ans}$$

$$\tau_B = \frac{1000(0.500)}{0.6666(1)}$$

$$= 750 \text{ psi} = 0.750 \text{ ksi} \qquad \textbf{Ans}$$

8–53. The tine ABC of the fork lift is subjected to a uniform distributed loading as shown. If it is pin connected at C and roller supported at B, determine the state of stress at points D and E, and show the results on differential elements. The tine is 3 in. wide and 0.5 in. thick.

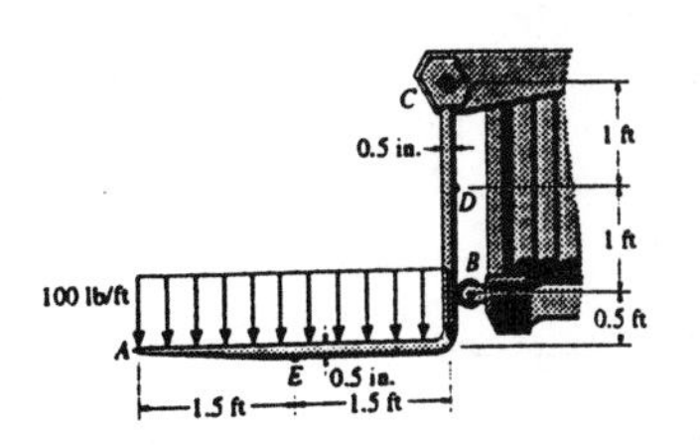

Support Reactions :

$$\circlearrowleft + \Sigma M_C = 0; \quad 300(1.5) - F_B(2) = 0 \quad F_B = 225 \text{ lb}$$

$$\xrightarrow{+} \Sigma F_x = 0; \quad C_x - 225 = 0 \quad C_x = 225 \text{ lb}$$

$$+\uparrow \Sigma F_y = 0; \quad C_y - 300 = 0 \quad C_y = 300 \text{ lb}$$

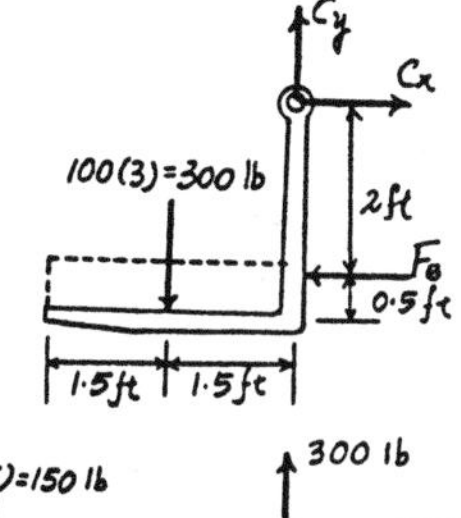

Internal Forces and Moment :

For point D

$$\xrightarrow{+} \Sigma F_x = 0; \quad 225 - V_D = 0 \quad V_D = 225 \text{ lb}$$

$$+\uparrow \Sigma F_y = 0; \quad 300 - N_D = 0 \quad N_D = 300 \text{ lb}$$

$$\circlearrowleft + \Sigma M_O = 0; \quad M_D - 225(1) = 0 \quad M_D = 225 \text{ lb} \cdot \text{ft}$$

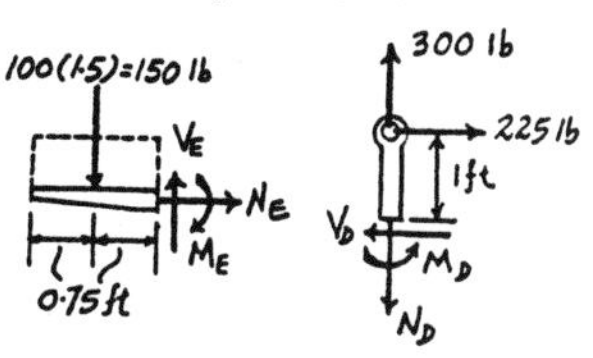

For Point E

$$\xrightarrow{+} \Sigma F_x = 0; \quad N_E = 0$$

$$+\uparrow \Sigma F_y = 0; \quad V_E - 150 = 0 \quad V_E = 150 \text{ lb}$$

$$\circlearrowleft + \Sigma M_O = 0; \quad 150(0.75) - M_E = 0 \quad M_E = 112.5 \text{ lb} \cdot \text{ft}$$

Section Properties :

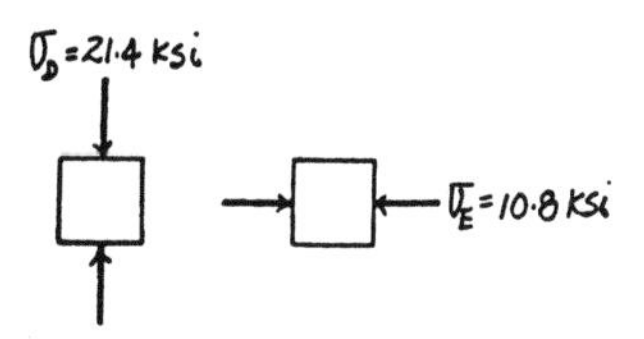

$$A = 3(0.5) = 1.50 \text{ in}^2$$

$$I = \frac{1}{12}(3)(0.5^3) = 0.03125 \text{ in}^4$$

$$Q_D = Q_E = 0$$

Normal Stress :

$$\sigma = \frac{N}{A} \pm \frac{My}{I}$$

$$\sigma_D = \frac{300}{1.5} - \frac{225(12)(0.25)}{0.03125}$$

$$= -21400 \text{ psi} = 21.4 \text{ ksi (C)} \qquad \textbf{Ans}$$

$$\sigma_E = 0 - \frac{112.5(12)(0.25)}{0.03125}$$

$$= -10800 \text{ psi} = 10.8 \text{ ksi (C)} \qquad \textbf{Ans}$$

Shear Stress : Since $Q_D = Q_E = 0$, then

$$\tau_D = \tau_E = 0 \qquad \textbf{Ans}$$

8–54. The strongback AB consists of a pipe that is used to lift the bundle of rods having a total mass of 3 Mg and center of mass at G. If the pipe has an outer diameter of 70 mm and a wall thickness of 10 mm, determine the state of stress acting at point C. Show the results on a differential volume element located at this point. Neglect the weight of the pipe.

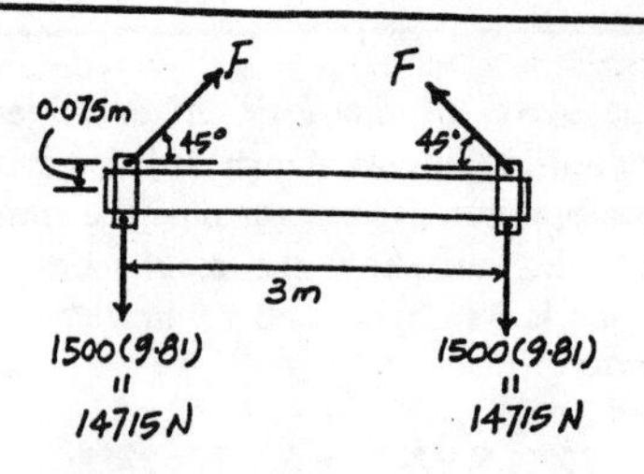

Support Reactions :

$$+\uparrow \Sigma F_y = 0; \quad 2F\sin 45° - 2(14\,715) = 0$$
$$F = 20\,810 \text{ N}$$

Internal Forces and Moment :

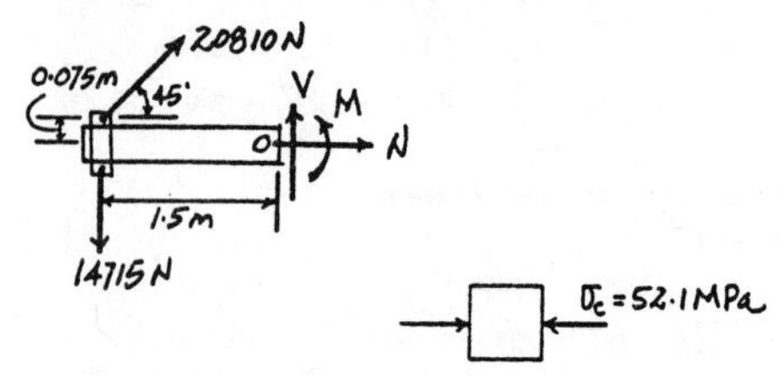

$$\overset{+}{\rightarrow} \Sigma F_x = 0; \quad 20\,810\cos 45° + N = 0 \qquad N = -14\,715 \text{ N}$$

$$+\uparrow \Sigma F_y = 0; \quad V + 20\,810\sin 45° - 14715 = 0 \qquad V = 0$$

$$\circlearrowleft + \Sigma M_O = 0; \quad M + 14\,715(1.5) - 20\,810\cos 45°(0.075) - 20\,810\sin 45°(1.5) = 0$$
$$M = 1103.625 \text{ N} \cdot \text{m}$$

Section Properties :

$$A = \pi\left(0.035^2 - 0.025^2\right) = 0.600\pi\left(10^{-3}\right) \text{ m}^2$$
$$I = \frac{\pi}{4}\left(0.035^4 - 0.025^4\right) = 0.2775\pi\left(10^{-6}\right) \text{ m}^4$$

Normal Stress :

$$\sigma = \frac{N}{A} \pm \frac{My}{I}$$

$$\sigma_C = \frac{-14\,715}{0.600\pi(10^{-3})} - \frac{1103.625(0.035)}{0.2775\pi(10^{-6})}$$
$$= -52.1 \text{ MPa} = 52.1 \text{ MPa (C)} \qquad \textbf{Ans}$$

Shear Stress : Since $V = 0$, then

$$\tau_C = 0 \qquad \textbf{Ans}$$

8–55. The strongback AB consists of a pipe that is used to lift the bundle of rods having a total mass of 3 Mg and center of mass at G. If the pipe has an outer diameter of 70 mm and a wall thickness of 10 mm, determine the state of stress acting at point D. Show the results on a differential volume element located at this point. Neglect the weight of the pipe.

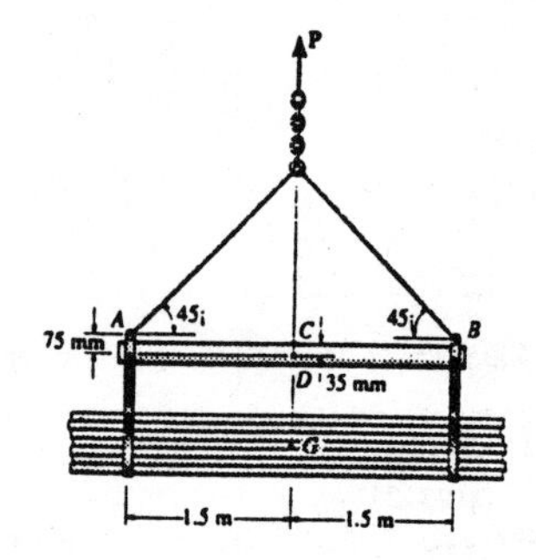

Support Reactions :

$$+\uparrow \Sigma F_y = 0; \quad 2F\sin 45° - 2(14715) = 0$$
$$F = 20\,810 \text{ N}$$

Internal Forces and Moment :

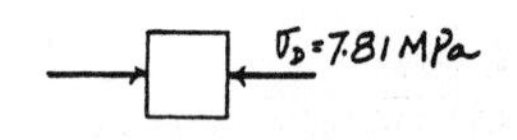

$$\overset{+}{\rightarrow} \Sigma F_x = 0; \quad 20\,810\cos 45° + N = 0 \qquad N = -14715 \text{ N}$$

$$+\uparrow \Sigma F_y = 0; \quad V + 20\,810\sin 45° - 14\,715 = 0 \qquad V = 0$$

$$\circlearrowleft + \Sigma M_O = 0; \quad M + 14\,715(1.5) - 20\,810\cos 45°(0.075) + 20\,810\sin 45°(1.5) = 0$$
$$M = 1103.625 \text{ N} \cdot \text{m}$$

Section Properties :

$$A = \pi\left(0.035^2 - 0.025^2\right) = 0.600\pi\left(10^{-3}\right) \text{ m}^2$$
$$I = \frac{\pi}{4}\left(0.035^4 - 0.025^4\right) = 0.2775\pi\left(10^{-6}\right) \text{ m}^4$$

Normal Stress :

$$\sigma = \frac{N}{A} \pm \frac{My}{I}$$

$$\sigma_D = \frac{-14\,715}{0.600\pi(10^{-3})} + \frac{1103.625(0)}{0.2775\pi(10^{-6})}$$
$$= -7.81 \text{ MPa} = 7.81 \text{ MPa (C)} \qquad \textbf{Ans}$$

Shear Stress : Since $V = 0$, then

$$\tau_D = 0 \qquad \textbf{Ans}$$

***8–56.** The member is subjected to the combination of distributed and concentrated loadings shown. Determine the state of stress at point A, and show the results on a differential volume element located at this point. *Hint:* Use Table 5-1.

Internal Forces and Moments :

$\Sigma F_y = 0;\quad V_y - 200 - 200 = 0 \quad V_y = 400\text{ N}$

$\Sigma F_z = 0;\quad V_z - 80.0 - 20.0 = 0 \quad V_z = 100\text{ N}$

$\Sigma M_x = 0;\quad T_x - 200(0.05) + 80.0(0.05) + 20.0(0.05) = 0$

$T_x = 5.00\text{ N}\cdot\text{m}$

$\Sigma M_y = 0;\quad M_y - 80.0(0.1) - 20.0(0.1333) = 0$

$M_y = 10.667\text{ N}\cdot\text{m}$

$\Sigma M_z = 0;\quad M_z + 200(0.2) + 200(0.1) = 0$

$M_z = -60.0\text{ N}\cdot\text{m}$

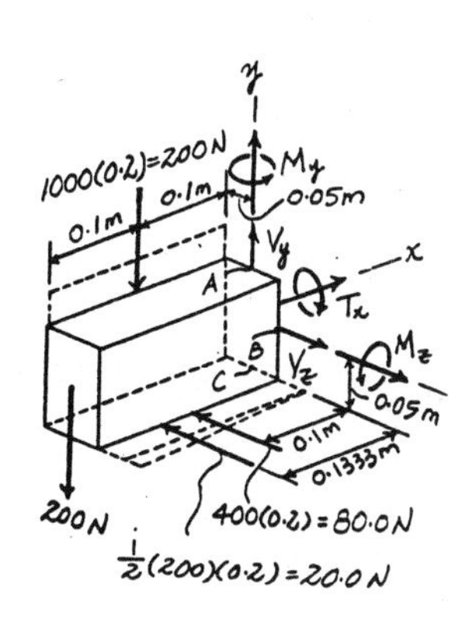

Section Properties :

$$A = 0.1(0.1) = 0.01\text{ m}^2$$

$$I_z = I_y = \frac{1}{12}(0.1)\left(0.1^3\right) = 8.333\left(10^{-6}\right)\text{ m}^4$$

$$(Q_B)_z = 0$$

$$(Q_B)_y = \bar{y}'A' = 0.025(0.05)(0.1) = 0.125\left(10^{-3}\right)\text{ m}^3$$

Normal Stress :

$$\sigma = -\frac{M_z y}{I_z} + \frac{M_y z}{I_y}$$

$$\sigma_B = -\frac{-60.0(0)}{8.333(10^{-6})} + \frac{10.667(0.05)}{8.333(10^{-6})}$$

$$= 64.0\text{ kPa (T)} \qquad \textbf{Ans}$$

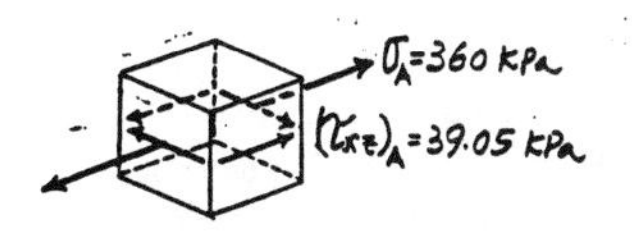

Shear Stress : The shear stress in the *y direction* is caused by the transverse shear stress in that direction and torsional shear stress. The transverse shear stress can be obtained using the shear formula, whereas the torsional shear stress formula for the square is given in Table 5 – 1 of the text.

$$(\tau_{xy})_B = \frac{V_y (Q_B)_y}{I_z t} - \frac{4.81 T_x}{a^3}$$

$$= \frac{400\left[0.125(10^{-3})\right]}{8.333(10^{-6})(0.1)} - \frac{4.81(5.00)}{0.1^3}$$

$$= 35.95\text{ kPa} \qquad \textbf{Ans}$$

The shear stress in the *z direction* is caused by transverse shear stress in that direction only. Since $(Q_B)_z = 0$, then

$$(\tau_{xz})_B = 0 \qquad \textbf{Ans}$$

8–57. The member is subjected to the combination of distributed and concentrated loadings shown. Determine the state of stress at point B, and show the results on a differential volume element located at this point. *Hint:* Use Table 5-1.

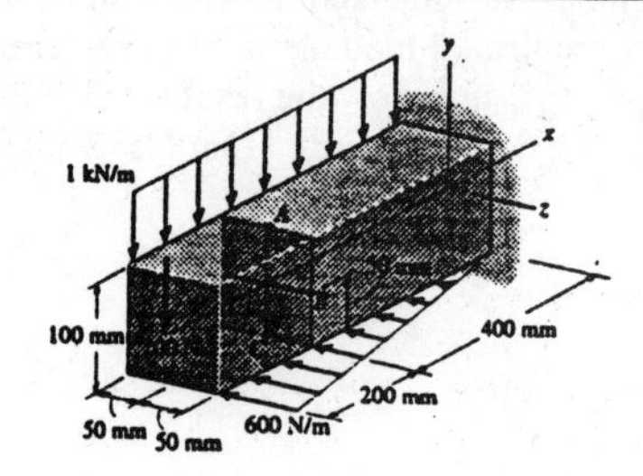

Internal Forces and Moments :

$$\Sigma F_y = 0; \quad V_y - 200 - 200 = 0 \quad V_y = 400 \text{ N}$$
$$\Sigma F_z = 0; \quad V_z - 80.0 - 20.0 = 0 \quad V_z = 100 \text{ N}$$
$$\Sigma M_x = 0; \quad T_x - 200(0.05) + 80.0(0.05) + 20.0(0.05) = 0$$
$$T_x = 5.00 \text{ N}\cdot\text{m}$$
$$\Sigma M_y = 0; \quad M_y - 80.0(0.1) - 20.0(0.1333) = 0$$
$$M_y = 10.667 \text{ N}\cdot\text{m}$$
$$\Sigma M_z = 0; \quad M_z + 200(0.2) + 200(0.1) = 0$$
$$M_z = -60.0 \text{ N}\cdot\text{m}$$

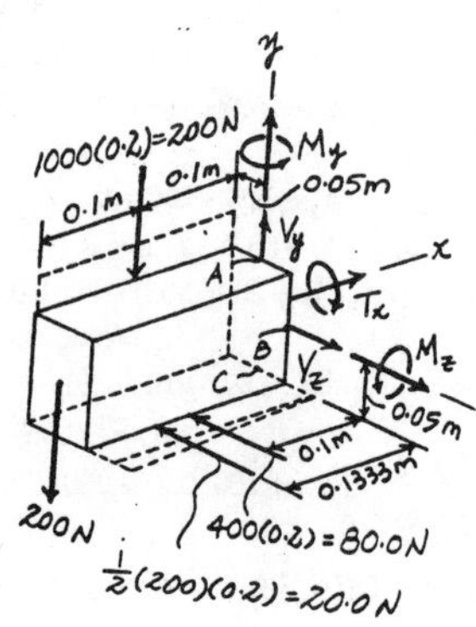

Section Properties :

$$A = 0.1(0.1) = 0.01 \text{ m}^2$$
$$I_z = I_y = \frac{1}{12}(0.1)\left(0.1^3\right) = 8.333\left(10^{-6}\right) \text{ m}^4$$
$$(Q_A)_y = 0$$
$$(Q_A)_z = \bar{y}'A' = 0.025(0.05)(0.1) = 0.125\left(10^{-3}\right) \text{ m}^3$$

Normal Stress :

$$\sigma = -\frac{M_z y}{I_z} + \frac{M_y z}{I_y}$$
$$\sigma_A = -\frac{-60.0(0.05)}{8.333(10^{-6})} + \frac{10.667(0)}{8.333(10^{-6})}$$
$$= 360 \text{ kPa (T)} \qquad \textbf{Ans}$$

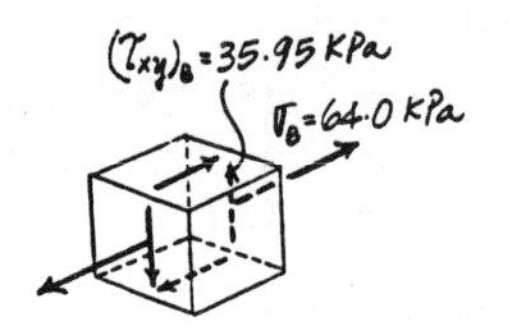

Shear Stress : The shear stress in the z *direction* is caused by the transverse shear stress in that direction and torsional shear stress. The transverse shear stress can be obtained using the shear formula, whereas the torsional shear stress formula for the square is given in Table 5 – 1 of the text.

$$(\tau_{xz})_A = \frac{4.81 T_x}{a^3} + \frac{V_z (Q_A)_z}{I_y t}$$
$$= \frac{4.81(5.00)}{0.1^3} + \frac{100\left[0.125(10^{-3})\right]}{8.333(10^{-6})(0.1)}$$
$$= 39.1 \text{ kPa} \qquad \textbf{Ans}$$

The shear stress in the y *direction* is caused by transverse shear stress in that direction only. Since $(Q_A)_y = 0$, then

$$(\tau_{xy})_A = 0 \qquad \textbf{Ans}$$

8–58. The member is subjected to the combination of distributed and concentrated loadings shown. Determine the state of stress at point *C*, and show the results on a differential volume element located at this point. *Hint:* Use Table 5-1.

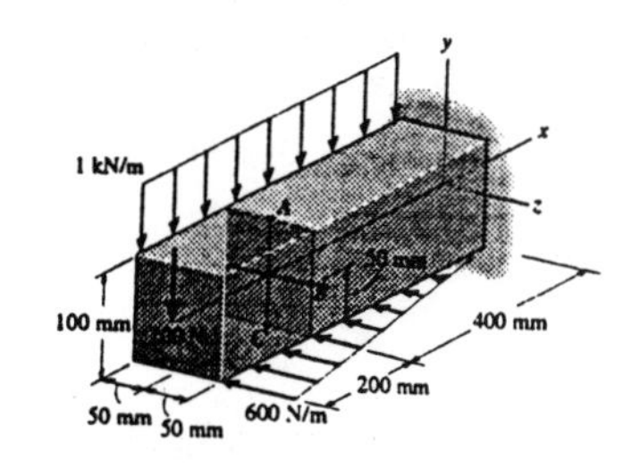

Internal Forces and Moments :

$$\Sigma F_y = 0; \quad V_y - 200 - 200 = 0 \quad V_y = 400\text{ N}$$
$$\Sigma F_z = 0; \quad V_z - 80.0 - 20.0 = 0 \quad V_z = 100\text{ N}$$
$$\Sigma M_x = 0; \quad T_x - 200(0.05) + 80.0(0.05) + 20.0(0.05) = 0$$
$$T_x = 5.00\text{ N}\cdot\text{m}$$
$$\Sigma M_y = 0; \quad M_y - 80.0(0.1) - 20.0(0.1333) = 0$$
$$M_y = 10.667\text{ N}\cdot\text{m}$$
$$\Sigma M_z = 0; \quad M_z + 200(0.2) + 200(0.1) = 0$$
$$M_z = -60.0\text{ N}\cdot\text{m}$$

Section Properties :

$$A = 0.1(0.1) = 0.01\text{ m}^2$$
$$I_z = I_y = \frac{1}{12}(0.1)\left(0.1^3\right) = 8.333\left(10^{-6}\right)\text{ m}^4$$
$$(Q_C)_y = 0$$
$$(Q_C)_z = \bar{y}'A' = 0.025(0.05)(0.1) = 0.125\left(10^{-3}\right)\text{ m}^3$$

Normal Stress :

$$\sigma = -\frac{M_z y}{I_z} + \frac{M_y z}{I_y}$$
$$\sigma_C = -\frac{-60.0(-0.05)}{8.333(10^{-6})} + \frac{10.667(0)}{8.333(10^{-6})}$$
$$= -360\text{ kPa} = 360\text{ kPa (C)} \qquad \textbf{Ans}$$

Shear Stress : The shear stress in the *z direction* is caused by transverse shear stress in that direction and torsional shear stress. The transverse shear stress can be obtained using the shear formula, whereas the torsional shear stress formula for the square section is given in Table 5 – 1 of the text.

$$(\tau_{xz})_C = \frac{V_z(Q_C)_z}{I_y t} - \frac{4.81T_x}{a^3}$$
$$= \frac{100\left[0.125(10^{-3})\right]}{8.333(10^{-6})(0.1)} - \frac{4.81(5.00)}{0.1^3}$$
$$= -9.05\text{ kPa} \qquad \textbf{Ans}$$

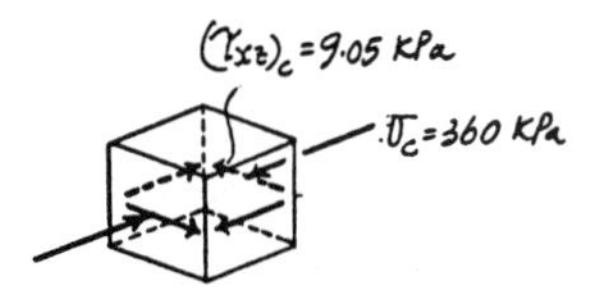

The shear stress in the *y direction* is caused by transverse shear stress in that direction only. Since $(Q_C)_y = 0$, then

$$(\tau_{xy})_C = 0 \qquad \textbf{Ans}$$

8–59. A post having the dimensions shown is subjected to the bearing load **P**. Specify the region to which this load can be applied without causing tensile stress to be developed at points A, B, C, and D.

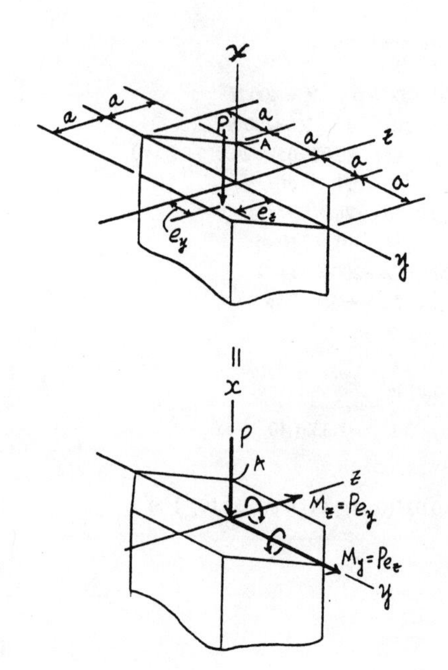

Equivalent Force System : As shown on FBD.

Section Properties :

$$A = 2a(2a) + 2\left[\frac{1}{2}(2a)\,a\right] = 6a^2$$

$$I_z = \frac{1}{12}(2a)(2a)^3 + 2\left[\frac{1}{36}(2a)\,a^3 + \frac{1}{2}(2a)\,a\left(a+\frac{a}{3}\right)^2\right]$$
$$= 5a^4$$

$$I_y = \frac{1}{12}(2a)(2a)^3 + 2\left[\frac{1}{36}(2a)\,a^3 + \frac{1}{2}(2a)\,a\left(\frac{a}{3}\right)^2\right]$$
$$= \frac{5}{3}a^4$$

Normal Stress :

$$\sigma = \frac{N}{A} - \frac{M_z y}{I_z} + \frac{M_y z}{I_y}$$
$$= \frac{-P}{6a^2} - \frac{Pe_y y}{5a^4} + \frac{Pe_z z}{\frac{5}{3}a^4}$$
$$= \frac{P}{30a^4}\left(-5a^2 - 6e_y y + 18e_z z\right)$$

At point A where $y = -a$ and $z = a$, we require $\sigma_A < 0$.

$$0 > \frac{P}{30a^4}\left[-5a^2 - 6(-a)\,e_y + 18(a)\,e_z\right]$$
$$0 > -5a + 6e_y + 18e_z$$
$$6e_y + 18e_z < 5a \qquad \textbf{Ans}$$

When $\quad e_z = 0, \qquad e_y < \frac{5}{6}a$

When $\quad e_y = 0, \qquad e_z < \frac{5}{18}a$

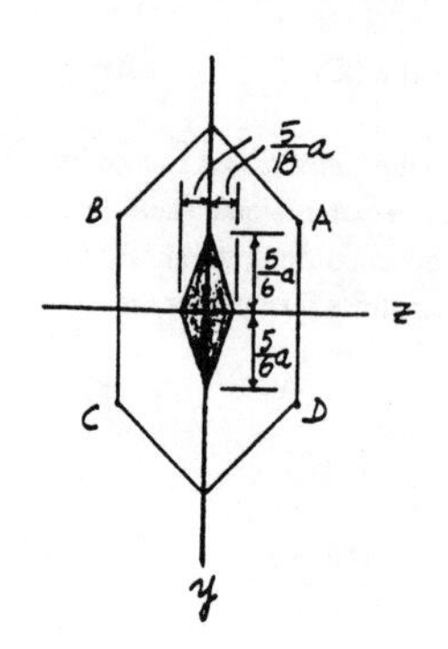

Repeat the same procedures for point B, C and D. The region where **P** can be applied without creating tensile stress at points A, B, C and D is shown shaded in the diagram.

***8–60.** The clamp exerts a compressive force of 500 lb on the block. Determine the maximum tensile and compressive stress at section *a–a*. Use the curved-beam formula to compute the bending stress.

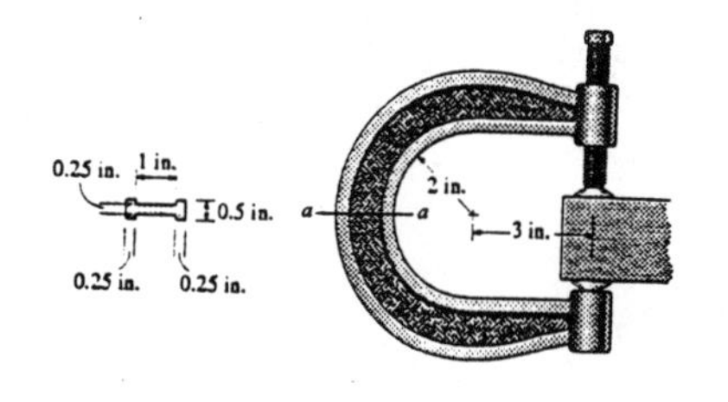

Section Properties :

$$\bar{r} = 2 + 0.75 = 2.75 \text{ in.}$$

$$\int_A \frac{dA}{r} = \Sigma b \ln \frac{r_2}{r_1}$$

$$= 0.5 \ln \frac{2.25}{2} + 0.25 \ln \frac{3.25}{2.25} + 0.5 \ln \frac{3.5}{3.25}$$

$$= 0.1878767 \text{ in.}$$

$$A = 2[0.5(0.25)] + 1(0.25) = 0.500 \text{ in}^2$$

$$R = \frac{A}{\int_A \frac{dA}{r}} = \frac{0.500}{0.1878767} = 2.6613199 \text{ in.}$$

$$\bar{r} - R = 2.75 - 2.6613199 = 0.088680 \text{ in.}$$

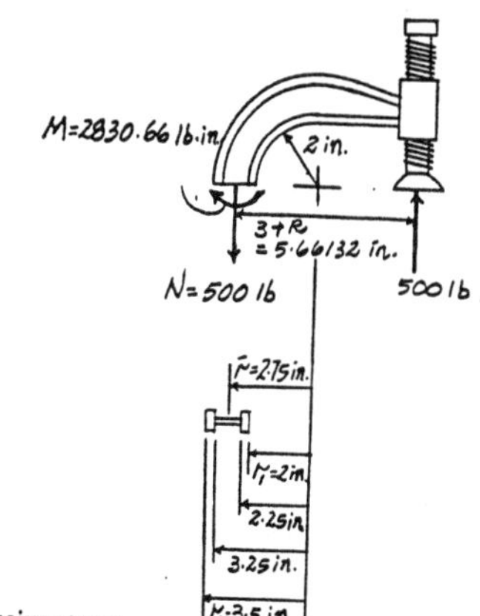

Internal Force and Moment : As shown on FBD. The internal moment must be computed about the neutral axis. $M = 2830.66$ lb · in. is positive since it tends to increase the beam's radius of curvature.

Normal Stress : Applying the curved - beam formula,
For tensile stress

$$(\sigma_t)_{\max} = \frac{N}{A} + \frac{M(R - r_1)}{Ar_1(\bar{r} - R)}$$

$$= \frac{500}{0.500} + \frac{2830.66(2.6613199 - 2)}{0.500(2)(0.088680)}$$

$$= 22109 \text{ psi} = 22.1 \text{ ksi (T)} \qquad \textbf{Ans}$$

For compressive stress

$$(\sigma_c)_{\max} = \frac{N}{A} + \frac{M(R - r_2)}{Ar_2(\bar{r} - R)}$$

$$= \frac{500}{0.500} + \frac{2830.66(2.6613199 - 3.5)}{0.500(3.5)(0.088680)}$$

$$= -14297 \text{ psi} = 14.3 \text{ ksi (C)} \qquad \textbf{Ans}$$

8–61. Determine the largest clamping force that can be exerted on the block if the allowable normal stress for the clamp is to be $\sigma_{\text{allow}} = 30$ ksi. Use the curved-beam formula to compute the bending stress.

Section Properties :

$$\bar{r} = 2 + 0.75 = 2.75 \text{ in.}$$

$$\int_A \frac{dA}{r} = \Sigma b \ln \frac{r_2}{r_1}$$

$$= 0.5 \ln \frac{2.25}{2} + 0.25 \ln \frac{3.25}{2.25} + 0.5 \ln \frac{3.5}{3.25}$$

$$= 0.1878767 \text{ in.}$$

$$A = 2[0.5(0.25)] + 1(0.25) = 0.500 \text{ in}^2$$

$$R = \frac{A}{\int_A \frac{dA}{r}} = \frac{0.500}{0.1878767} = 2.6613199 \text{ in.}$$

$$\bar{r} - R = 2.75 - 2.6613199 = 0.088680 \text{ in.}$$

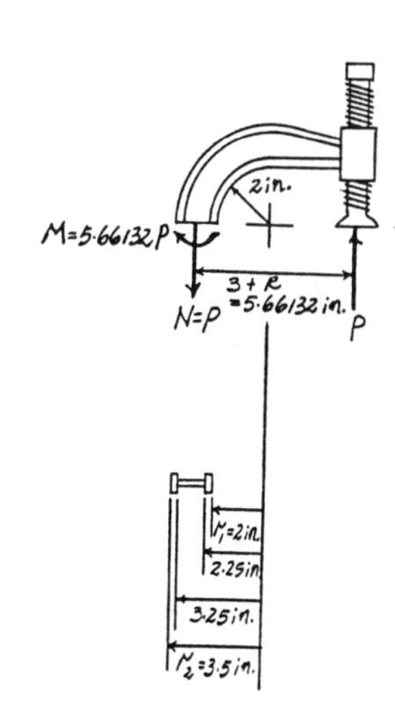

Internal Force and Moment : As shown on FBD. The internal moment must be computed about the neutral axi. $M = 5.66132P$ is positive since it tends to increase the beam's radius of curvature.

Allowable Normal Stress : Applying the curved - beam formul,
Assume compressive failure

$$(\sigma_c)_{\max} = \sigma_{\text{allow}} = \frac{N}{A} + \frac{M(R - r_2)}{Ar_2(\bar{r} - R)}$$

$$-30(10^3) = \frac{P}{0.500} + \frac{5.66132P(2.6613199 - 3.5)}{0.500(3.5)(0.088680)}$$

$$= 1049 \text{ lb}$$

Assume tensile failure

$$(\sigma_t)_{\max} = \sigma_{\text{allow}} = \frac{N}{A} + \frac{M(R - r_1)}{Ar_1(\bar{r} - R)}$$

$$30(10^3) = \frac{P}{0.500} + \frac{5.66132P(2.6613199 - 2)}{0.500(2)(0.088680)}$$

$$P = 678 \text{ lb} \quad (\textit{Controls!}) \qquad \textbf{Ans}$$

8–62. The eye is subjected to the force of 50 lb. Determine the maximum tensile and compressive stresses at section a–a. The cross section is circular and has a diameter of 0.25 in. Use the curved-beam formula to compute the bending stress.

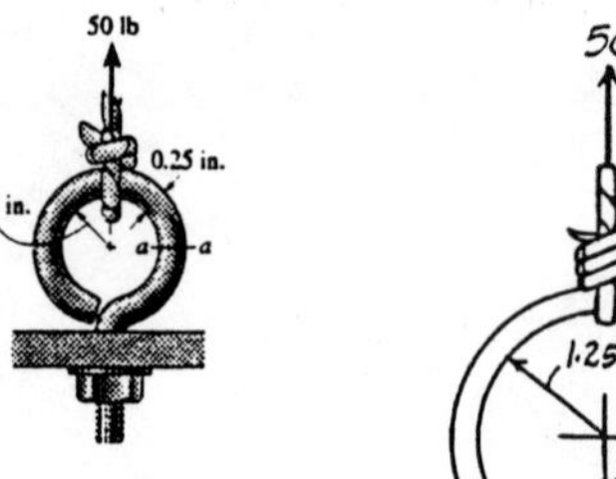

Section Properties :

$$\bar{r} = 1.25 + \frac{0.25}{2} = 1.375 \text{ in.}$$

$$\int_A \frac{dA}{r} = 2\pi\left(\bar{r} - \sqrt{\bar{r}^2 - c^2}\right)$$

$$= 2\pi\left(1.375 - \sqrt{1.375^2 - 0.125^2}\right)$$

$$= 0.035774 \text{ in.}$$

$$A = \pi\left(0.125^2\right) = 0.049087 \text{ in}^2$$

$$R = \frac{A}{\int_A \frac{dA}{r}} = \frac{0.049087}{0.035774} = 1.372153 \text{ in.}$$

$$\bar{r} - R = 1.375 - 1.372153 = 0.002847 \text{ in.}$$

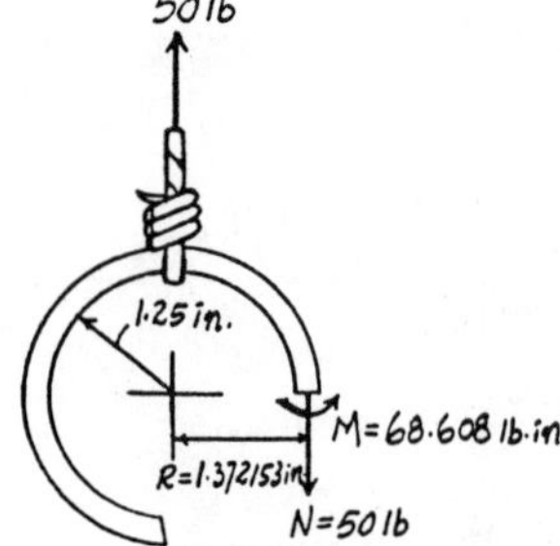

Internal Force andMoment : As shown on FBD. The internal moment must be computed about the neutral axis. $M = 68.608$ lb · in. is positive since it tends to increase the beam's radius of curvature.

Normal Stress : Applying the curved - beam formula,
For tensile stress

$$(\sigma_t)_{\max} = \frac{N}{A} + \frac{M(R - r_1)}{Ar_1(\bar{r} - R)}$$

$$= \frac{50.0}{0.049087} + \frac{68.608(1.372153 - 1.25)}{0.049087(1.25)(0.002847)}$$

$$= 48996 \text{ psi} = 49.0 \text{ ksi (T)} \qquad \textbf{Ans}$$

For compressive stress

$$(\sigma_c)_{\max} = \frac{N}{A} + \frac{M(R - r_2)}{Ar_2(\bar{r} - R)}$$

$$= \frac{50.0}{0.049087} + \frac{68.608(1.372153 - 1.50)}{0.049087(1.50)(0.002847)}$$

$$= -40826 \text{ psi} = 40.8 \text{ ksi (C)} \qquad \textbf{Ans}$$

8–63. Solve Prob. 8–62 if the cross section is square, having dimensions of 0.25 in. by 0.25 in.

Section Properties :

$$\bar{r} = 1.25 + \frac{0.25}{2} = 1.375 \text{ in.}$$

$$\int_A \frac{dA}{r} = b\ln\frac{r_2}{r_1} = 0.25\ln\frac{1.5}{1.25} = 0.045580 \text{ in.}$$

$$A = 0.25(0.25) = 0.0625 \text{ in}^2$$

$$R = \frac{A}{\int_A \frac{dA}{r}} = \frac{0.0625}{0.045580} = 1.371204 \text{ in.}$$

$$\bar{r} - R = 1.375 - 1.371204 = 0.003796 \text{ in.}$$

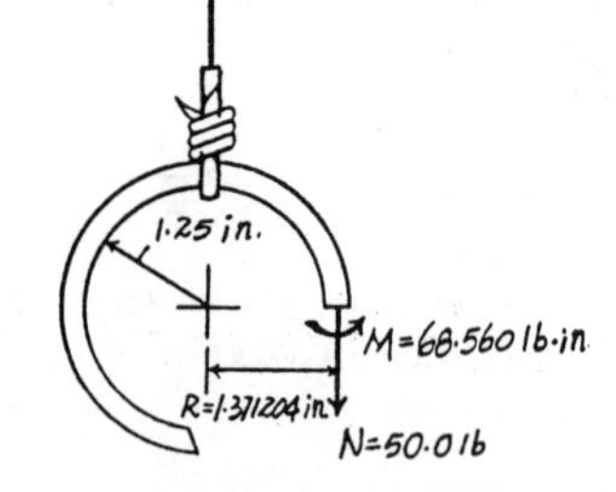

Internal Force and Moment : As shown on FBD. The internal moment must be computed about the neutral axis. $M = 68.560$ lb · in. is positive since it tends to increase the beam's radius of curvature.

Normal Stress : Applying the curved - beam formula,
For tensile stress

$$(\sigma_t)_{\max} = \frac{N}{A} + \frac{M(R - r_1)}{Ar_1(\bar{r} - R)}$$

$$= \frac{50.0}{0.0625} + \frac{68.560(1.371204 - 1.25)}{0.0625(1.25)(0.003796)}$$

$$= 28818 \text{ psi} = 28.8 \text{ ksi (T)} \qquad \textbf{Ans}$$

For Compressive stress

$$(\sigma_c)_{\max} = \frac{N}{A} + \frac{M(R - r_2)}{Ar_2(\bar{r} - R)}$$

$$= \frac{50.0}{0.0625} + \frac{68.560(1.371204 - 1.5)}{0.0625(1.5)(0.003796)}$$

$$= -24011 \text{ psi} = 24.0 \text{ ksi (C)} \qquad \textbf{Ans}$$

***8–64.** The man has a mass of 100 kg and center of mass at G. If he holds himself in the position shown, determine the maximum tensile and compressive stress developed in the curved bar at section a–a. He is supported uniformly by two bars, each having a diameter of 25 mm. Assume the floor is smooth.

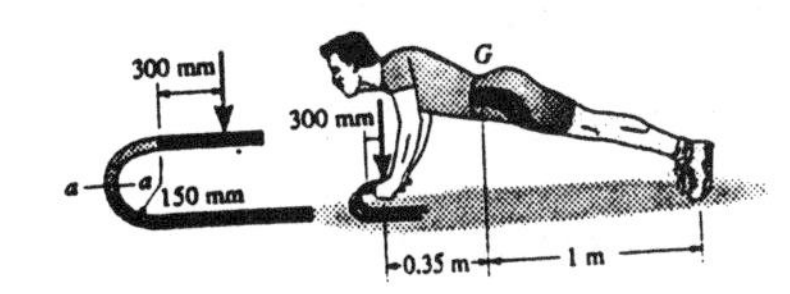

Equilibrium : For the man

$$\curvearrowleft + \Sigma M_B = 0; \quad 981(1) - 2F_A(1.35) = 0 \quad F_A = 363.33 \text{ N}$$

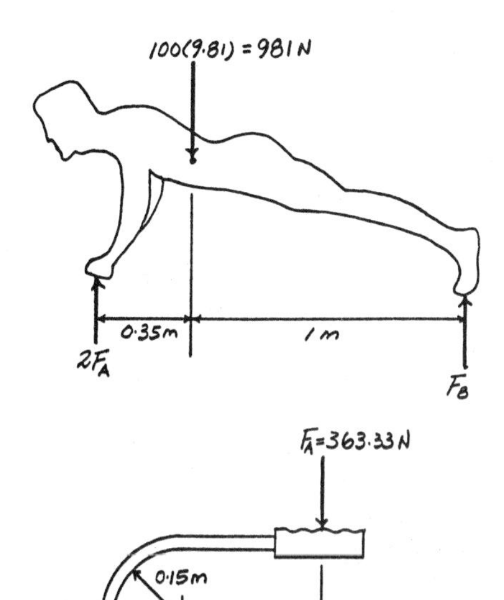

Section Properties :

$$\bar{r} = 0.15 + \frac{0.025}{2} = 0.1625 \text{ m}$$

$$\int_A \frac{dA}{r} = 2\pi\left(\bar{r} - \sqrt{\bar{r}^2 - c^2}\right)$$

$$= 2\pi\left(0.1625 - \sqrt{0.1625^2 - 0.0125^2}\right)$$

$$= 3.02524\left(10^{-3}\right) \text{ m}$$

$$A = \pi\left(0.0125^2\right) = 0.490874\left(10^{-3}\right) \text{ m}^2$$

$$R = \frac{A}{\int_A \frac{dA}{r}} = \frac{0.490874(10^{-3})}{3.02524(10^{-3})} = 0.162259 \text{ m}$$

$$\bar{r} - R = 0.1625 - 0.162259 = 0.240741\left(10^{-3}\right) \text{ m}$$

Internal Force and Moment : The internal moment must be computed about the neutral axis.

$$+\uparrow \Sigma F_y = 0; \quad -363.33 - N = 0 \quad N = -363.33 \text{ N}$$

$$\curvearrowleft + \Sigma M_O = 0; \quad -M - 363.33(0.462259) = 0 \quad M = -167.95 \text{ N}\cdot\text{m}$$

Normal Stress : Applying the curved - beam formula.

For tensile stress

$$(\sigma_t)_{max} = \frac{N}{A} + \frac{M(R - r_2)}{Ar_2(\bar{r} - R)}$$

$$= \frac{-363.33}{0.490874(10^{-3})} + \frac{-167.95(0.162259 - 0.175)}{0.490874(10^{-3})(0.175)0.240741(10^{-3})}$$

$= 103$ MPa(T) **Ans**

For compressive stress,

$$(\sigma_c)_{max} = \frac{N}{A} + \frac{M(R - r_1)}{Ar_1(\bar{r} - R)}$$

$$= \frac{-363.33}{0.490874(10^{-3})} + \frac{-167.95(0.162259 - 0.15)}{0.490874(10^{-3})(0.15)0.240741(10^{-3})}$$

$= -117$ MPa $= 117$ MPa(C) **Ans**

8–65. The hook is used to lift the force of 600 lb. Determine the maximum tensile and compressive stresses at section *a–a*. The cross section is circular and has a diameter of 1 in. Use the curved-beam formula to compute the bending stress.

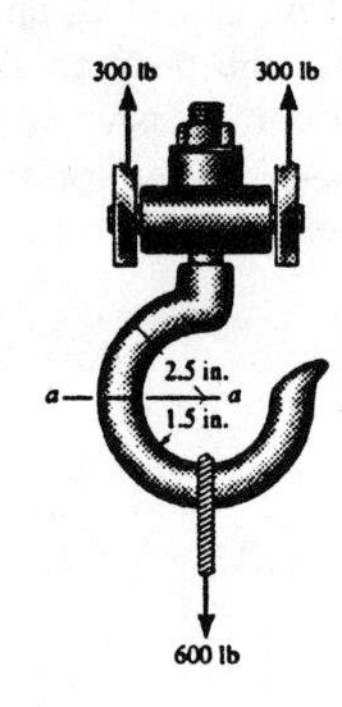

Section Properties :

$$\bar{r} = 1.5 + 0.5 = 2.00 \text{ in.}$$

$$\int_A \frac{dA}{r} = 2\pi\left(\bar{r} - \sqrt{\bar{r}^2 - c^2}\right)$$
$$= 2\pi\left(2.00 - \sqrt{2.00^2 - 0.5^2}\right)$$
$$= 0.399035 \text{ in.}$$

$$A = \pi\left(0.5^2\right) = 0.25\pi \text{ in}^2$$

$$R = \frac{A}{\int_A \frac{dA}{r}} = \frac{0.25\pi}{0.399035} = 1.968246 \text{ in.}$$

$$\bar{r} - R = 2 - 1.968246 = 0.031754 \text{ in.}$$

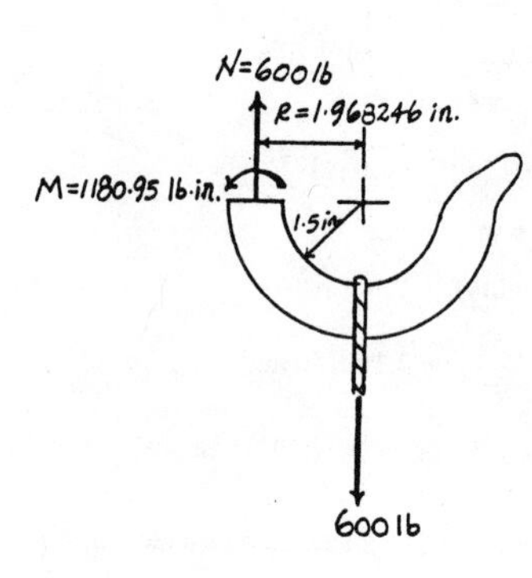

Internal Force and Moment : As shown on FBD. The internal moment must be computed about the neutral axis. $M = 1180.95$ lb · in. is positive since it tends to increase the beam's radius of curvature.

Normal Stress : Applying the curved - beam formula.
For tensile stress

$$(\sigma_t)_{\max} = \frac{N}{A} + \frac{M(R - r_1)}{A r_1 (\bar{r} - R)}$$
$$= \frac{600}{0.25\pi} + \frac{1180.95(1.968246 - 1.5)}{0.25\pi(1.5)(0.031754)}$$
$$= 15546 \text{ psi} = 15.5 \text{ ksi (T)} \qquad \textbf{Ans}$$

For compressive stress,

$$(\sigma_c)_{\max} = \frac{N}{A} + \frac{M(R - r_2)}{A r_2 (\bar{r} - R)}$$
$$= \frac{600}{0.25\pi} + \frac{1180.95(1.968246 - 2.5)}{0.25\pi(2.5)(0.031754)}$$
$$= -9308 \text{ psi} = 9.31 \text{ ksi (C)} \qquad \textbf{Ans}$$

8–66. The support is subjected to the compressive load **P**. Determine the absolute maximum and minimum normal stress acting in the material.

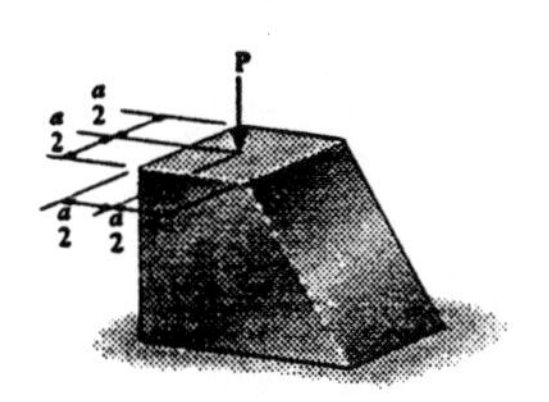

Section Properties :

$$w = a + x$$
$$A = a(a+x)$$
$$I = \frac{1}{12}(a)(a+x)^3 = \frac{a}{12}(a+x)^3$$

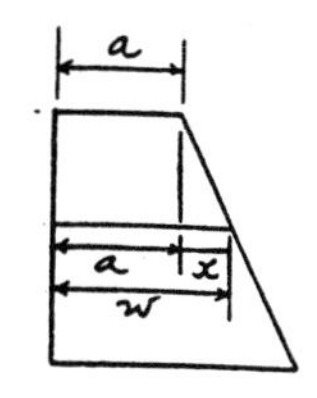

Internal Forces and Moment : As shown on FBD.

Normal Stress :

$$\sigma = \frac{N}{A} \pm \frac{Mc}{I}$$
$$= \frac{-P}{a(a+x)} \pm \frac{0.5Px\left[\frac{1}{2}(a+x)\right]}{\frac{a}{12}(a+x)^3}$$
$$= \frac{P}{a}\left[\frac{-1}{a+x} \pm \frac{3x}{(a+x)^2}\right]$$

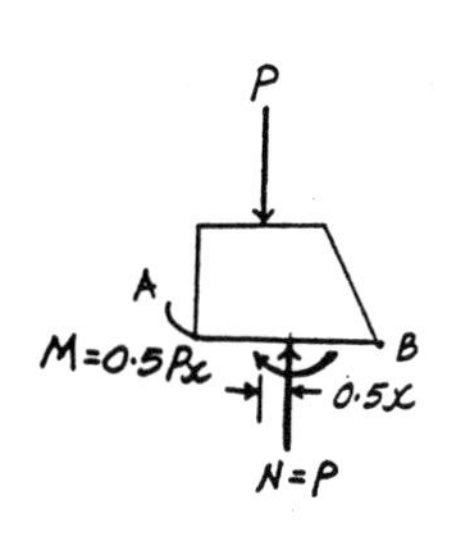

$$\sigma_A = -\frac{P}{a}\left[\frac{1}{a+x} + \frac{3x}{(a+x)^2}\right]$$
$$= -\frac{P}{a}\left[\frac{4x+a}{(a+x)^2}\right] \qquad [1]$$

$$\sigma_B = \frac{P}{a}\left[\frac{-1}{a+x} + \frac{3x}{(a+x)^2}\right]$$
$$= \frac{P}{a}\left[\frac{2x-a}{(a+x)^2}\right] \qquad [2]$$

In order to have maximum normal stress, $\frac{d\sigma_A}{dx} = 0$.

$$\frac{d\sigma_A}{dx} = -\frac{P}{a}\left[\frac{(a+x)^2(4) - (4x+a)(2)(a+x)(1)}{(a+x)^4}\right] = 0$$
$$-\frac{P}{a(a+x)^3}(2a - 4x) = 0$$

Since $\frac{P}{a(a+x)^3} \neq 0$, then

$$2a - 4x = 0 \qquad x = 0.500a$$

Substituting the result into Eq.[1] yields

$$\sigma_{max} = -\frac{P}{a}\left[\frac{4(0.500a)+a}{(a+0.5a)^2}\right]$$
$$= -\frac{1.33P}{a^2} = \frac{1.33P}{a^2} \text{ (C)} \qquad \textbf{Ans}$$

In order to have minimum normal stress, $\frac{d\sigma_B}{dx} = 0$.

$$\frac{d\sigma_B}{dx} = \frac{P}{a}\left[\frac{(a+x)^2(2) - (2x-a)(2)(a+x)(1)}{(a+x)^4}\right] = 0$$
$$\frac{P}{a(a+x)^3}(4a - 2x) = 0$$

Since $\frac{P}{a(a+x)^3} \neq 0$, then

$$4a - 2x = 0 \qquad x = 2a$$

Substituting the result into Eq.[2] yields

$$\sigma_{min} = \frac{P}{a}\left[\frac{2(2a)-a}{(a+2a)^2}\right] = \frac{P}{3a^2} \text{ (T)} \qquad \textbf{Ans}$$

8–67. The support is subjected to the compressive load **P**. Determine the maximum and minimum normal stress acting in the material.

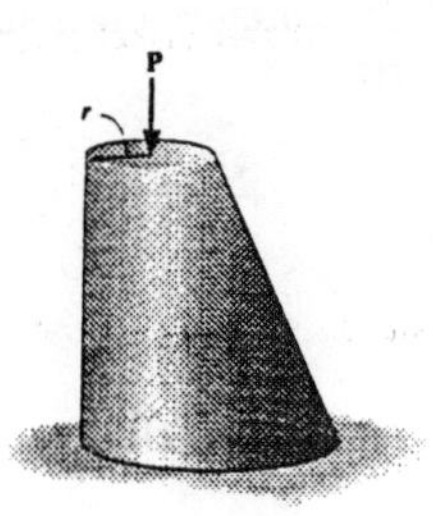

Section Properties :

$$d' = 2r + x$$
$$A = \pi(r+0.5x)^2$$
$$I = \frac{\pi}{4}(r+0.5x)^4$$

Internal Force and Moment : As shown on FBD.

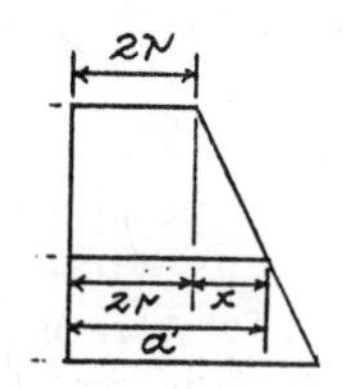

Normal Stress :

$$\sigma = \frac{N}{A} \pm \frac{Mc}{I}$$
$$= \frac{-P}{\pi(r+0.5x)^2} \pm \frac{0.5Px(r+0.5x)}{\frac{\pi}{4}(r+0.5x)^4}$$
$$= \frac{P}{\pi}\left[\frac{-1}{(r+0.5x)^2} \pm \frac{2x}{(r+0.5x)^3}\right]$$

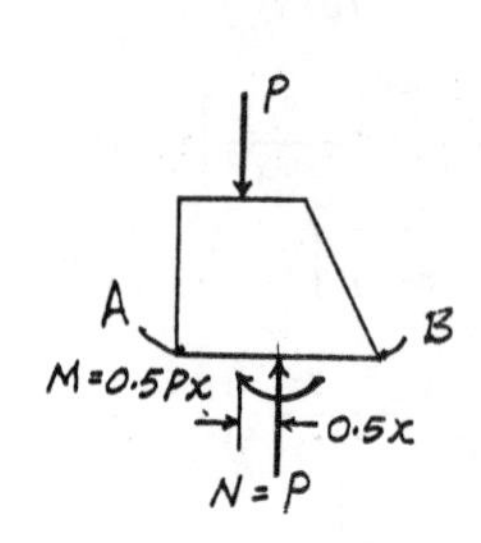

$$\sigma_A = -\frac{P}{\pi}\left[\frac{1}{(r+0.5x)^2} + \frac{2x}{(r+0.5x)^3}\right]$$
$$= -\frac{P}{\pi}\left[\frac{r+2.5x}{(r+0.5x)^3}\right] \qquad [1]$$

$$\sigma_B = \frac{P}{\pi}\left[\frac{-1}{(r+0.5x)^2} + \frac{2x}{(r+0.5x)^3}\right]$$
$$= \frac{P}{\pi}\left[\frac{1.5x-r}{(r+0.5x)^3}\right] \qquad [2]$$

In order to have maximum normal stress, $\frac{d\sigma_A}{dx} = 0$.

$$\frac{d\sigma_A}{dx} = -\frac{P}{\pi}\left[\frac{(r+0.5x)^3(2.5) - (r+2.5x)(3)(r+0.5x)^2(0.5)}{(r+0.5x)^6}\right] = 0$$
$$-\frac{P}{\pi(r+0.5x)^4}(r-2.5x) = 0$$

Since $\frac{P}{\pi(r+0.5x)^4} \neq 0$, then

$$r - 2.5x = 0 \qquad x = 0.400r$$

Substituting the result into Eq.[1] yields

$$\sigma_{max} = -\frac{P}{\pi}\left[\frac{r+2.5(0.400r)}{[r+0.5(0.400r)]^3}\right]$$
$$= -\frac{0.368P}{r^2} = \frac{0.368P}{r^2} \text{ (C)} \qquad \textbf{Ans}$$

In order to have minimum normal stress, $\frac{d\sigma_B}{dx} = 0$.

$$\frac{d\sigma_B}{dx} = \frac{P}{\pi}\left[\frac{(r+0.5x)^3(1.5) - (1.5x-r)(3)(r+0.5x)^2(0.5)}{(r+0.5x)^6}\right] = 0$$
$$\frac{P}{\pi(r+0.5x)^4}(3r-1.5x) = 0$$

Since $\frac{P}{\pi(r+0.5x)^4} \neq 0$, then

$$3r - 1.5x = 0 \qquad x = 2.00r$$

Substituting the result into Eq.[2] yields

$$\sigma_{min} = \frac{P}{\pi}\left[\frac{1.5(2.00r)-r}{[r+0.5(4.00r)]^3}\right] = \frac{0.0796P}{r^2} \text{ (T)} \qquad \textbf{Ans}$$

***8–68.** If $P = 15$ kN, plot the distribution of stress acting over the cross section a–a of the offset link.

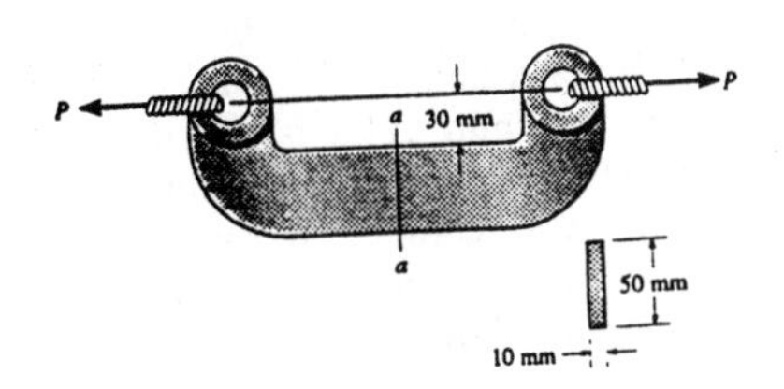

Internal Force and Moment : As shown on FBD.

Section Properties :

$$A = 0.01(0.05) = 0.500\left(10^{-3}\right)\ \text{m}^2$$

$$I = \frac{1}{12}(0.01)\left(0.05^3\right) = 0.104167\left(10^{-6}\right)\ \text{m}^4$$

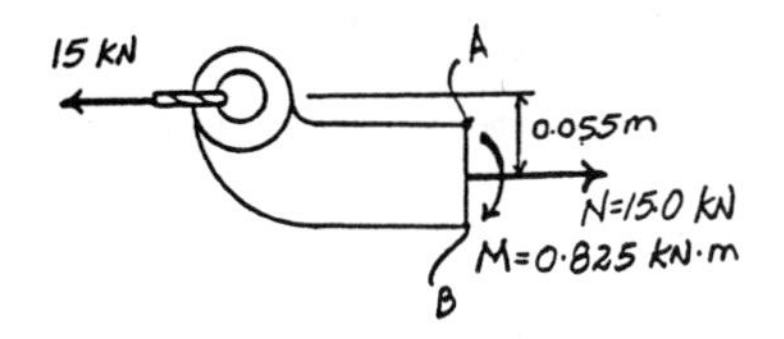

Normal Stress :

$$\sigma = \frac{N}{A} \pm \frac{Mc}{I}$$

$$= \frac{15.0(10^3)}{0.500(10^{-3})} \pm \frac{0.825(10^3)(0.025)}{0.104167(10^{-6})}$$

$$\sigma_A = 30.0 + 198 = 228\ \text{MPa (T)}$$

$$\sigma_B = 30.0 - 198 = -168\ \text{MPa} = 168\ \text{MPa (C)}$$

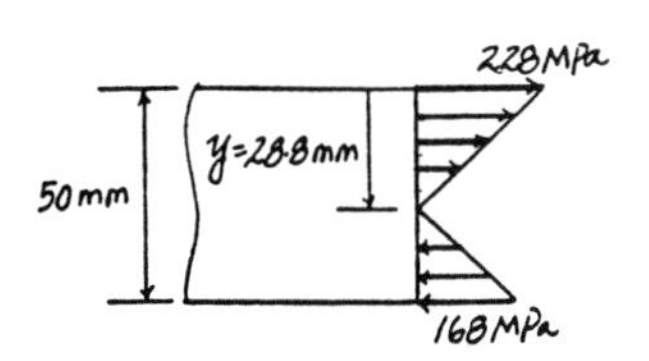

$$\frac{y}{228} = \frac{50-y}{168} \qquad y = 28.8\ \text{mm}$$

8–69. Determine the magnitude of the load **P** that will cause a maximum normal stress of $\sigma_{max} = 200$ MPa in the link at section a–a.

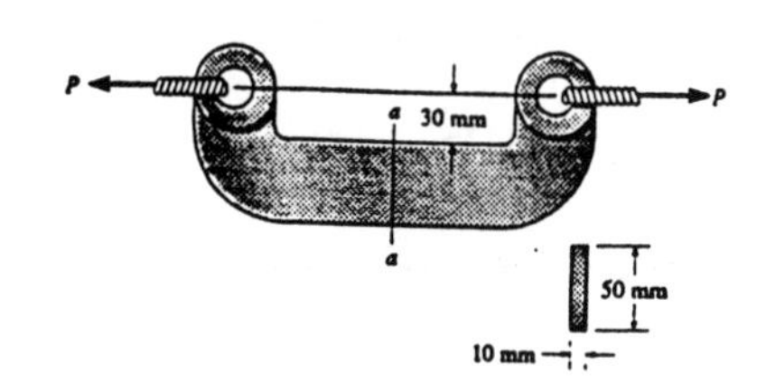

Internal Force and Moment : As shown on FBD.

Section Properties :

$$A = 0.01(0.05) = 0.500\left(10^{-3}\right)\ \text{m}^2$$

$$I = \frac{1}{12}(0.01)\left(0.05^3\right) = 0.104167\left(10^{-6}\right)\ \text{m}^4$$

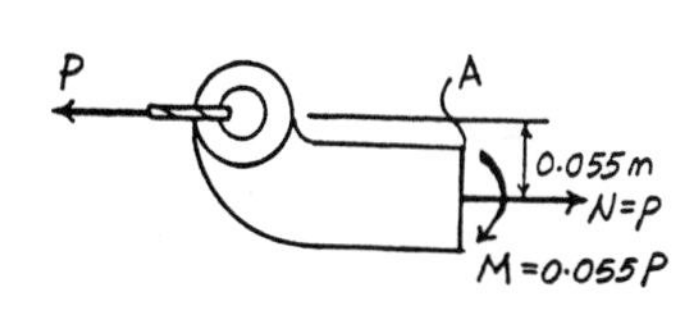

Allowable Normal Stress : The maximum normal stress occurs at point A.

$$\sigma_{max} = \sigma_{allow} = \frac{N}{A} + \frac{Mc}{I}$$

$$200\left(10^6\right) = \frac{P}{0.500(10^{-3})} + \frac{0.055P(0.025)}{0.104167(10^{-6})}.$$

$$P = 13158\ \text{N} = 13.2\ \text{kN} \qquad \textbf{Ans}$$

8–70. The cap on the cylindrical tank is bolted to the tank along the flanges. The tank has an inner diameter of 1.5 m and a wall thickness of 18 mm. If the largest normal stress is not to exceed 150 MPa, determine the maximum pressure the tank can sustain. Also, compute the number of bolts required to attach the cap to the tank if each bolt has a diameter of 20 mm. The allowable stress for the bolts is $(\sigma_{allow})_b = 180$ MPa.

Hoop Stress for Cylindrical Tank : Since $\frac{r}{t} = \frac{750}{18} = 41.6 > 10$, then *thin wall* analysis can be used. Applying Eq. 8 – 1

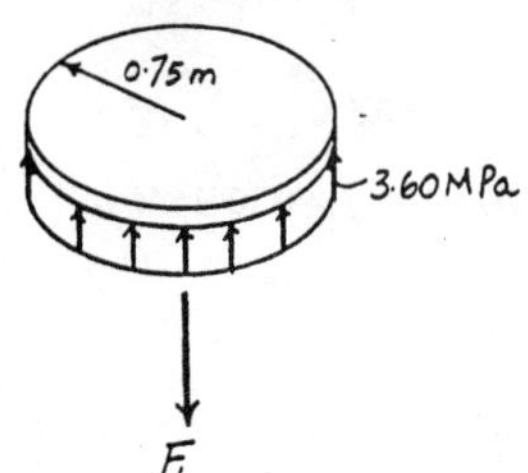

$$\sigma_1 = \sigma_{allow} = \frac{pr}{t}$$

$$150(10^6) = \frac{p(750)}{18}$$

$$p = 3.60 \text{ MPa} \qquad \textbf{Ans}$$

Force Equilibrium for the Cap :

$$+\uparrow \Sigma F_y = 0; \quad 3.60(10^6)\left[\pi(0.75^2)\right] - F_b = 0$$

$$F_b = 6.3617(10^6) \text{ N}$$

Allowable Normal Stress for Bolts :

$$(\sigma_{allow})_b = \frac{P}{A}$$

$$180(10^6) = \frac{6.3617(10^6)}{n\left[\frac{\pi}{4}(0.02^2)\right]}$$

$$n = 112.5$$

Use $n = 113$ bolts **Ans**

8–71. The cap on the cylindrical tank is bolted to the tank along the flanges. The tank has an inner diameter of 1.5 m and a wall thickness of 18 mm. If the pressure in the tank is $p = 1.20$ MPa, determine the force in the 16 bolts that are used to attach the cap to the tank. Also, specify the state of stress in the wall of the tank.

Hoop Stress for Cylindrical Tank : Since $\frac{r}{t} = \frac{750}{18} = 41.6 > 10$, then *thin wall* analysis can be used. Applying Eq. 8 – 1

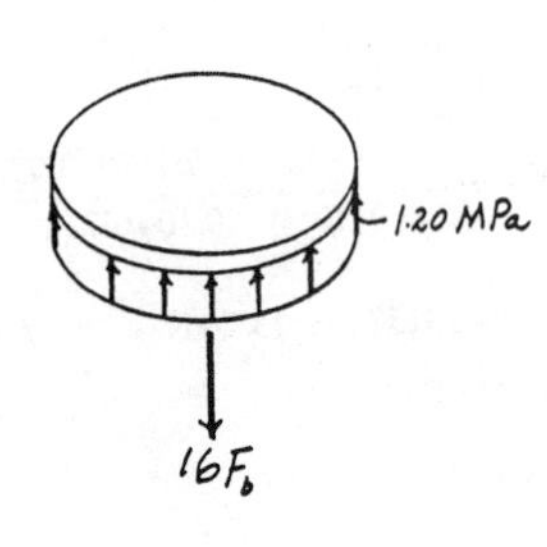

$$\sigma_1 = \frac{pr}{t} = \frac{1.20(10^6)(750)}{18} = 50.0 \text{ MPa} \qquad \textbf{Ans}$$

Longitudinal Stress for Cylindrical Tank :

$$\sigma_2 = \frac{pr}{2t} = \frac{1.20(10^6)(750)}{2(18)} = 25.0 \text{ MPa} \qquad \textbf{Ans}$$

Force Equilibrium for the Cap :

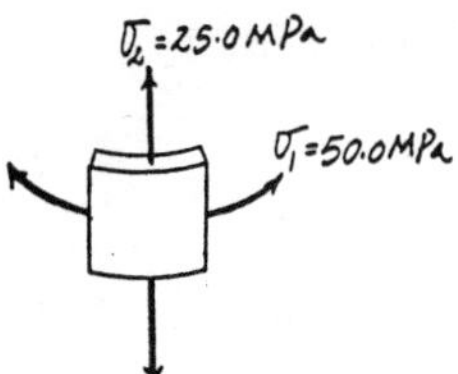

$$+\uparrow \Sigma F_y = 0; \quad 1.20(10^6)\left[\pi(0.75^2)\right] - 16F_b = 0$$

$$F_b = 132536 \text{ N} = 133 \text{ kN} \qquad \textbf{Ans}$$

***8–72.** The crowbar is used to pull out the nail at A. If a force of 8 lb is required, determine the stress components in the bar at points D and E. Show the results on a differential volume element located at each of these points. The bar has a circular cross section with a diameter of 0.5 in. No slipping occurs at B.

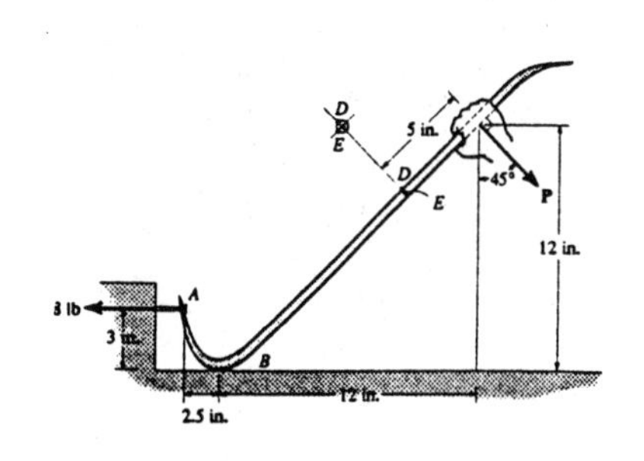

Support Reactions :

$$\circlearrowleft + \Sigma M_B = 0; \quad 8(3) - P(16.97) = 0 \qquad P = 1.414 \text{ lb}$$

Internal Forces and Moment :

$$+\Sigma F_{x'} = 0; \quad N = 0$$

$$+\Sigma F_{y'} = 0; \quad V - 1.414 = 0 \qquad V = 1.414 \text{ lb}$$

$$\circlearrowleft + \Sigma M_O = 0; \quad M - 1.414(5) = 0 \qquad M = 7.071 \text{ lb} \cdot \text{in.}$$

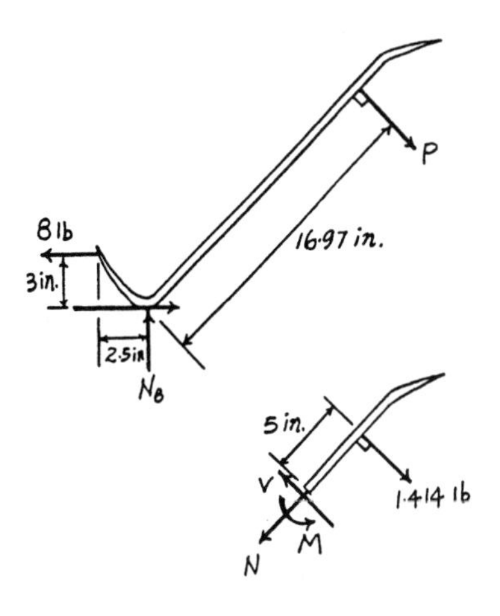

Section Properties :

$$A = \pi\left(0.25^2\right) = 0.0625\pi \text{ in}^2$$

$$I = \frac{\pi}{4}\left(0.25^4\right) = 0.9765625\pi\left(10^{-3}\right) \text{ in}^4$$

$$Q_D = 0$$

$$Q_E = \bar{y}'A' = \frac{4(0.25)}{3\pi}\left[\frac{1}{2}(\pi)\left(0.25^2\right)\right] = 0.0104167 \text{ m}^3$$

Normal Stress : Since $N = 0$, the normal stress is caused by bending stress only.

$$\sigma_D = \frac{Mc}{I} = \frac{7.071(0.25)}{0.9765625\pi(10^{-3})} = 576 \text{ psi (T)} \qquad \textbf{Ans}$$

$$\sigma_E = \frac{My}{I} = \frac{7.071(0)}{0.9765625\pi(10^{-3})} = 0 \qquad \textbf{Ans}$$

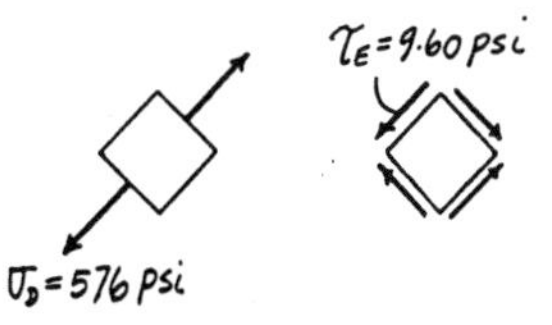

Shear Stress : Applying the shear formul, .

$$\tau_D = \frac{VQ_D}{It} = 0 \qquad \textbf{Ans}$$

$$\tau_E = \frac{VQ_E}{It} = \frac{1.414(0.0104167)}{0.9765625\pi(10^{-3})(0.5)} = 9.60 \text{ psi} \qquad \textbf{Ans}$$

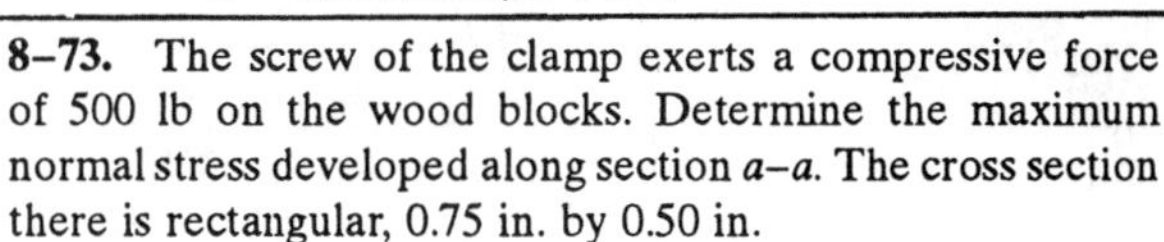

8–73. The screw of the clamp exerts a compressive force of 500 lb on the wood blocks. Determine the maximum normal stress developed along section a–a. The cross section there is rectangular, 0.75 in. by 0.50 in.

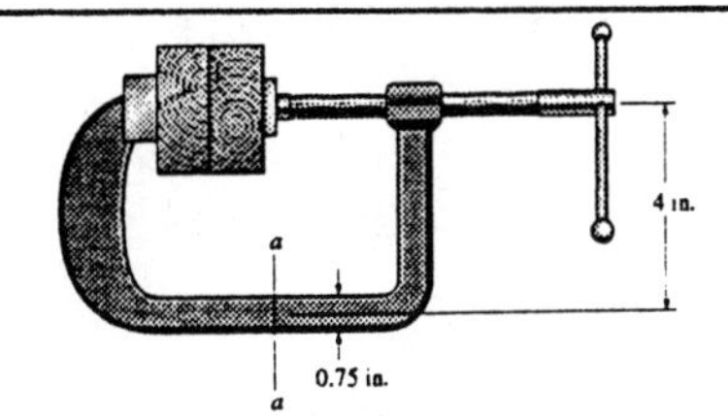

Internal Force and Moment : As shown on FBD.

Section Properties :

$$A = 0.5(0.75) = 0.375 \text{ in}^2$$

$$I = \frac{1}{12}(0.5)\left(0.75^3\right) = 0.017578 \text{ in}^4$$

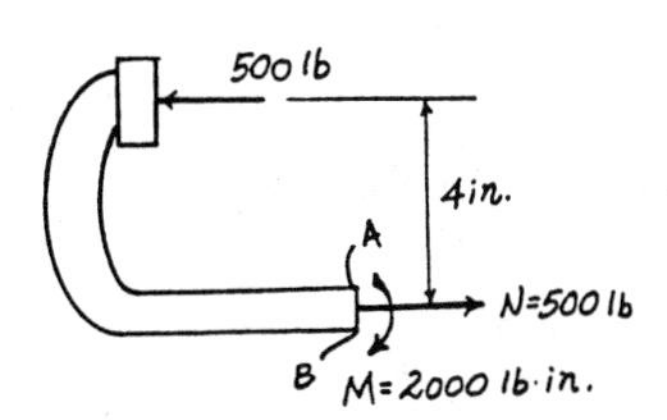

Maximum Normal Stress : Maximum normal stress occurs at point A.

$$\sigma_{\max} = \sigma_A = \frac{N}{A} + \frac{Mc}{I}$$

$$= \frac{500}{0.375} + \frac{2000(0.375)}{0.017578}$$

$$= 44000 \text{ psi} = 44.0 \text{ ksi (T)} \qquad \textbf{Ans}$$

8–74. The screw of the clamp exerts a compressive force of 500 lb on the wood blocks. Sketch the stress distribution along section a–a of the clamp. The cross section there is rectangular, 0.75 in. by 0.50 in.

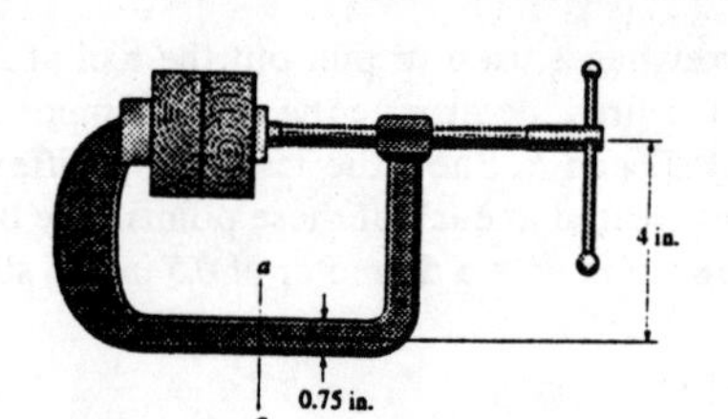

Internal Force and Moment : As shown on FBD.

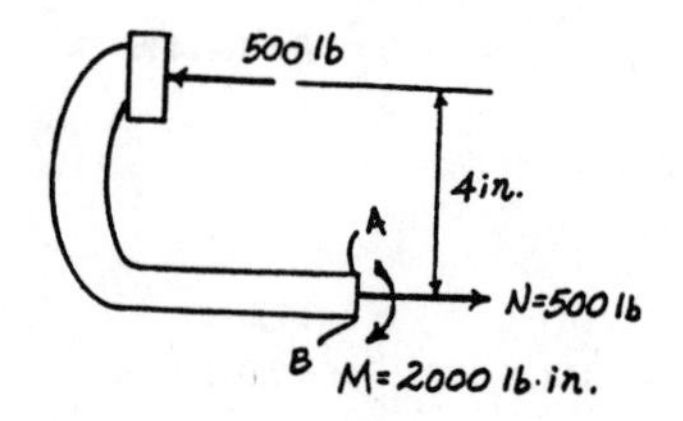

Section Properties :

$$A = 0.5(0.75) = 0.375 \text{ in}^2$$

$$I = \frac{1}{12}(0.5)\left(0.75^3\right) = 0.017578 \text{ in}^4$$

Maximum Normal Stress :

$$\sigma = \frac{N}{A} \pm \frac{Mc}{I}$$

$$= \frac{500}{0.375} \pm \frac{2000(0.375)}{0.017578}$$

$$\sigma_A = 1333.33 + 42666.67$$
$$= 44000 \text{ psi} = 44.0 \text{ ksi (T)}$$

$$\sigma_B = 1333.33 - 42666.67$$
$$= -41333 \text{ psi} = 41.3 \text{ ksi (C)}$$

$$\frac{y}{44000} = \frac{0.75 - y}{41333} \qquad y = 0.387 \text{ in.}$$

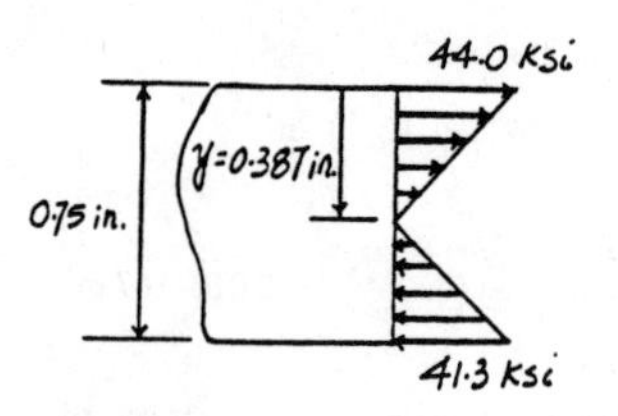

8–75. The clamp is made from members AB and AC, which are pin connected at A. If the compressive force at C and B is 180 N, determine the state of stress at point F, and indicate the results on a differential volume element. The screw DE is subjected only to a tensile force along its axis.

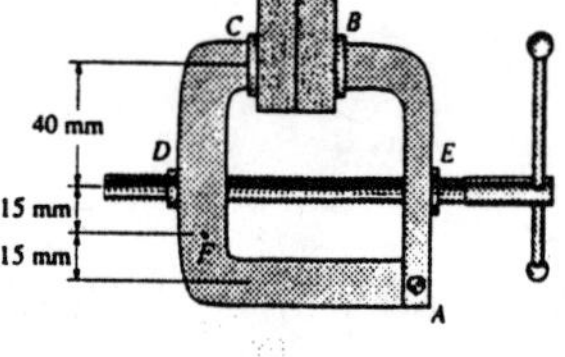

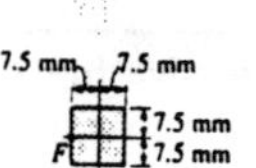

Support Reactions :

$$\circlearrowleft + \Sigma M_A = 0; \qquad 180(0.07) - T_{DE}(0.03) = 0$$
$$T_{DE} = 420 \text{ N}$$

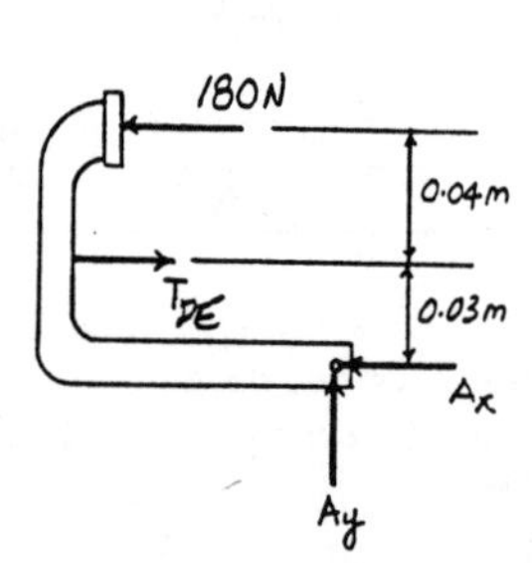

Internal Forces and Moment :

$$\xrightarrow{+} \Sigma F_x = 0; \qquad 420 - 180 - V = 0 \qquad V = 240 \text{ N}$$
$$+\uparrow \Sigma F_y = 0; \qquad N = 0$$
$$+\Sigma M_O = 0; \qquad 180(0.055) - 420(0.015) - M = 0$$
$$M = 3.60 \text{ N}\cdot\text{m}$$

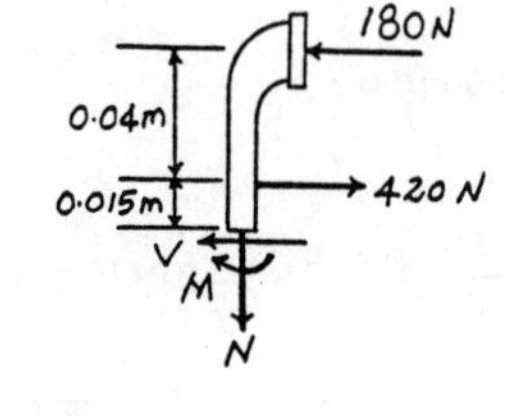

Section Properties :

$$A = 0.015(0.015) = 0.225\left(10^{-3}\right) \text{ m}^2$$

$$I = \frac{1}{12}(0.015)\left(0.015^3\right) = 4.21875\left(10^{-9}\right) \text{ m}^4$$

$$Q_F = 0$$

Normal Stress : Since $N = 0$, the normal stress is caused by bending stress only.

$$\sigma_F = \frac{Mc}{I} = \frac{3.60(0.0075)}{4.21875(10^{-9})} = 6.40 \text{ MPa (C)} \qquad \textbf{Ans}$$

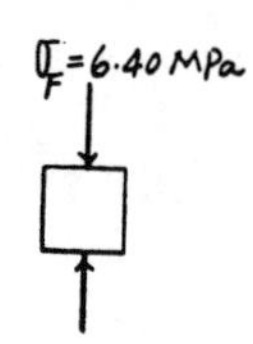

Shear Stress : Applying shear formula, we have

$$\tau_F = \frac{VQ_F}{It} = 0 \qquad \textbf{Ans}$$

***8–76.** The clamp is made from members AB and AC, which are pin-connected at A. If the compressive force at C and B is 180 N, determine the state of stress at point G, and indicate the results on a differential volume element. The screw DE is subjected only to a tensile force along its axis.

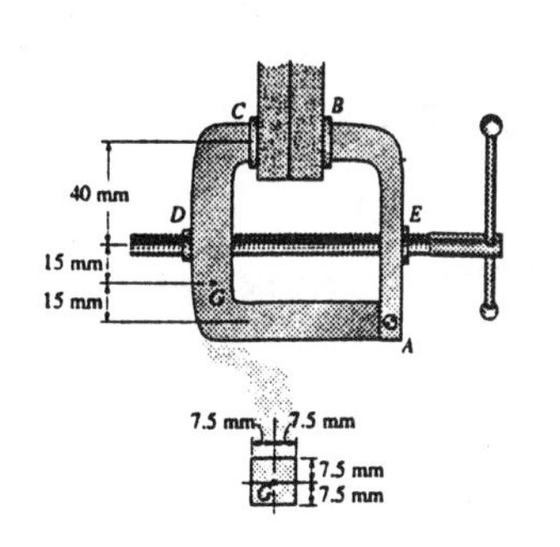

Support Reactions :

$$\curvearrowleft + \Sigma M_A = 0; \quad 180(0.07) - T_{DE}(0.03) = 0$$
$$T_{DE} = 420 \text{ N}$$

Internal Forces and Moment :

$$\xrightarrow{+} \Sigma F_x = 0; \quad 420 - 180 - V = 0 \quad V = 240 \text{ N}$$
$$+\uparrow \Sigma F_y = 0; \quad N = 0$$
$$\curvearrowleft + \Sigma M_O = 0; \quad 180(0.055) - 420(0.015) - M = 0$$
$$M = 3.60 \text{ N} \cdot \text{m}$$

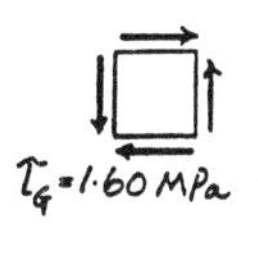

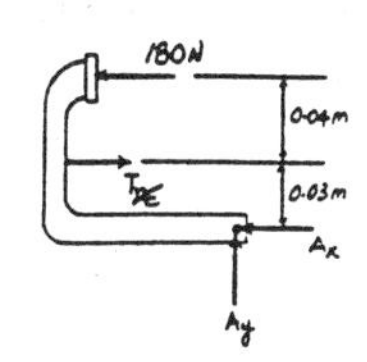

Section Properties :

$$A = 0.015(0.015) = 0.225\left(10^{-3}\right) \text{ m}^2$$
$$I = \frac{1}{12}(0.015)\left(0.015^3\right) = 4.21875\left(10^{-9}\right) \text{ m}^4$$
$$Q_G = \bar{y}'A' = 0.00375(0.0075)(0.015) = 0.421875\left(10^{-6}\right) \text{ m}^3$$

Normal Stress : Since $N = 0$, the normal stress is caused by bending stress only.

$$\sigma_G = \frac{My}{I} = \frac{3.60(0)}{4.21875(10^{-9})} = 0 \qquad \textbf{Ans}$$

Shear Stress : Applying shear formula, we have

$$\tau_G = \frac{VQ_G}{It} = \frac{240\left[0.421875(10^{-6})\right]}{4.21875(10^{-9})(0.015)} = 1.60 \text{ MPa} \qquad \textbf{Ans}$$

8–77. The wide-flange beam is subjected to the loading shown. Determine the state of stress at points A and B, and show the results on a differential volume element located at each of these points.

Support Reactions : As shown on FBD.

Internal Forces and Moment : As shown on FBD.

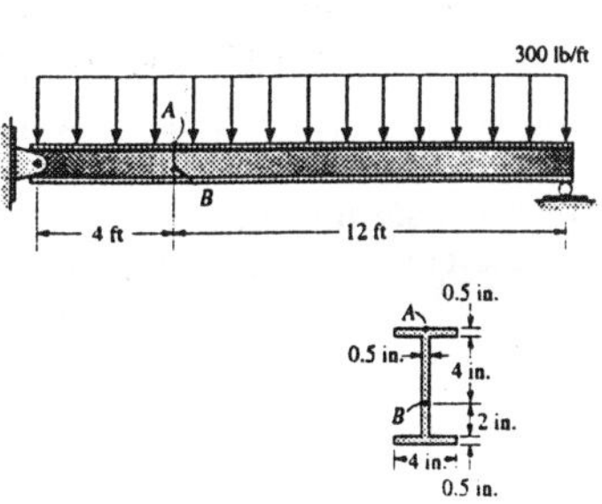

Section Properties :

$$A = 4(7) - 3.5(6) = 7.00 \text{ in}^2$$
$$I = \frac{1}{12}(4)\left(7^3\right) - \frac{1}{12}(3.5)\left(6^3\right) = 51.333 \text{ in}^4$$
$$Q_A = 0$$
$$Q_B = \Sigma\bar{y}'A' = 3.25(0.5)(4) + 2.00(0.5)(2) = 8.50 \text{ in}^3$$

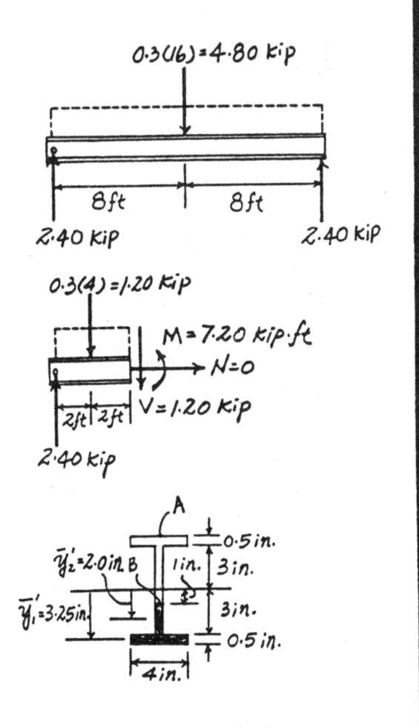

Normal Stress : Since $N = 0$, the normal stress is contributed by bending stress only.

$$\sigma_A = \frac{Mc}{I} = \frac{7.20(12)(3.5)}{51.333} = 5.89 \text{ ksi (C)} \qquad \textbf{Ans}$$
$$\sigma_B = \frac{My}{I} = \frac{7.20(12)(1)}{51.333} = 1.68 \text{ ksi (T)} \qquad \textbf{Ans}$$

Shear Stress : Applying the shear formula.

$$\tau_A = \frac{VQ_A}{It} = 0 \qquad \textbf{Ans}$$

$$\tau_B = \frac{VQ_B}{It} = \frac{1.20(8.50)}{51.333(0.5)} = 0.397 \text{ ksi} \qquad \textbf{Ans}$$

9–1. Prove that the sum of the normal stresses $\sigma_x + \sigma_y = \sigma_{x'} + \sigma_{y'}$ is constant.

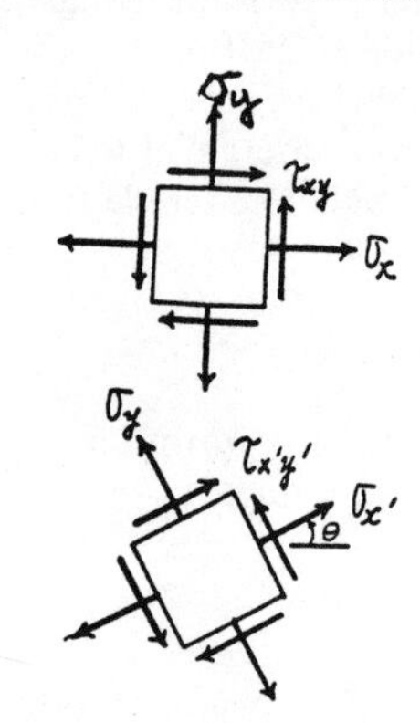

Stress Transformation Equations : Applying Eqs. 9 - 1 and 9 - 3 of the text.

$$\sigma_{x'} + \sigma_{y'} = \frac{\sigma_x + \sigma_y}{2} + \frac{\sigma_x - \sigma_y}{2}\cos 2\theta + \tau_{xy}\sin 2\theta$$
$$+ \frac{\sigma_x + \sigma_y}{2} - \frac{\sigma_x - \sigma_y}{2}\cos 2\theta - \tau_{xy}\sin 2\theta$$

$$\sigma_{x'} + \sigma_{y'} = \sigma_x + \sigma_y \qquad (Q.E.D.)$$

9–2. The state of stress at a point in a member is shown on the element. Determine the stress components acting on the inclined plane *AB*. Solve the problem using the method of equilibrium described in Sec. 9.1.

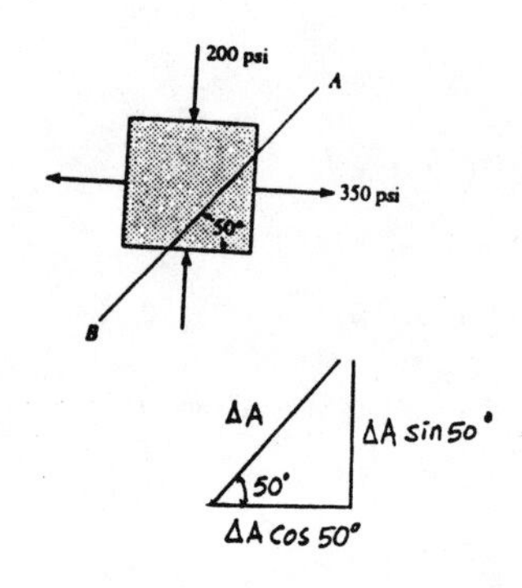

Force Equilibrium : For the sectioned element,

$$\nearrow + \Sigma F_{y'} = 0; \quad \Delta F_{y'} - 350(\Delta A \sin 50°)\sin 40° - 200(\Delta A \cos 50°)\sin 50° = 0$$

$$\Delta F_{y'} = 270.82\,\Delta A$$

$$\nwarrow + \Sigma F_{x'} = 0; \quad \Delta F_{x'} - 350(\Delta A \sin 50°)\cos 40° + 200(\Delta A \cos 50°)\cos 50° = 0$$

$$\Delta F_{x'} = 122.75\,\Delta A$$

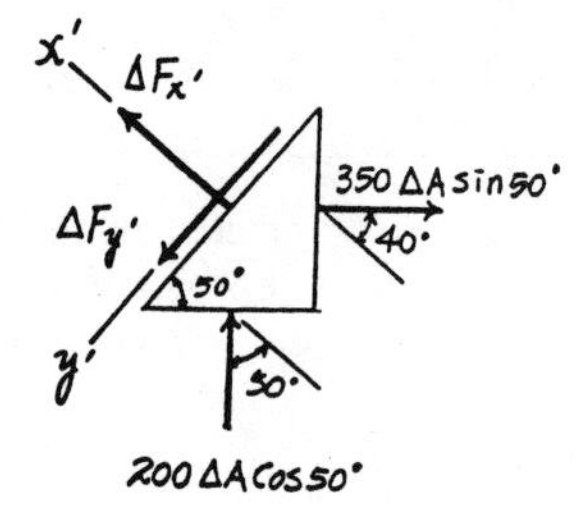

Normal and Shear Stress : For the inclined plane.

$$\sigma_{x'} = \lim_{\Delta A \to 0} \frac{\Delta F_{x'}}{\Delta A} = 123 \text{ psi} \qquad \textbf{Ans}$$

$$\tau_{x'y'} = \lim_{\Delta A \to 0} \frac{\Delta F_{y'}}{\Delta A} = 271 \text{ psi} \qquad \textbf{Ans}$$

9–3. Solve Prob. 9–2 using the stress-transformation equations developed in Sec. 9.2. Show the result on a sketch.

Normal and Shear Stress : In accordance with the established sign convention,

$$\theta = +140° \quad \sigma_x = 350 \text{ psi} \quad \sigma_y = -200 \text{ psi} \quad \tau_{xy} = 0$$

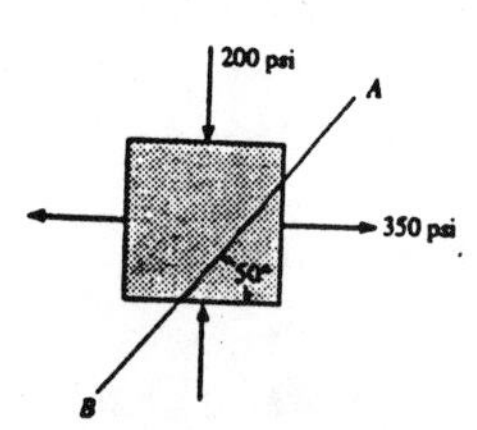

Stress Transformation Equations : Applying Eqs.9 – 1 and 9 – 2.

$$\sigma_{x'} = \frac{\sigma_x + \sigma_y}{2} + \frac{\sigma_x - \sigma_y}{2}\cos 2\theta + \tau_{xy}\sin 2\theta$$
$$= \frac{350 + (-200)}{2} + \frac{350 - (-200)}{2}\cos 280° + 0$$
$$= 123 \text{ psi} \qquad \textbf{Ans}$$

$$\tau_{x'y'} = -\frac{\sigma_x - \sigma_y}{2}\sin 2\theta + \tau_{xy}\cos 2\theta$$
$$= -\frac{350 - (-200)}{2}\sin 280° + 0$$
$$= 271 \text{ psi} \qquad \textbf{Ans}$$

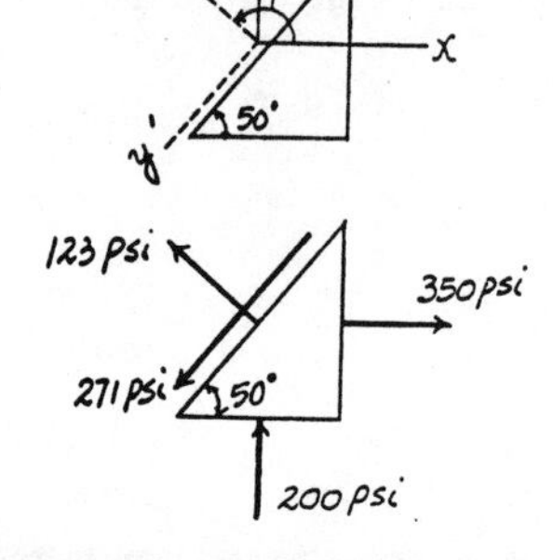

***9–4.** The state of stress at a point in a member is shown on the element. Determine the stress components acting on the inclined plane AB. Solve the problem using the method of equilibrium described in Sec. 9.1.

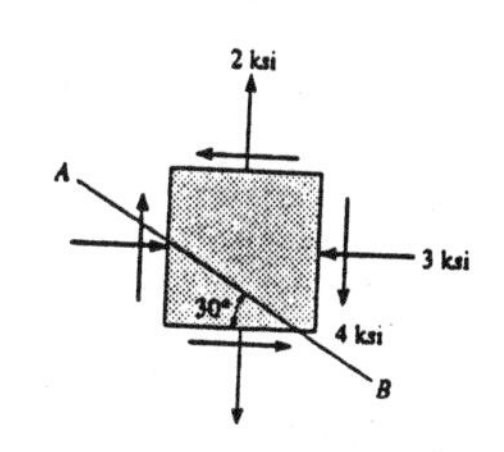

Force Equilibrium : For the sectioned element,

$\nearrow+\Sigma F_{y'} = 0;\quad \Delta F_{y'} - 3(\Delta A \sin 30°)\sin 60° + 4(\Delta A \sin 30°)\sin 30° - 2(\Delta A \cos 30°)\sin 30° - 4(\Delta A \cos 30°)\sin 60° = 0$

$$\Delta F_{y'} = 4.165\,\Delta A$$

$\nwarrow+\Sigma F_{x'} = 0;\quad \Delta F_{x'} + 3(\Delta A \sin 30°)\cos 60° + 4(\Delta A \sin 30°)\cos 30° - 2(\Delta A \cos 30°)\cos 30° + 4(\Delta A \cos 30°)\cos 60° = 0$

$$\Delta F_{x'} = -2.714\,\Delta A$$

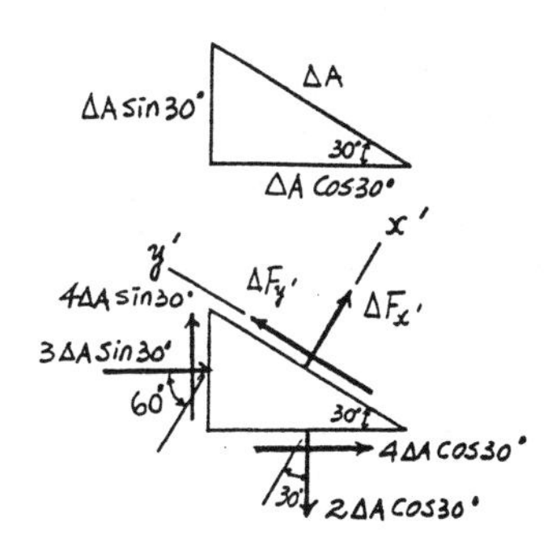

Normal and Shear Stress : For the inclined plane.

$$\sigma_{x'} = \lim_{\Delta A \to 0} \frac{\Delta F_{x'}}{\Delta A} = -2.71 \text{ ksi} \qquad \textbf{Ans}$$

$$\tau_{x'y'} = \lim_{\Delta A \to 0} \frac{\Delta F_{y'}}{\Delta A} = 4.17 \text{ ksi} \qquad \textbf{Ans}$$

Negative sign indicates that the sense of $\sigma_{x'}$ is opposite to that shown on FBD.

9–5. Solve Prob. 9–4 using the stress-transformation equations developed in Sec. 9.2. Show the result on a sketch.

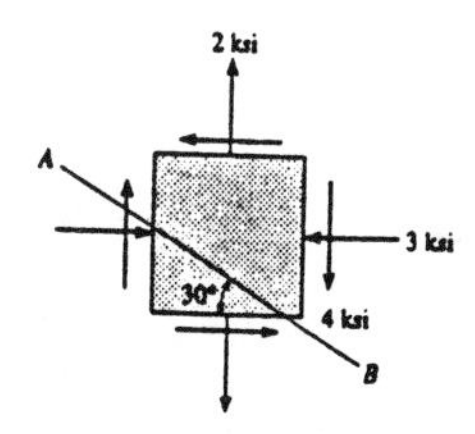

Normal and Shear Stress : In accordance with the established sign convention,

$\theta = +60° \qquad \sigma_x = -3 \text{ ksi} \qquad \sigma_y = 2 \text{ ksi} \qquad \tau_{xy} = -4 \text{ ksi}$

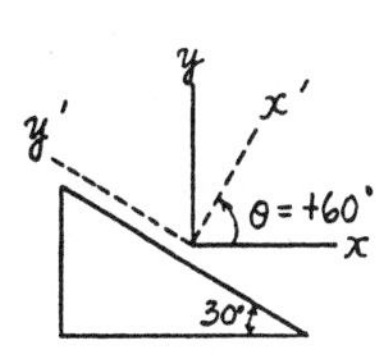

Stress Transformation Equations : Applying Eqs.9 – 1 and 9 – 2.

$$\sigma_{x'} = \frac{\sigma_x + \sigma_y}{2} + \frac{\sigma_x - \sigma_y}{2}\cos 2\theta + \tau_{xy}\sin 2\theta$$
$$= \frac{-3+2}{2} + \frac{-3-2}{2}\cos 120° + (-4\sin 120°)$$
$$= -2.71 \text{ ksi} \qquad \textbf{Ans}$$

$$\tau_{x'y'} = -\frac{\sigma_x - \sigma_y}{2}\sin 2\theta + \tau_{xy}\cos 2\theta$$
$$= -\frac{-3-2}{2}\sin 120° + (-4\cos 120°)$$
$$= 4.17 \text{ ksi} \qquad \textbf{Ans}$$

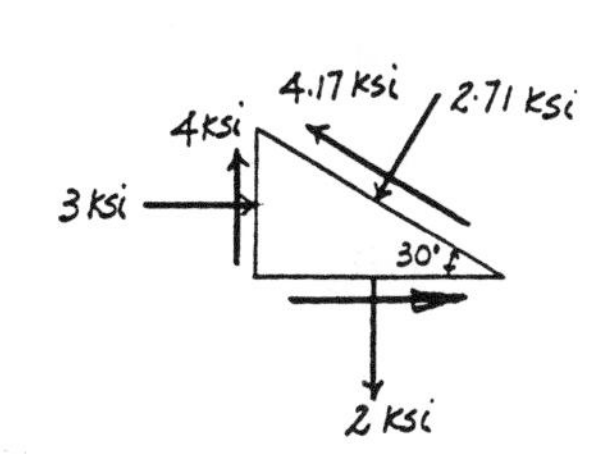

Negative sign indicates $\sigma_{x'}$ is a ***compressive*** stress

9–6. Determine the equivalent state of stress on an element if the element is oriented 30° counterclockwise from the element shown. Use the stress-transformation equations.

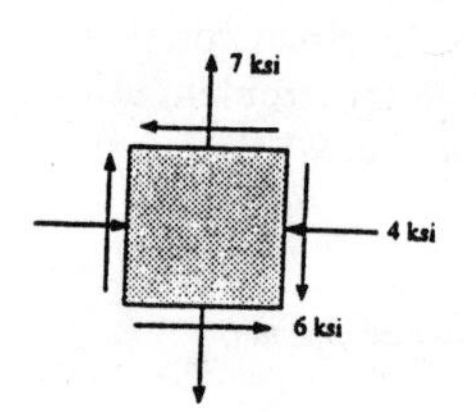

Normal and Shear Stress : In accordance with the established sign convention,

$$\theta = +30^\circ \quad \sigma_x = -4 \text{ ksi} \quad \sigma_y = 7 \text{ ksi} \quad \tau_{xy} = -6 \text{ ksi}$$

Stress Transformation Equations : Applying Eqs. 9–1, 9–2 and 9–3.

$$\sigma_{x'} = \frac{\sigma_x + \sigma_y}{2} + \frac{\sigma_x - \sigma_y}{2}\cos 2\theta + \tau_{xy}\sin 2\theta$$
$$= \frac{-4+7}{2} + \frac{-4-7}{2}\cos 60^\circ + (-6\sin 60^\circ)$$
$$= -6.45 \text{ ksi}$$

Ans

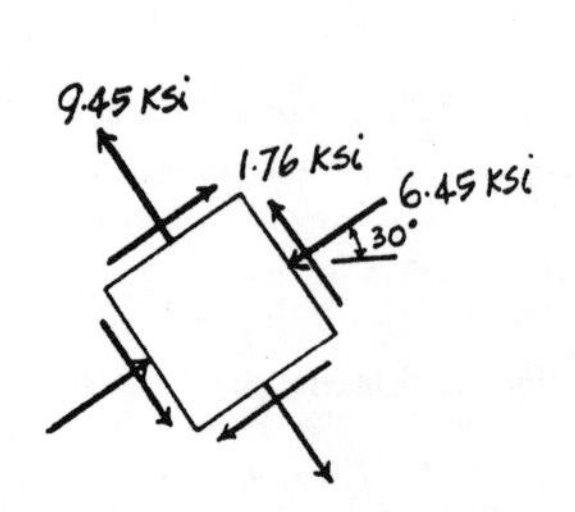

$$\sigma_{y'} = \frac{\sigma_x + \sigma_y}{2} - \frac{\sigma_x - \sigma_y}{2}\cos 2\theta - \tau_{xy}\sin 2\theta$$
$$= \frac{-4+7}{2} - \frac{-4-7}{2}\cos 60^\circ - (-6\sin 60^\circ)$$
$$= 9.45 \text{ ksi}$$

Ans

$$\tau_{x'y'} = -\frac{\sigma_x - \sigma_y}{2}\sin 2\theta + \tau_{xy}\cos 2\theta$$
$$= -\frac{-4-7}{2}\sin 60^\circ + (-6\cos 60^\circ)$$
$$= 1.76 \text{ ksi}$$

Ans

9–7. Determine the equivalent state of stress on an element if the element is oriented 60° clockwise from the element shown. Use the stress-transformation equations.

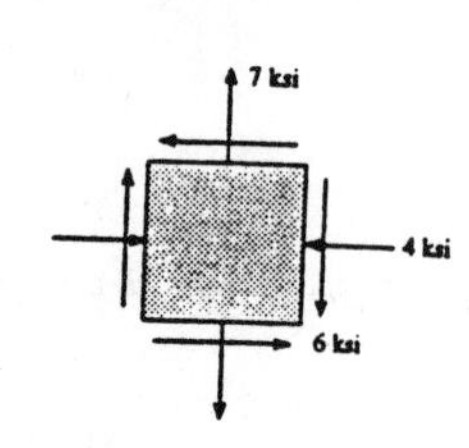

Normal and Shear Stress : In accordance with the established sign convention,

$$\theta = -60^\circ \quad \sigma_x = -4 \text{ ksi} \quad \sigma_y = 7 \text{ ksi} \quad \tau_{xy} = -6 \text{ ksi}$$

Stress Transformation Equations : Applying Eqs. 9–1, 9–2 and 9–3.

$$\sigma_{x'} = \frac{\sigma_x + \sigma_y}{2} + \frac{\sigma_x - \sigma_y}{2}\cos 2\theta + \tau_{xy}\sin 2\theta$$
$$= \frac{-4+7}{2} + \frac{-4-7}{2}\cos(-120^\circ) + [-6\sin(-120^\circ)]$$
$$= 9.45 \text{ ksi}$$

Ans

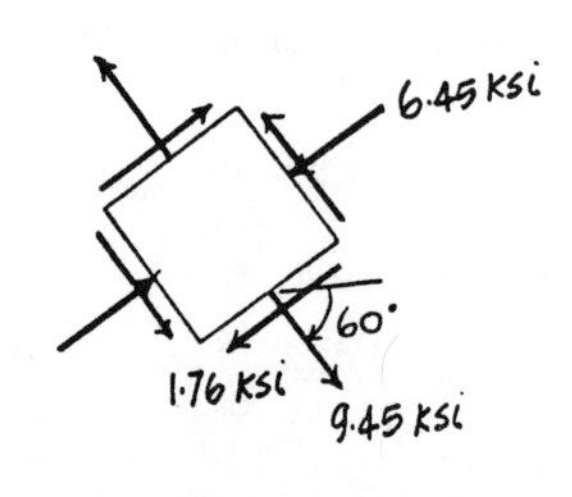

$$\sigma_{y'} = \frac{\sigma_x + \sigma_y}{2} - \frac{\sigma_x - \sigma_y}{2}\cos 2\theta - \tau_{xy}\sin 2\theta$$
$$= \frac{-4+7}{2} - \frac{-4-7}{2}\cos(-120^\circ) - [-6\sin(-120^\circ)]$$
$$= -6.45 \text{ ksi}$$

Ans

$$\tau_{x'y'} = -\frac{\sigma_x - \sigma_y}{2}\sin 2\theta + \tau_{xy}\cos 2\theta$$
$$= -\frac{-4-7}{2}\sin(-120^\circ) + [-6\cos(-120^\circ)]$$
$$= -1.76 \text{ ksi}$$

Ans

***9–8.** Determine the equivalent state of stress on an element if the element is oriented 60° clockwise from the element shown.

Normal and Shear Stress : In accordance with the established sign convention,

$$\theta = -60^\circ \qquad \sigma_x = 300 \text{ psi} \qquad \sigma_y = 0 \qquad \tau_{xy} = 120 \text{ psi}$$

Stress Transformation Equations : Applying Eqs. 9 – 1, 9 – 2 and 9 – 3.

$$\sigma_{x'} = \frac{\sigma_x + \sigma_y}{2} + \frac{\sigma_x - \sigma_y}{2}\cos 2\theta + \tau_{xy}\sin 2\theta$$

$$= \frac{300+0}{2} + \frac{300-0}{2}\cos(-120^\circ) + [120\sin(-120^\circ)]$$

$$= -28.9 \text{ psi}$$ **Ans**

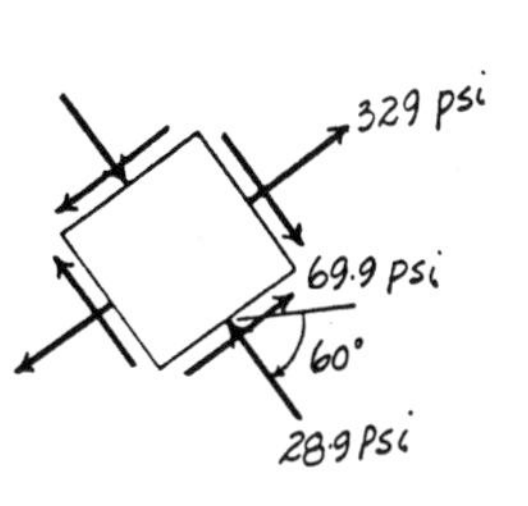

$$\sigma_{y'} = \frac{\sigma_x + \sigma_y}{2} - \frac{\sigma_x - \sigma_y}{2}\cos 2\theta - \tau_{xy}\sin 2\theta$$

$$= \frac{300+0}{2} - \frac{300-0}{2}\cos(-120^\circ) - [120\sin(-120^\circ)]$$

$$= 329 \text{ psi}$$ **Ans**

$$\tau_{x'y'} = -\frac{\sigma_x - \sigma_y}{2}\sin 2\theta + \tau_{xy}\cos 2\theta$$

$$= -\frac{300-0}{2}\sin(-120^\circ) + [120\cos(-120^\circ)]$$

$$= 69.9 \text{ psi}$$ **Ans**

9–9. Determine the equivalent state of stress on an element if the element is oriented 30° counterclockwise from the element shown.

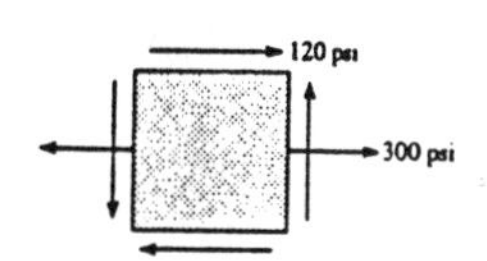

Normal and Shear Stress : In accordance with the established sign convention,

$$\theta = 30^\circ \qquad \sigma_x = 300 \text{ psi} \qquad \sigma_y = 0 \qquad \tau_{xy} = 120 \text{ psi}$$

Stress Tranformation Equations : Applying Eqs. 9 – 1, 9 – 2 and 9 – 3.

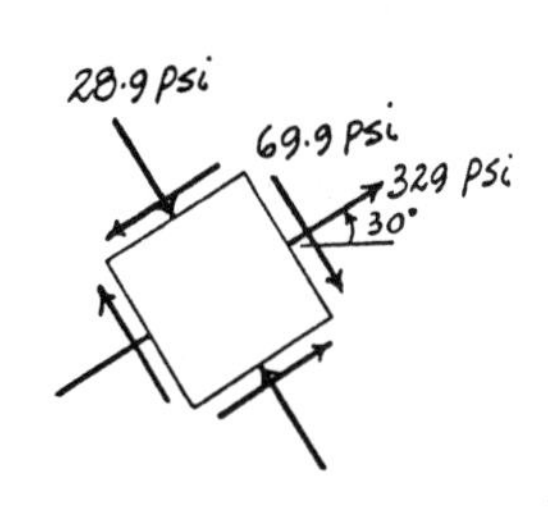

$$\sigma_{x'} = \frac{\sigma_x + \sigma_y}{2} + \frac{\sigma_x - \sigma_y}{2}\cos 2\theta + \tau_{xy}\sin 2\theta$$

$$= \frac{300+0}{2} + \frac{300-0}{2}\cos 60^\circ + 120\sin 60^\circ$$

$$= 329 \text{ psi}$$ **Ans**

$$\sigma_{y'} = \frac{\sigma_x + \sigma_y}{2} - \frac{\sigma_x - \sigma_y}{2}\cos 2\theta - \tau_{xy}\sin 2\theta$$

$$= \frac{300+0}{2} - \frac{300-0}{2}\cos 60^\circ - 120\sin 60^\circ$$

$$= -28.9 \text{ psi}$$ **Ans**

$$\tau_{x'y'} = -\frac{\sigma_x - \sigma_y}{2}\sin 2\theta + \tau_{xy}\cos 2\theta$$

$$= -\frac{300-0}{2}\sin 60^\circ + 120\cos 60^\circ$$

$$= -69.9 \text{ psi}$$ **Ans**

9–10. The state of stress at a point is shown on the element. Determine (a) the principal stresses and (b) the maximum in-plane shear stress and average normal stress at the point. Specify the orientation of the element in each case.

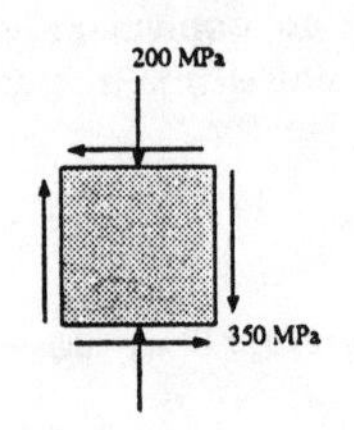

Normal and Shear Stress **:** In accordance with the established sign convention,

$$\sigma_x = 0 \qquad \sigma_y = -200 \text{ MPa} \qquad \tau_{xy} = -350 \text{ MPa}$$

a)

In - Plane Principal Stresses : Applying Eq. 9 – 5

$$\sigma_{1,2} = \frac{\sigma_x + \sigma_y}{2} \pm \sqrt{\left(\frac{\sigma_x - \sigma_y}{2}\right)^2 + \tau_{xy}^2}$$

$$= \frac{0 + (-200)}{2} \pm \sqrt{\left(\frac{0 - (-200)}{2}\right)^2 + (-350)^2}$$

$$= -100 \pm 364.0$$

$$\sigma_1 = 264 \text{ MPa} \qquad \sigma_2 = -464 \text{ MPa} \qquad \textbf{Ans}$$

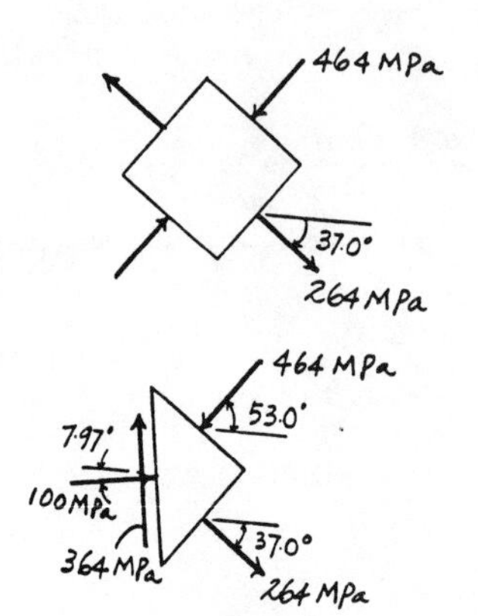

Orientation of Principal Plane : Applying Eq. 9 – 4,

$$\tan 2\theta_p = \frac{\tau_{xy}}{(\sigma_x - \sigma_y)/2} = \frac{-350}{[0 - (-200)]/2} = -3.50$$

$$\theta_p = -37.03° \quad \text{and} \quad 52.97°$$

Substituting the results into Eq. 9 - 1 with $\theta = -37.03°$ yields

$$\sigma_{x'} = \frac{\sigma_x + \sigma_y}{2} + \frac{\sigma_x - \sigma_y}{2}\cos 2\theta + \tau_{xy}\sin 2\theta$$

$$= \frac{0 + (-200)}{2} + \frac{0 - (-200)}{2}\cos(-74.06°) + (-350)\sin(-74.06°)$$

$$= 264.0 \text{ MPa} = \sigma_1$$

Hence,

$$\theta_{p_1} = -37.0° \qquad \theta_{p_2} = 53.0° \qquad \textbf{Ans}$$

b)

Maximum In - Plane Shear Stress : Applying Eq. 9 – 7,

$$\tau_{\substack{max \\ in\text{-}plane}} = \sqrt{\left(\frac{\sigma_x - \sigma_y}{2}\right)^2 + \tau_{xy}^2}$$

$$= \sqrt{\left(\frac{0 - (-200)}{2}\right)^2 + (-350)^2} = 364 \text{ MPa} \qquad \textbf{Ans}$$

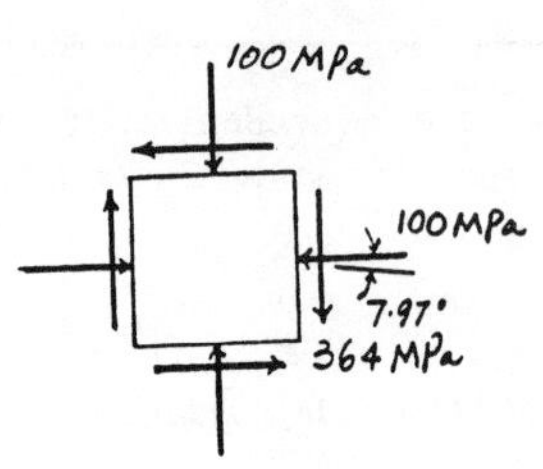

Orientation of the Plane for Maximum In - Plane Shear Stress **:** Applying Eq. 9 – 6,

$$\tan 2\theta_s = \frac{-(\sigma_x - \sigma_y)/2}{\tau_{xy}} = \frac{-[0 - (-200)]/2}{-350} = 0.2857$$

$$\theta_s = 7.97° \quad \text{and} \quad -82.0° \qquad \textbf{Ans}$$

By observation, in order to preserve equilibrium, $\tau_{\substack{max \\ in\text{-}plane}} = 364$ MPa has to act in the direction shown in the figure.

Average Normal Stress : Applying Eq. 9 – 8,

$$\sigma_{avg} = \frac{\sigma_x + \sigma_y}{2} = \frac{0 + (-200)}{2} = -100 \text{ MPa} \qquad \textbf{Ans}$$

9–11. The state of stress at a point is shown on the element. Determine (a) the principal stresses and (b) the maximum in-plane shear stress and average normal stress at the point. Specify the orientation of the element in each case.

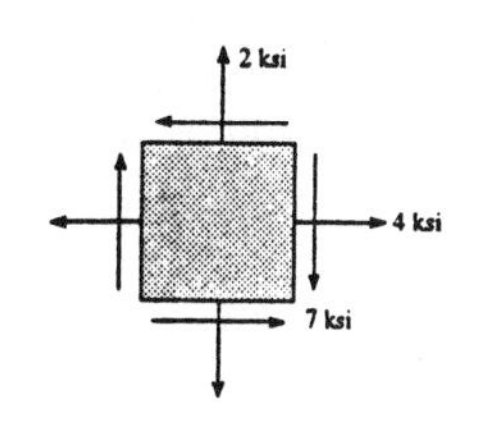

Normal and Shear Stress : In accordance with the established sign convention,

$$\sigma_x = 4 \text{ ksi} \qquad \sigma_y = 2 \text{ ksi} \qquad \tau_{xy} = -7 \text{ ksi}$$

a)

In-Plane Principal Stresses : Applying Eq.9 – 5,

$$\sigma_{1,2} = \frac{\sigma_x + \sigma_y}{2} \pm \sqrt{\left(\frac{\sigma_x - \sigma_y}{2}\right)^2 + \tau_{xy}^2}$$

$$= \frac{4+2}{2} \pm \sqrt{\left(\frac{4-2}{2}\right)^2 + (-7)^2}$$

$$= 3 \pm 7.071$$

$$\sigma_1 = 10.1 \text{ ksi} \qquad \sigma_2 = -4.07 \text{ ksi} \qquad \textbf{Ans}$$

Orientation of Principal Plane : Applying Eq.9 – 4,

$$\tan 2\theta_p = \frac{\tau_{xy}}{(\sigma_x - \sigma_y)/2} = \frac{-7}{(4-2)/2} = -7.00$$

$$\theta_p = -40.93^\circ \quad \text{and} \quad 49.07^\circ$$

Substituting the results into Eq.9 - 1 with $\theta = -40.93^\circ$ yields

$$\sigma_{x'} = \frac{\sigma_x + \sigma_y}{2} + \frac{\sigma_x - \sigma_y}{2}\cos 2\theta + \tau_{xy}\sin 2\theta$$

$$= \frac{4+2}{2} + \frac{4-2}{2}\cos(-81.86^\circ) + (-7)\sin(-81.86^\circ)$$

$$= 10.1 \text{ ksi} = \sigma_1$$

Hence,

$$\theta_{p_1} = -40.9^\circ \qquad \theta_{p_2} = 49.1^\circ \qquad \textbf{Ans}$$

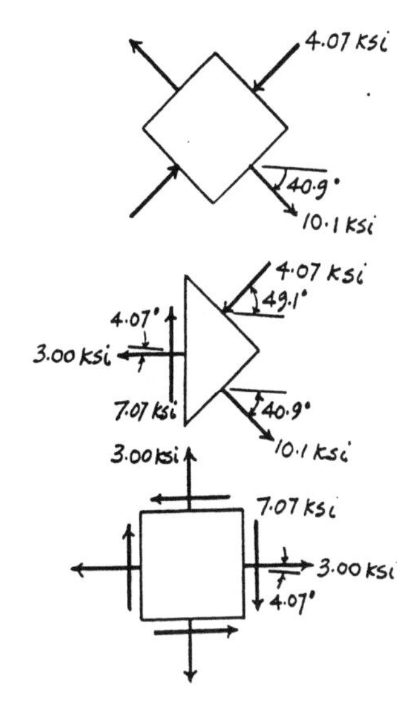

b)

Maximum In-Plane Shear Stress : Applying Eq.9 – 7,

$$\tau_{\substack{\max \\ \text{in-plane}}} = \sqrt{\left(\frac{\sigma_x - \sigma_y}{2}\right)^2 + \tau_{xy}^2}$$

$$= \sqrt{\left(\frac{4-2}{2}\right)^2 + (-7)^2} = 7.07 \text{ ksi} \qquad \textbf{Ans}$$

Orientation of the Plane for Maximum In-Plane Shear Stress : Applying Eq.9 – 6,

$$\tan 2\theta_s = \frac{-(\sigma_x - \sigma_y)/2}{\tau_{xy}} = \frac{-(4-2)/2}{-7} = 0.1429$$

$$\theta_s = 4.07^\circ \quad \text{and} \quad -85.9^\circ \qquad \textbf{Ans}$$

By observation, in order to preserve equilibrium, $\tau_{\substack{\max \\ \text{in-plane}}} = 7.07$ ksi has to act in the direction shown in the figure.

Average Normal Stress : Applying Eq.9 – 8,

$$\sigma_{avg} = \frac{\sigma_x + \sigma_y}{2} = \frac{4+2}{2} = 3.00 \text{ ksi} \qquad \textbf{Ans}$$

***9–12.** The state of stress at a point is shown on the element. Determine (a) the principal stresses and (b) the maximum in-plane shear stress and average normal stress at the point. Specify the orientation of the element in each case.

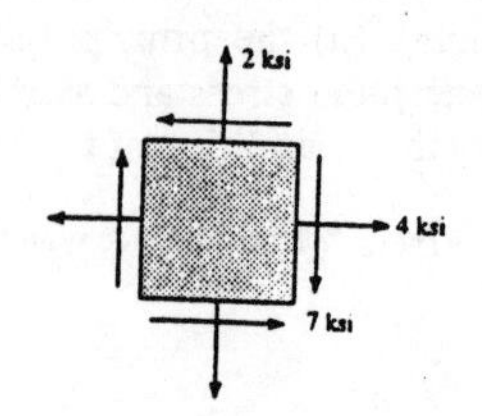

Normal and Shear Stress **:** In accordance with the established sign convention,

$$\sigma_x = 6 \text{ ksi} \qquad \sigma_y = 8 \text{ ksi} \qquad \tau_{xy} = -10 \text{ ksi}$$

a)

In - Plane Principal Stresses : Applying Eq. 9 – 5,

$$\sigma_{1,2} = \frac{\sigma_x + \sigma_y}{2} \pm \sqrt{\left(\frac{\sigma_x - \sigma_y}{2}\right)^2 + \tau_{xy}^2}$$

$$= \frac{6+8}{2} \pm \sqrt{\left(\frac{6-8}{2}\right)^2 + (-10)^2}$$

$$= 7.00 \pm 10.05$$

$$\sigma_1 = 17.0 \text{ ksi} \qquad \sigma_2 = -3.05 \text{ ksi} \qquad \textbf{Ans}$$

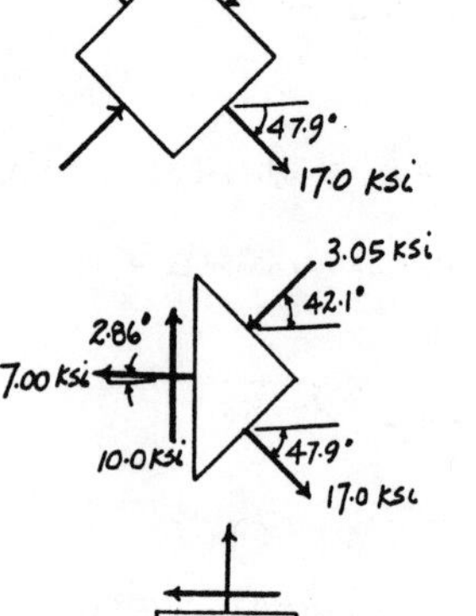

Orientation of Principal Plane : Applying Eq. 9 – 4,

$$\tan 2\theta_p = \frac{\tau_{xy}}{(\sigma_x - \sigma_y)/2} = \frac{-10}{(6-8)/2} = 10.00$$

$$\theta_p = 42.14° \quad \text{and} \quad -47.86°$$

Substituting the results into Eq. 9 - 1 with $\theta = 42.14°$ yields

$$\sigma_{x'} = \frac{\sigma_x + \sigma_y}{2} + \frac{\sigma_x - \sigma_y}{2} \cos 2\theta + \tau_{xy} \sin 2\theta$$

$$= \frac{6+8}{2} + \frac{6-8}{2} \cos 84.28° + (-10) \sin 84.28°$$

$$= -3.05 \text{ ksi} = \sigma_2$$

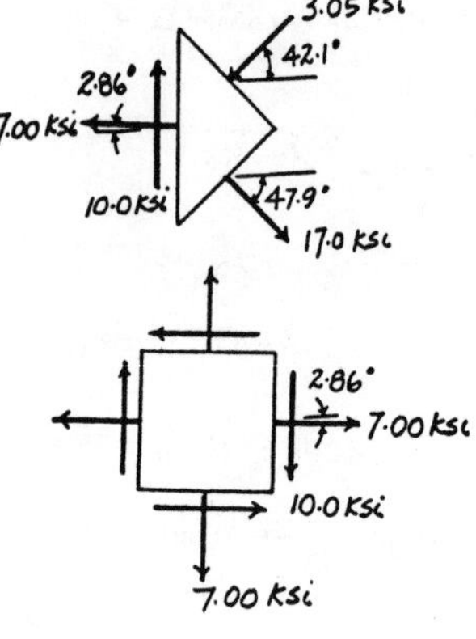

Hence,

$$\theta_{p_1} = -47.9° \qquad \theta_{p_2} = 42.1° \qquad \textbf{Ans}$$

b)

Maximum In - Plane Shear Stress : Applying Eq. 9 – 7,

$$\tau_{\substack{max \\ in\text{-}plane}} = \sqrt{\left(\frac{\sigma_x - \sigma_y}{2}\right)^2 + \tau_{xy}^2}$$

$$= \sqrt{\left(\frac{6-8}{2}\right)^2 + (-10)^2} = 10.0 \text{ ksi} \qquad \textbf{Ans}$$

Orientation of the Plane for Maximum In - Plane Shear Stress : Applying Eq. 9 – 6

$$\tan 2\theta_s = \frac{-(\sigma_x - \sigma_y)/2}{\tau_{xy}} = \frac{-(6-8)/2}{-10} = -0.100$$

$$\theta_s = -2.86° \quad \text{and} \quad 87.1° \qquad \textbf{Ans}$$

By observation, in order to preserve equilibrium, $\tau_{\substack{max \\ in\text{-}plane}} = 10.0$ ksi has to act in the direction shown in the figure.

Average Normal Stress : Applying Eq. 9 – 8,

$$\sigma_{avg} = \frac{\sigma_x + \sigma_y}{2} = \frac{8+6}{2} = 7.00 \text{ ksi} \qquad \textbf{Ans}$$

9–13. The state of stress at a point is shown on the element. Determine (a) the principal stresses and (b) the maximum in-plane shear stress and average normal stress on the element. Specify the orientation of the element in each case.

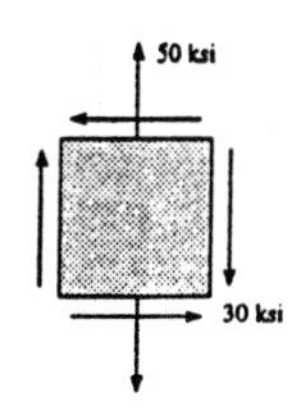

Normal and Shear Stress : In accordance with the established sign convention,

$$\sigma_x = 0 \qquad \sigma_y = 50 \text{ ksi} \qquad \tau_{xy} = -30 \text{ ksi}$$

a)

In - Plane Principal Stress : Applying Eq.9 – 5,

$$\sigma_{1,2} = \frac{\sigma_x + \sigma_y}{2} \pm \sqrt{\left(\frac{\sigma_x - \sigma_y}{2}\right)^2 + \tau_{xy}^2}$$

$$= \frac{0 + 50}{2} \pm \sqrt{\left(\frac{0 - 50}{2}\right)^2 + (-30)^2}$$

$$= 7.00 \pm 10.05$$

$$\sigma_1 = 64.1 \text{ ksi} \qquad \sigma_2 = -14.1 \text{ ksi} \qquad \textbf{Ans}$$

Orientation of Principal Plane : Applying Eq.9 – 4,

$$\tan 2\theta_p = \frac{\tau_{xy}}{(\sigma_x - \sigma_y)/2} = \frac{-30}{(0 - 50)/2} = 1.200$$

$$\theta_p = 25.10° \quad \text{and} \quad -64.90°$$

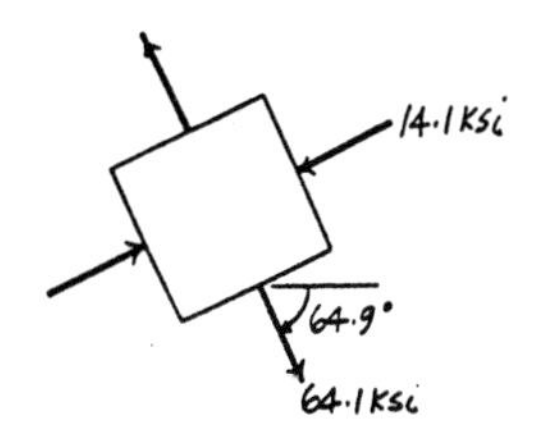

Substituting the results into Eq.9 - 1 with $\theta = 25.10°$ yields

$$\sigma_{x'} = \frac{\sigma_x + \sigma_y}{2} + \frac{\sigma_x - \sigma_y}{2}\cos 2\theta + \tau_{xy}\sin 2\theta$$

$$= \frac{0 + 50}{2} + \frac{0 - 50}{2}\cos 50.20° + (-30)\sin 50.20°$$

$$= -14.1 \text{ ksi} = \sigma_2$$

Hence,

$$\theta_{p_1} = -64.9° \qquad \theta_{p_2} = 25.1° \qquad \textbf{Ans}$$

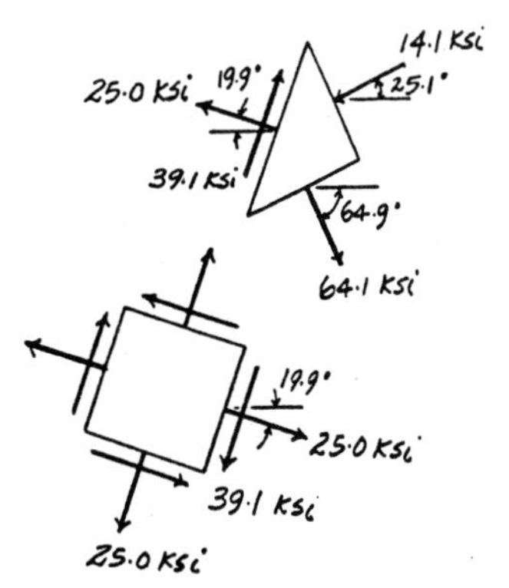

b)

Maximum In - Plane Shear Stress : Applying Eq.9 – 7,

$$\tau_{\substack{max \\ in\text{-}plane}} = \sqrt{\left(\frac{\sigma_x - \sigma_y}{2}\right)^2 + \tau_{xy}^2}$$

$$= \sqrt{\left(\frac{0 - 50}{2}\right)^2 + (-30)^2} = 39.1 \text{ ksi} \qquad \textbf{Ans}$$

Orientation of the Plane for Maximum In - Plane Shear Stress : Applying Eq.9 – 6,

$$\tan 2\theta_s = \frac{-(\sigma_x - \sigma_y)/2}{\tau_{xy}} = \frac{-(0 - 50)/2}{-30} = -0.8333$$

$$\theta_s = -19.9° \quad \text{and} \quad 70.1° \qquad \textbf{Ans}$$

By observation, in order to preserve equilibrium, $\tau_{\substack{max \\ in\text{-}plane}} = 39.1$ ksi has to act in the direction shown in the figure.

Average Normal Stress : Applying Eq.9 – 8,

$$\sigma_{avg} = \frac{\sigma_x + \sigma_y}{2} = \frac{0 + 50}{2} = 25.0 \text{ ksi} \qquad \textbf{Ans}$$

9–14. The state of stress at a point on the upper surface of the airplane wing is shown on the element. Determine (a) the principal stresses and (b) the maximum in-plane shear stress and average normal stress at the point. Specify the orientation of the element in each case.

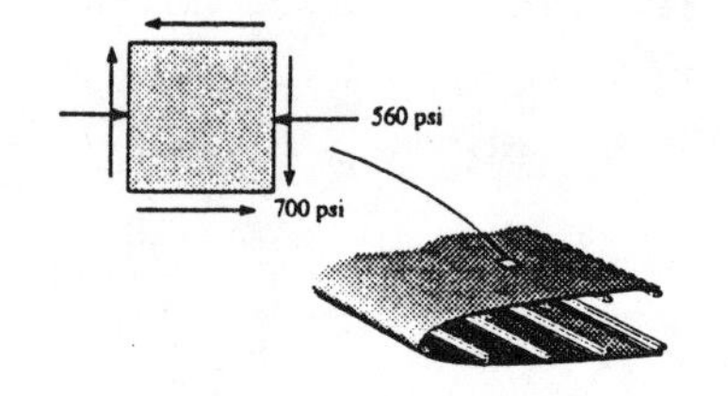

Normal and Shear Stress : In accordance with the established sign convention,

$$\sigma_x = -560 \text{ psi} \qquad \sigma_y = 0 \qquad \tau_{xy} = -700 \text{ psi}$$

a)

In-Plane Principal Stresses : Applying Eq.9–5

$$\sigma_{1,2} = \frac{\sigma_x + \sigma_y}{2} \pm \sqrt{\left(\frac{\sigma_x - \sigma_y}{2}\right)^2 + \tau_{xy}^2}$$

$$= \frac{-560+0}{2} \pm \sqrt{\left(\frac{-560-0}{2}\right)^2 + (-700)^2}$$

$$= -280 \pm 753.9$$

$$\sigma_1 = 474 \text{ psi} \qquad \sigma_2 = -1034 \text{ psi} \qquad \textbf{Ans}$$

Orientation of Principal Plane : Applying Eq.9–4.

$$\tan 2\theta_p = \frac{\tau_{xy}}{(\sigma_x - \sigma_y)/2} = \frac{-700}{(-560-0)/2} = 2.50$$

$$\theta_p = 34.10° \quad \text{and} \quad -55.90°$$

Substituting the results into Eq.9–1 with $\theta = 34.10°$ yields

$$\sigma_{x'} = \frac{\sigma_x + \sigma_y}{2} + \frac{\sigma_x - \sigma_y}{2}\cos 2\theta + \tau_{xy}\sin 2\theta$$

$$= \frac{-560+0}{2} + \frac{-560-0}{2}\cos 68.20° + (-700)\sin 68.20°$$

$$= -1034 \text{ psi} = \sigma_2$$

Hence,

$$\theta_{p_1} = -55.9° \qquad \theta_{p_2} = 34.1° \qquad \textbf{Ans}$$

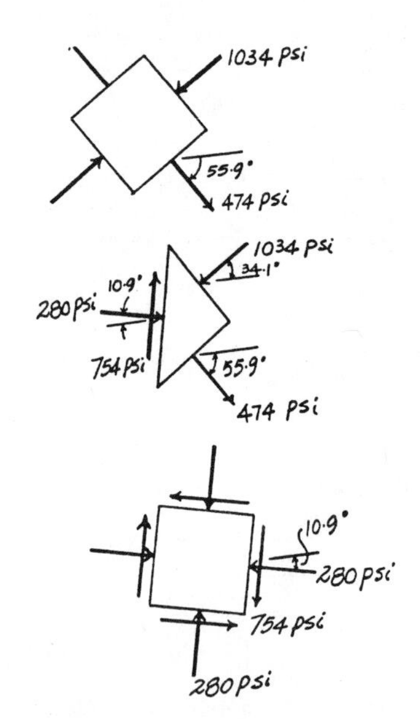

b)

Maximum In-Plane Shear Stress : Applying Eq.9–7

$$\tau_{\substack{max \\ in\text{-}plane}} = \sqrt{\left(\frac{\sigma_x - \sigma_y}{2}\right)^2 + \tau_{xy}^2}$$

$$= \sqrt{\left(\frac{-560-0}{2}\right)^2 + (-700)^2} = 754 \text{ psi} \qquad \textbf{Ans}$$

Orientation of the Plane for Maximum In-Plane Shear Stress : Applying Eq.9–6

$$\tan 2\theta_s = \frac{-(\sigma_x - \sigma_y)/2}{\tau_{xy}} = \frac{-(-560-0)/2}{-700} = -0.400$$

$$\theta_s = -10.9° \quad \text{and} \quad 79.1° \qquad \textbf{Ans}$$

By observation, in order to preserve equilibrium, $\tau_{\substack{max \\ in\text{-}plane}} = 754$ psi has to act in the direction shown in the figure.

Average Normal Stress : Applying Eq.9–8.

$$\sigma_{avg} = \frac{\sigma_x + \sigma_y}{2} = \frac{-560+0}{2} = -280 \text{ psi} \qquad \textbf{Ans}$$

9–15. The steel plate has a thickness of 10 mm and is subjected to the edge loading shown. Determine the maximum in-plane shear stress and the average normal stress developed in the steel.

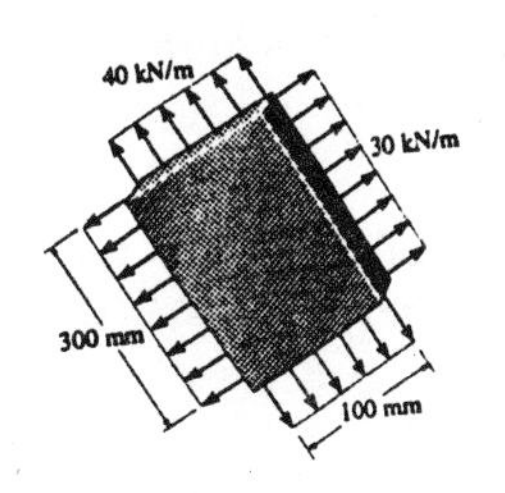

Normal and Shear Stress : In accordance with the established sign convention,

$$\sigma_x = \frac{30(10^3)}{0.01} = 3.00 \text{ MPa} \qquad \sigma_y = \frac{40(10^3)}{0.01} = 4.00 \text{ MPa}$$

$$\tau_{xy} = 0$$

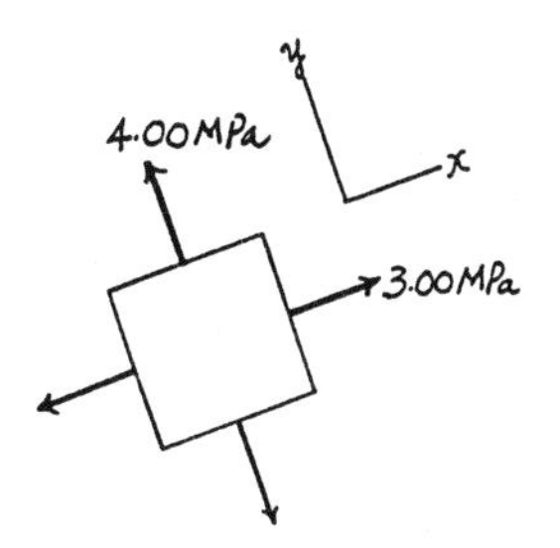

Maximum In-Plane Shear Stress : Applying Eq.9 – 7.

$$\tau_{\substack{\max \\ \text{in-plane}}} = \sqrt{\left(\frac{\sigma_x - \sigma_y}{2}\right)^2 + \tau_{xy}^2}$$

$$= \sqrt{\left(\frac{3.00 - 4.00}{2}\right)^2 + 0} = 0.500 \text{ MPa} \qquad \textbf{Ans}$$

Average Normal Stress : Applying Eq.9 – 8.

$$\sigma_{avg} = \frac{\sigma_x + \sigma_y}{2} = \frac{3.00 + 4.00}{2} = 3.50 \text{ MPa} \qquad \textbf{Ans}$$

***9–16.** The steel bar has a thickness of 0.5 in. and is subjected to the edge loading shown. Determine the principal stresses developed in the bar.

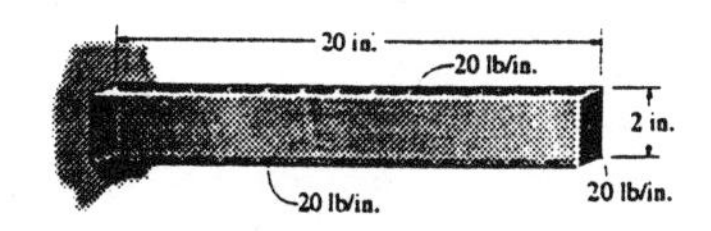

Normal and Shear Stress : In accordance with the established sign convention,

$$\sigma_x = 0 \qquad \sigma_y = 0 \qquad \tau_{xy} = \frac{20}{0.5} = 40.0 \text{ psi}$$

In-Plane Principal Stress : Applying Eq.9 – 5,

$$\sigma_{1,2} = \frac{\sigma_x + \sigma_y}{2} \pm \sqrt{\left(\frac{\sigma_x - \sigma_y}{2}\right)^2 + \tau_{xy}^2}$$

$$= 0 \pm \sqrt{0 + 40.0^2}$$

$$= 0 \pm 40.0$$

$$\sigma_1 = 40.0 \text{ psi} \qquad \sigma_2 = -40.0 \text{ psi} \qquad \textbf{Ans}$$

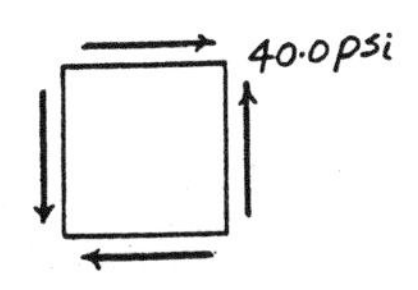

9-18. A point on a thin plate is subjected to the two successive states of stress shown. Determine the resultant state of stress represented on the element oriented as shown on the right.

200 MPa, 58 MPa, 350 MPa, 60°, 25°, σ_y, τ_{xy}

Stress Transformation Equations : Applying Eqs. 9-1, 9-2, and 9-3 to element (*a*) with $\theta = -30°$, $\sigma_{x'} = -200$ MPa, $\sigma_{y'} = -350$ MPa and $\tau_{x'y'} = 0$,

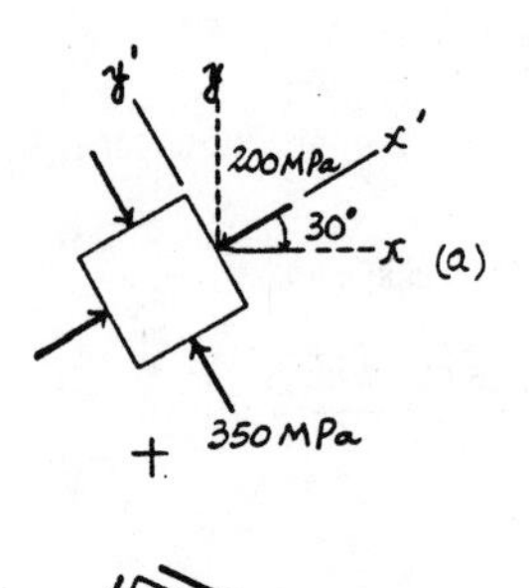

$$(\sigma_x)_a = \frac{\sigma_{x'} + \sigma_{y'}}{2} + \frac{\sigma_{x'} - \sigma_{y'}}{2}\cos 2\theta + \tau_{x'y'}\sin 2\theta$$

$$= \frac{-200 + (-350)}{2} + \frac{-200 - (-350)}{2}\cos(-60°) + 0$$

$$= -237.5 \text{ MPa}$$

$$(\sigma_y)_a = \frac{\sigma_{x'} + \sigma_{y'}}{2} - \frac{\sigma_{x'} - \sigma_{y'}}{2}\cos 2\theta - \tau_{x'y'}\sin 2\theta$$

$$= \frac{-200 + (-350)}{2} - \frac{-200 - (-350)}{2}\cos(-60°) - 0$$

$$= -312.5 \text{ MPa}$$

$$(\tau_{xy})_a = -\frac{\sigma_{x'} - \sigma_{y'}}{2}\sin 2\theta + \tau_{x'y'}\cos 2\theta$$

$$= -\frac{-200 - (-350)}{2}\sin(-60°) + 0$$

$$= 64.95 \text{ MPa}$$

For element (*b*), $\theta = 25°$, $\sigma_{x'} = \sigma_{y'} = 0$ and $\tau_{x'y'} = 58$ MPa,

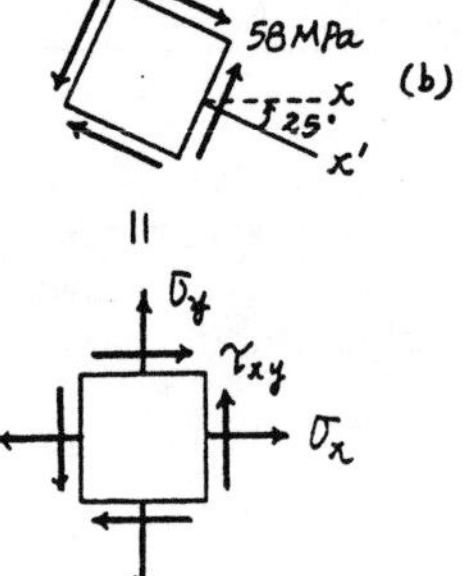

$$(\sigma_x)_b = \frac{\sigma_{x'} + \sigma_{y'}}{2} + \frac{\sigma_{x'} - \sigma_{y'}}{2}\cos 2\theta + \tau_{x'y'}\sin 2\theta$$

$$= 0 + 0 + 58\sin 50°$$

$$= 44.43 \text{ MPa}$$

$$(\sigma_y)_b = \frac{\sigma_{x'} + \sigma_{y'}}{2} - \frac{\sigma_{x'} - \sigma_{y'}}{2}\cos 2\theta - \tau_{x'y'}\sin 2\theta$$

$$= 0 - 0 - 58\sin 50°$$

$$= -44.43 \text{ MPa}$$

$$(\tau_{xy})_b = -\frac{\sigma_{x'} - \sigma_{y'}}{2}\sin 2\theta + \tau_{x'y'}\cos 2\theta$$

$$= -0 + 58\cos 50°$$

$$= 37.28 \text{ MPa}$$

Combining the stress components of two elements yields

$\sigma_x = (\sigma_x)_a + (\sigma_x)_b = -237.5 + 44.43 = -193$ MPa **Ans**

$\sigma_y = (\sigma_y)_a + (\sigma_y)_b = -312.5 - 44.43 = -357$ MPa **Ans**

$\tau_{xy} = (\tau_{xy})_a + (\tau_{xy})_b = 64.95 + 37.28 = 102$ MPa **Ans**

9–19. The stress along two planes at a point is indicated. Determine the normal stresses on plane b–b and the principal stresses.

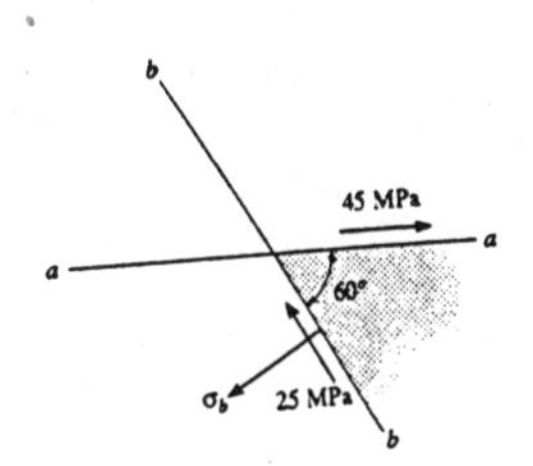

Stress Transformation Equations : Applying Eqs. 9 – 3 and 9 – 1 with $\theta = -150°$, $\sigma_y = 0$, $\tau_{xy} = 45$ MPa, $\tau_{x'y'} = -25$ MPa, and $\sigma_{x'} = \sigma_b$,

$$\tau_{x'y'} = -\frac{\sigma_x - \sigma_y}{2}\sin 2\theta + \tau_{xy}\cos 2\theta$$

$$-25 = -\frac{\sigma_x - 0}{2}\sin(-300°) + 45\cos(-300°)$$

$$\sigma_x = 109.70 \text{ MPa}$$

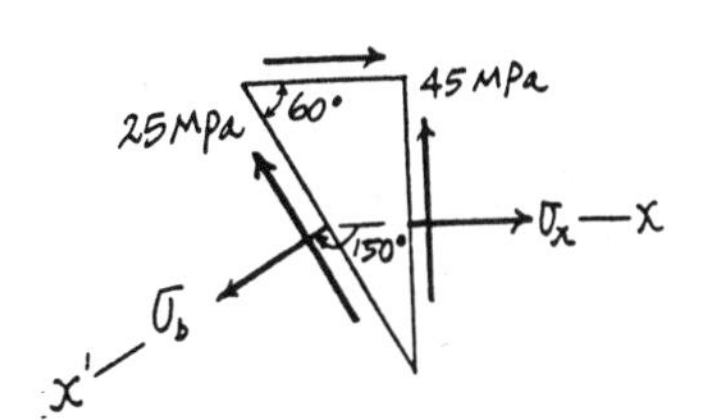

$$\sigma_{x'} = \frac{\sigma_x + \sigma_y}{2} + \frac{\sigma_x - \sigma_y}{2}\cos 2\theta + \tau_{xy}\sin 2\theta$$

$$\sigma_b = \frac{109.70 + 0}{2} + \frac{109.70 - 0}{2}\cos(-300°) + 45\sin(-300°)$$

$$= 121 \text{ MPa} \qquad \textbf{Ans}$$

In - Plane Principal Stress : Applying Eq. 9 – 5,

$$\sigma_{1,2} = \frac{\sigma_x + \sigma_y}{2} \pm \sqrt{\left(\frac{\sigma_x - \sigma_y}{2}\right)^2 + \tau_{xy}^2}$$

$$= \frac{109.70 + 0}{2} \pm \sqrt{\left(\frac{109.70 - 0}{2}\right)^2 + 45^2}$$

$$= 54.848 \pm 70.946$$

$$\sigma_1 = 126 \text{ MPa} \qquad \sigma_2 = -16.1 \text{ MPa} \qquad \textbf{Ans}$$

***9–20.** The stress acting on two planes at a point is indicated. Determine the normal stress σ_b and the principal stresses at the point.

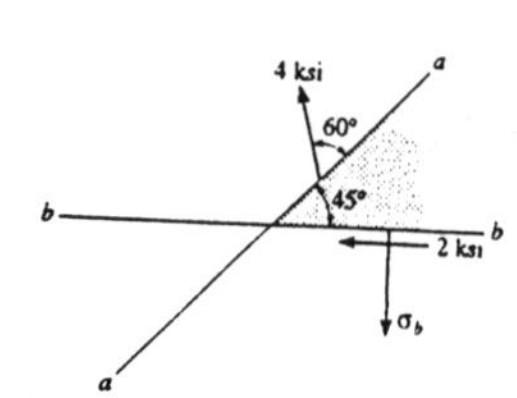

Stress Transformation Equations : Applying Eqs. 9 – 3 and 9 – 1 with $\theta = -135°$, $\sigma_y = 3.464$ ksi, $\tau_{xy} = 2.00$ ksi, $\tau_{x'y'} = -2$ ksi, and $\sigma_{x'} = \sigma_b$.,

$$\tau_{x'y'} = -\frac{\sigma_x - \sigma_y}{2}\sin 2\theta + \tau_{xy}\cos 2\theta$$

$$-2 = -\frac{\sigma_x - 3.464}{2}\sin(-270°) + 2\cos(-270°)$$

$$\sigma_x = 7.464 \text{ ksi}$$

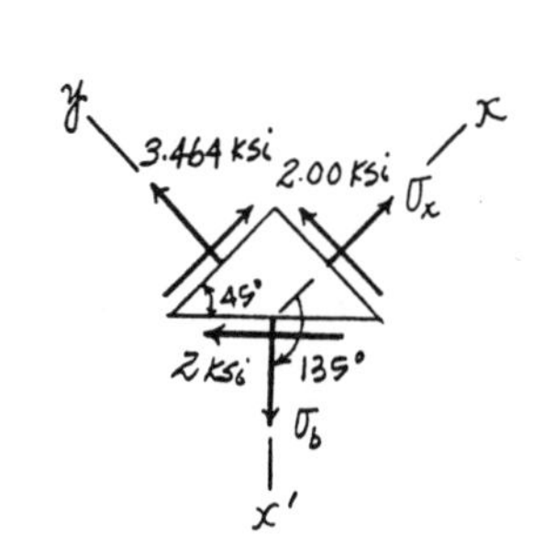

$$\sigma_{x'} = \frac{\sigma_x + \sigma_y}{2} + \frac{\sigma_x - \sigma_y}{2}\cos 2\theta + \tau_{xy}\sin 2\theta$$

$$\sigma_b = \frac{7.464 + 3.464}{2} + \frac{7.464 - 3.464}{2}\cos(-270°) + 2\sin(-270°)$$

$$= 7.46 \text{ ksi} \qquad \textbf{Ans}$$

In - Plane Principal Stress : Applying Eq. 9 – 5,

$$\sigma_{1,2} = \frac{\sigma_x + \sigma_y}{2} \pm \sqrt{\left(\frac{\sigma_x - \sigma_y}{2}\right)^2 + \tau_{xy}^2}$$

$$= \frac{7.464 + 3.464}{2} \pm \sqrt{\left(\frac{7.464 - 3.464}{2}\right)^2 + 2^2}$$

$$= 5.464 \pm 2.828$$

$$\sigma_1 = 8.29 \text{ ksi} \qquad \sigma_2 = 2.64 \text{ ksi} \qquad \textbf{Ans}$$

9–21. The wood beam is subjected to a distributed loading. If grains of wood in the beam at point A make an angle of 30° with the horizontal as shown, determine the normal and shear stresses that act perpendicular and parallel to the grains, respectively, due to the loading.

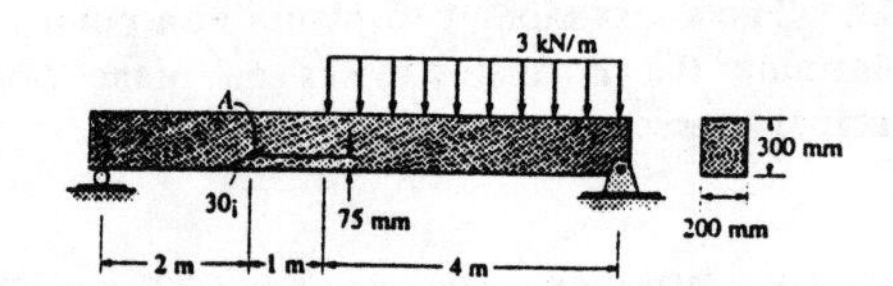

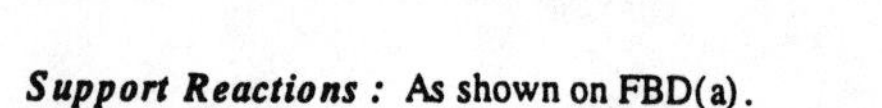

Support Reactions : As shown on FBD(a).

Internal Forces and Moment : As shown on FBD(b).

Section Properties :

$$I = \frac{1}{12}(0.2)\left(0.3^3\right) = 0.450\left(10^{-3}\right)\ \text{m}^4$$

$$Q_A = \bar{y}'A' = 0.1125(0.075)(0.2) = 1.6875\left(10^{-3}\right)\ \text{m}^3$$

Normal and Shear Stress : Applying the flexure and shear formulas.

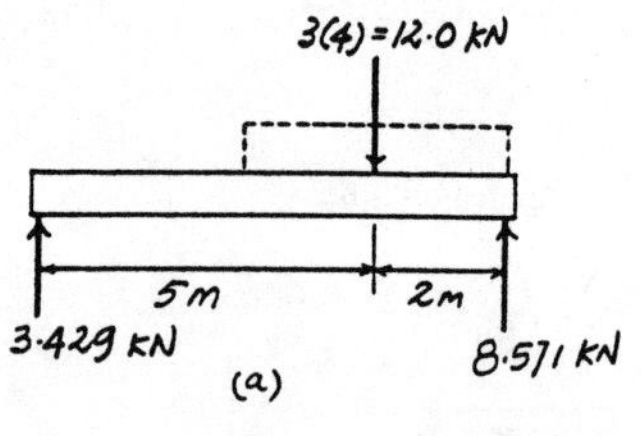

$$\sigma_A = -\frac{My}{I} = -\frac{6.857(10^3)(-0.075)}{0.450(10^{-3})} = 1.143\ \text{MPa}$$

$$\tau_A = \frac{VQ_A}{It} = \frac{3.429(10^3)\,1.6875(10^{-3})}{0.450(10^{-3})(0.2)} = 0.06429\ \text{MPa}$$

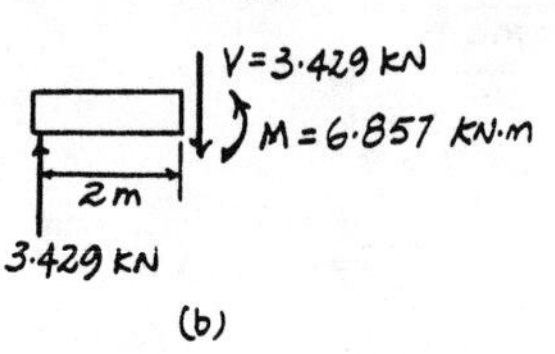

Stress Transformation Equations : Applying Eqs.9 – 1 and 9 – 2 with $\theta = 120°$, $\sigma_x = 1.143$ MPa, $\sigma_y = 0$, and $\tau_{xy} = -0.06429$ MPa,

$$\sigma_{x'} = \frac{\sigma_x + \sigma_y}{2} + \frac{\sigma_x - \sigma_y}{2}\cos 2\theta + \tau_{xy}\sin 2\theta$$

$$= \frac{1.143 + 0}{2} + \frac{1.143 - 0}{2}\cos 240° + (-0.06429\sin 240°)$$

$$= 0.341\ \text{MPa}$$ **Ans**

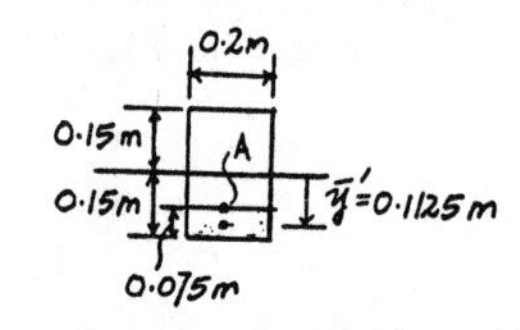

$$\tau_{x'y'} = -\frac{\sigma_x - \sigma_y}{2}\sin 2\theta + \tau_{xy}\cos 2\theta$$

$$= -\frac{1.143 - 0}{2}\sin 240° + (-0.06429\cos 240°)$$

$$= 0.527\ \text{MPa}$$ **Ans**

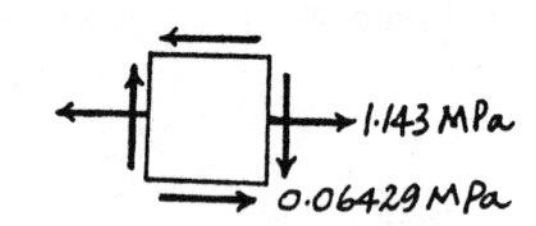

9–22. The wood beam is subjected to a distributed loading. Determine the principal stresses at point A and specify the orientation of the element.

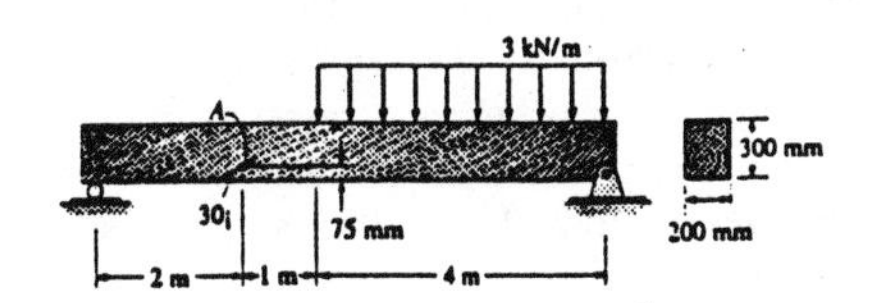

Support Reactions : As shown on FBD(a).
Internal Forces and Moment : As shown on FBD(b).

Section Properties :

$$I = \frac{1}{12}(0.2)\left(0.3^3\right) = 0.450\left(10^{-3}\right)\ \text{m}^4$$

$$Q_A = \bar{y}'A' = 0.1125(0.075)(0.2) = 1.6875\left(10^{-3}\right)\ \text{m}^3$$

Normal and Shear Stress : Applying the flexure and shear formul..

$$\sigma_A = -\frac{My}{I} = -\frac{6.857(10^3)(-0.075)}{0.450(10^{-3})} = 1.143\ \text{MPa}$$

$$\tau_A = \frac{VQ_A}{It} = \frac{3.429(10^3)\,1.6875(10^{-3})}{0.450(10^{-3})(0.2)} = 0.06429\ \text{MPa}$$

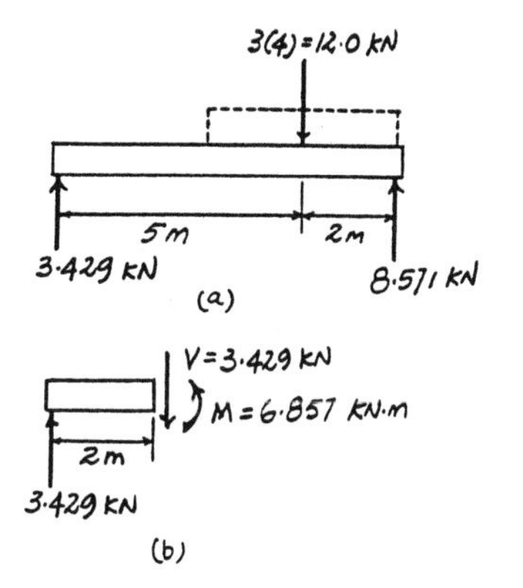

In-Plane Principal Stress : Applying Eq.9 – 5 with $\sigma_x = 1.143$ MPa, $\sigma_y = 0$ and $\tau_{xy} = -0.06429$ MPa.

$$\sigma_{1,2} = \frac{\sigma_x + \sigma_y}{2} \pm \sqrt{\left(\frac{\sigma_x - \sigma_y}{2}\right)^2 + \tau_{xy}^2}$$

$$= \frac{1.143 + 0}{2} \pm \sqrt{\left(\frac{1.143 - 0}{2}\right)^2 + (-0.06429)^2}$$

$$= 0.5714 \pm 0.057503$$

$\sigma_1 = 1.15$ MPa $\qquad \sigma_2 = -3.60$ kPa **Ans**

Orientation of Principal Plane : Applying Eq.9 – 4,

$$\tan 2\theta_p = \frac{\tau_{xy}}{\left(\sigma_x - \sigma_y\right)/2} = \frac{-0.06429}{(1.143 - 0)/2} = -0.1125$$

$$\theta_p = -3.209° \quad \text{and} \quad 86.79°$$

Substituting the results into Eq.9 - 1 with $\theta = -3.209°$ yields

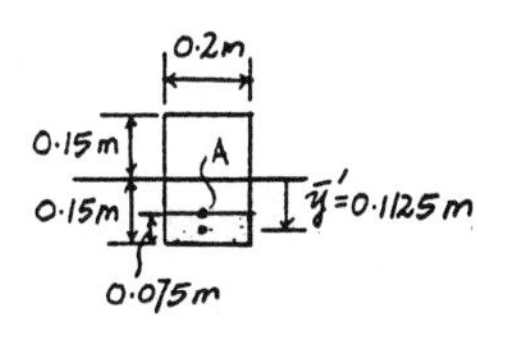

$$\sigma_{x'} = \frac{\sigma_x + \sigma_y}{2} + \frac{\sigma_x - \sigma_y}{2}\cos 2\theta + \tau_{xy}\sin 2\theta$$

$$= \frac{1.143 + 0}{2} + \frac{1.143 - 0}{2}\cos(-6.419°) + (-0.06429)\sin(-6.419°)$$

$$= 1.15\ \text{MPa} = \sigma_1$$

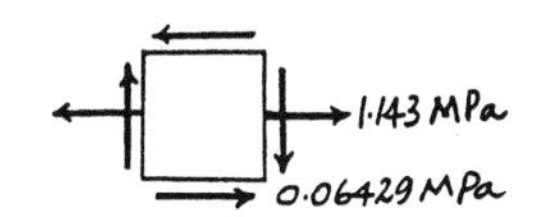

Hence,

$\theta_{p_1} = -3.21° \qquad \theta_{p_2} = 86.8°$ **Ans**

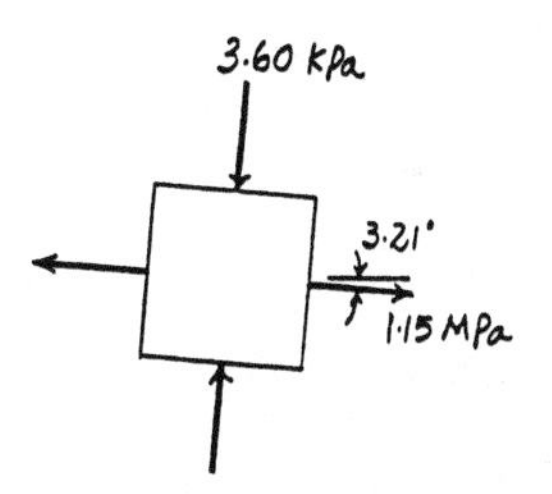

9–23. The clamp bears down on the smooth surface at E by tightening the bolt. If the tensile force in the bolt is 40 kN, determine the principal stresses at points A and B and show the results on elements located at each of these points. The cross-sectional area at A and B is shown in the adjacent figure.

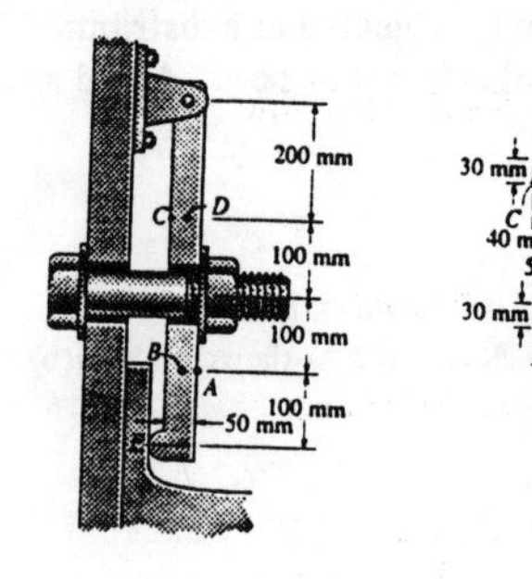

Support Reactions : As shown on FBD(a).

Internal Forces and Moment : As shown on FBD(b).

Section Properties :

$$I = \frac{1}{12}(0.03)(0.05^3) = 0.3125(10^{-6})\ \text{m}^4$$

$$Q_A = 0$$

$$Q_B = \bar{y}'A' = 0.0125(0.025)(0.03) = 9.375(10^{-6})\ \text{m}^3$$

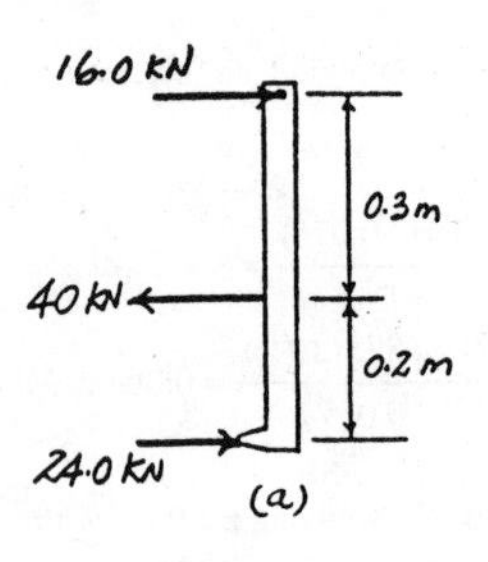

Normal Stress : Applying the flexure formula $\sigma = -\frac{My}{I}$,

$$\sigma_A = -\frac{2.40(10^3)(0.025)}{0.3125(10^{-6})} = -192\ \text{MPa}$$

$$\sigma_B = -\frac{2.40(10^3)(0)}{0.3125(10^{-6})} = 0$$

Shear Stress : Applying the shear formula $\tau = \frac{VQ}{It}$.

$$\tau_A = \frac{24.0(10^3)(0)}{0.3125(10^{-6})(0.03)} = 0$$

$$\tau_B = \frac{24.0(10^3)\left[9.375(10^{-6})\right]}{0.3125(10^{-6})(0.03)} = 24.0\ \text{MPa}$$

In - Plane Principal Stresses : $\sigma_x = 0$, $\sigma_y = -192$ MPa, and $\tau_{xy} = 0$ for point A. Since no shear stress acts on the element,

$$\sigma_1 = \sigma_x = 0 \qquad \textbf{Ans}$$

$$\sigma_2 = \sigma_y = -192\ \text{MPa} \qquad \textbf{Ans}$$

$\sigma_x = \sigma_y = 0$ and $\tau_{xy} = -24.0$ MPa for point B. Applying Eq. 9 – 5

$$\sigma_{1,2} = \frac{\sigma_x + \sigma_y}{2} \pm \sqrt{\left(\frac{\sigma_x - \sigma_y}{2}\right)^2 + \tau_{xy}^2}$$

$$= 0 \pm \sqrt{0 + (-24.0)^2}$$

$$= 0 \pm 24.0$$

$$\sigma_1 = 24.0 \qquad \sigma_2 = -24.0\ \text{MPa} \qquad \textbf{Ans}$$

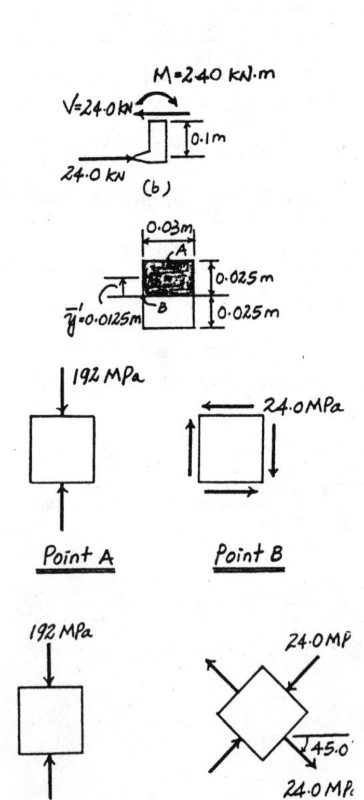

Orientation of Principal Plane : Applying Eq. 9 – 4 for point B,

$$\tan 2\theta_p = \frac{\tau_{xy}}{(\sigma_x - \sigma_y)/2} = \frac{-24.0}{0} = -\infty$$

$$\theta_p = -45.0^\circ \quad \text{and} \quad 45.0^\circ$$

Subsututing the results into Eq. 9 - 1 with $\theta = -45.0^\circ$ yields

$$\sigma_{x'} = \frac{\sigma_x + \sigma_y}{2} + \frac{\sigma_x - \sigma_y}{2}\cos 2\theta + \tau_{xy}\sin 2\theta$$

$$= 0 + 0 + [-24.0 \sin(-90.0^\circ)]$$

$$= 24.0\ \text{MPa} = \sigma_1$$

Hence,

$$\theta_{p_1} = -45.0^\circ \qquad \theta_{p_2} = 45.0^\circ \qquad \textbf{Ans}$$

***9–24.** Solve Prob. 9–23 for points C and D.

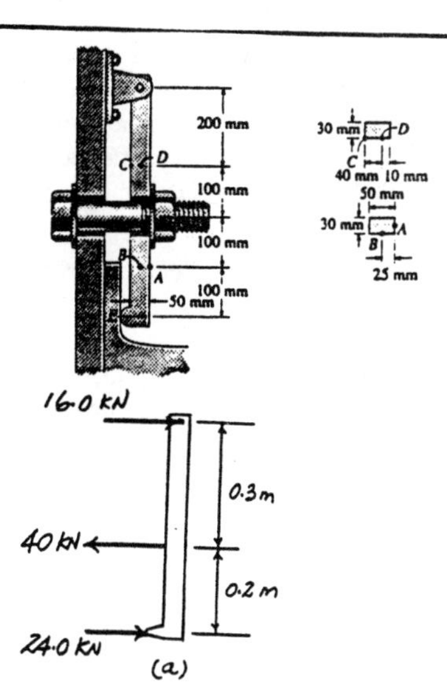

Support Reactions : As shown on FBD(a).

Internal Forces and Moment : As shown on FBD(b).

Section Properties :

$$I = \frac{1}{12}(0.03)(0.05^3) = 0.3125(10^{-6})\ \text{m}^4$$

$$Q_C = 0$$

$$Q_D = \bar{y}'A' = 0.02(0.01)(0.03) = 6.00(10^{-6})\ \text{m}^3$$

Normal Stresses : Applying the flexure formula $\sigma = -\frac{My}{I}$,

$$\sigma_C = -\frac{3.20(10^3)(-0.025)}{0.3125(10^{-6})} = 256\ \text{MPa}$$

$$\sigma_D = -\frac{3.20(10^3)(0.015)}{0.3125(10^{-6})} = -153.6\ \text{MPa}$$

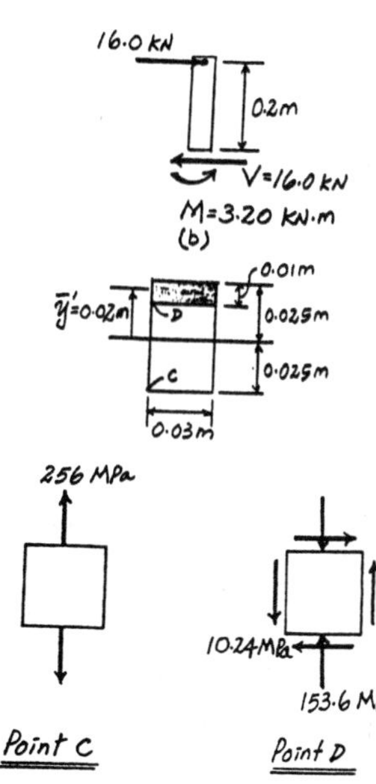

Shear Stress : Applying the shear formula $\tau = \frac{VQ}{It}$.

$$\tau_C = \frac{16.0(10^3)(0)}{0.3125(10^{-6})(0.03)} = 0$$

$$\tau_D = \frac{16.0(10^3)\left[6.00(10^{-6})\right]}{0.3125(10^{-6})(0.03)} = 10.24\ \text{MPa}$$

In-Plane Principal Stress : $\sigma_x = 0$, $\sigma_y = 256$ MPa, and $\tau_{xy} = 0$ for point C. Since no shear stress acts upon the element,

$$\sigma_1 = \sigma_y = 256\ \text{MPa} \qquad \textbf{Ans}$$

$$\sigma_2 = \sigma_x = 0 \qquad \textbf{Ans}$$

$\sigma_x = 0$, $\sigma_y = -153.6$ MPa, and $\tau_{xy} = 10.24$ MPa for point D. Applying Eq. 9 – 5,

$$\sigma_{1,2} = \frac{\sigma_x + \sigma_y}{2} \pm \sqrt{\left(\frac{\sigma_x - \sigma_y}{2}\right)^2 + \tau_{xy}^2}$$

$$= \frac{0 + (-153.6)}{2} \pm \sqrt{\left(\frac{0 - (-153.6)}{2}\right)^2 + 10.24^2}$$

$$= -76.8 \pm 77.48$$

$$\sigma_1 = 0.680\ \text{MPa} \qquad \sigma_2 = -154\ \text{MPa} \qquad \textbf{Ans}$$

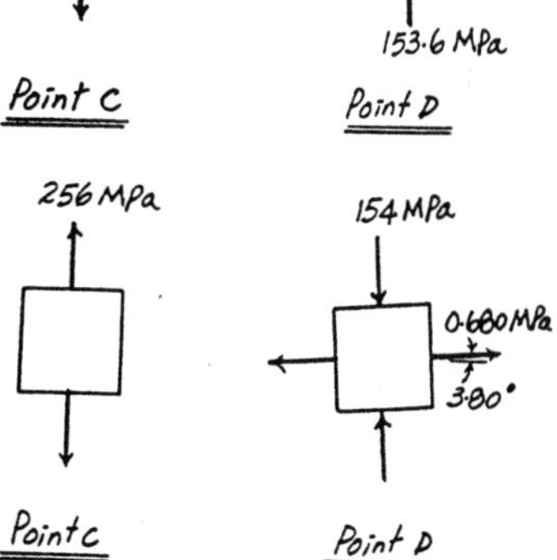

Orientation of Principal Plane : Applying Eq. 9 – 4 for point D,

$$\tan 2\theta_p = \frac{\tau_{xy}}{(\sigma_x - \sigma_y)/2} = \frac{10.24}{[0 - (-153.6)]/2} = 0.1333$$

$$\theta_p = 3.797° \quad \text{and} \quad -86.20°$$

Substituting the results into Eq. 9 – 1 with $\theta = 3.797°$ yields

$$\sigma_{x'} = \frac{\sigma_x + \sigma_y}{2} + \frac{\sigma_x - \sigma_y}{2}\cos 2\theta + \tau_{xy}\sin 2\theta$$

$$= \frac{0 + (-153.6)}{2} + \frac{0 - (-153.6)}{2}\cos 7.595° + 10.24 \sin 7.595°$$

$$= 0.678\ \text{MPa} = \sigma_1$$

Hence,

$$\theta_{p_1} = 3.80° \qquad \theta_{p_2} = -86.2° \qquad \textbf{Ans}$$

9–25. The clamp exerts a force of 150 lb on the boards at G. Determine the axial force in each screw, AB and CD, and then compute the principal stresses at points E and F. Show the results on properly oriented elements located at these points. The section through EF is rectangular and is 1 in. wide.

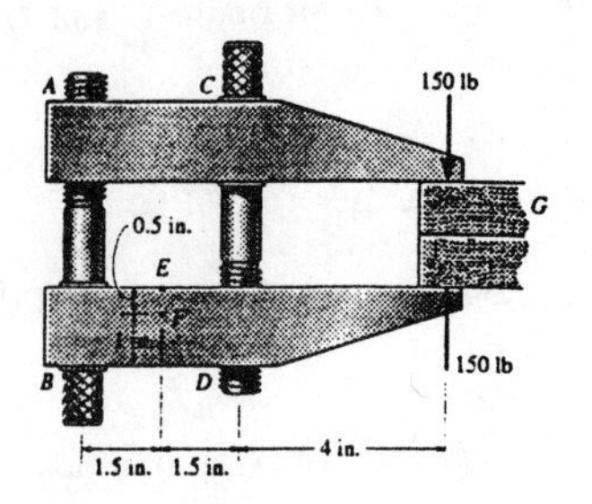

Support Reactions : FBD(a).

$$\curvearrowleft + \Sigma M_B = 0; \quad F_{CD}(3) - 150(7) = 0 \quad F_{CD} = 350 \text{ lb} \quad \textbf{Ans}$$

$$+\uparrow \Sigma F_y = 0; \quad 350 - 150 - F_{AB} = 0 \quad F_{AB} = 200 \text{ lb} \quad \textbf{Ans}$$

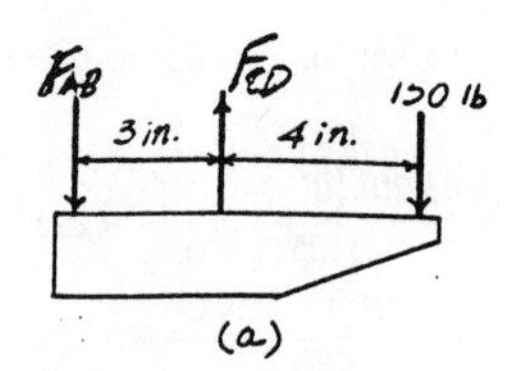

Internal Forces and Moment : As shown on FBD(b).

Section Properties :

$$I = \frac{1}{12}(1)(1.5^3) = 0.28125 \text{ in}^4$$

$$Q_E = 0$$

$$Q_F = \bar{y}'A' = 0.5(0.5)(1) = 0.250 \text{ in}^3$$

Normal Stresses : Applying the flexure formula $\sigma = -\frac{My}{I}$,

$$\sigma_E = -\frac{-300(0.75)}{0.28125} = 800 \text{ psi}$$

$$\sigma_F = -\frac{-300(0.25)}{0.28125} = 266.67 \text{ psi}$$

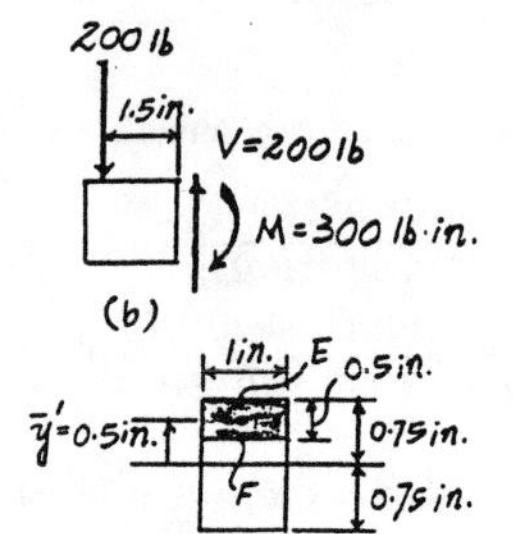

Shear Stress : Applying the shear formula $\tau = \frac{VQ}{It}$,

$$\tau_E = \frac{200(0)}{0.28125(1)} = 0$$

$$\tau_F = \frac{200(0.250)}{0.28125(1)} = 177.78 \text{ psi}$$

In-Plane Principal Stress : $\sigma_x = 800\text{psi}$, $\sigma_y = 0$ and $\tau_{xy} = 0$ for point E. Since no shear stress acts upon the element.

$$\sigma_1 = \sigma_x = 800 \text{ psi} \qquad \textbf{Ans}$$

$$\sigma_2 = \sigma_y = 0 \qquad \textbf{Ans}$$

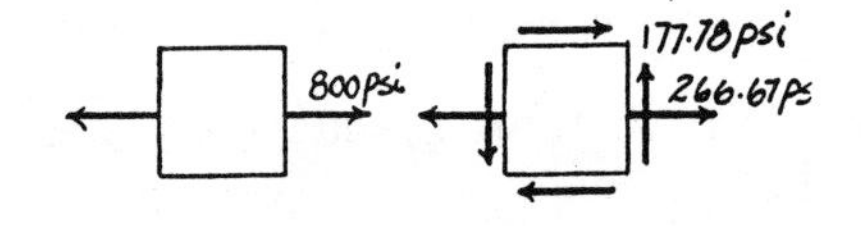

$\sigma_x = 266.67$ psi, $\sigma_y = 0$, and $\tau_{xy} = 177.78$ psi for point F. Applying Eq. 9–5

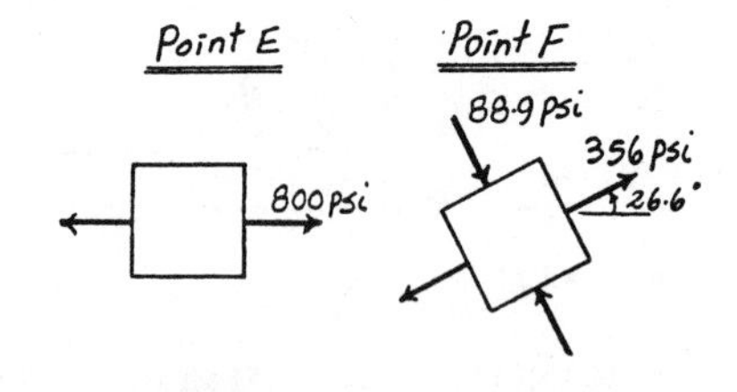

$$\sigma_{1,2} = \frac{\sigma_x + \sigma_y}{2} \pm \sqrt{\left(\frac{\sigma_x - \sigma_y}{2}\right)^2 + \tau_{xy}^2}$$

$$= \frac{266.67 + 0}{2} \pm \sqrt{\left(\frac{266.67 - 0}{2}\right)^2 + 177.78^2}$$

$$= 133.33 \pm 222.22$$

$$\sigma_1 = 356 \text{ psi} \qquad \sigma_2 = -88.9 \text{ psi} \qquad \textbf{Ans}$$

Orientation of Principal Plane : Applying Eq. 9–4 for point F,

$$\tan 2\theta_p = \frac{\tau_{xy}}{(\sigma_x - \sigma_y)/2} = \frac{177.78}{(266.67 - 0)/2} = 1.3333$$

$$\theta_p = 26.57° \quad \text{and} \quad -63.43°$$

Subsututing the results into Eq. 9 – 1 with $\theta = 26.57°$ yields

$$\sigma_{x'} = \frac{\sigma_x + \sigma_y}{2} + \frac{\sigma_x - \sigma_y}{2}\cos 2\theta + \tau_{xy}\sin 2\theta$$

$$= \frac{266.67 + 0}{2} + \frac{266.67 - 0}{2}\cos 53.13° + 177.78 \sin 53.13°$$

$$= 356 \text{ psi} = \sigma_1$$

Hence,

$$\theta_{p_1} = 26.6° \qquad \theta_{p_2} = -63.4° \qquad \textbf{Ans}$$

9–26. The T-beam is subjected to the distributed loading that is applied along its centerline. Determine the principal stresses at points A and B and show the results on elements located at each of these points.

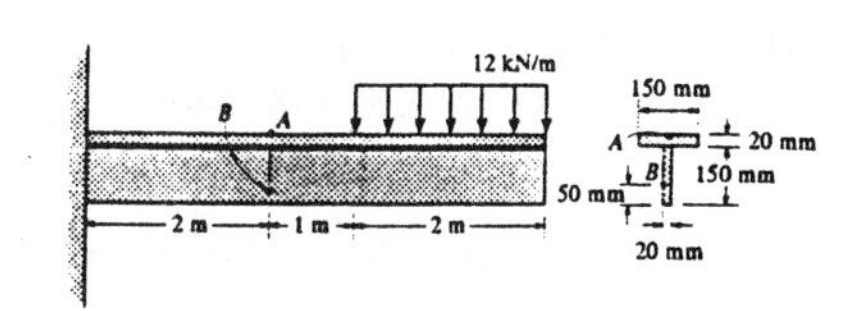

Internal Forces and Moment : As shown on FBD.

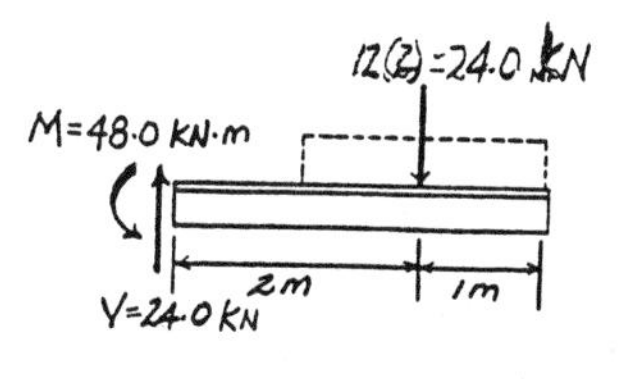

Section Properties :

$$\bar{y} = \frac{\Sigma \tilde{y}A}{\Sigma A} = \frac{0.01(0.02)(0.15)+0.095(0.15)(0.02)}{0.02(0.15)+0.15(0.02)} = 0.0525 \text{ m}$$

$$I = \frac{1}{12}(0.15)\left(0.02^3\right) + 0.15(0.02)(0.0525-0.01)^2 + \frac{1}{12}(0.02)\left(0.15^3\right) + 0.02(0.15)(0.095-0.0525)^2$$

$$= 16.5625\left(10^{-6}\right) \text{ m}^4$$

$$Q_A = 0$$

$$Q_B = \bar{y}'A' = 0.0925(0.05)(0.02) = 92.5\left(10^{-6}\right) \text{ m}^3$$

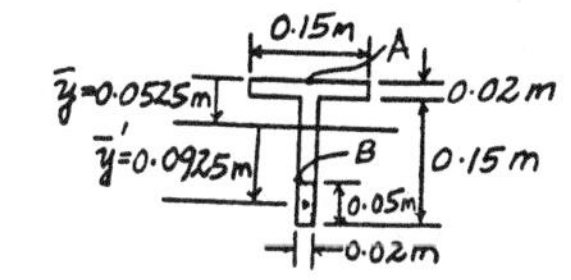

Normal Stress : Applying the flexure formula $\sigma = -\frac{My}{I}$.

$$\sigma_A = -\frac{-48.0(10^3)(0.0525)}{16.5625(10^{-6})} = 152.2 \text{ MPa}$$

$$\sigma_B = -\frac{-48.0(10^3)(-0.0675)}{16.5625(10^{-6})} = -195.6 \text{ MPa}$$

Shear Stress : Applying the shear formula $\tau = \frac{VQ}{It}$,

$$\tau_A = 0$$

$$\tau_B = \frac{24.0(10^3)\left[92.5(10^{-6})\right]}{16.5625(10^{-6})(0.02)} = 6.702 \text{ MPa}$$

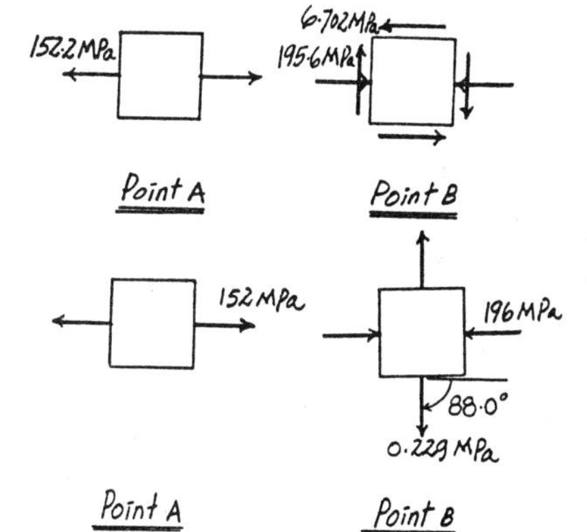

In-Plane Principal Stresses : $\sigma_x = 152.2$ MPa, $\sigma_y = 0$, and $\tau_{xy} = 0$ for point A. Since no shear stress acts on the element,

$$\sigma_1 = \sigma_x = 152 \text{ MPa} \qquad \textbf{Ans}$$

$$\sigma_2 = \sigma_y = 0 \qquad \textbf{Ans}$$

$\sigma_x = -195.6$ MPa, $\sigma_y = 0$ and $\tau_{xy} = -6.702$ MPa for point B. Applying Eq. 9 – 5,

$$\sigma_{1,2} = \frac{\sigma_x+\sigma_y}{2} \pm \sqrt{\left(\frac{\sigma_x-\sigma_y}{2}\right)^2 + \tau_{xy}^2}$$

$$= \frac{-195.6+0}{2} \pm \sqrt{\left(\frac{-195.6-0}{2}\right)^2 + (-6.702)^2}$$

$$= -97.811 \pm 98.041$$

$$\sigma_1 = 0.229 \text{ MPa} \qquad \sigma_2 = -196 \text{ MPa} \qquad \textbf{Ans}$$

Orientation of Principal Plane : Applying Eq. 9 – 4 for point B,

$$\tan 2\theta_p = \frac{\tau_{xy}}{(\sigma_x - \sigma_y)/2} = \frac{-6.702}{(-195.6-0)/2} = 0.06851$$

$$\theta_p = 1.960° \quad \text{and} \quad -88.04°$$

Substituting the results into Eq. 9 – 1 with $\theta = 1.960°$ yields

$$\sigma_{x'} = \frac{\sigma_x+\sigma_y}{2} + \frac{\sigma_x-\sigma_y}{2}\cos 2\theta + \tau_{xy}\sin 2\theta$$

$$= \frac{-195.6+0}{2} + \frac{-195.6-0}{2}\cos 3.920° + (-6.702\sin 3.920°)$$

$$= -196 \text{ MPa} = \sigma_2$$

Hence,

$$\theta_{p_1} = 88.0° \qquad \theta_{p_2} = 1.96° \qquad \textbf{Ans}$$

9–27. The bent rod has a diameter of 15 mm and is subjected to the force of 600 N. Determine the principal stresses and the maximum in-plane shear stress that are developed at point A and point B. Show the results on properly oriented elements located at these points.

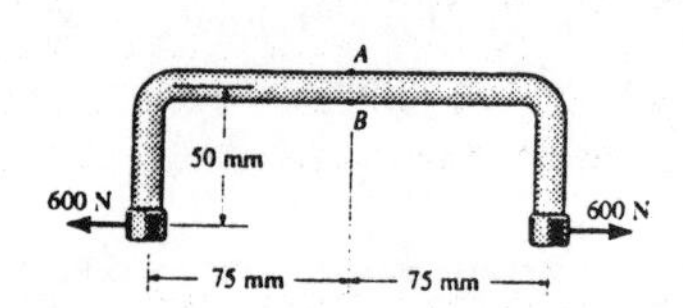

Internal Forces and Moment : As shown on FBD.

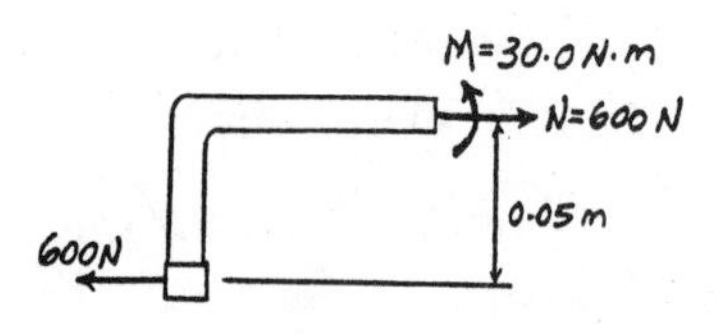

Section Properties :

$$A = \pi(0.0075^2) = 56.25\pi(10^{-6})\ \text{m}^2$$
$$I = \frac{\pi}{4}(0.0075^4) = 2.48505(10^{-9})\ \text{m}^4$$

Normal Stress :

$$\sigma = \frac{N}{A} \pm \frac{Mc}{I}$$
$$= \frac{600}{56.25\pi(10^{-6})} \pm \frac{30.0(0.0075)}{2.48505(10^{-9})}$$

$$\sigma_A = 3.395 - 90.541 = -87.146\ \text{MPa}$$
$$\sigma_B = 3.395 + 90.541 = 93.937\ \text{MPa}$$

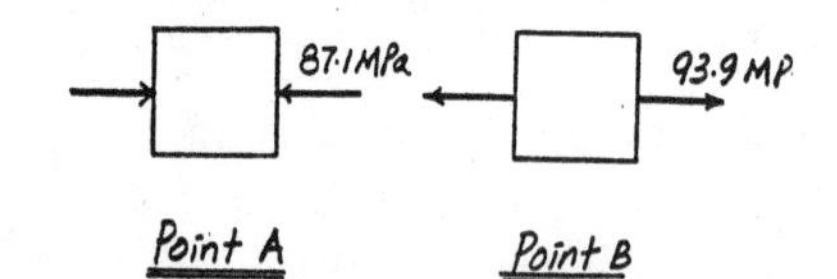

In-Plane Principal Stresses : $\sigma_x = -87.146$ MPa, $\sigma_y = 0$, and $\tau_{xy} = 0$ for point A. Since no shear stress acts on the element,

$$\sigma_1 = \sigma_y = 0 \qquad \textbf{Ans}$$
$$\sigma_2 = \sigma_x = -87.1\ \text{MPa} \qquad \textbf{Ans}$$

$\sigma_x = 93.937$ MPa, $\sigma_y = 0$ and $\tau_{xy} = 0$ for point B. Since no shear stress acts on the element,

$$\sigma_1 = \sigma_x = 93.9\ \text{MPa} \qquad \textbf{Ans}$$
$$\sigma_2 = \sigma_y = 0 \qquad \textbf{Ans}$$

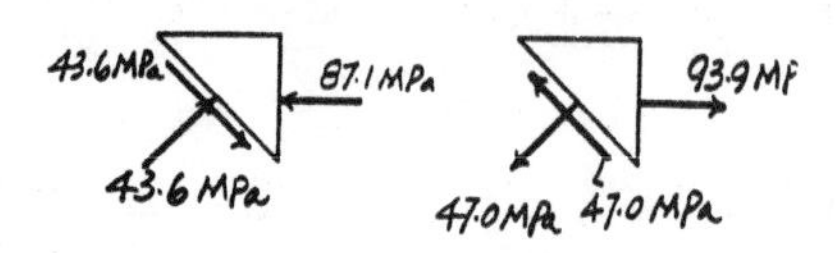

Maximum In-Plane Shear Stress : Applying Eq. 9–7 for point A,

$$\tau_{\substack{\max \\ \text{in-plane}}} = \sqrt{\left(\frac{\sigma_x - \sigma_y}{2}\right)^2 + \tau_{xy}^2}$$
$$= \sqrt{\left(\frac{-87.146 - 0}{2}\right)^2} + 0 = 43.6\ \text{MPa} \qquad \textbf{Ans}$$

Applying Eq. 9–7 for point B,

$$\tau_{\substack{\max \\ \text{in-plane}}} = \sqrt{\left(\frac{\sigma_x - \sigma_y}{2}\right)^2 + \tau_{xy}^2}$$
$$= \sqrt{\left(\frac{93.937 - 0}{2}\right)^2} + 0 = 47.0\ \text{MPa} \qquad \textbf{Ans}$$

Orientation of the Plane for Maximum In-Plane Shear Stress : Applying Eq. 9–6 for point A,

$$\tan 2\theta_s = \frac{-(\sigma_x - \sigma_y)/2}{\tau_{xy}} = \frac{-(-87.146 - 0)/2}{0} = \infty$$

$$\theta_s = 45.0^\circ \quad \text{and} \quad -45.0^\circ$$

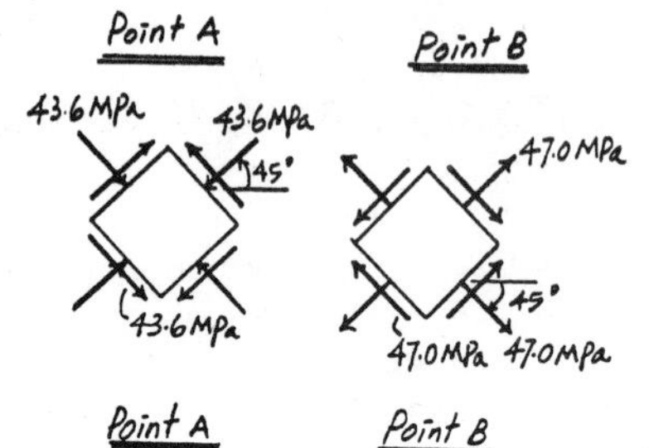

Applying Eq. 9–6 for point B,

$$\tan 2\theta_s = \frac{-(\sigma_x - \sigma_y)/2}{\tau_{xy}} = \frac{-(93.937 - 0)/2}{0} = -\infty$$

$$\theta_s = -45.0^\circ \quad \text{and} \quad 45.0^\circ$$

By observation, in order to preserve equilibrium, $\tau_{\substack{\max \\ \text{in-plane}}}$ has to act in the direction shown in the figure.

Average Normal Stress : Applying Eq. 9–8 for point A.

$$\sigma_{\text{avg}} = \frac{\sigma_x + \sigma_y}{2} = \frac{-87.146 + 0}{2} = -43.6\ \text{MPa}$$

Applying Eq. 9–8 for point B,

$$\sigma_{\text{avg}} = \frac{\sigma_x + \sigma_y}{2} = \frac{93.937 + 0}{2} = 47.0\ \text{MPa}$$

***9–28.** The beam has a rectangular cross section and is subjected to the loading shown. Determine the principal stresses and the maximum in-plane shear stress that are developed at point A and point B. Show the results on properly oriented elements located at these points.

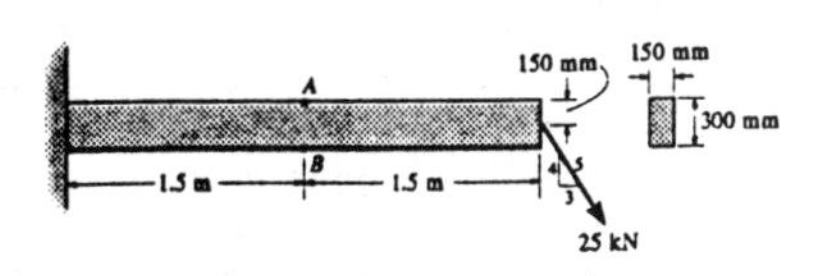

Internal Forces and Moment : As shown on FBD(a).

Section Properties :

$$A = 0.15(0.3) = 0.0450 \text{ m}^2$$

$$I = \frac{1}{12}(0.15)\left(0.3^3\right) = 0.3375\left(10^{-3}\right) \text{ m}^4$$

$$Q_A = Q_B = 0$$

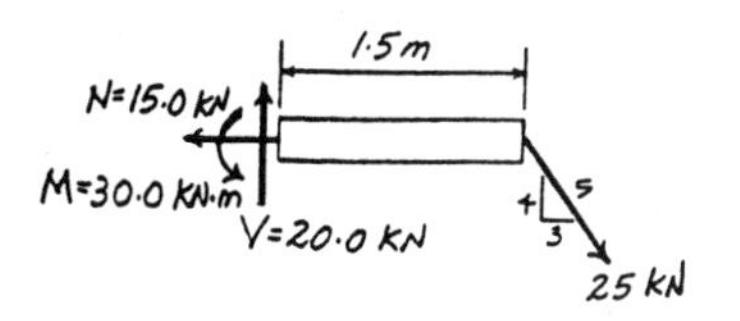

Normal Stresses :

$$\sigma = \frac{N}{A} \pm \frac{Mc}{I}$$

$$= \frac{15.0(10^3)}{0.0450} \pm \frac{30.0(10^3)(0.15)}{0.3375(10^{-3})}$$

$$\sigma_A = 0.3333 + 13.3333 = 13.67 \text{ MPa}$$

$$\sigma_B = 0.3333 - 13.3333 = -13.0 \text{ MPa}$$

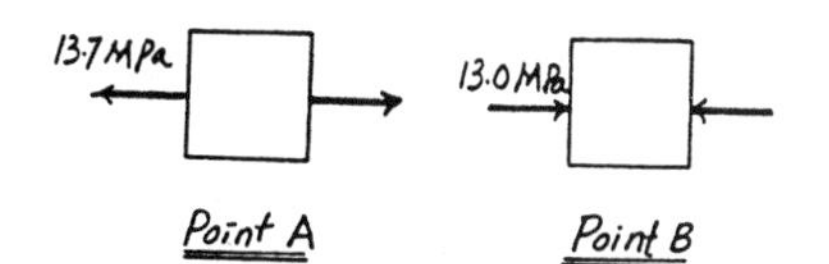

Shear Stresses : Since $Q_A = Q_B = 0$, hence, $\tau_A = \tau_B = 0$

In - Plane Principal Stress : $\sigma_x = 13.67$ MPa, $\sigma_y = 0$ and $\tau_{xy} = 0$ for point A. Since no shear stress acts on the element,

$$\sigma_1 = \sigma_x = 13.7 \text{ MPa} \qquad \textbf{Ans}$$

$$\sigma_2 = \sigma_y = 0 \qquad \textbf{Ans}$$

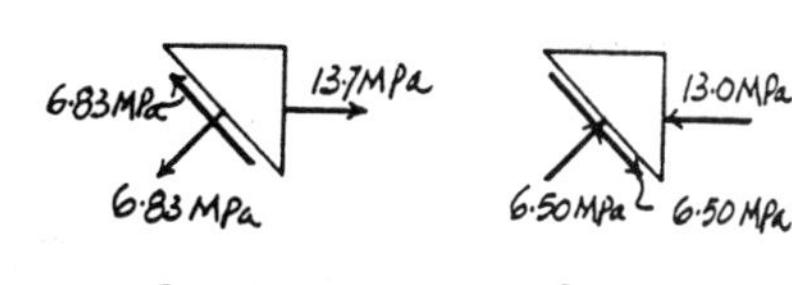

$\sigma_x = -13.0$ MPa, $\sigma_y = 0$ and $\tau_{xy} = 0$ for point B. Since no shear stress acts on the element,

$$\sigma_1 = \sigma_y = 0 \qquad \textbf{Ans}$$

$$\sigma_2 = \sigma_x = -13.0 \text{ MPa} \qquad \textbf{Ans}$$

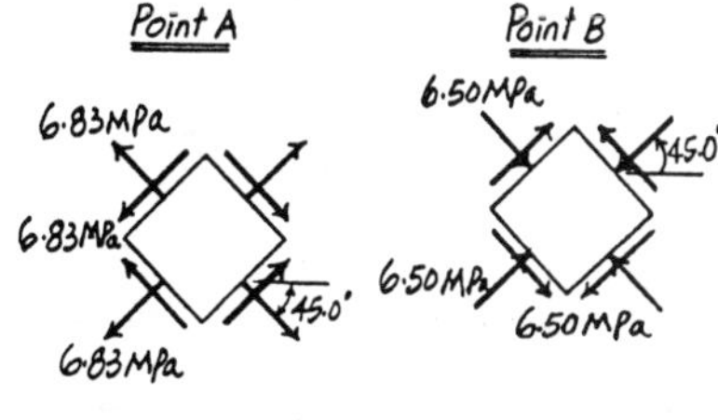

Maximum In - Plane Shear Stress : Applying Eq. 9 – 7 for point A,

$$\tau_{\substack{\max \\ \text{in-plane}}} = \sqrt{\left(\frac{\sigma_x - \sigma_y}{2}\right)^2 + \tau_{xy}^2}$$

$$= \sqrt{\left(\frac{13.67 - 0}{2}\right)^2 + 0} = 6.83 \text{ MPa} \qquad \textbf{Ans}$$

Applying Eq. 9 – 7 for point B.

$$\tau_{\substack{\max \\ \text{in-plane}}} = \sqrt{\left(\frac{\sigma_x - \sigma_y}{2}\right)^2 + \tau_{xy}^2}$$

$$= \sqrt{\left(\frac{-13.0 - 0}{2}\right)^2 + 0} = 6.50 \text{ MPa} \qquad \textbf{Ans}$$

Orientation of the Plane for Maximum In - Plane Shear Stress : Applying Eq. 9 – 6 for point A,

$$\tan 2\theta_s = \frac{-(\sigma_x - \sigma_y)/2}{\tau_{xy}} = \frac{-(13.67 - 0)/2}{0} = -\infty$$

$$\theta_s = -45.0° \quad \text{and} \quad 45.0°$$

Applying Eq. 9 – 6 for point B,

$$\tan 2\theta_s = \frac{-(\sigma_x - \sigma_y)/2}{\tau_{xy}} = \frac{-(-13.0 - 0)/2}{0} = \infty$$

$$\theta_s = 45.0° \quad \text{and} \quad -45.0°$$

By observation, in order to preserve equilibrium, $\tau_{\substack{\max \\ \text{in-plane}}}$ has to act in the direction shown in the figure.

Average Normal Stress : Applying Eq. 9 – 8 for point A.

$$\sigma_{\text{avg}} = \frac{\sigma_x + \sigma_y}{2} = \frac{13.67 + 0}{2} = 6.83 \text{ MPa}$$

Applying Eq. 9 – 8 for point B,

$$\sigma_{\text{avg}} = \frac{\sigma_x + \sigma_y}{2} = \frac{-13.0 + 0}{2} = -6.50 \text{ MPa}$$

9–29. The beam has a rectangular cross section and is subjected to the loadings shown. Determine the principal stresses and the maximum in-plane shear stress that are developed at point A and point B. These points are just to the left of the 2000-lb load. Show the results on properly oriented elements located at these points.

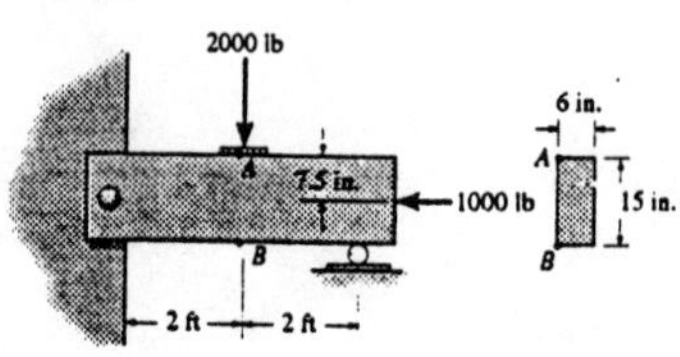

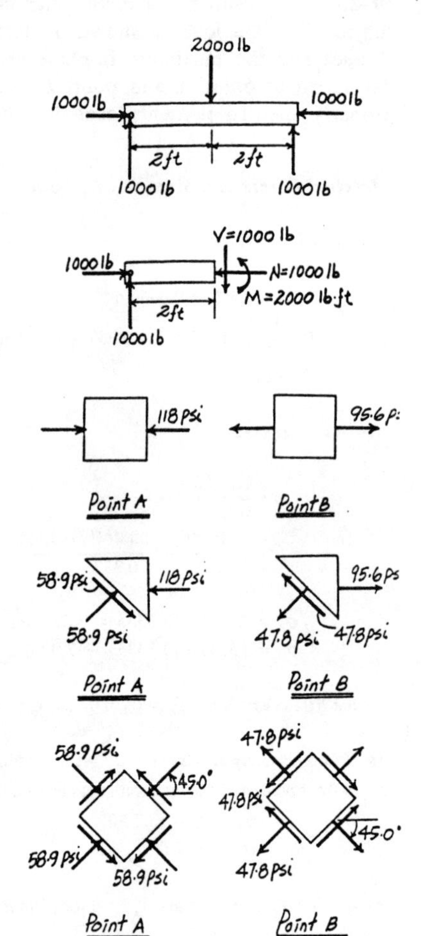

Support Reactions : As shown on FBD(a).

Internal Forces and Moment : As shown on FBD(b).

Section Properties :

$$A = 6(15) = 90.0 \text{ in}^2$$

$$I = \frac{1}{12}(6)(15^3) = 1687.5 \text{ in}^4$$

$$Q_A = Q_B = 0$$

Normal Stress :

$$\sigma = \frac{N}{A} \pm \frac{Mc}{I}$$

$$= \frac{-1000}{90.0} \pm \frac{2000(12)(7.5)}{1687.5}$$

$$\sigma_A = -11.11 - 106.67 = -117.78 \text{ psi}$$

$$\sigma_B = -11.11 + 106.67 = 95.56 \text{ psi}$$

Shear Stresses : Since $Q_A = Q_B = 0$, hence, $\tau_A = \tau_B = 0$

In-Plane Principal Stress : $\sigma_x = -117.78$ psi, $\sigma_y = 0$, and $\tau_{xy} = 0$ for point A. Since no shear stress acts upon the element,

$$\sigma_1 = \sigma_y = 0 \qquad \textbf{Ans}$$

$$\sigma_2 = \sigma_x = -118 \text{ psi} \qquad \textbf{Ans}$$

$\sigma_x = 95.56$ psi, $\sigma_y = 0$ and $\tau_{xy} = 0$ for point B. Since no shear stress acts upon the element,

$$\sigma_1 = \sigma_x = 95.6 \text{ psi} \qquad \textbf{Ans}$$

$$\sigma_2 = \sigma_y = 0 \qquad \textbf{Ans}$$

Maximum In-Plane Shear Stress : Applying Eq. 9–7 for point A,

$$\tau_{\substack{\max \\ \text{in-plane}}} = \sqrt{\left(\frac{\sigma_x - \sigma_y}{2}\right)^2 + \tau_{xy}^2}$$

$$= \sqrt{\left(\frac{-117.78 - 0}{2}\right)^2 + 0} = 58.9 \text{ psi} \qquad \textbf{Ans}$$

Applying Eq. 9–7 for Point B.

$$\tau_{\substack{\max \\ \text{in-plane}}} = \sqrt{\left(\frac{\sigma_x - \sigma_y}{2}\right)^2 + \tau_{xy}^2}$$

$$= \sqrt{\left(\frac{95.56 - 0}{2}\right)^2 + 0} = 47.8 \text{ psi} \qquad \textbf{Ans}$$

Orientation of the plane for Maximum In-Plane Shear Stress : Applying Eq. 9–6 for point A.

$$\tan 2\theta_s = \frac{-(\sigma_x - \sigma_y)/2}{\tau_{xy}} = \frac{-(-117.78 - 0)/2}{0} = \infty$$

$$\theta_s = 45.0° \quad \text{and} \quad -45.0°$$

Applying Eq. 9–6 for point B.

$$\tan 2\theta_s = \frac{-(\sigma_x - \sigma_y)/2}{\tau_{xy}} = \frac{-(95.56 - 0)/2}{0} = -\infty$$

$$\theta_s = -45.0° \quad \text{and} \quad 45.0°$$

By observation, in order to preserve equilibrium, $\tau_{\substack{\max \\ \text{in-plane}}}$ has to act in the direction shown in the figure.

Average Normal Stress : Applying Eq. 9–8 for point A.

$$\sigma_{\text{avg}} = \frac{\sigma_x + \sigma_y}{2} = \frac{-117.78 + 0}{2} = -58.9 \text{ psi}$$

Applying Eq. 9–8 for point B.

$$\sigma_{\text{avg}} = \frac{\sigma_x + \sigma_y}{2} = \frac{95.56 + 0}{2} = 47.8 \text{ psi}$$

9–30. The wide-flange beam is subjected to the loading shown. Determine the principal stress in the beam at point A and at point B. These points are located at the top and bottom of the web, respectively. Although it is not very accurate, use the shear formula to compute the shear stress.

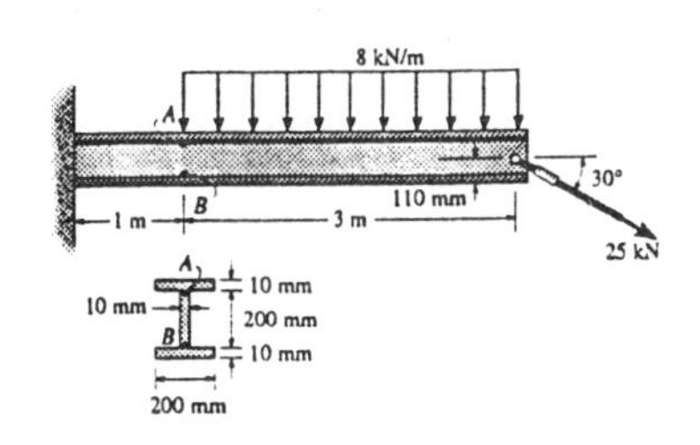

Internal Forces and Moment : As shown on FBD(a).

Section Properties :

$$A = 0.2(0.22) - 0.19(0.2) = 6.00\left(10^{-3}\right)\ \text{m}^2$$

$$I = \frac{1}{12}(0.2)\left(0.22^3\right) - \frac{1}{12}(0.19)\left(0.2^3\right) = 50.8\left(10^{-6}\right)\ \text{m}^4$$

$$Q_A = Q_B = \bar{y}'A' = 0.105(0.01)(0.2) = 0.210\left(10^{-3}\right)\ \text{m}^3$$

Normal Stress :

$$\sigma = \frac{N}{A} \pm \frac{My}{I}$$

$$= \frac{21.65(10^3)}{6.00(10^{-3})} \pm \frac{73.5(10^3)(0.1)}{50.8(10^{-6})}$$

$$\sigma_A = 3.608 + 144.685 = 148.3\ \text{MPa}$$

$$\sigma_B = 3.608 - 144.685 = -141.1\ \text{MPa}$$

Shear Stress : Applying the shear formula $\tau = \frac{VQ}{It}$,

$$\tau_A = \tau_B = \frac{36.5(10^3)\left[0.210(10^{-3})\right]}{50.8(10^{-6})(0.01)} = 15.09\ \text{MPa}$$

In - Plane Principal Stress : $\sigma_x = 148.3$ MPa, $\sigma_y = 0$, and $\tau_{xy} = -15.09$ MPa for point A. Applying Eq. 9 – 5,

$$\sigma_{1,2} = \frac{\sigma_x + \sigma_y}{2} \pm \sqrt{\left(\frac{\sigma_x - \sigma_y}{2}\right)^2 + \tau_{xy}^2}$$

$$= \frac{148.3 + 0}{2} \pm \sqrt{\left(\frac{148.3 - 0}{2}\right)^2 + (-15.09)^2}$$

$$= 81.381 \pm 82.768$$

$\sigma_1 = 150$ MPa $\qquad \sigma_2 = -1.52$ MPa $\qquad$ **Ans**

$\sigma_x = -141.1$ MPa, $\sigma_y = 0$, and $\tau_{xy} = -15.09$ MPa for point B. Applying Eq. 9 – 5,

$$\sigma_{1,2} = \frac{\sigma_x + \sigma_y}{2} \pm \sqrt{\left(\frac{\sigma_x - \sigma_y}{2}\right)^2 + \tau_{xy}^2}$$

$$= \frac{-141.1 + 0}{2} \pm \sqrt{\left(\frac{(-141.1) - 0}{2}\right)^2 + (-15.09)^2}$$

$$= -77.773 \pm 79.223$$

$\sigma_1 = 1.60$ MPa $\qquad \sigma_2 = -143$ MPa $\qquad$ **Ans**

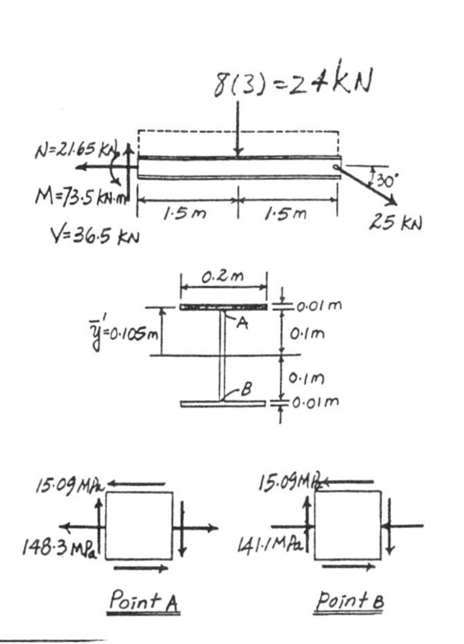

9–31. The shaft has a diameter d and is subjected to the loadings shown. Determine the principal stresses and the maximum in-plane shear stress that is developed at point A. The bearings only support vertical reactions.

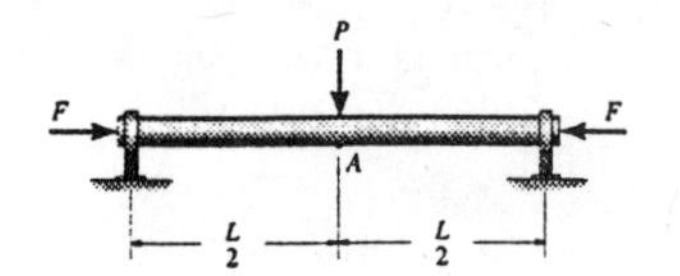

Support Reactions : As shown on FBD(a).

Internal Forces and Moment : As shown on FBD(b).

Section Properties :

$$A = \frac{\pi}{4}d^2 \qquad I = \frac{\pi}{4}\left(\frac{d}{2}\right)^4 = \frac{\pi}{64}d^4 \qquad Q_A = 0$$

Normal Stress :

$$\sigma = \frac{N}{A} \pm \frac{Mc}{I}$$

$$= \frac{-F}{\frac{\pi}{4}d^2} \pm \frac{\frac{PL}{4}\left(\frac{d}{2}\right)}{\frac{\pi}{64}d^4}$$

$$\sigma_A = \frac{4}{\pi d^2}\left(\frac{2PL}{d} - F\right)$$

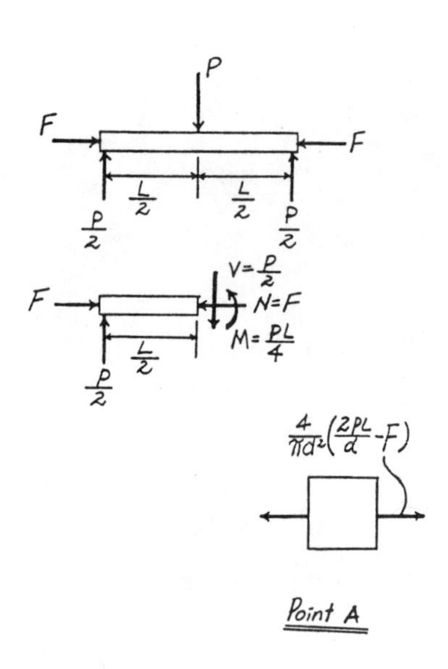

Shear Stress : Since $Q_A = 0$, $\tau_A = 0$

In - Plane Principal Stress : $\sigma_x = \frac{4}{\pi d^2}\left(\frac{2PL}{d} - F\right)$, $\sigma_y = 0$ and $\tau_{xy} = 0$ for point A. Since no shear stress acts on the element,

$\frac{4}{\pi d^2}\left(\frac{2PL}{d} - F\right)$

Point A

$$\sigma_1 = \sigma_x = \frac{4}{\pi d^2}\left(\frac{2PL}{d} - F\right) \qquad \textbf{Ans}$$

$$\sigma_2 = \sigma_y = 0 \qquad \textbf{Ans}$$

Maximum In - Plane Shear Stress : Applying Eq.9 – 7 for point A,

$$\tau_{\substack{\max \\ \text{in-plane}}} = \sqrt{\left(\frac{\sigma_x - \sigma_y}{2}\right)^2 + \tau_{xy}^2}$$

$$= \sqrt{\left(\frac{\frac{4}{\pi d^2}\left(\frac{2PL}{d} - F\right) - 0}{2}\right)^2 + 0}$$

$$= \frac{2}{\pi d^2}\left(\frac{2PL}{d} - F\right) \qquad \textbf{Ans}$$

***9–32.** The shaft has a diameter d and is subjected to the loadings shown. Determine the principal stresses and the maximum in-plane shear stress that is developed anywhere on the surface of the shaft.

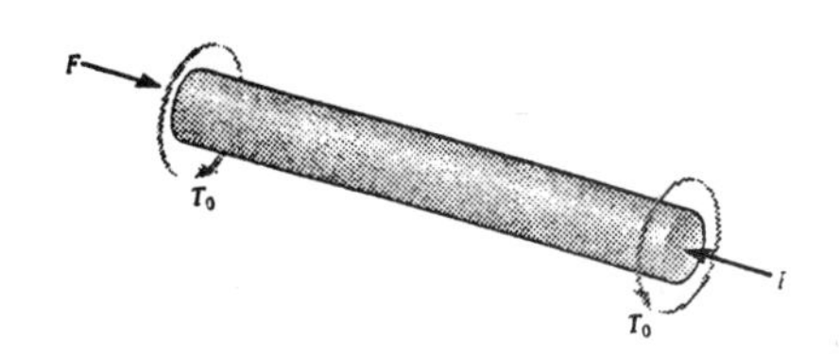

Internal Forces and Torque : As shown on FBD (a).

Section Properties :

$$A = \frac{\pi}{4}d^2 \qquad J = \frac{\pi}{2}\left(\frac{d}{2}\right)^4 = \frac{\pi}{32}d^4$$

Normal Stress :

$$\sigma = \frac{N}{A} = \frac{-F}{\frac{\pi}{4}d^2} = -\frac{4F}{\pi d^2}$$

Shear Stress : Applying the torsion formula,

$$\tau = \frac{Tc}{J} = \frac{T_0\left(\frac{d}{2}\right)}{\frac{\pi}{32}d^4} = \frac{16T_0}{\pi d^3}$$

In - Plane Principal Stresses : $\sigma_x = -\frac{4F}{\pi d^2}$, $\sigma_y = 0$, and

$\tau_{xy} = -\frac{16T_0}{\pi d^3}$ for any point on the shaft's surface. Applying Eq. 9 – 5,

$$\sigma_{1,2} = \frac{\sigma_x + \sigma_y}{2} \pm \sqrt{\left(\frac{\sigma_x - \sigma_y}{2}\right)^2 + \tau_{xy}^2}$$

$$= \frac{-\frac{4F}{\pi d^2} + 0}{2} \pm \sqrt{\left(\frac{-\frac{4F}{\pi d^2} - 0}{2}\right)^2 + \left(-\frac{16T_0}{\pi d^3}\right)^2}$$

$$= \frac{2}{\pi d^2}\left(-F \pm \sqrt{F^2 + \frac{64T_0^2}{d^2}}\right)$$

$$\sigma_1 = \frac{2}{\pi d^2}\left(-F + \sqrt{F^2 + \frac{64T_0^2}{d^2}}\right) \qquad \textbf{Ans}$$

$$\sigma_2 = -\frac{2}{\pi d^2}\left(F + \sqrt{F^2 + \frac{64T_0^2}{d^2}}\right) \qquad \textbf{Ans}$$

Maximum In - Plane Shear Stress : Applying Eq. 9 – 7,

$$\tau_{\substack{\max \\ \text{in-plane}}} = \sqrt{\left(\frac{\sigma_x - \sigma_y}{2}\right)^2 + \tau_{xy}^2}$$

$$= \sqrt{\left(\frac{-\frac{4F}{\pi d^2} - 0}{2}\right)^2 + \left(-\frac{16T_0}{\pi d^3}\right)^2}$$

$$= \frac{2}{\pi d^2}\sqrt{F^2 + \frac{64T_0^2}{d^2}} \qquad \textbf{Ans}$$

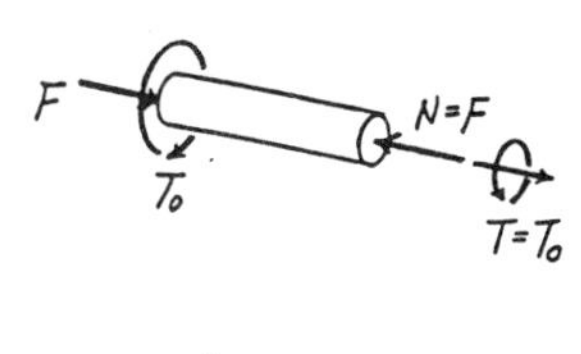

$\frac{4F}{\pi d^2}$

$\frac{16T_0}{\pi d^3}$

9–33. The rod has a diameter d and is subjected to the loadings shown. Determine the maximum normal stress and the maximum in-plane shear stress that is developed at point A and point B.

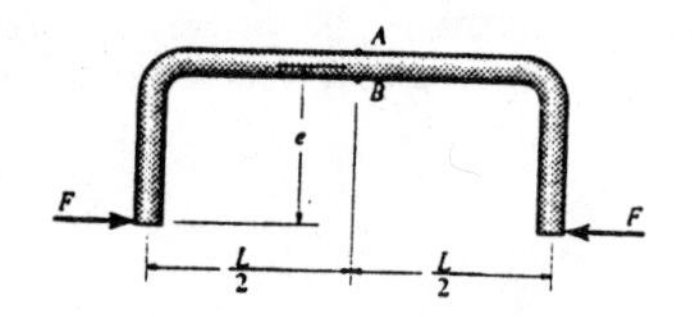

Internal Forces and Moment : As shown on FBD.
Section Properties :

$$A = \frac{\pi}{4}d^2 \qquad I = \frac{\pi}{4}\left(\frac{d}{2}\right)^4 = \frac{\pi}{64}d^4 \qquad Q_A = Q_B = 0$$

Normal Stress :

$$\sigma = \frac{N}{A} \pm \frac{Mc}{I}$$

$$= \frac{-F}{\frac{\pi}{4}d^2} \pm \frac{Fe\left(\frac{d}{2}\right)}{\frac{\pi}{64}d^4}$$

$$\sigma_A = \frac{4F}{\pi d^2}\left(\frac{8e}{d} - 1\right)$$

$$\sigma_B = -\frac{4F}{\pi d^2}\left(1 + \frac{8e}{d}\right)$$

Shear Stress : Since $Q_A = Q_B = 0$, $\quad \tau_A = \tau_B = 0$

In - Plane Principal Stress : $\sigma_x = \frac{4F}{\pi d^2}\left(\frac{8e}{d} - 1\right)$, $\sigma_y = 0$, and $\tau_{xy} = 0$ for point A. Since no shear stress acts on the element.

$$\sigma_1 = \sigma_x = \frac{4F}{\pi d^2}\left(\frac{8e}{d} - 1\right) \qquad \textbf{Ans}$$

$$\sigma_2 = \sigma_y = 0 \qquad \textbf{Ans}$$

$\sigma_x = -\frac{4F}{\pi d^2}\left(1 + \frac{8e}{d}\right)$, $\sigma_y = 0$, and $\tau_{xy} = 0$ for point B. Since no shear stress acts on the element,

$$\sigma_1 = \sigma_y = 0 \qquad \textbf{Ans}$$

$$\sigma_2 = \sigma_x = -\frac{4F}{\pi d^2}\left(1 + \frac{8e}{d}\right) \qquad \textbf{Ans}$$

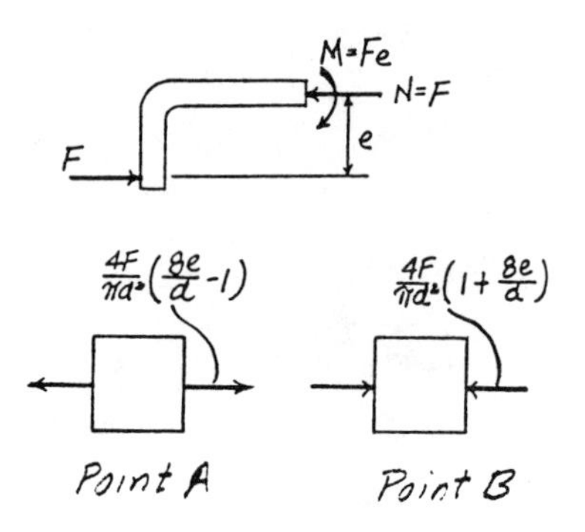

Maximum In - Plane Shear Stress : Applying Eq. 9 – 7 for point A.

$$\tau_{\substack{\max \\ \text{in-plane}}} = \sqrt{\left(\frac{\sigma_x - \sigma_y}{2}\right)^2 + \tau_{xy}^2}$$

$$= \sqrt{\left(\frac{\frac{4F}{\pi d^2}\left(\frac{8e}{d} - 1\right) - 0}{2}\right)^2 + 0}$$

$$= \frac{2F}{\pi d^2}\left(\frac{8e}{d} - 1\right) \qquad \textbf{Ans}$$

Applying Eq. 9 – 7 for point B,

$$\tau_{\substack{\max \\ \text{in-plane}}} = \sqrt{\left(\frac{\sigma_x - \sigma_y}{2}\right)^2 + \tau_{xy}^2}$$

$$= \sqrt{\left(\frac{-\frac{4F}{\pi d^2}\left(1 + \frac{8e}{d}\right) - 0}{2}\right)^2 + 0}$$

$$= \frac{2F}{\pi d^2}\left(1 + \frac{8e}{d}\right) \qquad \textbf{Ans}$$

9–34. The drill pipe has an outer diameter of 3 in., a wall thickness of 0.25 in. and a weight of 50 lb/ft. If it is subjected to a torque and axial load as shown, determine (a) the principal stresses and (b) the maximum in-plane shear stress at a point on its surface at section *a*.

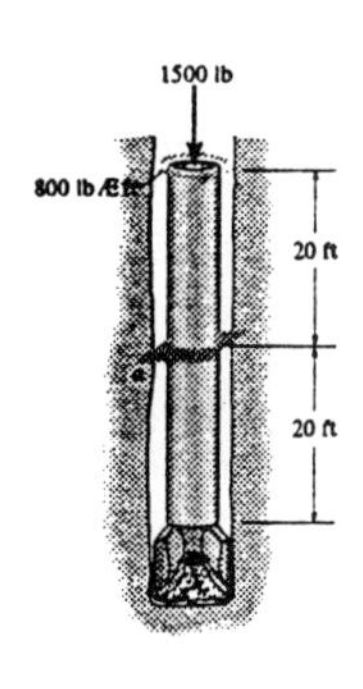

Internal Forces and Torque : As shown on FBD (a).

Section Properties :

$$A = \frac{\pi}{4}\left(3^2 - 2.5^2\right) = 0.6875\pi \text{ in}^2$$

$$J = \frac{\pi}{2}\left(1.5^4 - 1.25^4\right) = 4.1172 \text{ in}^4$$

Normal Stress :

$$\sigma = \frac{N}{A} = \frac{-2500}{0.6875\pi} = -1157.5 \text{ psi}$$

Shear Stress : Applying the torsion formula,

$$\tau = \frac{Tc}{J} = \frac{800(12)(1.5)}{4.1172} = 3497.5 \text{ psi}$$

a) ***In - Plane Principal Stresses :*** $\sigma_x = 0$, $\sigma_y = -1157.5$ psi and $\tau_{xy} = 3497.5$ psi for any point on the shaft's surface. Applying Eq.9 – 5,

$$\sigma_{1,2} = \frac{\sigma_x + \sigma_y}{2} \pm \sqrt{\left(\frac{\sigma_x - \sigma_y}{2}\right)^2 + \tau_{xy}^2}$$

$$= \frac{0 + (-1157.5)}{2} \pm \sqrt{\left(\frac{0 - (-1157.5)}{2}\right)^2 + (3497.5)^2}$$

$$= -578.75 \pm 3545.08$$

$\sigma_1 = 2966 \text{ psi} = 2.97 \text{ ksi}$ **Ans**

$\sigma_2 = -4124 \text{ psi} = -4.12 \text{ ksi}$ **Ans**

b) ***Maximum In - Plane Shear Stress :*** Applying Eq.9 –,7

$$\tau_{\substack{\max \\ \text{in-plane}}} = \sqrt{\left(\frac{\sigma_x - \sigma_y}{2}\right)^2 + \tau_{xy}^2}$$

$$= \sqrt{\left(\frac{0 - (-1157.5)}{2}\right)^2 + (3497.5)^2}$$

$$= 3545 \text{ psi} = 3.55 \text{ ksi}$$ **Ans**

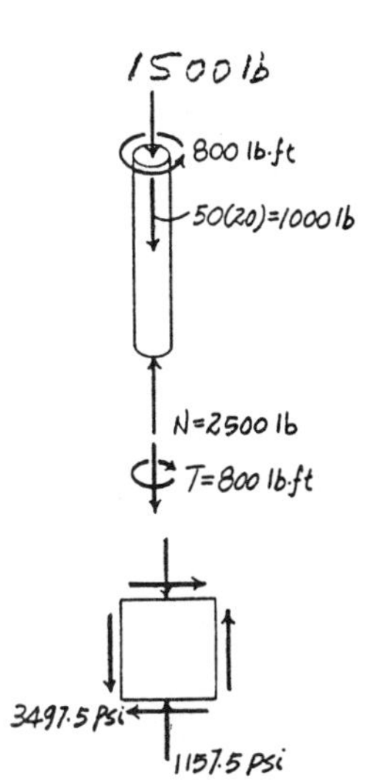

9–35. The internal loadings at a section of the beam are shown. Determine the principal stresses at point A. Also compute the maximum in-plane shear stress at this point.

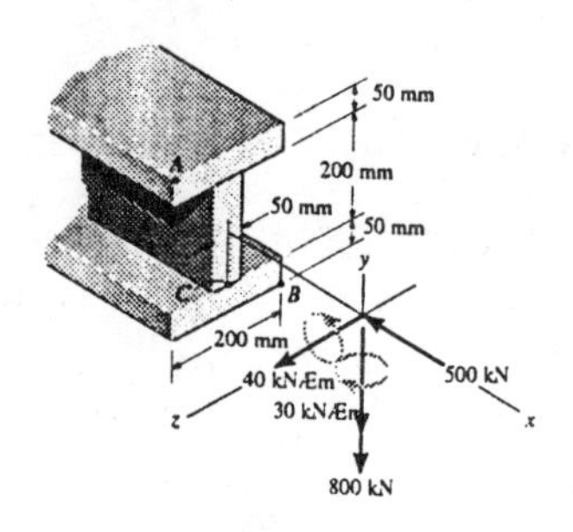

Section Properties :

$$A = 0.2(0.3) - 0.15(0.2) = 0.030 \text{ m}^4$$

$$I_z = \frac{1}{12}(0.2)\left(0.3^3\right) - \frac{1}{12}(0.15)\left(0.2^3\right) = 0.350\left(10^{-3}\right) \text{ m}^4$$

$$I_y = \frac{1}{12}(0.1)\left(0.2^3\right) + \frac{1}{12}(0.2)\left(0.05^3\right) = 68.75\left(10^{-6}\right) \text{ m}^4$$

$$(Q_A)_y = 0$$

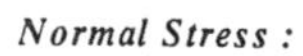

Normal Stress :

$$\sigma = \frac{N}{A} - \frac{M_z y}{I_z} + \frac{M_y z}{I_y}$$

$$\sigma_A = \frac{-500(10^3)}{0.030} - \frac{40(10^3)(0.15)}{0.350(10^{-3})} + \frac{-30(10^3)(0.1)}{68.75(10^{-6})}$$

$$= -77.45 \text{ MPa}$$

Shear Stress : Since $(Q_A)_y = 0$, $\quad \tau_A = 0$

In-Plane Principal Stresses : $\sigma_x = -77.45$ MPa, $\sigma_y = 0$, and $\tau_{xy} = 0$ for point A. Since no shear stress acts on the element,

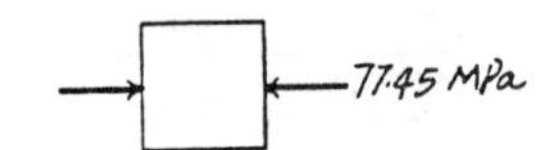

$$\sigma_1 = \sigma_y = 0 \qquad \textbf{Ans}$$

$$\sigma_2 = \sigma_x = -77.4 \text{ MPa} \qquad \textbf{Ans}$$

Maximum In-Plane Shear Stress : Applying Eq. 9–7,

$$\tau_{\substack{\max \\ \text{in-plane}}} = \sqrt{\left(\frac{\sigma_x - \sigma_y}{2}\right)^2 + \tau_{xy}^2}$$

$$= \sqrt{\left(\frac{-77.45 - 0}{2}\right)^2 + 0}$$

$$= 38.7 \text{ MPa} \qquad \textbf{Ans}$$

***9–36.** Solve Prob. 9–35 for point B.

Section Properties :

$$A = 0.2(0.3) - 0.15(0.2) = 0.030 \text{ m}^4$$

$$I_z = \frac{1}{12}(0.2)\left(0.3^3\right) - \frac{1}{12}(0.15)\left(0.2^3\right) = 0.350\left(10^{-3}\right) \text{ m}^4$$

$$I_y = \frac{1}{12}(0.1)\left(0.2^3\right) + \frac{1}{12}(0.2)\left(0.05^3\right) = 68.75\left(10^{-6}\right) \text{ m}^4$$

$$(Q_B)_y = 0$$

Normal Stress :

$$\sigma = \frac{N}{A} - \frac{M_z y}{I_z} + \frac{M_y z}{I_y}$$

$$\sigma_B = \frac{-500(10^3)}{0.030} - \frac{40(10^3)(-0.15)}{0.350(10^{-3})} + \frac{-30(10^3)(-0.1)}{68.75(10^{-6})}$$

$$= 44.11 \text{ MPa}$$

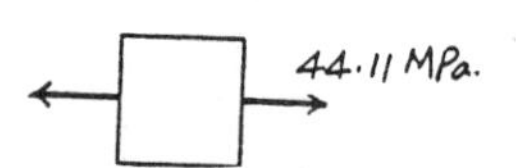

Shear Stress : Since $(Q_B)_y = 0$, $\quad \tau_B = 0$

In-Plane Principal Stress : $\sigma_x = 44.11$ MPa, $\sigma_y = 0$ and $\tau_{xy} = 0$ for point B. Since no shear stress acts on the element,

$$\sigma_1 = \sigma_x = 44.1 \text{ MPa} \qquad \textbf{Ans}$$

$$\sigma_2 = \sigma_y = 0 \qquad \textbf{Ans}$$

Maximum In-Plane Shear Stress : Applying Eq. 9–7,

$$\tau_{\substack{\max \\ \text{in-plane}}} = \sqrt{\left(\frac{\sigma_x - \sigma_y}{2}\right)^2 + \tau_{xy}^2}$$

$$= \sqrt{\left(\frac{44.11 - 0}{2}\right)^2 + 0}$$

$$= 22.1 \text{ MPa} \qquad \textbf{Ans}$$

9–37. Solve Prob. 9–35 for point C, located in the center on the bottom of the web.

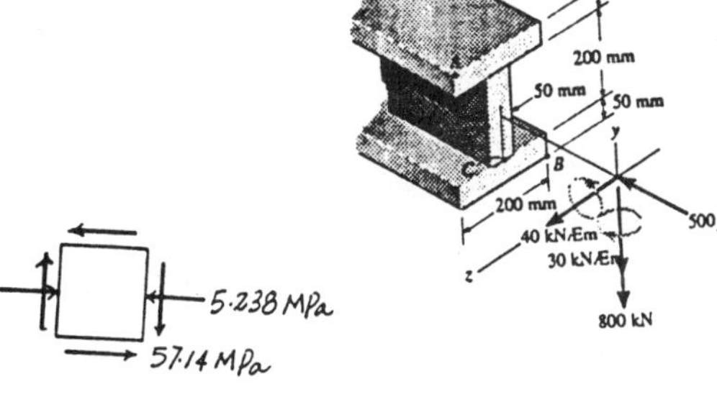

Section Properties :

$$A = 0.2(0.3) - 0.15(0.2) = 0.030 \text{ m}^4$$

$$I_z = \frac{1}{12}(0.2)\left(0.3^3\right) - \frac{1}{12}(0.15)\left(0.2^3\right) = 0.350\left(10^{-3}\right) \text{ m}^4$$

$$I_y = \frac{1}{12}(0.1)\left(0.2^3\right) + \frac{1}{12}(0.2)\left(0.05^3\right) = 68.75\left(10^{-6}\right) \text{ m}^4$$

$$(Q_C)_y = \bar{y}'A' = 0.125(0.05)(0.2) = 1.25\left(10^{-3}\right) \text{ m}^3$$

Normal Stress :

$$\sigma = \frac{N}{A} - \frac{M_z y}{I_z} + \frac{M_y z}{I_y}$$

$$\sigma_C = \frac{-500(10^3)}{0.030} - \frac{40(10^3)(-0.1)}{0.350(10^{-3})} + \frac{-30(10^3)(0)}{68.75(10^{-6})}$$

$$= -5.238 \text{ MPa}$$

Shear Stress : Applying the shear formula

$$\tau_C = \frac{V_y (Q_C)_y}{I_z t} = \frac{800(10^3)\left[1.25(10^{-3})\right]}{0.350(10^{-3})(0.05)} = 57.14 \text{ MPa}$$

In - Plane Principal Stress : $\sigma_x = -5.238$ MPa, $\sigma_y = 0$ and $\tau_{xy} = -57.14$ MPa for point C. Applying Eq. 9 – 5,

$$\sigma_{1,2} = \frac{\sigma_x + \sigma_y}{2} \pm \sqrt{\left(\frac{\sigma_x - \sigma_y}{2}\right)^2 + \tau_{xy}^2}$$

$$= \frac{-5.238 + 0}{2} \pm \sqrt{\left(\frac{-5.238 - 0}{2}\right)^2 + (-57.14)^2}$$

$$= -2.619 \pm 57.203$$

$\sigma_1 = 54.6$ MPa $\qquad \sigma_2 = -59.8$ MPa **Ans**

Maximum In - Plane Shear Stress : Applying Eq. 9 – 7,

$$\tau_{\substack{\text{max} \\ \text{in-plane}}} = \sqrt{\left(\frac{\sigma_x - \sigma_y}{2}\right)^2 + \tau_{xy}^2}$$

$$= \sqrt{\left(\frac{-5.238 - 0}{2}\right)^2 + (-57.14)^2}$$

$$= 57.2 \text{ MPa}$$

Ans

9–38. The internal loadings at a section of the beam consist of an axial force of 6 kip, a shear force of 12 kip, and a moment of 500 lb · ft. Determine the principal stresses at point A. Also compute the maximum in-plane shear stress at this point.

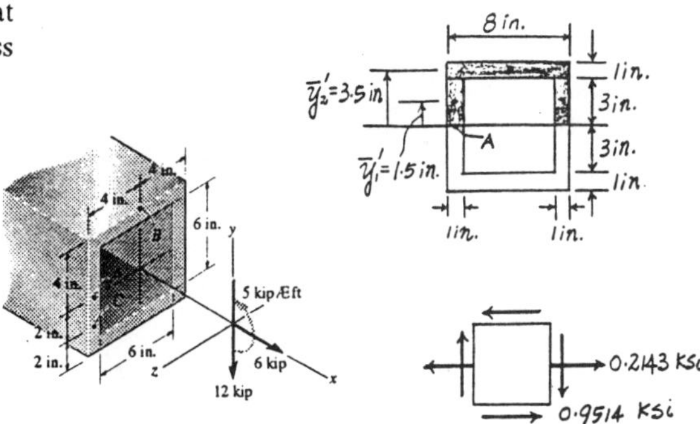

Section Properties :

$$A = 8(8) - 6(6) = 28.0 \text{ in}^2$$

$$I_z = \frac{1}{12}(8)\left(8^3\right) - \frac{1}{12}(6)\left(6^3\right) = 233.33 \text{ in}^4$$

$$(Q_A)_y = \Sigma\bar{y}'A' = 1.5(3)(2) + 3.5(1)(8) = 37.0 \text{ in}^3$$

Normal Stress :

$$\sigma = \frac{N}{A} - \frac{M_z y}{I_z}$$

$$\sigma_A = \frac{6}{28.0} - \frac{5(12)(0)}{233.33}$$

$$= 0.2143 \text{ ksi}$$

Shear Stress : Applying the shear formula,

$$\tau_A = \frac{V_y (Q_A)_y}{I_z t} = \frac{12(37.0)}{233.33(2)} = 0.9514 \text{ ksi}$$

In - Plane Principal Stress : $\sigma_x = 0.2143$ ksi, $\sigma_y = 0$ and $\tau_{xy} = -0.9514$ ksi for point A. Applying Eq. 9 – 5,

$$\sigma_{1,2} = \frac{\sigma_x + \sigma_y}{2} \pm \sqrt{\left(\frac{\sigma_x - \sigma_y}{2}\right)^2 + \tau_{xy}^2}$$

$$= \frac{0.2143 + 0}{2} \pm \sqrt{\left(\frac{0.2143 - 0}{2}\right)^2 + (-0.9514)^2}$$

$$= -2.619 \pm 57.203$$

$\sigma_1 = 1.06$ ksi $\qquad \sigma_2 = -0.850$ ksi **Ans**

Maximum In - Plane Shear Stress : Applying Eq. 9 – 7,

$$\tau_{\substack{\text{max} \\ \text{in-plane}}} = \sqrt{\left(\frac{\sigma_x - \sigma_y}{2}\right)^2 + \tau_{xy}^2}$$

$$= \sqrt{\left(\frac{0.2143 - 0}{2}\right)^2 + (-0.9514)^2}$$

$$= 0.957 \text{ ksi}$$

Ans

9–39. Solve Prob. 9–38 for point B.

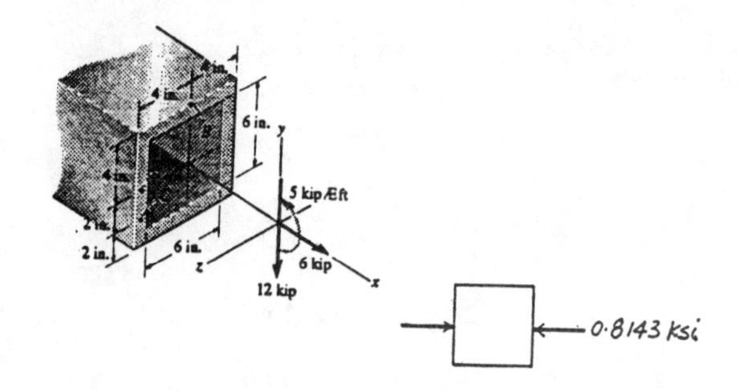

Section Properties :

$$A = 8(8) - 6(6) = 28.0 \text{ in}^2$$

$$I_z = \frac{1}{12}(8)(8^3) - \frac{1}{12}(6)(6^3) = 233.33 \text{ in}^4$$

$$(Q_B)_y = 0$$

Normal Stress :

$$\sigma = \frac{N}{A} - \frac{M_z y}{I_z}$$

$$\sigma_C = \frac{6}{28.0} - \frac{5(12)(4)}{233.33} = -0.8143 \text{ ksi}$$

Shear Stress : Since $(Q_B)_y = 0$, then $\tau_B = 0$

In-Plane Principal Stresses : $\sigma_x = -0.8143$ ksi, $\sigma_y = 0$ and $\tau_{xy} = 0$ for point B. Since no shear stress acts on the element,

$$\sigma_1 = \sigma_y = 0 \qquad \textbf{Ans}$$

$$\sigma_2 = \sigma_x = -0.814 \text{ ksi} \qquad \textbf{Ans}$$

Maximum In-Plane Shear Stress : Applying Eq. 9 – 7,

$$\tau_{\substack{\max \\ \text{in-plane}}} = \sqrt{\left(\frac{\sigma_x - \sigma_y}{2}\right)^2 + \tau_{xy}^2} = \sqrt{\left(\frac{-0.8143 - 0}{2}\right)^2 + 0} = 0.407 \text{ ksi} \qquad \textbf{Ans}$$

***9–40.** Solve Prob. 9–38 for point C.

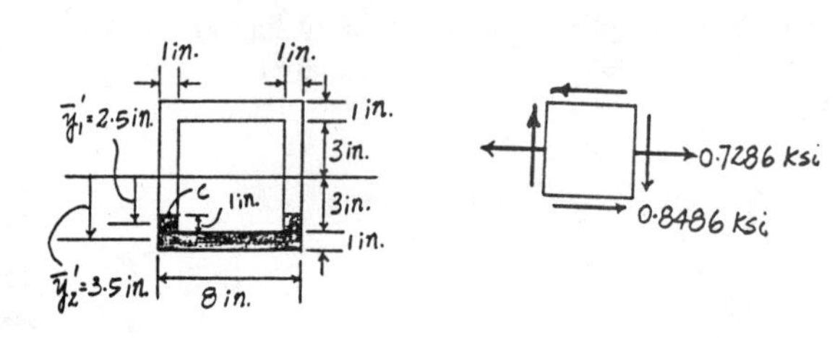

Section Properties :

$$A = 8(8) - 6(6) = 28.0 \text{ in}^2$$

$$I_z = \frac{1}{12}(8)(8^3) - \frac{1}{12}(6)(6^3) = 233.33 \text{ in}^4$$

$$(Q_C)_y = \Sigma \bar{y}'A' = 2.5(1)(2) + 3.5(1)(8) = 33.0 \text{ in}^3$$

Normal Stress :

$$\sigma = \frac{N}{A} - \frac{M_z y}{I_z}$$

$$\sigma_C = \frac{6}{28.0} - \frac{5(12)(-2)}{233.33} = 0.7286 \text{ ksi}$$

Shear Stress : Applying the shear formula,

$$\tau_C = \frac{V_y (Q_C)_y}{I_z t} = \frac{12(33.0)}{233.33(2)} = 0.8486 \text{ ksi}$$

In-Plane Principal Stresses : $\sigma_x = 0.7286$ ksi, $\sigma_y = 0$, and $\tau_{xy} = -0.8486$ ksi for point A. Applying Eq. 9 – 5,

$$\sigma_{1,2} = \frac{\sigma_x + \sigma_y}{2} \pm \sqrt{\left(\frac{\sigma_x - \sigma_y}{2}\right)^2 + \tau_{xy}^2} = \frac{0.7286 + 0}{2} \pm \sqrt{\left(\frac{0.7286 - 0}{2}\right)^2 + (-0.8486)^2} = 0.3643 \pm 0.9235$$

$$\sigma_1 = 1.29 \text{ ksi} \qquad \sigma_2 = -0.559 \text{ ksi} \qquad \textbf{Ans}$$

Maximum In-Plane Shear Stress : Applying Eq. 9 – 7,

$$\tau_{\substack{\max \\ \text{in-plane}}} = \sqrt{\left(\frac{\sigma_x - \sigma_y}{2}\right)^2 + \tau_{xy}^2} = \sqrt{\left(\frac{0.7286 - 0}{2}\right)^2 + (-0.8486)^2} = 0.923 \text{ ksi} \qquad \textbf{Ans}$$

9–41. The beam has a rectangular cross section and is subjected to the loadings shown. Determine the principal stresses that are developed at point A and point B, which are located just to the left of the 20-kN load. Show the results on elements located at these points.

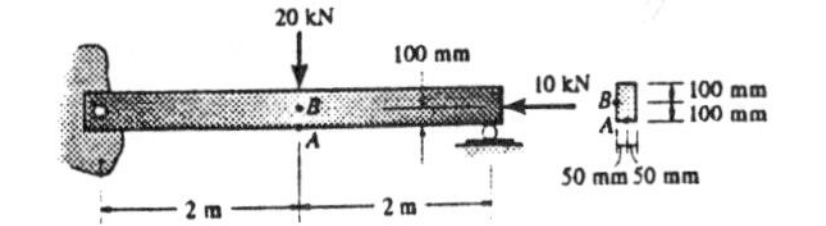

Internal Forces and Moment : As shown on FBD(b).

Section Properties :

$$A = 0.1(0.2) = 0.020\ \text{m}^2$$

$$I = \frac{1}{12}(0.1)\left(0.2^3\right) = 66.667\left(10^{-6}\right)\ \text{m}^4$$

$$Q_A = 0$$

$$Q_B = \bar{y}'A' = 0.05(0.1)(0.1) = 0.50\left(10^{-3}\right)\ \text{m}^3$$

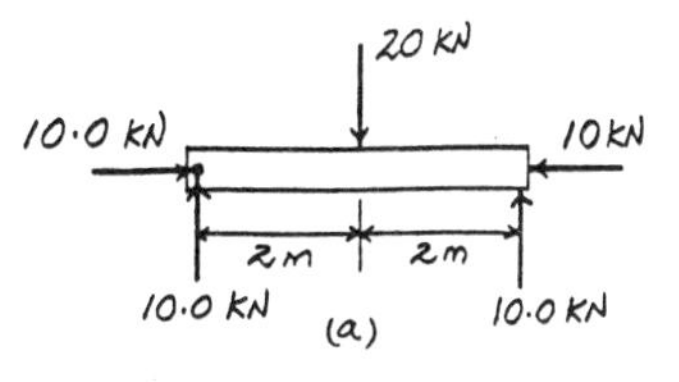

Normal Stresses :

$$\sigma = \frac{N}{A} \pm \frac{My}{I}$$

$$\sigma_A = \frac{-10.0(10^3)}{0.020} - \frac{20.0(10^3)(0.1)}{66.667(10^{-6})} = -30.5\ \text{MPa}$$

$$\sigma_B = \frac{-10.0(10^3)}{0.020} - \frac{20.0(10^3)(0)}{66.667(10^{-6})} = -0.500\ \text{MPa}$$

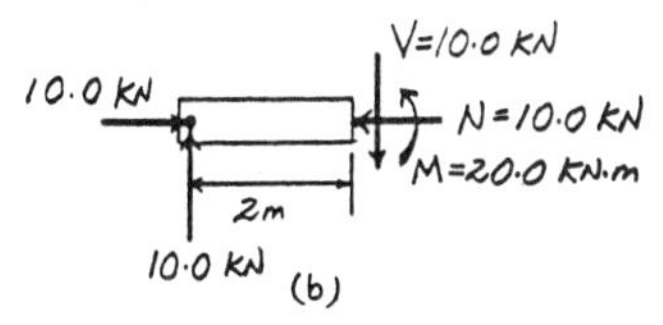

Shear Stress : Applying the shear formula $\tau = \frac{VQ}{It}$,

$$\tau_A = 0$$

$$\tau_B = \frac{10.0(10^3)\left[0.50(10^{-3})\right]}{66.667(10^{-6})(0.1)} = 0.750\ \text{MPa}$$

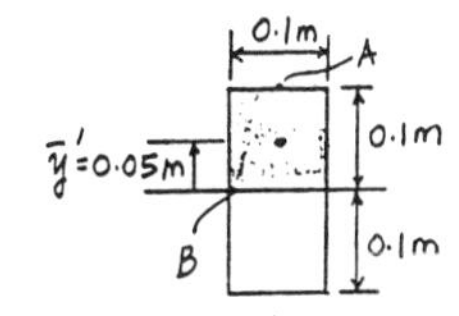

In - Plane Principal Stresses : $\sigma_x = -30.5$ MPa, $\sigma_y = 0$, and $\tau_{xy} = 0$ for point A. Since no shear stress acts on the element,

$$\sigma_1 = \sigma_y = 0 \qquad \textbf{Ans}$$

$$\sigma_2 = \sigma_x = -30.5\ \text{MPa} \qquad \textbf{Ans}$$

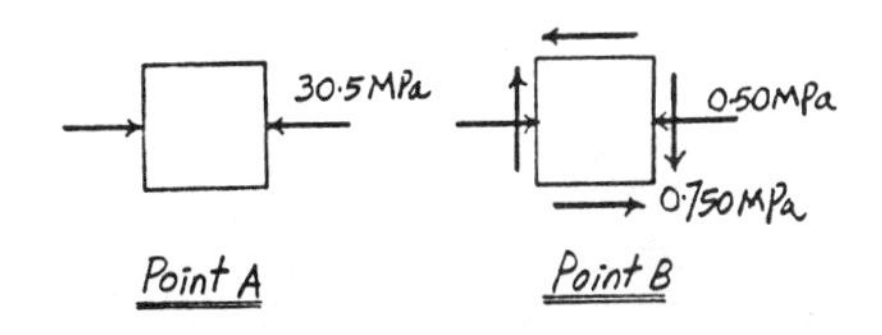

$\sigma_x = -0.500$ MPa, $\sigma_y = 0$ and $\tau_{xy} = -0.750$ MPa for point B. Applying Eq.9 – 5,

$$\sigma_{1,2} = \frac{\sigma_x + \sigma_y}{2} \pm \sqrt{\left(\frac{\sigma_x - \sigma_y}{2}\right)^2 + \tau_{xy}^2}$$

$$= \frac{-0.500 + 0}{2} \pm \sqrt{\left(\frac{-0.500 - 0}{2}\right)^2 + (-0.750)^2}$$

$$= -0.250 \pm 0.7906$$

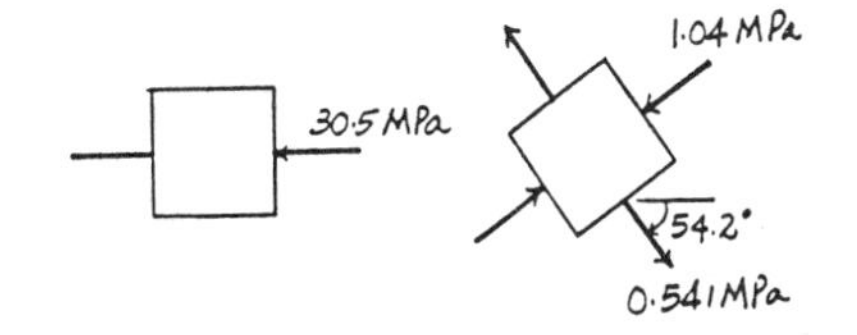

$$\sigma_1 = 0.541\ \text{MPa} \qquad \sigma_2 = -1.04\ \text{MPa} \qquad \textbf{Ans}$$

Orientation of Principal Plane : Applying Eq.9 – 4 for point B,

$$\tan 2\theta_p = \frac{\tau_{xy}}{(\sigma_x - \sigma_y)/2} = \frac{-0.750}{(-0.500 - 0)/2} = 3.000$$

$$\theta_p = 35.78° \quad \text{and} \quad -54.22°$$

Substituting the results into Eq.9 – 1 with $\theta = 35.78°$ yields

$$\sigma_{x'} = \frac{\sigma_x + \sigma_y}{2} + \frac{\sigma_x - \sigma_y}{2}\cos 2\theta + \tau_{xy}\sin 2\theta$$

$$= \frac{-0.500 + 0}{2} + \frac{-0.500 - 0}{2}\cos 71.56° + (-0.750\sin 71.56°)$$

$$= -1.04\ \text{MPa} = \sigma_2$$

Hence,

$$\theta_{p_1} = -54.2° \qquad \theta_{p_2} = 35.8° \qquad \textbf{Ans}$$

9–42. The solid propeller shaft on a ship extends outward from the hull. During operation it turns at $\omega = 15$ rad/s when the engine develops 900 kW of power. This causes a thrust of $F = 1.23$ MN on the shaft. If the shaft has an outer diameter of 250 mm, determine the principal stresses at any point located on the surface of the shaft.

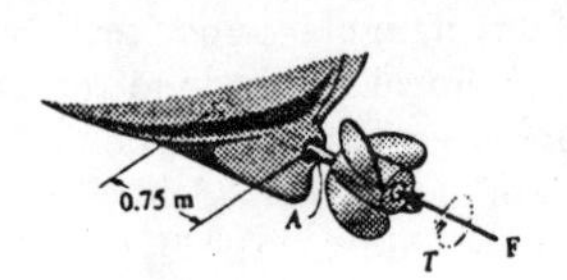

Power Transmission : Using the formula developed in Chapter 5,

$$P = 900 \text{ kW} = 0.900(10^6) \text{ N}\cdot\text{m/s}$$

$$T_0 = \frac{P}{\omega} = \frac{0.900(10^6)}{15} = 60.0(10^3) \text{ N}\cdot\text{m}$$

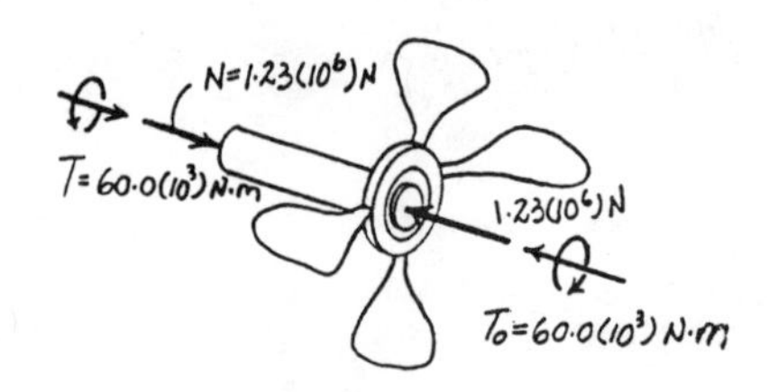

Internal Torque and Force : As shown on FBD.

Section Properties :

$$A = \frac{\pi}{4}(0.25^2) = 0.015625\pi \text{ m}^2$$

$$J = \frac{\pi}{2}(0.125^4) = 0.3835(10^{-3}) \text{ m}^4$$

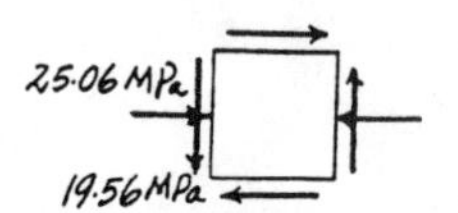

Normal Stress :

$$\sigma = \frac{N}{A} = \frac{-1.23(10^6)}{0.015625\pi} = -25.06 \text{ MPa}$$

Shear Stress : Applying the torsion formula.

$$\tau = \frac{Tc}{J} = \frac{60.0(10^3)(0.125)}{0.3835(10^{-3})} = 19.56 \text{ MPa}$$

In-Plane Principal Stresses : $\sigma_x = -25.06$ MPa, $\sigma_y = 0$ and $\tau_{xy} = 19.56$ MPa for any point on the shaft's surface. Applying Eq. 9 – 5,

$$\sigma_{1,2} = \frac{\sigma_x + \sigma_y}{2} \pm \sqrt{\left(\frac{\sigma_x - \sigma_y}{2}\right)^2 + \tau_{xy}^2}$$

$$= \frac{-25.06 + 0}{2} \pm \sqrt{\left(\frac{-25.06 - 0}{2}\right)^2 + (19.56)^2}$$

$$= -12.53 \pm 23.23$$

$\sigma_1 = 10.7$ MPa $\quad \sigma_2 = -35.8$ MPa **Ans**

9–43. The solid propeller shaft on a ship extends outward from the hull. During operation it turns at $\omega = 15$ rad/s when the engine develops 900 kW of power. This causes a thrust of $F = 1.23$ MN on the shaft. If the shaft has a diameter of 250 mm, determine the maximum in-plane shear stress at any point located on the surface of the shaft.

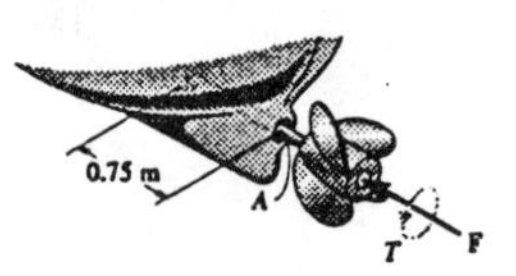

Power Transmission : Using the formula devloped in Chapter 5,

$$P = 900 \text{ kW} = 0.900(10^6) \text{ N}\cdot\text{m/s}$$

$$T_0 = \frac{P}{\omega} = \frac{0.900(10^6)}{15} = 60.0(10^3) \text{ N}\cdot\text{m}$$

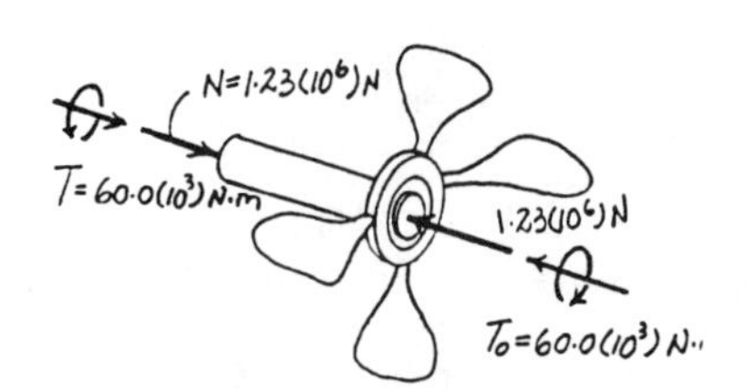

Internal Torque and Force : As shown on FBD.

Section Properties :

$$A = \frac{\pi}{4}(0.25^2) = 0.015625\pi \text{ m}^2$$

$$J = \frac{\pi}{2}(0.125^4) = 0.3835(10^{-3}) \text{ m}^4$$

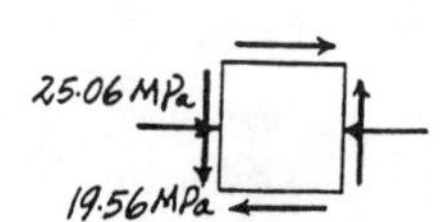

Normal Stress :

$$\sigma = \frac{N}{A} = \frac{-1.23(10^6)}{0.015625\pi} = -25.06 \text{ MPa}$$

Shear Stress : Applying the torsion formula,

$$\tau = \frac{Tc}{J} = \frac{60.0(10^3)(0.125)}{0.3835(10^{-3})} = 19.56 \text{ MPa}$$

Maximum In-Plane Shear Stress : $\sigma_x = -25.06$ MPa, $\sigma_y = 0$, and $\tau_{xy} = 19.56$ MPa for any point on the shaft's surface. Applying Eq. 9 – 7,

$$\tau_{\substack{\max \\ \text{in-plane}}} = \sqrt{\left(\frac{\sigma_x - \sigma_y}{2}\right)^2 + \tau_{xy}^2}$$

$$= \sqrt{\left(\frac{-25.06 - 0}{2}\right)^2 + (19.56)^2}$$

$$= 23.2 \text{ MPa}$$ **Ans**

***9–44.** The steel pipe has an inner diameter of 2.75 in. and an outer diameter of 3 in. If it is fixed at C and subjected to the horizontal 20–lb force acting on the handle of the pipe wrench at its end, determine the principal stresses in the pipe at point A which is located on the surface of the pipe.

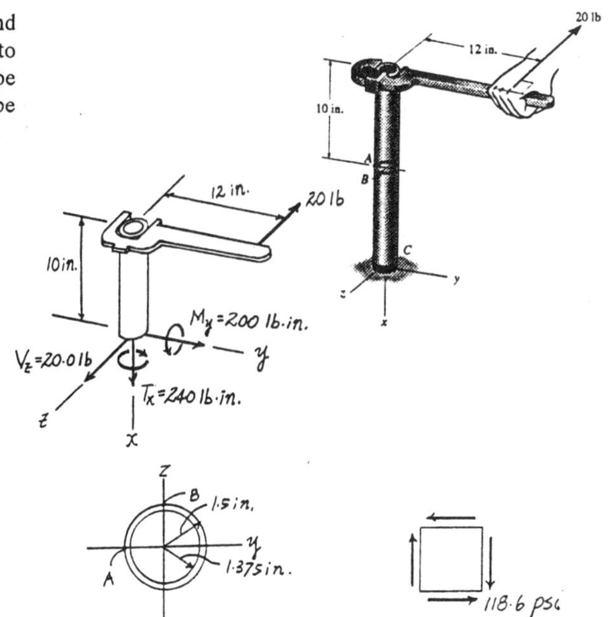

Internal Forces, Torque, and Moments : As shown on FBD.

Section Properties :

$$I = \frac{\pi}{4}\left(1.5^4 - 1.375^4\right) = 1.1687 \text{ in}^4$$

$$J = \frac{\pi}{2}\left(1.5^4 - 1.375^4\right) = 2.3374 \text{ in}^4$$

$$(Q_A)_z = \Sigma\bar{y}'A'$$

$$= \frac{4(1.5)}{3\pi}\left[\frac{1}{2}\pi\left(1.5^2\right)\right] - \frac{4(1.375)}{3\pi}\left[\frac{1}{2}\pi\left(1.375^2\right)\right]$$

$$= 0.51693 \text{ in}^3$$

Normal Stress : Applying the flexure formula $\sigma = \frac{M_y z}{I_y}$,

$$\sigma_A = \frac{200(0)}{1.1687} = 0$$

Shear Stress : The transverse shear stress in the z direction and the torsional shear stress can be obtained using shear formula and torsion formula, $\tau_V = \frac{VQ}{It}$ and $\tau_{twist} = \frac{T\rho}{J}$, respectively.

$$\tau_A = (\tau_V)_z - \tau_{twist}$$

$$= \frac{20.0(0.51693)}{1.1687(2)(0.125)} - \frac{240(1.5)}{2.3374}$$

$$= -118.6 \text{ psi}$$

In - Plane Principal Stress : $\sigma_x = 0$, $\sigma_z = 0$ and $\tau_{xz} = -118.6$ psi for point A. Applying Eq.9 – 5

$$\sigma_{1,2} = \frac{\sigma_x + \sigma_z}{2} \pm \sqrt{\left(\frac{\sigma_x - \sigma_z}{2}\right)^2 + \tau_{xz}^2}$$

$$= 0 \pm \sqrt{0 + (-118.6)^2}$$

$\sigma_1 = 119$ psi $\sigma_2 = -119$ psi **Ans**

9–45. Solve Prob. 9–44 for point B which is located on the surface of the pipe.

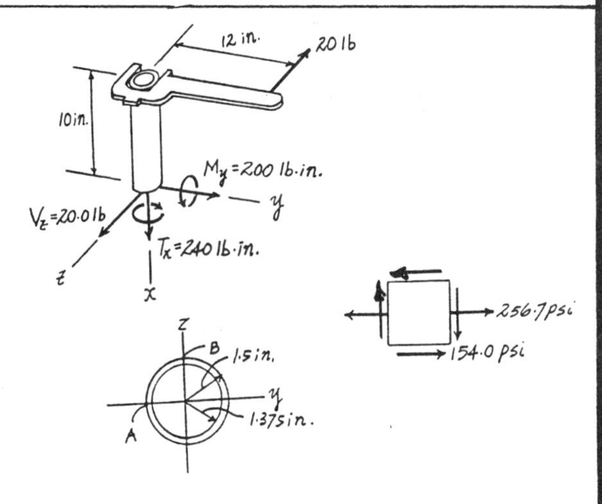

Internal Forces, Torque, and Moments : As shown on FBD.

Section Properties :

$$I = \frac{\pi}{4}\left(1.5^4 - 1.375^4\right) = 1.1687 \text{ in}^4$$

$$J = \frac{\pi}{2}\left(1.5^4 - 1.375^4\right) = 2.3374 \text{ in}^4$$

$$(Q_B)_z = 0$$

Normal Stress : Applying the flexure formula $\sigma = \frac{M_y z}{I_y}$,

$$\sigma_B = \frac{200(1.5)}{1.1687} = 256.7 \text{ psi}$$

Shear Stress : Torsional shear stress can be obtained using torsion formula, $\tau_{twist} = \frac{T\rho}{J}$.

$$\tau_B = \tau_{twist} = \frac{240(1.5)}{2.3374} = 154.0 \text{ psi}$$

In - Plane Principal Stress : $\sigma_x = 256.7$ psi, $\sigma_y = 0$, and $\tau_{xy} = -154.0$ psi for point B. Applying Eq.9 – 5,

$$\sigma_{1,2} = \frac{\sigma_x + \sigma_y}{2} \pm \sqrt{\left(\frac{\sigma_x - \sigma_y}{2}\right)^2 + \tau_{xy}^2}$$

$$= \frac{256.7 + 0}{2} \pm \sqrt{\left(\frac{256.7 - 0}{2}\right)^2 + (-154.0)^2}$$

$$= 128.35 \pm 200.49$$

$\sigma_1 = 329$ psi $\sigma_2 = -72.1$ psi **Ans**

9–46. The cantilevered beam is subjected to the load at its end. Determine the principal stresses in the beam at points A and B.

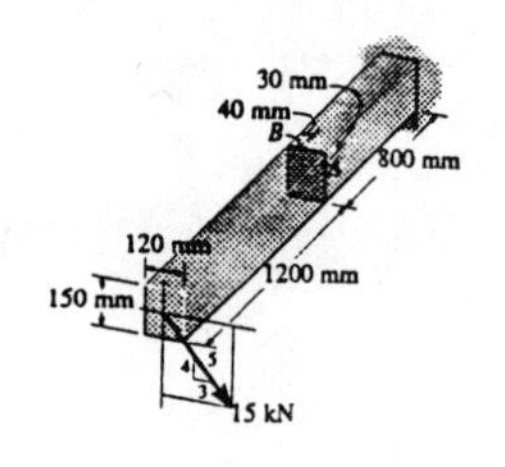

Internal Forces and Moment : As shown on FBD.

Section Properties :

$$I_z = \frac{1}{12}(0.12)\left(0.15^3\right) = 33.75\left(10^{-6}\right)\ \text{m}^4$$

$$I_y = \frac{1}{12}(0.15)\left(0.12^3\right) = 21.6\left(10^{-6}\right)\ \text{m}^4$$

$$(Q_A)_y = \bar{y}'A' = 0.06(0.03)(0.12) = 0.216\left(10^{-3}\right)\ \text{m}^3$$

$$(Q_A)_z = 0$$

$$(Q_B)_z = \bar{z}'A' = 0.04(0.04)(0.15) = 0.240\left(10^{-3}\right)\ \text{m}^3$$

$$(Q_B)_y = 0$$

Normal Stress :

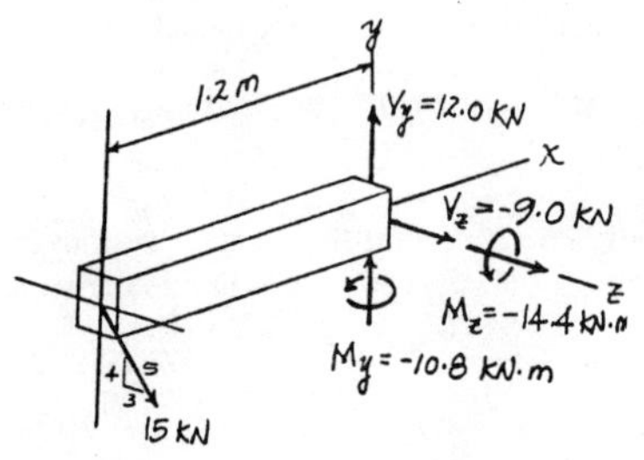

$$\sigma = -\frac{M_z y}{I_z} + \frac{M_y z}{I_y}$$

$$\sigma_A = -\frac{-14.4(10^3)(0.045)}{33.75(10^{-6})} + \frac{-10.8(10^3)(0.06)}{21.6(10^{-6})}$$
$$= -10.8\ \text{MPa}$$

$$\sigma_B = -\frac{-14.4(10^3)(0.075)}{33.75(10^{-6})} + \frac{-10.8(10^3)(-0.02)}{21.6(10^{-6})}$$
$$= 42.0\ \text{MPa}$$

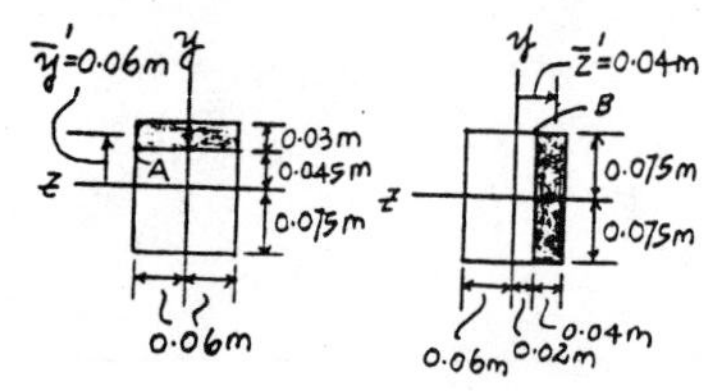

Shear Stress : Applyingthe shear formula

$$\tau_A = \frac{V_y (Q_A)_y}{I_z t} = \frac{12.0(10^3)\left[0.216(10^{-3})\right]}{33.75(10^{-6})(0.12)} = 0.640\ \text{MPa}$$

$$\tau_B = \frac{V_z (Q_B)_z}{I_y t} = \frac{-9.00(10^3)\left[0.240(10^{-3})\right]}{21.6(10^{-6})(0.15)} = -0.6667\ \text{MPa}$$

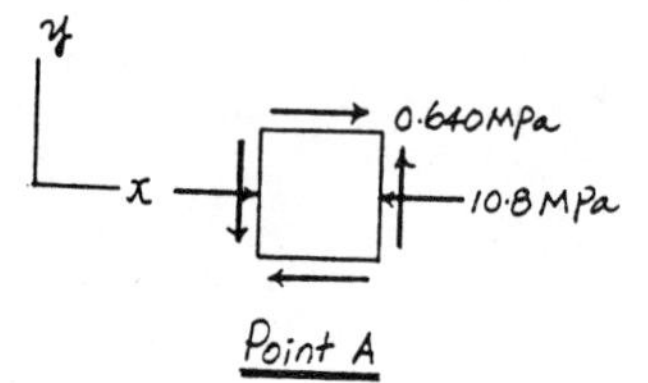

In - Plane Principal Stress : $\sigma_x = -10.8$ MPa, $\sigma_y = 0$ and $\tau_{xy} = 0.640$ MPa for point A. Applying Eq.9 – 5

$$\sigma_{1,2} = \frac{\sigma_x + \sigma_y}{2} \pm \sqrt{\left(\frac{\sigma_x - \sigma_y}{2}\right)^2 + \tau_{xy}^2}$$
$$= \frac{-10.8 + 0}{2} \pm \sqrt{\left(\frac{-10.8 - 0}{2}\right)^2 + 0.640^2}$$
$$= -5.40 \pm 5.4378$$

$\sigma_1 = 37.8$ kPa $\qquad \sigma_2 = -10.8$ MPa $\qquad$ **Ans**

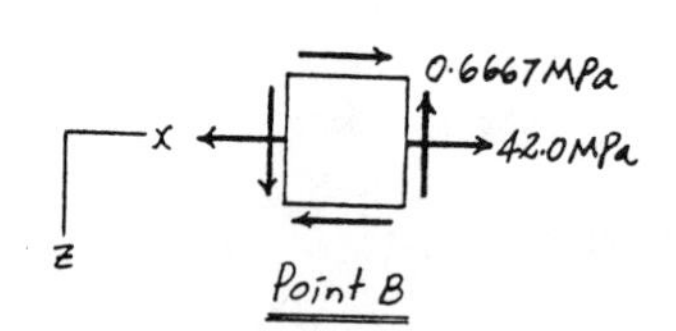

$\sigma_x = 42.0$ MPa, $\sigma_z = 0$, and $\tau_{xz} = 0.6667$ MPa for point B. Applying Eq.9 – 5

$$\sigma_{1,2} = \frac{\sigma_x + \sigma_y}{2} \pm \sqrt{\left(\frac{\sigma_x - \sigma_y}{2}\right)^2 + \tau_{xy}^2}$$
$$= \frac{42.0 + 0}{2} \pm \sqrt{\left(\frac{42.0 - 0}{2}\right)^2 + 0.6667^2}$$
$$= 21.0 \pm 21.0105$$

$\sigma_1 = 42.0$ MPa $\qquad \sigma_2 = -10.6$ kPa $\qquad$ **Ans**

9–49. Solve Prob. 9–3 using Mohr's circle.

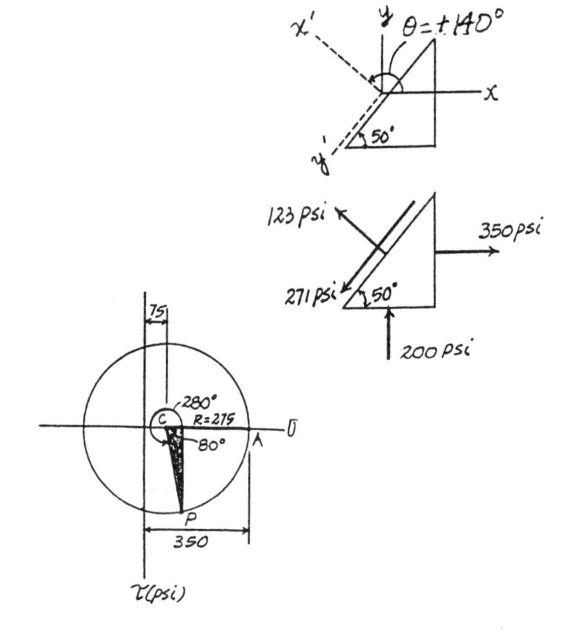

Construction of the Circle : In accordance with the sign convention, $\sigma_x = 350$ psi, $\sigma_y = -200$ psi, and $\tau_{xy} = 0$. Hence,

$$\sigma_{avg} = \frac{\sigma_x + \sigma_y}{2} = \frac{350 + (-200)}{2} = 75.0 \text{ psi}$$

The coordinates for reference points A and C are

$$A(350,\ 0) \qquad C(75.0,\ 0)$$

The radius of the circle is

$$R = 350 - 75.0 = 275 \text{ psi}$$

Stresses on the Inclined Plane : The normal and shear stress components $(\sigma_{x'}$ and $\tau_{x'y'})$ are represented by the coordinates of point P on the circle.

$$\sigma_{x'} = 75.0 + 275\cos 80° = 123 \text{ psi} \qquad \textbf{Ans}$$

$$\tau_{x'y'} = 275 \sin 80° = 271 \text{ psi} \qquad \textbf{Ans}$$

9–50. Solve Prob. 9–5 using Mohr's circle.

Construction of the Circle : In accordance with the sign convention, $\sigma_x = -3$ ksi, $\sigma_y = 2$ ksi, and $\tau_{xy} = -4$ ksi. Hence,

$$\sigma_{avg} = \frac{\sigma_x + \sigma_y}{2} = \frac{-3 + 2}{2} = -0.500 \text{ ksi}$$

The coordinates for reference points A and C are

$$A(-3,\ -4) \qquad C(-0.500,\ 0)$$

The radius of the circle is

$$R = \sqrt{(3 - 0.5)^2 + 4^2} = 4.717 \text{ ksi}$$

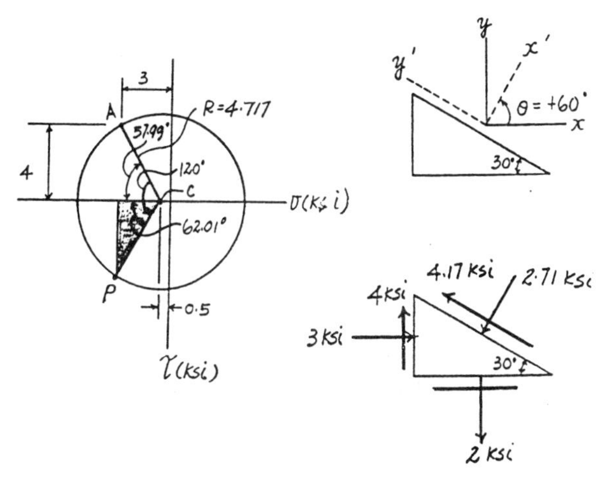

Stress on the Inclined Plane : The normal and shear stress components $(\sigma_{x'}$ and $\tau_{x'y'})$ are represented by the coordinates of point P on the circle.

$$\sigma_{x'} = -0.500 - 4.717\cos 62.01° = -2.71 \text{ ksi} \qquad \textbf{Ans}$$

$$\tau_{x'y'} = 4.717 \sin 62.01° = 4.17 \text{ ksi} \qquad \textbf{Ans}$$

9–51. Solve Prob. 9–6 using Mohr's circle.

Construction of the Circle : In accordance with the sign convention, $\sigma_x = -4$ ksi, $\sigma_y = 7$ ksi, and $\tau_{xy} = -6$ ksi. Hence,

$$\sigma_{avg} = \frac{\sigma_x + \sigma_y}{2} = \frac{-4+7}{2} = 1.50 \text{ ksi}$$

The coordinates for reference point A and C are

$$A(-4,\ -6) \qquad C(1.50,\ 0)$$

The radius of the circle is

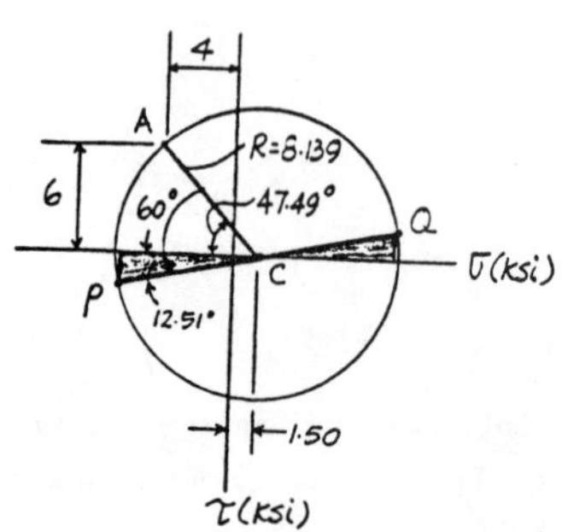

$$R = \sqrt{(4+1.50)^2 + 6^2} = 8.139 \text{ ksi}$$

Stress on The Rotated Element : The normal and shear stress components ($\sigma_{x'}$ and $\tau_{x'y'}$) are represented by the coordinates of point P on the circle. $\sigma_{y'}$ can be determined by calculating the coordinates of point Q on the circle.

$$\sigma_{x'} = 1.50 - 8.139\cos 12.51° = -6.45 \text{ ksi} \qquad \textbf{Ans}$$

$$\tau_{x'y'} = 8.139\sin 12.51° = 1.76 \text{ ksi} \qquad \textbf{Ans}$$

$$\sigma_{y'} = 1.50 + 8.139\cos 12.51° = 9.45 \text{ ksi} \qquad \textbf{Ans}$$

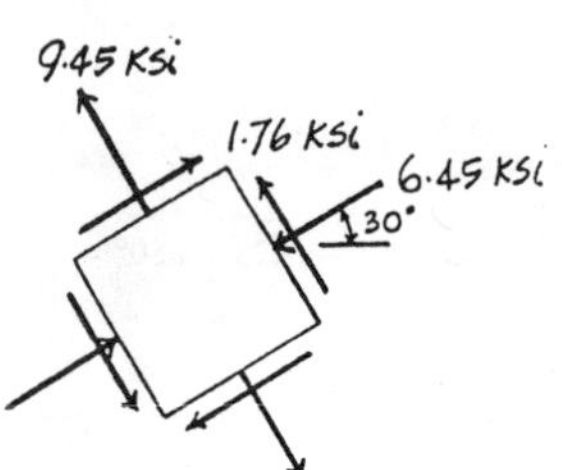

***9–52.** Solve Prob. 9–8 using Mohr's circle.

Construction of the Circle : In accordance with the sign convention, $\sigma_x = 300$ psi, $\sigma_y = 0$, and $\tau_{xy} = 120$ psi. Hence,

$$\sigma_{avg} = \frac{\sigma_x + \sigma_y}{2} = \frac{300+0}{2} = 150 \text{ psi}$$

The coordinates for reference point A and C are

$$A(300,\ 120) \qquad C(150,\ 0)$$

The radius of the circle is

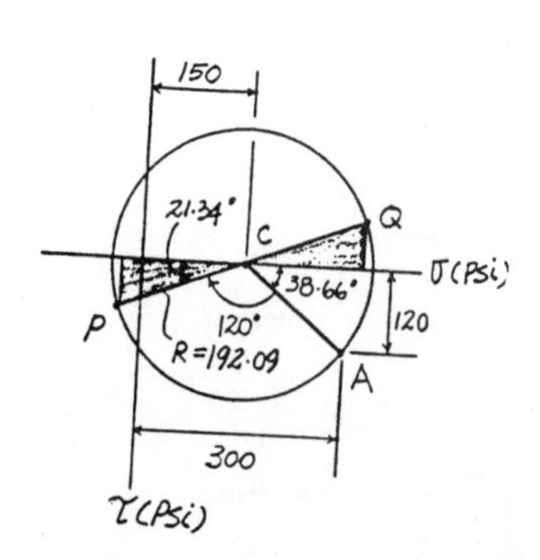

$$R = \sqrt{(300-150)^2 + 120^2} = 192.09 \text{ psi}$$

Stress on The Rotated Element : The normal and shear stress components ($\sigma_{x'}$ and $\tau_{x'y'}$) are represented by the coordinates of point P on the circle. $\sigma_{y'}$ can be determined by calculating the coordinates of point Q on the circle.

$$\sigma_{x'} = 150 - 192.09\cos 21.34° = -28.9 \text{ psi} \qquad \textbf{Ans}$$

$$\tau_{x'y'} = 192.09\sin 21.34° = 69.9 \text{ psi} \qquad \textbf{Ans}$$

$$\sigma_{y'} = 1.50 + 192.09\cos 21.34° = 329 \text{ psi} \qquad \textbf{Ans}$$

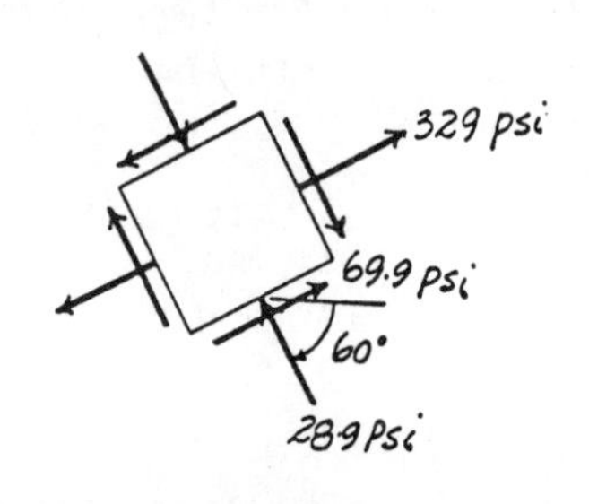

9–53. Solve Prob. 9–10 using Mohr's circle.

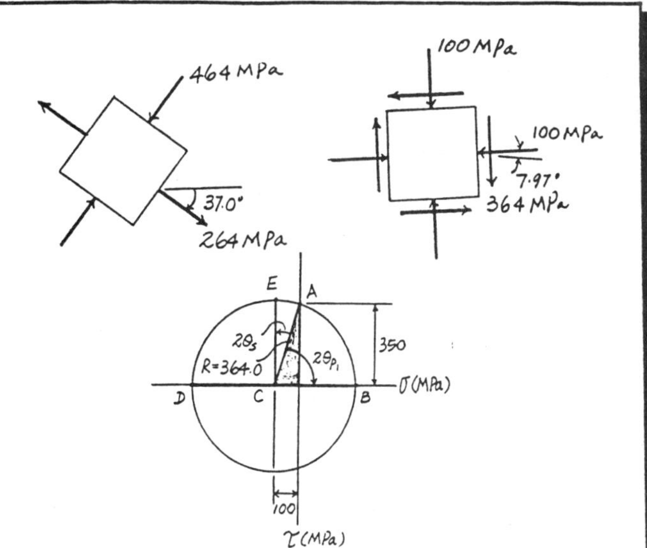

Construction of the Circle : In accordance with the sign convention, $\sigma_x = 0$, $\sigma_y = -200$ MPa, and $\tau_{xy} = -350$ MPa. Hence,

$$\sigma_{avg} = \frac{\sigma_x + \sigma_y}{2} = \frac{0 + (-200)}{2} = -100 \text{ MPa} \qquad \textbf{Ans}$$

The coordinates for reference point A and C are

$$A(0, -350) \qquad C(-100, 0)$$

The radius of the circle is

$$R = \sqrt{(0-100)^2 + 350^2} = 364.0 \text{ MPa}$$

a)

In-Plane Principal Stresses : The coordinates of points B and D represent σ_1 and σ_2, respectively.

$$\sigma_1 = -100 + 364.0 = 264 \text{ MPa} \qquad \textbf{Ans}$$
$$\sigma_2 = -100 - 364.0 = -464 \text{ MPa} \qquad \textbf{Ans}$$

Orientation of Principal Plane : From the circle

$$\tan 2\theta_{p_1} = \frac{350}{100} = 3.5000$$
$$\theta_{p_1} = 37.0° \ (\textit{Clockwise}) \qquad \textbf{Ans}$$

b)

Maximum In-Plane Shear Stress : Represented by the coordinates of point E on the circle.

$$\tau_{\substack{max \\ in\text{-}plane}} = -R = -364 \text{ MPa} \qquad \textbf{Ans}$$

Orientation of the Plane for Maximum In-Plane Shear Stress : From the circle

$$\tan 2\theta_s = \frac{100}{350} = 0.2857$$
$$\theta_s = 7.97° \ (\textit{Counterclockwise}) \qquad \textbf{Ans}$$

9–54. Solve Prob. 9–11 using Mohr's circle.

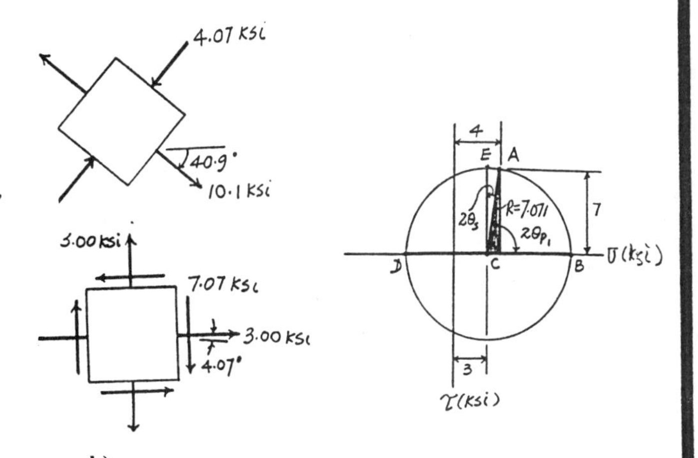

Construction of the Circle : In accordance with the sign convention, $\sigma_x = 4$ ksi, $\sigma_y = 2$ ksi, and $\tau_{xy} = -7$ ksi. Hence,

$$\sigma_{avg} = \frac{\sigma_x + \sigma_y}{2} = \frac{4 + 2}{2} = 3.00 \text{ ksi} \qquad \textbf{Ans}$$

The coordinates for reference point A and C are

$$A(4, -7) \qquad C(3, 0)$$

The radius of the circle is

$$R = \sqrt{(4-3)^2 + 7^2} = 7.071 \text{ ksi}$$

a)

In-Plane Principal Stresses : The coordinates of points B and D represent σ_1 and σ_2, respectively.

$$\sigma_1 = 3.00 + 7.071 = 10.1 \text{ ksi} \qquad \textbf{Ans}$$
$$\sigma_2 = 3.00 - 7.071 = -4.07 \text{ ksi} \qquad \textbf{Ans}$$

Orientation of Principal Plane : From the circle

$$\tan 2\theta_{p_1} = \frac{7}{4-3} = 3.5000$$
$$\theta_{p_1} = 40.9° \ (\textit{Clockwise}) \qquad \textbf{Ans}$$

b)

Maximum In-Plane Shear Stress : Represented by the coordinates of point E on the circle.

$$\tau_{\substack{max \\ in\text{-}plane}} = -R = -7.07 \text{ ksi} \qquad \textbf{Ans}$$

Orientation of the Plane for Maximum In-Plane Shear Stress : From the circle

$$\tan 2\theta_s = \frac{4-3}{7} = 0.1429$$
$$\theta_s = 4.07° \ (\textit{Counterclockwise}) \qquad \textbf{Ans}$$

9–55. Solve Prob. 9–13 using Mohr's circle.

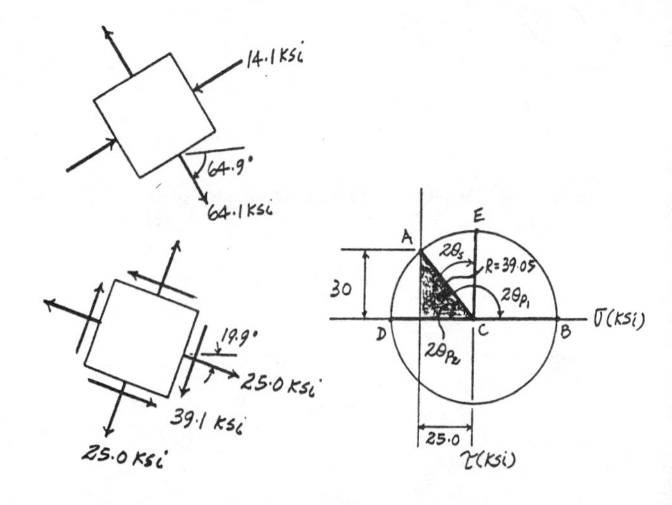

Construction of the Circle : In accordance with the sign convention, $\sigma_x = 0$, $\sigma_y = 50$ ksi, and $\tau_{xy} = -30$ ksi. Hence,

$$\sigma_{avg} = \frac{\sigma_x + \sigma_y}{2} = \frac{0+50}{2} = 25.0 \text{ ksi} \qquad \textbf{Ans}$$

The coordinates for reference points A and C are

$A(0, -30) \qquad C(25.0, 0)$

The radius of the circle is

$$R = \sqrt{(25.0-0)^2 + 30^2} = 39.05 \text{ ksi}$$

a)

In - Plane Principal Stresses : The coordinates of points B and D represent σ_1 and σ_2, respectively.

$$\sigma_1 = 25.0 + 39.05 = 64.1 \text{ ksi} \qquad \textbf{Ans}$$
$$\sigma_2 = 25.0 - 39.05 = -14.1 \text{ ksi} \qquad \textbf{Ans}$$

Orientation of Principal Plane : From the circle

$$\tan 2\theta_{p_2} = \frac{30}{25.0-0} = 1.200 \qquad 2\theta_{p_2} = 50.19°$$

$$2\theta_{p_1} = 180° - 2\theta_{p_2}$$
$$\theta_{p_1} = \frac{180° - 50.19°}{2} = 64.9° \ (\textit{Clockwise}) \qquad \textbf{Ans}$$

b)

Maximum In - Plane Shear Stress : Represented by the coordinates of point E on the circle.

$$\tau_{\substack{max \\ in\text{-}plane}} = -R = -39.1 \text{ ksi} \qquad \textbf{Ans}$$

Orientation of the Plane for Maximum In - Plane Shear Stress : From the circle

$$\tan 2\theta_s = \frac{25.0-0}{30} = 0.8333$$
$$\theta_s = 19.9° \ (\textit{Clockwise}) \qquad \textbf{Ans}$$

***9–56.** Solve Prob. 9–12 using Mohr's circle.

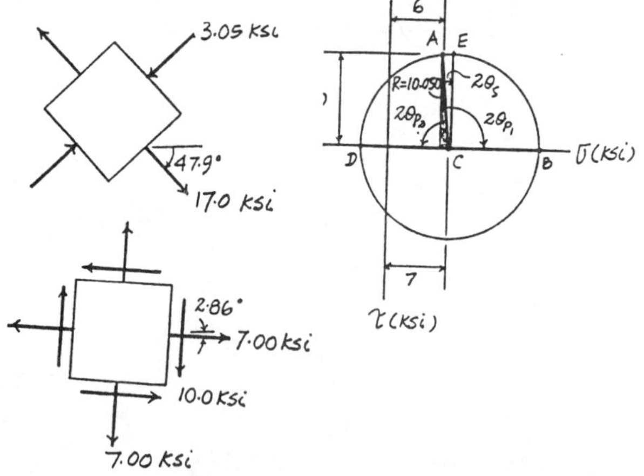

Construction of the Circle : In accordance with the sign convention, $\sigma_x = 6$ ksi, $\sigma_y = 8$ ksi, and $\tau_{xy} = -10$ ksi. Hence,

$$\sigma_{avg} = \frac{\sigma_x + \sigma_y}{2} = \frac{6+8}{2} = 7.00 \text{ ksi} \qquad \textbf{Ans}$$

The coordinates for reference points A and C are

$A(6, -10) \qquad C(7, 0)$

The radius of the circle is

$$R = \sqrt{(7-6)^2 + 10^2} = 10.050 \text{ ksi}$$

a)

In - Plane Principal Stresses : The coordinates of points B and D represent σ_1 and σ_2, respectively.

$$\sigma_1 = 7.00 + 10.050 = 17.0 \text{ ksi} \qquad \textbf{Ans}$$
$$\sigma_2 = 7.00 - 10.050 = -3.05 \text{ ksi} \qquad \textbf{Ans}$$

Orientation of Principal Plane : From the circle

$$\tan 2\theta_{p_2} = \frac{10}{7-6} = 10.0 \qquad 2\theta_{p_2} = 84.29°$$

$$2\theta_{p_1} = 180° - 2\theta_{p_2}$$
$$\theta_{p_1} = \frac{180° - 84.29°}{2} = 47.9° \ (\textit{Clockwise}) \qquad \textbf{Ans}$$

b)

Maximum In - Plane Shear Stress : Represented by the coordinates of point E on the circle.

$$\tau_{\substack{max \\ in\text{-}plane}} = -R = -10.0 \text{ ksi} \qquad \textbf{Ans}$$

Orientation of the Plane for Maximum In - Plane Shear Stress : From the circle

$$\tan 2\theta_s = \frac{7-6}{10} = 0.1000$$
$$\theta_s = 2.86° \ (\textit{Clockwise}) \qquad \textbf{Ans}$$

9–57. Solve Prob. 9–14 using Mohr's circle.

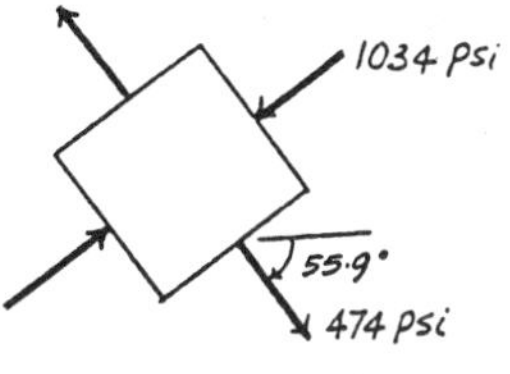

Construction of the Circle : In accordance with the sign convention, $\sigma_x = -560$ psi, $\sigma_y = 0$, and $\tau_{xy} = -700$ psi. Hence,

$$\sigma_{avg} = \frac{\sigma_x + \sigma_y}{2} = \frac{-560 + 0}{2} = -280 \text{ psi} \qquad \textbf{Ans}$$

The coordinates for reference points A and C are

$$A(-560, -700) \qquad C(-280, 0)$$

The radius of the circle is

$$R = \sqrt{(560 - 280)^2 + 700^2} = 753.9 \text{ psi}$$

a)

In - Plane Principal Stress : The coordinates of points B and D represent σ_1 and σ_2 respectively.

$$\sigma_1 = -280 + 753.9 = 474 \text{ psi} \qquad \textbf{Ans}$$

$$\sigma_2 = -280 - 753.9 = -1034 \text{ psi} \qquad \textbf{Ans}$$

Orientation of Principal Plane : From the circle

$$\tan 2\theta_{p_2} = \frac{700}{560 - 280} = 2.500 \qquad 2\theta_{p_2} = 68.20°$$

$$2\theta_{p_1} = 180° - 2\theta_{p_2}$$

$$\theta_{p_1} = \frac{180° - 68.20°}{2} = 55.9° \quad \textit{(Clockwise)} \qquad \textbf{Ans}$$

b)

Maximum In - Plane Shear Stress : Represented by the coordinates of point E on the circle.

$$\tau_{\substack{\max \\ \text{in-plane}}} = -R = -754 \text{ psi} \qquad \textbf{Ans}$$

Orientation of the Plane for Maximum In - Plane Shear Stress : From the circle

$$\tan 2\theta_s = \frac{560 - 280}{700} = 0.4000$$

$$\theta_s = 10.9° \quad \textit{(Clockwise)} \qquad \textbf{Ans}$$

9–58. Solve Prob. 9–15 using Mohr's circle.

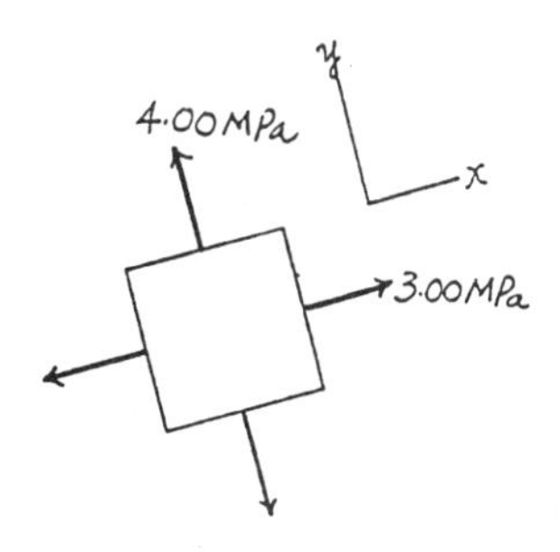

Construction of the Circle : In accordance with the sign convention, $\sigma_x = \frac{30(10^3)}{0.01} = 3.00$ MPa, $\sigma_y = \frac{40(10^3)}{0.01} = 4.00$ MPa, and $\tau_{xy} = 0$. Hence,

$$\sigma_{avg} = \frac{\sigma_x + \sigma_y}{2} = \frac{3.00 + 4.00}{2} = 3.50 \text{ MPa} \qquad \textbf{Ans}$$

The coordinates for reference points A and C are

$$A(3.00, 0) \qquad C(3.50, 0)$$

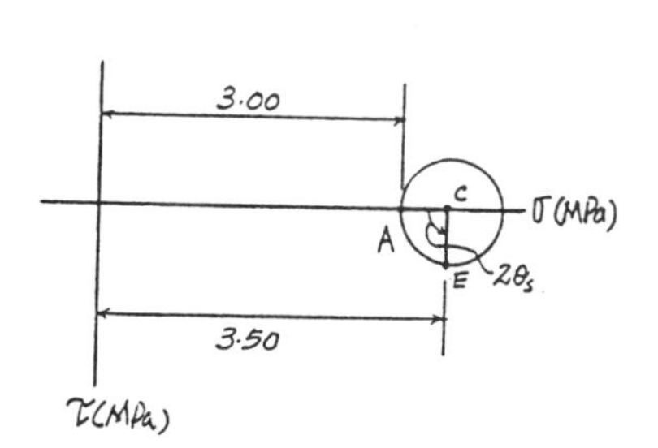

The radius of the circle is

$$R = 3.50 - 3.00 = 0.500 \text{ MPa}$$

Maximum In - Plane Shear Stress : Represented by the coordinates of point E on the circle.

$$\tau_{\substack{\max \\ \text{in-plane}}} = R = 0.500 \text{ MPa} \qquad \textbf{Ans}$$

9–59. Determine the equivalent state of stress if an element is oriented 20° clockwise from the element shown. Show the result on the element.

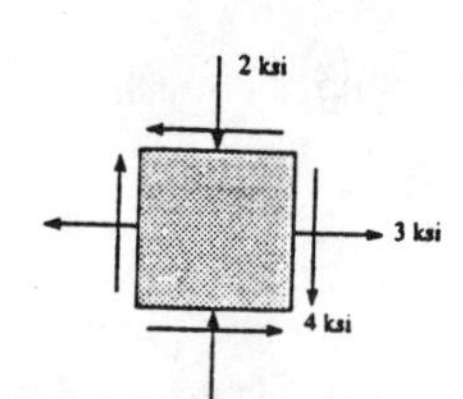

Construction of the Circle : In accordance with the sign convention, $\sigma_x = 3$ ksi, $\sigma_y = -2$ ksi and $\tau_{xy} = -4$ ksi. Hence,

$$\sigma_{avg} = \frac{\sigma_x + \sigma_y}{2} = \frac{3+(-2)}{2} = 0.500 \text{ ksi}$$

The coordinates for reference points A and C are

$$A(3,\ -4) \qquad C(0.500,\ 0)$$

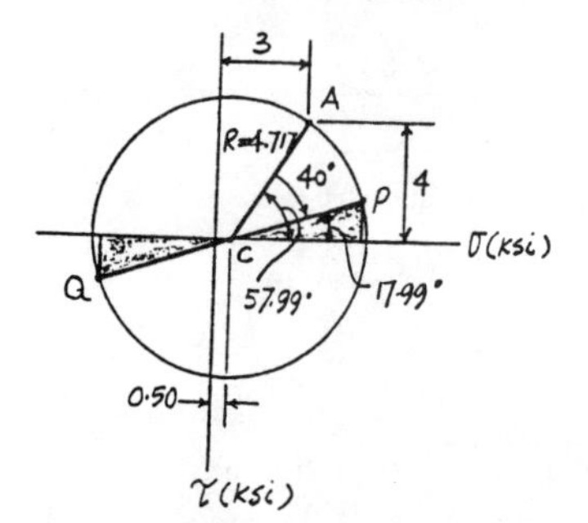

The radius of the circle is

$$R = \sqrt{(3-0.500)^2 + 4^2} = 4.717 \text{ ksi}$$

Stress on The Rotated Element : The normal and shear stress components ($\sigma_{x'}$ and $\tau_{x'y'}$) are represented by the coordinates of point P on the circle. $\sigma_{y'}$ can be determined by calculating the coordinates of point Q on the circle.

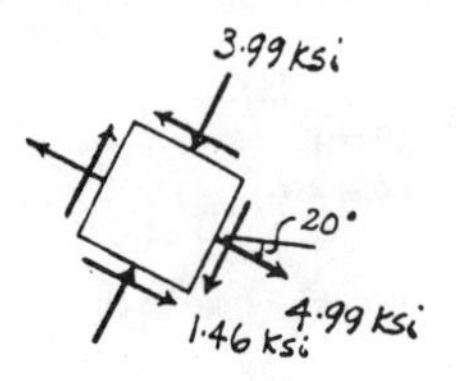

$\sigma_{x'} = 0.500 + 4.717\cos 17.99° = 4.99$ ksi **Ans**

$\tau_{x'y'} = -4.717\sin 17.99° = -1.46$ ksi **Ans**

$\sigma_{y'} = 0.500 - 4.717\cos 17.99° = -3.99$ ksi **Ans**

***9–60.** Determine the equivalent state of stress if an element is oriented 30° clockwise from the element shown. Show the result on the element.

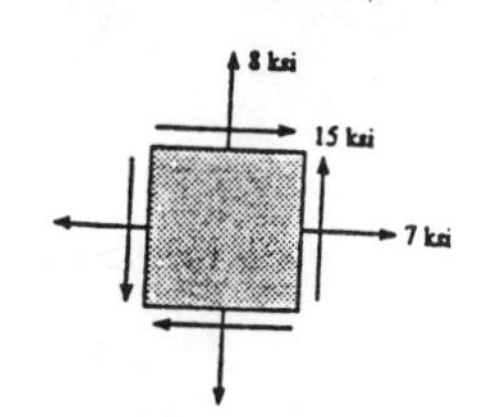

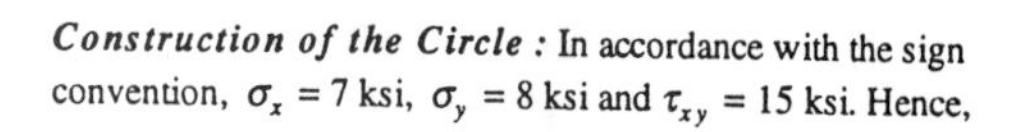

Construction of the Circle : In accordance with the sign convention, $\sigma_x = 7$ ksi, $\sigma_y = 8$ ksi and $\tau_{xy} = 15$ ksi. Hence,

$$\sigma_{avg} = \frac{\sigma_x + \sigma_y}{2} = \frac{7+8}{2} = 7.50 \text{ ksi}$$

The coordinates for reference points A and C are

$$A(7,\ 15) \qquad C(7.50,\ 0)$$

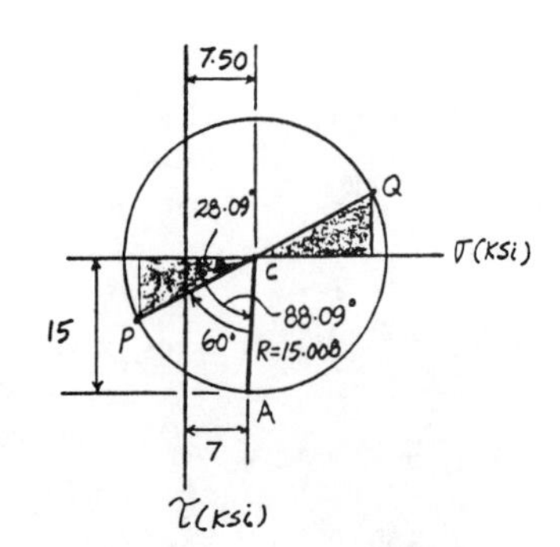

The radius of the circle is

$$R = \sqrt{(7.50-7)^2 + 15^2} = 15.008 \text{ ksi}$$

Stresses on The Rotated Element : The normal and shear stress components ($\sigma_{x'}$ and $\tau_{x'y'}$) are represented by the coordinates of point P on the circle. $\sigma_{y'}$ can be determined by calculating the coordinates of point Q on the circle.

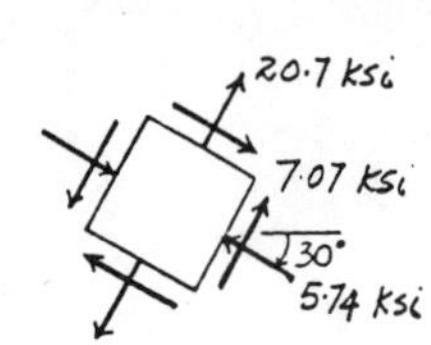

$\sigma_{x'} = 7.50 - 15.008\cos 28.09° = -5.74$ ksi **Ans**

$\tau_{x'y'} = 15.008\sin 28.09° = 7.07$ ksi **Ans**

$\sigma_{y'} = 7.50 + 15.008\cos 28.09° = 20.7$ ksi **Ans**

9–61. Determine (a) the principal stress and (b) the maximum in-plane shear stress and average normal stress. Specify the orientation of the element in each case.

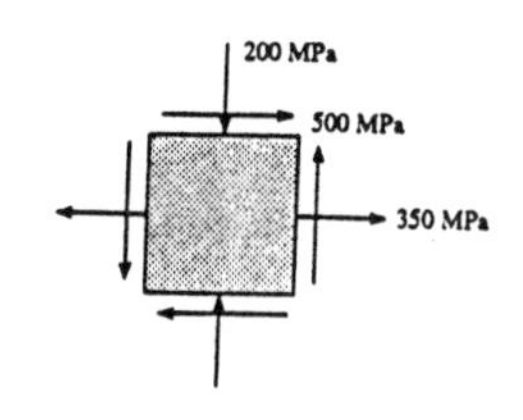

Construction of the Circle : In accordance with the sign convention, $\sigma_x = 350$ MPa, $\sigma_y = -200$ MPa, and $\tau_{xy} = 500$ MPa. Hence,

$$\sigma_{avg} = \frac{\sigma_x + \sigma_y}{2} = \frac{350 + (-200)}{2} = 75.0 \text{ MPa} \qquad \textbf{Ans}$$

The coordinates for reference points A and C are

$$A(350,\ 500) \qquad C(75.0,\ 0)$$

The radius of the circle is

$$R = \sqrt{(350 - 75.0)^2 + 500^2} = 570.64 \text{ MPa}$$

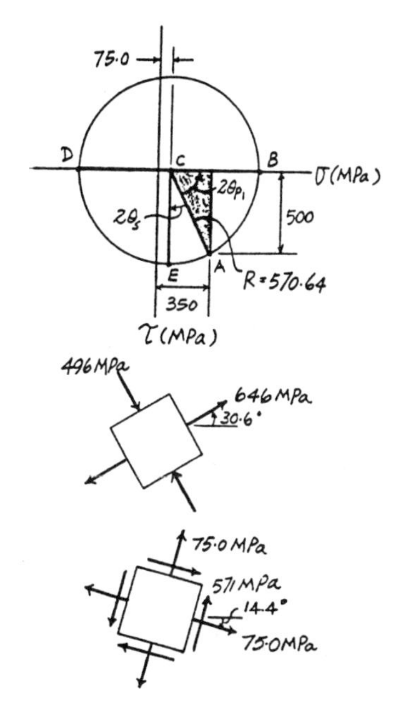

a)
In-Plane Principal Stresses : The coordinate of points B and D represent σ_1 and σ_2 respectively.

$$\sigma_1 = 75.0 + 570.64 = 646 \text{ MPa} \qquad \textbf{Ans}$$
$$\sigma_2 = 75.0 - 570.64 = -496 \text{ MPa} \qquad \textbf{Ans}$$

Orientation of Principal Plane : From the circle

$$\tan 2\theta_{p_1} = \frac{500}{350 - 75.0} = 3.5000$$
$$\theta_{p_1} = 30.6° \ (\textit{Counterclockwise}) \qquad \textbf{Ans}$$

b)
Maximum In-Plane Shear Stress : Represented by the coordinates of point E on the circle.

$$\tau_{\substack{max \\ in\text{-}plane}} = R = 571 \text{ MPa} \qquad \textbf{Ans}$$

Orientation of the Plane for Maximum In-Plane Shear Stress : From the circle

$$\tan 2\theta_s = \frac{350 - 75.0}{500} = 0.2857$$
$$\theta_s = 14.4° \ (\textit{Clockwise}) \qquad \textbf{Ans}$$

9–62. Determine (a) the principal stress and (b) the maximum in-plane shear stress and average normal stress. Specify the orientation of the element in each case.

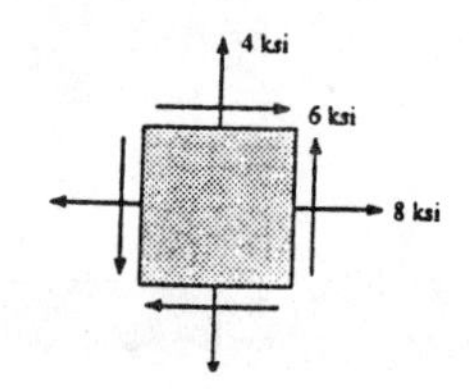

Construction of the Circle : In accordance with the sign convention, $\sigma_x = 8$ ksi, $\sigma_y = 4$ ksi, and $\tau_{xy} = 6$ ksi. Hence,

$$\sigma_{avg} = \frac{\sigma_x + \sigma_y}{2} = \frac{8+4}{2} = 6.00 \text{ ksi} \qquad \textbf{Ans}$$

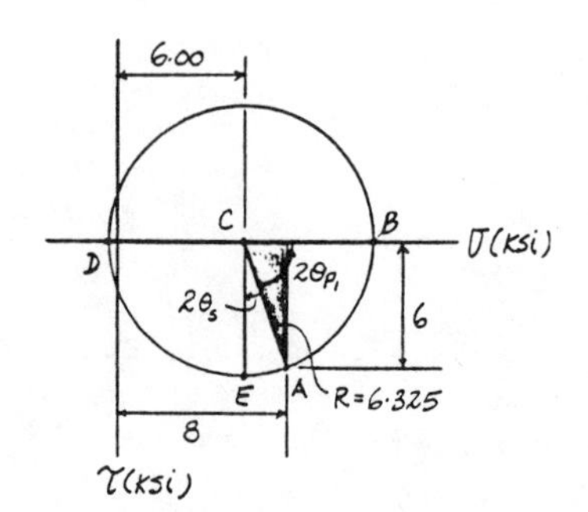

The coordinates for reference point A and C are

$$A(8,\ 6) \qquad C(6.00,\ 0)$$

The radius of the circle is

$$R = \sqrt{(8-6.00)^2 + 6^2} = 6.325 \text{ ksi}$$

a)

In - Plane Principal Stresses : The coordinates of points B and D represent σ_1 and σ_2, respectively.

$$\sigma_1 = 6.00 + 6.325 = 12.3 \text{ ksi} \qquad \textbf{Ans}$$

$$\sigma_2 = 6.00 - 6.325 = -0.325 \text{ ksi} \qquad \textbf{Ans}$$

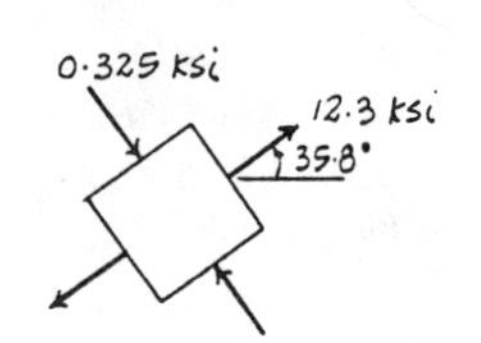

Orientation of Principal Plane : From the circle

$$\tan 2\theta_{p_1} = \frac{6}{8-6.00} = 3.00$$

$$\theta_{p_1} = 35.8° \ (\textit{Counterclockwise}) \qquad \textbf{Ans}$$

b)

Maximum In - Plane Shear Stress : Represented by the coordinates of point E on the circle.

$$\tau_{\substack{max \\ in\text{-}plane}} = R = 6.32 \text{ ksi} \qquad \textbf{Ans}$$

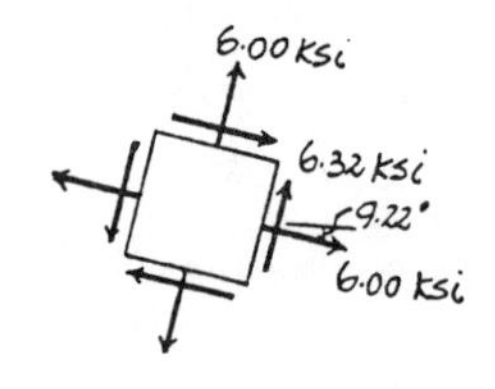

Orientation of the Plane for Maximum In - Plane Shear Stress : From the circle

$$\tan 2\theta_s = \frac{8-6.00}{6} = 0.3333$$

$$\theta_s = 9.22° \ (\textit{Clockwise}) \qquad \textbf{Ans}$$

9–63. Determine (a) the principal stress and (b) the maximum in-plane shear stress and average normal stress. Specify the orientation of the element in each case.

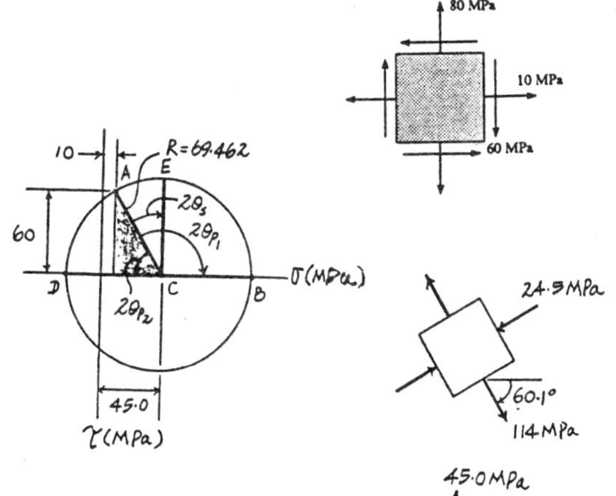

Construction of the Circle : In accordance with the sign convention, $\sigma_x = 10$ MPa, $\sigma_y = 80$ MPa and $\tau_{xy} = -60$ MPa. Hence,

$$\sigma_{avg} = \frac{\sigma_x + \sigma_y}{2} = \frac{10+80}{2} = 45.0 \text{ MPa} \qquad \textbf{Ans}$$

The coordinates for reference points A and C are

$$A(10,\ -60) \qquad C(45.0,\ 0)$$

The radius of circle is

$$R = \sqrt{(45.0-10)^2 + 60^2} = 69.462 \text{ MPa}$$

a)

In - Plane Principal Stress : The coordinate of points B and D represent σ_1 and σ_2 respectively.

$$\sigma_1 = 45.0 + 69.462 = 114 \text{ MPa} \qquad \textbf{Ans}$$
$$\sigma_2 = 45.0 - 570.64 = -24.5 \text{ MPa} \qquad \textbf{Ans}$$

Orientation of Principal Plane : From the circle

$$\tan 2\theta_{p_2} = \frac{60}{45.0-10} = 1.7143 \qquad 2\theta_{p_2} = 59.74$$

$$2\theta_{p_1} = 180° - 2\theta_{p_2}$$
$$\theta_{p_1} = \frac{180° - 59.74°}{2} = 60.1° \ \ (\textit{Clockwise}) \qquad \textbf{Ans}$$

b)

Maximum In - Plane Shear Stress : Represented by the coordinate of point E on the circle.

$$\tau_{\substack{max \\ in\text{-}plane}} = -R = -69.5 \text{ MPa} \qquad \textbf{Ans}$$

Orientation of the Plane for Maximum In - Plane Shear Stress : From the circle

$$\tan 2\theta_s = \frac{45.0-10}{60} = 0.5833$$
$$\theta_s = 15.1° \ \ (\textit{Clockwise}) \qquad \textbf{Ans}$$

***9–64.** The square steel plate has a thickness of 0.5 in. and is subjected to the edge loading shown. Determine the principal stresses developed in the steel.

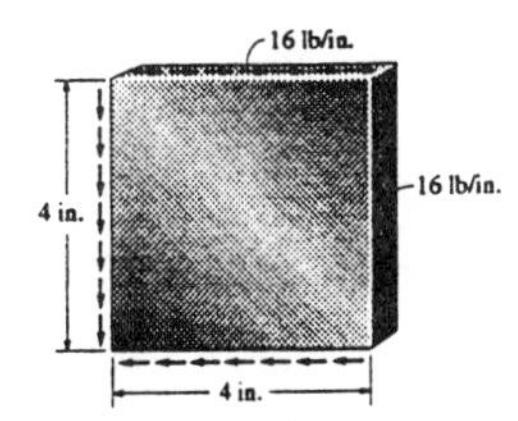

Construction of the Circle : In accordance with the sign convention, $\sigma_x = 0$, $\sigma_y = 0$, and $\tau_{xy} = \frac{16}{0.5} = 32.0$ psi. Hence,

$$\sigma_{avg} = \frac{\sigma_x + \sigma_y}{2} = \frac{0+0}{2} = 0$$

The coordinates for reference points A and C are

$$A(0,\ 32.0) \qquad C(0,\ 0)$$

The radius of the circle is

$$R = \sqrt{0 + 32.0^2} = 32.0 \text{ psi}$$

In - Plane Principal Stresses : The coordinates of points B and D represent σ_1 and σ_2, respectively.

$$\sigma_1 = 0 + 32.0 = 32.0 \text{ psi} \qquad \textbf{Ans}$$
$$\sigma_2 = 0 - 32.0 = -32.0 \text{ psi} \qquad \textbf{Ans}$$

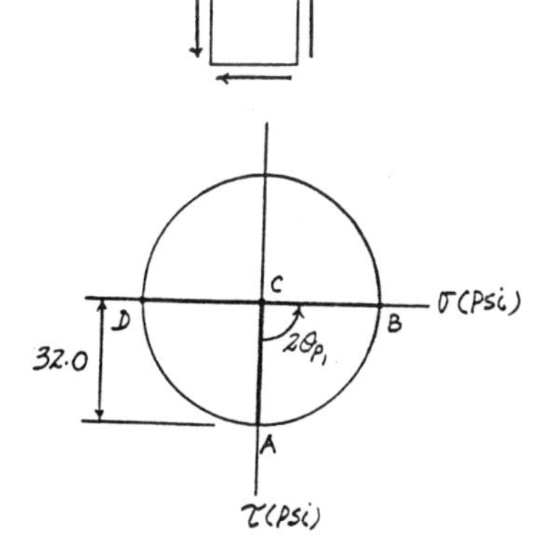

9–65. Determine (a) the principal stress and (b) the maximum in-plane shear stress and average normal stress. Specify the orientation of the element in each case.

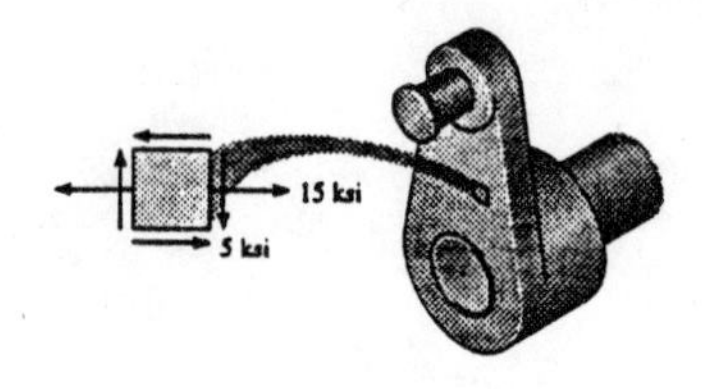

Construction of the Circle : In accordance with the sign convention, $\sigma_x = 15$ ksi, $\sigma_y = 0$ and $\tau_{xy} = -5$ ksi. Hence,

$$\sigma_{avg} = \frac{\sigma_x + \sigma_y}{2} = \frac{15+0}{2} = 7.50 \text{ ksi} \qquad \textbf{Ans}$$

The coordinates for reference point A and C are

$$A(15, \ -5) \qquad C(7.50, \ 0)$$

The radius of the circle is

$$R = \sqrt{(15-7.50)^2 + 5^2} = 9.014 \text{ ksi}$$

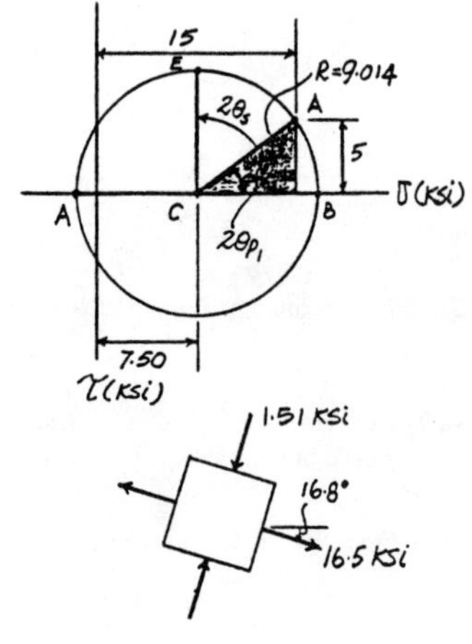

a)

In - Plane Principal Stress : The coordinates of points B and D represent σ_1 and σ_2, respectively.

$$\sigma_1 = 7.50 + 9.014 = 16.5 \text{ ksi} \qquad \textbf{Ans}$$
$$\sigma_2 = 7.50 - 9.014 = -1.51 \text{ ksi} \qquad \textbf{Ans}$$

Orientaion of Principal Plane : From the circle

$$\tan 2\theta_{p_1} = \frac{5}{15-7.50} = 0.6667$$
$$\theta_{p_1} = 16.8° \ (\textit{Clockwise}) \qquad \textbf{Ans}$$

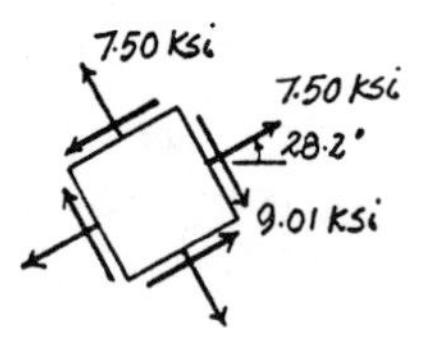

b)

Maximum In - Plane Shear Stress : Represented by the coordinates of point E on the circle.

$$\tau_{\substack{max \\ in\text{-}plane}} = -R = -9.01 \text{ ksi} \qquad \textbf{Ans}$$

Orientation of the Plane for Maximum In - Plane Shear Stress : From the circle

$$\tan 2\theta_s = \frac{15-7.50}{5} = 1.500$$
$$\theta_s = 28.2° \ (\textit{Counterclockwise}) \qquad \textbf{Ans}$$

9–66. Draw Mohr's circle that describes each of the following states of stress.

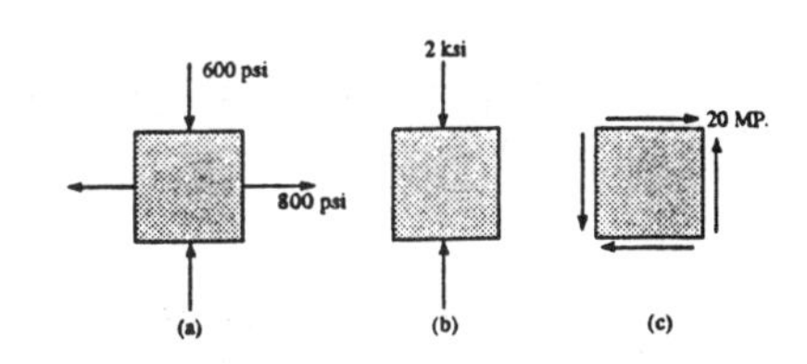

a) ***Construction of the Circle :*** In accordance with the sign convention, $\sigma_x = 800$ psi, $\sigma_y = -600$ psi, and $\tau_{xy} = 0$. Hence,

$$\sigma_{avg} = \frac{\sigma_x + \sigma_y}{2} = \frac{800 + (-600)}{2} = 100 \text{ psi}$$

The coordinates for reference point A and C are

$$A(800,\ 0) \qquad C(100,\ 0)$$

The radius of the circle is $\quad R = 800 - 100 = 700$ psi

b) ***Construction of the Circle :*** In accordance with the sign convention, $\sigma_x = 0$, $\sigma_y = -2$ ksi and $\tau_{xy} = 0$. Hence,

$$\sigma_{avg} = \frac{\sigma_x + \sigma_y}{2} = \frac{0 + (-2)}{2} = -1.00 \text{ ksi}$$

The coordinates for reference points A and C are

$$A(0,\ 0) \qquad C(-1.00,\ 0)$$

The radius of the circle is $\quad R = 1.00 - 0 = 1.00$ ksi

c) ***Construction of the Circle :*** In accordance with the sign convention $\sigma_x = \sigma_y = 0$ and $\tau_{xy} = 20$ MPa. Hence,

$$\sigma_{avg} = \frac{\sigma_x + \sigma_y}{2} = 0$$

The coordinates for reference points A and C are

$$A(0,\ 20) \qquad C(0,\ 0)$$

The radius of the circle is $\quad R = 20.0$ MPa

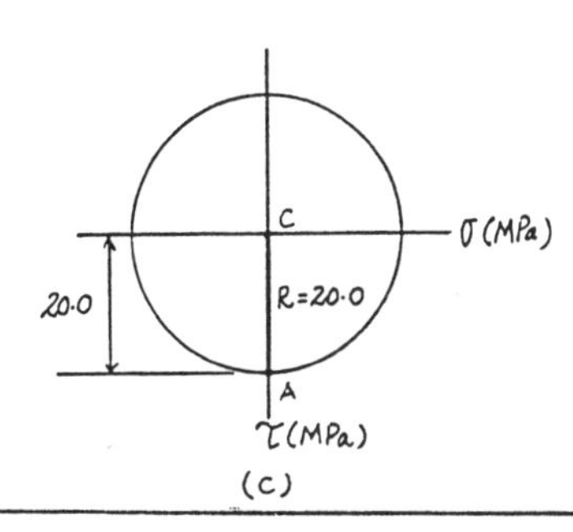

9–67. Draw Mohr's circle that describes each of the following states of stress.

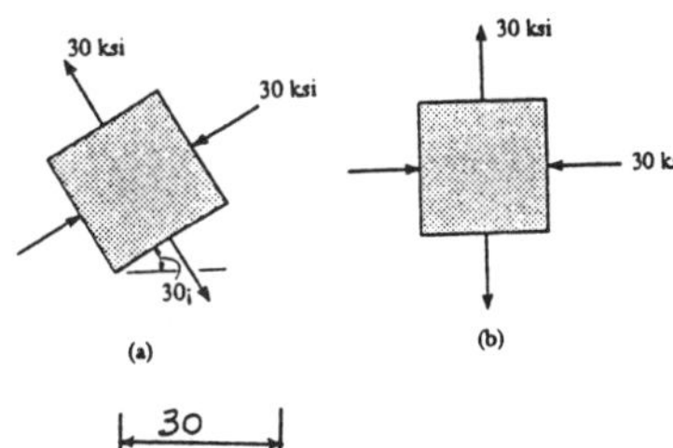

a, b) ***Construction of the Circle :*** In accordance with the sign convention, $\sigma_x = -30$ ksi, $\sigma_y = 30$ ksi, and $\tau_{xy} = 0$. Hence,

$$\sigma_{avg} = \frac{\sigma_x + \sigma_y}{2} = \frac{-30 + 30}{2} = 0$$

The coordinates for reference points A and C are

$$A(-30,\ 0) \qquad C(0,\ 0)$$

The radius of the circle is $R = 30 - 0 = 30.0$ ksi

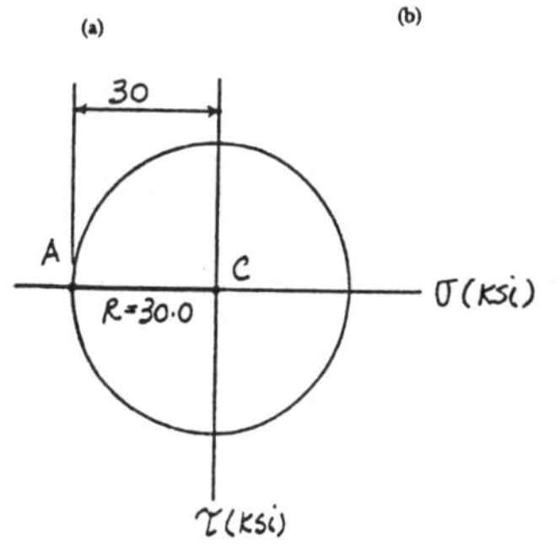

***9–68.** A point on a thin plate is subjected to two successive states of stress as shown. Determine the resulting state of stress with reference to an element oriented as shown at the right.

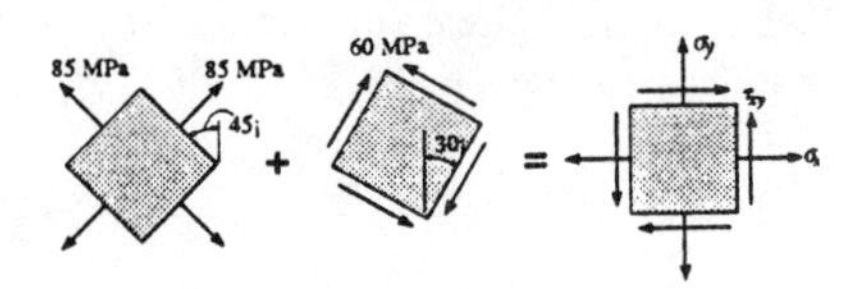

Construction of the Circle : In accordance with the sign convention, $\sigma_{x'} = 85$ MPa, $\sigma_{y'} = 85$ MPa and $\tau_{x'y'} = 0$.

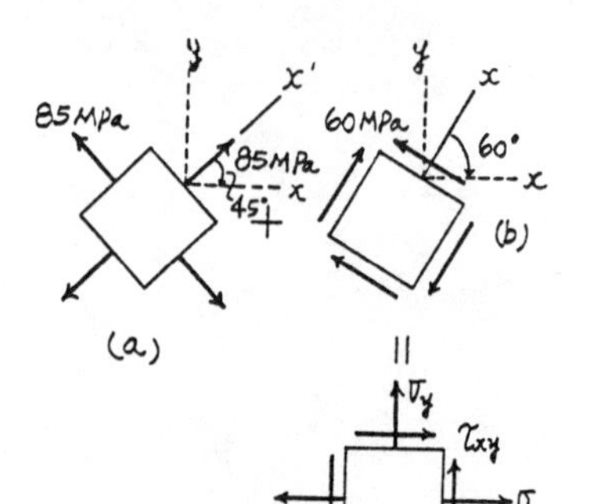

Hence,

$$\sigma_{avg} = \frac{\sigma_{x'} + \sigma_{y'}}{2} = \frac{85+85}{2} = 85.0 \text{ MPa}$$

The coordinates for reference points A and C are

$A(85, 0)$ $\quad C(85.0, 0)$

The radius ofthe circle is $R = 0$. Therefore, Mohr's circle is simply a dot at C. As the result, the state of stress is the same regardless of the orientation of the element.

$$(\sigma_x)_a = 85.0 \text{ MPa} \quad (\sigma_y)_a = 85.0 \text{ MPa} \quad (\tau_{xy})_a = 0$$

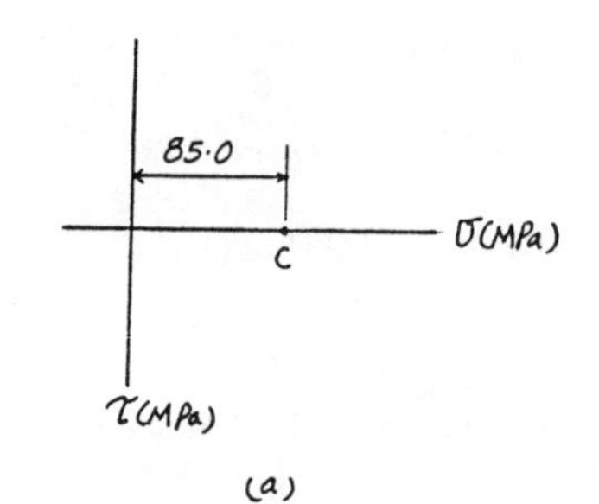

Construction of the Circle : In accordance with the sign convention, $\sigma_{x'} = \sigma_{y'} = 0$, and $\tau_{x'y'} = 60$ MPa for element (b). Hence,

$$\sigma_{avg} = \frac{\sigma_{x'} + \sigma_{y'}}{2} = 0$$

The coordinates for reference points A and C are

$A(0, 60)$ $\quad C(0, 0)$

The radius of the circle is $R = 60 - 0 = 60.0$ MPa

Stress on The Rotated Element : The normal and shear stress components $(\sigma_x)_b$, $(\tau_{xy})_b$ are represented by the coordinates of point P on the circle. $(\sigma_y)_b$ can be determined by calculating the coordinates of Q on the circle.

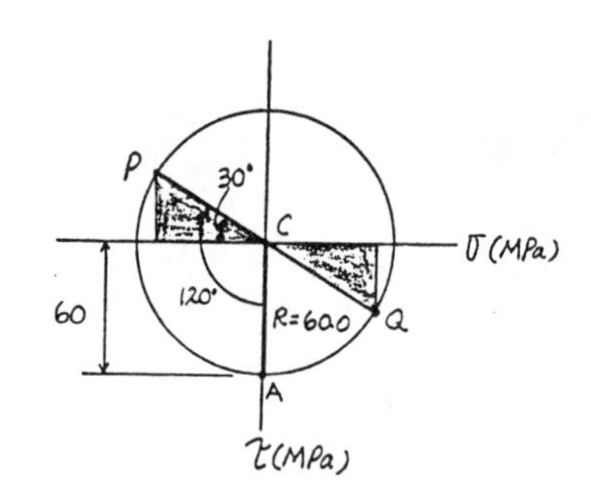

$$(\sigma_x)_b = 0 - 60.0\cos 30° = -51.96 \text{ MPa}$$
$$(\tau_{xy})_b = -60.0\sin 30° = -30.0 \text{ MPa}$$
$$(\sigma_y)_b = 0 + 60.0\cos 30° = 51.96 \text{ MPa}$$

Combine the stress components of the two elements.

$$\sigma_x = (\sigma_x)_a + (\sigma_x)_b = 85.0 + (-51.96) = 33.0 \text{ MPa} \qquad \textbf{Ans}$$
$$\sigma_y = (\sigma_y)_a + (\sigma_y)_b = 85.0 + 51.96 = 137 \text{ MPa} \qquad \textbf{Ans}$$
$$\tau_{xy} = (\tau_{xy})_a + (\tau_{xy})_b = 0 + (-30.0) = -30.0 \text{ MPa} \qquad \textbf{Ans}$$

9–69. Mohr's circle for the state of stress in Fig. 9–15*a* is shown in Fig. 9–15*b*. Show that the coordinates of point $P(\sigma_{x'}, \tau_{x'y'})$ on the circle gives the same value as the stress-transformation Eqs. 9–1 and 9–2.

Construction of the Circle : $\sigma_{avg} = \frac{\sigma_x + \sigma_y}{2}$ and the coordinates for reference point sA and C are

$$A(\sigma_x,\ \tau_{xy}) \qquad C\left(\frac{\sigma_x + \sigma_y}{2},\ 0\right)$$

The radius of the circle is

$$R = \sqrt{\left(\sigma_x - \frac{\sigma_x + \sigma_y}{2}\right)^2 + \tau_{xy}^2} = \sqrt{\left(\frac{\sigma_x - \sigma_y}{2}\right)^2 + \tau_{xy}^2}$$

Stress on The Rotated Element : The normal and shear stress components ($\sigma_{x'}$ and $\tau_{x'y'}$) are represented by the coordinates of point P on the circle. $\sigma_{y'}$ can be determined by calculating the coordinates of Q on the circle.

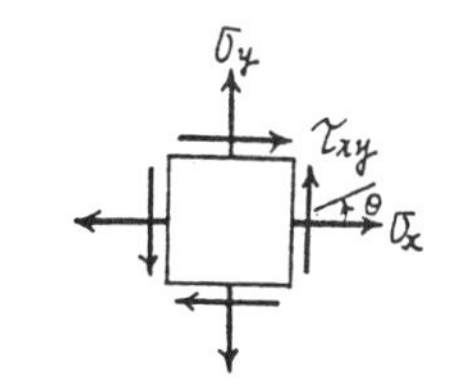

$$\sigma_{x'} = \frac{\sigma_x + \sigma_y}{2} + \left(\sqrt{\left(\frac{\sigma_x - \sigma_y}{2}\right)^2 + \tau_{xy}^2}\right)\cos\psi \qquad [1]$$

$$\tau_{x'y'} = \left(\sqrt{\left(\frac{\sigma_x - \sigma_y}{2}\right)^2 + \tau_{xy}^2}\right)\sin\psi \qquad [2]$$

However, $\psi = \phi - 2\theta$

$$\cos(\phi - 2\theta) = \cos\phi\cos 2\theta + \sin\phi\sin 2\theta \qquad [3]$$
$$\sin(\phi - 2\theta) = \sin\phi\cos 2\theta - \sin 2\theta\cos\phi \qquad [4]$$

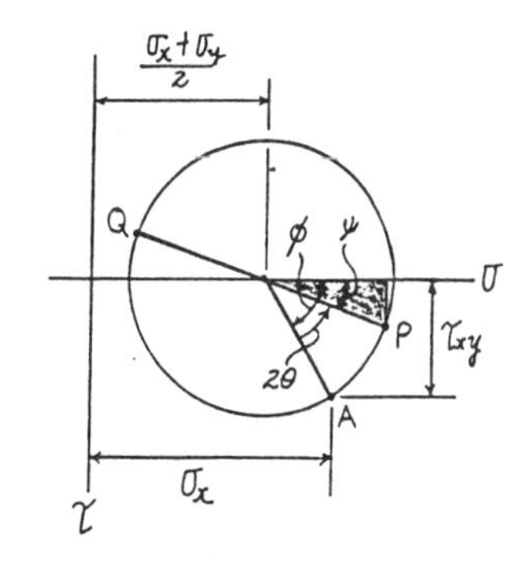

From the circle,

$$\cos\phi = \frac{\frac{\sigma_x - \sigma_y}{2}}{\sqrt{\left(\frac{\sigma_x - \sigma_y}{2}\right)^2 + \tau_{xy}^2}} = \frac{\sigma_x - \sigma_y}{2\sqrt{\left(\frac{\sigma_x - \sigma_y}{2}\right)^2 + \tau_{xy}^2}}$$

$$\sin\phi = \frac{\tau_{xy}}{\sqrt{\left(\frac{\sigma_x - \sigma_y}{2}\right)^2 + \tau_{xy}^2}} \qquad [6]$$

Substitute Eqs.[3] and [5] into [6] into [1] yield

$$\sigma_{x'} = \frac{\sigma_x + \sigma_y}{2} + \frac{\sigma_x - \sigma_y}{2}\cos 2\theta + \tau_{xy}\sin 2\theta \qquad (Q.E.D.)$$

Substitute Eq.[4] and [5] into [6] into [2] yield

$$\tau_{x'y'} = -\frac{\sigma_x - \sigma_y}{2}\sin 2\theta + \tau_{xy}\cos 2\theta \qquad (Q.E.D.)$$

9–70. The cantilevered rectangular bar is subjected to the force of 5 kip. Determine the principal stresses at point A.

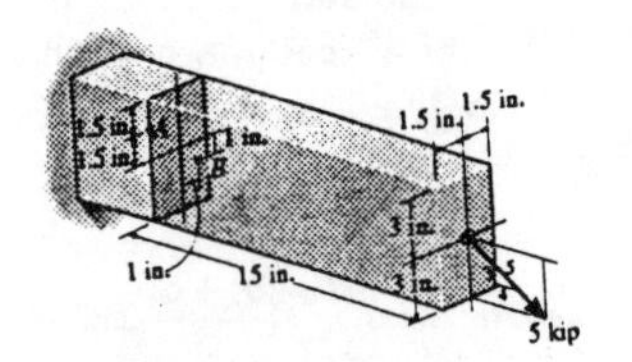

Internal Forces and Moment : As shown on FBD.

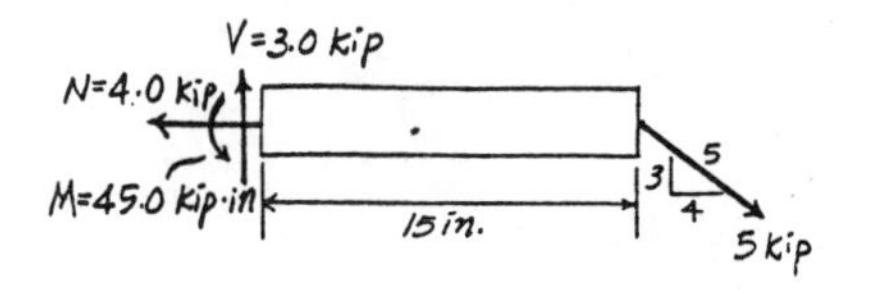

Section Properties :

$$A = 3(6) = 18.0 \text{ in}^2$$

$$I = \frac{1}{12}(3)\left(6^3\right) = 54.0 \text{ in}^4$$

$$Q_A = \bar{y}'A' = 2.25(1.5)(3) = 10.125 \text{ in}^3$$

Normal Stress :

$$\sigma = \frac{N}{A} \pm \frac{My}{I}$$

$$\sigma_A = \frac{4.00}{18.0} + \frac{45.0(1.5)}{54.0} = 1.4722 \text{ ksi}$$

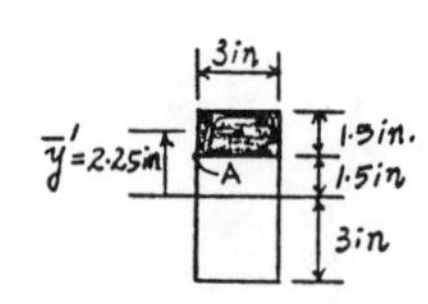

Shear Stress : Applying the shear formula $\tau = \frac{VQ}{It}$,

$$\tau_A = \frac{3.00(10.125)}{54.0(3)} = 0.1875 \text{ ksi}$$

Construction of the Circle : In accordance with the sign convention. $\sigma_x = 1.4722$ ksi, $\sigma_y = 0$, and $\tau_{xy} = -0.1875$ ksi. Hence,

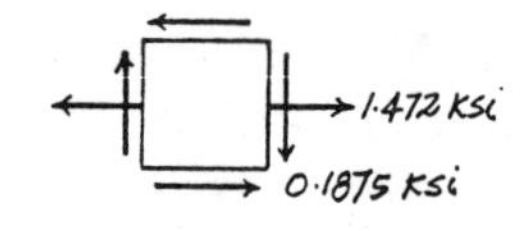

$$\sigma_{avg} = \frac{\sigma_x + \sigma_y}{2} = \frac{1.472 + 0}{2} = 0.7361 \text{ ksi}$$

The coordinates for reference points A and C are

$$A(1.4722,\ -0.1875) \qquad C(0.7361,\ 0)$$

The radius of the circle is

$$R = \sqrt{(1.4722 - 0.7361)^2 + 0.1875^2} = 0.7596 \text{ ksi}$$

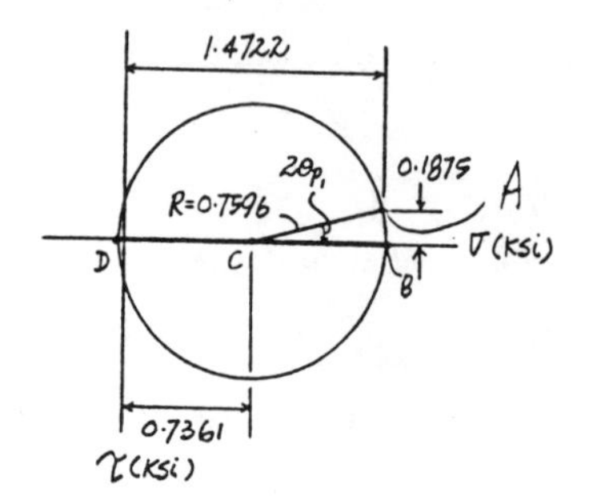

In - Plane Principal Stress : The coordinates of points B and D represent σ_1 and σ_2, respectively.

$$\sigma_1 = 0.7361 + 0.7596 = 1.50 \text{ ksi} \qquad \textbf{Ans}$$

$$\sigma_2 = 0.7361 - 0.7596 = -0.0235 \text{ ksi} \qquad \textbf{Ans}$$

9–71. Solve Prob. 9–70 for the principal stresses at point B.

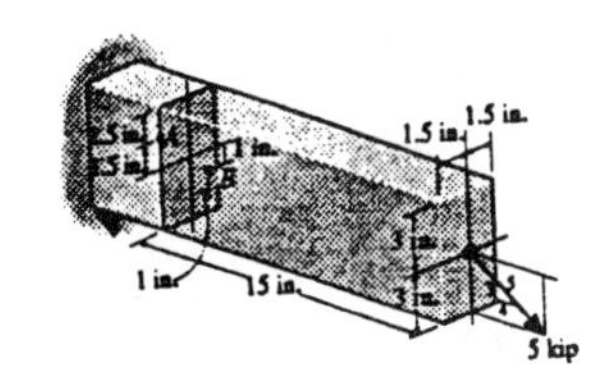

Internal Forces and Moment : As shown on FBD.

Section Properties :

$$A = 3(6) = 18.0 \text{ in}^2$$

$$I = \frac{1}{12}(3)(6^3) = 54.0 \text{ in}^4$$

$$Q_B = \bar{y}'A' = 2(2)(3) = 12.0 \text{ in}^3$$

Normal Stress :

$$\sigma = \frac{N}{A} \pm \frac{My}{I}$$

$$\sigma_B = \frac{4.00}{18.0} - \frac{45.0(1)}{54.0} = -0.6111 \text{ ksi}$$

Shear Stress : Applying the shear formula $\tau = \frac{VQ}{It}$.

$$\tau_B = \frac{3.00(12.0)}{54.0(3)} = 0.2222 \text{ ksi}$$

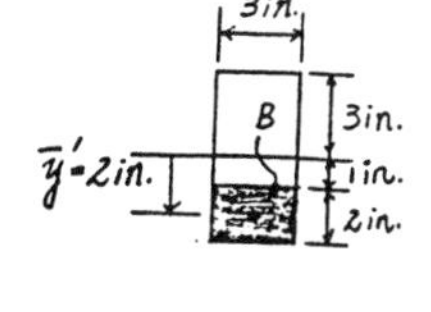

Construction of the Circle : In accordance with the sign convention, $\sigma_x = -0.6111$ ksi, $\sigma_y = 0$, and $\tau_{xy} = -0.2222$ ksi. Hence,

$$\sigma_{avg} = \frac{\sigma_x + \sigma_y}{2} = \frac{-0.6111 + 0}{2} = -0.3055 \text{ ksi}$$

The coordinates for reference points A and C are

$$A(-0.6111,\ -0.2222) \qquad C(-0.3055,\ 0)$$

The radius of the circle is

$$R = \sqrt{(0.6111 - 0.3055)^2 + 0.2222^2} = 0.3778 \text{ ksi}$$

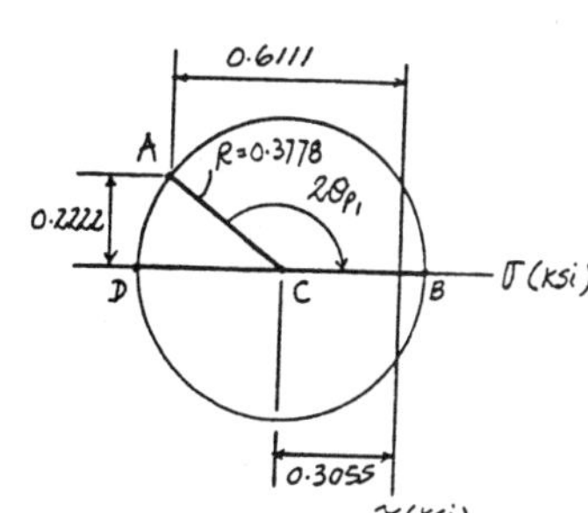
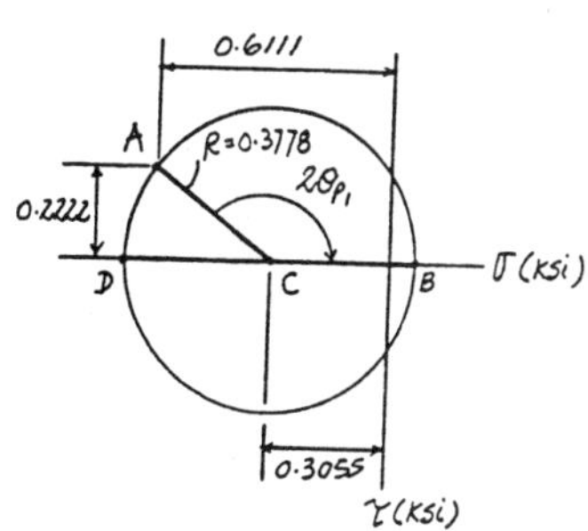

In-Plane Principal Stress : The coordinates of points B and D represent σ_1 and σ_2, respectively.

$$\sigma_1 = -0.3055 + 0.3778 = 0.0723 \text{ ksi} \qquad \textbf{Ans}$$

$$\sigma_2 = -0.3055 - 0.3778 = -0.683 \text{ ksi} \qquad \textbf{Ans}$$

***9–72.** The stair tread of the escalator is supported on two of its sides by the moving pin at A and the roller at B. If a man having a weight of 300 lb stands in the center of the tread, determine the principal stresses developed in the supporting truck on the cross section at point C. The stairs move at constant velocity.

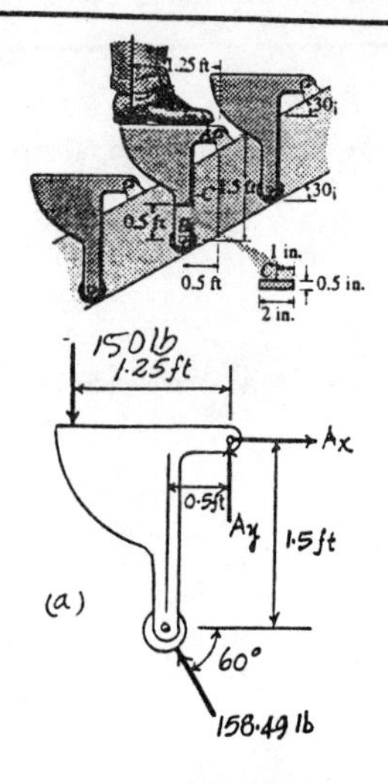

Support Reactions : As shown on FBD(a).

Internal Forces and Moment : As shown on FBD (b).

Section Properties :

$$A = 2(0.5) = 1.00 \text{ in}^2$$

$$I = \frac{1}{12}(0.5)\left(2^3\right) = 0.3333 \text{ in}^4$$

$$Q_B = \bar{y}'A' = 0.5(1)(0.5) = 0.250 \text{ in}^3$$

Normal Stress :

$$\sigma = \frac{N}{A} \pm \frac{My}{I}$$

$$\sigma_C = \frac{-137.26}{1.00} + \frac{475.48(0)}{0.3333} = -137.26 \text{ psi}$$

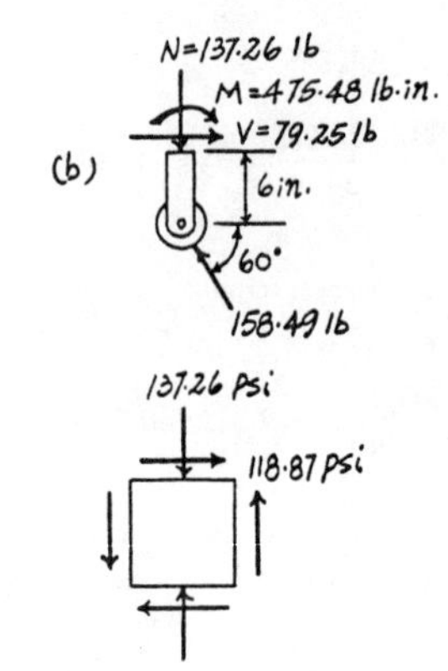

Shear Stress : Applying the shear formula $\tau = \frac{VQ}{It}$.

$$\tau_C = \frac{79.25(0.250)}{0.3333(0.5)} = 118.87 \text{ psi}$$

Construction of the Circle : In accordance with the sign convention, $\sigma_x = 0$, $\sigma_y = -137.26$ psi, and $\tau_{xy} = 118.87$ psi. Hence,

$$\sigma_{avg} = \frac{\sigma_x + \sigma_y}{2} = \frac{0 + (-137.26)}{2} = -68.63 \text{ psi}$$

The coordinates for reference points A and C are

$$A(0,\ 118.87) \qquad C(-68.63,\ 0)$$

The radius of the circle is

$$R = \sqrt{(68.63 - 0)^2 + 118.87^2} = 137.26 \text{ psi}$$

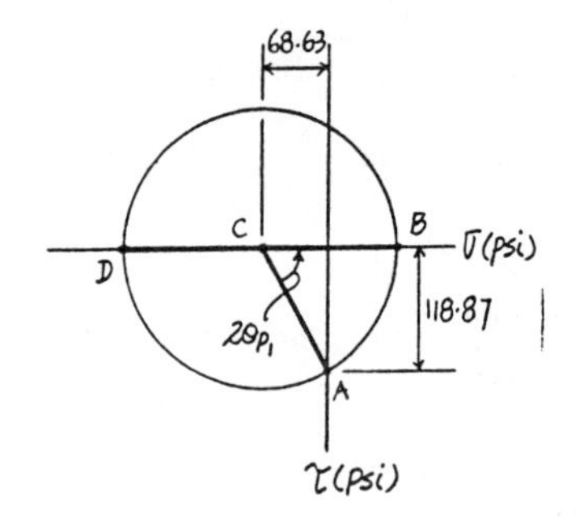

In - Plane Principal Stress : The coordinates of points B and D represent σ_1 and σ_2, respectively.

$$\sigma_1 = -68.63 + 137.26 = 68.6 \text{ psi} \qquad \textbf{Ans}$$

$$\sigma_2 = -68.63 - 137.26 = -206 \text{ psi} \qquad \textbf{Ans}$$

9–73. The bent rod has a diameter of 15 mm and is subjected to the force of 600 N. Determine the principal stresses and the maximum in-plane shear stress that are developed at point A and point B. Show the results on elements located at these points.

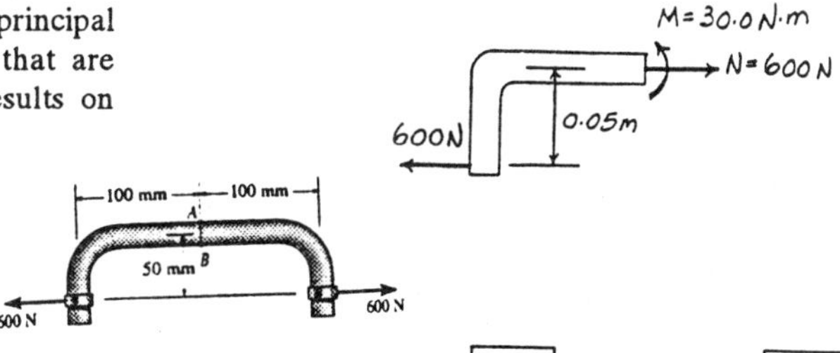

Section Properties :

$$A = \pi(0.0075^2) = 56.25\pi(10^{-6})\ \text{m}^2$$

$$I = \frac{\pi}{4}(0.0075^4) = 2.4850(10^{-9})\ \text{m}^4$$

$Q_A = Q_B = 0$

Stress :

$$\sigma = \frac{N}{A} \pm \frac{Mc}{I}$$

$$= \frac{600}{56.25\pi(10^{-6})} \pm \frac{30.0(0.0075)}{2.4850(10^{-9})}$$

$\sigma_A = 3.3953 - 90.5414 = -87.14$ MPa

$\sigma_B = 3.3953 + 90.5414 = 93.94$ MPa

$\tau_A = \tau_B = 0$ since $Q_A = Q_B = 0$

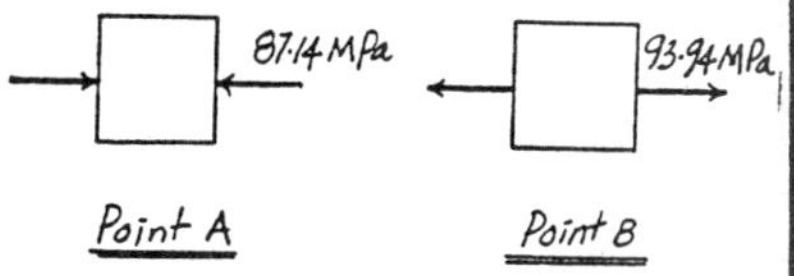

Construction of the Circle : In accordance with the sign convention, $\sigma_x = -87.14$ MPa, $\sigma_y = 0$, and $\tau_{xy} = 0$ for point A. Hence,

$$\sigma_{avg} = \frac{\sigma_x + \sigma_y}{2} = \frac{-87.14 + 0}{2} = -43.57 \text{ MPa}$$

The coordinates for reference points A and C are $A(-87.14, 0)$ $C(-43.57, 0)$.

The radius of the circle is $R = 87.14 - 43.57 = 43.57$ MPa

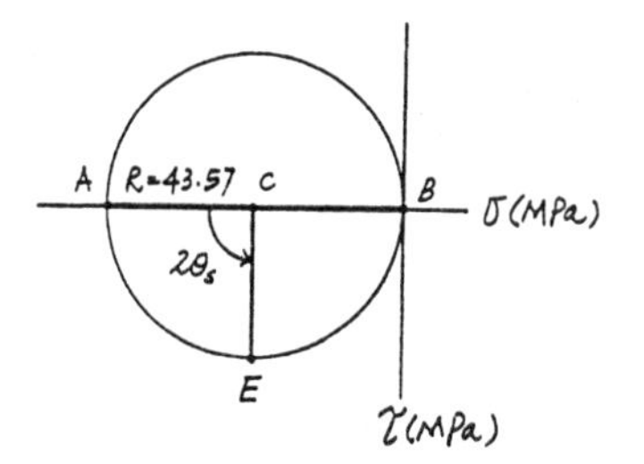

In-Plane Principal Stresses : The coordinates of points B and A represent σ_1 and σ_2, respectively.

$\sigma_1 = 0$ **Ans**

$\sigma_2 = -87.1$ MPa **Ans**

Maximum In-Plane Shear Stress : Represented by the coordinates of point E on the circle.

$\tau_{\substack{max \\ in\text{-}plane}} = R = 43.6$ MPa **Ans**

Orientation of the Plane for Maximum In-Plane Shear Stress : From the circle

$2\theta_s = 90°$ $\quad\theta_s = 45.0°$ (*Counterclockwise*) **Ans**

Construction of the Circle : In accordance with the sign convention, $\sigma_x = 93.94$ MPa, $\sigma_y = 0$, and $\tau_{xy} = 0$ for point B. Hence,

$$\sigma_{avg} = \frac{\sigma_x + \sigma_y}{2} = \frac{93.94 + 0}{2} = 46.97 \text{ MPa}$$

The coordinates for reference points A and C are $A(93.94, 0)$ $C(46.97, 0)$.

The radius of the circle is $R = 93.94 - 46.97 = 46.97$ MPa

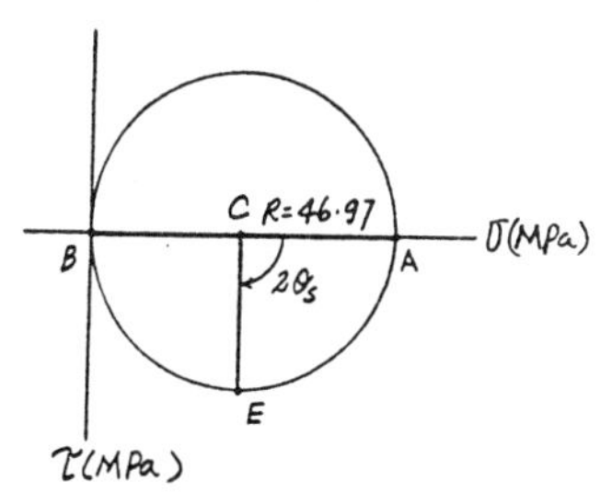

In-Plane Principal Stresses : The coordinates of points A and B represent σ_1 and σ_2, respectively.

$\sigma_1 = 93.9$ MPa **Ans**

$\sigma_2 = 0$ MPa **Ans**

Maximum In-Plane Shear Stress : Represented by point E on the circle.

$\tau_{\substack{max \\ in\text{-}plane}} = R = 47.0$ MPa **Ans**

Orientation of the Plane for Maximum In-Plane Shear Stress : From the circle

$2\theta_s = 90°$ $\quad\theta_s = 45.0°$ (*Clockwise*) **Ans**

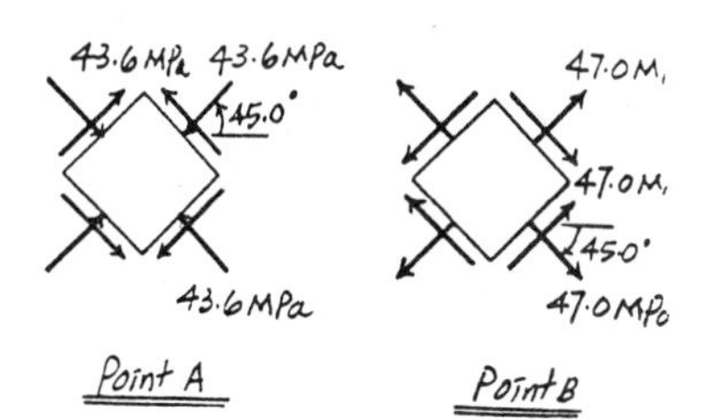

9–74. The beam has a rectangular cross section and is subjected to the loadings shown. Determine the principal stresses and the maximum in-plane shear stress that is developed at point A and point B. Show the results on elements located at these points.

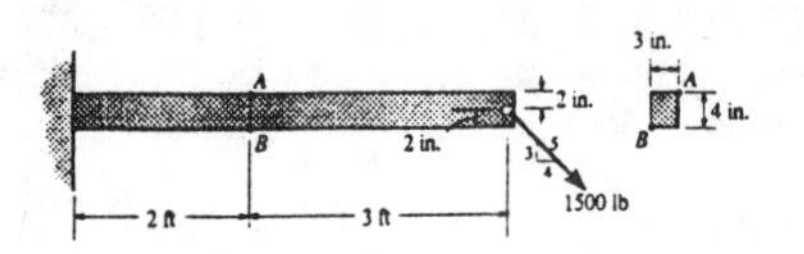

Internal Forces and Moment : As shown on FBD.

Section Properties :

$$A = 3(4) = 12.0 \text{ in}^2 \qquad I = \frac{1}{12}(3)\left(4^3\right) = 16.0 \text{ in}^4$$

$$Q_A = Q_B = 0$$

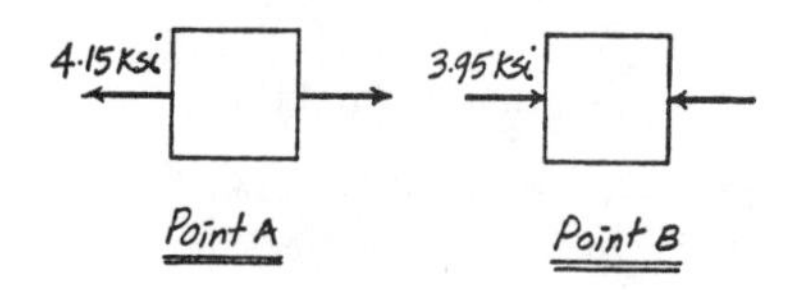

Normal Stress :

N = 1.20 kip, V = 0.90 kip, M = 32.4 kip·in, 36 in., 1.5 kip

$$\sigma = \frac{N}{A} \pm \frac{Mc}{I}$$
$$= \frac{1.20}{12} \pm \frac{32.4(2)}{16.0}$$

$$\sigma_A = 0.1 + 4.05 = 4.15 \text{ ksi}$$
$$\sigma_B = 0.1 - 4.05 = -3.95 \text{ ksi}$$

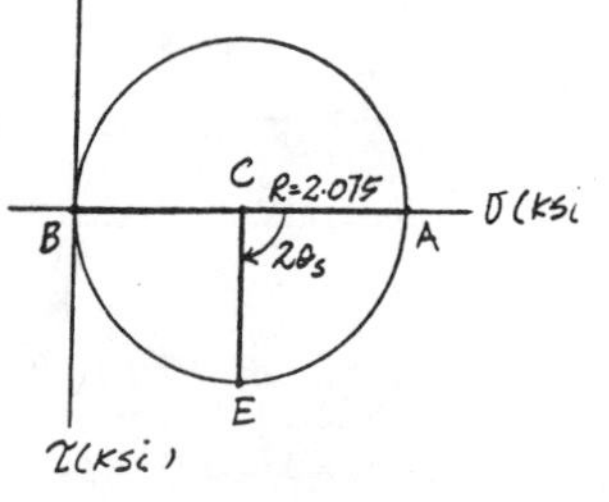

For Point A

Shear Stress : Since $Q_A = Q_B = 0$, then $\tau_A = \tau_B = 0$.

Construction of the Circle : In accordance with the sign convention, $\sigma_x = 4.15$ ksi, $\sigma_y = 0$, and $\tau_{xy} = 0$ for point A. Hence,

$$\sigma_{avg} = \frac{\sigma_x + \sigma_y}{2} = \frac{4.15 + 0}{2} = 2.075 \text{ ksi}$$

The coordinates for reference points A and C are $A(4.15,\ 0)$ $C(2.075,\ 0)$.

The radius of the circle is $R = 4.15 - 2.075 = 2.075$ ksi

For Point B

In-Plane Principal Stress : The coordinates of points A and B represent σ_1 and σ_2, respectively.

$$\sigma_1 = 4.15 \text{ ksi} \qquad \textbf{Ans}$$
$$\sigma_2 = 0 \qquad \textbf{Ans}$$

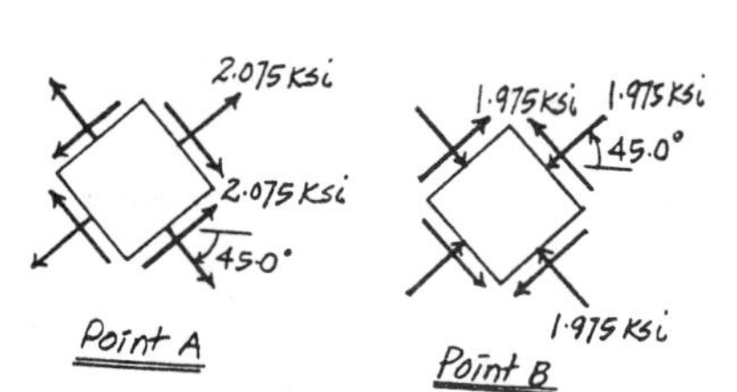

Maximum In-Plane Shear Stress : Represented by point E on the circle.

$$\tau_{\substack{max \\ in\text{-}plane}} = R = 2.075 \text{ ksi} \qquad \textbf{Ans}$$

Orientation of the Plane for Maximum In-Plane Shear Stress : From the circle

$$2\theta_s = 90^\circ \qquad \theta_s = 45.0^\circ \ (\textit{Clockwise}) \qquad \textbf{Ans}$$

Construction of the Circle : In accordance with the sign convention, $\sigma_x = -3.95$ ksi, $\sigma_y = 0$, and $\tau_{xy} = 0$ for point B. Hence,

$$\sigma_{avg} = \frac{\sigma_x + \sigma_y}{2} = \frac{-3.95 + 0}{2} = -1.975 \text{ ksi}$$

The coordinates for reference points A and C are $A(-3.95,\ 0)$ and $C(1.975,\ 0)$.

The radius ofthe circle is $R = 3.95 - 1.975 = 1.975$ ksi

In-Plane Principal Stress : The coordinates of points B and A represent σ_1 and σ_2, respectively.

$$\sigma_1 = 0 \qquad \textbf{Ans}$$
$$\sigma_2 = -3.95 \text{ ksi} \qquad \textbf{Ans}$$

Maximum In-Plane Shear Stress : Represented by the coordinates of point E on the circle.

$$\tau_{\substack{max \\ in\text{-}plane}} = R = 1.975 \text{ ksi} \qquad \textbf{Ans}$$

Orientation of the Plane for Maximum In-Plane Shear Stress : From the circle

$$2\theta_s = 90^\circ \qquad \theta_s = 45.0^\circ \ (\textit{Counterclockwise}) \qquad \textbf{Ans}$$

9–75. A spherical pressure vessel has an inner radius of 5 ft and a wall thickness of 0.5 in. Draw Mohr's circle for the state of stress at a point on the vessel and explain the significance of the result. The vessel is subjected to an internal pressure of 80 psi.

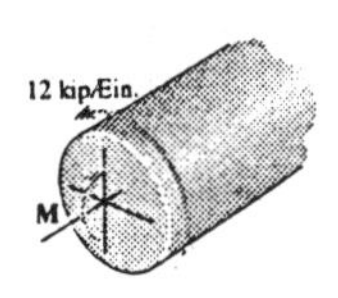

Spherical Vessels : Since $\frac{r}{t} = \frac{60}{0.5} = 120 > 10$, the *thin wall* analysis is valid. Applying Eq. 8 – 3

$$\sigma_x = \sigma_y = \frac{pr}{2t} = \frac{80(60)}{2(0.5)} = 4800 \text{ psi} = 4.80 \text{ ksi}$$

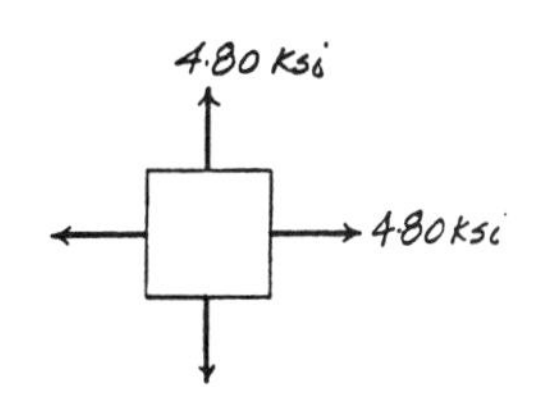

Construction of the Circle : In accordance with the sign convention, $\sigma_x = 4.80$ ksi, $\sigma_y = 4.80$ ksi, and $\tau_{xy} = 0$. Hence,

$$\sigma_{avg} = \frac{\sigma_x + \sigma_y}{2} = \frac{4.80 + 4.80}{2} = 4.80 \text{ ksi}$$

The coordinates for reference points A and C are

$$A(4.80,\ 0) \qquad C(4.80,\ 0)$$

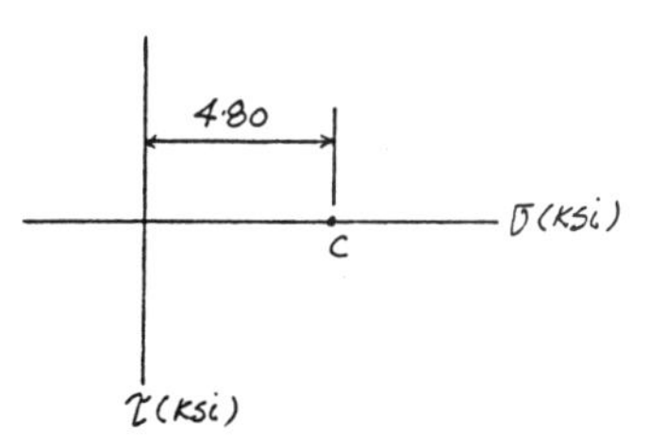

The radius of the circle is $R = 0$. Therefore, Mohr's circle is simply a dot at C. As a result, the state of stress is the **same** consisting of the same normal stress with zero shear stress **regardless** of the orientation of the element.

***9–76.** A rod has a circular cross section with a diameter of 2 in. It is subjected to a torque of 12 kip · in. and a bending moment **M**. If the greatest principal stress at the point of maximum flexural stress is 15 ksi, determine the magnitude of the bending moment.

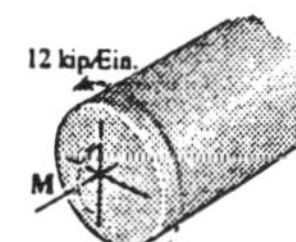

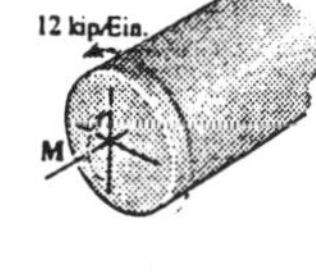

Section Properties :

$$I = \frac{\pi}{4}(1^4) = \frac{\pi}{4} \text{ in}^4 \qquad J = \frac{\pi}{2}(1^4) = \frac{\pi}{2} \text{ in}^4$$

Normal Stress : Applying the flexure formul,.

$$\sigma = \frac{Mc}{I} = \frac{M(1)}{\frac{\pi}{4}} = \frac{4M}{\pi}$$

Shear Stress : Applying the torsion formula,

$$\tau = \frac{Tc}{J} = \frac{12(1)}{\frac{\pi}{2}} = \frac{24}{\pi} \text{ ksi}$$

Construction of the Circle : In accordance with the sign convention, $\sigma_x = \frac{4M}{\pi}$, $\sigma_y = 0$, and $\tau_{xy} = -\frac{24}{\pi}$ ksi. Hence,

$$\sigma_{avg} = \frac{\sigma_x + \sigma_y}{2} = \frac{\frac{4M}{\pi} + 0}{2} = \frac{2M}{\pi}$$

The coordinates for reference points A and C are

$$A\left(\frac{4M}{\pi},\ -\frac{24}{\pi}\right) \qquad C\left(\frac{2M}{\pi},\ 0\right)$$

The radius of the circle is

$$R = \sqrt{\left(\frac{4M}{\pi} - \frac{2M}{\pi}\right)^2 + \left(\frac{24}{\pi}\right)^2} = \frac{2}{\pi}\sqrt{M^2 + 144}$$

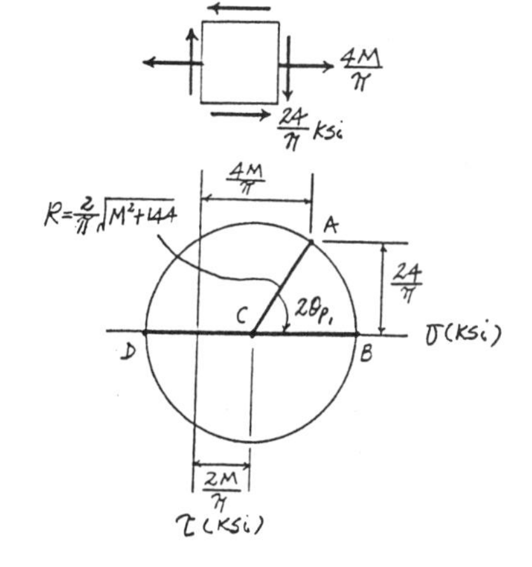

In - Plane Principal Stress : The coordinates of points B and D represent σ_1 and σ_2, respectively. Require $\sigma_1 = 15$ ksi

$$\sigma_1 = 15 = \frac{2M}{\pi} + \frac{2}{\pi}\sqrt{M^2 + 144}$$

$$M = 8.73 \text{ kip} \cdot \text{in} \qquad \textbf{Ans}$$

9–77. The beam is subjected to the two forces shown. Determine the principal stresses at point A.

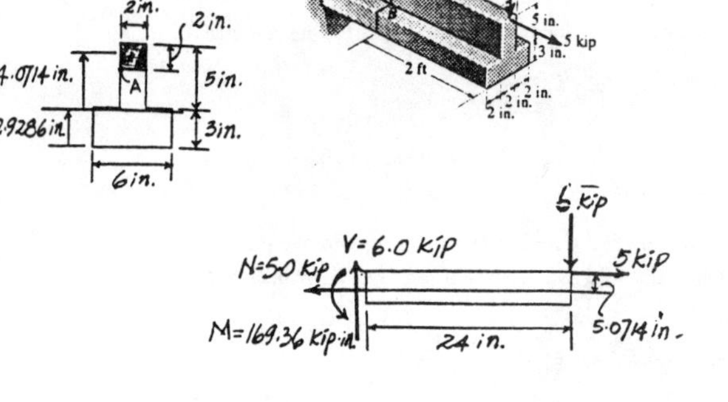

Section Properties :

$$A = 3(6) + 5(2) = 28.0 \text{ in}^2$$

$$\bar{y} = \frac{\Sigma \bar{y} A}{\Sigma A} = \frac{1.5(3)(6) + 5.5(5)(2)}{3(6) + 5(2)} = 2.9286 \text{ in.}$$

$$I = \frac{1}{12}(6)(3^3) + 6(3)(2.9286 - 1.5)^2 + \frac{1}{12}(2)(5^3) + 2(5)(5.5 - 2.9286)^2$$

$$= 137.19 \text{ in}^4$$

$$Q_A = \bar{y}'A' = 4.0714(2)(2) = 16.2857 \text{ in}^3$$

Internal Forces and Moment : As shown on FBD.

Normal Stress :

$$\sigma = \frac{N}{A} \pm \frac{My}{I}$$

$$\sigma_A = \frac{5.00}{28.0} + \frac{169.35(3.0714)}{137.19} = 3.970 \text{ ksi}$$

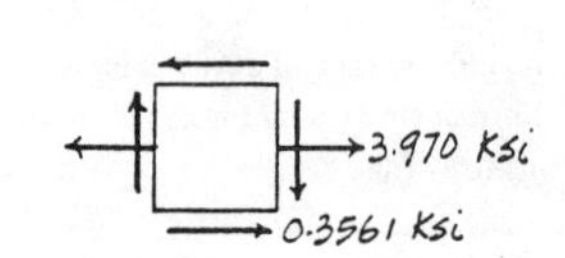

Shear Stress : Applying the shear formula,

$$\tau_A = \frac{VQ_A}{It} = \frac{6.00(16.2857)}{137.19(2)} = 0.3561 \text{ ksi}$$

Construction of the Circle : In accordance with the sign convention, $\sigma_x = 3.970$ ksi, $\sigma_y = 0$ and $\tau_{xy} = -0.3561$ ksi. Hence,

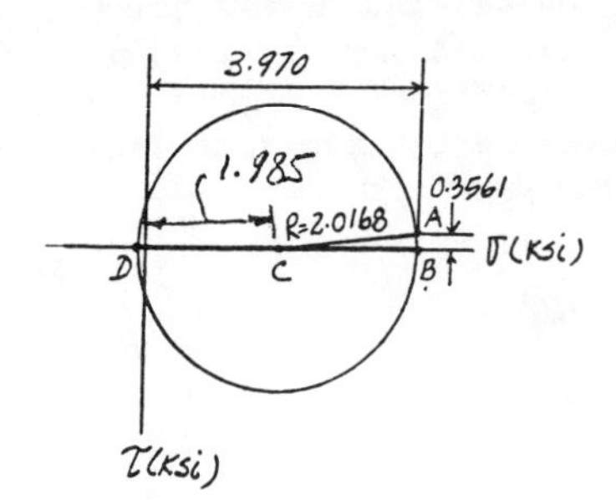

$$\sigma_{avg} = \frac{\sigma_x + \sigma_y}{2} = \frac{3.970 + 0}{2} = 1.985 \text{ ksi}$$

The coordinates for reference points A and C are

$$A(3.970,\ -0.3561) \qquad C(1.985,\ 0)$$

The radius of the circle is

$$R = \sqrt{(3.970 - 1.985)^2 + 0.3561^2} = 2.0168 \text{ ksi}$$

In - Plane Principal Stress : The coordinates of points B and D represent σ_1 and σ_2, respectively.

$$\sigma_1 = 1.985 + 2.0168 = 4.00 \text{ ksi} \qquad \textbf{Ans}$$

$$\sigma_2 = 1.985 - 2.0168 = -0.0317 \text{ ksi} \qquad \textbf{Ans}$$

9–78. The beam is subjected to the two forces shown. Determine the principal stresses at point B, which is located at the bottom of the vertical segment of the cross section.

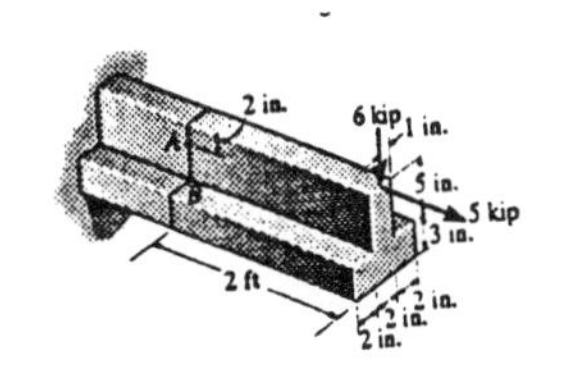

Section Properties :

$$A = 3(6) + 5(2) = 28.0 \text{ in}^2$$

$$\bar{y} = \frac{\Sigma \tilde{y} A}{\Sigma A} = \frac{1.5(3)(6) + 5.5(5)(2)}{3(6) + 5(2)} = 2.9286 \text{ in.}$$

$$I = \frac{1}{12}(6)(3^3) + 6(3)(2.9286 - 1.5)^2 + \frac{1}{12}(2)(5^3) + 2(5)(5.5 - 2.9286)^2$$

$$= 137.19 \text{ in}^4$$

$$Q_B = \bar{y}'A' = 2.5714(5)(2) = 25.7143 \text{ in}^3$$

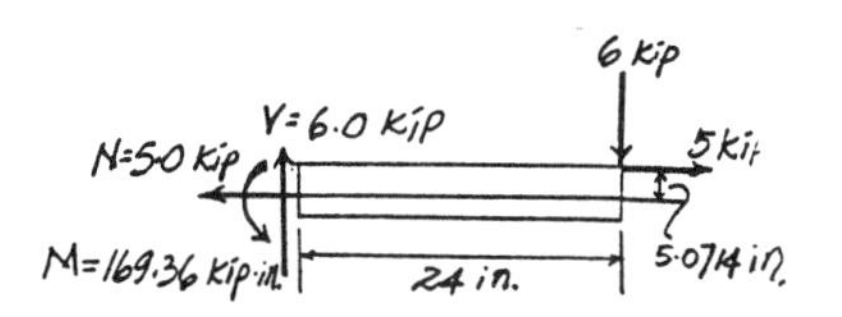

Internal Forces and Moment : As shown on FBD.

Normal Stress :

$$\sigma = \frac{N}{A} \pm \frac{My}{I}$$

$$\sigma_B = \frac{5.00}{28.0} + \frac{169.35(0.07143)}{137.19} = 0.2667 \text{ ksi}$$

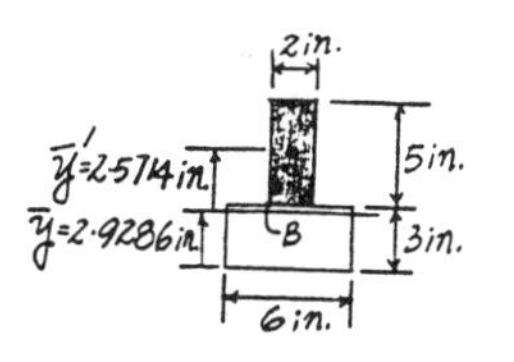

Shear Stress : Applying the shear formula,

$$\tau_B = \frac{VQ_B}{It} = \frac{6.00(25.7143)}{137.19(2)} = 0.5623 \text{ ksi}$$

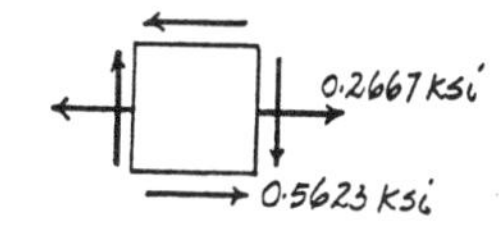

Construction of the Circle : In accordance with the sign convention, $\sigma_x = 0.2667$ ksi, $\sigma_y = 0$, and $\tau_{xy} = -0.5623$ ksi. Hence,

$$\sigma_{avg} = \frac{\sigma_x + \sigma_y}{2} = \frac{0.2667 + 0}{2} = 0.1334 \text{ ksi}$$

The coordinates for reference points A and C are

$$A(0.2667, \ -0.5623) \qquad C(0.1334, \ 0)$$

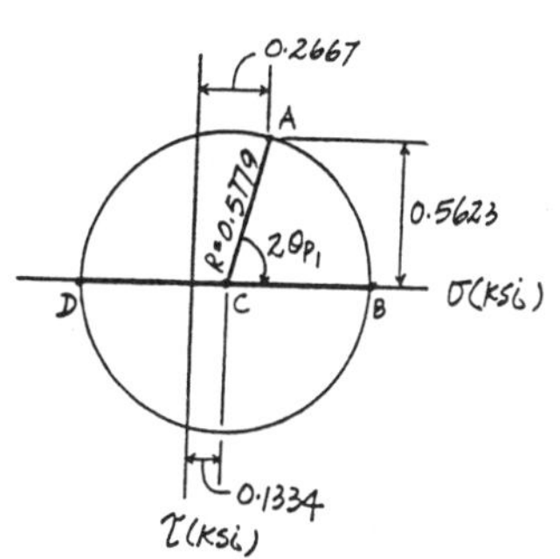

The radius of the circle is

$$R = \sqrt{(0.2667 - 0.1334)^2 + 0.5623^2} = 0.5779 \text{ ksi}$$

In - Plane Principal Stress : The coordinates of points B and D represent σ_1 and σ_2, respectively.

$$\sigma_1 = 0.1334 + 0.5779 = 0.711 \text{ ksi} \qquad \textbf{Ans}$$

$$\sigma_2 = 0.1334 - 0.5779 = -0.445 \text{ ksi} \qquad \textbf{Ans}$$

9–79. The pedal crank for a bicycle has the cross section shown. If it is fixed to the gear at B and does not rotate while subjected to a force of 75 lb, determine the principal stresses in the material on the cross section at point C.

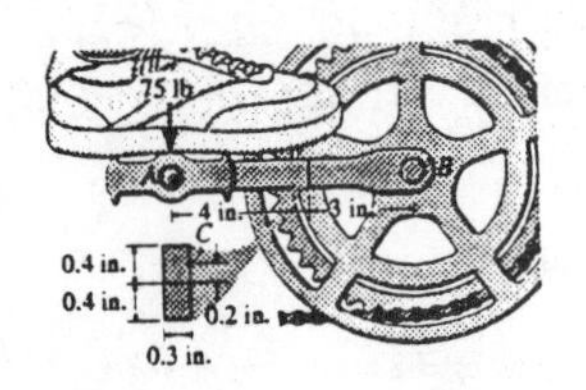

Internal Forces and Moment : As shown on FBD.

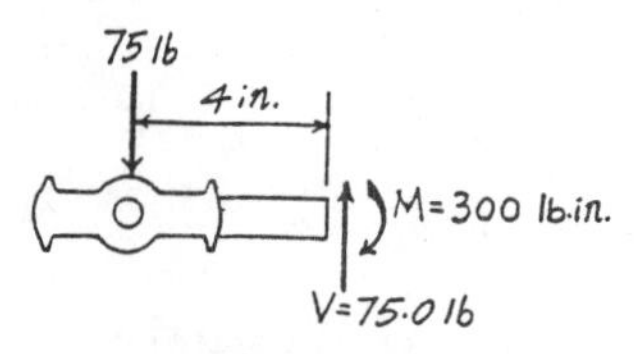

Section Properties :

$$I = \frac{1}{12}(0.3)\left(0.8^3\right) = 0.0128 \text{ in}^4$$

$$Q_C = \bar{y}'A' = 0.3(0.2)(0.3) = 0.0180 \text{ in}^3$$

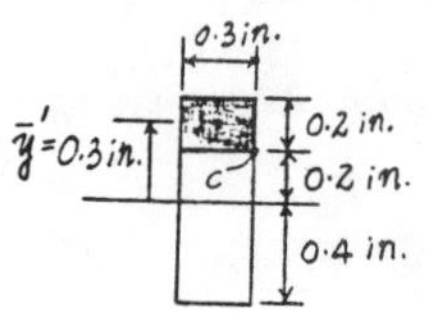

Normal Stress : Applying the flexure formula,

$$\sigma_C = -\frac{My}{I} = -\frac{-300(0.2)}{0.0128} = 4687.5 \text{ psi} = 4.6875 \text{ ksi}$$

Shear Stress : Applying the shear formula,

$$\tau_C = \frac{VQ_C}{It} = \frac{75.0(0.0180)}{0.0128(0.3)} = 351.6 \text{ psi} = 0.3516 \text{ ksi}$$

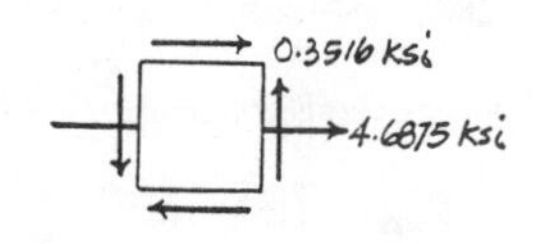

Construction of the Circle : In accordance with the sign convention, $\sigma_x = 4.6875$ ksi, $\sigma_y = 0$, and $\tau_{xy} = 0.3516$ ksi. Hence,

$$\sigma_{avg} = \frac{\sigma_x + \sigma_y}{2} = \frac{4.6875 + 0}{2} = 2.34375 \text{ ksi}$$

The coordinates for reference points A and C are

$$A(4.6875, 0.3516) \qquad C(2.34375, 0)$$

The radius of the circle is

$$R = \sqrt{(4.6875 - 2.34375)^2 + 0.3516^2} = 2.3670 \text{ ksi}$$

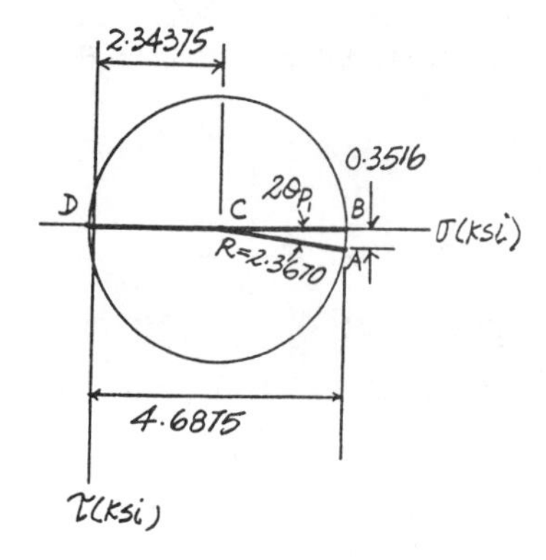

In - Plane Principal Stress : The coordinates of points B and D represent σ_1 and σ_2, respectively.

$$\sigma_1 = 2.34375 + 2.3670 = 4.71 \text{ ksi} \qquad \textbf{Ans}$$

$$\sigma_2 = 2.34375 - 2.3670 = -0.0262 \text{ ksi} \qquad \textbf{Ans}$$

***9–80.** The frame supports the distributed loading of 200 N/m. Determine the normal and shear stresses at point D that act perpendicular and parallel, respectively, to the grains. The grains at this point make an angle of 30° with the horizontal as shown.

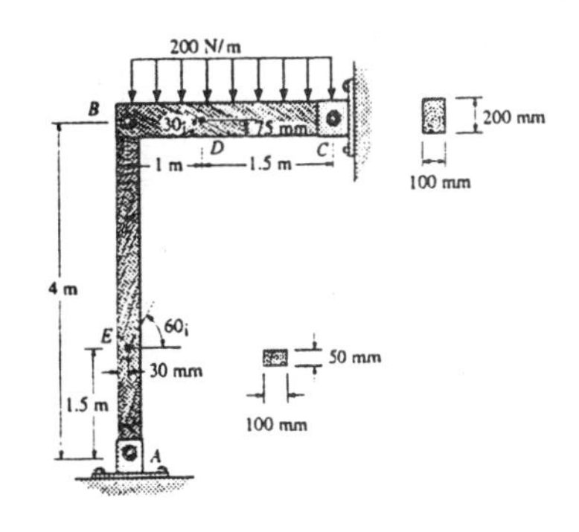

Support Reactions : As shown on FBD(a).

Internal Forces and Moment : As shown on FBD(b).

Section Properties :

$$I = \frac{1}{12}(0.1)\left(0.2^3\right) = 66.667\left(10^{-6}\right)\ \text{m}^4$$

$$Q_D = \bar{y}'A' = 0.0625(0.075)(0.1) = 0.46875\left(10^{-3}\right)\ \text{m}^3$$

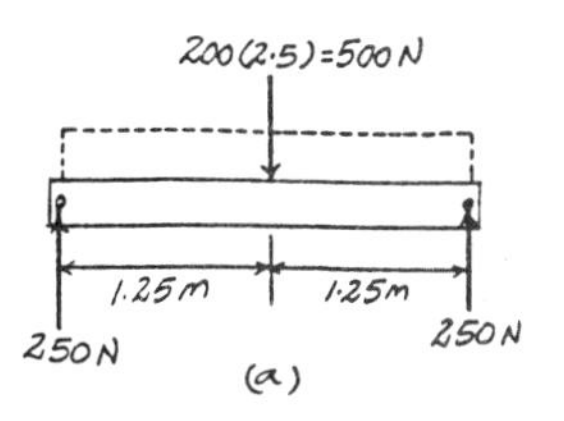

Normal Stre : Applying the flexure formula,

$$\sigma_D = -\frac{My}{I} = -\frac{150(-0.025)}{66.667(10^{-6})} = 56.25\ \text{kPa}$$

Shear Stress : Applying the shear formula,

$$\tau_D = \frac{VQ_D}{It} = \frac{50.0\left[0.46875(10^{-3})\right]}{66.667(10^{-6})(0.1)} = 3.516\ \text{kPa}$$

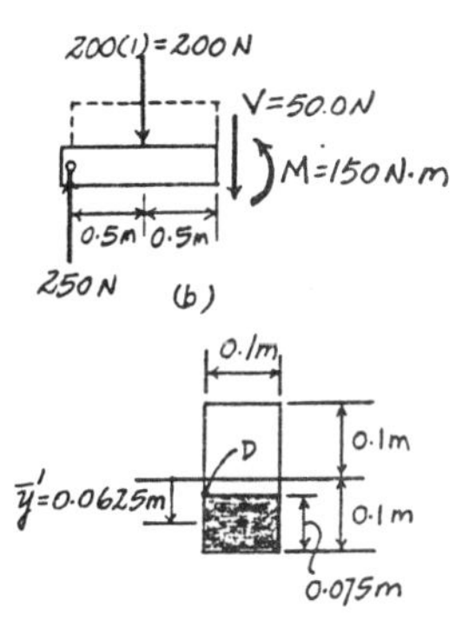

Construction of the Circle : In accordance to the established sign convention, $\sigma_x = 56.25$ kPa, $\sigma_y = 0$ and $\tau_{xy} = -3.516$ kPa. Hence,

$$\sigma_{avg} = \frac{\sigma_x + \sigma_y}{2} = \frac{56.25 + 0}{2} = 28.125\ \text{kPa}$$

The coordinates for reference point A and C are

$$A(56.25,\ -3.516) \qquad C(28.125,\ 0)$$

The radius of the circle is

$$R = \sqrt{(56.25 - 28.125)^2 + 3.516^2} = 28.3439\ \text{kPa}$$

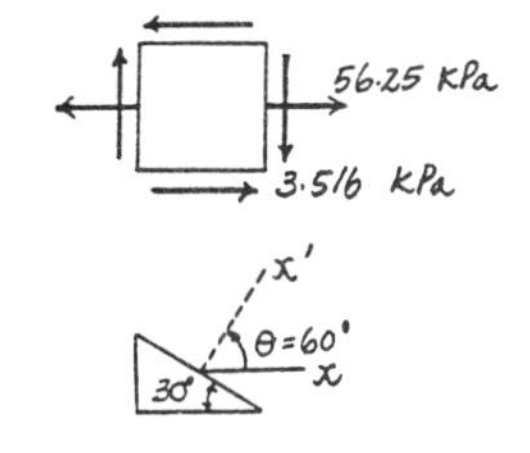

Stresses on The Rotated Element : The normal and shear stress components ($\sigma_{x'}$ and $\tau_{x'y'}$) are represented by the coordinates of point P on the circle. Here, $\theta = 60°$.

$$\sigma_{x'} = 28.125 - 28.3439\cos 52.875° = 11.0\ \text{kPa} \qquad \textbf{Ans}$$

$$\tau_{x'y'} = -28.3439\sin 52.875° = -22.6\ \text{kPa} \qquad \textbf{Ans}$$

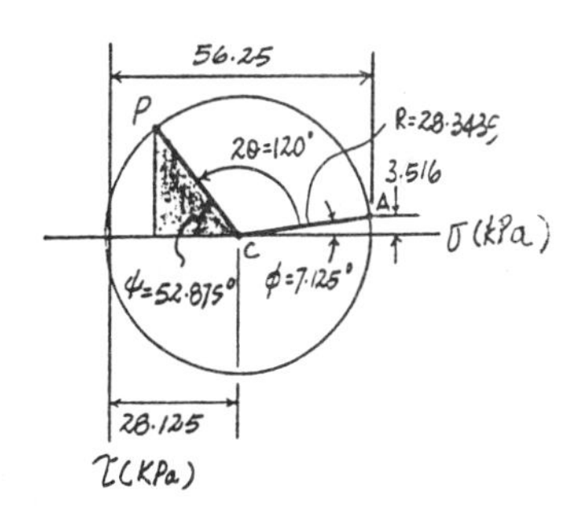

9–81. The frame supports the distributed loading of 200 N/m. Determine the normal and shear stresses at point E that act perpendicular and parallel, respectively, to the grains. The grains at this point make an angle of 60° with the horizontal as shown.

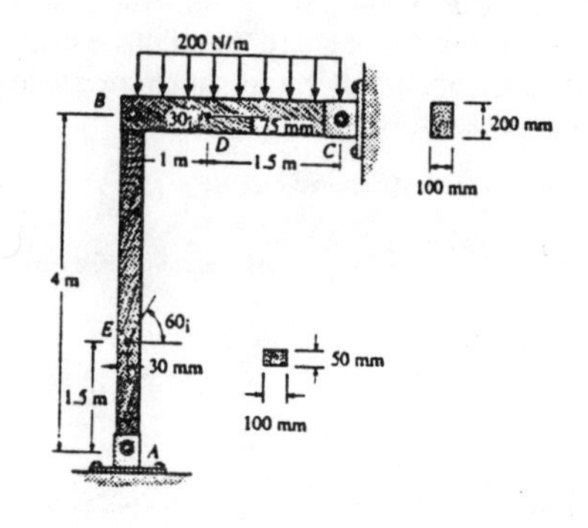

Support Reactions : As shown on FBD(a).

Internal Forces and Moment : As shown on FBD(b).

Section Properties :

$$A = 0.1(0.05) = 5.00\left(10^{-3}\right)\ \text{m}^2$$

Normal Stress :

$$\sigma_E = \frac{N}{A} = \frac{-250}{5.00(10^{-3})} = -50.0\ \text{kPa}$$

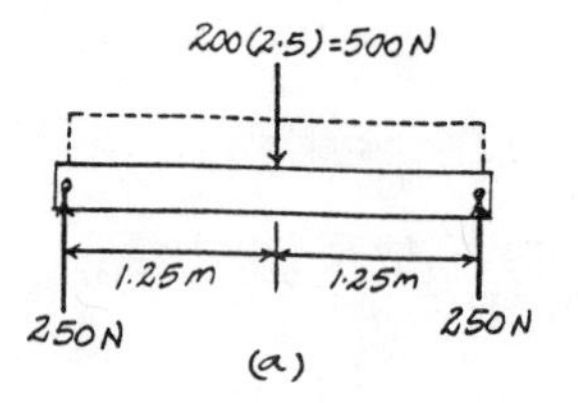

Construction of the Circle : In accordance with the sign convention, $\sigma_x = 0$, $\sigma_y = -50.0$ kPa, and $\tau_{xy} = 0$. Hence,

$$\sigma_{\text{avg}} = \frac{\sigma_x + \sigma_y}{2} = \frac{0 + (-50.0)}{2} = -25.0\ \text{kPa}$$

Ther coordinates for reference points A and C are

$$A(0,\ 0) \qquad C(-25.0,\ 0)$$

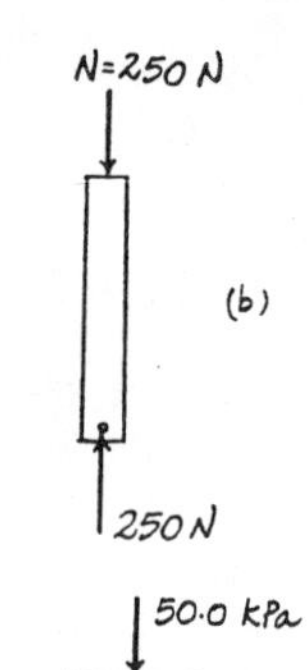

The radius of circle is $R = 25.0 - 0 = 25.0$ kPa

Stress on the Rotated Element : The normal and shear stress components ($\sigma_{x'}$ and $\tau_{x'y'}$) are represented by coordinates of point P on the circle. Here, $\theta = 150°$.

$$\sigma_{x'} = -25.0 + 25.0\cos 60° = -12.5\ \text{kPa} \qquad \textbf{Ans}$$

$$\tau_{x'y'} = 25.0\sin 60° = 21.7\ \text{kPa} \qquad \textbf{Ans}$$

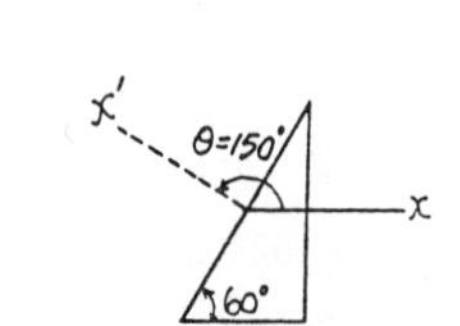

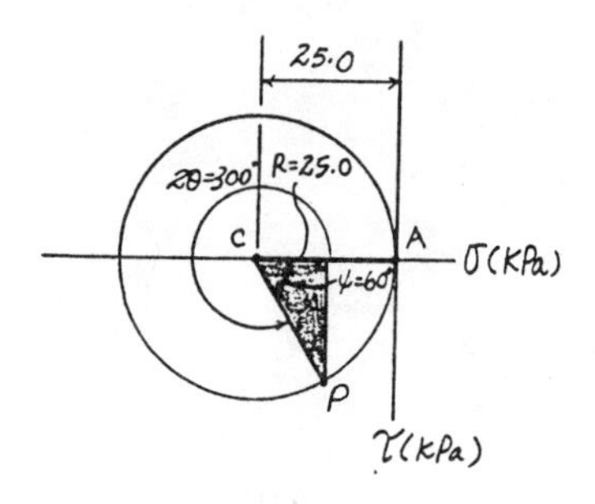

9–82. The thin-walled pipe has an inner diameter of 0.5 in. and a thickness of 0.025 in. If it is subjected to an internal pressure of 500 psi and the axial tension and torsional loadings shown, determine the principal stresses at a point on the surface of the pipe.

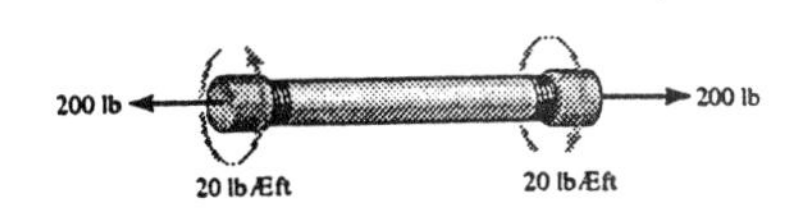

Section Properties :

$$A = \pi\left(0.275^2 - 0.25^2\right) = 0.013125\pi \text{ in}^2$$

$$J = \frac{\pi}{2}\left(0.275^4 - 0.25^4\right) = 2.84768\left(10^{-3}\right) \text{ in}^4$$

Normal Stress : Since $\frac{r}{t} = \frac{0.25}{0.025} = 10$, *thin wall* analysis is valid.

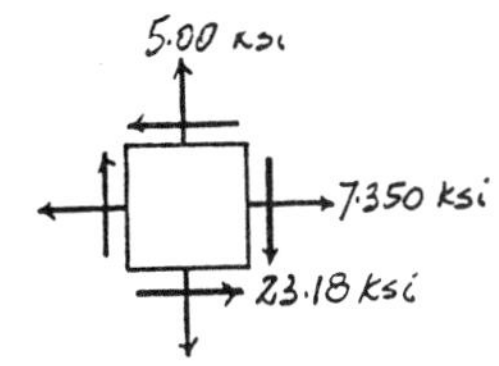

$$\sigma_{\text{long.}} = \frac{N}{A} + \frac{pr}{2t} = \frac{200}{0.013125\pi} + \frac{500(0.25)}{2(0.025)} = 7.350 \text{ ksi}$$

$$\sigma_{\text{hoop}} = \frac{pr}{t} = \frac{500(0.25)}{0.025} = 5.00 \text{ ksi}$$

Shear Stress : Applying the torsion formula.

$$\tau = \frac{Tc}{J} = \frac{20(12)(0.275)}{2.84768(10^{-3})} = 23.18 \text{ ksi}$$

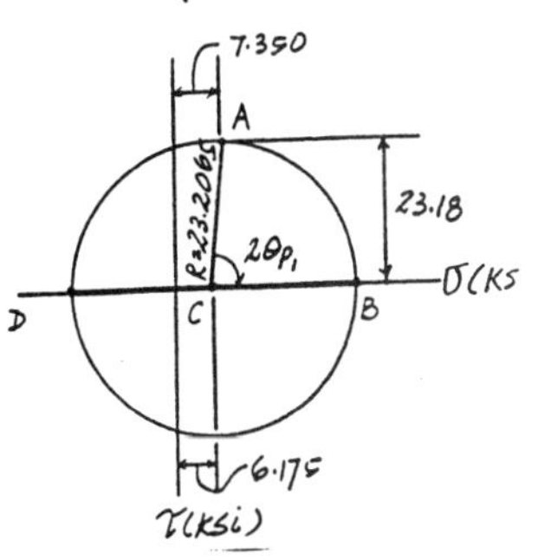

Construction of the Circle : In accordance with the sign convention $\sigma_x = 7.350$ ksi, $\sigma_y = 5.00$ ksi, and $\tau_{xy} = -23.18$ ksi. Hence,

$$\sigma_{\text{avg}} = \frac{\sigma_x + \sigma_y}{2} = \frac{7.350 + 5.00}{2} = 6.175 \text{ ksi}$$

The coordinates for reference points A and C are

$$A(7.350,\ -23.18) \qquad C(6.175,\ 0)$$

The radius of nthe circle is

$$R = \sqrt{(7.350 - 6.175)^2 + 23.18^2} = 23.2065 \text{ ksi}$$

In - Plane Principal Stress : The coordinates of points B and D represent σ_1 and σ_2, respectively.

$$\sigma_1 = 6.175 + 23.2065 = 29.4 \text{ ksi} \qquad \textbf{Ans}$$

$$\sigma_2 = 6.175 - 23.2065 = -17.0 \text{ ksi} \qquad \textbf{Ans}$$

9–83. The boom is used to support the 350-lb load. Determine the principal stresses acting in the boom at points A and B. The cross section is rectangular and has a height of 6 in. and a width of 3 in. The pulley has a radius of 0.5 ft.

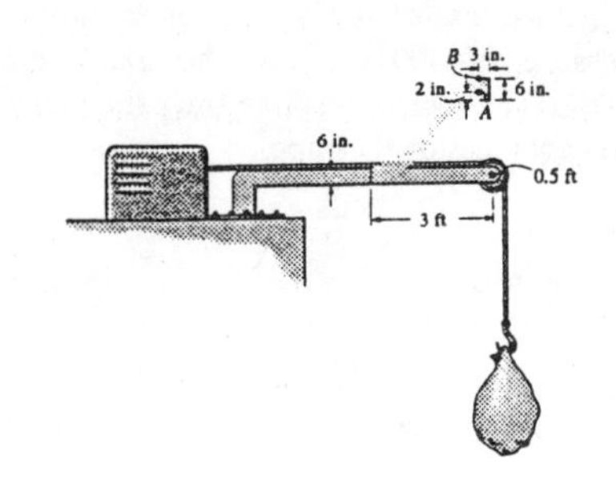

Internal Forces and Moment : As shown on FBD.

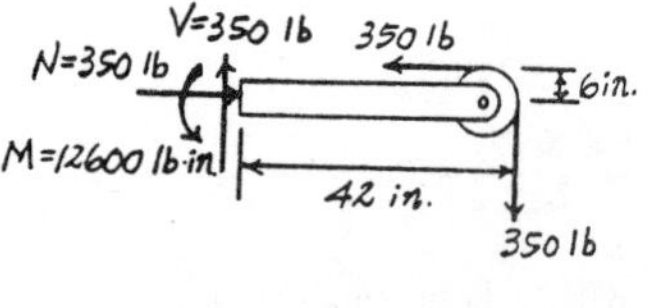

Section Properties :

$$A = 3(6) = 18.0 \text{ in}^2 \qquad I = \frac{1}{12}(3)\left(6^3\right) = 54.0 \text{ in}^3$$

$$Q_A = \bar{y}'A' = 2(2)(3) = 12.0 \text{ in}^3 \qquad Q_B = 0$$

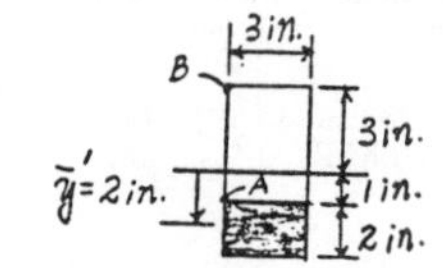

Normal Stress :

$$\sigma = \frac{N}{A} \pm \frac{My}{I}$$

$$\sigma_A = \frac{-350}{18.0} - \frac{12600(1)}{54.0} = -252.78 \text{ psi}$$

$$\sigma_B = \frac{-350}{18.0} + \frac{12600(3)}{54.0} = 680.56 \text{ psi}$$

Shear Stress : Applying the shear formula $\tau = \frac{VQ}{It}$.

$$\tau_A = \frac{VQ_A}{It} = \frac{350(12.0)}{54.0(3)} = 25.93 \text{ psi}$$

$$\tau_B = 0$$

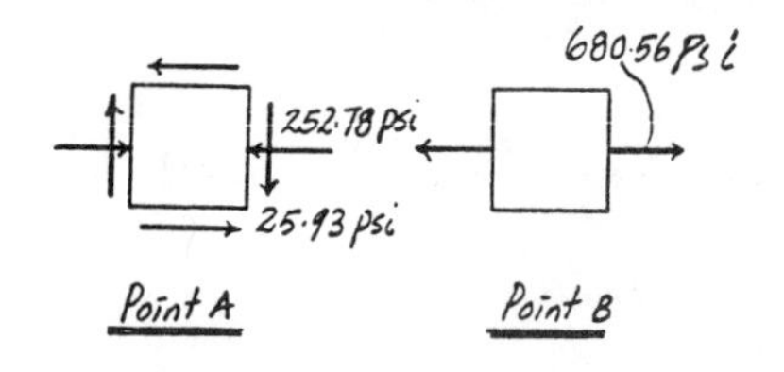

In - Plane Principal Stress : Since no shear stress acts on the element for point B, then

$$\sigma_1 = \sigma_x = 681 \text{ psi} \qquad \textbf{Ans}$$

$$\sigma_2 = \sigma_y = 0 \qquad \textbf{Ans}$$

Construction of the Circle : In accordance with the sign convention, $\sigma_x = -252.78$ psi, $\sigma_y = 0$, and $\tau_{xy} = -25.93$ psi for point A. Hence,

$$\sigma_{avg} = \frac{\sigma_x + \sigma_y}{2} = \frac{-252.78 + 0}{2} = -126.39 \text{ psi}$$

Coordinates for reference points A and C are

$$A(-252.78,\ -25.93) \qquad C(-126.39,\ 0)$$

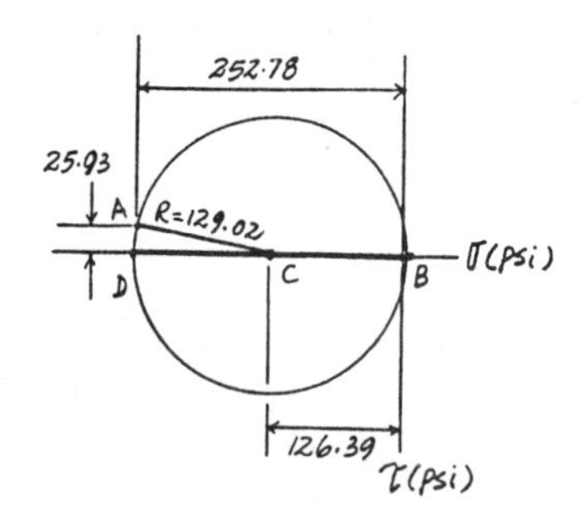

The radius ofthe circle is

$$R = \sqrt{(252.78 - 126.39)^2 + 25.93^2} = 129.02 \text{ psi}$$

In - Plane Principal Stress : The coordinates of points B and D represent σ_1 and σ_2, respectively.

$$\sigma_1 = -126.39 + 129.02 = 2.63 \text{ psi} \qquad \textbf{Ans}$$

$$\sigma_2 = -126.39 - 129.02 = -255 \text{ psi} \qquad \textbf{Ans}$$

***9–84.** The pipe has an inner radius of 25 mm and an outer radius of 27 mm. If it is subjected to an internal pressure of 8 MPa and a torsional moment of 500 N · m, determine the principal stresses and the maximum in-plane shear stress at point A, which lies on the pipe's outer surface.

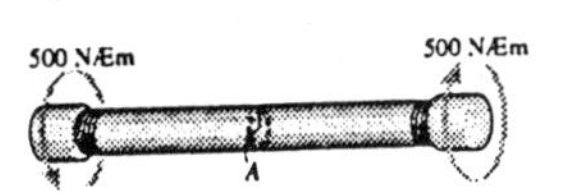

Section Properties :

$$A = \pi\left(0.027^2 - 0.025^2\right) = 0.104\pi\left(10^{-3}\right)\ \text{m}^2$$

$$J = \frac{\pi}{2}\left(0.027^4 - 0.025^4\right) = 70.408\pi\left(10^{-9}\right)\ \text{m}^4$$

Normal Stress : Since $\frac{r}{t} = \frac{25}{2} = 12.5 > 10$, the *thin wall* analysis for cylindrical pipe is valid.

$$\sigma_{\text{long.}} = \frac{pr}{2t} = \frac{8(25)}{2(2)} = 50.0\ \text{MPa}$$

$$\sigma_{\text{hoop}} = \frac{pr}{t} = \frac{8(25)}{2} = 100\ \text{MPa}$$

Shear Stress : Applying the torsion formula.

$$\tau = \frac{Tc}{J} = \frac{500(0.027)}{70.408\pi(10^{-9})} = 61.03\ \text{MPa}$$

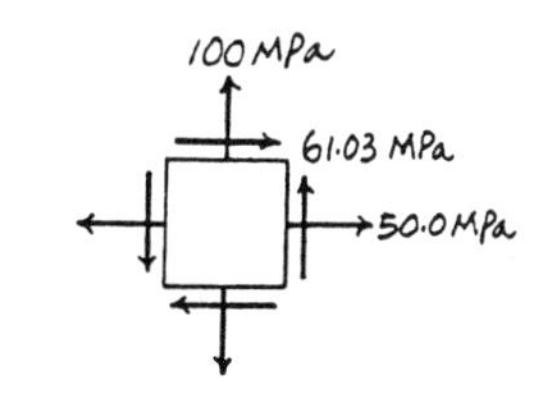

Construction of the Circle : In accordance with the sign convention, $\sigma_x = 50.0$ MPa, $\sigma_y = 100$ MPa, and $\tau_{xy} = 61.03$ MPa. Hence,

$$\sigma_{\text{avg}} = \frac{\sigma_x + \sigma_y}{2} = \frac{50.0 + 100}{2} = 75.0\ \text{MPa}$$

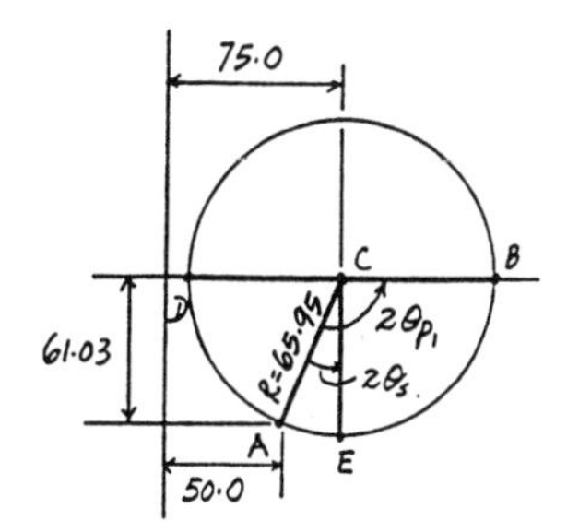

The coordinates for reference points A and C are

$$A(50.0,\ 61.03) \qquad C(75.0,\ 0)$$

The radius of the circle is $R = \sqrt{(75.0 - 50.0)^2 + 61.03^2} = 65.95$ MPa

In-Plane Principal Stress : The coordinates of points B and D represent σ_1 and σ_2, respectively.

$$\sigma_1 = 75.0 + 65.95 = 141\ \text{MPa} \qquad \textbf{Ans}$$

$$\sigma_2 = 75 - 65.95 = 9.05\ \text{MPa} \qquad \textbf{Ans}$$

Maximum In-Plane Shear Stress : Represented by the coordinate of point E on the circle.

$$\tau_{\substack{\max \\ \text{in-plane}}} = R = 66.0\ \text{MPa} \qquad \textbf{Ans}$$

9–85. Draw the three Mohr's circles that describe each of the following states of stress.

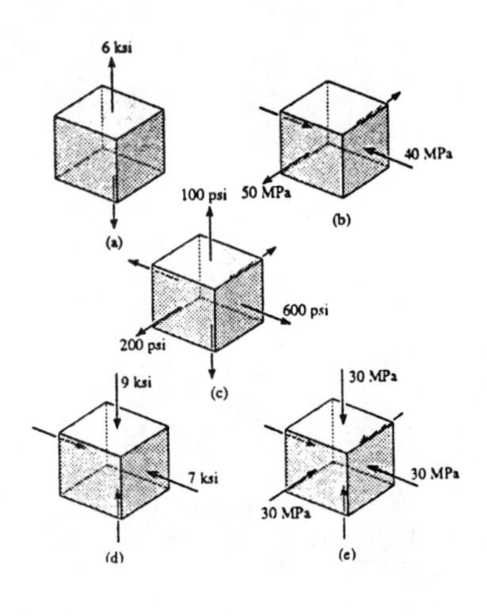

$\sigma_{int} = \sigma_{min} = 0$

σ (ksi)

c

$\sigma_{max} = 6$ ksi

τ (ksi)

(a)

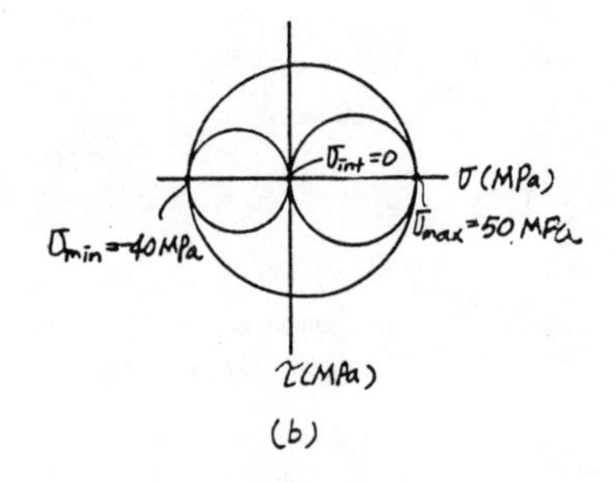

a) $\sigma_{max} = 6$ ksi $\quad \sigma_{int} = \sigma_{min} = 0$

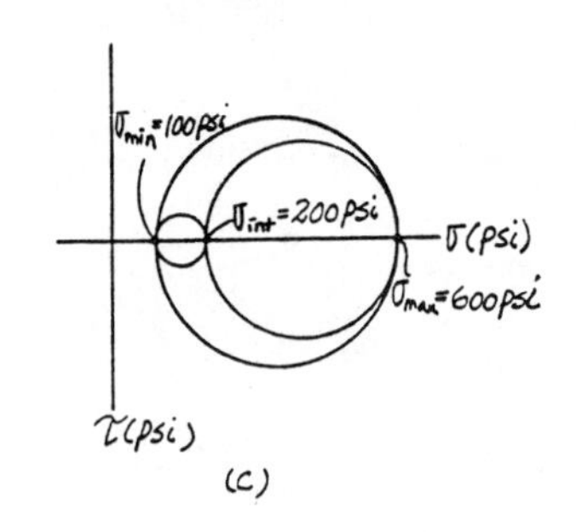

b) $\sigma_{max} = 50$ MPa $\quad \sigma_{int} = 0 \quad \sigma_{min} = -40$ MPa

c) $\sigma_{max} = 600$ psi $\quad \sigma_{int} = 200$ psi $\quad \sigma_{min} = 100$ psi

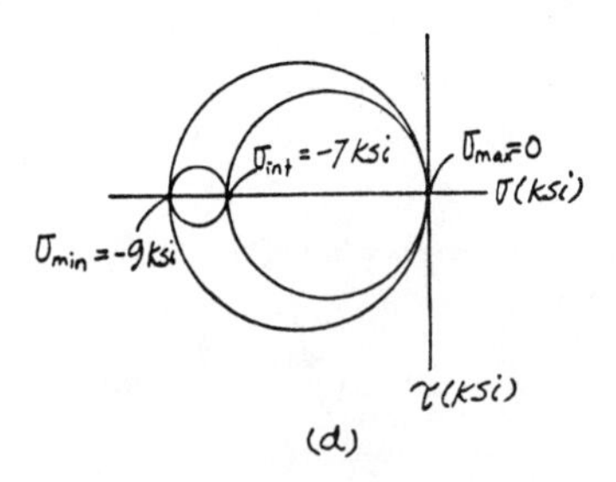

d) $\sigma_{max} = 0 \quad \sigma_{int} = -7$ ksi $\quad \sigma_{min} = -9$ ksi

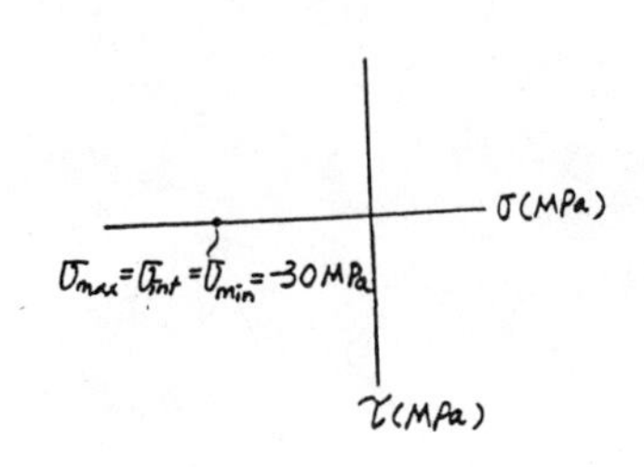

e) $\sigma_{max} = \sigma_{int} = \sigma_{min} = -30$ MPa

9–86. The principal stresses acting at a point in a body are shown. Draw the three Mohr's circles that describe this state of stress, and find the maximum in-plane shear stresses and associated average normal stresses for the x–y, y–z, and x–z planes. For each case, show the results on the element oriented in the appropriate direction.

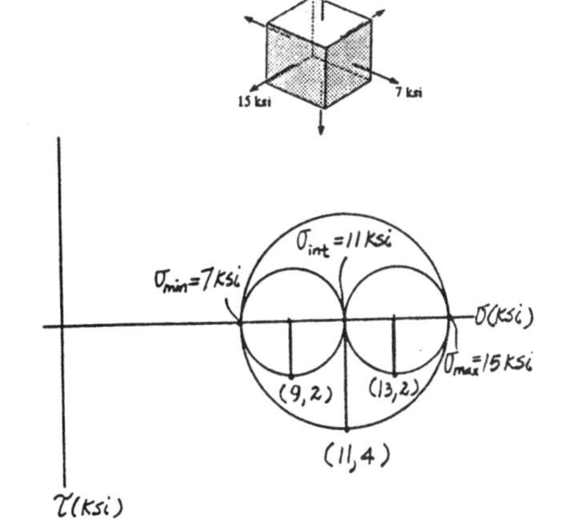

Three Mohr's Circles : $\sigma_{max} = 15$ ksi $\quad \sigma_{int} = 11$ ksi $\quad \sigma_{min} = 7$ ksi

For x - y Plane :

$\sigma_{avg} = 11$ ksi $\qquad \tau_{\substack{max \\ in\text{-}plane}} = 4$ ksi $\qquad$ **Ans**

For x - z Plane :

$\sigma_{avg} = 13$ ksi $\qquad \tau_{\substack{max \\ in\text{-}plane}} = 2$ ksi $\qquad$ **Ans**

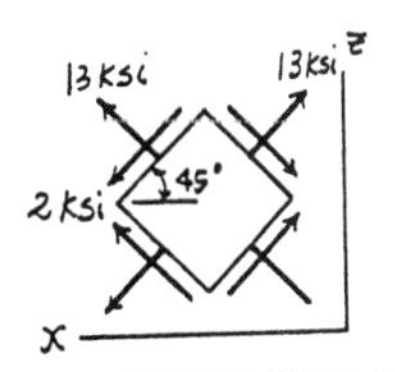

For y - z Plane :

$\sigma_{avg} = 9$ ksi $\qquad \tau_{\substack{max \\ in\text{-}plane}} = 2$ ksi $\qquad$ **Ans**

9–87. The stress at a point is shown on the element. Determine the principal stresses and the absolute maximum shear stress.

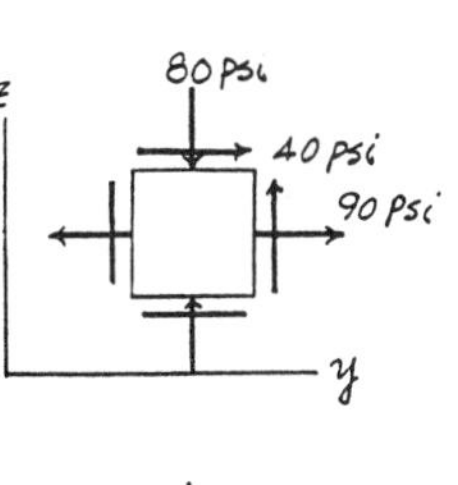

Construction of the Circle : Mohr's circle for the element in the $y-z$ plane is drawn first. In accordance with the sign convention, $\sigma_y = 90$ psi, $\sigma_z = -80$ psi, and $\tau_{yz} = 40$ psi. Hence,

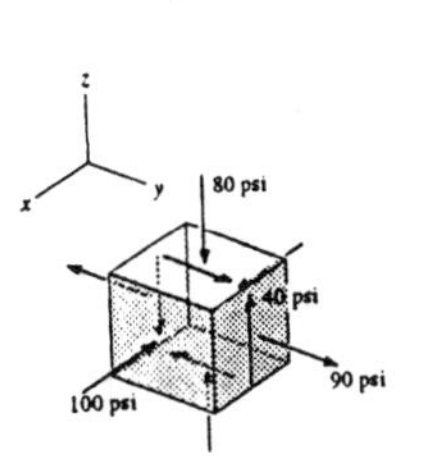

$$\sigma_{avg} = \frac{\sigma_y + \sigma_z}{2} = \frac{90 + (-80)}{2} = 5.00 \text{ psi}$$

The coordinates for reference points A and C are $A\,(90,\ 40)$ $C(5.00,\ 0)$.

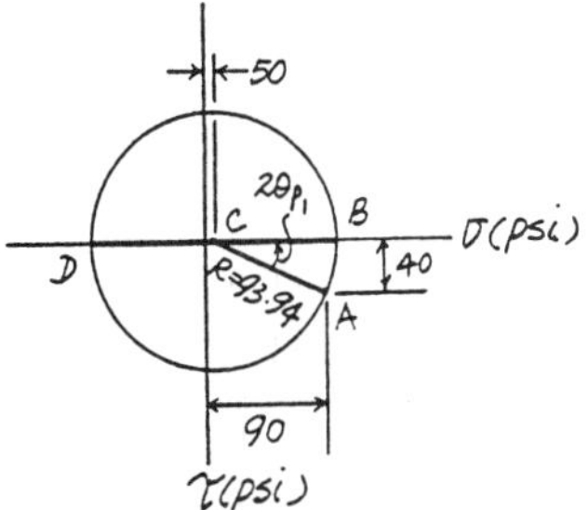

The radius of the circle is $R = \sqrt{(90 - 5.00)^2 + 40^2} = 93.94$ psi

In - Plane Principal Stress : The coordinates of points A and B represent σ_1 and σ_2, respectively.

$$\sigma_1 = 5.00 + 96.94 = 98.94 \text{ psi}$$
$$\sigma_2 = 5.00 - 96.94 = -88.94 \text{ psi}$$

Construction of Three Mohr's Circles : From the results obtained above,

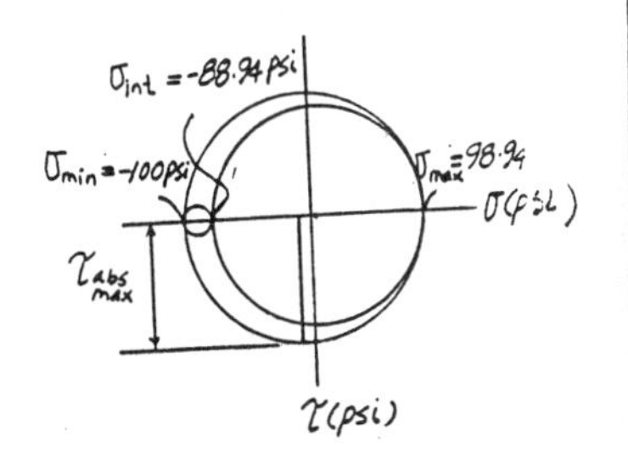

$\sigma_{max} = 98.9$ psi $\quad \sigma_{int} = -88.9$ psi $\quad \sigma_{min} = -100$ psi $\quad$ **Ans**

Absolute Maximum Shear Stress : From the three Mohr's circles

$$\tau_{\substack{abs \\ max}} = \frac{\sigma_{max} - \sigma_{min}}{2} = \frac{98.94 - (-100)}{2} = 99.5 \text{ psi} \qquad \textbf{Ans}$$

***9–88** The stress at a point is shown on the element. Determine the principal stresses and the absolute maximum shear stress.

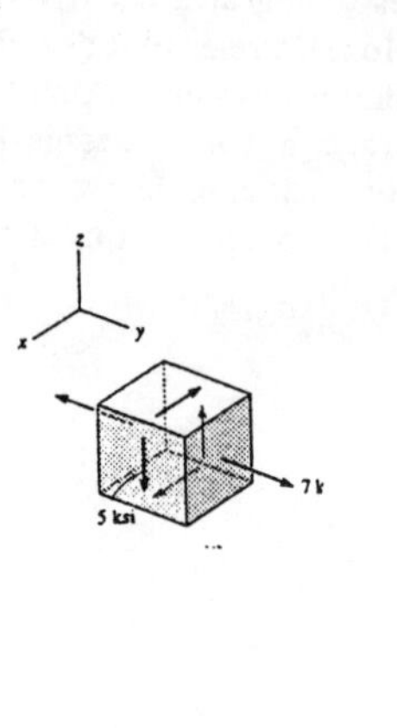

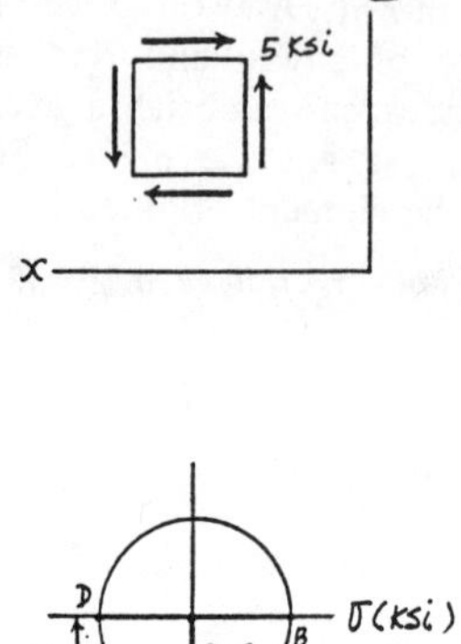

Construction of the Circle : Mohr's circle for the element in the $x-z$ plane is drawn first. In accordance with the sign convention, $\sigma_x = 0$, $\sigma_z = 0$, and $\tau_{xz} = 5$ ksi. Hence,

$$\sigma_{avg} = \frac{\sigma_y + \sigma_z}{2} = 0$$

The coordinates for reference points A and C are $A(0, 5)$ and $C(0, 0)$.

The radius of the circle is $R = 5.00$ ksi

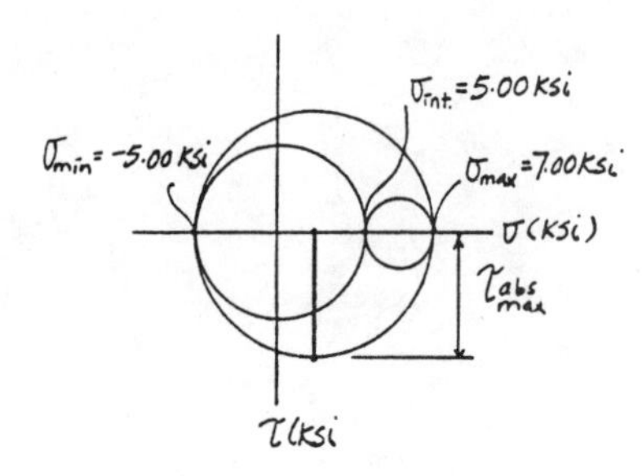

In - Plane Principal Stress : The coordinates of points A and B represent σ_1 and σ_2, respectively.

$$\sigma_1 = 0 + 5.00 = 5.00 \text{ ksi}$$
$$\sigma_2 = 0 - 5.00 = -5.00 \text{ ksi}$$

Construction of Three Mohr's Circles : From the results obtained above,

$\sigma_{max} = 7.00$ ksi $\quad \sigma_{int} = 5.00$ ksi $\quad \sigma_{min} = -5.00$ ksi **Ans**

Absolute Maximum Shear Stress : From the three Mohr's circles

$$\tau_{\substack{abs \\ max}} = \frac{\sigma_{max} - \sigma_{min}}{2} = \frac{7.00 - (-5.00)}{2} = 6.00 \text{ ksi} \quad \textbf{Ans}$$

9–89 The principal stresses acting at a point in a body are shown. Draw the three Mohr's circles that describe this state of stress, and find the maximum in-plane shear stresses and associated average normal stresses for the $x-y$, $y-z$, and $x-z$ planes. For each case, show the results on the element oriented in the appropriate direction.

Three Mohr's Circles : $\sigma_{max} = 40$ MPa $\quad \sigma_{int} = \sigma_{min} = -40$ MPa

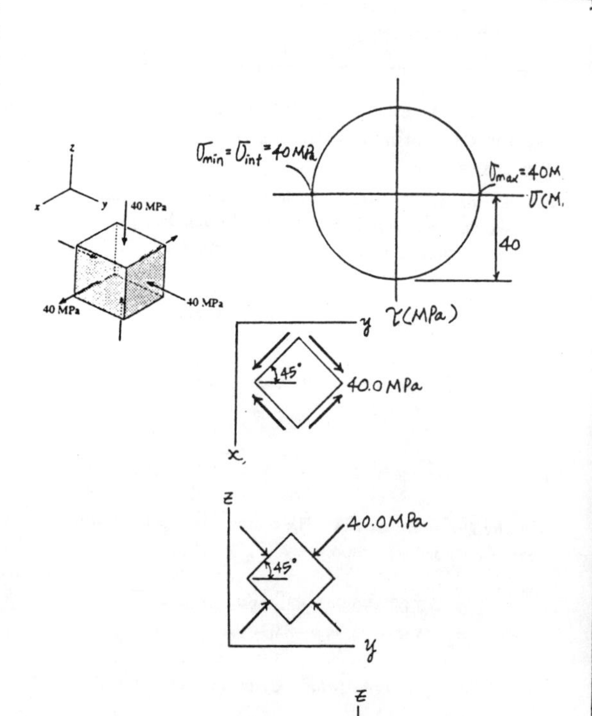

For x - y Plane :

$\sigma_{avg} = 0 \qquad \tau_{\substack{max \\ in\text{-}plane}} = 40.0$ MPa **Ans**

For y - z Plane :

$\sigma_{avg} = -40.0$ MPa $\qquad \tau_{\substack{max \\ in\text{-}plane}} = 0$ **Ans**

For x - z Plane :

$\sigma_{avg} = 0 \qquad \tau_{\substack{max \\ in\text{-}plane}} = 40.0$ MPa **Ans**

9–90. The solid shaft is subjected to a torque, bending moment, and shear force as shown. Determine the principal stresses acting at points A and B and the absolute maximum shear stress.

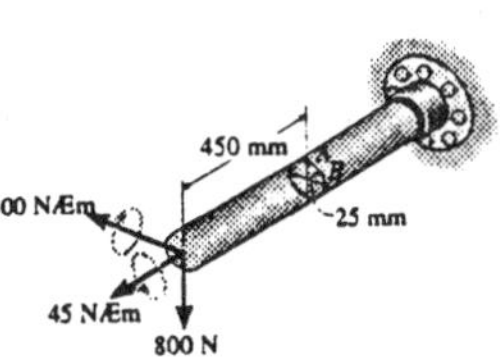

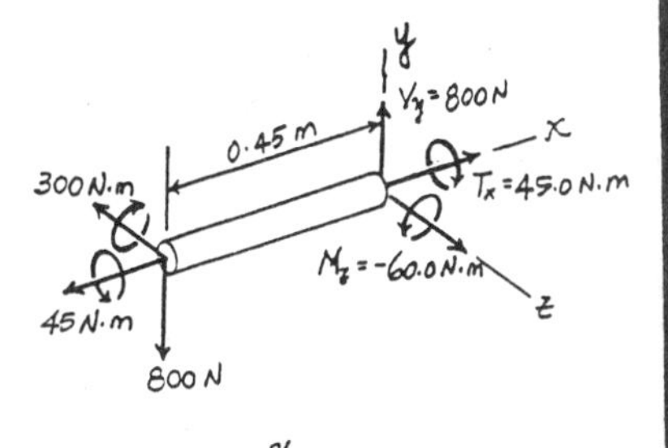

Internal Forces and Moment : As shown on FBD.

Section Properties :

$$I_z = \frac{\pi}{4}\left(0.025^4\right) = 0.306796\left(10^{-6}\right)\ \text{m}^4$$

$$J = \frac{\pi}{2}\left(0.025^4\right) = 0.613592\left(10^{-6}\right)\ \text{m}^4$$

$$(Q_A)_y = 0$$

$$(Q_B)_y = \bar{y}'A'$$

$$= \frac{4(0.025)}{3\pi}\left[\frac{1}{2}(\pi)\left(0.025^2\right)\right] = 10.417\left(10^{-6}\right)\ \text{m}^3$$

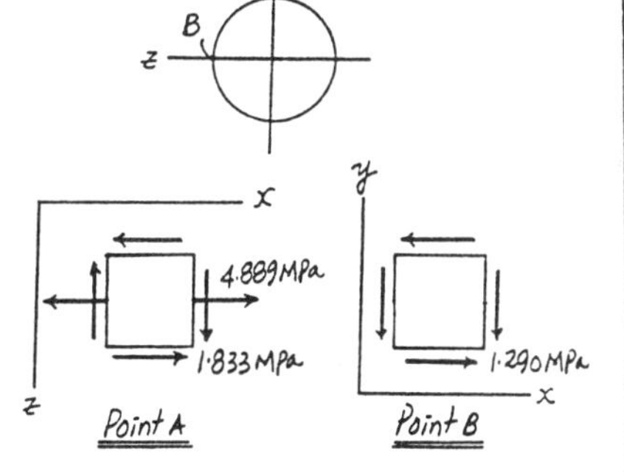

Normal Stress : Applying the flexure formula.

$$\sigma = -\frac{M_z y}{I_z}$$

$$\sigma_A = -\frac{-60.0(0.025)}{0.306796(10^{-6})} = 4.889\ \text{MPa}$$

$$\sigma_B = -\frac{-60.0(0)}{0.306796(10^{-6})} = 0$$

Shear Stress : Applying the torsion formula for point A,

$$\tau_A = \frac{Tc}{J} = \frac{45.0(0.025)}{0.613592(10^{-6})} = 1.833\ \text{MPa}$$

The transverse shear stress in the y direction and the torsional shear stress can be obtained using shear formula andtorsion formula, $\tau_V = \frac{VQ}{It}$ and $\tau_{\text{twist}} = \frac{T\rho}{J}$, respectively.

$$\tau_B = (\tau_V)_y - \tau_{\text{twist}}$$

$$= \frac{800\left[10.417(10^{-6})\right]}{0.306796(10^{-6})(0.05)} - \frac{45.0(0.025)}{0.613592(10^{-6})}$$

$$= -1.290\ \text{MPa}$$

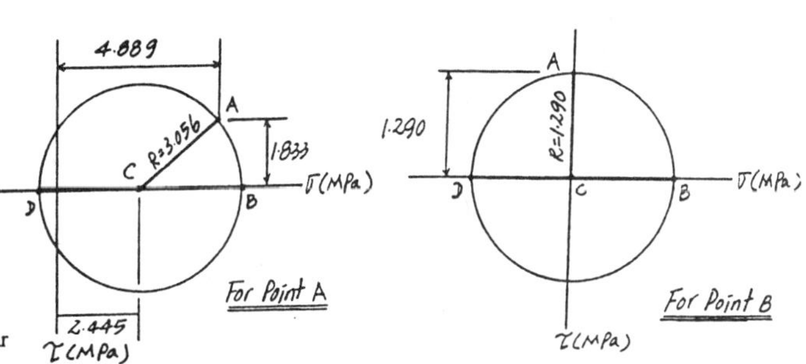

Construction of the Circle : $\sigma_x = 4.889$ MPa, $\sigma_z = 0$, and $\tau_{xz} = -1.833$ MPa for point A. Hence,

$$\sigma_{\text{avg}} = \frac{\sigma_x + \sigma_z}{2} = \frac{4.889 + 0}{2} = 2.445\ \text{MPa}$$

The coordinates for reference points A and C are $A(4.889,\ -1.833)$ and $C(2.445,\ 0)$.

The radius of the circle is

$$R = \sqrt{(4.889 - 2.445)^2 + 1.833^2} = 3.056\ \text{MPa}$$

$\sigma_x = \sigma_y = 0$ and $\tau_{xy} = -1.290$ MPa for point B. Hence,

$$\sigma_{\text{avg}} = \frac{\sigma_x + \sigma_z}{2} = 0$$

The coordinates for reference points A and C are $A(0,\ -1.290)$ and $C(0,\ 0)$.

The radius of the circle is $R = 1.290$ MPa

In - Plane Principal Stresses : The coordinates of points B and D represent σ_1 and σ_2, respectively. For point A

$$\sigma_1 = 2.445 + 3.056 = 5.50\ \text{MPa}$$

$$\sigma_2 = 2.445 - 3.056 = -0.611\ \text{MPa}$$

For point B,

$$\sigma_1 = 0 + 1.290 = 1.29\ \text{MPa}$$

$$\sigma_2 = 0 - 1.290 = -1.290\ \text{MPa}$$

Three Mohr's Circles : From the results obtained above, the principal stresses for point A are

$\sigma_{\max} = 5.50$ MPa $\quad \sigma_{\text{int}} = 0 \quad \sigma_{\min} = -0.611$ MPa **Ans**

And for point B

$\sigma_{\max} = 1.29$ MPa $\quad \sigma_{\text{int}} = 0 \quad \sigma_{\min} = -1.29$ MPa **Ans**

Absolute Maximum Shear Stress : For point A,

$$\tau_{\substack{\text{abs}\\ \max}} = \frac{\sigma_{\max} - \sigma_{\min}}{2} = \frac{5.50 - (-0.611)}{2} = 3.06\ \text{MPa}$$ **Ans**

For point B,

$$\tau_{\substack{\text{abs}\\ \max}} = \frac{\sigma_{\max} - \sigma_{\min}}{2} = \frac{1.29 - (-1.29)}{2} = 1.29\ \text{MPa}$$ **Ans**

9–91. Determine the principal stresses and the absolute maximum shear stress at point A on the frame. The cross-sectional area at this point is shown.

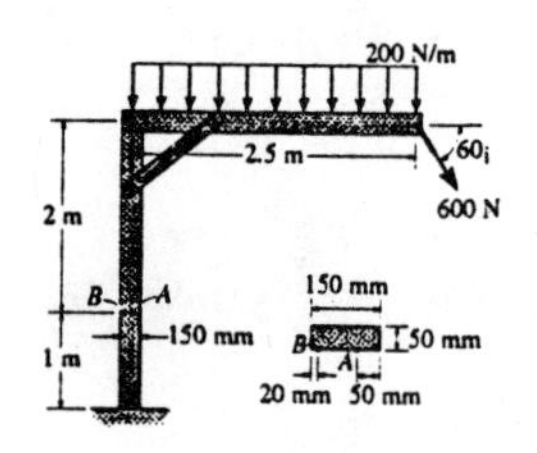

Internal Forces and Moment : As shown on FBD.

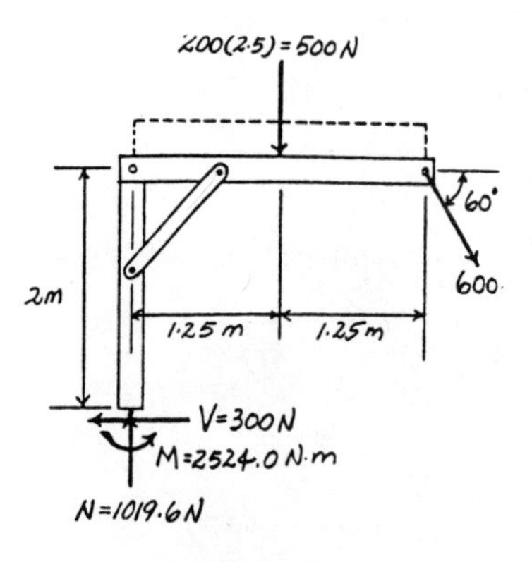

Section Properties :

$$A = 0.05(0.15) = 7.50(10^{-3})\ \text{m}^2$$

$$I = \frac{1}{12}(0.05)(0.15^3) = 14.0625(10^{-6})\ \text{m}^4$$

$$Q_A = \bar{x}'A' = 0.05(0.05)(0.05) = 0.125(10^{-3})\ \text{m}^3$$

Normal Stress :

$$\sigma = \frac{N}{A} \pm \frac{Mx}{I}$$

$$\sigma_A = \frac{-1019.6}{7.50(10^{-3})} - \frac{2524.0(0.025)}{14.0625(10^{-6})} = -4.623\ \text{MPa}$$

Shear Stress : Applying the shear formula,

$$\tau_A = \frac{VQ}{It} = \frac{300[0.125(10^{-3})]}{14.0625(10^{-6})(0.05)} = 0.0533\ \text{MPa}$$

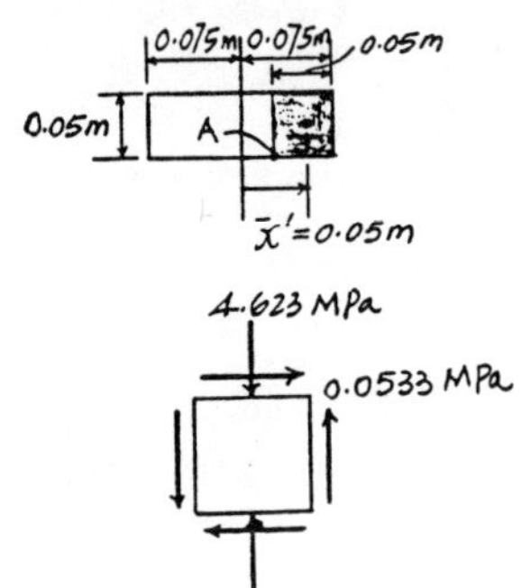

Construction of the Circle : $\sigma_x = 0$, $\sigma_y = -4.623$ MPa, and $\tau_{xy} = 0.0533$ MPa . Hence,

$$\sigma_{avg} = \frac{\sigma_x + \sigma_z}{2} = \frac{0 + (-4.623)}{2} = -2.3115\ \text{MPa}$$

The coordinates for reference points A and C are $A(0,\ 0.0533)$ and $C(-2.3115,\ 0)$.

The radius of the circle is $R = \sqrt{(2.3115 - 0)^2 + 0.0533^2} = 2.3122$ MPa

In - Plane Principal Stress : The coordinates of points B and D represent σ_1 and σ_2, respectively. For point A

$$\sigma_1 = -2.3115 + 2.3122 = 0.6152\ \text{kPa}$$

$$\sigma_2 = -2.3115 - 2.3122 = -4.624\ \text{MPa}$$

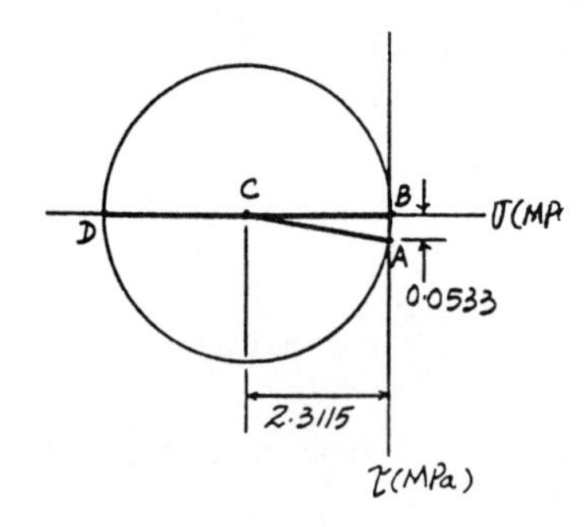

Three Mohr's Circles : From the results obtained above, the principal stresses are

$\sigma_{max} = 0.615$ kPa $\sigma_{int} = 0$ $\sigma_{min} = -4.62$ MPa **Ans**

Absolute Maximum Shear Stress :

$$\tau_{\substack{abs \\ max}} = \frac{\sigma_{max} - \sigma_{min}}{2} = \frac{0.6152(10^{-3}) - (-4.624)}{2} = 2.31\ \text{MPa} \quad \textbf{Ans}$$

***9–92.** Solve Prob. 9–91 for point B.

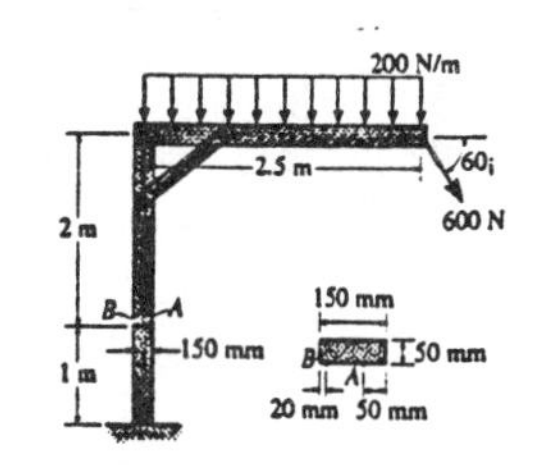

Internal Forces and Moment : As shown on FBD.

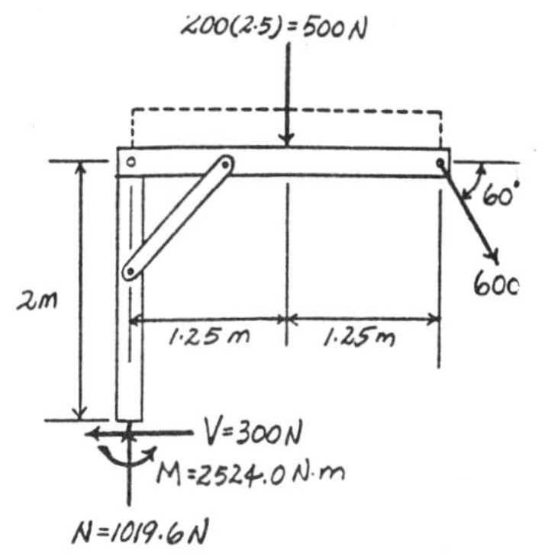

Section Properties :

$$A = 0.05(0.15) = 7.50(10^{-3})\ \text{m}^2$$

$$I = \frac{1}{12}(0.05)(0.15^3) = 14.0625(10^{-6})\ \text{m}^4$$

$$Q_B = \bar{x}'A' = 0.065(0.02)(0.05) = 65.0(10^{-6})\ \text{m}^3$$

Normal Stress :

$$\sigma = \frac{N}{A} \pm \frac{Mx}{I}$$

$$\sigma_B = \frac{-1019.6}{7.50(10^{-3})} + \frac{2524.0(0.055)}{14.0625(10^{-6})} = 9.736\ \text{MPa}$$

Shear Stress : Applying the shear formula,

$$\tau_A = \frac{VQ}{It} = \frac{300\left[65.0(10^{-6})\right]}{14.0625(10^{-6})(0.05)} = 0.02773\ \text{MPa}$$

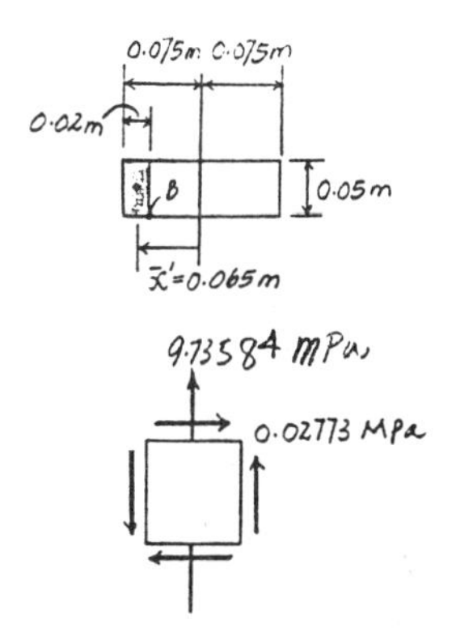

Construction of the Circle : $\sigma_x = 0$, $\sigma_y = 9.73584$ MPa, and $\tau_{xy} = 0.02773$ MPa. Hence,

$$\sigma_{avg} = \frac{\sigma_x + \sigma_z}{2} = \frac{0 + 9.73584}{2} = 4.86792\ \text{MPa}$$

The coordinates for reference points A and C are $A(0,\ 0.02773)$ and $C(4.86792,\ 0)$.

The radius of ther circle is $R = \sqrt{(4.86792 - 0)^2 + 0.02773^2} = 4.86800$ MPa

In - Plane Principal Stress : The coordinates of points B and D represent σ_1 and σ_2, respectively. For point A

$$\sigma_1 = 4.86792 + 4.86800 = 9.736\ \text{MPa}$$

$$\sigma_2 = 4.86792 - 4.86800 = -0.0790\ \text{kPa}$$

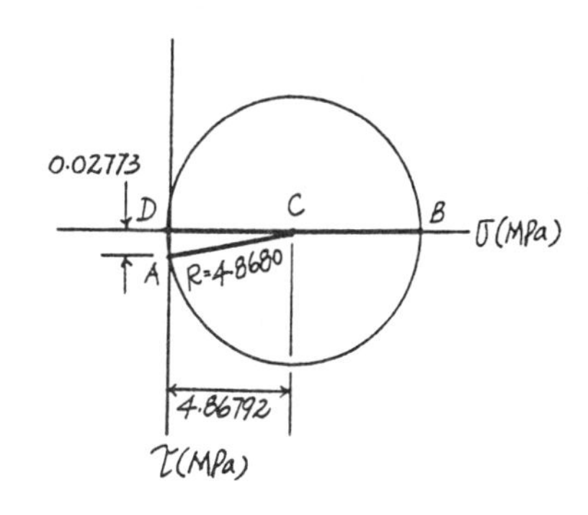

Three Mohr's Circles : From the results obtained above, the principal stresses are

$$\sigma_{max} = 9.74\ \text{MPa} \qquad \sigma_{int} = 0 \qquad \sigma_{min} = -0.0790\ \text{kPa} \qquad \textbf{Ans}$$

Absolute Maximum Shear Stress :

$$\tau_{\substack{abs \\ max}} = \frac{\sigma_{max} - \sigma_{min}}{2} = \frac{9.736 - \left[-0.0790(10^{-3})\right]}{2} = 4.87\ \text{MPa} \qquad \textbf{Ans}$$

9–93. The bolt is fixed to its support at C. If a force of 18 lb is applied to the wrench to tighten it, determine the principal stresses and the absolute maximum shear stress developed in the bolt shank at point A. Represent the results on an element located at this point. The shank has a diameter of 0.25 in.

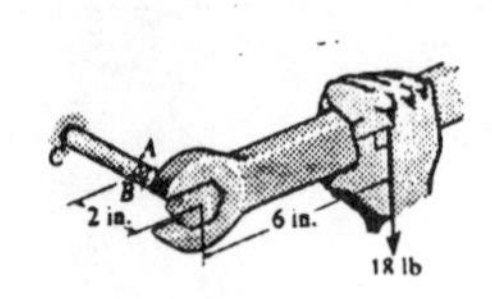

Internal Forces and Moment : As shown on FBD.

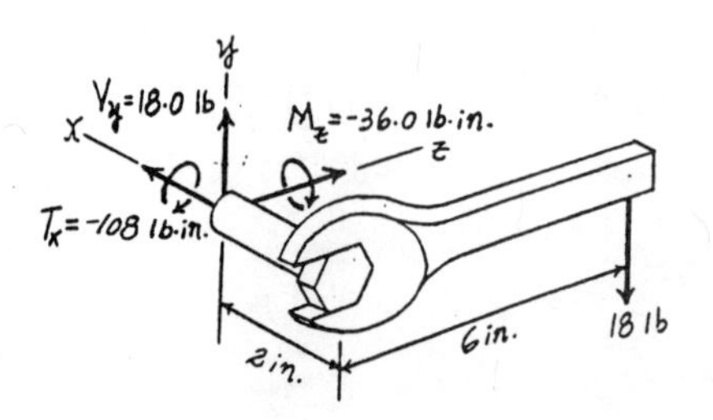

Section Properties :

$$I_z = \frac{\pi}{4}(0.125^4) = 0.191748(10^{-3})\ \text{in}^4$$

$$J = \frac{\pi}{2}(0.125^4) = 0.383495(10^{-3})\ \text{in}^4$$

$$Q_A = 0$$

Normal Stress : Applying the flexure formula,

$$\sigma_A = -\frac{M_z y}{I_z} = -\frac{-36.0(0.125)}{0.191748(10^{-3})} = 23.47\ \text{ksi}$$

Shear Stress : Applying the torsion formula,

$$\tau_A = \frac{T_x c}{J} = \frac{108(0.125)}{0.383495(10^{-3})} = 35.20\ \text{ksi}$$

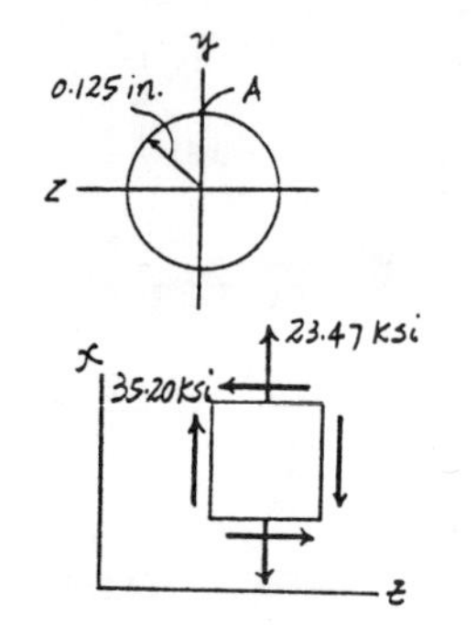

Construction of the Circle : $\sigma_z = 0$, $\sigma_x = 23.47$ ksi, and $\tau_{zx} = -35.20$ ksi. Hence,

$$\sigma_{avg} = \frac{\sigma_z + \sigma_x}{2} = \frac{0 + 23.47}{2} = 11.735\ \text{ksi}$$

The coordinates for reference points A and C are $A(0, -35.20)$ and $C(11.735, 0)$.

The radius of the circle is $R = \sqrt{(11.735 - 0)^2 + 35.20^2} = 37.11$ ksi

In - Plane Principal Stress (x - z) : The coordinates of points B and D represent σ_1 and σ_2, respectively.

$$\sigma_1 = 11.735 + 37.11 = 48.84\ \text{ksi}$$

$$\sigma_2 = 11.735 - 37.11 = -25.37\ \text{ksi}$$

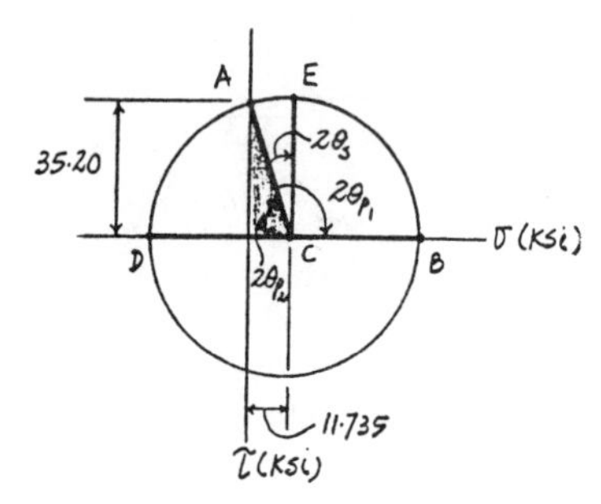

Orientaion of Principal Plane (x - z) : From the circle

$$\tan 2\theta_{p_2} = \frac{35.20}{11.735 - 0} = 3.00 \qquad 2\theta_{p_2} = 71.57°$$

$$2\theta_{p_1} = 180° - 2\theta_{p_2}$$

$$\theta_{p_1} = \frac{180° - 71.57°}{2} = 54.2°\ \textit{(Clockwise)}$$

Three Mohr's Circles : From the results obtained above, the principal stresses are

$$\sigma_{max} = 48.8\ \text{ksi} \qquad \sigma_{int} = 0 \qquad \sigma_{min} = -25.4\ \text{ksi} \qquad \textbf{Ans}$$

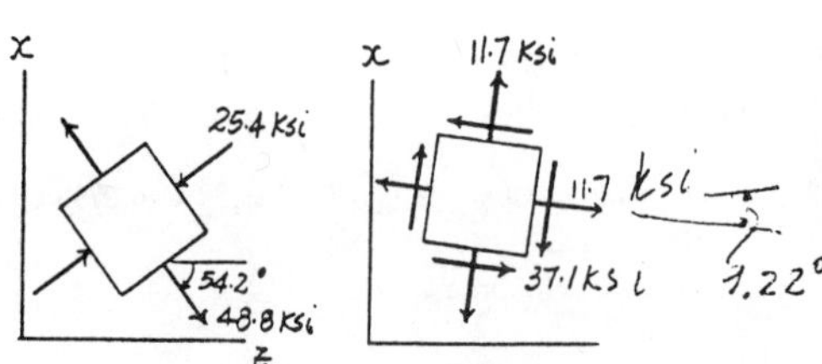

Absolute Maximum Shear Stress : The absolute maximum shear stress occurs within $x - z$ plane and the state of stress is represented by point E on the circle.

$$\tau_{\substack{abs \\ max}} = \frac{\sigma_{max} - \sigma_{min}}{2} = \frac{48.84 - (-25.37)}{2} = 37.1\ \text{ksi} \qquad \textbf{Ans}$$

And the orientation is

$$\tan 2\theta_s = \frac{11.735 - 0}{35.20} = 0.3333$$

$$\theta_s = 9.22°$$

9–94. Solve Prob. 9–93 for point B.

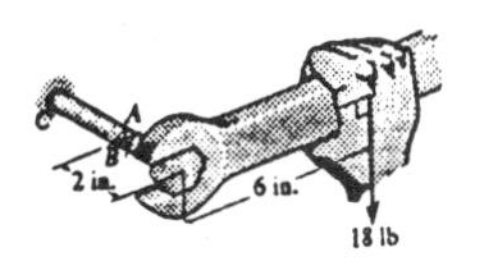

Internal Forces and Moment : As shown on FBD.

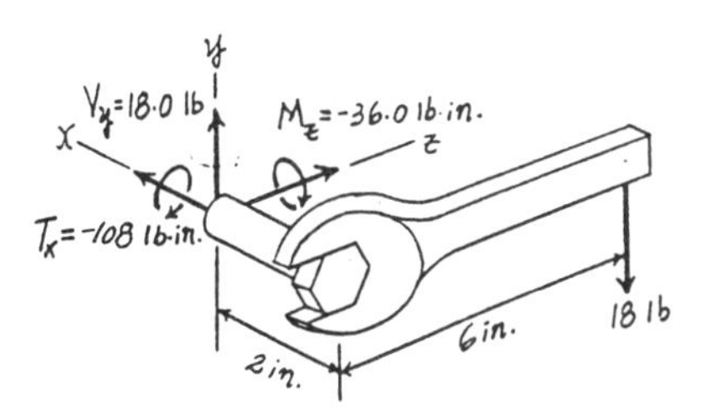

Section Properties :

$$I_z = \frac{\pi}{4}(0.125^4) = 0.191748(10^{-3})\ \text{in}^4$$

$$J = \frac{\pi}{2}(0.125^4) = 0.383495(10^{-3})\ \text{in}^4$$

$$Q_B = \frac{4(0.125)}{3\pi}\left[\frac{1}{2}(\pi)(0.125^2)\right] = 1.302083(10^{-3})\ \text{in}^4$$

Normal Stress : Applying the flexure formula,

$$\sigma_B = -\frac{M_z y}{I_z} = -\frac{-36.0(0)}{0.191748(10^{-3})} = 0$$

The transverse shear stress in the y direction andthe torsional shear stress can be obtained using shear formula andtorsion formula, $\tau_V = \frac{VQ}{It}$ $\tau_{twist} = \frac{T\rho}{J}$, respectively.

$$\tau_B = (\tau_V)_y - \tau_{twist}$$
$$= \frac{18.0[1.302083(10^{-3})]}{0.191748(10^{-3})(0.25)} - \frac{108(0.125)}{0.383495(10^{-3})}$$
$$= -34.71\ \text{ksi}$$

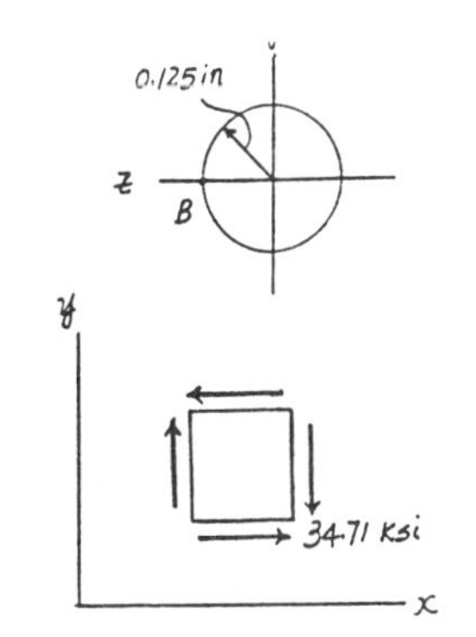

Construction of the Circle : $\sigma_x = \sigma_y = 0$, and $\tau_{xy} = -34.71$ ksi. Hence,

$$\sigma_{avg} = \frac{\sigma_z + \sigma_x}{2} = 0$$

The coordinates for reference points A and C are $A(0, -34.71)$ and $C(0, 0)$.

The radius of the circle is $R = 34.71$ ksi

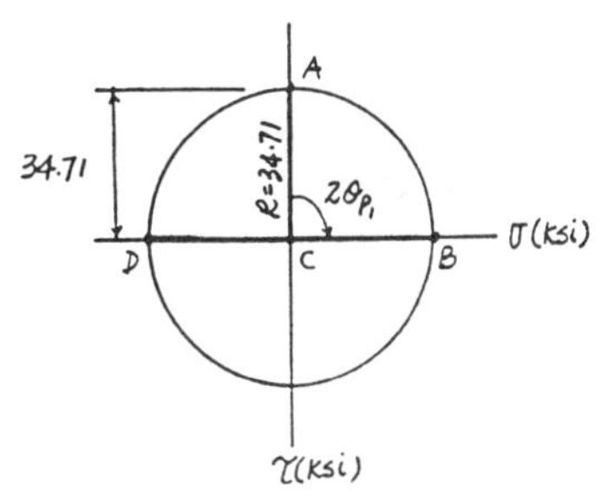

In-Plane Principal Stress (x - y) : The coordinates of points B and D represent σ_1 and σ_2, respectively.

$$\sigma_1 = 0 + 34.71 = 34.71\ \text{ksi}$$
$$\sigma_2 = 0 - 34.71 = -34.71\ \text{ksi}$$

Orientaion of Principal Plane (x - y) : From the circle

$$2\theta_{p_1} = 90° \qquad \theta_{p_1} = 45.0°\ (\textit{Clockwise})$$

Three Mohr's Circles : From the results obtained above, the principal stresses are

$$\sigma_{max} = 34.7\ \text{ksi} \qquad \sigma_{int} = 0 \qquad \sigma_{min} = -34.7\text{ksi} \qquad \textbf{Ans}$$

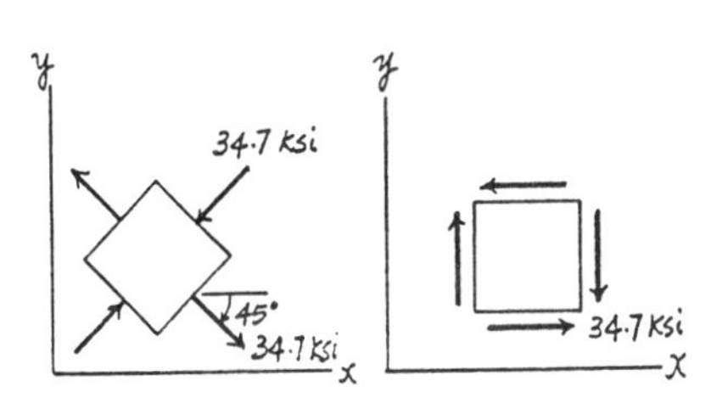

Absolute Maximum Shear Stress : The absolute maximum shear stress occurs within the $x-y$ plane and the state of stress is represented bypoint A on the circle.

$$\tau_{\substack{abs\\max}} = \frac{\sigma_{max} - \sigma_{min}}{2} = \frac{34.71 - (-34.71)}{2} = 34.7\ \text{ksi} \qquad \textbf{Ans}$$

9–95. The state of stress at a point is shown on the element. Determine (a) the principal stresses and (b) the maximum in-plane shear stress and the average normal stress at the point. Specify the orientation of the element in each case.

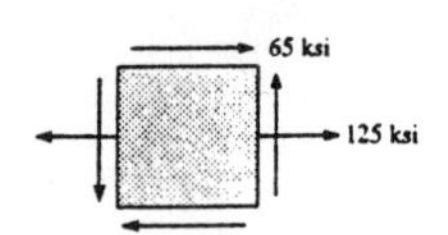

Construction of the Circle : In accordance with the sign convention, $\sigma_x = 125$ ksi, $\sigma_y = 0$, and $\tau_{xy} = 65$ ksi. Hence,

$$\sigma_{avg} = \frac{\sigma_x + \sigma_y}{2} = \frac{125+0}{2} = 62.5 \text{ ksi} \qquad \textbf{Ans}$$

The coordinates for reference points A and C are $A(125,\ 65)$ and $C(62.5,\ 0)$.

The radius of the circle is $R = \sqrt{(125-62.5)^2 + 65^2} = 90.173$ ksi

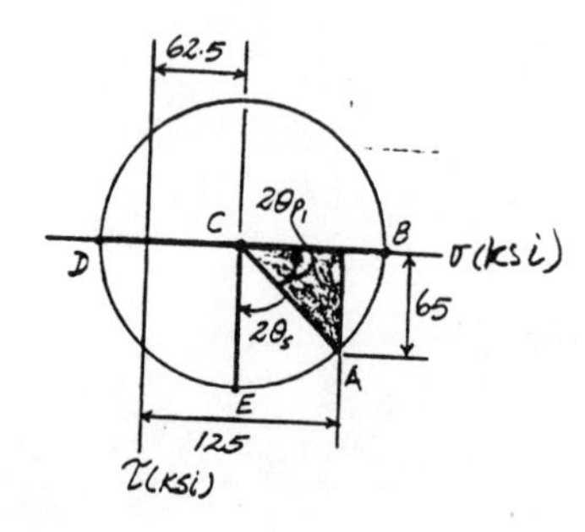

a)

In-Plane Principal Stress : The coordinates of points B and D represent σ_1 and σ_2, respectively.

$$\sigma_1 = 62.5 + 90.173 = 153 \text{ ksi} \qquad \textbf{Ans}$$
$$\sigma_2 = 62.5 - 90.173 = -27.7 \text{ ksi} \qquad \textbf{Ans}$$

Orientaion of Principal Plane : From the circle

$$\tan 2\theta_{p_2} = \frac{65}{125-62.5} = 1.04$$

$$\theta_{p_1} = 23.1° \ \ (\textbf{\textit{Counterclockwise}}) \qquad \textbf{Ans}$$

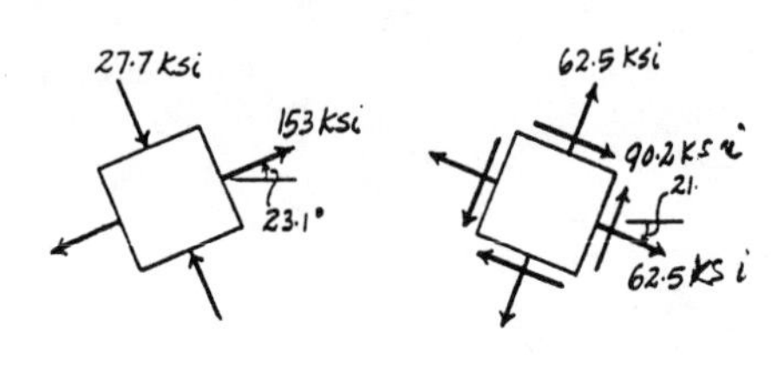

b)

Maximum In-Plane Shear Stress : Represented by the coordinates of point E on the circle.

$$\tau_{\substack{max \\ in\text{-}plane}} = R = 90.2 \text{ ksi} \qquad \textbf{Ans}$$

Orientation of the Plane for Maximum In-Plane Shear Stress : From the circle

$$\tan 2\theta_s = \frac{125-62.5}{65} = 0.9615$$
$$\theta_s = 21.9° \ \ (\textbf{\textit{Clockwise}}) \qquad \textbf{Ans}$$

***9–96.** Draw the three Mohr's circles for the state of stress shown, and determine the absolute maximum shear stress.

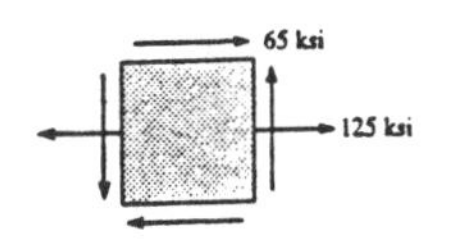

Construction of the Circle : In accordance to the sign convention, $\sigma_x = 125$ ksi, $\sigma_y = 0$, and $\tau_{xy} = 65$ ksi. Hence,

$$\sigma_{avg} = \frac{\sigma_x + \sigma_y}{2} = \frac{125 + 0}{2} = 62.5 \text{ ksi}$$

The coordinates for reference points A and C are $A(125, 65)$ and $C(62.5, 0)$.

The radius of the circle is $R = \sqrt{(125 - 62.5)^2 + 65^2} = 90.173$ ksi

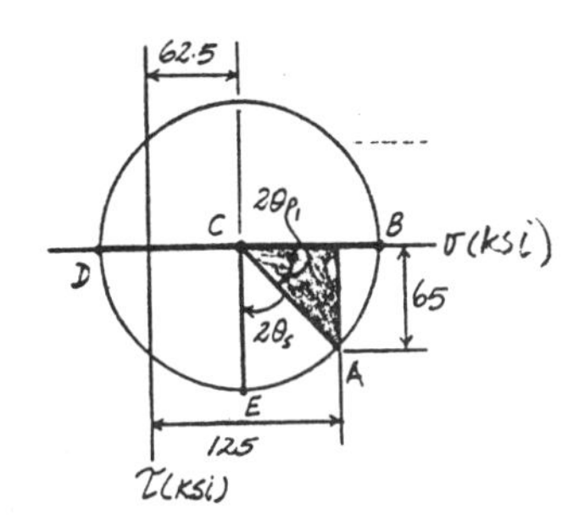

In - Plane Principal Stress : The coordinates of points B and D represent σ_1 and σ_2, respectively.

$$\sigma_1 = 62.5 + 90.173 = 152.67 \text{ ksi}$$
$$\sigma_2 = 62.5 - 90.173 = -27.67 \text{ ksi}$$

Three Mohr's Circles : From the results obtained above, the principal stresses are

$\sigma_{max} = 153$ ksi $\quad \sigma_{int} = 0 \quad \sigma_{min} = -27.7$ ksi **Ans**

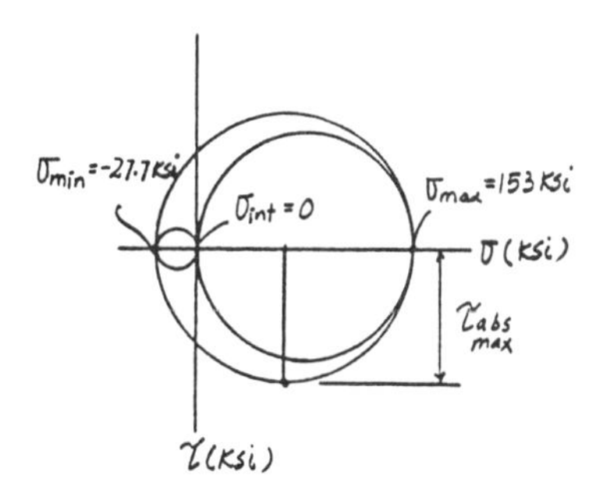

Absolute Maximum Shear Stress : The absolute maximum shear stress occurs within the $x-y$ plane.

$$\tau_{\substack{abs \\ max}} = \frac{\sigma_{max} - \sigma_{min}}{2} = \frac{152.67 - (-27.67)}{2} = 90.2 \text{ ksi} \quad \textbf{Ans}$$

9–97. The internal loadings on a cross section through the 6-in.-diameter drive shaft of a turbine consist of an axial force of 2500 lb, a bending moment of 800 lb · ft. and a torsional moment of 1500 lb · ft. Determine the principal stresses at point A. Also compute the maximum in-plane shear stress at this point.

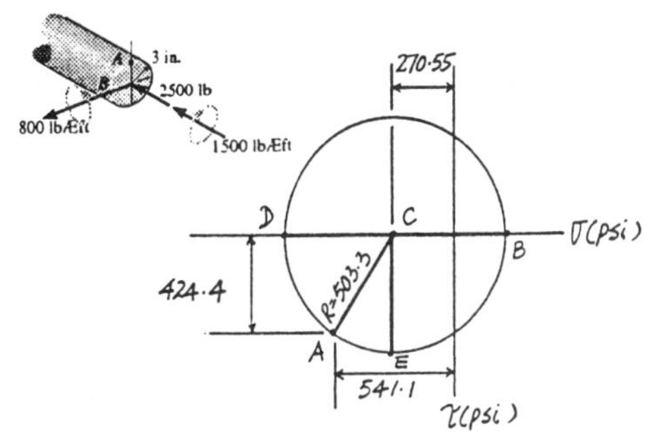

Section Properties :

$$A = \pi(3^2) = 9.0\pi \text{ in}^2 \qquad I = \frac{\pi}{4}(3^4) = 20.25\pi \text{ in}^4$$

$$J = \frac{\pi}{2}(3^4) = 40.5\pi \text{ in}^4$$

Normal Stress :

424.4 psi
541.1 psi

$$\sigma = \frac{N}{A} \pm \frac{My}{I}$$

$$\sigma_A = \frac{-2500}{9.0\pi} - \frac{800(12)(3)}{20.25\pi} = -541.1 \text{ psi}$$

Shear Stress : Applying the torsion formula,

$$\tau_A = \frac{Tc}{J} = \frac{1500(12)(3)}{40.5\pi} = 424.4 \text{ psi}$$

Construction of the Circle : In accordance with the sign convention, $\sigma_x = -541.1$ psi, $\sigma_y = 0$, and $\tau_{xy} = 424.4$ psi. Hence,

$$\sigma_{avg} = \frac{\sigma_x + \sigma_y}{2} = \frac{-541.1 + 0}{2} = -270.55 \text{ psi}$$

The coordinates for reference points A and C are $A(-541.1, 424.4)$ and $C(-270.55, 0)$.

The radius of the circle is

$$R = \sqrt{(541.1 - 270.55)^2 + 424.4} = 503.32 \text{ psi}$$

In - Plane Principal Stress : The coordinates of points B and D represent σ_1 and σ_2, respectively.

$\sigma_1 = -270.55 + 503.32 = 233$ psi **Ans**

$\sigma_2 = -270.55 - 503.32 = -774$ psi **Ans**

Maximum In - Plane Shear Stress : Represented by the coordinates of point E on the circle.

$\tau_{\substack{max \\ in\text{-}plane}} = R = 503$ psi **Ans**

9–98. The internal loadings at a cross section through the 6-in.-diameter drive shaft of a turbine consist of an axial force of 2500 lb, a bending moment of 800 lb · ft, and a torsional moment of 1500 lb · ft. Determine the principal stresses at point B. Also compute the maximum in-plane shear stress at this point.

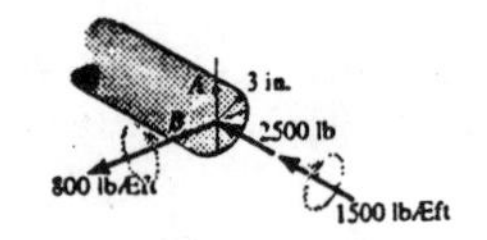

Section Properties :

$$A = \pi(3^2) = 9.0\pi \text{ in}^2 \qquad I = \frac{\pi}{4}(3^4) = 20.25\pi \text{ in}^4$$

$$J = \frac{\pi}{2}(3^4) = 40.5\pi \text{ in}^4$$

Normal Stress :

$$\sigma = \frac{N}{A} \pm \frac{My}{I}$$

$$\sigma_A = \frac{-2500}{9.0\pi} + \frac{800(12)(0)}{20.25\pi} = -88.42 \text{ psi}$$

Shear Stress : Applying the torsion formula,

$$\tau_A = \frac{Tc}{J} = \frac{1500(12)(3)}{40.5\pi} = 424.4 \text{ psi}$$

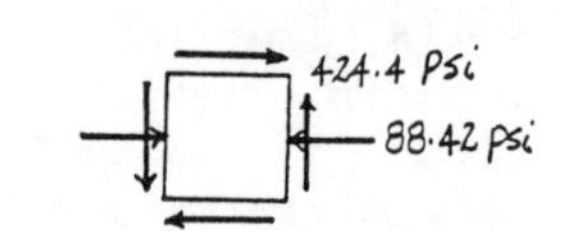

Construction of the Circle : In accordance with the sign convention, $\sigma_x = -88.42$ psi, $\sigma_y = 0$, and $\tau_{xy} = 424.4$ psi. Hence,

$$\sigma_{avg} = \frac{\sigma_x + \sigma_y}{2} = \frac{-88.42 + 0}{2} = -44.21 \text{ psi}$$

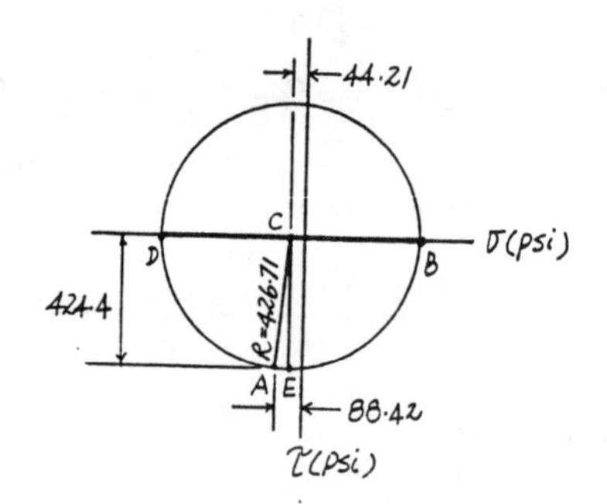

The coordinates for reference points A and C are $A(-88.42,\ 424.4)$ and $C(-44.21,\ 0)$.

The radius of the circle is

$$R = \sqrt{(88.42 - 44.21)^2 + 424.4} = 426.71 \text{ psi}$$

In - Plane Principal Stress : The coordinates of points B and D represent σ_1 and σ_2, respectively.

$$\sigma_1 = -44.21 + 426.71 = 382 \text{ psi} \qquad \textbf{Ans}$$

$$\sigma_2 = -44.21 - 426.71 = -471 \text{ psi} \qquad \textbf{Ans}$$

Maximum In - Plane Shear Stress : Represented by the coordinates of point E on the circle.

$$\tau_{\substack{\max \\ \text{in-plane}}} = R = 427 \text{ psi} \qquad \textbf{Ans}$$

9–99. A bar has a circular cross section with a diameter of 1 in. It is subjected to a torque and a bending moment. At the point of maximum bending stress the principal tensile stresses are 20 ksi and −10 ksi. Determine the torque and the bending moment.

Mohr's Circle : Under the given loading condition, $\sigma_y = 0$. From the circle,

$$\sigma_{avg} = \frac{\sigma_x + \sigma_y}{2}; \qquad 5 = \frac{\sigma_x + 0}{2} \qquad \sigma_x = 10.0 \text{ ksi}$$

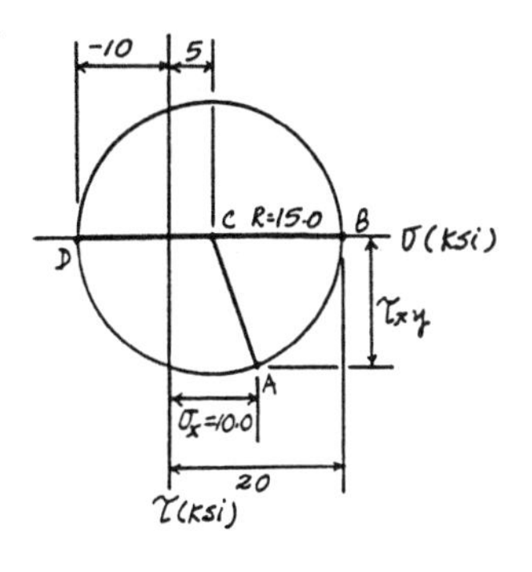

The radius of the circle is $R = 20 - 5 = 15.0$ ksi.

$$15.0 = \sqrt{(10.0-5)^2 + \tau_{xy}^2} \qquad \tau_{xy} = 14.14 \text{ ksi}$$

Normal Stress : Applying the flexure formula,

$$\sigma = \frac{Mc}{I}$$

$$10.0 = \frac{M(0.5)}{\frac{\pi}{4}(0.5^4)}$$

$$M = 0.9817 \text{ kip} \cdot \text{in} = 81.8 \text{ lb} \cdot \text{ft} \qquad \textbf{Ans}$$

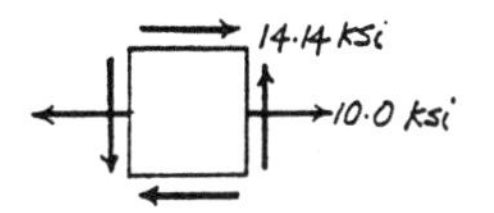

Shear Stress : Applying the torsion formula,

$$\tau = \frac{Tc}{J}$$

$$14.14 = \frac{T(0.5)}{\frac{\pi}{2}(0.5^4)}$$

$$T = 2.777 \text{ kip} \cdot \text{in} = 231 \text{ lb} \cdot \text{ft} \qquad \textbf{Ans}$$

***9–100.** The box beam is subjected to the loading shown. Determine the principal stresses in the beam at points A and B.

Support Reactions : As shown on FBD(a).

Internal Forces and Moment : As shown on FBD(b).

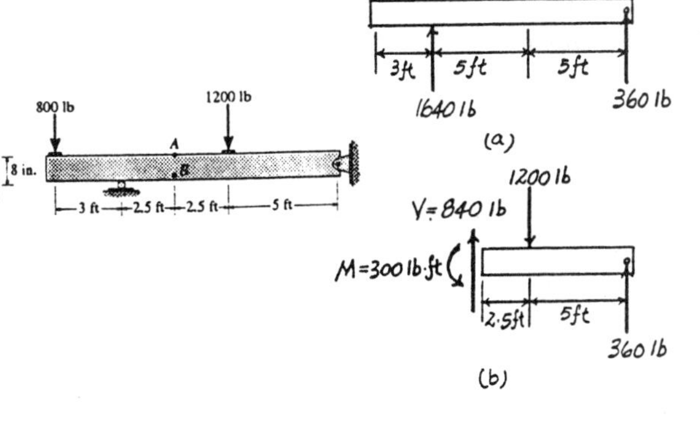

Section Properties :

$$I = \frac{1}{12}(8)(8^3) - \frac{1}{12}(6)(6^3) = 233.33 \text{ in}^4$$

$$Q_A = Q_B = 0$$

Normal Stress : Applying the flexure formula.

$$\sigma = -\frac{My}{I}$$

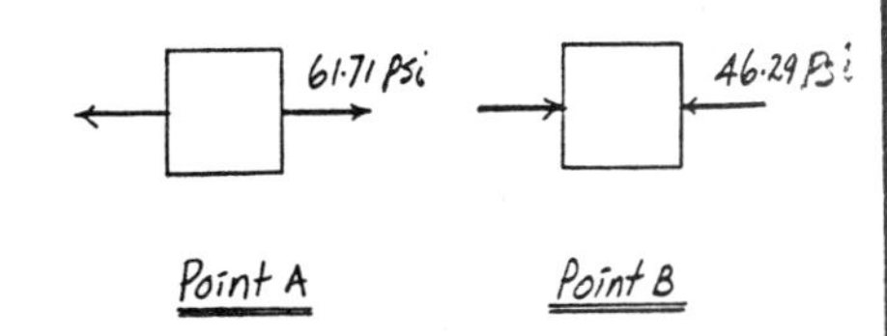

$$\sigma_A = -\frac{-300(12)(4)}{233.33} = 61.71 \text{ psi}$$

$$\sigma_B = -\frac{-300(12)(-3)}{233.33} = -46.29 \text{ psi}$$

Shear Stress : Since $Q_A = Q_B = 0$, then $\tau_A = \tau_B = 0$.

In - Plane Principal Stress : $\sigma_x = 61.71$ psi, $\sigma_y = 0$, and $\tau_{xy} = 0$ for point A. Since no shear stress acts on the element,

$$\sigma_1 = \sigma_x = 61.7 \text{ psi} \qquad \textbf{Ans}$$

$$\sigma_2 = \sigma_y = 0 \qquad \textbf{Ans}$$

$\sigma_x = -46.29$ psi, $\sigma_y = 0$, and $\tau_{xy} = 0$ for point B. Since no shear stress acts on the element,

$$\sigma_1 = \sigma_y = 0 \qquad \textbf{Ans}$$

$$\sigma_2 = \sigma_x = -46.3 \text{ psi} \qquad \textbf{Ans}$$

9–101. The state of stress at a point in a member is shown on the element. Determine the stress components acting on the inclined plane AB.

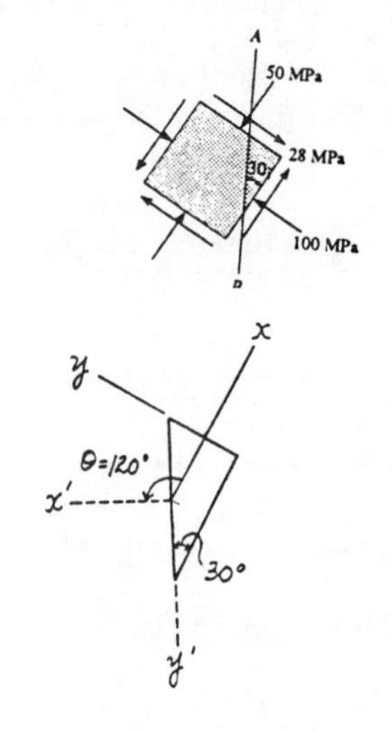

Construction of the Circle : In accordance with the sign convention, $\sigma_x = -50$ MPa, $\sigma_y = -100$ MPa, and $\tau_{xy} = -28$ MPa. Hence,

$$\sigma_{avg} = \frac{\sigma_x + \sigma_y}{2} = \frac{-50 + (-100)}{2} = -75.0 \text{ MPa}$$

The coordinates for reference points A and C are $A(-50,\ -28)$ and $C(-75.0,\ 0)$.

The radius of the circle is $R = \sqrt{(75.0 - 50)^2 + 28^2} = 37.54$ MPa.

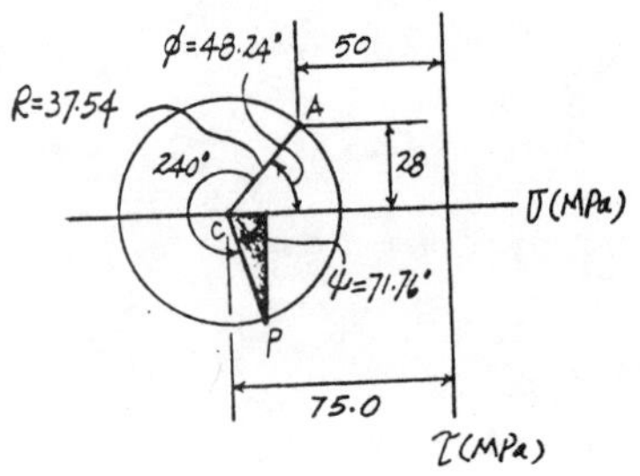

Stress on the Rotated Element : The normal and shear stress components ($\sigma_{x'}$ and $\tau_{x'y'}$) are represented by the coordinates of point P on the circle.

$$\sigma_{x'} = -75.0 + 37.54 \cos 71.76° = -63.3 \text{ MPa} \qquad \textbf{Ans}$$

$$\tau_{x'y'} = 37.54 \sin 71.76° = 35.7 \text{ MPa} \qquad \textbf{Ans}$$

9–102. The square steel plate has a thickness of 10 mm and is subjected to the edge loading shown. Determine the maximum in-plane shear stress and the average normal stress developed in the steel.

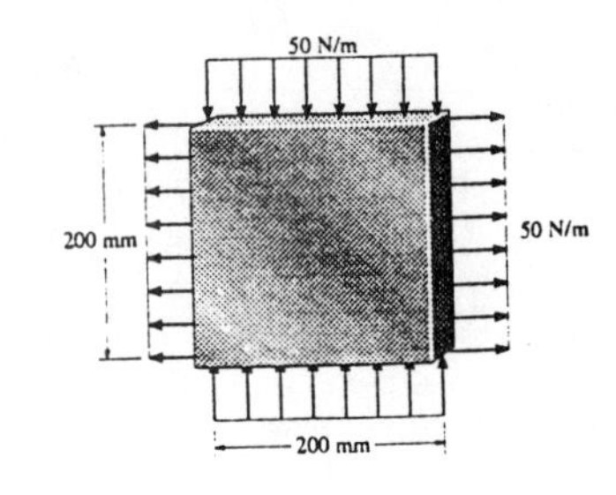

Construction of the Circle : In accordance with the sign convention, $\sigma_x = \frac{50}{0.01} = 5.00$ kPa, $\sigma_y = -\frac{50}{0.01} = -5.00$ kPa, and $\tau_{xy} = 0$. Hence,

$$\sigma_{avg} = \frac{\sigma_x + \sigma_y}{2} = \frac{5.00 + (-5.00)}{2} = 0 \qquad \textbf{Ans}$$

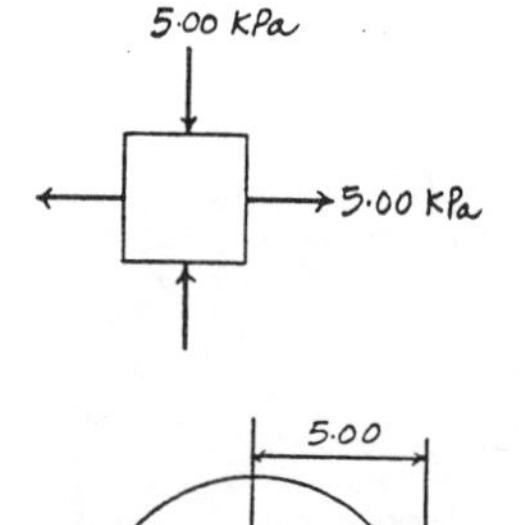

The coordinates for reference points A and C are $A(5.00,\ 0)$ and $C(0,\ 0)$.

The radius of the circle is $R = 5.00 - 0 = 5.00$ kPa

Maximum In - Plane Shear Stress : Represented by the coordinates of point E on the circle.

$$\tau_{\substack{max \\ in\text{-}plane}} = R = 5.00 \text{ kPa} \qquad \textbf{Ans}$$

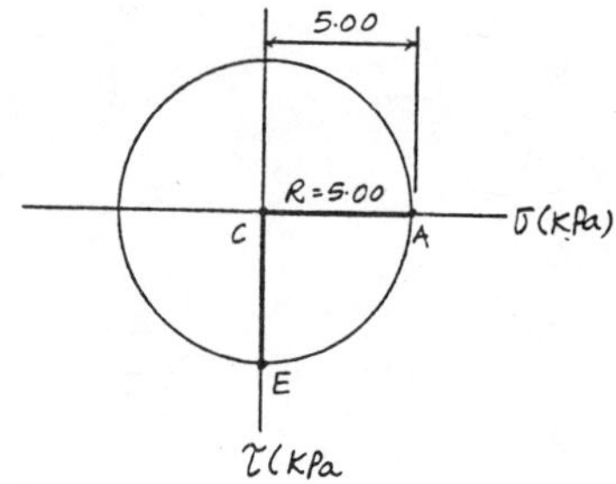

9–103. The square steel plate has a thickness of 10 mm and is subjected to the edge loading shown. Determine the principal stresses and the absolute maximum shear stress developed in the steel.

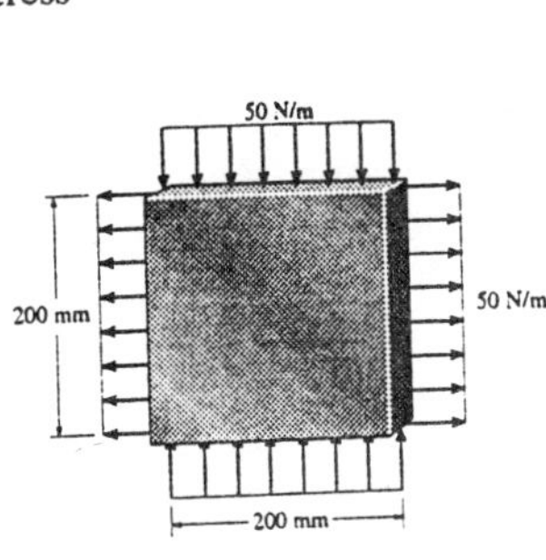

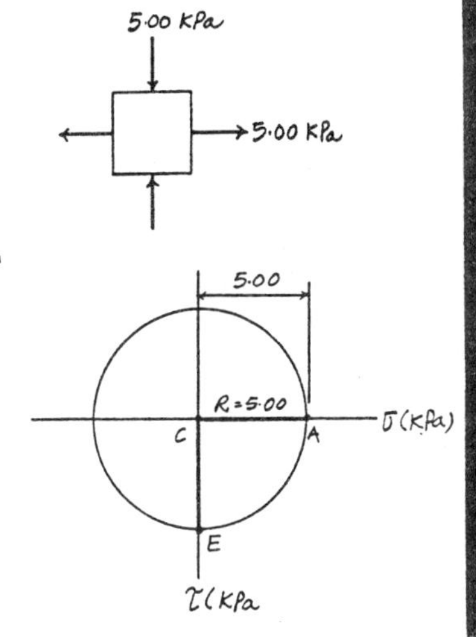

Construction of the Circle : In accordance with the sign convention, $\sigma_x = \dfrac{50}{0.01} = 5.00$ kPa, $\sigma_y = -\dfrac{50}{0.01} = -5.00$ kPa, and $\tau_{xy} = 0$. Hence,

$$\sigma_{avg} = \frac{\sigma_x + \sigma_y}{2} = \frac{5.00 + (-5.00)}{2} = 0$$

The coordinates for reference points A and C are $A(5.00,\ 0)$ and $C(0,\ 0)$.

The radius of the circle is $R = 5.00 - 0 = 5.00$ kPa. Then,

$$\sigma_1 = 5.00 \text{ kPa} \qquad \sigma_2 = -5.00 \text{ kPa}$$

Three Mohr's Circles : From the results obtained above, the principal stresses are

$$\sigma_{max} = 5.00 \text{ kPa} \qquad \sigma_{int} = 0 \qquad \sigma_{min} = -5.00 \text{kPa} \qquad \textbf{Ans}$$

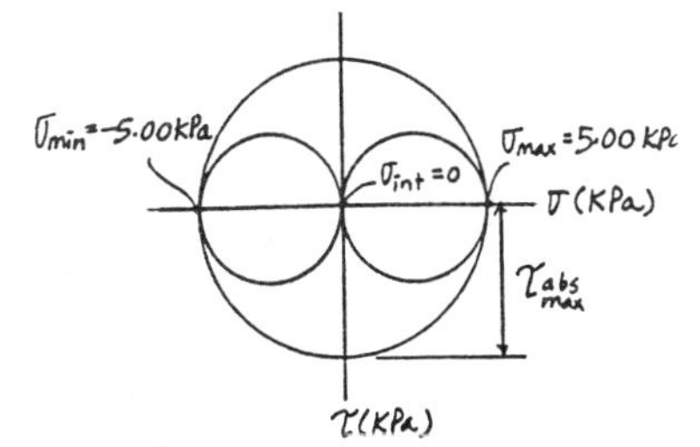

Absolute Maximum Shear Stress :

$$\tau_{\substack{abs \\ max}} = \frac{\sigma_{max} - \sigma_{min}}{2} = \frac{5.00 - (-5.00)}{2} = 5.00 \text{ kPa} \qquad \textbf{Ans}$$

***9–104.** The cylindrical pressure vessel has an inner radius of 1.25 m and a wall thickness of 15 mm. It is made from steel plates that are welded along a seam which makes an angle of 45 with the horizontal. Determine the normal and shear stress components along this seam if the vessel is subjected to an internal pressure of 3 MPa.

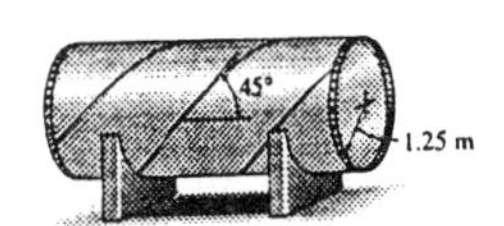

Normal Stress : Since $\dfrac{r}{t} = \dfrac{1250}{15} = 83.3 > 10$, *thin wall* analysis for a cylindrical pipe is valid.

$$\sigma_{long.} = \frac{pr}{2t} = \frac{3(1250)}{2(15)} = 125 \text{ MPa}$$

$$\sigma_{hoop} = \frac{pr}{t} = \frac{3(1250)}{15} = 250 \text{ MPa}$$

Construction of the Circle : In accordance with the sign convention, $\sigma_x = 125$ MPa, $\sigma_y = 250$ MPa, and $\tau_{xy} = 0$. Hence,

$$\sigma_{avg} = \frac{\sigma_x + \sigma_y}{2} = \frac{125 + 250}{2} = 187.5 \text{ MPa}$$

The coordinates for reference points A and C are $A(125,\ 0)$ and $C(187.5,\ 0)$.

The radius of the circle is $R = 187.5 - 125 = 62.5$ MPa

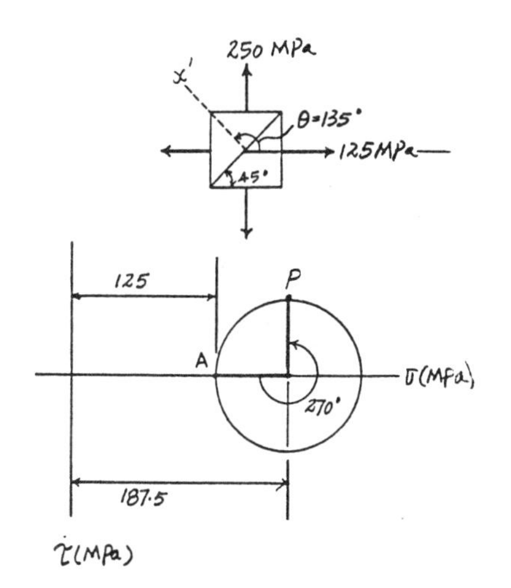

Stress on The Rotated Element : The normal and shear stress components ($\sigma_{x'}$ and $\tau_{x'y'}$) are represented by the coordinates of point P on the circle.

$$\sigma_{x'} = 187.5 \text{ MPa} \qquad \textbf{Ans}$$

$$\tau_{x'y'} = -62.5 \text{ MPa} \qquad \textbf{Ans}$$

10–1. Prove that the sum of the normal strains in perpendicular directions is constant.

Strain Transformation Equations : Applying Eqs. 10 – 5 and 10 – 7,

$$\varepsilon_{x'} = \frac{\varepsilon_x + \varepsilon_y}{2} + \frac{\varepsilon_x - \varepsilon_y}{2}\cos 2\theta + \frac{\gamma_{xy}}{2}\sin 2\theta \qquad [1]$$

$$\varepsilon_{y'} = \frac{\varepsilon_x + \varepsilon_y}{2} - \frac{\varepsilon_x - \varepsilon_y}{2}\cos 2\theta - \frac{\gamma_{xy}}{2}\sin 2\theta \qquad [2]$$

Eq. [1] + Eq. [2] yields :

$$\varepsilon_{x'} + \varepsilon_{y'} = \varepsilon_x + \varepsilon_y = \text{Constant} \qquad (\textit{Q.E.D.})$$

10–2. The state of strain at the point on the leaf of the caster assembly has components of $\epsilon_x = -400(10^{-6})$, $\epsilon_y = 860(10^{-6})$, and $\gamma_{xy} = 375(10^{-6})$. Use the strain-transformation equations to determine the equivalent in-plane strains on an element oriented at an angle of $\theta = 30°$ counterclockwise from the original position. Sketch the deformed element due to these strains within the x–y plane .

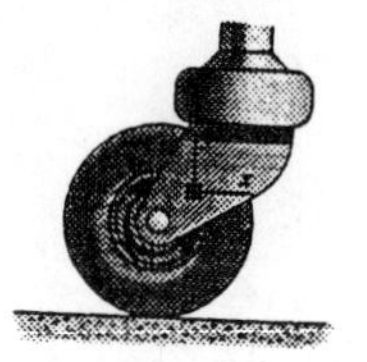

Normal Strain and Shear strain : In accordance with the sign convention,

$$\varepsilon_x = -400(10^{-6}) \qquad \varepsilon_y = 860(10^{-6}) \qquad \gamma_{xy} = 375(10^{-6})$$
$$\theta = +30°$$

Strain Transformation Equations : Applying Eqs. 10 – 5, 10 – 6, and 10 – 7,

$$\varepsilon_{x'} = \frac{\varepsilon_x + \varepsilon_y}{2} + \frac{\varepsilon_x - \varepsilon_y}{2}\cos 2\theta + \frac{\gamma_{xy}}{2}\sin 2\theta$$
$$= \left(\frac{-400+860}{2} + \frac{-400-860}{2}\cos 60° + \frac{375}{2}\sin 60°\right)(10^{-6})$$
$$= 77.4(10^{-6}) \qquad \textbf{Ans}$$

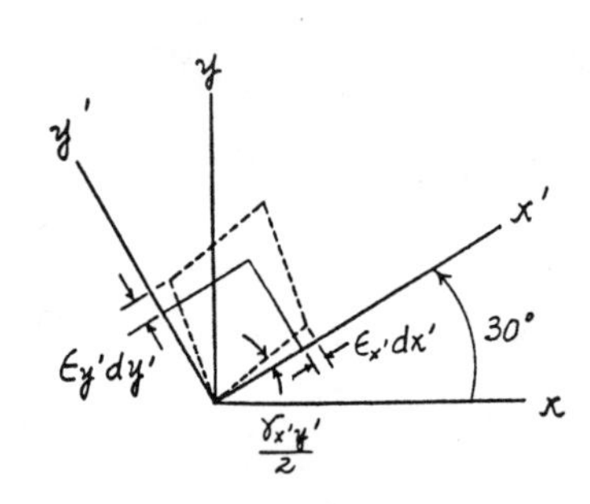

$$\frac{\gamma_{x'y'}}{2} = -\frac{\varepsilon_x - \varepsilon_y}{2}\sin 2\theta + \frac{\gamma_{xy}}{2}\cos 2\theta$$
$$\gamma_{x'y'} = [-(-400-860)\sin 60° + 375\cos 60°](10^{-6})$$
$$= 1279(10^{-6}) \qquad \textbf{Ans}$$

$$\varepsilon_{y'} = \frac{\varepsilon_x + \varepsilon_y}{2} - \frac{\varepsilon_x - \varepsilon_y}{2}\cos 2\theta - \frac{\gamma_{xy}}{2}\sin 2\theta$$
$$= \left(\frac{-400+860}{2} - \frac{-400-860}{2}\cos 60° - \frac{375}{2}\sin 60°\right)(10^{-6})$$
$$= 383(10^{-6}) \qquad \textbf{Ans}$$

10–3. The state of strain at the point on the pin leaf has components of $\epsilon_x = 200(10^{-6})$, $\epsilon_y = 180(10^{-6})$, and $\gamma_{xy} = -300(10^{-6})$. Use the strain-transformation equations and determine the equivalent in-plane strains on an element oriented at an angle of $\theta = 60°$ counterclockwise from the original position. Sketch the deformed element due to these strains within the x–y plane.

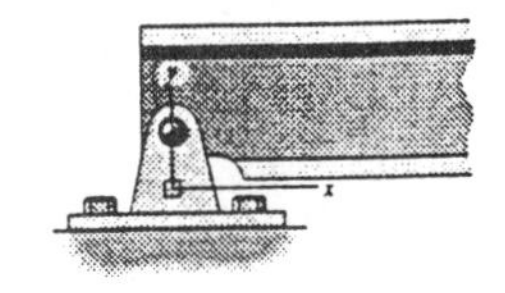

Normal Strain and Shear strain : In accordance with the sign convention,

$$\varepsilon_x = 200(10^{-6}) \qquad \varepsilon_y = 180(10^{-6}) \qquad \gamma_{xy} = -300(10^{-6})$$
$$\theta = +60°$$

Strain Transformation Equations : Applying Eqs. 10 – 5, 10 – 6, and 10 – 7,

$$\varepsilon_{x'} = \frac{\varepsilon_x + \varepsilon_y}{2} + \frac{\varepsilon_x - \varepsilon_y}{2}\cos 2\theta + \frac{\gamma_{xy}}{2}\sin 2\theta$$
$$= \left(\frac{200+180}{2} + \frac{200-180}{2}\cos 120° + \frac{-300}{2}\sin 120°\right)(10^{-6})$$
$$= 55.1(10^{-6}) \qquad \textbf{Ans}$$

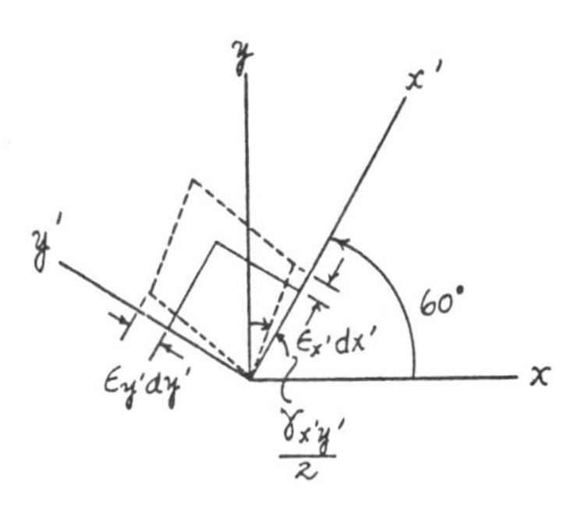

$$\frac{\gamma_{x'y'}}{2} = -\frac{\varepsilon_x - \varepsilon_y}{2}\sin 2\theta + \frac{\gamma_{xy}}{2}\cos 2\theta$$
$$\gamma_{x'y'} = [-(200-180)\sin 120° + (-300)\cos 120°](10^{-6})$$
$$= 133(10^{-6}) \qquad \textbf{Ans}$$

$$\varepsilon_{y'} = \frac{\varepsilon_x + \varepsilon_y}{2} - \frac{\varepsilon_x - \varepsilon_y}{2}\cos 2\theta - \frac{\gamma_{xy}}{2}\sin 2\theta$$
$$= \left(\frac{200+180}{2} - \frac{200-180}{2}\cos 120° - \frac{-300}{2}\sin 120°\right)(10^{-6})$$
$$= 325(10^{-6}) \qquad \textbf{Ans}$$

***10–4.** Solve Prob. 10–3 for an element oriented $\theta = 30°$ clockwise.

Normal Strain and Shear strain : In accordance with the sign convention,

$$\varepsilon_x = 200(10^{-6}) \qquad \varepsilon_y = 180(10^{-6}) \qquad \gamma_{xy} = -300(10^{-6})$$
$$\theta = -30°$$

Strain Transformation Equations : Applying Eqs. 10 – 5, 10 – 6, and 10 – 7,

$$\varepsilon_{x'} = \frac{\varepsilon_x + \varepsilon_y}{2} + \frac{\varepsilon_x - \varepsilon_y}{2}\cos 2\theta + \frac{\gamma_{xy}}{2}\sin 2\theta$$
$$= \left[\frac{200+180}{2} + \frac{200-180}{2}\cos(-60°) + \frac{-300}{2}\sin(-60°)\right](10^{-6})$$
$$= 325(10^{-6}) \qquad \textbf{Ans}$$

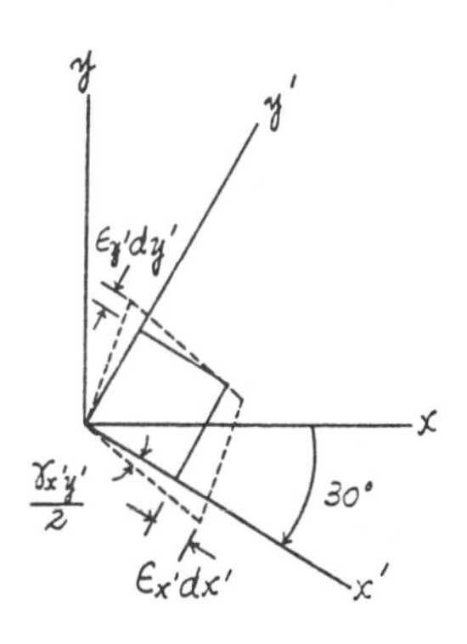

$$\frac{\gamma_{x'y'}}{2} = -\frac{\varepsilon_x - \varepsilon_y}{2}\sin 2\theta + \frac{\gamma_{xy}}{2}\cos 2\theta$$
$$\gamma_{x'y'} = [-(200-180)\sin(-60°) + (-300)\cos(-60°)](10^{-6})$$
$$= -133(10^{-6}) \qquad \textbf{Ans}$$

$$\varepsilon_{y'} = \frac{\varepsilon_x + \varepsilon_y}{2} - \frac{\varepsilon_x - \varepsilon_y}{2}\cos 2\theta - \frac{\gamma_{xy}}{2}\sin 2\theta$$
$$= \left(\frac{200+180}{2} - \frac{200-180}{2}\cos(-60°) - \frac{-300}{2}\sin(-60°)\right)(10^{-6})$$
$$= 55.1(10^{-6}) \qquad \textbf{Ans}$$

10–5. Due to the load **P**, the state of strain at the point on the bracket has components of $\epsilon_x = 500(10^{-6})$, $\epsilon_y = 350(10^{-6})$, and $\gamma_{xy} = -430(10^{-6})$. Use the strain-transformation equations to determine the equivalent in-plane strains on an element oriented at an angle of $\theta = 30°$ clockwise from the original position. Sketch the deformed element due to these strains within the x–y plane.

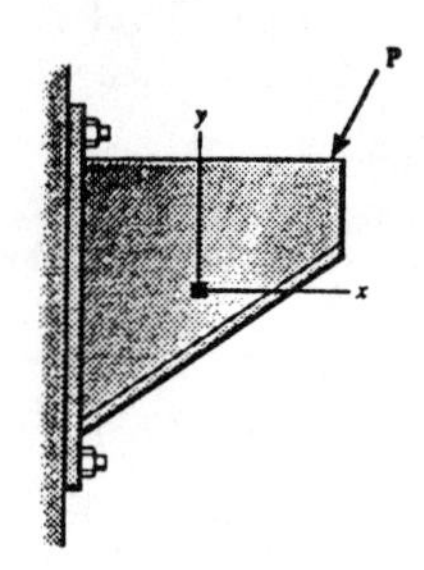

Normal Strain and Shear strain : In accordance with the sign convention,

$$\varepsilon_x = 500(10^{-6}) \qquad \varepsilon_y = 350(10^{-6}) \qquad \gamma_{xy} = -430(10^{-6})$$
$$\theta = -30°$$

Strain Transformation Equations : Applying Eqs. 10 – 5, 10 – 6, and 10 – 7,

$$\varepsilon_{x'} = \frac{\varepsilon_x + \varepsilon_y}{2} + \frac{\varepsilon_x - \varepsilon_y}{2}\cos 2\theta + \frac{\gamma_{xy}}{2}\sin 2\theta$$
$$= \left[\frac{500+350}{2} + \frac{500-350}{2}\cos(-60°) + \frac{-430}{2}\sin(-60°)\right](10^{-6})$$
$$= 649(10^{-6}) \qquad \textbf{Ans}$$

$$\frac{\gamma_{x'y'}}{2} = -\frac{\varepsilon_x - \varepsilon_y}{2}\sin 2\theta + \frac{\gamma_{xy}}{2}\cos 2\theta$$
$$\gamma_{x'y'} = [-(500-350)\sin(-60°) + (-430)\cos(-60°)](10^{-6})$$
$$= -85.1(10^{-6}) \qquad \textbf{Ans}$$

$$\varepsilon_{y'} = \frac{\varepsilon_x + \varepsilon_y}{2} - \frac{\varepsilon_x - \varepsilon_y}{2}\cos 2\theta - \frac{\gamma_{xy}}{2}\sin 2\theta$$
$$= \left(\frac{500+350}{2} - \frac{500-350}{2}\cos(-60°) - \frac{-430}{2}\sin(-60°)\right)(10^{-6})$$
$$= 201(10^{-6}) \qquad \textbf{Ans}$$

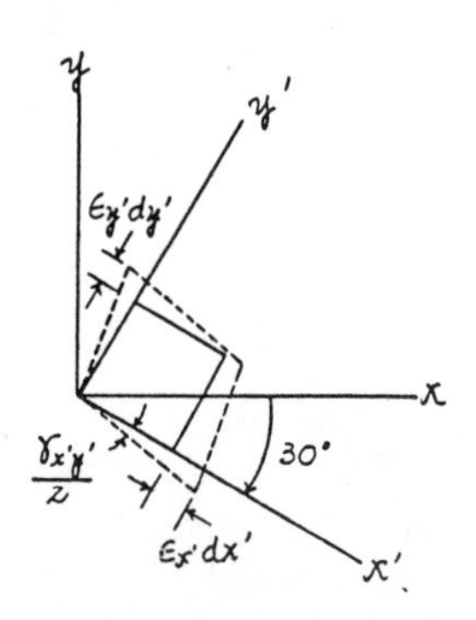

10–6. The state of strain at the point on the boom of the hydraulic engine crane has components of $\epsilon_x = 250(10^{-6})$, $\epsilon_y = 300(10^{-6})$, and $\gamma_{xy} = -180(10^{-6})$. Use the strain-transformation equations to determine (a) the in-plane principal strains and (b) the maximum in-plane shear strain and average normal strain. In each case specify the orientation of the element and show how the strains deform the element within the x–y plane.

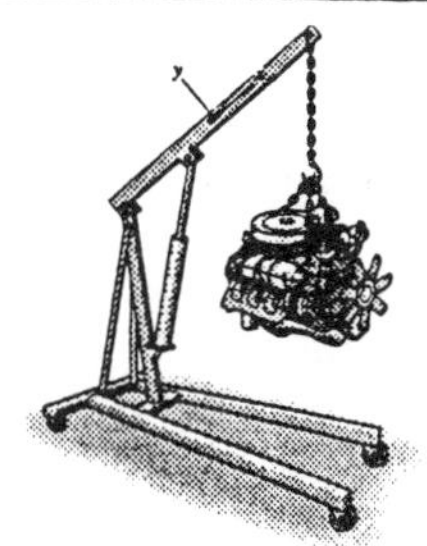

Normal Strain and Shear strain : In accordance with the sign convention,

$$\varepsilon_x = 250(10^{-6}) \qquad \varepsilon_y = 300(10^{-6}) \qquad \gamma_{xy} = -180(10^{-6})$$

a)

In - Plane Principal Strain : Applying Eq. 10–9,

$$\varepsilon_{1,2} = \frac{\varepsilon_x + \varepsilon_y}{2} \pm \sqrt{\left(\frac{\varepsilon_x - \varepsilon_y}{2}\right)^2 + \left(\frac{\gamma_{xy}}{2}\right)^2}$$

$$= \left[\frac{250+300}{2} \pm \sqrt{\left(\frac{250-300}{2}\right)^2 + \left(\frac{-180}{2}\right)^2}\right](10^{-6})$$

$$= 275 \pm 93.41$$

$$\varepsilon_1 = 368(10^{-6}) \qquad \varepsilon_2 = 182(10^{-6}) \qquad \textbf{Ans}$$

Orientation of Principal Strain : Applying Eq. 10–8,

$$\tan 2\theta_p = \frac{\gamma_{xy}}{\varepsilon_x - \varepsilon_y} = \frac{-180(10^{-6})}{(250-300)(10^{-6})} = 3.600$$

$$\theta_p = 37.24^\circ \quad \text{and} \quad -52.76^\circ$$

Use Eq. 10–5 to determine which principal strain deforms the element in the x' direction with $\theta = 37.24^\circ$.

$$\varepsilon_{x'} = \frac{\varepsilon_x + \varepsilon_y}{2} + \frac{\varepsilon_x - \varepsilon_y}{2}\cos 2\theta + \frac{\gamma_{xy}}{2}\sin 2\theta$$

$$= \left[\frac{250+300}{2} + \frac{250-300}{2}\cos 74.48^\circ + \frac{-180}{2}\sin 74.48^\circ\right](10^{-6})$$

$$= 182(10^{-6}) = \varepsilon_2$$

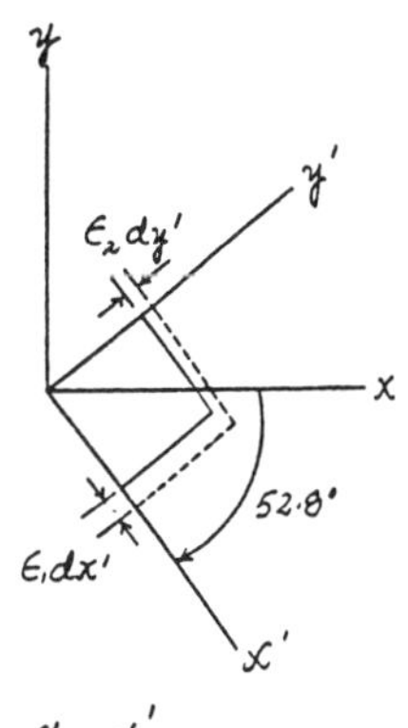

Hence, $\theta_{p_1} = -52.8^\circ$ and $\theta_{p_2} = 37.2^\circ$ **Ans**

b)

Maximum In - Plane Shear Strain : Applying Eq. 10–11,

$$\frac{\gamma_{\substack{\max \\ \text{in-plane}}}}{2} = \sqrt{\left(\frac{\varepsilon_x - \varepsilon_y}{2}\right)^2 + \left(\frac{\gamma_{xy}}{2}\right)^2}$$

$$\gamma_{\substack{\max \\ \text{in-plane}}} = 2\left[\sqrt{\left(\frac{250-300}{2}\right)^2 + \left(\frac{-180}{2}\right)^2}\right](10^{-6})$$

$$= 187(10^{-6}) \qquad \textbf{Ans}$$

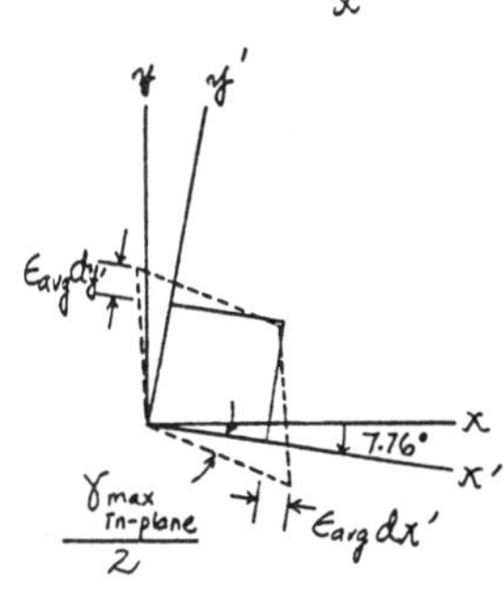

Orientation of Maximum In - Plane Shear Strain : Applying Eq. 10–10,

$$\tan 2\theta_s = -\frac{\varepsilon_x - \varepsilon_y}{\gamma_{xy}} = -\frac{250-300}{-180} = -0.2778$$

$$\theta_s = -7.76^\circ \quad \text{and} \quad 82.2^\circ \qquad \textbf{Ans}$$

The proper sign of $\gamma_{\substack{\max \\ \text{in-plane}}}$ can be determined by substituting $\theta = -7.76^\circ$ into Eq. 10–6.

$$\frac{\gamma_{x'y'}}{2} = -\frac{\varepsilon_x - \varepsilon_y}{2}\sin 2\theta + \frac{\gamma_{xy}}{2}\cos 2\theta$$

$$\gamma_{x'y'} = \{-[250-300]\sin(-15.52^\circ) + (-180)\cos(-15.52^\circ)\}(10^{-6})$$

$$= -187(10^{-6})$$

Average Normal Strain : Applying Eq. 10–12,

$$\varepsilon_{\text{avg}} = \frac{\varepsilon_x + \varepsilon_y}{2} = \left[\frac{250+300}{2}\right](10^{-6}) = 275(10^{-6}) \qquad \textbf{Ans}$$

10–7. The state of strain at the point on the vise-clamp has components of $\epsilon_x = 0$, $\epsilon_y = 280(10^{-6})$, and $\gamma_{xy} = 150(10^{-6})$. Use the strain-transformation equations to determine (a) the in-plane principal strains and (b) the maximum in-plane shear strain and average normal strain. In each case specify the orientation of the element and show how the strains deform the element within the x–y plane.

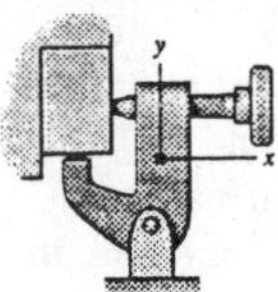

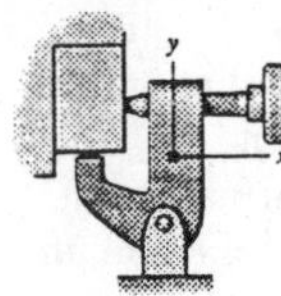

Normal Strain and Shear strain : In accordance with the sign convention,

$$\varepsilon_x = 0 \quad \varepsilon_y = 280(10^{-6}) \quad \gamma_{xy} = 150(10^{-6})$$

a)

In-Plane Principal Strain : Applying Eq. 10–9,

$$\varepsilon_{1,2} = \frac{\varepsilon_x + \varepsilon_y}{2} \pm \sqrt{\left(\frac{\varepsilon_x - \varepsilon_y}{2}\right)^2 + \left(\frac{\gamma_{xy}}{2}\right)^2}$$

$$= \left[\frac{0+280}{2} \pm \sqrt{\left(\frac{0-280}{2}\right)^2 + \left(\frac{150}{2}\right)^2}\right](10^{-6})$$

$$= 140 \pm 158.82$$

$$\varepsilon_1 = 299(10^{-6}) \qquad \varepsilon_2 = -18.8(10^{-6}) \qquad \textbf{Ans}$$

Orientation of Principal Strain : Applying Eq. 10–8,

$$\tan 2\theta_p = \frac{\gamma_{xy}}{\varepsilon_x - \varepsilon_y} = \frac{150(10^{-6})}{(0-280)(10^{-6})} = -0.5357$$

$$\theta_p = -14.09° \quad \text{and} \quad 75.91°$$

Use Eq. 10–5 to determine which principal strain deforms the element in the x' direction with $\theta = -14.09°$.

$$\varepsilon_{x'} = \frac{\varepsilon_x + \varepsilon_y}{2} + \frac{\varepsilon_x - \varepsilon_y}{2}\cos 2\theta + \frac{\gamma_{xy}}{2}\sin 2\theta$$

$$= \left[\frac{0+280}{2} + \frac{0-280}{2}\cos(-28.18°) + \frac{150}{2}\sin(-28.18°)\right](10^{-6})$$

$$= -18.8(10^{-6}) = \varepsilon_2$$

Hence, $\theta_{p_1} = 75.9°$ and $\theta_{p_2} = -14.1°$ **Ans**

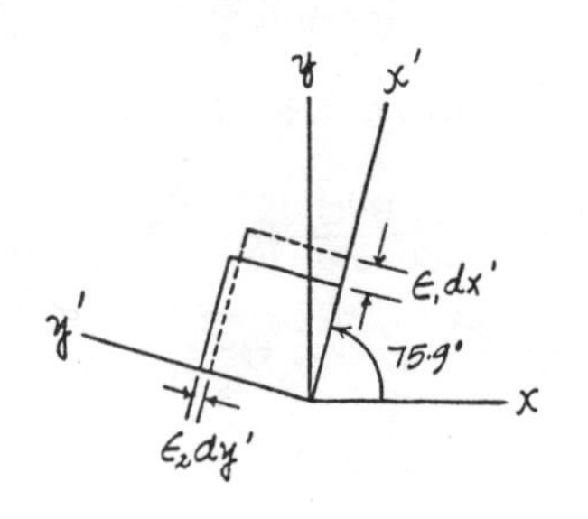

b)

Maximum In-Plane Shear Strain : Applying Eq. 10–11,

$$\frac{\gamma_{\substack{\max \\ \text{in-plane}}}}{2} = \sqrt{\left(\frac{\varepsilon_x - \varepsilon_y}{2}\right)^2 + \left(\frac{\gamma_{xy}}{2}\right)^2}$$

$$\gamma_{\substack{\max \\ \text{in-plane}}} = 2\left[\sqrt{\left(\frac{0-280}{2}\right)^2 + \left(\frac{150}{2}\right)^2}\right](10^{-6})$$

$$= 318(10^{-6}) \qquad \textbf{Ans}$$

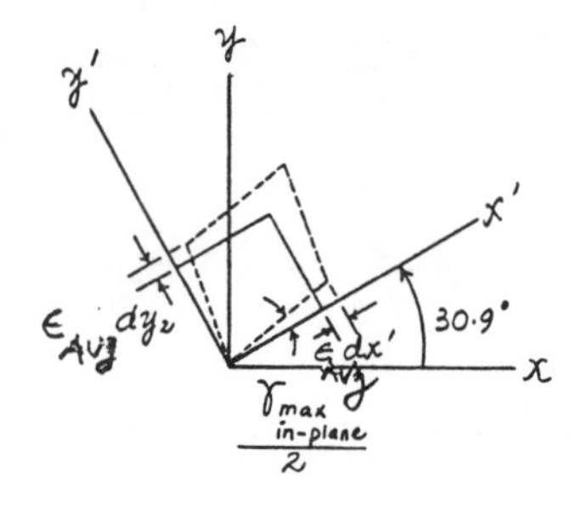

Orientation of Maximum In-Plane Shear Strain : Applying Eq. 10–10,

$$\tan 2\theta_s = -\frac{\varepsilon_x - \varepsilon_y}{\gamma_{xy}} = -\frac{0-280}{150} = 1.8667$$

$$\theta_s = 30.9° \quad \text{and} \quad -59.1° \qquad \textbf{Ans}$$

The proper sign of $\gamma_{\substack{\max \\ \text{in-plane}}}$ can be determined by substituting $\theta = 30.9°$ into Eq. 10–6,

$$\frac{\gamma_{x'y'}}{2} = -\frac{\varepsilon_x - \varepsilon_y}{2}\sin 2\theta + \frac{\gamma_{xy}}{2}\cos 2\theta$$

$$\gamma_{x'y'} = [-(0-280)\sin 61.8° + 150\cos 61.8°](10^{-6})$$

$$= 318(10^{-6})$$

Average Normal Strain : Applying Eq. 10–12,

$$\varepsilon_{avg} = \frac{\varepsilon_x + \varepsilon_y}{2} = \left[\frac{0+280}{2}\right](10^{-6}) = 140(10^{-6}) \qquad \textbf{Ans}$$

***10–8.** The state of strain at the point on the spanner wrench has components of $\epsilon_x = 260(10^{-6})$, $\epsilon_y = 320(10^{-6})$, and $\gamma_{xy} = 180(10^{-6})$. Use the strain-transformation equations to determine (a) the in-plane principal strains and (b) the maximum in-plane shear strain and average normal strain. In each case specify the orientation of the element and show how the strains deform the element within the x–y plane.

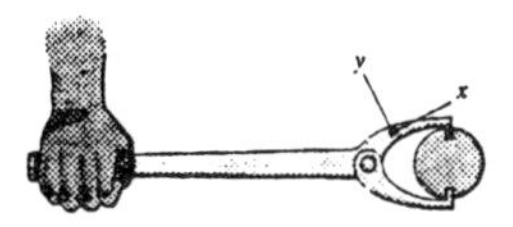

Normal Strain and Shear strain : In accordance with the sign convention,

$$\varepsilon_x = 260(10^{-6}) \qquad \varepsilon_y = 320(10^{-6}) \qquad \gamma_{xy} = 180(10^{-6})$$

a)

In-Plane Principal Strain : Applying Eq. 10–9,

$$\varepsilon_{1,2} = \frac{\varepsilon_x + \varepsilon_y}{2} \pm \sqrt{\left(\frac{\varepsilon_x - \varepsilon_y}{2}\right)^2 + \left(\frac{\gamma_{xy}}{2}\right)^2}$$

$$= \left[\frac{260+320}{2} \pm \sqrt{\left(\frac{260-320}{2}\right)^2 + \left(\frac{180}{2}\right)^2}\right](10^{-6})$$

$$= 290 \pm 94.87$$

$$\varepsilon_1 = 385(10^{-6}) \qquad \varepsilon_2 = 195(10^{-6}) \qquad \textbf{Ans}$$

Orientation of Principal Strain : Applying Eq. 10–8,

$$\tan 2\theta_p = \frac{\gamma_{xy}}{\varepsilon_x - \varepsilon_y} = \frac{180(10^{-6})}{(260-320)(10^{-6})} = -3.000$$

$$\theta_p = -35.78^\circ \quad \text{and} \quad 54.22^\circ$$

Use Eq. 10–5 to determine which principal strain deforms the element in the x' direction with $\theta = -35.78°$.

$$\varepsilon_{x'} = \frac{\varepsilon_x + \varepsilon_y}{2} + \frac{\varepsilon_x - \varepsilon_y}{2}\cos 2\theta + \frac{\gamma_{xy}}{2}\sin 2\theta$$

$$= \left[\frac{260+320}{2} + \frac{260-320}{2}\cos(-71.56^\circ) + \frac{180}{2}\sin(-71.56^\circ)\right](10^{-6})$$

$$= 195(10^{-6}) = \varepsilon_2$$

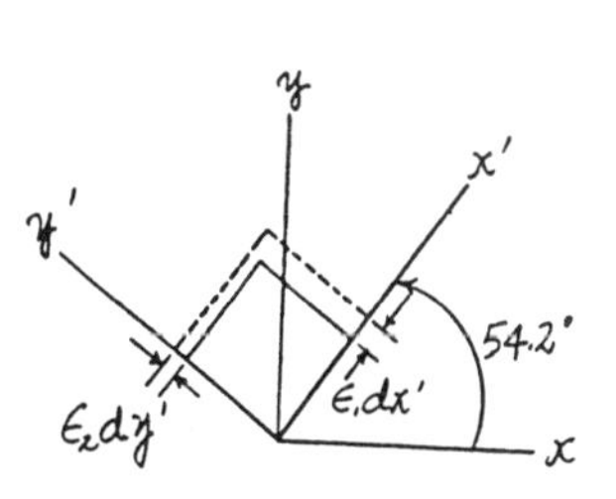

Hence, $\theta_{p_1} = 54.2^\circ$ and $\theta_{p_2} = -35.8^\circ$ **Ans**

b)

Maximum In-Plane Shear Strain : Applying Eq. 10–11,

$$\frac{\gamma_{\substack{max \\ in\text{-}plane}}}{2} = \sqrt{\left(\frac{\varepsilon_x - \varepsilon_y}{2}\right)^2 + \left(\frac{\gamma_{xy}}{2}\right)^2}$$

$$\gamma_{\substack{max \\ in\text{-}plane}} = 2\left[\sqrt{\left(\frac{260-320}{2}\right)^2 + \left(\frac{180}{2}\right)^2}\right](10^{-6})$$

$$= 190(10^{-6}) \qquad \textbf{Ans}$$

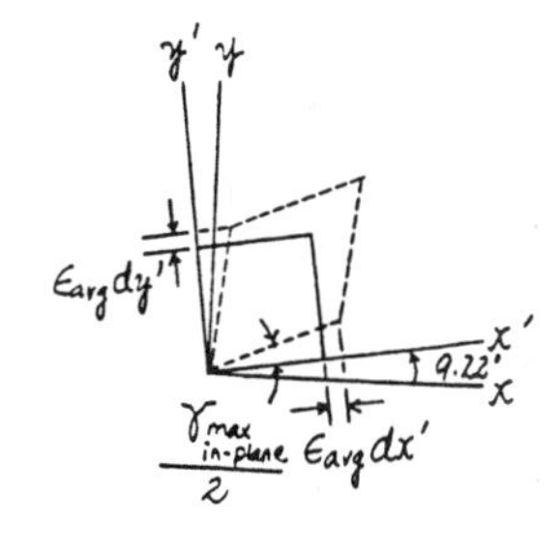

Orientation of Maximum In-Plane Shear Strain : Applying Eq. 10–10,

$$\tan 2\theta_s = -\frac{\varepsilon_x - \varepsilon_y}{\gamma_{xy}} = -\frac{260-320}{180} = 0.3333$$

$$\theta_s = 9.22^\circ \quad \text{and} \quad -80.8^\circ \qquad \textbf{Ans}$$

The proper sign of $\gamma_{\substack{max \\ in\text{-}plane}}$ can be determined by substituting $\theta = 9.22°$ into Eq. 10–6.

$$\frac{\gamma_{x'y'}}{2} = -\frac{\varepsilon_x - \varepsilon_y}{2}\sin 2\theta + \frac{\gamma_{xy}}{2}\cos 2\theta$$

$$\gamma_{x'y'} = [-(260-320)\sin 18.44^\circ + 180\cos 18.44^\circ](10^{-6})$$

$$= 190(10^{-6})$$

Average Normal Strain : Applying Eq. 10–12,

$$\varepsilon_{avg} = \frac{\varepsilon_x + \varepsilon_y}{2} = \left[\frac{260+320}{2}\right](10^{-6}) = 290(10^{-6}) \qquad \textbf{Ans}$$

10–9. The state of strain at the point on the fan blade has components of $\epsilon_x = 250(10^{-6})$, $\epsilon_{xy} = -450(10^{-6})$, and $\gamma_{xy} = -825(10^{-6})$. Use the strain-transformation equations to determine (a) the in-plane principal strains and (b) the maximum in-plane shear strain and average normal strain. In each case specify the orientation of the element and show how the strains deform the element within the x–y plane.

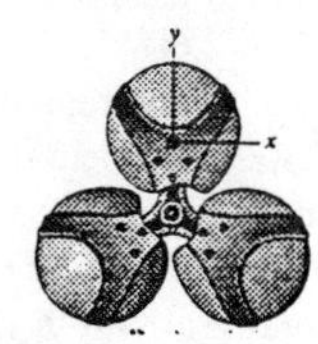

Normal Strain and Shear strain : In accordance with the sign convention,

$$\varepsilon_x = 250(10^{-6}) \qquad \varepsilon_y = -450(10^{-6}) \qquad \gamma_{xy} = -825(10^{-6})$$

a)

In - Plane Principal Strain : Applying Eq. 10 – 9,

$$\varepsilon_{1,2} = \frac{\varepsilon_x + \varepsilon_y}{2} \pm \sqrt{\left(\frac{\varepsilon_x - \varepsilon_y}{2}\right)^2 + \left(\frac{\gamma_{xy}}{2}\right)^2}$$

$$= \left[\frac{250 + (-450)}{2} \pm \sqrt{\left(\frac{250 - (-450)}{2}\right)^2 + \left(\frac{-825}{2}\right)^2}\right](10^{-6})$$

$$= -100 \pm 540.98$$

$$\varepsilon_1 = 441(10^{-6}) \qquad \varepsilon_2 = -641(10^{-6}) \qquad \textbf{Ans}$$

Orientation of Principal Strain : Applying Eq. 10 – 8,

$$\tan 2\theta_p = \frac{\gamma_{xy}}{\varepsilon_x - \varepsilon_y} = \frac{-825(10^{-6})}{[250 - (-450)](10^{-6})} = -1.1786$$

$$\theta_p = -24.84° \quad \text{and} \quad 65.16°$$

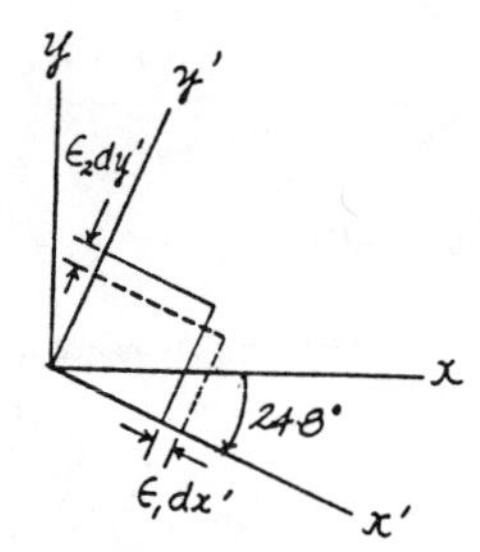

Use Eq. 10 – 5 to determine which principal strain deforms the element in the x' direction with $\theta = -24.84°$.

$$\varepsilon_{x'} = \frac{\varepsilon_x + \varepsilon_y}{2} + \frac{\varepsilon_x - \varepsilon_y}{2}\cos 2\theta + \frac{\gamma_{xy}}{2}\sin 2\theta$$

$$= \left[\frac{250 + (-450)}{2} + \frac{250 - (-450)}{2}\cos(-49.68°) + \frac{-825}{2}\sin(-49.68°)\right](10^{-6})$$

$$= 441(10^{-6}) = \varepsilon_1$$

Hence, $\theta_{p_1} = -24.8°$ and $\theta_{p_2} = 65.2°$ **Ans**

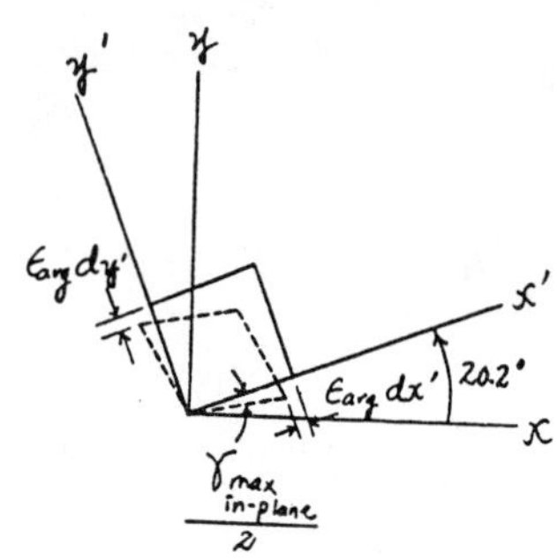

b)

Maximum In - Plane Shear Strain : Applying Eq. 10 – 11,

$$\frac{\gamma_{\substack{max \\ in-plane}}}{2} = \sqrt{\left(\frac{\varepsilon_x - \varepsilon_y}{2}\right)^2 + \left(\frac{\gamma_{xy}}{2}\right)^2}$$

$$\gamma_{\substack{max \\ in-plane}} = 2\left[\sqrt{\left(\frac{250 - (-450)}{2}\right)^2 + \left(\frac{-825}{2}\right)^2}\right](10^{-6})$$

$$= 1082(10^{-6}) \qquad \textbf{Ans}$$

Orientation of Maximum In - Plane Shear Strain : Applying Eq. 10 – 10,

$$\tan 2\theta_s = -\frac{\varepsilon_x - \varepsilon_y}{\gamma_{xy}} = -\frac{250 - (-450)}{-825} = 0.8485$$

$$\theta_s = 20.2° \quad \text{and} \quad -69.8° \qquad \textbf{Ans}$$

The proper sign of $\gamma_{\substack{max \\ in-plane}}$ can be determined by substituting $\theta = 20.2°$ into Eq. 10 – 6.

$$\frac{\gamma_{x'y'}}{2} = -\frac{\varepsilon_x - \varepsilon_y}{2}\sin 2\theta + \frac{\gamma_{xy}}{2}\cos 2\theta$$

$$\gamma_{x'y'} = \{-[250 - (-450)]\sin 40.3° + (-825)\cos 40.3°\}(10^{-6})$$

$$= -1082(10^{-6})$$

Average Normal Strain : Applying Eq. 10 – 12,

$$\varepsilon_{avg} = \frac{\varepsilon_x + \varepsilon_y}{2} = \left[\frac{250 + (-450)}{2}\right](10^{-6}) = -100(10^{-6}) \qquad \textbf{Ans}$$

10–10. A strain gauge is mounted on the 1-in.-diameter A-36 steel shaft in the manner shown. When the shaft is rotating with an angular velocity of $\omega = 1760$ rev/min, using a slip ring, the reading on the strain gauge is $\epsilon = 800(10^{-6})$. Determine the power output of the motor. Assume the shaft is only subjected to a torque.

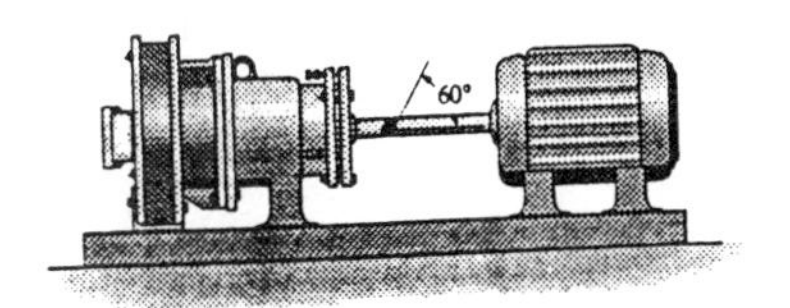

Strain Transformation Equations : For pure torsion $\varepsilon_x = \varepsilon_y = 0$. Applying Eq. 10 – 5 with $\theta = +60°$,

$$\varepsilon_{x'} = \frac{\varepsilon_x + \varepsilon_y}{2} + \frac{\varepsilon_x - \varepsilon_y}{2}\cos 2\theta + \frac{\gamma_{xy}}{2}\sin 2\theta$$

$$800(10^{-6}) = 0 + 0 + \frac{\gamma_{xy}}{2}\sin 120°$$

$$\gamma_{xy} = 1.8475(10^{-3})\ \text{rad}$$

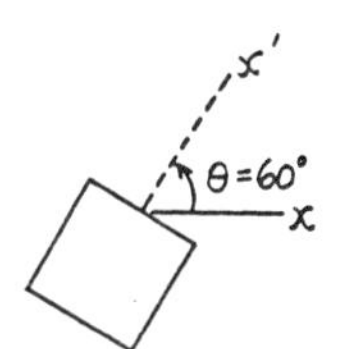

Shear Stress and Strain Relationship : Applying Hooke's law,

$$\tau = G\,\gamma_{xy} = 11.0(10^3)\,1.8475(10^{-3}) = 20.323\ \text{ksi}$$

Maximum Shear Stress : Applying the torsion formula.

$$\tau = \frac{Tc}{J}$$

$$20.323 = \frac{T(0.5)}{\frac{\pi}{2}(0.5^4)}$$

$$T = 3.9904\ \text{kip}\cdot\text{in} = 332.53\ \text{lb}\cdot\text{ft}$$

Power Transmission :

$$\omega = \left(1760\,\frac{\text{rev}}{\text{min}}\right)\left(\frac{1\ \text{min}}{60\ \text{sec}}\right)\left(\frac{2\pi\ \text{rad}}{1\ \text{rev}}\right) = 184.307\ \text{rad/s}$$

$$P = T\omega = 332.53(184.307)$$
$$= 61\,287.5\ \text{ft}\cdot\text{lb/s} = 111\ \text{hp} \qquad \textbf{Ans}$$

***10–12.** Solve Prob. 10–4 using Mohr's circle.

Construction of the Circle : In accordance with the sign convention, $\varepsilon_x = 200(10^{-6})$, $\varepsilon_y = 180(10^{-6})$, and $\frac{\gamma_{xy}}{2} = -150(10^{-6})$. Hence,

$$\varepsilon_{avg} = \frac{\varepsilon_x + \varepsilon_y}{2} = \left(\frac{200 + 180}{2}\right)(10^{-6}) = 190(10^{-6})$$

The coordinates for reference points A and C are

$$A(200,\ -150)(10^{-6}) \qquad C(190,\ 0)(10^{-6})$$

The radius of the circle is

$$R = \left(\sqrt{(200 - 190)^2 + 150^2}\right)(10^{-6}) = 150.33(10^{-6})$$

Strain on the Inclined Element : The normal and shear strain $\left(\varepsilon_{x'}\ \text{and}\ \frac{\gamma_{x'y'}}{2}\right)$ are represented by coordinates of point P on the circle. $\varepsilon_{y'}$ can be determined by calculating the coordinates of point Q on the circle.

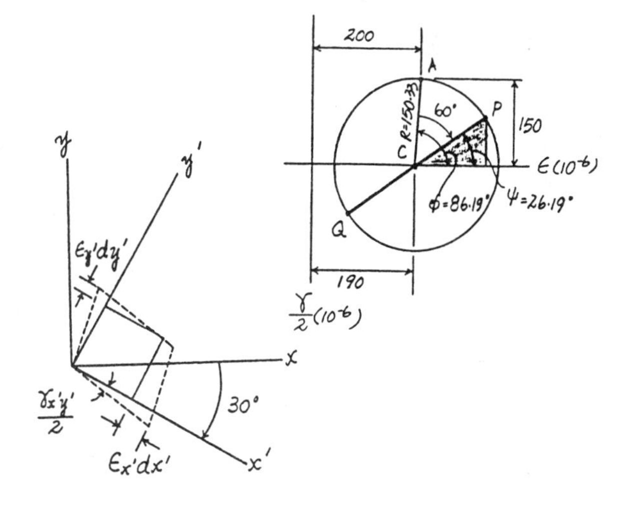

$$\varepsilon_{x'} = (190 + 150.33\cos 26.19°)(10^{-6}) = 325(10^{-6}) \qquad \textbf{Ans}$$

$$\frac{\gamma_{x'y'}}{2} = -(150.33\sin 26.19°)(10^{-6})$$

$$\gamma_{x'y'} = -133(10^{-6}) \qquad \textbf{Ans}$$

$$\varepsilon_{y'} = (190 - 150.33\cos 26.19°)(10^{-6}) = 55.1(10^{-6}) \qquad \textbf{Ans}$$

10–13. Solve Prob. 10–2 using Mohr's circle.

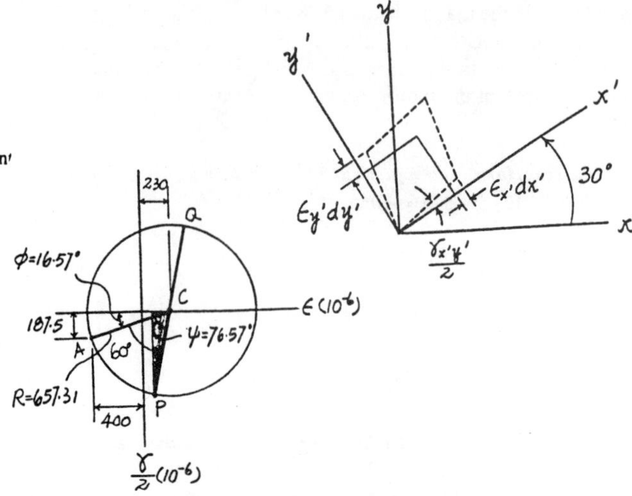

Construction of the Circle : In accordance with the sign conven' $\varepsilon_x = -400(10^{-6})$, $\varepsilon_y = 860(10^{-6})$ and $\frac{\gamma_{xy}}{2} = 187.5(10^{-6})$. Hence,

$$\varepsilon_{avg} = \frac{\varepsilon_x + \varepsilon_y}{2} = \left(\frac{-400 + 860}{2}\right)(10^{-6}) = 230(10^{-6})$$

The coordinates for reference points A and C are

$$A(-400,\ 187.5)(10^{-6}) \qquad C(230,\ 0)(10^{-6})$$

The radius of the circle is

$$R = \left(\sqrt{(400 + 230)^2 + 187.5^2}\right)(10^{-6}) = 657.31(10^{-6})$$

Strain on the Inclined Element : The normal and shear strain $\left(\varepsilon_{x'} \text{ and } \frac{\gamma_{x'y'}}{2}\right)$ are represented by the coordinates of point P on the circle. $\varepsilon_{y'}$ can be determined by calculating the coordinates of point Q on the circle.

$$\varepsilon_{x'} = (230 - 657.31\cos 76.57°)(10^{-6}) = 77.4(10^{-6}) \qquad \textbf{Ans}$$

$$\frac{\gamma_{x'y'}}{2} = (657.31\sin 76.57°)(10^{-6})$$

$$\gamma_{x'y'} = 1279(10^{-6}) \qquad \textbf{Ans}$$

$$\varepsilon_{y'} = (230 + 657.31\cos 76.57°)(10^{-6}) = 383(10^{-6}) \qquad \textbf{Ans}$$

10–14. Solve Prob. 10–3 using Mohr's circle.

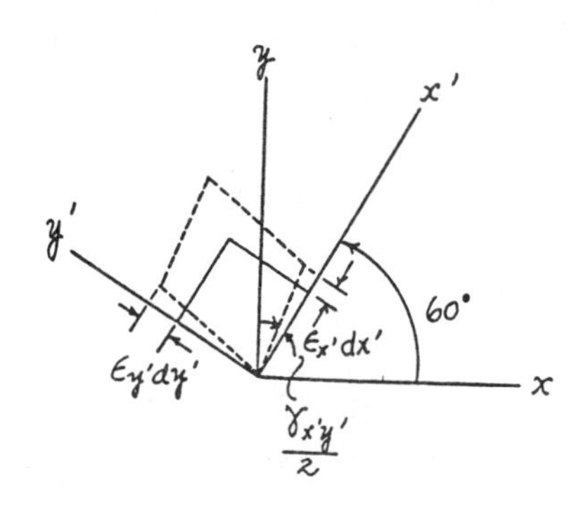

Construction of the Circle : In accordance with the sign convention, $\varepsilon_x = 200(10^{-6})$, $\varepsilon_y = 180(10^{-6})$, and $\frac{\gamma_{xy}}{2} = -150(10^{-6})$. Hence,

$$\varepsilon_{avg} = \frac{\varepsilon_x + \varepsilon_y}{2} = \left(\frac{200 + 180}{2}\right)(10^{-6}) = 190(10^{-6})$$

The coordinates for reference points A and C are

$$A(200,\ -150)(10^{-6}) \qquad C(190,\ 0)(10^{-6})$$

The radius of the circle is

$$R = \left(\sqrt{(200 - 190)^2 + 150^2}\right)(10^{-6}) = 150.33(10^{-6})$$

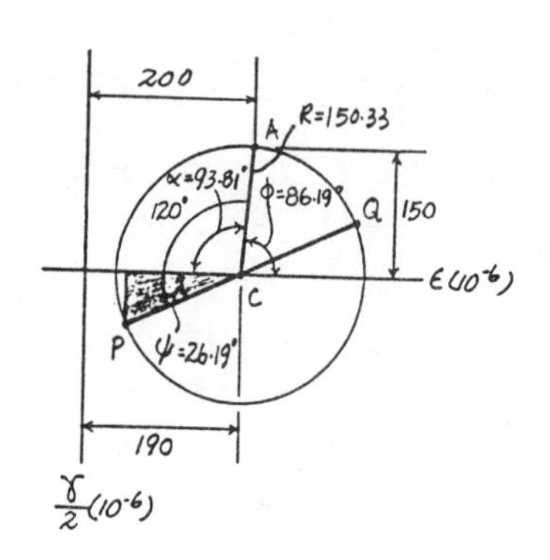

Strain on The Inclined Element : The normal and shear strain $\left(\varepsilon_{x'} \text{ and } \frac{\gamma_{x'y'}}{2}\right)$ are represented by coordinates of point P on the circle. $\varepsilon_{y'}$ can be determined by calculating the coordinates of point Q on the circle.

$$\varepsilon_{x'} = (190 - 150.33\cos 26.19°)(10^{-6}) = 55.1(10^{-6}) \qquad \textbf{Ans}$$

$$\frac{\gamma_{x'y'}}{2} = (150.33\sin 26.19°)(10^{-6})$$

$$\gamma_{x'y'} = 133(10^{-6}) \qquad \textbf{Ans}$$

$$\varepsilon_{y'} = (190 + 150.33\cos 26.19°)(10^{-6}) = 325(10^{-6}) \qquad \textbf{Ans}$$

10–15. Solve Prob. 10–5 using Mohr's circle.

Construction of the Circle : In accordance with the sign convention, $\varepsilon_x = 500(10^{-6})$, $\varepsilon_y = 350(10^{-6})$, and $\frac{\gamma_{xy}}{2} = -215(10^{-6})$. Hence,

$$\varepsilon_{avg} = \frac{\varepsilon_x + \varepsilon_y}{2} = \left(\frac{500+350}{2}\right)(10^{-6}) = 425(10^{-6})$$

and coordinates for reference points A and C are

$$A(500,\ -215)(10^{-6}) \qquad C(425,\ 0)(10^{-6})$$

The radius of the circle is

$$R = \left(\sqrt{(500-425)^2 + 215^2}\right)(10^{-6}) = 227.71(10^{-6})$$

Strain on the Inclined Element : The normal and shear strains $\left(\varepsilon_{x'} \text{ and } \frac{\gamma_{x'y'}}{2}\right)$ are represented by the coordinates of point P on the circle. $\varepsilon_{y'}$ can be determined by calculating the coordinates of point Q on the circle.

$$\varepsilon_{x'} = (425 + 227.71\cos 10.77°)(10^{-6}) = 649(10^{-6}) \qquad \textbf{Ans}$$

$$\frac{\gamma_{x'y'}}{2} = -(227.71\sin 10.77°)(10^{-6})$$

$$\gamma_{x'y'} = -85.1(10^{-6}) \qquad \textbf{Ans}$$

$$\varepsilon_{y'} = (425 - 227.71\cos 10.77°)(10^{-6}) = 201(10^{-6}) \qquad \textbf{Ans}$$

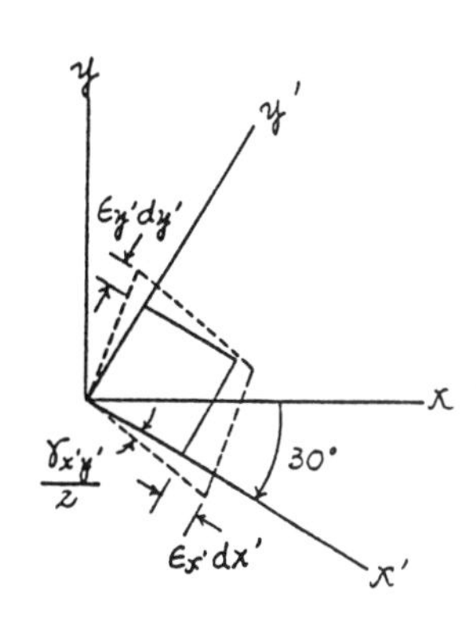

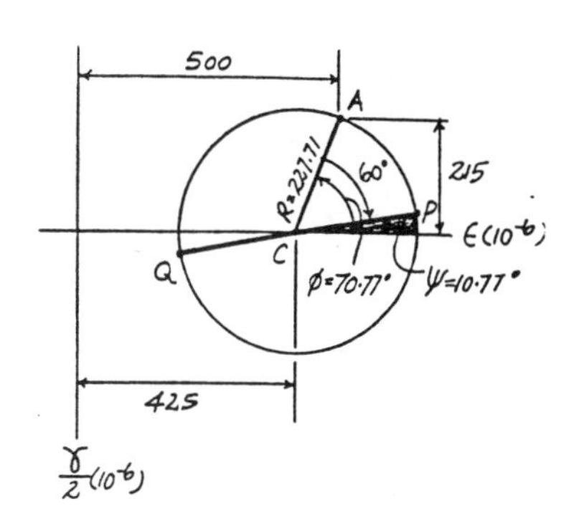

***10–16.** Solve Prob. 10–8 using Mohr's circle.

Construction of the Circle : In accordance with the sign convention, $\varepsilon_x = 260(10^{-6})$, $\varepsilon_y = 320(10^{-6})$, and $\frac{\gamma_{xy}}{2} = 90(10^{-6})$. Hence,

$$\varepsilon_{avg} = \frac{\varepsilon_x + \varepsilon_y}{2} = \left(\frac{260+320}{2}\right)(10^{-6}) = 290(10^{-6}) \qquad \textbf{Ans}$$

The coordinates for reference point A and C are

$$A(260, 90)(10^{-6}) \qquad C(290, 0)(10^{-6})$$

The radius of the circle is

$$R = \left(\sqrt{(290-260)^2 + 90^2}\right)(10^{-6}) = 94.868(10^{-6})$$

In - Plane Principal Strain : The coordinates of points B and D represent ε_1 and ε_2, respectively.

$$\varepsilon_1 = (290 + 94.868)(10^{-6}) = 385(10^{-6}) \qquad \textbf{Ans}$$

$$\varepsilon_2 = (290 - 94.868)(10^{-6}) = 195(10^{-6}) \qquad \textbf{Ans}$$

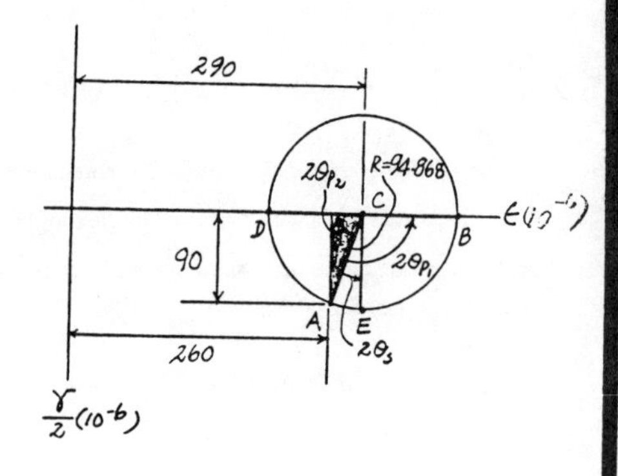

Orientation of Principal Strain : From the circle,

$$\tan 2\theta_{p_2} = \frac{90}{290-260} = 3.000 \qquad 2\theta_{p_2} = 71.57°$$

$$2\theta_{p_1} = 180° - 2\theta_{p_2}$$

$$\theta_{p_1} = \frac{180° - 71.57°}{2} = 54.2° \quad (\textit{Counterclockwise}) \qquad \textbf{Ans}$$

Maximum In - Plane Shear Strain : Represented by the coordinates of point E on the circle.

$$\frac{\gamma_{\substack{max\\in\text{-}plane}}}{2} = R = 94.868(10^{-6})$$

$$\gamma_{\substack{max\\in\text{-}plane}} = 190(10^{-6}) \qquad \textbf{Ans}$$

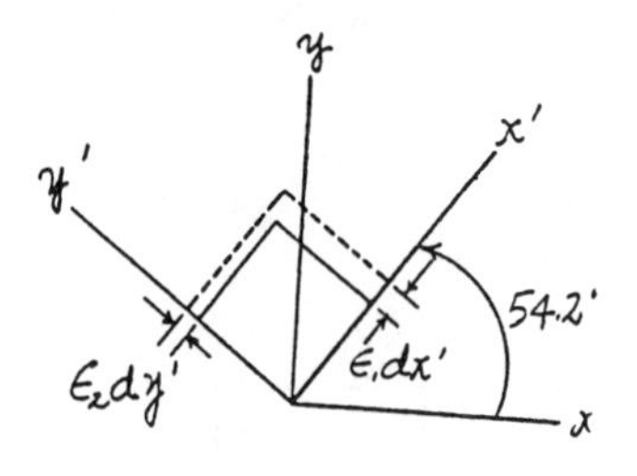

Orientation of Maximum In - Plane Shear Strain : From the cicle,

$$\tan 2\theta_s = \frac{290-260}{90} = 0.3333$$

$$\theta_s = 9.22° \quad (\textit{Counterclockwise}) \qquad \textbf{Ans}$$

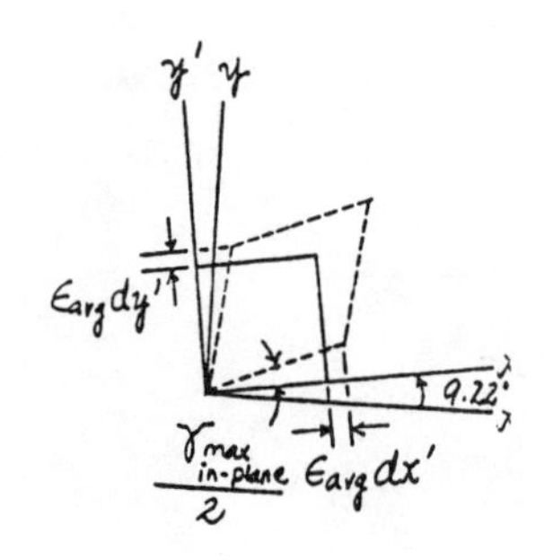

10–17. Solve Prob. 10–6 using Mohr's circle.

Construction of the Circle : In accordance with the sign convention, $\varepsilon_x = 250(10^{-6})$, $\varepsilon_y = 300(10^{-6})$, and $\frac{\gamma_{xy}}{2} = -90(10^{-6})$. Hence,

$$\varepsilon_{avg} = \frac{\varepsilon_x + \varepsilon_y}{2} = \left(\frac{250+300}{2}\right)(10^{-6}) = 275(10^{-6}) \qquad \textbf{Ans}$$

The coordinates for reference points A and C are

$$A(250,\ -90)(10^{-6}) \qquad C(275,\ 0)(10^{-6})$$

The radius of the circle is

$$R = \left(\sqrt{(275-250)^2 + 90^2}\right)(10^{-6}) = 93.408$$

In - Plane Principal Strain : The coordinates of points B and D represent ε_1 and ε_2, respectively.

$$\varepsilon_1 = (275+93.408)(10^{-6}) = 368(10^{-6}) \qquad \textbf{Ans}$$

$$\varepsilon_2 = (275-93.408)(10^{-6}) = 182(10^{-6}) \qquad \textbf{Ans}$$

Orientation of Principal Strain : From the circle,

$$\tan 2\theta_{p_2} = \frac{90}{275-250} = 3.600 \qquad 2\theta_{p_2} = 74.48°$$

$$2\theta_{p_1} = 180° - 2\theta_{p_2}$$

$$\theta_{p_1} = \frac{180° - 74.78°}{2} = 52.8° \ (\textit{Clockwise}) \qquad \textbf{Ans}$$

Maximum In - Plane Shear Strain : Represented by the coordinates of point E on the circle.

$$\frac{\gamma_{\substack{max \\ in\text{-}plane}}}{2} = -R = -93.408(10^{-6})$$

$$\gamma_{\substack{max \\ in\text{-}plane}} = -187(10^{-6}) \qquad \textbf{Ans}$$

Orientation of Maximum In - Plane Shear Strain : From the circle,

$$\tan 2\theta_s = \frac{275-250}{90} = 0.2778$$

$$\theta_s = 7.76° \ (\textit{Clockwise}) \qquad \textbf{Ans}$$

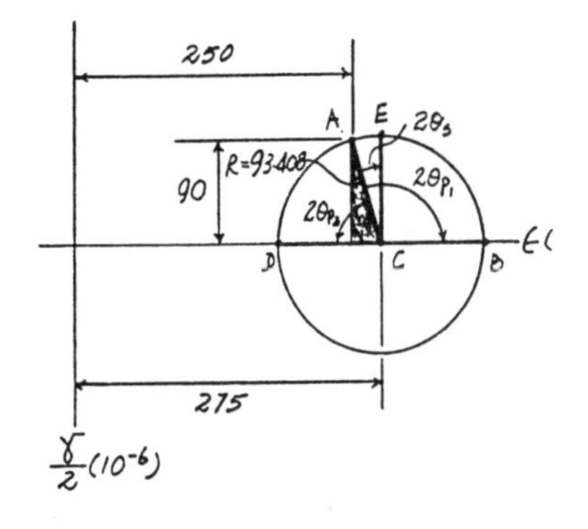

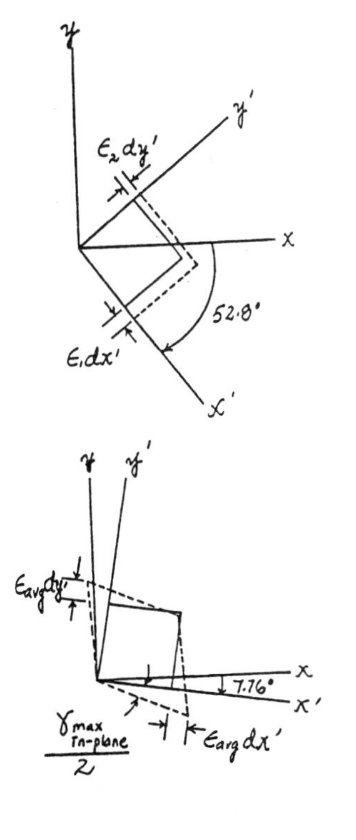

10–18. Solve Prob. 10–7 using Mohr's circle.

Construction of the Circle : In accordance with the sign convention, $\varepsilon_x = 0$, $\varepsilon_y = 280(10^{-6})$, and $\frac{\gamma_{xy}}{2} = 75(10^{-6})$. Hence,

$$\varepsilon_{avg} = \frac{\varepsilon_x + \varepsilon_y}{2} = \left(\frac{0+280}{2}\right)(10^{-6}) = 140(10^{-6}) \qquad \textbf{Ans}$$

The coordinates for reference points A and C are

$A(0,\ 75)(10^{-6}) \qquad C(140,\ 0)(10^{-6})$

The radius of the circle is

$$R = \left(\sqrt{(140-0)^2 + 75^2}\right)(10^{-6}) = 158.82(10^{-6})$$

In - Plane Principal Strain : The coordinates of points B and D represent ε_1 and ε_2 respectively.

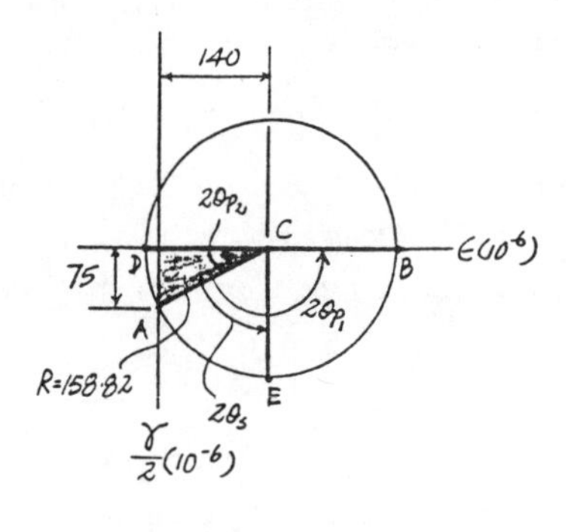

$$\varepsilon_1 = (140 + 158.82)(10^{-6}) = 299(10^{-6}) \qquad \textbf{Ans}$$

$$\varepsilon_2 = (140 - 158.82)(10^{-6}) = -18.8(10^{-6}) \qquad \textbf{Ans}$$

Orientation of Principal Strain : From the circle,

$$\tan 2\theta_{p_2} = \frac{75}{140-0} = 0.5357 \qquad 2\theta_{p_2} = 28.18°$$

$$2\theta_{p_1} = 180° - 2\theta_{p_2}$$

$$\theta_{p_1} = \frac{180° - 28.18°}{2} = 75.9° \quad (\textit{Counterclockwise}) \qquad \textbf{Ans}$$

Maximum In - Plane Shear Strain : Represented by the coordinates of point E on the circle.

$$\frac{\gamma_{\substack{max \\ in\text{-}plane}}}{2} = R = 158.82(10^{-6})$$

$$\gamma_{\substack{max \\ in\text{-}plane}} = 318(10^{-6}) \qquad \textbf{Ans}$$

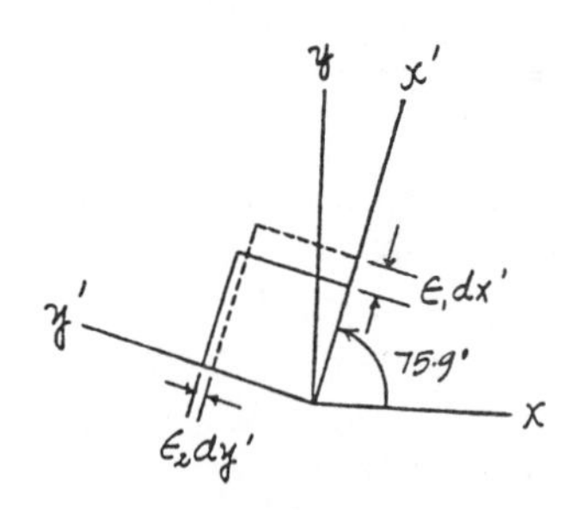

Orientation of Maximum In - Plane Shear Strain : From the circle,

$$\tan 2\theta_s = \frac{140-0}{75} = 1.8667$$

$$\theta_s = 30.9° \quad (\textit{Counterclockwise}) \qquad \textbf{Ans}$$

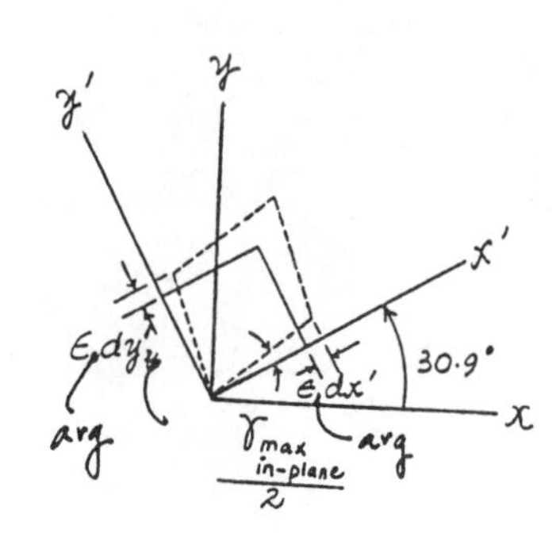

10–19. Solve Prob. 10–9 using Mohr's circle.

Construction of the Circle : In accordance with the sign convention, $\varepsilon_x = 250(10^{-6})$, $\varepsilon_y = -450(10^{-6})$ and $\frac{\gamma_{xy}}{2} = -412.5(10^{-6})$. Hence,

$$\varepsilon_{avg} = \frac{\varepsilon_x + \varepsilon_y}{2} = \left[\frac{250 + (-450)}{2}\right](10^{-6}) = -100(10^{-6}) \qquad \textbf{Ans}$$

The coordinates for reference points A and C are

$$A(250, -412.5)(10^{-6}) \qquad C(-100, 0)(10^{-6})$$

The radius of the circle is

$$R = \left(\sqrt{(250 + 100)^2 + 412.5^2}\right)(10^{-6}) = 540.98(10^{-6})$$

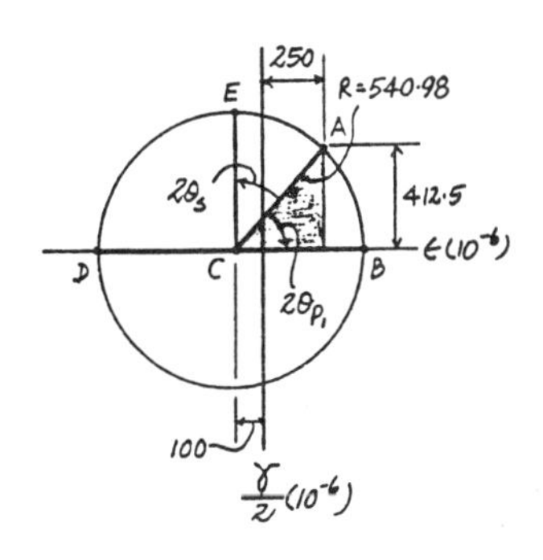

In - Plane Principal Strain : The coordinates of points B and D represent ε_1 and ε_2, respectively.

$$\varepsilon_1 = (-100 + 540.98)(10^{-6}) = 441(10^{-6}) \qquad \textbf{Ans}$$

$$\varepsilon_2 = (-100 - 540.98)(10^{-6}) = -641(10^{-6}) \qquad \textbf{Ans}$$

Orientation of Principal Strain : From the circle,

$$\tan 2\theta_{p_1} = \frac{412.5}{250 + 100} = 1.1786$$

$$\theta_{p_1} = 24.8° \,(\textit{Clockwise}) \qquad \textbf{Ans}$$

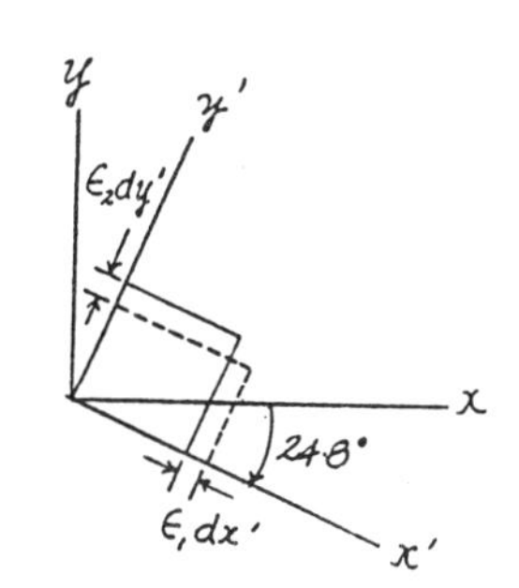

Maximum In - Plane Shear Strain : Represented by the coordinates of point E on the circle.

$$\frac{\gamma_{\substack{max \\ in\text{-}plane}}}{2} = -R = -540.98(10^{-6})$$

$$\gamma_{\substack{max \\ in\text{-}plane}} = -1082(10^{-6}) \qquad \textbf{Ans}$$

Orientation of Maximum In - Plane Shear Strain : From the cicle,

$$\tan 2\theta_s = \frac{250 + 100}{412.5} = 0.8485$$

$$\theta_s = 20.2° \,(\textit{Counterclockwise}) \qquad \textbf{Ans}$$

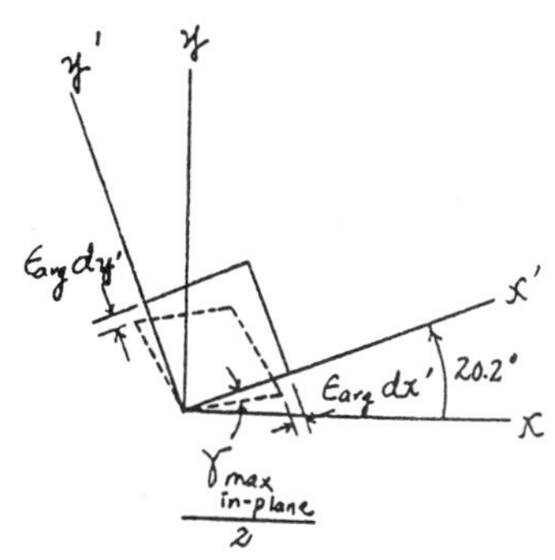

***10–20.** The strain at a point has components of $\epsilon_x = -480(10^{-6})$, $\epsilon_y = 300(10^{-6})$, $\gamma_{xy} = -650(10^{-6})$, and $\epsilon_z = 0$. Determine (a) the principal strains, (b) the maximum shear strain in the x–y plane. and (c) the absolute maximum shear strain.

Construction of the Circle (x - y Plane) : In accordance with the sign convention, $\varepsilon_x = -480(10^{-6})$, $\varepsilon_y = 300(10^{-6},)$ and $\frac{\gamma_{xy}}{2} = -325(10^{-6})$. Hence,

$$\varepsilon_{avg} = \frac{\varepsilon_x + \varepsilon_y}{2} = \left[\frac{-480+300}{2}\right](10^{-6}) = -90.0(10^{-6})$$

The coordinates for reference points A and C are

$$A(-480,\ -325)(10^{-6}) \qquad C(-90.0,\ 0)(10^{-6})$$

The radius of the circle is

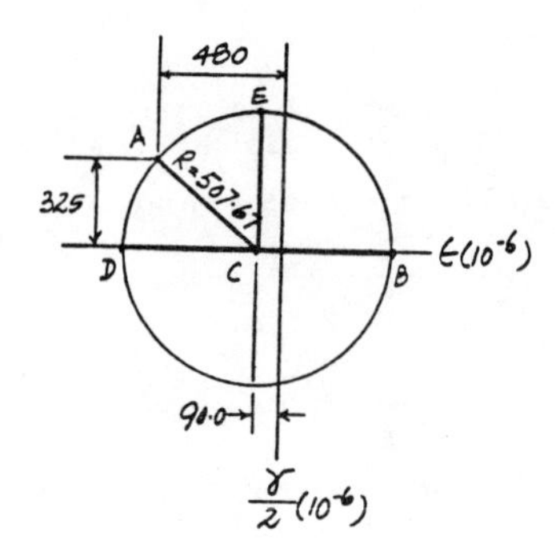

$$R = \left(\sqrt{(480-90)^2 + 325^2}\right)(10^{-6}) = 507.67(10^{-6})$$

In - Plane Principal Strain : The coordinates of points B and D represent ε_1 and ε_2, respectively.

$$\varepsilon_1 = (-90.0 + 507.67)(10^{-6}) = 417.67(10^{-6})$$

$$\varepsilon_2 = (-90.0 - 507.67)(10^{-6}) = -597.67$$

Maximum In - Plane Shear Strain (x - y Plane) : Represented by the coordinates of point E on the circle.

$$\frac{\gamma_{\substack{max\\in\text{-}plane}}}{2} = -R = -507.67(10^{-6})$$

$$\gamma_{\substack{max\\in\text{-}plane}} = -1015(10^{-6}) \qquad \textbf{Ans}$$

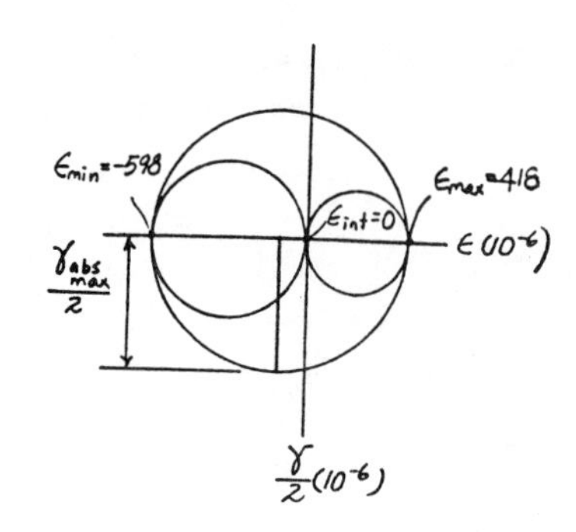

Three Mohr's Circles : From the results obtained above, the principal strains are

$$\varepsilon_{max} = 418(10^{-6}) \qquad \varepsilon_{int} = 0 \qquad \varepsilon_{min} = -598(10^{-6}) \qquad \textbf{Ans}$$

Absolute Maximum Shear Stress :

$$\gamma_{\substack{abs\\max}} = \varepsilon_{max} - \varepsilon_{min}$$

$$= [417.67 - (-597.67)](10^{-6}) = 1015(10^{-6}) \qquad \textbf{Ans}$$

10–21. The strain at a point has components of $\epsilon_x = -480(10^{-6})$, $\epsilon_y = 650(10^{-6})$, $\gamma_{xy} = 780(10^{-6})$, and $\epsilon_z = 0$. Determine (a) the principal strains, (b) the maximum shear strain in the x–y plane, and (c) the absolute maximum shear strain.

Construction of the Circle (x - y Plane) : In accordance with the sign convention, $\varepsilon_x = -480(10^{-6})$, $\varepsilon_y = 650(10^{-6})$, and $\frac{\gamma_{xy}}{2} = 390(10^{-6})$. Hence,

$$\varepsilon_{avg} = \frac{\varepsilon_x + \varepsilon_y}{2} = \left[\frac{-480+650}{2}\right](10^{-6}) = 85.0(10^{-6})$$

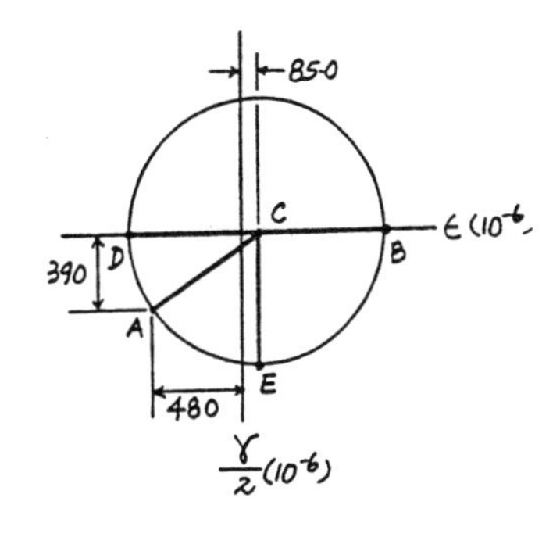

The coordinates for reference points A and C are

$$A(-480, 390)(10^{-6}) \qquad C(85.0, 0)(10^{-6})$$

The radius of the circle is

$$R = \left(\sqrt{(480+85.0)^2 + 390^2}\right)(10^{-6}) = 686.53(10^{-6})$$

In - Plane Principal Strain : The coordinates of points B and D represent ε_1 and ε_2, respectively.

$$\varepsilon_1 = (85.0 + 686.53)(10^{-6}) = 771.53(10^{-6})$$

$$\varepsilon_2 = (85 - 686.53)(10^{-6}) = -601.53(10^{-6})$$

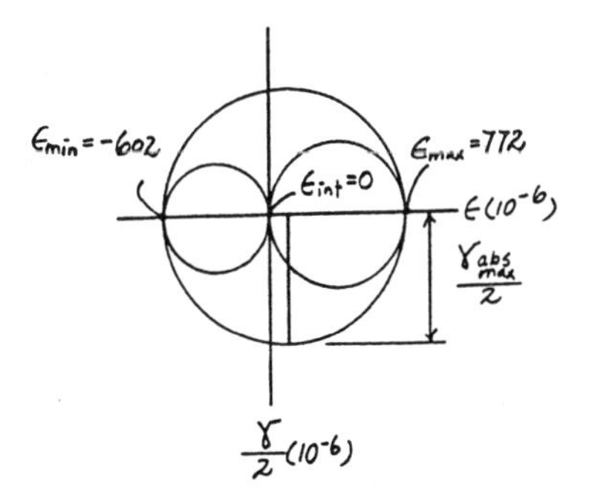

Maximum In - Plane Shear Strain (x - y Plane) : Represented by the coordinates of point E on the circle.

$$\frac{\gamma_{\substack{max \\ in\text{-}plane}}}{2} = R = 686.53(10^{-6})$$

$$\gamma_{\substack{max \\ in\text{-}plane}} = 1373(10^{-6}) \qquad \textbf{Ans}$$

Three Mohr's Circles : From the results obtained above, the principal strains are

$$\varepsilon_{max} = 772(10^{-6}) \qquad \varepsilon_{int} = 0 \qquad \varepsilon_{min} = -602(10^{-6}) \qquad \textbf{Ans}$$

Absolute Maximum Shear Stress :

$$\gamma_{\substack{abs \\ max}} = \varepsilon_{max} - \varepsilon_{min}$$

$$= [771.53 - (-601.53)](10^{-6}) = 1373(10^{-6}) \qquad \textbf{Ans}$$

10–22. The strain at a point has components of $\epsilon_x = 130(10^{-6})$, $\epsilon_y = 280(10^{-6})$, and $\gamma_{xy} = 75(10^{-6})$. and $\epsilon_z = 0$.Determine (a) the principal strains at the point, (b) the maximum in-plane shear strain in the x–y plane, and (c) the absolute maximum shear strain.

Construction of the Circle (x - y Plane) : In accordance with the sign convention, $\varepsilon_x = 130(10^{-6})$, $\varepsilon_y = 280(10^{-6})$, and $\frac{\gamma_{xy}}{2} = 37.5(10^{-6})$. Hence,

$$\varepsilon_{avg} = \frac{\varepsilon_x + \varepsilon_y}{2} = \left[\frac{130+280}{2}\right](10^{-6}) = 205(10^{-6})$$

The coordinates for reference points A and C are

$$A(130,\ 37.5)(10^{-6}) \qquad C(205,\ 0)(10^{-6})$$

The radius of the circle is

$$R = \left(\sqrt{(205-130)^2 + 37.5^2}\right)(10^{-6}) = 83.853(10^{-6})$$

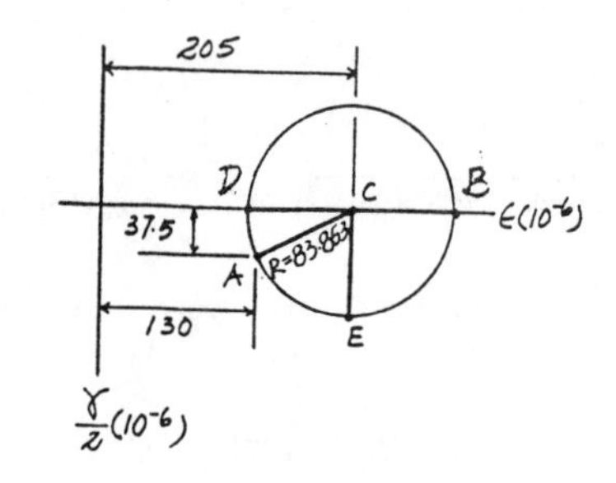

In - Plane Principal Strain : The coordinates of points B and D represent ε_1 and ε_2, respectively.

$$\varepsilon_1 = (205 + 83.853)(10^{-6}) = 288.85(10^{-6})$$

$$\varepsilon_2 = (205 - 83.853)(10^{-6}) = 121.15(10^{-6})$$

Maximum In - Plane Shear Strain (x - y Plane) : Represented by the coordinates of point E on the circle.

$$\frac{\gamma_{\substack{max \\ in\text{-}plane}}}{2} = R = 83.853(10^{-6})$$

$$\gamma_{\substack{max \\ in\text{-}plane}} = 168(10^{-6}) \qquad \textbf{Ans}$$

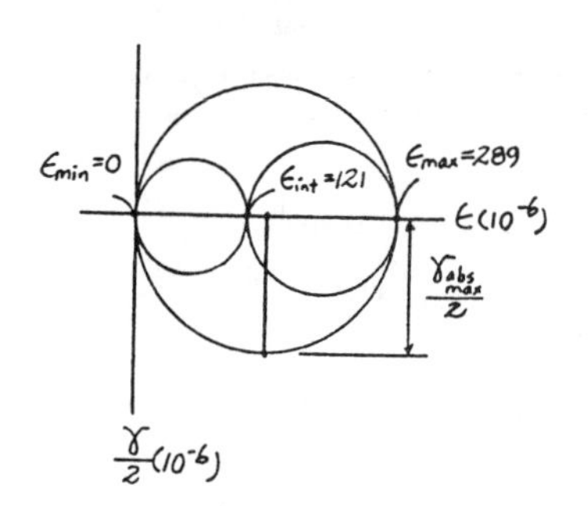

Three Mohr's Circles : From the results obtained above, the principal strains are

$$\varepsilon_{max} = 289(10^{-6}) \qquad \varepsilon_{int} = 121(10^{-6}) \qquad \varepsilon_{min} = 0 \qquad \textbf{Ans}$$

Absolute Maximum Shear Stress :

$$\gamma_{\substack{abs \\ max}} = \varepsilon_{max} - \varepsilon_{min}$$

$$= [288.85 - 0](10^{-6}) = 289(10^{-6}) \qquad \textbf{Ans}$$

10–23. The strain at a point has components of $\epsilon_x = 350(10^{-6})$, $\epsilon_y = -460(10^{-6})$, $\gamma_{xy} = -560(10^{-6})$, and $\epsilon_z = 0$. Determine (a) the principal strains at the point, (b) the maximum shear strain in the x–y plane, and (c) the absolute maximum shear strain.

Construction of the Circle (x - y Plane) : In accordance with the sign convention, $\varepsilon_x = 350(10^{-6})$, $\varepsilon_y = -460(10^{-6})$, and $\frac{\gamma_{xy}}{2} = -280(10^{-6})$. Hence,

$$\varepsilon_{avg} = \frac{\varepsilon_x + \varepsilon_y}{2} = \left[\frac{350 + (-460)}{2}\right](10^{-6}) = -55.0(10^{-6})$$

The coordinates for reference points A and C are

$$A(350,\ -280)(10^{-6}) \qquad C(-55.0,\ 0)(10^{-6})$$

The radius of the circle is

$$R = \left(\sqrt{(350 + 55.0)^2 + 280^2}\right)(10^{-6}) = 492.37(10^{-6})$$

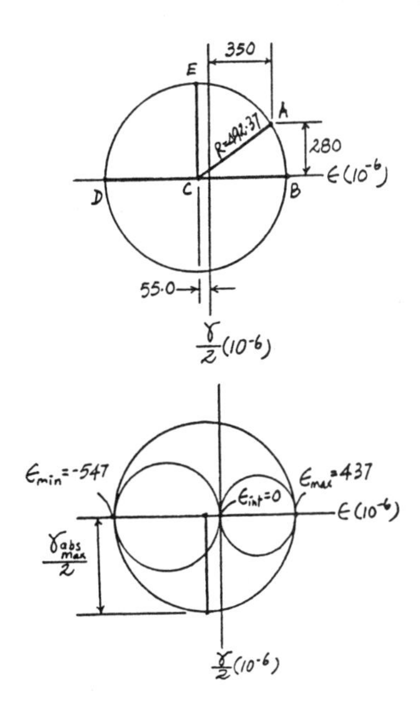

In - Plane Principal Strain : The coordinates of points B and D represent ε_1 and ε_2, respectively.

$$\varepsilon_1 = (-55.0 + 492.37)(10^{-6}) = 437.37(10^{-6})$$

$$\varepsilon_2 = (-55.0 - 492.37)(10^{-6}) = -547.37(10^{-6})$$

Maximum In - Plane Shear Strain (x - y Plane) : Represented by the coordinates of point E on the circle.

$$\frac{\gamma_{\substack{max \\ in\text{-}plane}}}{2} = -R = -492.37(10^{-6})$$

$$\gamma_{\substack{max \\ in\text{-}plane}} = -985(10^{-6}) \qquad \textbf{Ans}$$

Three Mohr's Circle : From the results obtained above, the principal strains are

$$\varepsilon_{max} = 437(10^{-6}) \qquad \varepsilon_{int} = 0 \qquad \varepsilon_{min} = -547(10^{-6}) \qquad \textbf{Ans}$$

Absolute Maximum Shear Stress :

$$\gamma_{\substack{abs \\ max}} = \varepsilon_{max} - \varepsilon_{min}$$

$$= [437.37 - (-547.37)](10^{-6}) = 985(10^{-6}) \qquad \textbf{Ans}$$

***10–24.** The strain at a point has components of $\epsilon_x = -520(10^{-6})$, $\epsilon_y = -350(10^{-6})$, $\gamma_{xy} = 720(10^{-6})$, and $\epsilon_z = 0$. Determine (a) the principal strains at the point, (b) the maximum shear strain in the $x-y$ plane, and (c) the absolute maximum shear strain.

Construction of the Circle (x - y Plane) : In accordance with the sign convention, $\varepsilon_x = -520(10^{-6})$, $\varepsilon_y = -350(10^{-6})$, and $\dfrac{\gamma_{xy}}{2} = 360(10^{-6})$. Hence,

$$\varepsilon_{avg} = \frac{\varepsilon_x + \varepsilon_y}{2} = \left[\frac{-520 + (-350)}{2}\right](10^{-6}) = -435(10^{-6})$$

The coordinates for reference points A and C are

$$A(-520,\ 360)(10^{-6}) \qquad C(-435,\ 0)(10^{-6})$$

The radius of the circle is

$$R = \left(\sqrt{(520-435)^2 + 360^2}\right)(10^{-6}) = 369.90(10^{-6})$$

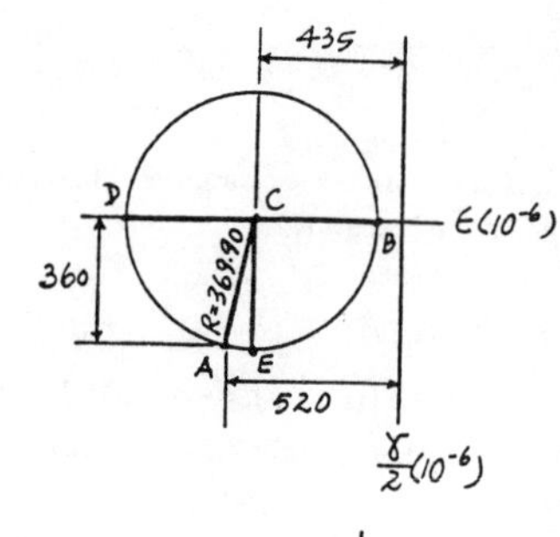

In - Plane Principal Strain : The coordinates of points B and D represent ε_1 and ε_2, respectively.

$$\varepsilon_1 = (-435 + 369.90)(10^{-6}) = -65.10(10^{-6})$$
$$\varepsilon_2 = (-435 - 369.90)(10^{-6}) = -804.90(10^{-6})$$

Maximum In - Plane Shear Strain (x - y Plane) : Represented by the coordinates of point E on the circle.

$$\frac{\gamma_{\substack{max \\ in\text{-}plane}}}{2} = R = 369.90(10^{-6})$$

$$\gamma_{\substack{max \\ in\text{-}plane}} = 740(10^{-6}) \qquad \textbf{Ans}$$

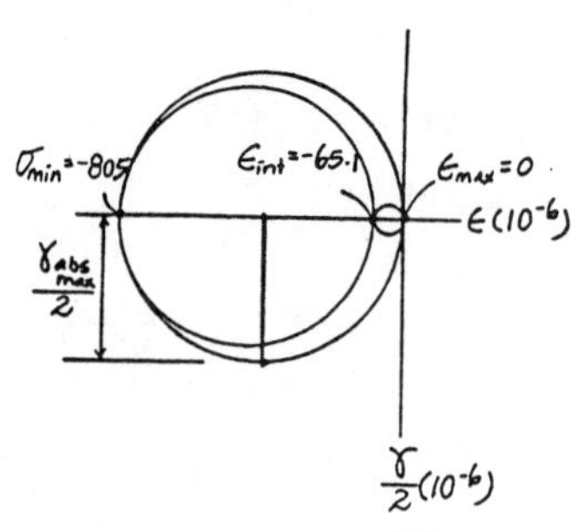

Three Mohr's Circles : From the results obtained above, the principal strains are

$$\varepsilon_{max} = 0 \qquad \varepsilon_{int} = -65.1(10^{-6}) \qquad \varepsilon_{min} = -805(10^{-6}) \qquad \textbf{Ans}$$

Absolute Maximum Shear Stress :

$$\gamma_{\substack{abs \\ max}} = \varepsilon_{max} - \varepsilon_{min}$$
$$= [0 - (-804.90)](10^{-6}) = 805(10^{-6}) \qquad \textbf{Ans}$$

10–25. The strain at a point has components of $\epsilon_x = 450(10^{-6})$, $\epsilon_y = 825(10^{-6})$, $\gamma_{xy} = 275(10^{-6})$, and $\epsilon_z = 0$. Determine (a) the principal strains at the point, (b) the maximum shear strain in the $x-y$ plane, and (c) the absolute maximum shear strain.

Construction of the Circle (x - y Plane) : In accordance with the sign convention, $\varepsilon_x = 450(10^{-6})$, $\varepsilon_y = 825(10^{-6})$ and $\frac{\gamma_{xy}}{2} = 137.5(10^{-6})$. Hence,

$$\varepsilon_{avg} = \frac{\varepsilon_x + \varepsilon_y}{2} = \left[\frac{450+825}{2}\right](10^{-6}) = 637.5(10^{-6})$$

The coordinates for reference points A and C are

$$A(450, 137.5)(10^{-6}) \qquad C(637.5, 0)(10^{-6})$$

The radius of the circle is

$$R = \left(\sqrt{(637.5-450)^2 + 137.5^2}\right)(10^{-6}) = 232.51(10^{-6})$$

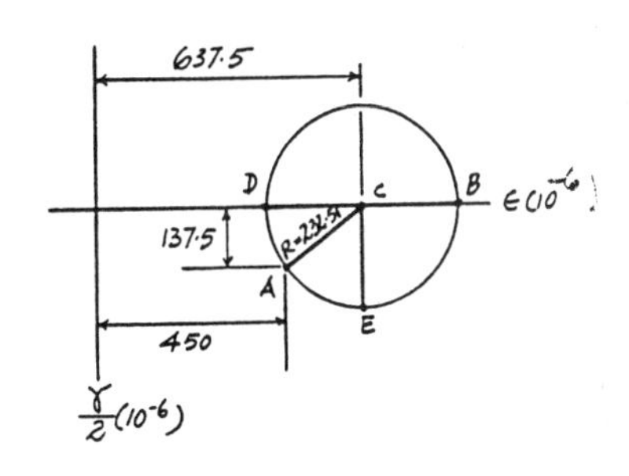

In - Plane Principal Strain : The coordinates of points B and D represent ε_1 and ε_2, respectively.

$$\varepsilon_1 = (637.5+232.51)(10^{-6}) = 870.01(10^{-6})$$
$$\varepsilon_2 = (637.5-232.51)(10^{-6}) = 404.99(10^{-6})$$

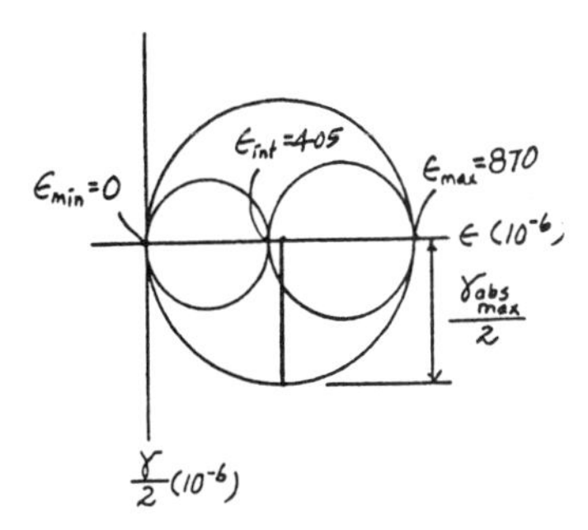

Maximum In - Plane Shear Strain (x - y Plane) : Represented by the coordinate of point E on the circle.

$$\frac{\gamma_{\substack{max \\ in\text{-}plane}}}{2} = R = 232.51(10^{-6})$$

$$\gamma_{\substack{max \\ in\text{-}plane}} = 465(10^{-6}) \qquad \textbf{Ans}$$

Three Mohr's Circle : From the results obtained above, the principal strains are

$$\varepsilon_{max} = 870(10^{-6}) \qquad \varepsilon_{int} = 405(10^{-6}) \qquad \varepsilon_{min} = 0 \qquad \textbf{Ans}$$

Absolute Maximum Shear Stress :

$$\gamma_{\substack{abs \\ max}} = \varepsilon_{max} - \varepsilon_{min}$$
$$= (870.01-0)(10^{-6}) = 870(10^{-6}) \qquad \textbf{Ans}$$

10–26. The state of strain at a point has components of $\epsilon_x = -400(10^{-6})$, $\epsilon_y = -200(10^{-6})$, $\gamma_{xy} = -250(10^{-6})$, and $\epsilon_z = 0$. Determine (a) the principal strains at the point, (b) the maximum shear strain in the x–y plane, and (c) the absolute maximum shear strain.

Construction of the Circle (x - y Plane) : In accordance with the sign convention, $\varepsilon_x = -400(10^{-6})$, $\varepsilon_y = -200(10^{-6})$, and $\dfrac{\gamma_{xy}}{2} = -125(10^{-6})$. Hence,

$$\varepsilon_{avg} = \frac{\varepsilon_x + \varepsilon_y}{2} = \left[\frac{-400 + (-200)}{2}\right](10^{-6}) = -300(10^{-6})$$

The coordinates for reference points A and C are

$$A(-400, -125)(10^{-6}) \qquad C(-300, 0)(10^{-6})$$

The radius of the circle is

$$R = \left(\sqrt{(400-300)^2 + 125^2}\right)(10^{-6}) = 160.08(10^{-6})$$

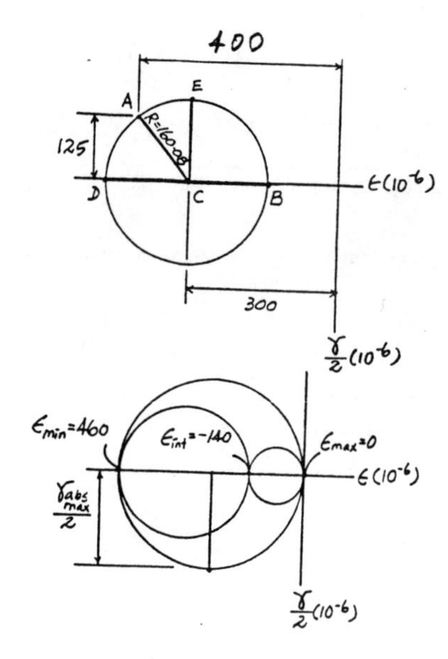

In - Plane Principal Strain : The coordinates of points B and D represent ε_1 and ε_2, respectively.

$$\varepsilon_1 = (-300 + 160.08)(10^{-6}) = -139.92(10^{-6})$$
$$\varepsilon_2 = (-300 - 160.08)(10^{-6}) = -460.08(10^{-6})$$

Maximum In - Plane Shear Strain (x - y Plane) : Represented by the coordinates of point E on the circle.

$$\frac{\gamma_{\substack{max \\ in\text{-}plane}}}{2} = -R = -160.08(10^{-6})$$

$$\gamma_{\substack{max \\ in\text{-}plane}} = -320(10^{-6}) \qquad \textbf{Ans}$$

Three Mohr's Circles : From the results obtained above, the principal strains are

$$\varepsilon_{max} = 0 \qquad \varepsilon_{int} = -140(10^{-6}) \qquad \varepsilon_{min} = -460(10^{-6}) \qquad \textbf{Ans}$$

Absolute Maximum Shear Stress :

$$\gamma_{\substack{abs \\ max}} = \varepsilon_{max} - \varepsilon_{min}$$
$$= [0 - (-460.08)](10^{-6}) = 460(10^{-6}) \qquad \textbf{Ans}$$

10–27. The 45° strain rosette is mounted on the surface of an aluminum plate. The following readings are obtained for each gauge: $\epsilon_a = 475(10^{-6})$, $\epsilon_b = 250(10^{-6})$, and $\epsilon_c = -360(10^{-6})$. Determine the in-plane principal strains.

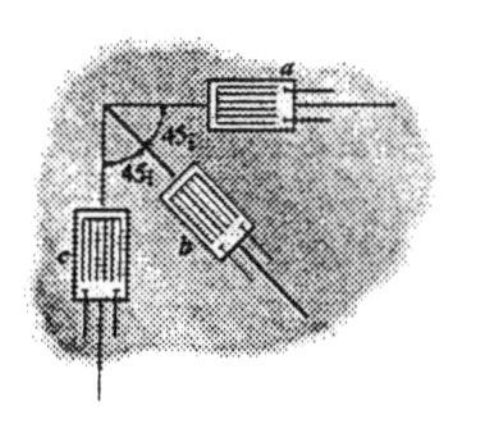

***Strain Rosettes* (45°) :** Applying the equations in the text with $\varepsilon_a = 475(10^{-6})$, $\varepsilon_b = 250(10^{-6})$, $\varepsilon_c = -360(10^{-6})$, $\theta_a = 0°$, $\theta_b = -45°$, and $\theta_c = -90°$,

$$475(10^{-6}) = \varepsilon_x \cos^2 0° + \varepsilon_y \sin^2 0° + \gamma_{xy} \sin 0° \cos 0°$$

$$\varepsilon_x = 475(10^{-6})$$

$$250(10^{-6}) = 475(10^{-6})\cos^2(-45°) + \varepsilon_y \sin^2(-45°) + \gamma_{xy}\sin(-45°)\cos(-45°)$$

$$250(10^{-6}) = 237.5(10^{-6}) + 0.5\,\varepsilon_y - 0.5\,\gamma_{xy}$$

$$0.5\varepsilon_y - 0.5\,\gamma_{xy} = 12.5(10^{-6}) \qquad [1]$$

$$-360(10^{-6}) = 475(10^{-6})\cos^2(-90°) + \varepsilon_y \sin^2(-90°) + \gamma_{xy}\sin(-90°)\cos(-90°)$$

$$\varepsilon_y = -360(10^{-6})$$

From Eq. [1], $\gamma_{xy} = -385(10^{-6})$

Therefore, $\varepsilon_x = 475(10^{-6})$ $\quad \varepsilon_y = -360(10^{-6})$ $\quad \gamma_{xy} = -385(10^{-6})$

***Construction of the Circle* :** With $\dfrac{\gamma_{xy}}{2} = -192.5(10^{-6})$ and

$$\varepsilon_{avg} = \frac{\varepsilon_x + \varepsilon_y}{2} = \left(\frac{475 + (-360)}{2}\right)(10^{-6}) = 57.5(10^{-6})$$

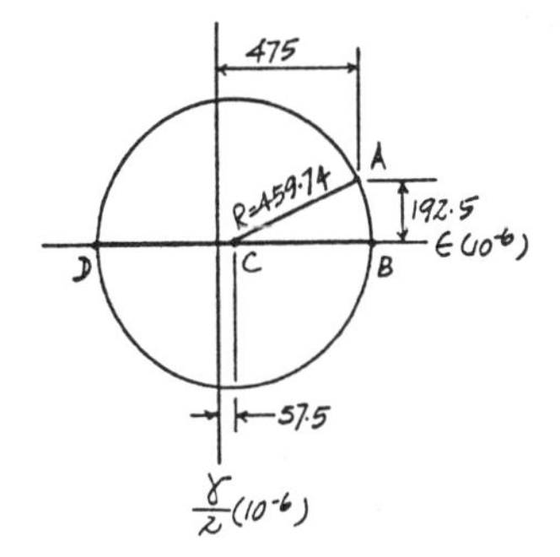

The coordinates for reference points A and C are

$$A(475,\ -192.5)(10^{-6}) \qquad C(57.5,\ 0)(10^{-6})$$

The radius of the circle is

$$R = \left(\sqrt{(475 - 57.5)^2 + 192.5^2}\right)(10^{-6}) = 459.74(10^{-6})$$

***In - Plane Principal Strain* :** The coordinates of points B and D represent ε_1 and ε_2, respectively.

$$\varepsilon_1 = (57.5 + 459.74)(10^{-6}) = 517(10^{-6}) \qquad \textbf{Ans}$$

$$\varepsilon_2 = (57.5 - 459.74)(10^{-6}) = -402(10^{-6}) \qquad \textbf{Ans}$$

***10–28.** The 45° strain rosette is mounted on a steel shaft. The following readings are obtained from each gauge: $\epsilon_a = 300(10^{-6})$, $\epsilon_b = 180(10^{-6})$, and $\epsilon_c = -250(10^{-6})$. Determine the in-plane principal strains and their orientation.

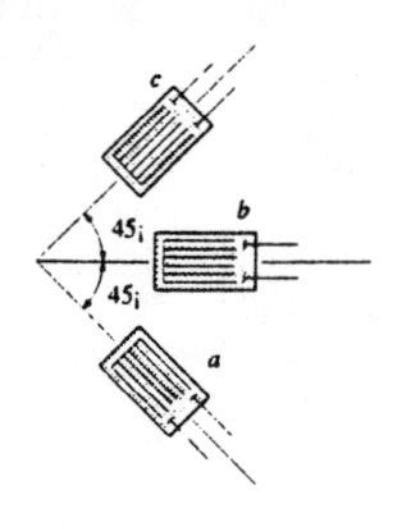

***Strain Rosettes* (45°) :** Applying the equations in the text with $\varepsilon_a = 300(10^{-6})$, $\varepsilon_b = 180(10^{-6})$, $\varepsilon_c = -250(10^{-6})$, $\theta_a = -45°$, $\theta_b = 0$, and $\theta_c = 45°$,

$$180(10^{-6}) = \varepsilon_x \cos^2 0° + \varepsilon_y \sin^2 0° + \gamma_{xy} \sin 0° \cos 0°$$

$$\varepsilon_x = 180(10^{-6})$$

$$300(10^{-6}) = 180(10^{-6}) \cos^2(-45°) + \varepsilon_y \sin^2(-45°) + \gamma_{xy} \sin(-45°)\cos(-45°)$$

$$210(10^{-6}) = 0.5\,\varepsilon_y - 0.5\,\gamma_{xy} \qquad [1]$$

$$-250(10^{-6}) = 180(10^{-6}) \cos^2 45° + \varepsilon_y \sin^2 45° + \gamma_{xy} \sin 45° \cos 45°$$

$$-340(10^{-6}) = 0.5\,\varepsilon_y + 0.5\,\gamma_{xy} \qquad [2]$$

Solving Eqs. [1] and [2] yields $\quad \varepsilon_y = -130(10^{-6}) \quad \gamma_{xy} = -550(10^{-6})$

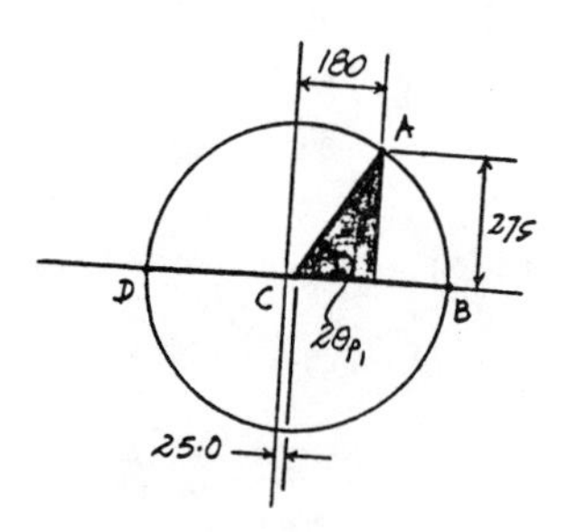

***Construction of the Circle* :** With $\varepsilon_x = 180(10^{-6})$, $\varepsilon_y = -130(10^{-6})$, and $\frac{\gamma_{xy}}{2} = -275(10^{-6})$.

$$\varepsilon_{avg} = \frac{\varepsilon_x + \varepsilon_y}{2} = \left[\frac{180 + (-130)}{2}\right](10^{-6}) = 25.0(10^{-6})$$

The coordinates for reference points A and C are

$$A(180,\ -275)(10^{-6}) \qquad C(25.0,\ 0)(10^{-6})$$

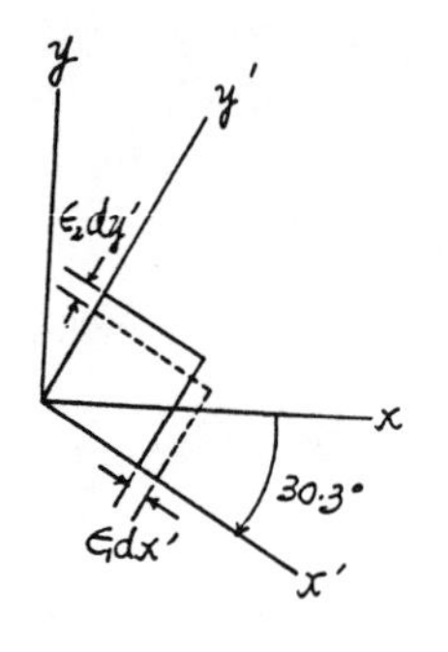

The radius of the circle is

$$R = \left(\sqrt{(180 - 25.0)^2 + 275^2}\right)(10^{-6}) = 315.67(10^{-6})$$

***In - Plane Principal Strain* :** The coordinates of points B and D represent ε_1 and ε_2, respectively.

$$\varepsilon_1 = (25.0 + 315.67)(10^{-6}) = 341(10^{-6}) \qquad \textbf{Ans}$$

$$\varepsilon_2 = (25.0 - 315.67)(10^{-6}) = -291(10^{-6}) \qquad \textbf{Ans}$$

***Orientation of Principal Strain* :** From the circle,

$$\tan 2\theta_{p_1} = \frac{275}{180 - 25.0} = 1.7742$$

$$\theta_{p_1} = 30.3° \ (\textit{Clockwise}) \qquad \textbf{Ans}$$

***10–29.** The 60° strain rosette is mounted on the surface of the bracket. The following readings are obtained for each gauge: $\epsilon_a = -780(10^{-6})$, $\epsilon_b = 400(10^{-6})$, and $\epsilon_c = 500(10^{-6})$. Determine (a) the principal strains and (b) the maximum in-plane shear strain and associated average normal strain. In each case show the deformed element due to these strains.

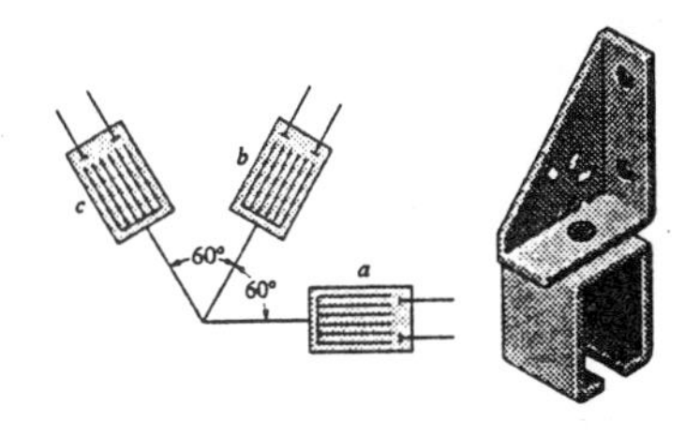

Strain Rosettes* (60°) *: Applying the equations in the text with $\varepsilon_a = -780(10^{-6})$, $\varepsilon_b = 400(10^{-6})$, $\varepsilon_c = 500(10^{-6})$, $\theta_a = 0°$, $\theta_b = 60°$, and $\theta_c = 120°$,

$$\varepsilon_x = \varepsilon_a = -780(10^{-6})$$

$$\begin{aligned}\varepsilon_y &= \frac{1}{3}(2\varepsilon_b + 2\varepsilon_c - \varepsilon_a)\\ &= \frac{1}{3}[2(400) + 2(500) - (-780)](10^{-6})\\ &= 860(10^{-6})\end{aligned}$$

$$\begin{aligned}\gamma_{xy} &= \frac{2}{\sqrt{3}}(\varepsilon_b - \varepsilon_c)\\ &= \frac{2}{\sqrt{3}}(400 - 500)(10^{-6})\\ &= -115.47(10^{-6})\end{aligned}$$

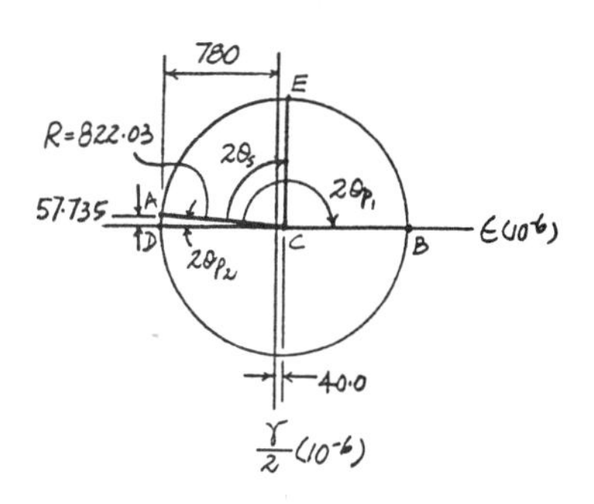

Construction of the Circle : With $\varepsilon_x = -780(10^{-6})$, $\varepsilon_y = 860(10^{-6})$, and $\frac{\gamma_{xy}}{2} = -57.735(10^{-6})$.

$$\varepsilon_{avg} = \frac{\varepsilon_x + \varepsilon_y}{2} = \left(\frac{-780 + 860}{2}\right)(10^{-6}) = 40.0(10^{-6}) \qquad \textbf{Ans}$$

The coordinates for reference points A and C are

$$A(-780, \ -57.735)(10^{-6}) \qquad C(40.0, \ 0)(10^{-6})$$

The radius of the circle is

$$R = \left(\sqrt{(780 + 40.0)^2 + 57.735^2}\right)(10^{-6}) = 822.03(10^{-6})$$

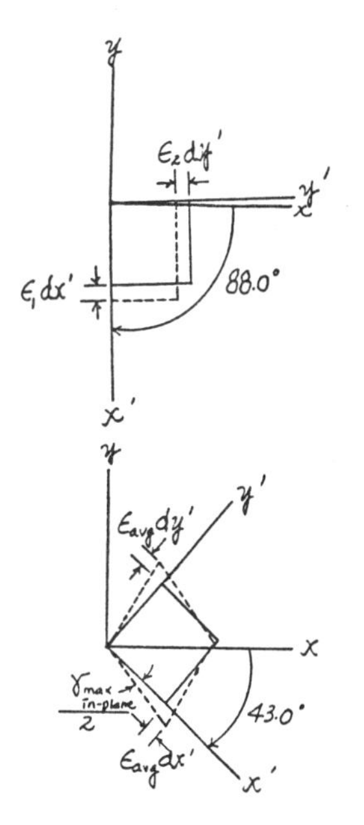

a)

In - Plane Principal Strain : The coordinates of points B and D represent ε_1 and ε_2, respectively.

$$\varepsilon_1 = (40.0 + 822.03)(10^{-6}) = 862(10^{-6}) \qquad \textbf{Ans}$$

$$\varepsilon_2 = (40.0 - 822.03)(10^{-6}) = -782(10^{-6}) \qquad \textbf{Ans}$$

Orientation of Principal Strain : From the circle,

$$\tan 2\theta_{p_2} = \frac{57.735}{780 + 40} = 0.07041 \qquad 2\theta_{p_2} = 4.03°$$

$$2\theta_{p_1} = 180° - 2\theta_{p_2}$$

$$\theta_{p_1} = \frac{180° - 4.03°}{2} = 88.0° \ (\textit{Clockwise}) \qquad \textbf{Ans}$$

b)

Maximum In - Plane Shear Strain : Represented by the coordinates of point E on the circle.

$$\frac{\gamma_{\substack{max \\ in\text{-}plane}}}{2} = -R = -822.03(10^{-6})$$

$$\gamma_{\substack{max \\ in\text{-}plane}} = -1644(10^{-6}) \qquad \textbf{Ans}$$

Orientation of Maximum In - Plane Shear Strain : From the circle,

$$\tan 2\theta_s = \frac{780 + 40}{57.735} = 14.2028$$

$$\theta_s = 43.0° \ (\textit{Clockwise}) \qquad \textbf{Ans}$$

10–30. The 45° strain rosette is mounted near the tooth of the wrench. The following readings are obtained for each gauge: $\epsilon_a = 800(10^{-6})$, $\epsilon_b = 520(10^{-6})$, and $\epsilon_c = -450(10^{-6})$. Determine (a) the in-plane principal strains and (b) the maximum in-plane shear strain and associated average normal strain. In each case show the deformed element due to these strains.

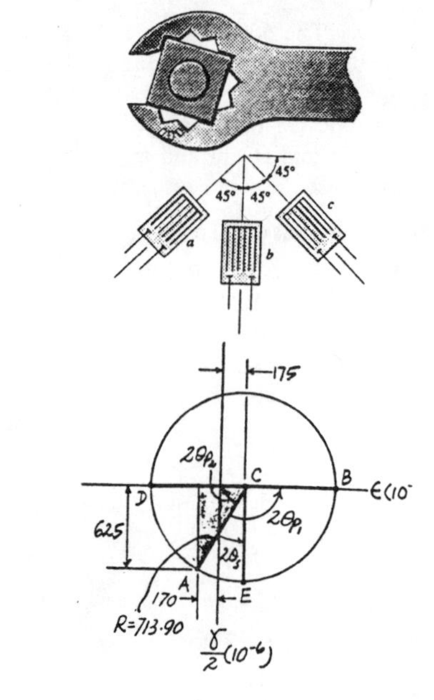

***Strain Rosettes* (45°) :** Applying the equations in the text with $\varepsilon_a = 800(10^{-6})$, $\varepsilon_b = 520(10^{-6})$, $\varepsilon_c = -450(10^{-6})$, $\theta_a = -135°$, $\theta_b = -90°$ and $\theta_c = -45°$,

$$520(10^{-6}) = \varepsilon_x \cos^2(-90°) + \varepsilon_y \sin^2(-90°) + \gamma_{xy} \sin(-90°)\cos(-90°)$$
$$\varepsilon_y = 520(10^{-6})$$

$$800(10^{-6}) = \varepsilon_x \cos^2(-135°) + 520(10^{-6})\sin^2(-135°) + \gamma_{xy}\sin(-135°)\cos(-135°)$$
$$540(10^{-6}) = 0.5\,\varepsilon_x + 0.5\,\gamma_{xy} \qquad [1]$$

$$-450(10^{-6}) = \varepsilon_x \cos^2(-45°) + 520(10^{-6})\sin^2(-45°) + \gamma_{xy}\sin(-45°)\cos(-45°)$$
$$-710(10^{-6}) = 0.5\,\varepsilon_x - 0.5\,\gamma_{xy} \qquad [2]$$

Solving Eqs. [1] and [2] yields $\quad \varepsilon_x = -170(10^{-6}) \quad \gamma_{xy} = 1250(10^{-6})$

***Construction of the Circle* :** With $\varepsilon_x = -170(10^{-6})$, $\varepsilon_y = 520(10^{-6})$, and $\frac{\gamma_{xy}}{2} = 625(10^{-6})$,

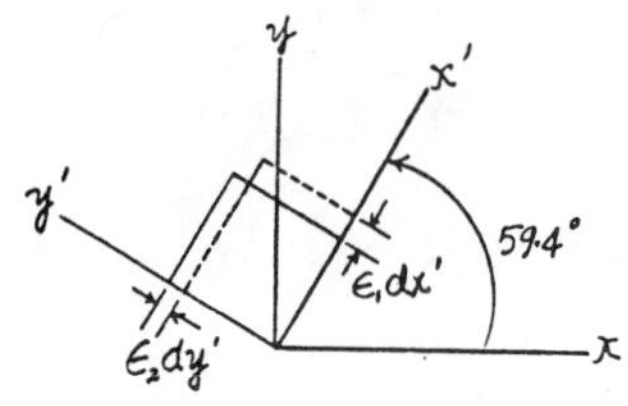

$$\varepsilon_{avg} = \frac{\varepsilon_x + \varepsilon_y}{2} = \left(\frac{-170+520}{2}\right)(10^{-6}) = 175(10^{-6}) \qquad \textbf{Ans}$$

The coordinates for reference points A and C are

$$A(-170,\ 625)(10^{-6}) \qquad C(175,\ 0)(10^{-6})$$

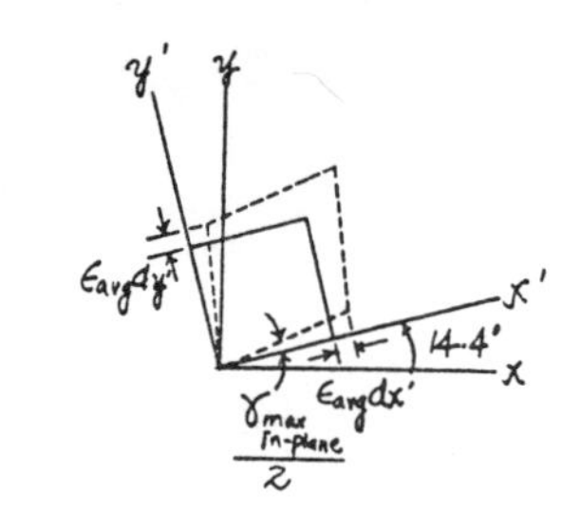

The radius of the circle is

$$R = \left(\sqrt{(170+175)^2 + 625^2}\right)(10^{-6}) = 713.90(10^{-6})$$

a)

***In-Plane Principal Strain* :** The coordinates of points B and D represent ε_1 and ε_2, respectively.

$$\varepsilon_1 = (175+713.90)(10^{-6}) = 889(10^{-6}) \qquad \textbf{Ans}$$
$$\varepsilon_2 = (175-713.90)(10^{-6}) = -539(10^{-6}) \qquad \textbf{Ans}$$

***Orientation of Principal Strain* :** From the circle,

$$\tan 2\theta_{p_2} = \frac{625}{170+175} = 1.8118 \qquad 2\theta_{p_2} = 61.10°$$

$$2\theta_{p_1} = 180° - 2\theta_{p_2}$$
$$\theta_{p_1} = \frac{180° - 61.10°}{2} = 59.4° \quad (\textit{Counterclockwise}) \qquad \textbf{Ans}$$

b)

***Maximum In-Plane Shear Strain* :** Represented by the coordinate of point E on the circle.

$$\frac{\gamma_{\substack{max \\ in\text{-}plane}}}{2} = R = 713.90(10^{-6})$$
$$\gamma_{\substack{max \\ in\text{-}plane}} = 1428(10^{-6}) \qquad \textbf{Ans}$$

***Orientation of Maximum In-Plane Shear Strain* :** From the circle,

$$\tan 2\theta_s = \frac{170+175}{625} = 0.552$$
$$\theta_s = 14.4° \quad (\textit{Counterclockwise}) \qquad \textbf{Ans}$$

10–31. The 60° strain rosette is mounted on a beam. The following readings are obtained from each gauge: $\epsilon_a = 150(10^{-6})$, $\epsilon_b = -330(10^{-6})$, and $\epsilon_c = 400(10^{-6})$. Determine (a) the in-plane principal strains and (b) the maximum in-plane shear strain and average normal strain. In each case show the deformed element due to these strains.

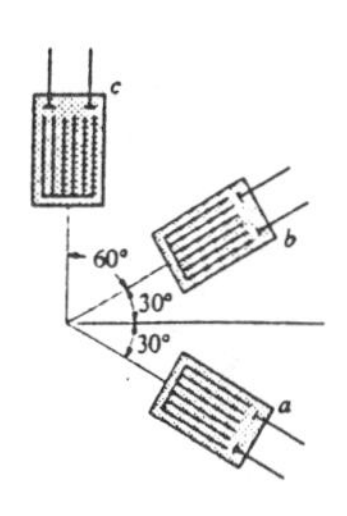

***Strain Rosettes* (60°) :** Applying the equations in the text with $\varepsilon_a = 150(10^{-6})$, $\varepsilon_b = -330(10^{-6})$, $\varepsilon_c = 400(10^{-6})$, $\theta_a = -30°$, $\theta_b = 30°$ and $\theta_c = 90°$,

$$400(10^{-6}) = \varepsilon_x \cos^2 90° + \varepsilon_y \sin^2 90° + \gamma_{xy} \sin 90° \cos 90°$$
$$\varepsilon_y = 400(10^{-6})$$

$$400(10^{-6}) = \varepsilon_x \cos^2 90° + \varepsilon_y \sin^2 90° + \gamma_{xy} \sin 90° \cos 90°$$
$$\varepsilon_y = 400(10^{-6})$$

$$150(10^{-6}) = \varepsilon_x \cos^2(-30°) + 400(10^{-6}) \sin^2(-30°) + \gamma_{xy} \sin(-30°)\cos(-30°)$$
$$50.0(10^{-6}) = 0.75\,\varepsilon_x - 0.4330\,\gamma_{xy} \qquad [1]$$

$$-330(10^{-6}) = \varepsilon_x \cos^2 30° + 400(10^{-6}) \sin^2 30° + \gamma_{xy} \sin 30° \cos 30°$$
$$-430(10^{-6}) = 0.75\,\varepsilon_x + 0.4330\,\gamma_{xy} \qquad [2]$$

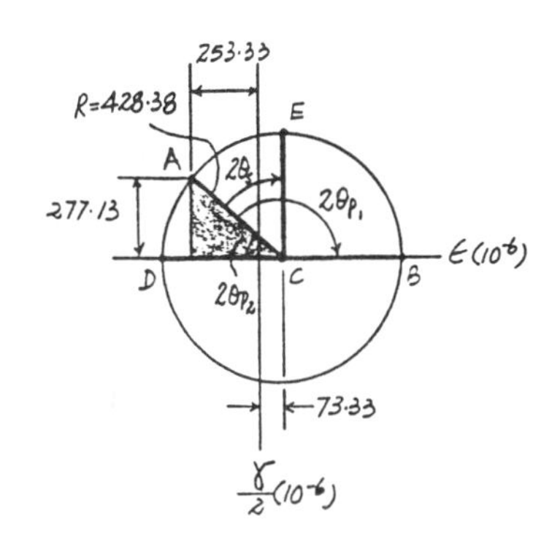

***Construction of the Circle* :** With $\varepsilon_x = -253.33(10^{-6})$, $\varepsilon_y = 400(10^{-6})$, and $\frac{\gamma_{xy}}{2} = -277.13(10^{-6})$

$$\varepsilon_{avg} = \frac{\varepsilon_x + \varepsilon_y}{2} = \left(\frac{-253.33 + 400}{2}\right)(10^{-6}) = 73.3(10^{-6}) \qquad \textbf{Ans}$$

Coordinates for reference points A and C are

$$A(-253.33,\ -277.13)(10^{-6}) \qquad C(73.33,\ 0)(10^{-6})$$

The radius of the circle is

$$R = \left(\sqrt{(253.33 + 73.33)^2 + 277.13^2}\right)(10^{-6}) = 428.38(10^{-6})$$

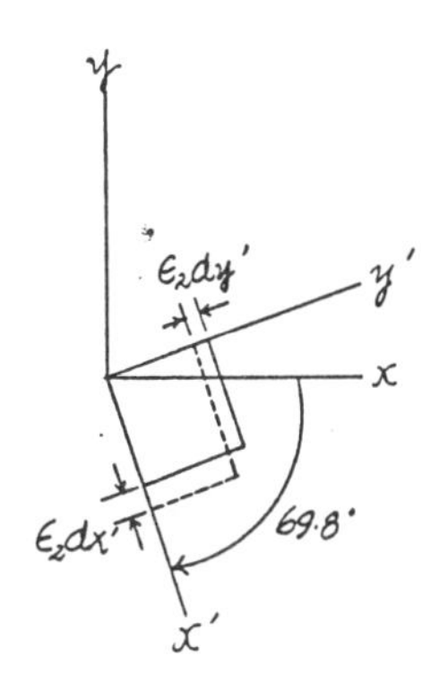

a)

***In-Plane Principal Strain* :** The coordinates of points B and D represent ε_1 and ε_2, respectively.

$$\varepsilon_1 = (73.33 + 428.38)(10^{-6}) = 502(10^{-6}) \qquad \textbf{Ans}$$
$$\varepsilon_2 = (73.33 - 428.38)(10^{-6}) = -355(10^{-6}) \qquad \textbf{Ans}$$

***Orientation of Principal Strain* :** From the circle,

$$\tan 2\theta_{p_2} = \frac{277.13}{253.33 + 73.33} = 0.8484 \qquad 2\theta_{p_2} = 40.31°$$

$$2\theta_{p_1} = 180° - 2\theta_{p_2}$$
$$\theta_{p_1} = \frac{180° - 40.31°}{2} = 69.8° \ (\textit{Clockwise}) \qquad \textbf{Ans}$$

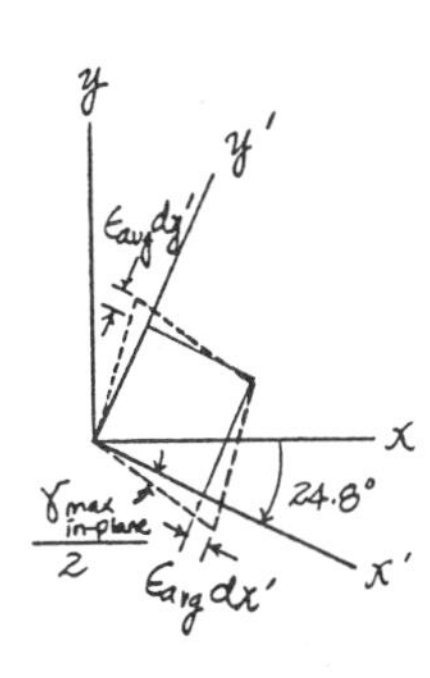

b)

***Maximum In-Plane Shear Strain* :** Represented by the coordinates of point E on the circle,

$$\frac{\gamma_{\substack{max \\ in\text{-}plane}}}{2} = -R = -428.38(10^{-6})$$
$$\gamma_{\substack{max \\ in\text{-}plane}} = -857(10^{-6}) \qquad \textbf{Ans}$$

***Orientation of Maximum In-Plane Shear Strain* :** From the circle.

$$\tan 2\theta_s = \frac{253.33 + 73.33}{277.13} = 1.1788$$

$$\theta_s = 24.8° \ (\textit{Clockwise}) \qquad \textbf{Ans}$$

10–33. For the case of plane stress, show that Hooke's law can be written as

$$\sigma_x = \frac{E}{(1-\nu^2)}(\epsilon_x + \nu\epsilon_y), \qquad \sigma_y = \frac{E}{(1-\nu^2)}(\epsilon_y + \nu\epsilon_x)$$

Generalized Hooke's Law : For plane stress, $\sigma_z = 0$. Applying Eq. 10–18,

$$\varepsilon_x = \frac{1}{E}(\sigma_x - \nu\sigma_y)$$

$$\nu E\varepsilon_x = (\sigma_x - \nu\sigma_y)\nu$$

$$\nu E\varepsilon_x = \nu\sigma_x - \nu^2\sigma_y \qquad [1]$$

$$\varepsilon_y = \frac{1}{E}(\sigma_y - \nu\sigma_x)$$

$$E\varepsilon_y = -\nu\sigma_x + \sigma_y \qquad [2]$$

Adding Eq. [1] and Eq. [2] yields,

$$\nu E\varepsilon_x + E\varepsilon_y = \sigma_y - \nu^2\sigma_y$$

$$\sigma_y = \frac{E}{1-\nu^2}(\nu\varepsilon_x + \varepsilon_y) \qquad (Q.E.D.)$$

Substituting σ_y into Eq. [2]

$$E\varepsilon_y = -\nu\sigma_x + \frac{E}{1-\nu^2}(\nu\varepsilon_x + \varepsilon_y)$$

$$\sigma_x = \frac{E(\nu\varepsilon_x + \varepsilon_y)}{\nu(1-\nu^2)} - \frac{E\varepsilon_y}{\nu}$$

$$= \frac{E\nu\varepsilon_x + E\varepsilon_y - E\varepsilon_y + E\varepsilon_y\nu^2}{\nu(1-\nu^2)}$$

$$= \frac{E}{1-\nu^2}(\varepsilon_x + \nu\varepsilon_y) \qquad (Q.E.D.)$$

10–34. Use Hooke's law, Eq. 10–18, to develop the strain-transformation equations, Eqs. 10–5 and 10–6, from the stress-transformation equations, Eqs. 9–1 and 9–2.

Stress Transformation Equations :

$$\sigma_{x'} = \frac{\sigma_x + \sigma_y}{2} + \frac{\sigma_x - \sigma_y}{2}\cos 2\theta + \tau_{xy}\sin 2\theta \qquad [1]$$

$$\tau_{x'y'} = -\frac{\sigma_x - \sigma_y}{2}\sin 2\theta + \tau_{xy}\cos 2\theta \qquad [2]$$

$$\sigma_{y'} = \frac{\sigma_x + \sigma_y}{2} - \frac{\sigma_x - \sigma_y}{2}\cos 2\theta - \tau_{xy}\sin 2\theta \qquad [3]$$

Generalized Hooke's Law : For plane stress, $\sigma_z = \tau_{xz} = \tau_{yz} = 0$. Applying Eq. 10–18, 10–19, and 10-20,

$$\varepsilon_x = \frac{\sigma_x}{E} - \frac{\nu\sigma_y}{E} \qquad [4]$$

$$\varepsilon_y = \frac{-\nu\sigma_x}{E} + \frac{\sigma_y}{E} \qquad [5]$$

$$\tau_{xy} = G\gamma_{xy} \qquad [6]$$

$$G = \frac{E}{2(1+\nu)} \qquad [7]$$

From Eqs. [4] and [5],

$$\varepsilon_x + \varepsilon_y = \frac{(1-\nu)}{E}(\sigma_x + \sigma_y) \qquad [8]$$

$$\varepsilon_x - \varepsilon_y = \frac{(1+\nu)}{E}(\sigma_x - \sigma_y) \qquad [9]$$

From Eqs. [6] and [7],

$$\tau_{xy} = \frac{E}{2(1+\nu)}\gamma_{xy} \qquad [10]$$

From Eq. [4],

$$\varepsilon_{x'} = \frac{\sigma_{x'}}{E} - \frac{\nu\sigma_{y'}}{E} \qquad [11]$$

Substituting Eqs. [1] and [3] into Eq. [11],

$$\varepsilon_{x'} = \frac{(1-\nu)(\sigma_x + \sigma_y)}{2E} + \frac{(1+\nu)(\sigma_x - \sigma_y)}{2E}\cos 2\theta + \frac{(1+\nu)\tau_{xy}\sin 2\theta}{E} \qquad [12]$$

By using Eqs. [8], [9], and [10] and substituting into Eq. [12],

$$\varepsilon_{x'} = \frac{\varepsilon_x + \varepsilon_y}{2} + \frac{\varepsilon_x - \varepsilon_y}{2}\cos 2\theta + \frac{\gamma_{xy}}{2}\sin 2\theta \qquad (Q.E.D.)$$

From Eq. [6],

$$\tau_{x'y'} = G\gamma_{x'y'} = \frac{E}{2(1+\nu)}\gamma_{x'y'} \qquad [13]$$

Substituting Eqs. [13], [6] and [9] into Eq. [2],

$$\frac{E}{2(1+\nu)}\gamma_{x'y'} = -\frac{E(\varepsilon_x - \varepsilon_y)}{2(1+\nu)}\sin 2\theta + \frac{E}{2(1+\nu)}\gamma_{xy}\cos 2\theta$$

$$\frac{\gamma_{x'y'}}{2} = -\frac{\varepsilon_x - \varepsilon_y}{2}\sin 2\theta + \frac{\gamma_{xy}}{2}\cos 2\theta \qquad (Q.E.D.)$$

10–35. Determine the bulk modulus for hard rubber if $E_r = 0.68(10^3)$ ksi and $\nu_r = 0.43$.

Bulk Modulus : Applying Eq. 10 – 25,

$$k = \frac{E}{3(1-2\nu)} = \frac{0.68(10^3)}{3[1-2(0.43)]} = 1.62\left(10^3\right) \text{ ksi} \qquad \textbf{Ans}$$

***10–36.** Determine the bulk modulus for gray cast iron if $E_{fe} = 14(10^3)$ ksi and $\nu_{fe} = 0.20$.

Bulk Modulus : Applying Eq. 10 – 25,

$$k = \frac{E}{3(1-2\nu)} = \frac{14(10^3)}{3[1-2(0.20)]} = 7.78\left(10^3\right) \text{ ksi} \qquad \textbf{Ans}$$

10–37. The rod is made of aluminum 2014-T6. If it is subjected to the tensile load of 700 N and has a diameter of 20 mm, determine the absolute maximum shear strain in the rod at a point on its surface.

Normal Stress : For uniaxial loading, $\sigma_y = \sigma_z = 0$.

$$\sigma_x = \frac{P}{A} = \frac{700}{\frac{\pi}{4}(0.02^2)} = 2.228 \text{ MPa}$$

Normal Strain : Applying the generalized Hooke's Law.

$$\begin{aligned} \varepsilon_x &= \frac{1}{E}\left[\sigma_x - \nu\left(\sigma_y + \sigma_z\right)\right] \\ &= \frac{1}{73.1(10^9)}\left[2.228\left(10^6\right) - 0\right] \\ &= 30.48\left(10^{-6}\right) \end{aligned}$$

$$\begin{aligned} \varepsilon_y &= \frac{1}{E}\left[\sigma_y - \nu(\sigma_x + \sigma_z)\right] \\ &= \frac{1}{73.1(10^9)}\left[0 - 0.35\left(2.228\left(10^6\right) + 0\right)\right] \\ &= -10.67\left(10^{-6}\right) \end{aligned}$$

$$\begin{aligned} \varepsilon_z &= \frac{1}{E}\left[\sigma_z - \nu\left(\sigma_x + \sigma_y\right)\right] \\ &= \frac{1}{73.1(10^9)}\left[0 - 0.35\left(2.228\left(10^6\right) + 0\right)\right] \\ &= -10.67\left(10^{-6}\right) \end{aligned}$$

Therefore,

$$\varepsilon_{max} = 30.48\left(10^{-6}\right) \qquad \varepsilon_{min} = -10.67\left(10^{-6}\right)$$

Absolute Maximum Shear Stress :

$$\begin{aligned} \gamma_{\substack{abs \\ max}} &= \varepsilon_{max} - \varepsilon_{min} \\ &= [30.48 - (-10.67)]\left(10^{-6}\right) = 41.1\left(10^{-6}\right) \qquad \textbf{Ans} \end{aligned}$$

10–38. The rod is made of aluminum 2014-T6. If it is subjected to the tensile load of 700 N and has a diameter of 20 mm, determine the principal strains at a point on the surface of the rod.

Normal Stress : For uniaxial loading, $\sigma_y = \sigma_z = 0$.

$$\sigma_x = \frac{P}{A} = \frac{700}{\frac{\pi}{4}(0.02^2)} = 2.228 \text{ MPa}$$

Normal Strains : Applying the generalized Hooke's Law,

$$\varepsilon_x = \frac{1}{E}\left[\sigma_x - \nu(\sigma_y + \sigma_z)\right]$$
$$= \frac{1}{73.1(10^9)}\left[2.228(10^6) - 0\right]$$
$$= 30.48(10^{-6})$$

$$\varepsilon_y = \frac{1}{E}\left[\sigma_y - \nu(\sigma_x + \sigma_z)\right]$$
$$= \frac{1}{73.1(10^9)}\left[0 - 0.35(2.228(10^6) + 0)\right]$$
$$= -10.67(10^{-6})$$

$$\varepsilon_z = \frac{1}{E}\left[\sigma_z - \nu(\sigma_x + \sigma_y)\right]$$
$$= \frac{1}{73.1(10^9)}\left[0 - 0.35(2.228(10^6) + 0)\right]$$
$$= -10.67(10^{-6})$$

Principal Strains : From the results obtained above,

$$\varepsilon_{max} = 30.5(10^{-6}) \qquad \varepsilon_{int} = \varepsilon_{min} = -10.7(10^{-6}) \qquad \textbf{Ans}$$

10–39. A uniform edge load of 500 lb/in. and 350 lb/in. is applied to the polystyrene specimen. If the specimen is originally square and has dimensions of $a = 2$ in., $b = 2$ in., and a thickness of $t = 0.25$ in., determine its new dimensions a', b', and t' after the load is applied. $E_p = 597(10^3)$ psi and $\nu_p = 0.25$.

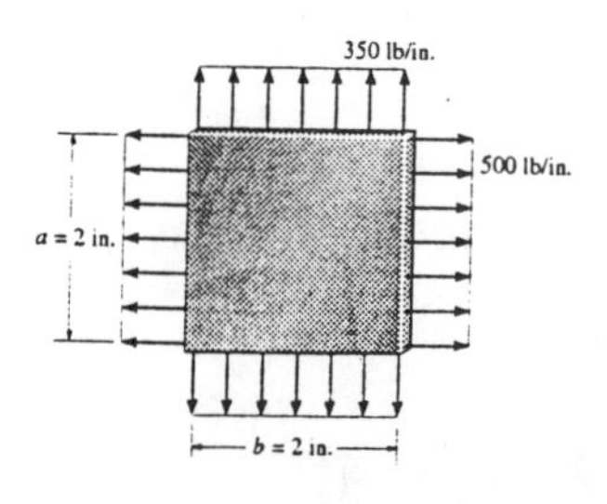

Normal Stresses : For plane stress, $\sigma_z = 0$.

$$\sigma_x = \frac{500}{0.25} = 2000 \text{ psi} \qquad \sigma_y = \frac{350}{0.25} = 1400 \text{ psi}$$

Normal Strains : Applying the generalized Hooke's Law,

$$\varepsilon_x = \frac{1}{E}\left[\sigma_x - \nu(\sigma_y + \sigma_z)\right]$$
$$= \frac{1}{597(10^3)}[2000 - 0.25(1400 + 0)]$$
$$= 2.7638(10^{-3})$$

$$\varepsilon_y = \frac{1}{E}\left[\sigma_y - \nu(\sigma_x + \sigma_z)\right]$$
$$= \frac{1}{597(10^3)}[1400 - 0.25(2000 + 0)]$$
$$= 1.5075(10^{-3})$$

$$\varepsilon_z = \frac{1}{E}\left[\sigma_z - \nu(\sigma_x + \sigma_y)\right]$$
$$= \frac{1}{597(10^3)}[0 - 0.25(2000 + 1400)]$$
$$= -1.4238(10^{-3})$$

The new dimensions for the new specimen are,

$$a' = 2 + 2\left[1.5075(10^{-3})\right] = 2.00302 \text{ in.} \qquad \textbf{Ans}$$
$$b' = 2 + 2\left[2.7638(10^{-3})\right] = 2.00553 \text{ in.} \qquad \textbf{Ans}$$
$$t' = 0.25 + 0.25\left[-1.4238(10^{-3})\right] = 0.24964 \text{ in.} \qquad \textbf{Ans}$$

***10–40.** The principal stresses at a point are shown. If the material is graphite for which $E_g = 800$ ksi and $\nu_g = 0.23$, determine the principal strains.

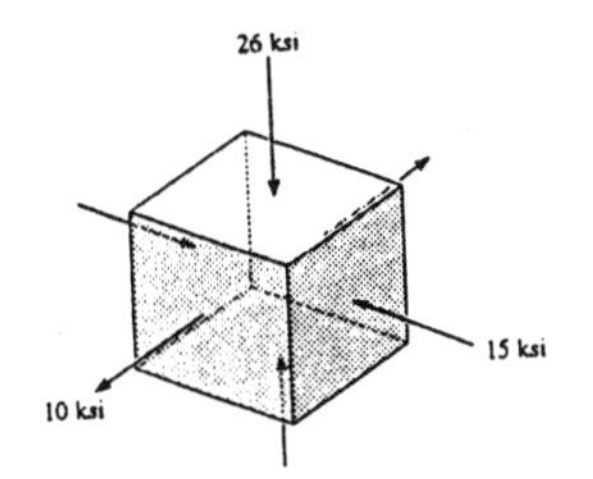

Normal Strains : Applying the generalized Hooke's Law with $\sigma_x = 10$ ksi, $\sigma_y = -15$ ksi, and $\sigma_z = -26$ ksi.

$$\varepsilon_x = \frac{1}{E}\left[\sigma_x - \nu\left(\sigma_y + \sigma_z\right)\right]$$
$$= \frac{1}{800}[10 - 0.23(-15 - 26)]$$
$$= 0.0242875$$

$$\varepsilon_y = \frac{1}{E}\left[\sigma_y - \nu\left(\sigma_x + \sigma_z\right)\right]$$
$$= \frac{1}{800}[-15 - 0.23(10 - 26)]$$
$$= -0.01415$$

$$\varepsilon_z = \frac{1}{E}\left[\sigma_z - \nu\left(\sigma_x + \sigma_y\right)\right]$$
$$= \frac{1}{800}[-26 - 0.23(10 - 15)]$$
$$= -0.0310625$$

Principal Strains : From the results obtained above,

$\varepsilon_{max} = 0.0243$ $\quad \varepsilon_{int} = -0.01415$ $\quad \varepsilon_{min} = -0.0311$ **Ans**

10–41. Determine the principal strains that occur at a point on a steel member where the principal stresses are $\sigma_{max} = 18$ ksi, $\sigma_{int} = 15$ ksi, $\sigma_{min} = -28$ ksi. $E_{st} = 29(10^3)$ ksi and $\nu_{st} = 0.3$.

Normal Strains : Applying the generalized Hooke's Law with $\sigma_{max} = 18$ ksi, $\sigma_{int} = 15$ ksi, and $\sigma_{min} = -28$ ksi.

$$\varepsilon_{max} = \frac{1}{E}\left[\sigma_{max} - \nu\left(\sigma_{int} + \sigma_{min}\right)\right]$$
$$= \frac{1}{29(10^3)}[18 - 0.3(15 - 28)]$$
$$= 0.755(10^{-3}) \qquad \textbf{Ans}$$

$$\varepsilon_{int} = \frac{1}{E}\left[\sigma_{int} - \nu\left(\sigma_{max} + \sigma_{min}\right)\right]$$
$$= \frac{1}{29(10^3)}[15 - 0.3(18 - 28)]$$
$$= 0.621(10^{-3}) \qquad \textbf{Ans}$$

$$\varepsilon_{min} = \frac{1}{E}\left[\sigma_{min} - \nu\left(\sigma_{max} + \sigma_{int}\right)\right]$$
$$= \frac{1}{29(10^3)}[-28 - 0.3(18 + 15)]$$
$$= -1.31(10^{-3}) \qquad \textbf{Ans}$$

10–42. A bar of plastic having a diameter of 0.5 in. is loaded in a tension machine, and it is determined that $\epsilon_x = 530(10^{-6})$ when the load is 80 lb. Determine the modulus of elasticity, E_p, and the dilatation, e_p, of the plastic. $\nu_p = 0.26$.

Normal Stresses : For uniaxial loading, $\sigma_y = \sigma_z = 0$.

$$\sigma_x = \frac{P}{A} = \frac{80}{\frac{\pi}{4}(0.5^2)} = 407.44 \text{ psi}$$

Normal Strains : Applying the generalized Hooke's Law,

$$\varepsilon_x = \frac{1}{E}\left[\sigma_x - \nu(\sigma_y + \sigma_z)\right]$$

$$530(10^{-6}) = \frac{1}{E_p}(407.44 - 0)$$

$$E_p = 768748 \text{ psi} = 0.769(10^{-3}) \text{ ksi} \quad \textbf{Ans}$$

Dilatation : Applying Eq. 10 – 23,

$$e = \frac{1-2\nu}{E}(\sigma_x + \sigma_y + \sigma_z)$$

$$= \frac{1-2(0.26)}{768748}(407.44 + 0 + 0)$$

$$= 0.254(10^{-3}) \quad \textbf{Ans}$$

10–43. A rod has a radius of 10 mm. If it is subjected to an axial load of 15 N such that the axial strain in the rod is $\epsilon_x = 2.75(10^{-6})$, determine the modulus of elasticity E and the change in its diameter. $\nu = 0.23$.

Normal Stresses : For uniaxial loading, $\sigma_y = \sigma_z = 0$.

$$\sigma_x = \frac{P}{A} = \frac{15}{\pi(0.01^2)} = 47.746 \text{ kPa}$$

Normal Strains : Applying the generalized Hooke's Law,

$$\varepsilon_x = \frac{1}{E}\left[\sigma_x - \nu(\sigma_y + \sigma_z)\right]$$

$$2.75(10^{-6}) = \frac{1}{E}\left[47.746(10^3) - 0\right]$$

$$E = 17.36 \text{ GPa} = 17.4 \text{ GPa} \quad \textbf{Ans}$$

$$\varepsilon_r = -\nu\varepsilon_x = -0.23(2.75)(10^{-6}) = -0.6325(10^{-6})$$

Then,

$$\delta d = \varepsilon_r d = -0.6325(10^{-6})(20) = 12.65(10^{-6}) \text{ mm} \quad \textbf{Ans}$$

***10–44.** The principal plane stresses and associated strains in a plane at a point are $\sigma_1 = 40$ ksi, $\sigma_2 = 25$ ksi, $\epsilon_1 = 1.15(10^{-3})$, and $\epsilon_2 = 0.450(10^{-3})$. If this is a case of plane stress, determine the modulus of elasticity and Poisson's ratio.

Normal Stresses : For plane stress, $\sigma_3 = 0$.

Normal Strains : Applying the generalized Hooke's Law,

$$\varepsilon_1 = \frac{1}{E}\left[\sigma_1 - \nu\,(\sigma_2 + \sigma_3)\right]$$

$$1.15(10^{-3}) = \frac{1}{E}[40 - \nu(25 + 0)]$$

$$1.15(10^{-3})\,E = 40 - 25\nu \qquad [1]$$

$$\varepsilon_2 = \frac{1}{E}\left[\sigma_2 - \nu\,(\sigma_1 + \sigma_3)\right]$$

$$0.450(10^{-3}) = \frac{1}{E}[25 - \nu(40 + 0)]$$

$$0.450(10^{-3})\,E = 25 - 40\nu \qquad [2]$$

Solving Eqs. [1] and [2] yields :

$\nu = 0.309$ $\qquad E = 28.1(10^3)$ ksi $\qquad$ **Ans**

10–45. From experiment, the principal strains in a plane at a point on a steel shell are $\epsilon_1 = 350(10^{-6})$ and $\epsilon_2 = -250(10^{-6})$. If $E_{st} = 200$ GPa and $\nu_{st} = 0.3$, determine the principal plane stresses in this plane.

Normal Stresses : For plane stress, $\sigma_3 = 0$.

Normal Strains : Applying the generalized Hooke's Law,

$$\varepsilon_1 = \frac{1}{E}\left[\sigma_1 - \nu\,(\sigma_2 + \sigma_3)\right]$$

$$350(10^{-6}) = \frac{1}{200(10^9)}[\sigma_1 - 0.3(\sigma_2 + 0)]$$

$$70(10^6) = \sigma_1 - 0.3\sigma_2 \qquad [1]$$

$$\varepsilon_2 = \frac{1}{E}\left[\sigma_2 - \nu\,(\sigma_1 + \sigma_3)\right]$$

$$-250(10^{-6}) = \frac{1}{200(10^9)}\left[\sigma_2 - 0.3(\sigma_1 + 0)\right]$$

$$-50(10^6) = \sigma_2 - 0.3\sigma_1 \qquad [2]$$

Solving Eqs. [1] and [2] yields :

$\sigma_1 = 60.4$ MPa $\qquad \sigma_2 = -31.9$ MPa $\qquad$ **Ans**

10–46. The spherical pressure vessel has an inner diameter of 2 m and a thickness of 10 mm. A strain gauge having a length of 20 mm is attached to it, and it is observed to increase in length by 0.012 mm when the vessel is pressurized. Determine the pressure causing this deformation, and find the maximum in-plane shear stress, and the absolute maximum shear stress at a point on the outer surface of the vessel, . The material is steel, for which E_{st} = 200 GPa and ν_{st} = 0.3.

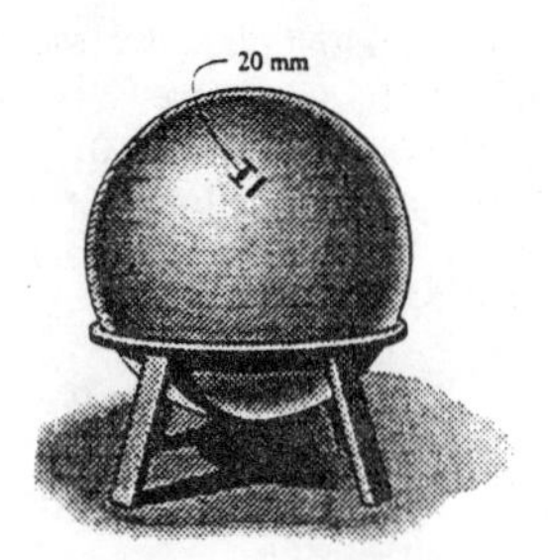

Normal Stresses : Since $\frac{r}{t} = \frac{1000}{10} = 100 > 10$, the *thin wall* analysis is valid to determine the normal stress in the wall of the spherical vessel. This is a plane stress problem where $\sigma_{min} = 0$ since there is no load acting on the outer surface of the wall.

$$\sigma_{max} = \sigma_{int} = \frac{pr}{2t} = \frac{p(1000)}{2(10)} = 50.0p \qquad [1]$$

Normal Strains : Applying the generalized Hooke's Law with $\varepsilon_{max} = \varepsilon_{int} = \frac{0.012}{20} = 0.600(10^{-3})$ mm/mm,

$$\varepsilon_{max} = \frac{1}{E}[\sigma_{max} - \nu(\sigma_{int} + \sigma_{min})]$$

$$0.600(10^{-3}) = \frac{1}{200(10^9)}[50.0p - 0.3(50.0p + 0)]$$

$$p = 3.4286 \text{ MPa} = 3.43 \text{ MPa} \qquad \textbf{Ans}$$

From Eq. [1] $\quad \sigma_{max} = \sigma_{int} = 50.0(3.4286) = 171.43$ MPa

Maximum In - Plane Shear Stress (Sphere's Surface) : Mohr's circle is simply a dot. As the result, the state of stress is the **same** consisting of two normal stresses with zero shear stress **regardless** of the orientation of the element.

$$\tau_{\substack{max \\ in\text{-}plane}} = 0 \qquad \textbf{Ans}$$

Absolute Maximum Shear Stress :

$$\tau_{\substack{abs \\ max}} = \frac{\sigma_{max} - \sigma_{min}}{2} = \frac{171.43 - 0}{2} = 85.7 \text{ MPa} \qquad \textbf{Ans}$$

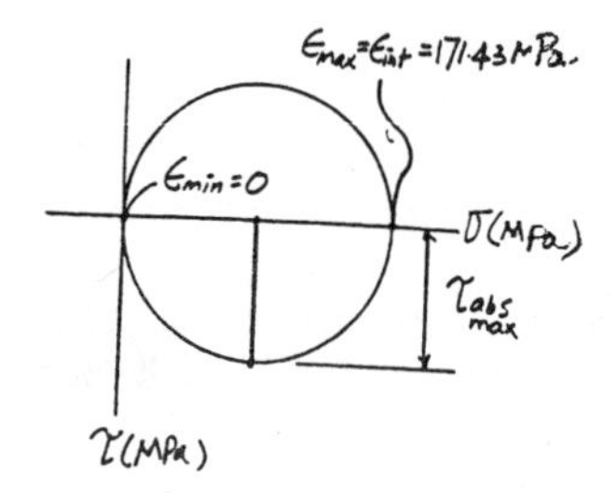

10–47. The principal strains in a plane, measured experimentally at a point on the aluminum fuselage of a jet aircraft, are $\epsilon_1 = 630(10^{-6})$ and $\epsilon_2 = 350(10^{-6})$. If this is a case of plane stress, determine the associated principal stresses at the point in the same plane. $E_{al} = 10(10^3)$ ksi and $\nu_{al} = 0.33$.

Normal Stresses : For plane stress, $\sigma_3 = 0$.

Normal Strains : Applying the generalized Hooke's Law.

$$\varepsilon_1 = \frac{1}{E}[\sigma_1 - \nu(\sigma_2 + \sigma_3)]$$

$$630(10^{-6}) = \frac{1}{10(10^3)}[\sigma_1 - 0.33(\sigma_2 + 0)]$$

$$6.30 = \sigma_1 - 0.33\sigma_2 \qquad [1]$$

$$\varepsilon_2 = \frac{1}{E}[\sigma_2 - \nu(\sigma_1 + \sigma_3)]$$

$$350(10^{-6}) = \frac{1}{10(10^3)}[\sigma_2 - 0.33(\sigma_1 + 0)]$$

$$3.50 = \sigma_2 - 0.33\sigma_1 \qquad [2]$$

Solving Eqs. [1] and [2] yields :

$$\sigma_1 = 8.37 \text{ ksi} \qquad \sigma_2 = 6.26 \text{ ksi} \qquad \textbf{Ans}$$

***10–48.** A single strain gauge, placed in the vertical plane on the outer surface and at an angle of 60° to the axis of the pipe, gives a reading at point A of $\epsilon_A = -250(10^{-6})$. Determine the vertical force P if the pipe has an outer diameter of 1 in. and an inner diameter of 0.6 in. The pipe is made of C86100 bronze.

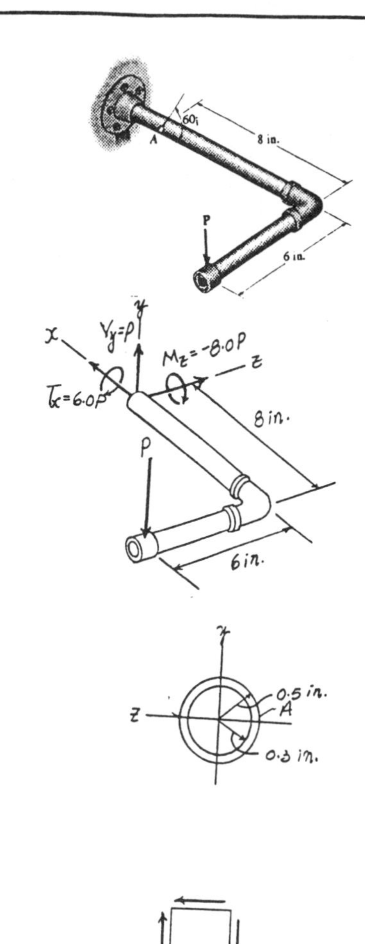

Internal Forces, Torque and Moments : As shown on FBD.

Section Properties :

$$I_z = I_y = \frac{\pi}{4}\left(0.5^4 - 0.3^4\right) = 0.0136\pi \text{ in}^4$$

$$J = \frac{\pi}{2}\left(0.5^4 - 0.3^4\right) = 0.0272\pi \text{ in}^4$$

$$(Q_A)_y = \Sigma \bar{y}' A'$$

$$= \frac{4(0.5)}{3\pi}\left[\frac{1}{2}\pi\left(0.5^2\right)\right] - \frac{4(0.3)}{3\pi}\left[\frac{1}{2}\pi\left(0.3^2\right)\right]$$

$$= 0.06533 \text{ in}^3$$

Normal Stress : Applying the flexure formula $\sigma = -\frac{M_z y}{I_z}$.

$$\sigma_A = -\frac{-8.00P(0)}{0.0136\pi} = 0$$

Shear Stress : The transverse shear stress in the y direction and the torsional shear stress can be obtained using shear formula and torsion formula, $\tau_V = \frac{VQ}{It}$ and $\tau_{twist} = \frac{Tc}{J}$, respectively.

$$(\tau_{xy})_A = (\tau_V)_y + \tau_{twist}$$

$$= \frac{p(0.06533)}{0.0136\pi(2)(0.2)} + \frac{6.00P(0.5)}{0.0272\pi}$$

$$= 38.93P$$

Strain Rosettes : For pure shear, $\varepsilon_x = \varepsilon_y = 0$. Applying Eq. 10 – 15 with $\varepsilon_b = -250\left(10^{-6}\right)$ and $\theta_b = 60°$,

$$-250\left(10^{-6}\right) = 0 + 0 + \gamma_{xy} \sin 60° \cos 60°$$

$$\gamma_{xy} = -577.35\left(10^{-6}\right)$$

Shear Stress and Strain Relationship : Applying Hooke's Law,

$$(\tau_{xy})_A = G\gamma_{xy}$$

$$-38.93P = 5.60\left(10^3\right)\left[-577.35\left(10^{-6}\right)\right]$$

$$P = 0.08305 \text{ kip} = 83.0 \text{ lb} \qquad \textbf{Ans}$$

10–49. A single strain gauge, placed in the vertical plane on the outer surface and at an angle of 60° to the axis of the pipe, gives a reading at point A of $\epsilon_A = -250(10^{-6})$. Determine the principal strains in the pipe at point A. The pipe has an outer diameter of 1 in. and an inner diameter of 0.6 in. and is made of C86100 bronze.

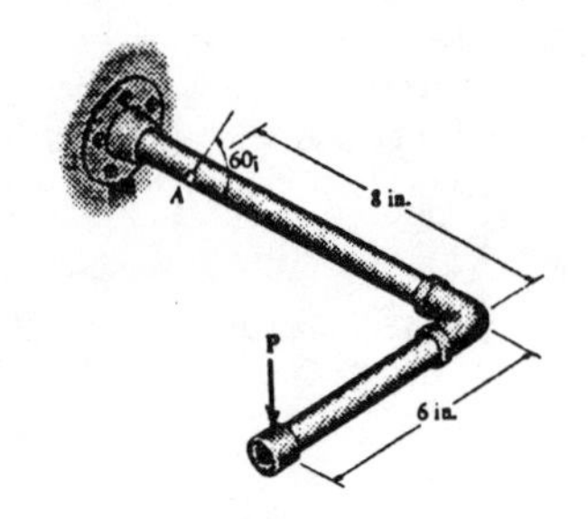

Internal Forces, Torque and Moments : As shown on FBD. By observation, this is a pure shear problem.

Strain Rosettes : For pure shear, $\varepsilon_x = \varepsilon_y = 0$. Applying Eq. 10 – 15 with $\varepsilon_b = 250(10^{-6})$ and $\theta_b = 60°$,

$$-250(10^{-6}) = 0 + 0 + \gamma_{xy} \sin 60° \cos 60°$$

$$\gamma_{xy} = -577.35(10^{-6})$$

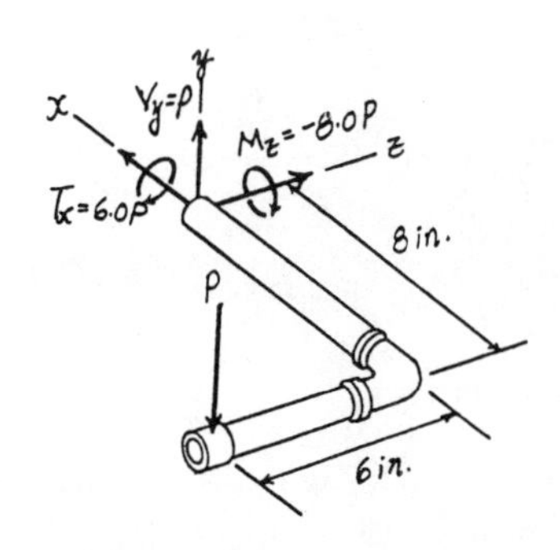

Construction of the Circle : In accordance with the sign convention.

$\varepsilon_x = \varepsilon_y = 0$ and $\frac{\gamma_{xy}}{2} = -288.675(10^{-6})$.

Hence,

$$\varepsilon_{avg} = \frac{\varepsilon_x + \varepsilon_y}{2} = 0$$

The coordinates for reference points A and C are

$$A(0, -288.675)(10^{-6}) \qquad C(0, 0)(10^{-6})$$

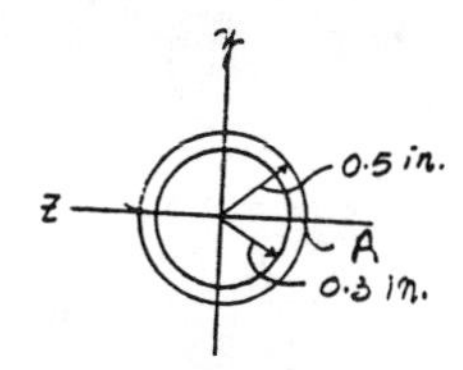

The radius of the circle is

$$R = \left(\sqrt{(0-0)^2 + 288.675^2}\right)(10^{-6}) = 288.675(10^{-6})$$

In - Plane Principal Strain : The coordinates of points B and D represent ε_1 and ε_2, respectively.

$$\varepsilon_1 = (0 + 288.675)(10^{-6}) = 288.675(10^{-6})$$

$$\varepsilon_2 = (0 - 288.675)(10^{-6}) = -288.675(10^{-6})$$

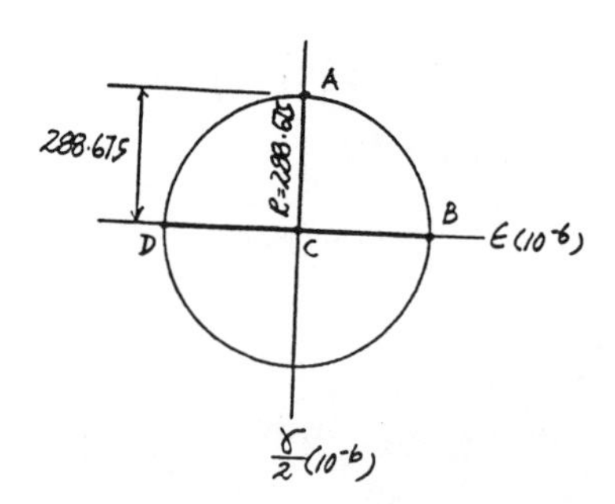

Principal Stress : Since $\sigma_x = \sigma_y = \sigma_z = 0$, then from the generalized Hooke's Law $\varepsilon_z = 0$. From the results obtained above, we have

$$\varepsilon_{max} = 289(10^{-6}) \qquad \varepsilon_{int} = 0 \qquad \varepsilon_{min} = -289(10^{-6}) \qquad \textbf{Ans}$$

10–50. The thin-walled cylindrical pressure vessel of inner radius r and thickness t is subjected to an internal pressure p. Determine the maximum x,y in-plane shear strain at point A on the outer surface. Also find the absolute maximum shear strain at A. The material properties are E and ν.

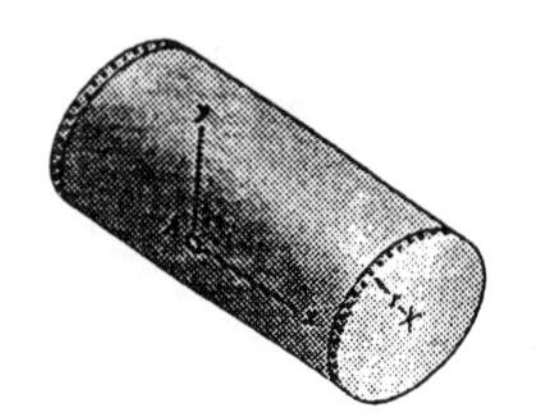

Normal Stress : This is a plane stress problem where $\sigma_{min} = 0$ since there is no load acting on the outer surface of the wall.

$$\sigma_{max} = \frac{pr}{t} \qquad \sigma_{int} = \frac{pr}{2t}$$

Normal Strain : Applying the generalized Hooke's Law.

$$\varepsilon_{max} = \frac{1}{E}[\sigma_{max} - \nu(\sigma_{int} + \sigma_{min})]$$
$$= \frac{1}{E}\left[\frac{pr}{t} - \nu\left(\frac{pr}{2t} + 0\right)\right]$$
$$= \frac{pr}{2tE}(2 - \nu)$$

$$\varepsilon_{int} = \frac{1}{E}[\sigma_{int} - \nu(\sigma_{max} + \sigma_{min})]$$
$$= \frac{1}{E}\left[\frac{pr}{2t} - \nu\left(\frac{pr}{t} + 0\right)\right]$$
$$= \frac{pr}{2tE}(1 - 2\nu)$$

$$\varepsilon_{min} = \frac{1}{E}[\sigma_{min} - \nu(\sigma_{max} + \sigma_{int})]$$
$$= \frac{1}{E}\left[0 - \nu\left(\frac{pr}{t} + \frac{pr}{2t}\right)\right]$$
$$= -\frac{3\nu pr}{2tE}$$

Maximum In-Plane Shear Strain (x-y Plane) : From the three Mohr's circles.

$$\gamma_{\substack{max \\ in\text{-}plane}} = \varepsilon_{max} - \varepsilon_{int}$$
$$= \frac{pr}{2tE}(2 - \nu) - \frac{pr}{2tE}(1 - 2\nu)$$
$$= \frac{pr}{2tE}(1 + \nu) \qquad \textbf{Ans}$$

Absolute Maximum Shear Strain : From the three Mohr's circles

$$\gamma_{\substack{abs \\ max}} = \varepsilon_{max} - \varepsilon_{min}$$
$$= \frac{pr}{2tE}(2 - \nu) - \left(-\frac{3\nu pr}{2tE}\right)$$
$$= \frac{pr}{tE}(1 + \nu) \qquad \textbf{Ans}$$

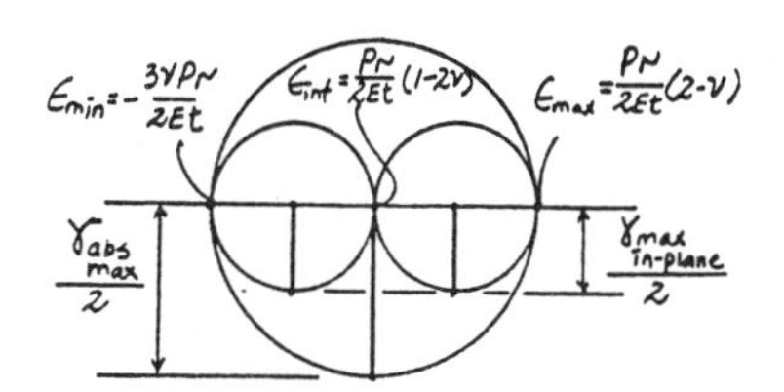

10–51. The strain gauge is placed on the surface of a thin-walled boiler as shown. If it is 0.20 in. long, determine the pressure in the boiler when the gauge elongates $0.08(10^{-3})$ in. The boiler has a thickness of 0.5 in. and inner diameter of 30 in., and it is made of 304 stainless steel. Also, determine the maximum x,y in-plane shear strain and the absolute maximum shear strain in the material.

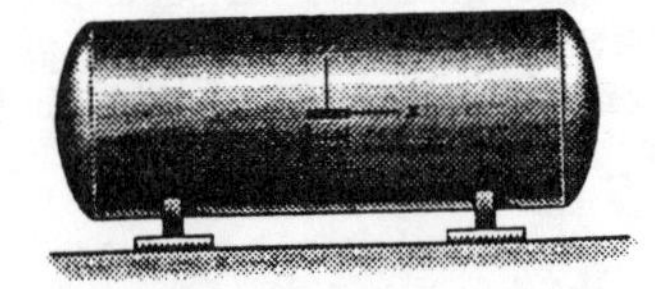

Principal Stress : Since $\frac{r}{t} = \frac{15}{0.5} = 30 > 10$, *thin wall* analysis can be used to determine the normal stress in the wall of the cylindrical vessel. This is a plane stress problem where $\sigma_{min} = 0$, since there is no load acting on the outer surface of the wall.

$$\sigma_{max} = \frac{pr}{t} = \frac{p(15)}{0.5} = 30.0p \qquad [1]$$

$$\sigma_{int} = \frac{pr}{2t} = \frac{p(15)}{2(0.5)} = 15.0p \qquad [2]$$

Principal Strain : Applying the generalized Hooke's Law with

$$\varepsilon_{max} = \frac{0.08(10^{-3})}{0.20} = 0.400(10^{-3}) \text{ in./in..}$$

$$\varepsilon_{max} = \frac{1}{E}\left[\sigma_{max} - \nu(\sigma_{int} + \sigma_{min})\right]$$

$$0.400(10^{-3}) = \frac{1}{28.0(10^3)}[30.0p - 0.27(15.0p + 0)]$$

$$p = 0.4316 \text{ ksi} = 0.432 \text{ ksi} \qquad \textbf{Ans}$$

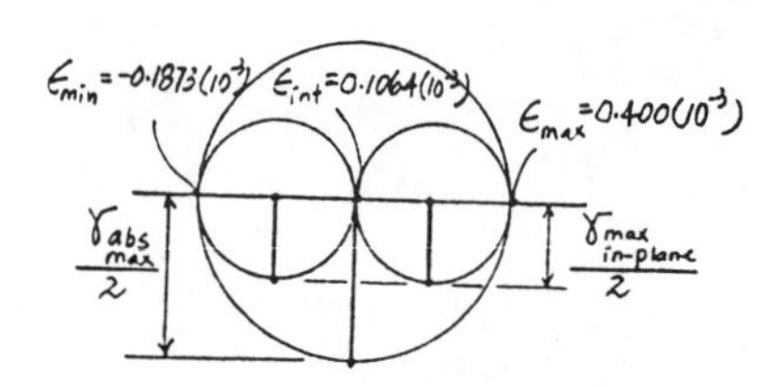

From Eq. [1] and [2] $\quad \sigma_{max} = 30.0(0.4316) = 12.95$ ksi

$\sigma_{int} = 15.0(0.4316) = 6.474$ ksi

$$\varepsilon_{int} = \frac{1}{E}\left[\sigma_{int} - \nu(\sigma_{max} + \sigma_{min})\right]$$

$$= \frac{1}{28.0(10^3)}[6.474 - 0.27(12.95 + 0)]$$

$$= 0.1064(10^{-3})$$

$$\varepsilon_{min} = \frac{1}{E}\left[\sigma_{min} - \nu(\sigma_{max} + \sigma_{int})\right]$$

$$= \frac{1}{28.0(10^3)}[0 - 0.27(12.95 + 6.474)]$$

$$= -0.1873(10^{-3})$$

Maximum In-Plane Shear Strain (x-y Plane) : From the three Mohr's circles,

$$\gamma_{\substack{max \\ in\text{-}plane}} = \varepsilon_{max} - \varepsilon_{int}$$

$$= 0.400(10^{-3}) - 0.1064(10^{-3})$$

$$= 0.294(10^{-3}) \qquad \textbf{Ans}$$

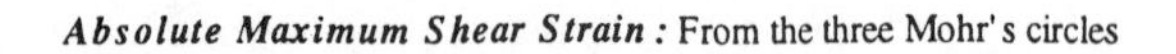

Absolute Maximum Shear Strain : From the three Mohr's circles

$$\gamma_{\substack{abs \\ max}} = \varepsilon_{max} - \varepsilon_{min}$$

$$= 0.400(10^{-3}) - \left[-0.1873(10^{-3})\right]$$

$$= 0.587(10^{-3}) \qquad \textbf{Ans}$$

***10–52.** The shaft has a radius of 15 mm and is made of L2 tool steel. Determine the torque T in the shaft if the two strain gauges, attached to the surface of the shaft, report strains of $\epsilon_{x'} = -45(10^{-6})$ and $\epsilon_{y'} = 45(10^{-6})$. Also, compute the strains acting in the x and y directions.

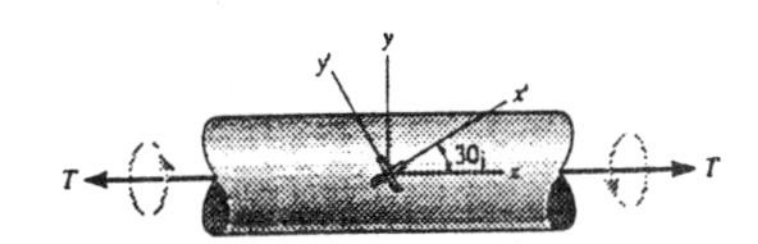

Shear Stress : The torsional shear stress can be obtained using the torsion formula, $\tau = \frac{Tc}{J}$.

$$\tau_{xy} = \frac{T(0.015)}{\frac{\pi}{2}(0.015^4)} = 188628T$$

Strain Rosettes : For pure shear, $\varepsilon_x = \varepsilon_y = 0$ **Ans**

Applying Eq. 10 – 15 with $\varepsilon_a = -45(10^{-6})$ and $\theta_a = 30°$,

$$-45(10^{-6}) = 0 + 0 + \gamma_{xy} \sin 30° \cos 30°$$

$$\gamma_{xy} = -103.92(10^{-6})$$

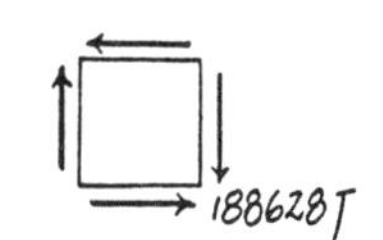

Or with $\varepsilon_b = 45(10^{-6})$ and $\theta_b = 120°$,

$$45(10^{-6}) = 0 + 0 + \gamma_{xy} \sin 120° \cos 120°$$

$$\gamma_{xy} = -103.92(10^{-6})$$

Shear Stress and Strain Relationship : Applying Hooke's Law.

$$\tau_{xy} = G\gamma_{xy}$$

$$-188628T = 78.0(10^9)\left[-103.92(10^{-6})\right]$$

$$T = 43.0 \text{ N} \cdot \text{m}$$ **Ans**

10–53. The shaft has a radius of 15 mm and is made of L2 tool steel. Determine the strains in the x' and y' directions if a torque of $T = 2$ kN · m is applied to the shaft.

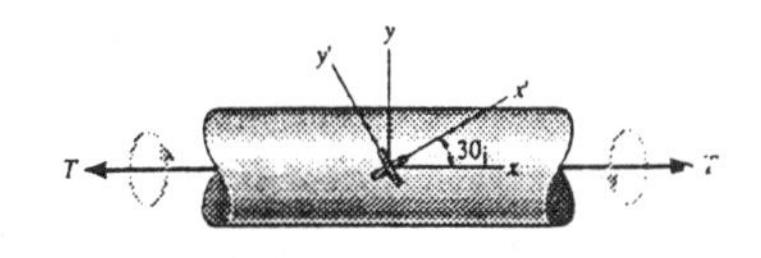

Shear Stress : The torsional shear stress can be obtained using the torsion formula, $\tau = \frac{Tc}{J}$.

$$\tau_{xy} = \frac{2(10^3)(0.015)}{\frac{\pi}{2}(0.015^4)} = 377.26 \text{ MPa}$$

Shear Stress and Strain Relationship : Applying Hooke's Law.

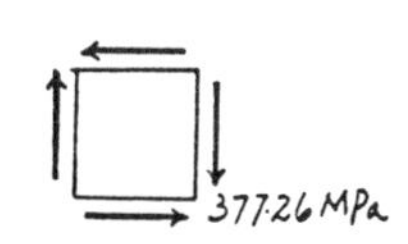

$$\tau_{xy} = G\gamma_{xy}$$

$$-377.26(10^6) = 78.0(10^9)\,\gamma_{xy}$$

$$\gamma_{xy} = -4.8366(10^{-3}) \text{ rad}$$

Strain Rosettes : For pure shear, $\varepsilon_x = \varepsilon_y = 0$. Applying Eq. 10 – 15 with $\varepsilon_a = \varepsilon_{x'}$ and $\theta_a = 30°$,

$$\varepsilon_{x'} = 0 + 0 + \left[-4.8366(10^{-3})\right] \sin 30° \cos 30°$$

$$= -2.09(10^{-3}) \text{ mm/mm}$$ **Ans**

Or with $\varepsilon_b = \varepsilon_{y'}$ and $\theta_b = 120°$,

$$\varepsilon_{y'} = 0 + 0 + \left[-4.8366(10^{-3})\right] \sin 120° \cos 120°$$

$$= 2.09(10^{-3}) \text{ mm/mm}$$ **Ans**

10–54. Determine the change in volume of the tapered plate when it is subjected to the axial load **P**. The material has a thickness t, a modulus of elasticity E, and Poisson's ratio is ν.

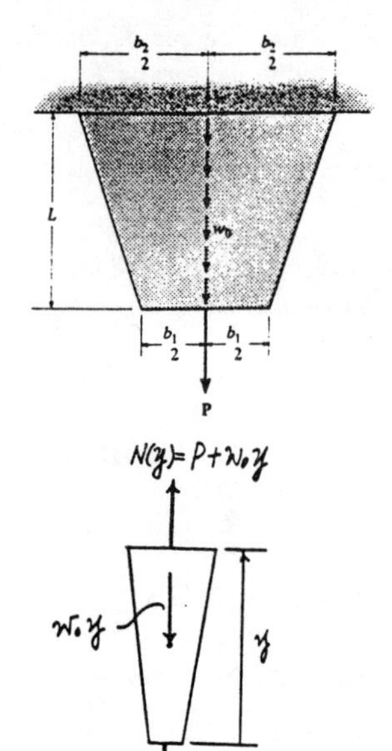

Normal Stress : The bar is subjected to uniaxial load. Therefore,

$\sigma_x = \sigma_z = 0$ and $\sigma_y = \dfrac{N(y)}{A(y)}$.

Dilatation : Applying Eq. 10 – 23.

$$\delta V = \frac{1-2\nu}{E}\int(\sigma_x + \sigma_y + \sigma_z)\,dV \qquad \text{However, } dV = A(y)\,dy$$

$$= \frac{1-2\nu}{E}\int \sigma_y\,dV$$

$$= \frac{1-2\nu}{E}\int_0^L \frac{N(y)}{A(y)}[A(y)\,dy]$$

$$= \frac{1-2\nu}{E}\int_0^L N(y)\,dy \qquad [1]$$

From the FBD, $N(y) = P + w_0 y$. Then, from Eq. [1]

$$\delta V = \frac{1-2\nu}{E}\int_0^L (P + w_0 y)\,dy$$

$$= \frac{1-2\nu}{E}\left[Py + \frac{w_0 y^2}{2}\right]\Bigg|_0^L$$

$$= \frac{1-2\nu}{2E}(2PL + w_0 L^2) \qquad \textbf{Ans}$$

10–55. The A-36 steel pipe is subjected to the axial loading of 60 kN. Determine the change in volume of the material after the load is applied.

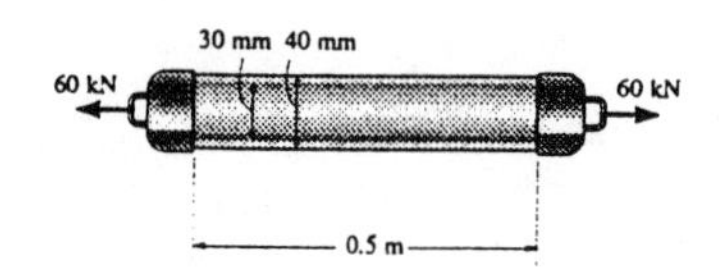

Normal Stress : The pipe is subjected to uniaxial load. Therefore,

$\sigma_y = \sigma_z = 0$ and $\sigma_x = \dfrac{N}{A}$.

Dilatation : Applying Eq. 10 – 23.

$$\frac{\delta V}{V} = \frac{1-2\nu}{E}(\sigma_x + \sigma_y + \sigma_z)$$

$$\frac{\delta V}{V} = \frac{1-2\nu}{E}\left(\frac{N}{A}\right)$$

$$\delta V = \frac{1-2\nu}{E}\left(\frac{N}{A}\right)V \qquad \text{However, } V = AL$$

$$\delta V = \left(\frac{1-2\nu}{E}\right)NL$$

$$= \left[\frac{1-2(0.32)}{200(10^9)}\right](60)(10^3)(0.5)$$

$$= 54.0(10^{-9})\ \text{m}^3 = 54.0\ \text{mm}^3 \qquad \textbf{Ans}$$

***10–56.** A thin-walled cylindrical pressure vessel has an inner radius r, thickness t, and length L. If it is subjected to an internal pressure p, show that the increase in its inner radius is $\delta r = pr^2(2 - \nu)/2Et$ and the increase in its length is $\delta L = pLr(1 - 2\nu)/2Et$. Using these results, show that the change in internal volume becomes $\delta V = \pi r^2(1 + \epsilon_1)^2(1 + \epsilon_2)L - \pi r^2 L$. Since ϵ_1 and ϵ_2 are small quantities, show further that the change in volume per unit volume, called *volumetric strain,* can be written as $\delta V/V = (pr/2Et)(5 - 4\nu)$.

Normal Stress : By ignoring the radial stress component, that is, $\sigma_3 = 0$, the problem becomes a plane stress problem.

$$\sigma_1 = \frac{pr}{t} \qquad \sigma_2 = \frac{pr}{2t}$$

Normal Strain : Applying the generalized Hooke's law,

$$\varepsilon_1 = \frac{1}{E}[\sigma_1 - \nu(\sigma_2 + \sigma_3)] = \frac{1}{E}\left(\frac{pr}{t} - \frac{\nu pr}{2t}\right) = \frac{pr}{2Et}(2 - \nu)$$

Therefore, $\delta r = \varepsilon_1 r = \frac{pr^2}{2Et}(2 - \nu)$ (Q.E.D.)

$$\varepsilon_2 = \frac{1}{E}[\sigma_2 - \nu(\sigma_1 + \sigma_3)] = \frac{1}{E}\left(\frac{pr}{2t} - \frac{\nu pr}{t}\right) = \frac{pr}{2Et}(1 - 2\nu)$$

Therefore, $\delta L = \varepsilon_2 L = \frac{pLr}{2Et}(1 - 2\nu)$ (Q.E.D.)

The volume of the cylindrical vessel before and after being pressurized is

$$V' = \pi(r + \varepsilon_1 r)^2(L + \varepsilon_2 L) \qquad V = \pi r^2 L$$

$$\delta V = V' - V = \pi r^2(1 + \varepsilon_1)^2(1 + \varepsilon_2)L - \pi r^2 L \qquad (Q.E.D.)$$

However, $(1 + \varepsilon_1)^2 = 1 + 2\varepsilon_1 + \cdots$ (neglect ε_1^2 term)

Therefore, $(1 + \varepsilon_1)^2(1 + \varepsilon_2) \approx (1 + 2\varepsilon_1)(1 + \varepsilon_2) \approx 1 + \varepsilon_2 + 2\varepsilon_1 + \cdots$ (neglect $\varepsilon_1\varepsilon_2$ term)

$$\frac{\delta V}{V} = 1 + \varepsilon_2 + 2\varepsilon_1 - 1 = \varepsilon_2 + 2\varepsilon_1$$

$$\frac{\delta V}{V} = \frac{pr}{2Et}(1 - 2\nu) + 2\left[\frac{pr}{2Et}(2 - \nu)\right] = \frac{pr}{2Et}(5 - 4\nu) \qquad (Q.E.D.)$$

10–57. A soft material is placed within the confines of a rigid cylinder which rests on a rigid support. Determine the factor by which the modulus of elasticity will be increased from not being confined when a load is applied. Take $\nu = 0.3$ for the material.

Normal Strain : Since the material is confined in a rigid cylinder, $\varepsilon_x = \varepsilon_y = 0$. Applying the generalized Hooke's Law,

$$\varepsilon_x = \frac{1}{E}[\sigma_x - \nu(\sigma_y + \sigma_z)]$$

$$0 = \sigma_x - \nu(\sigma_y + \sigma_z) \qquad [1]$$

$$\varepsilon_y = \frac{1}{E}[\sigma_y - \nu(\sigma_x + \sigma_z)]$$

$$0 = \sigma_y - \nu(\sigma_x + \sigma_z) \qquad [2]$$

Solving Eqs. [1] and [2] yields :

$$\sigma_x = \sigma_y = \frac{\nu}{1 - \nu}\sigma_z$$

Thus,

$$\varepsilon_z = \frac{1}{E}[\sigma_z - \nu(\sigma_x + \sigma_y)] = \frac{1}{E}\left[\sigma_z - \nu\left(\frac{\nu}{1-\nu}\sigma_z + \frac{\nu}{1-\nu}\sigma_z\right)\right] = \frac{\sigma_z}{E}\left[1 - \frac{2\nu^2}{1-\nu}\right] = \frac{\sigma_z}{E}\left[\frac{1 - \nu - 2\nu^2}{1 - \nu}\right] = \frac{\sigma_z}{E}\left[\frac{(1+\nu)(1-2\nu)}{1-\nu}\right]$$

Thus, when the material is not being confined and undergoes the same normal strain of ε_z, then the required modulus of elasticity is

$$E' = \frac{\sigma_z}{\varepsilon_z} = \frac{1 - \nu}{(1 - 2\nu)(1 + \nu)}E$$

The increased factor is $k = \frac{E'}{E} = \frac{1 - \nu}{(1 - 2\nu)(1 + \nu)} = \frac{1 - 0.3}{[1 - 2(0.3)](1 + 0.3)} = 1.35$ **Ans**

10–58. A thin-walled spherical pressure vessel having an inner radius r and thickness t is subjected to an internal pressure p. Show that the increase in volume within the vessel is $\delta V = (2p\pi r^4/Et)(1 - \nu)$. Use a small-strain analysis.

Normal Stress : By ignoring the radial stress component that is $\sigma_3 = 0$, the problem becomes a plane stress problem.

$$\sigma_1 = \sigma_2 = \frac{pr}{2t}$$

Normal Strain : Applying the generalized Hooke's law,

$$\varepsilon_1 = \frac{1}{E}\left[\sigma_1 - \nu(\sigma_2 + \sigma_3)\right] = \frac{1}{E}\left(\frac{pr}{2t} - \frac{\nu pr}{2t}\right) = \frac{pr}{2Et}(1-\nu)$$

The volume of the spherical vessel before and after being pressurized is

$$V' = \frac{4}{3}\pi(r + \varepsilon_1 r)^3 \qquad V = \frac{4}{3}\pi r^3$$

$$\delta V = V' - V = \frac{4}{3}\pi(r+\varepsilon_1 r)^3 - \frac{4}{3}\pi r^3 = \frac{4}{3}\pi r^3\left[(1+\varepsilon_1)^3 - 1\right] \qquad [1]$$

However, $(1+\varepsilon_1)^3 - 1 = 1 + 3\varepsilon_1 + \cdots - 1 = 3\varepsilon_1$
(neglect second order terms)

From Eq. [1]

$$\delta V = \frac{4}{3}\pi r^3(3\varepsilon_1) = 4\pi r^3\varepsilon_1 = 4\pi r^3\left[\frac{pr}{2Et}(1-\nu)\right] = \frac{2p\pi r^4}{Et}(1-\nu) \qquad (Q.E.D.)$$

10–59. The thin-walled cylindrical pressure vessel of inner radius r and thickness t is subjected to an internal pressure p. If the material constants are E and ν, determine the strains in the circumferential and longitudinal directions. Using these results, compute the increase in both the diameter and the length of a steel pressure vessel filled with air and having an internal gauge pressure of 20 MPa. The vessel is 2 m long and has an inner radius of 0.4 m and a thickness of 10 mm. $E_{st} = 200$ GPa, and $\nu_{st} = 0.3$.

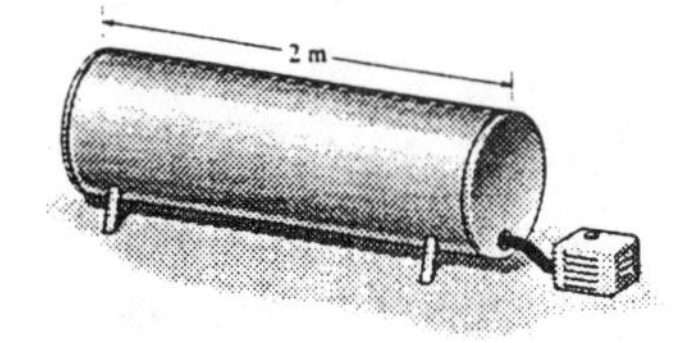

Normal Stress : By ignoring the radial stress component, that is, $\sigma_3 = 0$, the problem becomes a plane stress problem.

$$\sigma_1 = \frac{pr}{t} \qquad \sigma_2 = \frac{pr}{2t}$$

Normal Strain : Applying the generalized Hooke's law.

$$\varepsilon_1 = \frac{1}{E}\left[\sigma_1 - \nu(\sigma_2+\sigma_3)\right] = \frac{1}{E}\left(\frac{pr}{t} - \frac{\nu pr}{2t}\right) = \frac{pr}{2Et}(2-\nu) \qquad \textbf{Ans}$$

Therefore,

$$\delta d = 2\varepsilon_1 r = \frac{pr^2}{Et}(2-\nu) = \frac{20(10^6)(0.4^2)}{200(10^9)(0.01)}(2-0.3) = 2.72(10^{-3})\ \text{m} = 2.72\ \text{mm} \qquad \textbf{Ans}$$

$$\varepsilon_2 = \frac{1}{E}\left[\sigma_2 - \nu(\sigma_1+\sigma_3)\right] = \frac{1}{E}\left(\frac{pr}{2t} - \frac{\nu pr}{t}\right) = \frac{pr}{2Et}(1-2\nu) \qquad \textbf{Ans}$$

Therefore,

$$\delta L = \varepsilon_2 L = \frac{pLr}{2Et}(1-2\nu) = \frac{20(10^6)(2)(0.4)}{2(200)(10^9)(0.01)}[1-2(0.3)] = 1.60(10^{-3}) = 1.60\ \text{mm} \qquad \textbf{Ans}$$

***10–60.** Estimate the increase in volume of the tank in Prob. 10–59. *Suggestion:* Use the results of Prob. 10–56 as a check.

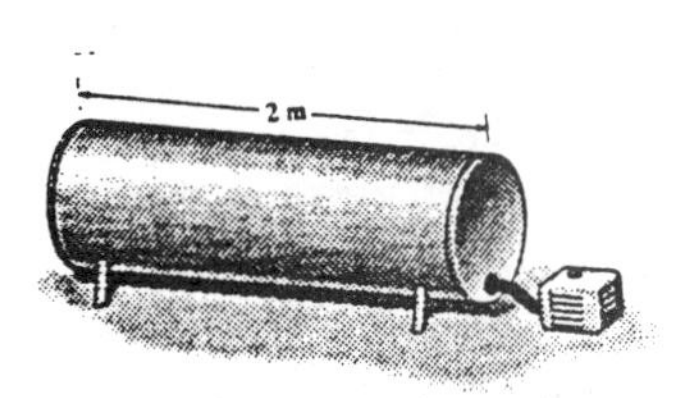

Normal Stress : By ignoring the radial stress component, that is, $\sigma_3 = 0$, the problem becomes a plane stress problem.

$$\sigma_1 = \frac{pr}{t} \qquad \sigma_2 = \frac{pr}{2t}$$

Normal Strain : Applying the generalized Hooke's law,

$$\varepsilon_1 = \frac{1}{E}[\sigma_1 - \nu(\sigma_2 + \sigma_3)] = \frac{1}{E}\left(\frac{pr}{t} - \frac{\nu pr}{2t}\right) = \frac{pr}{2Et}(2-\nu)$$

$$\varepsilon_2 = \frac{1}{E}[\sigma_2 - \nu(\sigma_1 + \sigma_3)] = \frac{1}{E}\left(\frac{pr}{2t} - \frac{\nu pr}{t}\right) = \frac{pr}{2Et}(1-2\nu)$$

The volume of the cylindrical vessel before and after being pressurized is

$$V' = \pi(r + \varepsilon_1 r)^2(L + \varepsilon_2 L) \qquad V = \pi r^2 L$$

$$\delta V = V' - V = \pi r^2(1+\varepsilon_1)^2(1+\varepsilon_2)L - \pi r^2 L = \pi r^2 L\left[(1+\varepsilon_1)^2(1+\varepsilon_2) - 1\right] \qquad [1]$$

However, $(1+\varepsilon_1)^2 = 1 + 2\varepsilon_1 + \cdots$ (neglect ε_1^2 term)

Therefore, $(1+\varepsilon_1)^2(1+\varepsilon_2) \approx (1+2\varepsilon_1)(1+\varepsilon_2) \approx 1 + \varepsilon_2 + 2\varepsilon_1 + \cdots$ (neglect $\varepsilon_1\varepsilon_2$ term)

From Eq.[1],

$$\delta V = \pi r^2 L(1 + \varepsilon_2 + 2\varepsilon_1 - 1) = \pi r^2 L(\varepsilon_2 + 2\varepsilon_1) = \pi r^2 L\left\{\frac{pr}{2Et}(1-2\nu) + 2\left[\frac{pr}{2Et}(2-\nu)\right]\right\}$$

$$= \frac{\pi p r^3 L}{2Et}(5-4\nu) = \frac{\pi(20)(10^6)(0.4^3)(2)}{2(200)(10^9)(0.01)}[5-4(0.3)] = 7.64(10^{-3})\ \text{m}^3 \qquad \textbf{Ans}$$

10–61. The smooth rigid-body cavity is filled with liquid 6061-T6 aluminum. When cooled it is 0.012 in. from the top of the cavity. If the top of the cavity is covered and the temperature is increased by 200°F, determine the stress components σ_x, σ_y, and σ_z in the aluminum. *Hint:* Use Eqs. 10–18 with an additional strain term of $\alpha\Delta T$ (Eq. 4–4).

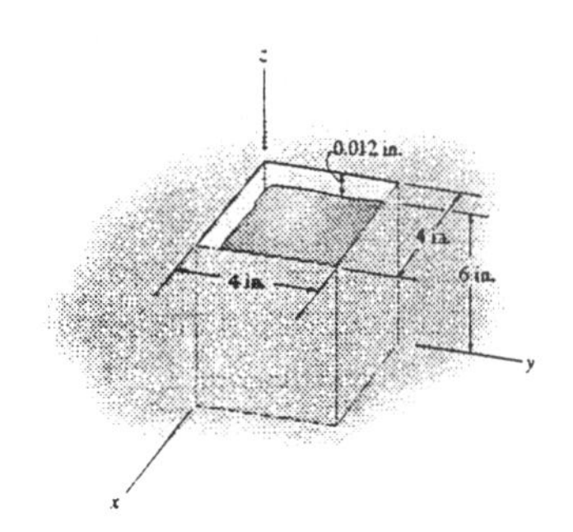

Normal Strains : Since the aluminum is confined at its sides by a rigid container and allowed to expand in the z direction, $\varepsilon_x = \varepsilon_y = 0$; whereas $\varepsilon_z = \frac{0.012}{6} = 0.002$. Applying the generalized Hooke's Law with the additional thermal strain,

$$\varepsilon_x = \frac{1}{E}\left[\sigma_x - \nu(\sigma_y + \sigma_z)\right] + \alpha\Delta T$$

$$0 = \frac{1}{10.0(10^3)}\left[\sigma_x - 0.35(\sigma_y + \sigma_z)\right] + 13.1(10^{-6})(200)$$

$$0 = \sigma_x - 0.35\sigma_y - 0.35\sigma_z + 26.2 \qquad [1]$$

$$\varepsilon_y = \frac{1}{E}[\sigma_y - \nu(\sigma_x + \sigma_z)] + \alpha\Delta T$$

$$0 = \frac{1}{10.0(10^3)}[\sigma_y - 0.35(\sigma_x + \sigma_z)] + 13.1(10^{-6})(200)$$

$$0 = \sigma_y - 0.35\sigma_x - 0.35\sigma_z + 26.2 \qquad [2]$$

$$\varepsilon_z = \frac{1}{E}\left[\sigma_z - \nu(\sigma_x + \sigma_y)\right] + \alpha\Delta T$$

$$0.002 = \frac{1}{10.0(10^3)}\left[\sigma_z - 0.35(\sigma_x + \sigma_y)\right] + 13.1(10^{-6})(200)$$

$$0 = \sigma_z - 0.35\sigma_x - 0.35\sigma_y + 6.20 \qquad [3]$$

Solving Eqs.[1], [2] and [3] yields :

$$\sigma_x = \sigma_y = -70.0\ \text{ksi} \qquad \sigma_z = -55.2\ \text{ksi} \qquad \textbf{Ans}$$

10–62. The smooth rigid-body cavity is filled with liquid 6061-T6 aluminum. When cooled it is 0.012 in. from the top of the cavity. If the top of the cavity is not covered and the temperature is increased by 200° F, determine the strain components ϵ_x, ϵ_y, and ϵ_z in the aluminum. *Hint:* Use Eqs. 10–18 with an additional strain term of $\alpha \Delta T$ (Eq. 4-4).

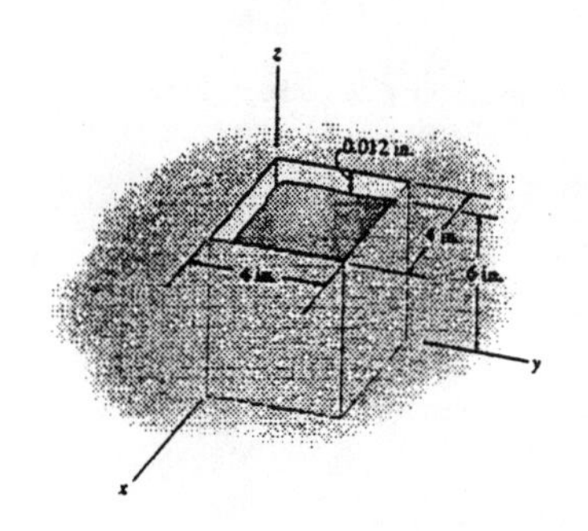

Normal Strains : Since the aluminum is confined at its sides by a rigid container , then

$$\varepsilon_x = \varepsilon_y = 0 \qquad \textbf{Ans}$$

and since it is not restrained in z *direction*, $\sigma_z = 0$. Applying the generalized Hooke's Law with the additional thermal strain,

$$\varepsilon_x = \frac{1}{E}\left[\sigma_x - \nu\left(\sigma_y + \sigma_z\right)\right] + \alpha \Delta T$$

$$0 = \frac{1}{10.0(10^3)}\left[\sigma_x - 0.35\left(\sigma_y + 0\right)\right] + 13.1\left(10^{-6}\right)(200)$$

$$0 = \sigma_x - 0.35\sigma_y + 26.2 \qquad [1]$$

$$\varepsilon_y = \frac{1}{E}\left[\sigma_y - \nu(\sigma_x + \sigma_z)\right] + \alpha \Delta T$$

$$0 = \frac{1}{10.0(10^3)}\left[\sigma_y - 0.35(\sigma_x + 0)\right] + 13.1\left(10^{-6}\right)(200)$$

$$0 = \sigma_y - 0.35\sigma_x + 26.2 \qquad [2]$$

Solving Eqs. [1] and [2] yields :

$$\sigma_x = \sigma_y = -40.31 \text{ ksi}$$

$$\varepsilon_z = \frac{1}{E}\left[\sigma_z - \nu\left(\sigma_x + \sigma_y\right)\right] + \alpha \Delta T$$

$$= \frac{1}{10.0(10^3)}\{0 - 0.35[-40.31 + (-40.31)]\} + 13.1\left(10^{-6}\right)(200)$$

$$= 5.44\left(10^{-3}\right) \qquad \textbf{Ans}$$

10–63. The block is fitted between the fixed supports. If the glued joint can resist a maximum shear stress of $\tau_{allow} = 2$ ksi, determine the temperature rise that will cause the joint to fail. Take $E = 10(10^3)$ ksi, $\nu = 0.2$, and $\alpha = 6.0\,(10^{-6})/°F$. *Hint:* Use Eqs. 10–18 with an additional strain term of $\alpha\Delta T$ (Eq. 4–4).

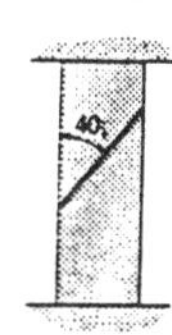

Normal Strain : Since the aluminum is confined along the *y direction* by the rigid frame, then $\varepsilon_y = 0$ and $\sigma_x = \sigma_z = 0$. Applying the generalized Hooke's Law with the additional thermal strain,

$$\varepsilon_y = \frac{1}{E}\left[\sigma_y - \nu(\sigma_x + \sigma_z)\right] + \alpha\Delta T$$

$$0 = \frac{1}{10.0(10^3)}\left[\sigma_y - 0.2(0+0)\right] + 6.0(10^{-6})(\Delta T)$$

$$\sigma_y = -0.06\Delta T$$

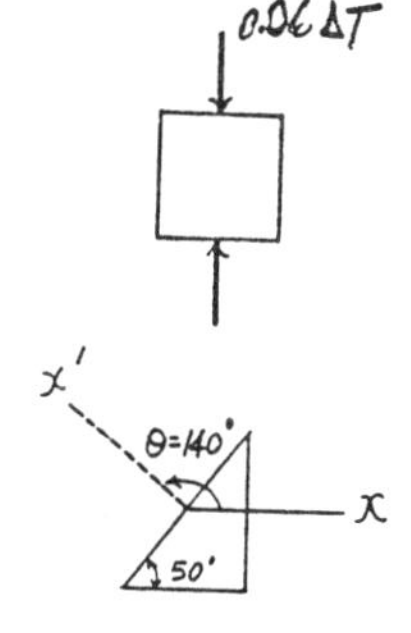

Construction of the Circle : In accordance with the sign convention, $\sigma_x = 0$, $\sigma_y = -0.06\Delta T$ and $\tau_{xy} = 0$. Hence,

$$\sigma_{avg} = \frac{\sigma_x + \sigma_y}{2} = \frac{0 + (-0.06\Delta T)}{2} = -0.03\Delta T$$

The coordinates for the reference points A and C are $A(0, 0)$ and $C(-0.03\Delta T,\ 0)$.

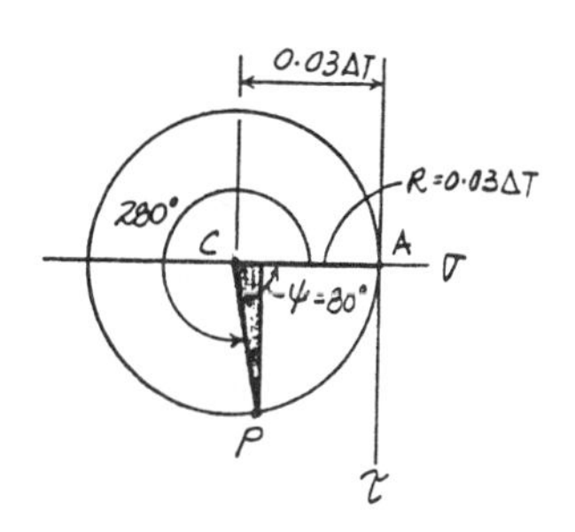

The radius of the circle is $R = \sqrt{(0 - 0.03\Delta T)^2 + 0} = 0.03\Delta T$

Stress on The Inclined Plane : The shear stress components $\tau_{x'y'}$ are represented by the coordinates of point P on the circle.

$$\tau_{x'y'} = 0.03\Delta T \sin 80° = 0.02954\Delta T$$

Allowable Shear Stress :

$$\tau_{allow} = \tau_{x'y'}$$

$$2 = 0.02954\Delta T$$

$$\Delta T = 67.7\ °F \qquad \textbf{Ans}$$

***10–64.** A material is subjected to plane stress. Express the maximum-distortion-energy theory of failure in terms of σ_x, σ_y, and τ_{xy}.

Maximum Distortion Energy Theory :

$$\sigma_1^2 - \sigma_1\sigma_2 + \sigma_2^2 = \sigma_Y^2 \qquad [1]$$

In-Plane Principal Stress : Using Eq. 9 – 5,

$$\sigma_{1,2} = \frac{\sigma_x + \sigma_y}{2} \pm \sqrt{\left(\frac{\sigma_x - \sigma_y}{2}\right)^2 + \tau_{xy}^2}$$

Let $a = \dfrac{\sigma_x + \sigma_y}{2}$ and $b = \sqrt{\left(\dfrac{\sigma_x - \sigma_y}{2}\right)^2 + \tau_{xy}^2}$

$$\sigma_1 = a + b \qquad \sigma_2 = a - b$$

$$\sigma_1^2 = a^2 + b^2 + 2ab \qquad \sigma_2^2 = a^2 + b^2 - 2ab$$

$$\sigma_1\sigma_2 = a^2 - b^2$$

Substituting the above results into Eq. [1],

$$a^2 + b^2 + 2ab - a^2 + b^2 + a^2 + b^2 - 2ab = \sigma_Y^2$$

$$a^2 + 3b^2 = \sigma_Y^2$$

$$\left(\frac{\sigma_x + \sigma_y}{2}\right)^2 + 3\left[\left(\frac{\sigma_x - \sigma_y}{2}\right)^2 + \tau_{xy}^2\right] = \sigma_Y^2$$

$$\sigma_x^2 + \sigma_y^2 - \sigma_x\sigma_y + 3\tau_{xy}^2 = \sigma_Y^2 \qquad \textbf{Ans}$$

10–65. A material is subjected to plane stress. Express the maximum-shear-stress theory of failure in terms of σ_x, σ_y, and τ_{xy}. Assume that the principal stresses are of different algebraic signs.

Maximum Shear Stress Theory :

$$|\sigma_1 - \sigma_2| = \sigma_Y \qquad [1]$$

In - Plane Principal Stress : Using Eq.9 – 5,

$$\sigma_{1,2} = \frac{\sigma_x + \sigma_y}{2} \pm \sqrt{\left(\frac{\sigma_x - \sigma_y}{2}\right)^2 + \tau_{xy}^2}$$

$$|\sigma_1 - \sigma_2| = 2\sqrt{\left(\frac{\sigma_x - \sigma_y}{2}\right)^2 + \tau_{xy}^2}$$

Substituting the results into Eq.[1] yields :

$$4\left[\left(\frac{\sigma_x - \sigma_y}{2}\right)^2 + \tau_{xy}^2\right] = \sigma_Y^2$$

$$(\sigma_x - \sigma_y)^2 + 4\,\tau_{xy}^2 = \sigma_Y^2 \qquad \textbf{Ans}$$

10–66. The components of plane stress at a critical point on an A-36 structural steel shell are shown. Determine if failure (yielding) has occurred on the basis of the maximum-shear-stress theory.

Normal and Shear Stress : In accordance with the sign convention,

$$\sigma_x = -75 \text{ MPa} \qquad \sigma_y = 125 \text{ MPa} \qquad \tau_{xy} = -80 \text{ MPa}$$

In - Plane Principal Stress : Applying Eq.9 – 5,

$$\sigma_{1,2} = \frac{\sigma_x + \sigma_y}{2} \pm \sqrt{\left(\frac{\sigma_x - \sigma_y}{2}\right)^2 + \tau_{xy}^2}$$

$$= \frac{-75 + 125}{2} \pm \sqrt{\left(\frac{-75 - 125}{2}\right)^2 + (-80)^2}$$

$$= 25.0 \pm 128.06$$

$$\sigma_1 = 153.06 \text{ MPa} \qquad \sigma_2 = -103.06 \text{ MPa}$$

Maximum Shear Stress Theory : σ_1 and σ_2 have opposite signs, so

$$|\sigma_1 - \sigma_2| = |153.06 - (-103.06)| = 256.12 \text{ MPa} > \sigma_Y$$

Based on the result obtained above, **the material yields according to the maximum shear stress theory.** **Ans**

10–67. The components of plane stress at a critical point on an A-36 structural steel shell are shown. Determine if failure (yielding) has occurred on the basis of the maximum-distortion-energy theory.

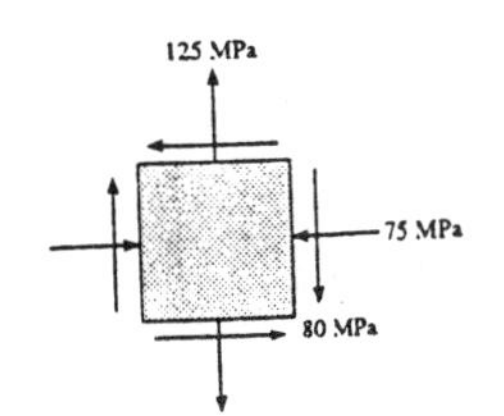

Normal and Shear Stress : In accordance with the sign convention,

$$\sigma_x = -75 \text{ MPa} \qquad \sigma_y = 125 \text{ MPa} \qquad \tau_{xy} = -80 \text{ MPa}$$

In-Plane Principal Stress : Applying Eq.9 – 5,

$$\sigma_{1,2} = \frac{\sigma_x + \sigma_y}{2} \pm \sqrt{\left(\frac{\sigma_x - \sigma_y}{2}\right)^2 + \tau_{xy}^2}$$
$$= \frac{-75 + 125}{2} \pm \sqrt{\left(\frac{-75 - 125}{2}\right)^2 + (-80)^2}$$
$$= 25.0 \pm 128.06$$

$$\sigma_1 = 153.06 \text{ MPa} \qquad \sigma_2 = -103.06 \text{ MPa}$$

Maximum Distortion Energy Theory :

$$\sigma_1^2 - \sigma_1\sigma_2 + \sigma_2^2 = \sigma_Y^2$$
$$153.06^2 - 153.06(-103.06) + (-103.06)^2 = 49825 < \sigma_Y^2 = 62500$$

Based on the result obtained above, **the material does not yield according to the maximum distortion energy theory.** **Ans**

***10–68.** The yield stress for a zirconium-magnesium alloy is $\sigma_Y = 15.3$ ksi. If a machine part is made of this material and a critical point in the material is subjected to in-plane principal stresses σ_1 and $\sigma_2 = -0.5\sigma_1$, determine the magnitude of σ_1 that will cause yielding according to the maximum-shear-stress theory.

Maximum Shear Stress Theory : σ_1 and $\sigma_2 = -0.5\sigma_1$ have opposite signs, so

$$|\sigma_1 - \sigma_2| = \sigma_Y$$
$$|\sigma_1 - (-0.5\sigma_1)| = 15.3$$
$$\sigma_1 = 10.2 \text{ ksi}$$ **Ans**

10–69. Solve Prob. 10–68 using the maximum-distortion-energy theory.

Maximum Distortion Energy Theory :

$$\sigma_1^2 - \sigma_1\sigma_2 + \sigma_2^2 = \sigma_Y^2$$
$$\sigma_1^2 - \sigma_1(-0.5\sigma_1) + (-0.5\sigma_1)^2 = 15.3^2$$

$$\sigma_1 = 11.6 \text{ ksi}$$ **Ans**

10–70. The yield stress for heat-treated beryllium copper is $\sigma_Y = 130$ ksi. If this material is subjected to plane stress and elastic failure occurs when one principal stress is 145 ksi, what is the smallest magnitude of the other principal stress? Use the maximum-distortion-energy theory.

Maximum Distortion Energy Theory : With $\sigma_1 = 145$ ksi,

$$\sigma_1^2 - \sigma_1\sigma_2 + \sigma_2^2 = \sigma_Y^2$$
$$145^2 - 145\sigma_2 + \sigma_2^2 = 130^2$$
$$\sigma_2^2 - 145\sigma_2 + 4125 = 0$$

$$\sigma_2 = \frac{-(-145) \pm \sqrt{(-145)^2 - 4(1)(4125)}}{2(1)}$$
$$= 72.5 \pm 33.634$$

Choose the smaller root, $\sigma_2 = 38.9$ ksi **Ans**

10–71. The yield stress for a plastic material is $\sigma_Y = 110$ MPa. If this material is subjected to plane stress and elastic failure occurs when one principal stress is 120 MPa, what is the smallest magnitude of the other principal stress? Use the maximum-distortion-energy theory.

Maximum Distortion Energy Theory : With $\sigma_1 = 120$ MPa,

$$\sigma_1^2 - \sigma_1\sigma_2 + \sigma_2^2 = \sigma_Y^2$$
$$120^2 - 120\sigma_2 + \sigma_2^2 = 110^2$$
$$\sigma_2^2 - 120\sigma_2 + 2300 = 0$$

$$\sigma_2 = \frac{-(-120) \pm \sqrt{(-120)^2 - 4(1)(2300)}}{2(1)}$$
$$= 60.0 \pm 36.056$$

Choose the smaller root, $\sigma_2 = 23.9$ MPa **Ans**

***10–72.** Solve Prob. 10–71 using the maximum-shear-stress theory. Both principal stresses have opposite signs.

Maximum Shear Stress Theory : $\sigma_1 = 120$ MPa and σ_2 have opposite signs, so

$$|\sigma_1 - \sigma_2| = \sigma_Y$$
$$|120 - (-\sigma_2)| = 110$$

$$\sigma_2 = -10.0 \text{ MPa}$$ **Ans**

10–73. The plate is made of Tobin bronze, which yields at $\sigma_Y = 25$ ksi. Using the maximum-shear-stress theory, determine the maximum tensile stress σ_x that can be applied to the plate if a tensile stress $\sigma_y = 0.75\sigma_x$ is also applied.

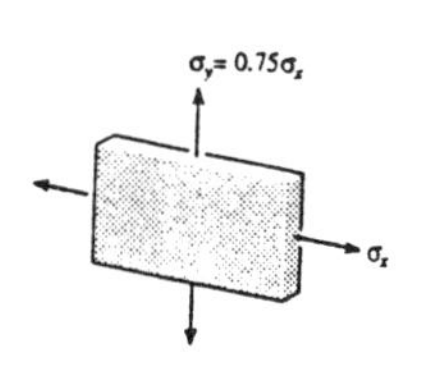

Maximum Shear Stress Theory : $\sigma_1 = \sigma_x$ and $\sigma_2 = 0.75\sigma_x$ have the same signs, so

$$|\sigma_2| = |0.75\sigma_x| = \sigma_Y$$
$$0.75\sigma_x = 25.0$$
$$\sigma_x = 33.3 \text{ ksi}$$

$$|\sigma_1| = |\sigma_x| = \sigma_Y$$
$$\sigma_x = 25.0 \text{ ksi} \quad (\textit{Controls!})$$

Ans

10–74. The plate is made of Tobin bronze, which yields at $\sigma_Y = 25$ ksi. Using the maximum-distortion-energy theory, determine the maximum tensile stress σ_x that can be applied to the plate if a tensile stress $\sigma_y = 0.75\sigma_x$ is also applied.

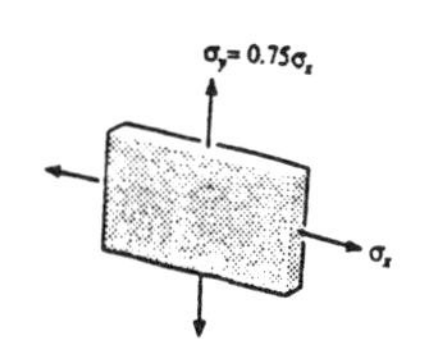

Maximum Distortion Energy Theory : With $\sigma_1 = \sigma_x$ and $\sigma_2 = 0.75\sigma_x$,

$$\sigma_1^2 - \sigma_1\sigma_2 + \sigma_2^2 = \sigma_Y^2$$
$$\sigma_x^2 - \sigma_x(0.75\sigma_x) + (0.75\sigma_x)^2 = 25^2$$

$$\sigma_x = 27.7 \text{ ksi}$$

Ans

10–75. A bar with a circular cross-sectional area is made of SAE 1045 carbon steel having a yield stress of $\sigma_Y = 150$ ksi. If the bar is subjected to a torque of 30 kip · in. and a bending moment of 56 kip · in., determine the required diameter of the bar according to the maximum-distortion-energy theory. Use a factor of safety of 2 with respect to yielding.

Normal and Shear Stresses : Applying the flexure and torsion formulas,

$$\sigma = \frac{Mc}{I} = \frac{56\left(\frac{d}{2}\right)}{\frac{\pi}{4}\left(\frac{d}{2}\right)^4} = \frac{1792}{\pi d^3}$$

$$\tau = \frac{Tc}{J} = \frac{30\left(\frac{d}{2}\right)}{\frac{\pi}{2}\left(\frac{d}{2}\right)^4} = \frac{480}{\pi d^3}$$

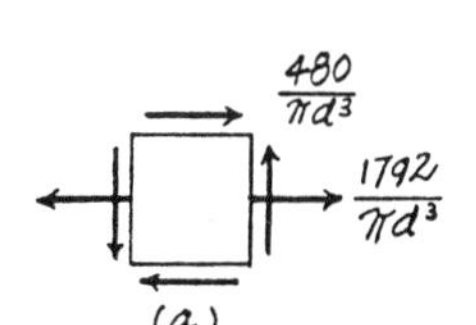

The critical state of stress is shown in Fig. (a) or (b), where

$$\sigma_x = \frac{1792}{\pi d^3} \qquad \sigma_y = 0 \qquad \tau_{xy} = \frac{480}{\pi d^3}$$

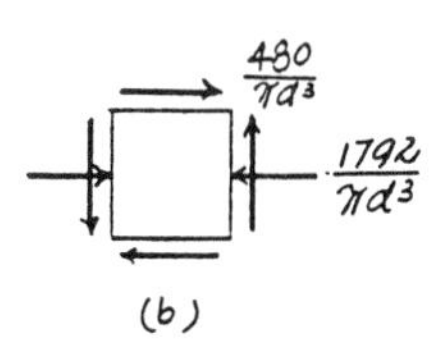

In - Plane Principal Stresses : Applying Eq. 9 – 5,

$$\sigma_{1,2} = \frac{\sigma_x + \sigma_y}{2} \pm \sqrt{\left(\frac{\sigma_x - \sigma_y}{2}\right)^2 + \tau_{xy}^2}$$

$$= \frac{\frac{1792}{\pi d^3} + 0}{2} \pm \sqrt{\left(\frac{\frac{1792}{\pi d^3} - 0}{2}\right)^2 + \left(\frac{480}{\pi d^3}\right)^2}$$

$$= \frac{896}{\pi d^3} \pm \frac{1016.47}{\pi d^3}$$

$$\sigma_1 = \frac{1912.47}{\pi d^3} \qquad \sigma_2 = -\frac{120.47}{\pi d^3}$$

Maximum Distortion Energy Theory :

$$\sigma_1^2 - \sigma_1\sigma_2 + \sigma_2^2 = \sigma_{\text{allow}}^2$$

$$\left(\frac{1912.47}{\pi d^3}\right)^2 - \left(\frac{1912.47}{\pi d^3}\right)\left(-\frac{120.47}{\pi d^3}\right) + \left(-\frac{120.47}{\pi d^3}\right)^2 = \left(\frac{150}{2}\right)^2$$

$$d = 2.03 \text{ in.}$$

Ans

***10–76.** A bar with a square cross-sectional area is made of a material having a yield stress of $\sigma_Y = 120$ ksi. If the bar is subjected to a bending moment of 75 kip · in., determine the required size of the bar according to the maximum-distortion-energy theory. Use a factor of safety of 1.5 with respect to yielding.

Normal and Shear Stress **:** Applying the flexure formula,

$$\sigma = \frac{Mc}{I} = \frac{75\left(\frac{a}{2}\right)}{\frac{1}{12}a^4} = \frac{450}{a^3}$$

In - Plane Principal Stress **:** Since no shear stress acts on the element.

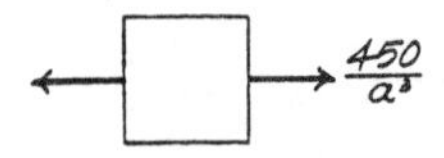

$$\sigma_1 = \sigma_x = \frac{450}{a^3} \qquad \sigma_2 = \sigma_y = 0$$

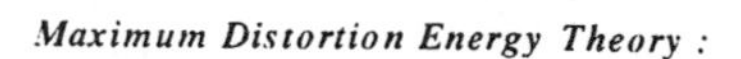

Maximum Distortion Energy Theory **:**

$$\sigma_1^2 - \sigma_1\sigma_2 + \sigma_2^2 = \sigma_{allow}^2$$

$$\left(\frac{450}{a^3}\right)^2 - 0 + 0 = \left(\frac{120}{1.5}\right)^2$$

$$a = 1.78 \text{ in.} \qquad \textbf{Ans}$$

10–77. Solve Prob. 10–76 using the maximum-shear-stress theory.

Normal and Shear Stress **:** Applying the flexure formula,

$$\sigma = \frac{Mc}{I} = \frac{75\left(\frac{a}{2}\right)}{\frac{1}{12}a^4} = \frac{450}{a^3}$$

In - Plane Principal Stress **:** Since no shear stress acts on the element.

$$\sigma_1 = \sigma_x = \frac{450}{a^3} \qquad \sigma_2 = \sigma_y = 0$$

Maximum Shear Stress Theory **:**

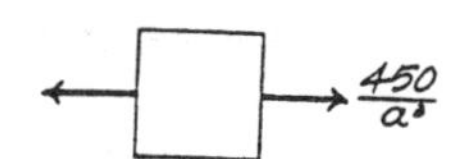

$$|\sigma_2| = 0 < \sigma_{allow} = \frac{120}{1.5} = 80.0 \text{ ksi } (O.K!)$$

$$|\sigma_1| = \sigma_{allow}$$

$$\frac{450}{a^3} = \frac{120}{1.5}$$

$$a = 1.78 \text{ in.} \qquad \textbf{Ans}$$

10–78. The principal plane stresses acting on a differential element are shown. If the material is machine steel having a yield stress of $\sigma_Y = 700$ MPa, determine the factor of safety with respect to yielding using the maximum-distortion-energy theory.

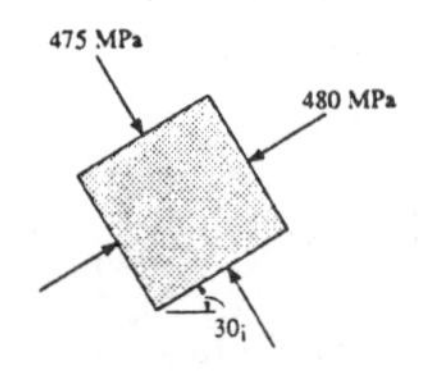

In - Plane Principal Stresses **:** Since no shear stress acts on the element.

$$\sigma_1 = \sigma_y = -475 \text{ MPa} \qquad \sigma_2 = \sigma_x = -480 \text{ MPa}$$

Maximum Distortion Energy Theory **:**

$$\sigma_1^2 - \sigma_1\sigma_2 + \sigma_2^2 = \sigma_{allow}^2$$

$$(-475)^2 - (-475)(-480) + (-480)^2 = \sigma_{allow}^2$$

$$\sigma_{allow} = 477.52 \text{ MPa}$$

The factor of safety is

$$F.S = \frac{\sigma_Y}{\sigma_{allow}} = \frac{700}{477.52} = 1.47 \qquad \textbf{Ans}$$

10–79. The state of plane stress at a critical point in an aluminum machine bracket is shown. If the yield stress for aluminum is $\sigma_Y = 60$ ksi, determine if yielding occurs using (a) the maximum-shear-stress theory, (b) the maximum-distortion-energy theory.

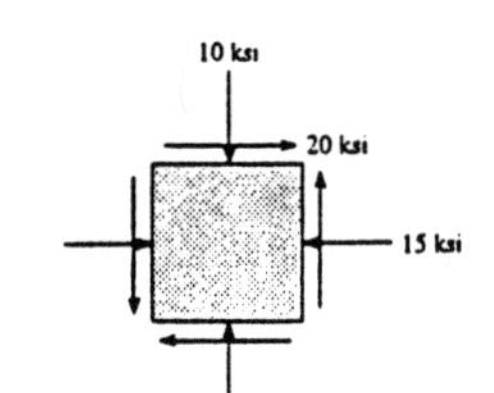

Normal and Shear Stress : In accordance with the sign convention,

$$\sigma_x = -15 \text{ ksi} \qquad \sigma_y = -10 \text{ ksi} \qquad \tau_{xy} = 20 \text{ ksi}$$

In-Plane Principal Stresses : Applying Eq. 9–5,

$$\sigma_{1,2} = \frac{\sigma_x + \sigma_y}{2} \pm \sqrt{\left(\frac{\sigma_x - \sigma_y}{2}\right)^2 + \tau_{xy}^2}$$
$$= \frac{-15 + (-10)}{2} \pm \sqrt{\left[\frac{-15 - (-10)}{2}\right]^2 + 20^2}$$
$$= -12.5 \pm 20.156$$

$$\sigma_1 = 7.656 \text{ ksi} \qquad \sigma_2 = -32.656 \text{ ksi}$$

Maximum Shear Stress Theory : σ_1 and σ_2 have opposite signs, so

$$|\sigma_1 - \sigma_2| = |7.656 - (-32.565)| = 40.31 \text{ ksi} < \sigma_Y$$

Based on the result obtained above, **the material does not yield according to the maximum shear stress theory.** **Ans**

Maximum Distortion Energy Theory :

$$\sigma_1^2 - \sigma_1\sigma_2 + \sigma_2^2 = \sigma_Y^2$$
$$7.656^2 - 7.656(-32.656) + (-32.656)^2 = 1375 < \sigma_Y^2 = 3600$$

Based on the result obtained above, **the material does not yield according to the maximum distortion energy theory.** **Ans**

***10–80.** If a shaft is made of nickel for which $\sigma_Y = 65$ ksi, determine the maximum torsional stress required to cause yielding using (a) the maximum-shear-stress theory and (b) the maximum-distortion-energy theory.

In-Plane Principal Stresses : For pure torsion, $\sigma_1 = \tau$ and $\sigma_2 = -\tau$.

a) *Maximum Shear Stress Theory* : σ_1 and σ_2 have opposite signs, so

$$|\sigma_1 - \sigma_2| = \sigma_Y$$
$$|\tau - (-\tau)| = 65$$

$$\tau = 32.5 \text{ ksi} \qquad \textbf{Ans}$$

b) *Maximum Distortion Energy Theory* :

$$\sigma_1^2 - \sigma_1\sigma_2 + \sigma_2^2 = \sigma_Y^2$$
$$\tau^2 - \tau(-\tau) + (-\tau)^2 = 65^2$$

$$\tau = 37.5 \text{ ksi} \qquad \textbf{Ans}$$

10–81. Derive an expression for an equivalent torque T_e that, if applied alone to a solid shaft with a circular cross section, would cause the same energy of distortion as the combination of an applied bending moment M and torque T.

In - Plane Principal Stress : In the case of pure torsion, $\sigma_1 = \tau_e$ and $\sigma_2 = -\tau_e$, where $\tau_e = \dfrac{T_e c}{J}$.

Energy of Distortion :

$$u_d = \frac{1+\nu}{3E}\left(\sigma_1^2 - \sigma_1\sigma_2 + \sigma_2^2\right)$$

$$(u_d)_1 = \frac{1+\nu}{3E}\left(\tau_e^2 - \tau_e(-\tau_e) + (-\tau_e)^2\right)$$

$$= \frac{1+\nu}{E}\tau_e^2$$

$$= \frac{1+\nu}{E}\left(\frac{T_e c}{J}\right)^2$$

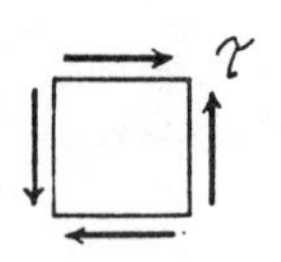

In - Plane Principal Stress : In the case of combined bending stress and torsional shear stress, $\sigma = \dfrac{Mc}{I}$ and $\tau = \dfrac{Tc}{J}$.

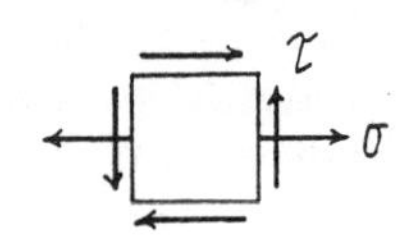

$$\sigma_{1,2} = \frac{\sigma_x + \sigma_y}{2} \pm \sqrt{\left(\frac{\sigma_x - \sigma_y}{2}\right)^2 + \tau_{xy}^2}$$

$$\sigma_{1,2} = \frac{\sigma + 0}{2} \pm \sqrt{\left(\frac{\sigma - 0}{2}\right)^2 + \tau^2}$$

$$\sigma_1 = \frac{\sigma}{2} + \sqrt{\frac{\sigma^2}{4} + \tau^2} \qquad \sigma_2 = \frac{\sigma}{2} - \sqrt{\frac{\sigma^2}{4} + \tau^2}$$

Energy of Distortion : Let $a = \dfrac{\sigma}{2}$ and $b = \sqrt{\dfrac{\sigma^2}{4} + \tau^2}$ then

$\sigma_1^2 = a^2 + b^2 + 2ab$, $\quad \sigma_1\sigma_2 = a^2 - b^2$, $\quad \sigma_2^2 = a^2 + b^2 - 2ab$.

and $\sigma_1^2 - \sigma_1\sigma_2 + \sigma_2^2 = 3b^2 + a^2$.

$$u_d = \frac{1+\nu}{3E}\left(\sigma_1^2 - \sigma_1\sigma_2 + \sigma_2^2\right)$$

$$(u_d)_2 = \frac{1+\nu}{3E}\left(3b^2 + a^2\right)$$

$$= \frac{1+\nu}{3E}\left(\sigma^2 + 3\tau^2\right)$$

$$= \frac{c^2(1+\nu)}{3E}\left(\frac{M^2}{I^2} + \frac{3T^2}{J^2}\right)$$

Require,

$$(u_d)_1 = (u_d)_2$$

$$\frac{1+\nu}{E}\left(\frac{T_e c}{J}\right)^2 = \frac{c^2(1+\nu)}{3E}\left(\frac{M^2}{I^2} + \frac{3T^2}{J^2}\right)$$

$$T_e = \sqrt{\frac{M^2 J^2}{3I^2} + T^2}$$

However, for the circular shaft $\dfrac{J^2}{I^2} = \dfrac{\left(\frac{\pi}{2}c^4\right)^2}{\left(\frac{\pi}{4}c^4\right)^2} = 4$, then

$$T_e = \sqrt{\frac{4}{3}M^2 + T^2} \qquad \textbf{Ans}$$

10–82. Derive an expression for an equivalent bending moment M_e that, if applied alone to a solid bar with a circular cross section, would cause the same maximum shear stress as the combination of an applied moment M and torque T. Assume that the principal stresses are of opposite algebraic signs.

In-Plane Principal Stress : In the case of pure bending, $\sigma_1 = \frac{M_e c}{I}$ and $\sigma_2 = 0$.

Maxiumum Shear Stress : σ_1 and σ_2 are assumed to have opposite signs.

$$\left(\tau_{\substack{abs\\max}}\right)_1 = \frac{\sigma_1 - \sigma_2}{2} = \frac{M_e c}{2I}$$

In-Plane Principal Stress : In the case of combined bending stress and torsional shear stress, $\sigma = \frac{Mc}{I}$ and $\tau = \frac{Tc}{J}$.

$$\sigma_{1,2} = \frac{\sigma_x + \sigma_y}{2} \pm \sqrt{\left(\frac{\sigma_x - \sigma_y}{2}\right)^2 + \tau_{xy}^2}$$

$$\sigma_{1,2} = \frac{\sigma + 0}{2} \pm \sqrt{\left(\frac{\sigma - 0}{2}\right)^2 + \tau^2}$$

$$\sigma_1 = \frac{\sigma}{2} + \sqrt{\frac{\sigma^2}{4} + \tau^2} \qquad \sigma_2 = \frac{\sigma}{2} - \sqrt{\frac{\sigma^2}{4} + \tau^2}$$

Maxiumum Shear Stress : σ_1 and σ_2 are assumed to have opposite signs.

$$\left(\tau_{\substack{abs\\max}}\right)_2 = \frac{\sigma_1 - \sigma_2}{2} = \sqrt{\frac{\sigma^2}{4} + \tau^2} = \sqrt{\frac{M^2c^2}{4I^2} + \frac{T^2c^2}{J^2}}$$

Requires,

$$\left(\tau_{\substack{abs\\max}}\right)_1 = \left(\tau_{\substack{abs\\max}}\right)_2$$

$$\frac{M_e c}{2I} = \sqrt{\frac{M^2c^2}{4I^2} + \frac{T^2c^2}{J^2}}$$

$$M_e = \sqrt{M^2 + \frac{4T^2I^2}{J^2}}$$

However, for circular shaft $\frac{I^2}{J^2} = \frac{\left(\frac{\pi}{4}c^4\right)^2}{\left(\frac{\pi}{2}c^4\right)^2} = \frac{1}{4}$, then

$$M_e = \sqrt{M^2 + T^2}$$ **Ans**

10–83. The state of stress acting at a critical point on the seat frame of an automobile during a crash is shown in the figure. Determine the smallest yield stress for a steel that can be selected for the member, based on the maximum-shear-stress theory.

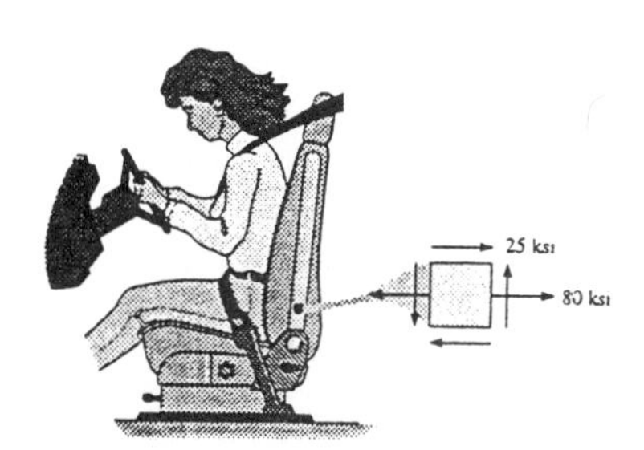

Normal and Shear Stress : In accordance with the sign convention,

$$\sigma_x = 80 \text{ ksi} \qquad \sigma_y = 0 \qquad \tau_{xy} = 25 \text{ ksi}$$

In-Plane Principal Stress : Applying Eq. 9 – 5,

$$\sigma_{1,2} = \frac{\sigma_x + \sigma_y}{2} \pm \sqrt{\left(\frac{\sigma_x - \sigma_y}{2}\right)^2 + \tau_{xy}^2}$$

$$= \frac{80 + 0}{2} \pm \sqrt{\left(\frac{80 - 0}{2}\right)^2 + 25^2}$$

$$= 40 \pm 47.170$$

$$\sigma_1 = 87.170 \text{ ksi} \qquad \sigma_2 = -7.170 \text{ ksi}$$

Maximum Shear Stress Theory : σ_1 and σ_2 have opposite signs, so

$$|\sigma_1 - \sigma_2| = \sigma_Y$$

$$|87.170 - (-7.170)| = \sigma_Y$$

$$\sigma_Y = 94.3 \text{ ksi}$$ **Ans**

***10–84.** Solve Prob. 10–83 using the maximum-distortion-energy theory.

Normal and Shear Stress : In accordance with the sign convention,

$$\sigma_x = 80 \text{ ksi} \qquad \sigma_y = 0 \qquad \tau_{xy} = 25 \text{ ksi}$$

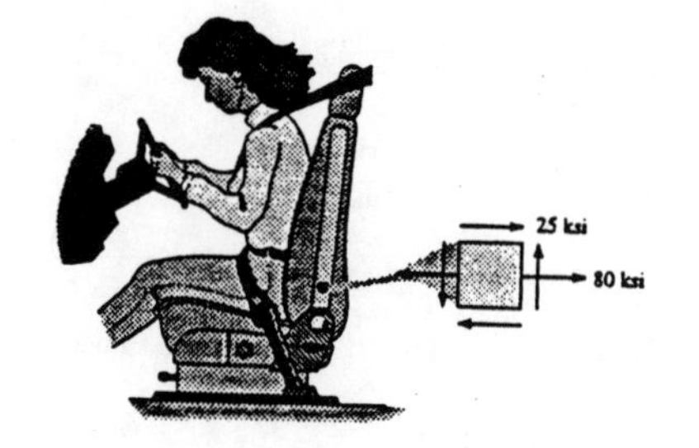

In-Plane Principal Stress : Applying Eq. 9–5,

$$\sigma_{1,2} = \frac{\sigma_x + \sigma_y}{2} \pm \sqrt{\left(\frac{\sigma_x - \sigma_y}{2}\right)^2 + \tau_{xy}^2}$$

$$= \frac{80+0}{2} \pm \sqrt{\left(\frac{80-0}{2}\right)^2 + 25^2}$$

$$= 40 \pm 47.170$$

$$\sigma_1 = 87.170 \text{ ksi} \qquad \sigma_2 = -7.170 \text{ ksi}$$

Maximum Distortion Energy Theory :

$$\sigma_1^2 - \sigma_1\sigma_2 + \sigma_2^2 = \sigma_Y^2$$

$$87.170^2 - 87.170(-7.170) + (-7.170)^2 = \sigma_Y^2$$

$$\sigma_Y = 91.0 \text{ ksi} \qquad \textbf{Ans}$$

10–85. If a machine part is made of titanium (Ti-6A1-4V) and a critical point in the material is subjected to plane stress, such that the principal stresses are σ_1 and $\sigma_2 = 0.5\sigma_1$, determine the magnitude of σ_1 in MPa that will cause yielding according to (a) the maximum-shear-stress theory, and (b) the maximum-distortion-energy theory.

a) ***Maximum Shear Stress Theory*** : σ_1 and $\sigma_2 = 0.5\sigma_1$ have the same signs, so

$$|\sigma_2| = |0.5\sigma_1| = \sigma_Y$$

$$0.5\sigma_1 = 924$$

$$\sigma_x = 1848 \text{ MPa}$$

$$|\sigma_1| = \sigma_Y$$

$$\sigma_1 = 924 \text{ MPa} \quad (\textit{Controls!}) \qquad \textbf{Ans}$$

b) ***Maximum Distortion Energy Theory*** : With σ_1 and $\sigma_2 = 0.5\sigma_1$,

$$\sigma_1^2 - \sigma_1\sigma_2 + \sigma_2^2 = \sigma_Y^2$$

$$\sigma_1^2 - \sigma_1(0.5\sigma_1) + (0.5\sigma_1)^2 = 924^2$$

$$\sigma_1 = 1067 \text{ MPa} \qquad \textbf{Ans}$$

10–86. An aluminum alloy 6061-T6 is to be used for a drive shaft such that it transmits 50 hp at 1800 rev/min. Using a factor of safety of F.S. = 2, with respect to yielding, determine the smallest-diameter shaft that can be selected based on the maximum-distortion-energy theory.

Internal Torque : Using the power transmission formula,

$$\omega = 1800\ \frac{\text{rev}}{\text{min}}\left(\frac{2\pi\ \text{rad}}{\text{rev}}\right)\frac{1\ \text{min}}{60\ \text{s}} = 60.0\pi\ \text{rad/s}$$

$$P = 50\ \text{hp}\left(\frac{550\ \text{ft}\cdot\text{lb/s}}{1\ \text{hp}}\right) = 27500\ \text{ft}\cdot\text{lb/s}$$

$$T = \frac{P}{\omega} = \frac{27500}{60.0\pi} = 145.89\ \text{lb}\cdot\text{ft}$$

Shear Stress : Applying the torsion formula,

$$\tau = \frac{Tc}{J} = \frac{145.89(12)\left(\frac{d}{2}\right)}{\frac{\pi}{2}\left(\frac{d}{2}\right)^4} = \frac{8916.26}{d^3}$$

In - Plane Principal Stress : In the case of pure torsion, $\sigma_1 = \frac{8916.26}{d^3}$ and $\sigma_2 = -\frac{8916.26}{d^3}$.

Maximum Distortion Energy Theory :

$$\sigma_1^2 - \sigma_1\sigma_2 + \sigma_2^2 = \sigma_{\text{allow}}^2$$

$$\left(\frac{8916.26}{d^3}\right)^2 - \left(\frac{8916.26}{d^3}\right)\left(-\frac{8916.26}{d^3}\right) + \left(-\frac{8916.26}{d^3}\right)^2 = \left[\frac{37.0(10^3)}{2}\right]^2$$

$$d = 0.942\ \text{in.} \qquad \textbf{Ans}$$

***10–87.** Solve Prob. 10–86 using the maximum-shear-stress theory.

Internal Torque : Using the power transmission formula,

$$\omega = 1800\ \frac{\text{rev}}{\text{min}}\left(\frac{2\pi\ \text{rad}}{\text{rev}}\right)\frac{1\ \text{min}}{60\ \text{s}} = 60.0\pi\ \text{rad/s}$$

$$P = 50\ \text{hp}\left(\frac{550\ \text{ft}\cdot\text{lb/s}}{1\ \text{hp}}\right) = 27500\ \text{ft}\cdot\text{lb/s}$$

$$T = \frac{P}{\omega} = \frac{27500}{60.0\pi} = 145.89\ \text{lb}\cdot\text{ft}$$

Shear Stress : Applying the torsion formula,

$$\tau = \frac{Tc}{J} = \frac{145.89(12)\left(\frac{d}{2}\right)}{\frac{\pi}{2}\left(\frac{d}{2}\right)^4} = \frac{8916.26}{d^3}$$

In - Plane Principal Stress : In the case of pure torsion, $\sigma_1 = \frac{8916.26}{d^3}$ and $\sigma_2 = -\frac{8916.26}{d^3}$.

Maximum Shear Stress Theory : σ_1 and σ_2 have opposite signs, so

$$|\sigma_1 - \sigma_2| = \sigma_{\text{allow}}$$

$$\left|\frac{8916.26}{d^3} - \left(-\frac{8916.26}{d^3}\right)\right| = \frac{37.0(10^3)}{2}$$

$$d = 0.988\ \text{in.} \qquad \textbf{Ans}$$

***10–88.** The element is subjected to the stresses shown. If $\sigma_Y = 50$ ksi, determine the factor of safety for this loading based on (a) the maximum-shear-stress theory and (b) the maximum-distortion-energy theory.

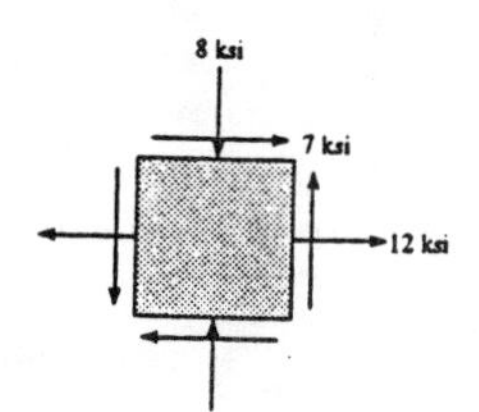

Normal and Shear Stress : In accordance with the sign convention,

$$\sigma_x = 12 \text{ ksi} \qquad \sigma_y = -8 \text{ ksi} \qquad \tau_{xy} = 7 \text{ ksi}$$

In - Plane Principal Stress : Applying Eq.9 – 5,

$$\sigma_{1,2} = \frac{\sigma_x + \sigma_y}{2} \pm \sqrt{\left(\frac{\sigma_x - \sigma_y}{2}\right)^2 + \tau_{xy}^2}$$

$$= \frac{12 + (-8)}{2} \pm \sqrt{\left[\frac{12 - (-8)}{2}\right]^2 + 7^2}$$

$$= 2.00 \pm 12.207$$

$$\sigma_1 = 14.207 \text{ ksi} \qquad \sigma_2 = -10.207 \text{ ksi}$$

a) ***Maximum Shear Stress Theory :*** σ_1 and σ_2 have opposite signs, so

$$|\sigma_1 - \sigma_2| = \sigma_{\text{allow}}$$

$$|14.207 - (-10.207)| = \sigma_{\text{allow}}$$

$$\sigma_{\text{allow}} = 24.414 \text{ ksi}$$

The factor of safety is

$$\text{F.S} = \frac{\sigma_Y}{\sigma_{\text{allow}}} = \frac{50}{24.414} = 2.05 \qquad \textbf{Ans}$$

b) ***Maximum Distortion Energy Theory :***

$$\sigma_1^2 - \sigma_1\sigma_2 + \sigma_2^2 = \sigma_{\text{allow}}^2$$

$$14.207^2 - 14.207(-10.207) + (-10.207)^2 = \sigma_{\text{allow}}^2$$

$$\sigma_{\text{allow}} = 21.237 \text{ ksi}$$

The factor of safety is

$$\text{F.S} = \frac{\sigma_Y}{\sigma_{\text{allow}}} = \frac{50}{21.237} = 2.35 \qquad \textbf{Ans}$$

10–89. An aluminum alloy 6061-T6 is to be used for a solid drive shaft such that it transmits 40 hp at 2400 rev/min. Using a factor of safety of F.S. = 2, with respect to yielding, determine the smallest-diameter shaft that can be selected based on the maximum-shear-stress theory.

Internal Torque : Using the power transmission formula,

$$\omega = 2400 \frac{\text{rev}}{\text{min}} \left(\frac{2\pi \text{ rad}}{\text{rev}}\right) \frac{1 \text{ min}}{60 \text{ s}} = 80.0\pi \text{ rad/s}$$

$$P = 40 \text{ hp} \left(\frac{550 \text{ ft} \cdot \text{lb/s}}{1 \text{ hp}}\right) = 22000 \text{ ft} \cdot \text{lb/s}$$

$$T = \frac{P}{\omega} = \frac{22000}{80.0\pi} = 87.54 \text{ lb} \cdot \text{ft}$$

Shear Stress : Applying the torsion formula,

$$\tau = \frac{Tc}{J} = \frac{87.54(12)\left(\frac{d}{2}\right)}{\frac{\pi}{2}\left(\frac{d}{2}\right)^4} = \frac{5349.76}{d^3}$$

In - Plane Principal Stress : In the case of pure torsion, $\sigma_1 = \frac{5349.76}{d^3}$ and $\sigma_2 = -\frac{5349.76}{d^3}$.

Maximum Shear Stress Theory : σ_1 and σ_2 have opposite signs, so

$$|\sigma_1 - \sigma_2| = \sigma_{\text{allow}}$$

$$\left|\frac{5349.76}{d^3} - \left(-\frac{5349.76}{d^3}\right)\right| = \frac{37.0(10^3)}{2}$$

$$d = 0.833 \text{ in.} \qquad \textbf{Ans}$$

10–90. Solve Prob. 10–89 using the maximum-distortion-energy theory.

Internal Torque : Using the power transmission formula,

$$\omega = 2400\ \frac{\text{rev}}{\text{min}}\left(\frac{2\pi\ \text{rad}}{\text{rev}}\right)\frac{1\ \text{min}}{60\ \text{s}} = 80.0\pi\ \text{rad/s}$$

$$P = 40\ \text{hp}\left(\frac{550\ \text{ft}\cdot\text{lb/s}}{1\ \text{hp}}\right) = 22000\ \text{ft}\cdot\text{lb/s}$$

$$T = \frac{P}{\omega} = \frac{22000}{80.0\pi} = 87.54\ \text{lb}\cdot\text{ft}$$

Shear Stress : Applying the torsion formula,

$$\tau = \frac{Tc}{J} = \frac{87.54(12)\left(\frac{d}{2}\right)}{\frac{\pi}{2}\left(\frac{d}{2}\right)^4} = \frac{5349.76}{d^3}$$

In - Plane Principal Stress : In the case of pure torsion, $\sigma_1 = \frac{5349.76}{d^3}$ and $\sigma_2 = -\frac{5349.76}{d^3}$.

Maximum Distortion Energy Theory :

$$\sigma_1^2 - \sigma_1\sigma_2 + \sigma_2^2 = \sigma_{\text{allow}}^2$$

$$\left(\frac{5349.76}{d^3}\right)^2 - \left(\frac{5349.76}{d^3}\right)\left(-\frac{5349.76}{d^3}\right) + \left(-\frac{5349.76}{d^3}\right)^2 = \left[\frac{37.0(10^3)}{2}\right]^2$$

$$d = 0.794\ \text{in.} \qquad \textbf{Ans}$$

10–91. The principal stresses acting at a point on a thin-walled cylindrical pressure vessel are $\sigma_1 = pr/t$, $\sigma_2 = pr/2t$, and $\sigma_3 = 0$. If the yield stress is σ_Y, determine the maximum value of p based on (a) the maximum-shear-stress theory and (b) the maximum-distortion-energy theory.

a) ***Maximum Shear Stress Theory :*** σ_1 and σ_2 have the same signs, then

$$|\sigma_2| = \sigma_Y \qquad \left|\frac{pr}{2t}\right| = \sigma_Y \qquad p = \frac{2t}{r}\sigma_Y$$

$$|\sigma_1| = \sigma_Y \qquad \left|\frac{pr}{t}\right| = \sigma_Y \qquad p = \frac{t}{r}\sigma_Y\ (\textit{Controls!}) \qquad \textbf{Ans}$$

b) ***Maximum Distortion Energy Theory :***

$$\sigma_1^2 - \sigma_1\sigma_2 + \sigma_2^2 = \sigma_Y^2$$

$$\left(\frac{pr}{t}\right)^2 - \left(\frac{pr}{t}\right)\left(\frac{pr}{2t}\right) + \left(\frac{pr}{2t}\right)^2 = \sigma_Y^2$$

$$p = \frac{2t}{\sqrt{3}r}\sigma_Y \qquad \textbf{Ans}$$

***10–92.** The state of stress acting at a critical point on a wrench is shown in the figure. Determine the smallest yield stress for steel that might be selected for the part, based on the maximum-distortion-energy theory.

Normal and Shear Stress : In accordance with the sign convention,

$$\sigma_x = 25 \text{ ksi} \qquad \sigma_y = 0 \qquad \tau_{xy} = 10 \text{ ksi}$$

In - Plane Principal Stress : Applying Eq.9 – 5,

$$\sigma_{1,2} = \frac{\sigma_x + \sigma_y}{2} \pm \sqrt{\left(\frac{\sigma_x - \sigma_y}{2}\right)^2 + \tau_{xy}^2}$$
$$= \frac{25+0}{2} \pm \sqrt{\left(\frac{25-0}{2}\right)^2 + 10^2}$$
$$= 12.5 \pm 16.008$$

$$\sigma_1 = 28.508 \text{ ksi} \qquad \sigma_2 = -3.508 \text{ ksi}$$

Maximum Distortion Energy Theory :

$$\sigma_1^2 - \sigma_1\sigma_2 + \sigma_2^2 = \sigma_Y^2$$
$$28.508^2 - 28.508(-3.508) + (-3.508)^2 = \sigma_Y^2$$

$$\sigma_Y = 30.4 \text{ ksi} \qquad \textbf{Ans}$$

10–93. The state of stress acting at a critical point on a wrench is shown in the figure. Determine the smallest yield stress for steel that might be selected for the part, based on the maximum-shear-stress theory.

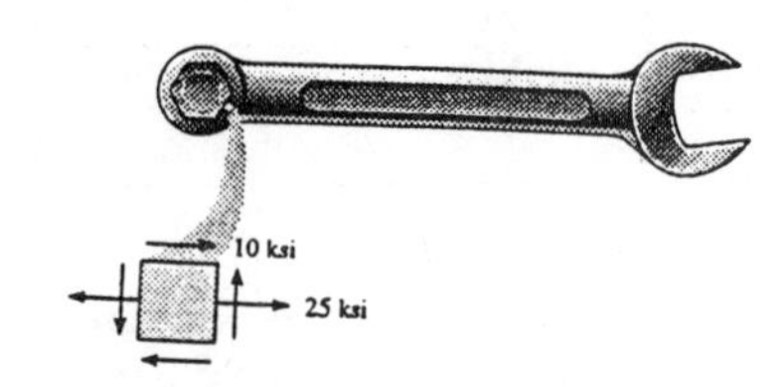

Normal and Shear Stresses : In accordance with the sign convention,

$$\sigma_x = 25 \text{ ksi} \qquad \sigma_y = 0 \qquad \tau_{xy} = 10 \text{ ksi}$$

In - Plane Principal Stress : Applying Eq.9 – 5

$$\sigma_{1,2} = \frac{\sigma_x + \sigma_y}{2} \pm \sqrt{\left(\frac{\sigma_x - \sigma_y}{2}\right)^2 + \tau_{xy}^2}$$
$$= \frac{25+0}{2} \pm \sqrt{\left(\frac{25-0}{2}\right)^2 + 10^2}$$
$$= 12.5 \pm 16.008$$

$$\sigma_1 = 28.508 \text{ ksi} \qquad \sigma_2 = -3.508 \text{ ksi}$$

Maximum Shear Stress Theory : σ_1 and σ_2 have opposite signs.

so

$$|\sigma_1 - \sigma_2| = \sigma_Y$$
$$|28.508 - (-3.508)| = \sigma_Y$$

$$\sigma_Y = 32.0 \text{ ksi} \qquad \textbf{Ans}$$

10–94. The internal loadings at a critical section along the steel drive shaft of a ship are calculated to be a torque of 2650 lb · ft, a bending moment of 2800 lb · ft, and an axial thrust of 3700 lb. If the yield points for tension and shear are $\sigma_Y = 100$ ksi and $\tau_Y = 50$ ksi, respectively, determine the required diameter of the shaft using the maximum-shear-stress theory.

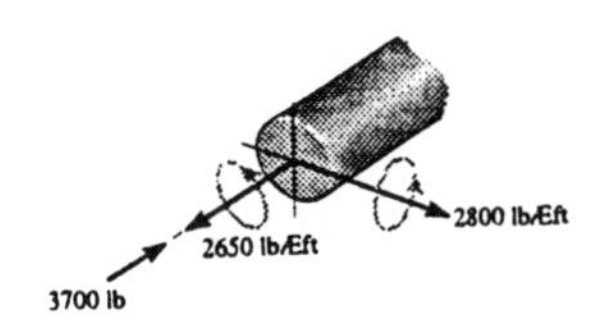

Normal Stress and Shear Stress : The critical point is at the bottom of the circular section.

$$\sigma = \frac{N}{A} \pm \frac{Mc}{I}$$

$$= \frac{-3.70}{\frac{\pi}{4}d^2} - \frac{2.80(12)\left(\frac{d}{2}\right)}{\frac{\pi}{4}\left(\frac{d}{2}\right)^4}$$

$$= -\left(\frac{14.8d + 1075.2}{\pi d^3}\right)$$

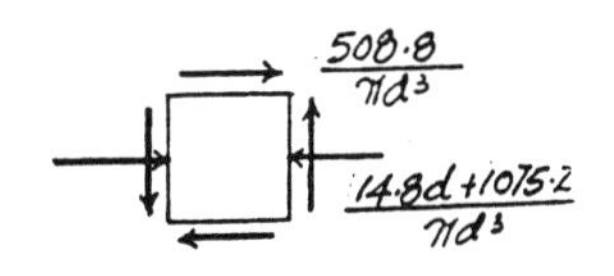

$$\tau = \frac{Tc}{J} = \frac{2.65(12)\left(\frac{d}{2}\right)}{\frac{\pi}{2}\left(\frac{d}{2}\right)^4} = \frac{508.8}{\pi d^3}$$

In - Plane Principal Stress : Applying Eq.9 – 5 with $\sigma_y = 0$, $\tau_{xy} = \frac{508.8}{\pi d^3}$, and $\sigma_x = -\left(\frac{14.8d + 1075.2}{\pi d^3}\right)$,

$$\sigma_{1,2} = \frac{\sigma_x + \sigma_y}{2} \pm \sqrt{\left(\frac{\sigma_x - \sigma_y}{2}\right)^2 + \tau_{xy}^2}$$

$$= -\left(\frac{7.40d + 537.6}{\pi d^3}\right) \pm \sqrt{\left(\frac{7.40d + 537.6}{\pi d^3}\right)^2 + \left(\frac{508.8}{\pi d^3}\right)^2} \qquad [1]$$

Maximum Shear Stress Theory : Assume σ_1 and σ_2 have opposite signs, then

$$|\sigma_1 - \sigma_2| = \sigma_Y$$

$$2\sqrt{\left(\frac{7.40d + 537.6}{\pi d^3}\right)^2 + \left(\frac{508.8}{\pi d^3}\right)^2} = 100$$

$$2500\pi^2 d^6 - 54.76d^2 - 7956.48d - 547891.2 = 0$$

Solving by trial and error,

$$d = 1.683361 \text{ in.} = 1.68 \text{ in.} \qquad \textbf{Ans}$$

From Eq.[1] $\quad \sigma_1 = 13.29$ ksi $\quad \sigma_2 = -86.71$ ksi.

The principal stresses σ_1 and σ_2 indeed have opposite signs as assumed. Therefore, the solution is valid.

10–95. The internal loadings at a critical section along the steel drive shaft of a ship are calculated to be a torque of 2650 lb · ft, a bending moment of 2800 lb · ft, and an axial thrust of 3700 lb. If the yield points for tension and shear are $\sigma_Y = 100$ ksi and $\tau_Y = 50$ ksi, respectively, determine the required diameter of the shaft using the maximum-distortion-energy theory.

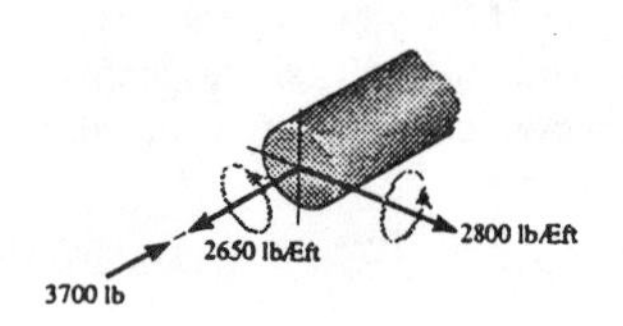

Normal Stress and Shear Stress : The critical point is at the bottom of the circular section.

$$\sigma = \frac{N}{A} \pm \frac{Mc}{I}$$

$$= \frac{-3.70}{\frac{\pi}{4}d^2} - \frac{2.80(12)\left(\frac{d}{2}\right)}{\frac{\pi}{4}\left(\frac{d}{2}\right)^4}$$

$$= -\left(\frac{14.8d + 1075.2}{\pi d^3}\right)$$

$$\tau = \frac{Tc}{J} = \frac{2.65(12)\left(\frac{d}{2}\right)}{\frac{\pi}{2}\left(\frac{d}{2}\right)^4} = \frac{508.8}{\pi d^3}$$

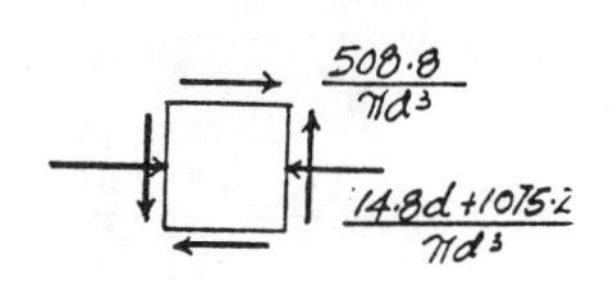

In - Plane Principal Stresses : Applying Eq.9 – 5 with $\sigma_y = 0$, $\tau_{xy} = \frac{508.8}{\pi d^3}$, and $\sigma_x = -\left(\frac{14.8d + 1075.2}{\pi d^3}\right)$.

$$\sigma_{1,2} = \frac{\sigma_x + \sigma_y}{2} \pm \sqrt{\left(\frac{\sigma_x - \sigma_y}{2}\right)^2 + \tau_{xy}^2}$$

$$= -\left(\frac{7.40d + 537.6}{\pi d^3}\right) \pm \sqrt{\left(\frac{7.40d + 537.6}{\pi d^3}\right)^2 + \left(\frac{508.8}{\pi d^3}\right)^2}$$

Maximum Distortion Energy Theory : Let $a = -\left(\frac{7.40d + 537.6}{\pi d^3}\right)$ and $b = \sqrt{\left(\frac{7.40d + 537.6}{\pi d^3}\right)^2 + \left(\frac{508.8}{\pi d^3}\right)^2}$, then $\sigma_1^2 = a^2 + b^2 + 2ab$, $\sigma_1 \sigma_2 = a^2 - b^2$, $\sigma_2^2 = a^2 + b^2 - 2ab$, and $\sigma_1^2 - \sigma_1\sigma_2 + \sigma_2^2 = 3b^2 + a^2$

$$\sigma_1^2 - \sigma_1\sigma_2 + \sigma_2^2 = \sigma_Y^2$$

$$3\left[\left(\frac{7.40d + 537.6}{\pi d^3}\right)^2 + \left(\frac{508.8}{\pi d^3}\right)^2\right] + \left(\frac{7.40d + 537.6}{\pi d^3}\right)^2 = 100^2$$

$$10000\pi^2 d^6 - 219.04d^2 - 31825.93d - 1932687.36 = 0$$

Solving by trial and error,

$d = 1.6491932$ in. $= 1.65$ in. **Ans**

***10–96.** The cast iron cylinder having a diameter of 100 mm is subjected to a torque of 600 N · m and an axial compressive force of 15 kN. Determine if it fails according to the maximum-normal-stress theory. The ultimate stress of the cast iron is $\sigma_{ult} = 170$ MPa.

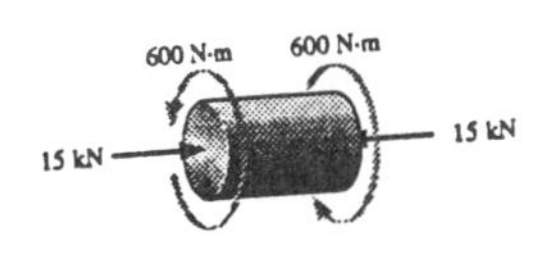

Normal Stress and Shear Stress :

$$\sigma = \frac{N}{A} = \frac{-15(10^3)}{\frac{\pi}{4}(0.1^2)} = -1.910 \text{ MPa}$$

$$\tau = \frac{Tc}{J} = \frac{600(0.05)}{\frac{\pi}{2}(0.05^4)} = 3.056 \text{ MPa}$$

In-Plane Principal Stress : Applying Eq. 9 – 5 with $\sigma_x = -1.910$ MPa, $\sigma_y = 0$, and $\tau_{xy} = 3.056$ MPa,

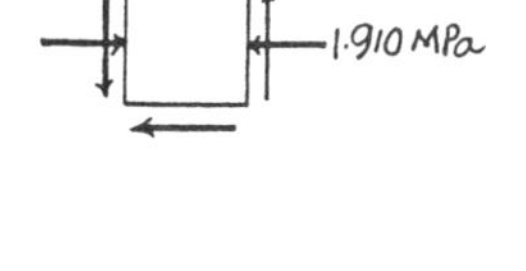

$$\sigma_{1,2} = \frac{\sigma_x + \sigma_y}{2} \pm \sqrt{\left(\frac{\sigma_x - \sigma_y}{2}\right)^2 + \tau_{xy}^2}$$

$$= \frac{-1.910 + 0}{2} \pm \sqrt{\left(\frac{-1.910 - 0}{2}\right)^2 + 3.056^2}$$

$$= -0.955 \pm 3.202$$

$\sigma_1 = 2.247$ MPa $\qquad \sigma_2 = -4.156$ MPa

Maximum Normal Stress Theory :

$$|\sigma_1| = 2.247 \text{ MPa} < \sigma_{ult} = 170 \text{ MPa}$$
$$|\sigma_2| = 4.156 \text{ MPa} < \sigma_{ult} = 170 \text{ MPa}$$

Based on the result obtained above, **the material does not fail according to the maximum normal stress theory.** **Ans**

10–97. Cast iron when tested in tension and compression has an ultimate strength of $(\sigma_{ult})_t = 280$ MPa and $(\sigma_{ult})_c = 420$ MPa, respectively. Also, when subjected to pure torsion it can sustain an ultimate shear stress of $\tau_{ult} = 168$ MPa. Plot the Mohr's circles for each case and establish the failure envelope. If a part made of this material is subjected to the state of plane stress shown, determine if it fails according to Mohr's failure criterion.

In-Plane Principal Stress : Applying Eq. 9 – 5 with $\sigma_x = 240$ MPa, $\sigma_y = -180$ MPa, and $\tau_{xy} = 70$ MPa,

$$\sigma_{1,2} = \frac{\sigma_x + \sigma_y}{2} \pm \sqrt{\left(\frac{\sigma_x - \sigma_y}{2}\right)^2 + \tau_{xy}^2}$$

$$= \frac{240 + (-180)}{2} \pm \sqrt{\left[\frac{240 - (-180)}{2}\right]^2 + 70^2}$$

$$= 30 \pm 221.36$$

$\sigma_1 = 251$ MPa $\qquad \sigma_2 = -191$ MPa

Mohr's Failure Criteria : As shown in Fig. (b), the coordinates of principal stress, which is represented by point A, are located outside the shaded region. Hence, **the material fails according to Mohr's failure criteria.** **Ans**

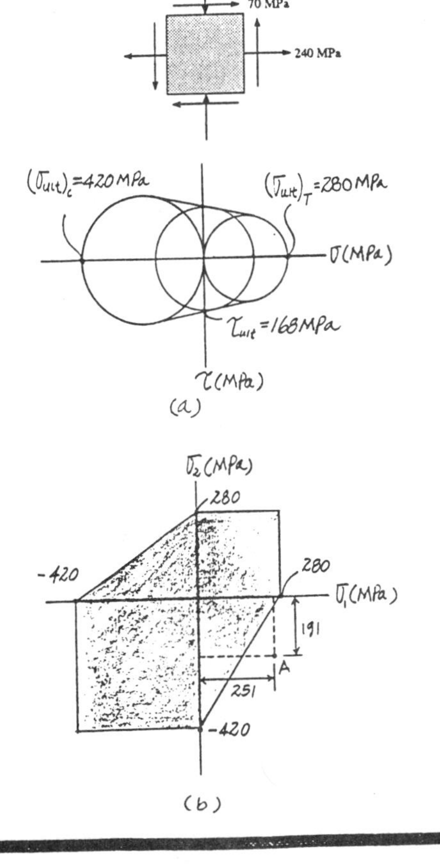

10–98. The state of strain at a point on the bearing has components of $\epsilon_x = 350(10^{-6})$, $\epsilon_y = -860(10^{-6})$, and $\gamma_{xy} = 250(10^{-6})$. Use the strain-transformation equations to determine the equivalent in-plane strains on an element oriented at an angle of $\theta = 45°$ clockwise from the original position. Sketch the deformed element within the x–y plane due to these strains.

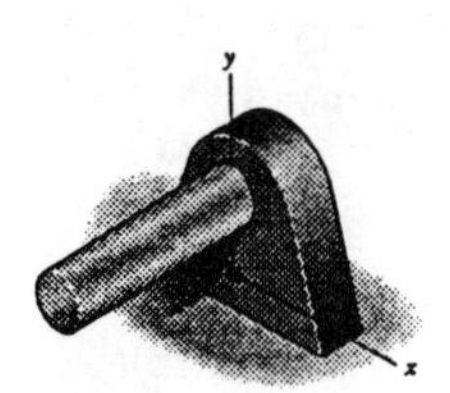

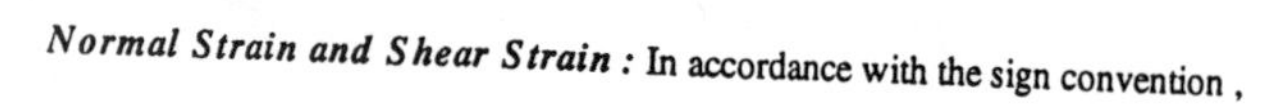

Normal Strain and Shear Strain : In accordance with the sign convention ,

$$\varepsilon_x = 350(10^{-6}) \qquad \varepsilon_y = -860(10^{-6}) \qquad \gamma_{xy} = 250(10^{-6})$$
$$\theta = -45°$$

Strain Transformation Equations : Applying Eq. 10 – 5, 10 – 6, and 10 – 7,

$$\varepsilon_{x'} = \frac{\varepsilon_x + \varepsilon_y}{2} + \frac{\varepsilon_x - \varepsilon_y}{2}\cos 2\theta + \frac{\gamma_{xy}}{2}\sin 2\theta$$
$$= \left[\frac{350+(-860)}{2} + \frac{350-(-860)}{2}\cos(-90°) + \frac{250}{2}\sin(-90°)\right](10^{-6})$$
$$= -380(10^{-6}) \qquad \textbf{Ans}$$

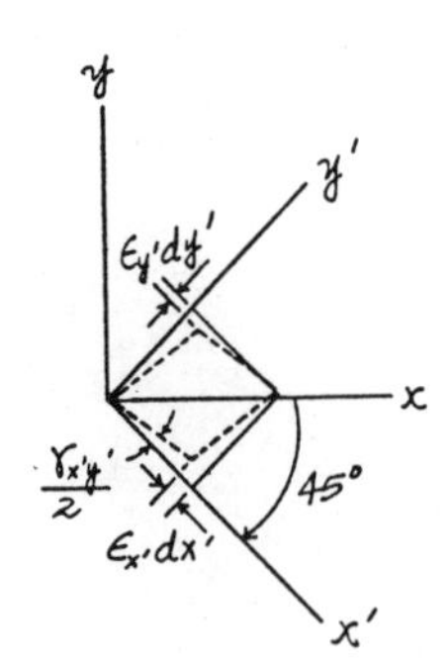

$$\frac{\gamma_{x'y'}}{2} = -\frac{\varepsilon_x - \varepsilon_y}{2}\sin 2\theta + \frac{\gamma_{xy}}{2}\cos 2\theta$$
$$\gamma_{x'y'} = \{-[350-(-860)]\sin(-90°) + 250\cos(-90°)\}(10^{-6})$$
$$= 1210(10^{-6}) \qquad \textbf{Ans}$$

$$\varepsilon_{y'} = \frac{\varepsilon_x + \varepsilon_y}{2} - \frac{\varepsilon_x - \varepsilon_y}{2}\cos 2\theta - \frac{\gamma_{xy}}{2}\sin 2\theta$$
$$= \left[\frac{350+(-860)}{2} - \frac{350-(-860)}{2}\cos(-90°) - \frac{250}{2}\sin(-90°)\right](10^{-6})$$
$$= -130(10^{-6}) \qquad \textbf{Ans}$$

10–99. Derive an expression for an equivalent bending moment M_e that if applied alone to a solid bar with a circular cross section would cause the same energy of distortion as the combination of an applied bending moment M and torque T.

In - Plane Principal Stress : In the case of pure bending, $\sigma_1 = \frac{M_e c}{I}$ and $\sigma_2 = 0$.

Energy of Distortion :

$$u_d = \frac{1+\nu}{3E}(\sigma_1^2 - \sigma_1\sigma_2 + \sigma_2^2)$$
$$(u_d)_1 = \frac{1+\nu}{3E}\left[\left(\frac{M_e c}{I}\right)^2 - 0 + 0\right]$$
$$= \frac{1+\nu}{3E}\left(\frac{M_e c}{I}\right)^2$$

In - Plane Principal Stress : In the case of combined bending stress and torsional shear stress, $\sigma = \frac{Mc}{I}$ and $\tau = \frac{Tc}{J}$.

$$\sigma_{1,2} = \frac{\sigma_x + \sigma_y}{2} \pm \sqrt{\left(\frac{\sigma_x - \sigma_y}{2}\right)^2 + \tau_{xy}^2}$$
$$\sigma_{1,2} = \frac{\sigma + 0}{2} \pm \sqrt{\left(\frac{\sigma - 0}{2}\right)^2 + \tau^2}$$
$$\sigma_1 = \frac{\sigma}{2} + \sqrt{\frac{\sigma^2}{4} + \tau^2} \qquad \sigma_2 = \frac{\sigma}{2} - \sqrt{\frac{\sigma^2}{4} + \tau^2}$$

Energy of Distortion : Let $a = \frac{\sigma}{2}$ and $b = \sqrt{\frac{\sigma^2}{4} + \tau^2}$, then $\sigma_1^2 = a^2 + b^2 + 2ab$, $\sigma_1\sigma_2 = a^2 - b^2$, $\sigma_2^2 = a^2 + b^2 - 2ab$, and $\sigma_1^2 - \sigma_1\sigma_2 + \sigma_2^2 = 3b^2 + a^2$

$$u_d = \frac{1+\nu}{3E}(\sigma_1^2 - \sigma_1\sigma_2 + \sigma_2^2)$$
$$(u_d)_2 = \frac{1+\nu}{3E}(3b^2 + a^2)$$
$$= \frac{1+\nu}{3E}(\sigma^2 + 3\tau^2)$$
$$= \frac{c^2(1+\nu)}{3E}\left(\frac{M^2}{I^2} + \frac{3T^2}{J^2}\right)$$

Require,

$$(u_d)_1 = (u_d)_2$$
$$\frac{1+\nu}{3E}\left(\frac{M_e c}{I}\right)^2 = \frac{c^2(1+\nu)}{3E}\left(\frac{M^2}{I^2} + \frac{3T^2}{J^2}\right)$$
$$M_e = \sqrt{M^2 + \frac{3T^2I^2}{J^2}}$$

However, for the circular shaft $\frac{I^2}{J^2} = \frac{\left(\frac{\pi}{4}c^4\right)^2}{\left(\frac{\pi}{2}c^4\right)^2} = \frac{1}{4}$, then

$$M_e = \sqrt{M^2 + \frac{3}{4}T^2} \qquad \textbf{Ans}$$

***10–100.** If a solid shaft having a diameter d is subjected to a torque **T** and moment **M,** show that by the maximum-shear-stress theory the maximum allowable shear stress is $\tau_{allow} = (16/\pi d^3)\sqrt{M^2 + T^2}$. Assume the principal stresses to be of opposite algebraic signs.

Section Properties :

$$I = \frac{\pi}{4}\left(\frac{d}{2}\right)^4 = \frac{\pi d^4}{64} \qquad J = \frac{\pi}{2}\left(\frac{d}{2}\right)^4 = \frac{\pi d^4}{32}$$

Normal Stress and Shear Stress : Applying the flexure and torsion formulas.

$$\sigma = \frac{Mc}{I} = \frac{M(\frac{d}{2})}{\frac{\pi d^4}{64}} = \frac{32M}{\pi d^3}$$

$$\tau = \frac{Tc}{J} = \frac{T(\frac{d}{2})}{\frac{\pi d^4}{32}} = \frac{16T}{\pi d^3}$$

In-Plane Principal Stress :

$$\sigma_{1,2} = \frac{\sigma_x + \sigma_y}{2} \pm \sqrt{\left(\frac{\sigma_x - \sigma_y}{2}\right)^2 + \tau_{xy}^2}$$

$$= \frac{16M}{\pi d^3} \pm \sqrt{\left(\frac{16M}{\pi d^3}\right)^2 + \left(\frac{16T}{\pi d^3}\right)^2}$$

$$= \frac{16M}{\pi d^3} \pm \frac{16}{\pi d^3}\sqrt{M^2 + T^2}$$

Maximum Shear Stress Theory : Assume σ_1 and σ_2 have opposite signs. Hence,

$$\tau_{allow} = \frac{|\sigma_1 - \sigma_2|}{2}$$

$$= \frac{2\left(\frac{16}{\pi d^3}\sqrt{M^2 + T^2}\right)}{2}$$

$$= \frac{16}{\pi d^3}\sqrt{M^2 + T^2} \qquad (Q.E.D.)$$

10–101. The state of strain at a point on the arm has components of $\epsilon_x = 250(10^{-6})$, $\epsilon_y = -450(10^{-6})$, $\gamma_{xy} = -825(10^{-6})$. Use the strain-transformation equations to determine (a) the in-plane principal strains and (b) the maximum in-plane shear strain and average normal strain. In each case specify the orientation of the element and show how the strains deform the element within the x–y plane.

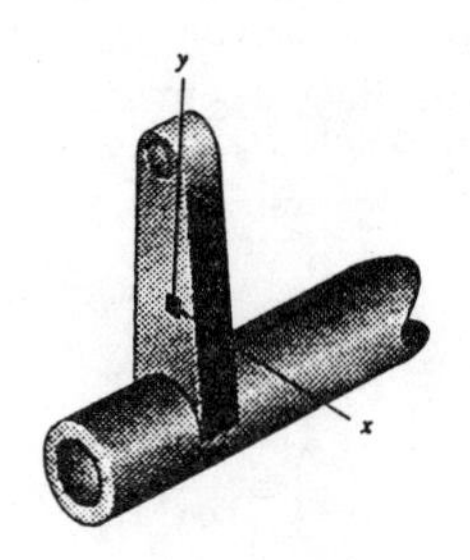

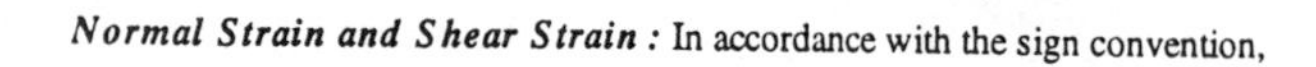

Normal Strain and Shear Strain : In accordance with the sign convention,

$$\varepsilon_x = 250(10^{-6}) \qquad \varepsilon_y = -450(10^{-6}) \qquad \gamma_{xy} = -825(10^{-6})$$

a)

In - Plane Principal Strain : Applying Eq. 10 – 9,

$$\varepsilon_{1,2} = \frac{\varepsilon_x + \varepsilon_y}{2} \pm \sqrt{\left(\frac{\varepsilon_x - \varepsilon_y}{2}\right)^2 + \left(\frac{\gamma_{xy}}{2}\right)^2}$$

$$= \left[\frac{250 + (-450)}{2} \pm \sqrt{\left[\frac{250 - (-450)}{2}\right]^2 + \left(\frac{-825}{2}\right)^2}\right](10^{-6})$$

$$= -100 \pm 540.98$$

$$\varepsilon_1 = 441(10^{-6}) \qquad \varepsilon_2 = -641(10^{-6}) \qquad \textbf{Ans}$$

Orientation of Principal Strain : Applying Eq. 10 – 8,

$$\tan 2\theta_p = \frac{\gamma_{xy}}{\varepsilon_x - \varepsilon_y} = \frac{-825(10^{-6})}{[250 - (-450)](10^{-6})} = -1.1786$$

$$\theta_p = -24.84° \quad \text{and} \quad 65.16°$$

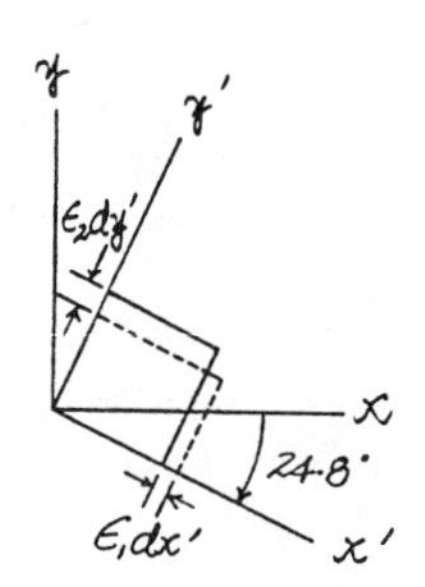

Use Eq. 10 – 5 to determine which principal strain deforms the element in the x' direction with $\theta = -24.84°$.

$$\varepsilon_{x'} = \frac{\varepsilon_x + \varepsilon_y}{2} + \frac{\varepsilon_x - \varepsilon_y}{2}\cos 2\theta + \frac{\gamma_{xy}}{2}\sin 2\theta$$

$$= \left[\frac{250 + (-450)}{2} + \frac{250 - (-450)}{2}\cos(-49.68°) + \frac{-825}{2}\sin(-49.68°)\right](10^{-6})$$

$$= 441(10^{-6}) = \varepsilon_1$$

Hence, $\theta_{p_1} = -24.8°$ and $\theta_{p_2} = 65.2°$ **Ans**

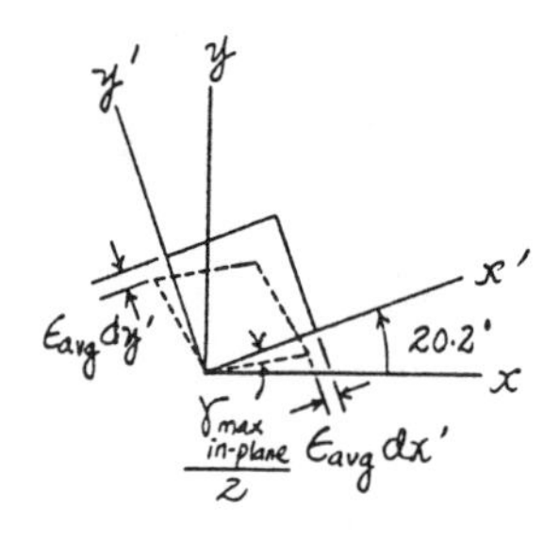

b)

Maximum In - Plane Shear Strain : Applying Eq. 10 – 11,

$$\frac{\gamma_{\substack{max \\ in\text{-}plane}}}{2} = \sqrt{\left(\frac{\varepsilon_x - \varepsilon_y}{2}\right)^2 + \left(\frac{\gamma_{xy}}{2}\right)^2}$$

$$\gamma_{\substack{max \\ in\text{-}plane}} = 2\left[\sqrt{\left[\frac{250 - (-450)}{2}\right]^2 + \left(\frac{-825}{2}\right)^2}\right](10^{-6})$$

$$= 1082(10^{-6}) = 1.08(10^{-3}) \qquad \textbf{Ans}$$

Orientation of Maximum In - Plane Shear Strain : Applying Eq. 10 – 10,

$$\tan 2\theta_s = -\frac{\varepsilon_x - \varepsilon_y}{\gamma_{xy}} = -\frac{250 - (-450)}{-825} = 0.8485$$

$$\theta_s = 20.2° \quad \text{and} \quad -69.8° \qquad \textbf{Ans}$$

The proper sign of $\gamma_{\substack{max \\ in\text{-}plane}}$ can be determined by substituting $\theta = 20.2°$ into Eq. 10 – 6.

$$\frac{\gamma_{x'y'}}{2} = -\frac{\varepsilon_x - \varepsilon_y}{2}\sin 2\theta + \frac{\gamma_{xy}}{2}\cos 2\theta$$

$$\gamma_{x'y'} = \{-[250 - (-450)]\sin 40.4° + (-825)\cos 40.4°\}(10^{-6})$$

$$= -1082(10^{-6})$$

Average Normal Strain : Applying Eq. 10 – 12,

$$\varepsilon_{avg} = \frac{\varepsilon_x + \varepsilon_y}{2} = \left[\frac{250 + (-450)}{2}\right](10^{-6}) = -100(10^{-6}) \qquad \textbf{Ans}$$

10–102. The principal stresses at a point are shown in the figure. If the material is nylon for which E_n = 2.5 GPa and ν_n = 0.4, determine the principal strains.

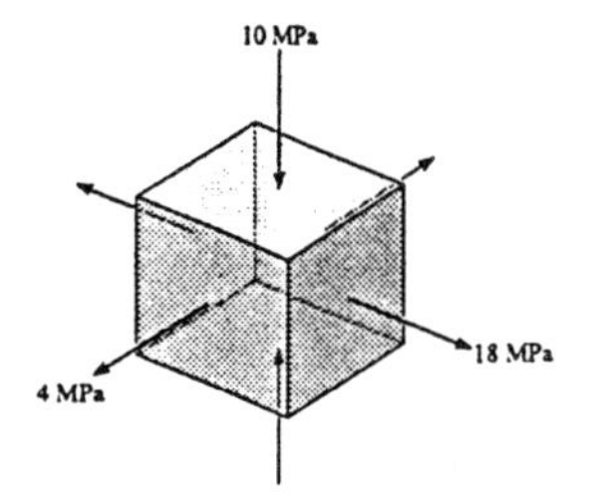

Normal Strain : Applying the generalized Hooke's Law with $\sigma_{max} = 18$ MPa, $\sigma_{int} = 4$ MPa, and $\sigma_{min} = -10$ MPa,

$$\varepsilon_{max} = \frac{1}{E}[\sigma_{max} - \nu(\sigma_{int} + \sigma_{min})]$$
$$= \frac{1}{2.5(10^9)}[18 - 0.4(4 - 10)](10^6)$$
$$= 8.16(10^{-3}) \qquad \textbf{Ans}$$

$$\varepsilon_{int} = \frac{1}{E}[\sigma_{int} - \nu(\sigma_{max} + \sigma_{min})]$$
$$= \frac{1}{2.5(10^9)}[4 - 0.4(18 - 10)](10^6)$$
$$= 0.320(10^{-3}) \qquad \textbf{Ans}$$

$$\varepsilon_{min} = \frac{1}{E}[\sigma_{min} - \nu(\sigma_{max} + \sigma_{int})]$$
$$= \frac{1}{2.5(10^9)}[-10 - 0.4(18 + 4)](10^6)$$
$$= -7.52(10^{-3}) \qquad \textbf{Ans}$$

10–103. A thin-walled spherical pressure vessel has an inner radius r, thickness t, and is subjected to an internal pressure p. If the material constants are E and ν, determine the strain in the circumferential direction in terms of the stated parameters.

Normal Stress : By ignoring the radial stress component, that is $\sigma_3 = 0$, the problem becomes a plane stress problem.

$$\sigma_1 = \sigma_2 = \frac{pr}{2t}$$

Normal Strain : Applying the generalized Hooke's law.

$$\varepsilon_1 = \frac{1}{E}[\sigma_1 - \nu(\sigma_2 + \sigma_3)]$$
$$= \frac{1}{E}\left(\frac{pr}{2t} - \frac{\nu pr}{2t}\right)$$
$$= \frac{pr}{2Et}(1 - \nu) \qquad \textbf{Ans}$$

***10–104.** Determine the bulk modulus for each of the following materials: (a) rubber, E_r = 0.4 ksi, ν_r = 0.48, and (b) glass, $E_g = 8(10^3)$ ksi, ν_g = 0.24.

Bulk Modulus : Applying Eq. 10 – 25,

$$k_r = \frac{E_r}{3(1 - 2\nu_r)} = \frac{0.4}{3[1 - 2(0.48)]} = 3.33 \text{ ksi} \qquad \textbf{Ans}$$

$$k_g = \frac{E_g}{3(1 - 2\nu_g)} = \frac{8(10^3)}{3[1 - 2(0.24)]} = 5.13(10^3) \text{ ksi} \qquad \textbf{Ans}$$

10–105. The 60° strain rosette is mounted on a beam. The following readings are obtained for each gauge: $\epsilon_a = 600(10^{-6})$, $\epsilon_b = -700(10^{-6})$, and $\epsilon_c = 350(10^{-6})$. Determine (a) the in-plane principal strains and (b) the maximum in-plane shear strain and average normal strain. In each case show the deformed element due to these strains.

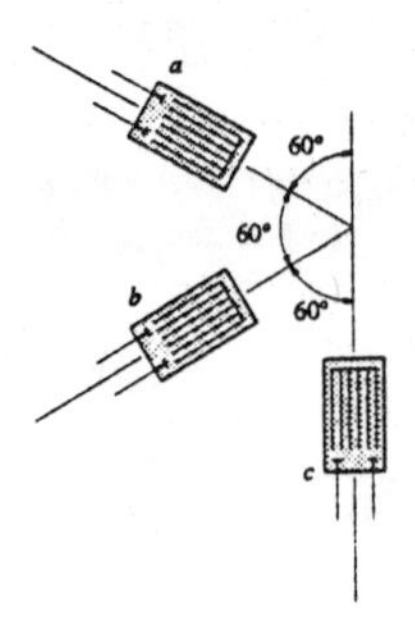

Strain Rosettes (60°) : Applying Eq. 10–15 with $\varepsilon_a = 600(10^{-6})$, $\varepsilon_b = -700(10^{-6})$, $\varepsilon_c = 350(10^{-6})$, $\theta_a = 150°$, $\theta_b = -150°$ and $\theta_c = -90°$,

$$350(10^{-6}) = \varepsilon_x \cos^2(-90°) + \varepsilon_y \sin^2(-90°) + \gamma_{xy} \sin(-90°) \cos(-90°)$$

$$\varepsilon_y = 350(10^{-6})$$

$$600(10^{-6}) = \varepsilon_x \cos^2 150° + 350(10^{-6}) \sin^2 150° + \gamma_{xy} \sin 150° \cos 150°$$

$$512.5(10^{-6}) = 0.75\,\varepsilon_x - 0.4330\,\gamma_{xy} \qquad [1]$$

$$-700(10^{-6}) = \varepsilon_x \cos^2(-150°) + 350(10^{-6}) \sin^2(-150°) + \gamma_{xy} \sin(-150°) \cos(-150°)$$

$$-787.5(10^{-6}) = 0.75\varepsilon_x + 0.4330\,\gamma_{xy} \qquad [2]$$

Solving Eq. [1] and [2] yields $\quad \varepsilon_x = -183.33(10^{-6}) \quad \gamma_{xy} = -1501.11(10^{-6})$

Construction of the Circle : With $\varepsilon_x = -183.33(10^{-6})$, $\varepsilon_y = 350(10^{-6})$, and $\frac{\gamma_{xy}}{2} = -750.56(10^{-6})$,

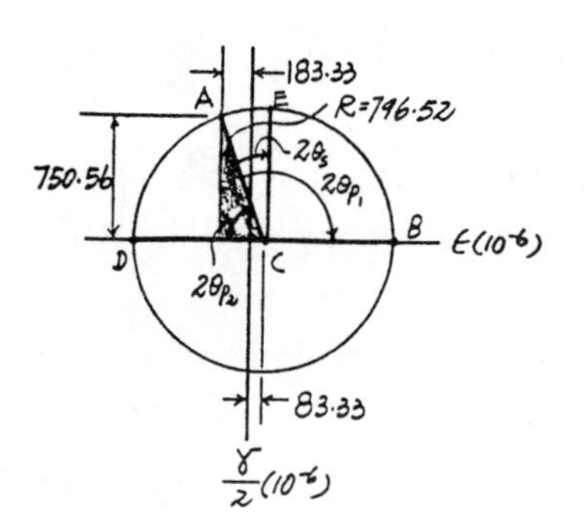

$$\varepsilon_{avg} = \frac{\varepsilon_x + \varepsilon_y}{2} = \left(\frac{-183.33 + 350}{2}\right)(10^{-6}) = 83.3(10^{-6}) \qquad \textbf{Ans}$$

The coordinates for reference points A and C are

$$A(-183.33,\ -750.56)(10^{-6}) \qquad C(83.33,\ 0)(10^{-6})$$

The radius of the circle is

$$R = \left(\sqrt{(183.33 + 83.33)^2 + 750.56^2}\right)(10^{-6}) = 796.52(10^{-6})$$

a)

In - Plane Principal Strain : The coordinates of points B and D represent ε_1 and ε_2, respectively.

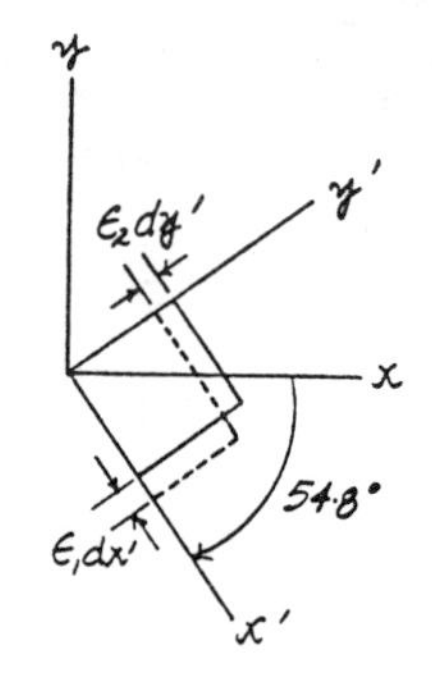

$$\varepsilon_1 = (83.33 + 796.52)(10^{-6}) = 880(10^{-6}) \qquad \textbf{Ans}$$

$$\varepsilon_2 = (83.33 - 796.52)(10^{-6}) = -713(10^{-6}) \qquad \textbf{Ans}$$

Orientation of Principal Strain : From the circle,

$$\tan 2\theta_{p_2} = \frac{750.56}{183.33 + 83.33} = 2.8145 \qquad 2\theta_{p_2} = 70.44°$$

$$2\theta_{p_1} = 180° - 2\theta_{p_2}$$

$$\theta_{p_1} = \frac{180° - 70.44°}{2} = 54.8° \quad (\textit{Clockwise}) \qquad \textbf{Ans}$$

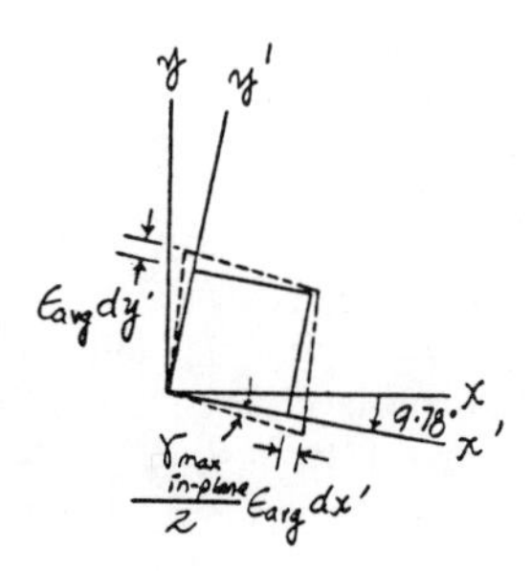

b)

Maximum In - Plane Shear Strain : Represented by the coordinates of point E on the circle.

$$\frac{\gamma_{\substack{max \\ in\text{-}plane}}}{2} = -R = -796.52(10^{-6})$$

$$\gamma_{\substack{max \\ in\text{-}plane}} = -1593(10^{-6}) \qquad \textbf{Ans}$$

Orientation of Maximum In - Plane Shear Strain : From the circle,

$$\tan 2\theta_s = \frac{183.33 + 83.33}{750.56} = 0.3553$$

$$\theta_s = 9.78° \quad (\textit{Clockwise}) \qquad \textbf{Ans}$$

11–1. The wooden beam has a rectangular cross section and is used to support a load of 1200 lb. If the allowable bending stress is $\sigma_{allow} = 2$ ksi and the allowable shear stress is $\tau_{allow} = 750$ psi, determine the height of the cross section to the nearest $\frac{1}{4}$ in. if it is to be rectangular and have a width of 3 in. Assume the supports at A and B only exert vertical reactions on the beam.

Bending Stress : From the moment diagram, $M_{max} = 2.88$ kip · ft. Assume bending controls the design. Applying the flexure formula.

$$\sigma_{allow} = \frac{M_{max}\,c}{I}$$

$$2 = \frac{2.88(12)\left(\frac{h}{2}\right)}{\frac{1}{12}(3)h^3}$$

$$h = 5.879 \text{ in.}$$

Use $\quad h = 6$ in. **Ans**

4ft 6ft
0.720 Kip 1.2 Kip 0.480 Kip

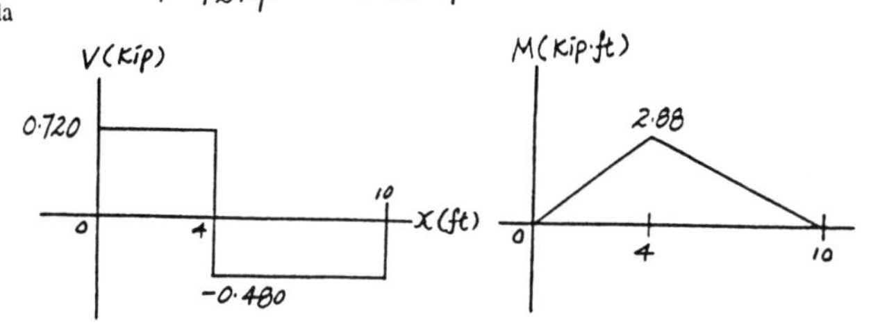

Shear Stress : Provide a shear stress check using the shear formula with $I = \frac{1}{12}(3)\left(6^3\right) = 54.0$ in^4 and $Q_{max} = 1.5(3)(3) = 13.5$ in^3. From the shear diagram, $V_{max} = 0.720$ kip.

$$\tau_{max} = \frac{V_{max}\,Q_{max}}{It} = \frac{0.720(13.5)}{54.0(3)} = 0.060 \text{ ksi} < \tau_{allow} = 0.750 \text{ ksi } (O.K!)$$

11–2. Solve Prob. 11–1 if the cross section has an unknown width but is to be square, i.e., $h = b$.

Bending Stress : From the moment diagram, $M_{max} = 2.88$ kip · ft. Assume bending controls the design. Applying the flexure formula,

$$\sigma_{allow} = \frac{M_{max}\,c}{I}$$

$$2 = \frac{2.88(12)\left(\frac{h}{2}\right)}{\frac{1}{12}h^4}$$

$$h = 4.698 \text{ in.}$$

Use $\quad h = 4\frac{3}{4}$ in. **Ans**

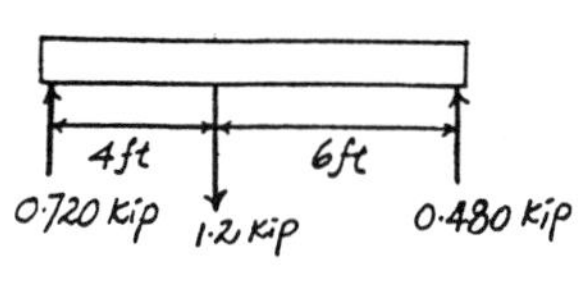

Shear Stress : Provide a shear stress check using the shear formula with $I = \frac{1}{12}\left(4.75^4\right) = 42.42$ in^4 and $Q_{max} = 1.1875(2.375)(4.75) = 13.40$ in^3. From the shear diagram, $V_{max} = 0.720$ kip.

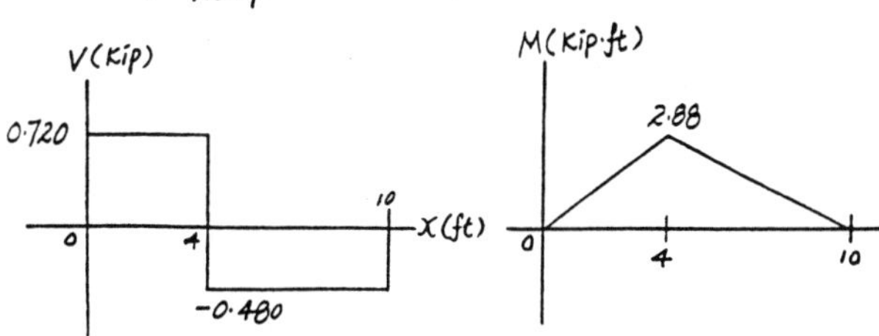

$$\tau_{max} = \frac{V_{max}\,Q_{max}}{It} = \frac{0.720(13.40)}{42.42(4.75)} = 0.0479 \text{ksi} < \tau_{allow} = 0.750 \text{ ksi } (O.K!)$$

11–3. The joists of a floor in a warehouse are to be selected using square timber beams made of oak. If each beam is to be designed to carry 90 lb/ft over a simply supported span of 25 ft, determine the dimension a of its square cross section to the nearest $\frac{1}{4}$ in. The allowable bending stress is $\sigma_{allow} = 4.5$ ksi and the allowable shear stress is $\tau_{allow} = 125$ psi.

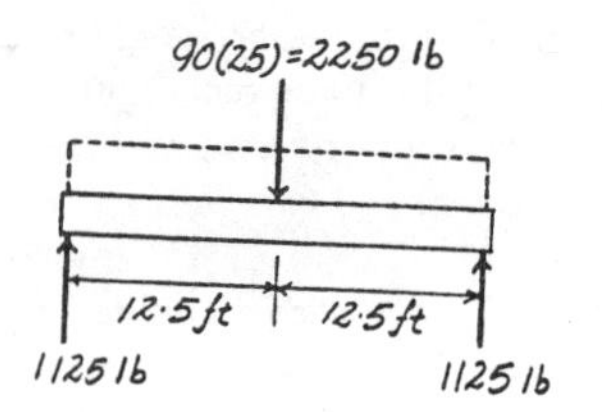

Bending Stress : From the moment diagram, $M_{max} = 7031.25$ lb · ft. Assume bending controls the design. Applying the flexure formula.

$$\sigma_{allow} = \frac{M_{max}\,c}{I}$$

$$4.5(10^3) = \frac{7031.25(12)\left(\frac{a}{2}\right)}{\frac{1}{12}a^4}$$

$$a = 4.827 \text{ in.}$$

Use $a = 5$ in. **Ans**

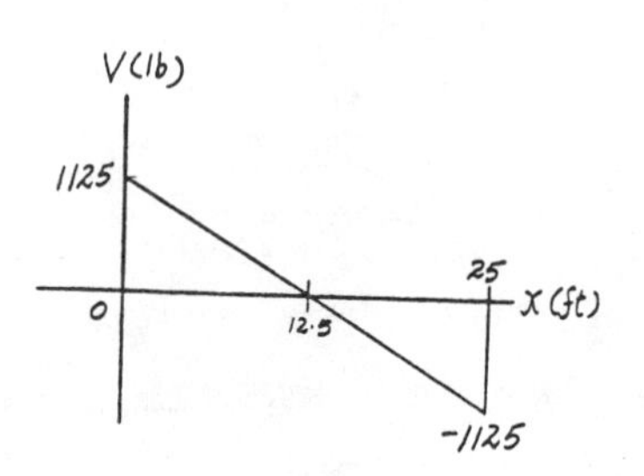

Shear Stress : Provide a shear stress check using the shear formula with $I = \frac{1}{12}(5^4) = 52.083$ in^4 and $Q_{max} = 1.25(2.5)(5) = 15.625$ in^3. From the shear diagram, $V_{max} = 1125$ lb.

$$\tau_{max} = \frac{V_{max}Q_{max}}{It}$$

$$= \frac{1125(15.625)}{52.083(5)}$$

$$= 67.5 \text{ psi} < \tau_{allow} = 125 \text{ psi } (O.K!)$$

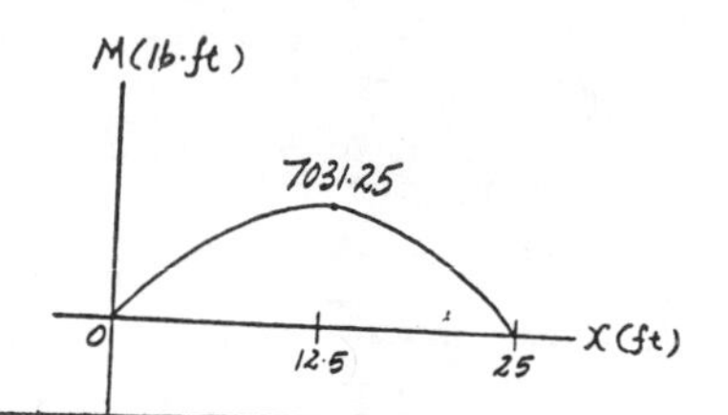

***11–4.** Select the lightest-weight steel wide-flange beam from Appendix *B* that will safely support the machine loading shown. The allowable bending stress is $\sigma_{allow} = 24$ ksi and the allowable shear stress is $\tau_{allow} = 14$ ksi.

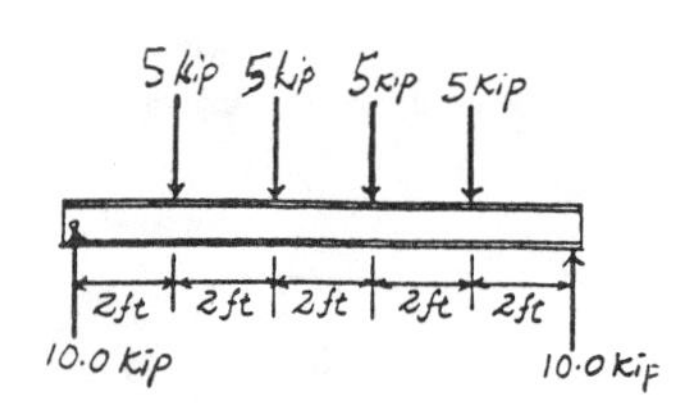

Bending Stress : From the moment diagram, $M_{max} = 30.0$ kip · ft. Assume bending controls the design. Applying the flexure formula.

$$S_{req'd} = \frac{M_{max}}{\sigma_{allow}}$$

$$= \frac{30.0(12)}{24} = 15.0 \text{ in}^3$$

Select W12×16 $\left(S_x = 17.1 \text{ in}^3,\ d = 11.99 \text{ in.},\ t_w = 0.220 \text{ in.}\right)$

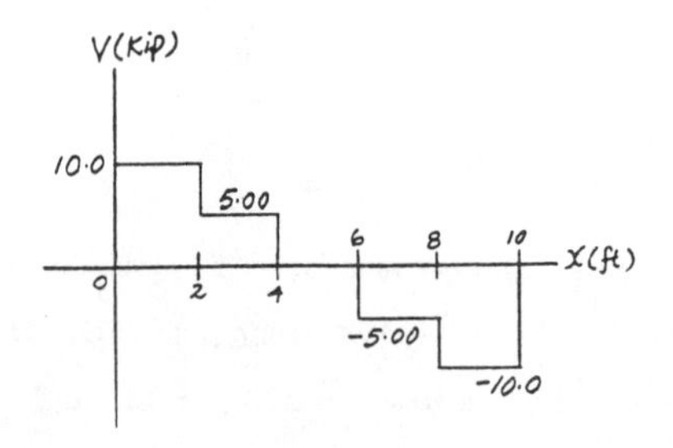

Shear Stress : Provide a shear stress check using $\tau = \frac{V}{t_w d}$ for the W12 × 16 wide-flange section. From the shear diagram, $V_{max} = 10.0$ kip

$$\tau_{max} = \frac{V_{max}}{t_w d}$$

$$= \frac{10.0}{0.220(11.99)}$$

$$= 3.79 \text{ ksi} < \tau_{allow} = 14 \text{ ksi } (O.K!)$$

Hence, **Use** W12×16 **Ans**

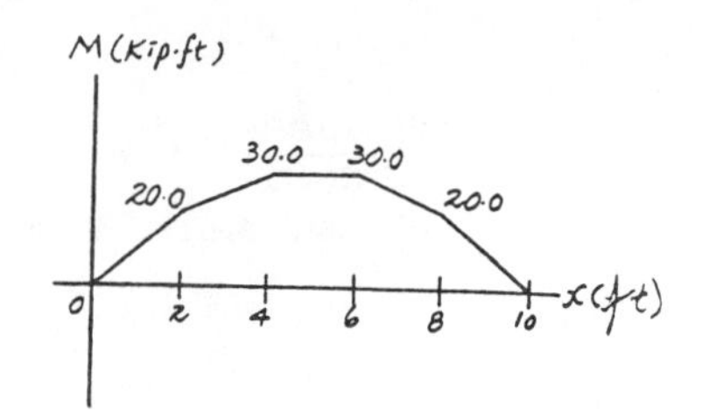

11–5. Select the lightest-weight steel wide-flange beam from Appendix *B* that will safely support the loading shown, where $w = 6$ kip/ft and $P = 5$ kip. The allowable bending stress is $\sigma_{allow} = 24$ ksi, and the allowable shear stress is $\tau_{allow} = 14$ ksi.

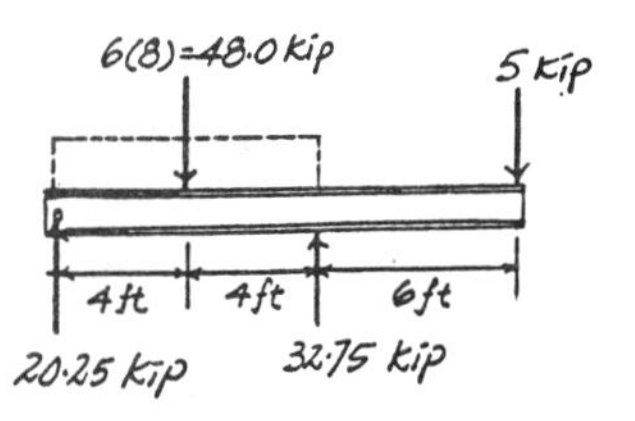

Bending Stress : From the moment diagram, $M_{max} = 34.17$ kip · ft. Assume bending controls the design. Applying the flexure formula,

$$S_{req'd} = \frac{M_{max}}{\sigma_{allow}} = \frac{34.17(12)}{24} = 17.09 \text{ in}^3$$

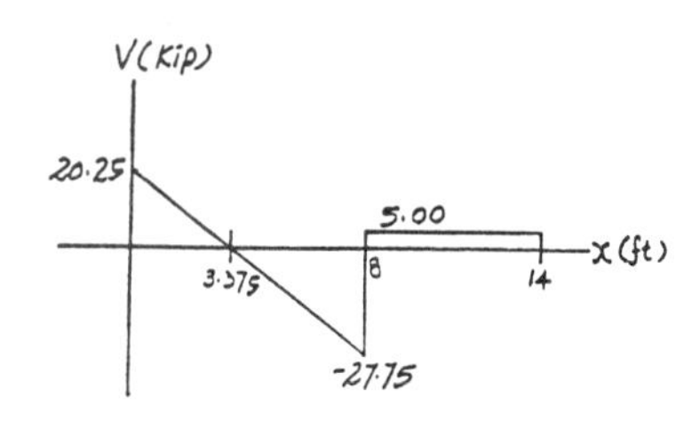

Select W12 × 16 $\left(S_x = 17.1 \text{ in}^3,\ d = 11.99 \text{ in.},\ t_w = 0.220 \text{ in.}\right)$

Shear Stress : Provide a shear stress check using $\tau = \dfrac{V}{t_w d}$ for the W12 × 16 wide-flange section. From the shear diagram, $V_{max} = 27.75$ kip

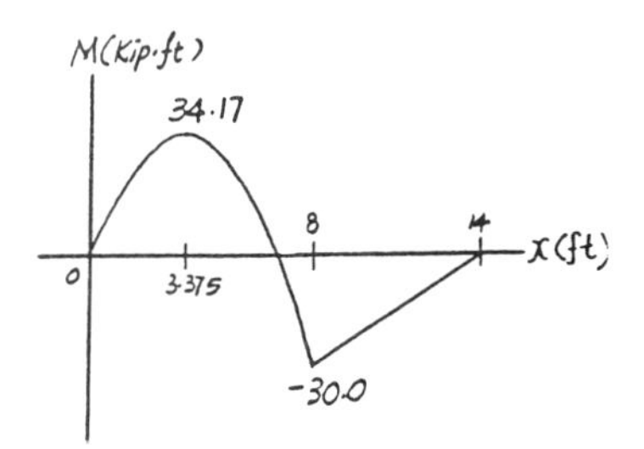

$$\tau_{max} = \frac{V_{max}}{t_w d} = \frac{27.75}{0.220(11.99)} = 10.52 \text{ ksi} < \tau_{allow} = 14 \text{ ksi} \quad (O.K!)$$

Hence, **Use** W12 × 16 **Ans**

11–6. Select the lightest-weight steel wide-flange beam having the shortest height from Appendix *B* that will safely support the loading shown, where $w = 0$ and $P = 10$ kip. The allowable bending stress is $\sigma_{allow} = 24$ ksi, and the allowable shear stress is $\tau_{allow} = 14$ ksi.

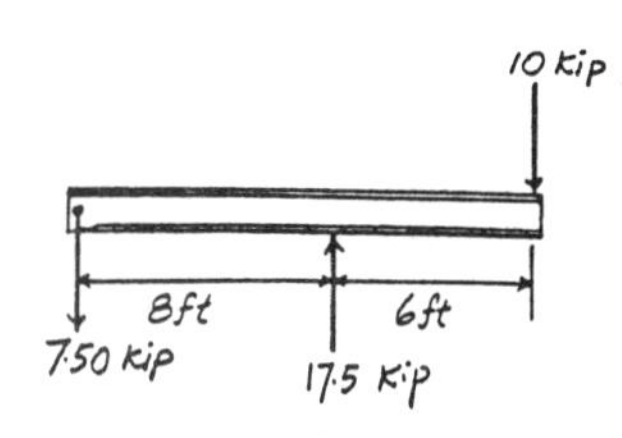

Bending Stress : From the moment diagram, $M_{max} = 60.0$ kip · ft. Assume bending controls the design. Applying the flexure formula,

$$S_{req'd} = \frac{M_{max}}{\sigma_{allow}} = \frac{60.0(12)}{24} = 30.0 \text{ in}^3$$

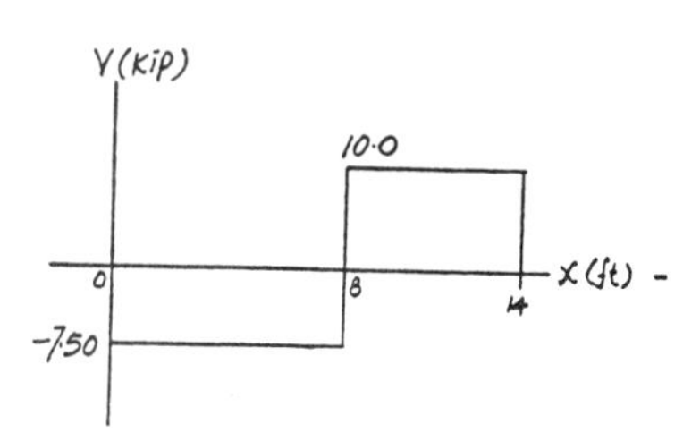

Three choices of wide flange section having the weight 26 lb/ft can be made. They are W12 × 26, W14 × 26, and W16 × 26. However, the shortest is the W12 × 26.

Select W12 × 26 $\left(S_x = 33.4 \text{ in}^3,\ d = 12.22 \text{ in.},\ t_w = 0.230 \text{ in.}\right)$

Shear Stress : Provide a shear stress check using $\tau = \dfrac{V}{t_w d}$ for the W12 × 26 wide-flange section. From the shear diagram, $V_{max} = 10.0$ kip.

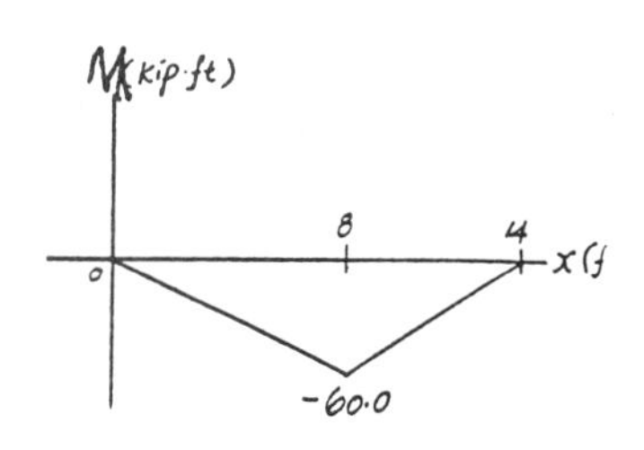

$$\tau_{max} = \frac{V_{max}}{t_w d} = \frac{10.0}{0.230(12.22)} = 3.56 \text{ ksi} < \tau_{allow} = 14 \text{ ksi} \quad (O.K!)$$

Hence, **Use** W12 × 26 **Ans**

11–7. Select the lightest-weight steel wide-flange beam from Appendix *B* that will safely suport the loading shown. The allowable bending stress is $\sigma_{allow} = 24$ ksi and the allowable shear stress is $\tau_{allow} = 14$ ksi.

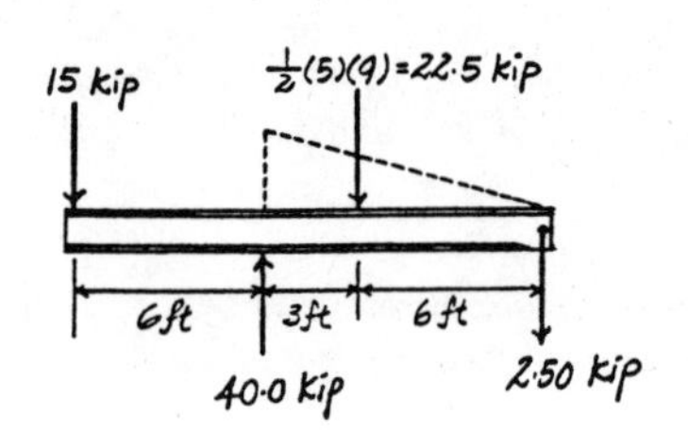

Bending Stress : From the moment diagram, $M_{max} = 90.0$ kip · ft. Assume bending controls the design. Applying the flexure formula.

$$S_{req'd} = \frac{M_{max}}{\sigma_{allow}} = \frac{90.0(12)}{24} = 45.0 \text{ in}^3$$

Select W16×31 $\left(S_x = 47.2 \text{ in}^3,\ d = 15.88 \text{ in.},\ t_w = 0.275 \text{ in.}\right)$

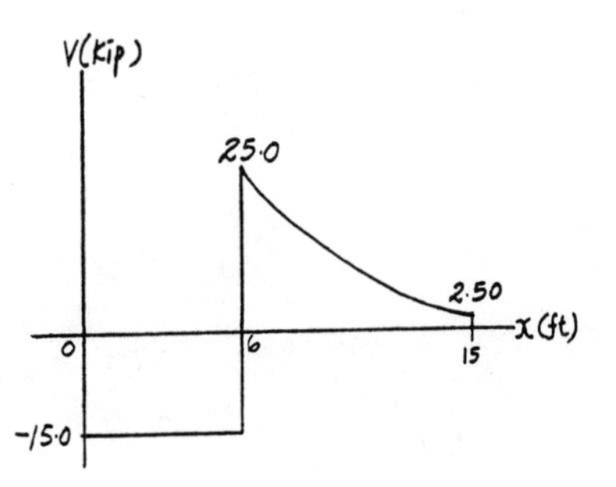

Shear Stress : Provide a shear stress check using $\tau = \dfrac{V}{t_w d}$ for the W16×31 wide flange section. From the shear diagram, $V_{max} = 25.0$ kip.

$$\tau_{max} = \frac{V_{max}}{t_w d} = \frac{25.0}{0.275(15.88)} = 5.72 \text{ ksi} < \tau_{allow} = 14 \text{ ksi } (O.K!)$$

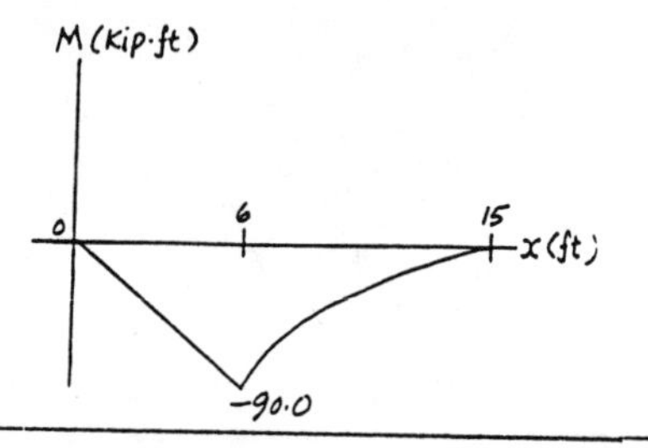

Hence, **Use** W16×31 **Ans**

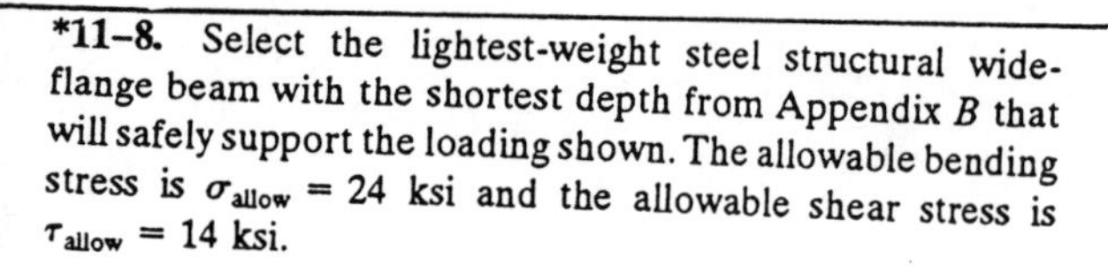

***11–8.** Select the lightest-weight steel structural wide-flange beam with the shortest depth from Appendix *B* that will safely support the loading shown. The allowable bending stress is $\sigma_{allow} = 24$ ksi and the allowable shear stress is $\tau_{allow} = 14$ ksi.

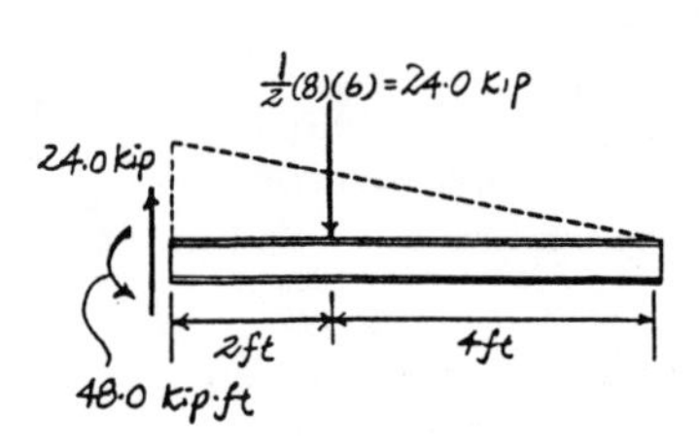

Bending Stress : From the moment diagram, $M_{max} = 48.0$ kip · ft. Assume bending controls the design. Applying the flexure formula.

$$S_{req'd} = \frac{M_{max}}{\sigma_{allow}} = \frac{48.0(12)}{24} = 24.0 \text{ in}^3$$

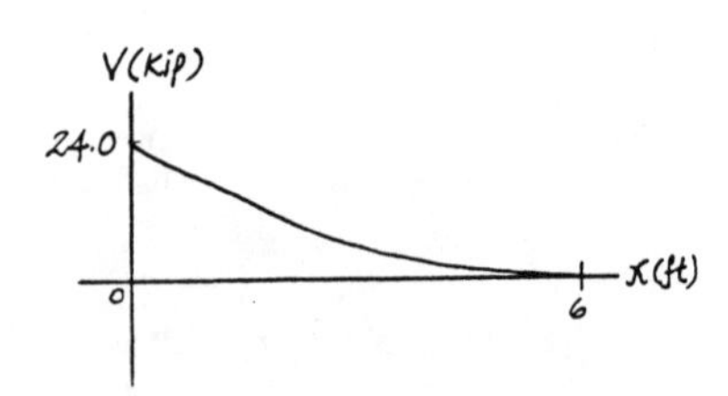

Two choices of wide flange section having the weight 22 lb/ft can be made. They are a W12×22 and W14×22 . However, the W12×22 is the shortest.

Select W12×22 $\left(S_x = 25.4 \text{ in}^3,\ d = 12.31 \text{ in.},\ t_w = 0.260 \text{ in.}\right)$

Shear Stress : Provide a shear stress check using $\tau = \dfrac{V}{t_w d}$ for the W12×22 wide-flange section. From the shear diagram, $V_{max} = 24.0$ kip.

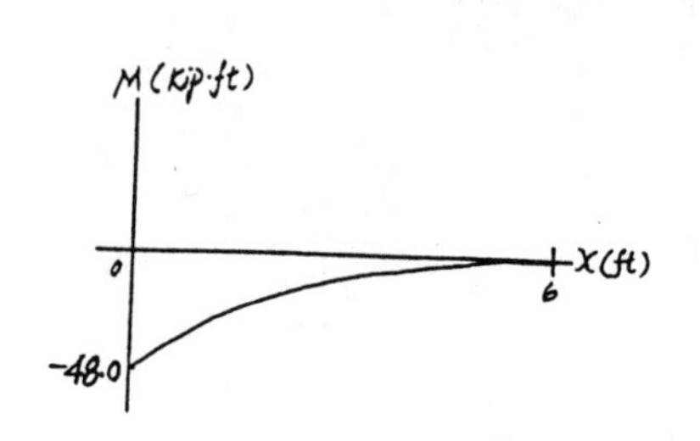

$$\tau_{max} = \frac{V_{max}}{t_w d} = \frac{24.0}{0.260(12.31)} = 7.50 \text{ ksi} < \tau_{allow} = 14 \text{ ksi } (O.K!)$$

Hence, **Use** W12×22 **Ans**

11–9. The simply supported beam is made of timber that has an allowable bending stress of $\sigma_{allow} = 6.5$ MPa and an allowable shear stress of $\tau_{allow} = 500$ kPa. Determine its dimensions if it is to be rectangular and have a height-to-width ratio of $h/b = 1.25$.

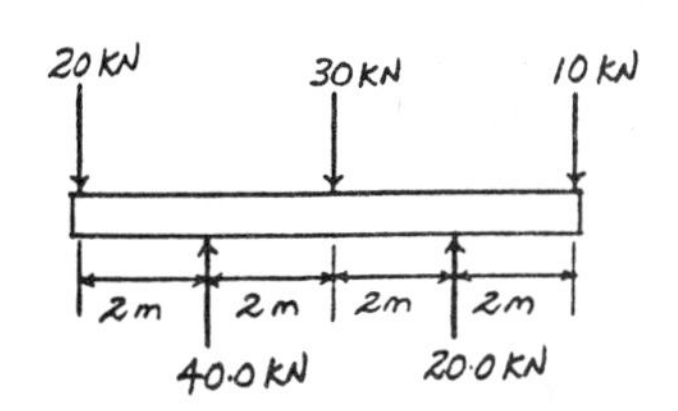

Bending Stress : From the moment diagram. $M_{max} = 40.0$ kN · m. Assume bending controls the design. Applying the flexure formula.

$$\sigma_{allow} = \frac{M_{max}\, c}{I}$$

$$6.5(10^6) = \frac{40.0(10^3)\left(\frac{1.25b}{2}\right)}{\frac{1}{12}(b)(1.25b)^3}$$

$$b = 0.2870 \text{ m} = 287 \text{ mm} \qquad \textbf{Ans}$$

Hence, $h = 1.25b = 359$ mm **Ans**

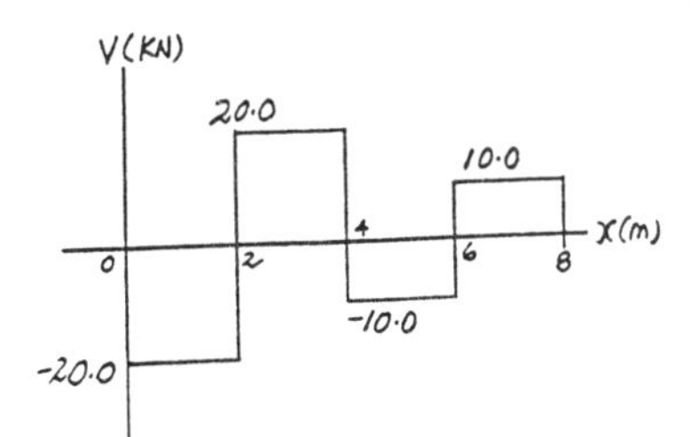

Shear Stress : Provide a shear stress check using the shear formula for a rectangular section. From the shear diagram, $V_{max} = 20.0$ kN.

$$\tau_{max} = \frac{3V_{max}}{2A}$$

$$= \frac{3[20.0(10^3)]}{2(0.2870)(1.25)(0.2870)}$$

$$= 291.4 \text{ kPa} < \tau_{allow} = 500 \text{ kPa } (O.K!)$$

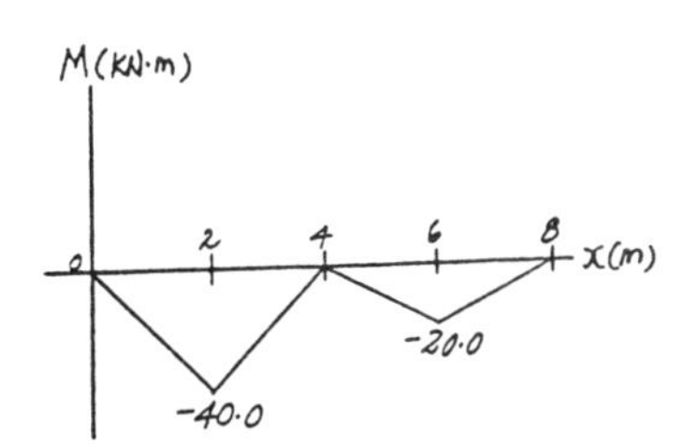

11–10. Solve Prob. 11–9 if the height-to-width ratio is to be $h/b = 1.5$.

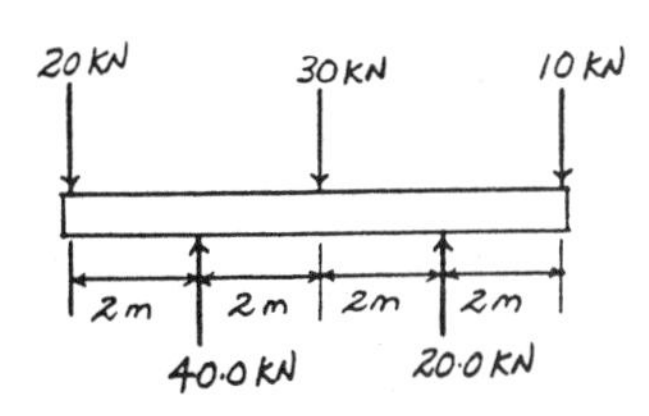

Bending Stress : From the moment diagram. $M_{max} = 40.0$ kN · m. Assume bending controls the design. Applying the flexure formula.

$$\sigma_{allow} = \frac{M_{max}\, c}{I}$$

$$6.5(10^6) = \frac{40.0(10^3)\left(\frac{1.5b}{2}\right)}{\frac{1}{12}(b)(1.5b)^3}$$

$$b = 0.2541 \text{ m} = 254 \text{ mm} \qquad \textbf{Ans}$$

Hence, $h = 1.5b = 381$ mm **Ans**

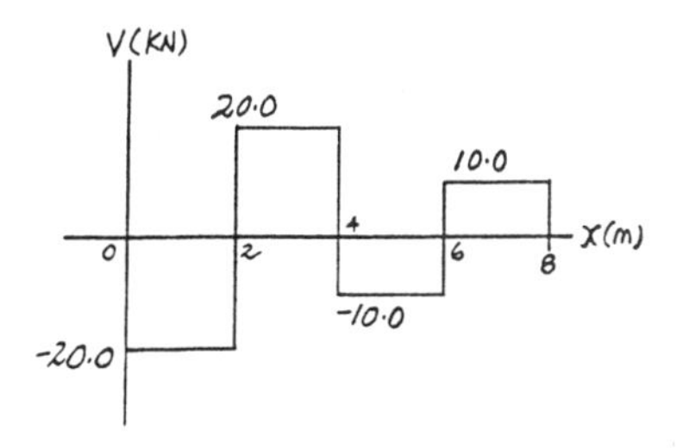

Shear Stress : Provide a shear stress check using the shear formula for a rectangular section. From the shear diagram, $V_{max} = 20.0$ kN.

$$\tau_{max} = \frac{3V_{max}}{2A}$$

$$= \frac{3[20.0(10^3)]}{2(0.2541)(1.5)(0.2541)}$$

$$= 309.7 \text{ kPa} < \tau_{allow} = 500 \text{ kPa } (O.K!)$$

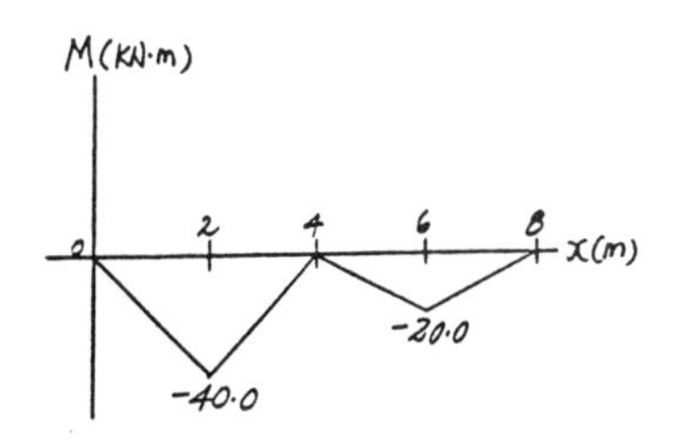

11–11. The beam is made of a ceramic material having an allowable bending stress of $\sigma_{allow} = 735$ psi and an allowable shear stress of $\tau_{allow} = 400$ psi. Determine the width b of the beam if the height $h = 2b$.

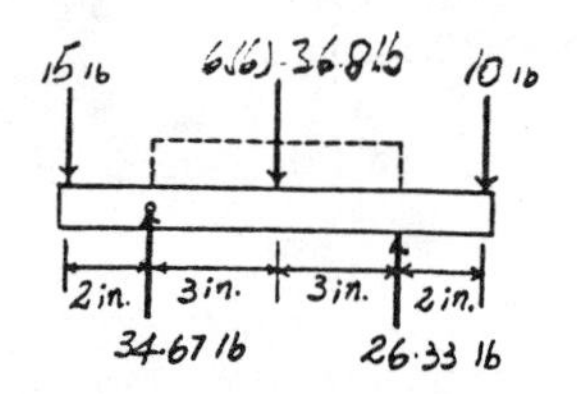

Bending Stress : From the moment diagram, $M_{max} = 30.0$ lb · in. Assume bending controls the design. Applying the flexure formula,

$$\sigma_{allow} = \frac{M_{max}\,c}{I}$$

$$735 = \frac{30.0\left(\frac{2b}{2}\right)}{\frac{1}{12}(b)(2b)^3}$$

$b = 0.3941$ in. $= 0.394$ in. **Ans**

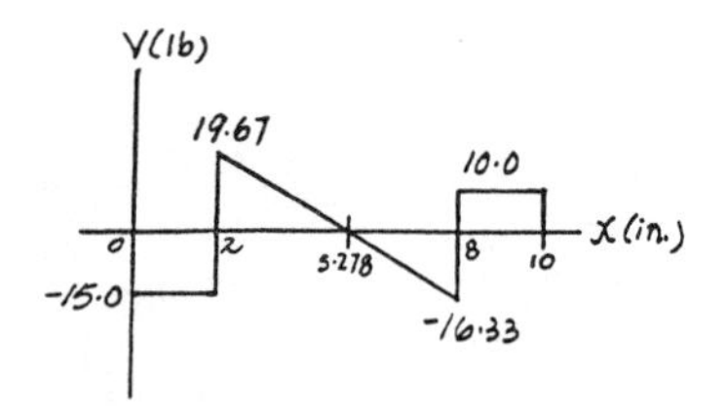

Shear Stress : Provide a shear stress check using the shear formula for a rectangular section. From the shear diagram, $V_{max} = 19.67$ lb.

$$\tau_{max} = \frac{3V_{max}}{2A}$$

$$= \frac{3(19.67)}{2(0.3941)(2)(0.3941)}$$

$= 94.95$ psi $< \tau_{allow} = 400$ psi $(O.K!)$

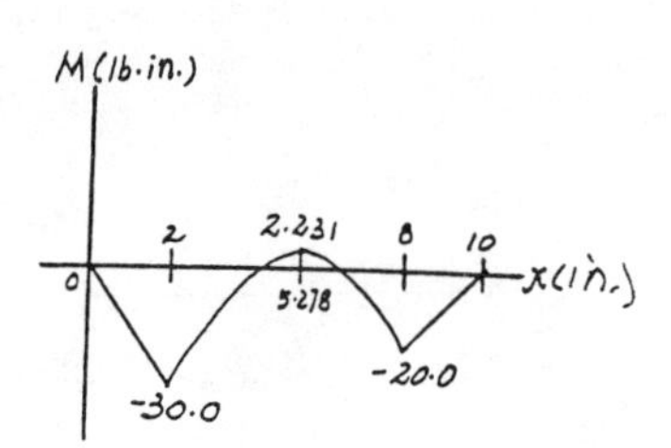

***11–12.** The beam is to be used to support the machine, which has a weight of 16 kip and a center of gravity at G. If the maximum bending stress is not to exceed $\sigma_{allow} = 22$ ksi, determine the required width b of the flanges. The supports at B and C are smooth.

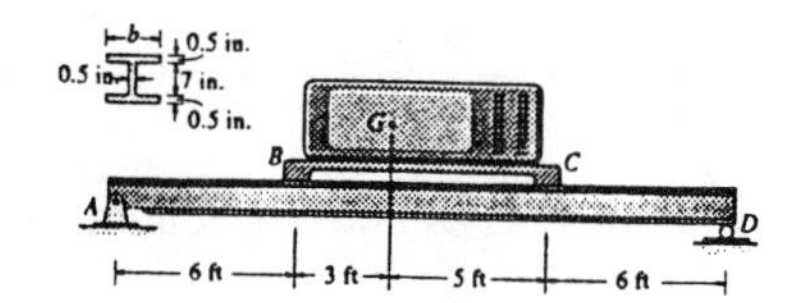

Section Properties :

$$I = \frac{1}{12}(b)\left(8^3\right) - \frac{1}{12}(b-0.5)\left(7^3\right) = 14.0833b + 14.2917$$

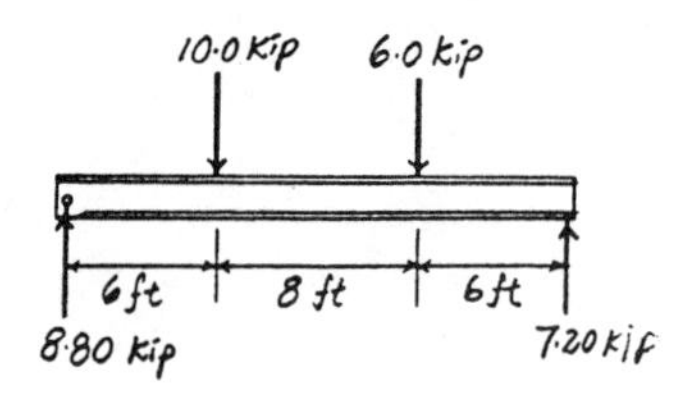

Bending Stress : From the moment diagram, $M_{max} = 52.8$ kip · ft. Applying the flexure formula,

$$\sigma_{allow} = \frac{M_{max}\,c}{I}$$

$$22 = \frac{52.8(12)(4)}{14.0833b + 14.2917}$$

$b = 7.165$ in. $= 7.17$ in. **Ans**

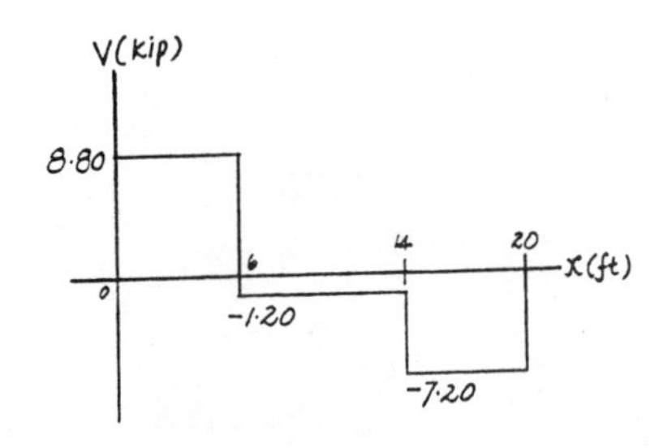

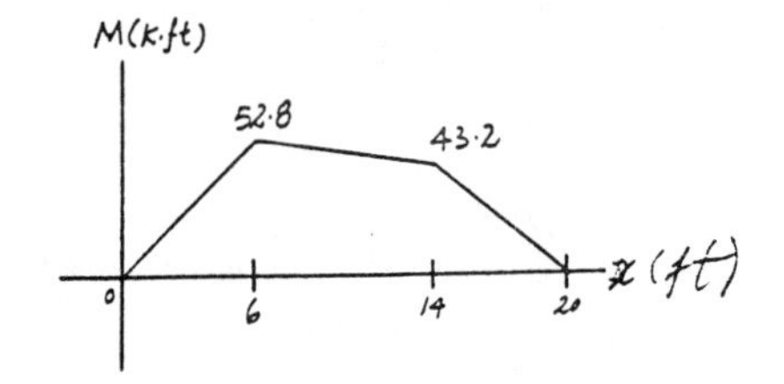

11–13. The beam has a flange width $b = 8$ in. If the maximum bending stress is not to exceed $\sigma_{allow} = 22$ ksi, determine the greatest weight of the machine that the beam can support. The center of gravity for the machine is at G, and the supports at B and C are smooth.

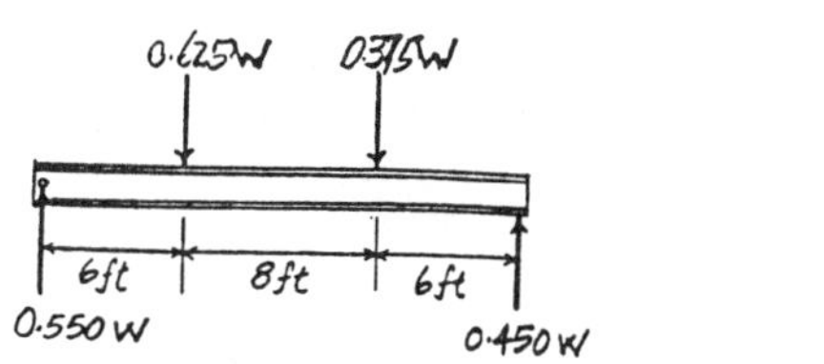

Section Property :

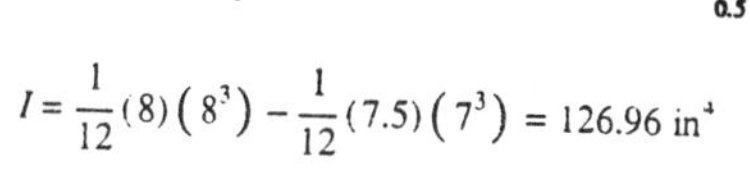

$$I = \frac{1}{12}(8)(8^3) - \frac{1}{12}(7.5)(7^3) = 126.96 \text{ in}^4$$

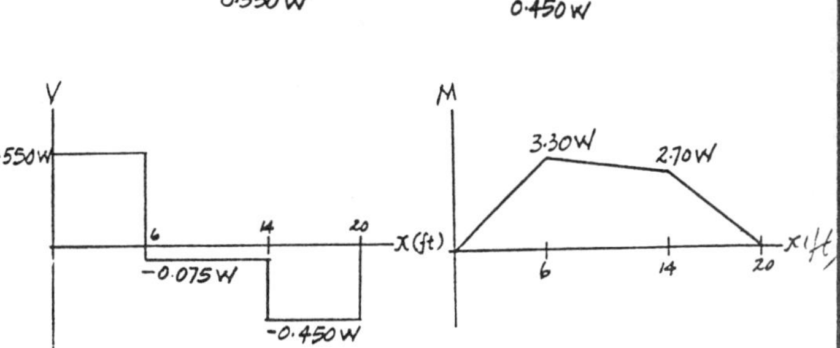

Bending Stress : From the moment diagram, $M_{max} = 3.30W$. Applying the flexure formula,

$$\sigma_{allow} = \frac{M_{max}c}{I}$$

$$22 = \frac{3.30W(12)(4)}{126.96}$$

$W = 17.63$ kip $= 17.6$ kip **Ans**

11–14. The compound beam is made from two sections, which are pinned together at B. Use Appendix B and select the light wide-flange beam that would be safe for each section if the allowable bending stress is $\sigma_{allow} = 24$ ksi and the allowable shear stress is $\tau_{allow} = 14$ ksi. The beam supports a pipe loading of 1200 lb and 1800 lb as shown.

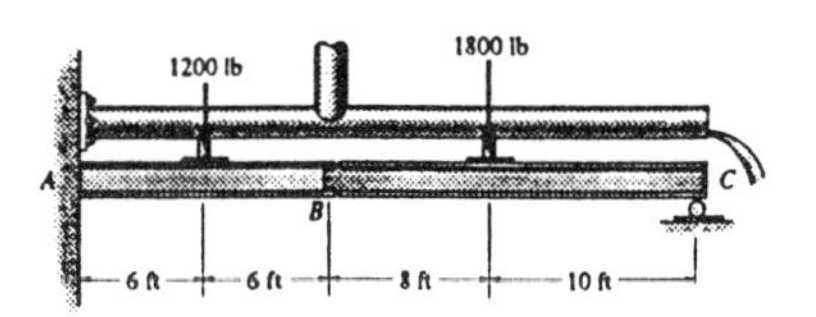

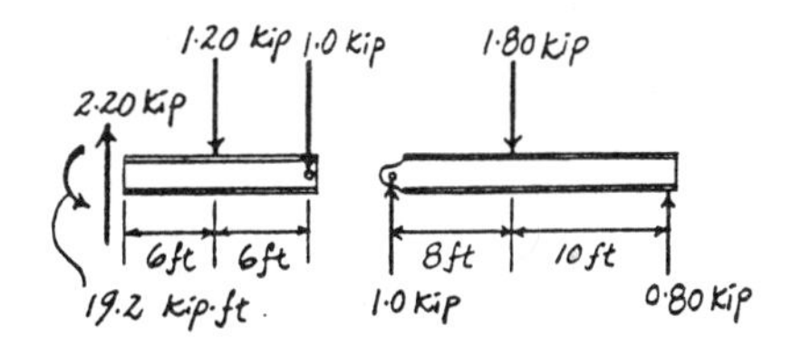

Bending Stress : From the moment diagram, $M_{max} = 19.2$ kip · ft for member AB. Assuming bending controls the design, applying the flexure formula.

$$S_{req'd} = \frac{M_{max}}{\sigma_{allow}} = \frac{19.2(12)}{24} = 9.60 \text{ in}^3$$

Select W10×12 $(S_x = 10.9 \text{ in}^3,\ d = 9.87 \text{ in.},\ t_w = 0.19 \text{ in.})$

For member BC, $M_{max} = 8.00$ kip · ft.

$$S_{req'd} = \frac{M_{max}}{\sigma_{allow}} = \frac{8.00(12)}{24} = 4.00 \text{ in}^3$$

Select W6×9 $(S_x = 5.56 \text{ in}^3,\ d = 5.90 \text{ in.},\ t_w = 0.17 \text{ in.})$

Shear Stress : Provide a shear stress check using $\tau = \frac{V}{t_w d}$ for the W10× 12 wide-flange section for member AB. From the shear diagram, $V_{max} = 2.20$ kip.

$$\tau_{max} = \frac{V_{max}}{t_w d} = \frac{2.20}{0.19(9.87)} = 1.17 \text{ ksi} < \tau_{allow} = 14 \text{ ksi } (O.K!)$$

Use W10×12 **Ans**

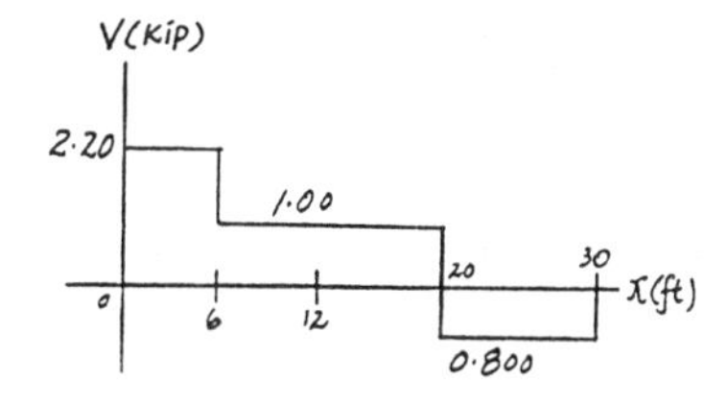

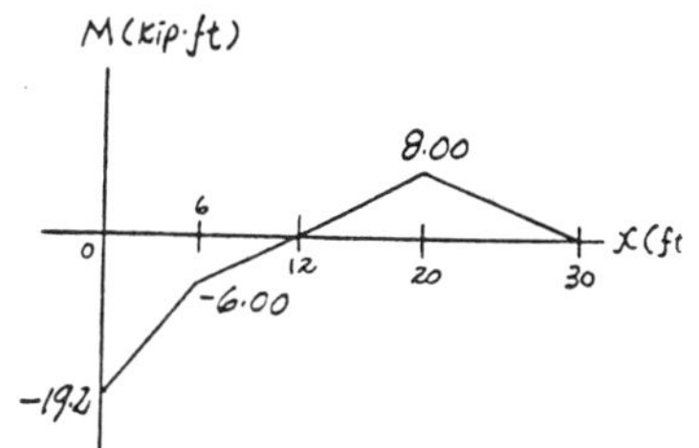

For member BC (W6×9), $V_{max} = 1.00$ kip.

$$\tau_{max} = \frac{V_{max}}{t_w d} = \frac{1.00}{0.17(5.90)} = 0.997 \text{ ksi} < \tau_{allow} = 14 \text{ ksi } (O.K!)$$

Hence, **Use** W6×9 **Ans**

11–15. Draw the shear and moment diagrams for the $W\,12 \times 14$ beam and check if the beam will safely support the loading. The allowable bending stress is $\sigma_{allow} = 22$ ksi and the allowable shear stress is $\tau_{allow} = 12$ ksi.

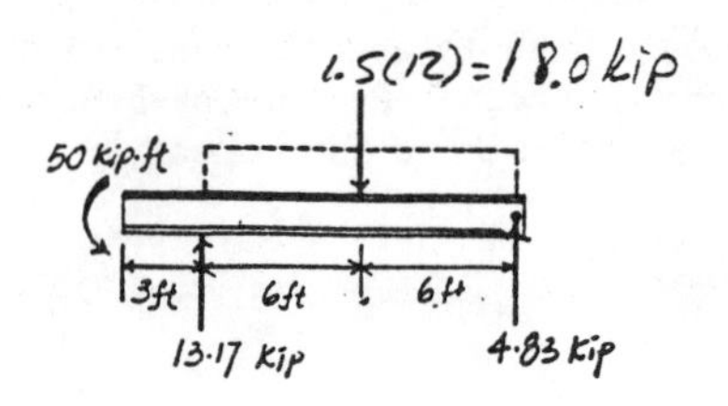

Bending Stress : From the moment diagram, $M_{max} = 50.0$ kip · ft.
Applying the flexure formula with $S = 14.9$ in^3 for a wide - flange section W12 × 14,

$$\sigma_{max} = \frac{M_{max}}{S}$$
$$= \frac{50.0(12)}{14.9} = 40.27 \text{ ksi} > \sigma_{allow} = 22 \text{ ksi } (\textit{No Good!})$$

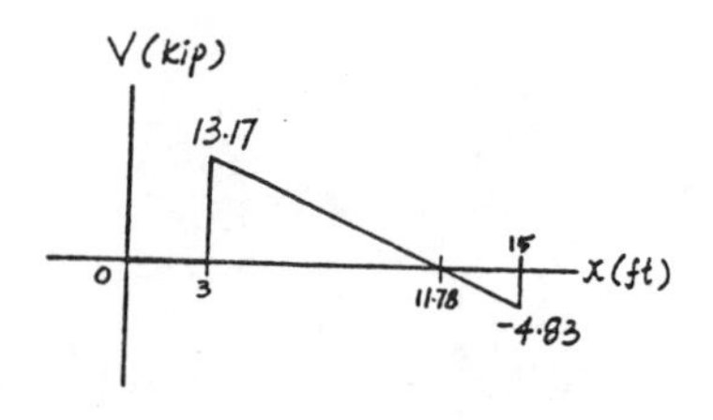

Shear Stress : From the shear diagram, $V_{max} = 13.17$ kip. Using $\tau = \frac{V}{t_w d}$ where $d = 11.91$ in. and $t_w = 0.20$ in. for W12 × 14 wide flange section.

$$\tau_{max} = \frac{V_{max}}{t_w d}$$
$$= \frac{13.17}{0.20(11.91)}$$
$$= 5.53 \text{ ksi} < \tau_{allow} = 12 \text{ ksi } (\textit{O.K!})$$

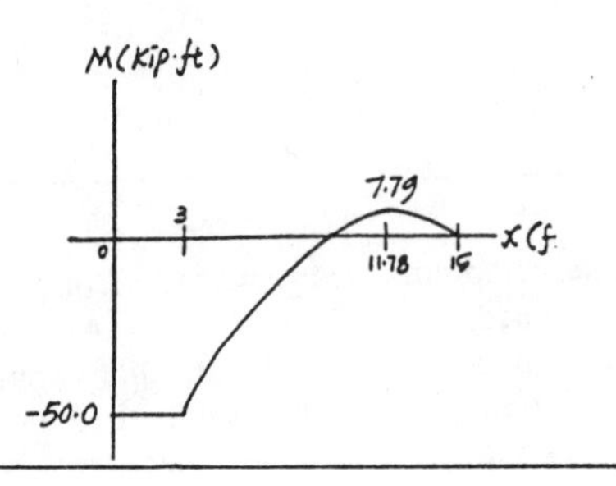

Hence, **the wide flange section W12 × 14 fails due to the bending stress and will not safely support the loading.** **Ans**

***11–16.** Select the lightest-weight steel wide-flange beam from Appendix *B* that will safely support the loading shown. The allowable bending stress is $\sigma_{allow} = 22$ ksi and the allowable shear stress is $\tau_{allow} = 12$ ksi.

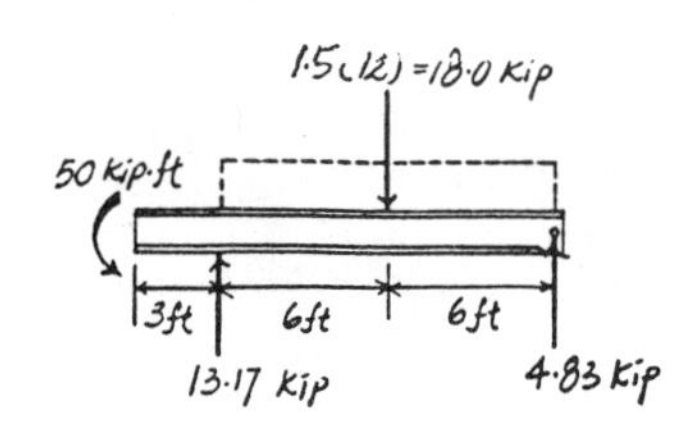

Bending Stress : From the moment diagram, $M_{max} = 50.0$ kip · ft.
Assume bending controls the design. Applying the flexure formula.

$$S_{req'd} = \frac{M_{max}}{\sigma_{allow}}$$
$$= \frac{50.0(12)}{22} = 27.27 \text{ in}^3$$

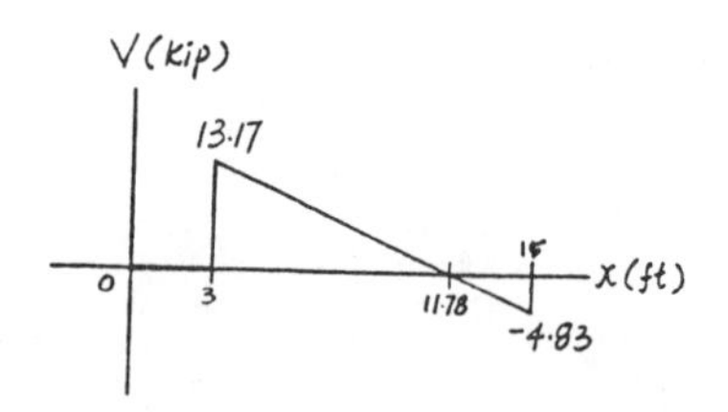

Select W14 × 22 $\left(S_x = 29.0 \text{ in}^3,\ d = 13.74 \text{ in.},\ t_w = 0.230 \text{ in.}\right)$

Shear Stress : Provide a shear stress check using $\tau = \frac{V}{t_w d}$ for the W14 × 22 wide - flange section. From the shear diagram, $V_{max} = 13.17$ kip

$$\tau_{max} = \frac{V_{max}}{t_w d}$$
$$= \frac{13.17}{0.230(13.74)}$$
$$= 4.17 \text{ ksi} < \tau_{allow} = 12 \text{ ksi } (\textit{O.K!})$$

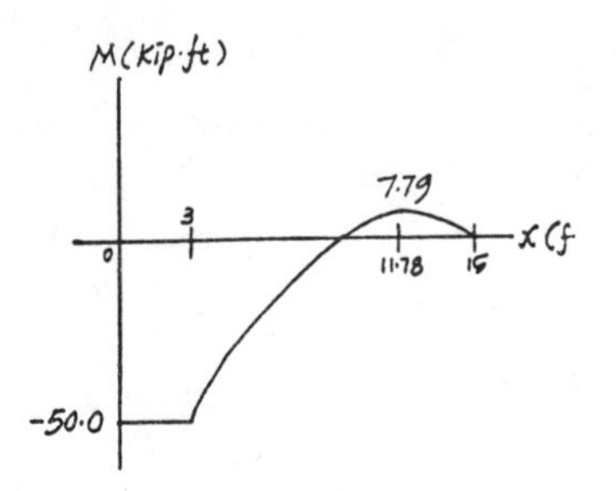

Hence, **Use** W14 × 22 **Ans**

11–17. Determine the smallest diameter rod that will safely support the loading shown. The allowable bending stress is $\sigma_{allow} = 167$ MPa and the allowable shear stress is $\tau_{allow} = 97$ MPa.

Bending Stress : From the moment diagram, $M_{max} = 24.375$ N·m. Assume bending controls the design. Applying the flexure formula,

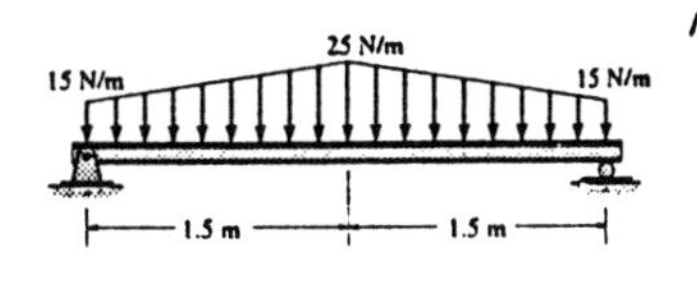

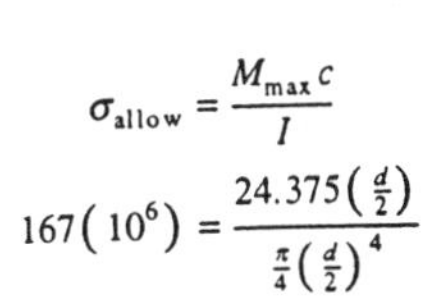

$$\sigma_{allow} = \frac{M_{max}\,c}{I}$$

$$167(10^6) = \frac{24.375\left(\frac{d}{2}\right)}{\frac{\pi}{4}\left(\frac{d}{2}\right)^4}$$

$$d = 0.01141 \text{ m} = 11.4 \text{ mm} \qquad \textbf{Ans}$$

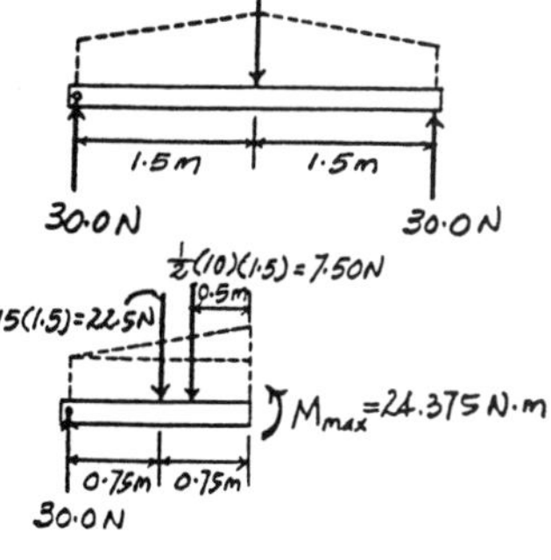

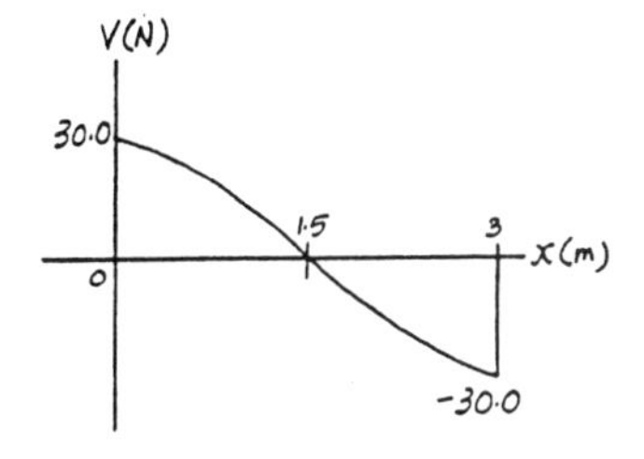

Shear Stress : Provide a shear stress check using the shear formula with

$$I = \frac{\pi}{4}\left(0.005706^4\right) = 0.8329\left(10^{-9}\right) \text{ m}^4$$

$$Q_{max} = \frac{4(0.005706)}{3\pi}\left[\frac{1}{2}(\pi)\left(0.005706^2\right)\right] = 0.1239\left(10^{-6}\right) \text{ m}^3$$

From the shear diagram, $V_{max} = 30.0$ N.

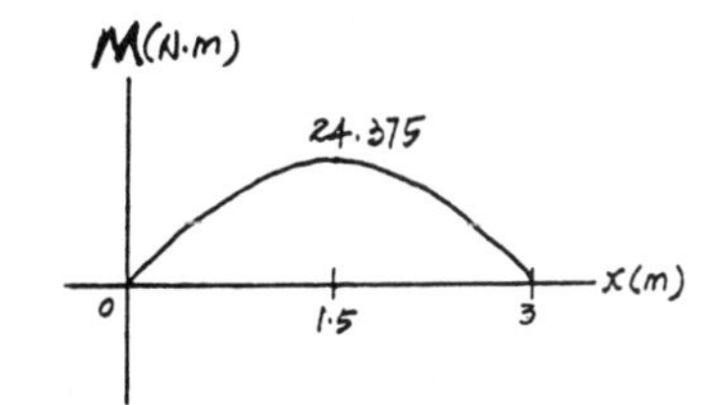

$$\tau_{max} = \frac{V_{max}\,Q_{max}}{It}$$

$$= \frac{30.0[0.1239(10^{-6})]}{0.8329(10^{-9})(0.01141)}$$

$$= 0.391 \text{ MPa} < \tau_{allow} = 97 \text{ MPa } (O.K!)$$

11–18. The pipe has an outside diameter of 15 mm. Determine the smallest inner diameter so that it will safely support the loading shown. The allowable bending stress is $\sigma_{allow} = 167$ MPa and the allowable shear stress is $\tau_{allow} = 97$ MPa.

Bending Stress : From the moment diagram, $M_{max} = 24.375$ N·m. Assume bending controls the design. Applying the flexure formula.

$$\sigma_{allow} = \frac{M_{max}\,c}{I}$$

$$167(10^6) = \frac{24.375(0.0075)}{\frac{\pi}{4}\left[0.0075^4 - \left(\frac{d_i}{2}\right)^4\right]}$$

$$d_i = 0.01297 \text{ m} = 13.0 \text{ mm} \qquad \textbf{Ans}$$

Shear Stress : Provide a shear stress check using the shear formula with

$$I = \frac{\pi}{4}\left(0.0075^4 - 0.006486^4\right) = 1.0947\left(10^{-9}\right) \text{ m}^4$$

$$Q_{max} = \frac{4(0.0075)}{3\pi}\left[\frac{1}{2}(\pi)\left(0.0075^2\right)\right] - \frac{4(0.006486)}{3\pi}\left[\frac{1}{2}(\pi)\left(0.006486^2\right)\right]$$

$$= 99.306\left(10^{-9}\right) \text{ m}^3$$

From the shear diagram, $V_{max} = 30.0$ N.

$$\tau_{max} = \frac{V_{max}\,Q_{max}}{It}$$

$$= \frac{30.0[99.306(10^{-9})]}{1.0947(10^{-9})(0.015-0.01297)}$$

$$= 1.34 \text{ MPa} < \tau_{allow} = 97 \text{ MPa } (O.K!)$$

11–19. The box beam has an allowable bending stress of $\sigma_{allow} = 10$ MPa and an allowable shear stress of $\tau_{allow} = 775$ kPa. Determine the maximum intensity w of the distributed loading that it can safely support. Also, determine the maximum safe nail spacing for each third of the length of the beam. Each nail can resist a shear force of 200 N.

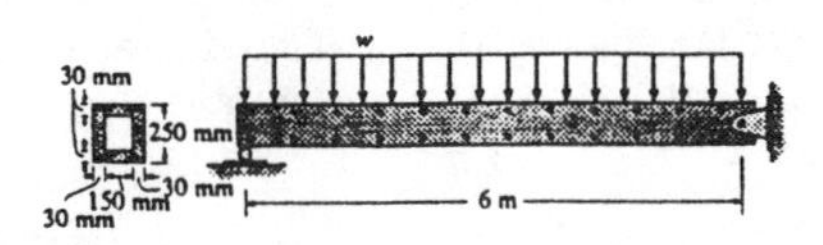

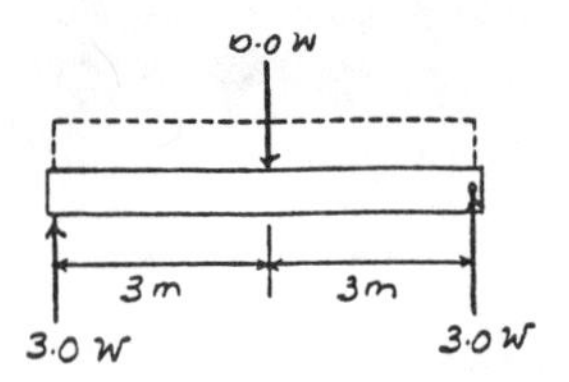

Section Properties :

$$I = \frac{1}{12}(0.21)(0.25^3) - \frac{1}{12}(0.15)(0.19^3) = 0.1877(10^{-3})\ \text{m}^4$$

$$Q_A = \bar{y}_1'A' = 0.11(0.03)(0.15) = 0.495(10^{-3})\ \text{m}^3$$

$$Q_{max} = \Sigma\bar{y}'A' = 0.11(0.03)(0.15) + 0.0625(0.125)(0.06) = 0.96375(10^{-3})\ \text{m}^3$$

Bending Stress : From the moment diagram, $M_{max} = 4.50w$. Assume bending controls the design. Applying the flexure formula.

$$\sigma_{allow} = \frac{M_{max}c}{I}$$

$$10(10^6) = \frac{4.50w(0.125)}{0.1877(10^{-3})}$$

$$w = 3336.9\ \text{N/m}$$

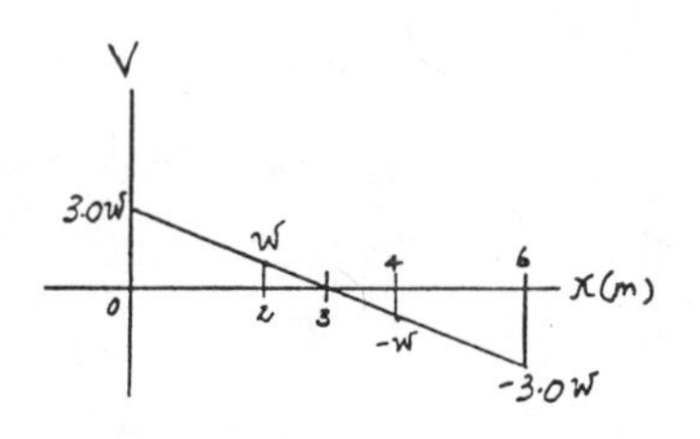

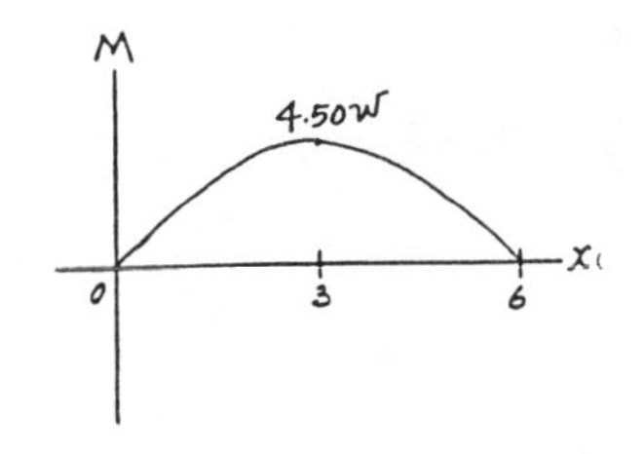

Shear Stress : Provide a shear stress check using the shear formula. From the shear diagram, $V_{max} = 3.00w = 10.01$ kN.

$$\tau_{max} = \frac{V_{max}Q_{max}}{It} = \frac{10.01(10^3)[0.96375(10^{-3})]}{0.1877(10^{-3})(0.06)}$$

$$= 857\ \text{kPa} > \tau_{allow} = 775\ \text{kPa}\ (\textit{No Good!})$$

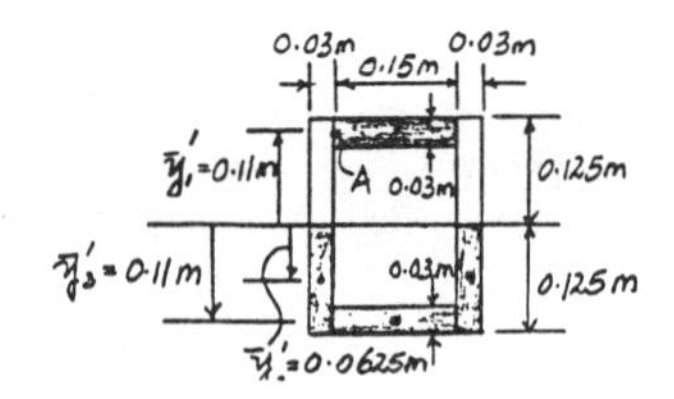

Hence, shear stress controls.

$$\tau_{allow} = \frac{V_{max}Q_{max}}{It}$$

$$775(10^3) = \frac{3.00w[0.96375(10^{-3})]}{0.1877(10^{-3})(0.06)}$$

$$w = 3018.8\ \text{N/m} = 3.02\ \text{kN/m} \qquad \textbf{Ans}$$

Shear Flow : Since there are two rows of nails, the allowable shear flow is $q = \frac{2(200)}{s} = \frac{400}{s}$.

For $0 \le x < 2$ m and $4\ \text{m} < x \le 6$ m, the design shear force is $V = 3.00w = 9056.3$ N.

$$q = \frac{VQ_A}{I}$$

$$\frac{400}{s} = \frac{9056.3[0.495(10^{-3})]}{0.1877(10^{-3})}$$

$$s = 0.01675\ \text{m} = 16.7\ \text{mm} \qquad \textbf{Ans}$$

For $2\text{m} < x < 4$ m, the design shear force is $V = w = 3018.8$ N.

$$q = \frac{VQ_A}{I}$$

$$\frac{400}{s} = \frac{3018.8[0.495(10^{-3})]}{0.1877(10^{-3})}$$

$$s = 0.05024\ \text{m} = 50.2\ \text{mm} \qquad \textbf{Ans}$$

***11–20.** The beam is constructed from two boards as shown. If each nail can support a shear force of 200 lb, determine the maximum spacing of the nails, s, s', and s'', to the nearest $\frac{1}{8}$ inch for regions AB, BC, and CD, respectively.

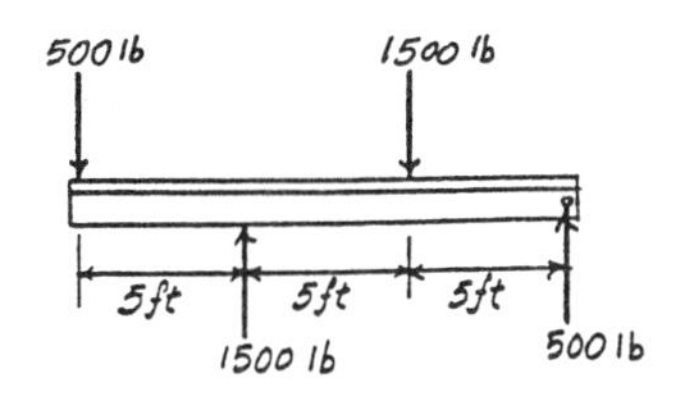

Section Properties :

$$\bar{y} = \frac{\Sigma \tilde{y}A}{\Sigma A} = \frac{0.5(8)(1)+4(6)(1)}{8(1)+6(1)} = 2.00 \text{ in.}$$

$$I = \frac{1}{12}(8)(1^3) + 8(1)(2-0.5)^2$$

$$\frac{1}{12}(1)(6^3) + 1(6)(4-2)^2$$

$$= 60.667 \text{ in}^4$$

$$Q_A = \bar{y}'A' = 1.5(1)(8) = 12.0 \text{ in}^3$$

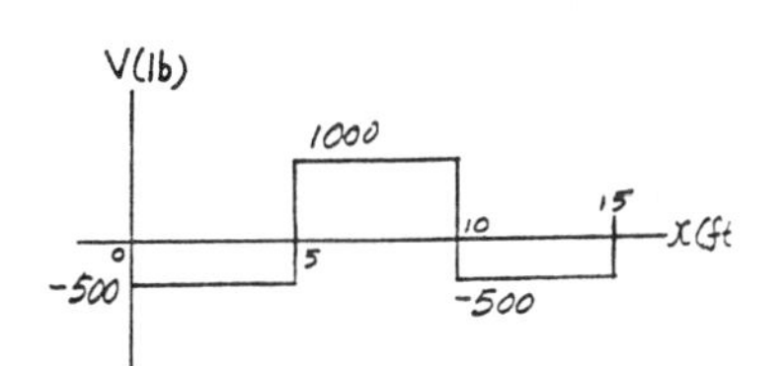

Shear Flow :
For $0 \le x < 5$ ft and $10 \text{ ft} < x \le 15$ ft (region AB and CD), the design shear force is $V = 500$ lb and the allowable shear flow is
$q = \frac{200}{s}$

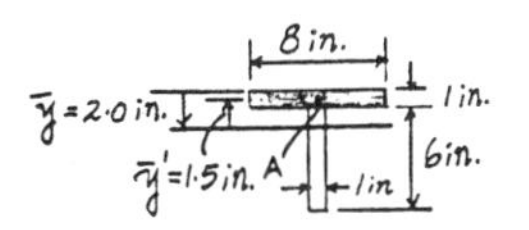

$$q = \frac{VQ_A}{I}$$

$$\frac{200}{s} = \frac{500(12.0)}{60.667}$$

$$s'' = s = 2.02 \text{ in}$$

Use $s'' = s = 2$ in. **Ans**

For $5 \text{ ft} < x < 10$ ft (region BC), the design shear force is $V = 1000$ lb and the allowable shear flow is $q = \frac{200}{s'}$

$$q = \frac{VQ_A}{I}$$

$$\frac{200}{s'} = \frac{1000(12.0)}{60.667}$$

$$s' = 1.01 \text{ in}$$

Use $s' = 1$ in. **Ans**

11–21. Determine the minimum width b of the beam to the nearest $\frac{1}{4}$ in. that will safely support the loading of $P = 8$ kip. The allowable bending stress is $\sigma_{\text{allow}} = 24$ ksi and the allowable shear stress is $\tau_{\text{allow}} = 15$ ksi.

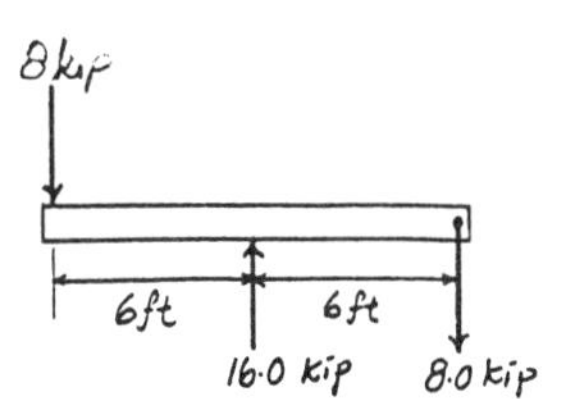

Bending Stress : From the moment diagram, $M_{\max} = 48.0$ kip · ft. Assume bending controls the design. Applying the flexure formula,

$$\sigma_{\text{allow}} = \frac{M_{\max}c}{I}$$

$$24 = \frac{48.0(12)(3)}{\frac{1}{12}(b)(6^3)}$$

$$b = 4.00 \text{ in.}$$

Ans

Shear Stress : Provide a shear stress check using the shear formula for a rectangular section. From the shear diagram, $V_{\max} = 8.00$ kip.

$$\tau_{\max} = \frac{3V_{\max}}{2A}$$

$$= \frac{3(8)}{2(4.00)(6)}$$

$$= 0.500 \text{ ksi} < \tau_{\text{allow}} = 15 \text{ ksi } (O.K!)$$

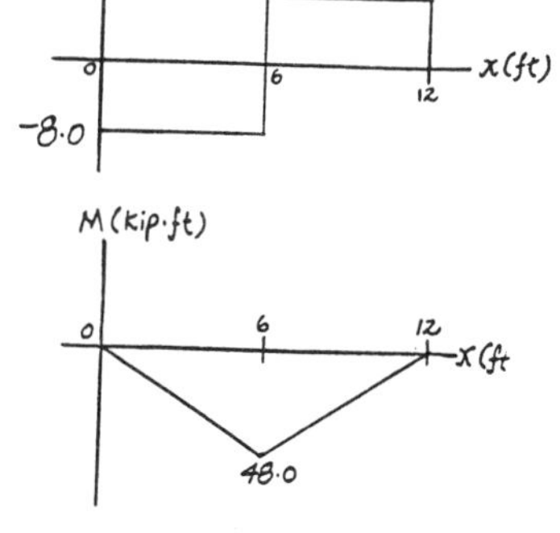

11–22. Solve Prob. 11–21 if P = 10 kip.

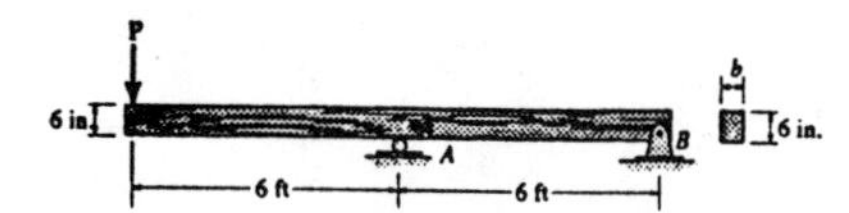

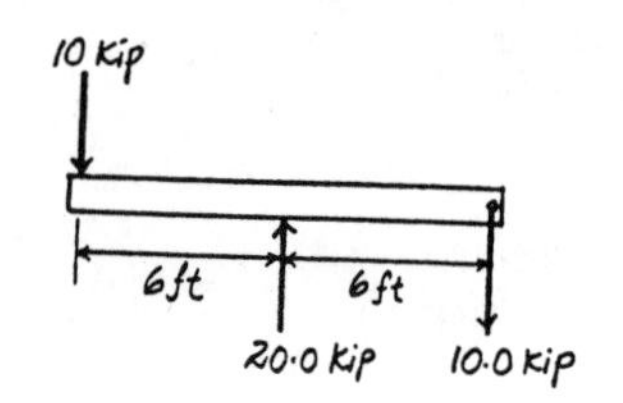

Bending Stress : From the moment diagram, M_{max} = 60.0 kip · ft. Assume bending controls the design. Applying the flexure formula,

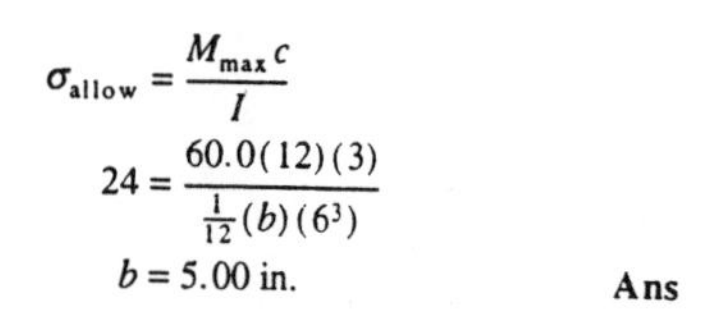

$$\sigma_{allow} = \frac{M_{max}c}{I}$$

$$24 = \frac{60.0(12)(3)}{\frac{1}{12}(b)(6^3)}$$

$$b = 5.00 \text{ in.} \qquad \text{Ans}$$

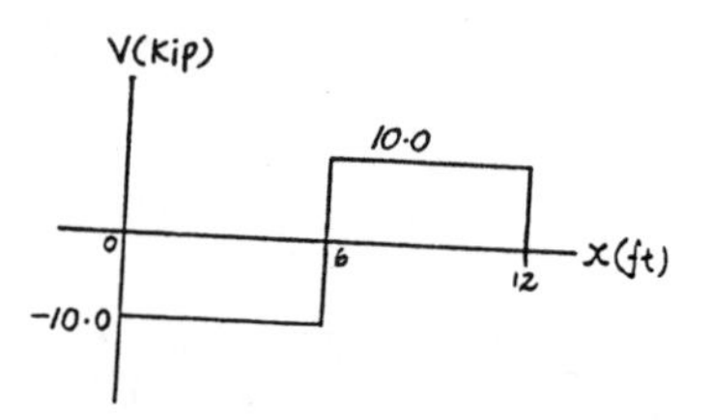

Shear Stress : Provide a shear stress check using the shear formula for a rectangular section. From the shear diagram, V_{max} = 10.0 kip.

$$\tau_{max} = \frac{3V_{max}}{2A} = \frac{3(10.0)}{2(5.00)(6)} = 0.500 \text{ ksi} < \tau_{allow} = 15 \text{ ksi } (O.K!)$$

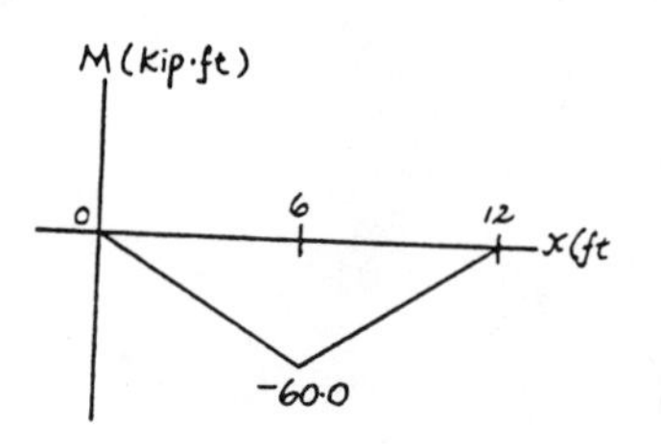

11–23. Draw the shear and moment diagrams for the shaft, and determine its required diameter to the nearest $\frac{1}{4}$ in. if σ_{allow} = 7 ksi and τ_{allow} = 3 ksi. The bearings at A and D exert only vertical reactions on the shaft. The loading is applied to the pulleys at B, C, and E. Take P = 110 lb.

Bending Stress : From the moment diagram, M_{max} = 1196.33 lb · in. Assume bending controls the design. Applying the flexure formula,

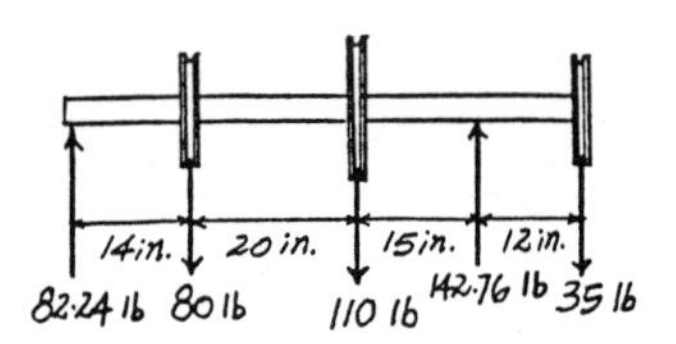

$$\sigma_{allow} = \frac{M_{max}c}{I}$$

$$7(10^3) = \frac{1196.33\left(\frac{d}{2}\right)}{\frac{\pi}{4}\left(\frac{d}{2}\right)^4}$$

$$d = 1.203 \text{ in.}$$

Use $\qquad d = 1\frac{1}{4}$ in. $\qquad$ **Ans**

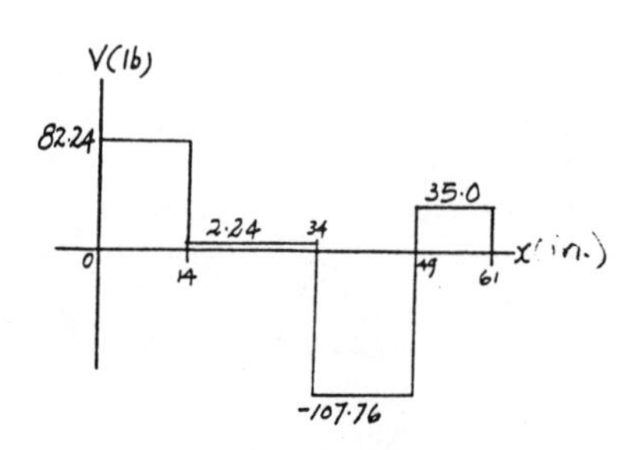

Shear Stress : Provide a shear stress check using the shear formula.

$$I = \frac{\pi}{4}(0.625^4) = 0.1198 \text{ in}^4$$

$$Q_{max} = \frac{4(0.625)}{3\pi}\left[\frac{1}{2}(\pi)(0.625^2)\right] = 0.1628 \text{ in}^3$$

From the shear diagram, V_{max} = 107.76 lb.

$$\tau_{max} = \frac{V_{max}Q_{max}}{It} = \frac{107.76(0.1628)}{0.1198(1.25)} = 117.1 \text{ psi} < \tau_{allow} = 3 \text{ ksi } (O.K!)$$

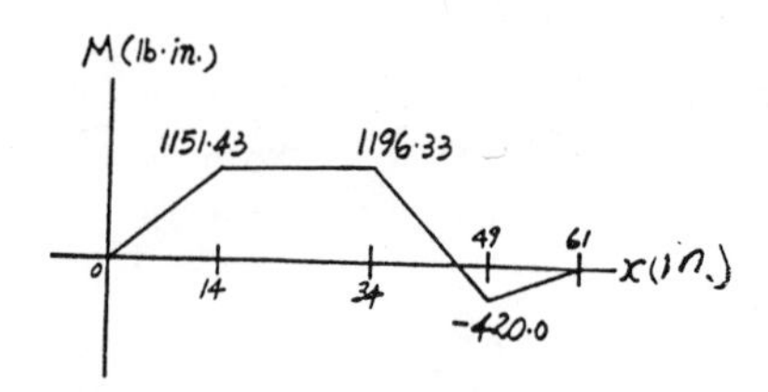

***11–24.** Draw the shear and moment diagrams for the shaft, and determine its required diameter to the nearest $\frac{1}{4}$ in. if $\sigma_{allow} = 7$ ksi and $\tau_{allow} = 3$ ksi. The bearings at A and D exert only vertical reactions on the shaft. The loading is applied to the pulleys at B, C, and E. Take $P = 80$ lb.

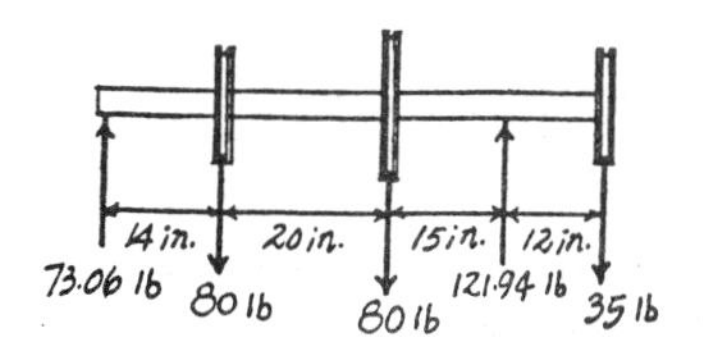

Bending Stress : From the moment diagram, $M_{max} = 1022.86$ lb · in. Assume bending controls the design. Applying the flexure formula,

$$\sigma_{allow} = \frac{M_{max}\,c}{I}$$

$$7(10^3) = \frac{1022.86\left(\frac{d}{2}\right)}{\frac{\pi}{4}\left(\frac{d}{2}\right)^4}$$

$$d = 1.142 \text{ in.}$$

Use $\quad d = 1\frac{1}{4}$ in. $\quad$ **Ans**

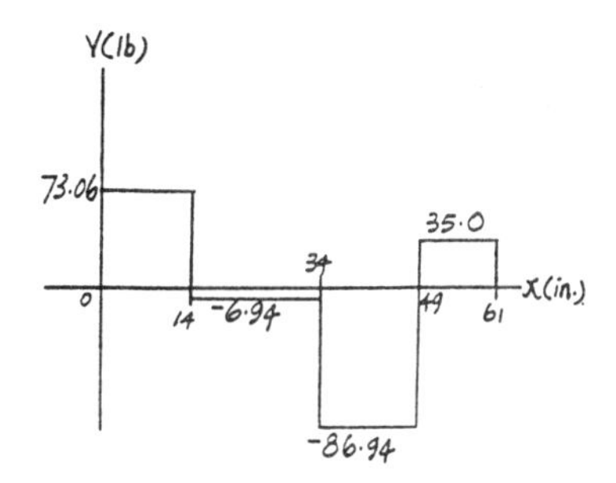

Shear Stress : Provide a shear stress check using the shear formula with

$$I = \frac{\pi}{4}\left(0.625^4\right) = 0.1198 \text{ in}^4$$

$$Q_{max} = \frac{4(0.625)}{3\pi}\left[\frac{1}{2}(\pi)\left(0.625^2\right)\right] = 0.1628 \text{ in}^3$$

From the shear diagram, $V_{max} = 86.94$ lb.

$$\tau_{max} = \frac{V_{max}\,Q_{max}}{It}$$

$$= \frac{86.94(0.1628)}{0.1198(1.25)}$$

$$= 94.46 \text{ psi} < \tau_{allow} = 3 \text{ ksi } (O.K!)$$

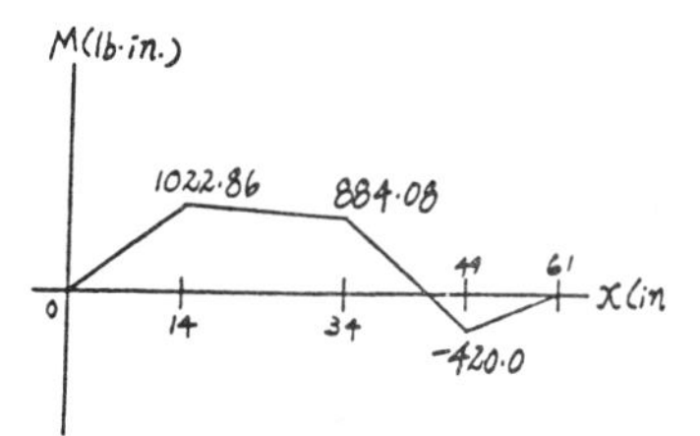

11–25. The brick wall exerts a uniform distributed load of 1.20 kip/ft on the beam. If the allowable bending stress is $\sigma_{allow} = 22$ ksi, determine the required width b of the flange to the nearest $\frac{1}{4}$ in.

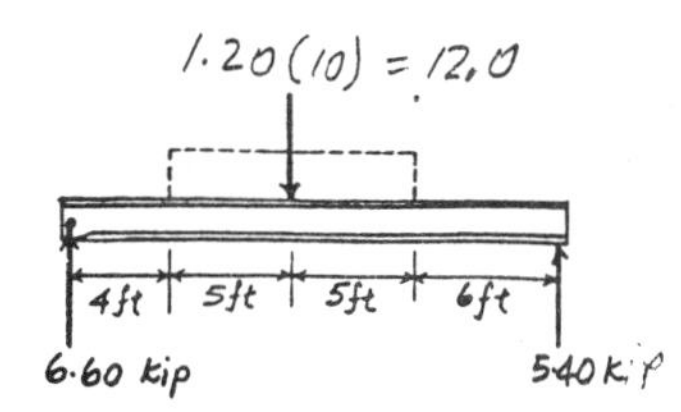

Section Property :

$$I = \frac{1}{12}(b)\left(10^3\right) - \frac{1}{12}(b-0.5)\left(9^3\right) = 22.583b + 30.375$$

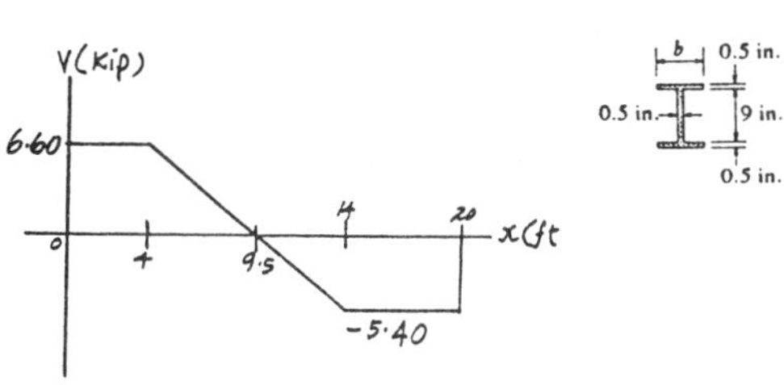

Bending Stress : From the moment diagram, $M_{max} = 44.55$ kip · ft.

$$\sigma_{allow} = \frac{M_{max}\,c}{I}$$

$$22 = \frac{44.55(12)(5)}{22.583b + 30.375}$$

$$b = 4.04 \text{ in.}$$

Use $\quad b = 4.25$ in. $\quad$ **Ans**

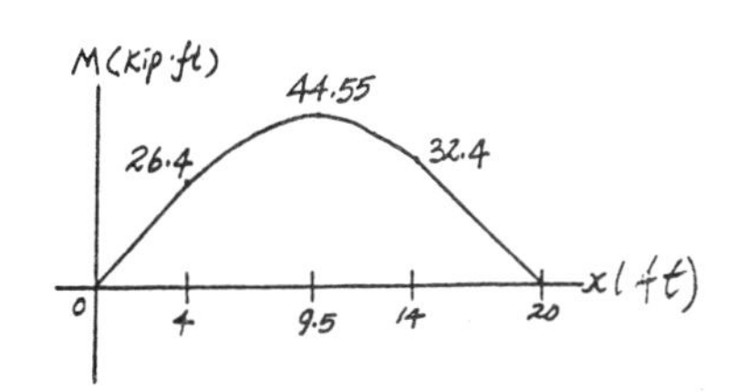

11–26. The brick wall exerts a uniform distributed load of 1.20 kip/ft on the beam. If the allowable bending stress is σ_{allow} = 22 ksi and the allowable shear stess is τ_{allow} = 12 ksi, select the lightest wide-flange section with the shortest depth from Appendix *B* that will safely support the load.

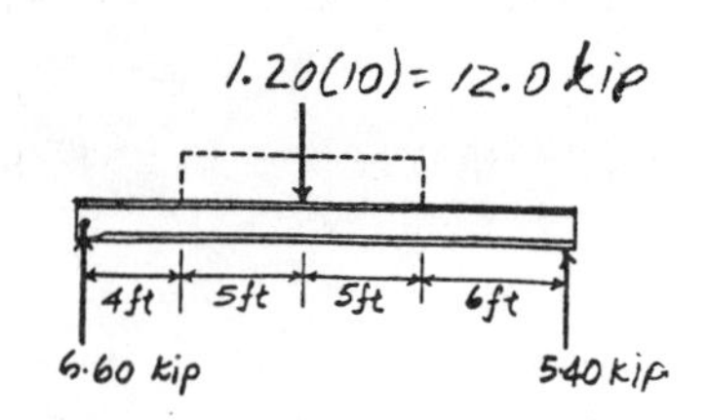

Bending Stress : From the moment diagram, M_{max} = 44.55 kip · ft. Assuming bending controls the design and applying the flexure formula,

$$S_{req'd} = \frac{M_{max}}{\sigma_{allow}}$$

$$= \frac{44.55(12)}{22} = 24.3 \text{ in}^3$$

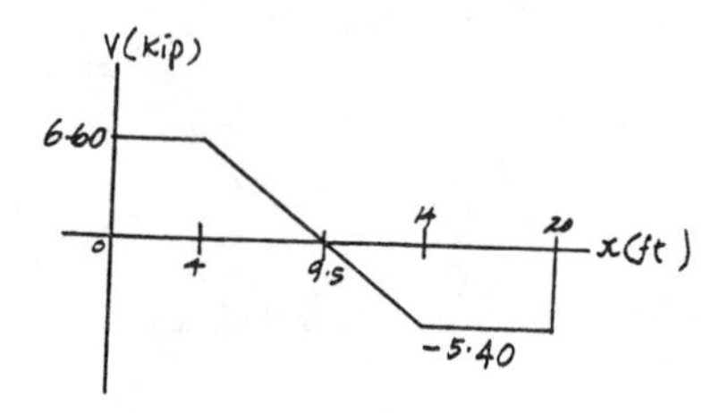

Two choices of wide flange section having the weight 22 lb/ft can be made. They are W12 × 22 and W14 × 22. However, W12 × 22 is the shortest.

Select W12×22 $\left(S_x = 25.4 \text{ in}^3,\ d = 12.31 \text{ in.},\ t_w = 0.260 \text{ in.}\right)$

Shear Stress : Provide a shear stress check using $\tau = \dfrac{V}{t_w d}$ for the W12 × 22 wide-flange section. From the shear diagram, V_{max} = 6.60 kip.

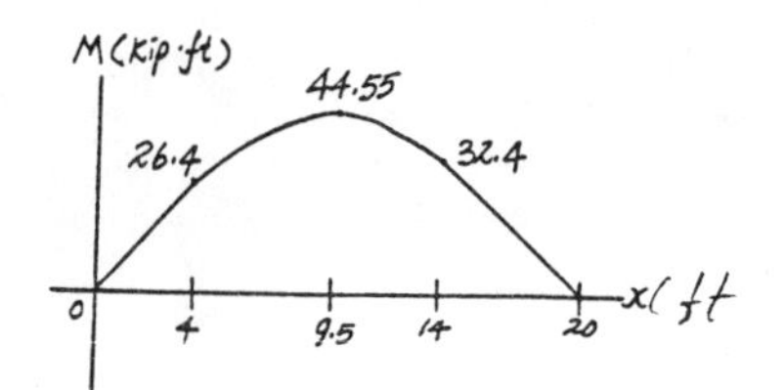

$$\tau_{max} = \frac{V_{max}}{t_w d}$$

$$= \frac{6.60}{0.260(12.31)}$$

$$= 2.06 \text{ ksi} < \tau_{allow} = 12 \text{ ksi } (O.K!)$$

Hence, **Use** W12×22 **Ans**

11–27. The simply supported joist is used in the construction of a floor for a building. In order to keep the floor low with respect to the sill beams *C* and *D*, the ends of the joists are notched as shown. If the allowable shear stress for the wood is τ_{allow} = 350 psi and the allowable bending stress is σ_{allow} = 1500 psi, determine the height *h* that will cause the beam to reach both allowable stresses at the same time. Also, what load *P* causes this to happen? Neglect the stress concentration at the notch.

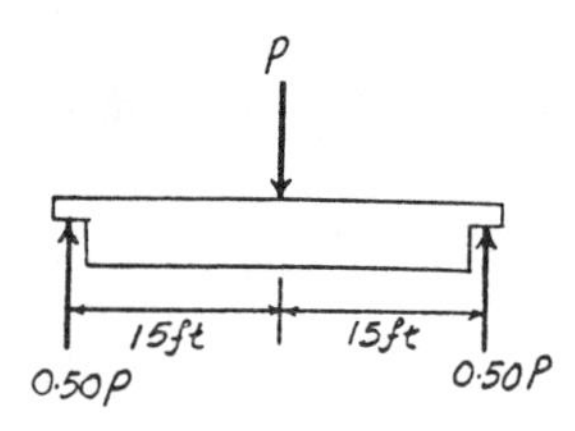

Bending Stress : From the moment diagram, M_{max} = 7.50*P*. Applying the flexure formula,

$$\sigma_{allow} = \frac{M_{max}\,c}{I}$$

$$1500 = \frac{7.50P(12)(5)}{\frac{1}{12}(2)(10^3)}$$

$$P = 555.56 \text{ lb} = 556 \text{ lb}$$ **Ans**

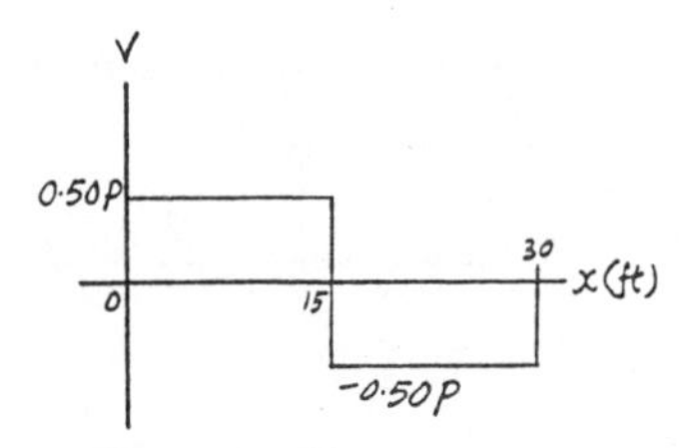

Shear Stress : From the shear diagram, V_{max} = 0.500*P* = 277.78 lb. The notch is the critical section. Using the shear formula for a rectangular section,

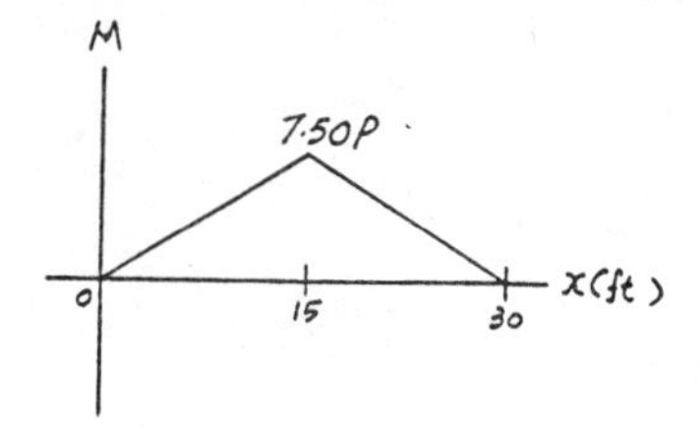

$$\tau_{allow} = \frac{3V_{max}}{2A}$$

$$350 = \frac{3(277.78)}{2(2)h}$$

$$h = 0.595 \text{ in.}$$ **Ans**

***11–28.** The simply supported joist is used in the construction of a floor for a building. In order to keep the floor low with respect to the sill beams C and D, the ends of the joists are notched as shown. If the allowable shear stress for the wood is $\tau_{allow} = 350$ psi and the allowable bending stress is $\sigma_{allow} = 1700$ psi, determine the smallest height h so that the beam will support a load of $P = 600$ lb. Also, will the entire joist safely support the load? Neglect the stress concentration at the notch.

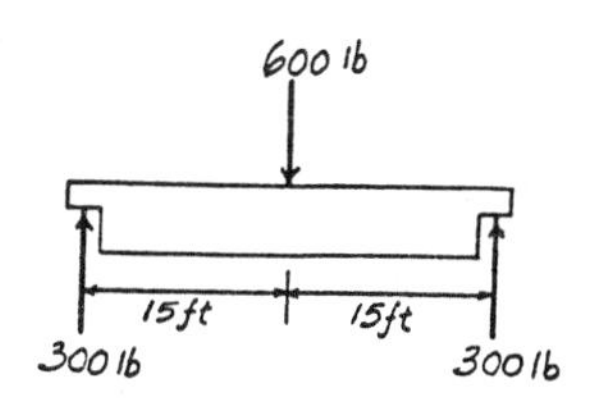

Shear Stress : From the shear diagram, $V_{max} = 300$ lb. The notch is the critical section. Using the shear formula for a rectangular section,

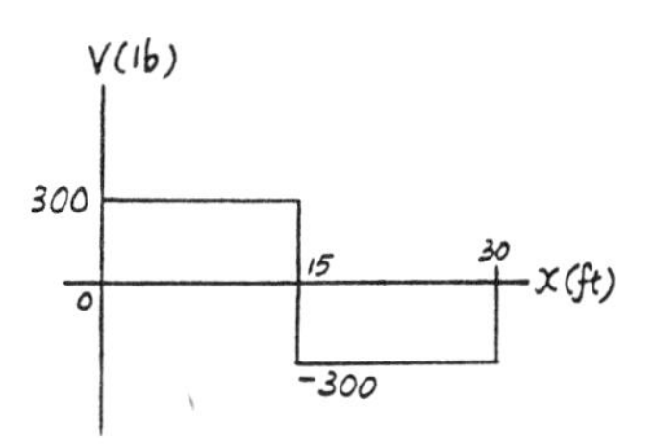

$$\tau_{allow} = \frac{3V_{max}}{2A}$$

$$350 = \frac{3(300)}{2(2)h}$$

$$h = 0.643 \text{ in.} \qquad \textbf{Ans}$$

Bending Stress : From the moment diagram, $M_{max} = 4500$ lb · ft. Applying the flexure formula,

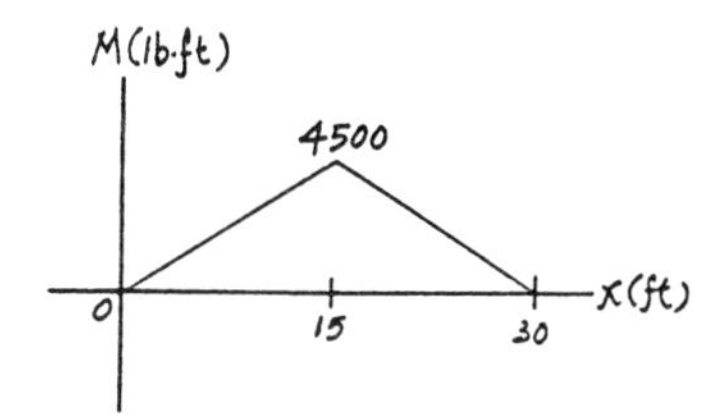

$$\sigma_{max} = \frac{M_{max}\,c}{I}$$

$$= \frac{4500(12)(5)}{\frac{1}{12}(2)(10^3)}$$

$$= 1620 \text{ psi} < \sigma_{allow} = 1700 \text{ psi } (O.K!)$$

Hence, **the joist is safe to support the load.** **Ans**

11–29. The tapered beam supports a concentrated force **P** at its center. If it is made from a plate that has a constant width b, determine the absolute maximum bending stress in the beam.

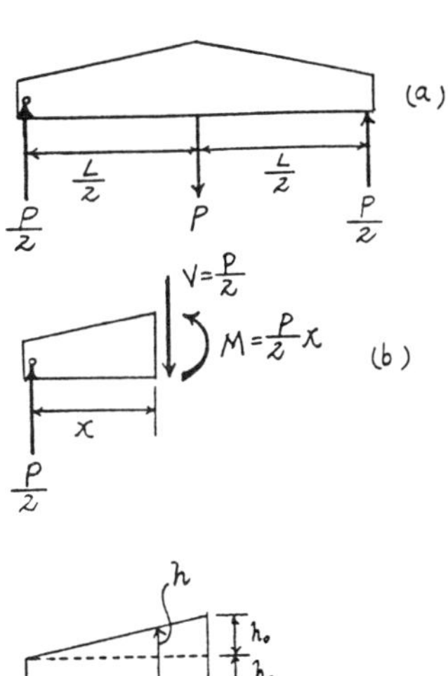

Section Properties :

$$\frac{h - h_0}{x} = \frac{h_0}{\frac{L}{2}} \qquad h = \frac{h_0}{L}(2x + L)$$

$$I = \frac{1}{12}(b)\left(\frac{h_0^3}{L^3}\right)(2x + L)^3$$

$$S = \frac{\frac{1}{12}(b)\left(\frac{h_0^3}{L^3}\right)(2x + L)^3}{\frac{h_0}{2L}(2x + L)} = \frac{bh_0^2}{6L^2}(2x + L)^2$$

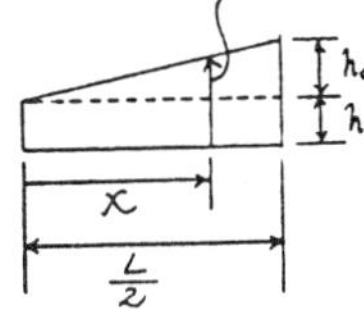

Bending Stress : Applying the flexure formula,

$$\sigma = \frac{M}{S} = \frac{\frac{Px}{2}}{\frac{bh_0^2}{6L^2}(2x + L)^2} = \frac{3PL^2x}{bh_0^2(2x + L)^2} \qquad [1]$$

In order to have the absolute maximum bending stress, $\frac{d\sigma}{dx} = 0$.

$$\frac{d\sigma}{dx} = \frac{3PL^2}{bh_0^2}\left[\frac{(2x+L)^2(1) - x(2)(2x+L)(2)}{(2x+L)^4}\right] = 0$$

$$x = \frac{L}{2}$$

Substituting $x = \frac{L}{2}$ into Eq. [1] yields

$$\sigma_{max} = \frac{3PL}{8bh_0^2} \qquad \textbf{Ans}$$

11–30. The beam is made from a plate having a constant thickness t and a width that varies as shown. If it supports a concentrated force **P** at its end, determine the absolute maximum bending stress in the beam and specify its location x.

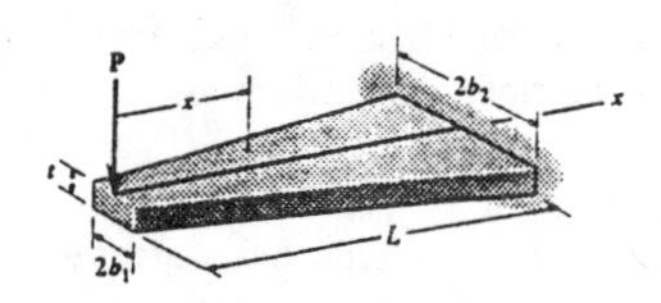

Section Properties :

$$\frac{b-b_1}{x}=\frac{b_2-b_1}{L} \qquad b=\frac{(b_2-b_1)x+b_1L}{L}$$

$$I=\frac{1}{12}(2b)t^3=\frac{1}{6}b\,t^3$$

$$S=\frac{I}{c}=\frac{1}{3}b\,t^2=\frac{t^2}{3L}[(b_2-b_1)x+b_1L]$$

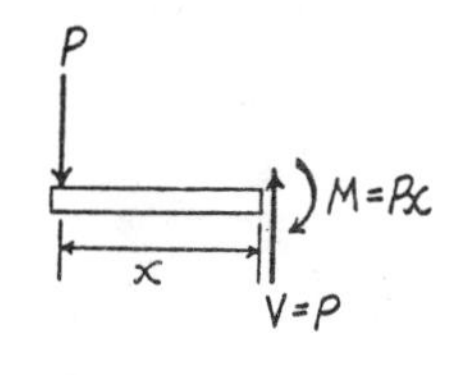

Bending Stress : Applying the flexure formula,

$$\sigma=\frac{M}{S}=\frac{3PLx}{t^2[(b_2-b_1)x+b_1L]} \qquad [1]$$

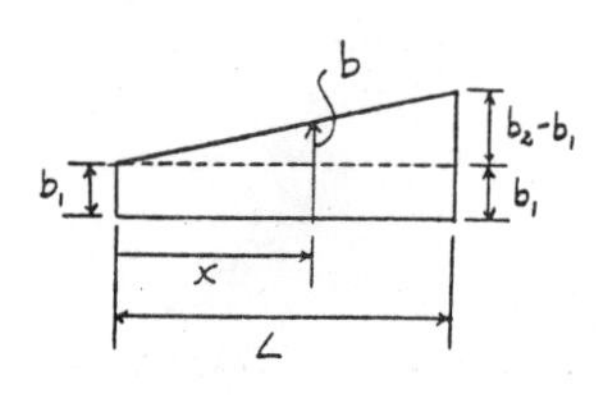

$$\frac{d\sigma}{dx}=\frac{3PL}{t^2}\left[\frac{(b_2-b_1)x+b_1L-x(b_2-b_1)}{[(b_2-b_1)x+b_1L]^2}\right]$$

$$=\frac{3PL}{t^2}\left[\frac{b_1L}{[(b_2-b_1)x+b_1L]^2}\right]$$

σ is an increasing function, therefore σ_{max} occurs at $x = L$. **Ans**

Substituting $x = L$ into Eq. [1] yields

$$\sigma_{max}=\frac{3PL}{t^2b_2} \qquad \textbf{Ans}$$

11–31. Determine the variation of the radius of the cantilevered beam that supports the uniform distributed load so that it has a constant maximum bending stress σ_{max} throughout its length.

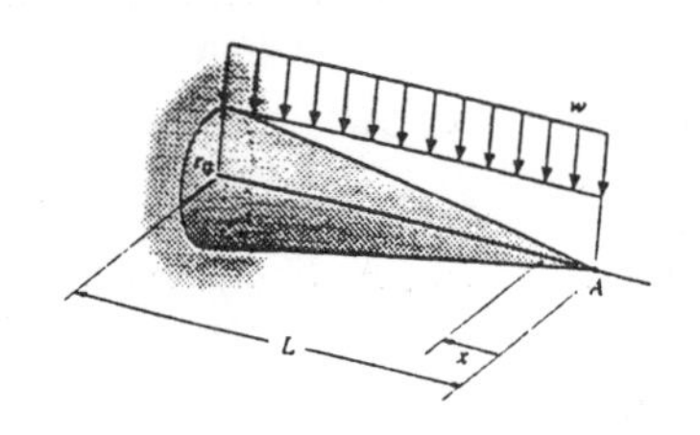

Moment Function : As shown on FBD.

Section Properties :

$$I=\frac{\pi}{4}r^4 \qquad S=\frac{I}{c}=\frac{\frac{\pi}{4}r^4}{r}=\frac{\pi}{4}r^3$$

Bending Stress : Applying the flexure formula.

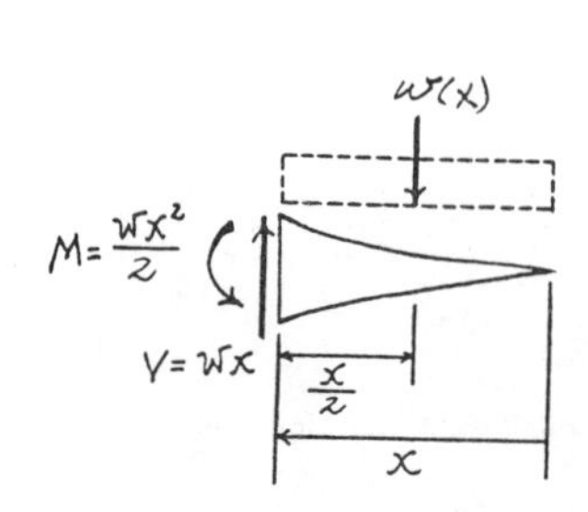

$$\sigma_{max}=\frac{M}{S}=\frac{\frac{wx^2}{2}}{\frac{\pi}{4}r^3}$$

$$\sigma_{max}=\frac{2wx^2}{\pi r^3} \qquad [1]$$

At $x = L$, $r = r_0$. From Eq. [1],

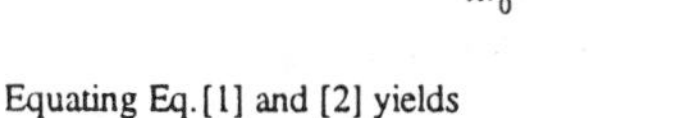

$$\sigma_{max}=\frac{2wL^2}{\pi r_0^3} \qquad [2]$$

Equating Eq. [1] and [2] yields

$$r^3=\frac{r_0^3}{L^2}x^2 \qquad \textbf{Ans}$$

***11–32.** The tapered cantilevered beam supports the concentrated force **P** at its end. Determine the absolute maximum bending stress in the beam.

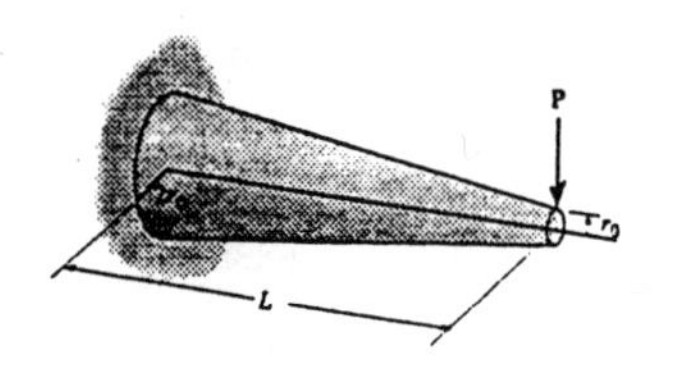

Section Properties :

$$\frac{r-r_0}{x}=\frac{r_0}{L} \qquad r=\frac{r_0}{L}(x+L)$$

$$I=\frac{\pi}{4}\left[\frac{r_0}{L}(x+L)\right]^4=\frac{\pi r_0^4}{4L^4}(x+L)^4$$

$$S=\frac{I}{c}=\frac{\frac{\pi r_0^4}{4L^4}(x+L)^4}{\frac{r_0}{L}(x+L)}=\frac{\pi r_0^3}{4L^3}(x+L)^3$$

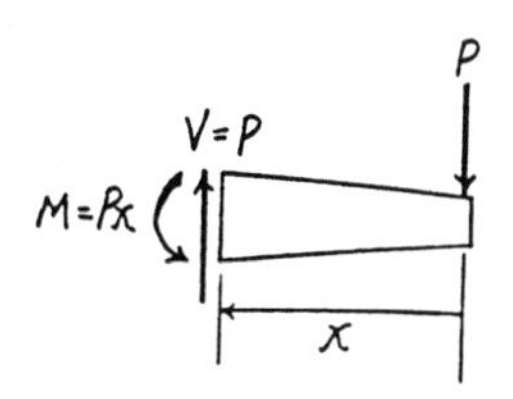

Bending Stress : Applying the flexure formula,

$$\sigma=\frac{M}{S}=\frac{Px}{\frac{\pi r_0^3}{4L^3}(x+L)^3}=\frac{4PL^3x}{\pi r_0^3(x+L)^3} \qquad [1]$$

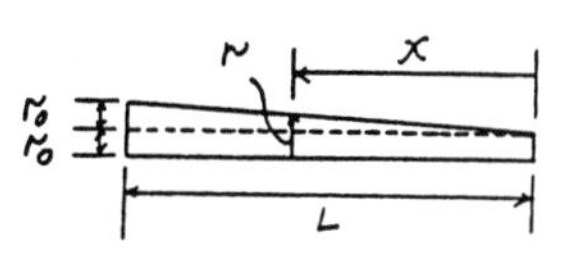

In order to have the absolute maximum bending stress, $\frac{d\sigma}{dx}=0$.

$$\frac{d\sigma}{dx}=\frac{4PL^3}{\pi r_0^3}\left[\frac{(x+L)^3(1)-x(3)(x+L)^2(1)}{(x+L)^6}\right]=0$$

$$x=\frac{L}{2}$$

Substituting $x=\frac{L}{2}$ into Eq. [1] yields

$$\sigma_{max}=\frac{16PL}{27\pi r_0^3} \qquad \textbf{Ans}$$

11–33. The beam is made into the shape of a frustum and has a diameter of 0.5 ft at A and a diameter of 1 ft at B. If it supports a force of 150 lb at A, determine the absolute maximum bending stress in the beam and specify its location x.

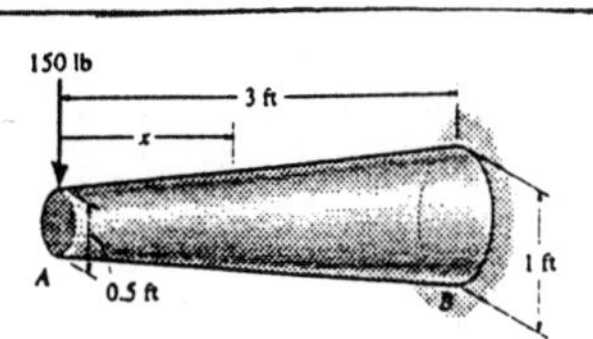

Section Properties :

$$\frac{r-3}{x}=\frac{3}{36} \qquad r=\frac{x+36}{12}$$

$$I=\frac{\pi}{4}\left(\frac{x+36}{12}\right)^4=\frac{\pi}{82944}(x+36)^4$$

$$S=\frac{\frac{\pi}{82944}(x+36)^4}{\frac{x+36}{12}}=\frac{\pi}{6912}(x+36)^3$$

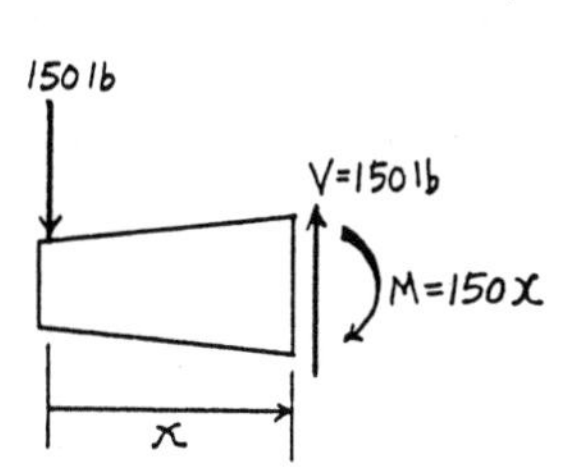

Bending Stress : Applying the flexure formula,

$$\sigma=\frac{M}{S}=\frac{150x}{\frac{\pi}{6912}(x+36)^3}=\frac{1036800x}{\pi(x+36)^3} \qquad [1]$$

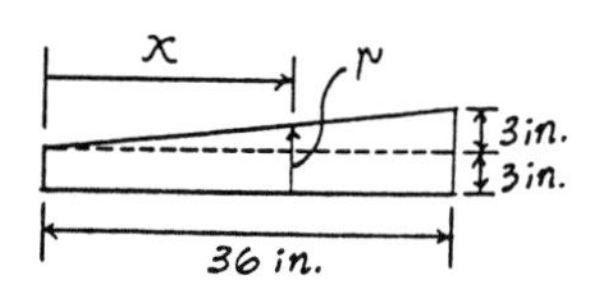

In order to have the absolute maximum bending stress, $\frac{d\sigma}{dx}=0$.

$$\frac{d\sigma}{dx}=\frac{1036800}{\pi}\left[\frac{(x+36)^3(1)-x(3)(x+36)^2(1)}{(x+36)^6}\right]=0$$

$$x=18.0\text{ in.}=1.50\text{ ft} \qquad \textbf{Ans}$$

Substituting $x=18.0$ in. into Eq. [1] yields

$$\sigma_{max}=\frac{1036800(18)}{\pi(18+36)^3}=37.7\text{ psi} \qquad \textbf{Ans}$$

11–34. The beam is made from a plate that has a constant thickness b. If it is simply supported and carries a uniform load w, determine the variation of its depth as a function of x so that it maintains a constant maximum bending stress σ_{allow} throughout its length.

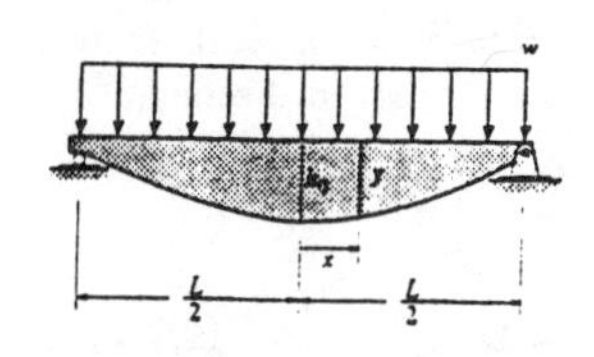

Moment Function : As shown on FBD(b).

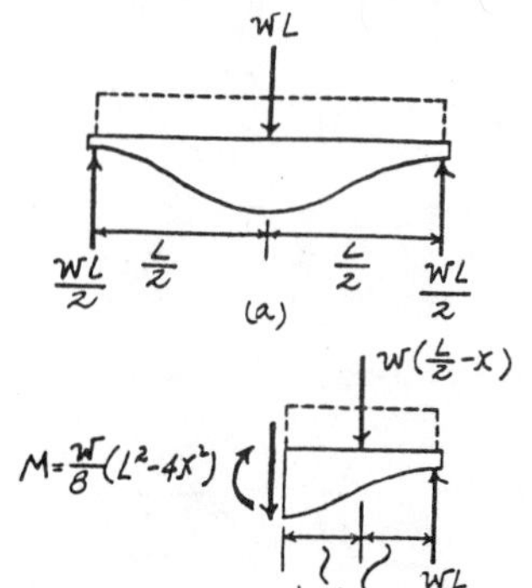

Section Properties :

$$I = \frac{1}{12}by^3 \qquad S = \frac{I}{c} = \frac{\frac{1}{12}by^3}{\frac{y}{2}} = \frac{1}{6}by^2$$

Bending Stress : Applying the flexure formula.

$$\sigma_{allow} = \frac{M}{S} = \frac{\frac{w}{8}(L^2 - 4x^2)}{\frac{1}{6}by^2}$$

$$\sigma_{allow} = \frac{3w(L^2 - 4x^2)}{4by^2} \qquad [1]$$

At $x = 0$, $y = h_0$. From Eq.[1].

$$\sigma_{allow} = \frac{3wL^2}{4bh_0^2} \qquad [2]$$

Equating Eq.[1] and [2] yields

$$y^2 = \frac{h_0^2}{L^2}(L^2 - 4x^2)$$

$$\frac{y^2}{h_0^2} + \frac{4x^2}{L^2} = 1 \qquad \textbf{Ans}$$

The beam has a **semi - elliptical** shape.

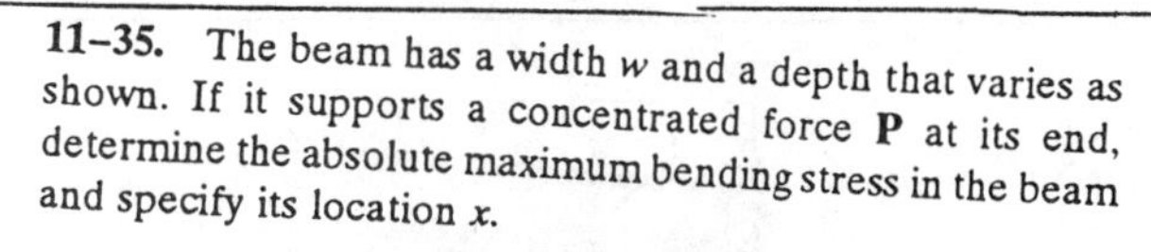
11–35. The beam has a width w and a depth that varies as shown. If it supports a concentrated force **P** at its end, determine the absolute maximum bending stress in the beam and specify its location x.

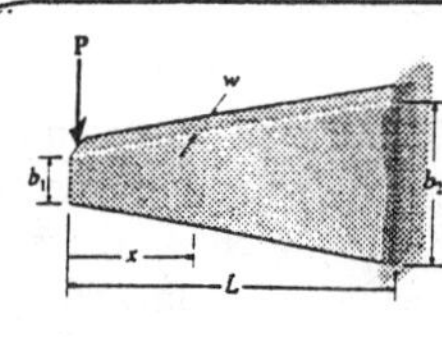

$$\frac{b - b_1}{x} = \frac{b_2 - b_1}{L} \qquad b = \frac{x(b_2 - b_1) + b_1 L}{L}$$

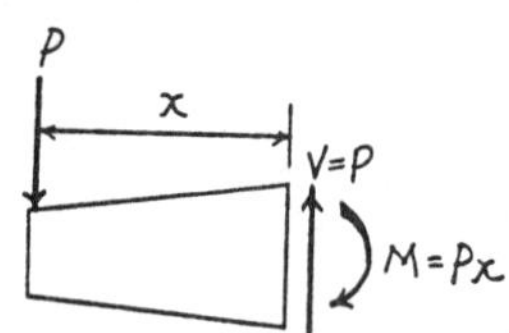

$$I = \frac{1}{12}wb^3$$

$$S = \frac{I}{c} = \frac{\frac{1}{12}wb^3}{\frac{b}{2}} = \frac{w}{6L^2}\left[x(b_2 - b_1) + b_1 L\right]^2$$

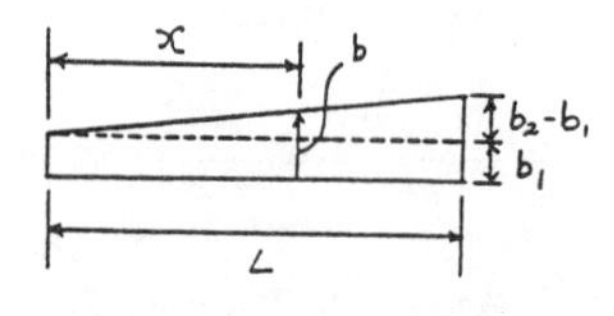

Bending Stress : Applying the flexure formula.

$$\sigma = \frac{M}{S} = \frac{Px}{\frac{w}{6L^2}\left[x(b_2 - b_1) + b_1 L\right]^2}$$

$$= \frac{6PL^2 x}{w\left[x(b_2 - b_1) + b_1 L\right]^2} \qquad [1]$$

In order to have the absolute maximum bending stress, $\frac{d\sigma}{dx} = 0$.

$$\frac{d\sigma}{dx} = \frac{6PL^2}{w}\left[\frac{\left[x(b_2 - b_1) + b_1 L\right]^2(1) - x(2)\left[x(b_2 - b_1) + b_1 L\right](b_2 - b_1)}{\left[x(b_2 - b_1) + b_1 L\right]^4}\right] = 0$$

$$x = \frac{b_1}{b_2 - b_1}L \qquad \textbf{Ans}$$

Substituting $x = \frac{b_1}{b_2 - b_1}L$ into Eq.[1] yields

$$\sigma_{max} = \frac{3PL}{2wb_1(b_2 - b_1)} \qquad \textbf{Ans}$$

***11–36.** The tapered simply supported beam supports the concentrated force **P** at its center. Determine the absolute maximum bending stress in the beam.

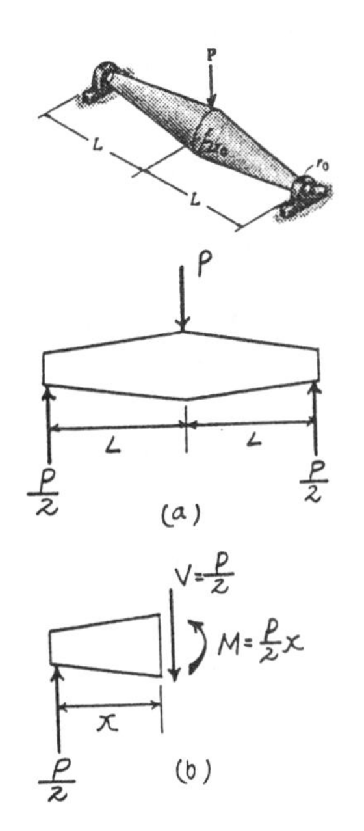

Moment Function : As shown on FBD(b).

Section Properties :

$$\frac{r-r_0}{x}=\frac{r_0}{L} \qquad r=\frac{r_0}{L}(x+L)$$

$$I=\frac{\pi}{4}\left[\frac{r_0}{L}(x+L)\right]^4=\frac{\pi r_0^4}{4L^4}(x+L)^4$$

$$S=\frac{I}{c}=\frac{\frac{\pi r_0^4}{4L^4}(x+L)^4}{\frac{r_0}{L}(x+L)}=\frac{\pi r_0^3}{4L^3}(x+L)^3$$

Bending Stress : Applying the flexure formula,

$$\sigma=\frac{M}{S}=\frac{\frac{P}{2}x}{\frac{\pi r_0^3}{4L^3}(x+L)^3}=\frac{2PL^3x}{\pi r_0^3(x+L)^3} \qquad [1]$$

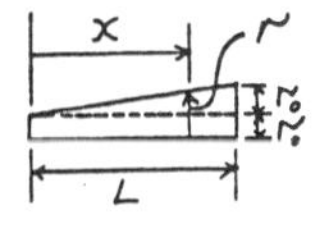

In order to have the absolute maximum bending stress, $\frac{d\sigma}{dx}=0.$

$$\frac{d\sigma}{dx}=\frac{2PL^3}{\pi r_0^3}\left[\frac{(x+L)^3(1)-x(3)(x+L)^2(1)}{(x+L)^6}\right]=0$$

$$x=\frac{L}{2}$$

Substitutung $x=\frac{L}{2}$ into Eq.[1] yields

$$\sigma_{\max}=\frac{8PL}{27\pi r_0^3} \qquad \textbf{Ans}$$

11–37. The motor at A is turning the shaft at a constant rate such that the torque **T** it develops creates horizontal reactions on the gears at C and D of 250 lb and 350 lb, respectively. The shaft is supported by a bearing in the motor at A and a bearing at B, which exert force components only in the x and z directions on the shaft. If the allowable shear stress for the shaft is $\tau_{\text{allow}}=10$ ksi, determine to the nearest $\frac{1}{8}$ in. the smallest diameter of the shaft that will support the loading. Use the maximum-shear-stress theory of failure.

Shaft Design : By observation, the critical section is located just to the left of gear C. Using the *maximum shear stress theory*,

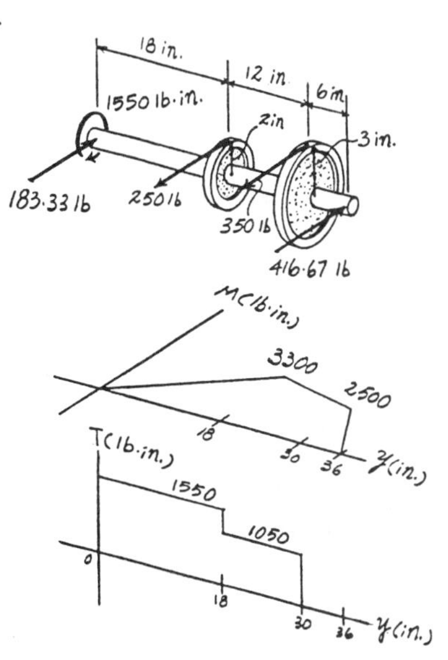

$$c=\left(\frac{2}{\pi\tau_{\text{allow}}}\sqrt{M^2+T^2}\right)^{\frac{1}{3}}$$

$$=\left(\frac{2}{\pi(10)(10^3)}\sqrt{3300^2+1550^2}\right)^{\frac{1}{3}}$$

$$=0.6146\text{ in.}$$

$$d=2c=2(0.6146)=1.229\text{ in.}$$

Use $\qquad d=1\frac{1}{4}$ in. $\qquad$ **Ans**

11–38. Solve Prob. 11–37 using the maximum-distortion-energy theory of failure with $\sigma_{allow} = 20$ ksi.

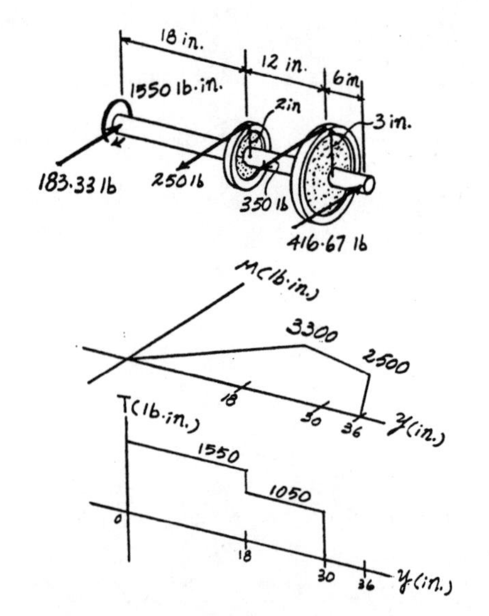

Support Reactions : As shown on FBD.

Torque and Moment Diagrams : As shown.

In-Plane Principal Stresses : Applying Eq. 9 – 5 with $\sigma_y = 0$,

$\sigma_x = \dfrac{Mc}{I} = \dfrac{4M}{\pi c^3}$, and $\tau_{xy} = \dfrac{Tc}{J} = \dfrac{2T}{\pi c^3}$,

$$\sigma_{1,2} = \frac{\sigma_x + \sigma_y}{2} \pm \sqrt{\left(\frac{\sigma_x - \sigma_y}{2}\right)^2 + \tau_{xy}^2}$$

$$= \frac{2M}{\pi c^3} \pm \sqrt{\left(\frac{2M}{\pi c^3}\right)^2 + \left(\frac{2T}{\pi c^3}\right)^2}$$

$$= \frac{2M}{\pi c^3} \pm \frac{2}{\pi c^3}\sqrt{M^2 + T^2}$$

Maximum Distortion Energy Theory : Let $a = \dfrac{2M}{\pi c^3}$ and $b = \dfrac{2}{\pi c^3}\sqrt{M^2+T^2}$, then $\sigma_1^2 = a^2 + b^2 + 2ab$, $\sigma_1\sigma_2 = a^2 - b^2$, $\sigma_2^2 = a^2 + b^2 - 2ab$, and $\sigma_1^2 - \sigma_1\sigma_2 + \sigma_2^2 = 3b^2 + a^2$.

$$\sigma_1^2 - \sigma_1\sigma_2 + \sigma_2^2 = \sigma_{allow}^2$$

$$3\left(\frac{2}{\pi c^3}\sqrt{M^2+T^2}\right)^2 + \left(\frac{2M}{\pi c^3}\right)^2 = \sigma_{allow}^2$$

$$c = \left[\frac{4}{\pi^2\sigma_{allow}^2}\left(4M^2 + 3T^2\right)\right]^{\frac{1}{6}}$$

Shaft Design : By observation, the critical section is located just to the left of gear *C*. Using the *maximum distortion energy theory*,

$$c = \left[\frac{4}{\pi^2\sigma_{allow}^2}\left(4M^2 + 3T^2\right)\right]^{\frac{1}{6}}$$

$$= \left\{\frac{4}{\pi^2[20(10^3)]^2}\left[4\left(3300^2\right) + 3\left(1550^2\right)\right]\right\}^{\frac{1}{6}}$$

$$= 0.6098 \text{ in.}$$

$$d = 2c = 2(0.6098) = 1.220 \text{ in.}$$

Use $\quad d = 1\frac{1}{4}$ in. $\quad$ **Ans**

11–39. The bearings at *A* and *D* exert only *y* and *z* components of force on the shaft. If $\tau_{allow} = 60$ MPa, determine to the nearest millimeter the smallest-diameter shaft that will support the loading. Use the maximum-shear-stress theory of failure.

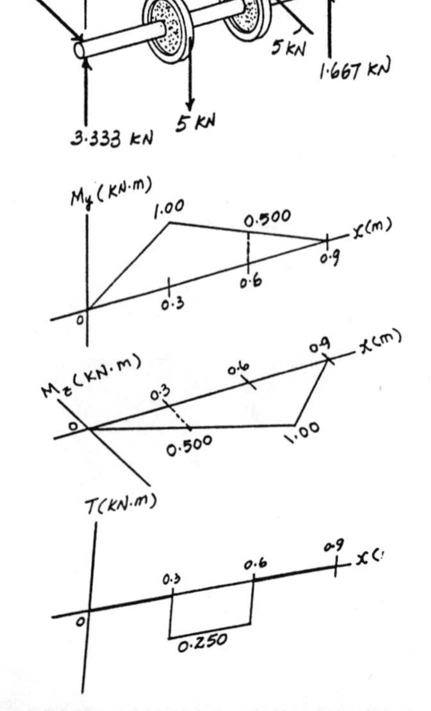

Shaft Design : By observation, the critical section is located just to the left of gear *C* and just to right of gear *B*, where $M = \sqrt{1.00^2 + 0.500^2} = 1.118$ kN · m and $T = 0.250$ kN · m. Using the *maximum shear stress theory*,

$$c = \left(\frac{2}{\pi\tau_{allow}}\sqrt{M^2+T^2}\right)^{\frac{1}{3}}$$

$$= \left[\frac{2}{\pi(60)(10^6)}\sqrt{[1.118(10^3)]^2 + [0.250(10^3)]^2}\right]^{\frac{1}{3}}$$

$$= 0.02299 \text{ m}$$

$$d = 2c = 2(0.02299) = 0.04599 \text{ m} = 45.99 \text{ mm}$$

Use $\quad d = 46$ mm $\quad$ **Ans**

***11–40.** Solve Prob. 11–39 using the maximum-distortion-energy theory of failure with $\sigma_{allow} = 180$ MPa.

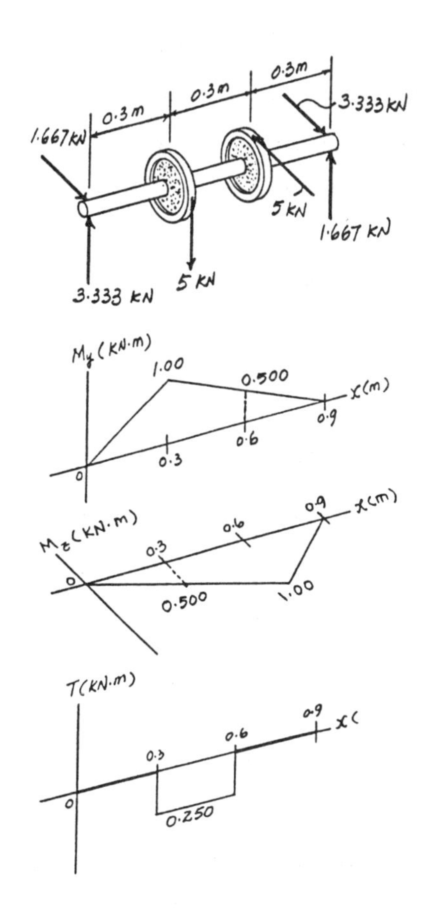

Torque and Moment Diagrams : As shown.

In-Plane Principal Stresses : Applying Eq. 9–5 with $\sigma_y = 0$,

$$\sigma_x = \frac{Mc}{I} = \frac{4M}{\pi c^3}\text{, and } \tau_{xy} = \frac{Tc}{J} = \frac{2T}{\pi c^3},$$

$$\begin{aligned}\sigma_{1,2} &= \frac{\sigma_x + \sigma_y}{2} \pm \sqrt{\left(\frac{\sigma_x - \sigma_y}{2}\right)^2 + \tau_{xy}^2}\\ &= \frac{2M}{\pi c^3} \pm \sqrt{\left(\frac{2M}{\pi c^3}\right)^2 + \left(\frac{2T}{\pi c^3}\right)^2}\\ &= \frac{2M}{\pi c^3} \pm \frac{2}{\pi c^3}\sqrt{M^2 + T^2}\end{aligned}$$

Maximum Distortion Energy Theory : Let $a = \frac{2M}{\pi c^3}$ and $b = \frac{2}{\pi c^3}\sqrt{M^2 + T^2}$ then $\sigma_1^2 = a^2 + b^2 + 2ab$, $\sigma_1\sigma_2 = a^2 - b^2$, $\sigma_2^2 = a^2 + b^2 - 2ab$ and $\sigma_1^2 - \sigma_1\sigma_2 + \sigma_2^2 = 3b^2 + a^2$.

$$\sigma_1^2 - \sigma_1\sigma_2 + \sigma_2^2 = \sigma_{allow}^2$$

$$3\left(\frac{2}{\pi c^3}\sqrt{M^2 + T^2}\right)^2 + \left(\frac{2M}{\pi c^3}\right)^2 = \sigma_{allow}^2$$

$$c = \left[\frac{4}{\pi^2\sigma_{allow}^2}\left(4M^2 + 3T^2\right)\right]^{\frac{1}{6}}$$

Shaft Design : By observation, the critical section is located just to the left of gear C and just to right of gear B, where $M = \sqrt{1.00^2 + 0.500^2} = 1.118$ kN · m and $T = 0.250$ kN · m. Using the *maximum distortion energy theory*,

$$\begin{aligned}c &= \left[\frac{4}{\pi^2\sigma_{allow}^2}\left(4M^2 + 3T^2\right)\right]^{\frac{1}{6}}\\ &= \left\{\frac{4}{\pi^2[180(10^6)]^2}\left[4(1118)^2 + 3(250)^2\right]\right\}^{\frac{1}{6}}\\ &= 0.02005 \text{ m}\end{aligned}$$

$$d = 2c = 2(0.02005) = 0.04009 \text{ m} = 40.09 \text{ mm}$$

Use $\qquad d = 41$ mm $\qquad$ **Ans**

11–41. The end gear connected to the shaft is subjected to the loading shown. If the bearings at A and B exert only y and z components of force on the shaft, determine the equilibrium torque T at gear C and then find the smallest diameter of the shaft to the nearest millimeter that will support the loading. Use the maximum-shear-stress theory of failure with $\tau_{\text{allow}} = 60$ MPa.

As shown on FBD, $T = 100$ N · m. **Ans**

Shaft Design : By observation, the critical section is located at support A, where $M = \sqrt{225^2 + 0} = 225$ N · m and $T = 150$ N · m. Using the *maximum shear stress theory*,

$$c = \left(\frac{2}{\pi \tau_{\text{allow}}}\sqrt{M^2 + T^2}\right)^{\frac{1}{3}}$$

$$= \left[\frac{2}{\pi(60)(10^6)}\sqrt{225^2 + 150^2}\right]^{\frac{1}{3}}$$

$$= 0.01421 \text{ m}$$

$$d = 2c = 2(0.01421) = 0.02842 \text{ m} = 28.42 \text{ mm}$$

Use $d = 29$ mm **Ans**

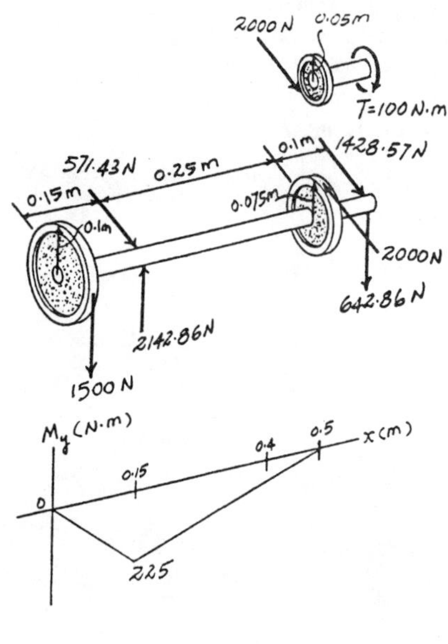

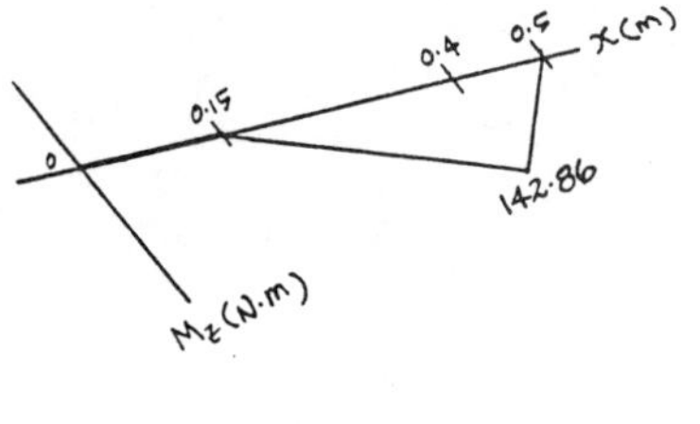

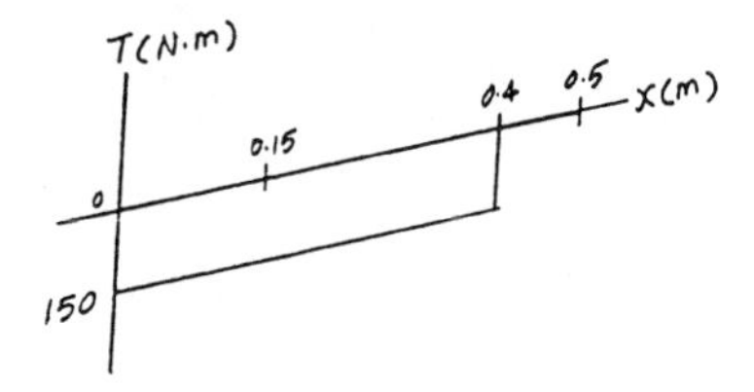

11–42. Solve Prob. 11–41 using the maximum-distortion-energy theory of failure with $\sigma_{\text{allow}} = 80$ MPa.

Torque and Moment Diagrams : As shown.

$T = 100$ N·m **Ans**

In - Plane Principal Stresses : Applying Eq. 9 – 5 with $\sigma_y = 0$,

$$\sigma_x = \frac{Mc}{I} = \frac{4M}{\pi c^3} \quad \text{and} \quad \tau_{xy} = \frac{Tc}{J} = \frac{2T}{\pi c^3}.$$

$$\sigma_{1,2} = \frac{\sigma_x + \sigma_y}{2} \pm \sqrt{\left(\frac{\sigma_x - \sigma_y}{2}\right)^2 + \tau_{xy}^2}$$

$$= \frac{2M}{\pi c^3} \pm \sqrt{\left(\frac{2M}{\pi c^3}\right)^2 + \left(\frac{2T}{\pi c^3}\right)^2}$$

$$= \frac{2M}{\pi c^3} \pm \frac{2}{\pi c^3}\sqrt{M^2 + T^2}$$

Maximum Distortion Energy Theory : Let $a = \dfrac{2M}{\pi c^3}$ and $b = \dfrac{2}{\pi c^3}\sqrt{M^2 + T^2}$ then $\sigma_1^2 = a^2 + b^2 + 2ab$, $\sigma_1 \sigma_2 = a^2 - b^2$, $\sigma_2^2 = a^2 + b^2 - 2ab$ and $\sigma_1^2 - \sigma_1 \sigma_2 + \sigma_2^2 = 3b^2 + a^2$.

$$\sigma_1^2 - \sigma_1 \sigma_2 + \sigma_2^2 = \sigma_{\text{allow}}^2$$

$$3\left(\frac{2}{\pi c^3}\sqrt{M^2 + T^2}\right)^2 + \left(\frac{2M}{\pi c^3}\right)^2 = \sigma_{\text{allow}}^2$$

$$c = \left[\frac{4}{\pi^2 \sigma_{\text{allow}}^2}\left(4M^2 + 3T^2\right)\right]^{\frac{1}{6}}$$

Shaft Design : By observation, the critical section is located at support A, where $M = \sqrt{225^2 + 0} = 225$ N · m and $T = 150$ N · m. Using the *maximum distortion energy theory*,

$$c = \left[\frac{4}{\pi^2 \sigma_{\text{allow}}^2}\left(4M^2 + 3T^2\right)\right]^{\frac{1}{6}}$$

$$= \left\{\frac{4}{\pi^2 [80(10^6)]^2}\left[4(225)^2 + 3(150)^2\right]\right\}^{\frac{1}{6}}$$

$$= 0.01605 \text{ m}$$

$$d = 2c = 2(0.01605) = 0.03210 \text{ m} = 32.1 \text{mm}$$

Use $d = 33$ mm **Ans**

11–43. The shaft is supported by bearings at A and B that exert force components only in the x and z directions on the shaft. If the allowable normal stress for the shaft is $\sigma_{allow} = 15$ ksi, determine to the nearest $\frac{1}{8}$ in. the smallest diameter of the shaft that will support the gear loading. Use the maximum-distortion-energy theory of failure.

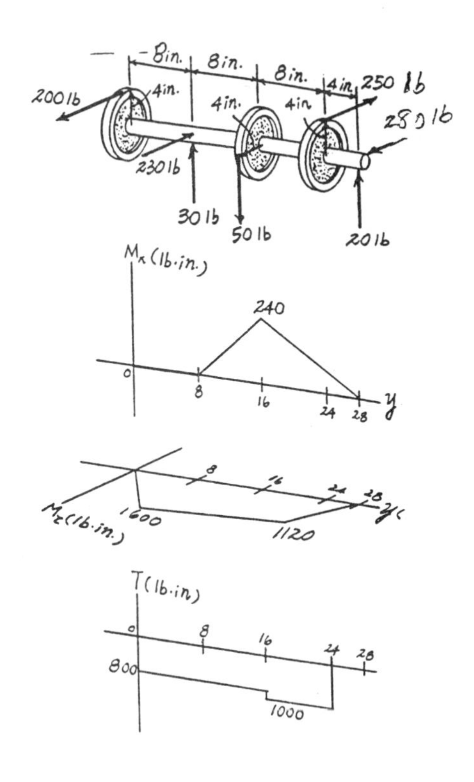

Torque and Moment Diagrams : As shown.

In - Plane Principal Stresses : Applying Eq. 9 – 5 with $\sigma_y = 0$,

$$\sigma_x = \frac{Mc}{I} = \frac{4M}{\pi c^3} \text{ and } \tau_{xy} = \frac{Tc}{J} = \frac{2T}{\pi c^3}.$$

$$\sigma_{1,2} = \frac{\sigma_x + \sigma_y}{2} \pm \sqrt{\left(\frac{\sigma_x - \sigma_y}{2}\right)^2 + \tau_{xy}^2}$$
$$= \frac{2M}{\pi c^3} \pm \sqrt{\left(\frac{2M}{\pi c^3}\right)^2 + \left(\frac{2T}{\pi c^3}\right)^2}$$
$$= \frac{2M}{\pi c^3} \pm \frac{2}{\pi c^3}\sqrt{M^2 + T^2}$$

Maximum Distortion Energy Theory : Let $a = \dfrac{2M}{\pi c^3}$ and $b = \dfrac{2}{\pi c^3}\sqrt{M^2+T^2}$, then $\sigma_1^2 = a^2 + b^2 + 2ab$. $\sigma_1\sigma_2 = a^2 - b^2$. $\sigma_2^2 = a^2 + b^2 - 2ab$, and $\sigma_1^2 - \sigma_1\sigma_2 + \sigma_2^2 = 3b^2 + a^2$.

$$\sigma_1^2 - \sigma_1\sigma_2 + \sigma_2^2 = \sigma_{allow}^2$$
$$3\left(\frac{2}{\pi c^3}\sqrt{M^2+T^2}\right)^2 + \left(\frac{2M}{\pi c^3}\right)^2 = \sigma_{allow}^2$$
$$c = \left[\frac{4}{\pi^2 \sigma_{allow}^2}\left(4M^2 + 3T^2\right)\right]^{\frac{1}{6}}$$

Shaft Design : By observation, the critical section is located at support A, where $M = \sqrt{1600^2 + 0} = 1600$ lb · in and $T = 800$ lb · in. Using the *maximum distortion energy theory*,

$$c = \left[\frac{4}{\pi^2 \sigma_{allow}^2}\left(4M^2 + 3T^2\right)\right]^{\frac{1}{6}}$$
$$= \left\{\frac{4}{\pi^2[15(10^3)]^2}\left[4(1600)^2 + 3(800)^2\right]\right\}^{\frac{1}{6}}$$
$$= 0.5290 \text{ in.}$$

$$d = 2c = 2(0.5290) = 1.058 \text{ in.}$$

Use $\quad d = 1\frac{1}{8}$ in. $\quad$ **Ans**

***11–44.** Solve Prob. 11–43 using the maximum-shear-stress theory of failure with $\tau_{allow} = 6$ ksi.

Shaft Design : By observation, the critical section is located at support A, where $M = \sqrt{1600^2 + 0} = 1600$ lb · in and $T = 800$ lb · in. Using the *maximum shear stress theory*,

$$c = \left(\frac{2}{\pi \tau_{allow}}\sqrt{M^2 + T^2}\right)^{\frac{1}{3}}$$
$$= \left[\frac{2}{\pi(6)(10^3)}\sqrt{1600^2 + 800^2}\right]^{\frac{1}{3}}$$
$$= 0.5747 \text{ in.}$$

$$d = 2c = 2(0.5747) = 1.149 \text{ in.}$$

Use $\quad d = 1\frac{1}{4}$ in. $\quad$ **Ans**

11–45. The shaft is supported on journal bearings that do not offer resistance to axial load. If the allowable normal stress for the shaft is $\sigma_{allow} = 80$ MPa, determine to the nearest millimeter the smallest diameter of the shaft that will support the loading. Use the maximum-distortion-energy theory of failure.

Torque and Moment Diagrams : As shown.

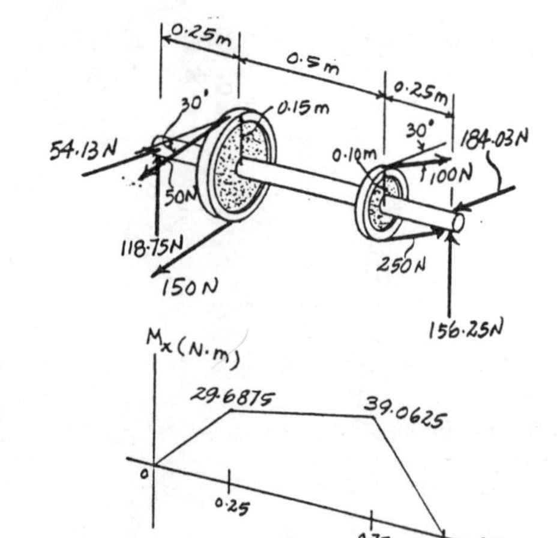

In - Plane Principal Stresses : Applying Eq.9 – 5 with $\sigma_y = 0$,

$$\sigma_x = \frac{Mc}{I} = \frac{4M}{\pi c^3}, \text{ and } \tau_{xy} = \frac{Tc}{J} = \frac{2T}{\pi c^3}.$$

$$\sigma_{1,2} = \frac{\sigma_x + \sigma_y}{2} \pm \sqrt{\left(\frac{\sigma_x - \sigma_y}{2}\right)^2 + \tau_{xy}^2}$$

$$= \frac{2M}{\pi c^3} \pm \sqrt{\left(\frac{2M}{\pi c^3}\right)^2 + \left(\frac{2T}{\pi c^3}\right)^2}$$

$$= \frac{2M}{\pi c^3} \pm \frac{2}{\pi c^3}\sqrt{M^2 + T^2}$$

Maximum Distortion Energy Theory : Let $a = \dfrac{2M}{\pi c^3}$ and $b = \dfrac{2}{\pi c^3}\sqrt{M^2+T^2}$, then $\sigma_1^2 = a^2 + b^2 + 2ab$, $\sigma_1\sigma_2 = a^2 - b^2$, $\sigma_2^2 = a^2 + b^2 - 2ab$, and $\sigma_1^2 - \sigma_1\sigma_2 + \sigma_2^2 = 3b^2 + a^2$.

$$\sigma_1^2 - \sigma_1\sigma_2 + \sigma_2^2 = \sigma_{allow}^2$$

$$3\left(\frac{2}{\pi c^3}\sqrt{M^2+T^2}\right)^2 + \left(\frac{2M}{\pi c^3}\right)^2 = \sigma_{allow}^2$$

$$c = \left[\frac{4}{\pi^2 \sigma_{allow}^2}\left(4M^2 + 3T^2\right)\right]^{\frac{1}{6}}$$

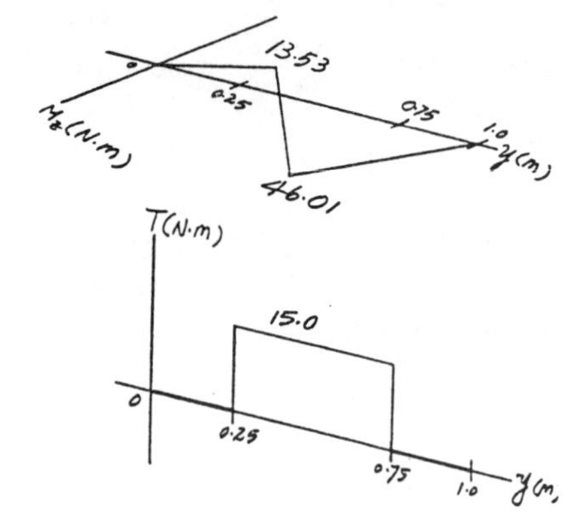

Shaft Design : By observation, the critical section is located just to the left of gear C, where $M = \sqrt{39.0625^2 + 46.01^2} = 60.354$ N · m and $T = 15.0$ N · m. Using the *maximum distortion energy theory*,

$$c = \left[\frac{4}{\pi^2 \sigma_{allow}^2}\left(4M^2 + 3T^2\right)\right]^{\frac{1}{6}}$$

$$= \left\{\frac{4}{\pi^2[80(10^6)]^2}\left[4(60.354)^2 + 3(15.0)^2\right]\right\}^{\frac{1}{6}}$$

$$= 0.009942 \text{ m}$$

$$d = 2c = 2(0.009942) = 0.01988 \text{ m} = 19.88 \text{ mm}$$

Use $d = 20$ mm **Ans**

11–46. The shaft is supported on journal bearings that do not offer resistance to axial load. If the allowable shear stress for the shaft is $\tau_{allow} = 35$ MPa, determine to the nearest millimeter the smallest diameter of the shaft that will support the loading. Use the maximum-shear-stress theory of failure.

Shaft Design : By observation, the critical section is located just to the left of gear C, where $M = \sqrt{39.0625^2 + 46.01^2} = 60.354$ N · m and $T = 15.0$ N · m. Using the *maximum shear stress theory*,

$$c = \left(\frac{2}{\pi \tau_{allow}}\sqrt{M^2 + T^2}\right)^{\frac{1}{3}}$$

$$= \left[\frac{2}{\pi(35)(10^6)}\sqrt{60.354^2 + 15.0^2}\right]^{\frac{1}{3}}$$

$$= 0.01042 \text{ m}$$

$$d = 2c = 2(0.01042) = 0.02084 \text{ m} = 20.84 \text{ mm}$$

Use $d = 21$ mm **Ans**

11–47. The bearings at A and B exert only x and z components of force on the steel shaft. Determine the shaft's diameter to the nearest millimeter so that it can resist the loadings of the gears without exceeding an allowable shear stress of $\tau_{allow} = 80$ MPa. Use the maximum-shear-stress theory of failure.

Shaft Design : By observation, the critical section is located just to the left of gear C, where $M = \sqrt{1250^2 + 250^2} = 1274.75$ N · m and $T = 375$ N · m. Using the *maximum shear stress theory*,

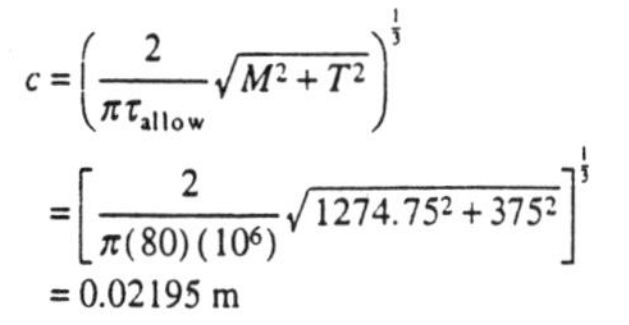

$$c = \left(\frac{2}{\pi \tau_{allow}}\sqrt{M^2 + T^2}\right)^{\frac{1}{3}}$$

$$= \left[\frac{2}{\pi(80)(10^6)}\sqrt{1274.75^2 + 375^2}\right]^{\frac{1}{3}}$$

$$= 0.02195 \text{ m}$$

$$d = 2c = 2(0.02195) = 0.04390 \text{ m} = 43.90 \text{ mm}$$

Use $d = 44$ mm **Ans**

***11–48.** The bearings at A and B exert only x and z components of force on the steel shaft. Determine the shaft's diameter to the nearest millimeter so that it can resist the loadings of the gears without exceeding an allowable normal stress of $\sigma_{allow} = 200$ MPa. Use the maximum-distortion-energy theory of failure.

Torque and Moment Diagrams : As shown.

In - Plane Principal Stresses : Applying Eq. 9 – 5 with $\sigma_y = 0$,

$$\sigma_x = \frac{Mc}{I} = \frac{4M}{\pi c^3}, \text{ and } \tau_{xy} = \frac{Tc}{J} = \frac{2T}{\pi c^3}.$$

$$\sigma_{1,2} = \frac{\sigma_x + \sigma_y}{2} \pm \sqrt{\left(\frac{\sigma_x - \sigma_y}{2}\right)^2 + \tau_{xy}^2}$$

$$= \frac{2M}{\pi c^3} \pm \sqrt{\left(\frac{2M}{\pi c^3}\right)^2 + \left(\frac{2T}{\pi c^3}\right)^2}$$

$$= \frac{2M}{\pi c^3} \pm \frac{2}{\pi c^3}\sqrt{M^2 + T^2}$$

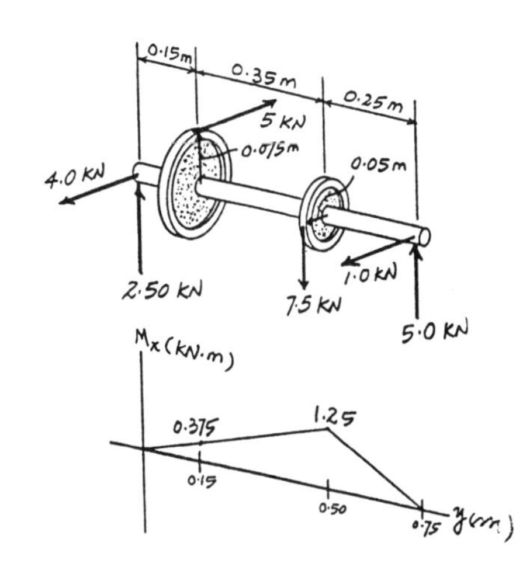

Maximum Distortion Energy Theory : Let $a = \frac{2M}{\pi c^3}$ and $b = \frac{2}{\pi c^3}\sqrt{M^2 + T^2}$, then $\sigma_1^2 = a^2 + b^2 + 2ab$, $\sigma_1 \sigma_2 = a^2 - b^2$, $\sigma_2^2 = a^2 + b^2 - 2ab$, and $\sigma_1^2 - \sigma_1 \sigma_2 + \sigma_2^2 = 3b^2 + a^2$.

$$\sigma_1^2 - \sigma_1 \sigma_2 + \sigma_2^2 = \sigma_{allow}^2$$

$$3\left(\frac{2}{\pi c^3}\sqrt{M^2 + T^2}\right)^2 + \left(\frac{2M}{\pi c^3}\right)^2 = \sigma_{allow}^2$$

$$c = \left[\frac{4}{\pi^2 \sigma_{allow}^2}\left(4M^2 + 3T^2\right)\right]^{\frac{1}{6}}$$

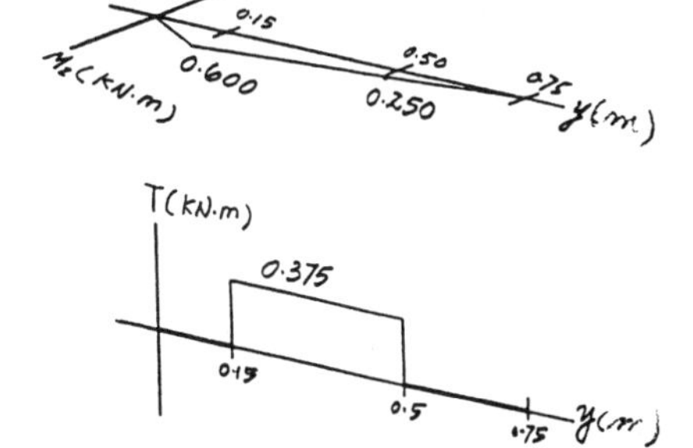

Shaft Design : By observation, the critical section is located just to the left of gear C, where $M = \sqrt{1250^2 + 250^2} = 1274.75$ N · m and $T = 375$ N · m. Using the *maximum distortion energy theory*,

$$c = \left[\frac{4}{\pi^2 \sigma_{allow}^2}\left(4M^2 + 3T^2\right)\right]^{\frac{1}{6}}$$

$$= \left\{\frac{4}{\pi^2 [200(10^6)]^2}\left[4(1274.75)^2 + 3(375)^2\right]\right\}^{\frac{1}{6}}$$

$$= 0.02031 \text{ m}$$

$$d = 2c = 2(0.02031) = 0.04062 \text{ m} = 40.62 \text{ mm}$$

Use $d = 41$ mm **Ans**

11–49. Draw the shear and moment diagrams for the shaft, and then determine its required diameter to the nearest millimeter if $\sigma_{allow} = 140$ MPa and $\tau_{allow} = 80$ MPa. The bearings at A and B exert only vertical reactions on the shaft.

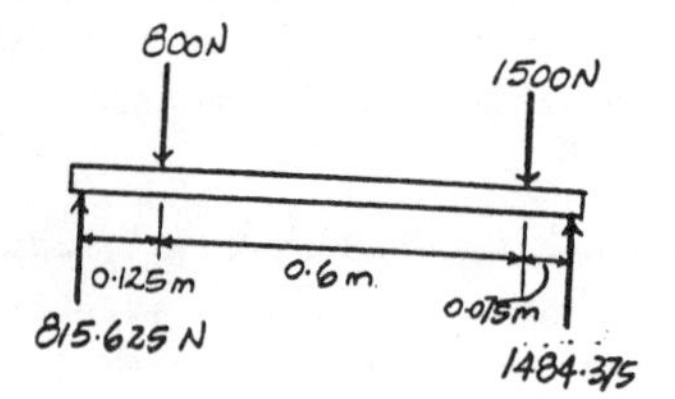

Bending Stress : From the moment diagram, $M_{max} = 111$ N·m. Assume bending controls the design. Applying the flexure formula,

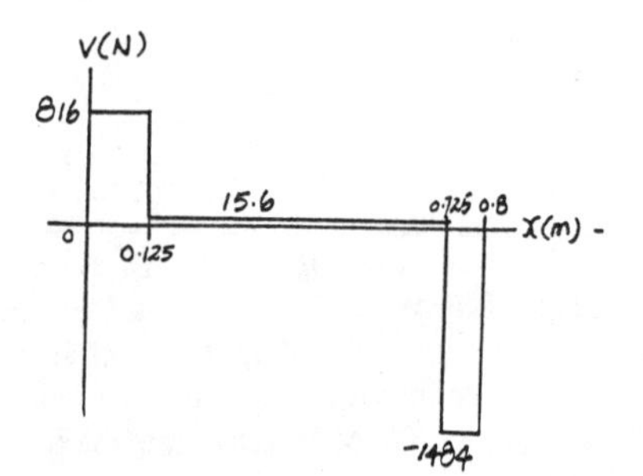

$$\sigma_{allow} = \frac{M_{max}\,c}{I}$$

$$140(10^6) = \frac{111\left(\frac{d}{2}\right)}{\frac{\pi}{4}\left(\frac{d}{2}\right)^4}$$

$$d = 0.02008 \text{ m} = 20.1 \text{ mm}$$

Use $d = 21$ mm **Ans**

Shear Stress : Provide a shear stress check using the shear formula with

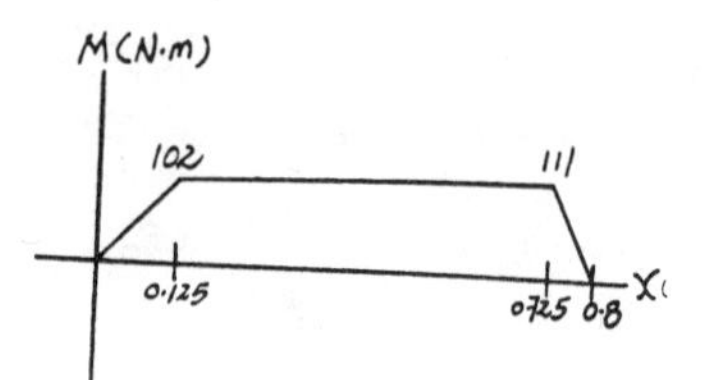

$$I = \frac{\pi}{4}\left(0.0105^4\right) = 9.5466\left(10^{-9}\right) \text{ m}^4$$

$$Q_{max} = \frac{4(0.0105)}{3\pi}\left[\frac{1}{2}(\pi)(0.0105)^2\right] = 0.77175\left(10^{-6}\right) \text{ m}^3$$

From the shear diagram, $V_{max} = 1484$ N.

$$\tau_{max} = \frac{V_{max}\,Q_{max}}{It}$$
$$= \frac{1484\left[0.77175(10^{-6})\right]}{9.5466(10^{-9})(0.021)}$$
$$= 5.71 \text{ MPa} < \tau_{allow} = 80 \text{ MPa } (O.K!)$$

11–50. Select the lightest-weight steel wide-flange beam from Appendix B that will safely support the loading shown. The allowable bending stress if $\sigma_{allow} = 22$ ksi, and the allowable shear stress is $\tau_{allow} = 12$ ksi.

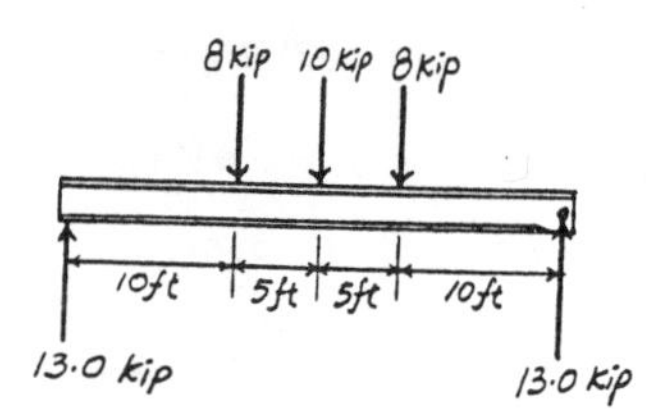

Bending Stress : From the moment diagram, $M_{max} = 155$ kip·ft. Assume bending controls the design. Applying the flexure formula,

$$S_{req'd} = \frac{M_{max}}{\sigma_{allow}}$$
$$= \frac{155(12)}{22} = 84.55 \text{ in}^3$$

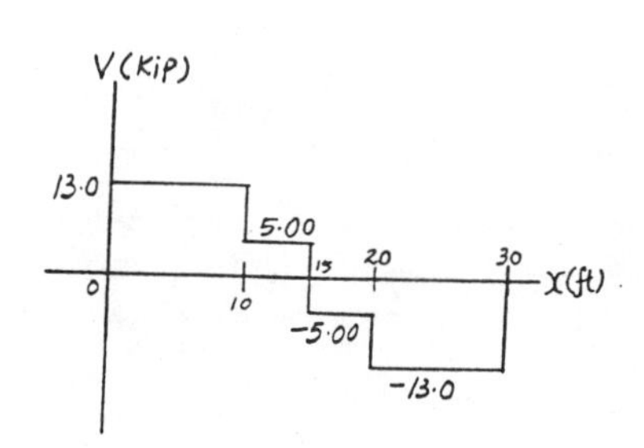

Select W18×50 $\left(S_x = 88.9 \text{ in}^3,\ d = 17.99 \text{ in.},\ t_w = 0.355 \text{ in.}\right)$

Shear Stress : Provide a shear stress check using $\tau = \frac{V}{t_w d}$ for a W18×50 wide-flange section. From the shear diagram, $V_{max} = 13.0$ kip.

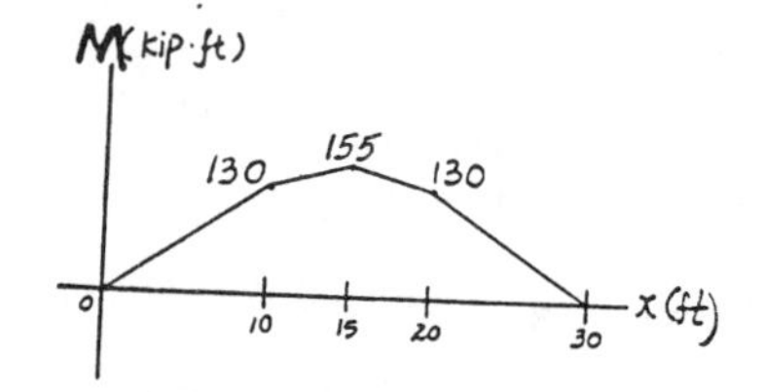

$$\tau_{max} = \frac{V_{max}}{t_w d}$$
$$= \frac{13.0}{0.355(17.99)}$$
$$= 2.04 \text{ ksi} < \tau_{allow} = 12 \text{ ksi } (O.K!)$$

Hence, **Use** W18×50 **Ans**

11–51. The beam is made in the shape of a frustum that has a diameter of 1 ft at A and a diameter of 2 ft at B. If it supports a couple moment of 8 kip · ft at its end, determine the absolute maximum bending stress in the beam and specify its location x.

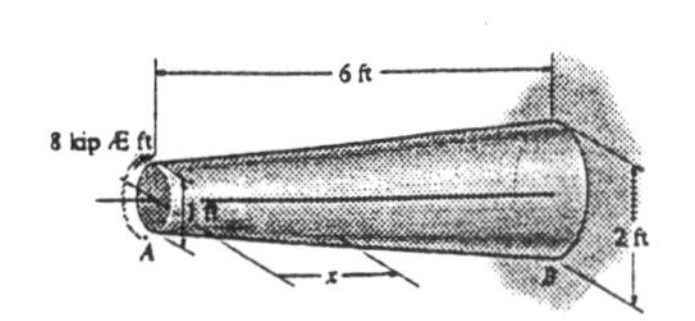

Section Properties :

$$\frac{r-6}{x} = \frac{6}{72} \qquad r = \frac{x+72}{12}$$

$$I = \frac{\pi}{4}\left(\frac{x+72}{12}\right)^4 = \frac{\pi}{82944}(x+72)^4$$

$$S = \frac{I}{c} = \frac{\frac{\pi}{82944}(x+72)^4}{\frac{x+72}{12}} = \frac{\pi}{6912}(x+72)^3$$

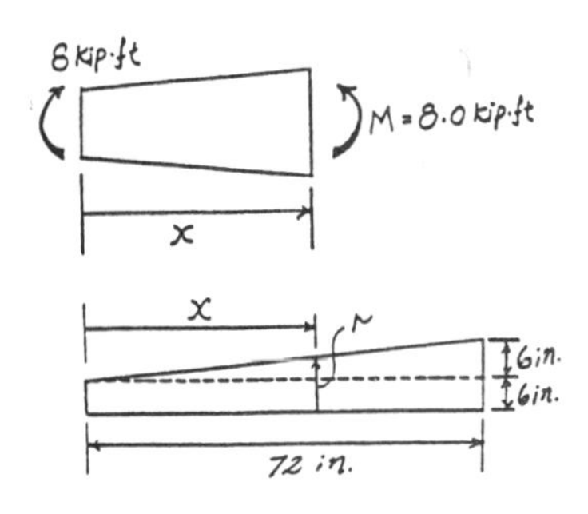

Bending Stress : Applying the flexure formula.

$$\sigma = \frac{M}{S} = \frac{8.00(12)}{\frac{\pi}{6912}(x+72)^3} = \frac{663552}{\pi(x+72)^3} \qquad [1]$$

Since σ is a decreasing function, σ_{max} occurs at $x = 0$ **Ans**

Substituting $x = 0$ into Eq.[1] yields

$$\sigma_{max} = \frac{663552}{\pi(0+72)^3} = 0.566\ ksi \qquad \textbf{Ans}$$

***11–52.** The beam is constructed from three boards as shown. If each nail can support a shear force of 50 lb, determine the maximum spacing of the nails, s, s', and s'', to he nearest $\frac{1}{8}$ in., for regions AB, BC, and CD, respectively.

5.4

Section Properties :

$$\bar{y} = \frac{\Sigma\tilde{y}A}{\Sigma A} = \frac{0.5(8)(1) + 4(6)(2)}{8(1) + 6(2)} = 2.60 \text{ in.}$$

$$I = \frac{1}{12}(8)(1^3) + 8(1)(2.60-0.5)^2$$

$$\frac{1}{12}(2)(6^3) + 2(6)(4-2.60)^2$$

$$= 9.\ \ 667 \text{ in}^4$$

$$Q_A = \bar{y}'A' = 2.10(1)(8) = 16.8 \text{ in}^3$$

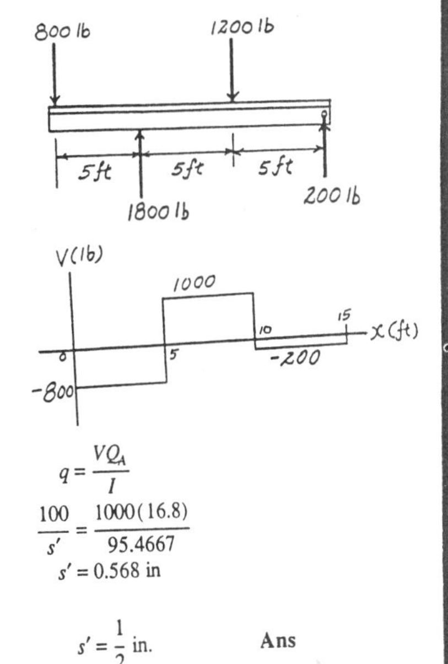

Shear Flow :

For $0 \le x < 5$ **ft** (region AB) , the design shear force is $V = 800$ lb.

Since there are two rows of nails, the allowable shear flow is $q = \frac{2(50)}{s} = \frac{100}{s}$.

$$q = \frac{VQ_A}{I}$$

$$\frac{100}{s} = \frac{800(16.8)}{95.4667}$$

$$s = 0.710 \text{ in}$$

Use $s = \frac{5}{8}$ in. **Ans**

For 5 **ft** $< x < 10$ **ft** (region BC), the design shear force is $V = 1000$ lb. Since there are two rows of nails, the allowable shear flow is $q = \frac{2(50)}{s'} = \frac{100}{s'}$.

$$q = \frac{VQ_A}{I}$$

$$\frac{100}{s'} = \frac{1000(16.8)}{95.4667}$$

$$s' = 0.568 \text{ in}$$

Use $s' = \frac{1}{2}$ in. **Ans**

For 10 **ft** $< x \le 15$ **ft** (region CD) , the design shear force is $V = 200$ lb. Since there are two rows of nails, the allowable shear flow is $q = \frac{2(50)}{s''} = \frac{100}{s''}$.

$$q = \frac{VQ_A}{I}$$

$$\frac{100}{s''} = \frac{200(16.8)}{95.4667}$$

$$s'' = 2.84 \text{ in}$$

Use $s'' = 2\frac{3}{4}$ in. **Ans**

11–53. Draw the shear and moment diagrams for the beam. Then select the lightest-weight steel wide-flange beam from Appendix *B* that will safely support the loading. Take $\sigma_{allow} = 22$ ksi, and $\tau_{allow} = 12$ ksi.

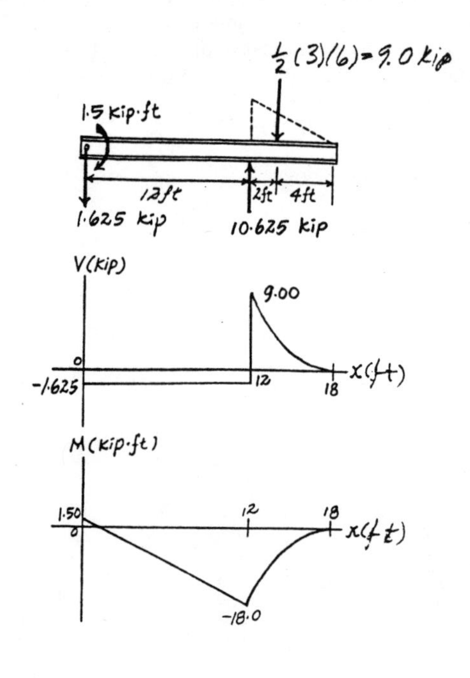

Bending Stress : From the moment diagram, $M_{max} = 18.0$ kip · ft. Assume bending controls the design. Applying the flexure formula,

$$S_{req'd} = \frac{M_{max}}{\sigma_{allow}} = \frac{18.0(12)}{22} = 9.82 \text{ in}^3$$

Select W10×12 $(S_x = 10.9 \text{ in}^3,\ d = 9.87 \text{ in.},\ t_w = 0.19 \text{ in.})$

Shear Stress : Provide a shear stress check using $\tau = \dfrac{V}{t_w d}$ for the W10×12 wide - flange section. From the shear diagram, $V_{max} = 9.00$ kip

$$\tau_{max} = \frac{V_{max}}{t_w d} = \frac{9.00}{0.19(9.87)} = 4.80 \text{ ksi} < \tau_{allow} = 12 \text{ ksi } (\textbf{\textit{O.K!}})$$

Hence, **Use** W10×12 **Ans**

11–54. Determine if the steel cantilevered T-beam can safely support the two loads of $P = 3$ kN if the allowable bending stress is $\sigma_{allow} = 170$ MPa and the allowable shear stress is $\tau_{allow} = 95$ MPa.

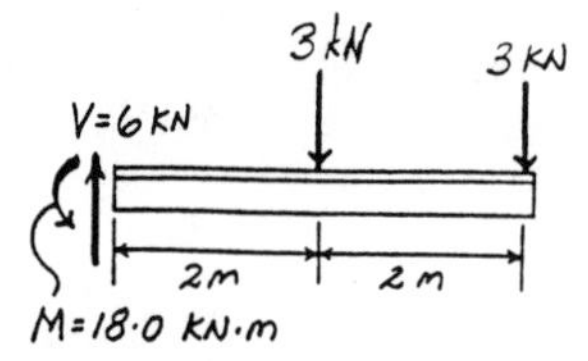

$$\bar{y} = \frac{\Sigma \tilde{y} A}{\Sigma A} = \frac{0.0075(0.015)(0.15) + 0.09(0.15)(0.015)}{0.015(0.15) + 0.15(0.015)} = 0.04875 \text{ m}$$

$$I = \frac{1}{12}(0.15)\left(0.015^3\right) + 0.15(0.015)(0.04875 - 0.0075)^2$$
$$\frac{1}{12}(0.015)\left(0.15^3\right) + 0.015(0.15)(0.09 - 0.04875)^2$$
$$= 11.9180\left(10^{-6}\right) \text{ m}^4$$

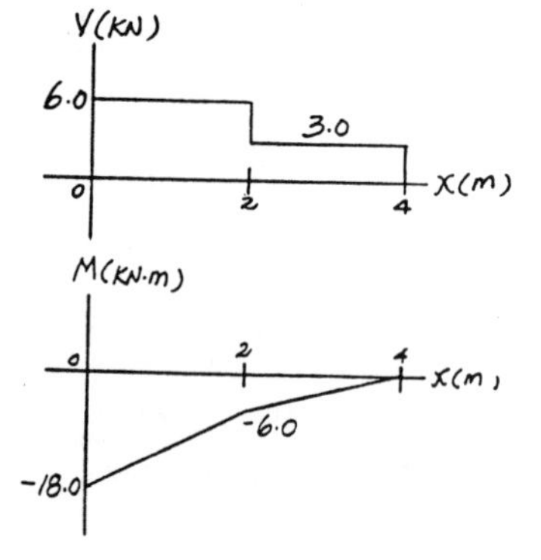

$$Q_{max} = \bar{y}'A' = 0.058125(0.11625)(0.015) = 0.10136\left(10^{-3}\right) \text{ m}^3$$

Bending Stress : From the moment diagram, $M_{max} = 18.0$ kN · m. Applying the flexure formula,

$$\sigma_{max} = \frac{M_{max}\,c}{I} = \frac{18.0(10^3)(0.11625)}{11.9180(10^{-6})} = 175.6 \text{ MPa} > \sigma_{allow} = 170 \text{ MPa } (\textbf{\textit{No GOOD!}})$$

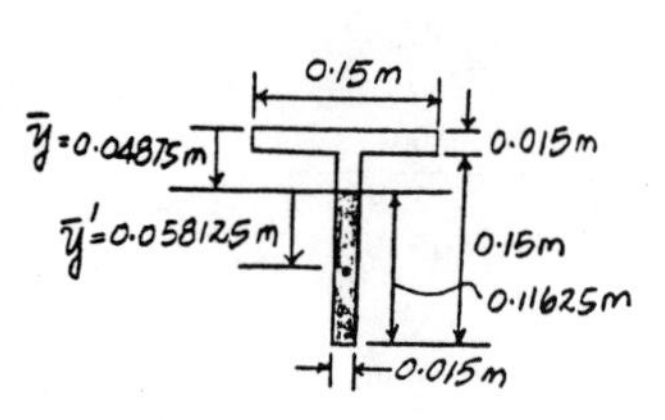

Shear Stress : From the shear diagram, $V_{max} = 6.00$ kN. Applying the shear formula,

$$\tau_{max} = \frac{V_{max}Q_{max}}{It} = \frac{6.00(10^3)\left[0.10136(10^{-3})\right]}{11.9180(10^{-6})(0.015)} = 3.40 \text{ MPa} < \tau_{allow} = 95 \text{ MPa } (\textbf{\textit{O.K!}})$$

Hence, **The T - beam fails due to the bending stress and will not safely support the loading.** **Ans**

11–55. The tapered beam supports a uniform distributed load w. If it is made from a plate and has a constant width b, determine the absolute maximum bending stress in the beam.

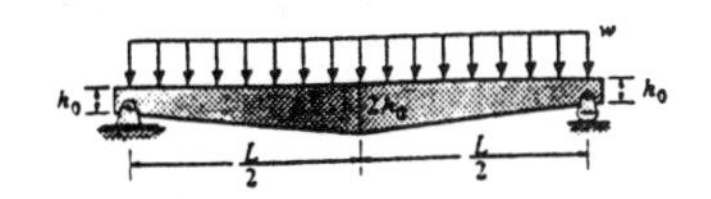

Support Reactions : As shown on FBD(a).

Moment Function : As shown on FBD(b).

Section Properties :

$$\frac{h-h_0}{x} = \frac{h_0}{\frac{L}{2}} \qquad h = \frac{h_0}{L}(2x+L)$$

$$I = \frac{1}{12}(b)\left(\frac{h_0^3}{L^3}\right)(2x+L)^3$$

$$S = \frac{\frac{1}{12}(b)\left(\frac{h_0^3}{L^3}\right)(2x+L)^3}{\frac{h_0}{2L}(2x+L)} = \frac{bh_0^2}{6L^2}(2x+L)^2$$

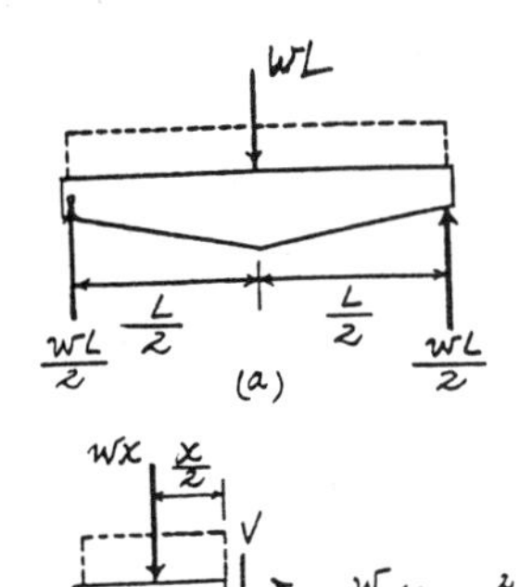

Bending Stress : Applying the flexure formula.

$$\sigma = \frac{M}{S} = \frac{\frac{w}{2}(Lx-x^2)}{\frac{bh_0^2}{6L^2}(2x+L)^2} = \frac{3wL^2(Lx-x^2)}{bh_0^2(2x+L)^2} \qquad [1]$$

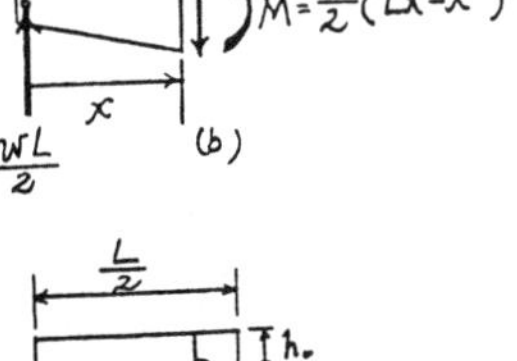

In order to have the absolute maximum bending stress, $\frac{d\sigma}{dx} = 0$.

$$\frac{d\sigma}{dx} = \frac{3wL^2}{bh_0^2}\left[\frac{(2x+L)^2(L-2x)-(Lx-x^2)(2)(2x+L)(2)}{(2x+L)^4}\right] = 0$$

$$x = \frac{L}{4}$$

Substituting $x = \frac{L}{4}$ into Eq.[1] yields

$$\sigma_{max} = \frac{wL^2}{4bh_0^2} \qquad \textbf{Ans}$$

***11–56.** The 10-mm-wide bracket is used to support a force of 70 N at its end A. Determine the absolute maximum bending stress in the bracket, and specify its location x.

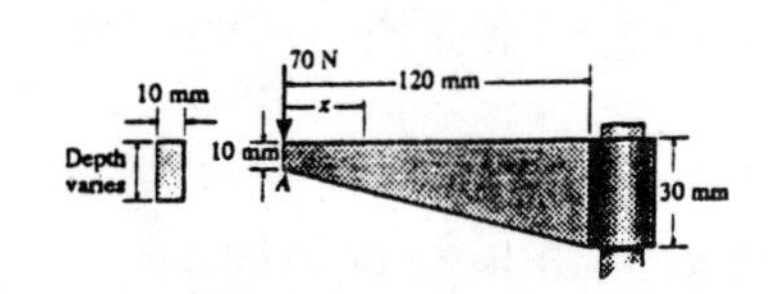

Moment Function : As shown on FBD.

Section Properties :

$$\frac{h-0.01}{x}=\frac{0.02}{0.12} \qquad h=0.1667x+0.01$$

$$I=\frac{1}{12}(0.01)(0.1667x+0.01)^3$$

$$S=\frac{I}{c}=\frac{\frac{1}{12}(0.01)(0.1667x+0.01)^3}{\frac{1}{2}(0.1667x+0.01)}$$
$$=0.001667(0.1667x+0.01)^2$$

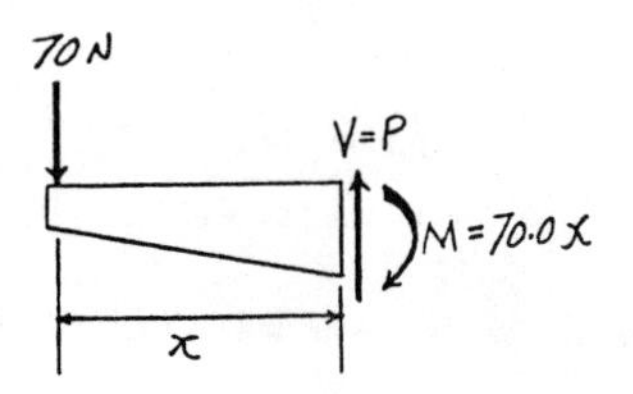

Bending Stress : Applying the flexure formula,

$$\sigma=\frac{M}{S}=\frac{70.0x}{0.001667(0.1667x+0.01)^2}$$
$$=\frac{42000x}{(0.1667x+0.01)^2} \qquad [1]$$

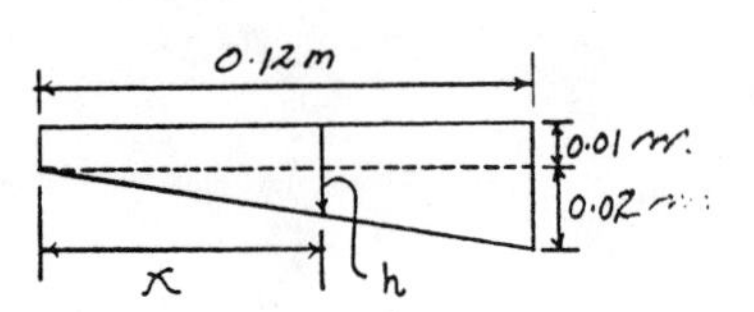

In order to have the absolute maximum bending stress, $\frac{d\sigma}{dx}=0$.

$$\frac{d\sigma}{dx}=42000\left[\frac{(0.1667x+0.01)^2(1)-x(2)(0.1667x+0.01)(0.1667)}{(0.1667x+0.01)^4}\right]=0$$

$$x=0.0600\ \text{m} \qquad \textbf{Ans}$$

Substituting $x=0.0600$ m into Eq. [1] yields

$$\sigma_{\max}=\frac{42000(0.0600)}{[0.1667(0.0600)+0.01]^2}=6.30\ \text{MPa} \qquad \textbf{Ans}$$

12–1. An A-36 steel strap having a thickness of 10 mm and a width of 20 mm is bent into a circular arc of radius ρ =10 m. Determine the maximum bending stress in the strap.

Moment - Curvature Relationship :

$$\frac{1}{\rho}=\frac{M}{EI} \quad \text{however,} \quad M=\frac{I}{c}\sigma$$

$$\frac{1}{\rho}=\frac{\frac{I}{c}\sigma}{EI}$$

$$\sigma=\frac{c}{\rho}E=\left(\frac{0.005}{10}\right)\left[200\left(10^{9}\right)\right]=100\ \text{MPa}$$

12–2. A picture is taken of a man performing a pole vault, and the minimum radius of curvature of the pole is estimated by measurement to be 4.5 m. If the pole is 40 mm in diameter and it is made of a glass-reinforced plastic for which E_g = 131 GPa, determine the maximum bending stress in the pole.

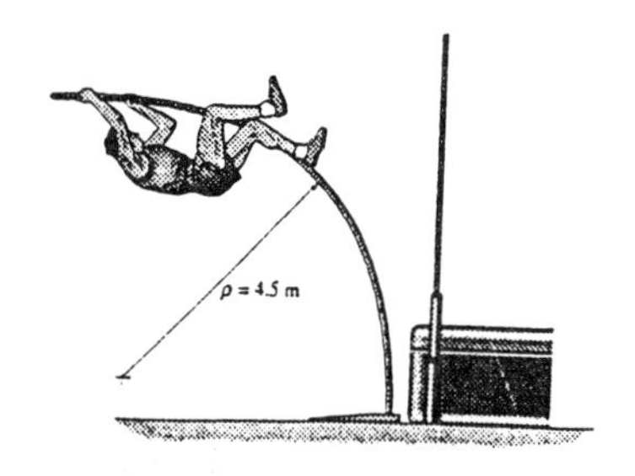

Moment - Curvature Relationship :

$$\frac{1}{\rho}=\frac{M}{EI} \quad \text{however,} \quad M=\frac{I}{c}\sigma$$

$$\frac{1}{\rho}=\frac{\frac{I}{c}\sigma}{EI}$$

$$\sigma=\frac{c}{\rho}E=\left(\frac{0.02}{4.5}\right)\left[131\left(10^{9}\right)\right]=582\ \text{MPa} \qquad \textbf{Ans}$$

12–3. Determine the equation of the elastic curve for the beam using the x coordinate that is valid for $0 \le x < L/2$. Specify the slope at A and the beam's maximum deflection. EI is constant.

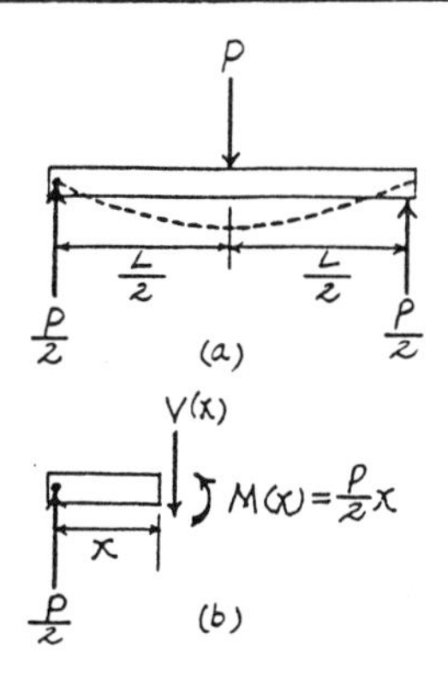

Support Reactions and Elastic Curve : As shown on FBD(a).

Moment Function : As shown on FBD(b).

Slope and Elastic Curve :

$$EI\frac{d^2v}{dx^2}=M(x)$$

$$EI\frac{d^2v}{dx^2}=\frac{P}{2}x$$

$$EI\frac{dv}{dx}=\frac{P}{4}x^2+C_1 \qquad [1]$$

$$EI\,v=\frac{P}{12}x^3+C_1x+C_2 \qquad [2]$$

Boundary Conditions : Due to symmetry, $\frac{dv}{dx}=0$ at $x=\frac{L}{2}$.

Also, $v=0$ at $x=0$.

From Eq.[1] $\quad 0=\frac{P}{4}\left(\frac{L}{2}\right)^2+C_1 \qquad C_1=-\frac{PL^2}{16}$

From Eq.[2] $\quad 0=0+0+C_2 \qquad C_2=0$

The Slope : Substitute the value of C_1 into Eq.[1],

$$\frac{dv}{dx}=\frac{P}{16EI}\left(4x^2-L^2\right)$$

$$\theta_A=\left.\frac{dv}{dx}\right|_{x=0}=-\frac{PL^2}{16EI} \qquad \textbf{Ans}$$

The negative sign indicates clockwise rotation.

The Elastic Curve : Substitute the values of C_1 and C_2 into Eq.[2],

$$v=\frac{Px}{48EI}\left(4x^2-3L^2\right) \qquad \textbf{Ans}$$

v_{max} occurs at $x=\frac{L}{2}$,

$$v_{max}=-\frac{PL^3}{48EI} \qquad \textbf{Ans}$$

The negative sign indicates downward displacement.

*12–4. Determine the equations of the elastic curve for the beam using the x_1 and x_2 coordinates. Specify the beam's maximum deflection. EI is constant.

Support Reactions and Elastic Curve : As shown on FBD(a).

Moment Function : As shown on FBD(b) and (c).

Slope and Elastic Curve :

$$EI\frac{d^2v}{dx^2} = M(x)$$

For $M(x_1) = -\frac{P}{2}x_1$,

$$EI\frac{d^2v_1}{dx_1^2} = -\frac{P}{2}x_1$$

$$EI\frac{dv_1}{dx_1} = -\frac{P}{4}x_1^2 + C_1 \qquad [1]$$

$$EI\,v_1 = -\frac{P}{12}x_1^3 + C_1x_1 + C_2 \qquad [2]$$

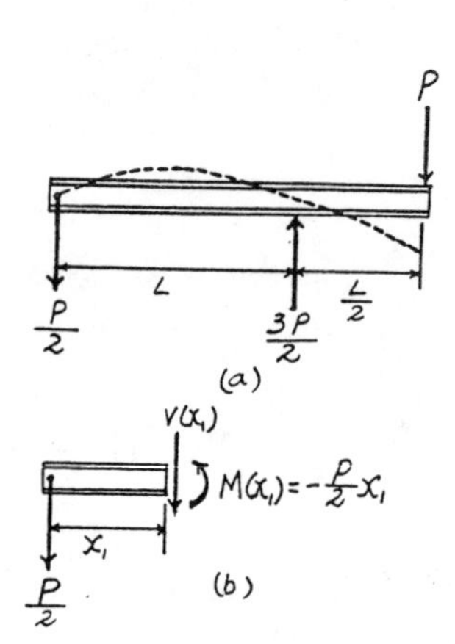

For $M(x_2) = -Px_2$,

$$EI\frac{d^2v_2}{dx_2^2} = -Px_2$$

$$EI\frac{dv_2}{dx_2} = -\frac{P}{2}x_2^2 + C_3 \qquad [3]$$

$$EI\,v_2 = -\frac{P}{6}x_2^3 + C_3x_2 + C_4 \qquad [4]$$

Boundary Conditions :

$v_1 = 0$ at $x_1 = 0$. From Eq.[2], $\qquad C_2 = 0$

$v_1 = 0$ at $x_1 = L$. From Eq.[2],

$$0 = -\frac{PL^3}{12} + C_1L \qquad C_1 = \frac{PL^2}{12}$$

$v_2 = 0$ at $x_2 = \frac{L}{2}$. From Eq. [4],

$$0 = -\frac{PL^3}{48} + \frac{L}{2}C_3 + C_4 \qquad [5]$$

Continuity Conditions :

At $x_1 = L$ and $x_2 = \frac{L}{2}$, $\quad \frac{dv_1}{dx_1} = -\frac{dv_2}{dx_2}$. From Eqs.[1] and [3],

$$-\frac{PL^2}{4} + \frac{PL^2}{12} = -\left(-\frac{PL^2}{8} + C_3\right) \qquad C_3 = \frac{7PL^2}{24}$$

From Eq.[5], $\qquad C_4 = -\frac{PL^3}{8}$

The Slope : Substitute the value of C_1 into Eq.[1],

$$\frac{dv_1}{dx_1} = \frac{P}{12EI}\left(L^2 - 3x_1^2\right)$$

$$\frac{dv_1}{dx_1} = 0 = \frac{P}{12EI}\left(L^2 - 3x_1^2\right) \qquad x_1 = \frac{L}{\sqrt{3}}$$

The Elastic Curve : Substitute the values of C_1, C_2, C_3, and C_4 into Eqs.[2] and [4], respectively,

$$v_1 = \frac{Px_1}{12EI}\left(-x_1^2 + L^2\right) \qquad \textbf{Ans}$$

$$v_D = v_1\big|_{x_1 = \frac{L}{\sqrt{3}}} = \frac{P\left(\frac{L}{\sqrt{3}}\right)}{12EI}\left(-\frac{L^2}{3} + L^2\right) = \frac{0.0321PL^3}{EI}$$

$$v_2 = \frac{P}{24EI}\left(-4x_2^3 + 7L^2x_2 - 3L^3\right) \qquad \textbf{Ans}$$

$$v_C = v_2\big|_{x_2 = 0} = -\frac{PL^3}{8EI}$$

Hence, $\qquad v_{max} = v_C = \frac{PL^3}{8EI} \qquad$ **Ans**

12-5. Determine the equations of the elastic curve for the beam using the x_1 and x_3 coordinates. Specify the beam's maximum deflection. EI is constant.

Support Reactions and Elastic Curve : As shown on FBD(a).

Moment Function : As shown on FBD(b) and (c).

Slope and Elastic Curve :

$$EI\frac{d^2v}{dx^2} = M(x)$$

For $M(x_1) = -\frac{P}{2}x_1$,

$$EI\frac{d^2v_1}{dx_1^2} = -\frac{P}{2}x_1$$

$$EI\frac{dv_1}{dx_1} = -\frac{P}{4}x_1^2 + C_1 \qquad [1]$$

$$EI\,v_1 = -\frac{P}{12}x_1^3 + C_1x_1 + C_2 \qquad [2]$$

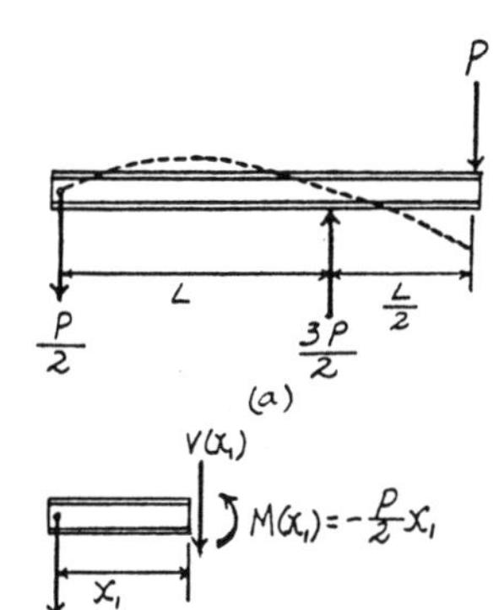

For $M(x_3) = Px_3 - \frac{3PL}{2}$,

$$EI\frac{d^2v_3}{dx_3^2} = Px_3 - \frac{3PL}{2}$$

$$EI\frac{dv_3}{dx_3} = \frac{P}{2}x_3^2 - \frac{3PL}{2}x_3 + C_3 \qquad [3]$$

$$EI\,v_3 = \frac{P}{6}x_3^3 - \frac{3PL}{4}x_3^2 + C_3x_3 + C_4 \qquad [4]$$

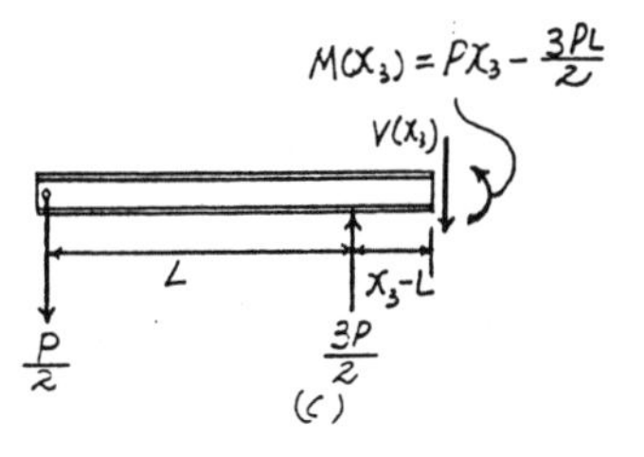

Boundary Conditions :

$v_1 = 0$ at $x_1 = 0$. From Eq.[2], $\quad C_2 = 0$

$v_1 = 0$ at $x_1 = L$. From Eq.[2],

$$0 = -\frac{PL^3}{12} + C_1L \qquad C_1 = \frac{PL^2}{12}$$

$v_3 = 0$ at $x_3 = L$ From Eq. [4],

$$0 = \frac{PL^3}{6} - \frac{3PL^3}{4} + C_3L + C_4$$

$$0 = -\frac{7PL^3}{12} + C_3L + C_4 \qquad [5]$$

Continuity Condition :

At $x_1 = x_3 = L$, $\quad \frac{dv_1}{dx_1} = \frac{dv_3}{dx_3}$. From Eqs.[1] and [3],

$$-\frac{PL^2}{4} + \frac{PL^2}{12} = \frac{PL^2}{2} - \frac{3PL^2}{2} + C_3 \qquad C_3 = \frac{5PL^2}{6}$$

From Eq. [5], $\quad C_4 = -\frac{PL^3}{4}$

The Slope : Substitute the value of C_1 into Eq.[1],

$$\frac{dv_1}{dx_1} = \frac{P}{12EI}\left(L^2 - 3x_1^2\right)$$

$$\frac{dv_1}{dx_1} = 0 = \frac{P}{12EI}\left(L^2 - 3x_1^2\right) \qquad x_1 = \frac{L}{\sqrt{3}}$$

The Elastic Curve : Substitute the values of C_1, C_2, C_3, and C_4 into Eqs.[2] and [4], respectively,

$$v_1 = \frac{Px_1}{12EI}\left(-x_1^2 + L^2\right) \qquad \textbf{Ans}$$

$$v_D = v_1\big|_{x_1 = \frac{L}{\sqrt{3}}} = \frac{P\left(\frac{L}{\sqrt{3}}\right)}{12EI}\left(-\frac{L^2}{3} + L^2\right) = \frac{0.0321PL^3}{EI}$$

$$v_3 = \frac{P}{12EI}\left(2x_3^3 - 9Lx_3^2 + 10L^2x_3 - 3L^3\right) \qquad \textbf{Ans}$$

$$v_C = v_3\big|_{x_3 = \frac{3}{2}L}$$

$$= \frac{P}{12EI}\left[2\left(\frac{3}{2}L\right)^3 - 9L\left(\frac{3}{2}L\right)^2 + 10L^2\left(\frac{3}{2}L\right) - 3L^3\right]$$

$$= -\frac{PL^3}{8EI}$$

Hence, $\quad v_{\max} = v_C = \frac{PL^3}{8EI} \qquad$ **Ans**

12–6. The shaft is supported at A by a journal bearing that exerts only vertical reactions on the shaft and at C by a thrust bearing that exerts horizontal and vertical reactions on the shaft. Determine the equations of the elastic curve using the coordinates x_1 and x_2. EI is constant.

Support Reactions and Elastic Curve : As shown on FBD(a).

Moment Function : As shown on FBD(b) and (c).

Slope and Elastic Curve :

$$EI\frac{d^2v}{dx^2} = M(x)$$

For $M(x_1) = -\frac{Pb}{a}x_1$,

$$EI\frac{d^2v_1}{dx_1^2} = -\frac{Pb}{a}x_1$$

$$EI\frac{dv_1}{dx_1} = -\frac{Pb}{2a}x_1^2 + C_1 \qquad [1]$$

$$EI\,v_1 = -\frac{Pb}{6a}x_1^3 + C_1x_1 + C_2 \qquad [2]$$

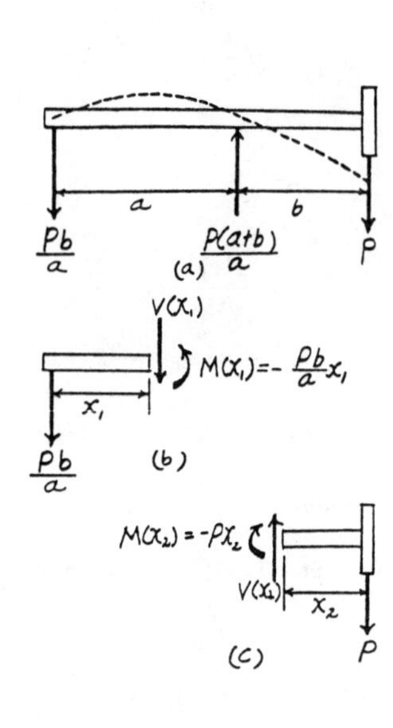

For $M(x_2) = -Px_2$,

$$EI\frac{d^2v_2}{dx_2^2} = -Px_2$$

$$EI\frac{dv_2}{dx_2} = -\frac{P}{2}x_2^2 + C_3 \qquad [3]$$

$$EI\,v_2 = -\frac{P}{6}x_2^3 + C_3x_2 + C_4 \qquad [4]$$

Boundary Conditions :

$v_1 = 0$ at $x_1 = 0$. From Eq.[2], $C_2 = 0$

$v_1 = 0$ at $x_1 = a$. From Eq.[2],

$$0 = -\frac{Pa^3b}{6a} + C_1(a) \qquad C_1 = \frac{Pab}{6}$$

$v_2 = 0$ at $x_2 = b$. From Eq. [4],

$$0 = -\frac{Pb^3}{6} + C_3(b) + C_4 \qquad [5]$$

Continuity Conditions :

At $x_1 = a$ and $x_2 = b$, $\frac{dv_1}{dx_1} = -\frac{dv_2}{dx_2}$. From Eqs.[1] and [3],

$$-\frac{Pab}{2} + \frac{Pab}{6} = -\left(-\frac{Pb^2}{2} + C_3\right) \qquad C_3 = \frac{Pb}{6}(2a + 3b)$$

From Eq.[5], $C_4 = -\frac{Pb^2}{3}(a+b)$

The Elastic Curve : Substitute the values of C_1, C_2, C_3, and C_4 into Eqs.[2] and [4], respectively,

$$v_1 = \frac{Pb}{6aEI}\left(-x_1^3 + a^2x_1\right) \qquad \textbf{Ans}$$

$$v_2 = \frac{P}{6EI}\left[-x_2^3 + b(2a+3b)x_2 - 2b^2(a+b)\right] \qquad \textbf{Ans}$$

12–7. Determine the equations of the elastic curve for the shaft using the x_1 and x_2 coordinates. Specify the slope at A and the deflection at the center of the shaft. EI is constant.

Support Reactions and Elastic Curve : As shown on FBD(a).

Moment Function : As shown on FBD(b) and (c).

Slope and Elastic Curve :

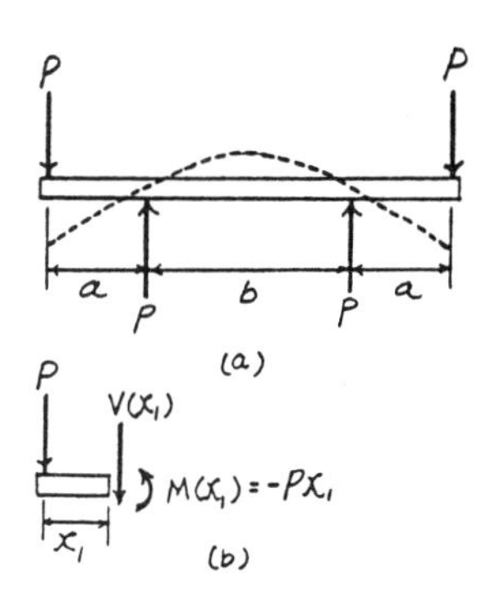

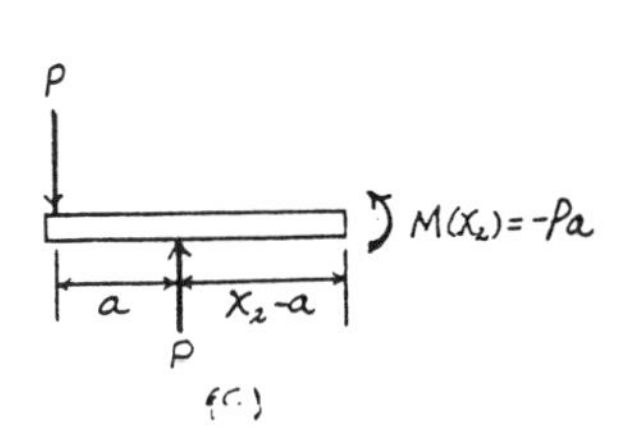

$$EI\frac{d^2 v}{dx^2} = M(x)$$

For $M(x_1) = -Px_1$,

$$EI\frac{d^2 v_1}{dx_1^2} = -Px_1$$

$$EI\frac{dv_1}{dx_1} = -\frac{P}{2}x_1^2 + C_1 \qquad [1]$$

$$EI\, v_1 = -\frac{P}{6}x_1^3 + C_1 x_1 + C_2 \qquad [2]$$

For $M(x_2) = -Pa$,

$$EI\frac{d^2 v_2}{dx_2^2} = -Pa$$

$$EI\frac{dv_2}{dx_2} = -Pax_2 + C_3 \qquad [3]$$

$$EI\, v_2 = -\frac{Pa}{2}x_2^2 + C_3 x_2 + C_4 \qquad [4]$$

Boundary Conditions :

$v_1 = 0$ at $x_1 = a$. From Eq.[2],

$$0 = -\frac{Pa^3}{6} + C_1 a + C_2 \qquad [5]$$

Due to symmetry, $\dfrac{dv_2}{dx_2} = 0$ at $x_2 = a + \dfrac{b}{2}$. From Eq.[3],

$$0 = -Pa\left(a + \frac{b}{2}\right) + C_3 \qquad C_3 = \frac{Pa}{2}(2a+b)$$

$v_2 = 0$ at $x_2 = a$. From Eq. [4],

$$0 = -\frac{Pa^3}{2} + \frac{Pa^2}{2}(2a+b) + C_4 \qquad C_4 = -\frac{Pa^2}{2}(a+b)$$

Continuity Condition :

At $x_1 = x_2 = a$, $\dfrac{dv_1}{dx_1} = \dfrac{dv_2}{dx_2}$. From Eqs.[1] and [3],

$$-\frac{Pa^2}{2} + C_1 = -Pa^2 + \frac{Pa}{2}(2a+b) \qquad C_1 = \frac{Pa}{2}(a+b)$$

From Eq. [5], $$C_2 = -\frac{Pa^2}{6}(2a+3b)$$

The Slope : Either Eq.[1] or [3] can be used. Substitute the value of C_1 into Eq.[1],

$$\frac{dv_1}{dx_1} = \frac{P}{2EI}\left[-x_1^2 + a(a+b)\right]$$

$$\theta_A = \left.\frac{dv_1}{dx_1}\right|_{x_1=a} = \frac{P}{2EI}\left[-a^2 + a(a+b)\right] = \frac{Pab}{2EI} \qquad \textbf{Ans}$$

The Elastic Curve : Substitute the values of C_1, C_2, C_3, and C_4 into Eqs.[2] and [4], respectively,

$$v_1 = \frac{P}{6EI}\left[-x_1^3 + 3a(a+b)x_1 - a^2(2a+3b)\right] \qquad \textbf{Ans}$$

$$v_2 = \frac{Pa}{2EI}\left[-x_2^2 + (2a+b)x_2 - a(a+b)\right] \qquad \textbf{Ans}$$

$$v_C = v_2\big|_{x_2=a+\frac{b}{2}}$$

$$= \frac{Pa}{2EI}\left[-\left(a+\frac{b}{2}\right)^2 + (2a+b)\left(a+\frac{b}{2}\right) - a(a+b)\right]$$

$$= \frac{Pab^2}{8EI} \qquad \textbf{Ans}$$

***12–8.** Determine the equations of the elastic curve for the shaft using the x_1 and x_3 coordinates. Specify the slope at A and the deflection at the center of the shaft. EI is constant.

Support Reactions and Elastic Curve : As shown on FBD(a).

Moment Function : As shown on FBD(b) and (c).

Slope and Elastic Curve :

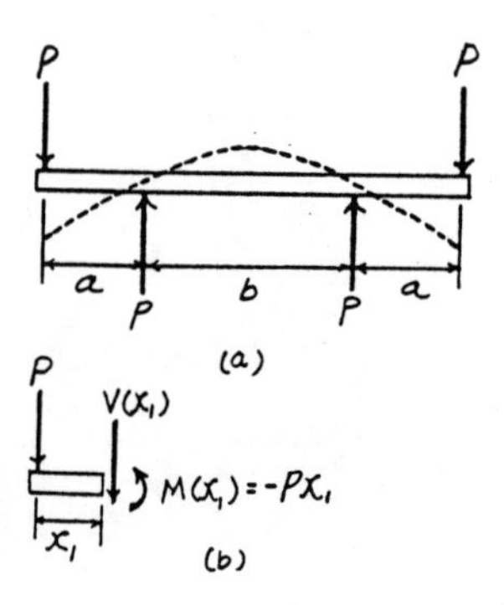

$$EI\frac{d^2v}{dx^2} = M(x)$$

For $M(x_1) = -Px_1$,

$$EI\frac{d^2v_1}{dx_1^2} = -Px_1$$

$$EI\frac{dv_1}{dx_1} = -\frac{P}{2}x_1^2 + C_1 \qquad [1]$$

$$EI\,v_1 = -\frac{P}{6}x_1^3 + C_1x_1 + C_2 \qquad [2]$$

For $M(x_3) = -Pa$,

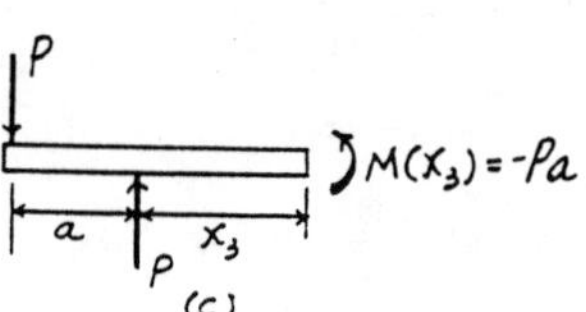

$$EI\frac{d^2v_3}{dx_3^2} = -Pa$$

$$EI\frac{dv_3}{dx_3} = -Pax_3 + C_3 \qquad [3]$$

$$EI\,v_3 = -\frac{Pa}{2}x_3^2 + C_3x_3 + C_4 \qquad [4]$$

Boundary Conditions :

$v_1 = 0$ at $x_1 = a$. From Eq.[2],

$$0 = -\frac{Pa^3}{6} + C_1a + C_2 \qquad [5]$$

Due to symmetry, $\frac{dv_3}{dx_3} = 0$ at $x_3 = \frac{b}{2}$. From Eq.[3]

$$0 = -Pa\left(\frac{b}{2}\right) + C_3 \qquad C_3 = \frac{Pab}{2}$$

$v_3 = 0$ at $x_3 = 0$ From Eq. [4], $C_4 = 0$

Continuity Condition :

At $x_1 = a$ and $x_3 = 0$, $\frac{dv_1}{dx_1} = \frac{dv_3}{dx_3}$. From Eqs.[1] and [3],

$$-\frac{Pa^2}{2} + C_1 = \frac{Pab}{2} \qquad C_1 = \frac{Pa}{2}(a+b)$$

From Eq. [5] $\qquad C_2 = -\frac{Pa^2}{6}(2a+3b)$

The Slope : Either Eq.[1] or [3] can be used. Substitute the value of C_1 into Eq.[1],

$$\frac{dv_1}{dx_1} = \frac{P}{2EI}\left[-x_1^2 + a(a+b)\right]$$

$$\theta_A = \left.\frac{dv_1}{dx_1}\right|_{x_1=a} = \frac{P}{2EI}\left[-a^2 + a(a+b)\right] = \frac{Pab}{2EI} \qquad \textbf{Ans}$$

The Elastic Curve : Substitute the values of C_1, C_2, C_3, and C_4 into Eqs.[2] and [4], respectively,

$$v_1 = \frac{P}{6EI}\left[-x_1^3 + 3a(a+b)x_1 - a^2(2a+3b)\right] \qquad \textbf{Ans}$$

$$v_3 = \frac{Pax_3}{2EI}(-x_3 + b) \qquad \textbf{Ans}$$

$$v_C = v_3\big|_{x_3=\frac{b}{2}} = \frac{Pa\left(\frac{b}{2}\right)}{2EI}\left(-\frac{b}{2} + b\right) = \frac{Pab^2}{8EI} \qquad \textbf{Ans}$$

12–9. The fence board weaves between the three smooth fixed posts, each of which has a diameter of 3 in. If the posts remain along the same line, determine the maximum bending stress in the board. The board has a width of 6 in. and a thickness of 0.5 in. $E = 1.60(10^3)$ ksi. Assume the deflection of each end of the board relative to its center is 3 in.

Support Reactions and Elastic Curve : As shown on FBD(a).

Moment Function : As shown on FBD(b).

Slope and Elastic Curve :

$$EI\frac{d^2v}{dx^2} = M(x)$$

$$EI\frac{d^2v}{dx^2} = \frac{P}{2}x$$

$$EI\frac{dv}{dx} = \frac{P}{4}x^2 + C_1 \qquad [1]$$

$$EI\,v = \frac{P}{12}x^3 + C_1x + C_2 \qquad [2]$$

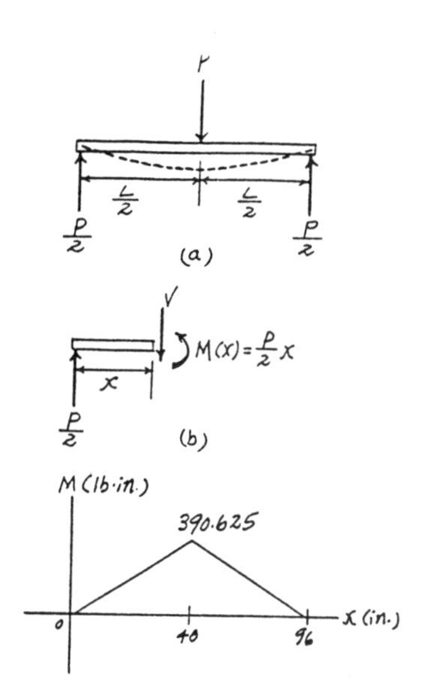

Boundary Conditions : Due to symmetry, $\frac{dv}{dx} = 0$ at $x = \frac{L}{2}$.
Also, $v = 0$ at $x = 0$.

From Eq.[1] $\quad 0 = \frac{P}{4}\left(\frac{L}{2}\right)^2 + C_1 \qquad C_1 = -\frac{PL^2}{16}$

From Eq.[2] $\quad 0 = 0 + 0 + C_2 \qquad C_2 = 0$

The Elastic Curve : Substitute the values of C_1 and C_2 into Eq.[2],

$$v = \frac{Px}{48EI}\left(4x^2 - 3L^2\right) \qquad [1]$$

Require at $x = 48$ in., $v = -3$ in. From Eq.[1],

$$-3 = \frac{P(48)}{48(1.60)(10^6)\left(\frac{1}{12}\right)(6)(0.5^3)}\left[4\left(48^2\right) - 3\left(96^2\right)\right]$$

$$P = 16.28 \text{ lb}$$

Maximum Bending Stress : From the moment diagram, the maximum moment is $M_{max} = 390.625$ lb · in. Applying the flexure formula,

$$\sigma_{max} = \frac{Mc}{I} = \frac{390.625(0.25)}{\frac{1}{12}(6)(0.5^3)} = 1562.5 \text{ psi} = 1.56 \text{ ksi} \qquad \textbf{Ans}$$

12–10. A beam torque wrench is used to tighten the nut on a bolt. If the dial indicates that a torque of 60 lb · ft is applied when the bolt is fully tightened, determine the force P acting at the handle and the distance s the needle moves along the scale. Assume only the portion AB of the beam distorts. The cross section is square having dimensions of 0.5 in. by 0.5 in. $E = 29(10^3)$ ksi.

Equations of Equilibrium : From FBD(a),

$\circlearrowleft + \Sigma M_A = 0; \quad 720 - P(18) = 0 \quad P = 40.0 \text{ lb}$ **Ans**

$+\uparrow \Sigma F_y = 0; \quad A_y - 40.0 = 0 \quad A_y = 40.0 \text{ lb}$

Moment Function : As shown on FBD(b).

Slope and Elastic Curve :

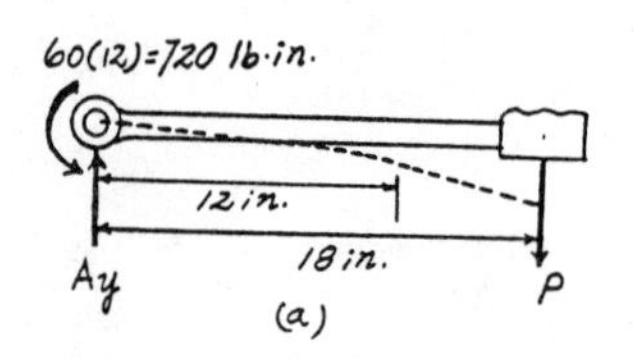

$$EI\frac{d^2v}{dx^2} = M(x)$$

$$EI\frac{d^2v}{dx^2} = 40.0x - 720$$

$$EI\frac{dv}{dx} = 20.0x^2 - 720x + C_1 \qquad [1]$$

$$EI\,v = 6.667x^3 - 360x^2 + C_1x + C_2 \qquad [2]$$

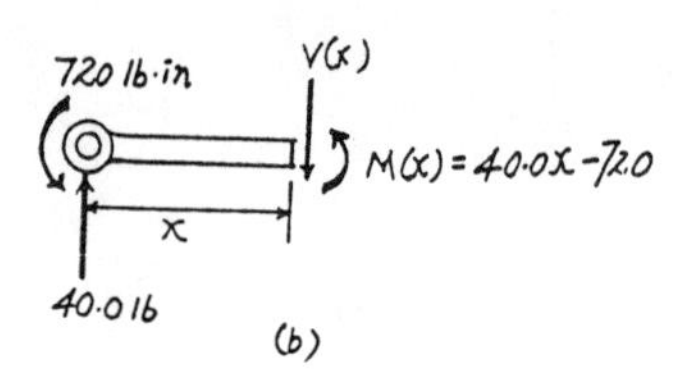

Boundary Conditions : $\frac{dv}{dx} = 0$ at $x = 0$ and $v = 0$ at $x = 0$.

From Eq.[1] $\quad 0 = 0 - 0 + C_1 \qquad C_1 = 0$

From Eq.[2] $\quad 0 = 0 - 0 + 0 + C_2 \qquad C_2 = 0$

The Elastic Curve : Substitute the values of C_1 and C_2 into Eq.[2],

$$v = \frac{1}{EI}\left(6.667x^3 - 360x^2\right) \qquad [1]$$

At $x = 12$ in., $v = -s$. From Eq.[1],

$$-s = \frac{1}{(29)(10^6)\left(\frac{1}{12}\right)(0.5)(0.5^3)}\left[6.667\left(12^3\right) - 360\left(12^2\right)\right]$$

$$s = 0.267 \text{ in.}$$ **Ans**

12–11. The shaft is supported at A by a journal bearing that exerts only vertical reactions on the shaft and at B by a thrust bearing that exerts horizontal and vertical reactions on the shaft. Draw the bending-moment diagram for the shaft and then, from this diagram, sketch the deflection or elastic curve for the shaft's centerline. Determine the equations of the elastic curve using the coordinates x_1 and x_2. EI is constant.

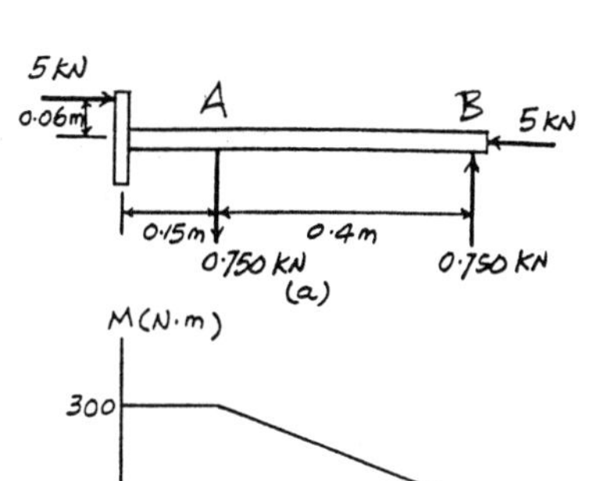

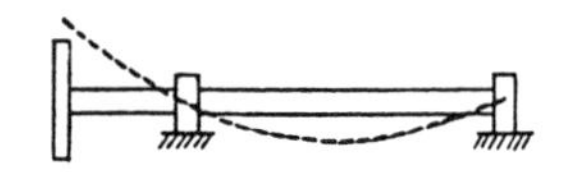

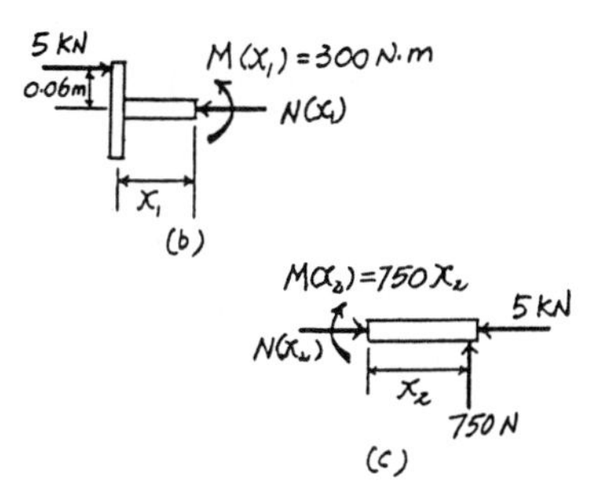

Elastic Curve : As shown.

Moment Function : As shown on FBD(b) and (c).

Slope and Elastic Curve :

$$EI\frac{d^2v}{dx^2} = M(x)$$

For $M(x_1) = 300\ \text{N}\cdot\text{m}$,

$$EI\frac{d^2v_1}{dx_1^2} = 300$$

$$EI\frac{dv_1}{dx_1} = 300x_1 + C_1 \qquad [1]$$

$$EI\,v_1 = 150x_1^2 + C_1x_1 + C_2 \qquad [2]$$

For $M(x_2) = 750x_2$,

$$EI\frac{d^2v_2}{dx_2^2} = 750x_2$$

$$EI\frac{dv_2}{dx_2} = 375x_2^2 + C_3 \qquad [3]$$

$$EI\,v_2 = 125x_2^3 + C_3x_2 + C_4 \qquad [4]$$

Boundary Conditions :

$v_1 = 0$ at $x_1 = 0.15$ m. From Eq.[2],

$$0 = 150\left(0.15^2\right) + C_1(0.15) + C_2 \qquad [5]$$

$v_2 = 0$ at $x_2 = 0$. From Eq.[4], $C_4 = 0$

$v_2 = 0$ at $x_2 = 0.4$ m. From Eq. [4],

$$0 = 125\left(0.4^3\right) + C_3(0.4) \qquad C_3 = -20.0$$

Continuity Condition :

At $x_1 = 0.15$ m and $x_2 = 0.4$ m, $\dfrac{dv_1}{dx_1} = -\dfrac{dv_2}{dx_2}$. From Eqs.[1] and [3],

$$300(0.15) + C_1 = -\left[375\left(0.4^2\right) - 20\right] \qquad C_1 = -85.0$$

From Eq. [5], $C_2 = 9.375$

The Elastic Curve : Substitute the values of C_1, C_2, C_3, and C_4 into Eqs.[2] and [4], respectively.

$$v_1 = \frac{1}{EI}\left(150x_1^2 - 85.0x_1 + 9.375\right)\ \text{N}\cdot\text{m}^3 \qquad \textbf{Ans}$$

$$v_2 = \frac{1}{EI}\left(125x_2^3 - 20.0x_2\right)\ \text{N}\cdot\text{m}^3 \qquad \textbf{Ans}$$

***12–12.** The A-36 steel beam has a depth of 10 in. and is subjected to a constant moment M_0, which causes the stress at the outer fibers to become $\sigma_Y = 36$ ksi. Determine the radius of curvature of the beam and the maximum slope and deflection.

Moment - Curvature Relationship :

$$\frac{1}{\rho} = \frac{M_0}{EI} \quad \text{however,} \quad M_0 = \frac{I}{c}\sigma_Y$$

$$\frac{1}{\rho} = \frac{\frac{I}{c}\sigma_Y}{EI}$$

$$\rho = \frac{Ec}{\sigma_Y} = \frac{29.0(10^3)(5)}{36} = 4027.78 \text{ in.} = 336 \text{ ft} \qquad \textbf{Ans}$$

Elastic Curve : As shown.

Moment Function : As shown on FBD.

Slope and Elastic Curve :

$$EI\frac{d^2v}{dx^2} = M(x)$$

$$EI\frac{d^2v}{dx^2} = -M_0$$

$$EI\frac{dv}{dx} = -M_0x + C_1 \qquad [1]$$

$$EI\,v = -\frac{M_0}{2}x^2 + C_1x + C_2 \qquad [2]$$

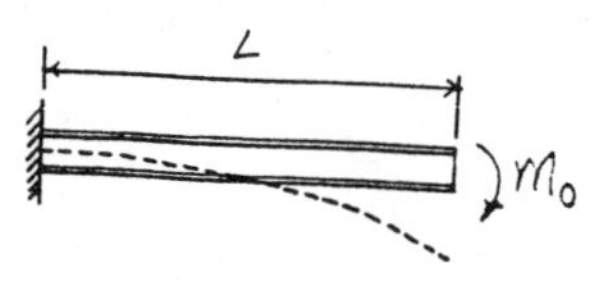

Boundary Conditions : $\frac{dv}{dx} = 0$ at $x = L$, and $v = 0$ at $x = L$.

From Eq. [1], $\quad 0 = -M_0L + C_1 \qquad C_1 = M_0L$

From Eq. [2], $\quad 0 = -\frac{M_0}{2}(L^2) + M_0L^2 + C_2 \qquad C_2 = -\frac{M_0L^2}{2}$

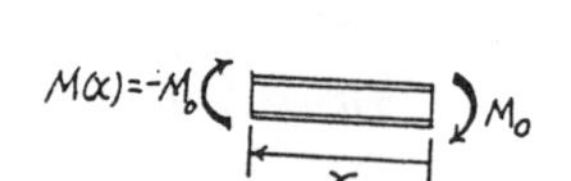

The Slope : Substitute the value of C_1 into Eq. [1],

$$\frac{dv}{dx} = \frac{M_0}{EI}(-x + L)$$

The maximum slope occurs at $x = 0$.

$$\theta_{max} = \left.\frac{dv}{dx}\right|_{x=0} = \frac{M_0L}{EI} \qquad \textbf{Ans}$$

The Elastic Curve : Substitute the values of C_1 and C_2 into Eq. [2],

$$v = \frac{M_0}{2EI}\left(-x^2 + 2Lx - L^2\right)$$

The maximum displacement occurs at $x = 0$.

$$v_{max} = -\frac{M_0L^2}{2EI} \qquad \textbf{Ans}$$

The negative sign indicates downward displacement.

12–13. Determine the elastic curve for the simply supported beam, which is subjected to the couple moment $\mathbf{M_0}$. Also, determine the maximum slope and the maximum deflection of the beam. EI is constant.

Support Reactions and Elastic Curve : As shown on FBD(a). The elastic curve is asymmetrical about the beam's center.

Moment Function : As shown on FBD(b).

Slope and Elastic Curve :

$$EI\frac{d^2 v}{dx^2} = M(x)$$

$$EI\frac{d^2 v}{dx^2} = \frac{M_0}{L}x$$

$$EI\frac{dv}{dx} = \frac{M_0}{2L}x^2 + C_1 \qquad [1]$$

$$EI\,v = \frac{M_0}{6L}x^3 + C_1 x + C_2 \qquad [2]$$

Boundary Conditions :

$v = 0$ at $x = 0$. From Eq.[2],

$$0 = 0 + 0 + C_2 \qquad C_2 = 0$$

Due to asymmetry, $v = 0$ at $x = \frac{L}{2}$. From Eq.[2],

$$0 = \frac{M_0}{6L}\left(\frac{L}{2}\right)^3 + C_1\left(\frac{L}{2}\right) \qquad C_1 = -\frac{M_0 L}{24}$$

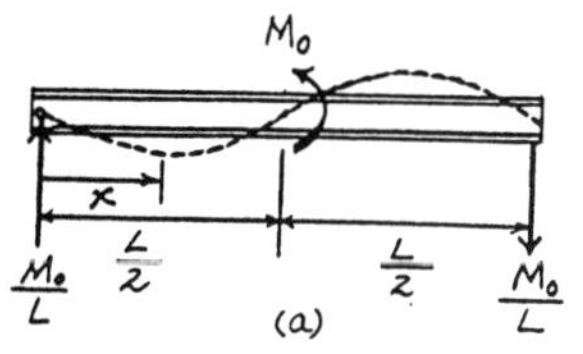

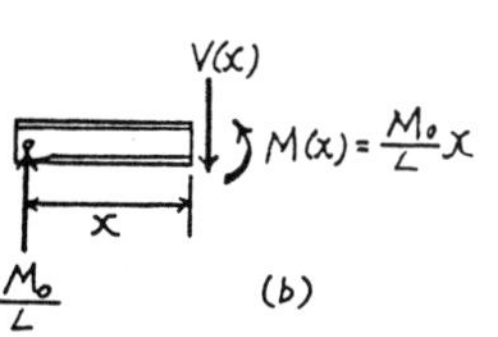

The Slope : Substitute the value of C_1 into Eq.[1],

$$\frac{dv}{dx} = \frac{M_0}{24LEI}\left(12x^2 - L^2\right)$$

$$\frac{dv}{dx} = 0 = \frac{M_0}{24LEI}\left(12x^2 - L^2\right) \qquad x = \frac{\sqrt{3}}{6}L$$

$$\theta_A = \left.\frac{dv}{dx}\right|_{x=0} = -\frac{M_0 L}{24EI}$$

$$\theta_{max} = \theta_C = \left.\frac{dv}{dx}\right|_{x=\frac{L}{2}} = \frac{M_0 L}{12EI} \qquad \textbf{Ans}$$

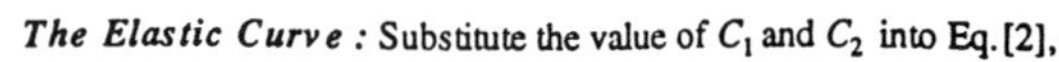

The Elastic Curve : Substitute the value of C_1 and C_2 into Eq.[2],

$$v = \frac{M_0}{24LEI}\left(4x^3 - L^2 x\right) \qquad \textbf{Ans}$$

v_{max} occurs at $x = \frac{\sqrt{3}}{6}L$,

$$v_{max} = -\frac{\sqrt{3}M_0 L^2}{216EI} \qquad \textbf{Ans}$$

The negative sign indicates downward displacement.

12–14. Determine the maximum slope and maximum deflection of the simply-supported beam which is subjected to the couple moment $\mathbf{M}_0$. EI is constant.

Support Reactions and Elastic Curve : As shown on FBD(a).

Moment Function : As shown on FBD(b).

Slope and Elastic Curve :

$$EI\frac{d^2v}{dx^2} = M(x)$$

$$EI\frac{d^2v}{dx^2} = \frac{M_0}{L}x$$

$$EI\frac{dv}{dx} = \frac{M_0}{2L}x^2 + C_1 \qquad [1]$$

$$EI\,v = \frac{M_0}{6L}x^3 + C_1x + C_2 \qquad [2]$$

Boundary Conditions :

$v = 0$ at $x = 0$. From Eq. [2],

$$0 = 0 + 0 + C_2 \qquad C_2 = 0$$

$v = 0$ at $x = L$. From Eq. [2],

$$0 = \frac{M_0}{6L}(L^3) + C_1(L) \qquad C_1 = -\frac{M_0L}{6}$$

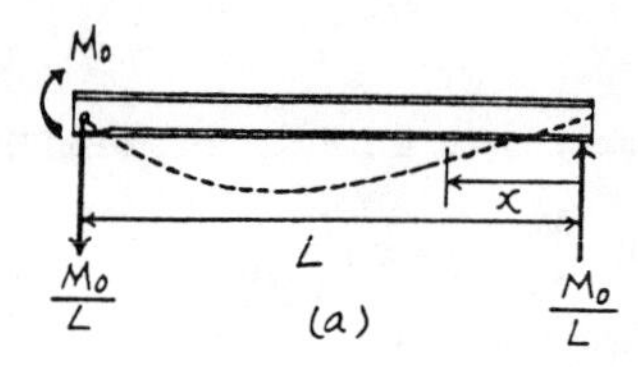

The Slope : Substitute the value of C_1 into Eq. [1],

$$\frac{dv}{dx} = \frac{M_0}{6LEI}(3x^2 - L^2)$$

$$\frac{dv}{dx} = 0 = \frac{M_0}{6LEI}(3x^2 - L^2) \qquad x = \frac{\sqrt{3}}{3}L$$

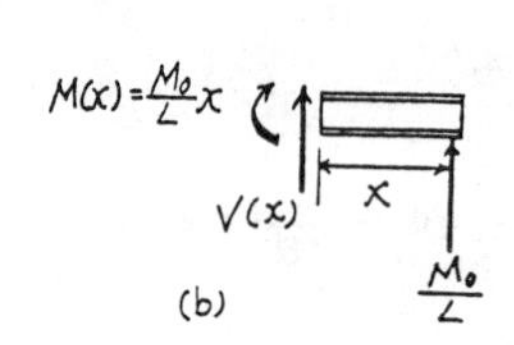

$$\theta_B = \left.\frac{dv}{dx}\right|_{x=0} = -\frac{M_0L}{6EI}$$

$$\theta_{max} = \theta_A = \left.\frac{dv}{dx}\right|_{x=L} = \frac{M_0L}{3EI} \qquad \textbf{Ans}$$

The Elastic Curve : Substituting the values of C_1 and C_2 into Eq. [2],

$$v = \frac{M_0}{6LEI}(x^3 - L^2x)$$

v_{max} occurs at $x = \frac{\sqrt{3}}{3}L$,

$$v_{max} = -\frac{\sqrt{3}M_0L^2}{27EI} \qquad \textbf{Ans}$$

The negative sign indicates downward displacement.

12–15. The shaft is supported at A by a journal bearing that exerts only vertical reactions on the shaft and at B by a thrust bearing that exerts horizontal and vertical reactions on the shaft. Draw the bending-moment diagram for the shaft and then, from this diagram, sketch the deflection or elastic curve for the shaft's centerline. Determine the equations of the elastic curve using the coordinates x_1 and x_2.

Elastic Curve : As shown.

Moment Function : As shown on FBD(b) and (c).

Slope and Elastic Curve :

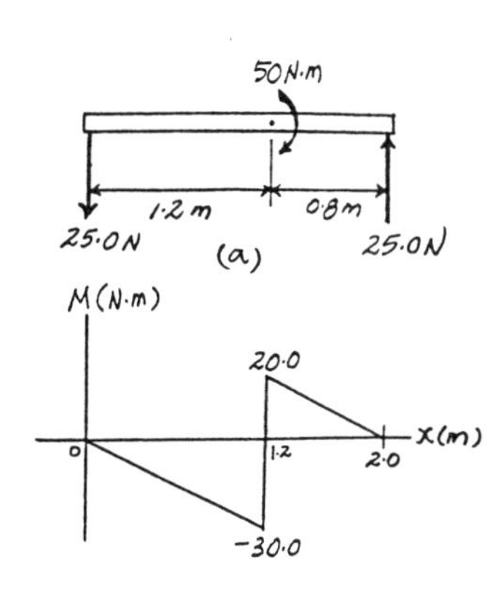

$$EI\frac{d^2v}{dx^2} = M(x)$$

For $M(x_1) = -25.0x_1$,

$$EI\frac{d^2v_1}{dx_1^2} = -25.0x_1$$

$$EI\frac{dv_1}{dx_1} = -12.5x_1^2 + C_1 \qquad [1]$$

$$EI\,v_1 = -4.1667x_1^3 + C_1x_1 + C_2 \qquad [2]$$

For $M(x_2) = 25.0x_2$,

$$EI\frac{d^2v_2}{dx_2^2} = 25.0x_2$$

$$EI\frac{dv_2}{dx_2} = 12.5x_2^2 + C_3 \qquad [3]$$

$$EI\,v_2 = 4.1667x_2^3 + C_3x_2 + C_4 \qquad [4]$$

Boundary Conditions :

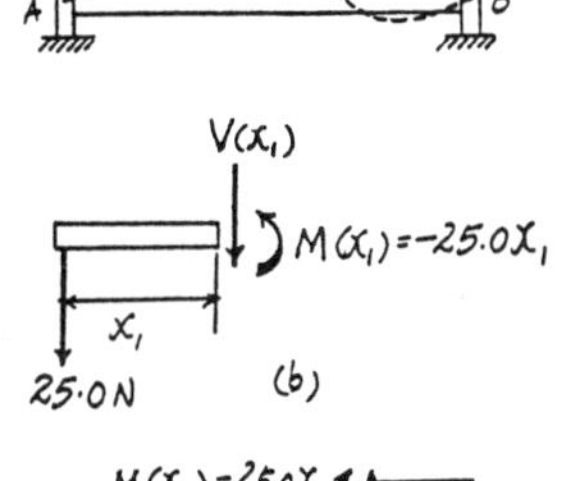

$v_1 = 0$ at $x_1 = 0$. From Eq.[2], $C_2 = 0$

$v_2 = 0$ at $x_2 = 0$. From Eq.[4], $C_4 = 0$

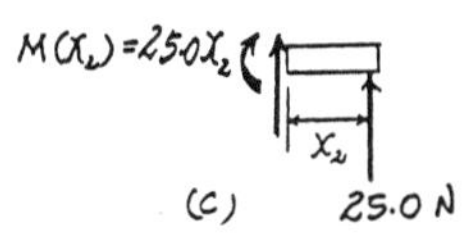

Continuity Conditions :

At $x_1 = 1.2$ m and $x_2 = 0.8$ m, $\frac{dv_1}{dx_1} = -\frac{dv_2}{dx_2}$. From Eqs.[1] and [3],

$$-12.5(1.2^2) + C_1 = -\left[12.5(0.8^2) + C_3\right]$$

$$C_1 + C_3 = 10.0 \qquad [5]$$

At $x_1 = 1.2$ m and $x_2 = 0.8$ m, $v_1 = v_2$. From Eqs.[2] and [4],

$$-4.1667(1.2^3) + C_1(1.2) = 4.1667(0.8^3) + C_3(0.8)$$

$$1.2C_1 - 0.8C_3 = 9.333 \qquad [6]$$

Solving Eqs.[5] and [6] yields,

$$C_1 = 8.667 \qquad C_3 = 1.333$$

The Elastic Curve : Substituting the values of C_1, C_2, C_3, and C_4 into Eqs.[2] and [4], respectively,

$$v_1 = \frac{1}{EI}\left(-4.17x_1^3 + 8.67x_1\right)\ \text{N}\cdot\text{m}^3 \qquad \textbf{Ans}$$

$$v_2 = \frac{1}{EI}\left(4.17x_2^3 + 1.33x_2\right)\ \text{N}\cdot\text{m}^3 \qquad \textbf{Ans}$$

***12–16.** Determine the equation of the elastic curve using the coordinate x, and specify the slope at point A and the deflection at point C. EI is constant.

Support Reactions and Elastic Curve : As shown on FBD(a).

Moment Function : As shown on FBD(b).

Slope and Elastic Curve :

$$EI\frac{d^2v}{dx^2} = M(x)$$

$$EI\frac{d^2v}{dx^2} = \frac{wL}{2}x - \frac{w}{2}x^2$$

$$EI\frac{dv}{dx} = \frac{wL}{4}x^2 - \frac{w}{6}x^3 + C_1 \qquad [1]$$

$$EI\,v = \frac{wL}{12}x^3 - \frac{w}{24}x^4 + C_1x + C_2 \qquad [2]$$

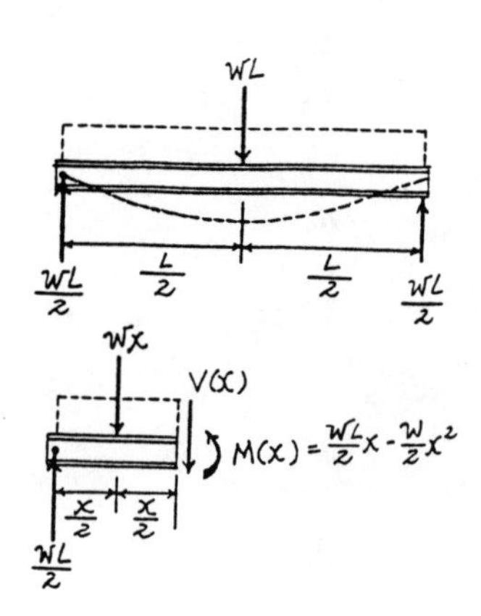

Boundary Conditions : Due to symmetry, $\frac{dv}{dx} = 0$ at $x = \frac{L}{2}$.
Also, $v = 0$ at $x = 0$.

From Eq.[1], $\quad 0 = \frac{wL}{4}\left(\frac{L}{2}\right)^2 - \frac{w}{6}\left(\frac{L}{2}\right)^3 + C_1 \qquad C_1 = -\frac{wL^3}{24}$

From Eq.[2], $\quad 0 = 0 + 0 + C_2 \qquad C_2 = 0$

The Slope : Substituting the value of C_1 into Eq.[1],

$$\frac{dv}{dx} = \frac{w}{24EI}\left(-4x^3 + 6Lx^2 - L^3\right)$$

$$\theta_A = \left.\frac{dv}{dx}\right|_{x=0} = -\frac{wL^3}{24EI} \qquad \textbf{Ans}$$

The negative sign indicates clockwise rotation.

The Elastic Curve : Substituting the values of C_1 and C_2 into Eq.[2],

$$v = \frac{wx}{24EI}\left(-x^3 + 2Lx^2 - L^3\right) \qquad \textbf{Ans}$$

$$v_C = v\,|_{x=\frac{L}{2}} = -\frac{5wL^4}{384EI} \qquad \textbf{Ans}$$

The negative sign indicates downward displacement.

12–17. Determine the equations of the elastic curve using the coordinates x_1 and x_2, and specify the slope at C and deflection at B. EI is constant.

Support Reactions and Elastic Curve : As shown on FBD(a).

Moment Function : As shown on FBD(b) and (c).

Slope and Elastic Curve :

$$EI\frac{d^2v}{dx^2} = M(x)$$

For $M(x_1) = wax_1 - \frac{3wa^2}{2}$,

$$EI\frac{d^2v_1}{dx_1^2} = wax_1 - \frac{3wa^2}{2}$$

$$EI\frac{dv_1}{dx_1} = \frac{wa}{2}x_1^2 - \frac{3wa^2}{2}x_1 + C_1 \qquad [1]$$

$$EI\,v_1 = \frac{wa}{6}x_1^3 - \frac{3wa^2}{4}x_1^2 + C_1x_1 + C_2 \qquad [2]$$

For $M(x_2) = -\frac{w}{2}x_2^2$,

$$EI\frac{d^2v_2}{dx_2^2} = -\frac{w}{2}x_2^2$$

$$EI\frac{dv_2}{dx_2} = -\frac{w}{6}x_2^3 + C_3 \qquad [3]$$

$$EI\,v_2 = -\frac{w}{24}x_2^4 + C_3x_2 + C_4 \qquad [4]$$

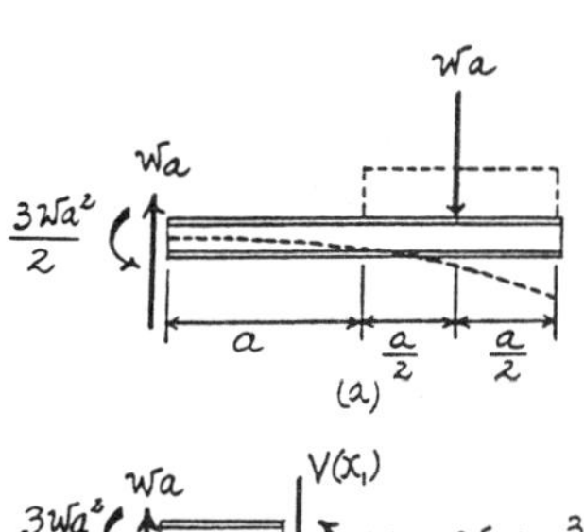

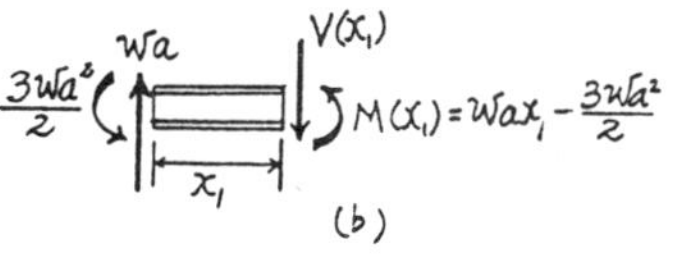

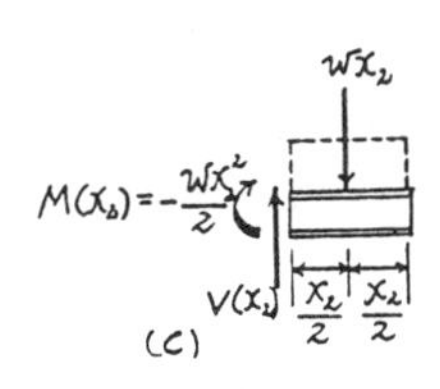

Boundary Conditions :

$\frac{dv_1}{dx_1} = 0$ at $x_1 = 0$. From Eq.[1], $C_1 = 0$

$v_1 = 0$ at $x_1 = 0$. From Eq.[2], $C_2 = 0$

Continuity Conditions :

At $x_1 = a$ and $x_2 = a$, $\frac{dv_1}{dx_1} = -\frac{dv_2}{dx_2}$. From Eqs.[1] and [3],

$$\frac{wa^3}{2} - \frac{3wa^3}{2} = -\left(-\frac{wa^3}{6} + C_3\right) \qquad C_3 = \frac{7wa^3}{6}$$

At $x_1 = a$ and $x_2 = a$, $v_1 = v_2$. From Eqs.[2] and [4],

$$\frac{wa^4}{6} - \frac{3wa^4}{4} = -\frac{wa^4}{24} + \frac{5wa^4}{6} + C_4 \qquad C_4 = -\frac{11wa^4}{8}$$

The Slope : Substituting into Eq.[1],

$$\frac{dv_1}{dx_1} = \frac{wax_1}{2EI}(x_1 - 3a)$$

$$\theta_C = \frac{dv_1}{dx_1}\bigg|_{x_1=a} = -\frac{wa^3}{EI} \qquad \textbf{Ans}$$

The Elastic Curve : Substituting the values of C_1, C_2, C_3, and C_4 into Eqs.[2] and [4], respectively,

$$v_1 = \frac{wax_1}{12EI}\left(2x_1^2 - 9ax_1\right) \qquad \textbf{Ans}$$

$$v_2 = \frac{w}{24EI}\left(-x_2^4 + 28a^3x_2 - 41a^4\right) \qquad \textbf{Ans}$$

$$v_B = v_2\big|_{x_2=0} = -\frac{41wa^4}{24EI} \qquad \textbf{Ans}$$

12–18. Determine the equations of the elastic curve using the coordinates x_1 and x_3, and specify the slope at B and deflection at C. EI is constant.

Support Reactions and Elastic Curve : As shown on FBD(a).
Moment Function : As shown on FBD(b) and (c).
Slope and Elastic Curve :

$$EI\frac{d^2v}{dx^2} = M(x)$$

For $M(x_1) = wax_1 - \dfrac{3wa^2}{2}$,

$$EI\frac{d^2v_1}{dx_1^2} = wax_1 - \frac{3wa^2}{2}$$

$$EI\frac{dv_1}{dx_1} = \frac{wa}{2}x_1^2 - \frac{3wa^2}{2}x_1 + C_1 \qquad [1]$$

$$EI\,v_1 = \frac{wa}{6}x_1^3 - \frac{3wa^2}{4}x_1^2 + C_1x_1 + C_2 \qquad [2]$$

For $M(x_3) = 2wax_3 - \dfrac{w}{2}x_3^2 - 2wa^2$,

$$EI\frac{d^2v_3}{dx_3^2} = 2wax_3 - \frac{w}{2}x_3^2 - 2wa^2$$

$$EI\frac{dv_3}{dx_3} = wax_3^2 - \frac{w}{6}x_3^3 - 2wa^2x_3 + C_3 \qquad [3]$$

$$EI\,v_3 = \frac{wa}{3}x_3^3 - \frac{w}{24}x_3^4 - wa^2x_3^2 + C_3x_3 + C_4 \qquad [4]$$

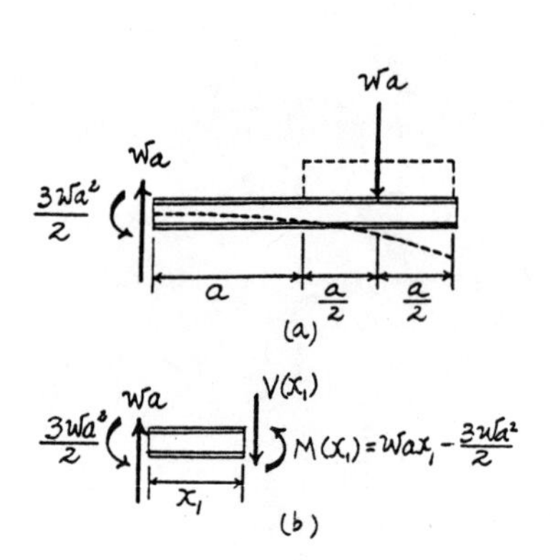

Boundary Conditions :

$\dfrac{dv_1}{dx_1} = 0$ at $x_1 = 0$. From Eq.[1], $C_1 = 0$

$v_1 = 0$ at $x_1 = 0$. From Eq.[2], $C_2 = 0$

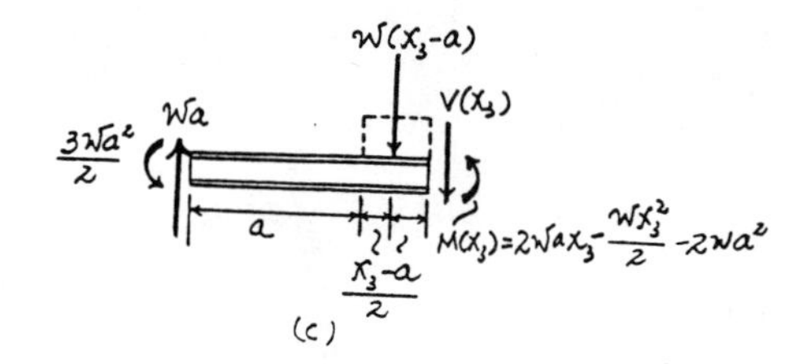

Continuity Conditions :

At $x_1 = a$ and $x_3 = a$, $\dfrac{dv_1}{dx_1} = \dfrac{dv_3}{dx_3}$. From Eqs.[1] and [3],

$$\frac{wa^3}{2} - \frac{3wa^3}{2} = wa^3 - \frac{wa^3}{6} - 2wa^3 + C_3 \qquad C_3 = \frac{wa^3}{6}$$

At $x_1 = a$ and $x_3 = a$, $v_1 = v_3$. From Eqs.[2] and [4],

$$\frac{wa^4}{6} - \frac{3wa^4}{4} = \frac{wa^4}{3} - \frac{wa^4}{24} - wa^4 + \frac{wa^4}{6} + C_4 \qquad C_4 = -\frac{wa^4}{24}$$

The Slope : Substituting the value of C_1 into Eq.[1],

$$\frac{dv_3}{dx_3} = \frac{w}{6EI}\left(6ax_3^2 - x_3^3 - 12a^2x_3 + a^3\right)$$

$$\theta_B = \left.\frac{dv_3}{dx_3}\right|_{x_3=2a} = -\frac{7wa^3}{6EI} \qquad \textbf{Ans}$$

The Elastic Curve : Substituting the values of C_1, C_2, C_3, and C_4 into Eqs.[2] and [4], respectively,

$$v_1 = \frac{wax_1}{12EI}\left(2x_1^2 - 9ax_1\right) \qquad \textbf{Ans}$$

$$v_C = v_1|_{x_1=a} = -\frac{7wa^4}{12EI} \qquad \textbf{Ans}$$

$$v_3 = \frac{w}{24EI}\left(-x_3^4 + 8ax_3^3 - 24a^2x_3^2 + 4a^3x_3 - a^4\right) \qquad \textbf{Ans}$$

12–19. Wooden posts used for a retaining wall have a diameter of 3 in. If the soil pressure along a post varies uniformly from zero at the top A to a maximum of 300 lb/ft at the bottom B, determine the slope and deflection at the top of the post. $E_w = 1.6(10^3)$ ksi.

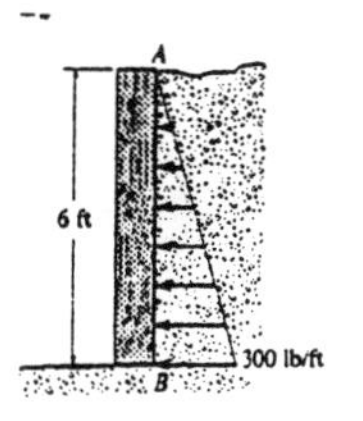

Moment Function : As shown on FBD.

Slope and Elastic Curve :

$$EI\frac{d^2v}{dy^2} = M(y)$$

$$EI\frac{d^2v}{dy^2} = -8.333y^3$$

$$EI\frac{dv}{dy} = -2.0833y^4 + C_1 \qquad [1]$$

$$EI\,v = -0.4167y^5 + C_1 y + C_2 \qquad [2]$$

Boundary Conditions : $\frac{dv}{dy} = 0$ at $y = 6$ ft and $v = 0$ at $y = 6$ ft.

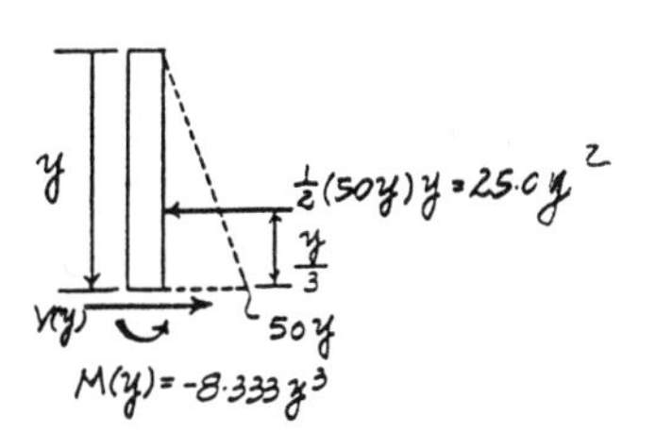

From Eq.[1], $\quad 0 = -2.0833(6^4) + C_1 \qquad C_1 = 2700$

From Eq.[2], $\quad 0 = -0.4167(6^5) + 2700(6) + C_2 \qquad C_2 = -12960$

The Slope : Substituting the value of C_1 into Eq.[1],

$$\frac{dv}{dy} = \left\{\frac{1}{EI}\left(-2.0833y^4 + 2700\right)\right\} \text{ lb}\cdot\text{ft}^2$$

$$\theta_A = \left.\frac{dv}{dy}\right|_{y=0} = \frac{2700 \text{ lb}\cdot\text{ft}^2}{EI}$$

$$= \frac{2700(144)}{1.6(10^6)\left(\frac{\pi}{4}\right)(1.5^4)}$$

$$= 0.0611 \text{ rad} \qquad \textbf{Ans}$$

The Elastic Curve : Substituting the values of C_1 and C_2 into Eq.[2],

$$v = \frac{1}{EI}\left\{\left(-0.4167y^5 + 2700y - 12960\right)\right\} \text{ lb}\cdot\text{ft}^3$$

$$v_A = v|_{y=0} = -\frac{12960 \text{ lb}\cdot\text{ft}^3}{EI}$$

$$= -\frac{12960(1728)}{1.6(10^6)\left(\frac{\pi}{4}\right)(1.5^4)}$$

$$= -3.52 \text{ in.} \qquad \textbf{Ans}$$

The negative sign indicates leftward displacement.

***12–20.** The beam is made of a material having a specific weight γ. Determine the displacement and slope at its end A due to its weight.

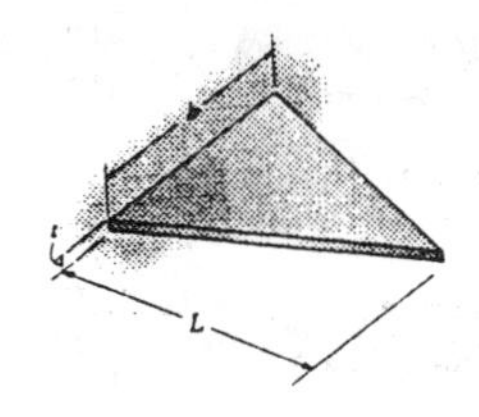

Section Properties :

$$b(x) = \frac{b}{L}x \qquad V(x) = \frac{1}{2}\left(\frac{b}{L}x\right)(x)(t) = \frac{bt}{2L}x^2$$

$$I(x) = \frac{1}{12}\left(\frac{b}{L}x\right)t^3 = \frac{bt^3}{12L}x$$

Moment Function : As shown on FBD.

Slope and Elastic Curve :

$$E\frac{d^2v}{dx^2} = \frac{M(x)}{I(x)}$$

$$E\frac{d^2v}{dx^2} = -\frac{\frac{bt\gamma}{6L}x^3}{\frac{bt^3}{12L}x} = -\frac{2\gamma}{t^2}x^2$$

$$E\frac{dv}{dx} = -\frac{2\gamma}{3t^2}x^3 + C_1 \qquad [1]$$

$$E\,v = -\frac{\gamma}{6t^2}x^4 + C_1x + C_2 \qquad [2]$$

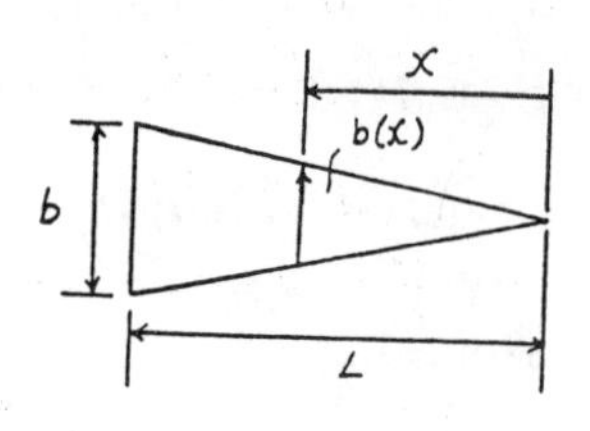

Boundary Conditions : $\frac{dv}{dx} = 0$ at $x = L$ and $v = 0$ at $x = L$.

From Eq. [1], $\quad 0 = -\frac{2\gamma}{3t^2}(L^3) + C_1 \qquad C_1 = \frac{2\gamma L^3}{3t^2}$

From Eq. [2], $\quad 0 = -\frac{\gamma}{6t^2}(L^4) + \left(\frac{2\gamma L^3}{3t^2}\right)(L) + C_2$

$$C_2 = -\frac{\gamma L^4}{2t^2}$$

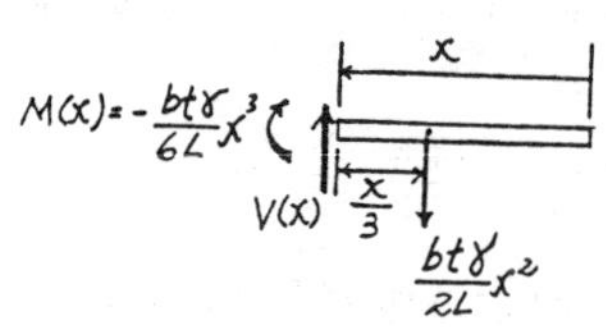

The Slope : Substituting the value of C_1 into Eq. [1],

$$\frac{dv}{dx} = \frac{2\gamma}{3t^2E}\left(-x^3 + L^3\right)$$

$$\theta_A = \left.\frac{dv}{dx}\right|_{x=0} = \frac{2\gamma L^3}{3t^2E} \qquad \textbf{Ans}$$

The Elastic Curve : Substituting the values of C_1 and C_2 into Eq. [2],

$$v = \frac{\gamma}{6t^2E}\left(-x^4 + 4L^3x - 3L^4\right)$$

$$v_A\big|_{x=0} = -\frac{\gamma L^4}{2t^2E} \qquad \textbf{Ans}$$

The negative sign indicates downward displacement.

12–21. The beam is made of a material having a specific weight γ. Determine the displacement and slope at its end A due to its weight.

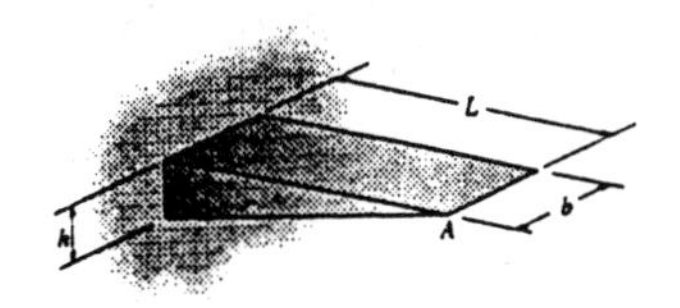

Section Properties :

$$h(x) = \frac{h}{L}x \qquad V(x) = \frac{1}{2}\left(\frac{h}{L}x\right)(x)(b) = \frac{bh}{2L}x^2$$

$$I(x) = \frac{1}{12}(b)\left(\frac{h}{L}x\right)^3 = \frac{bh^3}{12L^3}x^3$$

Moment Function : As shown on FBD.

Slope and Elastic Curve :

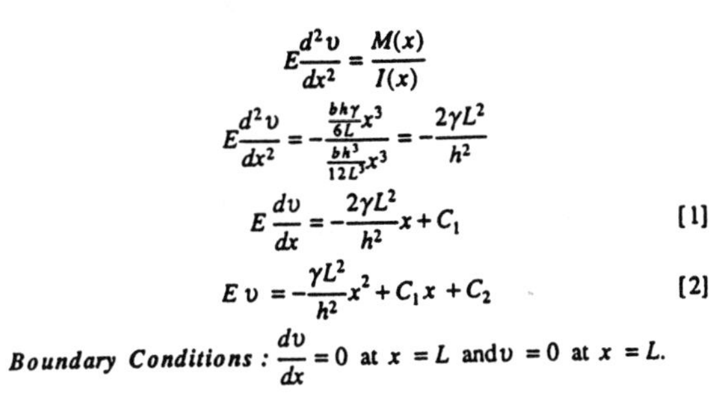

$$E\frac{d^2\upsilon}{dx^2} = \frac{M(x)}{I(x)}$$

$$E\frac{d^2\upsilon}{dx^2} = -\frac{\frac{bh\gamma}{6L}x^3}{\frac{bh^3}{12L^3}x^3} = -\frac{2\gamma L^2}{h^2}$$

$$E\frac{d\upsilon}{dx} = -\frac{2\gamma L^2}{h^2}x + C_1 \qquad [1]$$

$$E\,\upsilon = -\frac{\gamma L^2}{h^2}x^2 + C_1 x + C_2 \qquad [2]$$

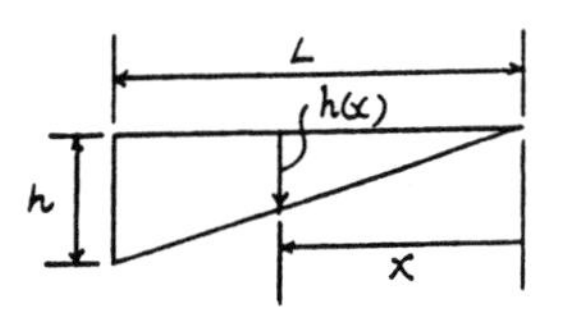

Boundary Conditions : $\frac{d\upsilon}{dx} = 0$ at $x = L$ and $\upsilon = 0$ at $x = L$.

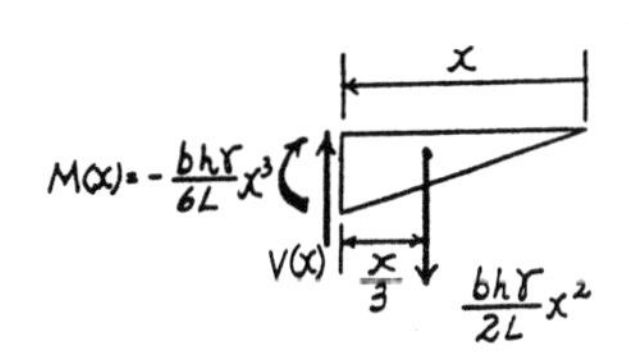

From Eq. [1], $\quad 0 = -\frac{2\gamma L^2}{h^2}(L) + C_1 \qquad C_1 = \frac{2\gamma L^3}{h^2}$

From Eq. [2], $\quad 0 = -\frac{\gamma L^2}{h^2}(L^2) + \frac{2\gamma L^3}{h^2}(L) + C_2 \qquad C_2 = -\frac{\gamma L^4}{h^2}$

The Slope : Substituting the value of C_1 into Eq. [1],

$$\frac{d\upsilon}{dx} = \frac{2\gamma L^2}{h^2 E}(-x + L)$$

$$\theta_A = \frac{d\upsilon}{dx}\bigg|_{x=0} = \frac{2\gamma L^3}{h^2 E} \qquad \textbf{Ans}$$

The Elastic Curve : Substituting the values of C_1 and C_2 into Eq. [2],

$$\upsilon = \frac{\gamma L^2}{h^2 E}\left(-x^2 + 2Lx - L^2\right)$$

$$\upsilon_A\,|_{x=0} = -\frac{\gamma L^4}{h^2 E} \qquad \textbf{Ans}$$

The negative sign indicates downward displacement.

12–22. The beam is made of a material having a specific weight of γ. Determine the displacement and slope at its end A due to its weight.

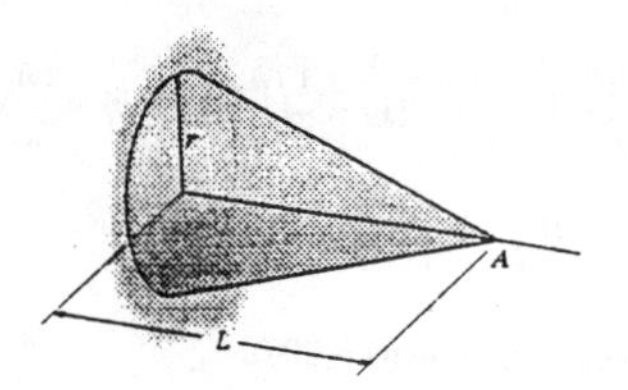

Section Properties :

$$r(x) = \frac{r}{L}x \qquad V(x) = \frac{\pi}{3}\left(\frac{r}{L}x\right)^2 x = \frac{\pi r^2}{3L^2}x^3$$

$$I(x) = \frac{\pi}{4}\left(\frac{r}{L}x\right)^4 = \frac{\pi r^4}{4L^4}x^4$$

Moment Function : As shown on FBD.

Slope and Elastic Curve :

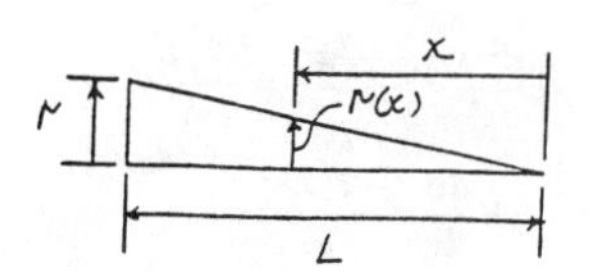

$$E\frac{d^2 v}{dx^2} = \frac{M(x)}{I(x)}$$

$$E\frac{d^2 v}{dx^2} = -\frac{\frac{\pi r^2\gamma}{9L^2}x^4}{\frac{\pi r^4}{4L^4}x^4} = -\frac{4\gamma L^2}{9r^2}$$

$$E\frac{dv}{dx} = -\frac{4\gamma L^2}{9r^2}x + C_1 \qquad [1]$$

$$E\,v = -\frac{2\gamma L^2}{9r^2}x^2 + C_1 x + C_2 \qquad [2]$$

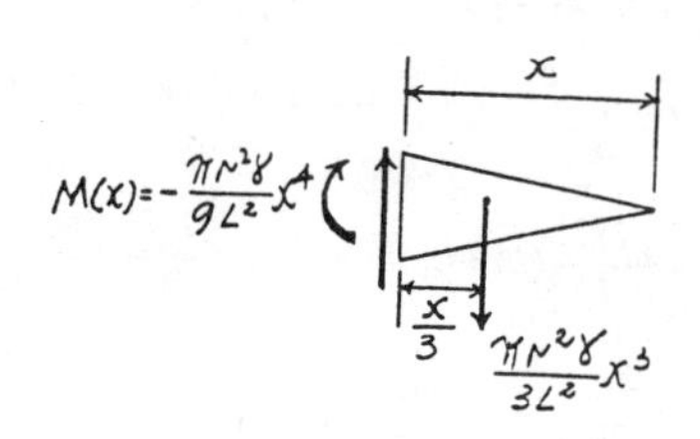

Boundary Conditions : $\frac{dv}{dx} = 0$ at $x = L$ and $v = 0$ at $x = L$.

From Eq.[1], $\quad 0 = -\frac{4\gamma L^2}{9r^2}(L) + C_1 \qquad C_1 = \frac{4\gamma L^3}{9r^2}$

From Eq.[2], $\quad 0 = -\frac{2\gamma L^2}{9r^2}\left(L^2\right) + \left(\frac{4\gamma L^3}{9r^2}\right)L + C_2 \qquad C_2 = -\frac{2\gamma L^4}{9r^2}$

The Slope : Substituting the value of C_1 into Eq.[1],

$$\frac{dv}{dx} = \frac{4\gamma L^2}{9r^2 E}(-x + L)$$

$$\theta_A = \left.\frac{dv}{dx}\right|_{x=0} = \frac{4\gamma L^3}{9r^2 E} \qquad \textbf{Ans}$$

The Elastic Curve : Substituting the values of C_1 and C_2 into Eq.[2],

$$v = \frac{2\gamma L^2}{9r^2 E}\left(-x^2 + 2Lx - L^2\right)$$

$$v_A\,|_{x=0} = -\frac{2\gamma L^4}{9r^2 E} \qquad \textbf{Ans}$$

The negative sign indicates downward displacement.

12–23. The leaf spring assembly is designed so that it is subjected to the same maximum stress throughout its length. If the plates of each leaf have a thickness t and can slide freely between each other, show that the spring must be in the form of a circular arc in order that the entire spring becomes flat when a large enough load **P** is applied. What is the maximum normal stress in the spring? Consider the spring to be made by cutting the n strips from the diamond-shaped plate of thickness t and width b. *Hint:* Show that the radius of curvature of the spring is constant.

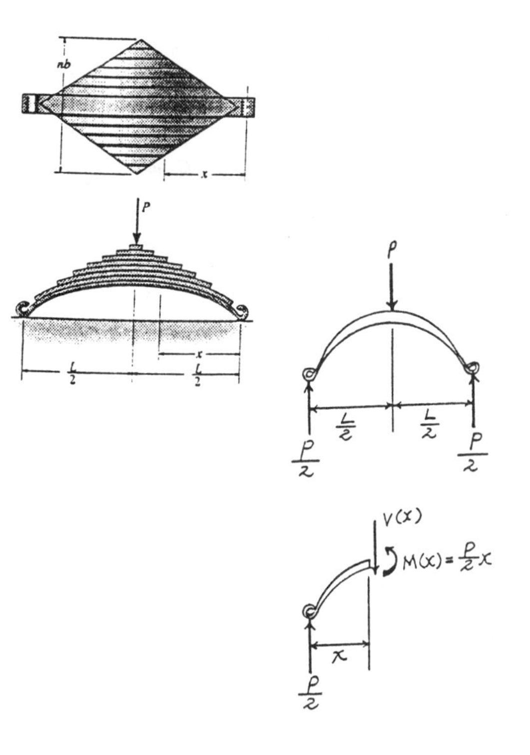

Section Properties : Since the plates can slide freely relative to each other, the plates resist the moment individually. At an arbitrary distance x from the support, the numbers of plates is $\frac{nx}{\frac{L}{2}} = \frac{2nx}{L}$. Hence,

$$I(x) = \frac{1}{12}\left(\frac{2nx}{L}\right)(b)\left(t^3\right) = \frac{nbt^3}{6L}x$$

Moment Function : As shown on FBD.

Bending Stress : Applying the flexure formula,

$$\sigma_{\max} = \frac{M(x)c}{I(x)} = \frac{\frac{Px}{2}\left(\frac{t}{2}\right)}{\frac{nbt^3}{6L}x} = \frac{3PL}{2nbt^2} \qquad \textbf{Ans}$$

Moment - Curvature Relationship :

$$\frac{1}{\rho} = \frac{M(x)}{EI(x)} = \frac{\frac{Px}{2}}{E\left(\frac{nbt^3}{6L}x\right)} = \frac{3PL}{nbt^3E} = \text{Constant} \quad (Q.E.D.)$$

***12–24.** The shaft supports the two pulley loads shown. Determine the equation of the elastic curve. The bearings at A and B exert only vertical reactions on the shaft. EI is constant.

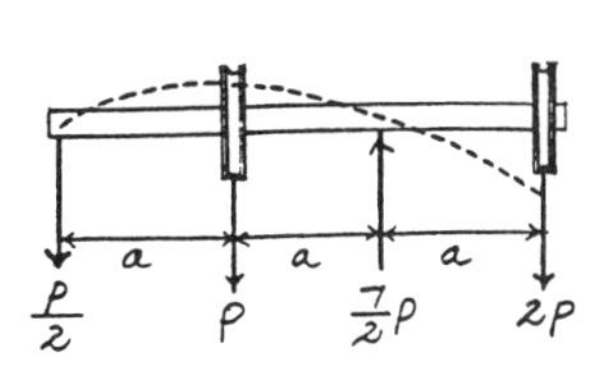

Moment Function : Using the discontinuity function,

$$M = -\frac{P}{2}<x-0> - P<x-a> - \left(-\frac{7}{2}P\right)<x-2a>$$
$$= -\frac{P}{2}x - P<x-a> + \frac{7}{2}P<x-2a>$$

Slope and Elastic Curve :

$$EI\frac{d^2v}{dx^2} = M$$

$$EI\frac{d^2v}{dx^2} = -\frac{P}{2}x - P<x-a> + \frac{7}{2}P<x-2a>$$

$$EI\frac{dv}{dx} = -\frac{P}{4}x^2 - \frac{P}{2}<x-a>^2 + \frac{7}{4}P<x-2a>^2 + C_1 \qquad [1]$$

$$EI\,v = -\frac{P}{12}x^3 - \frac{P}{6}<x-a>^3 + \frac{7}{12}P<x-2a>^3 + C_1x + C_2 \qquad [2]$$

Boundary Conditions :

$v = 0$ at $x = 0$. From Eq.[2], $C_2 = 0$

$v = 0$ at $x = 2a$. From Eq.[2],

$$0 = -\frac{P}{12}(2a)^3 - \frac{P}{6}(2a-a)^3 + 0 + C_1(2a) + 0$$

$$C_1 = \frac{5Pa^2}{12}$$

The Elastic Curve : Substituting the values of C_1 and C_2 into Eq.[2],

$$v = \frac{P}{12EI}\{-x^3 - 2<x-a>^3 + 7<x-2a>^3 + 5a^2x\} \qquad \textbf{Ans}$$

12–25. The shaft is supported at A by a journal bearing that exerts only vertical reactions on the shaft and at C by a thrust bearing that exerts horizontal and vertical reactions on the shaft. Determine the equation of the elastic curve. EI is constant.

Moment Function : Using discontinuity function,

$$M = -\frac{Pb}{a} < x-0 > - \left[-\frac{P(a+b)}{a}\right] < x-a >$$

$$= -\frac{Pb}{a}x + \frac{P(a+b)}{a} < x-a >$$

Slope and Elastic Curve :

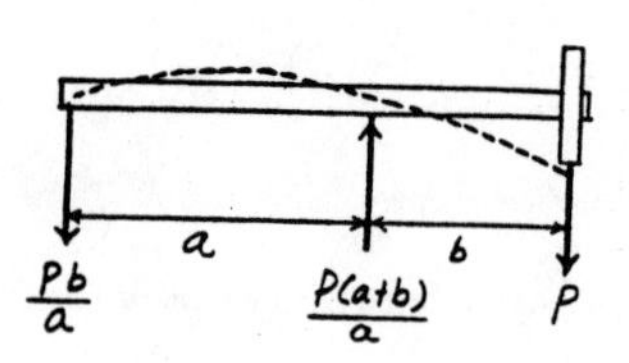

$$EI\frac{d^2v}{dx^2} = M$$

$$EI\frac{d^2v}{dx^2} = -\frac{Pb}{a}x + \frac{P(a+b)}{a} < x-a >$$

$$EI\frac{dv}{dx} = -\frac{Pb}{2a}x^2 + \frac{P(a+b)}{2a} < x-a >^2 + C_1 \qquad [1]$$

$$EI\,v = -\frac{Pb}{6a}x^3 + \frac{P(a+b)}{6a} < x-a >^3 + C_1x + C_2 \qquad [2]$$

Boundary Conditions :

$v = 0$ at $x = 0$. From Eq. [2], $\quad C_2 = 0$

$v = 0$ at $x = a$. From Eq. [2],

$$0 = -\frac{Pb}{6a}(a^3) + 0 + C_1 a \qquad C_1 = \frac{Pab}{6}$$

The Elastic Curve : Substituting the values of C_1 and C_2 into Eq. [2],

$$v = \frac{P}{6aEI}\{-bx^3 + (a+b) < x-a >^3 + a^2bx\} \qquad \textbf{Ans}$$

12–26. The shaft is made of 304 stainless steel and has a diameter of 15 mm. Determine its maximum deflection. The bearings at A and B exert only vertical reactions on the shaft.

Support Reactions and Elastic Curve : As shown on FBD.

Moment Function : Using the discontinuity function,

$$M = 300 < x-0 > -300 < x-0.2 > -300 < x-0.5 >$$
$$= 300x - 300 < x-0.2 > -300 < x-0.5 >$$

Slope and Elastic Curve :

$$EI\frac{d^2 v}{dx^2} = M$$

$$EI\frac{d^2 v}{dx^2} = 300x - 300 < x-0.2 > -300 < x-0.5 >$$

$$EI\frac{dv}{dx} = 150x^2 - 150 < x-0.2 >^2 - 150 < x-0.5 >^2 + C_1 \qquad [1]$$

$$EI\, v = 50x^3 - 50 < x-0.2 >^3 - 50 < x-0.5 >^3 + C_1 x + C_2 \qquad [2]$$

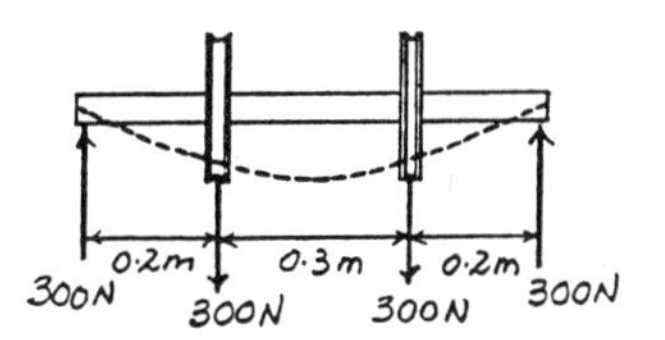

Boundary Conditions :

$v = 0$ at $x = 0$. From Eq.[2], $C_2 = 0$

$v = 0$ at $x = 0.7$ m. From Eq.[2],

$$0 = 50(0.7^3) - 50(0.7-0.2)^3 - 50(0.7-0.5)^3 + C_1(0.7)$$
$$C_1 = -15.0$$

The Elastic Curve : Substituting the value of C_1 and C_2 into Eq.[2],

$$v = \frac{1}{EI}\{50x^3 - 50 < x-0.2 >^3 - 50 < x-0.5 >^3 - 15.0x\} \text{ N}\cdot\text{m}^3$$

The maximum displacement occurs at the midspan of the shaft where $x = 0.35$ m.

$$v_{max} = -\frac{3.275 \text{ N}\cdot\text{m}^3}{EI}$$
$$= -\frac{3.275}{193(10^9)\left(\frac{\pi}{4}\right)(0.0075^4)}$$
$$= -0.006828 \text{ m} = 6.83 \text{ mm} \downarrow \qquad \textbf{Ans}$$

12–27. The shaft is made of 304 stainless steel and has a diameter of 15 mm. Determine the equation of the elastic curve and find the slopes at the bearings A and B. The bearings exert only vertical reactions on the shaft.

Support Reactions and Elastic Curve : As shown on FBD.

Moment Function : Using the discontinuity function,

$$M = 300 < x-0 > -300 < x-0.2 > -300 < x-0.5 >$$
$$= 300x - 300 < x-0.2 > -300 < x-0.5 >$$

Slope and Elastic Curve :

$$EI\frac{d^2 v}{dx^2} = M$$

$$EI\frac{d^2 v}{dx^2} = 300x - 300 < x-0.2 > -300 < x-0.5 >$$

$$EI\frac{dv}{dx} = 150x^2 - 150 < x-0.2 >^2 -150 < x-0.5 >^2 + C_1 \qquad [1]$$

$$EI\, v = 50x^3 - 50 < x-0.2 >^3 -50 < x-0.5 >^3 + C_1 x + C_2 \qquad [2]$$

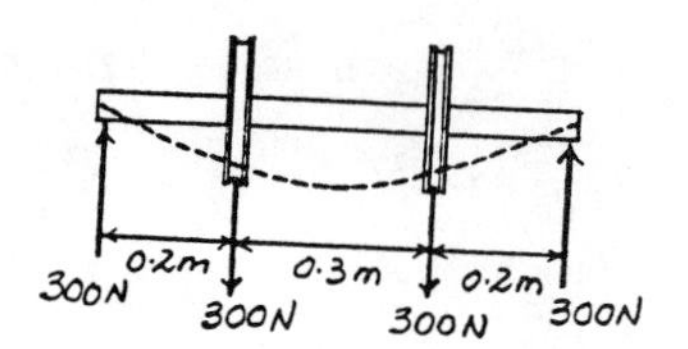

Boundary Conditions :

$v = 0$ at $x = 0$. From Eq. [2], $C_2 = 0$

$v = 0$ at $x = 0.7$ m. From Eq. [2],

$$0 = 50(0.7^3) - 50(0.7-0.2)^3 - 50(0.7-0.5)^3 + C_1(0.7)$$
$$C_1 = -15.0$$

The Slope : Substituting the value of C_1 into Eq. [1],

$$\frac{dv}{dx} = \frac{1}{EI}\{150x^2 - 150 < x-0.2 >^2 -150 < x-0.5 >^2 -15.0\}\ \text{N}\cdot\text{m}^2$$

Due to symmetrical loading and system,

$$\theta_B = \theta_A = \left.\frac{dv}{dx}\right|_{x=0} = \left|-\frac{15.0\ \text{N}\cdot\text{m}^2}{EI}\right|$$
$$= \frac{15.0}{193(10^9)\left(\frac{\pi}{4}\right)(0.0075^4)}$$
$$= 0.0313\ \text{rad} \qquad \textbf{Ans}$$

The Elastic Curve : Substituting the values of C_1 and C_2 into Eq. [2],

$$v = \frac{1}{EI}\{50x^3 - 50 < x-0.2 >^3 -50 < x-0.5 >^3 -15.0x\}\ \text{N}\cdot\text{m}^3 \qquad \textbf{Ans}$$

***12–28.** The beam is subjected to the loads shown. Determine the equation of the elastic curve. EI is constant.

Support Reactions and Elastic Curve : As shown on FBD.

Moment Function : Using the discontinuity function,

$$M = 2.50 < x - 0 > - 2 < x - 8 > - 4 < x - 16 >$$
$$= 2.50x - 2 < x - 8 > - 4 < x - 16 >$$

Slope and Elastic Curve :

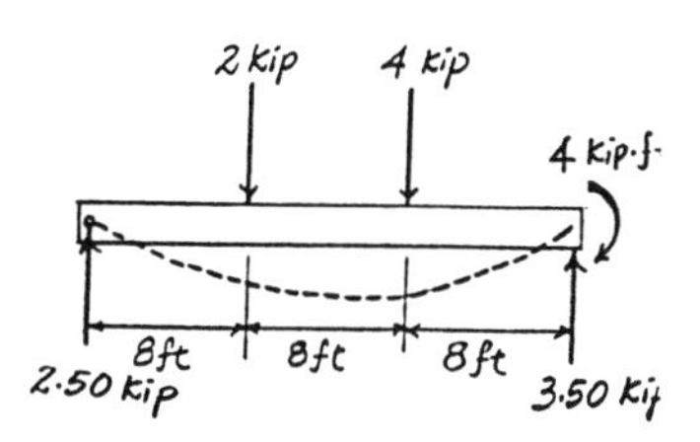

$$EI\frac{d^2\upsilon}{dx^2} = M$$

$$EI\frac{d^2\upsilon}{dx^2} = 2.50x - 2 < x - 8 > - 4 < x - 16 >$$

$$EI\frac{d\upsilon}{dx} = 1.25x^2 - < x - 8 >^2 - 2 < x - 16 >^2 + C_1 \qquad [1]$$

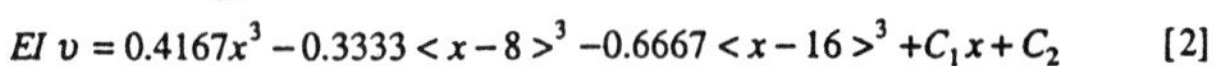

$$EI\,\upsilon = 0.4167x^3 - 0.3333 < x - 8 >^3 - 0.6667 < x - 16 >^3 + C_1 x + C_2 \qquad [2]$$

Boundary Conditions :

$\upsilon = 0$ at $x = 0$. From Eq.[2], $C_2 = 0$

$\upsilon = 0$ at $x = 24$ ft. From Eq.[2],

$$0 = 0.4167(24^3) - 0.3333(24-8)^3 - 0.6667(24-16)^3 + C_1(24)$$
$$C_1 = -168.89$$

The Elastic Curve : Substituting the values of C_1 and C_2 into Eq.[2],

$$\upsilon = \frac{1}{EI}\{0.417x^3 - 0.333 < x - 8 >^3 - 0.667 < x - 16 >^3 - 169x\}\ \text{kip}\cdot\text{ft}^3$$ **Ans**

12–29. The shaft supports the three pulley loads shown. Determine the equation of the elastic curve. The bearings at A and B exert only vertical reactions on the shaft. What is the maximum deflection? EI is constant.

Support Reactions and Elastic Curve : As shown on FBD.

Moment Function : Using the discontinuity function,

$$M = -P\langle x-0\rangle - \left(-\frac{3P}{2}\right)\langle x-a\rangle - P\langle x-2a\rangle - \left(-\frac{3P}{2}\right)\langle x-3a\rangle$$
$$= -Px + \frac{3P}{2}\langle x-a\rangle - P\langle x-2a\rangle + \frac{3P}{2}\langle x-3a\rangle$$

Slope and Elastic Curve :

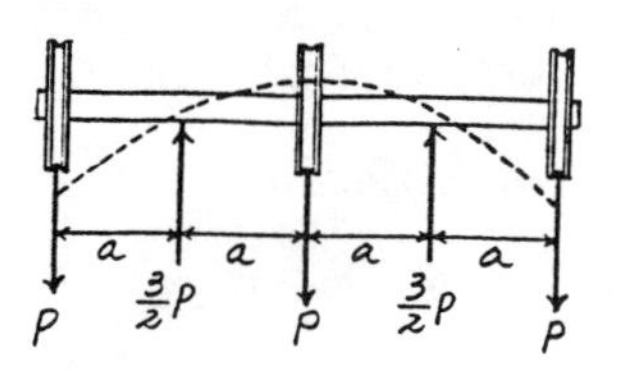

$$EI\frac{d^2v}{dx^2} = M$$

$$EI\frac{d^2v}{dx^2} = -Px + \frac{3P}{2}\langle x-a\rangle - P\langle x-2a\rangle + \frac{3P}{2}\langle x-3a\rangle$$

$$EI\frac{dv}{dx} = -\frac{P}{2}x^2 + \frac{3P}{4}\langle x-a\rangle^2 - \frac{P}{2}\langle x-2a\rangle^2 + \frac{3P}{4}\langle x-3a\rangle^2 + C_1 \qquad [1]$$

$$EI\,v = -\frac{P}{6}x^3 + \frac{P}{4}\langle x-a\rangle^3 - \frac{P}{6}\langle x-2a\rangle^3 + \frac{P}{4}\langle x-3a\rangle^3 + C_1x + C_2 \qquad [2]$$

Boundary Conditions :

Due to symmetry, $\frac{dv}{dx} = 0$ at $x = 2a$. From Eq. [1],

$$0 = -\frac{P}{2}(2a)^2 + \frac{3P}{4}(2a-a)^2 - 0 + 0 + C_1 \qquad C_1 = \frac{5Pa^2}{4}$$

$v = 0$ at $x = a$ From Eq. [2],

$$0 = -\frac{P}{6}(a^3) + 0 - 0 + 0 + \frac{5Pa^2}{4}(a) + C_2 \qquad C_2 = -\frac{13Pa^3}{12}$$

The Elastic Curve : Substituting the values of C_1 and C_2 into Eq. [2],

$$v = \frac{P}{12EI}\{-2x^3 + 3\langle x-a\rangle^3 - 2\langle x-2a\rangle^3 + 3\langle x-3a\rangle^3 + 15a^2x - 13a^3\} \quad \textbf{Ans}$$

$$v|_{x=2a} = \frac{P}{12EI}\{-2(2a)^3 + 3(2a-a)^3 - 0 + 0 + 15a^2(2a) - 13a^3\} = \frac{Pa^3}{3EI}$$

$$v_{\max} = v|_{x=0} = \frac{P}{12EI}\{-0 + 0 - 0 + 0 + 0 - 13a^3\} = -\frac{13Pa^3}{12EI} \qquad \textbf{Ans}$$

12–30. The shaft supports the three pulley loads shown. Determine the deflection of the shaft at its center and its slopes at the bearings A and B. The bearings exert only vertical reactions on the shaft. EI is constant.

Support Reactions and Elastic Curve : As shown on FBD.

Moment Function : Using the discontinuity function,

$$M = -P<x-0> - \left(-\frac{3P}{2}\right)<x-a> - P<x-2a> - \left(-\frac{3P}{2}\right)<x-3a>$$

$$= -Px + \frac{3P}{2}<x-a> - P<x-2a> + \frac{3P}{2}<x-3a>$$

Slope and Elastic Curve :

$$EI\frac{d^2v}{dx^2} = M$$

$$EI\frac{d^2v}{dx^2} = -Px + \frac{3P}{2}<x-a> - P<x-2a> + \frac{3P}{2}<x-3a>$$

$$EI\frac{dv}{dx} = -\frac{P}{2}x^2 + \frac{3P}{4}<x-a>^2 - \frac{P}{2}<x-2a>^2 + \frac{3P}{4}<x-3a>^2 + C_1 \qquad [1]$$

$$EI\,v = -\frac{P}{6}x^3 + \frac{P}{4}<x-a>^3 - \frac{P}{6}<x-2a>^3 + \frac{P}{4}<x-3a>^3 + C_1x + C_2 \qquad [2]$$

Boundary Conditions :

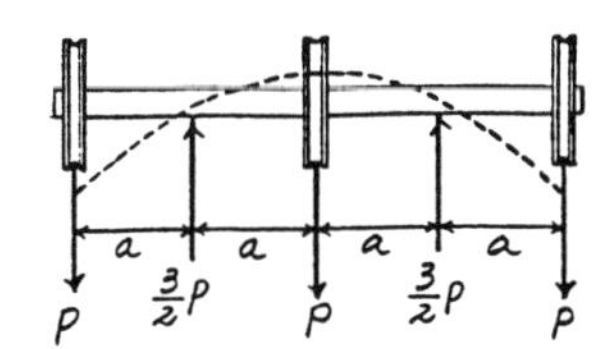

Due to symmetry, $\frac{dv}{dx} = 0$ at $x = 2a$ From Eq.[1],

$$0 = -\frac{P}{2}(2a)^2 + \frac{3P}{4}(2a-a)^2 - 0 + 0 + C_1 \qquad C_1 = \frac{5Pa^2}{4}$$

$v = 0$ at $x = a$ From Eq.[2],

$$0 = -\frac{P}{6}\left(a^3\right) + 0 - 0 + 0 + \frac{5Pa^2}{4}(a) + C_2 \qquad C_2 = -\frac{13Pa^3}{12}$$

The Slope : Substituting the value of C_1 into Eq.[1],

$$\frac{dv}{dx} = \frac{P}{4EI}\{-2x^2 + 3<x-a>^2 - 2<x-2a>^2 + 3<x-3a>^2 + 5a^2\}$$

Due to symmetrical loading and system,

$$\theta_B = \theta_A = \left.\frac{dv}{dx}\right|_{x=a} = \left|\frac{P}{4EI}\{-2a^2 + 0 - 0 + 0 + 5a^2\}\right| = \frac{3Pa^2}{4EI} \qquad \text{Ans}$$

The Elastic Curve : Substituting the values of C_1 and C_2 into Eq.[2],

$$v = \frac{P}{12EI}\{-2x^3 + 3<x-a>^3 - 2<x-2a>^3 + 3<x-3a>^3 + 15a^2x - 13a^3\}$$

$$v|_{x=2a} = \frac{P}{12EI}\{-2(2a)^3 + 3(2a-a)^3 - 0 + 0 + 15a^2(2a) - 13a^3\} = \frac{Pa^3}{3EI} \qquad \text{Ans}$$

12–31. The beam is subjected to the load shown. Determine the equation of the elastic curve. EI is constant.

Support Reactions and Elastic Curve : As shown on FBD.

Moment Function : Using discontinuity function,

$$M = 24.6 < x-0 > -1.5 < x-0 >^2 -(-1.5) < x-4 >^2 -50 < x-7 >$$
$$= 24.6x - 1.5x^2 + 1.5 < x-4 >^2 -50 < x-7 >$$

Slope and Elastic Curve :

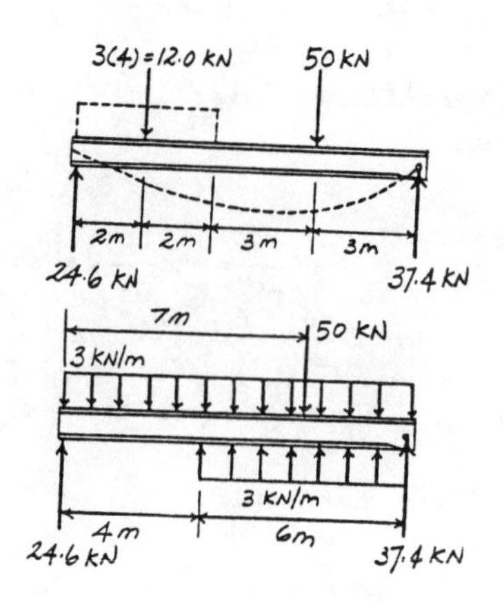

$$EI\frac{d^2v}{dx^2} = M$$
$$EI\frac{d^2v}{dx^2} = 24.6x - 1.5x^2 + 1.5 < x-4 >^2 -50 < x-7 >$$
$$EI\frac{dv}{dx} = 12.3x^2 - 0.5x^3 + 0.5 < x-4 >^3 -25 < x-7 >^2 + C_1 \quad [1]$$
$$EI\,v = 4.10x^3 - 0.125x^4 + 0.125 < x-4 >^4 -8.333 < x-7 >^3 + C_1x + C_2 \quad [2]$$

Boundary Conditions :

$v = 0$ at $x = 0$. From Eq. [2], $C_2 = 0$

$v = 0$ at $x = 10$ m. From Eq. [2],

$$0 = 4.10(10^3) - 0.125(10^4) + 0.125(10-4)^4 - 8.333(10-7)^3 + C_1(10)$$
$$C_1 = -278.7$$

The Elastic Curve : Substituting the values of C_1 and C_2 into Eq. [2],

$$v = \frac{1}{EI}\{4.10x^3 - 0.125x^4 + 0.125 < x-4 >^4 -8.33 < x-7 >^3 -279x\}\ \text{kN}\cdot\text{m}^3 \quad \textbf{Ans}$$

***12–32.** The beam is subjected to the load shown. Determine the deflection at $x = 7$ m and the slope at A. EI is constant.

Support Reactions and Elastic Curve : As shown on FBD.

Moment Function : Using the discontinuity function,

$$M = 24.6 < x-0 > -1.5 < x-0 >^2 -(-1.5) < x-4 >^2 -50 < x-7 >$$
$$= 24.6x - 1.5x^2 + 1.5 < x-4 >^2 -50 < x-7 >$$

Slope and Elastic Curve :

$$EI\frac{d^2 v}{dx^2} = M$$

$$EI\frac{d^2 v}{dx^2} = 24.6x - 1.5x^2 + 1.5 < x-4 >^2 -50 < x-7 >$$

$$EI\frac{dv}{dx} = 12.3x^2 - 0.5x^3 + 0.5 < x-4 >^3 -25 < x-7 >^2 + C_1 \qquad [1]$$

$$EI\, v = 4.10x^3 - 0.125x^4 + 0.125 < x-4 >^4 -8.333 < x-7 >^3 + C_1 x + C_2 \qquad [2]$$

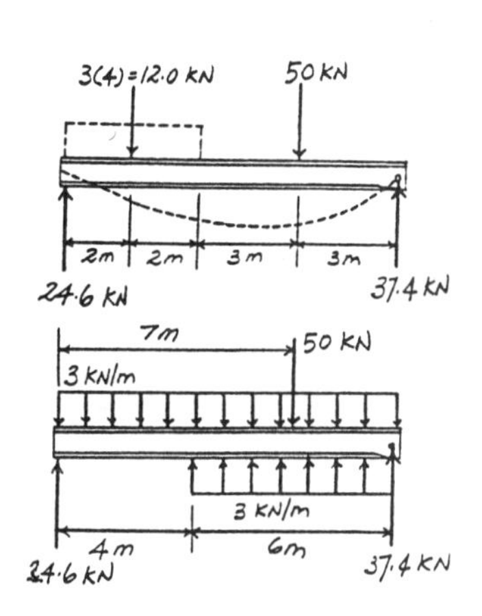

Boundary Conditions :

$v = 0$ at $x = 0$. From Eq.[2], $C_2 = 0$

$v = 0$ at $x = 10$ m. From Eq.[2],

$$0 = 4.10(10^3) - 0.125(10^4) + 0.125(10-4)^4 - 8.333(10-7)^3 + C_1(10)$$
$$C_1 = -278.7$$

The Slope : Substituting the value of C_1 into Eq.[1],

$$\frac{dv}{dx} = \frac{1}{EI}\{12.3x^2 - 0.5x^3 + 0.5 < x-4 >^3 -25 < x-7 >^2 -278.7\}\ \text{kN}\cdot\text{m}^2$$

$$\theta_A = \left.\frac{dv}{dx}\right|_{x=0} = \frac{1}{EI}\{0-0+0-0-278.7\} = -\frac{279\ \text{kN}\cdot\text{m}^2}{EI} \qquad \textbf{Ans}$$

The Elastic Curve : Substituting the values of C_1 and C_2 into Eq.[2],

$$v = \frac{1}{EI}\{4.10x^3 - 0.125x^4 + 0.125 < x-4 >^4 -8.33 < x-7 >^3 -278.7x\}\ \text{kN}\cdot\text{m}^3$$

$$v|_{x=7\text{m}} = \frac{1}{EI}\left\{4.10(7^3) - 0.125(7^4) + 0.125(7-4)^4 - 0 - 278.7(7)\right\}\ \text{kN}\cdot\text{m}^3$$
$$= -\frac{835\ \text{kN}\cdot\text{m}^3}{EI} \qquad \textbf{Ans}$$

12–33. The beam is subjected to the load shown. Determine the equation of the elastic curve. EI is constant.

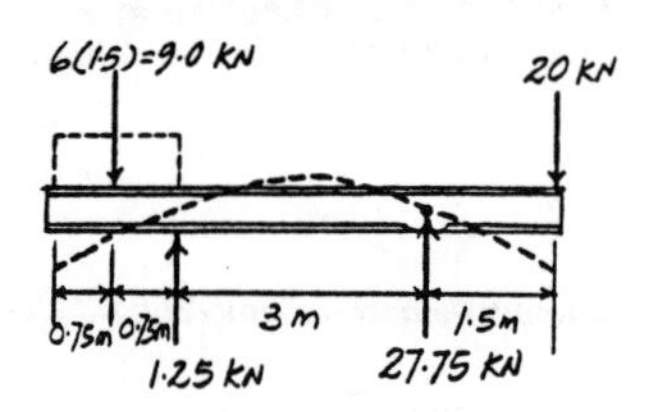

Support Reactions and Elastic Curve : As shown on FBD.

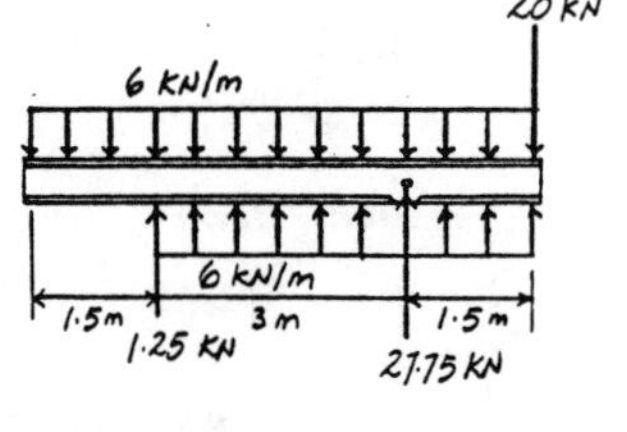

Moment Function : Using the discontinuity function,

$$M = -3 < x - 0 >^2 - (-3) < x - 1.5 >^2 - (-1.25) < x - 1.5 > - (-27.75) < x - 4.5 >$$
$$= -3x^2 + 3 < x - 1.5 >^2 + 1.25 < x - 1.5 > + 27.75 < x - 4.5 >$$

Slope and Elastic Curve :

$$EI \frac{d^2 \upsilon}{dx^2} = M$$

$$EI \frac{d^2 \upsilon}{dx^2} = -3x^2 + 3 < x - 1.5 >^2 + 1.25 < x - 1.5 > + 27.75 < x - 4.5 >$$

$$EI \frac{d\upsilon}{dx} = -x^3 + < x - 1.5 >^3 + 0.625 < x - 1.5 >^2 + 13.875 < x - 4.5 >^2 + C_1 \quad [1]$$

$$EI\, \upsilon = -0.25x^4 + 0.25 < x - 1.5 >^4 + 0.2083 < x - 1.5 >^3 + 4.625 < x - 4.5 >^3 + C_1 x + C_2 \quad [2]$$

Boundary Conditions :

$\upsilon = 0$ at $x = 1.5$ m. From Eq.[2],

$$0 = -0.25(1.5^4) + 0 + 0 + 0 + C_1(1.5) + C_2$$
$$0 = -1.265625 + 1.5C_1 + C_2 \quad [3]$$

$\upsilon = 0$ at $x = 4.5$ m. From Eq.[2],

$$0 = -0.25(4.5^4) + 0.25(4.5 - 1.5)^4 + 0.2083(4.5 - 1.5)^3 + 0 + C_1(4.5) + C_2$$
$$0 = -76.640625 + 4.5C_1 + C_2 = 0 \quad [4]$$

Solving Eqs.[3] and [4] yields,

$$C_1 = 25.125 \qquad C_2 = -36.421875$$

The Elastic Curve : Substituting the values of C_1 and C_2 into Eq.[2],

$$\upsilon = \frac{1}{EI}\{-0.25x^4 + 0.25 < x - 1.5 >^4 + 0.208 < x - 1.5 >^3 + 4.625 < x - 4.5 >^3 + 25.1x - 36.4\}\ \text{kN} \cdot \text{m}^3 \quad \text{Ans}$$

12–34. Determine the equation of the elastic curve. Specify the slope at A and the deflection at C. EI is constant.

Support Reactions and Elastic Curve : As shown on FBD.

Moment Function : Using the discontinuity function,

$$M = \frac{3}{4}wa < x-0 > -\frac{w}{2} < x-0 >^2 - \left(-\frac{w}{2}\right) < x-a >^2$$
$$= \frac{3wa}{4}x - \frac{w}{2}x^2 + \frac{w}{2} < x-a >^2$$

Slope and Elastic Curve :

$$EI\frac{d^2v}{dx^2} = M$$
$$EI\frac{d^2v}{dx^2} = \frac{3wa}{4}x - \frac{w}{2}x^2 + \frac{w}{2} < x-a >^2$$
$$EI\frac{dv}{dx} = \frac{3wa}{8}x^2 - \frac{w}{6}x^3 + \frac{w}{6} < x-a >^3 + C_1 \qquad [1]$$
$$EI\,v = \frac{wa}{8}x^3 - \frac{w}{24}x^4 + \frac{w}{24} < x-a >^4 + C_1x + C_2 \qquad [2]$$

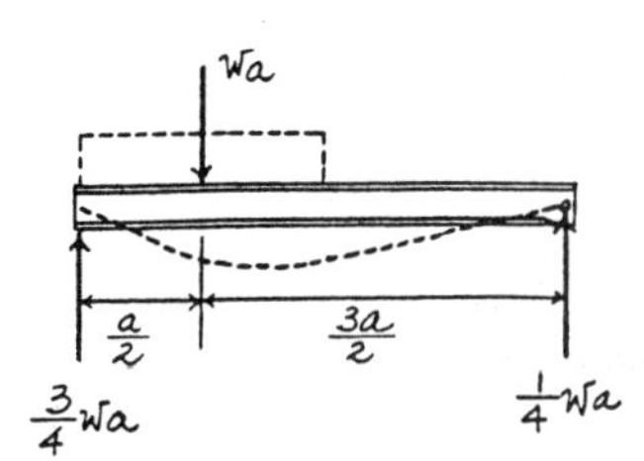

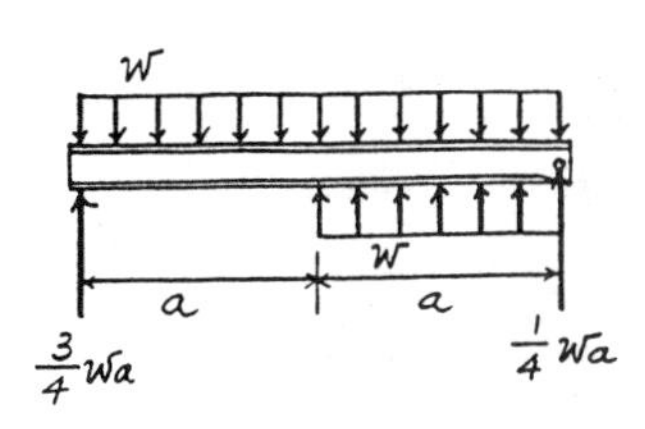

Boundary Conditions :

$v = 0$ at $x = 0$. From Eq.[2], $C_2 = 0$

$v = 0$ at $x = 2a$. From Eq.[2],

$$0 = \frac{wa}{8}(2a)^3 - \frac{w}{24}(2a)^4 + \frac{w}{24}(2a-a)^4 + C_1(2a)$$
$$C_1 = -\frac{3wa^3}{16}$$

The Slope : Substituting the value of C_1 into Eq.[1],

$$\frac{dv}{dx} = \frac{w}{48EI}\{18ax^2 - 8x^3 + 8 < x-a >^3 - 9a^3\}$$

$$\theta_A = \left.\frac{dv}{dx}\right|_{x=0} = \frac{w}{48EI}\{0-0+0-9a^3\} = -\frac{3wa^3}{16EI} \qquad \textbf{Ans}$$

The Elastic Curve : Substituting the values of C_1 and C_2 into Eq.[2],

$$v = \frac{w}{48EI}\{6ax^3 - 2x^4 + 2 < x-a >^4 - 9a^3x\} \qquad \textbf{Ans}$$

$$v_C = v|_{x=a} = \frac{w}{48EI}\{6a^4 - 2a^4 + 0 - 9a^4\} = -\frac{5wa^4}{48EI} \qquad \textbf{Ans}$$

12–35. Determine the equation of the elastic curve. Specify the slopes at A and B. EI is constant.

Support Reactions and Elastic Curve : As shown on FBD.

Moment Function : Using the discontinuity function,

$$M = \frac{3}{4}wa<x-0> - \frac{w}{2}<x-0>^2 - \left(-\frac{w}{2}\right)<x-a>^2$$
$$= \frac{3wa}{4}x - \frac{w}{2}x^2 + \frac{w}{2}<x-a>^2$$

Slope and Elastic Curve :

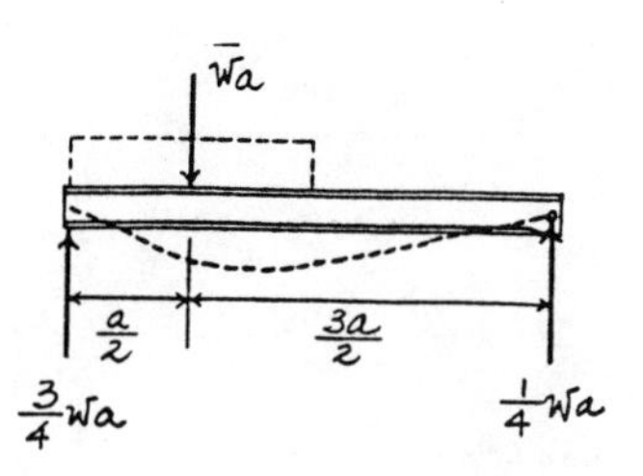

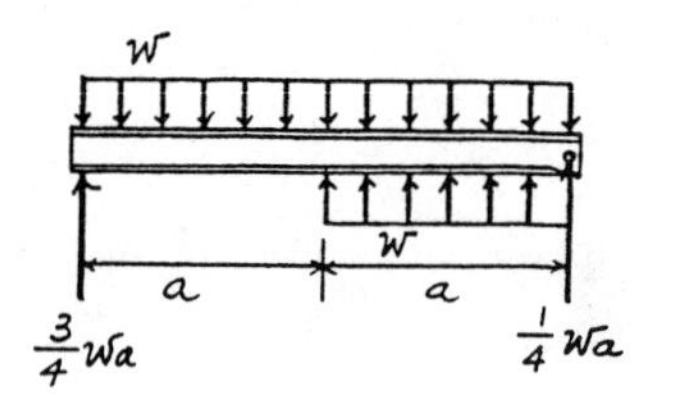

$$EI\frac{d^2v}{dx^2} = M$$
$$EI\frac{d^2v}{dx^2} = \frac{3wa}{4}x - \frac{w}{2}x^2 + \frac{w}{2}<x-a>^2$$
$$EI\frac{dv}{dx} = \frac{3wa}{8}x^2 - \frac{w}{6}x^3 + \frac{w}{6}<x-a>^3 + C_1 \qquad [1]$$
$$EI\,v = \frac{wa}{8}x^3 - \frac{w}{24}x^4 + \frac{w}{24}<x-a>^4 + C_1x + C_2 \qquad [2]$$

Boundary Conditions :

$v = 0$ at $x = 0$. From Eq. [2], $\qquad C_2 = 0$

$v = 0$ at $x = 2a$. From Eq. [2],

$$0 = \frac{wa}{8}(2a)^3 - \frac{w}{24}(2a)^4 + \frac{w}{24}(2a-a)^4 + C_1(2a)$$
$$C_1 = -\frac{3wa^3}{16}$$

The Slope : Substituting the value of C_1 into Eq. [1],

$$\frac{dv}{dx} = \frac{w}{48EI}\{18ax^2 - 8x^3 + 8<x-a>^3 - 9a^3\}$$

$$\theta_A = \left.\frac{dv}{dx}\right|_{x=0} = \frac{w}{48EI}\{0-0+0-9a^3\} = -\frac{3wa^3}{16EI} \qquad \textbf{Ans}$$

$$\theta_B = \left.\frac{dv}{dx}\right|_{x=2a} = \frac{w}{48EI}\left\{18a(2a)^2 - 8(2a)^3 + 8(2a-a)^3 - 9a^3\right\}$$
$$= \frac{7wa^3}{48EI} \qquad \textbf{Ans}$$

The Elastic Curve : Substituting the values of C_1 and C_2 into Eq. [2],

$$v = \frac{w}{48EI}\{6ax^3 - 2x^4 + 2<x-a>^4 - 9a^3x\} \qquad \textbf{Ans}$$

*12–36. The beam is subjected to the load shown. Determine the equation of the elastic curve. EI is constant.

Support Reactions and Elastic Curve : As shown on FBD.

Moment Function : Using the discontinuity function,

$$M = 66.75 < x-0 > -6 < x-0 >^2 -30 < x-3 >$$
$$= 66.75x - 6x^2 - 30 < x-3 >$$

Slope and Elastic Curve :

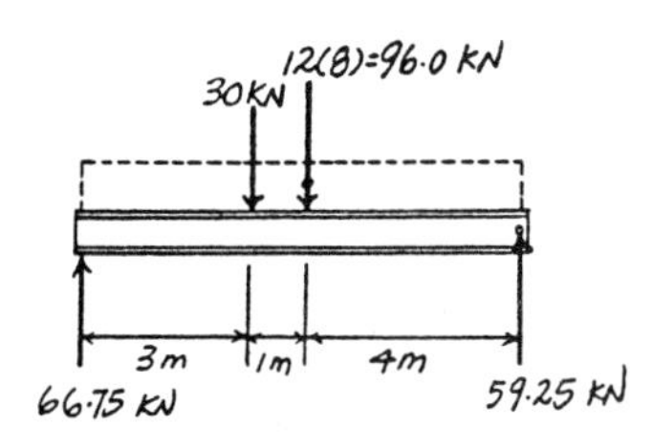

$$EI\frac{d^2 v}{dx^2} = M$$

$$EI\frac{d^2 v}{dx^2} = 66.75x - 6x^2 - 30 < x-3 >$$

$$EI\frac{dv}{dx} = 33.375x^2 - 2x^3 - 15 < x-3 >^2 + C_1 \qquad [1]$$

$$EI\, v = 11.125x^3 - 0.5x^4 - 5 < x-3 >^3 + C_1 x + C_2 \qquad [2]$$

Boundary Conditions :

$v = 0$ at $x = 0$. From Eq. [2], $C_2 = 0$

$v = 0$ at $x = 8$ m. From Eq. [2],

$$0 = 11.125(8^3) - 0.5(8^4) - 5(8-3)^3 + C_1(8)$$
$$C_1 = -377.875$$

The Elastic Curve : Substituting the values of C_1 and C_2 into Eq. [2],

$$v = \frac{1}{EI}\{11.1x^3 - 0.5x^4 - 5 < x-3 >^3 - 378x\}\ \text{kN}\cdot\text{m}^3 \qquad \textbf{Ans}$$

12–37. The beam is subjected to the load shown. Determine the slopes at A and B and the deflection at C. EI is constant.

Support Reactions and Elastic Curve : As shown on FBD.

Moment Function : Using the discontinuity function,

$$M = 66.75\langle x-0\rangle - 6\langle x-0\rangle^2 - 30\langle x-3\rangle$$
$$= 66.75x - 6x^2 - 30\langle x-3\rangle$$

Slope and Elastic Curve :

$$EI\frac{d^2\upsilon}{dx^2} = M$$

$$EI\frac{d^2\upsilon}{dx^2} = 66.75x - 6x^2 - 30\langle x-3\rangle$$

$$EI\frac{d\upsilon}{dx} = 33.375x^2 - 2x^3 - 15\langle x-3\rangle^2 + C_1 \qquad [1]$$

$$EI\,\upsilon = 11.125x^3 - 0.5x^4 - 5\langle x-3\rangle^3 + C_1x + C_2 \qquad [2]$$

Boundary Conditions :

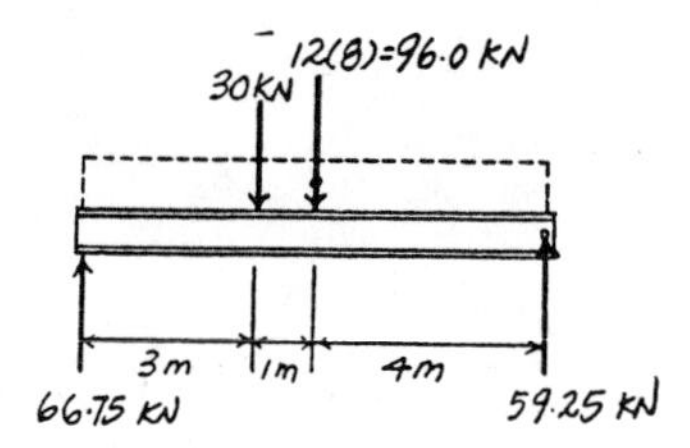

$\upsilon = 0$ at $x = 0$. From Eq. [2], $C_2 = 0$

$\upsilon = 0$ at $x = 8$ m. From Eq. [2],

$$0 = 11.125(8^3) - 0.5(8^4) - 5(8-3)^3 + C_1(8)$$
$$C_1 = -377.875$$

The Slope : Substituting the value of C_1 into Eq. [1],

$$\frac{d\upsilon}{dx} = \frac{1}{EI}\left\{33.375x^2 - 2x^3 - 15\langle x-3\rangle^2 - 377.875\right\}\ \text{kN}\cdot\text{m}^2$$

$$\theta_A = \left.\frac{d\upsilon}{dx}\right|_{x=0} = \frac{1}{EI}\{0-0-0-377.875\} = -\frac{378\ \text{kN}\cdot\text{m}^2}{EI} \qquad \textbf{Ans}$$

$$\theta_B = \left.\frac{d\upsilon}{dx}\right|_{x=8\text{m}}$$
$$= \frac{1}{EI}\left\{33.375(8^2) - 2(8^3) - 15(8-3)^2 - 377.875\right\}$$
$$= \frac{359\ \text{kN}\cdot\text{m}^2}{EI} \qquad \textbf{Ans}$$

The Elastic Curve : Substituting the values of C_1 and C_2 into Eq. [2],

$$\upsilon = \frac{1}{EI}\left\{11.125x^3 - 0.5x^4 - 5\langle x-3\rangle^3 - 377.875x\right\}\ \text{kN}\cdot\text{m}^3$$

$$\upsilon_C = \upsilon|_{x=3\text{m}} = \frac{1}{EI}\left\{11.125(3^3) - 0.5(3^4) - 0 - 377.875(3)\right\}$$
$$= -\frac{874\ \text{kN}\cdot\text{m}^3}{EI} \qquad \textbf{Ans}$$

12–38. The beam is subjected to the load shown. Determine the equations of the slope and elastic curve. *EI* is constant.

Support Reactions and Elastic Curve : As shown on FBD.

Moment Function : Using the discontinuity function,

$$M = 0.200 < x-0 > -\frac{1}{2}(2) < x-0 >^2 -\frac{1}{2}(-2) < x-5 >^2 -(-17.8) < x-5 >$$
$$= 0.200x - x^2 + < x-5 >^2 + 17.8 < x-5 >$$

Slope and Elastic Curve :

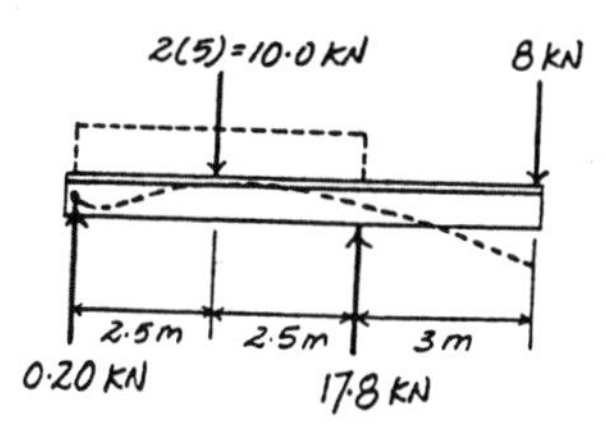

$$EI\frac{d^2v}{dx^2} = M$$
$$EI\frac{d^2v}{dx^2} = 0.200x - x^2 + < x-5 >^2 + 17.8 < x-5 >$$
$$EI\frac{dv}{dx} = 0.100x^2 - 0.3333x^3 + 0.3333 < x-5 >^3 + 8.90 < x-5 >^2 + C_1 \quad [1]$$
$$EI\,v = 0.03333x^3 - 0.08333x^4 + 0.08333 < x-5 >^4 + 2.9667 < x-5 >^3 + C_1x + C_2 \quad [2]$$

Boundary Conditions :

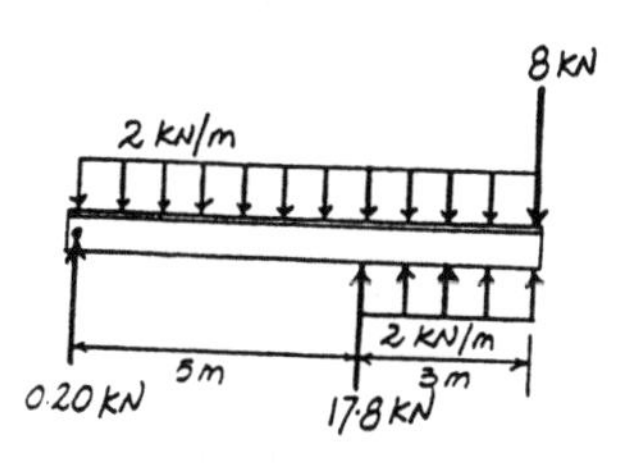

$v = 0$ at $x = 0$. From Eq.[2], $C_2 = 0$

$v = 0$ at $x = 5$ m. From Eq.[2],

$$0 = 0.03333(5^3) - 0.08333(5^4) + 0 + 0 + C_1(5)$$
$$C_1 = 9.5833$$

The Slope : Substituting the value of C_1 into Eq.[1],

$$\frac{dv}{dx} = \frac{1}{EI}\left\{0.100x^2 - 0.333x^3 + 0.333 < x-5 >^3 + 8.90 < x-5 >^2 + 9.58\right\} \text{ kN}\cdot\text{m}^2 \qquad \textbf{Ans}$$

The Elastic Curve : Substituting the values of C_1 and C_2 into Eq.[2],

$$v = \frac{1}{EI}\left\{0.0333x^3 - 0.0833x^4 + 0.0833 < x-5 >^4 + 2.97 < x-5 >^3 + 9.58x\right\} \text{ kN}\cdot\text{m}^3 \qquad \textbf{Ans}$$

12–39. The beam is subjected to the load shown. Determine the slope at A and the deflection at C. EI is constant.

Support Reactions and Elastic Curve : As shown on FBD.

Moment Function : Using the discontinuity function,

$$M = 0.200 < x-0 > -\frac{1}{2}(2) < x-0 >^2 -\frac{1}{2}(-2) < x-5 >^2 -(-17.8) < x-5 >$$

$$= 0.200x - x^2 + < x-5 >^2 + 17.8 < x-5 >$$

Slope and Elastic Curve :

$$EI\frac{d^2 v}{dx^2} = M$$

$$EI\frac{d^2 v}{dx^2} = 0.200x - x^2 + < x-5 >^2 + 17.8 < x-5 >$$

$$EI\frac{dv}{dx} = 0.100x^2 - 0.3333x^3 + 0.3333 < x-5 >^3 + 8.90 < x-5 >^2 + C_1 \qquad [1]$$

$$EI\, v = 0.03333x^3 - 0.08333x^4 + 0.08333 < x-5 >^4 + 2.9667 < x-5 >^3 + C_1 x + C_2 \qquad [2]$$

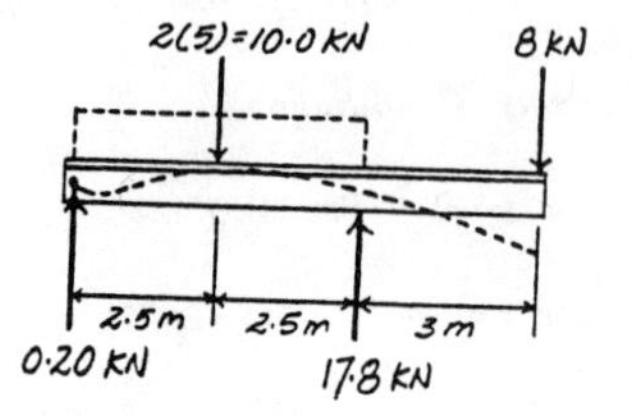

Boundary Conditions :

$v = 0$ at $x = 0$. From Eq.[2], $C_2 = 0$

$v = 0$ at $x = 5$ m. From Eq.[2],

$$0 = 0.03333(5^3) - 0.08333(5^4) + 0 + 0 + C_1(5)$$

$$C_1 = 9.5833$$

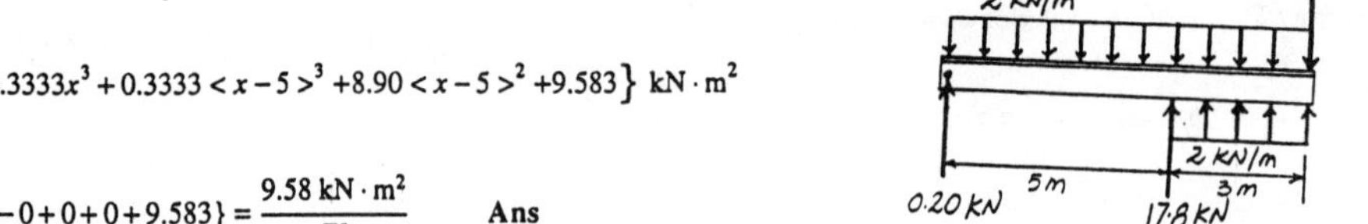

The Slope : Substituting the value of C_1 into Eq.[1],

$$\frac{dv}{dx} = \frac{1}{EI}\left\{0.100x^2 - 0.3333x^3 + 0.3333 < x-5 >^3 + 8.90 < x-5 >^2 + 9.583\right\} \text{ kN}\cdot\text{m}^2$$

$$\theta_A = \left.\frac{dv}{dx}\right|_{x=0} = \frac{1}{EI}\{0 - 0 + 0 + 0 + 9.583\} = \frac{9.58 \text{ kN}\cdot\text{m}^2}{EI} \qquad \textbf{Ans}$$

The Elastic Curve : Substituting the values of C_1 and C_2 into Eq.[2],

$$v = \frac{1}{EI}\left\{0.03333x^3 - 0.08333x^4 + 0.08333 < x-5 >^4 + 2.9667 < x-5 >^3 + 9.583x\right\} \text{ kN}\cdot\text{m}^3$$

$$v_C = v\big|_{x=8\text{m}}$$

$$= \frac{1}{EI}\left\{0.03333(8^3) - 0.08333(8^4) + 0.08333(8-5)^4 + 2.9667(8-5)^3 + 9.583(8)\right\}$$

$$= -\frac{161 \text{ kN}\cdot\text{m}^3}{EI} \qquad \textbf{Ans}$$

***12–40.** The beam is subjected to the load shown. Determine the equation of the elastic curve.

Support Reactions and Elastic Curve : As shown on FBD.

Moment Function : Using the discontinuity function,

$$M = -\frac{1}{2}(8) < x-0 >^2 - \frac{1}{6}\left(-\frac{8}{9}\right) < x-6 >^3 - (-88) < x-6 >$$

$$= -4x^2 + \frac{4}{27} < x-6 >^3 + 88 < x-6 >$$

Slope and Elastic Curve :

$$EI\frac{d^2 v}{dx^2} = M$$

$$EI\frac{d^2 v}{dx^2} = -4x^2 + \frac{4}{27} < x-6 >^3 + 88 < x-6 >$$

$$EI\frac{dv}{dx} = -\frac{4}{3}x^3 + \frac{1}{27} < x-6 >^4 + 44 < x-6 >^2 + C_1 \qquad [1]$$

$$EI\,v = -\frac{1}{3}x^4 + \frac{1}{135} < x-6 >^5 + \frac{44}{3} < x-6 >^3 + C_1 x + C_2 \qquad [2]$$

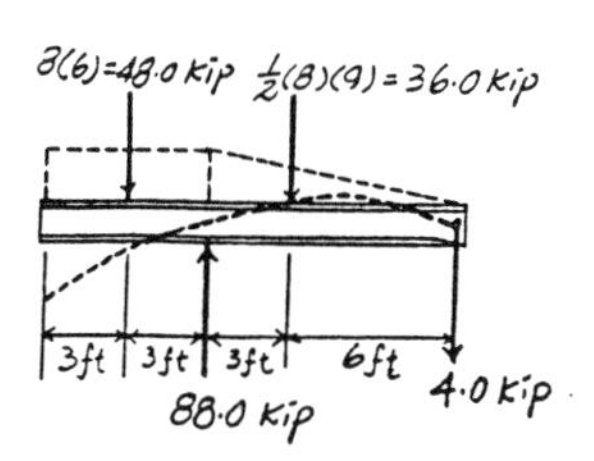

Boundary Conditions :

$v = 0$ at $x = 6$ ft. From Eq. [2],

$$0 = -\frac{1}{3}(6^4) + 0 + 0 + C_1(6) + C_2$$

$$432 = 6C_1 + C_2 \qquad [3]$$

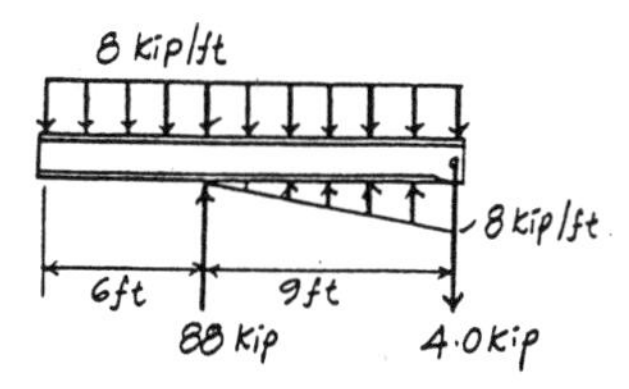

$v = 0$ at $x = 15$ ft. From Eq. [2],

$$0 = -\frac{1}{3}(15^4) + \frac{1}{135}(15-6)^5 + \frac{44}{3}(15-6)^3 + C_1(15) + C_2$$

$$5745.6 = 15C_1 + C_2 \qquad [4]$$

Solving Eqs. [3] and [4] yields,

$$C_1 = 590.4 \qquad C_2 = -3110.4$$

The Elastic Curve : Substituting the values of C_1 and C_2 into Eq. [2],

$$v = \frac{1}{EI}\left\{-0.333x^4 + 0.00741 < x-6 >^5 + 14.7 < x-6 >^3 + 590x - 3110\right\} \text{ kip}\cdot\text{ft}^3 \qquad \textbf{Ans}$$

12–41. Determine the deflection at C and the slope at A of the beam.

Support Reactions and Elastic Curve : As shown on FBD.

Moment Function : Using the discontinuity function,

$$M = -\frac{1}{2}(8) < x-0 >^2 - \frac{1}{6}\left(-\frac{8}{9}\right) < x-6 >^3 - (-88) < x-6 >$$

$$= -4x^2 + \frac{4}{27} < x-6 >^3 + 88 < x-6 >$$

Slope and Elastic Curve :

$$EI\frac{d^2 v}{dx^2} = M$$

$$EI\frac{d^2 v}{dx^2} = -4x^2 + \frac{4}{27} < x-6 >^3 + 88 < x-6 >$$

$$EI\frac{dv}{dx} = -\frac{4}{3}x^3 + \frac{1}{27} < x-6 >^4 + 44 < x-6 >^2 + C_1 \qquad [1]$$

$$EI\, v = -\frac{1}{3}x^4 + \frac{1}{135} < x-6 >^5 + \frac{44}{3} < x-6 >^3 + C_1 x + C_2 \qquad [2]$$

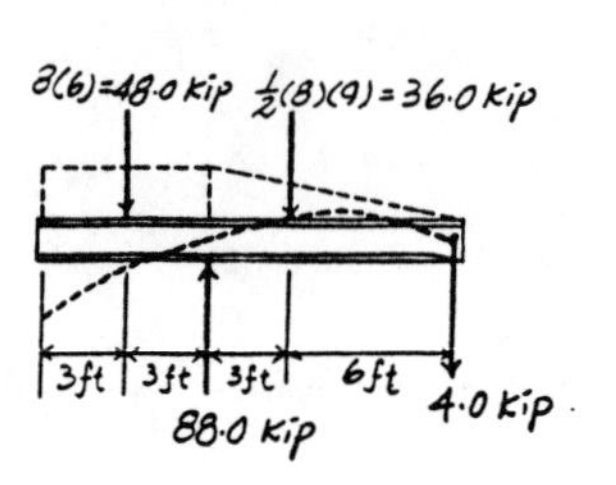

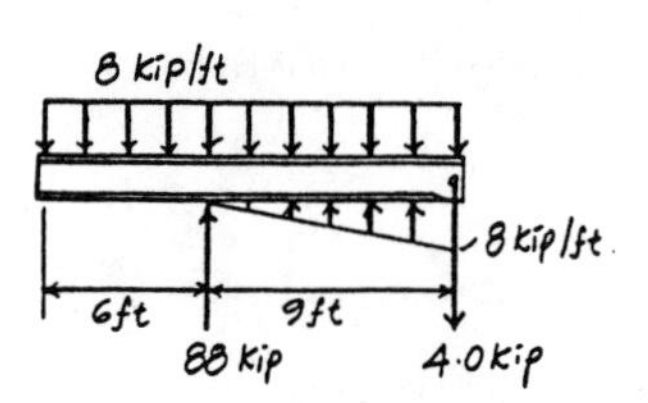

Boundary Conditions :

$v = 0$ at $x = 6$ ft. From Eq. [2],

$$0 = -\frac{1}{3}(6^4) + 0 + 0 + C_1(6) + C_2$$

$$432 = 6C_1 + C_2 \qquad [3]$$

$v = 0$ at $x = 15$ ft. From Eq. [2],

$$0 = -\frac{1}{3}(15^4) + \frac{1}{135}(15-6)^5 + \frac{44}{3}(15-6)^3 + C_1(15) + C_2$$

$$5745.6 = 15C_1 + C_2 \qquad [4]$$

Solving Eqs. [3] and [4] yields,

$$C_1 = 590.4 \qquad C_2 = -3110.4$$

The Slope : Substitute the value of C_1 into Eq. [1],

$$\frac{dv}{dx} = \frac{1}{EI}\left\{-\frac{4}{3}x^3 + \frac{1}{27} < x-6 >^4 + 44 < x-6 >^2 + 590.4\right\} \text{ kip}\cdot\text{ft}^2$$

$$\theta_A = \left.\frac{dv}{dx}\right|_{x=6\text{ft}} = \frac{1}{EI}\left\{-\frac{4}{3}(6^3) + 0 + 0 + 590.4\right\} = \frac{302 \text{ kip}\cdot\text{ft}^2}{EI} \qquad \textbf{Ans}$$

The Elastic Curve : Substitute the values of C_1 and C_2 into Eq. [2],

$$v = \frac{1}{EI}\left\{-\frac{1}{3}x^4 + \frac{1}{135} < x-6 >^5 + \frac{44}{3} < x-6 >^3 + 590.4x - 3110.4\right\} \text{ kip}\cdot\text{ft}^3$$

$$v_C = v|_{x=0} = \frac{1}{EI}\{-0+0+0+0-3110.4\}\text{kip}\cdot\text{ft}^3 = -\frac{3110 \text{ kip}\cdot\text{ft}^3}{EI} \qquad \textbf{Ans}$$

12–42. The 120-lb gymnast stands on the center of the simply supported balance beam. If the beam is made of wood and has the cross section shown, determine the maximum bending stress in the beam and its maximum deflection. The supports at A and B are assumed to be rigid. $E_w = 1.6(10^3)$ ksi.

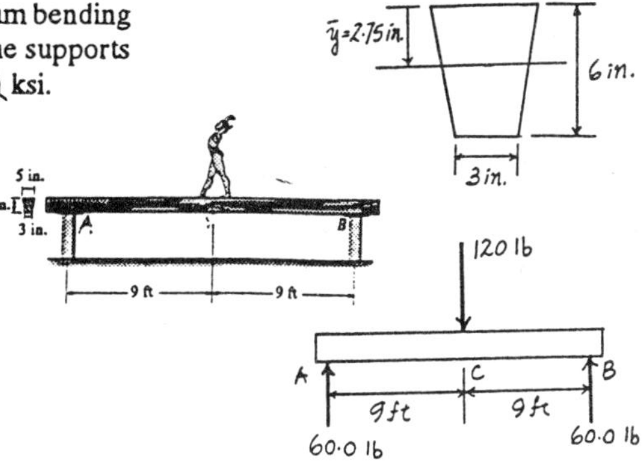

Section Properties :

$$\bar{y} = \frac{\Sigma \tilde{y} A}{\Sigma A} = \frac{3(3)(6) + 2\left(\frac{1}{2}\right)(2)(6)}{3(6) + \frac{1}{2}(2)(6)} = 2.75 \text{ in.}$$

$$I = \frac{1}{12}(3)\left(6^3\right) + 3(6)(3 - 2.75)^2 + \frac{1}{36}(2)\left(6^3\right) + \frac{1}{2}(2)(6)(2.75 - 2)^2 = 70.5 \text{ in}^4$$

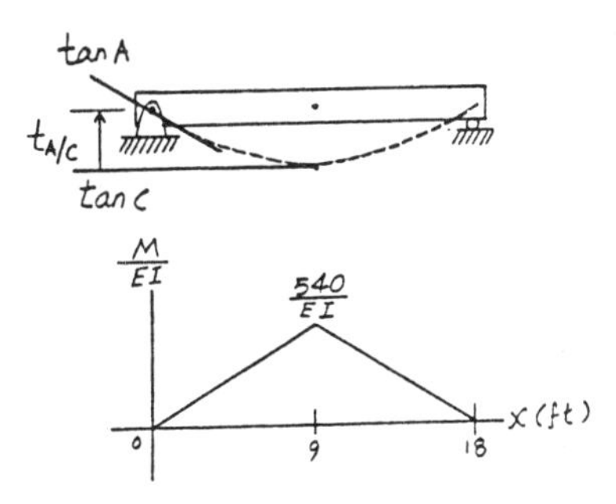

Support Reaction and Elastic Curve : As shown.

M/EI Diagram : As shown.

Moment - Area Theorems : Due to symmetry, the slope at midspan (point C) is zero. Hence,

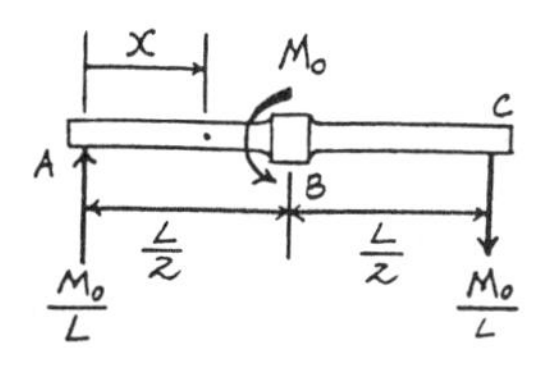

$$\Delta_{\max} = t_{A/C} = 6\left[\frac{1}{2}\left(\frac{540}{EI}\right)(9)\right] = \frac{14580 \text{ lb} \cdot \text{ft}^3}{EI} = \frac{14580(1728)}{1.60(10^6)(70.5)} = 0.223 \text{ in.} \downarrow \quad \textbf{Ans}$$

12–43. If the bearings exert only vertical reactions on the shaft, determine the slope at the bearings and the maximum deflection of the shaft. EI is constant.

Support Reactions and Elastic Curve : As shown.

M/EI Diagram : As shown.

Moment - Area Theorems :

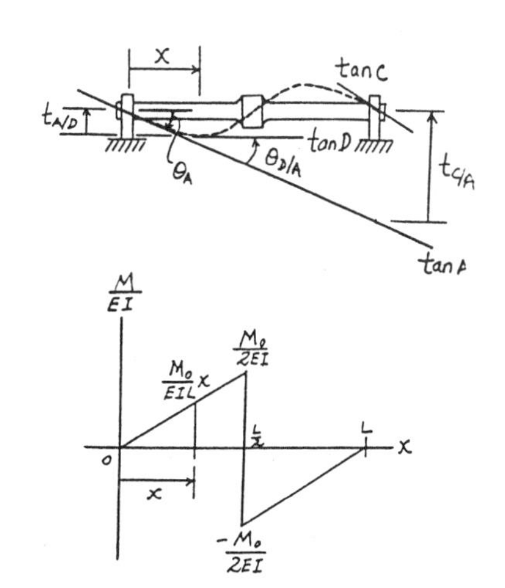

$$t_{C/A} = \frac{1}{2}\left(\frac{-M_0}{2EI}\right)\left(\frac{L}{2}\right)\left(\frac{L}{3}\right) + \frac{1}{2}\left(\frac{M_0}{2EI}\right)\left(\frac{L}{2}\right)\left(\frac{L}{2} + \frac{L}{6}\right) = \frac{M_0 L^2}{24EI}$$

$$\theta_A = \frac{|t_{C/A}|}{L} = \frac{\frac{M_0 L^2}{24EI}}{L} = \frac{M_0 L}{24EI} \quad \textbf{Ans}$$

In a similar manner,

$$\theta_C = \theta_A = \frac{M_0 L}{24EI} \quad \textbf{Ans}$$

The maximum displacement occurs at point D, where $\theta_D = 0$.

$$\theta_{D/A} = \frac{1}{2}\left(\frac{M_0}{EIL}x\right)(x) = \frac{M_0}{2EIL}x^2$$

$$\theta_D = \theta_A + \theta_{D/A}$$

$$0 = -\frac{M_0 L}{24EI} + \frac{M_0}{2EIL}x^2 \qquad x = \frac{\sqrt{3}}{6}L$$

The maximum displacement is,

$$\Delta_{\max} = t_{D/A} = \frac{1}{2}\left[\left(\frac{M_0}{EIL}\right)\left(\frac{\sqrt{3}}{6}L\right)\right]\left(\frac{\sqrt{3}}{6}L\right)\left(\frac{2}{3}\right)\left(\frac{\sqrt{3}}{6}L\right) = \frac{\sqrt{3}M_0 L^2}{216EI} \quad \textbf{Ans}$$

***12–44.** The shaft is supported by a journal bearing at A, which exerts only vertical reactions on the shaft, and by a thrust bearing at B, which exerts both horizontal and vertical reactions on the shaft. Determine the slope of the shaft at the bearings. EI is constant.

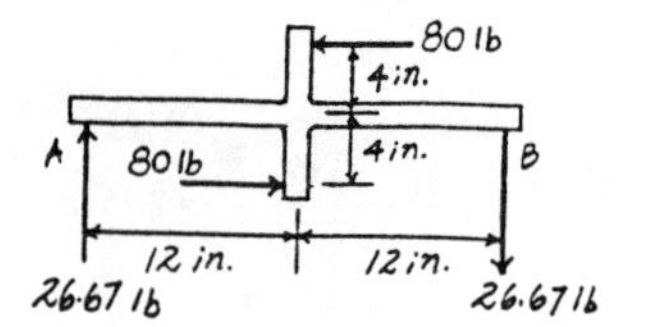

Support Reactions and Elastic Curve : As shown.

M/EI Diagram : As shown.

Moment - Area Theorems :

$$t_{B/A} = \frac{1}{2}\left(-\frac{320}{EI}\right)(12)(8) + \frac{1}{2}\left(\frac{320}{EI}\right)(12)(12+4)$$
$$= \frac{15360 \text{ lb} \cdot \text{in}^3}{EI}$$

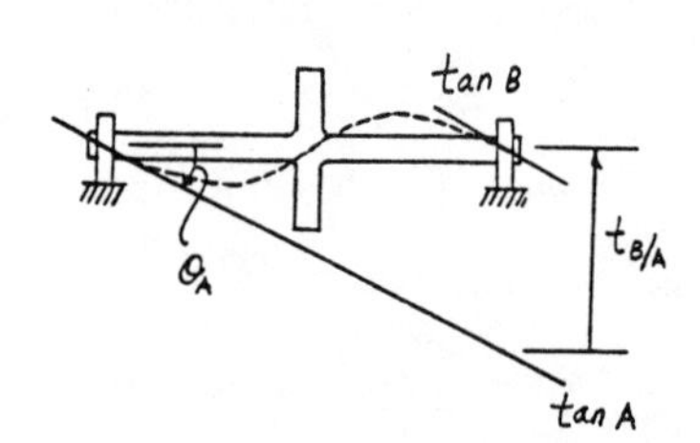

$$\theta_A = \frac{|t_{C/A}|}{L} = \frac{\frac{15360 \text{ lb} \cdot \text{in}^3}{EI}}{24 \text{ in.}} = \frac{640 \text{ lb} \cdot \text{in}^2}{EI} \qquad \textbf{Ans}$$

In a similar manner,

$$\theta_B = \theta_A = \frac{640 \text{ lb} \cdot \text{in}^2}{EI} \qquad \textbf{Ans}$$

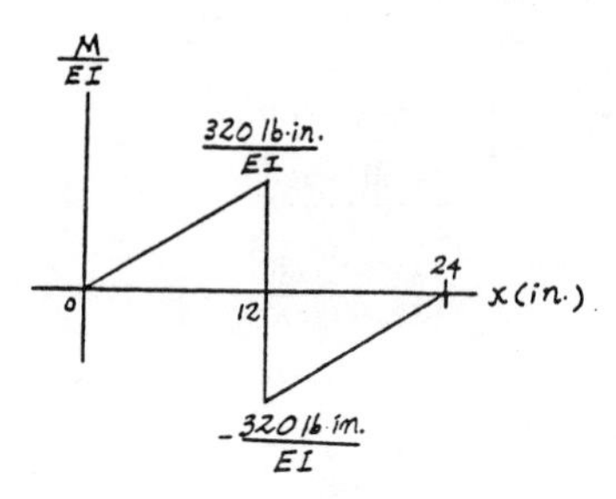

12–45. If the bearings at A and B exert only vertical reactions on the shaft, determine the slope at A. EI is constant.

M/EI Diagram : As shown.

Moment - Area Theorems :

$$t_{B/A} = \frac{1}{2}\left(-\frac{M_0}{2EI}\right)(a)\left(\frac{2a}{3}\right) + \frac{1}{2}\left(\frac{M_0}{2EI}\right)(a)\left(a+\frac{a}{3}\right)$$
$$+\frac{1}{2}\left(-\frac{M_0}{2EI}\right)(a)\left(2a+\frac{2a}{3}\right) + \frac{1}{2}\left(\frac{M_0}{2EI}\right)(a)\left(3a+\frac{a}{3}\right)$$
$$= \frac{M_0 a^2}{3EI}$$

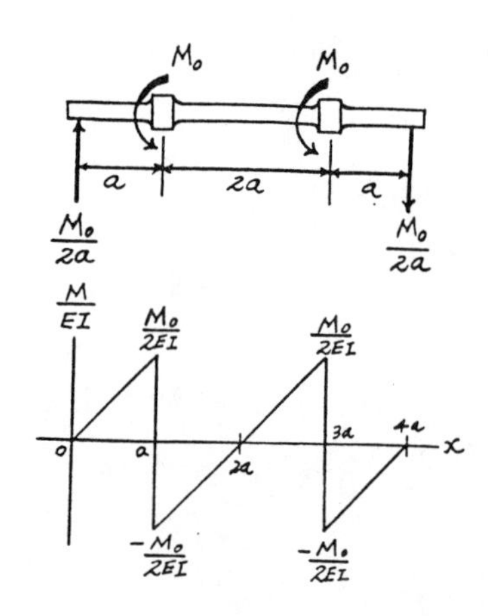

The slope at A is

$$\theta_A = \frac{|t_{B/A}|}{L} = \frac{\frac{M_0 a^2}{3EI}}{4a} = \frac{M_0 a}{12EI} \qquad \textbf{Ans}$$

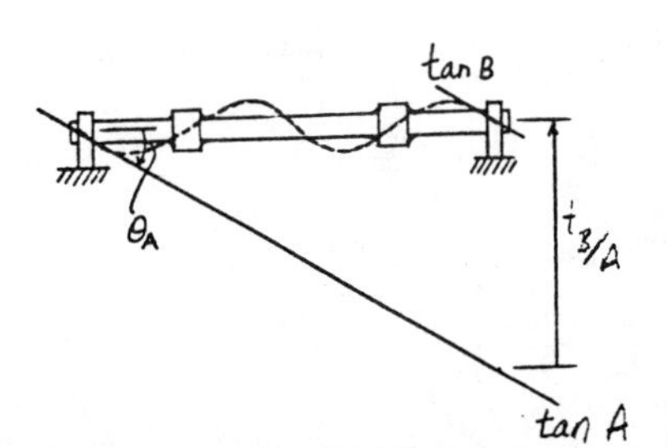

12–46. If the bearings at A and B exert only vertical reactions on the shaft, determine the slope at C. EI is constant.

Support Reactions and Elastic Curve : As shown.

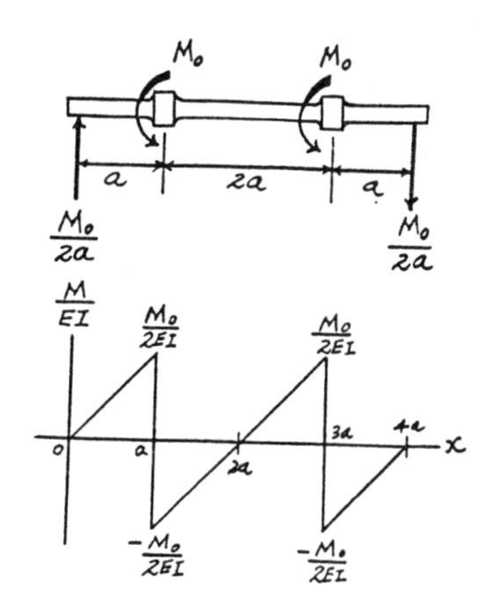

M/EI Diagram : As shown.

Moment - Area Theorems :

$$t_{B/A} = \frac{1}{2}\left(-\frac{M_0}{2EI}\right)(a)\left(\frac{2a}{3}\right)+\frac{1}{2}\left(\frac{M_0}{2EI}\right)(a)\left(a+\frac{a}{3}\right)$$
$$+\frac{1}{2}\left(-\frac{M_0}{2EI}\right)(a)\left(2a+\frac{2a}{3}\right)+\frac{1}{2}\left(\frac{M_0}{2EI}\right)(a)\left(3a+\frac{a}{3}\right)$$
$$= \frac{M_0 a^2}{3EI}$$

$$\theta_{C/A} = \frac{1}{2}\left(\frac{M_0}{2EI}\right)(a) = \frac{M_0 a}{4EI}$$

The slope at C is,

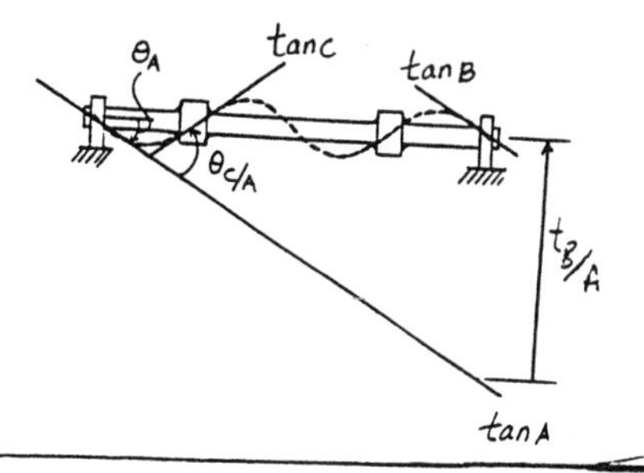

$$\theta_A = \frac{|t_{B/A}|}{L} = \frac{\frac{M_0 a^2}{3EI}}{4a} = \frac{M_0 a}{12EI}$$

$$\theta_C = \theta_A + \theta_{C/A}$$
$$= -\frac{M_0 a}{12EI}+\frac{M_0 a}{4EI} = \frac{M_0 a}{6EI} \qquad \textbf{Ans}$$

12–47. The shaft is subjected to the loading shown. If the bearings at A and B only exert vertical reactions on the shaft, determine the slope at A and the displacement at C. EI is constant.

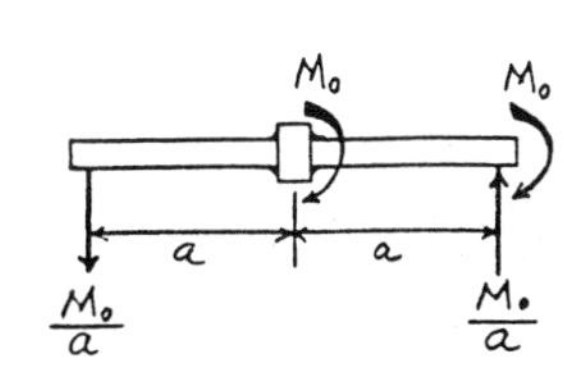

M/EI Diagram : As shown.

Moment - Area Theorems :

$$t_{B/A} = \frac{1}{2}\left(-\frac{M_0}{EI}\right)(a)\left(\frac{a}{3}\right)+\frac{1}{2}\left(-\frac{M_0}{EI}\right)(a)\left(a+\frac{a}{3}\right)$$
$$= -\frac{5M_0 a^2}{6EI}$$

$$t_{C/A} = \frac{1}{2}\left(-\frac{M_0}{EI}\right)(a)\left(\frac{a}{3}\right) = -\frac{M_0 a^2}{6EI}$$

The slope at A is

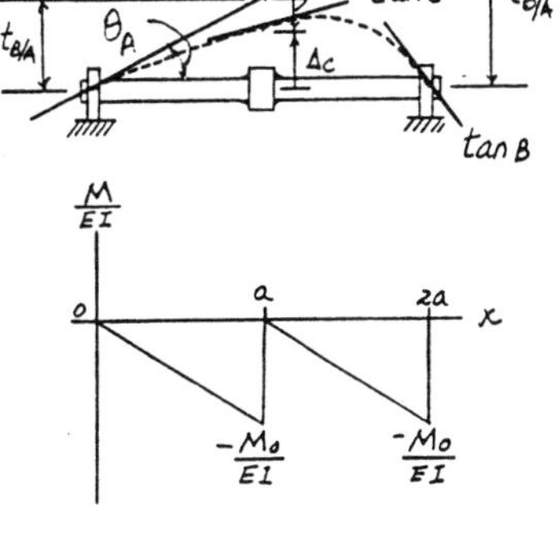

$$\theta_A = \frac{|t_{B/A}|}{L} = \frac{\frac{5M_0 a^2}{6EI}}{2a} = \frac{5M_0 a}{12EI} \qquad \textbf{Ans}$$

The displacement at C is,

$$\Delta_C = \left|\frac{1}{2}t_{B/A}\right| - |t_{C/A}|$$
$$= \frac{1}{2}\left(\frac{5M_0 a^2}{6EI}\right) - \frac{M_0 a^2}{6EI}$$
$$= \frac{M_0 a^2}{4EI} \uparrow \qquad \textbf{Ans}$$

***12–48.** The beam is subjected to the loading shown. Determine the slope at A and the displacement at C. Assume the support at A is a pin and B is a roller. EI is constant.

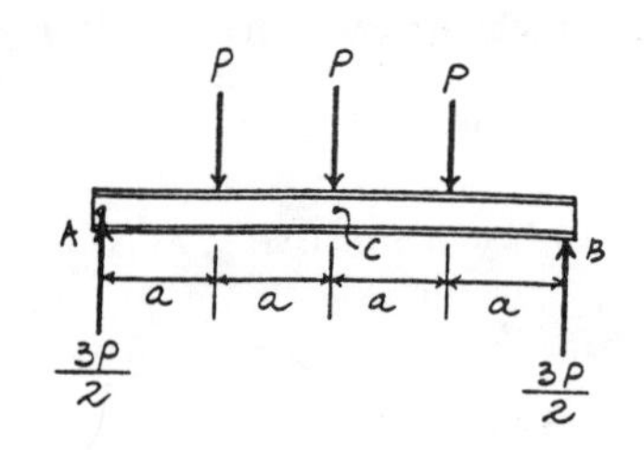

Support Reactions and Elastic Curve : As shown.

M/EI Diagram : As shown.

Moment - Area Theorems : Due to symmetry, the slope at midspan (point C) is zero. Hence the slope at A is

$$\theta_A = \theta_{A/C} = \frac{1}{2}\left(\frac{3Pa}{2EI}\right)(a) + \left(\frac{3Pa}{2EI}\right)(a) + \frac{1}{2}\left(\frac{Pa}{2EI}\right)(a)$$
$$= \frac{5Pa^2}{2EI} \qquad \textbf{Ans}$$

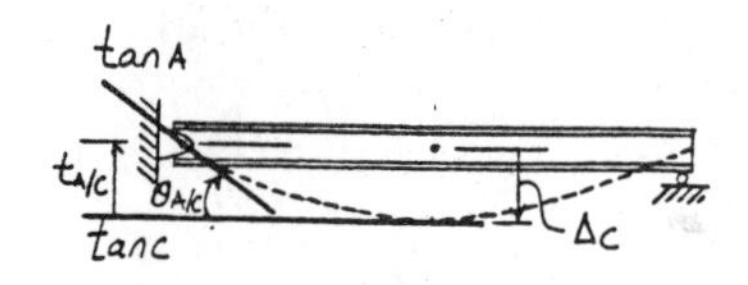

The displacement at C is

$$\Delta_C = t_{A/C} = \frac{1}{2}\left(\frac{3Pa}{2EI}\right)(a)\left(\frac{2a}{3}\right) + \left(\frac{3Pa}{2EI}\right)(a)\left(a + \frac{a}{2}\right)$$
$$+ \frac{1}{2}\left(\frac{Pa}{2EI}\right)(a)\left(a + \frac{2a}{3}\right)$$
$$= \frac{19Pa^3}{6EI} \downarrow \qquad \textbf{Ans}$$

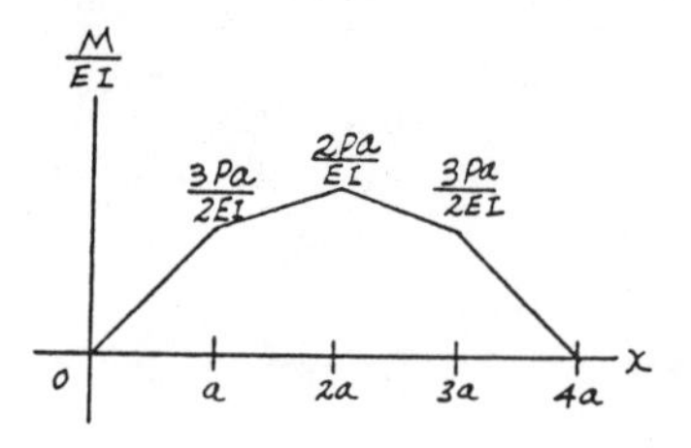

12–49. The bar is supported · : the roller constraint at C, which allows vertical displacem· · but resists axial load and moment. If the bar is subjecte to the loading shown, determine the slope and deflectio. at A. EI is constant.

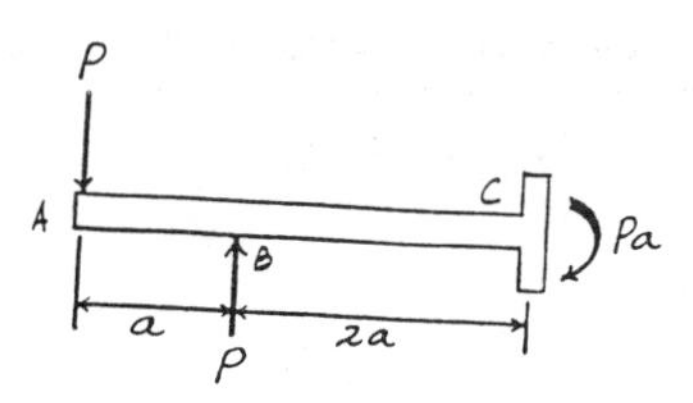

Support Reactions and Elastic Curve : As shown.

M/EI Diagram : As shown.

Moment - Area Theorems :

$$\theta_{A/C} = \left(-\frac{Pa}{EI}\right)(2a) + \frac{1}{2}\left(-\frac{Pa}{EI}\right)(a) = -\frac{5Pa^2}{2EI}$$
$$t_{B/C} = \left(-\frac{Pa}{EI}\right)(2a)\,(a) = -\frac{2Pa^3}{EI}$$
$$t_{A/C} = \left(-\frac{Pa}{EI}\right)(2a)\,(2a) + \frac{1}{2}\left(-\frac{Pa}{EI}\right)(a)\left(\frac{2}{3}a\right) = -\frac{13Pa^3}{3EI}$$

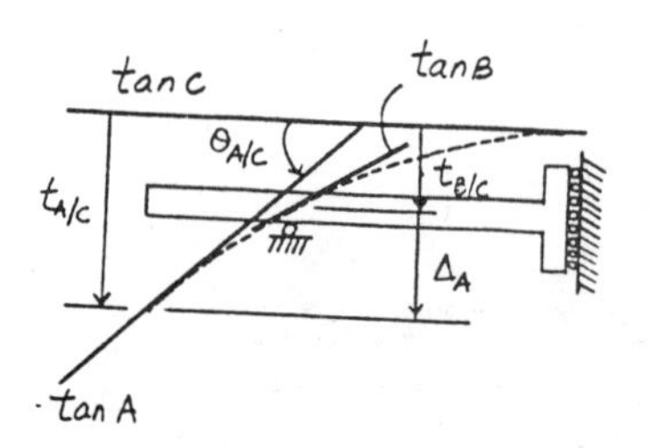

Due to the moment constraint, the slope at support C is zero. Hence, the slope at A is

$$\theta_A = |\theta_{A/C}| = \frac{5Pa^2}{2EI} \qquad \textbf{Ans}$$

and the displacement at A is

$$\Delta_A = |t_{A/C}| - |t_{B/C}|$$
$$= \frac{13Pa^3}{3EI} - \frac{2Pa^3}{EI} = \frac{7Pa^3}{3EI} \downarrow \qquad \textbf{Ans}$$

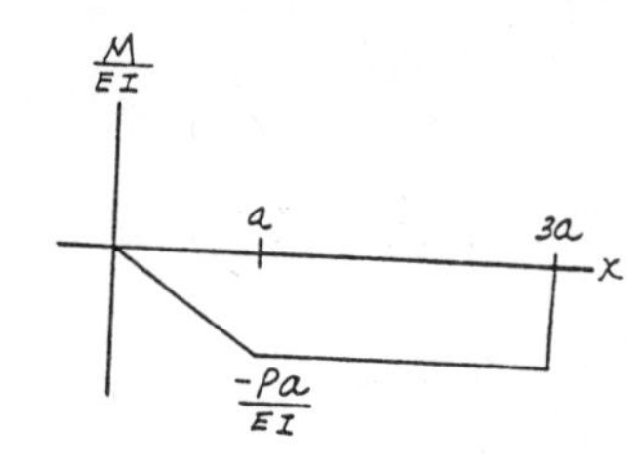

12–50. Determine the value of a so that the slope at A is equal to zero. EI is constant.

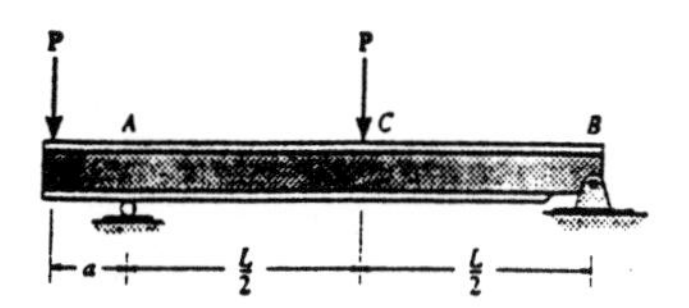

Moment-Area Theorems :

$$(\theta_A)_1 = (\theta_{A/C})_1 = \frac{1}{2}\left(\frac{PL}{4EI}\right)\left(\frac{L}{2}\right) = \frac{PL^2}{16EI}$$

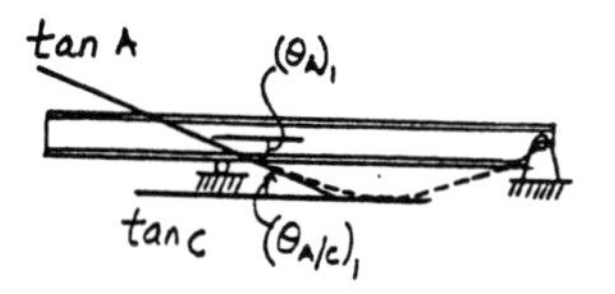

$$(t_{B/A})_2 = \frac{1}{2}\left(-\frac{pa}{EI}\right)(L)\left(\frac{2}{3}L\right) = -\frac{PaL^2}{3EI}$$

$$(\theta_A)_2 = \frac{|(t_{B/A})_2|}{L} = \frac{\frac{PaL^2}{3EI}}{L} = \frac{PaL}{3EI}$$

Require,

$$\theta_A = 0 = (\theta_A)_1 - (\theta_A)_2$$

$$0 = \frac{PL^2}{16EI} - \frac{PaL}{3EI}$$

$$a = \frac{3}{16}L \qquad \textbf{Ans}$$

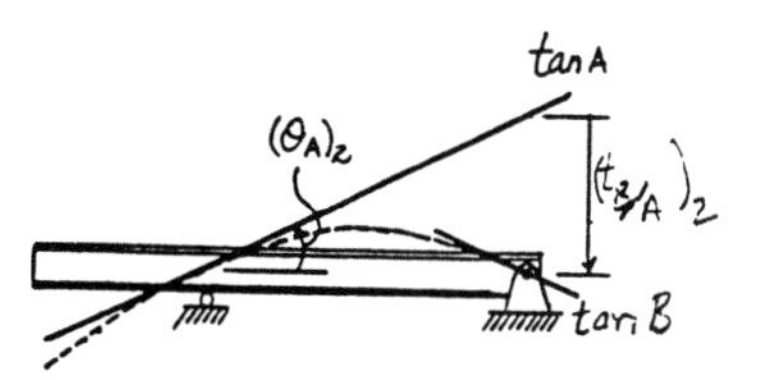

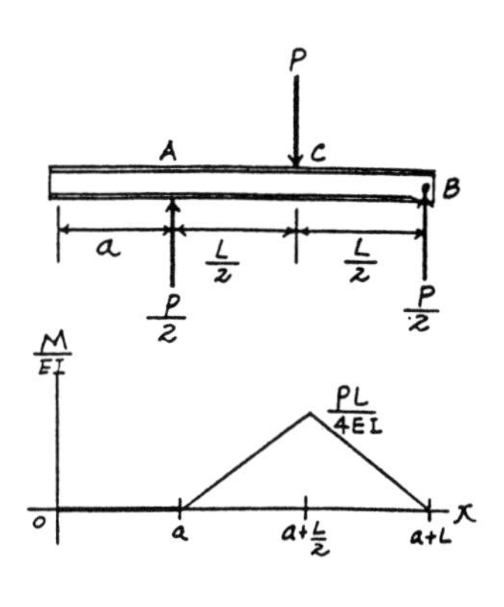

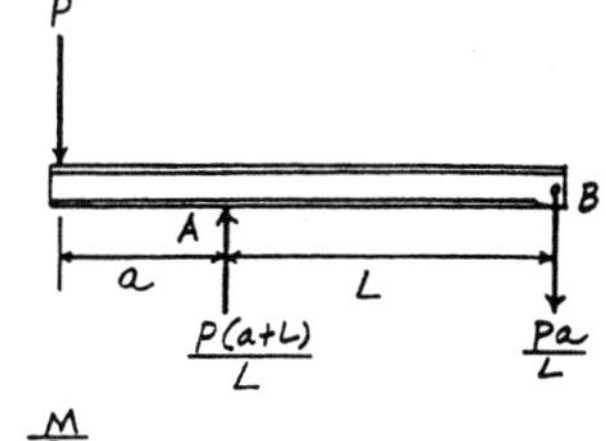

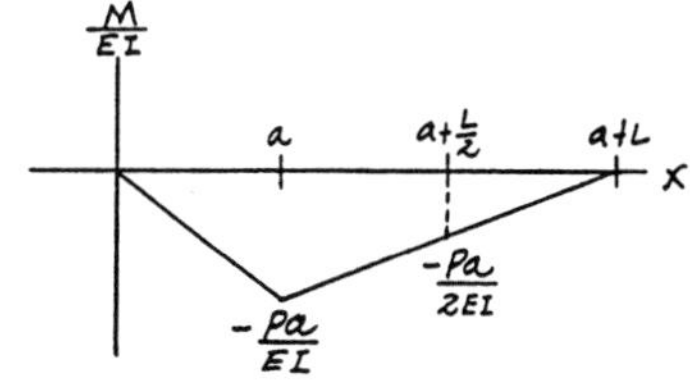

12–51. Determine the value of a so that the deflection at C is equal to zero. EI is constant.

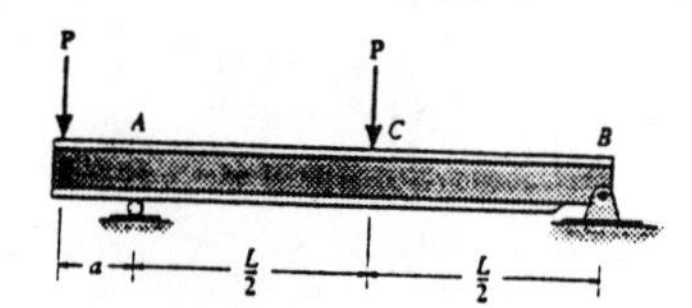

Moment - Area Theorems :

$$(\Delta_C)_1 = (t_{A/C})_1 = \frac{1}{2}\left(\frac{PL}{4EI}\right)\left(\frac{L}{2}\right)\left(\frac{L}{3}\right) = \frac{PL^3}{48EI}$$

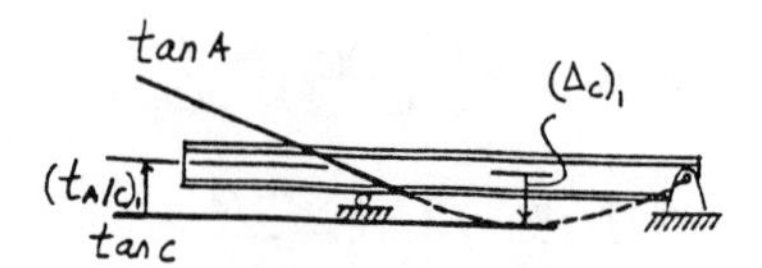

$$(t_{B/A})_2 = \frac{1}{2}\left(-\frac{pa}{EI}\right)(L)\left(\frac{2}{3}L\right) = -\frac{PaL^2}{3EI}$$

$$(t_{C/A})_2 = \left(-\frac{pa}{2EI}\right)\left(\frac{L}{2}\right)\left(\frac{L}{4}\right) + \frac{1}{2}\left(-\frac{pa}{2EI}\right)\left(\frac{L}{2}\right)\left(\frac{L}{3}\right) = -\frac{5PaL^2}{48EI}$$

$$(\Delta_C)_2 = \frac{1}{2}|(t_{B/A})_2| - |(t_{C/A})_2| = \frac{1}{2}\left(\frac{PaL^2}{3EI}\right) - \frac{5PaL^2}{48EI} = \frac{PaL^2}{16EI}$$

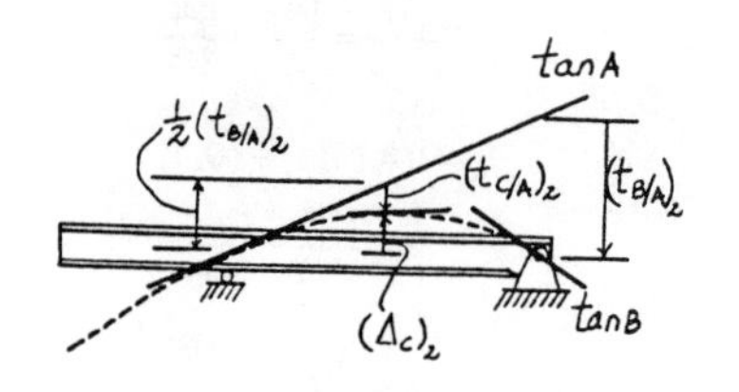

Require,

$$\Delta_C = 0 = (\Delta_C)_1 - (\Delta_C)_2$$

$$0 = \frac{PL^3}{48EI} - \frac{PaL^2}{16EI}$$

$$a = \frac{L}{3} \qquad \textbf{Ans}$$

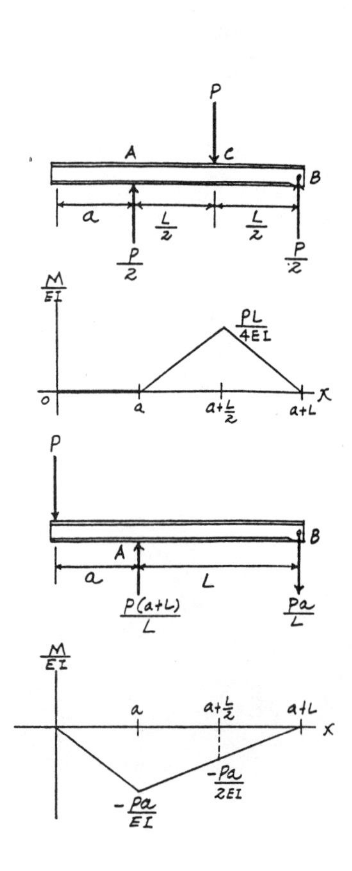

***12–52.** The beam is subjected to the two loads. Determine the slope and deflection at points A and B. EI is constant.

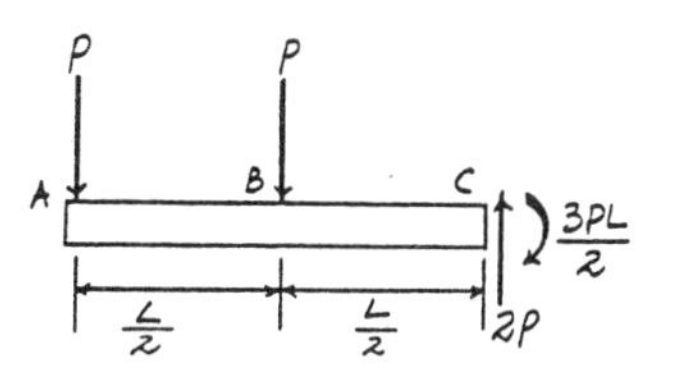

Moment - Area Theorems : The slope at support C is zero. The slopes at A and B are,

$$\theta_A = |\theta_{A/C}| = \frac{1}{2}\left(-\frac{PL}{2EI}\right)\left(\frac{L}{2}\right)+\left(-\frac{PL}{2EI}\right)\left(\frac{L}{2}\right)+\frac{1}{2}\left(-\frac{PL}{EI}\right)\left(\frac{L}{2}\right)$$
$$= \frac{5PL^2}{8EI} \qquad \textbf{Ans}$$

$$\theta_B = |\theta_{B/C}| = \left(-\frac{PL}{2EI}\right)\left(\frac{L}{2}\right)+\frac{1}{2}\left(-\frac{PL}{EI}\right)\left(\frac{L}{2}\right) = \frac{PL^2}{2EI} \qquad \textbf{Ans}$$

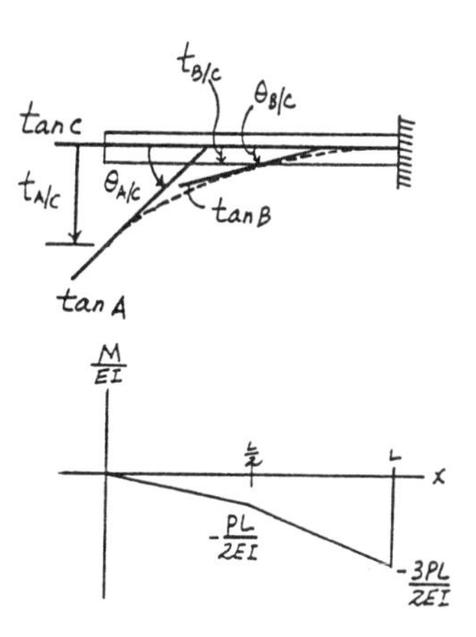

The displacements at A and B are,

$$\Delta_A = |t_{A/C}| = \frac{1}{2}\left(-\frac{PL}{2EI}\right)\left(\frac{L}{2}\right)\left(\frac{L}{3}\right)+\left(-\frac{PL}{2EI}\right)\left(\frac{L}{2}\right)\left(\frac{L}{2}+\frac{L}{4}\right)$$
$$+\frac{1}{2}\left(-\frac{PL}{EI}\right)\left(\frac{L}{2}\right)\left(\frac{L}{2}+\frac{L}{3}\right)$$
$$= \frac{7PL^3}{16EI} \downarrow \qquad \textbf{Ans}$$

$$\Delta_B = |t_{B/C}| = \left(-\frac{PL}{2EI}\right)\left(\frac{L}{2}\right)\left(\frac{L}{4}\right)+\frac{1}{2}\left(-\frac{PL}{EI}\right)\left(\frac{L}{2}\right)\left(\frac{L}{3}\right)$$
$$= \frac{7PL^3}{48EI} \downarrow \qquad \textbf{Ans}$$

12–53. The beam is made of a ceramic material. In order to obtain its modulus of elasticity, it is subjected to the elastic loading shown. If the moment of inertia is I and the beam has a measured maximum deflection Δ, determine E.

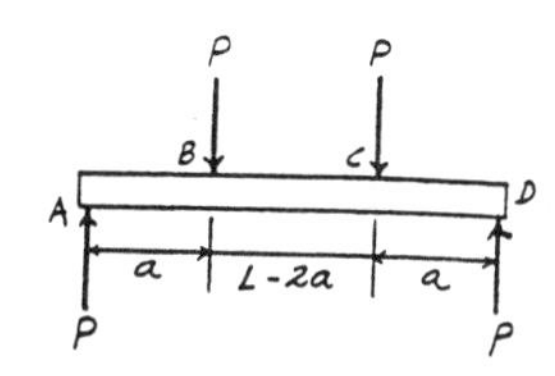

Moment - Area Theorems : Due to symmetry, the slope at midspan (point E) is zero. Hence the maximum displacement is,

$$\Delta_{max} = t_{A/E} = \left(\frac{Pa}{EI}\right)\left(\frac{L-2a}{2}\right)\left(a+\frac{L-2a}{4}\right)+\frac{1}{2}\left(\frac{Pa}{EI}\right)(a)\left(\frac{2}{3}a\right)$$
$$= \frac{Pa}{24EI}\left(3L^2-4a^2\right)$$

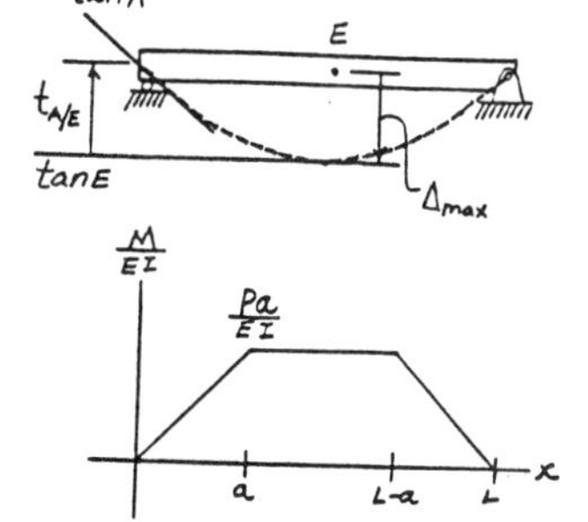

Require, $\Delta_{max} = \Delta$, then,

$$\Delta = \frac{Pa}{24EI}\left(3L^2-4a^2\right)$$

$$E = \frac{Pa}{24\Delta I}\left(3L^2-4a^2\right) \qquad \textbf{Ans}$$

12–54. At what distance a should the bearing supports at A and B be placed so that the deflection at the center of the shaft is equal to the deflection at its ends? The bearings exert only vertical reactions on the shaft. EI is constant.

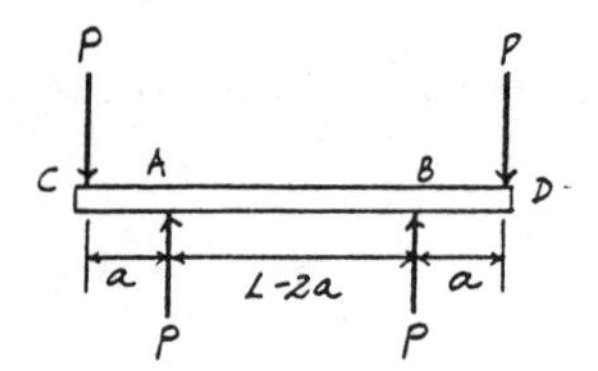

Support Reactions and Elastic Curve : As shown.

M/EI Diagram : As shown.

Moment-Area Theorems : Due to symmetry, the slope at midspan (point E) is zero.

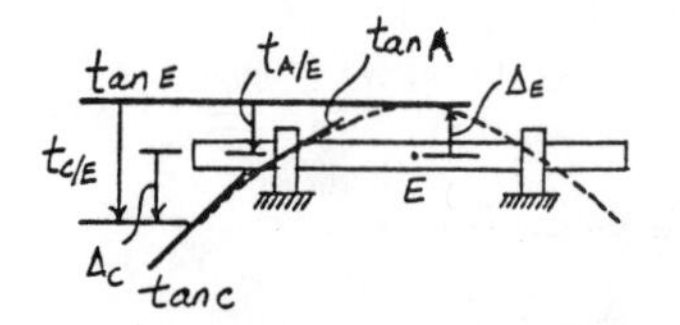

$$\Delta_E = |t_{A/E}| = \left(-\frac{Pa}{EI}\right)\left(\frac{L-2a}{2}\right)\left(\frac{L-2a}{4}\right) = \frac{Pa}{8EI}(L-2a)^2$$

$$t_{C/E} = \left(-\frac{Pa}{EI}\right)\left(\frac{L-2a}{2}\right)\left(a+\frac{L-2a}{4}\right)+\frac{1}{2}\left(-\frac{Pa}{EI}\right)(a)\left(\frac{2}{3}a\right)$$
$$= -\frac{Pa}{24EI}\left(3L^2-4a^2\right)$$

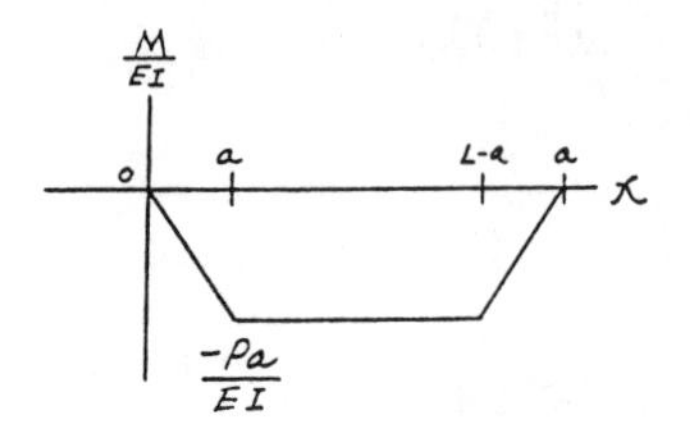

$$\Delta_C = |t_{C/E}| - |t_{A/E}|$$
$$= \frac{Pa}{24EI}\left(3L^2-4a^2\right) - \frac{Pa}{8EI}(L-2a)^2$$
$$= \frac{Pa^2}{6EI}(3L-4a)$$

Require, $\Delta_E = \Delta_C$, then,

$$\frac{Pa}{8EI}(L-2a)^2 = \frac{Pa^2}{6EI}(3L-4a)$$
$$28a^2 - 24aL + 3L^2 = 0$$

$$a = 0.152L \qquad \textbf{Ans}$$

12–55. Determine the slope of the 50-mm-diameter A-36 steel shaft at the bearings at A and B. The bearings exert only vertical reactions on the shaft.

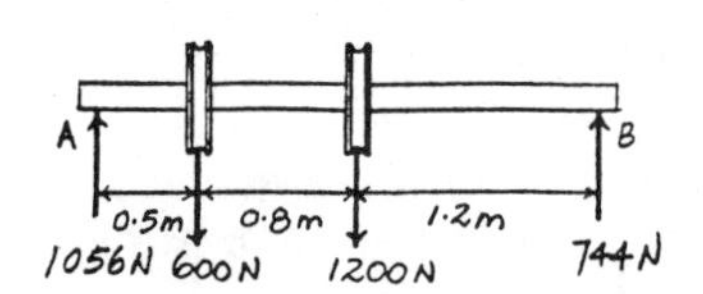

Support Reactions and Elastic Curve : As shown.

M/EI Diagram : As shown.

Moment-Area Theorems :

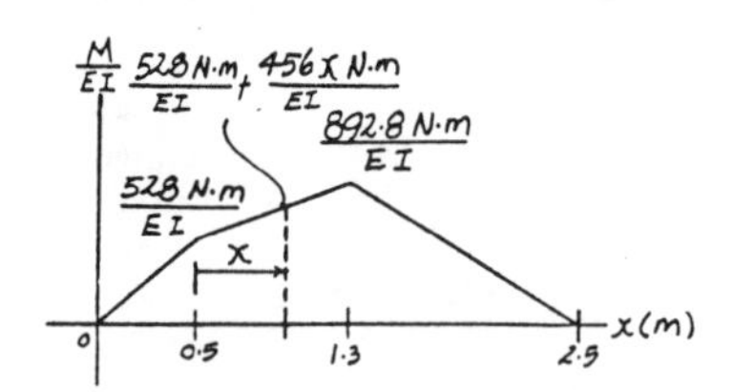

$$t_{B/A} = \frac{1}{2}\left(\frac{892.8}{EI}\right)(1.2)(0.8)+\frac{1}{2}\left(\frac{364.8}{EI}\right)(0.8)(1.4667)$$
$$+\left(\frac{528}{EI}\right)(0.8)(1.6)+\frac{1}{2}\left(\frac{528}{EI}\right)(0.5)(2.1667)$$
$$= \frac{1604.4\ \text{N}\cdot\text{m}^3}{EI}$$

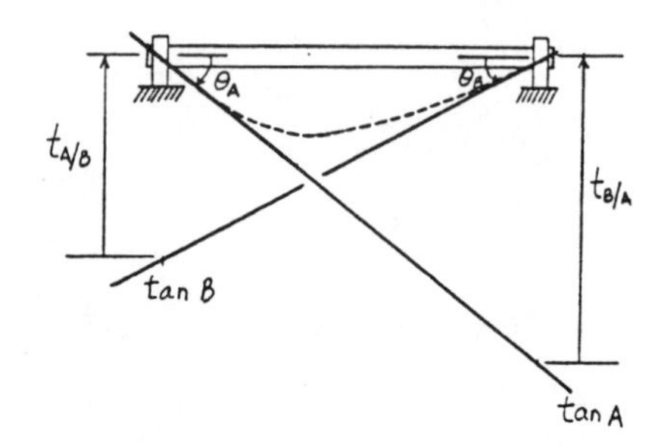

$$t_{A/B} = \frac{1}{2}\left(\frac{892.8}{EI}\right)(1.2)(1.7)+\frac{1}{2}\left(\frac{364.8}{EI}\right)(0.8)(1.0333)$$
$$+\left(\frac{528}{EI}\right)(0.8)(0.9)+\frac{1}{2}\left(\frac{528}{EI}\right)(0.5)(0.3333)$$
$$= \frac{1485.6\ \text{N}\cdot\text{m}^3}{EI}$$

The slopes at A and B are,

$$\theta_A = \frac{|t_{B/A}|}{L} = \frac{\frac{1604.4\ \text{N}\cdot\text{m}^3}{EI}}{2.5\ \text{m}}$$
$$= \frac{641.76\ \text{N}\cdot\text{m}^2}{EI}$$
$$= \frac{641.76}{200(10^9)\left(\frac{\pi}{4}\right)(0.025^4)} = 0.0105\ \text{rad} \qquad \textbf{Ans}$$

$$\theta_B = \frac{|t_{A/B}|}{L} = \frac{\frac{1485.6\ \text{N}\cdot\text{m}^3}{EI}}{2.5\ \text{m}}$$
$$= \frac{594.24\ \text{N}\cdot\text{m}^2}{EI}$$
$$= \frac{594.24}{200(10^9)\left(\frac{\pi}{4}\right)(0.025^4)} = 0.00968\ \text{rad} \qquad \textbf{Ans}$$

***12–56.** Determine the maximum deflection of the 50-mm-diameter A-36 steel shaft. It is supported by bearings at its ends A and B which only exert vertical reactions on the shaft.

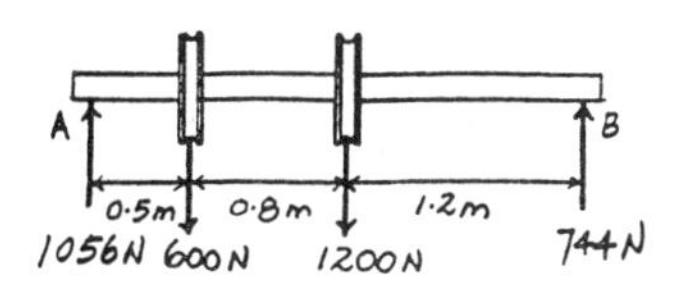

Moment - Area Theorems :

$$t_{B/A} = \frac{1}{2}\left(\frac{892.8}{EI}\right)(1.2)(0.8) + \frac{1}{2}\left(\frac{364.8}{EI}\right)(0.8)(1.4667)$$

$$+\left(\frac{528}{EI}\right)(0.8)(1.6) + \frac{1}{2}\left(\frac{528}{EI}\right)(0.5)(2.1667)$$

$$= \frac{1604.4\ \text{N}\cdot\text{m}^3}{EI}$$

$$\theta_A = \frac{|t_{B/A}|}{L} = \frac{\frac{1604.4\ \text{N}\cdot\text{m}^3}{EI}}{2.5\ \text{m}} = \frac{641.76\ \text{N}\cdot\text{m}^2}{EI}$$

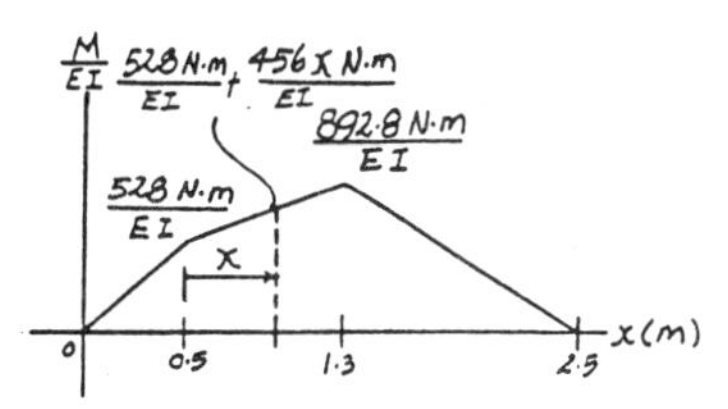

The maximum displacement occurs at point E, where $\theta_E = 0$.

$$\theta_{E/A} = \frac{1}{2}\left(\frac{528}{EI}\right)(0.5) + \left(\frac{528}{EI}\right)x + \frac{1}{2}\left(\frac{456}{EI}x\right)x$$

$$= \frac{1}{EI}\left(228x^2 + 528x + 132\right)$$

$$\theta_E = \theta_A + \theta_{E/A}$$

$$0 = -\frac{641.76}{EI} + \frac{1}{EI}\left(228x^2 + 528x + 132\right)$$

$$x = 0.7333\ \text{m} < 0.8\ \text{m}\quad (\textit{O.K!})$$

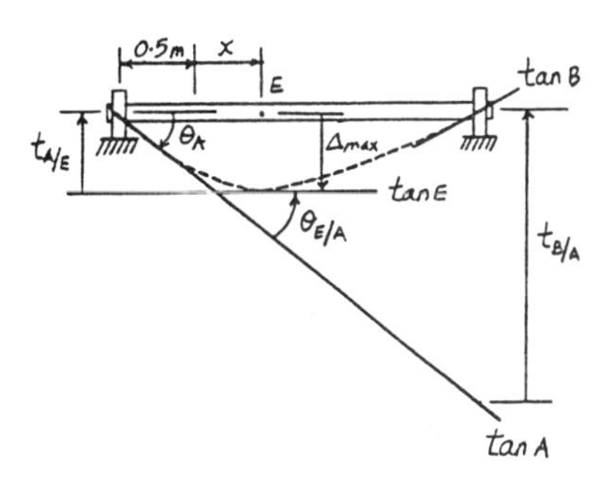

The maximum displacement is,

$$\Delta_{max} = |t_{A/E}| = \frac{1}{2}\left(\frac{528}{EI}\right)(0.5)(0.3333) + \left(\frac{528}{EI}\right)(0.7333)(0.8666)$$

$$+\frac{1}{2}\left(\frac{456}{EI}\right)\left(0.7333^2\right)(0.9888)$$

$$= \frac{500.76\ \text{N}\cdot\text{m}^3}{EI}$$

$$= \frac{500.76}{200(10^9)\left(\frac{\pi}{4}\right)(0.025^4)}$$

$$= 0.008161\ \text{m} = 8.16\ \text{mm}\downarrow \qquad \textbf{Ans}$$

12–57. Determine the slope of the 20-mm-diameter A-36 steel shaft at the bearings at A and B. The bearings exert only vertical forces on the shaft.

Moment - Area Theorems :

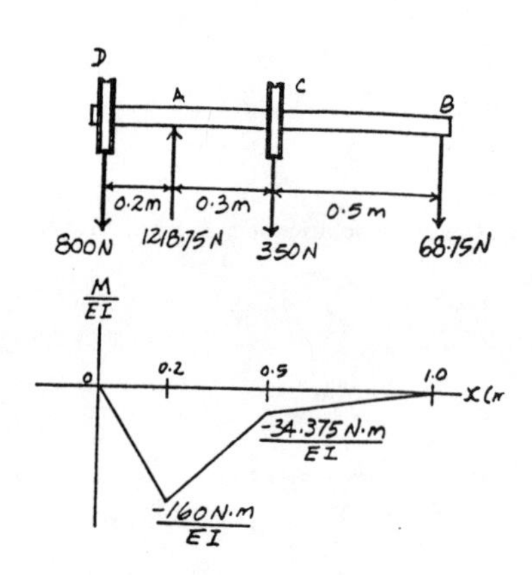

$$t_{B/A} = \frac{1}{2}\left(-\frac{34.375}{EI}\right)(0.5)(0.3333) + \frac{1}{2}\left(-\frac{125.625}{EI}\right)(0.3)(0.7) + \left(-\frac{34.375}{EI}\right)(0.3)(0.65)$$

$$= -\frac{22.75833 \text{ N}\cdot\text{m}^3}{EI}$$

$$t_{A/B} = \frac{1}{2}\left(-\frac{34.375}{EI}\right)(0.5)(0.4667) + \frac{1}{2}\left(-\frac{125.625}{EI}\right)(0.3)(0.1) + \left(-\frac{34.375}{EI}\right)(0.3)(0.15)$$

$$= -\frac{7.44167 \text{ N}\cdot\text{m}^3}{EI}$$

The slopes at A and B are,

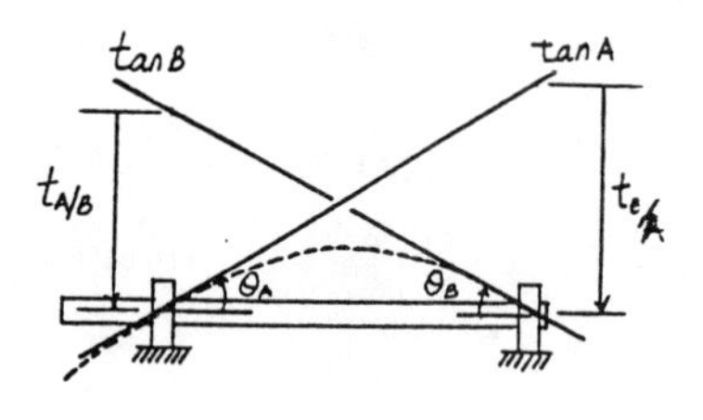

$$\theta_A = \frac{|t_{B/A}|}{L} = \frac{\frac{22.75833 \text{ N}\cdot\text{m}^3}{EI}}{0.8 \text{ m}}$$

$$= \frac{28.448 \text{ N}\cdot\text{m}^2}{EI}$$

$$= \frac{28.448}{200(10^9)\left(\frac{\pi}{4}\right)(0.01^4)} = 0.0181 \text{ rad} \qquad \textbf{Ans}$$

$$\theta_B = \frac{|t_{A/B}|}{L} = \frac{\frac{7.44167 \text{ N}\cdot\text{m}^3}{EI}}{0.8 \text{ m}}$$

$$= \frac{9.302 \text{ N}\cdot\text{m}^2}{EI}$$

$$= \frac{9.302}{200(10^9)\left(\frac{\pi}{4}\right)(0.01^4)} = 0.00592 \text{ rad} \qquad \textbf{Ans}$$

12–58. Determine the deflection of the 20-mm-diameter A-36 steel shaft at the pulley D. The bearings at A and B exert only vertical reactions on the shaft.

Support Reactions and Elastic Curve : As shown.

M/EI Diagram : As shown.

Moment - Area Theorems :

$$t_{D/B} = \frac{1}{2}\left(-\frac{34.375}{EI}\right)(0.5)(0.6667) + \frac{1}{2}\left(-\frac{125.625}{EI}\right)(0.3)(0.3) + \left(-\frac{34.375}{EI}\right)(0.3)(0.35) + \frac{1}{2}\left(-\frac{160}{EI}\right)(0.2)(0.1333)$$

$$= -\frac{17.125\ \mathrm{N \cdot m^3}}{EI}$$

$$t_{A/B} = \frac{1}{2}\left(-\frac{34.375}{EI}\right)(0.5)(0.4667) + \frac{1}{2}\left(-\frac{125.625}{EI}\right)(0.3)(0.1) + \left(-\frac{34.375}{EI}\right)(0.3)(0.15)$$

$$= -\frac{7.44167\ \mathrm{N \cdot m^3}}{EI}$$

The displacement at D is,

$$\Delta_D = |t_{D/B}| - |1.25t_{A/B}|$$

$$= \frac{17.125}{EI} - 1.25\left(\frac{7.44167}{EI}\right)$$

$$= \frac{7.823\ \mathrm{N \cdot m^3}}{EI}$$

$$= \frac{7.823}{200(10^9)\left(\frac{\pi}{4}\right)(0.01^4)}$$

$$= 0.00498\ \mathrm{m} = 4.98\ \mathrm{mm}\ \downarrow \qquad \textbf{Ans}$$

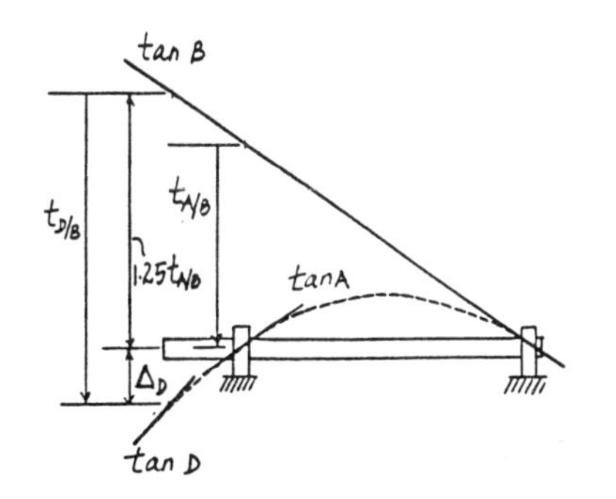

12–59. Determine the vertical deflection at the end A of the cantilevered beam due to the loading.

Support Reactions and Elastic Curve : As shown.

M/EI Diagram : As shown.

Moment - Area Theorems : The vertical displacement at A is,

$$(\Delta_A)_v = |t_{A/B}| = \left(\frac{3.897}{EI_x}\right)(15)(7.5) = \frac{438\ \mathrm{kip \cdot ft^3}}{EI_x} \qquad \textbf{Ans}$$

I_x is the moment of inertia about the x axis.

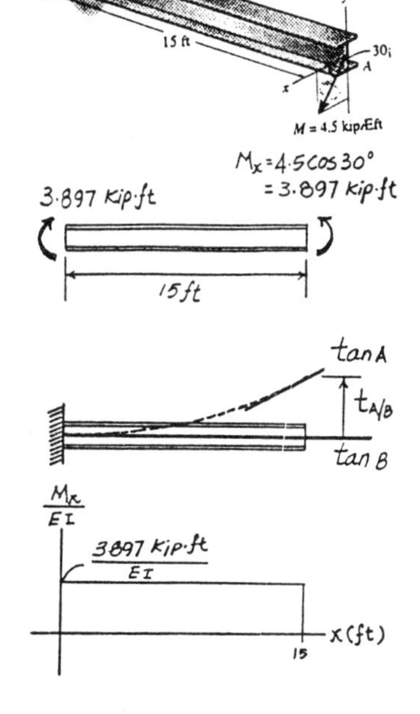

***12–60.** Determine the slope at B and the deflection at C. The bearings at A and B exert only vertical reactions on the shaft. EI is constant.

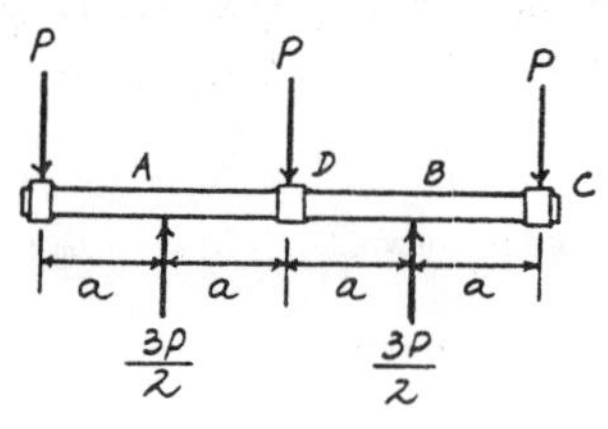

Moment - Area Theorems :

$$\theta_{B/D} = \frac{1}{2}\left(-\frac{Pa}{2EI}\right)(a) + \left(-\frac{Pa}{2EI}\right)(a) = -\frac{3Pa^2}{4EI}$$

$$t_{B/D} = \frac{1}{2}\left(-\frac{Pa}{2EI}\right)(a)\left(\frac{a}{3}\right) + \left(-\frac{Pa}{2EI}\right)(a)\left(\frac{a}{2}\right) = -\frac{Pa^3}{3EI}$$

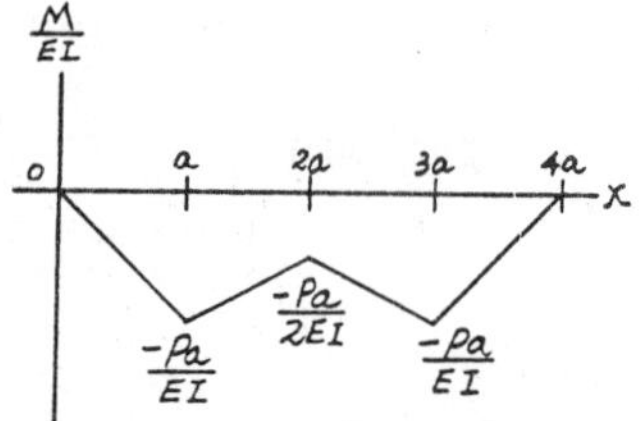

$$t_{C/D} = \frac{1}{2}\left(-\frac{Pa}{2EI}\right)(a)\left(a+\frac{a}{3}\right) + \left(-\frac{Pa}{2EI}\right)(a)\left(a+\frac{a}{2}\right) + \frac{1}{2}\left(-\frac{Pa}{EI}\right)(a)\left(\frac{2}{3}a\right)$$

$$= -\frac{17Pa^3}{12EI}$$

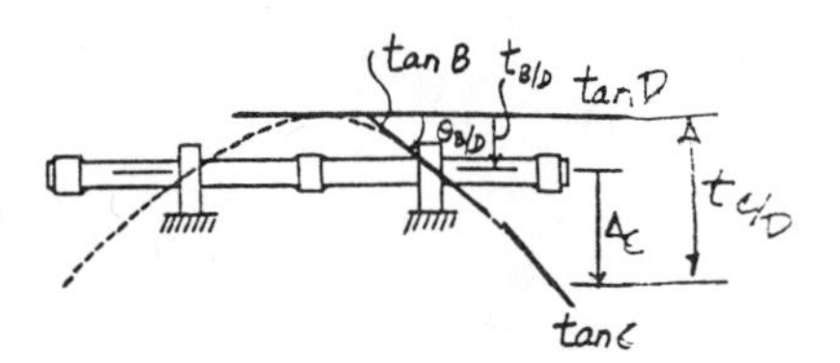

Due to symmetry, the slope at midspan (point D) is zero. Hence, the slope at B is

$$\theta_B = |\theta_{B/D}| = \frac{3Pa^2}{4EI} \qquad \textbf{Ans}$$

The displacement at C is

$$\Delta_C = |t_{C/D}| - |t_{B/D}|$$

$$= \frac{17Pa^3}{12EI} - \frac{Pa^3}{3EI}$$

$$= \frac{13Pa^3}{12EI} \downarrow \qquad \textbf{Ans}$$

12–61. Determine the deflection at D and the slope at C. The bearings at A and B exert only vertical reactions on the shaft. EI is constant.

Moment - Area Theorems :

$$\theta_{C/D} = \frac{1}{2}\left(-\frac{Pa}{2EI}\right)(a) + \left(-\frac{Pa}{2EI}\right)(a) + \frac{1}{2}\left(-\frac{Pa}{EI}\right)(a) = -\frac{5Pa^2}{4EI}$$

$$t_{B/D} = \frac{1}{2}\left(-\frac{Pa}{2EI}\right)(a)\left(\frac{a}{3}\right) + \left(-\frac{Pa}{2EI}\right)(a)\left(\frac{a}{2}\right) = -\frac{Pa^3}{3EI}$$

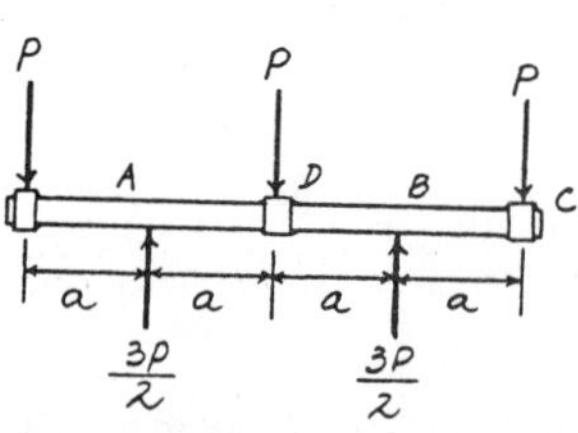

Due to symmetry, the slope at midspan (point D) is zero. Hence, the slope at C is

$$\theta_C = |\theta_{C/D}| = \frac{5Pa^2}{4EI} \qquad \textbf{Ans}$$

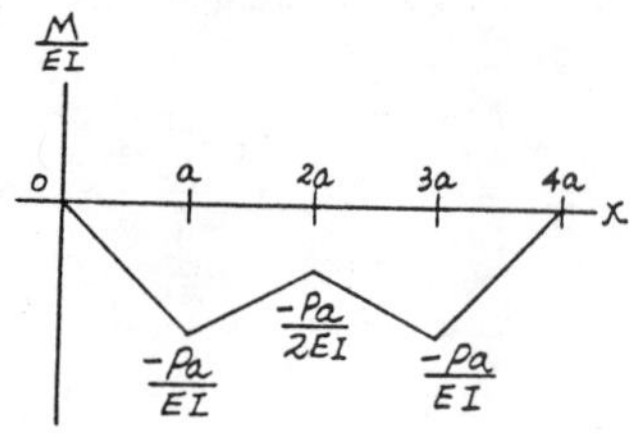

The displacement at D is

$$\Delta_D = |t_{B/D}| = \frac{Pa^3}{3EI} \uparrow \qquad \textbf{Ans}$$

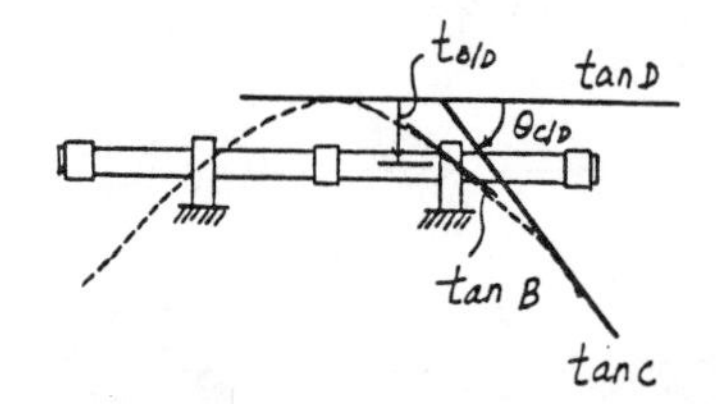

12–62. The simply supported shaft has a moment of inertia of $2I$ for region BC and a moment of inertia I for regions AB and CD. Determine the maximum deflection of the shaft due to the load $\mathbf{P}$. The modulus of elasticity is E.

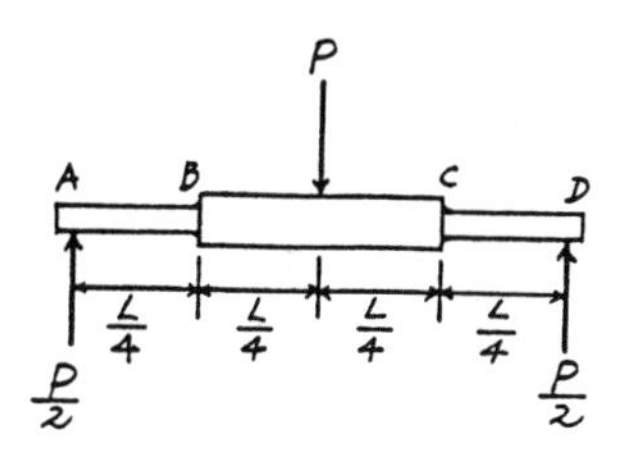

Support Reactions and Elastic Curve : As shown.

M/EI Diagram : As shown.

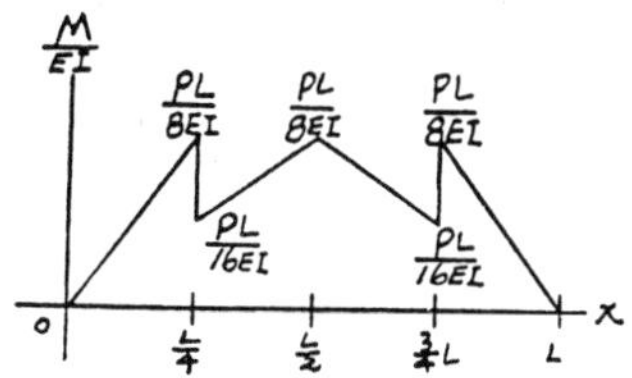

Moment - Area Theorems : Due to symmetry, the slope at midspan (point E) is zero. Hence,

$$\Delta_{max} = t_{A/E} = \frac{1}{2}\left(\frac{PL}{16EI}\right)\left(\frac{L}{4}\right)\left(\frac{L}{4}+\frac{L}{6}\right)+\left(\frac{PL}{16EI}\right)\left(\frac{L}{4}\right)\left(\frac{L}{4}+\frac{L}{8}\right) + \frac{1}{2}\left(\frac{PL}{8EI}\right)\left(\frac{L}{4}\right)\left(\frac{L}{6}\right)$$

$$= \frac{3PL^3}{256EI} \downarrow \qquad \textbf{Ans}$$

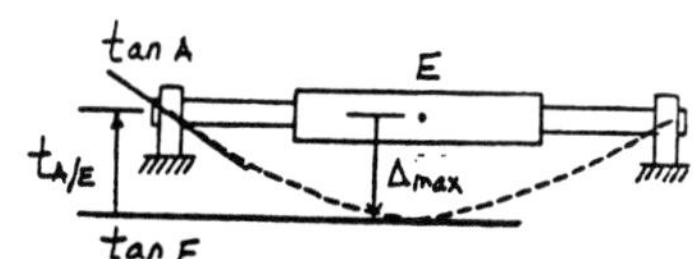

12–63. The simply supported shaft has a moment of inertia of $2I$ for region BC and a moment of inertia I for regions AB and CD. Determine the deflection at B and the slope at A of the shaft due to the load $\mathbf{P}$.

Support Reactions and Elastic Curve : As shown.

M/EI Diagram : As shown.

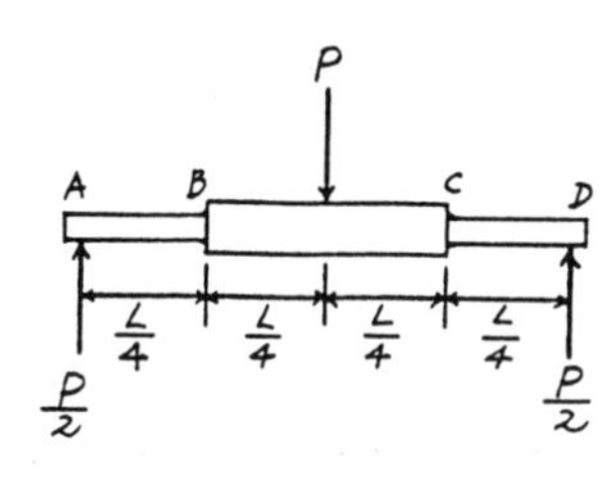

Moment - Area Theorems :

$$t_{A/E} = \frac{1}{2}\left(\frac{PL}{16EI}\right)\left(\frac{L}{4}\right)\left(\frac{L}{4}+\frac{L}{6}\right)+\left(\frac{PL}{16EI}\right)\left(\frac{L}{4}\right)\left(\frac{L}{4}+\frac{L}{8}\right) + \frac{1}{2}\left(\frac{PL}{8EI}\right)\left(\frac{L}{4}\right)\left(\frac{L}{6}\right)$$

$$= \frac{3PL^3}{256EI}$$

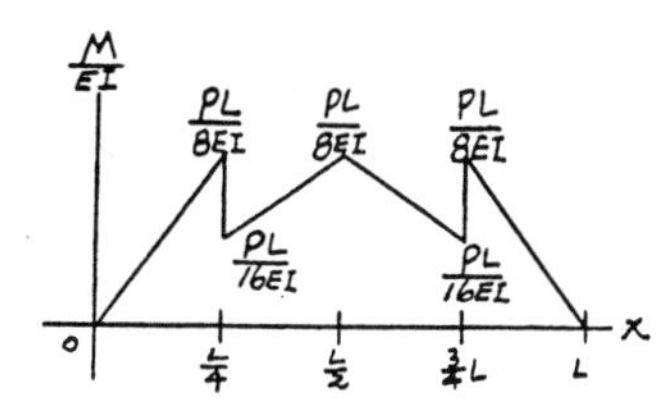

$$t_{B/E} = \frac{1}{2}\left(\frac{PL}{16EI}\right)\left(\frac{L}{4}\right)\left(\frac{L}{6}\right)+\left(\frac{PL}{16EI}\right)\left(\frac{L}{4}\right)\left(\frac{L}{8}\right) = \frac{5PL^3}{1536EI}$$

$$\theta_{A/E} = \frac{1}{2}\left(\frac{PL}{16EI}\right)\left(\frac{L}{4}\right)+\left(\frac{PL}{16EI}\right)\left(\frac{L}{4}\right)+\frac{1}{2}\left(\frac{PL}{8EI}\right)\left(\frac{L}{4}\right) = \frac{5PL^2}{128EI}$$

Due to symmetry, the slope at the midspan (point E) is zero. Hence, the slope at A is

$$\theta_A = |\theta_{A/E}| = \frac{5PL^2}{128EI} \qquad \textbf{Ans}$$

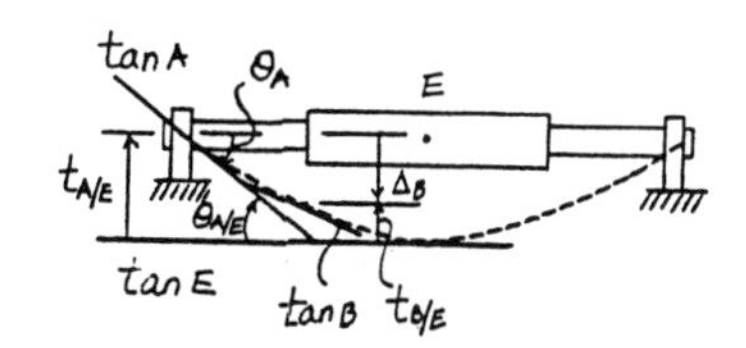

The displacement at D is

$$\Delta_D = |t_{A/E}| - |t_{B/E}| = \frac{3PL^3}{256EI} - \frac{5PL^3}{1536EI} = \frac{13PL^3}{1536EI} \qquad \textbf{Ans}$$

***12–64.** Determine the slope of the shaft at A and the deflection at D. EI is constant.

Support Reactions and Elastic Curve : As shown.

M/EI Diagram : As shown.

Moment-Area Theorems :

$$t_{B/A} = \frac{1}{2}\left(-\frac{Pa}{EI}\right)(a)\left(\frac{a}{3}\right) = -\frac{Pa^3}{6EI}$$

$$t_{D/A} = \frac{1}{2}\left(-\frac{Pa}{EI}\right)(a)\left(a+\frac{a}{3}\right)+\frac{1}{2}\left(-\frac{Pa}{EI}\right)(a)\left(\frac{2}{3}a\right) = \frac{Pa^3}{EI}$$

The slope at A is

$$\theta_A = \frac{|t_{B/A}|}{L} = \frac{\frac{Pa^3}{6EI}}{2a} = \frac{Pa^2}{12EI} \qquad \textbf{Ans}$$

The displacement at D is

$$\begin{aligned}\Delta_D &= |t_{D/A}| - \left|\frac{3}{2}t_{B/A}\right| \\ &= \frac{Pa^3}{EI} - \frac{3}{2}\left(\frac{Pa^3}{6EI}\right) \\ &= \frac{3Pa^3}{4EI}\end{aligned} \qquad \textbf{Ans}$$

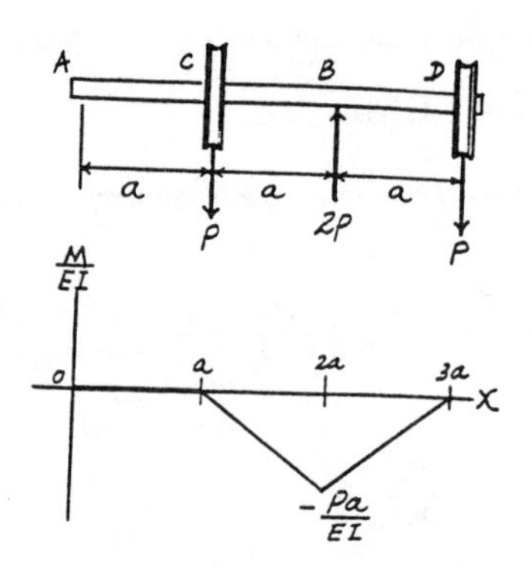

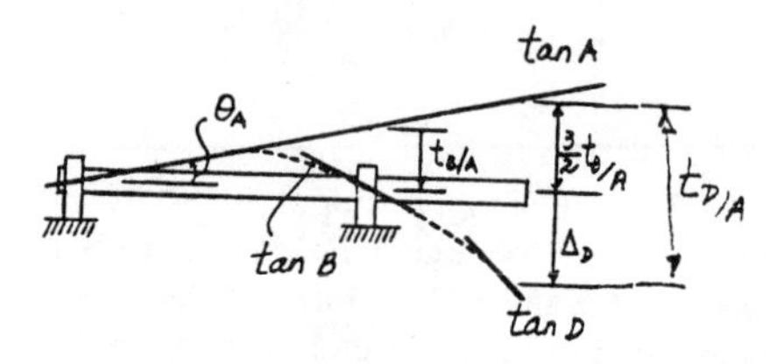

2–65. The beam is subjected to the load **P** as shown. Determine the magnitude of force **F** that must be applied at the end of the overhang C so that the deflection at C is zero. EI is constant.

Support Reactions and Elastic Curve : As shown.

M/EI Diagram : As shown.

Moment-Area Theorems :

$$t_{B/A} = \frac{1}{2}\left(\frac{Pa}{2EI}\right)(2a)(a)+\frac{1}{2}\left(-\frac{Fa}{EI}\right)(2a)\left(\frac{2}{3}a\right) = \frac{a^3}{6EI}(3P-4F)$$

$$\begin{aligned}t_{C/A} &= \frac{1}{2}\left(\frac{Pa}{2EI}\right)(2a)(a+a)+\frac{1}{2}\left(-\frac{Fa}{EI}\right)(2a)\left(a+\frac{2}{3}a\right) \\ &\quad+\frac{1}{2}\left(-\frac{Fa}{EI}\right)(a)\left(\frac{2}{3}a\right) \\ &= \frac{a^3}{EI}(P-2F)\end{aligned}$$

Require $\Delta_C = 0$, then

$$\begin{aligned}\Delta_C = 0 &= |t_{C/A}| - \left|\frac{3}{2}t_{B/A}\right| \\ 0 &= \frac{a^3}{EI}(P-2F) - \frac{3}{2}\left[\frac{a^3}{6EI}(3P-4F)\right] \\ F &= \frac{P}{4}\end{aligned} \qquad \textbf{Ans}$$

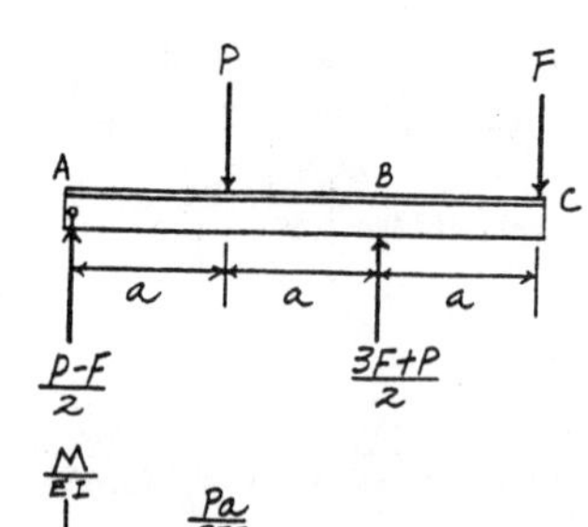

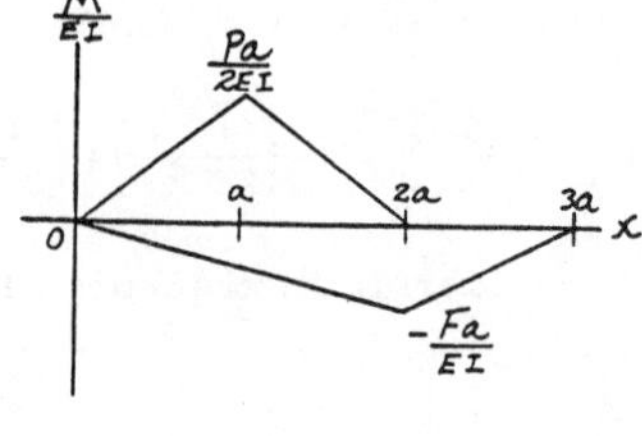

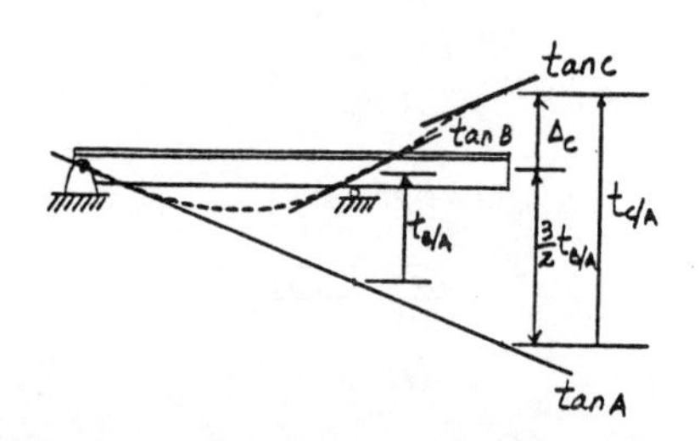

12–66. The beam is subjected to the load **P** as shown. If **F** = **P**, determine the deflection of the beam at D. EI is constant.

Support Reactions and Elastic Curve : As shown.

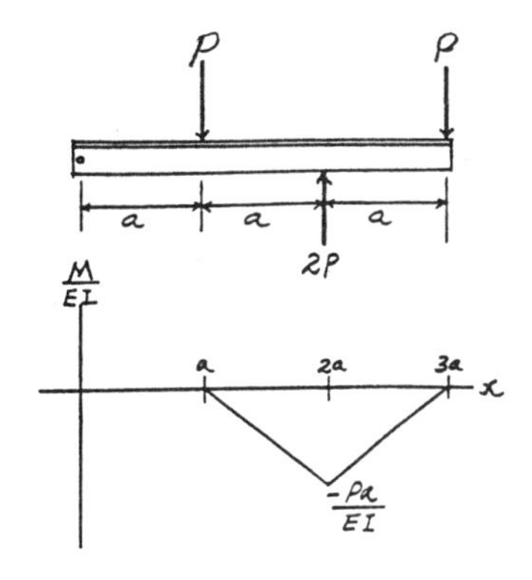

M/EI Diagram : As shown.

Moment - Area Theorems :

$$t_{B/A} = \frac{1}{2}\left(-\frac{Pa}{EI}\right)(a)\left(\frac{a}{3}\right) = -\frac{Pa^3}{6EI}$$

$$t_{D/A} = 0$$

The displacement at D is

$$\Delta_D = \frac{1}{2}|t_{B/A}| - |t_{D/A}|$$

$$= \frac{1}{2}\left(\frac{Pa^3}{6EI}\right) - 0$$

$$= \frac{Pa^3}{12EI} \uparrow \qquad \textbf{Ans}$$

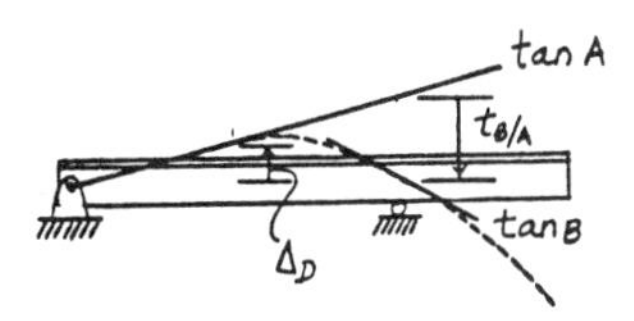

12–67. Determine the deflection at C, D, and E. The bearings at A and B exert only vertical reactions on the shaft. EI is constant.

Support Reactions and Elastic Curve : As shown.

M/EI Diagram : As shown.

Moment - Area Theorems :

$$t_{C/D} = \frac{1}{2}\left(\frac{Pa}{2EI}\right)(a)\left(\frac{2}{3}a\right) + \left(\frac{Pa}{EI}\right)(a)\left(\frac{a}{2}\right) = \frac{2Pa^3}{3EI}$$

$$t_{A/D} = \frac{1}{2}\left(\frac{Pa}{2EI}\right)(a)\left(a + \frac{2}{3}a\right) + \left(\frac{Pa}{EI}\right)(a)\left(a + \frac{a}{2}\right) + \frac{1}{2}\left(\frac{Pa}{EI}\right)(a)\left(\frac{2}{3}a\right)$$

$$= \frac{9Pa^3}{4EI}$$

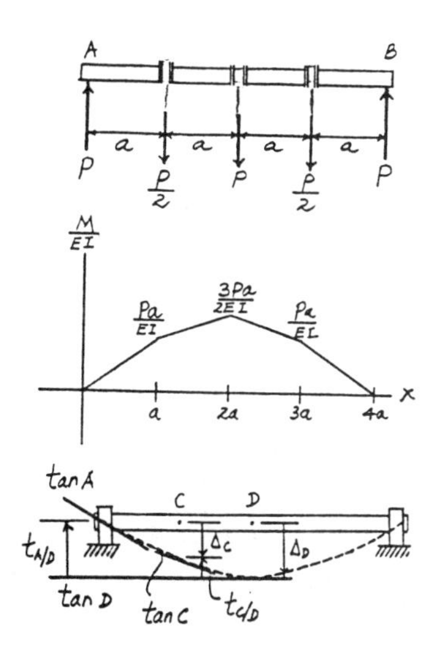

Due to symmetry, the slope at midspan (point D) is zero. Hence, the displacements at D and C(and E) are

$$\Delta_D = |t_{A/D}| = \frac{9Pa^3}{4EI} \downarrow \qquad \textbf{Ans}$$

$$\Delta_E = \Delta_C = |t_{A/D}| - |t_{C/D}|$$

$$= \frac{9Pa^3}{4EI} - \frac{2Pa^3}{3EI}$$

$$= \frac{19Pa^3}{12EI} \downarrow \qquad \textbf{Ans}$$

***12–68.** Determine the slope of the shaft at the bearings at A and B, which exert only vertical reactions on the shaft. EI is constant.

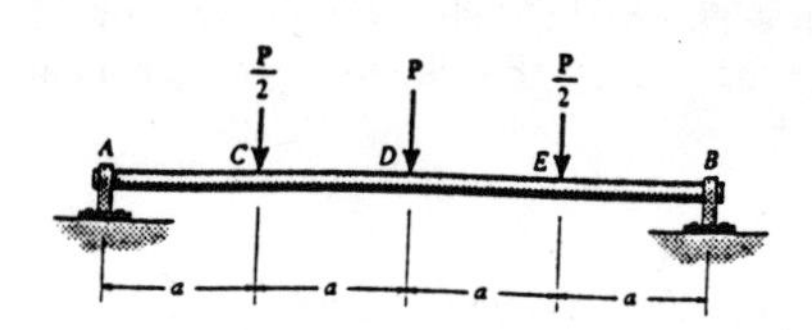

Support Reactions and Elastic Curve : As shown.

M/EI Diagram : As shown.

Moment - Area Theorems :

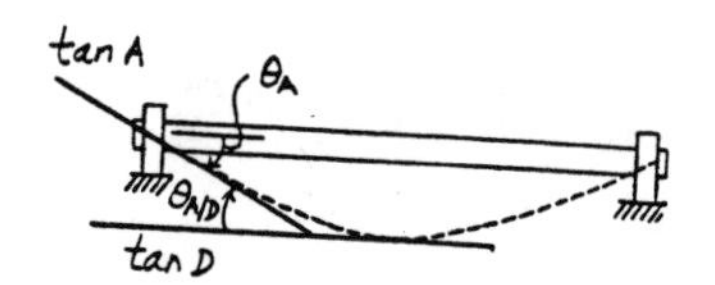

$$\theta_{A/D} = \frac{1}{2}\left(\frac{Pa}{2EI}\right)(a) + \left(\frac{Pa}{EI}\right)(a) + \frac{1}{2}\left(\frac{Pa}{EI}\right)(a) = \frac{7Pa^2}{4EI}$$

Due to symmetry, the slope at midspan (point D) is zero. Hence, the slopes at A and B are

$$\theta_B = \theta_A = |\theta_{A/D}| = \frac{7Pa^2}{4EI} \qquad \textbf{Ans}$$

12–69. The two force components act on the tire of the automobile as shown. The tire is fixed to the axle, which is supported by bearings at A and B. Determine the maximum deflection of the axle. Assume that the bearings resist only vertical loads. The thrust on the axle is resisted at C. The axle has a diameter of 1.25 in. and is made of A-36 steel. Neglect the effect of axial load on deflection.

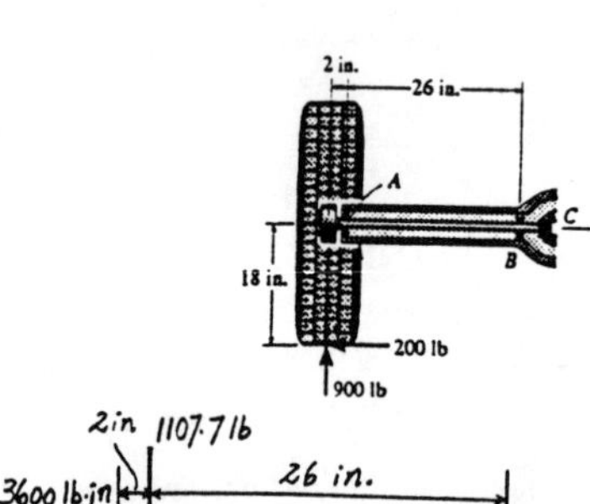

Support Reactions and Elastic Curve : As shown.

M/EI Diagram : As shown.

Moment - Area Theorems :

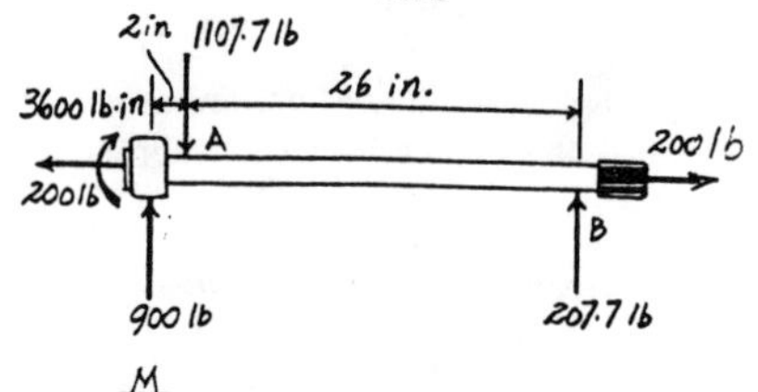

$$t_{A/B} = \frac{1}{2}\left(\frac{5400}{EI}\right)(26)\left(\frac{26}{3}\right) = \frac{608400 \text{ lb}\cdot\text{in}^3}{EI}$$

$$\theta_B = \frac{|t_{A/B}|}{L} = \frac{\frac{608400 \text{ lb}\cdot\text{in}^3}{EI}}{26 \text{ in.}} = \frac{23400 \text{ lb}\cdot\text{in}^2}{EI}$$

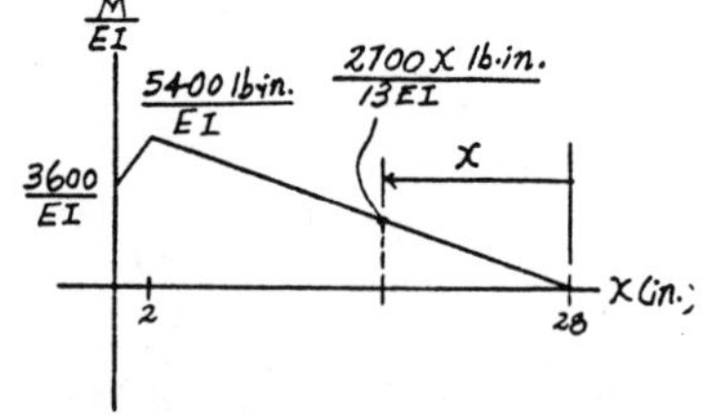

The maximum displacement occurs at point C, where $\theta_C = 0$.

$$\theta_{C/B} = \frac{1}{2}\left(\frac{2700}{13EI}x\right)(x) = \frac{103.846}{EI}x^2$$

$$\theta_C = \theta_B + \theta_{C/B}$$

$$0 = -\frac{23400}{EI} + \frac{103.846}{EI}x^2$$

$$x = 15.01 \text{ in.} < 26 \text{ in.} \quad (O.K!)$$

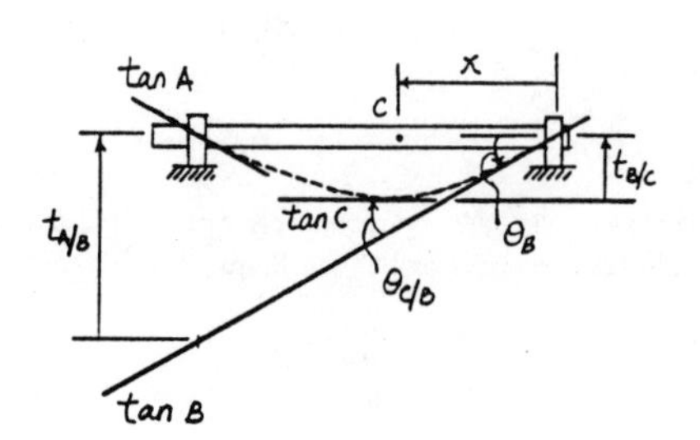

The maximum displacement is

$$\Delta_{max} = |t_{B/C}| = \frac{1}{2}\left(\frac{2700}{13EI}\right)(15.01^2)\left(\frac{2}{3}\right)(15.01)$$

$$= \frac{234173.27}{EI}$$

$$= \frac{234173.27}{29.0(10^6)\left(\frac{\pi}{4}\right)(0.625^4)}$$

$$= 0.0674 \text{ in.} \downarrow \qquad \textbf{Ans}$$

12–70. The cantilevered beam is subjected to the loading shown. Assume the support at A is fixed and determine the slope and displacement at C. EI is constant.

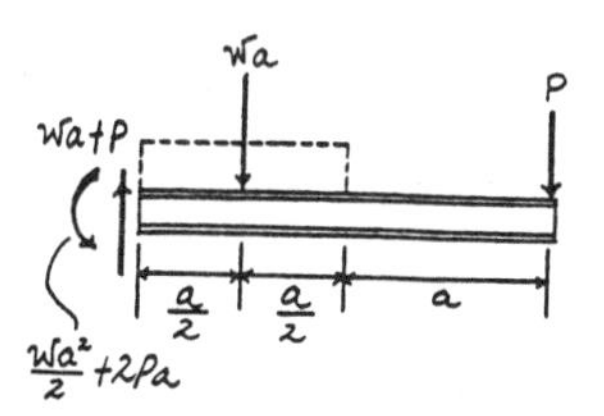

Support Reactions and Elastic Curve : As shown.

M/EI Diagrams : The M/EI diagrams for the uniform distributed load and concentrated load are drawn separately as shown.

Moment - Area Theorems : The slope at support A is zero. The slope at C is

$$\theta_C = |\theta_{C/A}| = \frac{1}{2}\left(-\frac{2Pa}{EI}\right)(2a) + \frac{1}{3}\left(-\frac{wa^2}{2EI}\right)(a)$$

$$= \frac{a^2}{6EI}(12P + wa) \qquad \textbf{Ans}$$

The displacement at C is

$$\Delta_C = |t_{C/A}| = \frac{1}{2}\left(-\frac{2Pa}{EI}\right)(2a)\left(\frac{4}{3}a\right) + \frac{1}{3}\left(-\frac{wa^2}{2EI}\right)(a)\left(a + \frac{3}{4}a\right)$$

$$= \frac{a^3}{24EI}(64P + 7wa) \downarrow \qquad \textbf{Ans}$$

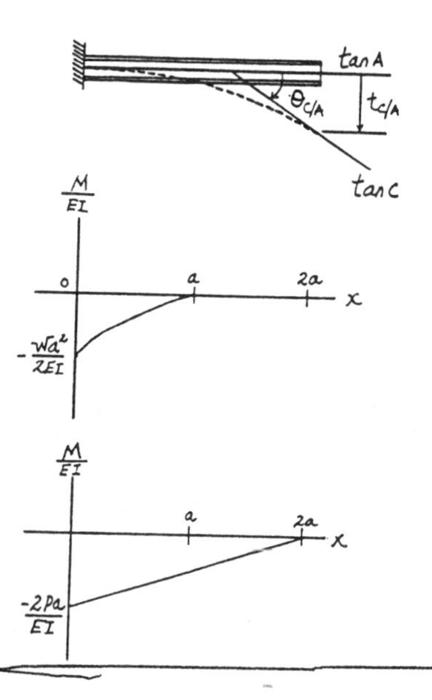

12–71. Determine the maximum deflection of the beam. EI is constant.

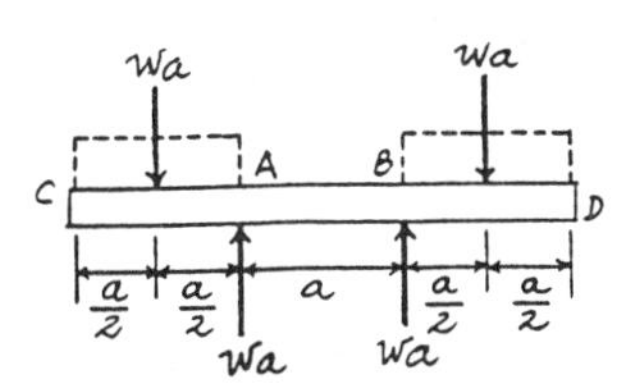

Support Reactions and Elastic Curve : As shown.

M/EI Diagram : As shown.

Moment - Area Theorems :

$$t_{A/E} = \left(-\frac{wa^2}{2EI}\right)\left(\frac{a}{2}\right)\left(\frac{a}{4}\right) = -\frac{wa^4}{16EI}$$

$$t_{C/E} = \left(-\frac{wa^2}{2EI}\right)\left(\frac{a}{2}\right)\left(a + \frac{a}{4}\right) + \frac{1}{3}\left(-\frac{wa^2}{2EI}\right)(a)\left(\frac{3a}{4}\right) = -\frac{7wa^4}{16EI}$$

Due to symmetry, the slope at midspan (point E) is zero. Hence, the maximum displacement occurs at E or C.

$$\Delta_E = |t_{A/E}| = \frac{wa^4}{16EI} \uparrow$$

$$\Delta_{max} = \Delta_C = |t_{C/E}| - |t_{A/E}|$$

$$= \frac{7wa^4}{16EI} - \frac{wa^4}{16EI}$$

$$= \frac{3wa^4}{8EI} \downarrow \qquad \textbf{Ans}$$

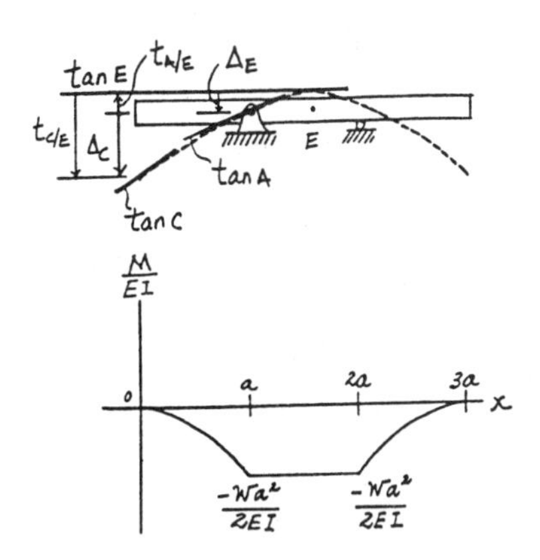

***12–72.** Determine the slope at B and the deflection at C. The member is an A-36 steel structural tee for which $I = 76.8\ \text{in}^4$.

Support Reactions and Elastic Curve : As shown.

M/EI Diagrams : The M/EI diagrams for the uniform distributed load and concentrated load are drawn separately as shown.

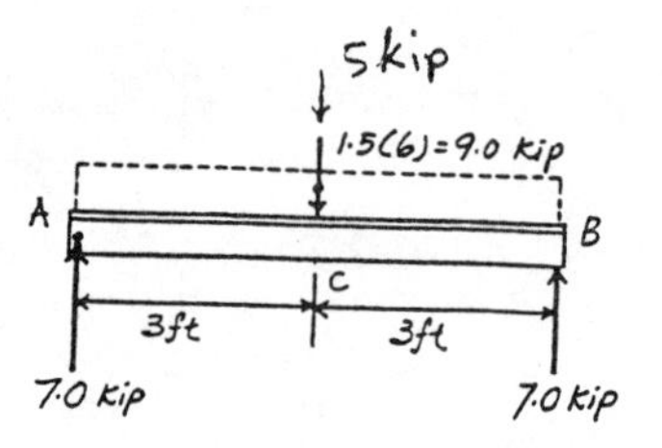

Moment - Area Theorems : Due to symmetry, the slope at midspan C is zero. Hence the slope at B is

$$\theta_B = |\theta_{B/C}| = \frac{1}{2}\left(\frac{7.50}{EI}\right)(3) + \frac{2}{3}\left(\frac{6.75}{EI}\right)(3)$$
$$= \frac{24.75\ \text{kip}\cdot\text{ft}^2}{EI}$$
$$= \frac{24.75(144)}{29.0(10^3)(76.8)}$$
$$= 0.00160\ \text{rad} \qquad \textbf{Ans}$$

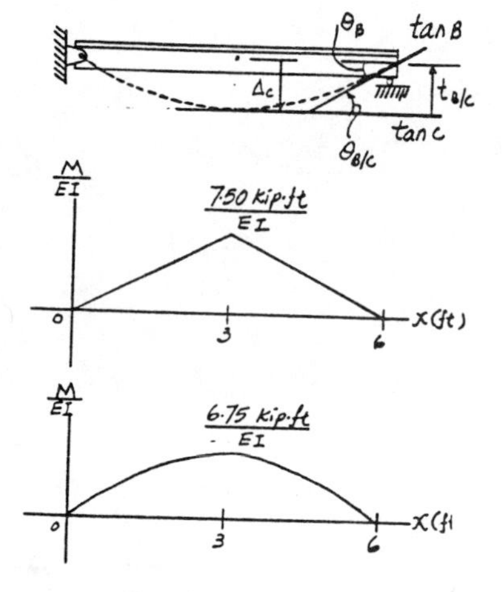

The dispacement at C is

$$\Delta_C = |t_{A/C}| = \frac{1}{2}\left(\frac{7.50}{EI}\right)(3)\left(\frac{2}{3}\right)(3) + \frac{2}{3}\left(\frac{6.75}{EI}\right)(3)\left(\frac{5}{8}\right)(3)$$
$$= \frac{47.8125\ \text{kip}\cdot\text{ft}^3}{EI}$$
$$= \frac{47.8125(1728)}{29.0(10^3)(76.8)}$$
$$= 0.0371\ \text{in.} \downarrow \qquad \textbf{Ans}$$

12–73. Determine the slope at C and deflection at B. EI is constant.

Support Reactions and Elastic Curve : As shown.

M/EI Diagram : As shown.

Moment - Area Theorems : The slope at support A is zero. The slope at C is

$$\theta_C = |\theta_{C/A}| = \frac{1}{2}\left(-\frac{wa^2}{EI}\right)(a) + \left(-\frac{wa^2}{2EI}\right)(a)$$
$$= \frac{wa^3}{EI} \qquad \textbf{Ans}$$

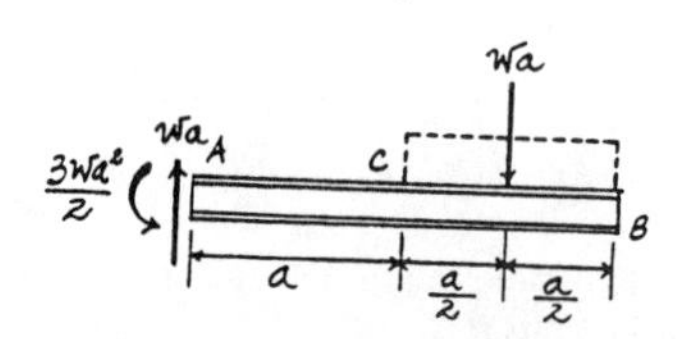

The displacement at B is

$$\Delta_B = |t_{B/A}| = \frac{1}{2}\left(-\frac{wa^2}{EI}\right)(a)\left(a+\frac{2}{3}a\right) + \left(-\frac{wa^2}{2EI}\right)(a)\left(a+\frac{a}{2}\right)$$
$$+\frac{1}{3}\left(-\frac{wa^2}{2EI}\right)(a)\left(\frac{3}{4}a\right)$$
$$= \frac{41wa^4}{24EI} \downarrow \qquad \textbf{Ans}$$

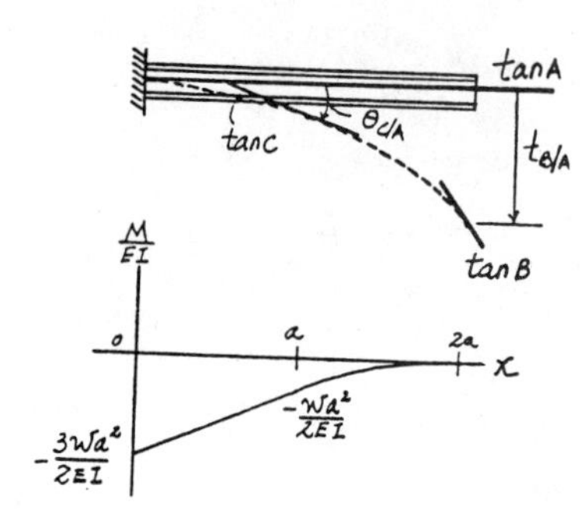

12–74. The $W8 \times 48$ cantilevered beam is made of steel and is subjected to the loading shown. Determine the deflection at its end A.

Elastic Curve : The elastic curves for the concentrated load and couple moment are drawn separately as shown.

Method of Superposition : Using the table in Appendix C, the required slope and displacement are

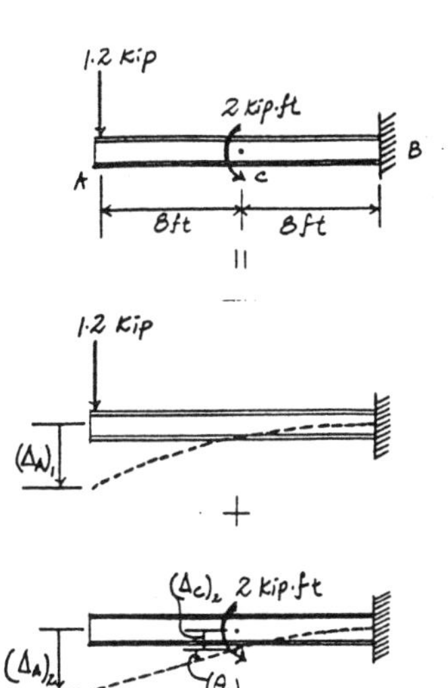

$$(\Delta_A)_1 = \frac{PL_{AB}^3}{3EI} = \frac{1.2(16^3)}{3EI} = \frac{1638.4 \text{ kip}\cdot\text{ft}^3}{EI} \downarrow$$

$$(\Delta_C)_2 = \frac{M_0 L_{BC}^2}{2EI} = \frac{2(8^2)}{2EI} = \frac{64.0 \text{ kip}\cdot\text{ft}^3}{EI}$$

$$(\theta_C)_2 = \frac{M_0 L_{BC}}{EI} = \frac{2(8)}{EI} = \frac{16.0 \text{ kip}\cdot\text{ft}^2}{EI}$$

$$(\Delta_A)_2 = (\Delta_C)_2 + (\theta_C)_2 L_{AC} = \frac{64.0}{EI} + \frac{16.0}{EI}(8) = \frac{192 \text{ kip}\cdot\text{ft}^3}{EI} \downarrow$$

The displacement at A is

$$\Delta_A = (\Delta_A)_1 + (\Delta_A)_2 = \frac{1638.4}{EI} + \frac{192}{EI} = \frac{1830.4 \text{ kip}\cdot\text{ft}^3}{EI} = \frac{1830.4(1728)}{29.0(10^3)(184)} = 0.593 \text{ in.} \downarrow \qquad \textbf{Ans}$$

12–75. The $W8 \times 48$ cantilevered beam is made of steel and subjected to the loading shown. Determine the deflection at C and the slope at A.

Elastic Curve : The elastic curves for the concentrated load and couple moment are drawn separately as shown.

Method of Superposition : Using the table in Appendix C, the required slope and displacement are

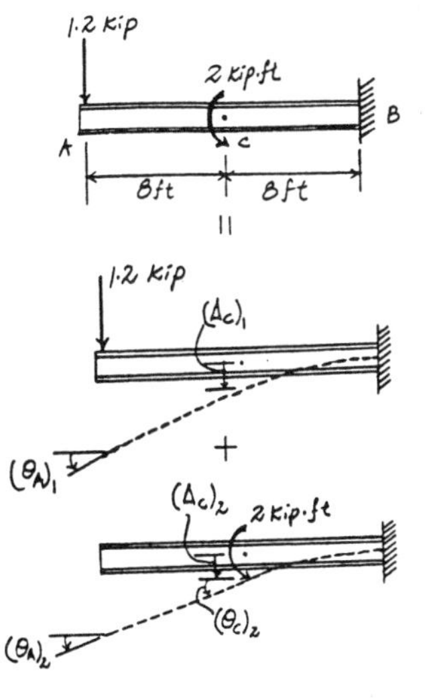

$$(\Delta_C)_1 = \frac{Px^2}{6EI}(3L_{AB} - x) = \frac{1.2(8^2)}{6EI}[3(16) - 8] = \frac{512 \text{ kip}\cdot\text{ft}^3}{EI} \downarrow$$

$$(\Delta_C)_2 = \frac{M_0 L_{BC}^2}{2EI} = \frac{2(8^2)}{2EI} = \frac{64.0 \text{ kip}\cdot\text{ft}^3}{EI} \downarrow$$

$$(\theta_A)_1 = \frac{PL_{AB}^2}{2EI} = \frac{1.2(16^2)}{2EI} = \frac{153.6 \text{ kip}\cdot\text{ft}^2}{EI}$$

$$(\theta_A)_2 = (\theta_C)_2 = \frac{M_0 L_{BC}}{EI} = \frac{2(8)}{EI} = \frac{16.0 \text{ kip}\cdot\text{ft}^2}{EI}$$

The slope at A is

$$\theta_A = (\theta_A)_1 + (\theta_A)_2 = \frac{153.6}{EI} + \frac{16.0}{EI} = \frac{169.6 \text{ kip}\cdot\text{ft}^2}{EI} = \frac{169.6(144)}{29.0(10^3)(184)} = 0.00458 \text{ rad} \qquad \textbf{Ans}$$

The displacement at C is

$$\Delta_C = (\Delta_C)_1 + (\Delta_C)_2 = \frac{512}{EI} + \frac{64.0}{EI} = \frac{576 \text{ kip}\cdot\text{ft}^3}{EI} = \frac{576(1728)}{29.0(10^3)(184)} = 0.187 \text{ in.} \downarrow \qquad \textbf{Ans}$$

***12–76.** The $W24 \times 104$ beam is made of A-36 steel and is subjected to the loading shown. Determine the deflection at its end A.

Elastic Curve : The elastic curves for the uniform distributed load and concentrated load are drawn separately as shown.

Method of Superposition : Using the table in Appendix C, the required displacements are

$$(\Delta_A)_1 = \frac{wL^4}{8EI} = \frac{5(15^4)}{8EI} = \frac{31640.625 \text{ kip}\cdot\text{ft}^3}{EI} \downarrow$$

$$(\Delta_A)_2 = \frac{PL^3}{3EI} = \frac{8(15^3)}{3EI} = \frac{9000 \text{ kip}\cdot\text{ft}^3}{EI} \downarrow$$

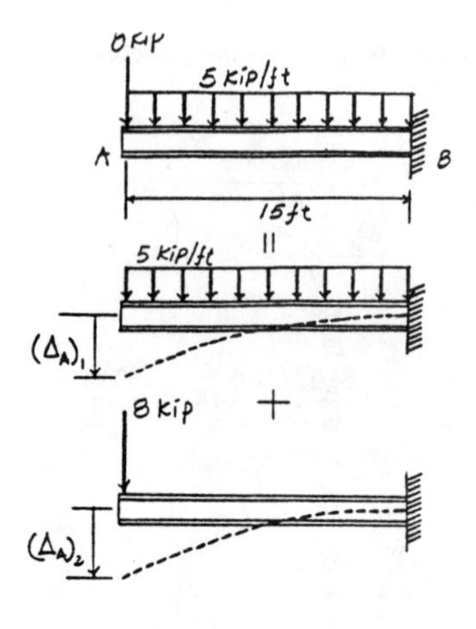

The displacement at A is

$$\begin{aligned}\Delta_A &= (\Delta_A)_1 + (\Delta_A)_2 \\ &= \frac{31640.625}{EI} + \frac{9000}{EI} \\ &= \frac{40640.625 \text{ kip}\cdot\text{ft}^3}{EI} \\ &= \frac{40640.625(1728)}{29.0(10^3)(3100)} = 0.781 \text{ in.} \downarrow \end{aligned}$$

Ans

12–77. Determine the moment M_0 in terms of the load P and dimension a so that the deflection at the center of the beam is zero. EI is constant.

Elastic Curve : The elastic curves for the concentrated load and couple moment are drawn separately as shown.

Method of Superposition : Using the table in Appendix C, the required slope and displacement are

$$(\Delta_C)_1 = \frac{Pa^3}{48EI} \downarrow$$

$$\begin{aligned}(\Delta_C)_2 = (\Delta_C)_3 &= \frac{M_0 x}{6EIL}(x^2 - 3Lx + 2L^2) \\ &= \frac{M_0\left(\frac{a}{2}\right)}{6EIa}\left[\left(\frac{a}{2}\right)^2 - 3(a)\left(\frac{a}{2}\right) + 2a^2\right] \\ &= \frac{M_0 a^2}{16EI} \uparrow\end{aligned}$$

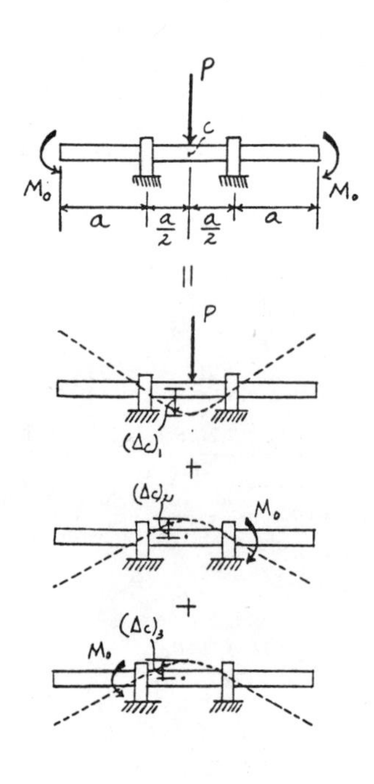

Require the displacement at C to equal zero.

$$(+\uparrow) \qquad \Delta_C = 0 = (\Delta_C)_1 + (\Delta_C)_2 + (\Delta_C)_3$$

$$0 = -\frac{Pa^3}{48EI} + \frac{M_0 a^2}{16EI} + \frac{M_0 a^2}{16EI}$$

$$M_0 = \frac{Pa}{6}$$

Ans

12–78. The 100-mm-diameter shaft is made of A-36 steel and is subjected to the loading shown. If the bearings at A and B only exert vertical reactions on the shaft, determine the slope at A and the deflection at C.

Elastic Curve : The elastic curves for the uniform distributed load and concentrated load are drawn separately as shown.

Method of Superposition : Using the table in Appendix C, the required slopes and displacements are

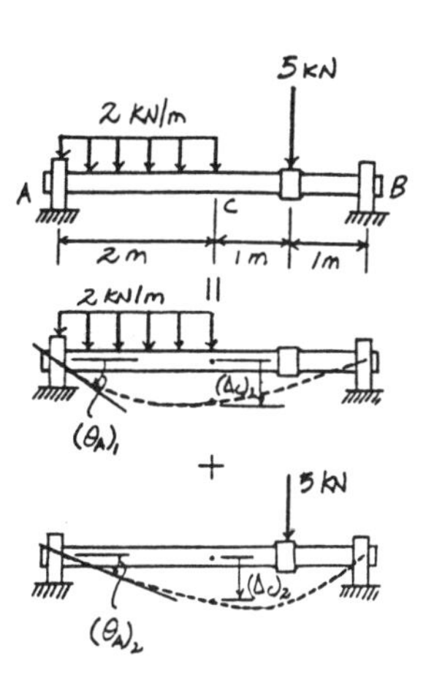

$$(\theta_A)_1 = \frac{3wL^3}{128EI} = \frac{3(2)(4^3)}{128EI} = \frac{3.00 \text{ kN}\cdot\text{m}^2}{EI}$$

$$(\theta_A)_2 = \frac{Pab(L+b)}{6EIL} = \frac{5(3)(1)(4+1)}{6EI(4)} = \frac{3.125 \text{ kN}\cdot\text{m}^2}{EI}$$

$$(\Delta_C)_1 = \frac{5wL^4}{768EI} = \frac{5(2)(4^4)}{768EI} = \frac{3.3333 \text{ kN}\cdot\text{m}^3}{EI} \downarrow$$

$$(\Delta_C)_2 = \frac{-Pbx}{6EIL}\left(L^2 - b^2 - x^2\right)$$

$$= \frac{-5(1)(2)}{6EI(4)}\left(4^2 - 1^2 - 2^2\right)$$

$$= -\frac{4.5833 \text{ kN}\cdot\text{m}^3}{EI} = \frac{4.5833 \text{ kN}\cdot\text{m}^3}{EI} \downarrow$$

The slope at A is

$$\theta_A = (\theta_A)_1 + (\theta_A)_2$$

$$= \frac{3.00}{EI} + \frac{3.125}{EI}$$

$$= \frac{6.125 \text{ kN}\cdot\text{m}^2}{EI}$$

$$= \frac{6.125(10^3)}{200(10^9)\left(\frac{\pi}{4}\right)(0.05^4)}$$

$$= 0.00624 \text{ rad}$$ **Ans**

The displacement at C is

$$\Delta_C = (\Delta_C)_1 + (\Delta_C)_2$$

$$= \frac{3.3333}{EI} + \frac{4.5833}{EI}$$

$$= \frac{7.9167 \text{ kN}\cdot\text{m}^3}{EI}$$

$$= \frac{7.9167(10^3)}{200(10^9)\left(\frac{\pi}{4}\right)(0.05^4)}$$

$$= 0.008064 \text{ m} = 8.06 \text{ mm} \downarrow$$ **Ans**

12–79. The relay switch consists of a thin metal strip or armature AB that is made of red brass C83400 and is attracted to the solenoid S by a magnetic field. Determine the smallest force F required to attract the armature at C in order that contact is made at the free end B. Also, what should the distance a be for this to occur? The armature is fixed at A and has a moment of inertia of $I = 0.18(10^{-12})$ m^4.

Elastic Curve : As shown.

Method of Superposition : Using the table in Appendix C, the required slopes and displacements are

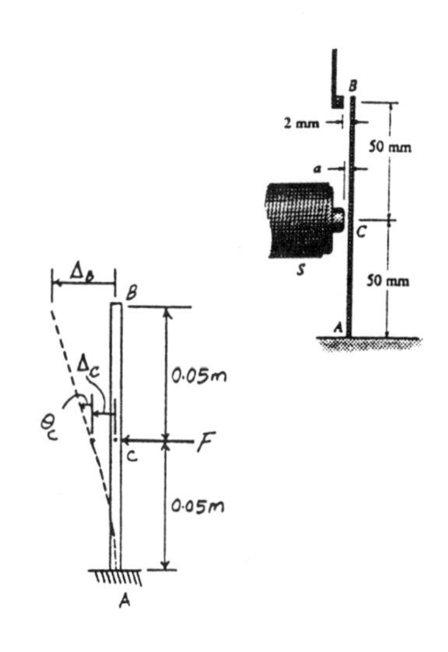

$$\theta_C = \frac{PL_{AC}^2}{2EI} = \frac{F(0.05^2)}{2EI} = \frac{0.00125F \text{ m}^2}{EI}$$

$$\Delta_C = \frac{PL_{AC}^3}{3EI} = \frac{F(0.05^3)}{3EI} = \frac{41.667(10^{-6})F \text{ m}^3}{EI} \qquad [1]$$

$$\Delta_B = \Delta_C + \theta_C L_{CB}$$

$$= \frac{41.667(10^{-6})F}{EI} + \frac{0.00125(10^{-6})F}{EI}(0.05)$$

$$= \frac{104.167(10^{-6})F \text{ m}^3}{EI} \qquad [2]$$

Requie the displacement $\Delta_B = 0.002$ m. From Eq.[2],

$$0.002 = \frac{104.167(10^{-6})F}{101(10^9)(0.18)(10^{-12})}$$

$$F = 0.349056 \text{ N} = 0.349 \text{ N}$$ **Ans**

From Eq.[1],

$$a = \Delta_C = \frac{41.667(10^{-6})(0.349056)}{101(10^9)(0.18)(10^{-12})}$$

$$= 0.800\left(10^{-3}\right) \text{ m} = 0.800 \text{ mm}$$ **Ans**

***12–80.** The two steel bars have a height of 1 in. and a width of 4 in. They are designed to act as a spring for the machine, which exerts a force of 4 kip on them at A and B. If all supports exert only vertical forces on the bars, determine the maximum deflection of the bottom bar. $E_{st} = 29(10^3)$ ksi.

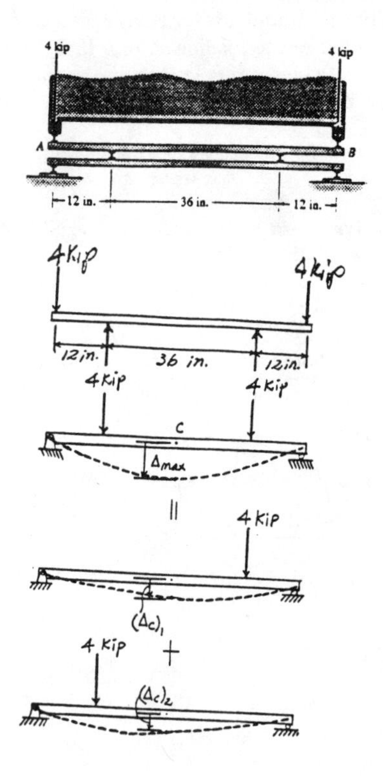

Method of Superposition : Due to the symmetry, the maximum displacement occurs at the midspan (point C). Using the table in Appendix C, the required displacements are

$$(\Delta_C)_1 = (\Delta_C)_2 = \frac{-Pbx}{6EIL}\left(L^2 - b^2 - x^2\right)$$
$$= \frac{-4(12)(30)}{6EI(60)}\left(60^2 - 12^2 - 30^2\right)$$
$$= -\frac{10224 \text{ kip}\cdot\text{in}^3}{EI} = \frac{10224 \text{ kip}\cdot\text{in}^3}{EI} \downarrow$$

The maximum displacement is

$$\Delta_{max} = (\Delta_C)_1 + (\Delta_C)_2$$
$$= \frac{10224}{EI} + \frac{10224}{EI}$$
$$= \frac{20448 \text{ kip}\cdot\text{in}^3}{EI}$$
$$= \frac{20\,448}{29.0(10^3)\left(\frac{1}{12}\right)(4)(1^3)} = 2.12 \text{ in.} \downarrow \qquad \textbf{Ans}$$

12–81. The shaft for an electric motor and generator supports the weight of the rotor, the armature, and the commutator. Determine the deflection at C of the shaft due to these loadings. The bearings exert vertical forces on the shaft at A and B. EI is constant.

Elastic Curve : The elastic curves for the two uniform distributed loads are drawn separately as shown.

Method of Superposition : Using the table in Appendix C, the required displacements are

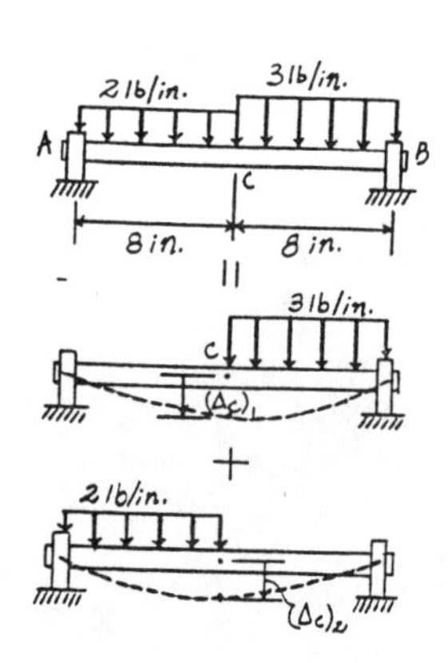

$$(\Delta_C)_1 = \frac{-5wL^4}{768EI} = \frac{-5(3)(16^4)}{768EI} = \frac{1280 \text{ lb}\cdot\text{in}^3}{EI} \downarrow$$
$$(\Delta_C)_2 = \frac{-5wL^4}{768EI} = \frac{-5(2)(16^4)}{768EI} = \frac{853.33 \text{ lb}\cdot\text{in}^3}{EI} \downarrow$$

The displacement at C is

$$\Delta_C = (\Delta_C)_1 + (\Delta_C)_2$$
$$= \frac{1280}{EI} + \frac{853.33}{EI}$$
$$= \frac{2133 \text{ lb}\cdot\text{in}^3}{EI} \downarrow \qquad \textbf{Ans}$$

12–82. The shaft for an electric motor and generator supports the weight of the rotor, the armature, and the commutator. Determine the slope at the bearings A and B due to these loadings. The bearings exert vertical forces on the shaft. EI is constant.

Elastic Curve : The elastic curves for the two uniform distributed loads are drawn separately as shown.

Method of Superposition : Using the table in Appendix C, the required slopes are

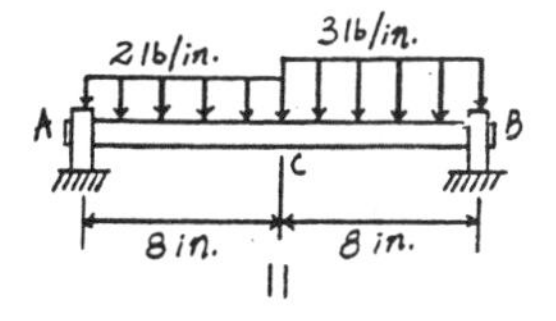

$$(\theta_A)_1 = \frac{7wL^3}{384EI} = \frac{7(3)(16^3)}{384EI} = \frac{224\ \text{lb}\cdot\text{in}^2}{EI}$$

$$(\theta_A)_2 = \frac{3wL^3}{128EI} = \frac{3(2)(16^3)}{128EI} = \frac{192\ \text{lb}\cdot\text{in}^2}{EI}$$

$$(\theta_B)_1 = \frac{3wL^3}{128EI} = \frac{3(3)(16^3)}{128EI} = \frac{288\ \text{lb}\cdot\text{in}^2}{EI}$$

$$(\theta_B)_2 = \frac{7wL^3}{384EI} = \frac{7(2)(16^3)}{384EI} = \frac{149.33\ \text{lb}\cdot\text{in}^2}{EI}$$

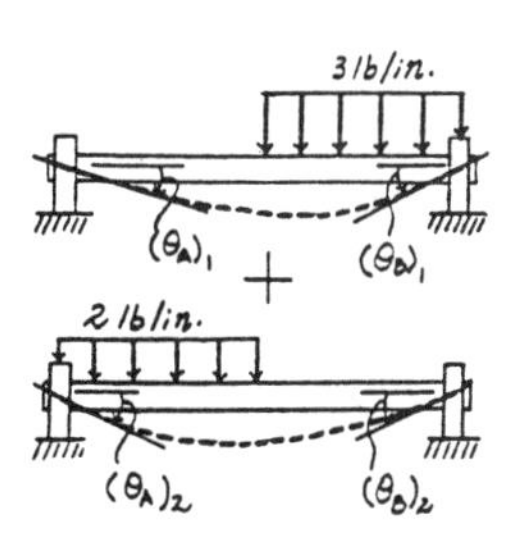

The slopes at A and B are

$$\theta_A = (\theta_A)_1 + (\theta_A)_2$$

$$= \frac{224}{EI} + \frac{192}{EI} = \frac{416\ \text{lb}\cdot\text{in}^2}{EI} \qquad \textbf{Ans}$$

$$\theta_B = (\theta_B)_1 + (\theta_B)_2$$

$$= \frac{288}{EI} + \frac{149.33}{EI} = \frac{437\ \text{lb}\cdot\text{in}^2}{EI} \qquad \textbf{Ans}$$

12–83. Determine the horizontal deflection at the end A of the bracket. Assume that the bracket is fixed supported at its base B, and neglect axial deformation of segment BC. EI is constant.

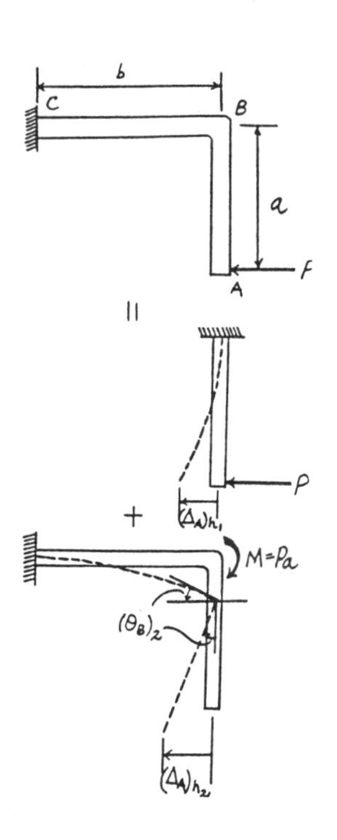

Elastic Curve : The elastic curves for the concentrated load and couple moment are drawn separately as shown.

Method of Superposition : Using the table in Appendix C, the required slopes and displacements are

$$(\Delta_A)_{h_1} = \frac{PL_{AB}^3}{3EI} = \frac{Pa^3}{3EI} \leftarrow$$

$$(\theta_B)_2 = \frac{M_0 L_{BC}}{EI} = \frac{Pab}{EI}$$

$$(\Delta_A)_{h_2} = (\theta_B)_2 L_{AB} = \frac{Pab}{EI}(a) = \frac{Pa^2 b}{EI} \leftarrow$$

The horizontal displacement at A is

$$(\Delta_A)_h = (\Delta_A)_{h_1} + (\Delta_A)_{h_2}$$

$$= \frac{Pa^3}{3EI} + \frac{Pa^2 b}{EI}$$

$$= \frac{Pa^2}{3EI}(a + 3b) \leftarrow \qquad \textbf{Ans}$$

***12–84.** Determine the slope at the end A of the bracket. Assume that the bracket is fixed supported at its base B, and neglect axial deformation of segment BC. EI is constant.

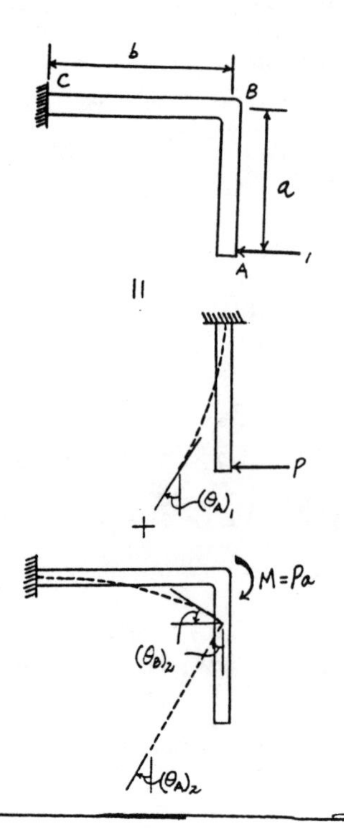

Elastic Curve : The elastic curves for the concentrated load and couple moment are drawn separately as shown.

Method of Superposition : Using the table in Appendix *C*, the required slopes are

$$(\theta_A)_1 = \frac{PL_{AB}^2}{2EI} = \frac{Pa^2}{2EI}$$

$$(\theta_A)_2 = (\theta_B)_2 = \frac{M_0 L_{BC}}{EI} = \frac{Pab}{EI}$$

The displacement at A is

$$\theta_A = (\theta_A)_1 + (\theta_A)_2 = \frac{Pa^2}{2EI} + \frac{Pab}{EI} = \frac{Pa}{2EI}(a + 2b) \qquad \textbf{Ans}$$

12–85. Determine the vertical deflection and slope at the end A of the bracket. Assume that the bracket is fixed supported at its base, and neglect the axial deformation of segment AB. EI is constant.

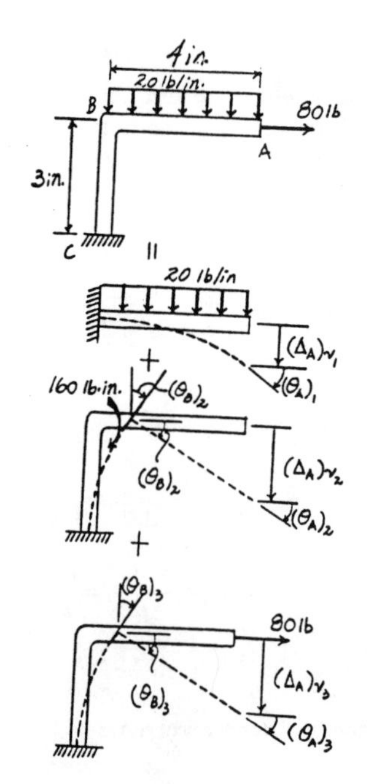

Elastic Curve : The elastic curves for the concentrated load, uniform distibuted load, and couple moment are drawn separately as shown.

Method of Superposition : Using the table in Appendix*C*, the required slopes and displacements are

$$(\theta_A)_1 = \frac{wL_{AB}^3}{6EI} = \frac{20(4^3)}{6EI} = \frac{213.33 \text{ lb}\cdot\text{in}^2}{EI}$$

$$(\theta_A)_2 = (\theta_B)_2 = \frac{M_0 L_{BC}}{EI} = \frac{160(3)}{EI} = \frac{480 \text{ lb}\cdot\text{in}^2}{EI}$$

$$(\theta_A)_3 = (\theta_B)_3 = \frac{PL_{BC}^2}{2EI} = \frac{80(3^2)}{2EI} = \frac{360 \text{ lb}\cdot\text{in}^2}{EI}$$

$$(\Delta_A)_{v_1} = \frac{wL_{AB}^4}{8EI} = \frac{20(4^4)}{8EI} = \frac{640 \text{ lb}\cdot\text{in}^3}{EI} \downarrow$$

$$(\Delta_A)_{v_2} = (\theta_B)_2 (L_{AB}) = \frac{480}{EI}(4) = \frac{1920 \text{ lb}\cdot\text{in}^3}{EI} \downarrow$$

$$(\Delta_A)_{v_3} = (\theta_B)_3 (L_{AB}) = \frac{360}{EI}(4) = \frac{1440 \text{ lb}\cdot\text{in}^3}{EI} \downarrow$$

The slope at A is

$$\theta_A = (\theta_A)_1 + (\theta_A)_2 + (\theta_A)_3 = \frac{213.33}{EI} + \frac{480}{EI} + \frac{360}{EI} = \frac{1053 \text{ lb}\cdot\text{in}^2}{EI} \qquad \textbf{Ans}$$

The vertical displacement at A is

$$(\Delta_A)_v = (\Delta_A)_{v_1} + (\Delta_A)_{v_2} (\Delta_A)_{v_3} = \frac{640}{EI} + \frac{1920}{EI} + \frac{1440}{EI} = \frac{4000 \text{ lb}\cdot\text{in}^3}{EI} \downarrow \qquad \textbf{Ans}$$

12–86. The W24 × 104 A-36 steel beam is used to support the uniform distributed load and a concentrated force which is applied at its end. If the force acts at an angle with the vertical as shown, determine the horizontal and vertical displacement at point A.

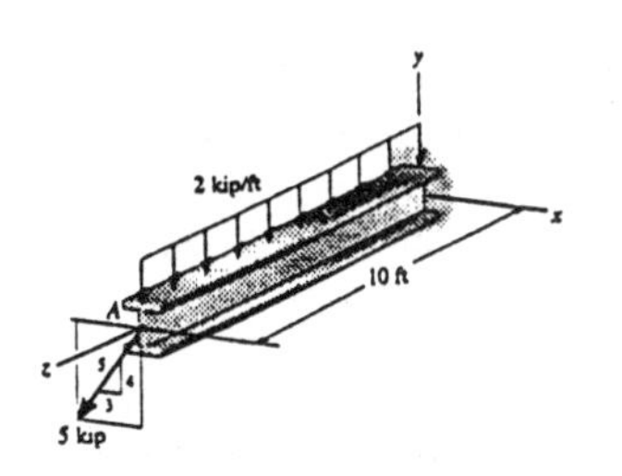

Method of Superposition : Using the table in Appendix C, the required vertical displacements are

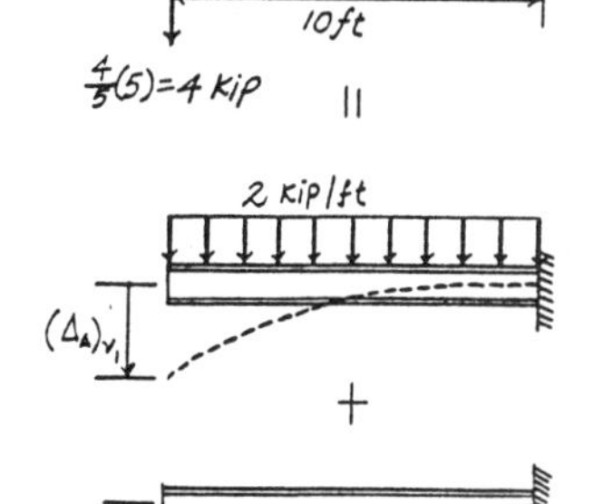

$$(\Delta_A)_{v_1} = \frac{wL^4}{8EI_x} = \frac{2(10^4)}{8EI_x} = \frac{2500\ \text{kip}\cdot\text{ft}^3}{EI_x}\ \downarrow$$

$$(\Delta_A)_{v_2} = \frac{P_yL^3}{3EI_x} = \frac{\frac{4}{5}(5)(10^3)}{3EI_x} = \frac{1333.33\ \text{kip}\cdot\text{ft}^3}{EI_x}\ \downarrow$$

The vertical displacement at A is

$$\begin{aligned}(\Delta_A)_v &= (\Delta_A)_{v_1} + (\Delta_A)_{v_2}\\ &= \frac{2500}{EI_x} + \frac{1333.33}{EI_x}\\ &= \frac{3833.33\ \text{kip}\cdot\text{ft}^3}{EI_x}\\ &= \frac{3833.33(1728)}{29.0(10^3)(3100)} = 0.0737\ \text{in.}\end{aligned}$$ **Ans**

2 kip/ft

$(\Delta_A)_{v_1}$

+

$(\Delta_A)_{v_2}$

4kip

The horizontal displacement at A is

$$(\Delta_A)_h = \frac{P_xL^3}{3EI_y} = \frac{\frac{3}{5}(5)(10^3)}{3EI_y} = \frac{1000\ \text{kip}\cdot\text{ft}^3}{EI_y} = \frac{1000(1728)}{29.0(10^3)(259)} = 0.230\ \text{in.}$$ **Ans**

12–87. The pipe assembly consists of two equal-sized pipes with flexibility stiffness EI and torsional stiffness GJ. Determine the vertical deflection at point A

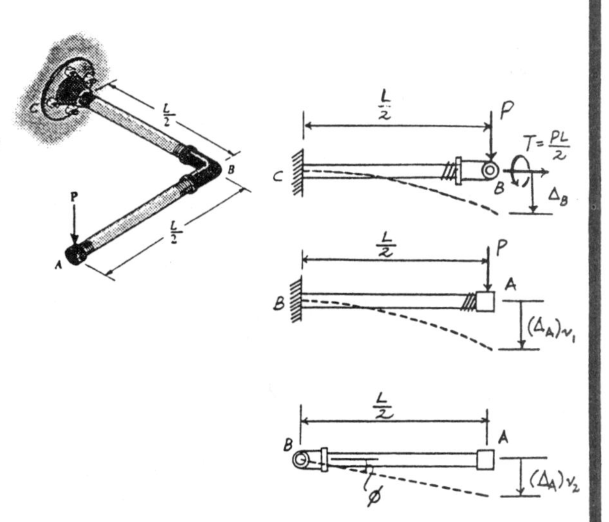

Elastic Curve : The elastic curves for segments BC and AB caused by the concentrated load and segment AB caused by torsion are drawn separately as shown.

Method of Superposition : Using the table in Appendix C, the required slopes and displacements are

$$\Delta_B = \frac{PL_{BC}^3}{3EI} = \frac{P\left(\frac{L}{2}\right)^3}{3EI} = \frac{PL^3}{24EI}\ \downarrow$$

$$(\Delta_A)_{v_1} = \frac{PL_{AB}^3}{3EI} = \frac{P\left(\frac{L}{2}\right)^3}{3EI} = \frac{PL^3}{24EI}\ \downarrow$$

The angle of twist for segment BC is

$$\phi = \frac{TL_{BC}}{JG} = \frac{\frac{PL}{2}\left(\frac{L}{2}\right)}{JG} = \frac{PL^2}{4JG}$$

Then, $$(\Delta_A)_{v_2} = \phi L_{AB} = \frac{PL^2}{4JG}\left(\frac{L}{2}\right) = \frac{PL^3}{8JG}\ \downarrow$$

The vertical displacement at A is

$$\begin{aligned}(\Delta_A)_v &= \Delta_B + (\Delta_A)_{v_1} + (\Delta_A)_{v_2}\\ &= \frac{PL^3}{24EI} + \frac{PL^3}{24EI} + \frac{PL^3}{8JG}\\ &= \frac{PL^3}{24}\left(\frac{2}{EI} + \frac{3}{JG}\right)\ \downarrow\end{aligned}$$ **Ans**

***12–88.** The assembly consists of a cantilevered beam *CB* and a simply supported beam *AB*. If each beam is made of A-36 steel and has a moment of inertia about its principal axis of $I_x = 118\ \text{in}^4$, determine the deflection at the center *D* of beam *BA*.

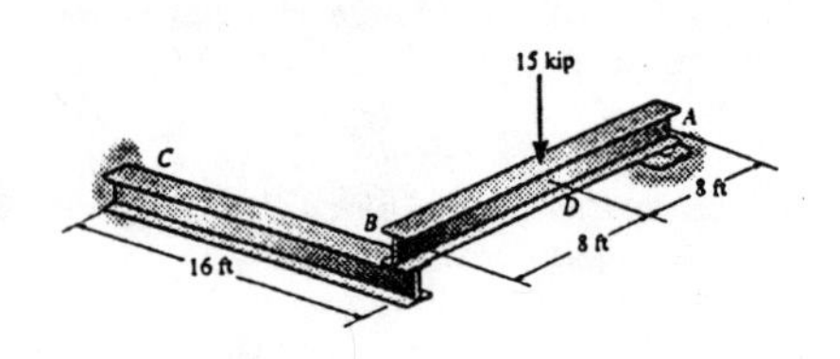

Method of Superposition : Using the table in Appendix C, the required slopes and displacements are

$$\Delta_B = \frac{PL_{BC}^3}{3EI} = \frac{7.50(16^3)}{3EI} = \frac{10240\ \text{kip}\cdot\text{ft}^3}{EI}\ \downarrow$$

$$(\Delta_D)_1 = \frac{PL_{AB}^3}{48EI} = \frac{15(16^3)}{48EI} = \frac{1280\ \text{kip}\cdot\text{ft}^3}{EI}\ \downarrow$$

$$(\Delta_D)_2 = \frac{1}{2}\Delta_B = \frac{5120\ \text{kip}\cdot\text{ft}^3}{EI}\ \downarrow$$

The vertical displacement at *A* is

$$\begin{aligned}\Delta_D &= (\Delta_D)_1 + (\Delta_D)_2\\ &= \frac{1280}{EI} + \frac{5120}{EI}\\ &= \frac{6400\ \text{kip}\cdot\text{ft}^3}{EI}\\ &= \frac{6400(1728)}{29.0(10^3)(118)} = 3.23\ \text{in.}\end{aligned}$$

Ans

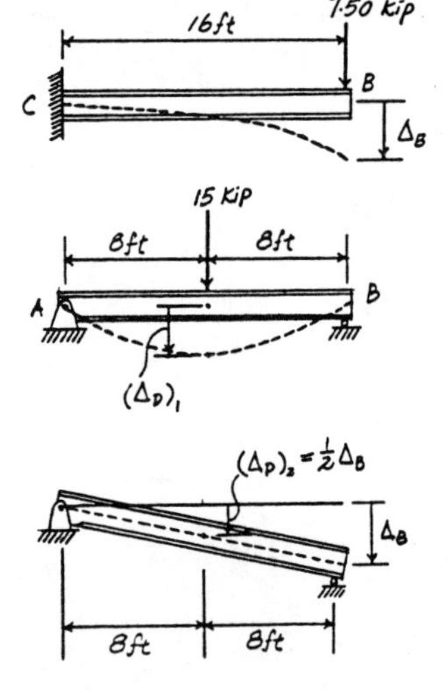

12–89. Determine the reactions at the supports A and B, then draw the shear and moment diagrams. EI is constant.

Support Reactions : FBD(a).

$$\xrightarrow{+} \Sigma F_x = 0; \qquad A_x = 0 \qquad \textbf{Ans}$$

$$+\uparrow \Sigma F_y = 0; \qquad A_y - B_y = 0 \qquad [1]$$

$$\circlearrowleft + \Sigma M_B = 0; \qquad M_0 - A_y L + M_B = 0 \qquad [2]$$

Moment Function : FBD(b)

$$\circlearrowleft + \Sigma M_{NA} = 0; \qquad M(x) + M_0 - A_y x = 0$$

$$M(x) = A_y x - M_0$$

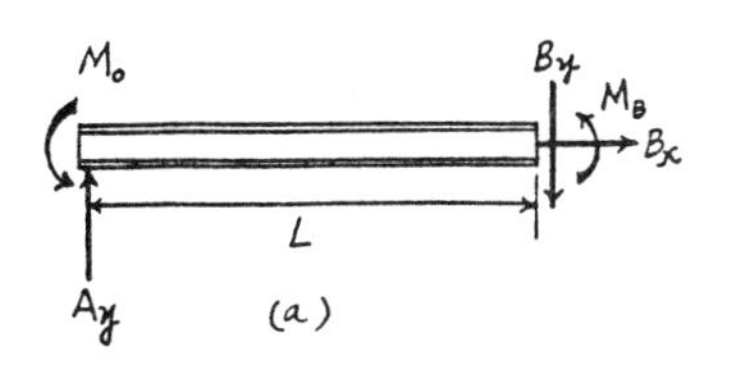

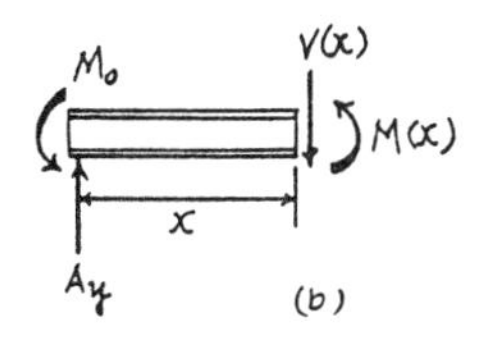

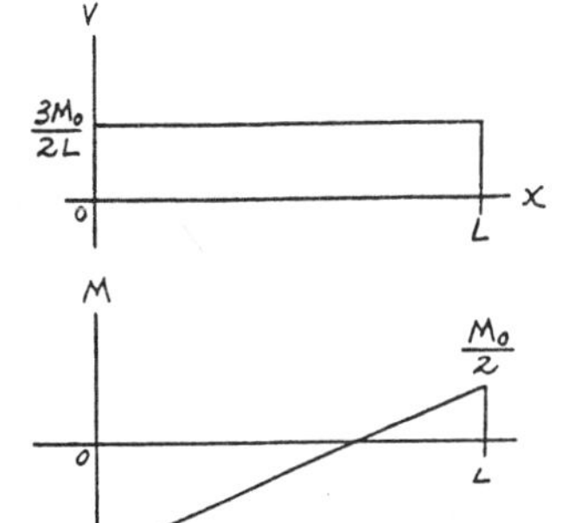

Slope and Elastic Curve :

$$EI\frac{d^2 v}{dx^2} = M(x)$$

$$EI\frac{d^2 v}{dx^2} = A_y x - M_0$$

$$EI\frac{dv}{dx} = \frac{A_y}{2}x^2 - M_0 x + C_1 \qquad [3]$$

$$EI\,v = \frac{A_y}{6}x^3 - \frac{M_0}{2}x^2 + C_1 x + C_2 \qquad [4]$$

Boundary Conditions :

At $x = 0$, $v = 0$. From Eq. [4], $\qquad C_2 = 0$

At $x = L$, $\frac{dv}{dx} = 0$. From Eq. [3],

$$0 = \frac{A_y L^2}{2} - M_0 L + C_1 \qquad [5]$$

At $x = L$, $v = 0$. From Eq. [4],

$$0 = \frac{A_y L^3}{6} - \frac{M_0 L^2}{2} + C_1 L \qquad [6]$$

Solving Eqs. [5] and [6] yields,

$$A_y = \frac{3M_0}{2L} \qquad \textbf{Ans}$$

$$C_1 = \frac{M_0 L}{4}$$

Substituting A_y into Eqs. [1] and [2] yields :

$$B_y = \frac{3M_0}{2L} \qquad M_B = \frac{M_0}{2} \qquad \textbf{Ans}$$

12–90. Determine the reactions at the supports, then draw the shear and moment diagram. EI is constant.

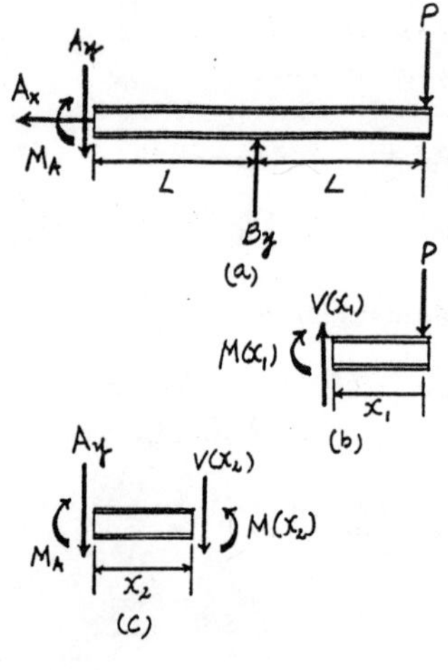

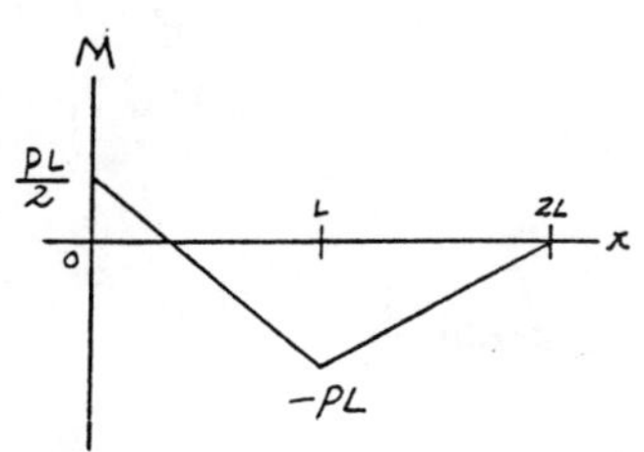

Support Reactions : FBD(a).

$\xrightarrow{+} \Sigma F_x = 0;\quad A_x = 0$ **Ans**

$+\uparrow \Sigma F_y = 0;\quad B_y - A_y - P = 0$ [1]

$\circlearrowleft + \Sigma M_B = 0;\quad A_y L - M_A - PL = 0$ [2]

Moment Functions : FBD(b) and (c).

$$M(x_1) = -Px_1$$
$$M(x_2) = M_A - A_y x_2$$

Slope and Elastic Curve :

$$EI\frac{d^2 v}{dx^2} = M(x)$$

For $M(x_1) = -Px_1$,

$$EI\frac{d^2 v_1}{dx_1^2} = -Px_1$$

$$EI\frac{dv_1}{dx_1} = -\frac{P}{2}x_1^2 + C_1 \qquad [3]$$

$$EI\, v_1 = -\frac{P}{6}x_1^3 + C_1 x_1 + C_2 \qquad [4]$$

For $M(x_2) = M_A - A_y x_2$,

$$EI\frac{d^2 v_2}{dx_2^2} = M_A - A_y x_2$$

$$EI\frac{dv_2}{dx_2} = M_A x_2 - \frac{A_y}{2}x_2^2 + C_3 \qquad [5]$$

$$EI\, v_2 = \frac{M_A}{2}x_2^2 - \frac{A_y}{6}x_2^3 + C_3 x_2 + C_4 \qquad [6]$$

Boundary Conditions :

$v_2 = 0$ at $x_2 = 0$. From Eq.[6], $C_4 = 0$

$\frac{dv_2}{dx_2} = 0$ at $x_2 = 0$. From Eq.[5], $C_3 = 0$

$v_2 = 0$ at $x_2 = L$. From Eq. [6],

$$0 = \frac{M_A L^2}{2} - \frac{A_y L^3}{6} \qquad [7]$$

Solving Eqs.[2] and [7] yields,

$$M_A = \frac{PL}{2} \qquad A_y = \frac{3P}{2} \qquad \textbf{Ans}$$

Substituting the value of A_y into Eq.[1],

$$B_y = \frac{5P}{2} \qquad \textbf{Ans}$$

Note : The other boundary and continuity conditions can be used to determine the constants C_1 and C_2 which are not needed here.

12–91. Determine the reactions at the supports A, B, and C; then draw the shear and moment diagrams. EI is constant.

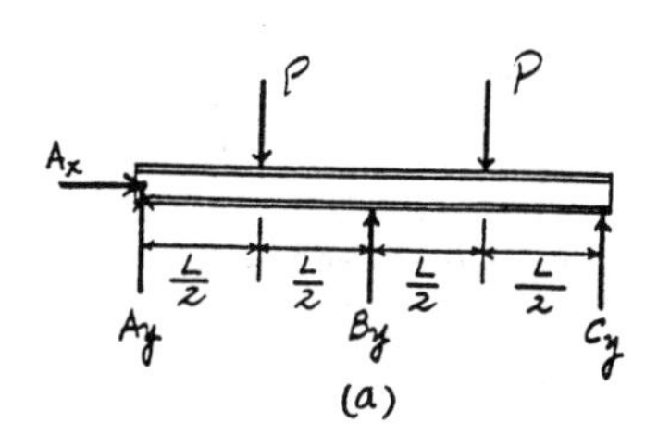

Support Reactions : FBD(a).

$$\xrightarrow{+} \Sigma F_x = 0; \qquad A_x = 0 \qquad \textbf{Ans}$$

$$+\uparrow \Sigma F_y = 0; \qquad A_y + B_y + C_y - 2P = 0 \qquad [1]$$

$$\circlearrowleft + \Sigma M_A = 0; \qquad B_y L + C_y(2L) - P\left(\frac{L}{2}\right) - P\left(\frac{3L}{2}\right) = 0 \qquad [2]$$

Moment Functions : FBD(b) and (c).

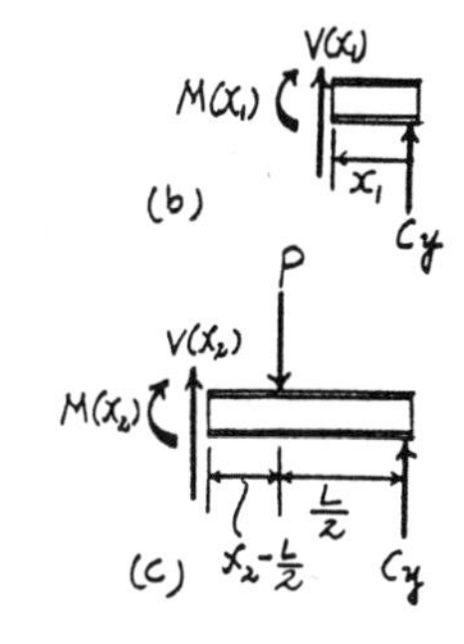

$$M(x_1) = C_y x_1$$

$$M(x_2) = C_y x_2 - P x_2 + \frac{PL}{2}$$

Slope and Elastic Curve :

$$EI\frac{d^2 v}{dx^2} = M(x)$$

For $M(x_1) = C_y x_1$,

$$EI\frac{d^2 v_1}{dx_1^2} = C_y x_1$$

$$EI\frac{dv_1}{dx_1} = \frac{C_y}{2}x_1^2 + C_1 \qquad [3]$$

$$EI\, v_1 = \frac{C_y}{6}x_1^3 + C_1 x_1 + C_2 \qquad [4]$$

For $M(x_2) = C_y x_2 - P x_2 + \frac{PL}{2}$,

$$EI\frac{d^2 v_2}{dx_2^2} = C_y x_2 - P x_2 + \frac{PL}{2}$$

$$EI\frac{dv_2}{dx_2} = \frac{C_y}{2}x_2^2 - \frac{P}{2}x_2^2 + \frac{PL}{2}x_2 + C_3 \qquad [5]$$

$$EI\, v_2 = \frac{C_y}{6}x_2^3 - \frac{P}{6}x_2^3 + \frac{PL}{4}x_2^2 + C_3 x_2 + C_4 \qquad [6]$$

Boundary Conditions :

$v_1 = 0$ at $x_1 = 0$. From Eq. [4], $\quad C_2 = 0$

Due to symmetry, $\frac{dv_2}{dx_2} = 0$ at $x_2 = L$. From Eq. [5],

$$0 = \frac{C_y L^2}{2} - \frac{PL^2}{2} + \frac{PL^2}{2} + C_3 \qquad C_3 = -\frac{C_y L^2}{2}$$

$v_2 = 0$ at $x_2 = L$. From Eq. [6],

$$0 = \frac{C_y L^3}{6} - \frac{PL^3}{6} + \frac{PL^3}{4} + \left(-\frac{C_y L^2}{2}\right)L + C_4$$

$$C_4 = \frac{C_y L^3}{3} - \frac{PL^3}{12}$$

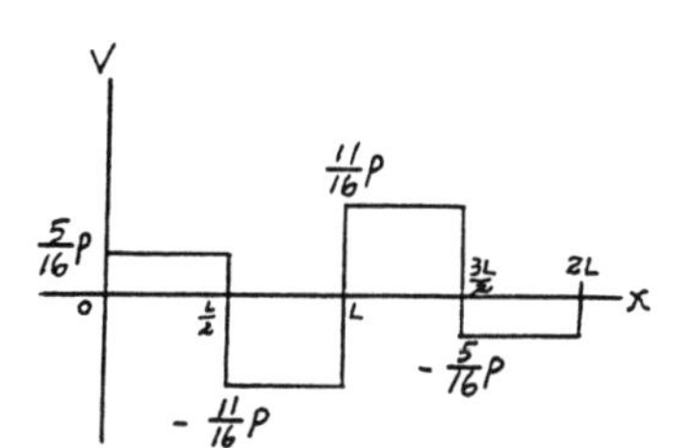

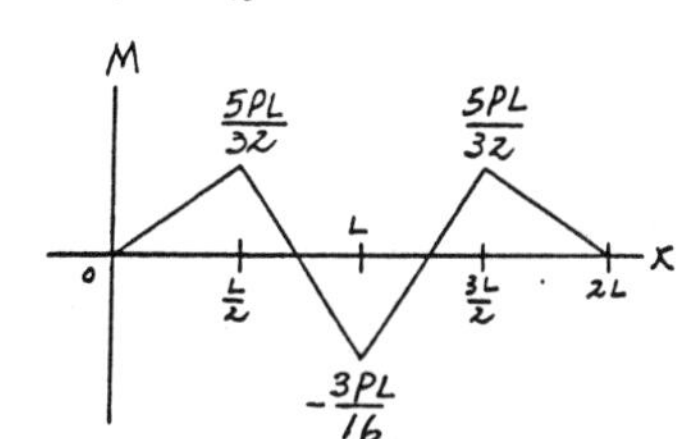

Continuity Conditions :

At $x_1 = x_2 = \frac{L}{2}$, $\quad \frac{dv_1}{dx_1} = \frac{dv_2}{dx_2}$. From Eqs. [3] and [5],

$$\frac{C_y}{2}\left(\frac{L}{2}\right)^2 + C_1 = \frac{C_y}{2}\left(\frac{L}{2}\right)^2 - \frac{P}{2}\left(\frac{L}{2}\right)^2 + \frac{PL}{2}\left(\frac{L}{2}\right) - \frac{C_y L^2}{2}$$

$$C_1 = \frac{PL^2}{8} - \frac{C_y L^2}{2}$$

At $x_1 = x_2 = \frac{L}{2}$, $\quad v_1 = v_2$. From Eqs. [4] and [6],

$$\frac{C_y}{6}\left(\frac{L}{2}\right)^3 + \left(\frac{PL^2}{8} - \frac{C_y L^2}{2}\right)\left(\frac{L}{2}\right)$$

$$= \frac{C_y}{6}\left(\frac{L}{2}\right)^3 - \frac{P}{6}\left(\frac{L}{2}\right)^3 + \frac{PL}{4}\left(\frac{L}{2}\right)^2 + \left(-\frac{C_y L^2}{2}\right)\left(\frac{L}{2}\right) + \frac{C_y L^3}{3} - \frac{PL^3}{12}$$

$$C_y = \frac{5}{16}P \qquad \textbf{Ans}$$

Substituting C_y into Eqs. [1] and [2],

$$B_y = \frac{11}{8}P \qquad A_y = \frac{5}{16}P \qquad \textbf{Ans}$$

***12–92.** Determine the moment reactions at the supports A and B. EI is constant.

Support Reactions : FBD(a).

$$\circlearrowleft + \Sigma M_B = 0; \quad Pa + P(L-a) + M_A - A_y L - M_B = 0$$

$$PL + M_A - A_y L - M_B = 0 \qquad [1]$$

Moment Functions : FBD(b) and (c).

$$M(x_1) = A_y x_1 - M_A$$
$$M(x_2) = A_y x_2 - P x_2 + Pa - M_A$$

Slope and Elastic Curve :

$$EI\frac{d^2 v}{dx^2} = M(x)$$

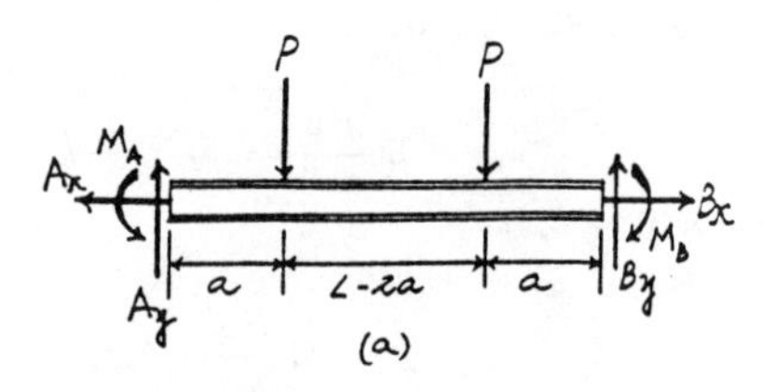

For $M(x_1) = A_y x_1 - M_A$,

$$EI\frac{d^2 v_1}{dx_1^2} = A_y x_1 - M_A$$

$$EI\frac{dv_1}{dx_1} = \frac{A_y}{2}x_1^2 - M_A x_1 + C_1 \qquad [2]$$

$$EI\, v_1 = \frac{A_y}{6}x_1^3 - \frac{M_A}{2}x_1^2 + C_1 x_1 + C_2 \qquad [3]$$

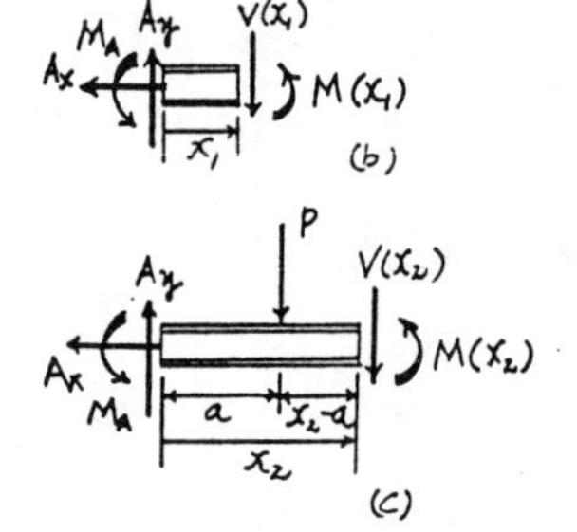

For $M(x_2) = A_y x_2 - P x_2 + Pa - M_A$,

$$EI\frac{d^2 v_2}{dx_2^2} = A_y x_2 - P x_2 + Pa - M_A$$

$$EI\frac{dv_2}{dx_2} = \frac{A_y}{2}x_2^2 - \frac{P}{2}x_2^2 + Pax_2 - M_A x_2 + C_3 \qquad [4]$$

$$EI\, v_2 = \frac{A_y}{6}x_2^3 - \frac{P}{6}x_2^3 + \frac{Pa}{2}x_2^2 - \frac{M_A}{2}x_2^2 + C_3 x_2 + C_4 \qquad [5]$$

Boundary Conditions :

$\frac{dv_1}{dx_1} = 0$ at $x_1 = 0$. From Eq.[2], $C_1 = 0$

$v_1 = 0$ at $x_1 = 0$. From Eq.[3], $C_2 = 0$

Due to symmetry, $\frac{dv_2}{dx_2} = 0$ at $x_2 = \frac{L}{2}$. From Eq.[4],

$$0 = \frac{A_y}{2}\left(\frac{L}{2}\right)^2 - \frac{P}{2}\left(\frac{L}{2}\right)^2 + Pa\left(\frac{L}{2}\right) - M_A\left(\frac{L}{2}\right) + C_3$$

$$C_3 = -\frac{A_y L^2}{8} + \frac{PL^2}{8} - \frac{PaL}{2} + \frac{M_A L}{2}$$

Due to symmetry, $\frac{dv_1}{dx_1} = -\frac{dv_2}{dx_2}$ at $x_1 = a$ and $x_2 = L - a$. From Eqs.[2] and [4],

$$\frac{A_y a^2}{2} - M_A a = -\frac{A_y}{2}(L-a)^2 + \frac{P}{2}(L-a)^2 - Pa(L-a)$$
$$+ M_A(L-a) + \frac{A_y L^2}{8} - \frac{PL^2}{8} + \frac{PaL}{2} - \frac{M_A L}{2}$$

$$-A_y a^2 - \frac{3A_y L^2}{8} + A_y aL + \frac{3PL^2}{8} - \frac{3PaL}{2} + \frac{3Pa^2}{2} + \frac{M_A}{2} = 0 \qquad [6]$$

Continuity Conditions :

At $x_1 = x_2 = a$, $\frac{dv_1}{dx_1} = \frac{dv_2}{dx_2}$. From Eqs.[2] and [4],

$$\frac{A_y a^2}{2} - M_A a$$
$$= \frac{A_y a^2}{2} - \frac{Pa^2}{2} + Pa^2 - M_A a - \frac{A_y L^2}{8} + \frac{PL^2}{8} - \frac{PaL}{2} + \frac{M_A L}{2}$$

$$\frac{Pa^2}{2} - \frac{A_y L^2}{8} + \frac{PL^2}{8} - \frac{PaL}{2} + \frac{M_A L}{2} = 0 \qquad [7]$$

Solving Eqs.[6] and [7] yields,

$$M_A = \frac{Pa}{L}(L-a) \qquad \textbf{Ans}$$
$$A_y = P$$

Substitute the value of M_A and A_y obtained into Eqs.[1],

$$M_B = \frac{Pa}{L}(L-a) \qquad \textbf{Ans}$$

12–93. Determine the value of a for which the maximum positive moment has the same magnitude as the maximum negative moment. EI is constant.

$$+\uparrow \Sigma F_y = 0; \quad A_y + B_y - P = 0 \qquad [1]$$

$$\circlearrowleft + \Sigma M_A = 0; \quad M_A + B_y L - Pa = 0 \qquad [2]$$

Moment Functions : FBD(b) and (c).

$$M(x_1) = B_y x_1$$

$$M(x_2) = B_y x_2 - Px_2 + PL - Pa$$

Slope and Elastic Curve :

$$EI\frac{d^2 v}{dx^2} = M(x)$$

For $M(x_1) = B_y x_1$,

$$EI\frac{d^2 v_1}{dx_1^2} = B_y x_1$$

$$EI\frac{dv_1}{dx_1} = \frac{B_y}{2}x_1^2 + C_1 \qquad [3]$$

$$EI\,v_1 = \frac{B_y}{6}x_1^3 + C_1 x_1 + C_2 \qquad [4]$$

For $M(x_2) = B_y x_2 - Px_2 + PL - Pa$,

$$EI\frac{d^2 v_2}{dx_2^2} = B_y x_2 - Px_2 + PL - Pa$$

$$EI\frac{dv_2}{dx_2} = \frac{B_y}{2}x_2^2 - \frac{P}{2}x_2^2 + PLx_2 - Pax_2 + C_3 \qquad [5]$$

$$EI\,v_2 = \frac{B_y}{6}x_2^3 - \frac{P}{6}x_2^3 + \frac{PL}{2}x_2^2 - \frac{Pa}{2}x_2^2 + C_3 x_2 + C_4 \qquad [6]$$

Boundary Conditions :

$v_1 = 0$ at $x_1 = 0$. From Eq.[4], $C_2 = 0$

$\dfrac{dv_2}{dx_2} = 0$ at $x_2 = L$. From Eq.[5]

$$0 = \frac{B_y L^2}{2} - \frac{PL^2}{2} + PL^2 - PaL + C_3$$

$$C_3 = -\frac{B_y L^2}{2} - \frac{PL^2}{2} + PaL$$

$v_2 = 0$ at $x_2 = L$. From Eq. [6],

$$0 = \frac{B_y L^3}{6} - \frac{PL^3}{6} + \frac{PL^3}{2} - \frac{PaL^2}{2} + \left(-\frac{B_y L^2}{2} - \frac{PL^2}{2} + PaL\right)L + C_4$$

$$C_4 = \frac{B_y L^3}{3} + \frac{PL^3}{6} - \frac{PaL^2}{2}$$

Continuity Conditions :

At $x_1 = x_2 = L - a$, $\dfrac{dv_1}{dx_1} = \dfrac{dv_2}{dx_2}$. From Eqs.[3] and [5],

$$\frac{B_y}{2}(L-a)^2 + C_1 = \frac{B_y}{2}(L-a)^2 - \frac{P}{2}(L-a)^2 + PL(L-a) - Pa(L-a) + \left(-\frac{B_y L^2}{2} - \frac{PL^2}{2} + PaL\right)$$

$$C_1 = \frac{Pa^2}{2} - \frac{B_y L^2}{2}$$

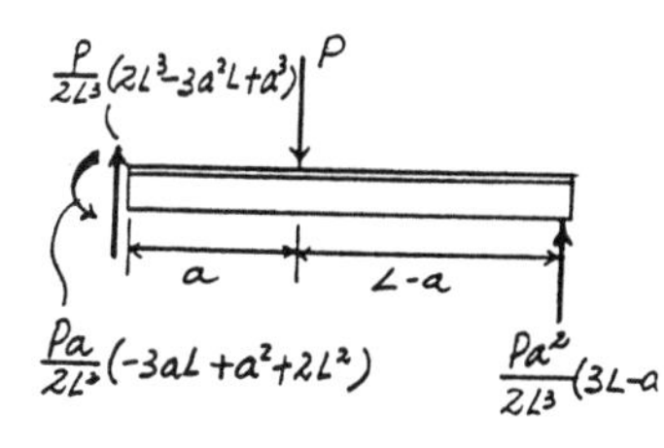

At $x_1 = x_2 = L - a$, $v_1 = v_2$. From Eqs.[4] and [6],

$$\frac{B_y}{6}(L-a)^3 + \left(\frac{Pa^2}{2} - \frac{B_y L^2}{2}\right)(L-a)$$

$$= \frac{B_y}{6}(L-a)^3 - \frac{P}{6}(L-a)^3 + \frac{PL}{2}(L-a)^2 - \frac{Pa}{2}(L-a)^2 + \left(-\frac{B_y L^2}{2} - \frac{PL^2}{2} + PaL\right)(L-a) + \frac{B_y L^3}{3} + \frac{PL^3}{6} - \frac{PaL^2}{2}$$

$$\frac{Pa^3}{6} - \frac{Pa^2 L}{2} + \frac{B_y L^3}{3} = 0$$

$$B_y = \frac{3Pa^2}{2L^2} - \frac{Pa^3}{2L^3} = \frac{Pa^2}{2L^3}(3L-a)$$

Substituting B_y into Eqs.[1] and [2], we have

$$A_y = \frac{P}{2L^3}\left(2L^3 - 3a^2 L + a^3\right)$$

$$M_A = \frac{Pa}{2L^2}\left(-3aL + a^2 + 2L^2\right)$$

Require $|M_{\max(+)}| = |M_{\max(-)}|$. From the moment diagram,

$$\frac{Pa^2}{2L^3}(3L-a)(L-a) = \frac{Pa}{2L^2}\left(-3aL + a^2 + 2L^2\right)$$

$$a^2 - 4aL + 2L^2 = 0$$

$$a = \left(2 - \sqrt{2}\right)L \qquad \textbf{Ans}$$

12–94. Determine the reactions at the supports, then draw the shear and moment dagrams. EI is constant.

Support Reactions : FBD(a).

$\xrightarrow{+} \Sigma F_x = 0;$ $\quad A_x = 0$ **Ans**

$+\uparrow \Sigma F_y = 0;$ $\quad A_y + B_y + C_y - w_0 L = 0$ [1]

$\circlearrowleft + \Sigma M_A = 0;$ $\quad B_y L + C_y(2L) - w_0 L(L) = 0$ [2]

Moment Function : FBD(b).

$\circlearrowleft + \Sigma M_{NA} = 0;$ $\quad -M(x) - \frac{1}{2}\left(\frac{w_0}{L}x\right)x\left(\frac{x}{3}\right) + C_y x = 0$

$$M(x) = C_y x - \frac{w_0}{6L}x^3$$

Slope and Elastic Curve :

$$EI\frac{d^2 v}{dx^2} = M(x)$$

$$EI\frac{d^2 v}{dx^2} = C_y x - \frac{w_0}{6L}x^3$$

$$EI\frac{dv}{dx} = \frac{C_y}{2}x^2 - \frac{w_0}{24L}x^4 + C_1 \quad [3]$$

$$EI\, v = \frac{C_y}{6}x^3 - \frac{w_0}{120L}x^5 + C_1 x + C_2 \quad [4]$$

Boundary Conditions :

At $x = 0$, $v = 0$. From Eq. [4], $\quad C_2 = 0$

Due to symmetry, $\frac{dv}{dx} = 0$ at $x = L$. From Eq. [3],

$$0 = \frac{C_y L^2}{2} - \frac{w_0 L^3}{24} + C_1$$

$$C_1 = -\frac{C_y L^2}{2} + \frac{w_0 L^3}{24}$$

At $x = L$, $v = 0$. From Eq. [4],

$$0 = \frac{C_y L^3}{6} - \frac{w_0 L^4}{120} + \left(-\frac{C_y L^2}{2} + \frac{w_0 L^3}{24}\right)L$$

$$C_y = \frac{w_0 L}{10} \quad \textbf{Ans}$$

Substituting C_y into Eqs. [1] and [2] yields :

$$B_y = \frac{4w_0 L}{5} \qquad A_y = \frac{w_0 L}{10} \quad \textbf{Ans}$$

Shear and Moment diagrams : The maximum span (positive) moment occurs when the shear force $V = 0$. From FBD (c),

$+\uparrow \Sigma F_y = 0;$ $\quad \frac{w_0 L}{10} - \frac{1}{2}\left(\frac{w_0}{L}x\right)x = 0$

$$x = \frac{\sqrt{5}}{5}L$$

$+\Sigma M_{NA} = 0;$ $\quad M + \frac{1}{2}\left(\frac{w_0}{L}x\right)(x)\left(\frac{x}{3}\right) - \frac{w_0 L}{10}(x) = 0$

$$M = \frac{w_0 L}{10}x - \frac{w_0}{6L}x^3$$

At $x = \frac{\sqrt{5}}{5}L$, $\quad M = \frac{\sqrt{5}w_0 L^2}{75}$

At $x = L$, $\quad M = -\frac{w_0 L^2}{15}$

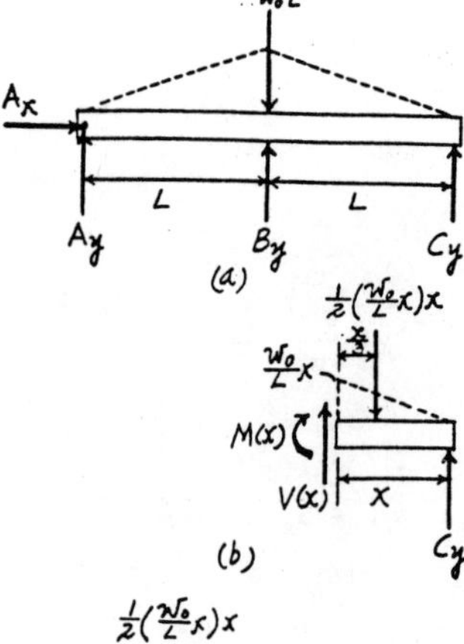

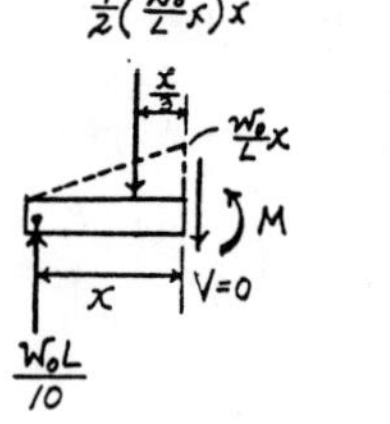

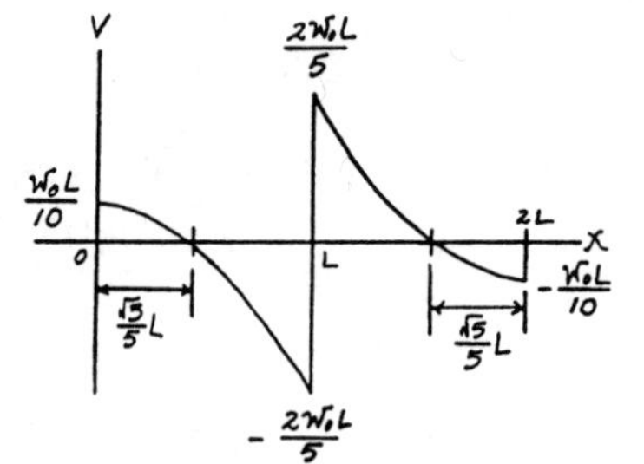

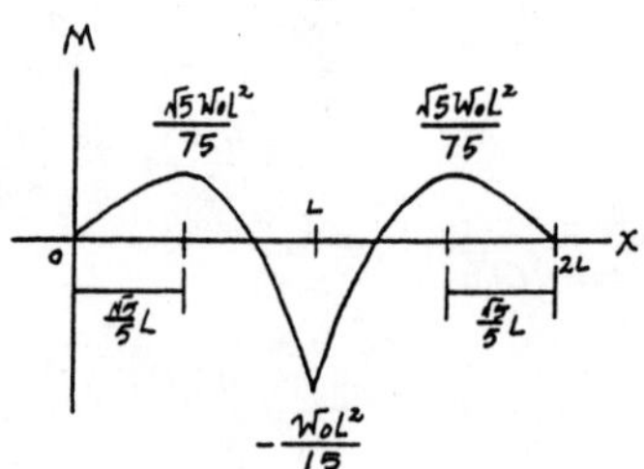

12–95. Determine the reactions at the supports A and B. EI is constant.

Support Reactions : FBD(a).

$$\xrightarrow{+} \Sigma F_x = 0; \qquad A_x = 0 \qquad \textbf{Ans}$$

$$+\uparrow \Sigma F_y = 0; \qquad A_y + B_y - \frac{w_0 L}{2} = 0 \qquad [1]$$

$$\curvearrowleft + \Sigma M_A = 0; \qquad B_y L + M_A - \frac{w_0 L}{2}\left(\frac{L}{3}\right) = 0 \qquad [2]$$

Moment Function : FBD(b).

$$\curvearrowleft + \Sigma M_{NA} = 0; \qquad -M(x) - \frac{1}{2}\left(\frac{w_0}{L}x\right)x\left(\frac{x}{3}\right) + B_y x = 0$$

$$M(x) = B_y x - \frac{w_0}{6L}x^3$$

Slope and Elastic Curve :

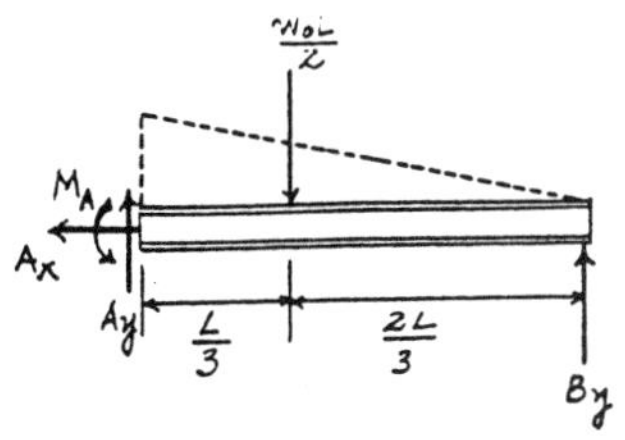

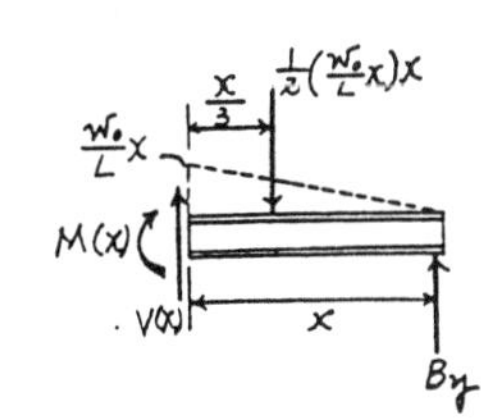

$$EI\frac{d^2 v}{dx^2} = M(x)$$

$$EI\frac{d^2 v}{dx^2} = B_y x - \frac{w_0}{6L}x^3$$

$$EI\frac{dv}{dx} = \frac{B_y}{2}x^2 - \frac{w_0}{24L}x^4 + C_1 \qquad [3]$$

$$EI\,v = \frac{B_y}{6}x^3 - \frac{w_0}{120L}x^5 + C_1 x + C_2 \qquad [4]$$

Boundary Conditions :

At $x = 0$, $v = 0$. From Eq.[4], $\qquad C_2 = 0$

At $x = L$, $\frac{dv}{dx} = 0$. From Eq. [3],

$$0 = \frac{B_y L^2}{2} - \frac{w_0 L^3}{24} + C_1$$

$$C_1 = -\frac{B_y L^2}{2} + \frac{w_0 L^3}{24}$$

At $x = L$, $v = 0$. From Eq. [4],

$$0 = \frac{B_y L^3}{6} - \frac{w_0 L^4}{120} + \left(-\frac{B_y L^2}{2} + \frac{w_0 L^3}{24}\right)L$$

$$B_y = \frac{w_0 L}{10} \qquad \textbf{Ans}$$

Substituting B_y into Eq. [1] and [2] yields,

$$A_y = \frac{2w_0 L}{5} \qquad M_A = \frac{w_0 L^2}{15} \qquad \textbf{Ans}$$

***12–96.** Determine the moment reactions at the supports A and B, then draw the shear and moment diagrams. EI is constant.

Support Reactions : FBD(a).

$$+\uparrow \Sigma F_y = 0; \quad A_y + B_y - wL = 0 \qquad [1]$$

$$\circlearrowleft + \Sigma M_B = 0; \quad -A_y L - M_B + M_A + wL\left(\frac{L}{2}\right) = 0 \qquad [2]$$

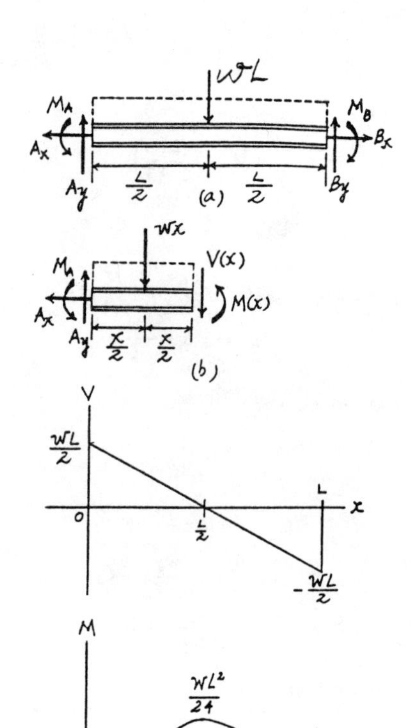

Moment Function : FBD(b).

$$\circlearrowleft + \Sigma M_{NA} = 0; \quad M(x) + wx\left(\frac{x}{2}\right) + M_A - A_y x = 0$$

$$M(x) = A_y x - \frac{w}{2}x^2 - M_A$$

Slope and Elastic Curve :

$$EI\frac{d^2 v}{dx^2} = M(x)$$

$$EI\frac{d^2 v}{dx^2} = A_y x - \frac{w}{2}x^2 - M_A$$

$$EI\frac{dv}{dx} = \frac{A_y}{2}x^2 - \frac{w}{6}x^3 - M_A x + C_1 \qquad [3]$$

$$EI\,v = \frac{A_y}{6}x^3 - \frac{w}{24}x^4 - \frac{M_A}{2}x^2 + C_1 x + C_2 \qquad [4]$$

Boundary Conditions :

At $x = 0$, $\frac{dv}{dx} = 0$ From Eq.[3], $C_1 = 0$

At $x = 0$, $v = 0$. From Eq.[4], $C_2 = 0$

At $x = L$, $\frac{dv}{dx} = 0$. From Eq. [3],

$$0 = \frac{A_y L^2}{2} - \frac{wL^3}{6} - M_A L \qquad [5]$$

At $x = L$, $v = 0$. From Eq. [4],

$$0 = \frac{A_y L^3}{6} - \frac{wL^4}{24} - \frac{M_A L^2}{2} \qquad [6]$$

Solving Eqs.[5] and [6] yields,

$$A_y = \frac{wL}{2}$$

$$M_A = \frac{wL^2}{12} \qquad \textbf{Ans}$$

Substituting A_y and M_A into Eqs. [1] and [2] yields,

$$B_y = \frac{wL}{2}$$

$$M_B = \frac{wL^2}{12} \qquad \textbf{Ans}$$

12–97. Determine the moment reactions at the supports A and B. EI is constant.

Support Reactions : FBD(a).

$$\circlearrowleft + \Sigma M_A = 0; \quad B_y L + M_A - M_B - \frac{w_0 L}{2}\left(\frac{L}{3}\right) = 0 \qquad [1]$$

Moment Function : FBD(b)

$$\circlearrowleft + \Sigma M_{NA} = 0; \quad -M(x) - \frac{1}{2}\left(\frac{w_0}{L}x\right)(x)\left(\frac{x}{3}\right) - M_B + B_y x = 0$$

$$M(x) = B_y x - \frac{w_0}{6L}x^3 - M_B$$

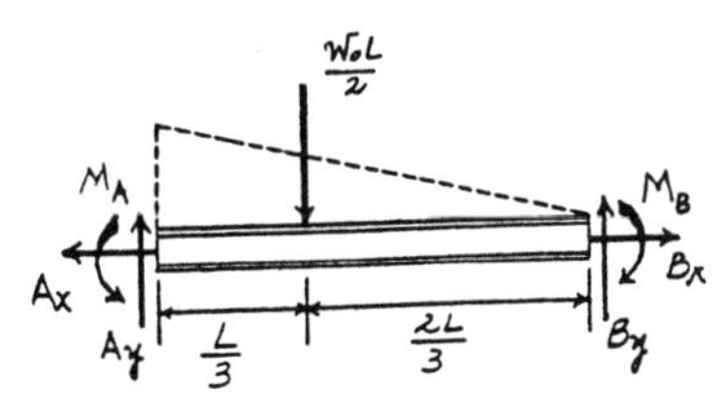

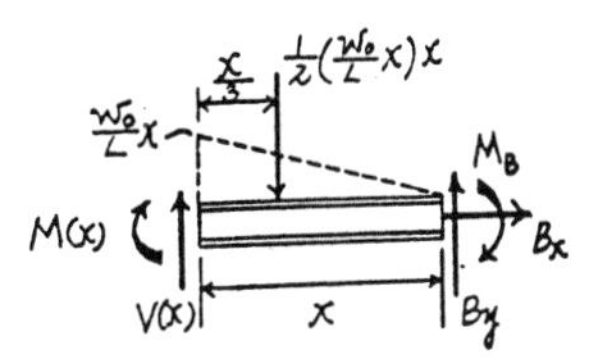

Slope and Elastic Curve :

$$EI\frac{d^2 v}{dx^2} = M(x)$$

$$EI\frac{d^2 v}{dx^2} = B_y x - \frac{w_0}{6L}x^3 - M_B$$

$$EI\frac{dv}{dx} = \frac{B_y}{2}x^2 - \frac{w_0}{24L}x^4 - M_B x + C_1 \qquad [2]$$

$$EI\, v = \frac{B_y}{6}x^3 - \frac{w_0}{120L}x^5 - \frac{M_B}{2}x^2 + C_1 x + C_2 \qquad [3]$$

Boundary Conditions :

At $x = 0$, $\frac{dv}{dx} = 0$ From Eq. [2], $C_1 = 0$

At $x = 0$, $v = 0$. From Eq. [3], $C_2 = 0$

At $x = L$, $\frac{dv}{dx} = 0$. From Eq. [2],

$$0 = \frac{B_y L^2}{2} - \frac{w_0 L^3}{24} - M_B L \qquad [4]$$

At $x = L$, $v = 0$. From Eq. [4],

$$0 = \frac{B_y L^3}{6} - \frac{w_0 L^4}{120} - \frac{M_B L^2}{2} \qquad [5]$$

Solving Eqs. [4] and [5] yields,

$$B_y = \frac{3w_0 L}{20}$$

$$M_B = \frac{w_0 L^2}{30} \qquad \textbf{Ans}$$

Substituting the value of B_y and M_B into Eq. [1] yields,

$$M_A = \frac{w_0 L^2}{20} \qquad \textbf{Ans}$$

12–98. Determine the reactions at the supports A and B, then draw the shear and moment diagrams. EI is constant.

Support Reaction **:** FBD(a).

$$\xrightarrow{+} \Sigma F_x = 0; \qquad A_x = 0 \qquad \textbf{Ans}$$

$$+\uparrow \Sigma F_y = 0; \qquad B_y - A_y = 0 \qquad [1]$$

$$\circlearrowleft + \Sigma M_A = 0; \qquad B_y L - M_A - M_0 = 0 \qquad [2]$$

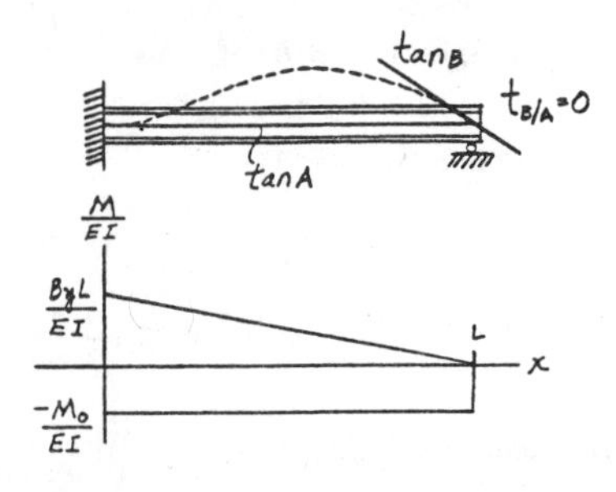

Elastic Curve **:** As shown.

M/EI Diagrams **:** M/EI diagrams for B_y and M_0 acting on a cantilever beam are shown.

Moment-Area Theorems **:** From the elastic curve, $t_{B/A} = 0$.

$$t_{B/A} = 0 = \frac{1}{2}\left(\frac{B_y L}{EI}\right)(L)\left(\frac{2}{3}L\right) + \left(-\frac{M_0}{EI}\right)(L)\left(\frac{L}{2}\right)$$

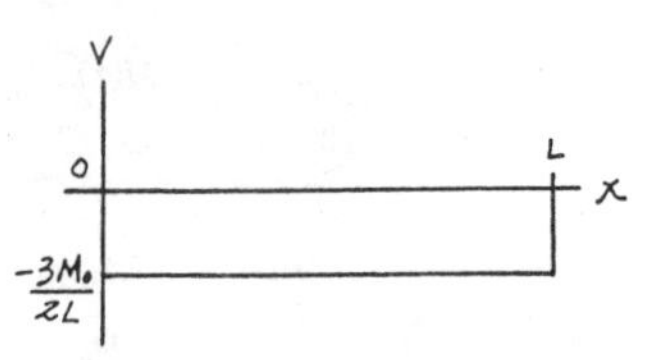

$$B_y = \frac{3M_0}{2L} \qquad \textbf{Ans}$$

Substituting the value of B_y into Eqs.[1] and [2] yields,

$$A_y = \frac{3M_0}{2L} \qquad M_A = \frac{M_0}{2} \qquad \textbf{Ans}$$

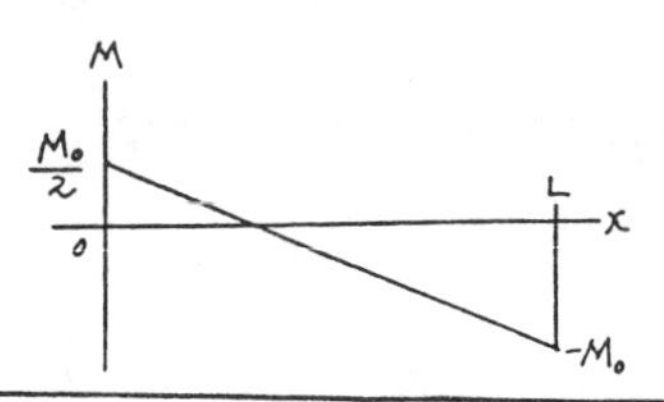

12–99. Determine the reactions at the supports A and B, then draw the shear and moment diagrams. EI is constant.

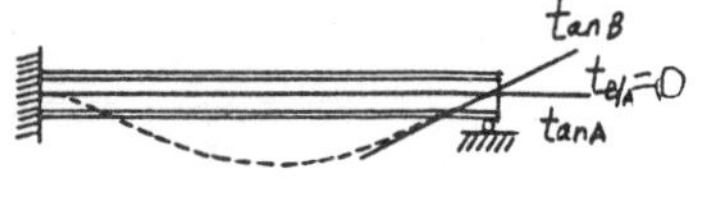

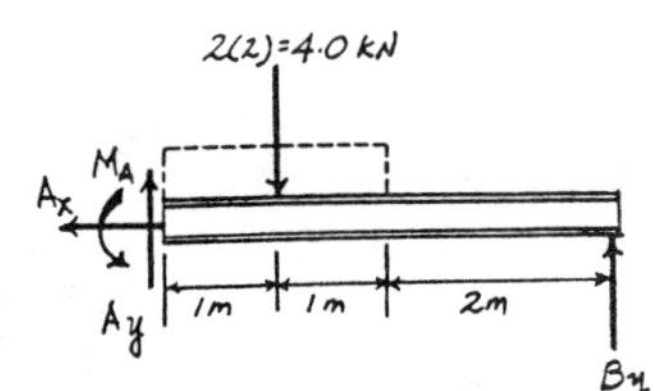

Support Reaction **:** FBD(a).

$$\xrightarrow{+} \Sigma F_x = 0; \qquad A_x = 0 \qquad \textbf{Ans}$$

$$+\uparrow \Sigma F_y = 0; \qquad B_y + A_y - 4.00 = 0 \qquad [1]$$

$$\circlearrowleft + \Sigma M_A = 0; \qquad B_y(4) + M_A - 4.00(1) = 0 \qquad [2]$$

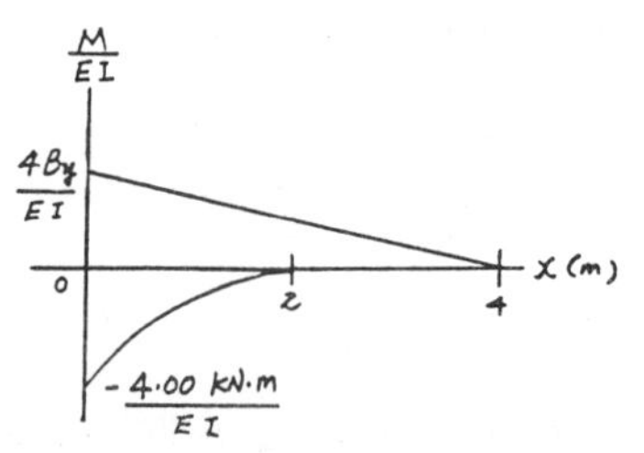

Elastic Curve **:** As shown.

M/EI Diagrams **:** M/EI diagrams for B_y and the uniform distributed load acting on a cantilever beam as shown.

Moment-Area Theorems **:** From the elastic curve, $t_{B/A} = 0$.

$$t_{B/A} = 0 = \frac{1}{2}\left(\frac{4B_y}{EI}\right)(4)\left(\frac{2}{3}\right)(4) + \frac{1}{3}\left(-\frac{4.00}{EI}\right)(2)\left[2 + \frac{3}{4}(2)\right]$$

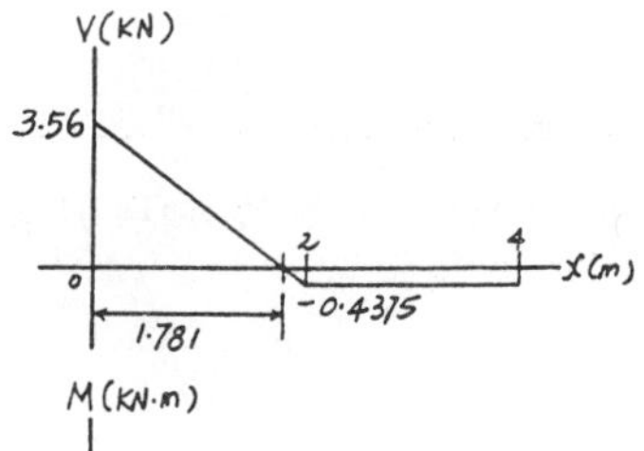

$$B_y = 0.4375 \text{ kN} \qquad \textbf{Ans}$$

Substituting the value of B_y into Eqs.[1] and [2] yields,

$$A_y = 3.56 \text{ kN} \qquad M_A = 2.25 \text{ kN}\cdot\text{m} \qquad \textbf{Ans}$$

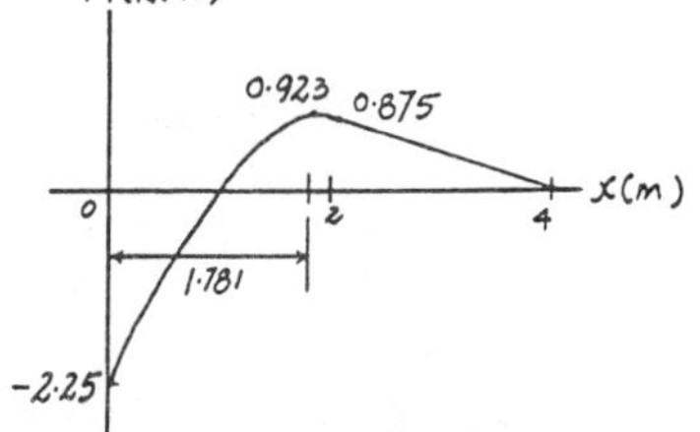

***12–100.** Determine the reactions at the supports, then draw the shear and moment diagrams. EI is constant.

Support Reaction : FBD(a).

$\xrightarrow{+} \Sigma F_x = 0;\qquad A_x = 0$ **Ans**

$+\uparrow \Sigma F_y = 0;\qquad B_y + A_y - 2P = 0$ [1]

$\curvearrowleft + \Sigma M_A = 0;\qquad B_y(3a) + M_A - P(a) - P(2a) = 0$ [2]

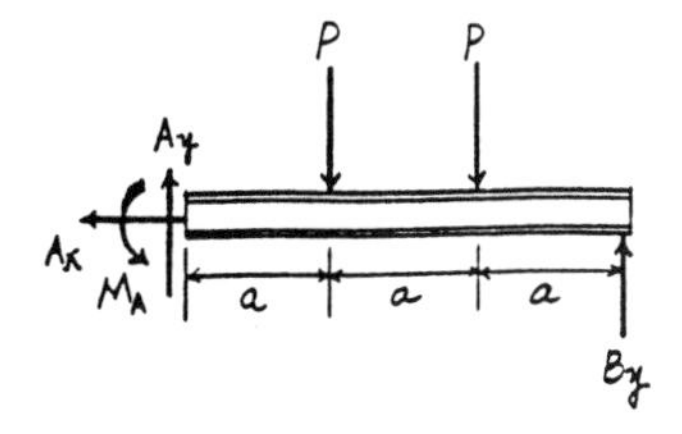

Elastic Curve : As shown.

M/EI Diagram : M/EI diagrams for B_y and P act on a cantilever beam as shown.

Moment-Area Theorems : From the elastic curve, $t_{B/A} = 0$.

$$t_{B/A} = 0 = \frac{1}{2}\left(\frac{3B_y a}{EI}\right)(3a)\left(\frac{2}{3}\right)(3a) + \frac{1}{2}\left(-\frac{2Pa}{EI}\right)(a)\left(2a+\frac{2}{3}a\right) + \left(-\frac{Pa}{EI}\right)(a)\left(2a+\frac{a}{2}\right) + \frac{1}{2}\left(-\frac{Pa}{EI}\right)(a)\left(a+\frac{2}{3}a\right)$$

$$B_y = \frac{2P}{3}$$ **Ans**

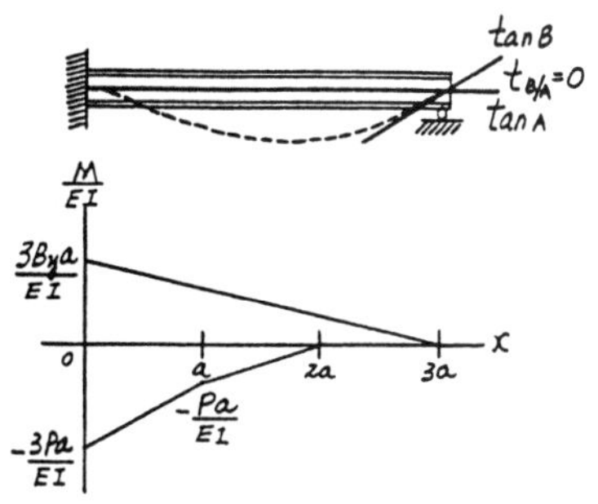

Substituting B_y into Eqs. [1] and [2] yields,

$$A_y = \frac{4P}{3}\qquad M_A = Pa$$ **Ans**

12–101. Determine the reactions at the supports, then draw the shear and moment diagrams. EI is constant.

Support Reaction : FBD(a).

$\xrightarrow{+} \Sigma F_x = 0;\qquad A_x = 0$ **Ans**

$+\uparrow \Sigma F_y = 0;\qquad -B_y + A_y = 0$ [1]

$\curvearrowleft + \Sigma M_A = 0;\qquad -B_y(3a) + M_A = 0$ [2]

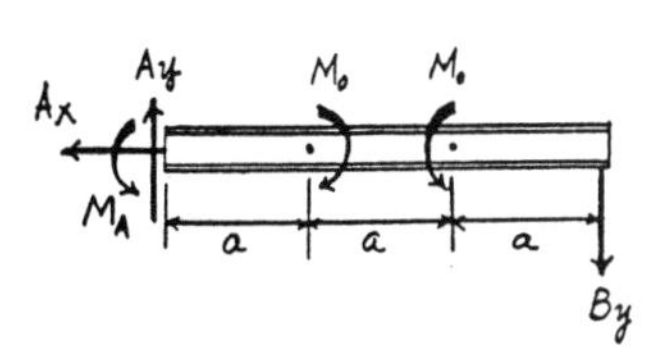

Elastic Curve : As shown.

M/EI Diagrams : M/EI diagrams for B_y and M_0 acting on a cantilever beam are drawn.

Moment-Area Theorems : From the elastic curve, $t_{B/A} = 0$.

$$t_{B/A} = 0 = \frac{1}{2}\left(-\frac{3B_y a}{EI}\right)(3a)\left(\frac{2}{3}\right)(3a) + \left(\frac{M_0}{EI}\right)(a)\left(a+\frac{a}{2}\right)$$

$$B_y = \frac{M_0}{6a}$$ **Ans**

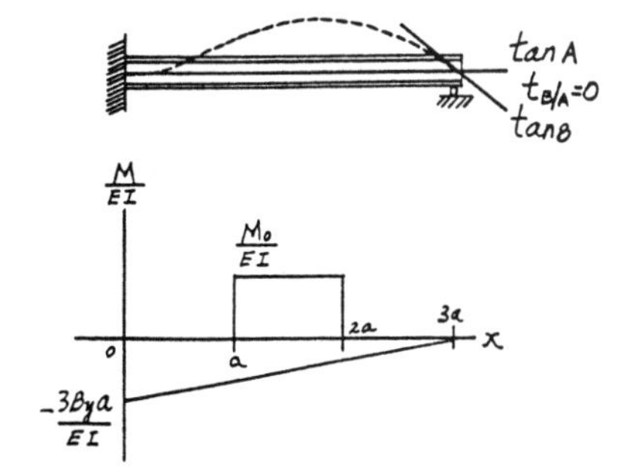

Substituting B_y into Eqs. [1] and [2] yields,

$$A_y = \frac{M_0}{6a}\qquad M_A = \frac{M_0}{2}$$ **Ans**

12–102. The rod is fixed at A, and the connection at B consists of a roller constraint which allows vertical displacement but resists axial load and moment. Determine the moment reactions at these supports. EI is constant.

Support Reaction : FBD(a).

$$\circlearrowleft + \Sigma M_A = 0; \qquad M_B + M_A - wL\left(\frac{L}{2}\right) = 0 \qquad [1]$$

Elastic Curve : As shown.

M/EI Diagrams : M/EI diagrams for M_B and the uniform distributed load act ing on a cantilever beam are shown.

Moment - Area Theorems : Since both tangents at A and B are horizontal (parallel), $\theta_{B/A} = 0$.

$$\theta_{B/A} = 0 = \left(\frac{M_B}{EI}\right)(L) + \frac{1}{3}\left(-\frac{wL^2}{2EI}\right)(L)$$

$$M_B = \frac{wL^2}{6} \qquad \textbf{Ans}$$

Substituting M_B into Eq.[1] ,

$$M_A = \frac{wL^2}{3} \qquad \textbf{Ans}$$

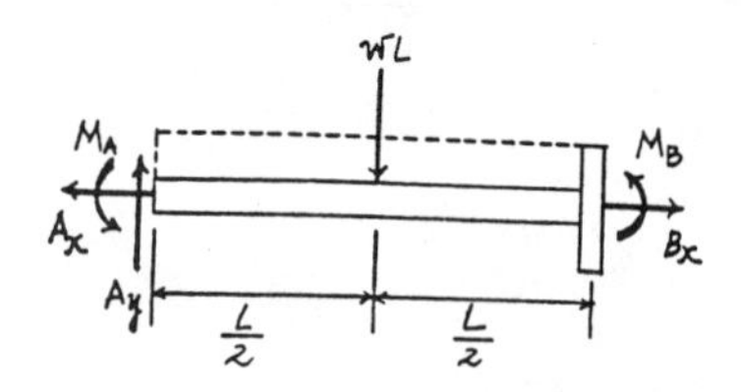

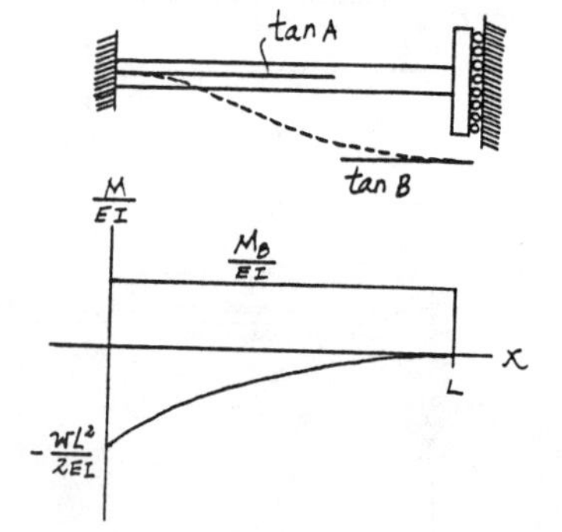

12–103. Determine the reactions at the supports, then draw the shear and moment diagrams. EI is constant. Support B is a thrust bearing.

Support Reaction : FBD(a).

$\xrightarrow{+} \Sigma F_x = 0; \qquad B_x = 0 \qquad$ **Ans**

$+\uparrow \Sigma F_y = 0; \qquad -A_y + B_y + C_y - P = 0 \qquad [1]$

$\circlearrowleft + \Sigma M_A = 0; \qquad B_y(L) + C_y(2L) - P\left(\frac{3L}{2}\right) = 0 \qquad [2]$

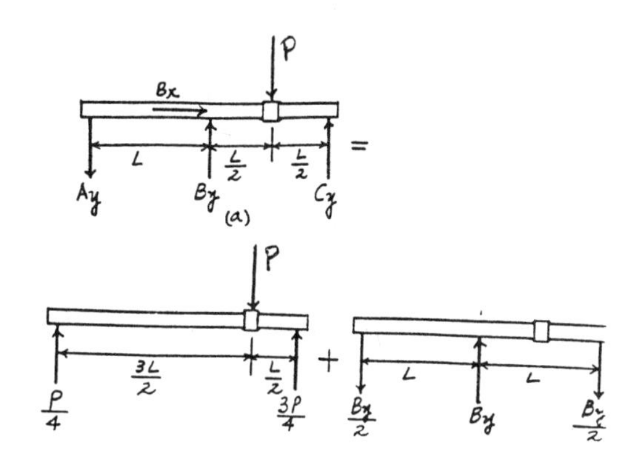

Elastic Curve : As shown.

M/EI Diagrams : M/EI diagrams for P and B_y acting on a simply supported beam are drawn separately.

Moment - Area Theorems :

$$(t_{A/C})_1 = \frac{1}{2}\left(\frac{3PL}{8EI}\right)\left(\frac{3L}{2}\right)\left(\frac{2}{3}\right)\left(\frac{3L}{2}\right) + \frac{1}{2}\left(\frac{3PL}{8EI}\right)\left(\frac{L}{2}\right)\left(\frac{3L}{2} + \frac{L}{6}\right) = \frac{7PL^3}{16EI}$$

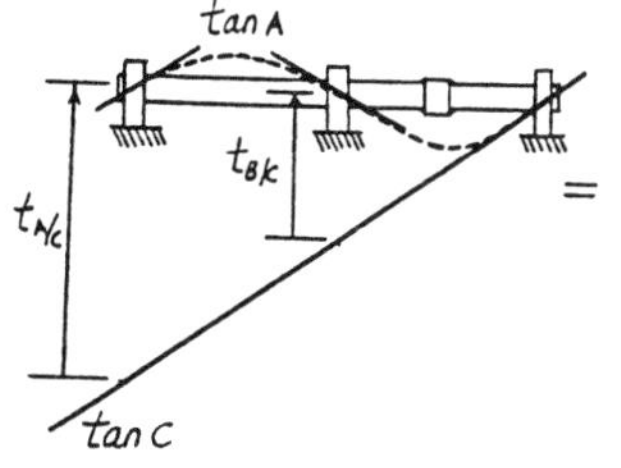

$$(t_{A/C})_2 = \frac{1}{2}\left(-\frac{B_yL}{2EI}\right)(2L)(L) = -\frac{B_yL^3}{2EI}$$

$$(t_{B/C})_1 = \frac{1}{2}\left(\frac{PL}{8EI}\right)\left(\frac{L}{2}\right)\left(\frac{2}{3}\right)\left(\frac{L}{2}\right) + \left(\frac{PL}{4EI}\right)\left(\frac{L}{2}\right)\left(\frac{L}{4}\right) + \frac{1}{2}\left(\frac{3PL}{8EI}\right)\left(\frac{L}{2}\right)\left(\frac{L}{2} + \frac{L}{6}\right) = \frac{5PL^3}{48EI}$$

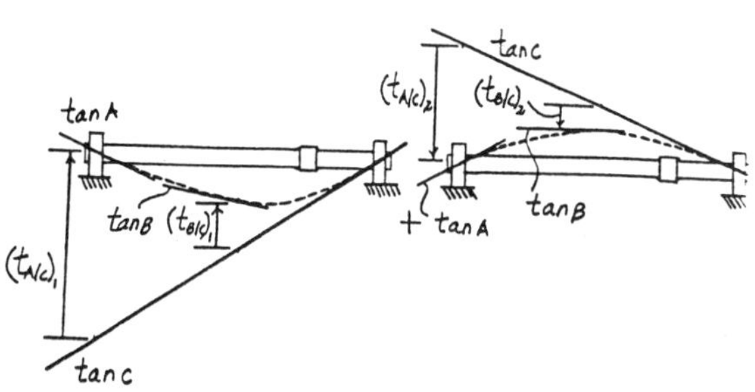

$$(t_{B/C})_2 = \frac{1}{2}\left(-\frac{B_yL}{2EI}\right)(L)\left(\frac{L}{3}\right) = -\frac{B_yL^3}{12EI}$$

$$t_{A/C} = (t_{A/C})_1 + (t_{A/C})_2 = \frac{7PL^3}{16EI} - \frac{B_yL^3}{2EI}$$

$$t_{B/C} = (t_{B/C})_1 + (t_{B/C})_2 = \frac{5PL^3}{48EI} - \frac{B_yL^3}{12EI}$$

From the elastic curve,

$$t_{A/C} = 2t_{B/C}$$

$$\frac{7PL^3}{16EI} - \frac{B_yL^3}{2EI} = 2\left(\frac{5PL^3}{48EI} - \frac{B_yL^3}{12EI}\right)$$

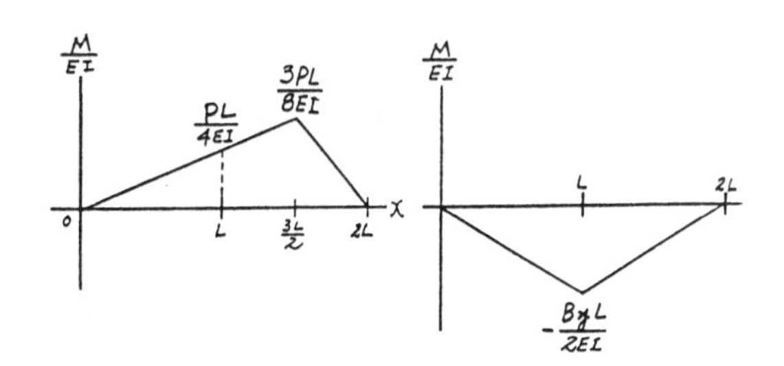

$$B_y = \frac{11P}{16} \qquad \textbf{Ans}$$

Substituting B_y into Eqs. [1] and [2] yields ,

$$C_y = \frac{13P}{32} \qquad A_y = \frac{3P}{32} \qquad \textbf{Ans}$$

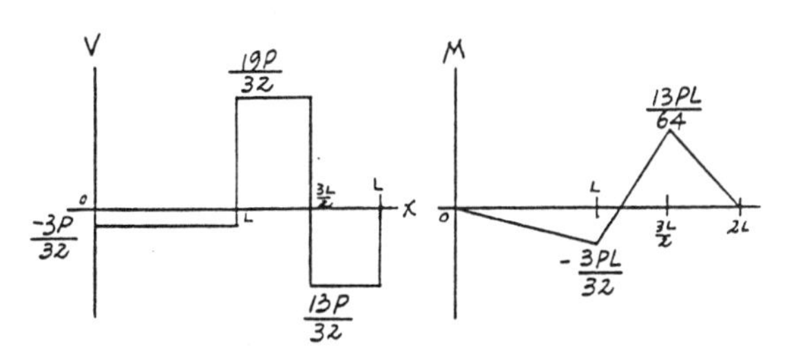

***12–104.** Determine the reactions at the supports, then draw the shear and moment diagrams. EI is constant. Support B is a thrust bearing.

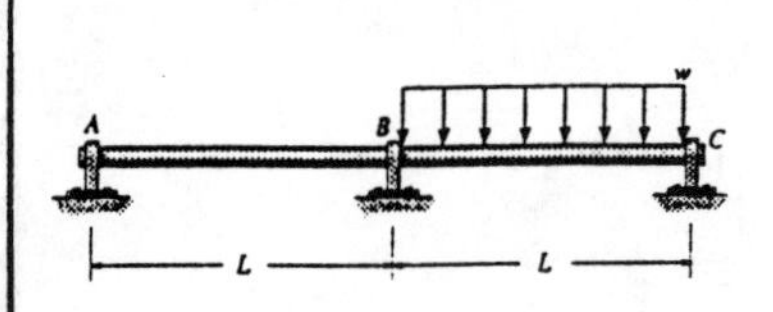

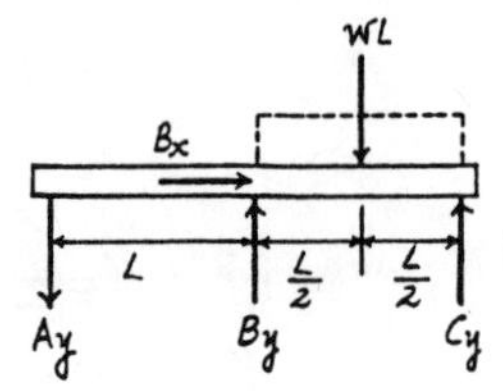

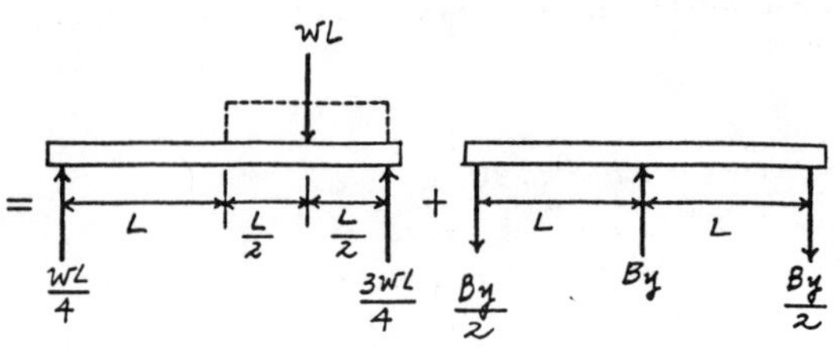

Support Reaction : FBD(a).

$$\xrightarrow{+} \Sigma F_x = 0; \qquad B_x = 0 \qquad \textbf{Ans}$$

$$+\uparrow \Sigma F_y = 0; \qquad -A_y + B_y + C_y - wL = 0 \qquad [1]$$

$$\circlearrowleft + \Sigma M_A = 0; \qquad B_y(L) + C_y(2L) - wL\left(\frac{3L}{2}\right) = 0 \qquad [2]$$

Elastic Curve : As shown.

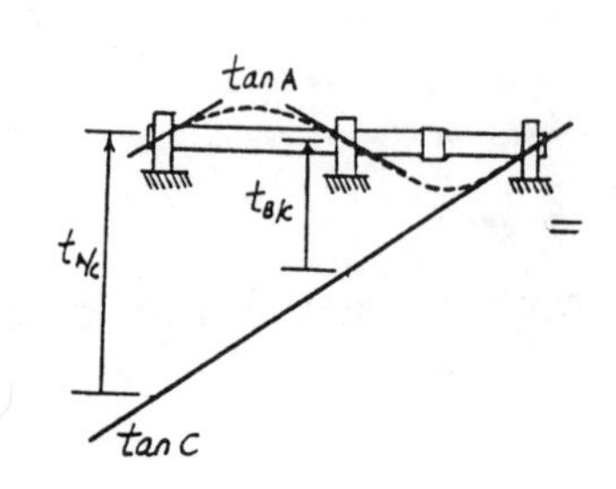

M/EI Diagrams : M/EI diagrams for the distributed load and B_y acting on a simply supported beam are drawn separately.

Moment - Area Theorems :

$$(t_{A/C})_1 = \frac{1}{2}\left(\frac{wL^2}{2EI}\right)(2L)\left(\frac{2}{3}\right)(2L) + \frac{1}{3}\left(-\frac{wL^2}{2EI}\right)(L)\left(L + \frac{3L}{4}\right)$$

$$= \frac{3wL^4}{8EI}$$

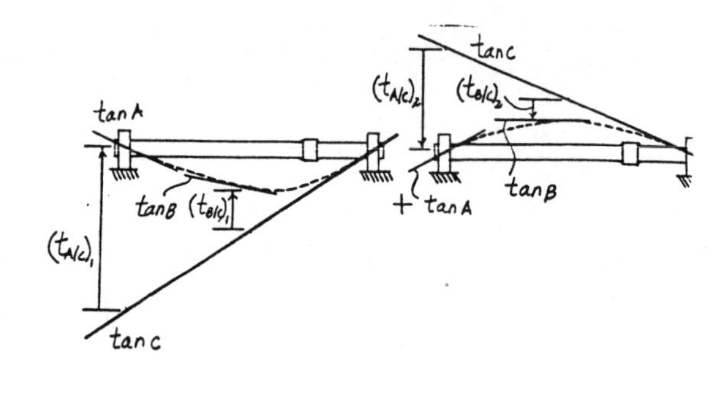

$$(t_{A/C})_2 = \frac{1}{2}\left(-\frac{B_y L}{2EI}\right)(2L)(L) = -\frac{B_y L^3}{2EI}$$

$$(t_{B/C})_1 = \frac{1}{2}\left(\frac{wL^2}{4EI}\right)(L)\left(\frac{2}{3}\right)(L) + \left(\frac{wL^2}{4EI}\right)(L)\left(\frac{L}{2}\right) + \frac{1}{3}\left(-\frac{wL^2}{2EI}\right)(L)\left(\frac{3L}{4}\right)$$

$$= \frac{wL^4}{12EI}$$

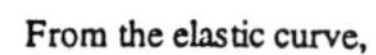

$$(t_{B/C})_2 = \frac{1}{2}\left(-\frac{B_y L}{2EI}\right)(L)\left(\frac{L}{3}\right) = -\frac{B_y L^3}{12EI}$$

$$t_{A/C} = (t_{A/C})_1 + (t_{A/C})_2 = \frac{3wL^4}{8EI} - \frac{B_y L^3}{2EI}$$

$$t_{B/C} = (t_{B/C})_1 + (t_{B/C})_2 = \frac{wL^4}{12EI} - \frac{B_y L^3}{12EI}$$

From the elastic curve,

$$t_{A/C} = 2t_{B/C}$$

$$\frac{3wL^4}{8EI} - \frac{B_y L^3}{2EI} = 2\left(\frac{wL^4}{12EI} - \frac{B_y L^3}{12EI}\right)$$

$$B_y = \frac{5wL}{8} \qquad \textbf{Ans}$$

Substituting B_y into Eqs. [1] and [2] yields ,

$$C_y = \frac{7wL}{16} \qquad A_y = \frac{wL}{16} \qquad \textbf{Ans}$$

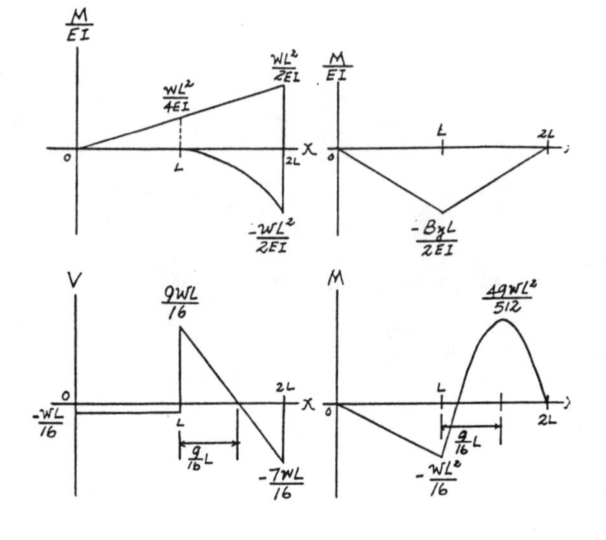

12–105. Determine the moment reactions at the supports A and B, then draw the shear and moment diagrams. EI is constant.

Support Reaction : FBD(a).

$$+\uparrow \Sigma F_y = 0; \quad A_y - B_y = 0 \qquad [1]$$

$$\circlearrowleft + \Sigma M_A = 0; \quad M_B + M_A + M_0 - B_y L = 0 \qquad [2]$$

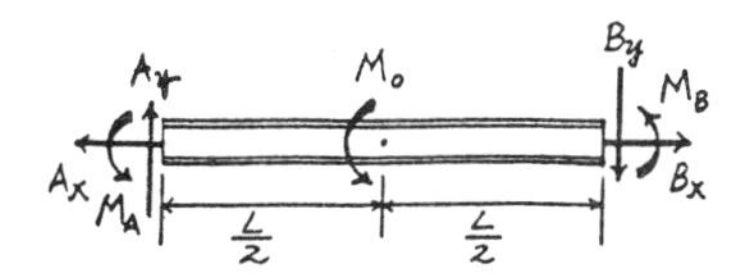

Elastic Curve : As shown.

M/EI Diagrams : M/EI diagrams for support reactions M_B, B_y and the couple moment M_0 act on a cantilever beam are drawn separately.

Moment - Area Theorems : Since both tangent at A and B are horizontal (parallel), $\theta_{B/A} = 0$.

$$\theta_{B/A} = 0 = \left(\frac{M_B}{EI}\right)(L) + \left(\frac{M_0}{EI}\right)\left(\frac{L}{2}\right) + \frac{1}{2}\left(-\frac{B_y L}{EI}\right)(L)$$

$$0 = 2M_B + M_0 - B_y L \qquad [3]$$

As shown on the elastic curve, $t_{B/A} = 0$

$$t_{B/A} = 0 = \left(\frac{M_B}{EI}\right)(L)\left(\frac{L}{2}\right) + \left(\frac{M_0}{EI}\right)\left(\frac{L}{2}\right)\left(\frac{L}{2} + \frac{L}{4}\right) + \frac{1}{2}\left(-\frac{B_y L}{EI}\right)(L)\left(\frac{2}{3}L\right)$$

$$0 = 12M_B + 9M_0 - 8B_y L \qquad [4]$$

Solving Eqs. [3] and [4] yields,

$$B_y = \frac{3M_0}{2L}$$

$$M_B = \frac{M_0}{4} \qquad \textbf{Ans}$$

Substituting M_B and B_y into Eqs. [1] and [2] yields ,

$$A_y = \frac{3M_0}{2L}$$

$$M_A = \frac{M_0}{4} \qquad \textbf{Ans}$$

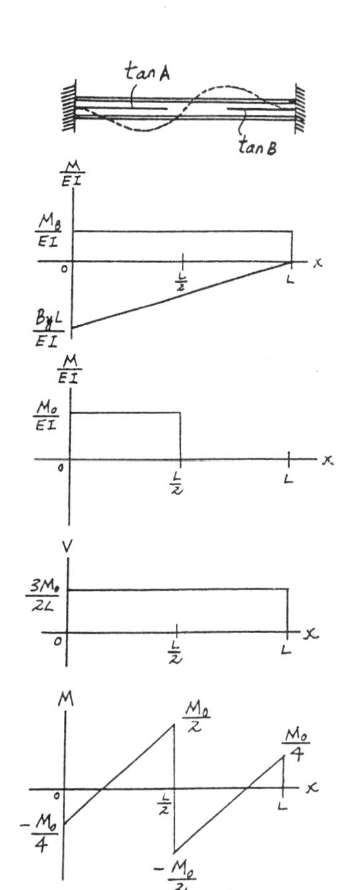

12–106. Determine the reactions at the supports A and B. EI is constant.

Support Reactions : FBD(a).

$$\xrightarrow{+} \Sigma F_x = 0; \qquad A_x = 0 \qquad \textbf{Ans}$$

$$+\uparrow \Sigma F_y = 0; \qquad A_y + B_y - \frac{w_0 L}{2} = 0 \qquad [1]$$

$$\circlearrowleft + \Sigma M_A = 0; \qquad B_y L + M_A - \frac{w_0 L}{2}\left(\frac{L}{3}\right) = 0 \qquad [2]$$

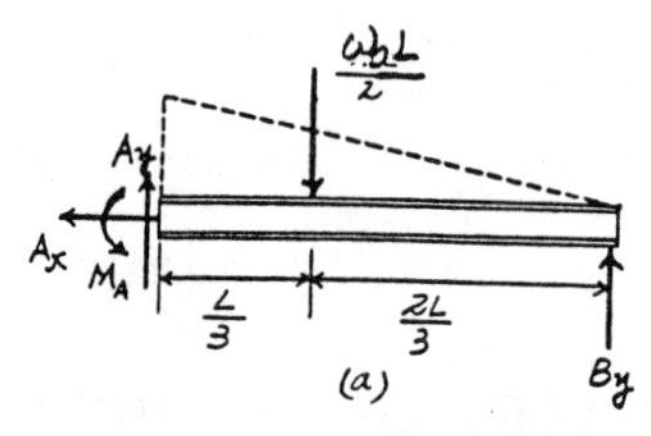

Method of Superposition : Using the table in Appendix C, the required displacements are

$$\upsilon_B{}' = \frac{w_0 L^4}{30EI} \downarrow \qquad \upsilon_B{}'' = \frac{B_y L^3}{3EI} \uparrow$$

The compatibility condition requires

$$(+\downarrow) \qquad 0 = \upsilon_B{}' + \upsilon_B{}''$$

$$0 = \frac{w_0 L^4}{30EI} + \left(-\frac{B_y L^3}{3EI}\right)$$

$$B_y = \frac{w_0 L}{10} \qquad \textbf{Ans}$$

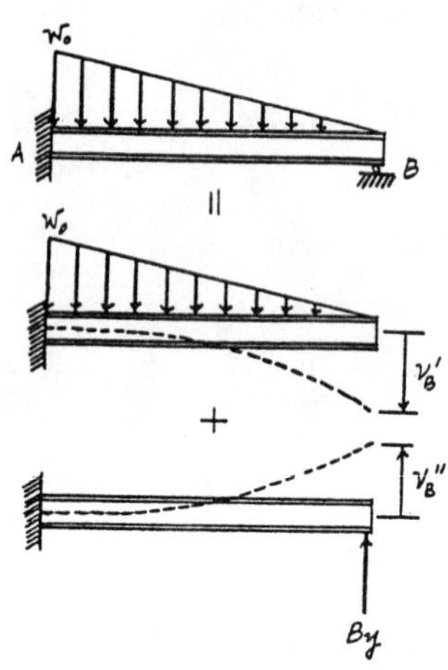

Substituting B_y into Eqs.[1] and [2] yields,

$$A_y = \frac{2w_0 L}{5} \qquad M_A = \frac{w_0 L^2}{15} \qquad \textbf{Ans}$$

12–107. Determine the reactions at the bearing supports A, B, and C of the shaft, then draw the shear and moment diagrams. EI is constant. Each bearing exerts only vertical reactions on the shaft.

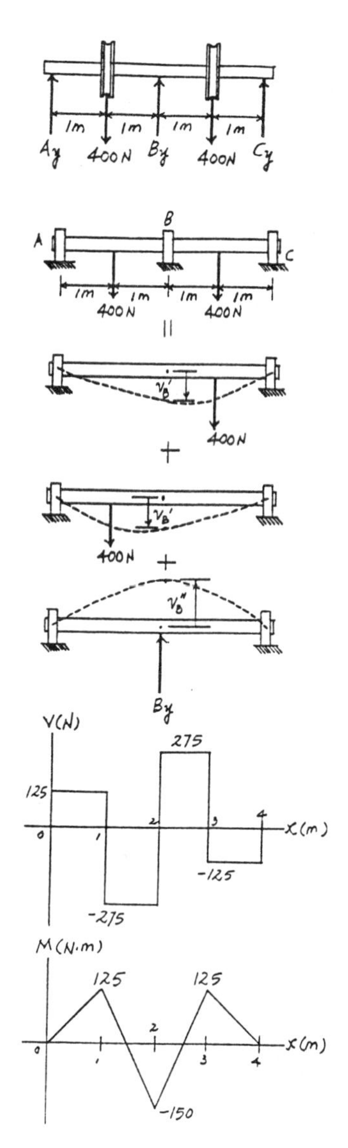

Support Reactions : FBD(a).

$$+\uparrow \Sigma F_y = 0; \qquad A_y + B_y + C_y - 800 = 0 \qquad [1]$$

$$\circlearrowleft + \Sigma M_A = 0; \qquad B_y(2) + C_y(4) - 400(1) - 400(3) = 0 \qquad [2]$$

Method of Superposition : Using the table in Appendix C, the required displacements are

$$\upsilon_B' = \frac{Pbx}{6EIL}\left(L^2 - b^2 - x^2\right)$$
$$= \frac{400(1)(2)}{6EI(4)}\left(4^2 - 1^2 - 2^2\right)$$
$$= \frac{366.67\ \text{N}\cdot\text{m}^3}{EI} \downarrow$$

$$\upsilon_B'' = \frac{PL^3}{48EI} = \frac{B_y(4^3)}{48EI} = \frac{1.3333B_y\ \text{m}^3}{EI} \uparrow$$

The compatibility condition requires

$$(+\downarrow) \qquad 0 = 2\upsilon_B' + \upsilon_B''$$
$$0 = 2\left(\frac{366.67}{EI}\right) + \left(-\frac{1.3333B_y}{EI}\right)$$

$$B_y = 550\ \text{N} \qquad \textbf{Ans}$$

Substituting B_y into Eqs.[1] and [2] yields,

$$A_y = 125\ \text{N} \qquad C_y = 125\ \text{N} \qquad \textbf{Ans}$$

***12–108.** Determine the reactions at the supports A, B, and C, then draw the shear and moment diagrams. EI is constant.

Support Reactions : FBD(a).

$$\xrightarrow{+} \Sigma F_x = 0; \qquad C_x = 0 \qquad \textbf{Ans}$$

$$+\uparrow \Sigma F_y = 0; \qquad A_y + B_y + C_y - 12 - 36.0 = 0 \qquad [1]$$

$$\curvearrowleft + \Sigma M_A = 0; \qquad B_y(12) + C_y(24) - 12(6) - 36.0(18) = 0 \qquad [2]$$

Method of Superposition : Using the table in Appendix C, the required displacements are

$$v_B' = \frac{5wL^4}{768EI} = \frac{5(3)(24^4)}{768EI} = \frac{6480 \text{ kip}\cdot\text{ft}^3}{EI} \downarrow$$

$$v_B'' = \frac{Pbx}{6EIL}\left(L^2 - b^2 - x^2\right)$$

$$= \frac{12(6)(12)}{6EI(24)}\left(24^2 - 6^2 - 12^2\right) = \frac{2376 \text{ kip}\cdot\text{ft}^3}{EI} \downarrow$$

$$v_B''' = \frac{PL^3}{48EI} = \frac{B_y(24^3)}{48EI} = \frac{288B_y \text{ ft}^3}{EI} \uparrow$$

The compatibility condition requires

$$(+\downarrow) \qquad 0 = v_B' + v_B'' + v_B'''$$

$$0 = \frac{6480}{EI} + \frac{2376}{EI} + \left(-\frac{288B_y}{EI}\right)$$

$$B_y = 30.75 \text{ kip} \qquad \textbf{Ans}$$

Substituting B_y into Eqs.[1] and [2] yields,

$$A_y = 2.625 \text{ kip} \qquad C_y = 14.625 \text{ kip} \qquad \textbf{Ans}$$

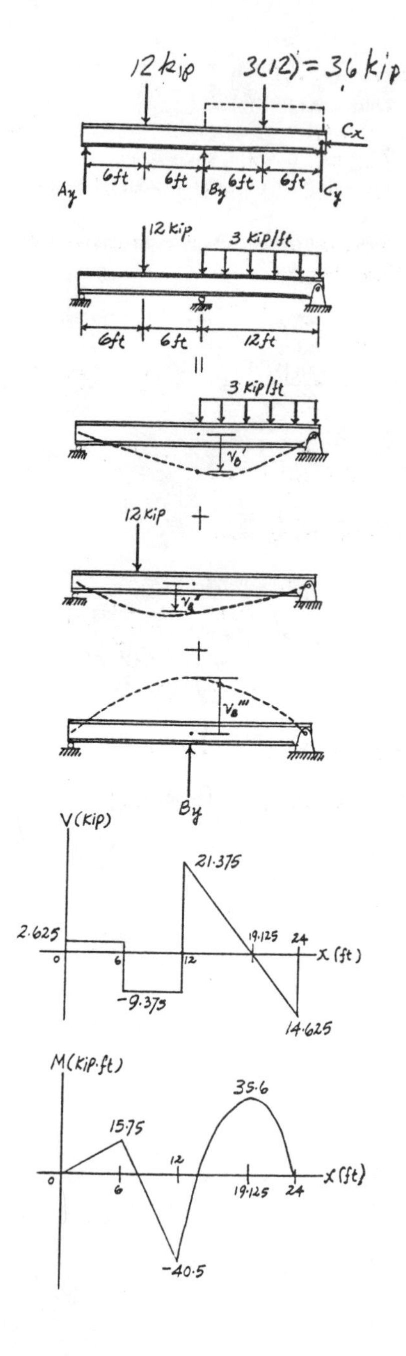

12–109. The beam is used to support the 20-kip loading. Determine the reactions at the supports. Assume A is fixed and B is a roller.

Support Reactions : FBD(a).

$\xrightarrow{+} \Sigma F_x = 0;\qquad A_x = 0$ **Ans**

$+\uparrow \Sigma F_y = 0;\qquad -A_y + B_y - 20 = 0$ [1]

$\circlearrowleft + \Sigma M_A = 0;\qquad B_y(8) - M_A - 20(12) = 0$ [2]

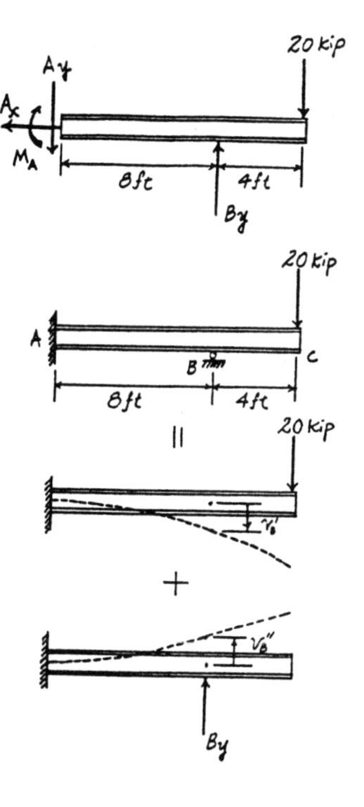

Method of Superposition : Using the table in Appendix C, the required displacements are

$$v_B' = \frac{Px^2}{6EI}(3L - x)$$

$$= \frac{20(8^2)}{6EI}[3(12) - 8] = \frac{5973.33 \text{ kip}\cdot\text{ft}^3}{EI} \downarrow$$

$$v_B'' = \frac{PL_{AB}^3}{3EI} = \frac{B_y(8^3)}{3EI} = \frac{170.67B_y \text{ ft}^3}{EI} \uparrow$$

The compatibility condition requires

$$(+\downarrow)\qquad 0 = v_B' + v_B''$$

$$0 = \frac{5973.33}{EI} + \left(-\frac{170.67B_y}{EI}\right)$$

$$B_y = 35.0 \text{ kip}$$ **Ans**

Substituting B_y into Eqs.[1] and [2] yields,

$A_y = 15.0$ kip $\qquad M_A = 40.0$ kip · ft **Ans**

12–110. Determine the reactions at the supports A and B, then draw the shear and moment diagrams. EI is constant.

Support Reactions : FBD(a).

$\overset{+}{\rightarrow}\Sigma F_x = 0;\qquad A_x = 0$ **Ans**

$+\uparrow \Sigma F_y = 0;\qquad -A_y + B_y - P = 0$ [1]

$(+\Sigma M_A = 0;\qquad B_y(L) - M_A - P(2L) = 0$ [2]

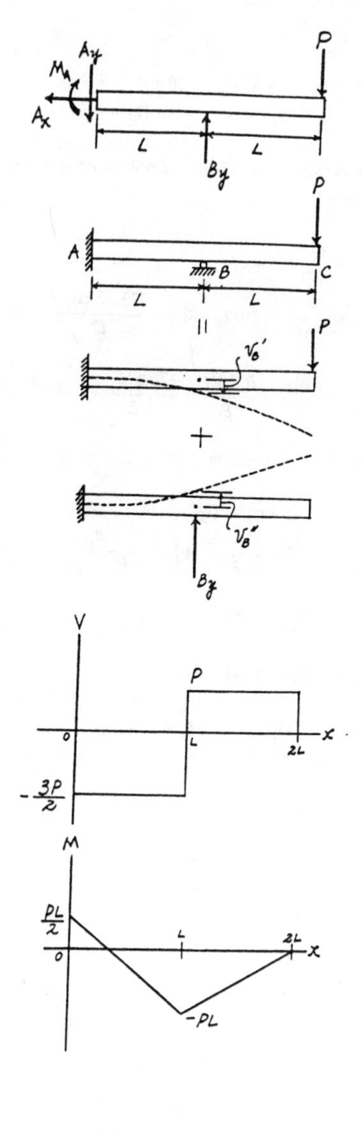

Method of Superposition : Using the table in Appendix C, the required displacements are

$$v_B' = \frac{Px^2}{6EI}(3L_{AC} - x)$$

$$= \frac{PL^2}{6EI}[3(2L) - L] = \frac{5PL^3}{6EI} \downarrow$$

$$v_B'' = \frac{PL_{AB}^3}{3EI} = \frac{B_y L^3}{3EI} \uparrow$$

The compatibility condition requires

$$(+\downarrow)\qquad 0 = v_B' + v_B''$$

$$0 = \frac{5PL^3}{6EI} + \left(-\frac{B_y L^3}{3EI}\right)$$

$$B_y = \frac{5P}{2}$$ **Ans**

Substituting B_y into Eqs. [1] and [2] yields,

$$A_y = \frac{3P}{2} \qquad M_A = \frac{PL}{2}$$ **Ans**

12–111. Determine the reactions at the supports A and B. EI is constant.

Support Reactions : FBD(a).

$\xrightarrow{+} \Sigma F_x = 0; \qquad A_x = 0$ **Ans**

$+\uparrow \Sigma F_y = 0; \qquad A_y + B_y - \frac{wL}{2} = 0$ [1]

$\curvearrowleft + \Sigma M_A = 0; \qquad B_y(L) + M_A - \left(\frac{wL}{2}\right)\left(\frac{L}{4}\right) = 0$ [2]

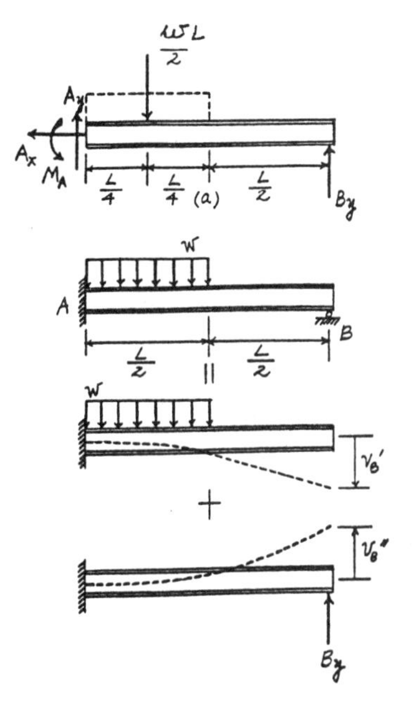

Method of Superposition : Using the table in appendix C, the required displacements are

$$v_B' = \frac{7wL^4}{384EI} \downarrow \qquad v_B'' = \frac{PL^3}{3EI} = \frac{B_yL^3}{3EI} \uparrow$$

The compatibility condition requires

$(+\downarrow)$

$$0 = v_B' + v_B''$$

$$0 = \frac{7wL^4}{384EI} + \left(-\frac{B_yL^3}{3EI}\right)$$

$$B_y = \frac{7wL}{128}$$ **Ans**

Substituting B_y into Eqs.[1] and [2] yields,

$$A_y = \frac{57wL}{128} \qquad M_A = \frac{9wL^2}{128}$$ **Ans**

***12–112.** Determine the reactions at support C. EI is constant for both beams.

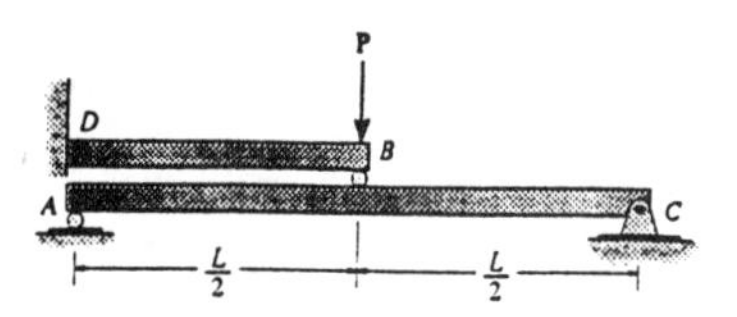

Support Reactions : FBD(a).

$\xrightarrow{+} \Sigma F_x = 0; \qquad C_x = 0$ **Ans**

$\curvearrowleft + \Sigma M_A = 0; \qquad C_y(L) - B_y\left(\frac{L}{2}\right) = 0$ [1]

Method of Superposition : Using the table in Appendix C, the required displacements are

$$v_B = \frac{PL^3}{48EI} = \frac{B_yL^3}{48EI} \downarrow$$

$$v_B' = \frac{PL_{BD}^3}{3EI} = \frac{P\left(\frac{L}{2}\right)^3}{3EI} = \frac{PL^3}{24EI} \downarrow$$

$$v_B'' = \frac{PL_{BD}^3}{3EI} = \frac{B_yL^3}{24EI} \uparrow$$

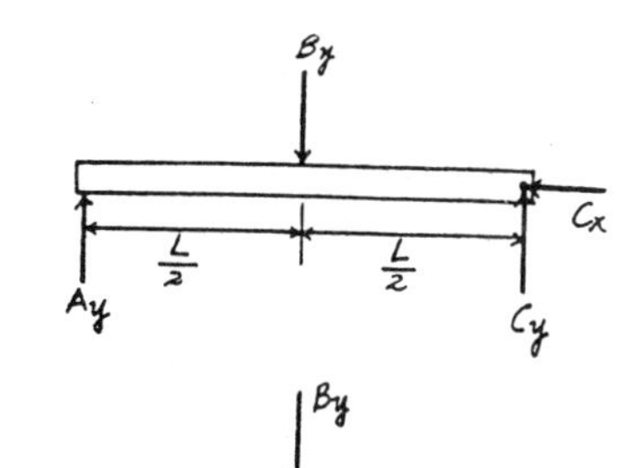

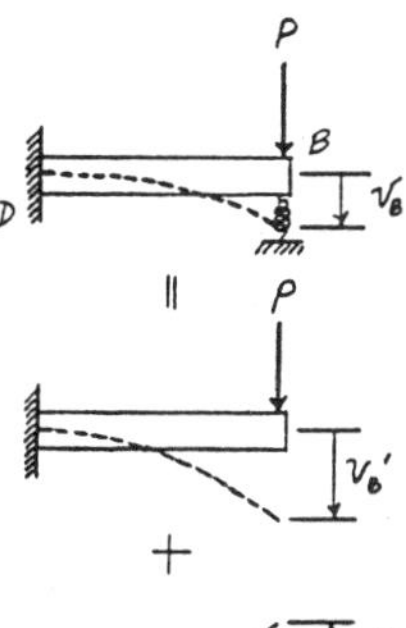

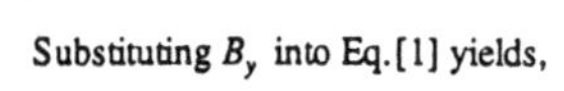

The compatibility condition requires

$(+\downarrow)$

$$v_B = v_B' + v_B''$$

$$\frac{B_yL^3}{48EI} = \frac{PL^3}{24EI} + \left(-\frac{B_yL^3}{24EI}\right)$$

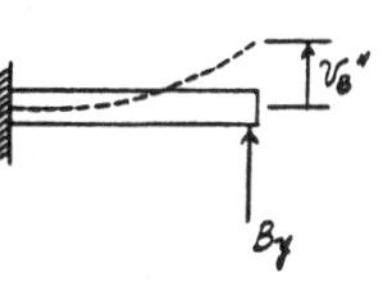

$$B_y = \frac{2P}{3}$$

Substituting B_y into Eq.[1] yields,

$$C_y = \frac{P}{3}$$ **Ans**

12–113. The assembly consists of a steel and an aluminum bar, each of which is 1 in. wide, fixed at its ends C and B, and pin connected to the *rigid* short link AD. If a vertical force of 150 lb is applied to the link as shown, determine the moments created at C and B. $E_{st} = 29(10^3)$ ksi, $E_{al} = 10(10^3)$ ksi.

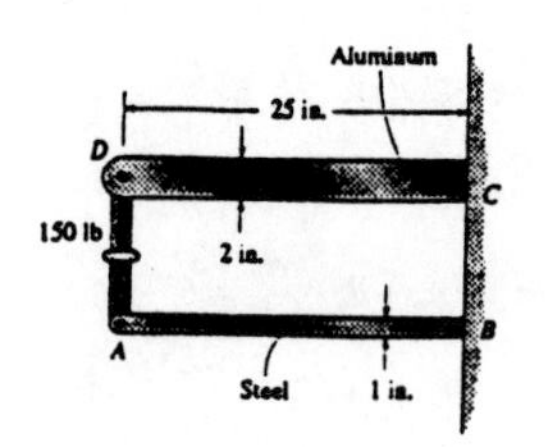

Support Reactions : From FBD(a),

$$+\uparrow \Sigma F_y = 0; \quad A_y + D_y - 150 = 0 \qquad [1]$$

From FBD(b),

$$\circlearrowleft + \Sigma M_C = 0; \quad D_y(25) - M_C = 0 \qquad [2]$$

From FBD(c),

$$\circlearrowleft + \Sigma M_B = 0; \quad A_y(25) - M_B = 0 \qquad [3]$$

Compatibility condition : Since link AD is rigid, $v_D = v_A$. Using the table in Appendix C,

$$v_D = \frac{PL^3}{3EI} = \frac{D_y(25^3)}{3(10)(10^6)\left(\frac{1}{12}\right)(1)(2^3)} = 0.78125\left(10^{-3}\right)D_y$$

$$v_A = \frac{PL^3}{3EI} = \frac{A_y(25^3)}{3(29)(10^6)\left(\frac{1}{12}\right)(1)(1^3)} = 0.002155A_y$$

$$v_D = v_A$$
$$0.78125\left(10^{-3}\right)D_y = 0.002155A_y$$
$$D_y = 2.7586A_y \qquad [4]$$

Solving Eqs. [1] and [4] yields,

$$A_y = 39.91 \text{ lb} \qquad D_y = 110.09 \text{ lb}$$

Substituting A_y and D_y into Eqs. [2] and [3] yields,

$$M_B = 0.998 \text{ kip}\cdot\text{in.} \qquad M_C = 2.75 \text{ kip}\cdot\text{in.} \qquad \textbf{Ans}$$

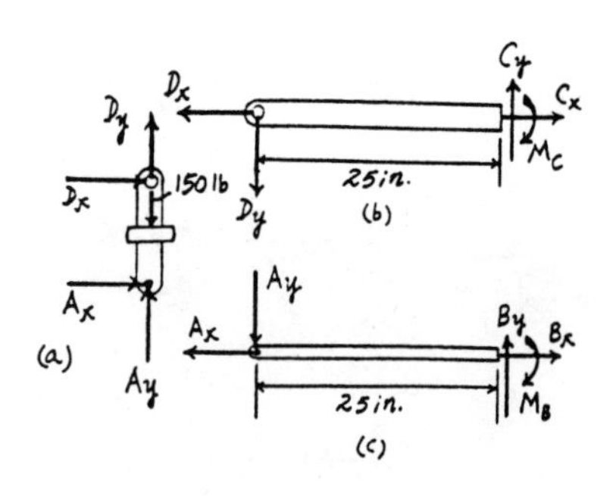

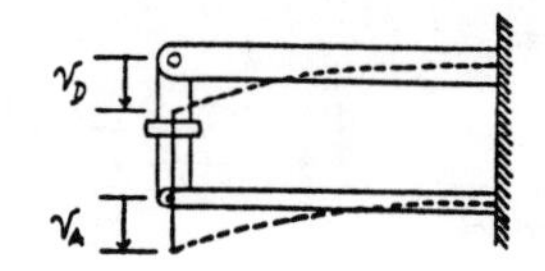

12–114. Determine the reactions at the supports, then draw the shear and moment diagrams. EI is constant.

Support Reactions : FBD(a).

$$\xrightarrow{+} \Sigma F_x = 0; \qquad A_x = 0 \qquad \textbf{Ans}$$

$$+\uparrow \Sigma F_y = 0; \qquad A_y + B_y + C_y - 2wL = 0 \qquad [1]$$

$$\circlearrowleft + \Sigma M_A = 0; \qquad B_y(L) + C_y(2L) - (2wL)(L) = 0 \qquad [2]$$

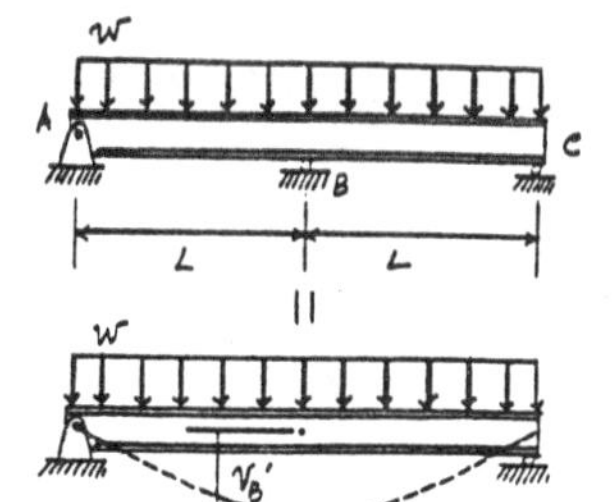

Method of Superposition : Using the table in Appendix C, the required displacements are

$$v_B' = \frac{5wL_{AC}^4}{384EI} = \frac{5w(2L)^4}{384EI} = \frac{5wL^4}{24EI} \downarrow$$

$$v_B'' = \frac{PL_{AC}^3}{48EI} = \frac{B_y(2L)^3}{48EI} = \frac{B_y L^3}{6EI} \uparrow$$

The compatibility condition requires

$$(+\downarrow) \qquad 0 = v_B' + v_B''$$

$$0 = \frac{5wL^4}{24EI} + \left(-\frac{B_y L^3}{6EI}\right)$$

$$B_y = \frac{5wL}{4} \qquad \textbf{Ans}$$

Substituting the value of B_y into Eqs.[1] and [2] yields,

$$C_y = A_y = \frac{3wL}{8} \qquad \textbf{Ans}$$

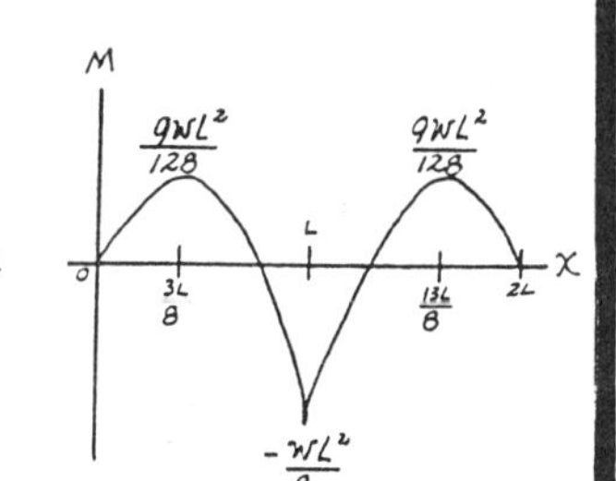

12–115. The beam is supported by a pin at A, a spring having a stiffness k at B, and a roller at C. Determine the force the spring exerts on the beam. EI is constant.

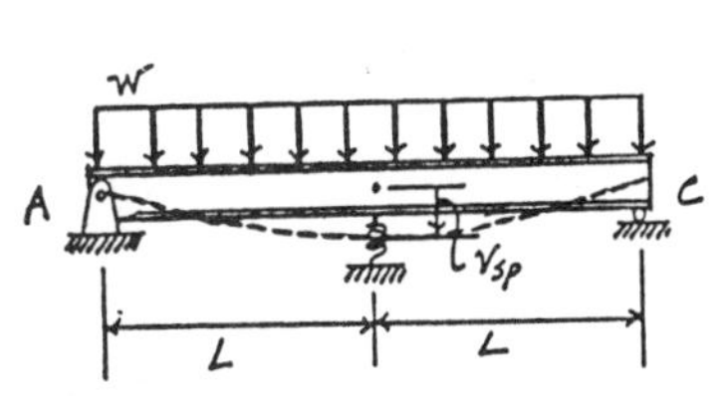

Method of Superposition : Using the table in appendix C, the required displacements are

$$v_B' = \frac{5wL_{AC}^4}{384EI} = \frac{5w(2L)^4}{384EI} = \frac{5wL^4}{24EI} \downarrow$$

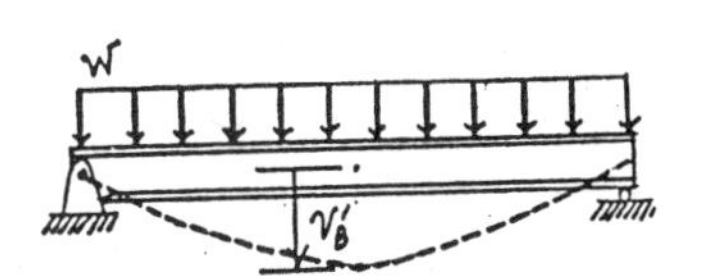

$$v_B'' = \frac{PL_{AC}^3}{48EI} = \frac{F_{sp}(2L)^3}{48EI} = \frac{F_{sp} L^3}{6EI} \uparrow$$

Using the spring formula, $\quad v_{sp} = \frac{F_{sp}}{k}$.

The compatibility condition requires

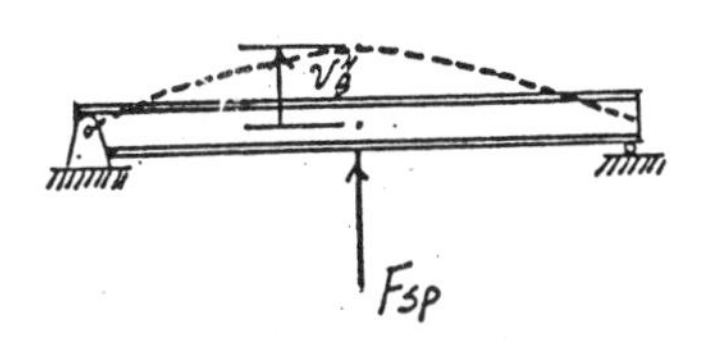

$$(+\downarrow) \qquad v_{sp} = v_B' + v_B''$$

$$\frac{F_{sp}}{k} = \frac{5wL^4}{24EI} + \left(-\frac{F_{sp} L^3}{6EI}\right)$$

$$F_{sp} = \frac{5wkL^4}{4(6EI + kL^3)} \qquad \textbf{Ans}$$

***12–116.** The beam has a moment of inertia I and is supported at its ends by a pin and roller and at its center by a rod having a cross-sectional area A and length L. Determine the tension in the rod when a uniform load w is placed on the beam. The modulus of elasticity is E.

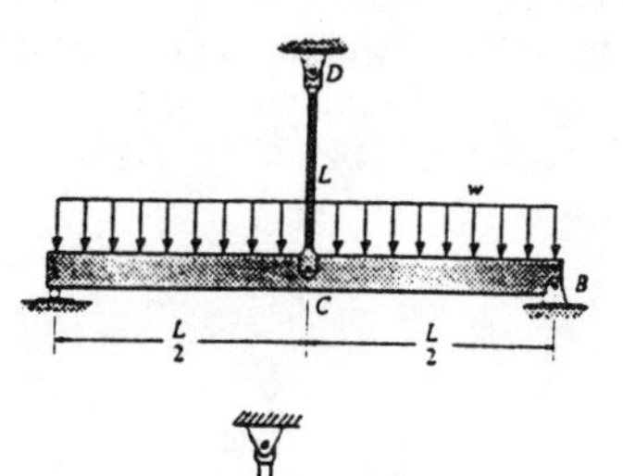

Method of Superposition : Using the table in Appendix C, the required displacements are

$$v_C' = \frac{5wL^4}{384EI} \downarrow \qquad v_C'' = \frac{PL^3}{48EI} = \frac{F_{CD}L^3}{48EI} \uparrow$$

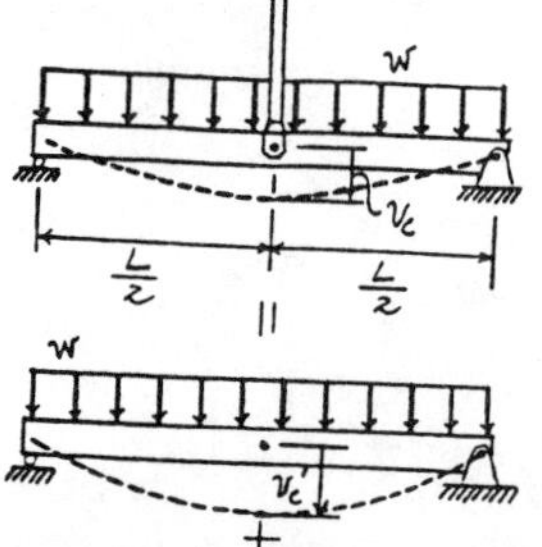

Using the axial force formula, $\quad v_C = \dfrac{F_{CD}L}{AE}$.

The compatibility condition requires

$$(+\downarrow) \qquad v_C = v_C' + v_C''$$

$$\frac{F_{CD}L}{AE} = \frac{5wL^4}{384EI} + \left(-\frac{F_{CD}L^3}{48EI}\right)$$

$$F_{CD} = \frac{5wAL^3}{8(48I + AL^2)} \qquad \textbf{Ans}$$

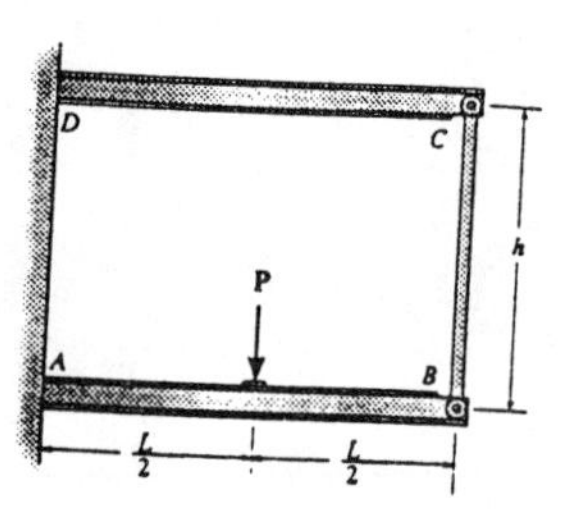

12–117. The two cantilevered beams, each having a moment of inertia I, are connected by a rod having a diameter d. Determine the tension developed in the rod if a force **P** is applied to the midpoint of the bottom beam. The modulus of elasticity for the material is E.

Method of Superposition : Using the table in Appendix C, the required displacements are

$$v_b = \frac{PL^3}{3EI} = \frac{F_{BC}L^3}{3EI} \downarrow \qquad v_B' = \frac{5PL^3}{48EI} \downarrow$$

$$v_B'' = \frac{PL^3}{3EI} = \frac{F_{BC}L^3}{3EI} \uparrow$$

Using the axial force formula, $\quad v_r = \dfrac{PL}{AE} = \dfrac{F_{BC}h}{\frac{\pi}{4}(d^2)E} = \dfrac{4F_{BC}h}{\pi d^2 E} \downarrow$

$$v_B = v_b + v_r = \frac{F_{BC}L^3}{3EI} + \frac{4F_{BC}h}{\pi d^2 E}$$

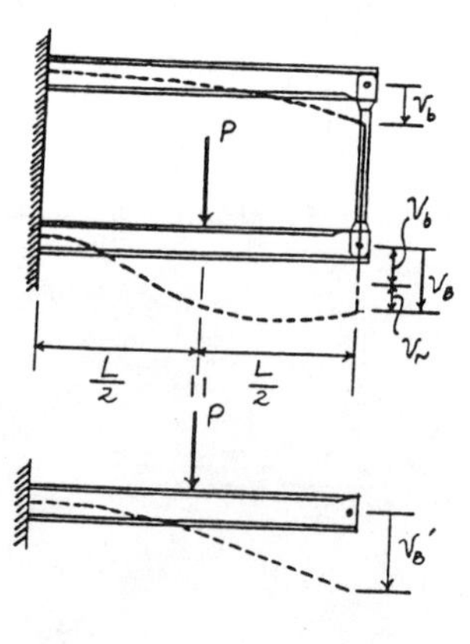

The compatibility condition requires

$$(+\downarrow) \qquad v_B = v_B' + v_C''$$

$$\frac{F_{BC}L^3}{3EI} + \frac{4F_{BC}h}{\pi d^2 E} = \frac{5PL^3}{48EI} + \left(-\frac{F_{BC}L^3}{3EI}\right)$$

$$F_{BC} = \frac{5\pi P d^2 L^3}{32(\pi d^2 L^3 + 6hI)} \qquad \textbf{Ans}$$

12–118. The beam AB has a moment of inertia $I = 475\ \text{in}^4$ and rests on the smooth supports at its ends. A 0.75-in.-diameter rod CD is welded to the center of the beam and to the fixed support at D. If the temperature of the rod is decreased by 150°F, determine the force developed in the rod. The beam and rod are both made of A-36 steel.

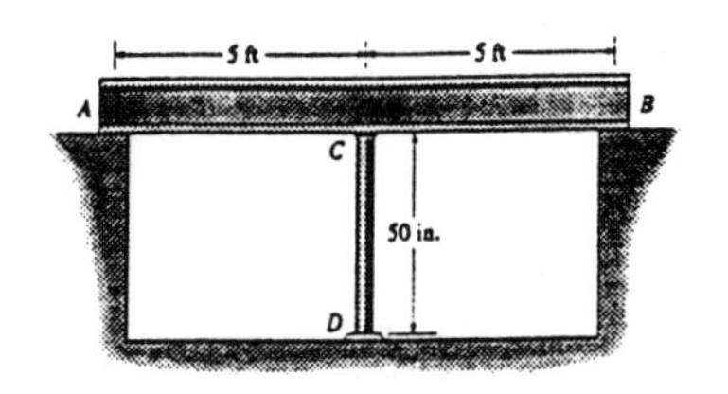

Method of Superposition : Using the table in Appendix C, the required displacements are

$$\upsilon_C = \frac{PL^3}{48EI} = \frac{F_{CD}(120^3)}{48(29)(10^3)(475)} = 0.002613F_{CD} \downarrow$$

Using the axial force formula,

$$\delta_F = \frac{PL}{AE} = \frac{F_{CD}(50)}{\frac{\pi}{4}(0.75^2)(29)(10^3)} = 0.003903F_{CD} \uparrow$$

The thermal contraction is,

$$\delta_T = \alpha\Delta TL = 6.5(10^{-6})(150)(50) = 0.04875\ \text{in.} \downarrow$$

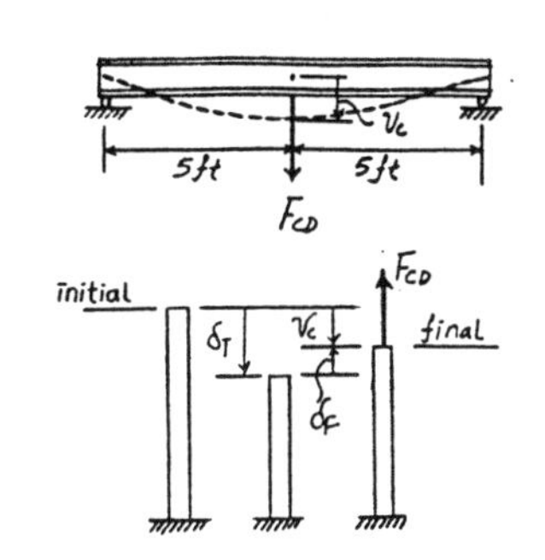

The compatibility condition requires

$$(+\downarrow) \qquad \upsilon_C = \delta_T + \delta_F$$
$$0.002613F_{CD} = 0.04875 + (-0.003903F_{CD})$$

$$F_{CD} = 7.48\ \text{kip} \qquad \textbf{Ans}$$

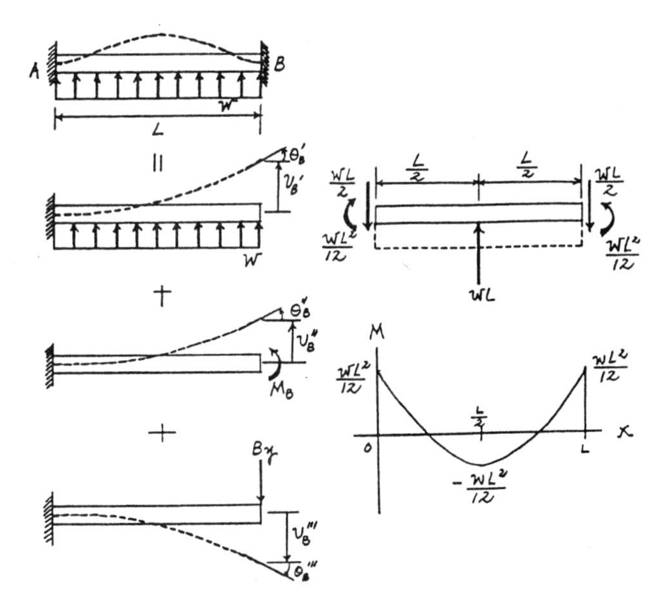

12–119. The box frame is subjected to a uniform distributed loading w along each of its sides. Determine the moment developed in each corner. Neglect the deflection due to axial load. EI is constant.

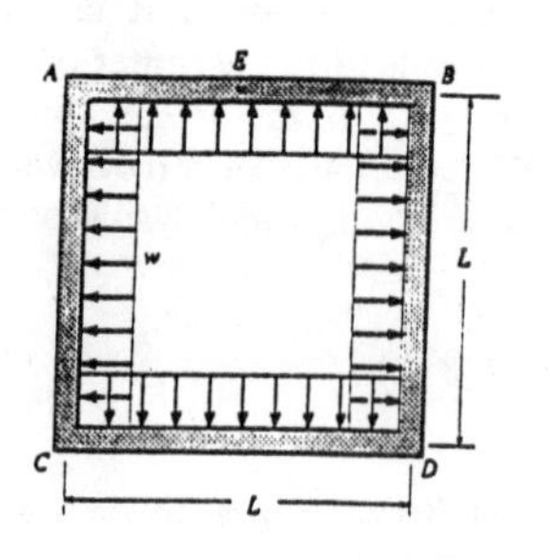

Elastic Curve : In order to maintain the right angle and zero slope (due to symmetrical loading) at the four corner joints, the box frame deformes into the shape shown when it is subjected to the internal uniform distributed load. Therefore, member AB of the frame can be modeled as a beam with both ends fixed.

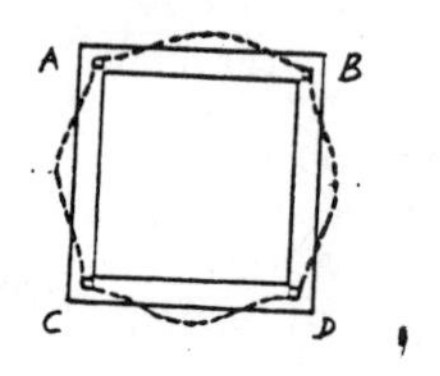

Method of Superposition : Using the table in Appendix C, the required displacements are

$$\theta_B' = \frac{wL^3}{6EI} \qquad \theta_B'' = \frac{M_B L}{EI} \qquad \theta_B''' = \frac{B_y L^2}{2EI}$$

$$\upsilon_B' = \frac{wL^4}{8EI} \uparrow \qquad \upsilon_B'' = \frac{M_B L^2}{2EI} \uparrow \qquad \upsilon_B''' = \frac{B_y L^3}{3EI} \downarrow$$

Compatibility conditions require,

$$0 = \theta_B' + \theta_B'' + \theta_B'''$$

$$0 = \frac{wL^3}{6EI} + \frac{M_B L}{EI} + \left(-\frac{B_y L^2}{2EI}\right)$$

$$0 = wL^2 + 6M_B - 3B_y L \qquad [1]$$

$$(+\uparrow) \qquad 0 = \upsilon_B' + \upsilon_B'' + \upsilon_B'''$$

$$0 = \frac{wL^4}{8EI} + \frac{M_B L^2}{2EI} + \left(-\frac{B_y L^3}{3EI}\right)$$

$$0 = 3wL^2 + 12M_B - 8B_y L \qquad [2]$$

Solving Eqs.[1] and [2] yields,

$$B_y = \frac{wL}{2}$$

$$M_B = \frac{wL^2}{12} \qquad \textbf{Ans}$$

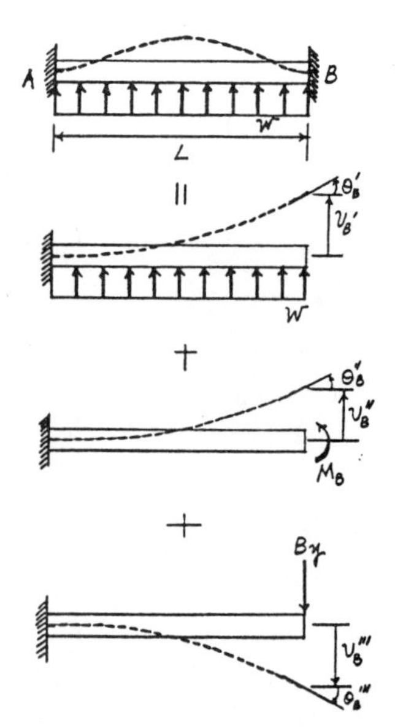

***12–120.** The rim on the flywheel has a thickness t, width b, and specific weight γ. If the flywheel is rotating at a constant rate of ω, determine the maximum moment developed in the rim. Assume that the spokes do not deform. *Hint*: Due to symmetry of the loading, the slope of the rim at each spoke is zero. Consider the radius to be sufficiently large so that the segment AB can be considered as a straight beam fixed at both ends and loaded with a uniform centrifugal force per unit length. Show that this force is $w = bt\gamma\omega^2 r/g$.

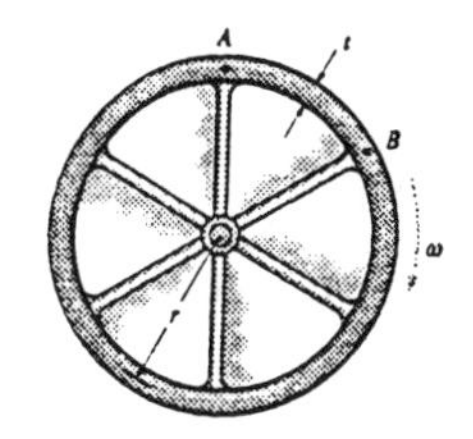

Centrifugal Force : The centrifugal force acting on a unit length of the rim rotating at a constant rate of ω is

$$w = m\omega^2 r = bt\left(\frac{\gamma}{g}\right)\omega^2 r = \frac{bt\gamma\omega^2 r}{g} \qquad (Q.E.D.)$$

Elastic Curve : Member AB of the rim is modeled as a straight beam with both of its ends fixed and subjected to a uniform centrifigal force w.
Method of Superposition : Using the table in Appendix C, the required displacements are

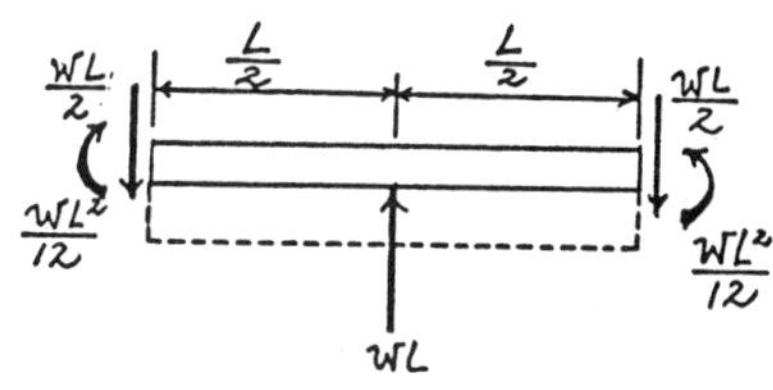

$$\theta_B{}' = \frac{wL^3}{6EI} \qquad \theta_B{}'' = \frac{M_B L}{EI} \qquad \theta_B{}''' = \frac{B_y L^2}{2EI}$$

$$\upsilon_B{}' = \frac{wL^4}{8EI}\uparrow \qquad \upsilon_B{}'' = \frac{M_B L^2}{2EI}\uparrow \qquad \upsilon_B{}''' = \frac{B_y L^3}{3EI}\downarrow$$

Compatibility requires,

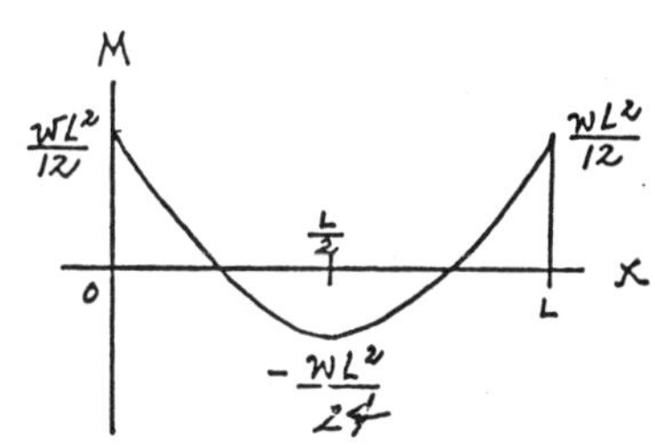

$$0 = \theta_B{}' + \theta_B{}'' + \theta_B{}'''$$

$$0 = \frac{wL^3}{6EI} + \frac{M_B L}{EI} + \left(-\frac{B_y L^2}{2EI}\right)$$

$$0 = wL^2 + 6M_B - 3B_y L \qquad [1]$$

$$(+\uparrow) \qquad 0 = \upsilon_B{}' + \upsilon_B{}'' + \upsilon_B{}'''$$

$$0 = \frac{wL^4}{8EI} + \frac{M_B L^2}{2EI} + \left(-\frac{B_y L^3}{3EI}\right)$$

$$0 = 3wL^2 + 12M_B - 8B_y L \qquad [2]$$

Solving Eqs. [1] and [2] yields,

$$B_y = \frac{wL}{2} \qquad M_B = \frac{wL^2}{12}$$

Due to symmetry, $A_y = \frac{wL}{2}$ $\qquad M_A = \frac{wL^2}{12}$

Maximum Moment : From the moment diagram, the maximum moment occurs at the two fixed end supports. With $w = \frac{bt\gamma\omega^2 r}{g}$ and $L = r\theta = \frac{\pi r}{3}$,

$$M_{\max} = \frac{wL^2}{12} = \frac{\frac{bt\gamma\omega^2 r}{g}\left(\frac{\pi r}{3}\right)^2}{12} = \frac{\pi^2 bt\gamma\omega^2 r^3}{108g} \qquad \textbf{Ans}$$

12-121. Use discontinuity functions to determine the equation of the elastic curve for the beam. Specify the slope and deflection at A. EI is constant.

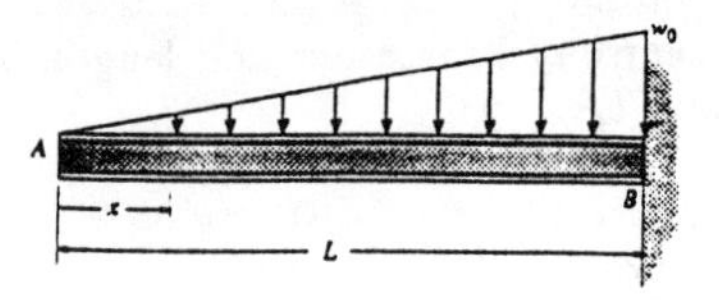

Support Reactions and Elastic Curve : As shown on FBD.
Moment Function : Using the discontinuity function,

$$M = -\frac{1}{6}\left(\frac{w_0}{L}\right) < x - 0 >^3 = -\frac{w_0}{6L}x^3$$

Slope and Elastic Curve :

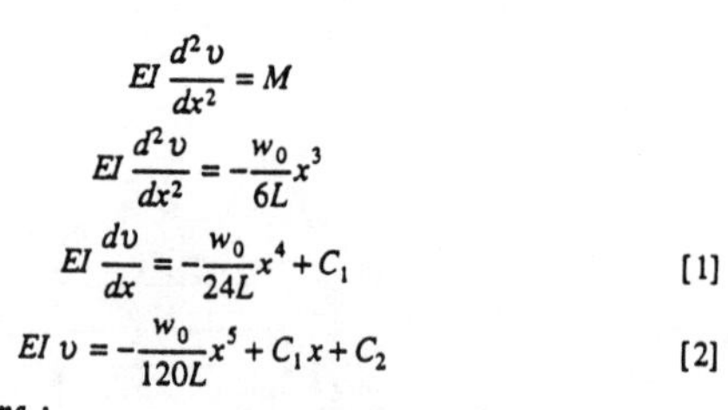

$$EI\frac{d^2 v}{dx^2} = M$$

$$EI\frac{d^2 v}{dx^2} = -\frac{w_0}{6L}x^3$$

$$EI\frac{dv}{dx} = -\frac{w_0}{24L}x^4 + C_1 \qquad [1]$$

$$EI\, v = -\frac{w_0}{120L}x^5 + C_1 x + C_2 \qquad [2]$$

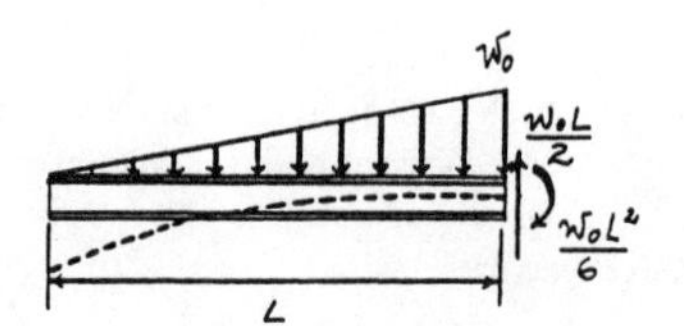

Boundary Conditions :

$\frac{dv}{dx} = 0$ at $x = L$. From Eq.[1],

$$0 = -\frac{w_0 L^3}{24} + C_1 \qquad C_1 = \frac{w_0 L^3}{24}$$

$v = 0$ at $x = L$. From Eq.[2],

$$0 = -\frac{w_0 L^4}{120} + \frac{w_0 L^4}{24} + C_2 \qquad C_2 = -\frac{w_0 L^4}{30}$$

Slope : Substituting C_1 into Eq.[1],

$$\frac{dv}{dx} = \frac{w_0}{24EIL}\left(-x^4 + L^4\right)$$

$$\theta_A = \frac{dv}{dx}\bigg|_{x=0} = \frac{w_0 L^3}{24EI} \qquad \textbf{Ans}$$

Elastic Curve : Substituting C_1 and C_2 into Eq.[2],

$$v = \frac{w_0}{120EIL}\left(-x^5 + 5L^4 x - 4L^5\right) \qquad \textbf{Ans}$$

$$v\big|_{x=0} = -\frac{w_0 L^4}{30EI} \qquad \textbf{Ans}$$

12–122. Use discontinuity functions to determine the slope at B and the deflection at C for the $W10 \times 45$. $E_{st} = 29(10^3)$ ksi.

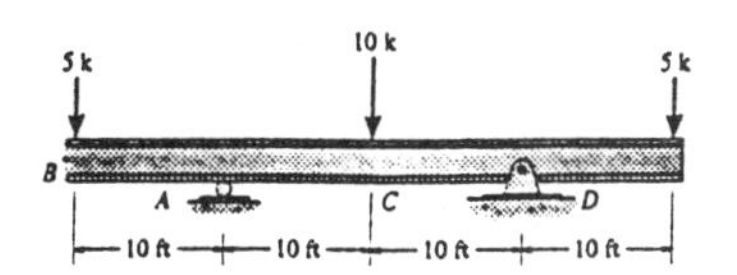

Support Reactions and Elastic Curve : As shown on FBD.

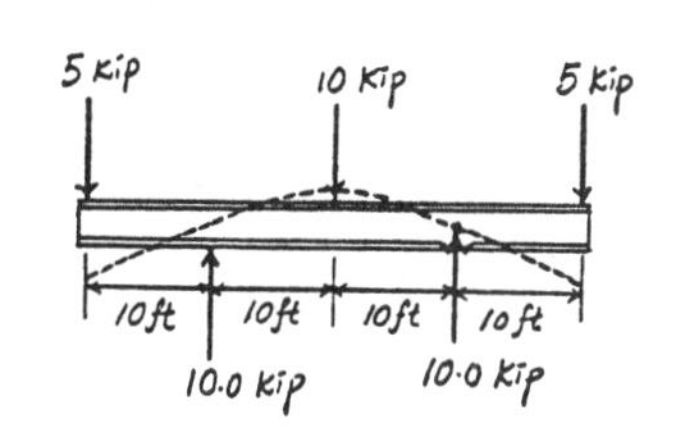

Moment Function : Using the discontinuity function,

$$M = -5 < x-0 > -(-10) < x-10 > -10 < x-20 > -(-10) < x-30 >$$
$$= -5 + 10 < x-10 > -10 < x-20 > +10 < x-30 >$$

Slope and Elastic Curve :

$$EI\frac{d^2 v}{dx^2} = M$$

$$EI\frac{d^2 v}{dx^2} = -5x + 10 < x-10 > -10 < x-20 > +10 < x-30 >$$

$$EI\frac{dv}{dx} = -\frac{5}{2}x^2 + 5 < x-10 >^2 -5 < x-20 >^2 +5 < x-30 >^2 + C_1 \quad [1]$$

$$EI\, v = -\frac{5}{6}x^3 + \frac{5}{3} < x-10 >^3 - \frac{5}{3} < x-20 >^3 + \frac{5}{3} < x-30 >^3 + C_1 x + C_2 \quad [2]$$

Boundary Conditions :

Due to symmetry, $\frac{dv}{dx} = 0$ at $x = 20$ ft. From Eq.[1],

$$0 = -\frac{5}{2}\left(20^2\right) + 5(20-10)^2 - 0 + 0 + C_1 \qquad C_1 = 500$$

$v = 0$ at $x = 10$ ft. From Eq.[2],

$$0 = -\frac{5}{6}\left(10^3\right) + 0 - 0 + 0 + 500(10) + C_2 \qquad C_2 = -\frac{12500}{3}$$

The Slope : Substituting C_1 into Eq.[1],

$$\frac{dv}{dx} = \frac{1}{EI}\left\{-\frac{5}{2}x^2 + 5 < x-10 >^2 -5 < x-20 >^2 +5 < x-30 >^2 +500\right\}$$

$$\theta_B = \left.\frac{dv}{dx}\right|_{x=0} = \frac{500 \text{ kip}\cdot\text{ft}^2}{EI} = \frac{500(144)}{29(10^3)(248)} = 0.0100 \text{ rad} \qquad \textbf{Ans}$$

Elastic Curve : Substituting C_1 and C_2 into Eq.[2],

$$v = \frac{1}{EI}\left\{-\frac{5}{6}x^3 + \frac{5}{3} < x-10 >^3 - \frac{5}{3} < x-20 >^3 + \frac{5}{3} < x-30 >^3 + 500x - \frac{12500}{3}\right\}$$

$$v_C = v|_{x=20\text{ft}} = \frac{1}{EI}\left\{-\frac{5}{6}\left(20^3\right) + \frac{5}{3}(20-10)^3 - 0 + 0 + 500(20) - \frac{12500}{3}\right\}$$
$$= \frac{2500 \text{ kip}\cdot\text{ft}^3}{3EI} = \frac{2500(1728)}{3(29)(10^3)(248)} = 0.200 \text{ in.} \qquad \textbf{Ans}$$

12–123. Solve Prob. 12–122 using the moment-area theorems.

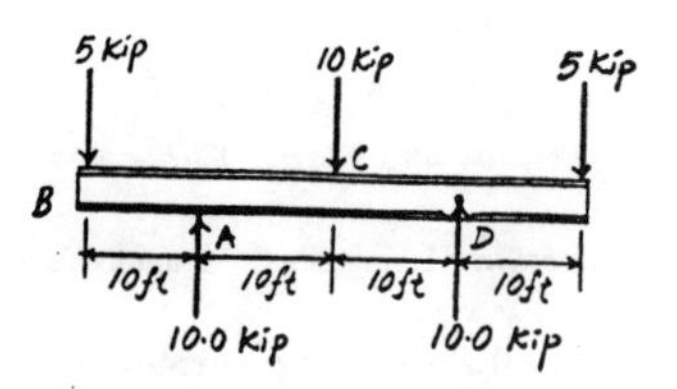

Support Reaction and Elastic Curve : As shown.

M/EI Diagram : As shown.

Moment-Area Theorems :

$$\theta_{B/C} = \frac{1}{2}\left(-\frac{50.0}{EI}\right)(20) = -\frac{500 \text{ kip}\cdot\text{ft}^2}{EI}$$

$$t_{A/C} = \frac{1}{2}\left(-\frac{50.0}{EI}\right)(10)\left(\frac{10}{3}\right) = -\frac{2500 \text{ kip}\cdot\text{ft}^3}{3EI}$$

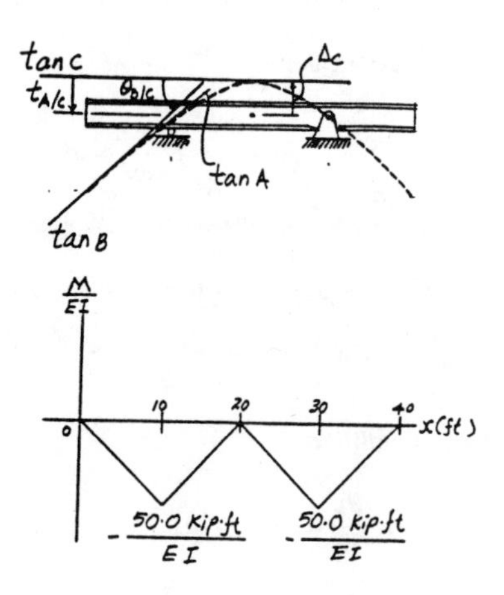

Due to symmetry, the slope at midspan (point C) is zero. Hence, the slope at B is

$$\theta_B = |\theta_{B/C}| = \frac{500 \text{ kip}\cdot\text{ft}^2}{EI} = \frac{500(144)}{29(10^3)(248)} = 0.0100 \text{ rad} \qquad \textbf{Ans}$$

The displacement at C is

$$\Delta_C = |t_{A/C}| = \frac{2500 \text{ kip}\cdot\text{ft}^3}{3EI} = \frac{2500(1728)}{3(29)(10^3)(248)} = 0.200 \text{ in.} \uparrow \qquad \textbf{Ans}$$

***12–124.** The wooden beam is subjected to the loading shown. Assume the support at A is a pin and B is a roller. Determine the slope at A and the displacement at C. Use the moment-area theorems. EI is constant.

Support Reaction and Elastic Curve : As shown.

M/EI Diagram : As shown.

Moment-Area Theorems :

$$t_{B/A} = \frac{1}{2}\left(-\frac{wa^2}{2EI}\right)(2a)\left(\frac{1}{3}\right)(2a) = -\frac{wa^4}{3EI}$$

$$t_{C/A} = \frac{1}{2}\left(-\frac{wa^2}{2EI}\right)(2a)\left(a+\frac{2}{3}a\right)+\frac{1}{3}\left(-\frac{wa^2}{2EI}\right)(a)\left(\frac{3a}{4}\right) = -\frac{23wa^4}{24EI}$$

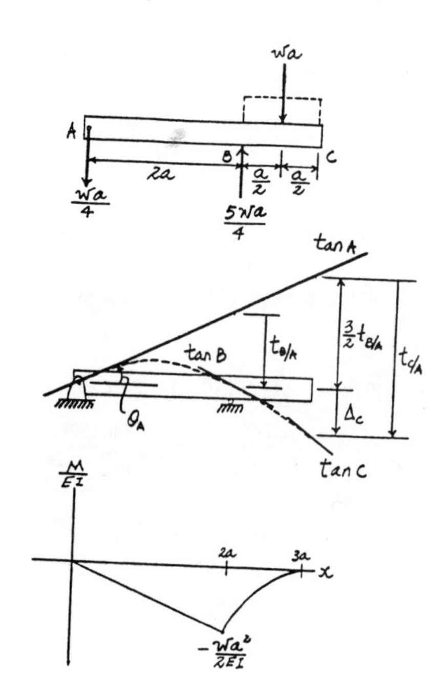

The slope at A is

$$\theta_A = \frac{|t_{B/A}|}{L_{AB}} = \frac{\frac{wa^4}{3EI}}{2a} = \frac{wa^3}{6EI} \qquad \textbf{Ans}$$

The displacement at C is

$$\Delta_C = |t_{C/A}| - \frac{3}{2}|t_{B/A}|$$
$$= \frac{23wa^4}{24EI} - \frac{3}{2}\left(\frac{wa^4}{3EI}\right)$$
$$= \frac{11wa^4}{24EI} \downarrow \qquad \textbf{Ans}$$

12–125. Determine the reactions at the supports. EI is constant. Use the method of superposition.

Support Reactions : FBD(a).

$$\xrightarrow{+} \Sigma F_x = 0; \qquad A_x = 0 \qquad \textbf{Ans}$$

$$+\uparrow \Sigma F_y = 0; \qquad A_y + B_y - \frac{w_0 L}{2} = 0 \qquad [1]$$

$$\curvearrowleft + \Sigma M_A = 0; \qquad B_y L + M_A - \frac{w_0 L}{2}\left(\frac{2}{3}L\right) = 0 \qquad [2]$$

Method of Superposition : Using the table in Appendix C, the required displacements are

$$\theta_A{}' = \frac{7 w_0 L^3}{360EI} \qquad \theta_A{}'' = \frac{M_A L}{3EI}$$

The compatibility condition requires

$$0 = \theta_A{}' + \theta_A{}''$$

$$0 = -\frac{7 w_0 L^3}{360EI} + \frac{M_A L}{3EI}$$

$$M_A = \frac{7 w_0 L^2}{120} \qquad \textbf{Ans}$$

Substituting M_A into Eqs. [1] and [2] yields,

$$B_y = \frac{11 w_0 L}{40} \qquad A_y = \frac{9 w_0 L}{40} \qquad \textbf{Ans}$$

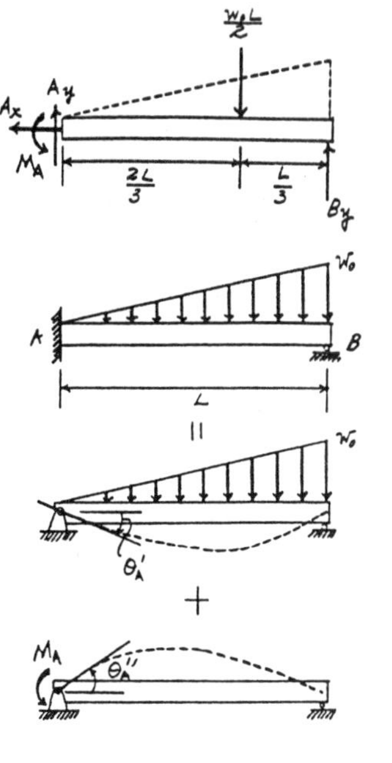

12–126. If the bearings at A and B exert only vertical reactions on the shaft, determine the slope at B and the deflection at C. Use the moment-area theorems. EI is constant.

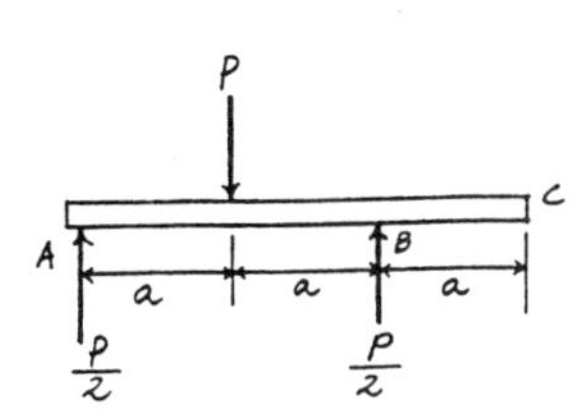

Support Reaction and Elastic Curve : As shown.

M/EI Diagram : As shown.

Moment - Area Theorems :

$$\theta_{B/D} = \frac{1}{2}\left(\frac{Pa}{2EI}\right)(a) = \frac{Pa^2}{4EI}$$

Due to symmetry, the slope at point D is zero. Hence, the slope at B is

$$\theta_B = |\theta_{B/D}| = \frac{Pa^2}{4EI} \qquad \textbf{Ans}$$

The displacement at C is

$$\Delta_C = \theta_B L_{BC} = \frac{Pa^2}{4EI}(a) = \frac{Pa^3}{4EI} \uparrow \qquad \textbf{Ans}$$

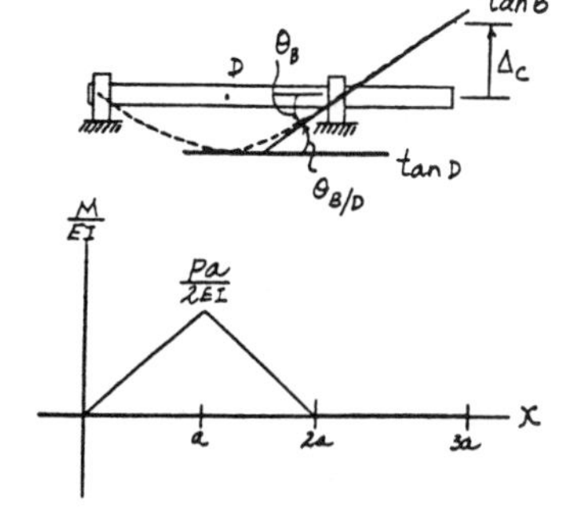

12–127. If the bearings at A and B exert only vertical reactions on the shaft, determine the slope at B and the deflection at C. Use the method of integration. EI is constant.

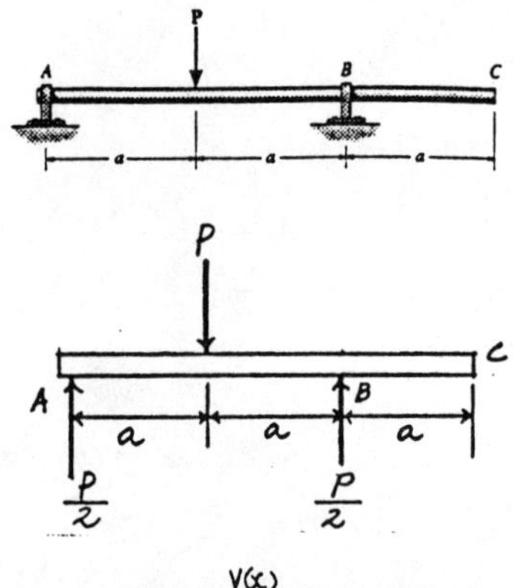

Slope and Elastic Curve :

$$EI\frac{d^2 v}{dx^2} = M(x)$$

$$EI\frac{d^2 v}{dx^2} = \frac{P}{2}x$$

$$EI\frac{dv}{dx} = \frac{P}{4}x^2 + C_1 \quad [1]$$

$$EI\, v = \frac{P}{12}x^3 + C_1 x + C_2 \quad [2]$$

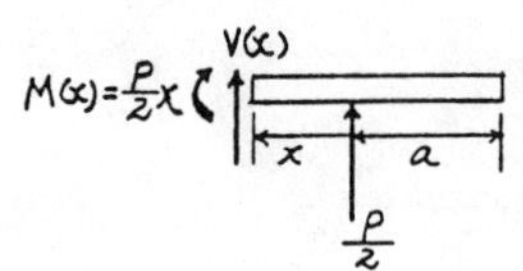

Boundary Conditions : Due to symmetry, $\frac{dv}{dx} = 0$ at $x = a$.

Also, $v = 0$ at $x = 0$.

From Eq. [1] $\quad 0 = \frac{P}{4}(a^2) + C_1 \qquad C_1 = -\frac{Pa^2}{4}$

From Eq. [2] $\quad 0 = 0 + 0 + C_2 \qquad C_2 = 0$

Slope : Substituting C_1 into Eq. [1],

$$\frac{dv}{dx} = \frac{P}{4EI}(x^2 - a^2)$$

$$\theta_B = \frac{dv}{dx}\bigg|_{x=0} = -\frac{Pa^2}{4EI} \qquad \textbf{Ans}$$

The displacement at C is

$$v_C = \theta_B L_{BC} = \frac{Pa^2}{4EI}(a) = \frac{Pa^3}{4EI} \uparrow \qquad \textbf{Ans}$$

***12–128.** The $W8 \times 24$ simply supported beam is subjected to the loading shown. Using the method of superposition, determine the deflection at its center C. The beam is made of A-36 steel.

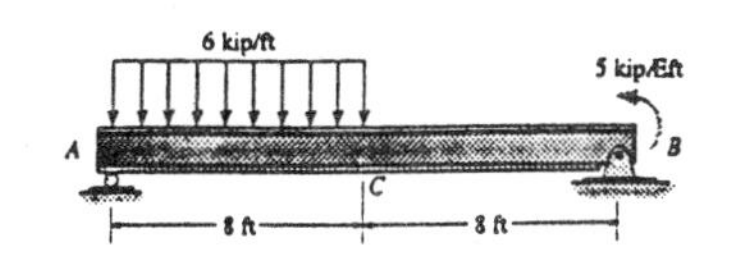

Elastic Curves : The elastic curves for the uniform distributed load and couple moment are drawn separately as shown.

Method of Superposition : Using the table in Appendix C, the required displacements are

$$(\Delta_C)_1 = \frac{-5wL^4}{768EI} = \frac{-5(6)(16^4)}{768EI} = \frac{2560 \text{ kip}\cdot\text{ft}^3}{EI} \downarrow$$

$$(\Delta_C)_2 = -\frac{M_0 x}{6EIL}(x^2 - 3Lx + 2L^2)$$

$$= -\frac{5(8)}{6EI(16)}\left[8^2 - 3(16)(8) + 2(16^2)\right]$$

$$= \frac{80 \text{ kip}\cdot\text{ft}^3}{EI} \downarrow$$

The displacement at C is

$$\Delta_C = (\Delta_C)_1 + (\Delta_C)_2$$

$$= \frac{2560}{EI} + \frac{80}{EI}$$

$$= \frac{2640 \text{ kip}\cdot\text{ft}^3}{EI}$$

$$= \frac{2640(1728)}{29(10^3)(82.8)} = 1.90 \text{ in.} \downarrow \qquad \textbf{Ans}$$

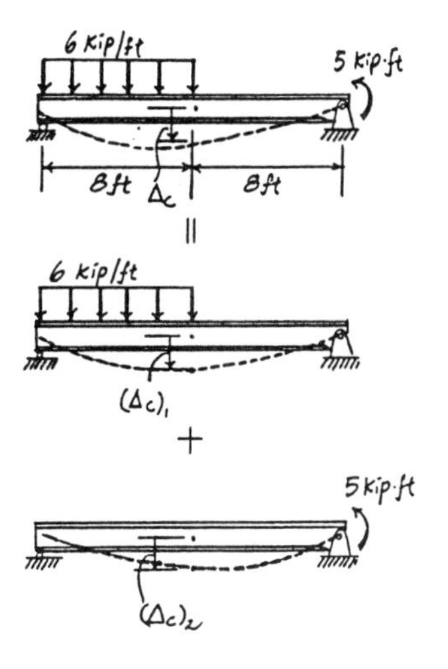

12–129. Determine the moment reactions at the supports A and B. Use the method of integration. EI is constant.

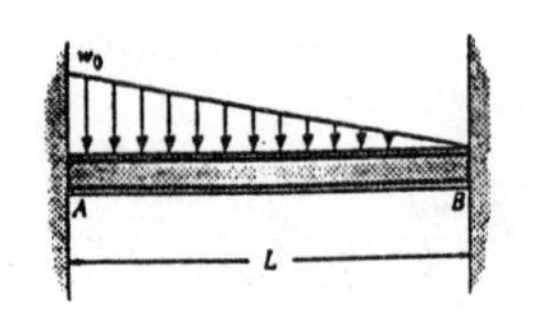

Support Reactions : FBD(a).

$$+\uparrow \Sigma F_y = 0; \quad A_y + B_y - \frac{w_0 L}{2} = 0 \qquad [1]$$

$$\curvearrowleft + \Sigma M_A = 0; \quad B_y L + M_A - M_B - \frac{w_0 L}{2}\left(\frac{L}{3}\right) = 0 \qquad [2]$$

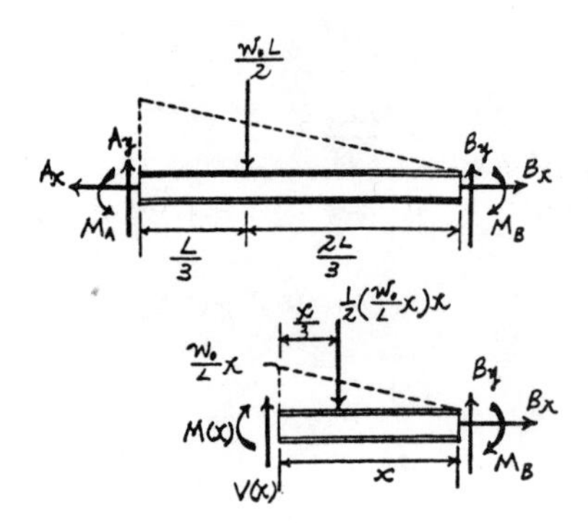

Moment Function : FBD(b).

$$\curvearrowleft + \Sigma M_{NA} = 0; \quad -M(x) - \frac{1}{2}\left(\frac{w_0}{L}x\right)x\left(\frac{x}{3}\right) - M_B + B_y x = 0$$

$$M(x) = B_y x - \frac{w_0}{6L}x^3 - M_B$$

Slope and Elastic Curve :

$$EI\frac{d^2 v}{dx^2} = M(x)$$

$$EI\frac{d^2 v}{dx^2} = B_y x - \frac{w_0}{6L}x^3 - M_B$$

$$EI\frac{dv}{dx} = \frac{B_y}{2}x^2 - \frac{w_0}{24L}x^4 - M_B x + C_1 \qquad [3]$$

$$EI\, v = \frac{B_y}{6}x^3 - \frac{w_0}{120L}x^5 - \frac{M_B}{2}x^2 + C_1 x + C_2 \qquad [4]$$

Boundary Conditions :

At $x = 0$, $\frac{dv}{dx} = 0$ From Eq.[3], $C_1 = 0$

At $x = 0$, $v = 0$. From Eq.[4] , $C_2 = 0$

At $x = L$, $\frac{dv}{dx} = 0$. From Eq. [3],

$$0 = \frac{B_y L^2}{2} - \frac{w_0 L^3}{24} - M_B L$$

$$0 = 12B_y L - w_0 L^2 - 24M_B \qquad [5]$$

At $x = L$, $v = 0$. From Eq. [4],

$$0 = \frac{B_y L^3}{6} - \frac{w_0 L^4}{120} - \frac{M_B L^2}{2}$$

$$0 = 20B_y L - w_0 L^2 - 60M_B \qquad [6]$$

Solving Eqs.[5] and [6] yields,

$$M_B = \frac{w_0 L^2}{30} \qquad \textbf{Ans}$$

$$B_y = \frac{3w_0 L}{20}$$

Substituting B_y and M_B into Eqs. [1] and [2] yields,

$$M_A = \frac{w_0 L^2}{20} \qquad \textbf{Ans}$$

$$A_y = \frac{7w_0 L}{20}$$

13–1. Determine the critical buckling load for the column. The material can be assumed rigid.

Equilibrium : The disturbing force F can be determine by summing moments about point A.

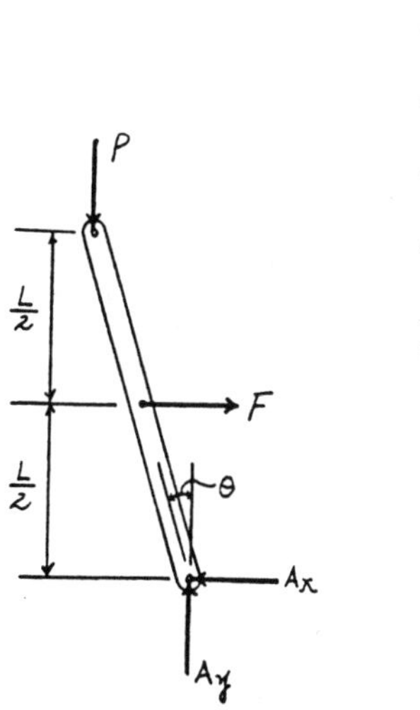

$$\circlearrowleft +\Sigma M_A = 0; \quad P(L\theta) - F\left(\frac{L}{2}\right) = 0$$

$$F = 2P\theta$$

Spring Formula : The restoring spring force F_s can be determine using spring formula $F_s = kx$.

$$F_s = k\left(\frac{L}{2}\theta\right) = \frac{kL\theta}{2}$$

Critical Buckling Load : For the mechanism to be on the verge of buckling, the disturbing force F must be equal to the restoring spring force F_s.

$$2P_{cr}\theta = \frac{kL\theta}{2}$$

$$P_{cr} = \frac{KL}{4} \qquad \textbf{Ans}$$

13–2. The rod is made from an A-36 steel rod. Determine the smallest diameter of the rod, to the nearest $\frac{1}{16}$ in., that will support the load of $P = 5$ kip without buckling. The ends are roller supported

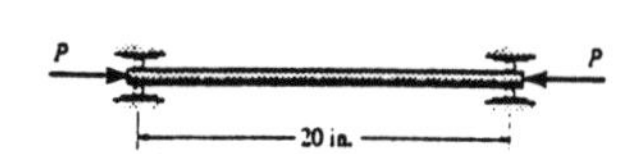

Critical Buckling Load : $I = \frac{\pi}{4}\left(\frac{d}{2}\right)^4 = \frac{\pi d^4}{64}$, $P_{cr} = 5$ kip and $K = 1$ for roller supported ends column. Applying *Euler's* formula,

$$P_{cr} = \frac{\pi^2 EI}{(KL)^2}$$

$$5 = \frac{\pi^2 (29)(10^3)\frac{\pi d^4}{64}}{[1(20)]^2}$$

$$d = 0.6142 \text{ in.}$$

Use $\qquad d = \frac{5}{8}$ in. $\qquad$ **Ans**

Critical Stress : Euler's formula is only valid if $\sigma_{cr} < \sigma_Y$.

$$\sigma_{cr} = \frac{P_{cr}}{A} = \frac{5}{\frac{\pi}{4}\left(\frac{5}{8}\right)^2} = 16.30 \text{ ksi} < \sigma_Y = 36 \text{ ksi } (O.K!)$$

13–3. The rod is made from a 1-in.-diameter steel rod. Determine the critical buckling load if the ends are roller supported. $E_{st} = 29(10^3)$ ksi, $\sigma_Y = 50$ ksi.

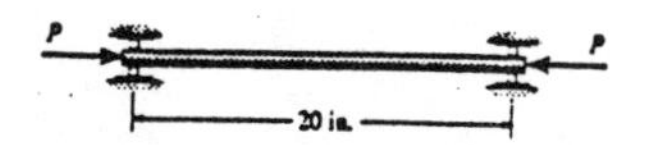

Critical Buckling Load : $I = \frac{\pi}{4}(0.5^4) = 0.015625\pi\ \text{in}^4$

and $K = 1$ for roller supported ends column. Applying *Euler's* formula,

$$P_{cr} = \frac{\pi^2 EI}{(KL)^2}$$
$$= \frac{\pi^2(29)(10^3)(0.015625\pi)}{[1(20)]^2}$$
$$= 35.12\ \text{kip} = 35.1\ \text{kip} \qquad \textbf{Ans}$$

Critical Stress : *Euler's* formula is only valid if $\sigma_{cr} < \sigma_Y$.

$$\sigma_{cr} = \frac{P_{cr}}{A} = \frac{35.12}{\frac{\pi}{4}(1^2)} = 44.72\ \text{ksi} < \sigma_Y = 50\ \text{ksi}\ (O.K!)$$

***13–4.** The W 10 × 45 is made of A-36 steel and is used as a column that has a length of 15 ft. If its ends are assumed pin supported and it is subjected to an axial load of 100 kip, determine the factor of safety with respect to buckling.

Critical Buckling Load : $I_y = 53.4\ \text{in}^4$ for a W10×45 wide flange section and $K = 1$ for pin supported ends column. Applying *Euler's* formula,

$$P_{cr} = \frac{\pi^2 EI}{(KL)^2}$$
$$= \frac{\pi^2(29)(10^3)(53.4)}{[1(15)(12)]^2}$$
$$= 471.73\ \text{kip}$$

Critical Stress : *Euler's* formula is only valid if $\sigma_{cr} < \sigma_Y$.
$A = 13.3\ \text{in}^2$ for the W10×45 wide-flange section.

$$\sigma_{cr} = \frac{P_{cr}}{A} = \frac{471.73}{13.3} = 35.47\ \text{ksi} < \sigma_Y = 36\ \text{ksi}\ (O.K!)$$

Factor of Safety :

$$\text{F.S} = \frac{P_{cr}}{P} = \frac{471.73}{100} = 4.72 \qquad \textbf{Ans}$$

13–5. The $W\,10 \times 45$ is made of A-36 steel and is used as a column that has a length of 15 ft. If the ends of the column are fixed supported, can the column support the critical load without yielding?

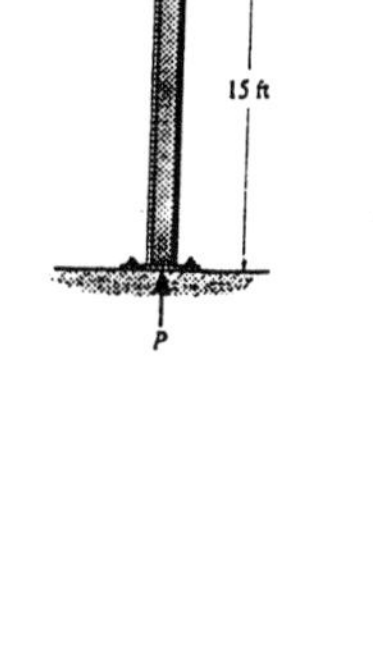

Critical Buckling Load : $I_y = 53.4\ \text{in}^4$ for a W10×45 wide flange section and $K = 0.5$ for fixed ends support column. Applying *Euler's* formula,

$$P_{cr} = \frac{\pi^2 EI}{(KL)^2}$$
$$= \frac{\pi^2(29)(10^3)(53.4)}{[0.5(15)(12)]^2}$$
$$= 1886.92\ \text{kip}$$

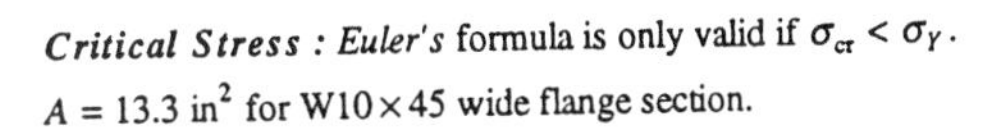

Critical Stress : *Euler's* formula is only valid if $\sigma_{cr} < \sigma_Y$.
$A = 13.3\ \text{in}^2$ for W10×45 wide flange section.

$$\sigma_{cr} = \frac{P_{cr}}{A} = \frac{1886.92}{13.3} = 141.87\ \text{ksi} > \sigma_Y = 36\ \text{ksi}\ (\textit{No!})\quad \textbf{Ans.}$$

The column will **yield** before the axial force achieves the critical load P_{cr} and th *Euler's* formula is not valid.

13–6. An A-36 steel column has a length of 4 m and is pinned at both ends. If the cross sectional area has the dimensions shown, determine the critical load.

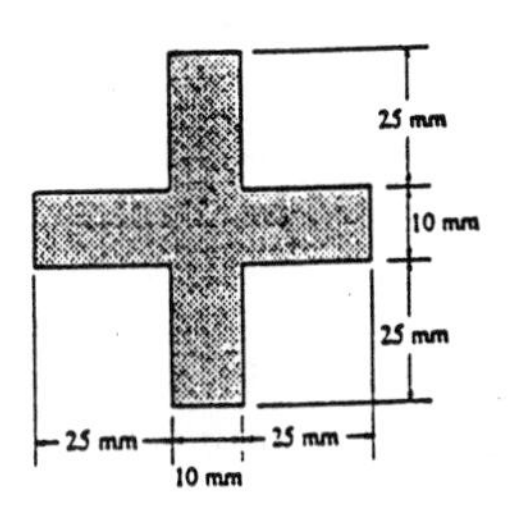

Section Properties :

$$A = 0.01(0.06) + 0.05(0.01) = 1.10\left(10^{-3}\right)\ \text{m}^2$$
$$I_x = I_y = \frac{1}{12}(0.01)\left(0.06^3\right) + \frac{1}{12}(0.05)\left(0.01^3\right) = 0.184167\left(10^{-6}\right)\ \text{m}^4$$

Critical Buckling Load : $K = 1$ for pin supported ends column. Applying *Euler's* formula,

$$P_{cr} = \frac{\pi^2 EI}{(KL)^2}$$
$$= \frac{\pi^2(200)(10^9)(0.184167)(10^{-6})}{[1(4)]^2}$$
$$= 22720.65\ \text{N} = 22.7\ \text{kN} \qquad \textbf{Ans}$$

Critical Stress : *Euler's* formula is only valid if $\sigma_{cr} < \sigma_Y$.

$$\sigma_{cr} = \frac{P_{cr}}{A} = \frac{22720.65}{1.10(10^{-3})} = 20.66\ \text{MPa} < \sigma_Y = 250\ \text{MPa}\ (\textit{O.K!})$$

13–7. Solve Prob. 13–6 if the column is fixed at its bottom and pinned at its top.

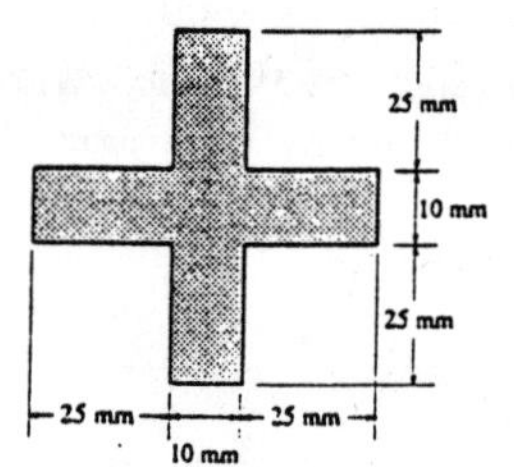

Section Properties :

$$A = 0.01(0.06) + 0.05(0.01) = 1.10\left(10^{-3}\right)\ \text{m}^2$$

$$I_x = I_y = \frac{1}{12}(0.01)\left(0.06^3\right) + \frac{1}{12}(0.05)\left(0.01^3\right) = 0.184167\left(10^{-6}\right)\ \text{m}^4$$

Critical Buckling Load : $K = 0.7$ for one end fixed and the other end pinned column. Applying *Euler's* formula,

$$P_{cr} = \frac{\pi^2 EI}{(KL)^2}$$

$$= \frac{\pi^2(200)(10^9)(0.184167)(10^{-6})}{[0.7(4)]^2}$$

$$= 46368.68\ \text{N} = 46.4\ \text{kN} \qquad \textbf{Ans}$$

Critical Stress : *Euler's* formula is only valid if $\sigma_{cr} < \sigma_Y$.

$$\sigma_{cr} = \frac{P_{cr}}{A} = \frac{46368.68}{1.10(10^{-3})} = 42.15\ \text{MPa} < \sigma_Y = 250\ \text{MPa}\ (O.K!)$$

***13–8.** The $W\,8 \times 67$ is used as a structural A-36 steel column that can be assumed fixed at its base and pinned at its top. Determine the largest axial force P that can be applied without causing it to buckle.

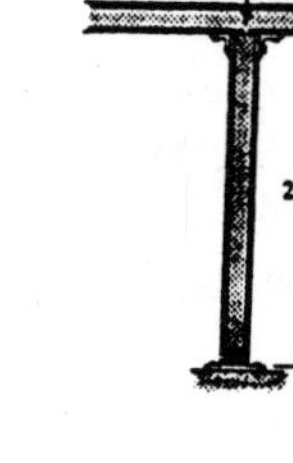

Critical Buckling Load : $I_y = 88.6\ \text{in}^4$ for a $W8 \times 67$ wide flange section and $K = 0.7$ for one end fixed and the other end pinned. Applying *Euler's* formula,

$$P_{cr} = \frac{\pi^2 EI}{(KL)^2}$$

$$= \frac{\pi^2(29)(10^3)(88.6)}{[0.7(25)(12)]^2}$$

$$= 575\ \text{kip} \qquad \textbf{Ans}$$

Critical Stress : *Euler's* formula is only valid if $\sigma_{cr} < \sigma_Y$. $A = 19.7\ \text{in}^2$ for a $W8 \times 67$ wide flange section.

$$\sigma_{cr} = \frac{P_{cr}}{A} = \frac{575.03}{19.7} = 29.19\ \text{ksi} < \sigma_Y = 36\ \text{ksi}\ (O.K!)$$

13–9. Solve Prob. 13–8 if the column is assumed fixed at its bottom and free at its top.

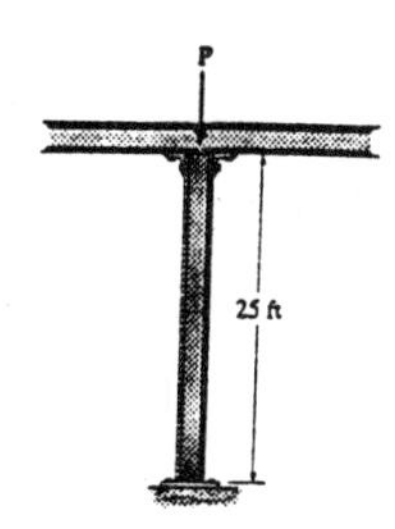

Critical Buckling Load : $I_y = 88.6\ \text{in}^4$ for a W8 × 67 wide flange section and $K = 2$ for one end fixed and the other end free. Applying *Euler's* formula,

$$P_{cr} = \frac{\pi^2 EI}{(KL)^2}$$
$$= \frac{\pi^2 (29)(10^3)(88.6)}{[2(25)(12)]^2}$$
$$= 70.4\ \text{kip} \qquad \textbf{Ans}$$

Critical Stress : *Euler's* formula is only valid if $\sigma_{cr} < \sigma_Y$.
$A = 19.7\ \text{in}^2$ for a W8 × 67 wide flange section.

$$\sigma_{cr} = \frac{P_{cr}}{A} = \frac{70.44}{19.7} = 3.58\ \text{ksi} < \sigma_Y = 36\ \text{ksi}\ (\boldsymbol{O.K!})$$

13–10. A steel column has a length of 9 m and is fixed both ends. If the cross-sectional area has the dim sions shown, determine the critical load. $E_{st} = 200$ G $\sigma_Y = 250$ MPa.

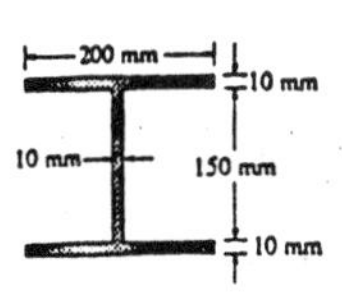

Section Properties :

$$A = 0.2(0.17) - 0.19(0.15) = 5.50(10^{-3})\ \text{m}^2$$
$$I_x = \frac{1}{12}(0.2)(0.17^3) - \frac{1}{12}(0.19)(0.15^3) = 28.44583(10^{-6})\ \text{m}^4$$
$$I_y = 2\left[\frac{1}{12}(0.01)(0.2^3)\right] + \frac{1}{12}(0.15)(0.01^3) = 13.34583(10^{-6})\ \text{m}^4 \quad (\textbf{\textit{Controls!}})$$

Critical Buckling Load : $K = 0.5$ for fixed support ends column. Applying *Euler's* formula,

$$P_{cr} = \frac{\pi^2 EI}{(KL)^2}$$
$$= \frac{\pi^2 (200)(10^9)(13.34583)(10^{-6})}{[0.5(9)]^2}$$
$$= 1300919\ \text{N} = 1.30\ \text{MN} \qquad \textbf{Ans}$$

Critical Stress : *Euler's* formula is only valid if $\sigma_{cr} < \sigma_Y$.

$$\sigma_{cr} = \frac{P_{cr}}{A} = \frac{1300919}{5.50(10^{-3})} = 236.53\ \text{MPa} < \sigma_Y = 250\ \text{MPa}\ (\boldsymbol{O.K!})$$

13–11. Solve Prob. 13–10 if the column is pinned at its top and bottom.

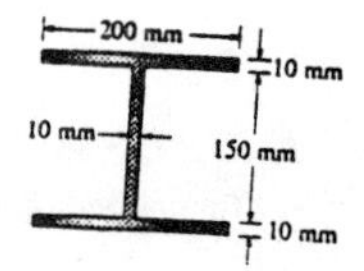

Section Properties :

$$A = 0.2(0.17) - 0.19(0.15) = 5.50(10^{-3})\ \text{m}^2$$

$$I_x = \frac{1}{12}(0.2)(0.17^3) - \frac{1}{12}(0.19)(0.15^3) = 28.44583(10^{-6})\ \text{m}^4$$

$$I_y = 2\left[\frac{1}{12}(0.01)(0.2^3)\right] + \frac{1}{12}(0.15)(0.01^3) = 13.34583(10^{-6})\ \text{m}^4 \quad (\textit{Controls !})$$

Critical Buckling Load : $K = 1$ for pin supported ends column. Applying *Euler's* formula,

$$P_{cr} = \frac{\pi^2 EI}{(KL)^2} = \frac{\pi^2(200)(10^9)(13.34583)(10^{-6})}{[1(9)]^2}$$

$$= 325229.87\ \text{N} = 325\ \text{kN} \qquad \textbf{Ans}$$

Critical Stress : *Euler's* formula is only valid if $\sigma_{cr} < \sigma_Y$.

$$\sigma_{cr} = \frac{P_{cr}}{A} = \frac{325229.87}{5.50(10^{-3})} = 59.13\ \text{MPa} < \sigma_Y = 250\ \text{MPa}\ (\textit{O.K !})$$

***13–12.** The $W\ 12 \times 87$ structural A-36 steel column has a length of 16 ft. If its bottom end is fixed supported while its top is free, and it is subjected to an axial load of $P = 275$ kip, determine the factor of safety with respect to buckling.

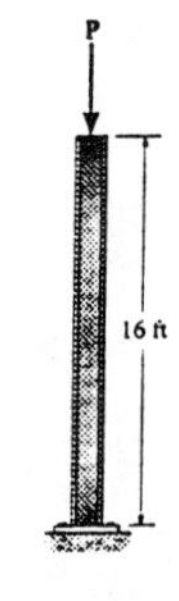

Critical Buckling Load : $I_y = 241\ \text{in}^4$ for a $W12 \times 87$ wide flange section and $K = 2$ for one end fixed and the other end free. Applying *Euler's* formula,

$$P_{cr} = \frac{\pi^2 EI}{(KL)^2} = \frac{\pi^2(29)(10^3)(241)}{[2(16)(12)]^2} = 467.79\ \text{kip}$$

Critical Stress : *Euler's* formula is only valid if $\sigma_{cr} < \sigma_Y$. $A = 25.6\ \text{in}^2$ for $W12 \times 87$ wide flange section.

$$\sigma_{cr} = \frac{P_{cr}}{A} = \frac{467.79}{25.6} = 18.27\ \text{ksi} < \sigma_Y = 36\ \text{ksi}\ (\textit{O.K !})$$

Factor of Safety :

$$\text{F.S} = \frac{P_{cr}}{P} = \frac{467.79}{275} = 1.70 \qquad \textbf{Ans}$$

13–13. The $W\,12 \times 87$ structural A-36 steel column has a length of 16 ft. If its bottom end is fixed supported while its top is free, determine the largest axial load it can support. Use a factor of safety with respect to buckling of F.S. = 1.75.

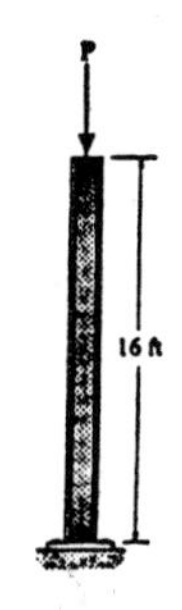

Critical Buckling Load : $I_y = 241\ \text{in}^4$ for a W12×87 wide flange section and $K = 2$ for one end fixed and the other end free. Applying *Euler's* formula,

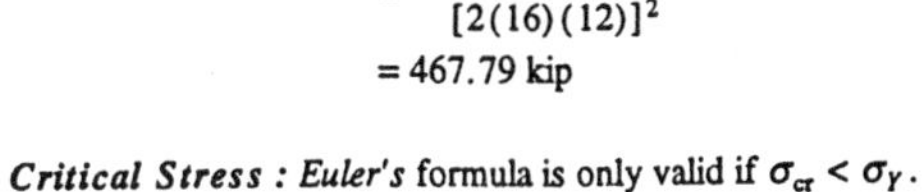

$$P_{cr} = \frac{\pi^2 EI}{(KL)^2}$$
$$= \frac{\pi^2(29)(10^3)(241)}{[2(16)(12)]^2}$$
$$= 467.79\ \text{kip}$$

Critical Stress : *Euler's* formula is only valid if $\sigma_{cr} < \sigma_Y$.
$A = 25.6\ \text{in}^2$ for a W12×87 wide flange section.

$$\sigma_{cr} = \frac{P_{cr}}{A} = \frac{467.79}{25.6} = 18.27\ \text{ksi} < \sigma_Y = 36\ \text{ksi}\ (\textbf{O.K!})$$

Factor of Safety :

$$\text{F.S} = \frac{P_{cr}}{P}$$
$$1.75 = \frac{467.79}{P}$$

$$P = 267\ \text{kip}$$ **Ans**

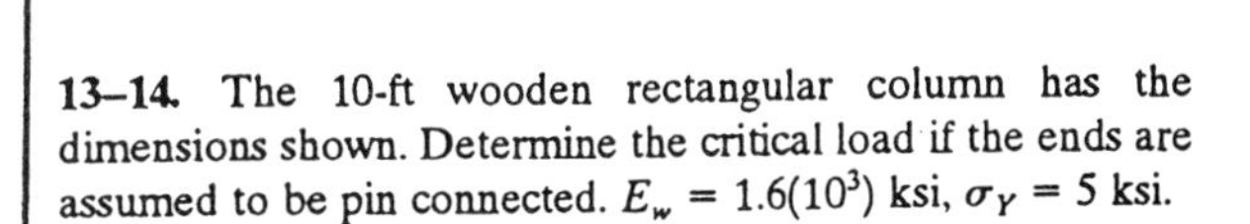

13–14. The 10-ft wooden rectangular column has the dimensions shown. Determine the critical load if the ends are assumed to be pin connected. $E_w = 1.6(10^3)$ ksi, $\sigma_Y = 5$ ksi.

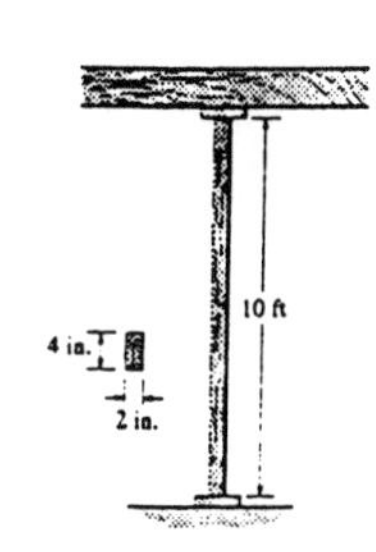

Section Properties :

$$A = 4(2) = 8.00\ \text{in}^2$$
$$I_x = \frac{1}{12}(2)(4^3) = 10.667\ \text{in}^4$$
$$I_y = \frac{1}{12}(4)(2^3) = 2.6667\ \text{in}^4\ (\textbf{Controls!})$$

Critical Buckling Load : $K = 1$ for pin supported ends column. Applying *Euler's* formula, .

$$P_{cr} = \frac{\pi^2 EI}{(KL)^2}$$
$$= \frac{\pi^2(1.6)(10^3)(2.6667)}{[1(10)(12)]^2}$$

$$= 2.924\ \text{kip} = 2.92\ \text{kip}$$ **Ans**

Critical Stress : *Euler's* formula is only valid if $\sigma_{cr} < \sigma_Y$.

$$\sigma_{cr} = \frac{P_{cr}}{A} = \frac{2.924}{8.00} = 0.3655\ \text{ksi} < \sigma_Y = 5\ \text{ksi}\ (\textbf{O.K!})$$

13–15. The 10-ft column has the dimensions shown. Determine the critical load if the bottom is fixed and the top is pinned. $E_w = 1.6(10^3)$ ksi, $\sigma_Y = 5$ ksi.

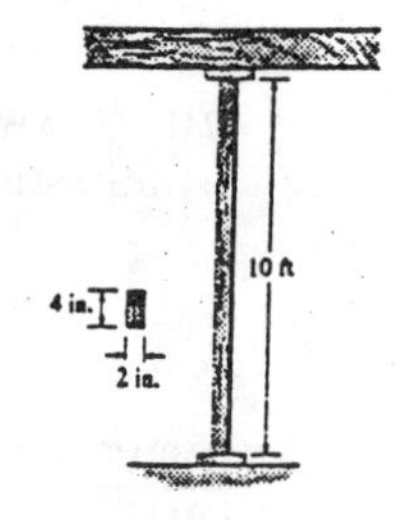

Section Properties :

$$A = 4(2) = 8.00 \text{ in}^2$$

$$I_x = \frac{1}{12}(2)(4^3) = 10.667 \text{ in}^4$$

$$I_y = \frac{1}{12}(4)(2^3) = 2.6667 \text{ in}^4 \textit{ (Controls!)}$$

Critical Buckling Load : $K = 0.7$ for column with one end fixed and the other end pinned. Applying *Euler's* formula,

$$P_{cr} = \frac{\pi^2 EI}{(KL)^2}$$

$$= \frac{\pi^2(1.6)(10^3)(2.6667)}{[0.7(10)(12)]^2}$$

$$= 5.968 \text{ kip} = 5.97 \text{ kip} \qquad \textbf{Ans}$$

Critical Stress : *Euler's* formula is only valid if $\sigma_{cr} < \sigma_Y$.

$$\sigma_{cr} = \frac{P_{cr}}{A} = \frac{5.968}{8.00} = 0.7460 \text{ ksi} < \sigma_Y = 5 \text{ ksi } (O.K!)$$

***13–16.** An L-2 steel link in a forging machine is pin connected to the forks at its ends as shown. Determine the maximum load P it can carry without buckling. Use a factor of safety with respect to buckling of F.S. = 1.75. Note from the figure on the left that the ends are pinned for buckling, whereas from the figure on the right the ends are fixed.

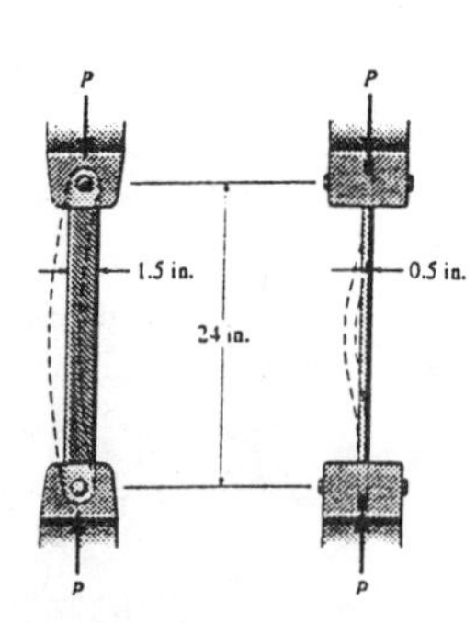

Section Properties :

$$A = 1.5(0.5) = 0.750 \text{ in}^2$$

$$I_x = \frac{1}{12}(0.5)(1.5^3) = 0.140625 \text{ in}^4$$

$$I_y = \frac{1}{12}(1.5)(0.5^3) = 0.015625 \text{ in}^4$$

Critical Buckling Load : With respect to the $x-x$ axis, $K = 1$ (column with both ends pinned). Applying *Euler's* formula,

$$P_{cr} = \frac{\pi^2 EI}{(KL)^2}$$

$$= \frac{\pi^2(29.0)(10^3)(0.140625)}{[1(24)]^2}$$

$$= 69.88 \text{ kip}$$

With respect to the $y-y$ axis, $K = 0.5$ (column with both ends fixed).

$$P_{cr} = \frac{\pi^2 EI}{(KL)^2}$$

$$= \frac{\pi^2(29.0)(10^3)(0.015625)}{[0.5(24)]^2}$$

$$= 31.06 \text{ kip} \qquad \textit{(Controls!)}$$

Critical Stress : *Euler's* formula is only valid if $\sigma_{cr} < \sigma_Y$.

$$\sigma_{cr} = \frac{P_{cr}}{A} = \frac{31.06}{0.75} = 41.41 \text{ ksi} < \sigma_Y = 102 \text{ ksi } (O.K!)$$

Factor of Safety :

$$\text{F.S} = \frac{P_{cr}}{P}$$

$$1.75 = \frac{31.06}{P}$$

$$P = 17.7 \text{ kip} \qquad \textbf{Ans}$$

13–17. The member has a symmetric cross section. Assuming that it is pin connected at its ends, determine the largest force it can support without buckling or yielding. The member is made of 2014-T6 aluminum.

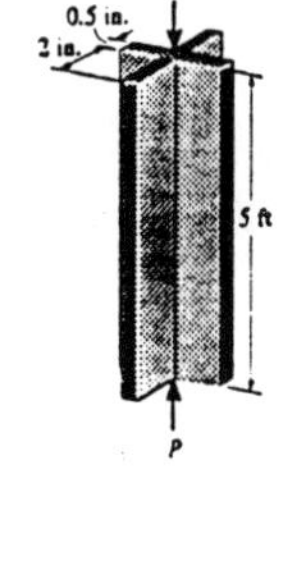

Section Properties :

$$A = 0.5(4.5) + 4(0.5) = 4.25 \text{ in}^2$$

$$I_x = I_y = \frac{1}{12}(0.5)\left(4.5^3\right) + \frac{1}{12}(4)\left(0.5^3\right) = 3.8385 \text{ in}^4$$

Critical Buckling Load : $K = 1$ for column with both ends pinned. Applying *Euler's* formula,

$$P_{cr} = \frac{\pi^2 EI}{(KL)^2} = \frac{\pi^2(10.6)(10^3)(3.8385)}{[1(5)(12)]^2}$$

$$= 111.55 \text{ kip} = 112 \text{ kip} \qquad \textbf{Ans}$$

Critical Stress : *Euler's* formula is only valid if $\sigma_{cr} < \sigma_Y$.

$$\sigma_{cr} = \frac{P_{cr}}{A} = \frac{111.55}{4.25} = 26.25 \text{ ksi} < \sigma_Y = 60 \text{ ksi } (\boldsymbol{O.K!})$$

13–18. Solve Prob. 13–17 if the column has fixed-connected ends.

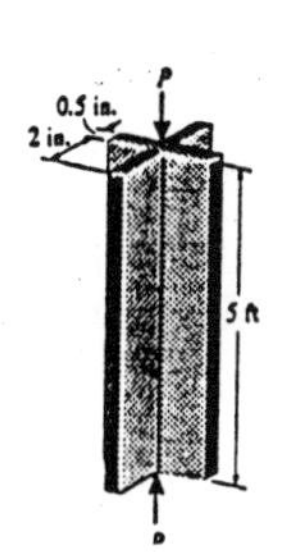

Section Properties :

$$A = 0.5(4.5) + 4(0.5) = 4.25 \text{ in}^2$$

$$I_x = I_y = \frac{1}{12}(0.5)\left(4.5^3\right) + \frac{1}{12}(4)\left(0.5^3\right) = 3.8385 \text{ in}^4$$

Critical Buckling Load : $K = 0.5$ for column with both ends fixed. Applying *Euler's* formula,

$$P_{cr} = \frac{\pi^2 EI}{(KL)^2} = \frac{\pi^2(10.6)(10^3)(3.8385)}{[0.5(5)(12)]^2}$$

$$= 446.20 \text{ kip}$$

Critical Stress : *Euler's* formula is only valid if $\sigma_{cr} < \sigma_Y$.

$$\sigma_{cr} = \frac{P_{cr}}{A} = \frac{446.20}{4.25} = 105.0 \text{ ksi} > \sigma_Y = 60 \text{ ksi } (\textit{No Good!})$$

Hence, **yielding** of the material controls.

$$P = \sigma_Y A = 60(4.25) = 255 \text{ kip} \qquad \textbf{Ans}$$

13–19. The A-36 steel pipe has an outer diameter of 2 in. and a thickness of 0.5 in. If it is held in place by a guywire, determine the largest vertical force P that can be applied without causing the pipe to buckle. Assume that the ends of the pipe are pin connected.

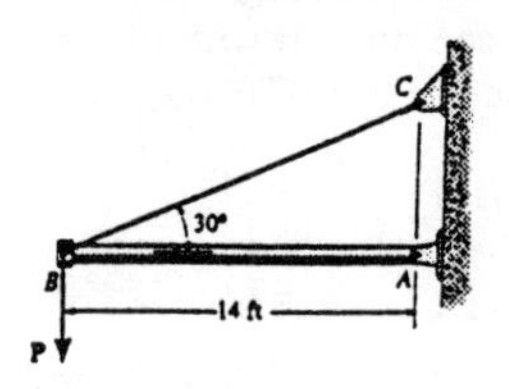

Member Forces : Use the method of joints,

$$+\uparrow \Sigma F_y = 0; \quad F_{BC}\sin 30° - P = 0 \qquad F_{BC} = 2.00P$$

$$\overset{+}{\rightarrow} \Sigma F_x = 0; \quad 2.00P\cos 30° - F_{AB} = 0 \qquad F_{AB} = 1.7321P$$

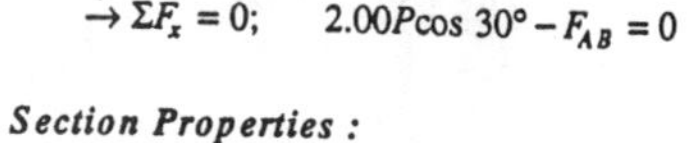

Section Properties :

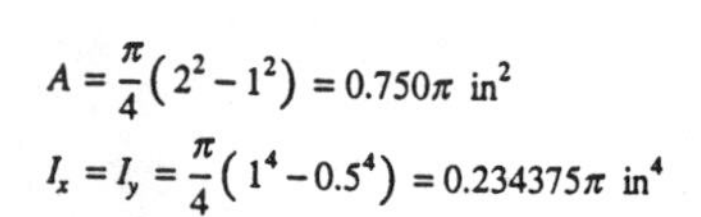

$$A = \frac{\pi}{4}\left(2^2 - 1^2\right) = 0.750\pi \text{ in}^2$$

$$I_x = I_y = \frac{\pi}{4}\left(1^4 - 0.5^4\right) = 0.234375\pi \text{ in}^4$$

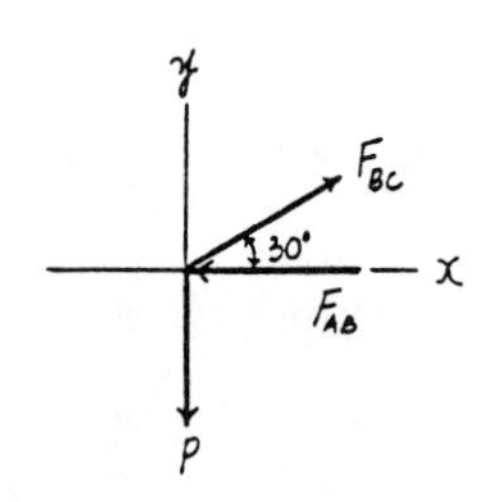

Critical Buckling Load : $K = 1$ for column with both ends pinned. Applying *Euler's* formula,

$$P_{cr} = \frac{\pi^2 EI}{(KL)^2}$$

$$1.7321P = \frac{\pi^2 (29)(10^3)(0.234375\pi)}{[1(14)(12)]^2}$$

$$P = 4.31 \text{ kip} \qquad \textbf{Ans}$$

Critical Stress : *Euler's* formula is only valid if $\sigma_{cr} < \sigma_Y$.

$$\sigma_{cr} = \frac{P_{cr}}{A} = \frac{1.7321(4.311)}{0.750\pi} = 3.169 \text{ ksi} < \sigma_Y = 36 \text{ ksi } (O.K!)$$

***13–20.** The A-36 steel pipe has an outer diameter of 2 in. If it is held in place by a guywire, determine its required inner diameter to the nearest $\frac{1}{8}$ in., so that it can support a maximum vertical load of $P = 4$ kip without causing the pipe to buckle. Assume the ends of the pipe are pin connected.

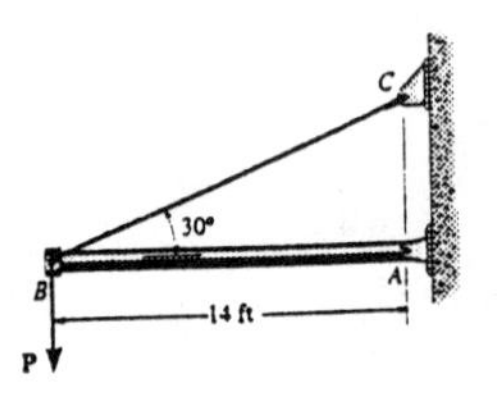

Member Forces : Use the method of joints,

$$+\uparrow \Sigma F_y = 0; \quad F_{BC}\sin 30° - 4 = 0 \qquad F_{BC} = 8.00 \text{ kip}$$

$$\overset{+}{\rightarrow} \Sigma F_x = 0; \quad 8.00\cos 30° - F_{AB} = 0 \qquad F_{AB} = 6.928 \text{ kip}$$

Section Properties :

$$A = \frac{\pi}{4}\left(2^2 - d_i^2\right) = \frac{\pi}{4}\left(4 - d_i^2\right)$$

$$I_x = I_y = \frac{\pi}{4}\left[1^4 - \left(\frac{d_i}{2}\right)^4\right] = \frac{\pi}{64}\left(16 - d_i^4\right)$$

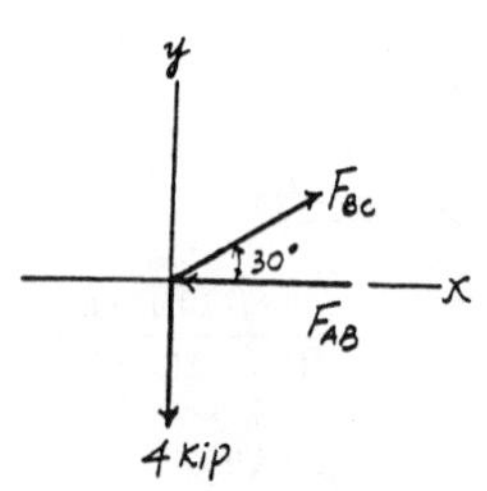

Critical Buckling Load : $K = 1$ for column with both ends pinned. Applying *Euler's* formula,

$$P_{cr} = \frac{\pi^2 EI}{(KL)^2}$$

$$6.928 = \frac{\pi^2 (29)(10^3)\left[\frac{\pi}{64}(16 - d_i^4)\right]}{[1(14)(12)]^2}$$

$$d_i = 1.201 \text{ in.}$$

Use $\quad d_i = 1\frac{1}{8}$ in. $\qquad$ **Ans**

Critical Stress : *Euler's* formula is only valid if $\sigma_{cr} < \sigma_Y$.

$$A = \frac{\pi}{4}\left[4 - \left(\frac{9}{8}\right)^2\right] = 2.1475 \text{ in}^2$$

$$\sigma_{cr} = \frac{P_{cr}}{A} = \frac{6.928}{2.1475} = 3.226 \text{ ksi} < \sigma_Y = 36 \text{ ksi } (O.K!)$$

13–21. The linkage is made using two A-36 steel rods, each having a circular cross section. Determine the diameter of each rod to the nearest $\frac{1}{8}$ in. that will support the 900-lb load. Assume that the rods are pin connected at their ends. Use a factor of safety with respect to buckling of F.S. = 1.8.

Member Forces : Use the method of joints,

$$\xrightarrow{+} \Sigma F_x = 0; \quad \frac{4}{5}F_{BA} - \frac{3}{5}F_{BC} = 0 \qquad [1]$$

$$+\uparrow \Sigma F_y = 0; \quad \frac{3}{5}F_{BA} + \frac{4}{5}F_{BC} - 900 = 0 \qquad [2]$$

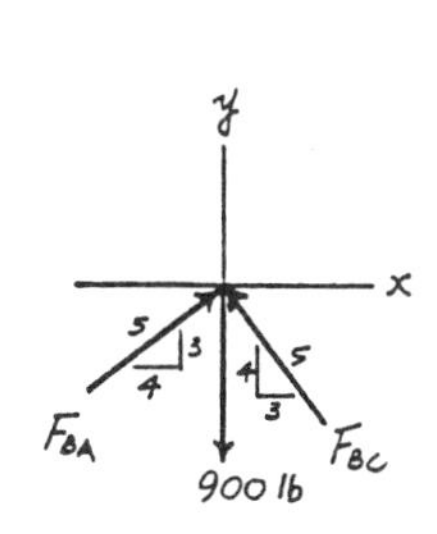

Solving Eqs. [1] and [2] yields,

$$F_{BC} = 720 \text{ lb} \qquad F_{BA} = 540 \text{ lb}$$

Critical Buckling Load : $K = 1$ for column with both ends pinned. Applying *Euler's* formula to member AB,

$$P_{cr} = 1.8F_{BA} = \frac{\pi^2 EI}{(KL_{AB})^2}$$

$$1.8(540) = \frac{\pi^2 (29)(10^6)\left(\frac{\pi}{64}d_{AB}^4\right)}{[1(20)(12)]^2}$$

$$d_{AB} = 1.413 \text{ in.}$$

Use $\quad d_{AB} = 1\frac{1}{2}$ in. $\quad$ **Ans**

For member BC,

$$P_{cr} = 1.8F_{BC} = \frac{\pi^2 EI}{(KL_{BC})^2}$$

$$1.8(720) = \frac{\pi^2 (29)(10^6)\left(\frac{\pi}{64}d_{BC}^4\right)}{[1(15)(12)]^2}$$

$$d_{BC} = 1.315 \text{ in.}$$

Use $\quad d_{BC} = 1\frac{3}{8}$ in. $\quad$ **Ans**

Critical Stress : *Euler's* formula is only valid if $\sigma_{cr} < \sigma_Y$.

$$(\sigma_{cr})_{AB} = \frac{P_{cr}}{A} = \frac{1.8(540)}{\frac{\pi}{4}(1.50^2)} = 550.0 \text{ psi} < \sigma_Y = 36 \text{ ksi } (\textit{O.K!})$$

$$(\sigma_{cr})_{BC} = \frac{P_{cr}}{A} = \frac{1.8(720)}{\frac{\pi}{4}(1.375^2)} = 872.8 \text{ psi} < \sigma_Y = 36 \text{ ksi } (\textit{O.K!})$$

13–22. The linkage is made using two A-36 steel rods, each having a circular cross section. If each rod has a diameter of $\frac{3}{4}$ in., determine the largest load it can support without causing any rod to buckle. Assume that the rods are pin connected at their ends.

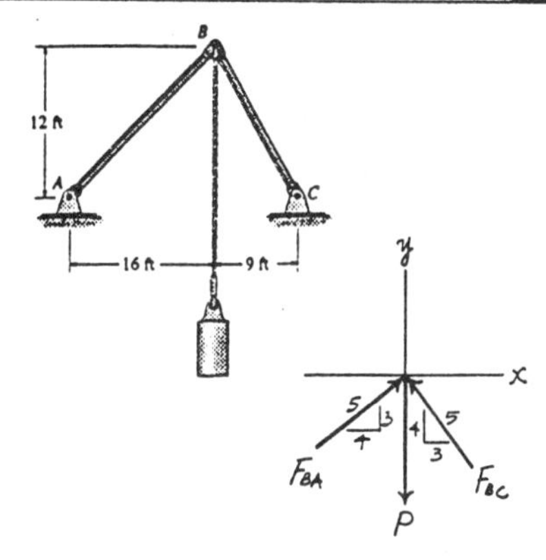

Member Forces : Use the method of joints,

$$\xrightarrow{+} \Sigma F_x = 0; \quad \frac{4}{5}F_{BA} - \frac{3}{5}F_{BC} = 0 \qquad [1]$$

$$+\uparrow \Sigma F_y = 0; \quad \frac{3}{5}F_{BA} + \frac{4}{5}F_{BC} - P = 0 \qquad [2]$$

Solving Eqs. [1] and [2] yields,

$$F_{BC} = 0.800P \qquad F_{BA} = 0.600P$$

Critical Buckling Load : $K = 1$ for column with both ends pinned Assume member AB buckles. Applying *Euler's* formula,

$$P_{cr} = F_{BA} = \frac{\pi^2 EI}{(KL_{AB})^2}$$

$$0.600P = \frac{\pi^2 (29)(10^6)\left[\frac{\pi}{4}(0.375^4)\right]}{[1(20)(12)]^2}$$

$$P = 128.6 \text{ lb} = 129 \text{ lb } (\textbf{\textit{Controls!}})$$

Assume member BC buckles,

$$P_{cr} = F_{BC} = \frac{\pi^2 EI}{(KL_{BC})^2}$$

$$0.800P = \frac{\pi^2 (29)(10^6)\left[\frac{\pi}{4}(0.375^4)\right]}{[1(15)(12)]^2}$$

$$P = 171.5 \text{ lb}$$

Critical Stress : *Euler's* formula is only valid if $\sigma_{cr} < \sigma_Y$.

$$(\sigma_{cr})_{BC} = \frac{P_{cr}}{A} = \frac{0.8(128.6)}{\frac{\pi}{4}(0.75^2)} = 232.9 \text{ psi} < \sigma_Y = 36 \text{ ksi } (\textit{O.K!})$$

13–23. Assuming the strut brace is pinned at its ends A and B, determine the maximum force P that can be applied to the end of member CD without causing the brace to buckle. The brace has a width of 1 in. and is made of aluminum 6061-T6.

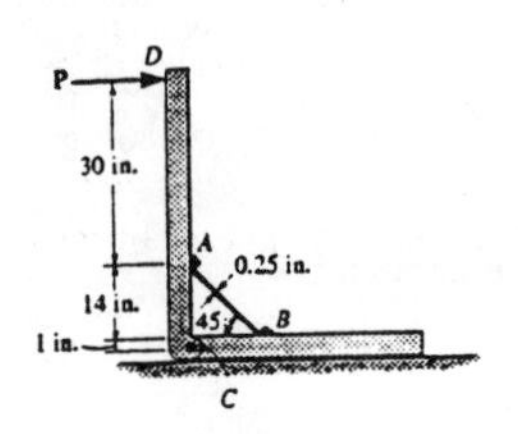

Support Reactions :

$$\circlearrowleft +\Sigma M_C = 0; \quad F_{AB}\cos 45°(15) - P(45) = 0$$
$$F_{AB} = 4.2426P$$

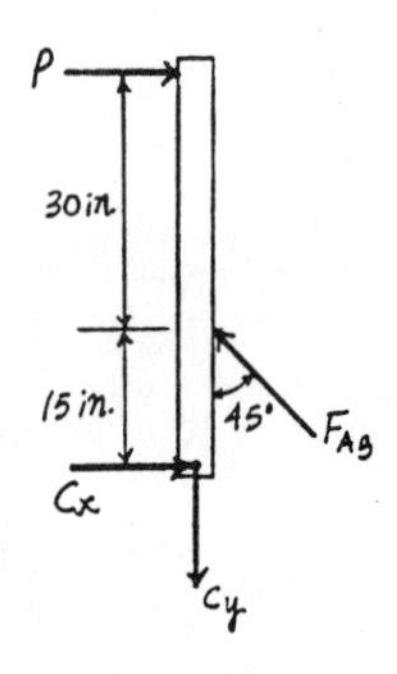

Section Properties :

$$L_{AB} = \frac{14}{\cos 45°} = 19.799 \text{ in.}$$
$$A = 0.25(1) = 0.250 \text{ in}^2$$
$$I = \frac{1}{12}(1)\left(0.25^3\right) = 1.302083\left(10^{-3}\right) \text{ in}^4$$

Critical Buckling Load : $K = 1$ for column with both ends pinned. Applying *Euler's* formula,

$$P_{cr} = F_{AB} = \frac{\pi^2 EI}{(KL_{AB})^2}$$
$$4.2426P = \frac{\pi^2(10)(10^6)\left[1.302083(10^{-3})\right]}{[1(19.799)]^2}$$
$$P = 77.27 \text{ lb} = 77.3 \text{ lb} \qquad \textbf{Ans}$$

Critical Stress : *Euler's* formula is only valid if $\sigma_{cr} < \sigma_Y$.

$$\sigma_{cr} = \frac{P_{cr}}{A} = \frac{4.2426(77.27)}{0.250} = 1311 \text{ psi} < \sigma_Y = 37 \text{ ksi } (\boldsymbol{O.K!})$$

***13–24.** Assume the strut brace is pinned at its ends A and B. If a force of $P = 100$ lb is applied to the end of member CD, determine the required width of the brace to the nearest $\frac{1}{8}$ in. to prevent its buckling. The brace is made of aluminum 6061-T6. Use a factor of safety with respect to buckling of F.S. = 2.25.

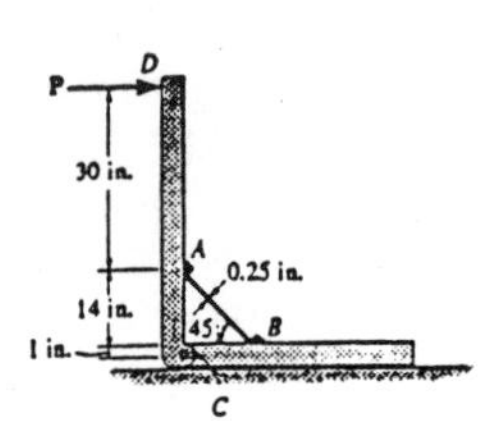

Support Reactions :

$$\circlearrowleft +\Sigma M_C = 0; \quad F_{AB}\cos 45°(15) - 100(45) = 0$$
$$F_{AB} = 424.26 \text{ lb}$$

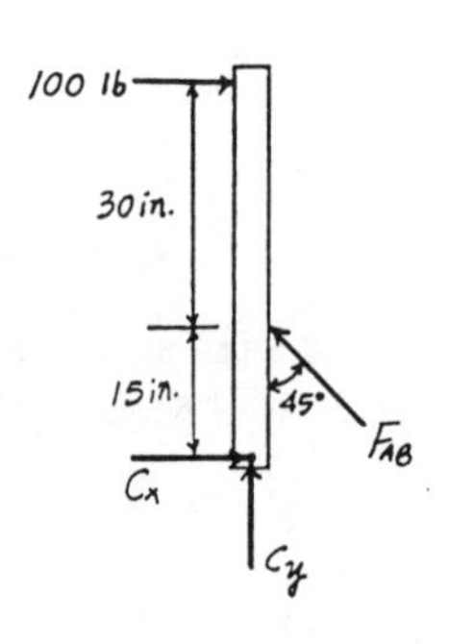

Section Properties :

$$L_{AB} = \frac{14}{\cos 45°} = 19.799 \text{ in.}$$
$$I = \frac{1}{12}(b)\left(0.25^3\right) = 1.302083\left(10^{-3}\right) b$$

Critical Buckling Load : $K = 1$ for column with both ends pinned. Applying *Euler's* formula,

$$P_{cr} = 2.25F_{AB} = \frac{\pi^2 EI}{(KL_{AB})^2}$$
$$2.25(424.26) = \frac{\pi^2(10)(10^6)\left[1.302083(10^{-3})b\right]}{[1(19.799)]^2}$$
$$b = 2.910 \text{ in.}$$

Use $\quad b = 3$ in. $\qquad$ **Ans**

Critical Stress : *Euler's* formula is only valid if $\sigma_{cr} < \sigma_Y$.

$$\sigma_{cr} = \frac{P_{cr}}{A} = \frac{2.25(424.26)}{(3)(0.25)} = 1273 \text{ psi} < \sigma_Y = 37 \text{ ksi } (\boldsymbol{O.K!})$$

13–25. The truss is made from A-36 steel bars, each of which has a circular cross section with a diameter of 1.5 in. Determine the maximum force P that can be applied without causing any of the members to buckle. The members are pin supported at their ends.

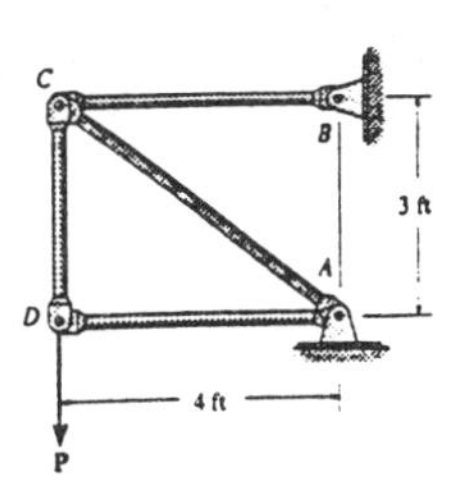

Member Forces : By observation of joint D, member AD is a zero - force member and member CD is subjected to a tension force P.

At **Joint** C,

$$+\uparrow \Sigma F_y = 0; \quad \frac{3}{5}F_{CA} - P = 0 \quad F_{CA} = 1.6667P \text{ (C)}$$

$$\xrightarrow{+} F_x = 0; \quad F_{CB} - \frac{4}{5}(1.6667P) = 0 \quad F_{CB} = 1.3333P \text{ (T)}$$

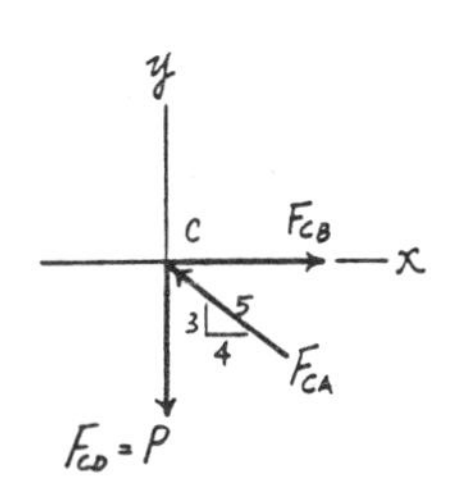

Section Properties :

$$L_{AC} = \sqrt{3^2 + 4^2} = 5.00 \text{ ft}$$

$$A = \frac{\pi}{4}\left(1.5^2\right) = 0.5625\pi \text{ in}^2$$

$$I = \frac{\pi}{4}\left(0.75^4\right) = 0.2485 \text{ in}^4$$

Critical Buckling Load : Only member AC is subjected to compressive force. $K = 1$ for a column with both ends pinned. Applying *Euler's* formula,

$$P_{cr} = F_{CA} = \frac{\pi^2 EI}{(KL_{AC})^2}$$

$$1.6667P = \frac{\pi^2(29)(10^3)(0.2485)}{[1(5.00)(12)]^2}$$

$$P = 11.85 \text{ kip} = 11.9 \text{ kip} \qquad \textbf{Ans}$$

Critical Stress : *Euler's* formula is only valid if $\sigma_{cr} < \sigma_Y$.

$$\sigma_{cr} = \frac{P_{cr}}{A} = \frac{1.6667(11.85)}{0.5625\pi} = 11.18 \text{ ksi} < \sigma_Y = 36 \text{ ksi } (O.K!)$$

13–26. The truss is made from A-36 steel bars, each of which has a circular cross section. If the applied load $P = 10$ kip, determine the diameter of member AC to the nearest $\frac{1}{8}$ in. that will prevent this member from buckling. The members are pin supported at their ends.

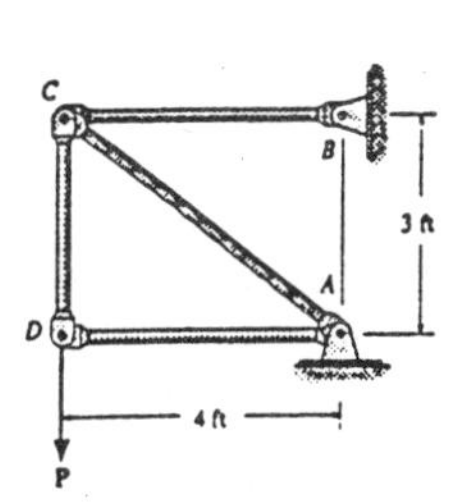

Member Forces : By observation of joint D, member AD is a zero force member and member CD is subjected to a tension of 10 kip.

At **Joint** C,

$$+\uparrow \Sigma F_y = 0; \quad \frac{3}{5}F_{CA} - 10 = 0 \quad F_{CA} = 16.67 \text{ kip(C)}$$

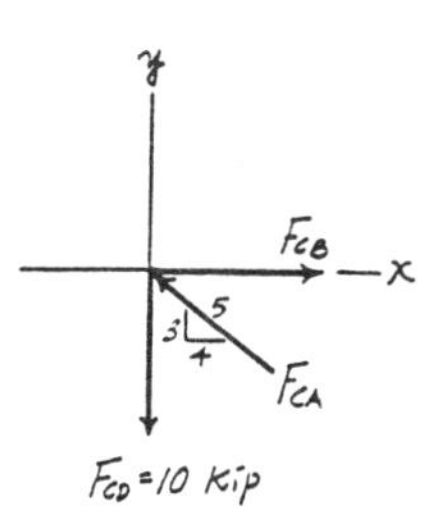

Section Properties :

$$L_{AC} = \sqrt{3^2 + 4^2} = 5.00 \text{ ft}$$

$$I = \frac{\pi}{4}\left(\frac{d}{2}\right)^4 = \frac{\pi d^4}{64}$$

Critical Buckling Load : $K = 1$ for a column with both ends pinned. Applying *Euler's* formula,

$$P_{cr} = F_{CA} = \frac{\pi^2 EI}{(KL_{AC})^2}$$

$$16.67 = \frac{\pi^2(29)(10^3)\left(\frac{\pi d^4}{64}\right)}{[1(5.00)(12)]^2}$$

$$d = 1.438 \text{ in.}$$

Use $\quad d = 1\frac{1}{2}$ in. $\qquad$ **Ans**

Critical Stress : *Euler's* formula is only valid if $\sigma_{cr} < \sigma_Y$.

$$\sigma_{cr} = \frac{P_{cr}}{A} = \frac{16.67}{\frac{\pi}{4}(1.50^2)} = 9.43 \text{ ksi} < \sigma_Y = 36 \text{ ksi } (O.K!)$$

13–27. The members of the truss are assumed to be pin connected. If member GF is an A-36 steel rod having a diameter of 2 in., determine the greatest magnitude of load **P** that can be supported by the truss without causing this member to buckle.

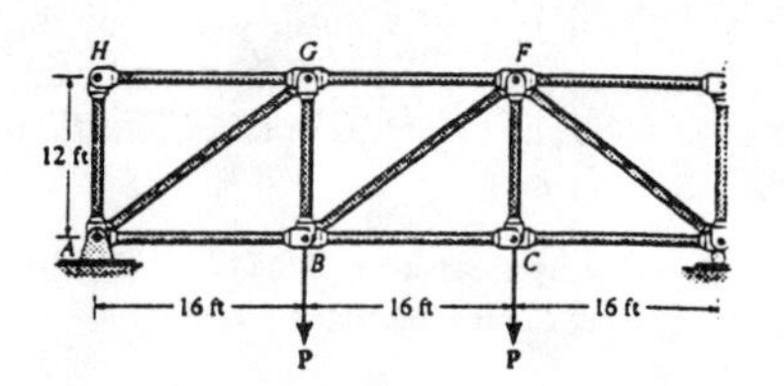

Support Reactions : As shown on FBD(a).

Member Forces : Use the method of sections [FBD(b)].

$$+\Sigma M_B = 0; \quad F_{GF}(12) - P(16) = 0 \quad F_{GF} = 1.3333P \text{ (C)}$$

Section Properties :

$$A = \frac{\pi}{4}(2^2) = \pi \text{ in}^2$$

$$I = \frac{\pi}{4}(1^4) = 0.250\pi \text{ in}^4$$

Critical Buckling Load : $K = 1$ for a column with both ends pinned. Applying *Euler's* formula,

$$P_{cr} = F_{GF} = \frac{\pi^2 EI}{(KL_{GF})^2}$$

$$1.3333P = \frac{\pi^2(29)(10^3)(0.250\pi)}{[1(16)(12)]^2}$$

$$P = 4.573 \text{ kip} = 4.57 \text{ kip} \qquad \textbf{Ans}$$

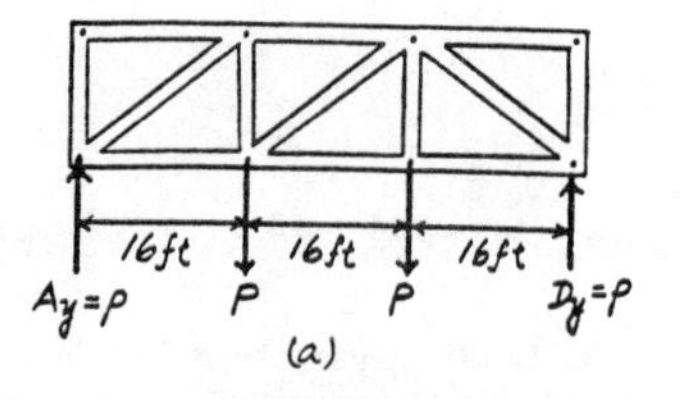

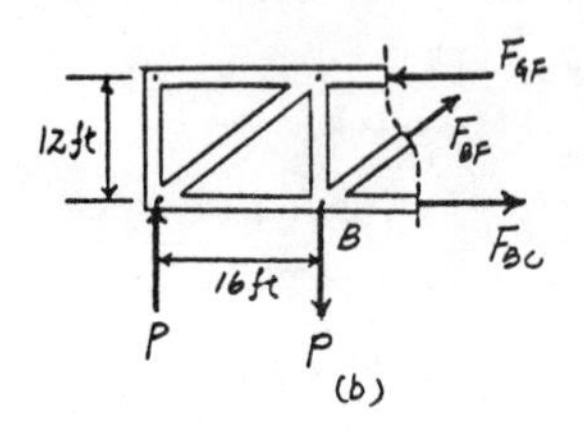

Critical Stress : *Euler's* formula is only valid if $\sigma_{cr} < \sigma_Y$.

$$\sigma_{cr} = \frac{P_{cr}}{A} = \frac{1.3333(4.573)}{\pi} = 1.94 \text{ ksi} < \sigma_Y = 36 \text{ ksi } (O.K!)$$

***13–28.** The members of the truss are assumed to be pin connected. If member AG is an A-36 steel rod having a diameter of 2 in., determine the greatest magnitude of load **P** that can be supported by the truss without causing this member to buckle.

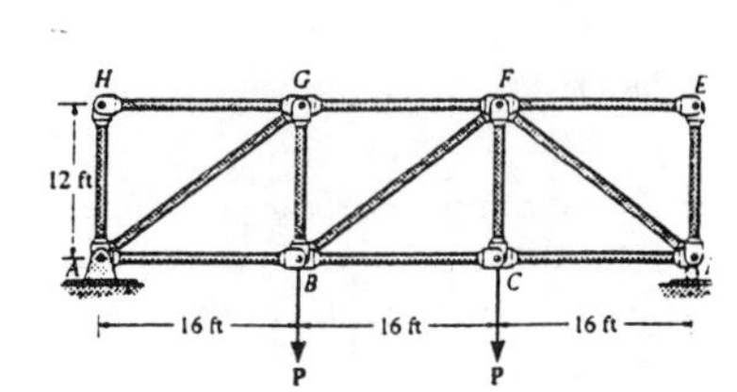

Support Reactions : As shown on FBD(a).

Member Forces : Use the method of joints [FBD(b)].

$$+\uparrow \Sigma F_y = 0; \quad P - \frac{3}{5}F_{AG} = 0 \quad F_{AG} = 1.6667P \text{ (C)}$$

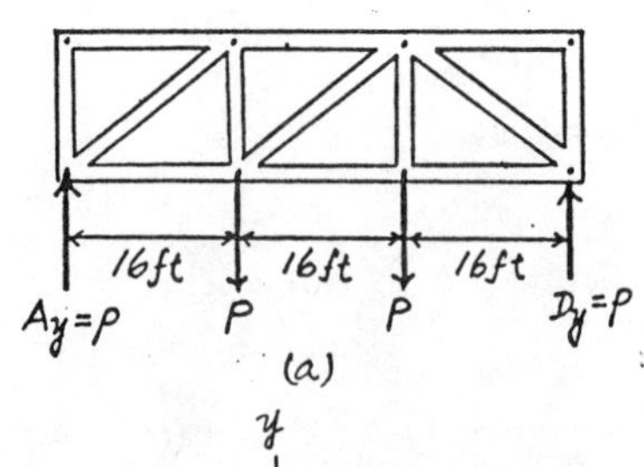

Section Properties :

$$L_{AG} = \sqrt{16^2 + 12^2} = 20.0 \text{ ft}$$

$$A = \frac{\pi}{4}(2^2) = \pi \text{ in}^2$$

$$I = \frac{\pi}{4}(1^4) = 0.250\pi \text{ in}^4$$

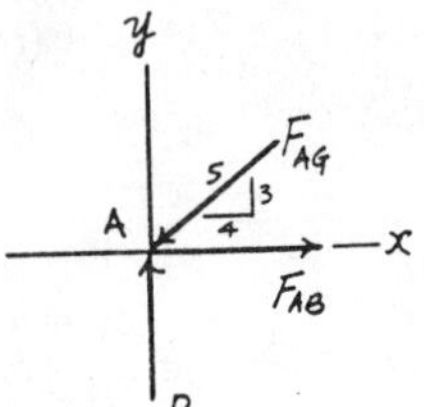

Critical Buckling Load : $K = 1$ for a column with both ends pinned. Applying *Euler's* formula,

$$P_{cr} = F_{GF} = \frac{\pi^2 EI}{(KL_{GF})^2}$$

$$1.6667P = \frac{\pi^2(29)(10^3)(0.250\pi)}{[1(20)(12)]^2}$$

$$P = 2.342 \text{ kip} = 2.34 \text{ kip} \qquad \textbf{Ans}$$

Critical Stress : *Euler's* formula is only valid if $\sigma_{cr} < \sigma_Y$.

$$\sigma_{cr} = \frac{P_{cr}}{A} = \frac{1.6667(2.342)}{\pi} = 1.24 \text{ ksi} < \sigma_Y = 36 \text{ ksi } (O.K!)$$

13–29. Determine the maximum force P that can be applied to the handle so that the A-36 steel control rod BC does not buckle. The rod has a diameter of 25 mm.

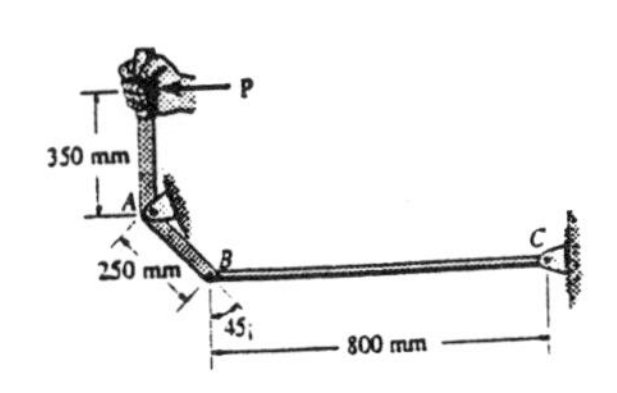

Support Reactions :

$$\curvearrowleft +\Sigma M_A = 0; \quad P(0.35) - F_{BC} \sin 45°(0.25) = 0$$
$$F_{BC} = 1.9799P$$

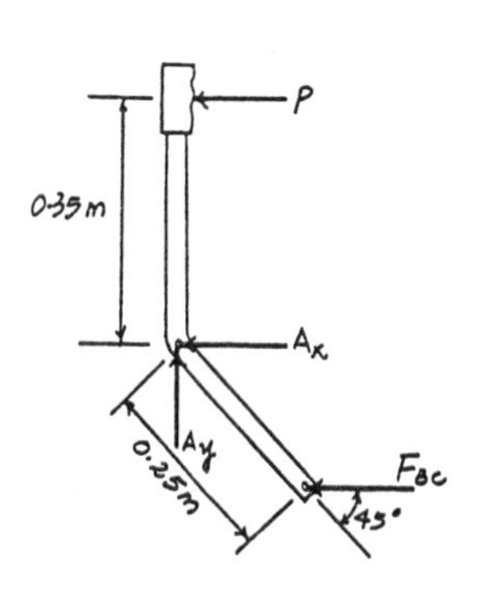

Section Properties :

$$A = \frac{\pi}{4}(0.025^2) = 0.15625(10^{-3})\,\pi \text{ m}^2$$
$$I = \frac{\pi}{4}(0.0125^4) = 19.17476(10^{-9}) \text{ m}^4$$

Critical Buckling Load : $K = 1$ for a column with both ends pinned. Applying *Euler's* formula,

$$P_{cr} = F_{BC} = \frac{\pi^2 EI}{(KL_{BC})^2}$$
$$1.9799P = \frac{\pi^2(200)(10^9)[19.17476(10^{-9})]}{[1(0.8)]^2}$$

$$P = 29\,870 \text{ N} = 29.9 \text{ kN} \qquad \textbf{Ans}$$

Critical Stress : *Euler's* formula is only valid if $\sigma_{cr} < \sigma_Y$.

$$\sigma_{cr} = \frac{P_{cr}}{A} = \frac{1.9799(29\,870)}{0.15625(10^{-3})\,\pi} = 120.5 \text{ MPa} < \sigma_Y = 250 \text{ MPa} \ (O.K!)$$

13–30. The rod of radius r is loosely pin supported at each end as shown when the temperature is T_1. Determine the increased temperature T_2 that will cause the rod to buckle. The coefficient of thermal expansion is α, and the modulus of elasticity is E. Assume the rod will not yield.

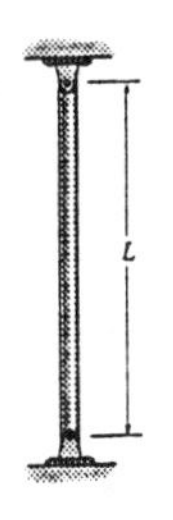

Compatibility Condition : This requires,

$$(+\downarrow) \qquad 0 = \delta_T + \delta_F$$
$$0 = \alpha(T_2 - T_1)L - \frac{FL}{\pi r^2 E}$$
$$F = \pi\alpha r^2 E(T_2 - T_1)$$

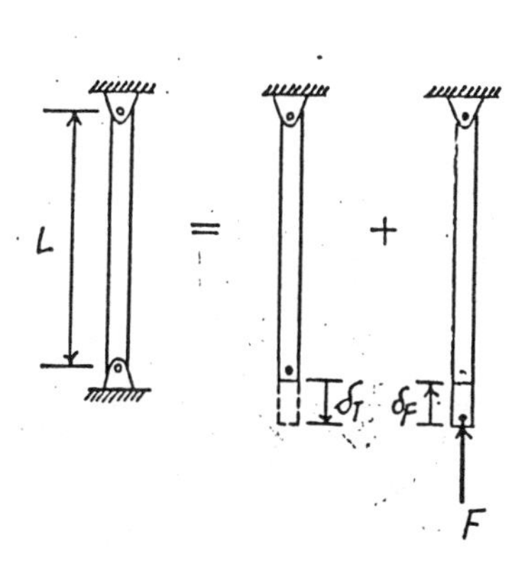

Critical Buckling Load : $K = 1$ for a column with both ends pinned. Applying *Euler's* formula,

$$P_{cr} = F = \frac{\pi^2 EI}{(KL)^2}$$
$$\pi\alpha r^2 E(T_2 - T_1) = \frac{\pi^2 E(\frac{\pi}{4}r^4)}{[1(L)]^2}$$

$$T_2 = \frac{\pi^2 r^2}{4\alpha L^2} + T_1 \qquad \textbf{Ans}$$

13–31. The steel pipe is constrained between the walls at A and B. If there is a gap of 3 mm at B when $T_1 = 4°C$, determine the temperature required to cause the pipe to become unstable and begin to buckle. The pipe has an outer diameter of 40 mm and a wall thickness of 10 mm. Assume that the collars at A and B provide fixed connections for the pipe. Neglect their size. $\alpha_{st} = 12(10^{-6})/°C$, $E_{st} = 200$ GPa, $\sigma_Y = 250$ MPa.

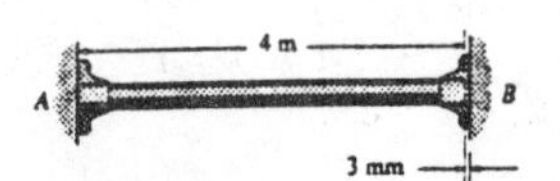

$$A = \frac{\pi}{4}\left(0.04^2 - 0.02^2\right) = 0.300\pi\left(10^{-3}\right)\ \text{m}^2$$
$$I = \frac{\pi}{4}\left(0.02^4 - 0.01^4\right) = 37.5\pi\left(10^{-9}\right)\ \text{m}^4$$

Compatibility Condition : This requires,

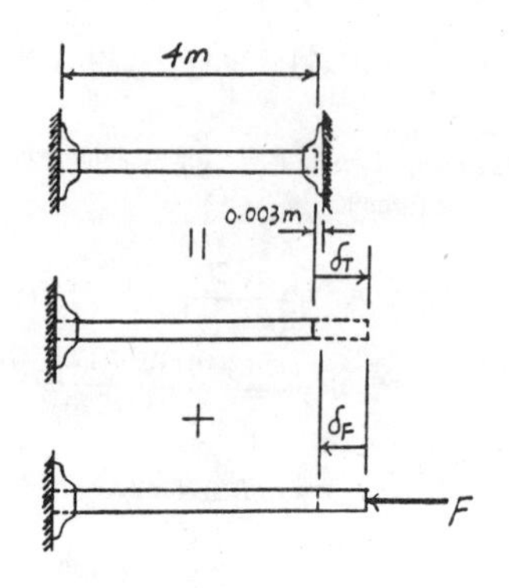

$$(\overset{+}{\rightarrow}) \qquad 0.003 = \delta_T + \delta_F$$
$$0.003 = 12\left(10^{-6}\right)(T_2 - 4)(4) - \frac{F(4)}{0.300\pi(10^{-3})(200)(10^9)}$$
$$F = 2261.95T_2 - 150\,419.46$$

Critical Buckling Load : $K = 0.5$ for column with both ends fixed. Applying *Euler's* formula,

$$P_{cr} = F = \frac{\pi^2 EI}{(KL)^2}$$
$$2261.95T_2 - 150\,419.46 = \frac{\pi^2(200)(10^9)\left[37.5\pi(10^{-9})\right]}{[0.5(4)]^2}$$

$$T_2 = 92.20\ °C = 92.2\ °C \qquad \textbf{Ans}$$

Critical Stress : *Euler's* formula is only valid if $\sigma_{cr} < \sigma_Y$.

$$P_{cr} = 2261.95(92.20) - 150\,419.46 = 58\,136.77\ \text{N}$$

$$\sigma_{cr} = \frac{P_{cr}}{A} = \frac{58\,136.77}{0.300\pi(10^{-3})} = 61.69\ \text{MPa} < \sigma_Y = 250\ \text{MPa}\ (O.K!)$$

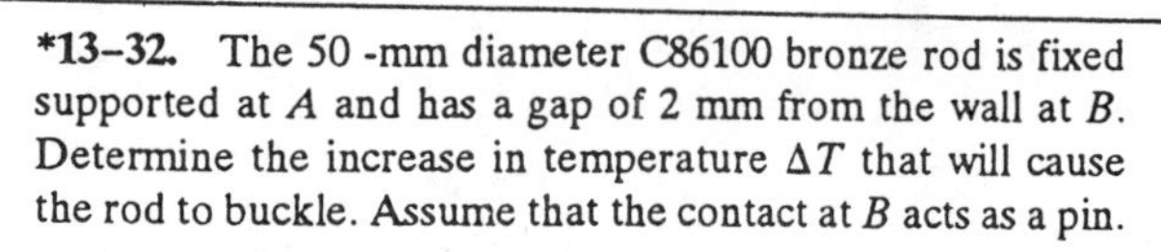

***13–32.** The 50 -mm diameter C86100 bronze rod is fixed supported at A and has a gap of 2 mm from the wall at B. Determine the increase in temperature ΔT that will cause the rod to buckle. Assume that the contact at B acts as a pin.

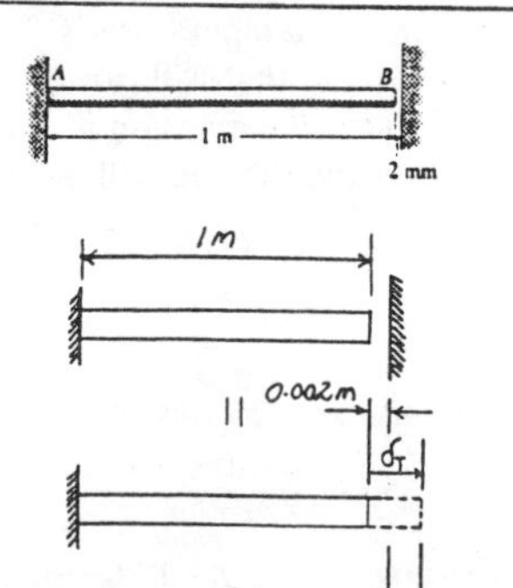

Section Properties :

$$A = \frac{\pi}{4}\left(0.05^2\right) = 0.625\pi\left(10^{-3}\right)\ \text{m}^2$$
$$I = \frac{\pi}{4}\left(0.025^4\right) = 97.65625\pi\left(10^{-9}\right)\ \text{m}^4$$

Compatibility Condition : This requires,

$$(\overset{+}{\rightarrow}) \qquad 0.002 = \delta_T + \delta_F$$
$$0.002 = 17\left(10^{-6}\right)(\Delta T)(1) - \frac{F(1)}{0.625\pi(10^{-3})(103)(10^9)}$$
$$F = 3438.08\Delta T - 404480.05$$

Critical Buckling Load : $K = 0.7$ for a column with one end fixed and the other end pinned. Applying *Euler's* formula,

$$P_{cr} = F = \frac{\pi^2 EI}{(KL)^2}$$
$$3438.08\Delta T - 404\,480.05 = \frac{\pi^2(103)(10^9)\left[97.65625\pi(10^{-9})\right]}{[0.7(1)]^2}$$

$$\Delta T = 302.78\ °C = 303\ °C \qquad \textbf{Ans}$$

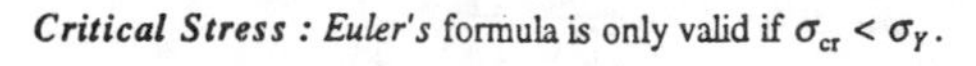

Critical Stress : *Euler's* formula is only valid if $\sigma_{cr} < \sigma_Y$.

$$P_{cr} = 3438.08(302.78) - 404\,480.05 = 636\,488.86\ \text{N}$$

$$\sigma_{cr} = \frac{P_{cr}}{A} = \frac{636\,488.86}{0.625\pi(10^{-3})} = 324.2\ \text{MPa} < \sigma_Y = 345\ \text{MPa}\ (O.K!)$$

13–33. The wood column has a length of 16 ft. It is fixed at its base A and pinned at its top C. Also, the two members are pin connected at B. If it has the cross section shown, determine the maximum axial force P that can be applied without causing it to buckle about the y–y axis. The column is braced so that it will not buckle about the x–x axis. $E_w = 1.6(10^3)$ ksi, $\sigma_Y = 8$ ksi.

Section Properties :

$$A = 4(2) = 8.00 \text{ in}^2$$

$$I_{y-y} = \frac{1}{12}(4)\left(2^3\right) = 2.6667 \text{ in}^4$$

Critical Buckling Load : For segment AB, $K = 0.7$ (column with one end pinned and the other end fixed). Applying *Euler's* formula,

$$P_{cr} = \frac{\pi^2 EI}{(KL_{AB})^2} = \frac{\pi^2(1.6)(10^3)(2.6667)}{[0.7(10)(12)]^2} = 5.968 \text{ kip} = 5.97 \text{ kip} \quad (\textit{Controls!}) \qquad \textbf{Ans}$$

For segment BC, $K = 1$ (column with both ends pinned).

$$P_{cr} = \frac{\pi^2 EI}{(KL)^2} = \frac{\pi^2(1.6)(10^3)(2.6667)}{[1(6)(12)]^2} = 8.123 \text{ kip}$$

Critical Stress : *Euler's* formula is only valid if $\sigma_{cr} < \sigma_Y$.

$$\sigma_{cr} = \frac{P_{cr}}{A} = \frac{5.968}{8} = 0.746 \text{ ksi} < \sigma_Y = 8 \text{ ksi } (O.K!)$$

13–34. The A-36 steel bar AB of the frame is pin connected at its ends. If $P = 30$ kip, determine the factor of safety with respect to buckling about the y–y axis due to the applied loading.

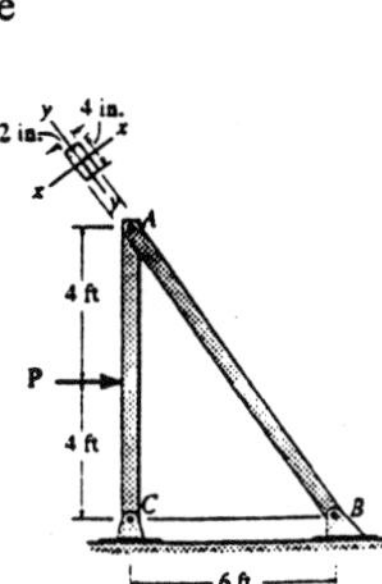

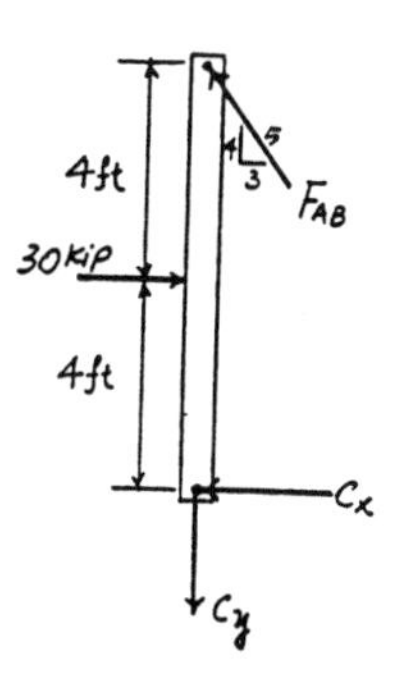

Support Reactions :

$$\circlearrowleft + \Sigma M_C = 0; \quad \frac{3}{5}F_{AB}(8) - 30(4) = 0 \qquad F_{AB} = 25.0 \text{ kip}$$

Section Properties :

$$L_{AB} = \sqrt{8^2 + 6^2} = 10\text{ft} = 120 \text{ in.}$$

$$A = 4(2) = 8.00 \text{ in}^2$$

$$I_y = \frac{1}{12}(4)\left(2^3\right) = 2.6667 \text{ in}^4$$

Critical Buckling Load : $K = 1$ for a column with both ends pinned. Applying *Euler's* formula,

$$P_{cr} = \frac{\pi^2 EI}{(KL_{AB})^2} = \frac{\pi^2(29)(10^3)(2.6667)}{[1(120)]^2} = 53.00 \text{ kip}$$

Critical Stress : *Euler's* formula is only valid if $\sigma_{cr} < \sigma_Y$.

$$\sigma_{cr} = \frac{P_{cr}}{A} = \frac{53.00}{8.00} = 6.625 \text{ ksi} < \sigma_Y = 36 \text{ ksi } (O.K!)$$

Factor of Safety :

$$\text{F.S} = \frac{P_{cr}}{F_{AB}} = \frac{53.00}{25.0} = 2.12 \qquad \textbf{Ans}$$

13–35. The A-36 steel bar AB of the frame is pin connected at its ends. Determine the largest load P that can be applied to the frame without causing it to buckle about the y-y axis.

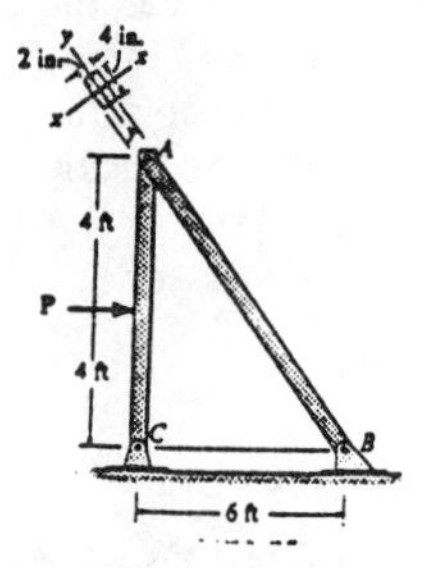

Support Reactions :

$$\curvearrowleft + \Sigma M_C = 0; \quad \frac{3}{5}F_{AB}(8) - P(4) = 0 \quad F_{AB} = 0.8333P$$

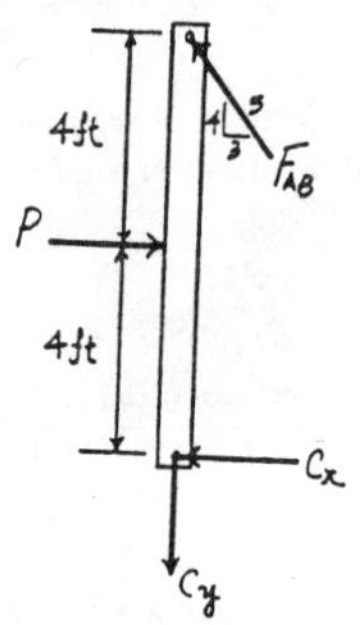

Section Properties :

$$L_{AB} = \sqrt{8^2 + 6^2} = 10\text{ft} = 120 \text{ in.}$$

$$A = 4(2) = 8.00 \text{ in}^2$$

$$I_y = \frac{1}{12}(4)\left(2^3\right) = 2.6667 \text{ in}^4$$

Critical Buckling Load : $K = 1$ for a column with both ends pinned. Applying *Euler's* formula,

$$P_{cr} = F_{AB} = \frac{\pi^2 EI}{(KL_{AB})^2}$$

$$0.8333P = \frac{\pi^2(29)(10^3)(2.6667)}{[1(120)]^2}$$

$$P = 63.60 \text{ kip} = 63.6 \text{ kip} \qquad \textbf{Ans}$$

Critical Stress : *Euler's* formula is only valid if $\sigma_{cr} < \sigma_Y$.

$$\sigma_{cr} = \frac{P_{cr}}{A} = \frac{0.8333(63.60)}{8.00} = 6.625 \text{ ksi} < \sigma_Y = 36 \text{ ksi } (O.K!)$$

***13–36.** Determine the maximum allowable load P that can be applied to member BC without causing member AB to buckle. Assume that AB is made of steel and is pinned at its ends for x–x axis buckling and fixed at its ends for y–y axis buckling. Use a factor of safety with respect to buckling of F.S. = 3. E_{st} = 200 GPa, σ_Y = 360 MPa.

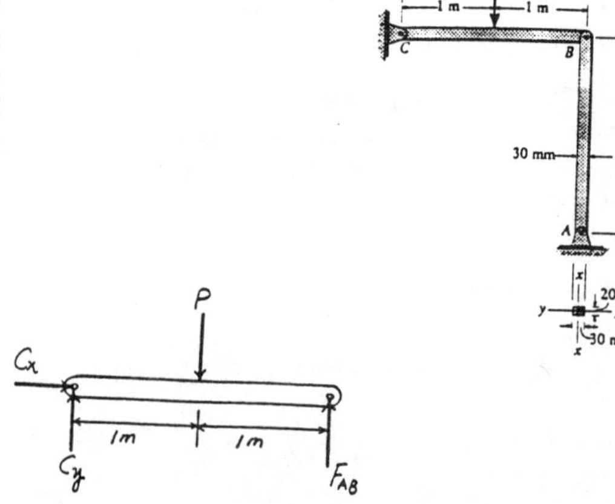

Support Reactions :

$$\curvearrowleft + \Sigma M_C = 0; \quad F_{AB}(2) - P(1) = 0 \quad F_{AB} = 0.500P$$

Section Properties :

$$A = 0.02(0.03) = 0.600\left(10^{-3}\right) \text{ m}^2$$

$$I_x = \frac{1}{12}(0.02)\left(0.03^3\right) = 45.0\left(10^{-9}\right) \text{ m}^4$$

$$I_y = \frac{1}{12}(0.03)\left(0.02^3\right) = 20.0\left(10^{-9}\right) \text{ m}^4$$

Critical Buckling Load : With respect to $x-x$ axis, $K = 1$ (column with both ends pinned). Applying *Euler's* formula,

$$P_{cr} = 3F_{AB} = \frac{\pi^2 EI}{(KL)^2}$$

$$3(0.500P) = \frac{\pi^2(200)(10^9)\left[45.0(10^{-9})\right]}{[1(2)]^2}$$

$$P = 14804.4 \text{ N} = 14.8 \text{ kN } (\textit{Controls!}) \qquad \textbf{Ans}$$

With respect to $y-y$ axis, $K = 0.5$ (column with both ends fixed).

$$P_{cr} = 3F_{AB} = \frac{\pi^2 EI}{(KL)^2}$$

$$3(0.500P) = \frac{\pi^2(200)(10^9)\left[20.0(10^{-9})\right]}{[0.5(2)]^2}$$

$$P = 26\,318.9 \text{ N}$$

Critical Stress : *Euler's* formula is only valid if $\sigma_{cr} < \sigma_Y$.

$$\sigma_{cr} = \frac{P_{cr}}{A} = \frac{3(0.5)(14\,804.4)}{0.600(10^{-3})} = 37.01 \text{ MPa} < \sigma_Y = 360 \text{ MPa } (O.K!)$$

13–37. Determine if the frame can support a load of $P = 20$ kN if the factor of safety with respect to buckling of member AB is F.S. = 3. Assume that AB is made of steel and is pinned at its ends for x–x axis buckling and fixed at its ends for y–y axis buckling. $E_{st} = 200$ GPa, $\sigma_Y = 360$ MPa.

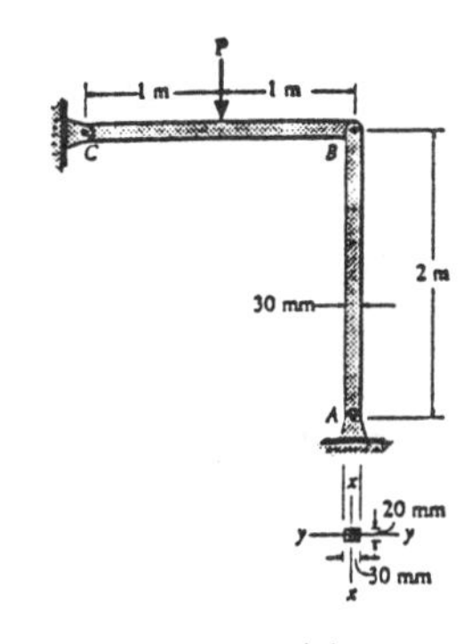

Support Reactions :

$$\circlearrowleft +\Sigma M_C = 0; \quad F_{AB}(2) - 20(1) = 0 \quad F_{AB} = 10.0 \text{ kN}$$

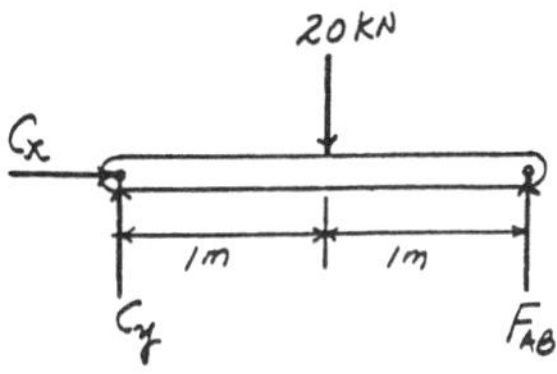

Section Properties :

$$A = 0.02(0.03) = 0.600(10^{-3}) \text{ m}^2$$

$$I_x = \frac{1}{12}(0.02)(0.03^3) = 45.0(10^{-9}) \text{ m}^4$$

$$I_y = \frac{1}{12}(0.03)(0.02^2) = 20.0(10^{-9}) \text{ m}^4$$

Critical Buckling Load : With respect to $x - x$ axis, $K = 1$ (column with both ends pinned) . Applying *Euler's* formula,

$$P_{cr} = \frac{\pi^2 EI}{(KL)^2}$$

$$= \frac{\pi^2(200)(10^9)[45.0(10^{-9})]}{[1(2)]^2}$$

$$= 22\,206.61 = 22.207 \text{ kN} \quad (\textit{Controls!})$$

With respect to $y - y$ axis, $K = 0.5$ (column with both ends fixed).

$$P_{cr} = \frac{\pi^2 EI}{(KL)^2}$$

$$= \frac{\pi^2(200)(10^9)[20.0(10^{-9})]}{[0.5(2)]^2}$$

$$= 39\,478.42 \text{ N}$$

Critical Stress : *Euler's* formula is only valid if $\sigma_{cr} < \sigma_Y$.

$$\sigma_{cr} = \frac{P_{cr}}{A} = \frac{32\,127.6}{5.00(10^{-3})} = 6.426 \text{ MPa} < \sigma_Y = 360 \text{ MPa} \quad (O.K!)$$

Factor of Safety : The required factor of safety is 3.

$$\text{F.S} = \frac{P_{cr}}{F_{AB}} = \frac{22.207}{10.0} = 2.22 < 3 \quad (\textit{No Good!})$$

Hence, the frame **cannot support the load** with the required F.S. **Ans**

13–38. The steel bar AC of the frame is pin connected at its ends. Determine the factor of safety with respect to buckling about the y–y axis due to the applied loading of $P = 15$ kN. $E_{st} = 200$ GPa, $\sigma_Y = 360$ MPa.

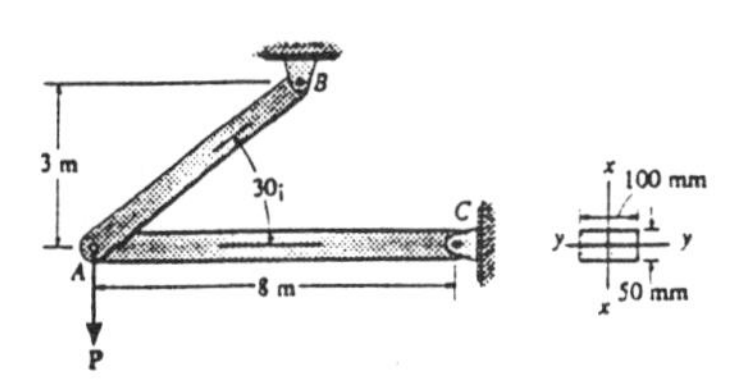

Member Forces : Use method of joints.

$$+\uparrow \Sigma F_y = 0; \quad F_{AB}\sin 30° - 15 = 0 \quad F_{AB} = 30.0 \text{ kN (T)}$$

$$\xrightarrow{+} \Sigma F_x = 0; \quad 30.0\cos 30° - F_{AC} = 0 \quad F_{AC} = 25.981 \text{ kN (C)}$$

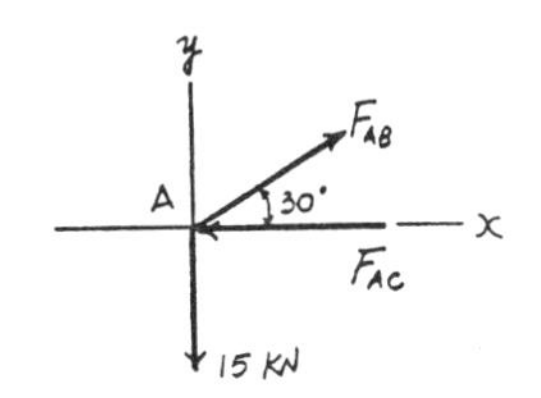

Section Properties :

$$A = 0.1(0.05) = 5.00(10^{-3}) \text{ m}^2$$

$$I_{y-y} = \frac{1}{12}(0.1)(0.05^3) = 1.04167(10^{-6}) \text{ m}^4$$

Critical Buckling Load : $K = 1$ for a column with both ends pinned. Applying *Euler's* formula,

$$P_{cr} = \frac{\pi^2 EI}{(KL)^2}$$

$$= \frac{\pi^2(200)(10^9)[1.04167(10^{-6})]}{[1(8)]^2}$$

$$= 32\,127.6 \text{ N} = 32.128 \text{ kN}$$

Critical Stress : *Euler's* formula is only valid if $\sigma_{cr} < \sigma_Y$.

$$\sigma_{cr} = \frac{P_{cr}}{A} = \frac{32\,127.6}{5.00(10^{-3})} = 6.426 \text{ MPa ksi} < \sigma_Y = 360 \text{ MPa} (O.K!)$$

Factor of Safety :

$$\text{F.S} = \frac{P_{cr}}{F_{AC}} = \frac{32.128}{25.981} = 1.24 \qquad \textbf{Ans}$$

13–39. The steel bar AC of the frame is pin connected at its ends. Determine if the frame can safely support a load of $P = 35$ kN without bar AC buckling about the y–y axis. $E_{st} = 200$ GPa, $\sigma_Y = 360$ MPa.

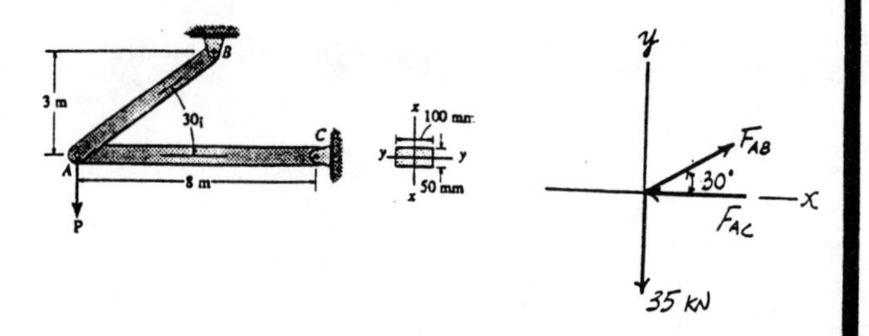

Member Forces : Use the method of joints.

$$+\uparrow \Sigma F_y = 0; \quad F_{AB}\sin 30^\circ - 35 = 0 \qquad F_{AB} = 70.0 \text{ kN (T)}$$

$$\xrightarrow{+} \Sigma F_x = 0; \quad 70.0\cos 30^\circ - F_{AC} = 0 \qquad F_{AC} = 60.622 \text{ kN (C)}$$

Section Properties :

$$A = 0.1(0.05) = 5.00(10^{-3}) \text{ m}^2$$

$$I_{y-y} = \frac{1}{12}(0.1)(0.05^3) = 1.04167(10^{-6}) \text{ m}^4$$

Critical Buckling Load : $K = 1$ for a column with both ends pinned. Applying *Euler's* formula,

$$P_{cr} = \frac{\pi^2 EI}{(KL)^2}$$

$$= \frac{\pi^2(200)(10^9)[1.04167(10^{-6})]}{[1(8)]^2}$$

$$= 32127.6 \text{ N} = 32.128 \text{ kN}$$

Critical Stress : *Euler's* formula is only valid if $\sigma_{cr} < \sigma_Y$.

$$\sigma_{cr} = \frac{P_{cr}}{A} = \frac{32127.6}{5.00(10^{-3})} = 6.426 \text{ MPa} < \sigma_Y = 360 \text{ MPa } (O.K!)$$

Factor of Safety :

$$\text{F.S} = \frac{P_{cr}}{F_{AC}} = \frac{32.128}{60.622} = 0.530 < 1 \ (\textit{No Good!})$$

Member AC buckles under the load, hence the frame **cannot safely support** the load. **Ans**

***13–40.** Determine the maximum distributed loading that can be applied to the wide–flange beam so that the brace CD does not buckle.The brace is an A-36 steel rod having a diameter of 50 mm.

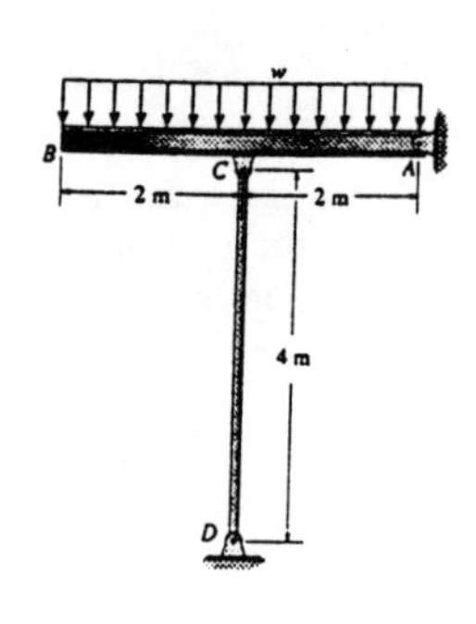

Support Reactions :

$$\circlearrowleft + \Sigma M_B = 0; \quad 4w(2) - F_{CD}(2) = 0 \qquad F_{CD} = 4.00w$$

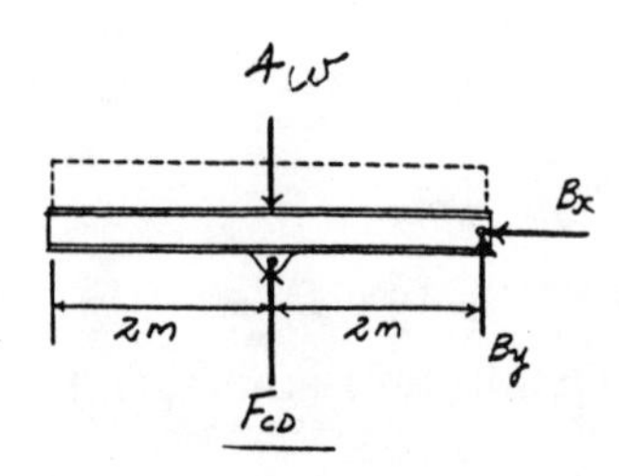

Section Properties :

$$A = \frac{\pi}{4}(0.05^2) = 0.625(10^{-3})\,\pi \text{ m}^2$$

$$I = \frac{\pi}{4}(0.025^4) = 97.65625(10^{-9})\,\pi \text{ m}^4$$

Critical Buckling Load : $K = 1$ for a column with both ends pinned. Applying *Euler's* formula,

$$P_{cr} = F_{CD} = \frac{\pi^2 EI}{(KL_{CD})^2}$$

$$4.00w = \frac{\pi^2(200)(10^9)[97.65625(10^{-9})\,\pi]}{[1(4)]^2}$$

$$= 9462.36 \text{ N/m} = 9.46 \text{ kN/m} \qquad \textbf{Ans}$$

Critical Stress : *Euler's* formula is only valid if $\sigma_{cr} < \sigma_Y$.

$$\sigma_{cr} = \frac{P_{cr}}{A} = \frac{4.00(9462.36)}{0.625(10^{-3})\,\pi} = 19.28 \text{ MPa} < \sigma_Y = 250 \text{ MPa } (O.K!)$$

13–41. The beam supports a distributed loading of w = 10 kN/m. Determine the smallest diameter of the A-36 steel brace CD so it does not buckle.

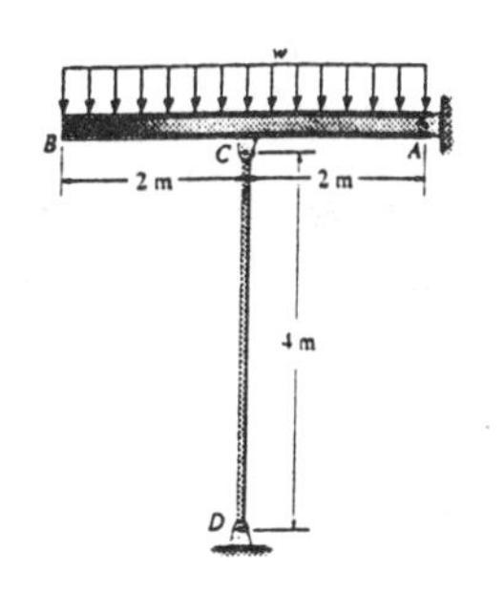

Support Reactions :

$$\circlearrowleft + \Sigma M_B = 0; \quad 40.0(2) - F_{CD}(2) = 0 \quad F_{CD} = 40.0 \text{ kN}$$

Section Properties :

$$A = \frac{\pi}{4}(d^2) \qquad I = \frac{\pi}{4}\left(\frac{d}{2}\right)^4 = \frac{\pi}{64}d^4$$

Critical Buckling Load : $K = 1$ for a column with both ends pinned. Appling *Euler's* formula,

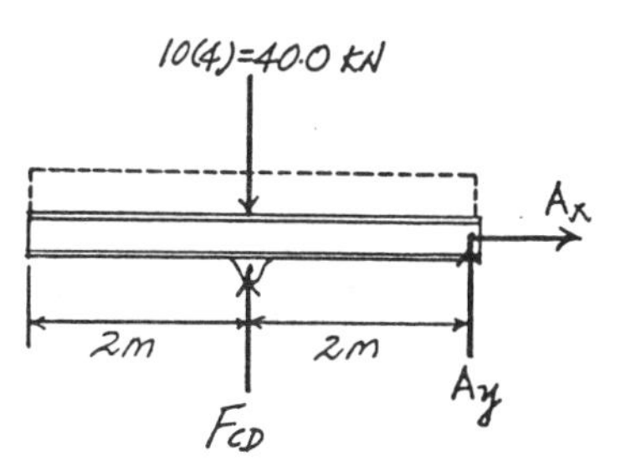

$$P_{cr} = F_{CD} = \frac{\pi^2 EI}{(KL_{CD})^2}$$

$$40.0(10^3) = \frac{\pi^2(200)(10^9)\left(\frac{\pi}{64}d^4\right)}{[1(4)]^2}$$

$$d = 0.0507 \text{ m} = 50.7 \text{ mm} \qquad \textbf{Ans}$$

Critical Stress : *Euler's* formula is only valid if $\sigma_{cr} < \sigma_Y$.

$$\sigma_{cr} = \frac{P_{cr}}{A} = \frac{40(10^3)}{\frac{\pi}{4}(0.0507^2)} = 19.82 \text{ MPa} < \sigma_Y = 250 \text{ MPa } (O.K!)$$

13–42. Consider an ideal column as in Fig. 13–12c, having both ends fixed. Show that the critical load on the column is given by $P_{cr} = 4\pi^2 EI/L^2$. *Hint:* Due to the vertical deflection of the top of the column, a constant moment $\mathbf{M'}$ will be developed at the supports. Show that $d^2v/dx^2 + (P/EI)v = M'/EI$. The solution is of the form $v = C_1 \sin(\sqrt{P/EI}x) + C_2 \cos(\sqrt{P/EI}x) + M'/P$.

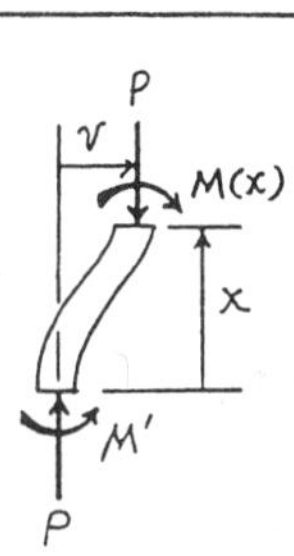

Moment Functions :

$$M(x) = M' - Pv$$

Differential Equation of The Elastic Curve :

$$EI\frac{d^2v}{dx^2} = M(x)$$

$$EI\frac{d^2v}{dx^2} = M' - Pv$$

$$\frac{d^2v}{dx^2} + \frac{P}{EI}v = \frac{M'}{EI} \qquad (Q.E.D.)$$

The solution of the above differential equation is of the form

$$v = C_1 \sin\left(\sqrt{\frac{P}{EI}}x\right) + C_2 \cos\left(\sqrt{\frac{P}{EI}}x\right) + \frac{M'}{P} \qquad [1]$$

and

$$\frac{dv}{dx} = C_1\sqrt{\frac{P}{EI}}\cos\left(\sqrt{\frac{P}{EI}}x\right) - C_2\sqrt{\frac{P}{EI}}\sin\left(\sqrt{\frac{P}{EI}}x\right) \qquad [2]$$

The integration constants can be determined from the boundary conditions.

Boundary Conditions :

At $x = 0$, $v = 0$. From Eq. [1], $C_2 = -\frac{M'}{P}$

At $x = 0$, $\frac{dv}{dx} = 0$. From Eq. [2], $C_1 = 0$

Elastic Curve :

$$v = \frac{M'}{P}\left[1 - \cos\left(\sqrt{\frac{P}{EI}}x\right)\right]$$

and

$$\frac{dv}{dx} = \frac{M'}{P}\sqrt{\frac{P}{EI}}\sin\left(\sqrt{\frac{P}{EI}}x\right)$$

However, due to symmetry $\frac{dv}{dx} = 0$ at $x = \frac{L}{2}$. Then,

$$\sin\left[\sqrt{\frac{P}{EI}}\left(\frac{L}{2}\right)\right] = 0 \quad \text{or} \quad \sqrt{\frac{P}{EI}}\left(\frac{L}{2}\right) = n\pi \quad \text{where } n = 1,2,3....$$

The smallest critical load occurs when $n = 1$.

$$P_{cr} = \frac{4\pi^2 EI}{L^2} \qquad (Q.E.D.)$$

13–43. Consider an ideal column as in Fig. 13–12*d*, having one end fixed and the other pinned. Show that the critical load on the column is given by $P_{cr} = 20.19\,EI/L^2$. *Hint:* Due to the vertical deflection at the top of the column, a constant moment **M′** will be developed at the fixed support and horizontal reactive forces **R′** will be developed at both supports. Show that $d^2v/dx^2 + (P/EI)v = (R'/EI)(L - x)$. The solution is of the form $v = C_1 \sin(\sqrt{P/EI}x) + C_2 \cos(\sqrt{P/EI}x) + (R'/P)(L - x)$. After application of the boundary conditions show that $\tan(\sqrt{P/EI}\,L) = \sqrt{P/EI}\,L$. Solve by trial and error for the smallest root.

Equilibrium : FBD(a).

Moment Functions : FBD(b).

$$M(x) = R'(L-x) - Pv$$

Differential Equation of The Elastic Curve :

$$EI\frac{d^2v}{dx^2} = M(x)$$

$$EI\frac{d^2v}{dx^2} = R'(L-x) - Pv$$

$$\frac{d^2v}{dx^2} + \frac{P}{EI}v = \frac{R'}{EI}(L-x)$$

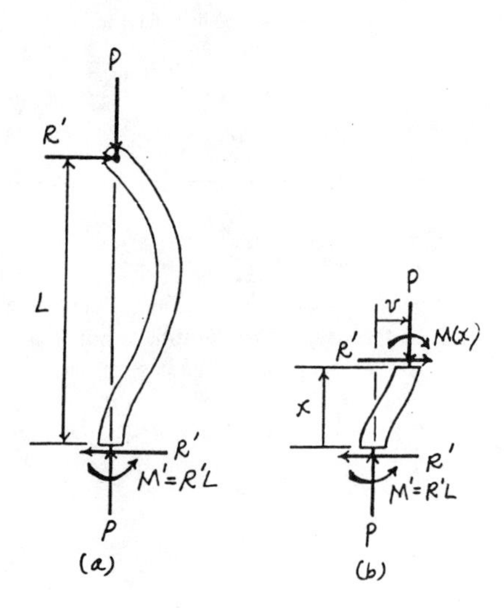

The solution of the above differential equation is of the form

$$v = C_1 \sin\left(\sqrt{\frac{P}{EI}}x\right) + C_2 \cos\left(\sqrt{\frac{P}{EI}}x\right) + \frac{R'}{P}(L-x) \qquad [1]$$

and

$$\frac{dv}{dx} = C_1\sqrt{\frac{P}{EI}}\cos\left(\sqrt{\frac{P}{EI}}x\right) - C_2\sqrt{\frac{P}{EI}}\sin\left(\sqrt{\frac{P}{EI}}x\right) - \frac{R'}{P} \qquad [2]$$

The integration constants can be determined from the boundary conditions.

Boundary Conditions :

At $x = 0$, $v = 0$. From Eq. [1], $\quad C_2 = -\frac{R'L}{P}$

At $x = 0$, $\frac{dv}{dx} = 0$. From Eq. [2], $\quad C_1 = \frac{R'}{P}\sqrt{\frac{EI}{P}}$

Elastic Curve :

$$v = \frac{R'}{P}\sqrt{\frac{EI}{P}}\sin\left(\sqrt{\frac{P}{EI}}x\right) - \frac{R'L}{P}\cos\left(\sqrt{\frac{P}{EI}}x\right) + \frac{R'}{P}(L-x)$$

$$= \frac{R'}{P}\left[\sqrt{\frac{EI}{P}}\sin\left(\sqrt{\frac{P}{EI}}x\right) - L\cos\left(\sqrt{\frac{P}{EI}}x\right) + (L-x)\right]$$

However, $v = 0$ at $x = L$. Then,

$$0 = \sqrt{\frac{EI}{P}}\sin\left(\sqrt{\frac{P}{EI}}L\right) - L\cos\left(\sqrt{\frac{P}{EI}}L\right)$$

$$\tan\left(\sqrt{\frac{P}{EI}}L\right) = \sqrt{\frac{P}{EI}}L \qquad (Q.E.D.)$$

By trial and error and choosing the smallest root, we have

$$\sqrt{\frac{P}{EI}}L = 4.49341$$

Then,

$$P_{cr} = \frac{20.19EI}{L^2} \qquad (Q.E.D.)$$

***13–44.** The ideal column is subjected to the force **F** at its midpoint and the axial load **P**. Determine the maximum displacement and the maximum moment in the column at midspan. EI is constant. *Hint:* Establish the differential equation for deflection Eq. 13–1. The general solution is $v = A \sin kx + B \cos kx - c^2x/k^2$, where $c^2 = F/2EI$, $k^2 = P/EI$.

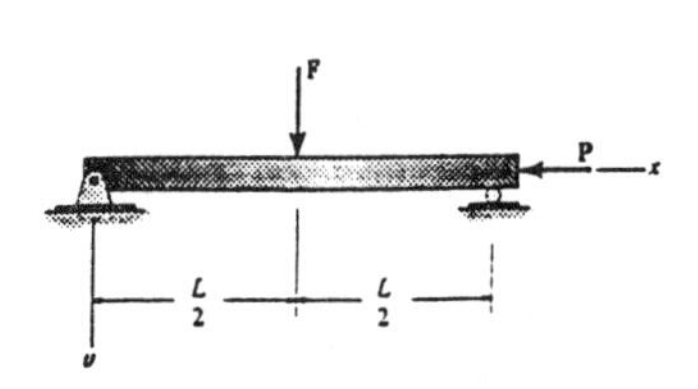

Moment Functions : FBD(b).

$$\circlearrowleft +\Sigma M_O = 0; \quad M(x) + \frac{F}{2}x + P(v) = 0$$

$$M(x) = -\frac{F}{2}x - Pv \qquad [1]$$

Differential Equation of The Elastic Curve :

$$EI\frac{d^2v}{dx^2} = M(x)$$

$$EI\frac{d^2v}{dx^2} = -\frac{F}{2}x - Pv$$

$$\frac{d^2v}{dx^2} + \frac{P}{EI}v = -\frac{F}{2EI}x$$

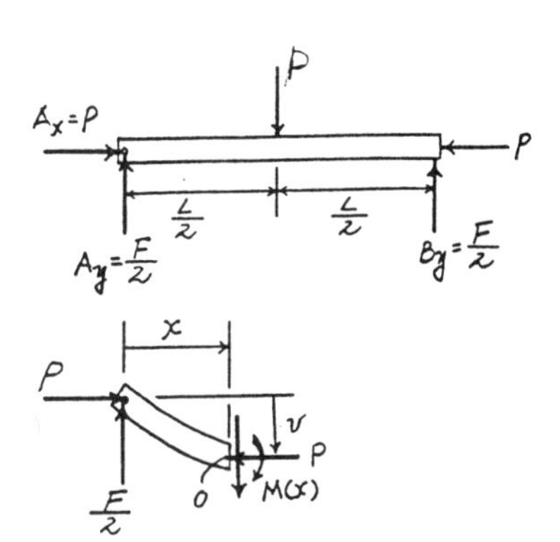

The solution of the above differential equation is of the form,

$$v = C_1 \sin\left(\sqrt{\frac{P}{EI}}x\right) + C_2 \cos\left(\sqrt{\frac{P}{EI}}x\right) - \frac{F}{2P}x \qquad [2]$$

and

$$\frac{dv}{dx} = C_1\sqrt{\frac{P}{EI}}\cos\left(\sqrt{\frac{P}{EI}}x\right) - C_2\sqrt{\frac{P}{EI}}\sin\left(\sqrt{\frac{P}{EI}}x\right) - \frac{F}{2P} \qquad [3]$$

The integration constants can be determined from the boundary conditions.

Boundary Conditions :

At $x = 0$, $v = 0$. From Eq. [2], $C_2 = 0$

At $x = \frac{L}{2}$, $\frac{dv}{dx} = 0$. From Eq. [3],

$$0 = C_1\sqrt{\frac{P}{EI}}\cos\left(\sqrt{\frac{P}{EI}}\frac{L}{2}\right) - \frac{F}{2P}$$

$$C_1 = \frac{F}{2P}\sqrt{\frac{EI}{P}}\sec\left(\sqrt{\frac{P}{EI}}\frac{L}{2}\right)$$

Elastic Curve :

$$v = \frac{F}{2P}\sqrt{\frac{EI}{P}}\sec\left(\sqrt{\frac{P}{EI}}\frac{L}{2}\right)\sin\left(\sqrt{\frac{P}{EI}}x\right) - \frac{F}{2P}x$$

$$= \frac{F}{2P}\left[\sqrt{\frac{EI}{P}}\sec\left(\sqrt{\frac{P}{EI}}\frac{L}{2}\right)\sin\left(\sqrt{\frac{P}{EI}}x\right) - x\right]$$

However, $v = v_{max}$ at $x = \frac{L}{2}$. Then,

$$v_{max} = \frac{F}{2P}\left[\sqrt{\frac{EI}{P}}\sec\left(\sqrt{\frac{P}{EI}}\frac{L}{2}\right)\sin\left(\sqrt{\frac{P}{EI}}\frac{L}{2}\right) - \frac{L}{2}\right]$$

$$= \frac{F}{2P}\left[\sqrt{\frac{EI}{P}}\tan\left(\sqrt{\frac{P}{EI}}\frac{L}{2}\right) - \frac{L}{2}\right] \qquad \textbf{Ans}$$

Maximum Moment : The maximum moment occurs at $x = \frac{L}{2}$. From Eq. [1],

$$M_{max} = -\frac{F}{2}\left(\frac{L}{2}\right) - Pv_{max}$$

$$= -\frac{FL}{4} - P\left\{\frac{F}{2P}\left[\sqrt{\frac{EI}{P}}\tan\left(\sqrt{\frac{P}{EI}}\frac{L}{2}\right) - \frac{L}{2}\right]\right\}$$

$$= -\frac{F}{2}\sqrt{\frac{EI}{P}}\tan\left(\sqrt{\frac{P}{EI}}\frac{L}{2}\right) \qquad \textbf{Ans}$$

13–45. The ideal column has a weight w (force/ length) and rests in the horizontal position when it is subjected to the axial load **P**. Determine the maximum displacement and the maxumum moment in the column at midspan. EI is constant. *Hint:* Establish the differential equation for deflection Eq. 13–1, with the origin at the midspan. The general solution is $v = A \sin kx + B \cos kx + \ldots$ where $k^2 = P/EI$.

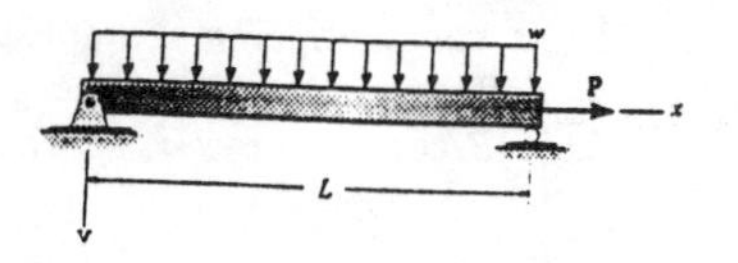

Moment Functions : FBD(b).

$$\curvearrowleft + \Sigma M_O = 0; \quad wx\left(\frac{x}{2}\right) - M(x) - \frac{wL}{2}x - Pv = 0$$

$$M(x) = \frac{w}{2}\left(x^2 - Lx\right) - Pv \qquad [1]$$

Differential Equation of The Elastic Curve :

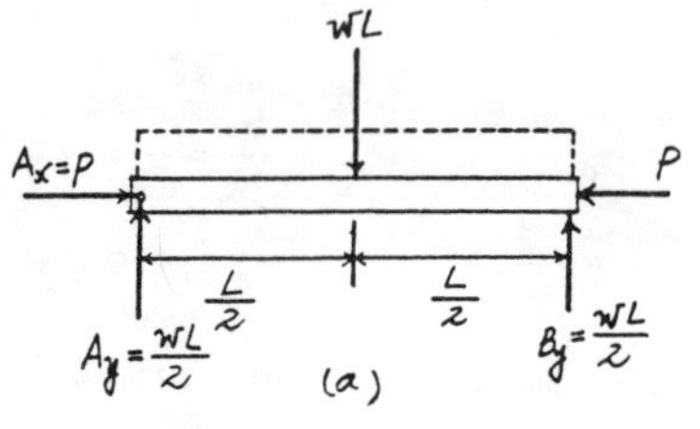

$$EI\frac{d^2 v}{dx^2} = M(x)$$

$$EI\frac{d^2 v}{dx^2} = \frac{w}{2}\left(x^2 - Lx\right) - Pv$$

$$\frac{d^2 v}{dx^2} + \frac{P}{EI}v = \frac{w}{2EI}\left(x^2 - Lx\right)$$

The solution of the above differential equation is of the form

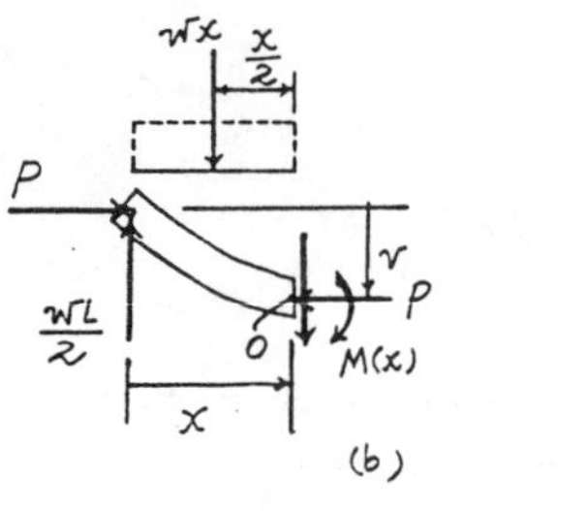

$$v = C_1 \sin\left(\sqrt{\frac{P}{EI}}x\right) + C_2 \cos\left(\sqrt{\frac{P}{EI}}x\right) + \frac{w}{2P}x^2 - \frac{wL}{2P}x - \frac{wEI}{P^2} \qquad [2]$$

and

$$\frac{dv}{dx} = C_1\sqrt{\frac{P}{EI}}\cos\left(\sqrt{\frac{P}{EI}}x\right) - C_2\sqrt{\frac{P}{EI}}\sin\left(\sqrt{\frac{P}{EI}}x\right) + \frac{w}{P}x - \frac{wL}{2P} \qquad [3]$$

The integration constants can be determined from the boundary conditions.

Boundary Conditions :

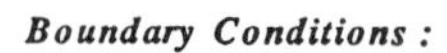

At $x = 0$, $v = 0$. From Eq. [2],

$$0 = C_2 - \frac{wEI}{P^2} \qquad C_2 = \frac{wEI}{P^2}$$

At $x = \frac{L}{2}$, $\frac{dv}{dx} = 0$. From Eq. [3],

$$0 = C_1\sqrt{\frac{P}{EI}}\cos\left(\sqrt{\frac{P}{EI}}\frac{L}{2}\right) - \frac{wEI}{P^2}\sqrt{\frac{P}{EI}}\sin\left(\sqrt{\frac{P}{EI}}\frac{L}{2}\right) + \frac{w}{P}\left(\frac{L}{2}\right) - \frac{wL}{2P}$$

$$C_1 = \frac{wEI}{P^2}\tan\left(\sqrt{\frac{P}{EI}}\frac{L}{2}\right)$$

Elastic Curve :

$$v = \frac{w}{P}\left[\frac{EI}{P}\tan\left(\sqrt{\frac{P}{EI}}\frac{L}{2}\right)\sin\left(\sqrt{\frac{P}{EI}}x\right) + \frac{EI}{P}\cos\left(\sqrt{\frac{P}{EI}}x\right) + \frac{x^2}{2} - \frac{L}{2}x - \frac{EI}{P}\right]$$

However, $v = v_{max}$ at $x = \frac{L}{2}$. Then,

$$v_{max} = \frac{w}{P}\left[\frac{EI}{P}\tan\left(\sqrt{\frac{P}{EI}}\frac{L}{2}\right)\sin\left(\sqrt{\frac{P}{EI}}\frac{L}{2}\right) + \frac{EI}{P}\cos\left(\sqrt{\frac{P}{EI}}\frac{L}{2}\right) - \frac{L^2}{8} - \frac{EI}{P}\right]$$

$$= \frac{wEI}{P^2}\left[\sec\left(\sqrt{\frac{P}{EI}}\frac{L}{2}\right) - \frac{PL^2}{8EI} - 1\right] \qquad \textbf{Ans}$$

Maximum Moment : The maximum moment occurs at $x = \frac{L}{2}$. From Eq. [1],

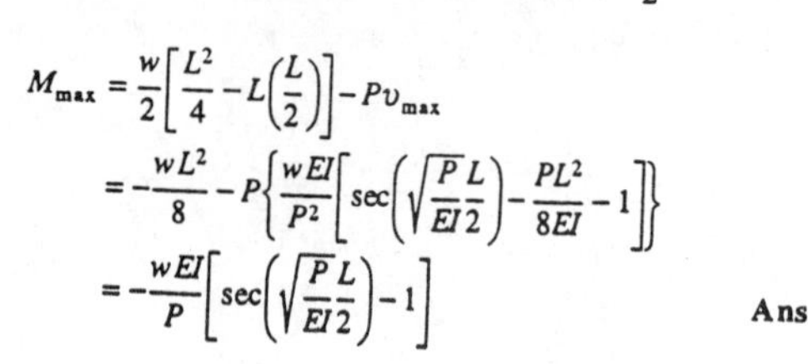

$$M_{max} = \frac{w}{2}\left[\frac{L^2}{4} - L\left(\frac{L}{2}\right)\right] - Pv_{max}$$

$$= -\frac{wL^2}{8} - P\left\{\frac{wEI}{P^2}\left[\sec\left(\sqrt{\frac{P}{EI}}\frac{L}{2}\right) - \frac{PL^2}{8EI} - 1\right]\right\}$$

$$= -\frac{wEI}{P}\left[\sec\left(\sqrt{\frac{P}{EI}}\frac{L}{2}\right) - 1\right] \qquad \textbf{Ans}$$

13–46. The wood column is assumed fixed at its base and pinned at its top. Determine the maximum eccentric load P that can be applied without causing the column to buckle or yield. $E_w = 1.8(10^3)$ ksi, $\sigma_Y = 8$ ksi.

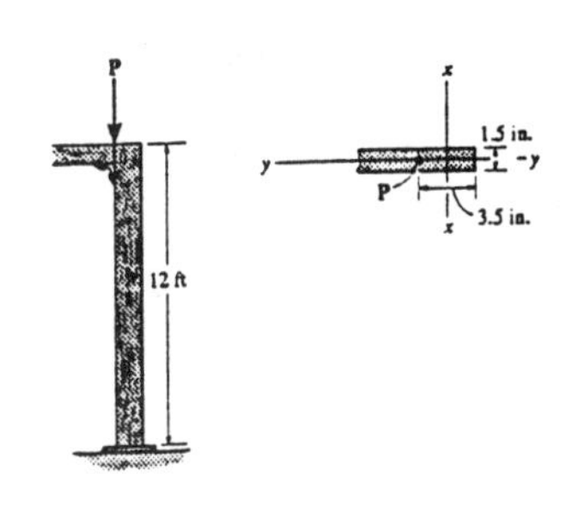

Section Properties :

$$A = 3.5(1.5) = 5.25 \text{ in}^2$$

$$I_x = \frac{1}{12}(1.5)\left(3.5^3\right) = 5.359375 \text{ in}^4$$

$$I_y = \frac{1}{12}(3.5)\left(1.5^3\right) = 0.984375 \text{ in}^4$$

$$r_x = \sqrt{\frac{I_x}{A}} = \sqrt{\frac{5.359375}{5.25}} = 1.01036 \text{ in.}$$

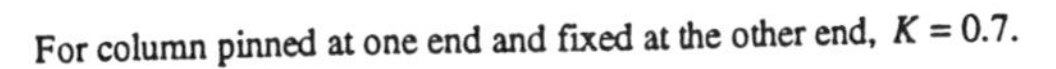

For column pinned at one end and fixed at the other end, $K = 0.7$.

$$(KL)_x = (KL)_y = 0.7(12)(12) = 100.8 \text{ in.}$$

Buckling About y - y Axis : Applying *Euler's* formula,

$$P = P_{cr} = \frac{\pi^2 E I_y}{(KL)_y^2}$$

$$= \frac{\pi^2(1.8)(10^3)(0.984375)}{100.8^2}$$

$$= 1.721 \text{ kip} = 1.72 \text{ kip} \qquad \textbf{Ans}$$

Critical Stress : *Euler's* formula is only valid if $\sigma_{cr} < \sigma_Y$.

$$\sigma_{cr} = \frac{P_{cr}}{A} = \frac{1.721}{5.25} = 0.328 \text{ ksi} < \sigma_Y = 8 \text{ ksi } (O.K!)$$

Yielding About x - x Axis : Applying the secant formula,

$$\sigma_{max} = \frac{P}{A}\left[1 + \frac{ec}{r_x^2}\sec\left(\frac{(KL)_x}{2r_x}\sqrt{\frac{P}{EA}}\right)\right]$$

$$= \frac{1.721}{5.25}\left[1 + \frac{1.75(1.75)}{1.01036^2}\sec\left(\frac{100.8}{2(1.01036)}\sqrt{\frac{1.721}{1.8(10^3)(5.25)}}\right)\right]$$

$$= 1.586 \text{ ksi} < \sigma_Y = 8 \text{ ksi } (O.K!)$$

13–47. The wood column is fixed at its base and fixed at its top. Determine the maximum eccentric load P that can be applied at its top without causing the column to buckle or yield. $E_w = 1.8(10^3)$ ksi, $\sigma_Y = 8$ ksi.

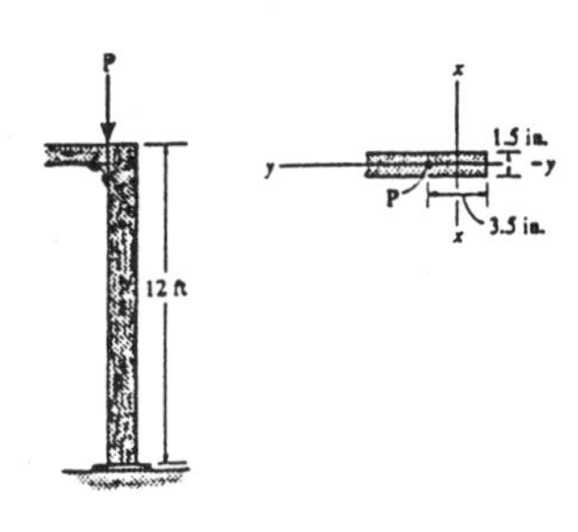

Section Properties :

$$A = 3.5(1.5) = 5.25 \text{ in}^2$$

$$I_x = \frac{1}{12}(1.5)\left(3.5^3\right) = 5.359375 \text{ in}^4$$

$$I_y = \frac{1}{12}(3.5)\left(1.5^3\right) = 0.984375 \text{ in}^4$$

$$r_x = \sqrt{\frac{I_x}{A}} = \sqrt{\frac{5.359375}{5.25}} = 1.01036 \text{ in.}$$

For a column fixed at both ends, $K = 0.5$.

$$(KL)_x = (KL)_y = 0.5(12)(12) = 72 \text{ in.}$$

Buckling About y - y Axis : Applying *Euler's* formula,

$$P = P_{cr} = \frac{\pi^2 E I_y}{(KL)_y^2}$$

$$= \frac{\pi^2(1.8)(10^3)(0.984375)}{72^2}$$

$$= 3.373 \text{ kip} = 3.37 \text{ kip} \qquad \textbf{Ans}$$

Critical Stress : *Euler's* formula is only valid if $\sigma_{cr} < \sigma_Y$.

$$\sigma_{cr} = \frac{P_{cr}}{A} = \frac{3.373}{5.25} = 0.643 \text{ ksi} < \sigma_Y = 8 \text{ ksi } (O.K!)$$

Yielding About x - x Axis : Applying the secant formula,

$$\sigma_{max} = \frac{P}{A}\left[1 + \frac{ec}{r_x^2}\sec\left(\frac{(KL)_x}{2r_x}\sqrt{\frac{P}{EA}}\right)\right]$$

$$= \frac{3.373}{5.25}\left[1 + \frac{1.75(1.75)}{1.01036^2}\sec\left(\frac{72}{2(1.01036)}\sqrt{\frac{3.373}{1.8(10^3)(5.25)}}\right)\right]$$

$$= 3.108 \text{ ksi} < \sigma_Y = 8 \text{ ksi } (O.K!)$$

***13–48.** The $W8 \times 48$ structural A-36 steel column is fixed at its bottom and free at its top. If it is subjected to the eccentric load of 75 kip, determine the factor of safety with respect to either the initiation of buckling or yielding.

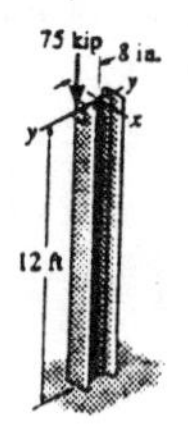

Section Properties : For a wide flange section W8×48,

$$A = 14.1 \text{ in}^2 \qquad r_x = 3.61 \text{ in.} \qquad I_y = 60.9 \text{ in}^4 \qquad d = 8.50 \text{ in.}$$

For a column fixed at one end and free at the other end, $K = 2$.

$$(KL)_y = (KL)_x = 2(12)(12) = 288 \text{ in.}$$

Buckling About y - y Axis : Applying *Euler's* formula,

$$P = P_{cr} = \frac{\pi^2 EI_y}{(KL)_y^2}$$
$$= \frac{\pi^2 (29.0)(10^3)(60.9)}{288^2}$$
$$= 210.15 \text{ kip}$$

Critical Stress : *Euler's* formula is only valid if $\sigma_{cr} < \sigma_Y$.

$$\sigma_{cr} = \frac{P_{cr}}{A} = \frac{210.15}{14.1} = 14.90 \text{ ksi} < \sigma_Y = 36 \text{ ksi} \ (O.K!)$$

Yielding About x - x Axis : Applying the secant formula,

$$\sigma_{max} = \frac{P_{max}}{A}\left[1 + \frac{ec}{r_x^2}\sec\left(\frac{(KL)_x}{2r_x}\sqrt{\frac{P_{max}}{EA}}\right)\right]$$

$$36 = \frac{P_{max}}{14.1}\left[1 + \frac{8\left(\frac{8.50}{2}\right)}{3.61^2}\sec\left(\frac{288}{2(3.61)}\sqrt{\frac{P_{max}}{29.0(10^3)(14.1)}}\right)\right]$$

$$36(14.1) = P_{max}\left(1 + 2.608943\sec 0.0623802\sqrt{P_{max}}\right)$$

Solving by trial and error,

$$P_{max} = 117.0 \text{ kip} \ (\textit{Controls!})$$

Factor of Safety :

$$\text{F.S.} = \frac{P_{max}}{P} = \frac{117.0}{75} = 1.56 \qquad \textbf{Ans}$$

13–49. The $W8 \times 48$ structural A-36 steel column is fixed at its bottom and pinned at its top. If it is subjected to the eccentric load of 75 kip, determine if the column fails by yielding. The column is braced so that it does not buckle about the y–y axis.

Section Properties : For the wide flange section W8×48,

$$A = 14.1 \text{ in}^2 \qquad r_x = 3.61 \text{ in.} \qquad d = 8.50 \text{ in.}$$

For a column fixed at one end and pinned at the other end, $K = 0.7$.

$$(KL)_x = 0.7(12)(12) = 100.8 \text{ in.}$$

Yielding About x - x Axis : Applying the secant formula,

$$\sigma_{max} = \frac{P}{A}\left[1 + \frac{ec}{r_x^2}\sec\left(\frac{(KL)_x}{2r_x}\sqrt{\frac{P}{EA}}\right)\right]$$
$$= \frac{75}{14.1}\left[1 + \frac{8\left(\frac{8.50}{2}\right)}{3.61^2}\sec\left(\frac{100.8}{2(3.61)}\sqrt{\frac{75}{29.0(10^3)(14.1)}}\right)\right]$$
$$= 19.45 \text{ ksi} < \sigma_Y = 36 \text{ ksi} \ (O.K!)$$

Hence, the column **does not fail by yielding.** **Ans**

13–50. A $W\,14 \times 30$ structural A-36 steel member is to be used as a pin-connected 20-ft column. Determine the maximum eccentric load P that can be applied so the column does not buckle or yield. Compare this value with an axial critical load P' applied through the centroid of the column.

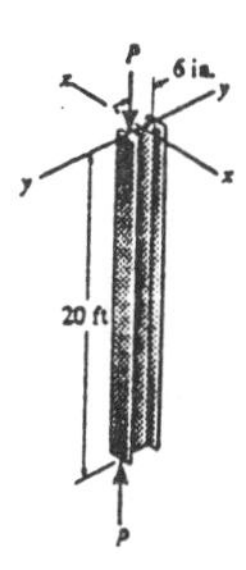

Section Properties : For a wide flange section W14×30,

$$A = 8.85\ \text{in}^2 \quad r_x = 5.73\ \text{in.} \quad I_y = 19.6\ \text{in}^4 \quad d = 13.84\ \text{in.}$$

For a column pinned at both ends, $K = 1$.

$$(KL)_y = (KL)_x = 1(20)(12) = 240\ \text{in.}$$

Buckling About y - y Axis : Applying *Euler's* formula,

$$P' = P = P_{cr} = \frac{\pi^2 EI_y}{(KL)_y^2}$$

$$= \frac{\pi^2(29.0)(10^3)(19.6)}{240^2}$$

$$= 97.39\ \text{kip} = 97.4\ \text{kip} \qquad \textbf{Ans}$$

Critical Stress : *Euler's* formula is only valid if $\sigma_{cr} < \sigma_Y$.

$$\sigma_{cr} = \frac{P_{cr}}{A} = \frac{97.39}{8.85} = 11.00\text{ksi} < \sigma_Y = 36\ \text{ksi}\ (O.K!)$$

Yielding About x - x Axis : Applying the secant formula,

$$\sigma_{max} = \frac{P}{A}\left[1 + \frac{ec}{r_x^2}\sec\left(\frac{(KL)_x}{2r_x}\sqrt{\frac{P}{EA}}\right)\right]$$

$$= \frac{97.39}{8.85}\left[1 + \frac{6\left(\frac{13.84}{2}\right)}{5.73^2}\sec\left(\frac{240}{2(5.73)}\sqrt{\frac{97.39}{29.0(10^3)(8.85)}}\right)\right]$$

$$= 26.17\ \text{ksi} < \sigma_Y = 36\ \text{ksi}\ \ (O.K!)$$

13–51. Solve Prob. 13–50 if the column is fixed connected at its ends.

Section Properties : For a wide flange section W14×30,

$$A = 8.85\ \text{in}^2 \quad r_x = 5.73\ \text{in.} \quad I_y = 19.6\ \text{in}^4 \quad d = 13.84\ \text{in.}$$

For a column fixed at both ends, $K = 0.5$.

$$(KL)_y = (KL)_x = 0.5(20)(12) = 120\ \text{in.}$$

Buckling About y - y Axis : Applying Euler's formula,

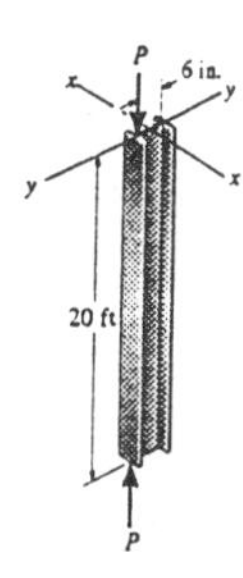

$$P = P_{cr} = \frac{\pi^2 EI_y}{(KL)_y^2}$$

$$= \frac{\pi^2(29.0)(10^3)(19.6)}{120^2}$$

$$= 389.58\ \text{kip}$$

Critical Stress : *Euler's* formula is only valid if $\sigma_{cr} < \sigma_Y$.

$$\sigma_{cr} = \frac{P_{cr}}{A} = \frac{389.58}{8.85} = 44.02\text{ksi} > \sigma_Y = 36\ \text{ksi}\ (\textit{No Good!})$$

Hence **yielding occurs!** Then,

$$P' = A\sigma_Y = 8.85(36) = 318.6\ \text{kip}$$

Yielding About x - x Axis : Applying the secant formula,

$$\sigma_{max} = \frac{P}{A}\left[1 + \frac{ec}{r_x^2}\sec\left(\frac{(KL)_x}{2r_x}\sqrt{\frac{P}{EA}}\right)\right]$$

$$36 = \frac{P}{8.85}\left[1 + \frac{6\left(\frac{13.84}{2}\right)}{5.73^2}\sec\left(\frac{120}{2(5.73)}\sqrt{\frac{P}{29.0(10^3)(8.85)}}\right)\right]$$

$$36(8.85) = P\left(1 + 1.264585\sec 0.020669\sqrt{P}\right)$$

Solving by trial and error,

$$P = 139\ \text{kip} \qquad \textbf{Ans}$$

The column would carry **2.29** $\left(\frac{P'}{P}\right)$ **times** more load if the load P acts concentrically. **Ans**

***13–52.** Solve Prob. 13–50 if the column is fixed at its bottom and free at its top.

Section Properties : For a wide flange section W14×30,

$A = 8.85\text{ in}^2 \quad r_x = 5.73\text{ in.} \quad I_y = 19.6\text{ in}^4 \quad d = 13.84\text{ in.}$

For a column fixed at one end and free at the other end, $K = 2$.

$$(KL)_y = (KL)_x = 2(20)(12) = 480\text{ in.}$$

Buckling About y - y Axis : Applying Euler's formula,

$$P' = P = P_{cr} = \frac{\pi^2 E I_y}{(KL)_y^2} = \frac{\pi^2(29.0)(10^3)(19.6)}{480^2} = 24.348\text{ kip} = 24.3\text{ kip} \qquad \textbf{Ans}$$

Critical Stress : *Euler's* formula is only valid if $\sigma_{cr} < \sigma_Y$.

$$\sigma_{cr} = \frac{P_{cr}}{A} = \frac{24.348}{8.85} = 2.75\text{ ksi} < \sigma_Y = 36\text{ ksi } (O.K!)$$

Yielding About x - x Axis : Applying the secant formula,

$$\sigma_{max} = \frac{P}{A}\left[1 + \frac{ec}{r_x^2}\sec\left(\frac{(KL)_x}{2r_x}\sqrt{\frac{P}{EA}}\right)\right]$$

$$= \frac{24.348}{8.85}\left[1 + \frac{6\left(\frac{13.84}{2}\right)}{5.73^2}\sec\left(\frac{480}{2(5.73)}\sqrt{\frac{24.348}{29.0(10^3)(8.85)}}\right)\right]$$

$$= 6.54\text{ ksi} < \sigma_Y = 36\text{ ksi } (O.K!)$$

13–53. The tube is made of copper and has an outer diameter of 35 mm and a wall thickness of 7 mm. Using a factor of safety with respect to buckling and yielding of F.S. = 2.5, determine the allowable eccentric load P. The tube is pin supported at its ends. $E_{cu} = 120$ GPa, $\sigma_Y = 750$ MPa.

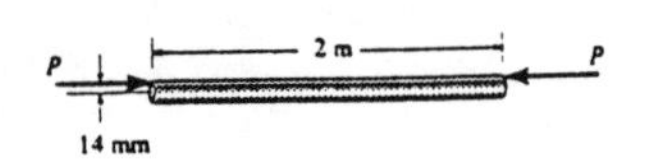

Section Properties :

$$A = \frac{\pi}{4}(0.035^2 - 0.021^2) = 0.61575(10^{-3})\text{ m}^2$$
$$I = \frac{\pi}{4}(0.0175^4 - 0.0105^4) = 64.1152(10^{-9})\text{ m}^4$$
$$r = \sqrt{\frac{I}{A}} = \sqrt{\frac{64.1152(10^{-9})}{0.61575(10^{-3})}} = 0.010204\text{ m}$$

For a column pinned at both ends, $K = 1$. Then $KL = 1(2) = 2$ m.

Buckling : Applying *Euler's* formula,

$$P_{max} = P_{cr} = \frac{\pi^2 EI}{(KL)^2} = \frac{\pi^2(120)(10^9)\left[64.1152(10^{-9})\right]}{2^2} = 18983.7\text{ N} = 18.98\text{ kN}$$

Critical Stress : *Euler's* formula is only valid if $\sigma_{cr} < \sigma_Y$.

$$\sigma_{cr} = \frac{P_{cr}}{A} = \frac{18983.7}{0.61575(10^{-3})} = 30.83\text{ MPa} < \sigma_Y = 750\text{ MPa}(O.K!)$$

Yielding : Applying the secant formula,

$$\sigma_{max} = \frac{P_{max}}{A}\left[1 + \frac{ec}{r^2}\sec\left(\frac{(KL)}{2r}\sqrt{\frac{P_{max}}{EA}}\right)\right]$$

$$750(10^6) = \frac{P_{max}}{0.61575(10^{-3})}\left[1 + \frac{0.014(0.0175)}{0.010204^2}\sec\left(\frac{2}{2(0.010204)}\sqrt{\frac{P_{max}}{120(10^9)[0.61575(10^{-3})]}}\right)\right]$$

$$750(10^6) = \frac{P_{max}}{0.61575(10^{-3})}\left(1 + 2.35294\sec 0.0114006\sqrt{P_{max}}\right)$$

Solving by trial and error,

$$P_{max} = 16\,885\text{ N} = 16.885\text{ kN} \quad (\textit{Controls!})$$

Factor of Safety :

$$P = \frac{P_{max}}{\text{F.S.}} = \frac{16.885}{2.5} = 6.75\text{ kN} \qquad \textbf{Ans}$$

13–54. The tube is made of copper and has an outer diameter of 35 mm and a wall thickness of 7 mm. Using a factor of safety with respect to buckling and yielding of F.S. = 2.5, determine the allowable eccentric load P that it can support without failure. The tube is fixed supported at its ends. E_{cu} = 120 GPa, σ_Y = 750 MPa.

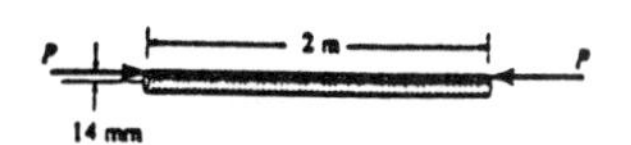

Section Properties :

$$A = \frac{\pi}{4}\left(0.035^2 - 0.021^2\right) = 0.61575\left(10^{-3}\right)\ \text{m}^2$$

$$I = \frac{\pi}{4}\left(0.0175^4 - 0.0105^4\right) = 64.1152\left(10^{-9}\right)\ \text{m}^4$$

$$r = \sqrt{\frac{I}{A}} = \sqrt{\frac{64.1152(10^{-9})}{0.61575(10^{-3})}} = 0.010204\ \text{m}$$

For a column fixed at both ends , $K = 0.5$. Then $KL = 0.5(2) = 1$ m.

Buckling : Applying *Euler's* formula,

$$P_{max} = P_{cr} = \frac{\pi^2 EI}{(KL)^2} = \frac{\pi^2(120)(10^9)\left[64.1152(10^{-9})\right]}{1^2} = 75\ 935.0\ \text{N} = 75.93\ \text{kN}$$

Critical Stress : *Euler's* formula is only valid if $\sigma_{cr} < \sigma_Y$.

$$\sigma_{cr} = \frac{P_{cr}}{A} = \frac{75\ 935.0}{0.61575(10^{-3})} = 123.3\ \text{MPa} < \sigma_Y = 750\ \text{MPa}(\textbf{\textit{O. K!}})$$

Yielding : Applying the secant formula,

$$\sigma_{max} = \frac{P_{max}}{A}\left[1 + \frac{ec}{r^2}\sec\left(\frac{(KL)}{2r}\sqrt{\frac{P_{max}}{EA}}\right)\right]$$

$$750\left(10^6\right) = \frac{P_{max}}{0.61575(10^{-3})}\left[1 + \frac{0.014(0.0175)}{0.010204^2}\sec\left(\frac{1}{2(0.010204)}\sqrt{\frac{P_{max}}{120(10^9)\left[0.61575(10^{-3})\right]}}\right)\right]$$

$$750\left(10^6\right) = \frac{P_{max}}{0.61575(10^{-3})}\left(1 + 2.35294\sec 5.70032\left(10^{-3}\right)\sqrt{P}\right)$$

Solving by trial and error,

$$P_{max} = 50\ 325\ \text{N} = 50.325\ \text{kN}\quad (\textbf{\textit{Controls!}})$$

Factor of Safety :

$$P = \frac{P_{max}}{\text{F.S.}} = \frac{50.325}{2.5} = 20.1\ \text{kN}$$ **Ans**

13–55. The brass rod is fixed at one end and free at the other end. If the eccentric load $P = 200$ kN is applied, determine the greatest allowable length L of the rod so that it does not buckle or yield. $E_{br} = 101$ GPa, $\sigma_Y = 69$ MPa.

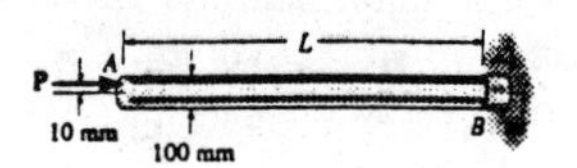

Section Properties :

$$A = \frac{\pi}{4}\left(0.1^2\right) = 2.50\left(10^{-3}\right)\pi \text{ m}^2$$

$$I = \frac{\pi}{4}\left(0.05^4\right) = 1.5625\left(10^{-6}\right)\pi \text{ m}^4$$

$$r = \sqrt{\frac{I}{A}} = \sqrt{\frac{1.5625(10^{-6})\pi}{2.50(10^{-3})\pi}} = 0.025 \text{ m}$$

For a column fixed at one end and free at the other end, $K = 2$. Then $KL = 2(L) = 2L$.

Buckling : Applying Euler's formula,

$$P_{cr} = \frac{\pi^2 EI}{(KL)^2}$$

$$200\left(10^3\right) = \frac{\pi^2(101)(10^9)\left[1.5625(10^{-6})\pi\right]}{(2L)^2}$$

$$L = 2.473 \text{ m}$$

Critical Stress : *Euler's* formula is only valid if $\sigma_{cr} < \sigma_Y$.

$$\sigma_{cr} = \frac{P_{cr}}{A} = \frac{200(10^3)}{2.50(10^{-3})\pi} = 25.46 \text{ MPa} < \sigma_Y = 69 \text{ MPa} (O.K!)$$

Yielding : Applying the secant formula,

$$\sigma_Y = \frac{P}{A}\left[1 + \frac{ec}{r^2}\sec\left(\frac{(KL)}{2r}\sqrt{\frac{P}{EA}}\right)\right]$$

$$69\left(10^6\right) = \frac{200(10^3)}{2.50(10^{-3})\pi}\left[1 + \frac{0.01(0.05)}{0.025^2}\sec\left(\frac{2L}{2(0.025)}\sqrt{\frac{200(10^3)}{101(10^9)[2.50(10^{-3})\pi]}}\right)\right]$$

$$69 = \frac{80}{\pi}(1 + 0.800\sec 0.635140L)$$

Solving by trial and error,

$$L = 1.7065 \text{ m} = 1.71 \text{ m} \quad (\textit{Controls!}) \qquad \text{Ans}$$

***13–56.** The brass rod is fixed at one end and free at the other end. If the length of the rod is $L = 2$ m, determine the greatest allowable load P that can be applied so that the rod does not buckle or yield. Also, determine the largest sidesway deflection of the rod due to the loading. $E_{br} = 101$ GPa, $\sigma_Y = 69$ MPa.

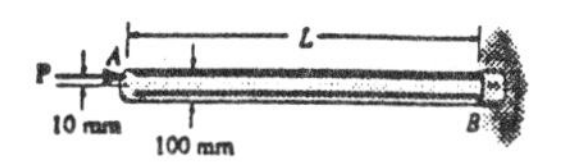

Section Properties :

$$A = \frac{\pi}{4}(0.1^2) = 2.50(10^{-3})\,\pi \text{ m}^2$$

$$I = \frac{\pi}{4}(0.05^4) = 1.5625(10^{-6})\,\pi \text{ m}^4$$

$$r = \sqrt{\frac{I}{A}} = \sqrt{\frac{1.5625(10^{-6})\,\pi}{2.50(10^{-3})\,\pi}} = 0.025 \text{ m}$$

For the column fixed at one end and free at the other end, $K = 2$. Then $KL = 2(2) = 4$ m.

Buckling *:* Applying *Euler's* formula,

$$P = P_{cr} = \frac{\pi^2 EI}{(KL)^2}$$
$$= \frac{\pi^2(101)(10^9)[1.5625(10^{-6})\,\pi]}{4^2}$$
$$= 305823.6 \text{ N} = 305.8 \text{ kN}$$

Critical Stress : *Euler's* formula is only valid if $\sigma_{cr} < \sigma_Y$.

$$\sigma_{cr} = \frac{P_{cr}}{A} = \frac{305823.6}{2.50(10^{-3})\,\pi} = 38.94 \text{ MPa} < \sigma_Y = 69 \text{ MPa} (O.K!)$$

Yielding *:* Applying the secant formula,

$$\sigma_Y = \frac{P}{A}\left[1 + \frac{ec}{r^2}\sec\left(\frac{(KL)}{2r}\sqrt{\frac{P}{EA}}\right)\right]$$

$$69(10^6) = \frac{P}{2.50(10^{-3})\,\pi}\left[1 + \frac{0.01(0.05)}{0.025^2}\sec\left(\frac{4}{2(0.025)}\sqrt{\frac{P}{101(10^9)[2.50(10^{-3})\,\pi]}}\right)\right]$$

$$69(10^6) = \frac{400P}{\pi}\left[1 + 0.800\sec 2.84043(10^{-3})\sqrt{P}\right]$$

Solving by trial and error,

$$P = 173700 \text{ N} = 174 \text{ kN} \quad (\textit{Controls!})$$ **Ans**

Maximum Displacement :

$$v_{max} = e\left[\sec\left(\sqrt{\frac{P}{EI}}\frac{KL}{2}\right) - 1\right]$$

$$= 0.01\left[\sec\left(\sqrt{\frac{173700}{101(10^9)[1.5625(10^{-6})\,\pi]}}\left(\frac{4}{2}\right)\right) - 1\right]$$

$$= 0.01650 \text{ m} = 16.5 \text{ mm}$$ **Ans**

13–57. The wood column is fixed at its base and can be assumed pin connected at its top. Determine the maximum eccentric load P that can be applied without causing the column to buckle or yield. $E_w = 1.8(10^3)$ ksi, $\sigma_Y = 8$ ksi.

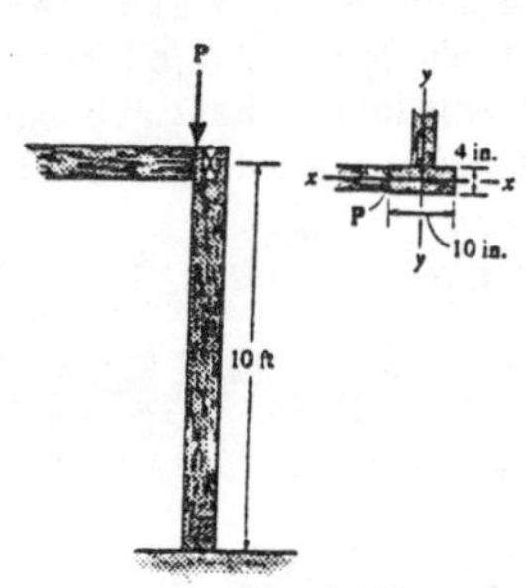

Section Properties :

$$A = 4(10) = 40.0 \text{ in}^2$$

$$I_x = \frac{1}{12}(4)\left(10^3\right) = 333.33 \text{ in}^4$$

$$I_y = \frac{1}{12}(10)\left(4^3\right) = 53.33 \text{ in}^4$$

$$r_x = \sqrt{\frac{I_x}{A}} = \sqrt{\frac{333.33}{40.0}} = 2.88675 \text{ in.}$$

For a column pinned at one end and fixed at the other end, $K = 0.7$.

$$(KL)_x = (KL)_y = 0.7(10)(12) = 84.0 \text{ in.}$$

Buckling About y - y Axis : Applying Euler's formula,

$$P = P_{cr} = \frac{\pi^2 EI_y}{(KL)_y^2}$$

$$= \frac{\pi^2(1.8)(10^3)(53.33)}{84.0^2}$$

$$= 134.28 \text{ kip} = 134 \text{ kip}$$

Critical Stress : *Euler's* formula is only valid if $\sigma_{cr} < \sigma_Y$.

$$\sigma_{cr} = \frac{P_{cr}}{A} = \frac{134.28}{40.0} = 3.357 \text{ ksi} < \sigma_Y = 8 \text{ ksi } (O.K!)$$

Yielding About x - x Axis : Applying the secant formula,

$$\sigma_{max} = \frac{P}{A}\left[1 + \frac{ec}{r_x^2}\sec\left(\frac{(KL)_x}{2r_x}\sqrt{\frac{P}{EA}}\right)\right]$$

$$8 = \frac{P}{40.0}\left[1 + \frac{5(5)}{2.88675^2}\sec\left(\frac{84.0}{2(2.88675)}\sqrt{\frac{P}{1.8(10^3)(40.0)}}\right)\right]$$

$$320.0 = P\left[1 + 3\sec 0.0542218\sqrt{P}\right]$$

Solving by trial and error,

$$P = 73.5 \text{ kip } (Controls!)$$ **Ans**

13–58. The wood column is fixed at its base and can be assumed fixed connected at its top. Determine the maximum eccentric load P that can be applied without causing the column to buckle or yield. $E_w = 1.8(10^3)$ ksi, $\sigma_Y = 8$ ksi.

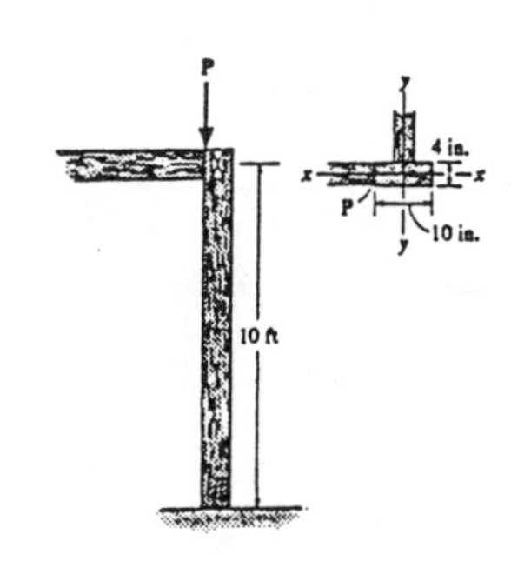

Section Properties :

$$A = 4(10) = 40.0 \text{ in}^2$$

$$I_x = \frac{1}{12}(4)\left(10^3\right) = 333.33 \text{ in}^4$$

$$I_y = \frac{1}{12}(10)\left(4^3\right) = 53.33 \text{ in}^4$$

$$r_x = \sqrt{\frac{I_x}{A}} = \sqrt{\frac{333.33}{40.0}} = 2.88675 \text{ in.}$$

For a column fixed at both ends, $K = 0.5$.

$$(KL)_x = (KL)_y = 0.5(10)(12) = 60.0 \text{ in.}$$

Buckling About y - y Axis : Applying *Euler's* formula,

$$P = P_{cr} = \frac{\pi^2 EI_y}{(KL)_y^2}$$

$$= \frac{\pi^2(1.8)(10^3)(53.33)}{60.0^2}$$

$$= 263.19 \text{ kip} = 263 \text{ kip}$$

Critical Stress : *Euler's* formula is only valid if $\sigma_{cr} < \sigma_Y$.

$$\sigma_{cr} = \frac{P_{cr}}{A} = \frac{263.19}{40.0} = 6.58 \text{ ksi} < \sigma_Y = 8 \text{ ksi } (O.K!)$$

Yielding About x - x Axis : Applying the secant formula,

$$\sigma_{max} = \frac{P}{A}\left[1 + \frac{ec}{r_x^2}\sec\left(\frac{(KL)_x}{2r_x}\sqrt{\frac{P}{EA}}\right)\right]$$

$$8 = \frac{P}{40.0}\left[1 + \frac{5(5)}{2.88675^2}\sec\left(\frac{60.0}{2(2.88675)}\sqrt{\frac{P}{1.8(10^3)(40.0)}}\right)\right]$$

$$320.0 = P\left[1 + 3\sec 0.0387298\sqrt{P}\right]$$

Solving by trial and error,

$$P = 76.6 \text{ kip } (Controls!)$$ **Ans**

13–59. Determine the load P required to cause the steel $W\ 12 \times 50$ structural A-36 steel column to fail either by buckling or by yielding. The column is fixed at its bottom and the cables at its top act as a pin to hold it.

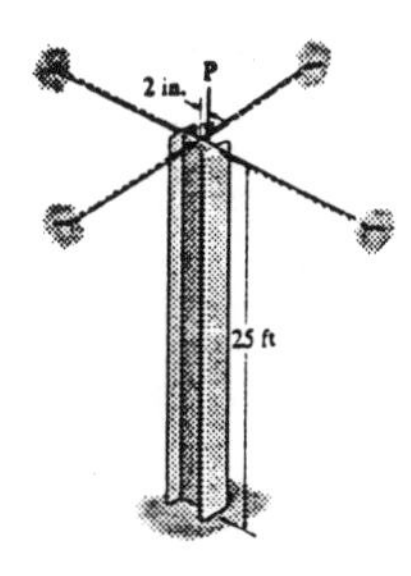

Section Properties : For the wide flange section W12×50,

$A = 14.7\ \text{in}^2 \quad r_x = 5.18\ \text{in.} \quad I_y = 56.3\ \text{in}^4 \quad d = 12.19\ \text{in.}$

For a column fixed at one end and pinned at the other end, $K = 0.7$.

$$(KL)_y = (KL)_x = 0.7(25)(12) = 210\ \text{in.}$$

Buckling About y - y Axis : Applying *Euler's* formula,

$$P = P_{cr} = \frac{\pi^2 EI_y}{(KL)_y^2} = \frac{\pi^2(29.0)(10^3)(56.3)}{210^2} = 365.40\ \text{kip}$$

Critical Stress : *Euler's* formula is only valid if $\sigma_{cr} < \sigma_Y$.

$$\sigma_{cr} = \frac{P_{cr}}{A} = \frac{365.4}{14.7} = 24.86\ \text{ksi} < \sigma_Y = 36\ \text{ksi}\ (O.K!)$$

Yielding About x - x Axis : Applying the secant formula,

$$\sigma_{max} = \frac{P}{A}\left[1 + \frac{ec}{r_x^2}\sec\left(\frac{(KL)_x}{2r_x}\sqrt{\frac{P}{EA}}\right)\right]$$

$$36 = \frac{P}{14.7}\left[1 + \frac{2\left(\frac{12.19}{2}\right)}{5.18^2}\sec\left(\frac{210}{2(5.18)}\sqrt{\frac{P}{29.0(10^3)(14.7)}}\right)\right]$$

$$36(14.7) = P\left(1 + 0.454302\sec 0.0310457\sqrt{P}\right)$$

Solving by trial and error,

$$P_{max} = 343.3\ \text{kip} = 343\ \text{kip}\ (\textit{Controls!})$$ **Ans**

***13–60.** Solve Prob. 13–59 if the column is a steel $W\ 12 \times 16$ section.

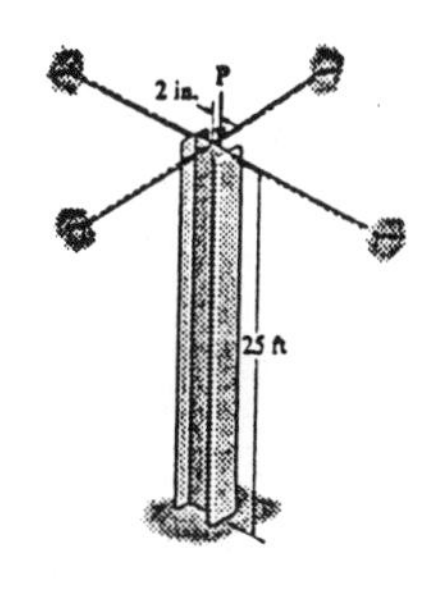

Section Properties : For the wide flange section W12×16,

$A = 4.71\ \text{in}^2 \quad r_x = 4.67\ \text{in.} \quad I_y = 2.82\ \text{in}^4 \quad d = 11.99\ \text{in.}$

For a column fixed at one end and pinned at the other end, $K = 0.7$.

$$(KL)_y = (KL)_x = 0.7(25)(12) = 210\ \text{in.}$$

Buckling About y - y Axis : Applying *Euler's* formula,

$$P = P_{cr} = \frac{\pi^2 EI_y}{(KL)_y^2} = \frac{\pi^2(29.0)(10^3)(2.82)}{210^2} = 18.30\ \text{kip} = 18.3\ \text{kip}$$ **Ans**

Critical Stress : *Euler's* formula is only valid if $\sigma_{cr} < \sigma_Y$.

$$\sigma_{cr} = \frac{P_{cr}}{A} = \frac{18.30}{4.71} = 3.89\ \text{ksi} < \sigma_Y = 36\ \text{ksi}\ (O.K!)$$

Yielding About x - x Axis : Applying the secant formula,

$$\sigma_{max} = \frac{P}{A}\left[1 + \frac{ec}{r_x^2}\sec\left(\frac{(KL)_x}{2r_x}\sqrt{\frac{P}{EA}}\right)\right]$$

$$= \frac{18.30}{4.71}\left[1 + \frac{2\left(\frac{11.99}{2}\right)}{4.67^2}\sec\left(\frac{210}{2(4.67)}\sqrt{\frac{18.30}{29.0(10^3)(4.71)}}\right)\right]$$

$$= 6.10\ \text{ksi} < \sigma_Y = 36\ \text{ksi}\ (O.K!)$$

13–61. The $W14 \times 53$ structural A-36 steel column is fixed at its base and free at its top. If $P = 75$ kip, determine the sidesway deflection at its top and the maximum stress in the column.

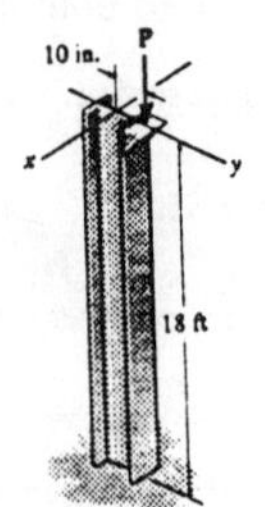

Section Properties : For the wide flange section W14×53,

$A = 15.6\text{ in}^2 \quad I_x = 541\text{ in}^4 \quad r_x = 5.89\text{ in.} \quad I_y = 57.7\text{ in}^4$
$d = 13.92\text{ in.}$

For a column fixed at one end and free at the other end, $K = 2$.

$$(KL)_y = (KL)_x = 2(18)(12) = 432\text{ in.}$$

Buckling About y - y Axis : Applying *Euler's* formula,

$$P_{cr} = \frac{\pi^2 EI_y}{(KL)_y^2} = \frac{\pi^2(29.0)(10^3)(57.7)}{432^2} = 88.49\text{ kip} > P = 75\text{ kip} \quad (O.K!)$$

Hence, the column **does not buckle** about the $y-y$ axis.

Critical Stress : *Euler's* formula is only valid if $\sigma_{cr} < \sigma_Y$.

$$\sigma_{cr} = \frac{P_{cr}}{A} = \frac{88.49}{15.6} = 5.67\text{ ksi} < \sigma_Y = 36\text{ ksi} \quad (O.K!)$$

Yielding About x - x Axis : Applying the secant formula,

$$\sigma_{max} = \frac{P}{A}\left[1 + \frac{ec}{r_x^2}\sec\left(\frac{(KL)_x}{2r_x}\sqrt{\frac{P}{EA}}\right)\right]$$

$$= \frac{75}{15.6}\left[1 + \frac{10\left(\frac{13.92}{2}\right)}{5.89^2}\sec\left(\frac{432}{2(5.89)}\sqrt{\frac{75}{29.0(10^3)(15.6)}}\right)\right]$$

$$= 15.6\text{ ksi} \qquad \textbf{Ans}$$

Since $\sigma_{max} < \sigma_Y = 36$ ksi, the column **does not yield.**

Maximum Displacement :

$$v_{max} = e\left[\sec\left(\sqrt{\frac{P}{EI}}\frac{KL}{2}\right) - 1\right] = 10\left[\sec\left(\sqrt{\frac{75}{29(10^3)(541)}}\left(\frac{432}{2}\right)\right) - 1\right]$$

$$= 1.23\text{ in.} \qquad \textbf{Ans}$$

13–62. The $W14 \times 53$ steel column is fixed at its base and free at its top. Determine the maximum eccentric load P that it can support without causing it to either buckle or yield. $E_{st} = 29(10^3)$ ksi, $\sigma_Y = 50$ ksi.

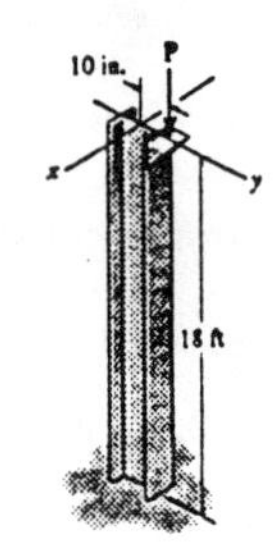

Section Properties : For a wide flange section W14×53,

$A = 15.6\text{ in}^2 \quad I_x = 541\text{ in}^4 \quad r_x = 5.89\text{ in.} \quad I_y = 57.7\text{ in}^4$
$d = 13.92\text{ in.}$

For a column fixed at one end and free at the other end, $K = 2$.

$$(KL)_y = (KL)_x = 2(18)(12) = 432\text{ in.}$$

Buckling About y - y Axis : Applying Euler's formula,

$$P = P_{cr} = \frac{\pi^2 EI_y}{(KL)_y^2} = \frac{\pi^2(29.0)(10^3)(57.7)}{432^2} = 88.49\text{ kip} = 88.5\text{ kip} \quad (\textbf{Control!}) \qquad \textbf{Ans}$$

Critical Stress : *Euler's* formula is only valid if $\sigma_{cr} < \sigma_Y$.

$$\sigma_{cr} = \frac{P_{cr}}{A} = \frac{88.49}{15.6} = 5.67\text{ ksi} < \sigma_Y = 50\text{ ksi} \quad (O.K!)$$

Yielding About x - x Axis : Applying the secant formula,

$$\sigma_{max} = \frac{P}{A}\left[1 + \frac{ec}{r_x^2}\sec\left(\frac{(KL)_x}{2r_x}\sqrt{\frac{P}{EA}}\right)\right]$$

$$= \frac{88.49}{15.6}\left[1 + \frac{10\left(\frac{13.92}{2}\right)}{5.89^2}\sec\left(\frac{432}{2(5.89)}\sqrt{\frac{88.49}{29.0(10^3)(15.6)}}\right)\right]$$

$$= 18.73\text{ ksi} < \sigma_Y = 50\text{ ksi} \quad (O.K!)$$

13–63. The wood column has a square cross section with dimensions 150 mm by 150 mm. It is fixed at its base and free at its top. Determine the load P that can be applied at $e = 200$ mm without causing the column to fail either by buckling or yielding. $E_w = 12$ GPa, $\sigma_Y = 55$ MPa

Section Properties :

$$A = 0.15(0.15) = 0.0225 \text{ m}^2$$

$$I = \frac{1}{12}(0.15)\left(0.15^3\right) = 42.1875\left(10^{-6}\right) \text{ m}^4$$

$$r = \sqrt{\frac{I}{A}} = \sqrt{\frac{42.1875(10^{-6})}{0.0225}} = 0.04330 \text{ m}$$

For a column fixed at one end and free at the other end, $K = 2$.

$$KL = 2(4) = 8 \text{ m}$$

Buckling About y - y Axis : Applying Euler's formula,

$$P = P_{cr} = \frac{\pi^2 EI}{(KL)^2}$$

$$= \frac{\pi^2(12)(10^9)[42.1875(10^{-6})]}{8^2}$$

$$= 78070 \text{ N} = 78.07 \text{ kN}$$

Critical Stress : *Euler's* formula is only valid if $\sigma_{cr} < \sigma_Y$.

$$\sigma_{cr} = \frac{P_{cr}}{A} = \frac{78070}{0.0225} = 3.47 \text{ MPa} < \sigma_Y = 55 \text{ MPa} (O.K!)$$

Yielding About x - x Axis : Applying the secant formula,

$$\sigma_{max} = \frac{P}{A}\left[1 + \frac{ec}{r_x^2}\sec\left(\frac{(KL)_x}{2r_x}\sqrt{\frac{P}{EA}}\right)\right]$$

$$55\left(10^6\right) = \frac{P}{0.0225}\left[1 + \frac{0.2(0.075)}{0.04330^2}\sec\left(\frac{8}{2(0.04330)}\sqrt{\frac{P}{12(10^9)(0.0225)}}\right)\right]$$

$$55\left(10^6\right) = \frac{P}{0.0225}\left[1 + 8\sec 5.621827\left(10^{-3}\right)\sqrt{P}\right]$$

Solving by trial and error,

$$P = 48510 \text{ N} = 48.5 \text{ kN} \quad (\textit{Controls!})$$ **Ans**

***13–64.** The steel column supports the two eccentric loadings. If it is assumed to be pinned at its top, fixed at the bottom, and braced against buckling about the y–y axis, determine the maximum deflection of the column and the maximum stress in the column. $E_{st} = 200$ GPa, $\sigma_Y = 360$ MPa.

Section Properties :

$$A = 0.12(0.1) - (0.1)(0.09) = 3.00\left(10^{-3}\right) \text{ m}^2$$

$$I_x = \frac{1}{12}(0.1)\left(0.12^3\right) - \frac{1}{12}(0.09)\left(0.1^3\right) = 6.90\left(10^{-6}\right) \text{ m}^4$$

$$r_x = \sqrt{\frac{I_x}{A}} = \sqrt{\frac{6.90(10^{-6})}{3.00(10^{-3})}} = 0.047958 \text{ m}$$

For a column fixed at one end and pinned at the other end, $K = 0.7$.

$$(KL)_x = 0.7(6) = 4.2 \text{ m}$$

The eccentricity of the two applied loads is,

$$e = \frac{130(0.12) - 50(0.08)}{180} = 0.06444 \text{ m}$$

Yielding About x - x Axis : Applying the secant formula,

$$\sigma_{max} = \frac{P}{A}\left[1 + \frac{ec}{r_x^2}\sec\left(\frac{(KL)_x}{2r_x}\sqrt{\frac{P}{EA}}\right)\right]$$

$$= \frac{180(10^3)}{3.00(10^{-3})}\left[1 + \frac{0.06444(0.06)}{0.047958^2}\sec\left(\frac{4.2}{2(0.047958)}\sqrt{\frac{180(10^3)}{200(10^9)(3.00)(10^{-3})}}\right)\right]$$

$$= 199 \text{ MPa}$$ **Ans**

Since $\sigma_{max} < \sigma_Y = 360$ MPa, the column **does not yield.**

Maximum Displacement :

$$v_{max} = e\left[\sec\left(\sqrt{\frac{P}{EI}}\frac{KL}{2}\right) - 1\right]$$

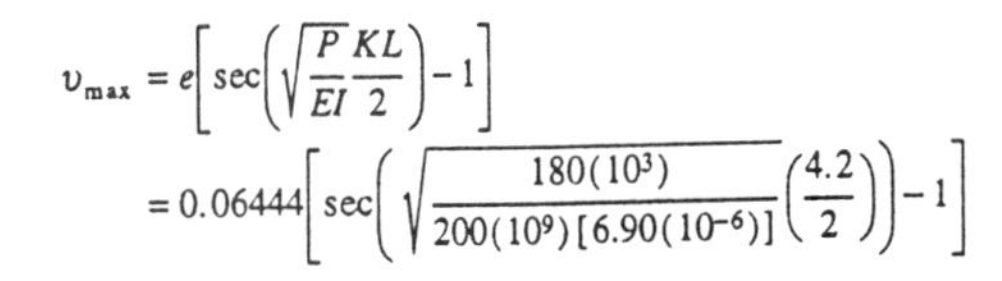

$$= 0.06444\left[\sec\left(\sqrt{\frac{180(10^3)}{200(10^9)[6.90(10^{-6})]}}\left(\frac{4.2}{2}\right)\right) - 1\right]$$

$$= 0.02433 \text{ m} = 24.3 \text{ mm}$$ **Ans**

13–65. The steel column supports the two eccentric loadings. If it is assumed to be fixed at its top and bottom, and braced against buckling about the y–y axis, determine the maximum deflection of the column and the maximum stress in the column. $E_{st} = 200$ GPa, $\sigma_Y = 360$ MPa.

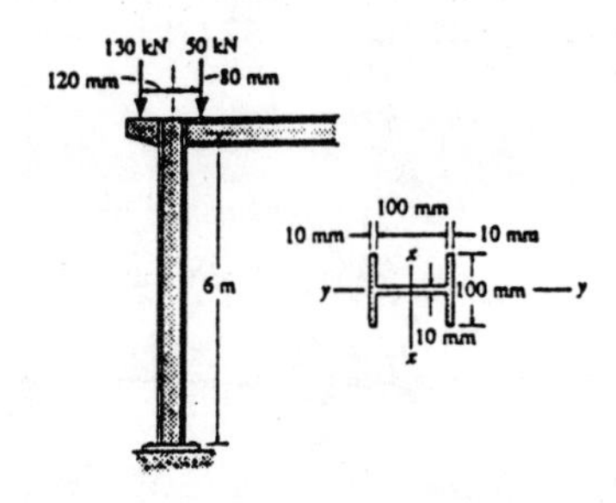

Section Properties :

$$A = 0.12(0.1) - (0.1)(0.09) = 3.00(10^{-3})\ \text{m}^2$$

$$I_x = \frac{1}{12}(0.1)(0.12^3) - \frac{1}{12}(0.09)(0.1^3) = 6.90(10^{-6})\ \text{m}^4$$

$$r_x = \sqrt{\frac{I_x}{A}} = \sqrt{\frac{6.90(10^{-6})}{3.00(10^{-3})}} = 0.047958\ \text{m}$$

For a column fixed at both ends, $K = 0.5$.

$$(KL)_x = 0.5(6) = 3.00\ \text{m}$$

The eccentricity of the two applied loads is,

$$e = \frac{130(0.12) - 50(0.08)}{180} = 0.06444\ \text{m}$$

Yielding About x - x Axis : Applying the secant formula,

$$\sigma_{\max} = \frac{P}{A}\left[1 + \frac{ec}{r_x^2}\sec\left(\frac{(KL)_x}{2r_x}\sqrt{\frac{P}{EA}}\right)\right]$$

$$= \frac{180(10^3)}{3.00(10^{-3})}\left[1 + \frac{0.06444(0.06)}{0.047958^2}\sec\left(\frac{3.00}{2(0.047958)}\sqrt{\frac{180(10^3)}{200(10^9)(3.00)(10^{-3})}}\right)\right]$$

$= 178$ MPa **Ans**

Since $\sigma_{\max} < \sigma_Y = 360$ MPa, the column **does not yield.**

Maximum Displacement :

$$v_{\max} = e\left[\sec\left(\sqrt{\frac{P}{EI}}\frac{KL}{2}\right) - 1\right]$$

$$= 0.06444\left[\sec\left(\sqrt{\frac{180(10^3)}{200(10^9)[6.90(10^{-6})]}}\left(\frac{3}{2}\right)\right) - 1\right]$$

$= 0.01077$ m $= 10.8$ mm **Ans**

13–66. A column of intermediate length buckles when the compressive stress is 40 ksi. If the slenderness ratio is 60, determine the tangent modulus.

Critical Stress : For inelastic buckling, apply *Engesser's* equation.

$$\sigma_{cr} = \frac{\pi^2 E_t}{\left(\frac{KL}{r}\right)^2} \quad \text{where} \quad \frac{KL}{r} = 60$$

$$40 = \frac{\pi^2 E_t}{60^2}$$

$E_t = 14.6(10^3)$ ksi **Ans**

13–67. The stress–strain diagram for a material can be approximated by the two line segments shown. If a bar having a diameter of 60 mm and a length of 2 m is made from this material, determine the critical load if both ends are pinned. Assume that the load acts through the axis of the bar. Use Engesser's equation.

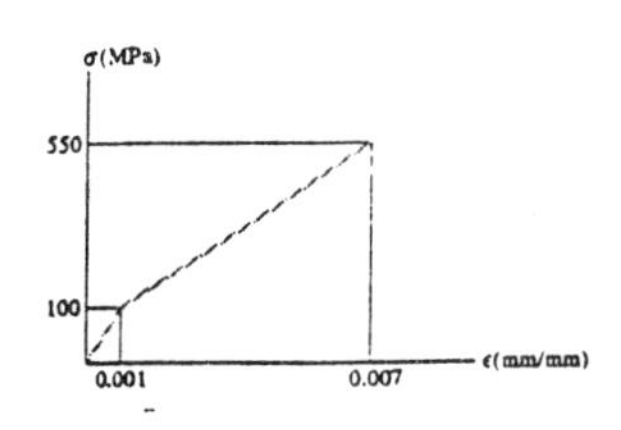

Section Properties :

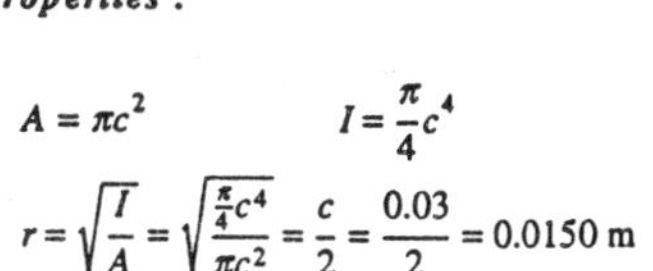

$$A = \pi c^2 \qquad I = \frac{\pi}{4}c^4$$

$$r = \sqrt{\frac{I}{A}} = \sqrt{\frac{\frac{\pi}{4}c^4}{\pi c^2}} = \frac{c}{2} = \frac{0.03}{2} = 0.0150\text{ m}$$

For a column with both of its ends pinned, $K = 1$.

$$\frac{KL}{r} = \frac{1(2)}{0.0150} = 133.33$$

Critical Stress : Applying *Engesser's equation,*

$$\sigma_{cr} = \frac{\pi^2 E_t}{\left(\frac{KL}{r}\right)^2} = \frac{\pi^2 E_t}{133.33^2} = 0.55517\left(10^{-3}\right)E_t \qquad [1]$$

From the stress – strain diagram, the *tangent moduli* are

$$(E_t)_1 = \frac{100(10^6)}{0.001} = 100\text{ GPa}$$

$$(E_t)_2 = \frac{(550-100)(10^6)}{0.007-0.001} = 75\text{ GPa}$$

Substitute $(E_t)_1 = 100$ GPa into Eq.[1], we have

$$\sigma_{cr} = 0.55517\left(10^{-3}\right)\left[100\left(10^9\right)\right]$$
$$= 55.52\text{ MPa} < 100\text{ MPa } (\textit{O.K!})$$

Therefore,

$$P_{cr} = \sigma_{cr}A$$
$$= 55.52\left(10^6\right)\left[\pi\left(0.03^2\right)\right]$$
$$= 156969\text{ N} = 157\text{ kN}$$

Ans

***13–68.** The stress–strain diagram for a material can be approximated by the two line segments shown. If a bar having a diameter of 60 mm and a length of 2 m is made from this material, determine the critical load provided the ends are fixed. Assume that the load acts through the axis of the bar. Use Engesser's equation.

Section Properties :

$$A = \pi c^2 \qquad I = \frac{\pi}{4}c^4$$

$$r = \sqrt{\frac{I}{A}} = \sqrt{\frac{\frac{\pi}{4}c^4}{\pi c^2}} = \frac{c}{2} = \frac{0.03}{2} = 0.0150\text{ m}$$

For a column with both of its ends pinned, $K = 0.5$.

$$\frac{KL}{r} = \frac{0.5(2)}{0.0150} = 66.667$$

Critical Stress : Applying *Engesser's* equation,

$$\sigma_{cr} = \frac{\pi^2 E_t}{\left(\frac{KL}{r}\right)^2} = \frac{\pi^2 E_t}{66.667^2} = 2.220661\left(10^{-3}\right)E_t \qquad [1]$$

From the stress – strain, the *tangent moduli* are

$$(E_t)_1 = \frac{100(10^6)}{0.001} = 100\text{ GPa}$$

$$(E_t)_2 = \frac{(550-100)(10^6)}{0.007-0.001} = 75\text{ GPa}$$

Substitute $(E_t)_1 = 100$ GPa into Eq.[1], we have

$$\sigma_{cr} = 2.220661\left(10^{-3}\right)\left[100\left(10^9\right)\right]$$
$$= 222.07\text{ MPa} > 100\text{ MPa } (\textit{No Good!})$$

Hence, inelastic buckling occurs. Substituting $(E_t)_2 = 75$ GPa into Eq.[1], we have

$$\sigma_{cr} = 2.220661\left(10^{-3}\right)\left[75\left(10^9\right)\right]$$
$$= 166.55\text{ MPa}$$

Since 100 MPa $< \sigma_{cr} <$ 550 MPa, the critical stress lies within the second region. Therefore,

$$P_{cr} = \sigma_{cr}A$$
$$= 166.55\left(10^6\right)\left[\pi\left(0.03^2\right)\right]$$
$$= 470908\text{ N} = 471\text{ kN}$$

Ans

13–69. The stress–strain diagram for a material can be approximated by the two line segments shown. If a bar having a diameter of 60 mm and length of 2 m is made from this material, determine the critical load provided one end is free and the other is fixed. Assume that the load acts through the axis of the bar. Use Engesser's equation.

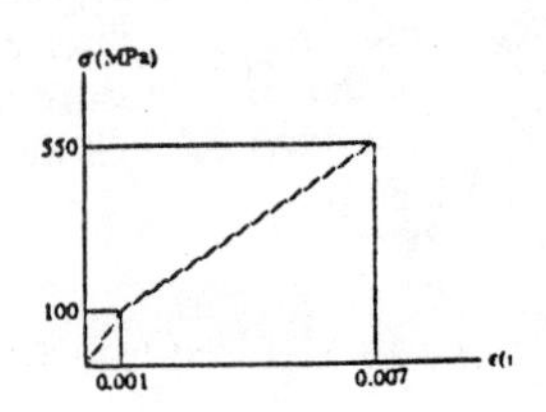

Section Properties :

$$A = \pi c^2 \qquad I = \frac{\pi}{4}c^4$$

$$r = \sqrt{\frac{I}{A}} = \sqrt{\frac{\frac{\pi}{4}c^4}{\pi c^2}} = \frac{c}{2} = \frac{0.03}{2} = 0.0150 \text{ m}$$

For a column with one end free and the other end fixed, $K = 2.0$.

$$\frac{KL}{r} = \frac{2.0(2)}{0.0150} = 266.7$$

Critical Stress : Applying *Engesser's equation,*

$$\sigma_{cr} = \frac{\pi^2 E_t}{\left(\frac{KL}{r}\right)^2} = \frac{\pi^2 E_t}{266.7^2} = 138.79\left(10^{-6}\right) E_t \qquad [1]$$

From the stress – strain, the *tangent moduli* are

$$(E_t)_1 = \frac{100(10^6)}{0.001} = 100 \text{ GPa}$$

$$(E_t)_2 = \frac{(550-100)(10^6)}{0.007-0.001} = 75 \text{ GPa}$$

Substitute $(E_t)_1 = 100$ GPa into Eq.[1], we have

$$\sigma_{cr} = 138.79\left(10^{-6}\right)\left[100\left(10^9\right)\right]$$
$$= 13.88 \text{ MPa} < 100 \text{ MPa } (\textit{OK!})$$

Thus,

$$P_{cr} = \sigma_{cr} A$$
$$= 13.88\left(10^6\right)\left[\pi\left(0.03^2\right)\right]$$
$$= 39\ 242 \text{ N} = 39.2 \text{ kN} \qquad \textbf{Ans}$$

13–70. Construct the buckling curve, P/A versus L/r, for a column that has a bilinear stress–strain curve in compression as shown. The column is pinned at its ends.

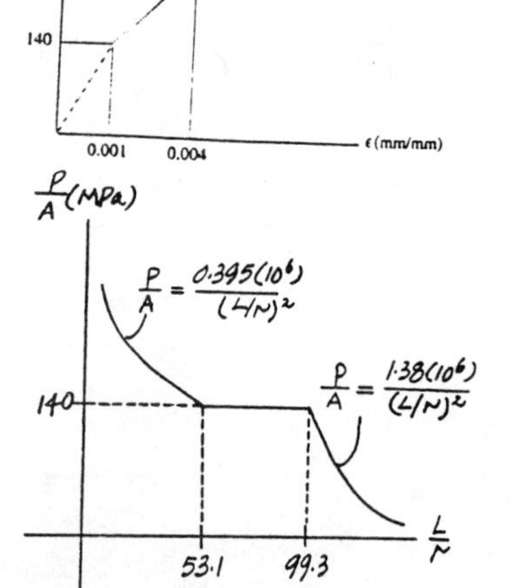

Tangent modulus : From the stress – strain diagram,

$$(E_t)_1 = \frac{140(10^6)}{0.001} = 140 \text{ GPa}$$

$$(E_t)_2 = \frac{(260-140)(10^6)}{0.004-0.001} = 40 \text{ GPa}$$

Critical Stress : Applying *Engesser's equation,*

$$\sigma_{cr} = \frac{P}{A} = \frac{\pi^2 E_t}{\left(\frac{L}{r}\right)^2} \qquad [1]$$

Substituting $(E_t)_1 = 140$ GPa into Eq.[1], we have

$$\frac{P}{A} = \frac{\pi^2\left[140(10^9)\right]}{\left(\frac{L}{r}\right)^2}$$

$$\frac{P}{A} = \frac{1.38(10^6)}{\left(\frac{L}{r}\right)^2} \text{ MPa}$$

When $\frac{P}{A} = 140$ MPa, $\frac{L}{r} = 99.3$

Substitute $(E_t)_2 = 40$ GPa into Eq.[1], we have

$$\frac{P}{A} = \frac{\pi^2\left[40(10^9)\right]}{\left(\frac{L}{r}\right)^2}$$

$$\frac{P}{A} = \frac{0.395(10^6)}{\left(\frac{L}{r}\right)^2} \text{ MPa}$$

When $\frac{P}{A} = 140$ MPa, $\frac{L}{r} = 53.1$

13–71. Determine the largest length of a structural A-36 steel rod if it is fixed supported and subjected to an axial load of 100 kN. The rod has a diameter of 50 mm. Use the AISC equations.

Section Properties :

$$A = \pi(0.025^2) = 0.625(10^{-3})\,\pi \text{ m}^2$$

$$I = \frac{\pi}{4}(0.025^4) = 97.65625(10^{-9})\,\pi \text{ m}^4$$

$$r = \sqrt{\frac{I}{A}} = \sqrt{\frac{97.65625(10^{-9})\pi}{0.625(10^{-3})\pi}} = 0.0125 \text{ m}$$

Slenderness Ratio : For a column fixed at both ends, $K = 0.5$. Thus,

$$\frac{KL}{r} = \frac{0.5L}{0.0125} = 40.0L$$

AISC Column Formula : Assume a *long* column.

$$\sigma_{allow} = \frac{12\pi^2 E}{23\left(\frac{KL}{r}\right)^2}$$

$$\frac{100(10^3)}{0.625(10^{-3})\pi} = \frac{12\pi^2[200(10^9)]}{23(40.0L)^2}$$

$$L = 3.555 \text{ m}$$

Here, $\frac{KL}{r} = 40.0(3.555) = 142.2$ and for A−36 steel, $\left(\frac{KL}{r}\right)_c = \sqrt{\frac{2\pi^2 E}{\sigma_Y}} = \sqrt{\frac{2\pi^2[200(10^9)]}{250(10^6)}} = 125.7$. Since $\left(\frac{KL}{r}\right)_c \le \frac{KL}{r} \le 200$, the assumption is correct. Thus,

$$L = 3.56 \text{ m} \qquad \textbf{Ans}$$

***13–72.** Determine the largest length of a $W\ 12 \times 45$ structural A-36 steel section if it is pin supported and subjected to an axial load of 200 kip. Use the AISC equations.

Section Properties : For a W12×45 wide flange section,

$$A = 13.2 \text{ in}^2 \qquad r_y = 1.94 \text{ in}$$

Slenderness Ratio : For a column pinned at both ends, $K = 1$. Thus,

$$\left(\frac{KL}{r}\right)_y = \frac{1(L)}{1.94} = 0.5155L$$

AISC Column Formula : Assume a *long* column.

$$\sigma_{allow} = \frac{12\pi^2 E}{23\left(\frac{KL}{r}\right)^2}$$

$$\frac{200}{13.2} = \frac{12\pi^2[29(10^3)]}{23(0.5155L)^2}$$

$$L = 192.6 \text{ in.}$$

Here, $\frac{KL}{r} = 0.5155(192.6) = 99.27$ and for A−36 steel, $\left(\frac{KL}{r}\right)_c = \sqrt{\frac{2\pi^2 E}{\sigma_Y}} = \sqrt{\frac{2\pi^2[29(10^3)]}{36}} = 126.1$. Since $\frac{KL}{r} < \left(\frac{KL}{r}\right)_c$, the assumption is not correct. Thus, the column is an *intermediate* column.

Applying Eq. 13-23,

$$\sigma_{allow} = \frac{\left[1 - \frac{(KL/r)^2}{2(KL/r)_c^2}\right]\sigma_Y}{\frac{5}{3} + \frac{3(KL/r)}{8(KL/r)_c} - \frac{(KL/r)^3}{8(KL/r)_c^3}}$$

$$\frac{200}{13.2} = \frac{\left[1 - \frac{(0.5155L)^2}{2(126.1^2)}\right](36)}{\frac{5}{3} + \frac{3(0.5155L)}{8(126.1)} - \frac{(0.5155L)^3}{8(126.1^3)}}$$

$$0 = 8.538213(10^{-9})L^3 - 19.851245(10^{-6})L^2 - 1.1532911(10^{-3})L + 0.709333$$

Solving by trial and error,

$$L = 158.73 \text{ in.} = 13.2 \text{ ft} \qquad \textbf{Ans}$$

13–73. Using the AISC equations, check if a $W\,6 \times 9$ structural A-36 steel column that is 10 ft long can support an axial load of 40 kip. The ends are fixed.

Section Properties : For a $W6\times9$ wide flange section,

$$A = 2.68\ \text{in}^2 \qquad r_y = 0.905\ \text{in}$$

Slenderness Ratio : For a column fixed at both ends, $K = 0.5$. Thus,

$$\left(\frac{KL}{r}\right)_y = \frac{0.5(10)(12)}{0.905} = 66.30$$

AISC Column Formula : For A–36 steel, $\left(\frac{KL}{r}\right)_c = \sqrt{\frac{2\pi^2 E}{\sigma_Y}}$ $= \sqrt{\frac{2\pi^2[29(10^3)]}{36}} = 126.1$. Since $\frac{KL}{r} < \left(\frac{KL}{r}\right)_c$, the column is an *intermediate* column. Applying Eq. 13–23,

$$\sigma_{\text{allow}} = \frac{\left[1 - \frac{(KL/r)^2}{2(KL/r)_c^2}\right]\sigma_Y}{\frac{5}{3} + \frac{3(KL/r)}{8(KL/r)_c} - \frac{(KL/r)^3}{8(KL/r)_c^3}}$$

$$= \frac{\left[1 - \frac{(66.30^2)}{2(126.1^2)}\right](36)}{\frac{5}{3} + \frac{3(66.30)}{8(126.1)} - \frac{(66.30^3)}{8(126.1^3)}}$$

$$= 16.809\ \text{ksi}$$

The allowable load is

$$P_{\text{allow}} = \sigma_{\text{allow}} A = 16.809(2.68) = 45.05\ \text{kip} > P = 40\ \text{kip} \quad (O.K!)$$

Thus, the column is **adequate**. **Ans**

13–74. Solve Prob. 13–73 if the ends are pin supported.

Section Properties : For a $W6\times9$ wide flange section,

$$A = 2.68\ \text{in}^2 \qquad r_y = 0.905\ \text{in}$$

Slenderness Ratio : For a column pinned at both ends, $K = 1$. Thus,

$$\left(\frac{KL}{r}\right)_y = \frac{1(10)(12)}{0.905} = 132.6$$

AISC Column Formula : For A–36 steel, $\left(\frac{KL}{r}\right)_c = \sqrt{\frac{2\pi^2 E}{\sigma_Y}}$ $= \sqrt{\frac{2\pi^2[29(10^3)]}{36}} = 126.1$. Since $\left(\frac{KL}{r}\right)_c \le \frac{KL}{r} \le 200$, the column is a *long* column. Applying Eq. 13–21,

$$\sigma_{\text{allow}} = \frac{12\pi^2 E}{23(KL/r)^2} = \frac{12\pi^2(29.0)(10^3)}{23(132.6^2)} = 8.493\ \text{ksi}$$

The allowable load is

$$P_{\text{allow}} = \sigma_{\text{allow}} A = 8.493(2.68) = 22.76\ \text{kip} < P = 40\ \text{kip} \quad (\textit{No Good!})$$

Thus, the column is **not adequate**. **Ans**

13–75. Using the AISC equations, select from Appendix *B* the lightest-weight structural A-36 steel column that is 30 ft long and supports an axial load of 200 kip. The ends are fixed.

Section Properties : Try a W8 × 48 wide flange section,

$$A = 14.1\text{ in}^2 \qquad r_y = 2.08\text{ in}$$

Slenderness Ratio : For a column fixed at both ends, $K = 0.5$. Thus,

$$\left(\frac{KL}{r}\right)_y = \frac{0.5(30)(12)}{2.08} = 86.54$$

AISC Column Formula : For A–36 steel, $\left(\frac{KL}{r}\right)_c = \sqrt{\frac{2\pi^2 E}{\sigma_Y}}$
$= \sqrt{\frac{2\pi^2[29(10^3)]}{36}} = 126.1$. Since $\frac{KL}{r} < \left(\frac{KL}{r}\right)_c$, the column is an *intermediate* column. Applying Eq. 13 – 23,

$$\sigma_{allow} = \frac{\left[1 - \frac{(KL/r)^2}{2(KL/r)_c^2}\right]\sigma_Y}{\frac{5}{3} + \frac{3(KL/r)}{8(KL/r)_c} - \frac{(KL/r)^3}{8(KL/r)_c^3}}$$

$$= \frac{\left[1 - \frac{(86.54^2)}{2(126.1^2)}\right](36)}{\frac{5}{3} + \frac{3(86.54)}{8(126.1)} - \frac{(86.54^3)}{8(126.1^3)}}$$

$$= 14.611\text{ ksi}$$

The allowable load is

$$P_{allow} = \sigma_{allow}A$$
$$= 14.611(14.1)$$
$$= 206\text{ kip} > P = 200\text{ kip} \quad (O.K!)$$

Thus, **Use** W8 × 48 **Ans**

***13–76.** Using the AISC equations, select from Appendix *B* the lightest-weight structural A-36 steel column that is 24 ft long and supports an axial load of 100 kip. The ends are fixed.

Section Properties : Try a W8 × 24 wide flange section,

$$A = 7.08\text{ in}^2 \qquad r_y = 1.61\text{ in}$$

Slenderness Ratio : For a column fixed at both ends, $K = 0.5$. Thus,

$$\left(\frac{KL}{r}\right)_y = \frac{0.5(24)(12)}{1.61} = 89.44$$

AISC Column Formula : For A–36 steel, $\left(\frac{KL}{r}\right)_c = \sqrt{\frac{2\pi^2 E}{\sigma_Y}}$
$= \sqrt{\frac{2\pi^2[29(10^3)]}{36}} = 126.1$. Since $\frac{KL}{r} < \left(\frac{KL}{r}\right)_c$, the column is an *intermediate* column. Applying Eq. 13 – 23,

$$\sigma_{allow} = \frac{\left[1 - \frac{(KL/r)^2}{2(KL/r)_c^2}\right]\sigma_Y}{\frac{5}{3} + \frac{3(KL/r)}{8(KL/r)_c} - \frac{(KL/r)^3}{8(KL/r)_c^3}}$$

$$= \frac{\left[1 - \frac{(89.44^2)}{2(126.1^2)}\right](36)}{\frac{5}{3} + \frac{3(89.44)}{8(126.1)} - \frac{(89.44^3)}{8(126.1^3)}}$$

$$= 14.271\text{ ksi}$$

The allowable load is

$$P_{allow} = \sigma_{allow}A$$
$$= 14.271(7.08)$$
$$= 101\text{ kip} > P = 100\text{ kip} \quad (O.K!)$$

Thus, **Use** W8 × 24 **Ans**

13–77. Determine the largest length of a $W\ 6 \times 16$ structural A-36 steel section if it is pin supported and subjected to an axial load of 70 kip. Use the AISC equations.

Section Properties : For a W6 × 16 wide flange section,

$$A = 4.74\ \text{in}^2 \qquad r_y = 0.966\ \text{in}$$

Slenderness Ratio : For a column pinned at both ends, $K = 1$. Thus,

$$\left(\frac{KL}{r}\right)_y = \frac{1(L)}{0.966} = 1.0352L$$

AISC Column Formula : Assume it is a *long* column.

$$\sigma_{\text{allow}} = \frac{12\pi^2 E}{23\left(\frac{KL}{r}\right)^2}$$

$$\frac{70}{4.74} = \frac{12\pi^2[29(10^3)]}{23(1.0352L)^2}$$

$$L = 97.14\ \text{in.}$$

Here, $\frac{KL}{r} = 1.0352(97.14) = 100.6$ and for A−36 steel, $\left(\frac{KL}{r}\right)_c = \sqrt{\frac{2\pi^2 E}{\sigma_Y}} = \sqrt{\frac{2\pi^2[29(10^3)]}{36}} = 126.1$. Since $\frac{KL}{r} < \left(\frac{KL}{r}\right)_c$, the assumption is not correct. Thus, the column is a *intermediate* column.

Applying Eq. 13 - 23,

$$\sigma_{\text{allow}} = \frac{\left[1 - \frac{(KL/r)^2}{2(KL/r)_c^2}\right]\sigma_Y}{\frac{5}{3} + \frac{3(KL/r)}{8(KL/r)_c} - \frac{(KL/r)^3}{8(KL/r)_c^3}}$$

$$\frac{70}{4.74} = \frac{\left[1 - \frac{(1.0352L)^2}{2(126.1^2)}\right](36)}{\frac{5}{3} + \frac{3(1.0352L)}{8(126.1)} - \frac{(1.0352L)^3}{8(126.1^3)}}$$

$$0 = 69.157737\left(10^{-9}\right)L^3 - 82.143521\left(10^{-6}\right)L^2 - 3.078517\left(10^{-3}\right)L + 0.771048$$

Solving by trial and error,

$$L = 82.2905\ \text{in.} = 6.86\ \text{ft} \qquad \textbf{Ans}$$

13–78. Determine the largest length of a $W\ 8 \times 48$ structural A-36 steel section if it is pin supported and subjected to an axial load of 55 kip. Use the AISC equations.

Section Properties : For a W8 × 48 wide flange section,

$$A = 14.1\ \text{in}^2 \qquad r_y = 2.08\ \text{in}$$

Slenderness Ratio : For a column pinned at both ends, $K = 1$. Thus,

$$\left(\frac{KL}{r}\right)_y = \frac{1(L)}{2.08} = 0.4808L$$

AISC Column Formula : Assume it is a *long* column.

$$\sigma_{\text{allow}} = \frac{12\pi^2 E}{23\left(\frac{KL}{r}\right)^2}$$

$$\frac{55}{14.1} = \frac{12\pi^2[29(10^3)]}{23(0.4808L)^2}$$

$$L = 407.0\ \text{in.}$$

Here, $\frac{KL}{r} = 0.4808(407.0) = 195.7$ and for A−36 steel, $\left(\frac{KL}{r}\right)_c = \sqrt{\frac{2\pi^2 E}{\sigma_Y}} = \sqrt{\frac{2\pi^2[29(10^3)]}{36}} = 126.1$. Since $\left(\frac{KL}{r}\right)_c \le \frac{KL}{r} \le 200$, the assumption is correct. Thus,

$$L = 407.0\ \text{in.} = 33.9\ \text{ft} \qquad \textbf{Ans}$$

13–79. Using the AISC equations, check if a column having the cross section shown can support an axial force of 1500 kN. The column has a length of 4 m, is made from A-36 steel, and its ends are pinned.

20 mm, 350 mm, 20 mm, 300 mm, 10 mm

Section Properties :

$$A = 0.3(0.35) - 0.29(0.31) = 0.0151\ \text{m}^2$$

$$I_y = \frac{1}{12}(0.04)(0.3^3) + \frac{1}{12}(0.31)(0.01^3) = 90.025833(10^{-6})\ \text{m}^4$$

$$r_y = \sqrt{\frac{I_y}{A}} = \sqrt{\frac{90.02583(10^{-6})}{0.0151}} = 0.077214\ \text{m}$$

Slenderness Ratio : For a column pinned at both ends, $K = 1$. Thus,

$$\left(\frac{KL}{r}\right)_y = \frac{1(4)}{0.077214} = 51.80$$

AISC Column Formula : For A–36 steel, $\left(\frac{KL}{r}\right)_c = \sqrt{\frac{2\pi^2 E}{\sigma_Y}}$

$= \sqrt{\frac{2\pi^2[200(10^9)]}{250(10^6)}} = 125.7$. Since $\frac{KL}{r} < \left(\frac{KL}{r}\right)_c$, the column is an *intermediate* column. Applying Eq. 13–23,

$$\sigma_{allow} = \frac{\left[1 - \frac{(KL/r)^2}{2(KL/r)_c^2}\right]\sigma_Y}{\frac{5}{3} + \frac{3(KL/r)}{8(KL/r)_c} - \frac{(KL/r)^3}{8(KL/r)_c^3}}$$

$$= \frac{\left[1 - \frac{(51.80^2)}{2(125.7^2)}\right](250)(10^6)}{\frac{5}{3} + \frac{3(51.80)}{8(125.7)} - \frac{(51.80^3)}{8(125.7^3)}}$$

$$= 126.2\ \text{MPa}$$

The allowable load is

$$P_{allow} = \sigma_{allow}A$$
$$= 126.2(10^6)(0.0151)$$
$$= 1906\ \text{kN} > P = 1500\ \text{kN} \quad (O.K!)$$

Thus, the column is **adequate**. **Ans**

***13–80.** Determine the largest length of a $W\ 8 \times 31$ structural A-36 steel column if it is to support an axial load of 35 kip. The ends are pinned.

Section Properties : For a W8 × 31 wide flange section,

$$A = 9.13\ \text{in}^2 \qquad r_y = 2.02\ \text{in}$$

Slenderness Ratio : For a column pinned at both ends, $K = 1$. Thus,

$$\left(\frac{KL}{r}\right)_y = \frac{1(L)}{2.02} = 0.49505L$$

AISC Column Formula : Assume a *long* column.

$$\sigma_{allow} = \frac{12\pi^2 E}{23\left(\frac{KL}{r}\right)^2}$$

$$\frac{35}{9.13} = \frac{12\pi^2[29(10^3)]}{23(0.49505L)^2}$$

$$L = 398.7\ \text{in.}$$

Here, $\frac{KL}{r} = 0.49505(398.7) = 197.4$ and for A–36 steel, $\left(\frac{KL}{r}\right)_c = \sqrt{\frac{2\pi^2 E}{\sigma_Y}} = \sqrt{\frac{2\pi^2[29(10^3)]}{36}} = 126.1$. Since $\left(\frac{KL}{r}\right)_c \le \frac{KL}{r} \le 200$, the assumption is correct. Thus,

$$L = 398.7\ \text{in.} = 33.2\ \text{ft}$$ **Ans**

13–81. Using the AISC equations, select from Appendix *B* the lightest-weight structural A-36 steel column that is 20 ft long and supports an axial load of 40 kip. The ends are pinned.

Section Properties : Try a W8×24 wide flange section,

$$A = 7.08 \text{ in}^2 \qquad r_y = 1.61 \text{ in}$$

Slenderness Ratio : For a column pinned at both ends, $K = 1$. Thus,

$$\left(\frac{KL}{r}\right)_y = \frac{1(20)(12)}{1.61} = 149.1$$

AISC Column Formula : For A−36 steel, $\left(\frac{KL}{r}\right)_c = \sqrt{\frac{2\pi^2 E}{\sigma_Y}}$ $= \sqrt{\frac{2\pi^2[29(10^3)]}{36}} = 126.1$. Since $\left(\frac{KL}{r}\right)_c < \frac{KL}{r} < 200$, the column is a *long* column. Applying Eq. 13−21,

$$\sigma_{allow} = \frac{12\pi^2 E}{23\left(\frac{KL}{r}\right)^2} = \frac{12\pi^2[29(10^3)]}{23(149.1^2)} = 6.720 \text{ ksi}$$

The allowable load is

$$P_{allow} = \sigma_{allow}A = 6.720(7.08) = 47.58 \text{ kip} > P = 40 \text{ kip} \quad (O.K!)$$

Thus, **Use** W8×24 **Ans**

13–82. Determine the largest length of a *W* 10 × 19 structural A-36 steel column if it is to support an axial load of 50 kip. The ends are fixed supported.

Section Properties : For a W10× 19 wide flange section,

$$A = 5.62 \text{ in}^2 \qquad r_y = 0.874 \text{ in}$$

Slenderness Ratio : For a column fixed at both ends, $K = 0.5$. Thus,

$$\left(\frac{KL}{r}\right)_y = \frac{0.5(L)}{0.874} = 0.57208L$$

AISC Column Formula : Assume it is a *long* column.

$$\sigma_{allow} = \frac{12\pi^2 E}{23\left(\frac{KL}{r}\right)^2}$$

$$\frac{50}{5.62} = \frac{12\pi^2[29(10^3)]}{23(0.57208L)^2}$$

$$L = 226.5 \text{ in.}$$

Here, $\frac{KL}{r} = 0.57208(226.5) = 129.6$ and for A−36 steel, $\left(\frac{KL}{r}\right)_c = \sqrt{\frac{2\pi^2 E}{\sigma_Y}} = \sqrt{\frac{2\pi^2[29(10^3)]}{36}} = 126.1$. Since $\left(\frac{KL}{r}\right)_c < \frac{KL}{r} < 200$, the assumption is correct. Thus,

$$L = 226.5 \text{ in.} = 18.9 \text{ ft} \qquad \textbf{Ans}$$

■13–83. Determine the largest length of a W 10 × 45 structural steel column if it is pin supported and subjected to an axial load of 290 kip. $E_{st} = 29(10^3)$ ksi, $\sigma_Y = 50$ ksi. Use the AISC equations.

Section Properties : For a W10×45 wide flange section,

$$A = 13.3 \text{ in}^2 \qquad r_y = 2.01 \text{ in}$$

Slenderness Ratio : For a column pinned at both ends, $K = 1$. Thus,

$$\left(\frac{KL}{r}\right)_y = \frac{1(L)}{2.01} = 0.49751L$$

AISC Column Formula : Assume a *long* column,

$$\sigma_{allow} = \frac{12\pi^2 E}{23\left(\frac{KL}{r}\right)^2}$$

$$\frac{290}{13.3} = \frac{12\pi^2[29(10^3)]}{23(0.49751L)^2}$$

$$L = 166.3 \text{ in.}$$

Here, $\frac{KL}{r} = 0.49751(166.3) = 82.76$ and for grade 50 steel, $\left(\frac{KL}{r}\right)_c = \sqrt{\frac{2\pi^2 E}{\sigma_Y}} = \sqrt{\frac{2\pi^2[29(10^3)]}{50}} = 107.0$. Since $\frac{KL}{r} < \left(\frac{KL}{r}\right)_c$, the assumption is not correct. Thus, the column is an *intermediate* column.

Applying Eq. 13-23,

$$\sigma_{allow} = \frac{\left[1 - \frac{(KL/r)^2}{2(KL/r)_c^2}\right]\sigma_Y}{\frac{5}{3} + \frac{3(KL/r)}{8(KL/r)_c} - \frac{(KL/r)^3}{8(KL/r)_c^3}}$$

$$\frac{290}{13.3} = \frac{\left[1 - \frac{(0.49751L)^2}{2(107.0^2)}\right](50)}{\frac{5}{3} + \frac{3(0.49751L)}{8(107.0)} - \frac{(0.49751L)^3}{8(107.0^3)}}$$

$$0 = 12.565658(10^{-9})L^3 - 24.788132(10^{-6})L^2 - 1.743638(10^{-3})L + 0.626437$$

Solving by trial and error,

$$L = 131.12 \text{ in.} = 10.9 \text{ ft} \qquad \textbf{Ans}$$

***13–84.** Determine the largest length of a W 14 × 43 structural A-36 steel column if it is pin supported and subjected to an axial load of 50 kip. Use the AISC equations.

Section Properties : For a W14×43 wide flange section,

$$A = 12.6 \text{ in}^2 \qquad r_y = 1.89 \text{ in}$$

Slenderness Ratio : For a column pinned at both ends, $K = 1$. Thus,

$$\left(\frac{KL}{r}\right)_y = \frac{1(L)}{1.89} = 0.52910L$$

AISC Column Formula : Assume it is a *long* column.

$$\sigma_{allow} = \frac{12\pi^2 E}{23\left(\frac{KL}{r}\right)^2}$$

$$\frac{50}{12.6} = \frac{12\pi^2[29(10^3)]}{23(0.52910L)^2}$$

$$L = 366.64 \text{ in.}$$

Here, $\frac{KL}{r} = 0.52910(366.64) = 194.0$ and for A−36 steel, $\left(\frac{KL}{r}\right)_c = \sqrt{\frac{2\pi^2 E}{\sigma_Y}} = \sqrt{\frac{2\pi^2[29(10^3)]}{36}} = 126.1$. Since $\left(\frac{KL}{r}\right)_c < \frac{KL}{r} < 200$, the assumption is correct. Thus,

$$L = 366.64 \text{ in.} = 30.6 \text{ ft} \qquad \textbf{Ans}$$

13–85. The bar is made of aluminum alloy 2014-T6. Determine its thickness b if its width is $5b$. Assume that it is pin connected at its ends.

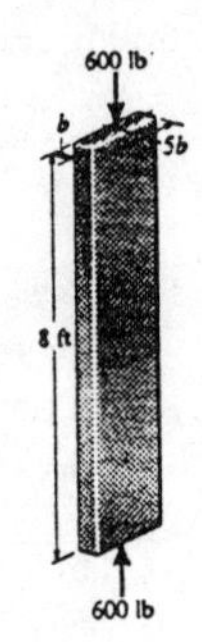

Section Properties :

$$A = b(5b) = 5b^2$$

$$I_y = \frac{1}{12}(5b)\left(b^3\right) = \frac{5}{12}b^4$$

$$r_y = \sqrt{\frac{I_y}{A}} = \sqrt{\frac{\frac{5}{12}b^4}{5b^2}} = \frac{\sqrt{3}}{6}b$$

Slenderness Ratio : For a column pinned at both ends, $K = 1$. Thus,

$$\left(\frac{KL}{r}\right)_y = \frac{1(8)(12)}{\frac{\sqrt{3}}{6}b} = \frac{332.55}{b}$$

Aluminum (2014 - T6 alloy) Column Formulas : Assume a *long* column and apply Eq. 13 - 26.

$$\sigma_{\text{allow}} = \frac{54\,000}{(KL/r)^2}$$

$$\frac{0.600}{5b^2} = \frac{54\,000}{\left(\frac{332.55}{b}\right)^2}$$

$$b = 0.7041 \text{ in.}$$

Here, $\frac{KL}{r} = \frac{332.55}{0.7041} = 472.3$. Since $\frac{KL}{r} > 55$, the assumption is correct.

Thus,

$b = 0.704$ in. **Ans**

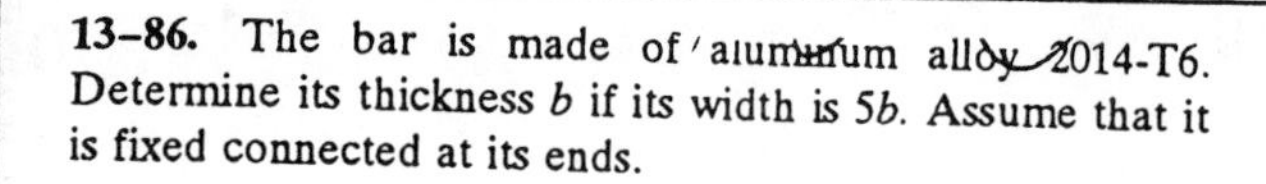

13–86. The bar is made of aluminum alloy 2014-T6. Determine its thickness b if its width is $5b$. Assume that it is fixed connected at its ends.

Section Properties :

$$A = b(5b) = 5b^2$$

$$I_y = \frac{1}{12}(5b)\left(b^3\right) = \frac{5}{12}b^4$$

$$r_y = \sqrt{\frac{I_y}{A}} = \sqrt{\frac{\frac{5}{12}b^4}{5b^2}} = \frac{\sqrt{3}}{6}b$$

Slenderness Ratio : For a column fixed at both ends, $K = 0.5$. Thus,

$$\left(\frac{KL}{r}\right)_y = \frac{0.5(8)(12)}{\frac{\sqrt{3}}{6}b} = \frac{166.28}{b}$$

Aluminum (2014 - T6 alloy) Column Formulas : Assume a *long* column and apply Eq. 13 - 26.

$$\sigma_{\text{allow}} = \frac{54\,000}{(KL/r)^2}$$

$$\frac{0.600}{5b^2} = \frac{54\,000}{\left(\frac{166.28}{b}\right)^2}$$

$$b = 0.4979 \text{ in.}$$

Here, $\frac{KL}{r} = \frac{166.28}{0.4979} = 334.0$. Since $\frac{KL}{r} > 55$, the assumption is correct.

Thus,

$b = 0.498$ in. **Ans**

13–87. The 2-in.-diameter rod is used to support an axial load of 8 kip. Determine its greatest allowable length L if it is made of 2014-T6 aluminum. Assume that the ends are pin connected.

Section Properties :

$$A = \pi\left(1^2\right) = \pi \text{ in}^2$$

$$I = \frac{\pi}{4}\left(1^4\right) = 0.25\pi \text{ in}^4$$

$$r = \sqrt{\frac{I}{A}} = \sqrt{\frac{0.25\pi}{\pi}} = 0.500 \text{ in.}$$

Slenderness Ratio : For a column pinned at both ends, $K = 1$. Thus,

$$\frac{KL}{r} = \frac{1(L)}{0.500} = 2.00L$$

Aluminum (2014 - T6 alloy) Column Formulas : Assume a *long* column and apply Eq. 13 - 26.

$$\sigma_{\text{allow}} = \frac{54\ 000}{(KL/r)^2}$$

$$\frac{8}{\pi} = \frac{54\ 000}{(2.00L)^2}$$

$$L = 72.81 \text{ in.}$$

Here, $\dfrac{KL}{r} = 2.00(72.81) = 145.6$. Since $\dfrac{KL}{r} > 55$, the assumption is correct. Thus,

$$L = 72.81 \text{ in.} = 6.07 \text{ ft} \qquad \textbf{Ans}$$

***13–88.** The 2-in.-diameter rod is used to support an axial load of 8 kip. Determine its greatest allowable length L if it is made of 2014-T6 aluminum. Assume that the ends are fixed connected.

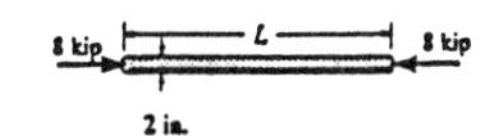

Section Properties :

$$A = \pi\left(1^2\right) = \pi \text{ in}^2$$

$$I = \frac{\pi}{4}\left(1^4\right) = 0.25\pi \text{ in}^4$$

$$r = \sqrt{\frac{I}{A}} = \sqrt{\frac{0.25\pi}{\pi}} = 0.500 \text{ in.}$$

Slenderness Ratio : For a column fixed at both ends, $K = 0.5$. Thus,

$$\left(\frac{KL}{r}\right)_y = \frac{0.5(L)}{0.500} = 1.00L$$

Aluminum (2014 - T6 alloy) Column Formulas : Assume a *long* column and apply Eq. 13 - 26.

$$\sigma_{\text{allow}} = \frac{54\ 000}{(KL/r)^2}$$

$$\frac{8}{\pi} = \frac{54\ 000}{(1.00L)^2}$$

$$L = 145.6 \text{ in.}$$

Here, $\dfrac{KL}{r} = 1.00(145.6) = 145.6$. Since $\dfrac{KL}{r} > 55$, the assumption is correct. Thus,

$$L = 145.6 \text{ in.} = 12.1 \text{ ft} \qquad \textbf{Ans}$$

13–89. A 5-ft-long rod is used in a machine to transmit an axial compressive load of 3 kip. Determine its diameter if it is pin connected at its ends and is made of a 2014-T6 aluminum alloy.

Section Properties :

$$A = \frac{\pi}{4}d^2 \qquad I = \frac{\pi}{4}\left(\frac{d}{2}\right)^4 = \frac{\pi}{64}d^4$$

$$r = \sqrt{\frac{I}{A}} = \sqrt{\frac{\frac{\pi}{64}d^4}{\frac{\pi}{4}d^2}} = \frac{d}{4}$$

Slenderness Ratio : For column pinned at both ends, $K = 1$. Thus,

$$\frac{KL}{r} = \frac{1(5)(12)}{\frac{d}{4}} = \frac{240}{d}$$

Aluminum (2014 - T6 alloy) Column Formulas : Assume a *long* column and apply Eq. 13 - 26.

$$\sigma_{\text{allow}} = \frac{54\,000}{(KL/r)^2}$$

$$\frac{3}{\frac{\pi}{4}d^2} = \frac{54\,000}{\left(\frac{240}{d}\right)^2}$$

$$d = 1.421 \text{ in.}$$

Here, $\frac{KL}{r} = \frac{240}{1.421} = 168.9$. Since $\frac{KL}{r} > 55$, the assumption is correct. Thus,

13–90. Solve Prob. 13–89 if the rod is fixed connected at its ends.

Section Properties :

$$A = \frac{\pi}{4}d^2 \qquad I = \frac{\pi}{4}\left(\frac{d}{2}\right)^4 = \frac{\pi}{64}d^4$$

$$r = \sqrt{\frac{I}{A}} = \sqrt{\frac{\frac{\pi}{64}d^4}{\frac{\pi}{4}d^2}} = \frac{d}{4}$$

Slenderness Ratio : For a column fixed at both ends, $K = 0.5$. Thus,

$$\frac{KL}{r} = \frac{0.5(5)(12)}{\frac{d}{4}} = \frac{120}{d}$$

Aluminum (2014 - T6 alloy) Column Formulas : Assume a *long* column and apply Eq. 13 - 26.

$$\sigma_{\text{allow}} = \frac{54\,000}{(KL/r)^2}$$

$$\frac{3}{\frac{\pi}{4}d^2} = \frac{54\,000}{\left(\frac{120}{d}\right)^2}$$

$$d = 1.005 \text{ in.}$$

Here, $\frac{KL}{r} = \frac{120}{1.005} = 119.4$. Since $\frac{KL}{r} > 55$, the assumption is correct. Thus,

$$d = 1.00 \text{ in.} \qquad \textbf{Ans}$$

13–91. The tube is 0.25 in. thick, is made of aluminum alloy 2014-T6 and is fixed at its bottom and pinned at its top. Determine the largest axial load that it can support.

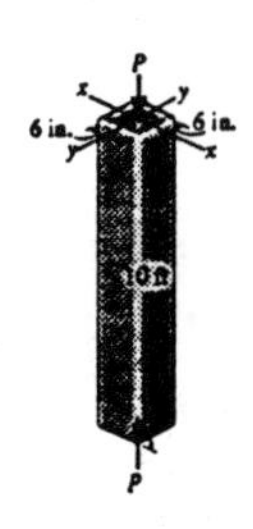

Section Properties :

$$A = 6(6) - 5.5(5.5) = 5.75 \text{ in}^2$$

$$I = \frac{1}{12}(6)(6^3) - \frac{1}{12}(5.5)(5.5^3) = 31.7448 \text{ in}^4$$

$$r = \sqrt{\frac{I}{A}} = \sqrt{\frac{31.7448}{5.75}} = 2.3496 \text{ in.}$$

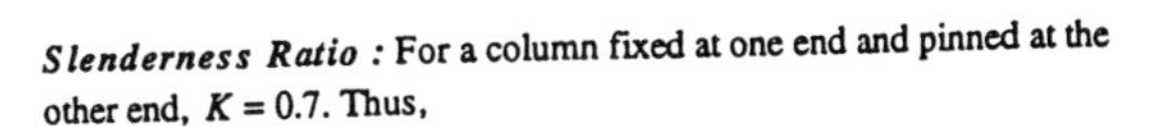

Slenderness Ratio : For a column fixed at one end and pinned at the other end, $K = 0.7$. Thus,

$$\frac{KL}{r} = \frac{0.7(10)(12)}{2.3496} = 35.75$$

Aluminium (2014 - T6 alloy) Column Formulas : Since $12 < \frac{KL}{r} < 55$, the column is classified as an *lintermediate* column. Applying Eq. 13 – 25,

$$\sigma_{allow} = \left[30.7 - 0.23\left(\frac{KL}{r}\right)\right] \text{ ksi}$$
$$= [30.7 - 0.23(35.75)]$$
$$= 22.48 \text{ ksi}$$

The allowable load is

$$P_{allow} = \sigma_{allow}A = 22.48(5.75) = 129 \text{ kip} \qquad \textbf{Ans}$$

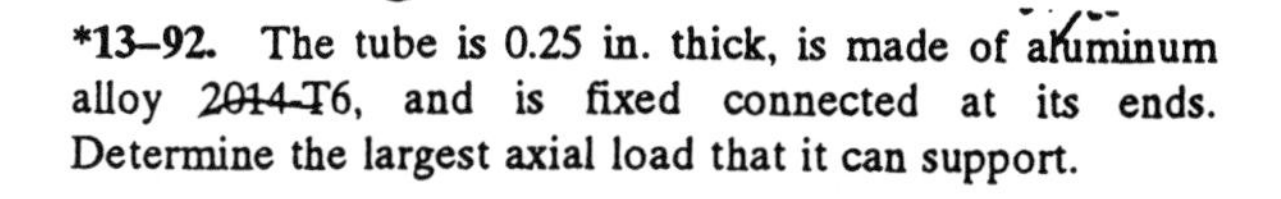

***13–92.** The tube is 0.25 in. thick, is made of aluminum alloy 2014-T6, and is fixed connected at its ends. Determine the largest axial load that it can support.

Section Properties :

$$A = 6(6) - 5.5(5.5) = 5.75 \text{ in}^2$$

$$I = \frac{1}{12}(6)(6^3) - \frac{1}{12}(5.5)(5.5^3) = 31.7448 \text{ in}^4$$

$$r = \sqrt{\frac{I}{A}} = \sqrt{\frac{31.7448}{5.75}} = 2.3496 \text{ in.}$$

Slenderness Ratio : For column fixed at both ends, $K = 0.5$. Thus,

$$\frac{KL}{r} = \frac{0.5(10)(12)}{2.3496} = 25.54$$

Aluminum (2014 - T6 alloy) Column Formulas : Since $12 < \frac{KL}{r} < 55$, the column is classified as an *intermediate* column. Applying Eq. 13 – 25,

$$\sigma_{allow} = \left[30.7 - 0.23\left(\frac{KL}{r}\right)\right] \text{ ksi}$$
$$= [30.7 - 0.23(25.54)]$$
$$= 24.83 \text{ ksi}$$

The allowable load is

$$P_{allow} = \sigma_{allow}A = 24.83(5.75) = 143 \text{ kip} \qquad \textbf{Ans}$$

13–93. The tube is 0.25 in. thick, is made of aluminum alloy 2014-T6, and is pin connected at its ends. Determine the largest axial load it can support.

Section Properties :

$$A = 6(6) - 5.5(5.5) = 5.75 \text{ in}^2$$

$$I = \frac{1}{12}(6)\left(6^3\right) - \frac{1}{12}(5.5)\left(5.5^3\right) = 31.7448 \text{ in}^4$$

$$r = \sqrt{\frac{I}{A}} = \sqrt{\frac{31.7448}{5.75}} = 2.3496 \text{ in.}$$

Slenderness Ratio : For a column pinned at both ends, $K = 1$. Thus,

$$\frac{KL}{r} = \frac{1(10)(12)}{2.3496} = 51.07$$

Aluminum (2014 - T6 alloy) Column Formulas : Since $12 < \frac{KL}{r} < 55$, the column is classified as an *intermediate* column. Applying Eq. 13 – 25,

$$\sigma_{\text{allow}} = \left[30.7 - 0.23\left(\frac{KL}{r}\right)\right] \text{ ksi}$$

$$= [30.7 - 0.23(51.07)]$$

$$= 18.95 \text{ ksi}$$

The allowable load is

$$P_{\text{allow}} = \sigma_{\text{allow}} A = 18.95(5.75) = 109 \text{ kip} \qquad \textbf{Ans}$$

13–94. A 6-ft-long rod is used in a machine to transmit an axial compressive load of 3 kip. Determine its diameter if it is fixed at one end and pinned at the other. The material is 2014-T6 aluminum alloy

Section Properties :

$$A = \frac{\pi}{4}d^2 \qquad I = \frac{\pi}{4}\left(\frac{d}{2}\right)^4 = \frac{\pi}{64}d^4$$

$$r = \sqrt{\frac{I}{A}} = \sqrt{\frac{\frac{\pi}{64}d^4}{\frac{\pi}{4}d^2}} = \frac{d}{4}$$

Slenderness Ratio : For a column pinned at one end and fixed at the other end, $K = 0.7$. Thus,

$$\frac{KL}{r} = \frac{0.7(6)(12)}{\frac{d}{4}} = \frac{201.6}{d}$$

Aluminum (2014 - T6 alloy) Column Formulas : Assume a *long* column and apply Eq. 13 - 26.

$$\sigma_{\text{allow}} = \frac{54\,000}{(KL/r)^2}$$

$$\frac{3}{\frac{\pi}{4}d^2} = \frac{54\,000}{\left(\frac{201.6}{d}\right)^2}$$

$$d = 1.302 \text{ in.}$$

Here, $\frac{KL}{r} = \frac{201.6}{1.302} = 154.8$. Since $\frac{KL}{r} > 55$, the assumption is correct. Thus,

$$d = 1.30 \text{ in.} \qquad \textbf{Ans}$$

13–95. The timber column has a square cross section and is assumed to be pin connected at its top and bottom. If it supports an axial load of 20 kip, determine its side dimensions a to the nearest $\frac{1}{2}$ in. Use the NFPA formulas.

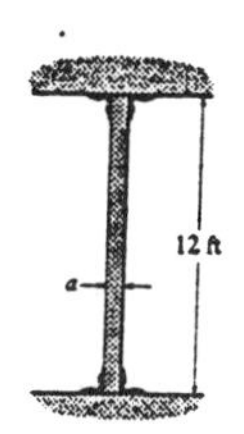

Slenderness Ratio : For column pinned at both ends , $K = 1$. Thus,

$$\frac{KL}{d} = \frac{1(12)(12)}{a} = \frac{144}{a}$$

NFPA Timber Column Formulas : Assume a *long* column and apply Eq. 13 – 29.

$$\sigma_{allow} = \frac{540}{(KL/d)^2} \text{ ksi}$$

$$\frac{20}{a^2} = \frac{540}{\left(\frac{144}{a}\right)^2}$$

$$a = 5.264 \text{ in.}$$

Here, $\frac{KL}{d} = \frac{144}{5.264} = 27.35$. Since $26 < \frac{KL}{d} < 50$, the assumption is correct. Thus,

Use $\quad a = 5\frac{1}{2}$ in. $\quad$ **Ans**

***13–96.** Solve Prob. 13–95 if the column is assumed to be fixed connected at its top and bottom.

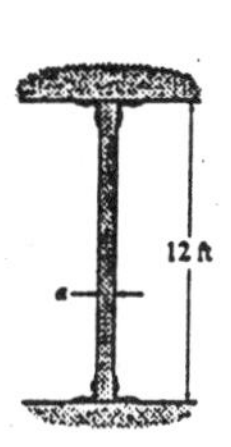

Slenderness Ratio : For column fixed at both ends , $K = 0.5$. Thus,

$$\frac{KL}{d} = \frac{0.5(12)(12)}{a} = \frac{72}{a}$$

NFPA Timber Column Formulas : Assume an *intermediate* column and apply Eq. 13 – 28.

$$\sigma_{allow} = 1.20\left[1 - \frac{1}{3}\left(\frac{KL/d}{26.0}\right)\right] \text{ ksi}$$

$$\frac{20}{a^2} = 1.20\left[1 - \frac{1}{3}\left(\frac{72/a}{26.0}\right)^2\right]$$

$$a = 4.384 \text{ in.}$$

Here, $\frac{KL}{d} = \frac{72}{4.384} = 16.42$. Since $11 < \frac{KL}{d} < 26$, the assumption is correct. Thus,

Use $\quad a = 4\frac{1}{2}$ in. $\quad$ **Ans**

13–97. The timber column has a length of 20 ft and is pin connected at its ends. Use the NFPA formulas to determine the largest axial force P that it can support.

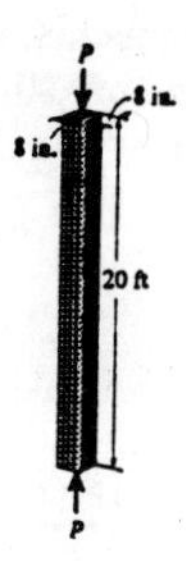

Slenderness Ratio : For column pinned at both ends, $K = 1$. Thus,

$$\frac{KL}{d} = \frac{1(20)(12)}{8} = 30$$

NFPA Timber Column Formulas : Since $26 < \frac{KL}{d} < 50$, it is a *long* column. Apply Eq. 13 – 29,

$$\sigma_{allow} = \frac{540}{(KL/d)^2} \text{ ksi}$$
$$= \frac{540}{30^2} = 0.600 \text{ ksi}$$

The largest axial force is

$$P = \sigma_{allow}A = 0.600[8(8)] = 38.4 \text{ kip} \qquad \textbf{Ans}$$

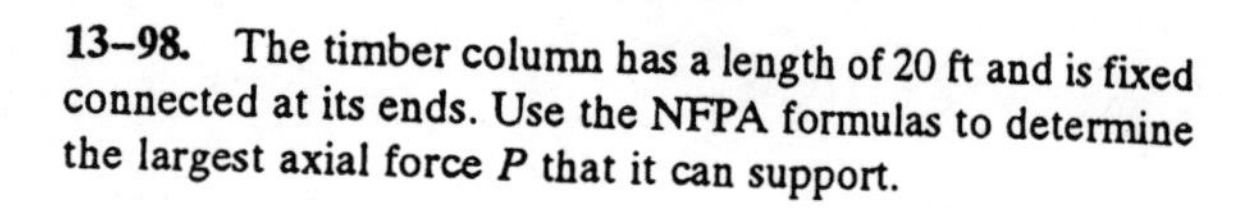

13–98. The timber column has a length of 20 ft and is fixed connected at its ends. Use the NFPA formulas to determine the largest axial force P that it can support.

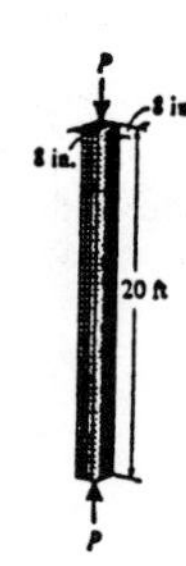

Slenderness Ratio : For a column fixed at both ends, $K = 0.5$. Thus,

$$\frac{KL}{d} = \frac{0.5(20)(12)}{8} = 15.0$$

NFPA Timber Column Formulas : Since $11 < \frac{KL}{d} < 26$, it is an *intermediate* column. Apply Eq. 13 – 28,

$$\sigma_{allow} = 1.20\left[1 - \frac{1}{3}\left(\frac{KL/d}{26.0}\right)\right] \text{ ksi}$$
$$= 1.20\left[1 - \frac{1}{3}\left(\frac{15.0}{26.0}\right)^2\right]$$
$$= 1.067 \text{ ksi}$$

The largest axial force is

$$P = \sigma_{allow}A = 1.067[8(8)] = 68.3 \text{ kip} \qquad \textbf{Ans}$$

13–99. The column is made of wood. It is fixed at its bottom and free at its top. Use the NFPA formulas to determine its greatest allowable length if it supports an axial load of $P = 2$ kip.

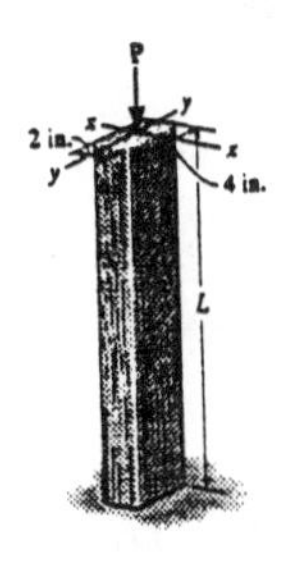

Slenderness Ratio : For a column fixed at one end and free at the other end , $K = 2$. Thus,

$$\frac{KL}{d} = \frac{2(L)}{2} = 1.00L$$

NFPA Timber Column Formulas : Assume a *long* column. Apply Eq. 13 – 29,

$$\sigma_{\text{allow}} = \frac{540}{(KL/d)^2}\ \text{ksi}$$

$$\frac{2}{2(4)} = \frac{540}{(1.00L)^2}$$

$$L = 46.48 \text{ in}$$

Here, $\frac{KL}{d} = 1.00(46.48) = 46.48$. Since $26 < \frac{KL}{d} < 50$, the assumption is correct. Thus,

$$L = 46.48 \text{ in.} = 3.87 \text{ ft} \qquad \textbf{Ans}$$

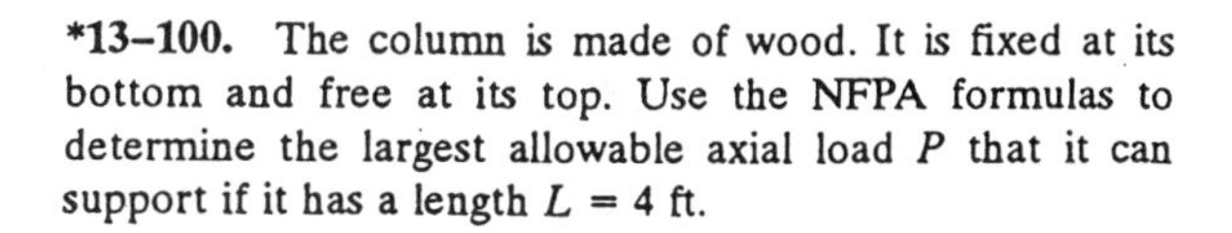
***13–100.** The column is made of wood. It is fixed at its bottom and free at its top. Use the NFPA formulas to determine the largest allowable axial load P that it can support if it has a length $L = 4$ ft.

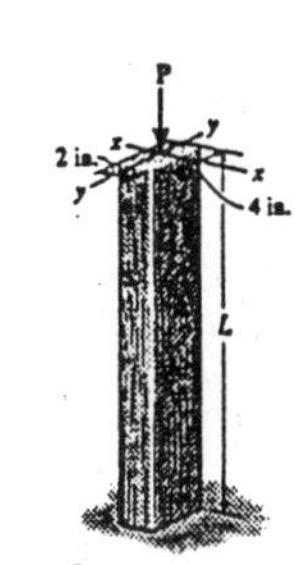

Slenderness Ratio : For a column fixed at one end and free at the other end , $K = 2$. Thus,

$$\frac{KL}{d} = \frac{2(4)(12)}{2} = 48.0$$

NFPA Timber Column Formulas : Since $26 < \frac{KL}{d} < 50$, it is a *long* column. Apply Eq. 13 – 29,

$$\sigma_{\text{allow}} = \frac{540}{(KL/d)^2}\ \text{ksi} = \frac{540}{48.0^2} = 0.234375 \text{ ksi}$$

The allowable axial force is

$$P_{\text{allow}} = \sigma_{\text{allow}}A = 0.234375[2(4)] = 1.875 \text{ kip} \qquad \textbf{Ans}$$

13–101. The $W\ 14 \times 53$ structural A-36 steel column supports an axial load of 80 kip in addition to an eccentric load P. Determine the maximum allowable value of P based on the AISC equations of Sec. 13.6 and Eq. 13–30. Assume the column is fixed at its base and at its top it is free to sway in the x–z plane while it is pinned in the y–z plane.

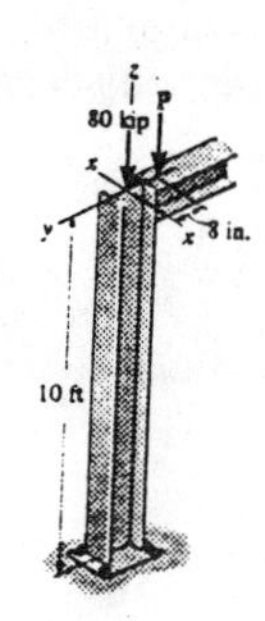

Section Properties : For a W14×53 wide flange section,

$A = 15.6\ \text{in}^2 \quad d = 13.92\ \text{in.} \quad I_x = 541\ \text{in}^4 \quad r_x = 5.89\ \text{in.}$
$r_y = 1.92\ \text{in.}$

Slenderness Ratio : By observation, the largest slenderness ratio is about $y-y$ axis. For a column fixed at one end and free at the other end, $K = 2$. Thus,

$$\left(\frac{KL}{r}\right)_y = \frac{2(12)(12)}{1.92} = 150$$

Allowable Stress : The allowable stress can be determined using *AISC Column Formulas*. For A – 36 steel, $\left(\frac{KL}{r}\right)_c = \sqrt{\frac{2\pi^2 E}{\sigma_Y}} = \sqrt{\frac{2\pi^2[29(10^3)]}{36}} = 126.1$. Since $\left(\frac{KL}{r}\right)_c \leq \frac{KL}{r} \leq 200$, the column is a *long* column. Applying Eq. 13 – 21,

$$\sigma_{\text{allow}} = \frac{12\pi^2 E}{23(KL/r)^2} = \frac{12\pi^2(29.0)(10^3)}{23(150^2)} = 6.637\ \text{ksi}$$

Maximum Stress : Bending is about $x-x$ axis. Applying Eq. 13 – 30, we have

$$\sigma_{\max} = \sigma_{\text{allow}} = \frac{P}{A} + \frac{Mc}{I}$$

$$6.637 = \frac{P + 80}{15.6} + \frac{P(10)\left(\frac{13.92}{2}\right)}{541}$$

$$P = 7.83\ \text{kip} \qquad \textbf{Ans}$$

13–102. The $W\ 12 \times 45$ structural A-36 steel column supports an axial load of 80 kip in addition to an eccentric load of $P = 60$ kip. Determine if the column fails based on the AISC equations of Sec. 13.6 and Eq. 13–30. Assume that the column is fixed at its base and at its top it is free to sway in the x–z plane while it is pinned in the y–z plane.

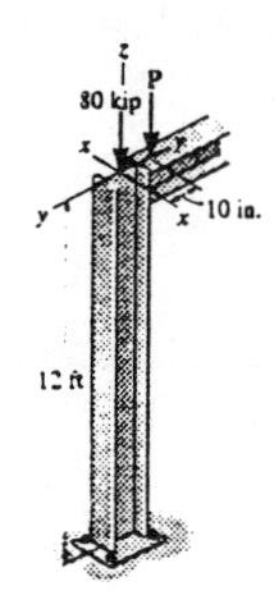

Section Properties : For a W12×45 wide flange section,

$A = 13.2\ \text{in}^2 \quad d = 12.06\ \text{in.} \quad I_x = 350\ \text{in}^4 \quad r_x = 5.15\ \text{in.}$
$r_y = 1.94\ \text{in.}$

Slenderness Ratio : By observation, the largest slenderness ratio is about $y-y$ axis. For a column fixed at one end and free at the other end, $K = 2$. Thus,

$$\left(\frac{KL}{r}\right)_y = \frac{2(12)(12)}{1.94} = 148.45$$

Allowable Stress : The allowable stress can be determined using *AISC Column Formulas*. For A – 36 steel, $\left(\frac{KL}{r}\right)_c = \sqrt{\frac{2\pi^2 E}{\sigma_Y}} = \sqrt{\frac{2\pi^2[29(10^3)]}{36}} = 126.1$. Since $\left(\frac{KL}{r}\right)_c \leq \frac{KL}{r} \leq 200$, the column is a *long* column. Applying Eq. 13 – 21,

$$\sigma_{\text{allow}} = \frac{12\pi^2 E}{23(KL/r)^2} = \frac{12\pi^2(29.0)(10^3)}{23(148.45^2)} = 6.776\ \text{ksi}$$

Maximum Stress : Bending is about $x-x$ axis. Applying Eq. 13 – 30, we have

$$\sigma_{\max} = \frac{P}{A} + \frac{Mc}{I} = \frac{140}{13.2} + \frac{60(10)\left(\frac{12.06}{2}\right)}{350} = 20.94\ \text{ksi}$$

Since $\sigma_{\max} > \sigma_{\text{allow}}$, the column **is not adequate.** **Ans**

13–103. The W 12 × 22 structural A-36 steel column is fixed at its bottom and free at its top. Determine the greatest eccentric load P that can be applied using Eq. 13–30 and the AISC equations of Sec. 13.6.

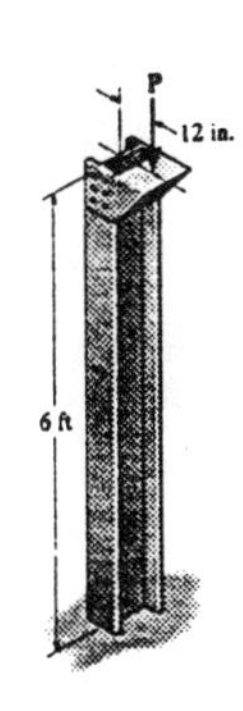

Section Properties : For a W12×22 wide flange section,

$$A = 6.48\ \text{in}^2 \qquad b_f = 4.030\ \text{in.} \qquad I_y = 4.66\ \text{in}^4 \qquad r_y = 0.847\ \text{in.}$$

Slenderness Ratio : By observation, the largest slenderness ratio is about $y-y$ axis. For a column fixed at one end and free at the other end, $K = 2$. Thus,

$$\left(\frac{KL}{r}\right)_y = \frac{2(6)(12)}{0.847} = 170.01$$

Allowable Stress : The allowable stress can be determined using *AISC Column Formulas*. For A−36 steel, $\left(\frac{KL}{r}\right)_c = \sqrt{\frac{2\pi^2 E}{\sigma_Y}} = \sqrt{\frac{2\pi^2[29(10^3)]}{36}} = 126.1$. Since $\left(\frac{KL}{r}\right)_c \leq \frac{KL}{r} \leq 200$, the column is a *long* column. Applying Eq. 13−21,

$$\sigma_{\text{allow}} = \frac{12\pi^2 E}{23(KL/r)^2} = \frac{12\pi^2(29.0)(10^3)}{23(170.01^2)} = 5.166\ \text{ksi}$$

Maximum Stress : Bending is about $y-y$ axis. Applying Eq. 13 − 30, we have

$$\sigma_{\max} = \sigma_{\text{allow}} = \frac{P}{A} + \frac{Mc}{I}$$

$$5.166 = \frac{P}{6.48} + \frac{P(12)\left(\frac{4.030}{2}\right)}{4.66}$$

$$P = 0.967\ \text{kip} \qquad \textbf{Ans}$$

***13–104.** The W 10 × 15 structural A-36 steel column is fixed at its bottom and free at its top. Determine the greatest eccentric load P that can be applied using Eq. 13–30 and the AISC equations of Sec. 13.6.

Section Properties : For a W10× 15 wide flange section,

$$A = 4.41\ \text{in}^2 \qquad b_f = 4.000\ \text{in.} \qquad r_y = 0.810\ \text{in.} \qquad I_y = 2.89\ \text{in}^4$$

Slenderness Ratio : By observation, the largest slenderness ratio is about $y-y$ axis. For a column fixed at one end and free at the other end, $K = 2$. Thus,

$$\left(\frac{KL}{r}\right)_y = \frac{2(6)(12)}{0.810} = 177.78$$

Allowable Stress : The allowable stress can be determined using *AISC Column Formulas*. For A−36 steel, $\left(\frac{KL}{r}\right)_c = \sqrt{\frac{2\pi^2 E}{\sigma_Y}} = \sqrt{\frac{2\pi^2[29(10^3)]}{36}} = 126.1$. Since $\left(\frac{KL}{r}\right)_c \leq \frac{KL}{r} \leq 200$, the column is a *long* column. Applying Eq. 13−21,

$$\sigma_{\text{allow}} = \frac{12\pi^2 E}{23(KL/r)^2} = \frac{12\pi^2(29.0)(10^3)}{23(177.78^2)} = 4.725\ \text{ksi}$$

Maximum Stress : Bending is about $y-y$ axis. Applying Eq. 13−30, we have

$$\sigma_{\max} = \sigma_{\text{allow}} = \frac{P}{A} + \frac{Mc}{I}$$

$$4.725 = \frac{P}{4.41} + \frac{P(12)\left(\frac{4.000}{2}\right)}{2.89}$$

$$P = 0.554\ \text{kip} \qquad \textbf{Ans}$$

13–105. The W 10 × 15 structural A-36 steel column is fixed at its bottom and free at its top. If it is subjected to a load of P = 2 kip, determine if it is safe based on the AISC equations of Sec. 13.6 and Eq. 13–30.

Section Properties : For a W10×15 wide flange section,

$A = 4.41\text{ in}^2 \quad b_f = 4.000\text{ in.} \quad r_y = 0.810\text{ in.} \quad I_y = 2.89\text{ in}^4$

Slenderness Ratio : By observation, the largest slenderness ratio is about $y-y$ axis. For a column fixed at one end and free at the other end, $K = 2$. Thus,

$$\left(\frac{KL}{r}\right)_y = \frac{2(6)(12)}{0.810} = 177.78$$

Allowable Stress : The allowable stress can be determined using *AISC Column Formulas*. For A–36 steel, $\left(\frac{KL}{r}\right)_c = \sqrt{\frac{2\pi^2 E}{\sigma_Y}}$ $= \sqrt{\frac{2\pi^2[29(10^3)]}{36}} = 126.1$. Since $\left(\frac{KL}{r}\right)_c \le \frac{KL}{r} \le 200$, the column is a *long* column. Applying Eq. 13–21,

$$\sigma_{allow} = \frac{12\pi^2 E}{23(KL/r)^2} = \frac{12\pi^2(29.0)(10^3)}{23(177.78^2)} = 4.725\text{ ksi}$$

Maximum Stress : Bending is about $y-y$ axis. Applying Eq. 13–30, we have

$$\sigma_{max} = \frac{P}{A} + \frac{Mc}{I} = \frac{2}{4.41} + \frac{2(12)\left(\frac{4.000}{2}\right)}{2.89} = 17.06\text{ ksi}$$

Since $\sigma_{max} > \sigma_{allow}$, the column **is not adequate.** **Ans**

13–106. The W 12 × 22 structural A-36 steel column is fixed at its bottom and free at its top. If it is subjected to a load of P = 4 kip, determine if it is safe based on the AISC equations of Sec. 13.6 and Eq. 13–30.

Section Properties : For a W12×22 wide flange section,

$A = 6.48\text{ in}^2 \quad b_f = 4.030\text{ in.} \quad I_y = 4.66\text{ in}^4 \quad r_y = 0.847\text{ in.}$

Slenderness Ratio : By observation, the largest slenderness ratio is about $y-y$ axis. For a column fixed at one end and free at the other end, $K = 2$. Thus,

$$\left(\frac{KL}{r}\right)_y = \frac{2(6)(12)}{0.847} = 170.01$$

Allowable Stress : The allowable stress can be determined using *AISC Column Formulas*. For A–36 steel, $\left(\frac{KL}{r}\right)_c = \sqrt{\frac{2\pi^2 E}{\sigma_Y}}$ $= \sqrt{\frac{2\pi^2[29(10^3)]}{36}} = 126.1$. Since $\left(\frac{KL}{r}\right)_c \le \frac{KL}{r} \le 200$, the column is a *long* column. Applying Eq. 13–21,

$$\sigma_{allow} = \frac{12\pi^2 E}{23(KL/r)^2} = \frac{12\pi^2(29.0)(10^3)}{23(170.01^2)} = 5.166\text{ ksi}$$

Maximum Stress : Bending is about $y-y$ axis. Applying Eq. 13–30, we have

$$\sigma_{max} = \frac{P}{A} + \frac{Mc}{I} = \frac{4}{6.48} + \frac{4(12)\left(\frac{4.030}{2}\right)}{4.66} = 21.37\text{ ksi}$$

Since $\sigma_{max} > \sigma_{allow}$, the column **is not adequate.** **Ans**

13–107. The $W\ 8 \times 15$ structural A-36 steel column is assumed to be pinned at its top and bottom. Determine the largest eccentric load P that can be applied using Eq. 13–30

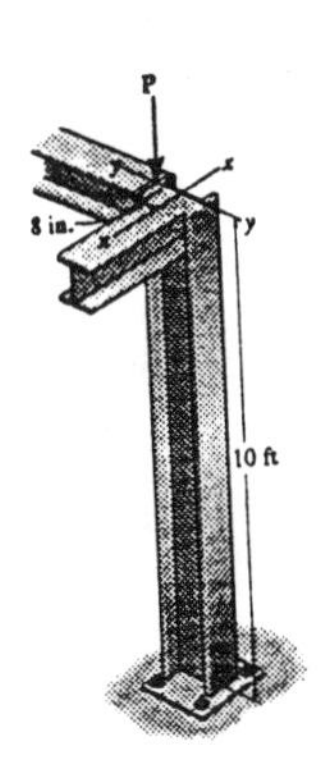

Section Properties : For a W8 × 15 wide flange section,

$A = 4.44\ \text{in}^2 \quad d = 8.11\ \text{in.} \quad I_x = 48.0\ \text{in}^4 \quad r_x = 3.29\ \text{in.}$
$r_y = 0.876\ \text{in.}$

Slenderness Ratio : By observation, the largest slenderness ratio is about $y-y$ axis. For a column pinned at both ends, $K = 1$. Thus,

$$\left(\frac{KL}{r}\right)_y = \frac{1(10)(12)}{0.876} = 137.0$$

Allowable Stress : The allowable stress can be determined using *AISC Column Formulas*. For A – 36 steel, $\left(\frac{KL}{r}\right)_c = \sqrt{\frac{2\pi^2 E}{\sigma_Y}}$ $= \sqrt{\frac{2\pi^2[29(10^3)]}{36}} = 126.1$. Since $\left(\frac{KL}{r}\right)_c \le \frac{KL}{r} \le 200$, the column is a *long* column. Applying Eq. 13 – 21,

$$\sigma_{\text{allow}} = \frac{12\pi^2 E}{23(KL/r)^2} = \frac{12\pi^2(29.0)(10^3)}{23(137.0^2)} = 7.958\ \text{ksi}$$

Maximum Stress : Bending is about $x-x$ axis. Applying Eq. 13 – 30, we have

$$\sigma_{\max} = \sigma_{\text{allow}} = \frac{P}{A} + \frac{Mc}{I}$$

$$7.958 = \frac{P}{4.44} + \frac{P(8)\left(\frac{8.11}{2}\right)}{48}$$

$$P = 8.83\ \text{kip} \qquad \textbf{Ans}$$

***13–108.** Solve Prob. 13–107 if the column is fixed at its top and bottom.

Section Properties : For a W8 × 15 wide flange section,

$A = 4.44\ \text{in}^2 \quad d = 8.11\ \text{in.} \quad I_x = 48.0\ \text{in}^4 \quad r_x = 3.29\ \text{in.}$
$r_y = 0.876\ \text{in.}$

Slenderness Ratio : By observation, the largest slenderness ratio is about $y-y$ axis. For a column fixed at both ends, $K = 0.5$. Thus,

$$\left(\frac{KL}{r}\right)_y = \frac{0.5(10)(12)}{0.876} = 68.49$$

Allowable Stress : The allowable stress can be determined using *AISC Column Formulas*. For A – 36 steel, $\left(\frac{KL}{r}\right)_c = \sqrt{\frac{2\pi^2 E}{\sigma_Y}}$ $= \sqrt{\frac{2\pi^2[29(10^3)]}{36}} = 126.1$. Since $\frac{KL}{r} < \left(\frac{KL}{r}\right)_c$, the column is an *intermediate* column. Applying Eq. 13 – 23,

$$\sigma_{\text{allow}} = \frac{\left[1 - \frac{(KL/r)^2}{2(KL/r)_c^2}\right]\sigma_Y}{\frac{5}{3} + \frac{3(KL/r)}{8(KL/r)_c} - \frac{(KL/r)^3}{8(KL/r)_c^3}} = \frac{\left[1 - \frac{(68.49^2)}{2(126.1^2)}\right](36)}{\frac{5}{3} + \frac{3(68.49)}{8(126.1)} - \frac{(68.49^3)}{8(126.1^3)}} = 16.586\ \text{ksi}$$

Maximum Stress : Bending is about $x-x$ axis. Applying Eq. 13 – 30, we have

$$\sigma_{\max} = \sigma_{\text{allow}} = \frac{P}{A} + \frac{Mc}{I}$$

$$16.586 = \frac{P}{4.44} + \frac{P(8)\left(\frac{8.11}{2}\right)}{48}$$

$$P = 18.4\ \text{kip} \qquad \textbf{Ans}$$

13–109. Solve Prob. 13–107 if the column is fixed at its bottom and pinned at its top.

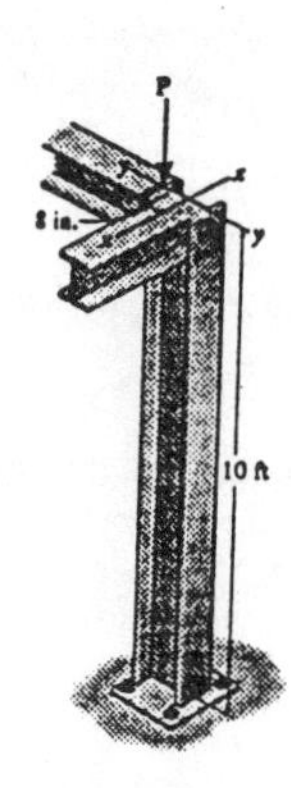

Section Properties : For a W8 × 15 wide flange section,

$$A = 4.44 \text{ in}^2 \quad d = 8.11 \text{ in.} \quad I_x = 48.0 \text{ in}^4 \quad r_x = 3.29 \text{ in.}$$
$$r_y = 0.876 \text{ in.}$$

Slenderness Ratio : By observation, the largest slenderness ratio is about $y-y$ axis. For a column fixed at both ends, $K = 0.7$. Thus,

$$\left(\frac{KL}{r}\right)_y = \frac{0.7(10)(12)}{0.876} = 95.89$$

Allowable Stress : The allowable stress can be determined using *AISC Column Formulas*. For A–36 steel, $\left(\frac{KL}{r}\right)_c = \sqrt{\frac{2\pi^2 E}{\sigma_Y}}$ $= \sqrt{\frac{2\pi^2[29(10^3)]}{36}} = 126.1$. Since $\frac{KL}{r} < \left(\frac{KL}{r}\right)_c$, the column is an *intermediate* column. Applying Eq. 13 – 23,

$$\sigma_{\text{allow}} = \frac{\left[1 - \frac{(KL/r)^2}{2(KL/r)_c^2}\right]\sigma_Y}{\frac{5}{3} + \frac{3(KL/r)}{8(KL/r)_c} - \frac{(KL/r)^3}{8(KL/r)_c^3}}$$
$$= \frac{\left[1 - \frac{(95.89^2)}{2(126.1^2)}\right](36)}{\frac{5}{3} + \frac{3(95.89)}{8(126.1)} - \frac{(95.89^3)}{8(126.1^3)}}$$
$$= 13.491 \text{ ksi}$$

Maximum Stress : Bending is about $x-x$ axis. Applying Eq. 13 – 30, we have

$$\sigma_{\max} = \sigma_{\text{allow}} = \frac{P}{A} + \frac{Mc}{I}$$
$$13.491 = \frac{P}{4.44} + \frac{P(8)\left(\frac{8.11}{2}\right)}{48}$$

$$P = 15.0 \text{ kip} \qquad \textbf{Ans}$$

13–110. A 20-ft-long column is made of aluminum alloy 2014-T6. If it is pinned at its top and bottom, and a compressive load **P** is applied at point *A*, determine the maximum allowable magnitude of **P** using the equations of Sec. 13.6 and Eq. 13–30.

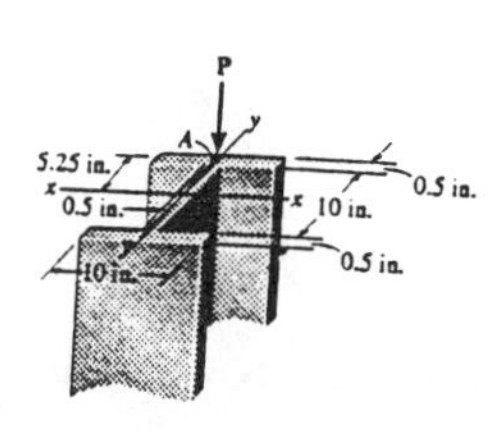

Section Properties :

$$A = 10(11) - 10(9.5) = 15.0 \text{ in}^2$$
$$I_x = \frac{1}{12}(10)(11^3) - \frac{1}{12}(9.5)(10^3) = 317.5 \text{ in}^4$$
$$I_y = \frac{1}{12}(1)(10^3) + \frac{1}{12}(10)(0.5^3) = 83.4375 \text{ in}^4$$
$$r_y = \sqrt{\frac{I_y}{A}} = \sqrt{\frac{83.4375}{15}} = 2.358 \text{ in.}$$

Slenderness Ratio : The largest slenderness ratio is about $y-y$ axis. For a column pinned at both ends, $K = 1.0$. Thus,

$$\left(\frac{KL}{r}\right)_y = \frac{1.0(20)(12)}{2.358} = 101.76$$

Allowable Stress : The allowable stress can be determined using the *aluminum (2014 - T6 alloy) column formulas*. Since $\frac{KL}{r} > 55$, the column is classified is a *long* column. Applying Eq. 13 – 26,

$$\sigma_{\text{allow}} = \left[\frac{54\,000}{(KL/r)^2}\right] \text{ksi}$$
$$= \frac{54\,000}{101.76^2}$$
$$= 5.215 \text{ ksi}$$

Maximum Stress : Bending is about $x-x$ axis. Applying Eq. 13 – 30, we have

$$\sigma_{\max} = \sigma_{\text{allow}} = \frac{P}{A} + \frac{Mc}{I}$$
$$5.215 = \frac{P}{15.0} + \frac{P(5.25)(5.5)}{317.5}$$

$$P = 33.1 \text{ kip} \qquad \textbf{Ans}$$

13–111. A 20-ft-long column is made of aluminum alloy 2014-T6. If it is pinned at its top and bottom, and a compressive load **P** is applied at point *A*, determine the

Sec. 13.6 and the interaction formula with $(\sigma_b)_{allow} = 20$ ksi.

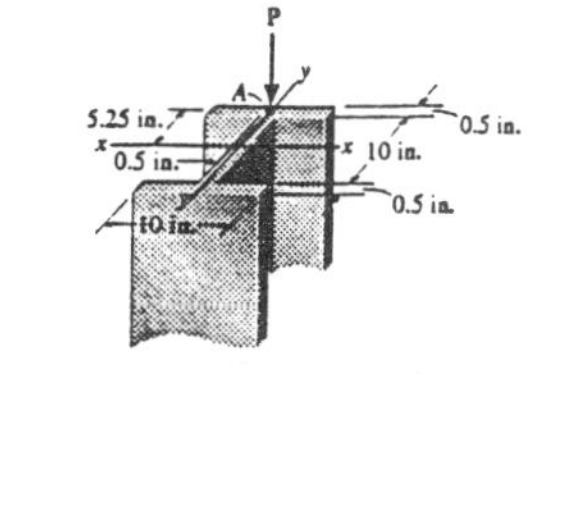

Section Properties :

$$A = 10(11) - 10(9.5) = 15.0 \text{ in}^2$$

$$I_x = \frac{1}{12}(10)\left(11^3\right) - \frac{1}{12}(9.5)\left(10^3\right) = 317.5 \text{ in}^4$$

$$I_y = \frac{1}{12}(1)\left(10^3\right) + \frac{1}{12}(10)\left(0.5^3\right) = 83.4375 \text{ in}^4$$

$$r_x = \sqrt{\frac{I_x}{A}} = \sqrt{\frac{317.5}{15}} = 4.601 \text{in.}$$

$$r_y = \sqrt{\frac{I_y}{A}} = \sqrt{\frac{83.4375}{15}} = 2.358 \text{ in.}$$

Slenderness Ratio : The largest slenderness ratio is about $y-y$ axis. For a column pinned at both ends, $K = 1.0$. Thus,

$$\left(\frac{KL}{r}\right)_y = \frac{1.0(20)(12)}{2.358} = 101.76$$

Allowable Axial Stress : The allowable stress can be determined using *aluminium (2014 - T6 alloy) column formulas*. Since $\frac{KL}{r} > 55$, the column is classified as a *long* column. Applying Eq. 13 – 26,

$$(\sigma_a)_{allow} = \left[\frac{54\,000}{(KL/r)^2}\right] \text{ksi} = \frac{54\,000}{101.76^2} = 5.215 \text{ ksi}$$

Interaction Formula : Bending is about $x-x$ axis. Applying Eq. 13 – 31, we have

$$\frac{P/A}{(\sigma_a)_{allow}} + \frac{Mc/Ar^2}{(\sigma_b)_{allow}} = 1$$

$$\frac{P/15.0}{5.215} + \frac{P(5.25)(5.5)/15.0(4.601^2)}{20} = 1$$

$$P = 57.7 \text{ kip} \qquad \textbf{Ans}$$

***13–112.** The 10-ft-long bar is made of aluminum alloy 2014-T6. If it is fixed at its bottom and pinned at the top, determine the maximum allowable eccentric load **P** that can be applied using the fomulas in Sec. 13.6 and Eq. 13–30.

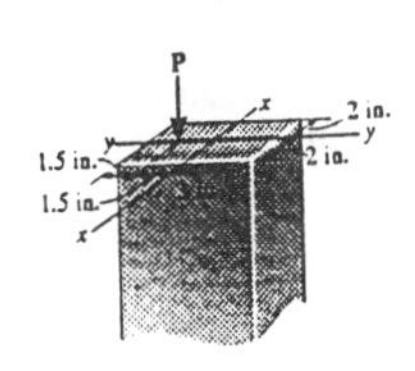

Section Properties :

$$A = 6(4) = 24.0 \text{ in}^2$$

$$I_x = \frac{1}{12}(4)\left(6^3\right) = 72.0 \text{ in}^4$$

$$I_y = \frac{1}{12}(6)\left(4^3\right) = 32.0 \text{ in}^4$$

$$r_y = \sqrt{\frac{I_y}{A}} = \sqrt{\frac{32.0}{24}} = 1.155 \text{ in.}$$

Slenderness Ratio : The largest slenderness ratio is about $y-y$ axis. For a column pinned at one end and fixed at the other end, $K = 0.7$. Thus,

$$\left(\frac{KL}{r}\right)_y = \frac{0.7(10)(12)}{1.155} = 72.75$$

Allowable Stress : The allowable stress can be determined using *aluminum (2014 - T6 alloy) column formulas*. Since $\frac{KL}{r} > 55$, the column is classified as a *long* column. Applying Eq. 13 – 26,

$$\sigma_{allow} = \left[\frac{54\,000}{(KL/r)^2}\right] \text{ksi} = \frac{54\,000}{72.75^2} = 10.204 \text{ ksi}$$

Maximum Stress : Bending is about $x-x$ axis. Applying Eq. 13 – 30, we have

$$\sigma_{max} = \sigma_{allow} = \frac{P}{A} + \frac{Mc}{I}$$

$$10.204 = \frac{P}{24.0} + \frac{P(1.5)(3)}{72.0}$$

$$P = 98.0 \text{ kip} \qquad \textbf{Ans}$$

13–113. The 10-ft-long bar is made of aluminum alloy 2014-T6. If it is fixed at its bottom and pinned at the top, determine the maximum allowable eccentric load **P** that can be applied using the equations of Sec. 13.6 and the interaction formula with $(\sigma_b)_{allow} = 18$ ksi.

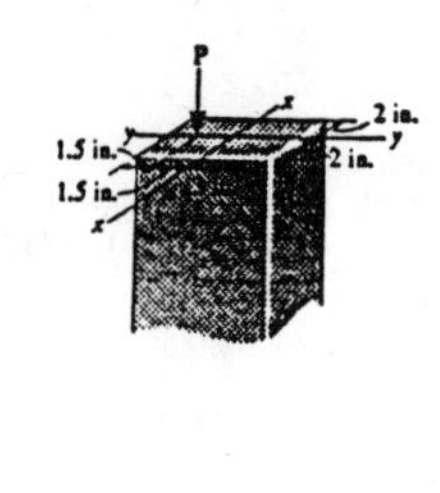

Section Properties :

$$A = 6(4) = 24.0 \text{ in}^2$$

$$I_x = \frac{1}{12}(4)(6^3) = 72.0 \text{ in}^4$$

$$I_y = \frac{1}{12}(6)(4^3) = 32.0 \text{ in}^4$$

$$r_x = \sqrt{\frac{I_x}{A}} = \sqrt{\frac{72.0}{24.0}} = 1.732 \text{ in.}$$

$$r_y = \sqrt{\frac{I_y}{A}} = \sqrt{\frac{32.0}{24.0}} = 1.155 \text{ in.}$$

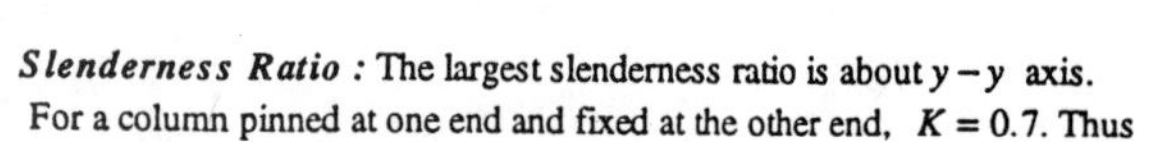

Slenderness Ratio : The largest slenderness ratio is about $y-y$ axis. For a column pinned at one end and fixed at the other end, $K = 0.7$. Thus

$$\left(\frac{KL}{r}\right)_y = \frac{0.7(10)(12)}{1.155} = 72.75$$

Allowable Stress : The allowable stress can be determined using *aluminum (2014 - T6 alloy) column formulas*. Since $\frac{KL}{r} > 55$, the column is classified as a *long* column. Applying Eq. 13 – 26,

$$(\sigma_a)_{allow} = \left[\frac{54\,000}{(KL/r)^2}\right] \text{ ksi}$$

$$= \frac{54\,000}{72.75^2}$$

$$= 10.204 \text{ ksi}$$

Interaction Formula : Bending is about $x-x$ axis. Applying Eq. 13 – 31, we have

$$\frac{P/A}{(\sigma_a)_{allow}} + \frac{Mc/Ar^2}{(\sigma_b)_{allow}} = 1$$

$$\frac{P/24.0}{10.204} + \frac{P(1.5)(3)/24.0(1.732^2)}{18} = 1$$

$$P = 132 \text{ kip} \qquad \textbf{Ans}$$

13–114. Using the NFPA equations of Sec. 13.6 and Eq. 13–30, determine the maximum allowable eccentric load P that can be applied to the wood column. Assume that the column is pinned at both its top and bottom.

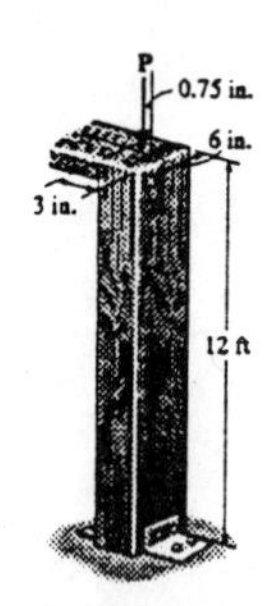

Section Properties :

$$A = 6(3) = 18.0 \text{ in}^2$$

$$I_y = \frac{1}{12}(6)(3^3) = 13.5 \text{ in}^4$$

Slenderness Ratio : For a column pinned at both ends, $K = 1.0$. Thus,

$$\left(\frac{KL}{d}\right)_y = \frac{1.0(12)(12)}{3} = 48.0$$

Allowable Stress : The allowable stress can be determined using *NFPA timber column formulas*. Since $26 < \frac{KL}{d} < 50$, it is a *long* column. Applying Eq. 13 – 29,

$$\sigma_{allow} = \frac{540}{(KL/d)^2} \text{ ksi}$$

$$= \frac{540}{48.0^2} = 0.234375 \text{ ksi}$$

Maximum Stress : Bending is about $y-y$ axis. Applying Eq. 13 – 30, we have

$$\sigma_{max} = \sigma_{allow} = \frac{P}{A} + \frac{Mc}{I}$$

$$0.234375 = \frac{P}{18.0} + \frac{P(0.75)(1.5)}{13.5}$$

$$P = 1.6875 \text{ kip} \qquad \textbf{Ans}$$

13–115. Using the NFPA equations of Sec. 13.6 and Eq. 13–30, determine the maximum allowable eccentric load P that can be applied to the wood column. Assume that the column is pinned at the top and fixed at the bottom.

Section Properties :

$$A = 6(3) = 18.0\ \text{in}^2$$

$$I_y = \frac{1}{12}(6)\left(3^3\right) = 13.5\ \text{in}^4$$

Slenderness Ratio : For a column pinned at one end and fixed at the other end, $K = 0.7$. Thus,

$$\left(\frac{KL}{d}\right)_y = \frac{0.7(12)(12)}{3} = 33.6$$

Allowable Stress : The allowable stress can be determined using *NFPA timber column formulas*. Since $26 < \frac{KL}{d} < 50$, it is a *long* column. Applying Eq. 13 – 29,

$$\sigma_{\text{allow}} = \frac{540}{(KL/d)^2}\ \text{ksi}$$

$$= \frac{540}{33.6^2} = 0.4783\ \text{ksi}$$

Maximum Stress : Bending is about $y-y$ axis. Applying Eq. 13 – 30, we have

$$\sigma_{\max} = \sigma_{\text{allow}} = \frac{P}{A} + \frac{Mc}{I}$$

$$0.4783 = \frac{P}{18.0} + \frac{P(0.75)(1.5)}{13.5}$$

$$P = 3.44\ \text{kip} \qquad \textbf{Ans}$$

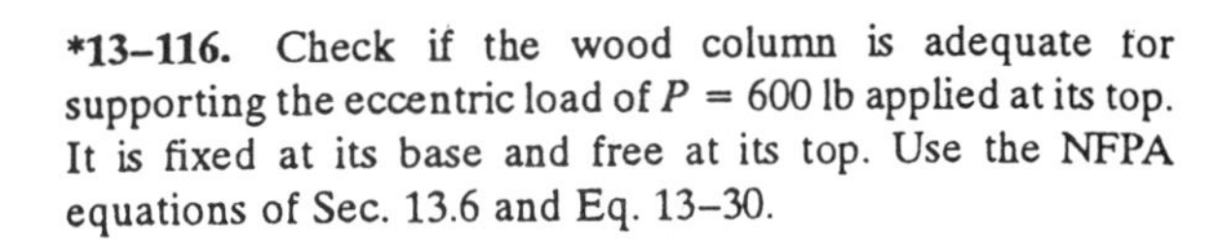
***13–116.** Check if the wood column is adequate for supporting the eccentric load of $P = 600$ lb applied at its top. It is fixed at its base and free at its top. Use the NFPA equations of Sec. 13.6 and Eq. 13–30.

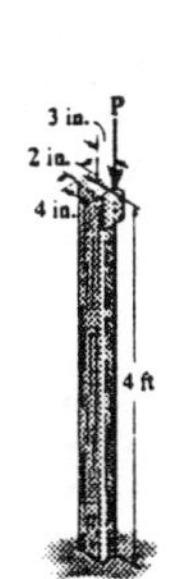

Section Properties :

$$A = 4(2) = 8.00\ \text{in}^2$$

$$I_x = \frac{1}{12}(2)\left(4^3\right) = 10.6667\ \text{in}^4$$

Slenderness Ratio : For a column fixed at one end and free at the other end, $K = 2$. Thus,

$$\left(\frac{KL}{d}\right)_y = \frac{2(4)(12)}{2} = 48.0$$

Allowable Stress : The allowable stress can be determined using *NFPA timber column formulas*. Since $26 < \frac{KL}{d} < 50$, it is a *long* column. Applying Eq. 13 – 29,

$$\sigma_{\text{allow}} = \frac{540}{(KL/d)^2}\ \text{ksi}$$

$$= \frac{540}{48.0^2} = 0.234375\ \text{ksi}$$

Maximum Stress : Bending is about $x-x$ axis. Applying Eq. 13 – 30, we have

$$\sigma_{\max} = \sigma_{\text{allow}} = \frac{P}{A} + \frac{Mc}{I}$$

$$0.234375 = \frac{P_{\text{allow}}}{8.00} + \frac{P_{\text{allow}}(3)(2)}{10.6667}$$

$$P_{\text{allow}} = 0.341\ \text{kip} = 341\ \text{lb}$$

Since $P_{\text{allow}} < P = 600$ lb, The column is **not adequate.** **Ans**

13–117. Determine the maximum allowable eccentric load P that can be applied to the wood column. The column is fixed at its base and free at its top. Use the NFPA equations of Sec. 13.6 and Eq. 13–30.

Section Properties :

$$A = 4(2) = 8.00 \text{ in}^2$$

$$I_x = \frac{1}{12}(2)\left(4^3\right) = 10.6667 \text{ in}^4$$

Slenderness Ratio : For a column fixed at one end and free at the other end, $K = 2$. Thus,

$$\left(\frac{KL}{d}\right)_y = \frac{2(4)(12)}{2} = 48.0$$

Allowable Stress : The allowable stress can be determined using *NFPA timber column formulas*. Since $26 < \dfrac{KL}{d} < 50$, it is a *long* column. Applying Eq. 13 – 29,

$$\sigma_{allow} = \frac{540}{(KL/d)^2} \text{ ksi}$$

$$= \frac{540}{48.0^2} = 0.234375 \text{ ksi}$$

Maximum Stress : Bending is about $x-x$ axis. Applying Eq. 13 – 30, we have

$$\sigma_{max} = \sigma_{allow} = \frac{P}{A} + \frac{Mc}{I}$$

$$0.234375 = \frac{P}{8.00} + \frac{P(3)(2)}{10.6667}$$

$$P = 0.341 \text{ kip} = 341 \text{ lb} \qquad \textbf{Ans}$$

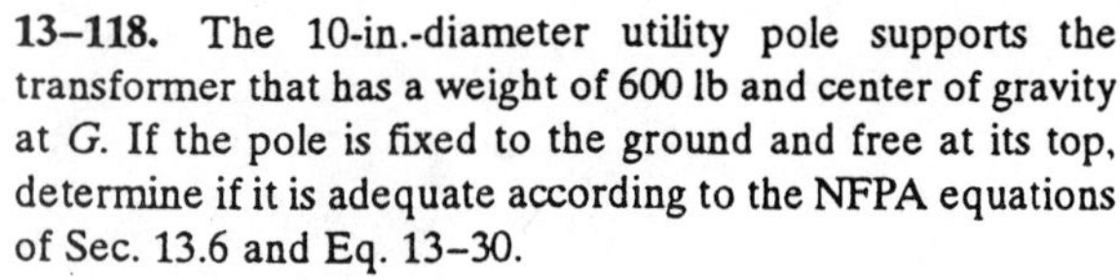
13–118. The 10-in.-diameter utility pole supports the transformer that has a weight of 600 lb and center of gravity at G. If the pole is fixed to the ground and free at its top, determine if it is adequate according to the NFPA equations of Sec. 13.6 and Eq. 13–30.

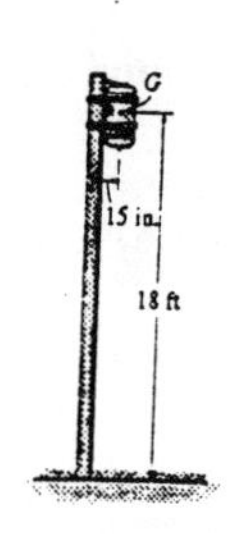

Section Properties :

$$A = \frac{\pi}{4}\left(10^2\right) = 25.0\pi \text{ in}^2$$

$$I = \frac{\pi}{4}\left(5^4\right) = 156.25\pi \text{ in}^4$$

Slenderness Ratio : For a column fixed at one end and free at the other end, $K = 2$. Thus,

$$\frac{KL}{d} = \frac{2(18)(12)}{10} = 43.2$$

Allowable Stress : The allowable stress can be determined using *NFPA timber column formulas*. Since $26 < \dfrac{KL}{d} < 50$, it is a *long* column. Apply Eq. 13 – 29,

$$\sigma_{allow} = \frac{540}{(KL/d)^2} \text{ ksi}$$

$$= \frac{540}{43.2^2} = 0.2894 \text{ ksi}$$

Maximum Stress : Applying Eq. 13 – 30, we have

$$\sigma_{max} = \sigma_{allow} = \frac{P}{A} + \frac{Mc}{I}$$

$$0.2894 = \frac{P_{allow}}{25.0\pi} + \frac{P_{allow}(15)(5)}{156.25\pi}$$

$$P_{allow} = 1.748 \text{ kip} = 1748 \text{ lb}$$

Since $P_{allow} > P = 600$ lb, The column is **adequate**. **Ans**

13–119. The wood column is 4 m long and is required to support the axial load of 25 kN. If the cross section is square, determine the dimension a of each of its sides using a factor of safety against buckling of F.S. = 2.5. The column is assumed to be pinned at its top and bottom. Use the Euler equation. E_w = 11 GPa, σ_Y = 10 MPa.

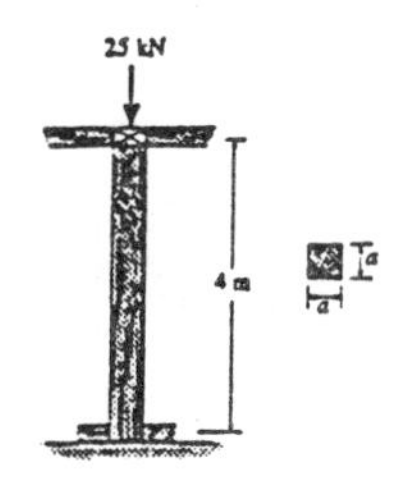

Critical Buckling Load : $I = \frac{1}{12}(a)\left(a^3\right) = \frac{a^4}{12}$,

$P_{cr} = (2.5)\,25 = 62.5$ kN and $K = 1$ for pin supported ends column.

Applying *Euler's* formula,

$$P_{cr} = \frac{\pi^2 EI}{(KL)^2}$$

$$62.5\left(10^3\right) = \frac{\pi^2(11)\left(10^9\right)\left(\frac{a^4}{12}\right)}{[1(4)]^2}$$

$$a = 0.1025 \text{ m} = 103 \text{ mm} \qquad \textbf{Ans}$$

Critical Stress : *Euler's* formula is only valid if $\sigma_{cr} < \sigma_Y$.

$$\sigma_{cr} = \frac{P_{cr}}{A} = \frac{62.5\left(10^3\right)}{0.1025(0.1025)} = 5.94 \text{ MPa} < \sigma_Y = 10 \text{ MPa} (O.K!)$$

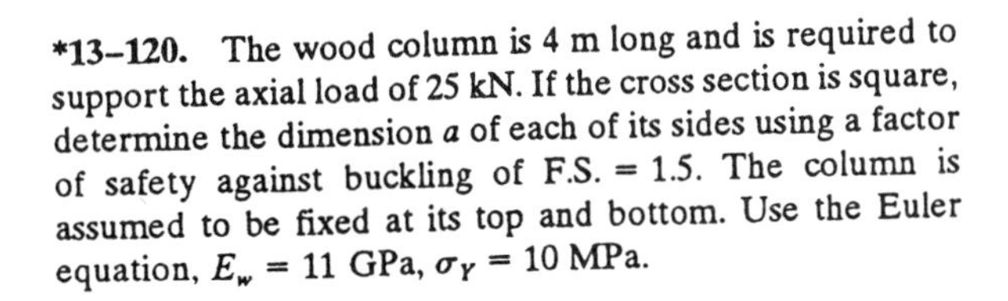

***13–120.** The wood column is 4 m long and is required to support the axial load of 25 kN. If the cross section is square, determine the dimension a of each of its sides using a factor of safety against buckling of F.S. = 1.5. The column is assumed to be fixed at its top and bottom. Use the Euler equation, E_w = 11 GPa, σ_Y = 10 MPa.

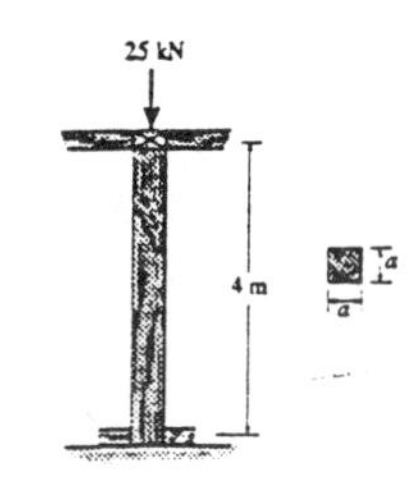

Critical Buckling Load : $I = \frac{1}{12}(a)\left(a^3\right) = \frac{a^4}{12}$,

$P_{cr} = (1.5)\,25 = 37.5$ kN and $K = 0.5$ for fix supported ends column.

Applying *Euler's* formula,

$$P_{cr} = \frac{\pi^2 EI}{(KL)^2}$$

$$37.5\left(10^3\right) = \frac{\pi^2(11)\left(10^9\right)\left(\frac{a^4}{12}\right)}{[0.5(4)]^2}$$

$$a = 0.06381 \text{ m} = 63.8 \text{ mm} \qquad \textbf{Ans}$$

Critical Stress : *Euler's* formula is only valid if $\sigma_{cr} < \sigma_Y$.

$$\sigma_{cr} = \frac{P_{cr}}{A} = \frac{37.5\left(10^3\right)}{0.06381(0.06381)} = 9.21 \text{ MPa} < \sigma_Y = 10 \text{ MPa } (O.K!)$$

13–121. The steel column is assumed to be pin connected at its top and bottom and braced against buckling about the y–y axis. If it is subjected to an axial load of 200 kN, determine the maximum moment M that can be applied to its ends without causing it to yield. E_{st} = 200 GPa, σ_Y = 250 MPa.

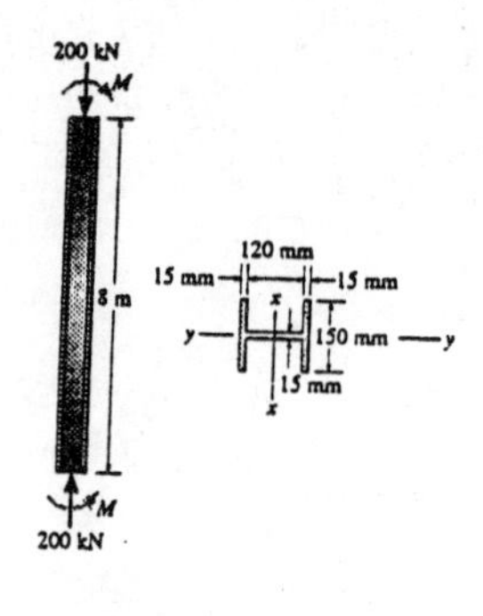

Section Properties :

$$A = 0.15(0.15) - 0.12(0.135) = 6.30\left(10^{-3}\right)\ \text{m}^2$$

$$I_x = \frac{1}{12}(0.15)\left(0.15^3\right) - \frac{1}{12}(0.135)\left(0.12^3\right)$$
$$= 22.7475\left(10^{-6}\right)\ \text{m}^4$$

$$r_x = \sqrt{\frac{I_x}{A}} = \sqrt{\frac{22.7475(10^{-6})}{6.30(10^{-3})}} = 0.060089\ \text{m}$$

For a column pinned at both ends, $K = 1$. Thus, $(KL)_x = 1(8) = 8$ m

Yielding About x - x Axis : Applying the secant formula,

$$\sigma_{\max} = \frac{P}{A}\left[1 + \frac{ec}{r_x^2}\sec\left(\frac{(KL)_x}{2r_x}\sqrt{\frac{P}{EA}}\right)\right]$$

$$250\left(10^6\right) = \frac{200(10^3)}{6.30(10^{-3})}\left[1 + \frac{e(0.075)}{0.060089^2}\sec\left(\frac{8}{2(0.060089)}\sqrt{\frac{200(10^3)}{200(10^9)[6.30(10^{-3})]}}\right)\right]$$

$$e = 0.2212\ \text{m}$$

Thus,

$$M = Pe = 200(0.2212) = 44.2\ \text{kN}\cdot\text{m} \qquad \textbf{Ans}$$

13–122. The steel column is assumed to be fixed connected at its top and bottom and braced against buckling about the y–y axis. If it is subjected to an axial load of 200 kN, determine the maximum moment M that can be applied to its ends without causing it to yield. E_{st} = 200 GPa, σ_Y = 250 MPa.

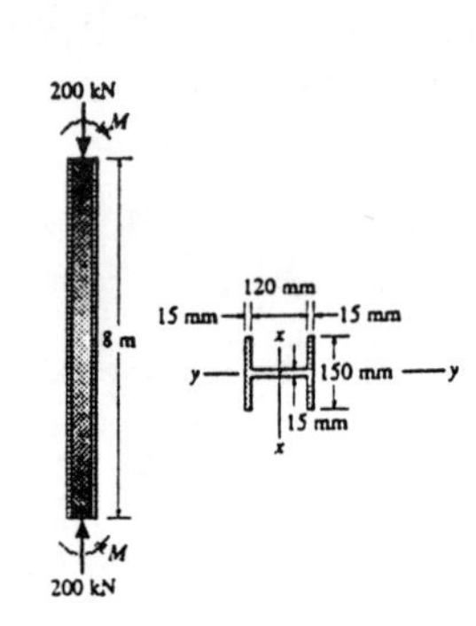

Section Properties :

$$A = 0.15(0.15) - 0.12(0.135) = 6.30\left(10^{-3}\right)\ \text{m}^2$$

$$I_x = \frac{1}{12}(0.15)\left(0.15^3\right) - \frac{1}{12}(0.135)\left(0.12^3\right)$$
$$= 22.7475\left(10^{-6}\right)\ \text{m}^4$$

$$r_x = \sqrt{\frac{I_x}{A}} = \sqrt{\frac{22.7475(10^{-6})}{6.30(10^{-3})}} = 0.060089\ \text{m}$$

For a column fixed at both ends, $K = 0.5$. Thus, $(KL)_x = 0.5(8) = 4$ m

Yielding About x - x Axis : Applying the secant formula,

$$\sigma_{\max} = \frac{P}{A}\left[1 + \frac{ec}{r_x^2}\sec\left(\frac{(KL)_x}{2r_x}\sqrt{\frac{P}{EA}}\right)\right]$$

$$250\left(10^6\right) = \frac{200(10^3)}{6.30(10^{-3})}\left[1 + \frac{e(0.075)}{0.060089^2}\sec\left(\frac{4}{2(0.060089)}\sqrt{\frac{200(10^3)}{200(10^9)[6.30(10^{-3})]}}\right)\right]$$

$$e = 0.3023\ \text{m}$$

Thus,

$$M = Pe = 200(0.3023) = 60.5\ \text{kN}\cdot\text{m} \qquad \textbf{Ans}$$

13–123. The steel bar AB has a rectangular cross section. If it is pin connected at its ends, determine the maximum allowable intensity w of the distributed load that can be applied to BC without causing bar AB to buckle. Use a factor of safety with respect to buckling of F.S. = 1.5. E_{st} = 200 GPa. σ_Y = 360 MPa.

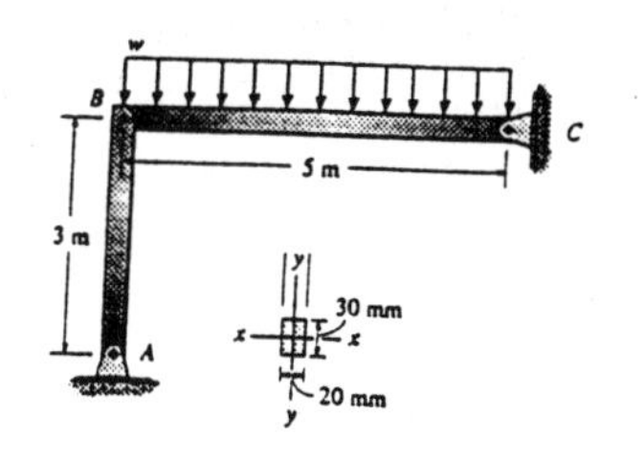

Support Reactions :

$$\circlearrowleft + \Sigma M_C = 0; \quad F_{AB}(5) - 5w(2.5) = 0 \quad F_{AB} = 2.50w$$

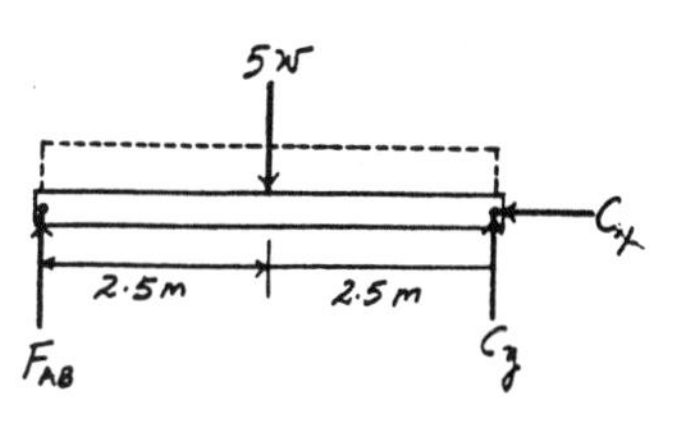

Section Properties :

$$A = 0.03(0.02) = 0.600\left(10^{-3}\right)\ \text{m}^2$$

$$I = \frac{1}{12}(0.03)\left(0.02^3\right) = 20.0\left(10^{-9}\right)\ \text{m}^4$$

Critical Buckling Load : $K = 1$ for a column with both ends pinned. Applying *Euler's* formula,

$$P_{cr} = 1.5F_{AB} = \frac{\pi^2 EI}{(KL_{AB})^2}$$

$$1.5(2.50w) = \frac{\pi^2(200)(10^9)\left[20.0(10^{-9})\right]}{[1(3)]^2}$$

$$w = 1169.7\ \text{N/m} = 1.17\ \text{kN/m} \qquad \textbf{Ans}$$

Critical Stress : *Euler's* formula is only valid if $\sigma_{cr} < \sigma_Y$.

$$\sigma_{cr} = \frac{P_{cr}}{A} = \frac{1.5(2.5)(1169.7)}{0.600(10^{-3})} = 7.31\ \text{MPa} < \sigma_Y = 360\ \text{MPa}\ (O.K!)$$

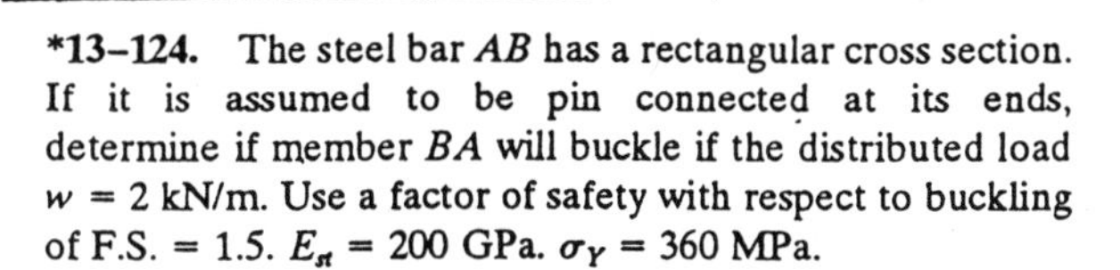

***13–124.** The steel bar AB has a rectangular cross section. If it is assumed to be pin connected at its ends, determine if member BA will buckle if the distributed load w = 2 kN/m. Use a factor of safety with respect to buckling of F.S. = 1.5. E_{st} = 200 GPa. σ_Y = 360 MPa.

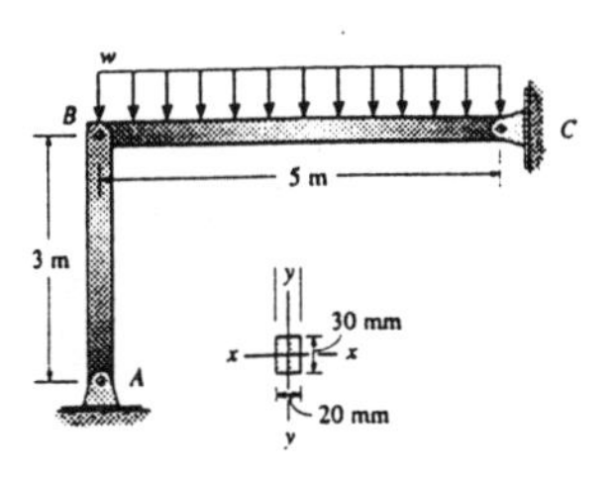

Support Reactions :

$$\circlearrowleft + \Sigma M_C = 0; \quad F_{AB}(5) - 10.0(2.5) = 0 \quad F_{AB} = 5.00\ \text{kN}$$

Section Properties :

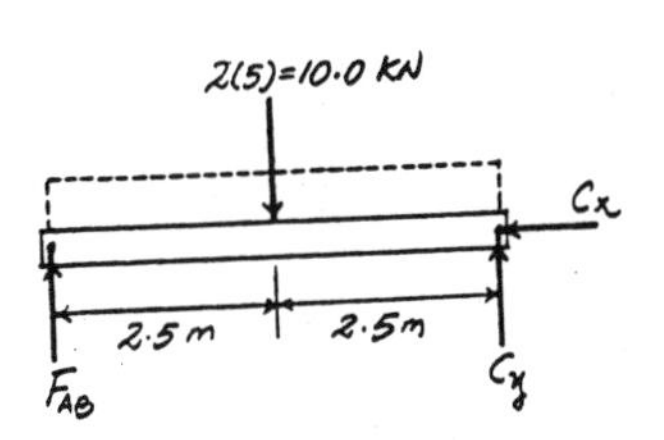

$$A = 0.03(0.02) = 0.600\left(10^{-3}\right)\ \text{m}^2$$

$$I = \frac{1}{12}(0.03)\left(0.02^3\right) = 20.0\left(10^{-9}\right)\ \text{m}^4$$

Critical Buckling Load : $K = 1$ for column with both ends pinned. Applying *Euler's* formula,

$$P_{cr} = \frac{\pi^2 EI}{(KL_{AB})^2} = \frac{\pi^2(200)(10^9)\left[20.0(10^{-9})\right]}{[1(3)]^2}$$

$$= 4386.5\ \text{N} = 4.386\ \text{kN}$$

Critical Stress : *Euler's* formula is only valid if $\sigma_{cr} < \sigma_Y$.

$$\sigma_{cr} = \frac{P_{cr}}{A} = \frac{4386.5}{0.600(10^{-3})} = 7.31\ \text{MPa} < \sigma_Y = 360\ \text{MPa}\ (O.K!)$$

Since $P_{cr} < 1.5F_{AB} = 7.50$ kN, member AB **will buckle**. **Ans**

13–125. Use the AISC equations and check if the $W\,6 \times 15$ A-36 steel column can support the axial load of 16 kip. The column is fixed at its base and free at its top.

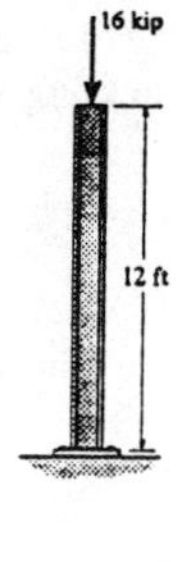

Section Properties : For a W6×15 wide flange section,

$$A = 4.43\text{ in}^2 \qquad r_y = 1.46\text{ in}$$

Slenderness Ratio : For a column fixed at one end and free at the other end, $K = 2$. Thus,

$$\left(\frac{KL}{r}\right)_y = \frac{2(12)(12)}{1.46} = 197.3$$

AISC Column Formula : For A−36 steel, $\left(\frac{KL}{r}\right)_c = \sqrt{\frac{2\pi^2 E}{\sigma_Y}}$ $= \sqrt{\frac{2\pi^2[29(10^3)]}{36}} = 126.1$. Since $\left(\frac{KL}{r}\right)_c \leq \frac{KL}{r} \leq 200$, the column is a *long* column. Applying Eq. 13−21,

$$\sigma_{allow} = \frac{12\pi^2 E}{23(KL/r)^2} = \frac{12\pi^2(29.0)(10^3)}{23(197.3^2)} = 3.838\text{ ksi}$$

The allowable load is

$$P_{allow} = \sigma_{allow}A = 3.838(4.43) = 17.00\text{ kip} > P = 16\text{ kip} \quad (O.K!)$$

Thus, the column **is adequate.** **Ans**

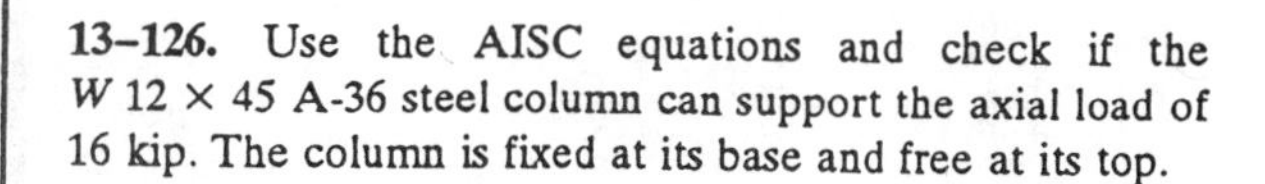

13–126. Use the AISC equations and check if the $W\,12 \times 45$ A-36 steel column can support the axial load of 16 kip. The column is fixed at its base and free at its top.

Section Properties : For a W12×45 wide flange section,

$$A = 13.2\text{ in}^2 \qquad r_y = 1.94\text{ in}$$

Slenderness Ratio : For a column fixed at one end and free at the other end, $K = 2$. Thus,

$$\left(\frac{KL}{r}\right)_y = \frac{2(12)(12)}{1.94} = 148.5$$

AISC Column Formula : For A−36 steel, $\left(\frac{KL}{r}\right)_c = \sqrt{\frac{2\pi^2 E}{\sigma_Y}}$ $= \sqrt{\frac{2\pi^2[29(10^3)]}{36}} = 126.1$. Since $\left(\frac{KL}{r}\right)_c \leq \frac{KL}{r} \leq 200$, the column is a *long* column. Applying Eq. 13−21,

$$\sigma_{allow} = \frac{12\pi^2 E}{23(KL/r)^2} = \frac{12\pi^2(29.0)(10^3)}{23(148.5^2)} = 6.776\text{ ksi}$$

The allowable load is

$$P_{allow} = \sigma_{allow}A = 6.776(13.2) = 89.44\text{ kip} > P = 16\text{ kip} \quad (O.K!)$$

Thus, the column **is adequate.** **Ans**

13–127. The A-36 steel bar AB has a square cross section. If it is pin connected at its ends, determine the maximum allowable load P that can be applied to the frame. Use a factor of safety with respect to buckling of F.S. = 2.

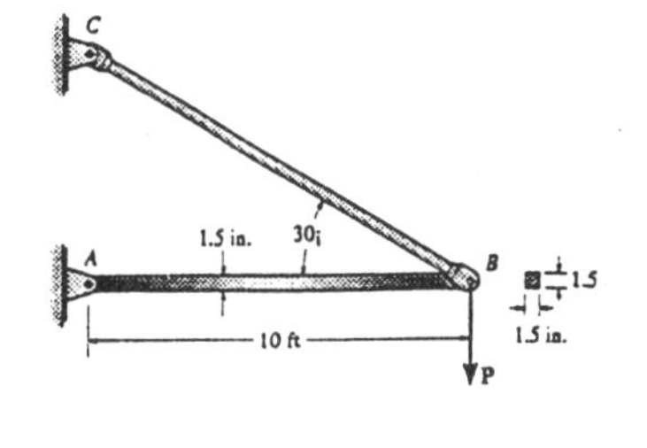

Member Forces : Apply method of joint at joint B.

$$+\uparrow \Sigma F_y = 0; \quad F_{BC}\sin 30° - P = 0 \quad F_{BC} = 2.00P \text{ (T)}$$

$$\xrightarrow{+} F_x = 0; \quad F_{AB} - 2.00P\cos 30° = 0 \quad F_{AB} = 1.732P \text{ (C)}$$

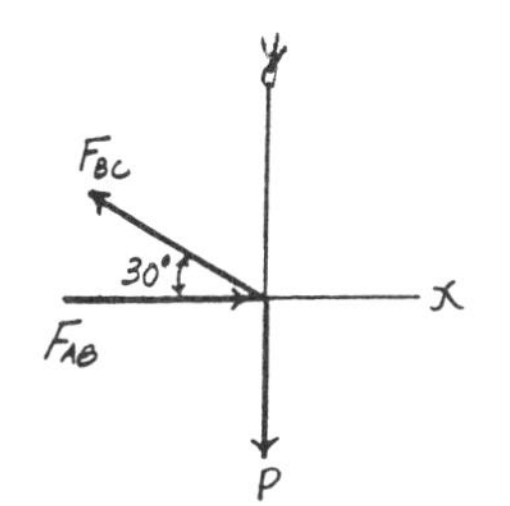

Section Properties :

$$A = 1.5(1.5) = 2.25 \text{ in}^2$$

$$I = \frac{1}{12}(1.5)\left(1.5^3\right) = 0.421875 \text{ in}^4$$

Critical Buckling Load : $K = 1$ for a column with both ends pinned. Applying *Euler's* formula,

$$P_{cr} = 2F_{AB} = \frac{\pi^2 EI}{(KL_{AB})^2}$$

$$2(1.732P) = \frac{\pi^2(29)(10^3)(0.421875)}{[1(10)(12)]^2}$$

$$P = 2.421 \text{ kip} = 2.42 \text{ kip} \qquad \textbf{Ans}$$

Critical Stress : *Euler's* formula is only valid if $\sigma_{cr} < \sigma_Y$.

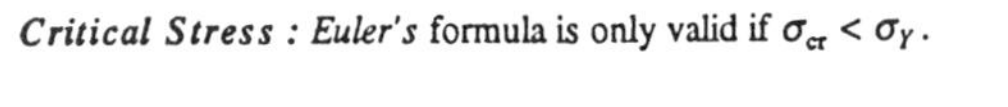

$$\sigma_{cr} = \frac{P_{cr}}{A} = \frac{2[1.732(2.421)]}{2.25} = 3.73 \text{ ksi} < \sigma_Y = 36 \text{ ksi } (O.K!)$$

***13–128.** The distributed loading is supported by two pin-connected columns, each having a solid circular cross section. If AB is made of aluminum and CD of steel, determine the required diameter of each column so that both will be on the verge of buckling at the same time. $E_{st} = 200$ GPa, $E_{al} = 70$ GPa, $(\sigma_Y)_{st} = 250$ MPa, $(\sigma_Y)_{al} = 100$ MPa.

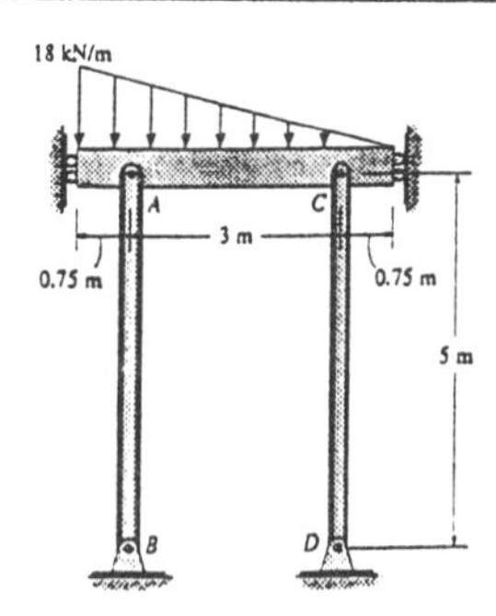

Support Reactions :

$$\circlearrowleft + \Sigma M_C = 0; \quad 40.5(2.25) - F_{AB}(3) = 0 \quad F_{AB} = 30.375 \text{ kN}$$

$$+\uparrow F_y = 0; \quad F_{CD} + 30.375 - 40.5 = 0 \quad F_{CD} = 10.125 \text{ kN}$$

Critical Buckling Load : $K = 1$ for column with both ends pinned. Applying *Euler's* formula to member AB,

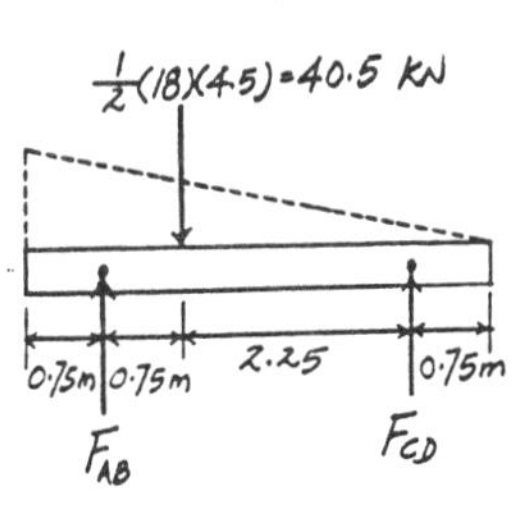

$$P_{cr} = F_{AB} = \frac{\pi^2 E_{al} I}{(KL_{AB})^2}$$

$$30.375\left(10^3\right) = \frac{\pi^2(70)(10^9)\left(\frac{\pi}{64}d_{AB}^4\right)}{[1(5)]^2}$$

$$d_{AB} = 0.06879 \text{ m} = 68.8 \text{ mm} \qquad \textbf{Ans}$$

For member BC,

$$P_{cr} = F_{CD} = \frac{\pi^2 E_{st} I}{(KL_{CD})^2}$$

$$10.125\left(10^3\right) = \frac{\pi^2(200)(10^9)\left(\frac{\pi}{64}d_{CD}^4\right)}{[1(5)]^2}$$

$$d_{CD} = 0.04020 \text{ m} = 40.2 \text{ mm} \qquad \textbf{Ans}$$

Critical Stress : *Euler's* formula is only valid if $\sigma_{cr} < \sigma_Y$.

$$(\sigma_{cr})_{AB} = \frac{P_{cr}}{A} = \frac{30.375(10^3)}{\frac{\pi}{4}(0.06879^2)} = 8.17 \text{ MPa} < (\sigma_Y)_{al} = 100 \text{ MPa } (O.K!)$$

$$(\sigma_{cr})_{CD} = \frac{P_{cr}}{A} = \frac{10.125(10^3)}{\frac{\pi}{4}(0.04020^2)} = 7.98 \text{ MPa} < (\sigma_Y)_{st} = 250 \text{ MPa } (O.K!)$$

14–1. A material is subjected to a general state of plane stress. Express the strain energy density in terms of the elastic constants E, G, and ν and the stress components σ_x, σ_y, and τ_{xy}.

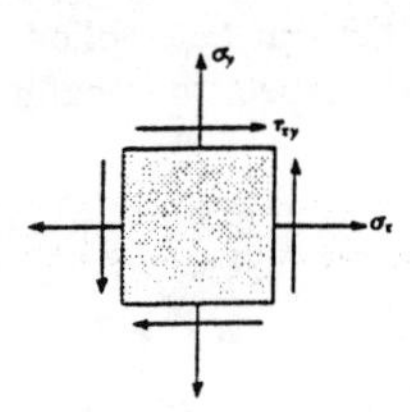

Strain Energy Due to Normal Stresses : We will consider the application of normal stresses on the element in two successive stages. For the first stage, we apply only σ_x on the element. Since σ_x is a constant, from Eq. 14 – 8, we have

$$(U_i)_1 = \int_V \frac{\sigma_x^2}{2E} dV = \frac{\sigma_x^2 V}{2E}$$

When σ_y is applied in the second stage, the normal strain ε_x will be strained by $\varepsilon_x' = -\nu\varepsilon_y = -\frac{\nu\sigma_y}{E}$. Therefore, the strain energy for the second stage is

$$(U_i)_2 = \int_V \left(\frac{\sigma_y^2}{2E} + \sigma_x \varepsilon_x'\right) dV$$

$$= \int_V \left[\frac{\sigma_y^2}{2E} + \sigma_x\left(-\frac{\nu\sigma_y}{E}\right)\right] dV$$

Since σ_x and σ_y are constants,

$$(U_i)_2 = \frac{V}{2E}\left(\sigma_y^2 - 2\nu\sigma_x\sigma_y\right)$$

Strain Energy Due to Shear Stress : The application of τ_{xy} does not strain the element in normal direction. Thus, from Eq. 14 – 11, we have

$$(U_i)_3 = \int_V \frac{\tau_{xy}^2}{2G} dV = \frac{\tau_{xy}^2 V}{2G}$$

The total strain energy is

$$U_i = (U_i)_1 + (U_i)_2 + (U_i)_3$$

$$= \frac{\sigma_x^2 V}{2E} + \frac{V}{2E}\left(\sigma_y^2 - 2\nu\sigma_x\sigma_y\right) + \frac{\tau_{xy}^2 V}{2G}$$

$$= \frac{V}{2E}\left(\sigma_x^2 + \sigma_y^2 - 2\nu\sigma_x\sigma_y\right) + \frac{\tau_{xy}^2 V}{2G}$$

and the strain energy density is

$$\frac{U_i}{V} = \frac{1}{2E}\left(\sigma_x^2 + \sigma_y^2 - 2\nu\sigma_x\sigma_y\right) + \frac{\tau_{xy}^2}{2G} \qquad \text{Ans}$$

14–2. The strain energy density must be the same whether the state of stress is represented by σ_x, σ_y, and τ_{xy}, or by the principal stresses σ_1 and σ_2. Equate the strain energy expressions for each of these two cases and show that $G = E/[2(1 + \nu)]$.

Strain Energy : The strain energy for an element with the same state of stress is the same regardless of the orientation. Applying Eq. 14 – 13 with $\sigma_z = 0$, we have

$$(U_i)_1 = \int_V \left[\frac{1}{2E}\left(\sigma_x^2 + \sigma_y^2\right) - \frac{\nu}{E}\sigma_x\sigma_y + \frac{\tau_{xy}^2}{2G}\right]dV$$

Since σ_x, σ_y and τ_{xy} are constant,

$$(U_i)_1 = \frac{V}{2E}\left(\sigma_x^2 + \sigma_y^2 - 2\nu\sigma_x\sigma_y\right) + \frac{\tau_{xy}^2 V}{2G} \qquad [1]$$

Applying Eq. 14 – 14 with $\sigma_3 = 0$, we have

$$(U_i)_2 = \int_V \left[\frac{1}{2E}\left(\sigma_1^2 + \sigma_2^2\right) - \frac{\nu}{E}\sigma_1\sigma_2\right]dV$$

Since σ_1 and σ_2 are constant,

$$(U_i)_2 = \frac{V}{2E}\left(\sigma_1^2 + \sigma_2^2 - 2\nu\sigma_1\sigma_2\right) \qquad [2]$$

However, $\sigma_{1,2} = \dfrac{\sigma_x + \sigma_y}{2} \pm \sqrt{\left(\dfrac{\sigma_x - \sigma_y}{2}\right)^2 + \tau_{xy}^2}$, then

$\sigma_1^2 + \sigma_2^2 = \sigma_x^2 + \sigma_y^2 + 2\tau_{xy}^2$ and $\sigma_1\sigma_2 = \sigma_x\sigma_y - \tau_{xy}^2$. Eq. [2] becomes

$$(U_i)_2 = \frac{V}{2E}\left[\sigma_x^2 + \sigma_y^2 + 2\tau_{xy}^2 - 2\nu\left(\sigma_x\sigma_y - \tau_{xy}^2\right)\right] \qquad [3]$$

Equating Eq. [1] and [3], we have

$$\frac{V}{2E}\left(\sigma_x^2 + \sigma_y^2 - 2\nu\sigma_x\sigma_y\right) + \frac{\tau_{xy}^2 V}{2G} = \frac{V}{2E}\left[\sigma_x^2 + \sigma_y^2 + 2\tau_{xy}^2 - 2\nu\left(\sigma_x\sigma_y - \tau_{xy}^2\right)\right]$$

Solving for G, we have

$$G = \frac{E}{2(1+\nu)} \quad (Q.E.D.)$$

14–3. If σ_x and σ_y are applied to an element, for what ratio of σ_x to σ_y is the strain energy a minimum?

Strain Energy : Applying Eq. 14 – 13 with $\sigma_z = \tau_{xy} = 0$, we have

$$U_i = \int_V \left[\frac{1}{2E}\left(\sigma_x^2 + \sigma_y^2\right) - \frac{\nu}{E}\sigma_x\sigma_y\right]dV$$

Since σ_x and σ_y are constant,

$$U_i = \frac{V}{2E}\left(\sigma_x^2 + \sigma_y^2 - 2\nu\sigma_x\sigma_y\right) = \frac{V\sigma_y^2}{2E}\left[\left(\frac{\sigma_x}{\sigma_y}\right)^2 - 2\nu\left(\frac{\sigma_x}{\sigma_y}\right)\right]$$

For strain energy U_i to be minimum, $\dfrac{dU_i}{d\left(\frac{\sigma_x}{\sigma_y}\right)} = 0$.

$$\frac{dU_i}{d\left(\frac{\sigma_x}{\sigma_y}\right)} = \frac{V\sigma_y^2}{2E}\left[2\left(\frac{\sigma_x}{\sigma_y}\right) - 2\nu\right] = 0$$

$$2\left(\frac{\sigma_x}{\sigma_y}\right) - 2\nu = 0$$

$$\frac{\sigma_x}{\sigma_y} = \nu \qquad \text{Ans}$$

Since $\dfrac{d^2U_i}{d\left(\frac{\sigma_x}{\sigma_y}\right)^2} = \dfrac{V\sigma_y^2}{E}$ is always a postive value, $\dfrac{\sigma_x}{\sigma_y} = \nu$ will indeed yield a minimum strain energy.

*14–4. Determine the torsional strain energy in the A-36 steel shaft. The shaft has a radius of 40 mm.

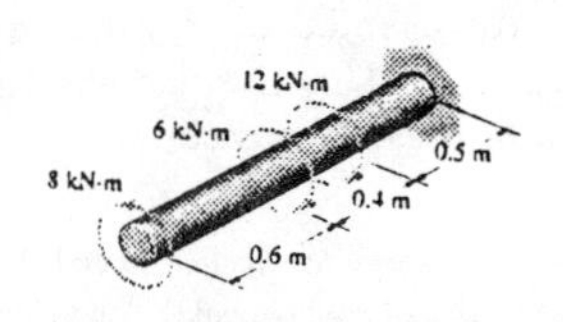

Internal Torsional Moment : As shown on FBD.

Torsional Strain Energy : With polar moment of inertia $J = \frac{\pi}{2}(0.04^4) = 1.28(10^{-6})\,\pi\ \text{m}^4$. Applying Eq. 14–22 gives

8 kN·m T=8.0 kN·m

8 kN·m 6 kN·m T=2.0 kN·m

8 kN·m 6 kN·m 12 kN·m T= −10.0 kN·m

$$U_i = \sum \frac{T^2 L}{2GJ}$$

$$= \frac{1}{2GJ}\left[8000^2(0.6) + 2000^2(0.4) + (-10000^2)(0.5)\right]$$

$$= \frac{45.0(10^6)\ \text{N}^2\cdot\text{m}^3}{GJ}$$

$$= \frac{45.0(10^6)}{75(10^9)[1.28(10^{-6})\pi]}$$

$$= 149\ \text{J}$$ **Ans**

14–5. Determine the bending strain energy in the beam. EI is constant.

Support Reactions : As shown on FBD(a).

Internal Moment Function : As shown on FBD(b).

Bending Strain Energy : Applying Eq. 14–17 gives

$$U_i = \int_0^L \frac{M^2 dx}{2EI}$$

$$= 2\left[\frac{1}{2EI}\int_0^{\frac{L}{2}} \left(\frac{P}{2}x\right)^2 dx\right]$$

$$= \frac{P^2}{4EI}\int_0^{\frac{L}{2}} x^2 dx$$

$$= \frac{P^2 L^3}{96EI}$$ **Ans**

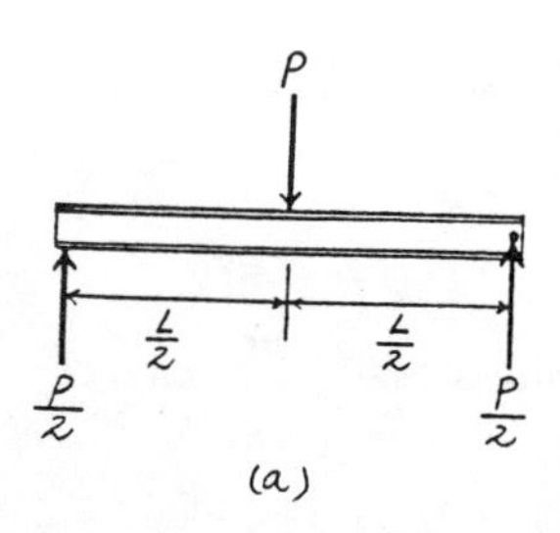

(a)

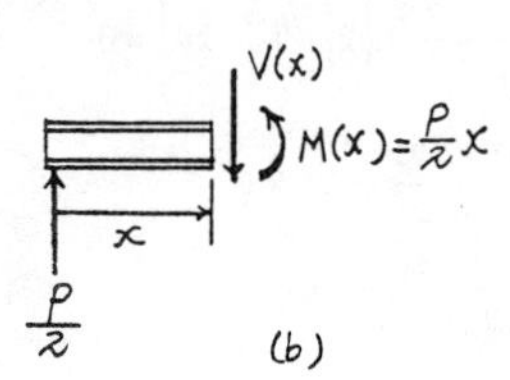

(b)

14–6. Determine the bending strain energy in the A-36 structural steel W 10 × 12 beam. Obtain the answer using the coordinates (a) x_1 and x_4, and (b) x_2 and x_3.

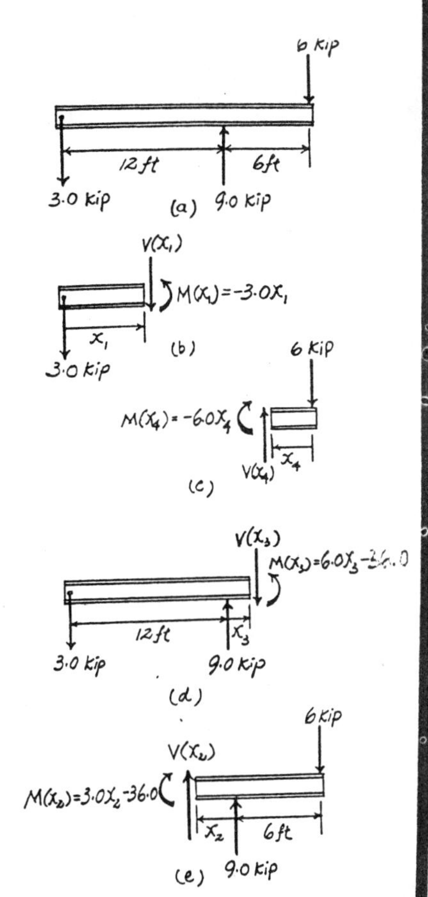

Support Reactions : As shown on FBD(a).

Internal Moment Function : As shown on FBD(b), (c), (d) and (e).

Bending Strain Energy : **a)** Using coordinates x_1 and x_4 and applying Eq. 14 – 17 gives

$$U_i = \int_0^L \frac{M^2 dx}{2EI}$$
$$= \frac{1}{2EI}\left[\int_0^{12\text{ft}} (-3.00x_1)^2 dx_1 + \int_0^{6\text{ft}} (-6.00x_4)^2 dx_4\right]$$
$$= \frac{1}{2EI}\left[\int_0^{12\text{ft}} 9.00x_1^2 dx_1 + \int_0^{6\text{ft}} 36.0x_4^2 dx_4\right]$$
$$= \frac{3888 \text{ kip}^2\cdot\text{ft}^3}{EI}$$

For W10 × 12 wide flange section, $I = 53.8$ in^4.

$$U_i = \frac{3888(12^3)}{29.0(10^3)(53.8)} = 4.306 \text{ in}\cdot\text{kip} = 359 \text{ ft}\cdot\text{lb} \qquad \textbf{Ans}$$

b) Using coordinates x_2 and x_3 and applying Eq. 14 – 17 gives

$$U_i = \int_0^L \frac{M^2 dx}{2EI}$$
$$= \frac{1}{2EI}\left[\int_0^{12\text{ft}} (3.00x_2 - 36.0)^2 dx_2 + \int_0^{6\text{ft}} (6.00x_3 - 36.0)^2 dx_3\right]$$
$$= \frac{1}{2EI}\left[\int_0^{12\text{ft}} \left(9.00x_2^2 - 216x + 1296\right) dx_2 + \int_0^{6\text{ft}} \left(36.0x_3^2 - 432x + 1296\right) dx_3\right]$$
$$= \frac{3888 \text{ kip}^2\cdot\text{ft}^3}{EI}$$

For W10 × 12 wide flange section, $I = 53.8$ in^4.

$$U_i = \frac{3888(12^3)}{29.0(10^3)(53.8)} = 4.306 \text{ in}\cdot\text{kip} = 359 \text{ ft}\cdot\text{lb} \qquad \textbf{Ans}$$

14–7. Determine the bending strain energy in the beam. EI is constant.

Support Reactions : As shown on FBD(a).

Internal Moment Function : As shown on FBD(b).

Bending Strain Energy : Applying Eq. 14 – 17 gives

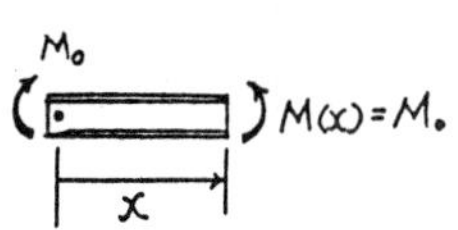

$$U_i = \int_0^L \frac{M^2 dx}{2EI}$$
$$= \frac{M_0}{2EI}\int_0^L dx$$
$$= \frac{M_0 L}{2EI} \qquad \textbf{Ans}$$

*14–8. Determine the bending strain energy in the simply supported beam due to a uniform load w. Solve the problem two ways. (a) Apply Eq. 14–17. (b) The load $w\,dx$ acting on the segment dx of the beam is displaced a distance y, where $y = w(-x^4 + 2Lx^3 - L^3x)/(24EI)$, the equation of the elastic curve. Hence the internal strain energy in the differential segment dx of the beam is equal to the external work, i.e., $dU_i = \frac{1}{2}(w\,dx)(-y)$. Integrate this equation to obtain the total strain energy in the beam. EI is constant.

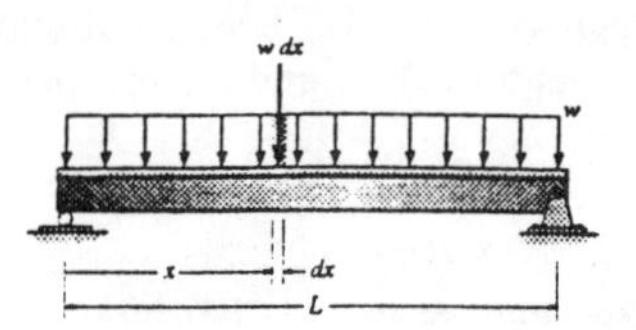

Support Reactions : As shown on FBD(a).

Internal Moment Function : As shown on FBD(b).

Bending Strain Energy : **a)** Applying Eq. 14 – 17 gives

$$U_i = \int_0^L \frac{M^2 dx}{2EI}$$
$$= \frac{1}{2EI}\left[\int_0^L \left[\frac{w}{2}\left(Lx - x^2\right)\right]^2 dx\right]$$
$$= \frac{w^2}{8EI}\left[\int_0^L \left(L^2x^2 + x^4 - 2Lx^3\right) dx\right]$$
$$= \frac{w^2L^5}{240EI} \qquad \textbf{Ans}$$

b) Integrating $dU_i = \frac{1}{2}(w\,dx)(-y)$

$$dU_i = \frac{1}{2}(w\,dx)\left[-\frac{w}{24EI}\left(-x^4 + 2Lx^3 - L^3x\right)\right]$$
$$dU_i = \frac{w^2}{48EI}\left(x^4 - 2Lx^3 + L^3x\right) dx$$

$$U_i = \frac{w^2}{48EI}\int_0^L \left(x^4 - 2Lx^3 + L^3x\right) dx$$
$$= \frac{w^2L^5}{240EI} \qquad \textbf{Ans}$$

14–9. Determine the bending strain energy in the cantilevered beam due to a uniform load w. Solve the problem two ways. (a) Apply Eq. 14–17. (b) The load $w\,dx$ acting on a segment dx of the beam is displaced a distance y, where $y = w(-x^4 + 4L^3x - 3L^4)/(24EI)$, the equation of the elastic curve. Hence the internal strain energy in the differential segment dx of the beam is equal to the external work, i.e., $dU_i = \frac{1}{2}(w\,dx)(-y)$. Integrate this equation to obtain the total strain energy in the beam. EI is constant.

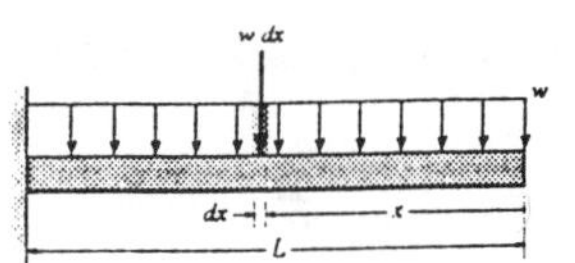

Internal Moment Function : As shown on FBD.

Bending Strain Energy : **a)** Applying Eq. 14 – 17 gives

$$U_i = \int_0^L \frac{M^2 dx}{2EI}$$
$$= \frac{1}{2EI}\left[\int_0^L \left[-\frac{w}{2}x^2\right]^2 dx\right]$$
$$= \frac{w^2}{8EI}\left[\int_0^L x^4 dx\right]$$
$$= \frac{w^2L^5}{40EI} \qquad \textbf{Ans}$$

b) Integrating $dU_i = \frac{1}{2}(w\,dx)(-y)$

$$dU_i = \frac{1}{2}(w\,dx)\left[-\frac{w}{24EI}\left(-x^4 + 4L^3x - 3L^4\right)\right]$$
$$dU_i = \frac{w^2}{48EI}\left(x^4 - 4L^3x + 3L^4\right) dx$$

$$U_i = \frac{w^2}{48EI}\int_0^L \left(x^4 - 4L^3x + 3L^4\right) dx$$
$$= \frac{w^2L^5}{40EI} \qquad \textbf{Ans}$$

14–10. Determine the bending strain energy in the beam. EI is constant.

Support Reactions : As shown on FBD(a).

Internal Moment Function : As shown on FBD(b).

Bending Strain Energy : Applying Eq. 14 – 17 gives

$$U_i = \int_0^L \frac{M^2 dx}{2EI}$$
$$= \frac{1}{2EI}\left[\int_0^L \left[\frac{w_0}{6L}\left(L^2x - x^3\right)\right]^2 dx\right]$$
$$= \frac{w_0^2}{72EIL^2}\left[\int_0^L \left(L^4x^2 + x^6 - 2L^2x^4\right) dx\right]$$
$$= \frac{w_0^2 L^5}{945EI} \quad \textbf{Ans}$$

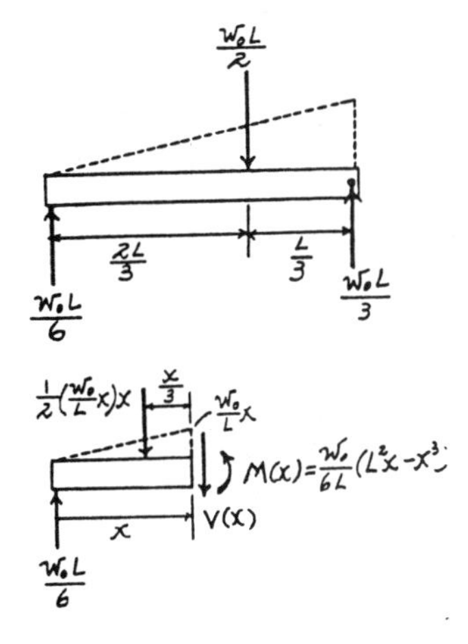

14–11. Determine the bending strain energy in the beam and the axial strain energy in each of the two rods. The beam is made of 2014-T6 aluminum and has a square cross section 50 mm by 50 mm. The rods are made of A-36 steel and have a circular cross section with a 20-mm diameter.

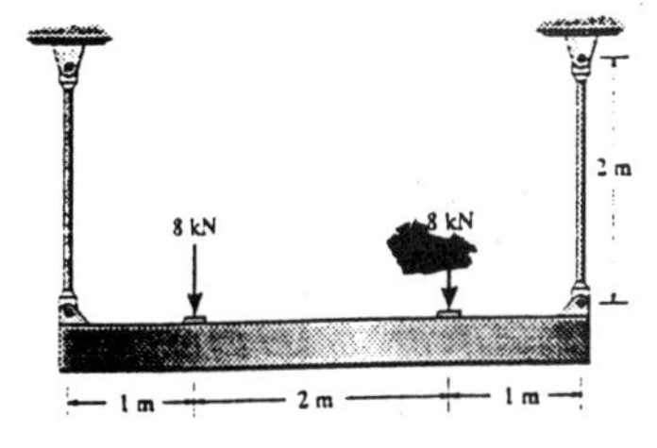

Support Reactions : As shown on FBD(a).

Internal Moment Function : As shown on FBD(b) and (c).

Axial Strain Energy : Applying Eq. 14 – 16 gives

$$(U_i)_a = \frac{N^2 L}{2AE}$$
$$= \frac{\left[8.00(10^3)\right]^2 (2)}{2AE}$$
$$= \frac{64.0(10^6)\ \text{N}^2 \cdot \text{m}}{AE}$$
$$= \frac{64.0(10^6)}{\frac{\pi}{4}(0.02^2)\left[200(10^9)\right]}$$
$$= 1.02\ \text{J} \quad \textbf{Ans}$$

Bending Strain Energy : Applying Eq. 14 – 17 gives

$$(U_i)_b = \int_0^L \frac{M^2 dx}{2EI}$$
$$= \frac{1}{2EI}\left[2\int_0^{1\,\text{m}} (8.00x_1)^2 dx_1 + \int_0^{2\,\text{m}} 8.00^2 dx_2\right]$$
$$= \frac{85.333\ \text{kN}^2 \cdot \text{m}^3}{EI}$$
$$= \frac{85.333(10^6)}{73.1(10^9)\left[\frac{1}{12}(0.05)(0.05^3)\right]}$$

$$= 2241.3\ \text{N} \cdot \text{m} = 2.24\ \text{kJ} \quad \textbf{Ans}$$

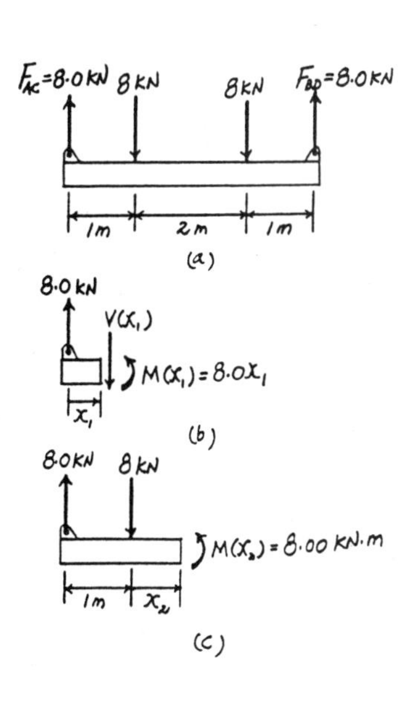

***14–12.** The beam shown is tapered along its width. If a force **P** is applied to its end, determine the strain energy in the beam and compare this result with that of a beam that has a constant rectangular cross section of width b and height h.

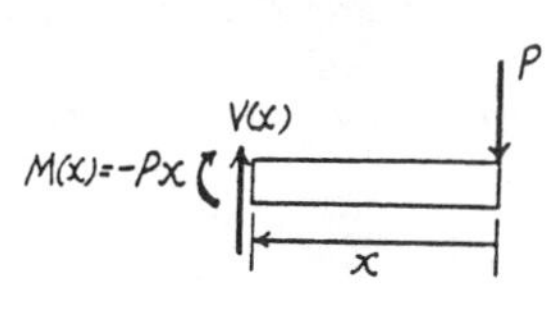

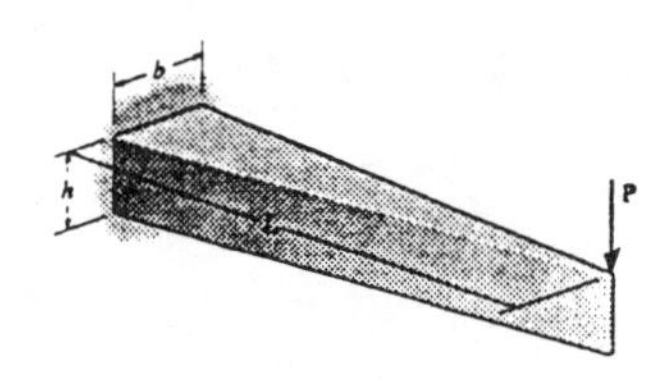

Moment of Inertia : For the beam with the uniform section,

$$I = \frac{bh^3}{12} = I_0$$

For the beam with the tapered section,

$$I = \frac{1}{12}\left(\frac{b}{L}x\right)\left(h^3\right) = \frac{bh^3}{12L}x = \frac{I_0}{L}x$$

Internal Moment Function : As shown on FBD.

Bending Strain Energy : For the beam with the tapered section, applying Eq. 14 – 17 gives

$$U_i = \int_0^L \frac{M^2 dx}{2EI}$$

$$= \frac{1}{2E}\int_0^L \frac{(-Px)^2}{\frac{I_0}{L}x}dx$$

$$= \frac{P^2L}{2EI_0}\int_0^L x\,dx$$

$$= \frac{P^2L^3}{4EI_0} = \frac{3P^2L^3}{bh^3E} \qquad \textbf{Ans}$$

For the beam with the uniform section,

$$U_i = \int_0^L \frac{M^2 dx}{2EI}$$

$$= \frac{1}{2EI_0}\int_0^L (-Px)^2 dx$$

$$= \frac{P^2L^3}{6EI_0}$$

The strain energy in the tapered beam is **1.5 times** as great as that in the beam having a uniform cross section. **Ans**

14–13. The bolt has a diameter of 10 mm, and the link AB has a rectangular cross section that is 12 mm wide by 7 mm thick. Determine the strain energy in the link due to bending and in the bolt due to axial force. The bolt is tightened so that it has a tension of 500 N. Both members are made of A-36 steel. Neglect the hole in the link.

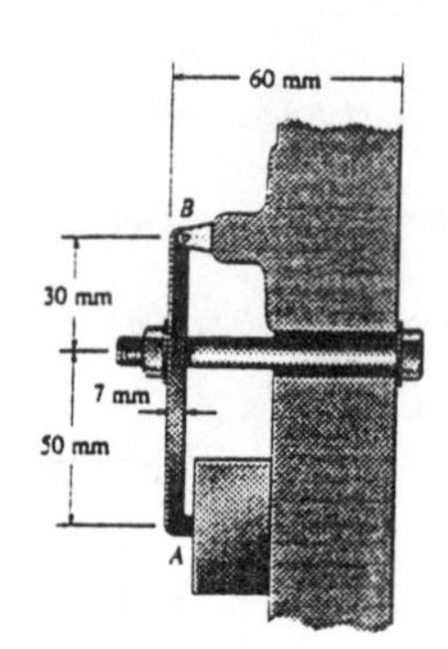

Axial Strain Energy : Applying Eq. 14 – 16 gives

$$(U_i)_a = \frac{N^2L}{2AE}$$

$$= \frac{500^2(0.06)}{2AE}$$

$$= \frac{7500\ \text{N}^2\cdot\text{m}}{AE}$$

$$= \frac{7500}{\frac{\pi}{4}(0.01^2)[200(10^9)]}$$

$$= 0.477\left(10^{-3}\right)\ \text{J} \qquad \textbf{Ans}$$

Bending Strain Energy : Applying Eq. 14 – 17 gives

$$(U_i)_b = \int_0^L \frac{M^2 dx}{2EI}$$

$$= \frac{1}{2EI}\left[\int_0^{0.05\text{m}}(187.5x_1)^2 dx_1 + \int_0^{0.03\text{m}}(312.5x_2)^2 dx_2\right]$$

$$= \frac{1.171875\ \text{N}^2\cdot\text{m}^3}{EI}$$

$$= \frac{1.171875}{200(10^9)\left[\frac{1}{12}(0.012)(0.007^3)\right]}$$

$$= 0.0171\ \text{J} \qquad \textbf{Ans}$$

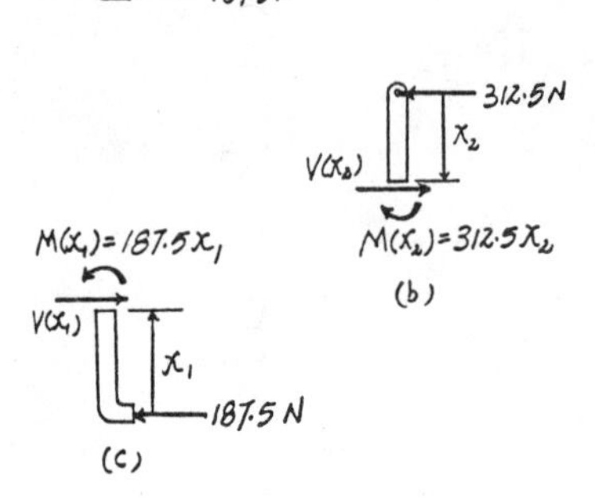

14–14. The steel beam is supported on two springs, each having a stiffness of $k = 8$ MN/m. Determine the strain energy in each of the springs and the bending strain energy in the beam. $E_{st} = 200$ GPa, $I = 5(10^6)$ mm^4.

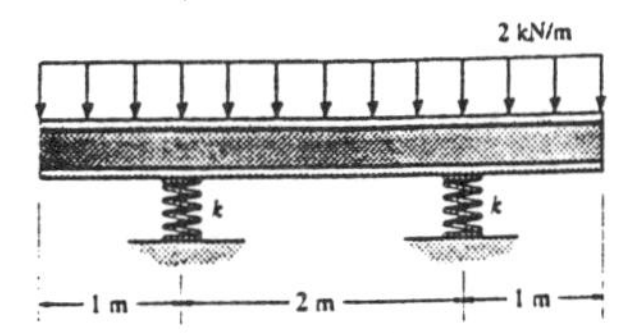

Spring Strain Energy : The spring deforms $\delta_{sp} = \frac{F_{sp}}{k} = \frac{4.00(10^3)}{8(10^6)}$ $= 0.500(10^{-3})$ m under the applied load.

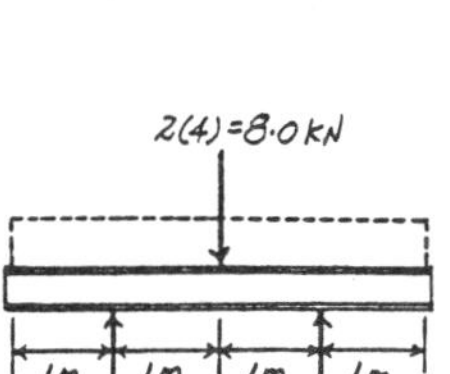

$$(U_i)_{sp} = \frac{1}{2}k\delta_{sp}^2$$
$$= \frac{1}{2}\left[8(10^6)\right]\left[0.500(10^{-3})\right]^2$$
$$= 1.00 \text{ J} \qquad \textbf{Ans}$$

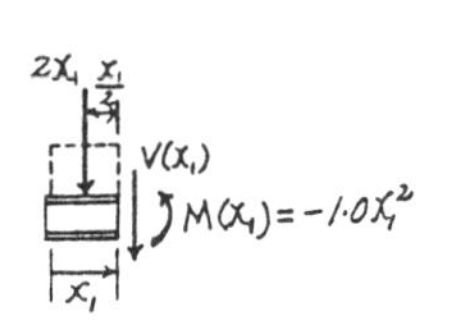

Bending Strain Energy : Applying Eq. 14 – 17 gives

$$(U_i)_b = \int_0^L \frac{M^2 dx}{2EI}$$
$$= \frac{1}{2EI}\left[2\int_0^{1\text{m}} \left(-1.00x_1^2\right)^2 dx_1 + \int_0^{2\text{m}} \left(2x_2 - x_2^2 - 1\right)^2 dx_2\right]$$
$$= \frac{0.400 \text{ kN}^2 \cdot \text{m}^3}{EI}$$
$$= \frac{0.400(10^6)}{200(10^9)\left[5(10^{-6})\right]}$$
$$= 0.400 \text{ J} \qquad \textbf{Ans}$$

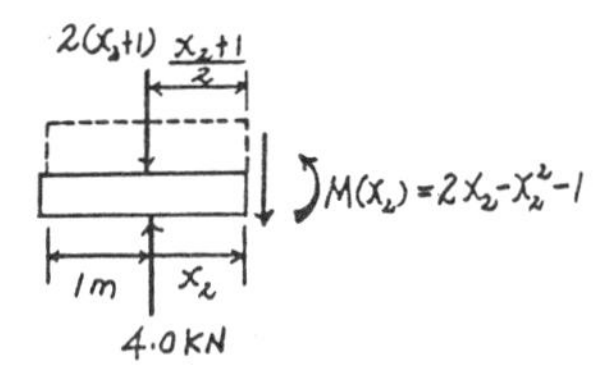

14–15. Determine the bending strain energy in the 2-in.-diameter A-36 steel rod due to the loading shown.

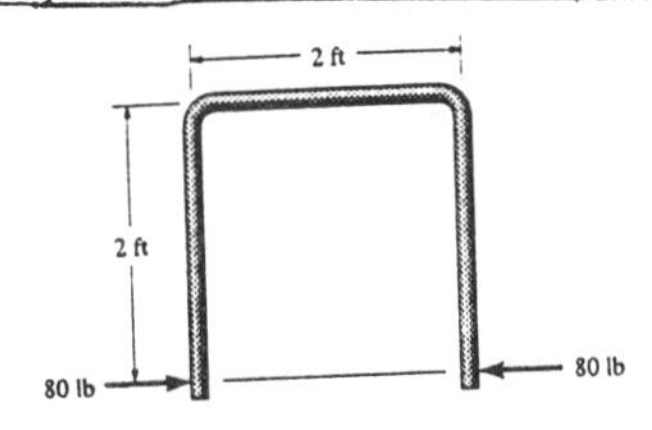

Internal Moment Function : As shown on FBD(a) and (b).

Bending Strain Energy : Applying Eq. 14 – 17 gives

$$(U_i)_b = \int_0^L \frac{M^2 dx}{2EI}$$
$$= \frac{1}{2EI}\left[2\int_0^{2\text{ft}} (80.0x_1)^2 dx_1 + \int_0^{2\text{ft}} 160^2 dx_2\right]$$
$$= \frac{42666.67 \text{ lb}^2 \cdot \text{ft}^3}{EI}$$
$$= \frac{42666.67(12^3)}{29.0(10^6)\left[\frac{\pi}{4}(1^4)\right]}$$
$$= 3.24 \text{ in.} \cdot \text{lb} \qquad \textbf{Ans}$$

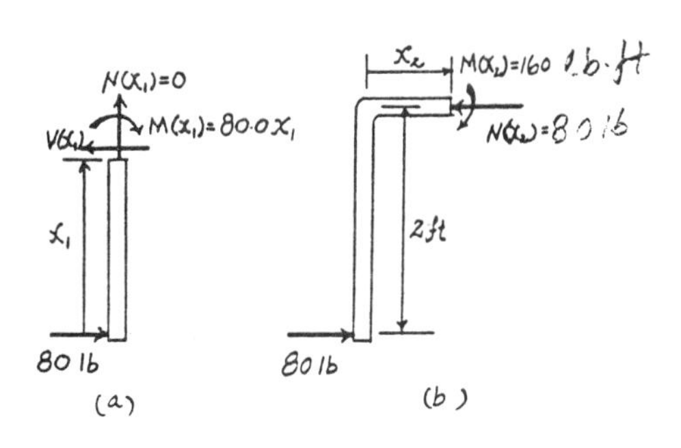

***14–16.** Determine the bending and axial strain energy in the 2-in.-diameter A-36 steel rod due to the loading shown.

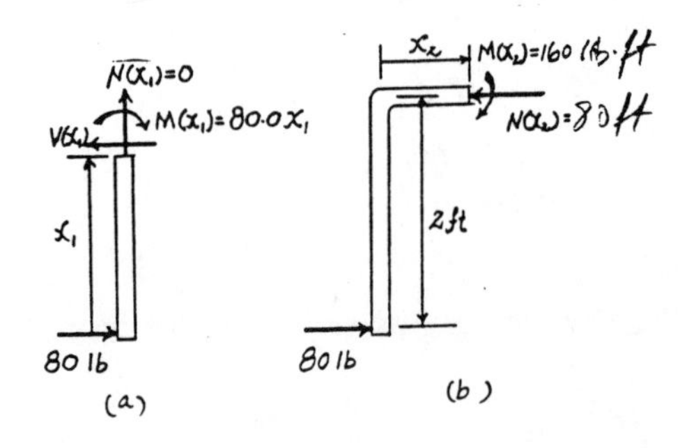

Internal Moment Function : As shown on FBD(a) and (b).

Axial Strain Energy : Applying Eq. 14 – 16 gives

$$(U_i)_a = \frac{N^2L}{2AE} = \frac{80^2(2)}{2AE} = \frac{6400\ \text{lb}^2\cdot\text{ft}}{AE} = \frac{6400(12)}{\frac{\pi}{4}(2^2)[29.0(10^6)]} = 0.843\left(10^{-3}\right)\ \text{in}\cdot\text{lb}$$ **Ans**

Bending Strain Energy : Applying Eq. 14 – 17 gives

$$(U_i)_b = \int_0^L \frac{M^2dx}{2EI} = \frac{1}{2EI}\left[2\int_0^{2\text{ft}}(80.0x_1)^2 dx_1 + \int_0^{2\text{ft}} 160^2 dx_2\right] = \frac{42666.67\ \text{lb}^2\cdot\text{ft}^3}{EI} = \frac{42666.67(12^3)}{29.0(10^6)\left[\frac{\pi}{4}(1^4)\right]}$$

$$= 3.24\ \text{in.}\cdot\text{lb}$$ **Ans**

14–17. The pipe lies in the horizontal plane. If it is subjected to a vertical force **P** at its end, determine the strain energy due to bending and torsion. Express the results in terms of the cross-sectional properties I and J and the material properties E and G.

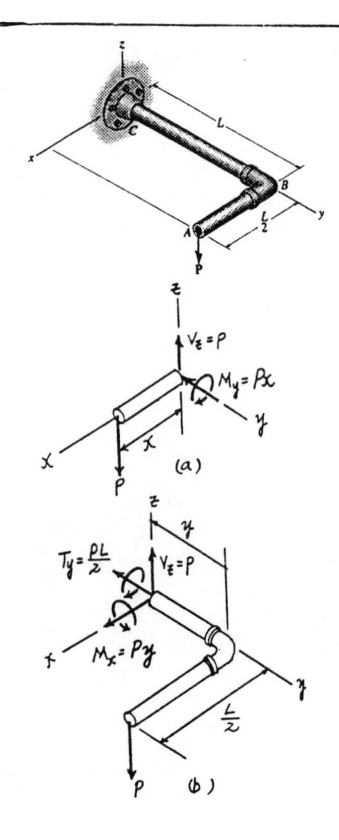

Internal Loading : As shown on FBD(a) and (b).

Strain Energy : Applying Eq. 14 – 17 and Eq. 14 – 21 gives

$$U_i = \int_0^L \frac{M^2dx}{2EI} + \int_0^L \frac{T^2}{2GJ}dx = \frac{1}{2EI}\left[\int_0^{\frac{L}{2}}(Px)^2 dx + \int_0^L (Py)^2 dy\right] + \frac{1}{2GJ}\int_0^L\left(\frac{PL}{2}\right)^2 dy$$

$$= \frac{P^2L^3}{16}\left(\frac{3}{EI} + \frac{2}{GJ}\right)$$ **Ans**

14–18. The A-36 steel bar consists of two segments, one of circular and one of square cross section. If it is subjected to the axial loading of P, determine the dimensions a of the square segment so that the strain energy within the square segment is the same as in the circular segment.

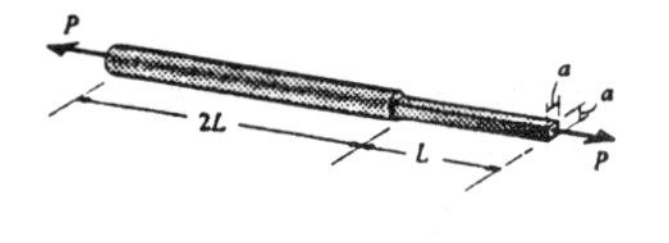

Axial Strain Energy : Applying Eq. 14 – 16 to the circular segment gives

$$(U_i)_c = \frac{N^2 L_c}{2AE} = \frac{P^2(2L)}{2(\pi r^2)E} = \frac{P^2 L}{\pi r^2 E}$$

Applying Eq. 14 – 16 to the square segment gives

$$(U_i)_s = \frac{N^2 L_s}{2AE} = \frac{P^2 L}{2(a^2)E} = \frac{P^2 L}{2a^2 E}$$

Require,

$$(U_i)_c = (U_i)_s$$

$$\frac{P^2 L}{\pi r^2 E} = \frac{P^2 L}{2a^2 E}$$

$$a = \sqrt{\frac{\pi}{2}}\, r \qquad \textbf{Ans}$$

14–19. Consider the thin-walled tube of Fig. 5–30. Use the formula for shear stress, $\tau_{avg} = T/2tA_m$, Eq. 5–18, and the general equation of shear strain energy, Eq. 14–11, to show that the twist of the tube is given by Eq. 5–20. *Hint:* Equate the work done by the torque T to the strain energy in the tube, determined from integrating the strain energy for a differential element, Fig. 14–4, over the volume of material.

Shear Strain Energy : Applying Eq. 14 – 11 with $\tau = \frac{T}{2tA_m}$, we have

$$U_i = \int_V \frac{\tau^2}{2G} dV$$

$$= \int_V \frac{T^2}{8t^2 A_m^2 G} dV$$

$$= \frac{T^2}{8A_m^2 G}\int_A \frac{dA}{t^2}\int_0^L dx$$

$$= \frac{T^2 L}{8A_m^2 G}\int_A \frac{dA}{t^2} \qquad \text{however, } dA = t\,ds$$

$$= \frac{T^2 L}{8A_m^2 G}\int \frac{ds}{t}$$

External Work : The external work done by torque T is

$$U_e = \frac{1}{2}T\phi$$

Require,

$$U_e = U_i$$

$$\frac{1}{2}T\phi = \frac{T^2 L}{8A_m^2 G}\int \frac{ds}{t}$$

$$\phi = \frac{TL}{4A_m^2 G}\int \frac{ds}{t} \qquad (Q.E.D.)$$

***14–20.** Determine the horizontal displacement of joint C. AE is constant.

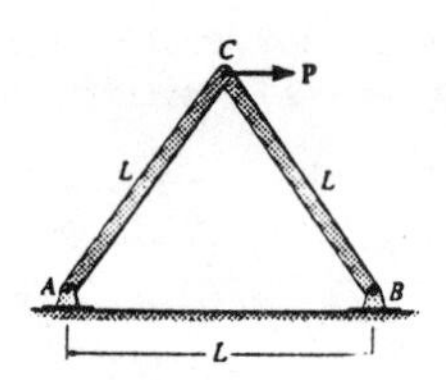

Member Forces : Applying the method of joints to C, we have

$$+\uparrow \Sigma F_y = 0; \quad F_{BC}\cos 30^\circ - F_{AC}\cos 30^\circ = 0 \qquad F_{BC} = F_{AC} = F$$

$$\xrightarrow{+} \Sigma F_x = 0; \quad P - 2F\sin 30^\circ = 0 \qquad F = P$$

Hence, $\quad F_{BC} = P \text{ (C)} \qquad F_{AC} = P \text{ (T)}$

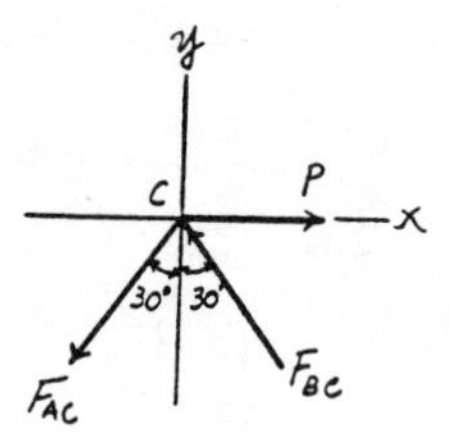

Axial Strain Energy : Applying Eq. 14–16, we have

$$U_i = \sum \frac{N^2 L}{2AE}$$
$$= \frac{1}{2AE}\left[P^2 L + (-P)^2 L\right]$$
$$= \frac{P^2 L}{AE}$$

External Work : The external work done by force P is

$$U_e = \frac{1}{2}P(\Delta_C)_h$$

Conservation of Energy :

$$U_e = U_i$$
$$\frac{1}{2}P(\Delta_C)_h = \frac{P^2 L}{AE}$$
$$(\Delta_C)_h = \frac{2PL}{AE} \qquad \textbf{Ans}$$

14–21. Determine the horizontal displacement of joint A. Each bar is made of A-36 steel and has a cross-sectional area of 1.5 in^2.

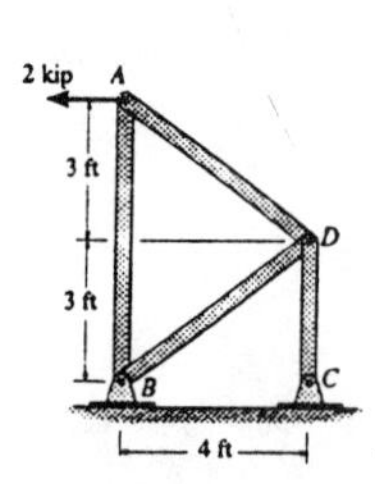

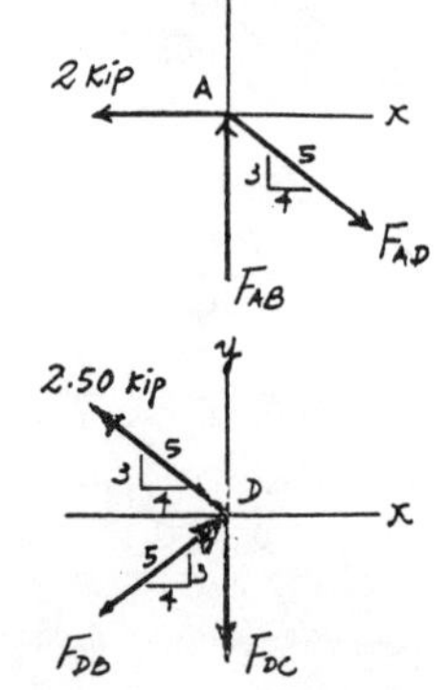

Member Forces : Applying the method of joints to joint at A, we have

$$\xrightarrow{+} \Sigma F_x = 0; \quad \frac{4}{5}F_{AD} - 2 = 0 \qquad F_{AD} = 2.50 \text{ kip (T)}$$

$$+\uparrow \Sigma F_y = 0; \quad F_{AB} - \frac{3}{5}(2.50) = 0 \qquad F_{AB} = 1.50 \text{ kip (C)}$$

At joint D

$$\xrightarrow{+} \Sigma F_x = 0; \quad \frac{4}{5}F_{DB} - \frac{4}{5}(2.50) = 0 \qquad F_{DB} = 2.50 \text{ kip (C)}$$

$$+\uparrow \Sigma F_y = 0; \quad \frac{3}{5}(2.50) + \frac{3}{5}(2.50) - F_{DC} = 0$$
$$F_{DC} = 3.00 \text{ kip (T)}$$

Axial Strain Energy : Applying Eq. 14–16, we have

$$U_i = \sum \frac{N^2 L}{2AE}$$
$$= \frac{1}{2AE}[2.50^2(5) + (-1.50)^2(6) + (-2.50)^2(5) + 3.00^2(3)]$$
$$= \frac{51.5 \text{ kip}^2\cdot\text{ft}}{AE}$$
$$= \frac{51.5(12)}{1.5[29.0(10^3)]} = 0.014207 \text{ in}\cdot\text{kip}$$

External Work : The external work done by 2 kip force is

$$U_e = \frac{1}{2}(2)(\Delta_A)_h = (\Delta_A)_h$$

Conservation of Energy :

$$U_e = U_i$$
$$(\Delta_A)_h = 0.014207$$
$$= 0.0142 \text{ in.} \qquad \textbf{Ans}$$

14–22. Determine the vertical displacement of joint D. AE is constant.

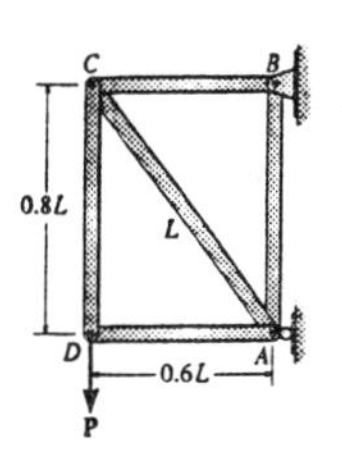

Member Forces : By inspetion of joint D, member AD is a zero force member and $F_{CD} = P$ (T). Applying the method of joints at C, we have

$+\uparrow \Sigma F_y = 0; \quad \frac{4}{5}F_{CA} - P = 0 \qquad F_{CA} = 1.25P \text{ (C)}$

$\xrightarrow{+} \Sigma F_x = 0; \quad F_{CB} - \frac{3}{5}(1.25P) = 0 \qquad F_{CB} = 0.750P \text{ (T)}$

At joint A

$+\uparrow \Sigma F_y = 0; \quad F_{AB} - \frac{4}{5}(1.25P) = 0 \qquad F_{BA} = 1.00P \text{ (T)}$

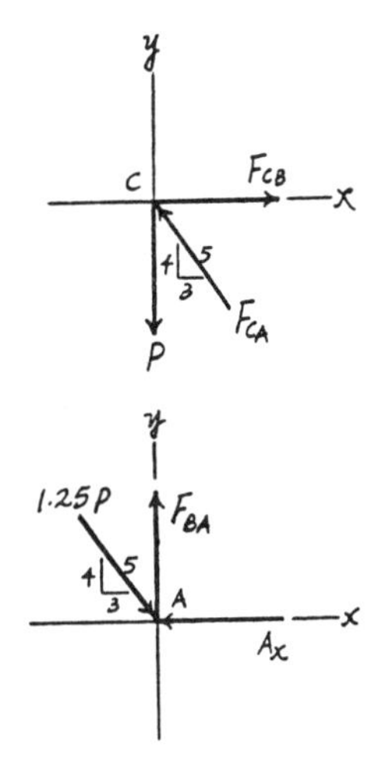

Axial Strain Energy : Applying Eq. 14 – 16, we have

$$U_i = \sum \frac{N^2L}{2AE}$$

$$= \frac{1}{2AE}[P^2(0.8L) + (-1.25P)^2(L) + (0.750P)^2(0.6L) + (1.00P)^2(0.8L)]$$

$$= \frac{1.750P^2L}{AE}$$

External Work : The external work done by force P is

$$U_e = \frac{1}{2}(P)(\Delta_D)_v$$

Conservation of Energy :

$$U_e = U_i$$

$$\frac{1}{2}(P)(\Delta_D)_v = \frac{1.750P^2L}{AE}$$

$$(\Delta_D)_v = \frac{3.50PL}{AE} \qquad \textbf{Ans}$$

14–23. The cantilevered beam has a rectangular cross-sectional area A, a moment of inertia I, and a modulus of elasticity E. If a load $\mathbf{P}$ acts at point B as shown, determine the displacement at B in the direction of $\mathbf{P}$, accounting for bending, axial force, and shear.

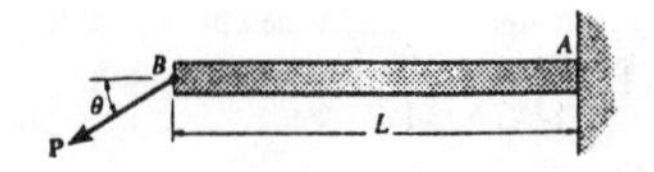

Strain Energy : Applying Eq. 14 – 15 , 14 – 17 and 14 - 19, we have

$$U_i = \int_0^L \frac{N^2 dx}{2AE} + \int_0^L \frac{M^2 dx}{2EI} + \int_0^L \frac{f_s V^2 dx}{2GA}$$

However, $f_s = \frac{6}{5}$ for a rectangular section.

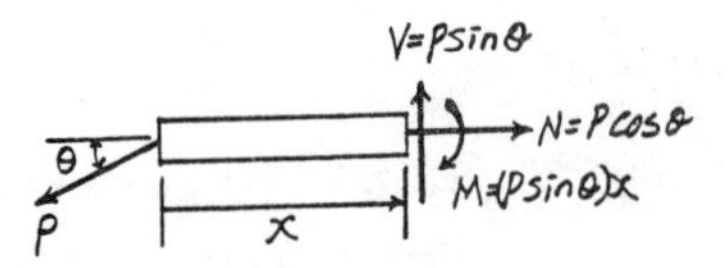

$$U_i = \int_0^L \frac{(P\cos\theta)^2 dx}{2AE} + \int_0^L \frac{[(P\sin\theta)x]^2 dx}{2EI} + \frac{6}{5}\int_0^L \frac{(P\sin\theta)^2 dx}{2GA}$$

$$= \frac{P^2 L}{30}\left(\frac{15\cos^2\theta}{AE} + \frac{5L^2\sin^2\theta}{EI} + \frac{18\sin^2\theta}{GA}\right)$$

External Work : The external work done by force P is

$$U_e = \frac{1}{2}(P)(\Delta_B)$$

Conservation of Energy : .

$$U_e = U_i$$

$$\frac{1}{2}(P)(\Delta_B) = \frac{P^2 L}{30}\left(\frac{15\cos^2\theta}{AE} + \frac{5L^2\sin^2\theta}{EI} + \frac{18\sin^2\theta}{GA}\right)$$

$$\Delta_B = \frac{PL}{15}\left(\frac{15\cos^2\theta}{AE} + \frac{5L^2\sin^2\theta}{EI} + \frac{18\sin^2\theta}{GA}\right) \qquad \textbf{Ans}$$

***14–24.** Determine the displacement of point B on the A-36 steel beam. $I = 250\ \text{in}^4$.

Moment Function : As shown on FBD.

Bending Strain Energy : Applying 14 – 17, we have

$$U_i = \int_0^L \frac{M^2 dx}{2EI}$$

$$= \frac{1}{2EI}\int_0^{15\text{ft}} (8.00x)^2 dx$$

$$= \frac{1}{2EI}\int_0^{15\text{ft}} 64.0x^2 dx$$

$$= \frac{36000\ \text{kip}^2 \cdot \text{ft}^3}{EI}$$

$$= \frac{36000(12^3)}{29.0(10^3)(250)} = 8.5804\ \text{in} \cdot \text{kip}$$

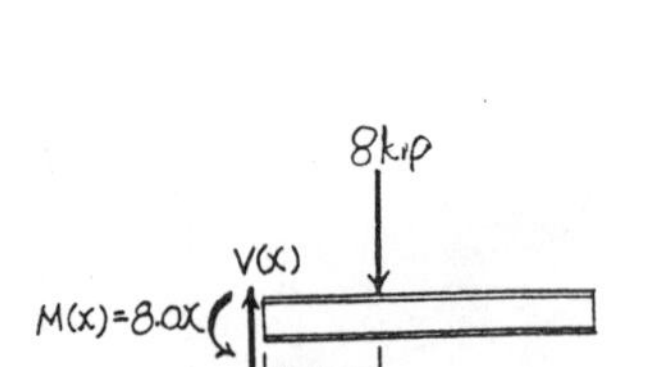

External Work : The external work done by 8 kip force is

$$U_e = \frac{1}{2}(8)(\Delta_B) = 4\Delta_B$$

Conservation of Energy :

$$U_e = U_i$$

$$4\Delta_B = 8.5804$$

$$\Delta_B = 2.15\ \text{in.} \qquad \textbf{Ans}$$

14–25. Determine the displacement of point B on the 2014-T6 aluminum beam.

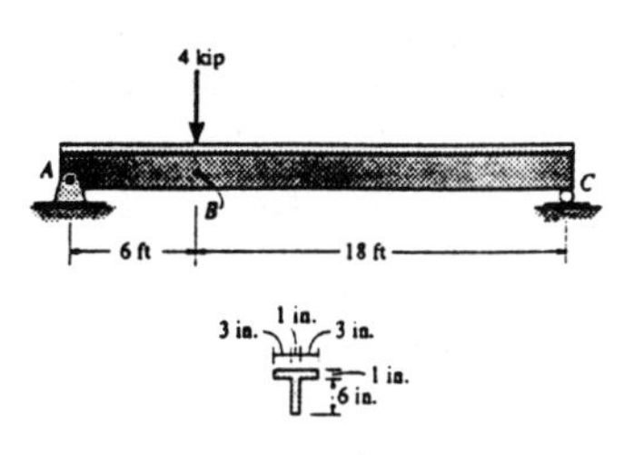

Section Properties :

$$\bar{y} = \frac{\Sigma \tilde{y} A}{\Sigma A} = \frac{0.5(1)(7) + 4(6)(1)}{1(7) + 6(1)} = 2.1154 \text{ in.}$$

$$I = \frac{1}{12}(7)\left(1^3\right) + 7(1)(2.1154 - 0.5)^2 + \frac{1}{12}(1)\left(6^3\right) + 1(6)(4 - 2.1154)^2$$
$$= 58.16 \text{ in}^4$$

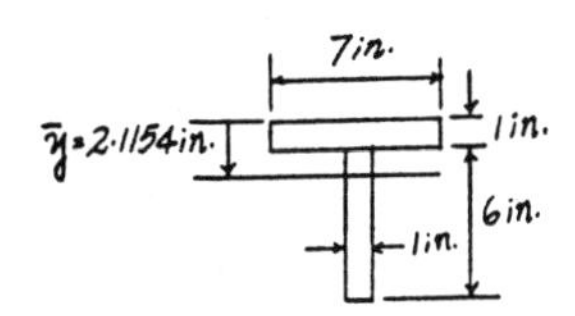

Support Reactions : As shown on FBD(a).

Moment Functions : As shown on FBD(b) and (c).

Bending Strain Energy : Applying 14 – 17, we have

$$U_i = \int_0^L \frac{M^2 dx}{2EI}$$
$$= \frac{1}{2EI}\left[\int_0^{6\text{ft}} (3.00x_1)^2 dx_1 + \int_0^{18\text{ft}} (1.00x_2)^2 dx_2\right]$$
$$= \frac{1296 \text{ kip}^2 \cdot \text{ft}^3}{EI}$$
$$= \frac{1296(12^3)}{10.6(10^3)(58.16)} = 3.6326 \text{ in} \cdot \text{kip}$$

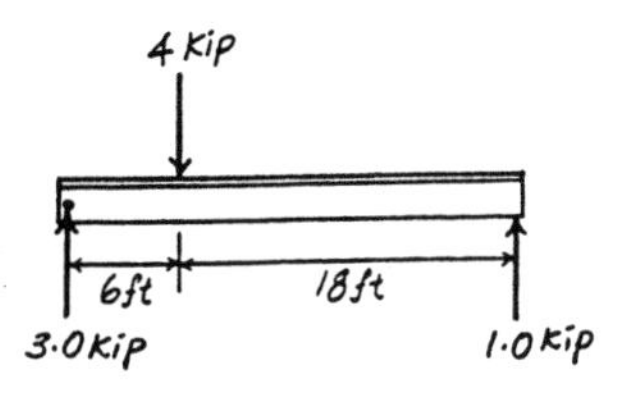

External Work : The external work done by 4 kip force is

$$U_e = \frac{1}{2}(4)(\Delta_B) = 2\Delta_B$$

Conservation of Energy :

$$U_e = U_i$$
$$2\Delta_B = 3.6326$$

$$\Delta_B = 1.82 \text{ in.} \qquad \textbf{Ans}$$

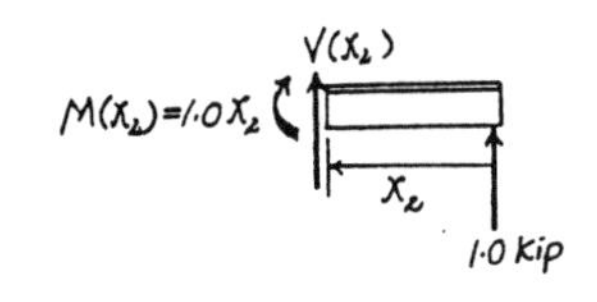

14–26. Determine the slope at point C of the A-36 steel beam. $I = 9.50(10^6)$ mm^4.

Support Reactions : As shown on FBD(a).

Moment Functions : As shown on FBD(b) and (c).

Bending Strain Energy : Applying 14 – 17, we have

$$U_i = \int_0^L \frac{M^2 dx}{2EI}$$

$$= \frac{1}{2EI}\left[\int_0^{4\,\text{m}} (-3.00x_1)^2 dx_1 + \int_0^{4\,\text{m}} (-12.0)^2 dx_2\right]$$

$$= \frac{384 \text{ kN}^2 \cdot \text{m}^3}{EI}$$

$$= \frac{384(10^6)}{200(10^9)[9.50(10^{-6})]} = 202.11 \text{ N}\cdot\text{m}$$

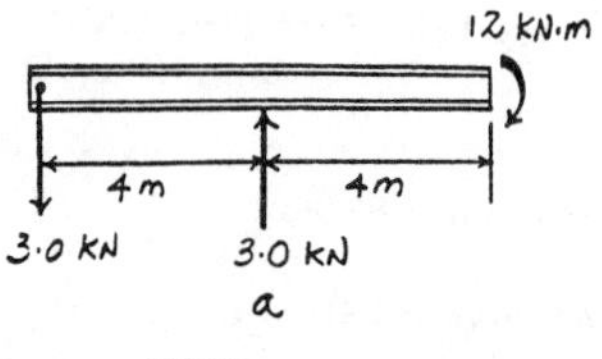

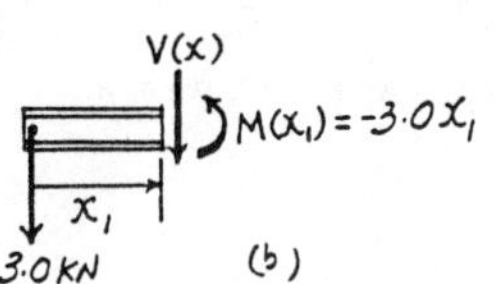

External Work : The external work done by 12 kN · m couple moment is

$$U_e = \frac{1}{2}\left[12(10^3)\right](\theta_C) = 6.00(10^3)\,\theta_C$$

Conservation of Energy :

$$U_e = U_i$$

$$6.00(10^3)\,\theta_C = 202.11$$

$$\theta_C = 0.0337 \text{ rad} = 1.93° \qquad \textbf{Ans}$$

14–27. The A-36 steel bars are pin connected at B. If each has a square cross section, determine the vertical displacement at B.

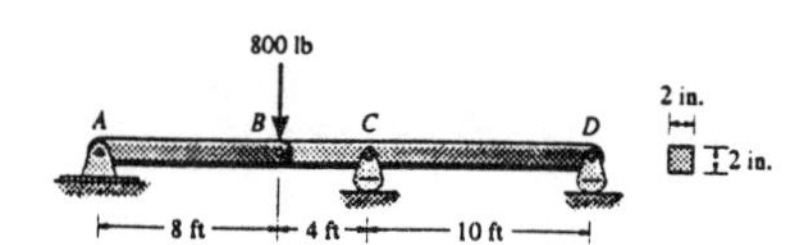

Support Reactions : As shown on FBD(a).

Moment Functions : As shown on FBD(b) and (c).

Bending Strain Energy : Applying 14 – 17, we have

$$U_i = \int_0^L \frac{M^2 dx}{2EI}$$

$$= \frac{1}{2EI}\left[\int_0^{4\,\text{ft}} (-800x_1)^2 dx_1 + \int_0^{10\,\text{ft}} (-320x_2)^2 dx_2\right]$$

$$= \frac{23.8933(10^6) \text{ lb}^2\cdot\text{ft}^3}{EI}$$

$$= \frac{23.8933(10^6)(12^3)}{29.0(10^6)\left[\frac{1}{12}(2)(2^3)\right]} = 1067.78 \text{ in}\cdot\text{lb}$$

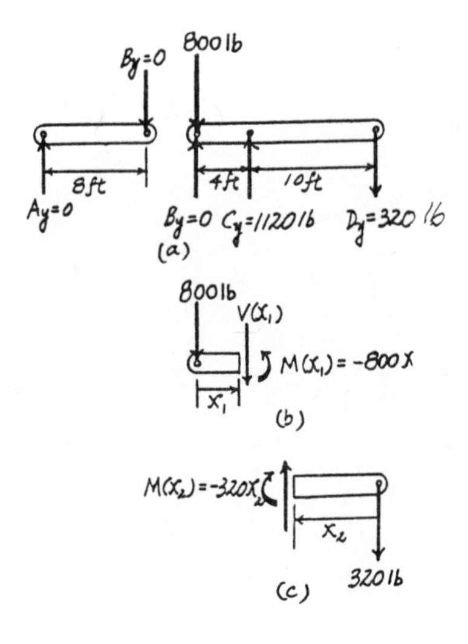

External Work : The external work done by 800 lb force is

$$U_e = \frac{1}{2}(800)(\Delta_B) = 400\Delta_B$$

Conservation of Energy :

$$U_e = U_i$$

$$400\Delta_B = 1067.78$$

$$\Delta_B = 2.67 \text{ in.} \qquad \textbf{Ans}$$

***14–28.** Determine the deflection of the beam at its center caused by shear. The shear modulus is G.

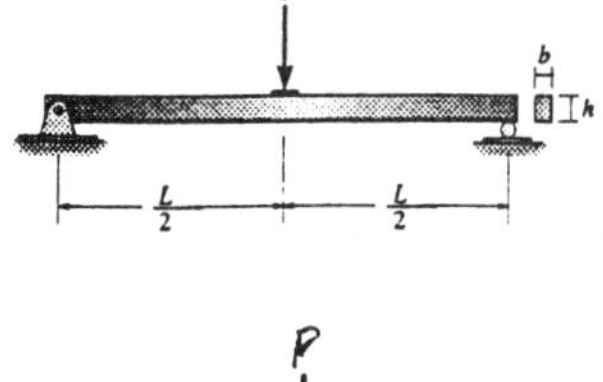

Support Reactions : As shown on FBD(a).

Shear Functions : As shown on FBD(b).

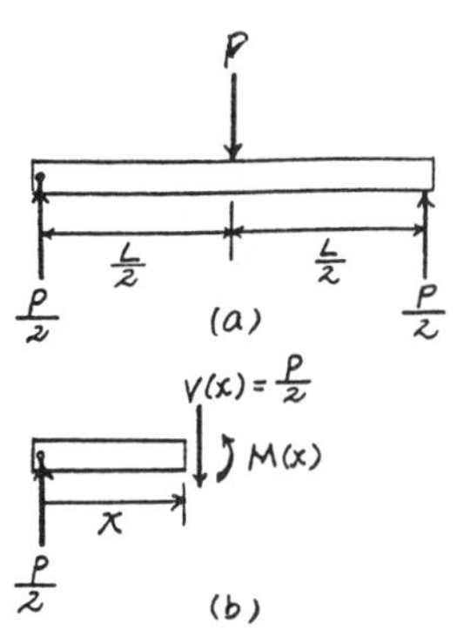

Shear Strain Energy : Applying 14 – 19 with $f_s = \frac{6}{5}$ for a rectangular section, we have

$$U_i = \int_0^L \frac{f_s V^2 dx}{2GA}$$

$$= \frac{1}{2bhG}\left[2\int_0^{\frac{L}{2}} \left(\frac{6}{5}\right)\left(\frac{P}{2}\right)^2 dx\right]$$

$$= \frac{3P^2L}{20bhG}$$

External Work : The external work done by force P is

$$U_e = \frac{1}{2}(P)\Delta$$

Conservation of Energy :

$$U_e = U_i$$

$$\frac{1}{2}(P)\Delta = \frac{3P^2L}{20bhG}$$

$$\Delta = \frac{3PL}{10bhG} \qquad \textbf{Ans}$$

14–29. Determine the deflection of the beam at its center caused by shear. The shear modulus is G.

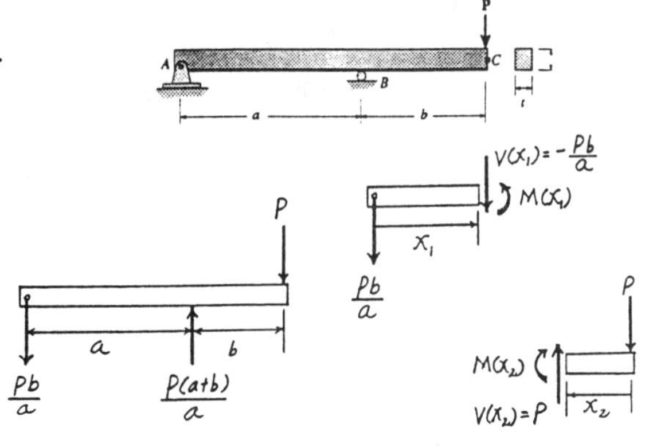

Support Reactions : As shown on FBD(a).

Shear Functions : As shown on FBD(b) and (c).

Shear Strain Energy : Applying 14 – 19 with $f_s = \frac{6}{5}$ for a rectangular section, we have

$$U_i = \int_0^L \frac{f_s V^2 dx}{2GA}$$

$$= \frac{1}{2thG}\left[\int_0^a \left(\frac{6}{5}\right)\left(-\frac{Pb}{a}\right)^2 dx_1 + \int_0^b \left(\frac{6}{5}\right)(P)^2 dx_2\right]$$

$$= \frac{3P^2b}{5thGa}(a+b)$$

External Work : The external work done by force P is

$$U_e = \frac{1}{2}(P)\Delta_C$$

Conservation of Energy :

$$U_e = U_i$$

$$\frac{1}{2}(P)\Delta_C = \frac{3P^2b}{5thGa}(a+b)$$

$$\Delta_C = \frac{6Pb}{5thGa}(a+b) \qquad \textbf{Ans}$$

14–30. The rod has a circular cross section with a moment of inertia I. If a vertical force **P** is applied at A, determine the vertical displacement at this point. Only consider the strain energy due to bending. The modulus of elasticity is E.

Bending Strain Energy : Applying 14 – 17 with $ds = r d\theta$, we have

$$U_i = \int_0^s \frac{M^2 ds}{2EI}$$
$$= \frac{1}{2EI}\int_0^{\pi} (Pr\sin\theta)^2 r d\theta$$
$$= \frac{P^2 r^3}{2EI}\int_0^{\pi} \sin^2\theta d\theta$$
$$= \frac{P^2 r^3}{4EI}\int_0^{\pi} (1-\cos 2\theta)\, d\theta$$
$$= \frac{\pi P^2 r^3}{4EI}$$

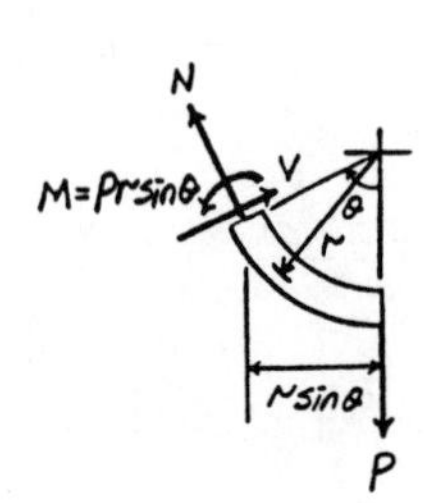

External Work : The external work done by force P is

$$U_e = \frac{1}{2}(P)(\Delta_A)$$

Conservation of Energy :

$$U_e = U_i$$
$$\frac{1}{2}(P)(\Delta_A) = \frac{\pi P^2 r^3}{4EI}$$

$$\Delta_A = \frac{\pi P r^3}{2EI} \qquad \textbf{Ans}$$

14–31. The coiled spring has n coils and is made of a material having a shear modulus G. Determine the stretch of the spring when it is subjected to the load **P**. Assume that the coils are close to each other so that $\theta \approx 0°$ and the deflection is caused entirely by the torsional stress in the coil.

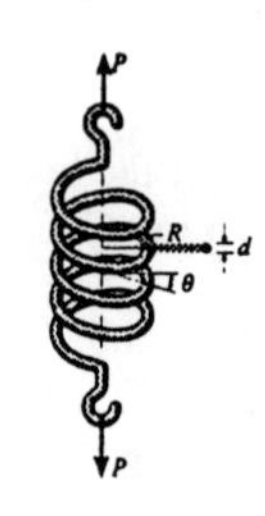

Bending Strain Energy : Applying 14 – 22, we have

$$U_i = \frac{T^2 L}{2GJ} = \frac{P^2 R^2 L}{2G\left[\frac{\pi}{32}(d^4)\right]} = \frac{16 P^2 R^2 L}{\pi d^4 G}$$

However, $L = n(2\pi R) = 2n\pi R$. Then

$$U_i = \frac{32 n P^2 R^3}{d^4 G}$$

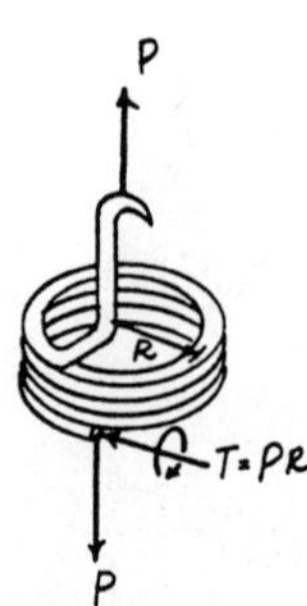

External Work : The external work done by force P is

$$U_e = \frac{1}{2}P\Delta$$

Conservation of Energy :

$$U_e = U_i$$
$$\frac{1}{2}P\Delta = \frac{32 n P^2 R^3}{d^4 G}$$

$$\Delta = \frac{64 n P R^3}{d^4 G} \qquad \textbf{Ans}$$

*14–32. A bar is 8 ft long and has a diameter of 0.5 in. If it is to be used to absorb energy in tension from an impact loading, determine the total amount of elastic energy that it can absorb if it is made of red brass C83400.

Elastic Strain Energy : The yielding axial force is $P_Y = \sigma_Y A$.
Applying Eq. 14 – 16, we have

$$U_i = \frac{N^2 L}{2AE} = \frac{(\sigma_Y A)^2 L}{2AE} = \frac{\sigma_Y^2 AL}{2E}$$

Substituting, we have

$$U_i = \frac{11.4^2\left[\frac{\pi}{4}(0.5^2)\right](8)(12)}{2[14.6(10^3)]} = 0.08389 \text{ in}\cdot\text{kip} = 6.99 \text{ ft}\cdot\text{lb} \quad \textbf{Ans}$$

14–33. Solve Prob. 14–32 if the bar is made of aluminum 2014-T6.

Elastic Strain Energy : The yielding axial force is $P_Y = \sigma_Y A$.
Applying Eq. 14 – 16, we have

$$U_i = \frac{N^2 L}{2AE} = \frac{(\sigma_Y A)^2 L}{2AE} = \frac{\sigma_Y^2 AL}{2E}$$

Substituting, we have

$$U_i = \frac{60^2\left[\frac{\pi}{4}(0.5^2)\right](8)(12)}{2[10.6(10^3)]} = 3.201 \text{ in}\cdot\text{kip} = 267 \text{ ft}\cdot\text{lb} \quad \textbf{Ans}$$

14–34. Determine the diameter of a red brass C83400 bar that is 8 ft long if it is to be used to absorb 800 ft · lb of energy in tension from an impact loading.

Elastic Strain Energy : The yielding axial force is $P_Y = \sigma_Y A$.
Applying Eq. 14 – 16, we have

$$U_i = \frac{N^2 L}{2AE} = \frac{(\sigma_Y A)^2 L}{2AE} = \frac{\sigma_Y^2 AL}{2E}$$

Substituting, we have

$$U_i = \frac{\sigma_Y^2 AL}{2E}$$

$$0.8(12) = \frac{11.4^2\left[\frac{\pi}{4}(d^2)\right](8)(12)}{2[14.6(10^3)]}$$

$$d = 5.35 \text{ in.} \quad \textbf{Ans}$$

14–35. The mass of 50 Mg is moving downward at $v = 0.5$ m/s when it is just over the top of the steel post having a length of 1 m and a cross-sectional area of 0.01 m^2. If the support is rigid, determine the maximum stress developed in the post and its maximum displacement. $E_{st} = 200$ GPa, $\sigma_Y = 600$ MPa.

Conservation of Energy : The equivalent spring constant for the post is $k = \frac{AE}{L} = \frac{0.01[200(10^9)]}{1} = 2.00(10^9)$ N/m.

$$U_e = U_i$$

$$\frac{1}{2}mv^2 + W\Delta_{max} = \frac{1}{2}k\Delta_{max}^2$$

$$\frac{1}{2}[50(10^3)](0.5^2) + [50(10^3)](9.81)\Delta_{max} = \frac{1}{2}[2.00(10^9)]\Delta_{max}^2$$

$$1.00(10^9)\Delta_{max}^2 - 490500\Delta_{max} - 6250 = 0$$

Solving for the positive root, we have

$$\Delta_{max} = 2.7573(10^{-3})\text{ m} = 2.76\text{ mm} \qquad \textbf{Ans}$$

Maximum Stress : The maximum axial force is $P_{max} = k\Delta_{max} = 2.00(10^9)[2.7573(10^{-3})] = 5514501$ N.

$$\sigma_{max} = \frac{P_{max}}{A} = \frac{5514501}{0.01} = 551\text{ MPa} \qquad \textbf{Ans}$$

Since $\sigma_{max} < \sigma_Y = 600$ MPa, the above analysis is valid.

***14–36.** Determine the speed v of the 50-Mg mass when it is just over the top of the steel post, if after impact the maximum stress developed in the post is 550 MPa. The post has a length of 1 m and a cross-sectional area of 0.01 m^2. Assume the support is rigid. $E_{st} = 200$ GPa, $\sigma_Y = 600$ MPa.

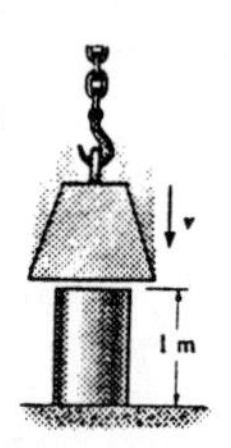

Maximum Axial Force :

$$P_{max} = \sigma_{max}A = 550(10^6)(0.01) = 5500\text{ kN}$$

Maximum Displacement : The equivalent spring constant for the post is $k = \frac{AE}{L} = \frac{0.01[200(10^9)]}{1} = 2.00(10^9)$ N/m.

$$\Delta_{max} = \frac{P_{max}}{k} = \frac{5500(10^3)}{2.00(10^9)} = 2.750(10^{-3})\text{ m}$$

Conservation of Energy :

$$U_e = U_i$$

$$\frac{1}{2}mv^2 + W\Delta_{max} = \frac{1}{2}k\Delta_{max}^2$$

$$\frac{1}{2}[50(10^3)](v^2) + [50(10^3)](9.81)[2.750(10^{-3})] = \frac{1}{2}[2.00(10^9)][2.750(10^{-3})]^2$$

$$v = 0.499\text{ m/s} \qquad \textbf{Ans}$$

14–37. The drop hammer of a pile driver has a mass of 150 kg. It is dropped from rest, 0.75 m from the top of a wooden pile having a diameter of 300 mm and a length of 6 m. Determine the maximum axial stress developed in the pile if the pile absorbs 70% of the impacting energy. Assume the material responds elastically. E_w = 11.5 GPa.

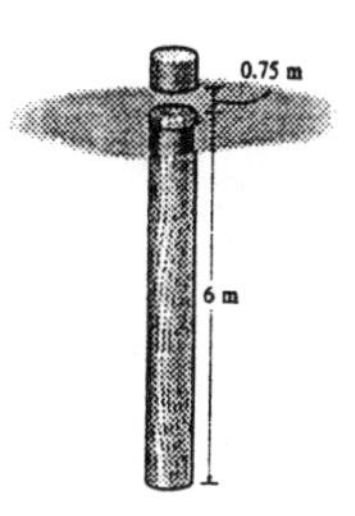

Conservation of Energy : The equivalent spring constant for the post is $k = \frac{AE}{L} = \frac{\frac{\pi}{4}(0.3^2)[11.5(10^9)]}{6} = 0.13548(10^9)$ N/m. Here, only 70% of the energy is conserved.

$$0.7U_e = U_i$$

$$0.7[W(h+\Delta_{max})] = \frac{1}{2}k\Delta_{max}^2$$

$$0.7[150(9.81)(0.75+\Delta_{max})] = \frac{1}{2}[0.13548(10^9)]\Delta_{max}^2$$

$$67.7406(10^6)\Delta_{max}^2 - 1030.05\Delta_{max} - 772.5375 = 0$$

Solving for the positive root, we have

$$\Delta_{max} = 3.3846(10^{-3}) \text{ m}$$

Maximum Stress : The maximum axial force is $P_{max} = k\Delta_{max}$ $= 0.13548(10^9)[3.3846(10^{-3})] = 458556$ N.

$$\sigma_{max} = \frac{P_{max}}{A} = \frac{458556}{\frac{\pi}{4}(0.3^3)} = 6.48 \text{ MPa} \qquad \textbf{Ans}$$

14–38. The A-36 steel bolt is required to absorb the energy of a 2-kg mass that falls h = 30 mm. If the bolt has a diameter of 4 mm, determine its required length L so the stress in the bolt does not exceed 150 MPa.

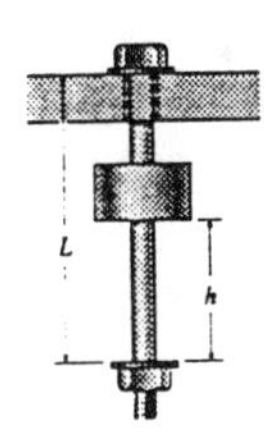

Maximum Stress : With $\Delta_{st} = \frac{WL}{AE} = \frac{2(9.81)(L)}{\frac{\pi}{4}(0.004^2)[200(10^9)]}$ $= 7.80655(10^{-6})L$ and $\sigma_{st} = \frac{W}{A} = \frac{2(9.81)}{\frac{\pi}{4}(0.004^2)} = 1.56131$ MPa, we have

$$\sigma_{max} = n\sigma_{st} \quad \text{where } n = 1+\sqrt{1+2\left(\frac{h}{\Delta_{st}}\right)}$$

$$150(10^6) = \left[1+\sqrt{1+2\left(\frac{0.03}{7.80655(10^{-6})L}\right)}\right][1.56131(10^6)]$$

$$L = 0.8504 \text{ m} = 850 \text{ mm} \qquad \textbf{Ans}$$

14–39. The A-36 steel bolt is required to absorb the energy of a 2-kg mass that falls $h = 30$ mm. If the bolt has a diameter of 4 mm and a length of $L = 200$ mm, determine if the stress in the bolt will exceed 175 MPa.

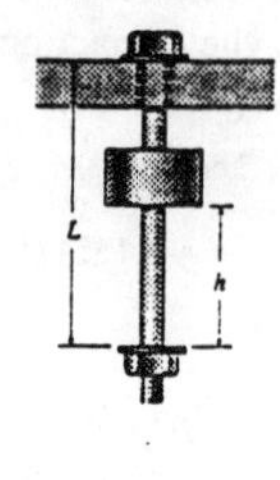

Maximum Stress : With

$$\Delta_{st} = \frac{WL}{AE} = \frac{2(9.81)(0.2)}{\frac{\pi}{4}(0.004^2)[200(10^9)]} = 1.56131(10^{-6})\text{ m}$$

$$\sigma_{st} = \frac{W}{A} = \frac{2(9.81)}{\frac{\pi}{4}(0.004^2)} = 1.56131\text{ MPa}$$

Applying Eq. 14 – 34, we have

$$n = 1 + \sqrt{1 + 2\left(\frac{h}{\Delta_{st}}\right)} = 1 + \sqrt{1 + 2\left(\frac{0.03}{1.56131(10^{-6})}\right)} = 197.04$$

Thus,

$$\sigma_{max} = n\sigma_{st} = 197.04(1.56131) = 307.6\text{ MPa}$$

Yes, σ_{max} exceeded 175 MPa. **Ans**

***14–40.** The A-36 steel bolt is required to absorb the energy of a 2-kg mass that falls along the 4-mm-diameter bolt shank that is $L = 150$ mm long. Determine the maximum height h of release so the stress in the bolt does not exceed 150 MPa.

Maximum Stress : With $\Delta_{st} = \frac{WL}{AE} = \frac{2(9.81)(0.15)}{\frac{\pi}{4}(0.004^2)[200(10^9)]}$

$= 1.17098(10^{-6})$ m and $\sigma_{st} = \frac{W}{A} = \frac{2(9.81)}{\frac{\pi}{4}(0.004^2)} = 1.56131$ MPa,

we have

$$\sigma_{max} = n\sigma_{st} \quad \text{where } n = 1 + \sqrt{1 + 2\left(\frac{h}{\Delta_{st}}\right)}$$

$$150(10^6) = \left[1 + \sqrt{1 + 2\left(\frac{h}{1.17098(10^{-6})}\right)}\right]\left[1.56131(10^6)\right]$$

$$h = 5.292(10^{-3})\text{ m} = 5.29\text{ mm}$$ **Ans**

14–41. The 50-lb weight is falling at 3 ft/s at the instant it is 2 ft above the spring and post assembly. Determine the maximum stress in the post if the spring has a stiffness of $k = 200$ kip/in. The post has a diameter of 3 in. and a modulus of elasticity of $E = 6.80(10^3)$ ksi. Assume the material will not yield.

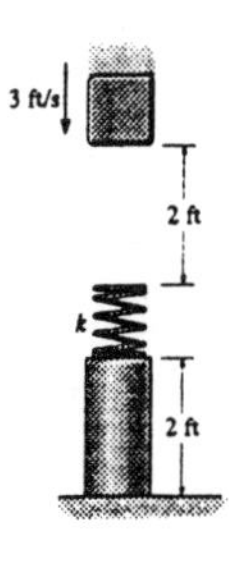

Equilibrium : This requires $F_{sp} = F_p$. Hence

$$k_{sp}\Delta_{sp} = k_p\Delta_p \quad \text{and} \quad \Delta_{sp} = \frac{k_p}{k_{sp}}\Delta_p \qquad [1]$$

Conservation of Energy : The equivalent sping constant for the post is $k_p = \frac{AE}{L} = \frac{\frac{\pi}{4}(3^2)\left[6.80(10^3)\right]}{2(12)} = 2.003\left(10^6\right)$ lb/in..

$$U_e = U_i$$

$$\frac{1}{2}mv^2 + W(h + \Delta_{max}) = \frac{1}{2}k_p\Delta_p^2 + \frac{1}{2}k_{sp}\Delta_{sp}^2 \qquad [2]$$

However, $\Delta_{max} = \Delta_p + \Delta_{sp}$. Then, Eq.[2] becomes

$$\frac{1}{2}mv^2 + W\left(h + \Delta_p + \Delta_{sp}\right) = \frac{1}{2}k_p\Delta_p^2 + \frac{1}{2}k_{sp}\Delta_{sp}^2 \qquad [3]$$

Substituting Eq.[1] into [3] yields

$$\frac{1}{2}mv^2 + W\left(h + \Delta_p + \frac{k_p}{k_{sp}}\Delta_p\right) = \frac{1}{2}k_p\Delta_p^2 + \frac{1}{2}\left(\frac{k_p^2}{k_{sp}}\Delta_p^2\right)$$

$$\frac{1}{2}\left(\frac{50}{32.2}\right)\left(3^2\right)(12) + 50\left[24 + \Delta_p + \frac{2.003(10^6)}{200(10^3)}\Delta_p\right]$$
$$= \frac{1}{2}\left[2.003\left(10^6\right)\right]\Delta_p^2 + \frac{1}{2}\left(\frac{\left[2.003(10^6)\right]^2}{200(10^3)}\right)\Delta_p^2$$

$$11.029\left(10^6\right)\Delta_p^2 - 550.69\Delta_p - 1283.85 = 0$$

Solving for positive root, we have

$$\Delta_p = 0.010814 \text{ in.}$$

Maximum Stress : The maximum axial force for the post is $P_{max} = k_p\Delta_p = 2.003\left(10^6\right)(0.010814) = 21.658$ kip.

$$\sigma_{max} = \frac{P_{max}}{A} = \frac{21.658}{\frac{\pi}{4}(3^2)} = 3.06 \text{ ksi} \qquad \textbf{Ans}$$

14–42. The collar has a mass of 5 kg and falls down the titanium Ti-6A1-4V bar. If the bar has a diameter of 20 mm, determine the maximum stress developed in the bar if the weight is (*a*) dropped from a height of $h = 1$ m, (*b*) released from a height $h \approx 0$, and (*c*) placed slowly on the flange at *A*.

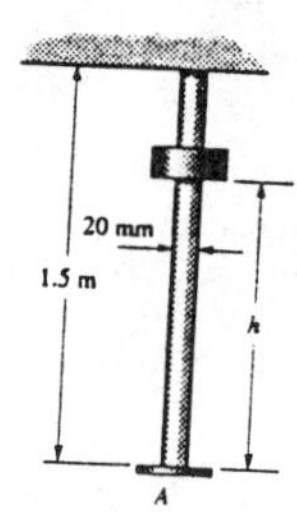

Maximum Stress : With $\Delta_{st} = \dfrac{WL}{AE} = \dfrac{5(9.81)(1.5)}{\frac{\pi}{4}(0.02^2)[120(10^9)]}$

$= 1.9516(10^{-6})$ m and $\sigma_{st} = \dfrac{W}{A} = \dfrac{5(9.81)}{\frac{\pi}{4}(0.02^2)} = 0.156131$ MPa and

Applying Eq. 14 – 34, we have

a)

$$n = 1+\sqrt{1+2\left(\frac{h}{\Delta_{st}}\right)} = 1+\sqrt{1+2\left(\frac{1}{1.9516(10^{-6})}\right)} = 1013.31$$

Thus,

$$\sigma_{max} = n\sigma_{st} = 1013.31(0.156131) = 158 \text{ MPa} \qquad \textbf{Ans}$$

b)

$$n = 1+\sqrt{1+2\left(\frac{h}{\Delta_{st}}\right)} = 1+\sqrt{1+2\left(\frac{0}{1.9516(10^{-6})}\right)} = 2$$

Thus,

$$\sigma_{max} = n\sigma_{st} = 2(0.156131) = 0.312 \text{ MPa} \qquad \textbf{Ans}$$

c)

$$\sigma_{max} = \sigma_{st} = 0.156 \text{ MPa} \qquad \textbf{Ans}$$

Since all of the $\sigma_{max} < \sigma_Y = 924$ MPa, the above analysis is valid.

14–43. The collar has a mass of 5 kg and falls down the titanium Ti-6A1-4V bar. If the bar has a diameter of 20 mm, determine if the weight can be released from rest at any point along the bar and not permanently damage the bar after striking the flange at *A*.

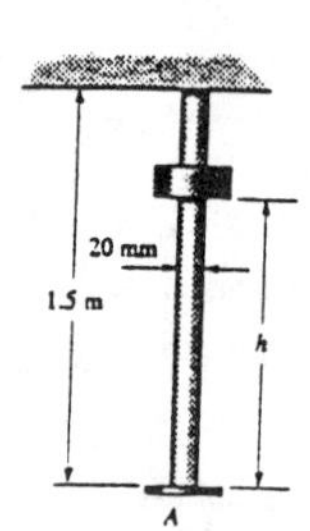

Maximum Stress : With $\Delta_{st} = \dfrac{WL}{AE} = \dfrac{5(9.81)(1.5)}{\frac{\pi}{4}(0.02^2)[120(10^9)]}$

$= 1.9516(10^{-6})$ m , $\sigma_{st} = \dfrac{W}{A} = \dfrac{5(9.81)}{\frac{\pi}{4}(0.02^2)} = 0.156131$ MPa and

$h = h_{max} = 1.5$ m. Applying Eq. 14 – 34 , we have

$$n = 1+\sqrt{1+2\left(\frac{h}{\Delta_{st}}\right)} = 1+\sqrt{1+2\left(\frac{1.5}{1.9516(10^{-6})}\right)} = 1240.83$$

Thus,

$$\sigma_{max} = n\sigma_{st} = 1240.83(0.156131) = 193.7 \text{ MPa}$$

Since $\sigma_{max} < \sigma_Y = 924$ MPa, the weight **can be released** from rest at any position along the bar without causing permanent damage to the bar. **Ans**

***14–44.** A cylinder having the dimensions shown is made of magnesium Am 1004-T61. If it is struck by a rigid block having a weight of 800 lb and traveling at 2 ft/s, determine the maximum stress in the cylinder. Neglect the mass of the cylinder.

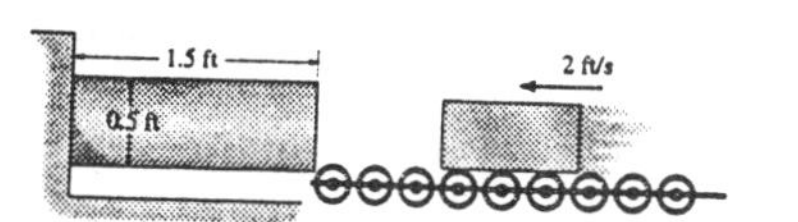

Conservation of Energy : The equivalent spring constant for the post is $k = \frac{AE}{L} = \frac{\frac{\pi}{4}(6^2)[6.48(10^6)]}{1.5(12)} = 10.1788(10^6)$ lb/in..

$$U_e = U_i$$

$$\frac{1}{2}mv^2 = \frac{1}{2}k\Delta_{max}^2$$

$$\left[\frac{1}{2}\left(\frac{800}{32.2}\right)(2^2)\right](12) = \frac{1}{2}\left[10.1788(10^6)\right]\Delta_{max}^2$$

$$\Delta_{max} = 0.01082 \text{ in.}$$

Maximum Stress : The maximum axial force is

$P_{max} = k\Delta_{max} = 10.1788(10^6)(0.01082) = 110175.5$ lb.

$$\sigma_{max} = \frac{P_{max}}{A} = \frac{110175.5}{\frac{\pi}{4}(6^2)} = 3897 \text{ psi} = 3.90 \text{ ksi} \qquad \textbf{Ans}$$

Since $\sigma_{max} < \sigma_Y = 22$ ksi, the above analysis is valid.

14–45. The sack of cement has a weight of 90 lb. If it is dropped from rest at a height of $h = 4$ ft onto the center of the W 10 × 39 structural steel A-36 beam, determine the maximum bending stress developed in the beam due to the impact. Also, what is the impact factor?

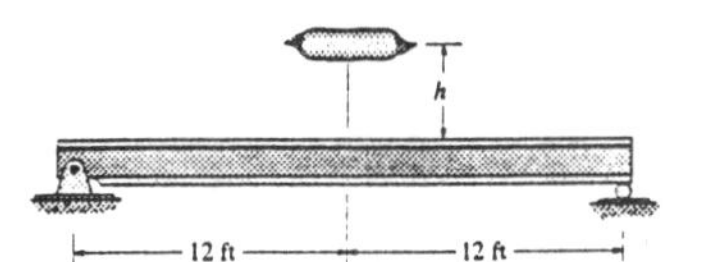

Impact Factor : From the table listed in Appendix C,

$$\Delta_{st} = \frac{PL^3}{48EI} = \frac{90[24(12)]^3}{48[29.0(10^6)](209)} = 7.3898(10^{-3}) \text{ in.}$$

Applying Eq. 14 – 34, we have

$$n = 1 + \sqrt{1 + 2\left(\frac{h}{\Delta_{st}}\right)}$$

$$= 1 + \sqrt{1 + 2\left(\frac{4(12)}{7.3898(10^{-3})}\right)}$$

$$= 114.98 = 115 \qquad \textbf{Ans}$$

Maximum Bending Stress : The maximum moment occurs at mid - span where $M_{max} = \frac{PL}{4} = \frac{90(24)(12)}{4} = 6480$ lb · in.

$$\sigma_{st} = \frac{M_{max}c}{I} = \frac{6480(9.92/2)}{209} = 153.78 \text{ psi}$$

Thus,

$$\sigma_{max} = n\sigma_{st} = 114.98(153.78) = 17.7 \text{ ksi} \qquad \textbf{Ans}$$

Since $\sigma_{max} < \sigma_Y = 36$ ksi, the above analysis is valid.

14–46. The sack of cement has a weight of 90 lb.lb. Determine the maximum height h from which it can bebe dropped from rest onto the center of the W 10 × 39;el structural steel A-36 beam so that the maximum bending to stress due to impact does not exceed 30 ksi.

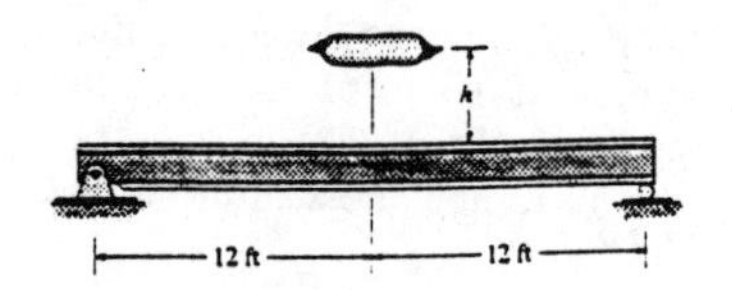

Maximum Bending Stress : The maximum moment occurs at mid-span where $M_{max} = \dfrac{PL}{4} = \dfrac{90(24)(12)}{4} = 6480 \text{ lb}\cdot\text{in.}$

$$\sigma_{st} = \frac{M_{max}\,c}{I} = \frac{6480(9.92/2)}{209} = 153.78 \text{ psi}$$

However,

$$\sigma_{max} = n\sigma_{st}$$
$$30\left(10^3\right) = n(153.78)$$
$$n = 195.08$$

Impact Factor : From the table listed in Appendix C,

$$\Delta_{st} = \frac{PL^3}{48EI} = \frac{90[24(12)]^3}{48[29.0(10^6)](209)} = 7.3898\left(10^{-3}\right) \text{ in.}$$

Applying Eq. 14–34, we have

$$n = 1 + \sqrt{1 + 2\left(\frac{h}{\Delta_{st}}\right)}$$
$$195.08 = 1 + \sqrt{1 + 2\left(\frac{h}{7.3898(10^{-3})}\right)}$$
$$h = 139.17 \text{ in.} = 11.6 \text{ ft} \qquad \textbf{Ans}$$

14–47. The weight of 175 lb is dropped from a height of 4 ft from the top of the A-36 steel beam. Determine the maximum deflection and maximum stress in the beam if the supporting springs at A and B each have a stiffness of k = 500 lb/in. The beam is 3 in. thick and 4 in. wide.

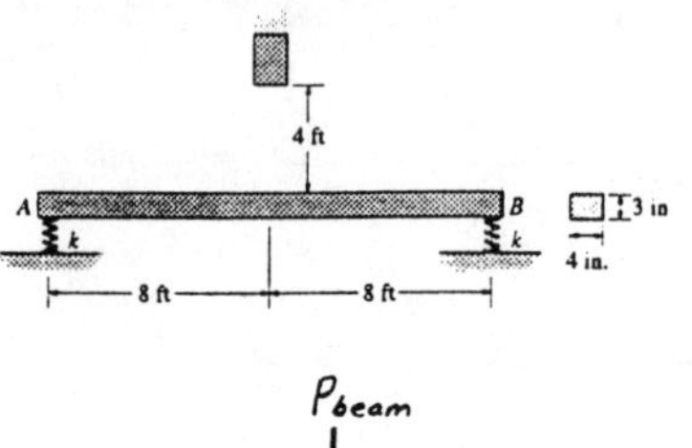

Equilibrium : This requires $F_{sp} = \dfrac{P_{beam}}{2}$. Then

$$k_{sp}\Delta_{sp} = \frac{k\Delta_{beam}}{2} \quad \text{or} \quad \Delta_{sp} = \frac{k}{2k_{sp}}\Delta_{beam} \qquad [1]$$

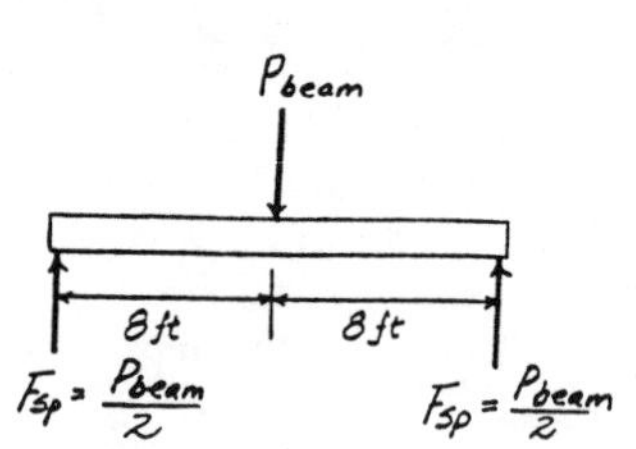

Conservation of Energy : The equivalent spring constant for the beam can be determined using the deflection table listed in the Appendix C.

$$k = \frac{48EI}{L^3} = \frac{48\left[29.0(10^6)\right]\left[\frac{1}{12}(4)(3^3)\right]}{[16(12)]^3} = 1770.02 \text{ lb/in.}$$

Thus,

$$U_e = U_i$$
$$W\left(h + \Delta_{sp} + \Delta_{beam}\right) = \frac{1}{2}k\Delta_{beam}^2 + 2\left(\frac{1}{2}k_{sp}\Delta_{sp}^2\right) \qquad [2]$$

Substituting Eq.[1] into [2] yields

$$W\left(h + \frac{k}{2k_{sp}}\Delta_{beam} + \Delta_{beam}\right) = \frac{1}{2}k\Delta_{beam}^2 + \frac{k^2}{4k_{sp}}\Delta_{beam}^2$$

$$175\left[4(12) + \frac{1770.02}{2(500)}\Delta_{beam} + \Delta_{beam}\right] = \frac{1}{2}(1770.02)\Delta_{beam}^2 + \frac{1770.02^2}{4(500)}\Delta_{beam}^2$$

Solving for the positive root, we have

$$\Delta_{beam} = 1.9526 \text{ in.}$$

Maximum Displacement : From Eq.[1], $\Delta_{sp} = \dfrac{1770.02}{2(500)}(1.9526) = 3.4561$ in..

$$\Delta_{max} = \Delta_{sp} + \Delta_{beam} = 3.4561 + 1.9526 = 5.41 \text{ in.} \qquad \textbf{Ans}$$

Maximum Stress : The maximum force on the beam is $P_{beam} = k\Delta_{beam}$ $= 1770.02(1.9526) = 3456.1$ lb. The maximum moment occurs at mid-span. $M_{max} = \dfrac{P_{beam}L}{4} = \dfrac{3456.1(16)(12)}{4} = 165893.3 \text{ lb}\cdot\text{in.}$

$$\sigma_{max} = \frac{M_{max}\,c}{I} = \frac{165893.3(1.5)}{\frac{1}{12}(4)(3^3)} = 26737 \text{ psi} = 27.6 \text{ ksi} \qquad \textbf{Ans}$$

Since $\sigma_{max} < \sigma_Y = 36$ ksi, the above analysis is valid.

*14–48. The weight of 175 lb is dropped from a height of 4 ft from the top of the A-36 steel beam. Determine the load factor n if the supporting springs at A and B each have a stiffness of $k = 300$ lb/in. The beam is 3 in. thick and 4 in. wide.

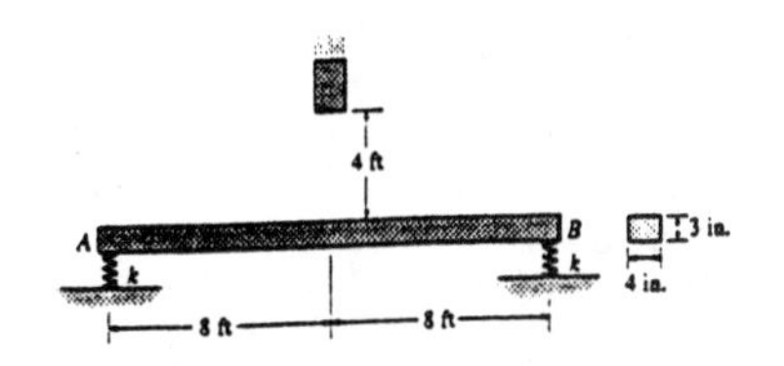

Equilibrium : This requires $F_{sp} = \dfrac{P_{beam}}{2}$. Then

$$k_{sp}\Delta_{sp} = \frac{k\Delta_{beam}}{2} \quad \text{or} \quad \Delta_{sp} = \frac{k}{2k_{sp}}\Delta_{beam} \qquad [1]$$

Conservation of Energy : The equivalent spring constant for the beam can be determined using the deflection table listed in the Appendix C.

$$k = \frac{48EI}{L^3} = \frac{48\left[29.0(10^6)\right]\left[\frac{1}{12}(4)(3^3)\right]}{[16(12)]^3} = 1770.02 \text{ lb/in.}$$

Thus,

$$U_e = U_i$$

$$W\left(h + \Delta_{sp} + \Delta_{beam}\right) = \frac{1}{2}k\Delta^2_{beam} + 2\left(\frac{1}{2}k_{sp}\Delta^2_{sp}\right) \qquad [2]$$

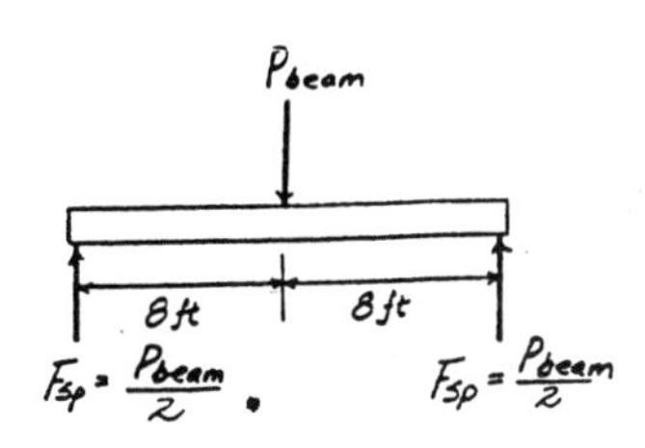

Substituting Eq.[1] into [2] yields

$$W\left(h + \frac{k}{2k_{sp}}\Delta_{beam} + \Delta_{beam}\right) = \frac{1}{2}k\Delta^2_{beam} + \frac{k^2}{4k_{sp}}\Delta^2_{beam}$$

$$175\left[4(12) + \frac{1770.02}{2(300)}\Delta_{beam} + \Delta_{beam}\right] = \frac{1}{2}(1770.02)\Delta^2_{beam} + \frac{1770.02^2}{4(300)}\Delta^2_{beam}$$

Solving for positive root, we have

$$\Delta_{beam} = 1.6521 \text{ in.}$$

Load Factor : The maximum force on the beam is $P_{beam} = k\Delta_{beam} = 1770.02(1.6521) = 2924.3$ lb.

$$n = \frac{P_{beam}}{W} = \frac{2924.3}{175} = 16.7 \qquad \textbf{Ans}$$

Maximum Stress : The maximum moment occurs at mid-span.

$$M_{max} = \frac{P_{beam}L}{4} = \frac{2924.3(16)(12)}{4} = 140367.2 \text{ lb}\cdot\text{in.}$$

$$\sigma_{max} = \frac{M_{max}c}{I} = \frac{140367.2(1.5)}{\frac{1}{12}(4)(3^3)} = 23.39 \text{ ksi} < \sigma_Y = 36 \text{ ksi } (\boldsymbol{O.K!})$$

14–49. The diver weighs 150 lb and while holding himself rigid strikes the end of a wooden diving board with a downward velocity of 4 ft/s when $h = 0$. Determine the maximum bending stress developed in the board. The board has a thickness of 1.5 in. and width of 1.5 ft. $E_w = 1.8(10^3)$ ksi. Assume the material does not yield.

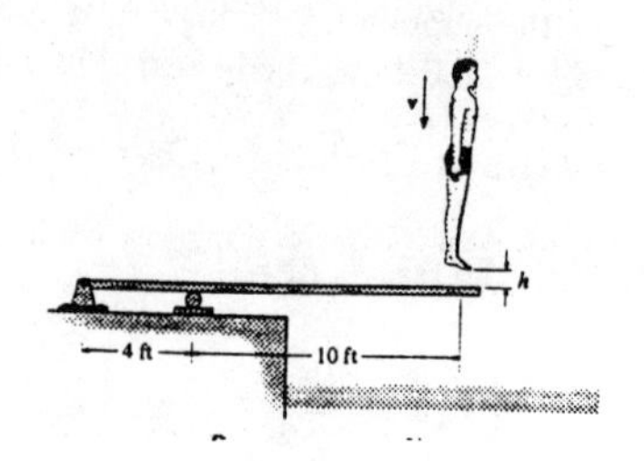

Static Displacment : The static displacement at the end of the diving board can be determined using the conservation of energy.

$$\frac{1}{2}P\Delta = \int_0^L \frac{M^2 dx}{2EI}$$

$$\frac{1}{2}(150)\,\Delta_{st} = \frac{1}{2EI}\left[\int_0^{4\text{ft}} (-375x_1)^2 dx_1 + \int_0^{10\text{ft}} (-150x_2)\,dx_2\right]$$

$$\Delta_{st} = \frac{70.0(10^3)\ \text{lb}\cdot\text{ft}^3}{EI}$$

$$= \frac{70.0(10^3)(12^3)}{1.8(10^6)\left[\frac{1}{12}(18)(1.5^3)\right]}$$

$$= 13.274\ \text{in.}$$

Conservation of Energy : The equivalent spring constant for the board is $k = \frac{W}{\Delta_{st}} = \frac{150}{13.274} = 11.30$ lb/in..

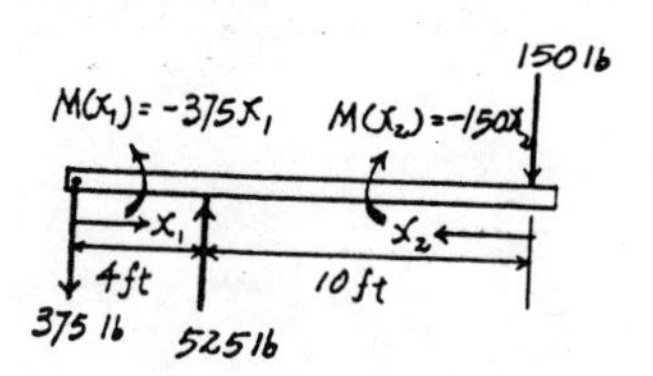

$$U_e = U_i$$

$$\frac{1}{2}mv^2 + W\Delta_{max} = \frac{1}{2}k\Delta_{max}^2$$

$$\left[\frac{1}{2}\left(\frac{150}{32.2}\right)(4^2)\right](12) + 150\Delta_{max} = \frac{1}{2}(11.30)\,\Delta_{max}^2$$

Solving for the positive root, we have

$$\Delta_{max} = 29.2538\ \text{in.}$$

Maximum Stress : The maximum force on the beam is $P_{max} = k\Delta_{max}$ $= 11.30(29.2538) = 330.57$ lb. The maximum moment occurs at the middle support. $M_{max} = 330.57(10)(12) = 39668.90$ lb · in.

$$\sigma_{max} = \frac{M_{max}c}{I} = \frac{39668.90(0.75)}{\frac{1}{12}(18)(1.5^3)} = 5877\ \text{psi} = 5.88\ \text{ksi} \qquad \textbf{Ans}$$

Note : The result will be somewhat inaccurate since the static displacement is so large.

14-50. The diver weighs 150 lb and while holding himself rigid strikes the end of the wooden diving board. Determine the maximum height h from which he can jump onto the board so that the maximum bending stress in the wood does not exceed 6 ksi. The board has a thickness of 1.5 in. and width of 1.5 ft. $E_w = 1.8(10^3)$ ksi.

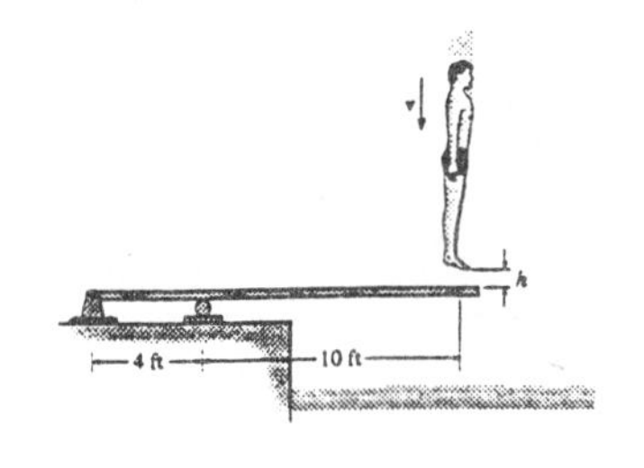

Static Displacment : The static displacement at the end of the diving board can be determined using the conservation of energy.

$$\frac{1}{2}P\Delta = \int_0^L \frac{M^2 dx}{2EI}$$

$$\frac{1}{2}(150)\Delta_{st} = \frac{1}{2EI}\left[\int_0^{4\text{ft}}(-375x_1)^2 dx_1 + \int_0^{10\text{ft}}(-150x_2)\,dx_2\right]$$

$$\Delta_{st} = \frac{70.0(10^3)\ \text{lb}\cdot\text{ft}^3}{EI}$$

$$= \frac{70.0(10^3)(12^3)}{1.8(10^6)\left[\frac{1}{12}(18)(1.5^3)\right]}$$

$$= 13.274\ \text{in.}$$

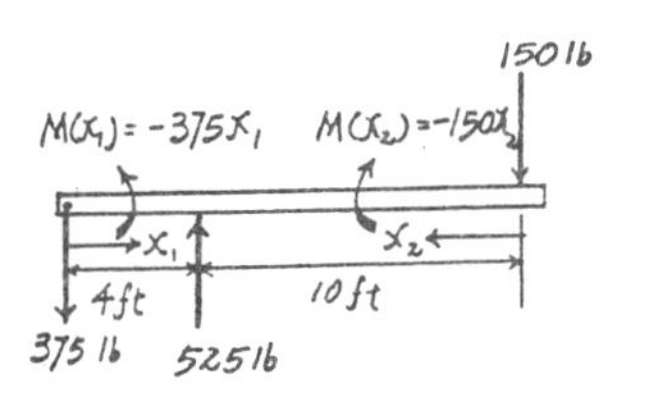

Maximum Stress : The maximum force on the beam is P_{max}. The maximum moment occurs at the middle support. $M_{max} = P_{max}(10)(12) = 120P_{max}$.

$$\sigma_{max} = \frac{M_{max}c}{I}$$

$$6(10^3) = \frac{120P_{max}(0.75)}{\frac{1}{12}(18)(1.5^3)}$$

$$P_{max} = 337.5\ \text{lb}$$

Conservation of Energy : The equivalent spring constant for the board is $k = \frac{W}{\Delta_{st}} = \frac{150}{13.274} = 11.30$ lb/in.. The maximum displacement at the end of the board is $\Delta_{max} = \frac{P_{max}}{k} = \frac{337.5}{11.30} = 29.867$ in.

$$U_e = U_i$$

$$W(h + \Delta_{max}) = \frac{1}{2}k\Delta_{max}^2$$

$$150(h + 29.867) = \frac{1}{2}(11.30)\left(29.867^2\right)$$

$$h = 3.73\ \text{in.} \qquad \textbf{Ans}$$

Note : The result will be somewhat inaccurate since the static displacement is so large.

14–51. The 75-lb block has a downward velocity of 2 ft/s when it is 3 ft from the top of the wooden beam. Determine the maximum bending stress in the beam due to the impact, and compute the maximum deflection of its end D. $E_w = 1.9(10^3)$ ksi. Assume the material will not yield.

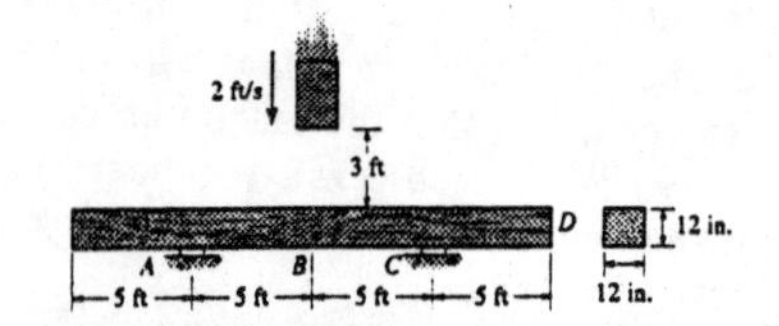

Conservation of Energy : The equivalent spring constant for the beam can be determined using the deflection table listed in Appendix C.

$$k = \frac{48EI}{L^3} = \frac{48\left[1.90(10^6)\right]\left[\frac{1}{12}(12)(12^3)\right]}{[10(12)]^3} = 91200 \text{ lb/in.}$$

Thus,

$$U_e = U_i$$

$$\frac{1}{2}mv^2 + W(h + \Delta_{max}) = \frac{1}{2}k\Delta_{max}^2$$

$$\left[\frac{1}{2}\left(\frac{75}{32.2}\right)(2^2)\right](12) + 75\left[3(12) + \Delta_{max}\right] = \frac{1}{2}(91200)\Delta_{max}^2$$

Solving for the positive root, we have

$$\Delta_{max} = 0.2467 \text{ in.}$$

Maximum Stress : The maximum force on the beam is $P_{max} = k\Delta_{max} = 91200(0.2467) = 22495.6$ lb $= 22.496$ kip. The maximum moment occurs at mid-span. $M_{max} = \frac{P_{max}L}{4} = \frac{22.496(10)(12)}{4} = 674.87$ kip · in.

$$\sigma_{max} = \frac{M_{max}c}{I} = \frac{674.87(6)}{\frac{1}{12}(12)(12^3)} = 2.34 \text{ ksi} \qquad \textbf{Ans}$$

Displacement : The maximum force on the beam is $P_{max} = k\Delta_{max} = 91200(0.2467) = 22495.6$ lb $= 22.496$ kip. From the deflection table listed on Appendix C, the slope at C is

$$\theta_C = \frac{P_{max}L^2}{16EI} = \frac{22.496[10(12)]^2}{16[1.9(10^3)]\left[\frac{1}{12}(12)(12^3)\right]} = 6.1665\left(10^{-3}\right) \text{ rad}$$

$$(\Delta_D)_{max} = \theta_C L_{CD} = 6.1665\left(10^{-3}\right)[5(12)] = 0.370 \text{ in.} \qquad \textbf{Ans}$$

***14–52.** The 75-lb block has a downward velocity of 2 ft/s when it is 3 ft from the top of the wood beam. Determine the maximum bending stress in the beam due to the impact, and compute the maximum deflection of point B. $E_w = 1.9(10^3)$ ksi.

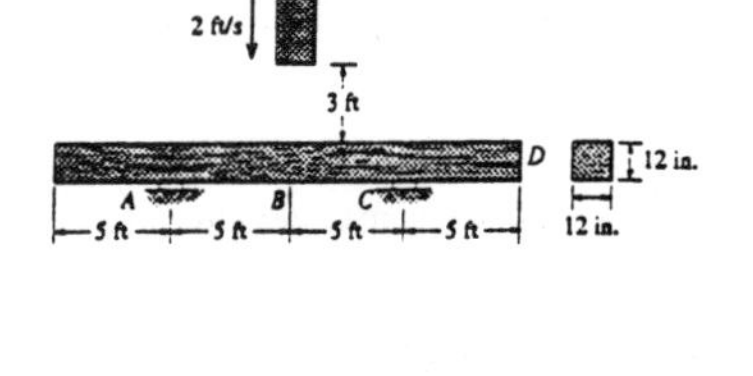

Conservation of Energy : The equivalent spring constant for the beam can be determined using the deflection table listed in the appendix C.

$$k = \frac{48EI}{L^3} = \frac{48[1.90(10^6)][\frac{1}{12}(12)(12^3)]}{[10(12)]^3} = 91200 \text{ lb/in.}$$

Thus,

$$U_e = U_i$$

$$\frac{1}{2}mv^2 + W(h + \Delta_{max}) = \frac{1}{2}k\Delta_{max}^2$$

$$\left[\frac{1}{2}\left(\frac{75}{32.2}\right)(2^2)\right](12) + 75[3(12) + \Delta_{max}] = \frac{1}{2}(91200)\Delta_{max}^2$$

Solving for the positive root, we have

$$\Delta_B = \Delta_{max} = 0.2467 \text{ in.} = 0.247 \text{ in.} \qquad \textbf{Ans}$$

Maximum Stress : The maximum force on the beam is $P_{max} = k\Delta_{max}$ $= 91200(0.2467) = 22495.6$ lb $= 22.496$ kip. The maximum moment occurs at mid-span. $M_{max} = \frac{P_{max}L}{4} = \frac{22.496(10)(12)}{4}$ $= 674.87$ kip · in.

$$\sigma_{max} = \frac{M_{max}c}{I} = \frac{674.87(6)}{\frac{1}{12}(12)(12^3)} = 2.34 \text{ ksi} \qquad \textbf{Ans}$$

14–53. Determine the maximum height h from which an 80-lb weight can be dropped onto the end of the A-36 steel W 6 × 12 beam without exceeding the maximum elastic stress.

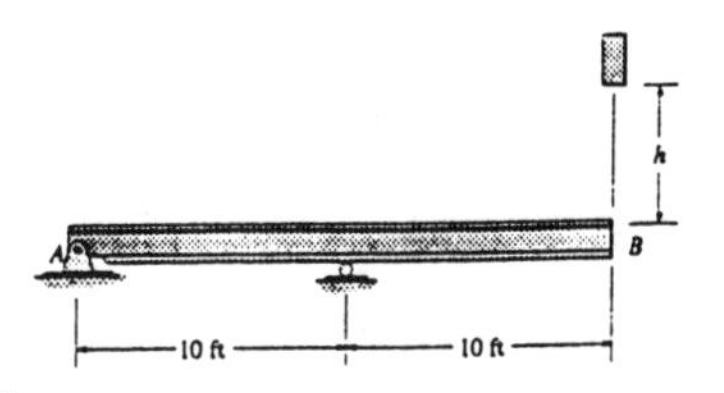

Static Displacment : The static displacement at the end of the beam can be determined using the conservation of energy.

$$\frac{1}{2}P\Delta = \int_0^L \frac{M^2 dx}{2EI}$$

$$\frac{1}{2}(80)\Delta_{st} = \frac{1}{2EI}\left[2\int_0^{10\text{ft}}(-80.0x)^2 dx\right]$$

$$\Delta_{st} = \frac{53.333(10^3)\text{ lb}\cdot\text{ft}^3}{EI}$$

$$= \frac{53.333(10^3)(12^3)}{29.0(10^6)(22.1)}$$

$$= 0.1438 \text{ in.}$$

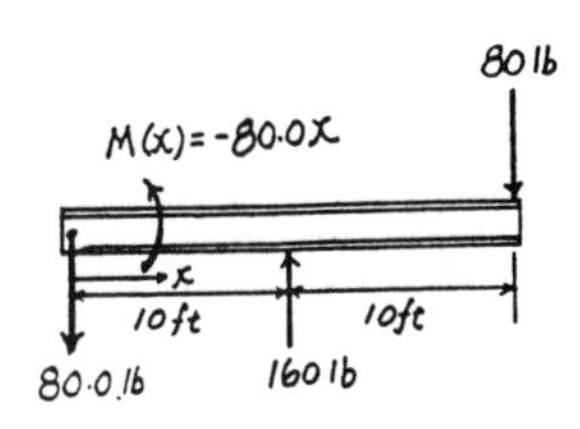

Maximum Stress : The maximum force on the beam is P_{max}. The maximum moment occurs at the middle support $M_{max} = P_{max}(10)(12)$ $= 120P_{max}$.

$$\sigma_{max} = \frac{M_{max}c}{I}$$

$$36(10^3) = \frac{120P_{max}\left(\frac{6.03}{2}\right)}{22.1}$$

$$P_{max} = 2199 \text{ lb}$$

Conservation of Energy : The equivalent spring constant for the beam is $k = \frac{W}{\Delta_{st}} = \frac{80}{0.1438} = 556.34$ lb/in. The maximum displacement at the end of the beam is $\Delta_{max} = \frac{P_{max}}{k} = \frac{2199}{556.34} = 3.9527$ in..

$$U_e = U_i$$

$$W(h + \Delta_{max}) = \frac{1}{2}k\Delta_{max}^2$$

$$80[h + 3.9527] = \frac{1}{2}(556.34)(3.9527^2)$$

$$h = 50.37 \text{ in.} = 4.20 \text{ ft} \qquad \textbf{Ans}$$

14–54. The 80-lb weight is dropped from rest at a height of $h = 4$ ft onto the end of the A-36 steel $W\,6 \times 12$ beam. Determine the maximum bending stress developed in the beam.

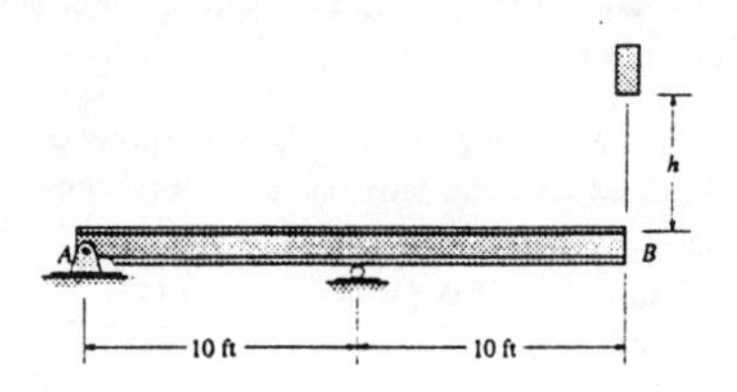

Static Displacment : The static displacement at the end of the beam can be determined using the conservation of energy method.

$$\frac{1}{2}P\Delta = \int_0^L \frac{M^2 dx}{2EI}$$

$$\frac{1}{2}(80)\Delta_{st} = \frac{1}{2EI}\left[2\int_0^{10\text{ft}}(-80.0x)^2\,dx\right]$$

$$\Delta_{st} = \frac{53.333(10^3)\ \text{lb}\cdot\text{ft}^3}{EI}$$

$$= \frac{53.333(10^3)(12^3)}{29.0(10^6)(22.1)}$$

$$= 0.1438\text{ in.}$$

Conservation of Energy : The equivalent spring constant for the beam is $k = \frac{W}{\Delta_{st}} = \frac{80}{0.1438} = 556.34$ lb/in..

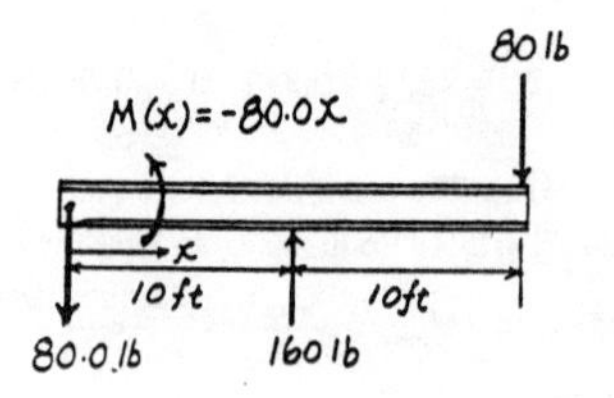

$$U_e = U_i$$

$$W(h+\Delta_{max}) = \frac{1}{2}k\Delta_{max}^2$$

$$80[4(12)+\Delta_{max}] = \frac{1}{2}(556.34)\Delta_{max}^2$$

Solving for the positive root, we have

$$\Delta_{max} = 3.862\text{ in.}$$

Maximum Stress : The maximum force on the beam is $P_{max} = k\Delta_{max} = 556.34(3.862) = 2148.6$ lb. The maximum moment occurs at the middle support. $M_{max} = 2148.6(10)(12) = 257830.9$ lb · in.

$$\sigma_{max} = \frac{M_{max}\,c}{I} = \frac{257830.9\left(\frac{6.03}{2}\right)}{22.1} = 35175\text{ psi} = 35.2\text{ ksi} \qquad \textbf{Ans}$$

Since $\sigma_{max} < \sigma_Y = 36$ ksi, the above analysis is valid.

14–55. The 100-lb block is falling at 2 ft/s when it is 3 ft above the spring. Determine the maximum deflection of the A-36 steel $W10 \times 12$ beam and the maximum bending stress in the beam due to the impact. The spring has a stiffness of $k = 1000$ lb/in.

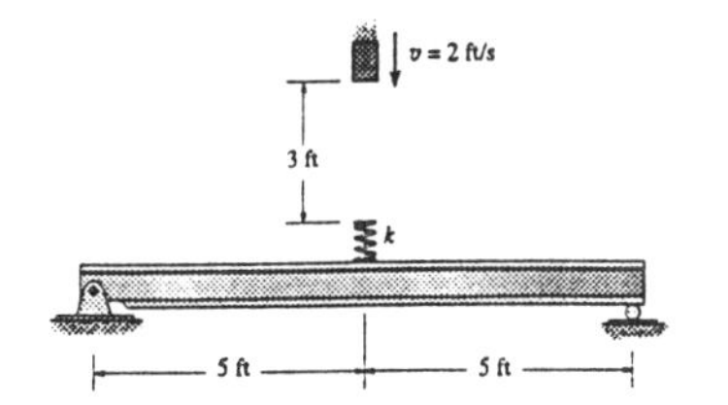

Equilibrium : This requires $F_{sp} = P_{beam}$. Then

$$k_{sp}\Delta_{sp} = k\Delta_{beam} \quad \text{or} \quad \Delta_{sp} = \frac{k}{k_{sp}}\Delta_{beam} \qquad [1]$$

Conservation of Energy : The equivalent spring constant for the beam can be determined using the deflection table listed in the Appendix C.

$$k = \frac{48EI}{L^3} = \frac{48[29.0(10^6)](53.8)}{[10(12)]^3} = 43338.89 \text{ lb/in.}$$

Thus,

$$U_e = U_i$$

$$\frac{1}{2}mv^2 + W\left(h + \Delta_{sp} + \Delta_{beam}\right) = \frac{1}{2}k\Delta_{beam}^2 + \frac{1}{2}k_{sp}\Delta_{sp}^2 \qquad [2]$$

Substituting Eq.[1] into [2] yields

$$\frac{1}{2}mv^2 + W\left(h + \frac{k}{k_{sp}}\Delta_{beam} + \Delta_{beam}\right) = \frac{1}{2}k\Delta_{beam}^2 + \frac{k^2}{2k_{sp}}\Delta_{beam}^2$$

$$\left[\frac{1}{2}\left(\frac{100}{32.2}\right)(2^2)\right](12) + 100\left[3(12) + \frac{43338.89}{1000}\Delta_{beam} + \Delta_{beam}\right]$$

$$= \frac{1}{2}(43338.89)\Delta_{beam}^2 + \frac{43338.89^2}{2(1000)}\Delta_{beam}^2$$

$$960799.1\Delta_{beam}^2 - 4433.89\Delta_{beam} - 3674.5 = 0$$

Solving for positive root, we have

$$\Delta_{beam} = 0.06419 \text{ in.} = 0.0642 \text{ in.} \qquad \textbf{Ans}$$

Maximum Stress : The maximum force on the beam is $P_{beam} = k\Delta_{beam}$ $= 43338.89(0.06419) = 2782.0$ lb. The maximum moment occurs at mid-span. $M_{max} = \frac{P_{beam}L}{4} = \frac{2782.0(10)(12)}{4} = 83461.1 \text{ lb} \cdot \text{in.}$

$$\sigma_{max} = \frac{M_{max}c}{I} = \frac{83461.1\left(\frac{9.87}{2}\right)}{53.8} = 7655.8 \text{ psi} = 7.66 \text{ ksi} \qquad \textbf{Ans}$$

Since $\sigma_{max} < \sigma_Y = 36$ ksi, the above analysis is valid.

***14–56.** The car bumper is made of polycarbonate-polybutylene terephthalate. If $E = 2.0$ GPa, determine the maximum deflection and maximum stress in the bumper if it strikes the rigid post when the car is coasting at $v = 0.75$ m/s. The car has a mass of 1.80 Mg, and the bumper can be considered simply supported on two spring supports connected to the rigid frame of the car. For the bumper take $I = 300(10^{6})$ mm^4, $c = 75$ mm, $\sigma_Y = 30$ MPa and $k = 1.5$ MN/m

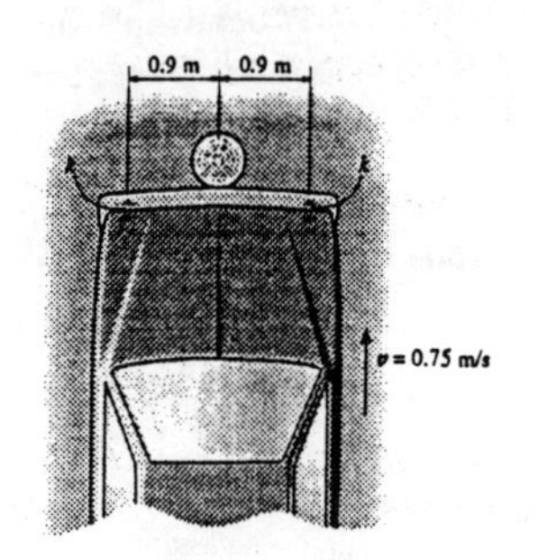

Equilibrium : This requires $F_{sp} = \frac{P_{beam}}{2}$. Then

$$k_{sp}\Delta_{sp} = \frac{k\Delta_{beam}}{2} \quad \text{or} \quad \Delta_{sp} = \frac{k}{2k_{sp}}\Delta_{beam} \qquad [1]$$

Conservation of Energy : The equivalent spring constant for the beam can be determined using the deflection table listed in the Appendix C.

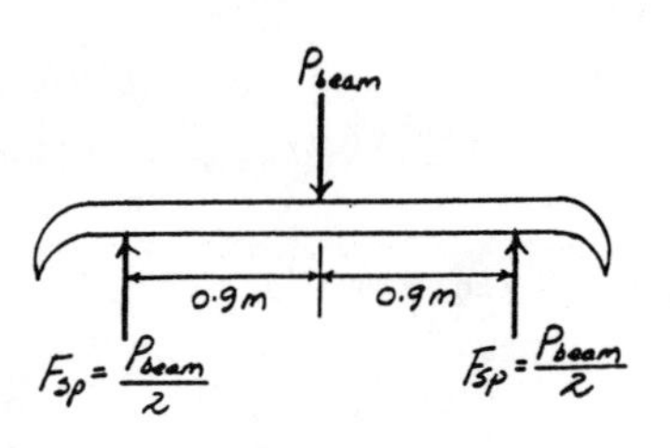

$$k = \frac{48EI}{L^3} = \frac{48[2(10^9)][300(10^{-6})]}{1.8^3} = 4\,938\,271.6 \text{ N/m}$$

Thus,

$$U_e = U_i$$

$$\frac{1}{2}mv^2 = \frac{1}{2}k\Delta^2_{beam} + 2\left(\frac{1}{2}k_{sp}\Delta^2_{sp}\right) \qquad [2]$$

Substitute Eq. [1] into [2] yields

$$\frac{1}{2}mv^2 = \frac{1}{2}k\Delta^2_{beam} + \frac{k^2}{4k_{sp}}\Delta^2_{beam}$$

$$\frac{1}{2}(1800)\left(0.75^2\right) = \frac{1}{2}(4\,93\,8271.6)\Delta^2_{beam} + \frac{(4\,93\,8271.6)^2}{4[1.5(10^6)]}\Delta^2_{beam}$$

$$\Delta_{beam} = 8.8025\left(10^{-3}\right) \text{ m}$$

Maximum Displacement : From Eq. [1], $\Delta_{sp} = \frac{4\,938\,271.6}{2[1.5(10^6)]}\left[8.8025\left(10^{-3}\right)\right]$ $= 0.014490$ m.

$$\Delta_{max} = \Delta_{sp} + \Delta_{beam}$$
$$= 0.014490 + 8.8025\left(10^{-3}\right)$$
$$= 0.02329 \text{ m} = 23.3 \text{ mm} \qquad \textbf{Ans}$$

Maximum Stress : The maximum force on the beam is $P_{beam} = k\Delta_{beam}$ $= 4\,938\,271.6\left[8.8025\left(10^{-3}\right)\right] = 43\,469.3$ N. The maximum moment occurs at mid - span. $M_{max} = \frac{P_{beam}L}{4} = \frac{43\,469.3(1.8)}{4} = 19\,561.2$ N · m.

$$\sigma_{max} = \frac{M_{max}c}{I} = \frac{19\,561.2(0.075)}{300(10^{-6})} = 4.89 \text{ MPa} \qquad \textbf{Ans}$$

Since $\sigma_{max} < \sigma_Y = 30$ MPa, the above analysis is valid.

14–57. The 150-lb weight has a velocity of 4 ft/s at a height of 4 ft from the top of the A-36 steel beam. Determine the maximum deflection and maximum bending stress in the beam if the supporting springs at A and B each have a stiffness of $k = 500$ lb/in.

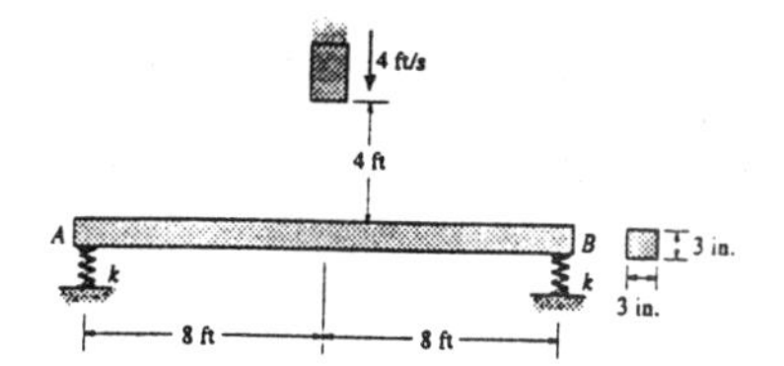

Equilibrium : This requires $F_{sp} = \dfrac{P_{beam}}{2}$. Then

$$k_{sp}\Delta_{sp} = \frac{k\Delta_{beam}}{2} \quad \text{or} \quad \Delta_{sp} = \frac{k}{2k_{sp}}\Delta_{beam} \qquad [1]$$

Conservation of Energy : The equivalent spring constant for the beam can be determined using the deflection table listed in Appendix C.

$$k = \frac{48EI}{L^3} = \frac{48\left[29.0(10^6)\right]\left[\frac{1}{12}(3)(3^3)\right]}{[16(12)]^3} = 1327.51 \text{ lb/in.}$$

Thus,

$$U_e = U_i$$

$$\frac{1}{2}mv^2 + W\left(h + \Delta_{sp} + \Delta_{beam}\right) = \frac{1}{2}k\Delta_{beam}^2 + 2\left(\frac{1}{2}k_{sp}\Delta_{sp}^2\right) \qquad [2]$$

Substituting Eq.[1] into [2] yields

$$\frac{1}{2}mv^2 + W\left(h + \frac{k}{2k_{sp}}\Delta_{beam} + \Delta_{beam}\right) = \frac{1}{2}k\Delta_{beam}^2 + \frac{k^2}{4k_{sp}}\Delta_{beam}^2$$

$$\left[\frac{1}{2}\left(\frac{150}{32.2}\right)\left(4^2\right)\right](12) + 150\left[4(12) + \frac{1327.51}{2(500)}\Delta_{beam} + \Delta_{beam}\right]$$

$$= \frac{1}{2}(1327.51)\Delta_{beam}^2 + \frac{1327.51^2}{4(500)}\Delta_{beam}^2$$

$$1544.90\Delta_{beam}^2 - 349.13\Delta_{beam} - 7647.20 = 0$$

Solving for the positive root, we have

$$\Delta_{beam} = 2.3407 \text{ in.}$$

Maximum Displacement : From Eq.[1], $\Delta_{sp} = \dfrac{1327.51}{2(500)}(2.3407) = 3.1073$ in.

$$\Delta_{max} = \Delta_{sp} + \Delta_{beam} = 3.1073 + 2.3407 = 5.45 \text{ in.} \qquad \textbf{Ans}$$

Maximum Stress : The maximum force on the beam is $P_{beam} = k\Delta_{beam} = 1327.51(2.3407) = 3107.3$ lb. The maximum moment occurs at mid-span. $M_{max} = \dfrac{P_{beam}L}{4} = \dfrac{3107.3(16)(12)}{4} = 149\,151.6$ lb · in.

$$\sigma_{max} = \frac{M_{max}c}{I} = \frac{149\,151.6(1.5)}{\frac{1}{12}(3)(3^3)} = 33\,145 \text{ psi} = 33.1 \text{ ksi} \qquad \textbf{Ans}$$

Since $\sigma_{max} < \sigma_Y = 36$ ksi, the above analysis is valid.

14-58. The 150-lb weight has a velocity of 4 ft/s at a height of 4 ft from the top of the A-36 steel beam. Determine the load factor n if the supporting springs at A and B each have a stiffness of $k = 300$ lb/in.

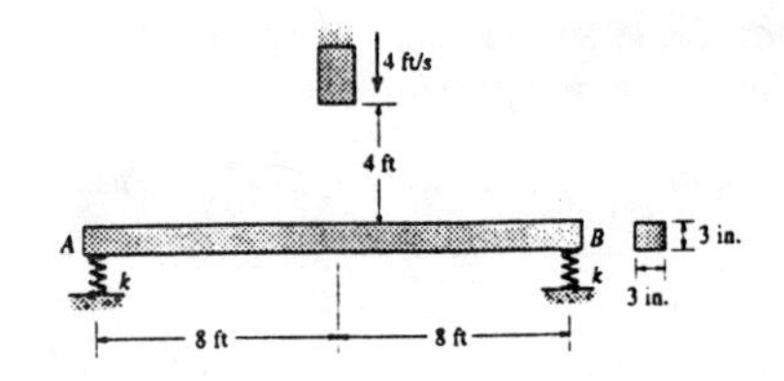

Equilibrium : This requires $F_{sp} = \frac{P_{beam}}{2}$. Then

$$k_{sp}\Delta_{sp} = \frac{k\Delta_{beam}}{2} \quad \text{or} \quad \Delta_{sp} = \frac{k}{2k_{sp}}\Delta_{beam} \qquad [1]$$

Conservation of Energy : The equivalent spring constant for the beam can be determined using the deflection table listed in Appendix C.

$$k = \frac{48EI}{L^3} = \frac{48\left[29.0(10^6)\right]\left[\frac{1}{12}(3)(3^3)\right]}{[16(12)]^3} = 1327.51 \text{ lb/in.}$$

Thus,

$$U_e = U_i$$

$$\frac{1}{2}mv^2 + W\left(h + \Delta_{sp} + \Delta_{beam}\right) = \frac{1}{2}k\Delta^2_{beam} + 2\left(\frac{1}{2}k_{sp}\Delta^2_{sp}\right) \qquad [2]$$

Substituting Eq.[1] into [2] yields

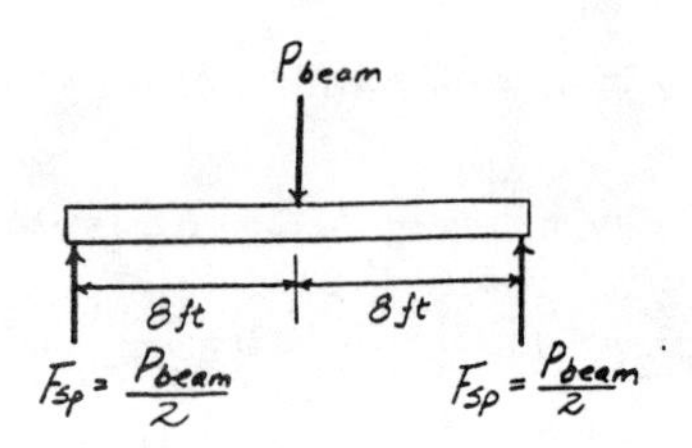

$$\frac{1}{2}mv^2 + W\left(h + \frac{k}{2k_{sp}}\Delta_{beam} + \Delta_{beam}\right) = \frac{1}{2}k\Delta^2_{beam} + \frac{k^2}{4k_{sp}}\Delta^2_{beam}$$

$$\left[\frac{1}{2}\left(\frac{150}{32.2}\right)(4^2)\right](12) + 150\left[4(12) + \frac{1327.51}{2(300)}\Delta_{beam} + \Delta_{beam}\right]$$

$$= \frac{1}{2}(1327.51)\Delta^2_{beam} + \frac{1327.51^2}{4(300)}\Delta^2_{beam}$$

$$2132.34\Delta^2_{beam} - 481.878\Delta_{beam} - 7647.20 = 0$$

Solving for positive root, we have

$$\Delta_{beam} = 2.0101 \text{ in.}$$

Load Factor : The maximum force on the beam is $P_{beam} = k\Delta_{beam} = 1327.51(2.0101) = 2668.5$ lb.

$$n = \frac{P_{beam}}{W} = \frac{2668.5}{150} = 17.8 \qquad \textbf{Ans}$$

Maximum Stress : The maximum moment occurs at mid-span.

$$M_{max} = \frac{P_{beam}L}{4} = \frac{2668.5(16)(12)}{4} = 128085.9 \text{ lb}\cdot\text{in.}$$

$$\sigma_{max} = \frac{M_{max}c}{I} = \frac{128\,085.9(1.5)}{\frac{1}{12}(3)(3^3)} = 28\,464 \text{ psi} = 28.5 \text{ ksi}$$

Since $\sigma_{max} < \sigma_Y = 36$ ksi, the above analysis is valid.

14-59. Determine the horizontal displacement of joint B. Each A-36 steel member has a cross-sectional area of 2 in^2.

Member Real Forces N : As shown on figure(a).

Member Virtual Forces n : As shown on figure(b).

Virtual - Work Equation : Applying Eq. 14–39, we have

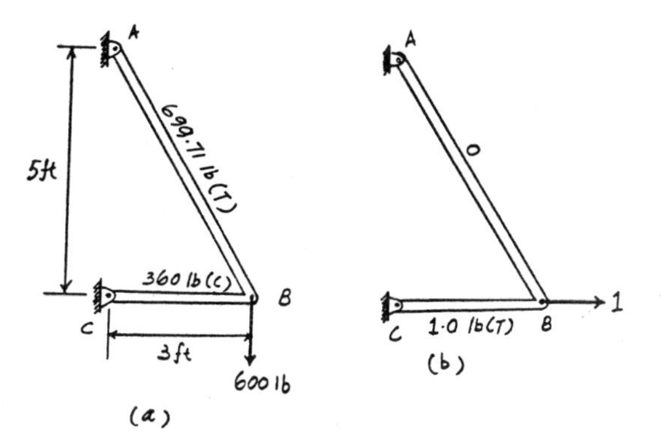

$$1 \cdot \Delta = \sum \frac{nNL}{AE}$$

$$1 \text{ lb} \cdot (\Delta_B)_h = \frac{1}{AE}[1.00(-360)(3)(12) + 0(699.71)(5.8310)(12)]$$

$$1 \text{ lb} \cdot (\Delta_B)_h = -\frac{12960 \text{ lb}^2 \cdot \text{in}}{AE}$$

$$(\Delta_B)_h = -\frac{12960}{2[29.0(10^6)]}$$

$$= -0.223(10^{-3}) \text{ in.} = 0.223(10^{-3}) \text{ in.} \leftarrow \qquad \textbf{Ans}$$

***14-60.** Determine the vertical displacement of joint B. Each A-36 steel member has a cross-sectional area of 2 in^2.

Member Real Forces N : As shown on figure(a).

Member Virtual Forces n : As shown on figure(b).

Virtual - Work Equation : Applying Eq. 14–39, we have

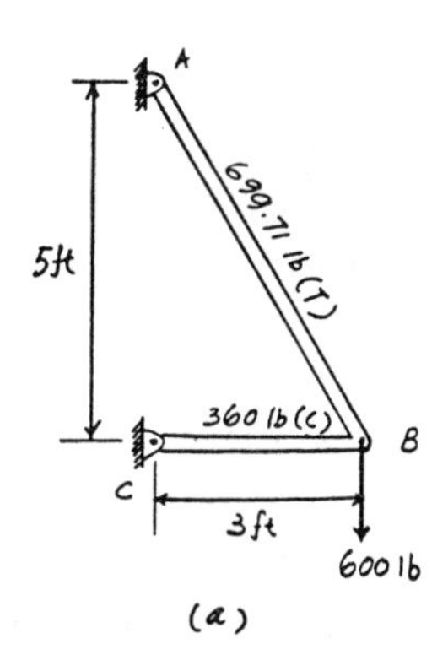

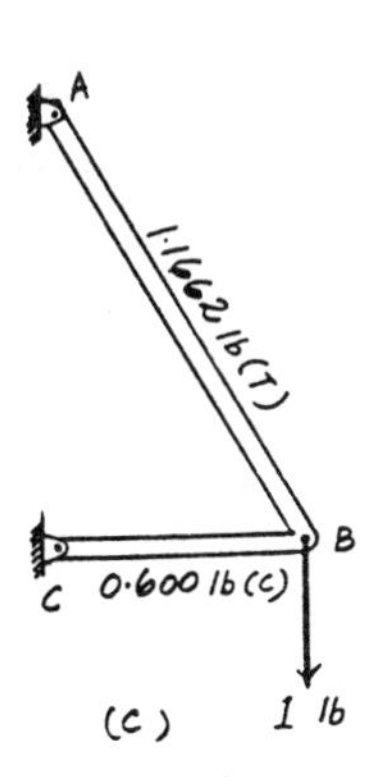

$$1 \cdot \Delta = \sum \frac{nNL}{AE}$$

$$1 \text{ lb} \cdot (\Delta_B)_v = \frac{1}{AE}[(-0.600)(-360)(3)(12) + 1.1662(699.71)(5.8310)(12)]$$

$$1 \text{ lb} \cdot (\Delta_B)_v = \frac{64872.68 \text{ lb}^2 \cdot \text{in}}{AE}$$

$$(\Delta_B)_v = \frac{64872.68}{2[29.0(10^6)]}$$

$$= 1.12(10^{-3}) \text{ in.} \downarrow \qquad \textbf{Ans}$$

14–61. Determine the horizontal displacement of point B. Each A-36 steel member has a cross-sectional area of 2 in^2.

Member Real Forces N : As shown on figure (a).

Member Virtual Forces n : As shown on figure (b).

Virtual - Work Equation : Applying Eq. 14 – 39, we have

$$1 \cdot \Delta = \sum \frac{nNL}{AE}$$

$$1 \text{ lb} \cdot (\Delta_B)_h = \frac{1}{AE}[0.8333(166.67)(10)(12) + (-0.8333)(-166.67)(10)(12) + 0.500(100)(12)(12)]$$

$$1 \text{ lb} \cdot (\Delta_B)_h = \frac{40533.33 \text{ lb}^2 \cdot \text{in}}{AE}$$

$$(\Delta_B)_h = \frac{40533.33}{2[29.0(10^6)]} = 0.699(10^{-3}) \text{ in.} \rightarrow \quad \textbf{Ans}$$

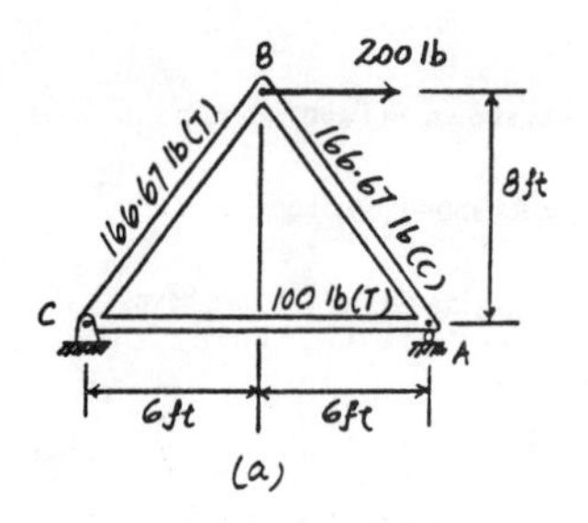

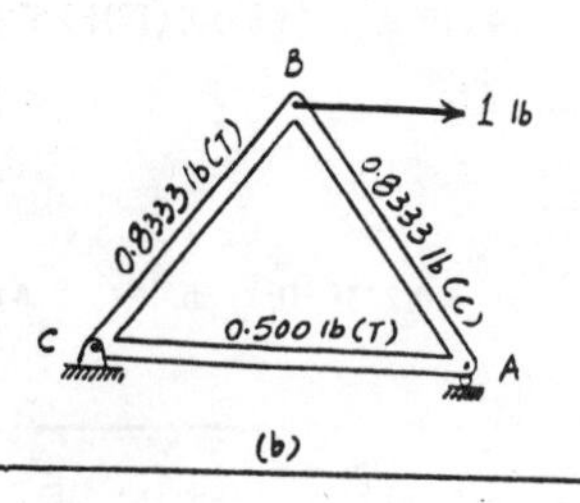

14–62. Determine the vertical displacement of point B. Each A-36 steel member has a cross-sectional area of 2 in^2.

Member Real Forces N : As shown on figure (a).

Member Virtual Forces n : As shown on figure (b).

Virtual - Work Equation : Applying Eq. 14 – 39, we have

$$1 \cdot \Delta = \sum \frac{nNL}{AE}$$

$$1 \text{ lb} \cdot (\Delta_B)_v = \frac{1}{AE}[(-0.625)(166.67)(10)(12) + (-0.625)(-166.67)(10)(12) + 0.375(100)(12)(12)]$$

$$1 \text{ lb} \cdot (\Delta_B)_v = \frac{5400 \text{ lb}^2 \cdot \text{in}}{AE}$$

$$(\Delta_B)_v = \frac{5400}{2[29.0(10^6)]} = 0.0931(10^{-3}) \text{ in.} \downarrow \quad \textbf{Ans}$$

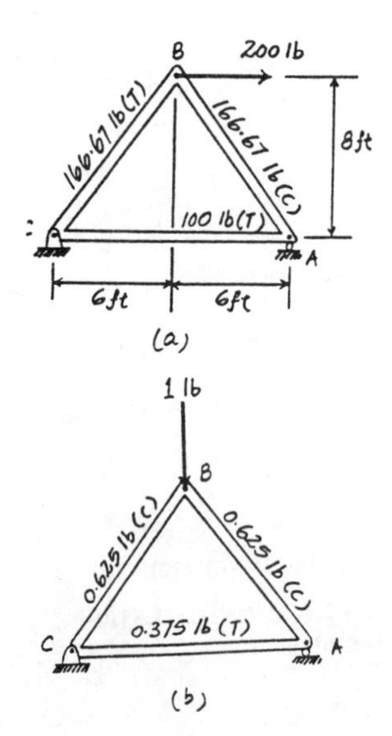

14–63. Determine the vertical displacement of point D. Each A-36 steel member has a cross-sectional area of 300 mm^2.

Member Real Forces N : As shown on figure (a).

Member Virtual Forces n : As shown on figure (b).

Virtual - Work Equation : Applying Eq. 14 – 39, we have

Member	n	N	L	nNL
AB	0.9014	$27.04(10^3)$	3.6056	$87.885(10^3)$
BC	0.9014	$27.04(10^3)$	3.6056	$87.885(10^3)$
CD	−0.750	$-22.5(10^3)$	3	$50.625(10^3)$
AD	−0.750	$-22.5(10^3)$	3	$50.625(10^3)$
BD	−1.00	$-30.0(10^3)$	2	$60.0(10^3)$
				$\sum 337.021(10^3)\ \text{N}^2\cdot\text{m}$

$$1\cdot\Delta = \sum\frac{nNL}{AE}$$

$$1\ \text{N}\cdot(\Delta_D)_v = \frac{337.021(10^3)\ \text{N}^2\cdot\text{m}}{AE}$$

$$(\Delta_D)_v = \frac{337.021(10^3)}{0.300(10^{-3})[200(10^9)]}$$

$$= 5.617(10^{-3})\ \text{m} = 5.62\ \text{mm}\downarrow \qquad \textbf{Ans}$$

***14–64.** Determine the vertical displacement of point B. Each A-36 steel member has a cross-sectional area of 300 mm^2.

Member Real Forces N : As shown on figure (a).

Member Virtual Forces n : As shown on figure (b).

Virtual - Work Equation : Applying Eq. 14 – 39, we have

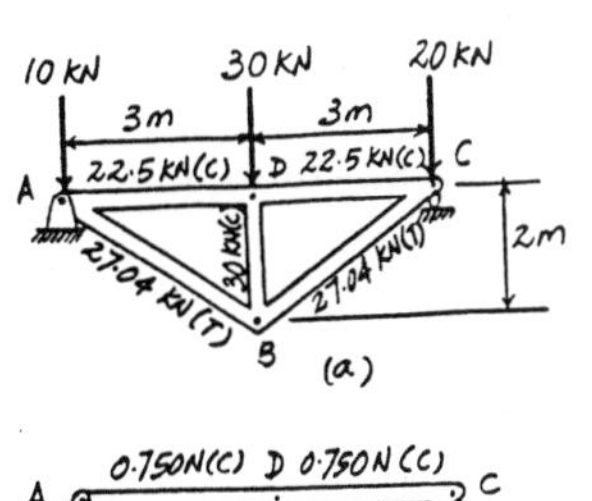

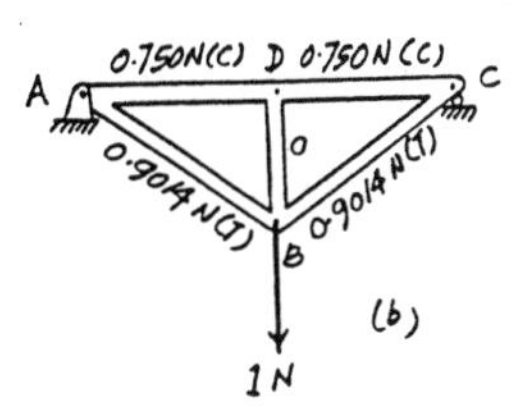

Member	n	N	L	nNL
AB	0.9014	$27.04(10^3)$	3.6056	$87.885(10^3)$
BC	0.9014	$27.04(10^3)$	3.6056	$87.885(10^3)$
CD	−0.750	$-22.5(10^3)$	3	$50.625(10^3)$
AD	−0.750	$-22.5(10^3)$	3	$50.625(10^3)$
BD	0	$-30.0(10^3)$	2	0
				$\sum 277.021(10^3)\ \text{N}^2\cdot\text{m}$

$$1\cdot\Delta = \sum\frac{nNL}{AE}$$

$$1\ \text{N}\cdot(\Delta_B)_v = \frac{277.021(10^3)\ \text{N}^2\cdot\text{m}}{AE}$$

$$(\Delta_B)_v = \frac{277.021(10^3)}{0.300(10^{-3})[200(10^9)]}$$

$$= 4.617(10^{-3})\ \text{m} = 4.62\ \text{mm}\downarrow \qquad \textbf{Ans}$$

14–65. Determine the horizontal displacement of point C. Each A-36 steel member has a cross-sectional area of 400 mm^2.

Member Real Forces N : As shown on figure(a).

Member Virtual Forces n : As shown on figure(b).

Virtual - Work Equation : Applying Eq. 14 – 39, we have

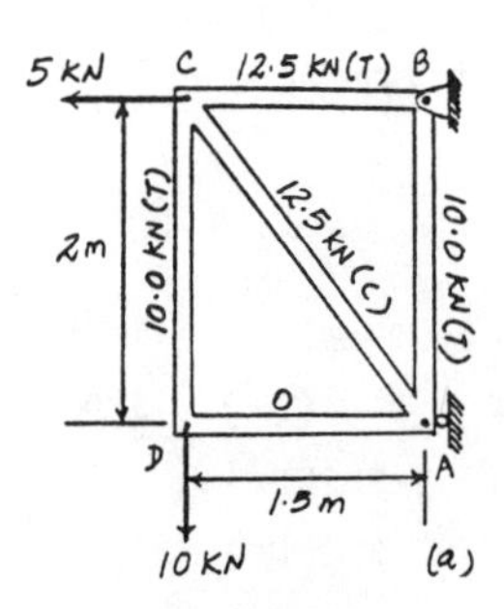

Member	n	N	L	nNL
AB	0	$10.0(10^3)$	2	0
BC	1.00	$12.5(10^3)$	1.5	$18.75(10^3)$
CD	0	$10.0(10^3)$	2	0
AD	0	0	1.5	0
AC	0	$-12.5(10^3)$	2.5	0
				$\sum 18.75(10^3)\ \text{N}^2\cdot\text{m}$

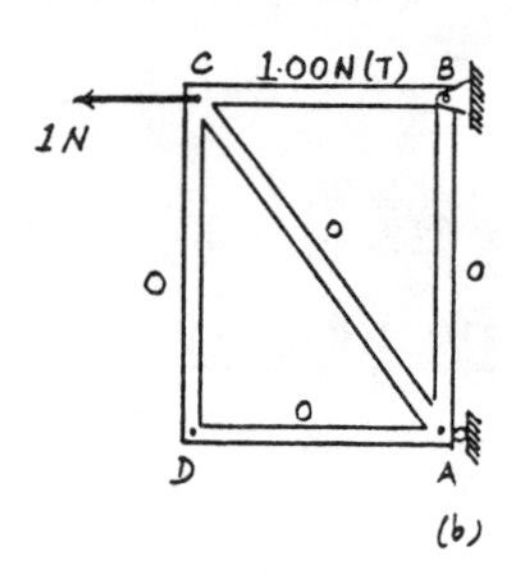

$$1\cdot\Delta = \sum\frac{nNL}{AE}$$

$$1\ \text{N}\cdot(\Delta_C)_h = \frac{18.75(10^3)\ \text{N}^2\cdot\text{m}}{AE}$$

$$(\Delta_C)_h = \frac{18.75(10^3)}{0.400(10^{-3})[200(10^9)]}$$

$$= 0.2344(10^{-3})\ \text{m} = 0.234\ \text{mm} \leftarrow \quad \textbf{Ans}$$

14–66. Determine the vertical displacement of point D. Each A-36 steel member has a cross-sectional area of 400 mm^2.

Member Real Forces N : As shown on figure(a).

Member Virtual Forces n : As shown on figure(b).

Virtual - Work Equation : Applying Eq. 14 – 39, we have

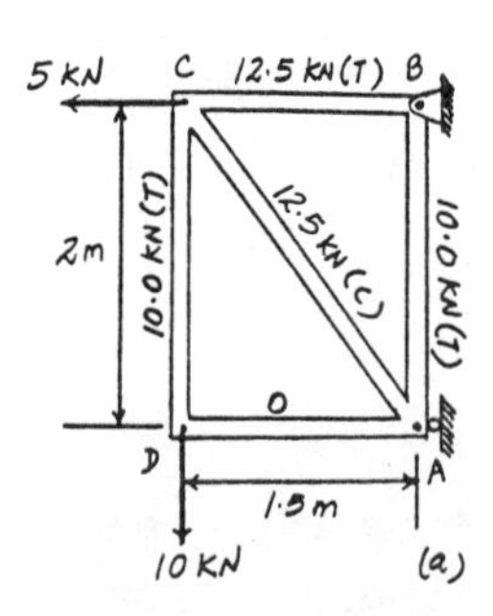

Member	n	N	L	nNL
AB	1.00	$10.0(10^3)$	2	$20.0(10^3)$
BC	0.750	$12.5(10^3)$	1.5	$14.0625(10^3)$
CD	1.00	$10.0(10^3)$	2	$20.0(10^3)$
AD	0	0	1.5	0
AC	-1.25	$-12.5(10^3)$	2.5	$39.0625(10^3)$
				$\sum 93.125(10^3)\ \text{N}^2\cdot\text{m}$

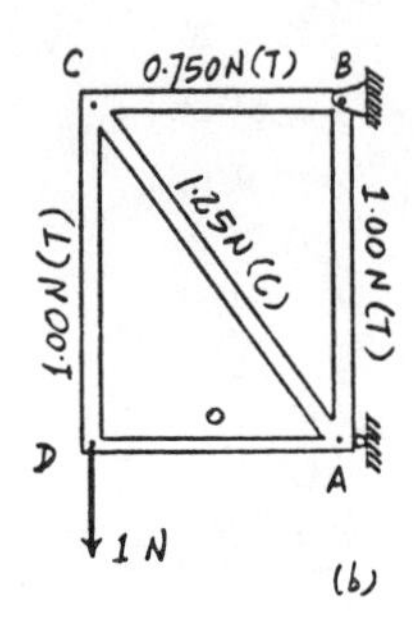

$$1\cdot\Delta = \sum\frac{nNL}{AE}$$

$$1\ \text{N}\cdot(\Delta_D)_v = \frac{93.125(10^3)\ \text{N}^2\cdot\text{m}}{AE}$$

$$(\Delta_D)_v = \frac{93.125(10^3)}{0.400(10^{-3})[200(10^9)]}$$

$$= 1.164(10^{-3})\ \text{m} = 1.16\ \text{mm} \downarrow \quad \textbf{Ans}$$

14–67. Determine the vertical displacement of point A. Each A-36 steel member has a cross-sectional area of 3 in^2.

Member Real Forces N : As shown on figure (a).

Member Virtual Forces n : As shown on figure (b).

Virtual - Work Equation : Applying Eq. 14 – 39, we have

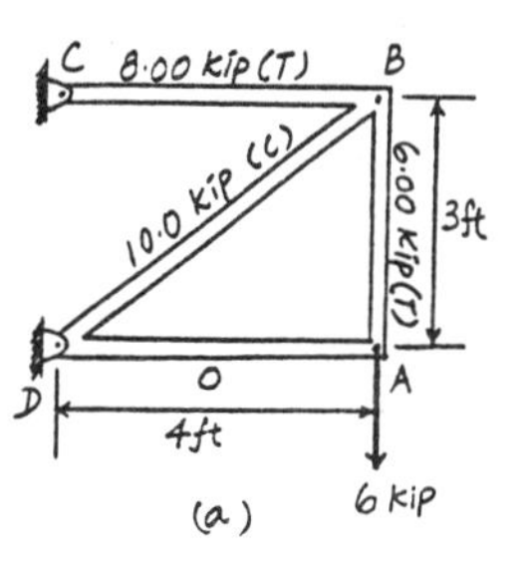

Member	*n*	*N*	*L*	*nNL*
AB	1.00	6.00	36	216
BC	1.333	8.00	48	512
AD	0	0	48	0
BD	− 1.667	− 10.0	60	1000
				$\sum$ 1728 kip² · in.

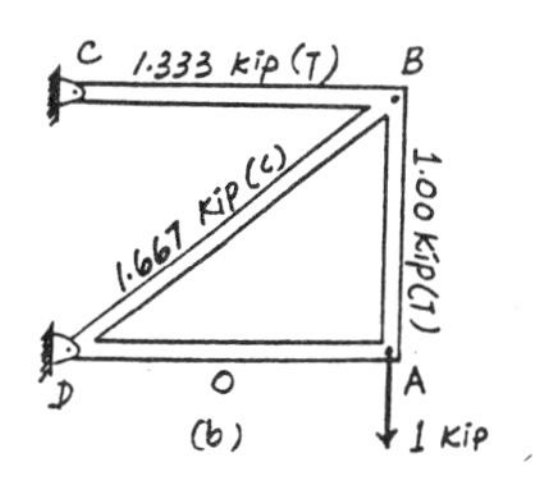

$$1 \cdot \Delta = \sum \frac{nNL}{AE}$$

$$1 \text{ kip} \cdot (\Delta_A)_v = \frac{1728 \text{ kip}^2 \cdot \text{in.}}{AE}$$

$$(\Delta_A)_v = \frac{1728}{3[29.0(10^3)]} = 0.0199 \text{ in.} \downarrow \quad \textbf{Ans}$$

***14–68.** Determine the vertical displacement of point B. Each A-36 steel member has a cross-sectional area of 3 in^2.

Member Real Forces N : As shown on figure (a).

Member Virtual Forces n : As shown on figure (b).

Virtual - Work Equation : Applying Eq. 14 – 39, we have

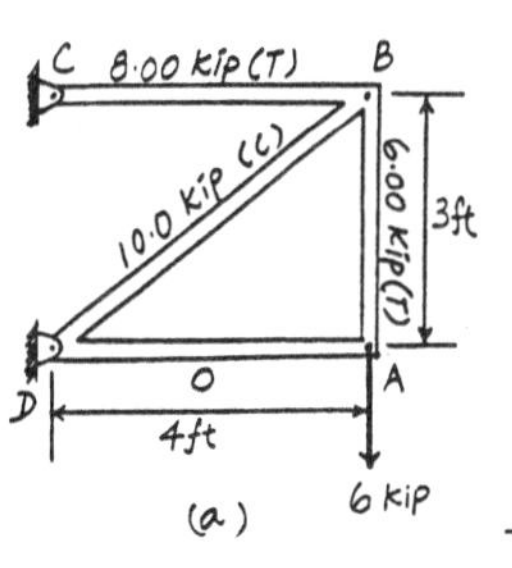

Member	*n*	*N*	*L*	*nNL*
AB	0	6.00	36	0
BC	1.333	8.00	48	512
AD	0	0	48	0
BD	− 1.667	− 10.0	60	1000
				$\sum$ 1512 kip² · in.

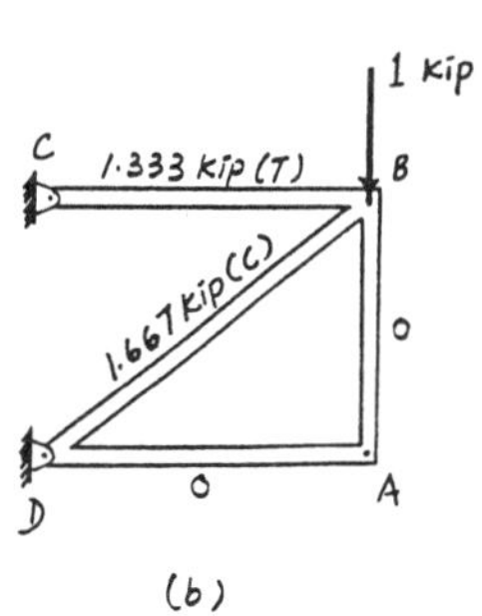

$$1 \cdot \Delta = \sum \frac{nNL}{AE}$$

$$1 \text{ kip} \cdot (\Delta_B)_v = \frac{1512 \text{ kip}^2 \cdot \text{in.}}{AE}$$

$$(\Delta_B)_v = \frac{1512}{3[29.0(10^3)]} = 0.0174 \text{ in.} \downarrow \quad \textbf{Ans}$$

14–69. Determine the vertical displacement of point A. Each A-36 steel member has a cross-sectional area of 400 mm^2.

Virtual - Work Equation : Applying Eq. 14 – 39, we have

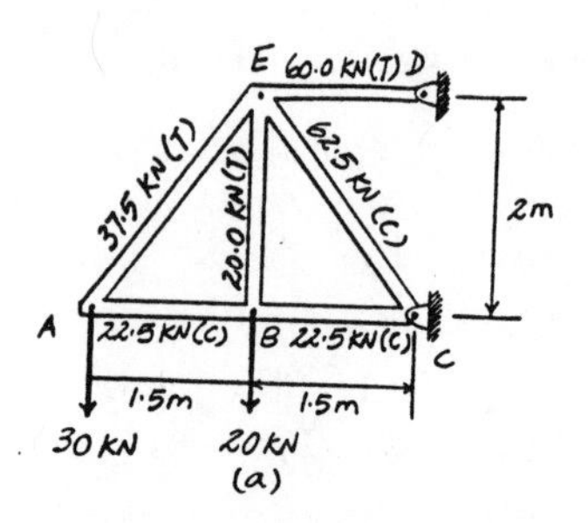

Member	*n*	*N*	*L*	*nNL*
AB	−0.750	$-22.5(10^3)$	1.5	$25.3125(10^3)$
BC	−0.750	$-22.5(10^3)$	1.5	$25.3125(10^3)$
AE	1.25	$37.5(10^3)$	2.5	$117.1875(10^3)$
CE	−1.25	$-62.5(10^3)$	2.5	$195.3125(10^3)$
BE	0	$20.0(10^3)$	2	0
DE	1.50	$60.0(10^3)$	1.5	$135.00(10^3)$
				$\sum 498.125(10^3)\ \text{N}^2\cdot\text{m}$

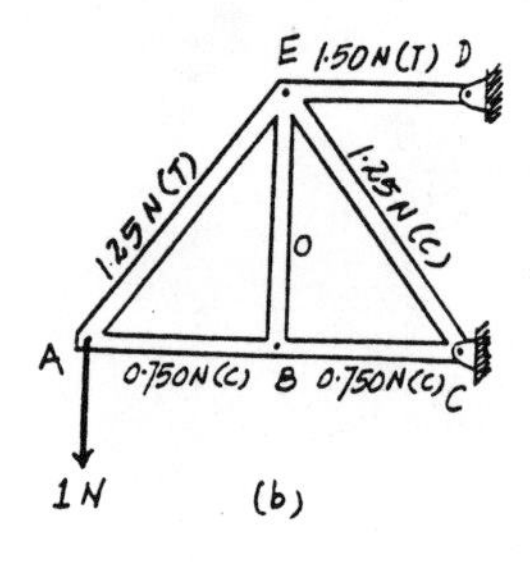

$$1\cdot\Delta = \sum \frac{nNL}{AE}$$

$$1\ \text{N}\cdot(\Delta_A)_v = \frac{498.125(10^3)\ \text{N}^2\cdot\text{m}}{AE}$$

$$(\Delta_A)_v = \frac{498.125(10^3)}{0.400(10^{-3})[200(10^9)]}$$

$$= 6.227(10^{-3})\ \text{m} = 6.23\ \text{mm}\ \downarrow \qquad \textbf{Ans}$$

14–70. Determine the vertical displacement of point B. Each A-36 steel member has a cross-sectional area of 400 mm^2.

Virtual - Work Equation : Applying Eq. 14 – 39, we have

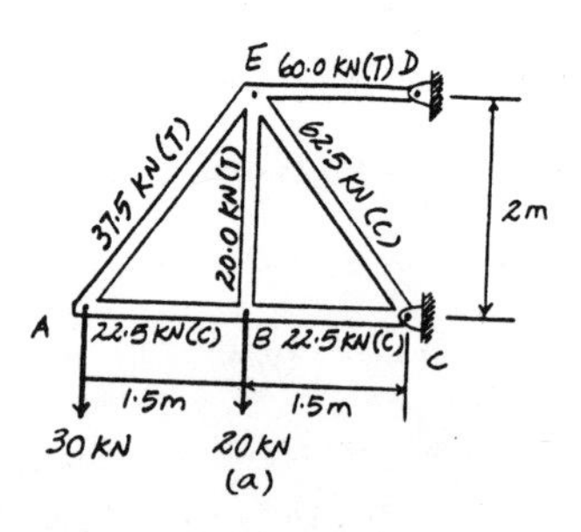

Member	*n*	*N*	*L*	*nNL*
AB	0	$-22.5(10^3)$	1.5	0
BC	0	$-22.5(10^3)$	1.5	0
AE	0	$37.5(10^3)$	2.5	0
CE	−1.25	$-62.5(10^3)$	2.5	$195.3125(10^3)$
BE	1.00	$20.0(10^3)$	2	$40.0(10^3)$
DE	0.750	$60.0(10^3)$	1.5	$67.5(10^3)$
				$\sum 302.8125(10^3)\ \text{N}^2\cdot\text{m}$

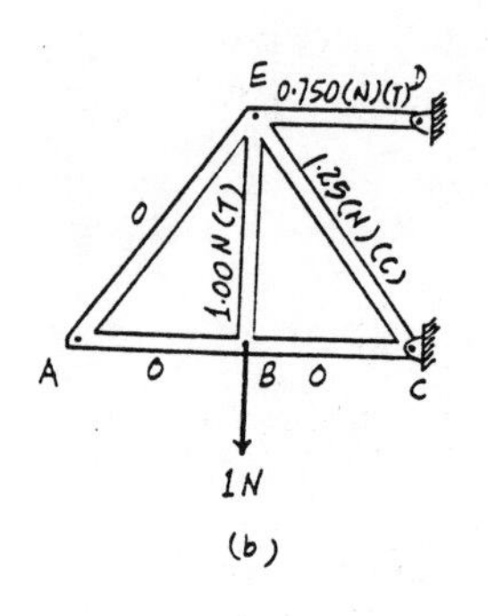

$$1\cdot\Delta = \sum \frac{nNL}{AE}$$

$$1\ \text{N}\cdot(\Delta_B)_v = \frac{302.8125(10^3)\ \text{N}^2\cdot\text{m}}{AE}$$

$$(\Delta_B)_v = \frac{302.8125(10^3)}{0.400(10^{-3})[200(10^9)]}$$

$$= 3.785(10^{-3})\ \text{m} = 3.79\ \text{mm}\ \downarrow \qquad \textbf{Ans}$$

14–71. Determine the horizontal displacement of point D. Each A-36 steel member has a cross-sectional area of 300 mm^2.

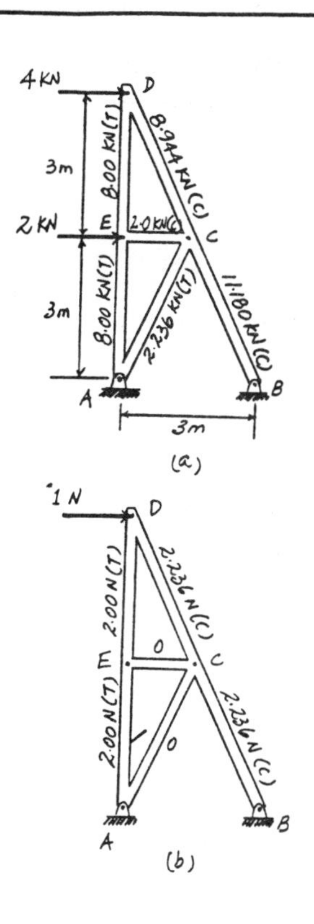

Virtual - Work Equation : Applying Eq. 14 – 39, we have

Member	*n*	*N*	*L*	*nNL*
AE	2.00	$8.00(10^3)$	3	$48.0(10^3)$
ED	2.00	$8.00(10^3)$	3	$48.0(10^3)$
CD	−2.236	$-8.944(10^3)$	3.354	$67.082(10^3)$
BC	−2.236	$-11.180(10^3)$	3.354	$83.853(10^3)$
CE	0	$-2.00(10^3)$	1.5	0
AC	0	$2.236(10^3)$	3.354	0
				$\sum 246.935(10^3)\ \text{N}^2\cdot\text{m}$

$$1\cdot\Delta = \sum\frac{nNL}{AE}$$

$$1\ \text{N}\cdot(\Delta_D)_h = \frac{246.935(10^3)\ \text{N}^2\cdot\text{m}}{AE}$$

$$(\Delta_D)_h = \frac{246.935(10^3)}{0.300(10^{-3})[200(10^9)]}$$

$$= 4.116(10^{-3})\ \text{m} = 4.12\ \text{mm} \rightarrow \qquad \textbf{Ans}$$

***14–72.** Determine the horizontal displacement of point E. Each A-36 steel member has a cross-sectional area of 300 mm^2.

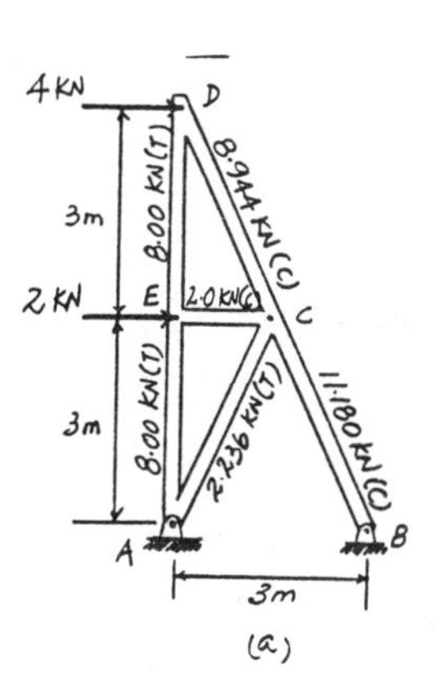

Virtual - Work Equation : Applying Eq. 14 – 39, we have

Member	*n*	*N*	*L*	*nNL*
AE	0	$8.00(10^3)$	3	0
ED	0	$8.00(10^3)$	3	0
CD	0	$-8.944(10^3)$	3.354	0
BC	−1.118	$-11.180(10^3)$	3.354	$41.926(10^3)$
CE	−1.00	$-2.00(10^3)$	1.5	$3.00(10^3)$
AC	1.118	$2.236(10^3)$	3.354	$8.385(10^3)$
				$\sum 53.312(10^3)\ \text{N}^2\cdot\text{m}$

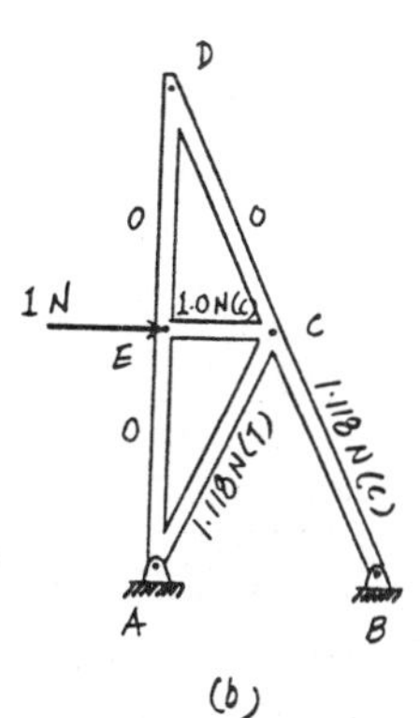

$$1\cdot\Delta = \sum\frac{nNL}{AE}$$

$$1\ \text{N}\cdot(\Delta_E)_h = \frac{53.312(10^3)\ \text{N}^2\cdot\text{m}}{AE}$$

$$(\Delta_E)_h = \frac{53.312(10^3)}{0.300(10^{-3})[200(10^9)]}$$

$$= 0.8885(10^{-3})\ \text{m} = 0.889\ \text{mm} \rightarrow \qquad \textbf{Ans}$$

14–73. Determine the vertical displacement of point B. Each A-36 steel member has a cross-sectional area of 4.5 in^2. $E_{st} = 29(10^3)$ ksi.

Virtual - Work Equation : Applying Eq. 14 – 39, we have

Member	n	N	L	nNL
AB	0.6667	3.333	96	213.33
BC	0.6667	3.333	96	213.33
CD	0	0	72	0
DE	0	0	96	0
EF	0	0	96	0
AF	0	0	72	0
AE	−0.8333	−4.167	120	416.67
CE	−0.8333	−4.167	120	416.67
BE	1.00	5.00	72	360.00
				Σ 1620 kip^2 · in.

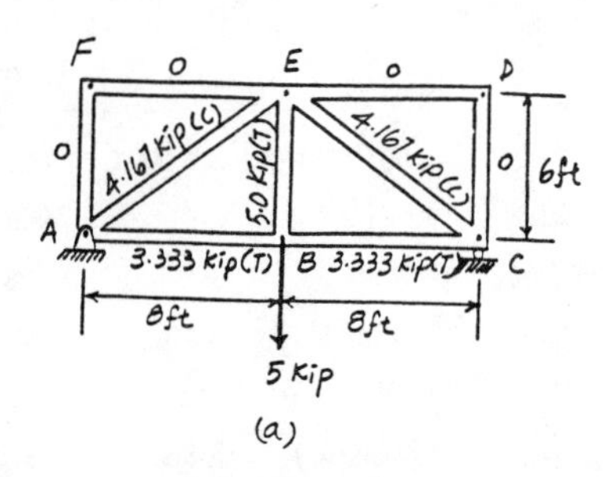

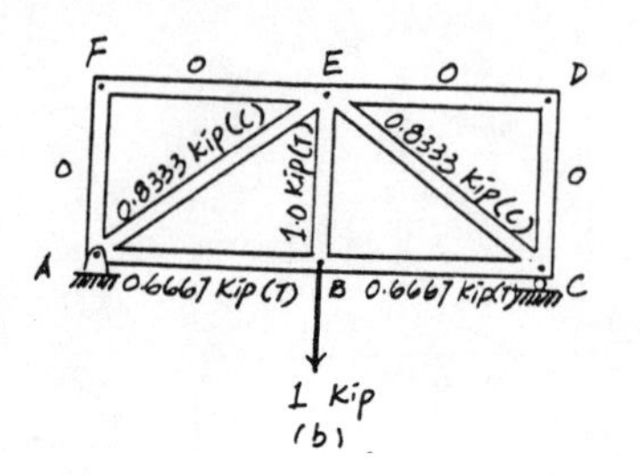

$$1 \cdot \Delta = \sum \frac{nNL}{AE}$$

$$1 \text{ kip} \cdot (\Delta_B)_v = \frac{1620 \text{ kip}^2 \cdot \text{in.}}{AE}$$

$$(\Delta_B)_v = \frac{1620}{4.5[29.0(10^3)]} = 0.0124 \text{ in.} \downarrow \qquad \textbf{Ans}$$

14–74. Determine the vertical displacement of point E. Each A-36 steel member has a cross-sectional area of 4.5 in^2.

Virtual - Work Equation : Applying Eq. 14 – 39, we have

Member	n	N	L	nNL
AB	0.6667	3.333	96	213.33
BC	0.6667	3.333	96	213.33
CD	0	0	72	0
DE	0	0	96	0
EF	0	0	96	0
AF	0	0	72	0
AE	−0.8333	−4.167	120	416.67
CE	−0.8333	−4.167	120	416.67
BE	0	5.00	72	0
				Σ 1260 kip^2 · in.

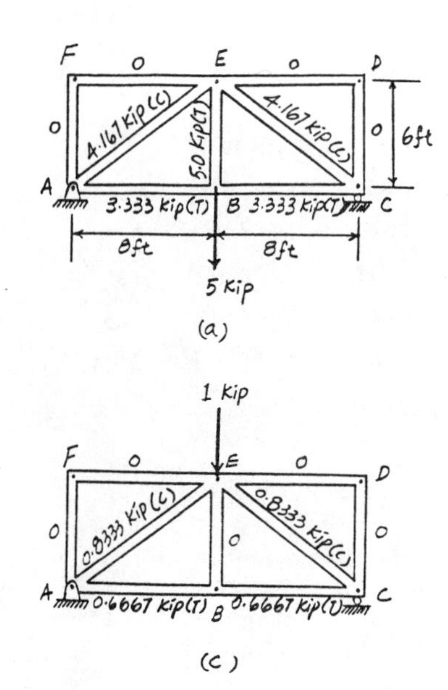

$$1 \cdot \Delta = \sum \frac{nNL}{AE}$$

$$1 \text{ kip} \cdot (\Delta_E)_v = \frac{1260 \text{ kip}^2 \cdot \text{in.}}{AE}$$

$$(\Delta_E)_v = \frac{1260}{4.5[29.0(10^3)]} = 0.00966 \text{ in.} \downarrow \qquad \textbf{Ans}$$

14–75. Determine the vertical displacement of joint C. Each A-36 steel member has a cross-sectional area of 4.5 in^2.

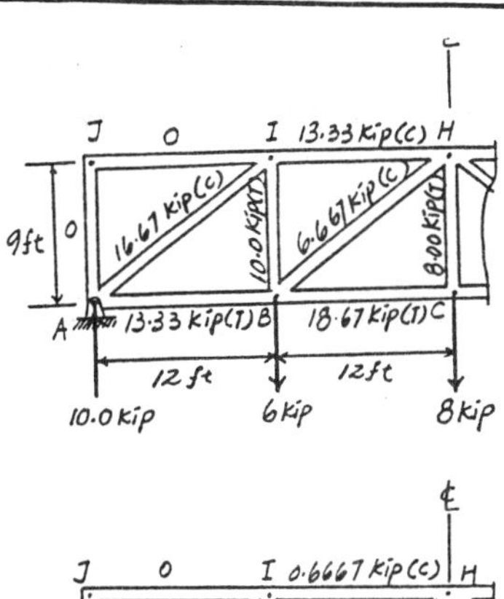

Virtual - Work Equation : Applying Eq. 14 – 39, we have

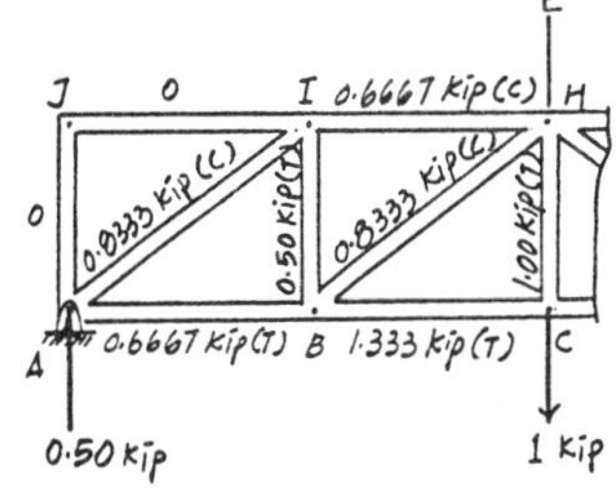

Member	*n*	*N*	*L*	*nNL*
AB	0.6667	13.33	144	1280.00
DE	0.6667	13.33	144	1280.00
BC	1.333	18.67	144	3584.00
CD	1.333	18.67	144	3584.00
AJ	0	0	108	0
EF	0	0	108	0
IJ	0	0	144	0
FG	0	0	144	0
HI	−0.6667	−13.33	144	1280.00
GH	−0.6667	−13.33	144	1280.00
AI	−0.8333	−16.67	180	2500.00
EG	−0.8333	−16.67	180	2500.00
BI	0.500	10.0	108	540.00
DG	0.500	10.0	108	540.00
BH	−0.8333	−6.667	180	1000.00
DH	−0.8333	−6.667	180	1000.00
CH	1.00	8.00	108	864.00
				$\sum$ 21232 kip^2 · in.

$$1 \cdot \Delta = \sum \frac{nNL}{AE}$$

$$1 \text{ kip} \cdot (\Delta_C)_v = \frac{21232 \text{ kip}^2 \cdot \text{in.}}{AE}$$

$$(\Delta_C)_v = \frac{21232}{4.5[29.0(10^3)]} = 0.163 \text{ in.} \downarrow \qquad \textbf{Ans}$$

***14–76.** Determine the vertical displacement of joint H. Each A-36 steel member has a cross-sectional area of 4.5 in^2.

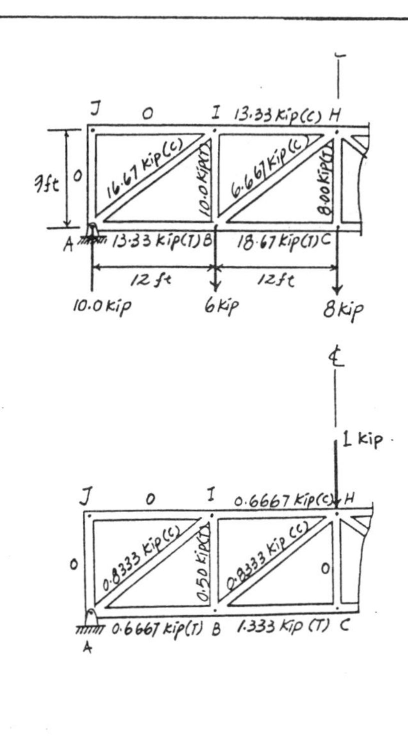

Virtual - Work Equation : Applying Eq. 14 – 39, we have

Member	*n*	*N*	*L*	*nNL*
AB	0.6667	13.33	144	1280.00
DE	0.6667	13.33	144	1280.00
BC	1.333	18.67	144	3584.00
CD	1.333	18.67	144	3584.00
AJ	0	0	108	0
EF	0	0	108	0
IJ	0	0	144	0
FG	0	0	144	0
HI	−0.6667	−13.33	144	1280.00
GH	−0.6667	−13.33	144	1280.00
AI	−0.8333	−16.67	180	2500.00
EG	−0.8333	−16.67	180	2500.00
BI	0.500	10.0	108	540.00
DG	0.500	10.0	108	540.00
BH	−0.8333	−6.667	180	1000.00
DH	−0.8333	−6.667	180	1000.00
CH	0	8.00	108	0
				$\sum$ 20368 kip^2 · in.

$$1 \cdot \Delta = \sum \frac{nNL}{AE}$$

$$1 \text{ kip} \cdot (\Delta_H)_v = \frac{20368 \text{ kip}^2 \cdot \text{in.}}{AE}$$

$$(\Delta_H)_v = \frac{20368}{4.5[29.0(10^3)]} = 0.156 \text{ in.} \downarrow \qquad \textbf{Ans}$$

14–77. Determine the displacement of point C and the slope at point B. EI is constant.

Real Moment Function $M(x)$ **:** As shown on figure(a).

Virtual Moment Functions $m(x)$ ***and*** $m_\theta(x)$ **:** As shown on figure(b) and (c).

***Virtual Work Equation* :** For the displacement at point C, apply Eq. 14–42.

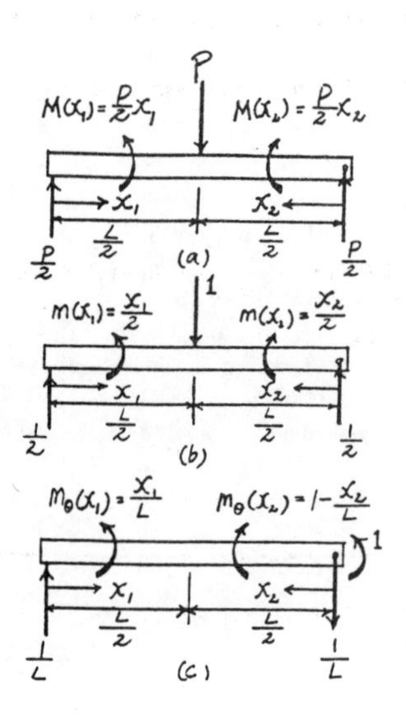

$$1\cdot\Delta=\int_0^L \frac{mM}{EI}dx$$

$$1\cdot\Delta_C=2\left[\frac{1}{EI}\int_0^{\frac{L}{2}}\left(\frac{x_1}{2}\right)\left(\frac{P}{2}x_1\right)dx_1\right]$$

$$\Delta_C=\frac{PL^3}{48EI}\ \downarrow \qquad \textbf{Ans}$$

For the slope at B, apply Eq. 14–43.

$$1\cdot\theta=\int_0^L \frac{m_\theta M}{EI}dx$$

$$1\cdot\theta_B=\frac{1}{EI}\left[\int_0^{\frac{L}{2}}\left(\frac{x_1}{L}\right)\left(\frac{P}{2}x_1\right)dx_1+\int_0^{\frac{L}{2}}\left(1-\frac{x_2}{L}\right)\left(\frac{P}{2}x_2\right)dx_2\right]$$

$$\theta_B=\frac{PL^2}{16EI} \qquad \textbf{Ans}$$

14–78. Determine the displacement of point C. EI is constant.

Real Moment Function $M(x)$ **:** As shown on figure(a).

Virtual Moment Functions $m(x)$ **:** As shown on figure(b).

***Virtual Work Equation* :** For the displacement at point C, apply Eq. 14–42.

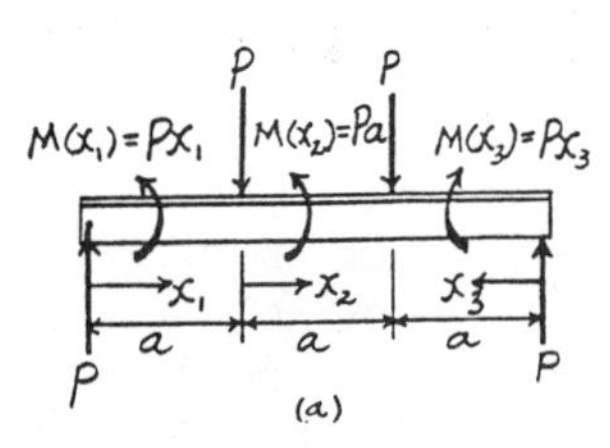

$$1\cdot\Delta=\int_0^L \frac{mM}{EI}dx$$

$$1\cdot\Delta_C=\frac{1}{EI}\int_0^a\left(\frac{2}{3}x_1\right)(Px_1)\,dx_1+\frac{1}{EI}\int_0^a\frac{1}{3}(2a-x_2)(Pa)\,dx_2+\frac{1}{EI}\int_0^a\left(\frac{x_3}{3}\right)(Px_3)\,dx_3$$

$$\Delta_C=\frac{5Pa^3}{6EI}\ \downarrow \qquad \textbf{Ans}$$

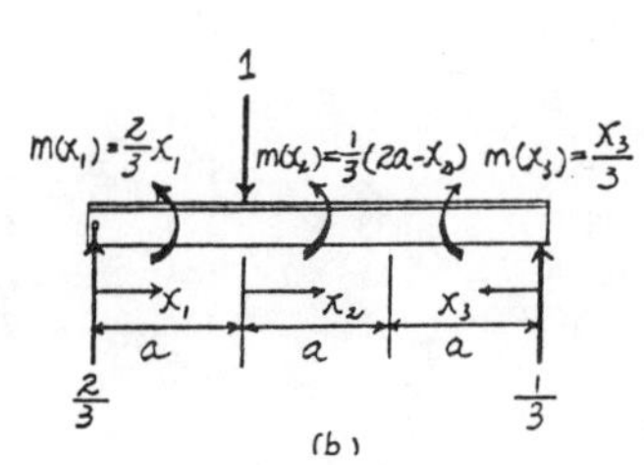

14–79. Determine the slope at point C. EI is constant.

Real Moment Function M(x) : As shown on figure(a).

Virtual Moment Functions $m_\theta(x)$: As shown on figure(b).

Virtual Work Equation : For the slope at point C, apply Eq. 14–43.

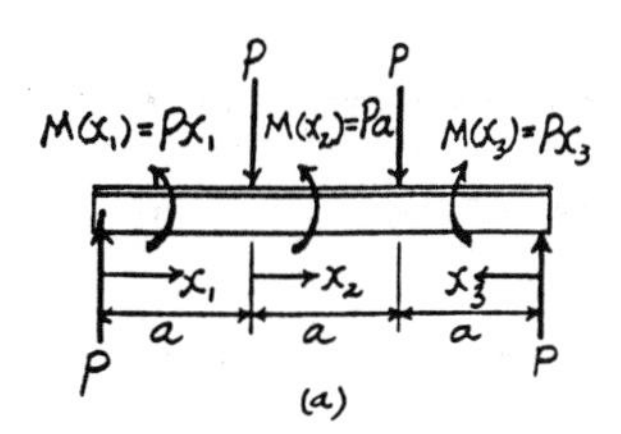

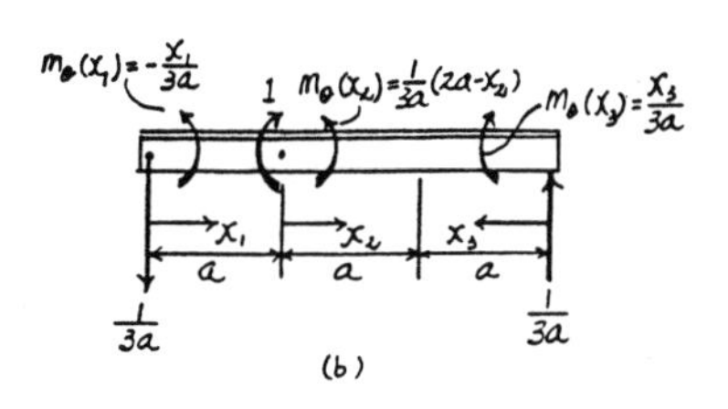

$$1 \cdot \theta = \int_0^L \frac{m_\theta M}{EI} dx$$

$$1 \cdot \theta_C = \frac{1}{EI}\int_0^a \left(-\frac{x_1}{3a}\right)(Px_1)\,dx_1 + \frac{1}{EI}\int_0^a \frac{1}{3a}(2a - x_2)(Pa)\,dx_2 + \frac{1}{EI}\int_0^a \left(\frac{x_3}{3a}\right)(Px_3)\,dx_3$$

$$\theta_C = \frac{Pa^2}{2EI} \qquad \textbf{Ans}$$

***14–80.** Determine the slope at point A. EI is constant.

Real Moment Function M(x) : As shown on figure(a).

Virtual Moment Functions $m_\theta(x)$: As shown on figure(b).

Virtual Work Equation : For the slope at point A, apply Eq. 14–43.

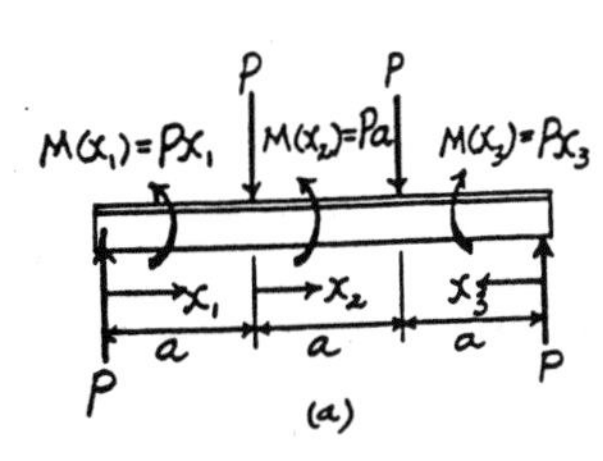

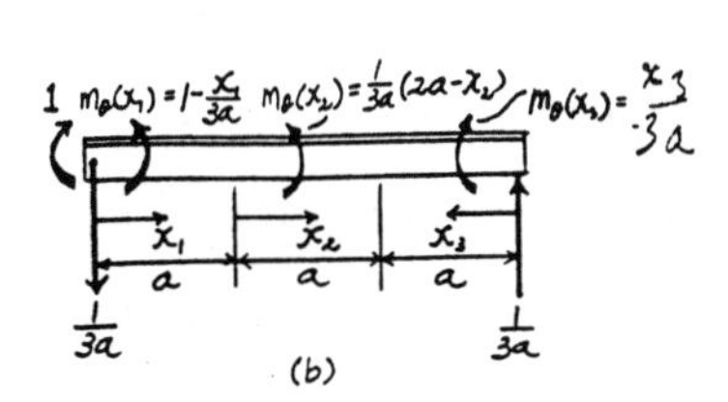

$$1 \cdot \theta = \int_0^L \frac{m_\theta M}{EI} dx$$

$$1 \cdot \theta_A = \frac{1}{EI}\int_0^a \left(1 - \frac{x_1}{3a}\right)(Px_1)\,dx_1 + \frac{1}{EI}\int_0^a \frac{1}{3a}(2a - x_2)(Pa)\,dx_2 + \frac{1}{EI}\int_0^a \left(\frac{x_3}{3a}\right)(Px_3)\,dx_3$$

$$\theta_A = \frac{Pa^2}{EI} \qquad \textbf{Ans}$$

14–81. Determine the slope of the 60-mm-diameter A-36 steel shaft at the bearing support A.

Real Moment Function $M(x)$ **:** As shown on figure(a).

Virtual Moment Functions $m_\theta(x)$ **:** As shown on figure(b).

Virtual Work Equation **:** For the slope at point A, apply Eq. 14 – 43.

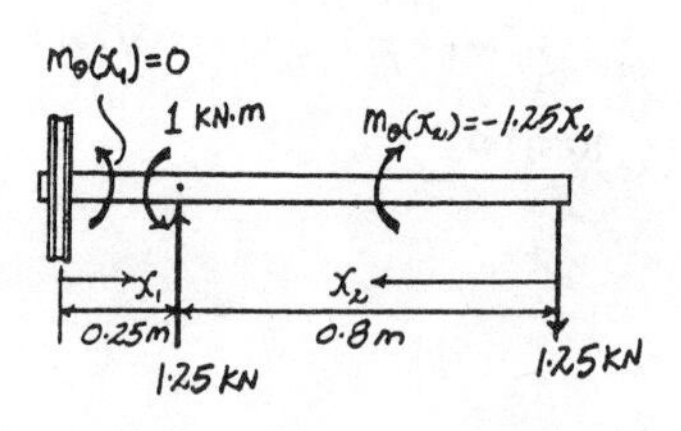

$$1\cdot\theta = \int_0^L \frac{m_\theta M}{EI}dx$$

$$1\ \text{kN}\cdot\text{m}\cdot\theta_A = \frac{1}{EI}\int_0^{0.25m}(0)(-24.0x_1)\,dx_1 + \frac{1}{EI}\int_0^{0.8m}(-1.25x_2)(-7.50x_2)\,dx_2$$

$$\theta_A = \frac{1.60\ \text{kN}\cdot\text{m}^2}{EI} = \frac{1.60(1000)}{200(10^9)\left[\frac{\pi}{4}(0.03^4)\right]}$$

$$= 0.0126\ \text{rad}$$ **Ans**

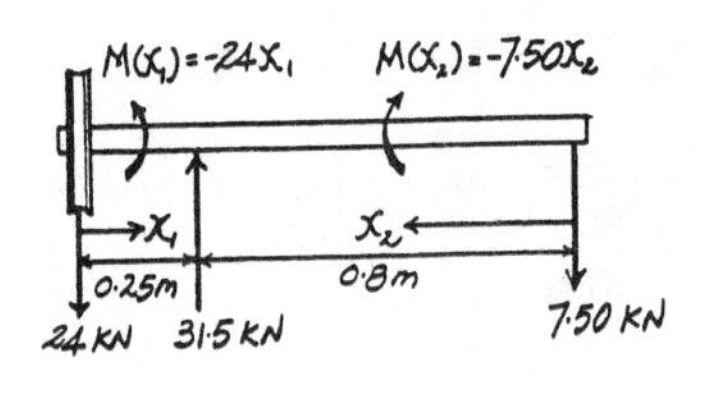

14–82. Determine the displacement at C of the 60-mm-diameter A-36 steel shaft.

Real Moment Function $M(x)$ **:** As shown on figure(a).

Virtual Moment Functions $m(x)$ **:** As shown on figure(b).

Virtual Work Equation **:** For the displacement at point C, apply Eq. 14 – 42.

$$1\cdot\Delta = \int_0^L \frac{mM}{EI}dx$$

$$1\ \text{kN}\cdot\Delta_C = \frac{1}{EI}\int_0^{0.25m}(-1.00x_1)(-24.0x_1)\,dx_1 + \frac{1}{EI}\int_0^{0.8m}(-0.3125x_2)(-7.50x_2)\,dx_2$$

$$\Delta_C = \frac{0.525\ \text{kN}\cdot\text{m}^3}{EI} = \frac{0.525(1000)}{200(10^9)\left[\frac{\pi}{4}(0.03^4)\right]}$$

$$= 4.126\left(10^{-3}\right)\ \text{m} = 4.13\ \text{mm}\ \downarrow$$ **Ans**

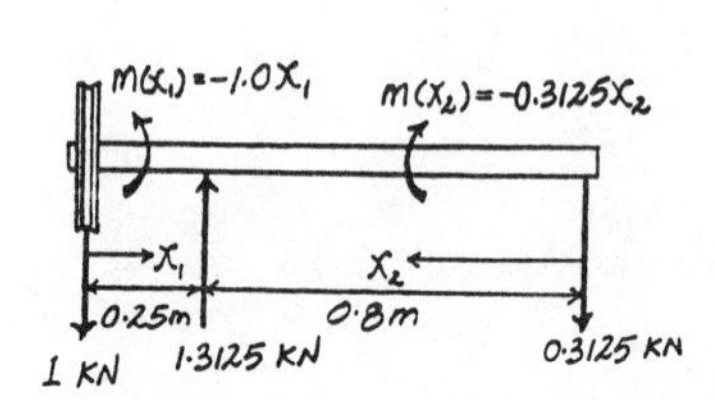

14–83. Determine the tilt of the pulley at C if the A-36 steel shaft has a 60-mm diameter.

Real Moment Function M(x) : As shown on figure(a).

Virtual Moment Functions $m_\theta(x)$: As shown on figure(b).

Virtual Work Equation : For the slope at point C, apply Eq. 14 – 43.

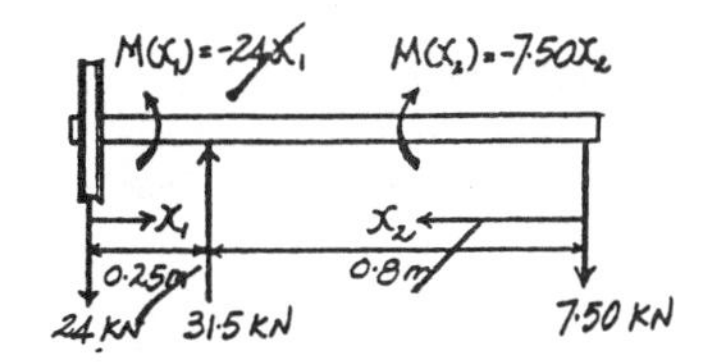

$$1 \cdot \theta = \int_0^L \frac{m_\theta M}{EI} dx$$

$$1 \text{ kN} \cdot \text{m} \cdot \theta_C = \frac{1}{EI} \int_0^{0.25m} (-1.00)(-24.0x_1)\, dx_1 + \frac{1}{EI} \int_0^{0.8m} (-1.25x_2)(-7.50x_2)\, dx_2$$

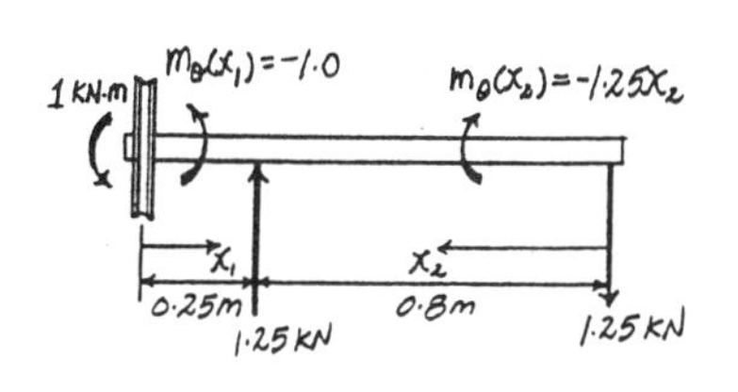

$$\theta_C = \frac{2.35 \text{ kN} \cdot \text{m}^2}{EI}$$

$$= \frac{2.35(1000)}{200(10^9)\left[\frac{\pi}{4}(0.03^4)\right]} = 0.0185 \text{ rad} \qquad \textbf{Ans}$$

***14–84.** Determine the displacement of collar B. The A-36 steel shaft has a diameter of 60 mm.

Real Moment Function M(x) : As shown on figure(a).

Virtual Moment Functions m(x) : As shown on figure(b).

Virtual Work Equation : For the displacement at point B, apply Eq. 14 – 42.

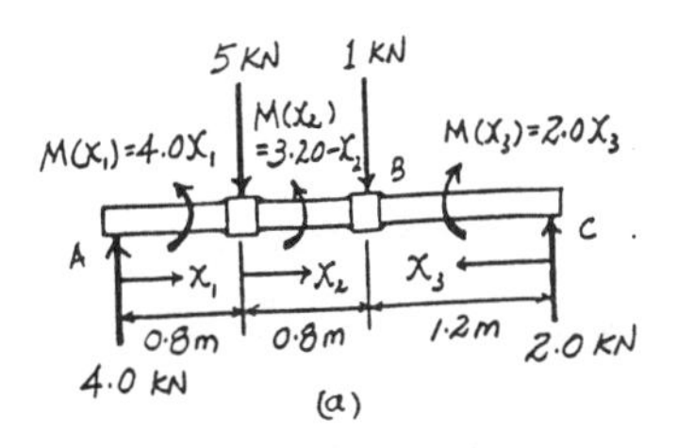

$$1 \cdot \Delta = \int_0^L \frac{mM}{EI} dx$$

$$1 \text{ kN} \cdot \Delta_B = \frac{1}{EI} \int_0^{0.8m} (0.4286x_1)(4.00x_1)\, dx_1 + \frac{1}{EI} \int_0^{0.8m} 0.4286(x_2 + 0.8)(3.20 - 1.00x_2)\, dx_2 + \frac{1}{EI} \int_0^{1.2m} (0.5714x_3)(2.00x_3)\, dx_3$$

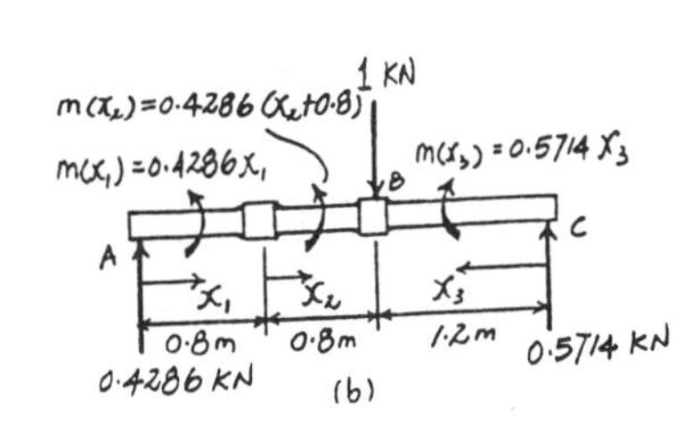

$$\Delta_B = \frac{2.0846 \text{ kN} \cdot \text{m}^3}{EI}$$

$$= \frac{2.0846(1000)}{200(10^9)\left[\frac{\pi}{4}(0.03^4)\right]}$$

$$= 0.01638 \text{ m} = 16.4 \text{ mm} \downarrow \qquad \textbf{Ans}$$

14–85. Determine the slope at C. The A-36 steel shaft has a diameter of 60 mm.

Real Moment Function M(x) : As shown on figure(a).

Virtual Moment Functions $m_\theta(x)$: As shown on figure(b).

Virtual Work Equation : For the slope at point C, apply Eq. 14–43.

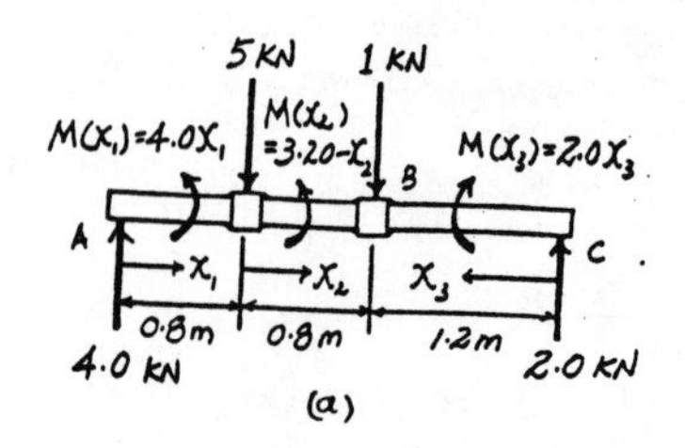

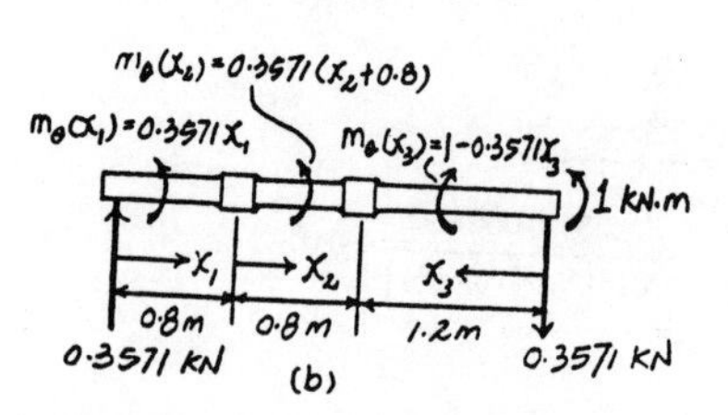

$$1 \cdot \theta = \int_0^L \frac{m_\theta M}{EI} dx$$

$$1 \text{ kN} \cdot \text{m} \cdot \theta_C = \frac{1}{EI}\int_0^{0.8m} (0.3571x_1)(4.00x_1)\,dx_1$$

$$+\frac{1}{EI}\int_0^{0.8m} 0.3571(x_2+0.8)(3.20-x_2)\,dx_2$$

$$+\frac{1}{EI}\int_0^{1.2m} (1-0.3571x_3)(2.00x_3)\,dx_3$$

$$\theta_C = \frac{2.2171 \text{ kN} \cdot \text{m}^2}{EI}$$

$$= \frac{2.2171(1000)}{200(10^9)\left[\frac{\pi}{4}(0.03^4)\right]} = 0.0174 \text{ rad} \qquad \textbf{Ans}$$

14–86. Determine the slope at A. The A-36 steel shaft has a diameter of 60 mm.

Real Moment Function M(x) : As shown on figure(a).

Virtual Moment Functions $m_\theta(x)$: As shown on figure(b).

Virtual Work Equation : For the slope at point C, apply Eq. 14–43.

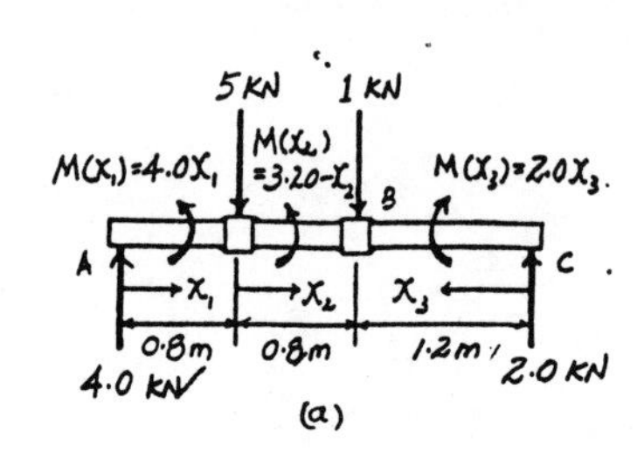

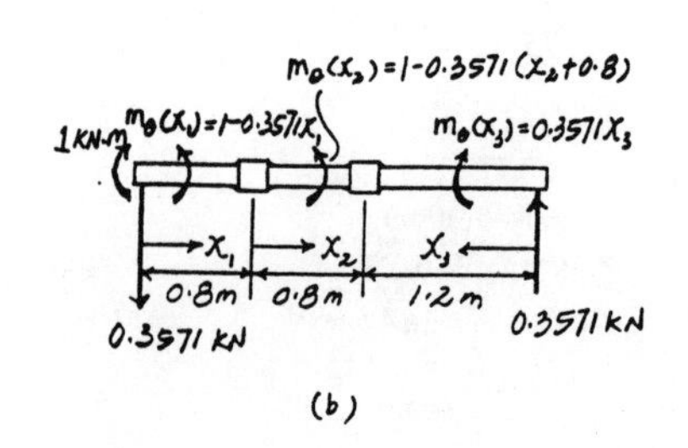

$$1 \cdot \theta = \int_0^L \frac{m_\theta M}{EI} dx$$

$$1 \text{ kN} \cdot \text{m} \cdot \theta_B = \frac{1}{EI}\int_0^{0.8m} (1-0.3571x_1)(4.00x_1)\,dx_1$$

$$+\frac{1}{EI}\int_0^{0.8m} [1-0.3571(x_2+0.8)](3.20-x_2)\,dx_2$$

$$+\frac{1}{EI}\int_0^{1.2m} (0.3571x_3)(2.00x_3)\,dx_3$$

$$\theta_B = \frac{2.7429 \text{ kN} \cdot \text{m}^2}{EI}$$

$$= \frac{2.7429(1000)}{200(10^9)\left[\frac{\pi}{4}(0.03^4)\right]} = 0.0216 \text{ rad} \qquad \textbf{Ans}$$

14–87. The A-36 steel beam has a moment of inertia of $I = 125(10^6)\ \text{mm}^4$. Determine the displacement at D.

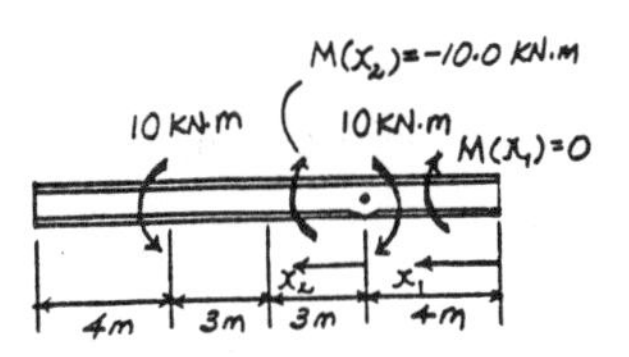

Real Moment Function $M(x)$: As shown on figure(a).

Virtual Moment Functions $m(x)$: As shown on figure(b).

Virtual Work Equation : For the displacement at point D, apply Eq. 14 – 42.

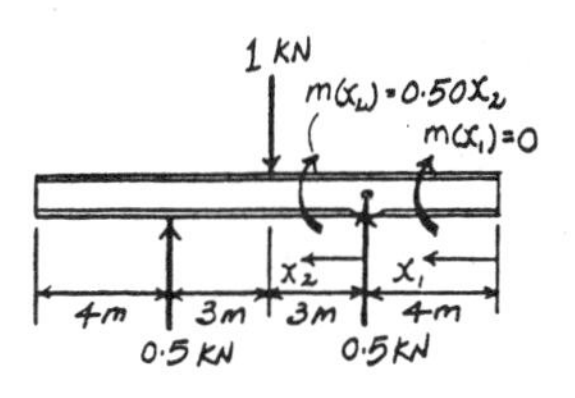

$$1 \cdot \Delta = \int_0^L \frac{mM}{EI} dx$$

$$1\ \text{kN} \cdot \Delta_D = 2\left[\frac{1}{EI}\int_0^{3m} (0.500x_2)(-10.0)\, dx_2\right]$$

$$\Delta_D = -\frac{45.0\ \text{kN} \cdot \text{m}^3}{EI}$$

$$= -\frac{45.0(1000)}{200(10^9)[125(10^{-6})]}$$

$$= -1.80(10^{-3})\ \text{m} = 1.80\ \text{mm} \uparrow$$ **Ans**

***14–88.** The A-36 steel beam has a moment of inertia of $I = 125(10^6)\ \text{mm}^4$. Determine the slope at E.

Real Moment Function $M(x)$: As shown on figure(a).

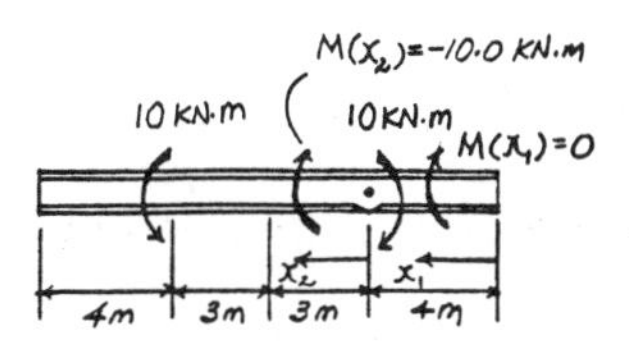

Virtual Moment Functions $m_\theta(x)$: As shown on figure(b).

Virtual Work Equation : For the slope at point E, apply Eq. 14 – 43.

$$1 \cdot \theta = \int_0^L \frac{m_\theta M}{EI} dx$$

$$1\ \text{kN} \cdot \text{m} \cdot \theta_E = \frac{1}{EI}\int_0^{4m} (-1.00)(0)\, dx_1 + \frac{1}{EI}\int_0^{6m}\left(\frac{x_2}{6} - 1\right)(-10.0)\, dx_2$$

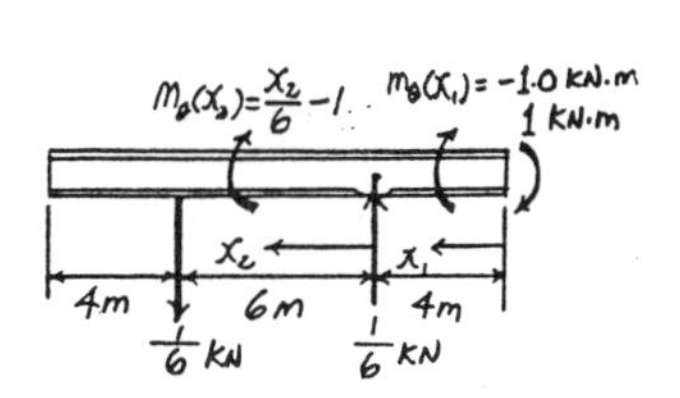

$$\theta_E = \frac{30.0\ \text{kN} \cdot \text{m}^2}{EI}$$

$$= \frac{30.0(1000)}{200(10^9)[125(10^{-6})]}$$

$$= 0.00120\ \text{rad}$$ **Ans**

14–89. The A-36 structural steel beam has a moment of inertia of $I = 125(10^6)$ mm^4. Determine the slope of the beam at B.

Real Moment Function $M(x)$: As shown on figure(a).

Virtual Moment Functions $m_\theta(x)$: As shown on figure(b).

Virtual Work Equation : For the slope at point B, apply Eq. 14 – 43.

$$1 \cdot \theta = \int_0^L \frac{m_\theta M}{EI} dx$$

$$1\ \text{kN} \cdot \text{m} \cdot \theta_B = \frac{1}{EI}\int_0^{6\text{m}} \left(-\frac{x}{6}\right)(-10.0)\, dx$$

$$\theta_B = \frac{30.0\ \text{kN} \cdot \text{m}^2}{EI}$$

$$= \frac{30.0(1000)}{200(10^9)[125(10^{-6})]}$$

$$= 1.20\left(10^{-3}\right)\ \text{rad} \qquad \textbf{Ans}$$

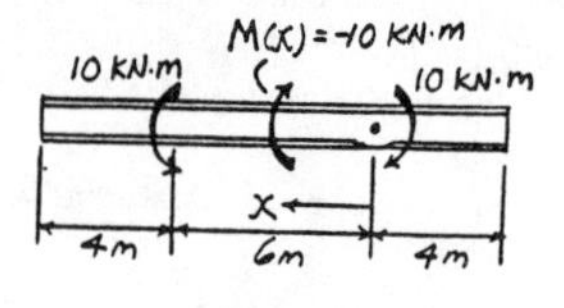

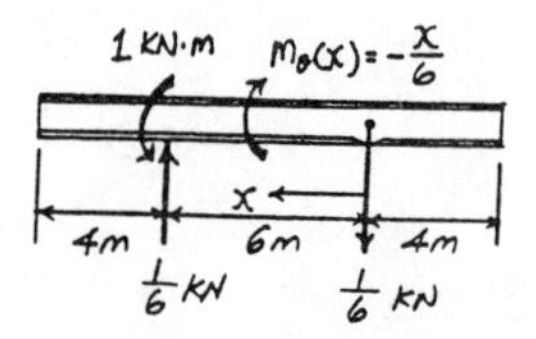

14–90. Determine the displacement at point B. The moment of inertia of the center portion DG of the shaft is $2I$, whereas the end segments AD and GC have a moment of inertia I. The modulus of elasticity for the material is E.

Real Moment Function $M(x)$: As shown on figure(a).

Virtual Moment Functions $m(x)$: As shown on figure(b).

Virtual Work Equation : For the displacement at point B, apply Eq. 14 – 42.

$$1 \cdot \Delta = \int_0^L \frac{mM}{EI} dx$$

$$1 \cdot \Delta_B = 2\left[\frac{1}{EI}\int_0^a \left(\frac{x_1}{2}\right)(wax_1)\, dx_1\right]$$

$$+2\left[\frac{1}{2EI}\int_0^a \frac{1}{2}(x_2 + a)\left[wa(a+x_2) - \frac{w}{2}x_2^2\right]dx_2\right]$$

$$\Delta_B = \frac{65wa^4}{48EI} \downarrow \qquad \textbf{Ans}$$

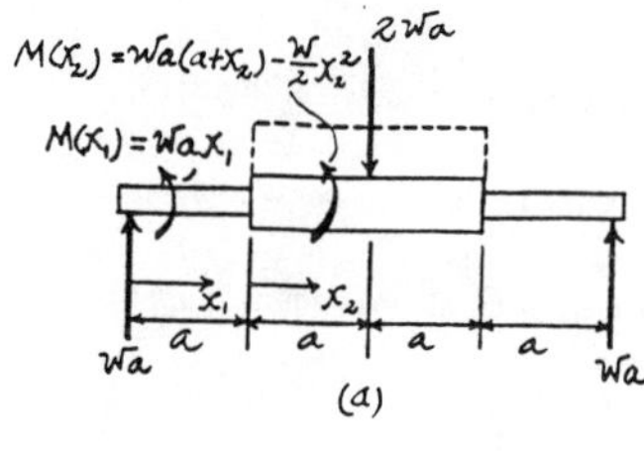

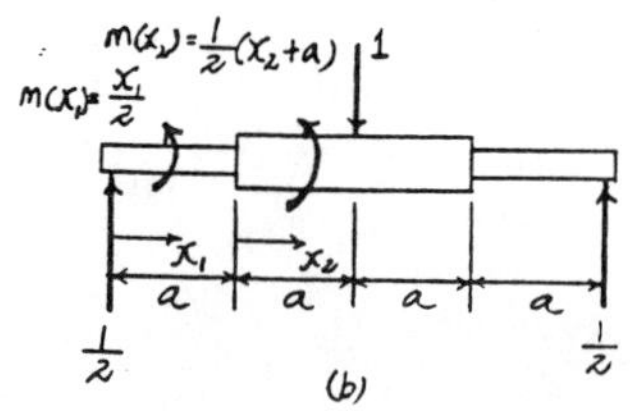

14–91. Determine the slope at A of the shaft in Prob. 14–90.

Real Moment Function $M(x)$: As shown on Figure(a).

Virtual Moment Functions $m_\theta(x)$: As shown on figure(b).

Virtual Work Equation : For the slope at point A, apply Eq. 14–43.

$$1 \cdot \theta = \int_0^L \frac{m_\theta M}{EI} dx$$

$$1 \cdot \theta_A = \frac{1}{EI}\int_0^a \left(1 - \frac{x_1}{4a}\right)(wax_1)\,dx_1 + \frac{1}{2EI}\int_0^{2a} \frac{1}{4a}(x_2 + a)\left[wa(a + x_2) - \frac{w}{2}x_2^2\right]dx_2 + \frac{1}{EI}\int_0^a \left(\frac{x_3}{4a}\right)(wax_3)\,dx_3$$

$$\theta_A = \frac{7wa^3}{6EI} \qquad \text{Ans}$$

***14–92.** Determine the displacement at D of the shaft in Prob. 14–90

Real Moment Function $M(x)$: As shown on figure(a).

Virtual Moment Functions $m_\theta(x)$: As shown on figure(b).

Virtual Work Equation : For the displacement at point D, apply Eq. 14–42.

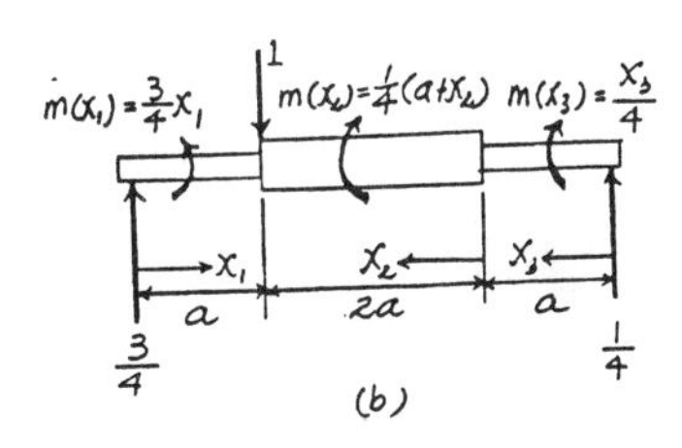

$$1 \cdot \Delta = \int_0^L \frac{mM}{EI} dx$$

$$1 \cdot \Delta_D = \frac{1}{EI}\int_0^a \left(\frac{3}{4}x_1\right)(wax_1)\,dx_1 + \frac{1}{2EI}\int_0^{2a} \frac{1}{4}(a + x_2)\left[wa(a + x_2) - \frac{w}{2}x_2^2\right]dx_2 + \frac{1}{EI}\int_0^a \left(\frac{x_3}{4}\right)(wax_3)\,dx_3$$

$$\Delta_D = \frac{wa^4}{EI} \downarrow \qquad \text{Ans}$$

14–93. Determine the displacement at C of the A-36 steel beam. $I = 70(10^6)$ mm^4.

Real Moment Function $M(x)$ **:** As shown on figure(a).

Virtual Moment Functions $m(x)$ **:** As shown on figure(b).

Virtual Work Equation **:** For the displacement at point C, apply Eq. 14–42.

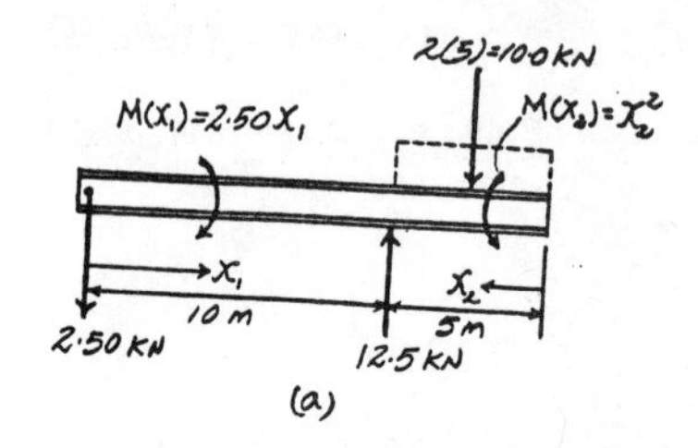

$$1 \cdot \Delta = \int_0^L \frac{mM}{EI} dx$$

$$1 \text{ kN} \cdot \Delta_C = \frac{1}{EI}\int_0^{10\text{m}} 0.500x_1\,(2.50x_1)\,dx_1 + \frac{1}{EI}\int_0^{5\text{m}} x_2\left(x_2^2\right)dx_2$$

$$\Delta_C = \frac{572.92 \text{ kN} \cdot \text{m}^3}{EI}$$

$$= \frac{572.92(1000)}{200(10^9)\,[70(10^{-6})]}$$

$$= 0.04092 \text{ m} = 40.9 \text{ mm} \downarrow \qquad \textbf{Ans}$$

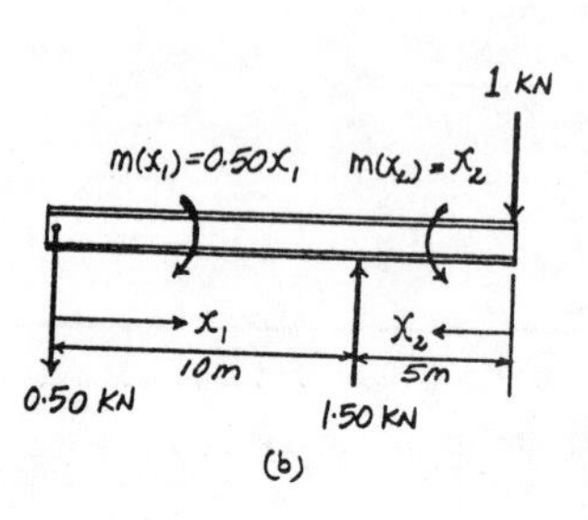

14–94. Determine the slope at A of the A-36 steel beam. $I = 70(10^6)$ mm^4.

Real Moment Function $M(x)$ **:** As shown on figure(a).

Virtual Moment Functions $m_\theta(x)$ **:** As shown on figure(b).

Virtual Work Equation **:** For the slope at point A, apply Eq. 14–43.

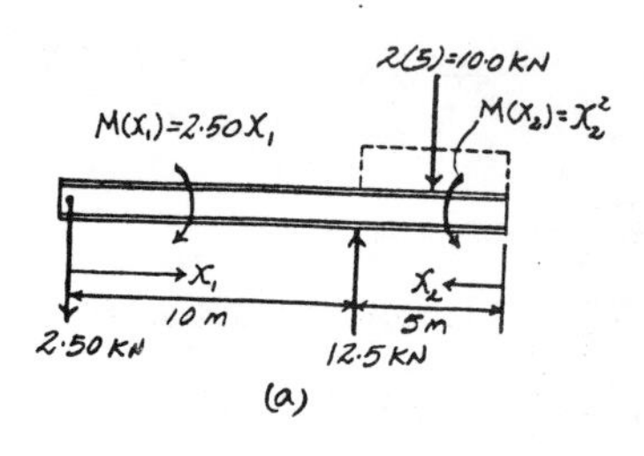

$$1 \cdot \theta = \int_0^L \frac{m_\theta M}{EI} dx$$

$$1 \text{ kN} \cdot \text{m} \cdot \theta_A = \frac{1}{EI}\int_0^{10\text{m}} (1 - 0.100x_1)(2.50x_1)\,dx_1 + \frac{1}{EI}\int_0^{5\text{m}} 0\left(1.00x_2^2\right)dx_2$$

$$\theta_A = \frac{41.667 \text{ kN} \cdot \text{m}^2}{EI}$$

$$= \frac{41.667(1000)}{200(10^9)\,[70(10^{-6})]} = 0.00298 \text{ rad} \qquad \textbf{Ans}$$

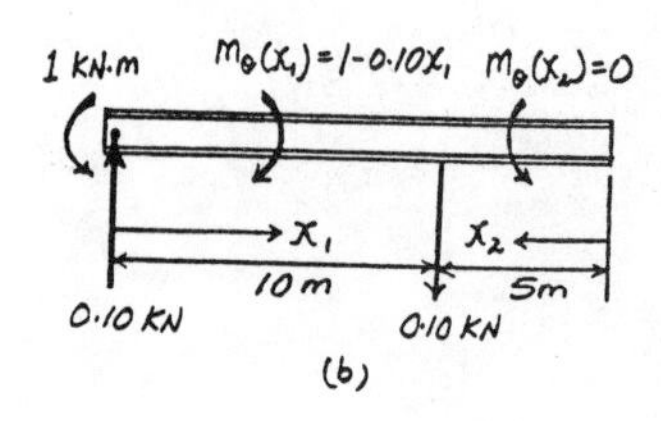

14–95. Determine the slope at B of the A-36 steel beam. $I = 70(10^6)\ \text{mm}^4$.

Real Moment Function $M(x)$ **:** As shown on figure(a).

Virtual Moment Functions $m_\theta(x)$ **:** As shown on figure(b).

***Virtual Work Equation* :** For the slope at point B, apply Eq. 14–43.

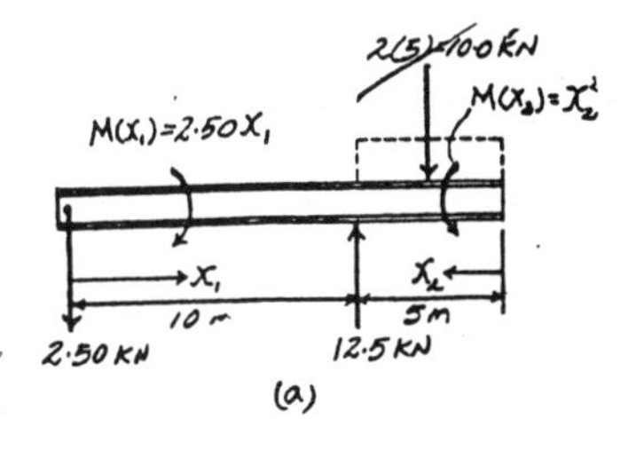

$$1 \cdot \theta = \int_0^L \frac{m_\theta M}{EI} dx$$

$$1\ \text{kN}\cdot\text{m}\cdot\theta_B = \frac{1}{EI}\int_0^{10\text{m}} 0.100x_1\,(2.50x_1)\,dx_1 + \frac{1}{EI}\int_0^{5\text{m}} 0\,(1.00x_2^2)\,dx_2$$

$$\theta_B = \frac{83.333\ \text{kN}\cdot\text{m}^2}{EI} = \frac{83.333(1000)}{200(10^9)\,[70(10^{-6})]} = 0.00595\ \text{rad} \qquad \textbf{Ans}$$

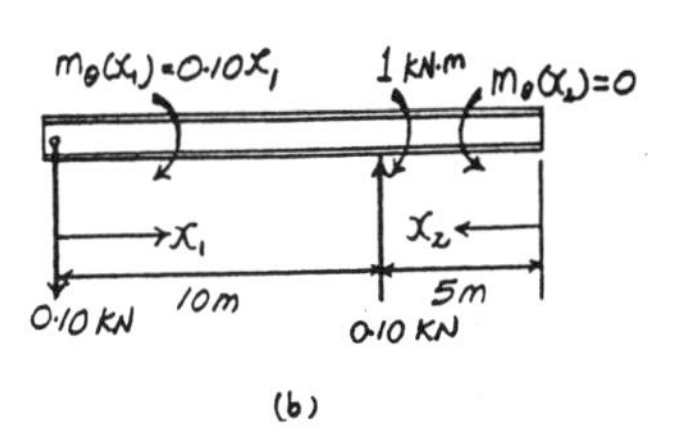

***14–96.** The beam is made of Douglas fir. Determine the displacement at A.

Real Moment Function $M(x)$ **:** As shown on figure(a).

Virtual Moment Functions $m(x)$ **:** As shown on figure(b).

***Virtual Work Equation* :** For the displacement at point A, apply Eq. 14–42.

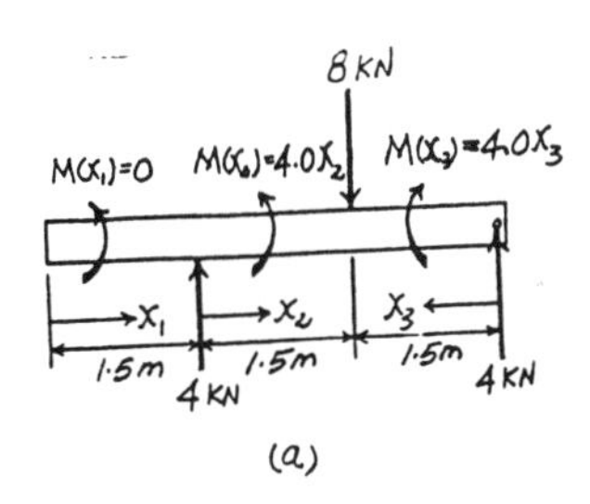

$$1 \cdot \Delta = \int_0^L \frac{mM}{EI} dx$$

$$1\ \text{kN}\cdot\Delta_A = 0 + \frac{1}{EI}\int_0^{1.5\text{m}} (0.500x_2 - 1.50)(4.00x_2)\,dx_2 + \frac{1}{EI}\int_0^{1.5\text{m}} (-0.500x_3)(4.00x_3)\,dx_3$$

$$\Delta_A = -\frac{6.75\ \text{kN}\cdot\text{m}^3}{EI} = -\frac{6.75(1000)}{13.1(10^9)\left[\frac{1}{12}(0.12)(0.18^3)\right]} = -8.835(10^{-3})\ \text{m} = 8.84\ \text{mm} \uparrow \qquad \textbf{Ans}$$

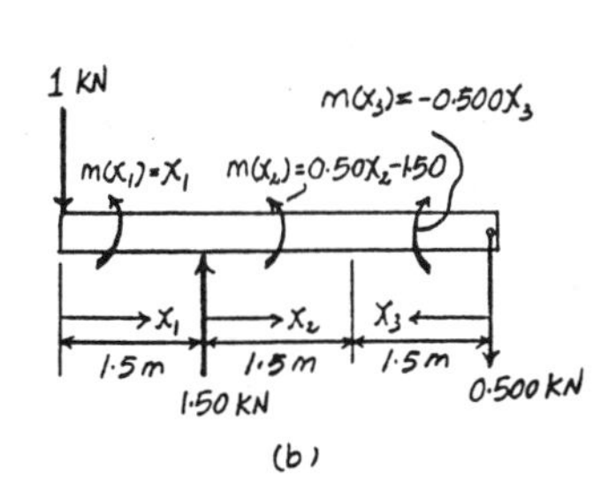

14–97. The beam is made of Douglas fir. Determine the slope at C..

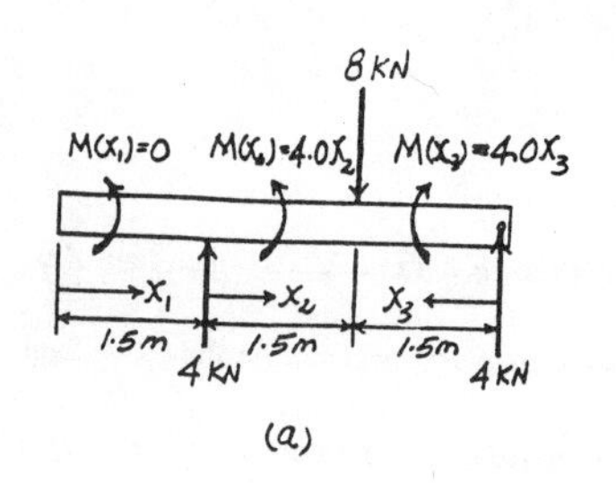

Virtual Work Equation : For the slope at point C, apply Eq. 14 – 43.

$$1 \cdot \theta = \int_0^L \frac{m_\theta M}{EI} dx$$

$$1 \text{ kN} \cdot \text{m} \cdot \theta_C = 0 + \frac{1}{EI}\int_0^{1.5\text{m}} (0.3333x_2)(4.00x_2)\,dx_2 + \frac{1}{EI}\int_0^{1.5\text{m}} (1 - 0.3333x_3)(4.00x_3)\,dx_3$$

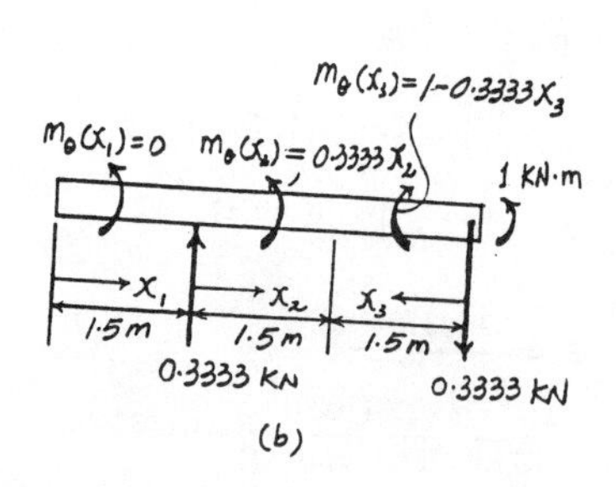

$$\theta_C = \frac{4.50 \text{ kN} \cdot \text{m}^3}{EI}$$

$$= -\frac{4.50(1000)}{13.1(10^9)\left[\frac{1}{12}(0.12)(0.18^3)\right]}$$

$$= 5.89\left(10^{-3}\right) \text{ rad} \qquad \textbf{Ans}$$

14–98. The beam is made of oak, for which E_o = 11 GPa. Determine the slope and displacement at A.

Virtual Work Equation : For the displacement at point A, apply Eq. 14 – 42.

$$1 \cdot \Delta = \int_0^L \frac{mM}{EI} dx$$

$$1 \text{ kN} \cdot \Delta_A = \frac{1}{EI}\int_0^{3\text{m}} x_1\left(\frac{2}{9}x_1^3\right)dx_1 + \frac{1}{EI}\int_0^{3\text{m}} (x_2+3)\left(2.00x_2^2 + 6.00x_2 + 6.00\right)dx_2$$

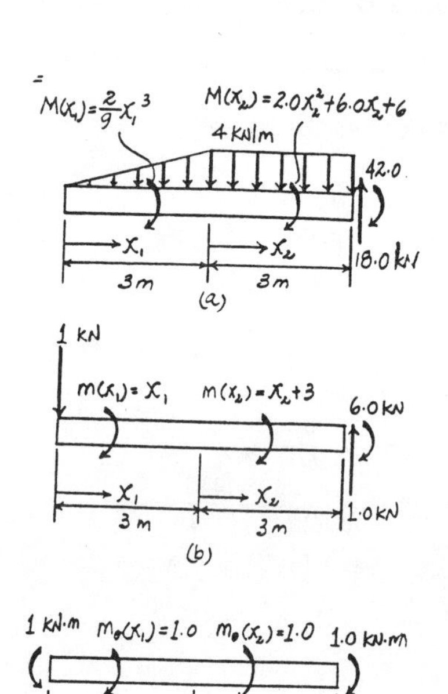

$$\Delta_A = \frac{321.3 \text{ kN} \cdot \text{m}^3}{EI}$$

$$= \frac{321.3(10^3)}{11(10^9)\left[\frac{1}{12}(0.2)(0.4^3)\right]}$$

$$= 0.02738 \text{ m} = 27.4 \text{ m} \downarrow \qquad \textbf{Ans}$$

For the slope at A, apply Eq. 14 – 43.

$$1 \cdot \theta = \int_0^L \frac{m_\theta M}{EI} dx$$

$$1 \text{ kN} \cdot \text{m} \cdot \theta_A = \frac{1}{EI}\int_0^{3\text{m}} 1.00\left(\frac{2}{9}x_1^3\right)dx_1 + \int_0^{3\text{m}} 1.00\left(2.00x_2^2 + 6.00x_2 + 6.00\right)dx_2$$

$$\theta_A = \frac{67.5 \text{ kN} \cdot \text{m}^2}{EI}$$

$$= \frac{67.5(1000)}{11(10^9)\left[\frac{1}{12}(0.2)(0.4^3)\right]}$$

$$= 5.75\left(10^{-3}\right) \text{ rad} \qquad \textbf{Ans}$$

14–99. Determine the displacement at C of the A-36 steel beam. $I = 36.9(10^6)\ \text{mm}^4$.

Real Moment Function M(x) : As shown on figure(a).

Virtual Moment Functions m(x) : As shown on figure(b).

Virtual Work Equation : For the displacement at point C, apply Eq. 14–42.

$$1\cdot\Delta = \int_0^L \frac{mM}{EI}dx$$

$$1\ \text{kN}\cdot\Delta_C = \frac{1}{EI}\int_0^{4\text{m}} 0.500x_1\left(2.00x_1^2 + 2.00x_1\right)dx_1 + \frac{1}{EI}\int_0^{2\text{m}} 1.0x_2(20.0x_2)\,dx_2$$

$$\Delta_C = \frac{138.67\ \text{kN}\cdot\text{m}^3}{EI} = \frac{138.67(1000)}{200(10^9)[36.9(10^{-6})]} = 0.01879\ \text{m} = 18.8\ \text{mm}\ \downarrow \qquad \textbf{Ans}$$

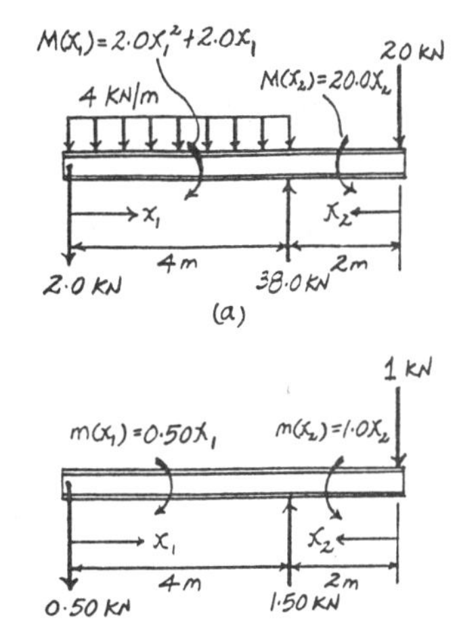

***14–100.** Determine the displacement at D of the A-36 steel beam. $I = 36.9(10^6)\ \text{mm}^4$.

Real Moment Function M(x) : As shown on Figure(a).

Virtual Moment Functions m(x) : As shown on figure(b).

Virtual Work Equation : For the displacement at point D, apply Eq. 14–42.

$$1\cdot\Delta = \int_0^L \frac{mM}{EI}dx$$

$$1\ \text{kN}\cdot\Delta_D = \frac{1}{EI}\int_0^{2\text{m}} (-0.500x_1)\left(2.00x_1^2 + 2.00x_1\right)dx_1 + \frac{1}{EI}\int_0^{2\text{m}} (-0.500x_2)\left(2.00x_2^2 - 18.0x_2 + 40.0\right)dx_2$$

$$\Delta_D = -\frac{26.667\ \text{kN}\cdot\text{m}^3}{EI} = -\frac{26.667(1000)}{200(10^9)[36.9(10^{-6})]} = -3.613\left(10^{-3}\right)\ \text{m} = 3.61\ \text{mm}\ \uparrow \qquad \textbf{Ans}$$

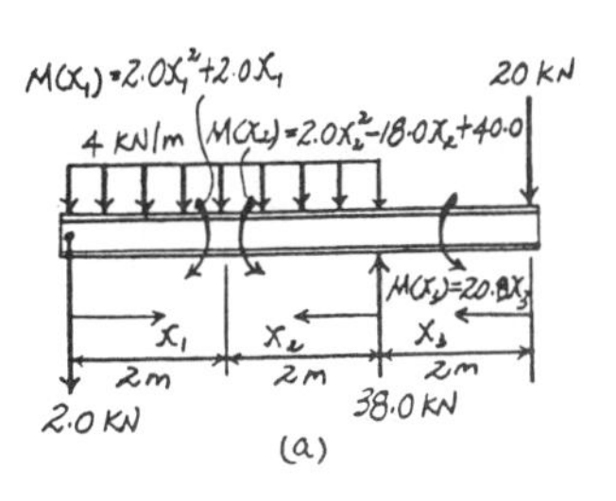

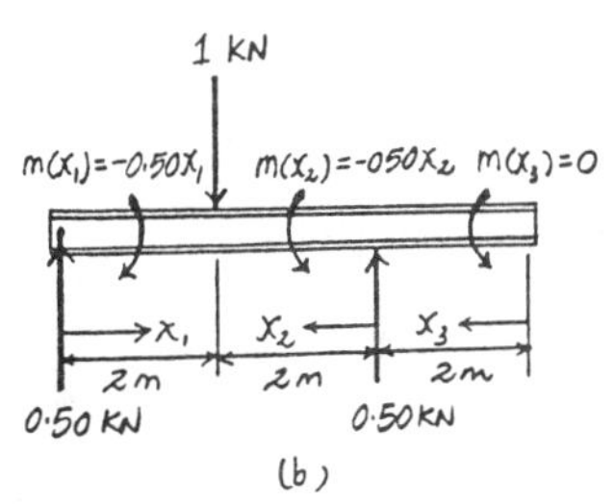

14–101. Beam AB has a square cross section of 100 mm by 100 mm. Bar CD has a diameter of 10 mm. If both members are made of A-36 steel, determine the vertical displacement of point B due to the loading of 10 kN.

Real Moment Function $M(x)$ **:** As shown on figure(a).

Virtual Moment Functions $m(x)$ **:** As shown on figure(b).

Virtual Work Equation **:** For the displacement at point B, combine Eq. 14–42 and Eq. 14–39.

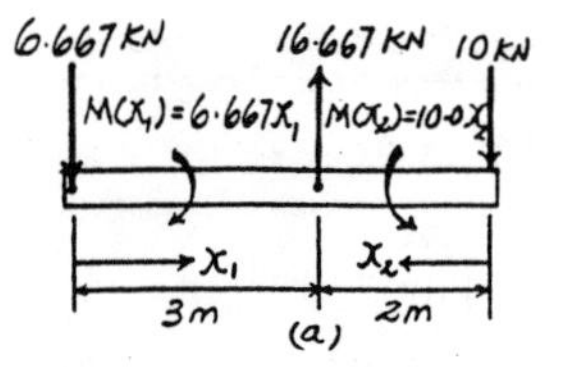

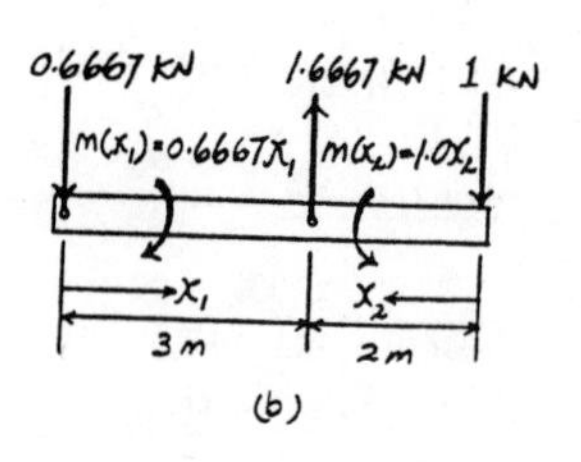

$$1\cdot\Delta=\int_0^L \frac{mM}{EI}dx+\frac{nNL}{AE}$$

$$1\text{ kN}\cdot\Delta_B=\frac{1}{EI}\int_0^{3\text{m}}(0.6667x_1)(6.667x_1)\,dx_1+\frac{1}{EI}\int_0^{2\text{m}}(1.00x_2)(10.0x_2)\,dx_2+\frac{1.667(16.667)(2)}{AE}$$

$$\Delta_B=\frac{66.667\text{ kN}\cdot\text{m}^3}{EI}+\frac{55.556\text{ kN}\cdot\text{m}}{AE}$$

$$=\frac{66.667(1000)}{200(10^9)\left[\frac{1}{12}(0.1)(0.1^3)\right]}+\frac{55.556(1000)}{\left[\frac{\pi}{4}(0.01^2)\right][200(10^9)]}$$

$$=0.04354\text{ m}=43.5\text{ mm}\downarrow \qquad \textbf{Ans}$$

14–102. Beam AB has a square cross section of 100 mm by 100 mm. Bar CD has a diameter of 10 mm. If both members are made of A-36 steel, determine the slope at A due to the loading of 10 kN.

Real Moment Function $M(x)$ **:** As shown on figure(a).

Virtual Moment Functions $m_\theta(x)$ **:** As shown on figure(b).

Virtual Work Equation **:** For the slope at point A, combine Eq. 14–42 and Eq. 14–39.

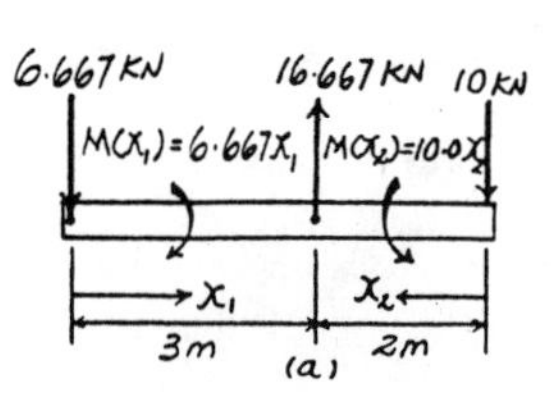

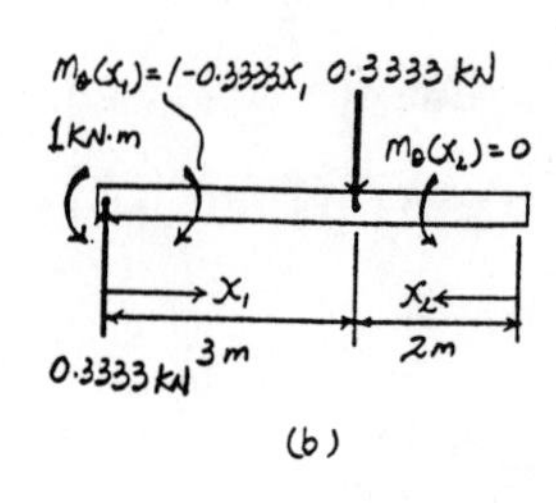

$$1\cdot\theta=\int_0^L \frac{m_\theta M}{EI}dx+\frac{nNL}{AE}$$

$$1\text{ kN}\cdot\text{m}\cdot\theta_A=\frac{1}{EI}\int_0^{3\text{m}}(1-0.3333x_1)(6.667x_1)\,dx_1+\frac{1}{EI}\int_0^{2\text{m}}0(10.0x_2)\,dx_2+\frac{(-0.3333)(16.667)(2)}{AE}$$

$$\theta_A=\frac{10.0\text{ kN}\cdot\text{m}^2}{EI}-\frac{11.111\text{ kN}}{AE}$$

$$=\frac{10.0(1000)}{200(10^9)\left[\frac{1}{12}(0.1)(0.1^3)\right]}-\frac{11.111(1000)}{\left[\frac{\pi}{4}(0.01^2)\right][200(10^9)]}$$

$$=0.00529\text{ rad} \qquad \textbf{Ans}$$

14–103. Bar ABC has a rectangular cross section of 300 mm by 100 mm. Attached rod DB has a diameter of 20 mm. If both members are made of A-36 steel, determine the vertical displacement of point C due to the loading. Consider only the effect of bending in ABC and axial force in DB.

Real Moment Function $M(x)$ **:** As shown on figure(a).

Virtual Moment Functions $m(x)$ **:** As shown on figure(b).

Virtual Work Equation **:** For the displacement at point C, combine Eq. 14–42 and Eq. 14–39.

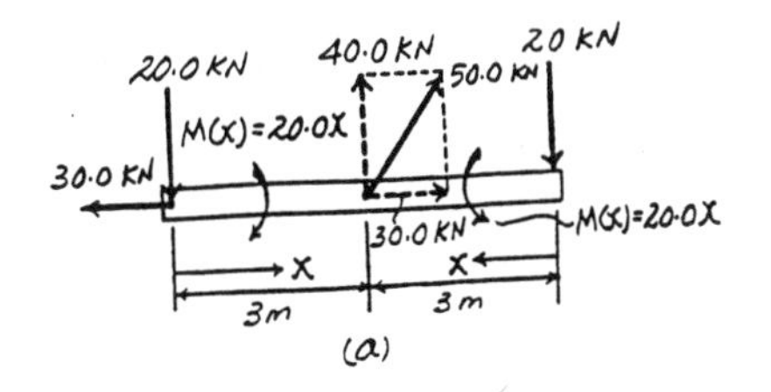

$$1 \cdot \Delta = \int_0^L \frac{mM}{EI} dx + \frac{nNL}{AE}$$

$$1\ \text{kN} \cdot \Delta_C = 2\left[\frac{1}{EI}\int_0^{3\text{m}} (1.00x)(20.0x)\,dx\right] + \frac{2.50(50.0)(5)}{AE}$$

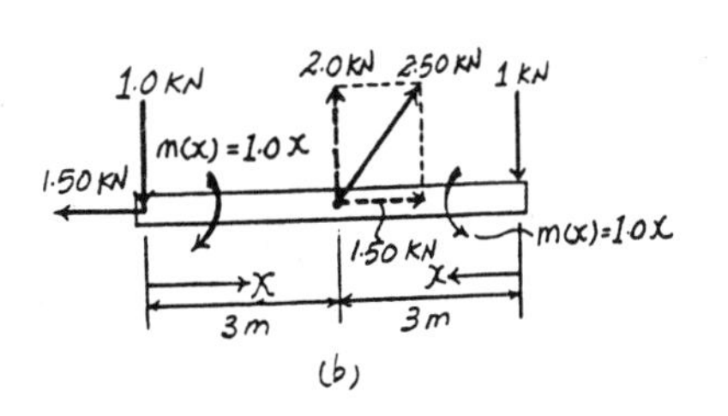

$$\Delta_C = \frac{360\ \text{kN} \cdot \text{m}^3}{EI} + \frac{625\ \text{kN} \cdot \text{m}}{AE}$$

$$= \frac{360(1000)}{200(10^9)\left[\frac{1}{12}(0.1)(0.3^3)\right]} + \frac{625(1000)}{\left[\frac{\pi}{4}(0.02^2)\right][200(10^9)]}$$

$$= 0.017947\ \text{m} = 17.9\ \text{mm} \downarrow \qquad \textbf{Ans}$$

***14–104.** Bar ABC has a rectangular cross section of 300 mm by 100 mm. Attached rod DB has a diameter of 20 mm. If both members are made of A-36 steel, determine the slope at A due to the loading. Consider only the effect of bending in ABC and axial force in DB.

Real Moment Function $M(x)$ **:** As shown on figure(a).

Virtual Moment Functions $m_\theta(x)$ **:** As shown on figure(b).

Virtual Work Equation **:** For the slope at point A, combine Eq. 14–43 and Eq. 14–39.

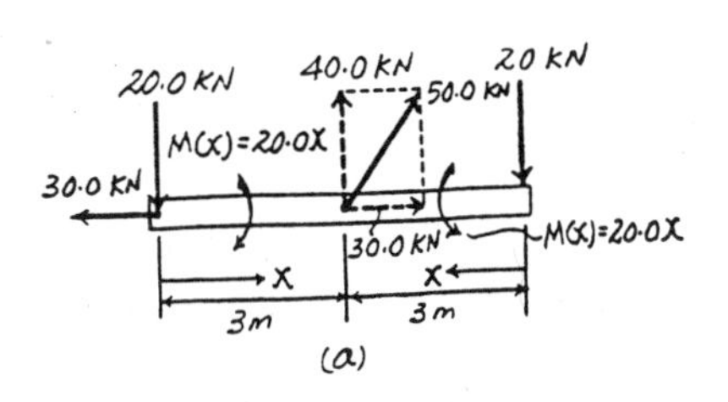

$$1 \cdot \theta = \int_0^L \frac{m_\theta M}{EI} dx + \frac{nNL}{AE}$$

$$1\ \text{kN} \cdot \text{m} \cdot \theta_A = \frac{1}{EI}\int_0^{3\text{m}} (1 - 0.3333x)(20.0x)\,dx + \frac{(-0.41667)(50.0)(5)}{AE}$$

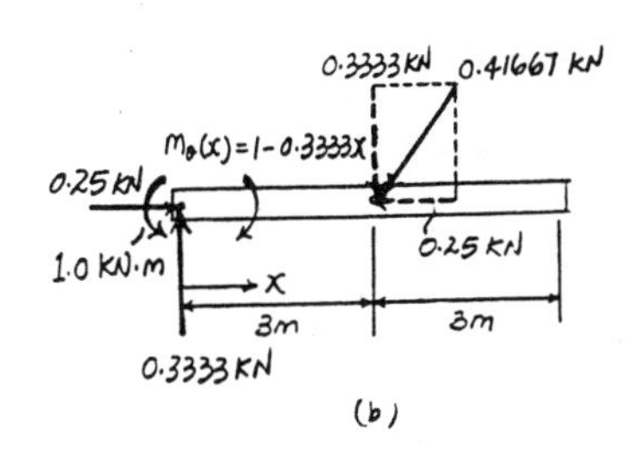

$$\theta_A = \frac{30.0\ \text{kN} \cdot \text{m}^2}{EI} - \frac{104.167\ \text{kN}}{AE}$$

$$= \frac{30.0(1000)}{200(10^9)\left[\frac{1}{12}(0.1)(0.3^3)\right]} - \frac{104.167(1000)}{\left[\frac{\pi}{4}(0.02^2)\right][200(10^9)]}$$

$$= -0.991\left(10^{-3}\right)\ \text{rad} = 0.991\left(10^{-3}\right)\ \text{rad} \qquad \textbf{Ans}$$

14–105. Bar ADC has a square cross section of 2 in. by 2 in. Attached rod CB has a diameter of 0.25 in. If both members are made of A-36 steel, determine the horizontal displacement of point C due to the 600-lb loading. Consider only the effect of bending in ADC and axial force in CB.

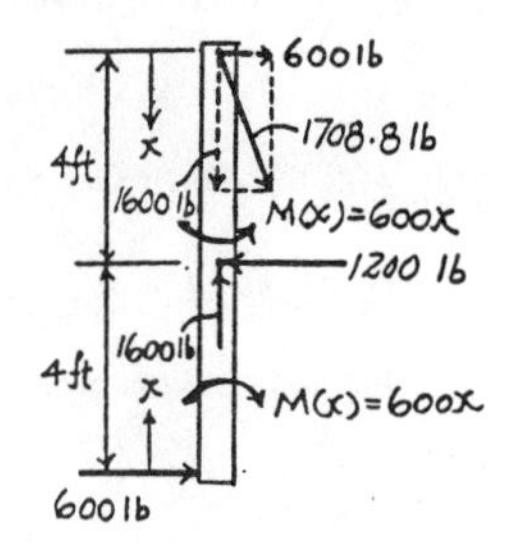

Real Moment Function M(x) : As shown on figure(a).

Virtual Moment Functions m(x) : As shown on figure(b).

Virtual Work Equation : For the displacement at point C, combine Eq. 14 – 42 and Eq. 14 – 39.

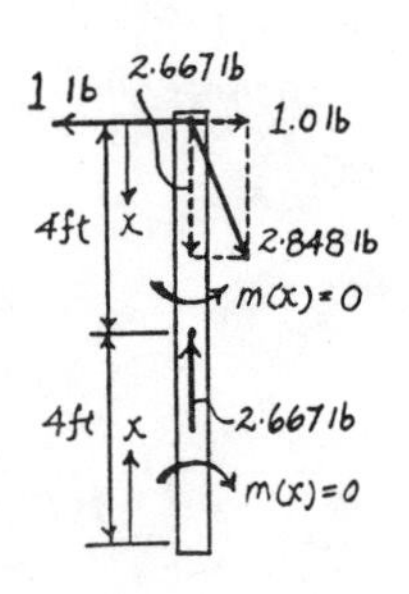

$$1 \cdot \Delta = \int_0^L \frac{mM}{EI} dx + \frac{nNL}{AE}$$

$$1 \text{ lb} \cdot \Delta_C = 2\left[\frac{1}{EI}\int_0^{4\text{ft}} (0)(600x)\, dx\right] + \frac{2.848(1708.8)[8.5440(12)]}{AE}$$

$$\Delta_C = 0 + \frac{498969.8 \text{ lb} \cdot \text{in.}}{AE}$$

$$= 0 + \frac{498969.8}{\left[\frac{\pi}{4}(0.25^2)\right]\left[29.0(10^6)\right]}$$

$$= 0.351 \text{ in.} \leftarrow \qquad \textbf{Ans}$$

14–106. Bar ADC has a square cross section of 2 in. by 2 in. Attached rod CB has a diameter of 0.25 in. If both members are made of A-36 steel, determine the horizontal displacement of point A due to the 600-lb loading. Consider only the effect of bending in ADC and axial force in CB.

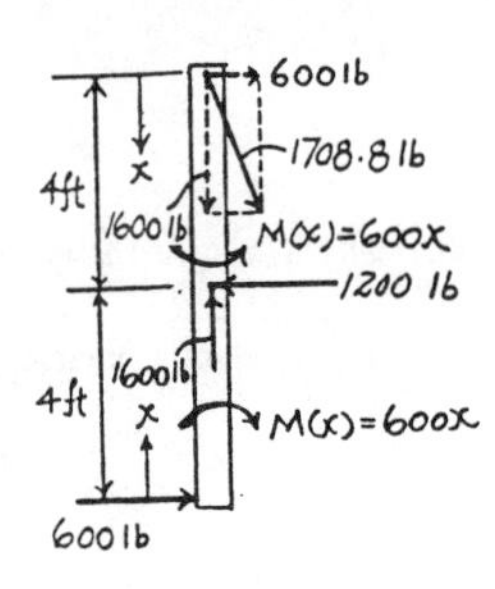

Real Moment Function M(x) : As shown on Figure(a).

Virtual Moment Functions m(x) : As shown on figure(b).

Virtual Work Equation : For the displacement at point A, combine Eq. 14 – 42 and Eq. 14 – 39.

$$1 \cdot \Delta = \int_0^L \frac{mM}{EI} dx + \frac{nNL}{AE}$$

$$1 \text{ lb} \cdot \Delta_A = 2\left[\frac{1}{EI}\int_0^{4\text{ft}} (1.00x)(600x)\, dx\right] + \frac{2.848(1708.8)[8.5440(12)]}{AE}$$

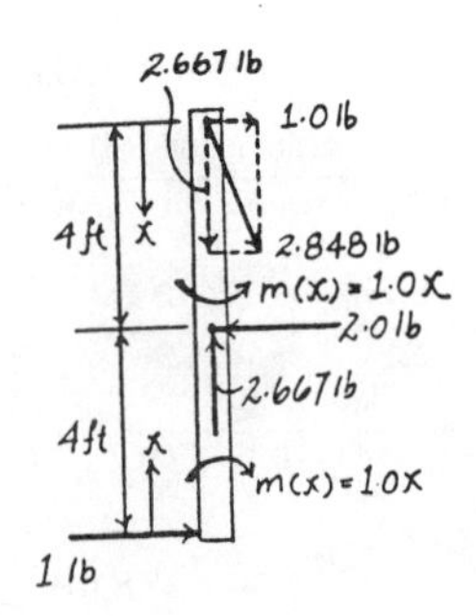

$$\Delta_A = \frac{25600 \text{ lb} \cdot \text{ft}^3}{EI} + \frac{498969.8 \text{ lb} \cdot \text{in.}}{AE}$$

$$= \frac{25600(12^3)}{29.0(10^6)\left[\frac{1}{12}(2)(2^3)\right]} + \frac{498969.8}{\left[\frac{\pi}{4}(0.25^2)\right]\left[29.0(10^6)\right]}$$

$$= 1.49 \text{ in.} \rightarrow \qquad \textbf{Ans}$$

14–107. Determine the horizontal displacement of point C. EI is constant. There is a fixed support at A. Consider only the effect of bending.

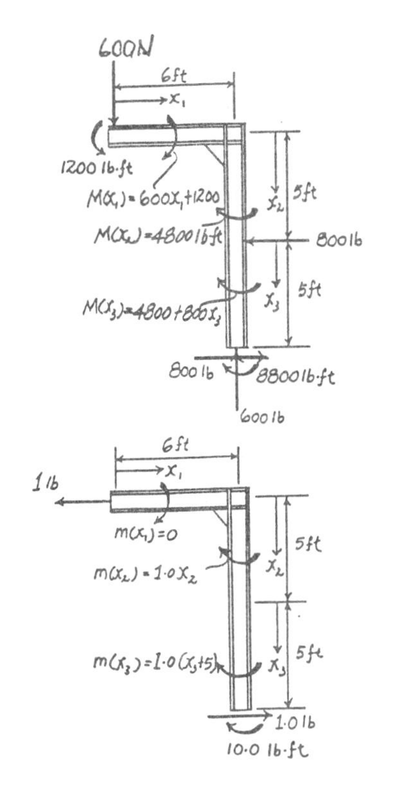

Real Moment Function $M(x)$ **:** As shown on figure(a).

Virtual Moment Functions $m(x)$ **:** As shown on figure(b).

Virtual Work Equation **:** For the horizontal displacement at point C, apply Eq. 14 – 42.

$$1 \cdot \Delta = \int_0^L \frac{mM}{EI} dx$$

$$1 \text{ lb} \cdot (\Delta_C)_h = 0 + \frac{1}{EI} \int_0^{5\text{ ft}} (1.00x_2)(4800)\, dx_2 + \frac{1}{EI} \int_0^{5\text{ ft}} 1.00(x_3 + 5)(4800 + 800x_3)\, dx_3$$

$$(\Delta_C)_h = \frac{323(10^3) \text{ lb} \cdot \text{ft}^3}{EI} \qquad \textbf{Ans}$$

***14–108.** Determine the horizontal displacement of point A on the angle bracket due to the concentrated force **P**. The bracket is fixed connected to its support. EI is constant. Consider only the effect of bending.

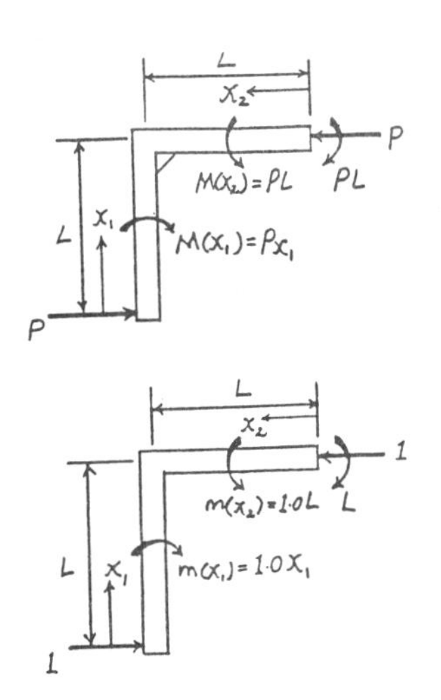

Real Moment Function $M(x)$ **:** As shown on figure(a).

Virtual Moment Functions $m(x)$ **:** As shown on figure(b).

Virtual Work Equation **:** For the horizontal displacement at point A, apply Eq. 14 – 42.

$$1 \cdot \Delta = \int_0^L \frac{mM}{EI} dx$$

$$1 \cdot (\Delta_A)_h = \frac{1}{EI} \int_0^L (1.00x_1)(Px_1)\, dx_1 + \frac{1}{EI} \int_0^L (1.00L)(PL)\, dx_2$$

$$(\Delta_A)_h = \frac{4PL^3}{3EI} \rightarrow \qquad \textbf{Ans}$$

14–109. The frame is made from two segments, each of length L and flexural stiffness EI. If it is subjected to the uniform distributed load, determine the vertical displacement of point C. Consider only the effect of bending.

Real Moment Function $M(x)$: As shown on figure(a).

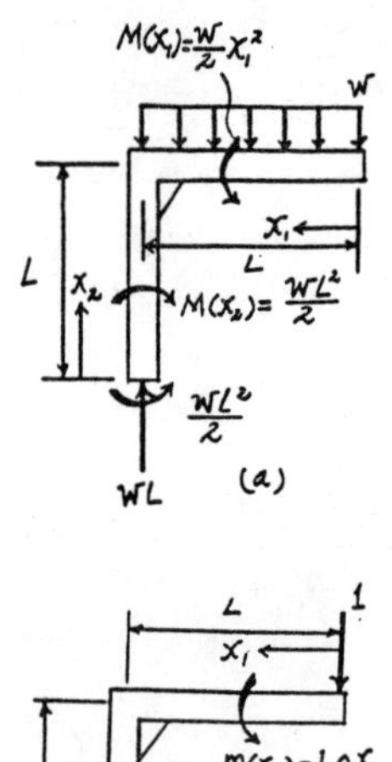

Virtual Moment Functions $m(x)$: As shown on figure(b).

Virtual Work Equation : For the vertical displacement at point C, apply Eq. 14–42.

$$1 \cdot \Delta = \int_0^L \frac{mM}{EI} dx$$

$$1 \cdot (\Delta_C)_v = \frac{1}{EI}\int_0^L (1.00x_1)\left(\frac{w}{2}x_1^2\right)dx_1 + \frac{1}{EI}\int_0^L (1.00L)\left(\frac{wL^2}{2}\right)dx_2$$

$$(\Delta_C)_v = \frac{5wL^4}{8EI} \downarrow \qquad \textbf{Ans}$$

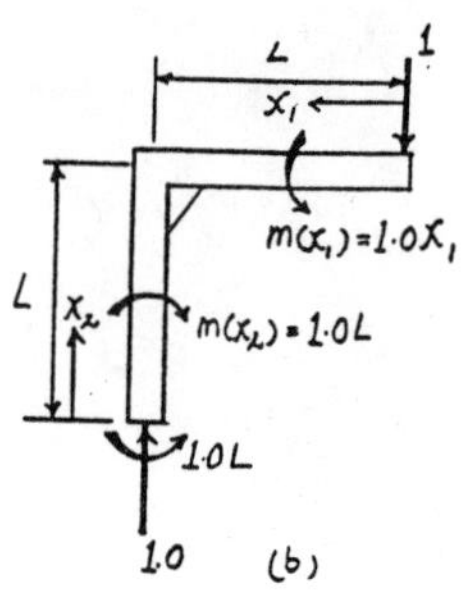

14–110. The frame is made from two segments, each of length L and flexural stiffness EI. If it is subjected to the uniform distributed load, determine the horizontal displacement of point B. Consider only the effect of bending.

Real Moment Function $M(x)$: As shown on figure(a).

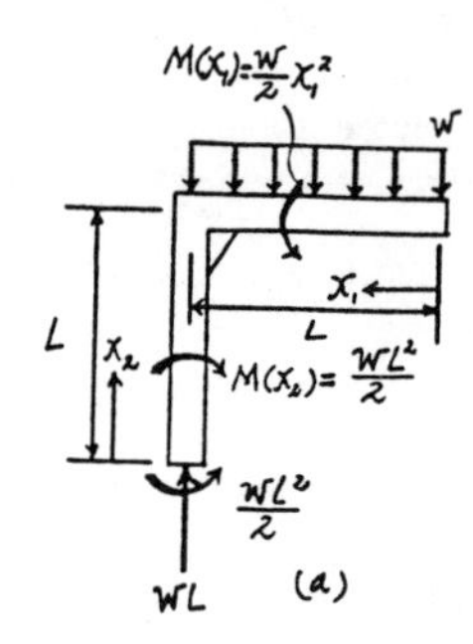

Virtual Moment Functions $m(x)$: As shown on figure(b).

Virtual Work Equation : For the horizontal displacement at point B, apply Eq. 14–42.

$$1 \cdot \Delta = \int_0^L \frac{mM}{EI} dx$$

$$1 \cdot (\Delta_B)_h = \frac{1}{EI}\int_0^L (0)\left(\frac{w}{2}x_1^2\right)dx_1 + \frac{1}{EI}\int_0^L (1.00L - 1.00x_2)\left(\frac{wL^2}{2}\right)dx_2$$

$$(\Delta_B)_h = \frac{wL^4}{4EI} \rightarrow \qquad \textbf{Ans}$$

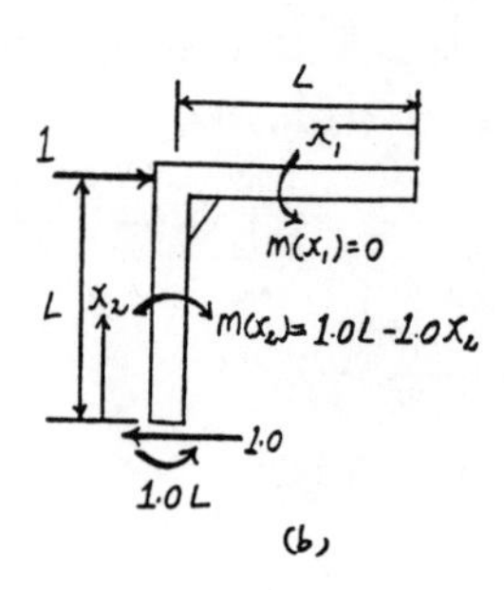

14–111. The ring rests on the rigid surface and is subjected to the vertical load **P**. Determine the vertical displacement at *B*. *EI* is constant.

Model : The ring can be modeled as a half ring as shown in figure(a).

Real Moment Function $M(x)$: As shown on figure(a).

Virtual Moment Functions $m(x)$ and $m_\theta(x)$: As shown on figure(b) and (c).

Virtual Work Equation : Due to symmetry, the slope at B remains horizontal, i.e., equal to zero. Applying Eq. 14 – 43, we have

$$1 \cdot \theta = \int_0^L \frac{m_\theta M}{EI} ds \qquad \text{Where } ds = r d\theta$$

$$1 \cdot \theta_B = 0 = \frac{1}{EI}\int_0^\pi 1.00\left(\frac{Pr}{2}\sin\theta - M_0\right) r d\theta$$

$$M_0 = \frac{Pr}{\pi}$$

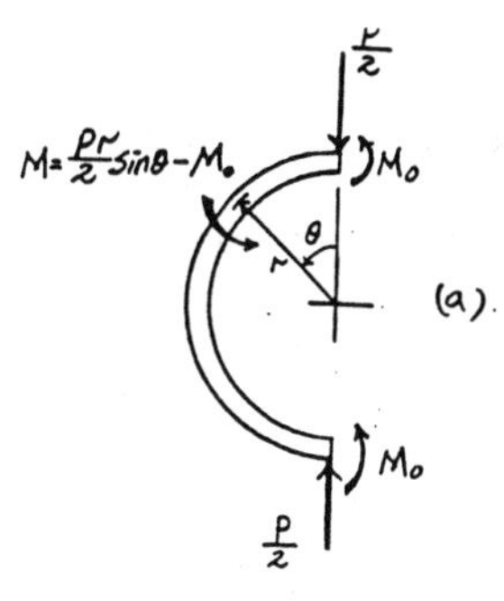

For the vertical displacement at B, apply Eq. 14 – 42.

$$1 \cdot \Delta = \int_0^L \frac{mM}{EI} ds$$

$$1 \cdot \Delta_B = \frac{1}{EI}\int_0^\pi (r\sin\theta)\left(\frac{Pr}{2}\sin\theta - \frac{Pr}{\pi}\right) r d\theta$$

$$= \frac{Pr^3}{2\pi EI}\int_0^\pi \left(\pi \sin^2\theta - 2\sin\theta\right) d\theta$$

$$= \frac{Pr^3}{4\pi EI}\int_0^\pi [\pi(1-\cos 2\theta) - 4\sin\theta]\, d\theta$$

$$\Delta_B = \frac{Pr^3}{4\pi EI}\left(\pi^2 - 8\right) \qquad \textbf{Ans}$$

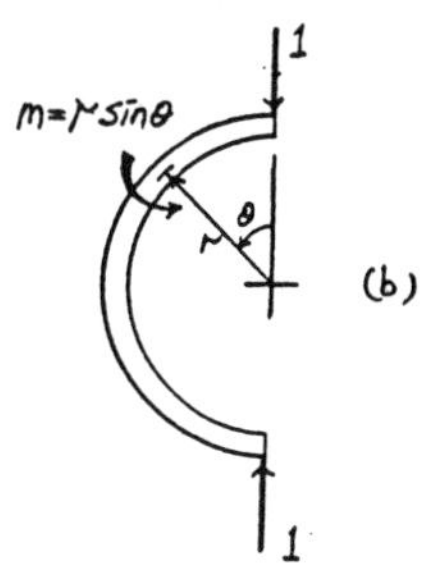

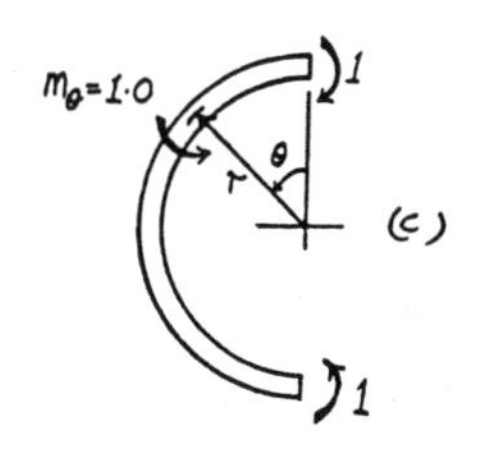

***14–112.** Solve Prob. 14–60 using Castigliano's theorem.

Member Forces N : Member forces due to external force **P** and external applied forces are shown on the figure.

Castigliano's Second Theorem : Applying Eq. 14 – 48, we have

Member	N	$\frac{\partial N}{\partial P}$	$N(P = 600 \text{ lb})$	L	$N\left(\frac{\partial N}{\partial P}\right)L$
AB	$1.1662P$	1.1662	699.71	5.8310	4758.06
BC	$-0.600P$	−0.600	−360.00	3	648.00
					$\sum$ 5406.06 lb · ft

$$\Delta = \sum N\left(\frac{\partial N}{\partial P}\right)\frac{L}{AE}$$

$$(\Delta_B)_v = \frac{5406.06 \text{ lb} \cdot \text{ft}}{AE}$$

$$= \frac{5406.06(12)}{2[29.0(10^6)]} = 1.12\left(10^{-3}\right) \text{ in.} \downarrow \qquad \textbf{Ans}$$

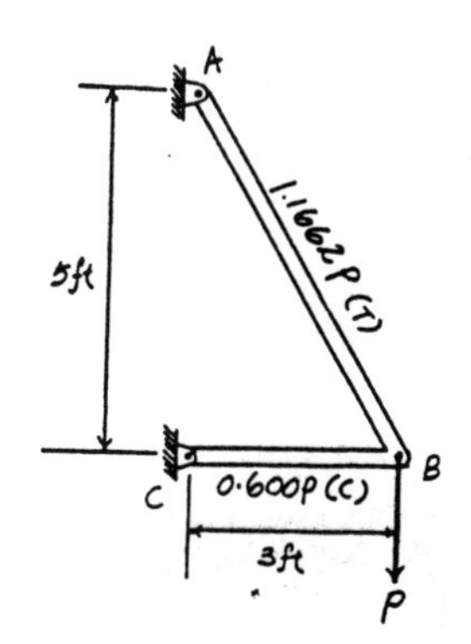

14–113. Solve Prob. 14–59 using Castigliano's theorem.

Member Forces N : Member forces due to external force **P** and external applied forces are shown on the figure.

Castigliano's Second Theorem : Applying Eq. 14 – 48, we have

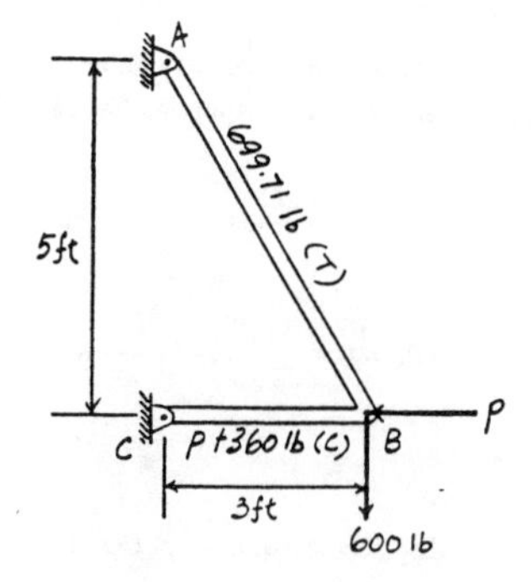

Member	N	$\frac{\partial N}{\partial P}$	$N(P=0)$	L	$N\left(\frac{\partial N}{\partial P}\right)L$
AB	699.71	0	699.71	5.8310	0
BC	$-(P+360)$	-1	-360.00	3	1080
					$\sum$ 1080 lb · ft

$$\Delta = \sum N\left(\frac{\partial N}{\partial P}\right)\frac{L}{AE}$$

$$(\Delta_B)_h = \frac{1080 \text{ lb} \cdot \text{ft}}{AE}$$

$$= \frac{1080(12)}{2[29.0(10^6)]} = 0.223\left(10^{-3}\right) \text{ in.} \leftarrow \qquad \textbf{Ans}$$

14–114. Solve Prob. 14–61 using Castigliano's theorem.

Member Forces N : Member forces due to external force **P** and external applied forces are shown on the figure.

Castigliano's Second Theorem : Applying Eq. 14 – 48, we have

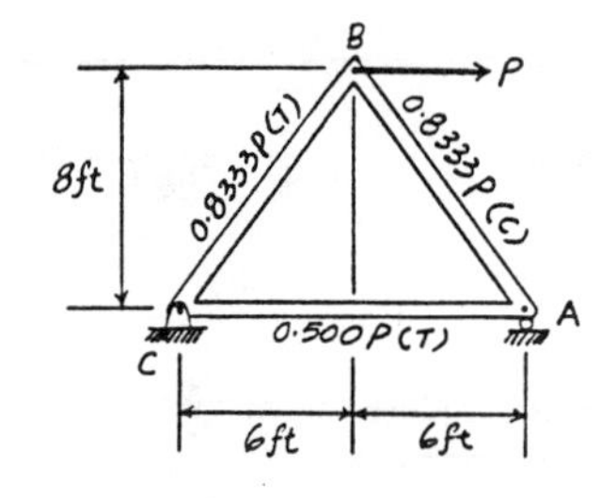

Member	N	$\frac{\partial N}{\partial P}$	$N(P=200\text{ lb})$	L	$N\left(\frac{\partial N}{\partial P}\right)L$
AB	$-0.8333P$	-0.8333	-166.67	10.0	1388.89
BC	$0.8333P$	0.8333	166.67	10.0	1388.89
AC	$0.500P$	0.500	100.00	12	600.00
					$\sum$ 3377.78 lb · ft

$$\Delta = \sum N\left(\frac{\partial N}{\partial P}\right)\frac{L}{AE}$$

$$(\Delta_B)_h = \frac{3377.78 \text{ lb} \cdot \text{ft}}{AE}$$

$$= \frac{3377.78(12)}{2[29.0(10^6)]} = 0.699\left(10^{-3}\right) \text{ in.} \rightarrow \qquad \textbf{Ans}$$

14–115. Solve Prob. 14–62 using Castigliano's theorem.

Member Forces N : Member forces due to external force **P** and external applied forces are shown on the figure.

Castigliano's Second Theorem : Applying Eq. 14 – 48, we have

Member	N	$\frac{\partial N}{\partial P}$	$N(P=0)$	L	$N\left(\frac{\partial N}{\partial P}\right)L$
AB	$-(0.625P+166.67)$	-0.625	-166.67	10.0	1041.67
BC	$-(0.625P-166.67)$	-0.625	166.67	10.0	-1041.67
AC	$0.375P+100$	0.375	100.00	12	450.00
					$\sum 450.00$ lb · ft

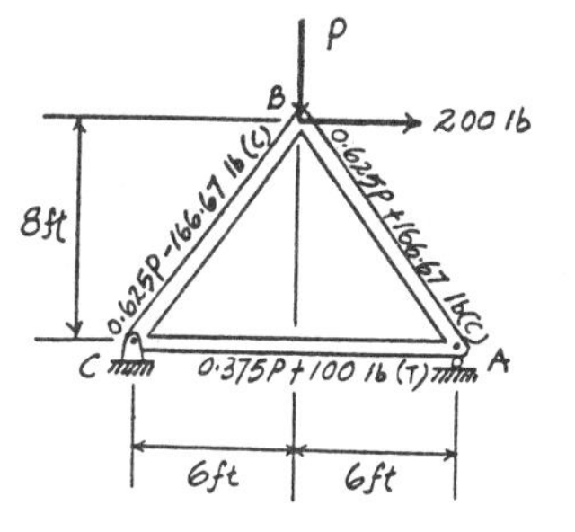

$$\Delta = \sum N\left(\frac{\partial N}{\partial P}\right)\frac{L}{AE}$$

$$(\Delta_B)_v = \frac{450.00 \text{ lb} \cdot \text{ft}}{AE}$$

$$= \frac{450(12)}{2[29.0(10^6)]} = 0.0931\left(10^{-3}\right) \text{ in.} \downarrow \qquad \textbf{Ans}$$

***14–116.** Solve Prob. 14–64 using Castigliano's theorem.

Member Forces N : Member forces due to external force **P** and external applied forces are shown on the figure.

Castigliano's Second Theorem : Applying Eq. 14 – 48, we have

Member	N	$\frac{\partial N}{\partial P}$	$N(P=0)$	L	$N\left(\frac{\partial N}{\partial P}\right)L$
AB	$0.9014P+27.04$	0.9014	27.04	3.6056	87.885
BC	$0.9014P+27.04$	0.9014	27.04	3.6056	87.885
CD	$-(0.750P+22.5)$	-0.750	-22.5	3	50.625
AD	$-(0.750P+22.5)$	-0.750	-22.5	3	50.625
BD	-30.0	0	-30.0	2	0
					$\sum 277.021$ kN · m

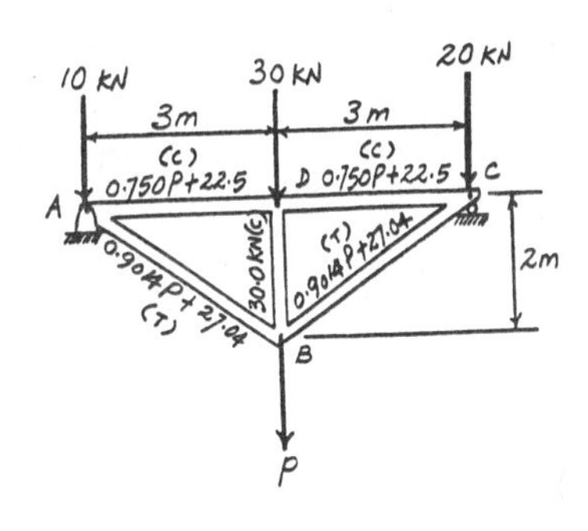

$$\Delta = \sum N\left(\frac{\partial N}{\partial P}\right)\frac{L}{AE}$$

$$(\Delta_B)_v = \frac{277.021 \text{ kN} \cdot \text{m}}{AE}$$

$$= \frac{277.021(1000)}{[0.3(10^{-3})][200(10^9)]}$$

$$= 4.617\left(10^{-3}\right) \text{ m} = 4.62 \text{ mm} \downarrow \qquad \textbf{Ans}$$

14–117. Solve Prob. 14–63 using Castigliano's theorem.

Member Forces N : Member forces due to external force **P** and external applied forces are shown on the figure.

Castigliano's Second Theorem : Applying Eq. 14 – 48, we have

Member	N	$\dfrac{\partial N}{\partial P}$	$N(P = 30\text{ kN})$	L	$N\left(\dfrac{\partial N}{\partial P}\right)L$
AB	$0.9014P$	0.9014	27.04	3.6056	87.885
BC	$0.9014P$	0.9014	27.04	3.6056	87.885
CD	$-0.750P$	−0.750	−22.5	3	50.625
AD	$-0.750P$	−0.750	−22.5	3	50.625
BD	$-1.00P$	−1.00	−30.0	2	60.00
					$\sum 337.021\text{ kN}\cdot\text{m}$

$$\Delta = \sum N\left(\frac{\partial N}{\partial P}\right)\frac{L}{AE}$$

$$(\Delta_D)_v = \frac{337.021\text{ kN}\cdot\text{m}}{AE}$$

$$= \frac{337.021(1000)}{[0.3(10^{-3})][200(10^{9})]}$$

$$= 5.617\left(10^{-3}\right)\text{ m} = 5.62\text{ mm} \downarrow \qquad \textbf{Ans}$$

14–118. Solve Prob. 14–65 using Castigliano's theorem.

Member Forces N : Member forces due to external force **P** and external applied forces are shown on the figure.

Castigliano's Second Theorem : Applying Eq. 14 – 48, we have

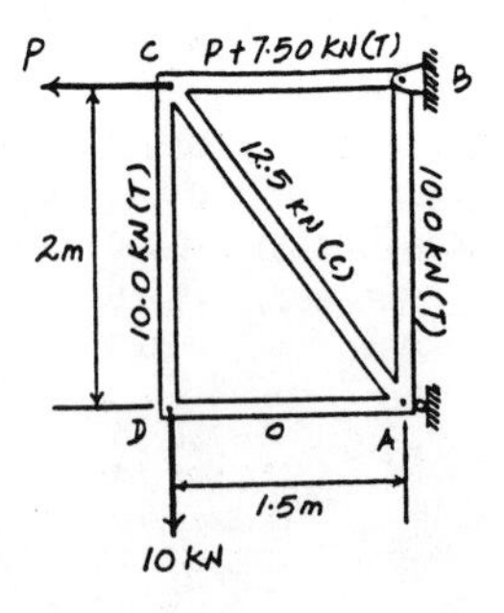

Member	N	$\dfrac{\partial N}{\partial P}$	$N(P = 5\text{ kN})$	L	$N\left(\dfrac{\partial N}{\partial P}\right)L$
AB	10.0	0	10.0	2	0
BC	$1.00P + 7.50$	1.00	12.5	1.5	18.75
CD	10.0	0	10.0	2	0
AD	0	0	0	1.5	0
AC	−12.5	0	−12.5	2.5	0
					$\sum 18.75\text{ kN}\cdot\text{m}$

$$\Delta = \sum N\left(\frac{\partial N}{\partial P}\right)\frac{L}{AE}$$

$$(\Delta_C)_h = \frac{18.75\text{ kN}\cdot\text{m}}{AE}$$

$$= \frac{18.75(10^{3})}{0.400(10^{-3})[200(10^{9})]}$$

$$= 0.2344\left(10^{-3}\right)\text{ m} = 0.234\text{ mm} \leftarrow \qquad \textbf{Ans}$$

14–119. Solve Prob. 14–66 using Castigliano's theorem.

Member Forces N : Member forces due to external force **P** and external applied forces are shown on the figure.

Castigliano's Second Theorem : Applying Eq. 14 – 48, we have

Member	N	$\dfrac{\partial N}{\partial P}$	$N(P = 10\text{ kN})$	L	$N\left(\dfrac{\partial N}{\partial P}\right)L$
AB	$1.00P$	1.00	10.0	2	20.00
BC	$0.750P+5.00$	0.750	12.5	1.5	14.0625
CD	$1.00P$	1.00	10.0	2	20.00
AD	0	0	0	1.5	0
AC	$-1.25P$	-1.25	-12.5	2.5	39.0625
					$\sum 93.125\text{ kN}\cdot\text{m}$

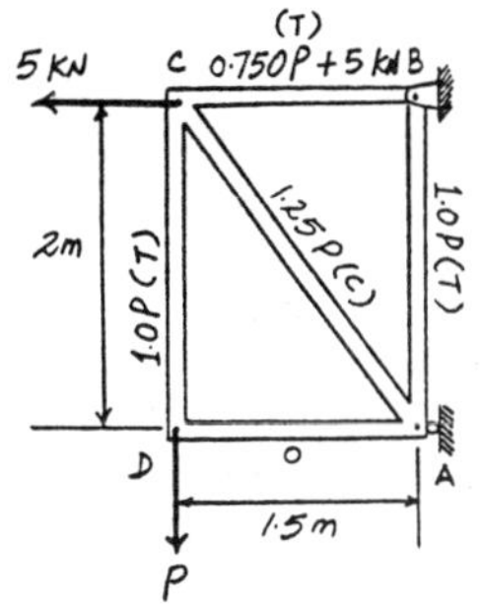

$$\Delta = \sum N\left(\frac{\partial N}{\partial P}\right)\frac{L}{AE}$$

$$(\Delta_D)_v = \frac{93.125\text{ kN}\cdot\text{m}}{AE}$$

$$= \frac{93.125(10^3)}{0.400(10^{-3})[200(10^9)]}$$

$$= 1.164\left(10^{-3}\right)\text{ m} = 1.16\text{ mm} \downarrow \qquad \textbf{Ans}$$

***14–120.** Solve Prob. 14–68 using Castigliano's theorem.

Member Forces N : Member forces due to external force **P** and external applied forces are shown on the figure.

Castigliano's Second Theorem : Applying Eq. 14 – 48, we have

Member	N	$\dfrac{\partial N}{\partial P}$	$N(P = 0)$	L	$N\left(\dfrac{\partial N}{\partial P}\right)L$
AB	6.00	0	6.00	36	0
BC	$1.333P+8.00$	1.333	8.00	48	512
AD	0	0	0	48	0
BD	$-(1.667P+10.0)$	-1.667	-10.0	60	1000
					$\sum 1512\text{ kip}\cdot\text{in.}$

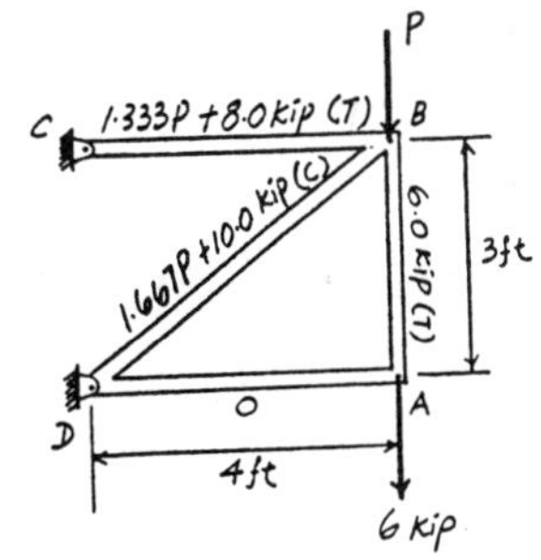

$$\Delta = \sum N\left(\frac{\partial N}{\partial P}\right)\frac{L}{AE}$$

$$(\Delta_B)_v = \frac{1512\text{ kip}\cdot\text{in.}}{AE}$$

$$= \frac{1512}{3[29.0(10^3)]} = 0.0174\text{ in.} \downarrow \qquad \textbf{Ans}$$

14–121. Solve Prob. 14–67 using Castigliano's theorem.

Member Forces N : Member forces due to external force **P** and external applied forces are shown on the figure.

Castigliano's Second Theorem : Applying Eq. 14 – 48, we have

Member	N	$\frac{\partial N}{\partial P}$	$N(P = 6\text{ kip})$	L	$N\left(\frac{\partial N}{\partial P}\right)L$
AB	$1.00P$	1.00	6.00	36	216
BC	$1.333P$	1.333	8.00	48	512
AD	0	0	0	48	0
BD	$-1.667P$	-1.667	-10.0	60	1000
					$\sum$ 1728 kip · in.

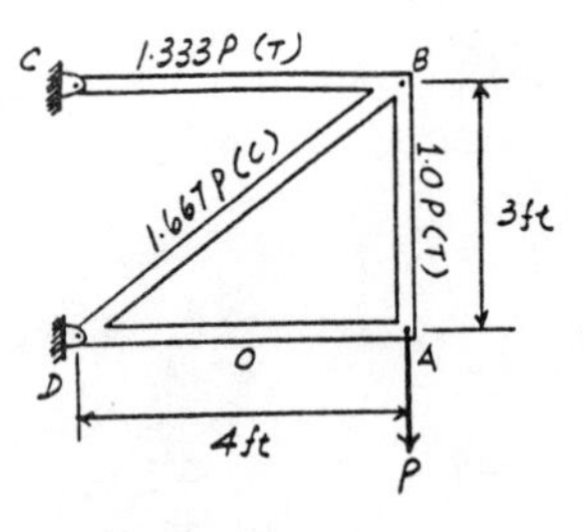

$$\Delta = \sum N\left(\frac{\partial N}{\partial P}\right)\frac{L}{AE}$$

$$(\Delta_A)_v = \frac{1728\text{ kip}\cdot\text{in.}}{AE}$$

$$= \frac{1728}{3[29.0(10^3)]} = 0.0199\text{ in.}\downarrow \qquad \textbf{Ans}$$

14–122. Solve Prob. 14–69 using Castigliano's theorem.

Member Forces N : Member forces due to external force **P** and external applied forces are shown on the figure.

Castigliano's Second Theorem : Applying Eq. 14 – 48, we have

Member	N	$\frac{\partial N}{\partial P}$	$N(P = 30\text{ kN})$	L	$N\left(\frac{\partial N}{\partial P}\right)L$
AB	$-0.750P$	-0.750	-22.5	1.5	25.3125
BC	$-0.750P$	-0.750	-22.5	1.5	25.3125
AE	$1.25P$	1.25	37.5	2.5	117.1875
CE	$-(1.25P + 25.0)$	-1.25	-62.5	2.5	195.3125
BE	20.0	0	20.0	2	0
DE	$1.50P + 15.0$	1.50	60.0	1.5	135.00
					$\sum$ 498.125 kN · m

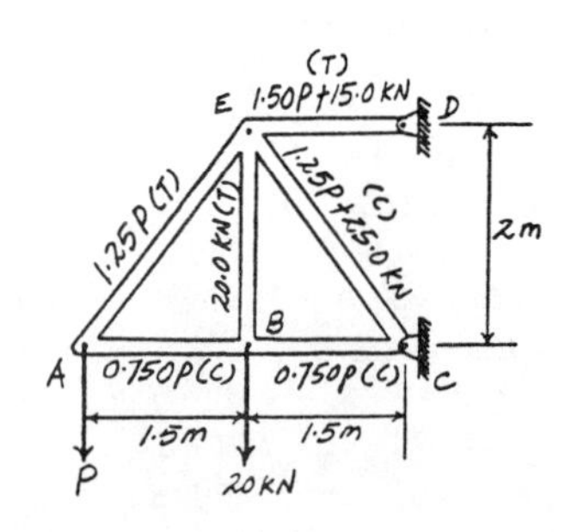

$$\Delta = \sum N\left(\frac{\partial N}{\partial P}\right)\frac{L}{AE}$$

$$(\Delta_A)_v = \frac{498.125\text{ kN}\cdot\text{m}}{AE}$$

$$= \frac{498.125(10^3)}{0.400(10^{-3})[200(10^9)]}$$

$$= 6.227\left(10^{-3}\right)\text{ m} = 6.23\text{ mm}\downarrow \qquad \textbf{Ans}$$

14–123. Solve Prob. 14–70 using Castigliano's theorem.

Member Forces N : Member forces due to external force **P** and external applied forces are shown on the figure.

Castigliano's Second Theorem : Applying Eq. 14 – 48, we have

Member	N	$\dfrac{\partial N}{\partial P}$	$N(P = 20\text{ kN})$	L	$N\left(\dfrac{\partial N}{\partial P}\right)L$
AB	-22.5	0	-22.5	1.5	0
BC	-22.5	0	-22.5	1.5	0
AE	37.5	0	37.5	2.5	0
CE	$-(1.25P+37.5)$	-1.25	-62.5	2.5	195.3125
BE	$1.00P$	1.00	20.0	2	40.00
DE	$0.750P+45$	0.750	60.0	1.5	67.50
					$\sum 302.8125\text{ kN}\cdot\text{m}$

$$\Delta = \sum N\left(\frac{\partial N}{\partial P}\right)\frac{L}{AE}$$

$$(\Delta_B)_v = \frac{302.8125\text{ kN}\cdot\text{m}}{AE}$$

$$= \frac{302.8125(10^3)}{0.400(10^{-3})[200(10^9)]}$$

$$= 3.785(10^{-3})\text{ m} = 3.79\text{ mm} \downarrow \qquad \textbf{Ans}$$

***14–124.** Solve Prob. 14–72 using Castigliano's theorem.

Member Forces N : Member forces due to external force **P** and external applied forces are shown on the figure.

Castigliano's Second Theorem : Applying Eq. 14 – 48, we have

Member	N	$\dfrac{\partial N}{\partial P}$	$N(P = 2\text{ kN})$	L	$N\left(\dfrac{\partial N}{\partial P}\right)L$
AE	8.00	0	8.00	3	0
ED	8.00	0	8.00	3	0
CD	-8.944	0	-8.944	3.354	0
BC	$-(1.118P+8.944)$	-1.118	-11.180	3.354	41.926
CE	$-1.00P$	-1.00	-2.00	1.5	3.00
AC	$1.118P$	1.118	2.236	3.354	8.385
					$\sum 53.312\text{ kN}\cdot\text{m}$

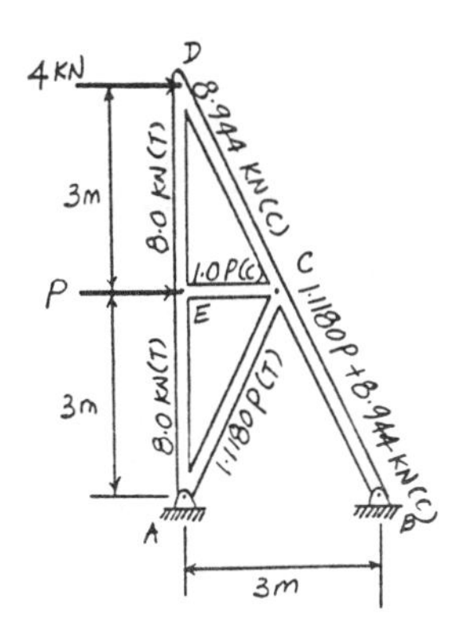

$$\Delta = \sum N\left(\frac{\partial N}{\partial P}\right)\frac{L}{AE}$$

$$(\Delta_E)_h = \frac{53.312\text{ kN}\cdot\text{m}}{AE}$$

$$= \frac{53.312(10^3)}{0.300(10^{-3})[200(10^9)]}$$

$$= 0.8885(10^{-3})\text{ m} = 0.889\text{ mm} \rightarrow \qquad \textbf{Ans}$$

14–125. Solve Prob. 14–71 using Castigliano's theorem.

Member Forces N : Member forces due to external force **P** and external applied forces are shown on the figure.

Castigliano's Second Theorem : Applying Eq. 14 – 48, we have

Member	N	$\frac{\partial N}{\partial P}$	$N(P = 4\ \text{kN})$	L	$N\left(\frac{\partial N}{\partial P}\right)L$
AE	$2.00P$	2.00	8.00	3	48.00
ED	$2.00P$	2.00	8.00	3	48.00
CD	$-2.236P$	-2.236	-8.944	3.354	67.082
BC	$-(2.236P+2.236)$	-2.236	-11.180	3.354	83.853
CE	-2.00	0	-2.00	1.5	0
AC	2.236	0	2.236	3.354	0
					$\sum 246.935\ \text{kN}\cdot\text{m}$

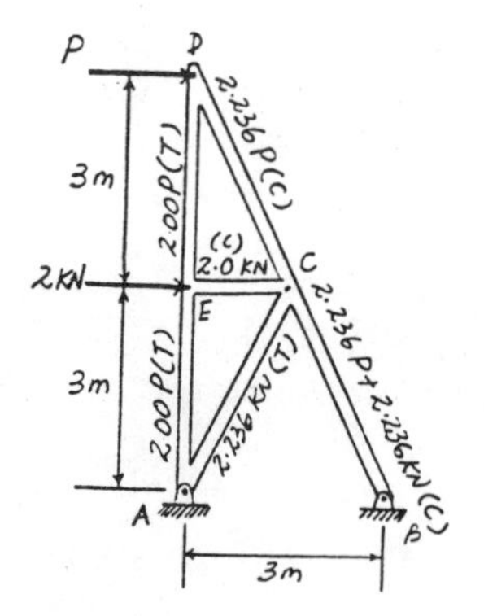

$$\Delta = \sum N\left(\frac{\partial N}{\partial P}\right)\frac{L}{AE}$$

$$(\Delta_D)_h = \frac{246.935\ \text{kN}\cdot\text{m}}{AE}$$

$$= \frac{246.935(10^3)}{0.300(10^{-3})[200(10^9)]}$$

$$= 4.116(10^{-3})\ \text{m} = 4.12\ \text{mm} \rightarrow \quad \textbf{Ans}$$

14–126. Solve Prob. 14–73 using Castigliano's theorem.

Member Forces N : Member forces due to external force **P** and external applied forces are shown on the figure.

Castigliano's Second Theorem : Applying Eq. 14 – 48, we have

Member	N	$\frac{\partial N}{\partial P}$	$N(P = 5\ \text{kip})$	L	$N\left(\frac{\partial N}{\partial P}\right)L$
AB	$0.6667P$	0.6667	3.333	96	213.33
BC	$0.6667P$	0.6667	3.333	96	213.33
CD	0	0	0	72	0
DE	0	0	0	96	0
EF	0	0	0	96	0
AF	0	0	0	72	0
AE	$-0.8333P$	-0.8333	-4.167	120	416.67
CE	$-0.8333P$	-0.8333	-4.167	120	416.67
BE	$1.00P$	1.00	5.00	72	360.00
					$\sum 1620\ \text{kip}\cdot\text{in.}$

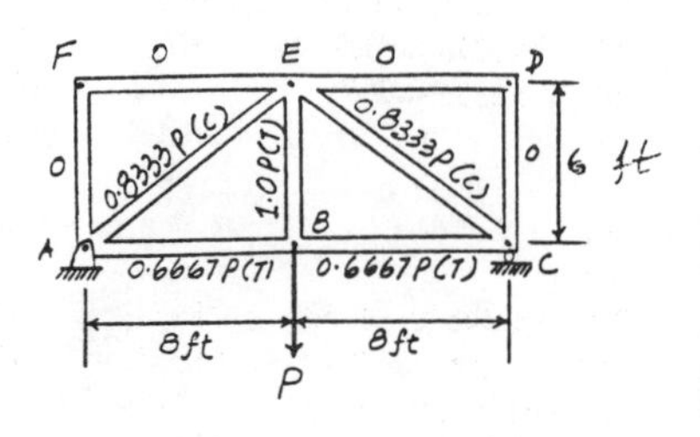

$$\Delta = \sum N\left(\frac{\partial N}{\partial P}\right)\frac{L}{AE}$$

$$(\Delta_B)_v = \frac{1620\ \text{kip}\cdot\text{in.}}{AE}$$

$$= \frac{1620}{4.5[29.0(10^3)]} = 0.0124\ \text{in.} \downarrow \quad \textbf{Ans}$$

14–127. Solve Prob. 14–74 using Castigliano's theorem.

Member Forces N : Member forces due to external force **P** and external applied forces are shown on the figure.

Castigliano's Second Theorem : Applying Eq. 14 – 48, we have

Member	N	$\frac{\partial N}{\partial P}$	$N(P=0)$	L	$N\left(\frac{\partial N}{\partial P}\right)L$
AB	$0.6667P+3.333$	0.6667	3.333	96	213.33
BC	$0.6667P+3.333$	0.6667	3.333	96	213.33
CD	0	0	0	72	0
DE	0	0	0	96	0
EF	0	0	0	96	0
AF	0	0	0	72	0
AE	$-(0.8333P+4.167)$	-0.8333	-4.167	120	416.67
CE	$-(0.8333P+4.167)$	-0.8333	-4.167	120	416.67
BE	5.0	0	5.00	72	0
					$\sum$ 1260 kip · in.

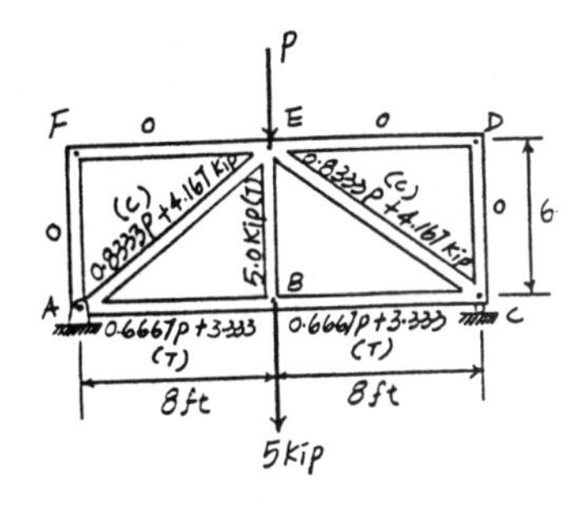

$$\Delta = \sum N\left(\frac{\partial N}{\partial P}\right)\frac{L}{AE}$$

$$(\Delta_E)_v = \frac{1260 \text{ kip} \cdot \text{in.}}{AE}$$

$$= \frac{1260}{4.5[29.0(10^3)]} = 0.00966 \text{ in.} \downarrow \qquad \textbf{Ans}$$

*14–128. Solve Prob. 14–76 using Castigliano's theorem.

Member Forces N : Member forces due to external force **P** and external applied forces are shown on the figure.

Castigliano's Second Theorem : Applying Eq. 14 – 48, we have

Member	N	$\frac{\partial N}{\partial P}$	$N(P=0)$	L	$N\left(\frac{\partial N}{\partial P}\right)L$
AB	$0.6667P+13.33$	0.6667	13.33	144	1280.00
DE	$0.6667P+13.33$	0.6667	13.33	144	1280.00
BC	$1.333P+18.67$	1.333	18.67	144	3584.00
CD	$1.333P+18.67$	1.333	18.67	144	3584.00
AJ	0	0	0	108	0
EF	0	0	0	108	0
IJ	0	0	0	144	0
FG	0	0	0	144	0
HI	$-(0.6667P+13.33)$	-0.6667	-13.33	144	1280.00
GH	$-(0.6667P+13.33)$	-0.6667	-13.33	144	1280.00
AI	$-(0.8333P+16.67)$	-0.8333	-16.67	180	2500.00
EG	$-(0.8333P+16.67)$	-0.8333	-16.67	180	2500.00
BI	$0.500P+10.0$	0.500	10.0	108	540.00
DG	$0.500P+10.0$	0.500	10.0	108	540.00
BH	$-(0.8333P+6.667)$	-0.8333	-6.667	180	1000.00
DH	$-(0.8333P+6.667)$	-0.8333	-6.667	180	1000.00
CH	8.00	0	8.00	108	0
					$\sum$ 20368 kip · in.

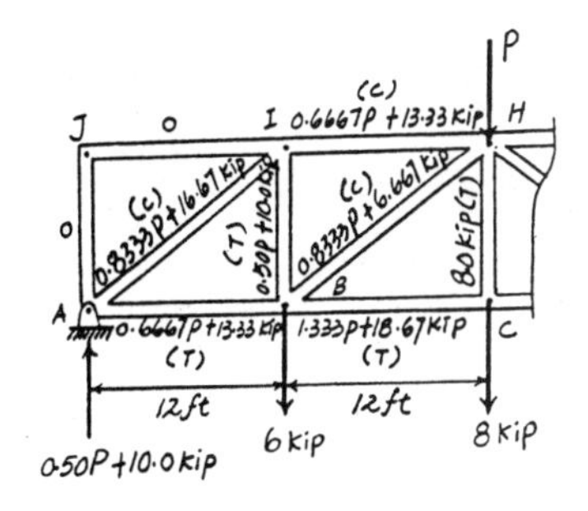

$$\Delta = \sum N\left(\frac{\partial N}{\partial P}\right)\frac{L}{AE}$$

$$(\Delta_H)_v = \frac{20368 \text{ kip} \cdot \text{in.}}{AE}$$

$$= \frac{20368}{4.5[29.0(10^3)]} = 0.156 \text{ in.} \downarrow \qquad \textbf{Ans}$$

14–129. Solve Prob. 14–75 using Castigliano's theorem.

Member Forces N : Member forces due to external force **P** and external applied forces are shown on the figure.

Castigliano's Second Theorem : Applying Eq. 14 – 48, we have

Member	N	$\frac{\partial N}{\partial P}$	$N(P = 8\text{ kip})$	L	$N\left(\frac{\partial N}{\partial P}\right)L$
AB	$0.6667P + 8.00$	0.6667	13.33	144	1280.00
DE	$0.6667P + 8.00$	0.6667	13.33	144	1280.00
BC	$1.333P + 8.00$	1.333	18.67	144	3584.00
CD	$1.333P + 8.00$	1.333	18.67	144	3584.00
AJ	0	0	0	108	0
EF	0	0	0	108	0
IJ	0	0	0	144	0
FG	0	0	0	144	0
HI	$-(0.6667P + 8.00)$	-0.6667	-13.33	144	1280.00
GH	$-(0.6667P + 8.00)$	-0.6667	-13.33	144	1280.00
AI	$-(0.8333P + 10.0)$	-0.8333	-16.67	180	2500.00
EG	$-(0.8333P + 10.0)$	-0.8333	-16.67	180	2500.00
BI	$0.500P + 6.00$	0.500	10.0	108	540.00
DG	$0.500P + 6.00$	0.500	10.0	108	540.00
BH	$-0.8333P$	-0.8333	-6.667	180	1000.00
DH	$-0.8333P$	-0.8333	-6.667	180	1000.00
CH	$1.00P$	1.00	8.00	108	864.00
					$\sum 21232\text{ kip}\cdot\text{in.}$

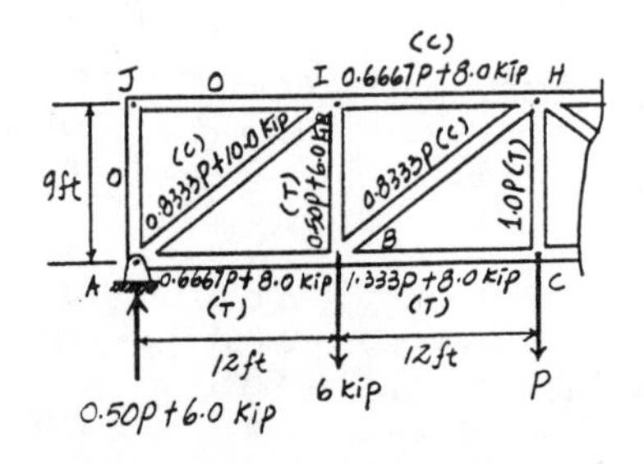

$$\Delta = \sum N\left(\frac{\partial N}{\partial P}\right)\frac{L}{AE}$$

$$(\Delta_C)_v = \frac{21232\text{ kip}\cdot\text{in.}}{AE}$$

$$= \frac{21232}{4.5[29.0(10^3)]} = 0.163\text{ in.}\downarrow \qquad \textbf{Ans}$$

14–130. Solve Prob. 14–77 using Castigliano's theorem.

Internal Moment Function M(x) : The internal moment function in terms of the load **P′** and couple moment **M′** and externally applied load are shown on figures (a) and (b), respectively.

Castigliano's Second Theorem : The displacement at C can be determined using Eq. 14 – 49 with $\frac{\partial M(x)}{\partial P'} = \frac{x}{2}$ and set $P' = P$.

$$\Delta = \int_0^L M\left(\frac{\partial M}{\partial P'}\right)\frac{dx}{EI}$$

$$\Delta_C = 2\left[\frac{1}{EI}\int_0^{\frac{L}{2}}\left(\frac{P}{2}x\right)\left(\frac{x}{2}\right)dx\right]$$

$$= \frac{PL^3}{48EI}\downarrow \qquad \textbf{Ans}$$

To determine the slope at B, we apply Eq. 14 – 50 with $\frac{\partial M(x_1)}{\partial M'} = \frac{x_1}{L}$, $\frac{\partial M(x_2)}{\partial M'} = 1 - \frac{x_2}{L}$ and setting $M' = 0$.

$$\theta = \int_0^L M\left(\frac{\partial M}{\partial M'}\right)\frac{dx}{EI}$$

$$\theta_B = \frac{1}{EI}\int_0^{\frac{L}{2}}\left(\frac{P}{2}x_1\right)\left(\frac{x_1}{L}\right)dx_1 + \frac{1}{EI}\int_0^{\frac{L}{2}}\left(\frac{P}{2}x_2\right)\left(1 - \frac{x_2}{L}\right)dx_2$$

$$= \frac{PL^2}{16EI} \qquad \textbf{Ans}$$

14–131. Solve Prob. 14–78 using Castigliano's theorem.

Internal Moment Function $M(x)$ **:** The internal moment function in terms of the load $\mathbf{P}'$ and externally applied load are shown on the figure.

Castigliano's Second Theorem **:** The displacement at C can be determined using Eq. 14 – 49 with $\dfrac{\partial M(x_1)}{\partial P'} = \dfrac{2}{3}x_1$, $\dfrac{\partial M(x_3)}{\partial P'} = \dfrac{x_3}{3}$, $\dfrac{\partial M(x_2)}{\partial P'} = \dfrac{1}{3}(2a - x_2)$ and setting $P' = P$.

$M(x_1) = \frac{1}{3}(P+2P')x_1$ $M(x_3) = \frac{1}{3}(2P+P')x_3$

$M(x_2) = \frac{1}{3}(2P'a + Pa + Px_2 - P'x_2)$

$$\Delta = \int_0^L M\left(\frac{\partial M}{\partial P'}\right)\frac{dx}{EI}$$

$$\Delta_C = \frac{1}{EI}\int_0^a (Px_1)\left(\frac{2}{3}x_1\right)dx_1 + \frac{1}{EI}\int_0^a (Pa)\left[\frac{1}{3}(2a - x_2)\right]dx_2 + \frac{1}{EI}\int_0^a (Px_3)\left(\frac{x_3}{3}\right)dx_3$$

$$= \frac{5Pa^3}{6EI} \downarrow \qquad \textbf{Ans}$$

***14–132.** Solve Prob. 14–84 using Castigliano's theorem.

Internal Moment Function $M(x)$ **:** The internal moment function in terms of the load $\mathbf{P}$ and external applied load are shown on the figure.

Castigliano's Second Theorem **:** The displacement at B can be determined using Eq. 14 – 49, with $\dfrac{\partial M(x_1)}{\partial P} = 0.4286x_1$, $\dfrac{\partial M(x_3)}{\partial P} = 0.5714x_3$, $\dfrac{\partial M(x_2)}{\partial P} = 0.3429 + 0.4286x_2$ and setting $P = 1$ kN.

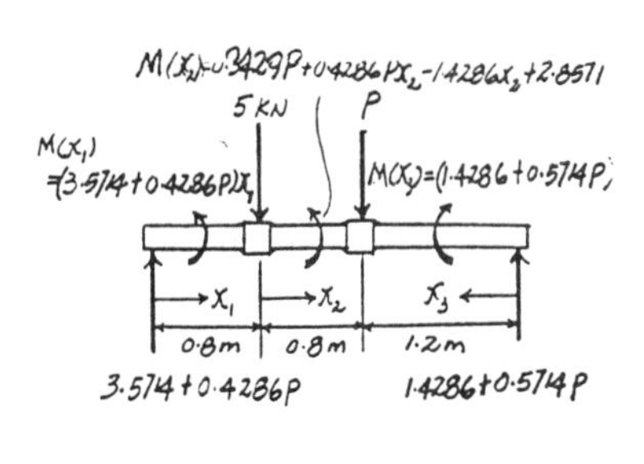

$$\Delta = \int_0^L M\left(\frac{\partial M}{\partial P}\right)\frac{dx}{EI}$$

$$\Delta_B = \frac{1}{EI}\int_0^{0.8\text{ m}} (4.00x_1)(0.4286x_1)\,dx_1 + \frac{1}{EI}\int_0^{0.8\text{ m}} (3.20 - 1.00x_2)(0.3429 + 0.4286x_2)\,dx_2 + \frac{1}{EI}\int_0^{1.2\text{ m}} (2.00x_3)(0.5714x_3)\,dx_3$$

$$= \frac{2.0846\ \text{kN}\cdot\text{m}^3}{EI}$$

$$= \frac{2.0846(1000)}{200(10^9)\left[\frac{\pi}{4}(0.03^4)\right]} = 0.01638\ \text{m} = 16.4\ \text{mm} \downarrow \qquad \textbf{Ans}$$

14–133. Solve Prob. 14–83 using Castigliano's theorem.

Internal Moment Function $M(x)$: The internal moment function in terms of the couple moment M' and external applied load are shown on the figure.

Castigliano's Second Theorem : The slope at C can be determined using Eq. 14 – 50 with $\frac{\partial M(x_1)}{\partial M'} = 1.00$, $\frac{\partial M(x_2)}{\partial M'} = 1.25x_2$ and setting $M' = 0$.

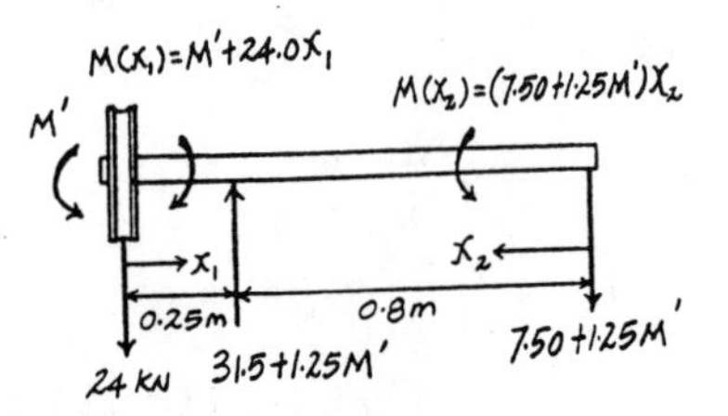

$$\theta = \int_0^L M\left(\frac{\partial M}{\partial M'}\right)\frac{dx}{EI}$$

$$\theta_C = \frac{1}{EI}\int_0^{0.25\,\text{m}} (24.0x_1)(1.00)\,dx_1 + \frac{1}{EI}\int_0^{0.8\,\text{m}} (7.50x_2)(1.25x_2)\,dx_2$$

$$= \frac{2.35\ \text{kN}\cdot\text{m}^2}{EI}$$

$$= \frac{2.35(10^3)}{200(10^9)\left[\frac{\pi}{4}(0.03^4)\right]} = 0.0185\ \text{rad} \qquad \textbf{Ans}$$

14–134. Solve Prob. 14–88 using Castigliano's theorem.

Internal Moment Function $M(x)$: The internal moment function in terms of the couple moment M' and externally applied load are shown on the figure.

Castigliano's Second Theorem : The slope at E can be determined using Eq. 14 – 50 with $\frac{\partial M(x_1)}{\partial M'} = \frac{x_1}{6}$, $\frac{\partial M(x_2)}{\partial M'} = 1.00$ and setting $M' = 0$.

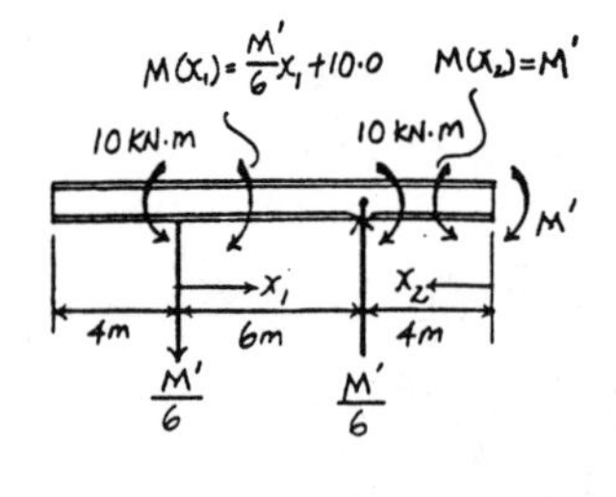

$$\theta = \int_0^L M\left(\frac{\partial M}{\partial M'}\right)\frac{dx}{EI}$$

$$\theta_E = \frac{1}{EI}\int_0^{6\text{m}} (10.0)\left(\frac{x_1}{6}\right)dx_1 + \frac{1}{EI}\int_0^{4\text{m}} (0)(-1.00)\,dx_2$$

$$= \frac{30.0\ \text{kN}\cdot\text{m}^2}{EI}$$

$$= \frac{30.0(10^3)}{200(10^9)[125(10^{-6})]} = 0.00120\ \text{rad} \qquad \textbf{Ans}$$

14–135. Solve Prob. 14–94 using Castigliano's theorem.

Internal Moment Function $M(x)$: The internal moment function in terms of the couple moment **M′** and the applied load are shown on the figure.

Castigliano's Second Theorem : The slope at A can be determined using Eq. 14 – 50 with $\dfrac{\partial M(x_1)}{\partial M'} = 1 - 0.100x_1$, $\dfrac{\partial M(x_2)}{\partial M'} = 0$ and setting $M' = 0$.

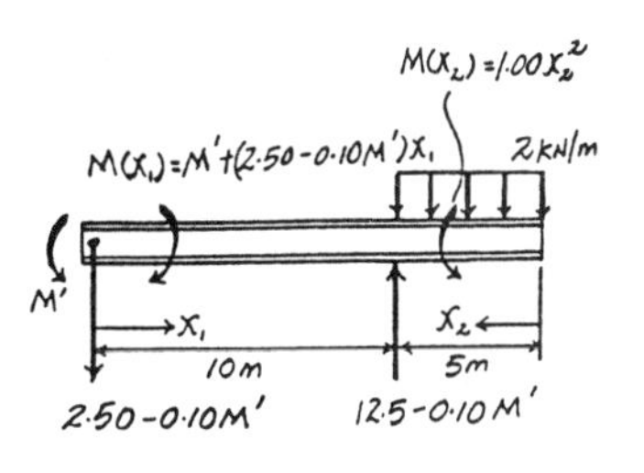

$$\theta = \int_0^L M\left(\frac{\partial M}{\partial M'}\right)\frac{dx}{EI}$$

$$\theta_A = \frac{1}{EI}\int_0^{10\,\text{m}} (2.50x_1)(1 - 0.100x_1)\,dx_1 + \frac{1}{EI}\int_0^{5\,\text{m}} (1.00x_2^2)(0)\,dx_2$$

$$= \frac{41.667\ \text{kN}\cdot\text{m}^2}{EI}$$

$$= \frac{41.667(10^3)}{200(10^9)[70(10^{-6})]} = 0.00298\ \text{rad} \qquad \textbf{Ans}$$

***14–136.** Solve Prob. 14–92 using Castigliano's theorem.

Internal Moment Function $M(x)$: The internal moment function in terms of the load **P** and external applied load are shown on the figure.

Castigliano's Second Theorem : The displacement at D can be determined using Eq. 14 – 49 with $\dfrac{\partial M(x_1)}{\partial P} = \dfrac{3}{4}x_1$, $\dfrac{\partial M(x_2)}{\partial P} = \dfrac{x_2}{4}$, $\dfrac{\partial M(x_3)}{\partial P} = \dfrac{1}{4}(a + x_3)$ and setting $P = 0$.

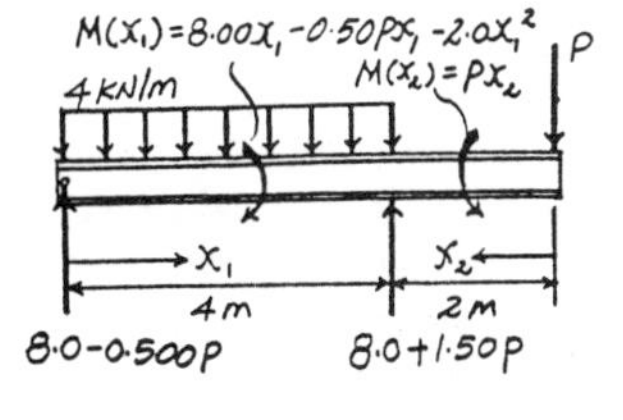

$$\Delta = \int_0^L M\left(\frac{\partial M}{\partial P}\right)\frac{dx}{EI}$$

$$\Delta_D = \frac{1}{EI}\int_0^a (wax_1)\left(\frac{3}{4}x_1\right)dx_1 + \frac{1}{EI}\int_0^a (wax_2)\left(\frac{x_2}{4}\right)dx_2 + \frac{1}{2EI}\int_0^{2a}\left(wa^2 + wax_3 - \frac{w}{2}x_3^2\right)\left[\frac{1}{4}(a + x_3)\right]dx_3$$

$$= \frac{wa^4}{EI} \downarrow \qquad \textbf{Ans}$$

14–137. Solve Prob. 14–102 using Castigliano's theorem.

Internal Axial Force N : The internal axial force for rod *CD* is indicated on the figure.

Internal Moment Function M(x) : The internal moment function in terms of the couple moment **M′** and external applied load are shown on the figure.

Castigliano's Second Theorem : The slope at *A* can be determined by combining Eq. 14 – 50 and Eq. 14 – 48 with $\frac{\partial M(x_1)}{\partial M'} = 1 - \frac{x_1}{3}$, $\frac{\partial M(x_2)}{\partial M'} = 0$, $\frac{\partial N}{\partial M'} = -\frac{1}{3}$ and setting $M' = 0$.

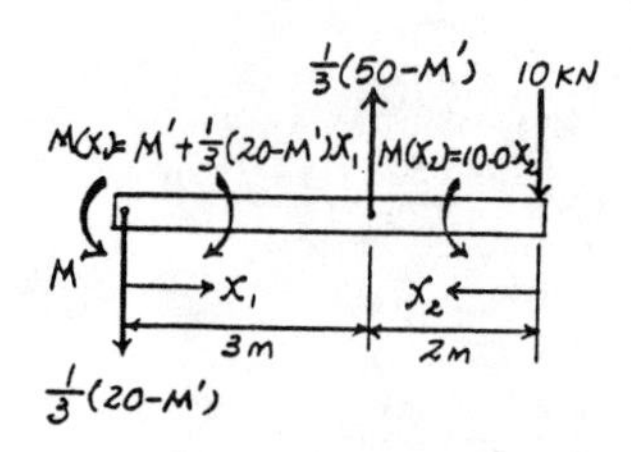

$$\theta = \int_0^L M\left(\frac{\partial M}{\partial M'}\right)\frac{dx}{EI} + N\left(\frac{\partial N}{\partial M'}\right)\frac{L}{AE}$$

$$\theta_A = \frac{1}{EI}\int_0^{3\,\text{m}}\left(\frac{20}{3}x_1\right)\left(1-\frac{x_1}{3}\right)dx_1 + 0 + \frac{50}{3}\left(-\frac{1}{3}\right)\frac{2}{AE}$$

$$= \frac{10.0\ \text{kN}\cdot\text{m}^2}{EI} - \frac{11.11\ \text{kN}}{AE}$$

$$= \frac{10.0(10^3)}{200(10^9)\left[\frac{1}{12}(0.1)(0.1^3)\right]} - \frac{11.11(10^3)}{\frac{\pi}{4}(0.01^2)\left[200(10^9)\right]}$$

$= 0.00529$ rad **Ans**

14–138. Solve Prob. 14–98 using Castigliano's theorem.

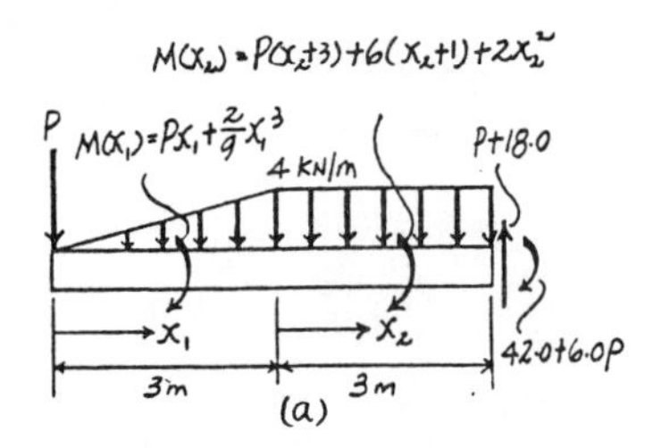

(a)

(b)

Internal Moment Function M(x) : The internal moment function in terms of the load **P** and couple moment **M′** and the external applied load are shown on figures (a) and (b), respectively.

Castigliano's Second Theorem : The displacement at *A* can be determined using Eq. 14 – 49 with $\frac{\partial M(x_1)}{\partial P} = x_1$, $\frac{\partial M(x_2)}{\partial P} = x_2 + 3$ and setting $P = 0$.

$$\Delta = \int_0^L M\left(\frac{\partial M}{\partial P'}\right)\frac{dx}{EI}$$

$$\Delta_A = \frac{1}{EI}\int_0^{3\,\text{m}}\left(\frac{2}{9}x_1^3\right)(x_1) + \frac{1}{EI}\int_0^{3\,\text{m}}\left[6(x_2+1)+2x_2^2\right](x_2+3)$$

$$= \frac{321.3\ \text{kN}\cdot\text{m}^3}{EI}$$

$$= \frac{321.3(10^3)}{11(10^9)\left[\frac{1}{12}(0.2)(0.4^3)\right]}$$

$= 0.02738$ m $= 27.4$ m ↓ **Ans**

To determine the slope at *A*, we apply Eq. 14 – 50 with $\frac{\partial M(x_1)}{\partial M'} = 1$, $\frac{\partial M(x_2)}{\partial M'} = 1$ and setting $M' = 0$.

$$\theta = \int_0^L M\left(\frac{\partial M}{\partial M'}\right)\frac{dx}{EI}$$

$$\theta_A = \frac{1}{EI}\int_0^{3\,\text{m}}\left(\frac{2}{9}x_1^3\right)(1) + \frac{1}{EI}\int_0^{3\,\text{m}}\left[6(x_2+1)+2x_2^2\right](1)$$

$$= \frac{67.5\ \text{kN}\cdot\text{m}^2}{EI}$$

$$= \frac{67.5(1000)}{11(10^9)\left[\frac{1}{12}(0.2)(0.4^3)\right]}$$

$= 5.75\left(10^{-3}\right)$ rad **Ans**

14–139. Solve Prob. 14–99 using Castigliano's theorem.

Internal Moment Function $M(x)$ **:** The internal moment function in terms of the load **P** and external applied load are shown on the figure.

***Castigliano's Second Theorem* :** The displacement at C can be determined using Eq. 14–49 with $\frac{\partial M(x_1)}{\partial P} = -0.500x_1$, $\frac{\partial M(x_2)}{\partial P} = 1.00x_2$ and setting $P = 20$ kN.

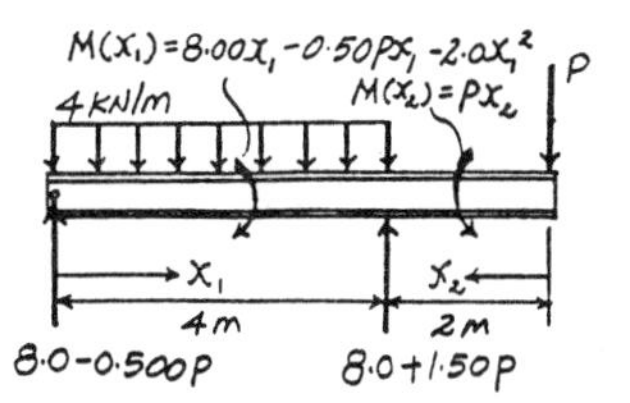

$$\Delta = \int_0^L M\left(\frac{\partial M}{\partial P}\right)\frac{dx}{EI}$$

$$\Delta_C = \frac{1}{EI}\int_0^{4\text{m}} \left(-2.00x_1 - 2.00x_1^2\right)(-0.500x_1)\,dx_1 + \frac{1}{EI}\int_0^{2\text{m}} (20.0x_2)(1.00x_2)\,dx_2$$

$$= \frac{138.67 \text{ kN}\cdot\text{m}^3}{EI}$$

$$= \frac{138.67(1000)}{200(10^9)[36.9(10^{-6})]}$$

$$= 0.01879 \text{ m} = 18.8 \text{ mm} \downarrow \qquad \textbf{Ans}$$

***14–140.** Solve Prob. 14–96 using Castigliano's theorem.

Internal Moment Function $M(x)$ **:** The internal moment function in terms of the load **P** and external applied load are shown on the figure.

***Castigliano's Second Theorem* :** The displacement at A can be determined using Eq. 14–49 with $\frac{\partial M(x_1)}{\partial P} = 1.00x_1$, $\frac{\partial M(x_2)}{\partial P} = 0.500x_2 - 1.50$, $\frac{\partial M(x_3)}{\partial P} = -0.500x_3$ and setting $P = 0$.

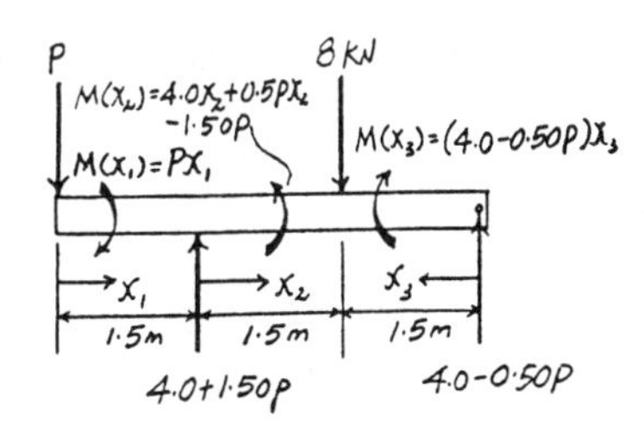

$$\Delta = \int_0^L M\left(\frac{\partial M}{\partial P}\right)\frac{dx}{EI}$$

$$\Delta_A = \frac{1}{EI}\int_0^{1.5\text{m}} (0)(1.00x_1)\,dx_1 + \frac{1}{EI}\int_0^{1.5\text{m}} (4.00x_2)(0.500x_2 - 1.50)\,dx_2 + \frac{1}{EI}\int_0^{1.5\text{m}} (4.00x_3)(-0.500x_3)\,dx_3$$

$$= -\frac{6.75 \text{ kN}\cdot\text{m}^3}{EI}$$

$$= -\frac{6.75(1000)}{13.1(10^9)\left[\frac{1}{12}(0.12)(0.18^3)\right]}$$

$$= -8.835\left(10^{-3}\right) \text{ m} = 8.84 \text{ mm} \uparrow \qquad \textbf{Ans}$$

14–141. Solve Prob. 14–103 using Castigliano's theorem.

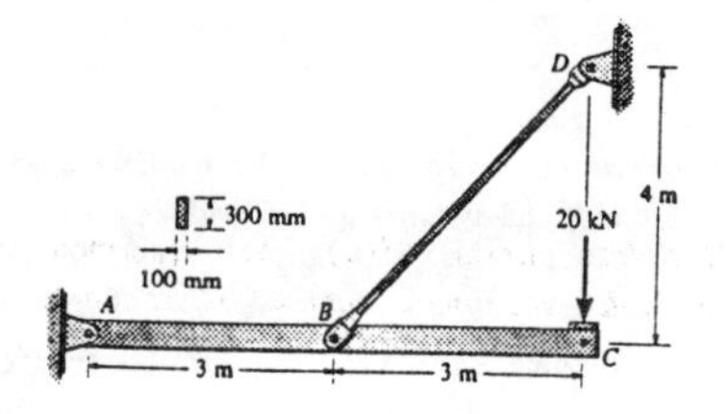

Internal Axial Force N : The component of the internal axial force for rod *BD* is indicated on the figure.

Internal Moment Function M(x) : The internal moment function in terms of force **P** and the external applied load are shown on the figure.

Castigliano's Second Theorem : The displacement at *C* can be determined by combining Eq. 14 – 49 and Eq. 14 – 48 with $\frac{\partial M(x)}{\partial P} = 1.00x$, $\frac{\partial N}{\partial P} = 2.50$ and setting $P = 20$ kN.

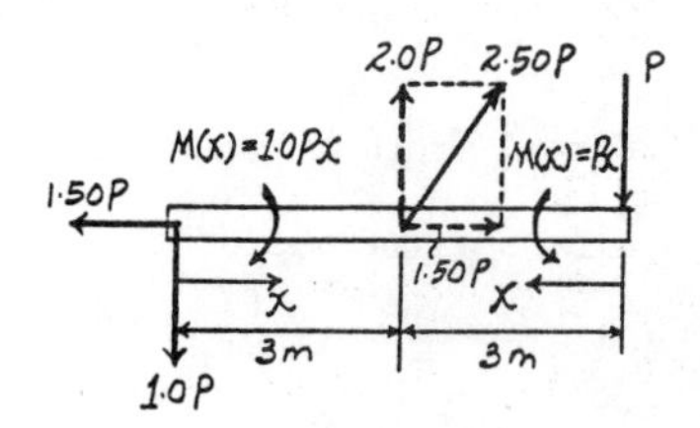

$$\Delta = \int_0^L M\left(\frac{\partial M}{\partial P}\right)\frac{dx}{EI} + N\left(\frac{\partial N}{\partial P}\right)\frac{L}{AE}$$

$$\Delta_C = 2\left[\frac{1}{EI}\int_0^{3\,\text{m}} (20.0x)(1.00x)\,dx\right] + \frac{50.0(2.50)(5)}{AE}$$

$$= \frac{360\ \text{kN}\cdot\text{m}^3}{EI} + \frac{625\ \text{kN}\cdot\text{m}}{AE}$$

$$= \frac{360(10^3)}{200(10^9)\left[\frac{1}{12}(0.1)(0.3^3)\right]} + \frac{625(10^3)}{\frac{\pi}{4}(0.02^2)\left[200(10^9)\right]}$$

$= 0.0179$ m $= 17.9$ mm ↓ **Ans**

14–142. Solve Prob. 14–106 using Castigliano's theorem.

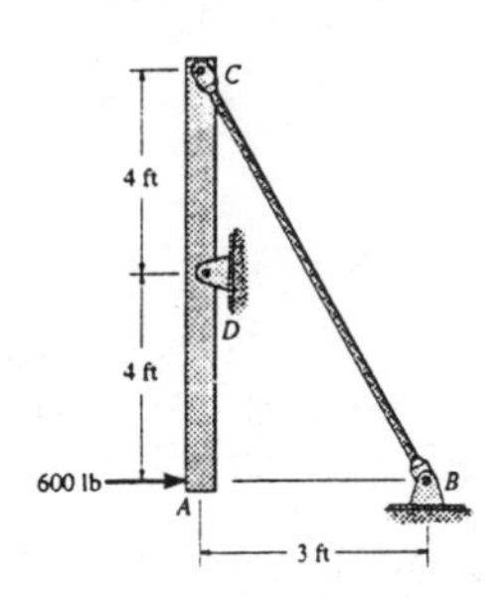

Internal Axial Force N : The internal axial force for rod *BC* is indicated on the figure.

Internal Moment Function M(x) : The internal moment function in terms of external force **P** and external applied load are shown on the figure.

Castigliano's Second Theorem : The displacement at *A* can be determined by combining Eq. 14 – 49 and Eq. 14 – 48 with $\frac{\partial M(x)}{\partial P} = 1.00x$, $\frac{\partial N}{\partial P} = 2.848$ and set $P = 600$ lb.

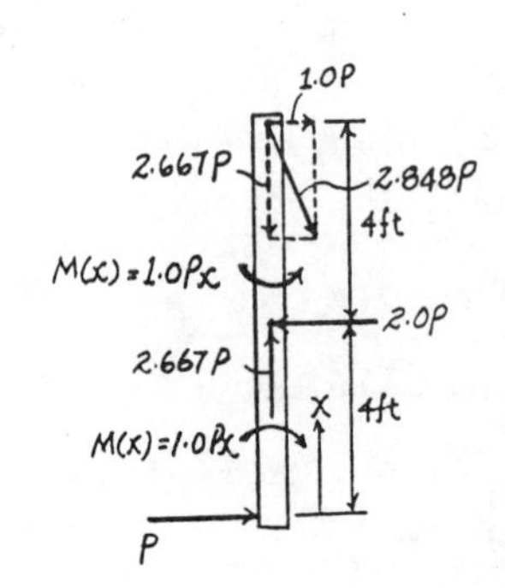

$$\Delta = \int_0^L M\left(\frac{\partial M}{\partial P}\right)\frac{dx}{EI} + N\left(\frac{\partial N}{\partial P}\right)\frac{L}{AE}$$

$$\Delta_A = 2\left[\frac{1}{EI}\int_0^{4\,\text{ft}} (600x)(1.00x)\,dx\right] + \frac{1708.8(2.848)(8.544)(12)}{AE}$$

$$= \frac{25600\ \text{lb}\cdot\text{ft}^3}{EI} + \frac{498969.8\ \text{lb}\cdot\text{in.}}{AE}$$

$$= \frac{25600(12^3)}{29.0(10^6)\left[\frac{1}{12}(2)(2^3)\right]} + \frac{498969.8}{\left[\frac{\pi}{4}(0.25^2)\right]\left[29.0(10^6)\right]}$$

$= 1.49$ in. → **Ans**

14–143. Solve Prob. 14–107 using Castigliano's theorem.

Internal Moment Function $M(x)$ **:** The internal moment function in terms of the load **P** and external applied load are shown on the figure.

***Castigliano's Second Theorem* :** The horizontal displacement at C can be determined using Eq. 14–49 with $\frac{\partial M(x_1)}{\partial P} = 0$, $\frac{\partial M(x_2)}{\partial P} = -1.00x_2$, $\frac{\partial M(x_3)}{\partial P} = 1.00x_3 - 10.0$ and setting $P = 0$.

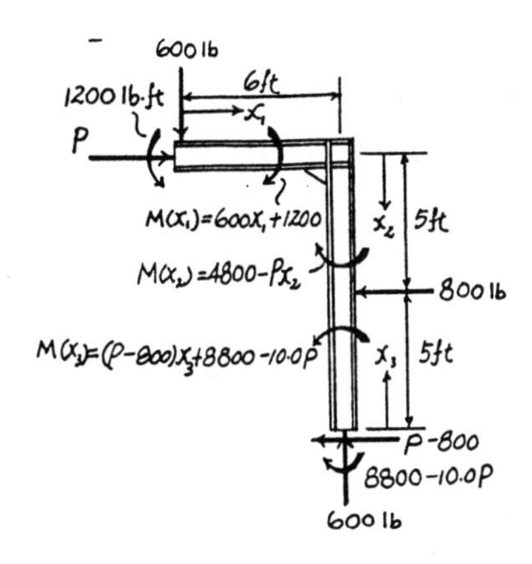

$$\Delta = \int_0^L M\left(\frac{\partial M}{\partial P}\right)\frac{dx}{EI}$$

$$(\Delta_C)_h = 0 + \frac{1}{EI}\int_0^{5\,\text{ft}} 4800(-1.00x_2)\,dx_2$$

$$+ \frac{1}{EI}\int_0^{5\,\text{ft}} (-800x_3 + 8800)(1.00x_3 - 10.0)\,dx_3$$

$$= -\frac{323333.33\ \text{lb}\cdot\text{ft}^3}{EI}$$

$$= \frac{323(10^3)\ \text{lb}\cdot\text{ft}^3}{EI} \leftarrow \qquad \textbf{Ans}$$

***14–144.** Solve Prob. 14–108 using Castigliano's theorem.

Internal Moment Function $M(x)$ **:** The internal moment function in terms of the load **P′** and external applied load are shown on the figure.

***Castigliano's Second Theorem* :** The horizontal displacement at A can be determined using Eq. 14–49 with $\frac{\partial M(x_1)}{\partial P'} = 1.00x_1$, $\frac{\partial M(x_2)}{\partial P'} = 1.00L$ and setting $P' = P$.

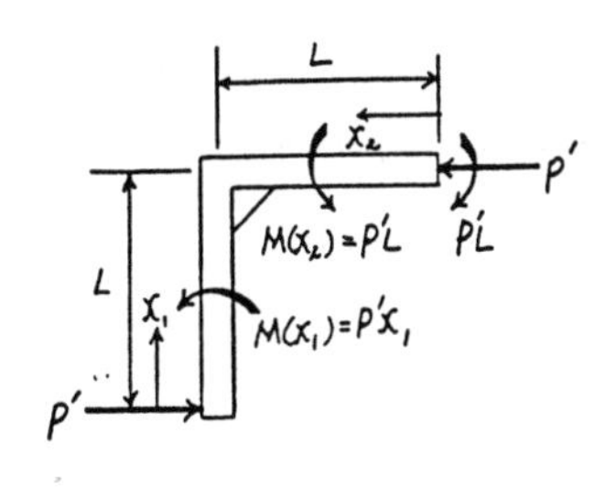

$$\Delta = \int_0^L M\left(\frac{\partial M}{\partial P}\right)\frac{dx}{EI}$$

$$(\Delta_A)_h = \frac{1}{EI}\int_0^L (Px_1)(1.00x_1)\,dx_1$$

$$+ \frac{1}{EI}\int_0^L (PL)(1.00L)\,dx_2$$

$$= \frac{4PL^3}{3EI} \rightarrow \qquad \textbf{Ans}$$

14–145. Solve Prob. 14–109 using Castigliano's theorem.

Internal Moment Function M(x) : The internal moment function in terms of the load **P** and external applied load are shown on the figure.

Castigliano's Second Theorem : The vertical displacement at C can be determined using Eq. 14 – 49 with $\frac{\partial M(x_1)}{\partial P} = 1.00x_1$, $\frac{\partial M(x_2)}{\partial P'} = 1.00L$ and setting $P = 0$.

$$\Delta = \int_0^L M\left(\frac{\partial M}{\partial P}\right)\frac{dx}{EI}$$

$$(\Delta_C)_v = \frac{1}{EI}\int_0^L \left(\frac{w}{2}x_1^2\right)(1.00x_1)\,dx_1 + \frac{1}{EI}\int_0^L \left(\frac{wL^2}{2}\right)(1.00L)\,dx_2$$

$$= \frac{5wL^4}{8EI} \downarrow \qquad \textbf{Ans}$$

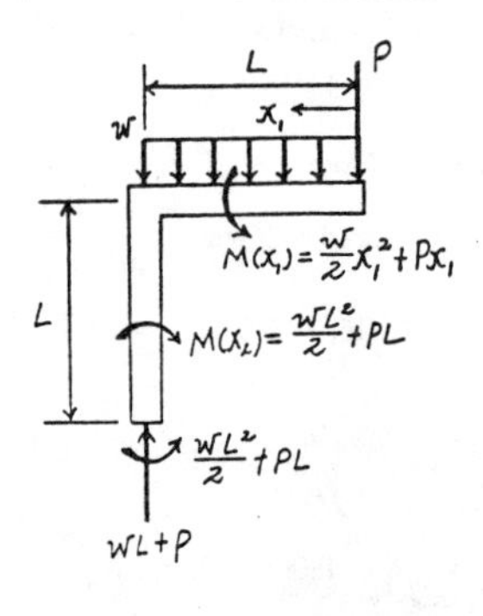

14–146. Solve Prob. 14–111 using Castigliano's theorem.

Model : The ring can be modeled as a half ring as shown in figure(a).

Internal Moment Function M(x) : The internal moment expressed in terms of the load **P** and couple moment **M'** and external applied load are shown on figures (b) and (c), respectively.

Castigliano's Second Theorem : Due to symmetry, the slope at B remain horizontal that is equal to zero. Applying Eq. 14 – 50 with $\frac{\partial M}{\partial M'} = -1.00$ and set $M' = M_0$, we have

$$\theta = \int_0^L M\left(\frac{\partial M}{\partial M'}\right)\frac{ds}{EI} \qquad \text{Where } ds = r\,d\theta$$

$$\theta_B = 0 = \frac{1}{EI}\int_0^\pi \left(\frac{Pr}{2}\sin\theta - M_0\right)(-1.00)\,r\,d\theta$$

$$M_0 = \frac{Pr}{\pi}$$

To determine the vertical diplacement at B, we apply Eq. 14 – 49 with $\frac{\partial M}{\partial P'} = r\sin\theta$, and setting $P' = \frac{P}{2}$.

$$\Delta = \int_0^L M\left(\frac{\partial M}{\partial P'}\right)\frac{ds}{EI}$$

$$(\Delta_B)_v = \frac{1}{EI}\int_0^\pi \left(\frac{Pr}{2}\sin\theta - \frac{Pr}{\pi}\right)(r\sin\theta)\,r\,d\theta$$

$$= \frac{Pr^3}{2\pi EI}\int_0^\pi \left(\pi\sin^2\theta - 2\sin\theta\right)d\theta$$

$$= \frac{Pr^3}{4\pi EI}\int_0^\pi [\pi(1-\cos 2\theta) - 4\sin\theta]\,d\theta$$

$$= \frac{Pr^3}{4\pi EI}(\pi^2 - 8) \qquad \textbf{Ans}$$

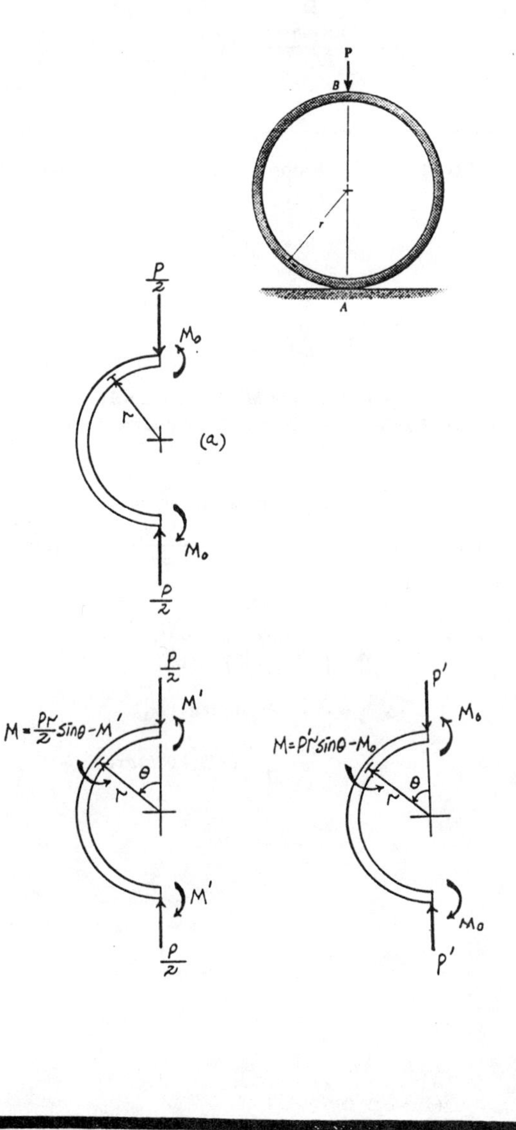

14–147. Use the method of virtual work to determine the slope of the beam at point *B*. *EI* is constant.

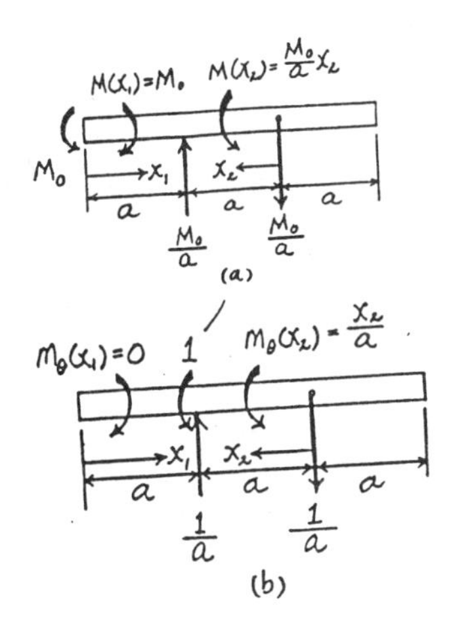

Real Moment Function $M(x)$ **:** As shown on Figure(a).

Virtual Moment Functions $m_\theta(x)$ **:** As shown on figure(b).

Virtual Work Equation **:** For the slope at point *B*, apply Eq. 14 – 43.

$$1 \cdot \theta = \int_0^L \frac{m_\theta M}{EI} dx$$

$$1 \cdot \theta_B = \frac{1}{EI} \int_0^a (0)(M_0)\, dx_1 + \frac{1}{EI} \int_0^a \left(\frac{x_2}{a}\right)\left(\frac{M_0}{a} x_2\right) dx_2$$

$$\theta_B = \frac{M_0 a}{3EI} \qquad \textbf{Ans}$$

***14–148.** Solve Prob. 14–147 using Castigliano's theorem.

Internal Moment Function $M(x)$ **:** The internal moment function in terms of the couple moment $\mathbf{M'}$ and external applied load are shown on the figure.

Castigliano's Second Theorem **:** The slope at *B* can be determined using Eq. 14 – 50 with $\frac{\partial M(x_1)}{\partial M'} = 0$, $\frac{\partial M(x_2)}{\partial M'} = \frac{x_2}{a}$ and setting $M' = 0$.

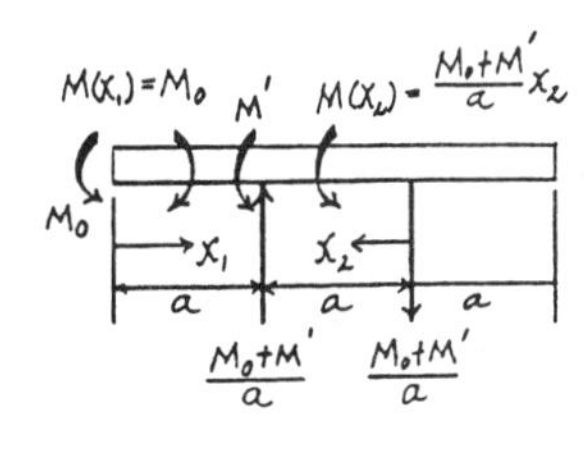

$$\theta = \int_0^L M \left(\frac{\partial M}{\partial M'}\right) \frac{dx}{EI}$$

$$\theta_B = \frac{1}{EI} \int_0^a (M_0)(0)\, dx_1 + \frac{1}{EI} \int_0^a \left(\frac{M_0}{a} x_2\right)\left(\frac{x_2}{a}\right) dx_2$$

$$= \frac{M_0 a}{3EI} \qquad \textbf{Ans}$$

14–149. Use the method of virtual work to determine the displacement at point A. EI is constant.

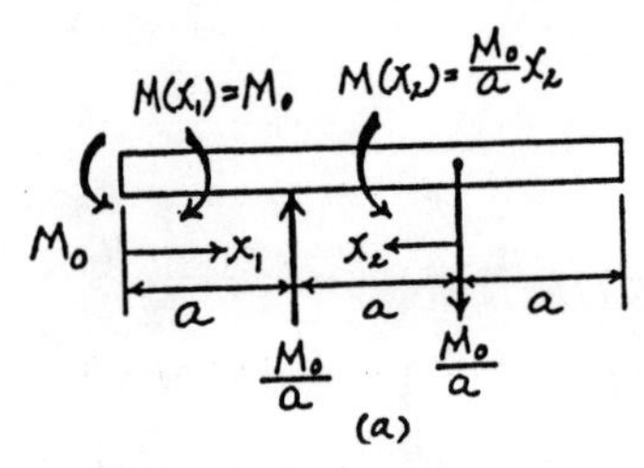

Real Moment Function $M(x)$: As shown on Figure(a).

Virtual Moment Functions $m(x)$: As shown on figure(b).

Virtual Work Equation : For the displacement at point A, apply Eq. 14 – 42.

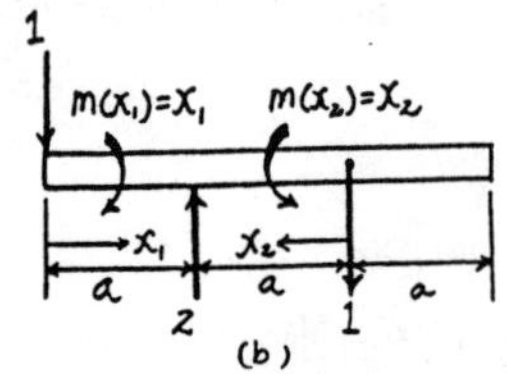

$$1 \cdot \Delta = \int_0^L \frac{mM}{EI} dx$$

$$1 \cdot \Delta_A = \frac{1}{EI}\int_0^a (x_1)(M_0)\, dx_1 + \frac{1}{EI}\int_0^a (x_2)\left(\frac{M_0}{a}x_2\right) dx_2$$

$$\Delta_A = \frac{5M_0 a^2}{6EI} \downarrow \qquad \textbf{Ans}$$

14–150. Solve Prob. 14–149 using Castigliano's theorem.

Internal Moment Function $M(x)$: The internal moment function in terms of the load **P** and the external applied load are shown on the figure.

Castigliano's Second Theorem : The displacement at A can be determined using Eq. 14 – 49 with $\frac{\partial M(x_1)}{\partial P} = x_1$, $\frac{\partial M(x_2)}{\partial P} = x_2$, and setting $P = 0$.

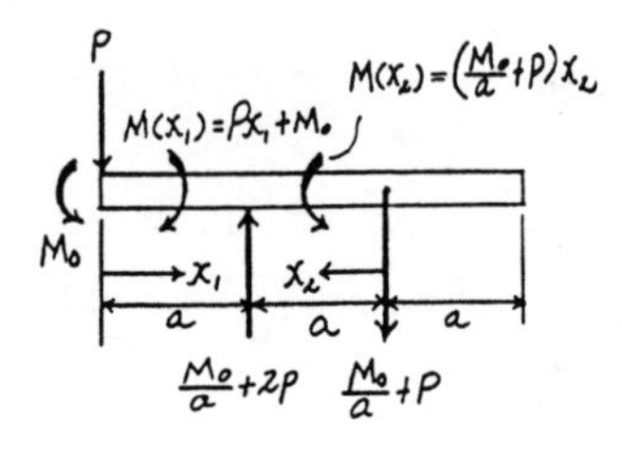

$$\Delta = \int_0^L M\left(\frac{\partial M}{\partial P}\right)\frac{dx}{EI}$$

$$\Delta_A = \frac{1}{EI}\int_0^a (M_0)(x_1)\, dx_1 + \frac{1}{EI}\int_0^a \left(\frac{M_0}{a}x_2\right)(x_2)\, dx_2$$

$$= \frac{5M_0 a^2}{6EI} \downarrow \qquad \textbf{Ans}$$

14–151. Determine the slope of point C of the A-36 steel beam. $I = 9.50(10^6)\ \text{mm}^4$. Use the method of virtual work.

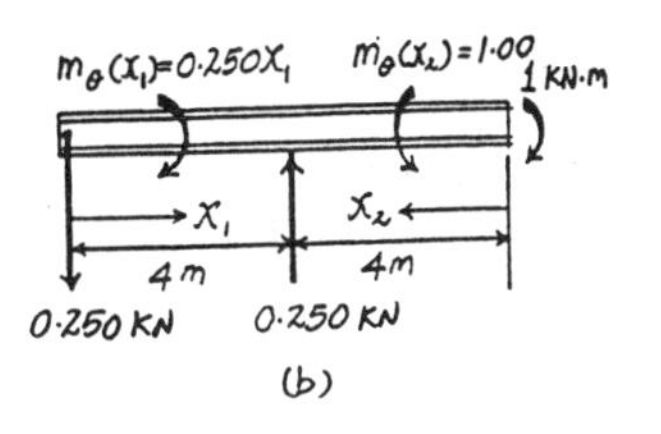

Real Moment Function $M(x)$ **:** As shown on figure(a).

Virtual Moment Functions $m_\theta(x)$ **:** As shown on figure(b).

***Virtual Work Equation* :** For the slope at point C, apply Eq. 14 – 43.

$$1 \cdot \theta = \int_0^L \frac{m_\theta M}{EI} dx$$

$$1\ \text{kN}\cdot\text{m}\cdot\theta_C = \frac{1}{EI}\int_0^{4\text{m}} (0.250x_1)(3.00x_1)\,dx_1 + \frac{1}{EI}\int_0^{4\text{m}} (1.00)(12.0)\,dx_2$$

$$\theta_C = \frac{64.0\ \text{kN}\cdot\text{m}^2}{EI} = \frac{64.0(1000)}{200(10^9)[9.50(10^{-6})]} = 0.0337\ \text{rad} \qquad \textbf{Ans}$$

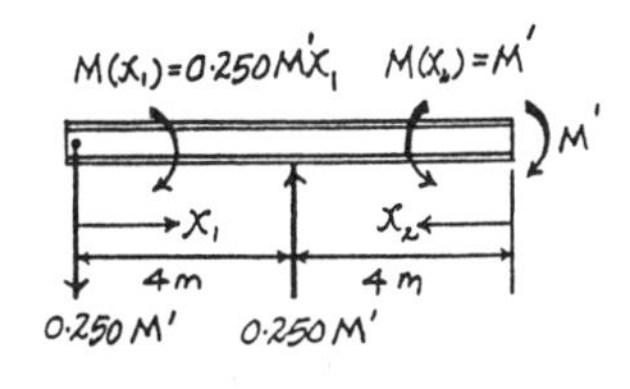

***14–152.** Solve Prob. 14–151 using Castigliano's theorem.

Internal Moment Function $M(x)$ **:** The internal moment function in terms of the couple moment $\mathbf{M}'$ and external applied load are shown on the figure.

***Castigliano's Second Theorem* :** The slope at C can be determined using Eq. 14 – 50 with $\frac{\partial M(x_1)}{\partial M'} = 0.250x_1$, $\frac{\partial M(x_2)}{\partial M'} = 1.00$ and setting $M' = 12\ \text{kN}\cdot\text{m}$.

M(x1)=0.250M'x1   M(x2)=M'

M'

x1   x2

4 m   4 m

0.250 M'   0.250 M'

$$\theta = \int_0^L M\left(\frac{\partial M}{\partial M'}\right)\frac{dx}{EI}$$

$$\theta_C = \frac{1}{EI}\int_0^{4\text{m}} (3.00x_1)(0.250x_1)\,dx_1 + \frac{1}{EI}\int_0^{4\text{m}} (12.00)(1.00)\,dx_2$$

$$= \frac{64.0\ \text{kN}\cdot\text{m}^2}{EI} = \frac{64.0(1000)}{200(10^9)[9.50(10^{-6})]} = 0.0337\ \text{rad} \qquad \textbf{Ans}$$

14–153. Determine the displacement at point C of the A-36 steel beam. $I = 9.50(10^6)\ \text{mm}^4$. Use the method of virtual work.

Real Moment Function M(x) : As shown on figure(a).

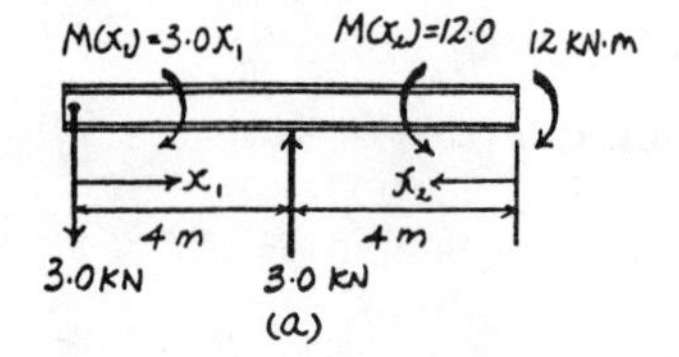

Virtual Moment Functions m(x) : As shown on figure(b).

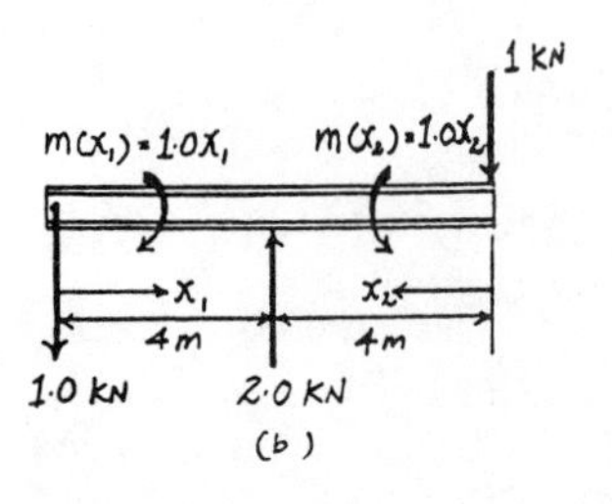

Virtual Work Equation : For the displacement at point C, apply Eq. 14 – 42.

$$1 \cdot \Delta = \int_0^L \frac{mM}{EI} dx$$

$$1\ \text{kN} \cdot \Delta_C = \frac{1}{EI}\int_0^{4\text{m}} (1.00x_1)(3.00x_1)\,dx_1 + \frac{1}{EI}\int_0^{4\text{m}} (1.00x_2)(12.0)\,dx_1$$

$$\Delta_C = \frac{160\ \text{kN} \cdot \text{m}^3}{EI} = \frac{160(1000)}{200(10^9)[9.50(10^{-6})]} = 0.08421\ \text{m} = 84.2\ \text{mm} \downarrow \qquad \textbf{Ans}$$

14–154. Solve Prob. 14–153 using Castigliano's theorem.

Internal Moment Function M(x) : The internal moment function in terms of the load **P** and external applied load are shown on the figure.

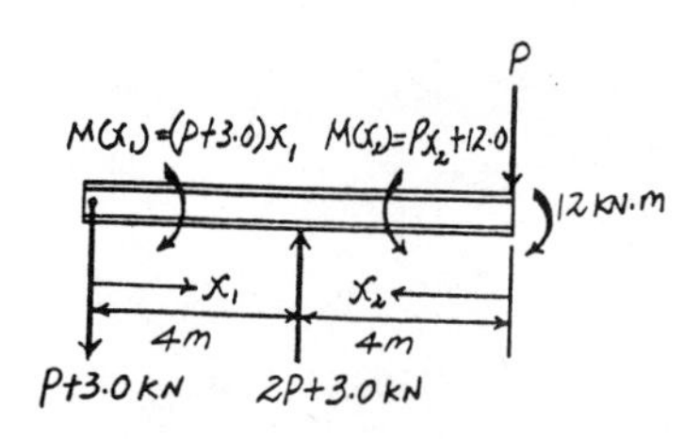

Castigliano's Second Theorem : The displacement at C can be determined using Eq. 14 – 49 with $\frac{\partial M(x_1)}{\partial P} = 1.00x_1$, $\frac{\partial M(x_2)}{\partial P} = 1.00x_2$, and setting $P = 0$.

$$\Delta = \int_0^L M\left(\frac{\partial M}{\partial P}\right)\frac{dx}{EI}$$

$$\Delta_C = \frac{1}{EI}\int_0^{4\text{m}} (3.00x_1)(1.00x_1)\,dx_1 + \frac{1}{EI}\int_0^{4\text{m}} (12.0)(1.00x_2)\,dx_2$$

$$= \frac{160\ \text{kN} \cdot \text{m}^3}{EI} = \frac{160(1000)}{200(10^9)[9.50(10^{-6})]} = 0.08421\ \text{m} = 84.2\ \text{mm} \downarrow \qquad \textbf{Ans}$$

14–155. Determine the strain energy in the *horizontal* curved bar due to torsion. There is a *vertical* force **P** acting at its end. JG is constant.

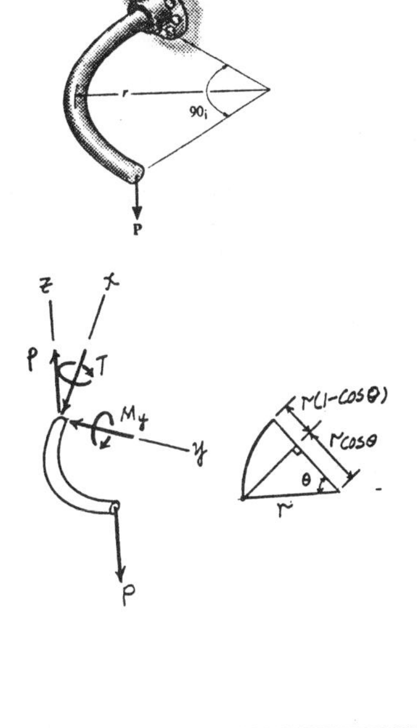

Internal Torque :

$$\Sigma M_x = 0; \quad P[r(1-\cos\theta)] - T = 0 \qquad T = Pr(1-\cos\theta)$$

Torsional Strain Energy : Here $ds = rd\theta$. Applying Eq. 14 – 22 gives

$$U_i = \int_0^L \frac{T^2}{2GJ} ds$$

$$= \frac{1}{2GJ}\int_0^{\frac{\pi}{2}} [Pr(1-\cos\theta)]^2 \, rd\theta$$

$$= \frac{P^2r^3}{2GJ}\int_0^{\frac{\pi}{2}} \left(\cos^2\theta - 2\cos\theta + 1\right) d\theta$$

$$= \frac{P^2r^3}{4GJ}\int_0^{\frac{\pi}{2}} (\cos 2\theta - 4\cos\theta + 3)\, d\theta$$

$$= \frac{P^2r^3}{8GJ}(3\pi - 8) \qquad \textbf{Ans}$$

***14–156.** Determine the total strain energy in the A-36 steel assembly. Consider the axial strain energy in the two 0.5-in.-diameter rods and the bending strain energy in the beam for which $I = 43.4 \text{ in}^4$.

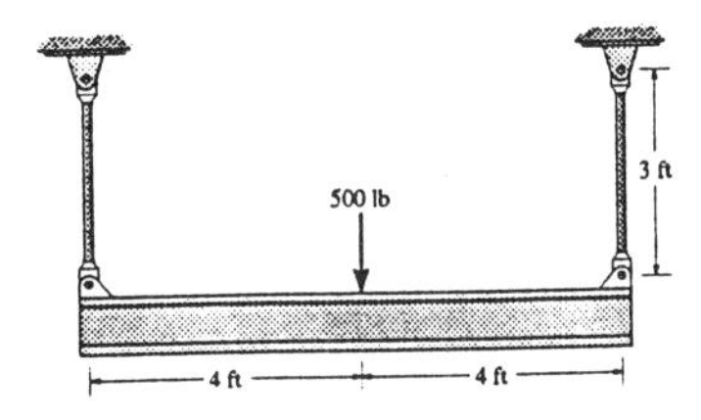

Support Reactions : As shown FBD(a).

Internal Moment Function : As shown on FBD(b).

Total Strain Energy : Combine Eq. 14 – 17 and Eq. 14 – 16 gives

$$(U_i)_T = \int_0^L \frac{M^2 dx}{2EI} + \frac{N^2 L}{2AE}$$

$$= 2\left[\frac{1}{2EI}\int_0^{4\text{ft}} (250x)^2 \, dx\right] + 2\left[\frac{250^2(3)}{2AE}\right]$$

$$= \frac{1.3333(10^6)\ \text{lb}^2\cdot\text{ft}^3}{EI} + \frac{0.1875(10^6)\ \text{lb}^2\cdot\text{ft}}{AE}$$

$$= \frac{1.3333(10^6)(12^3)}{29.0(10^6)(43.4)} + \frac{0.1875(10^6)\ (12)}{\frac{\pi}{4}(0.5^2)\,[29.0(10^6)]}$$

$$= 2.23 \text{ in}\cdot\text{lb} \qquad \textbf{Ans}$$

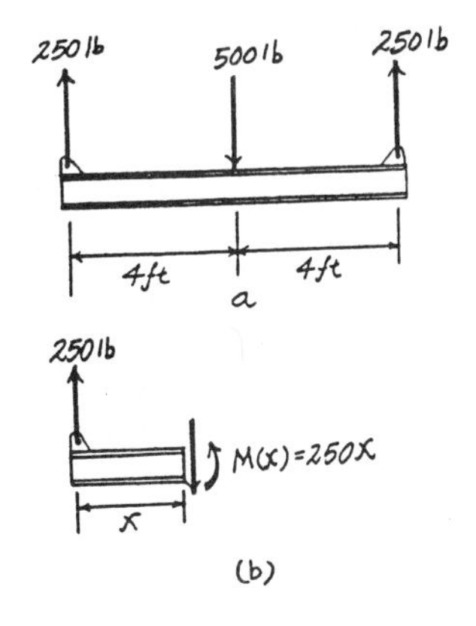

14–157. The $W\,10 \times 12$ beam is made of A-36 steel and is cantilevered from the wall at B. The spring mounted on the beam has a stiffness of $k = 1000$ lb/in. If a weight of 8 lb is dropped onto the spring from a height of $h = 3$ ft, determine the maximum bending stress developed in the beam.

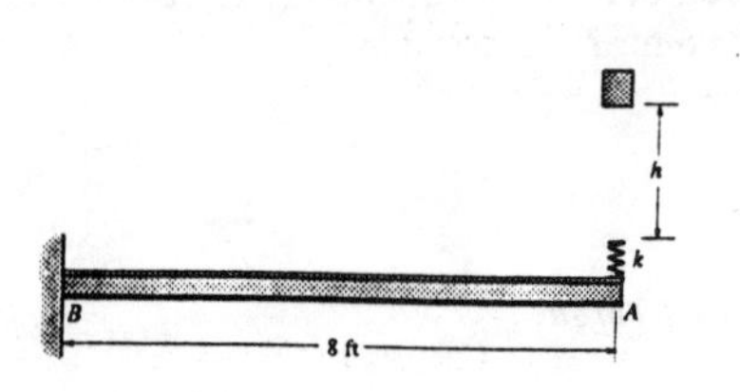

Equilibrium : This requires $F_{sp} = P_{beam}$. Then

$$k_{sp}\Delta_{sp} = k\Delta_{beam} \quad \text{or} \quad \Delta_{sp} = \frac{k}{k_{sp}}\Delta_{beam} \qquad [1]$$

Conservation of Energy : The equivalent spring constant for the beam can be determined using the deflection table listed in Appendix C.

$$k = \frac{3EI}{L^3} = \frac{3\left[29.0(10^6)\right](53.8)}{[8(12)]^3} = 5290.39 \text{ lb/in.}$$

Thus,

$$U_e = U_i$$

$$W\left(h + \Delta_{sp} + \Delta_{beam}\right) = \frac{1}{2}k\Delta_{beam}^2 + \frac{1}{2}k_{sp}\Delta_{sp}^2 \qquad [2]$$

Substituting Eq.[1] into [2] yields

$$W\left(h + \frac{k}{k_{sp}}\Delta_{beam} + \Delta_{beam}\right) = \frac{1}{2}k\Delta_{beam}^2 + \frac{k^2}{2k_{sp}}\Delta_{beam}^2$$

$$8\left[3(12) + \frac{5290.39}{1000}\Delta_{beam} + \Delta_{beam}\right] = \frac{1}{2}(5290.39)\Delta_{beam}^2 + \frac{5290.39^2}{2(1000)}\Delta_{beam}^2$$

$$16639.32\Delta_{beam}^2 - 50.323\Delta_{beam} - 288 = 0$$

Solving for positive root, we have

$$\Delta_{beam} = 0.13308 \text{ in.}$$

Maximum Stress : The maximum force on the beam is $P_{beam} = k\Delta_{beam} = 5290.39(0.13308) = 704.06$ lb. Then the maximum moment occurs at the fixed support B where $M_{max} = P_{beam}L = 704.06(8)(12) = 67589.5$ lb · in.

$$\sigma_{max} = \frac{M_{max}c}{I} = \frac{67589.5\left(\frac{9.87}{2}\right)}{53.8} = 6200 \text{ psi} = 6.20 \text{ ksi} \qquad \textbf{Ans}$$

Since $\sigma_{max} < \sigma_Y = 36$ ksi, the above analysis is valid.

A–1. Determine the distance $\bar{y}$ to the centroid C of the beam's cross-sectional area. The beam is symmetric with respect to the y axis.

Centroid :

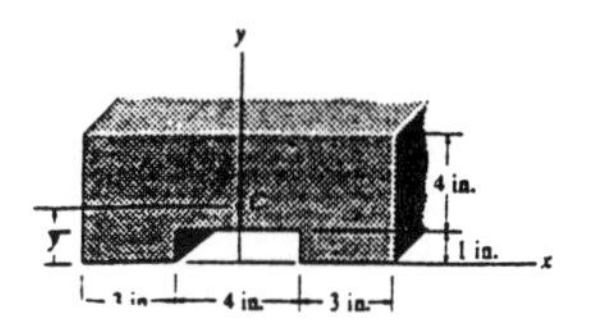

$$A = 2[3(5)] + 4(4) = 46.0 \text{ in}^2$$

$$\Sigma\tilde{y}A = 2[2.5(3)(5)] + 3(4)(4) = 123.0 \text{ in}^3$$

$$\bar{y} = \frac{\Sigma\tilde{y}A}{A} = \frac{123.0}{46.0} = 2.67 \text{ in.} \qquad \textbf{Ans}$$

A–2. Determine I_x and I_y for the beam's cross-sectional area.

Moment of Inertia :

$$I_x = 2\left[\frac{1}{12}(3)(5^3) + 3(5)(2.5^2)\right] + \frac{1}{12}(4)(4^3) + 4(4)(3^2)$$

$$= 415 \text{ in}^4. \qquad \textbf{Ans}$$

$$I_y = \frac{1}{12}(5)(10^3) - \frac{1}{12}(1)(4^3) = 411 \text{ in}^4 \qquad \textbf{Ans}$$

A–3. Determine the distance $\bar{y}$ to the centroid C of the beam's cross-sectional area, then find $\bar{I}_{x'}$.

Centroid :

$$A = 1(12) + 1(8) + 3(2) = 26.0 \text{ in}^2$$

$$\Sigma\tilde{y}A = 0.5(1)(12) + 4.5(1)(8) + 2.5(3)(2) = 57.0 \text{ in}^3$$

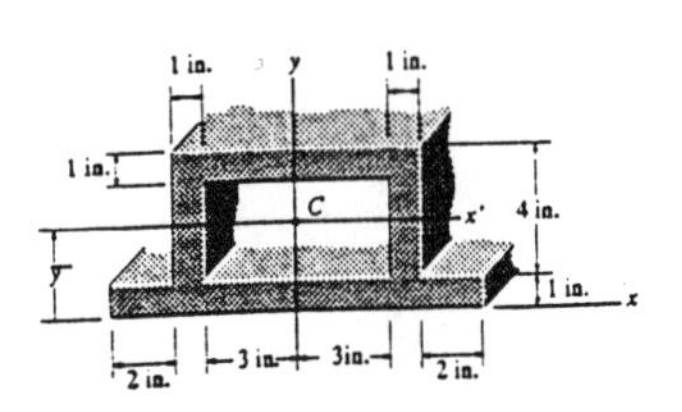

$$\bar{y} = \frac{\Sigma\tilde{y}A}{\Sigma A} = \frac{57.0}{26.0} = 2.1923 \text{ in.} = 2.19 \text{ in.} \qquad \textbf{Ans}$$

Moment of Inertia :

$$\bar{I}_{x'} = \frac{1}{12}(12)(1^3) + 12(1)(2.1923 - 0.5)^2 + \frac{1}{12}(8)(1^3) + 8(1)(4.5 - 2.1923)^2 + \frac{1}{12}(2)(3^3) + 2(3)(2.5 - 2.1923)^2$$

$$= 83.7 \text{ in}^4 \qquad \textbf{Ans}$$

***A–4.** Determine $\bar{I}_y$ for the beam's cross-sectional area.

Moment of Inertia :

$$\bar{I}_y = \frac{1}{12}(1)(12^3) + \frac{1}{12}(4)(8^3) - \frac{1}{12}(3)(6^3)$$
$$= 261 \text{ in}^4 \qquad \textbf{Ans}$$

A–5. Determine $\bar{y}$, which locates the centroid, and then find the moments of inertia $\bar{I}_{x'}$ and $\bar{I}_y$ for the T-beam.

Centroid :

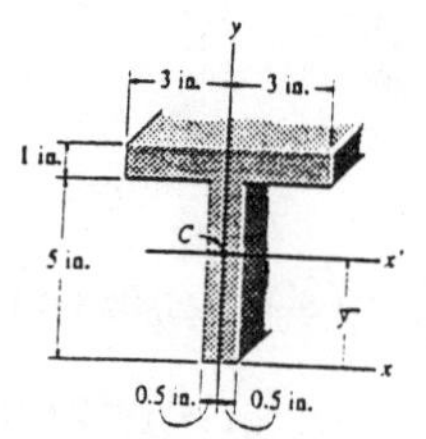

$$A = 1(6) + 5(1) = 11.0 \text{ in}^2$$
$$\Sigma\tilde{y}A = 2.5(5)(1) + 5.5(1)(6) = 45.5 \text{ in}^3$$

$$\bar{y} = \frac{\Sigma\tilde{y}A}{\Sigma A} = \frac{45.5}{11.0} = 4.1364 \text{ in.} = 4.14 \text{ in.} \qquad \textbf{Ans}$$

Moment of Inertia :

$$\bar{I}_{x'} = \frac{1}{12}(1)(5^3) + 1(5)(4.1364 - 2.5)^2 + \frac{1}{12}(6)(1^3) + 6(1)(5.5 - 4.1364)^2$$
$$= 35.5 \text{ in}^4 \qquad \textbf{Ans}$$

$$\bar{I}_y = \frac{1}{12}(1)(6^3) + \frac{1}{12}(5)(1^3) = 18.4 \text{ in}^4 \qquad \textbf{Ans}$$

A–6. Determine $\bar{y}$, which locates the centroid C, and then compute the moments of inertia $\bar{I}_{x'}$ and $\bar{I}_{y'}$ for the cross-sectional area.

Centroid :

$$A = 0.08(0.08) - 0.06(0.04) = 4.00(10^{-3}) \text{ m}^2$$
$$\Sigma\tilde{y}A = 0.04(0.08)(0.08) - 0.05(0.06)(0.04) = 0.1360(10^{-3}) \text{ m}^3$$

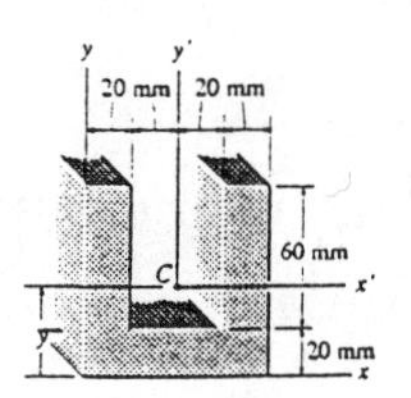

$$\bar{y} = \frac{\Sigma\tilde{y}A}{\Sigma A} = \frac{0.1360(10^{-3})}{4.00(10^{-3})} = 0.0340 \text{ m} = 34.0 \text{ mm} \qquad \textbf{Ans}$$

Moment of Inertia :

$$\bar{I}_{x'} = \frac{1}{12}(0.08)(0.08^3) + 0.08(0.08)(0.04 - 0.034)^2 - \left[\frac{1}{12}(0.04)(0.06^3) + 0.04(0.06)(0.05 - 0.034)^2\right]$$
$$= 2.31(10^{-6}) \text{ m}^4 = 2.31(10^6) \text{ mm}^4 \qquad \textbf{Ans}$$

$$\bar{I}_y = \frac{1}{12}(0.08)(0.08^3) - \frac{1}{12}(0.06)(0.04^3)$$
$$= 3.09(10^{-6}) \text{ m}^4 = 3.09(10^6) \text{ mm}^4 \qquad \textbf{Ans}$$

A–7. Determine the location $(\bar{x}, \bar{y})$ of the centroid C of the angle's cross-sectional area, then find the moments of inertia $\bar{I}_{x'}$ and $\bar{I}_{y'}$.

Centroid :

$$A = 3(1) + 1(3) = 6.00 \text{ in}^2$$
$$\Sigma\tilde{y}A = 0.5(3)(1) + 1.5(1)(3) = 6.00 \text{ in}^3$$
$$\Sigma\tilde{x}A = 0.5(3)(1) + 2.5(1)(3) = 9.00 \text{ in}^3$$

$$\bar{y} = \frac{\Sigma\tilde{y}A}{\Sigma A} = \frac{6.00}{6.00} = 1.00 \text{ in.} \qquad \textbf{Ans}$$
$$\bar{x} = \frac{\Sigma\tilde{x}A}{\Sigma A} = \frac{9.00}{6.00} = 1.50 \text{ in.} \qquad \textbf{Ans}$$

Moment of Inertia :

$$\bar{I}_{x'} = \frac{1}{12}(3)(1^3) + 3(1)(1.00 - 0.5)^2 + \frac{1}{12}(1)(3^3) + 1(3)(1.5 - 1.00)^2$$
$$= 4.00 \text{ in}^4 \qquad \textbf{Ans}$$

$$\bar{I}_{y'} = \frac{1}{12}(3)(1^3) + 3(1)(1.50 - 0.5)^2 + \frac{1}{12}(1)(3^3) + 1(3)(2.5 - 1.50)^2$$
$$= 8.50 \text{ in}^4 \qquad \textbf{Ans}$$

***A–8.** Determine the location $(\bar{x}, \bar{y})$ of the centroid C of the angle's cross-sectional area, then find the product of inertia $\bar{I}_{x'y'}$ with respect to the x' and y' axes.

Centroid :

$$A = 3(1) + 1(3) = 6.00 \text{ in}^2$$
$$\Sigma\tilde{y}A = 0.5(3)(1) + 1.5(1)(3) = 6.00 \text{ in}^3$$
$$\Sigma\tilde{x}A = 0.5(3)(1) + 2.5(1)(3) = 9.00 \text{ in}^3$$

$$\bar{y} = \frac{\Sigma\tilde{y}A}{\Sigma A} = \frac{6.00}{6.00} = 1.00 \text{ in.} \qquad \textbf{Ans}$$
$$\bar{x} = \frac{\Sigma\tilde{x}A}{\Sigma A} = \frac{9.00}{6.00} = 1.50 \text{ in.} \qquad \textbf{Ans}$$

Product of Inertia :

$$\bar{I}_{x'y'} = 3(1)(1)(-0.5) + 1(3)(-1)(0.5)$$
$$= -3.00 \text{ in}^4 \qquad \textbf{Ans}$$

A–9. Determine the location $(\bar{x}, \bar{y})$ of the centroid C of the angle's cross-sectional area, and then compute the moments of inertia $\bar{I}_{x'}$ and $\bar{I}_{y'}$.

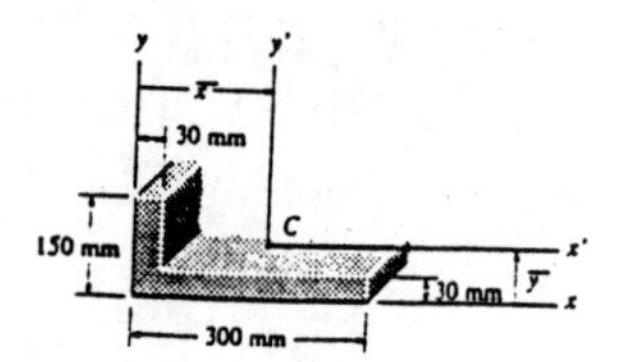

Centroid :

$$A = 270(30) + 30(150) = 12600 \text{ mm}^2$$
$$\Sigma\tilde{y}A = 15(270)(30) + 75(30)(150) = 459000 \text{ mm}^3$$
$$\Sigma\tilde{x}A = 165(30)(270) + 15(150)(30) = 1404000 \text{ mm}^3$$

$$\bar{y} = \frac{\Sigma\tilde{y}A}{\Sigma A} = \frac{459000}{12600} = 36.429 \text{ mm} = 36.4 \text{ mm} \qquad \textbf{Ans}$$
$$\bar{x} = \frac{\Sigma\tilde{x}A}{\Sigma A} = \frac{1404000}{12600} = 111.43 \text{ mm} = 111 \text{ mm} \qquad \textbf{Ans}$$

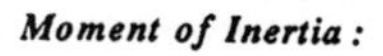

$$\bar{I}_{x'} = \frac{1}{12}(270)(30^3) + 270(30)(36.429 - 15)^2$$
$$+ \frac{1}{12}(30)(150^3) + 30(150)(75 - 36.429)^2$$
$$= 19.5(10^6) \text{ mm}^4 \qquad \textbf{Ans}$$

$$\bar{I}_{y'} = \frac{1}{12}(150)(30^3) + 150(30)(111.43 - 15)^2$$
$$+ \frac{1}{12}(30)(270^3) + 30(270)(165 - 111.43)^2$$
$$= 115(10^6) \text{ mm}^4 \qquad \textbf{Ans}$$

A–10. Determine the location $(\bar{x}, \bar{y})$ of the centroid C of the angle's cross-sectional area, and then compute the product of inertia $\bar{I}_{x'y'}$ with respect to the x' and y' axes.

Centroid :

$$A = 270(30) + 30(150) = 12600 \text{ mm}^2$$
$$\Sigma\tilde{y}A = 15(270)(30) + 75(30)(150) = 459000 \text{ mm}^3$$
$$\Sigma\tilde{x}A = 165(30)(270) + 15(150)(30) = 1404000 \text{ mm}^3$$

$$\bar{y} = \frac{\Sigma\tilde{y}A}{\Sigma A} = \frac{459000}{12600} = 36.429 \text{ mm} = 36.4 \text{ mm} \qquad \textbf{Ans}$$
$$\bar{x} = \frac{\Sigma\tilde{x}A}{\Sigma A} = \frac{1404000}{12600} = 111.43 \text{ mm} = 111 \text{ mm} \qquad \textbf{Ans}$$

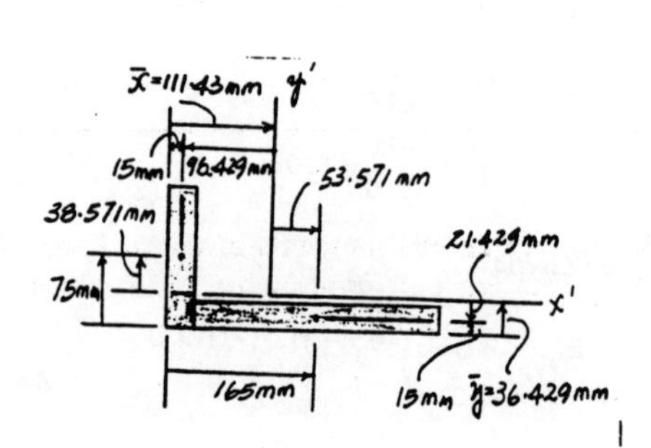

Product of Inertia :

$$\bar{I}_{x'y'} = 270(30)(53.571)(-21.429)$$
$$+ 150(30)(-96.429)(38.571)$$
$$= -26.1(10^6) \text{ mm}^4 \qquad \textbf{Ans}$$

A–11. Determine the location $(\bar{x}, \bar{y})$ of the channel's cross-sectional area, and then determine the moments of inertia $\bar{I}_{x'}$ and $\bar{I}_{y'}$.

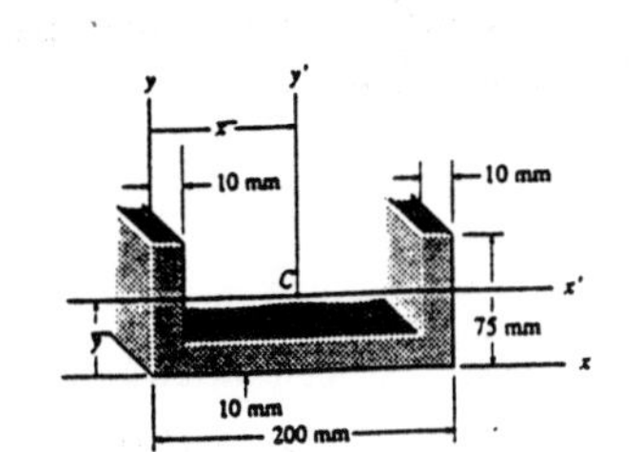

Centroid :

$$A = 180(10) + 20(75) = 3300 \text{ mm}^2$$
$$\Sigma\tilde{y}A = 5(180)(10) + 37.5(20)(75) = 65250 \text{ mm}^3$$

$$\bar{y} = \frac{\Sigma\tilde{y}A}{\Sigma A} = \frac{65250}{3300} = 19.773 \text{ mm} = 19.8 \text{ mm} \qquad \textbf{Ans}$$

Due to symmetry about the y' axis, $\bar{x} = \frac{200}{2} = 100 \text{ mm}$ **Ans**

Moment of Inertia :

$$\bar{I}_{x'} = \frac{1}{12}(180)(10^3) + 180(10)(19.773 - 5)^2 + \frac{1}{12}(20)(75^3) + 20(75)(37.5 - 19.773)^2$$
$$= 1.58(10^6) \text{ mm}^4 \qquad \textbf{Ans}$$

$$\bar{I}_{y'} = \frac{1}{12}(75)(200^3) - \frac{1}{12}(65)(180^3)$$
$$= 18.4(10^6) \text{ mm}^4 \qquad \textbf{Ans}$$

***A–12.** Determine the location $(\bar{x}, \bar{y})$ of the channel's cross-sectional area, and then determine the product of inertia $\bar{I}_{x'y'}$ with respect to the x' and y' axes.

Centroid :

$$A = 180(10) + 20(75) = 3300 \text{ mm}^2$$
$$\Sigma\tilde{y}A = 5(180)(10) + 37.5(20)(75) = 65250 \text{ mm}^3$$

$$\bar{y} = \frac{\Sigma\tilde{y}A}{\Sigma A} = \frac{65250}{3300} = 19.773 \text{ mm} = 19.8 \text{ mm} \qquad \textbf{Ans}$$

Due to symmetry about the y' axis, $\bar{x} = \frac{200}{2} = 100 \text{ mm}$ **Ans**

Product of Inertia : Due to symmetry about y' axis,

$$\bar{I}_{x'y'} = 0 \qquad \textbf{Ans}$$

A–13. Determine the moments of inertia $\bar{I}_x$ and $\bar{I}_y$ of the Z-section. The origin of coordinates is at the centroid C.

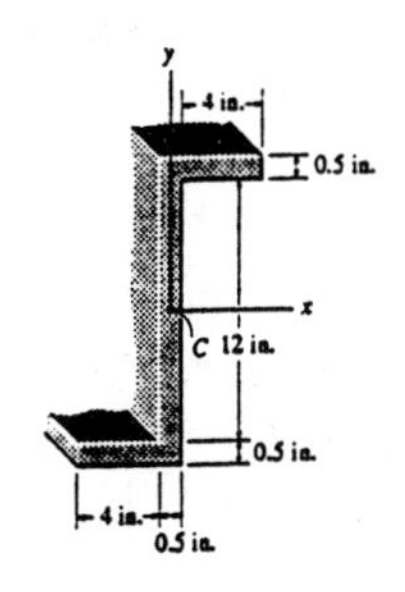

Moment of Inertia :

$$\bar{I}_x = \frac{1}{12}(0.5)(13^3) + 2\left[\frac{1}{12}(4)(0.5^3) + 4(0.5)(6.25^2)\right]$$
$$= 248 \text{ in}^4 \qquad \textbf{Ans}$$

$$\bar{I}_y = \frac{1}{12}(12.5)(0.5^3) + \frac{1}{12}(0.5)(8.5^3)$$
$$= 25.7 \text{ in}^4 \qquad \textbf{Ans}$$

A–14. Determine the product of inertia $\bar{I}_{xy}$ of the cross-sectional area of the Z-section. The origin of coordinates is at the centroid C.

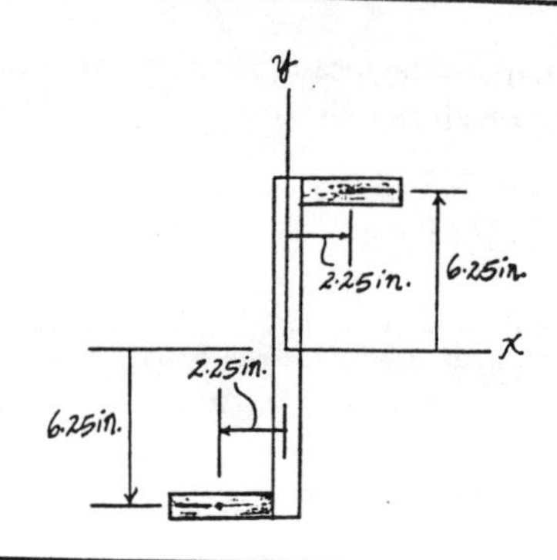

Product of Inertia :

$$\bar{I}_{xy} = 4(0.5)(2.25)(6.25) + 4(0.5)(-2.25)(-6.25)$$
$$= 56.25 \text{ in}^4 \qquad \textbf{Ans}$$

A–15. Determine the location $(\bar{x}, \bar{y})$ of the centroid C of the cross-sectional area, then determine the moments of inertia $\bar{I}_{x'}$ and $\bar{I}_{y'}$ with respect to the x' and y' axes that have their origin located at the centroid C.

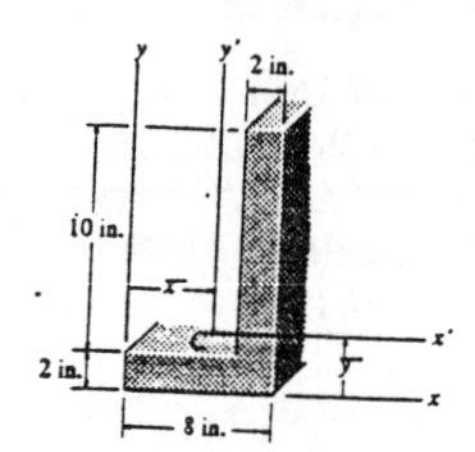

Centroid :

$$A = 6(2) + 2(12) = 36.0 \text{ in}^2$$
$$\Sigma\tilde{y}A = 1(6)(2) + 6(2)(12) = 156 \text{ in}^3$$
$$\Sigma\tilde{x}A = 3(2)(6) + 7(12)(2) = 204 \text{ in}^3$$

$$\bar{y} = \frac{\Sigma\tilde{y}A}{\Sigma A} = \frac{156}{36.0} = 4.33 \text{ in.} \qquad \textbf{Ans}$$
$$\bar{x} = \frac{\Sigma\tilde{x}A}{\Sigma A} = \frac{204}{36.0} = 5.67 \text{ in.} \qquad \textbf{Ans}$$

Moment of Inertia :

$$\bar{I}_{x'} = \frac{1}{12}(6)(2^3) + 6(2)(4.333-1)^2 + \frac{1}{12}(2)(12^3) + 2(12)(6-4.333)^2$$
$$= 492 \text{ in}^4 \qquad \textbf{Ans}$$

$$\bar{I}_{y'} = \frac{1}{12}(2)(6^3) + 2(6)(5.667-3)^2 + \frac{1}{12}(12)(2^3) + 12(2)(7-5.667)^2$$
$$= 172 \text{ in}^4 \qquad \textbf{Ans}$$

***A–16.** Determine the location $(\bar{x}, \bar{y})$ of the centroid C of the cross-sectional area, then determine the product of inertia $I_{x'y'}$ with respect to the x' and y' axes.

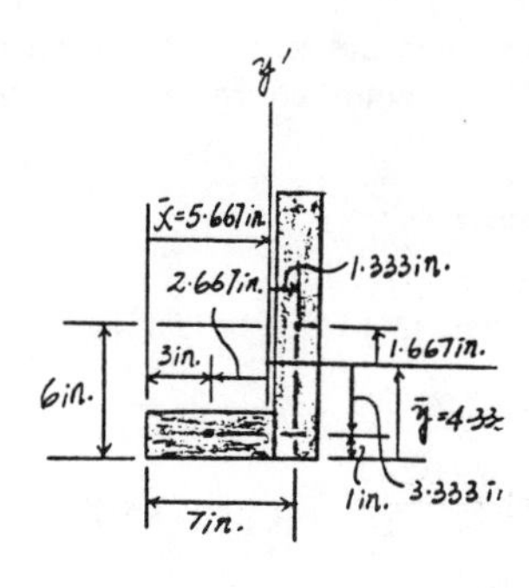

Centroid :

$$A = 6(2) + 2(12) = 36.0 \text{ in}^2$$
$$\Sigma\tilde{y}A = 1(6)(2) + 6(2)(12) = 156 \text{ in}^3$$
$$\Sigma\tilde{x}A = 3(2)(6) + 7(12)(2) = 204 \text{ in}^3$$

$$\bar{y} = \frac{\Sigma\tilde{y}A}{\Sigma A} = \frac{156}{36.0} = 4.33 \text{ in.} \qquad \textbf{Ans}$$
$$\bar{x} = \frac{\Sigma\tilde{x}A}{\Sigma A} = \frac{204}{36.0} = 5.67 \text{ in.} \qquad \textbf{Ans}$$

Product of Inertia :

$$I_{x'y'} = 12(2)(1.333)(1.667) + 6(2)(-2.667)(-3.333)$$
$$= 160 \text{ in}^4 \qquad \textbf{Ans}$$

A–17. Determine the product of inertia I_{xy} of the cross-sectional area with respect to the x and y axes.

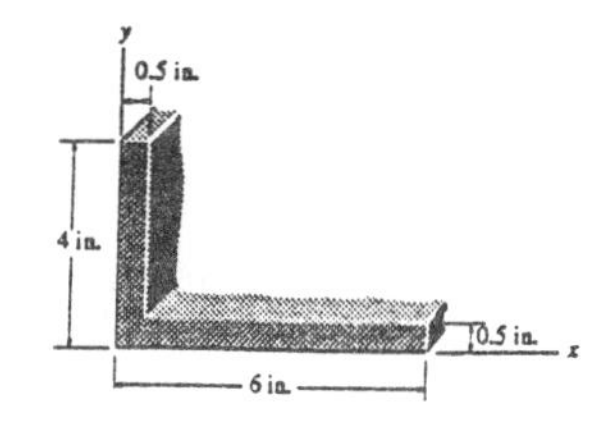

Product of Inertia :

$$\bar{I}_{xy} = 6(0.5)(3)(0.25) + 3.5(0.5)(0.25)(2.25)$$
$$= 3.23 \text{ in}^4 \qquad \textbf{Ans}$$

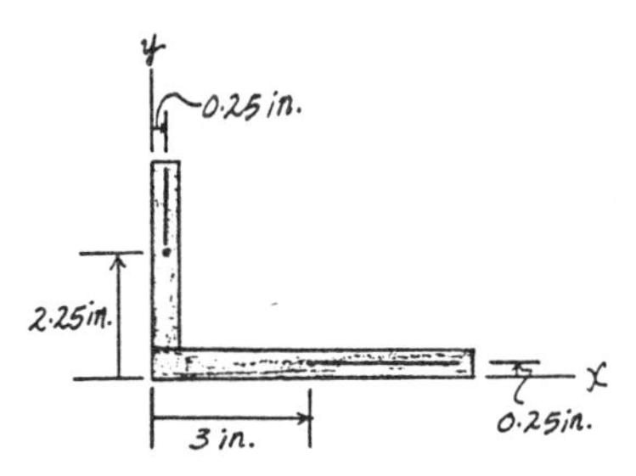

A–18. Compute the moments of inertia $I_{x'}$ and $I_{y'}$ and the product of inertia $I_{x'y'}$ of the cross-sectional area. Use the equations of Sec. A.4.

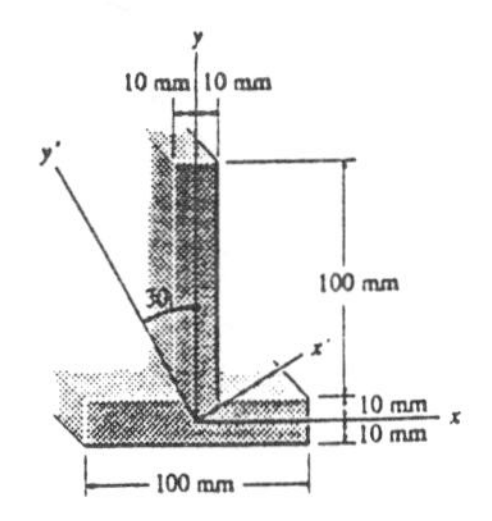

Moment of Inertia and Product of inertia about x and y axes :

$$I_x = \frac{1}{12}(100)\left(20^3\right) + \frac{1}{12}(20)\left(100^3\right) + 20(100)\left(60^2\right)$$
$$= 8.9333\left(10^6\right) \text{ mm}^4$$

$$I_y = \frac{1}{12}(100)\left(20^3\right) + \frac{1}{12}(20)\left(100^3\right) = 1.7333\left(10^6\right) \text{ mm}^4$$

Due to symmetry about y axis, $I_{xy} = 0$.

Moment of Inertia and Product of Inertia about x' and y' Axes :
Applying Eq. A – 10 with $\theta = 30°$ gives

$$I_{x'} = \frac{I_x + I_y}{2} + \frac{I_x - I_y}{2}\cos 2\theta - I_{xy}\sin 2\theta$$
$$= \left(\frac{8.9333 + 1.7333}{2} + \frac{8.9333 - 1.7333}{2}\cos 60° - 0\right)10^6$$
$$= 7.13\left(10^6\right) \text{ mm}^4 \qquad \textbf{Ans}$$

$$I_{y'} = \frac{I_x + I_y}{2} - \frac{I_x - I_y}{2}\cos 2\theta + I_{xy}\sin 2\theta$$
$$= \left(\frac{8.9333 + 1.7333}{2} - \frac{8.9333 - 1.7333}{2}\cos 60° + 0\right)10^6$$
$$= 3.53\left(10^6\right) \text{ mm}^4 \qquad \textbf{Ans}$$

$$I_{x'y'} = \frac{I_x - I_y}{2}\sin 2\theta + I_{xy}\cos 2\theta$$
$$= \left(\frac{8.9333 - 1.7333}{2}\sin 60° + 0\right)10^6$$
$$= 3.12\left(10^6\right) \text{ mm}^4 \qquad \textbf{Ans}$$

A–19. Determine the moments of inertia $I_{x'}$ and $I_{y'}$ and the product of inertia $I_{x'y'}$ for the rectangular area. The x' and axes pass through the centroid C. Use the equations of Sec. A.4.

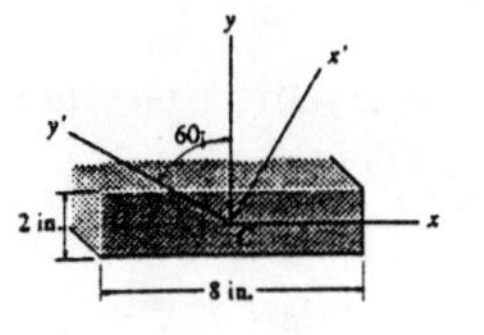

Moment of Inertia and Product of Inertia about x and y axes :

$$I_x = \frac{1}{12}(8)\left(2^3\right) = 5.333 \text{ in}^4$$

$$I_y = \frac{1}{12}(2)\left(8^3\right) = 85.33 \text{ in}^4$$

Due to symmetry , $\qquad I_{xy} = 0$

Moment of Inertia and Product of Inertia about x' and y' Axes :
Applying Eq. A– 10 with $\theta = 60°$ gives

$$\begin{aligned} I_{x'} &= \frac{I_x + I_y}{2} + \frac{I_x - I_y}{2}\cos 2\theta - I_{xy}\sin 2\theta \\ &= \frac{5.333 + 85.33}{2} + \frac{5.333 - 85.33}{2}\cos 120° - 0 \\ &= 65.3 \text{ in}^4 \end{aligned}$$

Ans

$$\begin{aligned} I_{y'} &= \frac{I_x + I_y}{2} - \frac{I_x - I_y}{2}\cos 2\theta + I_{xy}\sin 2\theta \\ &= \frac{5.333 + 85.33}{2} - \frac{5.333 - 85.33}{2}\cos 120° - 0 \\ &= 25.3 \text{ in}^4 \end{aligned}$$

Ans

$$\begin{aligned} I_{x'y'} &= \frac{I_x - I_y}{2}\sin 2\theta + I_{xy}\cos 2\theta \\ &= \frac{5.333 - 85.33}{2}\sin 120° + 0 \\ &= -34.6 \text{ in}^4 \end{aligned}$$

Ans

***A–20.** Determine the location $(\bar{x}, \bar{y})$ of the centroid C of the angle's cross-sectional area, then find the product of inertia $I_{x'y'}$. Use the equations of Sec. A.4.

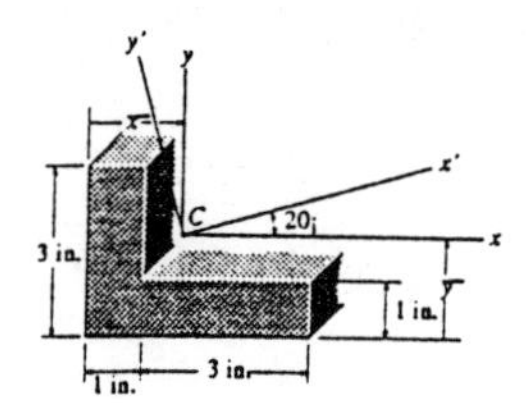

Centroid :

$$A = 3(1) + 1(3) = 6.00 \text{ in}^2$$
$$\Sigma\tilde{y}A = 0.5(3)(1) + 1.5(1)(3) = 6.00 \text{ in}^3$$
$$\Sigma\tilde{x}A = 0.5(3)(1) + 2.5(1)(3) = 9.00 \text{ in}^3$$

$$\bar{y} = \frac{\Sigma\tilde{y}A}{\Sigma A} = \frac{6.00}{6.00} = 1.00 \text{ in.}$$ **Ans**

$$\bar{x} = \frac{\Sigma\tilde{x}A}{\Sigma A} = \frac{9.00}{6.00} = 1.50 \text{ in.}$$ **Ans**

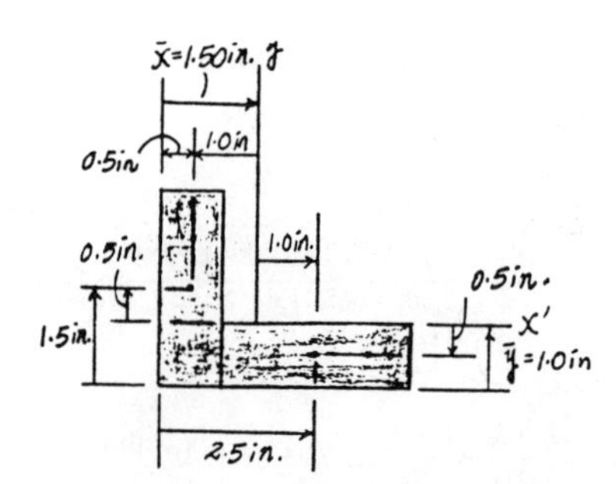

Moment of Inertia and Product of Inertia about x and y axes :

$$\begin{aligned} I_x &= \frac{1}{12}(3)\left(1^3\right) + 3(1)(1.00 - 0.5)^2 \\ &\quad + \frac{1}{12}(1)\left(3^3\right) + 1(3)(1.5 - 1.00)^2 \\ &= 4.00 \text{ in}^4 \end{aligned}$$

$$\begin{aligned} I_y &= \frac{1}{12}(3)\left(1^3\right) + 3(1)(1.50 - 0.5)^2 \\ &\quad + \frac{1}{12}(1)\left(3^3\right) + 1(3)(2.5 - 1.50)^2 \\ &= 8.50 \text{ in}^4 \end{aligned}$$

$$\begin{aligned} I_{xy} &= 3(1)(1)(-0.5) + 1(3)(-1)(0.5) \\ &= -3.00 \text{ in}^4 \end{aligned}$$

Product of Inertia About x' and y' Axes : Applying Eq. A– 10 with $\theta = 40°$ gives

$$\begin{aligned} I_{x'y'} &= \frac{I_x - I_y}{2}\sin 2\theta + I_{xy}\cos 2\theta \\ &= \frac{4.00 - 8.50}{2}\sin 40° + (-3.00)\cos 40° \\ &= -3.74 \text{ in}^4 \end{aligned}$$

Ans

A–21. Determine the location $(\bar{x}, \bar{y})$ of the centroid C of the angle's cross-sectional area, then find the principal moments of inertia and the orientation of the principal axes of inertia with respect to the x and y axes.

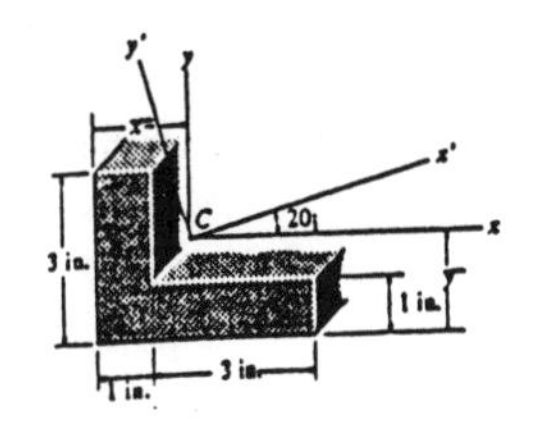

Centroid :

$$A = 3(1) + 1(3) = 6.00 \text{ in}^2$$
$$\Sigma \tilde{y} A = 0.5(3)(1) + 1.5(1)(3) = 6.00 \text{ in}^3$$
$$\Sigma \tilde{x} A = 0.5(3)(1) + 2.5(1)(3) = 9.00 \text{ in}^3$$

$$\bar{y} = \frac{\Sigma \tilde{y} A}{\Sigma A} = \frac{6.00}{6.00} = 1.00 \text{ in.} \qquad \textbf{Ans}$$
$$\bar{x} = \frac{\Sigma \tilde{x} A}{\Sigma A} = \frac{9.00}{6.00} = 1.50 \text{ in.} \qquad \textbf{Ans}$$

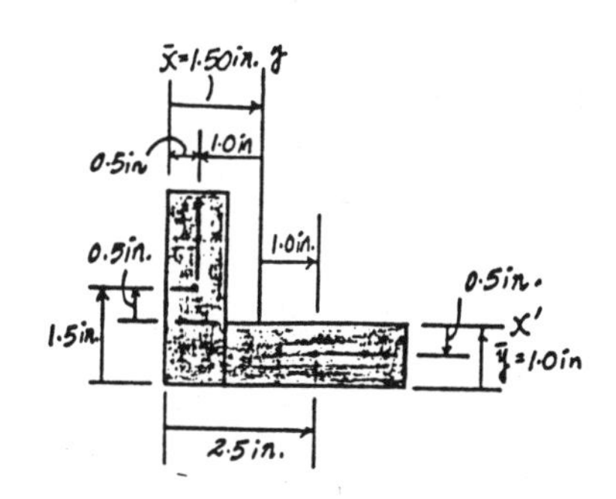

Moment of Inertia and Product of Inertia about x and y axes :

$$I_x = \frac{1}{12}(3)(1^3) + 3(1)(1.00 - 0.5)^2 + \frac{1}{12}(1)(3^3) + 1(3)(1.5 - 1.00)^2 = 4.00 \text{ in}^4$$

$$I_y = \frac{1}{12}(3)(1^3) + 3(1)(1.50 - 0.5)^2 + \frac{1}{12}(1)(3^3) + 1(3)(2.5 - 1.50)^2 = 8.50 \text{ in}^4$$

$$I_{xy} = 3(1)(1)(-0.5) + 1(3)(-1)(0.5) = -3.00 \text{ in}^4$$

Principal Moment of Inertia : Applying Eq. A – 12 gives

$$I_{\substack{\max \\ \min}} = \frac{I_x + I_y}{2} \pm \sqrt{\left(\frac{I_x - I_y}{2}\right)^2 + I_{xy}^2} = \frac{4.00 + 8.50}{2} \pm \sqrt{\left(\frac{4.00 - 8.50}{2}\right)^2 + (-3.00)^2}$$

$$I_{\max} = 10.0 \text{ in}^4 \qquad I_{\min} = 2.50 \text{ in}^4 \qquad \textbf{Ans}$$

Orientation of the Principal Axes of Inertia : Applying Eq. A – 11 gives

$$\tan 2\theta_p = \frac{-I_{xy}}{(I_x - I_y)/2} = \frac{-(-3.00)}{(4.00 - 8.50)/2} = -1.3333$$
$$2\theta_p = -53.13^\circ \quad \text{and} \quad 126.87^\circ$$

Substituting $2\theta_p = -53.13^\circ$ into Eq. A – 10, we have

$$I_{x'} = \frac{I_x + I_y}{2} + \frac{I_x - I_y}{2}\cos 2\theta - I_{xy}\sin 2\theta = \frac{4.00 + 8.50}{2} + \frac{4.00 - 8.50}{2}\cos(-53.13^\circ) - (-3.00)\sin(-53.13^\circ) = 2.50 \text{ in}^4 = I_{\min}$$

Thus, $\qquad \theta_{p_1} = 63.4^\circ \qquad \theta_{p_2} = -26.6^\circ \qquad$ **Ans**

A–22. Determine the moments of inertia $I_{x'}$ and $I_{y'}$ and the product of inertia $I_{x'y'}$ for the semicircular area.

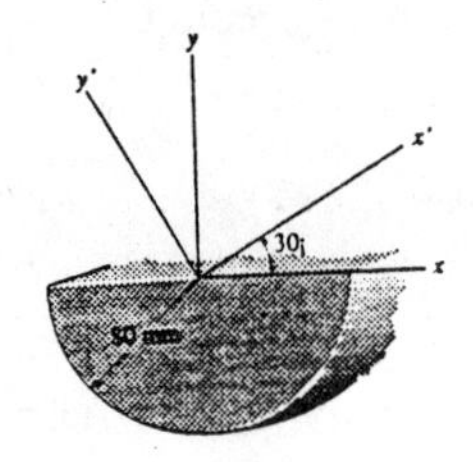

Moment of Inertia and Product of Inertia about x and y axes :

$$I_x = I_y = \frac{1}{2}\left[\frac{\pi}{4}\left(80^4\right)\right] = 5.12\pi\left(10^6\right)\ \text{mm}^4$$

Due to symmetry about y axis, $I_{xy} = 0$.

Moment of Inertia and Product of Inertia about x' and y' Axes :

Applying Eq. A – 10 with $\theta = 30°$ gives

$$I_{x'} = \frac{I_x + I_y}{2} + \frac{I_x - I_y}{2}\cos 2\theta - I_{xy}\sin 2\theta$$

$$= \left(\frac{5.12\pi + 5.12\pi}{2} + \frac{5.12\pi - 5.12\pi}{2}\cos 60° - 0\right)10^6$$

$$= 16.1\left(10^6\right)\ \text{mm}^4$$ **Ans**

$$I_{y'} = \frac{I_x + I_y}{2} - \frac{I_x - I_y}{2}\cos 2\theta + I_{xy}\sin 2\theta$$

$$= \left(\frac{5.12\pi + 5.12\pi}{2} - \frac{5.12\pi - 5.12\pi}{2}\cos 60° + 0\right)10^6$$

$$= 16.1\left(10^6\right)\ \text{mm}^4$$ **Ans**

$$I_{x'y'} = \frac{I_x - I_y}{2}\sin 2\theta + I_{xy}\cos 2\theta$$

$$= \left(\frac{5.12\pi - 5.12\pi}{2}\sin 60° + 0\right)10^6$$

$$= 0$$ **Ans**

A–23. Determine the principal moments of inertia and the orientation of the principal axes of inertia for the angle's cross-sectional area with respect to a set of principal axes that have their origin located at the centroid C. Use the equations developed in Sec. A.4.

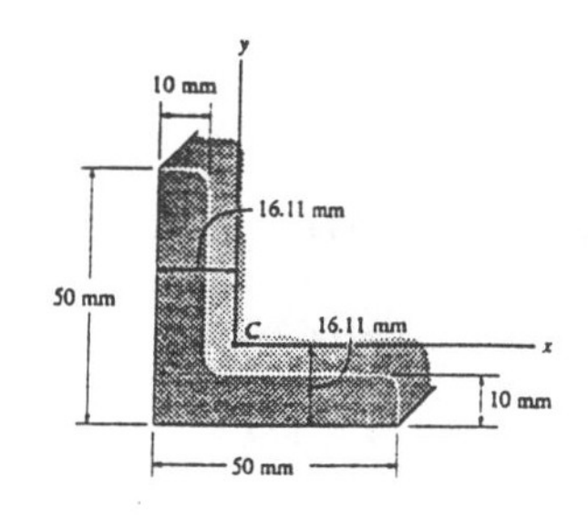

Moment of Inertia and Product of Inertia about x and y axes :

$$I_x = I_y = \frac{1}{12}(40)\left(10^3\right) + 40(10)(16.11-5)^2$$
$$+\frac{1}{12}(10)\left(50^3\right) + 10(50)(25-16.11)^2$$
$$= 196.39\left(10^3\right)\ \text{mm}^4$$

$$I_{xy} = 40(10)(13.89)(-11.11) + 50(10)(-11.11)(8.89)$$
$$= -111.11\left(10^3\right)\ \text{mm}^4$$

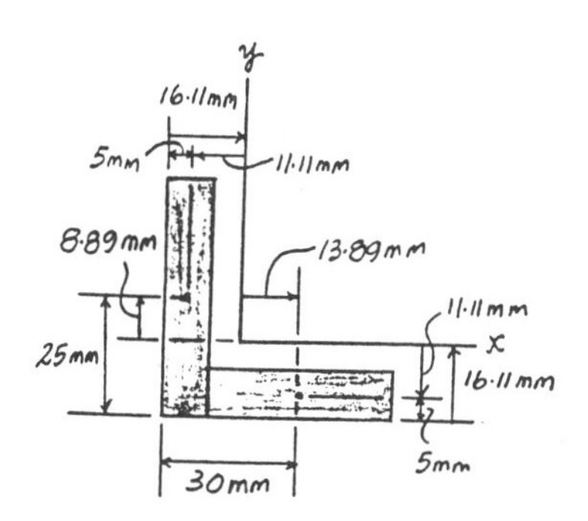

Principal Moment of Inertia : Applying Eq. A – 12 gives

$$I_{\substack{max\\min}} = \frac{I_x + I_y}{2} \pm \sqrt{\left(\frac{I_x - I_y}{2}\right)^2 + I_{xy}^2}$$
$$= \left[\frac{196.39 + 196.39}{2} \pm \sqrt{\left(\frac{196.39 - 196.39}{2}\right)^2 + (-111.11)^2}\,\right]10^3$$

$I_{max} = 308\left(10^3\right)\ \text{mm}^4 \qquad I_{min} = 85.3\left(10^3\right)\ \text{mm}^4$ **Ans**

Orientation of the Principal Axes of Inertia : Applying Eq. A – 11 gives

$$\tan 2\theta_p = \frac{-I_{xy}}{\left(I_x - I_y\right)/2} = \frac{-\left[-111.11(10^3)\right]}{(196.39 - 196.39)(10^3)/2} = \infty$$
$$2\theta_P = 90.0^\circ \quad \text{and} \quad -90.0^\circ$$

Substituting $2\theta_P = 90.0^\circ$ into Eq. A – 10, we have

$$I_{x'} = \frac{I_x + I_y}{2} + \frac{I_x - I_y}{2}\cos 2\theta - I_{xy}\sin 2\theta$$
$$= \left[\frac{196.39 + 196.39}{2} + \frac{196.39 - 196.39}{2}\cos 90.0^\circ - (-111.11)\sin 90.0^\circ\right]10^3$$
$$= 308\left(10^3\right)\ \text{mm}^4 = I_{max}$$

Thus, $\quad \theta_{p_1} = 45.0^\circ \qquad \theta_{p_2} = -45.0^\circ$ **Ans**

***A–24.** Solve Prob. A–23 using Mohr's circle.

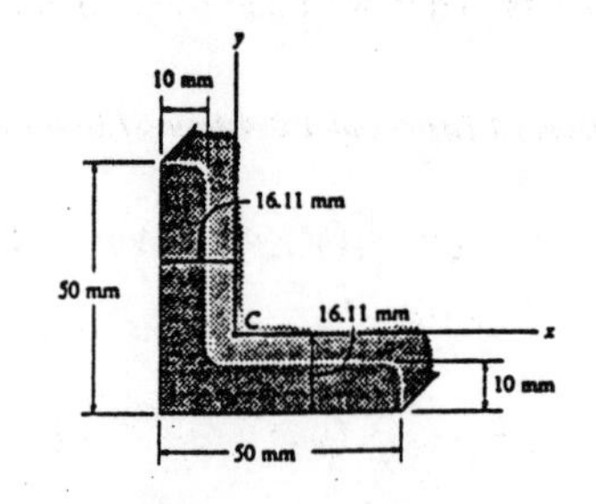

Moment of Inertia and Product of Inertia about x and y axes :

$$I_x = I_y = \frac{1}{12}(40)\left(10^3\right) + 40(10)(16.11 - 5)^2 + \frac{1}{12}(10)\left(50^3\right) + 10(50)(25 - 16.11)^2 = 196.39\left(10^3\right)\ \text{mm}^4$$

$$I_{xy} = 40(10)(13.89)(-11.11) + 50(10)(-11.11)(8.89) = -111.11\left(10^3\right)\ \text{mm}^4$$

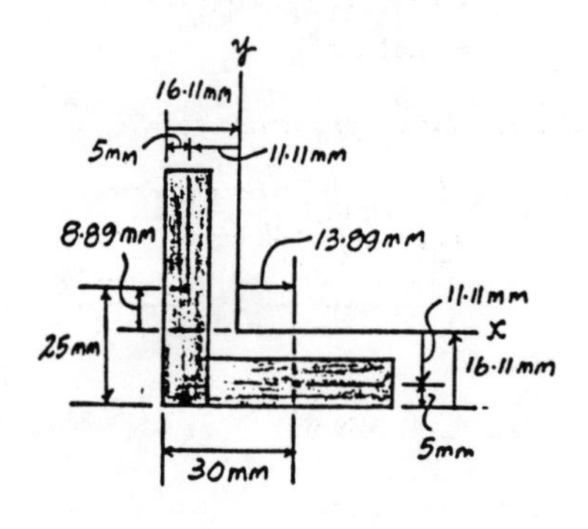

Circle Construction : Coordinates for reference point A and C are established first. Here, $\frac{I_x + I_y}{2} = \left(\frac{196.39 + 196.39}{2}\right)10^3 = 196.39\left(10^3\right)\ \text{mm}^4$. Hence,

$$A\ (196.39,\ -111.11)\left(10^3\right) \qquad C\ (196.39,\ 0)\left(10^3\right)$$

The radius of the circle is

$$R = \left(\sqrt{(196.39 - 196.39)^2 + 111.11^2}\right)\left(10^3\right) = 111.11\left(10^3\right)\ \text{mm}^4$$

Principal Moment of Inertia : The coordinates of points B and D represent I_{max} and I_{min} respectively.

$$I_{max} = (196.39 + 111.11)\left(10^3\right) = 308\left(10^3\right)\ \text{mm}^4 \qquad \textbf{Ans}$$

$$I_{min} = (196.39 - 111.11)\left(10^3\right) = 85.3\left(10^3\right)\ \text{mm}^4 \qquad \textbf{Ans}$$

Orientation of the Principal Axes of Inertia : From the circle,

$$2\theta_{P_1} = 90.0^\circ \qquad \theta_{P_1} = 45.0^\circ\ \textit{(Counterclockwise)} \qquad \textbf{Ans}$$

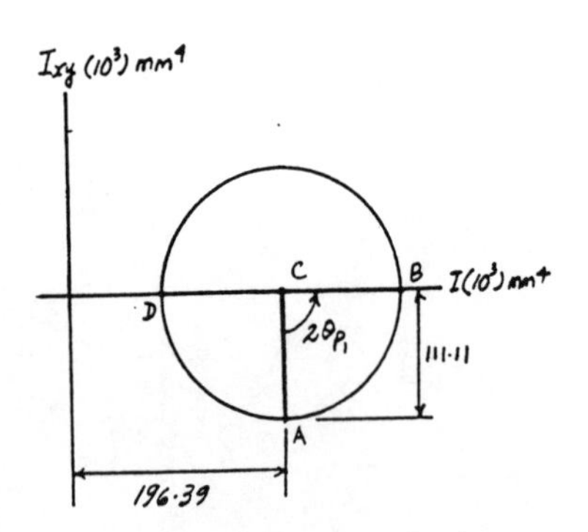

A–25. Determine the principal moments of inertia and the orientation of the principal axes of inertia of the cross-sectional area that have their origin located at the centroid C. Use the equations developed in Sec. A.4.

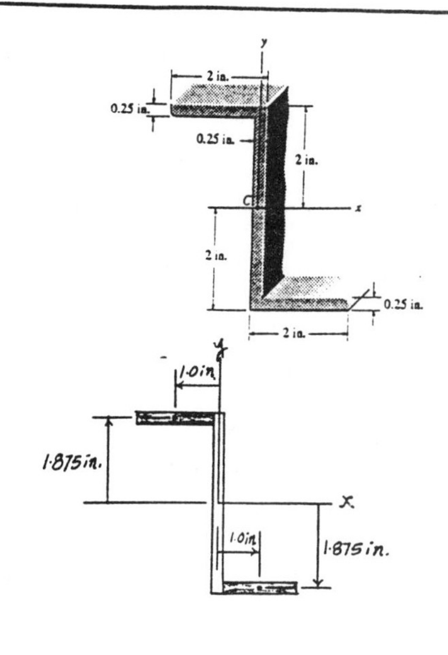

Moment of Inertia and Product of Inertia about x and y axes :

$$I_x = \frac{1}{12}(0.25)\left(4^3\right) + 2\left[\frac{1}{12}(1.75)\left(0.25^3\right) + 1.75(0.25)\left(1.875^2\right)\right]$$
$$= 4.4141 \text{ in}^4$$

$$I_y = \frac{1}{12}(3.75)\left(0.25^3\right) + \frac{1}{12}(0.25)\left(3.75^3\right) = 1.1035 \text{ in}^4$$

$$I_{xy} = 1.75(0.25)(-1.00)(1.875) + 1.75(0.25)(1.00)(-1.875)$$
$$= -1.6406 \text{ in}^4$$

Principal Moment of Inertia : Applying Eq. A – 12 gives

$$I_{\substack{\max \\ \min}} = \frac{I_x + I_y}{2} \pm \sqrt{\left(\frac{I_x - I_y}{2}\right)^2 + I_{xy}^2}$$
$$= \frac{4.4141 + 1.1035}{2} \pm \sqrt{\left(\frac{4.4141 - 1.1035}{2}\right)^2 + (-1.6406)^2}$$

$I_{\max} = 5.09 \text{ in}^4 \qquad I_{\min} = 0.428 \text{ in}^4$ **Ans**

Orientation of the Principal Axes of Inertia : Applying Eq. A – 11 gives

$$\tan 2\theta_p = \frac{-I_{xy}}{\left(I_x - I_y\right)/2} = \frac{-(-1.6406)}{(4.4141 - 1.1035)/2} = 1.06678$$
$$2\theta_p = 44.75^\circ \quad \text{and} \quad -135.25^\circ$$

Substitute $2\theta_p = 44.75^\circ$ into Eq. A – 10, we have

$$I_{x'} = \frac{I_x + I_y}{2} + \frac{I_x - I_y}{2}\cos 2\theta - I_{xy}\sin 2\theta$$
$$= \frac{4.4141 + 1.1035}{2} + \frac{4.4141 - 1.1035}{2}\cos 44.75^\circ - (-1.6406)\sin 44.75^\circ$$
$$= 5.09 \text{ in}^4 = I_{\max}$$

Thus, $\qquad \theta_{p_1} = 22.4^\circ \qquad \theta_{p_2} = -67.6^\circ$ **Ans**

A–25. Determine the principal moments of inertia and the orientation of the principal axes of inertia of the cross-sectional area that have their origin located at the centroid C. Use the equations developed in Sec. A.4.

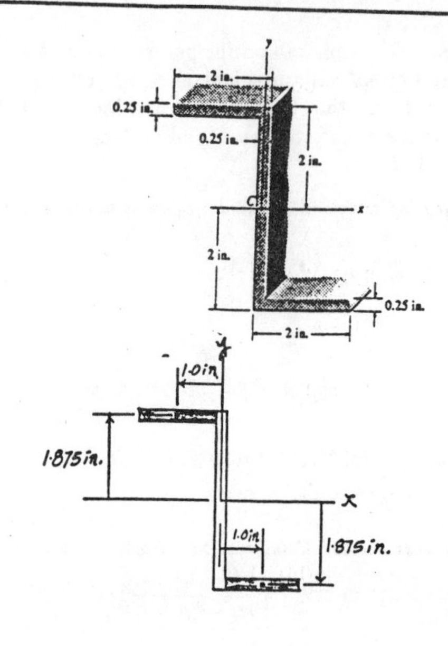

Moment of Inertia and Product of Inertia about x and y axes :

$$I_x = \frac{1}{12}(0.25)(4^3) + 2\left[\frac{1}{12}(1.75)(0.25^3) + 1.75(0.25)(1.875^2)\right]$$
$$= 4.4141 \text{ in}^4$$

$$I_y = \frac{1}{12}(3.75)(0.25^3) + \frac{1}{12}(0.25)(3.75^3) = 1.1035 \text{ in}^4$$

$$I_{xy} = 1.75(0.25)(-1.00)(1.875) + 1.75(0.25)(1.00)(-1.875)$$
$$= -1.6406 \text{ in}^4$$

Principal Moment of Inertia : Applying Eq. A–12 gives

$$I_{\substack{max \\ min}} = \frac{I_x + I_y}{2} \pm \sqrt{\left(\frac{I_x - I_y}{2}\right)^2 + I_{xy}^2}$$
$$= \frac{4.4141 + 1.1035}{2} \pm \sqrt{\left(\frac{4.4141 - 1.1035}{2}\right)^2 + (-1.6406)^2}$$

$I_{max} = 5.09 \text{ in}^4$ $\qquad$ $I_{min} = 0.428 \text{ in}^4$ $\qquad$ **Ans**

Orientation of the Principal Axes of Inertia : Applying Eq. A–11 gives

$$\tan 2\theta_p = \frac{-I_{xy}}{(I_x - I_y)/2} = \frac{-(-1.6406)}{(4.4141 - 1.1035)/2} = 1.06678$$
$$2\theta_p = 44.75° \quad \text{and} \quad -135.25°$$

Substitute $2\theta_p = 44.75°$ into Eq. A–10, we have

$$I_{x'} = \frac{I_x + I_y}{2} + \frac{I_x - I_y}{2}\cos 2\theta - I_{xy}\sin 2\theta$$
$$= \frac{4.4141 + 1.1035}{2} + \frac{4.4141 - 1.1035}{2}\cos 44.75° - (-1.6406)\sin 44.75°$$
$$= 5.09 \text{ in}^4 = I_{max}$$

Thus, $\qquad \theta_{p_1} = 22.4° \qquad \theta_{p_2} = -67.6°$ $\qquad$ **Ans**

A–26. Solve Prob. A–25 using Mohr's circle.

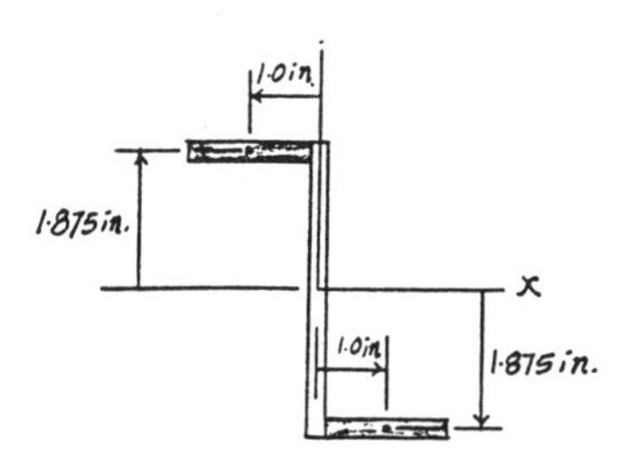

Moment of Inertia and Product of Inertia about x and y axes :

$$I_x = \frac{1}{12}(0.25)\left(4^3\right) + 2\left[\frac{1}{12}(1.75)\left(0.25^3\right) + 1.75(0.25)\left(1.875^2\right)\right]$$
$$= 4.4141 \text{ in}^4$$

$$I_y = \frac{1}{12}(3.75)\left(0.25^3\right) + \frac{1}{12}(0.25)\left(3.75^3\right) = 1.1035 \text{ in}^4$$

$$I_{xy} = 1.75(0.25)(-1.00)(1.875) + 1.75(0.25)(1.00)(-1.875)$$
$$= -1.6406 \text{ in}^4$$

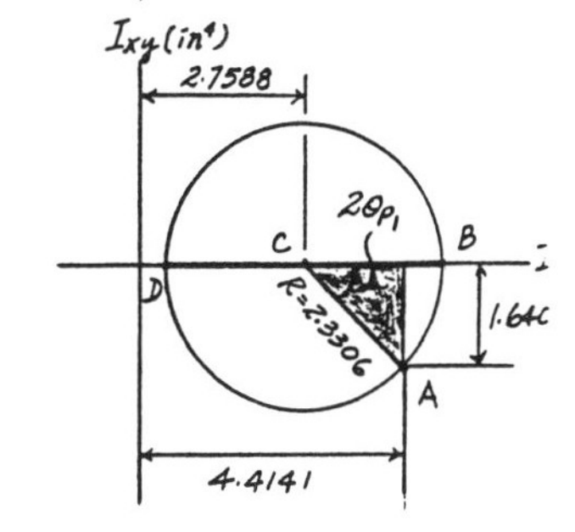

Circle Construction : Coordinates for reference point A and C are established first. Here, $\frac{I_x + I_y}{2} = \left(\frac{4.4141 + 1.1035}{2}\right) = 2.7588$ in^4. Hence,

$$A\ (4.4141,\ -1.6406) \qquad C\ (2.7588,\ 0)$$

The radius of the circle is

$$R = \sqrt{(4.4141 - 2.7588)^2 + 1.6406^2} = 2.3306 \text{ in}^4$$

Principal Moment of Inertia : The coordinates of points B and D represent $I_{\max}$ and $I_{\min}$ respectively.

$$I_{\max} = 2.7588 + 2.3306 = 5.09 \text{ in}^4 \qquad \textbf{Ans}$$
$$I_{\min} = 2.7588 - 2.3306 = 0.428 \text{ in}^4 \qquad \textbf{Ans}$$

Orientation of the Principal Axes of Inertia : From the circle,

$$\tan 2\theta_{p_1} = \frac{1.6406}{(4.4141 - 2.7588)} = 0.99115$$

$$\theta_{p_1} = 22.4^\circ \ (\textit{Counterclockwise}) \qquad \textbf{Ans}$$